Fundamentals of Electrochemistry

Fundamentals of Electrochemistry

Principles, Methods, Materials, Applications

Gunther Wittstock
with contributions by Bernd Speiser, William R. Heineman, Frank Marken, and Julia Witt

WILEY
WILEY-VCH GmbH

Author

Prof. Dr. Gunther Wittstock
Carl von Ossietzky University of Oldenburg
School for Mathematics and Science, Institute of Chemistry
Ammerländer Heerstrasse 114–118
26129 Oldenburg
Germany

Cover
Cover by Oleg Sklyar and Gunther Wittstock

Library of Congress Card No.: applied for

British Library Cataloguing-in-Publication Data
A catalogue record for this book is available from the British Library.

Bibliographic information published by the Deutsche Nationalbibliothek
The Deutsche Nationalbibliothek lists this publication in the Deutsche Nationalbibliografie; detailed bibliographic data are available on the Internet at <http://dnb.d-nb.de>.

Print ISBN: 978-3-527-34721-6
ePDF ISBN: 978-3-527-82555-4
ePub ISBN: 978-3-527-82556-1

Cover Design: Wiley
Typesetting: Straive, Chennai, India
Printing and Binding

Printed on acid-free paper
Printed and bound by CPI Group (UK) Ltd, Croydon, CR0 4YY
C9783527347216_250626

Contents

Symbols

Latin Symbols (Selection)

The quantities are listed with their typical units and the section where the quantity is introduced first.

A	area of the electrode [m^2, cm^2] without consideration of roughness, geometric electrode area	1.4
A	relaxation constant [$L^{1/2}$ $mol^{-1/2}$]	4.2.10
A	preexponential factor (frequency factor) in the Arrhenius equation [s^{-1}]	3.2.1
$\mathscr{A}$	optical absorbance	6.6.3
A'	area of the electrode [m^2, cm^2] with consideration of roughness	4.2.2
A'	area requirement per molecule (Chapter 13) [m^2]	13.1.7
A'	modified frequency factor [$cm\ s^{-1}$]	3.2.1
A_{ECSA}	electrochemical active surface area of an electrode [$m^2_{Pt}\ g^{-1}_{Pt}$]	12.4.1
a	general activity	2.2.1
a	normalized scan rate in cyclic voltammetry $a = nFv/(RT)$ [dimensionless]	6.2.4
a°	standard activity, e.g., $a^\circ_c = 1$ mol L^{-1}, $a^\circ_m = 1$ mol kg^{-1}, $a_x = 1$	2.2.1
$a_i^{\circ\alpha}$, a_i^α	standard activity (activity) of the ith compound in phase α	2.2.2
a_c	concentration activity [mol L^{-1}], also stated as sum formula in curved brackets {}	2.2.3
a_m	molality activity [mol kg^{-1}]	2.2.3
a_x	mole fraction activity [dimensionless]	2.2.3
a_{XY}	lateral interaction parameter between adsorbates X and Y [dimensionless]	13.4.4
$a_\pm$	mean activity [dimensionless]	2.4.2
B	electrophoretic constant [S cm^2 $mol^{-3/2}$ $L^{1/2}$]	4.2.10
b_i	molality of compound i [mol kg^{-1}]	2.2.3
b_i	adsorption coefficient of the compound i [L cm^{-2}]	13.4.2
C	capacitance [C V^{-1}, A s V^{-1}]	4.2.2
C	normalized concentration [dimensionless]	6.3.1
c	speed of light in vacuum (in the context of spectroscopy), $c = 2.9979 \times 10^8$ ms^{-1}	8.3
c	(amount) concentration [mol L^{-1}, mol cm^{-3}], also stated as sum formula of the compound in square brackets []	2.2.3

c^*	(amount) concentration [mol L^{-1}, mol cm^{-3}] in the bulk phase of the solution, also stated as sum formula of the compound in square brackets with asterisk []*	3.2.1
C_{dl}	double-layer capacitance [A s V^{-1}]	4.2.2
$C_{dl,A}$	area-related double layer capacitance [μF cm^{-2} = A s V^{-1} cm^{-2}]	8.10.3
$C^{d}_{dl,A}$	differential area-related double-layer capacitance [μF cm^{-2} = A s V^{-1} cm^{-2}]	8.10.3
$C^{i}_{dl,A}$	integral area-related double-layer capacitance [μF cm^{-2} = A s V^{-1} cm^{-2}]	8.10.3
c_{LOD}	limit of detection [mol L^{-1}]	5.3.10
C_r	C-rate [A/(A h) = h^{-1}]	15.2.2
$C_{sc,A}$	area-related capacitance of the space charge region [A s V^{-1} cm^{-2}]	10.2.2
C_s	serial capacitance [C]	7.3.3
C_s	compression modulus [N m^{-1}]	13.1.6
C_V	volume-related capacity "charge density" [A s cm^{-3}], [mA h cm^{-3} = A h L^{-1}]	15.2.1
D	diffusion coefficient [cm^2 s^{-1}]	4.4.2
D_+, D_-	diffusion coefficient for anions and cations [cm^2 s^{-1}]	4.4.2
d	diameter, distance, layer thickness [m]	8.5.2
D_{CT}	apparent charge transport coefficient [cm^2 s^{-1}]	13.4.5
D_f	diffusion coefficients within a film [cm^2 s^{-1}]	13.4.5
d_Q	thickness of a quartz resonator [mm]	8.8.1
D_S	density of states [cm^{-3}]	8.11.2
E	electrode potential [V]	2.1.2
$\dot{E}$	instantaneous electrode potential [V]	7.1
$\mathbf{E}$	electrical field strength [V m^{-1}]	4.2.2
$\mathbf{E}_p$	electrical field vector of a photon within the plane of incidence (parallel) [V m^{-1}]	8.2.1
$\mathbf{E}_s$	electrical field vector of a photon perpendicular to the plane of incidence (from German *senkrecht*) [V m^{-1}]	8.2.1
e	elementary charge, charge of an electron = $1.6021892 \times 10^{-19}$ A s	1.1.1
E	electrode potential [V]	2.1.1
E°	standard electrode potential [V]	2.2.2
$E^{\circ\prime}$	formal potential of an electrode or redox couple [V]	2.2.6
E_A	activation energy [J mol^{-1}]	3.2.1
E_{AC}	alternating-current component of a complex potential program [V]	5.3.9
$E^{\circ\prime}_{ads}$	formal potential of an immobilized/adsorbed redox system [V]	13.4.2
E_{aff}	electron affinity [eV]	10.1.3
E_B	binding energy of a core electron within an atom [eV]	8.5.1
E_{cor}	corrosion potential [V]	14.3.1
E_d	deposition potential [V]	5.3.10
E_d	detection potential [V]	5.4.1
E_{DC}	direct-current component of a complex potential program [V]	5.3.9
E_e	one-electron energy [eV]	8.11.2
E_{eff}	effective electrode potential [V]	7.2.2
E_{eq}	equilibrium electrode potential	3.2.3
E_F	Fermi energy, Fermi level [eV]	8.11.2

$E_{\mathrm{F,n}}$, $E_{\mathrm{F,p}}$	quasi-Fermi level for electrons and holes [eV]	10.1.5
E_{fb}	flat band potential [V]	10.2.2
E_{g}	band gap [eV]	10.1.2
E_{initial}	initial potential in voltammetry [V]	5.3.5
E_{ion}	ionization energy [eV]	10.1.3
E_{kin}	kinetic energy of emitted electrons [eV]	8.5.1
E_{NF}	Nernst factor [V], $E_{\mathrm{NF}} = RT/(nF) = 0.0257$ V for $T = 298$ K	2.2.2
E_{OCP}	open-circuit potential [V]	3.2.7
E_{P}	pitting potential [V]	14.3.2
E_{p}	electrostatic interaction energy [kJ mol^{-1}]	4.2.9
$E_{\mathrm{p/2}}$	potential at which half the peak current is reached	6.4.1
E_{p}, E_{pa}, E_{pc}	peak potential in voltammetry; anodic, cathodic peak potential [V]	5.3.5
E_{pas}	passivation potential [V]	14.3.2
$E_{\mathrm{P,I}}$	potential for inhibition of pitting corrosion [V]	14.5.3
$E_{\mathrm{P,n}}$	potential for stable nucleation of corrosion pits [V] [V]	14.5.3
$E_{\mathrm{P,r}}$	potential for repassivation of corrosion pits [V]	14.5.3
E_{pzc}	potential of zero charge [V]	5.3.5
E_{pzfc}	potential of zero free charge [V]	8.10.3
E_{pztc}	potential of zero total charge [V]	8.10.3
E_{S}	step potential [V]	5.3.8
E_{set}	set potential for an electronic feedback system [V]	5.7.1
E_{V}	valence band edge [eV]	10.1.3
E_{v}	vibrational energy of a harmonic oscillator [eV]	8.2
E_{vac}	energy of an electron in a field-free vacuum [eV]	10.1.3
E_{λ}	switching or vertex potential in cyclic voltammetry [V]	5.3.5
$E_{1/2}$	half-wave potential [V]	5.3.5
F	vibrational term $F = E_{\mathrm{v}}/hc$ [cm^{-1}]	8.2.1
F	molar Helmholtz free energy (as intensive quantity) [kJ mol^{-1}]	2.2.1
F	Faraday constant $= N_{\mathrm{A}} \cdot e = 96\,485.3$ A s mol^{-1} $\approx 100\,000$ A s mol^{-1}	1.3
$\mathbf{F}$	mechanical force (vector) [N]	8.11.5
$\mathbf{F}_{\mathrm{E}}$	acceleration force in an electric field [N]	4.2.2
$\mathbf{F}_{\mathrm{R}}$	friction force [N]	4.2.2
f	Helmholtz free energy (as extensive quantity) [kJ]	2.2.1
f	alternating current frequency [Hz = s^{-1}], circular frequency [s^{-1}], rotation per second	4.4.1
$f_{\pm}$	mean activity coefficient [dimensionless]	2.4.2
f_c	concentration activity coefficient [dimensionless]	2.2.3
f_m	molality activity coefficient [dimensionless]	2.2.3
f_x	mole fraction activity coefficient [dimensionless]	2.2.3
f_x	"safety factor" for specification of the size of a simulation domain in Section 6.3.1	6.3.1
f_{λ}	conductivity coefficient [dimensionless]	4.2.4
f_0	resonance frequency of a quartz resonator without deposited mass [s^{-1}]	8.8.1

G	gravitational force (weight) [N]	8.10.2
G	molar Gibbs free energy (as intensive quantity) [kJ mol^{-1}]	2.2.1
G	conductance [S, Ω^{-1}]	4.2.2
g	Gibbs free energy (as extensive quantity) [kJ]	2.2.1
g	gravitational acceleration on Earth g $\approx$ 9.81 m^2 s^{-1}	1.1.1
g	combined interaction parameter in the Frumkin isotherm [dimensionless]	13.4.4
G_{ideal}	molar Gibbs free energy of an ideal crystal [kJ mol^{-1}]	11.1.5
G_{real}	molar Gibbs free energy of a real crystal [kJ mol^{-1}]	11.1.5
H	molar enthalpy (as intensive quantity) [kJ mol^{-1}]	2.2.1
$H^{\neq}$	molar activation enthalpy (as intensive quantity) [kJ mol^{-1}]	3.2.1
h	enthalpy (as extensive quantity) [kJ]	2.2.1
h	Planck constant, $h = 6.626070\ 15 \times 10^{-34}$ J s	8.2
$\hbar$	reduced Planck constant $h/2\pi$	10.1.2
i	imaginary unit	7.1
i	index for the type of ion i	2.2.1
I	current [A]	1.1.1
I_0	exchange current [A]	3.2.2
I_0	current amplitude [A]	4.2.2
I_0	light intensity [W cm^{-2}]	10.2.6
$\dot{I}$	instantaneous current [A]	7.1
I_{a}, I_{c}	current of anodic and cathodic partial reaction [A]	3.2.2
I_{AC}	alternating current	8.13
I_{av}	mean current [A]	14.6.2
I_{be}	current within the bipolar electrode [A]	9.6.1
I_{bg}	background current [A]	6.2.2
I_{C}	capacitive current [A]	4.2.2
I_c	ionic strength on the basis of concentration [mol L^{-1}]	4.2.8
I_{cat}	catalytic current [A]	18.2.5
I_{D}	current at a disk of an RDE [A]	12.4.3
I_{d}	diffusion-limited current [A]	5.3.5
I_{d}, I_{c}	discharge and charging current [A]	15.2.3
I_{F}	faradaic current [A]	4.5.2
I_{f}	forward current in SQWV [A]	5.3.8
I_{i}	ion current in mass spectrometry [A]	12.4.5
I_j	general intensity of XP photoemission line from energy level j [counts] or [counts s^{-1}]	8.5.2
I_{K}	kinetically limited current [A]	12.4.2
I_{L}	Levich current, transport-limited current at an RDE [A]	12.4.2
I_m	ionic strength on the basis of molalities [mol kg^{-1}]	4.2.8
I_{p}	pulse current [A]	14.6.2
I_{p}, I_{pa}, I_{pc}	peak current in voltammetry; anodic, cathodic peak current [A]	5.3.5
I_{p}^{b}, I_{p}^{f}	peak currents of backward and forward scan in cyclic voltammetry [A]	6.2.2
I_{pas}	passivation current [A]	14.3.2
I_{ph}	photocurrent [A]	10.2.6

$I_{P,\max}$	current at the maximum power point [A]	10.2.6
I_R	current at the ring electrode of an RRDE [A]	12.4.3
I_r	reverse current in SQWV [A]	5.3.8
I_SC	short-circuit current [A]	10.3.2
$I_{\mathrm{T},\infty}$	probe current (tip) (SECM, ECSTM) [A]	8.12.2
I_tot	total current [A]	9.6.1
J	flux of the amount of substance [mol cm^{-2} s^{-1}]	4.4.2
J	nucleation rate [cm^{-2} s^{-1}]	9.2.3
j	index for the type of ion j	2.2.1
j	current density [A cm^{-2}]	1.4
j_0	exchange current density [A cm^{-2}]	3.2.3
$j_\mathrm{a}, j_\mathrm{c}$	current density for anodic and cathodic partial reactions [A cm^{-2}]	3.2.3
j_cor	corrosion current density [A m^{-2}]	14.3.1
j_pas	passivation current density [A cm^{-2}]	14.3.2
$\mathbf{K}$	translation vector in reciprocal crystal lattice [m^{-1}]	10.1.2
K	equilibrium constant, unit depends on the expression of the mass action law	2.2.3
k	rate constant, the unit depends on the reaction order and on whether it refers to a homogeneous or heterogeneous reaction. Indices refer to the number of the reaction.	3.2.1
$\mathbf{k}$	wave vector [cm^{-1}]	8.4
K_a	acid dissociation constant [mol L^{-1}]	2.2.2
k_app	apparent heterogeneous standard rate constant [cm s^{-1}]	5.6.6
k_a, k_c	heterogeneous rate constant for anodic and cathodic partial reaction [cm s^{-1}]	3.2.1
k_B	Boltzmann constant $= R/N_\mathrm{A} = 1.380649 \times 10^{-23}$ J K^{-1}	2.2.1
K_c	equilibrium constant of comproportionation [dimensionless]	6.4.1
k_eff	heterogeneous first-order rate constant at the sample in SECM experiments [cm s^{-1}]	8.12.2
k_F	force constant [N m^{-1}]	8.2
$\mathbf{k}_\mathrm{F}$	wave vector corresponding to the Fermi energy [cm^{-1}]	10.1.2
K_M	Michaelis–Menten constant [mol L^{-1}]	18.3.3
K'_M	apparent Michaelis–Menten constant [mol L^{-1}]	18.3.3
$K_\mathrm{M,I}$	potentiometric selectivity coefficient of ion M with respect to interfering ion I	5.2.4
k_n	nucleation rate constant [s^{-1}]	9.2.3
k_s°	standard rate constant of an immobilized redox system [s^{-1}]	13.4.3
K_SP	solubility product, unit depends on the stoichiometry, e.g., [mol^2 L^{-2}]	2.2.4
K_W	equilibrium constant for the autoprotolysis of water [mol^2 L^{-2}]	2.4.2
k_0	force constant of a freely suspended cantilever [N m^{-1}]	8.11.5
k°	heterogeneous standard (reaction) rate constant [cm s^{-1}]	3.2.2
L	normalized distance in SECM	8.12.1
l	length or distance [cm, m]	4.2.1
l_film	film thickness	13.4.7
L_h	hole diffusion length [cm]	10.2.6
l_pore	length of a pore [cm]	7.4.3

M_i	molar mass of the ith compound	2.4.1
m	mass [g]	1.1.1
m_e	mass of an electron [kg]	10.1.2
m_h	mass of a hole [kg]	10.1.2
m_i	mass of ith compound	2.4.1
m_r	reduced mass (vibrational spectroscopy) [kg]	8.2
m_{solv}	mass of the solvent [kg]	2.2.3
$m_{t,i}$	mass transport coefficient of species i [cm s^{-1}]	3.2.5
N	amount of objects [mol]	1.3
n	stoichiometric number of charges in an electrode reaction [dimensionless]	1.3
$\hat{n}$	complex refractive index [dimensionless], $\hat{n}(\omega) = n'(\omega) + in''(\omega)$	8.3
N_A	Avogadro number, amount of objects in one mole = 6.022140 76 × 10^{23} mol^{-1}	1.3
N_C	number of chemical compounds in the reaction system [dimensionless]	15.4
N_{CN}	coordination number at a specific crystal site [dimensionless]	11.1.5
n_{crit}	number of metal atoms in a nucleus of critical size [dimensionless]	9.2.2
$N_{d,eq}$	number of defects per amount of objects in equilibrium state [mol^{-1}]	11.1.5
N_e	number of electrons as amount of objects [mol]	10.1.2
n_e	number of valence electrons per space unit [cm^{-3}]	10.1.2
N_F	degrees of freedom (p, T, U)	15.4
N_i	amount of substance of the ith compound [mol]	2.2.2
N_p	number of current pulses [dimensionless]	14.6.2
N_{pore}	number of pores [dimensionless]	7.4.3
N_P	number of phases[dimensionless]	15.4
n_{solv}	number of solvent molecules replaced by one adsorbate molecule at the interface [dimensionless]	2.2.3
N^{α}	amount of substance in phase α	8.10.1
N_0	number of sites suitable for nucleation [cm^{-2}]	9.2.3
P	power [W] = [J s^{-1}] = [VA]	1.1.1
$\mathbf{p}$	translational momentum (vector)	10.1.2
p	pressure [Pa = N m^{-2}]	1.1.1
P_A	area-related power [W cm^{-2}]	2.4.4
p_i	partial pressure of the ith gas [Pa]	2.2.3
pI	isoelectric point of a protein	4.6.2
pK_a	negative decadic logarithm of the normalized acid dissociation constant	4.2.6
P_m	mass-specific power [W kg^{-1}]	15.2.1
pmf	protonmotive force [V]	18.1.1
P_V	volume-related power density [W cm^{-3}, W L^{-1})	15.2.1
$p°$	standard pressure, 100 000 Pa, older conventions 101 325 Pa = 760 Torr	2.2.3
Q	charge [C] = [A s]	1.1.1
Q	*Q*-factor (quality factor) of a resonator [dimensionless]	8.11.5
q	general inner coordinate in vibrational spectroscopy (bond angle or bond length)	8.2.1
q	exchanged heat [J]	2.4.4

$\mathbf{q}$	vector of all inner coordinates of a molecule (bond angles and bond lengths)	12.2.2
Q_{ads}	charge for the conversion of adsorbed species [C] = [A s]	6.5.3
Q_{CPE}	parameter of the constant phase element (CPE) [s^{γ} A V^{-1}], γ is a further parameter of the CPE	7.4.4
$Q_{CPE,A}$	area-related parameter of the constant phase element (CPE) [s^{γ} A V^{-1} cm^{-2}], γ is a further parameter of the CPE	7.4.4
Q_d, Q_c	charge for the discharge (d) and charge (c) [A s], [A h]	15.2.3
Q_{dl}	charge of the electric double layer [A s]	6.5.3
Q_m	specific capacity [1 A s g^{-1} = 0.278 mA h g^{-1} = 0.278 A h kg^{-1}]	15.3.1
Q_N	nominal capacity of a battery [A s], [A h]	15.2.1
$Q_{N,m}$	nominal specific capacity [1 A s g^{-1} = 0.278 mA h g^{-1} = 0.278 A h kg^{-1}]	15.2.1
$Q_{N,V}$	nominal volume-related capacity "charge density" [A s cm^{-3}], [mA h cm^{-3} = A h L^{-1}]	15.2.1
Q_{stored}	charge stored after charging a battery [A s], [A h]	15.2.3
$\mathbf{R}$	translation vector in a crystal lattice [m]	10.1.2
R	molar gas constant $R = 8.314\,462\,618\ J\,K^{-1}\ mol^{-1}$	2.2.2
R	ohmic resistance [Ω]	4.2.2
R	reflectivity [dimensionless]	8.4
$\mathbf{r}$	vector of three space coordinates	10.1.2
r	radius [m]	2.2.1
r	reaction rate; homogeneous [mol L^{-1} s^{-1}], heterogeneous [mol cm^{-2} s^{-1}]	1.4
R_C	resistance of a capacitor in an AC circuit [Ω]	4.2.2
r_{crit}	critical radius of a nucleus [nm]	9.2.2
R_{CT}	charge-transfer resistance [Ω]	3.2.5
r_D	Debye length [m]	4.2.8
R_E	electrolyte resistance [Ω]	4.2.2
r_E	radius of a contact electrodes on a quartz resonator [cm]	8.8.1
R_f	resistance in the electronic feedback loop of an operational amplifier [Ω]	5.7.1
R_g	normalized radius of the insulating sheath of a microelectrode probe in SECM [dimensionless]	8.12.1
r_i	hydrodynamic radius of an ion i [m]	4.2.1
R_{in}	input resistance of an electronic circuit [Ω]	5.7.1
$r_{kl,s}$, $r_{kl,p}$	amplitude ratio of reflected s- and p-polarized light [dimensionless]	8.3
r_{pore}	characteristic pore radius [cm]	7.4.3
r_Q	radius of a quartz resonator [cm]	8.8.1
R_S	normalized radius of the active sample in SECM [dimensionless]	8.12.1
R_s	serial resistance [Ω]	7.3.3
r_S	radius of the active electrode area of the sample in SECM	8.12.1
r_{sol}	radius of the solvation shell of an ion	4.3.2
r_T	radius of the microelectrode in SECM	8.12.1
R_{uc}	uncompensated electrolyte resistance [Ω]	5.3.1
$R_{0,l}$	specific pore resistance [Ω cm^{-1}] = $\rho/(\pi r_p^{\,2})$	7.4.3
R_0	half of the interelectrode spacing in a microelectrode array [cm]	5.6.6
r_0	exchange reaction rate; homogeneous [mol L^{-1} s^{-1}], heterogeneous [mol cm^{-2} s^{-1}]	3.2.1
r_0	radius of a microelectrode [cm]	5.6.6

s	entropy as extensive quantity [J K^{-1}]	2.2.1
S_i	molar entropy of compound i [J K^{-1} mol^{-1}]	2.2.4
$S_i°$	molar standard entropy of compound i [J K^{-1} mol^{-1}]	2.2.4
$S^{\neq}$	molar activation entropy (as intensive quantity) [J K^{-1} mol^{-1}]	3.2.1
T	absolute temperature [K]	2.2.1
t	time [s]	1.1.1
T_{b}	boiling point [K]	15.6.3
t_{c}	critical time [s]	5.6.4
t_{d}	lifetime of a drop of a dropping mercury electrode [s]	5.3.5
t_{d}, t_{c}	time for discharge (d) or charge (c) [s], [h]	15.2.3
T_{e}	transmission coefficient for a tunneling process [dimensionless]	8.11.1
T_{g}	glass transition temperature [K]	11.6.1
T_{i}	ignition temperature [K]	15.6.3
t_{initial}	initial time [s]	5.6.5
$t_{kl,s}$, $t_{kl,p}$	amplitude ratio of transmitted (refracted) s- and p-polarized light [dimensionless]	8.3
T_{m}	melting point [K]	15.6.3
t_{on}	pulse duration [ms]	14.6.2
t_{off}	resting time [mss]	14.6.2
T_{P}	throwing power [dimensionless]	9.3.3
t_{p}	time within a potential pulse [s]	5.3.9
t_+, t_-	transference number, ratio between the current carried by one kind of ion and the total ionic current [dimensionless]	2.3.3
U	voltage [V]	1.1.1
u	internal energy as extensive quantity [J]	2.2.1
U_{as}	asymmetry potential at a combination pH electrode	5.2.3
U_{bias}	bias voltage [V]	8.11.2
U_{cpd}	contact potential difference	8.13.5
$\langle U_{\mathrm{d}}\rangle$, $\langle U_{\mathrm{c}}\rangle$	mean voltage for discharge (d) and charge (c) [V]	15.2.3
u_{eo}	electroosmotic mobility [m^2 V^{-1} s^{-1}]	4.6.3
U_{eq}	equilibrium cell voltage [V]	2.1.2
u_{exp}	experimentally determined electroosmotic mobility [m^2 V^{-1} s^{-1}]	4.6.3
u_i	ion mobility of the ion i [m^2 s^{-1} V^{-1}]	4.2.1
U_{in}	general input voltage of a device [V]	5.7.1
U_{OCV}	open circuit voltage [V]	2.4.4
U_{onset}	onset voltage for an electrolysis [V]	2.1.1
U_{out}	general output voltage of a device [V]	5.7.1
$U_{P,\,\mathrm{max}}$	voltage at the maximum power point	10.2.6
$U°$	standard cell voltage [V]	2.2.2
U_0	voltage amplitude [V]	4.2.2
V	volume [L, cm^3]	1.1.1
V	potential energy [eV]	10.1.2
v	quantum number of the harmonic oscillator [dimensionless]	8.2
v	(potential) scan rate in voltammetry [V s^{-1}]	5.3.5

V_m	molar volume [L mol^{-1}, cm^3 mol^{-1}, ...]	9.2.4
v_Q	speed of sound in quartz [m s^{-1}]	8.8.1
W	work, energy [W h], [W s], [J]	1.1.1
W_a	Wagner number [dimensionless]	9.3.3
W_el	electric work [W s, J, N m, V A s]	2.2.1
$\mathbf{w}_\mathrm{eo}$	electroosmotic flow velocity [m s^{-1}]	4.6.3
$\mathbf{w}_i$	migration velocity [m s^{-1}], index i = kind of ion	4.2.1
w_i	mass fraction of compound i [dimensionless]	2.4.1
W_m	specific energy [W h kg^{-1}] = [mW h g^{-1}]	15.2.1
W_V	energy density [W h L^{-1}]	15.2.1
$\mathbf{w}_\mathrm{exp}$	experimentally determined migration velocity [m s^{-1}]	4.6.3
X	reactance [V A^{-1}]	7.1
X	normalized distance to the electrode [dimensionless]	6.3.1
x	space coordinate [m], e.g., distance to an electrode	2.2.1
$\{\mathrm{X}\}$	concentration activity of compound X [mol L^{-1}]	2.2.3
$[\mathrm{X}]$	concentration of compound X [mol L^{-1}]	2.2.3
$\dot{x}$	rate of thickness change of a work piece in corrosion tests [mm h^{-1}]	14.3.1
x_i	mole fraction of the ith compound [dimensionless]	2.2.3
$\mathbf{Y}$	admittance [A V^{-1}] = [Ω^{-1}]	7.3.2
$\mathbf{Z}$	impedance [Ω]	7.1
Z	absolute amount of the impedance [Ω]	7.3.1
$\mathbf{Z}_{0,l}$	specific interfacial impedance (related to width of a ring Δl) [Ω cm]	7.4.3
$\mathbf{Z}_A$	area-related interfacial impedance [Ω cm^2], determined at flat electrodes	7.4.3
$\mathbf{Z}_\mathrm{F}$	faradaic impedance [Ω]	7.3.3
z_i	ionic charge of the ion i [dimensionless], exactly the ratio between the charge of ion i [A s] and the charge of a proton [A s]	1.3
$\mathbf{Z}_\mathrm{pore}$	impedance of a pore [Ω]	7.4.3
Z_S	number of states within a volume element	10.1.2
$\mathbf{Z}_\mathrm{W}$	Warburg impedance [Ω]	7.3.3
z_+, z_-	ionic charges of cations and anions [dimensionless]	4.2.3

Greek Symbols (Selection)

The quantities are listed with their typical units and the section where the quantity is introduced first.

α	transfer coefficient [dimensionless]	3.2.2
α	degree of dissociation (dimensionless [0 … 1] or stated as %)	4.2.1
α	light absorption coefficient [cm^{-1}]	10.2.6
α, β	index of phase	2.2.1
β	decay constant in tunneling processes (STM) [m^{-1}]	8.11.2
β_i	adsorption equilibrium constant of substance i [mol^{-1} L]]	6.1.1
γ	exponent of a constant phase element (CPE) [$0 < \gamma < 1$]	7.4.4
γ	surface tension or interfacial tension [N m^{-1}]	9.2.2.
γ	duty cycle [dimensionless] in pulsed electrochemical machining	14.6.2
γ_i, γ°	interfacial concentration activity of species i and corresponding standard activity [mol cm^{-2}]	5.2.3
Γ_i	surface or interfacial excess concentration of substance i [mol cm^{-2}]	5.5.1
$\Gamma_{O,sa}$, $\Gamma_{r,sa}$	maximum interfacial excess concentration of compounds R and O [mol cm^{-2}]	13.4.2
Γ^*	total surface or interfacial excess concentration of a redox couple [mol cm^{-2}]	5.5.1
Δ	phase shift in ellipsometry [radian]	8.3
δ	thickness of the hydrodynamic boundary layer [cm]	4.7
δ	distance between two redox centers [cm]	4.7
$\Delta_{bg}g$	Gibbs free energy contribution due to bond breaks for one defect [kJ]	11.1.5
$\Delta_{bg}g_{EE}$	Gibbs free energy contribution due to breaking one element-element bond [kJ]	11.1.5
$\Delta_d g$	Gibbs free energy for formation of an individual defect [kJ]	11.1.5
$\Delta_d G$	molar Gibbs free energy for formation of defects in 1 mol of a pure compound [kJ mol^{-1}]	11.1.5
ΔE	amplitude of an AC voltage [V]	5.3.9
ΔE	potential step height in pulse voltammetry or staircase voltammetry [V]	5.3.6
ΔE_p	peak splitting, i.e., splitting of corresponding anodic and cathodic peak potentials in cyclic voltammetry [V]	5.3.5
$\Delta_F G_i$	molar Gibbs free energy of formation of compound i [kJ mol^{-1}]	2.2.4
$\Delta G^{\neq}$	molar Gibbs free energy of activation [kJ mol^{-1}]	3.2.1
$\Delta_R G$	molar Gibbs free reaction energy [kJ mol^{-1}]	2.2.2
$\Delta_R G^\circ$	molar standard Gibbs free reaction energy [kJ mol^{-1}]	2.2.2
$\Delta_F G^\circ$	molar standard Gibbs free energy of formation [kJ mol^{-1}]	2.2.4
$\Delta_{SS} G$	molar Gibbs free energy of super saturation [kJ mol^{-1}]	9.2.2
$\Delta_{vib} g$	Gibbs free energy contribution due to change of vibration frequencies at the defect location for one defect [kJ]	11.1.5
$\Delta H^{\neq}$	molar activation enthalpy [kJ mol^{-1}]	3.2.1
$\Delta_F H_i$	molar reaction enthalpy of formation of compound i [kJ mol^{-1}]	2.2.4
$\Delta_F H_i^\circ$	molar standard reaction enthalpy of formation of compound i [kJ mol^{-1}]	2.2.4
ΔI	current amplitude of alternating current [A]	7.1

Δl_{pore}	length of a pore segment [cm]	7.4.3
Δm	mass loss during corrosion, determined by weighting [g]	14.3.1
Δm_{theo}	mass loss during corrosion calculated from Faraday's law [g]	14.6.2
$\Delta_R H$	molar reaction enthalpy [kJ mol^{-1}]	2.2.2
$\Delta_R H°$	molar standard reaction enthalpy [kJ mol^{-1}]	2.2.2
$\Delta_R S$	molar reaction entropy [J K^{-1} mol^{-1}]	2.2.2
$\Delta_R S°$	molar standard reaction entropy [J K^{-1} mol^{-1}]	2.2.2
ΔS_{conf}	molar configuration enthalpy [J K^{-1} mol^{-1}]	11.1.5
$\Delta S^{\neq}$	molar activation entropy [J K^{-1} mol^{-1}]	3.2.1
Δx	thickness of the diffusion layer [cm], e.g., Nernst diffusion layer	5.3.5
$\Delta \varphi^{A}$	difference in Galvani potentials [V] between anode and solution	4.3.2
$\Delta \varphi^{\alpha,\beta}$	Galvani potential difference between phases α and β [V]	2.3
$\Delta \varphi^{\circ\alpha,\beta}$	standard Galvani potential difference [V] between phases α and β	2.3
$\Delta \varphi^{C}$	difference in Galvani potentials between cathode and solution	4.3.2
$\Delta \varphi_{\text{eq}}$	Galvani potential difference in equilibrium [V]	2.3
$\Delta \varphi^{\circ}_{\text{eq}}$	standard Galvani potential difference in equilibrium [V]	2.3
ε	yield [dimensionless], general efficiency [dimensionless], transfer coefficient [dimensionless]	1.4
	complex dielectric function [dimensionless]	8.3
ε_{be}	bipolar current efficiency [dimensionless]	9.6.1
ε_{C}	collection efficiency, transfer ratio [dimensionless]	12.4.3
ε_{C}	coulombic efficiency [dimensionless]	15.2.3
ε_{el}	energy efficiency [dimensionless]	15.2.3
$\varepsilon_{\text{F},i}$	current yield [dimensionless] or faradaic efficiency	1.4
$\hat{\varepsilon}_r$	relative dielectric constant [dimensionless]	2.2.1
ε_{V}	voltaic efficiency [dimensionless]	15.2.3
ε_0	vacuum electric permittivity $\varepsilon_0 = 8.8541878123 \times 10^{-12}$ A^2 s^4 m^{-3} kg^{-1} = $8.8541878123 \times 10^{-12}$ A s V^{-1} m^{-1} = $8.8541878123 \times 10^{-12}$ F m^{-1}	2.2.1
ζ	zeta-potential or surface potential [V]	4.6.3
η	overvoltage (of an electrochemical cell), overpotential (of an electrode) [V]	3.1.2, 3.2.3
η	viscosity [N s m^{-2}]	4.2.1
η_{T}	thermodynamic overpotential	12.3.2
θ	coverage [%]	5.6.5
θ	contact angle [°]	
θ	emission angle [°]	8.5.2
κ	specific conductivity [S cm^{-1}, Ω^{-1} cm^{-1}]	4.2.2
κ	normalized homogeneous rate constant $\kappa = RTk/(nFv)$ for first order and $\kappa = RTkc^*/(nFv)$ for second order [dimensionless]	6.2.5
κ	normalized heterogeneous rate constant (SECM) $\kappa = k_{\text{eff}}\, r_{\text{T}}/D$ [dimensionless]	8.12.2
κ	normalized rate constant of an immobilized redox system [dimensionless]	
Λ	molar conductivity [(Ω^{-1} cm^{-1})/mol^{-3}, S cm^2 mol^{-1}]	4.2.3
Λ	information depth in surface spectroscopies [m]	8.5.2
λ	wavelength or an electromagnetic radiation [nm] or [cm]	8.2.1

λ	partial discharge coefficient [dimensionless] during adsorption of a metal ion with charge number z^+, λz is the electrosorption valency	8.9.8
λ	reorganization energy in the framework of Marcus theory	12.2.2
λ	mean free path length of electrons in a solid [m]	8.5.2
λ	air-fuel ratio for an internal combustion engine [dimensionless]	5.2.8
Λ_{ephor}	decrease in molar conductivity by electrophoretic effect [$(\Omega^{-1}$ $cm^{-1})/(mol$ $cm^{-3})$, S cm^2 mol^{-1}]	4.2.10
Λ_{relax}	decrease in molar conductivity by relaxation effects [$(\Omega^{-1}$ $cm^{-1})/(mol$ $cm^{-3})$, S cm^2 mol^{-1}]	4.2.10
λ_i	molar conductivity for a single kind of ion i [S cm^2 mol^{-1}]	4.2.1
λ_i, λ_o	inner and outer reorganization energy [kJ mol^{-1}] or related to an individual molecule [eV]	12.2.2
Λ_0	molar conductivity at infinite dilution [$(\Omega^{-1}$ $cm^{-1})/(mol$ $cm^{-3})$, S cm^2 mol^{-1}]	4.2.4
$\lambda_{0,eq}$	equivalent conductivity	4.2.4
$\lambda_{0,eq,i}$	equivalent conductivity at infinite dilution for a single kind of ion i [S cm^2 mol^{-1}]	4.2.4
$\lambda_{0,i}$	molar conductivity at infinite dilution of a single kind of ion i [S cm^2 mol^{-1}]	4.2.4
$\boldsymbol{\mu}$	transition dipole moment in spectroscopy	8.2.1
μ	absolute value of the dipole moment of a molecule [A s m], usually given in Debye D; 1 D = 3.33564×10^{-30} A s m	8.2.1
μ	linear absorption coefficient of X-rays, [mm^{-1}]	8.5.3
μ_i	chemical potential of compound i [J mol^{-1}]	2.2.1
μ_i°	chemical standard potential of compound i [J mol^{-1}]	2.2.1
$\tilde{\mu}_i^\alpha$	electrochemical potential of compound i in phase α [J mol^{-1}]	2.2.1
$\tilde{\mu}_i^{\alpha\circ}$	electrochemical standard potential of compound i in the phase α [J mol^{-1}]	2.2.1
$\hat{\mu}_r$	complex magnetic permittivity number [dimensionless]	8.3
ν	frequency of electromagnetic radiation [Hz = s^{-1}] as circular frequency	8.2
$\tilde{\nu}$	wave number [cm^{-1}]	8.2.1
ν_D	mean velocity diffusion along one spatial coordinate [cm s^{-1}]	4.4.1
ν_i	signed stoichiometric coefficient (for reagents <0; for products >0) [dimensionless]	2.2.2
ν_n	nuclear frequency factor [Hz] = [s^{-1}]	12.2.3
ν_{osc}	frequency of an oscillator in vibrational spectroscopy [Hz] = [s^{-1}]	8.2
ν_+, ν_-	stoichiometric coefficients for cations and anions [dimensionless]	4.2.9
$\tilde{\nu}_d$	vibrational wavenumber for the neighboring atoms of the defect site [cm^{-1}]	11.1.5
ξ	[xi] reaction progression given as amount of objects, where objects is a count of completing a specific chemical reaction equation [mol]	2.2.2
Π	film pressure [N m^{-1}]	13.1.6
π	mathematical constant pi $\approx$ 3.141	5.3.8
ρ	mass density [g cm^{-3}] = [kg dm^{-3}] = [t m^{-3}]	2.4.1
ρ	specific resistance [Ω cm]	4.2.2
ρ_E	specific electrolyte resistance [Ω cm]	14.6.2
σ^M	excess charge density at a metal [A s cm^{-2}]	8.10.3
τ	time constant [s]	4.3.2

τ	step duration [s]	5.3.7
τ	drop time of a mercury electrode [s]	8.10.2
τ	transfer ratio during Langmuir–Blodgett transfer [dimensionless]	13.1.7
τ	degree of titration [dimensionless]	4.2.7
τ	normalized time in voltammetry $\tau = vt/E_{NF}$	6.3.1
Φ	work function [eV]	8.5.1
ϕ	inner potential or Galvani potential [V]	2.2.1
φ	phase shift in scanning force microscopy [radians]	8.11.5
φ^A	Galvani potential [V] within the anode material ($x \to \infty$)	4.3.2
φ^C	Galvani potential [V] within the cathode material ($x \to \infty$)	4.3.2
φ^S	Galvani potential [V] in the solution bulk ($x \to \infty$)	4.3.2
$\varphi_{x'=0}$	potential [V] at the outer Helmholtz plane	4.3.2
χ^α	surface potential [V] of a specific surface of phase α	2.2.1
Ψ	quantum mechanical wave function	8.11.2
Ψ	fugacity [Pa], i.e., the activity quantity for partial pressure of a gaseous compound	2.2.3
Ψ^α	outer potential or Volta potential [V] of the phase α	2.2.1
ψ	current function, normalized faradaic current [dimensionless]	5.3.8
ψ_T	current function, normalized microelectrode current in SECM	8.12.2
ψ	normalized heterogeneous standard reaction rate constant [dimensionless]	6.2.5
ω	angular frequency (omega = $2\pi f$) [s^{-1}], angle in radians per second	4.2.2
Ω	number of microstates [dimensionless; index ideal – for ideal crystal, index real for real crystal]	11.1.5

Important Abbreviations

Abbreviations and acronyms are listed with the section where they are introduced.

2D DIGE	two-dimensional differential gel electrophoresis	4.6.2
Ab	antibody	18.3.6
AC	alternating current	4.2.2
AChE	acetylcholinesterase	18.3.4
ACS	American Chemical Society	1.4
AC-SECM	SECM working modes using AC current	8.13.4
ACV	alternating current voltammetry	5.3.9
ADC	analog-digital converter	5.7.1
ADP	adenosine diphosphate	18.1
AES	Auger electron spectroscopy	8.5.1
AFC	alkaline fuel cell	2.4.4
AFM	atomic force microscopy	8.11.4
Ag	antigen	18.3.6
ALP	alkaline phosphatase	18.3.6
AO	atomic orbital	10.1.2
APT	atom probe tomography	14.5.7
APU	auxiliary power unit	16.4.4
aq	aqueous	1.2
ARXPS	angle resolved X-ray photoelectron spectroscopy	8.5.2
ASC	anode-supported cell	16.4
ATP	adenosine triphosphate	18.1
ATR	attenuated total reflection	6.6.3
Aux	auxiliary electrode (in a three-electrode setup)	5.3.1
BDD	boron-doped diamond	5.3.2
BET	Brunauer-Emmett-Teller (isotherm of gas adsorption)	15.9
BMS	battery management system	15.6.3
BOB	bisoxalylborate	11.4.2
CA	chronoamperometry	4.5.2
CB	conduction band (of a semiconductor)	10.1.2
CC	constant current (during charging of accumulators)	15.2.1
CCD	charge-coupled device	8.2.2
CE	capillary electrophoresis	4.6.3

CE	counter electrode (usually in a two-electrode setup)	5.3.1
CEI	cathode electrolyte interphase	15.6.4
CELT	confined etchant layer	8.12.7
CER	chlorine evolution reaction	12.5
CFME	carbon fiber microelectrode	5.6.5
cgs	unit system centimeter-gram-second	14.2
CHP	combined heat power	16.4.4
CL	catalyst layer	16.2.2
CMC	critical micelle concentration	4.6.3
CN	coordination number	11.1.3
CNT	carbon nanotube	5.3.2
COT	cyclooctatetraene	6.4.1
CPE	constant phase element	7.4.4
C-SFM	conductive scanning force microscopy	8.11.4
CT	charge transfer	13.4.5
CV	cyclic voltammetry	5.3.5
cyt	cytochrome	18.1.3
DA	dopamine	5.6.5
DAC	digital-analog converter	5.7.1
DC	direct current	1.2
DCECM	direct current electrochemical machining	14.6.2
DCM	dichloromethane	17.1.6
DDO	damped, driven oscillator	8.11.5
DDQ	2,3-dichloro-5,6-dicyano-1,4-benzoquinone	17.4.2
DEMS	differential electrochemical mass spectrometry	12.4.5
DES	deep eutectic solvent	11.4.3
DET	direct electron transfer	18.2.3
DFT	density functional theory	12.3.2
DHE	dynamic hydrogen electrode	2.4.1
DME	dropping mercury electrode	5.3.5
DMF	dimethyl formamide	5.3.2
DMFC	direct methanol fuel cell	2.4.4
DMSO	dimethyl sulfoxide	6.5.2
DNA	deoxyribonucleic acid	4.6.3
DOS	density of states	8.11.2
DPN	dip-pen nanolithography	5.3.7
DPV	differential pulse voltammetry	13.1.5
DSA	dimensionally stable anode	17.1.5
DSC	dye-sensitized solar cell	10.3.2
dsDNA	double strand deoxyribonucleic acid	18.3.7
EBSD	electron backscatter diffraction	8.13.3
EC	ethylene carbonate	15.6.3
ECM	electrochemical machining	14.6.2
ECS	The Electrochemical Society	1.4
ECSA	electrochemically active surface area	12.4.1
ECSFM	electrochemical scanning force microscopy	8.11.1

ECSTM	electrochemical scanning tunneling microscopy	8.11.1
EDLC	electrical double layer capacitors	15.9
EDTA	ethylenediaminetetraacetic acid	5.3.12
EDX	energy-dispersive X-ray analysis	8.5.1
EELS	electron energy loss spectroscopy	8.2.4
EIS	electrochemical impedance spectroscopy	7.
EMD	electrolytic manganese dioxide	15.4
ENA	electrochemical noise analysis	14.5.3
EQCM	electrochemical quartz crystal microbalance	8.8
EOF	electroosmotic flow	4.6.3
ESC	electrolyte-supported cell	16.4
EPR	electron paramagnetic resonance	6.6.3
ET	electron transfer	6
ETC	electron transfer chain	6.4.4
EWG	electron-withdrawing group	17.3.3
EXAFS	extended X-ray absorption fine structure	8.5.1
Fab	fragment of antigen binding	18.3.6
FAD	flavin adenine dinucleotide	18.2.1
FB	feedback mode (in SECM)	8.12.2
FC	fuel cell	16.1
Fc/Fc^+	ferrocene/ferrocenium	2.4.1
FEC	fluoroethylene carbonate	15.6.3
Fd	ferredoxin	18.1.3
FEM	finite element method	8.12.2
FFT	fast Fourier transform	6.5.6
FGT	functional group transformation	17.1.1
FMN	flavin mononucleotide	18.2.1
FR	Fermi resonance	8.2.1
FRA	frequency response analyzer	7.2.1
FSCV	fast scan cyclic voltammetry	5.6.5
FT	Fourier transform	6.5.6
FTO	fluorine-doped tin oxide	6.6.3
FWHM	full width at half maximum	5.5.1
g	gaseous	1.2
Gal	galactosidase	18.3.6
GC	generation-collection mode in SECM	8.12.4
GCE	glassy carbon electrode	5.3.2
GDE	gas diffusion electrode	15.7
GDH	glucose dehydrogenase	18.3.3
GDL	gas diffusion layer	15.1
GIC	graphite intercalation compound	15.6.3
GIXRD	grazing incidence X-ray diffraction	8.6
GMR	giant magnetoresistance	9.5.3
GOx	glucose oxidase	18.2.5
HAADF-STEM	high angular annular dark field scanning transmission electron microscopy	12.5.1

hcd	high current density	9.3.3
HER	hydrogen evolution reaction	3.2.3
HMDE	hanging mercury drop electrode	5.3.6
HOMO	highest occupied molecular orbital	10.1.2
HOPG	highly ordered pyrolytic graphite	8.13.5
HPLC	high-performance liquid chromatography	5.4
HREELS	High-resolution electron energy loss spectroscopy	8.2
HRP	horseradish peroxidase	18.3.6
HSAB	hard and soft acids and bases	9.4.3
HT PEMFC	high-temperature proton exchange membrane fuel cell	2.4.4
ICE	internal combustion engine	5.2.8
ICP	intrinsically conducting polymer	13.2.4
ICP-MS	inductively coupled plasma mass spectrometry	12.4.5
IEF	isoelectric focusing	4.6.2
IL TEM	identical location transmission electron microscopy	8.11.1
IMPS	intensity modulated photocurrent spectroscopy	10.2.6
IMVS	intensity modulated photovoltage spectroscopy	10.2.6
IR	infrared	6.6.3
IRRAS	infrared reflection absorption spectroscopy	8.2.1
ISE	International Society of Electrochemistry	6.
ISE	ion-selective electrode	5.2.2
ISS	ion scattering spectroscopy	8.6
ITO	indium tin oxide	6.6.3
IUPAC	International Union of Pure and Applied Electrochemistry	1.4
ITIES	interface between two immiscible electrolyte solutions	2.3.3
KPFM	Kelvin probe force microscopy	8.13.5
KPM	Kelvin probe microscopy	8.13.5
KKR	Kramers-Kronig relationship	7.2.1
l	liquid	1.2
LB	Langmuir-Blodgett (transfer)	13.1.7
LbL	layer-by-layer	13.2.1
LFM	lateral force microscopy	8.11.5
lcd	low current density	9.3.3
LIB	lithium-ion battery	15.3.1
LiBOB	lithium bis(oxalato)borate	15.6.3
LC	liquid condensed (phase)	13.1.7
LCM	lanthanum calcium manganite	16.4.2
LCO	lithium cobalt oxide	15.6.4
LE	liquid expanded (phase)	13.1.7
LEED	low energy electron diffraction	8.6
LEIS	localized electrochemical impedance spectroscopy	8.13.4
LFER	linear free energy relationship	12.3.2
LFM	lateral force microscopy	8.11.5
LFP	lithium ferrophosphate (lithium iron phosphate)	15.6.4
LHC	light harvesting complex	18.1.3
LHD	linear hydrodimer	17.3.3

LIB	lithium-ion battery	15.6
LiFSI	lithium (bisfluorosulfonyl)imide	15.6.3
LiTFSI	lithium bis(trifluoromethylsufonyl)imide	15.6.3
LIGA	from German *Lithographie – galvanische Abformung* (lithography – galvanic molding)	6.6.3
LMO	lithium manganese oxide	15.6.4
LNO	lithium nickel oxide	15.6.4
LOD	limit of detection	5.2.4
LS	Langmuir-Schaefer (transfer)	13.1.7
LSGM	$La_{1-x}Sr_xGa_{1-y}Mg_yO_{3-(x+y)/2}$	11.2.3
LSM	lanthanum strontium manganite	16.4.1
LSV	linear sweep voltammetry	5.3.5
LTO	Lithium titanium oxide	15.6.2
LUMO	lowest unoccupied molecular orbital	10.1.2
MCFC	molten carbonate fuel cell	2.4.4
MCP	microcontact printing	13.1.5
ME	microelectrode	8.12.1
MEA	membrane electrode assembly	2.4.4
Mem.	memory	5.7.1
MET	mediated electron transfer	18.2.3
MFE	mercury film electrode	5.3.10
MHCM	metal hexacyanometallate	13.3.2
MIC	microbiologically induced corrosion	14.5.5
MO	molecular orbital	10.1.2
MOF	metal organic framework (compounds)	13.3.2
MP	microprocessor	5.7.1
MPL	microporous layer in fuel cells	16.2.2
MQ	menaquinone	18.2.1
MS	mass spectrometry	12.4.5
MSC	metal-supported cell	16.4
μ-TAS	microstructured total analysis system	4.6.3
NaCMC	sodium carboxymethylcellulose	15.9.3
NAD^+	β-nicotinamide dinucleotide	18.2.1
NADH	reduced form of β-nicotinamide dinucleotide	18.2.1
$NADP^+$	β-nicotinamide adenine dinucleotide phosphate	18.2.1
NADPH	reduced form of β-nicotinamide adenine dinucleotide phosphate	18.2.1
NCA	nickel cobalt aluminum oxide	15.6.4
NCE	normal calomel electrode	2.4.1
NHE	normal hydrogen electrode	2.4.1
NHPI	*N*-hydroxypthalimide	17.4.2
Ni-MH	nickel metal hydride	15.5.3
NIR	near infrared (radiation)	6.6.3
NIST	National Institute of Standard and Technology, USA	5.2.3
NMC	nickel manganese cobalt oxide	15.6.4
NMP	N-methylpyrrolidone	15.3.1
NMR	nuclear magnetic resonance spectroscopy	6.6.2
NPOE	2-nitrophenyloctylether	5.2.6

NPV	normal pulse voltammetry	5.3.9
NTA	nitrilotriacetic acid	5.3.12
O	generic for oxidizing reagent of a redox couple	4.5.2
OA	operational amplifier	5.3.1
OCP	open circuit potential	3.2.7
OCV	open circuit voltage	10.3.2
ODC	oxygen depolarized electrode	17.1.5
ODE	ordinary differential equation	4.5.2
OER	oxygen evolution reaction	15.5.2
OMB	2,2′,3,3′,4,4′,5,5′-octamethylbipyrrole	6.5.7
OA	operational amplifier	5.7.1
ORR	oxygen reduction reaction	15.5.2
OTE	optically transparent electrode	6.6.3
OTTLE	optically transparent thin layer electrolysis (cell)	6.6.3
PAA	polyacrylic acid	13.2.1
PAD	pulsed amperometric detection	5.4.1
PAFC	phosphoric acid fuel cell	2.4.4
PANI	polyaniline	13.2.4
PAP	*p*-aminophenol	18.3.6
PAPG	*p*-aminophenyl-β-galactopyranoside	18.3.6
PBI	polybenzimidazol	16.3.2
PBS	phosphate buffer saline	5.3.3
PC	personnel computer	5.7.1
PC	propylene carbonate	15.6.3
Pc	plastocyanine	18.1.3
PCB	printed circuit board	9.4.3
PCR	polymerase chain reaction	4.6.3
PDDA	poly-diallyldimethyl-ammonium chloride	13.2.1
PDE	partial differential equation	4.5.1
PDMS	poly(dimethyl)siloxane	13.1.5
PE	poly(ethylene)	15.6.3
PECM	pulsed electrochemical machining	14.6.2
PEDOT	poly(diethoxythiophene)	13.2.4
PEEK	poly(etheretherketone)	5.3.4
PEG	poly(ethyleneglycol)	9.4.3
PEI	polyethyleneimine	13.2.1
PEMFC	polymer electrolyte membrane fuel cell or proton exchange membrane fuel cell	2.4.4
PFG-NMR	pulsed-field-gradient nuclear magnetic resonance	15.6.2
PGSE-NMR	pulse gradient spin echo nuclear magnetic resonance	6.6.2
PIGE	paraffine-impregnated graphite electrode	5.5.2
PINO	phthalimide-nitroxyl	17.4.2
PM IRRAS	polarization modulation infrared reflection absorption spectroscopy	8.2.1
PP	poly(propylene)	15.6.3
PPy	polypyrrole	13.2.4
PQ	plastoquinone	18.1.3

PQI	*p*-quinone imine	18.3.6
PQQ	pyrroloquinoline quinone	18.2.1
PS I, PS II	photosystem I, photosystem II	19.1.3
PSS	poly-styrene-sulfonate	13.2.1
PT	polythiophene	13.2.4
PTFE	poly(tetrafluoroethylene)	5.2.7
PVC	poly(vinylchloride)	5.2.6
PVDF	polyvinylidene difluoride	4.6.2
PVF	poly(vinyl ferrocene)	8.8.3
QRCE	combined quasi-reference and counter electrode	8.13.3
QRE	quasi-reference electrode	8.13.4
PZT	lead zirconium titanate	8.11.3
R	generic for reducing reagent of a redox couple	4.5.2
RAM	random access memory	11.2.3
RC	redox competition mode (of SECM)	8.12.5
RDE	rotating disk electrode	5.6.1
RDS	rate-determining step	3.2.6
RE	recognition element	18.3
Ref	reference electrode	5.3.1
RFB	redox flow battery	15.1
RHE	reversible hydrogen electrode	2.4.1
RNA	ribonucleic acid	18.3.7
RNS	reactive nitrogen species	5.6.5
ROS	reactive oxygen species	5.6.5
RPV	reverse pulse voltammetry	5.3.9
RRAM	resistive random access memory (RAM with switching of a cell resistance)	11.2.3
RRDE	rotating ring-disk electrode	12.4.2
RTIL	room-temperature ionic liquid	11.4.2
s	solid	1.2
sa	saturated	1.2
sol	solvated	1.2
SAM	self-assembled monolayer	13.1.1
SCE	saturated calomel electrode)	2.4.1
SCV	staircase-voltammetry	5.3.9
SDS	sodium dodecyl sulfate	4.6.2
SDS-PAGE	sodium dodecyl sulfate polyacrylamide gel electrophoresis	4.6.2
SECCM	scanning electrochemical cell microscopy	8.13.3
SECM	scanning electrochemical microscopy	5.6.4
SEI	solid electrolyte interphase	15.4
SEIRA	surface-enhanced infrared absorption	8.2.1
SEM	scanning electron microscopy	8.11.1
SERS	surface-enhance Raman spectroscopy	8.2.2
SFG	sum frequency generation	8.2.3
SFM	scanning force microscopy	8.11.4
SG/TC	sample-generation/tip-collection (mode in SECM)	8.12.4
SHE	standard hydrogen electrode	2.2.4

SHG	second harmonic generation	8.2.3
SICM	scanning ion-conductance microscopy	8.13.2
SIMS	secondary ion mass spectrometry	8.7
SI	surface interrogation mode of SECM	8.12.3
SLI	starter-lighter-ignition	15.5.1
SMDE	static mercury drop electrode	5.3.6
SNIFTIRS	subtractively normalized interfacial Fourier transform infrared spectroscopy	8.2.1
SoC	state of charge	15.2.3
SOFC	solid oxide fuel cell	2.4.4
SoH	state of health	15.2.4
SPECM	scanning photoelectrochemical microscopy	8.13.1
SPM	scanning probe microscopy	8.11.1
SPR	surface plasmon resonance	8.4
SPS	bis-3-sulfopropyl disulfide	9.4.3
SQWV	square wave-voltammetry	5.3.8
SRET	Scanning reference electrode technique	8.11.1
SSCE	saturated sodium chloride calomel electrode	2.4.1
ssDNA	single strand desoxyribonucleic acid	18.3.7
STM	scanning tunneling microscopy	4.3.3
STEM	scanning transmission electron microscopy	8.11.1
STS	scanning tunneling spectroscopy	8.11.2
SVET	scanning vibrating electrode technique	8.11.1
S&H	sample-and-hold amplifier	5.7.2
$TBAPF_6$	tetra-*n*-butylammonium hexafluorophosphate	5.3.2
TCO	transparent conducting oxide	6.6.3
TEAP	tetraethylammonium perchlorate	12.4.2
TEM	transmission electron microscopy	8.11.1
TEMPO	2,2,6,6-tetramethyl-piperidine-*N*-oxyl	17.4.2
Tf_2N	bis(trifluoromethylsulfonyl)imide	11.4.2
TFSI	bis(trifluoromethylsulfonyl)imide	11.4.2
TG/SC	tip-generation/sample collection (mode in SECM)	8.12.4
THF	tetrahydrofuran	6.5.3
TISAB	total ionic strength adjustment buffer	5.2.4
TMPD	N,N,N′,N′-tetramethylphenylenediamine	6.4.2
TSG	template-stripped gold	8.9.4
TSM	thickness shear mode	8.8.1
TSR	thermodynamically superfluous reaction	6.3.1
TSV	tip-substrate voltammetry	8.12.4
Tyr	tyrosinase	18.3.3
UA	urease	18.3.1
UHV	ultra-high vacuum	8.1
UPD	underpotential deposition	8.9.8
UPS	ultraviolet photoelectron spectroscopy	8.5.2
UQ	ubiquinone (coenzyme Q), subsequent number in subscript specifies the number of isoprenic units in the side chain (UQ_{10} = coenzyme Q_{10}).	18.1.2

USB	universal serial bus (in computer technology)	5.7.1
UV	ultraviolet (radiation)	6.6.3
VB	valence band (of semiconductors)	10.1.2
VC	vinylcarbonate	15.6.3
VEC	vinylethylene carbonate	15.6.3
VIA	vertical interconnect access	9.4.3
Vis	visible (radiation)	6.6.3
VRFB	vanadium redox flow battery	15.8
VRLA	valve-regulated lead acid cells	15.5.2
vs.	in comparison to or against (lat. *versus*)	2.4.1
WE	working electrode	5.3.1
YSZ	yttrium-stabilized zirconia	5.2.8
XANES	X-ray absorption near edge structure	8.5.1
XAS	X-ray absorption spectroscopy	8.5.1
XFA	X-ray fluorescence analysis	8.5.1
XPS	X-ray photoelectron spectroscopy	8.5.1
ZRA	zero-resistance ammeter	14.5.3
∅	diameter	4.5.2
□	free adsorption site on a surface	12.1.3

Fundamentals of Electrochemistry

Is Another Textbook on Electrochemistry Really Necessary?

When I grow up, I'm definitely not going to write a book,
certainly not voluntarily.

(Felix Wittstock, *2003)

Electrochemistry – Long Established, But Still Complicated?

As early as 1896, Wilhelm Ostwald published his textbook "*Electrochemistry – Its History and Teaching*" (translated title from German, Figure 0.1), in which he already referred to a long history at that time. For many decades, this textbook remained exemplary worldwide for teaching electrochemistry at universities with a focus on the thermodynamics of electrochemical reactions, electrolyte equilibria, and transport processes in solutions.

However, understanding modern applications of electrochemistry and even very simple experiments requires the *simultaneous* consideration of a whole range of complex concepts such as the kinetics of heterogeneous reactions, double layer, migration, diffusion, electrostatics, activation barriers, and catalysis, which cannot be reduced to thermodynamic consideration. What level should a course in electrochemistry start from? What basic topics need to be covered and in which sequence in order to make this subject accessible in a reasonable number of hours?

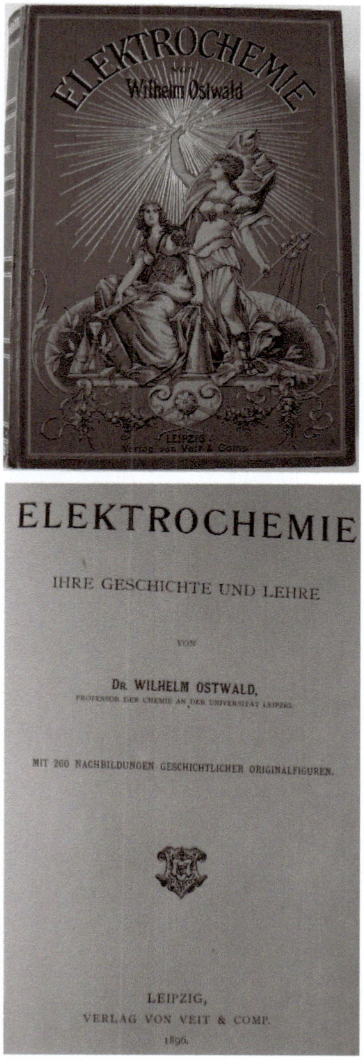

Figure 0.1 Cover and title page of the textbook by Wilhelm Ostwald [1], probably the first systematic presentation of electrochemistry in textbook format. Source: Reproduced with permission from Gerda and Klaus Tschira Stiftung, Heidelberg/Public domain.

A typical aspect of this dilemma is the concept of phase, which is central to thermodynamics and many introductory materials on electrochemistry. This is to some extent opposed to a molecular viewpoint that has become prevalent for the development of chemistry courses based on availability of exact structural data of molecules and solids from spectroscopy and diffraction experiments. This trend is spreading into many areas of biology and physics as well. As a result, today's students often struggle with the concept of (charged) phases, even though this approach has been followed by classic electrochemistry textbook over many decades.

In my experience, a systematic and consistently quantitative treatment of even the most important fundamentals of electrochemistry has a demotivating effect on a considerable fraction of students, as the supposed algebraic requirements risk covering up the core content of the course, especially at the beginning of a chemistry degree program. In contrast, an approach based primarily on the discussion of important current applications is likely to arouse spontaneous interest, but risks slipping into "infotainment" if a lack of time prevents the development of transferable basic knowledge and skills. However, I consider a solid foundation to be an essential prerequisite for the ability to carry out independent research and further self-guided training later on. The specific subject matter of such activities is naturally difficult to foresee during the first semesters at the university.

Bearing this difficulty in mind, many current textbooks have dedicated themselves to only one or a few areas of electrochemistry. It is, however, noticeable that almost all of these texts and even didactically structured review papers in scientific journals cannot achieve their goal without a renewed introduction to the basics, which often has to be extremely compact and without any real foundation. These, sometimes redundant introductions, also seem to be necessary in order to define which of the many competing terminologies and conventions the authors want to follow in their explanations.

Consequently, each new textbook on electrochemistry is an experiment to explore the sequence, in which basic concepts (also from other scientific disciplines) are covered and connected to the central concepts and experiments in electrochemistry, connected to the historically established (but currently not so popular) thermodynamic perspective, and connected to balance the coverage of basics and applications. Finally, a textbook has to guide the reader through the competing and simultaneously used conventions such as the sign of an electrolysis current, forming a distractive barrier for many students.

What Can You Expect from the Subject Area and This Book?

Beyond the specific communication problems, electrochemistry has been and remains an essential cross-sectional discipline in many areas of the economy and in very different fields of research. The attention paid to batteries in electric cars and several Nobel Prizes for electrochemical research are just a few indications of this. The decarbonization of the economy will not succeed without new contributions from electrochemistry, and personalized medicine at affordable prices will only be possible with electrochemical biosensors and assays. Although there are excellent texts on all these subjects, they cannot with sufficient clarity emphasize the relationships between the various subfields of electrochemistry and their common foundations. Furthermore, the rising importance of functional materials in electrochemistry has often not been systematically covered in textbook literature.

In this book, I aim for presenting electrochemistry as *one* unified field with many facets and interdisciplinary aspects, both in terms of fundamentals and applications. My favorite student Thomas [the Doubting] will help you and me in this endeavor by not being afraid to ask the supposedly simple questions.

Chapters 1–4 present essential concepts that are oriented toward the needs of students in a Bachelor's program or for students studying electrochemistry as a minor subject. We will start from simple experimental results and first anchor qualitative ideas. Central to this are graphic key schematics, which summarize essential concepts in a memorizable form. This is followed by the development of a detailed quantitative description of the facts, often using a limited number of examples. For this, we provide explanation of the used symbols with their typical units in the margin, for example, "I – Current [A]."

In Chapters 5–18, we will revisit most of the topics, but now deal in greater detail and depth with understanding important special cases in close relation to specific applications. Students should be enabled to use a consistent foundation of knowledge as a starting point for studying monographs and specialized review articles and to integrate the new knowledge into an understanding of the system – a level of proficiency that is suitable for the specialization phases at the end of a Bachelor's degree course or the beginning of a PhD program. It is also the level of Master's programs in those countries where the completion of a research master is a precondition for entry into a PhD program. Ideally, study of this book is accompanied by laboratory exercises in both the basic and advanced sections.

Lecturers can also develop application-oriented courses with very different emphases using just a selection of the chapters or sections. For this purpose, Chapters 1–4 can be combined with more advanced chapters or sections, for example,

- Batteries: Chapters 1–4, 7, 9, 15
- Electrocatalysis and fuel cells: Chapters 1–4, 7, 8, 11, 12, 16
- Electrochemical energy conversion: Chapters 1–4, 10, 11, 12, 15, 16
- Corrosion and surface treatment: Chapters 1–4, 7, 8, 9, 14
- Electrochemical analysis methods and biosensors: Chapters 1–4, 5, (7), 13, 18
- Bioelectrochemistry: Chapters 1–4, 5, 18
- Molecular electrochemistry: Chapters 1–4, 5, 6, 13, 17, (18)
- Electrochemical surface chemistry: Chapters 1–4, 8, 9, 12, (13)
- Electrochemical production processes: Chapters 1–4, 9, 17

In those course examples, other chapters may serve as a reference for looping around and refreshing specific aspects. Many cross-references between the chapters are intended to show interested students the systematic connections between the subfields of electrochemistry, perhaps even enticing them to read on without burdening other readers with an overloaded text. Such an approach cannot avoid making simplifications that bear the risk of causing misconceptions that hinder a deeper understanding of the interrelationships at a later time. Therefore, we will explicitly address pitfalls and common misconceptions and provide references to recommended further reading throughout the book. Some particularly interesting further aspects are presented in Excursions, which may be of interest to selected readers only.

The selection of specific content and applications is of course also subjected to preferences and abilities of the textbook author and the contributors. We strive for a balanced coverage of concepts, experimental methods, and important applications. Especially in the introductory chapters, we start from a molecular approach that is more familiar to chemistry students from other courses than the thermodynamic phase concept. The innovations of recent decades, such as the widespread use of personal computers to control, evaluate and simulate experiments, the use of single-crystal electrodes and microelectrodes, high-resolution microscopy data from electrode surfaces and the importance of functional materials, are already integrated into the introductory chapters. The limitations of an already rather extensive book require references to further specialized literature instead of covering these topics within the book. Unfortunately, the long history of electrochemistry cannot be covered in this book although its many anecdotes and the diverse interrelationships with neighboring disciplines such as physics, electrical engineering, and electrophysiology would provide ample material. Only where names of persons have become part of technical terms (e.g., Faraday constant), will you find at least their biographical data to place discoveries into a historical context. Otherwise, attribution of scientific discoveries to persons was omitted, as a selection would be mandatory and, in any case, would be subjective and thus controversial.

Precisely because electrochemistry has a long history as an established discipline, we want to explicitly point out unsolved, sometimes urgent research questions and their relevance for applications. These are questions that structure knowledge, initiate communication among students or scientists, bring the discipline to life, and thus, point toward the future. This applies both to science as a whole and to one's own learning progress. Therefore, the book is intended to promote a culture of questioning, even if one or two questions have to remain unanswered.

With this in mind, I wish you many Eureka moments, exciting discussions, and always new questions!

Gunther Wittstock

Historical and Current Textbooks

Like every textbook, this one builds on predecessors. Here, important mostly English textbooks are listed [2–29] focusing on electrochemistry as a whole or at least on a large part of it. Some German titles have been included as they and literature derived from them played a role in the formation of the textbook author and contributors [7]. An extensive list of early books on electrochemistry including textbooks and preuniversity educational materials has been compiled on the website of the Electrochemical Society under https://knowledge.electrochem.org/estir/old-books.htm (last access 04 Jan 2024).

References

1 W. Ostwald; *Elektrochemie: Ihre Geschichte und Lehre*, Verl. von Veit, Leipzig, 1896.

2 G. Kortüm, J. O'M. Bockris; *Textbook of Electrochemistry, Vol. 1 and 2*, Elsevier, New York, NY, 1951.

3 G. Kortüm; *Treatise on Electrochemistry*. 2nd ed., American Elsevier Publishing Co., New York, NY, 1965.

4 G. Milazzo; *Electrochemistry: Theoretical Principles and Applications*, Elsevier Publishing Co., New York, NY, 1965.

5 J. O'M. Bockris, A. K. N. Reddy; *Modern Electrochemistry, Vol. 1: Ionics*. 1st ed., Plenum Press, New York, 1970, 2nd ed. 1998; J. O'M. Bockris, A. K. N. Reddy, M. Gamboa-Aldeco; *Modern Electrochemistry: Vol. 2A Fundamentals of Electrodics*, Plenum Press, New York, 2000; J. O'M. Bockris, A. K. N. Reddy; *Modern Electrochemistry: Vol. 2B Electrodics in Chemistry, Engineering, Biology and Environmental Science*. Plenum Press, New York, 2000.

6 I. Friend; *The Chemistry of Electrode Processes*, Academic Press Inc., London, 1973.

7 G. Ackermann, W. Jugelt, H. D. Suschke, G. Werner; *Lehrwerk Chemie, Lehrbuch 5 – Elektrolytgleichgewichte und Elektrochemie*. 4th ed., VEB Deutscher Verlag für Grundstoffindustrie, Leipzig, 1974; and E.-G. Jäger, K. Schöne; *Lehrwerk Chemie, Arbeitsbuch 5 – Elektrolytgleichgewichte und Elektrochemie*. 3rd ed., VEB Deutscher Verlag für Grundstoffindustrie, Leipzig, 1977.

8 F. Hine; *Electrode Processes and Electrochemical Engineering*, Plenum Press, New York, NY, 1985.

9 J. Koryta, J. Dvořák, L. Kavan; *Principles of Electrochemistry*. 2nd ed., John Wiley & Sons, Chichester, 1993.

10 D. Pletcher, F. C. Walsh; *Industrial Electrochemistry*. 2nd ed., Springer, Dordrecht, 1993.

11 P. H. Rieger; *Electrochemistry*. 2nd ed., Springer Science + Media, Dordrecht, 1994.

12 C. M. A. Brett, A. M. O. Brett; *Electrochemistry: Principles, Methods, and Applications*, Oxford Science Publications, 1993.

13 P. T. Kissinger, W. R. Heineman; *Laboratory Techniques in Electroanalytical Chemistry*. 2nd ed., Marcel Dekker, New York, 1996.

14 K. B. Oldham, J. C. Myland; *Fundamentals of Electrochemical Science*, Academic Press, San Diego, 1998.

15 Southampton Electrochemistry Group; *Instrumental Methods in Electrochemistry*, Howard Publisher, Chichester, 2001.

16 J. Newman, K. E. Thomas-Alyea; *Electrochemical Systems*. 3rd ed., John Wiley & Sons, Hoboken, NJ, 2004.

17 H. H. Girault; *Analytical and Physical Electrochemistry*, EPFL Press, distributed by Marcel Dekker, New York, 2004.

18 V. S. Bagotsky; *Fundamentals of Electrochemistry*. 2nd ed., Wiley-Interscience, Hoboken, NJ, 2006.
19 C. H. Hamann, A. Hamnett, W. Vielstich; *Electrochemistry*. 2nd ed., Wiley-VCH, Weinheim, 2007.
20 C. G. Zoski; *Handbook of Electrochemistry*, Elsevier, Amsterdam, 2007.
21 D. Pletcher; *A First Course in Electrode Processes*. 2nd ed., RSC Publishing, Cambridge, 2009.
22 C. Lefrou, P. Fabry, J.-C. Poignet; *Electrochemistry – The Basics, With Examples*, Springer, Heidelberg, 2012.
23 K. B. Oldham, J. C. Myland, A. M. Bond; *Electrochemical Science and Technology: Fundamentals and Application*, Wiley, Chichester, 2012.
24 R. Holze; *Experimental Electrochemistry – A Laboratory Textbook*. 2nd ed., Wiley-VCH, Weinheim, 2019.
25 N. Eliaz, E. Gileadi; *Physical Electrochemistry: Fundamentals, Techniques and Applications*. 2nd ed., Wiley-VCH, Weinheim, 2019.
26 H. Baumgärtel; *Electrochemistry: A Guide for Newcomers*, de Gruyter, Berlin, 2020.
27 K. Jackowska, P. Krysiński; *Applied Electrochemistry*, de Gruyter, Berlin, 2020.
28 S. Petrovic; *Electrochemistry Crash Course for Engineers*, Springer, Cham, 2021.
29 A. J. Bard, L. R. Faulkner, H. S. White; *Electrochemical Methods: Fundamentals and Applications*. 3rd ed., John Wiley & Sons, New York, 2022.

Acknowledgments

I have been working on the German edition of this book since 2009 and would probably never have completed it without the support of Prof. Dr. Bernd Speiser, Eberhard-Karls-Universität Tübingen, and Dr. Julia Witt, then at the Carl von Ossietzky Universität Oldenburg, now at the Federal Institute for Materials Research and Testing. Chapter 6 by Bernd Speiser and the illustrations by Julia Witt (see below) are important building blocks of the book.

Apart from a number of factual corrections and few amendments (Sections 1.1.1, 5.5.2, 5.5.3, and 15.9), the textbook is a translation of the German original published in 2023. The translation was completed by Prof. Dr. Frank Marken (University of Bath, UK), Prof. William R. Heineman (University of Cincinnati, OH, USA), and myself with a contribution by B. Speiser to Chapter 6 between 2022 and 2024. For me personally, the exchange on all aspects of the manuscript with the contributors to the German and English editions was a stimulus, inspiration, and corrective, for which I am extremely grateful.

I would also like to thank many colleagues as well as current and former members of my research group for reviewing individual chapters, discussing technical issues, sending relevant literature, providing data, illustrations, and experimental instructions (listed below), as well as for their benevolent and supportive inquiries as to when the book would finally be finished. In particular, I would like to thank Prof. Dr. Ralf Rabus for his review of Excursion 4-4, Dr. Constanze Donner and Prof. Dr. Andreas Bund for their review of Chapter 9, Prof. Dr. Derck Schlettwein and Dr. Raffael Ruess for their comments on Chapter 10, Prof. Dr. Gerhard Hilt, Prof. Dr. Bernd Speiser and Dr. Robert Francke for their comments on and review of Chapter 17, and Dr. Ina Schmidt for her review of Chapters 1–4 and 9, especially from the perspective of students.

I would like to thank the publisher for their patience in waiting for the manuscript, their tolerance in terms of scope and their professionalism in converting the manuscript into the various publication formats.

Finally, my thanks go to my wife Marion and my children Alena and Felix, who often only saw me at my desk in the evenings and on weekends, but still were supportive.

Sources of Figures

I would like to thank the following persons and institutions (in alphabetical order) for illustrations or data for illustrations.

Prof. Dr. Helmut Baltruschat for the dataset for Figure 12.34.

Prof. Dr. Lothar Beyer, Mrs. Susanne Dietel, and Mrs. Katy Reimelt for their support in the search for W. Ostwald's textbook in good condition (Figure 0.1), which could finally be reproduced with the support of the Gerda and Klaus Tschira Stiftung. The foundation maintains the Wilhelm Ostwald Park in Großbothen, Ostwald's residence, where the book has been preserved over the decades.

Dr. Izabella Brand for spectra for Figure 8.4 and data for Figures 5.37 and 5.47
Dr. Heinz Bülter for templates for Figures 15.16b, 15.32, 15.33, and 15.35
Ritex GmbH, Bielefeld, Germany for Figure 4.24
Dr. Anita Cymann-Sachajdak for data for Figures 15.46 and Exercise 15.26
Dr. Praktik Das for data of Figures 15.2, 15.11, 15.12, and 15.16c
Prof. Dr. Christian Degrigny for Figure 14.23
Dr. Wilke Dononelli for Figures 8.54, 15.37–15.39
Dr. Carsten Dosche for Figures 8.22, 8.23, and 8.28
Company Metrohm DropSens, Llanera (Asturias), Spain for Figure 5.25
Dr. Johann Wilhelm Erning for Figure 14.26
Dr. Julian Lorenz for data of Figures 5.41, 12.19, 12.22, 12.24, 12.25, and 12.28
Dr. Victor Climent and Prof. Juan Feliu for Figure 8.61
Prof. Dr. Jürgen Heinze for Figure 13.47
Dr.-Ing. Jacob Jörissen† for Figures 17.5 and 17.7
Dr. Ludwig Kibler for Figures 8.55, 8.60, and 8.62
Dr. Lidiya Komsiyska for Figure 2.31d,e and data for Figures 15.13, 15.17, 15.29, and 15.41b
Prof. Dr. Robert Francke for Figure 17.36
Prof. Dr. Christophe Léger for the graphic file for Figure 18.26
Prof. Dr. Andreas Lesch for templates for Figures 14.43–14.46
Dr. Eva Mutoro for the color scheme, the pictograms for Thomas, Pitfall, and Key scheme and first drafts for Figures 1.1 to 1.6, 2.4 to 2.7, 2.12 to 2.16, 2.28, 2.31, 3.1b and the electrical double layer
Company PalmSens BV, Randhoeve, the Netherlands for Figure 5.90c
Prof. Dr. Ralf Rabus for Figure 4.61e
Prof. Dr. Richard Hanke-Rauschenbach for templates for Figures 16.17 and 16.18
Prof. Dr. Frank Uwe Renner for Figures 14.34 and 14.35
Nico Kuhlmann for data for Figures 5.37, 5.47, 5.68, and 14.20
Andreas Schank for Figures 6.51b and 6.52
Dr. Simon Schundelmeier for the data of Figures 6.2 and 6.8 and Tables 6.5 and 6.6 as well as for templates for most of the figures in Chapter 6
Prof. Dr. Wolfgang Schuhmann for Figures 18.35 and 18.36
Prof. Dr. Slawomir Sek for Figure 8.106
Prof. Dr. Peter Strasser for Figure 12.36
Prof. Dr. Jörg Weissmüller and Dr. Matthias Graf for Figure 14.33
Dr. Julia Witt for image editing, polishing existing illustrations and the index for Chapters 1–5, 14–16, as well as for the design and layout of Figures 2.10, 2.17, 2.19 to 2.23, 2.25 to 2.27, 2.32, 3.1a, 3.2 to 3.6, 3.9, Table 3.1, Figures 4.1, 4.3, 4.4, 4.6 to 4.13, 4.15 to 4.20, 4.22, 4.23, 4.25, 4.29 to 4.50, 4.56, 4.62 to 4.69, 4.71, 5.2 to 5.7, 5.10 to 5.14, 5.18, 5.19, 5.22, 5.24, 5.26 to 5.33, 5.48 to 5.58, 5.64 to 5.67, 5.82 to 5.85, 5.88, 14.3, 14.4, 14.6 to 14.11, 14.13, 14.15 to 14.17, 14.30, 15.3 to 15.10, 15.14, 15.15, 15.19, 15.20, 15.30, 15.34, 16.1, 16.2, 16.4, 16.7 to 16.14
Illustrations taken from the literature are acknowledged in the text.

Part I

Principles

1

The Electrode Reaction – The Core of Electrochemistry

In this chapter, we will compare electrochemical reactions with other chemical reactions in order to work out their special features. We will become familiar with basic terminology. This is followed by a discussion of the relationship between the amount of charge transferred in a process and the amount of product formed. We will identify the importance of agreed-upon conventions for the first time in this book. Conventions cannot be derived from experiments or basic principles of nature, but need to be agreed upon in a sensible way. There are cases of different committees reaching different agreements at different times and in different locations, which then led to ineffective communication, as we will see in the development of electrochemistry. The history of electrochemistry is very interesting and, in many ways, exemplary of the history of science. For reasons of brevity, we cannot unfold all aspects of the subject. We will, however, provide biographical footnotes when names have become part of technical terms, which will enable the reader to see the historical context.

Further Reading. A concise description of the conventions in electrochemistry, the set of algebraic symbols, and SI units is provided in a book by the International Union of Pure and Applied Electrochemistry (IUPAC, available also via internet) [1, 2]. The historical aspects of electrochemistry have been covered in some seminal books and articles [3–6]. A current reference source [7], a textbook [8], and Wikipedia [9] frequently provide further historical and biographical accounts of major scientific discoveries.

1.1 Electrical Current and Chemical Conversion in Electrochemical Cells

1.1.1 Charge, Current, and Voltage

Charge (or more precisely electrical charge) Q (coulomb [C][1]) is a property that describes how objects experience a force in an electromagnetic field. Charge (very much like mass) is one of the fundamental properties of subatomic particles such as electrons or protons. The charge of an electron is negative (by convention), and the charge of a proton is positive. Each electron and each proton have the same charge regardless of which state they are (e.g., bound in an atom or free). The charge of a neutron is zero; it is neither positively nor negatively charged but neutral, thus the name "neutron".

The charge of larger objects is calculated as the charge of all its protons minus the charge of all its electrons. The charge of a proton is constant, called the elementary

[1] Charles Augustin de Coulomb, 1736–1806, French physicist, founded the area of electrostatics.

charge $e \approx +1.602 \times 10^{-19}$ C; the charge of the electron is opposite $e^- \approx -1.602 \times 10^{-19}$ C. Thus, the charge of any body composed of N protons and M electrons must be an integral multiple of the elementary charge e, exactly $(N - M)e$. We see that the charge of an object can only assume certain discrete values. For instance, a charge of $\frac{3}{4}\,e$ is impossible. Such quantities are called quantized.

Charge is a conserved quantity, which means the charge of an isolated system cannot change. In another phrasing, the charge of a system can only change if charge is transferred across the system boundary. How this transfer happens is one of the central questions in electrochemistry.

Objects with positive and negative overall charge attract each other and objects with charges of equal sign (either both positive or both negative) repel each other. The force resulting from this interaction is called the coulombic[1] force.

We can see from the consideration above that charge cannot exist apart from material objects. We can also say some objects "carry a charge". This phrasing is the origin of the term "charge carrier" frequently used in this book. Of course, electrons and protons are charge carriers, but this term also includes more complex objects that we frequently encounter in chemistry and physics, such as positive and negative ions (Na^+, Cl^-) including complex molecular ions (SO_4^{2-}, acetate CH_3COO^-) or so-called quasi-particles. Quasi-particles are a common concept in physics that ascribes single particle properties (such a mass and charge) to states of matter that are much more complex than a physical particle. A typical example is a "hole" in a semiconductor. In the context of this book this refers to a semiconductor, which has one electron less than it has protons. In the ground state, this electron deficiency is localized in a binding orbital of one atom. However, it is possible that an electron from a similar orbital of a neighboring atoms moves to this initially vacant state. Thus, the vacant state, with it the electron deficiency, has moved to the neighboring atom. Instead of thinking of a deficiency that changes place, it is efficient to consider the deficient state as a mobile particle of the charge $+e$ in the semiconductor. We will make extensive use of this concept in Chapter 10.

An electric field is formed around electrically charged objects. If a charged object, for example, charge carriers move, they cause a flow of charge, called **(electric) current** (symbol I, unit ampere[2] [A]). Moving charges produce a magnetic field.

In addition to charge and current, we will use another quantity called **voltage** U [V]. It is the difference between two electrical potentials at location $\mathbf{r}_1$ and location $\mathbf{r}_2$. [In this book, bold upright symbols stand for a vector quantity. Here it is the location $\mathbf{r} = \vec{r} = \{x, y, z\}$.] This potential difference is related to the work per charge W/Q to move a small probe charge Q in the electric field between the locations $\mathbf{r}_1$ and $\mathbf{r}_2$.

Hydraulic models are sometimes used to illustrate the relation between the electrical quantities by means of a strongly simplified analogy that is more accessible to visual experience (Figure 1.1). The electric charge is represented by the liquid volume in the hydraulic model. With this definition, the electric current corresponds to the volume flow rate (Table 1.1). The voltage is a "tension" under which the charge is stored with respect to a second reservoir. The mechanical equivalent is the hydrostatic pressure difference of two reservoirs. After defining the relations in this way, further a relation can be derived for the stored energy W and the released power P of an ideal turbine and an ideal motor.

[2]André-Marie Ampère, 1775–1836, French physicist and mathematician, founder of electrodynamics, clarified the meaning of voltage and current.

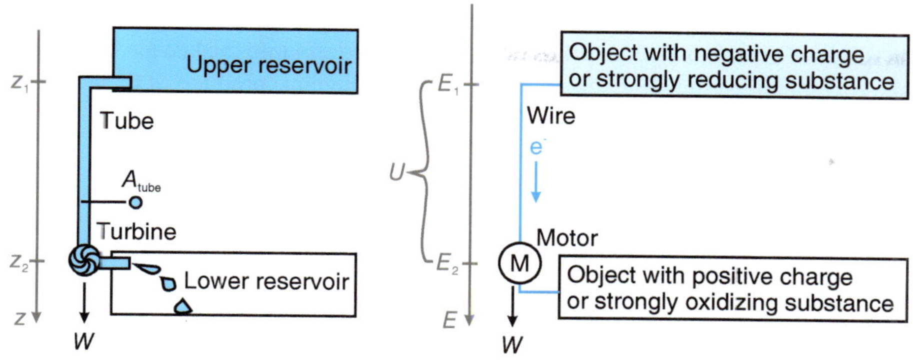

Figure 1.1 Analogy between mechanical energy stored in a water reservoir (left) and electrical energy obtained by moving charge between reservoirs of different electric potentials (right).

Table 1.1 Equation describing equivalent quantities in a hydropower station and an electrochemical cell. The equations are given with absolute values to avoid complication with sign definitions.

Quantity	Hydrostatic model	Electrochemical cell
Stored energy (or work required to fill the storage device) [W s] = [Nm]	$\vert W_{mech}\vert = mg\vert\Delta z\vert = V\rho g\vert\Delta z\vert$ $= Vp$ $= Pt$	$\vert W_{el}\vert = Pt$ $= UQ$ $= UIt$
Power [W]	$P = \mathrm{d}W/\mathrm{d}t$ $= \rho g\Delta z\frac{\mathrm{d}V}{\mathrm{d}t} = \rho g\Delta z\dot{V}$ $= p\dot{V}$	$P = \mathrm{d}W/\mathrm{d}t$ $= U\frac{\mathrm{d}Q}{\mathrm{d}t}$ $= UI$
"Tension" [Pa] or [V]	Hydrostatic pressure $p = F/A$ $= \frac{mg}{A_{tube}} = \frac{A_{tube}\rho gz}{A_{tube}}$ $= \rho gz$ Pressure difference $\Delta p = p_2 - p_1$ $= \rho g(z_2 - z_1)$ $= \rho g\Delta z$	Electric potential Voltage (potential difference) $U = \frac{\int_{r_1}^{r_2} F(Q)\,\mathrm{d}r}{Q} = E_2 - E_1$
Balanced quantity	Volume V	Charge Q
Current	Volume flow rate $\dot{V} = \mathrm{d}V/\mathrm{d}t$	Electric current ("charge flow rate") $I = \dot{Q} = \mathrm{d}Q/\mathrm{d}t$

Symbols: W – work/energy [W s] = [J] = [Nm], P – power [W], m – mass [kg], V – volume of the liquid [m^3], $\dot{V} = \mathrm{d}V/\mathrm{d}t$ – volume flow rate [m^3 s^{-1}], ρ – density of the liquid, g – local gravitational field of Earth [9.81 m^2 s^{-1}], p – pressure [Pa] = [N m^{-2}], $\Delta z = z_2 - z_1$ height difference between the reservoirs [m], Q – charge [C] = [A s], t = time [s], U – voltage [V].

1.1.2 Electrochemical Cells Convert Charge and Substances

Let us imagine the following experimental setup: a beaker contains two metal sheets of an inert metal, for instance, platinum (Figure 1.2). The metal sheets are connected to a direct current (DC) voltage source via a voltage divider that allows applying a fraction of the maximum output voltage (e.g., 50 V) to the two metal sheets. A current measuring instrument (called ampere meter) displays the resulting current. This could be a

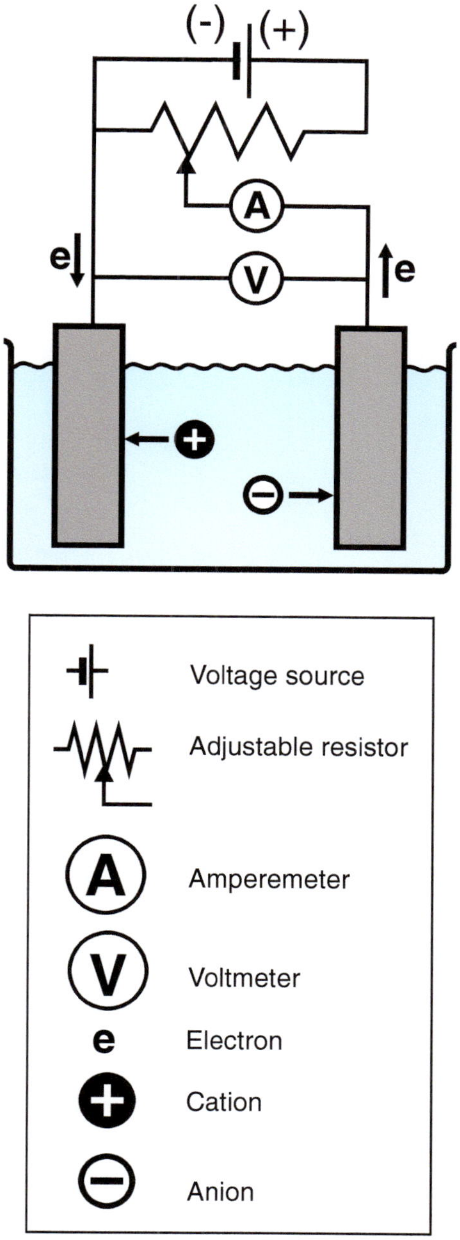

Figure 1.2 Basic experimental setup for an electrolysis cell.

multimeter or an electric light bulb connected in series. The metal sheets are immersed in different materials such as solids or liquids that we add into the beaker. We observe the current and at the same time the generation of chemical products at the interfaces of the metal sheets and the surrounding materials.

> We call the experimental setup with two metal sheets connected via an ion-conductor an **electrochemical cell**.

If we manipulate the metal sheets to touch a dry crystal of NaCl at both ends, we will not observe a current (a flow of charge). A similar observation is made when pouring clean water (e.g., deionized or doubly distilled water) into the beaker. However, if we dissolve the solid crystalline NaCl into the water, a current is observed as electrons start flowing in the wire. At the same time, gas bubbles rise from the two metal sheets to the surface of the liquid. Hydrogen gas is formed at the sheet connected to the negative pole of the voltage source. We could trap the gas in a test tube for analysis. At the other metal sheet, mainly chlorine gas is formed, revealed by its characteristic smell. *Caution, this electrolysis should be carried out for a very short time only because chlorine is toxic and hydrogen forms explosive gas mixtures with oxygen from air and with chlorine.*

Conceptually, this type of experimentation can be expanded by placing molten NaCl within a special crucible, employing diluted hydrochloric acid (HCl), or using aqueous sodium sulfate solution (Na_2SO_4). Electrochemical reactions in molten salts (Section 11.4.1) are often performed in salt mixtures in order to lower the melting temperature. Carbon rods are employed as current feeder electrodes because they are relatively stable against corrosion and show little passivation. Table 1.2 summarizes some of the observations made for the different experiments.

From this simple sequence of experiments, we can derive the following condition for a current in an electrochemical cell:

1) Current occurs only if there is a sufficient number of mobile charge carriers between the metal sheets to generate a measurable flow of charge. The aqueous solution of NaCl contains mobile Na^+ and Cl^- ions. It is therefore an ion-conductor. This is not the case for solid NaCl or pure H_2O.
2) A DC current is always coupled to chemical conversions at *both* metal sheets because the medium between the sheets is chemically converted. This process is called **electrolysis** (decomposition by electrical current).

   Let us consider these chemical conversions in more detail: Cl_2 is formed by oxidation of Cl^-, and H_2 is formed by reduction of H^+ in experiments (3) and (5) in Table 1.2.

$$2\,Cl^- \rightarrow Cl_2 + 2\,e^- \quad \text{(oxidation)} \tag{1.1}$$

$$2\,H^+ + 2\,e^- \rightarrow H_2 \quad \text{(reduction)} \tag{1.2}$$

Table 1.2 Reaction products for electrolysis in different electrolytes.

No.	Substance between metal sheets	Current?	Product at (+)	Product at (−)
1	H_2O pure	No	—	—
2	NaCl (solid)	No	—	—
3	NaCl in solution	Yes	Cl_2	H_2
4	NaCl melt	Yes	Cl_2	Na
5	HCl in solution	Yes	Cl_2	H_2
6	Na_2SO_4 in solution	Yes	O_2	H_2

Chlorine Cl_2 is formed by oxidation, and Na metal is formed by reduction in experiment (4).

$$2\,Cl^- \rightarrow Cl_2 + 2\,e^- \quad \text{(oxidation)} \tag{1.3}$$

$$2\,Na^+ + 2\,e^- \rightarrow 2\,Na \quad \text{(reduction)} \tag{1.4}$$

In experiment (6), the formation of O_2 (by oxidation) and of H_2 (by reduction) is observed.

$$2\,H_2O \rightarrow O_2 + 4\,H^+ + 4\,e^- \quad \text{(oxidation)} \tag{1.5}$$

$$4\,H_2O + 4\,e^- \rightarrow 2\,H_2 + 4\,OH^- \quad \text{(reduktion)} \tag{1.6}$$

3) Electrochemical cell reactions always comprise an **oxidation** reaction and a **reduction** reaction that occur at the *same time*, but in *different locations*.

The reaction that occurs on one of the current feeders (e.g., the Pt sheets) is called an **electrode reaction**.

We recognize that the mobile ionic species within the solution between the two metal sheets may be directly involved in the electrode reaction (example 4), but that this is not always the case (example 6). From these conditions for current flow through an electrochemical cell, two central questions for electrochemistry arise:

1) How does the **charge transport** proceed in detail in the ion-conducting medium (solution, membrane, ion-conducting solid)? (*ionics*)
2) How does the **charge transfer** occur across the interface between the metal sheet and the ion-conducting medium? (*electrodicss*)

1.2 Electrodes and Ions

In the nineteenth century, Faraday[3] realized that the interface between the current feeder and the electrolyte solution is the location at which the oxidation and reduction reactions proceed. He suggested calling this special interface the "electrode" and to use this terminology instead of the older term "pole". A contemporary phrasing of this principle reads:

An **electrode** is a phase boundary between *two electrically conductive phases* of different structure, at which electrochemical equilibria are attained or approached by *chemical conversions*. At least *one phase must be an ion-conductor*.

At least two electrodes are required to form an **electrochemical cell**. There are many different types of both electrodes and cells. Faraday was far ahead of his time. He identified the interfacial reaction as the unifying principle of all electrodes. This is all the more astonishing as several more decades had to pass until elemental particles and ions were finally discovered. Therefore, it is perhaps not so surprising that a colloquial meaning of the word "electrode" is still in use in addition to the exact scientific definition. If you ask your fellow students during the preparation of an experiment, according to Figure 1.2: "Could you please hand me the Pt electrode", she or he will give you the Pt sheet with an attached wire and a possibly existing casing of glass without any hesitation (Figure 1.3). The situation is indeed even more complicated

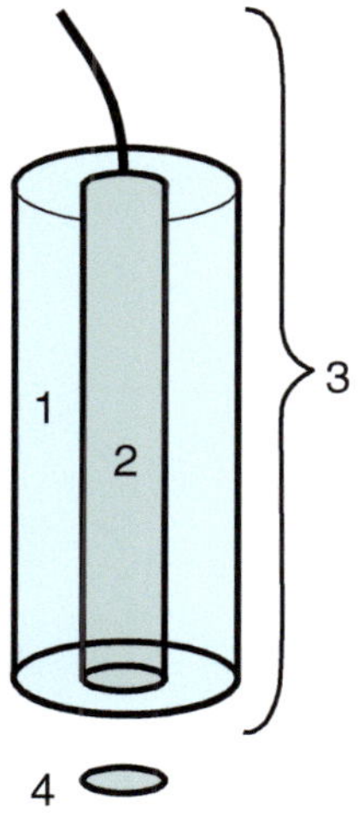

Figure 1.3 Colloquial and scientific meaning of the term "electrode". (1) Sheath or insulating casing, (2) metallic conductor, (3) colloquial meaning of the term electrode, and (4) electrochemical meaning of the term electrode according to Faraday (electrode area).

[3]Michael Faraday, 1791–1867, English physicist and chemist, made decisive discoveries in the fields of electricity and electromagnetism and discovered Faraday's laws named after him.

because one and the same scientific text may contain the term "electrode" in the meaning of Faraday's definition as well as in its colloquial understanding. For distinguishing those different nuances in the meaning of this central concept, further auxiliary terms are frequently used in electrochemical texts such as "electrode surface", "interface electrode-electrolyte", "electrode body", etc. There are many more terminology problems in electrochemistry. This results from the interdisciplinary nature of electrochemistry with different scientific disciplines providing important contributions in different historical periods together with their scientific terminology (chemistry with different subdisciplines, physics, electronics). As a result, a multitude of synonyms and terms with very similar but not completely identical meaning is in use, which makes the study of the electrochemical literature hard. This historical background is authoritatively explained in [6].

Q: What is the meaning of the term "phase" or "phase boundary" in Faraday's concept of an electrode?

A: In thermodynamics, "phase" refers to a compound or a mixture with spatially uniform properties. Phase boundaries are those interfaces at which properties change discontinuously. Examples: zinc powder (one phase), zinc chips suspended in water (two phases), zinc nitrate dissolved in water (one mixed phase), and water on a hotplate (70 °C at the bottom, 40 °C at the water–air boundary, but without a discontinuous change in properties, thus one phase).

Q: Are all electrochemical cells composed of exactly two electrodes?

A: No, electrochemical cells can contain more than two electrodes. We will discuss examples in Chapter 5.

According to Faraday, we call the electrode at which the **reduction** occurs ("electrons flow out of the electrode into the solution") the **cathode**. The **anode** is the electrode at which the oxidation happens ("electrons transfer from the solution into the electrode).

The liquid or the solid between the two electrodes in Figure 1.2 must be an ion-conductor. Liquids in which ions can move freely are obtained either by melting a salt or by dissolving ionizable salts, bases, or acids in a suitable solvent such as water. These compounds dissociate into positively and negatively charged species called **ions**. A negatively charged ion is called an **anion** because it migrates toward the anode during electrolysis ["(+) pole"]. A **cation** is a positively charged ion. It is attracted by the cathode [the "(−) pole" during an electrolysis]. An electrolyte is a material that contains mobile ions. It is not necessarily an aqueous solution. Electrolytes can also be formed with hydrophobic ions in organic solvents, ionic liquids, gel-like polymers, and some ionic solids, which become ion-conducting at elevated temperatures (solid electrolytes, see Chapter 11).

Q: Where do the expressions anode and cathode come from?

A: Faraday derived these terms (using expressions from the Greek language) from consideration of current direction, orientation of the magnetic field of the Earth, and the definition of current induction [6].

Practically, there is a large number of different types of electrodes, but we can distinguish two key classes: ion electrodes (Figure 1.4a) and redox electrodes (Figure 1.4b). In **ion electrodes**, it is an ion that transfers the charge across the phase boundary. If an electron crosses the phase boundary during the electrode reaction, the electrode is called a **redox electrode**.

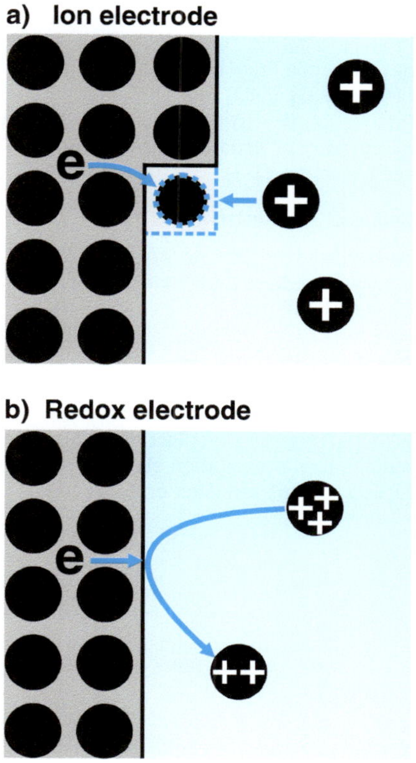

Figure 1.4 Distinction between a) an ion electrode (e.g., galvanic deposition of a metal) and b) a redox electrode (example reduction of Fe^{3+} to Fe^{2+} ions).

Within both classes, there are further subgroups. For a compact description of the electrode, a short-hand notation is used, in which chemical symbols describe the composition of the interacting phases. The boundary between two adjacent phases (e.g., solid|liquid) is indicated with a vertical bar. The description of the phase composition can be supplemented by providing the concentration or activity of the electrolyte solutions (1 mol L^{-1} KNO_3). The physical state is given in brackets (Figure 1.5a–d), where *s* stands for solid, *l* for liquid, *g* for gaseous, *aq* for an aqueous solution, and *sol* for a nonspecified solution. The abbreviation *sa* stands for a saturated solution.

Example Calculation 1.1 Provide the short notation of a silver wire immersed in a 0.1 mol L^{-1} $AgNO_3$ solution.

$$Ag(s)\,|\,0.1\ \text{mol}\ L^{-1}\ AgNO_3(aq)$$

Example Calculation 1.2 Provide the short notation of a silver wire covered by a layer of solid silver chloride and immersed in a 1 mol L^{-1} solution of NaCl.

$$Ag(s)\,|\,AgCl(s)\,|\,1\ \text{mol}\ L^{-1}\ NaCl(aq)$$

We should reflect for a moment and contrast heterogeneous electrochemical reaction with homogeneous charge transfer reactions in chemistry. Such reactions always require two reaction partners. In a homogeneous reaction, electrons or ions are transferred directly from a donating species to an accepting species. The reaction rate of such a transfer depends strongly on the distance and usually happens only over a few hundred picometers, that is, the length of a few chemical bonds. In an electrochemical reaction, one of the reaction partners is the electrode. Thus, the charge transfer step becomes a step in which an electron or an ion crosses the phase boundary (Section 3.1.1). The strong distance dependence of such a reaction step remains. Therefore, the reactants must be transported to the interface and often require very close contact with the electrode surface (Chapter 4).

The coupling of charge transfer processes and transport processes, such as the movement of an ion or molecule from one place to another, is a key feature of electrochemical reactions.

There are various types of ion electrodes (Figure 1.5e–j). If a metal is in contact with a solution of one of its salts, it is called an electrode of the first kind.

$$Ag \rightleftharpoons Ag^{+} + e^{-} \tag{1.7}$$

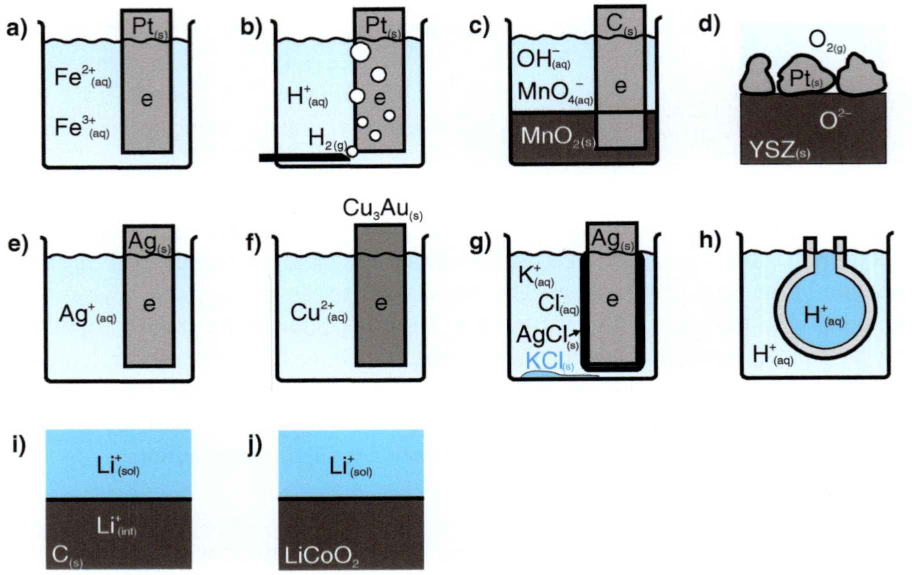

Figure 1.5 Examples of redox electrodes:

a) $Pt(s) \mid Fe^{2+}(aq), Fe^{3+}(aq)$;
b) gas electrode $Pt(s) \mid H_2(g), H^+(aq)$;
c) $C(s) \mid MnO_2(s), MnO_4^-(aq), OH^-(aq)$;
d) $Pt(s) \mid O_2(g), O^{2-}(s)$ (ionic conduction in the oxide lattice of yttria-stabilized zirconia, YSZ, see Excursion 11-1).

Examples of ion electrodes:

e) electrode of the 1st kind: $Ag(s) \mid Ag^+(aq)$;
f) alloy electrode of the 1st kind: $Cu_3Au(s) \mid Cu^{2+}(aq)$;
g) electrode of the 2nd kind: $Ag(s) \mid AgCl(s) \mid KCl(aq, sa)$;
h) glass electrode (Section 5.2.3);
i) graphite intercalation electrode: $[(C_6)_n][Li^+](s) \mid Li^+(sol)$ (Section 15.6); and
j) metal oxide intercalation electrode: $Li_xCoO_2(s) \mid Li^+(sol)$ (Section 15.6).

At an electrode of the second kind, the ion transfer is coupled to a further equilibrium reaction, most often a solubility equilibrium. For instance, a layer of solid AgCl is formed on a Ag wire during anodic dissolution in HCl solution. Two equilibria are attained simultaneously at two phase boundaries.

$$Ag \rightleftharpoons Ag^+ + e^- \tag{1.8}$$

$$Ag^+ + Cl^- \rightleftharpoons AgCl \tag{1.9}$$

The electrode reaction can be summarized as

$$Ag + Cl^- \rightleftharpoons AgCl + e^- \tag{1.10}$$

Very rarely, electrodes of the third kind are encountered. At such electrodes, the charge transfer is coupled to two further equilibria in solution. One example would be a metal ion electrode $M \mid M^{n+}$, at which the released metal ion M^{n+} is complexed by a ligand L. Another metal ion M'^{m+} in the solution competes with M^{n+} for the ligand. In this way, the electrode potential becomes dependent on M'^{m+}.

$$M \rightleftharpoons M^{n+} + ne^- \tag{1.11}$$

$$M^{n+} + L \rightleftharpoons ML^{n+} \tag{1.12}$$

$$M'^{m+} + L \rightleftharpoons M'L^{m+} \tag{1.13}$$

1.3 What Is the Relationship Between Degree of Chemical Conversion and Flow of Charge?

Faraday diligently investigated the relationship between current I and charge Q. For a constant current, the charge is the product of current and time t ($Q = I\,t$). For a current that changes with time (a transient current), we have to integrate the current over t and obtain **Faraday's First Law**.

Q – Charge [C = A s]
t – Time [s]
I – Current [A]

$$Q = \int_{t_1}^{t_2} I(t)\,\mathrm{d}t \tag{1.14}$$

The electrochemical reaction in Eq. (1.2) states that one electron reduces exactly one proton. Consequently, one mole of electrons can reduce one mole of protons. The required charge Q to convert a given amount of a material N, is described by Faraday's Second Law. Q is the sum of elementary charges transferred in the reaction (Section 2.2.4) and directly proportional to the amount N of converted substance. Usually, the symbol "n" is recommended for the amount of substance, but within the field of electrochemistry, the symbol n is strongly preferred for the stoichiometric number of charges transferred across the phase boundary per reaction turnover as written in the reaction equation. This book follows this common practice. For reactions like that in Eq. (1.11), $n = 2$ corresponds to the charges transferred per metal cation, for example, $Cu^{2+} + 2e^- \rightleftharpoons Cu$. However, for the hydrogen evolution reaction in Eq. (1.2) the stoichiometric number of electrons is $n = 2$ and thus different from the charge number $z = 1$ of H^+. The amounts of consumed reagents or formed products must be multiplied by the stoichiometric number of charges, the elementary charge e, and the number of objects in one mole (Avogadro's[4] constant N_A) to obtain the total charge.

$$Q = n \cdot N \cdot e \cdot N_A \tag{1.15}$$

$$Q = n \cdot N \cdot F \tag{1.16}$$

Q – Charge [C = A s]
N – Converted amount of substance [mol]
n – Stoichiometric number for charges as defined in the reaction equation [dimensionless]
e – Elementary charge = 1.602177×10^{-19} A s
N_A – Avogadro constant = 6.022140×10^{23} mol^{-1}
F – Faraday constant [A s mol^{-1}]

The natural constants e and N_A were not yet known in Faraday's time and could not be determined separately. However, the product $N_A \times e$ could be determined. This quantity is called the **Faraday constant** F. It is the electrical charge of 1 mol electrons expressed in the SI unit of the electric charge (1 C = 1 A s) [Eq. (1.17)].

Example Calculation 1.3 Calculation of the Faraday constant.

$$\begin{aligned} F &= e \times N_A \\ &= 1.602177 \times 10^{-19}\ \text{A s} \cdot 6.022140 \times 10^{23}\ \text{mol}^{-1} \\ &= 96485.34\ \text{A s} \cdot \text{mol}^{-1} \approx 100\,000\ \text{C mol}^{-1} \end{aligned} \tag{1.17}$$

1.4 What Is the Relationship Between Electrical Current and Electrochemical Reaction Rate?

A chemical process that occurs at the interface between two phases is called a **heterogeneous reaction**. Electrochemical reactions are important, but not the only examples of heterogeneous reactions. The reactions of gases at the surface of solids are heterogeneous. In contrast, **homogeneous reactions** occur in the bulk of a gas or a solution phase (Figure 1.6).

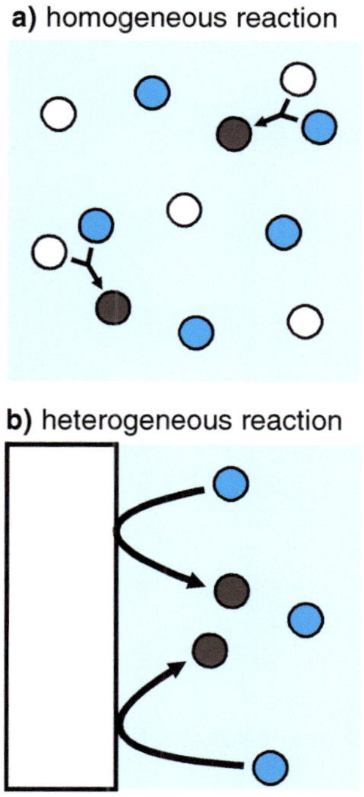

Figure 1.6 Comparison between a) a homogeneous and b) a heterogeneous chemical reaction.

[4] Lorenzo Romano Amedeo Carlo Avogadro, 1776–1856, Italian physicist and chemist, professor of mathematical physics at the University of Torino, lent his name also to the mineral Avogadrite and a crater on the Moon.

The reaction rate r for a homogeneous reaction is defined as the amount of product generated per unit time and per unit of the reaction volume V.

r – Homogeneous reaction rate [mol L^{-1} s^{-1}]
N – Converted amount of substance [mol]
t – Time [s]
V – Volume [L]
c – Concentration (of amount of substance) [mol L^{-1}]

$$r = \frac{1}{V} \cdot \frac{dN}{dt} = \frac{dc}{dt} \tag{1.18}$$

The reaction rate r for a heterogeneous reaction is obtained by relating the amount of product generated per unit time and per unit area A of the interface at which the reaction occurs.

r – Heterogeneous reaction rate [mol cm^{-2} s^{-1}]
A – Interfacial area [cm^2]

$$r = \frac{1}{A} \cdot \frac{dN}{dt} \tag{1.19}$$

From Faraday's First Law (1.14), one obtains for the current $I(t) = dQ/dt$. Replacing Q by the right side of Eq. (1.16) and considering n and F constant for a particular reaction yields the expression

Q – Charge [C = A s]
N – Converted amount of substance [mol]
n – Stoichiometric number of charges in the reaction [dimensionless]
F – Faraday constant [96 485 A s mol^{-1}]

$$I(t) = \frac{dQ}{dt} = \frac{d}{dt}(N \cdot n \cdot F) = n \cdot F \cdot \frac{dN}{dt} \tag{1.20}$$

Rearranging leads to

$$\frac{dN}{dt} = \frac{I(t)}{nF} \tag{1.21}$$

Inserting Eq. (1.21) into (1.19) provides an expression for the heterogeneous reaction rate

$$r = \frac{1}{A} \cdot \frac{I(t)}{nF} \tag{1.22}$$

We note as a second characteristic feature of electrochemical reactions, that the current is *proportional to the momentary reaction rate* at any given time. Thus, electrochemical systems offer the unique opportunity to measure reaction rates directly. An interruption of the current ($I = 0$) causes an instantaneous termination of the reaction, which offers a way for reaction control that is not easily available for other types of chemical systems. Reaction rate control via current is employed in electrosynthesis. In order to facilitate comparison between different experiments, the current is often related to the electrode area. This quantity is called the **current density** $j = I / A$. Thus, the reaction rate can be expressed as

r – Heterogeneous reaction rate [mol cm^{-2} s^{-1}]
n – Stoichiometric number of charges for the reaction [dimensionless]
F – Faraday constant = 96 485 A s mol^{-1}
j – Current density [A cm^{-2}]

$$r = \frac{1}{nF} \cdot j \tag{1.23}$$

Note that the choice of area A can be important here when surfaces are rough. Often, there are more than one electrode reaction occurring simultaneously at one electrode. In this case, the charge crossing the interface is distributed among several reaction channels. If there are m different electrode reactions, the current yield $\varepsilon_{\mathrm{F},l}$ describes the ratio between the current for reaction number l and the total current. The same quantity is also called faradaic efficiency or coulombic efficiency in different areas of electrochemistry.

$$\varepsilon_{\mathrm{F},l} = \frac{n_l F N_l}{\int I(t)\mathrm{d}t} = \frac{n_l F N_l}{\sum_{k=1}^{m} n_k F N_k} \tag{1.24}$$

$\varepsilon_{\mathrm{F},l}$ – Current yield or faradaic efficiency for the lth reaction (dimensionless, $0 \leq \varepsilon \leq 1$, speak "epsilon")
N – Number of formula conversions, subscript l is the number of the reaction [mol]
I – Current [A]
t – Time [s]

Q: What is the meaning of the subscripts F and l in $\varepsilon_{\mathrm{F},l}$? Can't we just drop them?

A: The subscript F stands for Michael Faraday, because we calculate the yield with reference to Faraday's Law. The subscript helps us to distinguish this particular yield from other measures of efficiency (Chapter 15; for energy efficiency of a battery, "energy yield" can be defined). The subscript F is printed nonitalicized because it is an abbreviation. In contrast, the symbol F, which you will find frequently in this book, stands for the Faraday constant. This is defined as the product of a number and a unit (F = 96 485.3 A s mol^{-1}). The subscript l refers to the numbering of the reactions at the electrode. It is printed italicized because it is a symbol (and not an abbreviation). One could, of course, drop the subscript if it is clear from the context which quantity is meant. The symbol η (eta) is also frequently used for the current yield. However, this symbol is also used to denote the overpotential or overvoltage (Chapter 3).

Signs are used to indicate the **direction of the current**. There are no clear-cut criteria to assign a specific current direction to the signs "+" or "−". In such a case, one must rely on an agreement (**convention**). Unfortunately, there are opposing or competing conventions in electrochemistry regarding the direction of current and other quantities. Both conventions exist in the contemporary literature in parallel. It is suggested to follow the convention of IUPAC (*International Union of Pure and Applied Chemistry*) [1], but electrochemists have to be aware of the conflicting conventions in the literature and be able to convert intuitively, especially when reading the older literature.

There are opposing sign conventions in electrochemistry.

According to the IUPAC definition, oxidative currents (anodic currents) are positive and reductive currents (cathodic currents) are negative (Figure 1.7). The direction of the current defined in this way corresponds to the physical motion of electrons in the metal wires and of anions in solution. In contrast, the technical current direction is positive toward the (−) pole of the circuit and thus opposite to the IUPAC definition. The American Chemical Society (ACS) and the Electrochemical Society (ECS) have often followed the definition of the technical current direction [2]. However, there are papers in major journals of those learned societies using both conventions even in the same issue! When reflecting about this and other nomenclature problems, one should always ask which rules are derived from experiments and therefore represent an objective fact (i.e., a fact that does not depend on the opinions of the observer) and which are resting on conventions. Conventions have frequently been changed and adapted in the history of science and could be changed again in the future.

Oxidation	Reduction
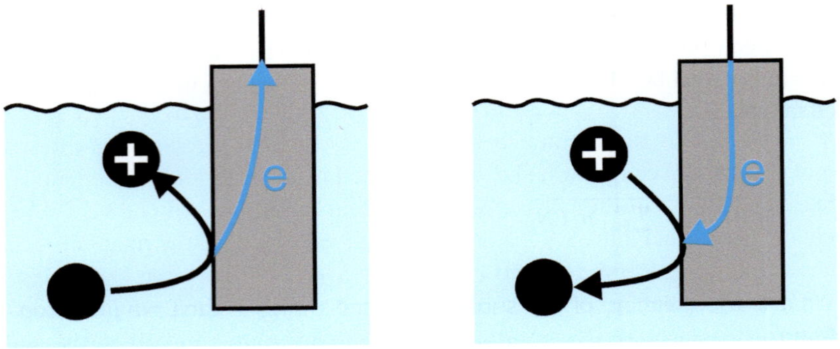	
• Electrons "flow" from the solution to the metal	• Electrons "flow" from the electrode metal into the solution
• *IUPAC:* positive electrochemical current direction, plotted upward in diagrams	• *IUPAC:* negative electrochemical current direction, plotted downward in diagrams
• *polarographic convention:* negative current direction, plotted downward in diagrams	• *polarographic convention:* positive current direction, plotted upward in diagrams
• negative technical current direction	• positive technical current direction

Figure 1.7 Definitions and explanation of the direction of current in electrochemical cells.

1.5 Galvanic Cells and Electrolytic Cells

The large variety of types of electrochemical cells can be divided into two main groups: An **electrolysis cell** is a cell in which a current and the associated electrochemical conversion are driven by an external voltage applied to the cell. Water electrolysis, from example 6 of Table 1.2. is a typical example. Electrical energy of the voltage source is converted into chemical energy (H_2 and O_2 contain much more Gibbs energy than H_2O at the same temperature). Electrolysis cells are applied in industry for the synthesis of commodity chemicals (Chapter 17), for the extraction and refinement of metals (Chapter 9), and in some chemical sensors (Sections 5.4 and 18.3.3).

In contrast, processes in electrochemical cells can be spontaneous. Energy-rich oxidants and reductants can be made to react spontaneously in an electrochemical cell. For instance, H_2 and O_2 react at the electrodes in a fuel cell, producing water and energy. The direct (homogeneous) reaction of the two gases occurs with an explosion over a wide range of the composition of the gaseous mixture. However, in a fuel cell, H_2 and O_2 react with each other in a controlled manner (i.e., with a controlled reaction rate). The electrons provided at the H_2 electrode are transmitted to the O_2 electrode via an external electrical circuit to produce electrical work. In such a cell, chemical energy is converted into electrical energy. These cells are called **galvanic cells**.[5] The reaction occurs spontaneously and is not driven. Important examples of galvanic cells are batteries (Section 2.4.4. and Chapter 15), fuel cells (Section 2.4.4 and Chapter 16),

[5]Luigi Galvani, 1737–1798, Italian physician, anatomist and physicist, professor of medicine in Bologna, famous for his work on electrophysiology with frog legs, lent his name to galvanic technology, the Galvani potential, galvanometer, and a crater on the moon.

and some types of chemical sensors (Section 5.1). Rechargeable batteries are galvanic cells during discharge and electrolysis cells during charging.

Of course, there is no *perpetuum mobile* in electrochemistry! For instance, the electrical energy used to drive a water electrolysis can by principle never be regained by bringing the formed H_2 und O_2 to a fuel cell for generating electricity. We will explore the energetics of electrochemical processes in detail in Chapters 2 and 3. We will treat the energy efficiency of electrochemical energy conversion devices in Chapter 15.

Figure 1.8 illustrates the relation between galvanic and electrolysis cells for the example of the Daniell cell.[6] A Cu sheet is immersed in the solution of a copper salt, for example $Cu(NO_3)_2$, in the left beaker. We call such an arrangement a **half-cell**. Electrochemical cells in Figure 1.8 consist of two half-cells each. In the right half-cells, a Zn sheet is placed in a zinc salt solution ($Zn(NO_3)_2$). The Cu and Zn electrodes in Figure 1.8 are immersed in different solutions. To distinguish them, the electrolyte around the anode is also called anolyte and the electrolyte around the cathode is called catholyte. The reactions in both half-cells commence only if the two metal sheets are connected by an electronic conductor (e.g., a metal wire). The electrolyte solutions in both half-cells also have to be connected electrically. A suitable connection could be based on a filter paper soaked in electrolyte solution, a porous glass disk (frit), or an ion-conducting membrane. In the laboratory, a salt bridge is frequently used. It consists of a glass tube terminated at both ends with a porous frit. The frits prevent the solution from leaking out of the tube, but allow ions to move to connect the two electrolyte reservoirs electrically. If the two metal sheets of the cell are connected via a voltmeter, the cell voltage can be determined. This voltage can also drive a small device. While electrical work is provided to the device, Cu is deposited and Zn is dissolved. If one connects a DC voltage source with the right polarity to the cell, a reversal of the reaction can be achieved: The noble Cu is dissolved and the nonnoble Zn is redeposited.

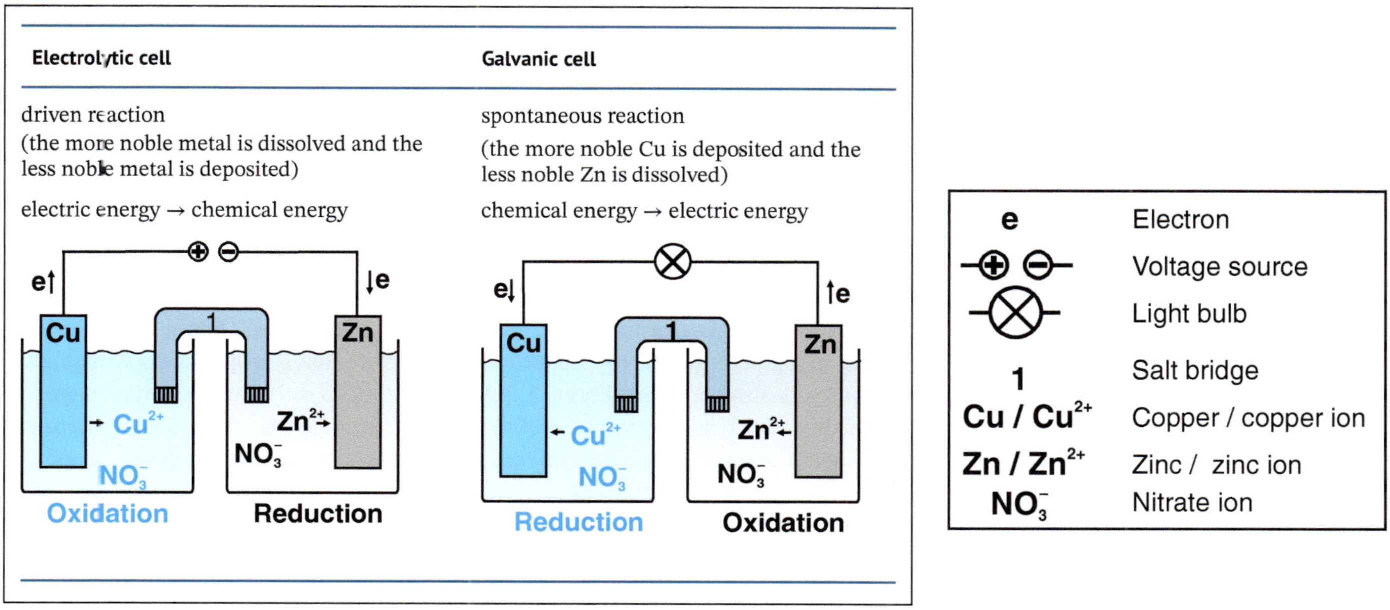

Figure 1.8 Comparison of electrolytic cell (left) and galvanic cell (right).

[6] John Frederic Daniell, 1790–1845, English scientist, professor of chemistry in London, the Daniell cell was widely used during the development of the telegraph system.

Q: It seemed to me that the (+) pole is always the anode and the (−) pole is always the cathode. Isn't this true?

A: It does not matter whether we consider a galvanic or an electrolysis cell; the anode is always the interface at which the oxidation happens and the cathode is associated with the reduction. This definition implies that the interface Zn | $Zn(NO_3)_2$ solution of the galvanic cell in Figure 1.8 is the anode, whereas it is the cathode in the electrolysis cell! It is recommended to avoid the terms (+) and (−) poles. In the context of battery technology, the electrode is called a cathode (anode) when the reduction (oxidation) occurs during the *discharge process*. Because this causes confusion in the context of rechargeable batteries, it is recommended to refer to the positive and the negative electrodes in this case.

By combining the shorthand notation for two electrodes from Figure 1.5 and salt bridge, we obtain

$$\mathrm{Cu} \mid \mathrm{Cu(NO_3)_2}(aq) \mid \mathrm{KNO_3}(aq) \mid \mathrm{Zn(NO_3)_2}(aq) \mid \mathrm{Zn} \mid \mathrm{Cu'}$$

The exact content of the salt bridge is not relevant (as long as no additional voltages are generated), and therefore, the salt bridge is often simplified and presented as a vertical double bar:

$$\mathrm{Cu} \mid 1\ \mathrm{mol\ L^{-1}\ Cu(NO_3)_2}(aq) \parallel 1\ \mathrm{mol\ L^{-1}\ Zn(NO_3)_2}(aq) \mid \mathrm{Zn} \mid \mathrm{Cu'}$$

If the cell contains sheets of different metals, we have to consider a wire connection based on a uniform material as current feeders. In the example, a Cu wire (Cu and Cu′) is connected to the Cu and Zn sheets. This creates an additional Zn | Cu interface. We will return to this problem in Chapter 2.

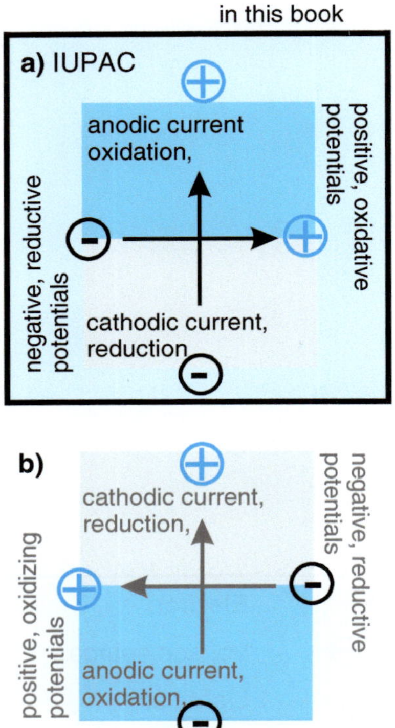

Figure 1.9 Conventions for plotting of currents and electrode potentials a) according to IUPAC and b) according to the polarographic convention. This book follows the convention in a), except for a few original figures taken from the literature.

1.6 Summary

Electrochemical reactions are redox reactions in which the oxidative and reductive partial reactions occur *simultaneously but spatially separated* at two electrodes: The oxidation proceeds at the anode, the reduction at the cathode. *Electrodes are interfaces* between structurally different, electrically conducting phases with at least one phase being an ion-conductor. The charge transmitted via the external electrical circuit equals the charge converted in the anodic and cathodic partial reaction $Q = nFN$. The current $I = \mathrm{d}Q/\mathrm{d}t$ is proportional to the heterogeneous reaction rate $I = nF\ \mathrm{d}N/\mathrm{d}t$ at any time. The current divided by the electrode area is the current density j. The sign of the current indicates the *direction of transport*. Oxidative (anodic) currents are positive and plotted upward according to the recommendation of IUPAC (Figure 1.9a). Positive and oxidizing potentials are plotted to the right. Note that the opposite *polarographic convention* is also common (Figure 1.9b). An *electrolysis cell* is used to drive chemical processes under the consumption of electrical energy (driven reaction). If chemical processes in an electrochemical cell occur spontaneously to generate electrical energy, we call the system a *galvanic cell*.

Test your knowledge

Exercise 1.1: Compile the definitions of scientific terms and physical quantities (with symbols and units) for: electrode, anode, cathode, electrode reaction, current density, Faraday constant, Avogadro constant, elementary charge, electrolysis cell, galvanic cell, half-cell, anodic current, cathodic current, ions, electrolyte, anion, cation. Provide an example and an algebraic formula for each term.

Exercise 1.2: A solution contains 0.4 mol L^{-1} potassium chloride and 0.3 mol L^{-1} magnesium chloride. Calculate the weight in grams of chloride ions contained in 200 mL of this solution.

Exercise 1.3: The electrolytic production of H_2 and O_2 is conducted with a current of 28 kA at electrodes with an area of 16 m^2 each. Calculate the current density and the reaction rate for the formation of molecular hydrogen in mol cm^{-2} s^{-1}.

Exercise 1.4: An electrolytic refinery plant for nickel is operated without interruption 24/7 with a constant current of 15 kA. Raw nickel is dissolved at the anode ($Ni \rightarrow Ni^{2+} + 2e^-$) and redeposited at the cathode ($Ni^{2+} + 2e^- \rightarrow Ni$).

a) How many tons can be purified in 16 days with an assumed faradaic yield of 100%?
b) What would be the length of an edge of a cube containing the amount of Ni purified in 16 days?

Exercise 1.5: For the electrolytic refinery of copper, copper is dissolved at the anode and redeposited at the cathode. In a pilot plant, an anode of 10 cm length, 10 cm width, and 1 cm thickness is used with a constant current of 10 A. The solution is stirred with 1500 revolutions per minute by a mechanical stirrer and kept at 40 °C. The energy for the pilot plant is supplied by a wind turbine with a specified output of 2 MW. The turbine runs under full load during the experiment. How long (hours and minutes) does this electrolysis take if the current yield is 90%.

Exercise 1.6: The chlorine gas obtained via chlorine-caustic electrolysis is filled into a steel cylinder of 100 L volume with a pressure of 100 MPa (at 298 K). One module of a large-scale electrolysis plant operates with 2.0 V cell voltage and at 200 kA current.

a) How long does it take to produce the chlorine for filling one steel cylinder?
b) How long would it take for a wind turbine of 2 MW specified power under full load to generate the electric energy required for the production of one cylinder of chlorine gas, provided that current yield is 100% and there are no energy conversion losses *outside* of the electrolytic cell (no losses in transformation, transmission, and rectification).
c) Discuss the consequences of the numerical results of a) and b) for using fluctuating sources of renewable electric for the described task.

Exercise 1.7: Provide a schematic drawing of the following electrochemical cell

$$Ag \mid AgCl \mid 1\ \text{mol}\ L^{-1}\ KCl \parallel 1\ \text{mol}\ L^{-1}\ KBr \mid AgBr \mid Ag$$

Exercise 1.8: Provide a shorthand notation of the electrochemical cell in Exercise 1.4 employing chloride as the anion in the electrolyte.

Exercise 1.9: Explain the fundamental differences between galvanic and electrolysis cells.

References

1 Convention, regulation and recommendation of IUPAC for electrochemistry and for other subject areas can be found in the internet http://goldbook.iupac.org/ The material is available as a printed book: IUPAC; A. D. McNaught, A. Wilkinson (Eds.); *Compendium of Chemical Terminology*. 2nd ed. (the "Gold Book"). Blackwell Scientific Publications, Oxford, 1997.

2 M. Ciobanu, J. P. Wilburn, M. L. Krim, D. E. Cliffel; Fundamentals, in *Handbook of Electrochemistry* (C. G. Zoski, Ed.), Elsevier, Amsterdam, 2007, pp. 3–8.

3 K. J. Laidler; The chemical history of a current. *Can. J. Chem.* **1997**, *75*, 1552–1565.

4 S. Ross; Faraday consults the scholars: The origins of the terms of electrochemistry. *Notes Rec. R. Soc. Lond.* **1961**, *16*:2, 187–220.

5 G. Inzelt; Milestones of the development of kinetics of electrode reactions. *J. Solid State Electrochem.* **2011**, *15*, 1373–1389.

6 P. Kurzweil; History. in *Encyclopedia of Electrochemical Power Sources* (J. Garche, C. Dyer, P. Mosley, Z. Ogumi, D. Rand, B. Scrosati, Eds.). 1st ed., Elsevier, Amsterdam, 2009. Available under https://doi.org/10.1016/B978-044452745-5.00007-1.

7 A. J. Bard, G. Inzelt, F. Scholz (Eds.); *Electrochemical Dictionary*. 2nd ed., Springer Verlag, Heidelberg, 2012.

8 K. Oldham, J. Myland, A. Bond; *Electrochemical Science and Technology – Fundamentals and Applications*, John Wiley & Sons, Chichester, 2012.

9 Wikipedia https://en.wikipedia.org/wiki/Main_Page.

2

Thermodynamics of Electrochemical Reactions

In Chapter 1, we have seen that charge transfer and the rate of product formation for electrochemical reactions are linked to elementary stoichiometric relationships. Given a specific electrolytic process either at the anode or at the cathode, there is a minimum applied voltage required for processes to commence and/or to go to completion. This minimum applied voltage requirement will be the topic addressed in this chapter. We will see that the applied voltage is directly linked to thermodynamic properties and the stoichiometry for the electrolytic reactions. Thermodynamic properties defined for each reaction provide a way to calculate energy requirements in electrochemical reactions. We will explore which electrochemical processes occur spontaneously (in galvanic cells) and which types of electrochemical reactions are driven with electrical energy (in electrolytic cells). As a further important consideration, the energetic efficiency of processes will be investigated based on thermodynamic arguments.

An important aspect of thermodynamics is the availability of tabulated data for many types of materials and processes. For a unified treatment of data, standard states and a unified potential scale with a zero point that is readily linked to practical laboratory reference electrodes are introduced. In fact, many parameters in thermodynamic tables have been measured with electrochemical techniques.

Thermodynamics is key to electrochemical processes in many areas, but it is especially important in electrochemical energy conversion and energy storage. With the tools provided in this chapter, it will be possible to calculate and explain the energy consumption/production in electrolytic cells (and in half-cells; see Section 2.4.4). After familiarizing ourselves with electrochemical energy conversion at the more fundamental level, the topics of batteries and fuel cells will be treated in more depth in Chapters 15 and 16, respectively.

Further Reading. This chapter builds on fundamental knowledge of chemical thermodynamics as developed in classic textbooks on physical chemistry [1–3]. Familiarization with the fundamentals of chemical thermodynamics is recommended and will be beneficial when following materials in this chapter.

2.1 Electrochemical Cell Voltage and Electrode Potentials

2.1.1 Definition of the Electrochemical Cell Voltage

It appears obvious that the conversion of chemical energy into electrical energy is linked to different types of chemical reactions producing different cell voltages and *vice versa*. We will attempt here to develop a scheme to predict the cell voltage exclusively based on the nature (thermodynamics) of the chemical reaction. This will include the conditions during the chemical reaction as well as stoichiometric factors.

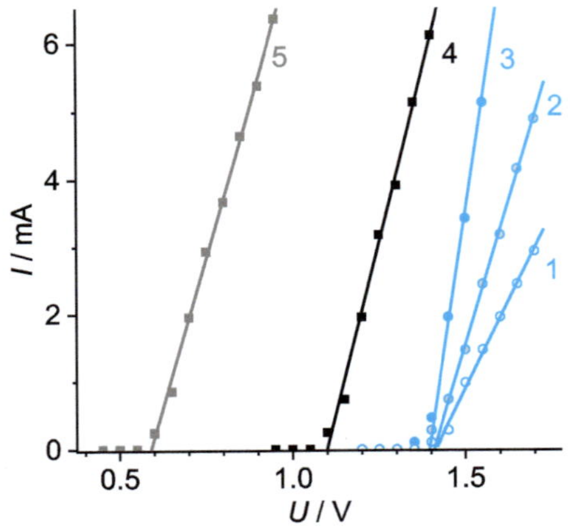

Figure 2.1 Current–voltage characteristics for a HCl, HBr, or HI electrolysis cell:

1) 0.05 M HCl, $U_{onset} = 1.40$ V
2) 0.1 M HCl, $U_{onset} = 1.40$ V
3) 0.2 M HCl, $U_{onset} = 1.40$ V
4) 0.1 M HBr, $U_{onset} = 1.10$ V
5) 0.1 M HI, $U_{onset} = 0.59$ V

The basic design of the electrolytic cell is shown in Figure 1.1, with a voltage source to apply a stable, adjustable voltage. When varying the applied voltage step-by-step and recording the resulting current, data in Figure 2.1 are generated. Curve 1 in blue shows data for aqueous HCl at low concentration. At a low applied voltage, essentially no current flows, but with an onset potential of 1.4 V, the current can be seen to increase. At this point, gas bubble formation at both cathode and anode can be attributed to hydrogen and chlorine production, respectively. When changing the electrolyte to aqueous HBr (curve 4) or to aqueous HI (curve 5), similar trends in current are observed, but the onset potentials are significantly lower. This suggests that the applied voltage at the onset point of the electrolytic reaction is directly linked to the chemical nature of processes at cathode and anode. An extrapolation to the onset potential is possible by fitting lines through the current data points. The intercept at the abscissa ($I = 0$) can be identified as the onset voltage U_{onset}. A comparison of data obtained at different HCl concentrations (curves 1 to 3) suggests that U_{onset} is *not* concentration dependent. However, the onset voltage can depend on electrode material and, more specifically, on the kinetic rate of electron transfer (see Section 3.2.3). The onset voltage is an important parameter for the design of electrolytic processes, but it is less appropriate for the determination of purely reaction-based thermodynamic parameters.

After discussing the electrolysis cell, a similar consideration of properties and applications of galvanic cell is instructive. In the following cell, defined as:

$$Cu' \mid Zn \mid Zn(NO_3)_2(aq) \parallel Cu(NO_3)_2(aq) \mid Cu$$

two copper wires are employed to measure the cell voltage (Figure 1.8). This **cell voltage** should be measured at zero current and is called an **open-circuit voltage**. Practically, this measurement can be performed with a high impedance voltmeter (to minimize residual current flow) or for higher precision with a compensation circuit or galvanic bridge. For compensation, a voltage of opposite polarity is applied to the cell, and the current is monitored with a sensitive ammeter. At the point of zero current, the exact voltage is determined. As a high-precision voltage source for this experiment, a very accurate temperature-calibrated voltage source can be employed, for example, based on a Weston cell.[7]

$$Hg \mid Hg, Cd \mid CdSO_4 \times 8/3\ H_2O(s) \mid CdSO_4(sa), Hg_2SO_4(s) \mid Hg'$$

A precision voltage bridge allows the relationship from standard cell to test cell to be determined under zero current conditions.

2.1.2 The Electrochemical Series

With the help of open-circuit voltage measurements, a wide range of electrochemical cells and chemical systems can be characterized. Ideally, the electrochemical cell is divided into two half-cells, and one of these half-cells is maintained whilst systematically changing the second one. The technical term for the half-cell chosen as a potential reference point is **reference electrode**. For the case of zero net current through the electrochemical cell and for equilibrium without any net chemical processes progressing, the measured voltage is the **equilibrium cell voltage** U_{eq}. Cell voltages

[7] Edward Weston (1850 Ostwestry, England – 1936 Montclair, NJ, USA), study of medicine, emigration to the United States, worked as chemist and entrepreneur.

determined versus a well-defined reference electrode are the **electrode potentials** E (Appendix, Table A-1). Every cell voltage for cells based on two half-cells can be expressed as the difference between two electrode potentials E_A and E_B of those two half-cells when determined with the same reference electrode. A systematic comparison of many half-cell voltages measured under well-defined standard conditions is called the galvanic series of half-cell reactions or **electrochemical series** (Figure 2.2). Solvents stabilize ions to a different degree, and therefore, the electrochemical series such as those tabulated in Appendix, Table A-1 are specific for one solvent.

Q: Why are there so many technical terms with voltage and potential? Aren't the open-circuit voltage and the equilibrium cell voltage the same?

A: For both the open-circuit voltage and the equilibrium cell voltage, zero current flow is imposed. The open-circuit voltage could be an equilibrium cell voltage, but doesn't have to be. Chemical processes within the cell can occur without affecting the external current, for example, a corrosion process in the cell could lead to continually changing concentration conditions (nonequilibrium). Furthermore, an anode process may be compensated for by *two* independent cathode processes (anodic metal dissolution may occur in conjunction with cathodic metal deposition and oxygen reduction simultaneously). As a result, the open-circuit voltage is associated with a mixed potential (Section 2.3.5). There are cases of irreversible reactions at phase boundaries contributing to a deviation of the open-circuit voltage from the equilibrium voltage (see Section 2.3.3, phase boundaries between two immiscible electrolyte solutions).

The cell voltage for a hypothetical galvanic cell U is obtained from the measured equilibrium cell voltages E_A and E_B.

$$U = E_A - E_B \tag{2.1}$$

U – Cell voltage [V]
E_A – Electrode potential of electrode A [V]
E_B – Electrode potential of electrode B [V]

In measurements, different types of reference electrodes can be employed (Section 2.4.1), but these can be calibrated based on the **standard hydrogen electrode** (SHE) as an agreed-upon reference point with $E_{SHE} = 0.0$ V (convention).

$$\text{Pt} \mid \text{H}_2 \text{ (100 000 Pa), H}^+(aq)\ (a_m(\text{H}^+) = 1 \text{ mol kg}^{-1})$$

The most recent IUPAC definition of the SHE uses the activity based on molality (see Section 2.2.3): Pt | H_2 (100 000 Pa), $H^+(aq)$ ($a_m H^+ = 1$ mol kg^{-1}). The resulting potential difference to a hydrogen electrode with $a_c = 1$ mol L^{-1} is very small under ambient conditions.

By definition, the half-cell potential for the standard hydrogen electrode is set to 0.0 V at all temperatures. This approach is consistent with the methodology introduced in chemical thermodynamics as will be explained below.

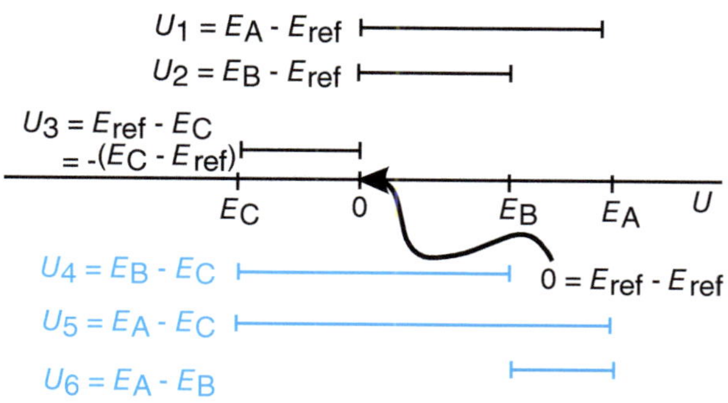

Figure 2.2 How to systematically develop the electrochemical series from open-circuit voltage data; black – measured open-circuit voltages, blue – predicted cell voltages.

We will then explore the link between half-cell voltages (or electrode potentials) and voltage contributions (Galvani voltages)[3] at each of the interfaces within an electrochemical cell.

2.2 What Is the Link from Cell Voltage to Thermodynamic Potentials?

2.2.1 The Electrochemical Potential

Note the ambiguity for some commonly used symbols: *u*, *U* – internal energy and cell voltage; *f*, *F* – activity coefficient, free energy, and Faraday constant.

In electrochemical processes, anode and cathode reactions are spatially separated, but linked by the flow of charges (both ions and electrons). This introduces the possibility of a contribution to the internal energy u (Figure 2.3) from the electrical energy term linked to charge transport through a potential gradient. In chemical thermodynamics, contributions to the internal energy are commonly expressed as a combination of an *extensive parameter* (system size- or mass-dependent, e.g., V or s) and an *intensive parameter* (system size- or mass-independent, e.g., p or T). The most practically relevant thermodynamic function is the **Gibbs free energy**[8] g. The Gibbs free energy is a function of pressure p, temperature T, and amount N_i of component i and represents the energy of a system that is available for conversion into work or other forms of energy under conditions of constant pressure. The Gibbs free energy $g = g(T, p, N_i)$ can be expressed as a total differential.

g – Gibbs free energy [J]
s – Entropy [J K^{-1}]
T – Absolute temperature [K]
p – Pressure [Pa = N m^{-2}]
N_i – Amount [mol]

$$\mathrm{d}g = -s\,\mathrm{d}T + V\,\mathrm{d}p + \sum_{i=1}^{n}\left(\frac{\partial g}{\partial N_i}\right)_{T,p,N_{j\neq 1}} \mathrm{d}N_i \tag{2.2}$$

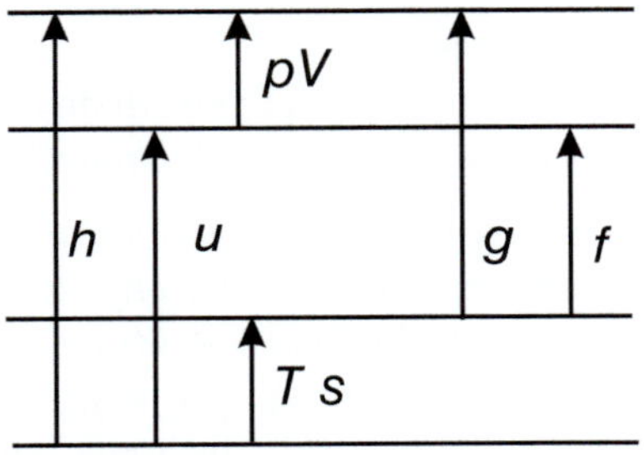

Figure 2.3 Relationship between thermodynamic functions. *h* – enthalpy, *u* – internal energy, *T s* – bound thermal energy, *p V* – external volume expansion work, *g* – Gibbs free energy, *f* – Helmholtz free energy, *T* – absolute temperature, *s* – entropy, *p* – pressure, and *V* – volume. Compared to textbooks on chemical thermodynamics [1–3].

From this total differential, we can see that at constant pressure ($p = const$) and constant temperature ($T = const$), changes in Gibbs free energy are linked only to changes in amount or composition in $N_1, \ldots N_n$ within the system. These are extensive parameters. The corresponding intensive parameter is the partial derivative of g for N_i for component i (at constant pressure, constant temperature, and keeping all other N_j constant, $j \neq i$). This partial differential defines the **chemical potential** μ_i of component i for the case of electrically neutral species.

$$(\mathrm{d}g)_{T,p,N_{j\neq 1}} = \left(\frac{\partial g}{\partial N_i}\right)_{T,p,N_{j\neq 1}} \mathrm{d}N_i = \mu_i\,\mathrm{d}N_i \tag{2.3}$$

Due to the stoichiometric link of product formation and starting material consumption, we have to consider all changes as a sum of partial differentials for each component (here for k components).

μ_i – Chemical potential of component i [J mol^{-1}]

$$(\mathrm{d}g)_{T,p} = \sum_{i=1}^{k}\left(\frac{\partial g}{\partial N_i}\right)_{T,p,n_j} \mathrm{d}N_i = \sum_{i=1}^{k}\mu_i\,\mathrm{d}N_i \tag{2.4}$$

Q: Some of this seems difficult to follow. We started discussing electrode potentials (in V), and now we discuss the chemical potential (in J mol^{-1}). Is there a link between these potentials and can the units be rationalized?

[8] Josiah Willard Gibbs, 1839–1903, US-American chemist, professor at Yale University, laid the foundations for thermodynamics (Gibbs energy, Gibbs–Duhem equation, Gibbs–Helmholtz equation) and contributed to mathematics. The Willard Gibbs Medal of the American Chemical Society and the Gibbs Lecture of the American Mathematical Society, a crater on the moon and an asteroid are named after him.

A: Indeed, this may seem confusing, but there is a rationale. Thermodynamic potentials are usually expressed as an energy (J mol^{-1}). The potential difference can be understood as a driving force for spontaneous processes to happen, and, for the case of a moving charge, the driving force would be the potential differences (or voltage, in V). The energy released for an elementary charge (say an electron) passing through a potential difference of 1 V can be calculated by multiplying the charge $e = 1.602 \times 10^{-19}$ A s with 1 V to give the energy 1.602×10^{-19} W s $= 1.602 \times 10^{-19}$ J = 1 eV. Therefore, the energy per particle in J (or in eV) is directly linked to the potential difference in V if we consider a single charge. If we consider a molar amount of charges, the energy needs to be multiplied by Avogadro's number to give $1\ \text{eV} \times N_A = 1.602 \times 10^{-19}\ \text{J} \cdot 6.022 \times 10^{23}\ \text{mol}^{-1} = 96.5\ \text{kJ mol}^{-1}$.

A more detailed derivation of thermodynamic potentials is beyond the scope of this chapter, but can be found in classic textbooks on chemical thermodynamics [1–3]. Some key results from chemical thermodynamics are summarized below.

Equation (2.4) has been derived for the case of neutral particles. In order to account for energies involving ionic species, we have to take into consideration the energy contributions for charges moving across interfaces and phase boundaries between charged phases. In order to describe the electrical energy term, we have to introduce the inner potential φ (intensive) of the phase and the amount of charge Q (extensive). For a more detailed consideration of the inner potential, see Excursion 2-1. The definition of the inner potential is based on the electrical work done W_{el} divided by the charge Q when the charge (defined for an infinitely small particle) is moved from a far-away field-free location under vacuum into a location within the phase. The charge Q is considered positive by convention.

$$\varphi = \lim_{Q \to 0} \frac{W_{el}}{Q} \tag{2.5}$$

φ – Inner potential [V]
W_{el} – Electrical work [V A s] = [J] = [Nm]
Q – Charge [C] = [A s]

The work term W_{el} can be understood as the electrical work associated with moving the charge between two locations.

$$\text{d}W_{el} = \varphi\, \text{d}Q \tag{2.6}$$

It seems entirely plausible that the addition of charged particles (electrons or ions) into a charged phase may occur uphill or downhill depending on the charge already present in this phase (Figure 2.4). In order to express the change in Gibbs free energy for our system dg, we have to add the change in electrical energy $\text{d}W_{el,}$ which equates to the inner potential multiplied by the change in charge dQ (see Eq. (2.4)).

$$(\text{d}g)_{T,p} = \sum_{i=1}^{n} \mu_i\, \text{d}N_i + \text{d}W_{el} = \sum_{i=1}^{n} \mu_i\, \text{d}N_i + \varphi \text{d}Q \tag{2.7}$$

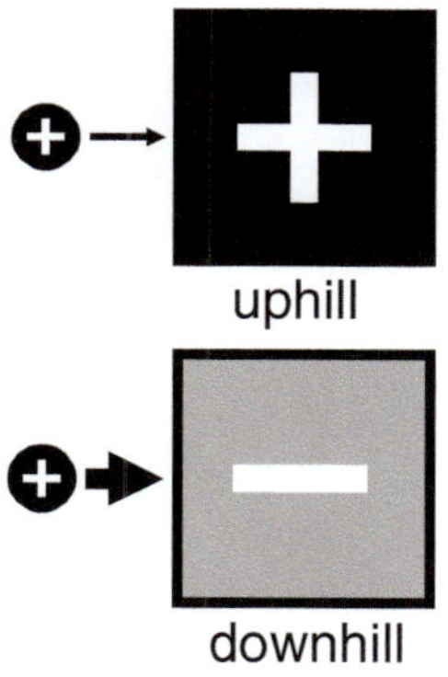

Figure 2.4 Illustration of electrical work needed to add a positively charged particle into a phase with either positive or negative inner potential.

From expression (2.7), it becomes apparent that the inner potential φ is given by the partial derivative of g for Q at constant T, p, and N_i. However, in electrochemical systems, any change in dQ has to be linked to stoichiometric changes in composition dN_i. We can develop a link from dQ to dN_i by defining the amount of one mol of charges as given in the Faraday constant F. One mol of singly charged particles represents the charge of $1\ \text{mol} \cdot 1\ F = 96\,485$ A s. For multiply charged particles, we introduce z_i as an integer describing the number of elementary charges per particle of type i (or the ionic charge). For cations, z_i is positive and for anions z_i is negative. With these definitions

N_i – Amount of component i [mol]
Q – Charge [C] = [A s]
F – Faraday constant = 96 485 A s mol^{-1}
z_i – Charge of the ion i (positive or negative; without units)

in place, we now express the change in charge dQ in terms of a product of ionic charge z_i, the Faraday constant F, and the change in amount of species i dN_i.

$$dQ = \sum_{i=1}^{n} z_i F dN_i \tag{2.8}$$

$\tilde{\mu}_i$ – Electrochemical potential [J mol^{-1}]
i – Index for components
α – Index for phases

Combining Eqs. (2.8) and (2.7) allows us to express the **electrochemical potential $\tilde{\mu}_i^\alpha = \mu_i + \varphi z_i F$ for phase α** in Eqs. (2.9) and (2.10) as a sum of chemical potential and an electrical term.

$$(dg)_{T,p} = \sum_{i=1}^{n} \mu_i dN_i + \varphi \sum_{i=1}^{n} z_i F dN_i \tag{2.9}$$

$$(dg)_{T,p} = \sum_{i=1}^{n} (\mu_i + \varphi z_i F) dN_i = \sum_{i=1}^{n} \tilde{\mu}_i dN_i \tag{2.10}$$

We obtained an expression for the change in Gibbs free energy dg for $p = const$ and $T = const$ as a function of changing amounts N_i for both neutral and charged particles. The comparison to the expression for neutral particles (2.4) suggests that Eq. (2.10) reverts to the simpler case for $z_i = 0$. For particles without charge, the chemical potential μ_i and the electrochemical potential $\tilde{\mu}_i$ are identical.

Electrochemical systems are composed of multiple phases, and therefore, we have to express the change in Gibbs free energy dg for multiphase systems. We introduce capital Roman numerals (I, II, III, IV, ...) for individual phases and small Greek letters (α, β, γ, ...; see Appendix, Table A-9) as variables for the phase index.

g – Gibbs free energy [J]
N_i – Amount of component i [mol]
z_i – Charge of ions i
μ_i^α – Chemical potential for component i in phase α
$\tilde{\mu}_i^\alpha$ – Electrochemical potential for component i in phase α [J mol^{-1}]
R – Universal gas constant [J K^{-1} mol^{-1}]
T – Absolute temperature [K]
a_i^α/a° – Activity of component i in phase α over standard activity for this component [dimensionless]

The inner potential φ is experienced equally for all charged particles/ions in a given phase. Therefore, the electrochemical potential is a property of the ion i in the phase α. It is given as the partial differential of the Gibbs free energy over N_i with constant $T, p, N_{j\neq i}$ in phase α.

$$\left(\frac{dg^\alpha}{dN_i}\right)_{T,p} = \tilde{\mu}_i^\alpha = \mu_i^\alpha + z_i F\varphi = \mu_i^{\alpha\circ} + RT\ln\left(\frac{a_i^\alpha}{a^\circ}\right) + z_i F\varphi. \tag{2.11}$$

A further consideration involves the formalism of chemical thermodynamics when activities are introduced. These describe nonideal behavior contributing to Gibbs free energy terms. For the chemical potential, the term μ_i° is then associated with a term $RT\ln(a_i^\alpha/a^\circ)$. Activities describe effective or corrected concentrations of a component, that is, their numerical values deviate somewhat from the corresponding concentrations. The use of activities in this book is linked to units, and therefore, division of the activity by a standard activity a_i^α/a° allows units to be formally removed in the logarithm (see Sections 2.1.2 and 2.2.1).

Excursion 2-1 Physical Analysis of Potentials in Electrochemistry

In this excursion, we will discuss the link between potentials discussed in physics textbooks and electrode potentials in electrochemistry (Figure 2.5). What are the connections and differences for parameters generally given in V, but associated with different physical phenomena?

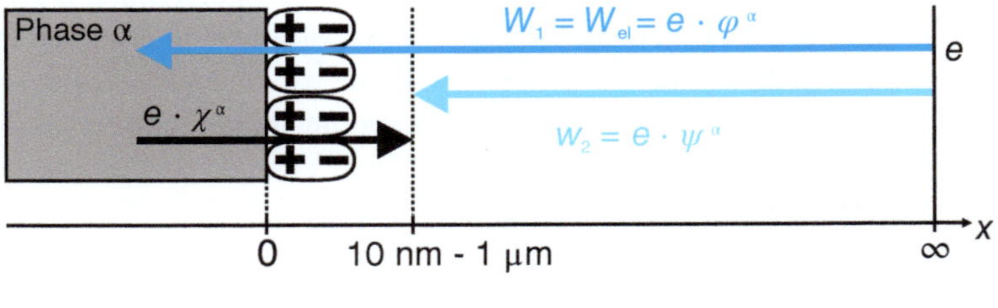

Figure 2.5 Potentials at phase boundaries.

We have encountered the inner potential φ (also called Galvani potential[5]). This has been linked to the presence of excess charges in a phase α (φ^α) that affect the transfer of a test charge from a field-free location in vacuum. There is no direct way to measure the inner potential. One would have to measure the electrical work w_1 for a test charge Q from a field-free vacuum to be transferred into the phase α. The inner potential is obtained by dividing w_1 by Q. In reality, the transfer of this charge would trigger chemical reactions that prohibit meaningful measurements.

The link from electrical potentials (in V) to thermodynamic potentials (in J) can be appreciated when considering the units. Every energetic aspect of the transfer of the test charge can be considered in J or in eV. The proportionality is based on the value of the elementary charge. The energy 1 eV is given by the kinetic energy gained by an electron when traversing a potential difference of 1 V in vacuum.

$1\ \text{eV} = 1.602 \times 10^{-19}\ \text{J}$
$1\ \text{eV} = 1.602 \times 10^{-19}\ \text{V A s}$
$1\ \text{eV} = 1.602 \times 10^{-19}\ \text{W s}$

It is interesting to explore the energy contributions to w_1. The electrostatic energy term is independent of the path of the charge from a field-free ($E = 0$) starting point in vacuum, and dependent only on the endpoint. The hypothetical endpoint is defined as closely as possible to the surface of phase α (without inducing any chemical interactions). This can be at a distance of 10–1000 nm. The potential at this point associated with the work term w_2 is called the outer potential Ψ^α (Psi) or also Volta potential[9]. The remaining transfer of the charge from outside to inside of phase α is associated with a work term or the corresponding potential defined as the surface potential χ^α (chi). This leads to expression (2.12).

$$\chi^\alpha = \varphi^\alpha - \Psi^\alpha \tag{2.12}$$

Ψ^α – Outer potential of phase α [V]
χ^α – Surface potential at a specific point outside of phase α [V]
φ^α – Inner potential of phase α [V]

The surface potential corresponds to the work required to move the charge through the complex charge distribution at the surface of phase α.

Surface potentials are generally present at all phase boundaries, for example, at the surface of metals. Metals can be envisaged to consist of positively charged metal cations surrounded by a negatively charged gas of electrons. The probability of finding electrons close to the surface decays exponentially and penetrates further out when compared to the location of the metal atoms. As a result, a net negative excess charge develops with a corresponding positive excess charge on the metal (Figure 2.6). This type of charge density distribution at the metal surface is called a surface dipole.

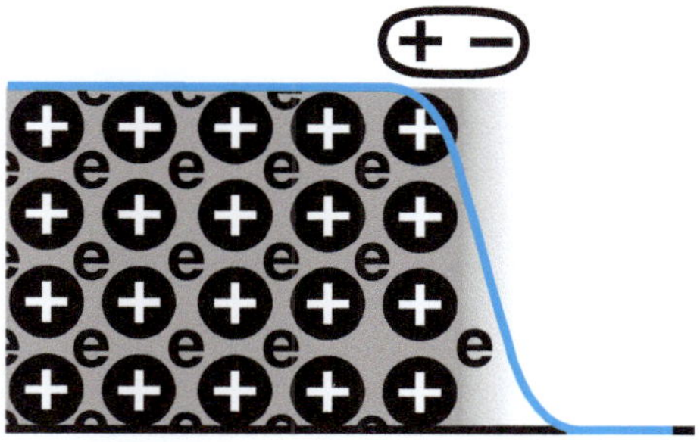

Figure 2.6 Formation of a surface dipole at the surface of a metal. The blue line indicates the electron density distribution.

The presence of the surface dipole (or surface potential) at metal phase boundaries has important implications, which will become clearer in later chapters. The following points are important:

1) The inner potential φ^α is a property of the phase, and it is path-independent (independent of the path of the charge moving from outside to inside).
2) The surface potential χ^α is surface structure-dependent and may vary, for example, for different crystal planes. Imagine a single crystal metal specimen (as employed in the study of single crystal electrocatalysis; Section 8.9). Each type of crystal plane will have a specific geometric pattern of atoms. As a result, the surface dipole due to the electron density distribution across the interface will also vary. χ^α is a property dependent on each specific surface of phase α.
3) If we consider χ^α surface specific, then the outer potential Ψ^α must also be surface specific [see Eq. (2.12)].

[9] Allesandro Giuseppe Antonio Anastasio Volta, 1745–1827, Italian physicist, professor of physics in Como and later in Pavia, developed Volta's pile experiment. This was crucial as the first constant voltage source. He also discovered the gas methane. Volta's pile basically is a zinc–air battery $Cu\,|\,O_2, H_2O\,||\,Zn^{2+}\,|\,Zn\,|\,Cu'$.

N_e – Number of electrons [mol]

Ψ – Outer potential [V]
r – Radius of the sphere [m]
d – Diameter of the sphere [m]
ε – Relative permittivity [dimensionless]
ε_0 – Permittivity of the free space [A s V^{-1} m^{-1}]

Electric potential of a sphere in a vacuum (see physics textbooks)
Q – Charge [C] = [A s]

N_A – Avogadro constant [mol^{-1}]
F – Faraday constant [A s mol^{-1}]

Example Calculation 2.1 Given an iron sphere of 1 cm diameter in vacuum, how many electrons are required to produce an outer potential of 1 V?

sought: N_e

known: $\Psi = 1$ V (as given)

$\varepsilon = 1$ (see physics textbooks)

$\varepsilon_0 = 8.854 \times 10^{-12}$ A s V^{-1} m^{-1}

$d = 1$ cm (as given)

$r = 0.5d = 0.5$ cm $= 5 \times 10^{-3}$ m

$$\Psi = \frac{1}{4\pi\varepsilon\varepsilon_0}\frac{Q}{r}$$

$$Q = 4\pi\varepsilon\varepsilon_0 r\Psi$$

$$Q = 4 \cdot 3.141 \cdot 1 \cdot 8.854 \times 10^{-12}\ \text{A s V}^{-1}\ \text{m}^{-1} \cdot 5 \times 10^{-3}\ \text{m} \cdot 1\ \text{V}$$

$$Q = 0.555 \times 10^{-12}\ \text{A s}$$

$$N_e = \frac{\text{Number of electrons}}{N_A} = \frac{Q}{F}$$

$$\text{Number of electrons} = \frac{Q}{F}N_A$$

$$= \frac{0.555 \times 10^{-12}\ \text{A s} \cdot 6.022 \times 10^{23}\ \text{mol}^{-1}}{96\,485\ \text{A s mol}^{-1}}$$

$$= 3.644 \times 10^5$$

In order to produce an outer potential of 1 V on an iron sphere of 1 cm diameter in a vacuum, we need 364 400 electrons. Is this a large number? It seems so, however, when related to the number of iron atoms of the sphere (or on the surface of the sphere), this is indeed a very small number.

Example Calculation 2.2 What would be the radius of an iron sphere in a vacuum, so that a single electron addition would result in a measurable change in outer potential? Here, we assume a measurable outer potential difference to correspond to 200 mV as a value significantly higher when compared to the thermal energy of the surroundings. [The energy of thermal noise can be estimated as $3k_BT$ or 76 meV.]

unknown: r

known: $\Psi = 200$ mV $= 0.2$ V

number of electrons = 1

$$\Psi = \frac{1}{4\pi\varepsilon\varepsilon_0}\frac{Q}{r} \qquad Q = \frac{F}{N_A}\ \text{(elemental charge)}$$

$$\Psi = \frac{1}{4\pi\varepsilon\varepsilon_0}\frac{F}{rN_A}$$

$$r = \frac{1}{4\pi\varepsilon\varepsilon_0}\frac{F}{\Psi N_A}$$

$$r = \frac{96\,485\ \text{A s mol}^{-1}}{4 \cdot 3.141 \cdot 8.854 \times 10^{-12}\text{A s V}^{-1}\,\text{m}^{-1} \cdot 6 \times 10^{23}\,\text{mol}^{-1} \cdot 0.2\,\text{V}}$$

$$r = 7.22 \times 10^{-9}\ \text{m} = 7.2\ \text{nm}$$

The iron sphere needs a radius of 7.2 nm or smaller. Given the diameter of atoms in the range of 0.1 to 0.3 nm, it seems entirely feasible to make such a nanosphere. Quantized charging of nanospheres has been experimentally studied and applications suggested, for example, in single-electron transistors.

In this case, we have chosen thermal noise consistent with room temperature conditions, but we could also consider going to much lower temperatures. How would the results change at temperatures of liquid helium (4 K)?

2.2.2 The Nernst Equation[10]

Coming back to the topic of thermodynamic description of electrochemical processes, we now turn our attention to the Daniell cell[6] as shown in Figure 1.4. The short notation of the cell is given as follows:

$$\text{Cu}'\ \ \text{Zn} \mid \text{Zn(NO}_3)_2(aq) \mid \text{KNO}_3(aq) \mid \text{Cu(NO}_3)_2(aq) \mid \text{Cu}$$

The cell reaction suggests the spontaneous plating of copper associated with the dissolution of the less noble zinc electrode.

$$\text{Cathode reaction: } Cu^{2+} + 2e^- \rightarrow Cu \tag{2.13}$$

$$\text{Anode reaction: } Zn \rightarrow Zn^{2+} + 2e^- \tag{2.14}$$

The overall cell reaction is obtained by adding Eqs. (2.13) and (2.14) together. Reaction components that occur on both sides ($2e^-$) are eliminated.

$$\text{Cell reaction: } Cu^{2+} + Zn \rightarrow Cu + Zn^{2+} \tag{2.15}$$

The fact that this cell reaction is spontaneous is shown easily by placing zinc powder into $Cu(NO_3)_2$ solution in a test tube. The reaction will go to almost completion with copper metal precipitating and Zn^{2+} dissolving. How does this compare to the galvanic cell? The cell is composed of Cu- and Zn-metal electrodes immersed in aqueous $Cu(NO_3)_2$ or $Zn(NO_3)_2$ solution, respectively. The electrodes are connected by wires, and the two electrolytic compartments are connected via a salt bridge (usually a gel with high concentration of aqueous electrolyte such as KCl). A high impedance voltmeter is placed between the two wires coming from the metal electrodes (Figure 2.7). As soon as the circuit is completed and the metals are immersed in solution, reactions (2.13) and (2.14) will occur. However, after a short period of time, the reactions stop and a constant voltage is obtained, indicative of the cell reaching equilibrium.

We can now interpret these observations based on considerations in Section 2.1.1. Every Cu^{2-} cation added to the copper lattice will add to the positive charge and therefore add to the inner potential. At the same time, the electrolyte solution containing $Cu(NO_3)_2$ is charged more negatively due to the loss of Cu^{2+} cations. We can identify a charge separation at the phase boundary due to Cu^{2+} cations transferring from solution

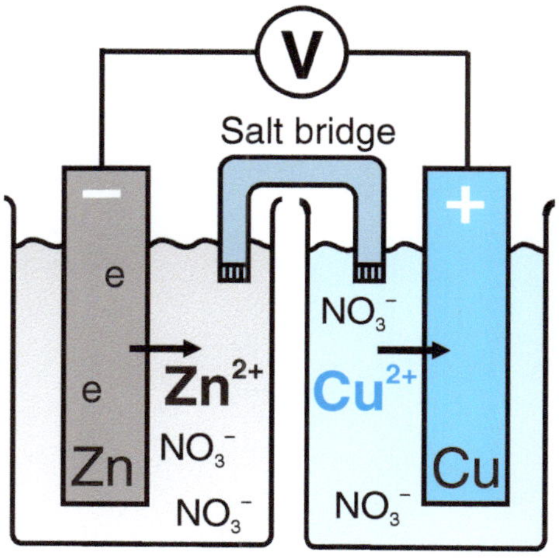

Figure 2.7 A galvanic cell: the Daniell cell.

[10]Walther Herrmann Nernst, 1864–1941, physical chemist, Nobel Prize 1920, worked in Leipzig, Göttingen, Berlin, and contributed to the Nernst equation, the third law of thermodynamics, the Nernst diffusion layer and treatment of phase equilibria. He obtained patents for light bulb designs and the Nernst glower.

to the metal phase. In Section 2.3, we will consider such charge transfer events in more microscopic detail. In the second compartment, every Zn^{2+} cation going into solution will leave 2 electrons behind in the zinc metal. This generates a net negative charge. The charging of copper and zinc both lead to a lower probability for Cu^{2+} to plate or for Zn^{2+} to dissolve. At a well-defined point in this process, a stable endpoint voltage is observed when all processes cease (although microscopically there can still be some dissolution and plating at each electrode). The system reaches a steady state, the net reaction (2.15) stops, and a well-defined potential difference, the **equilibrium cell voltage**, is detected between the two metal plates. The number of charges required to establish this equilibrium is low (see Example Calculation 2.1). Therefore, the amount of copper that is plated as well as the amount of zinc that dissolves are negligible. The charge transport between the two half-cells, occurring via electrolyte solution (ionic conduction via the salt bridge) also remains negligible.

The electrochemical equilibrium (similar to other chemical equilibria) has to be considered as a **dynamic equilibrium**: there is no net current flow and no net reaction, but there are continuing processes such as reductive Cu^{2+} discharge at the Cu electrode and Cu^{2+} dissolution due to copper oxidation. Oxidation and reduction occur simultaneously without resulting in a net change with time.

Next, we need to express the cell voltage at equilibrium based on thermodynamic properties. The equilibrium is established when the total differential for Gibbs free energy $g = g(p, T, N_i)$ is zero. We start with Eq. (2.2) and substitute in dg/dN from Eq. (2.11) to give Eq. (2.16).

g – Gibbs free energy [J]
s – Entropy [J K^{-1}]
T – Absolute temperature [K]
V – Volume [m^3]
p – Pressure [Pa] = [N m^{-2}]
$\tilde{\mu}_i^\alpha$ – Electrochemical potential of component i in phase α [J mol^{-1}]
N_i – Amount of component i [mol]

$$\mathrm{d}g = s\mathrm{d}T + V\mathrm{d}p + \sum_{i=1}^{n} \tilde{\mu}_i^\alpha \mathrm{d}N_i \tag{2.16}$$

For the Daniell cell (Figure 2.7), we assume that changes in amounts of components are based only on internal conversion without any externally added materials. The formal reaction progress coefficient ξ [mol] is the amount a given reaction runs and is defined as positive for a reaction formally going from left to right as written in the stoichiometric reaction equation. It is linked to the individual stoichiometric coefficients ν_i in Eq. (2.17). If the reaction proceeds 1 mol times, we can see that ν_i mol products are generated with i for products ($\nu_i > 0$) or for starting materials ($\nu_i < 0$).

ν_i – Stoichiometric coefficient for component i with positive sign for products and negative sign for reactants;
ξ – Formal reaction progress coefficient for a given reaction equation [mol].

$$\mathrm{d}N_i = \nu_i \,\mathrm{d}\xi \tag{2.17}$$

Q: Is it necessary to introduce all these symbols? Usually, they are small integers, and could these be employed without resort to all these symbols?

A: These more abstract equations may seem complicated, but they offer a single formalism for many different processes. Formal expressions like this are similar to computer programs that allow input data to be converted to output information. When using databases with many types of reactions and materials properties, the formal approach can provide clarity. Let's consider an example for the reaction equation associated with the Daniell cell, correctly written as spontaneous process: $Cu^{2+} + Zn \rightarrow Cu + Zn^{2+}$. We can write this formally with stoichiometric coefficients as in Eq. (2.18).

$$0 = (-1)\,Cu^{2+} + (-1)\,Zn + (+1)\,Cu + (+1)\,Zn^{2+} \tag{2.18}$$

The stoichiometric coefficients for $\{Cu^{2+}, Zn, Cu, Zn^{2+}\}$ are therefore $\{-1, -1, +1, +1\}$.

Under conditions of constant pressure ($\mathrm{d}p = 0$) and constant temperature ($\mathrm{d}T = 0$), we can rewrite Eq. (2.16).

$$(\mathrm{d}g)_{T,p} = \sum_{i=1}^{n} \nu_i \tilde{\mu}_i \mathrm{d}\xi = \left(\sum_{i=1}^{n} \nu_i \tilde{\mu}_i \right) \mathrm{d}\xi = \Delta_{\mathrm{R}} G \cdot \mathrm{d}\xi \tag{2.19}$$

The expression in brackets before $\mathrm{d}\xi$ can be identified as the change in molar Gibbs free energy or molar Gibbs energy of reaction, $\Delta_{\mathrm{R}}G$. Here, molar quantities such as G (in J mol^{-1}) are expressed in capital letters, and the corresponding extensive (mass-dependent) quantities are expressed with a small g (in J). A change in a property X for a given reaction equation or stoichiometry (with all other parameters kept constant) is denoted as $\Delta_{\mathrm{R}}X$. Having defined the change in molar Gibbs free energy, $\Delta_{\mathrm{R}}G$, for the electrolysis process, we can now distinguish three types of processes:

$\Delta_{\mathrm{R}}G$ – Molar Gibbs energy or reaction [J mol^{-1}]

$\Delta_{\mathrm{E}} G < 0$	Spontaneous reaction	Galvanic cell
$\Delta_{\mathrm{E}} G = 0$	Process in equilibrium	No net reaction in the cell
$\Delta_{\mathrm{E}} G > 0$	Reaction requires input of energy	Electrolytic cell

Next, we can express the theoretical maximum for the work done by the electrochemical cell (at constant temperature and pressure) for a very small (infinitesimal) amount of conversion as $\mathrm{d}g = \Delta_{\mathrm{R}}G\, \mathrm{d}\xi$. Here, the parameter $\mathrm{d}\xi$ suggests a very small amount of reaction without a significant overall change in composition. In a galvanic cell without external exchange of materials, the concentration of starting materials will go down very slightly, and the concentration of products will go up. If we allow more current to flow, and therefore, more reaction to occur, the molar Gibbs free energy change, $\Delta_{\mathrm{R}}G$, as well as other chemical potentials, may change.

The work done by the galvanic cell can be obtained as electrical work, $\mathrm{d}g = \mathrm{d}W_{\mathrm{el}}$. We can express the electrical work $\mathrm{d}W_{\mathrm{el}} = -U\mathrm{d}Q$ as work required for moving charge Q from the potential of one electrode to the potential of the other electrode. Based on Eq. (2.1), this difference in potential is equal to the difference in electrode potentials or $U = E_1 - E_2$. The amount of charge Q is linked to the reaction stoichiometry as indicated in Eq. (2.20).

$$\mathrm{d}Q = N_{\mathrm{A}}\, e\, n\, \mathrm{d}\xi = n\, F\, \mathrm{d}\xi. \tag{2.20}$$

Q – Charge [A s] = [C]
N_{A} – Avogadro number [mol^{-1}]
e – Elementary charge
n – Stoichiometric number of charges as defined in the reaction
F – Faraday constant

The parameter n here is an integer without units and describes the number of electrons taking part in the process (stoichiometry) as defined in the reaction equation. This parameter is the **electron stoichiometry**. Formally, n can be derived by adding up all charges exchanged by *one phase* (here α), taking into account stoichiometric factors ν_i and charges z_i:

$$n = \left| \sum_i \nu_i^{\alpha} z_i \right|. \tag{2.21}$$

n – Electron stoichiometry
z_i – Charge of ion i
ν_i – Stoichiometric coefficient of component i
α – Phase index
i – Ion index

We can now combine Eqs. (2.19) and (2.21) to give

$$(\mathrm{d}g)_{T,p} = \Delta_{\mathrm{R}}G\mathrm{d}\xi = \mathrm{d}W_{\mathrm{el}} = -nUF\mathrm{d}\xi \tag{2.22}$$

$$\Delta_{\mathrm{R}}G = -nFU_{\mathrm{eq}} \tag{2.23}$$

ξ – Formal reaction progress coefficient for a given reaction equation [mol]
W_{el} – Electrical work [J] = [V A s]
U_{eq} – Cell voltage in equilibrium [V]

Q: Why do we select the minus sign in this equation?

A: The minus sign originates from conventions that link the potential/energy scales and thermodynamic properties, but there is no fundamental law of nature involved. The key principles to remember in this context are (i) that in chemical thermodynamics, $\Delta_R G$ is negative for spontaneous reactions and (ii) in the galvanic cell, the parameter ξ is positive in the direction of the cell reaction as written. We therefore have to write the cell reaction correctly to represent a spontaneous process. This is what we did when writing Eq. (2.15) for the Daniell cell.

Based on the principles of chemical thermodynamics, we can now write the Gibbs–Helmholtz equation[11] to link the Gibbs free energy to enthalpy and entropy.

$\Delta_R H$ – Molar reaction enthalpy [J mol^{-1}]
$\Delta_R S$ – Molar reaction entropy [J K^{-1} mol^{-1}]
T – Absolute temperature [K]
$\Delta_R G^\circ$ – Molar Gibbs standard reaction energy [J mol^{-1}]
$\Delta_R H^\circ$ – Molar standard reaction enthalpy [J mol^{-1}]
$\Delta_R S^\circ$ – Molar standard reaction entropy [J K^{-1} mol^{-1}]

$$\Delta_R G = \Delta_R H - T\Delta_R S \tag{2.24}$$

A specific case of Eq. (2.24) is expressed when all components in the reaction are in their standard state (denoted by °), which refers to the appropriate phase and unit activity at 1 bar pressure and at a given temperature.

$$\Delta_R G^\circ = \Delta_R H^\circ - T\Delta_R S^\circ \tag{2.25}$$

Equation (2.25) offers convenient access to the molar Gibbs standard energy of reaction from tabulated chemical thermodynamic data. The parameter $\Delta_R G$ can be expressed based on a standard term and an activity-dependent term. Equation (2.26) shows these two terms. The standard term is obtained from tabulated values based on agreed-upon standard conditions. The activity-dependent term allows any variations in the experimental system to be expressed. This equation will also be the key to expressing the cell voltage for electrochemical cells.

R – Gas constant = 8.314 J mol^{-1} K^{-1} (obtained by multiplying the Boltzmann constant and the Avogadro constant)

$$\Delta_R G = \Delta_R G^\circ + RT\ln\prod_i \left(\frac{a_i}{a^\circ}\right)^{\nu_i}. \tag{2.26}$$

All components in the reaction are expressed in terms of their activities a_i as a measure of their effective presence. More details on the topic of activities will be provided in Section 2.2.3. Only for very low activities can we assume that the activity is equal to a concentration term. The standard activity a° is associated with the standard state of each component and effectively removes the units in the logarithmic term. Division by $-nF$ in Eq. (2.26) gives Eq. (2.27) and the cell voltage as defined in Eq. (2.1).

U – Equilibrium cell voltage [V]
U° – Standard equilibrium cell voltage [V]
R – Gas constant [J K^{-1} mol^{-1}]
a_i – Activity, for example, $a_{c,i}$ – activity of component i in terms of concentration [mol L^{-1}]
a° – Standard activity, for example, a_c° = 1 mol L^{-1}
ν_i – Stoichiometric coefficient with positive or negative sign

$$U_{eq} = U^\circ - \frac{RT}{nF}\ln\prod_i \left(\frac{a_i}{a^\circ}\right)^{\nu_i} \tag{2.27}$$

This equation is important and known as the **Nernst equation**[10]. In this equation, a standard term U° is associated with an activity-dependent term with stoichiometric coefficients ν_i, which are positive for products and negative for reactants. An alternative way of writing the Nernst equation is shown in Eq. (2.28), where products and

[11] Hermann Ludwig Ferdinand von Helmholtz, 1821–1894, German physiologist and physicist, educated as medical doctor, professor of physiology in Berlin, Königsberg, Bonn, and in Heidelberg, from 1870 professor of physics in Berlin, important contributions to physiology, optics, electrodynamics, mathematics, thermodynamics (see enthalpy = Helmholtz energy), and meteorology.

reactants are separated more clearly. In this case, the absolute values of the stoichiometric coefficients must be used.

$$U_{eq} = U^\circ - \frac{RT}{nF}\ln\frac{\prod_{\text{Products}}^{i}\left(\frac{a_i}{a^\circ}\right)^{|\nu_i|}}{\prod_{\text{Reagents}}^{j}\left(\frac{a_j}{a^\circ}\right)^{|\nu_j|}} \quad (2.28)$$

The multiplier $RT/(nF)$ in this expression is found very commonly and is often called **Nernst factor** E_{NF}. Very often, the decadic logarithm is employed rather than the natural logarithm, and an additional factor 2.3 = ln(10) is applied to yield Eq. (2.29).

$$U_{eq} = U^\circ - 2.3\frac{RT}{nF}\log\prod_i\left(\frac{a_i}{a^\circ}\right)^{\upsilon_i}. \quad (2.29)$$

$\ln(10) \approx 2.3$ or $e^{2.3} \approx 10$
therefore $K = 10^x = (e^{2.3})^x = (e^x)^{2.3}$ and
$\ln K = \ln(e^x)^{2.3} = 2.3x$
$\lg K = \lg 10^x = x$
and therefore $2.3 \lg K = \ln K$

By substituting in the gas constant $R = 8.314$ J K^{-1} mol^{-1} = 8.314 V A s K^{-1} mol^{-1}, the Faraday constant $F = 96\,485$ A s mol^{-1}, and the temperature $T = 298$ K = 25 °C, we obtain Eq. (2.30).

$$U_{eq} = U^\circ - \frac{0.059\,\text{V}}{n}\log\prod_i\left(\frac{a_i}{a^\circ}\right)^{\upsilon_i}(T = 298\ \text{K}). \quad (2.30)$$

This expression is of practical importance as it directly implies a link from activity (for each of the components i at the electrode surface) to the equilibrium cell voltage and *vice versa*.

The cell voltage at room temperature changes by 59 mV/n for a change in activity (or in concentration) by a factor of 10.

The equilibrium cell voltage is equal to the difference in two half-cell potentials $U_{eq} = E_2 - E_1$ [Eq. (2.1), Figure 2.2]. We select potential E_1 and E_2 to associate each activity with one of the two electrode processes. This can be illustrated for the case of the Daniell cell (Figure 2.7): the cell reaction is $Cu^{2+} + Zn \rightarrow Cu + Zn^{2+}$. Based on Eq. (2.27), we now write the equilibrium cell voltage, as shown in Eq. (2.31).

$$U_{eq} = U^\circ - \frac{RT}{2F}\ln\left[\left(\frac{a_{Cu^{(0)}}}{a^\circ_{Cu^{(0)}}}\right)^{(+1)}\left(\frac{a_{Zn^{2+}}}{a^\circ_{Zn^{2+}}}\right)^{(+1)}\left(\frac{a_{Zn^{(0)}}}{a^\circ_{Zn^{(0)}}}\right)^{(-1)}\left(\frac{a_{Cu^{2+}}}{a^\circ_{Cu^{2+}}}\right)^{(-1)}\right]. \quad (2.31)$$

In Figure (2.2), we illustrated that electrode potentials are always measured versus a reference electrode potential and by subtraction the reference electrode terms are eliminated. This is also the case for electrode processes under standard conditions, and therefore, we can write the standard equilibrium cell voltage U° based on the difference of two **standard electrode potentials** E°.

$$U^\circ = E_2^\circ - E_1^\circ \quad (2.32)$$

Standard electrode potentials are electrode potentials, as defined in Eq. (2.1), for the (sometimes hypothetical) case that measurements are performed with all components present in standard activity a° (as defined for each component). We select $E_2^\circ - E_1^\circ$ to make U° positive. That is, the less noble electrode potential is subtracted from the more noble electrode potential. Obviously, U° can be obtained only when the same agreed-upon reference electrode is employed for both half-cells. The electrodes also need to be at the same temperature. In Section 2.3.3, we will further explore complications linked to deviation in temperature. For the Daniell cell, we can write

$$U^\circ = E^\circ_{Cu|Cu^{2+}} - E^\circ_{Zn|Zn^{2+}}. \quad (2.33)$$

$\ln(a \cdot b) = \ln a + \ln b$
$\ln(a/b) = -\ln(b/a) = \ln[(b/a)^{(-1)}]$

Note the inversion of numerator and denominator for Zn and the negative signs before curly brackets and in the exponents.

E – Electrode potentials [V]
$E°$ – Standard electrode potentials [V]
R – Gas constant [8.314 J K^{-1} mol^{-1}]
T – Temperature [K]
F – Faraday constant [96 485 A s mol^{-1}]
a_i – Activity, for example, $a_{c,i}$ – concentration activity [mol L^{-1}]
$a°$ – Standard activity, for example, $a_c^{\circ} = 1$ mol L^{-1}

$-\ln(a/b) = +\ln(b/a)$

z – Charge of the metal ion and also charge transferred for reduction to metal ($= n$)

a_O/a_O° could be considered a normalized activity. Normalized parameters are important and commonly employed as they are free of units. They could appear abstract, but can be made plausible by considering the normalized parameter as relative to an agreed-upon standard value, here for the activity. A similar case can be seen for the charge of an ion. The parameter used to describe the charge of ions is dimensionless, but always relative to the charge of the proton. The authors recommend using and practicing the concept of dimensionless quantities because it is important in many situations in electrochemistry and elsewhere.

$|\nu_O|$ ($|\nu_R|$) – Absolute value of the stoichiometric coefficient for the oxidized (reduced) species
n – Electron stoichiometric number (dimensionless)

a_{Base} ($a_{\text{Säure}}$) – Activity of base (acid)
pK_a – Negative decadic logarithm of the acid dissociation constant

From Eq. (2.31), we can derive Eq. (2.34) by sorting the terms for each half-cell.

$$U = \left\{E^{\circ}_{\text{Cu|Cu}^{2+}} - \frac{RT}{2F}\ln\left[\left(\frac{a_{\text{Cu}^{(0)}}}{a^{\circ}_{\text{Cu}^{(0)}}}\right)^{(+1)}\left(\frac{a_{\text{Cu}^{2+}}}{a^{\circ}_{\text{Cu}^{2+}}}\right)^{(-1)}\right]\right\} - \left\{E^{\circ}_{\text{Zn|Zn}^{2+}} - \frac{RT}{2F}\ln\left[\left(\frac{a_{\text{Zn}^{(0)}}}{a^{\circ}_{\text{Zn}^{(0)}}}\right)^{(+1)}\left(\frac{a_{\text{Zn}^{2+}}}{a^{\circ}_{\text{Zn}^{2+}}}\right)^{(-1)}\right]\right\}. \tag{2.34}$$

Now, we can identify in Eq. (2.34) terms for each of the two electrodes in half-cells. We can write out formal expressions for $E_{\text{Cu/Cu}^{2+}}$ and for $E_{\text{Zn/Zn}^{2+}}$.

$$E_{\text{Cu|Cu}^{2+}} = E^{\circ}_{\text{Cu|Cu}^{2+}} - \frac{RT}{2F}\ln\left(\left(\frac{a_{\text{Cu}^{(0)}}}{a^{\circ}_{\text{Cu}^{(0)}}}\right)^{(+1)}\left(\frac{a^{\circ}_{\text{Cu}^{2+}}}{a_{\text{Cu}^{2+}}}\right)^{(+1)}\right), \tag{2.35}$$

$$E_{\text{Zn|Zn}^{2+}} = E^{\circ}_{\text{Zn|Zn}^{2+}} - \frac{RT}{2F}\ln\left(\left(\frac{a_{\text{Zn}^{(0)}}}{a^{\circ}_{\text{Zn}^{(0)}}}\right)^{(+1)}\left(\frac{a^{\circ}_{\text{Zn}^{2+}}}{a_{\text{Zn}^{2+}}}\right)^{(+1)}\right). \tag{2.36}$$

This result can be generalized for processes at metal electrodes in equilibrium with the corresponding metal cations.

$$E_{\text{M|M}^{z+}} = E^{\circ}_{\text{M|M}^{z+}} - \frac{RT}{zF}\ln\left(\frac{\frac{a_{\text{M}^{(0)}}}{a^{\circ}_{\text{M}^{(0)}}}}{\frac{a_{\text{M}^{z+}}}{a^{\circ}_{\text{M}^{z+}}}}\right) = E^{\circ}_{\text{M|M}^{z+}} + \frac{RT}{zF}\ln\left(\frac{\frac{a_{\text{M}^{z+}}}{a^{\circ}_{\text{M}^{z+}}}}{\frac{a_{\text{M}^{(0)}}}{a^{\circ}_{\text{M}^{(0)}}}}\right) \tag{2.37}$$

It is now apparent that for every redox process R $\rightleftharpoons$ O + n e$^-$ at an electrode, we can write out an equation analogous to the Nernst equation in (2.28). However, in this case, both products and reactants are present at a single electrode. In the derivation of this equation, sign conventions apply. As a rule of thumb, it is useful to remember that activity or concentration terms in the logarithm require oxidized forms of components in the numerator and reduced forms of components in the denominator if a plus sign is in front of the logarithm. Standard activity terms are included here, but are sometimes not shown explicitly, for example, when standard activity terms cancel or when working with dimensionless parameters. It is important to explicitly include standard activities a_O° or a_R° in cases where different units are employed (e.g., combining concentration and pressure) and to carry these forward throughout calculations. Unfortunately, this methodology can be prone to error.

$$E = E^{\circ} + \frac{RT}{nF}\ln\frac{\left(a_O/a_O^{\circ}\right)^{|\nu_O|}}{\left(a_R/a_R^{\circ}\right)^{|\nu_R|}} \tag{2.38}$$

There is a clear analogy from the Nernst Equation (2.38) to the Henderson–Hasselbalch[12] Equation (2.39). The former is associated with an electron transfer equilibrium, whereas the latter is associated with a proton transfer equilibrium.

$$\text{pH} = \text{p}K_a + \log\frac{a_{\text{Base}}}{a_{\text{Acid}}} \tag{2.39}$$

[12]Lawrence Joseph Henderson, 1878–1942, professor of biological chemistry at Harvard University; Karl Albert Hasselbalch, 1874–1962, Danish researcher who obtained a PhD in medicine and physiology, developed an understanding for the pH of blood.

The Nernst equation for electrode potentials describes the equilibration of electrons between two (or more) phases when oxidized and reduced species are both present at the interface (e.g., $Fe^{2+}(aq)$ and $Fe^{3+}(aq)$ both exchange e^- at a Pt metal electrode; or $Cu^{2+}(aq)$ cations in solution and $Cu^{(0)}$ in the lattice surrounded by an electron gas exchanging Cu^{2+} across the solid–liquid interface). The Henderson–Hasselbalch Equation (2.39) describes the interaction of a proton, to be in the system either as $H_3O^+(aq)$ or bound to a proton acceptor (Brønsted base) to yield the corresponding Brønsted acid.

2.2.3 Activities Describe Effective Content

In this section, practical examples will be discussed for equilibrium cell voltage calculations based on equations introduced in Section 2.1.2. We will consider each term of Eq. (2.27) in turn and start with the activity-dependent term.

Concentration and density are examples of amount-independent or intensive properties, whereas amount [mol], volume [cm^3, L], and mass [g, kg] are amount-dependent and therefore extensive properties. Concentration or density is derived by division of two extensive properties, for example, amount/volume or mass/volume. Activities a_i in Eq. (2.27) have to be considered as generalized and corrected concentrations. Generally, for types of mixtures (e.g., solids, liquids, electrolyte solutions, and gases) with interacting particles, we introduce activities to account for interactions that change the tendencies of particles to react. The activity coefficient f links the concentration parameter to the corresponding activity. For each type of mixture, a standard activity $a°$ is defined by convention.

For the case of electrolyte solutions, predominantly the **concentration** c_i for component i is employed and defined as follows:

$$c_i = \frac{N_i}{V_{sol}}. \tag{2.40}$$

c_i – Concentration of component i [mol L^{-1}]
V_{sol} – Volume of the solution [L]
N_i – Amount of component I [mol]

There are different styles of writing the concentration for a particular species with the help of subscripts or brackets, for example, $c_{Na^+} = c(Na^+) = [Na^+]$. The corresponding activity a_c can be written as $a_{Na^+} = a_{c,Na^+} = a_c(Na^+) = \{Na^+\}$. The choice of a particular presentation is often linked to avoiding complexity in notation. We can now introduce the activity coefficient f_c for the concentration.

$$a_c = c f_c \quad \text{or} \quad \{X\} = [X] f_c \tag{2.41}$$

a_c, $a_{c,X}$, $a_c(X)$, $\{X\}$ – Activity for concentration of X [mol L^{-1}]
f_c – Activity coefficient [dimensionless]
c, c_X, $c(X)$, $[X]$ – Concentration of X [mol L^{-1}]

The activity coefficient f_c is linked to intermolecular interactions, and therefore, itself is a function of the concentration c. Values for activity coefficients have to be measured by experiment (Section 2.4.2); in some cases, they can be found in tables in data collections [4–6], or we can employ interpolation tools in conjunction with tables (see Example Calculation 2.11). In the data section of this book (Appendix, Table A-5), activity coefficients are provided for exercises. The value of the activity coefficient approaches unity in the limit of very low concentrations (as interactions diminish). The corresponding case of $f_c = 1.0$ is called an **ideal solution**. The standard state for the electrolyte solution is defined as a solution with activity $a_c^\circ = 1$ mol L^{-1}. In this book, activity coefficients routinely refer to the case of concentrations, and therefore, we write f instead of f_c, avoiding the index c.

For high-precision experiments, working with the concentration parameter is associated with the disadvantage of volume changes with temperature. Heating of a solution will lead to expansion and therefore to a deviation in the concentration. This is the reason for introducing an alternative approach based on the amount of material per mass of solvent, the so-called **molality**. Molality is defined as follows:

$$b_i = \frac{\text{Amount of compound } i}{\text{Mass of the solvent}} = \frac{N_i}{m_{\text{solv}}}. \tag{2.42}$$

In the electrochemistry literature, molality is often represented by the symbol *m*, which can be easily confused with the symbol for mass. Therefore, we use a different symbol here. However, we stay with symbol *m* for molality when used as a subscript to activity or activity coefficient.

The activity associated with molality a_m and the corresponding activity coefficient f_m can be defined accordingly.

$$a_{m,i} = b_i f_{m,i}. \tag{2.43}$$

The standard state in this case has to be defined as the standard molality $b° = 1\,\text{mol kg}^{-1}$. Again, the activity coefficient will converge to unity for very low molalities when b converges toward zero (to give an ideal solution). Note the key difference that m_{solv} in Eq. (2.42) is linked to mass of the solvent (without solute), whereas V_{sol} in Eq. (2.40) is linked to the volume of the solution (containing both solvent and solute).

Q: How closely related are *molality* and *molarity*, and why are these terms associated with units? And how does this compare to the term *normality*?

A: The term *molality* refers to a physical quantity, and the term molarity refers to a unit [mol L^{-1}]. When stating the composition of a solution, similarity of the numerical value of the molality to that of *molarity* is possible only for a solvent with a density close to 1000 g L^{-1} (water). However, another important point is not to confuse a physical property such as the concentration, with expressions based on units such as *molarity*. Another example of a confusion of a property (e.g., the current I) and its unit (here A for ampere) can be found when current is referred to as "amperage". Physical properties represented by a symbol in italics are the product of a number and a unit denoted by a symbol without italics (e.g., $I = 100$ A = 0.1 kA). The physical property and its unit need to be clearly kept apart. This convention allows parameters such as the area of an electrode A to be clearly distinguished from the unit A for ampere [7]. A notable exception can be found in the property *voltage* (used throughout this book), describing an electrical tension, which is a property linked to the unit volt [V].

The physical property *normality* can be found in some older literature referring to the concentration of protons (e.g., 1 N H_2SO_4 = 0.5 M H_2SO_4 and 1 N HCl = 1 M HCl). The modern SI system recommends that terminology like *molarity* (for concentration) and *normality* be phased out. The very common abbreviation 1 M can be employed for 1 mol L^{-1} to represent the units for concentration, and both are correct.

b_i – Molality of component i [mol kg^{-1}]
N_i – Amount of component i [mol]
m_{solv} – Mass of solvent [kg]

For both solid solutions and mixtures of two liquids, the **mole fraction** x offers a convenient and dimensionless method of expressing the composition. The mole fraction takes on values between 0 and 1. The corresponding activity is the mole fraction activity a_x. The standard mole fraction is defined as the pure compound ($x = a_x^\circ = 1$). The corresponding mol fraction activity coefficient f_x converges toward unity for the pure compound.

$a_{m,i}$ – Activity for molality [mol kg^{-1}]
$f_{m,i}$ – Activity coefficient for molality [dimensionless]

$$x_i = \frac{N_i}{\sum_{j=1}^{k} N_j} \tag{2.44}$$

$$a_{x,i} = x_i f_{x,i} \tag{2.45}$$

x_i – Mole fraction for component i [dimensionless]
N_i, N_j – Amounts of components i and j [mol]

$a_{x,i}$ – Mole fraction activity [dimensionless]
$f_{x,i}$ – Mole fraction activity coefficient

The mole fraction is also readily applied to the solvent. For dilute aqueous electrolyte solutions, the mole fraction is essentially unity. For 1 L of water, we have approximately 55.6 mol H_2O. With the addition of typically 0.1 mol L^{-1} electrolyte, the deviation from unity for the mole fraction of H_2O is only 0.2%. When inspecting Eq. (2.11), we can also see that the ln (natural logarithm) in the concentration-dependent term makes results very insensitive to minor changes in $a(H_2O)$. As a consequence, $a(H_2O)$ is often assumed to be unity and eliminated from expressions. For example, in the definition of the acid dissociation constant K_a, contributions from $a(H_2O)$ are commonly ignored.

For processes in gaseous phase, a measure of composition is the **partial pressure** p_i for a gas component i. The partial pressure p_i can be expressed based on the mole fraction $p_i = x_i\, p$ with p being the total pressure. The corresponding activity is the so-called fugacity ψ. The standard state of a gas is defined as the pure ideal gas at standard pressure $p^\circ = 1$ bar $= 100\,000$ Pa. For this case, the fugacity coefficient is unity. In most electrochemical systems, the behavior of gases is assumed to be close to ideal so that the fugacity is often ignored, and partial pressures are used directly. There are notable exceptions, for example, when we use the cell voltage to measure the fugacity of a gas like chlorine.

Note that the symbol for the outer potential, for fugacity, and for the wave function in quantum mechanics are identical. These are entities that are easily distinguished from the context.

In older literature, the standard pressure is stated as $p = 1$ atm $= 101\,325$ Pa. The difference between the older $p = 101\,325$ Pa and the more recent definition of standard pressure as $p^\circ = 100\,000$ Pa leads to a potential shift in the hydrogen electrode potential (Section 2.4.1) of +0.169 mV. Therefore, this potential shift needs to be subtracted when employing older tabulated data for $p = 101\,325$ Pa. However, this is a minor shift in potential, and for most practical measurements, errors in potential readings may exceed 1 mV. Therefore, often this correction is negligible.

2.2.4 Calculating the Standard Cell Voltage

The standard cell voltage $U^\circ = U^\circ(T, p)$ in Eq. (2.28) is the equilibrium cell voltage for the case that all reaction components are in their standard state and present in their standard activity a°. This results in the logarithmic term in Eq. (2.28) going to zero with $a^\circ/a^\circ = 1$ Therefore, we have $U = U^\circ$. There are different approaches to the calculation of U°.

Calculations Based on Standard Electrode Potentials E°. There are extensive data collections for standard electrode potentials typically at 25 °C = 298.15 K at ambient pressure (100 000 Pa) (Appendix, Table A-1). U° can be obtained directly based on equation Gl. (2.32) by employing $U^\circ = E_2^\circ - E_1^\circ$.

Calculations Based on Tables of Molar Standard Gibbs Energies of Formation $\Delta_F G^\circ$. Molar standard Gibbs energies of formation provide data for the formation (from the elements in standard state) of a mol of a component under standard conditions (Appendix, Table A-9). Due to thermodynamic *functions of state* like the molar Gibbs free energy being *path-independent* (only depending on initial and final state), we can employ combinations of $\Delta_F G^\circ$ data to represent any chemical reaction. We combine the endpoint of the reaction (by expressing $\Delta_F G^\circ$ for products) with the starting point (by subtracting $\Delta_F G^\circ$ for the reactants) to give the molar standard Gibbs free energy for the reaction $\Delta_R G^\circ$. Equation (2.46) employs sign-dependent stoichiometric coefficients (negative for reactants and positive for products).

$$\Delta_R G^\circ = \sum_i \nu_i \Delta_F G_i^\circ. \tag{2.46}$$

$\Delta_R G^\circ$ – Molar standard Gibbs free energy of reaction [J mol^{-1}]

$\Delta_F G^\circ$ – Molar standard Gibbs free energy of formation [J mol^{-1}]

ν_i – Stoichiometric coefficient for component i, dimensionless, sign-dependent

From the calculated $\Delta_R G^\circ$, we then obtain the standard cell voltage with Eq. (2.23) $U^\circ = -\Delta_R G^\circ/(nF)$.

Calculations Based on Enthalpy of Formation and Entropy of Formation. For the case that $\Delta_F G^\circ$ data are not available, we can employ Eq. (2.25) to derive these data. We use tabulated values for the molar standard enthalpy of formation and the molar

$\Delta_R H^\circ$ – Molar standard enthalpy of reaction [J mol^{-1}]
$\Delta_F H^\circ$ – Molar standard enthalpy of formation [J mol^{-1}]
$\Delta_R S^\circ$ – Molar standard entropy of reaction [J K^{-1} mol^{-1}]
S_i° – Molar standard entropy [J K^{-1} mol^{-1}]

standard entropy to calculate $\Delta_R H^\circ$ and $\Delta_R S^\circ$. In this way, standard cell voltages are derived from predominantly calorimetric thermodynamic information.

$$\Delta_R H^\circ = \sum_i \nu_i \Delta_F H_i^\circ \tag{2.47}$$

$$\Delta_R S^\circ = \sum_i \nu_i S_i^\circ \tag{2.48}$$

This approach based on combining Eqs. (2.46)–(2.48) allows starting from enthalpies of formation and from entropies of formation to express the Gibbs energies and entropies for individual ions. However, a fundamental obstacle here is the need for a reference state, as solutions of purely cations or purely anions cannot be obtained. Here, by definition, the values for $\Delta_F G^\circ$, $\Delta_F H^\circ$, and S_i° for ions are given *relative* to those for protons in aqueous solution. By convention, the corresponding values for protons $H^+(aq)$ are set to zero.

$$\Delta_F G^\circ(H^+, aq) = 0, \tag{2.49}$$

$$\Delta_F H^\circ(H^+, aq) = 0, \tag{2.50}$$

$$S^\circ(H^+, aq) = 0. \tag{2.51}$$

This is an arbitrary choice, but consistent with the convention of selecting the standard cell potential for $H_2(g)/H^+(aq)$ as zero V. With these conventions in place, we are able to derive thermodynamic and standard cell voltage information for the formation or loss of ionic species. A systematic approach to formally writing out and analyzing standard cell voltages is now possible.

Example Calculation 2.3 Deriving the ion formation reactions for Cu^{2+} and for sulfate ions in aqueous solution as a starting point for the derivation of $\Delta_F G^\circ$, $\Delta_F H^\circ$, S_i°

before balancing with H^+/H_2: $Cu \rightarrow Cu^{2+} + 2e^-$
after balancing with H^+/H_2: $Cu(s) + 2\,H^+(aq) \rightarrow Cu^{2+}(aq) + H_2(g)$

before balancing with H^+/H_2: $S + 2\,O_2 + 2\,e^- \rightarrow SO_4^{2-}$
after balancing with H^+/H_2: $S(s) + 2\,O_2(g) + H_2(g) \rightarrow SO_4^{2-}(aq) + 2\,H^+(aq)$

By employing enthalpy and entropy to derive Gibbs free energy, we have accounted for primary temperature effects [Eq. (2.24)]. Therefore, values for $\Delta_R H^\circ$ and $\Delta_R S^\circ$ are relatively temperature-insensitive over small temperature ranges (and avoiding any phase transitions). However, temperature effects are not generally insignificant, and in some cases, the effect on $\Delta_F H^\circ$ and $\Delta_F S^\circ$ needs to be explicitly accounted for in terms of heat capacities. Further details can be found in textbooks for chemical thermodynamics [1–3].

2.2.5 Linking the Equilibrium Cell Voltage to Thermodynamic Functions of State

In this section, the rules and strategies for determining the equilibrium cell voltage U_{eq} will be put to the test. Examples for calculations will be provided to demonstrate the generality of the approach to equilibrium cell voltages based on thermodynamic properties.

Example Calculation 2.4 Effect of Concentration on the Equilibrium Cell Voltage for the Daniell Cell. We want to determine the equilibrium cell voltage for a Daniell cell at room temperature. We will assume that the copper electrode is immersed in aqueous 0.1 mol L^{-1} $CuSO_4$ solution and the zinc electrode is immersed in 2 mol L^{-1} $ZnSO_4$ solution.

Approach:
We employ tables for standard electrode potentials
Cell reaction: $Cu^{2+}(aq) + Zn(s) \rightarrow Cu(s) + Zn^{2+}(aq)$ ($n = 2$)
Determine U_{eq}
Known:

$$[Cu^{2+}] = 0.1 \text{ mol L}^{-1}$$
$$f_c(Cu^{2+}) = 0.150 \text{ (from Table A-5)}$$
$$[Zn^{2+}] = 2 \text{ mol L}^{-1}$$
$$f_c(Zn^{2+}) = 0.035 \text{ (from Table A-5)}$$
$$E^\circ(Cu/Cu^{2+}) = +0.3400 \text{ V (from Table A-1)}$$
$$E^\circ(Zn/Zn^{2+}) = +0.7000 \text{ V (from Table A-1)}$$
$$T = 298 \text{ K (we assume a temperature of 25 °C)}$$
$$n = 2$$
$$F = 96\,485 \text{ A s mol}^{-1}$$

$$U = U^\circ - \frac{RT}{nF}\ln\prod_i \frac{a_i^{\nu_i}}{a^{\circ\nu_i}} = [E_1^\circ - E_2^\circ] - \frac{RT}{nF}\ln\prod_i \frac{c_i^{\nu_i} f_{c,i}^{\nu_i}}{a^{\circ\nu_i}} \tag{2.52}$$

$$U = [E^\circ(Cu/Cu^{2+}) - E^\circ(Zn/Zn^{2+})] - \frac{RT}{2F}\ln\frac{[Zn^{2+}]\cdot f_c(Zn^{2+})\cdot\{Cu^{(0)}\}}{\{Zn^{(0)}\}\cdot[Cu^{2+}]\cdot f_c(Cu^{2+})} \tag{2.53}$$

Note that here the activity of pure solid phases Cu(*s*) and Zn(*s*) can be set to unity (Section 2.2.1).

$$U = [0.3400\,\text{V} - 0.7000\,\text{V}] - \frac{8.314\,\text{J mol}^{-1}\,\text{K}^{-1} \times 298\,\text{K}}{2 \times 96\,485\,\text{A s mol}^{-1}}\ln\frac{2\,\text{mol L}^{-1} \times 0.035}{0.1\,\text{mol L}^{-1} \times 0.150} \tag{2.54}$$

1 J = 1 W s = 1 V A s

1 J/1 A s = 1 V

$$U = 1.1000\text{ V} - 0.01977\text{ V} \tag{2.55}$$

$$U = 1.0802\text{ V} \tag{2.56}$$

The equilibrium cell voltage for the Daniell cell under the specified conditions is 1.08 V.

◢

Example Calculation 2.5 The Equilibrium Cell Voltage for the Daniell Cell as a Function of Temperature. We will derive the equilibrium cell voltage for the Daniell cell (Example Calculation 2.4) for the temperatures 50 and 70 °C. This will be done by employing standard molar enthalpy and standard molar entropy data and Eq. (2.23).

Sought: $U(T_2)$, $U(T_3)$
Known:

$$\Delta_F H(Cu^{2+}) = -64.77 \text{ kJ mol}^{-1} \text{ (from Table A-9)}$$
$$S(Cu^{2+}) = -99.6 \text{ J K}^{-1}\text{ mol}^{-1} \text{ (from Table A-9)}$$

$$S(\mathrm{Cu}) = 33.150\ \mathrm{J\ K^{-1}\ mol^{-1}}\ \text{(from Table A-9)}$$

$$\Delta_F H(\mathrm{Zn^{2+}}) = -153.89\ \mathrm{kJ\ mol^{-1}}\ \text{(from Table A-9)}$$

$$S(\mathrm{Zn^{2+}}) = -112.1\ \mathrm{J\ K^{-1}\ mol^{-1}}\ \text{(from Table A-9)}$$

$$S(\mathrm{Zn}) = 41.63\ \mathrm{J\ K^{-1}\ mol^{-1}}\ \text{(from Table A-9)}$$

For simplicity, activity coefficients from Example Calculation 2.4 are assumed to be independent of temperature.

$$T_1 = 298\ \mathrm{K}\ \text{(assumed as room temperature)}$$

$$T_2 = 323\ \mathrm{K}$$

$$T_3 = 343\ \mathrm{K}$$

$$F = 96\,485\ \mathrm{A\ s\ mol^{-1}}$$

Cell reaction: $Cu^{2+}(aq) + Zn(s) \rightarrow Cu(s) + Zn^{2+}(aq)$ $(n = 2)$

$$U = U^\circ - \frac{RT}{nF} \ln \prod_i \frac{a_i^{\nu_i}}{a^{\circ \nu_i}} \tag{2.57}$$

Here, both the standard cell voltage and the concentration-dependent term are temperature-dependent. We can express the temperature effect on the standard cell voltage based on the analogous expression for the molar standard reaction Gibbs free energy $\Delta_R G^\circ$ as in Eqs. (2.23) and (2.25).

$$U^\circ = -\frac{\Delta_R G^\circ}{nF} = -\frac{\Delta_R H^\circ}{nF} + T\frac{\Delta_R S^\circ}{nF} \tag{2.58}$$

The effect of temperature can be expressed as the partial derivative of the standard cell voltage for temperature (at constant pressure). We apply the product rule to give Eq. (2.59).

$$\left(\frac{\partial U^\circ}{\partial T}\right)_p = \frac{-1}{nF}\left(\frac{\partial \Delta_R G^\circ}{\partial T}\right)_p = \frac{-1}{nF}\left(\frac{\partial \Delta_R H^\circ}{\partial T}\right)_p + \frac{\Delta_R S^\circ}{nF}\left(\frac{\partial T}{\partial T}\right) + \frac{T}{nF}\left(\frac{\partial \Delta_R S^\circ}{\partial T}\right) \tag{2.59}$$

When considering the temperature range from 298 to 333 K, there are no phase transitions, and we can treat $\Delta_R H$ and $\Delta_R S$ as essentially constant. Therefore, the derivatives $\partial \Delta_R H / \partial T$ and $\partial \Delta_R S / \partial T$ can be assumed to be zero.

$$\left(\frac{\partial U^\circ}{\partial T}\right)_p = \frac{\Delta_R S^\circ}{nF}. \tag{2.60}$$

The equilibrium cell voltage for the Daniell cell was determined in Example Calculation 2.4 for $T_1 = 298$ K, and based on this, we can now obtain the standard equilibrium voltage for other temperatures employing the linear correction in Eq. (2.60) as shown in Figure 2.8. We assume that $\Delta_R S$ remains constant.

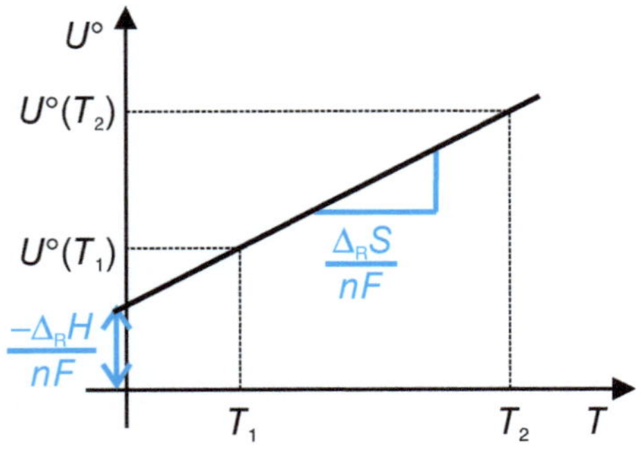

Figure 2.8 Determination of $\Delta_R H^\circ$ and $\Delta_R S^\circ$ from standard equilibrium cell voltage measurements.

$$U(T_2) = U^\circ(T_1) + \frac{\Delta_R S^\circ}{nF}(T_2 - T_1) - \frac{RT_2}{nF} \ln \prod_i \frac{c_i^{\nu_i} f_i^{\nu_i}}{a^{\circ \nu_i}} \tag{2.61}$$

$$\Delta_R S^\circ = \sum_i \nu_i S_i^\circ = (+1)S^\circ(\mathrm{Zn^{2+}})\,(+1)S^\circ(\mathrm{Cu})\,(-1)S^\circ(\mathrm{Zn})\,(-1)S^\circ(\mathrm{Cu^{2+}})$$

$$= [-112.1 + 33.15 - (-99.6) - 41.63]\ \mathrm{J\ K^{-1}\ mol^{-1}}$$

$$= -20.98\ \mathrm{J\ K^{-1}\ mol^{-1}} = -20.98\ \mathrm{V\ A\ s\ K^{-1}\ mol^{-1}}$$

We can now substitute this result into Eq. (2.55) to give

$$\begin{aligned} U(323\,\text{K}) &= 1.1000\,\text{V} + \frac{-20.98\,\text{V A s K}^{-1}\,\text{mol}^{-1}}{2 \cdot 96\,485\,\text{A s mol}^{-1}}(323 - 298\,\text{K}) \\ &\quad - \frac{8.314\,\text{A s V mol}^{-1}\,\text{K}^{-1} \cdot 323\,\text{K}}{2 \times 96\,485\,\text{A s mol}^{-1}} \ln \frac{2\,\text{mol L}^{-1} \cdot 0.035}{0.1\,\text{mol L}^{-1} \cdot 0.150} \\ &= 1.1000\,\text{V} + (-0.00272\,\text{V}) - 0.02143\,\text{V} \\ U(323\,\text{K}) &= 1.0758\,\text{V} \end{aligned}$$

$$\begin{aligned} U(343\,\text{K}) &= 1.1000\,\text{V} + \frac{-20.98\,\text{V A s K}^{-1}\,\text{mol}^{-1}}{2 \cdot 96\,485\,\text{A s mol}^{-1}}(343 - 298\,\text{K}) \\ &\quad - \frac{8.314\,\text{J mol}^{-1}\,\text{K}^{-1} \cdot 343\,\text{K}}{2 \times 96\,485\,\text{A s mol}^{-1}} \ln \frac{2\,\text{mol L}^{-1} \cdot 0.035}{0.1\,\text{mol L}^{-1} \cdot 0.150} \\ &= 1.1000\,\text{V} + (-0.00228\,\text{V}) - 0.02276\,\text{V} \\ U(343\,\text{K}) &= 1.0723\,\text{V} \end{aligned}$$

The equilibrium cell voltage for the Daniell cell is 1.076 and 1.072 V at 50 and 70 °C, respectively. Therefore, the effect of temperature on the Daniell cell is relatively minor, with about –0.61 mV K^{-1}.

◢

Example Calculation 2.6 Electrode Potential for an Electrode of the Second Kind (Involving a Precipitation Equilibrium). What would be the equilibrium electrode potential for an Ag|AgCl electrode immersed in aqueous 1 mol L^{-1} NaCl solution at 25 °C versus a standard hydrogen electrode?

Determine U

Known:

$[\text{NaCl}] = 1\ \text{mol L}^{-1}$

$f_c(\text{NaCl}) = 0.657$ (from Table A-5)

$E°(\text{Ag}|\text{Ag}^+) = +0.80\ \text{V}$ (from Table A-1)

$E(\text{SHE}) = 0$ (by definition)

$n = 2$ (as written in Eq. (2.62))

$pK_{SP}(\text{AgCl}) = 9.75$ (from Table A-3)

$K_{SP}(\text{AgCl}) = a^{\circ 2}\, 10^{-pK_L(\text{AgCl})} = 1\ \text{mol}^2\ \text{L}^{-2}\ 10^{-8.88} = 1.778\ 10^{-10}\ \text{mol}^2\ \text{L}^{-2}$

K_{SP}(AgCl) – Solubility product for AgCl 1.778×10^{-10} mol^2 L^{-2} (from Table A-3)

We can consider the dissolution of a precipitate of a poorly soluble salt $M_nX_m \rightleftharpoons n\,M^{(+)} + m\,X^{(-)}$.

The corresponding equilibrium constant can be written as follows:

$$K_{SP} = \frac{\{M\}^n\{X\}^m}{\{M_nX_m\}} = \{M\}^n\{X\}^m.$$

Formally, the activity of the solid phase in the denominator can be assumed to be unity (mole fraction of a pure solid compound) to give the expression for the solubility product.

$$H_2 + 2\,Ag^+ \rightleftharpoons 2\,H^+ + 2\,Ag \qquad (2.62)$$

$$U = E(\text{Ag}|\text{AgCl}\,|\,1\ \text{mol L}^{-1}\ \text{NaCl}) - E(\text{SHE}) \qquad (2.63)$$

$$U = E°(\text{Ag}|\text{Ag}^+) + \frac{RT}{2F}\ln\frac{\{\text{Ag}^+\}^2}{a_c^\circ\{\text{Ag}^{(0)}\}} \qquad (2.64)$$

We can use the solubility product $K_S = \{Ag^+\}\{Cl^-\}$ to express the silver cation activity $\{Ag^+\} = K_{SP}/\{Cl^-\}$. The mole fraction activity of pure solid silver in Eq. (2.64) is unity. We employ the transformation $\ln a^2 = 2 \ln a$ to cancel out 2 in the denominator and the exponent. From this transformation, it becomes clear that the expression for the electrode potential is independent of the way the reaction is formally written, for example, we could also employ the equation of the cell reaction with other stoichiometric coefficients, for example, ½ H_2 + $Ag^+ \rightleftharpoons H^+$ + Ag.

$$U = E°(\text{Ag}|\text{Ag}^+) + \frac{RT}{F}\ln\frac{K_{SP}}{a_c^\circ\{\text{Cl}^-\}} \qquad (2.65)$$

We substitute $\{Cl^-\} = f_c\,[Cl^-]$ to give

$$U = E^\circ(\mathrm{Ag|Ag^+}) + \frac{RT}{F}\ln\frac{K_{SP}}{a_c^\circ f_c\,[\mathrm{Cl^-}]} \tag{2.66}$$

$$U = +0.80\,\mathrm{V} + \frac{8.314\,\mathrm{V\,A\,s\,mol^{-1}K^{-1}} \cdot 298\,\mathrm{K}}{96\,485\,\mathrm{A\,s\,mol^{-1}}}\ln\frac{1.778\times10^{-10}\,\mathrm{mol^2\,L^{-2}}}{1\,\mathrm{mol\,L^{-1}}\times 1\,\mathrm{mol\,L^{-1}}\times 0.657} \tag{2.67}$$

$$U = E_2 = 0.234\,\mathrm{V}.$$

The electrode potential for the silver–silver chloride electrode in aqueous 1 mol L^{-1} NaCl solution is 0.234 V.

◢

"Condensed phase" is a term describing solid and liquid states.

"Fluid phase" is a term describing liquids, supercritical liquids, gases, and plasmas. Fluid phases will uniformly fill the volume of a container.

Example Calculation 2.7 Equilibrium Cell Voltage for a Fuel Cell. Effects of pressure on the equilibrium cell voltage are usually negligible due to the negligible compressibility of liquids or solids (condensed phases). In contrast, effects of pressure on electrochemical processes involving gases need to be taken into account. Here, we consider a fuel cell operating at 1 bar pressure and at 25 °C with hydrogen and air. In good approximation, we can employ 0.2 bar as the partial pressure of oxygen.

$2\,H_2 + O_2 \rightleftharpoons 2\,H_2O$

Known:

$$p(\mathrm{H_2}) = p^\circ$$
$$p(\mathrm{O_2}) = 0.2\,p^\circ$$
$$T = 25\,^\circ\mathrm{C} = 298\,\mathrm{K}$$
$$E^\circ(\mathrm{H_2}) = 0 \text{ (by convention)}$$
$$E^\circ(\mathrm{O_2/H_2O}) = 1.23\,\mathrm{V} \text{ (see Table A-1)}$$
$$n = 4$$

Sought: U

Combine Eqs. (2.1) and (2.38) to give

$$U = \left(E^\circ(\mathrm{H_2O\,|\,O_2}) + \frac{RT}{4F}\ln\frac{\frac{p_{O_2}}{p^\circ}\left(\frac{\{H^+\}}{a^\circ}\right)^4}{\left(\frac{x(H_2O)}{a_x^\circ}\right)^2}\right) - \left(E^\circ(\mathrm{H_2\,|\,H^+}) + \frac{RT}{4F}\ln\frac{\left(\frac{\{H^+\}}{a^\circ}\right)^4}{\left(\frac{p_{H_2}}{p^\circ}\right)^2}\right) \tag{2.68}$$

$$U = [E^\circ(\mathrm{H_2O\,|\,O_2}) - E^\circ(\mathrm{H_2\,|\,H^+})] + \left[\frac{RT}{4F}\ln\left(\frac{p_{O_2}}{p^\circ}\frac{p_{H_2}^2}{p^{\circ 2}}\right)\right] \tag{2.69}$$

$$U = [1.23\,\mathrm{V} - 0] + \left[\frac{8.314\,\mathrm{J\,mol^{-1}\,K^{-1}}298\,\mathrm{K}}{4\times 96\,485\,\mathrm{A\,s\,mol^{-1}}}\ln\left(\frac{0.2p^\circ}{p^\circ}\frac{p^{\circ 2}}{p^{\circ 2}}\right)\right] \tag{2.70}$$

$x(H_2O) \approx 1 = a_x^\circ$ for water in aqueous solution

$$U = 1.23\,\mathrm{V} + 0.0064\,\mathrm{V}\ln 0.2 \tag{2.71}$$

$$U = 1.2197\,\mathrm{V}$$

Therefore, the equilibrium cell voltage for this fuel cell is 1.22 V. This value is only slightly lower than the standard electrode potential difference for O_2/H_2 because the partial pressure for oxygen is less than 1 bar.

◢

2.2.6 Formal Potentials

The treatment above has demonstrated that working with standard electrode potentials requires the knowledge of activity coefficients, which are often unknown. Furthermore, in many systems, the concentrations of other species in solution (complexing agents, ions involved in ion pairing, H^+/OH^- involved in the electrode reaction) depart substantially from thermodynamically defined standard states with $a_c = 1$ mol L^{-1}. A good example would be buffers with the ionic strength and the pH of biological cells that are used for the characterization of biological redox systems (Section 18.2) because these reaction conditions are much more relevant for the understanding of biological redox systems than their behavior at the standard state (e.g., pH 0). **Formal potentials** help to tabulate such measurements by avoiding the need to deal with activity coefficients.

The formal potential is the half-cell equilibrium potential when the oxidized and reduced species are present in a concentration ratio that satisfies the condition $(c_O/c^\circ)^{|\nu_O|} : (c_R/c^\circ)^{|\nu_R|} = 1$. For the most common case $|\nu_O| = |\nu_R|$, this condition simply means $c_O = c_R$. Starting from Eq. (2.38), activity-dependent term can be separated into a term for activity coefficients and concentrations. Afterwards, the term with the activity coefficients and possible further terms for pH, concentration of ligands, etc. can be lumped together with the standard electrode potential to leave the concentration-dependent term in the Nernst Equation (2.72).

E – Electrode potential [V]
E° – Standard electrode potential [V]
$E^{\circ\prime}$ – Formal potential [V]
a_O (a_R) – *Activity* of the oxidized (reduced) species
c_O (c_R) – *Concentration* of the oxidized (reduced) species
f_O (f_R) – Activity coefficients of the oxidized (reduced) species
a_O° (a_R°), c_O° (c_R°) – Corresponding standard activity or concentration of the oxidized (reduced) species
$|\nu_O|$ ($|\nu_R|$) – Absolute value of the stoichiometric coefficient of the oxidized (reduced) species
R – Gas constant [J K^{-1} mol^{-1}]
T – Absolute temperature [K]
F – Faraday constant [A s mol^{-1}]
n – Stoichiometric number of electrons [dimensionless]

$$\begin{aligned} E &= E^\circ + \frac{RT}{nF}\ln\frac{(a_O/a_O^\circ)^{|\nu_O|}}{(a_R/a_R^\circ)^{|\nu_R|}} \\ &= E^\circ + \frac{RT}{nF}\ln\frac{(f_O)^{|\nu_O|}}{(f_R)^{|\nu_R|}} + \frac{RT}{nF}\ln\frac{(c_O/a_O^\circ)^{|\nu_O|}}{(c_R/a_R^\circ)^{|\nu_R|}} \\ &= E^{\circ\prime} + \frac{RT}{nF}\ln\frac{(c_O/a_O^\circ)^{|\nu_O|}}{(c_R/a_R^\circ)^{|\nu_R|}} \end{aligned} \quad (2.72)$$

Formal potentials can be regarded as a modified standard electrode potential, which is more convenient for many experimental works because accurate concentrations are much easier to establish than activities in most cases.

Activity coefficients depend on ionic strength and solvents. Because the formal potentials differ from the corresponding standard electrode potentials by the term with the activity coefficients (set in blue in Eq. (2.72)), formal potentials depend on the particular electrolyte solution in which they were determined.

Furthermore, ion pairing or formation of complexes between the metal ions and anions contained in the electrolyte can establish a large number of simultaneous solution equilibria, for which usually not all details are readily accessible (e.g., different stages of complexation and ion pairing). Formal potentials offer a convenient way to cope with this situation by using the *overall* concentration of oxidized forms and the *overall* concentration of the reduced forms in the concentration-dependent term in Eq. (2.72). In most cases, the overall concentrations will be the concentrations for which the solutions have been prepared by weighing defined amounts of pure substances and filling up to a defined volume of a volumetric flask with an electrolyte solution. In other words, the total concentration is the concentration that would result when the weighted substance would dissolve without entering solution equilibria.

2.3 Cell Voltages Are Associated with Charge Separation at Interfaces – The Galvani Voltage

When discussing the separation of charges at the interface of solid (metal) and liquid (electrolyte solution) phases in Section 2.2.1, we developed a picture to describe the cell voltage and the electrode potential. Here, we analyze the components of the cell voltage. In particular, we dissect the charge separation in processes at phase boundaries. A boundary between phases α and β is considered with charged particles (with index i) that cross this phase boundary. Each particle is characterized by an electrochemical potential $\tilde{\mu}_i^\alpha$, $\tilde{\mu}_i^\beta$, and the associated charge z_i. In equilibrium, the sum can be expressed over all charged particles at the boundary $\alpha|\beta$ with net zero charge transfer.

Index for phase: small Greek letter or Roman numeral, superscript.

Index for a species: small Latin letter or Arabic numeral, subscript.

$$\left(\frac{\partial g}{\partial \xi}\right)_{T,p} = \Delta_R G = \sum_i \nu_i^\alpha \tilde{\mu}_i^\alpha + \sum_j \nu_j^\beta \tilde{\mu}_j^\beta = 0. \tag{2.73}$$

$\tilde{\mu}_i^\alpha$, $(\tilde{\mu}_j^\beta)$ – Electrochemical potential of component i (j) [J mol^{-1}] in phase α (β)

ν_i – Stoichiometric coefficient (sign sensitive) for component i [dimensionless]

ξ – Amount corresponding to one reaction stoichiometry [mol]

$\Delta_R G$ – Partial molar Gibbs free energy [J mol^{-1}]

We need to express the voltage across this phase boundary in equilibrium without net current ($I = 0$). We insert the expression for the electrochemical potential (2.11) for $\tilde{\mu}_i^\alpha$ and $\tilde{\mu}_i^\beta$, rearrange, and separate out the difference in inner potentials (Galvani potentials). The difference in Galvani potentials is the **equilibrium Galvani voltage** $\Delta\varphi_{eq}$:

$$0 = \sum \nu_i^\alpha \mu_i^\alpha + \sum \nu_i^\alpha z_i F \varphi^\alpha + \sum \nu_i^\beta \mu_i^\beta + \sum \nu_i^\beta z_i F \varphi^\beta \tag{2.74}$$

$$0 = \sum \nu_i^\alpha \mu_i^\alpha + \sum \nu_i^\beta \mu_i^\beta + \sum \nu_i^\alpha z_i F \varphi^\alpha + \sum \nu_i^\beta z_i F \varphi^\beta \tag{2.75}$$

Substitute the stoichiometry of charges

$$n = -\sum \nu_i^\alpha z_i = \sum \nu_i^\beta z_i$$

$$0 = \sum \nu_i^\alpha \mu_i^\alpha + \sum \nu_i^\beta \mu_i^\beta - nF\,(\varphi^\alpha - \varphi^\beta) \tag{2.76}$$

$$0 = \sum_{\text{Phase }\alpha,\beta} \nu_i \mu_i - nF\,(\varphi^\alpha - \varphi^\beta) \tag{2.77}$$

$$\Delta\varphi_{eq}^{\alpha,\beta} = (\varphi^\alpha - \varphi^\beta)_{eq} = \frac{\sum_{\text{Phase }\alpha,\beta} \nu_i \mu_i}{n \cdot F}. \tag{2.78}$$

The chemical potential part of the expression is dissected into a standard term and an activity-dependent term [1–3]:

$$\Delta\varphi_{eq}^{\alpha,\beta} = \frac{\sum_{\text{Phase }\alpha,\beta} \nu_i \left(\mu_i^\circ + RT \ln \frac{a_i}{a^\circ}\right)}{nF} \tag{2.79}$$

$$\Delta\varphi_{eq}^{\alpha,\beta} = \frac{\sum_{\text{Phase }\alpha,\beta} \nu_i \mu_i^\circ}{zF} + \frac{RT}{nF} \sum_{\text{Phase }\alpha,\beta} \nu_i \ln \frac{a_i}{a^\circ} \tag{2.80}$$

$$\Delta\varphi_{eq}^{\alpha,\beta} = \frac{\sum_{\text{Phase }\alpha,\beta} \nu_i \mu_i^\circ}{nF} + \frac{RT}{nF} \ln \prod_{\text{Phase }\alpha,\beta} \frac{a_i^{\nu_i}}{a^{\circ\nu_i}} = \Delta\varphi_{eq}^{\circ\alpha,\beta} + \frac{RT}{nF} \ln \prod_{\text{Phase }\alpha,\beta} \left(\frac{a_i}{a^\circ}\right)^{\nu_i}. \tag{2.81}$$

As a result, we can now identify the algebraic structure of the Nernst equation in Eq. (2.81). Next, we will apply Eq. (2.81) for specific cases of charge transfer at phase boundaries.

2.3.1 The Galvani Voltage for Electrodes with Ion Transfer

A mechanistic picture can be developed for the case of the elementary processes at a copper electrode immersed in $Cu(NO_3)_2$ solution, $Cu|Cu(NO_3)_2(aq)$. In copper metal (phase I), copper atoms share their valence electrons that form a mobile free electron gas. Due to charge neutrality, there is initially no significant excess charge. A charge transfer process can now be considered as transfer of Cu^{2+} cations from the metal lattice into the solution phase (phase II). This is illustrated in Figure 2.9 for the case of a singly charged metal cation.

$$Cu^{2+}(I) \rightleftharpoons Cu^{2+}(II) \tag{2.82}$$

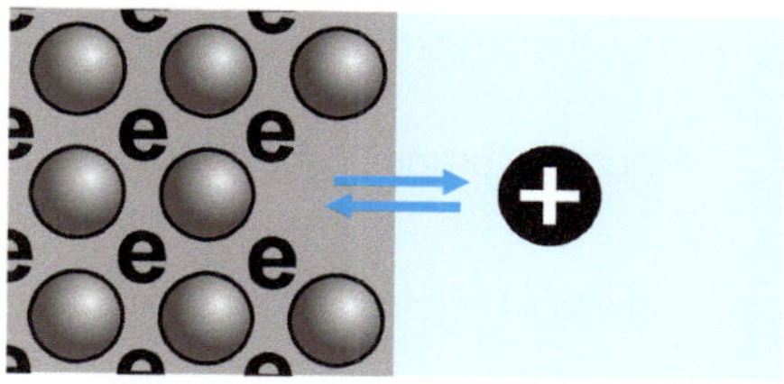

Figure 2.9 Schematic showing the separation of charges at a metal–electrolyte phase boundary due to cation transfer.

In this interfacial process, electrons remain in the free electron gas in the bulk metal. The electrode reaction can be formally written as

$$Cu(I) \rightleftharpoons Cu^{2+}(II) + 2\,e^-(I) \tag{2.83}$$

Electrode processes associated with ion release into the electrolyte without restriction on the activity of the ion in solution are called electrode of the first kind. This is in contrast to electrodes of the second kind (see below) with an insoluble salt in the electrolyte fixing the ion activity during release. In the expression for the Galvani potential, we have to include a term for copper metal, $Cu^{(0)}$. This is associated with activity 1.0 because the composition is stated as mole fraction of the pure Cu metal ($a_x = 1$). The activity of the electrons in Eq. (2.83) is not eliminated, but expressed as part of the standard potential. In Excursion 2-1, it was shown that only a few excess charges are sufficient for a measurable change in the potential of the metal phase. Therefore, the electrostatic charging of a metal will change the inner or Galvani potential (φ) without significantly changing the concentration or free electrons in the metal phase. We can write Eq. (2.84).

$$\Delta\varphi_{eq}^{I,II} = \Delta\varphi_{eq}^{\circ I,II} + \frac{RT}{2F}\ln\frac{\{Cu^{2+}\}}{a^\circ}\frac{1}{a_{x,Cu^{(0)}}}. \tag{2.84}$$

$\Delta\varphi_{eq}^{I,II}$ – Equilibrium Galvani voltage
$\Delta\varphi_{eq}^{\circ I,II}$ – Standard equilibrium Galvani voltage
$\{Cu^{2+}\}$ – Activity of the oxidized species [mol L^{-1}]
a° – Standard activity of a solution [1 mol L^{-1}]
$a_{x,Cu}$ – Mole fraction activity of copper in the metal [dimensionless]
R – Gas constant
T – Absolute temperature [K]
F – Faraday constant
$n = 2$

If the copper metal electrode is made from pure copper, the activity is $a_{x,Cu} = 1$, and Eq. (2.84) is obtained.

$$\Delta\varphi_{eq}^{I\,II} = \Delta\varphi_{eq}^{\circ I,II} + \frac{RT}{2F}\ln\frac{\{Cu^{2+}\}}{a^\circ}. \tag{2.85}$$

In contrast, if the metallic phase is based on an alloy (alloy electrode), then $a_{x,Cu} \neq 1$ and Eq. (2.84) cannot be simplified. In this case, each component of the alloy may have to be considered.

For electrode processes with ion transfer, the interfacial charge transfer can be associated with an additional equilibrium process. For the case of insoluble salts in electrolyte media, this effectively fixes the ion activity of the transferring ion, and we consider these to be electrodes of the second kind. This is commonly exploited, for example, for silver metal cations (Ag^+), forming a highly insoluble precipitate (AgCl) with chloride anions (Cl^-). As a result, we have to consider three phases with phase I, the solid metal; phase II, the solid AgCl directly on the metallic Ag; and phase III, the electrolyte solution. The resulting silver–silver chloride electrode (Figure 2.10) is a commonly used reference electrode. The relevant reactions with their phases labelled are:

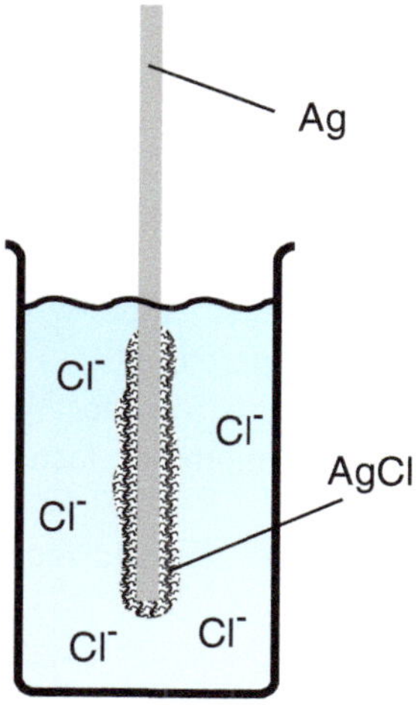

Figure 2.10 Schematic drawing of the silver–silver chloride electrode (electrode of second kind).

$$Ag\,(I) \rightleftharpoons Ag^+\,(III) + e^-(I) \tag{2.86}$$

$$Ag^+\,(III) + Cl^-\,(III) \rightleftharpoons AgCl\,(II) \tag{2.87}$$

$$Ag\,(I) + Cl^-\,(III) \rightleftharpoons AgCl\,(II) + e^-\,(I). \tag{2.88}$$

In order to express the Galvani voltage, we can use the expression derived for electrodes of the first kind and then substitute in the expression for the solubility product $K_{SP}(AgCl) = \{Ag^+\}^{(1)}\{Cl^-\}^{(1)}$, replacing the metal cation activity.

$$\Delta\varphi_{eq}^{I,III} = \Delta\varphi_{eq,1^{st}\,kind}^{\circ I,III} + \frac{RT}{F}\ln\frac{\{Ag^+\}}{a^\circ} \tag{2.89}$$

a° in the numerator and $a^{\circ 2}$ in the denominator are standard activities (1 mol L^{-1} and 1 mol^2 L^{-2});
K_{SP} – Solubility product [mol^2 L^{-2}].

$+\ln(a/b) = -\ln(b/a)$

$$\Delta\varphi_{eq}^{I,III} = \Delta\varphi_{eq,1^{st}\,kind}^{\circ I,III} + \frac{RT}{F}\ln\frac{K_{SP}(AgCl)\,a^\circ}{\{Cl^-\}\ a^{\circ 2}} \tag{2.90}$$

$$\Delta\varphi_{eq}^{I,III} = \varphi_{eq,\,1.Art}^{\circ I,III} + \frac{RT}{F}\ln\frac{K_L(AgCl)}{a^{\circ 2}} + \frac{RT}{F}\ln\frac{a^\circ}{\{Cl^-\}}. \tag{2.91}$$

It is now possible to combine the constant from the activity-dependent term with the standard term. With an additional sign correction, Eq. (2.92) is obtained. Note that the Galvani voltage for the metal now depends solely on the concentration of an electrolyte anion that is not involved in the original charge transfer process. For a high chloride anion activity, the Galvani voltage will be stable and suitable as a reference voltage.

$$\Delta\varphi_{eq}^{I,III} = \varphi_{eq,\,2.Art}^{\circ I,III} - \frac{RT}{F}\ln\frac{\{Cl^-\}}{a^\circ}. \tag{2.92}$$

Note that processes with ion transfer are not limited to metal–electrolyte phase boundaries. Closely related types of processes are observed for glass electrodes (Section 5.2.3) and for insertion electrodes (Section 15.6.2).

2.3.2 The Galvani Voltage for Electrodes with Electron Transfer

In contrast to electrode processes with ion transfer, we can identify another broad class of electrode processes based on electron transfer or **redox electrode processes**. Electrons are transferred from an electrically conducting phase (metal, semiconductor, conducting polymer, etc.) into an ion-conducting phase. Formally, the electron from phase I is in equilibrium with the reduced form (R) and the oxidized form (O) in phase II.

$$\nu_R R(II) \rightleftharpoons e^-(I) + \nu_O O(II) \tag{2.93}$$

An analytical approach to this process is to dissect it into two processes: in the solution phase, the reduced species R is in equilibrium with the oxidized form O and the electron or "R(II) $\rightleftharpoons$ O(II) + e(II)". Next, at the interface, the electron can transfer from solution phase to metallic conductor phase or "e$^-$ (II) $\rightleftharpoons$ e$^-$ (I)". This corresponds to an oxidation. For the reduction, the sequence of steps is reversed. The thermodynamic analysis is based on initial state and end state only and is path-independent. Therefore, the mechanism does not require any consideration. The expression for the Galvani voltage for the redox couple R/O (with both R and O in the electrolyte phase with standard activities $a_O^\circ = a_R^\circ = 1$ mol L^{-1}) is equivalent to that for the corresponding electrode potential (2.94).

$\Delta\varphi_{eq}^{I,II}$ – Equilibrium Galvani voltage
$\Delta\varphi_{eq}^{\circ I,II}$ – Standard equilibrium Galvani voltage
R – Gas constant [J K^{-1} mol^{-1}]
T – Absolute temperature [K]
F – Faraday constant [A s mol^{-1}]
n – Electron stoichiometric factor [dimensionless]
a_O (a_R) – Activity of the oxidized (reduced) species [mol L^{-1}]
$|\nu_O|$ ($|\nu_R|$) – Absolute stoichiometric coefficient for the oxidized (reduced) species

$$\Delta\varphi_{eq}^{I,II} = \Delta\varphi_{eq}^{\circ I,II} + \frac{RT}{nF}\ln\frac{\{O\}^{|\nu_O|}}{\{R\}^{|\nu_R|}}. \tag{2.94}$$

Note that formally, the chemical potential for the electrons in phase I is considered constant and summed into the standard term.

There is a broad variety of redox electrode processes with or without coupled chemical reaction steps. The simplest case associated (apparently) only with charge transfer (the so-called E-mechanism) is often observed for metal ions or metal complexes ($Fe^{2+} \rightleftharpoons Fe^{3+} + e^-$, $Ce^{3+} \rightleftharpoons Ce^{4+} + e^-$, $[Ru(NH_3)_6]^{2+} \rightleftharpoons [Ru(NH_3)_6]^{3+} + e^-$). Processes with coupled chemical equilibria are often associated with solution species or with the solvent and protolysis products from the solvent (e.g., H_2O, H^+, and OH^-). In order to factor in this additional chemical equilibrium, we have to add the reaction partners into the formal reaction equation and into the corresponding activity-dependent term in the Nernst equation. This consideration is important for many biological redox processes involving protons. The equilibrium electrode potential becomes pH-dependent. Here, the example of the quinone/hydroquinone redox system will be considered as typical for biological redox systems (Figure 2.11). Quinones exhibit a wide range of molecular structures naturally associated with hydrophobicity, with the acid–base constant, and with the pH-dependent redox potential. In equations, we use Q (quinone) and H_2Q (hydroquinone).

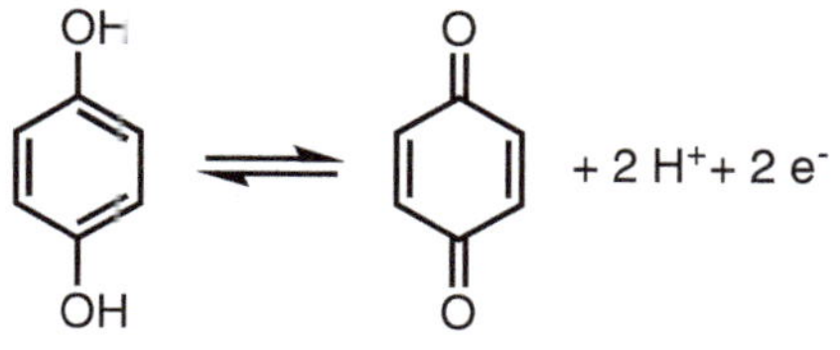

(2.95)

H₂Q **Q**

$$\Delta\varphi_{eq}^{I,II} = \Delta\varphi_{eq}^{\circ I,II} + \frac{RT}{2F}\ln\frac{\{Q\}\{H^+\}^2}{\{H_2Q\}a^{\circ 2}} \tag{2.96}$$

$$\Delta\varphi_{eq}^{I,II} = \Delta\varphi_{eq}^{\circ I,II} - \frac{2.3RT}{F}\mathrm{pH} + \frac{2.3RT}{2F}\log\frac{\{Q\}}{\{H_2Q\}} \tag{2.97}$$

In many biological processes, reactions occur at 37 °C = 310 K, which leads to a factor $2.3E_{NF} = 2.3\,RT/(nF) = 0.0614$ V. This leads to a pH-dependent Galvani voltage of

$$\Delta\varphi_{eq}^{I,II} = \Delta\varphi_{eq}^{\circ I,II} - 0.0614\,\mathrm{V}\cdot\mathrm{pH} + 0.0307\,\mathrm{V}\cdot\log\frac{\{Q\}}{\{H_2Q\}} \tag{2.98}$$

For redox processes involving gaseous reaction partners (gas electrodes), we noted in Section 2.2.3 that the fugacity of the gas needs to be considered in the activity-dependent term of the Nernst equation. Usually, at ambient conditions, the gas can be considered close to ideal and fugacity effects ignored. Here, the case of the hydrogen electrode is considered with partial pressure p_i and with standard activity p°. The reaction associated with the hydrogen electrode is

$$H_2(II) \rightleftharpoons 2\,H^+(III) + 2\,e^-(I). \tag{2.99}$$

The corresponding equilibrium Galvani voltage is pH- and pressure-dependent and formally given in Eq. (2.98).

$$\Delta\varphi_{eq}^{I,III} = \Delta\varphi_{eq}^{\circ I,III} + \frac{RT}{2F}\ln\left(\frac{\{H^+\}^2}{a^{\circ 2}}\frac{p^\circ}{p(H_2)}\right) \tag{2.100}$$

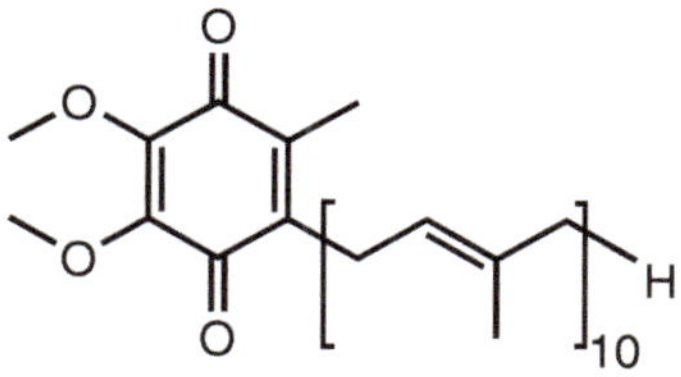

a) Ubiquinone, carrier for electrons and protons in respiration

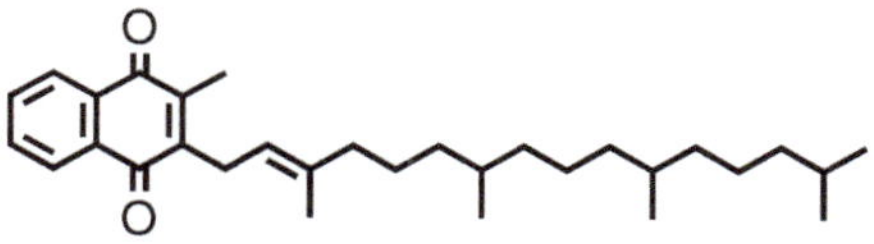

b) Vitamin K_1 (α-phylloquinone) is essential in the biosynthesis of blood clotting factors.

Figure 2.11 Molecular structures for biological redox systems with a quinone redox center.

$\mathrm{pH} = -\log(\{H^+\}/a^\circ)$

$2.3 \log a \approx \ln a$

2.3.3 Galvani Voltage at Phase Boundaries Without Electrodes

Galvani Voltage at Junctions between Electron Conductors. Galvani voltages are observed across the interface of electron conductors (metals, semiconductors, conducting polymers) when the chemical potential of the electrons during transfer changes. We illustrate this case for the junction of two different metals. The transfer of the electron across the phase boundary is a step in the cell reaction. The corresponding Galvani voltage contributes to the cell voltage. When two metal wires are combined (see phases I and II in Figure 2.12), a flow of electrons occurs to balance out the chemical potentials of electrons in phases I and II. The change in electron activity in the metallic conductor remains insignificant and can be ignored (compare Example Calculation 2.1). Therefore, chemical potentials for electrons in metals are commonly assumed to be constant without an activity-dependent term.

$$\Delta\varphi_{eq}^{I,II} = (\varphi^{I} - \varphi^{II})_{eq} = \frac{\mu^{\circ,I}(e^-) - \mu^{\circ,II}(e^-)}{F} = \Delta\varphi_{eq}^{\circ I,II} \tag{2.101}$$

a)

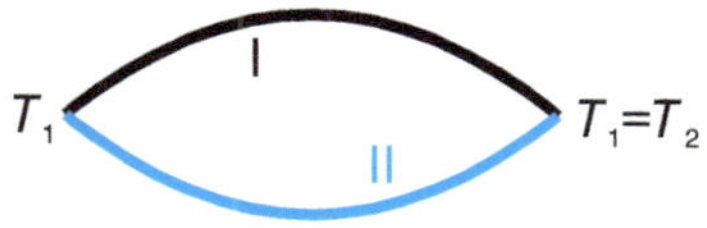

b)

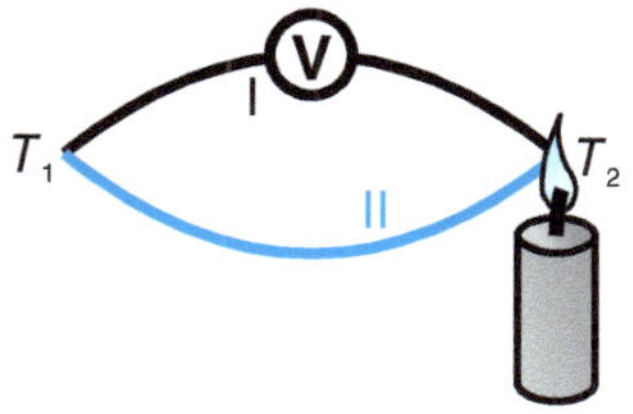

Figure 2.12 Galvani voltage for two-electron conductors. a) for $T_1 = T_2$ the voltage is zero; b) for $T_1 < T_2$ a voltage is generated due to a difference in interfacial voltages (thermoelectric voltage).

In the circuit in Figure 2.12, the sum of Galvani voltages is measured ($\Delta\varphi_{eq}^{I,II} + \Delta\varphi_{eq}^{II,I}$) as zero in the absence of a temperature difference. Due to the temperature at the two phase boundaries being the same, the two voltages $\Delta\varphi_{eq}^{I,II} = -\Delta\varphi_{eq}^{II,I}$ balance out to net zero. However, the chemical potential of electrons in metals is temperature-dependent in a way that is metal-dependent. Heating up one of the contact points therefore leads to $\Delta\varphi_{eq}^{I,II}(T_1) \neq -\Delta\varphi_{eq}^{II,I}(T_2)$ and a thermoelectric voltage is observed. Such metal–metal contact point voltages can affect equilibrium cell voltage data for other types of circuits and cells, and therefore, unintended effects of temperature gradients on experiments need careful consideration.

Galvani Voltage at Junctions between two Immiscible Ion-conducting Electrolytes. ITIES (interface between two immiscible electrolyte solutions, Figure 2.13) provides a liquid–liquid phase boundary at which cation and anion partitioning equilibria occur. In contrast to the partitioning of neutral molecules across this phase boundary, the partitioning of ions is associated with transport of charges and with the formation of an equilibrium voltage. The Galvani voltage resulting from this ion partitioning process leads to a situation in which net charge transfer ceases and a dynamic equilibrium is reached (zero net current). Accordingly, we can drive ion transfer reactions by intentionally applying an external voltage. The ability for ions to transfer across the liquid–liquid phase boundary can be attributed to different phenomena:

1) Hydrophobic anions (ClO_4^-, PF_6^-, BF_4^-, I^-, etc.) exhibit some degree of solubility in organic solvents and may therefore partition across the water–organic phase boundary (Figure 2.13a). As a result, the organic phase acquires a net negative charge, leaving an excess positive charge in the aqueous phase.
2) Metal cations commonly exhibit a high charge density (charge per volume) and stronger interaction with water dipoles. This lowers the tendency of cations to partition into the organic solvent phase. However, a complexing agent in the organic phase can be effective in shielding the high cation charge density and in triggering partitioning into the organic phase (Figure 2.13b). Very effective complexing agents for cations are crown ethers. The partitioning of cations into the organic phase causes a net positive charge and a negative excess charge in the aqueous phase. These types of Galvani voltages for specific cations or anions are exploited in analytical potentiometric electrodes and sensors (Section 5.2.6).

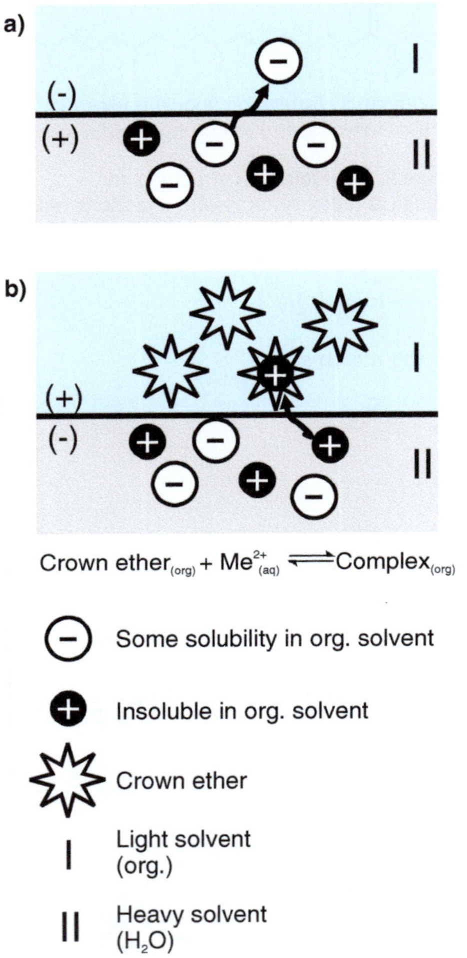

Figure 2.13 Distribution of ions at an ITIES. a) Hydrophobic anions show a tendency to partition into the organic solvent phase; b) the addition of a complexing agent (e.g., a crown ether) causes transfer of metal cations into the organic solution phase. This phenomenon is exploited in *phase transfer catalysis*.

The investigation of ion transfer processes at ITIES is a special subject of electrochemistry for which many methods have been adapted that are also used for the study of electrodes with solid–liquid interfaces. The interested reader is referred to specialized reviews and textbook chapters [8–10].

Galvani Voltage between two Miscible Ion-conducting Electrolytes. For two electrolyte solutions with differing ion concentrations, formally a gradient in the electrochemical potentials $\tilde{\mu}_i^{\alpha}$ and $\tilde{\mu}_i^{\beta}$ exists. The difference in $\tilde{\mu}_i^{\alpha}$ and $\tilde{\mu}_i^{\beta}$ introduces a driving force and as a response, transport of ionic species will occur in the form of irreversible transport (mainly via diffusion, Section 4.3). This kind of phenomenon is linked to a nonequilibrium process; electrochemical cells with this kind of gradient or interface remain in a nonequilibrium state as long as the activity gradient is present. Electrochemical cells of this type are cells with liquid junction at which an activity gradient exits. Such cells are practically very common, especially in two-compartment systems separated by a membrane or frit. The separator (membrane, diaphragm, salt bridge, etc.) prevents mixing of solutions but allows diffusion. The diffusion process then leads to the diffusion potential.

Despite the ongoing transport processes, rather constant potentials are observed over prolonged periods of time. Note that this case corresponds to a nonequilibrium potential in contrast to the equilibrium potentials discussed above. A property that remains constant with time without being in equilibrium is called a stationary or steady-state property. The Galvani voltage associated with ion transport in nonequilibrium is called a **diffusion potential** (or sometimes liquid junction potential). In order to calculate and analyze stationary diffusion potentials, there are approximate analytical solutions (equations) or computer simulation tools to explicitly treat the irreversible nature of these processes [11].

In order to illustrate the fundamental principles of diffusion potentials, Figure 2.14 shows a typical nonequilibrium scenario: on the left-hand side, a higher concentration of hydrochloric acid is present. Both diffusion of H^+ and of Cl^- will occur from left to right driven by the electrochemical potentials. However, proton diffusion is an order of magnitude faster in water compared to chloride diffusion (Section 4.2.6). This results in a build-up of excess positive charge in the right-hand side compartment and a corresponding negative excess charge in the left-hand side compartment. The resulting voltage slows down proton transport and speeds up chloride transport until a stationary state is reached. The final result is a time-independent (apparently constant) stationary diffusion potential across the separator.

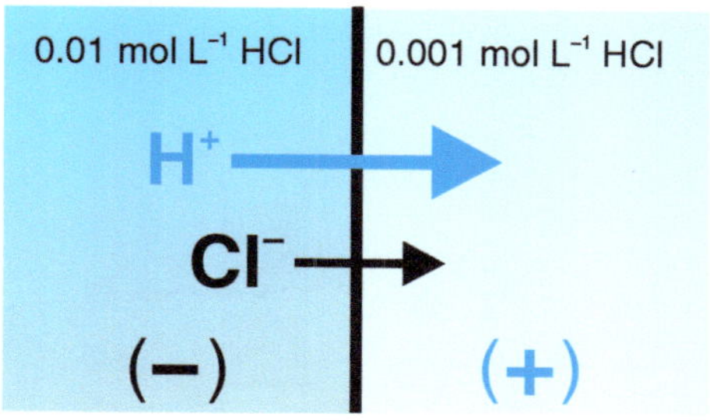

Figure 2.14 Illustration of the mechanism leading to diffusion potentials under conditions of electrochemical potential gradients. Arrows represent net fluxes of H^+ and Cl^-.

In practical experiments with electrochemical cells, it is important to minimize (or at least stabilize) effects from diffusion potentials. The following recommendations can help in practical work:

- Implementation of salt bridges (rather than direct junctions) to employ conditions under which diffusion potentials at both ends of the salt bridge partially compensate.
- Use of high concentration electrolyte media in the salt bridge to make diffusion of ions from the salt bridge out into the two half-cells the dominant process.
- Use of equal transference electrolyte media for the salt bridge. The transference number describes the relative diffusivity, and some typical values are given in Table 2.1. For a 1–1 electrolyte, the transference numbers t_+ and t_- add up to 1.0 (Section 4.2.7).

Table 2.1 Electrolytes for salt bridges with transference numbers.

Electrolyte	t_+	t_-
KCl	0.49	0.51
KNO_3	0.51	0.49
CsCl	0.5025	0.4975

In order to measure the effect of the electrolyte concentration in the salt bridge, it is possible to insert two reference electrodes (e.g., two Hg|Hg_2Cl_2 electrodes) into both sides of the cell shown in Figure 2.15. The left half-cell contains 0.1 mol L^{-1} HCl, and the right half-cell contains 0.1 mol L^{-1} KCl, and a KCl-containing salt bridge provides the ion-conducting connection. Due to the essentially identical chloride activities $\{Cl^-\}$ in both compartments of the cell, the measured cell voltage is dominated by the diffusion potentials in and out of the salt bridge. Varying the concentration [KCl] in the salt bridge then leads to a clear trend as summarized in Table 2.2. Only with a high KCl concentration in the salt bridge will the K^+ and Cl^- contributions dominate and suppress the diffusion potential effect.

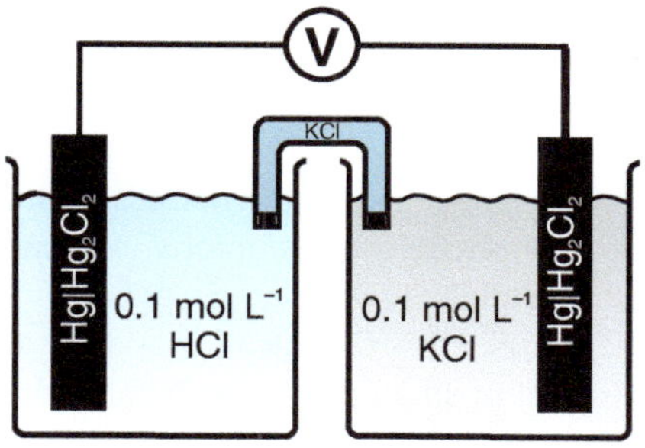

Figure 2.15 Measurement of diffusion potentials as a function of KCl concentration in the salt bridge.

Table 2.2 Effect of the KCl concentration in a salt bridge on the diffusion potential.

[KCl]/(mol L^{-1})	U/mV
0.1	27
0.2	20
0.5	13
1.0	8.4
2.5	3.4
3.5	1.1
4.2 (saturated)	<1

Source: [12] original data from H. A. Fales, W. C. Vosburgh, *J. Am. Chem. Soc.* **1918**, *40*, 1291; E. A. Guggenheim, *J. Am. Chem. Soc.* **1930**, *52*, 1315; A. L. Ferguson, K. van Lente, R. Hitchens, *J. Am. Chem. Soc.* **1931**, *54*, 1285.

From these considerations, it is apparent that diffusion potentials are very common. Experiments to determine electrode potentials or standard equilibrium voltages always carry systematic errors due to diffusion potentials. It is prudent to take into account an uncertainty in potential of ±5 mV due to diffusion potentials. Stationary cell voltages with zero net current that contain contributions from nonequilibrium processes (e.g., diffusion potentials or mixed potentials, Section 2.3.5) are generally called **open-circuit voltage** (U_{OCV}). The open-circuit voltage is equivalent to the equilibrium cell voltage in the absence of irreversible processes (e.g., no concentration gradients and no corrosion).

Galvani Voltages in Membranes. Membranes can be defined as thin films of porous materials (e.g., polymers, ceramics, glasses, or biological lipid structures that self-assemble). Membranes are permeable to different types of ions. For a membrane separating two different electrolyte solutions, the permeation of ions will result in Galvani voltages. In order to understand these Galvani voltages, we have to consider the permeability of the membrane material.

Membranes with coarse pores (e.g., a glass frit and diaphragms) allow diffusion of all ions but stop convective mixing of the two electrolyte solutions. For this type of membrane, the nonequilibrium conditions result in diffusion potentials equivalent to those between two miscible electrolyte media (Figure 2.14).

Membranes with fine pores (e.g., glass membranes, solid state electrolyte membranes, and ion exchange membranes) only allow very slow diffusion for one or several types of ions. In this case, the diffusion potentials are negligible. In this case, equilibrium Galvani voltages according to Eq. (2.81) develop at both membrane-electrolyte interfaces. Ions rapidly partition across the interfaces, and an equilibrium is reached with excess charges resulting in a Galvani voltage. In cases where ion diffusion within the membrane is fast (or the membrane is very thin), both nonequilibrium diffusion potentials and equilibrium ion-partitioning effects may occur simultaneously.

A further technically important type of membrane is **semipermeable membrane**. Semipermeable membranes are based on very small pores with either cation or anion permeability. Localized electric fields within the pores can restrict access to only one type of ion. Biological membranes are an example of those membranes containing highly specific ion channels that are permeable only to one ion type. Many biological tissues allow semipermeable ion transport (e.g., bladder wall tissue and digestive tract wall tissue). Larger multiply charged macromolecules (polyelectrolytes based on polyanions or polycations with charges along a polymer backbone) or colloid particles (particles with diameter less than 1 μm, often carrying surface charges) are repelled by these tissues. However, the smaller counteranions that accompany the colloid can pass through the tissue. The resulting equilibrium Galvani voltage is called a Donnan potential[13] or membrane voltage.

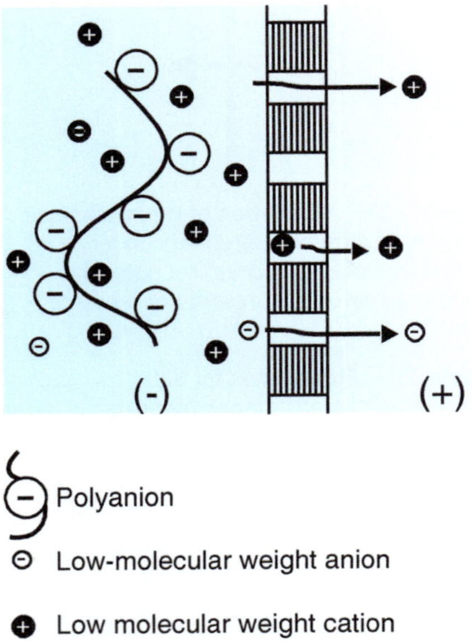

Figure 2.16 Donnan potential at semipermeable membrane.

An illustration of the case of a semipermeable membrane is shown in Figure 2.16. A macromolecular polyanion cannot pass the membrane, but smaller anions or cations, or neutral species (e.g., H_2O) can pass. In equilibrium, the electrochemical potentials of all three mobile species (cation, anion, and neutral species) in phase I and in phase II must be equal. By considering the charged species only, we can explore the formation of a net voltage across the membrane. Based on Eq. (2.11), we obtain expressions for the Donnan potentials.

$\tilde{\mu}_+^{\mathrm{I}}$, $\tilde{\mu}_+^{\mathrm{II}}$, $\tilde{\mu}_-^{\mathrm{I}}$, $\tilde{\mu}_-^{\mathrm{II}}$ – Electrochemical potential of small cations/anions in phase I or II [J mol^{-1}]

$\mu_+^{\mathrm{I}\circ}, \mu_+^{\mathrm{II}\circ}, \mu_-^{\mathrm{I}\circ}, \mu_-^{\mathrm{II}\circ}$ – Chemical standard potential of small cations/anions in phase I or II [J mol^{-1}]

z_+, z_- – Integer charge with sign for cations or anions

F – Faraday constant [A s mol^{-1}]

R – Gas constant [J K^{-1} mol^{-1}]

T – Absolute temperature [K]

$$\tilde{\mu}_+^{\mathrm{I}} = \tilde{\mu}_+^{\mathrm{II}} \tag{2.102}$$

$$\mu_+^{\mathrm{I},\circ} + RT\ln\frac{a_+^{\mathrm{I}}}{a^\circ} + z_+F\varphi^{\mathrm{I}} = \mu_+^{\mathrm{II},\circ} + RT\ln\frac{a_+^{\mathrm{II}}}{a^\circ} + z_+F\varphi^{\mathrm{II}} \tag{2.103}$$

$$\tilde{\mu}_-^{\mathrm{I}} = \tilde{\mu}_-^{\mathrm{II}} \tag{2.104}$$

$$\mu_-^{\mathrm{I},\circ} + RT\ln\frac{a_-^{\mathrm{I}}}{a^\circ} + z_-F\varphi^{\mathrm{I}} = \mu_-^{\mathrm{II},\circ} + RT\ln\frac{a_-^{\mathrm{II}}}{a^\circ} + z_-F\varphi^{\mathrm{II}} \tag{2.105}$$

The equilibrium conditions derived for each charge-carrying species (2.103) and (2.105) need to be solved simultaneously. For the case of the same solvent being employed on both sides of the membrane, the standard chemical potentials for ions in phase I and in phase II are identical, and we obtain Eq. (2.103).

$$\Delta\varphi_{\mathrm{eq}}^{\mathrm{I,II}} = (\varphi^{\mathrm{I}} - \varphi^{\mathrm{II}})_{\mathrm{eq}} = \frac{RT}{z_-F}\ln\frac{a_-^{\mathrm{II}}}{a_-^{\mathrm{I}}} = \frac{RT}{z_+F}\ln\frac{a_+^{\mathrm{II}}}{a_+^{\mathrm{I}}}$$
$$= \frac{RT}{F}\ln\left(\frac{a_-^{\mathrm{II}}}{a_-^{\mathrm{I}}}\right)^{\frac{1}{z_-}} = \frac{RT}{F}\ln\left(\frac{a_+^{\mathrm{II}}}{a_+^{\mathrm{I}}}\right)^{\frac{1}{z_+}}. \tag{2.106}$$

[13] Frederick George Donnan, 1870–1956, Irish chemist, worked at the University College London and held a chair of physical chemistry at the University of Liverpool, founding member of the Faraday Society.

Based on this expression, the term (2.107)

$$\frac{F\Delta\varphi_{\text{eq}}^{\text{I,II}}}{RT} = \ln\left(\frac{a_-^{\text{II}}}{a_-^{\text{I}}}\right)^{\frac{1}{z_-}} = \ln\left(\frac{a_+^{\text{II}}}{a_+^{\text{I}}}\right)^{\frac{1}{z_+}} \tag{2.107}$$

is essentially an equilibrium constant and known as the Donnan distribution coefficient.

In Excursion 2-1, we discovered that only a few ions (relative to the overall number of ions) are sufficient to produce significant potential differences at phase boundaries. A similar analysis applies in the case of semipermeable membranes. Therefore, there are only very small deviations from the condition of electroneutrality for phases I and II even when a measurable Donnan potential has been established.

2.3.4 Combining Galvani Voltages to Give the Cell Voltage

Next, the cell voltage for the Daniell cell (Figure 2.17) is developed based on Galvani voltages for each phase boundary. Consequences for the practical approach to cell voltage measurements are considered. The phases are numbered with capital Roman numerals from left to right. Phase I′ denotes the copper wire connected to the zinc electrode and to the voltmeter. The sum of all Galvani voltages allows the cell voltage to be calculated (Figure 2.18).

$$U = \Delta\varphi_{\text{eq}}^{\text{I,II}} + \Delta\varphi_{\text{eq}}^{\text{II,III}} + \Delta\varphi_{\text{eq}}^{\text{III,IV}} + \Delta\varphi_{\text{eq}}^{\text{IV,V}} + \Delta\varphi_{\text{eq}}^{\text{V,I'}} \tag{2.108}$$

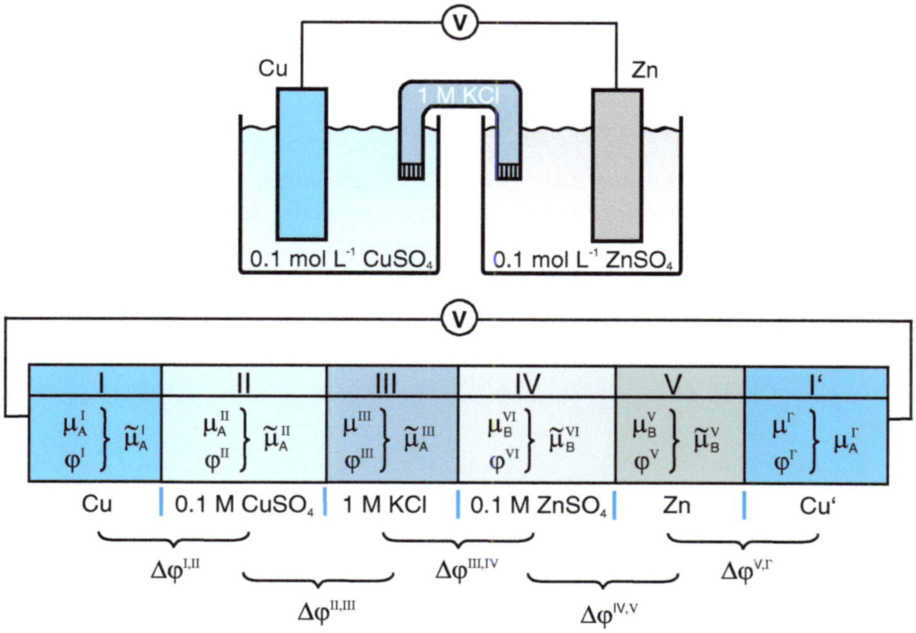

Figure 2.17 Schematic dissection of the Daniell cell into 6 separate phases. Phase I′ corresponds to the copper wire connected to the Zn electrode.

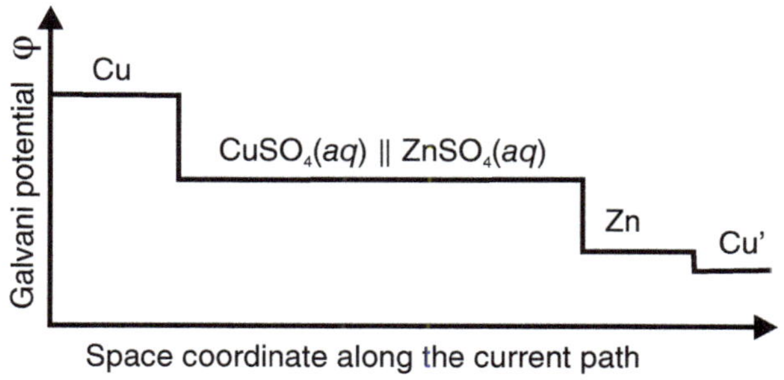

Figure 2.18 Potential profile along the current path in the Daniell cell. Under equilibrium conditions, there is no potential gradient within any of the phases. Nonequilibrium conditions are associated with potential gradients in one or more phases. Note that there are additional diffusion potentials at the salt bridge, which are neglected here due to their small values.

In this case, the salt bridge introduces diffusion potentials $\Delta\varphi^{\mathrm{II,III}}$ and $\Delta\varphi^{\mathrm{III,IV}}$, which prevent true equilibrium conditions from being reached. The choice of the electrolyte in the salt bridge (Table 2.2) is essential to minimize deviations from the thermodynamic equilibrium cell voltage and the measured open-circuit voltage. Based on Table 2.2, we have to expect deviations of up to 8 mV, which is acceptable for many practical applications but not for precision measurements. Precision measurements are important, for example, for the preparation of thermodynamic data tables and for the most accurate analytical measurements. Let us assume that

$$\Delta\varphi^{\mathrm{II,III}} \approx -\Delta\varphi^{\mathrm{III,IV}}, \tag{2.109}$$

so that the cell voltage is expressed as in Eq. (2.110).

$$U = \Delta\varphi_{\mathrm{eq}}^{\circ\mathrm{Cu|Cu^{2+}}} + \frac{RT}{2F}\ln\frac{\{\mathrm{Cu^{2+}}\}}{a^\circ} - \left(\Delta\varphi_{\mathrm{eq}}^{\circ\mathrm{Zn|Zn^{2+}}} + \frac{RT}{2F}\ln\frac{\{\mathrm{Zn^{2+}}\}}{a^\circ}\right) + \Delta\varphi_{\mathrm{eq}}^{\mathrm{Zn|Cu}}. \tag{2.110}$$

The parameter $\Delta\varphi^{\mathrm{V,I'}}$ in Eq. (2.108) expresses the Galvani voltage between Zn and Cu metal. The negative sign for the $\mathrm{Zn|Zn^{2+}}$ term is associated with the inversion of the sequence of solid and liquid phases. Equation (2.110) can be rearranged to Eq. (2.111).

$$U = \left(\Delta\varphi_{\mathrm{eq}}^{\circ,\mathrm{Cu|Cu^{2+}}} - \Delta\varphi_{\mathrm{eq}}^{\circ,\mathrm{Zn|Zn^{2+}}} + \Delta\varphi_{\mathrm{eq}}^{\circ,\mathrm{Zn|Cu}}\right) + \frac{RT}{2F}\ln\frac{\{\mathrm{Cu^{2+}}\}}{\{\mathrm{Zn^{2+}}\}}. \tag{2.111}$$

Expression (2.111) can now be compared to Eq. (2.34) derived above for the cell voltage given by two electrode potentials expressed with an identical reference potential. Rewriting $\{\mathrm{Cu^{2+}}\} = a_{\mathrm{Cu^{2+}}}$ and $\{\mathrm{Zn^{2+}}\} = a_{\mathrm{Zn^{2+}}}$ and considering $a^\circ_{\mathrm{Zn^{2+}}} = a^\circ_{\mathrm{Cu^{2+}}}$ as well as $a_{\mathrm{Cu^{(0)}}}/a^\circ_{\mathrm{Cu^{(0)}}} = a_{\mathrm{Zn^{(0)}}}/a^\circ_{\mathrm{Zn^{(0)}}} = 1$ for pure Cu and Zn, Eq. (2.34) simplifies to

$$U = \left(E^\circ_{\mathrm{Cu|Cu^{2+}}} - E^\circ_{\mathrm{Zn|Zn^{2+}}}\right) + \frac{RT}{2F}\ln\frac{\{\mathrm{Cu^{2+}}\}}{\{\mathrm{Zn^{2+}}\}}. \tag{2.34}$$

The activity-dependent term is identical in Eqs. (2.34) and (2.111). However, the standard term is clearly different. This shows that Galvani potentials are *not* identical to electrode potentials! The definition of the standard electrode potential requires that the contribution from the Galvani potential for metal–metal contacts be considered for both of the two standard electrode potentials E°. To do this, we use the metal wire employed in the reference electrode, for example, platinum.

$$\Delta\varphi_{\mathrm{eq}}^{\mathrm{Zn|Cu}} = (\varphi^{Zn} - \varphi^{Cu})_{\mathrm{eq}} = (\varphi^{Zn} - \varphi^{Pt})_{\mathrm{eq}} - (\varphi^{Cu} - \varphi^{Pt})_{\mathrm{eq}} = \Delta\varphi_{\mathrm{eq}}^{\mathrm{Zn|Pt}} - \Delta\varphi_{\mathrm{eq}}^{\mathrm{Cu|Pt}} \tag{2.112}$$

This allows the two standard electrode potentials to be identified and expressed based on Galvani potentials as $E^\circ_{\mathrm{Cu|Cu^{2+}}} = \Delta\varphi_{\mathrm{eq}}^{\circ\mathrm{Cu|Cu^{2+}}} - \Delta\varphi_{\mathrm{eq}}^{\mathrm{Cu|Pt}}$ and $E^\circ_{\mathrm{Zn|Zn^{2+}}} = \Delta\varphi_{\mathrm{eq}}^{\circ,\mathrm{Zn\,|\,Zn^{2+}}} - \Delta\varphi_{\mathrm{eq}}^{\circ,\mathrm{Zn\,|\,Pt}}$. When changing the reference electrode, only the distribution of the term $\Delta\varphi_{\mathrm{eq}}^{\circ,\mathrm{Zn\,|\,Cu}}$ across the two electrode potentials is affected without a change in the difference of the electrode potentials, that is, the calculated cell voltage.

Q: Given that the equilibrium cell voltage needs to be measured with the same metal wire connections, would the results of the measurement change when we change a copper wire to a silver wire?

A: The corresponding cell can be formally written as

$\mathrm{Ag|Cu|0.1\ mol\ L^{-1}\ CuSO_4||0.1\ mol\ L^{-1}\ ZnSO_4|Zn|Ag'}$

In this case, a new metal–metal contact term needs to be considered, and Eq. (2.111) has to be written as Eq. (2.113).

$$U = \left(\Delta\varphi_{eq}^{o,Ag|Cu} \Delta\varphi_{eq}^{o,Cu|Cu^{2+}} - \Delta\varphi_{eq}^{o,Zn|Zn^{2+}} + \Delta\varphi_{eq}^{o,Zn|Ag} \right) + \frac{RT}{2F} \ln \frac{\{Cu^{2+}\}}{\{Zn^{2+}\}}. \quad (2.113)$$

Closer inspection shows that

$$\begin{aligned} \Delta\varphi_{eq}^{o\ Ag|Cu} + \Delta\varphi_{eq}^{o,Zn|Ag} &= (\varphi^{Ag} - \varphi^{Cu})_{eq} + (\varphi^{Zn} - \varphi^{Ag})_{eq} \\ &= (\varphi^{Zn} - \varphi^{Cu})_{eq} \\ &= \Delta\varphi_{eq}^{o,Zn|Cu} \end{aligned} \quad (2.114)$$

and therefore, the two equations transform into each other under the condition that *the contact points Ag|Cu and Zn|Ag are at the same temperature.* This can be assumed valid for most experiments, but more care is required for cells with temperature gradients.

Q: What would happen if we change the wire connection from copper or zinc electrodes in the Daniell cell to the measurement device?

A: Under the condition that there are no temperature gradients, each new metal introduced as a wire will introduce two new interfaces, and the contributions from Galvani potentials at these two interfaces should cancel out. The net cell voltage measurement remains unaffected.

2.3.5 Electrodes with Mixed Potentials

In Sections 2.3.1 and 2.3.2, we discussed reactions that occur at the interface of electrode and electrolyte. The equilibrium of processes was shown to be quantitatively related to an electrode potential or a cell voltage. However, equilibrium conditions at an electrode may involve more than one reaction. For the case of **multireaction electrodes**, a combination of oxidation and reduction processes needs to be considered. For example, for a silver wire immersed in an aqueous $AgNO_3$ solution exposed to air, the following reactions can contribute to the equilibrium voltage:

$$Ag \rightleftharpoons Ag^{+} + e^{-} \quad (2.115)$$

$$O_2 + 4\,e^{-} + 4\,H^{+} \rightarrow 2\,H_2O \quad (2.116)$$

$$O_2 + 2\,e^{-} + 2\,H^{+} \rightarrow H_2O_2 \quad (2.117)$$

When measuring the open-circuit voltage of an electrochemical cell, there is no net current externally to the cell. Reactions (2.116) and (2.117) are chemically irreversible, that is, these processes only occur in a forward direction. Given that externally $I = 0$, all electrons required for processes (2.116) and (2.117) have to be generated by the oxidation of Ag in Eq. (2.115). Oxygen in solution is reduced while silver metal is gradually oxidized to $Ag^{+}(aq)$. Therefore, the electrode potential may be constant and the electrode in a steady state, but the electrode is *not* in equilibrium in spite of $I = 0$. In this case, the electrode potential is described as mixed potential. The value of the mixed potential at $I = 0$ is not governed by thermodynamic equilibrium conditions, but instead depends on the rates for the oxygen reduction versus the rate of silver oxidation. These kinetic rates are again dependent on solution conditions. In many cases, one of the two processes responsible for the mixed potential has a much faster rate and is therefore dominating the potential. This process is called the potential-determining process. However, there are also important cases in which the mixed potential departs very significantly from the equilibrium potentials of a metal electrode. This is particularly important in the field of corrosion. Excursion 14-2 describes an impressive example of such a situation. We will return to the prediction of mixed potentials, but only after further study of the topics of *electrode kinetics* and *transport processes.*

2.4 Applications of Cell Voltage Measurements and of Galvanic Cells

Experimental aspects and applications of galvanic cells are important in the context of understanding thermodynamic concepts and their application. A more detailed analysis of applied aspects of galvanic cells will require further study of materials in Chapters 3 and 4. Here, an introductory level survey about applications of galvanic cells will be provided based on equilibrium thermodynamics.

2.4.1 Reference Electrodes

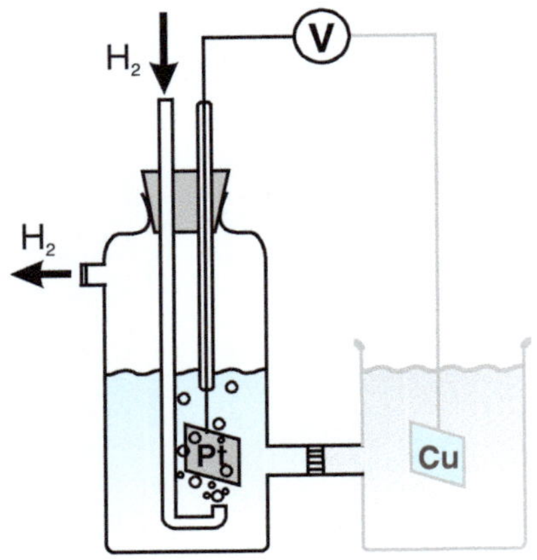

Figure 2.19 Construction of a standard hydrogen electrode.

The Standard Hydrogen Electrode. As discussed in Section 2.1.2, the electrochemical potential scale has been defined with 0.0 V versus the standard hydrogen electrode (SHE). This is the definition for all temperatures, although usually data for 298 K are discussed.

$$\text{Pt} \mid \text{H}_2\ (100\,000\ \text{Pa}), \text{H}^+(aq)\ (a_\text{m}(\text{H}^+) = 1\ \text{mol kg}^{-1})$$

Occasionally, the term normal hydrogen electrode (NHE) is employed, referring to a *concentration* of 1 mol L^{-1} protons. This is an outdated approach, and therefore, NHE should not be used anymore according to IUPAC.

Figure 2.19 illustrates the experimental design of the SHE. A platinum foil or gauze electrode is connected via platinum wire. The platinum wire is readily sealed into glass to seal the apparatus. A high surface area is achieved by electrodeposition of "platinum black" onto the platinum foil (the rate of the electrolytic reaction is linked to the active area, and therefore, highly dispersed platinum is employed; the black color is a result of small platinum particles absorbing the light). Hydrogen gas (purified) is supplied from a pressurized cylinder. A continuous flow of hydrogen allows bubbles to interact with the platinum electrode surface. Note the hazards of working with hydrogen gas in the presence of oxygen/air and the need to safely capture and dispose of excess hydrogen.

The hydrogen employed for the standard hydrogen electrode contains isotopes (e.g., natural abundance of deuterium of 0.015%) and the equilibrium ratio of *ortho*- to *para*-hydrogen.

The pressure must be p = 100 000 Pa, although in older literature a value of p = 101 325 Pa H_2 pressure was often used. To convert, a voltage of 0.169 mV needs to be subtracted to convert to standard pressure. This is often negligible compared to further errors of typically 1 mV or higher.

The platinum electrode is immersed in hydrochloric acid with proton molality activity of 1 mol kg^{-1}. That is, the proton molality multiplied by the activity coefficient has to be unity. A problem arising at this point is linked to the fact that only the average activity coefficient for protons and chloride is known/accessible. This will be the topic of a later section on the determination of the activity coefficient for ions. Average activity coefficients for hydrochloric acid are tabulated. However, tables provide average activity coefficients only for selected values of concentration or molality. The use of these tabulated values will be explained next.

Example Calculation 2.8 Calculating Concentrations from Molalities $c = f(b)$

In tables, activities are commonly listed in terms of molalities. How can we use this information to derive the activity coefficient for a given concentration? What is the error introduced when using the numerical value from molality for concentration?

The molality b_i [Eq. (2.42)] and the concentration c_i [Eq. (2.40)] are linked by the density ρ of the solution.

ρ – Density of the solution, temperature-dependent, concentration-dependent [g cm^{-3}] = [g mL^{-1}] = [kg L^{-1}]
m_sol – Mass of the solution [kg]
m_i – Mass of component i [kg]

$$\rho = \frac{\text{Mass of the solution}}{\text{Volume of the solution}} = \frac{m_\text{sol}}{V_\text{sol}} = \frac{m_\text{solv} + \sum_{i=1}^{n} m_i}{V_\text{sol}} \tag{2.118}$$

We can rearrange this equation to derive the expression for the volume of the solution.

m_{solv} – Mass of solvent [kg]
V_{sol} – Volume of solution [L]

N_i – Amount of component i [mol]
M_i – Molar mass of component i [g mol^{-1}]
b_i – Molality of component i [mol kg^{-1}]

$$V_{sol} = \frac{m_{solv} + \sum_{i=1}^{n} m_i}{\rho} = \frac{m_{solv} + \sum_{i=1}^{n} N_i M_i}{\rho} = \frac{m_{solv} + \sum_{i=1}^{n} b_i m_{solv} M_i}{\rho} = \frac{m_{solv}\left(1 + \sum_{i=1}^{n} b_i M_i\right)}{\rho} \tag{2.119}$$

We can now substitute (2.119) into Eq. (2.40) to give

$$c_i = \frac{N_i}{V_{sol}} = \frac{b_i m_{solv} \rho}{m_{solv}\left(1 + \sum_{i=1}^{n} b_i M_i\right)} = \frac{b_i \rho}{1 + \sum_{i=1}^{n} b_i M_i}. \tag{2.120}$$

Note the units: the concentration c in mol L^{-1} is obtained from the density ρ in g cm^{-3} = kg L^{-1} and the molar mass of components M_i in kg mol^{-1}.

Example Calculation 2.9 Calculating Molalities from Concentrations $b = f(c)$

In a similar way, we can derive the expression for calculating the molality b_i from concentrations c_i.

$$b_i = \frac{N_i}{m_{solv}} = \frac{c_i V_{Lsg}}{m_{sol} - \sum_{i=1}^{n} m_i} = \frac{c_i V_{Lsg}}{\rho V_{sol} - \sum_{i=1}^{n} c_i V_{sol} M_i} = \frac{c_i}{\rho - \sum_{i=1}^{n} c_i M_i} \tag{2.121}$$

Example Calculation 2.10 Calculating Molalities from Mass Fractions (or Mass Percentage) $b = f(m\text{-}\%)$

For tabulated values of density or activity coefficients given in terms of mass fraction $w_1 = m_1/m_{sol}$, we can calculate the molality of the dissolved component b_1 based on the following expression.

w_1 – Mass fraction for component 1 [dimensionless]
m_1 – Mass of component 1 [kg, g]
m_{solv} – Mass of solvent [kg, g]
m_{sol} – Mass of solution [kg, g]

$$w_1 = \frac{m_1}{m_{sol}} = \frac{m_1}{m_{solv} + m_1}$$

$$w_1 m_{solv} + w_1 m_1 = m_1$$

$$m_1 = \frac{w_1 m_{solv}}{1 - w_1} \tag{2.122}$$

N_1 – Amount of component 1 [mol]
M_1 – Molar mass of component 1 [g mol^{-1}]

$$N_1 = \frac{m_1}{M_1} = \frac{w_1 m_{solv}}{(1 - w_1) M_1} \tag{2.123}$$

$$b_1 = \frac{N_1}{m_{solv}} = \frac{w_1}{(1 - w_1) M_1}. \tag{2.124}$$

Note that the units of molality in mol kg^{-1} require that the molar mass is given in kg mol^{-1}.

Example Calculation 2.11 Interpolation of Density Values from Tabulated Values $\rho = f(b)$

Commonly, density data is tabulated as a function of mass fraction as in Table 2.3. In order to derive density values in between tabulated values, we could employ the closest value in the table, or more accurately, interpolate the density by assuming a

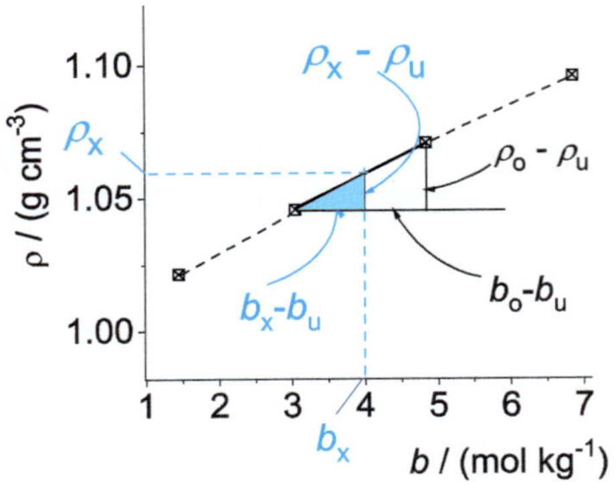

Figure 2.20 Densities as a function of molality corresponding to data in Table 2.3. In order to interpolate values that are not listed, we can employ a linear interpolation approach. The desired density ρ_x for molality b_x can be obtained from the listed values (b_1, r_1) and (b_2, r_2) by linear interpolation:
$\rho_x = \rho_1 + (\rho_x - \rho_1)$
$\rho_x = \rho_1 + [(\rho_2 - \rho_1)/(b_2 - b_1)] \cdot b_x - b_1$

Table 2.3 Density of aqueous HCl solution at 25 °C as a function of mass fraction of HCl [4]. The corresponding molalities were calculated based on Eq. (2.124) and using the molecular weight M(HCl) = 36.5 g mol^{-1}.

w(HCl)/%	b(HCl)/mol kg^{-1}	ρ(298 K)/g cm^{-3}
0	0	0.9970 (H_2O)
2	0.55912	1.0076
5	1.44196	1.0216
10	3.04414	1.0457
15	4.83481	1.0707
20	6.84931	1.0957
25	9.13242	1.1211
30	11.7417	1.1463
35	14.7523	1.1705

linear relationship, as shown in Figure 2.20. For cases of nonlinear functional dependence (or generally for higher precision), interpolation with polynomials of high order is possible [13]. These methods are commonly implemented in spreadsheet software or in database systems.

◢

Example Calculation 2.12 Conversion of Molality to Concentration

Next, data for molality in Table 2.4 will be converted and compared to concentration. First, for the molality b, the corresponding density ρ needs to be interpolated (Example Calculation 2.11). We employ a linear interpolation (although higher polynomials could be employed) and demonstrate in Table 2.4 that the effects remain mostly small for low concentrations (usually within experimental error).

In column 4 of Table 2.4, we can see that numerical values for b and c deviate by less than 1% for concentrations of less than 0.2 mol L^{-1}. However, for higher concentrations, significant changes in numerical values are observed.

◢

Table 2.4 Conversion of molalities b into concentrations c.

b/(mol kg^{-1})	ρ/(g cm^{-3})[a)]	c/(mol L^{-1})	Δ/%[b)]
0 (H_2O)	0.9970	0	0
0.0001	0.99704889	9.97045E−05	−0.295
0.0002	0.99705077	0.000199409	−0.296
0.0005	0.99705644	0.000498519	−0.296
0.001	0.99706587	0.000997029	−0.297
0.002	0.99708475	0.001994024	−0.299
0.005	0.99714137	0.004984797	−0.304
0.01	0.99723574	0.009968719	−0.313
0.02	0.99742448	0.019933938	−0.330
0.05	0.9979907	0.049808634	−0.383

Table 2.4 (Continued)

b/(mol kg^{-1})	ρ/(g cm^{-3})[a]	c/(mol L^{-1})	Δ/%[b]
0.1	1.00031914	0.099668125	−0.332
0.2	1.00190494	0.198928808	−0.536
0.5	1.00666235	0.494310017	−1.138
1	1.01459137	0.978862872	−2.114
1.5	1.02247302	1.454097685	−3.060
2	1.02999403	1.919839758	−4.008
3	1.04503605	2.825694579	−5.810
4	1.059045	3.696492147	−7.588

[a] Densities interpolated based on data in Table 2.3.
[b] Percent deviation of numerical values of c from b given as $100c/(\text{mol L}^{-1}) - b/(\text{mol kg}^{-1}))/b/(\text{mol kg}^{-1})$.

Example Calculation 2.13 Effect of Activity Coefficients

What is the electrode potential for a hydrogen electrode at standard pressure and at 25 °C, given a solution containing a concentration of $c = 1$ mol L^{-1}?

Sought: E

Known:

$$[\text{HCl}] = 1\ \text{mol L}^{-1}$$

$$T = 25\ °\text{C} = 298.15\ \text{K}$$

$$E°(H_2/H^+) = 0 \quad \text{(by Definition)}$$

$$R = 8.314\ \text{J K}^{-1}\ \text{mol}^{-1} = 8.314\ \text{V A s K}^{-1}\ \text{mol}^{-1}$$

$$F = 96\,485\ \text{A s mol}^{-1}$$

$$2\,H^- + 2\,e^- \rightleftharpoons H_2 \quad (n = 2)$$

$$E = E°(H_2|H^+) + \frac{RT}{2F}\ln\frac{\left(\frac{\{H^+\}}{a°}\right)^2}{\left(\frac{p(H_2)}{p°}\right)} \tag{2.125}$$

With $p(H_2) = p°$ and $\{H^+\} = [H^+]f_\pm$ we obtain

$$E = E°(H_2|H^+) + \frac{RT}{F}\ln\frac{[H^+]f_\pm}{a°}$$
$$= E°(H_2|H^+) + \frac{RT}{F}\ln\frac{[H^+]}{a°} + \frac{RT}{F}\ln f_\pm. \tag{2.126}$$

Here, all values are known apart from $f_\pm$. If we select unit activity coefficients ($f_\pm = 1$), the calculation results in

$$E = 0\,\text{V} + \frac{8.314\,\text{A V s mol}^{-1}\,\text{K}^{-1}\cdot 298.15\,\text{K}}{96\,485\,\text{A s mol}^{-1}}\ln\frac{1\ \text{mol L}^{-1}}{1\ \text{mol L}^{-1}}$$
$$= 0\,\text{V} + 0.02569\,\text{V}\ln(1) = 0.$$

ln 1 = 0

Next, we introduce the activity coefficient, which is commonly tabulated for molality b (Table 2.5). From Table 2.5, we obtain $f_\pm = 0.8090$ for $b = 1$ mol kg^{-1}. When ignoring the difference between c and b, Eq. (2.126) gives a potential of −5 mV. This difference is clearly measurable, and therefore, assuming $f = 1$ is not appropriate for accurate

Table 2.5 Data for the average activity coefficients for HCl at 298 K.

b/(mol kg^{-1})	c/ (mol L^{-1})[a]	$f_{\pm}$
0.0001	9.97045E−05	0.9891
0.0002	0.00019941	0.9845
0.0005	0.00049852	0.9752
0.001	0.00099703	0.9656
0.002	0.00199402	0.9521
0.005	0.00498480	0.9285
0.01	0.00996872	0.9048
0.02	0.01993394	0.8755
0.05	0.04980863	0.8304
0.1	0.09966812	0.7964
0.2	0.19892881	0.7667
0.5	0.49431002	0.7571
1	0.97886287	0.8090
1.5	1.45409768	0.8962
2	1.91983976	1.009
3	2.82569458	1.316
4	3.69649215	1.762

[a] The concentration data c were calculated employing Eq. (2.120) with interpolated densities from Table 2.4.
Source: Data from [6].

experiments. However, experiments employing a salt bridge may result in experimental errors of similar magnitude (Table 2.2).

$$E = 0\,\text{V} + 0.02569\,\text{V}\ln(1) + 0.02569\,\text{V}\ln(0.8090)$$
$$= 0.02569\,\text{V}\cdot(-0.2119) = -0.005446\,\text{V}$$

For a more accurate calculation, we take into account the difference between b and c. The value for f for $c = 1.0000$ mol L^{-1} is obtained by linear interpolation for tabulated values of $c_1 = 0.97886$ mol L^{-1} and $c_2 = 1.45409$ mol L^{-1} (Figure 2.20):

$$f(c_x) = f_u + \frac{f_o - f_u}{c_o - c_u}(c_x - c_u)$$
$$= 0.809 + \frac{0.8962 - 0.809}{(1.45410 - 0.97886)\,\text{mol}\,\text{L}^{-1}}(1.00000 - 0.97886)\,\text{mol}\,\text{L}^{-1}$$
$$= 0.81288$$

Substitution of this value into Eq. (2.126) gives

$$E = 0\,\text{V} + 0.02569\,\text{V}\ln(1) + 0.02569\,\text{V}\ln(0.81288)$$
$$= 0.02569\,\text{V}\cdot(-0.2072) = -0.005323\,\text{V}.$$

The correction compared to the previously calculated potential is very minor. Substituting c for b does not lead to significant error ($\Delta E = [5.446 - 5.323]$ mV = 0.1235 mV) in this concentration range and temperature range. This difference will be important only in precision measurements. The functional dependence of f on c is nonlinear, and therefore, further gains in accuracy are possible by employing nonlinear interpolation tools. However, for most laboratory measurements, errors in measured values will be higher when compared to these corrections.

◢

The Reversible Hydrogen Electrode (RHE). This type of commercially available hydrogen electrode is based on a platinum flag or gauze immersed in an aqueous solution without the requirement of $a(H^+) = 1$ mol kg^{-1}. Any hazards arising from the use of hydrogen gas can be minimized by locally producing hydrogen gas by preelectrolysis and by trapping the hydrogen gas around the platinum electrode (Figure 2.21). In this way, the platinum electrode is simultaneously in contact with the solution phase and with $H_2(g)$. Small effects of local pressure on hydrogen activity can be ignored.

The use of hydrogen reference electrodes (SHE or RHE) in general-purpose laboratory applications is problematic due to hazards associated with hydrogen gas as well as problems maintaining highly catalytic platinum surfaces. Hydrogen reference electrodes are essential in high precision measurements; however, a range of other types of reference electrodes have been developed for general-purpose use. Very commonly used reference electrodes are **electrodes of the second kind** (Table 2.6). Other types of reference electrodes can be found in the literature [14–16].

Other Reference Electrodes. A reference electrode very commonly employed for general-purpose work in the past is the saturated calomel electrode (SCE, Figure 2.22a). A droplet of liquid Hg is coated with highly insoluble Hg_2Cl_2 (calomel). The solution phase contains saturated KCl, with chloride being the potential defining species in solution. The potential of the saturated calomel electrode remains stable due to excess chloride in the solution, providing a reservoir of Cl^- anions during production or consumption of Hg_2Cl_2. Solid KCl can be added to stabilize the chloride activity by ensuring saturation.

Due to some chloride leaching out from this type of reference electrode, it is often (e.g., in electrocatalysis) desirable to employ Cl^--free electrodes. The sulfate-based

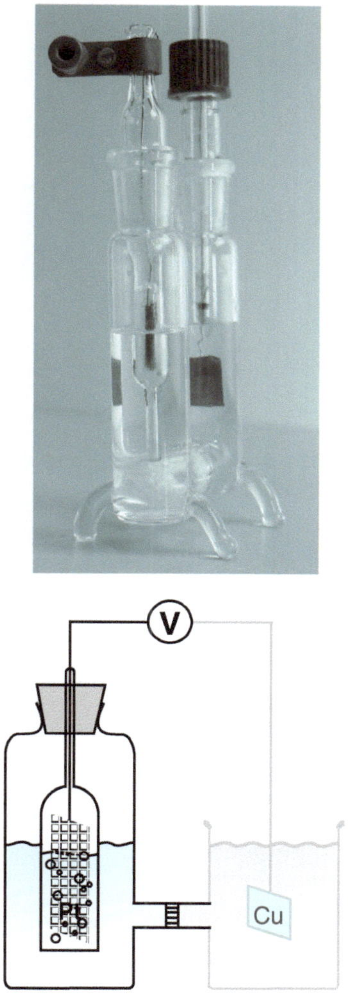

Figure 2.21 Construction of a reversible hydrogen electrode with a trapped hydrogen reservoir.

Table 2.6 Electrodes of the second kind with application as a reference electrode. The potential is given relative to the potential of the SHE at 25 °C under standard conditions. Each reference electrode will exhibit a distinct temperature dependence in potential, and therefore, values for conversion of reference electrode scale will differ at other temperatures.

Half-cell	*E*/V vs. SHE
Pt ∣ H_2 (100 kPa), $H^+(aq)$ ($a(H^+) = 1$ mol L^{-1}) (SHE)	0.00000000
Pt ∣ H_2 (100 kPa), $H^+(aq)$ ($c(H^+) = 1$ mol L^{-1}) (NHE)	−0.006
Ag ∣ AgCl KCl(*aq*, *sa*)	+0.197
Ag ∣ AgCl KCl(*aq*, $c = 3$ mol L^{-1})	+0.210
Ag ∣ AgCl KCl(*aq*, $c = 1$ mol L^{-1})	+0.236
Ag ∣ AgCl ∣ KCl(*aq*, $c = 0.1$ mol L^{-1})	+0.290
Ag ∣ AgCl ∣ NaCl(*aq*, $c = 3$ mol L^{-1})	+0.209
Hg ∣ Hg_2Cl_2 ∣ KCl(*aq*, *sa*) (SCE, saturated calomel electrode)	+0.241
Hg ∣ Hg_2Cl_2 ∣ KCl(*aq*, $c = 1$ mol L^{-1}) (NCE, normal calomel electrode)	+0.280
Hg ∣ Hg_2Cl_2 ∣ KCl(*aq*, $c = 0.1$ mol L^{-1})	+0.334
Hg ∣ Hg_2Cl_2 ∣ NaCl(*aq*, *sa*) (SSCE, sodium saturated calomel electrode)	+0.236
Hg ∣ Hg_2SO_4 ∣ K_2SO_4(*aq*, *sa*)	+0.640
Hg ∣ Hg_2SO_4 ∣ H_2SO_4(*aq*, $c = 0.5$ mol L^{-1})	+0.679
Hg ∣ HgO ∣ KOH(*aq*, $c = 1$ mol L^{-1})	+0.107
Hg ∣ HgO ∣ NaOH(*aq*, $c = 0.1$ mol L^{-1})	+0.165
Cu ∣ $CuSO_4$ ∣ $CuSO_4$(*sa*) for corrosion test	+0.320
Pb ∣ $PbSO_4$ ∣ K_2SO_4(*sa*)	−0.277

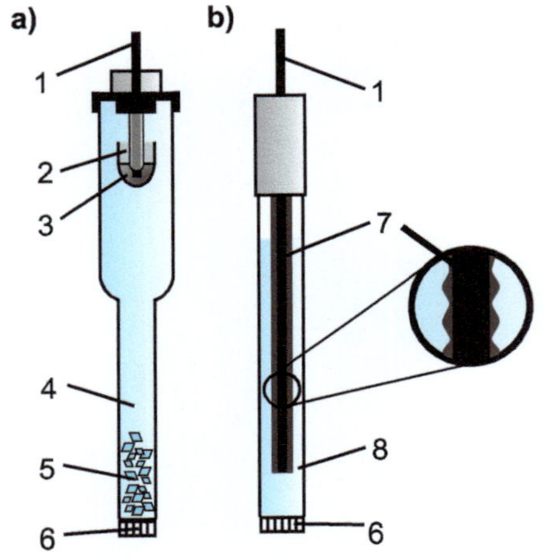

Figure 2.22 Design of a) an SCE and b) an Ag|AgCl reference electrode.

1) Connection wire,
2) Pt wire,
3) Hg-droplet with Hg_2Cl_2 (calomel) coating,
4) saturated KCl solution,
5) solid KCl,
6) diaphragm or ion-conducting glass (Vycor) to connect to the sample solution,
7) Ag wire coated with solid AgCl, and
8) KCl solution with defined activity (e.g., 1 mol L^{-1} KCl or 3 mol L^{-1} KCl).

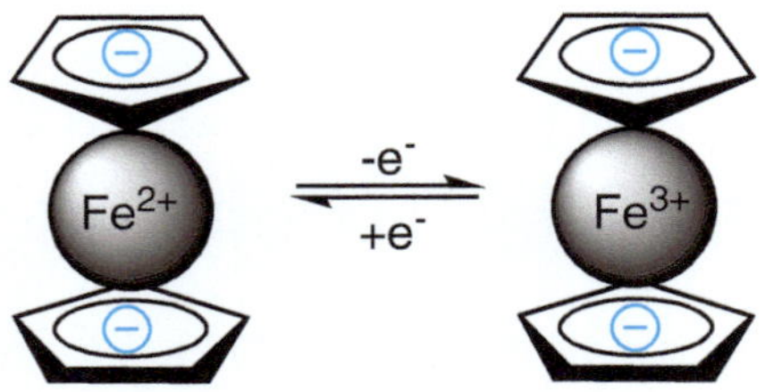

Figure 2.23 Molecular structure of ferrocene (Fc) and ferrocenium (Fc^+).

$Hg|Hg_2SO_4|K_2SO_4(sa)$ electrode offers an alternative. In alkaline electrolyte media, the Hg|HgO reference electrode can be employed.

Commercial availability of mercury-containing reference electrodes is now restricted because of mercury toxicity, and often silver-based reference electrodes can be employed instead, for example, Ag|AgCl|KCl(*sa*) (Figure 2.22b). Practical problems with these and similar reference electrodes often arise when the frit is clogged up. Therefore, commercial reference electrodes often contain lower concentrations of KCl (e.g., 1 mol L^{-1} or 3 mol L^{-1}). Particular problems arise in perchlorate-containing sample solutions due to K^+ and ClO_4^- forming poorly soluble precipitates. In this case, commercial reference electrodes based on NaCl electrolyte solution are available.

Reference electrodes based on electrodes of the second kind show specific temperature effects (versus SHE) and are therefore linked to a temperature function. This function, often provided as a polynomial, allows the effect of temperature on the reference potential versus SHE to be calculated. The fundamental reasons for the temperature function are linked to the entropy of the cell reaction, $\Delta_R S \neq 0$ [Eq. (2.58)], and the considerable temperature effect on solubility products that define the potential for electrodes of the second kind.

Special types of reference electrodes are, for example, the dynamic reference electrodes. These are operated with a well-defined current density to result in a constant, well-defined nonequilibrium potential. A practical example is the dynamic hydrogen electrode (DHE) [17].

A reference electrode based on hydrogen insertion rather than gaseous hydrogen is the palladium hydride electrode. Palladium is known to undergo relatively fast hydrogen insertion and diffusion into/through the metal lattice. The electrochemical reduction of H^+, for example, in 1 M $HClO_4$ solution leads to hydride materials PdH_x and a phase transformation (from α to β) as a function of hydride loading [18]. Both phases are present in an equilibrium for $0.03 < x < 0.57$, and this phase equilibrium leads to a constant electrode potential of 0.05 V vs. SHE [19] making the PdH_x electrode suitable as a reference electrode.

Potential Reference for Organic Electrolyte Solutions. For electrochemical cells based on organic solvents aqueous reference electrodes are often employed, but this can lead to contamination with water and salt. An alternative is to employ a silver wire immersed in the organic solvent (e.g., acetonitrile) with a silver salt (e.g., $AgClO_4$) and a suitable electrolyte salt dissolved. A porous frit can be used to separate the reference electrode electrolyte and the electrolytic cell compartment (Section 4.1) or better, a further electrolyte compartment with another frit is added to minimize contamination of the electrolytic cell (Figure 5.24). The additional diffusion potentials introduced for each frit can be substantial, and a calibration of the reference electrode potential with a well-known redox probe can be necessary. In voltammetric measurements (Section 5.3.5), the ferrocene/ferrocenium redox system (Figure 2.23) offers a convenient molecular system as an internal potential reference. The ferrocene/ferrocenium (Fc/Fc^+) potential is at approximately 0.40 V vs. SHE [20] and has been observed to be only weakly dependent on the type of organic solvent employed. The IUPAC recommendation is to provide redox potentials in organic solution versus ferrocene/ferrocenium (Fc/Fc^+) [21].

Q: Practically, how can redox potentials be compared to standard potentials?

A: Standard electrode potentials $E°$ (or more correctly formal potentials $E°'$, Section 3.2.2) are obtained in cyclic voltammetric measurements (Section 5.3.5, Figure 5.29). The reference electrode in the organic solution can be based on $Ag|Ag^+$ (0.01 mol L^{-1} $AgClO_4$ + 0.1 mol L^{-1} $[(H_9C_4)_4N]^+[PF_6]^-$ in an organic solvent, for example, CH_3CN (Figure 5.24). The resulting diffusion potentials are acceptable when employing a calibration, for example, with the well-known molecular redox system Fc/Fc^+ (Figure 2.23). The internal calibration procedure is most commonly employed. Initially, the electrolyte solution in the electrolytic cell contains a redox system either in the oxidized (O) or in the reduced (R) form, together with a supporting electrolyte. The formal potential $E^{\circ\prime}$ is determined for the reaction $R \rightleftharpoons O + ne^-$ employing the $Ag|Ag^+$ reference electrode [$E^{\circ\prime}$(R/O) versus $Ag|Ag^+$]. After completion of the measurements, a small amount of ferrocene (Fc; typically 1 mmol L^{-1}) is added directly into the same electrolyte solution. The additional voltammetric signal associated with ferrocene allows the determination of the formal potential $E^{\circ\prime}$(Fc/Fc^+ versus $Ag|Ag^+$). The calibrated standard electrode potential (versus Fc/Fc^+) is then obtained as follows:

$$E^{\circ\prime}(\mathrm{R/O\ vs.\ Fc/Fc^+}) = E^{\circ\prime}(\mathrm{R/O\ vs.\ Ag|Ag^+}) - E^{\circ\prime}(\mathrm{Fc/Fc^+\ vs.\ Ag|Ag^+}) \quad (2.127)$$

A more visual illustration of this calibration process is given in Figure 2.24.

In rare cases, the internal calibration process is complicated, for example, by reaction of ferrocene or ferrocenium with the investigated redox system. As a result, the voltammetric response for R/O or for Fc/Fc^+ can be distorted and calibration affected. In this case, it is possible to employ external calibration of the potential scale. Initially, a solution of 1 mmol L^{-1} Fc is investigated in the organic solvent with supporting electrolyte. This allows determination of the formal potential of ferrocene E°(Fc/Fc^+ vs. $Ag|Ag^+$). This measurement also allows the experimental system, the potential window, and the electrode area to be assessed. With appropriate equipment, it is possible to estimate the uncompensated resistance in the measurement cell (to then apply a correction, Section 5.3.1 and Excursion 6-1). Typical standard potentials (or formal potentials) are E°(Fc/Fc^+ vs. $Ag|Ag^+$) = −0.211 V in dichloromethane and +0.162 V in tetrahydrofuran. Next, the electrolytic cell is cleaned, and a new experiment is started with the O/R redox system, in the *same* solvent, with the *same* supporting electrolyte, at the *same* electrolyte concentration, and with the *same* measurement cell geometry.

The calibration of the reference $Ag|Ag^+$ electrode has to be repeated every time the solution with $AgClO_4$ is renewed. This is recommended once a month due to some inherent chemical instability. The calibration also needs to be renewed when changing the solvent or the electrolyte used in experiments, or when changing electrolyte concentration or the temperature of the electrolytic cell [22].

Example Calculation 2.14 Conversion of Electrode Potentials from one Type of Reference Electrode to Another

The equilibrium potential for a redox system (R/O) was measured as −0.150 V vs. SCE. What would be the corresponding equilibrium potential versus the SHE reference electrode?

Sought: E_{SHE}(R/O)

Known:

$$E_{\mathrm{SCE}}(\mathrm{R/O}) = -0.150\ \mathrm{V\ vs.\ SCE}$$

$$E_{\mathrm{SHE}}(\mathrm{SCE}) = +0.241\ \mathrm{V\ vs.\ SHE\ (Table\ 2.6)}$$

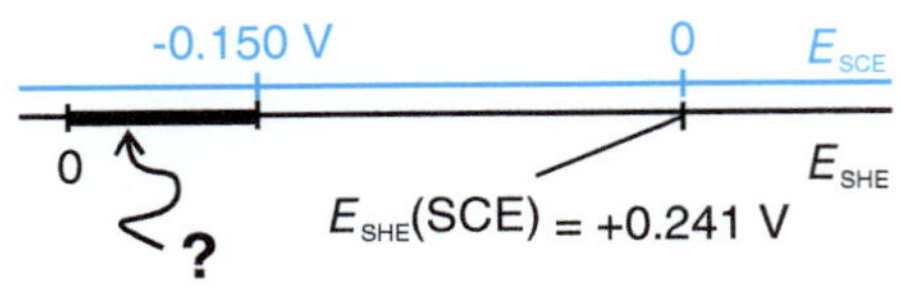

Figure 2.24 Illustration of the conversion from SCE potential scale to SHE potential scale.

It is convenient to illustrate the potential scale conversion with a line plot (Figure 2.24). The mathematical expression for the potential scale conversion is

$$\begin{aligned} E_{\mathrm{SHE}}(\mathrm{R/O}) &= E_{\mathrm{SCE}}(\mathrm{R/O}) + E_{\mathrm{SHE}}(\mathrm{SCE}) \\ &= -0.150\ \mathrm{V} + 0.241\ \mathrm{V} \\ &= +0.091\ \mathrm{V} \end{aligned}$$

The following procedure is recommended for converting the potential scale for reference electrode A to reference electrode B: The measured potential versus A is added to the measured potential of reference A versus B. To avoid confusion, it is always a good idea to make a sketch of a number line (Figure 2.24).

◢

2.4.2 Determination of Thermodynamic Properties

Determination of Standard Electrode Potentials. For the experimental determination of standard electrode potentials with high precision, a high-precision voltmeter and a carefully designed electrochemical cell are necessary. The electrochemical cell needs to be in equilibrium (i.e., mixed potentials and diffusion potentials have to be eliminated). This means that an electrochemical cell without salt bridge must be used (Section 2.3.3). To give an example, we consider the standard electrode potential for an electrode of the second kind.

U_{eq} – Equilibrium cell voltage [V]

$E_{\mathrm{Ag|AgCl}}$ – Equilibrium electrode potential for the Ag|AgCl electrode [V]

$E_{\mathrm{H_2|H^+}}$ – Equilibrium electrode potential for the Pt|H_2|HCl electrode [V]

$$\mathrm{Ag} \mid \mathrm{AgCl} \mid \mathrm{HCl}(aq) \mid \mathrm{H_2} \mid \mathrm{Pt} \mid \mathrm{Ag'}. \tag{2.128}$$

The equilibrium cell voltage is obtained from the difference in electrode potentials.

$$U_{\mathrm{eq}} = E_{\mathrm{Ag|AgCl}} - E_{\mathrm{H_2|H^+}} \tag{2.129}$$

For the Ag|AgCl electrode, we can substitute in the solubility product for AgCl (analogously to Eqs. (2.89)–(2.91)):

$a°$ – Standard activity (1 mol L^{-1}), employed to cancel out units

{X} – Concentration activity for component X [mol L^{-1}]

K_{SP} – Solubility product for silver chloride AgCl, $K_{SP}(\mathrm{AgCl}) = \{\mathrm{Ag^+}\}\{\mathrm{Cl^-}\}$ [mol² L^{-2}]

$$E_{\mathrm{Ag|AgCl}} = E^{\circ}_{\mathrm{Ag|Ag^+}} + \frac{RT}{F}\ln\frac{\{\mathrm{Ag^+}\}}{a^{\circ}} \tag{2.130}$$

$$E_{\mathrm{Ag|AgCl}} = E^{\circ}_{\mathrm{Ag|Ag^+}} + \frac{RT}{F}\ln\frac{K_{\mathrm{SP}}(\mathrm{AgCl})}{a^{\circ 2}} - \frac{RT}{F}\ln\frac{\{\mathrm{Cl^-}\}}{a^{\circ}}. \tag{2.131}$$

For the hydrogen electrode $0.5\ \mathrm{H_2} \rightleftharpoons \mathrm{H^+} + \mathrm{e^-}$, we can express the electrode potential as a function of HCl activity.

$$E_{\mathrm{H_2|H^+}} = E^{\circ}_{\mathrm{H_2|H^+}} + \frac{RT}{F}\ln\frac{\{\mathrm{H^+}\}}{a^{\circ}} - \frac{RT}{F}\ln\sqrt{\frac{p(\mathrm{H_2})}{p^{\circ}}}. \tag{2.132}$$

When performing the experiment under standard pressure $p(\mathrm{H_2}) = p°$, the hydrogen pressure-dependent term in Eq. (2.132) is removed. The standard potential for the hydrogen electrode $E^{\circ}_{\mathrm{H_2|H^+}}$ is by definition equal to zero. The standard electrode potential for the Ag|Ag^+ electrode in Eq. (2.131) contains the solubility product K_S(AgCl). The expression simplifies for the standard electrode potential for the Ag|AgCl electrode. Combining Eqs. (2.131) and (2.132) with Eq. (2.129) results in

$$U = E^{\circ}_{\mathrm{Ag|AgCl}} - \frac{RT}{F}\ln\frac{\{\mathrm{Cl^-}\}}{a^{\circ}} - \frac{RT}{F}\ln\frac{\{\mathrm{H^+}\}}{a^{\circ}}. \tag{2.133}$$

Activity parameters for $\{\mathrm{Cl^-}\}$ and $\{\mathrm{H^+}\}$ cannot be determined individually and are therefore replaced with the average ion activity $a_{\pm}$, which is determined as the geometric average ($a_{\pm,\mathrm{HCl}} = \sqrt{\{\mathrm{H^+}\}\{\mathrm{Cl^-}\}}$). Therefore, Eq. (2.133) can be written as follows:

$\bar{x} = \sqrt[n]{x_1 \cdot x_2 \cdot \ldots \cdot x_n}$

$2 \ln a = \ln a^2$

$$U = E^{\circ}_{\mathrm{Ag|AgCl}} - \frac{2RT}{F}\ln\frac{a_{\pm,\mathrm{HCl}}}{a^{\circ}}. \tag{2.134}$$

Separating the concentration [HCl] and the mean or average activity coefficient $f_\pm$ then gives

[HCl] – Concentration of HCl in mol L^{-1}
$f_\pm$ – Mean activity coefficient

$$U = E^\circ_{Ag|AgCl} - \frac{2RT}{F}\ln\frac{[HCl]}{a^\circ} - \frac{2RT}{F}\ln f_\pm. \tag{2.135}$$

In Eq. (2.135), U is accessible by experiment and [HCl] is known. The average activity coefficient $f_\pm$ is unknown. However, we know that $f_\pm$ converges to unity when [HCl] converges to zero. We want to determine $E^\circ_{Ag|AgCl}$. Therefore, all known or measurable parameters are collected on the left-hand side of the equation with unknown parameters collected on the right-hand side.

$$U + \frac{2RT}{F}\ln\frac{[HCl]}{a^\circ} = E^\circ_{Ag|AgCl} - \frac{2RT}{F}\ln f_\pm. \tag{2.136}$$

Experimentally, the equilibrium cell voltage U is now determined as a function of the concentration [HCl]. Empirically observed and supported by theory of electrolytes (Section 4.2.8) is a correlation of $\ln f_\pm = A\sqrt{c/c^\circ}$. Therefore, a plot of the left-hand side terms in Eq. (2.136) versus $\sqrt{[HCl]}$ results in a straight line plot [Eq. (2.137), Figure 2.25]. It is now possible to extrapolate to the limit at [HCl] = 0 to obtain the required standard potential $E^\circ_{Ag|AgCl}$ for $f_\pm$ converging to unity with $\ln(1) = 0$.

$$\underbrace{U + \frac{2RT}{F}\ln\frac{[HCl]}{a^\circ}}_{y} = \underbrace{E^\circ_{Ag|AgCl}}_{b_0} - \underbrace{\frac{2RT}{F}A}_{b_1}\ \underbrace{\sqrt{[HCl]/c^\circ}}_{x}. \tag{2.137}$$

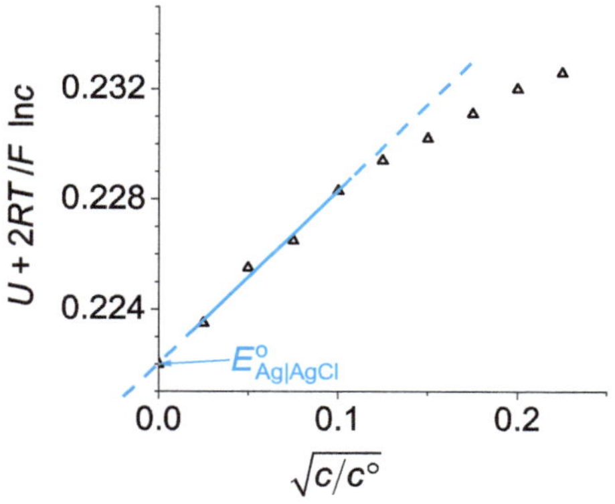

Figure 2.25 Extrapolation based on Eq. (2.137).

Determination of Activity Coefficients. Once the standard electrode potential $E^\circ_{Ag|AgCl}$ has been determined, we can rearrange Eq. (2.136) for any concentration $[HCl] > 0$ to give the average activity coefficient based on the measured equilibrium cell voltage.

$$\ln f_\pm = \frac{F}{2RT}\left[E^\circ_{Ag|AgCl} - U - \frac{2RT}{F}\ln\frac{[HCl]}{a^\circ}\right] \tag{2.138}$$

Determination of the Solubility Product. The solubility product K_{SP} for AgCl can be obtained in high precision from equilibrium cell voltage measurements. Both are required: the standard electrode potential for the $Ag|Ag^+$ electrode and the standard electrode potential for the Ag|AgCl electrode. Based on Eq. (2.131) and for the standard state ($\{Cl^-\} = a^\circ$), we obtain

$$\ln\frac{K_{SP}(AgCl)}{a^{\circ 2}} = \frac{F}{RT}\left(E_{Ag|AgCl} - E^\circ_{Ag|Ag^+}\right) \tag{2.139}$$

The pK_{SP} value (which is often found in data tables) corresponds to the negative decadic logarithm of the K_{SP}. We can obtain this value directly from the natural logarithm by dividing by $\ln(10) \approx 2.3$.

$2.3 \log x \approx \ln x$, due to $e^{2.3} = \exp(2.3) \approx 10$

$$pK_{SP} = -\log\frac{K_{SP}}{a^{\circ 2}} = -\frac{1}{2.3}\ln\frac{K_{SP}}{a^{\circ 2}} \tag{2.140}$$

Determination of the Ion Product for Water. Water undergoes autoprotolysis to result in a constant background level of protons and hydroxide anions.

$$H_2O + H_2O \rightleftharpoons H_3O^+ + OH^- \tag{2.141}$$

The equilibrium constant for this reaction can be written as follows:

K_W – Equilibrium constant for the autoprotolysis of water

$$K_W = \frac{a_{H_3O^+}a_{OH^-}}{a^2_{H_2O}}. \tag{2.142}$$

We replace the general activities in Eq. (2.142) with concentration activities for H_3O^+ and OH^-. We simplify the expression by writing H^+ instead of H_3O^+. According to the convention, the activity of the solvent (water) has to be stated as the mole fraction activity a_x [Eqs. (2.44) and (2.45)], which is very close to one in the limit for very low concentrations of H^+ and OH^-. This simplifies Eq. (2.142) to a product of activities, the so-called **ion product of water**:

$$K_W = \{H^+\}\{OH^-\}. \tag{2.143}$$

We can now rearrange Eq. (2.143) to give $\{H^+\}$ and substitute into Eq. (2.133). By expanding the concentration-dependent term with a°/a°, we obtain

$$U = E^\circ_{Ag|AgCl} - \frac{RT}{F}\ln\frac{\{Cl^-\}}{a^\circ} + \frac{RT}{F}\ln\frac{\{OH^-\}}{a^\circ} - \frac{RT}{F}\ln\frac{K_W}{a^{\circ 2}}. \tag{2.144}$$

Next, we replace activities $\{X\} = [X]f_x$ and collect all known or measurable parameters on the left:

$$U - E^\circ_{Ag|AgCl} + \frac{RT}{F}\ln\frac{[Cl^-]}{[OH^-]} = -\frac{RT}{F}\ln\frac{K_W}{a^{\circ 2}} - \frac{RT}{F}\ln\frac{f_{Cl^-}}{f_{OH^-}}. \tag{2.145}$$

Measurements have to be performed in the following electrochemical cell:

$$Ag \mid AgCl \mid x\ \text{mol}\ L^{-1}\ KCl, x\ \text{mol}\ L^{-1}\ KOH \mid H_2\ (p(H_2) = p^\circ) \mid Pt \mid Ag.$$

For very low concentrations of KCl and KOH, the activity coefficients f_{Cl^-} and f_{OH^-} converge to the same value, which allows the last term in Eq. (2.145) to be eliminated. Now, K_W can be directly measured to give 1.01×10^{-14} mol^2 L^{-2} at 298 K. Based on 55.6 mol H_2O in 1 L of water, we can determine the neutral point concentrations as 1×10^{-7} mol L^{-1} H_3O^+ and 1×10^{-7} mol L^{-1} OH^-. The experiment can be further refined by changing the ratio of $[OH^-]$ to $[Cl^-]$. Measurements are then analyzed based on the ionic strength $I_c = \frac{1}{2}\sum c_i z_i^2$ (Section 4.2.8). For very low concentrations of KCl and KOH, the ionic strength also becomes very small and, correspondingly, the activity coefficients f_{Cl^-} and f_{OH^-} in Eq. (2.145) converge toward unity and allow calculation of K_W.

I_c is the ionic strength and not a current! Here, the index "c" refers to the concentration and must therefore be italicized. We obtain I_c from the concentrations c_i. In contrast, often the index "c" is not italicized (e.g., for I_c as a symbol for the cathodic current, where c is an abbreviation and not a physical parameter).

2.4.3 Chemical Analysis with Potentiometry

Electrode potentials and equilibrium cell voltages are dependent on the activity of ions and molecules that participate in the electrode reaction. Therefore, potentiometric sensors can be designed based on electrode potentials, for example, the Ag|AgCl electrode based on Eq. (2.131) can be employed to determine the activity of Cl^- ions in solution. When the Cl^- activity is calculated from a cell voltage based on Eq. (2.131), the method is called direct potentiometry. The accuracy of this methodology is critically dependent on the stability of the reference electrode. Selectivity for specific types of ions can often be achieved with ion-selective membranes that restrict the ionic transport to specific ions. Ion-selective electrodes are often employed in the endpoint determination during titration. In this case, the potentiometric signal can be used to determine the amount of a component from the change in equilibrium potential during titration (potentiometric titration). For potentiometric titration, a broader range of electrodes can be employed beyond the type explained in the context of Eq. (2.131). Often, the selectivity of the electrode is limited, or additional diffusion potentials are present, without severely limiting the accuracy of the titration because the selectivity of the titration is based on the reaction between the analyte and the titrant. Combining an ion-selective electrode with a reference electrode into the *same* physical device gives chemical sensors. The "pH electrode" is a good example. In fact, it contains the indicator electrode and another reference electrode. When an enzyme is incorporated into the sensor mechanism, high selectivity can be achieved, and the device is called a biosensor (Figure 18.27). The lambda-probe in automotive combustion engine control is a potentiometric sensor operating at high temperature with a solid ion-conducting oxide (yttria-stabilized zirconia, YSZ) electrolyte membrane. We will explore potentiometry in more detail in Section 5.2.

2.4.4 Electrochemical Energy Conversion Systems: An Overview

Applications of electrochemical energy conversion. Galvanic cells convert chemical energy into electrical energy. Before the invention of the dynamo, galvanic cells provided the only technical source of electrical power (e.g., Volta's pile[9]). In this context, electrochemical cells were optimized for the generation and storage of electrical energy. With electricity slowly replacing combustion engines, there is now renewed interest in electrochemical energy storage. Electrical energy storage is essential in different sectors with different requirements in terms of power and cost: power supplies for cars, ships, and planes; and power supplies for mobile technologies such as electrical tools, laptop computers, cameras, mobile phones, or in medical equipment such as sensors, diagnostic equipment, therapeutic equipment (hearing aids, heart pace makers). The electrical power required in electric cars or in electric forklifts is supplied, for example, by a galvanic cell with special technology designed for high energy density, high power density, or high safety. Energy supply in remote locations without access to power lines is an important application (remote villages, autonomous instrumentation or sensors, military technology, space technologies). Galvanic cells are employed for the emergency power supply in hospitals and for critically important computer and communication installations.

In the future, an area of considerable importance is load leveling in electrical power supply networks. Electrical power needs to be supplied in times of high demand and stored in times of low demand. With an increasing contribution from wind and solar energy to the domestic electricity supply (e.g., in 2019 in Germany, wind and solar energy contributed 53% of electricity and 17% of overall energy consumption [23]) fluctuations in supply due to intermittent power generation from solar or wind need to be avoided. Under conditions of strong wind during night (with weak consumer demand for power), there is already an oversupply, and energy needs to be stored effectively. The control of electricity networks requires management of both over-supply and under-supply of electrical power, which could be addressed by large-scale electrochemical power storage. Installations of this type are under development (2016) [24].

Types of electrochemical energy conversion devices. Electrochemical energy conversion systems are based on electrode reactions for power production. In the device, there are separate paths for ion transport and for electron transport connected at the electrode surface. Currently, the most important energy conversion devices are **batteries** and **fuel cells**. Both are galvanic cells based on the electrochemical conversion of chemical energy into electrical energy. Batteries are closed devices, that is, the oxidant and reductant are contained in an enclosed unit together with electrodes and other components. Fuel cells are open systems (thermodynamically) due to a continuous external supply of gaseous reagents (O_2) and gaseous or liquid fuels (H_2, CH_4, etc.). Reaction products such as H_2O and CO_2 are released.

Batteries are based on galvanic cell networks in series or in parallel to give the desired output voltage and/or current. The reactive cell components (oxidant and reductant) are referred to as active mass. Many older types of batteries are designed for a single discharge/use and are referred to as primary batteries. Reusable batteries based on active materials that allow recharging by electrolysis are called secondary batteries (synonymous with accumulator or rechargeable battery).

The economically dominant battery system is still the **lead–acid battery** (Figure 2.26), which is commonly employed in automotive on-board electronic applications (e.g., ignition, lighting, and electronics). In its charged state, the lead–acid battery contains the active materials PbO_2 as oxidant and Pb as reductant. The aqueous electrolyte is based on sulfuric acid, which is consumed during the discharge process. Equation (2.146) describes the discharge process based on PbO_2 being reduced (with electrons from an external source). This is the cathodic process. The PbO_2 cathode takes on a positive charge. The resulting $PbSO_4$ is relatively insoluble and deposits as a crystalline film onto the cathode.

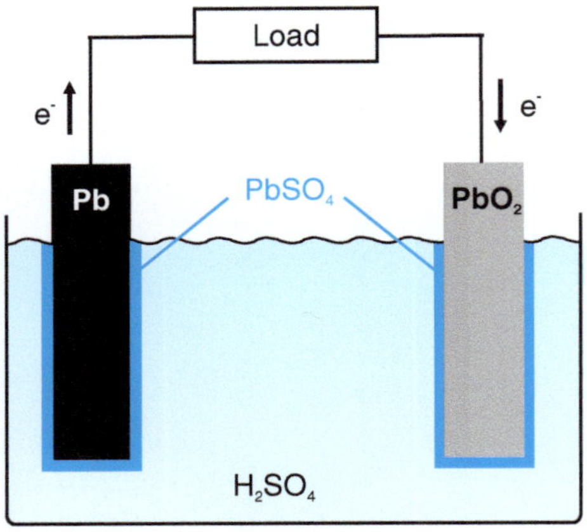

Figure 2.26 Schematic of a single element in the lead–acid battery based on Pb anode and PbO_2 cathode.

$$PbO_2 + H_2SO_4 + 2\,H^+ + 2\,e^- \rightarrow PbSO_4 + 2\,H_2O \tag{2.146}$$

The corresponding process at the Pb anode is shown in Eq. (2.147) and linked to release of electrons (going into the external circuit). The formation of Pb^{2+} is an oxidation, and therefore, the Pb electrode is called the anode. In the presence of SO_4^{2-} anions, precipitation of insoluble $PbSO_4$ occurs at the anode. The Pb anode takes on a negative charge.

$$Pb + H_2SO_4 \rightarrow PbSO_4 + 2\,H^+ + 2\,e^- \tag{2.147}$$

The overall cell reaction can be expressed as the spontaneous process given in Eq. (2.148).

$$PbO_2 + Pb + 2\,H_2SO_4 \rightarrow 2\,PbSO_4 + 2\,H_2O. \tag{2.148}$$

The process is reversed during charging when an external voltage is applied to the cell. During charging, the $Pb|PbSO_4$ electrode is the cathode, and the $PbO_2|PbSO_4$ electrode is the anode. Although there is a switch of anode/cathode assignment when comparing discharging and charging processes, it has become conventional for battery systems to always assign anode/cathode consistent with the processes during discharging. For the lead–acid battery, this leads to Pb being the anode and PbO_2 being the cathode.

Example Calculation 2.15 Determining the Cell Voltage of a Lead–Acid Battery

The cell reaction for the lead–acid battery during discharging is

$$Pb(s) + PbO_2(s) + 4\,H^+ + 2\,SO_4^{2-} \rightarrow 2\,PbSO_4(s) + 2\,H_2O\ (n = 2)$$

Determine: U

There are two approaches to solve this problem. We can initially obtain the thermodynamic cell potential from tabulated $\Delta_B G^\circ$ data for the reactants. We have to assume that the activity of all components is 1.0. This assumption is justified for pure solid phase components (by definition of the mole fraction activity) but obviously wrong for solution species. Due to the concentration of H_2SO_4 continuously decreasing during discharging, we can accept the resulting smaller error at lower concentrations.

$$\Delta_F G^\circ(Pb) = 0 \text{ (by definition)}$$

$$\Delta_F G^\circ(PbO_2) = -217.33\ \text{kJ mol}^{-1}\ [a]$$

$$\Delta_F G^\circ(H^+) = 0 \text{ (by definition)}$$

$$\Delta_F G^\circ\left(SO_4^{2-}\right) = -744.53\ \text{kJ mol}^{-1}\ [a]$$

$$\Delta_F G^\circ(PbSO_4) = -193.89\ \text{kcal mol}^{-1} = -811.23\ \text{kJ mol}^{-1}\ [b]$$

$$\Delta_F G^\circ(H_2O) = -237.13\ \text{kJ mol}^{-1}\ [a]$$

[a] Atkins, Physical Chemistry, 2nd ed. Wiley
[b] Handbook of Chemistry and Physics, CRC

$$U^\circ = \Delta_R G^\circ/(nF)$$

$$\Delta_R G^\circ = \sum_i \nu_i \, \Delta_F G_i^\circ$$

$$\begin{aligned}
\Delta_R G^\circ &= 2 \cdot \Delta_F G^\circ(\mathrm{PbSO_4}) + 2 \cdot \Delta_F G^\circ(\mathrm{H_2O}) - 2 \cdot \Delta_F G^\circ\left(\mathrm{SO_4^{2-}}\right) \\
&\quad -4 \cdot \Delta_F G^\circ(\mathrm{H^+}) - \Delta_F G^\circ(\mathrm{PbO_2}) - \Delta_F G^\circ(\mathrm{Pb}) \\
&= 2 \cdot (-811.23\ \mathrm{kJ\ mol^{-1}}) + 2 \cdot (-237.13\ \mathrm{kJ\ mol^{-1}}) \\
&\quad -2 \cdot (-744.53\ \mathrm{kJ\ mol^{-1}}) - (-217.33\ \mathrm{kJ\ mol^{-1}}) \\
&= -390.33\ \mathrm{kJ\ mol^{-1}}
\end{aligned}$$

$$\begin{aligned}
U^\circ &= -390.33 \times 10^3\ \mathrm{V\ A\ s\ mol^{-1}}/(2 \cdot 96\,485\ \mathrm{A\ s\ mol^{-1}}) \\
&= 2.023\ \mathrm{V}
\end{aligned}$$

After evaluation of the standard equilibrium cell potential, we can introduce the effects of concentrations and activity coefficients by employing the Nernst equation:

$$U_{eq} = U^\circ - \frac{RT}{nF} \ln \prod_i \left(\frac{a_i}{a^\circ}\right)^{\nu_i}.$$

Here, for the case of a spontaneous reaction, reaction products are in the numerator ($\nu_i > 0$) and reactants are in the denominator ($\nu_i < 0$). Due to sign inversion and using the correct concentration terms, we obtain

$$U_{eq} = U^\circ + \frac{RT}{\cancel{2}F} \ln \frac{a_x^{\circ\cancel{2}} a_x(\mathrm{Pb}) a_x(\mathrm{PbO_2}) \{\mathrm{H^+}\}^{\cancel{4}2} \{\mathrm{SO_4^{2-}}\}^{\cancel{2}}}{a_x(\mathrm{H_2O})^{\cancel{2}} a_x(\mathrm{PbSO_4})^2 a_c^{\circ\cancel{6}3}}.$$

Mole fraction activities for pure solids (in blue) are 1.0. The mole fraction activity for water is very close to 1.0 due to water being present in excess. We cancel exponents and insert $T = 298$ K, select the concentration activity based on the product of concentration of H_2SO_4 (2 mol L^{-1}) and average activity coefficient (0.128), and determine the multiplier $RT/F = 8.314\ \mathrm{V \cdot A \cdot s \cdot mol^{-1}} \cdot 298\ \mathrm{K} / 96\,485\ \mathrm{A\ s\ mol^{-1}} = 0.0256\ \mathrm{V}$.

$$\begin{aligned}
U_{eq} &= U^\circ + \frac{RT}{F} \ln \frac{[\mathrm{H^+}]^2 \left[\mathrm{SO_4^{2-}}\right]}{c^{\circ 3}} + \frac{RT}{F} \ln f_{c,\pm}^3 \\
&= 2.023\,\mathrm{V} + 0.0256\ \mathrm{V} \cdot \ln((2+2)^2 \cdot 2) + 0.0256\ \mathrm{V} \cdot \ln(0.128)^3 \\
&= 2.023\,\mathrm{V} + 0.08872\ \mathrm{V} - 0.1578\ \mathrm{V} \\
&= 1.954\ \mathrm{V}
\end{aligned} \tag{2.149}$$

An alternative approach to solving this problem with further insights is possible when determining the individual electrode potentials for anode and cathode. The cell voltage is then obtained as the difference between the electrode potentials. For the PbO_2 electrode, we can express the reaction for charging:

$$\mathrm{Pb^{2+} + 2\,H_2O \rightleftharpoons PbO_2 + 4\,H^+ + 2\,e^-}, E^\circ = 1.46\ \mathrm{V} \tag{2.150}$$

This reaction is associated with the solubility product for $PbSO_4$.

$$\mathrm{PbSO_4 \rightleftharpoons Pb^{2+} + SO_4^{2-}} \tag{2.151}$$

The electrode potential is obtained for the mole fraction activity 1.0 for solid phase components and with substitution of the solubility product.

$$
\begin{aligned}
E_1 &= E^\circ(\mathrm{Pb^{2+}|PbO_2}) + \frac{RT}{2F}\ln\frac{a_x(\mathrm{PbO_2})\,a_x^{\circ 2}\{\mathrm{H^+}\}^4}{a_x^\circ a_x(\mathrm{H_2O})^2\{\mathrm{Pb^{2+}}\}\,a_c^{\circ 3}} \\
&= E^\circ + \frac{RT}{2F}\ln\frac{\{\mathrm{H^+}\}^4\{\mathrm{SO_4^{2-}}\}}{K_\mathrm{L}(\mathrm{PbSO_4})\,a_c^{\circ 3}} \\
&= E^\circ + \frac{RT}{2F}\ln\frac{[\mathrm{H^+}]^4[\mathrm{SO_4^{2-}}]}{K_\mathrm{L}(\mathrm{PbSO_4})\,a_c^{\circ 3}} + \frac{RT}{2F}\ln f_{c,\pm}^5 \\
&= 1.46\ \mathrm{V} + 0.01284\ \mathrm{V}\cdot\ln\frac{4^4\cdot 2}{1.585\times 10^{-8}} + 0.01284\ \mathrm{V}\cdot\ln(0.01284)^5 \\
&= 1.46\ \mathrm{V} + 0.3106\ \mathrm{V} - 0.13196\ \mathrm{V} \\
&= 1.639\ \mathrm{V}
\end{aligned}
$$

Similarly, for the $\mathrm{Pb|PbSO_4}$ electrode, we can write the electrode reaction

$$\mathrm{Pb} \rightleftharpoons \mathrm{Pb^{2+}} + 2\mathrm{e^-} \tag{2.152}$$

$$\mathrm{PbSO_4} \rightleftharpoons \mathrm{Pb^{2+}} + \mathrm{SO_4^{2-}} \tag{2.153}$$

$$
\begin{aligned}
E_2 &= E^\circ(\mathrm{Pb|Pb^{2+}}) + \frac{RT}{2F}\ln\frac{\{\mathrm{Pb^{2+}}\}a_x^{\circ 1}}{a_c^\circ a_x(\mathrm{Pb})} \\
&= E^\circ + \frac{RT}{2F}\ln\frac{K_\mathrm{SP}(\mathrm{PbSO_4})}{\{\mathrm{SO_4^{2-}}\}} \\
&= E^\circ + \frac{RT}{2F}\ln\frac{K_\mathrm{SP}(\mathrm{PbSO_4})[\mathrm{H^+}]^4}{[\mathrm{SO_4^{2-}}]c^\circ} + \frac{RT}{2F}\ln\frac{1}{f_{c,\pm}} \\
&= -0.1263\ \mathrm{V} + 0.01284\ \mathrm{V}\cdot\ln\frac{1.585\times 10^{-8}\ \mathrm{mol\ L^{-1}}}{2\ \mathrm{mol\ L^{-1}}\cdot 1\ \mathrm{mol\ L^{-1}}} + 0.01284\ \mathrm{V}\cdot\ln\frac{1}{0.01284} \\
&= -0.1263\ \mathrm{V} + (-0.2306) + 0.0263\ \mathrm{V} \\
&= -0.330\ \mathrm{V}
\end{aligned} \tag{2.154}
$$

The overall cell voltage is then obtained as follows:

$$
\begin{aligned}
U &= E_1 - E_2 \\
&= 1.639\ \mathrm{V} - (-0.330\ \mathrm{V}) \\
&= 1.969\ \mathrm{V}
\end{aligned} \tag{2.155}
$$

Both approaches to the calculation provide consistent results indicating an equilibrium cell voltage of 2 V for a single galvanic cell. The typical 12 V output for a car battery is based on connecting 6 galvanic cells in series. The cell voltages add up.

◢

In a more realistic case of a real lead–acid battery, the concentration of sulfuric acid in the fully charged state is ~4.5 mol L^{-1}. We also need to introduce hydrogen sulfate (HSO_4^-) as a predominant species in solution. However, the overall result of a calculation remains similar with an equilibrium voltage only some mV higher than calculated in Example Calculation 2.15.

Of interest here is that the electrode potential for the $Pb|PbSO_4$ electrode is considerably more negative when compared to the hydrogen electrode. Thermodynamically, the production of H_2 gas is expected. The electrode potential for the $PbSO_4|PbO_2$ electrode is more positive when compared to the thermodynamic potential for oxygen gas evolution. Thermodynamically, the production of oxygen is expected. The production of oxygen and hydrogen would be highly dangerous. Here, kinetic effects are responsible for stopping gas evolution reactions and for maintaining the charge state of the lead–acid battery. Kinetic phenomena will be discussed in Section 3.2 and again in Chapter 15.

The rechargeability of batteries is dependent on complex factors, including the chemical reversibility and also the nature of morphological changes during discharge (e.g., both Pb and PbO_2 must redeposit). Lithium metal offers a low-weight material with very negative redox potential $Li\,|\,Li^+$ for applications as an anode material in nonaqueous solvents. However, Li is known to redeposit during charging not as a smooth film but rather in the form of dendrites (Figure 15.4). A dendritic deposit tends to grow toward the opposite electrode, sometimes piercing the separator and leading to short-circuit currents. The short-circuit current can be considerable, and a thermal runaway process may then lead to fire or even explosion of galvanic cells (the active mass components in Li batteries are rich in energy, and organic solvents tend to be combustible!). Considerable research effort goes into suppression of lithium dendrite growth.

In order to minimize such technical problems, it is now customary to employ lithium insertion electrodes in **Li-ion batteries**, rather than lithium metal. The most prominent anode material is based on chemically treated graphite, where graphitic layers are able to intercalate Li^+ ions [Figure 2.27; Eq. (2.156)]. Charge compensation occurs by negatively charged electrons entering the graphitic layers together with positive Li^+. During lithium insertion and release, this type of graphitic anode exhibits only minor changes in external shape. A corresponding cathode can be designed based on transitional metal oxides with the ability to undergo a similarly facile insertion and release of Li^+ ions, but at a much more positive electrode potential [Eq. (2.156)]. In transition metal oxides, the Li^+ ions are incorporated into lattice positions. In this type of Li-ion battery, charging and discharging are accompanied by Li^+ ions moving between the anode and the cathode. This mechanism can be compared to the movement of a rocking chair (rocking chair cell). The electrons are transferred from the current collectors to the host lattices (graphite or $Li_{1-x}CoO_2$).

$$\{LiC_6\} \rightleftharpoons \{C_6\} + Li^+ + e^- \tag{2.156}$$

$$Li_{1-x}Co^{III,IV}O_2 + xLi + xe^- \rightleftharpoons LiCo^{III}O_2 \tag{2.157}$$

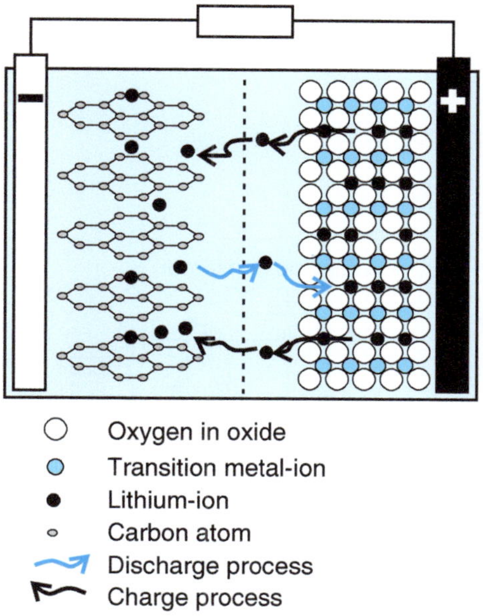

Figure 2.27 Schematic showing a Li-ion battery with negative graphite electrode (the anode during discharge) and with positive transition metal oxide electrode (the cathode during discharge).

Electrodes with concomitant electron transfer and ion transfer into the bulk of the electrode materials are called **insertion electrodes**.

Chapter 15 will be dedicated to the topic of batteries in more depth.

In **Fuel cells**, the anode reaction is based on a liquid or gaseous fuel feed. Depending on the fuel, operational conditions, and device design, there are a plethora of fuel cell types (Figure 2.28). Solid oxide fuel cells (SOFCs) operate at high temperatures and require solid ion-conductors. In contrast, polymer electrolyte membrane fuel cells (PEMFCs) operate in an aqueous membrane environment at lower temperature, but require much better catalysts.

A very common type of fuel cell is the polymer electrolyte membrane fuel cell (PEMFC). A key component in this device is the cation-conducting membrane. The

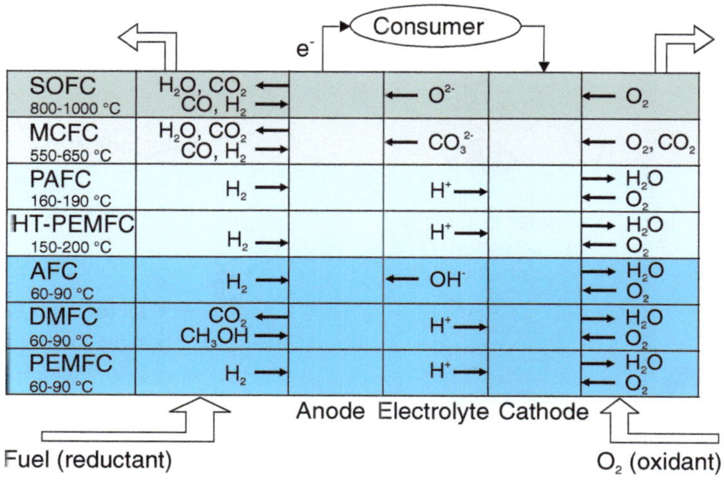

SOFC – Solid oxide fuel cell
MCFC – Molten carbonate fuel cell
PAFC – Phosphoric acid fuel cell
HT-PEMFC – High-temperature polymer electrolyte membrane fuel cell
AFC – Alkaline fuel cell
DMFC – Direct methanol fuel cell
PEMFC – Polymer electrolyte membrane fuel cell

Figure 2.28 Fuel cell types are listed according to operational temperature. Arrows indicate starting materials and products that diffuse in/out of the porous gas electrodes. Arrows in the membrane represent the transport of mobile ionic species.

operational temperature in PEMFCs is typically 80 °C, with higher temperatures causing problems with water transport/loss and membrane resistivity. The anode reaction is hydrogen oxidation.

$$H_2 \rightarrow 2\,H^+ + 2\,e^- \tag{2.158}$$

The cathode reaction is the 4-electron oxygen reduction to water.

$$O_2 + 4\,H^+ + 4\,e^- \rightarrow 2\,H_2O \tag{2.159}$$

A summary of the flux of materials and energy is shown in Figure 2.29.

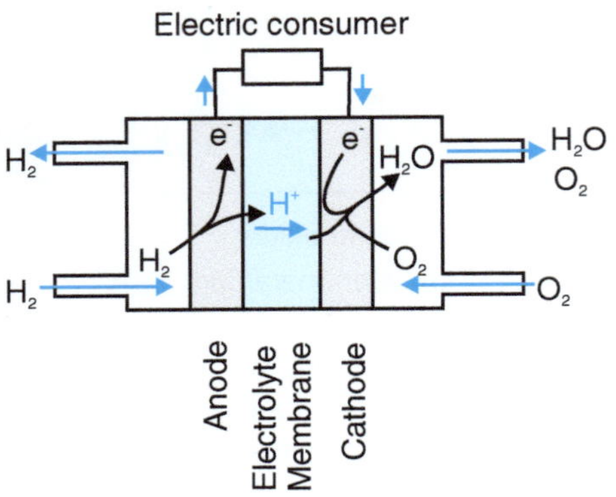

Figure 2.29 Transport of materials and energy in a PEMFC.

The proton-conducting membrane is important for the transport of ionic charge from anode to cathode. Most commonly employed are perfluorinated polymer systems with sulfonic acid groups in side chains (Figure 2.30). The sulfonic acid groups dissociate to give a well-defined number of fixed anionic sulfonate groups. Polymers with immobilized charges are called ionomers. The perfluorinated polymer backbone is hydrophobic. Hydrophilic water channels are formed by self-assembly when the hydrophilic sulfonate groups arrange to minimize the interaction with the perfluorinated backbone and maximize the interaction with water during swelling of the material. These channels are cation-conducting with negatively charged walls (Figure 2.30). The commercially most successful membrane material of this type is Nafion (trademarked by the company DuPont).

A multiscale illustration of the highly integrated design of a fuel cell is shown in Figure 2.31. Processes occur at different length scales, and therefore, optimization of reactions and fluxes needs to include all length scales. The electrodes in the fuel cell are graphitic, coated with platinum nanoparticles as catalyst (Figure 2.31e). The carbon-platinum catalyst hybrid material needs to be combined with Nafion and placed directly onto the proton-conducting membrane. The combination of ion conduction and electron conduction directly at the catalyst particle ensures effective charge transfer at the catalyst surface. This part of the fuel cell is called the

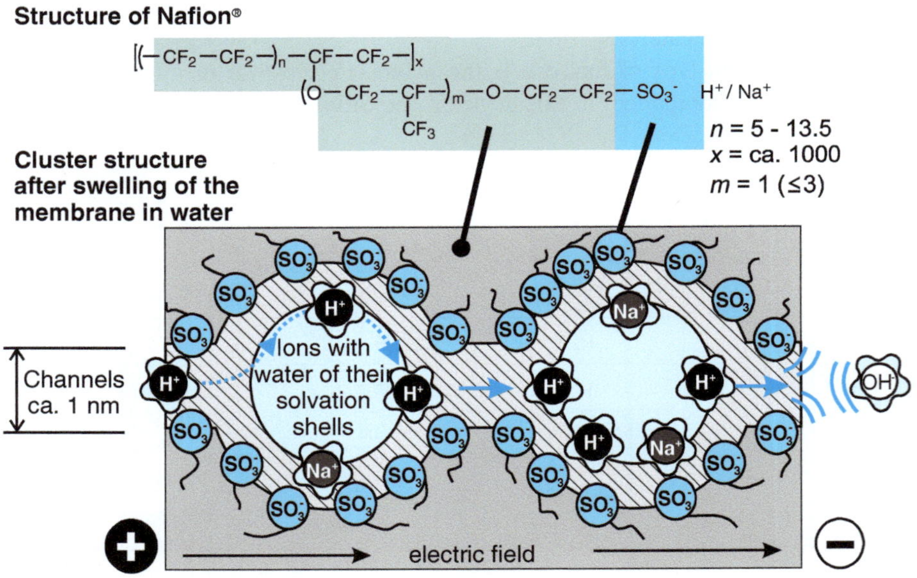

Figure 2.30 Schematic of the micropore structure in Nafion with highly charged hydrophilic channels and pore walls.

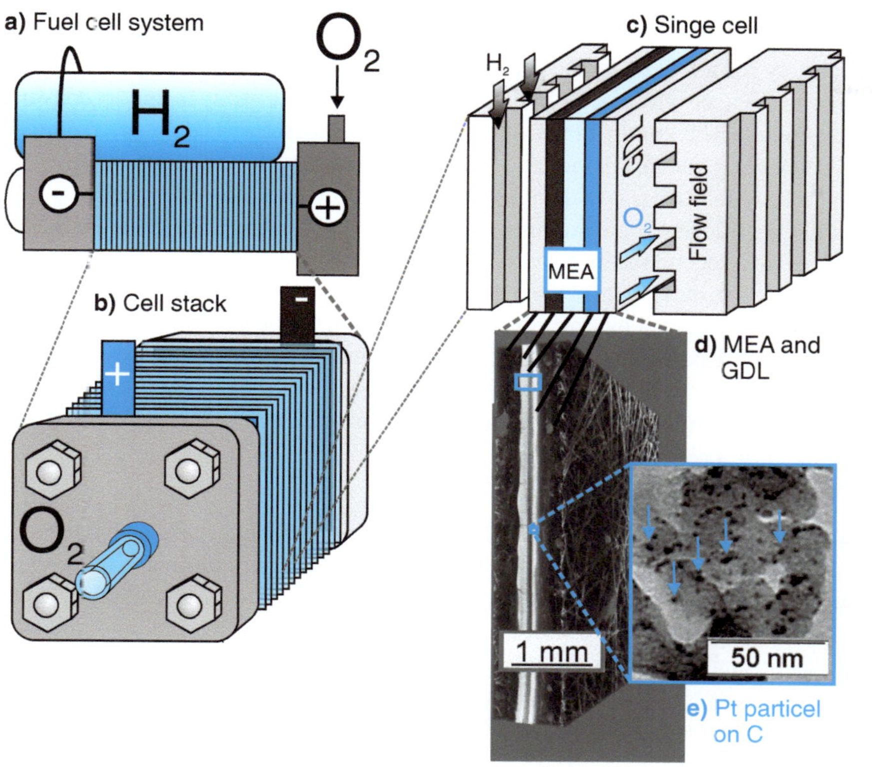

Figure 2.31 System components of a PEMFC on different length scales.

a) Fuel cell system with external units for hydrogen storage, gas purification, and gas humidification.
b) Fuel cell stack containing individual fuel cells in series.
c) Schematic of a single fuel cell unit.
d) X-ray tomographic view of a membrane electrode assembly (anode–membrane electrolyte–cathode) with gas diffusion layer (GDL).
e) Transmission electron micrograph of platinum nanoparticle catalyst (blue arrows) on a carbon substrate.

Panels d) and e) were provided by the German Aerospace Institute of Networked Energy Systems Oldenburg.

membrane electrode assembly (MEA, Figure 2.31d). In contact with the membrane-electrode assembly is a porous carbon film to conduct electrons and to allow gaseous reactants and products to access the catalyst. This is the gas diffusion layer (GDL, Figure 2.31d). Electrical contacts are then achieved with carbon or metal plates (Figure 2.31c). These also need channel structures, called flow fields, to allow transport of gaseous reagents/products. Individual electrochemical cell units are then assembled into stacks (Figure 2.31b). Each anode is in direct electrical contact with the next cathode, resulting in a bipolar plate structure.

Fuel cells theoretically allow energy conversion efficiencies of up to 80%. This number states the fraction of the chemical energy converted to electrical energy (Example Calculation 2.16). In contrast to heat engines, the theoretical energy conversion efficiency of a fuel cell does not depend on the use of high temperatures. Practical efficiencies are substantially lower (Chapter 16).

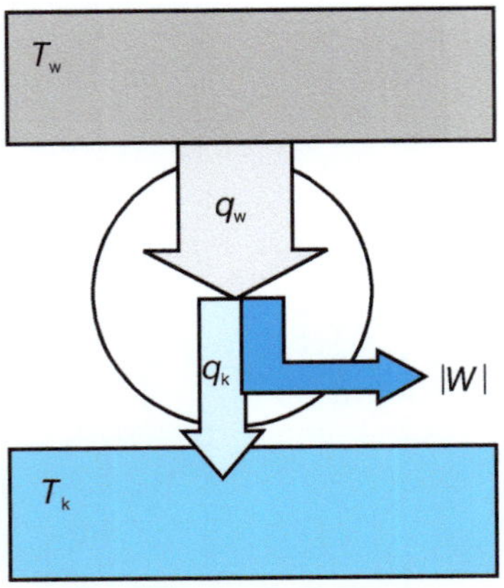

Figure 2.32 Schematic of a heat engine with heat energy flowing from high temperature to low temperature while producing mechanical work |W|. The width of the arrows represents the amount of energy converted.

Example Calculation 2.16 Comparison of the Theoretical Energy Conversion Efficiency for a Heat Engine and for a Fuel Cell (PEMFC).

A heat engine (see white circle in Figure 2.32) represents a group of energy conversion devices where heat energy flow from high to low temperature is employed to produce useful work (e.g., steam engines and combustion engines). By taking an amount of heat q_h ($q_h > 0$, by definition as taken up by the machine) from a reservoir with temperature T_h, and then transferring an amount of heat q_c ($q_c < 0$, moving from the machine to the reservoir) to a second reservoir with lower temperature T_c, a certain amount of useful work W ($W < 0$, moving from the machine to the outside) can be performed. This ideal heat engine, envisaged by Carnot[14] and conceptualized by Clausius,[15] allows mathematical analysis based on the entropy ΔS. Every energy conversion process requires an element of irreversibility and therefore energy loss, similar to an energy tax to be paid for the conversion. For the overall system, we can express the change in entropy as

$$\Delta S = -\frac{|q_w|}{T_w} + \frac{|q_c|}{T_c}. \tag{2.160}$$

We use absolute values for q_h and q_c. *Fundamentally*, heat generated by a chemical reaction in the hot reservoir can only be partially converted into mechanical energy. Every energy conversion process is associated with an loss in useful energy. The second law of thermodynamics states that for a closed system $\Delta S \geq 0$. Only for a (hypothetical) reversible process $\Delta S = 0$ can the maximum amount of work be performed with $W_{max} = |q_h| - |q_{c,min}|$. In this case Eq. (2.160) suggests $|q_h|/T_h = |q_{c,min}|/T_c$ and $|q_{c,min}| = (T_c/T_h)\,|q_h|$. Temperatures here refer to absolute temperatures [K]. The maximum energy conversion efficiency ε can now be expressed as

$$\varepsilon_{max} = \frac{|W_{max}|}{|q_w|} = \frac{|q_w| - |q_{c,min}|}{|q_w|} = \frac{|q_w| - \frac{T_c}{T_w}|q_w|}{|q_w|} = 1 - \frac{T_c}{T_w}. \tag{2.161}$$

For practical examples, let us consider (i) a machine with a hot reservoir at 1000 °C and a cold reservoir at 300 °C (e.g., a gasoline or diesel combustion engine) and (ii) a machine with hot reservoir at 200 °C and a cold reservoir at 100 °C (e.g., a steam engine). For these two machines, the *maximum theoretical* energy conversion efficiencies based on Eq. (2.161) are

	Combustion engine	Steam engine
θ_h/°C	1000	200
θ_c/°C	300	100
T_h/K	1273	473
T_c/K	573	373
ε_{max}	0.54	0.21
ε_{max} /%	54	21

[14] Nicolas Léonard Sadi Carnot, 1796–1832, French mechanical engineer, military scientist, and physicist, envisaged the need for better energy conversion efficiency in steam engines in his book *Reflections on the Motive Power of Fire*.

[15] Rudolf Julius Emanuel Clausius, 1822–1888, German physicist and mathematician, considered one of the central founders of the science of thermodynamics, first stated the basic ideas of the second law of thermodynamics, introduced the concept of entropy in 1865.

The fuel cell is not based on heat flow. However, we can express the energy conversion efficiency more generally based on the electrical work that can be done W_{el} and employing Eq. (2.23).

$$W_{el} = nFU_{eq} = -\Delta_R G. \tag{2.162}$$

The molar reaction enthalpy $\Delta_R H$ corresponds to the heat released during reaction of 1 mol of reactant under conditions of constant pressure and constant temperature. This allows us to express the energy efficiency for a process based on electrical power generation.

$$\epsilon_{el,\max} = \frac{|W_{el,\max}|}{|\Delta_R H|} = \frac{|\Delta_R G|}{|\Delta_R H|} = \frac{|-nFU_{eq}|}{|\Delta_R H|} = \frac{|-U_{eq}|}{|\Delta_R H|/_{nF}}. \tag{2.163}$$

The equilibrium cell voltage U_{eq} (at $I = 0$) can be combined with tabulated thermodynamic data for $\Delta_R H$ (using nF to express the corresponding voltage) to give the maximum theoretical energy conversion efficiency for fuel cells. Alternatively, we can employ the reaction entropy $\Delta_R S$ instead of the equilibrium cell voltage U_{eq}.

$$\varepsilon_{el,\max} = \frac{|W_{el,\max}|}{|\Delta_R H|} = \frac{|\Delta_R G|}{|\Delta_R H|} = \frac{|\Delta_R H - T\Delta_R S|}{|\Delta_R H|} = \left|1 - \frac{T\Delta_R S}{\Delta_R H}\right| \tag{2.164}$$

Substitution into this equation allows the theoretical energy conversion efficiency to be calculated for a fuel cell operating under standard conditions (reactants at 1 bar pressure, H_2O in liquid state, and $T = 25\ °C = 298$ K).

Reaction	$H_{2(g)}$	+	$0.5\ O_{2(g)}$	→	$H_2O_{(l)}$
ν_i	−1		−0.5		+1
$\Delta_B H°/(kJ\ mol^{-1})$	0		0		−285.83
$S°/(kJ\ K^{-1}\ mol^{-1})$	0.1306		0.205		0.07008

Using Eqs. (2.47), (2.48), and (2.25), we obtain

$$\Delta_R H° = \sum_i \nu_i \Delta_F H_i° = -285.83\,\text{kJ mol}^{-1}$$

$$\Delta_R S° = \sum_i \nu_i S_i° = -0.163\ \text{kJ K}^{-1}\,\text{mol}^{-1}$$

The PEMFC has a typical operational temperature of 80 °C (353 K), and $\Delta_R H$ and $\Delta_R S$ are assumed to remain approximately constant between 25 °C and 80 °C (there are no phase transitions in this temperature range).

$$\begin{aligned}\Delta_R G° &= \Delta_R H° - T\Delta_R S°\\ &= -285.53\ \text{kJ mol}^{-1} - 298\ \text{K} \cdot (-0.163\ \text{kJ K}^{-1}\ \text{mol}^{-1})\\ &= -228.28\ \text{kJ mol}^{-1}\end{aligned}$$

$$\begin{aligned}\eta_{el,\max} &= \frac{|\Delta_R G|}{|\Delta_R H|}\\ &= \frac{228.28\,\text{kJ mol}^{-1}}{285.53\,\text{kJ mol}^{-1}}\\ &= 0.7995\end{aligned}$$

Therefore, the *maximum theoretical* energy conversion efficiency for a PEMFC is close to 80%. This compares to typically 54% for an *ideal* combustion engine!

As an exercise, transfer these calculations into a data spreadsheet to then systematically explore how these results change with reaction conditions and temperature (T).

In spite of the high theoretical energy conversion efficiency, PEMFCs have not replaced traditional combustion engines. There are a number of reasons for this. Both anode and cathode require platinum, a noble metal catalyst with limited supply and high cost. Annually, only 268 t of platinum (2010) are mined, with an additional 80 t (ca. 30%) provided by recycling [25]. The cost of membranes is also considerable. Highly purified gaseous hydrogen and air are required to avoid catalyst poisoning. Production of hydrogen is currently still dominated by reforming from natural gas (mainly CH_4), and the resulting mixture of gases has to undergo multistage purification for hydrogen to be of sufficient quality for PEMFC applications. A particular problem is carbon monoxide (CO) gas, which strongly adsorbs to the platinum catalyst surface sites to suppress reactions with both H_2 and O_2. The purification processes increase costs and lower the energy efficiency for fuel cells. Hydrogen from water electrolysis is very pure but still expensive, although production from sustainable electricity (wind, solar) may change this in the future. Storage of hydrogen gas will require the development of new technologies and processes.

Q: What is the difference when comparing adsorption and absorption?

A: Adsorption describes the binding of a molecule at an interface, for example, at the surface of a solid powder. Absorption describes the transfer of molecules into another phase, for example, absorption of radiation or absorption of a gas into a liquid.

The problem of CO adsorption and poisoning of fuel cell catalysts can be overcome by going to higher operational temperatures. Solid oxide fuel cells (SOFC) have been developed to allow both CO and H_2 to be used as fuels. At these higher operational temperatures, water is gaseous, and many polymer-based membrane materials would degrade or thermolyze. Therefore, ion-conducting oxides are employed. At temperatures of 700 °C and higher, many transition metal oxides allow oxide dianion (O^{2-}) diffusion due to lattice vacancies to provide ion-conducting membranes for fuel cells. These types of SOFCs provide constant power and have commercial success in stationary applications. However, for smaller-scale applications and with fluctuations in power output/demand, there are disadvantages. To reach operational conditions, the SOFC needs to be heated and fluctuations during on/off operations are associated with thermal tension leading to mechanical cell degradation. Avoiding thermal tension is possible but only with expensive materials, which again lowers current economic impact.

What are the current–voltage characteristics of fuel cells during operation? Figure 2.33 shows the voltage U of a fuel cell as a function of the current density j. This kind of data plot provides key operational parameters for fuel cells or batteries. Measurements require that resistive loads are varied externally between anode and cathode to give current I and voltage U. The current I is divided by the area A of the electrode to give the current density $j = I/A$. The corresponding equilibrium voltage of the fuel cell (Example Calculation 2.7) is indicated as a horizontal line. This voltage is proportional to the molar Gibbs reaction energy $\Delta_R G$, which is a measure of the

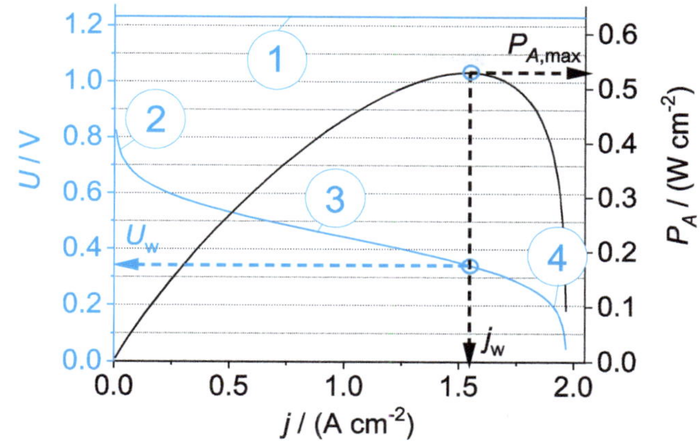

Figure 2.33 Voltage–current density plot (characteristic curve) for a fuel cell. On the left is the voltage measured for the cell, and on the right, the power density $P_A = j\,U$ (calculated with a spreadsheet).

1) Calculated equilibrium cell voltage.
2) Loss of cell voltage due to activation of electron transfer reactions at anode and/or cathode.
3) Further ohmic losses of cell voltage due to transport of charges in the membrane ($U = R\,I$).
4) Collapse of the cell voltage due to limitations in transport of reagents.

driving force. The equilibrium cell voltage at zero current or open-circuit voltage U_{OCV} (OCV) is a mixed potential and lower compared to the equilibrium voltage obtained in Example Calculation 2.7.

For all current densities ($j > 0$), the cell voltage for the fuel cell is always lower compared to the calculated equilibrium voltages. This is linked to the energy conversion being detrimentally affected by internal losses (resistive losses, gradients in reagents, slow charge transfer, or slow chemical reactions). The current density in electrochemical cells is linked to the reaction rates during electrode reactions (Chapter 1). In order to reach a desired current density (reaction rate), some of the energy driving the reaction is lost in a sequence of reaction and transport steps. A more detailed discussion of these reaction steps will be possible in Chapter 3. It will be shown that loss of voltage compared to U_{OCV} at low current densities is associated with the activation of electrochemical reactions. At slightly higher current densities, transport of charges in polymer membranes consumes additional energy and thereby lowers cell voltages, and at very high current densities, transport of reagents to electrodes becomes rate-limiting and most energy-consuming.

For each point in the plot in Figure 2.33, we can multiply current density j [A cm^{-2}] and voltage U [V] to give the power density P_A [W cm^{-2}]. The power density is shown as a plot with secondary ordinate on the right. The highest power output consistent with the optimum energy conversion in the fuel cell is indicated as $P_{A,max}$, with j_w and $U_w(j_w)$ defining the operation point.

From this analysis, it seems obvious that the use of electrochemical thermodynamics provides the foundation for understanding the equilibrium cell voltage for fuel cells or batteries. However, further insight into these energy conversion systems during operation with $j > 0$ will require consideration of the rate of reactions and of transport steps in these cells. These will be topics in Chapters 3 and 4. In Chapters 15 and 16, we then reexamine the technical and materials aspects of electrochemical energy conversion systems in more detail.

2.5 Summary

The equilibrium cell voltage is directly proportional to the molar Gibbs free energy $\Delta_R G$ for the cell reaction. The equilibrium cell voltage is dependent on concentrations of reactants, the temperature, and the partial pressure of gaseous reactants. The effect of concentration on the cell voltage is expressed in the Nernst equation. In order to compare the potential for different types of electrodes, we can determine the equilibrium cell voltage with a defined reference electrode. The equilibrium cell voltages determined versus a reference electrode are called electrode potentials. The electrochemical series is a table of electrode potentials put in order. The difference between two electrode potentials allows the theoretical cell voltage to be calculated. By convention, the potential of the standard hydrogen electrode (SHE) has been selected as the reference point with 0.0 V vs. SHE in the electrochemical potential scale. In practical laboratory experiments, a wider range of reference electrodes is employed. Measurements with different types of reference electrodes are readily converted to the SHE potential scale.

The equilibrium voltage of an electrochemical cell is composed from additive contributions (Galvani voltages) of a sequence of interfaces along the current path at which charge is transferred. A Galvani voltage is the difference between the Galvani potentials (inner potentials) of two adjacent phases. In electrochemical cells, charge separation not only occurs at electrodes but also at contacts between wires of different metals, at contacts of miscible liquid electrolytes (cell with salt bridge, diffusion potentials), and in the presence of parallel electrode reactions (mixed potentials). The latter are examples of nonequilibrium processes that contribute to the cell voltage. For precision measurements of equilibrium cell voltages, the net current must be zero, $I = 0$, and nonequilibrium processes must be eliminated. Equilibrium cell voltages provide experimental access to standard electrode potentials, thermodynamic properties, activity coefficients, and equilibrium constants.

Important applications of galvanic cells are in potentiometric sensors for environmental and medical applications, as well as in electrochemical energy storage and energy conversion for batteries or for fuel cells. Fuel cells are energy conversion devices with high energy conversion efficiency. They generate electrical energy directly from chemical energy either at ambient temperature (PEMFC) or at elevated temperatures (SOFC), to overcome limitations by catalyst poisoning. In contrast, combustion engines produce a heat gradient, which then allows conversion into useful work within the limits of the second law of thermodynamics, ultimately leading to lower efficiencies. The cell voltages provided by batteries and fuel cells under current load are always lower compared to the open circuit potential (and compared to the theoretical equilibrium cell voltage) due to voltage losses for energizing internal processes in the device.

Test your knowledge

Exercise 2.1: Sketch and label the components of the classic Daniell cell. Write down the electrode reactions that occur **spontaneously** in the half-cells. Calculate the standard cell voltage with $E°(Zn|Zn^{2+}) = -0.763$ V and $E°(Cu|Cu^{2+}) = 0.337$ V.

Exercise 2.2: In a galvanic cell at 30 °C, an open-circuit potential of 0.117 V is measured. Calculate the Gibbs free energy $\Delta_R G$ (the stoichiometric number of electrons is $n = 2$). Calculate the reaction enthalpy $\Delta_R H$ given that the reaction entropy is $\Delta_R S = -21$ J mol^{-1} K^{-1}.

Exercise 2.3: How can the reaction Gibbs free energy $\Delta_R G$ be expressed by the reaction enthalpy $\Delta_R H$ and the reaction entropy $\Delta_R S$? Calculate $\Delta_R S$ for a reaction with $\Delta_R G = -58.5$ kJ mol^{-1} and $\Delta_R H = -90.7$ kJ mol^{-1} (at room temperature, with a pressure of $p = 720$ mbar).

Exercise 2.4: For a galvanic cell with a cell reaction with $\Delta_R S = 73.33$ J mol^{-1} K^{-1} at 24 °C, an equilibrium cell voltage of 90.1 mV ($n = 2$) is measured. Determine $\Delta_R H$? What would be the equilibrium cell voltage at 10 °C? Assume that $\Delta_R H$ and $\Delta_R S$ are temperature-independent.

Exercise 2.5: The open-circuit voltage (U_{OCV}) for a galvanic cell Cu|$CuSO_4$||3 M KCl|AgCl|Ag|Cu′ is determined as a function of copper sulfate concentration:

$c(Cu^{2+})/(mol\ L^{-1})$	U_{OCV}/mV
10^{-4}	4.5
10^{-3}	24.2
10^{-2}	48.1
10^{-1}	66.9
1	87.9

Plot the open-circuit cell voltages versus decadic logarithm of $[Cu^{2+}]$ and use linear regression to express the equation that best represents these data (employ the correct units for slope and intercept). Use the equation to determine the corresponding concentration $[Cu^{2+}]$ for a measured value of $U_{OCV} = +43.8$ mV.

Exercise 2.6:
a) Write down the Nernst equation for a Cu electrode in aqueous $CuSO_4$ and define all parameters (with correct units).
b) Given an electrode potential of $E = 286$ mV at 15 °C, calculate the concentration $[Cu^{2+}]$ Beware of the difference between activity and concentration. ($E° = 337$ mV vs. SHE, $f_{\pm} = 0.16$).

Exercise 2.7: Write down the Nernst equation (a) for the silver electrode and (b) for the silver–silver chloride electrode. Next, calculate the potential (c) for a silver electrode in 0.01 mol L^{-1} silver nitrate solution and (d) for the silver–silver chloride electrode in 0.01 mol L^{-1} KCl solution.

Exercise 2.8:
a) Calculate the equilibrium cell voltage for a fuel cell operating with air and pure hydrogen at 1 bar and at 60 °C.
b) In operation (under load), this fuel cell exhibits a voltage of 0.75 V and a current of 1.234 A. Determine the energy conversion efficiency under these conditions.

Exercise 2.9: Sketch the design of a standard hydrogen electrode: Pt|H_2(1000.00 hPa)|HCl.

Exercise 2.10: Consider 50 mL of a solution containing 0.1 mol L^{-1} CH_3COO^- and 0.01 mol L^{-1} CH_3COOH. Assume that all activity coefficients are 1.0 and $pK_a(CH_3COOH) = 4.75$.

a) Calculate the pH value.
b) What would be the change in pH upon addition of 10 mL 0.1 mol L^{-1} NaOH solution?

Exercise 2.11: The potential of a Hg|Hg_2Cl_2|KCl electrode in 1 mol L^{-1} KCl solution is 0.2801 V versus standard hydrogen electrode. Calculate the solubility product for Hg_2Cl_2. (The activity coefficient for Cl^- in 1 M KCl is approximately 0.6).

Exercise 2.12: The cell Pt|H_2 (*g*)|HCl(*aq*)|AgCl(*s*)|Ag provides an equilibrium cell voltage of +0.322 V at 25 °C. Calculate the pH for the electrolyte solution. Assume that $a(H^+) = a(Cl^-)$ and $f_\pm = 1$.

Exercise 2.13: Calculate for the following half-cell reactions the standard potentials versus standard hydrogen electrode from thermodynamic data given in the Appendix, Tables A-10 and A-11.

a) $CO(g) + H_2O(l) \rightarrow CO_2(g) + 2H^+ + 2e^-$
b) $CH_4(g) + 2H_2O(l) \rightarrow CO_2(g) + 8H^+ + 8e^-$

Exercise 2.14: Calculate the standard potential for the Cu|Cu^+ electrode given $E°(Cu|Cu^{2+}) = 0.346$ V and $E°(Cu^+|Cu^{2+}) = 0.170$ V.

Exercise 2.15: Calculate the difference in electrode potential for SHE and NHE.

References

1 P. W. Atkins, J. de Paula; *Physical Chemistry*. 8th ed., Oxford University Press, Oxford, 2006, Chapters 2 and 3.
2 H. Kuhn, D. H. Waldeck, H.-D. Försterling; *Principles of Physical Chemistry: Structure of Atoms, Molecules and Solids*, John Wiley & Sons, Hoboken, 2025, Chapters 16–19.
3 W. Moore; *Physical Chemistry*, Prentice-Hall, New Jersey, 1972, Chapters 2 and 3.
4 E. Lax, (Hrsg.); *Taschenbuch für Chemiker und Physiker*, Spinger, Berlin, 1967, pp. 809–814.
5 D. Dobos; *Electrochemical Data – A Handbook for Electrochemists in Industry and Universities*, Elsevier, Amsterdam, 1975.
6 B. E. Conway, *Electrochemical Data*, Greenwood Press, Westport, 1952.
7 *Compendium of Chemical Terminology*. 2nd ed. (the "Gold Book"). Compiled by A. D. McNaught and A. Wilkinson. Blackwell Scientific Publications, Oxford, 1997. http://goldbook.iupac.org/, IUPAC.
8 A. J. Bard, L. R. Faulkner, H. S. White; *Electrochemical Methods – Fundamentals and Applications*. 3rd ed., Wiley, Hoboken, 2022, pp. 339–344.
9 H. H. Girault; Electrochemistry at liquid–liquid interfaces. in *Electroanalytical Chemistry, Vol. 23* (A. J. Bard, C. G. Zoski, Eds.), CRC Press, Boca Raton, 2010, pp. 1–104.
10 Y. Shao; Electrochemistry at liquid/liquid interfaces. in *Handbook of Electrochemistry* (C. G. Zoski, Ed.), Elsevier, Amsterdam, 2007, pp. 785–809.
11 E. J. F. Dickinson, J. G. Limon-Petersen, R. G. Compton; The electroneutrality approximation in electrochemistry. *J. Solid State Electrochem.* **2011**, *15*, 1335–1345.

12 A. J. Bard, L. R. Faulkner, H. S. White; *Electrochemical Methods – Fundamentals and Applications*. 3rd ed., Wiley, Hoboken, 2022, p. 100.

13 D. M. Hirst; *Mathematics for Chemists*, Macmillan Press Ltd., London, 1976.

14 D. J. G. Ives, G. J. Janz (Eds.); *Reference Electrodes – Theory and Practice*, Academic Press, New York, 1961.

15 G. Inzelt, A. Lewenstam, F. Scholz; *Handbook of Reference Electrodes*, Springer, Berlin, 2013.

16 https://www.edaq.com/wiki/Reference_Electrode_Potentials (last access 21 Jan 2026)

17 T. J. Smith, K. J. Stevenson; Reference electrodes. in *Handbook of Electrochemistry*, Z. Zoski (Hrsg.), Elsevier, Amsterdam, 2007, p. 85.

18 H. C. Jamieson, G. C. Weathley, F. D. Manchester; The $\beta \rightarrow \alpha$ phase transformation in palladium–hydrogen alloys, *J. Less Common Met.* **1976**, *50*, 85–102.

19 J. V. Dobson; Potentials of the palladium hydride reference electrode between 25 °C and 195 °C. *J. Electroanal. Chem.* **1972**, *35*, 129–135.

20 A. M. Bond, E. A. McLennan, R. S. Stojanovic, F. G. Thomas; Assessment of conditions under which the oxidation of ferrocene can be used as a standard voltammetric reference process in aqueous media. *Anal. Chem.* **1987**, *59*, 2853–2860.

21 G. Gritzner, J. Kůta; Recommendations on reporting electrode potentials in nonaqueous solvents. *Pure Appl. Chem.* **1984**, *56*, 461–466.

22 A. A. J. Torriero, S. W. Feldberg, J. Zhang, A. N. Simonov, A. M. Bond; On choosing a reference redox system for electrochemical measurements: a cautionary tale. *J. Solid State Electrochem.* **2013**, *17*, 3021–3026.

23 Working Group Renewable Energy Statistics (AGEE-Statt) from 18 Dec 2020, available via https://www.umweltbundesamt.de/themen/klima-energie/erneuerbare-energien/erneuerbare-energien-in-zahlen#uberblick (last access 06 Feb 2021).

24 https://en.wikipedia.org/wiki/Battery_storage_power_station (last access 25 May 2021).

25 D. R. Wilburn; Global Exploration and Production Capacity for Platinum-Group Metals From 1995 Through 2015. Scientific Investigation Reports 2012-5164. US Department of the Interior and U.S. Geological Survey, Reston, Virginia 2012. https://pubs.usgs.gov/publication/sir20125164 (last access 10 Aug 2024).

3

Electrochemical Kinetics – Homogeneous and Heterogeneous Processes

When considering important applications of electrochemistry such as fuel cells, we saw in Chapter 2 that the properties of electrochemical cells cannot be explained solely by the thermodynamics of the cell reaction (the energy balance). We also have to investigate the kinetics of the electrode reaction (the reaction rates). We will do this in Chapter 3 by first taking a closer look at the substeps of electrochemical reactions. There is always a series (or better, a network) of steps involved in electrode reactions, but each electrode reaction must have at least one charge transfer step, that is, the transition of a charged particle across the phase boundary. This charge transfer step is coupled to charge transport processes, which will be considered in more detail in Chapter 4.

In Chapter 3, we explore the quantitative kinetic relationship between the rate of charge transfer at an interface and the applied electrode potential. It will become clear that the rates of electrode reactions are in some cases very sensitive and in other cases insensitive to the type of electrode material and the nature of the surface. Processes that depend strongly on the electrode material are often associated with the need to form adsorbed intermediates. When adsorbed reaction intermediates are formed, the bond between surface and reactant/adsorbent should not be too strong (stopping further reactions) and not too weak (leading to low surface concentrations). This relationship is often illustrated in a *volcano plot*. The adsorption energy has to be balanced to achieve sufficient activation without blocking the electrode surface sites with low energy intermediates. Finally, products need to desorb to free up the catalytic sites.

Further Reading. Here, simple reactions have been selected for educational purposes. Many practically relevant types of reactions can be dissected in a similar way and there are authoritative textbooks and monographs to provide more in-depth coverage [1–7]. Volcano plots are commonly employed in the discussion of the kinetics of electrocatalytic reactions, but have to be used with caution [8]. This topic will be discussed in more detail in Chapter 12.

3.1 Coupling Between Mass Transport and Interfacial Reactions

Mass transport denotes the movement of molecules or ions from one location to another in the same phase. This process is also frequently denoted as mass transfer. Charge transfer across an interface is coupled to mass transport to (i) allow reactants and products to interact with the interface and (ii) to maintain electroneutrality when the transport of charges at both sides of the interface is coupled together.

3.1.1 What Are the Important Steps During Electrode Reactions with Interfacial Electron or Ion Transfer?

In Chapter 2, we introduced thermodynamic principles to quantitatively derive the equilibrium cell voltage for galvanic cells under open-circuit conditions. For the case of the fuel cell, it became apparent that the flow of charge strongly impacts the cell voltage. The energy conversion process was shown to be linked to losses, for example, due to activation of individual reaction steps in electrode reactions. This loss of energy is observed as resistance or loss of cell voltage and is addressed here. The processes associated with energy losses will be dissected and discussed quantitatively in subsequent chapters.

Electrode reactions with ion transfer can be dissected into a sequence of elementary processes. The interfacial **charge transfer** (CT) can occur only if ions approach the interface to a distance smaller than a solvent molecule. Consider a solvated metal cation as shown in Figure 3.1a. Once transported and adsorbed to the electrode surface, several reaction steps are required to remove ligand molecules and to integrate the metal cation into the metal lattice. For charge transfer reaction to continue, transport processes for ions are necessary to bring other solvated metal cations to the interface.

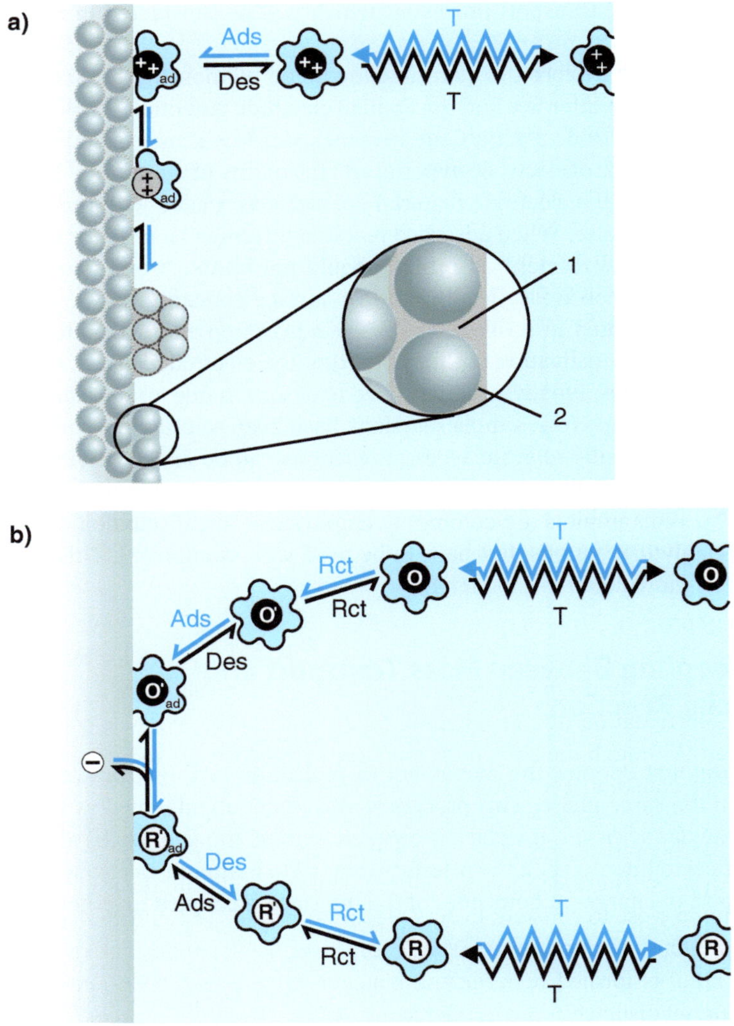

Figure 3.1 Schematic for exemplary electrode reactions associated with a) ion transfer and b) electron transfer.

Notes:
1) electrons in the electron gas in the metals;
2) metal atoms at lattice positions of the metal;

T – Mass transport;
Rct – Chemical reaction steps (homogeneous) preceding or following the charge transfer;
Ads/Des – Adsorption/desorption.

For electrodes with electron transfer, similar elementary reaction steps can be identified. Here electrons are transferred across the interface, a process that is called a heterogeneous electron transfer and can be described by quantum mechanical tunneling. This process is strongly distance-dependent. Hence, reactant molecules and ions must be transported toward and away from the electrode. Additional heterogeneous or homogeneous reaction steps are often associated with the heterogeneous electron transfer. They may be protonation/deprotonation, complexation, or dimerization steps which occur either as preceding or as follow-up reaction steps (Figure 3.1b). For a large group of electrode reactions (but not for all), we have to consider adsorption of reactants and desorption of products due to stronger interactions with the electrode surface during electron transfer.

A broad and useful distinction of two types of interfacial electron transfer processes has emerged based on the strength of interaction with the electrode surface before and after electron transfer. By analogy to the terminology employed in homogeneous redox chemistry of transition metal complexes, it is customary to distinguish outer and inner coordination sphere processes (Figure 3.2). For inner sphere processes, the metal cation directly (covalently or coordinatively) interacts with the electrode surface during electron transfer. Similar interactions are possible for other types of molecular redox species. For outer sphere processes, the solvation shell with oriented solvent dipoles remains intact during electron tunneling (Figure 3.2).

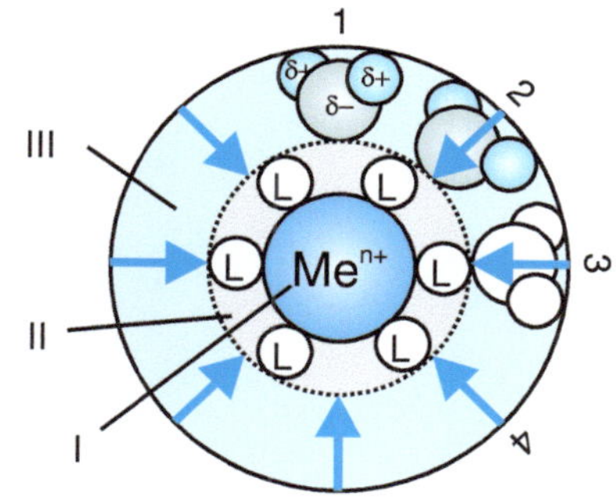

Figure 3.2 Schematic of a solvated metal cation (I). A shell of six ligands L surround the cation to form an inner coordination sphere (II). The metal complex $[MeL_6]^{n+}$ is solvated by H_2O dipoles that are oriented according to the local electric field (here $[MeL_6]^{n+}$ is assumed positive).

Numbers 1–4 indicate solvent dipoles as 1) space-filling calotte model with partial charges, 2) and 3) as calotte model with dipole vector, and 4) as dipole vector representing the solvent molecule.

Q Could electrons be transferred to be solvated and then transported to react with solution species?

A: Electron transfer between molecules and ions in solution or between an electrode surface and molecules/ions usually occurs up to a distance of 1.2 nm (equivalent to the length of 10 covalent C—H bonds). If there are no molecules with accessible energy levels for electrons to transfer, there is a possibility for transfer directly into the solvent (giving solvated electrons), for example in aqueous media at potentials of −2.87 V versus SHE. This is an extremely negative potential where reactions such as hydrogen evolution are likely. Short-lived solvated electrons in homogeneous solution can be produced with high-energy ionizing radiation. Once generated, the free electron is solvated within 1 ps (= 10^{-12} s) by $2H_2O$ molecules. Within 1 ms (at pH 7), the solvated electron reacts with a proton to give an H atom [9]. Solvated electrons in water are very short-lived and reactive toward oxidants. Solvated electrons in other types of solvents such as liquid ammonia (NH_3) form stable blue (or metallic at higher concentration) solutions with applications, for example, in organic Birch electrosynthesis [10].

Outer sphere reactions are based on electron transfer to the central ion, which changes its charge number (Table 3.1). The coordination to the ligands stays essentially unchanged, and solvent molecules in the outer sphere may undergo some degree of reorientation due to the change in the local electric field after transfer of the electron. Although the compositions of the solvation shell and inner coordination shell remain constant, the bond distances are likely to change after changing the redox state. Examples for fast outer sphere redox systems that are considered to be close to the theoretical model are the oxidation of ferrocene and the reduction of $[Ru(NH_3)_6]^{3+}$. For these systems, atomic bond distances within the inner coordination sphere remain close to constant during electron transfer. The rate of electron transfer for these systems is often insensitive to the type of electrode material because no chemical bond is formed between the reactant and the electrode material, and the electron tunnel distance controls the reaction rate.

Table 3.1 Comparison of outer sphere and inner sphere reactions. Outer sphere reactions occur with the solvent layer at the interface/around the reactant intact (electrode 1) and solvent layer 2) indicated). In contrast, for inner sphere reactions, the solvent layer has been removed 3), and a stronger interaction occurs between reactant and electrode surfaces.

	Outer sphere reaction	Inner sphere reaction
Homogeneous reaction	$[Co(NH_3)_6]^{3+} + [Cr(bpy)_3]^{2+} \rightarrow [Co(NH_3)_6]^{2+} + [Cr(bpy)_3]^{3+}$	$[Co(NH_3)_5Cl]^{2+} + [Cr(H_2O)_6]^{2+} \rightarrow [(NH_3)_5CoClCr(H_2O)_5]^{4+}$
Heterogeneous reaction	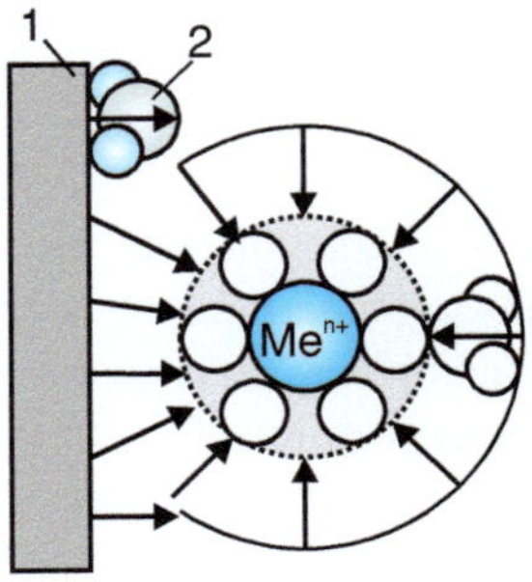	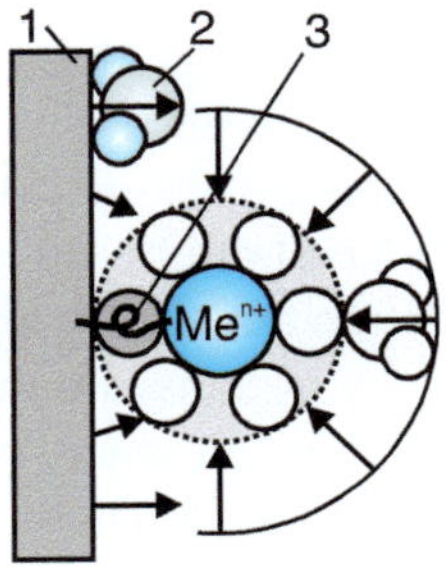
Examples	Ferrocene/ferrocenium $[Ru(NH_3)_6]^{2+/3+}$	$[Cr(H_2O)_6]^{2+/3+}$ at Hg-electrode in presence of Cl^- or Br^-

Inner sphere reactions are based on electron transfer reactions that occur via strong reactant–electrode interaction (replacing the solvent layer) and lead to significant changes in covalent bonds of the reactant (Table 3.1). Processes may involve adsorption at the electrode or binding via a bridging ligand between the redox center and the electrode. The fact that significant changes in covalent bonds occur and positions of atoms relative to each other change leads to slower reaction rates. Due to strong interfacial interactions, inner sphere reactions tend to be strongly sensitive to the nature of the electrode material.

3.1.2 Electrode Reactions Occur via Multiple Elementary Steps

Electrode reactions can be dissected into charge transfer, transport of reactants, and transport of products. Further steps may occur such as adsorption and heterogeneous/homogeneous chemical steps.

Multistep reactions are called "complex" in chemical kinetics. In this sense, the transport of reactants, the charge transfer step, and the transport of products can be considered as a sequence of coupled elementary reaction steps. For such a reaction scheme, the slowest elementary step of the sequence will determine the overall reaction rate. Thus, the slowest step is called the **rate-determining step** (RDS). Although some of the reaction steps are fast, the rate-determining step will create a bottleneck for the reaction rate and limit the overall conversion. This principle is illustrated in Figure 3.3 based on the analogy of the sequential reaction steps to a bucket brigade transporting water. When developing a new electrode process, for instance, in electrocatalysis, it is often necessary to identify the rate-determining step with the aim to improve its kinetics by variation in electrode materials subsequently.

The transport processes in electrochemical systems can be divided into **diffusion** (macroscopic transport resulting from random [brownian] motion of individual particles), **migration** (electric field-induced drift of charged particles [ions]), and

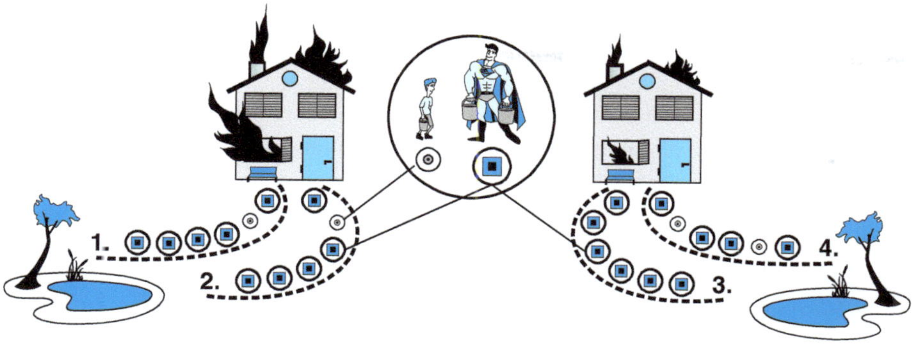

Figure 3.3 Illustration of the rate-determining step by analogy to a bucket brigade. The weakest link in the chain limits the overall efficiency. With two weaker elements (symbolized by junior volunteers) distributed over both chains of (fully professional) heroes, the transport of water is limited in both chains (left). When both weaker links are in the same chain, the net water transport rate will be significantly higher (right). Chain 4 will perform similarly to chains 1 or 2, but chain 3 will be significantly faster. For the water transport rates expressed as volume flow rate $\dot{V}i$ [L s^{-1}], we can state $\dot{V}1 + \dot{V}2 \ll \dot{V}3 + \dot{V}4$.

convection (flow or volume transport of the solution phase). These processes will be considered in more depth in Chapter 4.

A further approach to dissect the electrochemical cell can be based on electrical circuit analysis. The electrochemical cell can be considered as a network of elements such as electric resistors (more generally, impedances) and capacitors that reproduce the time dependence of the current output after applying a potential input signal. Every process is associated with a specific resistor (impedance), which is in some cases a serial or parallel network by itself (see for instance, Figure 4.9 or Chapter 7). For a sequence of resistors, the overall resistance is given by the sum of all resistors in series. There is a voltage drop at each resistor (process) which can be considered as the driving force for the associated process. The sum of all voltage drops can be regarded as the driving force that is necessary in order to achieve a *specified* current through the electrochemical cell. While a very small current will flow in an electrical circuit already at infinitesimally small voltages, a specified current can only be obtained at significantly higher applied voltages. In an electrochemical context, this additional voltage is called an **overpotential** η. This term is applied here for both, the voltage of a complete cell and the potential of an individual electrode. Returning to a thermodynamic view on cell reaction, the overpotential is the cell voltage during current flow ($|I| > 0$) with respect to the equilibrium cell voltage, U_{eq}.

$$\eta = U(|I| > 0) - U_{eq} \tag{3.1}$$

η – Overpotential [V]
$U(|I| > 0)$ – Cell voltage during current flow [V]
U_{eq} – Equilibrium cell voltage [V]

Q Is there a difference between open-circuit cell potential and equilibrium cell potential?

A: The equilibrium cell voltage U_{eq} is observed for the case that the oxidation and the corresponding reduction processes occur at the same rate. In contrast, the open-circuit cell voltage $U(I = 0)$ can also be measured and defined for nonequilibrium conditions and for irreversible processes. In this case, the currents due to oxidation and reduction processes balance to net zero, but there are multiple processes and the electrode potential has to be considered a mixed potential (e.g., during corrosion; see Section 3.2.7). For the case that there are only one anodic reaction and one corresponding cathodic reaction at an electrode, the open-circuit potential of an electrode or the open-circuit voltage of an electrochemical cell becomes identical to the equilibrium electrode potential or the equilibrium cell voltage.

The equilibrium cell voltage represents the thermodynamic energy difference between starting materials and products. In order to drive an electrolysis reaction in an electrolytic cell, the applied potential has to be higher than the equilibrium cell voltage, and therefore, the overpotential is higher than zero ($\eta > 0$). For galvanic cells (batteries), the flow of current results in a loss of voltage. The cell voltage is lower compared to the open-circuit cell voltage ($\eta < 0$).

Overpotentials are associated with each part of the reaction sequence and are named accordingly: charge transfer overpotential (due to the limited rate of interfacial charge transfer), diffusion overpotential (due to diffusional transport), reaction overpotential (due to limited rate of preceding or follow-up chemical reactions), or nucleation overpotential (due to slow formation of nuclei of critical size in phase formation processes).

3.2 Kinetics of Charge Transfer

The link from current (or current density) to the applied potential (or voltage) not only describes the electrochemical cell as a part of an electrical circuit but also provides direct insight into microscopic processes (the reaction rate) at the electrode–electrolyte interface [compare with Eq. (1.19)]. As early as 1905, Tafel[16] reported the observation that the logarithm of the current is proportional to the applied overpotential. The functional dependency, commonly observed for many electrode processes, can be expressed as

η – Overpotential [V]
$\log|j|$ – Logarithm of the absolute value of the current density (valid for both anodic and cathodic currents)
a, b – Empirical constants

$$\log|j| = a + b\,\eta \tag{3.2}$$

Equation (3.2) is based on purely empirical observations and has historically not been derived from fundamental laws. However, the link to fundamental laws can be established and the behavior in Eq. (3.2) rationalized. Here, the Tafel equation will be derived formally based on equations for kinetics. In this way, the physical phenomena responsible for the logarithmic relationship between overpotential and current can be understood based on molecular processes.

3.2.1 Homogeneous Equilibrium Reactions and Temperature Effects on Reaction Rates

Initially, we will consider a homogeneous model reaction based on compound A converting to compound B and vice versa in an elementary unimolecular reaction with a rate constant k_1 forward and a rate constant k_{-1} backward [Eq. (3.3)].

k_1, k_{-1} – Rate constants (forward and backward) for homogeneous first order reactions [s^{-1}]

$$\mathrm{A} \underset{k_{-1}}{\overset{k_1}{\rightleftharpoons}} \mathrm{B} \tag{3.3}$$

An elementary reaction refers to a chemical process with a *single* transition state (in contrast to multistep reactions). There are many cases of chemical reactions written formally as in Eq. (3.3) with multiple elementary steps (and more complex rate laws),

[16] Julius Tafel, 1862–1918, German chemist, Professor of Chemistry in Würzburg, known for his contributions such as the Tafel equation, the hydrogen coulometer in electrochemistry, and the Tafel rearrangement in organic chemistry. One mechanism in hydrogen evolution has been named Tafel mechanism in his honor.

and these are therefore unsuitable for this discussion. The reaction rate for the forward reaction $r_{h,1}$ is

$$r_{h,1} = k_1[A]. \quad (3.4)$$

For the back reaction, the reaction rate $r_{h,-1}$ is

$$r_{h,-1} = k_{-1}[B] \quad (3.5)$$

The net reaction rate is given by

$$r_h = r_{h,1} - r_{h,-1} = k_1[A] - k_{-1}[B] \quad (3.6)$$

At equilibrium, the net reaction rate approaches zero ($r_h = 0$) although both forward and back reactions continue at the microscopic scale (dynamic equilibrium). The corresponding reaction rate at equilibrium r_0 is the exchange reaction rate.

$$r_h = 0 = k_1[A]_{eq} - k_{-1}[B]_{eq} \quad (3.7)$$

$$k_1[A]_{eq} = k_{-1}[B]_{eq} = r_0 \quad (3.8)$$

$$K = \frac{[B]_{eq}}{[A]_{eq}} = \frac{k_1}{k_{-1}} \quad (3.9)$$

Kinetic reaction rate constants are associated with temperature effects primarily due to the population of the transition state. Empirically, it was observed that the logarithm of the rate constant is proportional to $1/T$ (**Arrhenius equation**[17]):

$$k = A \exp\left[\frac{-E_A}{RT}\right] \quad (3.10)$$

The term in the exponent has to be dimensionless and reflects the ratio of the kinetic energy barrier (E_A) and the thermal energy in the surroundings (RT). For surroundings with high thermal energy (high T), the term $-E_A/RT$ converges to zero, letting the exponential factor approach unity. Therefore, the rate constant k converges at high T to the frequency factor A. The value A corresponds to the frequency factor, that is, the limit of the reaction rate constant without any energy barrier. The value E_A corresponds to the activation energy. For conditions when E_A/RT is higher than unity ($E_A \gg$ thermal energy), only a small fraction of the molecules is able to populate the energy level E_A required for the reaction to proceed (see Boltzmann distribution[18] of energies in textbooks on chemical thermodynamics).

The empirically derived Arrhenius equation can be put on proper theoretical foundations, and for some types of reactions (in particular gas phase reactions), a very detailed molecular description emerges. Quantum mechanics can be used to calculate the internal energy of a molecular system and a reaction partner as a function of atomic distances and bond angles ($x_1, \ldots, x_n$). A plot of the calculated energies as a function of nuclear coordinates provides an energy landscape $E_{pot} = f(x_1, \ldots, x_n)$. For example, a diatomic molecule AB reacts with atom C according to Eq. (3.11), and parameters x_1 and x_2 for the energy landscape can be introduced.

$$\mathrm{A{-}B + C \rightleftharpoons [A\cdots B\cdots C]^{\neq} \rightleftharpoons A + B{-}C} \quad (3.11)$$

r_h – Homogeneous reaction rate
$r_{h,i}$ – Homogeneous reaction rate for elementary reaction step i
K – Equilibrium constant
[A], [B] – Concentrations of compounds A and B [mol L^{-1}]

$[A]_{eq}$, $[B]_{eq}$ – Concentrations of compounds A and B at equilibrium [mol L^{-1}]
r_0 – Exchange reaction rate [mol L^{-1} s^{-1}]

k – Kinetic reaction rate constant for a first order reaction [s^{-1}]
A – Frequency factor [s^{-1}]
E_A – Activation energy [J mol^{-1}]
R – Gas constant [8.314 J K^{-1} mol^{-1}]
T – Absolute temperature [K]

[17] Svante August Arrhenius, 1859–1927, Swedish physico-chemist, Nobel Prize 1903 for the theory of electrolyte dissociation.
[18] Ludwig Eduard Boltzmann, 1844–1906, Austrian physicist, professor in Graz and in Vienna, pioneering work in statistical thermodynamics and the understanding of entropy.

The distance between atoms A and B (r_{AB}) is employed as coordinate x_1 and the distance between atoms B and C (r_{BC}) is employed as coordinate x_2. The calculated internal energy for this system (A, B, C) is presented as a function of x_1 and x_2 to give a two-dimensional energy landscape with hills and valleys. There are local minima corresponding to the compounds AB and BC. Both are linked through a valley and a pass (saddle point) following the low-energy route from AB to BC. This low-energy path is known as the **reaction coordinate**. Figure 3.4a illustrates the reaction coordinate for this type of simplified three-atom reaction. Commonly energy landscapes are presented in a more simplified form directly as a function of the reaction coordinate. Any atomic configurations that deviate from the low-energy path are neglected. The resulting two-dimensional plot is shown in Figure 3.4b. This choice of plot will be useful in the discussion of electrode reactions below.

k – Reaction rate constant [s^{-1}]
A – Frequency factor [s^{-1}]
$\Delta H^{\neq}$ – Molar activation enthalpy [J mol^{-1}]

A' – Modified frequency factor [s^{-1}]
$\Delta S^{\neq}$ – Molar activation entropy [J K^{-1} mol^{-1}]

$\Delta G^{\neq}$ – Molar Gibbs free energy of activation [J mol^{-1}]

The point of maximum energy in Figure 3.4b corresponds to the saddle point in Figure 3.4a. The configuration of atoms A, B, and C at this point is known as the **transition state** and indicated with superscript ≠. The energy barrier for the reaction pathway can be expressed as change in molar internal energy, $\Delta U^{\neq}$. This is the internal energy change associated with 1 mol AB + C reacting to give the transition state. In condensed phase systems, volume changes during reactions are negligible and, therefore, $\Delta H^{\neq} = \Delta U^{\neq} + (pV)^{\neq} \approx \Delta U^{\neq}$. We can rewrite Eq. (3.10) based on the molar enthalpy for the formation of the transition state.

! Here, the symbol U is employed for the molar internal energy. Do not confuse with cell voltage U elsewhere in the book.

$$k = A \exp\left[\frac{-\Delta H^{\neq}}{RT}\right] \tag{3.12}$$

The frequency factor A can now be dissected into a modified factor A' and the factor $\exp(\Delta S^{\neq}/R)$, containing $\Delta S^{\neq}$, the molar entropy for the formation of the transition state (the activation entropy).

$$k = A' \exp\left[\frac{\Delta S^{\neq}}{R}\right] \exp\left[\frac{-\Delta H^{\neq}}{RT}\right] \tag{3.13}$$

Employing the Gibbs–Helmholtz Equation (2.24), we can now use the expression $-\Delta G^{\neq} = -\Delta H^{\neq} + T\Delta S^{\neq}$ to express the rate constant as a function of Gibbs free energy of activation.

$$k = A' \exp\left[\frac{-(\Delta H^{\neq} - T\Delta S^{\neq})}{RT}\right] = A' \exp\left[\frac{-\Delta G^{\neq}}{RT}\right] \tag{3.14}$$

This equation, derived from the Arrhenius equation, will be helpful due to the link from Gibbs free energy to the cell voltage for an electrochemical cell, $\Delta G = -nFU$ [Eq. (2.23), note the use of U for the cell voltage]. It will be possible to express the reaction rate constant for electrode reactions as a function of electrode potential.

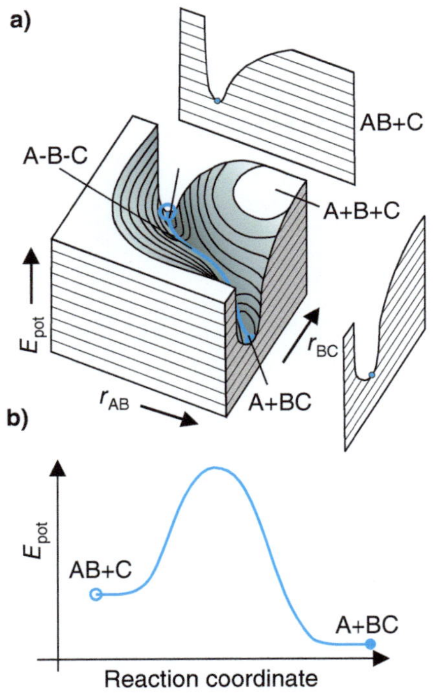

Figure 3.4 Illustration of the potential energy surface for reaction (3.11). a) Energy landscape for $E_{pot} = f(r_{AB}, r_{BC})$ and b) a plot of E_{pot} along the reaction coordinate (see blue line in the r_{AB}, r_{BC} plane in a). A more realistic or complex path for the reaction coordinate would include vibrational behavior for AB and BC.

3.2.2 Derivation of the Butler–Volmer Equation

Next, it will be shown how the empirical expression suggested by Tafel is linked theoretically to the effect of the electrode potential on the reaction rate. The theory will provide further insights into how the phenomena at the microscopic or molecular scale cause macroscopic scale changes in current flow. The discussion of reaction rates in Section 3.2.1 will be expanded to include electrochemical reactions. The reaction rate for heterogeneous electrode reactions is given by the amount of product being formed as a function of time and *electrode area* [see Eq. (1.19)]. Based on Eq. (1.23), we can convert the reaction rate into current density. For a hypothetical electrochemical reaction, we introduce the rate constants k_a and k_c for the elementary oxidation (anodic) and reduction (cathodic) reactions.

$$O + e^- \underset{k_a}{\overset{k_c}{\rightleftharpoons}} R \tag{3.15}$$

The reaction rates can now be expressed as first order processes for the cathode and the anode.

$$r_c = \frac{-I_c}{nFA} = k_c c_O(x = 0, t) \tag{3.16}$$

$$r_a = \frac{I_a}{nFA} = k_a c_R(x = 0, t) \tag{3.17}$$

The rate law for electrochemical reactions is expressed based on the concentration c of the oxidized form (c_O) or the reduced form (c_R). The units for the rate constants k_c and k_a are given in length per time (m s^{-1} in the SI system or cm s^{-1} in the cgs system). Often in electrochemical processes, the concentration in the vicinity of the electrode is dependent on distance x to the surface and on time t. In Eqs. (3.16) and (3.17), the *concentration at the electrode surface* ($x = 0$) is required, which only in ideal situations (for extremely fast mass transport), can be assumed to be identical to the *bulk concentration*. The bulk concentration is indicated with an asterisk symbol (c_O^*, c_R^*).

The net current through the external circuit can now be expressed keeping in mind the sign convention for anodic (+) and cathodic (−) currents.

$$\begin{aligned} I &= |I_a| - |I_c| \\ &= nFA[k_a c_R(0, t) - k_c c_O(0, t)] \end{aligned} \tag{3.18}$$

For k_a and k_c, an expression analogous to Eq. (3.4) can be inserted to demonstrate a strong dependence of the rate constants, and therefore, the net current on the electrode potential. The effect of the electrode potential can be illustrated with a plot of Gibbs free energy of activation versus reaction coordinate for the processes $O + e^- \rightarrow R$ and $R \rightarrow O + e^-$ (Figure 3.5). A Na^+ cation solvated in organic solvent is assumed to undergo reduction at a liquid mercury electrode. The reaction coordinate can be envisaged in terms of the distance from the solvent–mercury interface. Upon approaching the interface, the Na^+ cation changes the solvation layer, which is associated with a change in Gibbs free energy. Once the Na^+ cation is transferred across the solvent–mercury interface, an electron must be supplied through the external circuit to keep the charge of the electrode constant. This causes an external current. Once the Na atom is completely surrounded by mercury and inside the mercury phase, the Gibbs free energy G remains constant (Figure 3.5).

For the case that the externally applied electrode potential E is equal to the equilibrium electrode potential E_{eq}, there is no net current ($I = 0$). If we assume the same bulk concentration for O and R ($c_O^* = c_R^*$), the activation barriers for reduction and oxidation reactions are equal (Figure 3.5a). For an electrode potential other than E_{eq}, the rates (and activation barriers) for elementary reactions change. At more positive electrode potentials, the oxidation process dominates (the activation barrier for oxidation is lowered), and for more negative electrode potentials, the reduction process dominates (the activation barrier for reduction is lowered). An illustration of the activation barriers as a function of applied potential is shown in Figure 3.5b and c.

For a typical electron transfer reaction such as the one-electron transfer reaction from $[Fe(CN)_6]^{3-}$ to $[Fe(CN)_6]^{4-}$, the average bond distance from the Fe atom to the C atom is strongly linked to the reaction coordinate. The higher positive charge on

k_a, (k_c) – Heterogeneous rate constant for a first-order reaction for the anodic (cathodic) elementary reaction [cm s^{-1}]

r_a, (r_c) – Reaction rate for the anodic (cathodic) reaction [mol cm^{-2} s^{-1}]
I_a, (I_c) – Anodic (cathodic) current component [A], the sign for cathodic current is negative by definition
n – Stoichiometric number of electrons
F – Faraday constant [A s mol^{-1}]
$c_R(x = 0, t)$ [$c_O(x = 0, t)$] – Concentration of the reduced (index R) [oxidized (index O)] redox species at distance $x = 0$ from the electrode at time t [mol cm^{-3}]
A – Area of the electrode [cm^{-2}]

Figure 3.5 Plot of the molar Gibbs free energy versus reaction coordinate for the electrode Na(Hg)|$Na^+ + e^-$.

a) Reaction in equilibrium (the activation barriers for anodic and for cathodic reactions are equal);
b) $E > E_{eq}$, the oxidation dominates (lower activation barrier for oxidation); and
c) $E < E_{eq}$, the reduction dominates (lower activation barrier for reduction).

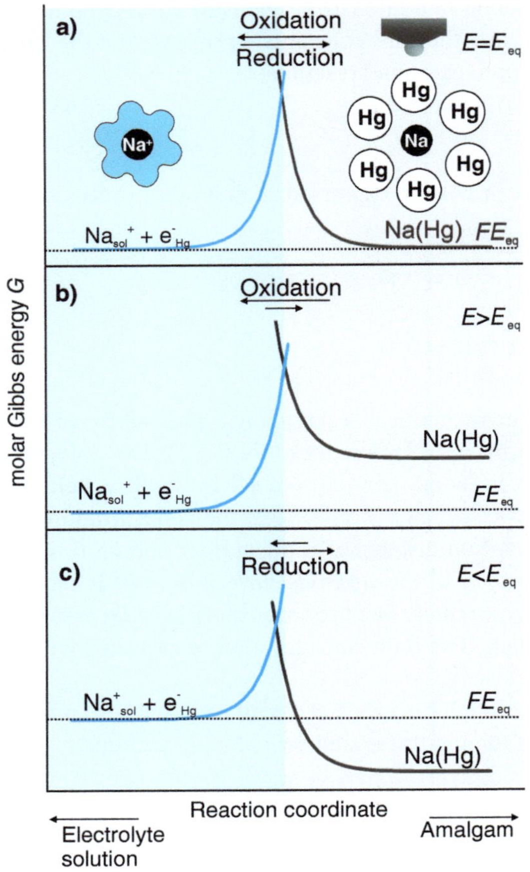

Fe(III) compared to Fe(II) causes a shorter bond distance to the cyanide ligands. As a chemical bond, the Fe–C bond is in first approximation consistent with the harmonic oscillator model (for more detail see textbooks for physical chemistry or spectroscopy), and therefore, the change in Gibbs free energy G with distance is mathematically represented by a parabola. A parabola can be drawn for Fe(II) and for Fe(III) at two different points along the reaction coordinate. These two parabolas intersect at a point consistent with the molar Gibbs free energy for activation. We assume that the bulk concentration for the oxidized form and the reduced form are equal ($c_O^* = c_R^*$). In this case, Eq. (3.18) suggests a net current of zero for $k_a(E) = k_c(E)$. In this situation, the electrode potential equals the formal potential $E^{\circ\prime}$ (Section 2.2.6).

Due to the formal potential being very convenient, we will base the next steps in the derivation on $E^{\circ\prime}$ and on the case of a single electron transfer ($n = 1$). At the electrode potential $E^{\circ\prime}$, the Gibbs free energy of activation is equal for the anodic and the cathodic elementary reactions, $\Delta G_a^{\neq\circ} = \Delta G_c^{\neq\circ}$. The symbol for the standard is employed here to denote this special situation. This point of equal activation Gibbs energies is indicated with a black curve in Figure 3.6. A change in electrode potential by $\Delta E = E - E^{\circ\prime}$ causes a lower Gibbs free energy for the electrons in the metal ($-F\Delta E$), and therefore, a lowering of the parabola for the oxidized state in the reaction coordinate (gray line in Figure 3.6). The Gibbs free energies of O and R in solution are not affected. Only the potential gradient in the electrical double layer close to the electrode surface (Section 4.3) changes, with no potential changes in the bulk solution.

The main effect of the potential on the reaction coordinate is associated with a lowering of the activation energy for the reaction $R \rightarrow O + e^-$. In contrast, the activation energy for the opposite reaction $O + e^- \rightarrow R$ increases. The bottom part of Figure 3.6

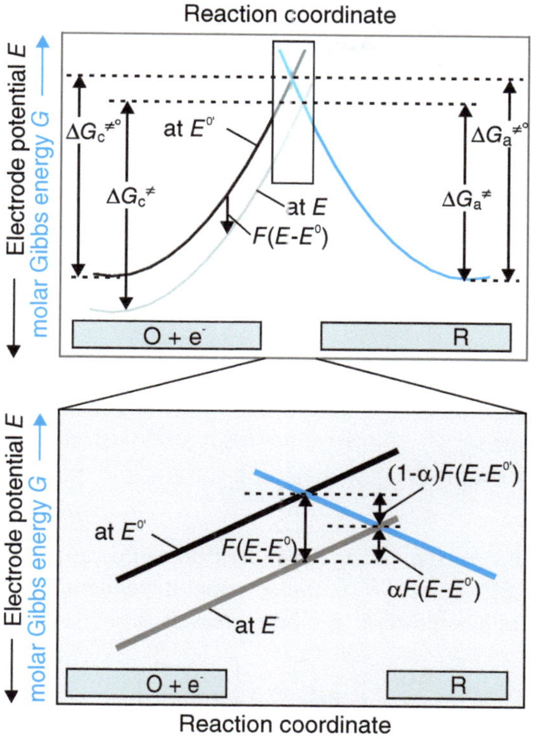

Figure 3.6 The effect of applied potential on the reaction coordinate (top). Magnified and linearized plot of the region close to the transition state (bottom).

focuses on the transition state. In this region the intersection of parabolas can be simplified to the intersection of two lines. In this illustration it can be seen that for a Gibbs free energy change of $-F\Delta E$ the activation barrier for the oxidation only changes by a fraction $(1-\alpha)$ with $0 \leq \alpha \leq 1$. The dimensionless coefficient α is the **transfer coefficient** (synonymous with symmetry coefficient). We can now express the molar Gibbs free energy of activation for the oxidation.

$$\Delta G_a^{\neq} = \Delta G_a^{\neq\circ} - (1-\alpha)F(E - E^{\circ\prime}). \tag{3.19}$$

α – Transfer coefficient [dimensionless]

Correspondingly, a shift in Gibbs free energy by $-F(E - E^{\circ\prime})$ increases the Gibbs free energy of activation for the reduction by $\alpha F(E - E^{\circ\prime})$. Sometimes $1-\alpha$ is replaced by β in the literature with $\alpha + \beta = 1$ or symbols for anodic and cathodic transfer coefficients are denoted as $\alpha_a + \alpha_c = 1$. Further symbols need to be introduced for multielectron processes. Here, the symbol α for the transfer coefficient is sufficient. The Gibbs free energy of activation for the cathodic reaction can be written as

$$\begin{aligned}\Delta G_c^{\neq} &= \Delta G_c^{\neq\circ} - [-F(E - E^{\circ\prime})] - (1-\alpha)F(E - E^{\circ\prime}) \\ &= \Delta G_c^{\neq\circ} + \alpha F(E - E^{\circ\prime}).\end{aligned} \tag{3.20}$$

We can now substitute for $\Delta G_c^{\neq}$ in Eq. (3.14) to give Eqs. (3.21) and (3.22) for the reduction.

$$k_c = A_c' \exp\left[\frac{-\Delta G_c^{\neq}}{RT}\right] = A_c' \exp\left[\frac{-\left[\Delta G_c^{\neq\circ} + \alpha F(E - E^{\circ\prime})\right]}{RT}\right]. \tag{3.21}$$

$$k_c = A_c' \exp\left[\frac{-\Delta G_c^{\neq\circ}}{RT}\right] \cdot \exp\left[\frac{-\alpha F(E - E^{\circ\prime})}{RT}\right] \tag{3.22}$$

n – Stoichiometric number of electrons
F – Faraday constant [A s mol^{-1}]
α – Transfer coefficient [dimensionless]
A_a', A_c' – Modified frequency factor for anodic and cathodic reactions [cm s^{-1}]
E – Electrode potential [V]
$E^{\circ\prime}$ – Formal potential [V]
$\Delta G_a^{\neq}$, $\Delta G_c^{\neq}$ – Molar Gibbs free energy of activation for cathodic and anodic elementary reactions [J mol^{-1}]

An analogous treatment yields Eq. (3.23) for the oxidation.

$$k_a = A'_a \exp\left[\frac{-\Delta G_a^{\neq\circ}}{RT}\right] \cdot \exp\left[\frac{+(1-\alpha)F(E-E^{\circ\prime})}{RT}\right] \tag{3.23}$$

It is now possible to define a standard rate constant for heterogeneous electron transfer. For the special case of equal concentrations of oxidized and reduced species ($c_O^* = c_R^*$) and for equilibrium at $E = E^{\circ\prime}$ [Eq. (2.72)], the rates for anodic and cathodic reactions become equal, $r_a = r_c$. Substitution into Eq. (3.18) results in $k_c c_O^* = k_a c_R^*$. With bulk concentrations $c_O^* = c_R^*$ at the potential $E^{\circ\prime}$ we obtain $k_c(E^{\circ\prime}) = k_a(E^{\circ\prime}) = k^\circ$.

At the formal potential $E^{\circ\prime}$, the heterogeneous rate constants for charge transfer, anodic or cathodic, take on the same value, which is the **heterogeneous standard reaction rate constant** k°. The heterogeneous standard reaction rate constant is specific for redox couple R/O, the electrode surface, and the solution composition.

The standard rate constant k° is dependent on temperature but not on electrode potential. For $E = E^{\circ\prime}$ ($E - E^{\circ\prime} = 0$), the potential-dependent terms in Eqs. (3.22) and (3.23) become unity. Therefore, k° can be expressed as

k° – Heterogeneous standard reaction rate constant [cm s^{-1}]
T – Absolute temperature [K]
A'_a, A'_c – Modified frequency factor for anodic and cathodic reactions [cm s^{-1}]

$$k^\circ(T) = A'_c \exp\left[\frac{-\Delta G_c^{\neq\circ}}{RT}\right] = A'_a \exp\left[\frac{-\Delta G_a^{\neq\circ}}{RT}\right]. \tag{3.24}$$

The expressions for the potential-dependent rate constants for reduction k_c and for oxidation k_a can be rewritten with Eqs. (3.22)–(3.24) to give Eqs. (3.25) and (3.26).

$$k_c = k^\circ(T) \cdot \exp\left[\frac{-\alpha F(E-E^{\circ\prime})}{RT}\right], \tag{3.25}$$

$$k_a = k^\circ(T) \cdot \exp\left[\frac{+(1-\alpha)F(E-E^{\circ\prime})}{RT}\right]. \tag{3.26}$$

Substitution of the expressions for the rate constants into the expression for the net current [Eq. (3.18)] gives Eq. (3.27). Note that for the case of net current flow (nonequilibrium) concentration gradients will develop at the electrode surface, and therefore, the surface concentration can deviate from the bulk concentration ($c_R(0,t) \neq c_R^*$).

$c_R(0,t), c_O(0,t)$ – Concentrations of reduced and oxidized species at the electrode surface ($x = 0$) at time t [mol cm^{-3}]

$$I = nFAk^\circ \left\{ c_R(0,t) \exp\left[\frac{+(1-\alpha)F(E-E^{\circ\prime})}{RT}\right] - c_O(0,t) \exp\left[\frac{-\alpha F(E-E^{\circ\prime})}{RT}\right] \right\} \tag{3.27}$$

This is the **Butler–Volmer equation.**[19] The term Butler–Volmer equation is sometimes associated with other similar equations that always link current to potential in an exponential function.

The net current is given as the sum of anodic (+) and cathodic (−) currents. Both are exponentially dependent on the potential difference to the formal potential ($E - E^{\circ\prime}$)

[19] John Alfred Valentine Butler, 1899–1977, English physicochemist and biophysicist; Max Volmer, 1885–1965, German physicochemist, worked in Leipzig, Berlin, and Hamburg, known also from the Stern–Volmer equation for fluorescence quenching. Important contributions to the development of the Butler–Volmer equation were made by T. Erdey-Grúz, 1902–1976, Hungarian professor of physical chemistry, President of the Hungarian Academy of Sciences 1970–1976 [11].

(Figure 3.7). For $E > E^{\circ\prime}$, the oxidation process dominates, and the current for reduction becomes negligible. For $E < E^{\circ\prime}$, the reduction process dominates, and the current for oxidation becomes negligible. At the formal potential $E = E^{\circ\prime}$ cathodic and anodic current components cancel out. Because we have assumed equal concentrations for species R and O, a special case arises with the formal potential $E^{\circ\prime}$ being equal to the equilibrium potential E_{eq}. Generally, for different concentrations of oxidized form O and reduced form R, the equilibrium potential E_{eq} (the point of zero net current) will be different from $E^{\circ\prime}$.

The Butler–Volmer equation can be rewritten for specific cases as will be shown below. It is important to note that the elementary charge transfer at the electrode surface is only one step in a sequence of reaction steps in electrochemical processes (Figure 3.1). Therefore, the observed current can be lower than that predicted by Eq. (3.27) (e.g., with mass transport limitations). The net current cannot be higher than that predicted by the Butler–Volmer equation.

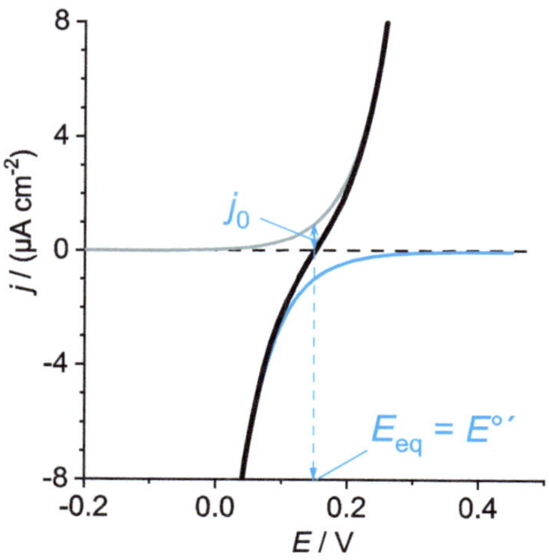

Figure 3.7 Plot of the anodic (gray) and the cathodic (blue) current components. The sum or net current is shown in black. Parameters: $E^{\circ\prime} = 0.15$ V, $\alpha = 0.5$, $n = 1$, $k^\circ = 1 \times 10^{-5}$ cm s^{-1} (exchange current density $j_0 = 0.96$ µA cm^{-2}, Section 3.2.3), $T = 298$ K, $c_R(0,t) = c_O(0,t) = 0.001$ mol L^{-1}.

3.2.3 Parameters in the Butler–Volmer Equation

It is instructive to analyze the components of the Butler–Volmer equation. The term $RT/F = 0.0257$ V ($n = 1$, $T = 298$ K) is the thermodynamic Nernst factor, E_{NF}, as introduced in the discussion of the Nernst Equation (2.28).

Q: I have read that the Nernst factor at room temperature is $59/n$ mV with n referring to the number of electrons. Why is the value above different from 59 mV?

A: This is only an apparent difference that is due to the difference between the natural and the decadic logarithm. Often the decadic logarithm is convenient. For conversion from natural to decadic logarithm we can use $a = e^x = 10^y = (e^{2.3})^y = e^{2.3 \cdot y}$ and $\ln a = x = 2.3 \cdot y$ (therefore, $2.3 \cdot 25.7$ mV $= 59$ mV). Here, we employ the abbreviation for the Nernst factor $E_{NF} = RT/(nF)$ due to its common use. The Nernst factor can also be helpful in unit conversions or in quantity calculus.

Inspection of Eq. (3.27) suggests that the difference in electrode potential versus $E^{\circ\prime}$ relative to E_{NF} is important (and dimensionless). The potential difference is relative to E_{NF} in a similar way as the activation energy is relative to the thermal energy of the surroundings RT in the Arrhenius equation. The current I in the Butler–Volmer equation is often normalized with respect to the electrode area and given as the current density $j = I/A$. With these adjustments, the Butler–Volmer equation can be written as in Eq. (3.28).

$$j = nFk^\circ \left\{ c_R(0,t) \exp\left[\frac{+(1-\alpha)(E - E^{\circ\prime})}{E_{NF}}\right] - c_O(0,t) \exp\left[\frac{-\alpha(E - E^{\circ\prime})}{E_{NF}}\right] \right\} \quad (3.28)$$

$E_{NF} = RT/(nF)$ – Nernst factor [V]

A spreadsheet calculation document provided as supplementary material for download is very useful for familiarization with the Bulter–Volmer equation. Effects from each parameter in Eq. (3.28) on the net current can be assessed.

The **exchange current density** j_0 describes the absolute value of the anodic $|j_a|$ or cathodic $|j_c|$ current density at equilibrium, which is for zero net current density with $j = |j_a| - |j_c| = 0$. The exchange current density is always positive. Anodic and cathodic currents follow different sign conventions. Here, we define cathodic current as negative and anodic current as positive (see IUPAC recommendations, Figure 1.9a).

At the point, where $j = |j_a| - |j_c| = 0$, the equilibrium potential E_{eq} is applied to the electrode. This equilibrium potential E_{eq} can be calculated with the Nernst Equation (3.29). Due to the simplicity of the reaction considered here, the standard activities in the concentration term cancel out. With $n = 1$ and under equilibrium conditions with zero net current, bulk concentrations can be employed with $c_R(0, t) = c_R^*$ and $c_O(0, t) = c_O^*$.

$$E_{eq} = E^{\circ\prime} + E_{NF} \ln \frac{c_O^*}{c_R^*} \tag{3.29}$$

Rearranging this equation gives the ratio of concentrations for oxidized and reduced species O and R.

$$\frac{c_O^*}{c_R^*} = \exp\left[\frac{E_{eq} - E^{\circ\prime}}{E_{NF}}\right] \tag{3.30}$$

From Eq. (3.28) under equilibrium conditions ($E = E_{eq}, j = 0$), the exchange current density is obtained with $j_0 = j_a = |j_c|$. Substitution of Eq. (3.30) into the cathodic current term in Eq. (3.28) provides the cathodic current density in equilibrium.

$e^{-ax} = (e^x)^{-a} = (e^{[1/x]})^a$

$$j_0 = |j_c(E_{eq})| = \left| nFk^\circ \left(-c_O^*\right) \exp\left[\frac{-\alpha(E_{eq} - E^{\circ\prime})}{E_{NF}}\right] \right| = nFk^\circ c_O^* \left(\frac{c_O^*}{c_R^*}\right)^{-\alpha}. \tag{3.31}$$

$$j_0 = nFk^\circ c_O^{*\,(1-\alpha)}\, c_R^{*\,\alpha} \tag{3.32}$$

Only for the special case of $c = c_O^* = c_R^*$ can we simplify to Eq. (3.33).

$$j_0 = nFck^\circ \tag{3.33}$$

The exchange current density j_0 provides kinetic information for an electrochemical reaction equivalent to that in the heterogeneous standard reaction rate constant k°.

The **current–overpotential curve** can now be obtained by dividing Eq. (3.28) by Eq. (3.32).

$$\begin{aligned} \frac{I}{I_0} = \frac{j}{j_0} &= \frac{nFk^\circ}{nFk^\circ c_O^{*\,(1-\alpha)}\, c_R^{*\,\alpha}} \left\{ c_R(0,t) \exp\left[\frac{+(1-\alpha)(E - E^{\circ\prime})}{E_{NF}}\right] \right. \\ &\quad \left. - c_O(0,t) \exp\left[\frac{-\alpha(E - E^{\circ\prime})}{E_{NF}}\right] \right\} \\ &= \frac{c_R(0,t)}{c_R^*} \left(\frac{c_O^*}{c_R^*}\right)^{-(1-\alpha)} \exp\left[\frac{+(1-\alpha)(E - E^{\circ\prime})}{E_{NF}}\right] \\ &\quad - \frac{c_O(0,t)}{c_O^*} \left(\frac{c_O^*}{c_R^*}\right)^{\alpha} \exp\left[\frac{-\alpha(E - E^{\circ\prime})}{E_{NF}}\right]. \end{aligned} \tag{3.34}$$

With Eq. (3.30), we can derive

$$\left(\frac{c_O^*}{c_R^*}\right)^{-(1-\alpha)} = \exp\left[\frac{E_{eq} - E^{\circ\prime}}{E_{NF}}\right]^{-(1-\alpha)} = \exp\left[\frac{(1-\alpha)(E^{\circ\prime} - E_{eq})}{E_{NF}}\right] \tag{3.35}$$

and

$$\left(\frac{c_O^*}{c_R^*}\right)^{\alpha} = \exp\left[\frac{E_{eq} - E^{\circ\prime}}{E_{NF}}\right]^{\alpha} = \exp\left[\frac{-\alpha(E^{\circ\prime} - E_{eq})}{E_{NF}}\right]. \tag{3.36}$$

Substitution of Eqs. (3.35) and (3.36) into Eq. (3.34) followed by simplification of the exponential terms gives Eq. (3.37).

$$\frac{I}{I_0} = \frac{c_R(0,t)}{c_R^*}\exp\left[\frac{+(1-\alpha)(E-E^{\circ\prime}+E^{\circ\prime}-E_{eq})}{E_{NF}}\right] - \frac{c_O(0,t)}{c_O^*}\exp\left[\frac{-\alpha(E-E^{\circ\prime}+E^{\circ\prime}-E_{eq})}{E_{NF}}\right]$$
$$= \frac{c_R(0,t)}{c_R^*}\exp\left[\frac{+(1-\alpha)(E-E_{eq})}{E_{NF}}\right] - \frac{c_O(0,t)}{c_O^*}\exp\left[\frac{-\alpha(E-E_{eq})}{E_{NF}}\right]. \quad (3.37)$$

Here, a new potential difference $E - E_{eq}$ appears in the exponential term. This is the difference between the electrode potential and the equilibrium potential at given bulk concentrations c_O^* and c_R^*. This potential difference is the **overpotential** $\eta = E - E_{eq}$. This allows further simplification of the Butler–Volmer equation in (3.38).

$$j = j_0\left\{\frac{c_R(0,t)}{c_R^*}\exp\left[\frac{+(1-\alpha)\eta}{E_{NF}}\right] - \frac{c_O(0,t)}{c_O^*}\exp\left[\frac{-\alpha\eta}{E_{NF}}\right]\right\} \quad (3.38)$$

j, j_0 – Current density, exchange current density [A cm^{-2}]
η – Overpotential [V]
α – Transfer coefficient [dimensionless]
$c_R(0, t)$, [$c_O(0, t)$] – Concentration of the reduced [oxidized] species at the electrode surface ($x = 0$) at time t [mol L^{-1}]
c_R^* (c_O^*) – Bulk concentration of the reduced (oxidized) species [mol L^{-1}]

Figure 3.8 shows a plot of Eq. (3.38) for different exchange current densities (i.e., different heterogeneous standard rate constants $k°$) for $c_O^* = c_R^* =$ const. For a smaller exchange current density, the current close to the equilibrium potential appears lower and and the slope smaller around E_{eq}. This is consistent with a high fraction of the electrical energy being consumed for the activation of the charge transfer. Substantial currents are observed only for potentials that are significantly different from those predicted by equilibrium thermodynamics (Chapter 2). The plots in Figure 3.8 also suggest that for reactions that are very slow close to the equilibrium potential, we can select a higher electrode potential to drive the reaction much faster. In contrast to most chemical reactions, where the activation barrier is overcome by heating, the activation barrier for slow electrochemical processes can be overcome by changing the applied potential at room temperature.

With a high value for $k°$, exchange currents are high even at equilibrium. In order to achieve a net current density lower than j_0, the overpotential η can be small, just offsetting the equilibrium enough to achieve the current. Due to the high anodic and cathodic current densities, the change in potential will affect the ratio of concentrations of oxidized and reduced species *at the electrode surface*. The change in concentrations will be rapid (instantaneous), and therefore, the concentrations will remain consistent with those predicted by the Nernst Equation (3.30). The concentration ratio at the electrode surface can be expressed as:

$$\frac{c_O(x=0,t)}{c_R(x=0,t)} = \exp\left[\frac{E-E^{\circ\prime}}{E_{NF}}\right] \quad (3.39)$$

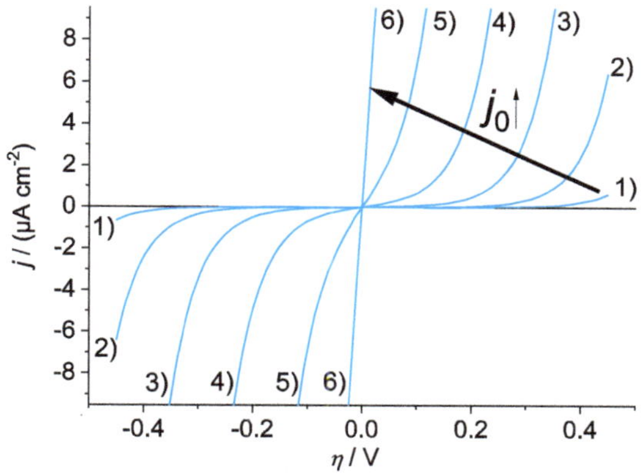

Figure 3.8 Plots of the current density versus overpotential according to the Butler–Volmer equation (without consideration of mass transport limitations); $c_R(0,t)/c_R^* = c_O(0,t)/c_O^* = 1$, $c_R^* = c_O^*$, $\alpha = 0.5$ and different j_0 in µA cm^{-2}: 1) 10^{-4} 2) 10^{-3}, 3) 10^{-2}, 4) 10^{-1}, 5) 1, 6) 10.

This type of fast electrode reaction maintains nernstian conditions and is therefore called an electrochemically **reversible reaction** or a **nernstian reaction**. The nernstian conditions are retained independent of how fast the potential at the electrode surface is changed. Note, that for all real electrode reactions, there is a time domain (at short times) for which nernstian conditions are not maintained and the electrode reaction appears to become irreversible. The terminology of **electrochemically irreversible reactions** refers to processes with small values of $k°$. Alternatively, the concentrations of O or R can also depart from the prediction of Eq. (3.30) due to a coupled homogeneous reaction consuming O or R. Such reactions are called **chemically irreversible reaction**. The occurrence of coupled homogeneous following reactions is very common in organic electrochemistry, and we will discuss experimental methods for a detailed mechanistic analysis and the determination of the involved rate constants in Section 6.2.

In case of an electrochemically irreversible reaction (small $k°$), exchange current densities are small, too. A high overpotential η is required to achieve significant net currents. A high overpotential is associated with a substantial change in concentrations at the surface, which cannot be achieved quickly. The conditions at the electrode surface deviate from the nernstian case.

In the context of electrochemistry, "electrochemically reversible" or "electrochemically irreversible" refers to the kinetics of the heterogeneous reaction. In contrast, "chemically reversible" and "chemically irreversible" refer to the presence of homogeneous chemical reaction steps coupled to the charge transfer.

For a specific electrode reaction, the exchange current densities (and therefore the heterogeneous standard rate constants) can vary substantially on different electrode materials. For the intensely studied hydrogen evolution reaction (HER), different types of electrode materials can be arranged to give volcano plots. The logarithm of j_0 (or $k°$) is plotted versus the metal–hydrogen bond energy for each metal (M) (Figure 3.9). The shape of the resulting plot is reminiscent to the outline of a volcano, thus the name. The highest exchange current densities can be observed for M—H bond energies of approximately 240 kJ mol^{-1}. The metal with the highest exchange current density is platinum (Pt). The range of observed exchange current densities is enormous, with ten orders of magnitude. When comparing the exchange current density for Pt (10^{-3} A cm^{-2}) with a supersonic plane (2300 km h^{-1}), then the exchange current density for mercury Hg (10^{-12} A cm^{-2}) corresponds to only 8.2 mm h^{-1}, much

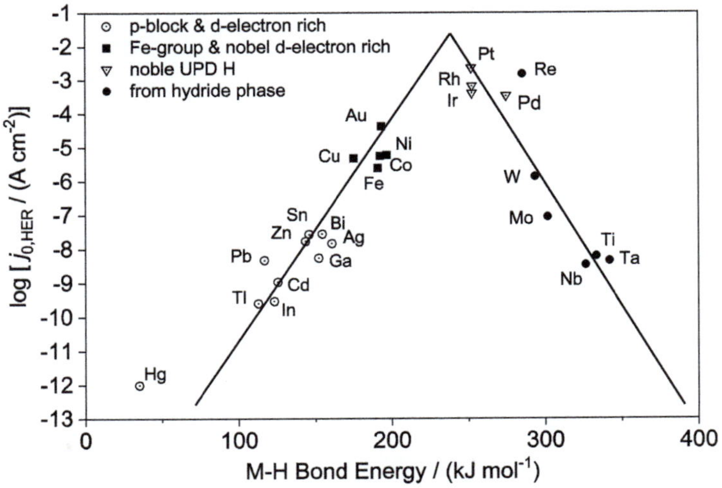

Figure 3.9 Volcano plot for the hydrogen evolution reaction. Source: Adapted from [12–17].

slower than a garden snail. Qualitatively, the volcano plot can be readily explained. The hydrogen evolution reaction is catalytic and involves surface intermediates. The key reaction intermediate is a hydrogen atom bound to the metal surface, M–H. With a weak M–H bond, there is not enough M–H formed to allow a fast reaction. For a strong M–H bond, the reaction intermediate is too stable to form products. The catalyst surface saturates, and the reaction is slow. This trade-off based on adsorption energies for intermediates is the basis for the Sabatier principle[20] in heterogeneous catalysis. This principle states that the bond energy between the catalyst and the key intermediate has to be just right to give optimum reaction rates.

The volcano plot (Figure 3.9) is a key illustration referred to repeatedly in this textbook. The volcano plot can be used to explain why some nonnoble metals can be electroplated without significant hydrogen evolution or why noble metals are so important in fuel cell catalysis. The use of volcano plots can also lead to systematic exploration, for example, when using alloys in electrocatalysis.

Although offering a powerful tool, volcano plots need to be used with care and a couple of warnings may be appropriate [8].

1) A critical attempt to verify the effect of the adsorption energy for hydrogen on the hydrogen evolution kinetics would require that we vary the adsorption energy *without changing any other parameters*. For metals, a change in the adsorption enthalpy $\Delta_{ads}H$ is also linked to a change in work function ϕ, in the sublimation energy of the metal, the potential of zero charge E_{pzc}, in the free electron density, in the crystal structure, in the M–M bond distances etc. Some of these parameters follow very similar trends, for example, ϕ, $\Delta_{ads}H$, and E_{pzc} are linearly correlated. Similar plots could be constructed without a necessarily causal link.
2) On the right-hand side of the plot in Figure 3.9 are metals, which are known to form very stable oxide layers. Measurements of the hydrogen evolution kinetics are therefore the kinetics of an oxide-covered surface for which the M–H bond energy is not important.

The numerical correlation needs to be interpreted with care as causal links from the data to the underlying physicochemical processes may be complex. Scaling laws couple kinetic parameters to a range of thermodynamic parameters even if there is no direct causal link to the observed reaction.

3.2.4 Interpretation of the Transfer Coefficient α

Figure 3.6 shows that the transfer coefficient becomes $\alpha = 0.5$ for the case where the intersecting lines for Gibbs energies for starting materials and for products exhibit similar slopes (symmetry). Indeed, α is commonly observed to be close to 0.5 or in the range $0.3 < \alpha < 0.7$. Important exceptions are observed for semiconductor electrodes where α is either close to 1 or close to 0 (i.e., either the oxidation or the reduction is strongly favored, see Chapter 10). The transfer coefficient α is lower than 0.5 when the curve $G^{\neq} = f$(reaction coordinate) for the oxidation $R \rightarrow O + e^-$ is steeper compared to the curve for the reduction. This leads to the nuclear coordinates of the atoms in the transition state being closer to those of the reduced state and less similar to those of the oxidized state. The electrode potential more strongly affects the activation barrier for oxidation.

For $\alpha > 0.5$, the reduction $O + e^- \rightarrow R$ is associated with the steeper curve for $G^{\neq} = f$(reaction coordinate). The electrode potential has a stronger effect on the activation barrier for the reduction. The structure of the transition state is more similar to the oxidized form.

The consequences of changing the transfer coefficient α on the current density–potential curves are shown in Figure 3.10. With $\alpha = 0.5$, the curve is symmetric, and anodic or cathodic currents increase in a similar fashion with potential difference

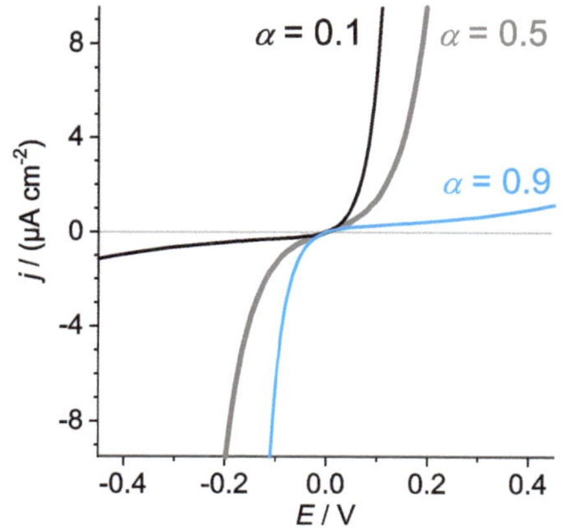

Figure 3.10 Effect of the transfer coefficient α on the current density–electrode potential curve. α is 0.1 (—— black), 0.5 (solid, blue) and 0.9 (—— blue). Other parameters $E^{\circ\prime} = 0$ V, $T = 298$ K, $j_0 = 0.2$ µA cm^{-2}.

[20] Paul Sabatier, 1854–1941, French chemist, Professor for Chemistry in Toulouse, Nobel Prize for Chemistry 1912 together with Victor Grignard for contributions to the hydrogenation of organic compounds in the presence of metal catalysts.

positive or negative, respectively. With $\alpha = 0.1$, the oxidation produces high currents, but the currents for reduction remain small. With $\alpha = 0.9$, the opposite trend is observed. The effect of α on the symmetry of the intersecting Gibbs free energy curves in the vicinity of the transition state is the reason for the expression "symmetry coefficient" employed synonymously with "transfer coefficient". In the following discussion, we assume $\alpha = 0.5$ if not stated otherwise.

3.2.5 Limiting Cases Derived from the Butler–Volmer Equation

The Butler–Volmer Equation with Mass Transport Limitation. Based on Figure 3.8, one could assume that the rate of electrochemical processes can be increased simply by increasing the overpotential. However, Figure 3.1 shows that this is not the case. In a sequence of elementary processes there is always one rate-limiting step. This may be the charge transfer step or, for example, the mass transport process. Commonly, mass transport-limited processes are observed for sufficiently high overpotentials. We can derive an expression for the transition from purely charge transfer kinetic controlled to purely mass transport-controlled current densities by realizing that the sum of the inverse of the limiting current densities equals the inverse of the resulting total current density (the lower current density is always limiting; for a more comprehensive explanation, see the discussion below Eqs. (12.78)–(12.80). By taking into account both the mass transport-limiting current densities for anodic and cathodic processes, $j_{1,a}$ and $j_{1,c}$ (index 1 for limiting), we can obtain the current density–overpotential curve in Eq. (3.40).

$j_{1,a}$, $j_{1,c}$ – Anodic and cathodic limiting current densities [A cm^{-2}]

$$j = \frac{-\exp\left(\frac{-\alpha\eta}{E_{NF}}\right) + \exp\left(\frac{(1-\alpha)\eta}{E_{NF}}\right)}{\frac{1}{j_0} - \frac{1}{j_{1,c}}\exp\left(\frac{-\alpha\eta}{E_{NF}}\right) + \frac{1}{j_{1,a}}\exp\left(\frac{(1-\alpha)\eta}{E_{NF}}\right)} \tag{3.40}$$

Plots of Eq. (3.40) are shown in Figure 3.11. In this example, the current for high positive overpotentials $|\eta| > 0.3$ V converges to a constant limit. This is the anodic mass transport limit. In the potential range $0.1\text{ V} < |\eta| < 0.3$ V, the net current is dominated by the anodic branch of the Butler–Volmer expression. Below 0.1 V both anodic and cathodic components of the current are important until −0.1 at which point the net current becomes dominated by the cathodic branch of the Butler–Volmer expression. An Excel spreadsheet in the download materials allows the effect of different parameters in Eq. (3.40) to be explored. The plateau current for $|\eta| > 0.3$ V is called a diffusion-limited current (see Section 4.4.2). The blue curve in Figure 3.11 is based on concentrations $c_R^* = c_O^*$ and the assumption of equal diffusion coefficients leading to equal anodic and cathodic diffusion-limited current densities ($j_{l,a} = -j_{l,c}$). For the case of $c_O^* = 0.1\, c_R^*$, the cathodic diffusion-limited current density $j_{l,c}$ is only 10% of the original value (black curve). Now the equilibrium potential E_{eq} (i.e., the potential with $j = |j_a| - |j_c| = 0$) is also shifted by $2.3 \cdot E_{NF} = 59$ mV in the negative direction (black curve in Figure 3.11).

Q: How can Eq. (3.40) be derived to explain the curves shown in Figure 3.11?

A: The derivation of Eq. (3.40) requires material covered in Chapter 4 and is therefore only indicated at this point [1]. For the case of mass transport limitations, an approximation can be employed based on the current being linearly dependent on the mass transport coefficient m_t [cm s^{-1}] and on the concentration difference between the electrode surface and the bulk.

$$j = nFm_t\left[c_O^* - c_O(0,t)\right]$$

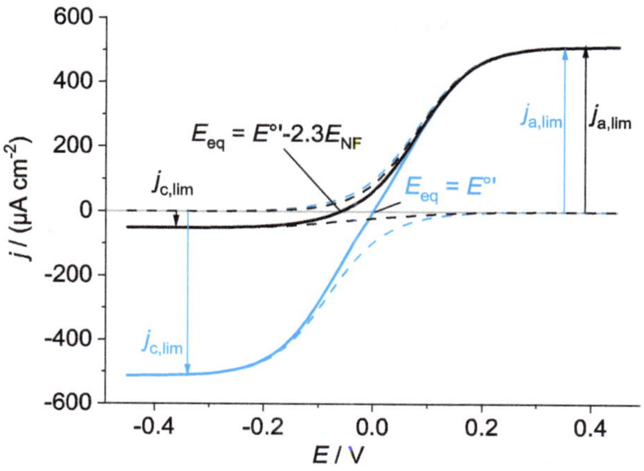

Figure 3.11 Current density–potential curve taking into account mass transport limitations; parameters $E^{\circ\prime} = 0$ V, $m = 2 \times 10^{-4}$ cm s^{-1}, $k^\circ = 0.001$ cm s^{-1}, $\alpha = 0.5$, $T = 298$ K, with different concentrations: 1) $c_O^* = c_R^* = 1$ mM (blue, solid); 2) $c_R^* = 1$ mM, $c_O^* = 0.1$ mM (black). Dashed lines indicate the anodic and cathodic current branches.

The current density reaches a limit for $c_O(0, t)$ 0. The diffusion-limited current density for the cathodic process is therefore

$$j_{1,c} = nFm_t\left(c_O^* - 0\right) = nFm_t c_O^*$$

The ratio of current density to limiting current density is

$$\frac{j}{j_{1,c}} = \frac{c_O^* - c_O(0,t)}{c_O^*} = 1 - \frac{c_O(0,t)}{c_O^*}$$

Therefore,

$$\frac{c_O(0,t)}{c_O^*} = 1 - \frac{j}{j_{1,c}} \text{ and correspondingly } \frac{c_R(0,t)}{c_R^*} = 1 - \frac{j}{j_{1,a}}.$$

Substitution of this result into Eq. (3.37) gives:

$$\frac{j}{j_0} = \left(1 - \frac{j}{j_{1,a}}\right)\exp\left[\frac{(1-\alpha)(E - E_{eq})}{E_{NF}}\right] - \left(1 - \frac{j}{j_{1,c}}\right)\exp\left[\frac{-\alpha(E - E_{eq})}{E_{NF}}\right]$$

$$\frac{j}{j_0} - \frac{j}{j_{1,c}}\exp\left[\frac{-\alpha(E - E_{eq})}{E_{NF}}\right] + \frac{j}{j_{1,a}}\exp\left[\frac{+(1-\alpha)(E - E_{eq})}{E_{NF}}\right]$$

$$= \exp\left[\frac{(1-\alpha)(E - E_{eq})}{E_{NF}}\right] - \exp\left[\frac{-\alpha(E - E_{eq})}{E_{NF}}\right]$$

$$j\left\{\frac{1}{j_0} - \frac{1}{j_{1,c}}\exp\left[\frac{-\alpha(E - E_{eq})}{E_{NF}}\right] + \frac{1}{j_{1,a}}\exp\left[\frac{+(1-\alpha)(E - E_{eq})}{E_{NF}}\right]\right\} =$$

$$\exp\left[\frac{(1-\alpha)(E - E_{eq})}{E_{NF}}\right] - \exp\left[\frac{-\alpha(E - E_{eq})}{E_{NF}}\right]$$

$$j = \frac{\exp\left[\frac{(1-\alpha)(E - E_{eq})}{E_{NF}}\right] - \exp\left[\frac{-\alpha(E - E_{eq})}{E_{NF}}\right]}{\left\{\frac{1}{j_0} - \frac{1}{j_{1,c}}\exp\left[\frac{-\alpha(E - E_{eq})}{E_{NF}}\right] + \frac{1}{j_{1,a}}\exp\left[\frac{+(1-\alpha)(E - E_{eq})}{E_{NF}}\right]\right\}}$$

Division by the expression in curly brackets yields Eq. (3.40).

The **limiting case of low overpotentials** is readily accessible experimentally in the range of low net current densities (e.g., limiting the current density to less than 10% of the lowest diffusion-limited current density $j_{1,a}$ or $j_{1,c}$. Stirring the solution or rotating the electrode can be used to increase the limits. Under these conditions, the currents are limited purely by charge transfer kinetics, and the ratio of surface to bulk concentration approaches unity, $c_R(0, t)/c_R^* \approx 1$ and $c_O(0, t)/c_O^* \approx 1$. Under these conditions, the Butler–Volmer equation can be simplified to Eq. (3.41).

$$j = j_0\left\{\exp\left[\frac{+(1-\alpha)\eta}{E_{NF}}\right] - \exp\left[\frac{-\alpha\eta}{E_{NF}}\right]\right\} \tag{3.41}$$

j, j_0 –Current density, exchange current density [A cm^{-2}]
η – Overpotential [V]
E_{NF} – Nernst factor [V]
α – Transfer coefficient [dimensionless]

For very small overpotentials η (small with respect to E_{NF}), the exponential function can be expressed as a Taylor series[21] at the point $x = -\alpha\eta/E_{NF} = 0$. As a result, a linear relationship is found between the current density and the overpotential, that is, an *ohmic* dependency. The corresponding resistance is the charge transfer resistance R_{CT}.

$$j \approx j_0 \left\{ 1 + \frac{+(1-\alpha)\eta}{E_{NF}} - \left(1 + \frac{-\alpha\eta}{E_{NF}} \right) \right\} \approx j_0 \frac{\eta}{E_{NF}} \tag{3.42}$$

Q: What is the Taylor series?

A: The Taylor series offers a convenient method of approximating a function close to a known value $f(x_0)$, the source point. The function is replaced with a polynomial expressed as an infinite series to give the desired value in the limit $f(x)$ with $n \to \infty$. For values very close to the known value at x_0, the Taylor series can be terminated for $n = 1$ to give very convenient expressions. The Taylor series is developed at the point x_0 with $\Delta x = x - x_0$:

$$f(x) = f(x_0) + \sum_{n=1}^{\infty} \frac{1}{n!} f^{(n)}(x_0)(\Delta x)^n$$

Here, $f(x)$ is any function. For the specific case of the exponential function e^x at the point x_0 0, we can write

$$f(x) = f'(x) = f''(x) = f^{(n)}(x) = e^x$$

$$f'(0) = f'(0) = e^0 = 1$$

$$\Delta x = x - x_0 = x - 0 = x$$

$$e^x \approx 1 + \frac{1}{1!} \cdot 1 \cdot x = 1 + x$$

The variable x represents the term in the exponent in Eq. (3.41). The Taylor series needs to be applied twice to simplify Eq. (3.41) to Eq. (3.42).

Equation (3.42) shows that a linear current density to voltage relationship results for small enough overpotentials. The resistance is the charge transfer resistance R_{CT}. We can derive the value by substituting $\eta = E_{NF}$ and $j = j_0$ (or $I = I_0$):

$$R_{CT} = \frac{\eta}{I} = \frac{E_{NF}}{I_0} = \frac{RT}{nFI_0} \tag{3.43}$$

A recording of current versus overpotential at very small overpotentials will give a straight-line plot with slope $R_{CT}{}^{-1}$. Similarly, a small amplitude ($\Delta U \leq 10\,\text{mV}$) low frequency alternating current (AC) measurement will give R_{CT}. This then allows I_0 to be determined and with A, $c_O^* = c_R^*$ and Eq. (3.33), $k°$ can be calculated. A classic tool for this measurement is electrochemical impedance spectroscopy (Chapter 7).

In the **limit of high overpotentials** (without mass transport limitation), one of the two exponential terms in Eq. (3.41) can be considered insignificant (Figure 3.7). For the case of $E \ll E_{eq}$ only the cathodic part of the current contributes to the total current, and the anodic current becomes negligible. The equation simplifies to

$$j = -j_0 \exp\left[\frac{-\alpha\eta}{E_{NF}}\right] \tag{3.44}$$

By taking the logarithm of the absolute current density, this equation can be linearized to $\ln|j| = f(\eta)$ or $\eta = f(\ln|j|)$. This is the equation employed at the beginning

j, j_0 – Current density, exchange current density [A cm^{-2}]
η – Overpotential [V]
E_{NF} – Nernst factor [V]
α – Transfer coefficient [dimensionless]

[21] Brook Taylor, 1685–1731, English mathematician.

of this chapter consistent with the empirically derived Tafel plot [Eq. (3.2)]. For the case of a *simple charge transfer* and cathodic currents, we can express the link from the Butler–Volmer equation to the Tafel line.

$$\ln|j| = \ln j_0 + \frac{-\alpha}{E_{NF}}\eta = \ln j_0 + \frac{-\alpha F}{RT}\eta \tag{3.45}$$

$$|\eta| = \frac{E_{NF}}{+\alpha}\ln j_0 - \frac{E_{NF}}{+\alpha}\ln|j| = \frac{RT}{\alpha F}\ln j_0 - \frac{RT}{\alpha F}\ln|j| \tag{3.46}$$

For the case of the anodic current, analogous expressions are obtained by replacing $(-\alpha)$ in Eqs. (3.44)–(3.46) by $(1-\alpha)$.

It is instructive to define the conditions for either only the anodic or only the cathodic branch in the Butler–Volmer equation determines the overall current. If we assume a minimum factor of 100 between the two current components, we can use Eq. (3.47) to derive the needed overpotential of $\eta \leq -0.118$ V for reduction. Analogously, we obtain $\eta \geq +0.118$ V for an oxidation (or generally $|\eta| \geq 0.118$ V).

$$\frac{\exp\left(\frac{-\alpha\eta}{E_{NF}}\right)}{\exp\left(\frac{(1-\alpha)\eta}{E_{NF}}\right)} = \exp\left(\frac{-\eta}{E_{NF}}\right) \geq 100; \quad \eta \leq -\frac{RT}{F}\ln(100) \tag{3.47}$$

Figure 3.12 shows a set of Tafel plots derived from data in Figure 3.8. The linear region is indicated with blue lines, and clear deviations from linearity are seen for $|\eta| < 0.118$ V. For the case of high exchange current density (curve 6), the linear region of the Tafel slope will be difficult to measure experimentally due to mass transport limitation cutting off the current. Even at low overpotentials of $|\eta| = 0.118$ V significant mass transport limitations are expected as shown in Figure 3.11. As a conclusion, linear Tafel plots require (i) a sufficiently wide potential range for measurements, (ii) relatively slow charge transfer kinetics (low exchange current densities), and (iii) a high rate of mass transport during measurements (e.g., employing rotating disk electrode methods).

If we extrapolate from the linear region to the intersection at $\eta = 0$, Eq. (3.45) suggests that we can obtain $\ln j_0$ and thus the exchange current density j_0. The slope of lines is $-\alpha F/(RT)$ for the cathodic branch and $+(1-\alpha)F/(RT)$ for the anodic branch. When using the decadic logarithm (log) with base 10 as shown in Figure 3.12, we can obtain j_0 conveniently. For this type of plot the slope for the cathodic branch is $-\alpha F/(2.3\ RT)$ and the slope for the anodic branch is $(1-\alpha)F/(2.3\ RT)$.

Typical areas where Tafel plots are crucially important are in corrosion science and in studies of electrocatalysis using rotating disk electrodes (see Chapters 12, 14).

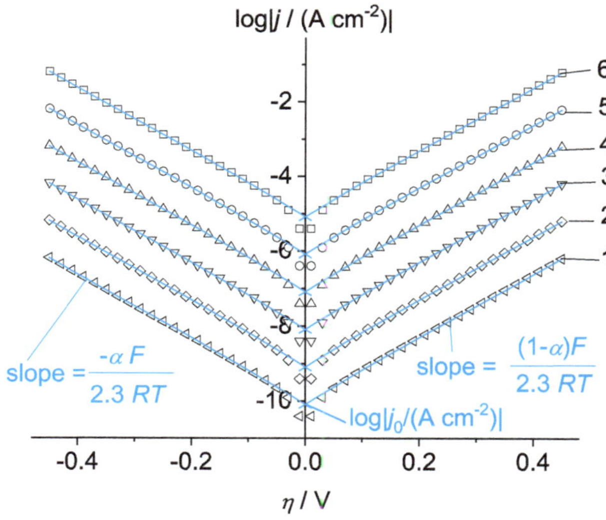

Figure 3.12 Tafel plot for data in Figure 3.8. Data points shown as symbols with a linear trend line (in blue). Trend lines intersect at zero overpotential to give log j_0. The slope gives the transfer coefficient α. Note that current densities are converted from μA cm^{-2} in Figure 3.8 into A cm^{-2}.

3.2.6 Can We Employ the Butler–Volmer Equation for Multistep Electrode Reactions?

In Section 3.1.1, we have indicated that electrode reactions can be dissected into elementary processes. Even for concerted processes, there is always a single electron transferred at a time. Each elementary electron transfer process can be assigned a potential-dependent rate law based on the formalism developed in Section 3.2.2. Each sequential charge transfer process (index *i*) is associated with parameters k_i°, α_i, and $E_i^{\circ\prime}$. There are intermediates S_i at the electrode to consider with a surface concentration of $c_{S_i}(x = 0, t)$. Furthermore, there is complexity introduced by additional homogeneous chemical reaction steps and the resulting intermediates in the vicinity of the electrode surface. The formation of intermediates may not be potential dependent but will affect the evolution of reactant and product concentrations at the electrode surface. This overwhelming complexity of processes and parameters makes a full discussion of multistep charge transfer reactions in the context of the Butler–Volmer equation unwarranted and ultimately futile. There is no algorithm or approach to assign and reliably quantify the mechanism for an arbitrary type of electrode reaction. On the other hand, the mechanism is often the key to improved performance and new technologies. This lack of general theory leads to complexity in the analysis of electrochemical data (or generally in heterogeneous kinetics), but also to an element of adventure and satisfaction when solving kinetic puzzles.

What is the best way of approaching complex kinetic problems? There are cases or reaction mechanisms that allow simplifying assumptions so that some of the complexity can be removed. For example, for electrode reactions with multiple electron transfer steps, one of these steps can be assumed to be rate determining (i.e., significantly slower compared to the other electron transfer steps; RDS = rate-determining step). With all other steps sufficiently fast to remain in equilibrium at all times, we can apply the Nernst equation to express concentrations as a function of electrode potential (i.e., fast electron transfer reaction steps are *reversible*, see Section 3.2.3). We can employ expressions for the ratio of intermediate concentrations analogously to Eq. (3.39). For chemical reaction steps reversible conditions can be assumed, and equilibrium expressions for acid–base equilibria, complexation equilibria, and adsorption–desorption equilibria employed to express concentrations at the electrode surface. Furthermore, the validity of the assumption of equilibrium conditions needs to be considered in the context of the time domain of the experiment. For faster experiments these conditions are more likely to become invalid.

Q: Could we determine the concentration of intermediates experimentally?

A: Techniques for the determination of some reaction intermediates exist, but there is no simple or generally applicable method. Some techniques can interfere with the reaction and lead therefore to more complexity. Techniques for *in situ* monitoring of reaction intermediates will be discussed in Chapters 6 and 8.

Based on data in a Tafel plot, there is no reliable method to unequivocally conclude the reaction mechanism or to obtain constants for the elementary reactions.

The mathematical analysis based on one rate-determining step and otherwise reversible conditions allows complex reaction types to be dissected into individual cases with distinct Butler–Volmer expressions. Within a given range of applied potentials, the theoretically predicted Tafel slope may fit the experimental data, which may indicate that the assumed mechanism could be valid. Note that Tafel slope information can be complex and associated with electrode processes more complex than the classic case in Figure 3.12. Interpretation of data requires considerable care. The derivation of cases of Butler–Volmer equations can be found in classic literature and textbooks [1–7]. Specific cases will be treated in later chapters. This discussion will be linked to specific experimental tools and methods, and it will require the consideration of mass transport effects treated in Chapter 4. We need to consider experimental methods based on their time domain. This will provide us with a powerful tool to dissect complex electrode processes.

The field of electrochemistry dedicated to mechanisms and the rate of electrochemical reactions is **electrochemical reaction kinetics**.

Q: Why not use the term "electrokinetics"?

A: Electrokinetics is an area of electrochemistry dedicated to transport processes in electrolyte and membrane materials with applied electric fields (see Section 4.1).

3.2.7 Multiple Redox Systems Present at the Electrode Surface Lead to Mixed Potentials

In Section 2.3.5, a silver electrode immersed in an oxygen-containing electrolyte solution was used as an example of a multiple redox system developing a mixed potential. Mixed potentials are observed in cases where there is no equilibrium established and where two processes (one oxidation and one reduction) occur simultaneously in a steady state. Here, the case of simultaneous interfacial redox reactions is discussed based on (i) silver oxidation/reduction and (ii) the four-electron reduction of O_2 to H_2O. The two-electron reduction of O_2 to H_2O_2 may also be important in this context, but effects are assumed to be negligible.

$$Ag \rightleftharpoons Ag^+ + e^-, E^\circ = 0.800\ V \quad (3.48)$$

$$O_2 + 4\ e^- + 4\ H^+ \rightarrow 2\ H_2O, E^\circ = 1.229\ V \quad (3.49)$$

Figure 3.13 illustrates the current–electrode potential relationships for individual current components and the net current.

The equilibrium potential of the silver electrode E_{eq} is obtained by adding the current components for silver oxidation (upper dashed black line) and for silver reduction (lower dashed black line in Figure 3.13). In the absence of oxygen, this potential is consistent with equilibrium (solid black line, intersection at $I = 0$). However, in the presence of oxygen, the zero-current crossing of the net current (blue solid line) is consistent with the open-circuit potential E_{OCP} (i.e., the potential with $I = 0$), which is *not* an equilibrium potential. At E_{OCP}, a chemical reaction based on concomitant oxygen

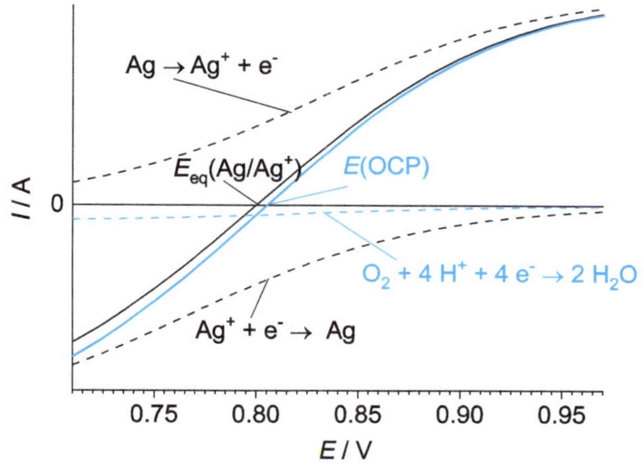

Figure 3.13 Illustration of the development of a mixed potential for a silver electrode immersed in oxygen-containing electrolyte. Current components for the oxidation of silver (dashed, black), the reduction of Ag^+ (dashed, black), and the reduction of O_2 (dashed, blue) are shown in conjunction with net currents in the absence of oxygen (solid, black) and in the presence of oxygen (solid, blue).

reduction and silver oxidation continues to occur despite a zero current in the external circuit. This is the case of a mixed potential.

> When more than one redox pair contributes to the open-circuit potential E_{OCP}, then a **mixed potential** develops, at which a nonequilibrium process (continuing chemical reactions) occurs at the electrode despite a zero net current.

In the example in Figure 3.13, the value for E_{OCP} deviates only slightly from that for E_{eq}. This is linked to the exchange current density for Ag/Ag^+ being much higher compared to that for oxygen reduction. In this case, the Ag/Ag^+ redox couple is the **potential-determining redox system**. Similar situations with mixed potentials are encountered for ion-selective electrodes (Section 5.2). Large deviations between E_{OCP} and E_{eq} may occur in corrosion processes (Chapter 14).

3.2.8 What Are the Limitations of the Butler–Volmer Formalism?

When formally deriving the Butler–Volmer equation, important factors such as the molecular structure of reactants or the composition and structure of the electrode surface have been ignored. Therefore, the Butler–Volmer equation cannot be employed to predict the standard rate constant for heterogeneous electron transfer $k°$ for unknown reactions. However, once $k°$ and α have been determined experimentally, then Eq. (3.27) allows current–voltage relationships to be calculated. The Butler–Volmer formalism allows an empirical description of interfacial redox reactions with the elementary charge transfer step being rate limiting.

Q: What is the relevance of the term "empirical" here? The equations look "theory-loaded" to me?

A: A theory is called "empirical" when all predictions made by this theory are based on measured parameters (e.g., j_0). For example, the prediction of j as a function of electrode potential requires knowledge of j_0 from experiment without providing insight into the origin of j_0 being large or small. Another example of an empirical relationship is the Arrhenius equation.

When discussing the hydrogen evolution electrode reaction kinetics (which was fundamentally developed in work by Erdey-Grúz and Volmer 1930 [18]) at different types of electrode materials, it is apparent that the exchange current density j_0 (or the corresponding heterogeneous standard rate constant $k°$) varies by up to nine orders of magnitude (Figure 3.9). For this extreme range of rates, there is no explanation or interpretation based on the Butler–Volmer equation. Similarly, with known parameters j_0 or $k°$ for one redox system, there is no way to extrapolate to the kinetic parameters for another redox system. This limitation is not linked to complexity in equations (as in the case in Section 3.2.6), but is fundamentally linked to the lack of consideration of the important role of molecular structure and electrode surface structure in electron transfer.

A theory linking kinetic parameters to the molecular structure of molecules and surfaces is called a "microscopic" theory. Based on molecular structure information, this kind of theory can predict and explain the magnitude of kinetic parameters. Computational methods in theoretical chemistry are in rapid development and ambitious aims like the prediction of the rate of surface catalytic charge transfer reactions are becoming more realistic given recent progress [19, 20]. Two important approaches to microscopic theories are based on Marcus theory[22] and the formalism developed by Gerischer.[23] These will be introduced here at a qualitative level because they offer orientation regarding the extreme diversity of kinetic phenomena in electrode reactions.

In order to explore **Marcus theory** we can start with a transition metal complex, for example, $[Fe(H_2O)_6]^{3+}$ (Figure 3.2) with coordinative bonds to ligands (here water). In the ground vibrational state these ligands undergo a vibrational oscillation of the Fe–O bond length (at the vibrational zero-point energy). The harmonic oscillator model suggests an energy state with parabolic dependency on Fe–O bond distance. These oscillations are accessible in vibrational spectroscopy. A change in the redox state of the central cation ($Fe^{3+} + e^- \rightarrow Fe^{2+}$) affects the equilibrium bond distance. The transfer of an electron produces Fe^{2+} and weakens the Fe–O bond. In both $[Fe(H_2O)_6]^{3+}$ and $[Fe(H_2O)_6]^{2+}$, vibrational energy is associated with a parabolic function with the energy minimum for these two complexes at different Fe–O bond lengths. The two parabolas intersect at a point where the Fe–O bond length is longer than that for Fe(III) and shorter than that for Fe(II). This intersection point can be identified (approximately) as the transition state with both redox states for $[Fe(H_2O)_6]^{3+}$ and for $[Fe(H_2O)_6]^{2+}$ having the same energy. At this point in the reaction coordinate, the electron is equally likely to stay in the electrode or to tunnel out (without energy transfer) into the metal complex. The electron transfer is orders of magnitude faster when compared to any vibrational bond distance changes (see Franck–Condon principle[24,25], in textbooks for spectroscopy), which decouples electron and nuclear motion. The distance in terms of energy from the parabola minima for $[Fe(H_2O)_6]^{3+}$ and $[Fe(H_2O)_6]^{2+}$ to the parabola intersection point approximately represents the activation energy for the charge transfer reaction. Metal complexes with a small change in energy from the parabola minima to the intersection point (this is called the reorganization energy) are kinetically fast (the activation energy is low). Large transition metal complexes or metal centers in redox enzymes often exhibit very fast electron transfer (small changes in bond distances lead to a low reorganization energy). However, technically important processes involving small molecules (e.g., H_2, O_2) are often kinetically slow (large changes in bond distances lead to a high reorganization energy). Larger molecular systems have a higher degree of freedom in terms of charge compensation mechanisms (a broader saddle point) to lower the activation energy. The intersection point of the two parabolas (in linear approximation) allows the transfer coefficient α to be estimated and explained based on molecular properties such as vibrational energies.

The **Gerischer formalism or Gerischer–Marcus theory** focuses on the effects introduced by the electrode material and is particularly powerful for the description of electron transfer processes at semiconductor electrodes (Chapter 10). For the electrode material, the density of electronic states is represented in the energy density diagram

---

[22] Rudolph A. Marcus, born 1923, US-American physicochemist, Professor of Chemistry in Brooklyn, Urbana and from 1978 at California Institute of Technology in Pasadena, Nobel Prize in Chemistry 1992.

[23] Heinz Gerischer, 1919–1994, German chemist, worked in Leipzig, Berlin, Stuttgart, Munich, 1969–1987 Director of the Fritz-Haber Institute of the Max-Planck Society in Berlin.

[24] James Franck, 1882–1964, German chemist and physicist, oversaw the use of chemical warfare agents in World War I, heavily injured by chemical warfare agents, 1920–1933 professor of physics in Göttingen, 1933 emigration into the United States, professor of physics and physical chemistry in Baltimore, in Copenhagen, and in Chicago, Nobel Prize 1925 for the Franck–Hertz experiment.

[25] Edward Uhler Condon, 1902–1974, US American physicist at the Westinghouse Electric Company, Corning Glass Works, various national research institutions in connection with the Manhattan Project, and finally as a professor at Washington University in St. Louis and the University of Colorado at Boulder. The Slater–Condon rules of quantum mechanics and a lunar crater are also named after him.

(the energy of the electron on the y-axis and the number of states or density of states on the x-axis; see density of states in Excursion 10-1). In this diagram, the variation of energy (due to molecular vibration) for the oxidized and the reduced forms of the solution redox system are superimposed [see Figure 10.19]). The electron transfer requires oxidized and reduced forms of the solution species to be on the same energy level. The probability of an electron transfer scales with the overlap of high density of states in the solid and high probability of the momentary energy of the dissolved reactant in solution. This formalism combines microscopic information from reduced and oxidized species in solution with the energy state density of the electrode material.

3.3 Summary

Electrochemical reactions are always multistep or complex reactions based on at least the elementary charge transfer step and mass transport processes, but are often coupled to further types of elementary reaction steps. These reaction steps are usually considered to be sequential but can also occur in parallel (i.e., represented by a network).

In cases when the heterogeneous charge transfer is the rate-limiting reaction step, the reaction rate for the anodic and the cathodic components of the reaction (i.e., the current components) is exponentially dependent on overpotential. The overpotential η is the difference between the electrode potential E and the (thermodynamically determined) equilibrium potential of the electrode E_{eq}.

For the kinetics of heterogeneous charge transfer, a key parameter is the exchange current density j_0. The exchange current density is proportional to the heterogeneous standard reaction rate constant $k°$. Both, j_0 and $k°$ are interchangeable in the Butler–Volmer formalism. Both parameters are temperature-dependent and describe the ability of an electrode to transfer charge without activation (at equilibrium). Both, j_0 and $k°$ can vary over many orders of magnitude. For low overpotentials η, both anodic and cathodic current components contribute to the net current. In this potential region, a linear relationship is observed for net current and η. The inverse of the current–overpotential slope gives the charge transfer resistance R_{CT}. An increase in the overpotential η into the nonlinear region is linked to only one current component, anodic or cathodic, dominating. Even for very low j_0 (or $k°$), reaction rates can be very high, when the overpotential η is increased to lower the activation barrier. The high-η potential region is consistent with the Tafel region with a linear dependence of the logarithm of net current on electrode potential (or better η).

A second important parameter in heterogeneous charge transfer kinetics is the transfer coefficient α. This parameter describes how close (or far) the transition state for the charge transfer is to the structure of the reactant or product on the reaction coordinate. For many reactions α is approximately 0.5 (i.e., the transition state is halfway between the minima for oxidized and reduced equilibrium states on the reaction coordinate). Deviations from $\alpha \approx 0.5$ are observed for electron transfer-induced bond breaking processes and for processes at semiconductor electrodes (often $\alpha \approx 1$ or $\alpha \approx 0$).

When analyzing kinetic data for a specific electrode reaction, it is important to consider mass transport effects depending on how the experiment is designed. For large overpotentials, mass transport can be rate limiting, and experiments can be designed to avoid mass transport limitations. In addition, the charge transfer reaction is likely to be based on a sequence of elementary reaction steps (with or without additional charge transfer steps). Each case of a rate-limiting charge transfer step can be expressed in the form of a Butler–Volmer equation. Practically, this approach soon becomes difficult and complex with many unknown kinetic parameters. In some cases, limiting cases for the Butler–Volmer equation can be expressed assuming one clearly defined rate-limiting step and assuming equilibrium conditions for all other elementary steps (chemical and electrochemical). These Butler–Volmer equations for ideal reactions can then be compared to experimental data. In this way, a mechanistic hypothesis can be confirmed but not proven.

All kinetic discussions based on the Butler–Volmer equation provide empirical relationships based on experimentally determined (empirical) parameters j_0, $k°$, and α. Based on these parameters, currents can be predicted for different electrode potentials, concentrations, and temperatures. The prediction of values for j_0, $k°$, and α requires more advanced theories that take into account molecular and electrode structure.

The Butler–Volmer equation for electrochemical reaction kinetics is as important and significant as the Arrhenius equation for chemical reaction kinetics.

Test your knowledge

Exercise 3.1: Sketch the elementary reaction steps for electrode reactions (a) with ion transfer and (b) with electron transfer.

Exercise 3.2: (a) Calculate the net current density for the standard hydrogen electrode ($j_0 = 1$ mA cm^{-2}, $\alpha = 0.5$) with an applied overpotential of 30 mV. [Note that this is a process with consecutive electron transfers with the first electron transfer being rate limiting]

(b) For the same type of electrode process, the solution pH is adjusted to 2 without a change in hydrogen fugacity. Calculate the exchange current density for this concentration ratio. Calculate the net current density for a 30 mV overpotential.

Exercise 3.3: The diagram below shows the current–voltage plot for Butler–Volmer kinetics without mass transport limitation and for $\alpha = 0.5$, $n = 1$, $T = 298$ K, $j° = 1$ μA cm^{-2} and $c^*_O = c^*_R = 0.001$ mol L^{-1}.

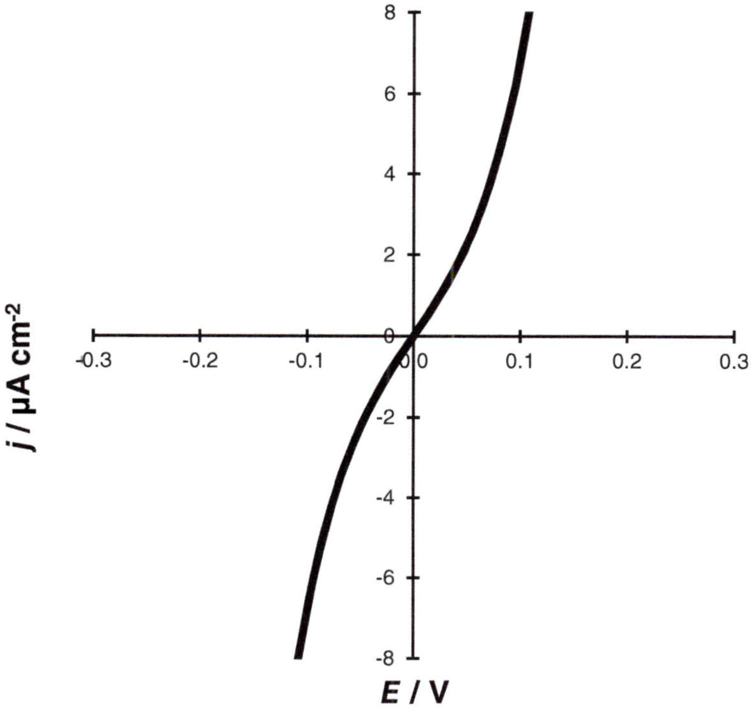

Sketch changes to this current–voltage curve for

a) $\alpha = 0.2$?
b) $j^\circ = 0.1\ \mu A\ cm^{-2}$?
c) $n = 2$?
d) $c_O^* = 0.1\ mol\ L^{-1}$?
e) The case where limitation from mass transport occurs?

The sketches should be by hand and qualitative to indicate only significant trends.

Exercise 3.4: Now attempt Exercise 3.3 with a computer spreadsheet analysis. The parameters α, j°, n, and c_O^* should be freely variable with superimposed plots.

Exercise 3.5: Calculate the heterogeneous standard rate constant k_0 for the redox reaction $[Fe(CN)_6]^{3-} + e^- \rightleftharpoons [Fe(CN)_6]^{4-}$ assuming $c_O(x = 0) = c_R(x = 0) = 2\ mmol\ L^{-1}$ (exchange current density $j_0 = 2.0\ mA\ cm^{-2}$, transfer coefficient $\alpha = 0.5$).

Exercise 3.6: Calculate the net current density j for an exchange current density j_0 of a) $10^{-6}\ A\ cm^{-2}$ or b) $10^{-3}\ A\ cm^{-2}$ and an overpotential of

1) 0.01 V
2) 0.1 V
3) 0.3 V?

In this calculation, use $\alpha = 0.5$, $n = 1$, $c_O^* = c_R^* = 1\ mol\ L^{-1}$, and $T = 298\ K$ without any mass transport limitations.

Exercise 3.7: A Tafel plot $\log(|I|/[A\ cm^{-2}]) = f(E)$ shows a slope of $8.5\ V^{-1}$. Calculate the transfer coefficient α for $n = 1$.

Exercise 3.8: **(difficult, requires material from Chapter 4)**: A platinum electrode with $A = 2\ cm^2$ and a counter electrode are immersed in a solution containing $1\ mmol\ L^{-1}$ $[Fe(H_2O)_6]^{2+}$ and $1\ mmol\ L^{-1}$ $[Fe(H_2O)_6]^{3+}$ and $1\ mol\ L^{-1}$ KNO_3 as electrolyte acidified to pH 2. An AC signal with an effective voltage of 10 mV and a frequency 80 Hz is applied. The effective AC current is determined as 40.94 μA. In contrast, in pure $1\ mol\ L^{-1}$ KNO_3 solution without $Fe^{3+/2+}$ ions, the AC current is 2.00 μA. Calculate

a) the charge transfer resistance
b) the exchange current
c) the exchange current density
d) the heterogeneous standard reaction rate constant
e) the peak values for AC voltage and AC currents.

Estimate the concentration of $[Fe(H_2O)_6]^{2+}$ and $[Fe(H_2O)_6]^{3+}$ at a time 30 s after switching off the AC voltage, given that the experiment occurred over a period of 10 s.

Note: In $1\ mol\ L^{-1}$ KNO_3 electrolyte solution, the AC voltage causes charging/discharging of the electrical double layer without any net reaction in solution. This charging/discharging process can be assumed to be the same in the presence of low concentrations of $Fe^{3+/2+}$ ions.

Exercise 3.9: Data were obtained for a Tafel plot based on

$$j_a = j^\circ \exp\left[\frac{(1-\alpha)nF}{RT}(E - E^\circ)\right]$$

From the linear region of the Tafel plot, the following data points were selected:

$E - E^\circ$ / mV	j_a / (μA cm^{-2})
100	1.883
200	7.626

Calculate the exchange current density j° and the transfer coefficient α for $n = 1$. Assume that the current densities were measured (a) at 298 K and (b) at 315 K.

Exercise 3.10: An electrolysis experiment is performed with 0.1 mol L^{-1} $CdSO_4$ solution (pH 7) with a current density of 1 mA cm^{-2} with a cadmium cathode and a platinum anode at 60 °C. The two electrodes have areas of 1 m^2 with an interelectrode gap of 25 cm. Calculate the Cd^{2+} concentration at the time when hydrogen evolution commences at the cathode. The overpotential for hydrogen evolution on cadmium is −0.6 V. Assume that all activity coefficients are 1.0.

References

1 A. J. Bard, J. R. Faulkner, H. S. White; *Electrochemical Methods – Fundamentals and Application.* 3rd ed., John Wiley & Sons, New York, 2022, pp. 121–181.

2 J. M. Bockris, A. K. N. Reddy, M. Gambola-Aldeco; *Modern Electrochemistry, Vol. 2A: Fundamentals of Electrodics.* 2nd ed. Plenum Publishers, New York, 2000, Ch. 7.

3 W. J. Albery; *Electrode Kinetics*, Clarendon, Oxford, 1975.

4 K. J. Vetter; *Electrochemical Kinetics*, Academic, New York, 1967, Ch. 2–3.

5 I. Rubinstein, *Physical Electrochemistry*, Marcel Dekker, New York, 1995.

6 E. Gileadi; *Physical Electrochemistry: Fundamentals, Techniques and Applications.* 2nd ed., Wiley-VCH, Weinheim, 2019, Ch. 4–7.

7 T. Eredy-Grúz; *Kinetics of Electrode Processes*, Wiley-Interscience, New York, 1972, Ch. 4.

8 P. Quainc, F. Juarez, E. Santos, W. Schmickler; Volcano plots in hydrogen electrocatalysis – Uses and abuses, *Beilstein J. Nanotechnol.* **2014**, *5*, 846–854.

9 A. J. Bard, G. Inzelt, F. Scholz (Eds.); *Electrochemical Dictionary*, 2nd ed., Springer Verlag, Berlin, 2012, p. 463, entry "hydrated electron".

10 R. Stockman via YouTube: https://www.youtube.com/watch?v=PTT6qPKIijc (posted on 2.5.2014, last access 30.10.2016).

11 A. J. Bard, G. Inzelt, F. Scholz (Eds.); *Electrochemical Dictionary*, 2nd ed., Springer Verlag, Berlin, 2012, pp. 89–91, 352, 954, entries "Butler", "Butler–Volmer equation", "Eredey–Grúz", "Volmer".

12 S. Trasatti; Work function, electronegativity, and electrochemical behaviour of metals: III. Electrolytic hydrogen evolution in acid solutions, *J. Electroanal. Chem.* **1972**, *39*, 163–184.

13 B. E. Conway, G. Jerkiewicz; Relation of energies and coverages of underpotential and overpotential deposited H at Pt and other metals to the 'volcano curve' for cathodic H_2 evolution kinetics, *Electrochim. Acta* **2000**, *45*, 4075–4083.

14 C. H. Hamann, A. Hamnett, W. Vielstich; *Electrochemistry.* 2nd ed., Wiley-VCH, Weinheim, 2007, p. 222.

15 H. Geiger, K. Scheel; *Handbuch der Physik*, *Vol. 21*, Verlag Von Julius Springer, Berlin, 1929.

16 J. O. M. Bockris, A. K. N. Reddy, M. Gambola-Aldeco; *Modern Electrochemistry, Vol. 2A: Fundamentals of Electrodics*. 2nd ed. Plenum Publishers, New York, 2000, p. 1285.

17 K.-H. Trostmann; *Korrosion - Ursachen und Vermeidung*, Wiley-VCH, 2001, p. 47.

18 T. Eredey-Grúz, M. Volmer; Zur Theorie der Wasserstoffüberspannung. *Z. Phys. Chem.* **1930**, *A150*, 203.

19 K. H. Bevan, M. S. Hossain, A. Iqbal, Z. Wang; Exploring bridges between quantum transport and electrochemistry. I, *J. Phys. Chem. C* **2016**, *120*, 179–187.

20 M. S. Hossain, K. H. Bevan; Exploring bridges between quantum transport and electrochemistry. II. A theoretical study of redox-active monolayers, *J. Phys. Chem. C* **2016**, *120*, 188–194.

4

Mass Transport and Boundary Layers

Very important in electrochemical reactions are the associated transport processes. Charge transfer reactions occur only in the vicinity of the interface. Therefore, the reactants must be transported toward the interface. Simultaneously, the charge must be transported in the electrolyte between the two electrodes by movement of ions.

This chapter deals with the most important mechanisms of mass transport. These are diffusion, migration, and convection. For diffusion and migration, we will derive quantitative relationships between current and transport of particles. This is important for identifying rate-limiting conditions in electrode reactions and for the realization of a number of electroanalytical applications. We will treat these types of transport mechanisms individually and explore their importance in classical electrochemical experiments. Experimental conditions can be chosen to approach the limits of either (i) eliminating mass transport effects or (ii) working under complete mass transport control with well-defined mathematical equations predicting the outcome and overall conversion. At the end of this chapter, we will introduce electrophoretic separation as an example for exploiting several concomitantly proceeding mechanisms of mass transport. As a conclusion of this chapter, we summarize and compare all transport processes within the framework of the Nernst–Planck transport equation.

The conversion of solution species at the electrode surface causes a deviation of the local concentration *close to* the electrode surface from that in the bulk solution. As a result of this concentration gradient, transport occurs from the bulk toward the surface. We will see that each transport mechanism can be associated with a boundary layer (diffusion layer, electrical double layer, and hydrodynamic boundary layer).

Further Reading. Comprehensive coverage of migration phenomena in electrolyte solutions has been given by Bockris [1]. Collections of transport data for a range of compounds and ions useful for laboratory experiments are available [2–4]. Here, we can only provide some illustrative examples of solving partial differential equations for mass transport. For further examples, the reader is directed to more specialized electrochemistry textbooks [1, 5] or monographs [6–8]. These also include modern numerical procedures for solving systems of equations computationally. The significance of such numerical procedures for the understanding of experimental methods in electroanalysis will be highlighted several times in Chapters 6 and 8. We will cover the charge transport in solid electrolytes in Chapter 11.

Coupled transport processes are an important subject in the field of thermodynamics of irreversible processes, in which transport processes are part of a unified theory together with other irreversible processes (such as heat transport) [9, 10]. Laboratory experiments with quantitative describable transport processes provided a central challenge in the history of electrochemistry, for example, leading to the explanation of ion conduction processes in solution [11].

4.1 Survey and Terminology

We had already emphasized the role of transport processes for electrode reactions in Figure 3.1. Reactants must reach the electrode surface to undergo charge transfer within the tunneling distance. Current flow in the electrolyte can only be realized by the transport of ions, which have a significantly higher mass compared to electrons in metals, and therefore they move more slowly.

F: What does significantly higher mass mean here?
A: You may calculate the ratios between charge and mass for electrons (e/m_e) and for a Na^+ ion [$e/M(Na)/N_A$]. Compare both ratios. They should differ by orders of magnitude.

Due to transport processes and charge transfer processes, concentrations of reactants in the vicinity of the interface deviate from those in the undisturbed parts of the solution, for which we use the term bulk solution. In general, we distinguish three mechanisms for mass transport: migration, diffusion, and convection (Figure 4.1, left column). Common synonyms for migration are drift or ion conductance.

Migration is the directed movement of solvated ions of an electrolyte solution within an electric field.

Migration is the cause of electrical conductivity of electrolyte solutions.

Diffusion is the random thermal movement of individual ions or molecules relative to neighboring particles.

Diffusion occurs continuously in solution and leads to the equilibration of concentration differences.

Convection refers to the movement of ions or molecules together with neighboring particles of a volume element of a fluidic phase (liquid, gaseous, or supercritical).

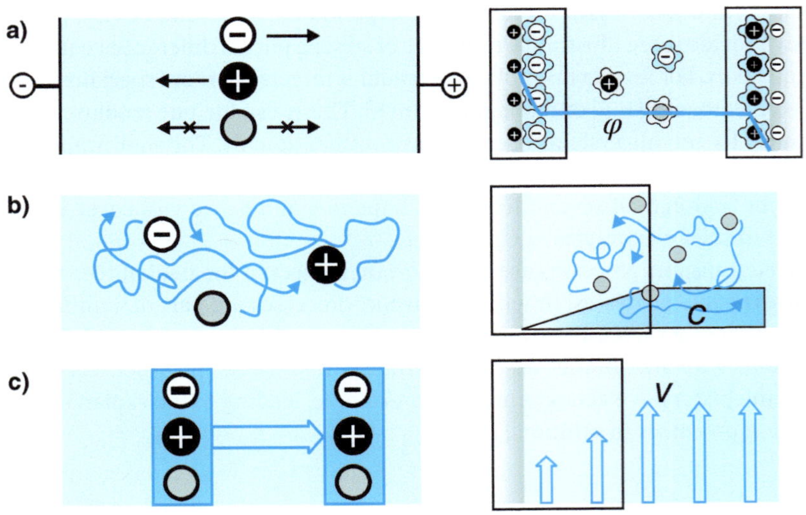

Figure 4.1 Schematics for transport processes and the corresponding boundary layers, in which state variables are changed compared to those in the bulk solution.

a) Migration and electrical double layer with the local variation in the inner potential;
b) Diffusion and diffusion layer with local variation in concentration;
c) Convection and hydrodynamic boundary layer with local variation in flow velocity.

The schemes are not to scale!

This phenomenon is observed, for instance, when an electrolyte solution streams through a tube. The solvent molecules and the ions of one volume element jointly change their position relative to the wall of the tube.

There is a boundary layer associated with each of these transport processes at the electrode surface (Figure 4.1, right column): The **electrical double layer**, the **diffusion layer**, and the **hydrodynamic boundary layer**. In these boundary layers, at least one intensive property of the liquid phase deviates from that in the solution bulk. The different transport mechanisms and the corresponding boundary layers can and often will occur *concomitantly* in electrochemical cells. However, most laboratory electrochemical experiments are designed in a way that transport processes limit the overall conversion (and thus the current) either not at all or in a well-predictable way. For instance, laboratory experiments are often conducted in quiescent solution (no convection) or with exactly defined solution flow (controlled convection). An inert supporting electrolyte is often added to the solution. The supporting electrolyte can be an acid, a base, or a salt that dissociates into ions but does not react at the electrodes. It is provided in large excess compared to the reactants in the electrode reaction. Therefore, most of the ionic current in the electrolyte solution is carried by the ions of the supporting electrolyte rather than by the reactants or products of the electrode reaction. As a consequence, the reactants in the electrode reaction do not migrate appreciably, and diffusion remains as the only mechanism by which the reactants are transported to/from the electrode surface. *In these cases*, we may consider the mechanisms of mass transport as independent from each other, and as such, we will initially treat them here.

4.2 Migration

4.2.1 Structure of the Electrolyte Phase and Ion Mobility

The liquid electrolyte phase consists of anions, cations, and solvent molecules (e.g., H_2O). The solvent molecules are dipoles. This means that there is a permanently asymmetric distribution of charge within the molecule with distinct negative and positive regions. Close to a cation, the negatively charged region of the dipole will turn toward the positive charge of the cation (in a H_2O molecule, the negative charge center is located close to the O-atom, Figure 4.2a). To represent dipolar molecules, we use vector arrows (see Section 3.1.1) pointing from the positive to the negative charge center. These oriented solvent molecules form a solvation shell around the cation. In the solvation shell, the arrangement of the solvent molecules deviates from the random orientation in the solution bulk (Figure 4.2c). Without further calculations, we can hypothesize that the orientation of H_2O molecules is influenced in more than one molecular layer around the cation. With increasing distance from the cation, the electric field strength induced by the cation decreases, and thermal motion of the solvent molecules may overcome the electrostatic forces that act to align the solvent dipoles. As a consequence, the orientation of the dipolar molecules does not affect the entire solution but only a relatively small spherical volume element around each ion.

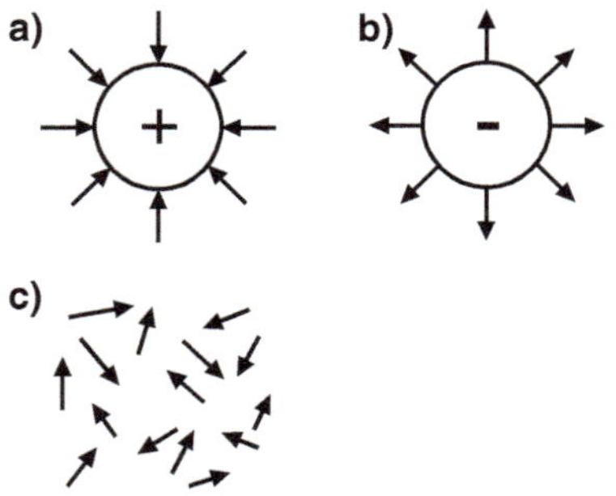

Figure 4.2 Orientation of solvent molecule dipoles.

a) around a cation,
b) around an anion, and
c) random arrangement of H_2O molecule dipoles in the solution bulk without ions.

A corresponding solvation layer is also formed around anions in which the positive charge centers of the solvent dipoles orient toward the negative charge of the anions (Figure 4.2b). When anions or cations move, they drag their solvation shells with them. The solution phase is overall electrically neutral because the charge of all cations is compensated by the charge of all anions.

It is helpful to distinguish between **strong and weak electrolytes**. These attributes refer to the tendency of compounds in a polar solvent to dissociate into cations and anions. Strong electrolytes dissociate completely into their respective ions.

Examples are the most important inorganic acids (mineral acids), bases, and many salts (e.g., many nitrates). Weak electrolytes dissociate incompletely. Many organic acids, bases, and their salts belong to this group. Acetic acid is a prototypical* example of a weak electrolyte. For a concentration of 0.1 mol L^{-1} acetic acid in water (6 g in 1 L solution), the resulting solution contains oxonium cations (H_3O^+), acetate anions (CH_3COO^-), water molecules, and nondissociated molecules of acetic acid (CH_3COOH) in exactly defined proportions (here we neglect the auto-protolysis of water, which causes the occurrence of trace amounts of OH^-, see Section 2.4.2). The fraction of dissociated molecules of acetic acid with respect to the total concentration c_0 of acetic acid (0.1 mol L^{-1}) is called the **degree of dissociation** α.

*prototypical = a typical and well-known example for a larger group of individual phenomena or substances

$$\alpha = \frac{c_{\text{dissociated}}}{c_0} \tag{4.1}$$

α - Degree of dissociation (dimensionless, [0 … 1] or stated in %)

Now, we will estimate how fast an ion moves and how large the hydration shell is. We consider a solvated ion i in the presence of an electric field between two parallel plate electrodes (Figure 4.3). The strength of the electric field **E** is calculated from the voltage U between the two electrodes and their distance l.

$$\mathbf{E} = \frac{U}{l} \tag{4.2}$$

E - Electric field strength [V m^{-1}], symbol **E** as vectoral quantity with a bold symbol
U - Voltage [V]
l - Distance [m]

As a charged particle, the ion is accelerated by the electric field. The acceleration force is obtained by the charge number z_i (−1, −2, −3, etc.), the elemental charge e and **E**.

$$\mathbf{F}_{\mathrm{E}} = z_i e\mathbf{E} \tag{4.3}$$

$\mathbf{F}_{\mathrm{E}}$ - Acceleration force within an electric field [N] = [Nm m^{-1}] = [A s V m^{-1}]
e - Elemental charge = 1.6×10^{-19} A s
z - Charge number of an ion [dimensionless]

In a viscous solvent, the movement of the particle is decelerated according to Stokes' law.[26] The retarding friction force $\mathbf{F}_{\mathrm{R}}$ is obtained from the viscosity of the solvent η (eta), from the migration velocity $\mathbf{w}_i$ of ion i and from the hydrodynamic radius r_i of the ion. The hydrodynamic radius is the radius of a sphere, which would cause the same retarding force in the considered solvent as the solvated ion i:

$$\mathbf{F}_{\mathrm{R}} = 6\pi r_i \eta \mathbf{w}_i. \tag{4.4}$$

$\mathbf{F}_{\mathrm{R}}$ - Retarding force [N] = [m N s m^{-2} m s^{-1}]
r_i - Hydrodynamic radius of the ion [m]
η -Viscosity [N s m^{-2}]
$\mathbf{w}_i$ - Migration velocity [m s^{-1}].

Please note that retarding friction force grows proportional to the translation velocity. You can experience this fact yourself, for example, during cycling or skiing.

When a uniform movement is reached at the constant velocity $\mathbf{w}_i$, the acceleration force and the retarding force are balanced. For simplification, we avoid an explicit consideration of the vectorial character of some quantities in Eqs. (4.3) and (4.4), and consider only the amount of those vectorial quantities:

$$|\mathbf{F}_{\mathrm{E}}| = |\mathbf{F}_{\mathrm{R}}|$$
$$z_i e\mathbf{E} = 6\pi r_i \eta \mathbf{w}_i. \tag{4.5}$$

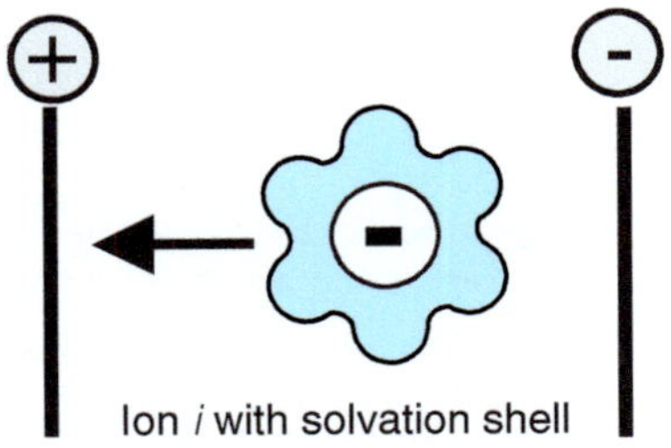

Figure 4.3 Acceleration of an anion within an electric field between two plate electrodes.

Rearranging for the migration velocity yields

$$\mathbf{w}_i = \frac{z_i e\mathbf{E}}{6\pi r_i \eta} \tag{4.6}$$

The migration velocity of an ion i is proportional to the electric field strength **E** and is thus determined by the setup of our experiment. For tabulating data specific to an ion, one would prefer finding a quantity that depends only on intensive substance properties independent of the experiment. The quantity ion mobility u_i of the ion i fulfills

[26] Georg Gabriel Stokes, 1819–1903, Irish mathematician and physicist.

this requirement. The ion mobility is the ratio of migration velocity $\mathbf{w}_i$ and the applied electric field strength $\mathbf{E}$:

$$u_i = \frac{\mathbf{w}_i}{\mathbf{E}} = \frac{z_i e \mathbf{E}}{6\pi r_i \eta}\frac{1}{\mathbf{E}} = \frac{z_i e}{6\pi r_i \eta} \quad (4.7)$$

u_i – Ion mobility [$m^2\ s^{-1}\ V^{-1}$]

The new quantity u_i only depends on natural constants and ionic properties (z_i, r_i). In addition, the solvent viscosity η is important. A temperature dependence is not immediately obvious from Eq. (4.7). Because η generally decreases with increasing temperature, ion mobilities also increase with temperature.

For colored ions (e.g., MnO_4^- or CrO_4^{2-}), the migration velocity can be directly observed and measured in a U-tube apparatus (Figure 4.4), from which ion-specific u_i and r_i can be calculated as in Example Calculation 4.1.

Ion mobilities depend on temperature, although the temperature T does not appear in Eq. (4.7)!

Example Calculation 4.1 The boundary between permanganate solution and the supernatant solution in the experiment shown in Figure 4.4 moved by 8 mm toward the anode within 25 min after application of a voltage of 100 V. The distance between the electrodes was 0.50 m. Calculate the ion mobility [$cm^2\ V^{-1}\ s^{-1}$], the molar conductivity and the hydrodynamic radius of the permanganate ion. The viscosity of the solution was 0.89×10^{-3} N s m^{-2} at 25 °C.

Unknown: u_i, r_i, λ_i

Given:

$$U = 100\ \text{V}$$

$$l = 0.5\ \text{m} = 50\ \text{cm}$$

$$t = 25\ \text{min} = 1500\ \text{s}$$

$$s = 8\ \text{mm} = 0.8\ \text{cm}$$

$$z_i = 1\ (\text{MnO}_4^-)$$

$$\eta = 0.89 \times 10^{-3}\ \text{N s m}^{-2}$$

$$|\mathbf{E}| = \frac{U}{l},\ |\mathbf{w}_i| = \frac{s}{t}$$

$$u_i = \frac{|w_i|}{|\mathbf{E}|} = \frac{s}{t}\cdot\frac{l}{U}$$

$$u_i = \frac{0.8\ \text{cm}}{1500\ \text{s}}\cdot\frac{50\ \text{cm}}{100\ \text{V}}$$

$$u_i = 2.66 \times 10^{-4}\ \text{cm}^2\ \text{s}^{-1}\ \text{V}^{-1}$$

$$\lambda_i = z_i\, e\, N_A\, u_i$$

$$\lambda_i = z_i F u_i$$

$$\lambda_i = 1 \cdot 96485\ \text{A s mol}^{-1} \cdot 2.66 \times 10^{-4}\ \text{cm}^2\ \text{s}^{-1}\ \text{V}^{-1}$$

$$\lambda_i = 25.7\ (\text{A/V})\ \text{cm}^2\ \text{mol}^{-1} = 25.7\ \Omega^{-1}\ \text{cm}^2\ \text{mol}^{-1}$$

$$r_i = \frac{z_i e}{6\pi\eta u_i}$$

$$= \frac{1 \cdot 1.602 \times 10^{-19}\text{A s}}{6\pi \cdot 0.89 \times 10^{-3}\ \text{N s m}^{-2} \cdot 2.66 \times 10^{-8}\ \text{m}^2\ \text{s}^{-1}\ \text{V}^{-1}}$$

$$= 2.57 \times 10^{-10}\ \text{m}$$

$$r_i = 358\ \text{pm}$$

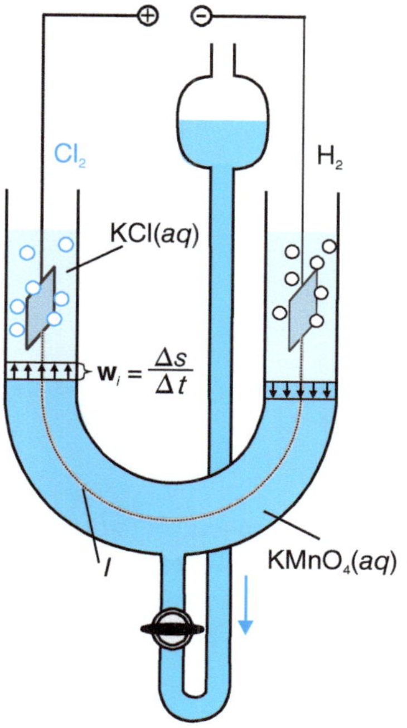

Figure 4.4 The moving boundary experiment for direct observation of the migration velocity of a colored ion. For a better displacement of the KCl solution by the permanganate solution and to establish a clearer separation between them, the density of the permanganate solution is increased by addition of urea. The distance l between the two electrodes must be measured along the current path.

The ion mobility amounts to 2.66×10^{-4} cm^2 s^{-1} V^{-1}, the molar conductivity is 25.7 Ω cm^2 mol^{-1} and the hydrodynamic radius is 358 pm.

The permanganate ion has a tetrahedral arrangement of O-atom surrounding the central Mn atom with a crystallographic Mn–O bond distance of 161 pm. As expected, the hydrodynamic radius (radius of the solvation shell) is significantly larger than the Mn-O bond distance.

4.2.2 Electrical Quantities for the Characterization of Charge Carrier Mobility

The movement of electrons in metallic conductors can be related to the quantities current I, voltage U, and resistance R. We call the quotient U/I the resistance. We can calculate the ohmic resistance[27] from the tabulated data for specific resistance ρ, the length l, and the cross-sectional area A of the conductor.

Q: The term ohmic resistance seems to me like unnecessary doubling, like in "free gift" or in "hot water heater", doesn't it?

A: No, "ohmic resistance" is not an avoidable pleonasm. The attribute is sensible because there are electrical resistors with quite different (nonohmic) properties. For instance, a capacitor exhibits different resistive properties in an electrical alternating current (AC) circuit (Chapter 7). The electrical double layer behaves similarly to a capacitor within an electrical circuit. Thus, we find resistive elements in electrochemical cells that do not necessarily behave like an ohmic resistor.

R – Resistance [Ω]
I – Current [A]
U – Voltage [V]

$$R = \frac{U}{I} \tag{4.8}$$

l – Length of a conductor [cm]
A – Cross-sectional area [cm^2]
ρ – Specific resistance [Ω cm^1]

$$R = \rho\frac{l}{A} \tag{4.9}$$

Another important quantity is the **conductivity** G defined as the reciprocal of the resistance. We can consider the **specific conductivity** κ (kappa) as the electric conductivity provided by a cube of the solution with 1 cm edge length.

G – Conductivity, 1 siemens[28] = 1 S = 1 Ω^{-1}
κ – Specific conductivity, [S cm^{-1}] = [Ω^{-1} cm^{-1}]

$$G = \frac{1}{R} = \frac{I}{U} = \kappa\frac{A}{l} \tag{4.10}$$

$$\kappa = \frac{1}{\rho} = \frac{1}{R}\frac{l}{A} \tag{4.11}$$

We will now consider how we can experimentally determine the conductivity of an electrolyte solution. Initially, we note that a direct current (DC) experiment is likely to fail in this respect.

Q: Well, this task shouldn't be that hard. We set up an electrolysis cell, apply a fixed voltage, and measure the current. That's it, isn't it?

A: Unfortunately, it is not that easy. We had already considered the electrolysis of HCl solution at the beginning of Chapter 2 (Figure 4.5). We noted that

---

27 Georg Simon Ohm, 1789–1854, German physicist.

28 Werner von Siemens, 1816–1892, German electrical engineer and entrepreneur, discovered the electrodynamic principle. He founded a small workshop for communication devices that developed into Siemens AG.

we initially measure (almost) no current at small voltages. Only beyond a threshold voltage (the electrolysis voltage), the current starts to rise with further increase in the voltage applied to the cell. Hence, there is *no* linear relationship between U and I. An electrochemical cell does *not* behave like an ohmic resistor in a DC circuit. In addition, the current often depends strongly on the material of the electrodes. That is, for measuring properties of the solution bulk, we have to carefully eliminate effects from electrodes.

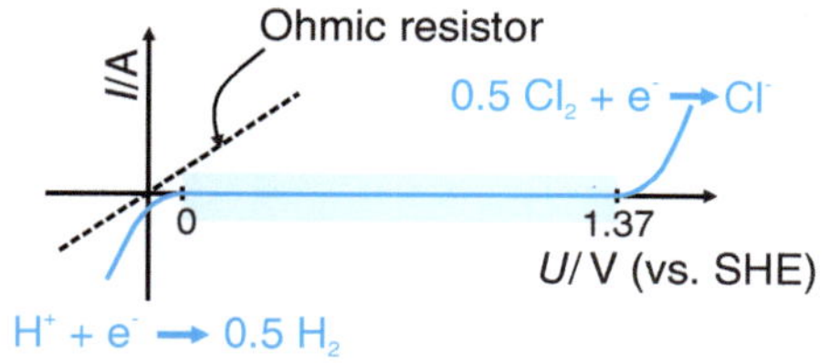

Figure 4.5 Comparison of I–U characteristics of an electrolysis cell (—) filled with 1 mol L^{-1} HCl solution and an ohmic resistor (- - -). There is no electrolysis in the shaded region of the blue curve, although a voltage is applied. We say: "The electrode is polarized."

At a voltage below the electrolysis voltage, a current is recorded that is small but not negligible. What happens at those voltages at the two electrodes at microscopic level? The ions are attracted to the oppositely charged electrodes, but they cannot react due to insufficient thermodynamic driving force for the reaction. At the surface of each electrode, a thin layer of charge-compensating ions builds up (Figure 4.6). This layer is called the **electrical double layer** because it is composed of two regions with opposite excess charges, that is, a layer with accumulated ions to compensate charges in the metallic electrode due to a surplus or deficit of free (mobile) electrons. The double layer behaves electrically similar to a capacitor. A capacitor is often imagined as a parallel plate capacitor. In the electrical double layer, one plate is the metallic electrode material with free electrons, and the second plate is represented by the ion-conducting solution with excess charges close to the electrode surface. The capacitance of the capacitor is calculated from charge Q divided by voltage U.

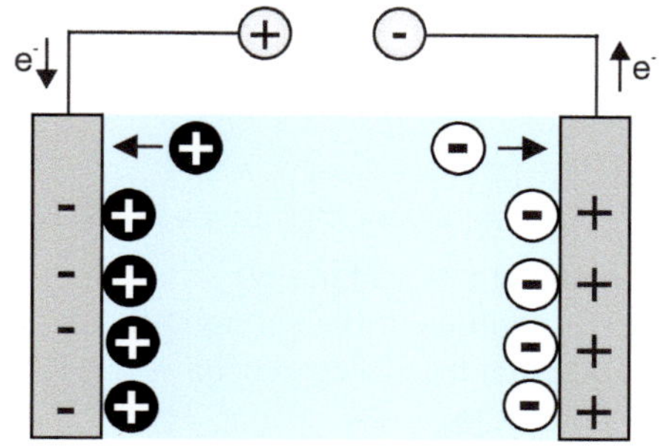

Figure 4.6 Evolution of a capacitive current in an electrochemical cell.

$$C = \frac{Q}{U} \tag{4.12}$$

The charge stored in the capacitor is

$$Q = C\,U \tag{4.13}$$

C – Capacitance [C V^{-1}] = [A s V^{-1}]
Q – Charge [A s]
U – Voltage [V]

Q: Isn't the charge neutrality violated in the double layer?
A: This is indeed the case. Only if we sum up the charges on the electrode material and the ionic excess charge in the solution close to the electrode will the charges be balanced. After learning about homogenous redox reactions in a General Chemistry course, where the conservation of charge is an important principle, we may not be used to the concept that there is charge conservation, but charges can, of course, be separated in the real world. There are numerous experiments in physics demonstrating how charges are separated and recombined (often in spectacular fashion).

We will return to the structure of the electrical double layer in Sections 4.3, 8.10, and 15.9. For now, it is sufficient to realize that the excess of ionic charges shields the excess charge of mobile electrons in the metallic conductor of the electrode. When the charge is completely shielded, the current ceases, and the ions in the bulk solution "do not feel" the electric field anymore. However, when the charge on the electrode surface is changed by varying the electrode potential, ions are again directionally transported until the ionic excess charge in the double layer has adjusted. An ionic current in the electrolyte solution is observed during the time required for restructuring of the electrical double layer. Because the ionic current is caused by a change in the charge of a capacitor (the electrical double layer), we call this current component capacitive current I_C as opposed to the faradaic current I_F from the chemical conversion (e.g., during electrolysis) at the electrode (see Chapter 1).

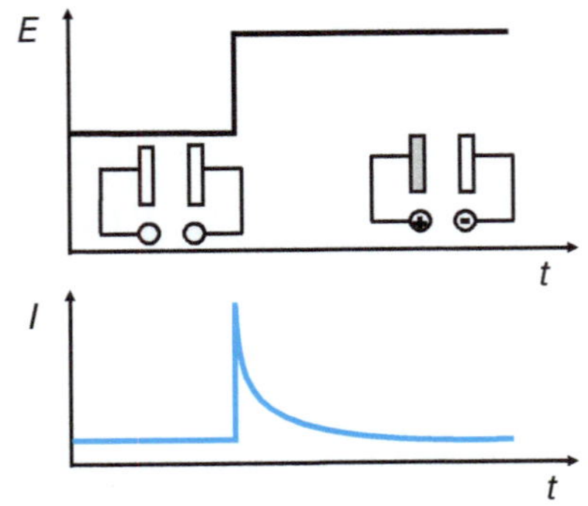

Figure 4.7 Capacitive current transient after applying a potential step.

The capacitive current in electrochemical cells results from a *change in* the charge in the electrical double layer. It occurs *without* an accompanying chemical conversion (Figure 4.6).

After a potential step is applied to an electrical double layer (or a capacitor), an electrical current is recorded for a short period of time (Figure 4.7). We also say: "A transient current is observed." The current after the potential step decays exponentially according to Eq. (4.14). The equation describes the current response after a potential step is applied to an RC element (a resistor and capacitor in series). Here, the capacitor is caused by the capacitance of the electrical double layer C_{dl}, and the ohmic resistor is due to the resistance R_E of the bulk electrolyte. The product of R_E and C_{dl} has the dimension of time and is therefore called the **time constant** of the electrochemical cell.

ΔE – Change in the electrode potential [V]
R_E – Electrolyte resistance [Ω]
t – Time [s]
C_{dl} – Double layer capacitance [A s V^{-1}]

$$I = \frac{\Delta E}{R_E} \exp\left(\frac{-t}{R_E C_{dl}}\right) \tag{4.14}$$

When applying an AC voltage instead of a DC voltage [Eq. (4.15)] to the electrical double layer (or generally to a capacitor), the sign of the voltage changes *before* the double layer is fully charged. In this case, an alternating current [Eq. (4.16)] is passing through the electrochemical cell, although *no* electrolysis takes place (Figure 4.8).

U – Momentary voltage [V]
U_0 – Amplitude of the voltage [V]
ω – Angular frequency (omega = $2\pi f$) [s^{-1}]
I – Current [A]
I_0 – Amplitude of the current [A]
$\pi/2$ – Phase shift (90°)

$$U = U_0 \sin(\omega t) \tag{4.15}$$

$$I_C = I_0 \sin\left(\omega t - \frac{\pi}{2}\right) \tag{4.16}$$

The resistance (or better the "impedance") of a capacitor in an AC circuit amounts to

R_C – Resistance of a capacitor in an AC circuit [V A^{-1}] = [Ω]
C – Capacitance [A s V^{-1}]

$$R_C = \frac{1}{\omega C}. \tag{4.17}$$

The capacitance of a parallel plate capacitor can be calculated according to Eq. (4.18).

ε_0 – Dielectric permittivity of free space = 8.8542×10^{-12} A s V^{-1} m^{-1}
ε_r – (Relative) permittivity of the medium between the plates [dimensionless]
A' – Area of the electrodes considering their roughness [m^2]
d – Distance between the electrodes [m]

$$C = \varepsilon_0 \varepsilon_r \frac{A'}{d} \tag{4.18}$$

There are a number of synonyms commonly used for ε_0 and ε_r. The dielectric permittivity of free space ε_0 is also called vacuum permittivity, absolute dielectric permittivity of vacuum, electric constant, or distributed capacitance of the vacuum. The relative permittivity ε_r is also called the dielectric constant. For the electrical double layer, A' in Eq. (4.18) corresponds to the electrode area and d to the distance between the metal surface and the mean distance of the excess ionic charges within the double layer. We denote the area as A' in order to indicate that we refer to the true electrode area, that is, the electrode area under consideration of the roughness.

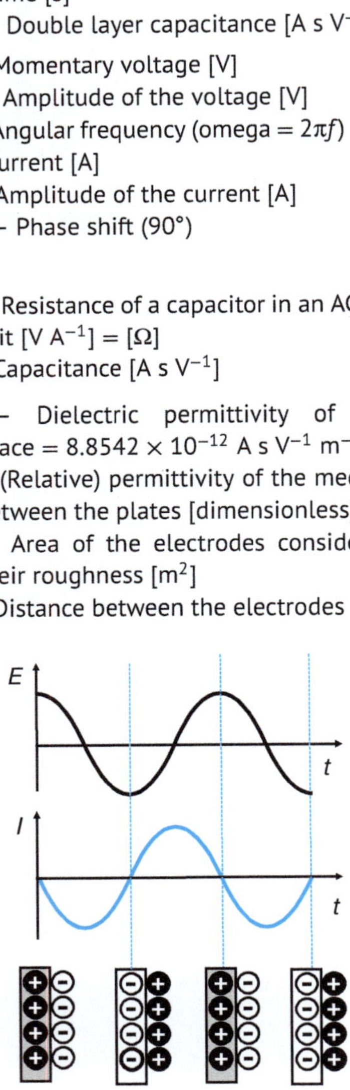

Figure 4.8 Capacitive current (blue) when an AC voltage (black) is applied. Maximum voltage points are consistent with maximum change in current.

With these initial clarifications, we return to our actual aim, the determination of the conductivity (or resistance) due to the movement of ions in the electrolyte solution in the electrochemical cell. An electrochemical cell is composed of at least two electrodes and an electrolyte between them. We now devise an electronic circuit containing electric components (ohmic resistors, capacitors, electric coils) that behave electrically like the electrochemical cell under study. Such circuits are called **equivalent circuits**. Figure 4.9 shows a very simple version of an equivalent circuit that is already sufficient to illustrate several principles. The two shaded areas, one for each electrode, contain a parallel circuit of a (potential-dependent) ohmic resistor $R_{CT}(E)$ for the charge transfer and a capacitor C_{dl} for the electrical double layer. The faradaic current flows via R_{CT}, and the capacitive current via C_{dl}. The two shaded areas are linked via the electrolyte resistance R_E, which we would like to measure. When measuring the resistance via

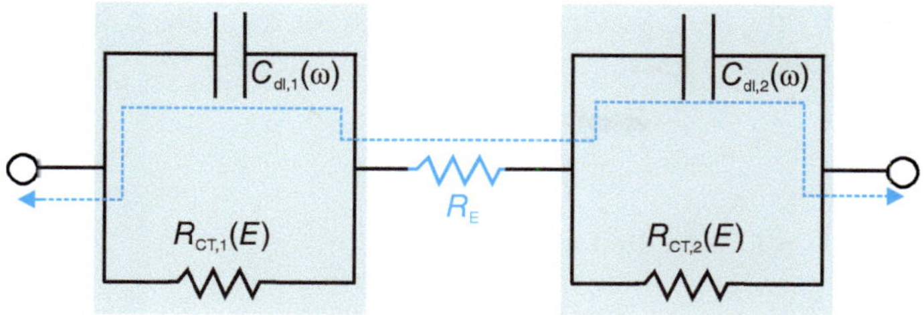

Figure 4.9 Equivalent circuit for an electrochemical cell. The current path during a conductivity measurement (without charge transfer) is indicated by the blue dashed line.

the two terminal contact points with an applied current, we can only access the overall resistance of the entire circuit (corresponding to the entire electrochemical cell). We have to avoid consecutive and parallel elements interfering with the measurement. Thus, we have to take care that only R_E significantly contributes to the overall resistance of the electrochemical cell.

Because the faradaic currents do not directly depend on charge carrier mobility in the bulk electrolyte, we aim to exclude any influence of the electrochemical reactions. Therefore, we apply an AC voltage to the cell with an amplitude that is sufficiently low to avoid driving any unwanted electrochemical conversion (5–10 mV). In this case, $R_{CT}(E)$ values approach infinity, and therefore no current passes through the resistors $R_{CT,1}(E)$ and $R_{CT,2}(E)$. Thus, all the current through the cell is of a capacitive nature and flows via $C_{dl,1}$ and $C_{dl,2}$. Next, we need to take care that R_C for electrodes 1 and 2, according to Eq. (4.17), is much smaller than R_E. We fulfill this requirement by selecting a sufficiently high angular frequency ω for the applied AC voltage. In addition, we increase the true electrode area A' (while maintaining the same geometric electrode area A) by depositing finely dispersed Pt onto the Pt sheet electrodes ("platinized" platinum, platinum black). There is a defined distance between the two electrodes. In the simplest and most instructive case, we use electrodes with a geometric electrode area $A = 1\ \text{cm}^2$ and a distance of 1 cm (Figure 4.10). In this case, the numerical value of the measured conductivity G equals the numerical value of the specific conductivity κ (the conductivity of a cube of the electrolyte solution with edge length of 1 cm). The manufacturing of such measurement cells is possible with only limited precision. Therefore, one settles with cells that have roughly those dimensions, and a correction factor (the cell constant) is determined for each individual cell.

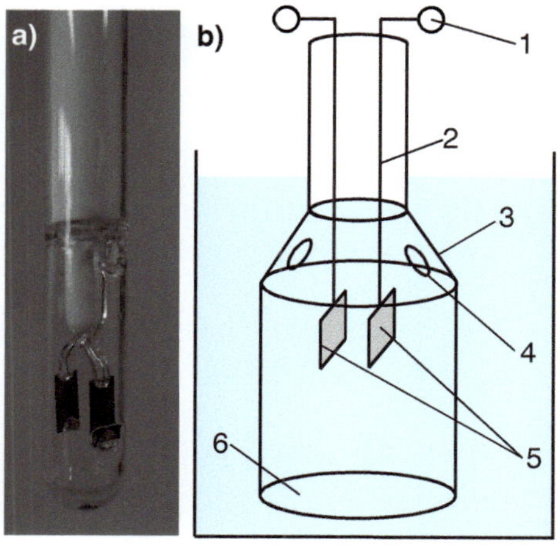

Figure 4.10 Setup of a conductivity cell. a) Photograph of a conductivity cell; b) schematic drawing, 1) connection to the conductivity meter, 2) connecting wires, 3) protective cover, 4) opening for releasing air and solution during immersion and stirring, 5) platinized Pt sheet electrodes, 6) opening for influx of solution during immersion.

$$\text{cell constant } [\text{cm}^{-1}] = \frac{l}{A} \tag{4.19}$$

In contrast to A' in Eq. (4.18), the geometric area A is used in Eq. (4.19). This is the area without consideration of the roughness, because it has a negligible effect on the enclosed solution volume.

The determination of the resistance of the conductivity cell is carried out with a commercial conductivity meter or by means of Wheatstone's bridge circuit[29] (Figure 4.11). In the bridge circuit, the variable resistor is changed until there is no current between nodes 1 and 2. This process is called compensation and can be followed very precisely with a galvanometer or an oscilloscope. For the partial currents across the resistors R_1, R_2, R_x, and R_N, the following conditions hold after compensation: $I_1 = I_2$ and $I_N = I_x$. This is because no current branches off from the upper or the lower branches at nodes 1 and 2. The voltage between nodes 1 and 2 must be zero once the compensation

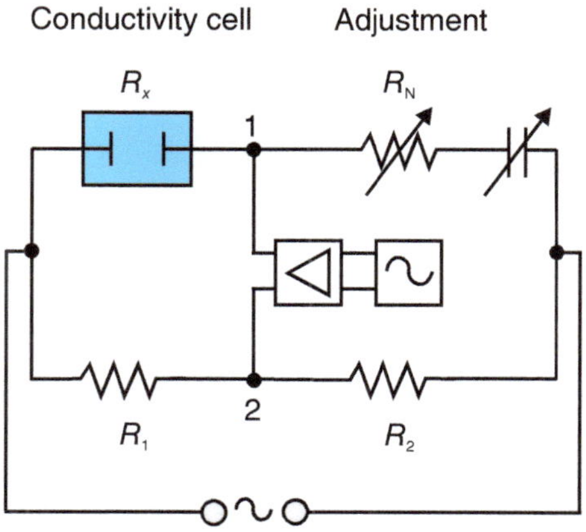

Figure 4.11 Wheatstone's bridge circuit with fixed resistors R_1 and R_2 for setting the measurement range and the variable resistor R_N for fine adjustment. The conductivity cell has the unknown resistance R_x. The nodes are indicated by 1 and 2.

[29] Sir Charles Wheatstone, 1802–1875, English physicist, popularized the circuit for the determination of resistors, which had been published originally by Samuel Hunter Christie, 1784–1865, English mathematician, and natural scientist in 1833. Wheatstone realized the practical importance of this circuit.

is successful. With this, the following relations hold for the voltage drops U_1, U_2, U_N, and U_x over the respective resistors:

$$\frac{U_x}{U_N} = \frac{U_1}{U_2} \tag{4.20}$$

For each resistor $U = I\,R$ and therefore

$$\frac{I_x R_x}{I_N R_N} = \frac{I_1 R_1}{I_2 R_2} \tag{4.21}$$

Due to the compensation condition $I_1 = I_2$ and $I_N = I_x$, the resistors must obey the following relation:

$$\frac{R_x}{R_N} = \frac{R_1}{R_2}$$
$$R_x = \frac{R_1}{R_2} R_N. \tag{4.22}$$

4.2.3 The Relationship Between Ion Mobility and Conductivity

There are many relationships between microscopic and macroscopic quantities in physics and in chemistry. Here, we will derive the relationship between the mobility of individual ions and the conductivity of a cubic centimeter of an electrolyte solution. One particular advantage of taking this educational example is the simple realization of the corresponding experiments in a laboratory course. In the history of chemistry, such experiments were milestones for the development of our understanding of atoms, ions, and electrolyte solutions.

1) We consider the movement of ions along one space coordinate perpendicular to the surface of the electrodes in a conductivity cell. The movement of the ions is connected with a transport of charge.
2) For simplicity, we consider a salt $Me_{\nu+}X_{\nu-}$ with a concentration given as concentration of formula units $c = c(Me_{\nu+}X_{\nu-})$. The salt dissociates completely in the solution:

$$Me_{\nu+}X_{\nu-} \rightleftharpoons \nu_+ Me^{z+} + \nu_- X^{z-} \tag{4.23}$$

3) The total current I is composed of the partial currents carried by the cations (I_+) and the anions (I_-). Each of the partial currents can be expressed as a quotient of the transported charge and the measurement time.

$$I = I_+ + I_-$$
$$= \frac{Q_+}{t} + \frac{Q_-}{t} \tag{4.24}$$

4) We consider a section of the bulk electrolyte solution next to an imagined charge counting plane A (Figure 4.12). The cation moves with the migration velocity w_+ from left to right, the anions with w_- from right to left.

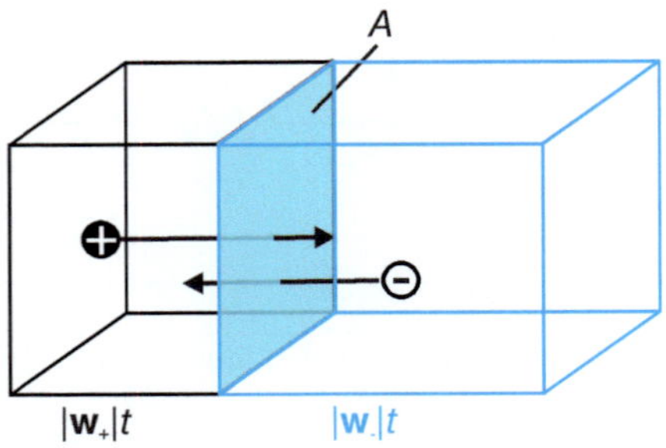

Figure 4.12 Schematic visualization of the movement of cations and anions through the counting plane A during the time t.

During the time t, the ions move the distance $x = \mathbf{w}\,t$. Consequently, all cations that reside inside the left wire-frame box in Figure 4.12 will pass through the counting plane within the observation time t. The analogous consideration holds for the anions within the box on the right side of the counting plane. We obtain the amounts of cations and anions that pass the counting plane by multiplying the volumes of the two boxes (V_+, V_-) with the respective concentrations (c_+, c_-) of cations and anions.

Multiplication of the object numbers with the charge per mole of the ions (z_+eN_A, z_-eN_A) yields the transported partial charges from Eq. (4.24).

$$I = \frac{V_+c_+z_+eN_A}{t} + \frac{V_-c_-z_-eN_A}{t} \quad (4.25)$$

V_+, V_- – Volumes of the boxes in Figure 4.12 [L]
c_+, c_- – Concentrations of cations and anions [mol L^{-1}]
z_+, z_- – Charge numbers [dimensionless]

The volume of the counting boxes is $V = A\,\mathbf{w}\,t$. The ion concentrations are the products of the concentration of formula units and the stoichiometric coefficients.

$$I = \frac{Aw_+t\,\nu_+c\,z_+eN_A}{t} + \frac{Aw_-t\,\nu_-c\,z_-eN_A}{t} \quad (4.26)$$

A – Area of the counting plane [cm^2]
t – Observation time [s]
e – Elemental charge [A s]
N_A – Avogadro number [mol^{-1}]

We substitute the Faraday constant $F = eN_A$, reduce by t, and factor out A, F, and c:

$$I = A\,F\,c\,[\nu_+z_+\mathbf{w}_+ + \nu_-z_-\mathbf{w}_-] \quad (4.27)$$

ν_+, ν_- – Stoichiometric coefficients of anions and cations [dimensionless]
c – Concentration of formula unit [mol L^{-1}]
$\mathbf{w}_i$ – Migration velocity of ion i [m s^{-1}]

Equation (4.27) is not really practical because the migration velocities $\mathbf{w}$ strongly depend on the cell geometry and the applied voltage. Therefore, it does not reveal substance-specific quantities. In order to rearrange the relation, we consider Eqs. (4.2) and (4.7) and form a "unity factor" [Eq. (4.28)].

$$1 = \frac{U}{l}\frac{u_+}{|\mathbf{w}_+|} = \frac{U}{l}\frac{u_-}{|\mathbf{w}_-|} \quad (4.28)$$

$\mathbf{E} = \frac{\mathbf{w}_i}{u_i}$, $\mathbf{E} = \frac{U}{l}$
$\mathbf{w}_i$ – Migration velocity [m s^{-1}], Index i = kind of ions
u_i – Ion mobility [m^2 s^{-1} V^{-1}]
$\mathbf{E}$ – Electric field strength [V m^{-1}]
U – Voltage [V]
l – Distance [m]

We multiply the summands in the square brackets of Eq. (4.27) with this special unity factor:

$$I = \frac{AFcU}{l}\left[\nu_+z_+|\mathbf{w}_+|\frac{u_+}{|\mathbf{w}_+|} + \nu_-z_-|\mathbf{w}_-|\frac{u_-}{|\mathbf{w}_-|}\right] = \frac{AFc}{l}[\nu_+z_+u_+ + \nu_-z_-u_-]U. \quad (4.29)$$

$\kappa = \frac{1}{R}\,\frac{l}{A}$

A comparison of the coefficients in Eq. (4.29) with Ohm's law[27], $R = U/I$ or $I = U/R$, yields that the term before U in Eq. (4.29) must be equal to $1/R$. If we then consider the influence of the electrode area A and the distance l between the electrodes [i.e., the cell constant, Eq. (4.19)], we obtain the specific conductivity κ [Ω^{-1} cm^{-1}] [Eq. (4.11)] as intensive quantity that we can determine experimentally using a calibrated conductivity cell.

$$\kappa = F\,c\,[\nu_+\,z_+\,u_+ + \nu_-\,z_-\,u_-] \quad (4.30)$$

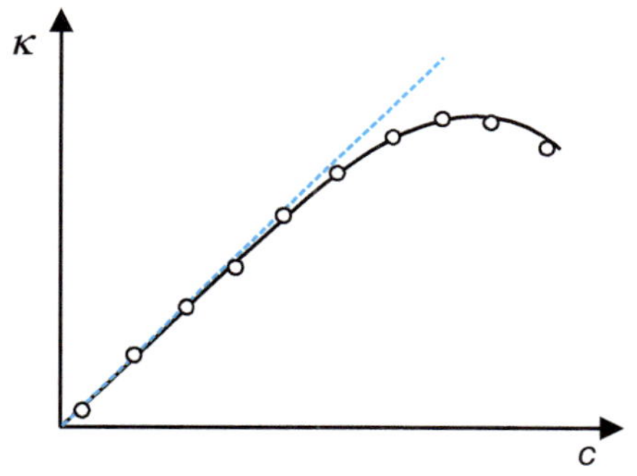

Figure 4.13 Plot of specific conductivity data versus concentration of KCl. The dashed line shows the expected behavior for $u_+ = const$ and $u_- = const$.

According to Eq. (4.30), we expect a straight line plot of the specific conductivity versus concentration of the salt (Figure 4.13) with constant u_+ and u_-. However, from Figure 4.13, we recognize that the specific conductivity is indeed a function of concentration. Also, the ion mobilities u_+ and u_- are themselves functions of concentration. The influence of the ion mobilities becomes even more evident if we form the quotient of κ and c, a quantity called **molar conductivity** Λ.

$$\Lambda = \frac{\kappa}{c} = [\nu_+\;z_+\,u_+F + \nu_-\;z_-\;u_-F] \quad (4.31)$$

A plot of molar conductivity over the concentration for strong and weak electrolytes is shown in Figure 4.14. The selected examples are 1–1 electrolytes because one formula unit yields one singly charged anion and a singly charged cation. A 2–1 electrolyte would dissociate into a doubly charged cation and two singly charged anions.

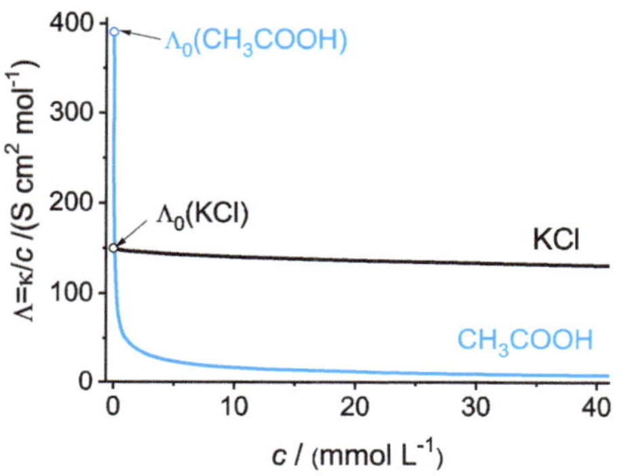

Figure 4.14 Plots of the molar conductivity versus concentrations of KCl and acetic acid as examples for a strong electrolyte (KCl) and a weak electrolyte (CH_3COOH).

$$KCl \rightleftharpoons K^+ + Cl^- \quad (4.32)$$

$$CH_3COOH + H_2O \rightleftharpoons CH_3COO^- + H_3O^+ \quad (4.33)$$

4.2.4 Why Are Molar Conductivities of Strong Electrolytes Concentration Dependent?

For strong electrolytes, which are completely dissociated, Λ slightly increases with dilution (at lower concentration) as shown in Figure 4.14. The directed movement of a cation is retarded by the surrounding anions and *vice versa*. The higher the total ion concentration, the stronger is the mutual retardation of the rate of migration. If we extrapolate values of $\Lambda(c)$ toward $c = 0$ (at "infinite dilution"), we obtain a value which depends for a given solvent only on the kind of ions and their stoichiometric coefficient in the considered compound. We call this value molar conductivity at infinite dilution Λ_0 or limiting conductivity [Eq. (4.34)].

Λ – Molar conductivity $[(\Omega^{-1}\ \mathrm{cm}^{-1})/(\mathrm{mol\ cm}^{-3})] = [\mathrm{S\ cm}^2\ \mathrm{mol}^{-1}]$
κ – Specific conductivity $[\Omega^{-1}\ \mathrm{cm}^{-1}]$

$$\lim_{c \to 0} \Lambda = \Lambda_0 \tag{4.34}$$

Equation (4.7) predicts that in the limiting case, u for a specific ion depends only on the ratio between ionic charge and hydrodynamic radius. According to Eq. (4.31), this also holds for Λ_0 at constant viscosity of the solvent (which depends on the nature of interactions in the solvent and the temperature). At high electrolyte concentrations, many solvent molecules are part of the solvation shell of ions. Those solvent molecules are not free to move anymore, unlike the unbound solvent molecules. This limits their ability to screen charges (e.g., at electrodes in the conductivity cell), which means a decrease in the relative permittivity ε_r of the solvent. At very high concentrations, ion pairs are formed in solution. These are due to transient association of an anion and a cation surrounded by a shared solvation shell (Figure 4.15). In this state, the ion pair is overall electrically neutral and thus does not contribute to the conductivity. Therefore, Λ decreases.

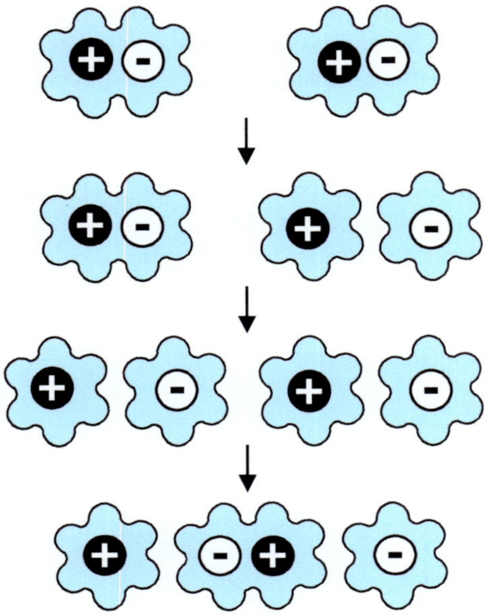

Figure 4.15 Schematic illustration of the formation and decay of ion pairs.

We use the **conductivity coefficients** f_Λ to describe the molar conductivity at arbitrary concentrations, analogous to the use of the activity coefficients in Chapter 2.

$$f_\Lambda = \frac{\Lambda(c)}{\Lambda_0} \tag{4.35}$$

f_λ – Conductivity coefficient [dimensionless]
Λ – Molar conductivity $[\mathrm{S\ cm}^2\ \mathrm{mol}^{-1}]$
Λ_0 – Molar conductivity at infinite solution $[\mathrm{S\ cm}^2\ \mathrm{mol}^{-1}]$

Kohlrausch[30] found empirically a square-root relationship for the concentration dependence of molar conductivities, the "square-root equation" of Kohlrausch.

$$\Lambda(c) = \Lambda_0 - const\sqrt{c} \tag{4.36}$$

The numerical value of *const* depends on the charge number of the ions. To summarize, the specific conductivity of a strong electrolyte (degree of dissociation $\alpha = 1$ or 100%) is obtained by

$$\kappa = c\Lambda(c) = c\Lambda_0 f_\Lambda \tag{4.37}$$

Table 4.1 Molar conductivity at infinite dilution of some salts (aqueous solution, 25 °C).

Salt	$\Lambda_0/(\mathrm{S\ cm}^2\ \mathrm{mol}^{-1})$
NaCl	126.5
KCl	149.9
$NaNO_3$	121.5
KNO_3	144.9

4.2.5 The Law of Independent Migration of Ions

Equation (4.36) facilitates the determination of conductivity data for a large number of salts. Kohlrausch noted that the difference in Λ_0 values between Na and K salts with a common anion is constant. From the data in Table 4.1, we can form the difference $\Lambda_0(\mathrm{KCl}) - \Lambda_0(\mathrm{NaCl}) = 23.4\ \mathrm{S\ cm}^2\ \mathrm{mol}^{-1}$, which equals the difference $\Lambda_0(\mathrm{KNO_3}) - \Lambda_0(\mathrm{NaNO_3}) = 23.4\ \mathrm{S\ cm}^2\ \mathrm{mol}^{-1}$. This observation is explained straightforwardly if we assume that each type of ion in the solution contributes independently to the overall conductivity. For the case of an infinitely diluted solution, this is entirely plausible because the dilution precludes ion–ion interactions.

Convince yourself that the reasoning exemplified for salts with a common anion also holds for chlorides and nitrates with a common cation. For this purpose, compare $\Lambda_0(\mathrm{NaCl}) - \Lambda_0(\mathrm{NaNO_3})$ to $\Lambda_0(\mathrm{KCl}) - \Lambda_0(\mathrm{KNO_3})$.

[30] Friedrich Wilhelm Georg Kohlrausch, 1840–1910, German physicist, professor in Göttingen, Zurich, Würzburg, president of the German Physical-Technical Reichs Institut Berlin.

Law of independent migration of ions: In an infinitely diluted solution, the ions move independently of each other in an electric field.

If indeed the ions contribute individually to the overall conductivity, it also makes sense to define conductivity quantities for individual ion types (rather than for salts, acids, and bases only). We denote those quantities by a lower-case lambda, for example, λ_+, λ_-. For a 1–1 electrolyte, the total conductivity would be $\Lambda_0 = \lambda_+ + \lambda_-$. In principle, we could determine λ_+ and λ_- by first determining Λ_0 of a salt and then obtaining λ from direct observation of migration velocity according to Figure 4.4. However, there is a much more precise method known as Hittorf's method[33], which we deal with in Section 4.2.7. It also has the advantage that it works with colorless ions. In the older literature, you will frequently encounter the term equivalent conductivity of ion i at infinite dilution $\lambda_{\mathrm{eq},i}$, for which an additional normalization to the charge number z_i of the ion is made:

Make sure you understand in each case, what the authors of a publication describe by the symbol λ, molar or equivalent conductivities!

$$\lambda_{0\mathrm{eq},i} = \frac{\lambda_{0,i}}{z_i} \tag{4.38}$$

$\lambda_{0,\mathrm{eq},i}$ – Equivalent conductivity at infinite dilution for the ion i [S cm^2 mol^{-1}]

$\lambda_{0,i}$ – Molar conductivity at infinite dilution for the ion i [S cm^2 mol^{-1}]

z_i – Charge number of the ion i [dimensionless]

The calculation of conductivities of electrolyte solutions is made by means of Eq. (4.37) using tabulated data of $\lambda_{0,i}$ or $\lambda_{\mathrm{eq},i}$ at a defined temperature and in a specific solvent, at the concentration c, or tabulated values for $f_\lambda(c)$ from published data collections [3, 4] or electronic databases. Of course, the conductivity can also be obtained from tabulated $\kappa(c, T)$ data for specific common electrolyte solutions [2, 3]. For concentration without $f_\lambda(c)$ or κ data available, the values must be interpolated from the nearest tabulated values analogous to Example Calculation 2.11.

4.2.6 Why Does the Molar Conductivity of Weak Electrolytes Depend Strongly on Concentration?

The molar conductivity Λ at a concentration of $c_0 = 0.1$ mol L^{-1} is much smaller for the weak electrolyte acetic acid compared to that of a strong electrolyte, such as KCl solution at the same concentration. In fact, this is not surprising at all because the degree of dissociation of 0.1 mol L^{-1} aqueous solution of acetic acid amounts to about 1% only. That means 99% of the acetic acid molecules are dissolved in water as neutral molecules, CH_3COOH, and thus cannot contribute to the ionic conductivity. For the solution of acetic acid in water, the concentration of ions is given by Eq. (4.1).

$$c_{\mathrm{dissociated}} = [\mathrm{H}^+] = [\mathrm{ac}] = \alpha c_0 \tag{4.39}$$

[H$^+$] – Concentration of oxonium ions H_3O^+ [mol L^{-1}]

[ac] – Concentration of acetate ions [mol L^{-1}]

c_0 – Total concentration of the weighted acetic acid [mol L^{-1}]

α – Degree of dissociation [dimensionless]

Using this relationship in the expression for the mass action law of the protonation equilibrium with an acid dissociation constant K_a yields an expression known as Ostwald's dilution law[31]:

$$K_\mathrm{a} = \frac{\{\mathrm{H}^+\}\{\mathrm{ac}\}}{\{\mathrm{Hac}\}} \approx \frac{[\mathrm{H}^+][\mathrm{ac}]}{[\mathrm{Hac}]} = \frac{\alpha c_0 \cdot \alpha c_0}{(1-\alpha)c_0} c_0 = \frac{\alpha^2}{(1-\alpha)} c_0 \tag{4.40}$$

K_a – Acid dissociation constant [mol L^{-1}]

Because K_a is a constant, α must increase steeply for $c_0 \to 0$. We can now calculate the specific conductivity assuming $f_\lambda \approx 1$. Assuming $f_\lambda \approx 1$ is reasonable because there is an overall very low number of dissociated acid molecules and thus low concentrations

[31] Friedrich Wilhelm Ostwald, 1853 (Riga)–1932 (Großbothen near Leipzig), German chemist, Nobel Prize 1909 for contributions to catalysis, author of an influential textbook about the history of electrochemistry (see Figure 0.1), further terms: "Ostwald ripening."

of oxonium ions and acetate ions. Interionic interactions (that would cause f_λ to depart from unity) are negligible in this situation.

$$\kappa = c\,\Lambda(c) = c\,\Lambda_0\alpha = c\,(\lambda_{0,+} + \lambda_{0,-})\alpha \tag{4.41}$$

From Eq. (4.41) follows immediately

$$\alpha = \frac{\Lambda(c)}{\Lambda_0} \tag{4.42}$$

Inserting this expression into Eq. (4.40) yields

$$K_a = \frac{\Lambda^2}{\Lambda_0^2\left(1-\frac{\Lambda}{\Lambda_0}\right)}c_0 = \frac{\Lambda^2}{\left(\Lambda_0^2 - \Lambda_0\Lambda\right)}c_0 = \frac{\Lambda^2}{(\Lambda_0 - \Lambda)\Lambda_0}c_0 \tag{4.43}$$

At low concentrations, the dissociation is almost complete ($\alpha \approx 1$) and thus $\Lambda_0 \approx \Lambda$. We can now replace Λ^2 by ${\Lambda_0}^2$ in the numerator of Eq. (4.43), reducing the fraction and isolate Λ:

$$\Lambda = \Lambda_0 - \frac{\Lambda_0}{K_a}c_0 = \Lambda_0 - const\ c_0 \tag{4.44}$$

For low degrees of dissociation (i.e., high concentrations), we find $\Lambda \ll \Lambda_0$ and thus can substitute $\Lambda_0 - \Lambda \approx \Lambda_0$ in the denominator to obtain

$$\Lambda = \Lambda_0\sqrt{\frac{K_a}{c_0}} \tag{4.45}$$

We can now convincingly explain that the strong concentration dependence of Λ for small concentrations of weak electrolytes is caused by the steep change in the degree of dissociation. Figure 4.14 shows that at low concentrations Λ(Hac) even surpasses Λ(KCl)! We can make this observation for all aqueous electrolyte solutions in which H^+ and OH^- contribute to the conductivity. These ions possess an **extraordinarily high conductivity** in protic solvents that is based on a prototropic transport mechanism, which is frequently called Grotthuss[32] mechanism. While a K^+ ion must physically move with all its mass between two points in space to carry charge between these locations, a H^+ ion can attach to a nonbonding electron pair of a water cluster $(H_2O)_n$ $(1 \leq n \leq 5)$ in liquid water (Figure 4.16a). Through shifting of electron density (with much lower mass), another proton at the opposite side of the water cluster can be released. In this way, the charge is translocated over a much larger distance than the distance by which the nuclear coordinates of the proton are moved. There are corresponding considerations for OH^- ions (Figure 4.16b).

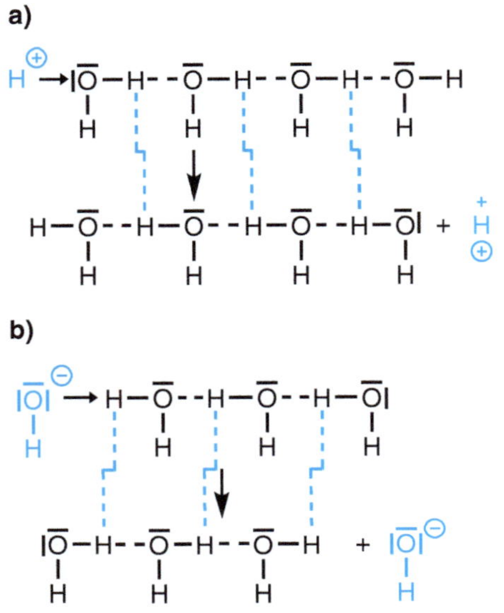

Figure 4.16 Mechanism of the extraordinarily high conductivity of a) protons and b) hydroxide ions. The charge (blue) is transported over a larger distance from left to right than the change in nuclear coordinates.

Example Calculation 4.2 An acid HB with a molar conductivity at infinite dilution of $\Lambda_0 = 246\ S\ cm^2\ mol^{-1}$ and a dissociation according to $HB \rightleftharpoons H^+ + B^-$ was investigated in pure water with a conductivity of 0.0032 mS. The cell constant was $0.9812\ cm^{-1}$. The following values were measured.

$c/(mol\ L^{-1})$	G/mS
0.005	0.132
0.010	0.183
0.050	0.630

[32]Christian Johann Dietrich Grotthuss, 1785 (Geddutz, today located in Lithuania)–1822, Lithuanian chemist, first description of a jump mechanism of the electrical conductivity in 1805. At that time, electrons and ions had not been discovered yet! Consequently, details of Grotthuss' theory cannot be in agreement with our contemporary understanding of structure and dynamics of electrolyte solutions.

Calculate for all concentrations the specific conductivity of the acid solution, the molar conductivity of the solution, the degree of dissociation, the acid dissociation constant K_a, and the equilibrium exponent $pK_a = -\log(K_a/a°)$. Use a spreadsheet application.

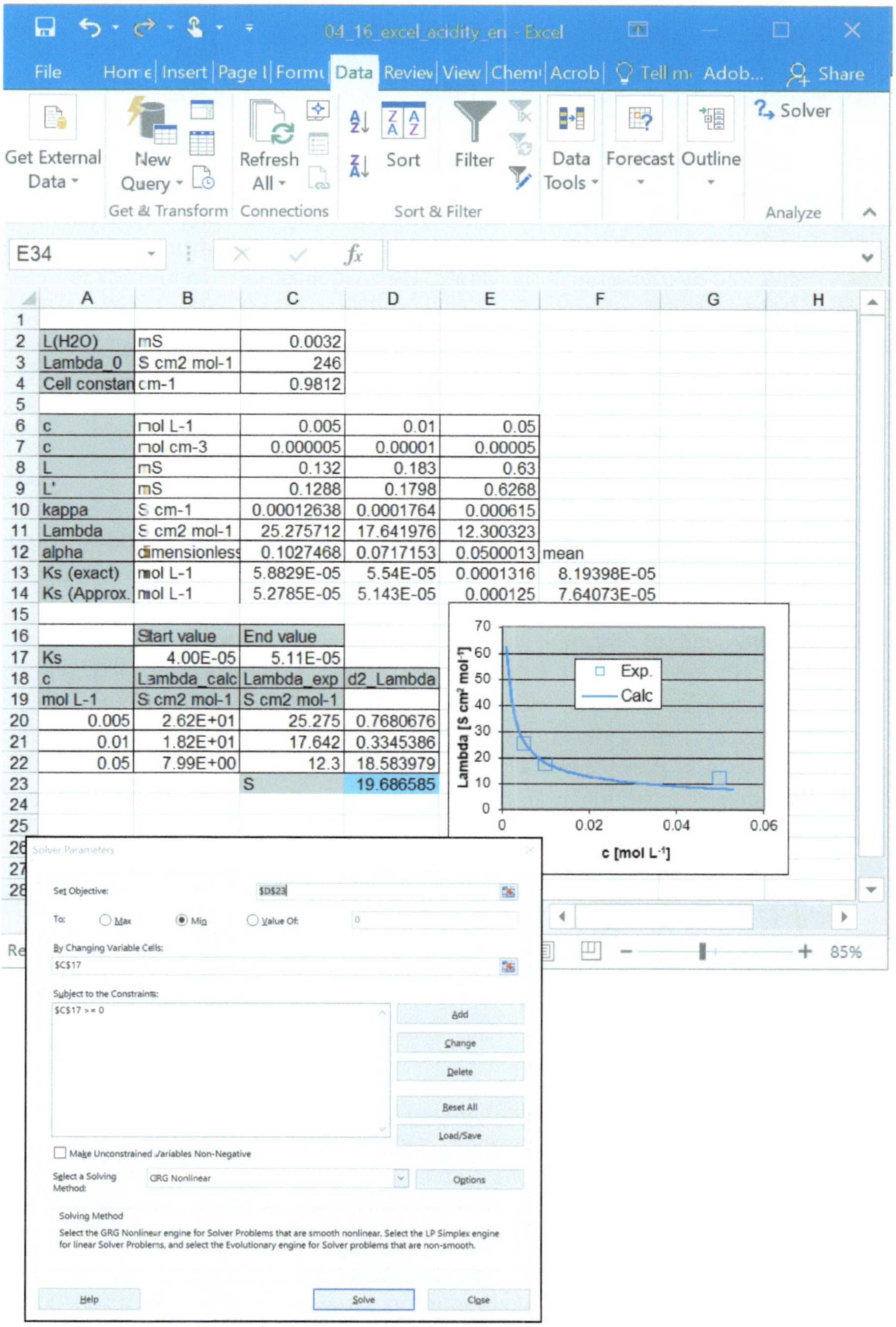

Figure 4.17 Section of a Microsoft Excel spreadsheet for solving the exercise. The pop-up window shows the settings of the Solver. The line in the diagram was calculated with Eq. (4.52) for c values between 0.001 mol L^{-1} and 0.055 mol L^{-1} with increments of 1×10^{-6} mol L^{-1} and connected by a line in the plot. K_a is optimized to 5.51×10^{-5} mol L^{-1}. Similar functionalities are available with other spreadsheet programs.

First, we subtract the conductivity of the pure water from the measured values. From these corrected conductivities and the cell constant, we obtain the specific conductivity. This value and the concentration yield the molar conductivity, which provides us directly with the degree of dissociation α.

$$G' \text{ [mS]} = G \text{ [mS]} - G(\text{H}_2\text{O}) \text{ [mS]} \tag{4.46}$$

$$\kappa \text{ [S cm}^{-1}] = 0.001 \cdot G' \text{ [mS]} \cdot \text{cell constant [cm}^{-1}] \tag{4.47}$$

$$c \text{ [mol cm}^{-3}] = \frac{0.001 \text{ mol cm}^{-3}}{1 \text{ mol L}^{-1}} c \text{ [mol L}^{-1}] \tag{4.48}$$

$$\Lambda = \kappa \text{ [S cm}^{-1}]/c \text{ [mol cm}^{-3}] \tag{4.49}$$

$$\alpha = \Lambda/\Lambda_0 \tag{4.50}$$

The calculation of K_a can be executed for each measurement according to Eqs. (4.51) or (4.43) followed by averaging of the values. Here we must use c in mol L^{-1} in order to obtain K_a in common units.

$$K_a = \frac{\Lambda^2}{(\Lambda_0 - \Lambda)\Lambda_0} c_0 \tag{4.43}$$

Approximate solution:

$$K_a \approx \frac{\Lambda^2}{\Lambda_0^2} c_0 \tag{4.51}$$

Alternatively, we may use a spreadsheet program for a nonlinear regression analysis. We express $\Lambda(c)$ as a formula and search for the K_a value, for which the best agreement between measured and calculated values is found. First, we rearrange Eq. (4.40), then apply the quadratic formula solution to replace α according to Eq. (4.42) and obtain Eq. (4.52). As shown in Eq. (4.53), we form the sum of the squared deviations S_{R^2} between measured values Λ_i and values $\Lambda_{\text{calc}}(c_i)$ calculated by means of Eq. (4.52). Now, we vary K_a systematically until S_{R^2} reaches a minimum. In the program MS Excel, this process can be performed by using the tool "Solver". The Solver may require separate installation or activation as an Add-On. The optimized value of K_a is located in cell C17 in Figure 4.17.

$$\Lambda(c) = \alpha(c)\Lambda_0 = -\frac{\Lambda_0 K_a}{2c_0} + \Lambda_0 \sqrt{\left(\frac{K_a}{2c_0}\right)^2 + \frac{K_a}{c_0}} \tag{4.52}$$

$$S_{R^2} = \sum_{n=1}^{3} (\Lambda_{calc}(c_i) - \Lambda_i)^2 \tag{4.53}$$

In the general equation for the calculation of specific conductivities [Eq. (4.37)], we have to consider the influence of c, α, and f_λ. Only for limiting cases, the equation can be simplified by using approximations such as $\alpha = 1$. Further parameters such as the temperature or type of solvent could be important (Figure 4.18). The viscosity of the solution decreases exponentially with increasing temperature. Other solvent properties that change with temperature may counteract this effect. For instance, κ is generally increases with temperature in aqueous solution. However, κ decreases above 90 °C because of a decrease in the relative permittivity ε_r.

An extraordinarily high conductivity is observed only for H^+ and OH^- in protic solvents (H_2O > MeOH > EtOH, Figure 4.18). In acetone (a nonprotic solvent) the conductivity of H^+ is insignificantly higher compared to that of other singly charged cations.

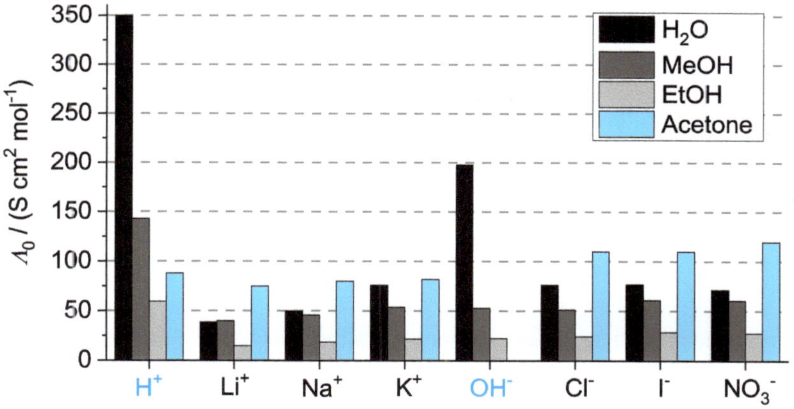

Figure 4.18 Molar conductivities at infinite dilution for selected ions in water, methanol, ethanol, and acetone at 25 °C. Source: Plot prepared by the book authors using data from E.-G. Jäger, K. Schöne; *Elektrochemie und Elektrolytgleichgewichte*, Lehrwerk Chemie, Arbeitsbuch 5. Deutscher Verlag für Grundstoffindustrie, Leipzig, 1977, p. 272.

Further Reading. The scheme in Figure 4.16 is not only geometrically but also conceptually strongly simplified. In fact, the processes are considerably more complicated, and the differences in the conduction mechanisms between H^+ and OH^- are larger than suggested by Figure 4.16. This is demonstrated by a number of more detailed experimental findings [12] as well as by molecular dynamics calculations [13, 14].

4.2.7 Transference Numbers Describe the Contribution of Ions to the Total Current

We would like to consider now which *fraction* of the total current is transported by each type of ions, for example, cations and anions. In 1–1 electrolyte solution, there are only anions and cations as mobile charge carriers. Therefore, the total current I is the sum of the partial currents carried by anions I_- or cations I_+.

$$I = I_+ + I_- \tag{4.54}$$

Q: This statement seems trivial, doesn't it? What else should be the case?

A: Well, the statement is less trivial than it seems on first sight. In solid-state electrolytes (Chapter 11), in which ions may move between different lattice positions at elevated temperatures, electron conduction may also contribute. Such situations are called "mixed conductivity". It is very important, for instance, in active battery materials (Chapter 15) or for passive layers (Chapter 14) in the context of corrosion. Here, we would like to consider only the much simpler case of aqueous electrolyte solutions.

We define the **transference number** as the fraction of the total current carried by one type of ion. For the solution of a single salt, this is

$$t_+ = \frac{I_+}{I} \quad \text{and} \quad t_- = \frac{I_-}{I}. \tag{4.55}$$

I_+ – Current transported by cations [A]
I_- – Current transported by anions [A]
I – Total current [A]
t_+, t_- – Fractions of the total current carried by one ion type (transference number) [dimensionless]

Furthermore, all transference numbers sum up to unity. For the solution of a single 1–1 salt, this is

$$t_+ + t_- = 1 \tag{4.56}$$

Using these relations, we can express the partial currents as

$$I_+ = t_+ I \text{ and } I_- = t_- I \tag{4.57}$$

ν_-, ν_+ - Stoichiometric coefficients [dimensionless].

Inserting molar conductivities and stoichiometric coefficients, we obtain for t_+ in the solution of a *single* salt

$$t_+ = \frac{\nu_+\Lambda_+}{\nu_+\Lambda_+ + \nu_-\Lambda_-} = \frac{\nu_+\Lambda_+}{\Lambda} \tag{4.58}$$

In the presence of more than two types of ion, we obtain the transference number t_i of the ion i

$$t_i = \frac{c_i\Lambda_i}{\sum_j c_j\Lambda_j}. \tag{4.59}$$

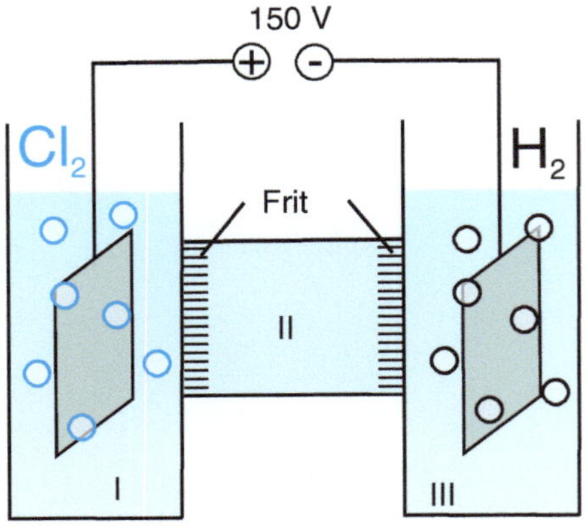

Figure 4.19 Scheme of a divided electrolysis cell for the determination of transference numbers. Within each compartment, fast mixing is ensured. I) Anode compartment, II) central compartment, and III) cathode compartment.

Transference numbers are experimentally determined by the method of Hittorf[33] in an electrolysis cell, which is divided by two diaphragms into an anode compartment, a central compartment, and a cathode compartment (Figure 4.19). We will discuss the experiment using HCl solution as an example. The transference number of the cation (H^+) shall be

$$t_+ = \frac{\Lambda(H^+)}{\Lambda(H^+) + \Lambda(Cl^-)} = 0.8 \tag{4.60}$$

$$t_- = 1 - t_+ = 0.2 \tag{4.61}$$

Reaction (4.62) proceeds at the anode.

$$Cl^- \rightarrow \tfrac{1}{2}\,Cl_2 + e^- \ (n = 1) \tag{4.62}$$

The reduction of oxonium ions occurs at the cathode.

$$H^+ + e^- \rightarrow \tfrac{1}{2}\,H_2 \ (n = 1) \tag{4.63}$$

We would like to understand the consequences of electrolysis and migration using HCl electrolysis as a thought experiment (Figure 4.20). Initially, there are 10 equivalents of object amount in each of the three compartments (e.g., 10 mmol). Each symbol

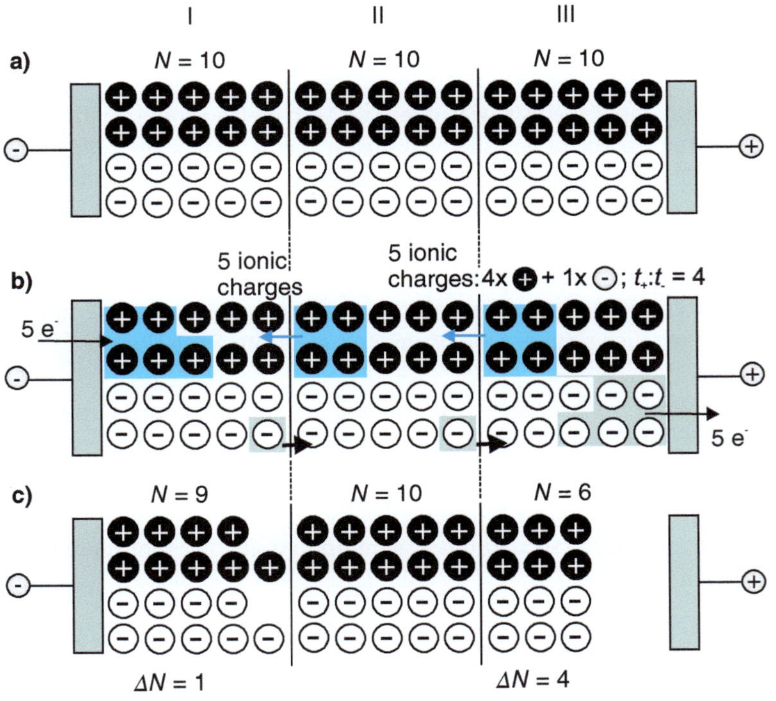

Figure 4.20 Change in the amounts of ions with $t_+ = 0.8$ and $t_- = 0.2$ during the electrolysis of HCl in the cathode compartment (I), the central compartment (II), and the anode compartment (III) of an electrolysis cell according to Hittorf (schematically);

a) before electrolysis;
b) reaction of 5 charge equivalents;
c) after the reaction.

[33] Johann Wilhelm Hittorf, 1824–1914, German physical chemist, professor of chemistry and physics in Münster from 1889, worked on electrolyte theory and cathode rays.

for an ion in Figure 4.20a stands for one of those equivalents. During the electrolysis, 5 charge equivalents (here 5 mmol · F) shall be transferred at anode and cathode. This is indicated by 5 electrons in Figure 4.20b. Those 5 charge equivalents cause the conversion (consumption N_{rct}) of 5 cation equivalents in the cathode compartment as well as of 5 anion equivalents in the anode compartment. The charge equivalents must also be transported by ion migration across the diaphragms. According to the transference numbers $N_{rct} = \Delta N_+ + \Delta N_- = t_+ N_{rct} + t_- N_{rct}$, the cations transport $\Delta N_+ = 0.8 \cdot 5 = 4$ charge equivalents and the anions $\Delta N_- = 0.2 \cdot 5 = 1$ charge equivalent. If we sum up all charge equivalents consumed, transported in, and transported out for each compartment and for each type of ion in a balance sheet (Figure 4.20c), it becomes evident that there is an imbalance. The faster of the two ions provides a better resupply of the ions consumed in the reaction. In the cathode compartment (consumption of cations), the total of substance equivalents decreases by 1, and in the anode compartment by 4 (consumption of the slower anions). There are no changes in the object numbers in the central compartment due to equal incoming and outgoing currents.

We can determine the amounts of ions in the anode and cathode compartments (N_a and N_c) by potentiometric titration. The ratio of the observed *changes in* the amounts $\Delta N_c / \Delta N_a$ equals the reciprocal ratio of the corresponding transference numbers t_-/t_+. The more mobile ion compensates more effectively for the consumption of the respective ion in the electrode reaction. Therefore, the observed change in the number of cations in the cathode compartment is smaller. Note that charge neutrality is maintained in each of the three compartments.

$$\frac{\Delta N_c}{\Delta N_a} = \frac{-N_{rct} + t_+ N_{rct}}{-N_{rct} + t_- N_{rct}} = \frac{N_{rct}(t_+ - 1)}{N_{rct}(t_- - 1)} = \frac{-t_-}{t_- - 1} = \frac{1}{4} = \frac{t_-}{t_+} \tag{4.64}$$

$\Delta N_c, \Delta N_a$ – Changes in the amount of ions in the cathode and anode compartments [mol]

N_{rct} – Consumed amount of ions in the electrolysis at anode and cathode for $n = 1$ [mol]

t_+, t_- – Fraction of the total current carried by one type of ion (transference number) [dimensionless] $t_+ + t_- = 1$

This approach yields experimental transference number $t_- = 0.2$ and with Eq. (4.56) $t_+ = 0.8$.

By a combination of measured molar conductivities of salts and transference numbers of ion types or measured ionic mobilities (for the case of colored ions), individual molar conductivities λ and ion mobilities $u = \lambda/(zF)$ can be derived and tabulated. For KCl, this yields $\lambda_0(K^+) = t_+ \Lambda_0(KCl)$ and $\lambda_0(Cl^-) = t_- \Lambda_0(KCl)$.

As shown in Table 4.2, transference numbers can span a wide range. In some salt solution, for example, in KCl solution, the ionic current is almost equally divided between anions and cations ($t_+ \approx t_-$). This is the reason for KCl being particularly suitable for salt bridges (cf. Figure 2.15). In contrast, acids and bases have very different t_+ and t_- values due to the effect of the extraordinarily high conductivity of H^+ and OH^-. We can assume that t_+ and t_- are in good approximation concentration-independent. That is, the molar conductivity does change with concentration, but the *ratio* of the conductivity of anions and cations does not.

Table 4.2 Transference numbers of selected electrolytes at 25 °C.

Electrolyte	t_+	$t_- = 1 - t_+$
KCl	0.4906	0.5094
NH_4Cl	0.4909	0.5091
HCl	0.821	0.179
KOH	0.274	0.726
NaCl	0.3962	0.6038
$NaOOCCH_3$	0.5507	0.4493
$KOOCCH_3$	0.6427	0.3573
$CuSO_4$	0.375	0.625

The shaded cells show examples in which the extraordinarily high conductivity of H^+ and OH^- plays a role.
Source: Adapted from Landolt–Börnstein; *Zahlenwerte und Funktionen*, Vol. 2, part 7, Springer Verlag, Berlin 1960; and C. H. Hamann, W. Vielstich; *Elektrochemie*, 3rd ed., Wiley-VCH, Weinheim, 2003, p. 29.

An extreme case of transference is found in ion-exchange membranes. We had briefly introduced Nafion membranes in the context of fuel cells in Figure 2.30. The material Nafion consists of perfluorinated and extremely hydrophobic polymer chains. The side chains end in a very hydrophilic sulfonic acid group. These contrasting properties within one and the same polymer molecule cause self-assembly into a conformation in which the sulfonic acid groups line a water-filled pore. The perfluorinated chains form a hydrophobic matrix. Ions can move within the water-filled pores of about 1 nm in diameter. However, only the cations are mobile because the anions are the sulfonate groups covalently linked to the polymer backbone. In such membranes, the transference numbers are $t_+ \approx 1$ and correspondingly $t_- \approx 0$.

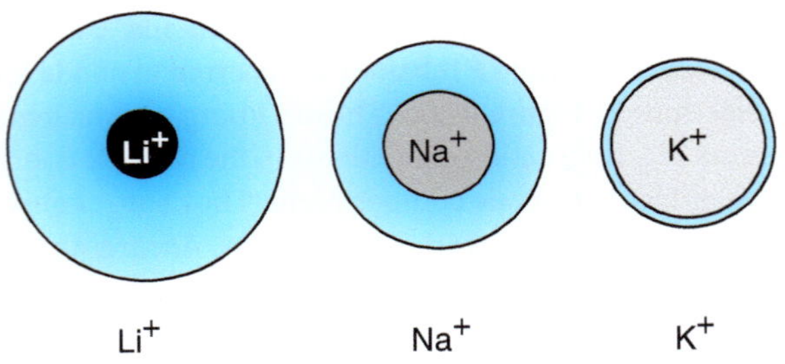

Figure 4.21 Schematic illustration of the relationship between crystallographic ionic radii, charge density, and hydrodynamic ionic radii.

Table 4.3 Molar conductivity at infinite dilution and equivalent conductivities at infinite dilution for important ions at 25 °C.

Ion	λ_0/ (Ω^{-1} mol^{-1} cm^2)	λ_0/z/ (Ω^{-1} mol^{-1} cm^2)
H^+	349.8	349.8
OH^-	197.0	197.0
Li^+	38.7	38.7
Na^+	50.1	50.1
K^+	73.5	73.5
Rb^+	77.5	77.5
Cs^+	77.0	77.0
$NH_4{}^+$	73.7	73.7
Ag^+	62.2	62.2
Be^{2+}	90.0	45.0
Mg^{2+}	106.0	53.0
Ca^{2+}	119.6	59.8
Sr^{2+}	118.8	59.4
Ba^{2+}	126.4	63.2
$CrO_4{}^{2-}$	166.0	83.0
$MnO_4{}^-$	62.8	62.8
F^-	55.4	55.4
Cl^-	76.4	76.4
Br^-	78.1	78.1
I^-	76.5	76.5
$NO_3{}^-$	71.5	71.5
$SO_4{}^{2-}$	161.6	80.8
$ClO_4{}^-$	67.4	67.4
Acetate	40.9	40.9
Benzoate	32.4	32.4

Source: Data from R. Altmann, G. Brandes, O. Regen, J. Schneider; *Chemisch-technische Stoffwerte – Eine Datensammlung*. 2nd ed., Deutscher Verlag für Grundstoffindustrie, Leipzig, 1987, p. 82.

Table 4.3 provides a comparison of ion conductivities at infinite dilution for several ion types. For such comparisons, it is practical to express the data as equivalent conductivities $\lambda_{eq,0} = \lambda_0/z$, which are obtained by dividing the molar quantities by the charge number z of the ion. H^+ and OH^- stand out due to the extraordinarily high conductivity. Surprisingly, the comparison between the alkali metal ions leads to the sequence $\lambda_0(Cs^+) \approx \lambda_0(Rb^+) > \lambda_0(K^+) > \lambda_0(Na^+) > \lambda_0(Li^+)$. The reason for this sequence is the higher hydrodynamic radius of Li^+ compared to all other alkali metal ions. But why does the trend of the hydrodynamic ionic radii run opposite to the trend of the crystallographic ionic radii? We can answer this question by considering Figure 4.21. All alkali metal cations have the charge number +1. This charge is distributed on a much smaller sphere in the case of Li^+ compared to K^+ due to Li^+ having a much smaller crystallographic ionic radius than K^+. (Keep in mind the r^3-dependence of the volume.) Therefore, the charge density (i.e., charge per volume) is much higher for Li^+. The charge density is indicated by the gray scale of the ions in Figure 4.21. Due to the higher charge density, Li^+ forms a much stronger electric field, which causes an orientation of more water molecule dipoles. Therefore, the thickness of the solvation shell is higher for ions with a larger charge density (blue in Figure 4.21). The stronger influence on the solvent molecules overcompensates the smaller crystallographic radius. A similar trend is found when comparing the data for the alkaline earth metal cations. The NH_4^+ ion has approximately the same size as K^+ and also a similar molar conductivity. The divalent ions have a larger molar conductivity due to the charge number +2 or −2, yet the molar conductivity is *clearly less than the double* of the molar conductivity of the alkali metal ions of the same period in the periodic table. The divalent cations have a higher charge density and thus a larger hydrodynamic radius, leading to a lower ionic mobility.

Excursion 4-1 Application of Conductivity Measurements

Quality Assurance of Ultrapure Water. We have already noted that all ions in solution contribute to the conductivity. A conductivity measurement, or conductometry, is highly sensitive, that is, the conductivity signal changes significantly upon very small changes in ion concentration. However, the measurement is not selective at all. An increased concentration of arbitrary types of ions causes a similar signal increase. When the presence of any ions is linked to a loss in water quality, **conductometry can be used as an analytical method**, for instance, for service water. You may have met a conductometric device at the ultrapure water facility in your laboratory. The quality of deionized water is often specified by its specific resistivity ρ.

$$\rho = \frac{1}{\kappa} = \left(\sum_{i=1}^{2} c_i \lambda_{0,i} f_{\lambda,i}\right)^{-1} \tag{4.65}$$

Absolutely pure water at pH 7 and 25 °C contains some ions $[H^+] = [OH^-] = 1 \times 10^{-7}$ mol L^{-1} due to the auto-protolysis of water. At such low ionic strengths, ions do not interact, thus $f_{\lambda,i} = 1$. With the data from Table 4.3, we obtain:

$$\begin{aligned}\rho &= \left(10^{-7}\,\text{mol L}^{-1} \cdot 349.8\,\Omega^{-1}\,\text{mol cm}^2 \right.\\ &\quad \left. + 10^{-7}\,\text{mol L}^{-1} \cdot 197.0\,\Omega^{-1}\,\text{mol cm}^2\right)^{-1}\\ &= \left(10^{-10}\,\text{mol cm}^{-3} \cdot 349.8\,\Omega^{-1}\,\text{mol cm}^2 \right.\\ &\quad \left. + 10^{-10}\,\text{mol cm}^{-3} \cdot 197.0\,\Omega^{-1}\,\text{mol cm}^2\right)^{-1}\\ &= \left(5.47 \times 10^{-8}\,\Omega^{-1}\,\text{cm}^{-1}\right)^{-1}\\ &= 1.828 \times 10^{7}\,\Omega\,\text{cm} = 18.3\,\text{M}\Omega\,\text{cm}\end{aligned} \tag{4.66}$$

A conductometric measurement can also be instrumental in determining the salt content in a solution in which the relative contents of the different types of ions do not vary significantly. An example is the salt content of near-shore seawater, which may be diluted by freshwater from rivers and rainfall. The relative ion concentrations remain fairly constant. The freshwater/seawater ratio can be determined, but the effects of temperature on molar conductivities must be compensated for.

Conductometric Indication of the Endpoint in Titrations. If a solution contains more than one ion type, a titration can be employed, followed by conductometry in order to find the endpoint (endpoint indication). In this application, the absolute value of κ is not really of interest, but rather the *change in* κ is monitored as the titrant is added. Reactions in the solution are revealed by changes in conductivity. The selectivity of the method does depend on the selectivity of the reaction of the titrant with the analyte ions but not on the selectivity of the conductometric probe itself. Figure 4.22 shows the result for an acid–base titration according to reaction (4.67).

$$H^+ + Cl^- + (Na^+ + OH^-) \rightarrow H_2O + Na^+ + Cl^- \tag{4.67}$$

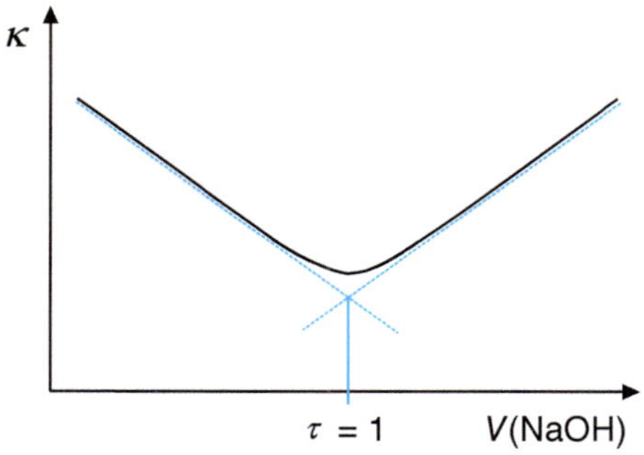

Figure 4.22 Titration curve for the titration of HCl with NaOH with conductometric endpoint indication.

Before the equivalence point ($\tau < 1$), one H^+ ion of HCl in the sample with very high molar conductivity is replaced by Na^+ of NaOH titrant with just ordinary molar conductivity (Table 4.3). Therefore, the specific conductivity of the reaction mixture decreases almost linearly as titrant is added. After the equivalence point ($\tau > 1$), the added ions do not react and the specific conductivity of the solution increases almost linearly with further addition of NaOH solution. The titration graph is curved near the equivalence point. That is why the titration endpoint is determined at the intersection point of the two regression lines through the linear branches of the titration curves for either $\tau < 1$ or $\tau > 1$. The equivalence point is the volume that must be added to complete the reaction. The endpoint is the volume determined practically by the indication method. Ideally, the two should be equal but for some indication methods, there can be a systematic deviation (compare to the problem of selecting an appropriate indicator for optical detection in acid–base titrations).

Q: What is the meaning of $\tau < 1$?

A: The symbol τ (tau) stands for the dimensionless degree of titration, that is, the relative conversion. If exactly the stoichiometric amount of titrant has been added to convert all the analyte in the sample (here H^+), the equivalence point is reached, corresponding to the degree of titration $\tau = 1$. The degree of titration can be calculated from the ratio of added volume of the titrant and the equivalence volume

$$\tau = V(\text{NaOH})/V_{\text{eq}}(\text{NaOH}) \tag{4.68}$$

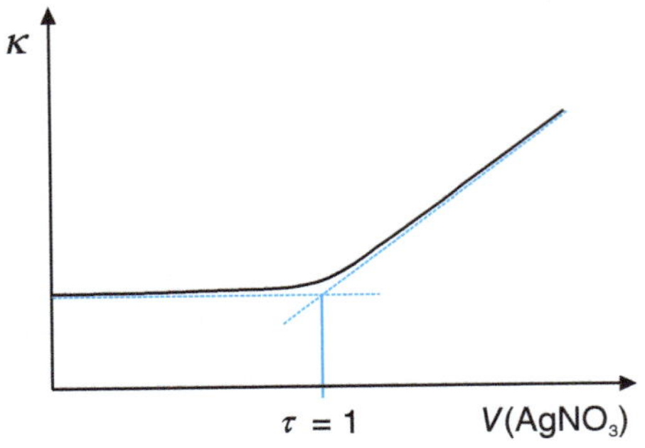

Figure 4.23 Titration curve for the precipitation titration of NaCl with $AgNO_3$ and conductometric detection of the endpoint.

Figure 4.24 Impermeability test of condoms by means of a conductivity measurement between an inner electrode and an outer electrode in an immersion bath.
Source: Photograph provided by Ritex GmbH, Bielefeld, Germany.

This quantity enables a unified description of titrations independently of the particular amount of analyte in the sample or the particular concentration of the titrant.

The use of conductometric endpoint indication is not limited to acid–base titrations. Figure 4.23 shows the conductometric monitoring of a precipitation titration for the determination of chloride according to reaction (4.69).

$$Na^+ + Cl^- + (Ag^+ + NO_3^-) \rightarrow AgCl \downarrow + Na^+ + NO_3^- \tag{4.69}$$

For ($\tau < 1$), the conductivity of the consumed Cl^- and the added NO_3^- are not very different (Table 4.3). Therefore, the specific conductivity is initially roughly constant. Only after passing the equivalence point ($\tau > 1$), the conductivity increases approximately linearly with titrant because the ions stop reacting (4.69). In this case, a conductometric indication of the endpoint is particularly important due to optical methods failing in a sample that is turbid due to precipitated AgCl.

Conductometric Leakage Testing. Here is another simple, transferable, and potentially life-saving application of conductometry: the impermeability test of condoms. From each production batch, a random sample is filled with an electrolyte solution and an inner electrode. The electrolyte-filled condom is immersed in an electrolyte bath with the outer electrode (Figure 4.24). For an impermeable condom, there is no ionic conductivity between the inner electrode and the outer electrode [15]. The available ions do not find a pathway across the polymer film of the condom. The substantially larger viruses, bacteria, or sperm cells would also not be able to pass the barrier. The test procedure is contactless and can be performed automatically. Any leakages cause an electrical signal and imply production problems. In an analogous way, a leakage test can also be conducted at other barrier layers.

4.2.8 Interactions Between Ions

The reason for the nonideal properties of electrolyte solutions, evident from activity coefficients $f \neq 1$ and conductivity coefficients $f_\lambda \neq 1$, is interionic interactions. In this section, we will quantify the extent of these interactions. Empirical findings of Lewis[34] and Randall[35] showed that the activity coefficient of an ion type depends on the concentration of all ions in solution, whereby ions with a charge +2 (e.g., 1 mol L^{-1} Ca^{2+}) cause a larger deviation from the ideal behavior than twice the concentration of singly charged ions (e.g., 2 mol L^{-1} Na^+). To express the effective total ion concentration, the ionic strength I_c has been introduced.

$$I_c = \frac{1}{2}\sum_{i=1}^{n} c_i\ z_i^2 \tag{4.70}$$

Q: What is the meaning of the subscript c in the symbol I_c?
A: The subscript c indicates that the ionic strength refers to the concentration, more exactly, the amount of ions per volume of solution. For precision measurement, the quantity molality b (often with symbol m) is preferred. From this quantity, a corresponding ionic strength I_m can be calculated. The textbook author is also particularly happy that the subscripts help avoiding confusion with the symbol I for current ☺.

[34]Gilbert Newton Lewis, 1875–1946, US-American physicochemist, professor at MIT and at the University of California in Berkley, important contribution to the theory of chemical bonds (Lewis formula), acid–base theory (Lewis acids), the concept of activities, and phosphorescence of organic molecules.
[35]Merle Randall, 1888–1950, US-American physicochemist, professor at the University of California in Berkley.

Example Calculation 4.3 Calculate the ionic strength of a solution that contains 1×10^{-2} mol L^{-1} NaCl and 1×10^{-2} mol L^{-1} Na_2SO_4.

unknown: I_c

known:

$$[Na^+] = 1 \times 10^{-2}\,\text{mol}\,L^{-1} + 2 \times 10^{-2}\,\text{mol}\,L^{-1}$$

$$[Cl^-] = 1 \times 10^{-2}\,\text{mol}\,L^{-1}$$

$$[SO_4^{2-}] = 1 \times 10^{-2}\,\text{mol}\,L^{-1}$$

$$I_c = \frac{1}{2}\sum_{i=1}^{n} c_i\, z_i^2$$

$$= \frac{1}{2}(3 \times 10^{-2}\,\text{mol}\,L^{-1} \cdot 1^2 + 1 \times 10^{-2}\,\text{mol}\,L^{-1} \cdot 1^2 + 1 \times 10^{-2}\,\text{mol}\,L^{-1} \cdot 2^2)$$

$$= \frac{1}{2}(3 \cdot 1 + 1 \cdot 1 + 1 \cdot 2^2) \times 10^{-2}\,\text{mol}\,L^{-1}$$

$$I_c = 4 \times 10^{-2}\,\text{mol}\,L^{-1}$$

The ionic strength is 4×10^{-2} mol L^{-1} and we realize that the sulfate anions provide the largest contribution of all ion types due to the charge number −2.

◢

Furthermore, we must realize that activities and activity coefficients can principally not be determined for single type of ions because in any experimental system we can only add cations and anions together. Therefore, we have resorted to mean activities and mean activity coefficients, and mean concentrations. These mean values are based on the geometric mean of the respective single ion quantities. For a strong electrolyte $Me_{\nu+}X_{\nu-}$, the concentration c_0 of the dissolved formula units provides the quantities according to Eqs. (4.71) to (4.73).

$$a_{\pm} = \sqrt[\nu_+ + \nu_-]{a_+^{\nu_+} \cdot a_-^{\nu_-}} \tag{4.71}$$

$$f_{\pm} = \sqrt[\nu_+ + \nu_-]{f_+^{\nu_+} \cdot f_-^{\nu_-}} \tag{4.72}$$

$$c_{\pm} = \sqrt[\nu_+ + \nu_-]{c_+^{\nu_+} \cdot c_-^{\nu_-}} = \sqrt[\nu_+ + \nu_-]{(\nu_+ c_0)_+^{\nu_+} \cdot (\nu_- c_0)_-^{\nu_-}}$$
$$= c_0 \cdot \sqrt[\nu_+ + \nu_-]{(\nu_+)_+^{\nu_+} \cdot (\nu_-)_-^{\nu_-}} \tag{4.73}$$

Example Calculation 4.4 Provide the formula for the calculation of the mean concentration of $ZnCl_2$.

$$c_{\pm} = c_0 \cdot \sqrt[1+2]{(1)_+^1 \cdot (2)_-^2}$$
$$= c_0 \sqrt[3]{4} \approx 1.59\, c_0$$

◢

Example Calculation 4.5 Provide the formula for the calculation of the mean concentration of $Al_2(SO_4)_3$.

$$c_{\pm} = c_0 \cdot \sqrt[2+3]{(2)^2 \cdot (3)^3}$$
$$= c_0 \sqrt[5]{4 \cdot 27} \approx 2.55\, c_0$$

◢

Example Calculation 4.6 Calculate the mean activity of 0.1 mol L^{-1} Na_2SO_4 with $f_\pm = 0.453$ (from tabulated data).

unknown: $a_\pm$

known:

$$c_0 = 1 \times 10^{-1}\ \mathrm{mol\ L^{-1}}$$

$$f_\pm = 0.453$$

$$a_\pm = \sqrt[2+1]{a_+^2 \cdot a_-^1}$$

$$= \sqrt[3]{(f_\pm c_+)^2 \cdot (f_\pm c_-)}$$

$$= f_\pm \sqrt[3]{(2c_0)^2 \cdot (1c_0)^1}$$

$$= f_\pm c_0 \sqrt[3]{2^2 \cdot 1^1}$$

$$\approx 0.453 \cdot 1.587 \cdot 0.1\ \mathrm{mol\ L^{-1}}$$

$$\approx 0.07191\ \mathrm{mol\ L^{-1}}$$

The mean activity is 0.072 mol L^{-1}.

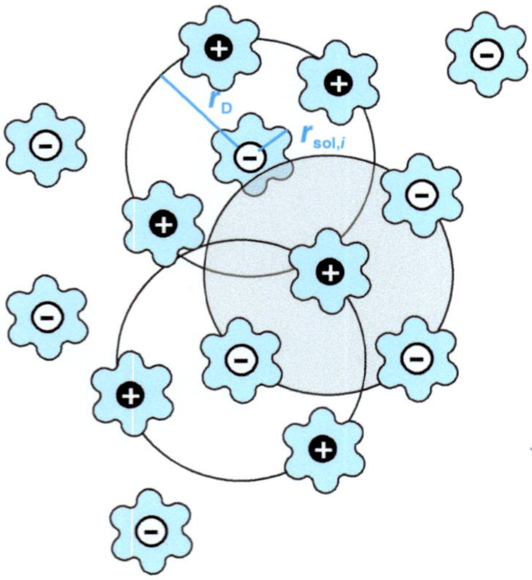

Figure 4.25 Schematic illustration of the positions of ions in an electrolyte solution for a moment in time. The ion cloud is indicated for selected ions by circles.

How can we interpret the decrease in the effective concentration (i.e., of the activity a) in the mass action law compared to the analytical concentration c? Let us imagine, we could use a magic camera to take a picture of the momentary positions of ions in a solution of a completely dissociated (strong) electrolyte. A schematically simplified depiction of such an image is given in Figure 4.25. Consider the surroundings of a single cation that is emphasized by the gray circle. Although the movement of the ions in the absence of an electric field is random, there are more anions in the immediate vicinity of a cation (in the example, the ratio is 4:1). Analogous considerations apply to all ions in Figure 4.25. Every ion can be considered as the central ion of an ion cloud of opposite charge surrounding it. At the same time, every ion is also part of the ion cloud of another neighboring ion.

The ion clouds decrease the effective concentration. Before an ion can undergo a chemical reaction, the surrounding ion cloud must be broken up under consumption of energy, that is, the electrostatic attraction between the central ions and the ion cloud must be overcome. **The Debye–Hückel theory**[36,37] attempts an estimation of the extent of nonideality in diluted electrolyte solutions based on ion–ion interactions. The theory considers the ions as point charge carriers that do not have a volume and that are not polarizable. The interaction between ions is assumed to be exclusively of electrostatic nature. Chemical interactions are neglected to keep the system relatively simple. In such a system, the ions should spontaneously order into a crystal if there were no random thermal fluctuations that counteract the ordering by crystallization. The thermal energy is assumed to be large compared to electrostatic attraction and is described by a Boltzmann[18] distribution of kinetic energy. These assumptions are strictly valid only in dilute electrolyte solutions. We can further assume that the

[36] Peter Joseph Wilhelm Debye, 1884–1966, Dutch physicist, professor in Zurich, Utrecht, Göttingen, Leipzig, Kaiser Wilhelm Institut Berlin, Cornell University Ithaca, Nobel Prize 1936, worked on X-ray diffraction (Debye–Scherrer method), on the theory of electrolyte solutions (Debye length, Debye–Hückel theory, Debye relaxation time).

[37] Erich Armand Joseph Hückel, 1896–1980, German mathematician and physicist, contributions to the theory of electrolyte solutions and molecular orbital theory (Hückel MO).

relative permittivity of the electrolyte solution ε_r does not differ significantly from that of pure water ($\varepsilon_r \approx \varepsilon_r(H_2O)$).

Due to space constraints, we omit a complete derivation and instead refer to classical textbooks of Physical Chemistry [16, 17]. As a result, we obtain an expression for the radius of the ion cloud r_D (Debye length)

$$r_D = \sqrt{\frac{\varepsilon_0 k_B}{2e^2 N_A}} \cdot \sqrt{\frac{\varepsilon_r T}{I_c}}. \tag{4.74}$$

r_D – Radius of an ion cloud given as Debye length [m]

$k_B = R/N_A = 1.381 \times 10^{-23}$ J K^{-1} – Boltzmann constant

$e = F/N_A = 1.602 \times 10^{-19}$ A s – Elemental charge

$N_A = 6.022 \times 10^{23}$ mol^{-1} – Avogadro constant

$\varepsilon_0 = 8.854 \times 10^{-12}$ A^2 s^4 m^{-3} kg^{-1} – Permittivity of vacuum

ε_r – Relative permittivity of the solvent [dimensionless]

T – Temperature [K]

I_c – Ionic strength [mol L^{-1}], it is convenient to recalculate the quantity to mol m^{-3} before inserting it in the formula

The quantity r_D refers to the radius of the shell around a central ion at which the opposite charges have shielded the local electric potential to $1/e = \exp(-1)$ or 36.8%. This distance represents the region that contains most of the oppositely charged excess ions (the ion cloud), shielding the charge of the central ion.

Example Calculation 4.7 Derive from Eq. (4.74) an expression $r_D = k_1 \cdot (\varepsilon_r T/I_c)^{1/2}$. Obtain k_1 for water at 25 °C.

We need to pay attention to the compatibility of units. The natural constants are typically given in the SI units ([m] for length), while we are used to giving ionic strength in [mol L^{-1}] = [mol dm^{-3}]. When converting the units, we use the relationship 1 J = 1 Nm = 1 (kg m s^{-2}) m = 1 kg m^2 s^{-2}.

$$r_D = \sqrt{\frac{8.854\times10^{-12}\mathrm{A^2\,s^4\,m^{-3}\,kg^{-1}} \cdot 1.381\times10^{-23}\,\mathrm{kg\,m^2\,s^{-2}}}{2(1.602\times10^{-19}\mathrm{As})^2 \cdot 6.022\times10^{23}\,\mathrm{mol^{-1}}}} \cdot \sqrt{\frac{\varepsilon_r T}{I_c}}$$

$$= \sqrt{3.868\times10^{-21}} \cdot \sqrt{\frac{\mathrm{m^{-1}K^{-1}}}{\mathrm{mol^{-1}}}} \cdot \sqrt{\frac{\varepsilon_r T}{I_c}}$$

$$= 6.219\times10^{-11} \cdot \sqrt{\frac{\mathrm{m^{-1}K^{-1}}}{\mathrm{mol^{-1}}}} \cdot \sqrt{\frac{\varepsilon_r T}{I_c}}$$

This is the first part of the solution. The units in the middle term of the solution show that *i*) we should provide I_c in mol m^{-3} in order to reduce the units in a sensible way, and *ii*) we obtain r_D [m]. There is nothing wrong with providing I_c in 1 mol L^{-1} = 1 mol dm^{-3} = 1000 mol m^{-3}. However, for repeated use, it is more convenient to adapt our equation in such a way that we can use I_c in the usual units [mol L^{-1}] = [mol dm^{-3}]. The treatment of the units may be a bit confusing due to the square root. In turn, this provides an opportunity to test a suitable "trick". We arrange the length units in a way that they form a unity factor, for example, $1 = \frac{1\ \mathrm{m}}{10\ \mathrm{dm}}$. This expression we can multiply by any term in the equation (i.e., expanding the fractional expression). This is helpful at two places in this expression. We use the method to transform the unit m^{-1} in the first root, which provides r_D [dm], which then should further transform, for instance, to [m], [cm], or [nm]. Even better would be to integrate the transformation into the formula:

$$r_D = 6.219\times10^{-11} \cdot \sqrt{\frac{1\ \mathrm{m}}{10\ \mathrm{dm}}\frac{\mathrm{m^{-1}\,K^{-1}}}{\mathrm{mol^{-1}}}} \cdot \sqrt{\frac{\varepsilon_r T}{I_c}\frac{1\ \mathrm{m}}{10\ \mathrm{dm}}}$$

$$= 6.219\times10^{-11}\frac{1}{10\sqrt{10}}\mathrm{m}\sqrt{\frac{\mathrm{K^{-1}}}{\mathrm{dm^3\,mol^{-1}}}} \cdot \sqrt{\frac{\varepsilon_r T}{I_c}}$$

$$= 1.967\times10^{-12}\mathrm{m}\sqrt{\frac{\mathrm{mol}}{\mathrm{L\cdot K}}} \cdot \sqrt{\frac{\varepsilon_r T}{I_c}} \tag{4.75}$$

This formula is often quoted for r_D, unfortunately, without a definition of the units of the factor preceding the root. There are also variations of this expression with a supposedly dimensionless prefactor that provides a correct result for r_D [cm] *only* under the condition that I_c [mol dm^{-3}] is used without any indication of this convention in the expression itself. Because preferences for units are different from subject area to subject area (electrochemistry is an interdisciplinary field!) and the preferences also change over time, the textbook author strongly discourages the use of calculus formulas with hidden dimensions. They can provide correct results only if used with units according to an "agreement". This agreement is *not* evident from the formula itself, and all too often it is never recorded at all.

In scientifically correctly typed texts, the difference between symbols (e.g., m for mass) and units (e.g., m for meter) is made clear by typing symbols (and only symbols) in italics and units in nonitalicized fonts. Using Eq. (4.75), we can take I_c [mol L^{-1}] and T [K] and obtain r_{D} [m]. Finally, we can treat the frequently encountered special case of an aqueous solution at room temperature ($T = 25\,°\mathrm{C} \approx 298$ K and $\varepsilon_{\mathrm{r}} = 78.54$):

$$\begin{aligned} r_{\mathrm{D}} &= 1.967 \times 10^{-12}\,\mathrm{m}\sqrt{\frac{\mathrm{mol}}{\mathrm{L \cdot K}}} \cdot \sqrt{\frac{78.54 \cdot 298\,\mathrm{K}}{I_c}} \\ &= 3.0 \times 10^{-10}\,\mathrm{m} \cdot \sqrt{\frac{\mathrm{mol}}{\mathrm{L}}} \cdot \sqrt{\frac{1}{I_c}} \end{aligned} \tag{4.76}$$

◢

4.2.9 Activity Coefficients

Activity coefficients are associated with the electrical work required to introduce an additional ion into a given electrolyte solution. Conceptually, we first place a neutral particle into the solution requiring no electrical work. Second, we charge up the particle to the ionic charge $z_i \cdot e$. The electrical work required for this process is divided into two parts. One contribution is required to build up the electric field around the particle. This work would also be required in an infinitely dilute solution (without ion–ion interaction) and thus corresponds to the model of an ideal solution. The second contribution is linked to ion–ion interactions. If there are other ions already present in the solution, the added ion immediately interacts with the ion cloud. The energy of electrostatic interaction is

$$E_{\mathrm{p}} = -\frac{N_{\mathrm{A}} z_i^2 e^2}{8\pi\varepsilon_0\varepsilon_{\mathrm{r}} r_{\mathrm{D}}} \tag{4.77}$$

E_{p} usually stabilizes the ionic charge. Thus, charging up the additional ion requires slightly less energy compared to that for the ideal solution. It is exactly this stabilizing contribution that causes the deviation between ideal and real behavior of electrolyte solutions described by the activity coefficients f_i. We obtain an expression for f_i from the following relationship:

$$\begin{aligned} \mu_i^{\mathrm{real}} - \mu_i^{\mathrm{ideal}} &= E_{\mathrm{p}} = RT \ln f_i \\ &= -\frac{N_{\mathrm{A}} z_i^2 e^2}{8\pi\varepsilon_0\varepsilon_{\mathrm{r}}} \sqrt{\frac{2e^2 N_{\mathrm{A}} I_c}{\varepsilon_0 k_{\mathrm{B}} \varepsilon_{\mathrm{r}} T}} \end{aligned} \tag{4.78}$$

Isolating $\ln f_i$, substituting $R = k_{\mathrm{B}} \cdot N_{\mathrm{A}}$ and sorting natural constants and prefactors yields

$$\begin{aligned} \ln f_i &= -\frac{z_i^2 e^2}{8\pi\,\varepsilon_0\varepsilon_{\mathrm{r}} k_{\mathrm{B}} T} \sqrt{\frac{2e^2 N_{\mathrm{A}} I_c}{\varepsilon_0\varepsilon_{\mathrm{r}} k_{\mathrm{B}} T}} \\ &= -\frac{1}{8\pi}\left(\frac{e^2}{\varepsilon_0 k_{\mathrm{B}}}\right)^{3/2} (2N_{\mathrm{A}})^{1/2} z_i^2 \left(\frac{1}{\varepsilon_{\mathrm{r}} T}\right)^{3/2} (I_c)^{1/2} \\ &= -4.1989 \times 10^6\,\mathrm{dm}^{3/2}\,\mathrm{mol}^{-1/2}\,\mathrm{K}^{3/2}\, z_i^2 \left(\frac{1}{\varepsilon_{\mathrm{r}} T}\right)^{3/2} (I_c)^{1/2} \end{aligned} \tag{4.79}$$

or expressed with the decadic logarithm

$$\log f_i = -\frac{4.1989\times 10^6}{2.3}\,\mathrm{dm}^{3/2}\,\mathrm{mol}^{-1/2}\,\mathrm{K}^{3/2}\,z_i^2\left(\frac{1}{\varepsilon_\mathrm{r}T}\right)^{3/2}(I_c)^{1/2}. \tag{4.80}$$

$\log x = \log(\mathrm{e}^{\ln x}) = \ln x \cdot \log \mathrm{e} \approx \ln x/2.3$

For water at $T = 298$ K ($\varepsilon_\mathrm{r} = 78.54$), one obtains

$$\begin{aligned}\log f_i &= -A(\varepsilon_\mathrm{r}, T)z_i^2\sqrt{I_c}\\ &= -0.5095\ \mathrm{L}^{1/2}\,\mathrm{mol}^{-1/2}\,z_i^2\sqrt{I_c}\end{aligned} \tag{4.81}$$

$A(\varepsilon_\mathrm{r},T)$ – Numerical constant.

or

$$\ln f_i = -1.1717\ \mathrm{L}^{1/2}\,\mathrm{mol}^{-1/2}\,z_i^2\sqrt{I_c}. \tag{4.82}$$

The single ion activity coefficients depend on the ionic strength I_c, the solvent (via ε_r), the temperature, and the absolute amount of the charge of the considered ion, but not on the chemical identity of the considered ion. For all ions with the same ionic charge, f_i should be equal for a given solvent, T and I_c. For a comparison to experimental data in Figure 4.26, we have to calculate the mean activity coefficients $f_\pm$ according to Eq. (4.72). Experimentally accessible are only the mean activity coefficients $f_\pm$ (Section 2.4.2).

$$\begin{aligned}\log f_\pm &= \frac{1}{\nu_+ + \nu_-}(\nu_+ \log f_+ + \nu_- \log f_-)\\ &= \frac{1}{\nu_+ + \nu_-}\left(\nu_+ - A(\varepsilon_\mathrm{r}, T)z_+^2\sqrt{I_c} - \nu_-\lg A(\varepsilon_\mathrm{r}, T)z_-^2\sqrt{I_c}\right)\\ &= \frac{-A(\varepsilon_\mathrm{r}, T)}{\nu_+ + \nu_-}\left(\nu_+ z_+^2 + \nu_- z_-^2\right)\sqrt{I_c}\end{aligned} \tag{4.83}$$

From the electroneutrality condition ($z_+ > 0$, $z_- < 0$), we obtain

$$\nu_+ z_+ + \nu_- z_- = 0 \tag{4.84}$$

Multiplication with z_+ or z_- yields

$$\nu_+ z_-{}^2 + \nu_- z_-\,z_+ = 0 \tag{4.85}$$

$$\nu_+ z_-\,z_- + \nu_- z_-{}^2 = 0 \tag{4.86}$$

Addition of Eqs. Gl. (4.85) and (4.86) yields after rearrangement

$$\nu_+ z_+{}^2 + \nu_- z_-{}^2 = -\nu_+ z_+\,z_- - \nu_- z_-\,z_+ = -(\nu_+ + \nu_-)\,z_+\,z_- \tag{4.87}$$

Inserting this expression into Eq. (4.83) gives Eq. (4.88), which is often written with an explicit negative sign and absolute values of the charge numbers of the ions $|z_i| > 0$ and with the numerical constant $A(\varepsilon_\mathrm{r},T) > 0$.

$$\log f_\pm = z_+ z_- A(\varepsilon_\mathrm{r}, T)\sqrt{I_c} = -|z_+ z_-|A(\varepsilon_\mathrm{r}, T)\sqrt{I_c} \tag{4.88}$$

This expression agrees with the experimental findings insofar as the logarithm of the activity coefficient in dilute electrolyte solution at constant temperature is proportional to the charge number and the ionic strength. Now, we are able to explain the origin of the proportionality constant. For $I_c \to 0$ we find $\log f_\pm \to 0$ and thus $f_\pm \to 1$. For $I_c > 0$, it follows $\log f_\pm < 0$ and thus $0 < f_\pm < 1$. Ions with higher charge number provide a larger contribution to nonideality compared to twice the concentration of singly charged ions. Ions with higher charge numbers influence the prefactor in Eq. (4.88) *and* enter into the calculation of I_c according to Eq. (4.70).

The comparison to experimentally determined mean activity coefficients in Figure 4.26 reveals that a good agreement is found only for very low ionic strengths $I_c < 0.01$ mol L^{-1}. At ionic strengths that are used in many laboratory experiments,

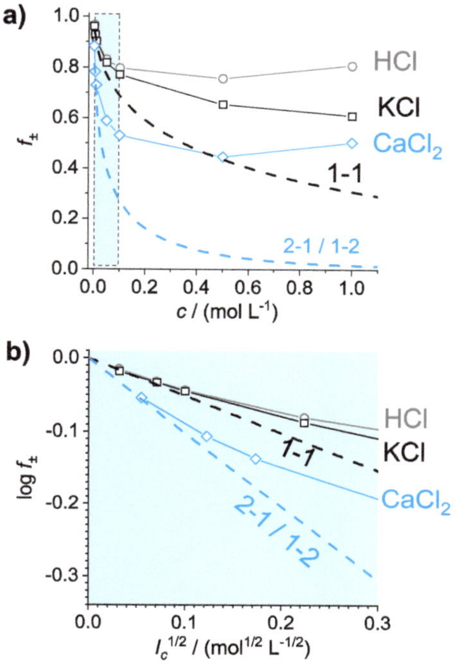

Figure 4.26 Mean activity coefficients of selected strong electrolytes a) as a function of concentrations and b) as linearized plot according to Eq. (4.81) for low concentrations corresponding to the dotted rectangle in a) for a 1–1 electrolyte.

The predictions from Eq. (4.81) are plotted as dashed lines. The solid lines connect the experimental values from one data set to enhance the clarity (*a guide to the eye*). They are not calculated or measured intermediate values. Values for experimental activity coefficients were taken from [4].

Eq. (4.81) is not valid. The range of validity can be somewhat expanded by introduction of further correction terms (extended Debye–Hückel theory). The real importance of Eq. (4.81) is not so much the precise calculation of $f_{\pm}$, but more the *prediction* of trends for the limiting cases of low concentrations. We had already used this insight (somewhat unconsciously) for the extrapolation procedure in the determination of thermodynamic quantities in Section 2.4.2.

Severe deviations between experimental data and Eq. (4.81) are observed for highly concentrated electrolyte solutions (Figure 4.27). So far, the calculation of activity coefficients has escaped a general, practically relevant theoretical treatment. Instead, tabulated experimental data are used from charts and increasingly from electronic databases, which offer integrated sophisticated algorithms for interpolation of values (Example Calculation 2.11).

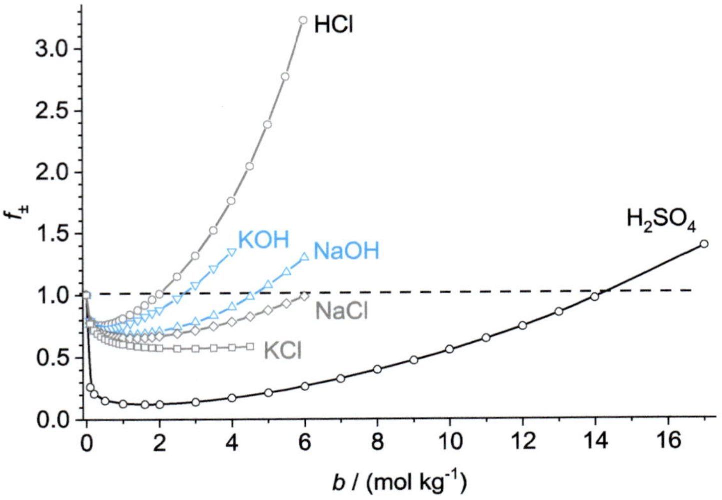

Figure 4.27 Dependence of mean activity coefficients in concentrated electrolyte solutions. The tabulated values from [4] are shown as symbols. The connecting lines are only guiding the eye.

Q: Why is the *x*-axis in Figure 4.27 labeled with *b*?

A: In general, solution contents in exact thermodynamic experiments are tabulated in molalities (Section 2.2.3). While the numerical values of molality and (object number) concentration do not differ significantly for dilute solutions (Example Calculation 2.12), this does not hold for concentrated electrolyte solutions. Therefore, molalities were chosen for the abscissa ("*x*-axis") in Figure 4.27.

Activity coefficients also show a temperature dependence. Such values have already been measured and tabulated for the most important electrolyte solutions. Figure 4.28 shows an example of temperature and concentration dependencies in sulfuric acid. Similar trends are observed for many types of electrolyte solutions.

4.2.10 Conductivity Coefficients

The existence of ion clouds is responsible for the concentration dependency of the molar conductivity Λ and the existence of the conductivity coefficients f_{Λ}. Although the definition of f_{Λ} is reminiscent of the definition of the activity coefficient f and both are caused by the ion clouds, there are fundamental differences between the two quantities. The activity coefficient describes nonidealities of an equilibrium, while the conductivity coefficient summarizes the effect of nonidealities on a nonequilibrium process.

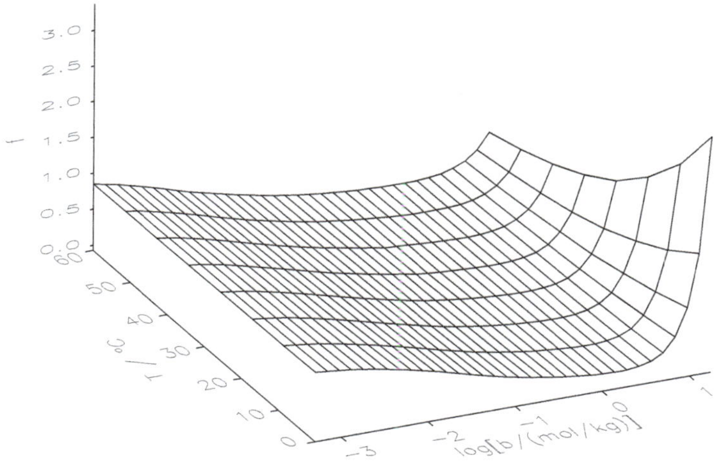

Figure 4.28 Dependency of the mean activity coefficient $f_{\pm}$ for sulfuric acid in water on the molality b (note the logarithmic axis!) and the temperature. The data are interpolated from tabulated data in [4].

The migration of a central ion in an electric field disturbs the symmetry of the ion cloud (Figure 4.29) until a new arrangement of the ions is established. However, the rearrangement of the ions cannot be infinitely fast, and therefore, a retarding force will act on the central ion in the electric field for the duration of such rearrangement (the relaxation time). This decreases the translation velocity of the central ion compared to the situation in an infinitely diluted solution. This effect is called the **relaxation effect,** which leads to decrease in the molar conductivity by Λ_{relax}. The relaxation effect depends on the radius of the ion cloud r_D (Eq. 4.74) and the charge of the ions. In addition, the relaxation effect becomes more important if the ion moves faster. The velocities of ions can be compared by their molar conductivity at infinite dilution (see the relationship between ion mobility and conductivity, Section 4.2.3).

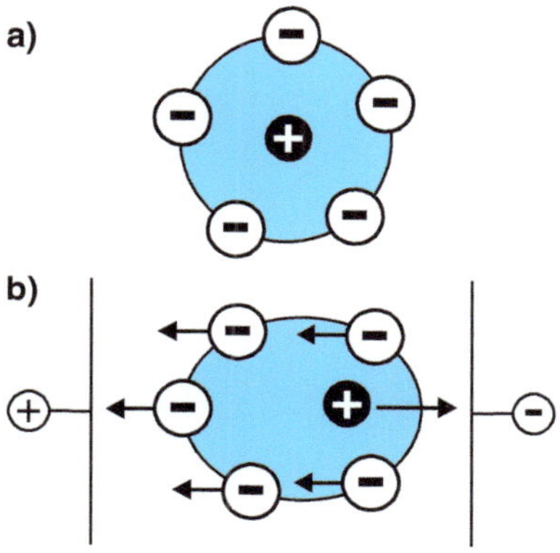

Figure 4.29 Schematic describing the relaxation effect. a) Ion cloud without external field; b) ion cloud in an external electric field.

Furthermore, the central ion does not move in a resting solvent environment. It rather moves in opposite direction to the ions of its ion cloud, dragging their solvation shells with them. The solvent movement associated with the migration of the ion cloud causes a streaming of the local solvent environment of the central ion opposite to its own migration. This situation diminishes the effective migration velocity of the central ion. This effect is the electrophoretic effect Λ_{ephor}. The electrophoretic effect depends on the charge numbers of the ions, the radius of the ion cloud and according to Stokes' law[26] Eq. (4.4) on the viscosity η of the solution. The molar conductivity is obtained from

$$\Lambda = \Lambda_0 - \Lambda_{relax} - \Lambda_{ephor} \tag{4.89}$$

Note that Λ_0 by itself is a function of the solvent (η and ε), as shown in Figure 4.18. The constants A and B in Eqs. (4.91) and (4.93) are derived within the Debye–Hückel–Onsager[38] theory under similar assumptions as those made for the derivation of the Debye–Hückel theory, with r_D according to Eq. (4.74) and with the auxiliary quantity q according to Eq. (4.90).

$$q = \frac{|z_+ z_-|}{|z_+| + |z_-|} \frac{\lambda_{0,+} + \lambda_{0,-}}{|z_+|\lambda_{0,+} + |z_-|\lambda_{0,-}} \tag{4.90}$$

Λ – Molar conductivity [S cm^2 mol^{-1}]
Λ_0 – Molar conductivity at infinite dilution [S cm^2 mol^{-1}]
Λ_{relax} – Decrease in conductivity due to relaxation [S cm^2 mol^{-1}]
Λ_{ephor} – Decrease in conductivity due to electrophoretic effect [S cm^2 mol^{-1}]

$|z_+|$, $|z_-|$ – Absolute value of the charge numbers of anion and cation [dimensionless]
$\lambda_{0,+}$, $\lambda_{0,-}$ – Single ion molar conductivity at infinite dilution for the anion and the cation [S cm^2 mol^{-1}]

[38] Lars Onsager, 1903 Christiana, Norway – 1976, US-American physicochemist, Nobel Prize for chemistry in 1968 for the discovery of reciprocity relationship as a basis for the thermodynamics of irreversible processes.

This yields the relaxation effect

r_D – Radius of the ionic cloud given as the Debye length [m]
ε_0 – Permittivity of vacuum [A^2 s^4 m^{-3} kg^{-1}]
k_B – Boltzmann constant [J K^{-1}]
e – Elemental charge [A s]
N_A – Avogadro constant [mol^{-1}]
ε_r – Relative permittivity of the solvent [dimensionless]
T – Temperature [K]
I_c – Ionic strength [mol L^{-1}]
A – Relaxation constant [$L^{1/2}$ $mol^{-1/2}$]

$$\begin{aligned}
\Lambda_{\text{relax}} &= \frac{e^2}{2k_B}\frac{1}{4\pi\varepsilon_0\varepsilon_r T}\frac{|z_+z_-|\,q}{1+\sqrt{q}}\frac{\Lambda_0}{r_D} \\
&= \frac{e^2}{2k_B}\sqrt{\frac{2e^2N_A}{\varepsilon_0 k_B}}\frac{1}{4\pi\varepsilon_0}\frac{|z_+z_-|\,q}{1+\sqrt{q}}\cdot\left(\frac{1}{\varepsilon_r T}\right)^{3/2}\sqrt{I_c}\Lambda_0 \\
&= \frac{(1.602\times10^{-19}\text{A s})^2}{2\cdot1.381\times10^{-23}\text{J K}^{-1}}\frac{1}{6.219\times10^{-11}}\left(\frac{\text{m K}}{\text{mol}}\right)^{1/2}\frac{1}{4\pi\cdot8.854\times10^{-12}\text{A}^2\,\text{s}^4\,\text{m}^{-3}\,\text{kg}^{-1}} \\
&\qquad \frac{|z_+z_-|\,q}{1+\sqrt{q}}\cdot\left(\frac{1}{\varepsilon_r T}\right)^{3/2}\sqrt{I_c}\Lambda_0 \\
&= 8.861\times10^4\left(\frac{\text{K}^3\,\text{m}^3}{\text{mol}}\right)^{1/2}\frac{|z_+z_-|\,q}{1+\sqrt{q}}\cdot\left(\frac{1}{\varepsilon_r T}\right)^{3/2}\sqrt{I_c}\Lambda_0 \\
&= A(\varepsilon,T)\Lambda_0\sqrt{I_c}.
\end{aligned} \tag{4.91}$$

For an aqueous 1–1 electrolyte, setting in $q = ½$, $\varepsilon_r = 78.54$ and $T = 298$ K, as well as considering the transformation of the concentration unit with $1 = (10\text{ dm})^3/(1\text{ m})^3 = 1000\text{ dm}^3/\text{m}^3$ yields

$$\Lambda_{\text{relax}} = 0.23\left(\frac{\text{dm}^3}{\text{mol}}\right)^{1/2}\Lambda_0\sqrt{I_c}. \tag{4.92}$$

The electrophoretic effect is calculated according to Eq. (4.93).

B – Electrophoretic constant [S cm^2 $mol^{-3/2}$ $L^{1/2}$]

$$\begin{aligned}
\Lambda_{\text{ephor}} &= \frac{e^2N_A}{6\pi}\cdot\frac{|z_+|+|z_-|}{\eta}\cdot\frac{1}{r_D} \\
&= \frac{e^2N_A}{6\pi}\cdot\sqrt{\frac{2e^2N_A}{\varepsilon_0 k_B}}\cdot\frac{|z_+|+|z_-|}{\eta\sqrt{\varepsilon_r T}}\cdot\sqrt{I_c} \\
&= \frac{(1.602\times10^{-19}\text{As})^2\cdot6.02\times10^{23}\text{mol}^{-1}}{6\pi}\cdot\frac{1}{6.219\times10^{-11}} \\
&\quad\times\left(\frac{\text{m K}}{\text{mol}}\right)^{1/2}\cdot\frac{|z_+|+|z_-|}{\eta\sqrt{\varepsilon_r T}}\cdot\sqrt{I_c} \\
&= 1.31\times10^{-5}\frac{\text{A}^2\text{s}^2\text{m}^{1/2}\text{K}^{1/2}}{\text{mol}^{3/2}}\cdot\frac{|z_+|+|z_-|}{\eta\sqrt{\varepsilon_r T}}\sqrt{I_c} \\
&= B(\varepsilon,T,\eta)\sqrt{I_c}
\end{aligned} \tag{4.93}$$

1 Nm = 1 kg m^2 s^{-2} = 1 V · A · s
1 kg = 1 V A s^3 m^{-2}
1 S = 1 Ω^{-1} = 1 A V^{-1}

For an aqueous 1–1 electrolyte, setting in $|z_+| + |z_-| = 2$, $\varepsilon_r = 78.54$, $\eta = 0.89 \times 10^{-3}$ kg m^{-1} s^{-1} at $T = 298$ K and conversion of concentration units with $1 = 1000\text{ dm}^3/\text{m}^3$ as well as conversion of units yields

$$\begin{aligned}
\Lambda_{\text{ephor}} &= 1.92\times10^{-4}\frac{\text{A}^2\text{s}^3\text{m}^{7/2}}{\text{mol}^{3/2}\text{kg}}\sqrt{I_c} \\
&= 6.07\times10^{-3}\frac{\text{A}^2\text{s}^3\text{m}^2}{\text{VAs}^3\,\text{mol}}\left(\frac{\text{dm}^3}{\text{mol}}\right)^{1/2}\sqrt{I_c} \\
&= 60.7\,\Omega^{-1}\text{cm}^2\,\text{mol}^{-1}\left(\frac{\text{L}}{\text{mol}}\right)^{1/2}\sqrt{I_c}.
\end{aligned} \tag{4.94}$$

From Eqs. (4.91) and (4.93), the molar conductivity is

$$\Lambda = \Lambda_0 - A(T,\varepsilon)\Lambda_0\sqrt{I_c} - B(T,\varepsilon,\eta)\sqrt{I_c}. \tag{4.95}$$

With a deviation of a few percent, the equations for 1–1 electrolytes are applicable up to concentrations of 0.1 mol L^{-1}. For electrolytes with ions of higher charge numbers, deviations between Eq. (4.95) and experimental values become evident already for significantly lower concentrations. For a particular electrolyte, Λ_0 is a constant, and c and I_c are proportional to each other. This can be used to simplify Eq. (4.95), which then takes the form of the empirically discovered square-root Eq. (4.36) of Kohlrausch.

$$\Lambda = \Lambda_0 - k_1\sqrt{I_c} - k_2\sqrt{I_c} = \Lambda_0 - (k_1 + k_2)\sqrt{I_c} = \Lambda_0 - k_3'\sqrt{c}$$

For the conductivity coefficient, we obtain from Eq. (4.95)

$$f_\Lambda = \frac{\Lambda}{\Lambda_0} = 1 - \left[A(T,\varepsilon) - \frac{B(T,\varepsilon,\eta)}{\Lambda_0}\right]\sqrt{I_c}. \tag{4.96}$$

For a more careful consideration of weak electrolytes, we ought to consider the concentration-dependent degree of dissociation $\alpha = f(c)$. This leads to a modification of Eq. (4.95):

$$\Lambda = \alpha\left\{\Lambda_0 - [A(T,\varepsilon)\Lambda_0 - B(T,\varepsilon,\eta)]\sqrt{\alpha c}\right\}. \tag{4.97}$$

An in-depth consideration and comparison between experiments and calculations for different electrolyte solutions can be found in [1].

4.3 The Electrical Double Layer

4.3.1 The Electrical Properties of the Double Layer

We have already considered some important electrical properties of the electrical double layer in Section 4.2.2. We assumed two electrodes immersed in an electrolyte solution. If we apply a small voltage between the two electrodes, there is only a small capacitive current rapidly decaying to zero (Figure 4.5). We can imagine this current as the charging current of two parallel-plate capacitor in series. For each capacitor, one of the plates corresponds to the metal surface of the electrode, and the other plate is at the position of the oppositely charged electrolyte ions that have approached the metal surface as closely as possible. Such a capacitor would have the capacitance

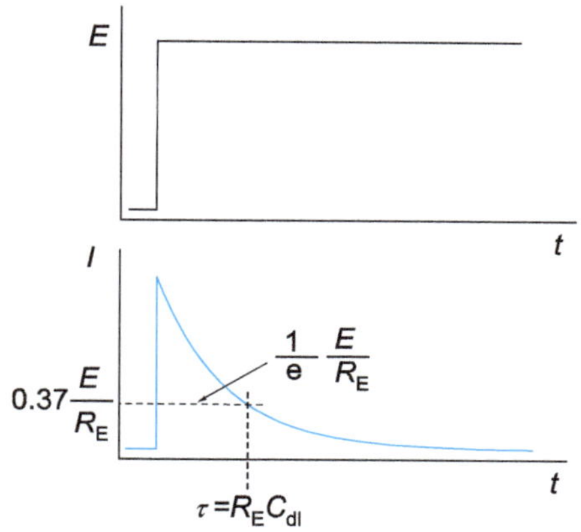

Figure 4.30 Current transient at an electrode after a *potential step* in the absence of an electrochemical conversion.

$$C = \varepsilon_0\,\varepsilon_r\frac{A'}{d} \tag{4.98}$$

ε_0 – Permittivity of vacuum = 8.8542 × 10^{-12} A s V^{-1} m^{-1}
ε_r – Relative permittivity of the medium between the plates [dimensionless]
A' – Area of the electrode [m^2]
d – Distance between the electrode surface and the center of gravity of those electrolyte ions that approached the electrode surface most closely [m]

When the electrolysis voltage of the solution is reached (1.37 V in Figure 4.5), a reaction current (faradaic current, electrolysis current) starts to flow, which significantly exceeds the charging current of the double layer. We had considered a simplified equivalent circuit for an electrochemical cell in Figure 4.9. For the investigation of the capacitance of the electrical double layer, we work in a potential range, in which there are *no* charge transfer reactions across the metal–solution interface and thus $R_{CT} \to \infty$. In addition, we must ensure that the result is influenced by the properties of one electrode only. The simplest way to achieve this is to select the electrode of interest with a much smaller area A' than the counter electrode. We can then determine the double layer capacitance C_{dl} either from a transient experiment (Figure 4.30) or from the current caused by applying a potential sweep (Figure 4.31). In a pulse experiment, the capacitive current decays according to Eq. (4.99).

$$I_C = \frac{\Delta E}{R_E}\cdot\exp\left[\frac{-t}{R_E C_{dl}}\right] \tag{4.99}$$

I_C – Capacitive current [A]
ΔE – Potential jump [V]
R_E – Electrolyte resistance [Ω]
t – Time after potential step [s]
C_{dl} – Double layer capacitance [A s V^{-1}] = [C V^{-1}]

The current at the beginning of a potential sweep is described by Eq. (4.100).

$$I_C = vC_{dl}\left[1 - \exp\left(\frac{-t}{R_E C_{dl}}\right)\right] \tag{4.100}$$

v – Potential scan rate [V s^{-1}]

For $t \gg R_E C_{dl}$, the current assumes a constant value $I_C = \nu C_{dl}$, from which C_{dl} can be determined conveniently with cyclic voltammetry (Section 5.3.5). In this procedure, potential ranges should be excluded in which electron transfer reactions of dissolved or surface-bound reactants occur (e.g., see the peaks in the black curves in Figure 4.32b). At a solid electrode, the nature of the corresponding processes is not always immediately clear, especially if they may result from adsorbed impurities, surface-oxidized metals, or carbon materials.

In all these procedures, C_{dl} is proportional to the true area or microscopic area A' of the electrode, that is, the area under consideration of the roughness. The roughness is much larger than the dimensions of ions and solvent molecules (or the Debye length), and thus, the rough electrode can accommodate more ions than a smooth electrode. The area A' must be distinguished from the geometric electrode area A, in which the microscopic area is projected on a plane perpendicular to the surface normal. Therefore, the relationship $A \leq A'$ always holds (Figure 4.33). The geometric area A was to be used for the consideration of electrolyte conductivity, and we will also encounter it again in the description of diffusion-controlled reactions in Chapter 5. If one is interested in the determination of interfacial concentrations $\Lambda = N/A'$ [mol cm^{-2}], the microscopic electrode area A' is needed, that is, the area with consideration of roughness. The roughness factor is defined as $r = A'/A$. Typical values for a well-polished metal electrode are 2–3 but can also be much higher.

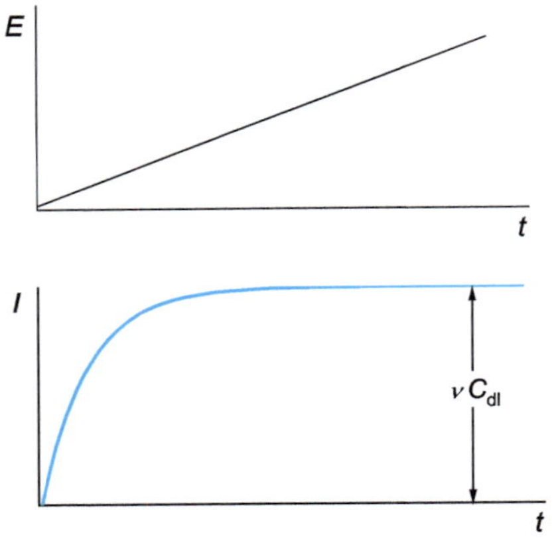

Figure 4.31 Current transient at an electrode during a *potential sweep* in the absence of an electrochemical conversion.

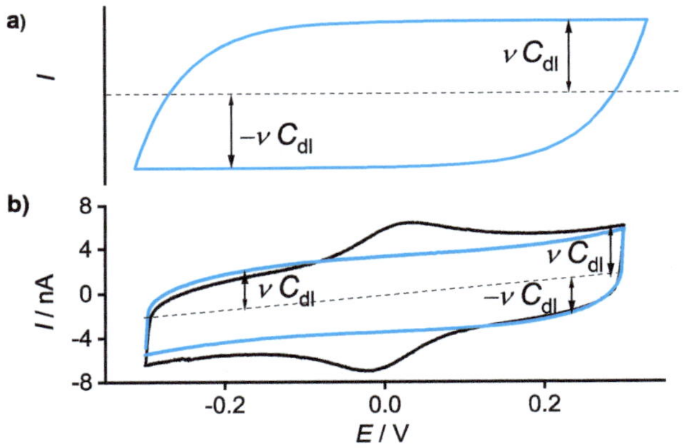

Figure 4.32 Determination of the double layer capacitance from cyclic voltammograms. The potential scan rate is $v = dE/dt$.

a) Potential range in which there are *no* electrochemical reactions;
b) Example of a real cyclic voltammogram (black), in which there is a superposition of double layer charging current and a current from the conversion of surface groups of a carbon electrode (also called pseudocapacitance). The estimation of the capacitive current is shown in blue. (Data have been kindly provided by S. Dongmo, Carl von Ossietzky Universität Oldenburg.)

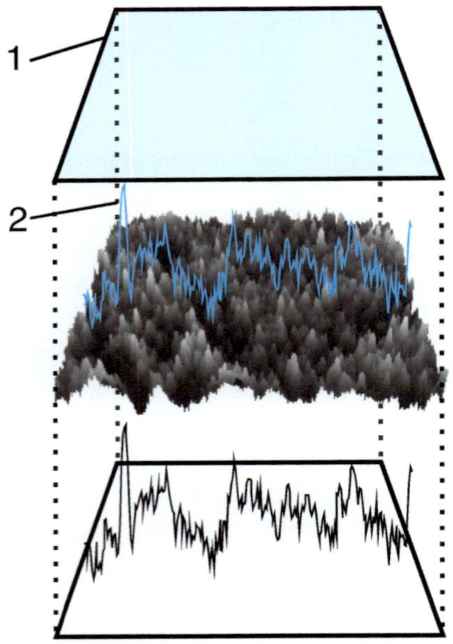

Figure 4.33 Relationship between the geometric area *A* (1, blue) and true (microscopic) electrode area *A′* (2, gray) as a three-dimensional representation or as a cross-sectional profile.

4.3.2 The Molecular Structure of the Electrical Double Layer

Returning to the analogy between the electrical double layer and the parallel plate capacitor, we may ask for the smallest possible distance between a charged metal surface and oppositely charged ions located in front of it but not reacting with or at the metal surface. This is the radius of the ions r_{sol}, including their solvation shell. This arrangement of ions is also called **compact or rigid double layer** or Helmholtz layer[11]). The plane passing through the center of ionic excess charge of the compact layer is called the outer Helmholtz plane. Figure 4.34 shows a schematic depiction of the situation in which only the excess charges in the double layer are colored. For point-like charges in the double layer (this is equivalent to the assumption that the ions do not possess an own volume), the solution of the Poisson equation[39] yields a linear change in the potential between the surface of the electrode and the Helmholtz plane (region I in Figure 4.34, more details on solving the equation can be found in physics textbooks).

The assumption of the compact double layer cannot be sufficient. If the charge density on the solid electrode is too high, one monolayer of oppositely charged ions would not be sufficient to compensate for the charge, because the ions have a volume (contrary to the assumption made in deriving the model). We have already seen during the treatment of the Debye–Hückel theory in Section 4.2.8 that random thermal motions of the ions act against a rigid arrangement of ions. This causes some ions to "break out" of the compact layer (Figure 4.35). Accordingly, the excess charge on the electrolyte side of the double layer must be "smeared out" to some extent. Expressed more correctly, the excess charge must occur as a diffuse space charge (region II in Figure 4.35). This layer is called the **diffuse layer** or Gouy–Chapman layer[40,41].

The potential distribution of this arrangement in Figure 4.35 can be derived in the following way: We consider a space coordinate x perpendicular to the electrode surface,

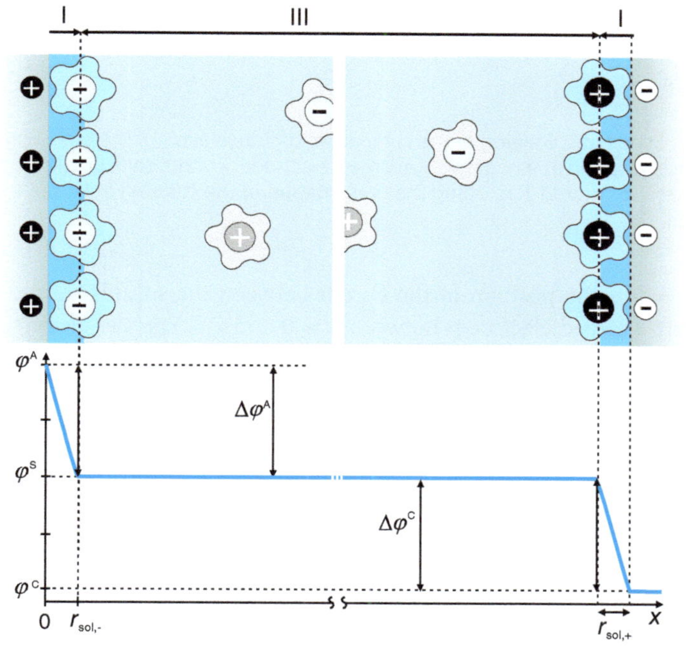

Figure 4.34 Schematic depiction of the compact double layer at a positively charged and a negatively charged electrode. For the sake of clarity, only the excess charges are colored. In the solution bulk (region III), the solvated ions are found in their stoichiometric concentration, and their charges compensate each other. They are indicated in gray. The scheme does not represent the solution concentrations realistically.

[39] Siméon Denis Poisson, 1781–1840, French physicist and mathematician, Poisson equation, Poisson crater on the Moon, Poisson number in the science on strength of materials, and Poisson bracket as mathematical operator in differential equations.

[40] Luis George Gouy, 1854–1926, French physical chemist, professor in Lyon, worked on electrical double layer, electrocapillarity, Brownian molecular motion, optical interference, and magnetic effects (Gouy balance).

[41] David Leonard Chapman, 1869–1958, English chemist, professor at Manchester and at Oxford.

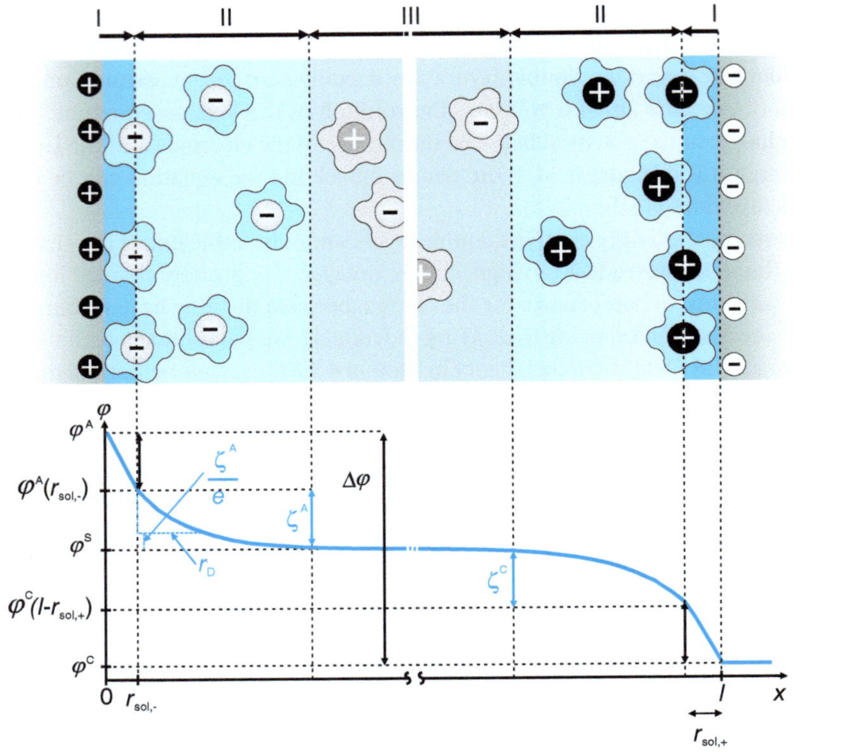

Figure 4.35 Schematic drawing of the compact and diffuse parts of the electrical double layer at a positively and at a negatively charged electrode. For the sake of clarity, only the excess charges are colored. Regarding the presentation of the solution bulk, see explanation in Figure 4.34.

in which $x = 0$ is the position of the border between the solution and the metal of the positively charged electrode. Between $x = 0$ and $x = r_{sol}$, the electrical potential $\varphi(x)$ decreases linearly to $\varphi(r_{sol})$. The further decay from $\varphi(r_{sol})$ to the value φ^S in the solution bulk is defined by a differential Eq. (4.101) similar to the treatment of the ionic cloud. In order to solve the differential equation, it is useful to introduce a transformed coordinate $x' = x - r_{sol}$ perpendicular to the electrode surface. It is the distance of a point in solution to the outer Helmholtz plane.

r_D – Debye length [m], Gl. (4.74)
φ^S – Potential [V] in the solution bulk ($x \to \infty$)
$\varphi_{x'=0}$ – Potential [V] at the outer Helmholtz plane
x' – Space coordinate [m] perpendicular to the electrode surface with the origin at the position of the outer Helmholtz plane

$$\frac{d^2\varphi(x')}{dx'^2} = \left(\frac{1}{r_D}\right)^2 (\varphi_{x'=0} - \varphi^S) \tag{4.101}$$

The solution of this differential equation, with consideration of the boundary conditions, is the potential profile according to Eq. (4.102). (Please, check by yourself that Eq. (4.102) will yield $\varphi = \varphi_{x'=0}$ for $x' = 0$ and $\varphi = \varphi^S$ for $x' = \infty$.)

$$\varphi(x') = \varphi^S + (\varphi_{x'=0} - \varphi^S)\exp\left(\frac{-x'}{r_D}\right) \tag{4.102}$$

We realize that the quantity r_D is a characteristic decay length (this is the Debye length). It is a decay constant in space, somewhat analogous to the decay lifetime of a reagent in a reaction with first-order kinetics. If $x' = r_D$, the exponential term in Eq. (4.102) takes the value $e^{-1} = 1/e \approx 0.37$. At distances $x' \geq r_D$, the potential difference between the position of the outer Helmholtz plane and the solution bulk decays to a fraction of 37% of its value. Consequently, 63% must have decayed already at shorter distances $x' < r_D$. If we consider r_D as a measure for the thickness of the electrical double layer, we can make an estimation for a 1–1 electrolyte at different ionic strengths using Eq. (4.76). The results are summarized in Table 4.4. We realize that there is, first, a strong concentration dependence and, second, that for typical concentrations of about 10^{-1} mol L^{-1}, the diffuse double layer has a thickness of slightly less than 1 nm. This layer thickness is thus not much thicker than an individual ion and about the size of a medium-sized organic molecule! Evidently, the violation of charge neutrality is restricted to a very small zone in the electrolyte solution. For the accuracy of those estimates, the same restrictions apply as those we have discussed for the Debye–Hückel theory.

Table 4.4 Thickness of the diffuse electrical double layer (estimated as the Debye length) of a completely dissociated 1–1 electrolyte at 298 K for different ionic strengths.

I_c/(mol L^{-1})	r_D/m	r_D/nm
10^{-4}	3.0×10^{-8}	30
10^{-3}	9.5×10^{-9}	9.5
10^{-2}	3.0×10^{-9}	3.0
10^{-1}	9.5×10^{-10}	0.95

The potential difference $\varphi_{x'=0} - \varphi^S$ is also called **ζ-potential** (zeta-potential). We will meet this quantity in various aspects of electrochemistry and other subjects. We should note that electrical double layers occur at *all* charged solids immersed in electrolyte solutions, including at the interface between insulators and electrolyte solutions (.). In the case of insulators, the surface charge is determined by chemical reactions of surface groups, for instance, dissociation of Si–OH group or the attachment of ions from the solution to the surface. The surface charge of an insulator can only be manipulated by the composition of the solution, for instance, by pH and the kind and concentration of other ions but not via an external voltage source as in the case of an electrode–electrolyte interface.

ζ-Potentials have a pivotal importance for the stability of colloidal solutions, the adhesion of bacteria to surfaces, and the emergence of the electroosmotic flow in electrophoresis (Section 4.6).

Overall, the potential difference between anode and solution bulk is the sum of the potential differences in the compact and diffuse parts of the electrical double layer. In Figure 4.35, the potential profiles of anode and cathode have been drawn symmetrically with respect to φ^S. However, this is not valid for all electrodes, solutions, and cell potentials.

4.3.3 How Does the Adsorption of Ions or Molecules Change the Structure and the Properties of the Electrical Double Layer?

So far, we have considered the enrichment or depletion of ions close to electrode surface as caused exclusively by electrostatic attraction between oppositely charged bodies. Consequently, it does not matter if a positive excess charge on a metal electrode is compensated by an iodide or by a perchlorate (ClO_4^-) ion. The Debye-Hückel formalism does not provide any clues which of the available anions in solution will be preferentially found close to the interface. Therefore, this mechanism of enrichment of ions at the outer Helmholtz plane is called a "nonspecific adsorption".

So far, the large variety of chemical interactions has been blanked out. However, chemical interaction could potentially lead to an enrichment of ions at a metal surface. We can imagine a metal as a regular (periodic) arrangement of metal cations held together by a free electron gas moving in between the cations (Figure 4.36a). A short glimpse in every textbook or elementary inorganic chemistry shows that metal cations form coordinative bonds with halogenide and pseudohalogenide ions (Cl^-, Br^-, I^-, CN^-, SCN^-, OCN^-, $S_2O_3^-$), yielding, for instance, octahedral complexes (Figure 4.36b). Knowing this, we have to expect that there are also specific reactions between anions in solution and cations within the metal, such as the formation of a coordinative bond at the surface of a metal (Figure 4.36c). In order for such reaction to happen, the anions must release some of the surrounding solvent molecules and electron density must be transferred between the metal and the anion. All the ions that are enriched at the metal surface under partial loss of the solvation shell are called "specifically adsorbed ions".

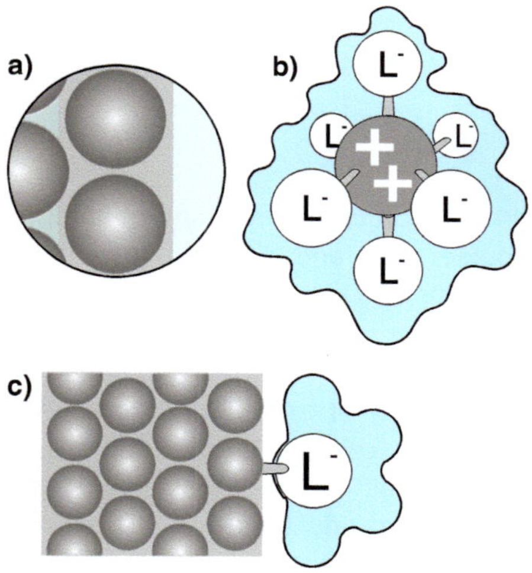

Figure 4.36 Schematic illustration of a) the electron gas model of a metal with mobile valence electron between the metal atoms (indicated by the gray area), b) a transition metal complex made of a metal cation and six negatively charged ligands (e.g., $[Fe(CN)_6]^{4-}$), and c) the specific adsorption of complexing anions to the metal surface.

This term is typically used for anions only. In Section 8.9.8 we will deal with the phenomenon of underpotential deposition of metal cations that we could consider as the analogy to specific adsorption of anions.

When considering specifically adsorbed anions and nonspecifically adsorbed ions of both charges, the structure of the electrical double layer would be as shown in Figure 4.37. The specifically adsorbed ions have partially stripped their solvation shell and, therefore, the center of charges of those ions is closer to the electrode surface than for nonspecifically adsorbed ions. The part of the double layer containing the specifically adsorbed anions is called inner Helmholtz layer. The plane through the center of charge of the specifically adsorbed anions is called the inner Helmholtz plane.

We can summarize the parts of the electrical double layer, as shown in Figure 4.38. The compact part of the electrical double layer always contains the outer Helmholtz layer, in which cations or anions with their solvation shell are held directly at the

Due to the adsorption of anion at the negatively charged cathode, the potential may have a local minimum very close to the electrode surface. In this model, the exact potential distribution depends on the specific kind of metal and ions, and open-circuit potential or externally applied potential. The scheme shows only one of several possible situations.

Unfortunately, the terms diffuse double layer and diffusion layer sound quite similar but refer to quite different things. We will consider diffusion layers in Section 4.5. They are formed as a consequence of concentration changes of neutral or charged species during electrolysis. The thickness is approximately 30–500 μm and thus much larger than the diffuse part of the electrical double layer with 1–30 nm. The ratio between the thicknesses of both types of boundary layers is about 10 000!!

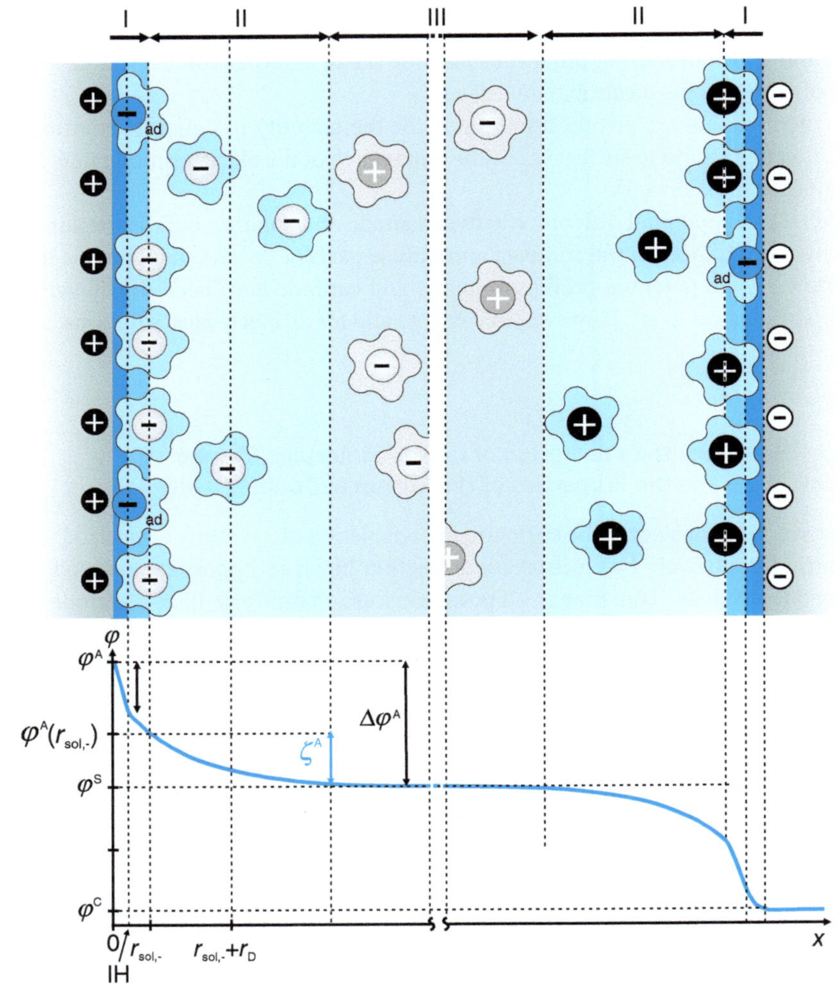

Figure 4.37 Schematic of the electrical double layer with specifically adsorbed anions in the inner Helmholtz layer of both electrodes and nonspecifically adsorbed ions in the outer Helmholtz layer and in the diffuse double layer. For the sake of clarity, only the excess ionic charges are colored, and only details of the anode are labelled. Regarding the representation of the solution bulk, see explanation in Figure 4.34.

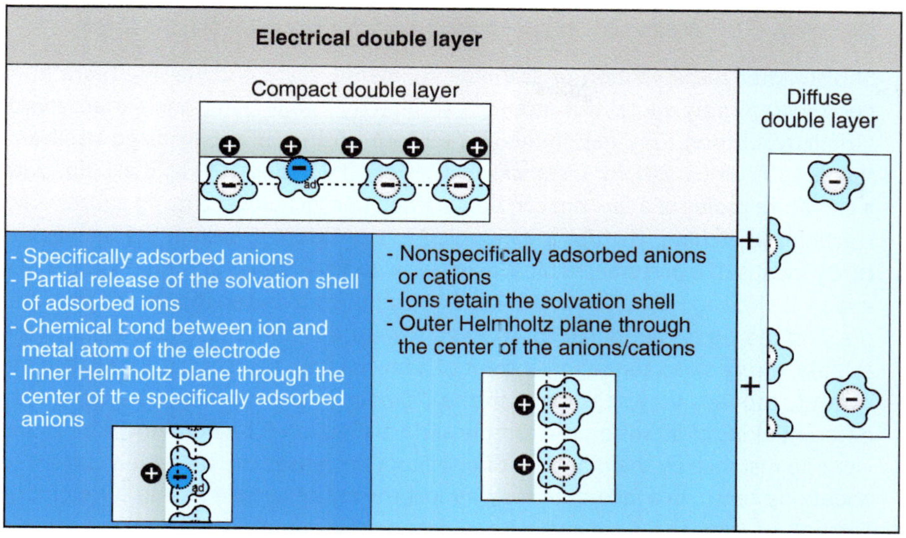

Figure 4.38 Components of the electrical double layer.

electrode surface by electrostatic force (i.e., chemically nonspecific). If there are complexing anions present in solution, they may be bound coordinatively at the electrode surface (chemically specific), whereby they lose part of their solvation shell. If the chemical interaction is very strong, it may even happen that an anion binds to a negatively charged metal electrode. The layer with chemically bound anions is called the inner Helmholtz layer and is also part of the compact double layer. In addition to the compact double layer, there is the diffuse double layer with an excess of those ions that are oppositely charged compared to the electrode. These ions follow a random Brownian motion due to the surrounding thermal energy.

In addition to ions, neutral molecules can also adsorb to the electrode surface. In aqueous solution, many organic compounds are not very soluble. The organic molecules preferentially bind at the interfaces of the liquid (liquid–gas interface, walls of the container, and the liquid–electrode interface) because their interaction with solvent molecules is minimized in this way. Such molecules are also called surface-active molecules or surfactants. Surfactants are particularly effective in binding to interfaces if one part of the molecule is hydrophilic and the other part of the molecule is hydrophobic (amphiphilic molecules). Such structures favor a directional arrangement of the molecule at the interface. In many cases, the neutral molecules "insert" between the electrode and the ions of the outer Helmholtz layer. The structure of such layers has been studied particularly well for adsorbed *n*-alkane thiolates $H_3C-(CH_2)_n-S^-$ on gold electrodes. The unbranched alkyl chains form spontaneously a highly ordered layer, in which the alkyl chains are aligned parallel, similar to timber logs in a wood pile. They form a barrier that is impermeable to ions and solvent molecules (Figure 4.39). Therefore, the minimum distance is increased between the electrode and the center of charge of the ions in the outer Helmholtz layer. According to Eq. (4.29), the increase in the distance d between the charges causes a decrease in the capacitance. In addition, the medium between the charges changes from water with a high ε_r to a hydrocarbon with low ε_r, which also contributes to the decrease in the double layer capacitance. The capacitance decrease upon adsorption of neutral compounds can be exploited for sensitive analytical methods and sensors (see tensammetry, Section 5.3.9, Figure 5.45).

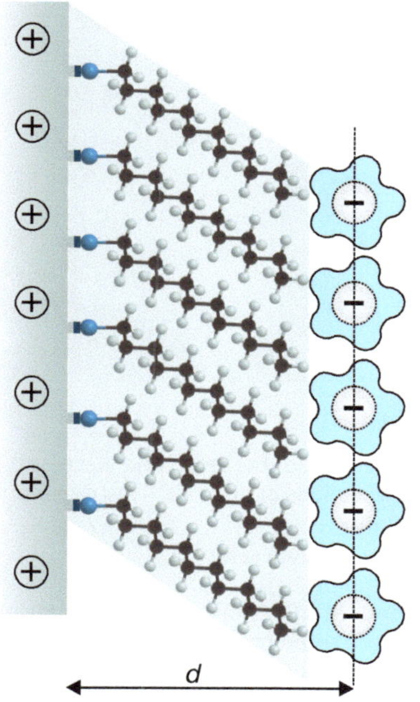

Figure 4.39 Effect of neutral adsorbed molecules on the structure and properties of the electrical double layer. Here, the special case of a self-assembled monolayer of alkanethiolates on gold is shown.

$$C = \varepsilon_0 \varepsilon_r \frac{A'}{d}$$

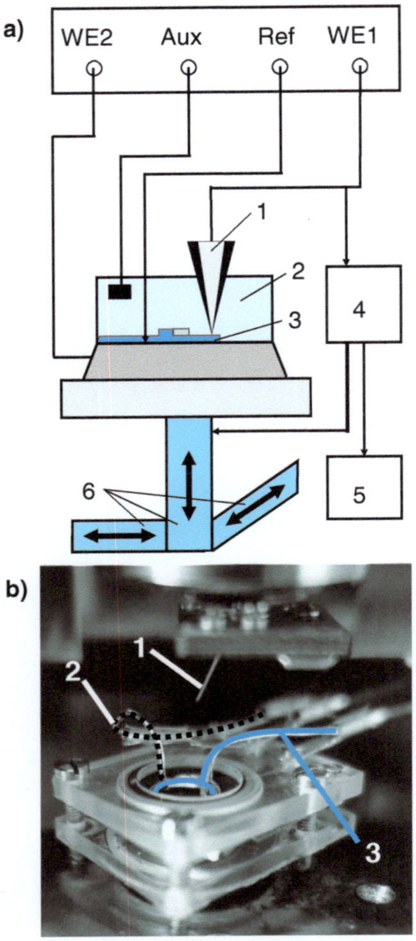

Figure 4.40

a) Schematic of the setup and electrical connections of an ECSTM instrument;
   1) tip,
   2) electrolyte reservoir,
   3) sample,
   4) electronic feedback circuit for maintaining the setpoint,
   5) recording of the vertical tip displacement of the *z*-actuator,
   6) piezomechanical positioning system.

b) Photograph of the ECSTM cell. The labelled components are
   1) the tip,
   2) the auxiliary electrode with connecting wire, and
   3) the reference electrode.

The sample has approximately the size of a 1-cent coin and forms the bottom of the beaker-like liquid reservoir. The schematic in a) is not to scale.

Excursion 4-2 Atomically Resolved Images of the Electrical Double Layer

Shortly after the invention of scanning tunneling microscopy (STM), there have been increasingly successful attempts to directly image electrode surfaces with atomic resolution. One requirement is a sharp tip that is not oxidized itself and that is sufficiently stiff, for instance, an alloy of 90% Pt and 10% Ir. Sharp tips with a curvature radius of a few nm can be obtained by mechanical cutting and/or electrochemical etching. The tips are mounted to a positioning system that moves the tip by means of a piezomechanical actuator with very high precision (Figure 4.40a and b). If a voltage (bias, about 0.1 V to 1 V) is applied between the tip and the sample, and the tip is cautiously approached toward a distance of about 1 nm to the sample, a very small tunneling current of a few nA can be recorded. The tunneling current responds very sensitively to the changes in the working distance and to different kinds of atoms on the sample surface (Section 8.11.2). An increase in the working distance by the diameter of one atom decreases the tunneling current to about one-tenth. In a favorable case, almost the entire current passes through one protruding atom at the tip and a few atoms on the sample. The current is set to a desired reference value (setpoint), and the tip is moved in a zigzag fashion laterally above the sample surface. An electronic feedback loop continuously corrects the vertical position of the tip in such a way that the current remains on the setpoint even if the tip moves over step edges and protruding atoms. The required expansion of the piezoelectric actuator in the vertical direction (z) is translated into a gray-scale or false-color image and plotted as a function of the horizontal coordinates (x, y). This basic experiment can also be used in the case when the sample is an electrode surface (ECSTM, electrochemical scanning tunneling microscopy).

Usual laboratory electrodes are polycrystalline. They consist of only one kind of atom but are made up of individual entangled crystallites, which are oriented in different directions in space (Figure 8.47). In order to achieve atomically resolved images, the current must always pass through the same atom of the tip, which is usually achieved only if the sample is atomically flat. Such experiments are realized with specially prepared single crystal electrodes (Section 8.9), in which the entire macroscopic electrode is formed from one single crystal. Many metals such as Pt, Au, Ag, and Cu crystallize in a face-centered cubic (fcc) structure (Figure 4.41). The surface of a macroscopic crystal can, in principle, be formed by different planes of the crystal. These planes are called lattice planes. Their direction with respect to the unit cell is stated by a number triplet (h k l). The numbers are the coordinates of a vector in reciprocal space that stands perpendicular to the lattice plane and whose length corresponds to the distance between two neighboring parallel lattice planes (Section 8.9.1). Thus, a notion like Ag(1 1 1) indicates a specific arrangement of Ag atoms at the surface. Most often, the surfaces with low index numbers are formed. Three typical examples are shown in Figure 4.41. From these examples, it becomes clear that the lattice planes are not all parallel to the faces of a cubic unit cell. The Ag(1 1 1) surface shows a hexagonal arrangement of Ag atoms within the plane, which is the arrangement expected from the closest packing of spheres.

At such surfaces, the structure of the double layer can be investigated with atomic resolution as a function of the applied electrode potential and the composition of the electrolyte solution. Atoms that diffuse rapidly on the surface are not imaged because they change their position between two subsequent passages of the tip. This applies to all nonspecifically adsorbed ions. In contrast, specifically adsorbed ions move much more slowly and can be imaged. Depending on the relative size with respect to the metal atoms and the exact energetics of the bound

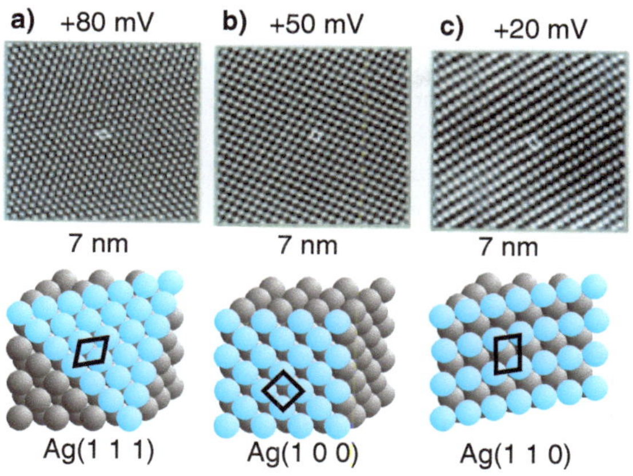

Figure 4.41 Low index lattice planes of a cubic face-centered crystal of metal atoms and ECSTM images of the corresponding surfaces. Source: Reproduced with permission from M. A. Schneeweiss, D. M. Kolb; Das Rastertunnelmikroskop in der Elektrochemie, *Chem. Unserer Zeit* **2000**, *34*, 72–83. Copyright 2000, Wiley-VCH.

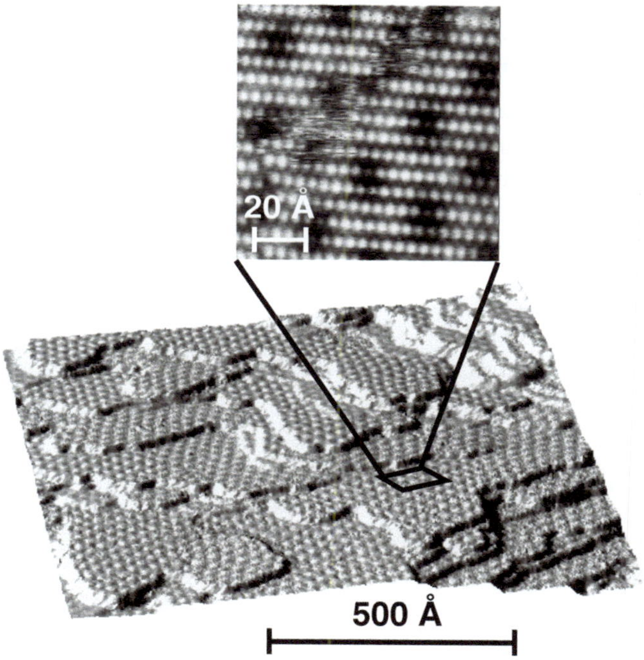

Figure 4.42 ECSTM image of a Cu(1 1 1) electrode in H_2SO_4. The bright spots in the upper panel are an ordered structure of adsorbed HSO_4^- ions (and not the topmost layer of Cu atoms). The structure also shows another periodicity of larger repeat length (moiré pattern, dark hexagonally arranged regions). The moiré pattern can also be observed in larger image frames, in which atomically flat terraces with monoatomic step height are visible. The moiré pattern is caused by a rearrangement of the topmost layer of Cu atoms that is induced by the adsorbate binding. For an overview describing the multitude of phenomena of the potential-dependent (!) adsorption of ions onto metal electrodes, the reader is referred to a review paper [18] and Chapter 8. Source: Reproduced with permission from O. M. Magnussen, R. J. Behm; Atomic-scale processes in Cu corrosion and corrosion inhibition, *Mater. Res. Bull.* **1999**, *24*, 16, Copyright 1999, Material Research Society.

ions, the adsorbate may form a regular two-dimensional lattice on the atomically ordered electrode. Such structures are called superstructure (Figure 4.42).

With very fast ECSTM instruments (video ECSTM), one can even record the movement of specifically adsorbed anions and arrange the image frames as a movie sequence with 10–30 frames per second. This requires a dilution of the specifically adsorbed ions on the surface [19].

ECSTM images also allow the investigation of potential-dependent and composition-dependent surface reconstruction. When a surface reconstruction is present, the atomic arrangement at the surface deviates from the arrangement of atoms in the undisturbed bulk of the crystal. This partially offsets the energy loss by the missing coordination partners at the surface of the metal crystal. A detailed consideration of this process will be provided in Section 8.9.2 or can be found in a review paper [18].

4.4 Diffusion

Diffusion is the random movement of particles (molecules and ions). By this movement, an individual particle may reach all locations within a fluidic phase. The term "fluidic phase" is a generalization of gaseous, liquid, and supercritical phases. Diffusion may also occur in solid phases. We will first look at how fast an individual particle will move from a starting point (the source) to any given location in the phase. Next, we will see how the random movement of many particles in a fluid leads to a directional transport from locations with high concentration to locations with low concentration.

4.4.1 Random Movement of a Single Particle

Let us turn first to the movement of a single particle. The line connecting the position of one particle at different times is called a trajectory. We can divide a trajectory into many small linear segments between the changes in direction in the random movement. The reason for the changes in direction is collisions with other particles or the container wall. In liquids, the linear segments can be thought of as the location change traveled by one place exchange with a neighboring solvent molecule. In order to illustrate the process, we use a macroscopic analogy, that is, a model. Traditionally, there is some popularity among students for the random walk, or more vividly, the drunken sailor (Figure 4.43). Imagine a sailor after a strenuous journey who has spent an evening

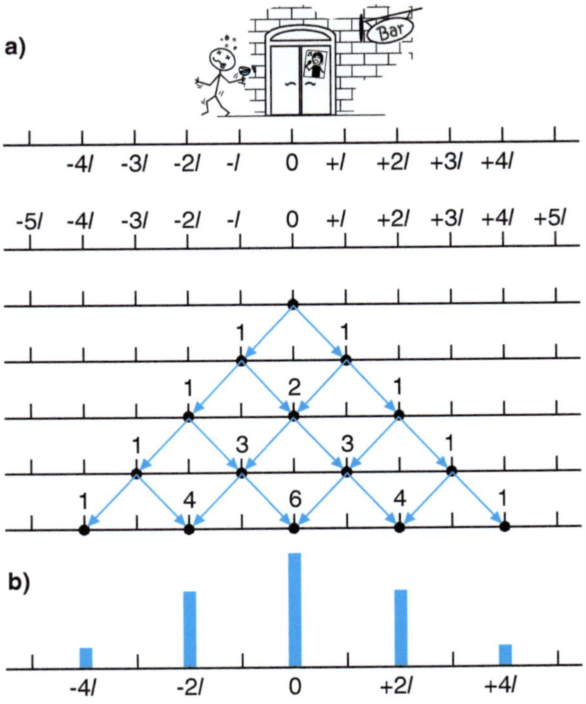

Figure 4.43 Principle of the random walk along one dimension in space; a) trajectories for the first four steps; b) bar chart for the probability of a specific endpoint of those trajectories for the first four steps.

drinking in a harbor bar. After leaving the bar (the source), he aims for new destinations on the other side of the driveway. Due to the adaptation to seaway and despite some balance problems, the sailor executes exactly one step within a fixed time interval (the steptime τ, e.g., $\tau = 2$ s). These steps shall have a constant step length in the direction of the other side of the driveway. However, each step is always connected to a lateral sidestep of constant length l in direction of the driveway. After the time in the bar, the sailor makes the sidesteps in random sequence but with constant length and frequency, either to the left *or* the right. Now we can ask ourselves, at which point on the other side of the road the sailor will eventually end up?

Because the movement of the sailor is thought to be random, we must phrase our question as a probability problem. We seek the probability P to find the sailor after m steps, that is, after a time $t = m\,\tau$ $(m = 0, 1, 2, 3, \ldots)$ at a coordinate x_i along the driveway. We place the zero position at the source (i.e., the door of the harbor bar). Should all side steps lead toward the left side, the final position would be $-m\,l$. If all steps were made to the right, the position would be $+m\,l$.

The endpoints of all possible trajectories of Figure 4.43 have been collected in Table 4.5. Column 1 shows the time in units of the step time. Because of $t = m\,\tau$ this is also equal to the number of steps (Have you noted the equivalence of m to a normalized time?). Column 2 shows the total number N of the existing trajectories (number of possible outcomes). For $t = 0$, this corresponds to the position of the

Table 4.5 Possible positions of a one-dimensional random walker after a finite step number m.

1	2	3	4	5	6	7
t	N	x_i	N_i [d)]	$<x>$	$\sum N_i x_i^2$	$\langle x^2 \rangle$
0	$1 = 2^0$	0	1	0	0	0
$1\,\tau$	$2 = 2^1$	$-1\,l$	1	0	$2\,l^2$	l^2
		$+1\,l$	1			
$2\,\tau$	$4 = 2^2$	$-2\,l$	1	0	$8\,l^2$	$2\,l^2$
		0	2			
		$+2\,l$	1			
$3\,\tau$	$8 = 2^3$	$-3\,l$	1	0	$24\,l^2$	$3\,l^2$
		$-1\,l$	3			
		$+1\,l$	3			
		$+3\,l$	1			
$4\,\tau$	$16 = 2^4$	$-4\,l$	1	0	$64\,l^2$	$4\,l^2$
		$-2\,l$	4			
		0	6			
		$+2\,l$	4			
		$+4\,l$	1			
$m\,\tau$	$N = 2^m$	$\{-m, -m+2, -m+4, \ldots, +m\}$	$\frac{m!}{r!(m-r)!}$	0	$m\,N\,l^2$	$m\,l^2$

Column 1) Total time in multiples of the step time
Column 2) Total number of all possible trajectories
Column 3) Possible values for the displacement
Column 4) Weighting factors, the number of trajectories ending at the specific position x_i
Column 5) Arithmetic mean of the displacement. This is the mean of the possible displacements multiplied with their weighting factors, $\langle x \rangle = \frac{1}{N} \sum N_i \cdot x_i$
Column 6) Sum of the squared displacements, sum of the weighted squared displacements
Column 7) Arithmetic mean of column 6), $\langle x^2 \rangle = \frac{1}{N} \sum N_i x_i^2$

source, hence $N = 1$. Column 3 lists the possible positions, and column 4 the number of trajectories ending at each position (number of realizations). Next, we look for a quantity that describes the mean lateral displacement of the random walker. We form the arithmetic mean of the x values at the end of the trajectories for all times in column 5. For the arithmetic mean, we use the notation with angle brackets $\langle\ \rangle$ around the considered quantity. When we form the arithmetic mean, negative and positive deviations from the source coordinate (0) just balance each other, and the mean will be zero for all times. Instead, it is more informative if we first square the positions and then sum up. All squares are positive regardless of the sign of the original lateral displacement, and therefore, negative and positive positions do not balance each other out anymore. Finally, we form the arithmetic mean of the squared displacements by dividing by the number N of all possible trajectories. In the last row of the table, we collect the formulas for arbitrary step numbers $m \geq 0$. In order to find an algebraic expression for column 4, we note the lateral displacement with the step number m and an integer auxiliary parameter $0 \leq r \leq m$ according to Eq. (4.103).

m – Step number [dimensionless]
r – Auxiliary integer parameter [0, …, m]
l – Step length [m]
x – Displacement (in the direction of the driveway) [m]

$$x_r = (-m + 2\,r)\,l \tag{4.103}$$

We recognize that column 4 holds the binomial coefficients for which we can look up an expression in mathematics textbooks if needed. The probability $P(m, r)$ to find the random walker after m steps at position x_r is expressed by Eq. (4.104).

$$\begin{aligned} P(m, r) &= \text{Normalization factor} \cdot \text{Weighting factor} \\ &= \frac{1}{N} \cdot \text{Binomial coefficient} \\ &= \frac{1}{2^m} \cdot \frac{m!}{r!(m-r)!} \end{aligned} \tag{4.104}$$

Let us examine the mean squared displacement $\langle x^2 \rangle$ a bit more in detail. The factor m was the number of steps in our example. We may consider it also as the ratio between the total time t and the step time τ. Step time and step length are inherent properties (parameters) of our random walker, which do not change during the "experiment". If we assemble all the constants in one prefactor [Eq. (4.105)], we can immediately realize that there must be a proportionality between the mean squared displacement $\langle x^2 \rangle$ and the total time t. In addition, we can consider the step time τ [s] as the reciprocal of the step frequency f [Hz] = [s^{-1}], that is, the number of steps per second.

ω – Step frequency [Hz] = [s^{-1}]

$$\langle x^2 \rangle = m l^2 = \frac{t}{\tau} l^2 = \frac{l^2}{\tau} t = \omega l^2 t \tag{4.105}$$

The square root of the mean squared displacements is a measure of the mean distance and is obtained from Eq. (4.106).

$$\sqrt{\langle x^2 \rangle} = \sqrt{\omega l^2 t} = const\sqrt{t} \tag{4.106}$$

It remains to be seen how we can determine the prefactor of the model parameters because we do not know ω and l and have no means to estimate them with reasonable assumptions for a molecular random walker. We will be able to make this step only at the end of the next section.

Let us return to the metaphor of the drunken sailor and try to estimate the velocity at which the sailor will be displaced along the direction of the driveway. We form the average velocity from the square root of the mean squared displacements and the observation time according to Eq. (4.107), whereby it becomes obvious that the displacement velocity decreases with increasing duration of the observation!

v_D – Mean diffusion velocity along one space coordinate [m s^{-1}]

$$v_D = \frac{\sqrt{\langle x^2 \rangle}}{t} = const\frac{\sqrt{t}}{t} = const\frac{1}{\sqrt{t}} \tag{4.107}$$

There is also a very nice model of inanimate nature, the Galton Board[42] (Figure 4.44). Identify the quantities l, τ, ω, x_i, $\langle x^2 \rangle$, which we have used above, for the example of the Galton Board. There are very nice animations on the web visualizing the operation of the Galton Board [20].

Each ion or molecule in a solution behaves like a random walker. This process is also called self-diffusion. Next, we will see how the random movement of many particles causes a directional transport.

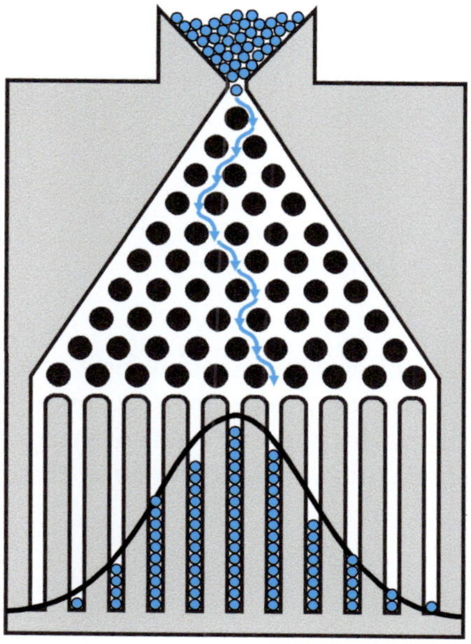

Figure 4.44 Schematic of the Galton board after passage of many spheres through the apparatus.

4.4.2 Mass Transport by Diffusion – Fick's First Law

How does directional transport in an ensemble of many particles result from the random movement of each individual particle? We can imagine the particles as dye molecules. If the solution of the dye is completely mixed at the beginning of the experiment, that is, the concentration of the dye $c(x, t = 0)$ is equal in all volume elements, the random walk of individual dye molecules will only lead to statistical fluctuations in the concentration within volume elements. If the volume elements are sufficiently large and the concentration is high (which is almost always the case for real chemical systems), these statistical fluctuations are too small to measure analytically.

Q: Why is the statement restricted by "almost"?

A: Indeed, there are fluorescence experiments and also some electrochemical experiments, in which the random movement of individual molecules leads to measurable signals. Restrictions of a general textbook do not permit a detailed consideration here. However, you may conduct a simple estimation under which conditions such random fluctuations might be observed. Imagine an extremely thin liquid channel with 100 μm length, 10 μm width, and a thickness of 10 nm. At which concentration would the particle number in the channel be, on average, 1.5? What would this mean for an experiment with these particles? What does this, in turn, mean for "ordinary" electrochemical experiments? Further reading [21].

The situation is different if there are different starting concentrations in the volume elements at the beginning of the experiment. We consider the process again along one space dimension, for example, along the direction of the central axis of a tube. The semester clock in Figure 4.46 is a visual realization of such an experiment, in which a colored ion spreads within the entire liquid volume of a large measuring cylinder during an academic semester. We consider a counting plane that spans the cylinder perpendicular to the symmetry axis of the tube (Figure 4.45). At a distance of a step length l of the molecule, we imagine further planes on the left and right of the counting plane. Left of the counting plane, there shall be a particle number $N(x)$ [mol], and on the right side of the counting plane, there shall be the particle number $N(x + l)$.

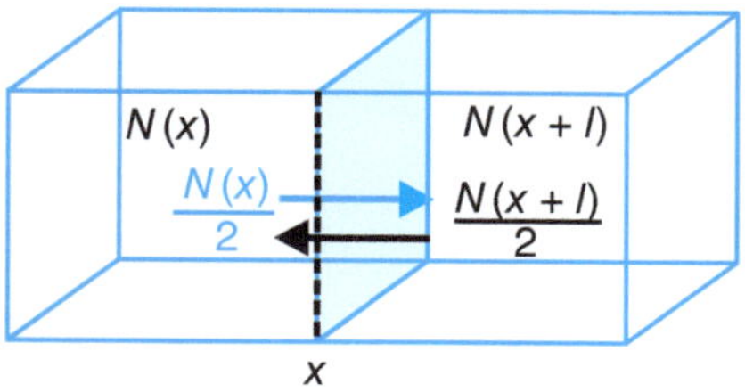

Figure 4.45 Particle flux through a counting plane along a concentration gradient.

Within the step time τ, half of the $N(x)$ particles on the left of the counting plane will move to the right and thereby pass the counting plane. The other half moves to the left and leaves the observation window. On the right side of the counting plane, half of the particles move to the left (across the counting plane) and the other half to the right (out of the observation window). Making a balance sheet of the particles passing the counting plane coming from the left and from the right side, division by the size A of the counting plane and the step time τ (observation time) yields the (particle flux) $dN/(A\,dt)$. The symbol J [mol cm^{-2} s^{-1}] is often used for this differential coefficient. J depends on the position x along the tube direction and the observation time t.

[42] Francis Galton, 1822–1911, British natural scientist and writer.

J – Particle flux [mol cm^{-2} s^{-1}]
N – Amount of particles [mol]
A – Area [cm^2]
τ – Step time [s]

$$\frac{1}{A}\cdot\frac{\partial N}{\partial t}=J(x,t)=\frac{1}{A}\frac{\frac{N(x)}{2}-\frac{N(x+l)}{2}}{\tau} \quad (4.108)$$

After multiplication of the right side of Eq. (4.108) with $1 = l^2/l^2$ and both sides with (−1), as well as introduction of the (particle) concentration $c = N/V = N/(A \cdot l)$, we obtain

l – Step length [cm]
c – Concentration [mol cm^{-3}]

$$-J(x,t)=\frac{l^2}{2\tau}\frac{\frac{N(x+l)}{Al}-\frac{N(x)}{Al}}{l}=\frac{l^2}{2\tau}\frac{c(x+l)-c(x)}{l} \quad (4.109)$$

*Derivatives with respect to time specify the rate of change in a quantity, for example, reaction rate = rate of change in concentration, translation rate = change in position.

The term on the right in Eq. (4.108) is a difference coefficient, with a limiting value for $l \to 0$ giving the partial differential coefficient $\partial c\,(x, t)/\partial x$. In general, partial derivatives with respect to a space coordinate are called gradients, which we can imagine as a slope, here a "slope" of the concentration, or expressed exactly as a concentration gradient* (Figure 4.47). Gradients can be visualized as vectors.

In addition, we realize that the prefactor $l^2/(2\tau)$ only contains constants. This proportionality constant is called the diffusion coefficient D. The negative sign indicates that a higher concentration at the right side of the counting plane (i.e., a positive slope of the concentration profile) causes a particle flux toward the left, that is opposite to the direction of the gradient. With these considerations, we obtain **Fick's first law**[43], Eq. (4.110). The effective transport that results from random movement of an ensemble of particles is also called transport diffusion.

$$J(x,t)=-D\frac{\partial c(x,t)}{\partial x} \quad (4.110)$$

We can now specify the proportionality constant in Eqs. (4.105) and (4.106). Because $D = l^2/(2\tau)$ according to Eq. (4.109), the proportionality constants are related as $l^2/\tau = \omega l^2 = 2D$. With this substitution, Eq. (4.106) turns out to be the well-known **Einstein–Smoluchowski equation**[44,45] (4.111), which links the mean squared displacement of an individual random walker to the macroscopic diffusion coefficient of an ensemble of particles.

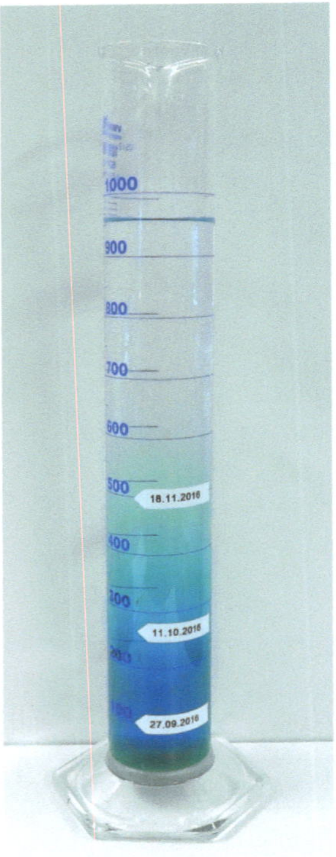

Figure 4.46 The semester clock shows the passage of an academic semester by the spread of a colored salt solution in a water-filled measuring cylinder. The calendar dates mark the shift of the zone with a just perceptible coloration.

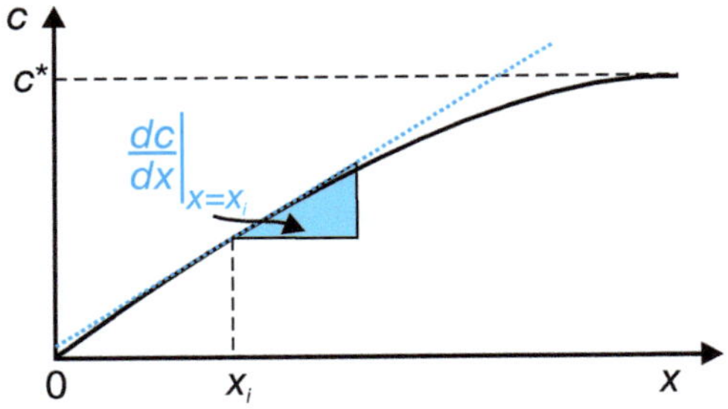

Figure 4.47 Schematic representation of a concentration profile, that is, a plot of local concentration c as a function of the location x. Generally, $x = 0$ refers to the position of the electrode–electrolyte interface. The gradient at the point $x = x_i$ corresponds to the slope of the tangent at the position $x = x_i$.

[43] Adolf Eugen Fick, 1829–1901, German physician, professor of physiology in Würzburg, stated Fick's laws for the first time in 1852 based on experimental observations.
[44] Albert Einstein, 1879–1955, German-Jewish physicist, Nobel Prize 1921 for the explanation of the photoelectric effect, professor in Prague, Zurich, at the Kaiser Wilhelm Institut in Berlin and after his emigration 1933 at Princeton, MA, USA. Within the area of electrochemistry, three equations carry his name: Nernst–Einstein equation, Stokes–Einstein equation, and Einstein–Smochulowski equation.
[45] Marian Smochulowski, 1872–1917, Polish physicist, professor in Vienna, L'viv, Cracow. Well-known are two equations for the electroosmotic flow in capillaries and the maximum bimolecular reaction rate in solution.

$$\sqrt{\langle x^2 \rangle} = \sqrt{2Dt} \qquad (4.111)$$

D – Diffusion coefficient [cm^2 s^{-1}]

Diffusion coefficients of different ions in aqueous solutions do not differ a lot from each other. A typical value that is well suited for rough calculations is $D = 5 \times 10^{-6}$ cm^2 s^{-1} = 5×10^{-10} m^2 s^{-1}. Using this value, we can estimate how fast and how far an ion or molecule departs from the source point within a given observation time (Table 4.6). We will use these results for qualitative estimations at several places in the book.

Diffusion coefficients are typically stated in the unit cm^2 s^{-1} in the field of electrochemistry. Therefore, it is useful to convert concentration units before calculations into the cgs system (centimeter, gram, second). Of course, it would also be possible to convert the diffusion coefficients to dm^2 s^{-1} in order to be compatible with the quantity of solution concentration mol L^{-1}. However, this is uncommon. Recently, the usage of complex simulation programs (for instance, COMSOL Multiphysics) has seen a steep increase in applications, not only in electrochemistry. These software packages sometimes require the input of parameters strictly in SI units (m^2 s^{-1}), even if this is otherwise not common.

Table 4.6 Square root of the mean squared displacements according to Eq. (4.111) and mean diffusion velocity according to Eq. (4.107) of a dissolved ion or molecule with $D = 5 \times 10^{-6}$ cm^2 s^{-1} along one space coordinate x.

t	Δx/cm	Δx/μm	v_D/(μm s^{-1})
0.1 μs	1×10^{-6}	0.01	1×10^{5}
10 μs	1×10^{-5}	0.1	1×10^{4}
1 ms	1×10^{-4}	1	1×10^{3}
0.1 s	1×10^{-3}	10	100
10 s	1×10^{-2}	100	10

Similar considerations of a random walker in two or three space dimensions yield expressions for the square root of the mean squared displacement of $(4Dt)^{1/2}$ and $(6Dt)^{1/2}$, respectively.

In addition, we would like to turn our attention to a few systematic relations between the diffusion coefficient (i.e., the random movement in the field-free space), the size of the ion, and the mobility of a dissolved ion in an electric field. A systematic treatment can be found in [1]. The **Stokes–Einstein relationship**[26,44] (4.112) connects the diffusion coefficient D of a particle i with the viscosity η of the solution and the radius of the particle r_i.

$$D_i = \frac{k_B T}{6\pi \eta r_i} \qquad (4.112)$$

k_B – Boltzmann constant [N m K^{-1}] = [J K^{-1}]
T – Temperature [K]
r_i – Hydrodynamic radius of the particle i [m]
η – Viscosity [N s m^{-2}]

The **Einstein Equation** (4.113) relates the diffusion coefficient of an ion to the corresponding ion mobility.

$$D_i = \frac{RTu_i}{|z_i|F} = \frac{k_B T u_i}{|z_i|e} \qquad (4.113)$$

R – Gas constant [J K^{-1} mol^{-1}]
F – Faraday constant [A s mol^{-1}]
z_i – Charge number of the ion [dimensionless]
u_i – Ion mobility [m^2 s^{-1} V^{-1}]

In combination with Eq. (4.31), we obtain a relationship between the molar conductivity at infinite dilution and the diffusion coefficients, the charges of ions, and the stoichiometric coefficients of the ions in a salt, the **Nernst–Einstein Equation**[10,44] (4.114).

$$\Lambda_0 = \frac{F^2}{RT}\left[\nu_+ z_+^2 D_+ + \nu_- z_-^2 D_-\right] \qquad (4.114)$$

ν_+, ν_- – Stoichiometric coefficients for anions and cations of a salt [dimensionless]
z_+, z_- – Charge number of cation and anion [dimensionless]
D_+, D_- – Diffusion coefficients of cation and anion [cm^2 s^{-1}]

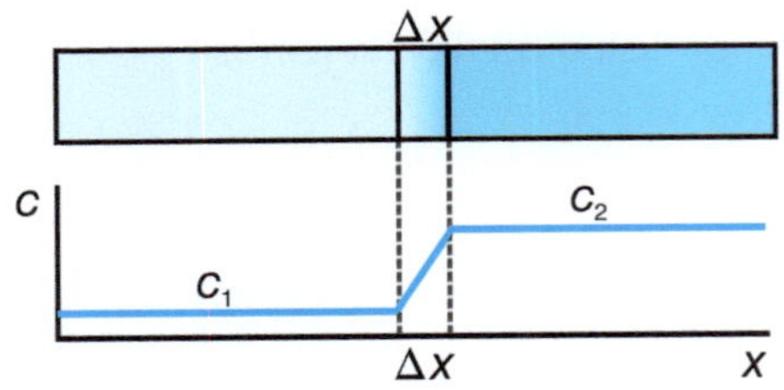

Figure 4.48 Schematic illustration of the problem in Example Calculation 4.8.

Example Calculation 4.8 A diaphragm of 2 mm thickness separates two equally large, well-mixed containers with glucose solution. The concentrations are $c_1(\text{glucose}) = 0.01\ \text{mol L}^{-1}$ and $c_2(\text{glucose}) = 0.1\ \text{mol L}^{-1}$. Calculate the diffusion coefficient if the steady state diffusional flux is $J = 3.15 \times 10^{-9}\ \text{mol cm}^{-2}\ \text{s}^{-1}$. The concentration profile is shown schematically in Figure 4.48.

unknown: D

known:

$$c_1 = 0.01\ \text{mol L}^{-1} = 1 \times 10^{-5}\ \text{mol cm}^{-3}$$

$$c_2 = 0.1\ \text{mol L}^{-1} = 10 \times 10^{-5}\ \text{mol cm}^{-3}$$

$$\Delta x = 2\ \text{mm} = 0.2\ \text{cm}$$

$$J(x,t) = -D\frac{\partial c(x,t)}{\partial x}$$

(Because of the linear change in the local concentration with x, the simplification $\partial c/\partial x \approx \Delta c/\Delta x$ is possible.)

$$\begin{aligned} D &= \left|\frac{J(x,t)\Delta x}{\Delta c}\right| \\ &= \frac{3.15 \times 10^{-9}\ \text{mol cm}^{-2}\ \text{s}^{-1} \times 0.2\ \text{cm}}{(0.1 - 0.01) \times 10^{-3}\ \text{mol cm}^{-3}} \\ D &= 7 \times 10^{-6}\ \text{cm}^2\ \text{s}^{-1} \\ &= 7 \times 10^{-6}\ \text{cm}^2\ \text{s}^{-1}\frac{1\ \text{m}}{100\ \text{cm}}\frac{1\ \text{m}}{100\ \text{cm}} \\ &= 7 \times 10^{-10}\ \text{m}^2\ \text{s}^{-1} \end{aligned}$$

The diffusion coefficient in the diaphragm is $7 \times 10^{-6}\ \text{cm}^2\ \text{s}^{-1}$ or $7 \times 10^{-10}\ \text{m}^2\ \text{s}^{-1}$.

4.5 The Diffusion Layer

4.5.1 Change in Concentration Profiles with Time–Fick's Second Law

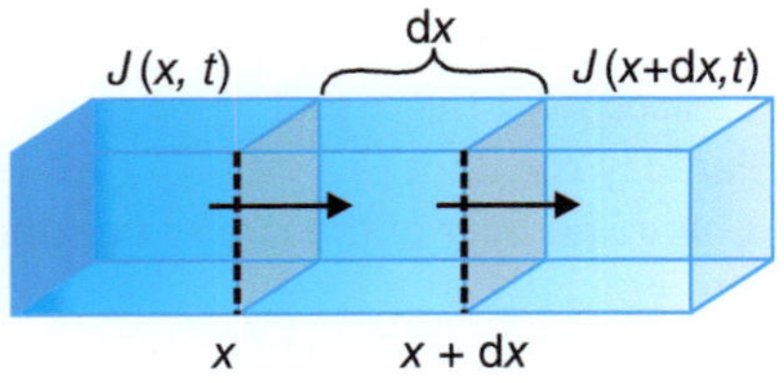

Figure 4.49 Concentration gradient in a tube with two counting planes with a mutual distance dx.

The transport of ions and molecules within a concentration gradient causes a change in the local concentration in solution. How quickly will the concentration at a specific location change with time within a concentration gradient? To answer this question, we seek an expression for the partial differential quotient $\partial c(x,t)/\partial t$. We simplify the problem by considering diffusion along one space dimension x only, for instance, within a tube. We count the flux at two planes perpendicular to the symmetry axis of the tube with a distance $\mathrm{d}x$ (Figure 4.49). We consider the balance for the fluxes J entering and leaving the volume element with $V = A\ \mathrm{d}x$. The rate of concentration change is generally given by

$$\frac{\partial c(x,t)}{\partial t} = \frac{1}{A\mathrm{d}x} \cdot \frac{\mathrm{d}N}{\mathrm{d}t}. \tag{4.115}$$

According to Eq. (4.108), we can substitute the rate at which the particle number is changing at each plane by the fluxes passing the counting plane.

$$\frac{\partial c(x,t)}{\partial t} = \frac{1}{\mathrm{d}x} \cdot [J(x,t) - J(x + \mathrm{d}x, t)] \tag{4.116}$$

Because J is a continuous function, we can replace the expression $J(x + \mathrm{d}x)$, that is the diffusional flux at a point located very close to the observation point x, by a Taylor series expansion[21] with respect to the dependence of J on the location x. The series is terminated after the first (linear) term [Eq. (4.117)].

Taylor series expansion of a function in the source point x_0 with $\Delta x = x - x_0$

$$f(x) = f(x_0) + \sum_{n=1}^{\infty} \frac{1}{n!} f^{(n)}(x_0)(\Delta x)^n$$

$$J(x + \mathrm{d}x, t) = J(x, t) + \frac{\partial J(x, t)}{\partial x} \cdot \mathrm{d}x \tag{4.117}$$

Inserting Eq. (4.117) into Eq. (4.116) yields

$$\frac{\partial c(x, t)}{\partial t} = \frac{J(x, t) - J(x, t) - \dfrac{\partial J(x, t)}{\partial x} \cdot \mathrm{d}x}{\mathrm{d}x} = -\frac{\partial J(x, t)}{\partial x} \tag{4.118}$$

Next, we replace $J(x, t)$ using Fick's first law [Gl. (4.110)]

$$\frac{\partial c(x, t)}{\partial t} = -\frac{\partial}{\partial x}\left[-D\frac{\partial c(x, t)}{\partial x}\right] \tag{4.119}$$

and obtain Fick's second law [Gl. (4.120)] after some more simplifying.

$$\frac{\partial c(x, t)}{\partial t} = D\frac{\partial^2 c(x, t)}{\partial x^2} \tag{4.120}$$

Fick's second law is a classic example of a partial differential equation (PDE). The solutions of a PDE are functions $c = f(x, t)$. These are called **concentration profiles** in this textbook. The solution of PDEs is demanding and requires initial and boundary conditions. We can only exemplify this approach in the subsequent sections for selected, relatively simple problems (Excursion 4-3).

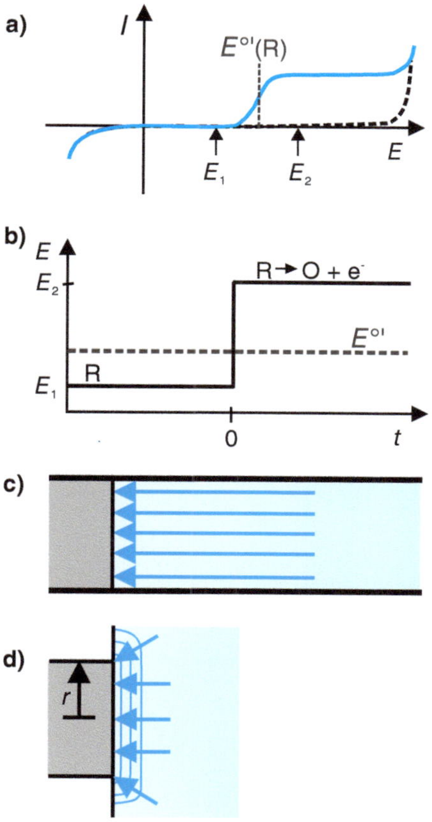

Figure 4.50 Schematic illustration of chronoamperometry. a) Steady-state current–potential curve at a rotating electrode in a pure electrolyte solution (background, – –) and after addition of an oxidizable compound R (—) for the selection of the potentials E_1 and E_2; b) potential program in chronoamperometry; c) cross-sectional view on the diffusion layer at a recessed electrode at the bottom of a solution-filled cylinder. The vectors indicate the concentration gradients; d) cross-sectional view of the diffusion layer above an electrode forming an inlaid disk in an insulator in quiescent solution. For this electrode, the situation shown in c) is a good approximation as long as the radius r of the electrode is much larger than the square root of the mean squared displacement for the diffusion of R. The diffusion layers are indicated in this example by iso-concentration lines. The additional current caused by diffusion to the edge of the electrode is called "edge effect".

4.5.2 Current Transients and Concentration Profiles During Potential Step Experiments (Chronoamperometry)

Chronoamperometry (CA) is a switch-on experiment in which the change in current with time (i.e., the current transient) is recorded after a potential step. It is one of the most basic electrochemical experiments that is especially suited to investigate the mass transport in detail. We will come back to this type of experiment in different parts of this textbook. For the following consideration, we assume that the solution contains a neutral oxidizable compound. A good example would be ferrocene (Figure 2.23). In the schematic depiction of the experiment in Figure 4.50, this compound is abbreviated and generalized with "R" (for *reduced* redox form). In addition, the solution contains a background electrolyte (or supporting electrolyte). This could be a salt (e.g., $[(H_3C{-}CH_2{-}CH_2{-}CH_2)_4N^+][PF_6^-]$) that is soluble in the solvent (here, for instance, acetonitrile $H_3C{-}C{\equiv}N$). The supporting electrolyte is present in much higher concentration than R and does *not* react at the electrodes at the considered potentials. The electrode potential is switched between two values (Figure 4.50b). At the starting potential E_1, neither R nor the solvent nor the electrolyte can be oxidized. The potential E_2 is selected from the inspection of a current–potential curve (Figure 4.50a, called a voltammogram) in such a way that R is oxidized under diffusion-controlled conditions [$E_2 \gg E^{\circ\prime}(\mathrm{R/O})$], but solvent and supporting electrolyte are not converted. After a resting time at potential E_1, the potential is stepped to E_2 at the time $t = 0$. For this purpose, a potentiostat is used (Section 5.3.1). We would like to calculate the faradaic current (electrolysis current) for each time $t > 0$. We already know that the potential step leads to redistribution of charge in the electrical double layer and an associated capacitive current I_C that decays with e^{-kt} [Eq. (4.99)]. We do not want to consider this current contribution here and, instead, we calculate only the current for the electrolysis I_F. Experimentally, I_F is obtained approximately as the difference of the chronoamperograms in the presence and absence of R in the solution.

We assume that all molecules of R that reach the electrode are instantaneously converted ($k_a \to \infty$). In this situation, the charge transfer does not limit the current, and therefore diffusion will be limiting (Figure 4.50c). The electrolysis current (charge per time) can be calculated according to Eq. (4.121) as the product of the diffusional flux [Fick's first law, Eq. (4.110)] directly at the electrode surface ($x = 0$), the geometric electrode area A, the number n of transferred electrons per molecule of R, and the conversion factor of 1 mol electrons to electrical units (F). Both sides of Eq. (4.121) are given as absolute amounts because the signs of I and J have different meanings. The sign of I indicates whether it is an anodic or cathodic current. The negative sign of J shows that the mass transport occurs in the opposite direction of a positive concentration gradient. The notation of Eq. (4.121) with vertical bars for the absolute values of I and J is not used very often in electrochemistry but is implied in the majority of cases. We will also omit them in the subsequent treatment and consider the signs separately.

Please note the different sign conventions for the current direction, see Figure 1.9.

$$|I(t)| = \left| -nFAD\left(\frac{\partial c_R(x,t)}{\partial x}\right)_{x=0} \right| \tag{4.121}$$

In order to calculate $I(t)$, we need $c_R(x, t)$, which we obtain as the solution of the partial differential Eq. (4.120), that is, Fick's second law. We impose the initial condition that the concentration of R is equal to the bulk concentration c_R^* everywhere in solution before the potential step at $t = 0$.

$$\text{Initial condition: } c_R(x, 0) = c_R^* \tag{4.122}$$

We consider a solution volume that is sufficiently large so that the concentration $c_R(x, t)$ at the outer parts of the simulation domain will not deviate from c_R^*. That is, the diffusion layers remain smaller than our simulation domain. Homogeneous reactions that could change $c_R(x, t)$ are excluded.

$$\text{Boundary condition 1 in } x\text{: } \lim_{x \to \infty} c_R(x, t) = c_R^* \tag{4.123}$$

Because we study a diffusion-controlled reaction, a second boundary condition [Eq. (4.124)] is imposed. Due to the very fast electrode reaction, the concentration of the reactant R at the electrode surface ($x = 0$) instantaneously drops to zero when the potential E_2 is applied. That is for $t > 0$.

$$\text{Boundary condition 2 in } x\text{: } c_R(0, t) = 0; t > 0 \tag{4.124}$$

The solution of this PDE is possible analytically by using the Laplace transform[46] (Excursion 4-3). The solutions are concentration profiles $c_R(x, t)$ according to Eq. (4.125). A selection of them is shown in Figure (4.52) for different times. The zone (in space) in which the local concentration deviates from c_R^* is called the diffusion layer because the nonzero concentration gradient in this zone drives transport diffusion.

$$c_R(x, t) = c_R^* \operatorname{erf}\left[\frac{x}{2\sqrt{D_R t}}\right] \tag{4.125}$$

[46] Pierre-Simon Laplace, 1749–1827, French mathematician, physicist, and astronomer, Laplace transform.

Q: What is the meaning of erf(x).

A: erf stand for error function or Gauss' error function.[47] We can explain it with variables v and u (selected to avoid confusion with the meaning of x in the main text). The error function is defined by Eq. (4.126).

$$\mathrm{erf}(u) \equiv \frac{2}{\sqrt{\pi}} \int_0^u \mathrm{e}^{-v^2} \mathrm{d}v. \tag{4.126}$$

This integral approaches unity for $u \to \infty$. Often, the error function complement is needed as well [Eq. (4.127)].

$$\mathrm{erfc}(u) \equiv 1 - \mathrm{erf}(u) = \frac{2}{\sqrt{\pi}} \int_u^\infty \mathrm{e}^{-v^2} \mathrm{d}v \tag{4.127}$$

The values of both integrals are tabulated. They are also accessible by a Maclaurin series expansion[48] for small x. The Maclaurin series expansion can be considered as a Taylor series expansion[21] in the source point $x = 0$.

$$\mathrm{erfc}(u) = \frac{2}{\sqrt{\pi}} \left[u - \frac{u^3}{3} + \frac{u^5}{5 \cdot 2!} - \frac{u^7}{7 \cdot 3!} + \ldots \right] \approx \frac{2u}{\sqrt{\pi}}. \tag{4.128}$$

The derivative of the error function in Eq. (4.125) with respect to u yields

$$\frac{\mathrm{d}}{\mathrm{d}x} \mathrm{erf} \left[\frac{x}{2\sqrt{D_\mathrm{R} t}} \right] = \frac{1}{\sqrt{\pi}\sqrt{D_\mathrm{R} t}}. \tag{4.129}$$

After derivatization and insertion in Eq. (4.121), we obtain the time-dependent (transient) diffusion-limited current according to Eq. (4.130).

$$I(t) = \frac{nFA\sqrt{D_\mathrm{R}}}{\sqrt{\pi}} c_\mathrm{R}^* \frac{1}{\sqrt{t}} \tag{4.130}$$

This equation is very important and is called the **Cottrell equation**[49]. It facilitates the calculation of the *maximally possible current* in an electrochemical system. This current is always proportional to c_R^*. This proportionality is the basis for the use of electrochemical methods for concentration analysis (Section 5.3). In addition, we see that the current decays with a dependence on the square root of time that is characteristic of diffusion processes (Figure 4.51). This is despite the fact that the concentration in the bulk phase does not change at all (Figure 4.52).

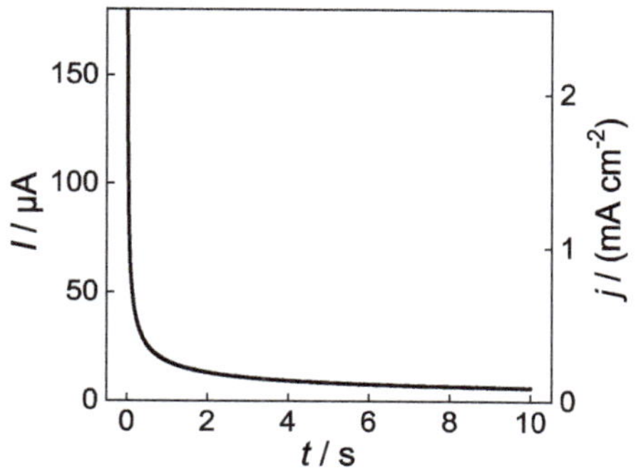

Figure 4.51 Experimental transient diffusion-limited current after a potential step for planar diffusion to a flat electrode with radius $r = 1.5$ mm; $c = 2$ mmol L^{-1} ferrocenylmethanol in 0.1 mol L^{-1} Na_2SO_4.

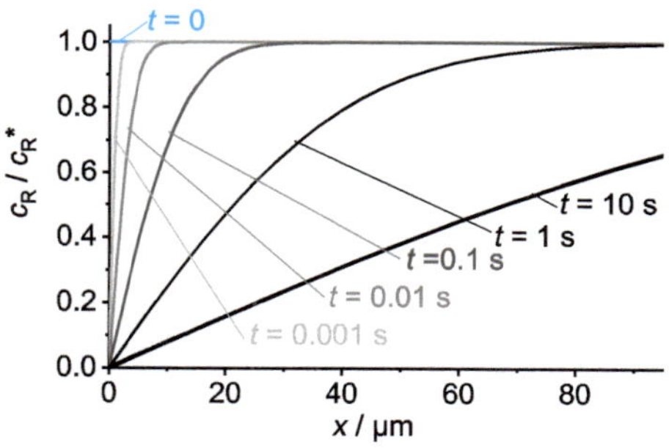

Figure 4.52 Calculated concentration profiles at different times t_i. To allow application to any initial concentration, the ordinate axis is given as normalized concentration $c(x,t)/c^*$; $D = 5 \times 10^{-6}$ cm^2 s^{-1}.

[47] Carl Friedrich Gauss (German spelling Gauß), 1777–1855, German mathematician, professor, and director for the astronomical observatory of the University of Göttingen, contributed to theory of numbers, astronomy, and land surveying. In this context, he developed important calculus algorithms that are still massively used, for instance, for regression analysis and statistics. His name is also used for instance in Gauss' elimination methods for matrices, Gauß–Jordan method for matrices, Gauss' law of error propagation, Gauss' integral theorem, Gauss' normal distribution, Gauss' plane for complex numbers (Chapter 7), and Gauss–Newton method for solving nonlinear equations (used in this book for fitting spectra in Chapter 8).

[48] Colin Maclaurin, 1698–1746, Scottish mathematician, geodesist, and geographer.

[49] Frederik Gardner Cottrell, 1877–1948, US-American chemist, developed the Cottrell equation during his PhD research in Leipzig, Germany. Professor of chemistry in Berkley. Patented a device for electrostatic deposition of particles from gas streams (still used today), later Director of the US Bureau of Mines and Director of the Fixed Nitrogen Research Laboratory of the US Department of Agriculture.

The reason for the decaying current is not the complete consumption of R, but the fact that the diffusion layers expand with time. The diffusional transport must cover larger and larger distances, thus decreasing the absolute value of the concentration gradient and making the transport less effective (Figure 4.52).

Unfortunately, the technical terms "diffuse double layer" and "diffusion layer" sound very similar, but refer to completely different phenomena. The diffuse double layer is part of the electrical double layer that also exists in the case of a chemical equilibrium at an electrode surface, that is, in the absence of a net reaction. Within the diffuse double layer, ions are enriched or depleted compared to the solution bulk to compensate for the charge excess on the electrode. Depending on ionic strength, the diffuse double layer extends 1–30 nm into the solution. A diffusion layer is formed only as a consequence of a net consumption of a substance during electrolysis and is always linked to a nonequilibrium state.

Chronoamperograms are recorded in the laboratory practice in order to characterize diffusion processes. Besides the case of **planar diffusion** treated here, there are other electrode geometries (microdisk, hemisphere, sphere, band, cone) for which the solution of Fick's second law is known (Section 5.6). Chronoamperometry at disk electrodes can be used for the determination of diffusion coefficients. In the treatment above, we considered only concentration gradients perpendicular to the electrodes as found at a recessed electrode in Figure 4.50c. For a disk electrode (Figure 4.50d), there is an enhanced transport toward the perimeter of the disk because of an additional diffusional flux parallel to the electrodes surface. This additional contribution causes an increase in the current, called edge effect, at a disk compared to a recessed disk of the same diameter. In many practical situations, the Cottrell equation derived for the situation in Figure 4.50c is also a good approximation for the disk electrode in Figure 4.50d because the edge effect is negligibly small for a sufficiently large electrode (Ø > 1 mm). For data evaluation, the current is plotted versus $1/t^{1/2}$ (Figure 4.53), yielding a Cottrell[49] plot. The data point recorded just after the potential step are located on the right-hand side of the chart, and the data point at the end of the experiment at the left-hand side, that is, *opposite* to the reading direction in English language. At known concentration of R and known geometric electrode area, the diffusion coefficient is calculated from the slope of the curve.

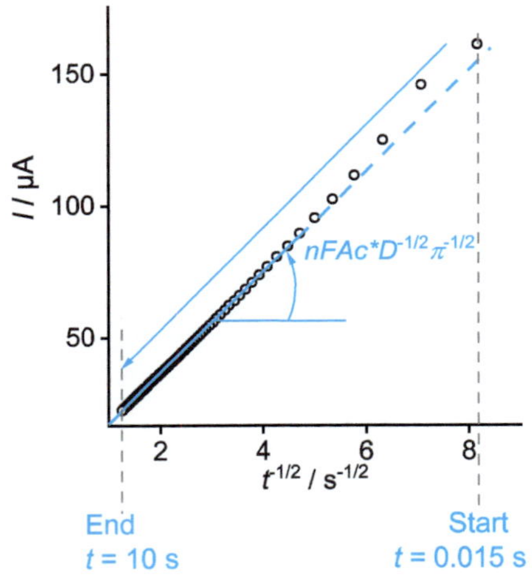

Figure 4.53 Linearized plot of the experimental chronoamperogram of Figure 4.51 (Cottrell plot).

Some limitations of chronoamperometry should be kept in mind. First, the data points immediately after the potential step may still be influenced by the capacitive current that has not been taken into account when deriving Eq. (4.130). As a consequence, the first three data points in Figure 4.53 are clearly above the regression line. For calculation, those points should not be used. The duration during which the capacitive current contributes significantly to the total current in the electrolytic cell depends on the time constant $C_{dl}R_{uc}$ of the cell. C_{dl} is the double layer capacitance and R_{uc} is the part of the solution resistance between working and auxiliary electrode that is not compensated by the potentiostat (Section 5.3.1). If fast processes are to be studied (and thus discarding the first data points is not an option), C_{dl} can be minimized by the use of microelectrodes (Section 5.6). Second, a potentiostat cannot apply a potential step infinitely fast (or instantaneously) because it behaves like an RC element itself. Third, the sharp rise of the current may also be a challenge for the recording device and may not be correctly recorded. Fourth, waiting very long is not a good option because convective effects commence in quiescent solution after approximately 20 s (natural convection). Further problems are caused by local heating or the emergence of density gradients in the diffusion layer as a consequence of the electrolysis reaction.

So far, we have not discussed how the Cottrell equation and many other equations describing electrochemical signals are derived. Usually, this requires solving (systems of) partial differential equations (PDEs). This is possible with a pen and a piece of paper only for relatively simple cases such as in Excursion 4-3. For most practical purposes, dedicated numerical computer programs are used to solve the problem by approximating the PDEs and solving the problem for discrete time steps and on a grid in space. The programs are so advanced that the user only needs to correctly specify the problem and does not need to program the solution. However, the accuracy of the solution must be checked. For such a purpose, analytical solution as the one specified in Excursion 4-3 can be used. It is therefore important that the user can differentiate between approximate and exact solutions of PDEs when working in this field.

Excursion 4-3 Solution of Partial Differential Equations by Means of the Laplace Transform Technique

You may already know mathematical transforms of functions from other areas of chemistry or physics. For instance, periodic processes can be treated very efficiently using the Fourier transform,[50] by which a signal is "translated" from the time domain into the frequency domain. For the treatment of switch-on processes, and as such we can consider a chronoamperometric experiment; the Laplace transform[46] is as useful as the Fourier transform is for periodic processes. Both transforms have in common that the initial function is mapped from a **domain** according to a set of rules onto an **image**. The calculation is solved in the image, and the solution is transformed back to the domain. A very simple example of this idea is the execution of a multiplication task by means of a slide rule (Figure 4.54).

The application of the Laplace transform can be illustrated using the derivation of the Cottrell equation. For a more in-depth coverage, the reader is referred to textbooks of mathematics [22, 23]. Using the Laplace transform, we convert the problem in question into an easier-to-solve problem in the image, solve the problem, and apply the inverse transform in order to convert the solution to the preimage. In this way, the ordinary differential equation (ODE) in the domain becomes an algebraic equation in the image, and a partial differential equation (PDE) in the domain (e.g., Fick's second law) becomes an ODE in the image.

The Laplace transform is described by the operator L. Generally, a mathematical operator is a calculus instruction applied to the expression right of the operator, called operand. For instance, the term $\partial/\partial x$ in Eq. (4.119) is an operator meaning "differentiate the following term with respect to x."

The function we seek in the domain is called $F(t)$. The transformed function (in the image) is $f(s)$. For better distinction, the function $f(s)$ is additionally labelled by an overbar. For our specific problem, $c(t)$ is an example of a general function $F(t)$ in the domain and becomes $\bar{c}(s)$ in the image. The Laplace transform is defined as

$$\bar{f}(s) \equiv \mathfrak{L}\{F(t)\} \equiv \int_0^\infty \exp(-st)F(t)\mathrm{d}t. \tag{4.131}$$

The integral expression, which we need in order to assemble the right-hand side of Eq. (4.119), is already known for many practically relevant cases and can

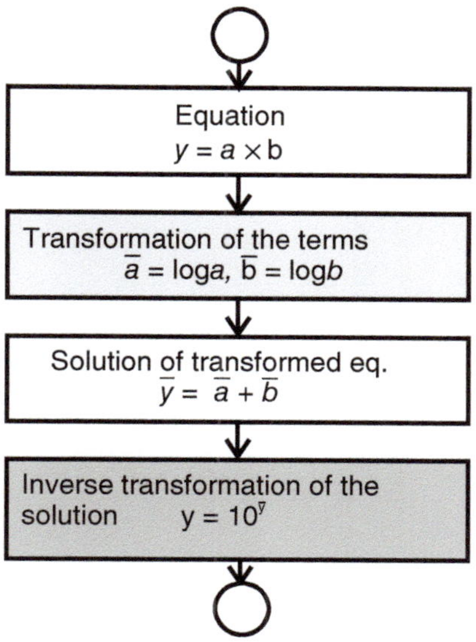

Figure 4.54 Illustration of a transformation for the solution of a multiplication exercise. We would like to multiply two large numbers $y = a \cdot b$ without using any electronic calculator or computer program. Instead of a direct multiplication, we form the decadic logarithms of a and b. Because of log $(a \cdot b) = \log a + \log b$, we can add the logarithms and obtain the solution of the original exercise after inverse transformation of our intermediate result $y = 10^{(\log a + \log b)}$. The flow chart illustrates the algorithm.
Slide rules with logarithmic scales worked by this principle and were widely used as a routine tool by engineers and scientists till the 1970s. The main author of this book used it to solve the high school leaving examination, which did not cause persistent damage to him. Golden Gate Bridge (and every other older bridge) was constructed by this tool as well.

[50] Jean Baptiste Joseph Fourier, 1768–1830, French mathematician and physicist.

Table 4.7 Subset of the table of transforms [23] with the expressions needed for deriving the Cottrell equation.

	in the domain $F(t)$	in the image $f(s)$
1	$const$	$\frac{const}{s}$
2	$\operatorname{erfc}\left[\frac{x}{2\sqrt{const \cdot t}}\right]$	$\frac{1}{s}\exp\left[-\sqrt{\frac{s}{const}}x\right]$
3	$\frac{1}{\sqrt{\pi t}}$	$\frac{1}{\sqrt{s}}$

be looked up in so-called "tables of transforms" [23]. Textbooks of electrochemistry also contain specialized tables of transforms and example solutions [24, 25]. Table 4.7 compiles the expressions that we need for deriving the Cottrell equation.

The following arithmetic rules apply to the transformed functions in the image:

1) Linearity with arbitrary constants a and b.

$$\mathfrak{L}\{a \cdot F(t) + b \cdot G(t)\} = a \cdot f(s) + b \cdot g(s) \tag{4.132}$$

2) For time derivatives, the following rules apply (providing the usefulness for solving ODEs and PDEs):

$$\mathfrak{L}\left\{\frac{dF(t)}{dt}\right\} = \mathfrak{L}\{F'(t)\} = s \cdot f(s) - F(0) \tag{4.133}$$

$$\mathfrak{L}\left\{\frac{d^2F(t)}{dt^2}\right\} = \mathfrak{L}\{F''(t)\} = s^2 \cdot f(s) - s \cdot F(0) - F'(0) \tag{4.134}$$

$$\mathfrak{L}\{F^{(n)}(t)\} = s^n \cdot f(s) - s^{(n-1)} \cdot F(0) - s^{(n-2)} \cdot F'(0) - \ldots - s^0 \cdot F^{(n-1)}(0) \tag{4.135}$$

Concerning derivatives with respect to other variables, the Laplace transform is invariant.

$$\mathfrak{L}\left\{\frac{dF(x,t)}{dx}\right\} = \frac{f(x,s)}{dx} \tag{4.136}$$

3) Transform of integrals

$$\mathfrak{L}\left\{\int_0^t F(x,t)dx\right\} = \frac{1}{s}f(x,s) \tag{4.137}$$

4) Multiplication with an exponential function

$$\mathfrak{L}\{\exp(at)F(x,t)\} = f(s-a) \tag{4.138}$$

5) Inverse transform (L^{-1}) of products

$$\mathfrak{L}^{-1}\{f(s) \cdot g(s)\} = F(t) * G(t) = \int_0^t F(t-\tau) \cdot G(t)d\tau \tag{4.139}$$

The expression $F(t) * G(t)$ is called a **convolution integral**.

We return to the Cottrell equation and transform Fick's second law (4.120) using Eqs. (4.133) and (4.136). The result is an ODE for $\bar{c}(x,s)$ in x. Subsequently, we always consider only the concentration of the reactant, be it R (for oxidations) or O (for reductions). Therefore, we can omit the index R or O, which helps us increase the clarity of the notation in the subsequent section.

$$\mathfrak{L}\left\{\frac{\partial c(x,t)}{\partial t}\right\} = \mathfrak{L}\left\{D\frac{\partial^2 c(x,t)}{\partial x^2}\right\}$$

$$s \cdot \bar{c}(x,s) - c^* = D\frac{d^2\bar{c}(x,s)}{dx^2}$$

$$\frac{d^2\bar{c}(x,s)}{dx^2} - \frac{s}{D}\bar{c}(x,s) = \frac{c^*}{D} \tag{4.140}$$

Equation (4.140) is an inhomogeneous ODE of second order. Its solution is the sum of the general solution $\bar{c}_h(x,s)$ of the corresponding homogeneous ODE and of one particular solution $\bar{c}_p(x,s)$ of the inhomogeneous ODE. The corresponding homogeneous ODE is Eq. (4.141).

$$\frac{d^2\bar{c}(x,s)}{dx^2} - \frac{s}{D}\bar{c}(x,s) = 0 \tag{4.141}$$

Its general solution is Eq. (4.142).

$$\bar{c}_h(x,s) = A(s)\cdot\exp(-kx) + B(s)\cdot\exp(+kx),\quad \text{with } k=\sqrt{\frac{s}{D}} \tag{4.142}$$

The solution of the homogeneous ODE $y''(x) - my(x) = 0$ is $y(x) = Ae^{-kx} + Be^{kx}$ with $m = k^2$ or $k = \sqrt{m}$.

A particular solution of the ODE (4.140) would be

$$\bar{c}_p(x,s) = \frac{c^*}{s}\ \text{for all } x \tag{4.143}$$

With this, we obtain the solution of Eq. (4.140) as the sum of the general solution and one particular solution [Eq. (4.144)].

$$\begin{aligned}\bar{c}(x,s) &= \bar{c}_p(x,s) + \bar{c}_h(x,s)\\ &= \frac{c^*}{s} + A(s)\cdot\exp(-kx) + B(s)\cdot\exp(+kx),\quad \text{with } k=\sqrt{\frac{s}{D}}\end{aligned} \tag{4.144}$$

We still have to specify the unknown functions $A(s)$ and $B(s)$. In order to do so, we transform the boundary condition 1 [Eq. (4.123)] according to line 1 of Table 4.7:

$$\begin{aligned}\lim_{x\to\infty}\mathfrak{L}\{c(x,t)\} &= \mathfrak{L}\{c^*\}\\ \lim_{x\to\infty}\bar{c}(x,s) &= \frac{c^*}{s}\end{aligned} \tag{4.145}$$

Because of $k > 0$ (otherwise we cannot calculate the square root within the range of real numbers), the boundary conditions (4.145) can only be satisfied if $B(s) = 0$ in Eq. (4.144).

$$\bar{c}(x,s) = \frac{c^*}{s} + A(s)\cdot\exp(-kx),\quad \text{with } k=\sqrt{\frac{s}{D}}. \tag{4.146}$$

The function $A(s)$ is obtained after transforming the boundary condition 2 $c_R(0,t) = 0; t > 0$:

$$\begin{aligned}\mathfrak{L}\{c(0,t)\} &= \mathfrak{L}\{0\}\\ \bar{c}(0,s) &= \frac{0}{s} = 0\end{aligned} \tag{4.147}$$

Inserting $x = 0$ into Eq. (4.146) gives the exponential term as $\exp(0) = 1$. The entire expression becomes zero only if $A(s) = -c^*/s$. Using the substitution $k = (s/D)^{1/2}$, the solution of the ODE in the image range is given in Eq. (4.148).

$$\bar{c}(x,s) = \frac{c^*}{s} - \frac{c^*}{s}\cdot\exp\left(-\sqrt{\frac{s}{D}}\cdot x\right) \tag{4.148}$$

The inverse transform using lines 1 and 2 of Table 4.7 in connection with the arithmetic rule (4.132) yields the concentration profile in Eq. (4.149).

$$\begin{aligned}\mathfrak{L}^{-1}\{\bar{c}(x,s)\} &= \mathfrak{L}^{-1}\left\{\frac{c^*}{s}\right\} - \mathfrak{L}^{-1}\left\{\frac{c^*}{s}\cdot\exp\left(-\sqrt{\frac{s}{D}}\cdot x\right)\right\}\\ c(x,t) &= c^* - c^*\mathrm{erfc}\left[\frac{x}{2\sqrt{Dt}}\right]\\ &= c^*\left\{1-\left(1-\mathrm{erf}\left[\frac{x}{2\sqrt{Dt}}\right]\right)\right\}\\ &= c^*\mathrm{erf}\left[\frac{x}{2\sqrt{Dt}}\right]\end{aligned} \tag{4.149}$$

We obtain the time-dependent current by transforming Eq. (4.121) into the image [Eq. (4.150)], insert the derivative of Eq. (4.148) with respect to x [Eq. (4.151) to

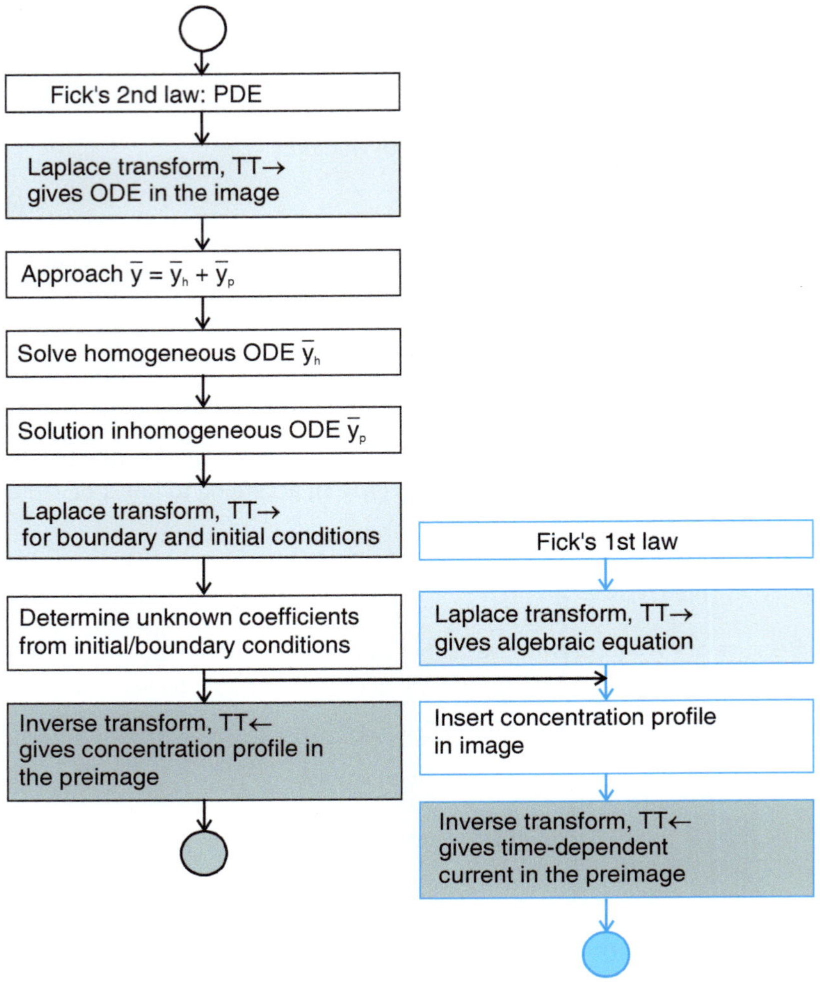

Figure 4.55 Scheme for solving the partial differential equation by means of the Laplace transform technique. (TT – table of transforms; →, ← indicate the reading direction in the table of transforms)

(4.153)] and apply the inverse transform (line 3 of Table 4.7) to the result using the arithmetic rule (4.132) to give Eq. (4.154).

$$\mathfrak{L}\{I(t)\} = \mathfrak{L}\left\{\left|-nFAD\frac{\partial c(x,t)}{\partial x}\right|_{x=0}\right|\right\}$$

$$\bar{I}(s) = nFAD\frac{\partial \bar{c}(x,s)}{\partial x}\bigg|_{x=0} \tag{4.150}$$

$$\left(\frac{\partial \bar{c}(x,s)}{\partial x}\right) = 0 - \frac{c^*}{s}\cdot\exp\left(\sqrt{\frac{s}{D}}\cdot x\right)\cdot\sqrt{\frac{s}{D}} \tag{4.151}$$

$$\left(\frac{\partial \bar{c}(x,s)}{\partial x}\right)_{x=0} = \frac{c^*}{s}\cdot\sqrt{\frac{s}{D}} = \frac{c^*}{\sqrt{sD}} \tag{4.152}$$

$$\bar{I}(s) = nFAD\frac{c^*}{\sqrt{sD}} \tag{4.153}$$

$$\mathfrak{L}^{-1}\{\bar{I}(s)\} = \mathfrak{L}^{-1}\left\{nFA\sqrt{D}c^*\frac{1}{\sqrt{s}}\right\}$$

$$I(t) = nFA\sqrt{D}c^*\frac{1}{\sqrt{\pi t}} \tag{4.154}$$

Equation (4.154) is finally the exact solution of Fick's law for a chronoamperometric experiment at a disk electrode. The entire procedure is summarized in Figure 4.55.

For more complicated geometries, analytical solutions for partial differential equations become very challenging. In these cases, approximate numerical solutions are used instead. A number of program packages have been developed in the previous years, which allow solving such problems without coding work. We will describe an example in Excursion 8-4.

4.6 Combined Transport Processes During Electrophoresis

So far, we have considered each transport mechanism as an isolated phenomenon. However, this is relevant only in specifically designed situations and is an exception rather than the rule. In the subsequent section, we will explore the simultaneous occurrence of different transport mechanisms.

4.6.1 Basic Principles

Electrophoretic separation experiments provide an excellent example to illustrate the simultaneous action of different transport mechanisms. Such experiments have a pivotal importance for the separation of proteins and nucleic acids (DNA and RNA) in biochemistry. In this section, we can only consider the electrochemical aspects of these methods. Further information with more practical details and applications is available in specialized textbooks and monographs [26, 27].

Electrophoresis refers to the separation of a mixture of dissolved compounds based on differences in the migration velocity. Electrophoretic methods belong to the larger group of **electrokinetic separation techniques**. A typical laboratory experiment is shown in Figure 4.56. Here, we consider this experimental system as an electrochemical cell. There are two separate half-cells, in which water electrolysis is performed at Pt wires. At the anode (+), O_2 is formed, and H_2 emerges at the cathode (−). Both half-cells are connected by a "salt bridge", which is in fact the most important part of

Most proteins do not have any or only a weak own color. Therefore, the bands are not or only faintly visible during the separation process. The clearly discernible bands in Figures 4.60 and 4.61e are only obtained after a staining process conducted after the separation. In the textbook figure, the bands are shown in high contrast to explain the principle.

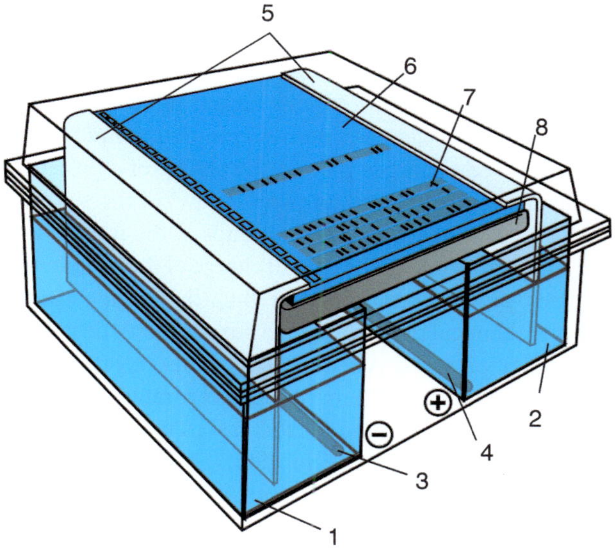

Figure 4.56 Laboratory experiment for electrophoresis. The apparatus contains the following major parts: 1) buffer reservoir for the cathode, 2) buffer reservoir for the anode, 3) cathode (Pt wire), 4) anode (Pt wire), 5) buffer-soaked fleece paper as ion-conducting connection to the separation gel, 6) separation gel, 7) separated bands of biomolecules, and 8) plate to support the separation gel often with water cooling to remove Joule's heat. Space-saving vertical arrangements are also very common.

the system. There, the mixture is separated into its components. The "salt bridge" is formed by a gel with a thickness of about 0.5 mm called a separation gel. The pores of the gel hold an aqueous buffer with freely moving low-molecular-weight ions between the loosely arranged polymer chains of the gel. The ends of the gel are connected to the anodic or cathodic half-cell by buffer-soaked filter or fleece papers. In contrast to an electrochemical cell designed for the most effective conversion of reagents to products with controlled potentials, here, a very high fraction of the voltage applied to the electrophoresis apparatus drops purposefully within the separation gel. This causes Joule heating[51] $W = I\,\Delta U\,t$, which may lead to a deformation of the gel during the separation process (typically 30 min to 2 h). Heating accelerates band broadening by diffusion of the components. In order to counteract the heating, the separation gel is actively cooled, in most cases by means of an electrically isolated, water-cooled metal block below the separation gel. The large voltage drop is a prerequisite for the formation of a strong electric field ($\mathbf{E} = U/l$) driving the separation process.

Q: The experimental system seems to me very similar to that of thin layer chromatography, doesn't it?

A: Yes, that's right. Most textbooks treat electrophoresis as one of many separation techniques (like chromatography and field flow fractionation). The separation gel in Figure 4.56 will hold an inner electropherogram analogous to a thin layer chromatogram because the separated components are kept at specific positions within the gel. In contrast, the separated components in capillary electrophoresis will pass the detector and leave the capillary one after the other. This is an outer electropherogram, which is analogous to the chromatogram in capillary gas chromatography of high-pressure liquid chromatography. Figures of merit such as resolution, limit of detection, and separation time can be defined and applied very similarly to procedures in chromatography. In this section, we emphasize the electrochemical principles of electrophoresis and the relationship to the transport mechanisms. This is not intended to replace a separate and thorough study of separation techniques.

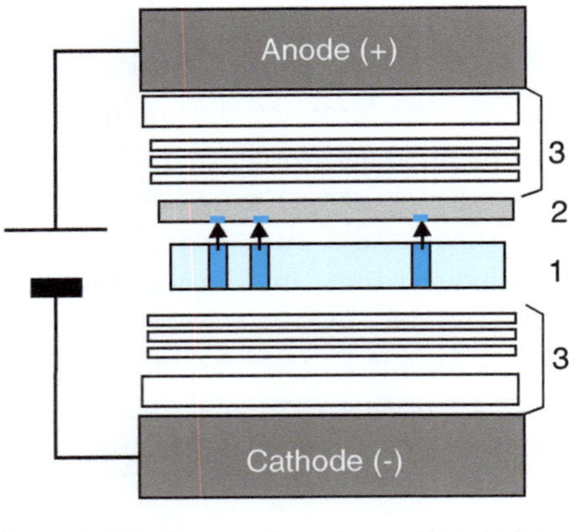

Figure 4.57 Schematic cross-sectional view of an electroblotting setup in exploded view. The proteins separated in the gel (1) are transferred to a PVDF membrane (2) by the electric field between cathode and anode. Buffer-soaked filter or fleece paper serves as an electrical contact.

4.6.2 Methods of Protein Electrophoresis

Electrophoresis of protein mixtures can be carried out in different approaches. During **native electrophoresis**, the proteins migrate according to their charge-to-size ratio at different speeds in the separation gel. Because proteins are polyelectrolytes, that is, macromolecules with several charged groups, the charge of a protein depends on the kind of amino acids making up the protein and the pH of the separation buffer. The pH determines which of the acidic and basic groups of the protein are protonated or deprotonated and thus contribute to the overall charge of the protein molecule. During native electrophoresis, proteins retain their original three-dimensional structure and also the associated biochemical properties, for example, catalytic properties or binding properties toward other proteins. After the end of the electrophoresis, the bands broaden quickly due to diffusion. Therefore, the separated proteins must be fixed. Electroblotting is an important method for this task. The separation gel is placed on a foil of nitrocellulose or poly(vinylidene fluoride) (PVDF, $[-CH_2-CF_2-]_n$). Similar to a slice of cheese in a sandwich (Figure 4.57), the arrangement is first placed between layers of filter paper and wetted with a transfer buffer. The entire stack is then

---

[51] James Prescott Joule, 1818–1889, English physicist, described the first law of thermodynamics, especially the equivalence between mechanical work and heat as well as the heat released during the flow of an electrical current across a resistor. His name occurs in the unit of energy Joule J, Joule's heat, the Joule–Thompson effect, the Joule effect of magnetostriction, and the Joule process (ideal circular process).

mounted between two plate electrodes. The electric field between the plates causes a movement of the protein from the gel onto the membrane. On the membrane, they stick. Because most proteins are not colored, the proteins on the membrane must be stained. The dye Coomassie Brilliant Blue or a silver stain can be employed. The proteins are first treated with an Ag^+-containing solution. The proteins retain Ag^+ ions. After discarding the supernatant solution, the retained Ag^+ ion will be reduced by a chemical reductant to $Ag^{(0)}$ in form of nanoparticles in the protein bands. The nanoparticles give the protein bands a gray to black coloration that is more intense with a higher protein content. The foil with the separated proteins is then photographically documented. The separated proteins can also be punched out and subjected to further biochemical tests, such as enzymatic reactions producing a color as indication of enzymatic activity. Mass-spectrometric analysis can be applied after partial trypsin digestion. The proteins can also be precipitated by a specific antibody.

Very often, **SDS-PAGE** methods are employed. The abbreviation stands for sodium dodecyl sulfate, $[C_{12}H_{13}{-}OSO_3^-]$ $[Na^+]$ and poly(acrylamide) gel electrophoresis. The gel is cast from two precursor solutions with different densities and different contents of crosslinking monomers. This results in an adjustable gradient of pore sizes in the gel. SDS is an anionic surfactant. When SDS comes into contact with a protein, the hydrophobic parts of SDS attach noncovalently to the hydrophobic parts of the protein (Figure 4.58). This weakens the hydrophobic interactions within the protein molecule that are responsible for stabilizing the three-dimensional protein structure. After weakening this structure-stabilizing intramolecular interaction, the proteins denature usually connected with a loss of their biochemical functions. The protein-SDS adduct assumes a negative net charge. Due to the uniform charge and the weakened intramolecular interactions, the entire protein migrates toward the anode (+). Depending on the protein size, it is held back at a specific position by the separation gel. The mesh size becomes increasingly narrow along the separation track to separate protein sizes. If a mixture of proteins with known sizes runs in a parallel track as a molecular weight standard, the covered migration distance can be used for an estimation of the molar mass of the separated proteins in the sample.

The **isoelectric focusing** (IEF) is carried out in a gel containing a gradient of pH. Such gradients are obtained when polyelectrolytes with different contents of weakly acidic and weakly basic functional groups are separated in an electric field (Figure 4.59b). These polyelectrolytes constitute the separation buffer for the subsequent protein electrophoresis. As long as the voltage is applied, the pH gradient remains stable over the duration of the experiment. There are also methods based on chemically fixed gradients. The mixture of proteins is then applied approximately in the middle of the separation gel. Proteins with a negative net charge migrate toward the anode (+), while proteins with a positive net charge migrate toward the cathode (−). Because of the stable pH gradient, the proteins pass through regions of different pH. This, in turn, constantly changes the protonation state of the proteins and thus their net charge. If the local pH of the separation gel is equal to the isoelectric point pI of the protein, the migration process comes to an end (Figure 4.59d). The **isoelectric point pI** is the pH at which the net charge of an amino acid or an entire protein is zero. The band broadening by diffusion only plays a minor role in IEF because the proteins are always pulled back to the point in the separation gel at which the pH equals the pI (therefore the name isoelectric *focusing*, Figure 4.59d). Isoelectric focusing can be used as a fingerprint method for identifying different kinds of animal origins in meat extracts. Samples of homogenized muscle tissue are separated and compared to samples of certified origin (Figure 4.60).

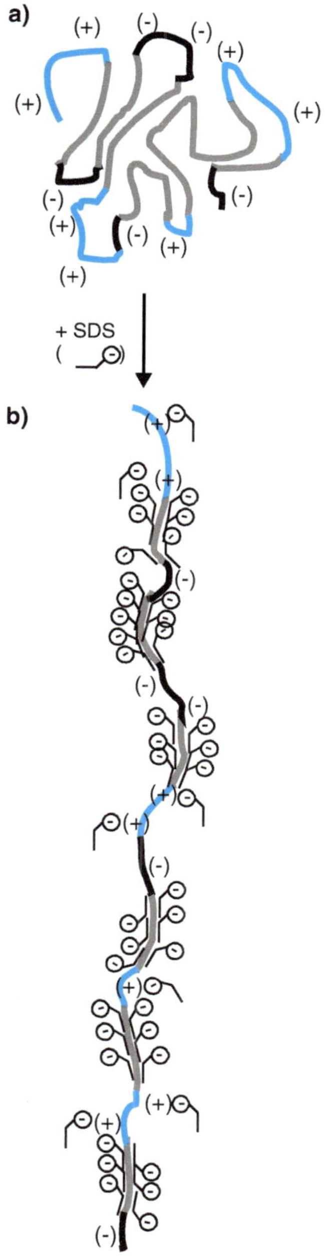

Figure 4.58 Schematic image of the interaction between a protein and SDS. In the native conformation (a), the hydrophobic regions of the protein are localized in the inner parts, and the charged regions (blue/black) are turned toward the periphery. The globular structure is stabilized by hydrophobic interactions between the hydrophobic components of the protein. b) Attachment of the hydrocarbon chains of SDS to the hydrophobic parts of the protein (gray), which assumes an overall negative charge. This strongly overcompensates for any net charge of the native protein. The loss of stabilizing intramolecular interactions and the repulsion of the multitude of negative charges of SDS cause a stretching of the entire protein.

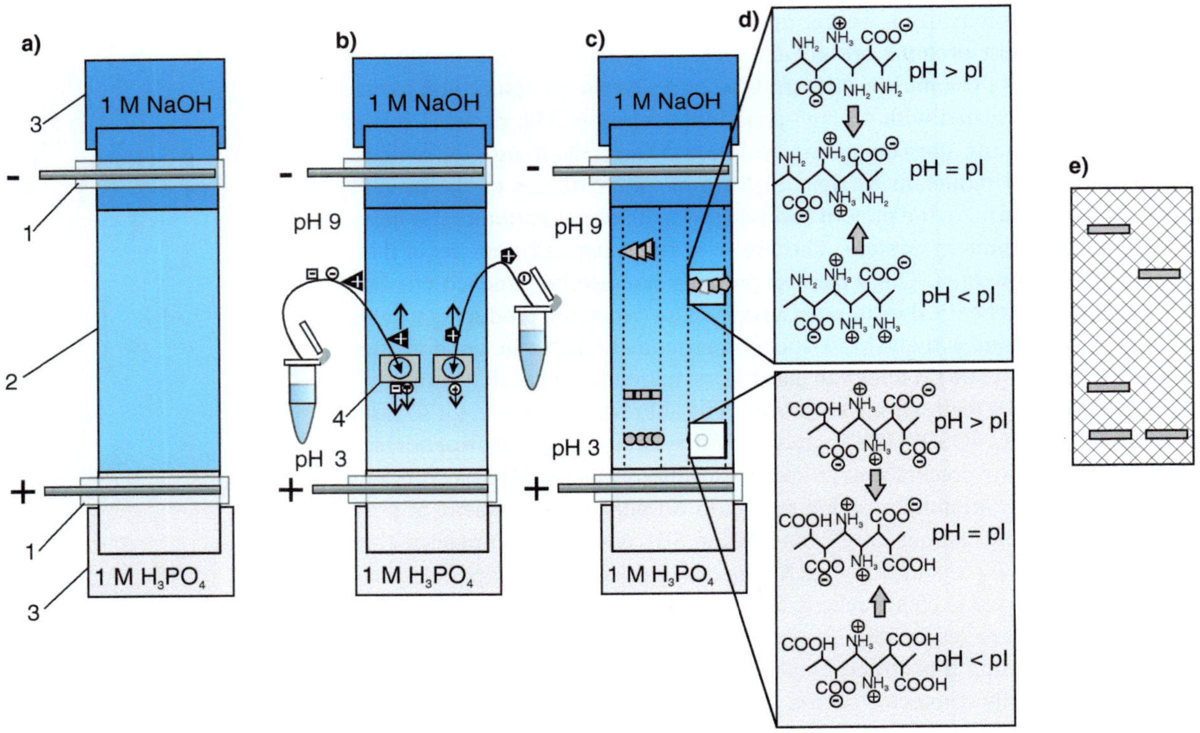

Figure 4.59 Principle of isoelectrical focusing (top view). a) The gel with a mixture of polyelectrolytes having different pI values and high buffer capacity is cast and placed between two filter papers. b) In a preelectrolysis, the polyelectrolyte components are "sorted", thereby establishing a pH gradient within the separation gel. Next, the sample is dosed into the middle of the gel via a sample application reservoir. Depending on the protonation state, the proteins of the mixture carry a different net charge and move with different speeds and directions in the electric field. c) The proteins reach the location in the separation gel at which the pI of the proteins is equal to the local pH in the separation gel. The net charge of the proteins reduces to zero, and the proteins do not migrate anymore. d) The enlarged schematic view shows how acidic or basic the proteins that diffuse away from their band are pulled back to the location with pI = pH. e) The separation gel is treated to fix and stain the proteins in the gel. This is followed by documentation and by quantification.

1) Electrodes with fleece paper (in a)
2) Mixtures of polyelectrolytes (in a)
3) Buffer reservoir (in a)
4) Reservoir for sample application (in b).

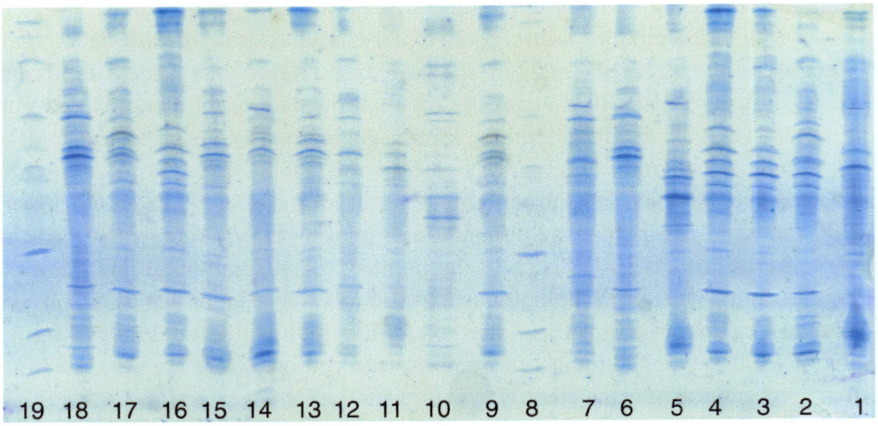

Figure 4.60 What's on the plate today? IEF separation of meat extracts. The origins of the meat in the samples 16, 17, and 18 are sought. The following reference samples have been applied and separated: 1 – rabbit, 2 – roe deer (*Capreolus capreolus*), 3 and 4 – beef, 5 – domestic pig, 6 – turkey, 7 – pigeon, 8 – pI marker, 9 – red deer (*Cervus elaphus*), 10 – chicken, 11 – European hare (*Lepus europaeus*), 12 – Muscovy duck, 13 – mutton, 14 – lamb, 15 – sheep, and 19 – pI marker. The comparison is made by comparing the bands of the samples with the band patterns of the reference samples like a fingerprint.

Answer: 16 = beef, 17 = red deer, 18 = turkey

Excursion 4-4 Proteomics with Two-Dimensional Electrophoresis

In **two-dimensional electrophoresis**, two different electrophoretic principles are combined, for example, isoelectrical focusing and SDS-PAGE. A complex protein mixture is then first separated by IEF according to the isoelectric point. Still, many proteins may have a very similar pI. Therefore, a second separation is performed in perpendicular direction to the first separation using SDS-PAGE, which separates out the mixture according to the molecular weight *M* (Figure 4.61). Separations of highly complex protein mixtures are very important in proteomics applications. Proteomics refers to the analysis of the entire momentary protein content of cells, tissues, and organisms as a function of development stage, health status, or metabolic state. The example of such a 2D gel in Figure 4.61e shows how complex the protein composition of a cell extract can be. The identification of the proteins is usually achieved by mass spectrometry in combination with bioinformatics. A complete assignment of each spot in the gel is very demanding and often not relevant. Instead, it is much more interesting to identify, for instance, proteins with an abundance that differs very strongly between healthy and tumor cells. Another question that can be addressed is the activation of certain metabolic pathways in bacteria

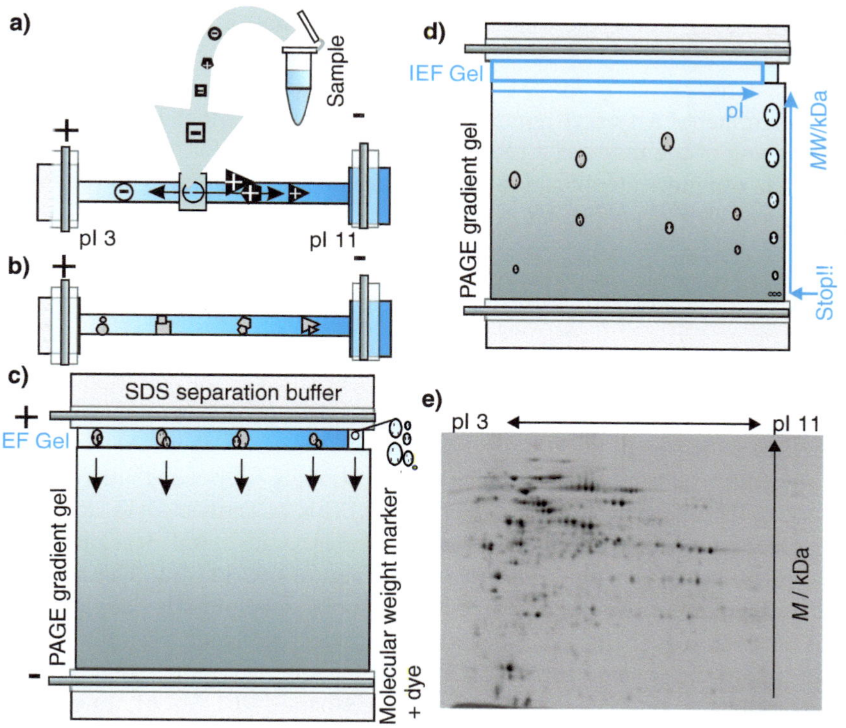

Figure 4.61 Schematic of two-dimensional electrophoresis in top view. a) Start of the IEF; b) end of the IEF, the proteins are separated according to their pI. Here, it is assumed that each band in the IEF gel contains two proteins of equal pI but different size. c) Chemical reduction of the disulfide bridges, denaturation of the proteins, noncovalent attachment of SDS, and positioning of the pretreated IEF gel strip between the start of a PAGE gradient gel and an SDS separation buffer reservoir with fleece paper. A small section loaded with proteins of known *M* and of a low-molecular-weight dye is also added. d) Separation by SDS-PAGE according to *M*. The electrophoresis is stopped when the low-molecular-weight dye has reached the positive electrode. After the separation, the proteins must be fixed and stained. e) Example of a 2D gel of an extract of marine bacteria. Source: Photo (e) by R. Rabus, Carl von Osseitzky University of Oldenburg.

The schematic is not to scale! The protein molecules are extremely enlarged in size. The spots in the separation gel are composed of many protein molecules, not of a single molecule.

depending on the available carbon-containing substrates. The activation of certain metabolic pathways is associated with characteristic concentration changes in enzymes.

Quantitative two-dimensional differential gel electrophoresis (**2D DIGE**) is an elegant solution to the problems posed above. Three protein extracts are employed: (1) the sample to be analyzed, (2) a reference sample (e.g., the extract of a healthy cell), and (3) a mixture of (1) and (2) as an internal standard. The samples are labelled with three *different* dyes. These dyes are clearly distinct in their excitation and emission spectra. The charge of the dyes and their coupling chemistry are adjusted in a way that it alters the pI of the labelled proteins only marginally. Now the *three samples are mixed* (!!) and subjected to a 2D electrophoresis as schematically shown in Figure 4.61a–d. The fluorescence of the separated and labelled proteins is observed and documented. Initially, the wavelength range suitable for Sample 3, that is, the mixture of unknown sample and reference, is used. Image 3 shows the positions of all protein spots in the gel. Next, the correct wavelength range is employed to excite the dye of Sample 1 to yield Image 1. Subsequently, the wavelength range suitable for Sample 2 is used to yield Image 2. After recording the three images, a computer program is used to compare spot by spot the normalized intensities of the spots in Images 1 and 2 with those of Image 3 (mixture of unknown sample and reference). In those cases, where the intensities in Images 1 and 2 differ from each other, a different abundance of this particular protein in the sample and the reference was detected. Only for this protein, an identification is performed by mass spectrometry. A typical number is five spots per gel, that is, a very significant focusing of the task and a reduction in time and costs for the analysis compared to the complete identification of all spots. Further explanation on the methodical basis of 2D electrophoresis in proteomics can be found in [28].

4.6.3 Capillary Electrophoresis

The electrophoretic techniques exhibit some fundamental limitations. They are manually demanding and generally not very reproducible; the detection of intensities is semiquantitative at best, and Joule heating leads to band broadening, which prevents a very high separation performance. Therefore, there are many attempts to integrate electrophoretic principles into strongly automated instruments as it is common for other analytical techniques. This is connected to an expectation to provide quantitative electrophoretic processes. A technical development in this direction is **capillary electrophoresis** (CE) (Figure 4.62) [27–32]. This technique has gained high importance for the separation of polar, medium-sized organic molecules and for DNA sequencing. The separation is carried out inside a fused silica capillary with a polyimide-coated surface. The inner diameter is between 25 μm and 75 μm. The capillary is conditioned by rinsing with methanol, HCl, NaOH, water, and finally the selected separation buffer. The pretreatment removes residual contaminants from previous separation runs and establishes a defined state of the inner capillary wall. The chemical state of the inner capillary wall is very important for the reproducibility of the migration time and separation (*vide infra*). The capillary can have a length of 1 m, and it connects two buffer reservoirs. Due to the small diameter and round geometry, the capillary enables a very efficient removal of Joule heat, which solves a fundamental problem of the classical gel electrophoresis. In addition, the detection of the separated compounds is possible with an instrumental method similar to those employed in chromatography. Thus, CE can be performed in an automated, parallelized, and miniaturized manner.

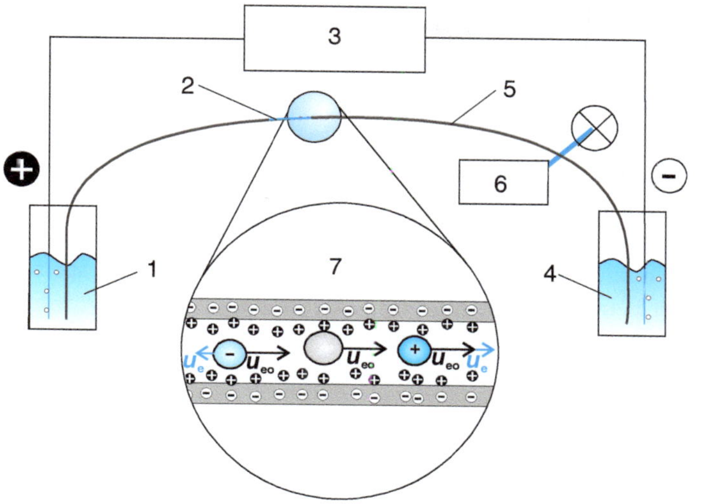

Figure 4.62 Schematic illustration of a CE system,

1) Buffer reservoir for inlet;
2) Sample solution;
3) Voltage source;
4) Buffer reservoir for outlet;
5) Capillary;
6) Detector;
7) Enlarged depiction of processes inside the capillary.

The scheme is not to scale!

Between the platinum wires immersed in the buffer reservoirs, a high voltage of 30 000 V is applied. This causes a continuous electrolysis in the buffer reservoirs. In this experimental system, the current is limited by the migration of ions through the capillary between the two buffer reservoirs.

To load the sample into the capillary, the inlet end is removed for a short time from the buffer reservoir with the separation buffer and immersed in a vial with the sample mixture. A small plug of the sample solution is loaded into the capillary. This is achieved by applying an overpressure to the atmosphere above the sample vial for a short time (or by lifting the sample vial with the capillary end versus the outlet). This kind of sample injection is called a hydrodynamic injection. All components of the sample are transferred into the capillary with the same efficiency. Alternatively, the electrokinetic injection can be used, in which the migration between the two electrodes is used to drive ions into the inlet of the separation capillary. Because the ion mobilities are different for each ion, there is a different probability for different kinds of ions to enter the capillary and thus the separation process.

After the injection, the inlet of the capillary is returned to the reservoir with the separation buffer, and the actual separation is started by application of a high voltage between the two buffer reservoirs. Under the influence of a constant electric field, the ions contained in the sample migrate with different velocities and reach the detector after different times. The detector records the analyte concentration as a function of separation time, similar to the processes in capillary chromatography (outer electropherogram).

In the literature about capillary electrophoresis, the symbol μ is often used for the ion mobility. We refrain from doing so in this textbook in order to avoid confusion with the symbol of the chemical potential.

As explained with the help of Eq. (4.7), an ion assumes a constant migration velocity $\mathbf{w}_i$, whereby cations move toward the capillary outlet and anions in the direction of the capillary inlet.

$$\mathbf{w}_i = u_i\mathbf{E} = \frac{z_i e}{6\pi r_i \eta}\mathbf{E} \tag{4.155}$$

u_i – Ion mobility [$m^2\ s^{-1}\ V^{-1}$]
r_i – Hydrodynamic radius of the ion i [m]
η – Viscosity of the separation buffer [$N\ s\ m^{-2}$]
$\mathbf{w}_i$ – Migration velocity [$m\ s^{-1}$]
z_i – Effective) ionic charge of ion i
e – Elemental charge
$\mathbf{E}$ – Electric field strength [$V\ m^{-1}$]

According to Eq. (4.155), the separation of ions during electrophoresis occurs in the order of their charge-to-radius ratio (z_ie/r_i). However, many analytes are part of equilibrium reactions (protonation, complexation, partitioning, supramolecular association), in which the charge or the radius can change. In these cases, the effective charge (or the average charge) or the effective radius is of importance. We want to demonstrate this for acetic acid as an example of a weak acid ($pK_a = 4.75$). If the pH of the separation buffer is much smaller than 4.75, all molecules of acetic acid HAc are protonated and would not migrate in the electric field. If the pH is much larger than 4.75, all molecules are deprotonated and occur as single negatively charged acetate ion Ac^-, which migrates according to the tabulated molar ionic conductivity at infinite dilution corrected by the conductivity coefficient. However, if the pH of the separation buffer is exactly 4.75, half of the molecules are protonated acetic acid HAc, and half of the molecules are deprotonated acetate ions Ac^- (Figure 4.63). Let us imagine for a moment that the electric field causes migration of the acetate ions only and thus separates them from neutral acetic acid. The separated protonation stages HAc and Ac^- would almost instantaneously equilibrate according to the mass action law, so that at each location in the column, there is a 1:1 mixture of Hac and Ac^-. This is because the buffer capacity of the separation buffer is much larger than the concentration of the analytes in the sample, and thus dictates the constant pH within the capillary all the time. (This is different from the consideration of a pure weak acid in Figure 4.14). The fast reestablishment of the protonation equilibrium makes it impossible to separate HAc and Ac^- into two bands. Instead, each molecule of acetic acid will be found statistically half of the separation time in the protonated state ($z_i = 0$) and half of the time in the deprotonated state ($z_i = -1$). Migration occurs only in the deprotonated state. The effective velocity of HAc/Ac^- will therefore be only roughly half of the velocity predicted for the acetate ion from tabulated ionic conductivities and ionic mobilities.

If the analytical task is the separation of various weak acids that show almost no difference in the ionic mobilities of the corresponding anions, but have different pK_a values, the task is solved by cleverly selecting the pH of the separation buffer in such

Figure 4.63 Thought experiment for the comparison of the mobility of the anion of a strong acid (Cl^-) and a weak acid partially deprotonated in the separation buffer (CH_3COOH/CH_3COO^-) in an electrophoretic experiment. a) Starting state, b) movement of anions toward the anode, c) equilibration of the protonation states, a′) repeated separation, etc.

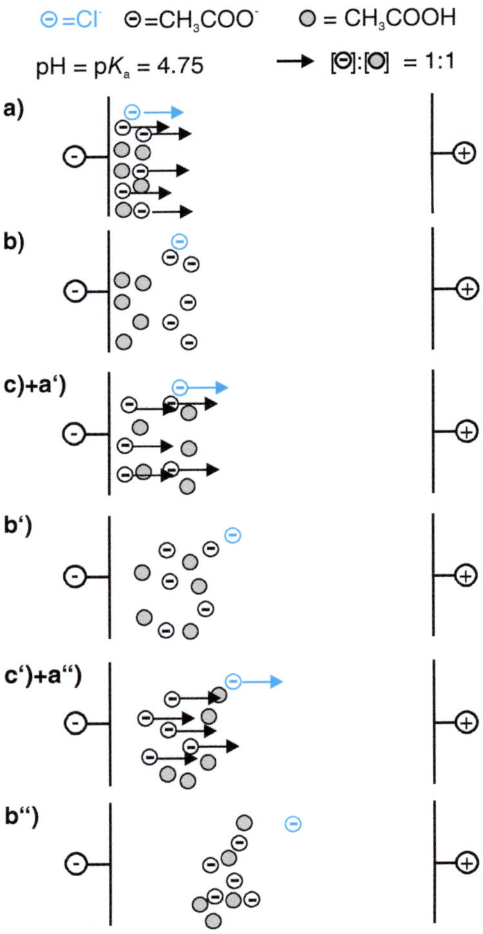

The thought experiment differs from reality because the proton transfer reactions are very fast compared to the electrophoretic mobilities. A delayed equilibration, as suggested by the schematic, does *not* occur. Therefore, the band does not experience an extra broadening above the value predicted from the prolonged overall separation time.

a way that the different weak acids have a different degree of dissociation and thus a different effective charge that is then exploited for the separation.

A further possibility is a separation by effective radii. The radius r_i appearing in Eq. (4.155) is likewise an effective hydrodynamic radius, which can be modified by complexation equilibria with a weakly binding ligand that facilitates the separation of cations. The weakly binding ligands cause the cations of the sample to occur at different times in the form of the complex or as water-solvated cations. Both forms have different radii and, depending on the charge of the ligand, also different charges which can be exploited for cations that otherwise have the same charge and almost the same radius (thus roughly the same ionic mobility when solvated in water).

In addition to migration, the electroosmotic flow (EOF) plays an important role in CE. The inner wall of the fused silica tube is covered by silanol groups (Si–OH), which are partially deprotonated above pH 2 ($Si{-}OH + H_2O \rightleftharpoons SiO^- + H_3O^+$). The capillary wall is therefore negatively charged under those conditions. The charged wall and the solution layer adjacent to the capillary wall form an electrical double layer. It is composed of a rigid and a diffuse part. The cations in the rigid part of the double layer neutralize only a fraction of the negative charge of the capillary wall. The remaining charge on the capillary wall is compensated by an excess of cations in the diffuse part of the double layer. In an electric field *parallel* to the surface of the capillary wall, the cations in the diffuse part of the double layer are accelerated toward the cathode and drag their solvation shell with them. The movement of the excess of cations along the entire perimeter of the capillary causes a net movement of the entire liquid plug in the capillary (Figure 4.64) which carries all ions and neutral molecules with it. This movement is called the electroosmotic flow. Its velocity $\mathbf{w}_{eo}$ is electric field strength dependent and given by Eq. (4.156).

$$\mathbf{w}_{eo} = u_{eo}\mathbf{E} = \frac{\varepsilon_r \varepsilon_0 \xi}{\eta}\mathbf{E}. \tag{4.156}$$

An electrical double layer is also formed at charged solid–liquid interfaces of electrical insulators (Section 4.3).

$\mathbf{w}_{eo}$ – Velocity of the EOF [m s^{-1}]
u_{eo} – Electroosmotic mobility [m^2 V^{-1} s^{-1}]
$\mathbf{E}$ – Electric field strength [V m^{-1}]
ε_0 – Permittivity of vacuum = 8.8542×10^{-12} A s V^{-1} m^{-1}
ε_r – Relative permittivity of the medium between the electrodes [dimensionless]
ζ – Zeta-potential or surface potential [V]
η – Viscosity of the separation buffer [N s m^{-2}]

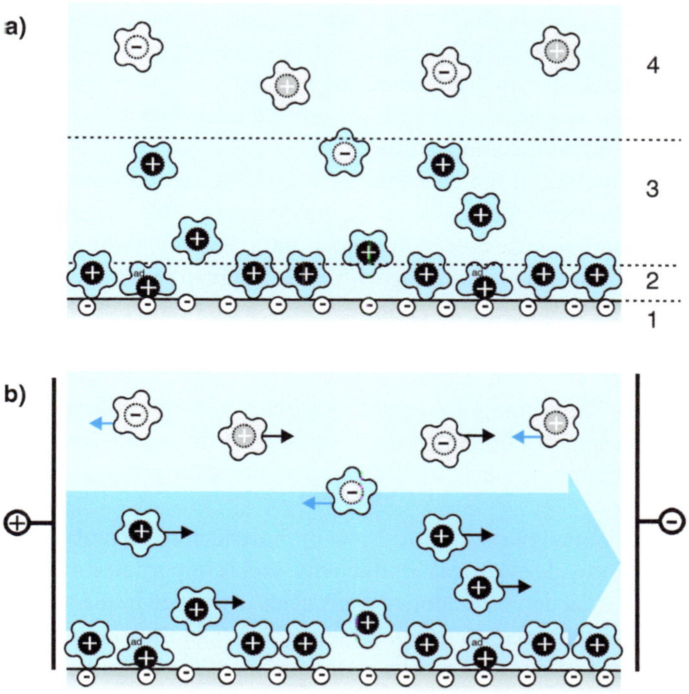

Figure 4.64 Illustration of the emergence of the electroosmotic flow at a charged capillary wall. a) electrical double layer; 1) deprotonated silanol group, 2) rigid part of the double layer, 3) diffuse part of the double layer, and 4) bulk phase; b) After turning on the electric field, the ions move with their solvation shell toward the oppositely charged electrode. Due to the excess of cations in the diffuse part of the double layer, a hydrodynamic flux of the entire solution plug is observed.

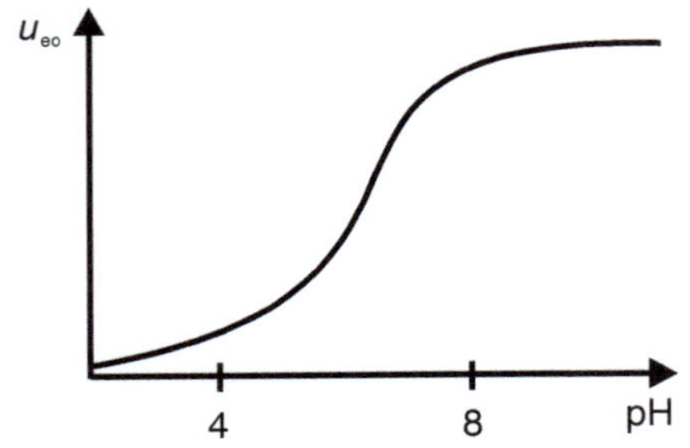

Figure 4.65 Change in EOF with the pH of the separation buffer. The curve is reminiscent of a titration curve because u_{eo} depends on the charge on the capillary wall, which in turn is determined by the protolysis equilibrium of the capillary surface.

The larger the pH of the separation buffer, the larger is the fraction of deprotonated silanol groups on the inner capillary wall. An increase in the charge density on the capillary wall increases the ζ-potential and thus the EOF (Figure 4.65). The ionic strength of the separation buffer may influence the thickness of the electric double layer. The smaller the total ionic strength, the larger is the EOF. The experimentally observed velocity $\mathbf{w}_{exp}$ can be calculated from the time t, a substance needs to move the distance l_d from the capillary inlet to the detector.

$$\mathbf{w}_{exp} = \frac{l_d}{t} \tag{4.157}$$

According to Eqs. (4.155) and (4.156), the velocity contributions of migration and of the electroosmotic flow are both proportional to the electric field strength **E**. We can thus combine the components into an experimental mobility for each compound. The electric field strength is obtained according to Eq. (4.2) from the separation voltage U and the total capillary length l_t (from anode to cathode).

$$\mathbf{w}_{exp} = u_{exp}\mathbf{E} = u_{exp}\frac{U}{l_t} \tag{4.158}$$

Equating Eqs. (4.157) and (4.158) and subsequent rearrangement gives an expression for the calculation of the experimental mobility from a particular electrophoresis experiment.

$$u_{exp} = \frac{l_t\ l_d}{Ut} \tag{4.159}$$

Table 4.8 Overall mobility resulting from combining the effects of ion migration and EOF. The positive direction is defined as toward the capillary outlet (–).

Particle	Experimental mobility
Cations	$u_{exp,i} = u_{eo} + u_i$
Neutral molecules	$u_{exp,neutral} = u_{eo}$
Anions	$u_{exp,i} = u_{eo} - u_i$

The experimentally observed mobility u_{exp} is a result of the vectorial addition of the electrophoretic mobility u_e and the electroosmotic mobility u_{eo} (Table 4.8). Cations migrate in the direction of the EOF. Their experimental mobility is thus faster than the EOF. Neutral molecules move with the velocity of the EOF. For anions, the electrophoretic mobility and the EOF point in opposite directions. By choosing a separation buffer with a suitable pH, it can be ensured that the EOF has a larger absolute value than u_e of the anions. In this case, the overall movement of the anions occurs in the direction of the EOF, albeit with a velocity that is smaller than that of neutral molecules. Thus, all types of components pass the detector close to the outlet of the capillary, first cations, then neutral molecules, and lastly the anions. Between different types of cations and anions, a separation is achieved according to the ratio of effective charge to effective hydrodynamic radius.

For the determination of the concentration of the separated compounds, a detector is employed similar to those used in chromatography. UV detection, conductivity detection, and amperometric detection are common. For demanding analytical tasks, laser-induced fluorescence or coupling with a mass spectrometer are further options. For the determination of unknown concentrations, a calibration plot is established by measuring reference samples with known content of the analyte. For ions that do not exhibit light absorption in the accessible wavelength range, indirect detection can be applied. An ion with a strong absorption (e.g., chromate, phthalate, and benzoate) is added to the separation buffer. In the zone of the sample, a decrease in the optical absorption is found because sample components drive out the absorbing buffer components.

The separation buffer plays a critical role in an effective separation. It must dissolve the sample, provide electrical conductivity, and is important for establishing the EOF by causing partial deprotonation of weak acids contained in the sample, and thus enables their separation according to pK_a values. The separation according to the ratio of effective charge and effective radius is not sufficient for complex mixtures. Especially neutral molecules would not be separated by this mechanism. However, the separation of neutral molecules can be achieved by buffer additives, such as anionic surfactants. If they are added in a concentration exceeding the critical micelle concentration (CMC), the surfactant exists in an equilibrium between micelles and molecularly dispersed surfactant molecules. The micelles are negatively charged as are the anionic surfactants forming them. They migrate toward the anode. The EOF can be adjusted by the pH in such a way that the net movement of micelles occurs toward the cathode.

If the sample plug contains neutral molecules, they enter a partitioning equilibrium between the solution and the hydrophobic internal part of the micelles (Figure 4.66). If the neutral molecules are dissolved in the solution, their mobility corresponds to the electroosmotic flow u_{eo}. If they are located in the inner part of the micelle, their transport toward the outlet is slower with a mobility $u_{eo} - u_e$. Because the exchange between the solution and the internal part of the micelles is very rapid, each kind of neutral molecule moves with *one* effective velocity that is between that of the EOF and that of the negatively charged micelles. It depends on the ratio of residence times of the neutral molecule inside the micelle and in the solution. Because this ratio depends on details of the chemical structure of the neutral molecule, this principle facilitates the separation of neutral molecules.

In this case, the separation is based on a partitioning equilibrium of a neutral molecule between the solution and the inner part of the micelles (multiplicative partitioning). In addition, the solution and the micelles move with different velocities toward the detector. This mechanism is very similar to chromatography. The micelles correspond to the stationary phase and retard the movement of neutral molecules with respect to the buffer as mobile phase. Therefore, this separation is also called micellar electrokinetic chromatography.

Further additives to the buffer are, for example, complexing agents which can improve the separation of cations, cyclodextrins (cyclic carbohydrates), or other chiral compounds as chiral auxiliary compounds in the separation of enantiomers. Polar organic solvents are often added in order to enhance the solubility of organic compounds. Other additives are employed to modify the inner capillary walls, for instance in order to reverse the EOF or suppress the adsorption of proteins to the capillary walls.

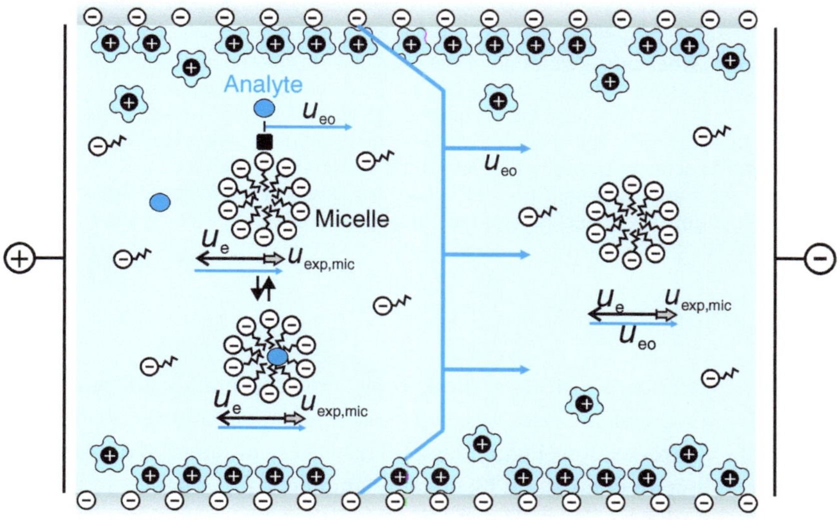

Figure 4.66 Partitioning of neutral molecules (blue dots) of a sample between the separation buffer and the inner part of a micelle. The negatively charged micelles have a lower total mobility than the neutral molecules in solution and therefore retain the neutral molecules of the sample compared to the EOF. (The flow velocity profile is indicated by a blue line).

The scheme is not to scale!

Excursion 4-5 Capillary Electrophoresis in DNA Sequencing

The CE played a very important role in instrumental **DNA sequencing**. Individual section of DNA can be replicated multiple times from a start sequence (primer) by the polymerase chain reaction (PCR) until the amount is sufficiently high for a chemical analysis. During the replication, DNA oligomers of different lengths are formed, all of them carrying a fluorescence dye at one of the ends. Four different dyes are used, each connected to one of the bases (A, T, C, and G) that is the last base in the particular oligomer. The dyes can be distinguished by the emission wavelength.

The capillary is loaded with a physical gel. It consists of polymer chains that are loosely coiled and form a sieve through which longer DNA oligomers move more slowly than shorter ones. Therefore, the shorter DNA oligomers reach the detector before the longer oligomers. Now, one only needs to read four channels for the four dyes marking the last base of each oligomeric chain and obtains directly the sequence of bases in the DNA section (Figure 4.67). The automated and highly parallel execution of this process was one of the technological requirements for the success of the Human Genome Project [33].

Of course, there are multiple copies of oligomers of every chain length! For the sake of clarity, this scheme shows only one representative molecule for all molecules of the same length and the same dye label.

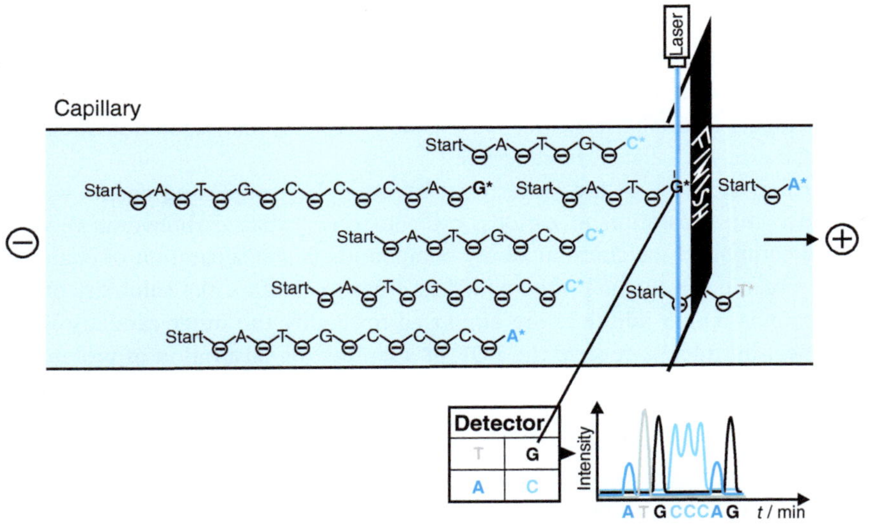

Figure 4.67 Schematic illustration of the principle of DNA sequencing by CE. The DNA sections are replicated starting from a primer (Start) and are terminated by a dye-labelled base (A*, G*, T*, C*). Each base has its unique dye label. The detector records a time series for each color ("trace"). The overlay of all four traces allows a direct reading of the base sequence. The scheme shows only a very small section from such a CE run. For the industrial scale application of this method, several levels of optimization were required based on the gels, the detection and labelling techniques, and based on the chemicals applied in the detection.

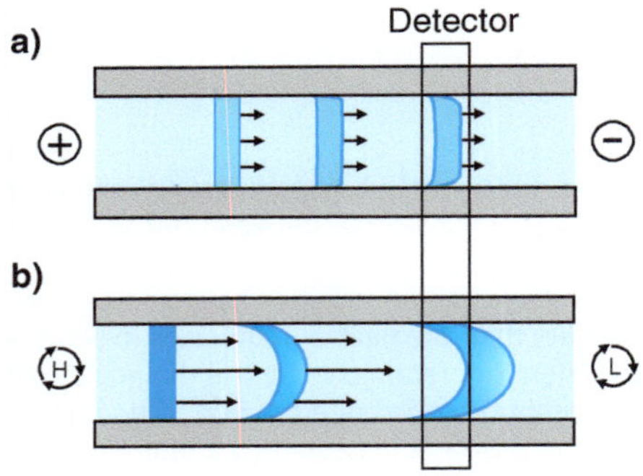

Figure 4.68 Formation of a plug flow streaming in CE as compared to a laminar hydrodynamic tube flow with a parabolic flow profile of linear streaming velocities along a pressure gradient from high (H) to low (L) pressure.

The schematic is not to scale!

The EOF contributes decisively to the extremely high separation efficiency of CE. It causes a plug flow, which causes nearly the same linear streaming velocity over the entire capillary cross section (Figure 4.68a). This is very different to a hydrodynamic laminar streaming process caused by pressure gradients, for instance, in liquid chromatography (LC, Figure 4.68b). Therefore, peaks broaden less in CE compared to LC.

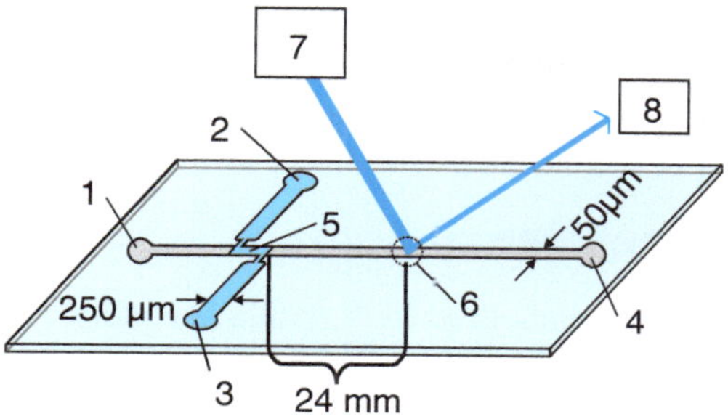

Figure 4.69 Illustration of the experimental system for electrophoresis on a chip. Voltage between ports 2 and 3: sample is loaded into the "injection zone" (5). Voltage between ports 1 and 4: separation of the sample segment between the two side arms (5) into its components and detection at the end of the separation distance (6). Fluorescence detection is used in this example with an excitation light source (7) and a detector behind an optical filter (8).

The EOF can be used as an absolutely pulsation-free pumping method. An example of this is the transfer of an electrophoresis experiment onto a microchip [34, 35]. Such chips are a step toward the vision to integrate an entire chemical laboratory onto a microchip (Lab on the Chip), also called micro total analysis system (μTAS) [36]. The chip contains a channel in a shape that is remotely similar to a Christian cross (Figure 4.69). However, a closer inspection reveals that the confluences of the two short channel arms to the long channel are slightly offset. The part between the confluences of the shorter arms is used as "injection loop". In the first step of a separation, the sample is dosed into port 2. Afterward, application of a voltage between ports 2 and 3 causes a filling of the channels between 2 and 3 by the resulting EOF. This also fills the "injection loop". For the actual separation, a voltage is applied between ports 1 and 4. The solution plug between the two side-arms is separated into its components, which are detected at the end of the separation distance by a suitable detector. The high electric field strength (U/l) facilitates very efficient separations within a short period of time. The dimensions of the device can be significantly reduced, causing a decrease in analysis time and solvent consumption (Figure 4.70).

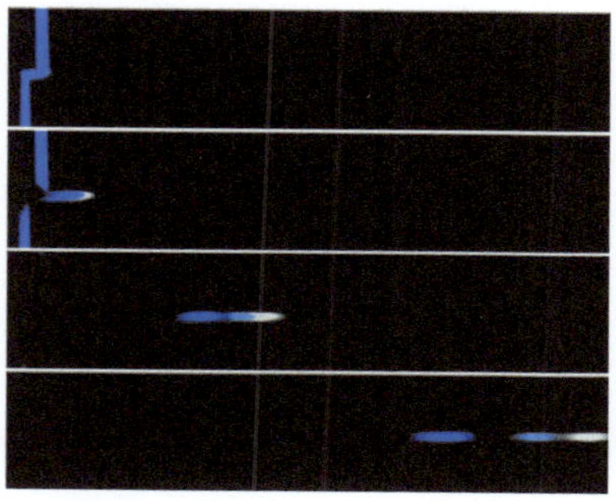

Figure 4.70 Separation of three fluorescence dyes of rhodamine type within a separation distance of 3 mm and within less than 5 s. Source: Reproduced with permission from S. Götz, U. Karst, *Sensors Actuators B* **2005**, *123*, 622–627, Copyright 2005, Elsevier.

4.7 Combined Transport Processes in the Nernst–Planck Transport Equation

The transport of chemical compounds toward interfaces is a central feature of electrochemical reactions. In all cases when the space coordinates of neutral molecules shift, their local concentration (dN/dV) changes. We can define a **molecular flux** J_i [mol cm^{-2} s^{-1}], for each kind of molecule and ion i. If the space coordinates of ions are shifted, **charge will be moved concomitantly with the change in concentration**. In electrolyte solution, the (electrical) current density is the sum of the (signed) ionic current densities for all kinds of ions. The ionic current density is linked to the molecular fluxes J_i by the ionic charges z_i and the Faraday constant F.

$$j\,[\mathrm{A\ cm^{-2}}] = F\,[\mathrm{A\ s\ mol^{-1}}]\sum_i z_i J_i\,[\mathrm{mol\ cm^{-2}\ s^{-1}}] \tag{4.160}$$

This is an important but also particularly simple special case of a more general treatment, because the electrolyte solution is a pure ionic conductor, and electrons cannot migrate over longer distances. We will see later in Chapter 11 that solid electrolytes may exhibit a mixed electronic and ionic conductivity.

Transport processes move a chemical system **toward the equilibrium state**, which is characterized by equality of the electrochemical potentials $\tilde{\mu}_i^\alpha$ of all ions and molecules with index i in the participating phases with index α. It is calculated according to Eq. (4.161).

$$\tilde{\mu}_i^\alpha = \mu_i^{\alpha\circ} + RT\ln\frac{a_i}{a^\circ} + z_i F\phi^\alpha \tag{4.161}$$

Gradients are partial derivatives of an intensive quantity with respect to space coordinates. Gradients can cause general transport processes (mass transport, heat transport, ...), which are treated within the framework of irreversible thermodynamics [9, 10]. Gradients can be denoted as shown in Eq. (4.162).

$$\frac{\partial y}{\partial x} = \frac{\partial}{\partial x} y = \text{grad}(y) \tag{4.162}$$

The operator grad is often symbolized by ∇ (nabla operator). For three-dimensional Cartesian coordinates with the unity vectors $\mathbf{e}_x$, $\mathbf{e}_y$, and $\mathbf{e}_z$, it takes the form of Eq. (4.163).

$$\nabla = \mathbf{e}_x \frac{\partial}{\partial x} + \mathbf{e}_y \frac{\partial}{\partial y} + \mathbf{e}_z \frac{\partial}{\partial z} \tag{4.163}$$

The specific form of the operator changes with the type of coordinate system such as spherical coordinates or cylinder coordinates.

The driving force for mass transport is the gradient in the electrochemical potential. In this chapter, we discussed transport processes along only one space dimension x. Of course, this is a simplification and generally we should recognize the transport as a vector **J** with all possible orientations in all spatial dimensions. As an illustration, we provide the description for one dimension below, but give the description for the three-dimensional case in the margins in operator notation. For the linear mass transport, Eq. (4.164) holds.

$$\mathbf{J}_i = -\frac{c_i D_i}{RT} \cdot \nabla \tilde{\mu}_i \tag{4.165}$$

$$J_i = -\frac{D_i c_i}{RT} \cdot \frac{\partial \tilde{\mu}_i}{\partial x} \tag{4.164}$$

Inserting Eq. (4.161) with the approximation $c_i \approx a_i$ yields Eq. (4.166), which holds for quiescent solutions or ion-conducting solids.

$$\mathbf{J}_i = -D_i \nabla c_i - \frac{z_i F}{RT} D_i c_i \nabla \phi^\alpha \tag{4.167}$$

$$\begin{aligned} J_i &= -D_i c_i \frac{\partial}{\partial x} ln \frac{c_i}{c^\circ} - \frac{D_i c_i z_i F}{RT} \frac{\partial \phi^\alpha(x)}{\partial x} \\ &= -D_i c_i \frac{c^\circ}{c_i} \frac{\partial \frac{c_i}{c^\circ}}{\partial x} - \frac{z_i F}{RT} D_i c_i \frac{\partial \phi^\alpha(x)}{\partial x} \\ &= -D_i \frac{\partial c_i}{\partial x} - \frac{z_i F}{RT} D_i c_i \frac{\partial \phi^\alpha(x)}{\partial x} \end{aligned} \tag{4.166}$$

The additional consideration of convective transport, which we obtain as a product of the local concentration and the linear flow velocity $v(x)$ [cm s^{-1}] at location x yields the general **Nernst–Planck transport equation**[10,52] (4.168).

$$\mathbf{J}_i = -D_i \nabla c_i - \frac{z_i F}{RT} D_i c_i \nabla \varphi^\alpha + c_i \mathbf{v} \tag{4.169}$$

$$J_i = -D_i \frac{\partial c_i}{\partial x} - \frac{z_i F}{RT} D_i c_i \frac{\partial \varphi^\alpha(x)}{\partial x} + c_i v(x) \tag{4.168}$$

Table 4.9 Survey of transport processes and corresponding boundary layers.

Prozess	Boundary layer	Thickness
Migration	Electrical double layer	0.9–30 nm
Diffusion	Diffusion layer	1–300 μm
Convection	Hydrodynamic boundary layer	10–40 μm

There is a **characteristic boundary** layer associated with each transport mechanism or transport process, in which the intensive properties deviate from those in the solution bulk (Table 4.9).

In many electrochemical systems, the different transport processes occur concomitantly and are coupled, that is, the overall effect is not a simple superposition of each of the three processes. This makes a quantitative description much more difficult. Therefore, one often uses "tricks" aimed at simplifying the situation to make it more tractable. An example is the addition of a supporting electrolyte in excess of the reactants of the electrochemical reaction. The supporting electrolyte is not converted in the electrochemical cell but conducts most of the ionic current due to

[52] Max Karl Ernst Ludwig Planck, 1858–1947, German physicist, Nobel Prize 1918, professor in Kiel and Berlin, president of the Kaiser Wilhelm Society, known for the description of the irradiation of black bodies and Planck's constant h.

the concentration excess over the reactants. Consequently, the transport of reactants follows Fick's laws of diffusion in good approximation.

In systems with convection, the solution in front of the electrode is constantly renewed. However, this does not hold for all parts of the hydrodynamic boundary layer. In this layer, immediately adjacent to the electrode surface, the solution rests (no slip condition), and the transport occurs in the presence of a supporting electrolyte by diffusion only. The thickness of the diffusion layer within the hydrodynamic boundary layer can be approximated by extrapolating the linear part of the concentration profile close to the electrode surface (Figure 4.71). The distance of the intersection of this line with the horizontal line $c = c^*$ indicates the thickness δ of the diffusion layer with an assumed sharp transition between the resting solution layer close to the electrode surface and a completely mixed solution in the rest of the volume. This construction is a popular simplification of the real situation and is called the **Nernst diffusion layer**. Because the exact thickness of the diffusion coefficient D and δ are rarely known precisely, a mass transport coefficient $m_{t,i}$ [cm s^{-1}] is defined that summarizes both quantities.

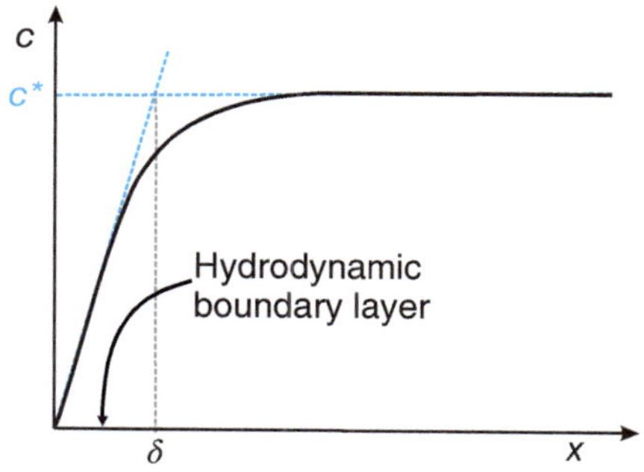

Figure 4.71 Concentration profile (solid line) and construction of Nernst's diffusion layer extending to the intersection between the tangent at the concentration profile at $x = 0$ and the horizontal line $c = c^*$.

$$J_i = -D_i\left(\frac{\partial c_i}{\partial x}\right)_{x=0} \approx -D_i\left(\frac{c_i^* - c_i(x=0)}{\delta}\right) = m_{t,i}\left(c_i^* - c_i(x=0)\right) \tag{4.170}$$

The coefficient $m_{t,i}$ has an analogous meaning (and the same units) as a rate constant for an interfacial chemical reaction. For a kinetically controlled reaction, if only one partial reaction is of significance ($|\eta| \geq -0.118$ V), the transport can easily keep up with the chemical transformation, and the local concentration close to the interface barely deviates from that in the solution bulk $c(x=0) \approx c^*$. According to Eq. (3.17), the current is then described by Eq. (4.171).

$$i = nFA \cdot k(E) \cdot c_i^* \tag{4.171}$$

For a diffusion-controlled reaction, the concentration of educts at the electrode is $c(x=0) \approx 0$, because the reaction is fast enough that all reactant molecules that reach the electrode are immediately converted. From Eq. (4.170), we obtain the faradaic current in Eq. (4.172).

$$i = nFA \cdot m_{t,i} \cdot c_i^* \tag{4.172}$$

Q: So far, you have talked about 409 equations. If Chapters 1–4 are meant to be just the basics, how many more equations are out there in the rest of the book?

A: Indeed, the first part is leaning toward algebraic equations. However, three more, hopefully exciting parts of the book are in front of us, in which we will frequently refer to those basics. Perhaps the deeper meaning of some equations will become clear to you if you consider the basics from the perspective of an advanced and application-oriented topic. For this purpose, the initial chapters of the book have been equipped with an "equation apparatus" that is consistent in itself and should help you in *understanding* fundamental aspects. Understanding takes time. The book author and all contributors to the German and English editions can assure you that it also took some time for us to absorb the content of those chapters that are central to almost all areas of electrochemistry. The fun part became more evident when working on special aspects of the subject. The following parts of the book treat important measurement techniques (Chapters 5–8), then material aspects in electrochemistry (Chapters 9–13), and finally application-oriented topics such as energy storage and conversion, corrosion, electrosynthesis, and bioelectrochemistry (Chapters 14–18). In all those chapters, we will make connections to the content of Chapters 1–4. So, please, continue accompanying us with your critical questions!

Test your knowledge

Exercise 4.1: Calculate the ionic strength of a solution containing 1×10^{-2} mol L^{-1} NaCl and 1×10^{-3} mol L^{-1} Na_2SO_4.

Exercise 4.2: Calculate the ionic strength of a solution containing 2×10^{-1} mol L^{-1} Na_2SO_4 and 10^{-2} mol L^{-1} $K_4[Fe(CN)_6]$.

Exercise 4.3: Calculate the mean activity of 0.1 mol L^{-1} Na_2SO_4 ($f_\pm = 0.453$).

Exercise 4.4: An aqueous 0.02 mol L^{-1} solution of KCl has a molar conductivity $\Lambda = 138.3$ cm^2 S^1 mol^{-1} at 298 K. In a conductivity cell, a resistance of 74.85 Ω was measured. Determine the cell constant in [cm^{-1}] of the used conductivity cell.

Exercise 4.5: Complete the following table for the area of solution conductivity.

Quantity(ies) to determine	Functional relation	Condition met for (strong/weak electrolytes)
Λ, Λ_0	$\Lambda = \Lambda_0 f_\Lambda$	
	$\alpha =$	
Λ	$\Lambda = f_\Lambda \Lambda_0$	
	$\alpha =$	
Λ	$\Lambda = \alpha\Lambda_0$	
	$f_\Lambda =$	
α	$\alpha = \dfrac{\Lambda}{\Lambda_0}$	
	$f_\Lambda =$	

Exercise 4.6: Draw a schematic of the electrical double layer. Indicate especially the following components: oriented water dipoles, water dipoles without preferred orientation, specifically adsorbed ions, solvated excess charges, inner and outer Helmholtz planes.

Exercise 4.7: Calculate the extension of the diffuse part of the electrical double layer as $3\,r_D$ for 1 mol L^{-1} Na_2SO_4, 0.1 mol L^{-1} Na_2SO_4, 0.01 mol L^{-1} Na_2SO_4, 0.001 mol L^{-1} Na_2SO_4 as well as for 1 mol L^{-1} $NaNO_3$, 0.1 mol L^{-1} $NaNO_3$, 0.01 mol L^{-1} $NaNO_3$, 0.001 mol L^{-1} $NaNO_3$ at 298 K. (Pay particular attention to proper handling of the units.)

Exercise 4.8: Calculate the equilibrium cell voltage of the cell Ag|$AgNO_3$||$NiCl_2$|Ni, if the Ag|$AgNO_3$ half cell contains 1.8 mg $AgNO_3$ dissolved in 250 mL H_2O. Calculate the activity coefficients $f_\pm(AgNO_3)$ needed according to the theory of Debye and Hückel. Further data are $M(AgNO_3) = 169.87$ g mol^{-1}, $E°(Ni|Ni^{2+}) = -0.23$ V vs. SHE, $a(Ni^{2+}) = 1$ mol L^{-1}.

Exercise 4.9: A specific conductivity of 2.570×10^{-5} S cm^{-1} was measured for a 2×10^{-4} mol L^{-1} solution of acetic acid. Determine the degree of dissociation for acetic acid in this solution. Further data are $\lambda°(H^+) = 349.6$ S cm^2 mol^{-1}, $\lambda°(Ac^-) = 40.9$ S cm^2 mol^{-1}.

Exercise 4.10: The molar conductivities of NaCl and KCl at infinite dilution in $cm^2\ \Omega^{-1}\ mol^{-1}$ are 126.45 and 149.86, respectively. The ion mobility of K^+ equals $76.1 \times 10^{-5}\ cm^2\ V^{-1}\ s^{-1}$ under those conditions. Calculate the ratio of the hydrodynamic radii of both cations.

Exercise 4.11: You titrate phosphoric acid with sodium hydroxide solution. Draw schematically the expected titration curve and mark the equivalent points and the pK_a values. Note the reaction equations for all dissociation steps.

Exercise 4.12: The limiting value of the ion mobilities of MnO_4^- equals $65 \times 10^{-5}\ cm^2\ V^{-1}\ s^{-1}$ in diluted solutions.

a) Calculate the molar conductivity at infinite dilution for MnO_4^-.
b) Calculate the ion mobility, the molar conductivity, and the transference numbers if the molar conductivity at infinite dilution of $KMnO_4$ was determined as $136.3\ cm^2\ \Omega^{-1}\ mol^{-1}$.

Exercise 4.13: A diaphragm of 1 mm thickness separates two equally sized, well-mixed reservoirs containing two $CuSO_4$ solutions with $c_1(CuSO_4) = 0.08\ mol\ L^{-1}$ and $c_2(CuSO_4) = 0.12\ mol\ L^{-1}$.

a) Calculate the diffusion coefficient if the molar diffusional flux is
$J = \frac{1}{A} \cdot \frac{dN}{dt} = 2.256 \cdot 10^{-9}\ mol\ cm^{-2}\ s^{-1}$.
b) How large should the area of the diaphragm be to allow a diffusion of 5 mg $CuSO_4$ from one reservoir to the other reservoir within 5 min?

References

1 J. O'M. Bockris, A. K. N. Reddy; *Modern Electrochemistry, Vol. 1 – Ionics*. 2nd ed., Plenum Press, 1998, Chapter 4.
2 D. Dobos; *Electrochemical Data – A Handbook for Electrochemists in Industry and Universities*, Elsevier, Amsterdam, 1975.
3 J. R. Rumble, D. R. Lide, T. J. Bruno; *CRC Handbook of Chemistry and Physics*. 100th ed., CRC Press, Taylor and Francis Group, Boca Raton, 2019, pp. 5–90 to 5–96.
4 B. E. Conway; *Electrochemical Data*, Greenwood Press, Westport, 1952.
5 A. J. Bard, L. R. Faulkner, H. S. White; *Electrochemical Methods, Fundamentals and Applications*. 3rd ed., Wiley, New York, 2022.
6 J. S. Vrentas, C. M. Vrentas; *Diffusion and Mass Transfer*, CRC Press, Boca Raton, 2013.
7 D. Britz; *Digital Simulation in Electrochemistry*. 3rd ed., Springer Verlag, Berlin, 2005 (*Lect. Notes Phys.* **2005**, *666*, 1–338).
8 B. Speiser; Numerical simulation of electrochemical experiments: Recent advances in methodology. in *Electroanalytical Chemistry, Vol. 19* (A. J. Bard, I. Rubinstein, Eds.), Marcel Dekker, New York, 1996, pp. 1–108.
9 D. Kondepudi, I. Prigogine; *Modern Thermodynamics: From Heat Engines to Dissipative Structures*. 2nd ed., Wiley, New York, 2014.
10 Y. Demirel; *Nonequilibrium Thermodynamics: Transport and Rate Processes in Physical, Chemical and Biological Systems*. 3rd ed., Elsevier, Amsterdam, 2013.
11 A. S. Shukla, T. P. Kumar; Pillars of modern electrochemistry: A brief history. in *Electrochemistry Encyclopedia*, (Z. Nagy, Ed.), The Electrochemical Society (ECS), 2008, https://knowledge.electrochem.org/encycl/art-p05-pillars-of-ec.htm (last access 25 Jan 2024).
12 N. Agmon; The Grotthuss mechanism. *Chem. Phys. Lett.* **1995**, *244*, 456–462.

13 M. Tuckerman, K. Laasonen, M. Sprik, M. Parrinello; Ab initio molecular dynamics simulation of the solvation and transport of hydronium and hydroxyl ions in water. *J. Chem. Phys.* **1995**, *103*, 150–161.

14 M. E. Tuckerman, D. Marx, M. Parrinello; The nature and transport mechanism of hydrated hydroxide ions in aqueous solution. *Nature* **2002**, *417*, 925–429.

15 World Health Organization; Working document QAS/19.807, July 2019. https://cdn.who.int/media/docs/default-source/medicines/norms-and-standards/current-projects/qas19-807-condom-quality-assurance.pdf?sfvrsn=d6300a87_2 (last access 26 April 2021).

16 T. Engel, P. Reid; *Physical Chemistry*, Pearson Education, San Francisco, 2006, pp. 230–234.

17 J. O'M. Bockris, A. K. N. Reddy; *Modern Electrochemistry, Vol. 1.* 2nd ed., Plenum Press, New York, 1998, pp. 225–293.

18 O. M. Magnussen; Ordered anion adlayers on metal electrode surfaces. *Chem. Rev.* **2002**, *102*, 679–726.

19 T. Tansel, O. Magnussen; Video STM studies of adsorbate diffusion at electrochemical interfaces. *Phys. Rev. Lett.* **2006**, *96*, 026101/1–4.

20 https://en.wikipedia.org/wiki/Galton_board (last access 23 Nov 2024).

21 S. G. Lemay, S. Kang, K. Mattwig, P. S. Singh; Single-molecule electrochemistry: Present status and outlook. *Acc. Chem. Res.* **2013**, *46*, 369–377.

22 J. S. Ventras, C. M. Ventras; *Diffusion and Mass Transfer*, CRC Press, Boca Raton, 2013, pp. 204–212.

23 K. R. Riley, M. P. Hobeson, S. J. Bence; *Mathematical Methods for Physics and Engineering*. 3rd ed., Cambridge University Press, Cambridge, 2024, pp. 453–459.

24 J. O'M. Bockris, A. K. N. Reddy; *Modern Electrochemistry, Vol. 1 – Ionics*. 2nd ed., Plenum Press, 1998, p. 380.

25 A. J. Bard, L. R. Faulkner, H. S. White; *Electrochemical Methods, Fundamentals and Applications*. 3rd ed., Wiley, New York, 2022, pp. 967–976.

26 R. Westermeier; *Electrophoresis in Practise*. 5th ed., Wiley-VCH, Weinheim, 2016.

27 D. C. Harris, C. A. Lucy; *Quantitative Chemical Analysis*. 10th ed., W. H. Freeman & Co, 2019, pp. 893–912.

28 L. Wöhlbrand, K. Trautwein, R. Rabus; Proteomic tools for environmental microbiology – A roadmap from sample preparation to protein identification and quantification. *Proteomics*. **2013**, *13*, 2700–2730.

29 H. Engelhardt, W. Beck, T. Schmitt; *Capillary Electrophoresis – Methods and Potentials*, Springer, Berlin, 2012.

30 M. G. Khaledi (Ed.); *High-performance Capillary Electrophoresis – Theory, Techniques, and Application*, John Wiley & Sons, New York, 1998.

31 J. P. Landers (Ed.); *Handbook of Capillary Electrophoresis*. 2nd ed., CRC Press, Boca Raton, 1997.

32 R. Weinberger; *Practical Capillary Electrophoresis*. 2nd ed., Academic Press, San Diego, 2000.

33 N. J. Dovichi, J. Zhan; How capillary electrophoresis sequenced the human genome. *Angew. Chem. Int. Ed.* **2000**, *39*, 4463–4468.

34 A. S. Rathore, A. Guttman; *Electrokinetic Phenomena – Principles and Application in Analytical Chemistry and Microchip Technology*, Marcel Dekker, New York, 2004.

35 H. H. Girault; *Analytical and Physical Electrochemistry*. EPFL Press, distributed by Marcel Dekker, Lausanne, 2004, pp. 229–256.

36 A. Manz, H. Becker (Eds.); *Microsystem Technology in Chemistry and Life Science*, Springer, Berlin, 1999.

Part II

Experimental Methods in Electrochemistry

5

Electroanalytical Methods and Electrochemical Sensors

This chapter is devoted to the application of electrochemical principles to chemical analysis. This refers to the determination of the identity and concentration of a (dissolved or gaseous) compound, called the **analyte**. We will consider a variety of methods, including titration with instrumental determination of the end point, measurement of pH by means of a glass electrode, polarography, and modern methods based on voltammetry. Historically, the former constitutes one of the first instrumental analytical methods. In a wider sense, electroanalytical methods also comprise methods for the elucidation of reaction mechanisms and characterization of material properties, which will be treated in more detail in subsequent chapters. A few related methods, such as voltammetry at rotating ring disc electrodes, will be discussed in separate chapters in the context of specific applications (e.g., electrocatalysis).

Since the 1980s, the practical importance of electrochemical methods has decreased for routine analysis due to atomic spectroscopy and mass-spectrometry offering advantages in broad applications, limit of detection, and selectivity. However, electrochemical methods remain important in a few niche applications, especially for mobile or low-cost sensing. Recent research focuses on using electroanalysis in disposable sensors, in miniaturized analysis systems (lab-on-a-chip), for DNA sequencing, and for biochemical assays. In these areas, electrochemical methods can be very powerful because electrochemical cells can be *miniaturized* without concomitant loss in sensitivity. This is in contrast to optical methods, in which the absorbance depends linearly on the layer thickness the light beam has passed through. Thus, knowledge of electroanalytical methods is key for shaping the future development of electrochemical sensors.

Further Reading. The scope of this textbook mandates the omission of some of the derivations of equations presented here. These can be found in more specialized textbooks [1–4] and in the literature resources listed in each section. Furthermore, we can only discuss the signal generation and data treatment within an analytical process. The other important parts of an analytical process (Table 5.1) should be studied by consulting a textbook on analytical chemistry [5–7].

Table 5.1 Steps of an analytical process.

Steps
• Sampling
• Sample preparation
• Signal generation
• Signal analysis
• Problem-related interpretation
• Quality management

5.1 The Family Tree of Electroanalytical Methods

There are many electroanalytical methods, exploiting almost all quantitative relationships discussed before in Chapters 1–4. Here, we discuss them in the context of their utility for solving analytical problems. The large variety of methods paired with a nonuniform nomenclature may confuse the novice (and sometimes even experienced electrochemists). This textbook cannot resolve this situation, but we will try to use consistent nomenclature and a limited set of acronyms.

Q: What are acronyms?

A: An acronym is an abbreviation of a longer term and is used by itself like a word. Examples are differential pulse voltammetry → DPV or light amplification by stimulated emission of radiation → LASER or, more commonly, laser.

Electrochemical methods can be categorized and subsequently arranged in a family tree (Figure 5.1). First, we divide the methods into two groups depending on whether the signal is limited by processes at the electrode interface or by processes in the solution bulk. For the categorization, it is important which process *limits* or *controls* the measured signal. Processes at interfaces and methods for their investigation fall under the umbrella of *electrodics*, while limiting processes in the bulk and the corresponding measurement methods (e.g., conductometry in Excursion 4-1) fall under *ionics*.

A further categorization of the interface-controlled methods is possible by asking whether a significant current passes through the interface during the measurement. Measurements in a currentless state ($I = 0$) are called **static methods**, while methods requiring a significant current ($I \neq 0$) are termed **dynamic methods**.

For further categorization, we consider the signal–response relationship. Many methods rely on applying a specific excitation signal (e.g., a linear potential ramp) to the electrochemical cell and recording the resulting response of the system. There is a plethora of different excitations signals, of which only a few will be considered. For a deeper coverage of the topic, the reader is referred to classical textbooks [1]. Execution of the methods in quiescent solution (no agitation or convection) or under

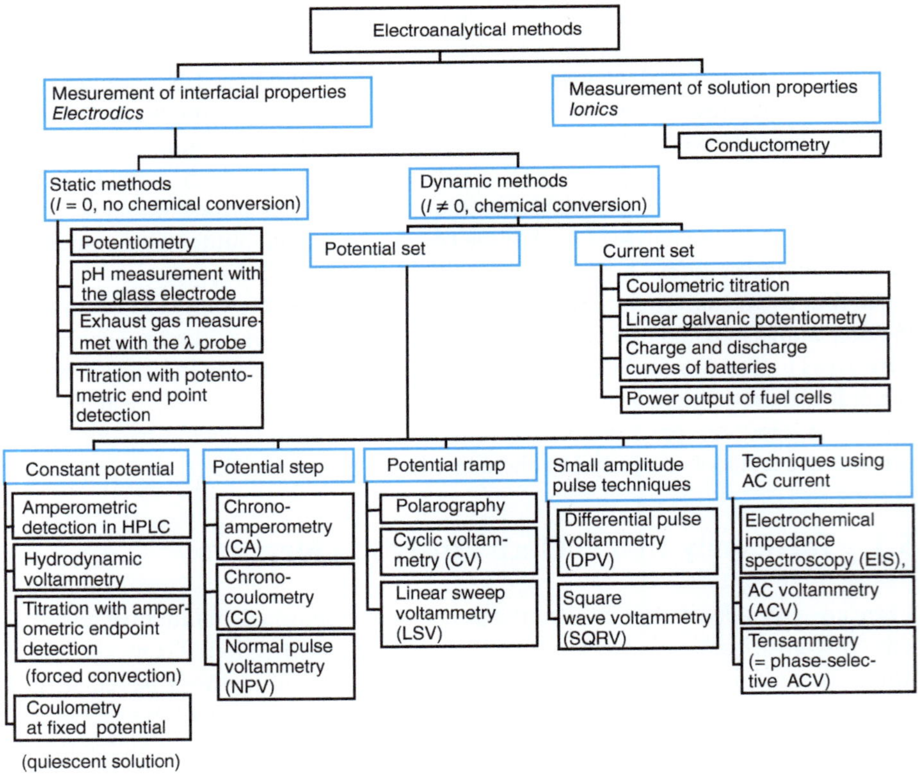

Figure 5.1 Family tree of electroanalytical methods. The criteria for categorizing are in blue boxes.

well-defined convection is a further criterion for categorizing the methods. Separation methods based on electrochemical principles have been the topic of Section 4.6 and are omitted from Figure 5.1. Impedance spectroscopy is the topic of Chapter 7. Galvanodynamic methods are covered in Excursion 15-3 in the context of batteries.

5.2 Potentiometry

Potentiometry usually refers to the measurement of the open-circuit potential of an indicator (or "sensing") electrode with respect to a reference electrode in order to determine the concentration of a specific analyte. As we discussed in detail in Chapter 2 in connection with the Nernst Equation (2.27), potentiometry actually determines activities. Although concentration and activities have the same unit mol L^{-1}, they are not identical, and we may regard activities as effective (or modified) concentrations. Below we will discuss examples in which either the effective concentration is of direct relevance (pH), or special procedures must be employed to obtain the concentration from a measurement of activities (concentration of K^+ in blood).

Instead of potentiometry, the measurement of electrode potentials is sometimes called voltametry (with only one "m" and similar to amperometry for the measurements of currents) in older literature. This terminology is discouraged because it is likely to cause confusion with a similar name, "voltammetry" (with double "mm") (Section 5.3). Voltammetry refers to the recording of potential [V] and current [A], that is, the measurement of a current–voltage curve or volt-ampere-metry.

5.2.1 When Is an Open-Circuit Voltage Selective for Only One Kind of Ion?

The open-circuit voltage for analytical purposes is measured versus a reference electrode with a constant half-cell potential independent of the composition of the sample. The signal sensitive to the analyte is provided by an indicator electrode, which is an ion-selective electrode. The measured potential depends on the concentration activity of the ion type M $a_c = \{M\}$ according to the Nernst Equation (2.37). If more than one type of ion is present in the sample, there can be competition. Previously, we have seen the example of a silver electrode where more than one electrode reaction may proceed concomitantly at a metal electrode surface (metal dissolution, metal deposition, oxygen reduction reaction). The potential measured at such an electrode is a mixed potential by nature (Section 2.3.5). Although Eq. (2.37) must not be used to calculate a_c from the open-circuit potential of such electrodes, analytical determinations of solution compositions are still possible if the electrode is used for end-point indication in a titration. In this case, the composition is obtained based on the consumption of the titrant and the stoichiometry of the reaction (Section 2.4.3). Thus, titration methods belong to the most accurate analytical methods and represent **absolute analytical methods** or synonymously primary methods (i.e., methods that do not need a calibration and can be traced to accepted metrological standards). Because those procedures are manually demanding (labor, time, reagent consumption), often there is a need for reagentless alternatives. In the case of ion-selective electrodes, a reagentless measurement is achieved by the combination of two concepts:

- Use of a membrane that allows the transfer of only one type of ion.
- Calibration of the electrode over the range of expected sample compositions.

Calibration is generally used in situations in which the quantitative relationship between a measured quantity (here electrode potential E) and the measured composition (concentration c or activity a) is known in principle but is too complex or affected by further parameters in such a way that the use of a simple formula is unwarranted. Instead, the signal is measured for a range of standard solutions with precisely known concentrations of the analyte to give a calibration curve. Such methods are called **relative analytical methods**. Most analytical methods are indeed relative methods.

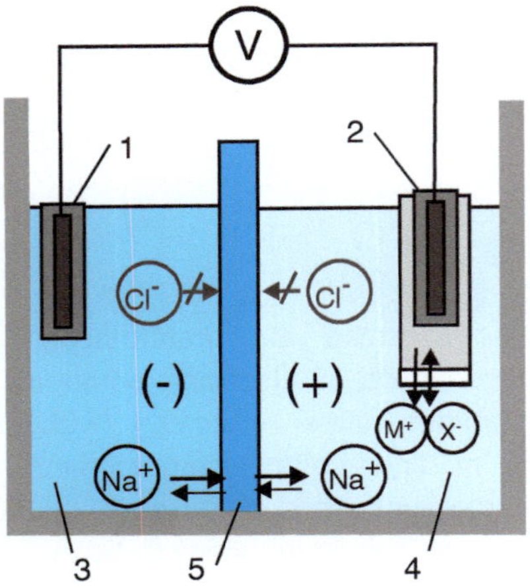

Figure 5.2 Formation of a potential at a Na^+-selective membrane placed between two reference electrodes.

1) Inner reference electrode;
2) Outer reference electrode;
3) Inner electrolyte solution (reference solution, phase I);
4) Outer electrolyte solution (sample with unknown concentration of the analyte ion Na^+, phase II);
5) Na^+-selective membrane.

$\tilde{\mu}_M^I$, $\tilde{\mu}_M^{II}$ – Electrochemical potential of the transferable cation in phase I or II [J mol^{-1}];

$\mu_M^{I^\circ}$, $\mu_M^{II^\circ}$ – Chemical standard potential of the transferable cation in phase I or II [J mol^{-1}];

z_M – Sign-dependent charge number of the cation M;

a_M^I, a_M^{II} – Activity of the cation M in phase I or II [mol L^{-1}];

a° – Standard activity for electrolyte solution [1 mol L^{-1}];

φ^I, φ^{II} – Inner potential (Galvani potential) of phase I or II [V];

F – Faraday constant [A s mol^{-1}];

T – Temperature [K];

R – Universal gas constant [J K^{-1} mol^{-1}].

$a_M = \{M\} = c\, f_c = [M]\, f_c$ – Activity of the cation M in the sample ($a_M{}^{II}$)

5.2.2 Membrane Electrodes

In Section 2.3.2, we have investigated how charge separation occurs at a membrane and how this causes the formation of a Galvani potential difference across the membrane. An ideal ion-selective membrane allows the transfer of only one kind of ion. Such a membrane separates the sample with an unknown concentration of the analyte ion from a reference solution with a *constant* concentration of the analyte ion.

The sample and the reference solution are contacted via the inner and outer reference electrodes (Figure 5.2). The diaphragm at the outer reference electrode allows nonspecific transport of ions between the filling solution of the reference electrode and the sample, but protects the reference electrode by preventing a rapid convective mixing of both solutions. Because complex analytes (blood, sea water, milk, beer, landfill leachate, ...) always contain different kinds of ions, the ion-selective membrane not only has to separate anions from cations but also discriminate between different kinds of cations, even those with the same charge number (e.g., Na^+, K^+, and NH_4^+)! The inner reference electrode, the inner solution, and the ion-selective membrane are often combined in one device (Figure 5.3) called **ion-selective electrode (ISE)**. It corresponds to the left half-cell with membrane in Figure 5.2.

The Galvani potential difference $\Delta\varphi_{eq}^{I,II}$ across the ion-selective membrane is obtained for one transferable kind of ion M from the equality of the electrochemical potentials $\tilde{\mu}_M^K$ for the ion M in the reference solution (phase I with constant a_M^I) and the sample solution (phase II) according to Eqs. (2.101), (2.102), and (2.106).

$$\tilde{\mu}_M^I = \tilde{\mu}_M^{II} \tag{5.1}$$

$$\mu_M^{I,\circ} + RT\ \ln\frac{a_M^I}{a^\circ} + z_M F\varphi^I = \mu_M^{II,\circ} + RT\ \ln\frac{a_M^{II}}{a^\circ} + z_M F\varphi^{II} \tag{5.2}$$

$$\Delta\varphi_{eq}^{I,II} = \frac{RT}{z_M F}\ \ln\frac{a_M^{II}}{a_M^I} = -\frac{RT}{z_M F}\ \ln\frac{a_M^I}{a^\circ} + \frac{RT}{z_M F}\ \ln\frac{a_M^{II}}{a^\circ} = const + \frac{RT}{z_M F}\ \ln\frac{a_M^{II}}{a^\circ} \tag{5.3}$$

This Galvani potential difference makes up the most important contribution to the overall cell voltage U. Other smaller voltage contributions (e.g., the diffusion potential, Section 2.3.3) at the outer reference electrode also contribute to the overall cell voltage. They must be kept small and, more importantly, constant. If this is achieved, the constant contributions to the cell voltage are summed up in the constant term of the calibration function of an ion-selective electrode:

$$U = const' + \frac{RT}{z_M F}\ln\frac{\{M\}}{a^\circ} = const' + 2.3\frac{RT}{z_M F}\log\frac{\{M\}}{a^\circ}. \tag{5.4}$$

We expect from Eq. (5.4) that the cell voltage will change at room temperature by 59 mV if the activity {M} of a singly charged ion is altered by one order of magnitude. Starting from Figure 5.2, we may note a few challenges that must be met when developing or using an ISE:

- Contamination of the sample by ions that leach out of the ion-selective membrane or the outer reference electrode. This will be especially important at low concentrations of the analyte ion.
- Formation of a potential difference due to diffusion at the diaphragm of the outer reference electrode. This will be especially important at low ionic strength of the sample.
- Incomplete selectivity of the membrane. This will be important at high concentrations of an interfering ion that is structurally similar to the analyte ion M.

- Constancy of temperature between calibration and measurement.
- Low sensitivity (signal change per change in activity) for multiply charged ions ($z_M > 1$).
- Retarded equilibration (i.e., slow response time) due to transport processes in solution or within the membrane.
- Logarithmic relationship between signal and activity, and thus different sensitivity in different concentration ranges.

Q: The cell shown in Figure 5.3 will always have a diffusion potential difference at the diaphragm of the reference electrode. The open-circuit voltage of the entire cell should steadily change as long as equilibration of the concentrations is underway between the sample and the inner filling solution of the reference electrode. How can we then relate the signal U recorded with such a cell to the distribution processes at the membrane–sample interface only?

A: The answer to this important question comprises two aspects. The two processes (equilibration at the reference electrode and ion distribution at the membrane–sample interface) occur on very different time scales. Within the reading of the signal from the ISE, the diffusion potential difference at the reference electrode reaches an almost steady-state value, that is, this contribution to the overall cell voltage does not change significantly and can be taken into account by a calibration procedure. However, such slow processes determine other practically relevant properties of ISE such as lifetime or the need for recalibration.

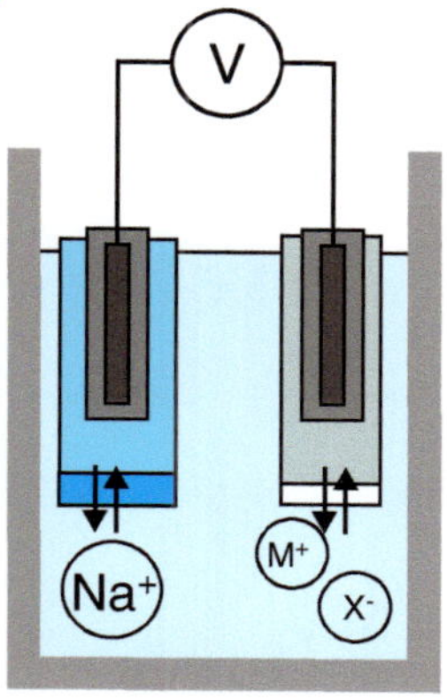

Figure 5.3 Arrangement of the electrochemical cell from Figure 5.2 as an indicator electrode with ion-selective membrane on the left and an external reference electrode on the right. Indicator electrode and reference electrode are often combined in one physical body (called combination electrode). An example is given in Figure 5.5.

Ion-selective membranes can be made from different materials, which are often grouped as shown in Figure 5.4. We will discuss a few general aspects for the example of glass electrodes in Section 5.2.3, provide an overview about the analytical characteristics in Section 5.2.4, and cover solid-state membranes and liquid membranes in Sections 5.2.5 and 5.2.6.

5.2.3 Glass Electrodes

The glass electrode should actually be called glass membrane electrode. The pH electrode can serve as a prototypical sensor with an extremely large dynamic range, a high selectivity, and a low temperature dependence. The dynamic range comprises the concentration span from the lowest to the highest concentration that can be determined. For the pH electrode, this is roughly $\{H^+\}$ between 10^{-14} and 1 mol L^{-1}. The processes

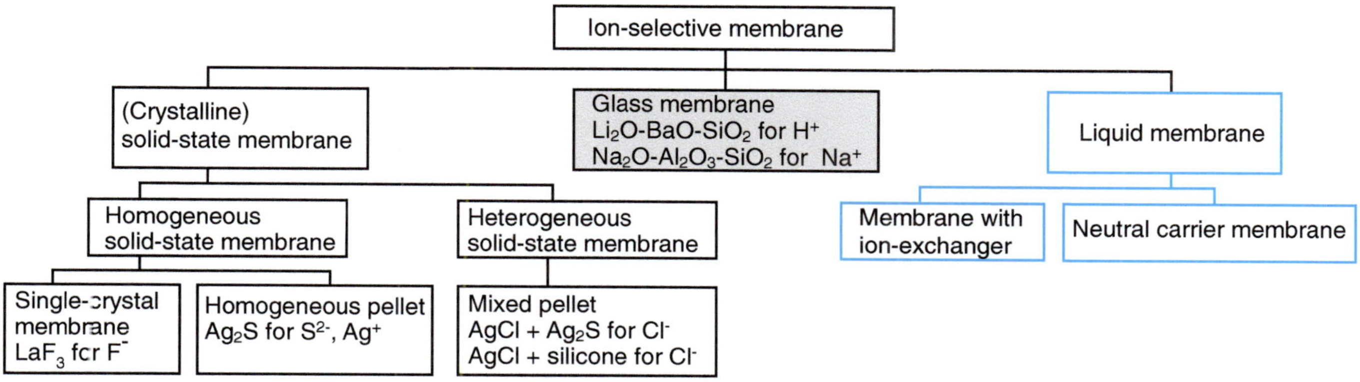

Figure 5.4 Classification of ion-selective membranes with examples for various analytes.

responsible for forming the signal occur in a hydrated surface layer of the glass membrane with a total thickness of only about 50–1000 μm. Glass is formed by an irregular network of silicate groups that enclose cavities as a host for cations. These cations have a specific mobility between the cavities in the silicate network. The movement occurs along gradients of the electrochemical potential within an electric field formed as a consequence of charge separation processes at the interfaces such as dissociation of surface functional groups, adsorption, or ion-exchange.

Different glass compositions allow the glass membrane to be tuned for the sensing of different analytes, such as (H^+/Na^+, Figure 5.4), and for different analytical tasks. Measurements of pH in the range of pH 12–14 face the problem that the activity of H^+ is orders of magnitude smaller than the activity of Na^+, which is typically contained in the solution due to the addition of NaOH as a base. Common pH electrodes yield an error because at this excess of Na^+, Na^+ is incorporated into the silicate network instead of H^+, giving rise to the "alkali error". A glass membrane with a high content of Li_2O causes smaller cavities for the cation and thus further discriminates the larger Na^+ ion against the smaller H^+ ion, which extends the applicability up to pH 14. Such membranes also have a lower electrical resistance, facilitating the manufacturing of relatively thick and thus mechanically stable membranes. The disadvantage of a high Li content is the poor processability, preventing, for instance, the use of glass blowing for shaping the electrode. Furthermore, the lifetime of the membrane is reduced, and the susceptibility to corrosion is increased. The use at elevated temperatures (e.g., thermal sterilization for measurement in cell cultures) is not possible. Thus, there are good reasons for the continuing use of conventional pH electrodes with a reduced dynamic range of pH 0–12 but increased robustness and larger temperature range.

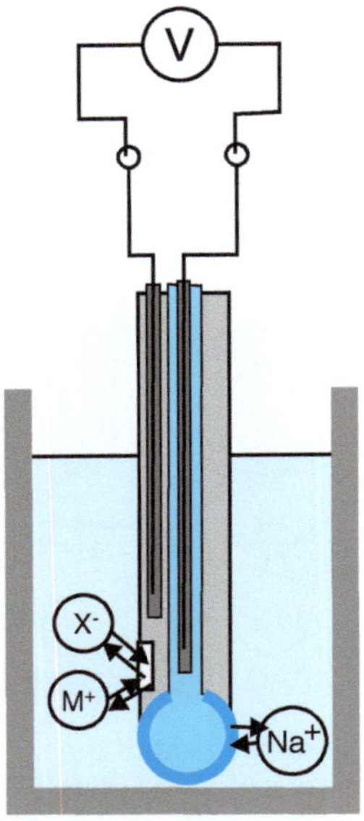

Figure 5.5 Schematic of a combination pH electrode containing the ion-selective electrode and the reference electrode within one device.

Glass electrodes are available in different shapes. In a typical laboratory electrode, the indicator electrode and the reference electrode are combined into one device with several compartments. They are called a combination pH electrode (Figure 5.5). Besides the rod-like arrangements shown in Figure 5.5, there are pointed shapes or completely flat probes for measurements on wet surfaces. The design of the diaphragm of the outer reference electrode is important for the functionality. There are several technical realizations for this, such as the use of porous glasses or ceramics, ground joints, polymers, or simple "holes" in various shapes.

Q: Recently, I looked at the connector plug of a pH electrode to the pH meter. There were more than two wires – why?

A: Commercial combination pH electrodes usually contain a temperature sensor, for instance, a precise resistor from micrometer-sized Pt structures (Pt-100 element), which changes its ohmic resistance of about 100 Ω in a predictable way so that the temperature can be calculated from the precise value of resistance of the Pt-100 element. Due to the temperature measurement, the accuracy of the glass electrode can be warranted over a larger *T*-interval during calibration and measurement. The need to connect the temperature sensor explains the occurrence of more than two wires in the cable of a typical combination pH electrode.

Although pH measurement using a glass membrane electrode is arguably the most frequently performed chemical analysis, there is no consensus about the precise mechanism of the formation of the potential difference across the glass membrane [8] because several processes are superimposed in a complex manner [9, 10]. For a simplified description, two dissociation equilibria at the surface must be considered:

Index (*s*) for the solvated interfacial layer (solid), index (*aq*) for the aqueous electrolyte phase.

$$\equiv\text{SiOH}(s) + \text{H}_2\text{O}(aq) \rightleftharpoons \equiv\text{SiO}^-(s) + \text{H}_3\text{O}^+(aq) \tag{5.5}$$

$$\equiv\text{SiOH}(s) + \text{OH}^-(aq) \rightleftharpoons \equiv\text{SiO}^-(s) + \text{H}_2\text{O}(aq) \tag{5.6}$$

Starting from the equilibrium condition, that is, the equality of the electrochemical potential for the transferring ion, the separation in the standard term, the concentration-dependent term, and the contribution of the Galvani potentials of both phases provides Eq. (5.7) for reaction (5.5)

$$\mu_{\text{SiOH}}^{(s),\circ} + RT\ \ln\frac{\gamma_{\text{SiOH}}}{\gamma^\circ} + F\varphi^{(s)} = \mu_{\text{H}^+}^{(aq),\circ} + \mu_{\text{SiO}^-}^{(s),\circ} + RT\ \ln\frac{\{\text{H}^+\}\gamma_{\text{SiO}^-}}{a^\circ\gamma^\circ} + F\varphi^{(aq)}, \quad (5.7)$$

γ, γ° – Interfacial concentration activity and the corresponding standard activity [mol cm^{-2}]

Here, the ionic charge $z_{\text{H}^+} = 1$ and the simplified notation "H^+" $\equiv H_3O^+$ are used. The activity of the silanol groups in the hydrated interfacial layer of the glass membrane is expressed by the interfacial concentration Γ_{SiOH} [mol cm^{-2}] and the corresponding interfacial concentration activity γ_{SiOH} [mol cm^{-2}]. The standard terms can be combined into an acid dissociation constant K_a'. It has a similar meaning for the acidity of the hydrated layer as K_a has for the acidity of an acid in an aqueous solution.

$$\varphi^{(s)} - \varphi^{(aq)} = -\frac{RT}{F}K_\text{a}' + \frac{RT}{F}\ \ln\frac{\{\text{H}^+\}\gamma_{\text{SiO}^-}}{a^\circ\gamma_{\text{SiOH}}} \quad (5.8)$$

$$K_\text{a}' = -\frac{\mu_{\text{H}^+}^{(aq),\circ} + \mu_{\text{SiO}^-}^{(s),\circ} - \mu_{\text{SiOH}}^{(s),\circ}}{RT} \quad (5.9)$$

K_a' – Acid dissociation constant for silanol groups at the surface of the glass membrane

From the reaction (5.6), we obtain for the processes in alkaline solution:

$$\varphi^{(s)} - \varphi^{(aq)} = \frac{RT}{F}K_\text{b}' + \frac{RT}{F}\ \ln\frac{a^\circ\gamma_{\text{SiO}^-}}{\{\text{OH}^-\}\gamma_{\text{SiOH}}} \quad (5.10)$$

$$K_\text{b}' = -\frac{\mu_{\text{SiOH}}^{(s),\circ} - \mu_{\text{SiO}^-}^{(s),\circ} + \mu_{\text{OH}^-}^{(aq),\circ}}{RT} \quad (5.11)$$

K_b' – Base association constant of protons to silicate groups at the surface of the glass membrane

Replacing $\{\text{OH}^-\}$ with the self-dissociation constant of water K_W and the proton activity $\{\text{H}^+\}$, we obtain the potential dependence for the entire pH range [Eq. (5.12)]. Although Eq. (5.12) is identical to Eq. (5.8), the importance of reaction (5.6) shall be emphasized here. Otherwise, it would not be possible to achieve a sufficiently high exchange current density at the glass membrane to retain protons as the potential-determining kind of ion in alkaline solution.

$$K_\text{W} = K_\text{a}'K_\text{b}' = \frac{\{\text{H}^+\}\gamma_{\text{SiO}^-}}{\gamma_{\text{SiOH}}}e^{\left(\frac{-F\Delta\phi}{RT}\right)}\frac{\{\text{OH}^-\}\gamma_{\text{SiOH}}}{\gamma_{\text{SiO}^-}}e^{\left(\frac{F\Delta\phi}{RT}\right)} = \{\text{H}^+\}\{\text{OH}^-\}$$

$$\varphi^{(s)} - \varphi^{(aq)} = -\frac{RT}{F}\frac{K_\text{W}}{K_\text{b}'} + \frac{RT}{F}\ \ln\frac{\{\text{H}^+\}\gamma_{\text{SiO}^-}}{a^\circ\gamma_{\text{SiOH}}} = -\frac{RT}{F}K_\text{a}' + \frac{RT}{F}\ \ln\frac{\{\text{H}^+\}\gamma_{\text{SiO}^-}}{a^\circ\gamma_{\text{SiOH}}} \quad (5.12)$$

Completely analogous to the treatment of the cell voltage in Section 2.3.4, we can add up the Galvani potential differences for each interface within the combination pH electrode to obtain the open-circuit voltage of the cell:

$$U = \varphi^\text{I} - \varphi^{\text{I}'} = \Delta\varphi_\text{eq}^{\text{I,II}} + \varphi_\text{eq}^{\text{II,III}} + \ldots + \Delta\varphi_\text{eq}^{\text{VIII,IX}} + \Delta\varphi_\text{eq}^{\text{IX,I}'} \quad (5.13)$$

I – Ag wire of the inner reference electrode
II – AgCl at the inner reference electrode
III – 3 mol L^{-1} KCl, buffered, inner filling of the indicator electrode with known and constant pH
IV – Inner surface of the glass membrane
V – Glass membrane
VI – Outer surface of the glass membrane
VII – Sample with unknown $\{\text{H}^+\}$
VIII – 3 mol L^{-1} KCl of the outer reference electrode
IX – AgCl of the outer reference electrode
I′ – Ag wire of the outer reference electrode

Due to the symmetrical design of the cell, some Galvani potential differences balance in Eq. (5.13), these are $\Delta\varphi_\text{eq}^{\text{I,II}} = -\Delta\varphi_\text{eq}^{\text{IX,I}'}$, $\varphi_\text{eq}^{\text{II,III}} = -\Delta\varphi_\text{eq}^{\text{VIII,IX}}$, $\Delta\varphi_\text{eq}^{\text{IV,V}} \approx -\Delta\varphi_\text{eq}^{\text{V,VI}}$. The remaining contributions originate at the interfaces of the glass membrane-sample (with unknown pH) and glass membrane-inner solution (with known pH):

$$\begin{aligned} U &= \Delta\varphi_\text{eq}^{\text{III,IV}} + \Delta\varphi_\text{eq}^{\text{VI,VII}} = -\Delta\varphi_\text{eq}^{\text{IV,III}} + \Delta\varphi_\text{eq}^{\text{VI,VII}} \\ &= +\frac{RT}{F}K_\text{a}' - \frac{RT}{F}\ \ln\frac{\{\text{H}^+\}^{\text{III}}\gamma_{\text{SiO}^-}^{\text{IV}}}{a^\circ\gamma_{\text{SiOH}}^{\text{IV}}} - \frac{RT}{F}K_\text{a}' + \frac{RT}{F}\ \ln\frac{\{\text{H}^+\}^{\text{VII}}\gamma_{\text{SiO}^-}^{\text{VI}}}{a^\circ\gamma_{\text{SiOH}}^{\text{VI}}} \\ &= \frac{RT}{F}\ \ln\frac{\{\text{H}^+\}^{\text{VII}}\gamma_{\text{SiOH}}^{\text{IV}}\gamma_{\text{SiO}^-}^{\text{VI}}}{\{\text{H}^+\}^{\text{III}}\gamma_{\text{SiO}^-}^{\text{IV}}\gamma_{\text{SiOH}}^{\text{VI}}} \end{aligned} \quad (5.14)$$

The concentrations of the protonated and deprotonated silanol groups on both sides of the glass membrane are roughly constant because the dissociation of only a few groups causes a substantial change in the potential, which counteracts further dissociation. We have already illustrated this important point in Excursion 2-1. Conversion to the decadic logarithm and taking 298 K as room temperature provides the so-called theoretical pH characteristics of the electrode [Eq. (5.15)].

pH = $-\log(\{H^+\}/a°$

$$U = \frac{2.3RT}{F}\log\frac{\{H^+\}^{VII}}{a°} - \frac{2.3RT}{F}\log\frac{\{H^+\}^{III}}{a°} = 0.059\ V(pH^{III} - pH^{VII}), \quad T = 298\ K \tag{5.15}$$

In reality, the pH electrode contains a few unavoidable asymmetries, such as the diffusion potential at the diaphragm. Such contributions are summed up as an asymmetry potential U_{as}. In addition, the experimentally determined slope of the pH characteristics is often slightly smaller than that of the theoretical pH characteristics of $S = 2.3\ RT/F = 59$ mV (at 298 K) by a factor $\alpha < 1$. With these considerations, we can write the experimental pH characteristics of the combination pH electrode as:

$$U = \varphi^{inner} - \varphi^{outer} = U_{as} + \alpha S(pH^{inner} - pH^{outer}) = const - \alpha\, 0.059\ V\, pH^{outer}, \quad T = 298\ K \tag{5.16}$$

Recent studies have shown that α can reach at best values of 0.995–0.999, even for an ideal glass membrane electrode, because a change in the activity of oxonium ions (or sodium ions) is always associated with changes in the activities of other reaction partners in the equilibrium [10]. In addition, aging of the glass membrane causes a further decrease in α.

Because all these factors are not precisely known for a particular glass membrane electrode and changes occur slowly over time, pH electrodes require calibration before use. Modern pH meters contain a microprocessor and a small display by which the user is guided through such a calibration process. The calibration relies on the known pH of standard buffer systems, which are distributed together with the combination pH electrode and the pH meter (Table 5.2). They are similar in composition to buffer solutions developed at the National Institute of Standards and Technology (NIST, USA). Those buffer systems can be easily traced to a primary standard, they have good (but not infinite!!) shelf life, and they show a low temperature dependence in their pH [11]. The calibration buffers should be selected in such a way that they encompass the expected pH of the sample.

Table 5.2 Typical calibration buffer systems.

pH	Buffer
1.68	Potassium oxalate
4.01	Potassium hydrogen phthalate
6.86	KH_2PO_4/Na_2HPO_4
9.18	Borax

An acetate buffer shall be prepared and checked with pH of 5.0. Suitable calibration buffer systems would be those with pH 6.86 and 4.01.

5.2.4 Selectivity and Limit of Detection for ISEs

The two-point calibration described above is sufficient for many routine tasks, but allows the determination of pH with an accuracy of only one decimal place at best(!). While this is sufficient for many tasks, precision measurement or measurements in solvents other than H_2O [12, 13], quality control or development of new ion-selective electrodes require the determination of the calibration plot over the widest possible activity range of the analyte ion M in different matrices (i.e., in the presence of various interfering ions I and other matrix components). We will discuss the effect of an interfering ion I using the example of a Na^+-selective glass membrane electrode (thus, the analyte ion M is Na^+). Na-selective glasses are obtained by adding oxides of trivalent metals (Al_2O_3) to Na-rich glasses. From an analytical point of view, it is important that these electrodes function in the presence of interfering ions, I ≡ H^+ and I ≡ K^+.

The pH is usually kept constant by buffering. Thus, the remaining selectivity issue is the discrimination against K^+ with respect to Na^+.

The behavior of an ISE in the presence of the analyte ion M and several interfering ions I is described by the Nikolskij–Eisenman Equation[53] (5.17). More complex treatments of this problem are summarized in Ref. [14].

$$U = U_{as} + \alpha S \lg\left(\frac{\{M\}}{a^\circ} + \sum_{I=Ions} K_{M,I}\left(\frac{\{I\}}{a^\circ}\right)^{z_M/z_I}\right) \tag{5.17}$$

A corresponding calibration plot is shown in Figure 5.6. Note that the abscissa shows the *activity* (and not the concentration) of the analyte ion M. In the case that the ISE needs to be used for the determination of concentration, a solution with a high concentration of noninterfering ions is added to the sample. This addition causes the ionic strength to rise to a very high value and therefore determines the activity coefficient of the analyte ion M, which is present at much lower concentration. Only if the activity coefficient of the analyte ion is constant over the entire measurement range will there be a proportionality between the cell voltage and the concentration. Such an additive is called TISAB (total ionic strength buffer). These additives are commercially available.

The concentration-dependent term in Eq. (5.17) contains both {M} and a sum of the (normalized) activities of the interfering ions {I} and a weighting factor $K_{M,I}$. The weighting factor is called potentiometric selectivity coefficient. If M and I have different ionic charges z_M and z_I, their ratio must be considered in the exponential term for each interfering ion. However, it has been shown that Eq. (5.17) does not adequately describe the behavior of the electrodes in such situations. Consequently, more complicated equations have been proposed for those situations [15]. For the special case of the interference of K^+ with the determination of Na^+, Eq. (5.17) can be simplified to Eq. (5.18).

$$U = U_{as} + \alpha S \lg\left(\frac{\{Na^+\}}{a^\circ} + K_{Na^+,K^+}\frac{\{K^+\}}{a^\circ}\right) \tag{5.18}$$

The graph of this function is shown in Figure 5.6 with annotations.

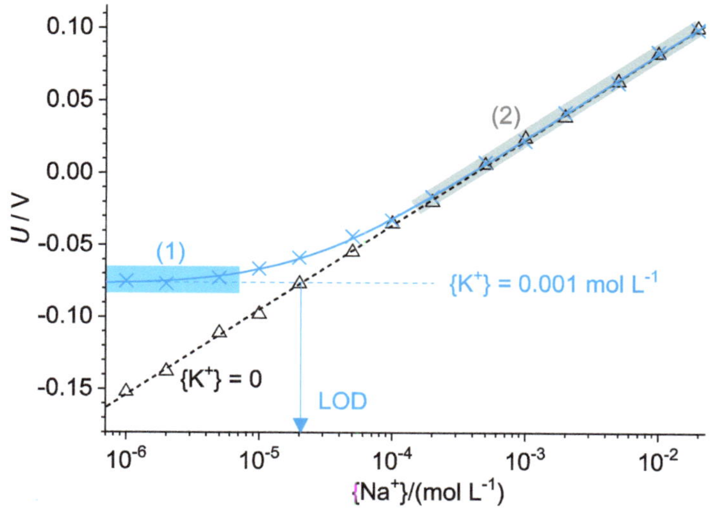

Figure 5.6 Calibration plot of a Na^+-sensitive glass membrane electrode. The black curve shows the calibration in the absence of interfering ions. The blue curve was measured with solutions containing K^+ with an activity of 10^{-3} mol L^{-1} K^+. $K_{Na^+,K^+} = 0.02$, $U_{as} = 0.2$ V. LOD is the limit of detection.

[53] Boris Petrovich Nikolskij, 1900–1990, Russian physico-chemist and radiochemist, professor, and academician in Leningrad (St. Petersburg); George Eisenman, professor of physiology in Salt Lake City and Los Angeles.

The potential of the electrode is determined by the interfering ion I if $K_{M,I}\{I\} > \{M\}$. This part of the calibration plot is highlighted by the blue rectangle (1) in Figure 5.6. In contrast, the potential depends only on the analyte ion M in the part of the calibration plot marked by the gray rectangle (2). This part is the working range in the presence of the interfering ion I. It is very much limited toward smaller activities of M compared to solutions that contain only the analyte ion M and no interfering ion (black curve in Figure 5.6). Typical limits of detection in the absence of interfering ions are just below 10^{-6} mol L^{-1}. We will treat some ISEs with exceptionally low limits of detection as well as new developments in the area of potentiometry in subsequent sections.

Example Calculation 5.1 Determine the selectivity coefficient K_{Na^+,K^+} from the blue calibration plot in Figure 5.6.

Solution: Extrapolation of the rising straight part of the calibration plot and of the horizontal part of the plot in the presence of K^+ leads to an intercept. The activity belonging to the intersection of the two straight lines corresponds to the limit of detection (LOD) $\{Na^+\}_{LOD}$ in the presence of a particular activity of the interfering ion $\{K^+\} = 0.001$ mol L^{-1}. At this activity $\{Na^+\}$, the analyte ion, and the interfering ion have the same effect on the potential of the ISE. The exact solution can be obtained by first determining the linear regression functions using a spreadsheet application. This gives

I: $U = 0.200 + 0.059\ \text{V} \log(\{Na^+\}/a°)$
II: $U' = -0.077$ V

From $U = U'$, one obtains by rearranging $\log(\{Na^+\}/a°) = (-0.077\ \text{V} - 0.2\ \text{V})/0.059\ \text{V} = -4.695$. This yields $\{Na^+\}_{LOD} = 10^{-4.695}a° = 2.0 \times 10^{-5}$ mol L^{-1}. From $\{Na^+\} = K_{Na^+,K^+}\{K^+\}$ at this point, we obtain the selectivity coefficient

$$K_{Na^+,K^+} = \frac{\{Na^+\}_{LOD}}{\{K^+\}^{z_{Na^+}/z_{K^+}}} = \frac{2 \times 10^{-5}\,\text{mol}\,\text{L}^{-1}}{1 \times 10^{-3}\,\text{mol}\,\text{L}^{-1}} = 0.02 \tag{5.19}$$

In another IUPAC-recommended method, one first determines the open-circuit voltage U of the ISE versus the reference electrode in a 0.1 mol L^{-1} solution of the analyte ion M, and subsequently in another solution with 0.1 mol L^{-1} of the interfering ion I. From the two values, one obtains the selectivity coefficient by Eq. (5.20).

$$K_{M,I} = \exp\left[z_M F \frac{U(I) - U(M)}{RT}\right] \tag{5.20}$$

Numerical values of K_{Na^+,K^+} determined by the two methods may differ substantially from each other. Therefore, the method of determining the selectivity coefficients must be exactly specified. Reference [16] is recommended as a more detailed review of the characterization of ISE selectivity.

Example Calculation 5.2 What is the maximum permissible selectivity coefficient K_{Na^+,K^+} for a Na^+-ISE for the determination of Na^+-activity in blood when the interference from K^+ should remain below 1%?

The normal range of $\{Na^+\}$ in blood is (1.02–1.12) mol L^{-1}. The normal range for $\{K^+\}$ is (2.6–3.7) × 10^{-3} mol L^{-1}. Analogously to Example Calculation 5.1, we set 1% of the lower limit (LL) of the working range equal to the maximum (UL, upper limit) weighted activity of K^+.

$$0.01\,\{Na^+\}_{LL} = K_{Na^+,K^+}\{K^+\}_{UL}^{z_{Na^+}/z_{K^+}} \quad (z_{Na^+} = z_{K^+})$$

$$K_{Na^+,K^+} = 0.01\frac{\{Na^+\}_{LL}}{\{K^+\}_{UL}}$$
$$= 0.01\frac{1\times 10^{-1}\ \text{mol L}^{-1}}{3.7\times 10^{-3}\ \text{mol L}^{-1}}$$
$$= 0.27 \tag{5.21}$$

The potentiometric selectivity coefficient for the determination of Na^+ with respect to K^+ in blood must be equal to or below 0.27.

5.2.5 Solid-State Membrane Electrodes

A solid-state membrane electrode with an extremely good selectivity can be obtained from slices of a LaF_3 single crystal. Doping with Eu^{2+} increases the solid-state conductivity of F^- in the crystal to be effective even at room temperature. Because only F^- ions can be incorporated or ejected at the interface crystal-electrolyte solution, these electrodes can have extreme selectivities. The analytical samples are buffered before the analysis, typically to pH 5–6. Otherwise, HF would be formed in acidic solution, which does not interact with the membrane. In alkaline solutions, the equilibration may be disturbed by the formation of $La(OH)_3$.

$$LaF_3 + H^+ \rightleftharpoons LaF_2^+\ HF \tag{5.22}$$

$$LaF_3 + 3\ OH^- \rightleftharpoons La(OH)_3 + 3\ F^- \tag{5.23}$$

In addition, the buffer contains ions for buffering the ionic strength and a complexing agent like citrate in order to mask triply charged cations that tend to form complexes with F^-. The only significant disadvantage of the LaF_3-membrane electrode is the relatively high cost of polished slices of the high-quality single crystal LaF_3.

Other solid-state membranes are made from pellets of salts of low solubility. Very often, silver sulfide (Ag_2S) is used because it exhibits a high Ag^+ conductivity (Chapter 11). The resulting electrode can be used for the determination of Ag^+ or S^{2-}. [Due to the low solubility product, both ions cannot occur simultaneously at high concentrations (activities).] The only significant interfering ion is Hg^{2+}. In order to remove adsorbates or deposited salts, the electrode can be regenerated by polishing.

Mixed pellets of Ag_2S (for increasing the solid-state conductivity) and another salt of low solubility such as CuS, AgCl, AgBr, or AgI can be used as the membrane in ISEs for Cu^{2+}, Cl^-, Br^- or I^-. The selectivities are strictly coupled to the solubility product of the employed salt of low solubility. The anion with the lowest solubility product will be detected most sensitively. Therefore, an initially Cl^--sensitive electrode will be converted into a I^--sensitive ISE upon prolonged storage in an I^--containing solution. Disturbances, especially in biological samples (blood, milk, ...), are caused by the adsorption of sulfur-containing proteins and peptides or by complexation of Ag^+ in the membrane by $-NH_2$ and $-SH$ groups of proteins in the sample. The addition of binders such as silicone is possible in order to improve the mechanical properties and the processability of the materials for the pellets.

Solubility of Ag salts (from low to high solubility product) $S^{2-} < I^- < Br^- < Cl^-$.

5.2.6 Liquid Membrane Electrodes

Liquid membrane electrodes are available for more than 50 analytes [17, 18]. They are used for direct potentiometry, for instance, in automated analyzers in large-scale clinical laboratories. In most cases, the "liquid membrane" is not a freely floating liquid but is mechanically stabilized within a polymeric network or microscopic porous network. Polymers with a particularly low glass transition temperature are used as well. The mobile polymer chains can be considered as a liquid environment for the other membrane components. The membrane is always hydrophobic, and the charge separation at the membrane–sample interface is based on a distribution equilibrium of the analyte ion between the aqueous solution of the sample and the hydrophobic liquid membrane.

Excursion 5-1 Making of a Simple Liquid Membrane Electrode

A simple liquid membrane ISE can be made as a laboratory exercise [19]. The membrane cocktail (1 g) is dissolved in 10 mL of tetrahydrofuran in a closed glass vial. The cocktail is a mixture that could, for instance, consist of 66% 2-nitrophenyloctylether (NPOE) as plasticizer, 33% poly(vinyl chloride) (PVC) of medium or high molecular weight, and 1% of the ion-selective host molecule and a strongly lipophilic salt. The viscous mixture is poured into a round mold. Very suitable for this purpose are O-rings with a diameter of 3 cm on a glass plate. The molds are covered by a filter paper and a glass plate in order to allow the slow evaporation of the solvent within about 48 h. After hardening, the membrane with an approximate thickness of 200 μm is pulled off the glass plate together with the O-ring. A scalpel is used to remove the O-ring. Several smaller membrane patches can be cut from the membrane with a laboratory cork drill or a hole punch. Those patches are then glued to the end of a PVC tube of approximately 1 cm in diameter. A glue can be obtained by dissolving some PVC in cyclohexanone. After a test of the tightness of the seal via a resistance measurement (see Figure 4.24), the inner electrolyte and the inner reference electrode (Ag|AgCl) are placed into the PVC tube. The inner electrolyte contains the analyte ion with a concentration that is within the expected concentration range of the analyte ion in the sample. In addition, the inner electrolyte contains (10^{-3}–10^{-2}) mol L^{-1} Cl^- for the Ag|AgCl electrode. The Ag|AgCl electrode itself can be prepared from an Ag wire by immersion into 0.1 mol L^{-1} HCl and application of 0.5 mA cm^{-2} in the galvanostatic mode. The formation of a brown coating indicates the formation of a dense AgCl layer on the Ag wire. The ion-selective component is commercially available. Tetraphenylborates are suitable lipophilic salts for analyte cations, while tetraalkylammonium ions in combination with the analyte anion are a good choice for analyte anions. The electrode is calibrated as described above. After testing for possible interfering ions, it can be used for the determination of unknown activities (concentrations) or to indicate the end point in titrations.

In the case of liquid membrane electrodes with ion-exchanger, the selective distribution of the analyte ion occurs between an ion pair with a hydrophobic counter ion within the membrane and the dissolved ion in the aqueous sample. The ion exchanger can be dissolved in the membrane or covalently anchored to the polymer backbone. The membrane with ion exchanger responds to all oppositely charged counter ions. If a salt of a hydrophobic cation and a hydrophilic anion is dissolved in the cocktail for the hydrophobic membrane, the hydrophilic anion can be exchanged while the hydrophobic cation remains inside the membrane. The selectivity of the membrane is achieved during a conditioning process, that is, storage of the membrane in a solution containing a high concentration of the analyte ion. This causes the ion-exchange process for the hydrophilic anion and yields a membrane that is selective for the anion used during the conditioning step. However, such a membrane will respond to all anions, whereby the selectivity follows the Hofmeister series[54] (Table 5.3). Hydrophobic anions are preferred over hydrophilic anions.

Table 5.3 The Hofmeister series.

hydrophobic
organic anions, ReO_4^-, ClO_4^-, IO_4^-, BF_4^-, salicylate, SCN^-, I^-, $HCrO_4^-$, ClO_3^-, NO_3^-, Br^-, NO_2^-, Cl^-, SO_4^{2-}, $H_2PO_4^-$
hydrophilic

[54] Franz Hofmeister, 1850 (Prague)–1922 (Würzburg), Bohemian biochemist, professor in Prague and in Strasbourg, recognized for suggestion of protein structure and amide bonds.

Whether such an electrode is suitable for a particular analytical problem depends on the composition of the matrix of the sample, which should not contain anions that are more hydrophobic than the analyte anion. Occasionally, the sequence of selectivity coefficients does not follow exactly the Hofmeister series when the hydrophobic interfering anion is present in very low concentrations ($<10^{-4}$ mol L^{-1}). Although the interfering ion is preferably incorporated into the membrane, the solution is depleted of the interfering ion in the vicinity of the membrane and thus the adverse effect is smaller than expected from a purely (thermodynamic) equilibrium consideration (see diffusion layer in Section 4.5.2, Figure 4.50). For cations, there is a sequence of hydrophobicity analogous to the Hofmeister series (Table 5.4).

Table 5.4 Hydrophobicity of cations.

hydrophobic
organic cations, Li^+, Na^+, K^+, NH_4^+, Rb^+, Cs^+
hydrophilic

Interesting applications of liquid membrane electrodes are for the determination of charged pharmaceuticals and surfactants (hydrophobic ions) in formulations, that is, in mixtures in which the contents of other ions are in a narrow and well-known range.

Another more unconventional application is the detection of the drug heparin and its antidote protamine [20]. Heparin is a multiply sulfated oligomeric glycosaminoglycan (a polyanionic carbohydrate) that acts as a blood anticoagulant, which is used for different medical indications and during surgical interventions. The effect can be neutralized by protaminesulfate, an arginine-rich protein (a polycation). Suitable liquid membrane electrodes can be used to determine very low amounts of those pharmaceuticals. The signal is based on mass transport-limited extraction of the poly-ions into the membrane rather than by an equilibration between membrane and solution. Therefore, a long-term exposition of the membrane to those ions must be avoided to retain its capacity for extraction.

The limitations of the hydrophobicity series of membranes with ion exchangers can be avoided by the use of neutral carrier membranes. These are liquid membranes with an electrically neutral ionophore (carrier), that is, an organic compound forming a supramolecular host-guest complex with the analyte ion. The name "neutral carrier" indicates their ability for selective transport of (charged) ions across biomimetic membranes that have a very hydrophobic inner part and otherwise repel any charged and hydrophilic species. The ionophore typically contains a hydrophilic binding pocket for the analyte ion with a set of coordinating atoms from the ionophore. The nature and geometry of these determine the selectivity for a specific ion. The binding pocket is surrounded by hydrophobic groups that ensure the retention of the carrier within the hydrophobic matrix. From a chemical point of view, carriers usually contain structural motifs of crown ethers (Figure 5.7), of thioamides, basket-like calixarene, metalloporphyrines, or phthalocyanines. A wide range of carriers and and compatible membrane

biomimetic = artificially made but inspired by the structure or functionality of a biological example

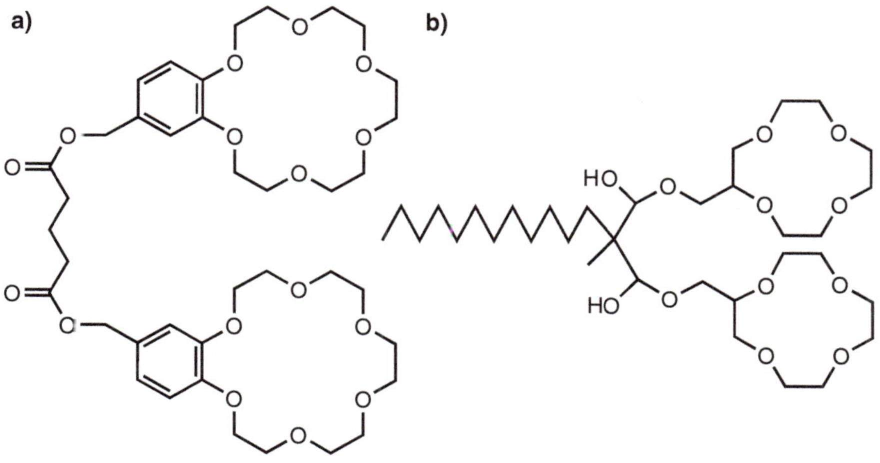

Figure 5.7 Molecular structure of crown ethers. a) Bis[(benzo-15-crown-5)-4′-ylmethyl]-pimelate for the enrichment of K^+ and b) bis[(12-crown-4-methyl]-dodecylmethyl-malonate for the enrichment of Na^+.

components can be found in reviews [18–21]. The membrane cocktail must contain a lipophilic ion exchanger in order to provide a counterion for the cation-loaded and thus charged carrier. The required compositional adjustment of the components is performed mostly by empirical approaches.

ISEs with ionophores are applied for the determination of intracellular ion activities, for which the membrane cocktail is loaded into the tip of a finely pulled capillary pipette (the diameter of the orifice should be about 1 μm). After inserting the inner solution and the inner reference electrode, the capillary electrode is inserted into the biological cell under an optical microscope. The potential of the electrode is measured versus an external reference electrode located outside the biological cell (extracellular). When calculating the ion activities from the measured voltage, the membrane potential difference of the biological cell (to be measured separately, Section 18.4) must be taken into account.

The topic of ISEs had already been regarded as mature, before new discoveries caused intensified research efforts since 1998. Of central importance for this development was the insight that the limits of detection, which are typically 10^{-7} to 10^{-6} mol L^{-1} for conventional ISEs, are often not caused by an interfering ion but by leaching of the analyte ion out of the membrane cocktail into the sample solution. If the membrane contains the analyte ion in an approximate concentration of 10^{-2} mol L^{-1}, a change in the analyte concentration [M] within the membrane by only 1% suffices to augment the concentration in the sample (near the membrane) by 10^{-7} mol L^{-1} compared to the bulk phase of the sample. This is just the concentration range at the limit of detection for many ISEs. Subsequently, several methods were developed to extend the working range of ISEs toward lower concentrations. The methods comprise the decrease in [M] of the inner filling solution, the suppression of transport processes within the membrane by covalently anchoring the ionophore to the polymer backbone of the membrane polymer, reducing the concentration of the ion exchanger, using thicker membranes, and increasing mass transport in the sample by stirring during the measurement. Those measures opened the area of trace analysis to ISEs such as the determination of lead and other heavy metals in drinking water. When comparing the results of ISEs to data provided by other analytical methods, one should keep in mind that ISEs determine activities rather than total concentrations of the analyte. This particular feature can also be exploited for speciation analysis. An authoritative summary of the new developments can be found in [22], and a more extended discussion including further developments is given in [23].

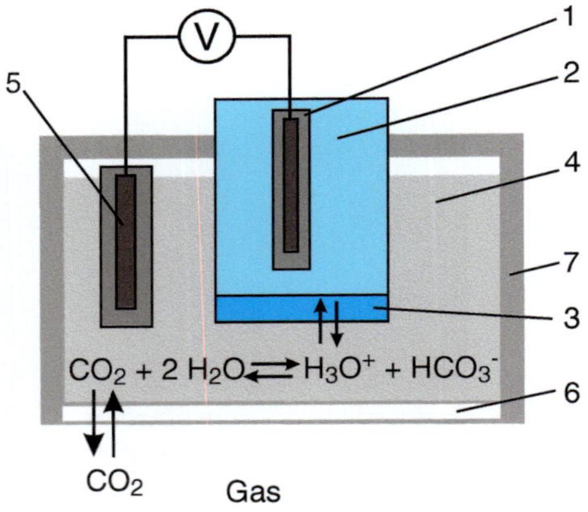

Figure 5.8 Schematic of a potentiometric gas sensor for CO_2 based on an ISE.

1) Inner reference electrode;
2) Inner solution of the ISE;
3) Ion-selective membrane;
4) inner solution of the gas sensor $NaHCO_3$/NaCl electrolyte;
5) Outer reference electrode;
6) Gas-permeable membrane;
7) Outer case of the entire sensor (corresponds to the cell body of the galvanic cell).

The schematic is not to scale!

5.2.7 Potentiometric Gas Sensors

Based on the principles of ISEs for dissolved ions, potentiometric gas sensors can be constructed for the determination of gas content in gas mixtures. Those sensors find very broad application, for example, in blood gas analyzers. The measurements exploit the protolysis equilibrium of CO_2 and are based on the detection of the resulting ions.

The schematic in Figure 5.8 applies the color scheme used before in Figure 5.3 to demarcate the different sensor components. The galvanic cell comprises the ISE and the external reference electrode, both housed in *one* casing (7). The inner solution of the sensor (4, corresponds to the inner solution of the external reference electrode in Figure 5.3) is connected via a gas-permeable membrane (6), for example, from silicon or porous poly(tetrafluoroethylene) (PTFE, Teflon) to the gaseous sample. The gap between the ion-selective membrane (3) and the gas-permeable membrane (5) is strongly exaggerated in Figure 5.8. In fact, both membranes are placed directly on top of each other, and the liquid film exists only in the pores of the gas-permeable membrane. In this way, a short response time can be achieved for the sensor.

The following distribution and protolysis equilibria are simultaneously established:

$$CO_2(gas) \rightleftharpoons CO_2(pore) \tag{5.24}$$

$$CO_2(pore) \rightleftharpoons CO_2(inner\ solution) \tag{5.25}$$

$$CO_2(inner\ solution) + 2H_2O \rightleftharpoons H_3O^+ + HCO_3^- \tag{5.26}$$

From the protolysis equilibrium (5.26) and the established distribution equilibrium, one obtains Eq. (5.27) using the shorthand $H^+ \equiv H_3O^+$.

$$const = \frac{\{H^+\}\{HCO_3^-\}}{p_{CO_2(g)}} \tag{5.27}$$

Because of the inner solution with a high concentration of bicarbonate (ca. 0.1 mol L^{-1}), $\{HCO_3^-\}$ remains roughly constant even if some CO_2 from the gaseous sample is hydrolyzed. Under these conditions, the oxonium ion activity is directly proportional to the CO_2 partial pressure in the gas atmosphere.

$$\{H^+\} = p_{CO_2(g)} \frac{const}{\{HCO_3^-\}} = const' p_{CO_2(g)} \tag{5.28}$$

Insertion into Eq. (5.16) gives Eq. (5.29).

$$\begin{aligned} U &= const + \alpha S \log\{H^+\} \\ &= const'' + \alpha S \log p_{CO_2(g)} \end{aligned} \tag{5.29}$$

The potentiometric gas sensor responds to the CO_2 partial pressure in a logarithmic fashion like a pH electrode does to $\{H^+\}$. The slope of the linear calibration plot is consistent with the Nernst factor (59 mV per decade at 298 K). Due to Henry's law[55] $p_{CO_2(g)} = K\{CO_2\}$, the partial pressure of the gas is also proportional to the dissolved gas concentration. Thus, the sensor can be used for both the determination of the partial pressure in a gas mixture and the determination of activities of dissolved gases in liquids. Note that gases are neutral species, and therefore, the numerical value of the activities is very similar to that of the concentration. The hydrophobic gas-permeable membrane prevents other ions from passing through the membrane and exerting an influence on the signal. However, other acidic or alkaline gases as well as organic acids and bases might pass through the membrane (in neutral state) and subsequently alter the pH of the inner solution of the sensor and thus the readout of the CO_2 sensor.

Sensors for other gases (those reacting as acids or as bases) can be constructed by adding the corresponding base or acid to the inner solution of the sensor.

$$SO_2(g) + 2H_2O \rightleftharpoons H_3O^+ + HCO_3^- \tag{5.30}$$

$$NO_2(g) + 2H_2O \rightleftharpoons NO_3^- + NO_2^- + H_3O^+ \tag{5.31}$$

$$CH_3COOH(g) + H_2O \rightleftharpoons CH_3COO^- + H_3O^+ \tag{5.32}$$

$$NH_3(g) + H_2O \rightleftharpoons NH_4^+ + OH^- \tag{5.33}$$

[55]William Henry (1774–1836) English physician and chemist.

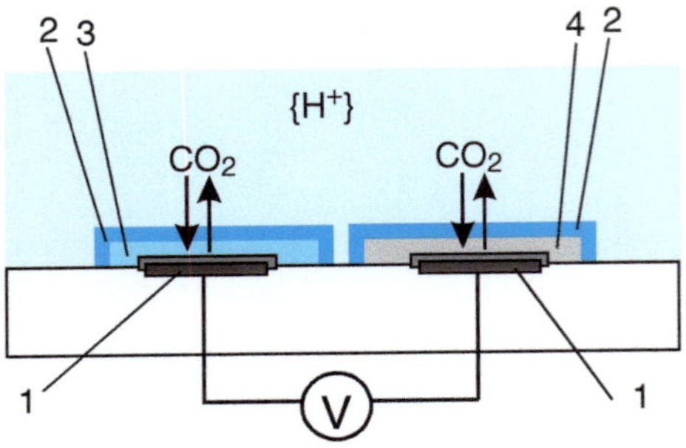

Figure 5.9 Schematic of a differential CO_2 blood gas sensor.

1) Equal reference electrodes;
2) equal gas permeable, pH-sensitive polymer membranes;
3) buffered internal solution;
4) unbuffered $NaHCO_3$/NaCl-internal solution.

The schematic is not to scale!

An increase in selectivity is possible if the ionized form of the gas is determined by an ISE instead of by $\{H^+\}$. For the measurement of p_{CO_2} in clinical blood gas analyzers, a differential measurement is used between two almost identical ISEs with a gas-permeable and a pH-sensitive membrane (Figure 5.9) [18]. The difference between the two electrodes is the composition of the inner solution. One side contains an unbuffered $NaHCO_3$/NaCl solution, while the other inner solution has a high buffer capacity with respect to pH. A change in the pH of the sample does not cause a voltage difference between the two electrodes because the pH-sensitive membranes of both electrodes respond to it in the same way. However, if CO_2 passes through the membrane and hydrolyzes in the inner solution, only the pH of the unbuffered $NaHCO_3$/NaCl solution changes, while the pH of the buffered solution does not change, and neither does the potential of the electrode. The sensor is produced cost-effectively in a planar setup on cheap carrier materials suitable for mass production.

Potentiometric biosensors can be made for a number of charged and neutral biological molecules [18]. In order to detect a molecule in a matrix with many structurally similar molecules, one can use enzymes as biological recognition elements.

Q: What are enzymes?

A: Enzymes are also called ferments or biocatalysts. Chemically, they are mostly proteins formed by a linear sequence of coupled natural amino acids, a polypeptide. Enzymes may contain several polypeptide chains and, in addition, a reactive group (prosthetic group), for example, a coordinated transition metal ion. Enzymes accelerate reactions.

As an example, we consider the enzyme urease that accelerates the hydrolysis of urea $H_2N–CO–NH_2$ [Eq. (5.34)].

$$H_2N\text{–}CO\text{–}NH_2 + 3\,H_2O \xrightarrow{\text{Urease}} 2\,NH_4^+ + CO_2 + 2\,OH^- \tag{5.34}$$

The more urea is contained in the sample, the more NH_4^+ and OH^- will be produced per unit of time. A steady state is quickly reached in which the production of NH_4^+ and OH^- by reaction (5.34) at the sensor surface is balanced by the transport rate of those ions away from the sensor surface by diffusion or forced convection. In a simple version suitable for a lab course experiment, the resulting steady-state concentrations can be followed by a combination pH electrode (Figure 18.27) or, with more selectivity in complex matrices, by an NH_4^+-sensitive glass membrane electrode or a polymer membrane electrode. The analytical properties of such a sensor are determined by the properties of the ISE as well as by the kinetic properties of the enzyme.

5.2.8 The Lambda Probe (λ-Probe)

The most frequently used potentiometric sensor is the lambda probe in motor vehicles for the determination of the oxygen content in exhaust gases from internal combustion engines (ICEs). In an ICE, hydrocarbons with the generic sum formula C_xH_y are oxidized by molecular oxygen from air. Complete combustion requires a stoichiometric amount of O_2, which can be obtained from Eq. (5.35).

$$C_xH_y + (x + y/4)\,O_2 \rightarrow x\,CO_2 + y/2\,H_2O \tag{5.35}$$

Quantitatively, the mixture is characterized by the air–fuel ratio λ [Eq. (5.36)].

$$\lambda = \frac{\left(\frac{m_{air}}{m_{gasoline}}\right)_{measured}}{\left(\frac{m_{air}}{m_{gasoline}}\right)_{stoichiom}} \tag{5.36}$$

m_{air} – Mass of air
$m_{gasoline}$ – Mass of gasoline
$()_{measured}$ – Measured mass ratio
$()_{stoichiom}$ – Stoichiometric mass ratio

The lambda number is $\lambda = 1$ if air is mixed stoichiometrically with the fuel (ca. 14.5 kg air and 1 kg gasoline). In a "rich" mixture, there is an excess of fuel and a deficit of O_2 ($\lambda < 1$). A "lean" mixture has an excess of oxygen ($\lambda > 1$).

The readout of the lambda probe is used to regulate the air intake of the carburetor in such a way that lambda always remains close to unity. Despite this measurement, toxic side products such as CO, NO, and incompletely oxidized hydrocarbons are formed in the combustion process and are further decomposed in the three-way catalytic converter. The lambda probe can be regarded as a complete electrochemical cell. Different from the cells discussed so far, it contains a high-temperature solid electrolyte (Figure 5.10) instead of the liquid electrolyte that we have encountered so far in other examples.

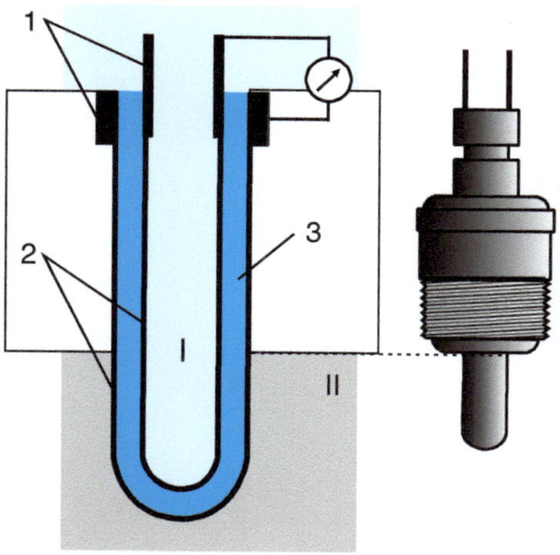

Figure 5.10 Schematic of a potentiometric lambda probe.

(I) – Air as reference gas;
(II) – exhaust gas stream;

1) electrical contact of the inner and outer coatings;
2) inner and outer Pt electrodes;
3) solid electrolyte.

The lambda probe is used in different geometrical shapes. In the classical version, a hollow cylinder with one closed and one open end from yttrium-stabilized zirconia (ZrO_2, or YSZ) is inserted with the closed end into the hot exhaust gas stream. Doped ZrO_2 shows a selective ionic conductivity for O^{2-} (for basics of ion conduction in solids, see Chapter 11). The inner side of the cylinder is in contact with air that serves as a reference mixture, similar to the inner solution of a pH electrode. The inner and outer sides of the zirconia cylinder are coated with an electronically conductive, porous Pt layer. This serves as an electrical connection and as catalyst for both the reduction of O_2 to O^{2-} and the oxidation of O^{2-} to O_2.

If the inner and outer conductive coatings of the zirconia layer are connected to a voltmeter, a galvanic cell is formed, here a differential concentration cell:

$$O_2, p_{ref}(O_2) \,|\, Pt \,|\, ZrO_2 \,|\, Pt \,|\, O_2, p_m(O_2) \tag{5.37}$$

$p_{ref}(O_2)$ – Oxygen partial pressure in the reference gas (air) [Pa]

$p_m(O_2)$ – Oxygen partial pressure in the sample (exhaust gas stream) [Pa]

The following reaction occurs at the three-phase boundary Pt|air|ZrO_2:

$$O_2(g) + 4e^-(Pt) \rightleftharpoons 2O^{2-}(ZrO_2) \tag{5.38}$$

The oxide ions are directly incorporated into O-sites of the ZrO_2 lattice or released from there. The equilibrium voltage of the cell is obtained from the Nernst Equation (5.39) where the temperature is especially high:

$$U = E_m - E_{ref} = \frac{RT}{4F} \ln \frac{p_m(O_2)}{p_{ref}(O_2)} \tag{5.39}$$

Assuming a constant O_2 partial pressure in air, one obtains the calibration function of the sensor:

$$U = \text{const} + \frac{RT}{4F} \ln \frac{p_m(O_2)}{p°} \tag{5.40}$$

$p°$ – Standard pressure (1 bar = 100 000 Pa)

If the oxygen content in the air-fuel mixtures is varied systematically, the lambda probe records a "titration curve" for the consumption of the fuel by air (Figure 5.11). The signal changes very strongly next to the equivalence point ($\lambda = 1$). This makes it difficult to regulate exactly at this point. In addition, the Nernst factor exhibits a temperature dependence, which limits the accuracy of the measurement due to changing temperature gradients between the exhaust gas stream and the outside air. Therefore, modern lambda probes work preferably in the amperometric mode (Figure 5.12). By applying a constant voltage between the two porous Pt layers, oxygen is reduced and incorporated into the ZrO_2 lattice at one Pt electrode, whereas O^{2-} is transported through the lattice of the solid electrolyte to be oxidized at the other Pt electrode and then released as O_2 into the adjacent gas atmosphere. An artificial diffusion barrier on the side of the tail gas stream minimizes the effect of different convective transport conditions in different operational stages of the motor. An electrical heating circuit allows the sensor to reach its operating temperature rapidly. A housing with holes for gas intake provides protection against mechanical damage.

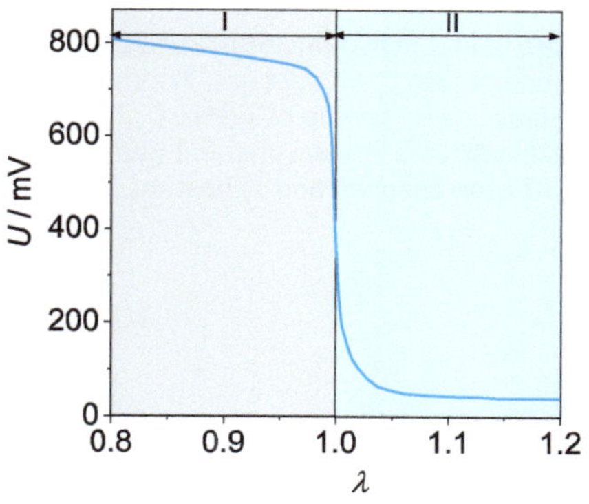

Figure 5.11 Response characteristic of a potentiometric lambda probe.

1) rich mixture (excess of fuel);
2) lean mixture (excess of O_2).

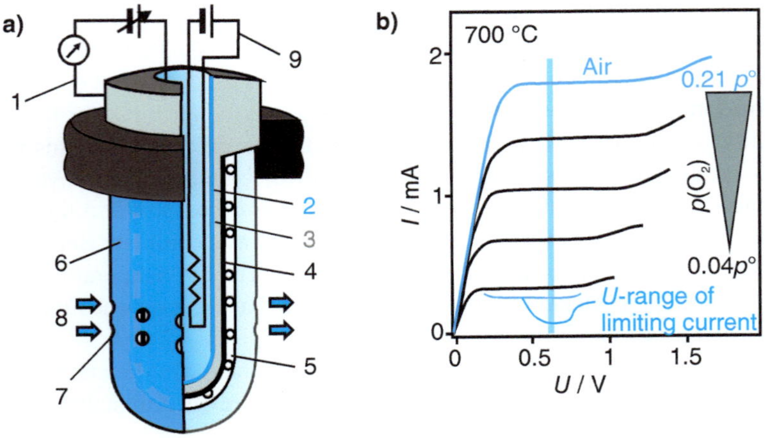

Figure 5.12 Amperometric lambda probe.

a) **Schematic;** 1) Measurement circuit, 2) anode (Pt), 3) solid electrolyte (doped ZrO_2), 4) cathode (Pt), 5) porous layer as diffusion barrier, 6) housing, 7) opening for 8) gas intake from the exhaust gas stream, and 9) heating circuit;

b) **Voltammograms (schematic)** of the lambda probe for different oxygen partial pressures; the vertical blue line shows the normally applied constant cell voltage for amperometric operation.

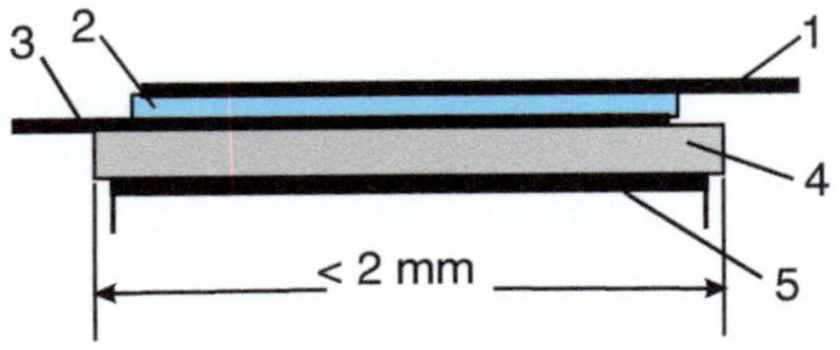

Figure 5.13 Lambda probe in the shape of a thin film sensor; 1) anode (Pt), 2) thin film of the solid electrolyte, 3) cathode (Pt), 4) substrate/diffusion barrier (porous Al_2O_3), and 5) resistive heating (Pt).

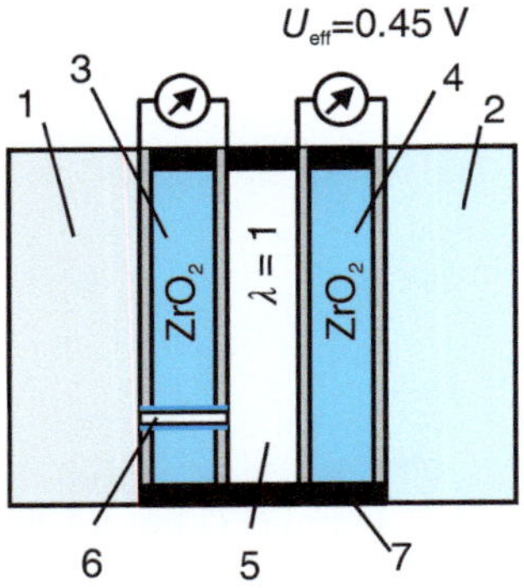

Figure 5.14 Schematic of a broad-band lambda probe; 1) exhaust gas, 2) air as reference gas, 3) pump cell ($Pt|ZrO_2|Pt$), 4) galvanic cell, 5) measurement gap, 6) diffusion channel, and 7) heating.

The sensor can also be produced in a thin layer version (Figure 5.13) on a planar ceramic substrate. This broad-band lambda probe (Figure 5.14) facilitates the adjustment of the working point over a wider range and has applications in gasoline direct fuel injection engines and Diesel engines. The probe contains two electrochemical cells in a planar design, which are separated from each other by a small, gas-filled measurement gap. The measurement gap is in contact with the exhaust gas stream via a diffusion channel. One cell, the pump cell, works like an amperometric sensor and continuously pumps oxygen into or out of the measurement gap in such a way that $\lambda = 1$ is always maintained inside the gas-filled measurement gap. The reference cell operates potentiometrically between the measurement gap with $\lambda = 1$ and the outside air and provides $U = 0.45$ V. Should the oxygen content in the exhaust gas stream deviate from $\lambda = 1$, an equilibration occurs between the tail gas and the gas mixture in the measurement gap through the diffusion channel. The pump current must equalize incoming or outgoing diffusional O_2 flux in the measurement gap. Thus, the value of the required pump current correlates with the deviation from $\lambda = 1$ in the tail gas stream.

5.2.9 Summary for Potentiometry

1) Potentiometric measurements are suitable for direct determination of ions, that is, charged species. The ion to be determined is called analyte, and all other ions and molecules in the sample are regarded as the matrix.
2) Real samples usually contain several different ions. Therefore, it is imperative that electrodes be constructed in such a way that a charge transfer occurs with essentially only one type of ion, the analyte. This is achieved by the application of chemically selective membranes. The chemical nature of this selective membrane can vary widely, as well as the construction details of the electrodes. Those electrodes are called ion-selective electrodes. The pH electrode is an example of a frequently applied ISE.

3) An ISE must be combined with a reference electrode to give a galvanic cell. This also applies when the ISE and the reference electrode are located within one housing or device as is the case for the combination pH electrode. The potential of the reference electrode must not depend on the composition of the sample.
4) The filling solution of the reference electrode is typically separated from the sample by a diaphragm, a porous glass connector, or similar material. This results in a cell with an activity gradient at the junction between the half-cells. The measured open-circuit voltage deviates from a true equilibrium voltage by the diffusion potential formed at the junction. The measured signal is therefore not a true equilibrium voltage despite the fact that it is (almost) constant with time.
5) The measured cell voltage is proportional to the logarithm of the activity of the analyte ion. The voltage change is 59 mV for a change in the activity for a singly charged analyte ion at 25 °C; for doubly charged ions, it is 59/2 mV = 29.5 mV. Therefore, ISEs are inherently more sensitive to singly charged analyte ions than to ions with a higher charge, all other things being equal.
6) An ISE can be applied with different measurement approaches. The determination of activities and concentrations from measured open-circuit voltages (after calibration) can only be accurate if a truly selective membrane is employed *and* no time-dependent nonequilibrium processes (diffusion potentials at the external reference electrode) contribute to the cell voltage. In this case, one obtains the activity (i.e., an effective concentration) as measure for the composition. This approach is called direct potentiometry. The measurement of activity (instead of concentration) is sometimes desired (as in case of pH); however, more often the concentration of the analyte is sought. This can be enabled with appropriate calibration by measuring the open-circuit voltage in solutions of known analyte concentration. From the voltage-concentration data, a calibration plot is constructed. The activity coefficients depend on the concentration of *all* ions in the sample, that is, also on the ions of the matrix. Thus, this approach is only promising if the concentration of all ions is high and constant. Under those conditions, variation of the analyte ion concentration does not cause a change in the activity coefficient. This condition can be met by intentional addition of a high concentration of noninterfering ions (TISAB) to the sample before the measurement.
7) Titrations with indication of the end point by an ISE represent very accurate analytical methods (potentiometric titration). A stepwise change in the open-circuit voltage occurs right at the equivalence point (Figure 5.15). The concentration of the analyte is calculated from the *consumption* (V) of the standard solution until the point where the voltage jump occurs using the stoichiometric relationship of the titration reaction. Less well-controlled factors such as diffusion potentials, formation of mixed potentials, or changing activity coefficients may change the value of the recorded cell voltage, but they do *not* influence, at which added volume the reaction between analyte and standard solution is completed, and the point at which the potential jump occurs. Titration methods are absolute analytical methods (no calibration needed) and belong to the most accurate procedures because they can be traced to primary standards.

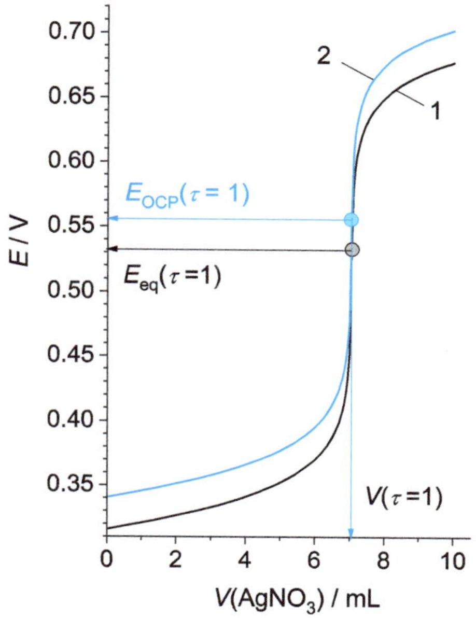

Figure 5.15 Precipitation titration of chloride ions in milk with 0.1 mol L^{-1} $AgNO_3$ solution and an Ag electrode as the indicator electrode. Curve 1 is the theoretically calculated equilibrium potentials based on the standard electrode potential of an Ag|Ag^+ electrode. Curve 2 is calculated taking into account the formation of a mixed potential at the silver electrode. Both curves exhibit a potential jump after the addition of the *same volume* of $AgNO_3$ solution. However, the particular *potential recordings differ* between Curves 1 and 2.

5.3 Dynamic Electrochemical Methods

Dynamic electrochemical methods are those that are conducted with a nonzero current. The electrochemical cells are thus electrolysis cells that are supplied with electrical energy by a potentiostat or galvanostat.

5.3.1 Potentiostats and Galvanostats

Potentiostats and galvanostats are electrical power sources that supply electrolysis cells with electrical energy in a *well-defined way*. A potentiostat keeps the potential of one of the two electrodes constant with respect to a reference electrode. A galvanostat forces a set current through the electrochemical cell and measures the voltage required to achieve the set current. Modern advanced laboratory devices allow switching between both operational modes. Many electrochemical methods not only use a fixed electrode potential or a fixed current, but also apply rather complex excitation signals to the electrochemical cell and follow its response. Modern instruments can be used without detailed knowledge of their inner setup and functionality. However, special measurements, the location and fixing of malfunctions, or the assembly of complex experiments require a basic understanding of how potentiostats and galvanostats work. Particular caution is recommended in the following situations:

- Measurement of very small currents (<10 nA) or very large currents (>1A)
- Very rapidly changing excitation signals
- Large solution resistance (e.g., organic electrolyte solutions)
- Parallel operation of several working electrodes in the same solution
- Investigation of surface-bound redox systems

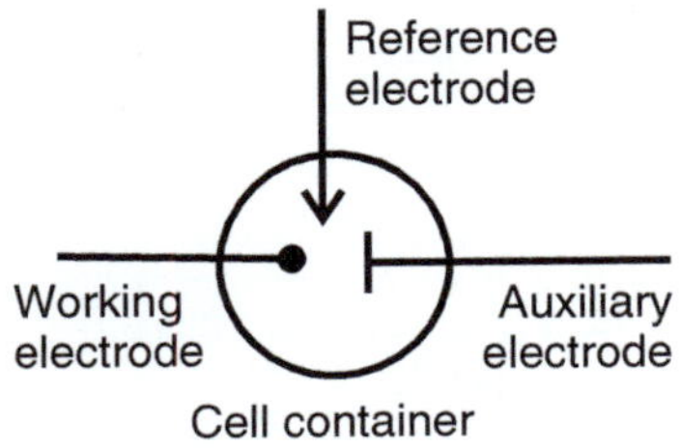

Figure 5.16 Symbolic depiction of an electrochemical three-electrode cell.

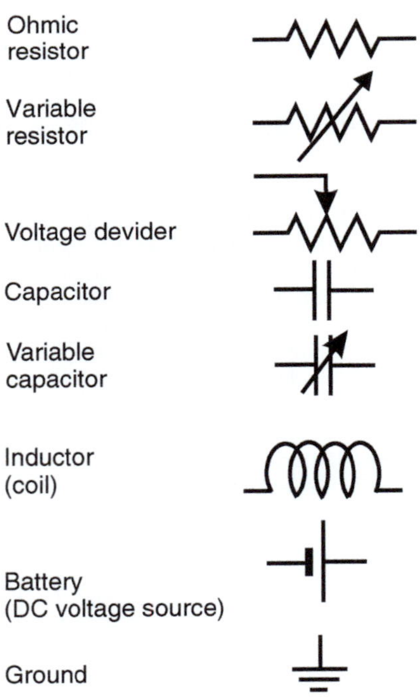

Figure 5.17 Symbols for elements of electrical circuits.

We discuss the basic electronic circuits with operational amplifiers (OAs) and their application to electrochemical instrumentation in Section 5.7. Here, we are concerned mainly with potentiostatic three-electrode experiments. The three-electrode setup in Figure 5.16 is often preferred because it provides clear information about the reactions at *one* of the two current-carrying electrodes without interfering influence of the second electrode. The potentiostatic working mode provides larger flexibility with respect to designing excitation signals. Keeping the current instead of the potential constant is instrumentally simpler and preferred for high current densities (preparative electrolysis) and for special problems in laboratory practice such as charging and discharging of batteries.

In a three-electrode configuration (Figure 5.16), the reaction to be studied occurs at the working electrode (WE). The potential is adjusted with respect to a reference electrode (Ref). The current loop is closed via an auxiliary electrode (Aux). This electrode is also called counter electrode (CE). However, it is advisable to reserve this term for the second electrode in a two-electrode setup. Symbols of further electrical elements used in this book are collected in Figure 5.17.

In order to understand the features of a potentiostatic experiment, we initially imagine a two-electrode setup of a working electrode and a counter (auxiliary) electrode. The externally supplied electrolysis voltage drops at different points along the current path: within the electrical cables between the electrodes and the voltage source, across the electrical double layers of the two electrodes, and within the electrolyte solution. A schematic of the potential along the current path is shown in Figure 5.18. It is important to realize that the voltage drop in the metallic cables is usually negligibly small (except for very high currents in large-scale technical electrolysis cells). Within the electrolyte, the potential drop depends on the current I and the solution resistance R_E between working electrode and counter electrode ($U_{drop} = IR_E$). For an electrolyte resistance of $R_E = 1\ \Omega$ and a current $I = 100\ \mu A$, U_{drop} reaches 0.1 mV only, which is insignificant compared to other factors like diffusion potentials (Figure 5.18a). This conclusion changes drastically when passing higher currents through the same cell: For $I = 100$ mA, U_{drop} amounts to 0.1 V, a value that is not acceptable for most measurements. In addition to the influence of U_{drop}, the major problem for the use of two-electrode cells for laboratory studies is the fact that we cannot easily tell how the remaining voltage drop is distributed between the working electrode and the counter electrode.

The potentiostat uses an electronic feedback circuit in order to apply the desired potential E_{set} to the working electrode independent of the potential required for the reaction at the auxiliary electrode (and therefore independent of the total cell voltage $U = E_{WE} - E_{Aux}$). For this purpose, the potentiostat must adjust the voltage drop between the working electrode and the electrolyte solution. This is accomplished by means of the reference electrode, which is the third electrode in the three-electrode setup. There is essentially no current passing through the reference electrode. The reference electrode records the momentary potential E_{WE} somewhere in the electrolyte solution between the working electrode and the auxiliary electrode. In the currentless state, E_{WE} is equal to the electrode potential (Section 2.1.2, Figure 5.18a). The desired potential difference E_{set} between working and reference electrode is provided to the potentiostat. The actually *measured* potential E_{WE} is continuously compared to E_{set}. If $E_{WE} < E_{set}$, the potentiostat increases the total cell voltage U between the working electrode and the auxiliary electrode until $E_{WE} = E_{set}$ is reached. The resulting current between the working electrode and the auxiliary electrode is measured as the response of the cell. In the case of $E_{WE} = E_{set}$, E_{set} equals the sum of the following voltage contributions:

- Galvani potential difference between the cable and the electrode material of the WE.
- Galvani potential difference between the WE and the electrolyte solution just outside of the electrical double layer.
- Potential drop in solution between WE and reference electrode determined by the overall current and the electrolyte resistance.
- Constant Galvani potential difference between the electrolyte solution and the inner solution of the reference electrode. (This contribution is constant because there is no current through the reference electrode.)

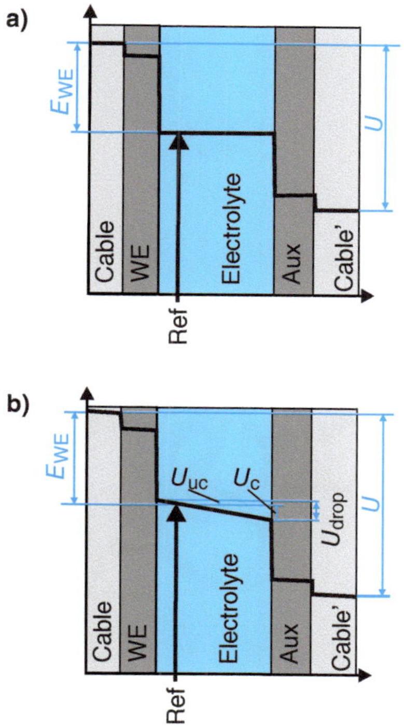

Figure 5.18 Potential–distance plot inside an electrochemical cell; a) at equilibrium and b) with current flow. The location of the reference electrode is symbolized by a black arrow.

The so-called uncompensated resistance (uncompensated resistance, R_{uc}) causes an uncompensated potential drop $U_{uc} = I\,R_{uc}$. $I\,R_{uc}$ is also called ohmic drop or *IR* drop. The electrode potential deviates from E_{set} by $I\,R_{uc}$. The potential drop through the remaining fraction of the solution resistance $U_c = I\,(R_E - R_{uc})$ and the Galvani potential difference at the auxiliary electrode $\Delta\varphi_{Aux}$ are compensated for completely (Figure 5.18b). This principle also works if E_{set} is changed over time. There are various excitation signals $E_{set} = f(t)$ that are purposefully applied for solving different scientific questions. The measured current, the response, is converted by the potentiostat to a voltage $U_{out} = kI$ that is proportional to the measured current. The voltage U_{out} can be directed to an x, y-recorder, an oscilloscope, an analog–digital converter, or any other recording unit (Section 5.7).

In many analytical experiments, the effect of the uncompensated potential drop $U_{uc} = IR_u$ is very small (see examples before the key icon). However, this does not apply to experiments in less conductive organic electrolyte solutions (Excursion 6-1) or for the use of large currents (at large electrodes or at high concentrations). The general guideline for mitigating the adverse effects of U_{uc} is the sequence 1) avoid, 2) compensate, and 3) correct. Avoiding a significant U_{uc} involves the selection of an appropriate electrolyte and a cell design with particular attention to the relative positions of Ref and WE (impacting R_E), and choosing an analyte concentration and electrode area (impacting I).

Modern potentiostats offer technical means for further compensation of U_{uc} that will be discussed in Section 5.7.2 and are used for nonsteady state voltammograms as discussed in Chapters 5 and 6.

Numerical correction methods are applied to the obtained data *after* the measurement. This is commonly used in steady-state current–voltage curves. These are curves in which the recorded current does not depend on what potential was applied before the particular reading. An example is the voltammetry with rotating disk electrodes for the characterization of electrocatalysts in Figure 12.25.

5.3.2 Electrodes for Voltammetry

The measurement cell for cyclic voltammetry experiments usually contains a working electrode, a reference electrode, and an auxiliary electrode. Inert electrodes are used for the working and auxiliary electrodes. These electrodes are made from materials that do not dissolve/react in the potential range used for the voltammetric experiment. Instead, they can provide or accept electrons by electron transfer across the solid–liquid interface. They can also serve as the substrate or support onto which other metals can be deposited/dissolved. The most important electrode materials are platinum, gold, mercury, silver, and glassy carbon (abbreviated below as GCE for glassy carbon electrode). Glassy carbon has a low porosity and is produced by heating aromatic polymers under exclusion of air. It consists of intertwined graphitic regions of a mean thickness L_c and lateral extension L_a (Figure 5.19). The intertwined graphite bands cause the unusual hardness of the material. After polishing, the surface of a GCE glistens similarly to darkly colored glass. More recently introduced electrode materials are boron-doped diamond (BDD) [25] and carbon nanomaterials such as carbon nanotubes (CNTs) and graphene (one or a few layers of sp^2-hybridized carbon atoms), as well as carbon black that can be processed with a binder and further functional agents for electrodes, especially for sensors [26].

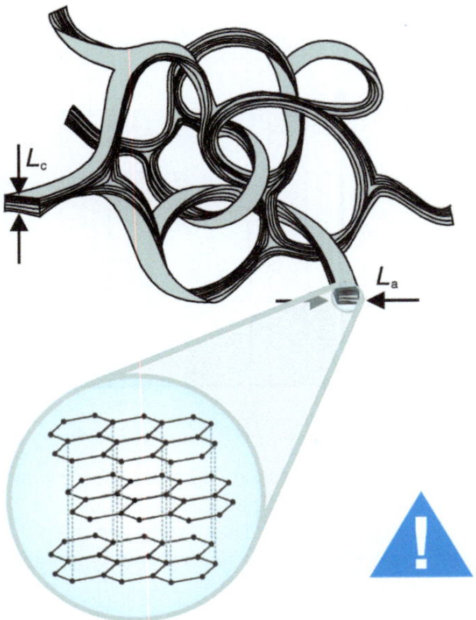

Figure 5.19 Structure of glassy carbon, schematic made by the author according to [24].

The active electrode material of the working electrode is often encased in an insulating electrode body (usually in the shape of a cylinder, see Figure 1.2). The insulating electrode body allows the electrolyte solution to contact only the pretreated end face (an inlaid disk). Solid electrode materials must be polished or otherwise pretreated (e.g., powder processing for screen-printed electrodes) in order to establish a reproducible initial state of the surface. This is a decisive prerequisite for quantitative work in electrochemistry! There are many different pretreatment methods for solid electrodes, which are selected according to the electrode material, but also taking into account the solution in which previous or subsequent work is conducted. A comprehensive survey of pretreatment methods for carbon electrodes is given in [27]. Original literature sources often provide detailed information about the source of the electrode material and any pretreatments. In Section 8.9.4, we will discuss a few methods especially suited for single-crystal electrodes. They are important for fundamental studies, but do not play a role in concentration determination treated in this chapter.

A mechanical polishing step is almost always part of the electrode pretreatment. A polishing agent Al_2O_3 powder (alumina) is used, which is a very hard material commercially available in different grain sizes. A spatula tip of the polishing agent is placed on a polishing felt and wetted with double-distilled water. This prevents dust formation (dusts of this grain size are always a health risk; therefore, never polish with dry powders!).

Figure 5.20 Polishing an analytical electrode with an aqueous suspension of alumina (Al_2O_3) on a polishing felt along an imagined trace of an "eight".

The end face of the electrode is immersed in the suspension and moved under slight pressure similar to writing the numeral 8 (Figure 5.20). The motion avoids polishing marks having a preferential direction on the finished surface of the electrode. Instead, the electrode is alternately moved "left–right" and "top–down" over the abrasive material. Tilting of the electrode should be avoided, which is aided by a simple holder (Figure 5.21). Avoiding canting is especially important for the preparation of test specimens, single-crystal electrodes, and electrodes without an insulating body. The polishing is usually conducted for about 2 min. Afterward, the polishing agent is removed by a gentle jet of doubly-distilled water and wiped with a lint-free paper tissue ("Kim-wipe"). Alternatively, the polishing agent can also be removed by immersion in a water-filled beaker inserted in a water-filled ultrasonic bath. Subsequently, the procedure is repeated with the next smaller grain size of the polishing agent. Common grain sizes are 5 μm, 1 μm, 0.3 μm, and 0.05 μm. If an electrode is used for the first time or after very strong contamination, all grain sizes are to be used in order to achieve a sufficient removal of contaminated materials and smoothing, while use of 0.3 μm and 0.05 μm is usually sufficient for regeneration of electrodes between two similar measurements.

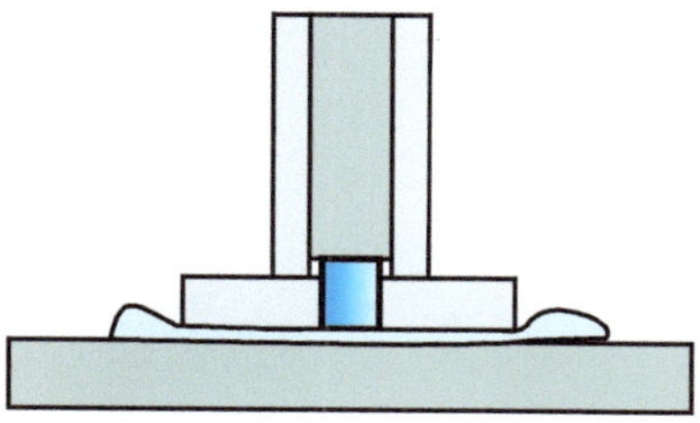

Figure 5.21 Cross-sectional view of an electrode holder that prevents tilting of the electrode during polishing. The electrode body is highlighted in blue.

In many cases, the mechanical polishing is followed by an electrochemical pre-treatment, also called electrochemical polishing, which we will discuss only for the frequently used electrode material Pt. The procedure is conducted in chloride-free, more exactly halogenide-free, solution because halogenide ions specifically adsorb on Pt surfaces and etch Pt surfaces. Such undesirable ions may be released from commonly used reference electrodes, for example, Ag|AgCl|KCl(*sa*). Thus, the use of a chloride-free reference electrode is recommended here. A suitable electrolyte is 0.5 mol L^{-1} H_2SO_4. A continuous sawtooth voltage profile is applied to the cell. The switching potentials of the voltage profile are selected in such a way that the negative limit is in the region of the commencing hydrogen evolution and the positive limit is at the foot of the signal from the oxygen evolution reaction. Under these conditions, the typical cyclic voltammogram of a Pt electrode is observed (Figure 5.22). Below 0.25 V vs. RHE, the adsorption and desorption of hydrogen on Pt is seen, while the formation of a thin oxide/hydroxide film occurs at potentials positive of 0.7 V vs. RHE. The rise of cathodic current at 0.0 V vs. RHE and of anodic current at 1.45 V vs. RHE is caused by the onset of H_2 and O_2 evolution reactions, respectively. The charge represented by the H-adsorption and H-desorption peaks is indicative of free adsorption sites. The more free adsorption sites are available per surface area, the cleaner the electrode surface is. The combined H-adsorption and H-desorption peaks of a clean Pt electrode represent a charge of 210 × 10^{-6} A s cm^{-2} (or 210 μC cm^{-2}, [28]). A larger charge is caused by roughening of the electrode during operation, especially when surface oxides are formed and reduced, as shown in Figure 5.22.

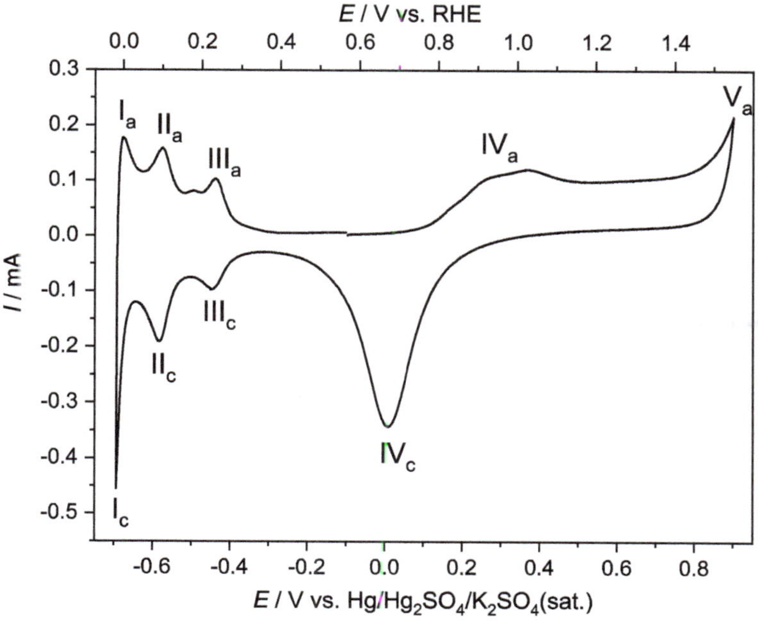

Figure 5.22 Cyclic voltammogram of a Pt electrode in 0.5 mol L^{-1} H_2SO_4. Peak assignment:
I_c/I_a – H_2 evolution/H_2 oxidation;
II_c/II_a – H adsorption/H desorption at terraces with predominantly (110) structure;
III_c/III_a – H adsorption/H desorption at steps of (111) planes and at (100) terraces in the vicinity of steps;
IV_a – surface oxidation, formation of Pt–OH and PtO;
IV_c – reduction of surface oxides;
V_a – O_2 evolution.

Roughening can also be caused by mechanical polishing or special pretreatments such as platinizing (electrochemical deposition of sponge-like Pt structures on a flat Pt electrode). The exact assignment of peaks II–IV is complicated and requires comparison to corresponding cyclic voltammograms of single-crystal electrodes and the consideration of stepped surfaces [29] (Sections 8.9, 12.4.1).

Among the working electrode materials, mercury has exceptional properties. This metal is liquid at room temperature but is immiscible with water and organic solvents. Thus, a new and perfectly smooth surface can be created simply by the formation of a new drop that forms the interface to the electrolyte solution. This constitutes a magnificent advantage in comparison to the partially complicated and time-consuming pretreatment procedures for solid electrodes. Furthermore, Hg has a very high overvoltage for the hydrogen evolution reaction. This means that the reaction from H^+ to H_2 proceeds unrecognizably slowly at potentials at which this reaction is expected to occur from a thermodynamic point of view (Figure 3.9). This offers the opportunity to reduce metal cations with very negative standard redox potentials at Hg, without interference from overlapping H^+ reduction. After reduction of the metal cations to atoms, most metals form amalgams with metallic Hg. Amalgams are alloys of mercury. At high Hg concentrations, amalgams are liquid themselves. For some metals, the potentials for the formation of amalgams are significantly less negative than for the metal deposition onto itself [cf. Eq. (5.41) and (5.42)].

$$K^+ + e^- \rightleftharpoons K(s, \text{on}\,K) \quad E^\circ = -2.94\ \text{V} \tag{5.41}$$

$$K^+ + e^- + Hg \rightleftharpoons K(Hg) \quad E^\circ = -1.98\ \text{V} \tag{5.42}$$

Therefore, very electronegative alkali metal ions can be reduced at Hg electrodes in neutral and alkaline solutions without significant disturbance by H_2 evolution. Hg electrodes are especially suited for the investigation of reductions (Excursion 5-2). They are much less useful for the study of oxidation reactions because they dissolve electrolytically at moderately positive potentials. The useful potential range is further restricted by the presence of complexing ligands (e.g., Cl^-) that make mercury easier to oxidize. Metals can be enriched in a mercury drop by amalgam formation over a prolonged electrolysis time and then oxidized back to metal ions in a short time. This preconcentration step is exploited in stripping methods (Section 5.3.10). The amalgam formation also greatly reduces problems with nucleation phenomena.

These significant advantages of mercury as an electrode material have to be weighed against the high toxicity of metallic mercury and all its compounds as a severe disadvantage for routine work [30]. The toxicity increases from metallic mercury to mercury cations and metalorganic mercury compounds. The latter not only have an extremely high toxicity, but they also enrich along the food chain. Therefore, the use of mercury and all its compounds is restricted or even completely prohibited. The regulatory measures are not uniform in different countries. As an example, the use of Hg as an electrode material is completely prohibited in Sweden, whereas in Germany it is still permissible and considered unavoidable for certain tasks and safe if conducted properly [31]. The substitution of Hg electrodes for analytical procedures is an active field of research and has brought positive results for specific applications only, for example, the use of Bi electrodes [32]. In any case, the use of Hg should be limited as much as possible. Used mercury metal must be recycled via the supplier of the mercury electrode material.

Depending on the electrode material and the electrolyte solution, different potential ranges are applicable in which neither the electrode material nor the electrolyte components are electrolyzed (Figure 5.23).

Electrodes of the 2nd kind are preferred as reference electrodes for voltammetric experiments because of their simple handling (Table 2.6). Criteria for selection are the avoidance of contamination of the sample by constituents of the inner solution of

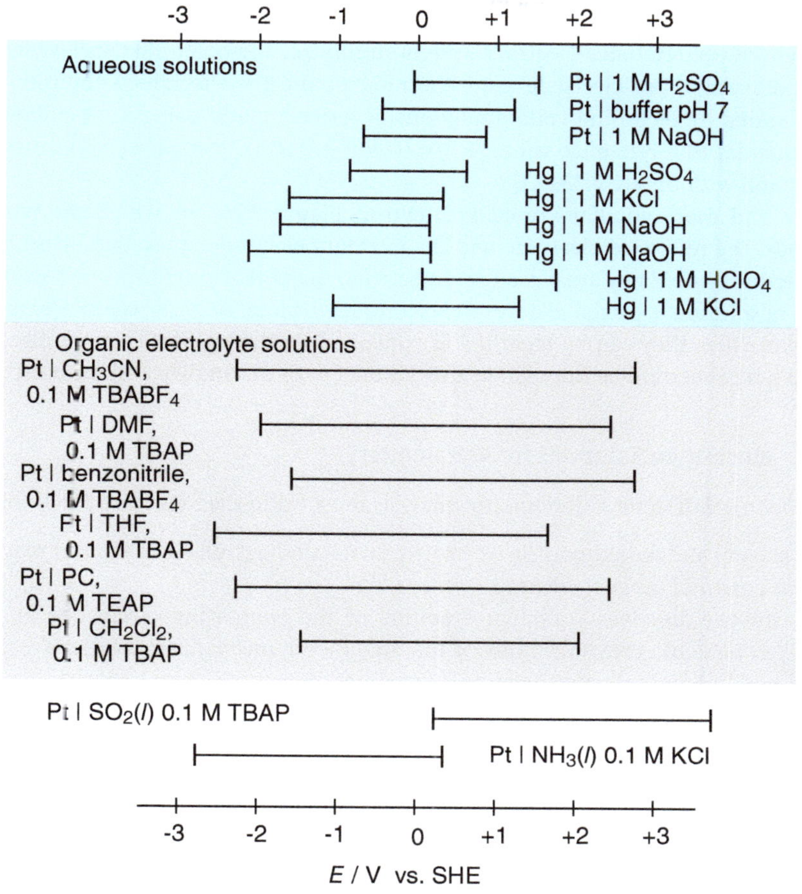

Figure 5.23 Usable potential ranges of important analytical electrodes in typical solvents. Values were taken from [1] and converted to SHE. TBABF is tetra-*n*-butylammonium-tetrafluoroborate; TBAP is tetra-*n*-butylammonium-perchlorate; TEAP is tetraethylammonium-perchlorate.

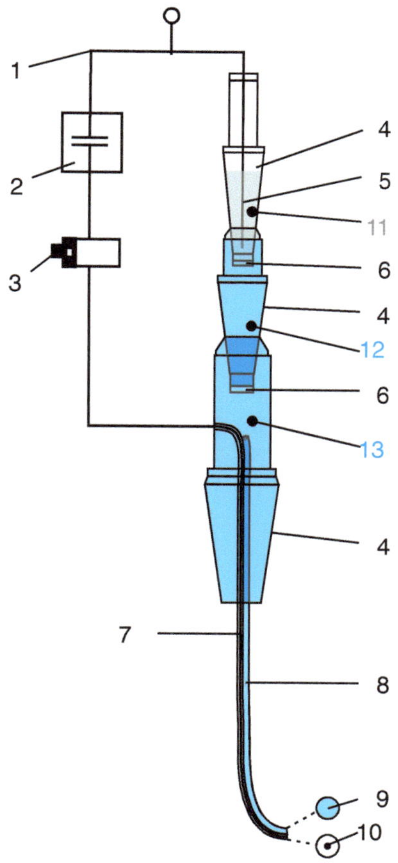

Figure 5.24 Reference electrode with parallel running capacitively coupled Pt wire for work in organic electrolyte solutions, after [42].

1) Cu wire;
2) Capacitor (0.01 μF);
3) Clamp;
4) Ground glass joint;
5) Ag wire;
6) Glass frit;
7) Pt wire sealed in glass;
8) Haber–Luggin capillary;[56,57]
9) Terminal of the Luggin capillary (cross section);
10) Terminal of the Pt wire (cross section); exemplary solution compositions for use in 0.1 mol L^{-1} tetrabutylammonium hexafluorophosphate ($TBAPF_6$) in dimethylformamide (DMF);
11) 0.01 M $AgClO_4$;
12) 0.1 M $TBAPF_6$ in DMF;
13) 0.1 M $TBAPF_6$ in DMF.

the reference electrodes. In this view, chloride-free reference electrodes are important. Fillings with NaCl (instead of KCl) are used to avoid the precipitation of $KClO_4$ during work in perchlorate-containing samples. While the selection of reference electrodes is not difficult for aqueous samples, further considerations are necessary for organic electrolyte solutions. Long-term stable reference electrodes are not available for those systems that facilitate high potential scan rates without inflicting significant noise on the recorded voltammetric curves. The reference electrode shown in Figure 5.24 has

[56] Fritz Haber, 1868–1934, German-Jewish chemist, Nobel Price 1912 for contribution to the ammonia synthesis, played a major role in the conception of gas as a chemical warfare agent during World War I, important contribution such as the Born–Haber cyclic process for the determination of crystal energies, worked on pH glass electrodes, quinone/hydroquinone electrode, founding director of the Kaiser Wilhelm Institute of Physical Chemistry and Electrochemistry in Berlin, today the Fritz Haber Institute of the Max Planck Society. In 1933, Fritz Haber resigned from his director position and emigrated to Cambridge, UK, in response to the take-over by the Nazi party in Germany and the subsequent dismissal of Jewish scientists from public institutions. He passed away in Basel, Switzerland, on the journey to Rehovot (today in Israel), where he was about to accept the position of the founding head of a research center, the Weizman Institute. A contemporary survey of the controversial reflection on Haber can be found in B. Friedrich's Fritz Haber at one hundred fifty: Evolving views of and on a German Jewish Patriot. *Bunsenmaganzin* **2019**, *21*, 129–143.

[57] Hans Luggin, 1863–1899, Austrian physicist, the setup of the reference electrode was jointly developed by Luggin and Haber during their common time in Karlsruhe, Germany, where they prepared their habilitations.

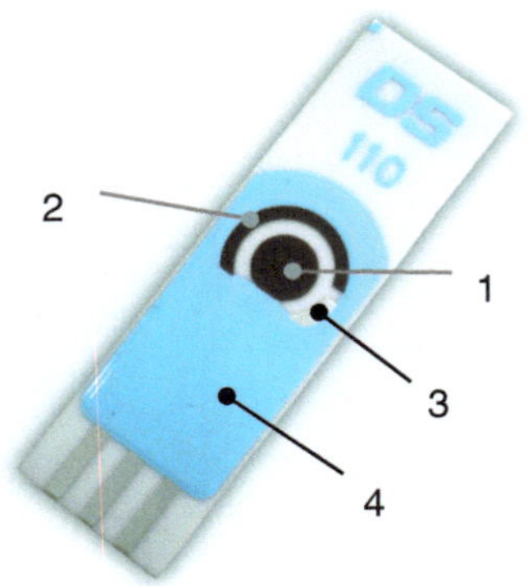

Figure 5.25 Example of an electrochemical cell made by screen printing:

1) WE,
2) Aux,
3) Ref,
4) Cover;

Source: Reproduced with permission from Dropsense; Llanero, Spain.

proven useful in those situations. A capacitively coupled Pt wire parallel to the current path of the reference electrode system minimizes the potential oscillations. The exact potential of this electrode is not thermodynamically determined and thus must be calibrated by recording voltammograms of ferrocene and using the obtained formal potential as a reference value [$E^{\circ\prime}$ = (0.400 ± 0.005) V vs. SHE] as discussed in connection with Figure 2.23 [43].

More and more popular are cells shown in Figure 5.25, for which the working electrode, the reference electrode and the auxiliary electrode are screen-printed on a ceramic substrate. They are directly connected to the potentiostat by a connector, and a drop of solution is placed to cover the electrode surfaces. As these electrodes are not very expensive, they can be regarded as consumables/disposables that are discarded should a regeneration be impossible after contact with the analyte and its matrix.

5.3.3 Electrolyte Solutions for Voltammetry

Electrolyte solution for voltammetric analysis must fulfill the following conditions:

- The solvent and the supporting electrolyte do not undergo electron transfer reactions in the potential range to be used.
- The solvent dissolves sufficient amounts of the supporting electrolyte and the analyte. Sensible concentrations of the analyte for mechanistic studies are about 10^{-3} mol L^{-1} and for concentration determinations, usually much lower.
- The supporting electrolyte ensures the conductivity of the liquid phase. The supporting electrolyte exhibits such a high ionic conductivity that a charged analyte is transported to the electrode essentially by diffusion only (and not by migration). This is usually the case if the contribution of the supporting electrolyte to the ionic conductivity is 50 to 100 times larger than that of an analyte ion.

Naturally, there is a large variety of solutions that meet these conditions. However, a clever choice of a specific solution can greatly facilitate a voltammetric experiment (or cause the opposite effect). Let us consider some additional requirements for a typical application. For the trace determination of heavy metals, it is of particular importance to keep the blank value of the analyte low by excluding sources of contamination. Double-distilled water or water from an ion-exchange system is employed, ideally followed by UV irradiation to destroy (mineralize) organic compounds that could adsorb to the electrode surface and hinder (or inhibit) the electrode reaction. Instead of an electrolyte salt, suprapure HNO_3 is used. Protons of the acid provide a particularly large contribution to the ionic conductivity. Therefore, lower amounts of the acid must be added to reach the desired conductivity. This also means potentially lower amounts of trace impurities that are added to the system together with the electrolyte. By using an acidic solution, the deposition of metal ions as hydroxides or their adsorption to the walls of the cell container is prevented.

Q: What is a blank value?
A: The blank value is the concentration of the analyte that is introduced to the sample by added reagents and contact with sample vials between the sampling steps and the actual measurement. That means this concentration of the analyte was *not* in the sample before sampling.

Solutions for bioelectrochemical studies must be buffered in order to maintain a constant pH despite electrode processes that consume or release protons. Often, a buffer named PBS is used. This abbreviation stands for phosphate-buffered saline and is a phosphate buffer of pH 7.4 containing the physiological concentration of

NaCl [8.8 mmol L^{-1} (1.556 g L^{-1}) $Na_2HPO_4 \times 2\ H_2O$ + 1.2 mmol L^{-1} (0.163 g L^{-1}) KH_2PO_4 + 150 mmol L^{-1} (8.775 g L^{-1}) NaCl]. A buffer with a physiological pH is very good nutrient solution for microorganisms that disturb subsequent electrochemical measurement. Therefore, nonsterile storage of buffers is possible only for a very short time, especially with proteins or other carbon sources in the buffer. The durability can be prolonged by addition of 0.3 mmol L^{-1} (0.2 g L^{-1}) sodium azide NaN_3.

Completely dissolve the salt in a small volume of double-distilled H_2O, then fill up to one liter, perhaps adjusting the pH with 0.1 mol L^{-1} HCl or 0.1 mol L^{-1} NaOH.

Ready-to-use solutions in ampules are available commercially.

Q: Why does NaN_3 prevents the growth of microorganisms?

A: The N_3^- ion is a pseudo-halogenide and, similarly to CN^-, it permanently blocks the oxygen binding site of the enzyme cytochrome *c* oxidase in the respiratory chain of microorganisms. Such solutions must therefore be disposed of as pollutants.

For the mechanistic study of organic and element-organic compounds, organic solvents are often employed. Otherwise, sensitive compounds/intermediates would directly react with trace water. The solvents must be purified by special procedures because traces of water are usually contained even in highly pure commercial solvents. Traces of water and other trace impurities may significantly limit the usable potential range or react with electrochemically formed reaction intermediates. The commercial purity "p.a." (lat. *per analysi*, analytical grade) is often insufficient for voltammetric experiments. Organic solvents can only dissolve hydrophobic salts. Therefore, commonly used electrolyte salts are tetraalkyl-ammonium perchlorates (beware of perchlorate explosion hazards!!), hexafluorophosphates, and tetrafluoroborates. More detailed recommendations can be found in [33–37] for the selection and purification of solvents and electrolyte salts.

5.3.4 Cell Constructions for Voltammetry

Special cells are used for voltammetry that accommodate three electrodes, enable deoxygenation when necessary, and prevent sample contamination. Vials for electrochemical cells should be inert with respect to the solvent. As in many other areas of chemistry, glass is the material of choice for most of the reaction mixtures. There are a few exceptions, though. For extreme trace analysis, vials from special polymers are better because they do not adsorb material onto the surface and do not contain heavy metals. Alkaline solution can cause glass corrosion, and the dissolved silicates may adsorb on the electrode surfaces. Alternatives for such cases are containers of fluorinates polymers [poly-(tetrafluoroethylene), (Teflon™); poly(chlorotrifluoroethylene) (until 1995 Kel-F™, Neoflon™, Aclar™)], or poly(ether etherketone) (PEEK).

For simple experiments, the container is a beaker holding 10–50 mL of solution with a detachable lid. The lid contains openings for the insertion of the three electrodes and allows rapid solution exchange and cleaning of the cell. A stream of inert gas can be passed into the cell for deaerating the solution (Figure 5.31 in Excursion 5-2). The purging gas must leave the cell via an exhaust path. Inlet and outlet can be combined in one glass device. The inert gas can be highly purified nitrogen or argon. Ar is heavier than air and thus forms a blanket over the solution–gas interface, helping to limit the contact to atmospheric O_2.

Q: How long must the gas be bubbled into the cell?

A: The O_2 concentration in solution is determined by Henry's law[55], an equilibrium relationship. If the oxygen-containing atmosphere above the solution is depleted of O_2, then O_2 dissolved in solution is released into the gas atmosphere to reestablish the gas dissolution equilibrium.

The continuous gas stream out of the cell transports O_2 away from the solution–gas interface, thus causing more release of O_2 from the solution. Bubbling the inert gas into the solution promotes a convective circulation of the solution, which accelerates the gas exchange by bringing O_2-rich volume elements of the solution to the solution–gas interface. It takes about 3–5 min for a gas exchange in an aqueous solution with not too strict requirements. Using a glass frit that disperses the gas as tiny bubbles greatly increases the gas–liquid interfacial area and speeds up the deoxygenation process. Organic electrolyte solutions can dissolve much more O_2 than aqueous solutions. Thus, it usually takes longer to deaerate these solutions (15–30 min).

In order to prevent evaporation of volatile solvents into the gas stream, the gas stream can be saturated with solvent vapor in a small gas wash bottle before entering the cell. The gas wash bottle is filled with the same electrolyte solution as used in the cell.

The variation in temperature can be important for mechanistic investigations. In such cases, double-walled or "jacketed" containers are used that have an inlet and an outlet through which a cooling/heating liquid can circulate between the cell and a cryostat/thermostat. Investigations in organic solvents require strict shielding against moisture and O_2 from the air. This can be achieved by custom glass apparatuses [38–41]. The cell may contain a built-in adsorption column, over which the electrolyte solution is passed once before the compound under study is added. In case of extreme requirements, one should keep in mind that polymer tubing is not completely impermeable to O_2. Therefore, small but detectable amounts of O_2 can enter a gas stream between the high-pressure gas cylinder and the electrochemical cell. Thin metal tubing and gas adsorption cartridges installed directly before the gas stream enters the cell can be a solution in such situations. Under some circumstances, working in a glovebox is desirable. For experiments with newly synthesized compounds, often available only in minute amounts, it is advantageous to reduce the solution volume. Pear-shaped flasks can be used for that purpose, which are equipped with several conical ground glass joints for insertion of the electrodes and the gas supply.

Q: What does "small amount" mean in this context?
A: Please calculate the required mass of a compound under study for 5 mL of a 1 mmol L^{-1} solution if the analyte has an assumed molar mass of 230 g mol^{-1}.

5.3.5 Cyclic Voltammetry

Cyclic voltammetry (CV) is conducted in a three-electrode cell with a triangular voltage ramp as the excitation signal $E(t)$ (Figure 5.26a). The initial potential is typically selected such that the analyte is stable. The potential is changed with the (potential) scan rate v

$$v = \frac{dE}{dt} \tag{5.43}$$

to the switching potential E_λ (or vertex potential), where the scan direction is reversed and the potential is scanned back to the initial potential. The switching potential is typically selected in a potential range in which the analyte is converted under diffusion-controlled condition (see below) and the solvent and electrolyte ions are not electrolyzed. More complex potential programs are possible and may involve multiple potential cycles or a rest time at specific potentials. The scan rate is a very important

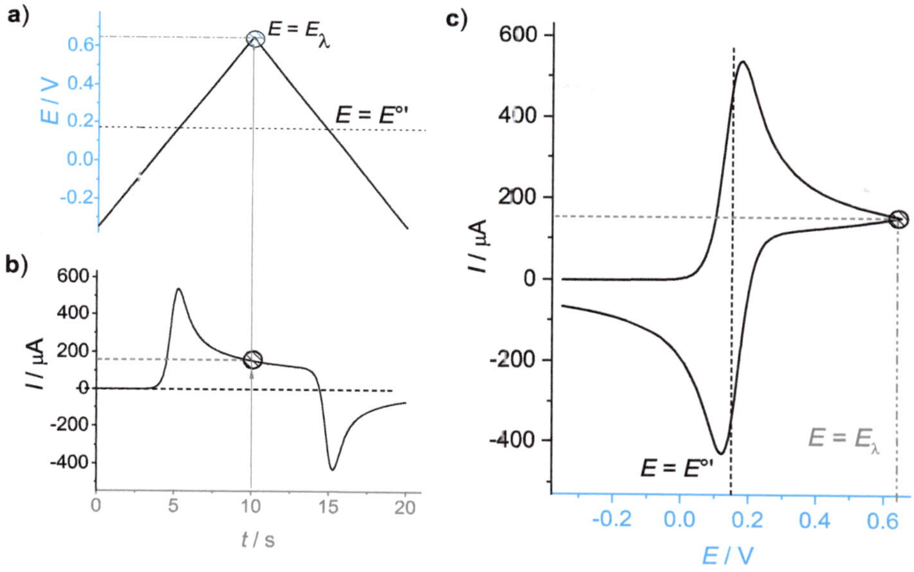

Figure 5.26 a) Excitation signal; b) response signal; and c) typical plot in cyclic voltammetry (cyclic voltammogram). The construction of the cyclic voltammogral(E) from $E(t)$ and $I(t)$ is exemplified for the switching potential E_λ.

parameter in CV because it adjusts the timescale within which reaction steps must cause a measurable change in the response in order to be detected.

As a response, the current $I(t)$ is recorded (Figure 5.26b). Usually, the data are plotted as $I = I(\mathrm{E})$, which leads to the typical "duck" shape of cyclic voltammograms in Figure 5.26c. According to the IUPAC convention, positive potentials are plotted to the right, and anodic currents (oxidation) upward (pay attention to the zero point of the current axis!). This convention is not uniformly applied in the scientific literature and is often not even within one journal. The reader must reassure herself/himself in each case, which convention the authors have applied (Figure 1.9).

The opposite convention is also widespread, especially in the USA. In this convention, negative potentials are plotted toward the right and cathodic currents (reductions) upward (Figure 1.9). This was common practice in polarography, in which mercury electrodes were used primarily for the study of reduction reactions (of metal ions). The resulting plots were graphs that increased in reading direction (Excursion 5-2).

We will use the example in Figure 5.26 to understand how the characteristic shape of a voltammogram evolves, because CV is an important technique, and at the same time, it is very instructive for understanding the interplay of interfacial kinetics and mass transport. The voltammogram in Figure 5.26 was recorded for an oxidation reaction. In order to make the evolution of this plot more evident, Figure 5.27 shows a sequence of "snapshots" at different times t during the potential scan. For each snapshot, the right column holds a plot of the local concentration of the reduced redox form R (reactant) and the oxidized redox form O (product) versus the distance x to the electrode surface. These are so-called concentration–distance profiles $c_\mathrm{R}(x)$ and $c_\mathrm{O}(x)$. O and R are used here as generic symbols for the oxidized and reduced forms of *one* redox couple. The working solution initially contains about 1 mmol L^{-1} of R and 100 mmol L^{-1} of the supporting electrolyte, which provides the conductivity to the working solution and reduces the ohmic drop to a negligible value. Because the supporting electrolyte is present in large excess compared to R, R itself is transported to the electrode surface mainly by diffusion (and not by migration).

The formal potential of the redox couple R/O in Figure 5.27 is $E^{\circ\prime} = +0.15$ V. The potential scan is initiated at E_initial, which is negative of $E^{\circ\prime}$. The rate of oxidation of R to O is negligible at this potential because the thermodynamic driving force for this transformation is not sufficient. Therefore, there is almost no faradaic current ($I \approx 0$, Figure 5.27a). R is available everywhere in the cell at the initial concentration $c = 1$ mmol L^{-1}. The product O has not been formed yet, and therefore, the blue curve

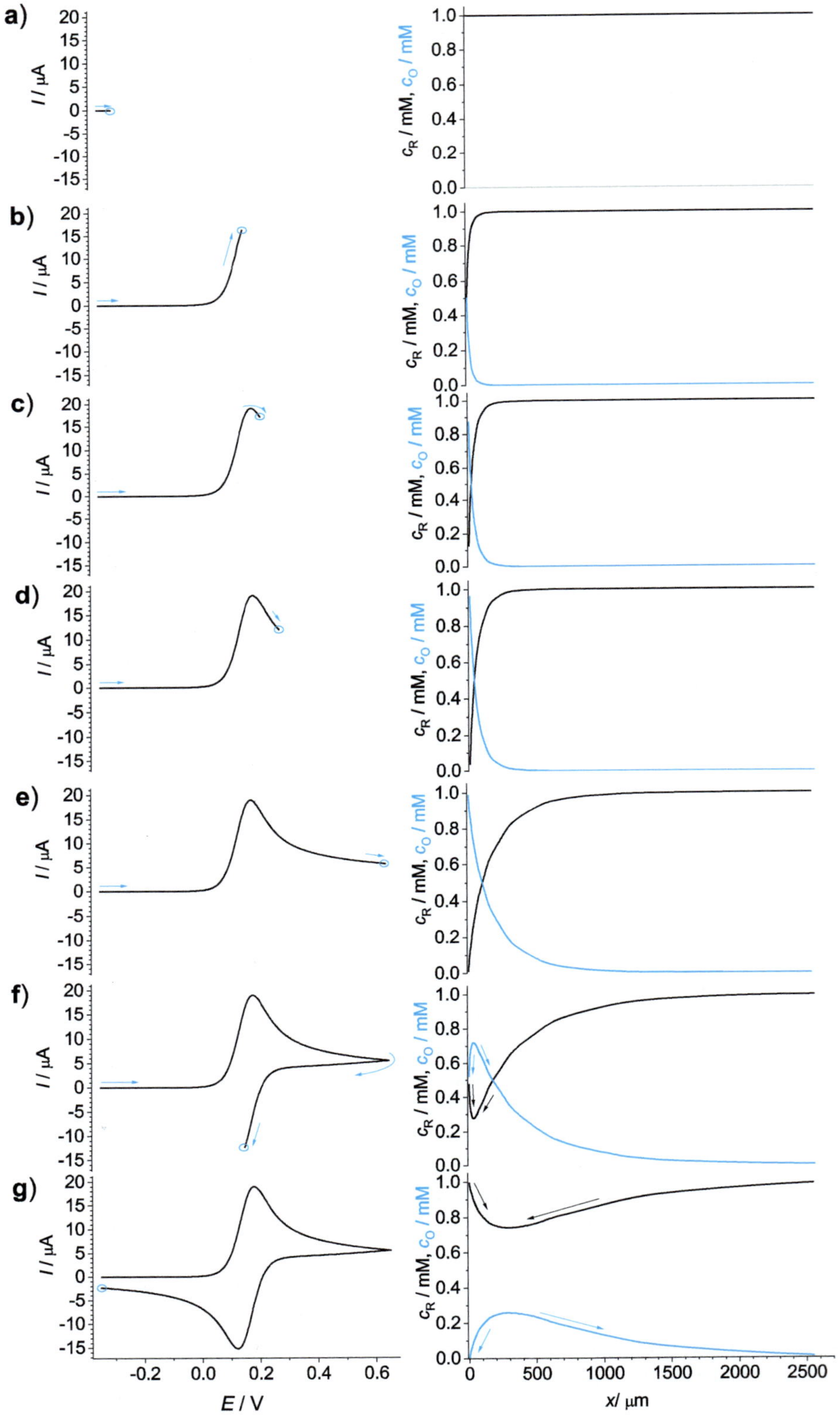

Figure 5.27 Relationship between the shape of a cyclic voltammogram and the local concentrations of R and O in the vicinity of the electrode surface. Calculated curves for $E^{\circ\prime} = 0.15$ V, $c_R^* = 1$ mmol L^{-1}, $D_R = D_O = 1 \times 10^{-5}$ cm^2 s^{-1}, $v = 0.1$ V s^{-1} and $A = 0.071$ cm^2. (a) to (g) Development of the voltammetric curve and momentary concentration as function of the distance to the electrode surface at the potentials E / V of a) −0.350; b) 0.150; c) 0.209; d) 0.268; e) 0.650; f) 0.150; and g) −0.350 during the potential scan.

for its concentration superimposes with the abscissa. The potential is increased and approaches $E^{\circ\prime}$. At the electrode, the concentration ratio of O and R is governed by the Nernst equation:

$$\frac{c_O(x=0)}{c_R(x=0)} = e^{\left[\frac{nF}{RT}(E(t) - E^{\circ\prime})\right]} \tag{5.44}$$

We call a redox system a **reversible redox system** or a **nernstian system** if the establishment of the concentration ratio according to Eq. (5.44) takes place instantaneously. This is the case if the charge transfer reaction is *fast*, and there is *no chemical following reaction* consuming the reaction product.

We will discuss the different voltammetric techniques first using the example of a nernstian system. In Chapter 6, we will consider consecutively more and more complications associated with charge transfer kinetics at the electrode surface or with following chemical reactions.

If the applied potential is $E(t) = E^{\circ\prime}$, the ratio $c_O/c_R = 1$ at $x = 0$, and due to the stoichiometric relationship, $c_O = c_R = 0.5\ \text{mmol L}^{-1}$ (Figure 5.27b). Now, the concentration at the electrode surface $c_R(x = 0)$ deviates from the concentration c_R^* in the solution bulk. A concentration gradient is formed that drives the diffusion of R from the solution bulk to the electrode surface. The steeper the concentration gradient, the more molecules of R per unit time diffuse to the electrode, and the larger the faradaic current will be. The larger the difference $E(t) - E^{\circ\prime}$ is, the larger the concentration ratio c_O/c_R will be. Table 5.5 illustrates the idea. In this simplified example, we require that the sum of c_R and c_O be equal to the initial concentration c_R^* at any arbitrary position in the solution. Actually, this is only fulfilled if $D_R = D_O$. The case of $D_R \neq D_O$ affects concentration gradients and, therefore, the shape of the voltammogram.

The concentration $c_R(x = 0)$ changes very rapidly in the potential range around $E^{\circ\prime}$, resulting in a steep concentration gradient and a correspondingly high current. Upon further increase in $E(t)$ well beyond $E^{\circ\prime}$, the ratio (quotient) $c_O(x = 0)/c_R(x = 0)$ increases further, but the concentration *difference* from the solution bulk changes substantially only around the formal potential but only marginally for $E(t) \gg E^{\circ\prime}$ (Table 5.5). If the concentration difference remains roughly constant, we expect a decrease in current with time because the diffusion layer expands (Figure 5.27c–e), as we have discussed before with the example of chronoamperometry (Figure 4.52).

The potential scan direction (direction of change in the applied potential) is inverted at the switching potential E_λ (Figure 5.27e). The oxidation current continues to decay steadily, which is particularly visible in Figure 5.25f. As $E(t)$ approaches $E^{\circ\prime}$, the reduction commences. The oxidation products O formed in the first half cycle are reduced

Table 5.5 Concentration ratio for a redox couple with $n = 1$ at the electrode surface while recording a cyclic voltammogram of a reversible redox system at 298 K, $D_R = D_O$. R is initially present in the bulk solution.

Potential $E(t) - E^{\circ\prime}$/V	Concentration ratio at the electrode surface $c_O(x=0)/c_R(x=0)$	Concentration at the electrode surface $c_R(x=0)$	Concentration difference from the solution bulk $c_R^* - c_R(x=0)$
−0.118	1:100	0.99 c_R^*	0.01 c_R^*
−0.059	1:10	0.91 c_R^*	0.09 c_R^*
0	1:1	0.50 c_R^*	0.50 c_R^*
+0.059	10:1	0.09 c_R^*	0.91 c_R^*
+0.118	100:1	0.01 c_R^*	0.99 c_R^*

back to R. O is available only in the vicinity of the electrode and is shown as blue curves in the concentration profiles in Figure 5.27. The evolution of the cathodic current peak I_{pc} at the peak potential E_{pc} can be explained in an analogous way to that for the oxidation peak current. The concentration profiles in the reverse potential scan show a maximum (O) or a minimum (R). The arrows indicate the directions in which O and R are transported by diffusion. At the end of the potential scan, the initial concentrations are *not* completely restored. Therefore, there is still a sizable cathodic current at the end of the potential cycle. In this situation, we can imagine the diffusion layer in front of the electrode as a "concentration scar" comparable to a "memory" linked to the just completed experiment. The "concentration scar" has to be reequilibrated with the bulk solution concentration by stirring (short stirring with a magnetic stir bar, bubble agitation with the stream of inert gas, shaking of the cell container) before the recording of another voltammogram can be started.

Q: You mentioned the need for stirring before recording the next voltammogram. However, a significantly larger charge is converted for the oxidation than for the reduction in Figure 5.26b. Shouldn't one discard the solution each time a cyclic voltammogram has been recorded because the concentration c_R must be smaller than the initial concentration of R once a voltammogram has been recorded?

A: Indeed, c_R decreases, but the decrease after equilibration of the "concentration scar" is so small that it is almost impossible to measure it. The following calculation can illustrate this. Let us assume the electrochemical cell contains 10 mL of solution with 0.001 mol L^{-1} $[Fe(CN)_6]^{4-}$. An electrolysis is conducted for 2 min with 100 μA ($[Fe(CN)_6]^{4-} \rightarrow [Fe(CN)_6]^{3-} + e^-$). What fraction of the initially contained $[Fe(CN)_6]^{4-}$ ions have been converted?

Unknown: $\Delta N/N_0$

Given:

$$c_0 = 0.001 \text{ mol L}^{-1}$$
$$V = 10 \text{ mL} = 0.01 \text{ L}$$
$$I = 100 \text{ μA}$$
$$t = 2 \text{ min} = 120 \text{ s}$$
$$n = 1$$
$$Q = nF\Delta N = I\,t$$
$$N_0 = c_0\,V$$

$$\frac{\Delta N}{N_0} = \frac{It}{nFc_0V}$$
$$= \frac{100\times10^{-6}\text{ A}\cdot120\text{ s}}{1\cdot96485\text{ A s mol}^{-1}\cdot10^{-3}\text{ mol L}^{-1}\cdot0.01\text{ L}}$$
$$= 0.0124 = 1.24\%$$

We can approximate the shape of the voltammetric curve in the first half cycle between $E_{initial}$ and the switching potential E_λ with a combination of a Butler–Volmer curve for $E < E^{\circ\prime}$ and a chronoamperogram for $E > E^{\circ\prime}$ (Figure 5.28). In the transition between the two curves, a maximum occurs, the anodic peak current I_{pa}. Please note that the current after the peak does *not* decay to the baseline. This causes the specific shape that is reminiscent of a duck's bill.

The exact shape of the function $I(E)$ of a cyclic voltammogram cannot be given as an analytical function, but can be calculated numerically (Section 6.3.1), by calculating the current pointwise from Fick's laws and from the initial and boundary conditions.

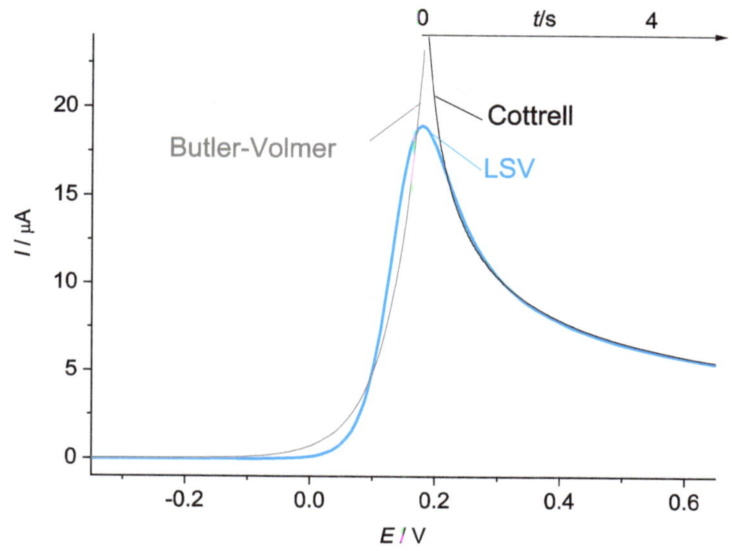

Figure 5.28 Emergence of the curve shape of the first half cycle in a cyclic voltammogram (i.e., linear sweep voltammetry) with $v = 0.1\ \text{V s}^{-1}$ with superposition of a Butler–Volmer curve [Eq. (3.44) calculated with $c_R(t = 0) = c^*$ and $c_O(t = 0) = 0$] and a Cottrell curve [Eq. (4.130)]. The smaller of these two theoretical contributions determines the experimentally observed net current.

In the case of a simple cyclic voltammogram, the boundary conditions would be a statement about the concentrations of R and O at the electrode surface according to Eq. (5.44) and the undisturbed initial concentrations in the bulk phase, $c_R = c_R^*$ and $c_O = 0$ (i.e , the concentrations at quasi-infinite distance to the electrode surface). Computer programs for the stepwise calculation of voltammetric curves (digital simulation) are often provided as part of commercial computer-controlled potentiostats and are also separately available commercially or as open-source application (Excursion 6-2). For the purpose of a laboratory course, even spreadsheet applications [44] can be used to explore the influence of different parameters on such curves. Data in Figures 5.26 and 5.27 have been calculated in such a way. Even without complete comprehension of the underlying numerical principles, "playing" with such programs is of great benefit for understanding the signal–response relationship, especially in conjunction with a real practical experiment in the laboratory.

A wealth of information can be extracted from cyclic voltammograms about the thermodynamics and kinetics of an electrochemical reaction, as well as about the concentration of the reacting substances. As a first step, four characteristic quantities are extracted from the voltammogram: the anodic and cathodic peak currents (I_{pa}, I_{pc}) and the anodic and cathodic peak potentials (E_{pa}, E_{pc}) (Figure 5.29). Based on these parameters, further quantities can be evaluated (Chapter 6).

The **formal potential** is obtained as the arithmetic mean of E_{pa} and E_{pc} (assuming $D_R = D_O$).

The potential difference between the anodic and cathodic peak potentials, the so-called peak separation ΔE_p, is important. This parameter assumes a constant value for electrochemically reversible systems. The value does *not* depend on the potential scan rate v. The exact value is $\Delta E_p = 57$ mV for a reversible one-electron system ($n = 1$) and a quasi-infinite potential difference between the switching potential E_λ and the

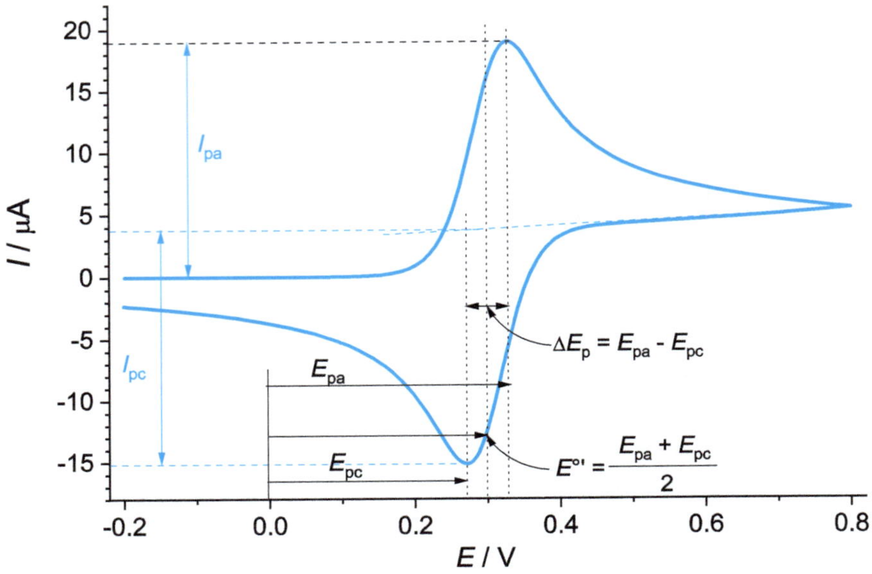

Figure 5.29 Important characteristic parameters of a cyclic voltammogram that can be used as diagnostic criteria. Calculated curves for $E°' = 0.15$ V, $c_R^* = 1$ mmol L^{-1}, $D_R = D_O = 1 \times 10^{-5}$ cm^2 s^{-1}, $v = 0.1$ V s^{-1} and $A = 0.071$ cm^2, $k° \to \infty$.

Table 5.6 Dependence of ΔE_p on the potential difference between the peak potential E_p and the switching potential E_λ [45].

$n\,|E_\lambda - E_{pa}|$/mV	$n\,\Delta E_p$/mV
∞	57.0
272	57.8
172	58.3
122	59.2
72	60.5

Those considerations are valid in analogous form if the oxidized redox form of a redox system is provided and a reduction is studied.

anodic peak potential E_{pa}. The exact value of ΔE_p is slightly larger for smaller values of $E_\lambda - E_{pa}$ (Table 5.6). For systems with slower electron transfer (quasireversible systems or irreversible systems), ΔE_p increases when the experiment is conducted with a faster scan rate v (Figure 6.31). The decisive observation that tells us whether a redox system can be considered as reversible is not so much the exact value of ΔE_p (which can be determined with limited accuracy) but the constancy of ΔE_p upon increase in v. ΔE_p also depends on the number of transferred electrons n. For $n = 2$, ΔE_p is halved compared to a redox system with $n = 1$. We will often find similar n-dependent relations in electrochemistry. Therefore, criteria based on the peak separation are mostly tabulated for $n\,\Delta E_p$ so that they can be applied to a wider range of systems using the correct value of n.

The faradaic **currents** at every point in the voltammogram **are proportional to the concentration** c_R^* of the studied compound. This follows from the proportional augmentation of all concentration gradients with c_R^* (Figure 5.27).

After proper calibration, one could use the peak current for concentration determination. Note that there are other voltammetric techniques that are usually much better for concentration determinations. (Sections 5.3.6, 5.3.7).

The **peak currents** in voltammograms must be measured against an appropriate **baseline**.

Figure 5.29 shows that this baseline is the extrapolation of the curve before the onset of the anodic peak. In this specific example, the capacitive current is so small that the baseline essentially overlaps with the line $I = 0$. However, this is often not the case! If there are two redox systems present or surface redox processes (Figure 5.22), or if experiments are conducted at high scan rates (high capacitive currents, Excursion 5-2), a nonzero baseline has to be extrapolated from the current trace before the peak onset. Control programs for modern electrochemical workstations sometimes propose automatically selected baselines, but these can be wrong. In these cases, an electrochemically well-versed person is needed to decide on a proper baseline.

The baseline problem becomes more evident when we consider the curve for the second half cycle of the voltammogram, the back scan or reverse scan. The reading against the zero line ($I = 0$) yields a peak current that is considerably smaller than the anodic peak current in the first half cycle – a result that would appear unrealistic for a reversible system. Furthermore, the peak current defined in this way strongly depends on the value of the switching potential. A more reasonable approach for establishing the baseline of the reverse scan becomes evident when looking at the currents in the $I = f(t)$ plot in Figure 5.26b. We should elongate the curve between $t = 6$ s and $t = 14$ s and use *this curve* as the baseline. However, this is not so simple because the baseline is curved. Fortunately, we already know that the curved baseline results from the growth of the diffusion layer (Figure 5.27d and e) and that the current can be approximated by the Cottrell equation for potentials $E > E_{pa}$ in the first half cycle (Figure 5.28). We extend the curve according to the Cottrell equation beyond E_λ. This line represents the current that would result if we were to hold the potential at E_λ and continue to record the current as a function of time. This current $I(t)$ would be plotted against a hypothetical E that would result if the potential scan had been continued in the positive scan direction during the hold time (Figure 5.30, dashed blue line, $v = dE/dt$, $\Delta E = v\Delta t$, $E = const + v\Delta t$). If this line is "mirrored" at an imagined vertical line at $E = E_\lambda$ (gray line in Figure 5.30), we obtain a suitable baseline for the determination of I_{pc} (dashed black line).

The peak current I_{pc} determined in this way equals I_{pa} for a reversible system.

Deviations from the ratio $I_{pc}/I_{pa} = 1$ are an indication of a more complex reaction mechanism. For instance, the species formed in the first half cycle can be consumed in a homogeneous chemical reaction and would not be available for a reduction in the subsequent back scan (Section 6.4.3).

Because the construction of the baseline is time-consuming in practice, a recommendation of Nicholson [46] is frequently followed, which provides directly the *ratio* I_{pc}/I_{pa} from Eq. (5.45) using the uncorrected cathodic peak current $I_{pc,0}$, the current

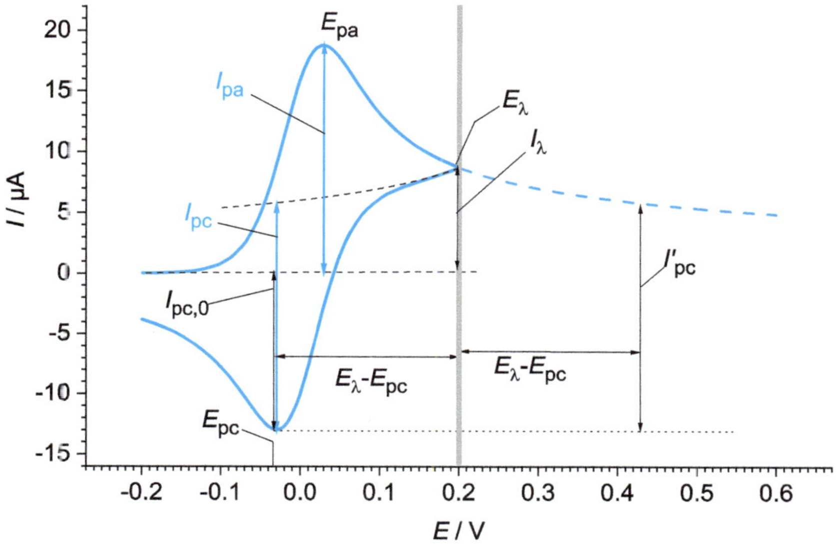

Figure 5.30 Determination of the peak current I_{pc} in the reverse scan of a cyclic voltammogram. The considerations also hold in an analogous manner if the oxidized form O is provided in solution and the experiment starts with reduction followed by reoxidation. Calculated curves for $E^{\circ\prime} = 0.0$ V, $c_R^* = 1$ mmol L^{-1}, $D_R = D_O = 1 \times 10^{-5}$ cm^2 s^{-1}, $v = 0.1$ V s^{-1}, and $A = 0.071$ cm^2.

at the switching potential I_λ, and I_{pa}. The *absolute current values* have to be used in Eq. (5.45). Of course, this can also be applied in an analogous manner if the solution initially contains O, which is reduced in the first half cycle.

$$\frac{|I_{pc}|}{|I_{pa}|} = \frac{|I_{pc,0}|}{|I_{pa}|} + \frac{0.485\,|I_\lambda|}{|I_{pa}|} + 0.086 \tag{5.45}$$

The peak current not only depends on the concentration, but also on the scan rate v. This is not so obvious as the dependence on c^*_R. We would like to postpone a quantitative treatment to Section 6.2.4 but do not want to omit an instructive qualitative view of the problem. We consider the development of the local concentrations between the time corresponding to Figure 5.27c and d. The local concentrations are depicted as concentration profiles $c = f(x)$, which we can imagine as "snapshots" of the local concentrations in the diffusion layer.

Q: Can one actually measure such "snapshots" and how do we know that they actually look like they are shown in Figure 5.27?

A: The concentration profiles in Figure 5.27 were calculated by numerically solving the partial differential equations. Programs for this task are widely available (Excursion 6-2), and a simple version can be downloaded from the Supporting Information of Ref. [44]. This allows one to retrace the approximation given here even more exactly. With more numerical effort, the calculations can be made extremely accurate. Experimentally, the local concentrations can be measured by means of microelectrodes positioned in the vicinity of a larger electrode [47]. Optical measurements with fluorescing products of electrochemical reactions [48] through a confocal microscope provide direct experimental observation of local concentrations in thin solution layers [49]. Practically, the situation is even a bit more complicated because "natural convection" (caused by local density fluctuations or vibrations) cannot be completely eliminated [50].

The difference between Figures 5.27c and d corresponds to a potential range of 0.1 V in the cyclic voltammogram or to a time of 1 s, if the experiment was performed with $v = 0.1$ V s^{-1}. In the concentration profiles in Figure 5.27c, the concentration at the electrode surface $c_R(x = 0)$ is almost zero. Thus, the current is a diffusion-limited current. The diffusion-limited current is proportional to the concentration gradient dc/dx at the electrode surface. In the next concentration profile (Figure 5.27d), $c_R(x = 0)$ has decreased a little bit further, and thus the concentration difference to the bulk phase $\Delta c = c^*_R - c_R(x = 0)$ has increased slightly. However, the current significantly decreases between Figure 5.27c and d because the diffusion layer has expanded according to Fick's second law within the second that passed between the two "snapshots". The expansion of the diffusion layer has a stronger effect on the gradient $dc/dx \approx \Delta c/\Delta x$ than the minor increase in Δc. This is the reason for the reduction of the gradient (and thus the current). Let us imagine now that we repeat the experiment but with a slower scan rate of $v = 0.01$ V s^{-1}. We take the "snapshots" of the concentration profiles at the same potentials. The passage of a potential range of 0.1 V takes 10 s when using $v = 0.01$ V s^{-1}. This allows much more time for the expansion of the diffusion layer. We can approximate the resulting concentration gradient as $\Delta c/\Delta x$ and use $\Delta c \approx c^*_R - 0 = c^*_R$. We find the thickness of the diffusion layer approximately as the square root of the mean squared displacement that we have looked at when deriving Fick's second law in Section 4.4.2:

$$\Delta x = \sqrt{\langle \Delta x^2 \rangle} = \sqrt{2Dt} = const\,\sqrt{\Delta E/v} \tag{5.46}$$

We obtain Eq. (5.47) as an approximation for the dependence of the concentration gradient (and the current) on v.

$$I \propto \left.\frac{\mathrm{d}c}{\mathrm{d}x}\right|_{x=0} \approx \frac{c_R^* - 0}{\Delta x} \approx \frac{c_R^*}{const'}\sqrt{v} \tag{5.47}$$

> The currents in cyclic voltammograms with a reversible redox system are proportional to the square root of the potential scan rate v.

The detailed quantitative relationship is covered in Section 6.2.4.

Applications. Experimental observation deviating from this expectation can be an indication of different mass transport conditions that occur for redox systems that are chemically bound to the electrode surface (Section 5.5.1), a hemispherical diffusion layer at microelectrodes (Section 5.6.4), diffusion layers with constant thickness caused by convection in solution (Section 12.4.2), or catalytic regeneration of R by a homogeneous reaction in solution (Section 18.2.5). Also note that the use of continuous potential cycling in voltammetry will lead to results that are different from those discussed here (Figure 6.39).

Besides characterization of diffusion behavior, the most important application of CV is certainly the study of redox states and electrochemical reaction mechanisms of molecular compounds, whose detailed coverage is the topic of Chapter 6. Extensive practical instructions for beginners, including experimental exercises, are compiled in [51].

In some cases, all necessary information can be obtained by just recording the first half scan (forward scan) of a cyclic voltammogram and omitting the reverse scan. This experiment is called **linear sweep voltammetry (LSV)**. The diagnostic criteria such as $I \propto c$ and $I \propto v^{1/2}$ are of course also valid if only the first half-cycle of a voltammogram is recorded. A historically very important method is polarography, which can be considered as LSV at a dropping mercury electrode. Although the method has retained practical significance only for very special applications, the following Excursion provides an overview of this method because many modern voltammetric methods for concentration analysis have been developed out of polarography. An understanding of the corresponding original literature is almost impossible without some appreciation of the background upon which the development of modern voltammetry took place.

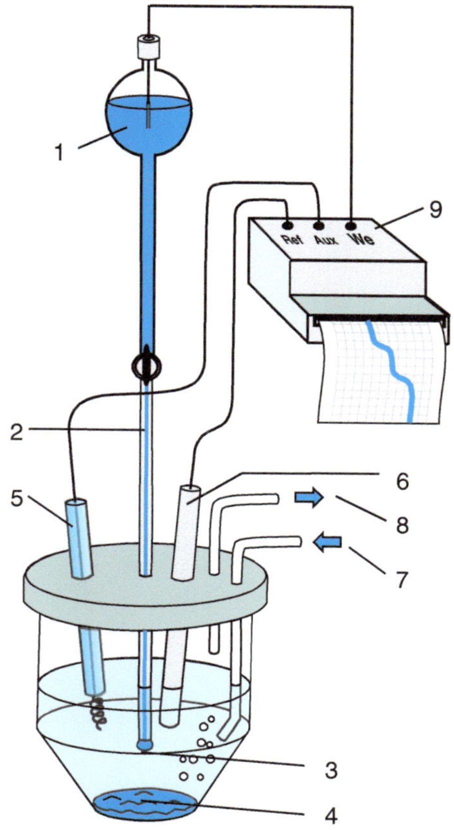

Figure 5.31 Schematic of a setup for polarography.

1) Hg reservoir;
2) Capillary with stopper;
3) Growing Hg drop;
4) Used Hg;
5) Auxiliary electrode (Pt or GCE);
6) Reference electrode;
7) Ar inlet;
8) Ar outlet;
9) Potentiostat with $x-t$ recorder for recording the data.

Excursion 5-2 Classic Polarography

Liquid mercury is used as the electrode material for the working electrode in classical polarography developed by Heyrovský[58] [52, 53]. It constantly flows out of a thin capillary with an inner diameter of about 50 μm. Hg is resupplied from a reservoir that holds enough Hg that the consumption during an analysis does not alter the filling height significantly (Figure 5.31). The flow rate is determined by the height difference between the liquid level in the reservoir and the end of the capillary, which can be conveniently adjusted with the help of a scale attached next to the capillary. A hanging drop of Hg forms at the end of the capillary. The drop grows steadily from the constant volume flow of Hg through the capillary. The drop surface is in contact with the electrolyte solution and forms the electrode area. When the droplet diameter has reached about 1 mm, the droplet weight

[58] Jaroslav Heyrovský, 1890–1967, Czech physical chemist, professor in Prague, Nobel Prize 1959 for the development of polarography.

cannot be counterbalanced by the surface tension of Hg, and the drop falls off the capillary. Immediately afterward, the growth of a new drop begins. The available electrode area changes periodically every 0.5–10 s. This type of electrode is called dropping mercury electrode (DME).

The main advantages and disadvantages of Hg as an electrode material have already been discussed in Section 5.3.2. The DME consumes unnecessarily large amounts of Hg that should be disposed of properly and used only in specialized laboratories (e.g., for the measurement of adsorption properties by means of drop rate variation, see Section 8.10.2). In addition to the disadvantages rooted in the toxicity of Hg, the DME also comes with disadvantageous analytical properties. Due to the constantly changing electrode area, the current changes periodically. We can distinguish two causes of a current flow at an electrode. The current from an electrochemical reaction is called faradaic current (less frequently reaction current or electrolysis current). Only this current component depends on the kind and concentration of the analyte and is calculated and plotted in Figure 5.27. The faradaic current is the useful signal. It depends on the electrode area. Overlaid on the faradaic current is the current component caused by charging the electrical double layer when either the potential and/or the electrode area is changed. This current is called capacitive current. It is an unavoidable interfering signal that is usually independent of the kind and concentration of the analyte. The common aim for the development of voltammetric methods was to maximize the ratio between the useful signal (faradaic current) and the interfering signal (capacitive current). The capacitive current can be observed when recording a linear sweep voltammogram at a DME in a solution of the supporting electrolyte (i.e., without addition of the analyte). The resulting current is a background current for the actual measurement. Figure 5.32 shows the periodic current oscillations resulting from the changing electrode area that are also clearly observable in the background current and make the compensation of the background current even more difficult.

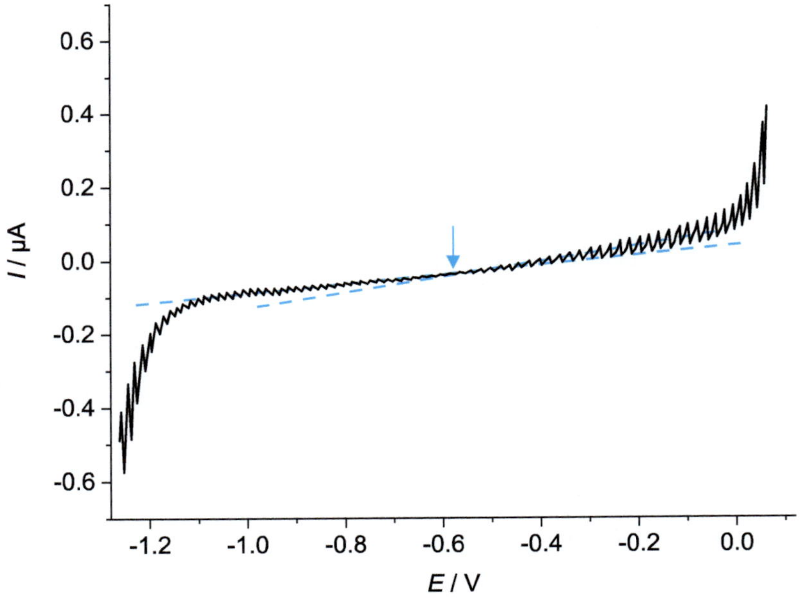

Figure 5.32 Background current of a 0.1 mol L^{-1} HCl solution. The oxidation current for $E > 0$ V results from the electrolytic dissolution of the Hg drop. The current at $E < -1.2$ V marks the onset of H_2 evolution. The arrow indicates the potential of zero charge E_{pzc} (electrocapillary maximum, see Section 8.10.3). The reference electrode is an SCE. The dashed lines emphasize the change in slope at E_{pzc}. Source: Data set from L. Meites, *Polarographic Techniques*, 2nd ed., John Wiley and Sons, New York, 1965, p. 101. Plot redrawn and adapted to the sign convention in this book.

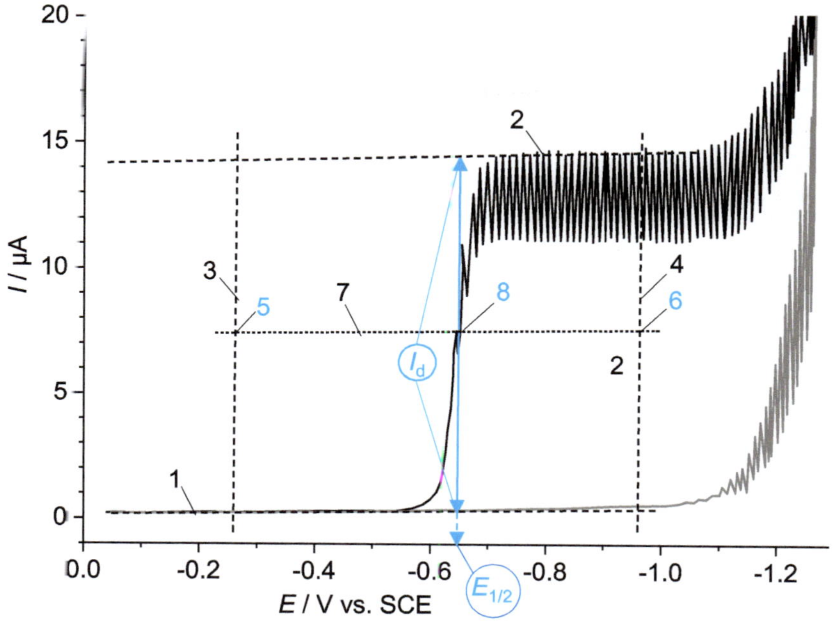

Figure 5.33 Determination of $E_{1/2}$ and I_d for a polarogram of 0.5 mmol L^{-1} Cd^{2+} + 1 mol L^{-1} HCl (black curve). The background current measured in the pure supporting electrolyte solution (1 mol L^{-1} HCl) is also shown as a gray curve. First, tangents are constructed for the measured curve before (1) and after (2) the polarographic step. They are extended over the entire potential range. Two vertical auxiliary lines (3 and 4) are placed clearly before and after the polarographic step. The midpoints (5 and 6) of the vertical lines are determined and connected by another auxiliary line (7). This line intersects with the measured curve at point 8. The coordinates of point 8 on the potential axis define $E_{1/2}$. The vertical distance between the tangents 1 and 2 at $E = E_{1/2}$ is I_d. Source: Measured data from *Polarographic Techniques*, 2nd ed., John Wiley and Sons, New York, 1965, p. 3. Auxiliary lines and annotations by the book author.

This is an example of the classical polarographic sign convention: reductive currents are positive and plotted upward, and negative potentials are plotted to the right.

The faradaic current is observed only in the presence of the analyte that can be reduced or oxidized. Figure 5.33 shows the polarogram when Cd^{2+} is present in HCl solution as supporting electrolyte. The reduction of Cd^{2+} causes the current at $E < -0.6$ V. The analyte was frequently called depolarizer in the context of polarography because its reduction current is observed in a potential range in which the supporting electrolyte does not show any significant current, which means at potentials at which the electrode is polarized. Two features of the curve are important for data evaluation: The half-wave potential $E_{1/2}$ and the diffusion-limited current I_d. $E_{1/2}$ refers to the potential at which the current reaches just half of the diffusion-limited current. I_d is taken "vertically" between the current peaks in the presence and absence of the Cd^{2+}. The graphical determination of I_d and $E_{1/2}$ is explained in the caption of Figure 5.33.

Q: Why does Figure 5.33 show a step-like curve and the first half-cycle of the cyclic voltammogram in Figure 5.27e a peak despite the fact that both are a linear-sweep voltammogram of a fast (reversible) redox system?

A: At the DME, a drop falls off periodically. Thereby, the diffusion layer is stirred and removed with the falling drop. The formation of a new drop starts in a solution of the undisturbed bulk concentration of the analyte. The extension of the diffusion layer is limited by the lifetime of a Hg drop at the orifice of the capillary in Figure 5.33, while it is the time for the potential scan from $E_{1/2}$ to the switching potential $t = (E_\lambda - E_{1/2})/v$ in Figure 5.27. Step-shaped voltammograms are also recorded in other cells with forced convection, for instance, at a rotating disk electrode (Section 12.4.2).

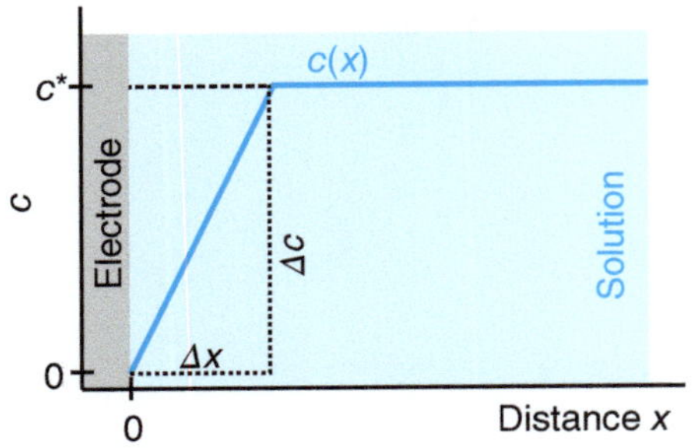

Figure 5.34 Plot of the local concentration for a diffusion-controlled conversion with $c(x = 0) = 0$. A popular simplification is the Nernst diffusion layer, for which a linear change in concentration with distance is assumed for the diffusion layer. The actual course of $c(x)$ is shown in Figure 4.52.

How does the diffusion-limited current I_d (or diffusion-controlled current) depend on the bulk concentration c_O^*? Generally, the diffusion-limited current is proportional to the concentration gradient of the analyte at the electrode surface.

$$I = nFAD_O \left.\frac{\partial c_O^*}{\partial x}\right|_{x=0} \tag{5.48}$$

Because we are interested in the diffusion-limited current I_d, we consider the situation where there are no metal cations O remaining at $x = 0$ (the reduction rate at the electrode surface is so fast that all analyte species O are immediately reduced). In addition, we assume that the diffusion layer always reaches the same thickness Δx at the end of a drop lifetime t_d (Figure 5.34). This is a sensible assumption because the drop falls off the capillary orifice in regular intervals. Therefore, the diffusion layer always has an identical time span for expansion. Furthermore, we simplify the concentration profile as linear. This approximation is called **Nernst diffusion layer**[10] (compare Figure 5.34 to Figures 4.52 and 4.71). Under these assumptions, we can replace the partial differential quotient in Eq. (5.48) with a difference quotient to arrive at Eq. (5.49).

$$I_d = nFAD_O \frac{c_O^* - 0}{\Delta x} = kc_O^*, \quad k = nFAD_O/\Delta x = nFAm_{t,O} \tag{5.49}$$

The half-wave potential characterizes which ion causes the signal. We can rationalize this with a simple consideration. The following relations hold for the current at each point of the curve

$$I = k\,\left[c_O^* - c_O(x = 0)\right], \; k = nFAD_O/\Delta x = nFAm_{t,O} \tag{5.50}$$

$$= I_d - \mathrm{k}\,c_O(x = 0).$$

Isolation of $c_O(x = 0)$ provides

$$c_O(x = 0) = \frac{I_d - I}{k}. \tag{5.51}$$

Next, we consider in a similar way the concentration of the product R, which would be the metal atoms dissolved in the liquid Hg droplet. For the polarographic analysis of organic compounds, R would usually be the reduced compound dissolved in the electrolyte solution. At the beginning of the droplet formation, the concentration is zero and only builds up in the same way as O is reduced ($O + ne^- \rightarrow R$). Therefore, we can express the current using the concentration gradient for R in an analogous way to Eqs. (5.48) and (5.49). The proportionality coefficient k' has a different value because of different D_R and Δx.

$$I = k'\,c_R(x = 0) \tag{5.52}$$

Rearranging provides an expression for $c_R(x = 0)$:

$$c_R(x = 0) = \frac{I}{k'}, \quad k' = nFA\frac{D_R}{\Delta x'} = nFAm_{t,R} \tag{5.53}$$

The concentrations of R and O at the electrode surface can be inserted into the Nernst Equation (5.54) when assuming that the metal cations are reduced rapidly (reversible system). Indeed, that is the case for a number of important analyte ions such as Pb^{2+} and Cd^{2+}.

$$E = E^{\circ\prime} + \frac{RT}{nF}\ln\frac{c_O(x = 0)}{c_R(x = 0)}$$

$$= E^{\circ\prime} + \frac{RT}{nF}\ln\left(\frac{I_d - I}{I}\frac{k'}{k}\right) \tag{5.54}$$

This general expression can be written for specific potential and current values. By definition, the current I is exactly half of I_d at the half-wave potential $E_{1/2}$. Setting the relationship for this specific data pair into (5.54) yields the simplified Eq. (5.55).

$$E_{1/2} = E^{\circ\prime} + \frac{RT}{nF} \ln \left(\frac{I_d - ^1\!/_2 I_d}{^1\!/_2 I_d} \frac{k'}{k} \right) = E^{\circ\prime} + \frac{RT}{nF} \ln \frac{k'}{k} \quad (5.55)$$

Because k and k' are constants, we realize that the half-wave potential $E_{1/2}$ has a fixed difference to the formal electrode potential $E^{\circ\prime}$ and therefore also to the standard electrode potential E°. We take the expressions for the two proportionality constants $k = nFAD_O/\Delta x$ and $k' = nFAD_R/\Delta x'$ from Eqs. (5.50) and (5.53). Using Eq. (5.46) as an expression of the extension of the diffusion layers of O and R at the end of the drop life, we obtain Eq. (5.56).

$$k = nFA\frac{D_O}{\sqrt{2D_O t_d}} = \frac{nFA}{\sqrt{2t_d}}\sqrt{D_O}$$
$$k' = nFA\frac{D_R}{\sqrt{2D_R t_d}} = \frac{nFA}{\sqrt{2t_d}}\sqrt{D_R} \quad (5.56)$$

Inserting those expressions into Eq. (5.55) gives the most frequently cited relationship (5.57) between $E_{1/2}$ and $E^{\circ\prime}$. This equation applies generally for cyclic voltammetry with planar diffusion and $D_R \neq D_O$.

$$E_{1/2} = E^{\circ\prime} + \frac{RT}{nF} \ln \sqrt{\frac{D_R}{D_O}} \quad (5.57)$$

Diffusion coefficients in liquid phase do not deviate very much from each other. For example, the diffusion coefficients are 7.15×10^{-6} cm^2 s^{-1} for Cd^{2+} at 25 °C in 0.1 mol L^{-1} KCl solution and 18.9×10^{-6} cm^2 s^{-1} for Cd atoms in mercury amalgam [55]. The square root is calculated from the ratio of the numbers, and then the logarithm is formed. Thus, the numerical value of $E_{1/2}$ does not deviate significantly from $E^{\circ\prime}$ for many systems.

Summary: The half-wave potential allows a qualitative determination of the type of analyte, and the diffusion-limited current is proportional to the bulk concentration of the analyte.

5.3.6 Modern Mercury Electrodes and Pulse Programs

Today, the mercury electrode presented in Excursion 5-2 is rarely used for concentration analysis. The periodically growing electrode area causes oscillating capacitive currents that form a background signal, which makes the evaluation of smaller faradaic currents difficult. The consumption and the free flow of toxicologically worrying mercury are a concern for occupational safety and a possible source of contamination of the environment. Modern voltammetric methods have been enabled by two innovations: The introduction of improved constructions of the mercury electrode and more complex pulse programs, which facilitate a better resolution of the signal and improve the ratio between the faradaic current and the capacitive current [53, 54]. The curves in the pulse programs are typically realized by a sequence of discrete measurement points (as opposed to a continuous curve in classical polarography). Modern Hg electrodes consist of a sealed mercury reservoir, which is connected to the capillary by a needle valve. The valve is opened and closed pneumatically while always maintaining an electrical connection to the mercury column in the capillary.

By means of this valve, the flow of mercury can be coordinated with the measurement program. A small electromechanically operated hammer can dislodge the Hg drop from the end of the capillary before reaching its maximum size.

With such an apparatus, two types of electrodes are employed based on this setup: the static mercury drop electrode (SMDE) and the hanging mercury drop electrode (HMDE). At the SMDE, a new drop of Hg is generated with reproducible size at each measurement point within a voltammogram. Only after the valve is closed and the drop has stopped growing is the current recorded. Afterward, the potential is increased, the old droplet is dislodged, and a new droplet is formed, and so forth. In this way, the surface area of the drop has a static (constant) size during the current recording.

When using an HMDE, the droplet is formed only once before recording a voltammogram, so that the entire voltammogram is recorded with a single droplet of constant size. Before recording the next voltammogram, the droplet is dislodged and renewed. (In fact, several drops are usually dislodged and formed to ensure that the mercury close to the orifice of the capillary does not become part of the drop that is used in the next voltammogram. (The Hg close to the capillary orifice may have been contaminated by reaction products that diffuse into the Hg column inside the capillary.) Then, a new drop is formed by shortly opening the needle valve at the top end of the capillary. The HMDE is a prerequisite for voltammetric methods with an integrated preconcentration step (stripping voltammetry, Section 5.3.10).

Although many pulse methods have originally been developed with the idea of synchronizing the pulse cycle with the droplet renewal of an SMDE, it turned out that they can also be used advantageously at the HMDE as well as with other types of solid electrodes.

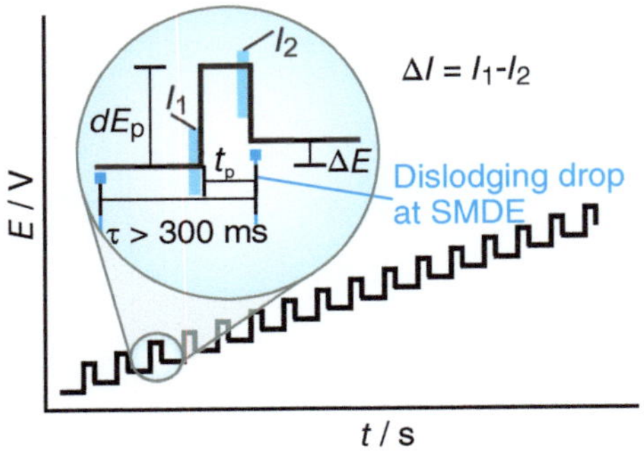

Figure 5.35 Potential excitation signal for differential pulse voltammetry. The dislodgement indicated in the figure is done only with an SMDE. The potential excitation signal can also be carried out without renewal of the electrode surface at solid electrodes or at an HMDE.

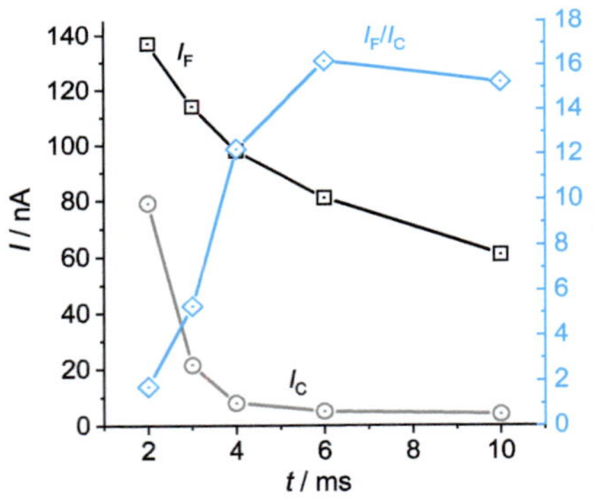

Figure 5.36 Comparison of the transients of the capacitive current I_C and the faradaic current I_F after a potential step. I_C can be determined experimentally in a solution of the supporting electrolyte without analyte. After the addition of a redox-active analyte (e.g., a metal cation), the total current I_T can be measured. The faradaic current is obtained from $I_F = I_T - I_C$. Source: Data from U. Schröder, H. Kahlert; *Practical Voltammetry*. Metrohm, Herisau, 2002 [56].

5.3.7 Differential Pulse Voltammetry

The term differential pulse voltammetry (DPV) refers to the potential program, not to the electrodes used, because the method can be used advantageously not only with Hg but also with solid electrodes, for instance, made from Au, Pt, or glassy carbon. The basic potential program (Figure 5.35) is a staircase ramp, in which each step is applied for 0.5 to 2 s (τ). The step height ΔE amounts to 5 to 20 mV. At the end of each step, a difference pulse with a pulse amplitude dE_p of 5 to 50 mV (in the direction of the potential ramp) is added relative to the step potential for 5 to 100 ms. For each step, the current is read twice, giving the values I_1 and I_2. The first reading I_1 is made just before, and the second reading I_2 is at the end of the difference pulse. If DPV is conducted at an SMDE, the Hg drop is dislodged after reading I_2 followed by formation of a new drop of defined size. The falling drop stirs the solution and equalizes the local concentration in such a way that the new drop is formed in a solution of undisturbed analyte concentration.

Because the current reading is delayed with respect to the drop formation, both the faradaic current I_F [Eq. (4.130)] as well as the capacitive currents I_C [Eq. (4.99)] have decayed. However, the decay of I_C is much steeper than that of I_F (Figure 5.36). Because I_C is an (interfering) background signal for concentration determinations, pulse voltammetry improves the signal-to-background ratio. This also reduces the noise of the background signal, which ultimately leads to significantly improved limits of detection compared to CV or classical polarography.

The DP voltammogram is a plot of the difference current $\Delta I = I_2 - I_1$ as a function of the step potential. Figure 5.37 shows an example of the simultaneous determination of three different metal cations. Compared to the step-shaped polarograms in Figure 5.33 or the "duck bill" shape of the cyclic voltammograms in Figure 5.28, the difference current ΔI decays almost to zero after passing through the peak (Figure 5.37).

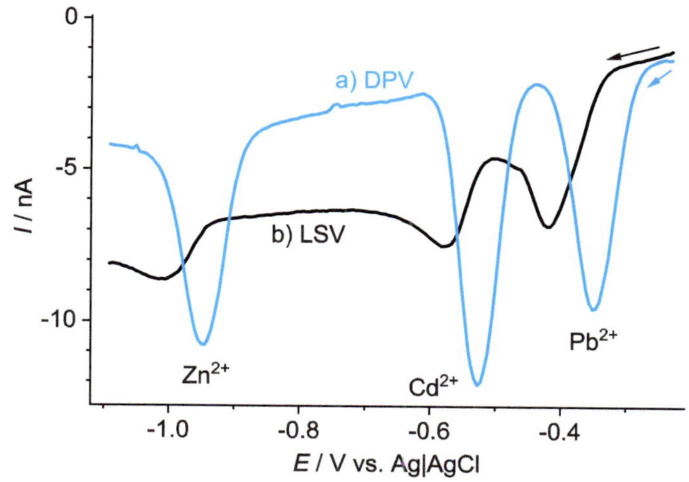

Figure 5.37 Comparison of voltammograms of 1 µmol L^{-1} Pb^{2+}, 1 µmol L^{-1} Cd^{2+}, and 1 µmol L^{-1} Zn^{2+} with a) DPV at a SMDE and b) LSV at a HMDE, $v = 0.02$ V s^{-1}. DPV pulse amplitude = 50 mV. Source: Data provided by Izabella Brand and Nico Roth, University of Oldenburg.

We can explain this with the example of the reduction of a metal cation (e.g., reduction of Pb^{2+}). If the step potential is positive of $E_{1/2}$ [Eq. (5.57)], no reduction is possible neither at the step potential nor during the potential pulse, thus the currents are similar $I_1 \approx I_2 \approx 0$, and the difference current $\Delta I \approx 0$ is negligible. If the step potential is negative of $E_{1/2}$, a diffusion-controlled electrolysis will occur at the step potential as well as during the potential pulse. The electrolysis during the step potential causes the formation of a diffusion layer that is not significantly changed during the short difference pulse. This leads again to $I_2 \approx I_1$ and thus $\Delta I \approx 0$. A significant difference between I_1 and I_2 will occur only at potentials around $E_{1/2}$. The difference will be largest when the step potential before the pulse is $\mathrm{d}E_\mathrm{p}/2$ positive of $E_{1/2}$, and during the pulse, the potential is $\mathrm{d}E_\mathrm{p}/2$ negative of $E_{1/2}$. Thus, the voltammogram is similar to the first derivative of a polarogram. However, it is clearly not exactly the first derivative! One difference is the position of the maximum in the DP voltammogram, which is found at $E_{max} = E_{1/2} - \mathrm{d}E_\mathrm{p}/2$ [Eq. (5.58)], while the first derivative of a polarogram has a maximum at $E_{1/2}$.

The peak of a DP voltammogram for a reversible system is found at

$$E_{max} = E^{\circ\prime} + \frac{RT}{nF}\ln\sqrt{\frac{D_R}{D_O}} - \frac{\mathrm{d}E_\mathrm{p}}{2} = E_{1/2} - \frac{\mathrm{d}E_\mathrm{p}}{2} \tag{5.58}$$

The peak current for a reversible system equals

$$\Delta I_{max} = \frac{nFA\sqrt{D_O}}{\sqrt{\pi t_\mathrm{p}}} \cdot \frac{1-\sigma}{1+\sigma} \cdot c_O^* \tag{5.59}$$

t_p – Pulse duration [s] (Figure 5.35)

in which σ is the exponential function of half the pulse amplitude in units of the Nernst factor [Eq. (5.60)].

$$\sigma = \exp\left(\frac{\mathrm{d}E_\mathrm{p}}{2E_{NF}}\right) = \exp\left(\frac{nF}{RT}\cdot\frac{\mathrm{d}E_\mathrm{p}}{2}\right) \tag{5.60}$$

Eq. (5.59) is not used for concentration determination. Instead, this is made after a calibration. This means DPV is used as a *relative method of analysis* because the underlying parameters are not sufficiently well known in practical situations. For instance, it is often unknown if the investigated system really behaves like a reversible system (fast charge transfer without follow-up reactions). Components of the matrix of a real sample could adsorb to the electrode surface and inhibit (slow down) the charge transfer. Furthermore, A and D_O are often not known with sufficiently high precision, and their determination would be more cumbersome than making a calibration plot. We will demonstrate the evaluation of voltammetric data for concentration determination in Section 5.3.10.

The analytical advantages of DPV compared to polarography or CV are based on the suppressed capacitive currents and therefore the improved limits of detection. The peak-shaped signals are also easier to evaluate because ΔI_{max} is easily identified and the construction of the baseline is simplified (Figure 5.37). DPV also improves the separation of closely spaced signals on the potential axis in the case of mixtures. For these reasons, DPV is the voltammetric method most frequently applied for concentration determination.

5.3.8 Square Wave Voltammetry

Square wave voltammetry (SQWV) uses an excitation signal composed of rectangular potential pulses in and reverse direction to the potential scan direction.

The entire potential program is executed at a solid electrode or in the case of Hg at an HMDE.

There are several other methods that were introduced as square wave voltammetry or polarography to the literature but refer to different procedures from those covered here. Most of them have lost importance but may appear in a literature search. Among these is the idea to apply SQWV to a dropping Hg electrode as square wave polarography. There is a similar technique that works with very small pulse heights (Barker square wave voltammetry[59]), which we may conceptually regard as an alternating current voltammetry with rectangular excitation signal [59]. In older literature, the method covered here is called Osteryoung square wave voltammetry (OSQWV)[60,61].

An extensive survey of this method is provided by a review [57, 58]. We will cover only a few essentials. The pulse program in Figure 5.38 may initially look like a special case of DPV. However, evaluation of the data follows an entirely different approach. It is therefore sensible to think of the excitation signal as a staircase potential of step potential E_S and step duration τ, on which a rectangular voltage of the frequency $\nu = 1/\tau$ is superimposed. The rectangular voltage is formed by a forward and reverse pulse (with respect to the scan direction of the staircase voltage). Forward and reverse pulses have the same duration $t_p = \tau/2$ and the same absolute value dE_p. Similar to the DPV, the current is read shortly before each pulse in each cycle. After a complete period of the rectangular voltage, the step potential is incremented by ΔE_S, and the next cycle of the rectangular voltage is applied. An important difference to DPV is the duration of the pulses and, therefore, the timescale that is probed by an SQWV experiment. The following settings can be recommended as typical (n is the stoichiometric number of transferred electrons) [57]:

$$\Delta E_S = 10\ \text{mV}/n \approx (2\text{–}10)\ \text{mV}$$

$$dE_p = 50\ \text{mV}/n \approx (15\text{–}50)\ \text{mV}$$

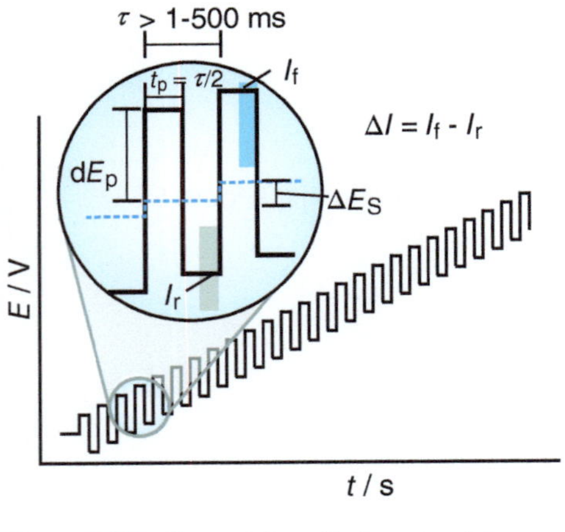

Figure 5.38 Potential excitation signal for SQWV.

The pulse duration can be varied over two orders of magnitude. The recommended range is about

$$t_p = (1\text{–}500)\ \text{ms and thus } f = (1\text{–}500)\ \text{Hz.}$$

The resulting scan rate for the step potential E_S is thus

$$v_{max} = \Delta E_S/\Delta t = 0.01\ \text{V}/(0.002\ \text{s}) = 5\ \text{V s}^{-1} \text{ and}$$

$$v_{min} = 0.01\ \text{V}/(1\ \text{s}) = 0.01\ \text{V s}^{-1}$$

Running an SQWV over a potential range of 1 V takes only between $t_{min} = |E_{end} - E_{initial}|/v = 1\ \text{V}/(5\ \text{V/s}) = 0.2\ \text{s}$ and $t_{max} = 1\ \text{V}/(0.01\ \text{V/s}) = 100\ \text{s}$. Thus, SQWV is a rapid method compared to some other voltammetric methods.

In contrast to DPV, not only the difference current ΔI, but also the forward currents I_f and reverse currents I_r, which are recorded at the end of the respective rectangular

[59] Geoffrey Cecil Barker, 1915–2000; British electrochemist.
[60] Robert Allen Osteryoung, 1927–2004, US-American electrochemist.
[61] Janet G. Osteryoung, US-American electrochemist. R.A.O and J.G.O. worked with others at the Colorado State University, State University of New York at Buffalo, and the North Carolina State University.

pulses, are stored permanently. As a result, a SQW voltammogram consists of three data sets $\Delta\bar{I}(E_S)$, $I_f(E_S)$, and $I_r(E_S)$. Because SQWV always requires a computer-controlled potentiostat, storage of the data is not a problem.

For the theoretical treatment of SQWV data, we must consider that the diffusion layer is not reset periodically by the droplet fall at the SMDE. The concentration profile at the end of a drop life is rather the initial condition for the subsequent pulse period. Usually, a superposition principle is used to treat the staircase program and the rectangular wave separately. For the mathematical treatment, we refer to the literature [1, 57, 60, 61] and only consider the basics semiquantitatively. The working solution initially contains only the reduced form R of a reversible (this means fast) redox couple R/O. Initially, E_S is much smaller than $E_{1/2}$, and no electrolysis will occur during the pulses. I_f, I_r, and thus ΔI values are approximately zero. When E_S is approaching $E_{1/2}$ (and thus approaching $E^{\circ\prime}$), the rate of oxidation will increase during the forward pulse and I_f will assume significant values. Within one potential pulse, the current decays according to the Cottrell Equation (4.130). Figure 5.39 shows the current transient and the location of the points at which I_f and I_r are recorded. The ordinate of Figure 5.39a

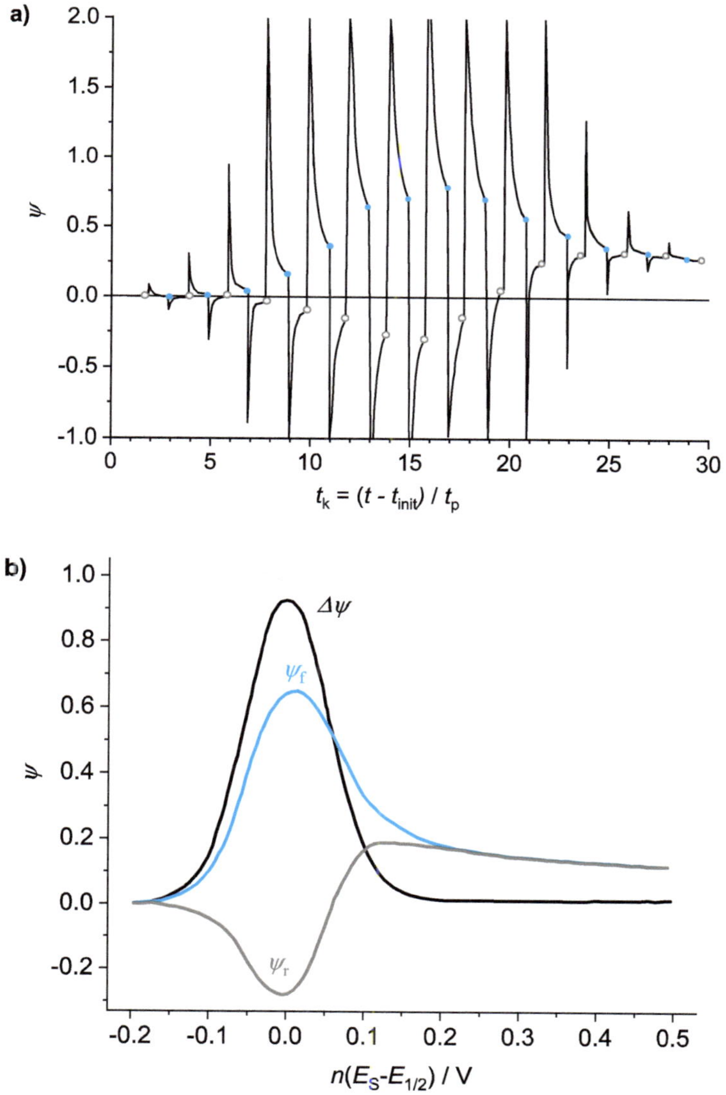

Figure 5.39 a) Current response signal and b) plotted signals for SQWV. ψ_f (blue) is the forward current function measured at the end of the forward pulse taken at the blue dots in (a), ψ_r (gray) is the current function of the reverse pulse taken at the open circles in (a), and $\Delta\psi$ (black) is the difference $\psi_f - \psi_r$. Source: Reproduced with permission from J. G. Osteryoung, J. J. O'Dea; Square-Wave Voltammetry. in *Electroanalytical Chemistry, Vol. 14* (A. J. Bard, Ed.), Marcel Dekker, New York **1986**, pp. 209–308. Copyright 1986, Marcel Dekker. Coloration and adaptation of the sign convention by the textbook author.

is a current function. Such current functions are often used for a comprehensive theoretical treatment of voltammetric methods because they facilitate generalization of the results. Current functions are proportional to the measured current. We can think of the current density $j = I/A$ as a kind of current function. When deriving the Cottrell Equation (4.130), we have already learned that the current is not only proportional to the electrode area but also to the bulk concentration, to the square root of the diffusion coefficient, and to additional factors. Thus, there are, in principle, many ways of forming a current function. When looking for a *sensible* way of forming a current function, one tries to identify a dimensionless expression that reflects all influences. For SQWV, the current function ψ is

ψ – Current function [dimensionless]
t_p – Pulse duration [s]
D – Diffusion coefficient of the analyte [$cm^2\ s^{-1}$]
c^* – Undisturbed bulk concentration of the analyte [$mol\ L^{-1}$]
A – Geometric electrode area [cm^2]

$$\psi = I\frac{\sqrt{\pi}}{nFA\sqrt{D}}\frac{\sqrt{t_p}}{c^*} \tag{5.61}$$

The time is also stated in a dimensionless quantity t_k, according to Eq. (6.62). It is the number of half-waves within the experiment. At integral values of the dimensionless time, the current I is recorded (Figure 5.39a) and plotted versus E (Figure 5.39b).

t – Time [s]
t_{init} – Initial time of the SQWV scans [s]
k – Index of the measurement point

$$t_k = \frac{t - t_{init}}{t_p} \tag{5.62}$$

The majority of the oxidation products formed during the forward pulse are reduced during the reverse pulse, and I_r reaches significant values with opposite sign to I_f. Consequently, ΔI asumes absolute values that are larger than either the absolute values of I_f or I_r. This is the main reason for the sensitivity advantage of SQWV when compared to DPV.

If E_S is positive to $E_{1/2}$, the oxidation also prevails during the reverse pulse. The oxidation is then only somewhat slower compared to the process during the forward pulse. The difference in reaction rate vanishes as E_S increases because eventually the reaction will be diffusion-controlled during both the forward and the reverse pulse. In this situation, ΔI decays to zero.

The peak-shaped signal ΔI can be used advantageously for concentration determination. The maximum difference current ΔI_{max} occurs at $E_S = E_{1/2}$ (Eq. 5.57). ΔI_{max} is proportional to the concentration and follows the expression in Eq. (5.63) for a reversible system.

$$\Delta I_{max} = \frac{nFA\sqrt{D_R}}{\sqrt{\pi t_p}} \cdot c_R^* \cdot \Delta\psi_{max} \tag{5.63}$$

ΔI_{max} can also be expressed as a function of frequency $f = (2t_p)^{-1}$ of the rectangular voltage:

$$\Delta I_{max} = \frac{nFA\sqrt{D_R}\cdot\sqrt{2f}}{\sqrt{\pi}} \cdot c_R^* \cdot \Delta\psi_{max}\cdot \tag{5.64}$$

Table 5.7 Dimensionless peak currents $\Delta\psi_{max}$ for $n\Delta E_S = 10$ mV.

$n\ dE_p$/mV	$\Delta\psi_{max}$
0	0.0437
10	0.2726
20	0.4845
50	0.9281
100	1.1675

Source: Data from [57].

Next to the bulk concentration of R, we recognize in Eq. (5.63) a Cottrell term (similar to Eq. (5.59) for DPV) with the pulse duration t_p as characteristic time and a factor $\Delta\psi_{max}$. This factor is a dimensionless, normalized current factor as used frequently for the mathematical treatment of voltammetric methods [Eq. (5.61)]. The current in physical units can be obtained by multiplying the current function by the contained constant factors and applying the appropriate sign convention [×(+1) or ×(−1)]. Values of ψ can be found in extensive tables in [57] from which excerpts are reproduced in [1]. Table 5.7 illustrates the influence of the pulse height dE_p on the signal for a recommended step height $\Delta E_s = 10\ mV/n$. A pulse height of $dE_p = 0$ corresponds to staircase voltammetry. The signal increases significantly up to $dE_p = 50$ mV. A further increase in dE_p causes only a slight additional increase but comes with a lower resolution of partially overlapping signals. The replacement of the pulse time t_p in Eq. (5.63) by the frequency f of the rectangular voltage makes clear the relationship to alternating current (AC) voltammetry, in which the signal height is proportional to the square root of the frequency [cf. to Eq. (5.71)].

SQWV is particularly sensitive to reversible immobilized (= surface-bound) redox systems. These are systems in which the reactants R and O are covalently or adsorptively bound to the electrode surface (Section 5.3.10). In addition to the application of the effect in adsorptive stripping voltammetry, there is an important application for the detection of redox-active labels in biosensors and biochips (Sections 18.3.6 and 18.3.7). In this case, the experimenter may choose the best-suited redox system as a label and, as such, can select a particularly fast redox system, which makes the most efficient use of the sensitivity advantage of SQWV over other voltammetric methods. In contrast to DPV, I_f and I_r give two curves as shown in Figure 5.39b. Together they are reminiscent of the shape of a cyclic voltammogram (Figure 5.26). Indeed, the reason for this curve shape is similar to that discussed for CV before. Similarly, to the use of CV (Chapter 6), the I_f and I_r curves can also be used for mechanistic studies, because their shapes change in a characteristic manner when complex reaction mechanisms occur with preceding and subsequent homogeneous reaction steps or finite kinetics of the electron transfer step. For this purpose, the measured curves can be compared to simulated curves that have been calculated under the assumption of specific details of the reaction mechanism. The theory for the calculation of such voltammograms is well developed for various typical scenarios. Introductions and examples can be found in [61–63].

5.3.9 Survey of Further Excitation Signals and Their Application

The number of excitation methods in voltammetry is almost inexhaustible. Many of these are less important for the novice or for practical applications. However, new electrochemical workstations and their control software provide a sometimes confusing variety of excitation signals because these are relatively easy to implement. Therefore, we would like to provide here an overview of the most important excitation signals as well as a critical assessment of their application. More specialized textbooks [1, 61, 64] can be consulted for a more in-depth coverage.

Staircase Voltammetry (SCV). Figure 5.40 shows the excitation signal of staircase voltammetry. The potential is incremented in steps, and the current is read out at constant potential. In terms of the DME, the pulses in the excitation signal can be synchronized with the drop time in order to limit the effect of the changing electrode area. When using a solid electrode or an HMDE, peak-shaped signals are obtained that are very similar to LSV or CV. One idea for the application of SCV aims to increase the ratio of faradaic current to capacitive current, by recording the current after the capacitive current has already appreciably decayed. According to Eq. (4.14), one must wait at least $R_E C_{cl}$ (the time constant of the electrochemical cell). The potential steps must be relatively long. Therefore, the suppression of capacitive currents with SCV is of limited practical value.

However, staircase voltammetry is important. Modern potentiostats operate on the basis of microprocessors and digital–analog converters (Section 5.7.1). They can realize a linear potential ramp only as a sequence of very small steps in a staircase, that is, by SCV or cyclic staircase voltammetry. The scan rate of the intended cyclic voltammogram $v = dE/dt$ is then controlled by the corresponding staircase voltammogram as $v = \Delta E/t_p$. The software often allows the setting of the step height and the interval t_s within t_p, during which the current recording is made. Depending on the mechanism and rate of the electrode reaction, t_s should be between 0.25 t_p and 0.5 t_p in order

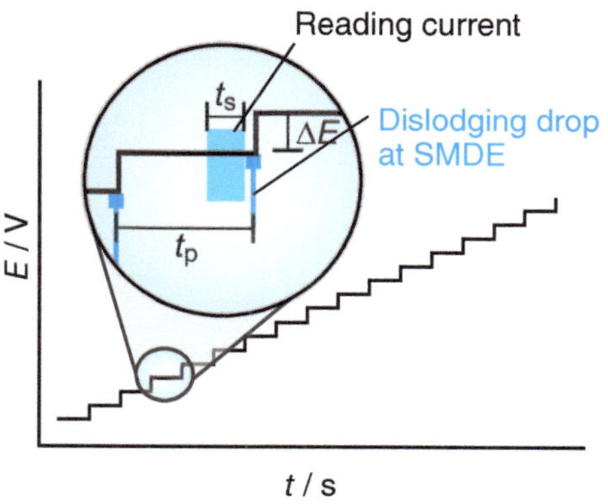

Figure 5.40 Excitation signal for staircase voltammetry. The potential program can not only be executed at the SMDE as noted in the scheme, but also at solid electrodes or at the HMDE without droplet removal (and surface renewal).

to record a voltammogram by SCV that is comparable to the analog cyclic voltammogram of the same system. This interval can also be shifted within t_p. Probably more practical is the approach to make the step height ΔE very small (this means 1–2 mV). Under these conditions, CV and SCV yield indistinguishable voltammetric data for diffusing redox systems.

The desired equivalence of CV and SCV is often *not* fulfilled for immobilized redox systems.

However, caution must be exercised if immobilized redox systems are studied! Because the current decay with time is not governed by the Cottrell Equation (4.130) (there is no diffusion for these systems), the faradaic current can decrease very rapidly after the potential step. If the data recording sets in too late, the immobilized systems have already reached the new equilibrium concentration, the recorded current will be small, and the surface concentrations of redox species may be underestimated dramatically. Figure 5.41a shows cyclic voltammograms of polycrystalline Pt in sulfuric acid solution, a system studied very commonly. A true linear ramp (blue line) is compared to cyclic staircase voltammetry data with different step heights (gray to black). While the peak for the reduction of Pt surface oxides at 0.75 V does not depend on the details of the potential program, the signals for the hydrogen adsorption and desorption between 0.05 V and 0.35 V change significantly with the potential step height under otherwise identical conditions. Even at a step as small as 1 mV (light blue curve), the amount of adsorbed hydrogen would be dramatically underestimated.

In order to mitigate against this problem, advanced potentiostats often contain an additional ramp generator that generates a true linear potential ramp for CV and LSV (blue curve in Figure 5.41b). Another approach is the continuous recording of the current over t_p and averaging. The user must ensure the suitability of a particular algorithm by reference measurements with well-known systems, perhaps also using different types of instruments. This problem is especially encountered in electrochemical surface science (Chapter 8), where the charge and eventually the surface concentration Γ are obtained by integration of currents (an absolute method). For electrochemical sensors, the bulk solute concentration c^* can be obtained despite these difficulties using calibration procedures (a relative method).

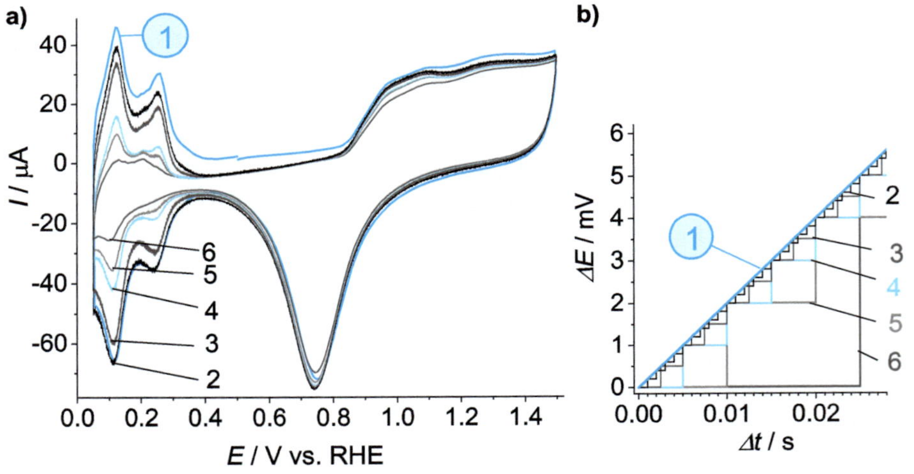

Figure 5.41 Influence of the step height on the appearance of the voltammograms of immobilized redox systems. a) Cyclic voltammogram and cyclic staircase voltammograms of polycrystalline Pt (A_{geom} = 0.0707 cm^2) in 0.5 mol L^{-1} H_2SO_4, deaerated with Ar, v = 0.2 V s^{-1}; b) excitation signals 1) true linear potential ramp (CV, blue) and staircase potential excitation signals (gray to black) with step heights of 2) 0.2 mV, 3) 0.5 mV, 4) 1 mV, 5) 2 mV, 6) 5 mV. Experimental data were provided by Julian Lorenz, DLR Institute of Networked Energy Systems, Oldenburg. The voltammograms have been recorded with potentiostats from different manufacturers.

Normal Pulse Voltammetry (NPV). In this technique, potential pulses of increasing height are added sequentially to a base potential at which no electrode reaction occurs. The height of the pulses is incremented from pulse to pulse by ΔE (Figure 5.42). If the pulse potential is sufficient to drive an electrochemical reaction, the resulting current $I(t)$ during the pulse time t_p represents a chronoamperogram from which a current value is stored at a predetermined time t_s (sampling time). If the pulse potential can cause a diffusion-controlled reaction, the current follows the Cottrell Equation (4.130). In order to ensure that this is also the case for subsequent pulses, the initial concentration must be recovered at the electrode surface between the pulses, that is, the initial concentration must be reestablished (the "concentration scar" must be smoothened). This is easily achieved at an SMDE by dislodging the Hg drop and forming a new one. At solid electrodes, the renewal of the concentration close to the electrode is realized during a waiting time t_w, during which a short stirring phase may facilitate the process, for instance, at a rotating disk electrode.

If the local concentrations can be reset to the initial conditions between the pulses, a sigmoidal (step-like) voltammogram is obtained. The half-wave potential is identical to that in polarography. It is linked to the formal potential $E^{\circ\prime}$ by Eq. (5.57). The height of the step is defined via the Cottrell Equation (5.65).

$$I_{\text{lim,NPV}} = \frac{nFAc_R^* \sqrt{D_R}}{\sqrt{\pi t_s}} \tag{5.65}$$

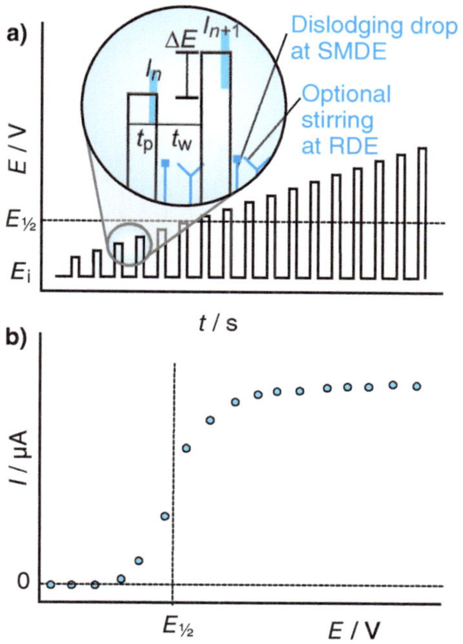

Figure 5.42 a) Excitation signal and b) schematic voltammogram for normal pulse voltammetry. Initially, the solution contains only the reduced form that is then oxidized at the electrode.

Because the limiting current in NPV does not depend on the kinetics of the electron transfer reaction, the method can be used for determining the number n of transferred electrons if concentration and diffusion coefficient are known. The method can also be very helpful for reactions accompanied by film formation on the electrode because the electrode is held at reaction potentials for a short time only. This helps to limit the influence of unwanted side reactions and follow-up processes on the reaction under study.

The so-called slope analysis provides an estimate for the reversibility (fast and invertible) of an electrode reaction and can be used for NPV and some other voltammetric techniques with sigmoidal waves. The required condition set out in Eq. (3.39) states that the local concentrations directly at the electrode surface must always obey the Nernst equation. Analogous to the derivation in Excursion 5-2, a reversible system will follow Eq. (5.66) for NPV.

$$\begin{aligned} E &= E_{1/2} + \frac{RT}{nF} \ln \frac{I_{\text{lim,NPV}} - I(E)}{I(E)} \\ &= E_{1/2} + \frac{2.3RT}{nF} \log \frac{I_{\text{lim,NPV}} - I(E)}{I(E)} \end{aligned} \tag{5.66}$$

A plot of E versus $\log[(I_{\text{lim,NPV}} - I)/I]$ gives a line with a slope of 59 mV/n at 25 °C (298 K).

Reverse Pulse Voltammetry (RPV). This method is suitable for studying the products of an electrode reaction in mechanistic investigations [65]. Similar to NPV, the excitation signal is a sequence of potential pulses of increasing heights. However, the initial situation for each pulse is completely different. The base potential E_b is selected in order to cause a diffusion-controlled reaction. Each pulse starts after preelectrolysis when the solution close to the electrode has been depleted in the reactants and enriched in products similar to the concentration profiles in Figure 4.52. Starting from this situation, potential pulses are applied with the duration t_p and increasing heights that eventually reach the potential required for the reverse reaction

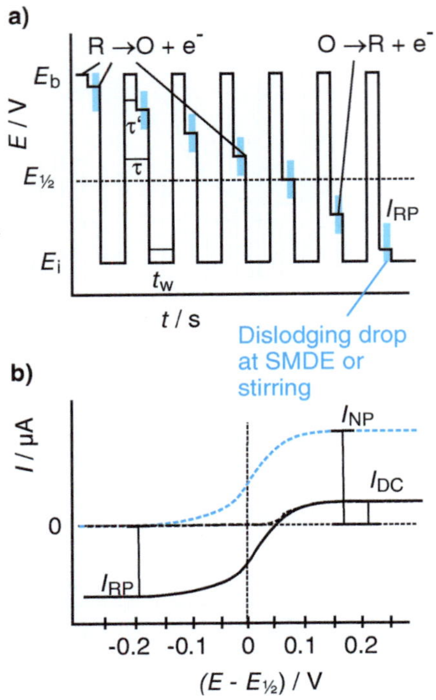

Figure 5.43 a) Excitation signal and b) schematic current response for reverse pulse voltammetry for a reversible electrode reaction (black). The response of an NPV experiment is shown for comparison as a blue dashed line. The RPV response of a chemically irreversible system is given as a black dashed line. I_{DC} is the diffusion-controlled current at the base potential. The solution initially contains the reduced form in this example.

to proceed (Figure 5.43). If the pulses exceed this potential, the compound formed at E_b will react back to the reactants, provided that the compound has not been consumed in a homogeneous chemical follow-up reaction. Consequently, the method is used mainly for verification of chemical reversibility (Section 3.2.3).

The initial concentration conditions can be reset at an SMDE by the falling droplet. At solid electrodes, an additional potential step at the potential E_i is necessary. There is no electrode reaction at E_i, allowing the reactant concentration at the electrode to recover to its initial value. This process can be accelerated by temporarily switching on a stirrer.

For evaluation of the data, the experimental current plateaus from the reverse pulses are compared to the calculated current for a reversible system. Using the relationship $\tau = \tau' + t_p$ the plateau current $I_{\text{lim,DC}}$ is obtained by a modified Cottrell Equation (5.67) for the potentials close to E_b, for which the diffusion-controlled preelectrolysis also occurs during the reverse pulse.

$$|I_{\text{lim,DC}}| = \frac{nFAc_R^*\sqrt{D_R}}{\sqrt{\pi}\sqrt{\tau}} \tag{5.67}$$

If the back reaction occurs during the reverse pulse at a diffusion-controlled rate, the current is given by Eq. (5.68).

$$|I_{\text{lim,RP}}| = \frac{nFAc_R^*\sqrt{D_R}}{\sqrt{\pi}}\left[\frac{1}{\sqrt{\tau}} - \frac{1}{\sqrt{t_p}}\right] \tag{5.68}$$

The comparison to the corresponding NPV experiment gives Eq. (5.69).

$$|I_{\text{lim,NPV}}| = |I_{\text{lim,DC}}| + |I_{\text{lim,RPV}}| \tag{5.69}$$

The absolute values of the currents must be used in Eq. (5.69) regardless of the current convention. Specifically, $I_{\text{lim,DC}}$ and $I_{\text{lim,RPV}}$ will have an opposite sign.

Alternating Current Voltammetry (ACV). Alternating current voltammetry uses a combination of a linear potential ramp E_{DC} (often approximated by a staircase profile on microprocessor-controlled potentiostats) and an AC voltage E_{AC} with a small amplitude ΔE of 5–20 mV.

$$\begin{aligned} E(t) &= E_{DC} + E_{AC} \\ &= E_1 + v \cdot t + \Delta E_{AC}\cos(\omega t) \end{aligned} \tag{5.70}$$

The frequency of the AC voltage is within the range of 10–600 Hz (Figure 5.44). There are mainly three fields of application for this technique:

- One observes the change in the current response as a function of the applied frequency (often in an even wider range than stated above) at constant E_{DC}. This technique is called electrochemical impedance spectroscopy, and we devote the entire Chapter 7 to it.
- ACV is conducted with constant frequency and changing E_{DC} in a solution containing a redox system. This provides a very sensitive detection for reversible redox systems. Mechanistic studies are possible for quasireversible and irreversible redox systems.
- At constant frequency and changing E_{DC}, the differential double layer capacitance can be determined very sensitively in a solution without a redox system. This can be applied to concentration determination of surface-active compounds that adsorb to the electrode surface and change the double layer capacity. This method is also called tensammetry.

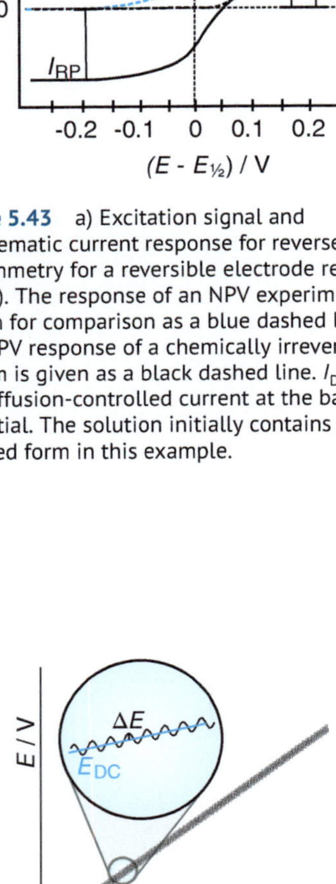

Figure 5.44 Excitation signal of AC voltammetry.

All consideration of ACV starts from the assumption that the rate of change is so different for the DC voltage and for the AC voltage that we can consider both components as decoupled. This will hold when the scan rate of the DC voltage $v = dE_{DC}/dt$ is much smaller than the rate of change in the AC voltage $\Delta E_{AC} \cdot \omega$. The redox systems may be divided into reversible/quasireversible or irreversible with respect to the DC voltage and with respect to the AC voltage. As we have already discussed in the context of SQWV, a peak-shaped signal results for redox systems that are reversible with respect to the AC voltage. The peak height is proportional to the concentration and the square root of the frequency (Eq. (5.71)). The peak height decreases considerably when the system reacts slowly (quasireversible or irreversible with respect to ω). This enables the detection of reversible systems by ACV or SQWV in the presence of an irreversible system. This can be used to detect analytes in the presence of dissolved oxygen without deaerating the solution, which considerably speeds up and simplifies the entire procedure [66, 67].

$$\Delta I_{max} = \frac{n^2F^2A\sqrt{D_O} \cdot \sqrt{\omega}\Delta E_{AC}}{4RT} \cdot c_O^* \tag{5.71}$$

ΔI_{max} – Peak current [A]
ΔE_{AC} – Amplitude of the AC voltage [V]
ω – Angular frequency = $2\pi f$

For quasireversible systems, a number of physicochemical parameters can be obtained by ACV; for example, the symmetry factor α can be determined very sensitively [1]. ACV can be performed particularly sensitively by means of a lock-in amplifier, which offers the opportunity to select the phase angle. Capacitive current dominates at a phase angle of 90°, and faradaic current prevails at smaller phase angles at which the capacitive current is largely suppressed.

In a solution that does not contain redox-active components, the change in the capacitive current with the DC electrode potential can be followed. In this situation, the current is determined by the differential double layer capacitance C_{dl}, which is diminished when neutral organic compounds adsorb to the electrode (Figure 4.39). If the DC potential is scanned over the range in which the electrode material and the electrolyte are stable (Figure 5.45), the drop in C_{dl} can be seen in the potential range between −0.4 and −1.5 V, in which a surfactant (here a long-chain alcohol) is adsorbed on the electrode surface. This variation of phase-selective ACV is called tensammetry. The decrease in C_{dl} is proportional to the *interfacial* concentration and can be used to measure adsorption isotherms of surface-active compounds that are *not* redox-active [56, 68]. The potential range in which adsorption occurs is flanked by two peaks in the C_{dl}–E_{DC} plot. At the potentials of the peaks, desorption occurs. During this process, the charge must flow toward the interface momentarily, which would normally accumulate at the bare interface continuously during the potential sweep. The heights of the maxima depend on the interfacial concentration of the surfactant but are also influenced by other experimental parameters. We will consider the thermodynamics of adsorption in more detail in Section 8.10.

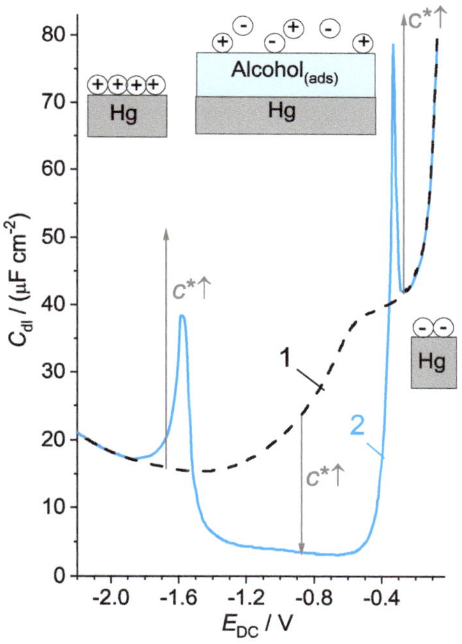

Figure 5.45 Tensammetric investigation at 1) a bare Hg in 0.5 mol L^{-1} Na_2SO_4 solution and 2) the same electrode after addition of *n*-heptanol to the electrolyte. The reference electrode is Hg|Hg_2Cl_2|1 mol L^{-1} KCl. The arrows indicate the change in the curve with increasing solution concentration of *n*-heptanol. Source: Adapted from [69].

5.3.10 Methods with Integrated Preconcentration (Stripping Methods)

The limit of detection of voltammetric methods can be lowered (improved) by 2 to 3 orders of magnitude by means of an integrated preconcentration step. This enables analysis of ultratrace concentrations (down to about 10^{-10} mol L^{-1}) with comparably simple instrumentation. This instrumentation is portable and may be used on board research vessels, buoys, monitoring stations, or in mobile laboratories [70, 71]. The extreme sensitivity requires careful compliance with the general principles of trace analysis. This includes the avoidance of any sources of contamination. The samples may, for instance, take up trace elements during storage in sample vials (heavy metals from ordinary glass) or during pipetting (Cd from polymers or from consumables

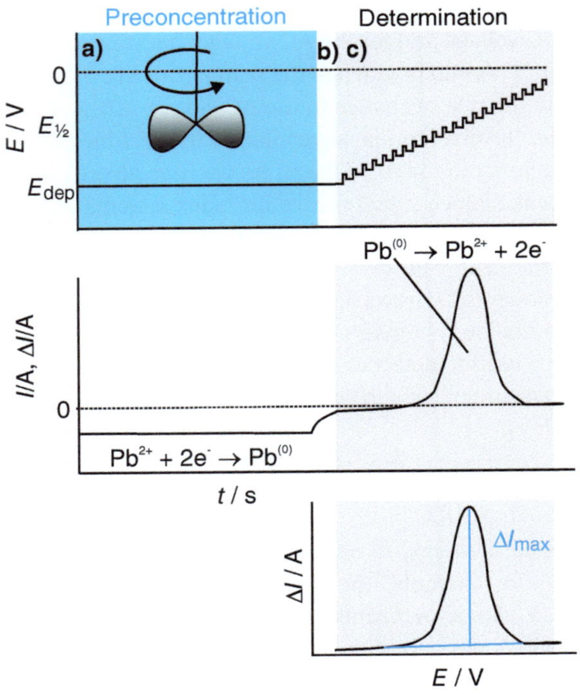

Figure 5.46 Principle of anodic stripping voltammetry using the example of Pb^{2+} determination.

Top: Excitation signal for a) preconcentration by electrolytic deposition of Pb^{2+} at $E_d \ll E_{½}(Pb/Pb^{2+}) \approx -1.0$ V under defined and constant stirring of the solution, for instance, with a rotating cylinder; b) Resting period after stirring under continuous deposition; c) determination step with DPV in the direction of positive potentials.

Middle: Current–time response signal. The plot shows the DC current in the preconcentration step and the current difference in the determination step (not to scale!).

Bottom: Plot of the resulting stripping voltammogram. Only the difference in current ΔI is plotted over the step potential during the determination step. The peak current is measured against a baseline and used for further evaluation.

for sample handling). They may also lose trace metal ions originally contained in the sample by adsorption to the wall of the sample container. Because of the overall small amount of the analyte in the sample, a small amount of substances can cause a significant relative change in solution concentration.

Anodic Stripping Voltammetry. This is the most frequently applied analytical stripping method. Typical analytes are metal ions that are electrolytically deposited as an amalgam in the drop of an HMDE under potential control and while stirring the solution during a preconcentration step (Figure 46a).

$$Me^{n+} + ne^- + Hg \rightarrow Me(Hg) \tag{5.72}$$

If the electrolysis is conducted for 10–300 s in a sample volume of 5–25 mL, only a small fraction of the analyte contained in the sample is actually converted. This may be quite different when working in microcells (e.g., for the determination of heavy metals in tear drops). In a volume of 0.1 mL, complete electrolysis is already achieved after 15 min. Prolonging the electrolysis time does not make sense. At the end of the preconcentration step, the concentration of the analyte atoms in the Hg drop is much larger than the previous concentration of metal ions in the original sample. This is the reason for the enormous change in sensitivity and the resulting lower limits of detection. Before the end of the preconcentration, the convection is stopped (Figure 46b) in order to conduct the subsequent determination step in quiescent solution. The electrode reaction during the determination step (Figure 5.47),

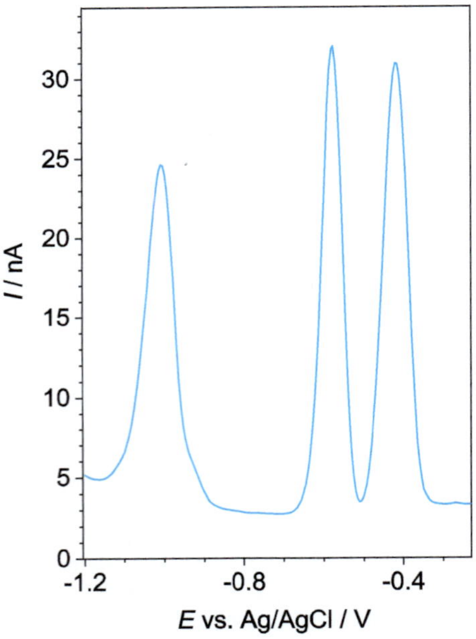

Figure 5.47 Anodic stripping voltammogram of 3×10^{-8} mol L^{-1} Pb^{2+}, 3×10^{-8} mol L^{-1} Zn^{2+}, 3×10^{-8} mol L^{-1} Cd^{2+} in 0.1 mol L^{-1} KNO_3 at a HMDE after electrolytic preconcentration at $E_d = -1.2$ V for 60 s. Source: Data were supplied by Izabella Brand and Nico Roth, University of Oldenburg.

$$Me(Hg) \rightarrow Me^{n+} + ne^- + Hg \tag{5.73}$$

can be followed by LSV, SCV, DPV, SQWV, or ACV (Figure 5.46c). Subsequently, the drop of the HMDE is dislodged, and the next determination is carried out at a new Hg drop. An example for the determination of heavy metals with DPV in the detection step is given in Figure 5.47.

Q: If there is something like anodic stripping voltammetry, is there also cathodic stripping voltammetry?

A: Yes, there is. In cathodic stripping voltammetry, anions (e.g., chloride as a trace component in suprapure chemicals) are preconcentrated as a salt layer of low solubility on the surface of the electrode, for instance,

$$Ag^{(0)} - Cl^- \rightarrow AgCl + e^-.$$

The reductive dissolution occurs in the determination step according to

$$AgCl + e^- \rightarrow Ag^{(0)} + Cl^-.$$

The mercury film electrode (MFE) is commonly used for stripping analysis. An MFE is prepared by electrolytic deposition of metallic Hg on glassy carbon or paraffin-impregnated graphite electrodes by reduction of Hg^{2+} during the preconcentration step. The analysis solution is spiked with 10^{-5} mol L^{-1} to 10^{-4} mol L^{-1} Hg^{2+}. The Hg layer of an MFE will be only about 10 nm thick, but the amount of deposited metals from the sample is roughly equal to that deposited at an HMDE. The enrichment factor and thus the sensitivity are therefore higher at an MFE. This advantage is offset by the need to handle Hg^{2+} with the additional need to take care of the finely dispersed $Hg^{(0)}$ of the MFE and the remaining Hg^{2+} in the sample solution. Hg^{2+} is significantly more toxic than metallic Hg. For elements that cannot be determined at Hg electrodes (Ag, Au, Hg), solid electrodes (Au, Pt, C) are employed. Unfortunately, they work less reliably than Hg electrodes because it is more difficult to renew the surface reproducibly. Reproducibility is one of the decisive prerequisites for low detection limits. For a linear calibration function $y = S \cdot c + b$ between the signal y, the analyte concentration c, and the blank value b, the sensitivity is given by $S = \partial y / \partial c$. The limit of detection (LOD) c_{LOD} is the concentration at which the signal y exceeds the standard deviation of the blank value σ_b by a factor of 3 ($3\sigma_b = S \cdot c_{LOD} \Rightarrow c_{LOD} = 3\sigma_b / S$). A high reproducibility means a small σ_b, which, along with a high sensitivity S, positively influences the limit of detection, which means pushing c_{LOD} to lower values [5–7].

The combination of electrolytic deposition and electrolytic dissolution allows preconcentration for metal ions that form amalgams. However, some metal ions form intermetallic compounds upon reduction at a Hg electrode. Many anions are difficult to preconcentrate [70]. Table 5.8 provides an overview of processes that are applied for preconcentration of such analytes. The adsorptive accumulation of Ni^{2+} as its dimethylglyoxime complex, followed by the oxidative conversion of the ligand during the determination step, is an example of adsorptive stripping voltammetry. Some anions, for example, halogenides and CN^-, can be accumulated on metal electrodes as salts of low solubility (see Q/A box above) from trace concentrations in solution. The reductive decomposition of salt provides the signal in the detection step, thus the term cathodic stripping voltammetry. Very carefully developed and proven analytical procedures can be found in Application Bulletins and Application Notes on the website of major instrumentation manufacturers (or they can be requested) [72, 73]. These protocols provide treasure troves of practical experiences and help avoid beginner's mistakes. Note that the combination of preconcentration and determination steps is only the electrochemical steps of the analytical procedure. There are many more steps, starting with taking a sample and including almost always a digestion or extraction step [74]. The details of sample preparation depend on the matrix and the concentration. The application examples from device manufacturers usually contain complete details for the sample preparation and analysis.

Table 5.8 Combinations of preconcentration steps and determination steps for stripping voltammetry with representative examples.

Reduction to a metal			
a) Formation of an amalgam	$M^{n+} + ne^- \rightarrow M^0(Hg)$	$M^0 \rightarrow M^{n+} + ne^-$ (anodic stripping voltammetry)	a) Zn, Cd, Bi, Pb, Tl (HMDE)
b) Deposition of a metal film	$M^{n+} + ne^- \rightarrow M^0$		b) Au, Ag, Hg (carbon electrode)
Adsorptive preconcentration of coordination complexes or complexing agents	$M^{n+} + k\text{HL} \rightarrow ML_k(\text{sol}) + m\,H^+$ $MeL_{k(\text{sol})} \rightarrow ML_{k(\text{ads})}$	$ML_k \rightarrow ML_k^{\ +} + e^-$ (oxidation of the metal ion or the ligand, adsorptive stripping voltammetry)	Ni^{2+} + dimethylglyoxime
	$L_{(\text{sol})} \rightarrow L_{(\text{ads})}$ $M^{n+} + L_{(\text{ads})} \rightarrow ML_k^{\ n+}{}_{(\text{ads})}$		Zr, U + phosphoric acid ester
Formation of precipitates of low solubility by electrochemical oxidation without reagent addition	$M^{n+} + (k+1)\,OH^- \rightarrow Me(OH)_{k+1} + e^-$	$M(OH)_{n+1} + e^- \rightarrow M^{n+} + (k+1)\,OH^-$	Mn, Tl, Ce, Pb
Formation of precipitates of low solubility *with* reagent addition			
a) by electrochemical oxidation	$M^{n+} + (k+1)\,HL \rightarrow ML_{k+1} + e^- + (n+1)\,H^+$	$ML_{k+1} + e^- \rightarrow M^{n+} + (k+1)\,L^-$	Co^{2+} + α-nitroso-β-naphthol
b) by electrochemical reduction	$M^{(n+1)+} + e^- + k\,HL \rightarrow ML_k + n\,H^+$	$ML_k \rightarrow M^{(n+1)+} + e^- + k\,L^-$	Sn^{4+} + diethyl dithiophosphate
Precipitation of anions through formation of salts of low solubility after oxidation of the electrode material	$M^{(0)} + X^{n-} \rightarrow MX + ne^-$	$MX + ne^- \rightarrow M^{(0)} + X^{n-}$ (cathodic stripping voltammetry)	S^{2-}, Cl^-, CN^-, Br^-, organic thiols, M = Ag, Hg
Deposition by coelectrolysis of solution components	$M_1^{n+}, M_2^{n'+} + (n+n')\,e^- \rightarrow M_1^0M_2^0$	$M_1^0M_2^0 + e^- \rightarrow M_1^0 + M_2^{\ -}$ (cathodic stripping voltammetry)	M_1 = Cu M_2 = 1 Se, Te (analyte)

Source: Adapted from [71].

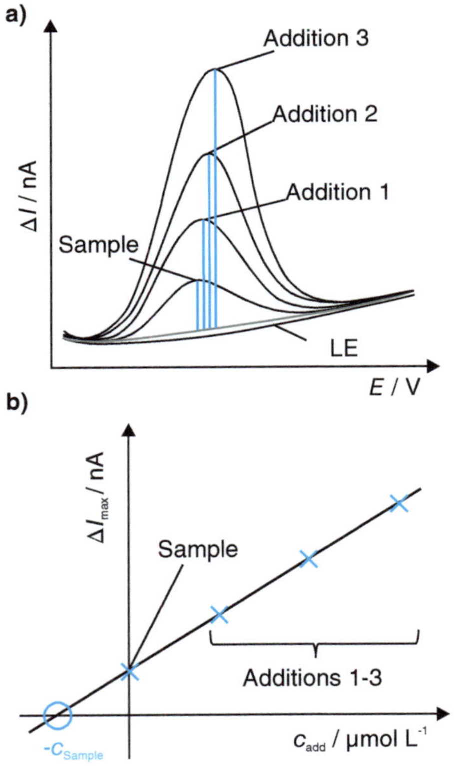

Figure 5.48 Concentration determination by means of standard addition from voltammetric curves; a) voltammetric curves of the sample and after additions of standard solution; b) plot of peak height and determination of the unknown concentration of the analyte in the sample.

5.3.11 The Use of Standard Additions to Mitigate Matrix Effects

For the determination of ΔI_{max}, the peak in voltammetric curves is measured versus a baseline (Figure 5.48a). From data pairs (c^*, ΔI_{max}), a calibration plot can be constructed to relate measured ΔI_{max} values from samples to the concentration. One of the important disadvantages of voltammetric concentration determination is the relatively strong dependence of the signal on the matrix. Components of the matrix may be complexed metal ions that are then not accessible for preconcentration. Substances that are not redox active by themselves may adsorb to the electrode surface, form a thin film that slows down (i.e., inhibits) the rate at which the analyte ions react. Furthermore, the diffusion coefficient D_O may change with the concentration of matrix components. This is the reason why a digestion procedure is usually used that completely mineralizes the organic compounds of the sample (this means oxidizes them to CO_2 and H_2O) or at least degrades them to small hydrophilic compounds [74]. Despite this effort, it is usually impossible to exclude matrix effects completely.

Because the matrix components themselves may vary in concentration, for instance, in environmental samples, **standard additions** are often employed as a calibration routine, taking into account the possible matrix effects. First, a DP voltammogram is recorded from the digested sample (Figure 5.48a). Next, small volumes of a standard solution are added with a precisely *known* concentration of the analyte. After each addition, the measurement is repeated, and the peak current is measured versus the baseline. The additions are made in such a way that the concentration and the signal roughly double, triple, and quadruple compared to that of the sample. Subsequently, the ΔI_{max} values are plotted against the *added* concentration c^*_{add} calculated using Eq. (5.74).

$$c^*_{add} = \frac{V_{std}\, c_{std}}{V_{tot}} = \frac{V_{std}\, c_{std}}{V_0 + V_{std}} \tag{5.74}$$

c^*_{add} – Added concentration of the standard in the measurement solution [mol L^{-1}]
c_{std} – Concentration of the analyte in the stock solution of the standard [mol L^{-1}]
V_{std} – Volume of the added standard solution [L]
V_{tot} – Total volume = volume of the sample + V_{std} [L]
V_0 – Initial volume of the sample solution [L]

ΔI_{max} of the sample before the standard addition is plotted at $c^*_{add} = 0$. A regression line is determined and extrapolated toward $\Delta I_{max} = 0$ (Figure 5.48b). The absolute value of the concentration at the intersection of abscissa ("*x*-axis") and the regression line is the sought concentration of the analyte in the sample.

5.3.12 Speciation by Means of Voltammetric Methods

The term **speciation** refers to the analysis of binding states, that is, the analytical differentiation of ions of one and the same element in different binding states. This is very important due to the largely different toxicity and bioavailability of ions and compounds of one and the same element. For instance, the toxicity of mercury compounds increases in the sequence $Hg^{(0)} < Hg^{2+} <$ organic mercury compounds. CrO_4^- is much more toxic than Cr^{3+}. Most organisms can uptake only Fe^{2+}. Therefore, primary production of aquatic ecosystems may be limited by Fe^{2+} despite abundance of Fe^{3+}. Bioavailability is also influenced by complexation. Because the half-wave potential (formal potential) is specific for a compound (and not for an element, as in atomic spectroscopy), voltammetric methods are especially suited for speciation analysis. Let us discuss first how complexation changes the half-wave potential or formal potential. We consider a complexation equilibrium of the metal ion M^{m+} with k ligands L.

$$M^{m+} + k\,L \rightleftharpoons ML_k{}^{m+} \tag{5.75}$$

The charge of the ligand is not stated in order to keep the notation simple. The general treatment would not be changed if a charged ligand (e.g., CN^-, $S_2O_3^{2-}$) were used. The tendency of complexation is quantified by the overall stability constant β of the complex.

$$\beta = \frac{\{ML_k{}^{m+}\}}{\{M^{m+}\}\{L\}^k} = \frac{[ML_k{}^{m+}]}{[M^{m+}][L]^k}\frac{f_{ML}}{f_M f_L^k} \tag{5.76}$$

Exemplary values can be found in Table A-4 of the Appendix. The prefix "overall" expresses that we summarize over all steps for a stepwise complexation reaction. For the expression of the formal potential, we need a similar expression using concentrations instead of activities in the mass action term. The corresponding modified constant is called **conditional overall stability constant** β'.

$$\beta' = \beta\frac{f_M f_L^k}{f_{ML}} = \frac{[ML_k{}^{m+}]}{[M^{m+}][L]^k} \tag{5.77}$$

Rearranging Eq. (5.77) gives the concentration of free (noncomplexed) metal ions.

$$[M^{m+}] = \frac{[ML_k^{m+}]}{\beta'[L]^k} \tag{5.78}$$

Inserting this expression into the equation of the potential of a metal ion electrode [Eq. (2.37)] gives a modified Nernst equation.

$$E = E^{\circ\prime}{}_{M(Hg)|M^{n+}} + \frac{RT}{nF}\ln\frac{[M^{m+}]}{[M(Hg)]} = E^{\circ\prime}{}_{M(Hg)|M^{n+}} + \frac{RT}{nF}\ln\frac{[ML_k^{m+}]}{[M(Hg)]\beta'[L]^k} \tag{5.79}$$

Rearranging provides Eq. (5.80).

$$E = E^{\circ\prime}{}_{\mathrm{M(Hg)|M}^{n+}} + \frac{RT}{nF}\ln\frac{1}{\beta'[\mathrm{L}]^k} + \frac{RT}{nF}\ln\frac{[\mathrm{ML}_k^{m+}]}{[\mathrm{M(Hg)}]} \tag{5.80}$$

According to Eq. (5.57), the half-wave potential $E_{1/2}$ is within a fixed, small distance of $E^{\circ\prime}$. The half-wave potential of a voltammetric curve in the presence of an excess of a complexing ligand shifts relative to the half-wave potential without any ligand (more precisely with only the ligand H_2O present) according to Eq. (5.81).

$$E'_{1/2} = E_{1/2} - \frac{RT}{nF}\ln\beta'(c^\circ)^k - k\frac{RT}{nF}\ln\frac{[\mathrm{L}]}{c^\circ} \tag{5.81}$$

$E_{1/2}$ – Half-wave potential in the absence of the ligand

The standard concentration $c^\circ = 1\ \mathrm{mol\,L^{-1}}$ serves the management of the units, because we can calculate the logarithm only from real numbers but not from a quantity with a unit. In the preceding equations, the units have cancelled.

Using the decadic logarithm and the tabulation temperature of 298 K, we obtain Eq. (5.82).

$$E'_{1/2} = E_{1/2} - \frac{0.059\ \mathrm{V}}{n}\log\beta'(c^\circ)^k - k\frac{0.059\ \mathrm{V}}{n}\log\frac{[\mathrm{L}]}{c^\circ} \tag{5.82}$$

The position of the voltammetric signals is shifted in the negative direction in the presence of the ligand. The stronger the ligand binds (larger β') and the greater the ligand concentration excess, the larger the shift. The dominating effect is caused by the term $\log\beta'$. In the presence of a strong ligand, the metal appears less noble (the dissolution occurs already at more negative potentials than in a noncomplexing electrolyte solution). We may regard Eq. (5.82) as a straight-line equation with $\log([\mathrm{L}]/c^\circ)$ as the argument ("x"). From a measurement in the absence of the ligand, we obtain $E_{1/2}$. Further measurements with different [L] (sensible variation on a logarithmic scale!) provide values ($\log([\mathrm{L}]/c^\circ$, $E'_{1/2}$) for a regression line. Its intercept with the ordinate ("y-axis") is $\log\beta'$ and the slope can be used to determine the number k of the complexing ligands.

Voltammetry is not restricted to the determination of the most toxic metal ions in drinking water, process water, cooling water, or wastewater. The concentration and binding strength of possible ligands are also of interest. Some ligands are not easy to detect, especially at trace concentrations, and may have considerable adverse effects when released into the environment. Therefore, there is a need to monitor especially the industrially used chelating ligands ethylenediamine tetraacetic acid (EDTA) and nitrilotriacetic acid (NTA). They cannot be determined satisfactorily by optical methods. In contrast, electrochemical methods are well-suited and have been adopted for standardized analytical protocols [75]. An excess of Bi^{3+}, a heavy metal ion of relatively low toxicity, is added to a sample that may contain NTA or EDTA. The addition is made until all NTA and EDTA are converted to the Bi complexes $[Bi(edta)]^-$ or $[Bi(nta)(H_2O)_3]$ with an excess of Bi^{3+} remaining uncomplexed. Acidification of the sample to pH 2 with HNO_3 ensures that only the 1 : 1 complex of Bi^{3+} and NTA is formed. A DP voltammogram is recorded between +0.1 V and −0.7 V vs. Ag|AgCl|KCl(*sa*) as shown in Figure 5.49. Peak 1 at −0.05 V results from noncomplexed Bi^{3+}. EDTA ($\log\beta'$ = 22.8) is an even stronger complexing agent than NTA ($\log\beta'$ = 18.5). Therefore, peak 2 with the lower shift at −0.25 V results from the NTA complex, and peak 3 at −0.48 V is assigned to the EDTA complex. The concentration is determined by standard additions, in which a standard solution with known concentrations of NTA and EDTA is employed. Even better is the use of a standard solution in which the entire amount of the complexing agents is available as their Bi complexes because this ensures the required excess of noncomplexed Bi^{3+} in the measurement solution without further additions. After standard addition, the peaks for $[Bi(edta)]^-$ or [Bi(nta)] grow. The original concentrations $[NTA]_0$ and $[EDTA]_0$ are obtained from a standard addition experiment in a similar way as shown in Figure 5.48.

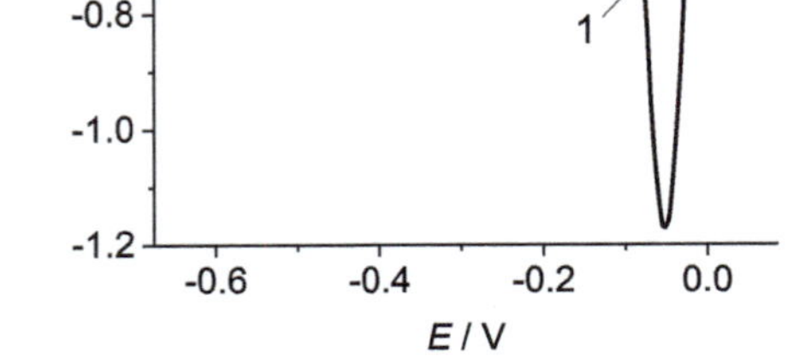

Figure 5.49 DP voltammogram of 3 μmol L^{-1} EDTA and 4 μmol L^{-1} NTA after addition of 20 μmol L^{-1} Bi^{3+}.

Samples of surface water often contain a mixture of organic compounds without precisely known molecular structures but with the ability to complex metal ions. Depending on sampling location, this can be humic substances, amino acids, etc.

These organic compounds influence the cycle of substances of metal ions, which is important for risk assessments with respect to toxic metal ions but also for the bioavailability of essential metal ions. The complexing capacity of a water sample is a sum parameter characterizing the ability of the sample to bind metal ions.

For the determination of the complexing capacity, one determines up to which (trace) concentration the mixture of complexing agents in the water sample is able to mask the trace determination of Pb^{2+} by stripping voltammetry using a deposition potential $E_{dep} = -0.9$ V [56]. By setting E_{dep}, we fix the minimum binding strength (otherwise described by $\log \beta'$) a ligand must have in order to be "counted" for the sum parameter [Eq. (5.82)]. The water sample (without digestion step) is spiked with small concentrations of Pb^{2+} followed by recording stripping voltammograms. After the first additions, no significant stripping peak is recorded for the dissolution of deposited Pb because the ligands in the sample bind Pb^{2+} and prevent it from being electrodeposited at E_{dep} as the corresponding amalgam (Figure 5.50). If the added Pb^{2+} concentration exceeds the concentration of strong complexing compounds in the sample, some Pb^{2+} will be strongly bound to ligands in the sample, but the remaining Pb^{2+} can only be weakly coordinated. During the subsequent preconcentration step, weakly coordinated Pb^{2+} is reduced at the HMDE and gives a stripping peak, which then grows linearly with each further addition of Pb^{2+}. The complexing capacity is the concentration at the intersection of the horizontal and the linearly increasing part of the plot in Figure 5.50.

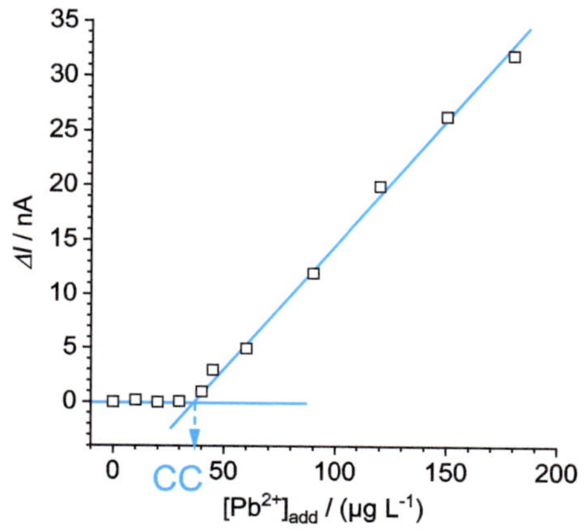

Figure 5.50 Principle of the determination of the complexing capacity (CC) of water. A DP stripping voltammogram is initially recorded from the sample without addition of a supporting electrolyte (surface water has a sufficient conductivity due to the natural salt content). Next, further stripping voltammograms are recorded after addition of small concentrations of a Pb^{2+} stock solution. The concentration $[Pb^{2+}]_{add}$, at which the complexing capacity of the sample is exhausted, is the concentration value of the intersection of the two linear parts of the curve.

5.4 Amperometric Sensors

Amperometry, that is, the measurement of a faradaic current after application of a constant potential, is not very selective because all redox-active compounds that can be converted at the applied potential will contribute to the signal. A chemically selective sensor can only be obtained when combining amperometry with another chemically selective principle. We will discuss here three important examples of such a combination. An amperometric determination can be carried out after a separation by high-performance liquid chromatography (HPLC). A nonselective determination can also be used for end point indication of a titration (as discussed before for ion-selective electrodes), and finally, a membrane can be used for a selective passage of only a few compounds to the electrode surface (e.g., gases).

In general, the amperometric signal is proportional to the particle flux to the surface of the amperometric sensor. The particle flux is connected by Fick's first law to the bulk concentration c^* of the analyte. Note the difference to potentiometric sensors, where there is a proportionality to the logarithm of the activity and thus different sensitivities $\partial\text{signal}/\partial c$ in different parts of the calibration plot.

Q: We have discussed extensively chronoamperometry (CA) in Chapter 4. The concentration appears in the Cottrell Equation (4.130). Can't we use this equation as well?

A: Yes, this is possible. However, the solution in the experiment discussed in Chapter 4 contained only *one* redox-active compound of *known* bulk concentration. In this situation, we could apply CA in order to characterize the diffusion process. During the application of sensors, the sample typically contains several redox-active compounds with unknown concentrations. They may also depart from simple planar diffusion due to specific geometries of the sensor. Therefore, we use the term CA in this textbook only in cases where we follow the change in current with time after a potential step, but not if the current changes with time because of changing concentration c^*, for instance, at the end of a chromatographic column.

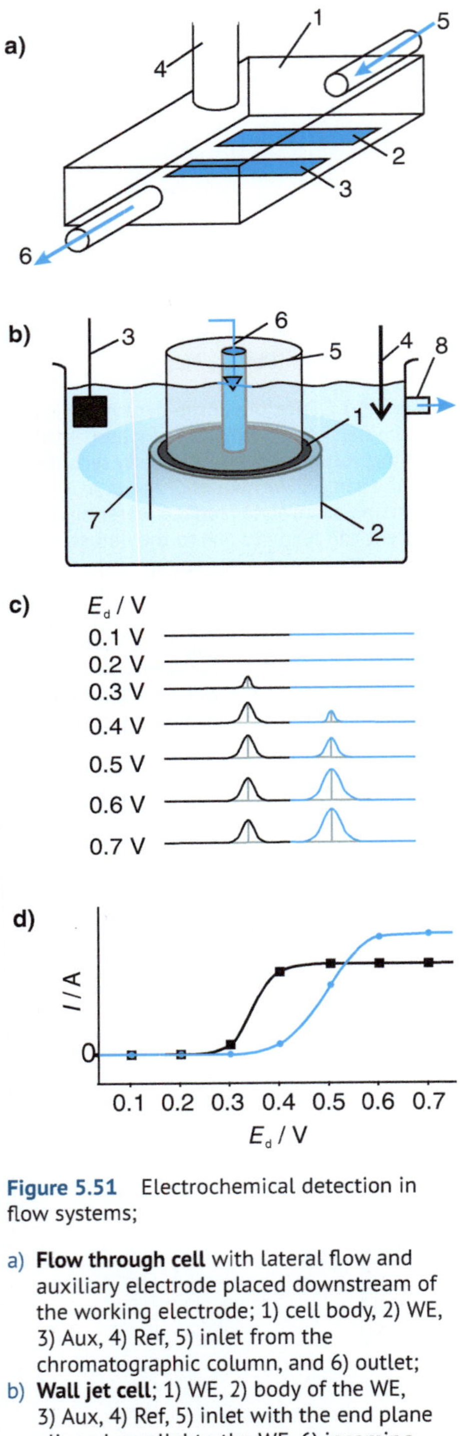

Figure 5.51 Electrochemical detection in flow systems;

a) **Flow through cell** with lateral flow and auxiliary electrode placed downstream of the working electrode; 1) cell body, 2) WE, 3) Aux, 4) Ref, 5) inlet from the chromatographic column, and 6) outlet;
b) **Wall jet cell**; 1) WE, 2) body of the WE, 3) Aux, 4) Ref, 5) inlet with the end plane aligned parallel to the WE, 6) incoming solution, 7) solution emerging from the detection slit, mixing with the background electrolyte, and 8) outlet;
c) **Chromatograms** for different E_d;
d) **Hydrodynamic voltammograms** obtained from the peak areas in (c).

5.4.1 Amperometric Detection in Flow Systems

For amperometric detection, the working electrode is embedded into the wall of the electrochemical cell through which the electrolyte solution passes (flow cell). Such cells are commercially available. Reference and auxiliary electrodes are often placed downstream of the working electrode in order to avoid contamination of the working electrode with products of the reaction at the auxiliary electrode (Figure 5.51a,b). The inlet of the flow cell is connected to the outlet of a chromatographic column or a flow injection system.

Frequently, the detection potential E_d has to be determined experimentally in the specific setup, because the components of a mixture are not available individually, or because a voltage drop in the flow cell causes small shifts of the potential compared to a standard laboratory cell. E_d is determined by hydrodynamic voltammetry, in which E_d is incremented by a constant amount, for example, 50 or 100 mV. For each E_d, a chromatogram is recorded for the same sample composition (Figure 5.51c). The peak area is determined for each compound and plotted separately for each compound against E_d. The resulting plot in Figure 5.51d is called hydrodynamic voltammogram, which is similar to that of normal pulse voltammetry in Figure 5.42. The most suitable detection potential can be selected by exploiting selectivity effects to some degree.

There are a number of variations to this approach [76]. The liquid stream can be directed perpendicular to the surface of a disk electrode that is placed together with the other electrodes within a larger liquid reservoir. When the incoming sample stream contains a segment with a dissolved redox-active compound, it will be detected when it impinges on the electrode surface and then be swept away and immediately diluted in the large liquid reservoir. This arrangement is called a wall-jet cell (Figure 5.51b).

Some detectors contain porous electrodes through which the analyte solution streams. In such an arrangement, a much larger fraction of the analyte molecules can be converted, similar to bulk electrolysis. Such detectors are also called coulometric detectors.

For many analytical samples, repeated recording of chromatograms may *not* yield reproducible signals at the electrochemical detector. This can be caused by a thin film electrodeposited on the working electrode. The film is caused by reaction products from the electrochemical reaction (typical for samples containing compounds with phenolic or alcoholic OH-groups). Radical intermediates can form and cross-link with other compounds in the sample, yielding insoluble polymers that deposit on the working electrode and eventually inhibit the detection (autoinhibition). The effect on analytical determination can be mitigated by pulsed amperometric detection (PAD), during which E_d is only applied for a short time (100 ms) during which the current is recorded (Figure 5.52b). In intervals, a potential is applied in which no electrolysis occurs and no reaction products are formed. After a delay time, E_d is applied again, and a further point of the chromatogram is recorded. Even if only one point per second is stored, this will often be sufficient for a correct recording of the chromatogram. The pulse programs can be refined by incorporation of cleaning steps, during which the deposited compounds are reductively or oxidatively removed (Figure 5.52c). In other cases, compounds of the sample mixture can be adsorptively preconcentrated before the detection pulse.

5.4.2 Amperometric Detection During Titrations

The selectivity of amperometric titration results from the stoichiometry of the redox reaction and *not* from the electrode process. Amperometry is merely used to follow the concentration of the added standard solution. The formation of diffusion layers

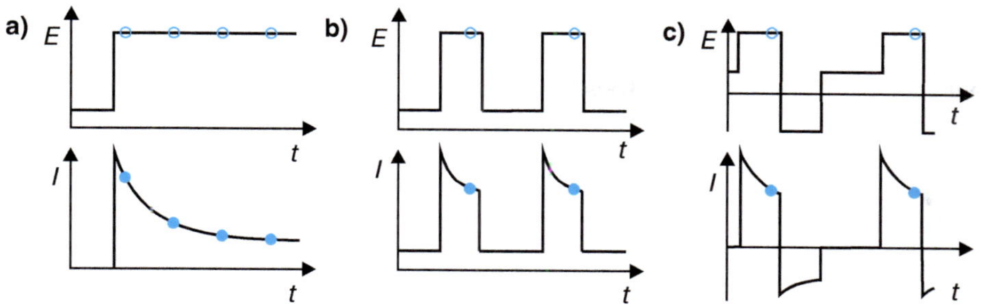

Figure 5.52 Potential and current transients for amperometric detection in flow systems for constant analyte concentration: a) constant potential and current decay due to autoinhibition; b) pulsed amperometric detection; and c) pulsed amperometric detection with cleaning pulse. The recording of the current values for the chromatogram is indicated by blue dots.

and current decay according to the Cottrell equation are avoided by imposing forced convection, for example, by application of a rotating electrode (Section 12.4.2). One of the most important analytical procedures is the determination of water content by Karl Fischer titration[62] [77, 78] and end point indication by means of bipotentiometric detection.

Equations (5.83) and (5.84) are the central reactions for Karl Fischer titration. A 1 : 1 stoichiometry between I_2 and H_2O is achieved by optimization of the reagent mixture.

$$CH_3OH + SO_2 + RN \rightleftharpoons [RNH]^+[CH_3\text{–}OSO_2]^- \tag{5.83}$$

Components in black are components of the working solution.

$$H_2O + I_2 + 2\,RN + [RNH]^+[CH_3\text{–}OSO_2]^- \rightleftharpoons [RNH]^+[CH_3\text{–}OSO_3]^- + 2\,[RNH]^+ + 2\,I^- \tag{5.84}$$

The standard solution is the Karl Fischer reagent containing SO_2, I_2, and a solvent mixture (e.g., methanol, diethyleneglycol-monomethylether $CH_3\text{–}O\text{–}[CH_2\text{–}CH_2\text{–}O]_2\text{–}OH$, and additional solvents depending on the type of sample) and a base RN such as imidazole or diethanolamine (Figure 5.53). The adjustment of the pH within the range 5.7 is important for the chemical reaction to yield only *one* product and to have an exact stoichiometry. The end point of the titration is reached when I_2 remains in the solution and is not consumed anymore.

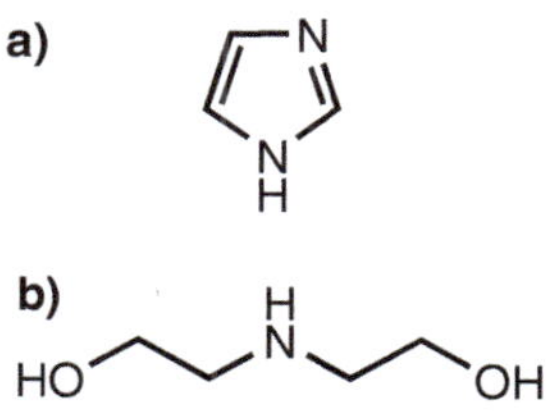

Figure 5.53 Structure of imidazole and diethanolamine.

Instead of storing a relatively unstable solution of I_2, I_2 can be produced by electrolysis directly in the reaction vessel (Figure 5.54). The cell contains *two* electrolysis cells. The cell shown in the left part of Figure 5.54 is called the generator cell, and the cell in the right part of the figure is called the detector cell. The generator is composed of two Pt electrodes, from which the anode contacts the sample solution and produces I_2 while recording the charge consumed in this process. This is particularly straightforward in a galvanostatic operation mode at constant current. In this case, the charge and thus the generated amount of I_2 are proportional to the electrolysis time. The cathode of this cell is located in another compartment and separated by a membrane permeable to ions only. During the electrolysis, one makes use of the fact that I^-/I_2 shows a facile electron transfer reaction on Pt while SO_2/H_2SO_4 reacts sluggishly. This allows selective oxidation of I^- at Pt in the presence of SO_2. I_2 produced at the anode of the generator cell is immediately consumed in the homogeneous reaction (5.84) as long as H_2O is available in the measurement cell.

The detection is carried out at two Pt electrodes that extend as parallel wires out of the glass shaft. A constant current of about 10 μA is applied between them (galvanostatic working mode). At the beginning of the titration, I_2 is formed at one of the Pt wires but in such a low amount that the stoichiometry is almost unaffected. H^+ reduction occurs at the other Pt wire. The resulting cell voltage is similar to that in Figure 2.1, curve 5.

---

[62] Karl Albert Otto Franz Fischer, 1901–1958, German chemist, developed the procedure at the analytical laboratory of Edeleanu Ltd.

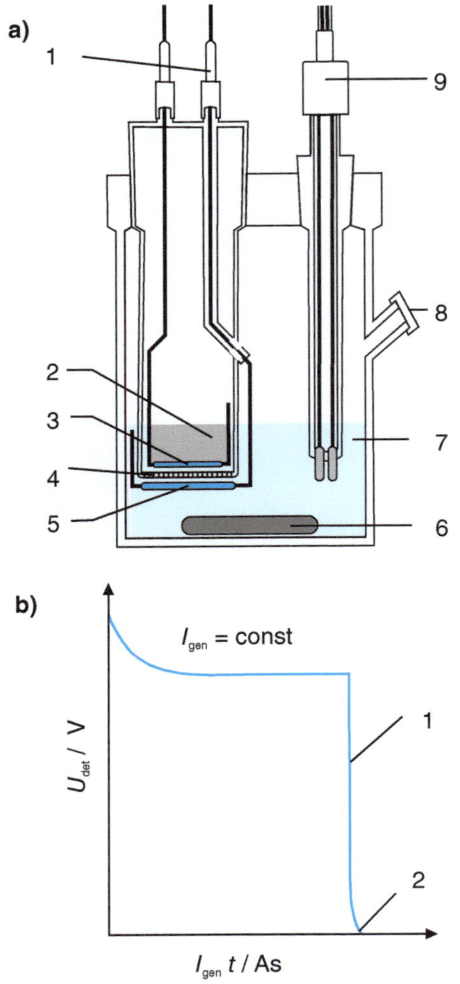

Figure 5.54 Coulometric Karl Fischer titration.

a) **Measurement cell**; 1) electrical connection for the generator cell for production of I_2, 2) electrolyte solution for the cathode compartment, 3) cathode of the generator, 4) ion-permeable membrane, 5) anode of the generator (I_2 formation), 6) magnetic stirring bar, 7) solution of the anode compartment with reagents to which the sample is injected through the septum (8), and 9) cell of two Pt electrodes for detection of the end point.

b) **Schematic titration curve**; 1) voltage drop at the end point and 2) low remaining voltage for the reaction of the system $I_2/2I^-$.

After complete consumption of H_2O in the working solution, I_2 production continues in the generator cell and causes I_2 to accumulate in the measurement solution. This changes the reaction in the detector cell: I_2 formation continues as the reaction at one Pt wire while the other Pt wire changes to I_2 reduction. The required electrolysis voltage drops drastically because I_2/I^- is a facile redox couple at Pt electrodes. The drop in electrolysis voltage is used as a very sensitive indicator (dead stop indication).

A similar experimental system can also be used for other iodometric titrations. For practical applications, one first titrates traces of H_2O that may be contained in the reagent solution. Next, a precisely measured mass or volume of the sample is added through the septum to the reagent solution, and a second titration experiment is carried out in the *same* solution. This method is routinely applied to measure the (residual) water content in many products during quality assurance and in research comprising such a wide range as processed food, cosmetics, pharmaceutical products, or organic solvents.

5.4.3 The Clark Electrode

The Clark electrode[63] is used for the determination of the concentration of dissolved oxygen O_2 in aqueous solution, especially in biological samples such as blood. The Clark electrode is, in fact, a complete two-electrode cell consisting of an electrolyte reservoir in which a Pt working electrode and an Ag|AgCl counter electrode are immersed (Figure 5.55a). The Ag|AgCl electrode functions as a combined reference and auxiliary electrode. The electrodes and the inner electrolyte reservoir are separated from the sample solution by a thin hydrophobic membrane. The membrane is

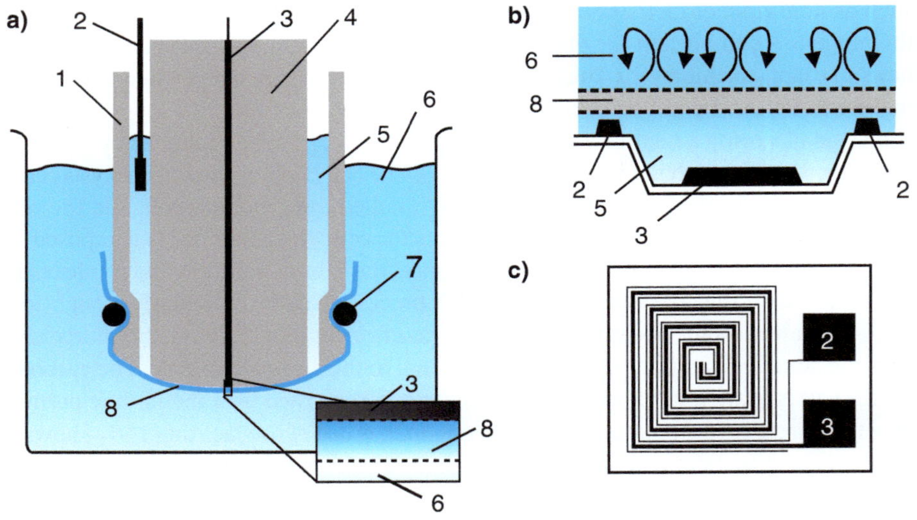

Figure 5.55 Design of a Clark electrode.

a) **Classical design**; 1) cell body, 2) Ag|AgCl counter electrode, 3) Pt-WE (O_2 reduction), 4) sheath of the Pt electrode, 5) inner solution reservoir, 6) sample solution, 7) O ring, and 8) O_2 permeable membrane; the inset shows an enlargement of the layering immediately in front of the working electrode.

b) and c) **Miniaturized Clark electrode**; b) cross section (numbers refer to the same components as in (a); c) top view on the planar sensor.

[63]Leland Charles Clark, 1918–2005, US American biochemist, professor at the University of Alabama and the Cincinnati Children's Hospital Research Foundation, also demonstrated the first glucose sensor and made important research contributions to artificial blood substitutes.

permeable to molecular oxygen and constitutes a diffusion barrier. Within the sample, O_2 is transported mainly by convection, but within the membrane, the transport occurs exclusively by diffusion with a constant thickness of the diffusion layer, that is, the thickness of the membrane. The membrane prevents the access of other potentially interfering constituents of the sample that cannot pass through the membrane and thus provides the exceptional selectivity of this very important sensor. The membrane also protects the electrodes from contact with proteins or other macromolecular components of the sample matrix, which would otherwise adsorb on the electrode surface and inhibit electron transfer reactions.

There are many different variations of Clark electrodes. Figure 5.55b, for instance, shows a miniaturized, planar design. The sensor can be miniaturized in such a way that it becomes suitable as an *in vivo* sensor that can be implanted into patients or laboratory animals and enables monitoring of the blood oxygen level, for example, during surgical procedures using a heart–lung machine. The sensor is operated with a constant voltage of −0.6 V in the two-electrode arrangement, causing the following processes:

$$\text{Transfer through the membrane: } O_{2,\text{sample}} \rightleftharpoons O_{2,\text{inner\_sol}} \tag{5.85}$$

$$\text{Pt working electrode: } O_2 + 4\,H^+ + 4\,e^- \rightleftharpoons 2\,H_2O \tag{5.86}$$

$$\text{Ag/AgCl counter electrode: } 4\,Ag + 4\,Cl^- \rightleftharpoons 4\,AgCl + 4\,e^- \tag{5.87}$$

The oxygen concentration is usually stated as partial pressure $p(O_2)$. This value refers to the partial pressure of a gas atmosphere that would be in equilibrium with the dissolved content of oxygen according to Henry's law. Often, a two-point calibration in an oxygen-free and an oxygen-saturated solution suffices for the calibration.

Q: How large is the equilibrium concentration of dissolved O_2 in an aqueous solution under standard pressure?

A: The exact value also depends on temperature and other constituents of the aqueous solution. As a rule of thumb, one may memorize that the ambient equilibrium concentration amounts to $c^*(O_2) \approx 0.27\ \text{mmol L}^{-1}$ at room temperature.

Surveys about other types of amperometric sensors can be found, for example, in Ref. [79].

5.5 Voltammetry Under Special Mass. Transport Conditions – Reactants Bound to the Electrode

5.5.1 Thin Layer Voltammetry and Voltammetry of Immobilized Redox Systems

When discussing cyclic voltammetry in Section 5.3.5, we assumed that mass transport of reagents occurs by diffusion only. In experiments, this situation is achieved by adding an excess of a supporting electrolyte. The electrolyte does not react at the electrode (within a given potential range) but enables charge transport within the electrolyte solution. In other words, the transference number of the redox-active component (if it is ionic) approaches zero. In addition, we assumed quiescent solution, that is, the absence of convection. In addition to these restrictions, we have only considered planar diffusion in which the vectors representing the concentration gradients are parallel to each other and perpendicular with respect to the electrode surface (Figure 4.50). Furthermore, the solution volume is so large that the diffusion layer never reaches the container wall opposite to the electrode surface. As a consequence, the bulk solution can always resupply reagents to the electrode, although mass transport to the electrode becomes less effective with time [Gl. (4.130)]. The diffusion of reagents to an electrode surface from an infinitely large solution volume is called a **semi-infinite diffusion**. The diffusion process with flat iso-concentration lines is also called planar diffusion, which we consider as the "default case" for voltammetric methods. In this section, we go beyond the default case and consider other important cases and their applications.

The presence of microconvection leads to deviation from semi-infinite diffusion, and this occurs in the absence of intended forced convection (e.g., stirring, rotating electrode, and streaming of the solution toward the electrode in a flow cell). Microconvection is caused by density gradients (and fluctuations) that form as a result of an electrochemical conversion and induces convection in the solution volume close to the electrode. It takes a certain time between the onset of the electrolysis and the onset of microconvection, during which a quiescent solution is found close to the working electrode. This fact is the reason for the recommendation to restrict the recording of chronoamperograms for the characterization of diffusion processes to about 10 s (Section 4.5.2). There are extensive studies available about the behavior in the (gravity-induced) transition region between a truly quiescent solution and the onset of microconvection, which is in fact very important for amperometric sensors [50].

In practical experiments and contrary to the model of semi-infinite diffusion, the electrolyte may be a thin solution layer, a thin gel layer or an ion-conducting membrane. In these cases, the diffusion layer will grow across the entire electrolyte volume and cause an exhaustive electrolysis within the considered timescale because there is no infinite solution reservoir for resupply of the reagents. Recording voltammetric data under these conditions is called thin-layer voltammetry.

With reagents trapped in a thin film, they cannot move away from the electrode surface, and they are considered immobilized redox systems. A special case is an electrode with redox-active reagents bound by adsorption or covalently directly to the electrode surface. We have discussed adsorptive stripping voltammetry as an application without considering further implications of this situation. We will encounter voltammogramms of immobilized redox systems or thin layer voltammograms during the study of electrocatalytic reactions (Figure 5.22), the use of optically transparent thin-layer electrodes for spectroelectrochemistry (Figure 18.19), the study of electrochromic systems, or during the investigation of immobilizes redox proteins and their use in sensors or biofuel cells (Chapter. 18).

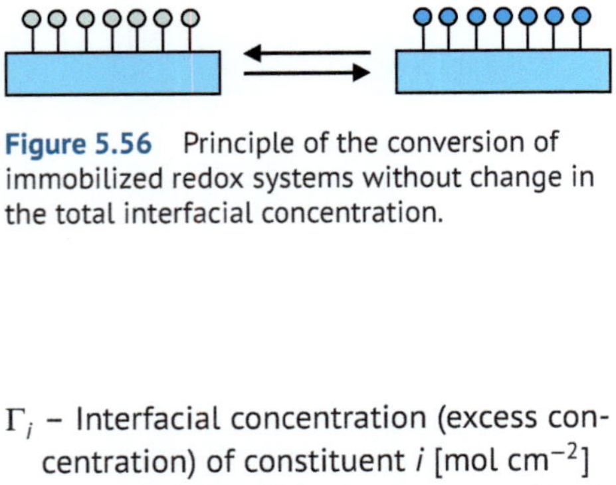

Figure 5.56 Principle of the conversion of immobilized redox systems without change in the total interfacial concentration.

It is instructive to consider first a case that is simplified as far as possible (cf. Section 5.3.5). We assume that the electrode is completely covered by a monolayer of the reduced form of the redox couple R_{ads}/O_{ads} (Figure 5.56), the oxidation product O_{ads} remains adsorbed to the electrode surface, and the solution does not contain any redox-active compounds. The electrode reaction can then be written as

$$R_{ads} \rightleftharpoons O_{ads} + e^- \tag{5.88}$$

The composition of such systems is described by the interfacial concentration Γ_i ($i = $ R, O) of component *i*.

Γ_i – Interfacial concentration (excess concentration) of constituent *i* [mol cm^{-2}]
Γ^* – Total interfacial concentration of both redox forms R/O [mol cm^{-2}]
N_i – Adsorbed amount of the constituents *I*
A – Area of the electrode [cm^2]

$$\Gamma_i = \frac{N_i}{A} \tag{5.89}$$

If there are no adsorption or desorption processes during the conversion and with stoichiometric coefficients of one, the total interfacial concentration of R_{ads} and O_{ads} remains constant according to Eq. (5.90) during the reaction.

$$\Gamma_R(t) + \Gamma_O(t) = const = \Gamma^* \tag{5.90}$$

In addition, the reaction is assumed to be reversible. That means, regardless of how fast the electrode potential may change, the ratio of the interfacial concentrations Γ_O/Γ_R will always follow the Nernst Equation (5.91).

E – Electrode potential [V]
$E^{\circ\prime}_{ads}$ – Formal potential of R_{ads}/O_{ads} [V]
E_{NF} – Nernst factor = $RT/(nF)$
T – Absolute temperature [K]
R – Universal gas constant [J K^{-1} mol^{-1}]
n – Stoichiometric number of electrons [dimensionless]
F – Faraday constant [A s mol^{-1}]

$$\frac{\Gamma_O}{\Gamma_R} = \exp\left[\frac{E - E^{\circ\prime}_{ads}}{E_{NF}}\right] \tag{5.91}$$

Q: Why do we use concentration here instead of the corresponding activities?
A: Indeed, we should use the interfacial activities γ_R and γ_O ($\gamma_i = f_i\Gamma_i$) for an exact treatment. However, the activity coefficients are difficult to determine. We will see later that numerical values of interfacial concentrations and interfacial activities may depart from each other just as volume concentrations and activities do. Here we look at an idealized system. Very important for the ideality of a system ($f = 1$) is how well R_{ads} and O_{ads} "mix" with each other at the interface when R_{ads} and O_{ads} occur simultaneously at the electrode during the reaction. Here, we assume an ideal mixture, which means the interaction between adsorbed R_{ads} with a neighboring R_{ads} is equal to the interaction between R_{ads}–O_{ads} and O_{ads}–O_{ads}. This assumption is implicitly contained in Eq. (5.91) and is one of the main simplifications in this treatment.

The relationship between the change in Γ_O and Γ_R on one side and the current on the other side follows from Faraday's first law as shown in Eq. (5.92).

$$I = \frac{dQ}{dt} = \frac{d(nFN)}{dt} = nFA\frac{d\Gamma_O}{dt} = -nFA\frac{d\Gamma_R}{dt} \tag{5.92}$$

The investigation of such a system by cyclic voltammetry, implying a linear change in potential with time $E(t) = E_i + vt$ gives the response as shown in Figure 5.57. Equation (5.93) gives the current $I(E)$ without capacitive currents.

$$I = nFA\Gamma^* \frac{v}{E_{NF}} \frac{\exp\left(\frac{E - E^{\circ\prime}_{ads}}{E_{NF}}\right)}{\left[1 + \exp\left(\frac{E - E^{\circ\prime}_{ads}}{E_{NF}}\right)\right]^2} \tag{5.93}$$

v – Scan rate [V/s]
t – Time [s]

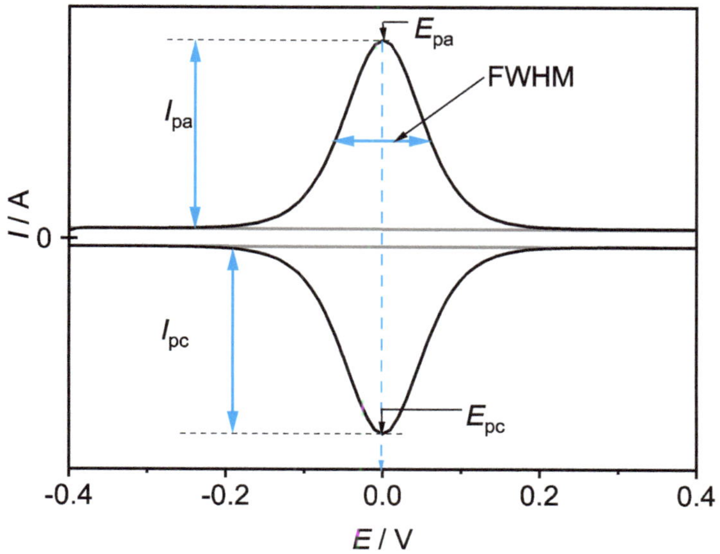

Figure 5.57 Cyclic voltammogram of an ideal immobilized nernstian redox system with $E^{\circ\prime}_{ads} = 0$ V and consideration of a capacitive current (gray line).

The peak current and the peak potential are given by Eqs. (5.94) and (5.95).

$$I_p = \frac{n^2F^2}{4RT}vA\Gamma^*_R \tag{5.94}$$

$$E_p = E^{\circ\prime}_{ads} \tag{5.95}$$

For comparison, we found the following features for cyclic voltammetry of a nernstian system with semi-infinite diffusion:

- Peak separation of E_{pa} and E_{pc} by 59 mV/n (298 K)
- Decay of current after the peak with $1/t^{1/2}$, $I = 0$ is not reached
- $I_p \propto v^{1/2}$.

For the ideal system, the peaks are symmetric to the potential axis. After the maximum, the current decays to the baseline because the redox-active species is completely consumed and no resupply occurs from the solution. Finally, I_p (and the current for each potential) is proportional to v. These features are suitable to identify immobilized redox systems (in contrast to diffusing redox systems) from their voltammetric response.

Because the entire immobilized amount of R_{ads} is converted in one half-cycle, the total interfacial concentration Γ^* can be calculated from the charge (corrected for double layer charge). For calculating the charge from the voltammogram, the area under the $I(E)$ curve must be corrected using v. Because of $v = dE/dt$, the differential time increment can be expressed as $dt = dE/v$. With this substitution, the charge Q is obtained by Eq. (5.96).

$$Q = \int_{t_1}^{t_2} I_F(t)\mathrm{d}t = \int_{E(t_1)}^{E(t_2)} I_F(E)\frac{1}{v}\mathrm{d}E = \frac{1}{v}\int_{E(t_1)}^{E(t_2)} I_F(E)\mathrm{d}E. \tag{5.96}$$

With this, Γ^* is obtained from the integral in Eq. (5.97), in which the integral corresponds to the area under the curve of one half-cycle in Figure 5.57 with respect to the baseline (gray).

$$\Gamma^* = \frac{Q}{nFA} = \frac{1}{nFAv}\int_{E(t_1)}^{E(t_2)} I_F(E)\mathrm{d}E$$
$$= \frac{1}{nFAv}\int_{E(t_1)}^{E(t_2)} I(E) - I_C(E)\mathrm{d}E \tag{5.97}$$

The peak width at half maximum (FWHM) amounts to the value given in Eq. (5.98).

$$\mathrm{FWHM} = 3.53\frac{RT}{nF} = \frac{90.6\,\mathrm{mV}}{n} \quad (298\ \mathrm{K}). \tag{5.98}$$

Important deviations from the case of an ideal surface redox system will be considered in Section 13.4, including unequally strong adsorption of R and O, existence of energetically distinct adsorption sites, slow electron transfer, and different lateral interactions between adsorbed R and O.

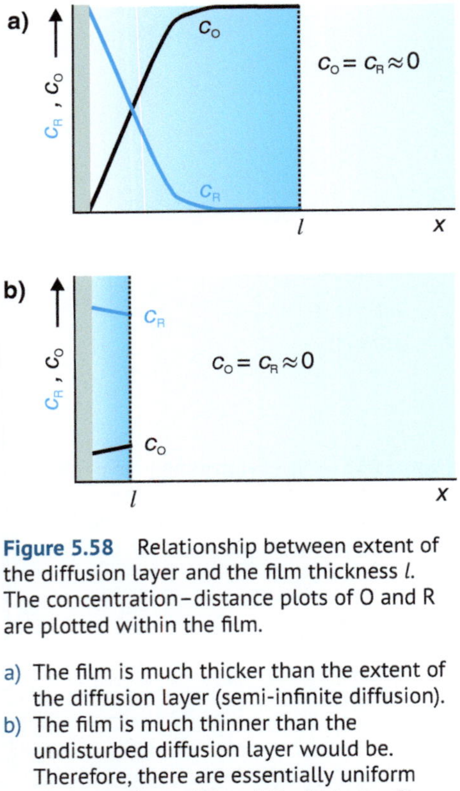

Figure 5.58 Relationship between extent of the diffusion layer and the film thickness l. The concentration–distance plots of O and R are plotted within the film.

a) The film is much thicker than the extent of the diffusion layer (semi-infinite diffusion).
b) The film is much thinner than the undisturbed diffusion layer would be. Therefore, there are essentially uniform concentrations of R and O within the film.

For a film of finite thickness l (instead of a monolayer on the electrode surface), the volume concentration has to be used instead of the interfacial concentration. The corresponding relations are $N_i = A\Gamma_i = c_i V = c_i A l$. The observed behavior qualitatively depends on the timescale of the experiment (v) relative to the film thickness (l) and the diffusion coefficient of R and O *within the film* (Figure 5.58). If the experiment is fast, the diffusion is slow and/or the film is thick, the diffusion layer does not advance to the film–solution interface (Figure 5.58a). The corresponding cyclic voltammogram will be that expected for semi-infinite diffusion. If the diffusion layer traverses rapidly through the entire film (Figure 5.58b), the resulting voltammogram will be similar to that of a monolayer. Of course, intermediate cases are possible and can intentionally be employed in experiments by variation of v for one and the same system.

5.5.2 Voltammetry of Immobilized Microparticles

This type of voltammetry uses directly a finely ground powder of a solid as the sample without previously dissolving it. This has the advantage of preserving the chemical state of the sample material, which can be minerals, alloys, and organic solids, as well as their solid-state chemical degradation products. The sensitivity to the chemical state is a tremendous advantage compared to solution voltammetry (Section 5.3) after digestion of the solid (transfer of the contained metal species to their solvated ions). However, the complexity of the possible reactions of the solid particle represents a formidable challenge. The analysis of the voltammograms of such solids is very instructive for the understanding of modified electrodes (Sections 13.2–13.4) and active materials in batteries (Sections 15.3.1, 15.6.2, and 15.6.4).

The analysis is often conducted with an inert electrode, that is, an electrode that ideally does not give a signal in the potential range of interest. Particularly attractive is the paraffin-impregnated graphite electrode (PIGE) that can be produced by adding a polycrystalline and porous graphite rod into molten paraffin in a closed container and applying a low vacuum to promote the release of gas bubbles from the pores of the carbon material. When returning to ambient pressure, the pore space in the graphite rod is filled with paraffin [80]. This avoids the permeation of the electrolyte solution into the pore of the base electrode in later steps of the experiment, which would cause high capacitive currents and contamination of the base electrode. A PIGE can be polished by simply rubbing it on pieces of (different) papers with a final polishing step on weighing paper. A small amount of the powdered sample is placed on the weighing paper, and the polished electrode area is rubbed with the powder. Only a *very small amount* of powder should be transferred to the PIGE (typically only visible under an optical microscope). Nanograms to micrograms of the sample powder are sufficient to generate an analytically useful signal; application of more than necessary quantities increases the overall current and thus causes distortion of voltammograms by ohmic drop (details in Section 5.3.1, 5.7.2, and Excursion 6-1). The soft paraffin in the pores of the PIGE helps retain the sample particles at the electrode surface. In some cases, the solid sample can also be transferred from its original location (e.g., film of corrosion products on macroscopic objects or minerals) to the base electrode by mechanical abrasion. Thus, the method is sometimes called abrasive stripping voltammetry. Other materials such as carbon–binder composites, highly ordered pyrolytic graphite, pencil lead (i.e., graphite), boron-doped diamond electrodes, and glassy carbon have also been used as base electrodes [80].

The electrode with the attached particles is then used as a working electrode in a potentiostatic three-electrode setup in an electrolyte solution, in which the *solid powder sample does not dissolve*. The usual voltammetric techniques such as LSV, CV, DPV, and SQWV can be applied (Section 5.3). Depending on the chemical nature of the sample, a negative-going potential scan or a positive-going potential scan is initiated; very often, both directions are explored with an intermittent preparation of a new sample.

Figure 5.59 shows a schematic view of the particles on the base electrode forming a **three-phase electrode** comprising the base electrode, the solid sample material, and the liquid electrolyte solution. The blue lines around the "footprint" of the particles indicate the locations where all three phases are in contact with each other, the **three-phase boundary** (synonym: **triple phase boundary**). There are many reaction mechanisms that can occur in such systems, depending on the chemical nature of the solid sample and the liquid electrolyte solution. Metallic particles can simply dissolve electrolytically, provided the metal cations are well soluble in the electrolyte solution. Sometimes metallic particles also cause a significant catalytic current, for example, for the hydrogen evolution reaction or the oxygen reduction reaction that would not occur at the unmodified base electrode at the applied potential. Such reactions may be important in the clarification of degradation mechanisms in complex composites found in historical objects of art (different inorganic pigments, complex organic binders).

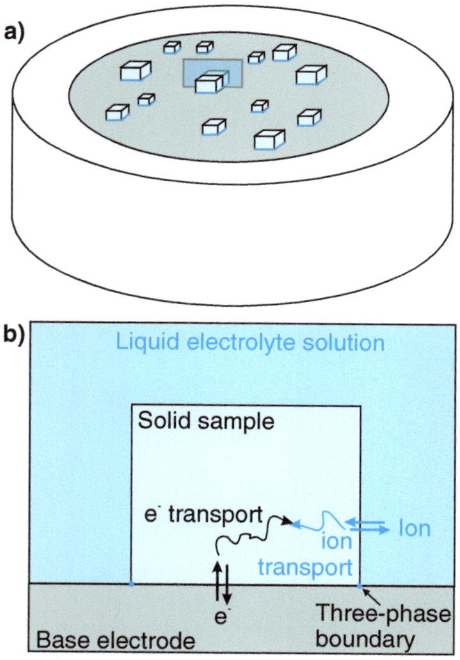

Figure 5.59 Schematic representation of a solid particle attached to a base electrode. a) Indication of the three-phase boundary as blue lines; b) cross-sectional view with indication of ion-transfer, electron transfer, ion-transport, and electron transport processes for an insertion electrode.

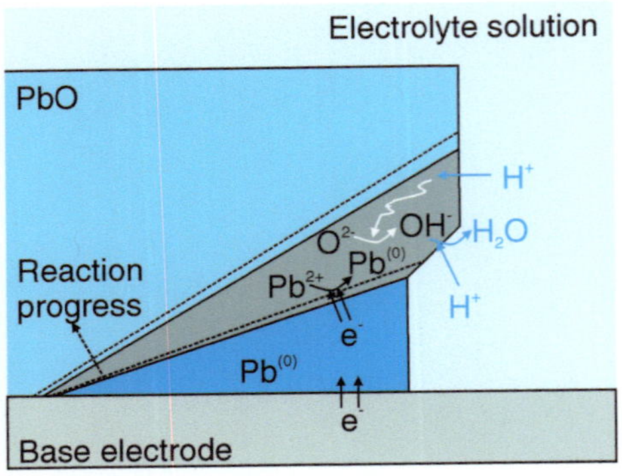

Figure 5.60 Schematic representation of the reduction of a metal oxide (PbO) to a metal (Pb) with two solid phases of sample and product and the fluxes of electrons and ions. An interphase is indicated in which the reaction happens, and mixed conductivity occurs.

Of particular interest are reactions in which the sample is electrochemically *converted to another solid material* attached to the base electrode. For instance, metal oxides can be reduced to metals. This process requires electrons supplied from the base electrode and protons provided by the electrolyte solution, for example,

$$PbO(s) + 2\,e^-(base) + 2\,H^+(aq) \rightarrow Pb(s) + H_2O(aq) \tag{5.99}$$

Protons and electrons are available at the three-phase boundary and, thus, the reaction commences there (Figure 5.60). Metallic lead forms a new, electron-conducting phase. After the onset of the reaction and before reaching complete conversion, the electrode comprises four phases. The details on how the reaction front propagates through the solid material depend on the relative rates at which electrons and ions can be transported through the solids. An amorphous *interphase* may emerge between the solid phases of the sample and the product with different transport properties, which complicates the quantitative description and prediction of the reaction. A complete conversion of the solid particle requires that the sample particle has **mixed conductivity**, that is, electronic conductivity *and* ionic conductivity. Detailed mechanisms of electron transport are covered in Section 13.4.5. The slower of the two processes usually determines the shape of the voltammograms.

However, there are also solid materials where the solid sample and its conversion product form a homogeneous solid mixture, called solid solution. Many metal hexacyanoferrates, such as Prussian Blue $KFe^{III}[Fe^{II}(CN)_6]$ (Figure 13.52a), are examples of this class of material. The material can be oxidized and reduced.

$$KFe^{III}[Fe^{II}(CN)_6](s) \rightleftharpoons Fe^{III}[Fe^{III}(CN)_6](s) + e^-(base) + K^+(aq) \tag{5.100}$$

$$KFe^{III}[Fe^{II}(CN)_6](s) + e^-(base) + K^+(aq) \rightleftharpoons K_2Fe^{II}[Fe^{II}(CN)_6](s) \tag{5.101}$$

The transferred electrons cause valence changes of the C- and N-coordinated $Fe^{2+/3+}$ ions. The charge of the solid is balanced by the insertion or release of K^+ ions from the crystal lattice. Within the solid, K^+ ions can be transported in channels formed between the $Fe^{2+/3+}$ ions and the bridging cyanide ligands. During the transfer of K^+ ions, the crystal lattice is preserved, and thus both oxidation states of the sample can occur in a single solid phase with gradients in the local content of Fe^{2+}, Fe^{3+}, and K^+. Such electrodes are called **insertion electrodes** (synonym: intercalation electrode). Electrochemical conversions of solids and especially insertion electrodes have tremendous importance in the area of batteries (Chapter 15). Films of intrinsically conductive polymers are another example of practically important insertion electrodes that will be discussed in Section 13.2.4.

A qualitative analysis aims for the identification of solid compounds in a sample, for instance, revealing the identity of pigments within a complex multicomponent powder. After the transfer of particles to the base electrode, only redox-active compounds contribute to the voltammetric signals. The signals appear at a potential that is characteristic of the observed redox reaction and can be compared to the peak potentials recorded from reference samples with known compounds that are suspected to occur in the sample.

A quantitative analysis must take into account that there is a significant variation in how many particles are transferred to the base electrode in each preparation run. An internal standard can be used to account for this variation. For instance, a known amount of AgCl particles is added to the sample. When recording the voltammogram, the ratio between the peak height of the signal from the analyte and signal for the reduction of the internal standard $AgCl + e^- \rightarrow Ag + Cl^-$ should be proportional to the product of stoichiometric numbers of electrons n_i and amounts N_i of analyte and internal standard. Examples are given in [81].

The evaluation of the electrochemical conversion of solid-state materials can be facilitated by coupling the voltammetry of particles with techniques that provide complementary structural information (Chapter 8, [80]).

Further Reading. The monograph [80] can be recommended for a comprehensive overview of the electrochemistry of solid particles and small droplets of redox-active liquids immiscible with the electrolyte solution. Particularly interesting and instructive applications have evolved in the analysis of pigments, ceramics, glasses and metals in historical, artistic, and archeological objects, including the changes/degradation of such materials over time. The voltammetric analysis can contribute to tracing the origin of materials and objects, dating, as well as evaluation of restoration and preservation efforts [81]. The concept of three-phase electrodes is also of paramount importance in the area of batteries, supercapacitors (Chapter 15), fuel cells, and electrolyzers (Chapter 16).

5.5.3 Nanoobject Collision Electrochemistry

In recent years, another approach toward the analysis of solid particles and other nanosized objects has seen rapid advances. The sample is a stable suspension of nanoparticles of metals, oxides, sulfides, droplets of redox-active liquids, supramolecular aggregates (e.g., vesicles with redox-active cargo), and so on. Typically, these entities are substantially smaller than the particles studied in Section 5.5.2. The base electrode is a microelectrode with a typical radius between 0.1 and 10 μm (Section 5.6). The small size of the base electrode reduces the capacitive background currents. In addition, the electrode is so small that the average time between subsequent collisions of particles at reasonable concentrations becomes so large that they can be recorded as sudden changes in current that are well resolved on the time axis. The approach facilitates single entity electrochemistry. Often, specialized potentiostats with very low time constants are required to record such rapid signal changes without artifacts [82, 83].

Only in the moment when a nanoparticle collides with the base electrode, a three-phase electrode is formed that enables the reaction discussed in Section 5.5.2 to occur, depending on the chemical nature of the particle and the electrolyte solution. For a single particle recording, a constant potential is applied to the base electrode, and the current is followed with high time resolution (chronoamperometry). Three main approaches have been successful:

1) The suspended particles are insulating and redox inactive. A redox-active molecular compound (e.g., ferrocene derivatives, $[Fe(CN)_6]^{4-/3-}$) is added as a redox probe. The microelectrode is held at a potential that causes the diffusion-controlled oxidation of the redox probe, which gives a steady-state current at the base microelectrode due to the hemispherical diffusion (see Section 5.6). When the redox-inactive analyte nanoparticles collide with and stick to the base electrode surface, the electrode area available for the probe reaction is partially blocked, and thus the resulting current is stepwise decremented (Figure 5.61a). Each current decrease corresponds to the arrival of an individual analyte nanoparticle [84].
2) The analyte nanoparticle is redox active and is immediately converted upon collision, causing a short current spike for only a few milliseconds. Each current spike corresponds to the arrival of a single analyte nanoparticle [85]. Integration of each current spike provides a charge that is related to the amount of material converted in a single collision event (Figure 5.61b). This amount may be directly linked to the size of the nanoparticle, provided that the particles were completely converted during a single collision event. Instead of nanoparticles, stochastic collisions of nanodroplets of redox-active liquids or vesicles with redox-active cargo can be followed [86].

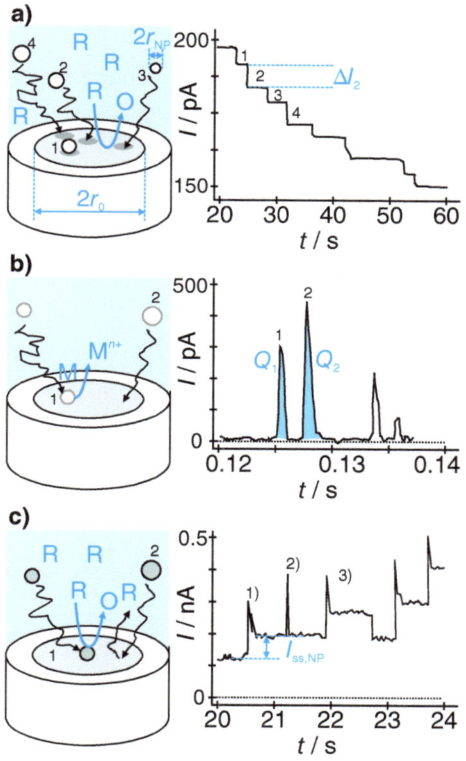

Figure 5.61 Schematic chronoamperometric responses for nanoparticle impact voltammetry. a) Blocking insulating analyte nanoparticles; b) electrolytic conversion of redox-active nanoparticles; c) electrocatalytically active nanoparticles. The qualitatively different behavior of nanoparticles is illustrated in (c) for 1) particle permanently sticking to the base electrode and maintaining catalytic activity, 2) collision and immediate detachment, and 3) collision, sticking, and detachment.

3) The analyte nanoparticle has electrocatalytic activity. It is not converted at the potential of the base electrode, but it causes an electrocatalytic reaction of a species provided in solution at its surface (Figure 5.61c). The electrocatalytic reaction does not occur at the base electrode without attached analyte particles. Each arrival of a nanoparticle increments the current [87]. If particles detach, the current may also decrease again toward the baseline. Several electrocatalytic reactions, for example, hydrogen evolution reaction, hydrazine (N_2H_4) oxidation, water oxidation, $NaBH_4$ oxidation, oxygen reduction, and H_2O_2 reduction, have been used as electrocatalytic reactions with different nanoparticle electrocatalysts to be analyzed. This approach can be expanded to reversible and quasireversible probe reactions if the base electrode is blocked for the probe reaction by a nanometer-thin insulating layer. When a metal nanoparticle lands on the blocked base electrode, a tunneling electrode is formed: The probe system is oxidized or reduced at the nanoparticles, and electrons are exchanged between the base electrode and the nanoparticles by a tunneling process [88].

f_{NP} – Collision frequency of nanoparticles with electrode [s^{-1}]
D_{NP} – Diffusion coefficient of the nanoparticle [$cm^2\ s^{-1}$]
c^* – Molar concentration of nanoparticles [mo L^{-1}]
r_0 – Radius of the disk-shaped microelectrode [cm]
N_A – Avogadro number [mol^{-1}]
Δt – Observation time [s]
σ^2 – Variance [dimensionless]

Nanoparticle Concentration and Collision Frequency. The frequency f_{NP} by which nanoparticles of a concentration c^* collide with an electrode is given by the relationship [89]

$$f_{NP} = 4\, D_{NP}\, c^*\, r_0\, N_A \tag{5.102}$$

Within a time interval Δt, there will be $N_{NP}(\Delta t)$ collisions

$$N_{NP}(\Delta t) = f_{NP}\, \Delta t = 4\, D_{NP}\, c^*\, r_0\, N_A\, \Delta t \tag{5.103}$$

With variances $\sigma^2[N_{NP}(\Delta t)] = (4\, D_{NP}\, c^*\, r_0\, N_A\, \Delta t)$ due to the random motion of discrete particles. The fluctuation is $\sigma[N_{NP}(\Delta t)] = (4\, D_{NP}\, c^*\, r_0\, N_A\, \Delta t)^{1/2}$ around the expected mean value $N_{NP}(\Delta t)$. Thus, the ratio between the fluctuation and the mean value is $\sqrt{N_{NP}(\Delta t)}$: $N_{NP}(\Delta t) = 1/\sqrt{N_{NP}(\Delta t)}$. This ratio increases dramatically with small $N_{NP}(\Delta t)$, caused by either small f_{NP} or small observation time Δt. D_{NP} can be estimated from the Stokes–Einstein relationship[26,44] [Eq. (4.112)].

Example Calculation 5.3 Calculate the expected relative error in a concentration determination by single impact experiments of nanoparticles with a collision frequency of 0.1 s^{-1}, 1 s^{-1}, and 10 s^{-1} over observation intervals from 1 s, 10 s, and 100 s. The expected error is $\sigma[N_{NP}(\Delta t)]$, and the mean value is $N_{NP}(\Delta t)$.

sought: $\Delta x/\langle x\rangle$

given: $f_{NP} = [0.1\ s^{-1}, 1\ s^{-1}, 10\ s^{-1}]$

$\Delta t = [1\ s, 10\ s, 100\ s]$

The numerical solution is detailed for one example with the expected largest relative error. The remaining examples are solved in a spreadsheet application.

$$\langle x\rangle = f_{NP}\, \Delta t$$

$$\Delta x = \sqrt{\langle x\rangle} = \sqrt{f_{NP}\, \Delta t}$$

$$\frac{\Delta x}{\langle x\rangle} = \frac{\sqrt{f_{NP}\, \Delta t}}{f_{NP}\, \Delta t} = \frac{1}{\sqrt{f_{NP}\, \Delta t}}$$

$$= \frac{1}{\sqrt{0.1\, s^{-1} \cdot 1\, s}}$$

$$= 3.16 = 316\%$$

The tabulated results are

f_{NP}/s^{-1}	$\Delta x/\langle x\rangle$		
	$\Delta t = 1$ s	$\Delta t = 10$ s	$\Delta t = 100$ s
0.1	316%	100%	31.6%
1	100%	31.6%	10%
10	31.6%	10%	3.2%

In the selected example, an observation time of 100 s would allow a reasonable estimation of concentration only for collision frequencies ≥1 s^{-1}.

Size Determination. In the case of a blocking nanoparticle (Figure 5.61a), the ratio of the current drop ΔI and the total steady state current I_{ss} before the first nanoparticle collision is approximately equal to the ratio of the cross section of the particle πr_{NP}^2 and the area of the base electrode πr_0^2 [89].

$$\frac{\Delta I}{I_{ss}} = \frac{\pi r_{NP}^2}{\pi r_0^2} \quad (5.104)$$

Rearranging yields Eq. (5.105).

$$r_{NP} = r_0\sqrt{\frac{\Delta I}{I_{ss}}} \quad (5.105)$$

ΔI – Current drop [A] in Figure 5.61a
I_{ss} – Steady-state electrolysis current of the redox probe at the base microdisk electrode before nanoparticle impact ($I_{SS} = 4nFDr_0c_R^*$);
r_{NP} – Radius of the nanoparticle [m]
r_0 – Radius of the base microelectrode [m]

Eqs. (5.104) and (5.105) are approximations because the microdisk base electrode shows a nonuniform current density distribution: a particle landing close to the circumference of the base electrode causes a larger ΔI than a particle of the same size landing in the center of the base electrode. Another hidden approximation is in the assumption of a constant value for D_{NP} close to the base electrode surface.

For a complete electrolysis of spherical nanoparticles colliding with the base electrode, the electrolysis charge of a single current spike Q_i is related to the number of atoms (m_{NP}/M) in the nanoparticle and thus to its volume ($V_{NP} = 4\pi r_{NP}^3/3$) and mass ($m_{NP} = \rho V_{NP}$) [89].

$$Q_i = nF\frac{\rho}{M}\frac{4}{3}\pi r_{NP}^3 S \quad (5.106)$$

Rearranging yields

$$r_{NP} = \sqrt[3]{\frac{3Q_iM}{4\pi nF\rho}} \quad (5.107)$$

Q_i – Charge of the ith current spike [A s] = [C] in Figure 5.61b from integration of current over time
M – Molar mass [g mol^{-1}]
ρ – Density of the nanoparticle [g cm^{-3}]
n – Stoichiometric number of electrons for conversion of the nanoparticle in Eqs. (5.106) and (5.107) [dimensionless]
F – Faraday constant [96 485 A s mol^{-1}]
n – Stoichiometric number of electrons for conversion of R/O in Eq. (5.108) [dimensionless]
D_R – Diffusion coefficient of reagent for electrocatalytic reaction [cm^2 s^{-1}]
c_R^* – Bulk concentration of the reagent for electrocatalytic reaction [mol cm^{-3}]

For collisions of electrocatalytically active particles, the catalytic current can be estimated under the condition that the probe reaction is diffusion controlled, the nanoparticle remains attached to the electrode surface, it does not deform, and the nanoparticles act as independent microelectrodes (cf. Figure 5.84). In this case, the current of an individual nanoparticle equals the current at a spherical microelectrode (Table 5.9) diminished by a factor of ln 2 ≈ 0.693 due to the diffusional shielding by the base electrode [89].

$$I_{ss,NP} = (\ln 2)4\pi nFD_Rc_R^*r_{NP} \quad (5.108)$$

It should be noted that the chronoamperometric response of particles in all mechanisms can be very complicated, depending on the sticking behavior of the observed nanoparticles [82, 90, 91]. This is exemplified schematically in Figure 5.61c for a particle sticking permanently to the surface and maintaining activity, a particle that collides and immediately detaches, and a particle that sticks and later detaches. The sticking coefficient may depend on the pretreatment or surface modification of the base electrode as well as on the chemical nature of the ligand shell typically found around nanoparticles to prevent their spontaneous aggregation [82].

Further Reading. The area of single entity electrochemistry is a rapidly growing research field that is covered in a number of reviews and book chapters that also include other approaches such as optical detections and perspectives for applications [89, 92–102].

5.6 Voltammetry Under Special Mass Transport Conditions – Microelectrodes

5.6.1 Steady-State Currents

According to Eq. (4.130) and many similar equations for voltammetry, the current is proportional to the electrode area. Already in Chapter 1, we have defined current density $j = I/A$ in order to compare the faradaic current with respect to the intensive properties of the participating phases (c, D). Although very small electrodes have already been applied before World War II for concentration determination in biomedical research, it was not until 1980 that it became clear that the simple proportionality between current I and electrode area A is not followed when the size of the electrode falls below about 30 μm (or below the extent of the diffusion layer) in at least one spatial dimension. Although the absolute currents can be very small (pA to nA), they exceed the prediction of the Cottrell Equation (4.130) by a wide margin. In other words, the current density is very large at these electrodes.

Unfortunately, the nomenclature in electrochemistry is not uniform, even for relatively recent terms. Historically, the term "microelectrode" referred to an analytical laboratory electrode such as a microdisk electrode of 1–3 mm diameter and distinguished those electrodes from a "macroelectrode" as used in (industrial) electrolysis processes. The particular properties of electrodes with radii below about 30 μm were summarized under the term "ultramicroelectrodes (UME)" [1, 103]. In this terminology, the properties not only depend on the size but rather on size relative to the timescale of the experiment (*vide infra*). With the emergence of electrodes of nanometer size, the term "nanoelectrode" became popular in the literature. ("Nanodes" was not broadly adopted). Nanoelectrode was complemented by "microelectrode" with the micrometer size of the electrode in mind. Finally, the development led to the currently valid IUPAC recommendation [104], to which this textbook adheres. Unfortunately, the emphasis on the particular behavior of small electrodes is lost together with the term "ultramicroelectrode", which not only depends on the size but also on the timescale of the experiment.

The increased current densities at microelectrodes are exemplified in Figure 5.62 in comparison to conventional electrodes under otherwise identical conditions.

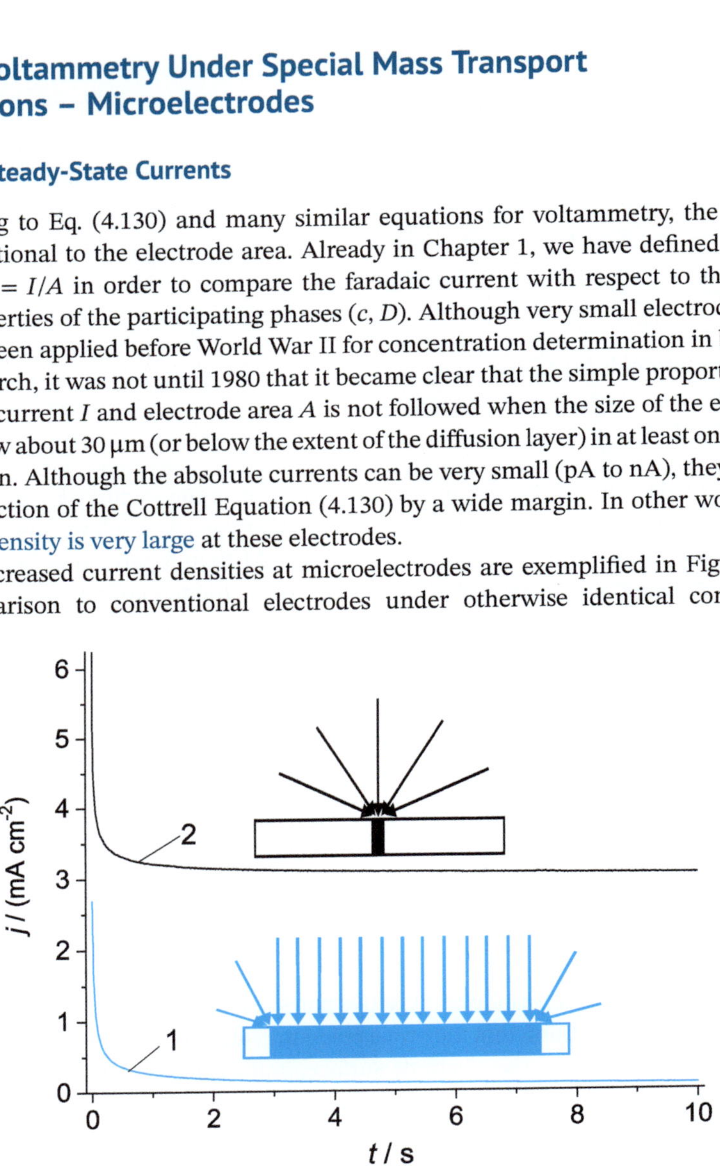

Figure 5.62 Chronoamperograms at disk-shaped electrodes in 2 mmol L^{-1} solution of ferrocene methanol in 0.1 mol L^{-1} phosphate buffer.
Curve 1) conventional electrode with a diameter of 3 mm (blue);
Curve 2) microelectrode with a diameter of 10 μm (black).
The ordinate is the current density. The absolute currents of curve 2 are, of course, much smaller than those of curve 1!

In addition, the measurement shows that the current at the microelectrode quickly approaches a steady-state current, that is, a current that does not change with time. For such electrodes, the term microelectrode was established. They exhibit fundamentally different properties from the electrodes that we have dealt with so far. Contrary to the steady-state current at microelectrodes, the Cottrell equation predicts that the current continues to decay with $1/t^{1/2}$ because the diffusion layer expands. This is observed even when the bulk phase of the solution contains an unchanged reactant concentration c^* (Section 4.5.2). The special properties of microelectrodes are readily observed in cyclic voltammograms because of the qualitatively different shape of the curves. For conventional electrodes, the cyclic voltammogram has a peak with a drawn-out flank at higher potentials (shape of a duck bill). In contrast, cyclic voltammograms at microelectrodes yield a sigmoidal shape (shape of a stretched character "s", Figure 5.63, curve 2). The current plateau at 0.5 V shows again a steady-state current that we also noted in the chronoamperogram in Figure 5.62, curve 2.

Here, it is instructive to compare this microelectrode behavior to the rotating disk electrode (RDE), for which steady-state currents (sigmoidal) are caused by convective mixing of the electrolyte solution outside of the hydrodynamic boundary layer in solution. This prevents the growth of the diffusion layer. Due to the relevance of these electrodes for the characterization of electrocatalysts, we discuss procedures with rotating disk electrodes in Chapter 12.

Microelectrodes have enabled many new analytical methods since the 1980s (Excursions 5-5, 5-6, Section 8.11). Below, we will identify the particular features of microelectrodes using qualitative consideration, then introduce a few important equations, and finally derive further properties. Subsequently, we will consider the behavior of ensembles or arrays of several microelectrodes. We will encounter situations in which a certain qualitative feature (such as the appearance of sigmoidally shaped cyclic voltammograms) not only depends on the size of the microelectrode but on the size in relationship to the timescale of the experiment (expressed by the scan rate v) and the mass transport in the liquid phase (characterized by the diffusion coefficient D). Due to its exemplary nature, we will discuss this relationship in depth.

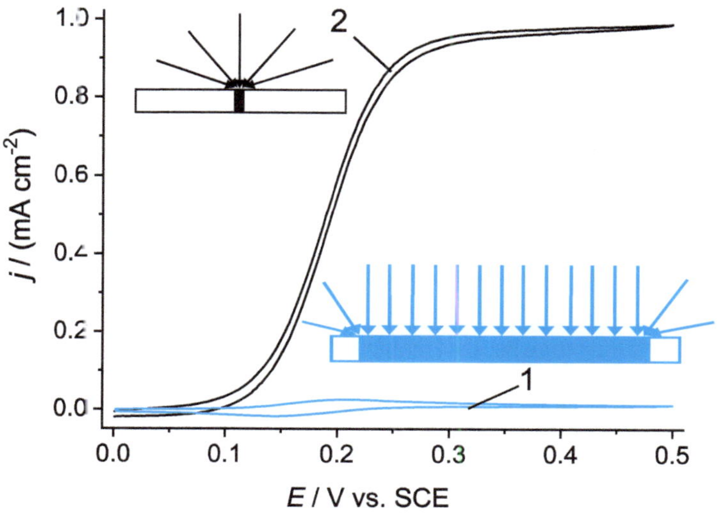

Figure 5.63 Cyclic voltammograms at $v = 0.01\ V\ s^{-1}$ at disk-shaped electrodes in 0.5 mmol L^{-1} solution of ferrocene methanol in 0.1 mol L^{-1} phosphate buffer.
Curve 1) conventional electrode with a diameter of 3 mm (blue);
Curve 2) microelectrode with a diameter of 10 μm (black).
The ordinate is the current density. The absolute currents of curve 2 are, of course, much smaller than those of curve 1!

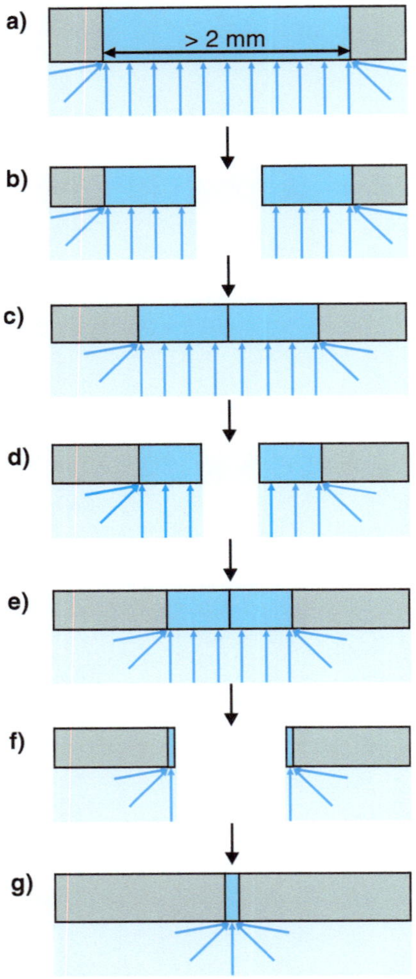

Figure 5.64 Schematic cross section of electrodes illustrating the transition from a macroelectrode to a microelectrode by removal of the central part of the disk, followed by shrinkage of the arrangement to reestablish the disk geometry with a smaller area. Eventually, the disk becomes so small that the edge effect is the dominating contribution to mass transport.

5.6.2 Why Are There Steady-State Currents?

For the subsequent discussion, we assume that the solution contains the reduced form R of a reversible redox couple and an excess of supporting electrolyte. If we carry out a chronoamperometry experiment at a conventional disk electrode, a diffusion layer emerges with the concentration gradient directed perpendicular toward the electrode surface, and the iso-concentration lines are parallel to the electrode surface (Figure 5.64a, planar diffusion). Only at the edge of the electrode, the iso-concentration lines are curved, and the concentration gradients have a component parallel to the electrode surface. Therefore, more R per unit area is transported to the edge of the electrode than to the center of the electrode. Thus, the disk electrode belongs to the category of nonuniformly accessible electrodes. For a millimeter-sized electrode, the edge effect will cause a small increase in the current compared to the predictions of the Cottrell Equation (4.130). For this reason, we have used a recessed electrode to derive the Cottrell equation in Chapter 4. The active area of the recessed electrode forms the bottom of a solution-filled cylinder, so that the side walls exclude the edge effect (Figure 4.50c).

Let us imagine we were in possession of a magic hole puncher, which we use to remove the central area of the disk electrode (Figure 5.64b) and "heal" the hole by shrinkage of the assembly to a new disk with smaller area (Figure 5.64c). The edge effect contributes a larger fraction to the mass transport compared to the situation in Figure 5.64a. This procedure can be repeated until the edge effect becomes the dominating form of the mass transport and iso-concentration lines are strongly curved and arranged approximately hemispherically around the center of the electrode (Figure 5.64d to g).

It should be clear that the Cottrell Equation (4.130) does not apply in the situation shown in Figure 5.64g. Now, R is transported from a three-dimensional volume to a one-dimensional (dot-like) electrode. For simplicity, we assume it to be hemispherical with the radius r_0 (inset in Figure 5.65). This makes it plausible that this geometry leads to a steady-state current. (Although this is not yet a proof!). Let us assume that the iso-concentration surfaces are hemispherical shells (similar to the shells of an onion). We now consider segments of a sphere which are projected from inside to outside on the hemispherical shell layers (Figure 5.66). By definition, a steady-state current does not change with time. In other words, the same number of species R reaches the electrode in each time interval. In order to maintain such a flux, the same number of species R must cross each segment of the different hemispheres. Because the area of each segment (and each hemisphere) decreases with decreasing distance to the electrode, the diffusional flux $J = \partial N/(A\partial t)$ [Eq. (4.108)] increases. Because the flux is given by Fick's first law [Eq. (4.110)] as $J = -D\partial c/\partial z$ and the diffusion coefficient is a constant, the concentration gradient $\partial c/\partial z$ increases significantly as the electrode is approached (Figure 5.65). The gradient stays approximately constant with time immediately at the electrode at $(z - r_0) = 0$. After some time, the diffusion layer shows almost negligible growth even for a long time because an incremental increase in the thickness increases enormously the *volume* of solution that is supplying R to the electrode ($V \propto r^3$). This is in strong contrast to the analogous plot for a macroelectrode in Figure 4.50.

The steady-state diffusion-limited current I_{ss} (index ss for steady state) is obtained for a disk-shaped microelectrode embedded in a quasi-infinitely large insulating plane and exposed to a quasi-infinitely large solution volume according to Eq. (5.109).

$$I_{ss} = 4nFDr_0c^* \tag{5.109}$$

D – Diffusion coefficient [cm^2 s^{-1}]

r_0 – Radius [cm], convert to sensible units, for example, 10 μm = 0.001 cm

c^* – Concentration [mol cm^{-3}], convert to consistent units, for example 1 mmol L^{-1} = 10^{-6} mol cm^{-3}

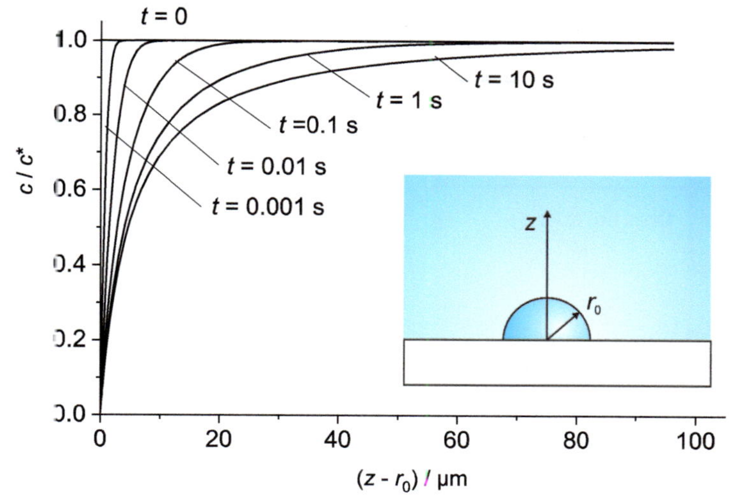

Figure 5.65 Concentration profiles around a hemispherical microelectrode with $r_0 = 5$ µm, $D = 5 \times 10^{-6}$ cm^2 s^{-1}. The profiles are strongly bent and lead to an increased transport in the immediate vicinity of the electrode.

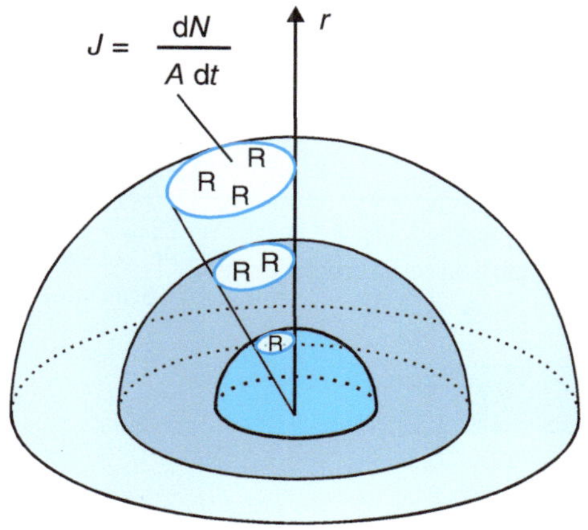

Figure 5.66 Diffusion through segments of sphere surfaces with different distances to the center of a microelectrode. After establishing a steady-state current, the same number of molecules R diffuse per unit time across the sphere segments, although the segments with smaller distance to the electrode have a much smaller area. The diffusive flux (particles per area per time) increases with decreasing distance to the electrode.

This equation is the exact analytical solution of a partial differential equation [105]. Because the microdisk electrode is a nonuniformly accessible electrode, the mathematical solution of the partial differential equation is demanding.

Example Calculation 5.4 Estimate the steady-state diffusion-limited current for a 1 mmol L^{-1} solution of a redox system with $n = 1$ and a (typical) diffusion coefficient $D = 5 \times 10^{-6}$ cm^2 s^{-1} at a microdisk electrode with 10 µm diameter.

sought: I_{ss}

given: $D = 5 \times 10^{-6}\ \text{cm}^2\ \text{s}^{-1}$

$c^* = 1\ \text{mmol L}^{-1} = 10^{-3}\ \text{mol L}^{-1} = 1 \times 10^{-6}\ \text{mol cm}^{-3}$

$r_0 = 5\ \mu\text{m} = 5 \times 10^{-4}\ \text{cm}$

$F \approx 100\,000\ \text{A s mol}^{-1}$ ("rounded" value, good enough for an estimate)

$$I_{ss} = 4\,nFDr_0c^*$$
$$= 4 \cdot 1 \cdot 10^5\ \text{A s mol}^{-1} \cdot 5 \times 10^{-6}\ \text{cm}^2\ \text{s}^{-1} \cdot 5 \times 10^{-4}\ \text{cm} \cdot 1 \times 10^{-6}\ \text{mol cm}^{-3}$$
$$= 1 \times 10^{-9}\ \text{A} = 1\ \text{nA}$$

This approximate value is very suitable for a rule of thumb that is very useful for practical work. At a microelectrode with a diameter of 10 µm in a solution with $c^* = 1$ mmol L^{-1} of the redox-active compound with $D = 5 \times 10^{-6}$ cm^2 s^{-1} (a typical value for aqueous solutions), one can expect a current of about 1 nA. Using proportionalities for r_0 and c^* from Eq. (5.109), diffusion-limited currents for other radii and concentrations can be approximated. In daily laboratory routines, this is useful, for instance, to adjust suitable current ranges of potentiostats.

◢

Please note that Eq. (5.109) does not contain the electrode area (for this, the term r_0^2 should appear somewhere in the equation). The current is proportional to the circumference ($2\pi r_0$) of the disk electrode, rather than to the electrode area. At the

Table 5.9 Overview of different microelectrode geometries: sketch of the geometry, the differential equation in a suitable coordinate system, and the expression of steady-state current.

Geometry	Fick's 2nd law for electrode geometry and coordinate system	Steady-state diffusion-limited current	Symbol	Comment
Planar diffusion (for comparison to macroelectrode) 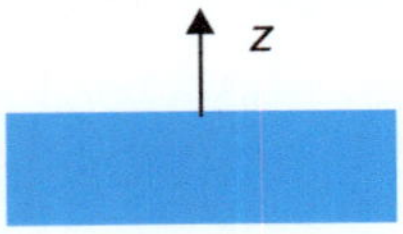	$\frac{\partial c}{\partial t} = D\frac{\partial^2 c}{\partial z^2}$ (Cartesian coordinate system)	Does not exist	z – Vertical distance	For better comparison, the space coordinate is called z here
Microdisk 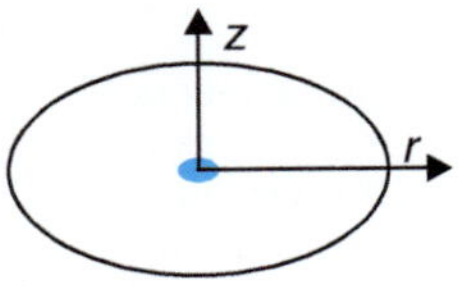	$\frac{\partial c}{\partial t} = D\left(\frac{\partial^2 c}{\partial r^2} + \frac{1}{r}\frac{\partial c}{\partial r} + \frac{\partial^2 c}{\partial z^2}\right)$ Cylinder coordinates	$I_d = 4nFDr_0c^*$	r_0 – Radius of the microelectrode	This is the exact solution for an infinitely thick shaft of the electrode [105]. For finite thickness of the shaft, the numerical prefactor becomes slightly larger [106].
Band 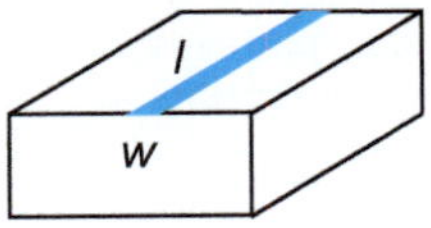	$\frac{\partial c}{\partial t} = D\left(\frac{\partial^2 c}{\partial x^2} + \frac{\partial^2 c}{\partial z^2}\right)$ (Cartesian coordinate system)	$I_d = \frac{2\pi nFDlc^*}{\ln\frac{64Dt}{w^2}}$	t – Time l – Length of the band w – Width of the band	Approximate solution
Cylindrical electrode (microwire) 	$\frac{\partial c}{\partial t} = D\left(\frac{\partial^2 c}{\partial r^2} + \frac{1}{r}\frac{\partial c}{\partial r}\right)$ (Cylinder coordinates)	$I_d = \frac{4\pi nFD_0lc^*}{\ln\frac{4Dt}{r_0^2}}$ $I_d = \frac{2\pi nFDlc^*}{\ln\frac{4Dt}{r_0^2}}$	t – Time l – Length of the cylinder r_0 – cylinder radius	Approximate solution for freely standing cylinder Approximate solution for half embedded cylinder
Microsphere (spherical electrode) 	$\frac{\partial c}{\partial t} = D\left(\frac{\partial^2 c}{\partial r^2} + \frac{2}{r}\frac{\partial c}{\partial r}\right)$ (Spherical coordinates)	$I_d = 4\pi nFDr_0c^*$		[107]
Microhemisphere 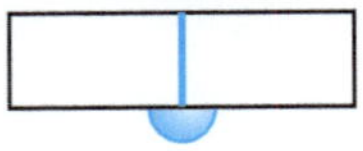	$\frac{\partial c}{\partial t} = D\left(\frac{\partial^2 c}{\partial r^2} + \frac{1}{r}\frac{\partial c}{\partial r}\right)$ (Spherical coordinates)	$I_d = 2\pi nFDr_0c^*$		Half the area of the spherical electrode [107].
Microcone tip 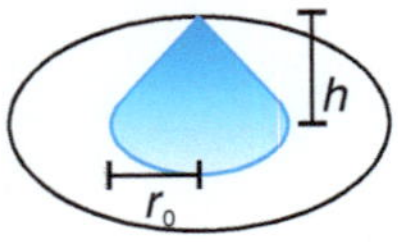	$\frac{\partial c}{\partial t} = D\left(\frac{\partial^2 c}{\partial r^2} + \frac{1}{r}\frac{\partial c}{\partial r} + \frac{\partial^2 c}{\partial z^2}\right)$ (Cylinder coordinates)	$I_d = 4nFDc^*r_0 \left(1 + 0.360661\left(\frac{h}{r_0}\right)^{1.14466}\right)$	h – Height of the cone r_0 – Radius of the cone	Analytical solution for $0 < h < 3r_0$ from time-dependent simulation [106].
Microring 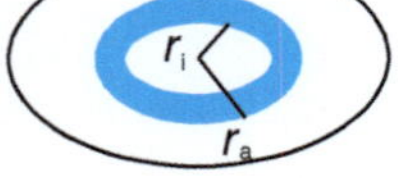	$\frac{\partial c}{\partial t} = D\left(\frac{\partial^2 c}{\partial r^2} + \frac{1}{r}\frac{\partial c}{\partial r} + \frac{\partial^2 c}{\partial z^2}\right)$ (Cylinder coordinates)	$I_d = \frac{\pi^2(r_a + r_i)nFDc^*}{\ln\frac{16(r_a + r_i)}{(r_a - r_i)}}$	r_a – Outer radius r_i – Inner radius	Approximate solution for $t \to \infty$

circumference, the local current density is highest at the microdisk electrode. Whereas the current density $j = I/A$ is usually independent of the electrode area A, j increases at microelectrodes with decreasing radius and area $A = \pi\, r_0^{\,2}$ [Eq. (5.110)]. This constitutes an important difference compared to macroscopic electrodes.

$$j_{\rm ss} = \frac{I_{\rm ss}}{A} = \frac{4nFDr_0c^*}{\pi r_0^2} = \frac{4nFD_0c^*}{\pi r} \tag{5.110}$$

Microelectrodes are not necessarily of disk geometry. There is a large variety of shapes that are preferred for specific applications or are a result of particular preparation routines (Table 5.9). It is only important that at least one dimension of the electrode is smaller than the diffusion length in typical experiments. This dimension is called the critical dimension. This is, for instance, the micrometer-sized radius of a microwire electrode, which may have an appreciable length.

5.6.3 How Does the Current Evolve Before Reaching the Steady State?

In the previous section, we have considered the steady-state current at microelectrodes. Now we want to derive a mathematical relationship for the time-dependent current immediately after a potential step and before attaining the steady-state current. We select a spherical microelectrode as an example because the resulting equations are comprehensible. We must seek the solution to Fick's 2nd law in spherical coordinates and with boundary and initial conditions.

$$\frac{\partial c(r,t)}{\partial t} = D\left(\frac{\partial^2 c(r,t)}{\partial r^2} + \frac{2}{r}\frac{\partial c(r,t)}{\partial r}\right) \tag{5.111}$$

$$c(r, t = 0) = c^* \; (r > r_0) \tag{5.112}$$

$$c(r, t > 0) = 0 \tag{5.113}$$

$$\lim_{r\to\infty} c(r,t) = c^* \tag{5.114}$$

The analytical solution is Eq. (5.115) [107]

$$I(t) = nFAD\left[\frac{1}{\sqrt{\pi Dt}} + \frac{1}{r_0}\right] \tag{5.115}$$

After multiplying the factor by the expression in brackets and reducing by $D^{1/2}$, the first term resembles the Cottrell Equation (4.130). That is, the planar diffusion case for which the current approaches zero for $t \to \infty$. It describes the part of the current that changes before establishing the steady state. In addition, there is a second term $nFAD/r_0$, which is time-independent, and this describes the steady-state current. (Inserting $A = 4\pi r_0^{\,2}$ and reducing by r_0 yields the equation in Table 5.9). As a result, we see in a quantitative manner what we have already made plausible by qualitative reasoning at the beginning of this Section.

Equation (5.115) facilitates an estimation of how long it takes to establish steady state at a microelectrode. We select the time at which the current contribution of planar diffusion has decreased to below 10% of the constant steady-state diffusion-limited current [Eq. (5.116)]. Table 5.10 shows that this time $t_{\rm ss}$ varies strongly with the size of the microelectrode. For a typical electrode with a diameter of 10 µm, the steady-state current is established after a little more than a second. According to Eq. (5.117), we could expect a steady-state current for every electrode size if we were only prepared to wait long enough. However, the value of $t_{\rm ss}$ for an electrode with a diameter of 1 mm

Table 5.10 Time t_{ss} to establish a steady state current according to Eq. (5.117) at a spherical electrode for a diffusion-controlled conversion, $D = 5 \times 10^{-6}$ cm^2 s^{-1}.

$\varnothing = 2\,r_0$	1 mm	100 μm	25 μm	10 μm	1 μm	100 nm	10 nm
t_{ss}	15,900 s	159 s	9.95 s	1.59 s	15.9 ms	159 μs	1.59 μs

is so large (15,900 s = 4:25 h) that microconvection due to density gradients caused by the electrochemical conversion or Joule heating would contribute significantly to the mass transport. In this case, I cannot be described by Eq. (5.115). At the other end of the size domain, nanometer-sized electrodes are expected to provide a steady state current within only a few μs.

$$\frac{nFAD}{\sqrt{\pi D t_{ss}}} < 0.1\frac{nFAD}{r_0} \tag{5.116}$$

$$t_{ss} > \frac{r_0^2}{0.1^2\,\pi D} \tag{5.117}$$

Microdisk electrodes are practically more important than spherical electrodes; however, their mathematical treatment is demanding due to the nonuniform current density distribution.

Q: What does nonuniform current density distribution mean?

A: We have seen in the discussion above, especially in Figure 5.64, that the current density at a disk electrode is highest at the edge and lowest in the center. Although the current density is overall increased by hemispherical diffusion, the current density is not equal at all locations on the disk. This is especially evident immediately after the potential step, when the thin diffusion layer causes a mostly planar diffusion to the disk except for the edge, where lateral diffusion makes an additional contribution to the transport of reagent to the electrode. In contrast, spherical or hemispherical electrodes exhibit the same current density at each point of their surface for all times after a potential step, that is, they possess a uniform current density distribution.

The time dependence of the current at a microdisk electrode cannot be given by an analytical (exact) solution of the corresponding partial differential equation. The Cottrell Equation (4.130) describes the current during a very short period immediately after the potential step when planar diffusion is dominating. For very long times, the steady-state current is dominating according to Eq. (5.109). For the transition region, there are approximate equations in common use, for example, Eq. (5.118) [108] or Eq. (5.119) [109].

$$I(t) = nF\pi D r_0 c^* \left[1 + \frac{r_0}{\sqrt{\pi D t}} + \left(\frac{4}{\pi} - 1\right)\exp\left(\frac{-0.39115\,r_0}{\sqrt{Dt}}\right)\right] \tag{5.118}$$

$$I(t) = nF\pi D r_0 c^* \begin{cases} \dfrac{1}{\sqrt{\pi\theta}} + 1 + \sqrt{\dfrac{\theta}{4\pi}} - \dfrac{3}{25}\theta + \dfrac{3}{226}\theta^{3/2} & \theta \le 1.281 \\[2ex] \dfrac{4}{\pi} + \dfrac{8}{\sqrt{\pi^5\theta}} + \dfrac{25}{2792}\theta^{-3/2} - \dfrac{\theta^{-5/2}}{3880} - \dfrac{\theta^{-7/2}}{4500} & \theta \ge 1.281 \end{cases} \tag{5.119}$$

Here θ is a normalized dimensionless time

$$\theta = \frac{Dt}{r_0^2}. \tag{5.120}$$

Equation (5.119) is slightly more accurate than Eq. (5.118), which is of relevance only for very strict requirements.

Q: Which strict requirements could these possibly be?

A: We will discuss in Chapter 6 and in Section 8.12 the important role of digital simulation, which can be executed with commercial software packages such as COMSOL or implemented in Matlab. When setting up a simulation environment, it is sensible to test the coding of the problem and the technical parameters such as mesh sizes and time steps with a problem for which the exact solution is known. This "true" solution is ideally an analytical solution of the partial differential equation. One compares the values according to the known solution with the output of the simulation software. The output of the simulation software always remains an approximation because the space and in most cases the time have to be discretized, which means split up into small space elements and time steps. The solution to the problem is calculated between these very small time and space steps by strongly simplified procedures, for example, linear approximations.

Most general considerations obtained from the treatment of a spherical electrode can be qualitatively transferred to a microdisk electrode because a molecule that is far away from the electrode surface (e.g., 10 r_0) encounters an almost identical diffusion field in both cases. However, it must be kept in mind that the details differ somewhat because the current density at a microdisc electrode is larger at the edge than in the center, while it is completely uniform over the surface for a spherical or hemispherical electrode. Such electrodes are also called **uniformly accessible electrodes**.

For quantitative experimental analysis, a chronoamperogram is recorded in a solution of the redox system in a supporting electrolyte (faradaic current and capacitive current) and then another chronoamperogram in the supporting electrolyte without the redox system (capacitive current only). By subtraction of the second chronoamperogram from the first one, the background-corrected current $I'(t)$ (faradaic current only) is obtained. The current $I'(t)$ can be plotted in dimensionless form, which is derived from Eq. (5.118). The time is plotted as the reciprocal square root of the dimensionless time τ on the abscissa [Eq. (5.121)]. The current function $f(\tau)$ is the current normalized with the steady-state diffusion-limited current of a coplanarly embedded microdisk electrode [Eq. (5.122)].

$$\tau = \frac{4Dt}{r_0^2} \tag{5.121}$$

$$f(\tau) = \frac{I'(t)}{4nFDr_0c^*} \tag{5.122}$$

The experimental current function is compared to the theoretical current function from Eq. (5.118).

$$f(\tau) \approx 0.7854 + 0.8862\frac{1}{\sqrt{\tau}} + 0.2146\exp\left(-0.7823\frac{1}{\sqrt{\tau}}\right) \tag{5.123}$$

Electrode surfaces that do not have a circular disk shape, but are a square or an ellipse, can be treated approximately by using the equations for a disk that has the same area as the electrode in question [110].

A common complication is recessed electrodes, at which the electrochemically active area is not in a plane with the surrounding insulating shaft (for example, due to unequal polishing rates; Figure 5.67a). Instead, the electrode is recessed by the

a)

b)

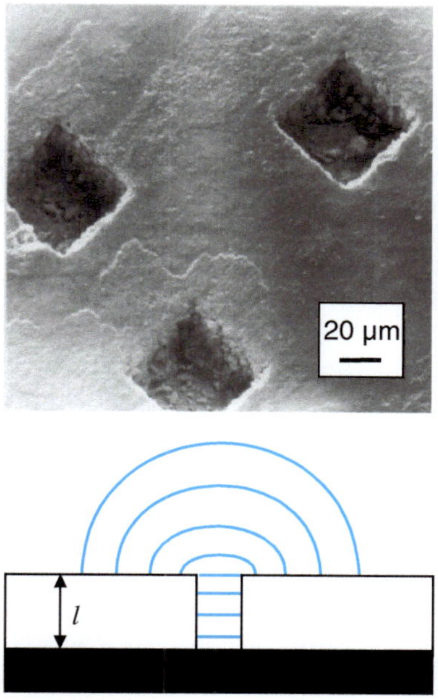

Figure 5.67 Geometry of a recessed microelectrode; a) electron micrograph; b) schematics of the cross section of the recessed electrode with iso-concentration lines in the steady state during diffusion-controlled conversion. Source: Reproduced with permission from [111], Copyright 1998, Wiley-VCH.

thickness l of a cover layer (Figure 5.67b). The steady-state diffusion-limited current $I_{ss,recess}$ is smaller than that ($I_{ss,disk}$) of an electrode of equal size that is coplanarly embedded in its insulating shaft. In the steady state, a fraction of the concentration difference between the electrode surface and the solution bulk decays within the approximately hemispherical diffusion layer, and another fraction decays within a linear concentration gradient inside the pore, through the insulating cover layer. The ratio χ between the steady-state currents of a coplanarly embedded microdisk electrode and a recessed microelectrode of the same size can be derived [Gl. (5.124)].

$$\chi = \frac{I_{ss,disk}}{I_{ss,recess}} = 1 + \frac{4l}{\pi r_0} \tag{5.124}$$

The steady-state current of the recessed electrode is thus comparable to that of a coplanarly embedded microelectrode at the concentration c^*/χ. Thus, the recess causes a decrease in the sensitivity, which becomes larger the larger the aspect ratio of the pore-shaped recess (aspect ratio = pore depth/pore diameter).

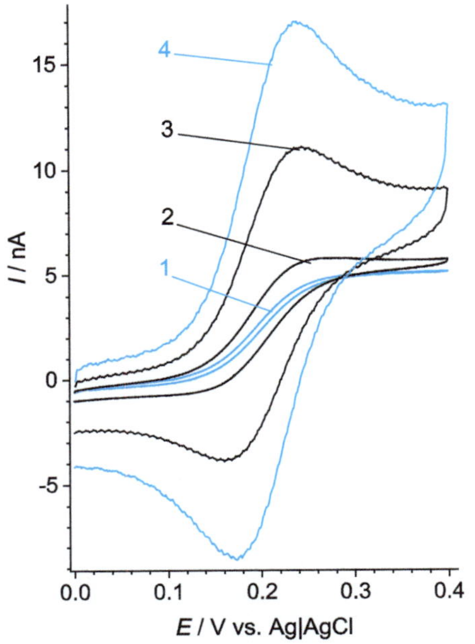

Figure 5.68 Cyclic voltammograms at a microdisk electrode with a diameter of 25 μm in a solution of 0.5 mmol L^{-1} hydroxymethylferrocene in 0.1 mol L^{-1} Na_2SO_4 using v of 1) 4 mV s^{-1}, 2) 40 mV s^{-1}, 3) 400 mV s^{-1}, 4) 1000 mV s^{-1}. Source: Data were provided by Nico Roth, Universität Oldenburg.

5.6.4 Cyclic Voltammetry at Microdisk Electrodes

Based on the discussion of chronoamperometry, you may already suspect that the diffusion layers at a microdisk electrode during cyclic voltammetry should depend on the timescale of the experiment, controlled by the scan rate v. If the experiment is conducted with a very fast scan rate, then the diffusion layer remains very thin and may not exceed the size of the microelectrode. In this situation, the mass transport is controlled by planar diffusion (similar to the case of a macroscopic electrode). The resulting voltammogram is peak-shaped as shown in Figure 5.68, curve 4. If the experiment is performed with a lower scan rate (i.e., on a longer timescale), a hemispherical diffusion layer is formed around the microelectrode, which allows the recording of a steady-state current. The result is a sigmoidal curve shape such as in curves 1 and 2 of Figure 5.68. In contrast to chronoamperometry, the user can diagnose the diffusion case simply by looking at the shape of a cyclic voltammogram to identify whether hemispherical diffusion prevails and a steady-state current is reached. The step height of the sigmoidal voltammograms corresponds to the steady-state diffusion-controlled current from Table 5.9 and is independent of v. [For a microdisk electrode, insertion of Eq. (5.109) for the steady-state current leads to Eq. (5.125)]. The I–E curves of both half-cycles are superimposed for the steady state, except for a small capacitive current. There is no peak in the reverse scan because hemispherical diffusion transports the reaction products away from the electrode so quickly that they cannot be reduced in the back scan.

E_{NF} - Nernst factor = $RT/(nF)$

$$I(E) = \frac{4nFDc^* r_0}{1 + \exp\left[\frac{E - E^{\circ\prime}}{E_{NF}}\right]} \tag{5.125}$$

If v is increased stepwise, a small peak at the beginning of the step is formed that grows with further incrementation of v, until the cyclic voltammogram resembles that of a macroelectrode (Figure 5.68, curve 4), which also includes the appearance of a peak in the back scan. In this case, the potential scan is so fast (with respect to diffusion) that the oxidized species formed in the forward scan have not diffused away from the electrode when the potential has already reached values that cause the rereduction. The peak current in this range of v values is proportional to the square root of v.

Next, the transition from sigmoidal to peak-shaped voltammogram is considered in more detail. We will need dimensionless parameters. Here, the dimensionless parameter is formed from a substance property (D), a parameter of the experiment characterizing the timescale (v), and the geometry of the electrode represented by its radius (r_0). We form the ratio $r_0/(2Dt_c)^{1/2}$ based on the size of the microelectrode r_0 and the distance molecules diffuse on average $(2Dt_c)^{1/2}$ along one space dimension during the critical time t_c. The critical time t_c is the time it takes to increment/decrement the electrode potential by the Nernst factor E_{NF} using the scan rate v of the experiment. We can consider the time t_c as the time it takes to "switch on" diffusion. During this time, the concentration ratio is changed from 1:e to 1:1 or from 1:1 to e:1.

$$t_c = \frac{E_{NF}}{v} = \frac{RT}{nFv}. \quad (5.126)$$

t_c – Critical time for the experiment [s]

If the ratio $r_0/(2Dt_c)^{1/2} \ll 1$, the cyclic voltammogram will exhibit a sigmoidal shape, whereas it will be peak-shaped for $r_0/(2Dt_c)^{1/2} \gg 1$.

Example Calculation 5.5 How fast must a cyclic voltammogram be recorded to obtain a peak-shaped or a sigmoidal curve for a microelectrode with $\varnothing = 2r_0 = 25\ \mu m$ immersed in an aqueous solution?

We calculate $x = r_0/(2Dt_c)^{1/2}$ as a criterion for the decision. To be certain we have a "safety margin", that is, we use $x \leq 0.5$ for a sigmoidal voltammogram and $x \geq 2$ for a peak-shaped voltammogram. For the calculation, we need to assume a diffusion coefficient that is not provided in the problem. For many aqueous and organic electrolyte solutions, the diffusion coefficients are actually quite similar. In addition, the diffusion coefficient enters the calculation only as a square root. We arrive at a sensible estimate if we take a reasonable value $D = 5 \times 10^{-6}\ cm^2\ s^{-1}$. (Of course, this does not hold anymore when the viscosity is very high and thus D is very small, as found in polymer gels or ionic liquids.)

Unknown: v_c for $x = 0.5$ and $x = 2$

Given: $r_0 = 12.5\ \mu m = 0.00125\ cm$

$D = 5 \times 10^{-6}\ cm^2\ s^{-1}$

$E_{NF} = RT/(nF) = 0.0257\ V$

($n = 1$, $T = 298$ K, room temperature if not stated otherwise)

$$x = \frac{r_0}{\sqrt{2Dt_c}} = \frac{r_0}{\sqrt{2D\frac{E_{NF}}{v_c}}} = \frac{r_0\sqrt{v_c}}{\sqrt{2DE_{NF}}}$$

$$v_c = \frac{2DE_{NF}x^2}{r_0^2}$$

$$v_{c,ss} = \frac{0.0257\ V \cdot 2 \cdot 5 \times 10^{-6}\ cm^2\ s^{-1}(0.5)^2}{(0.00125\ cm)^2}$$

$$v_{c,ss} = 0.04112\ V\ s^{-1} \approx 0.04\ V\ s^{-1}$$

$$v_{c,peak} = \frac{0.0257\ V \cdot 2 \cdot 5 \times 10^{-6}\ cm^2\ s^{-1}(2)^2}{(0.00125\ cm)^2}$$

$$v_{c,peak} = 0.66\ V\ s^{-1}$$

We expect a sigmoidal voltammogram for $v \leq 0.04\ V\ s^{-1}$ and a peak-shaped cyclic voltammogram for $v \geq 0.66\ V\ s^{-1}$. It has been somewhat arbitrary how we selected the margin or transition region. Figure 5.68 demonstrates how the peak-shaped voltammogram develops gradually. An exact calculation of the cyclic voltammograms in the transition region is available in Ref. [131].

A lot of information can be obtained from the slope of steady-state voltammograms. For example, kinetic reversibility can be assessed for a new electrode

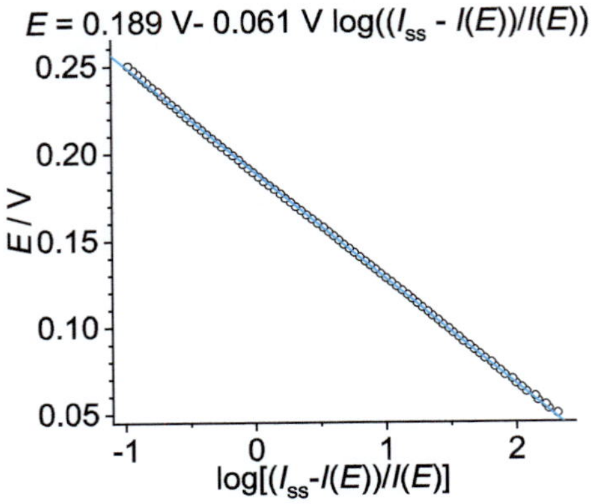

Figure 5.69 Slope analysis for the sigmoidal cyclic voltammogram of a microdisk electrode from Figure 5.63, curve 2 in the potential range $0.05 < E/V < 0.25$. The first half cycle was used for the evaluation.

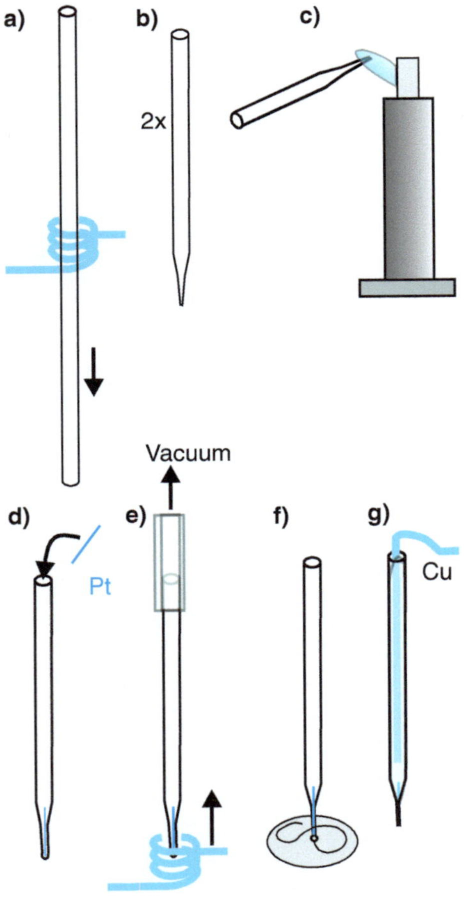

Figure 5.70 Fabrication of microelectrodes; a) pulling of the capillary; b and c) closing the pulled ends in a flame; d) inserting the Pt microwire; e) sealing the wire under vacuum; f) exposure of the cross section of the Pt wire by grinding and polishing; g) making the electrical contact of the Pt wire with the silver-containing epoxy glue and a Cu wire.

material or for new redox systems by slope analysis. Starting from the linearization in Eq. (5.127), a spreadsheet application is used to calculate the new data pairs ($\log[(I_{ss} - I_i)/I_i]$, E_i) from the measured data (E_i, I_i) and the diffusion-controlled current I_d.

$$E = E_{1/2} + 2.3\frac{RT}{nF}\log\frac{I_{ss} - I(E)}{I(E)} = E_{1/2} + E_{NF}\ln\frac{I_{ss} - I(E)}{I(E)} \tag{5.127}$$

From the plot in Figure 5.69, one calculates the slope. For $T = 298$ K and $n = 1$, the absolute value should be 0.059 V = 59 mV (for log) or 0.0257 V = 25.7 mV when using the natural logarithm. Smaller values are indicative of a redox couple with $n = 2$; larger slopes point to sluggish electrode kinetics.

Excursion 5-3 Fabrication of Microdisk Electrodes

There are many different procedures for the fabrication of Pt microdisk electrodes. The basic principle is the sealing of a Pt microwire into glass that has a similar thermal expansion coefficient to that of Pt. A glass capillary (100 mm length, 1.5 mm outer diameter, 0.375 mm wall thickness) is cleaned in ethanol, dried, and mounted centrically into the heating coil of a capillary puller. After switching on the heating current, the position may require readjustment because the thermal expansion of the heating coil may slightly shift its center (Figure 5.70a). Once the glass is soft, both ends are pulled apart, recovered from the capillary puller (Figure 5.70b) and heated briefly in a small gas flame in order to close the open capillary (Figure 5.70c). The Pt wire of 10 μm or 25 μm diameter is cut to a length of 1 cm and cleaned in ethanol. The cleaning step is important to obtain a good seal after melting the wire into glass. The wires are inserted into the open end of the capillary by means of a tweezer and then moved down to the closed end by gently tapping the tweezer against the outside of the capillary (Figure 5.70d). A stereomicroscope of low magnification can aid in following the process precisely. Subsequently, the capillary is mounted again centrally into the capillary puller slightly above the heating coil. The open end of the capillary is connected to a laboratory vacuum pump. While pumping down, the glass capillary loaded with the Pt wire is slowly lowered down about 5 mm into the glowing heating coil. When the glass softens, it encloses the Pt wire and embeds it tightly (Figure 5.70e).

Next, the cross section of the sealed wire is exposed by grinding on a grade 1200 grinding paper. Depending on the overall thickness of the capillary, the electrode can be held by hand, but micro-polishing equipment is more advisable (Figure 5.70f). Caution: Always wet the grinding paper in order to prevent the release of micrometer-sized dust. For a special requirement, the thickness of the insulating sheath around the Pt wire can be thinned down by placing the electrode at a tilt angle on the rotating polishing wheel while rotating the electrode around its own axis. After exposing the cross section, the pretreatment is continued by polishing in a suspension of alumina (Al_2O_3) in water. Usually, different grain sizes, such as 0.3 μm and 0.05 μm, are used one after the other.

A thin copper wire of 0.5 mm diameter is used for electrical contact after removal of the insulating polymer layer. One end is placed into silver-containing epoxy glue. The creamy glue is placed into the open end of the capillary until the Cu wire contacts the rear end of the sealed Pt wire. After thermally curing the glue according to the manufacturer's instructions, the electrode is ready for electrochemical testing (Figure 5.70 g). Further instructions for the fabrication of microelectrodes are available in the literature [113–116].

Excursion 5-4 Fabrication of Nanoelectrodes

The procedures presented in Excursion 5-3 can only be used for electrodes with a diameter equal to the diameter of the Pt microwire. The diameter of commercial Pt wires has a lower limit of 10 μm because smaller wires cannot be produced by wire drawing. Smaller electrodes can be obtained from Wollaston wires (Pt core, Ag shell). The Ag shell is etched at the lower end of the wire, exposing the Pt core. This part of the wire is then sealed into glass [117]. The cleaning of these wires is important and tricky. The insertion and handling of the etched fragile wire is something for manually talented persons with a high tolerance of frustration. Another way is to start from wires with a diameter of 25 μm and thin them by electrochemically etching. These wires are then sealed into glass [118]. The conical end of the wire has the consequence that the radius r_0 of the electrode will change with each polishing step.

A popular procedure uses a laser-heated capillary puller (laser puller) for electrophysiology to seal a 25 μm diameter Pt wire into a quartz capillary (Figure 5.71). This is achieved by placing the Pt wire roughly in the middle of the capillary that is mounted horizontally in the laser puller, applying vacuum to both open ends of the capillary, and mechanically blocking the pulling mechanisms of the instrument with a steel clamp. After sealing the Pt wire, the assembly is heated again, causing the quartz to soften while also Pt becomes liquid. In this state, quartz and Pt are pulled together, yielding electrodes with very small diameters that can even be polished. Detailed procedures can be found in [119]. When using the instrument settings from the literature, one should keep in mind that the optimum parameters vary for each individual device and may also require adjustment over the lifetime of the instrument.

Another elegant experimental access to nanoelectrodes starts from a drawn open capillary that is relatively easy to obtain with orifices of nanometer dimensions [120]. Such a capillary is inserted concentrically into another larger capillary. A fuel gas is flushed into the larger end of the pulled capillary. At the same time, an inert gas is flowing in countercurrent in the larger capillary (Figure 5.72) in order to dilute the fuel gas once it exits from the capillary orifice. The assembly is heated from the outside by a heating coil or the flame of a gas burner until the outer capillary is slightly glowing. At this temperature, the fuel gas will thermally decompose to carbon and hydrogen. Carbon is deposited as a black layer on the inner wall of the capillary. The flow of the fuel gas stops automatically when the inner capillary is completely clogged by carbon. Other materials can be electrodeposited on the nanometer-sized carbon electrode [120].

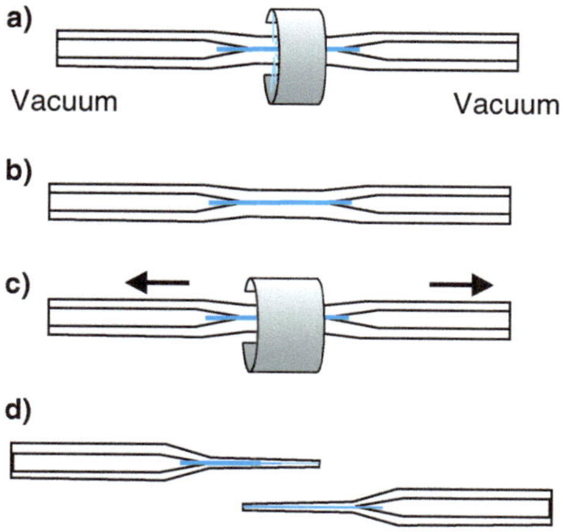

Figure 5.71 Fabrication of nanoelectrodes. a) Seal Pt wire in quartz capillary, but do not pull yet (!); b) check the correct sealing; c) heat again slowly so that Pt and quartz soften, then pull; d) the process yields two Pt electrodes. (Subsequent polishing and contacting are not shown.)

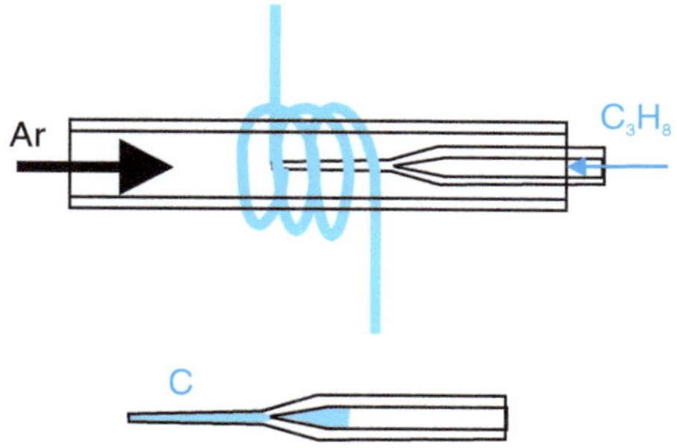

Figure 5.72 Fabrication of nanoelectrodes by thermal decomposition of propane or ethyne in the inner capillary. (Subsequent polishing and contacting are not shown.)

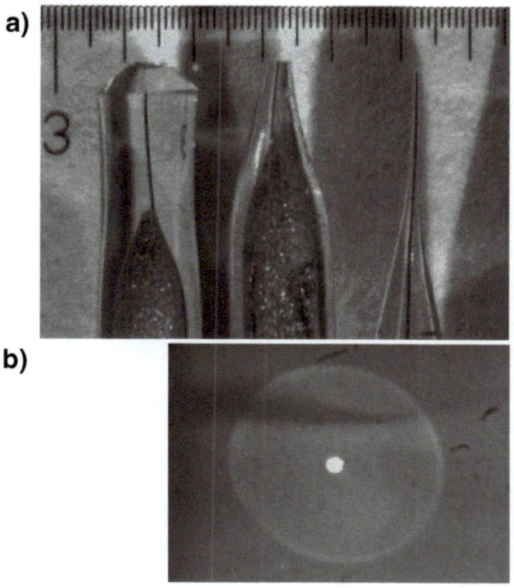

Figure 5.73 Optical inspection of the microelectrode; a) side view; b) top view.

Excursion 5-5 Characterization of Microelectrodes

Microelectrodes have to be tested for their functionality, either directly after fabrication or again later during usage when unexpected behavior is encountered. Such a functionality test includes several steps.

Optical Inspection. The sealed wire is inspected under a light microscope after focusing (Figure 5.73a). If the interface between glass and wire is blurred or gas bubbles are visible, the electrode is to be rejected. This method also allows one to confirm that the wire is located concentrically in the insulating sheath (Figure 5.73b). For special applications (Section 8.12), the diameter of the insulating sheath in the plane of the electrode can be measured.

Potential Cycling in H_2SO_4 Solution. Potential cycling the microelectrode in 0.5 mol L^{-1} H_2SO_4 allows one to confirm the proper electrical contact and cleanliness of the electrode surface. For noble metal electrodes, it is advisable to use a

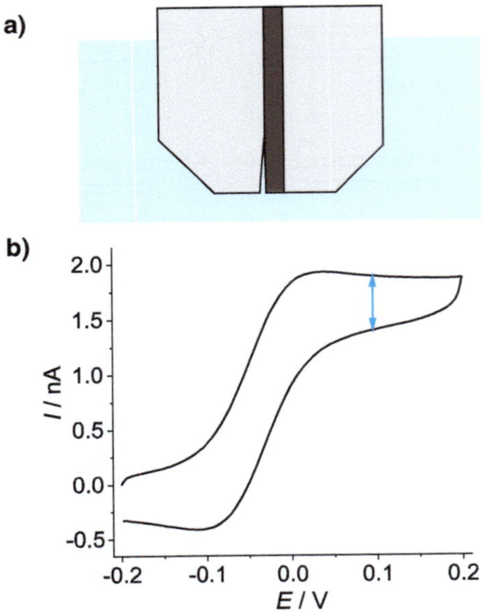

Figure 5.74 Microelectrode with enlarged capacitive current; a) cross section with a possible defect at the seal close to the exposed disk electrode; b) cyclic voltammogram ($v = 0.05$ V s^{-1}) in 1 mmol L^{-1} tetracyanoquinodimethane +1 mol L^{-1} $LiClO_4$ + dimethylsulfoxide.

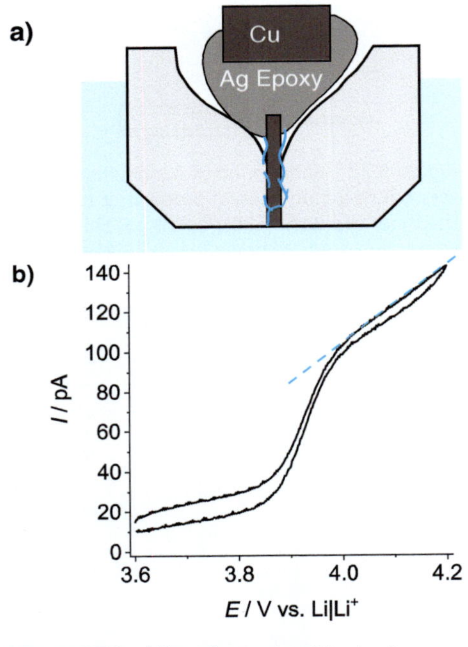

Figure 5.75 Microelectrode with ohmic contact between the contacting wire and the electrolyte solution; a) schematic cross section with indicated hairline cracks in the glass sheath that are often not visible in optical microscopy and are shown in an enlarged way; b) cyclic voltammogram ($v = 0.05$ V s^{-1}) of 1 mmol L^{-1} 2,5-di-tert-butyl-1,4-dimethoxybenzol +1 mol L^{-1} $LiPF_6$ + ethylene carbonate/propylene carbonate.

chloride-free solution and reference electrode, for example, Hg|Hg_2SO_4|$KHSO_4$(*sa*). All subsequent potentials in this Excursion refer to this reference electrode. The limits of the potential ranges are placed within the onset of H_2 and O_2 evolution. For Pt, this would be for the inspection of the electrode using $E_{\text{initial}} = -0.61$ V, $E_\lambda = +1.0$ V, $v = 0.1$ V s^{-1}. Should the voltammogram indicate a contamination, potential cycling can also be used for cleaning the surface. A typical procedure would be 50 cycles with $E_{\text{initial}} = -0.81$ V, $E_\lambda = +1.6$ V, $v = 0.1$ V s^{-1}. Next, another inspection is made. The cycling is terminated at a potential of −0.21 V, where the potential is kept for 2 min. After this procedure, the surface voltammogram of a clean Pt surface should be obtained (Figure 5.22). The resolution of the peaks for microelectrodes is often not as good as in Figure 5.22. Electrodes must be rejected if one of the following conditions is met.

- No current is measured even at very high sensitivities. In this case, the electrical contact failed, or the contact may have been broken due to vibration during polishing.
- A voltammogram similar to Figure 5.22 is obtained, but it is overlaid with a straight tilted line. Sometimes this feature evolves gradually or only after prolonged experimentation. The reason is usually a bad seal of the wire. Electrolyte solution may penetrate through microfissures to the macroscopic contacting wire and form an additional ohmic contact between the electrolyte solution and the contacting wire.

Diffusion-Controlled Current. If the electrode has passed these tests, it is removed from the solution, rinsed carefully with water, and transferred into a solution of a reversible redox couple. A cyclic voltammogram is recorded with 10 mV s^{-1} and should exhibit a sigmoidal shape as in Figure 5.63, curve 2.

Quality defects are:

- A large capacitive current (Figure 5.74). This results from an incomplete sealing process or when the metal–glass seal is damaged during the polishing process. The area of the metal–solution interface in the slit between the wire and the glass sheath contributes to the capacitive current, but the faradaic current is not increased significantly over that of a perfectly embedded disk.
- A slow rise in the current toward the plateau value is indicative of slow electron transfer that could be caused by coverage of the electrode surface by adsorbates. If the current continues to rise instead of staying constant in the potential region, in which a plateau is expected (Figure 5.75), an ohmic current between the contacting wire and the solution contributes to the current as was discussed already for the test in sulfuric acid solution. Because the microscopic fissures may be filled slowly by the electrolyte solution, it is possible that this effect occurs only after prolonged usage and disappears temporarily during rigorous drying, but returns soon upon contact with the electrolyte solution. Such electrodes are not suitable for quantitative work.

The apparent electrode radius r_{app} can be calculated from Eq. (5.128). The apparent radius is the radius of a perfect microdisk electrode that would give the same diffusion-controlled current as the electrode under study.

$$r_{\text{app}} = \frac{I_d}{4nFDc^*} \tag{5.128}$$

When using this equation, one should be sure to use compatible units (Example Calculation 5.4). Deviation between the real size and r_{app} can be caused by recessed electrodes, a protruding microelectrode, or by an oval instead of a circular cross section of the wire. These geometric deviations only influence specific measurements and therefore can only be further diagnosed using such techniques (e.g., by approach curves using scanning electrochemical microscopy [SECM] or by scanning force microscopy).

Storage. The Pt microelectrodes are stored in a box for protection from dust in the air. Slotted foam pads have proven useful as spacer and protection against mechanical damage while moving the storage box.

5.6.5 Analytical Features and Application of Microelectrodes

Microelectrodes have paved the way to entirely new applications in electroanalytical chemistry because they exhibit fundamentally different characteristics from their macroscopic counterparts [1, 103, 112].

Improved Ratio of Desired Signals to Interfering Signals. If a microelectrode is operated under steady-state diffusion-controlled conditions, the faradaic current $I_F = I_d$ is proportional to the radius of the electrode. Here, the faradaic current is the desired signal and the capacitive current is the interfering signal, which is proportional to the scan rate v and the double layer capacitance C_{dl} (which in turn is proportional to the electrode area $A' \approx \pi r_0^2$). According to Eq. (4.100), I_C is approximated by $I_C = vC_{dl}$, where $C_{dl}/A' \approx 20\ \mu F\ cm^{-2}$ for typical noble metal electrodes in nonadsorbing electrolytes. The important ratio I_F/I_C is proportional to r_0^{-1}, thus, it increases with decreasing electrode radius. Note that for very small microelectrodes, "stray capacitance" can increase the capacitive current.

For voltammetric pulse techniques, the capacitive current decays exponentially with $t/(R_E C_{dl})$ [Eq. (4.99)]. For macroscopic electrodes, the faradaic current decreases according to the Cottrell equation with $t^{-1/2}$. However, the decay of the faradaic current at microelectrodes is much lower, especially if the factor Dt_{pulse}/r_0^2 is large. Under those conditions, voltammetric pulse methods can reach especially high ratios of I_F/I_C.

Small Time Constants and Fast Voltammetry. The time constant of an electrochemical cell is controlled by C_{dl} and the electrolyte resistance R_E according to Eq. (5.129).

$$\tau = R_E C_{dl} \tag{5.129}$$

The time constant of microelectrodes is very small due to the small electrode area and the resulting low C_{dl}. According to Eq. (4.99), establishing a certain potential at an electrode requires at least a time of 3τ. In other words, changes in the potential drop across the double layer cannot be realized faster than given by τ (Table 5.11), independent of the scan rate the potentiostat applies to the contacting wires. This is a very important limitation for the study of fast electrochemical processes. Due to their small time constant, microelectrodes are especially well suited for voltammetry with high scan rates v.

Table 5.11 Time for charging the electrical double layer [3τ according to Eq. (5.129)] for disk electrodes with different diameters; $R_E = 1000\ \Omega$, area-specific double layer capacitance $C_{dl,A} = 2 \times 10^{-5}\ A\ s\ V^{-1}$.

$\varnothing = 2\,r_0$	3 mm	1 mm	100 μm	25 μm	10 μm
τ	4.2 ms	470 μs	4.7 μs	300 ns	47 ns

Low-Voltage Drop and Measurements in Solution of Low Conductivity. We have already discussed the potential distribution in electrochemical cells in Section 5.3.1. This made it clear that there will be a deviation between the desired potential and the effective potential of $(I_C + I_F)R_E$. Miniaturization causes a reduction of I_C as well as I_F, thus the potential drop is decreased. This effect can be exploited by using microelectrodes for measurements in solutions of low conductivity (high specific resistance). This can be solvents without deliberate addition of supporting electrolytes. Ubiquitous impurities and reaction products of the electrolysis are then carrying the charge in the liquid phase. Measurements in solvents with a low dielectric constant are possible as well. In these solvents, the supporting electrolyte is poorly soluble or forms (nonconducting) ion pairs. For example, measurements have been reported in benzene, toluene, and hexane. Further details and quantitative considerations are collected in [121].

Small Size of Microelectrodes and Detection in Constrained Volumes. The small size of microelectrodes facilitates their use for investigation in tiny spaces, a feature that is especially used for the investigation of biological cells. We call such studies *in vivo* if the microelectrode is implanted in a living organism and if it provides concentration measurements from inside the organism. The small size is a prerequisite to conduct the measurements without a significant impact on the life processes themselves (Excursion 5-6). Measurements are called *in vitro* if cultured cells are investigated outside of an organism. Such experiments allow one to mimic an intercellular cleft by a biological cell and a microelectrode positioned very close to it. Such arrangements are used to investigate the mechanisms of exocytosis (Excursion 5-7). Nanoelectrodes have been used successfully to penetrate individual cells. The puncture site in the biomembrane often seals spontaneously against the side wall of the nanoelectrode, allowing the study of redox reaction inside the sealed interior of a cell [122]. Finally, we should note the use of microelectrodes as probes in scanning electrochemical microscopy (SECM) (Section 8.12).

High Current Density and Independence from Convection. High current densities are achieved at microelectrodes compared to those at macroelectrodes in the same solution (Figure 5.63). This is a consequence of the small size of the diffusion layer and the strongly curved concentration profile close to the microelectrode (cf. Figure 5.65 to Figure 4.52). If macroelectrodes are used in stirred solution, the concentration differences in solution will be equalized by convection except for the hydrodynamic boundary layer (Section 4.6). Within this layer, the solution "sticks" to the electrode and its insulating surrounding. The thickness of the hydrodynamic boundary layer determines the Nernst diffusion layer. At microelectrodes, the thickness of the solution layer in which most of the concentration difference from the bulk solution is found is typically thinner than the hydrodynamic boundary layer. Therefore, convection in the solution bulk does not affect the current significantly. This feature is used in hand-held sensors, where the signal should not depend on the muscle tremor of the user. Scanning electrochemical microscopy (Section 8.12) makes use of this feature as well. Otherwise, the current of the scanning microelectrode would change very much compared to the resting electrode (Section 8.12).

Excursion 5-6 Neurotransmitter Concentration in the Brain During Drug Consumption

The human brain is made up of $10^{10}-10^{12}$ nerve cells (neurons, Figure 5.76). Each neuron is composed of a cell body (3) containing the cell nucleus (2), a branched net of dendrites (1), and the axon (4), whose end branches out to an extension called axon terminals (5). Signals are received at the dendrites, passed to the cell body (possibly processed) and from there passed as an action potential along

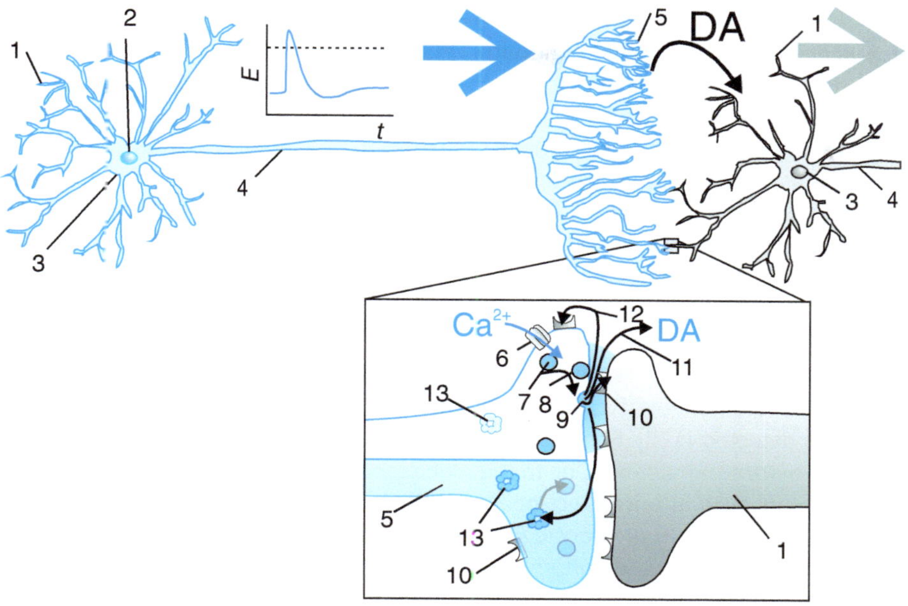

Figure 5.76 Excitation conduction from the presynaptic cell (blue) to the postsynaptic cell (gray). 1) Dendrite, 2) cell nucleus, 3) cell body, 4) axon, 5) axon terminal in enlarged view. When an action potential arrives, Ca^{2+} diffuses through the ion channels (6) into the cell and triggers, via a signal cascade, the fusion of dopamine (DA)-filled vesicles (7) with the cell membrane (8). The content of the vesicles is discharged into the synaptic cleft (exocytosis, 9). DA can be bound by a postsynaptic receptor (10), which finally causes the build-up of an action potential in the postsynaptic cell. The remaining DA can diffuse into the intracellular fluid (11), be bound by a receptor of the presynaptic cell (autoreceptor, 12), or be transported by a DA transporter (13) to the inside of the presynaptic cell. There, the neurotransmitter is chemically converted or stored again in a vesicle.

(see inset) the axon to the axon terminals [123]. The transmittance of an excitation between different neurons must overcome the synaptic cleft (enlarged inset). The synaptic cleft is filled with extracellular fluid, has a thickness of 1–100 nm, and separates the presynaptic cell (from where the excitation signal arrives) from the postsynaptic cell (receiving the excitation signal). Transmittance is made by specific compounds called neurotransmitters. Neurons are distinguished by the neurotransmitters used by them and correspond to their basic function in the brain. Among these neurotransmitters, catecholamines [dopamine (DA), norepinephrine, epinephrine] and serotonin can be detected voltammetrically or amperometrically [124].

The voltammetric microsensors for DA consist of a carbon fiber with a diameter of 7 μm that extends by 150 μm from an insulating fused silica capillary with a diameter of 90 μm and an approximate length of 1 cm (Figure 5.77). The detection is based on the oxidation of DA according to Eq. (5.130). The calibration of the sensor is performed externally in a flow-through apparatus in the concentration range of 50–1000 nmol L^{-1} [125].

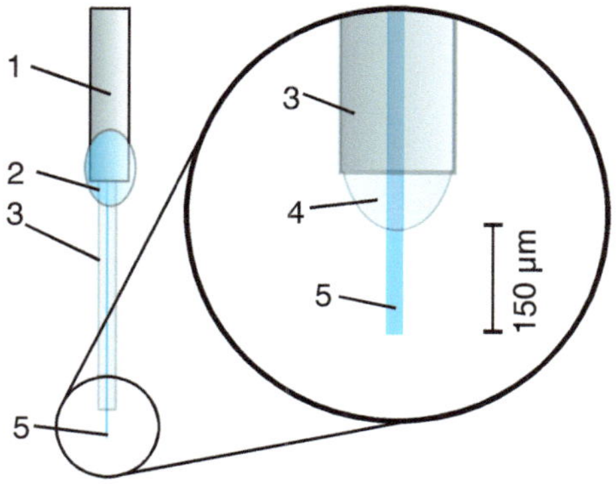

Figure 5.77 Design of a DA sensor for implantation in the brain of a rat; 1) electrode contact pin; 2) conductive connection with Ag-containing epoxy glue mechanically reinforced by nonconductive epoxy glue; 3) fused silica capillary ∅ 90 μm; filling with epoxy glue; 5) carbon fiber, ∅ 7 μm. Source: Adapted from [125].

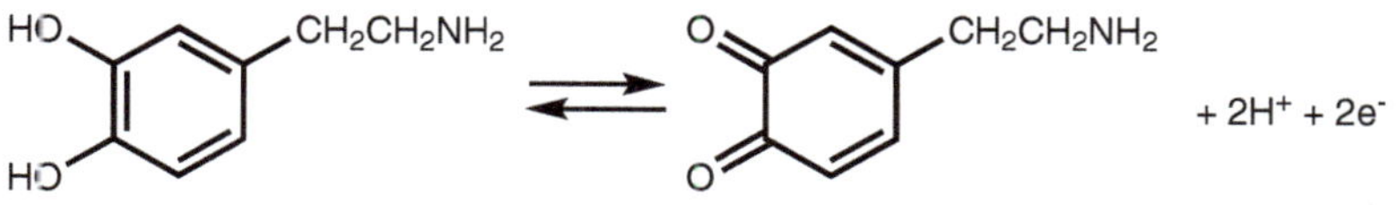

(5.130)

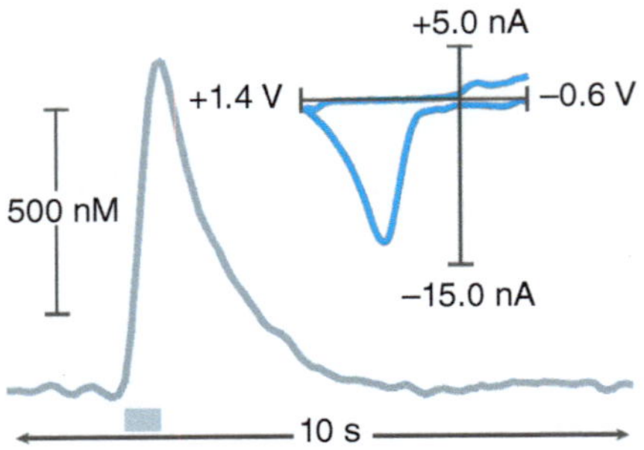

Figure 5.78 Increase in [DA] (gray) by electrical stimulation within the time window marked in gray above the abscissa. The FSCV signal (blue) shows the signal at the end of the stimulus and agrees with that of DA in buffer solutions. Note the use of the polarographic sign convention: Oxidation currents are negative in the FSCV trace, and plotted downward, oxidizing potentials are positive and plotted toward the left side. Source: Reproduced with permission from [127], Copyright 2003, Springer Nature.

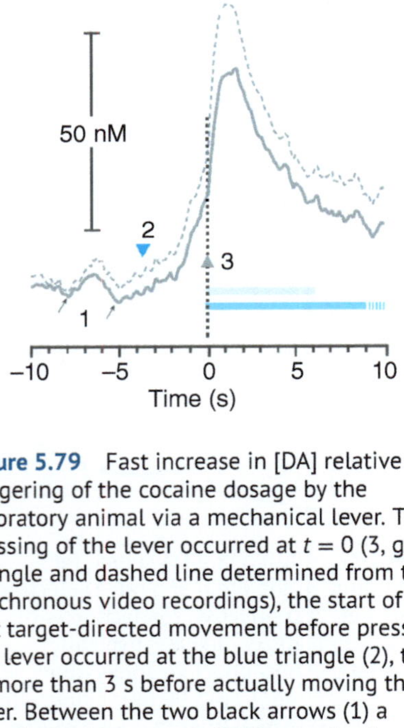

Figure 5.79 Fast increase in [DA] relative to triggering of the cocaine dosage by the laboratory animal via a mechanical lever. The pressing of the lever occurred at $t = 0$ (3, gray triangle and dashed line determined from the synchronous video recordings), the start of the last target-directed movement before pressing the lever occurred at the blue triangle (2), that is, more than 3 s before actually moving the lever. Between the two black arrows (1) a time-limited increase in [DA] precedes the action. Shown are averaged transients $[DA] = f(t)$ of 6 laboratory animals. Source: Reproduced with permission from [127] Copyright 2003, Springer Nature.

The sensors are so tiny that they can be implanted into the brain of rats and, subsequently, facilitate monitoring DA concentrations [DA] in the nucleus accumbens [126], a region of the forebrain rich in DA, for up to 25 days without significant changes in the rat's behavior. A large part of the DA-containing neurons can be activated synchronously, upon which they react by dopamine secretion within 200 ms. It is assumed that the resulting increased DA concentrations are the chemical basis of the neuronal reward system, which plays a central role in the development of addiction. More weight could be added to that hypothesis by time-dependent $[DA] = f(t)$ recordings.

The correct placement of the sensors in the nucleus accumbens can be verified by an electrical stimulation leading to an increase in [DA]. CV data obtained between −0.6 V and +1.4 V vs. Ag|AgCl with a scan rate of $v = 400\ V\ s^{-1}$ (FSCV, fast-scan cyclic voltammetry, 100 voltammograms per second!) could prove that indeed DA was detected [127]. For this purpose, a few FSCV curves are recorded in the tissue before stimulation. Due to the high scan rate, the curve shows high capacitive currents even at microelectrodes and additional signatures of other redox-active compounds. They are considered a background signal, which is subtracted after stimulation from the voltammograms recorded. The resulting difference signal (Figure 5.78, right) can be compared to the FSCV signal of DA in buffer solution outside of the test animal. A plot of the peak currents of all FSCV curves yields a chronoamperogram (Figure 5.78, left), from which the function $[DA] = f(t)$ can be obtained by calibration.

The test animals had been trained before implantation of the sensor to trigger a cocaine injection upon appearance of the lever by pressing the lever. This action also caused an optical signal (as an environmental stimulus). It was already known that [DA] increased upon cocaine administration because cocaine retards the reuptake of DA by the neurons. However, it was very surprising that the rise in [DA] commenced as much as 3–4 s *before* the actual injection at the moment when the animals initiated a targeted move toward the lever (Figure 5.79, (2)). Even before that, a short increase in [DA] could be measured ((1) in between the two arrows). The hypothesis that this short rise in DA levels triggered the desire for a new cocaine dosage could be substantiated as follows: When an electrical stimulation was used to cause temporarily increased [DA] levels, the time sequence by which different laboratory animals triggered cocaine injections could be synchronized. In a control group without electrical stimulation, the test animals operated the lever without a time pattern. Furthermore, it was shown that an environmental stimulus, which had been associated with cocaine administration, caused similar changes in DA concentration as a real injection, however, only for those animals for which the real cocaine administration and the environmental stimulus had been synchronized during the training phase.

Those studies were preceded by more than 25 years of intensive electrochemical research and development before sensors, instruments, and potential programs became available. More importantly, the integration of the electrochemical procedures with standards of biological experimentation had been developed to a level that biologically relevant conclusions could be derived from the recordings. In any case, such experiments are only meaningful against the background of many control experiments that challenge the electrochemical methods under *in vivo* conditions as well as the biological hypothesis, for instance, by blocking biological functions pharmacologically.

Excursion 5-7 *In Vitro* Investigation of Single Exocytosis Events

Neurotransmitters are stored in membrane-coated vesicles of the presynaptic cell close to the axon terminals (7 in the inset of Figure 5.76). If the action potential arrives at the axon terminal, a complex signal chain initiates the fusion of the vesicle with the cell membrane (8). As a consequence, neurotransmitters are released into the synaptic cleft (exocytosis, 9 in Figure 5.76). Once the neurotransmitters have reached the postsynaptic cell and if there is a corresponding receptor (10 in Figure 5.76), the permeability of the postsynaptic membrane changes for Na^+ and other small ions, which causes a change in the membrane potential. The magnitude of the change in membrane potential depends on the amount of bound neurotransmitters. After integration of signals of hundreds of synapses by the postsynaptic cell, the fate of the excitation is decided by either generating a new action potential in the postsynaptic cell or suppressing the transmittance of the signal.

In competition with the binding at the postsynaptic membrane, neurotransmitters can also be taken up by autoreceptors of the presynaptic cell (12 in Figure 5.76) or via DA transporters (13). After recovery by the presynaptic cell, DA is metabolically converted or stored again in vesicles. DA can also emerge by diffusion from the synaptic cleft. Therefore, the transient DA concentration in nerve tissue is a key experimental parameter to uncover the mechanisms of excitation transmission under healthy and pathological conditions, as well as for the mode of action of pharmaceuticals or illicit drugs.

An experimental system that mimics the transient concentrations within an individual synapse consists of a biological cell that grows outside of an organism in a cell culture medium and then adheres to a transparent surface. By means of an optical microscope and a micropositioning system, a microelectrode, with tilted cross section with respect to the symmetric axis of the carbon fiber (CFME, carbon fiber microelectrode), is positioned above the cell in such a way that the small distance between the cell and the CFME corresponds to the synaptic cleft (Figure 5.80). The CFME consists of a similar material as explained in Excursion 5-5, but has an insulating sheath to minimize background current.

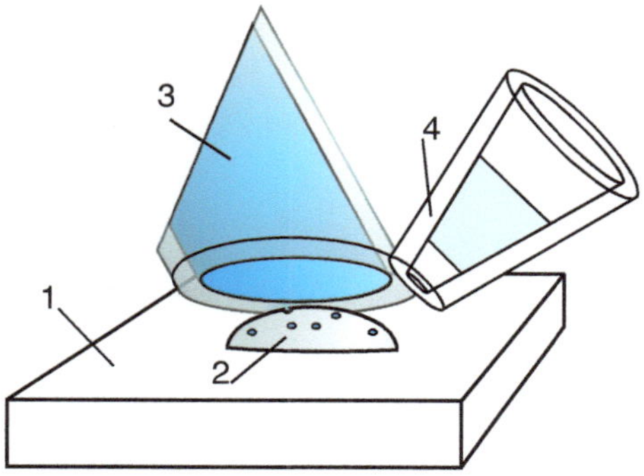

Figure 5.80 Experimental system for monitoring individual exocytosis events. 1) Bottom of the cell culture plate; 2) PC12 cell with DA-loaded vesicles (greatly enlarged in the schematic); 3) CFME; 4) micropipette with solution for stimulation of the cell.

The direct investigation of neurons from mammals is difficult due to the low amount of neurotransmitters present. Their vesicles measure only 50 nm in diameter and contain 1000–3000 dopamine molecules each [128]. For this reason, other cell lines (called cell models) with larger vesicles are used, which behave similarly to neurons with respect to exocytosis. Most studies have been performed with the PC12 cell line (pheochromocytoma cells of rat) that can be grown readily in cell culture – an important difference to many neuronal cells. PC12 cells release acetylcholine, dopamine, and norepinephrine. They also take up those neurotransmitters. The availability of different ion channels enables experiments in which precisely dosed chemical stimulants cause exocytosis events [128].

Because the cross section of the CFME is much larger than an individual vesicle and the distance between the cell surface and the CFME is very small, almost all released dopamine molecules will be detected in an amperometric experiment (Figure 5.81) at the CFME (collection efficiency ≈ 100%). From the charge of one background-corrected current spike, the amount of dopamine in one vesicle can be determined by Faraday's law:

$$N_{DA} = \frac{Q_F}{nF} = \frac{1}{nF}\left[\int_{t_{initial}}^{t_{end}} I(t)dt - \frac{I(t_{initial}) + I(t_{end})}{2}(t_{end} - t_{initial})\right] \qquad (5.131)$$

N_{DA} – Amount of DA [mol]

Q_F – Charge from the faradaic current for DA conversion [A s]

F – Faraday constant [A s mol^{-1}]

n – Number of transferred electrons (2 for DA)

$t_{initial}$ – Time immediately before current rise [s]

t_{end} – Time, at which the current has declined back to the baseline [s]

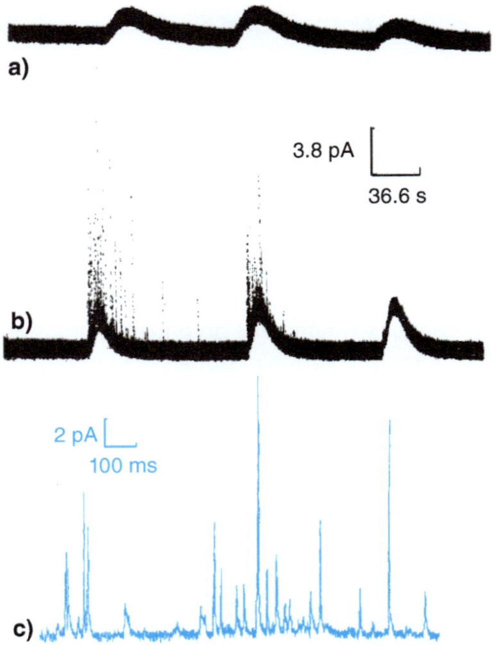

Figure 5.81 Amperograms from a CFME with 5 μm diameter at +0.65 V (versus SSCE).

a) Change in the current at the CFME during three consecutive dosages of 30 nL of a 1 mmol L^{-1} nicotine solution as stimulant. The CFME is located at some distance (200 μm) from the PC12 cell. The neurotransmitters escape detection due to the rapid dilution in the culture medium. The recorded signal is only a variation of the background current. The broad trace reflects the noise during the recording.
b) Like (a) but the CFME is located directly above the PC12 cell. During the dosages, sharp current spikes are recorded for individual exocytosis events on top of the signal background.
c) Enlarged view over 1 s from b). Individual exocytosis events are more recognizable.

Source: T. K. Chen, G. Luo, A. G. Ewing; Amperometric Monitoring of Stimulated Catecholamine Release from Rat Pheochromocytoma (PC12) Cells at the Zeptomole Level. *Anal. Chem.* **1994**, *66*, 3031–3035. Copyright 1994, American Chemical Society.

The determined amounts of N_{DA} are between 190 and 1290 zmol (zeptomol = 10^{-21} mol). They are *not* normally distributed; 85% of the vesicles contain less than 330 zmol or 2×10^5 molecules [128]. Because many exocytosis experiments are recorded during one experiment (Figure 5.81), statistical data evaluation can describe the frequency with which the events occur as well as N_{DA} per vesicle in control experiments and experiments after pharmacological manipulation [128]. Similar approaches are used for other molecules that are short-lived intermediates in biological cells. Reactive oxygen species (ROS) and reactive nitrogen species (RNS) belong to this group [129].

5.6.6 Arrays of Microelectrodes

Q: What is an **array**?

A: The term array refers to a planar arrangement of several separated, but equal objects on a common support or substrate. For microelectrode arrays, the objects are the individual microelectrodes, and the support is the insulating sheath or housing. Individually addressable electrode arrays contain electrodes that can be individually connected to a potentiostat. In contrast, parallel arrays contain microelectrodes that are all connected to only one potentiostat and thus share the same potential.

One of the major disadvantages of microelectrodes is the relatively small current. Therefore, measurements must be carried out in a grounded Faraday cage and require a dedicated potentiostat. This is one reason why the advantageous properties of microelectrodes cannot be easily transferred to robust sensors for use outside of dedicated laboratories.

One option to overcome this drawback is to use N_{ME} equal microdisk electrodes in parallel. That is, connecting them jointly as a working electrode to one potentiostat. All microelectrodes will have the same potential, and the potentiostat will record the sum of the currents of all individual microelectrodes. We expect a current that is N_{ME}-times higher than the current at an individual microelectrode and thus can be brought into an easily accessible current range. This approach will be successful only if the diffusion layers of the individual microelectrodes do *not* overlap during the experiment. Calculated iso-concentration surfaces in the steady state (Figure 5.82) show that the local concentration departs from c^* by only 10% at a distance of $6.4r_0$ and only by 1% at $64r_0$ [105]. Only if the interelectrode distance ($2R_0$) between two neighboring microelectrodes of the array is larger than $2 \times 6.5\ r_0$ can we assume that the array behaves indeed like the sum of N_{ME} microelectrodes (Figure 5.83). This case is referred to as diffusionally independent microelectrodes.

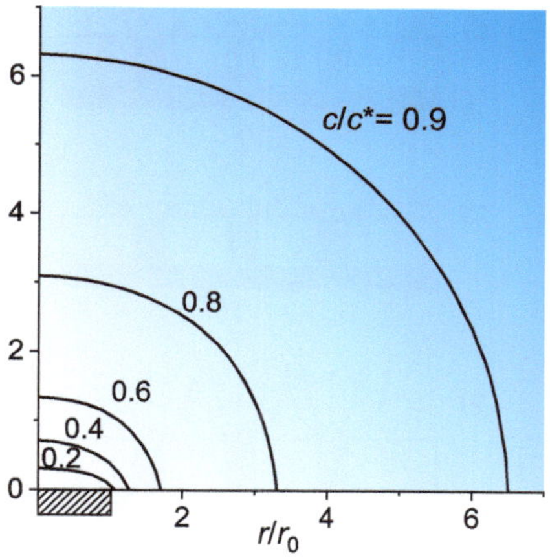

Figure 5.82 Cross section through isoconcentration surfaces above the center of a single microdisk electrode in the steady state. The surfaces are formed by rotation around the vertical (*z*) axis of the coordinate system. Source: The calculation of isoconcentration lines is based on the equations in [105].

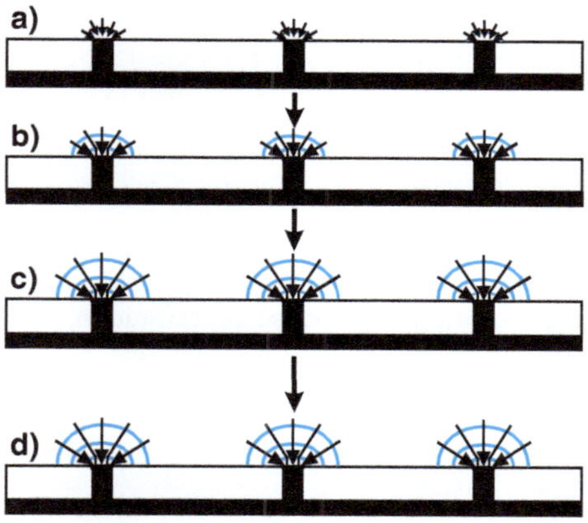

Figure 5.83 Development of diffusion layers in a microelectrode array of diffusionally independent electrodes after the onset of a diffusion-controlled electrolysis.

a) Planar diffusion above individual electrodes;

b–d) Hemispherical diffusion above each microelectrode at different times.

For small interelectrode distances, the behavior depends on the timescale considered. For very short times, the diffusion layers above each electrode are much smaller than the radius of the microelectrode (Figure 5.84a). There is planar diffusion and the current follows the Cottrell Equation (4.130), in which the active area of all individual electrodes must be set for $A = N_{ME}\pi r_0^2$. Later hemispherical diffusion layers are formed (Figure 5.84b). The current will be N_{ME}-fold larger than the current predicted by Eq. (5.118) for an individual microelectrode. Even later, the diffusion layers of the individual microelectrodes start overlapping (Figure 5.84c and d). The individual microelectrodes compete for the reactants. This phenomenon is called shielding. The current falls below the prediction of N_{ME}-fold Eq. (5.118). When the diffusion layers have merged completely, as in Figure 5.84d, planar diffusion controls the current. The current follows the Cottrell equation but now with A being the area of the entire array, which is the area of the active electrodes plus the area of the insulating parts between the electrodes. In other words, at long enough time, such an array provides a current that is equal to a current of a macroscopic electrode of the same total area!

These predictions have been confirmed for gold nanoelectrodes that had been deposited in track etch membranes as templates [130]. Cyclic voltammograms are peak-shaped due to the planar diffusion. The arrays still have favorable features for analytical applications because there is a very low capacitive current (=interfering signal). This translates into an improvement in the limits of detection by a factor of 1000!

This example illustrates again that almost everything is relative in electrochemistry. Therefore, it is very important to understand which area A has to be used in the different equations. In the example, it is the area of the gold electrodes $N_{ME}\pi r_0^2$ for short times, and at longer times it is the total area of the array, that is, $N_{ME}\pi R_0^2$, with R_0 being half of the center–center distance of the microelectrodes, and for the calculation of the capacitive current, the product of $N_{ME}\pi r_0^2$ and the roughness factor of the gold electrodes! We have already discussed the difference between geometric and microscopic electrode area (Figure 4.33). We realize now that the decision of which area is relevant in a particular situation may also depend on the timescale under consideration in the specific experiment (Figure 5.85).

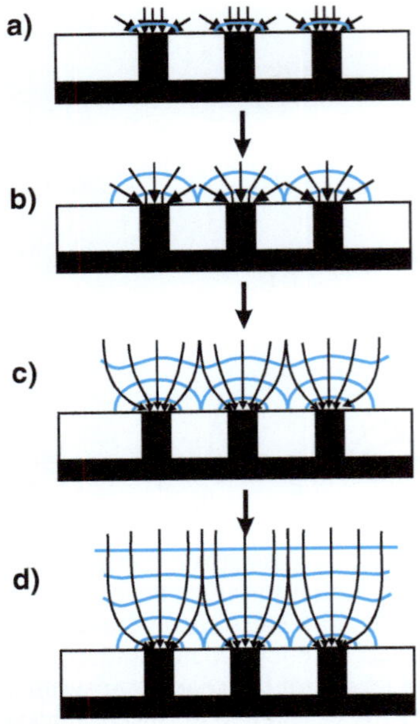

Figure 5.84 Development of diffusion layers in a microelectrode array with diffusive shielding after the start of a diffusion-controlled electrolysis.
a) Planar diffusion above individual electrodes;
b) Hemispherical diffusion above each microelectrode;
c) Merging of the diffusion layers;
d) Planar diffusion with respect to the total area of the array.
Blue – cut through the iso-concentration surfaces, thin black lines indicate the concentration gradient. Schematic is not to scale.

t_c – Critical time for CV = $RT/(nFv)$
R_0 – Half of the interelectrode spacing [cm]
D – Diffusion coefficient [$cm^2\ s^{-1}$]
$k°$ – Heterogeneous standard rate constant at the uncovered electrode [$cm\ s^{-1}$]

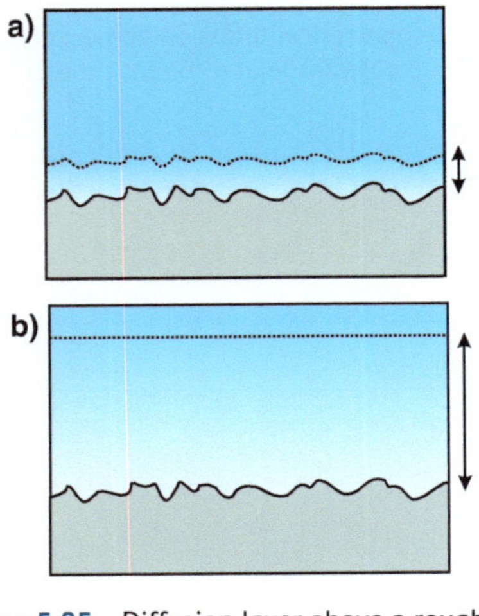

Figure 5.85 Diffusion layer above a rough electrode at a) short and b) long time.

The behavior of microelectrode arrays is of fundamental importance also for other common situations in electrochemistry, such as nanoparticles of catalysts in fuel cells, and for defects in insulating coatings. Therefore, we will discuss the implications of arrays a bit more in detail. A hexagonal arrangement of microelectrodes, as shown in Figure 5.86, is best suited for a quantitative treatment due to its symmetry. For further mathematical treatment, we use the approximation $\pi/(2 \cdot 3^{1/2}) = 0.906 \approx 1$. We can regard the insulating surroundings of the microelectrode as a coverage. The fraction of the covered area is the coverage θ, which should be $\theta > 0.9$. The fraction of the active electrode surface $(1-\theta)$ and the inactive parts θ of the array are:

$$(1-\theta) = \frac{3 \cdot \frac{1}{6}\pi r_0^2}{\frac{1}{2}(2R_0) \cdot \frac{1}{2}\sqrt{3}(2R_0)} = \frac{\frac{1}{2}\pi r_0^2}{\sqrt{3}R_0^2} = \frac{\pi r_0^2}{2\sqrt{3}R_0^2} \approx \frac{r_0^2}{R_0^2} \tag{5.132}$$

$$\theta \approx 1 - \frac{r_0^2}{R_0^2} \tag{5.133}$$

The shape of a cyclic voltammogram recorded at such a microelectrode array depends on θ, the apparent heterogeneous standard rate constant $k°_{app}$ and the timescale of the experiment given by the scan rate v, expressed here as the critical time t_c according to Eq. (5.126). The quite different cyclic voltammograms of a quasi-reversible redox system can be summarized in a zone diagram (Figure 5.87). The axes of the diagram are two dimensionless quantities describing the geometry-dependence and the predominantly kinetic dependence. Lines symbolize borders around zones of the diagram with qualitatively similar voltammograms. Those zones are separated from each other by transition regions. The geometry dependence is captured by the dimensionless parameter $\log(K\lambda^{1/2})$:

$$\log(K\sqrt{\lambda}) = \log\left(\sqrt{1-\theta}\frac{\sqrt{t_c D}}{0.6R_0}\right). \tag{5.134}$$

The predominantly kinetic dependence is given by the ratio Λ of the heterogeneous standard rate constant $k°$ at the uncovered electrode and the "diffusion rate" expressed as $(D/t_c)^{1/2}$:

$$\log[\Lambda(1-\theta)] = \log(1-\theta) + \log\left(k°\sqrt{\frac{t_c}{D}}\right). \tag{5.135}$$

If $\log(K\lambda^{1/2})$ exceeds one, a wave-shaped cyclic voltammogram is recorded. If additionally $\log[\Lambda(1-\theta)]$ assumes large values, the voltammogram is similar to a reversible voltammogram at a macroelectrode. At smaller values of $\log[\Lambda(1-\theta)]$, it changes to quasireversible and finally electrochemically irreversible voltammograms. The peak splitting between anodic and cathodic peaks of quasireversible voltammograms can be used to determine $k°_{app}$. If the heterogeneous standard rate constant of the uncovered electrode is known, the coverage can be estimated with the linear model in Eq. (5.136).

$$k°_{app} = k°(1-\theta) \tag{5.136}$$

An exemplary analysis of a microelectrode array with complex geometry can be found in [111], together with further background information and literature.

Band electrodes have analytical importance. They are typically combined in a comb-like array. If two of these "combs" are moved into each other, an interdigitated electrode array is formed (Figure 5.88). If different potentials are applied to both "combs", a redox-active molecule can be oxidized at one band and be reduced again

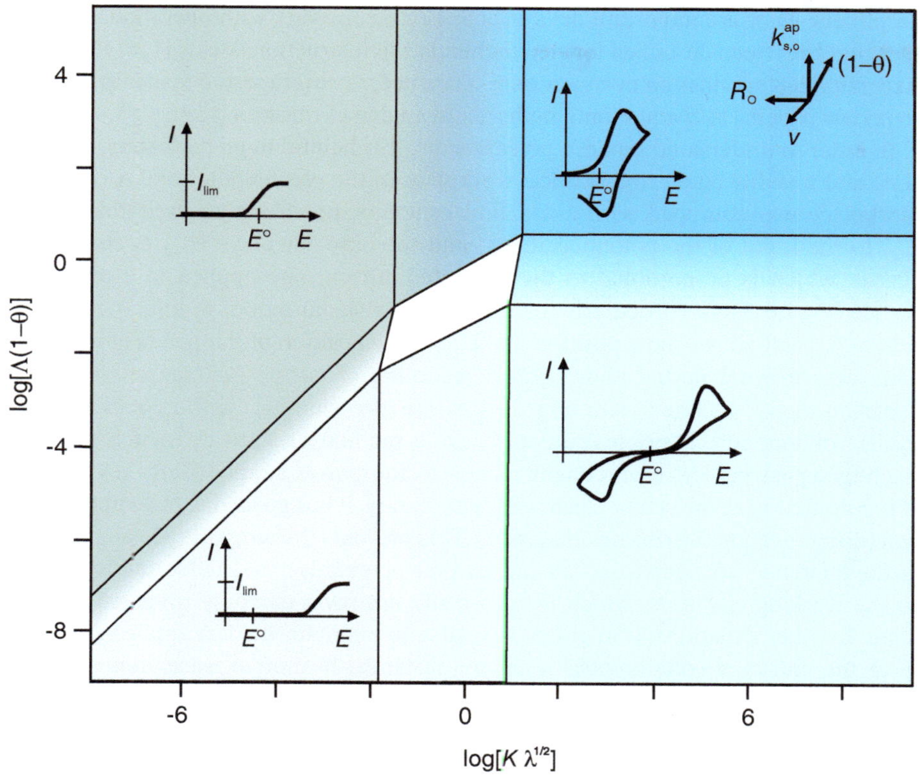

Figure 5.87 Zone diagram for the appearance of cyclic voltammograms at partially covered electrodes. Source: Reproduced with permission from C. Amatore, J.-M. Saveant, D. Tessier; Charge transfer at partially blocked surfaces: A model for the case of microscopic active and inactive sites. *J. Electroanal. Chem.* **1983**, *147*, 39, Copyright 1983, Elsevier.

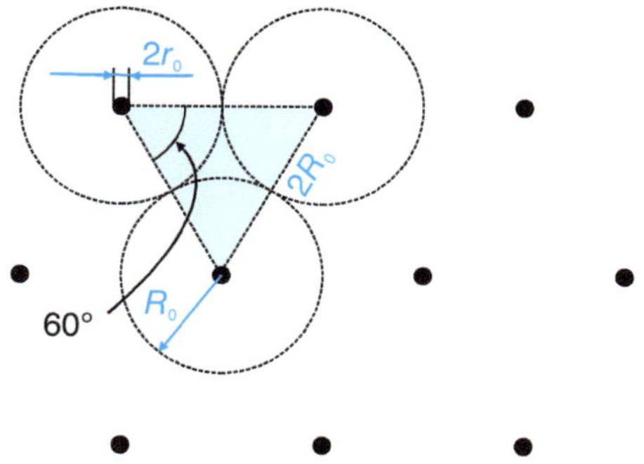

Figure 5.86 Top view of a hexagonal microelectrode array with the radius r_0 of individual microelectrodes and the interelectrode distance $2R_0$.

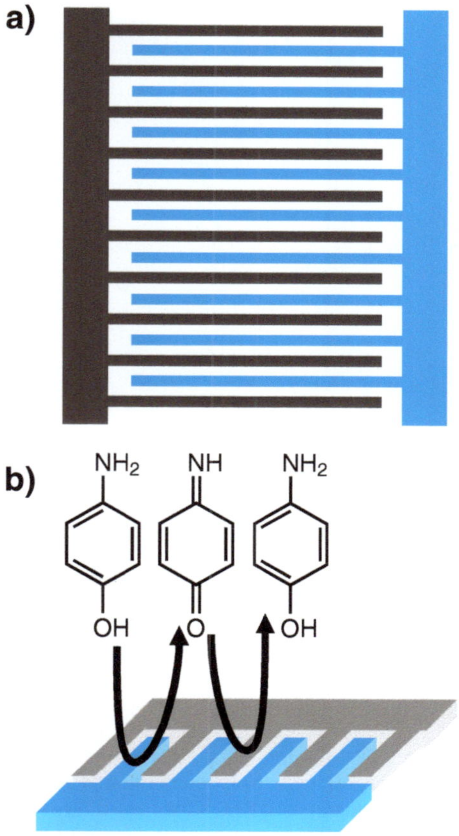

Figure 5.88 Principle of signal amplification and sensitivity enhancement at interdigitated arrays; a) top view; b) amplification by redox cycling between the band electrodes.

at the neighboring band. If the bands are closely spaced, there will be multiple oxidations and reductions for each molecule, causing a charge transport between both band arrays. This phenomenon is called redox cycling or redox shuttling. Significant enhancements of the sensitivity of amperometric or voltammetric detection methods have been demonstrated using miniaturized interdigitated arrays produced by means of technologies from the semiconductor industry [132, 133].

5.7 Foundations of Electrochemical Instrumentation

5.7.1 Modern Voltammetric Workstations

Today, commercially available potentiostats and their software allow their usage without detailed knowledge of electronic circuitry and electronics. On the one hand, this is extremely useful and, on the other hand, it is temptingly dangerous as illustrated by a metaphor from motorsports: driving a car is certainly possible without knowing a lot about the working principle of an engine, however a race car pilot who aims for getting all out of his vehicle and uses it close to the technical limits, should better know many details about the technology at her/his disposal. Correspondingly, electrochemists may come into this position in order to get the most out of the available equipment or to plan and justify the purchase of new dedicated instrumentation. It is then time to pose questions concerning the working principles of electrochemical instrumentation. Central to the operation of potentiostats is the functioning of operational amplifiers. They are important devices of analog electronics, with operational principles that have been described for electrochemists in instructional texts [1, 6, 134–136]. Excursion 5-8 provides an overview here. For a deeper coverage, the reader is referred to review papers [137] and monographs about analog electronics [138].

In order to understand modern potentiostats, it is helpful to go back some steps in time and imagine an electrochemical workplace in the precomputer era. A complete workplace was equipped with a function generator providing an excitation signal E_{set} to the input of the potentiostat. E_{set} and the output voltage U_{out} of the potentiostat, which is proportional to the measured current, are applied to a recording device (Figure 5.89). Particularly instructive is the signal output to an x–y-recorder, where E_{set} influences the x-position and U_{out} the y-position of the pen (Figure 5.89). This experimental design allows direct recording of current–voltage curves (thus "volt-am-metry"). Other recording devices are two-channel oscilloscopes for fast cyclic voltammetry. Many instruments have a manual or software switch for connecting/separating the electrochemical cell to the power supply (Cell on/Cell off). If manipulation at the cell or electrodes is necessary, it is a good idea to switch off the cell (disconnecting the counter electrode). This prevents the temporary loss of contact to the reference electrode from causing extreme potentials to be inadvertently applied to the working electrode, which in turn could destroy a carefully prepared surface state. In other situations, it might be advisable to maintain the last applied potential after the end of a voltammetric scan; for instance, if another measurement shall directly follow as part of a complex experimental sequence. Often, the instrumentation control software has an option to control whether the cell should be switched off after termination of an experiment.

Modern workstations are typically computer-controlled, for which two concepts exist:

1) The computer may use a digital–analog converter (DAC) to generate the excitation signal $E_{set} = f(t)$ and an analog–digital converter (ADC) to read U_{out} (Figure 5.90a). The potential ramp is replaced by a staircase function in E_{set}. Provided the steps of the staircase are small, this usually has little effect on the recorded data (Figure 5.41b). Workstations with computer control of the measurements usually offer advantages due to their flexibility with respect to defining excitation signals. The recorded data can be saved and effortlessly displayed as a plot. A direct linkage to data treatment routines and numerical simulations is implemented in many commercial devices. Besides these advantageous properties, there are a few exceptions for which analog measurements are better suited. One example is the study of surface-bound redox systems (Figure 5.41a).

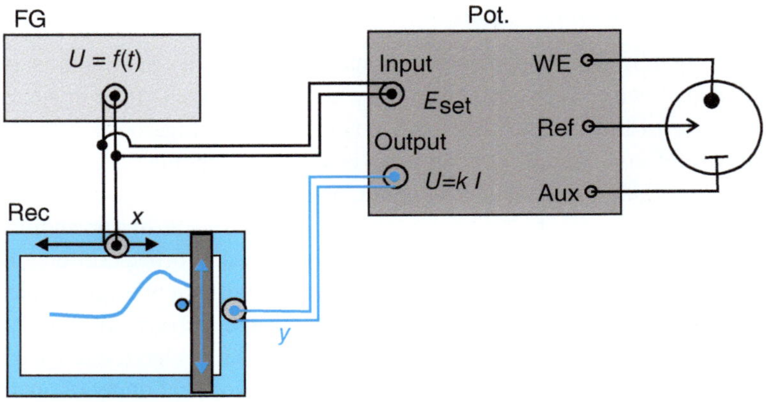

Figure 5.89 Components of a voltammetric workplace; FG is the function generator, Pot is the potentiostat, and Rec is the recording device exemplified here by an x–y-recorder. The cables connecting FG, Pot, and Rec are shielded cables, in which a voltage is applied between the inner conductor and the shielding.

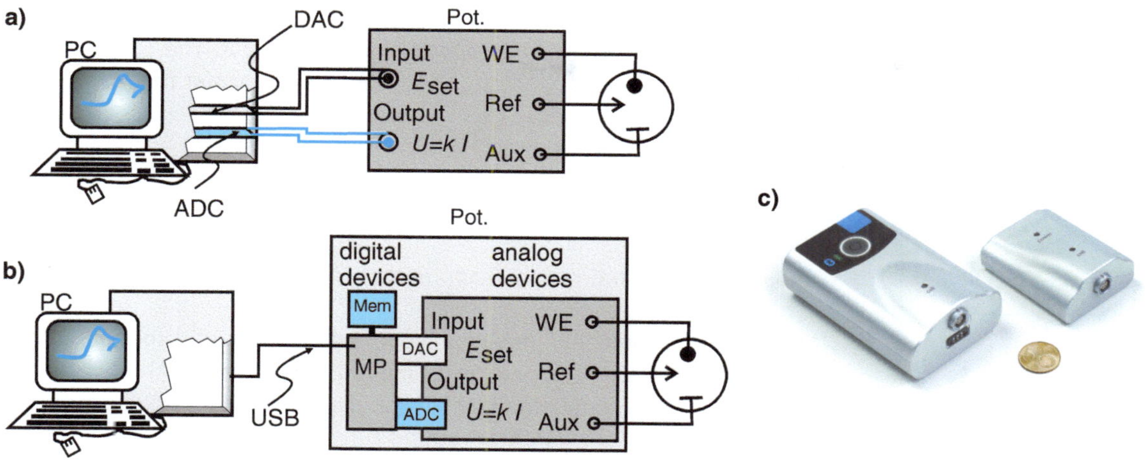

Figure 5.90 Components of computer-controlled potentiostats: a) a PC with an ADC–DAC unit takes over the role of the function generator and recording device. The actual potentiostat is still an analog electronic device. b) A PC is used only for communication with the user. The actual measurement is controlled by a microprocessor (MP) with ADC and DAC units inside the potentiostat (digital potentiostat). c) Examples of a miniaturized potentiostat. Panel c) Reproduced with permission from PalmSense BV, Randhoeve, NL.

2) More recent instrumentation even employs two computers in one workplace. One personal computer (PC) is operated by the user. The PC transmits the instruction for the measurements by USB or Ethernet connection to a microprocessor (MP) located in the housing of the potentiostat (Figure 5.90b). The MP controls the measurements and stores the data in an intermediate memory (Mem). The advantage of this concept is the independence of the MP from the ill-defined time control of PC operating systems. The recorded data are transferred from the temporary memory to the PC as data packages, displayed graphically on the PC, and stored on a hard drive. Contrary to the schematic sketch in Figure 5.90b, devices with microprocessors can be miniaturized to a large extent, which offers them new application areas (Figure 5.90c).

Potentiostats for Special Applications. Important special types of potentiostats are potentiostats for two independently controlled working electrodes (bipotentiostats), which are required, for instance, to operate ring–disk electrodes and interdigitated array electrodes. Multipotentiostats can be used for the parallel investigations of several samples, for example, in corrosion testing or for the characterization of sensors over prolonged time. Some monopotentiostats have a fourth connecting cable to the measurement cell. This is often connected to the working electrode together with the actual WE cable and allows for compensating the voltage drop inside the electrode and electrical cables at higher currents used in electrosynthesis. These two electrodes can be separated to allow 4-electrode membrane studies.

Excursion 5-8 Using Operational Amplifiers

Operational amplifiers (OAs) are integrated electronic devices that can be used for many amplifier and controller applications. The name originates from the original use in analog computers. OAs are the central devices of potentiostats and other measurement instruments. Although their structural shape is most closely to a cuboid (Figure 5.91a), at which 8 or more contact pins are available, the graphical symbol in Figure 5.91b shows only three electrical connections corresponding to the signal lines. The connections for the supply voltage of +15 V and −15 V have

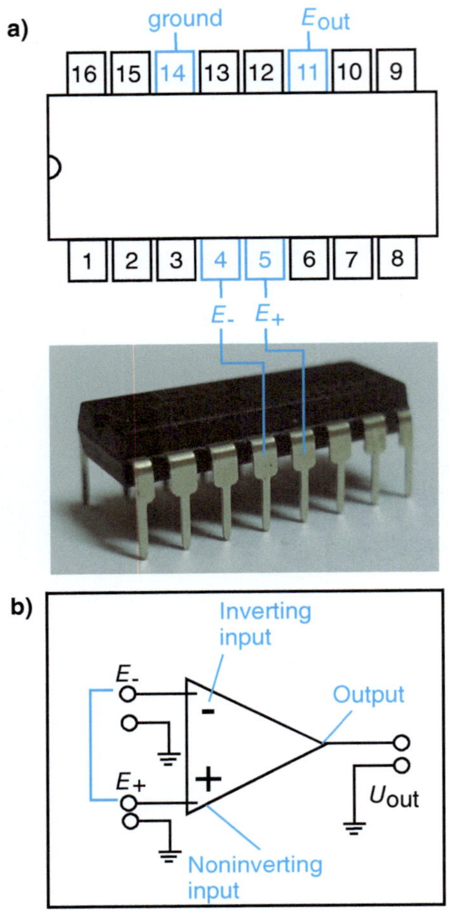

Figure 5.91 Construction and function of an operational amplifier.

a) Wiring of the signal lines and a photograph of an OA;
b) Simplified wiring scheme of an operational amplifier. The feeding lines of the power supply are omitted in a) and b) for clarity.

$$U_{out} = -\frac{R_f}{R_{in}} U_{in}$$

been omitted for clarity. The connections relevant for the signal processing are an inverting input (–), a noninverting input (+), and an output. The output voltage U_{out} results from the input voltage $U_{in} = E_+ - E_-$ and a transfer functionG, which is defined by the wiring [Eq. (5.137)].

$$U_{out} = G\,(E_+ - E_-) = G\,U_{in} \tag{5.137}$$

The most important properties of OAs are

1) Without any additional wiring, G approaches infinity (real open-loop gains are 10^4 to 10^8).
2) OAs exhibit an almost infinitely high input impedance (real values are 10^5 bis 10^{13} Ω). Therefore, OAs are suitable for measuring a voltage without drawing a significant current via their inputs that would change the object under study.
3) The output can deliver an arbitrary output current within the limits determined by the power supply. This means the output impedance approaches zero (real values are ±10 mA at ±14 V power supply).
4) If there is no input voltage, there is also no voltage at the output. This specification is often not precisely met. OAs may therefore feature a nulling mechanism to adjust this offset voltage to zero. The compensating voltage is the voltage U_{in} for which $U_{out} = 0$.
5) The OA responds to a change in U_{in} within 10 μs. This property determines the bandwidth of the OA, that is the statement, which frequencies can still be processed. Some experiments require even faster potential changes, which constitutes a formidable challenge for the selection and combination of electronic components.

OAs can be used to realize several basic circuits that transform the signal, similar to common arithmetic operations [139]. Some of those that are important for electrochemical instrumentation are listed below.

Inverter with Gain. The circuit in Figure 5.92 is used to increase or decrease a voltage while inverting its sign. The change in voltage is controlled by the ratio of resistors, which multiplies the input voltage by $-R_f/R_{in}$. The input current is given by $I_{in} = U_{in}/R_{in}$ and the current in the feedback branch is given by $I_f = U_f/R_f$. The sum of all currents into the summing point S must be zero. Because OAs have an almost infinitely high input impedance, almost no current passes through the inverting input. The remaining currents must add to zero, thus $I_{in} = -I_f$. The potential of the summation point will be kept by the feedback resistor close to the ground potential (0 V, virtual ground). With $I_{in} = U_{in}/R_{in}$, $I_f = U_f/R_f$, and $I_{in} = -I_f$, the transfer function given in the margin of the page is obtained.

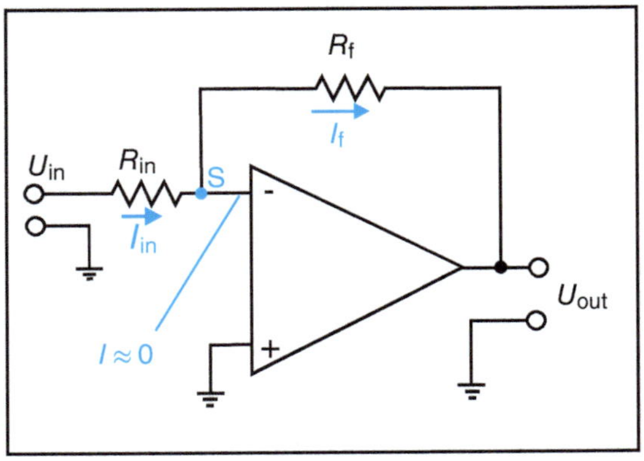

Figure 5.92 Inverting amplifier with gain equal to the ratio $-R_f/R_{in}$.

Adder. The circuit in Figure 5.93 adds several input voltages to a potentiostat and can generate complex signals, for instance, a slowly changing potential ramp and a superimposed sine wave for the measurement of the potential-dependent double layer capacitance. The transfer function is derived by balancing the currents at S under consideration that S is at virtual ground. $I_f = -(I_1 + I_2 + I_3) = U_{out}/R_f = -(U_1/R_1 + U_2/R_2 + U_3/R_3)$.

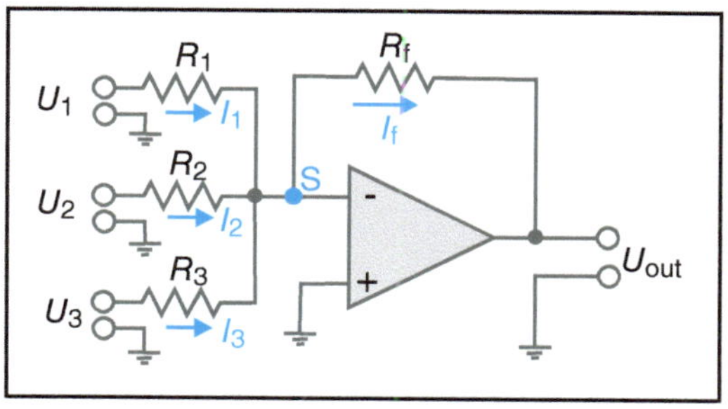

Figure 5.93 Circuit diagram of an adder with three input voltages.

$$U_{out} = -\left[\frac{R_f}{R_1}U_1 + \frac{R_f}{R_2}U_2 + \frac{R_f}{R_3}U_3\right]$$

or

$$U_{out} = -\sum_{i=1}^{n} U_i,$$

if $R_f = R_1 = R_2 = \ldots$

Current Follower. Analogous to the inverter with gain, the current follower in Figure 5.94 is governed by the relationship $I_{in} = -I_f = -U_f/R_g \approx -U_{out}/R_f$.

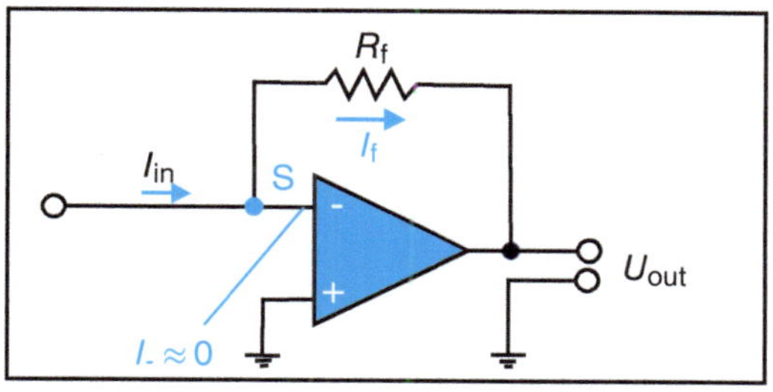

Figure 5.94 Circuit diagram of a current follower.

$$U_{out} = -R_f I_{in}$$

Voltage Follower. The circuit in Figure 5.95 provides an output voltage that is exactly equal to the input voltage. It does not change the input voltage. Due to the high input impedance of the OA, the object under study is not changed, even if U_{out} is used to drive a load, for example, moving the pen in an x–y-recorder. Without the voltage follower, the voltage at the source would collapse when a current is drawn. Circuits of this type are needed for potential measurement at ion-selective electrodes.

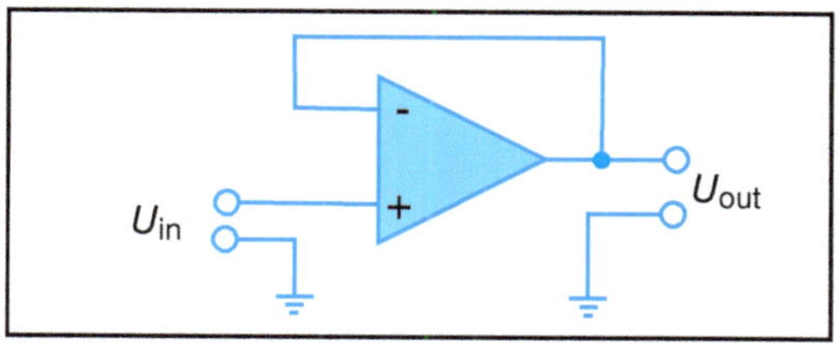

Figure 5.95 Circuit diagram of a voltage follower.

$$U_{out} = U_{in}$$

These basic types of OA circuits can be combined to construct circuits for more complex electrochemical devices. Please note that we cover here the basic principles of electrochemical instrumentation only. Further reading in [134, 139, 140] leads to even more specialized literature and discusses opportunities and limitations of specific circuits in more detail.

Galvanostat. A circuit for a galvanostat (Figure 5.96) is obtained straightforwardly by combining the inverter with gain amplifier circuit with a follower amplifier circuit and by replacing the feedback resistor for OA1 with the electrochemical cell. The OA1 forces a current through the cell that is necessary to compensate $I_i = U_{in}/R_i$ at the point S. Adding a reference electrode provides the opportunity to measure its potential against electrical ground using the voltage follower (OA2). Because the WE is connected without a resistor to S and the potential of S is close to ground, the following relationship holds $U_{out} = E_{ref} - 0 = -(0 - E_{ref}) = -(E_{we} - E_{ref})$. Further circuit elements might be necessary to enable high currents.

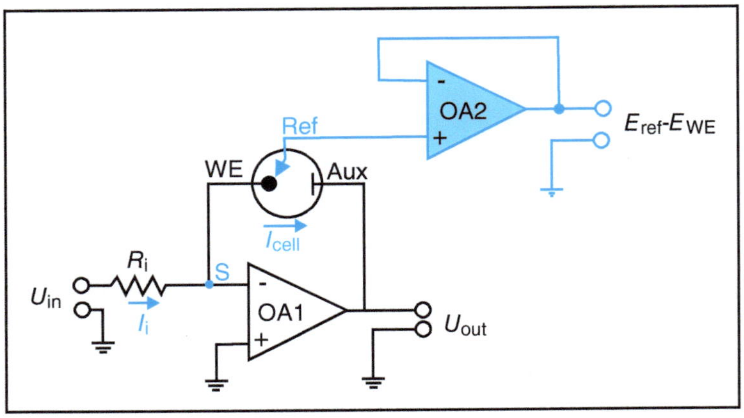

Figure 5.96 Circuit diagram of a simple galvanostat. The color shading highlights the combination of an inverter with gain and a voltage follower.

Potentiostat. The development of a potentiostat circuit is shown here in two steps [140]. First, Figure 5.97 shows how OA1 provides the addition of three input voltages, which could be a constant DC voltage, a linear voltage ramp, and a superimposed sine wave. Such an excitation signal is needed for AC voltammetry to set the initial potential for a voltammogram and combine a linear scan with an alternating voltage for the voltammogram. The WE is held on the ground. The feedback loop from Ref to the point S keeps the potential of Ref on the value set by $U_1 + U_2 + U_3$ with respect to ground (and thus with respect to the WE as well).

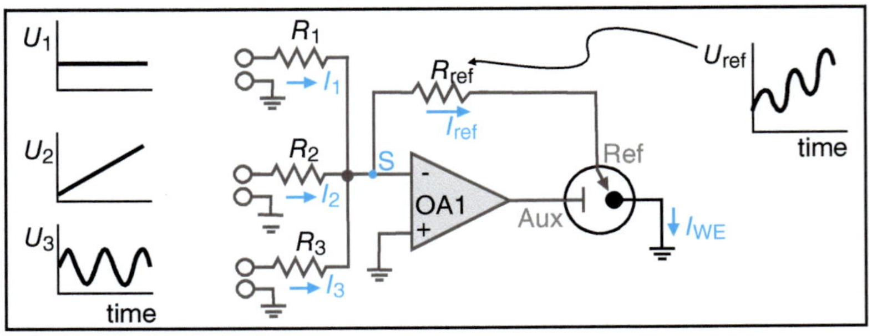

Figure 5.97 Circuit diagram of an adder in combination with an electrochemical three-electrode cell.

Potentiostat with Current Measurement. A refinement of the potentiostat in Figure 5.98 is made by a current follower OA2 to convert the current through the working electrode into a current-proportional output voltage $U_2 = I_{WE}R_n$. The voltage can be read by a plotter or an AD converter. Using the switch SW,

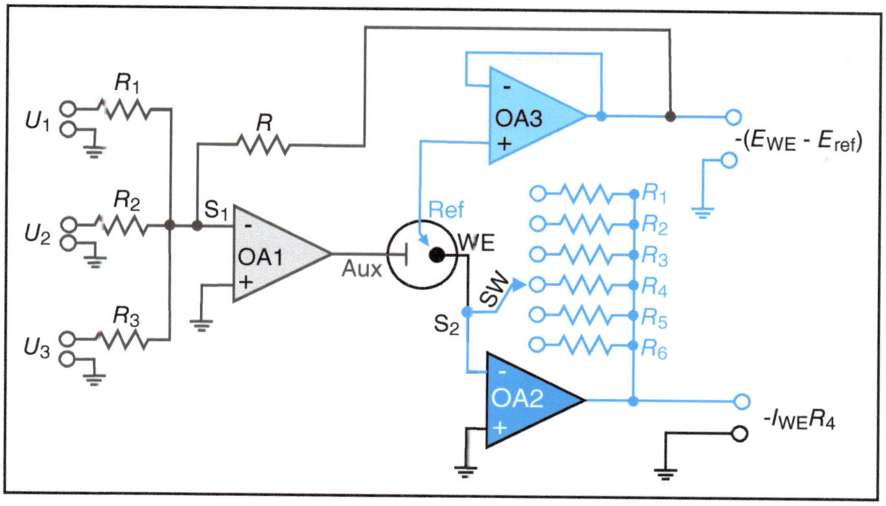

Figure 5.98 Circuit diagram of a simple potentiostat made from an adder (gray), a voltage follower (light blue), and a current follower (blue) with different sensitivities or current ranges. The electrochemical cell is a three-electrode cell.

one out of several resistors R_n can be selected to provide different current ranges. The voltage follower OA3 prevents current from flowing through the reference electrode, which would distort the exact potential measurement. To this end, the resistor R must be placed between OA3 and S1 to complete the adder circuit. Further options (not shown) would be amplifiers to enlarge the current and voltage range.

5.7.2 Instrumental Compensation of Ohmic Drop

As we have seen in Figure 5.18b, the use of a potentiostat and a three-electrode arrangement corrects for most of the ohmic voltage drop between the working electrode and the auxiliary electrode. Figure 5.99 shows this situation as an electrical circuit diagram. The reference electrode is supposed to measure the solution potential E_{ref} right in front of the working electrode. The potentiostat compares this measured potential with the set potential E_{set} and adjusts the potential of the auxiliary electrode (and thus the cell voltage) to maintain the condition $E_{set} = E_{ref}$. When a current I is passed between the WE and the Aux, an ohmic drop $U_{uc} = I \cdot R_{uc}$ occurs due to the unavoidable electrolyte

U_{uc} – Ohmic voltage drop [V]

R_{uc} – Uncompensated electrolyte resistance [Ω]

R_c – Compensated electrolyte resistance [Ω]

$R_E = R_c + R_{uc}$ – electrolyte resistance between WE and Aux [Ω];

R_{ct} – Charge transfer resistance at the WE [Ω]

C_{dl} – Double layer capacitance at the WE [F] = [A s/V]

$R_{ct,Aux}$ – Charge transfer resistance at the WE [Ω]

$C_{dl,Aux}$ – Double layer capacitance at the Aux [F] = [A s/V]

I – Cell current [A] between WE and Aux

E_{set} – Desired electrode potential

t_b – Time base [s]

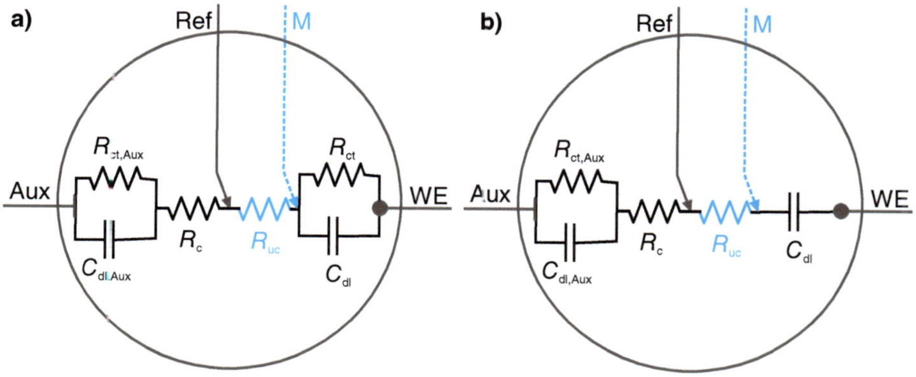

Figure 5.99 Simplified circuit diagram of an electrochemical three-electrode cell; a) at a working electrode potential that causes a faradaic current over R_{ct} at the WE and b) at working electrode potentials where no faradaic current occurs.

resistance between the surface of the working electrode and the position of the reference electrode. The potentiostat still controls the cell voltage between WE and Aux to maintain $E_{set} = E_{ref}$, but the experimenter aims for the control of the potential drop between WE and point M. This potential drop cannot be measured directly but can be estimated by different methods. Using this measurement, E_{set} can be incremented by $U_{uc} = I \cdot R_{uc}$ to meet the condition $E_{set} = E_{Ref} = E_{WE} + U_{uc}$. There are different instrumental methods for approximate determination of U_{uc} and incrementation of E_{set}. Due to their practical importance, we discuss the positive feedback compensation and the current interrupt method.

It is important to realize that all compensation methods cannot completely avoid the problem of ohmic drop and usually require adjustment by the user (details later).

Before attempting to compensate R_{uc}, all efforts should be made to minimize U_{uc} (see also Section 5.3.1) by increasing the conductivity of the electrolyte (higher concentration of the supporting electrolyte will decrease R_{uc}) and a small overall current (smaller electrode area, smaller concentration of reagent, smaller scan rates in the case of voltammetry will decrease U_{uc} at constant R_{uc}). There are situations in which the avoidance strategies will not suffice. For instance, an increase in the electrolyte concentration in corrosion studies would not only change R_{uc} but, more significantly, may considerably alter the corrosion mechanisms and the corrosion rate (see, for example, Section 14.5.3). Likewise, a decrease in reagent concentration in a plating study not only decreases the current but may also change the morphology of the obtained layers (see, for instance, Figure 9.17). Another example is a mechanistic study by cyclic voltammetry, where quantitative parameters are extracted from peak potentials (Excursion 6-1).

Positive-feedback IR_{uc} compensation is based on measuring R_{uc} in the cell setup prior to the actual experiment, for example, by means of an impedance measurement at high frequency (Figure 7.12). Figure 5.100 shows the potentiostat from Figure 5.98 with an additional feedback loop that supplies a fraction x of the current-proportional output voltage ($U_{out} = I_{WE}R_4$) to the adder (OA1). The fraction x is set with the adjustable resistor R_{fc}.

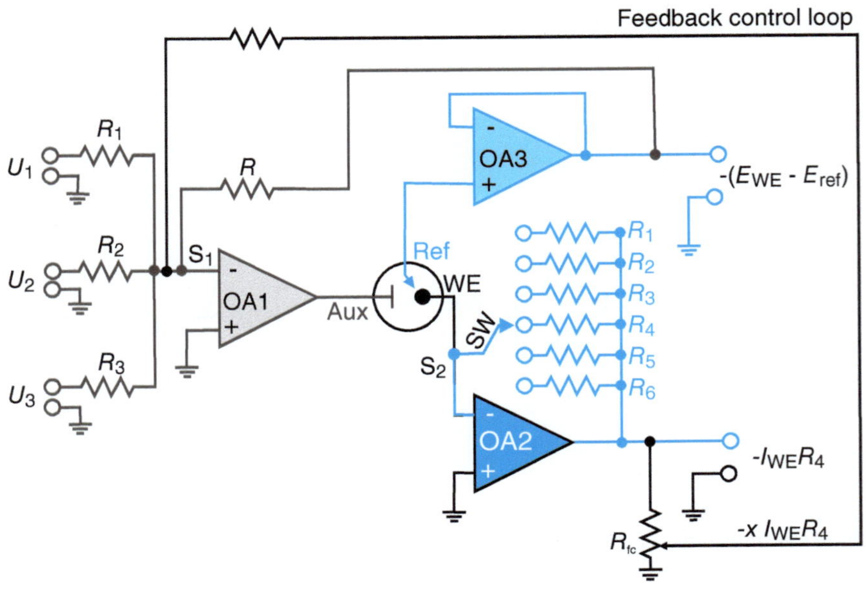

Figure 5.100 Circuit diagram of the potentiostat from Figure 5.98 with an additional control loop (in black) that applies a fraction x of the output voltage $U_{out} = I_{WE}R_4$ to the inverting input of the adder. The value of x (0 (x) < 1) is determined by the position of the adjustable resistor R_{fc}.

The positive-feedback compensation method has the advantage of working correctly even during rapid changes in E_{set} and the resulting current in voltammetry, provided that R_{uc} remains constant. The size of R_{uc} that can be compensated is restricted to the effective resistances in the current follower. The method fails when the measurement range of the current follower is automatically switched during the experiment, an approach often used in corrosion studies (Chapter 14), but strongly discouraged in voltammetry of surface-bound or molecular species treated in Chapters 6, 12, and 13.

It is generally impossible to completely correct for R_{uc} because the electronics and the electrochemical cell cause a very small delay in the current response, and thus, the R_{uc} correction applied to the adder OA1. This can cause the potentiostat to oscillate. Typically, the experimenter aims to correct as much of R_{uc} as possible, but to stay just below the limit where the potentiostat would start to oscillate. An empirical approach is to first set R_{uc} to 100% and, if the potentiostat starts to oscillate, reduce the compensation to 80–90% of R_{uc}. This bears the risk that strong oscillations are induced that might damage a (pretreated) electrode surface. Alternatively, one may start with compensation of 70% of R_{uc} and increase until the potentiostat starts to oscillate, followed by a slight reduction in the compensation. Modern computer-controlled instruments may contain an analog–digital converter after R_{fc}, and, after processing, provide U_{uc} via a digital–analog converter to the adder OA1. Such systems facilitate advanced routines to determine and set R_{uc}. Typically, R_{uc} is measured in a potential range where no faradaic reaction occurs at the WE by impedance measurement over a range of (high) frequencies or by monitoring the transient current response upon a small potential step such as 50 mV. With the measured R_{uc}, the feedback control loop can be adjusted automatically by observing first signs of oscillation.

The current interrupt methodexploits the fact that the $U_{uc} = R_{uc} \cdot I$ instantaneously drops to zero when the current is instrumentally interrupted for a short time (0.1–1 ms). The difference in the potential reading just before and after the current interrupt provides U_{uc}. The current interrupt is repeated several times per second and thus allows dynamic changes in U_{uc} to be followed. The main advantage of the current interrupt method is the independence of prior knowledge of R_{uc} and the possibility to work when R_{uc} changes during the experiment, and even when current ranges change automatically during one experiment. Such situations are encountered in corrosion studies where the dissolution of a metallic sample may cause a significant increase in ionic strength in the vicinity of the sample and thus a change in R_{uc}.

A realization of this principle with analog instrumentation is shown in Figure 5.101. There are three clock counters that fire a trigger at the same frequency as the current interrupt, but they are slightly offset in time. Upon a timer event by Clock 1, the voltage $(E_{WE} - E_{Ref})$ measured by OA3 is stored in the sample-and-hold amplifier S&H1. An S&H amplifier keeps the output voltage at the input voltage at the time of the clock event until the next timer event causes a new input voltage to be applied to the output. Clock 2 causes the analog switch AS to interrupt the connection to the auxiliary electrode. After a short delay, Clock 3 causes the recording of $-(E_{WE} - E_{Ref})$ when $I = 0$ at S&H2. The output from S&H1 and S&H2 is fed to the adder OA1. Since an inverter OA4 is applied to the output of S&H2, it is actually the difference in E_{WE} before and after the current interrupt that is used to increment the set voltage atOA1.

However, the measured difference in E_{WE} before and after the current interrupt equals U_{uc} only when E_{WE} remains constant after the current interrupt, as indicated by the dashed blue line in Figure 5.102. If the potential E_{WE} is sufficient for faradaic reactions, the double layer of the WE is slowly discharged via R_{ct} in Figure 5.99a, causing an exponential decay in $-(E_{WE} - E_{Ref})$ with time (black curves in Figure 5.102). Depending on the timing of Clock 3, U_{uc} would be overestimated.

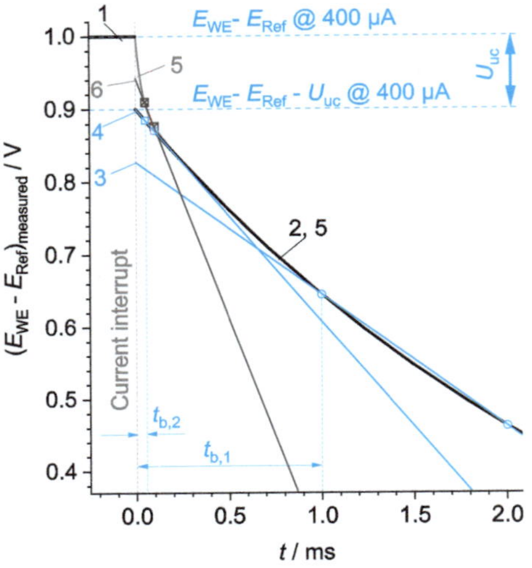

Figure 5.102 Estimation of U_{uc} by means of the current interrupt method using the potential difference between the terminals of the working and reference electrodes in a potentiostatic three-electrode cell, calculated for a working electrode with $C_{dl} = 1$ μF, $R_{CT} = 3000$ Ω, $R_{uc} = 200$ Ω. 1) Voltage before current interrupt at 400 μA; 2) decay of the measured voltage (black); 3) estimation of U_{uc} from curve 2 by extrapolation of readings with$t_{b,1} = 1$ ms (blue, overestimation); 4) estimation of U_{uc} from curve 2 by extrapolation from readings with $t_{b,2} = 0.05$ ms (blue, correct estimation); 5) decay of voltage with an additional cable capacitance of 350 pF discharged over the entire solution resistance $R_E = 100$ kΩ (gray); 6) estimation of U_{uc} from curve 5 by extrapolation from readings with $t_{b,2} = 0.05$ ms (gray, underestimation). Source: Figure inspired by data in [142].

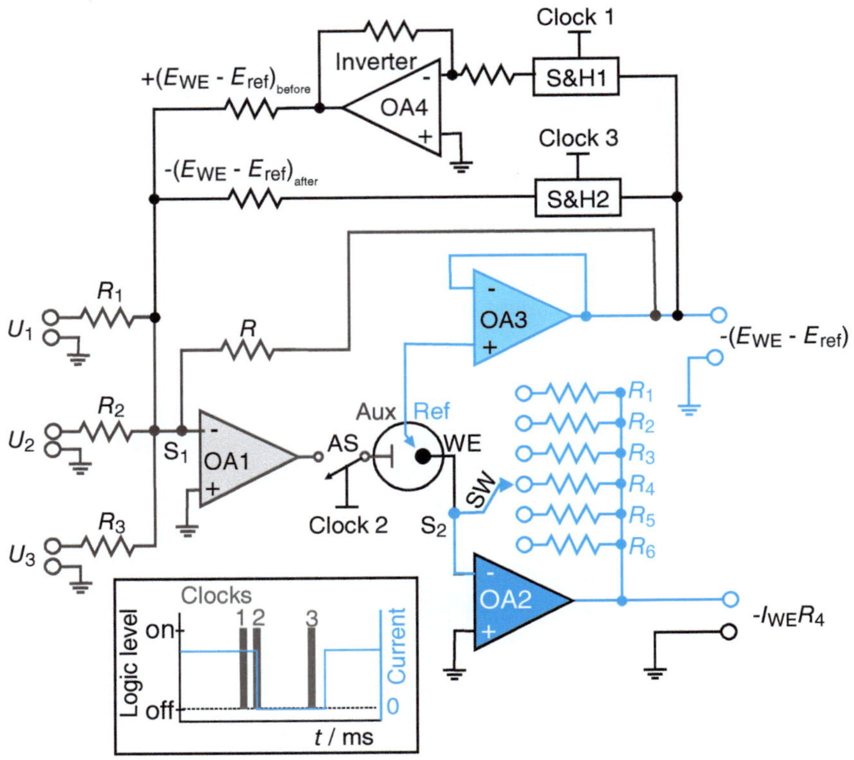

Figure 5.101 Circuit diagram of the potentiostat from Figure 5.98 with IR_{uc} compensation by the current interrupt method. The inset shows the timing events of clocks 1, 2, and 3 and the interrupt of the current. AS is the analog switch; S&H is the sample-and-hold amplifier.

Of course, the procedure can be refined by modern digital equipment. For instance, $-(E_{WE} - E_{Ref})$ can be read with a time base t_b multiple times. The data points $-(E_{WE} - E_{Ref})_i(t_{b,i})$ can be used to extrapolate to the time of the current interrupt ($t = 0$ in Figure 5.102). Even this procedure can only provide an estimation of U_{uc} and the accuracy may depend strongly on the tuning of t_b. For a large t_b, the curvature of the exponential curve causes an overestimation of U_{uc} (point 3 in Figure 5.102). U_{uc} becomes more accurate with decreasing t_b (point 4 in Figure 5.102). However, the capacitance of the auxiliary electrode cable may cause another exponential decay of the measured potential with a very small time constant rather than the presumed instantaneous drop (curve 5 in Figure 5.102). If t_b is selected to be within this first exponential decay, U_{uc} is underestimated (point 6 in Figure 5.102). As we can see, the accuracy of the compensation depends critically on the adjustment of t_b. A guideline for the selection of t_b is $t_b < 0.1\, R_{ct}C_{dl}$. Generally, R_{uc} must be below a threshold set by the charge transfer resistance R_{ct} of the working electrode. The limited frequency of the current interrupt limits the method to relatively small potential scan rates.

As a user, one should first of all recognize the effects of R_{uc} on a particular experiment and select the most appropriate compensation method from the possibilities offered by commercial instrumentation, keeping in mind the limitations and tuning options briefly discussed above. For instance, the current interrupt method is suitable for measurements with a slow change of E_{set} with a switch of the current ranges (as in many corrosion studies), but less suitable for voltammetric studies of organic reaction mechanisms with rapid changes in E_{set} (Chapter 6). Generally, optimum results can only be obtained after experimenting and fine-tuning the settings of the compensation method.

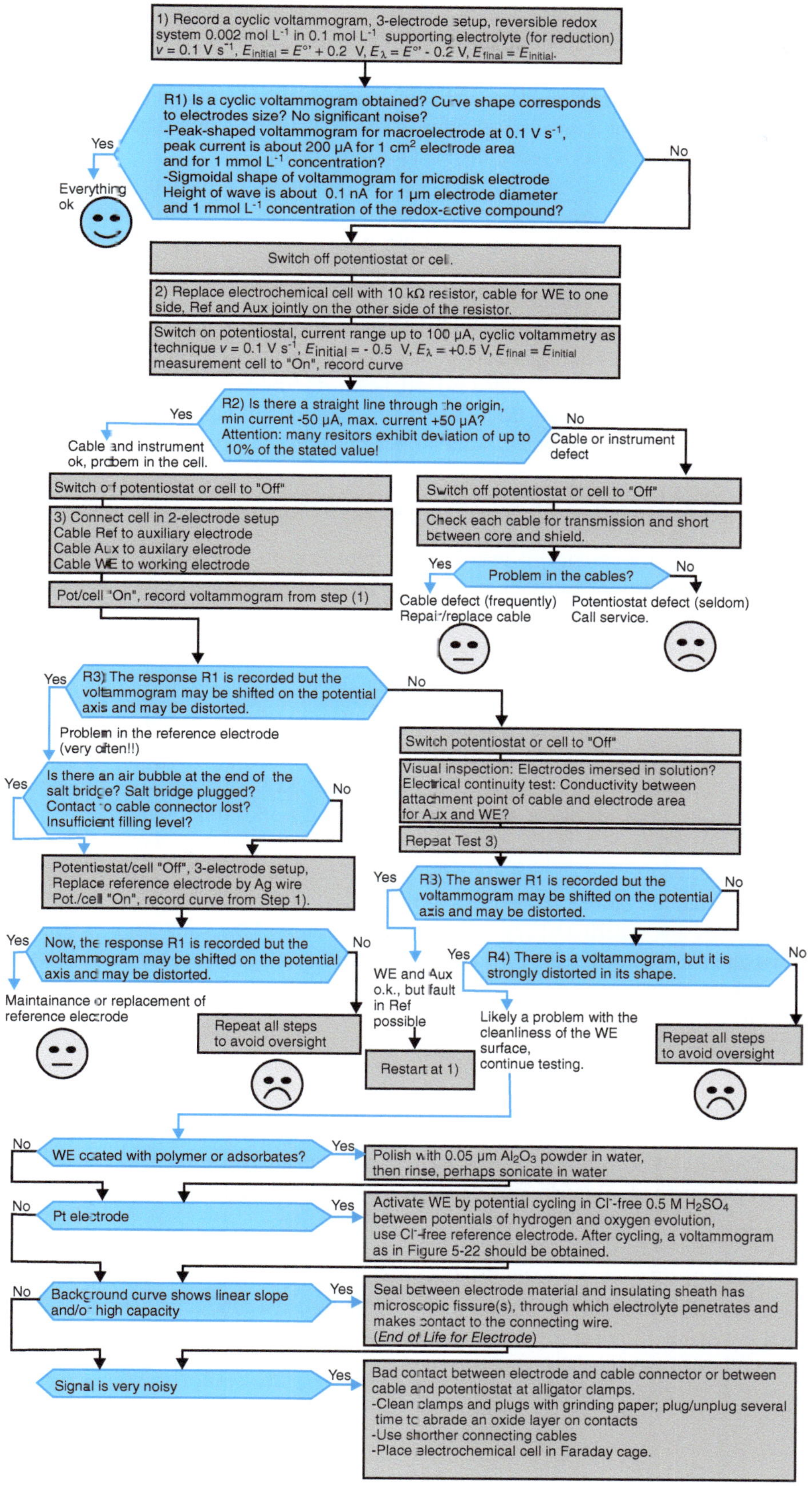

Figure 5.103 Flow chart for troubleshooting and for narrowing the cause of faults in electrochemical experiments.

Another fundamental limitation for any IR_{uc} compensation method is a variation in U_{uc} across the working electrode surface, which makes attempts to compensate futile. Especially for porous electrodes, considerable errors can be introduced by attempting IR_{uc} compensation. One should know the cell setup and the specific operation principle of the potentiostat for a sensible selection of the particular method. The consultation of further literature [141] is recommended for those cases.

Numerical correction methods, that is, computational correction of the potential *after* completing the measurement, can only be recommended for stationary current–voltage curves as found, for example, in LSV with rotating disk electrodes for the characterization of electrocatalysts (see Figure 12.25 for an example).

Further Reading. A more detailed entry with quantitative considerations can be found in the application notes of manufacturers, such as [142]. A more extensive coverage of the electronic circuit elements and their relationships to the electronic property of electrochemical cells are provided in [140, 143, 144]. The particular challenge of correcting the ohmic drop in very fast voltammetric experiments has been developed with dedicated instrumentation and is summarized in [145].

5.7.3 Troubleshooting and Eliminating Faults in Potentiostatic Systems

In daily laboratory practice, it may happen from time to time that electrochemical workstations seem not to function properly or to provide meaningless recordings. There are several potential reasons for this, and each requires different measures to restore the functionality of the system. Figure 5.103 shows a flow chart for narrowing down the reason for the malfunctioning of the experiment (based on recommendations in [140]). The chart assumes that a three-electrode configuration is used with WE, Ref, and Aux. The user should prepare a test solution. Very suitable are reversible redox couples with a 50-fold (or higher) excess of the supporting electrolyte, for instance, 2 mmol L^{-1} $[Ru(NH_3)_6]Cl_3$ in 0.1 mol L^{-1} KCl. This redox system can be reduced at −0.19 V (vs. SCE). At the electrochemical workstation, the method cyclic voltammetry (Section 5.3.5) is selected with the following parameters $E_{initial} = 0$ V (vs. SCE), $E_\lambda = -0.5$ V, $E_{final} = 0$ V, $v = 0.1$ V s^{-1}. If everything works, a voltammogram similar to that in Figure 5.26b with an initial reduction should be recorded.

Failure of this experiment suggests a fault. The cause of the fault can be narrowed down and fixed following the flow chart in Figure 5.103. Because there can be several faults at the same time, one should work through the flow chart multiple times until the anticipated cyclic voltammogram is recorded.

Test your knowledge

Exercise 5.1: Define and explain the terms absolute analytical method and relative analytical method. Provide two examples of electroanalytical methods for each category.

Exercise 5.2: Why is a conductivity measurement not suitable for determining the sulfate concentration in industrial wastewater? Why can a conductivity measurement then be used to monitor the status of a laboratory deionization system for purifying water ("Millipore water")?

Exercise 5.3: Justify the allocation of the following methods to either “absolute analytical methods” or “relative analytical methods”:

- pH measurement with a glass electrode;
- Potentiometric titration of Cl^- with Ag^+ using an Ag electrode for end point indication;
- Pb^{2+} determination by anodic stripping voltammetry;
- Electrogravimetry for the determination of Cu^{2+};
- Blood glucose measurement with an enzyme electrode;
- Conductometric monitoring of the quality of double-distilled water.

Exercise 5.4: **a)** Sketch the formation of a mixed potential at a metal–ion electrode (e.g., Ag|Ag$^-$). **b)** Why can such an electrode, which is not very selective, be used for a very precise potentiometric endpoint detection?

Exercise 5.5: Sketch schematically a combination pH electrode. Label the components.

Exercise 5.6: During a potentiometric titration of a strong acid with a strong base using a combination pH electrode for end point detection (inner solution with pH 7.0), the following voltages were recorded at different times during the titration:

$$E_1 = +0.2366 \text{ V}$$
$$E_2 = +0.0296 \text{ V}$$
$$E_3 = 0.0000 \text{ V}$$
$$E_4 = -0.1775 \text{ V}$$

Calculate the pH in the titration mixture for each of the given voltages. Assume an asymmetry voltage of zero, an ideal slope of the calibration plot, and a room temperature of 298 K.

Exercise 5.7: Sketch the relationship of signal and concentration for an ion-selective electrode (schematic plot with labelled axes). Which effects limit the working range toward small analyte concentrations? Which effects limit the working range toward high concentrations?

Exercise 5.8: What is the meaning of TISAB in potentiometry? Why is a matrix-adapted calibration required when direct potentiometry is to be used for the determination of concentrations?

Exercise 5.9: Provide three examples of analytes that can be determined by potentiometry. Specify the membrane material and the class of membrane that enables the construction of a suitable ion-selective electrode for thistask.

Exercise 5.10: Matrix effects are to be expected when potentiometric ion-selective electrodes are used for samples that contain ions other than just the analyte ion. Assume the ion-selective electrode shall be used for the direct potentiometric determination of concentrations. Name and explain two possible mechanisms by which ions of the sample matrix can influence the signal.

Why can titrations with end point indication by an ion-selective electrode provide very exact results even in the presence of other ions in the matrix?

Exercise 5.11: Provide a rough sketch with exact labelling of the axes analogous to Figures 5.26 and 5.27, but for the investigation of a reversible redox couple for which the oxidized form O is initially present in solution. The formal potential is −0.6 V. How must the experimental parameters be changed compared to the experiment in Figures 5.26 and 5.27?

Exercise 5.12: Explain the following abbreviations in a sentence and/or a schematic sketch: DME, SMDE, andHMDE.

Exercise 5.13: Name and explain two advantages of differential pulse voltammetry at an SMDE compared to DC polarography at aDME.

Exercise 5.14: The diffusion coefficients at 25 °C are 7.2×10^{-6} cm^2 s^{-1} for Cd^{2+} in 0.1 mol L^{-1} KCl solution and 1.9×10^{-5} cm^2 s^{-1} for Cd atoms in mercury amalgam. What is the potential difference between the formal potential $E^{\circ\prime}(Cd/Cd^{2+})$ and the polarographic half-wave potential $E_{1/2}(Cd/Cd^{2+})$? (task is related to Excursion 5-2.)

Exercise 5.15: Provide a schematic plot of the excitation signal for differential pulse voltammetry and the expected shape of the voltammogram. What can be concluded from the height of the voltammogram? What can be concluded from the position of the signal on the potential axis? When and why are such complicated excitation signals preferred over a simple linear potential ramp?

Exercise 5.16: Compare the equation for the current function of SQWV with the Cottrell equation. List similarities and differences.

Exercise 5.17: Why does a polarogram for the deposition of Pb at a dropping mercury electrode look like a step, but it looks like one half-cycle of a cyclic voltammogram ("duck bill") for the deposition of Pb at a HMDE in unstirred solution?

Exercise 5.18: The graph below shows the excitation signal of a voltammetric technique. Name the technique. Provide a sketch of the resulting voltammogram recorded by this method using an SMDE as working electrode (WE), SCE as reference electrode (Ref), and glassy carbon as auxiliary electrode (Aux) in 0.1 mol L^{-1} HCl solution with 5×10^{-6} mol L^{-1} Cd^{2+}. $E^{\circ\prime}(Cd^{2+}|Cd^0) = -0.403$ V vs. SHE.

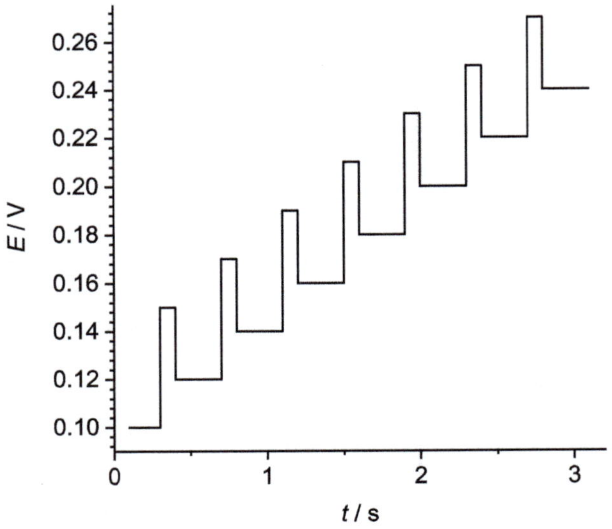

Exercise 5.19: Explain the principle of anodic stripping voltammetry. Which analytes can be determined by this method? Why must organic compounds in the sample be destroyed by a digestion process, even if they do not exhibit any redox reactions in the potential range used for the determination?

Exercise 5.20: Provide sketches and explain the procedures for the use of calibration plots and the use of standard additions. Why are standard additions often preferred over calibration plots for the analysis of environmental samples?

Exercise 5.21: A water sample was digested, and 25 mL was transferred into the voltammetric cell. A stripping voltammogram was recorded for the determination of Pb^{2+} followed by three standard additions of 25 µL of a $Pb(NO_3)_2$ stock solution of 4000 µg L^{-1} Pb^{2+}.

Calculate the added concentration after each standard addition using $V_{ges} = V_0 + V_{add}$ and without volume correction ($V_{ges} \approx V_0$).

Prepare a rough sketch of how the standard addition plot should look when the sample contains 3 µg L^{-1} Pb^{2+}.

Exercise 5.22: The analysis of Exercise 5.21 is repeated with another 25 mL aliquot of the digested water sample from Exercise 5.21. However, the stock solution now contains only 400 µg L^{-1} Pb^{2+}, and therefore, the three additions are made with 250 µL stock solution.

Calculate the added concentration after each standard addition using $V_{tot} = V_0 + V_{add}$ and without volume correction ($V_{tot} \approx V_0$).

Discuss advantages and disadvantages of the concentrations of the stock solutions in Exercises 5.21 and 5.22. For which concentration range would both stock solutions be well suited?

Exercise 5.23: Calculate the mass of $Pb(NO_3)_2$ that must be weighed to prepare 1 L of the stock solution in Exercise 5.21.

Exercise 5.24: Assign the signals in Figure 5.47 to the different metal ions using a comparison to Figure 5.37. Estimate the sensitivity of the determination by DPV at the SMDE and by anodic stripping voltammetry at the HMDE. How would the sensitivity change when the deposition time is doubled in anodic stripping voltammetry? What is the relationship between sensitivity and limit of detection?

Exercise 5.25: The signals shown below were recorded during the stripping voltammetry of Pb^{2+} using standard additions as specified below the graph. Determine graphically, using the second diagram, the concentration of the sample and **state the result in one sentence**. You may ignore the volume increase by the addition of 0.1 mL stock solution. Challenge: Which deviation in the final result is caused by this simplifying assumption? How does the deviation scales relative to other measurement uncertainties in this analysis?

Curve (1) Voltammogram of the 25 mL sample

Curve (2) Voltammogram of sample with addition of 50 µL of 1 mmol L^{-1} Pb^{2+} stock solution

Curve (3) Voltammogram of sample with addition of 75 µL of 1 mmol L^{-1} Pb^{2+} stock solution

Curve (4) Voltammogram of sample with addition of 100 µL of 1 mmol L^{-1} Pb^{2+} stock solution

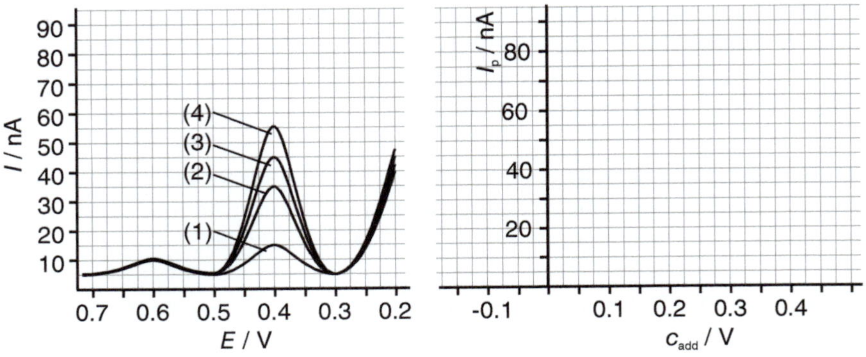

Exercise 5.26: The result of a voltammetric investigation of water was a concentration of (50 ± 7) μg L^{-1} Pb^{2+} and of (40 ± 8) μg L^{-1} Cd^{2+}. Which of the two metal ions is contained in the sample in higher (substance quantity) concentration? Provide the result and a justification with equations.

Exercise 5.27: What is the complexing capacity of natural waters? Describe the key points and, with a schematic sketch, how this parameter can be determined by electrochemical methods.

Exercise 5.28: Explain the procedure to calibrate a Clark sensor for molecular oxygen.

Exercise 5.29: Calculate the cyclic voltammogram of the adsorbed redox system in Figure 5.57 using a spreadsheet application. Arrange the spreadsheet in such a way that the formal potential and the interfacial concentration of the redox system can be easily changed.

Use an analytical expression for the curves, and in addition, calculate the curve by stepwise calculation.

Determine the total concentration of the redox system by stepwise integration of the curve obtained by the analytical expression and by integration of the data obtained by stepwise calculation. Compare the results with the total concentration that was the input for the calculation of the curves.

Exercise 5.30: How large is the steady-state diffusion-controlled current for a 1 mmol L^{-1} solution of a redox system with $n = 1$ and a diffusion coefficient of $D = 5 \times 10^{-6}$ cm^2 s^{-1} at a microdisk electrode with a diameter of 25 μm? Calculate the current in convenient units and also strictly in SI units (meter, kilogram, second).

Exercise 5.31: Square-shaped microelectrodes with an edge length of 15 μm have been produced by photolithography. Calculate the radius of the equivalent microdisk that has the same area and would produce approximately the same current.

Exercise 5.32: An analog potentiostat provides a current-proportional voltage as output between −2.0 V and + 2.0 V for each current range. The current range is 2.5 μA/V. The voltage measured at the signal output is 1.5 V.

a) What is the value of the current at the working electrode?
b) State the current range in the unit volt per microampere.

Exercise 5.33: An analog potentiostat provides a current-proportional voltage as output between −2.0 V and + 2.0 V for each current range. This voltage is measured with an analog–digital converter. The converter available has a range from −10.00 V bis +10.00 V. Suggest an amplifier circuit that amplifies the current-proportional output voltage of the potentiostat in such a way that the available range of the analog–digital is used optimally.

Give for a current range of 2.50 μA/V the corresponding value after the amplification described above.

Refine the circuit diagram to provide a means for a fine adjustment of the amplifier. How would you conduct such a fine adjustment experimentally?

Exercise 5.34: An analog potentiostat provides a current-proportional voltage as output between −10.0 V and + 10.0 V for each current range. This voltage is measured by means of an analog–digital converter. The converter available has a range from −2.000 V bis +2.000 V. Suggest a circuit that scales the current-proportional output voltage of the potentiostat in such a way that the available range of the analog–digital converter is used optimally. Give for the current range of 2.50 μA/V the corresponding range after the scaling described above.

Refine the circuit diagram to provide a means for a fine adjustment of the scaler. How would you conduct such a fine adjustment experimentally?

Exercise 5.35: An analog potentiostat provides a current-proportional voltage as output between −2.0 V and + 2.0 V for each current range. This voltage is measured by means of an analog–digital converter. The converter available has a range from 0.0 V to +5.0 V.

Suggest a circuit that adjusts the current-proportional output voltage of the potentiostat (bipolar) to the available voltage range of the analog–digital converter (unipolar).

Refine the circuit to allow a fine adjustment of the amplification. How would you proceed experimentally to make the fine adjustment?

References

1 A. J. Bard, L. F. Faulkner, H. S. White; *Electrochemical Methods. Fundamentals and Applications*, 3rd ed. John Wiley & Sons, New York, 2022.

2 P. T. Kissinger, W. R. Heineman (Eds.); *Laboratory Techniques in Electroanalytical Chemistry*, 2nd ed., Marcel Dekker, New York, 1996.

3 F. Scholz (Ed.); *Electroanalytical Methods – Guide to Experiments and Application*, 2nd ed., Springer, Berlin, 2010.

4 P. S. Monk; *Fundamentals of Electroanalytical Chemistry*, John Wiley & Sons, Chichester, 2001.

5 D. C. Harris; *Quantitative Chemical Analysis*, 9th ed. W. H. Freeman & Company, New York, NY, 2016.

6 D. A. Skoog, H. J. James, S. R. Crouch; *Principles of Instrumental Analysis*. 7th ed., Cengage Learning, Boston, 2017.

7 J. W. Robinson, E. M. Skelly, G. M. Frame II; *Instrumental Analytical Chemistry: An Introduction*, CRC Press, Oxon, 2021.

8 R. G. Bates; *Determination of pH: Theory and Practice*, John Wiley & Sons, New York, 1964.

9 F. G. K. Baucke, Further insight into the dissociation mechanism of glass electrodes. The response in heavy water. *J. Phys. Chem. B* **1998**, *102*, 4835–4841.

10 A. J. Bard, G. Inzelt, F. Scholz (Eds.); *Electrochemical Dictionary*, 2nd ed. Springer, Berlin, 2012, entry "glass electrode," "glass electrode – dissociative mechanism"

11 K. W. Pratt; Longitudinal meta-analysis of NIST pH standard reference materials: A complement to pH key comparisons. *Anal. Bioanal. Chem.* **2015**, *407*, 3193–3203.

12 H. Galster; *pH Measurement: Fundamentals, Methods, Applications, Instrumentation*. VCH, Weinheim, 1991.

13 J. Barbosa, D. Barrón; S. Butí; Chromatographic behaviour of ionizable compounds in liquid chromatography. Part I, pH scale, pK_a and pH_S values for standard buffers in tetrahydrofuran-water. *Anal. Chim. Acta* **1999**, *389*, 31–42.

14 D. G. Hall; Ion-selective membrane electrodes: A general limiting treatment of interference effects. *J. Phys. Chem.* **1996**, *100*, 7230–7236.

15 M. Nägele, E. Bakker, E. Pretsch; General description of the simultaneous response of potentiometric ionophore-based sensors to ions of different charge. *Anal. Chem.* **1999**, *71*, 1041–1048.

16 E. Bakker, E. Pretsch, P. Bühlmann; Selectivity of potentiometric sensors. *Anal. Chem.* **2000**, *72*, 1127–1133.

17 P. Bühlmann, E. Pretsch, E. Bakker; Carrier-based ion-selective electrodes and bulk optodes. 2. ionophores for potentiometric and optical sensors. *Chem. Rev.* **1998**, *98*, 1593–1687.

18 E. Bakker, M. E. Meyerhoff; Ion-selective electrodes for measurements in biological fluids, in *Encyclopedia of Electrochemistry, Vol. 9*, (G. S. Wilson, A. J. Bard, M. Stratmann, Eds.) Wiley-VCH, Weinheim 2002, pp. 277–307.

19 K. Cammann (Ed.), *Instrumentelle Analytische Chemie – Verfahren, Anwendungen, Qualitätssicherung*. Spektrum Akademischer Verlag, Heidelberg, 2001, pp. 7–17.

20 M. E. Meyerhoff, B. Fu, E. Bakker, J. H. Yun, V. C. Yang; Polyion-sensitive membrane electrodes for biomedical analysis, *Anal. Chem.* **1996**, *68*, 168A–175A.

21 E. Bakker, P. Bühlmann, E. Pretsch; Carrier-based ion-selective electrodes and bulk optodes. 1. General characteristics. *Chem. Rev.* **1997**, *97*, 3083–3132.

22 E. Bakker, E. Pretsch; The new wave of ion-selective electrodes. *Anal. Chem.* **2002**, *74*, 420A–426A.

23 E. Bakker, E. Pretsch; Advances in potentiometry. in *Electroanalytical Chemistry, Vol. 24*, (A. J. Bard, C. G. Zoski, Eds.), CRC-Press, Boca Raton, 2012, pp. 1–73.

24 G. M. Jenkins, K. Kawamura; Structure of glassy carbon. *Nature*, **1971**, *231*, 175–176.

25 G. M. Swain; Electrically conducting diamond thin films: Advanced electrode materials for electrochemical technologies. in *Electroanalytical Chemistry, Vol. 22*, (A. J. Bard, I. Rubinstein, Eds.); Marcel Dekker, New York, 2004, pp. 182–277.

26 R. L. McCreery, Advanced carbon electrode materials for molecular electrochemistry. *Chem. Rev.* **2008**, *108*, 2646–2687.

27 R. L. McCreery, K. K. Cline; Carbon electrodes. in *Laboratory Techniques in Electroanalytical Chemistry*. 2nd ed. (P. T. Kissinger, W. R. Heineman, Eds.), Marcel Dekker, New York, 1996, pp. 293–332.

28 S. Trasatti, O.A. Petrii; Real surface area measurement in electrochemistry. *Pure Appl. Chem.* **1991**, *63*, 711–734.

29 J. Solla-Gullón, P. Rodríguez, E. Herrero, A. Aldaz, J. M. Feliu; Surface characterization of platinum electrodes. *Phys. Chem. Chem. Phys.* **2008**, *10*, 1359–1373.

30 L. A. Broussard, C. A. Hammett-Stabler, R. E. Winecker, J. D. Ropero-Miller; The Toxicology of Mercury. *Lab. Med.* **2002**, *33*, 614. http://labmed.ascpjournals.org/content/33/8/614.full.pdf (last access 19 Mar 2016).

31 T. Navrátil, I. Švancara, K. Mrázová, K. Nováková, I. Šestáková, M. Heyrovský; Mercury and mercury electrodes: The ultimate battle for naked existence. in *Sensing in Electroanalysis, Vol. 6*, (K. Kalcher, R. Metelka, I. Švancara, K. Vytřas, Eds.), University Press Centre, Pardubize, 2011, pp. 23–53. https://hdl.handle.net/10195/42533 (last access 25 Jan 2026).

32 J. Wang; Stripping analysis at Bismuth electrodes: A review. *Electroanalysis* **2005**, *17*, 1341–1346.

33 D. T. Sawyer, J. L. Roberts, Jr.; *Experimental Electrochemistry for Chemists*. Wiley, New York, 1974.

34 C. K. Mann; Nonaqueous solvents for electrochemical use. in *Electroanalytical Chemistry, Vol. 3*, (A. J. Bard Ed.), Marcel Dekker, New York, 1969, pp. 57–134.

35 J. Jörissen, B. Speiser; Preparative electrolysis on the laboratory scale. in *Organic Electrochemistry*, 5th ed. (O. Hammerich, B. Speiser, Eds.), CRC Press, Boca Raton, 2016, Appendices pp. 300–330.

36 K. Izutsu; *Electrochemistry in Nonaqueous Solvents*. Wiley-VCH, Weinheim, 2002, pp. 287–311.

37 S. Creager; Solvents and supporting electrolytes. in *Handbook of Electrochemistry*. (C. G. Zosky, Ed.), Elsevier, Amsterdam, 2007, pp. 57–72.

38 H. Kiesele; Electrochemical cell for voltammetry, coulometry, and synthesis in superdry media. *Anal. Chem.* **1981**, *53*, 1952–1954.

39 D. T. Sawyer, J. L. Roberts, Jr.; *Experimental Electrochemistry for Chemists*. John Wiley & Sons, New York, 1974, pp. 117–165.

40 F. M. Hawkridge; Electrochemical cells. in *Laboratory Techniques in Electroanalytical Chemistry*, 2nd ed., (W. R. Heineman, P. T. Kissinger, Eds.), Marcel Dekker, New York, 1996, pp. 813–853.

41 S. Chen; Practical electrochemical cells. in *Handbook of Electrochemistry*. (C. G. Zosky, Ed.), Elsevier, Amsterdam, 2007, pp. 33–56.

42 B. Gollas, B. Kraus, B. Speiser, H. Stahl; Design of a single unit Haber–Luggin capillary/dual reference-electrode system. *Curr. Sep.* **1994**, *13*, 42–44.

43 A. M. Bond, E. A. McLennan, R. S. Stojanovic, F. G. Thomas; Assessment of conditions under which the oxidation of ferrocene can be used as a standard voltammetric reference process in aqueous media. *Anal. Chem.* **1987**, *59*, 2853–2860.

44 J. H. Brown; Analysis of two redox couples in a series: An expanded experiment to introduce undergraduate students to cyclic voltammetry and electrochemical simulations. *J. Chem. Educ.* **2016**, *93*, 1326–1329.

45 R. S. Nicholson, I. Shain; Theory of stationary electrode polarography. Single scan and cyclic methods applied to reversible, irreversible, and kinetic systems. *Anal. Chem.* **1964**, *36*, 706–723.

46 R. S. Nicholson; Theory and application of cyclic voltammetry for measurement of electrode reaction kinetics. *Anal. Chem.* **1965**, *37*, 1351–1355.

47 R. C. Engstrom, T. Meaney, R. Tople, R. M. Wightman; Spatiotemporal description of the diffusion layer with a microelectrode probe. *Anal. Chem.* **1987**, *59*, 2005–2010.

48 R. C. Engstrom, S. Ghaffrari, H. Qu; Fluorescence imaging of electrode solution interfacial processes. *Anal. Chem.* **1992**, *64*, 2525–2529.

49 P. Schwager, S. Dongmo, D. Fenske, G. Wittstock; Reactive oxygen species formed in organic lithium–oxygen batteries. *Phys. Chem. Chem. Phys.* **2016**, *18*, 10774–10780.

50 N. Baltes, L. Thounin, C. Amatore, J. Heinze; Imaging concentration profiles of Redox-active species with nanometric amperometric probes: Effect of natural convection on transport at microdisk electrodes. *Angew. Chem. Int. Ed. Engl.* **2004**, *43*, 1431–1435.

51 N. Elgrishi, K. J. Rountree, B. D. McCarthy, E. S. Rountree, T. T. Eisenhart; J. I. Dempsey; A practical beginner's guide to cyclic voltammetry. *J. Chem. Educ.* **2015**, *95*, 197–206.

52 L. Meites; *Polarographic Techniques*. 2nd ed., John Wiley & Sons, New, 1955.

53 A. M. Bond; *Modern Polarographic Methods in Analytical Chemistry.* Marcel Dekker, New York, 1980.

54 Z. Galus; Mercury electrodes. in *Laboratory Techniques in Electroanalytical Chemistry*, 2nd ed., (W. R. Heineman, P. T. Kissinger, Eds.), Marcel Dekker, New York, 1996, pp. 443–468.

55 K. B. Oldham, J. C. Myland; *Fundamentals of Electrochemistry*, Academic Press, London, 1994, p. 233.

56 U. Schröder, H. Kahlert, *Practical Voltammetry*. Metrohm, Herisau, 2002, https://www.metrohm.com/en_gb/products/8/1085/81085083.html (last access 22 Feb 2025).

57 J. G. Osteryoung, J. J. O'Dea; Square-wave voltammetry. in *Electroanalytical Chemistry, Vol. 14*, (A. J. Bard, Ed.), Marcel Dekker, New York **1986**, pp. 209–308.

58 J. G. Osteryoung, R. A. Osteryoung; Square wave voltammetry. *Anal. Chem.* **1985**, *57*, 101A–110A.

59 G. C. Barker, J. L. Jenkins; Square-wave polarography. *Analyst* **1952**, *77*, 685–696.

60 H. H. Girault; *Analytical and Physical Electrochemistry*, EPFL Press Lausanne, Marcel Dekker, New York, 2004, pp. 319–331.

61 Á. Molina, J. González; *Pulse Volatmmetry in Physical Electrochemistry and Electroanalysis: Theory and Application*, Springer, Cham, 2016.

62 M. Lovrić; Square-wave voltammetry. in *Electroanalytical Methods*, 2nd ed. (F. Scholz, Ed.), Springer Verlag, Berlin, 2010, pp. 121–145.

63 V. Mirceski, S. Skrzypek, L. Stojanov; Square-wave voltammetry. *ChemTexts* **2018**, *4*, 17.

64 Z. Stojek; Pulse voltammetry. in *Electroanalytical Methods*, 2nd ed. (F. Scholz, Ed.), Springer Verlag, Berlin, 2010, pp. 107–119.

65 J. Osteryoung, E. Kirowa-Eisner; Reverse pulse polarography. *Anal. Chem.* **1980**, *52*, 62–66.

66 J. Zhang, S.-X. Guo, A. M. Bond, F. Marken; Large amplitude fourier transformed high-harmonic alternating current cyclic voltammetry: Kinetic discrimination of interfering faradaic processes at glassy carbon and at boron-doped diamond electrodes. *Anal. Chem.* **2004**, *76*, 3619–3629.

67 A. Economou, P. R. Fielden; Square wave adsorptive stripping voltammetry on mercury film electrodes, *Anal. Chim. Acta* **1993**, *273*, 27–34.

68 B. Breyer. S. Hacobian; Tensammetry: A method of investigating surface phenomena by alternating current measurements. *Aust. J. Sci. Res. Ser. B* **1952**, *A5*, 500–520.

69 D. C. Grahame; The electrical double layer and the theory of electrocapillarity. *Chem. Rev.* **1947**, *41*, 441–501.

70 J. Wang; Stripping analysis. in *Encyclopedia of Electrochemistry, Vol. 3. Instrumentation and Electroanalytical Chemistry* (A. J. Bard, M. Stratmann, P. R. Unwin, Eds.) Wiley-VCH, Weinheim, 2007, pp. 125–133.

71 G. Henze, R. Neeb; *Elektrochemische Analytik*; Springer Verlag, Berlin, 1986, p. 108.

72 http://www.metrohm.com/en/applications/#methods=3DC44D9FB5154E97A3213 4A30CAF0C1E (11 May 2015).

73 http://www.princetonappliedresearch.com/Literature/index.aspx (last access 11 May 2015).

74 Sample Preparation Techniques in Voltammetric Trace Analysis (8.018.5003, P. A. Bruttel, J. Schäfer, Eds.), Metrohm AG, https://www.metrohm.com/en/products-overview/27970010 (last access 23 Oct 2021).

75 Metrohm; Polarographic determination of nitrilotriacetic acid (NTA) and ethylenediaminetetraacetic acid (EDTA) according to DIN 38413 part 5, accessible via https://www.metrohm.com/en/applications/AB-076.

76 S. M. Lunte, C. E. Lunte, P. T. Kissinger; Electrochemical detection in liquid chromatography and capillary electrophoresis. in *Laboratory Techniques in Electroanalytical Chemistry*, 2nd ed., (W. R. Heineman, P. T. Kissinger, Eds.), Marcel Dekker, New York, 1996, pp. 813–853.

77 Honeywell, Hydranal-Handbook, https://lab.honeywell.com/en/search/resources-search?subcat=Hydranal%20Manuals (last access 23 Oct 2021).

78 E. Scholz, *Karl Fischer Titration: Determination of Water*, Springer Verlag, Berlin, 1984.

79 P. Gründler, *Chemical Sensors – An Introduction for Scientists and Engineers*, Springer, Heidelberg, 2007, p. 166.

80 F. Scholz, U. Schröder, R. Gulaboski, A. Doménech-Carbó; *Electrochemistry of Immobilized Particles and Droplets: Experiments with Three-Phase Electrodes*. 2nd ed., Springer, Cham, 2015.

81 A. Doménech-Carbó, M. T. Doménech-Carbó; Electrochemistry for cultural heritage. in *Monographs in Electrochemistry* (F. Scholz, L. Péter, Eds.), Springer, Cham, 2023.

82 C.-H. Chen, E. R. Ravenhill, D. Momotenko, Y.-R. Kim, S. C. S. Lai, P. R. Unwin; Impact of surface chemistry on nanoparticle electrode interactions in the electrochemical detection of nanoparticle collisions. *Langmuir* **2015**, *31*, 11 11932–11942.

83 D. A. Robinson, A. A. Edwards, H. Ren, H. S. White; Effects of instrumental filters on electrochemical measurements of single-nanoparticle collision dynamics. *ChemElectroChem* **2018**, *5*, 3059–3067.

84 E. Laborda, A. Molina, C. Batchelor-McAuley, R. G. Compton; Individual detection and characterization of non-electrocatalytic, redox-inactive particles in solution by using electrochemistry. *ChemElectroChem* **2018**, *5*, 410–417.

85 E. Kätelhön, E. E. L. Tanner, C. Batchelor-McAuley, R. G. Compton; Destructive nano-impacts: What information can be extracted from spike shapes? *Electrochim. Acta* **2016**, *199*, 297–305.

86 B.-K. Kim, A. Boika, J. Kim, J. E. Dick, A. J. Bard; Characterizing emulsions by observation of single droplet collisions—Atoliter electrochemical reactors. *J. Am. Chem. Soc.* **2014**, *136*, 4849–4852.

87 X. Xiao, A. J. Bard; Observing single nanoparticle collisions at an ultramicroelectrode by electrocatalytic amplification. *J. Am. Chem. Soc.* **2007**, *129*, 9610–9612.

88 J. Kim, B.-K. Kim, S. K. Cho, A. J. Bard; Tunneling ultramicroelectrode: Nanoelectrodes and nanoparticle collisions. *J. Am. Chem. Soc.* **2014**, *136*, 8173–8176.

89 A. J. Bard, L. R. Faulkner, H. S. White; *Electrochemical Methods. Fundamentals and Applications*. 3rd ed., John Wiley & Sons, Hoboken, 2022, pp. 851–875.

90 Y.-G. Zhou, N. V. Rees, R. G. Compton; Electrode-nanoparticle collisions: The measurement of the sticking coefficient of silver nanoparticles on a glassy carbon electrode. *Chem. Phys. Lett.* **2011**, *514*, 291–293.

91 S. M. Oja, D. A. Robinson, N. J. Vitti, M. A. Edwards, Y. Liu, H. S. White, B. Zhang; Observation of multipeak collision behavior during electro-oxidation of single Ag nanoparticles. *J. Am. Chem. Soc.* **2017**, *139*, 708–718.

92 H. Zhou, J. H. Park, F.-R. F. Fan, A. J. Bard; Observation of single metal nanoparticle collisions by open circuit (mixed) potential changes at an ultramicroelectrode. *J. Am. Chem. Soc.* **2012**, *134*, 13212–13215.

93 S. E. F. Kleijn, S. C. S. Lai, M. T. M. Koper, P. R. Unwin; Electrochemistry of nanoparticles; *Angew. Chem. Int. Ed.* **2014**, *53*, 3558–3586.

94 J. Kim, J. E. Dick, A. J. Bard; Advanced electrodeposition of individual metal clusters electrodeposited atom by atom to nanometer by nanometer. *Acc. Chem. Res.* **2016**, *49*, 2587–2595.

95 K. J. Stevenson, K. Tschulik; A materials driven approach for understanding single entity nano impact electrochemistry. *Curr. Opin. Electrochem.* **2017**, *6*, 38–45.

96 C. Bentley, M. Kang, P. R. Unwin; Nanoscale surface structure–activity in electrochemistry and electrocatalysis. *J. Am. Chem. Soc.* **2019**, *141*, 2179–2193.

97 F. T. Patrice, K. Qiu, Y.-L. Ying, Y.-T. Long; Single nanoparticle electrochemistry. *Annu. Rev. Anal. Chem.* **2019**, *12*, 347–370.

98 Y.-Y. Peng, R.-C. Qian, M. E. Hafez, Y.-T. Long; Stochastic collision nanoelectrochemistry: A review of recent developments. *ChemElectroChem* **2017**, *4*, 977–985.

99 L. A. Baker; Perspective and prospectus on single-entity lectrochemistry. *J. Am. Chem. Soc.* **2018**, *140*, 15 15549–15559.

100 S. V. Sokolov, S. Eloul, E. Kätelhön, C. Batchelor-Auley, R. G. Compton; Electrode-particle impacts: A user guide. *Phys. Chem. Chem. Phys.* **2018**, *19*, 28–43.

101 Z. Deng, C. Renault; Detection of individual insulating entities by electrochemical blocking. *Curr. Opin. Electrochem.* **2021**, *25*, 10619.

102 M. Jaugstetter, N. Blanc, M. Kratz, K. Tschulik; Electrochemistry under confinement. *Chem. Soc. Rev.* **2022**, *51*, 2491–2543.

103 J. Heinze; Ultramicroelectrodes in electrochemistry. *Angew. Chem. Int. Ed.* **1993**, *32*, 1268–1288.

104 K. Stulik, C. Amatore, K. Holub, V. Marecek, W. Kutner; Microelectrodes. Definitions, characterization and application (Technical report). *Pure Appl. Chem.* **2000**, *72*, 1483–1492.

105 Y. Saito; A theoretical study on the diffusion current at the stationary electrodes of circular and narrow band types. *Rev. Polarogr.* **1968**, *15*, 177–187.

106 C. G. Zoski, M. V. Mirkin; Steady-state limiting currents at finite conical microelectrodes. *Anal. Chem.* **2002**, *74*, 1986–1992.

107 D. MacGillavry, E. K. Rideal; On the theory of limiting currents. I. Polarographic limiting currents. *Recl. Trav. Chim. Pays-Bas* **1937**, *56*, 1013–1021.

108 D. Shoup, A. Szabo; Chronoamperometric current at finite disk electrodes. *J. Electroanal. Chem. Interfacial Electrochem.* **1982**, 140, 237–245.

109 P. J. Mahon, K. B. Oldham; Diffusion-controlled chronoamperometry at a disk electrode. *Anal. Chem.* **2005**, *77*, 6100–6101.

110 R. Kelly, R. M. Wightman; Bevelled carbon-fiber ultramicroelectrodes. *Anal. Chim. Acta* **1986**, *187*, 79–87.

111 G. Wittstock; B. Gründig; B. Stehlitz, K. Zimmer; Evaluation of microelectrode arrays for amperometric detection by scanning electrochemical microscopy. *Electroanalysis* **1998**, *10*, 526–531.

112 R. J. Forster; Microelectrodes — Retrospect and prospect. in *Encyclopedia of Electrochemistry, Vol. 3. Instrumentation and Electroanalytical Chemistry* (A. J Bard, M. Stratmann, P. Unwin, Eds.), Wiley-VCH, Weinheim, 2003, pp. 160–195.

113 A. C. Michael, R. M. Wightman; Microelectrodes. in *Laboratory Techniques in Electroanalytical Chemistry*, 2nd ed. (P. T. Kissinger, W. R. Heineman, Eds.), Marcel Dekker, New York, 1996, pp. 369–372.

114 R. J. Forster, T. E. Keyes, J. Mauzeroll, R. J. LeSuer, B. Liu, R. Tel-Vered, A. Kucernak, C. Demaille, K. B. Holt, Y. Lee, B. Zhang, G. Wang, H. S. White; Ultramicroelectrodes. in *Handbook of Electrochemistry.* (C. G. Zosky, Ed.), Elsevier, Amsterdam, 2007, pp. 155–260.

115 F.-R. F. Fan; Preparation of tips for scanning electrochemical microscopy. in *Scanning Electrochemical Microscopy*, 2nd ed. (A. J. Bard, M. V. Mirkin, Eds.), CRC Press, Boca Raton, 2012, pp. 25–52.

116 C. G. Zoski; Ultramicroelectrodes: Design, fabrication, and characterization. *Electroanalysis* **2002**, *14*, 1041–1051.

117 C. D. Baer, N. J. Stone, D. A. Sweigart; Fabrication of platinum-disk ultramicroelectrodes. *Anal. Chem.* **1988**, *60*, 188–191.

118 H. Shiku, T. Takeda, H. Yamada, T. Matsue, I. Uchida; Microfabrication and characterization of diaphorase-patterned surfaces by scanning electrochemical microscopy. *Anal. Chem.* **1995**, *67*, 312–317.

119 M. A. Mezour, M. Morin, J. Mauzeroll; Fabrication and characterization of laser pulled platinum microelectrodes with controlled geometry. *Anal. Chem.* **2011**, *83*, 2378–2382.

120 P. Actis, S. Tokar, J. Clausmeyer, B. Babkinejad, S. Mikhaleva, R. Cornut, Y. Takahashi, A. López Córdoba, P. Novak, A. I. Sevchuk, J. A. Dougan, S. G. Karzarian, P. V. Gorelkin, A. S. Erofeev, I. V. Yaminsky, P. R. Unwin, W. Schuhmann, D. Klenerman, D. A. Rusakov, E. V. Sviderskaya, Y. E. Korchev; Electrochemical nanoprobes for single-cell analysis. *ACS Nano* **2014**, *8*, 875–884.

121 R. M. Wightman, D. O. Wipf, Voltammetry at Ultramicroelectrodes. in *Electroanalytical Chemistry, Vol. 15*, (A. J. Bard, Ed.), Marcel Dekker, 1989, pp. 267–353.

122 Y. Wang, J.-M. Noel, J. Velmurugan, W. Nogala, M. V. Mirkin, C. Lu, M. G. Collignon, F. Lemaitre, C. A. Amatore; Nanoelectrodes for determination of reactive oxygen and reactive nitrogen species inside murine macrophages. *Proc. Natl. Acad. Sci. USA* **2012**, *109*, 11534–11539.

123 https://en.wikipedia.org/wiki/Action_potential (last access 23 Oct 2021).

124 R. A. Clark, S. E. Zerby, A. G. Ewing; Electrochemistry in neuronal microenvironments. in *Electroanalytical Chemistry, Vol. 20*, (A. J. Bard, I. Rubinstein, Eds.), Marcel Dekker, New York, 1998, pp. 227–294.

125 J. J. Clark, S. G. Sandberg, M. J. Wanat, J. O. Gan, E. A. Horne, A. S. Hart, C. A. Akers, J. G. Parker, I. Wilhuhn, V. Martinez, S. B. Evans, N. Stella, P. E. M. Phillips; Chronic microsensors for longitudinal, subsecond dopamine detection in behaving animals. *Nat. Methods*, **2010**, *2*, 126–129.

126 https://en.wikipedia.org/wiki/Nucleus_accumbens (last access 23 Oct 2021).

127 P. E. M. Phillips, G. D. Stuber, M. L. A. V. Heien, R. M. Wightman, R. M. Carrel; Subsecond dopamine release promotes cocaine seeking. *Nature* **2003**, *422*, 614–618.

128 L. A. Sombers, A. G. Ewing; Electrochemical monitoring of exocystosis from individual PC12 cells in culture; in *Electroanalytical Methods for Biological Materials* (A. Braitjer-Toth, J. Q. Chambers, Eds.), Marcel Dekker, New York, 2002, pp. 279–327.

129 C. Amatore, S. Arbault, C. Bouton, K. Coffi, J.-C. Drapier, H. Ghandour, Y. Tong; Monitoring in real time with a microelectrode the release of reactive oxygen and nitrogen species by a single macrophage stimulated by its membrane mechanical depolarization. *ChemBioChem* **2006**, *7*, 653–661.

130 C. R. Menon, C. R. Martin; Fabrication and evaluation of nanoelectrode ensembles. *Anal. Chem.* **1995**, *67*, 1920–1928.

131 A. C. Michael, R. M. Wightman, C. A. Amatore; Microdisk electrodes: Part I. Digital simulation with a conformal map. *J. Electroanal. Chem.* **1989**, *267*, 33.

132 K. Aoki, M. Morita, O. Niwa, H. Tabei; Quantitative analysis of reversible diffusion-controlled currents of redox soluble species at interdigitated array electrodes under steady-state conditions. *J. Electroanal. Chem. Interfacial Electrochem.* **1988**, *256*, 269–282.

133 U. Wollenberger, M. Paeschke, R. Hintsche; Interdigitated array microelectrodes for the determination of enzyme activities. *Analyst* **1994**, *119*, 1245–1249.

134 P. T. Kissinger, Introduction to analog instrumentation, in *Laboratory Techniques in Electroanalytical Chemistry*, 2nd ed. (W. R. Heineman, P. T Kissinger, Eds.), Marcel Dekker, New York, 1996, pp. 165–194.

135 C. N. Reilley; IV. Instrumentation Based on Operational Amplifiers (Part I). *J. Chem. Educ.* **1962**, *39*, A853.

136 C. N. Reilley; IV. Instrumentation based on operational amplifiers (Part II). *J. Chem. Educ.* **1962**, *39*, A933.

137 D. Wipf; Analog and digital instrumentation. in *Encyclopedia of Electrochemistry, Vol. 3. Instrumentation and Electroanalytical Chemistry* (A. J Bard, M. Stratmann, P. Unwin, Eds.), Wiley-VCH, Weinheim, 2003, pp. 24–50.

138 P. Horrowitz, W. Hill; *The Art of Electronics*. 2nd ed., Cambridge University Press, Cambridge, 1989.

139 D. A. Skoog, H. J. James, S. R. Crouch; *Principles of Instrumental Analysis*. 7th ed., Cengage Learning, Boston, 2017, Ch. 2–4, pp. 23–118.

140 A. J. Bard, L. F. Faulkner, H. S. White; *Electrochemical Methods. Fundamentals and Applications*, 3rd ed., John Wiley & Sons, New York, 2022, pp. 721–754.

141 W. Zheng; *iR* compensation for electrocatlytic studies: Considerations and recommendations. *ACS Energy Lett.* **2023**, *8*, 1952–1958.

142 Gamry Instruments; *Understanding IR Compensation. Application Note Rev. 3 (16 Jan 2018), Gamry Instruments, Warminster* 2018. Available under: https://www.gamry.com/application-notes/instrumentation/understanding-ir-compensation/ (last access 4 Oct 2024).

143 D. Britz; IR elimination in electrochemical cells. *J. Electroanal. Chem.* **1978**, *88*, 309–352.

144 D. K. Roe; Overcoming solution resistance with stability and grace in potentiostatic circuits. in *Laboratory Techniques in Electroanalytical Chemistry*, 2nd ed., (P. T. Kissinger, W. R. Heineman, Eds.), Marcel Dekker, New York, 1996, pp. 195–235.

145 C. Amatore, E. Maisonhaute, G. Simoneau; Ultrafast cyclic voltammetry: Performing in the few megavolts per second range without ohmic drop. *Electrochem. Commun.* **2000**, *2*, 81–84.

6

Investigation of Electrochemical Reaction Mechanisms in Liquid Electrolytes

Electrode reactions are heterogeneous chemical reactions and, thus, involve several individual steps, which include transport processes between the electrode and the bulk of the solution (Figure 3.1). In Chapters 2 and 3, we dealt with thermodynamics (change in reactant energy, equilibria) and kinetics (time-dependent change in concentration and current, reaction pathways, and rate-limiting steps) of electrode reactions. Now we want to discuss the details of the reaction mechanisms of dissolved redox-active molecules. This involves a description of the processes at the molecular level in the form of models. In addition to the steps of transport and electron transfer (ET), which have already been discussed in detail, we must now also consider reaction steps in which molecules undergo *chemical* transformations that occur without further charge transfer at the electrode, including equilibria preceding the ET or following reactions of the energy-rich products of the ET. We can assign a transition state to each of the reaction steps, and each step leads to an intermediate or the final product. These concepts emphasize the *chemical* aspects of electrochemistry, the "doing chemistry with electrodes" [1] because an electrochemical ET is linked to sometimes complex molecular chemical reactions.

The molecular interpretation of electrode reactions forms an essential part of the field of "molecular electrochemistry". The reactions triggered by an ET at the electrode can be used for electrosynthesis (Chapter 17), where products are isolated. The elucidation of electrochemical reaction mechanisms plays a central role in the optimization of such syntheses. Mechanistic studies of electrochemical reactions are also of great importance in many areas of inorganic, organic, and organometallic chemistry, since redox-active molecules are relevant to all of these contexts. Furthermore, material properties that have been determined electrochemically support applications of new organic molecules for solar cells, organometallic catalysts, or battery materials, where the stability of the resulting ET products is of particular interest. The uniform basic principles that apply in all of these areas illustrate the "cross-sectional character" of electrochemistry.

The International Society of Electrochemistry (ISE) defines the field of molecular electrochemistry as follows: "Structural and mechanistic aspects of electrode processes of inorganic, metallorganic and organic substances; synthetic applications" [2]. One of the seven divisions of the ISE is dedicated to this subfield of electrochemistry.

After defining the most relevant terms in Section 6.1, we will discuss how a likely reaction mechanism can be derived (Section 6.2) and verified (Section 6.3). In this context, the simulation of electrochemical processes plays an important role. Among the variety of reaction mechanisms described in the literature, we consider some relevant examples in detail (Section 6.4) before we turn to the determination of physicochemical parameters of compounds, materials, and reaction steps in Section 6.5. The use of complementary methods and the combination of spectroscopic and electroanalytical techniques (spectroelectrochemistry) is often necessary to provide unambiguous results (Section 6.6).

☆ Sections 6.1 to 6.6.2 have been contributed by Prof. Dr. Bernd Speiser, University of Tübingen, Germany.

Before studying this chapter, you should have an understanding of the voltammetric methods discussed in Chapter 5, especially Section 5.3.5 on cyclic voltammetry (CV).

Further Reading. The textbook [3] can serve as a reference work on the application of electroanalytical methods. A textbook series specifically addressing the fundamentals and applications of voltammetry is also available [4–6]. A large number of detailed analyses of relevant mechanisms can be found in [7], while the monograph [8] demonstrates the breadth of application of molecular electrochemistry in organic chemistry.

6.1 Reaction Steps of an Electrode Reaction Mechanism

6.1.1 What Types of Processes Need to Be Considered?

In general, we are dealing with the following individual steps in an electrochemical reaction mechanism (key schematic in Figure 3.1b):

1) Transport, mostly diffusion of reactants, intermediates, and products. (In the context of this Chapter, we prevent migration from playing a role in the transport of ionic reactants by adding a supporting electrolyte. By working in quiescent solutions, we eliminate convection as a transport mechanism – however, exceptions are possible).
2) Electron transfer. *At least one* step of the overall mechanism involves the transfer of electrons between the electrode and a reactant, that is, the passage of electrons across the phase boundary between the electrode and the electrolyte solution. This step is also called heterogeneous ET or heterogeneous redox reaction in order to distinguish it from ET steps between molecules dissolved in the liquid electrolyte.
3) Chemical reaction steps, namely preceding reactions that form the redox-active species from an often redox-inactive precursor, or following reactions that are initiated by the ET.
4) Adsorption (specific accumulation of species at the phase boundary).

The combination of the individual steps comprises the electrode reaction.

The chemical reaction steps can be almost any reaction from the whole variety of reactions in organic, inorganic, and organometallic chemistry. Bond formation, bond breaking, or a simple ET are called primitive reaction steps. In contrast, a concerted reaction step involves *several* primitive changes of the molecule in a single transition state. All these reaction steps occur at some distance from the electrode–electrolyte interface, that is, in the solution phase. We therefore call these conversions homogeneous reactions, even if they take place near the electrode surface within the diffusion layer. As a special case, the homogeneous reaction can also be an ET that occurs between freely diffusing molecules (homogeneous ET). Reactions of adsorbed molecules play a particularly important role in electrocatalysis and will be discussed in Chapter 13.

Depending on the type of redox-active molecules, serial and parallel combinations of reaction steps provide an almost unlimited number of possible mechanisms. Therefore, we need to answer the following questions for each reaction:

1) Which mechanism occurs in a specific case, that is, which individual steps are involved and in which sequence do they proceed?
2) How do we characterize the individual steps?

We refer to the answer to the first question as "qualitative" analysis and the answer to the second question as "quantitative" analysis of the process (Table 6.1). We obtain suitable experimental data using the electroanalytical methods discussed in Chapter 5,

Table 6.1 Examples of questions occurring in qualitative and quantitative analysis of electrochemical reaction mechanisms.

Qualitative analysis	Quantitative analysis
– Is the analyte redox active? – Which steps control the rate of the process? – How many electron transfers take place? – What are the intermediates? – Which kinetic order do the individual steps follow? – Are products of the ETs stable? – Are new redox-active molecules formed?	– At which potentials do oxidations and reductions occur? – How fast are transport processes, ETs, and chemical reactions? – Which equilibrium positions are characteristic?
→ Reaction mechanism (model)	→ Physical properties and reaction parameters

with CV (Section 5.3.5) being particularly important. Our discussion will therefore mostly refer to this and similar instrumental techniques.

The aim of the qualitative analysis is to establish a plausible sequence of reaction steps, from the initial reagent(s) to a stable final product, and to define the steps combined in this mechanistic model. The result forms the basis for the quantitative analysis in which we determine characteristic quantities such as formal potentials, rate constants, and equilibrium constants (with numerical value and unit!) for as many individual steps as possible. Which quantities these are exactly depends on the physicochemical details of the steps (Table 6.2). The results lead to an interpretation in terms of physicochemical properties or reaction parameters.

Table 6.2 Typical relationships and corresponding intrinsic quantities for electrochemical reaction steps.

Reaction step	Typical relationship	Intrinsic quantities [typical unit] to be determined
Electron transfer	Nernst Equation (3.29)	$E^{\circ\prime}$ [V],
	Butler–Volmer Equation (3.27)	heterogeneous standard reaction rate constant k° [cm s^{-1}], α (dimensionless)
Transport (diffusion)	Fick's laws, Eqs. (4.110) and (4.120)	D [cm^2 s^{-1}]
Homogeneous reaction	Kinetic rate laws, e.g., Eq. (3.4)	homogeneous rate constant k [s^{-1}], [L mol^{-1} s^{-1}] or others,
	Law of mass action, Eq. (3.9)	Equilibrium constant K, often dimensionless
Adsorption	Isotherms (Section 8.10.4)	maximum interfacial concentration Γ_{max} [mol cm^{-2}], adsorption equilibrium constant β [mol^{-1} L].

For practical consideration, it is sensible to distinguish between parameters that are determined by the chemical system under study and those that can be varied by the experimenter to explore the behavior of the system.

We refer to properties determined by the "electrochemical system" as "**intrinsic quantities**" (belonging to the chemical and electrochemical processes) or **reaction parameters**. This includes the cell with electrodes and all chemical components of the electrolyte (solvent, supporting electrolyte, analyte, additives).

Quantities influencing the course of an electrochemical experiment from the outside of the electrochemical system can typically be controlled by the experimenter. They are called **"extrinsic quantities"** or **"control parameters"**. Examples are temperature and the rate at which the applied potential is changed.

Table 6.3 Examples of the notation for electrode reaction mechanisms.

Mechanism	Naming
$A + e^- \rightleftharpoons B$ ET differentiated by reversibility	E_{rev} or E_{qr}
$A + e^- \rightleftharpoons B$ $B \rightarrow C$ ET with irreversible following reaction	EC or E$\underline{C}$
$A \rightleftharpoons B$ $B + e^- \rightleftharpoons C$ Preceding chemical equilibrium	CE or $\underline{\underline{C}}$E
$A + e^- \rightleftharpoons B$ $B \rightarrow C$ $C \rightleftharpoons D + e^-$ Two ETs linked by an irreversible chemical reaction, with two different directions of charge transfer	$\vec{E}C\overleftarrow{E}$

6.1.2 Nomenclature for Electrode Reaction Mechanisms

For the effective, yet compact naming of mechanisms in molecular electrochemistry, an IUPAC recommendation exists [9]. However, this is not strictly applied and often extended, for example, to express specific reaction details. Therefore, we describe here only some generally and frequently used symbols from this nomenclature (Table 6.3).

ET reactions at the electrode are marked with the symbol E, chemical reaction steps with C. In case the chemical reaction is a dimerization, the symbol D is also used. Mass transport is not included in the nomenclature. Sequential reaction steps are written with an appropriate sequence of symbols. Additionally, it has been recommended to use a "+" between two reaction symbols for sequential processes and omission of the "+" to indicate concerted processes [9]. However, this is rarely followed in the literature. Parallel reaction sequences can be indicated by parentheses.

The basic symbols E and C are adapted in several ways. For example, the reversibility, quasireversibility, or irreversibility of an ET is specified by subscript indices ("rev", "qr", or "irr"). However, it is also suggested that the symbol be written with *one* (irreversible) or *two* (reversible) underscore(s) [10]. The reaction order can be encoded by subscript numbers. Subscripts have also been used to indicate which species react with each other, where the number is assigned to the species in the order of their occurrence in the reaction sequence. Examples can be found in Tables 6.17 and 6.18. The direction of an ET is indicated only in special cases, for example, when the overall reaction contains both an oxidation step *and* a reduction step. For this purpose, we use arrows pointing to the right or left *above* the symbol E. Sometimes, reaction steps that are *not* rate-determining are written in lower case.

Closer qualifications of a C-step are made with DISP for disproportionations or SET (SET) for a homogeneous (solution) ET. Some complex mechanisms are referred to in a figurative way, such as "square", "ladder", or "fence" schemes (examples in Section 6.4.5). Finally, a symbol with the suffix "ads" indicates that the corresponding reaction step takes place in the adsorbed state.

6.2 From the Experimental Voltammogram to a Model Mechanism

When discussing CV in this chapter, we mean *staircase* cyclic voltammetry in most cases (SCV, Section 5.3.9). SCV can be regarded as a digital implementation of CV and is found in computer-controlled electrochemical workstations. The use of SCV instead of CV is justified since both methods give analogous results for diffusing reactants under practical experimental conditions, that is, in the *absence of adsorption phenomena*. A detailed comparative discussion can be found in [11, 12].

6.2.1 How Should Experiments Be Conducted?

The most important method for elucidating reaction mechanisms is certainly CV (Section 5.3.5). CV may be supplemented by other dynamic techniques such as chronoamperometry (CA), which is also used in the variant of chronocoulometry. In the latter case, the charge $Q = \int_0^t I(t)\mathrm{d}t$ is recorded instead of the current I, that is, the integral is plotted and analyzed as a function of time (for details, see Section 6.5.3). In addition, double potential step CA is of interest. Here, the potential is not only stepped at the beginning of the experiment to a value where the redox process is initiated (as in CA), but also switched to another constant value after a certain time. Thus, the initially generated products of ET can be detected. Other complementary techniques will be discussed in Section 6.6.

In CV and other electrochemical methods that we use to elucidate reaction mechanisms, the three quantities E, I, and t are closely interrelated. In the example of CV (Section 5.3.5), the time-dependent potential of the electrode $E = f(t)$ (Figure 5.26a) is used as the excitation signal of electrochemical processes ("excitation function" or input function). The resulting "response" (or output function) of the experimental arrangement is the time-dependent current $I = f(t)$. The experimental trace is conventionally represented as a cyclic voltammogram $I = f(E)$.

Excitation is understood here in the sense of "stimulation" of a chemical reaction, not in the sense of "electronic excitation", that is, a change in the electronic structure without transfer of charge (as in spectroscopy).

Mostly, cells with a three-electrode arrangement are used. Only the processes at the working electrode (WE, Section 5.3.1) are considered. In molecular electrochemistry, not only aqueous electrolytes but also organic electrolyte systems are used to ensure sufficient solubility, and to exclude the reaction of reactive intermediates with the proton donor and nucleophile water (Section 5.3.3). We refer to the redox-active chemical compound to be investigated as the analyte (investigated substance, substrate). The analyte concentration is often in the range 10^{-5} mol L^{-1}. High demands are therefore put on the purity of solvent and supporting electrolyte, since these are present in a very large excess over the analyte (details in Sections 6.5.6 and 17.1.6 with references).

One general approach toward deciphering the reaction mechanism is a systematic variation of the extrinsic quantities of our experiment and the observation of the resulting changes, which will depend in a specific way on the reaction mechanism in place. Apart from selecting the experimental setup and the identity of the analyte, we can control (set) method-dependent quantities such as the potential range investigated, the potential scan rate v in CV, or the duration of the potential step in double potential step CA. In addition, there are control parameters generally relevant for chemical reactions, such as the temperature T and the analyte concentration c^* (Section 6.5.6). The experiment can also be influenced systematically by the addition of such reagents that react with the primary products of the ET (or with the starting compound), and by the concentration of reagents.

We emphasize the importance of those quantities which define the timescale of the experiment, for example, v in CV. We use the experimental timescale to define a window for the rates of the observable (chemical) reaction steps. If the timescale of the experiment is too short (v too large), slow processes can only proceed incompletely or not at all and, therefore, remain hidden (Section 3.2.1). Conversely, if the experimental timescale is too long (v too small), intermediates may react before we are able to detect them. This already shows that the variation of the timescale is of paramount importance for the analysis of reaction mechanisms (Section 6.5).

While purely electrochemical techniques are well suited for obtaining mechanistic, thermodynamic, and kinetic information, it should be emphasized already at this point that they generally cannot provide **direct information about the structure of reagents and products.**

Structural properties can be inferred only by spectroscopic methods, whereby direct coupling of spectroscopy and electrochemical measurements in *a single* experiment is particularly effective (spectroelectrochemistry, Section 6.6.3). Further approaches to deepen the understanding are the comparison of results for systematically varied reagent molecules or the isolation of electrode reaction products and their subsequent structural analysis. Recently, the latter has also become feasible in miniaturized form (Section 6.6.1 and [13]).

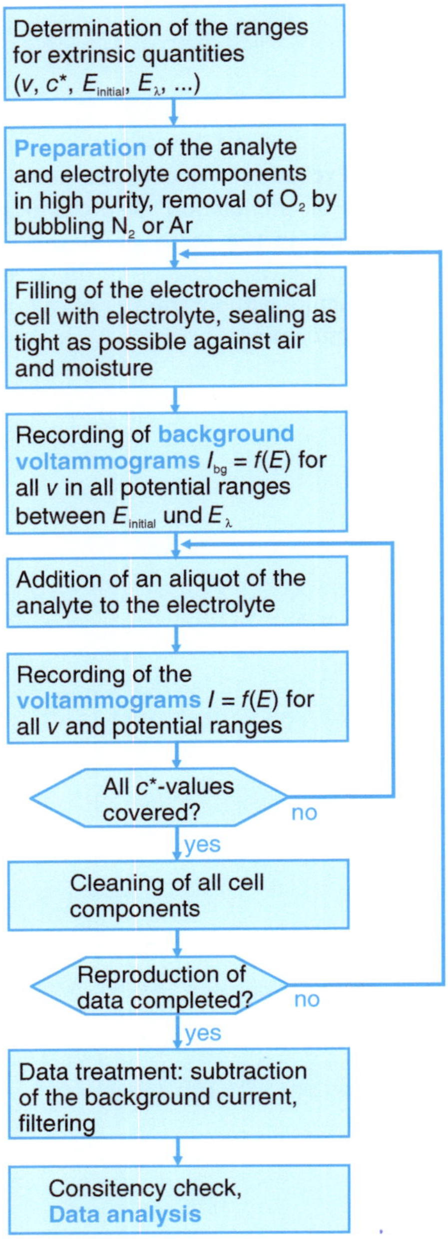

Figure 6.1 Exemplary flow chart for dynamic electrochemical experiments using the example of CV.

Background subtraction is possible only if the analyte does not cause any change in the background (capacitive) current through adsorption of electrolyte components or deposition of reaction products on the electrode ("fouling"). In most cases, the software of the electrochemical workstation also requires identical potential values for all data points of the two voltammograms to be processed.

6.2.2 How to Design a Set of Experiments

For the analysis of reaction mechanisms and the determination of physicochemical parameters, we need high-quality data. Here, we provide some guidelines for planning voltammetric experiments so that we can derive qualitative information about the reaction mechanism, determine physicochemical parameters, and detect systematic deviations between a hypothetical mechanism and the reaction under study. We will return to a more detailed coverage of this topic at the end of the chapter (Section 6.5.6).

A reasonable experimental workflow is shown in Figure 6.1 as a flow chart for CV. We assume that the potential range between the initial and the switching potentials E_{initial} and E_λ in which the analyte is redox active has been determined in previous exploratory voltammograms. Such experiments probe the entire accessible "electrochemical window" of the electrolyte (see Section 6.5.6) and do not have to meet the highest quality requirements yet. Apart from the potential range, the experimenter also defines a relevant concentration range for the chemical components of the system.

After preparing the electrolyte components (analyte, solvent, supporting electrolyte, any additives) and the cell (with all electrodes), we first record background voltammograms in the absence of analyte. Background currents $I_{\text{bg}} = f(E)$ are later subtracted from the voltammograms measured in the presence of the analyte. The same potential ranges and the same values of v must be applied for the background voltammograms and the later measurements in the presence of the analyte. Since we want to cover a wide range of v, it is advised to vary v in 1–2–5 steps (0.1, 0.2, 0.5 V s^{-1} etc.) over several orders of magnitude. It may be useful to select different potential ranges, for example, to separate *one* out of two stepwise ETs (wave clipping, Sections 6.5.6 and 6.5.7). This is exemplified in [14] for the investigation of benzophenone reduction.

After recording all background voltammograms, an initial aliquot of the analyte is added. Now, we record voltammograms (for all v) containing background current *and* a faradaic current component originating from the electrolysis of the analyte. This is then repeated after further analyte additions to the electrolyte. The resulting voltammograms can be arranged as a matrix of experiments at different, defined combinations of v and c^*.

For an independent repetition of the experiment, the cell is emptied. The cell body, electrodes, and other components are carefully cleaned. Then the cell is refilled with electrolyte, and all measurements are repeated for the pure electrolyte and after the addition of the analyte.

After completing all desired measurements, the obtained data are postprocessed: We subtract background current curves $I_{\text{bg}} = f(E)$ from the curves $I = f(E)$ recorded in the presence of analyte for matching control parameters (v, potential range, T, ...) and, if necessary, filter out spurious frequencies (Section 6.5.6). The resulting data set is checked for consistency and analyzed according to the techniques described in Sections 6.2.3 to 6.2.5 and 6.5.

6.2.3 How to Define and Identify Reversibility in the Context of CV

In Section 5.3.5, we have already discussed the most important experimental quantities that we can extract directly from a cyclic voltammogram, namely peak potentials E_{p} and peak currents I_{p}. We will use these quantities as well as diagnostic criteria derived from them first to test the consistency of the data, and then to identify basic elements of the presumed (hypothetical) reaction mechanism.

A diagnostic criterion makes a qualitative or quantitative statement about the change in voltammograms (or other electrochemical measurements) when changing one or more control parameters (one or more of the extrinsic quantities).

It often turns out that one of the experimental quantities or some derived quantity remains constant or has a simple relationship with extrinsic quantities. An example of such a derived quantity is the difference in peak potentials ΔE_p in CV obtained from the peak potentials E_{pa} and E_{pc} of a peak couple resulting from *one* simple redox process.

$$\Delta E_p = |E_{pa} - E_{pc}| \tag{6.1}$$

ΔE_p – Difference in peak potentials of a redox couple in CV [V]
E_{pa}, E_{pc} – Peak potentials [V]; c for cathodic, a for anodic

As follows from Eq. (5.45), we easily calculate the ratio of the peak currents, typically by taking the ratio of the *absolute* current values in the reverse scan (numerator) and the forward scan (denominator).

For the investigation of an oxidation reaction, the peak current ratio would be $|I_{pc}|/|I_{pa}|$. Of course, there are also many important reactions, where the first ET is a reduction. The peak current ratio would then be $|I_{pa}|/|I_{pc}|$. This situation is often summarized in the literature by using the term $|I_p^b|/|I_p^f|$ [b = back (or reverse) scan; f = forward].

The peak currents are usually not read against the zero current line but against a baseline that is constructed as explained in Figure 5.30.

Many relationships for oxidation and reduction reactions can be easily converted into each other. For brevity and clarity, we almost always assume in the following that we first oxidize the analyte in the forward scan and then reduce it in the reverse scan.

One must therefore *always* first check the formulas in this Chapter for the direction of reaction and adjust them if necessary before applying them to your task.

Peak current ratios can also be calculated between the currents of two redox processes not directly linked as forward and reverse ETs, for example, for the two peaks in the forward scan when two successive ETs occur in the EE mechanism (Figure 6.3a).

The peak current itself can be transformed into a normalized form $I_p/(c^* \sqrt{v})$ according to Eq. (5.47). This yields a new quantity in which the dependence on concentration (linear) and scan rate (root) has been eliminated. These two control parameters are easy to vary in practice and are, therefore, particularly valuable for experiments (Section 6.2.1 and Figure 6.1).

For the case of a reversible ET process already discussed in Section 5.3.5, Table 6.4 gives a summary of how the experimental and derived quantities behave as diagnostic criteria. This set of four relations is used to determine whether a process under investigation is classified as electrochemically and/or chemically reversible.

Table 6.4 Diagnostic criteria for a reversible ET (mechanism E_{rev}).

$E_p = const$, in particular $E_p \neq f(v, c)$	(6.2)
$\Delta E_p = 58\ \text{mV}/n$	(6.3)
$I_p/(c^* \sqrt{v}) = const$, in particular $I_p/(c^* \sqrt{v}) \neq f(v, c)$	(6.4)
$I_p^b/I_p^f = 1.0$ of course $I_p^b/I_p^f \neq f(v, c)$	(6.5)

Electrochemical reversibility means that the ET is fast with respect to diffusion, that is, the current is limited by the diffusion process. We also say "the electrochemical process is diffusion controlled" or "the process is diffusion limited". This is diagnosed in particular by the independence of ΔE_p and $I_p/(c^* \sqrt{v})$ on v and c^*.

Chemical reversibility is associated with the absence of any following reaction that removes the product of the ET reaction. Therefore, this product is available for the reverse ET reaction, although only a fraction of the molecules will return to the electrode as dictated by the diffusion processes (see the discussion on the "concentration scar" in Section 5.3.5). This property is checked by the value of $|I_p^b|/|I_p^f|$, where a value of unity indicates chemical reversibility.

Deviations from the diagnostic criteria provide valuable information about differences between the investigated process and the simple reversible model. Such criteria not only exist for the reversible ET process but also for more complex mechanistic models.

For an electrochemically reversible process, *both* criteria for the peak potential difference *and* the current must be fulfilled (Section 6.4.1).

Excursion 6-1 Why and How Do We Detect and Minimize Ohmic Losses?

The diagnostic criteria that refer to reversibility or its absence are only valid for data that are not compromised. In real experiments, data may be contaminated with artifacts, which render the criteria unusable or cause erroneous interpretations in the worst case.

One such experimental artifact is the *IR* drop (or ohmic loss, ohmic drop; Section 5.3.1). In particular, the interpretation of cyclic voltammograms requires the detection and minimization of *IR* drop and, thus, the systematic control of *IR* drop contributions to data quality and consistency. From Eq. (6.3), we see that ideally the quantity ΔE_p is independent of v and c^* for a reversible system. For most other systems, we can assume that ΔE_p is independent of at least c^* if determined without artifacts. However, this is not true for the dependence on v, as already seen for the case of quasireversibility (Sections 5.3.5 and 6.4.1) for which ΔE_p increases with v.

I – Current [A].

R_{uc} – Uncompensated resistance [Ω] between working and reference electrodes (Section 5.3.1).

Experimental data for ΔE_p are, however, additionally influenced by the *IR* drop. An uncompensated ohmic drop causes an increase in ΔE_p with *v and* c^*, which may hamper the application of ΔE_p as a diagnostic criterion.

What are the reasons for ohmic drop effects on the peak potential difference? If a current passes through the working and counter electrodes, a voltage drop $U_{drop} = I\ R_E$ is created (Section 5.3.1). As can be seen in Figure 5.18, even in a three-electrode cell configuration controlled by a potentiostat, an uncompensated resistance R_{uc} exists between the working and the reference electrodes (Ref) and, thus, the potential that the potentiostat applies to the WE deviates from the desired value by $I\ R_{uc}$ (Section 5.3.1). With increasing currents, the product $I\ R_{uc}$ will also increase (assuming that R_{uc} is constant and positive). As a result, the potentiostatic control error U_{uc} becomes larger, and we observe a distortion of the voltammogram. The error reaches a maximum at the peak, $I_p\ R_{uc}$, where the current is at a maximum. Anodic peaks shift to higher (i.e., more positive or less negative) potentials since the positive current I_p is multiplied by R_{uc}. Likewise, cathodic peaks with a negative peak current shift to lower (i.e., less positive or more negative) potentials. In summary, ΔE_p increases with R_{uc}.

See the sign convention for currents in Section 1.4.

Upon an increase in scan rate v, the current increases according to Eq. (5.47) with the square root, and so does the distortion of the current-potential curve. Unfortunately, the resulting increase in ΔE_p with v qualitatively resembles the increase caused by a slow ET step (see Sections 5.3.5 and 6.4.1; quasireversibility). It is thus impossible to distinguish between the presence of an ohmic drop and quasireversibility based on the scan rate dependence of ΔE_p alone.

On the other hand, $I \cdot R_{uc}$ (and thus $I_p\ R_{uc}$) is directly proportional to the concentration [Eq. (5.47)]. The control error and, as a result, ΔE_p increase linearly with c^*. This holds strictly for reversible, quasireversible, and irreversible systems without chemical steps, and also for many, but not all, other mechanisms. The criterion $\Delta E_p \neq f(c^*)$ may then be used to check the absence of significant *IR* drop contributions. If an appreciable effect of the *IR* drop is detected in such experiments with a variation in c^* [$\Delta E_p = f(c^*)$], one needs to mitigate the effects of U_{uc} by avoiding $I\ R_{uc}$ (proper positioning of the reference electrode, reduction in peak currents, increase in electrolyte conductance), or use instrumental compensation (Section 5.7.2).

If we *expect* a concentration dependence of ΔE_p for a certain mechanism based on the reaction steps, this type of analysis is, of course, not applicable.

The optimal position of the reference electrode should be selected close to the WE (Section 5.3.1), however, without shielding the transport processes. If the reference electrode is constructed as a Haber–Luggin[56,57] capillary (Figure 5.24), the capillary tip should be about 1–2 mm from the WE surface. Peak currents can be reduced by lowering the concentration of the analyte, the electrode surface area, or the scan rate. Of course, this is limited by the experimental facilities and the details of the experiment being conducted. The conductance of the electrolyte increases if we increase the concentration of the supporting electrolyte. In the case of organic electrolytes, we might have to consider an alternative solvent to increase the solubility of the supporting electrolyte or its dissociation.

As an example, Table 6.5 shows experimental ΔE_p values in cyclic voltammograms of ferrocene (Figure 2.23) in dichloromethane as a solvent with 0.1 mol L^{-1} tetra-*n*-butylammonium hexafluorophosphate (NBu_4PF_6) as supporting electrolyte. This salt is widely used in organic electrochemistry. In such an electrolyte, typical values of $R_{uc} = 1\,k\Omega$ are found for a distance of 1–2 mm between the tip of the reference electrode and the WE. For a WE with a geometric area of about 0.065 cm^2 and ferrocene concentrations of about 0.1 mmol L^{-1}, peak currents of several μA result. From Ohm's law, significant errors follow, $U_{uc} = IR_{uc} \approx 10\,\mu A \cdot 10^3\,\Omega = 10\,mV$.

In fact, the ΔE_p values in Table 6.5 (without applying any measure to control the *IR* drop) increase with concentration, especially strongly at large scan rates. If, on the other hand, the ohmic drop is compensated for in the experiment (Section 5.7.2) by the positive feedback technique, data as in Table 6.6 are obtained for the same system. The compensation of IR_{uc} was limited to 80% to avoid electronic instability of the potentiostat circuit. Even so, the increase in ΔE_p with concentration is almost eliminated in Table 6.6 and, thus, the *IR* drop artifact is essentially absent. ΔE_p increases slightly only at larger *v*.

In practice, the ΔE_p values even at small *v* are often slightly above the value suggested from the theory of CV [15] (58 mV/*n*, with $n = 1$ for the oxidation of ferrocene). This is explained by the use of the SCV technique and depends on details of the measurement (Section 5.3.9). For the data in Tables 6.5 and 6.6, $t_s = 0.25\,t_p$ was used, that is, the current was recorded at a point in time that was at $\frac{1}{4}$ of the step duration t_p, which is an optimal value for nearly reversible ETs without a subsequent reaction.

An additional problem is caused by the fact that the *IR* drop may attain different values at different locations along the electrode surface (e.g., for a circular disk) as a result of locally different current densities (edge effect, Figure 4.50d). The

Table 6.5 Peak potential differences ΔE_p in cyclic voltammograms of ferrocene + 0.1 mol L^{-1} NBu_4PF_6 + dichloromethane *without* compensation of ohmic drop.

	c^*/(mmol L^{-1})		
$v/(V\,s^{-1})$	0.074 ΔE_p/mV	0.18 ΔE_p/mV	0.25 ΔE_p/mV
0.05	61	64	67
0.1	62	66	72
0.2	63	72	78
0.5	74	82	91
1.0	76	92	105
2.0	83	104	121
5.0	100	132	156
10.0	119	159	192

Table 6.6 Peak potential differences ΔE_p in cyclic voltammograms of ferrocene + 0.1 mol L^{-1} NBu_4PF_6 + dichloromethane *with* successfully applied compensation (80%) of ohmic drop.

	c^*/(mmol L^{-1})		
	0.074	0.18	0.25
v/(V s^{-1})	ΔE_p/mV	ΔE_p/mV	ΔE_p/mV
0.05	58	58	59
0.1	58	59	61
0.2	58	61	63
0.5	58	63	65
1.0	65	70	75
2.0	70	73	81
5.0	82	87	91
10.0	88	97	107

nonuniformity of IR_{uc} hampers satisfactory compensation. The sum of all these effects results in the deviations from theory that are still observable in Table 6.6. These values, however, represent a realistic situation and the quality of data that can be reasonably obtained.

The concentration dependence of ΔE_p can also be used in simple cases for subsequent correction of the IR drop [16]. However, this procedure is rather complex and therefore not generally recommended.

6.2.4 Diagnostic Criteria Indicate Building Blocks of Electrochemical Model Mechanisms

We will now use diagnostic criteria to check which individual steps in a mechanism are important and how they relate to each other. In doing so, we will repeatedly use normalized quantities, which link intrinsic and extrinsic quantities with each other and with natural constants. Let us consider a reoccurring example: We have already introduced the Nernst factor $E_{NF} = RT/(nF)$ as an abbreviation in Section 2.2.2. It has the following physical meaning: If an electrode is in equilibrium with a redox system $R \rightleftharpoons O + ne^-$ and if the electrode potential is changed by E_{NF}, then the activity ratio $\{O\}/\{R\}$ must change by the eulerian number e ≈ 2.718. We can take E_{NF} [V] as an alternative unit for the voltage, and use it to derive further quantities. We have already discussed the importance of the scan rate v [V s^{-1}] in CV for the analysis of reaction mechanisms. However, scanning the potential has different effects depending on the stoichiometric number of charges n and the temperature T. We can combine v, n, and T in a new quantity, a,

$$a = \frac{v}{E_{NF}} = \frac{nF}{RT} v \tag{6.6}$$

which has the unit s^{-1} and indicates how many times per second the Nernst factor is scanned in an experiment. The following sections contain some formulas with a particularly large number of symbols. If one combines them in a sensible way, the underlying principles are more easily recognized. We occasionally show the more extended equations in addition to the more compact forms because one may need both of them to understand the literature.

Diffusing Redox Systems. To verify that the experimental signal arises indeed from freely diffusing ions or molecules, we use equations for the current in a cyclic voltammogram or a chronoamperogram. The following applies to the peak current in CV under diffusion control [15]:

$$
\begin{aligned}
I_p &= 0.4463 nFAc^* \sqrt{Da} \\
&= 0.4463 nFAc^* \sqrt{\frac{DnFv}{RT}} \\
&= 0.4463 n^{3/2} F^{3/2} Ac^* [Dv/(RT)]^{1/2}
\end{aligned}
\tag{6.7}
$$

Compare for yourself which notation of Eq. (6.7) is more intuitive for you. The numerical constant in Eq. (6.7) follows from calculations which use the Laplace transformation[46] (Excursion 4-3) or the algorithms in Section 6.3.1 and combine a slow, rate-determining diffusion with a fast (= reversible) ET step.

Q: I have seen other equations for the peak currents of reversible voltammograms in the literature. Is there a reason for this?

A: Equation (6.7) resulted from the work of Nicholson and Shain on the theory of cyclic voltammetry [15]. However, the proportionality between I_p and $\sqrt{v}$ had been recognized in even earlier work. Inserting into Eq. (6.7) $T = 298$ K, $R = 8.314$ V A s K^{-1} mol^{-1}, and $F = 96\,485$ A s mol^{-1}, you obtain

$$
I_p = 2.69 \times 10^5 \, \frac{\text{A s mol}^{-1}}{\sqrt{\text{V}}} n^{3/2} A c^* D^{1/2} v^{1/2} \tag{6.8}
$$

This equation is often given without the units of the prefactor as

$$
I_p = (2.69 \times 10^5)\, n^{3/2} A c^* D^{1/2} v^{1/2} \tag{6.9}
$$

where A must be inserted in cm^2, D in cm^2 s^{-1}, c^* in mol cm^{-3} and v in V s^{-1}. In this book, we usually avoid equations that give correct results only with a specific set of units because you would have to memorize the prescribed units in addition to the equation, without this being apparent from the algebraic expression itself. The book authors generally and strongly discourage the use of such equations because the preference for certain units is continuously changing (but the underlying physics is not!!). For example, in the context of simulation environments, there is currently a trend to use quantities exclusively and strictly only with SI base units, for example, D in m^2 s^{-1}. Eq. (6.7) still gives a correct result then (as long as the units are converted and harmonized correctly). Doing the same for Eq. (6.9) is potentially troublesome because implicit and then sensible conventions about the units made in the past are not evident, may not be reasonable anymore, and may not be deduced easily from the context. Generally, the term Randles–Ševčík equation[64,65] refers to the proportionality between I_p and $\sqrt{v}$.

Note also the sign of I_p: Strictly speaking, Eq. (6.7), as given here, is valid only for an oxidation. For a reduction, we use a negative sign for the current

[64] John Edward Brough Randles, 1912–1998, professor in Birmingham, developed, among other things, the first potentiostat with which voltammograms could be recorded (J. E. B. Randles, *Trans. Faraday Soc.* **1948**, 44, 322–327). He also developed the Randles circuit for interpreting electrochemical impedance spectra.

[65] Augustin Ševčík, 1926–2006, Czech electrochemist, developed the equation as part of his Ph.D. thesis; A. Ševčík, *Coll. Czech. Chem. Commun.* **1948**, 13, 349–377.

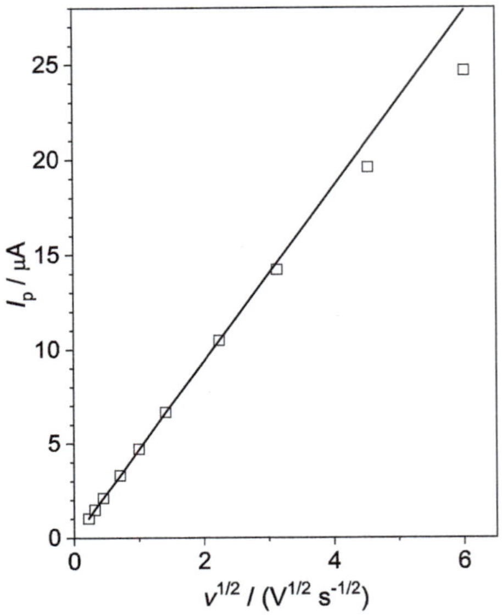

Figure 6.2 Anodic peak currents I_p as a function of $\sqrt{v}$ in cyclic voltammograms of 0.06 mmol L^{-1} ferrocene + 0.1 mol L^{-1} $NBu_4PF_6 + CH_2Cl_2$; A = 0.062 cm^2.

(by definition). More exactly, we should write Eq. (6.7) as "$|I_p|$ = ..." which, however, is very uncommon. This "sloppiness" brings up another semantic problem. The statement "The peak current increases as v increases." is not unambiguous. This is because the numerical value of the current becomes smaller for a reduction peak (negative number with larger magnitude, for example, $-3 > -5$, but $|-3| < |-5|$). Again, we would have to say more accurately, "The absolute value of the peak current increases as v increases." These formulations inflate texts without really communicating content. Therefore, we only occasionally refer in this book explicitly to the absolute value of the peak currents to avoid this uncertainty. "Decreasing currents" therefore means decreasing absolute currents for a reduction.

Table 6.7 EC mechanism.

First step is oxidation	
$A \rightleftharpoons B + e^-$	(6.10)
(identical to $A - e^- \rightleftharpoons B$)	
$B \rightarrow C$	
First step is reduction	
$A + e^- \rightleftharpoons B$	(6.11)
$B \rightarrow C$	
General formulation	
$A \pm e^- \rightleftharpoons B$	
$B \rightarrow C$	(6.12)

We formulate the electrochemical step in Eq. (6.12) such that it can be understood as an oxidation *or* a reduction: "±" indicates either electron uptake or electron release. Unless otherwise indicated, we select as examples cyclic voltammograms in which analyte A is *oxidized* first. In the case of a reduction as the first step, voltammograms would appear as "rotated by 180°".

Equations (6.7) and (6.9) have in common that the current is proportional to $\sqrt{v}$. Therefore, a plot of I_p versus $\sqrt{v}$ gives a straight line (Figure 6.2). At small scan rates up to $v \approx 5\,V\,s^{-1}$ ($\sqrt{v} = 2.25\,V^{1/2}\,s^{-1/2}$), the data points in this example follow exactly the straight line according to Eq. (6.7). Thus, under the experimental conditions and at these time scales, diffusing particles undergo ET at the electrode. At larger v, the current deviates from the straight line *to smaller values*. An explanation follows from the relative rates of the subprocesses: At small v, the process is electrochemically reversible, that is, the ET step is very fast and diffusion, as the slowest substep, determines the total conversion and the current. Transport is the rate-determining step in the sequence of steps of the electrode reaction (Figure 3.1b). As v increases, the diffusion rate also increases because the concentration profiles at the electrode become steeper. Diffusive transport is then accelerated according to Fick's first law [Eq. (4.110)]. At $v > 5\,V\,s^{-1}$, diffusion becomes so fast that the rate of the ET step [Butler–Volmer Equation (3.27)] becomes significant ("quasireversible" process). Now, diffusional transport and ET are of comparable rates. The deviation of the peak currents from the straight line to smaller values is consistent with the fact that now the ET limits the current, and therefore not all the reactant molecules delivered by diffusion are immediately converted at the electrode. The Randles–Ševčík Equation (6.7) no longer applies. Only in the extreme case of very slow ETs ("irreversible" process, Section 6.4.1) do we observe linearity between I_p and $\sqrt{v}$ again, but with a smaller proportionality factor than in Eq. (6.7). This range is not reached in Figure 6.2 because the studied ferrocene/ferrocenium (Fc/Fc^+) redox system is characterized by a relatively fast ET reaction.

Similarly, CA and a plot according to the Cottrell Equation (4.130) and Figure 4.53, which also describes a diffusion-controlled process, can be used to qualitatively verify reversibility. Deviations from reversible ET behavior result in nonlinear Cottrell plots.

Immobilized Redox Systems. As we saw in Eq. (5.93), the ideal faradaic current response of an immobilized redox system in CV is proportional to v. In such a case, plotting I_p against v would yield a straight line, but plotting against $\sqrt{v}$ would not. This is, therefore, a diagnostic criterion to check whether the redox-active particles diffuse or are immobilized. Note that many other factors may influence the peak current in immobilized systems. This will be considered in more detail in Chapter 13 in the context of modified electrodes.

Redox Systems with Following Reactions. The "chemical irreversibility" mentioned in Section 6.2.3 or the presence of only partially reversible following reactions can be checked by analysis of the peak current ratio $|I_p^b|/|I_p^f|$ in cyclic voltammograms. This criterion takes on a value of unity for cases where a subsequent reaction is *not* relevant.

We consider the EC mechanism in Table 6.7. The starting compound A is converted electrochemically to B. As long as the rate of the following reaction to product C is negligible, $|I_p^b|/|I_p^f|$ remains equal to one, as in the case of the reversible ET. However, if the product B of the redox step reacts in a significant amount, the peak current ratio drops below one because the amount of B generated is no longer fully available for the reverse ET to form A. From $|I_p^b|/|I_p^f| < 1$, we then conclude the presence of such an irreversible following reaction. We find here an interplay between the rate of the following reaction and the timescale of the experiment, which is defined by v. A short timescale (large v) hides the following reaction, the system appears chemically more reversible, and $|I_p^b|/|I_p^f$ approaches unity. In contrast, the following reaction has a stronger effect at a long timescale (small v) and $|I_p^b|/|I_p^f|$ has smaller values.

Redox System with Multiple Electron Transfers. We will learn more about diagnostic criteria in Section 6.4, where we discuss the most important models for electrode reactions in liquid electrolytes. In the present context, we now consider the complications arising when multiple ETs are linked, either as in the EE mechanism (Table 6.8) or by intermediate chemical reaction steps as in the ECE mechanism (Table 6.9).

In simple cases of the **EE mechanism** (Table 6.8), each ET results in a separate signal as in Figure 6.3a. This is the case when the formal potentials $E_1^{\circ\prime}$ and $E_2^{\circ\prime}$ of the two ET steps are separated enough (i.e., by more than 200 mV). This situation is often observed because the transfer of the second electron is usually thermodynamically less favorable (energetically more expensive) than the transfer of the first electron. In a vacuum, a difference of about 5 V is estimated for the introduction of the first versus the second charge to a neutral reagent molecule. This value, however, is reduced in an electrolyte by solvation effects to a few 100 mV [17] ("normal potential order"). We denote the difference in formal potentials as $\Delta E^{\circ\prime}$. The cyclic voltammogram is then the sum of the two virtual voltammograms for the two one-electron transfers at $E_1^{\circ\prime}$ and $E_2^{\circ\prime}$. The diagnostic criteria of the E_{rev} mechanism apply to both peak couples (Table 6.4, reversible ET). For example, the peak potential differences are easily obtained – as in the case of a single ET.

When determining the peak currents in such cyclic voltammograms, we must note that – except for the "first peak" (I) – a suitable baseline must be selected for all other signals. This is *not* the line of zero current! The process is analogous to Figures 5.29 and 5.30. For the second peak (II), the extrapolated decreasing current of the first peak forms the baseline (Figure 6.3). Then, the ratio of the two peak currents during the forward scan is $|I_p^f(\text{II})|/|I_p^f(\text{I})| = 1$ (Figure 6.3a). The Randles–Ševčík Equation (6.7) is strictly valid only for the peak current of the first peak $I_p^f(\text{I})$. For $I_p^f(\text{II})$, it can only be used if the zero line is correctly defined as above. The typical peak current ratio for the first peak couple cannot be determined because of the lack of a suitable baseline for the reverse peak.

In the **ECE mechanism** (Table 6.9), the relative size of the peaks depends on the position of the equilibrium B $\rightleftharpoons$ C (characterized by the equilibrium constant K) and how fast it is established (rate constant k). Examples can be found in [18]. More details will be discussed in Section 6.4.4.

Let us now consider the peak current ratio $|I_p^f(\text{II})|/|I_p^f(\text{I})|$ in Figure 6.3b. The equilibrium B $\rightleftharpoons$ C is assumed to be shifted far to the side of C, because K is much larger than one in this example. The size of the second peak in the cyclic voltammogram depends on the amount of C produced. As a result of the relatively slow rate of conversion of B to C, little C is formed during the course of the experiment. Therefore, $I_p^f(\text{II})$ is much smaller than $I_p^f(\text{I})$ and thus $|I_p^f(\text{II})|/|I_p^f(\text{I})| < 1$. Another feature in the voltammogram of Figure 6.3b is the small reduction peak of the first peak couple, at about 0.3 V. This is because species B, which would be reduced here, is removed by the equilibrium B $\rightleftharpoons$ C that is shifted to C. Thus, only a few molecules of B remain available for reduction.

Table 6.8 EE Mechanism.

First step is oxidation	
$A \rightleftharpoons B + e^-$	
(identical to $A - e^- \rightleftharpoons B$)	
$B \rightleftharpoons C + e^-$	
(identical to $B - e^- \rightleftharpoons C$)	(6.15)
First step is reduction	
$A + e^- \rightleftharpoons B$	
$B + e^- \rightleftharpoons C$	(6.16)
General formulation	
$A \pm e^- \rightleftharpoons B$	
$B \pm e^- \rightleftharpoons C$	(6.17)

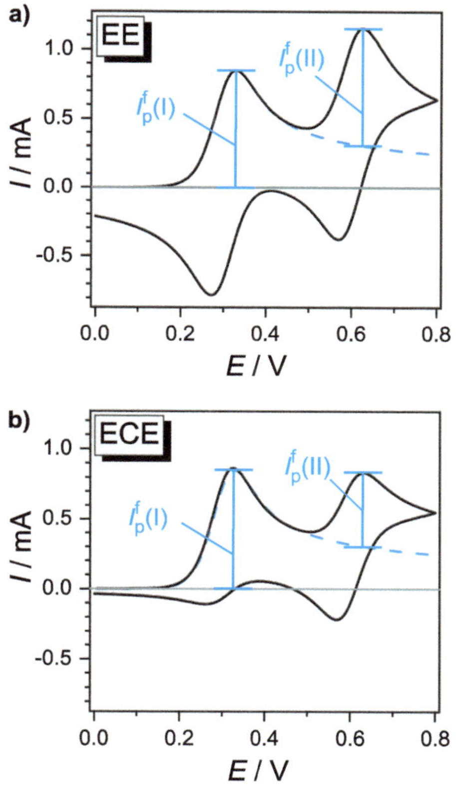

Figure 6.3 Cyclic voltammograms of two-electron transfers at $v = 1\ \text{V s}^{-1}$ according to a) the EE and b) the ECE mechanism. $E_1^{\circ\prime} = 0.3$ V, $E_2^{\circ\prime} = 0.6$ V, $c^* = 1\ \text{mmol L}^{-1}$; for ECE, $K = 10^5$, $k = 8\ \text{s}^{-1}$ (definitions of symbols in text). The dashed curves are for a reversible single ET with $E_1^{\circ\prime}$ and define the baseline for peak current determination of peak II; both ETs are assumed to be reversible.

Table 6.9 ECE mechanism.

First step is oxidation	
$A \rightleftharpoons B + e^-$	
(identical to $A - e^- \rightleftharpoons B$)	
$B \rightleftharpoons C$	
$C \rightleftharpoons D + e^-$	
(identical to $C - e^- \rightleftharpoons D$)	(6.18)
First step is reduction	
$A + e^- \rightleftharpoons B$	
$B \rightleftharpoons C$	
$C + e^- \rightleftharpoons D$	(6.19)
General formulation	
$A \pm e^- \rightleftharpoons B$	
$B \rightleftharpoons C$	
$C \pm e^- \rightleftharpoons D$	(6.20)

ψ – Normalized heterogeneous standard reaction rate constant [20] [dimensionless]

$k°$ – Standard heterogeneous reaction rate constant, $k(E = E°')$ [$\mathrm{cm\,s^{-1}}$]

$D = D_R = D_O$ – Diffusion coefficient of oxidized and reduced redox forms [$\mathrm{cm^2\,s^{-1}}$]

v – Potential scan rate [$\mathrm{V\,s^{-1}}$]

π – Mathematical constant pi, ratio between circumference and diameter of a circle [dimensionless]

For $D_O \neq D_R$, the expression

$$\psi = \frac{1}{\sqrt{\pi D_O^{1-\alpha} D_R^{\alpha} a}} k°$$

holds if the forward reaction is an oxidation. $(1 - \alpha)$ is the transfer coefficient for the anodic partial reaction.

In Ref. [21], the parameters are denoted as λ (here κ) and $\Lambda = k°/\sqrt{aD}$ (instead of ψ here). Λ and ψ are similar, but not identical ($\Lambda = \pi^{1/2}\,\psi$).

6.2.5 Zone Diagrams

The simple examples in Section 6.2.4 already demonstrate that the experimental curves, for example, cyclic voltammograms, resulting from certain reaction mechanisms behave in extreme cases as we would expect from simpler mechanisms. For example, the I–E curves for an EC mechanism at large scan rates turn into those of a simple, chemically reversible ET process (E mechanism), which is even electrochemically reversible if the rate constant of the ET is suitably high. This transition into simpler limiting cases can be illustrated with so-called zone diagrams, which are most often plotted in two dimensions, although three-dimensional examples have been constructed for more complex situations [19].

For a two-dimensional zone diagram, we identify two parameters, and we consider their effects on the shape of the curve. In more complex cases, several parameters can sometimes be combined, often at the expense of clarity. For the EC mechanism as an example, we can limit the discussion to two relevant parameters, κ and ψ. It is convenient to link the reaction parameters to our control parameters, for example, the potential scan rate v. A given value of the resulting normalized quantity can result from many combinations of the reaction parameter (e.g., $k°$) and the control parameter (e.g., v). From a rate constant k [$\mathrm{s^{-1}}$] of a homogeneous first-order reaction, we obtain

$$\kappa = \frac{k}{a} = E_{NF}\frac{k}{v} = \frac{RT}{nF}\frac{k}{v} \tag{6.13}$$

From a rate constant k [$\mathrm{L\,mol^{-1}\,s^{-1}}$] of a second-order homogeneous reaction, we choose Eq. (6.14).

$$\kappa = \frac{kc^*}{a} = E_{NF}\frac{kc^*}{v} = \frac{RT}{nF}\frac{kc^*}{v} \tag{6.14}$$

When normalizing the standard rate constant $k°$ [$\mathrm{cm\,s^{-1}}$] of a heterogeneous ET, characteristics of the transport step must be included in the form of the diffusion coefficient D. For simplification, we assume that the diffusion coefficient of the oxidized form D_O and that of the reduced form D_R are equal ($D = D_O = D_R$).

$$\psi = \frac{k°}{\sqrt{\pi D a}} = \frac{k°}{\sqrt{\pi D \frac{v}{E_{NF}}}} = \sqrt{\frac{RT}{\pi D n F v}} \cdot k° \tag{6.21}$$

The normalized quantities illustrate the interplay between the reaction and the control parameters. They also play an important role in simulation (Section 6.3.1) and in the discussion of reaction models (Section 6.4).

To construct a two-dimensional zone diagram, we create a rectangular coordinate system with two (often logarithmic) axes for the two normalized quantities used. Figure 6.4 shows this for the EC mechanism, Eq. (6.12).

The kinetics of the two reaction steps are defined by the dimensioned rate constants $k°$ (ET) and k (homogeneous first-order following reaction assumed to be irreversible). Accordingly, we use κ and ψ as the normalized, dimensionless quantities (defined above). The axes are logarithmic. Each axis tick corresponds to a 10-fold increase in κ or ψ. In total, seven zones are distinguished for the EC mechanism. In each zone, there is a different diagnostic criterion for one particular characteristic quantity. In Figure 6.4, it is the peak potential E_p of the forward scan. The different criteria are connected to different combinations of the values for the normalized quantities κ and ψ.

We start the discussion in the upper left corner of the diagram, where the rate of the following reaction is slow (κ small) and the rate constant of the ET is large (ψ large), This zone is called "diffusion only" (DO), and the peak potential in the voltammograms behaves as in a simple reversible system: Diffusion control prevails and the following

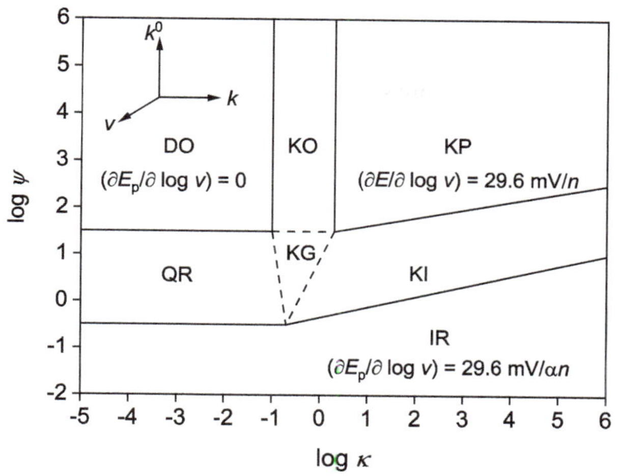

Figure 6.4 Zone diagram for the EC mechanism based on the peak potential E_p for the forward scan (edited, after [21]). The zones are:

DO – diffusion only (diffusion-limited case)
QR – quasireversible case
IR – electrochemically irreversible case
KP – "pure kinetic conditions"
KI – transition from KP to the electrochemically irreversible case
KO – transition from KP to the diffusion-limited case
KG – "kinetic general", all processes have comparable rates

reaction has no effect. The peak potential is then independent of v, which is expressed by $(\partial E_p/\partial \log v) = 0$ as the diagnostic criterion.

Let us now explore the behavior when the ET rate constant becomes smaller, meaning $k°$ and thus ψ decrease. As in zone DO, the following reaction may not have a significant effect because κ remains small. This means, we "walk" down the left side of the zone diagram from top to bottom through the zone of the quasireversible case (QR) into the zone of the electrochemically irreversible case (IR). Here, the rate of ET completely controls the overall reaction rate. This zone is associated with slow ET, and it covers the complete range of κ: Even if there is a chemical following process, this is obscured by slow ET. According to the kinetic bottleneck principle, an acceleration of the following reaction (κ becomes larger) goes unnoticed, because only molecules or ions previously formed in the rate-controlling ET step can react. Within the zone IR, a different diagnostic criterion (Figure 6.4) applies to the variation of E_p with v, which is characteristic of irreversible ETs: E_p shifts by 29.6 mV/(α n) for a 10-fold increase in v (one tick mark on the logarithmic axis; α is the transfer coefficient, Section 3.2.2; the numerical value is valid for 25 °C).

The exact position of the boundary lines between the zones depends on how precisely we can determine the characteristic quantity, here E_p. For peak potentials, 2 mV is often assumed as the experimental uncertainty [21].

When going from zone DO to the right, the parameter κ will increase toward faster homogeneous kinetics of the following reaction. The zone KP represents "pure kinetic conditions" [21], where ET and homogeneous reaction steps are extremely fast and a small "quasi-stationary" or "steady-state" concentration of B is established (top right of the zone diagram). Here, we find a characteristic shift in peak potential with the scan rate of $(\partial E_p/\partial \log v) = 29.6$ mV/n, which defines the zone.

The steady-state approximation (also called the Bodenstein[66] principle) is often used to simplify expressions in chemical kinetics. It states that for an intermediate (here B) whose concentration c_B is small compared to the sum of the reactant and product concentrations ($c_A + c_C$), the temporal change is negligible. Thus, $\partial c_B/\partial t = 0$ [22] and B is approximately at steady-state (quasi-stationary).

The KO and KI zones represent transition zones where kinetic control changes, in KO from diffusion control, in KI from ET control, respectively, to the quasi-steady state (KP).

Finally, there is a range "kinetic general" (KG) in which all reaction steps (diffusion, ET, following reaction) have comparable rates. Here, E_p is affected in a complex way by all steps.

The effect of the kinetic constants $k°$ and k is obvious when comparing different real systems. When changing the chemical system, that is, typically the chemical compound to be investigated (the analyte), we move in the zone diagram *vertically*

[66]Max Ernst August Bodenstein, 1871–1942, German physical chemist and one of the founders of chemical kinetics.

(variation of $k°$) and/or *horizontally* (variation of k). With the control parameter v, we can push a given molecular system (fixed values for $k°$ and k) from one zone to another. Since v affects both ψ and κ for given $k°$ and k, we move *diagonally* within the zone diagram when changing v. In the graph, this is symbolized by correspondingly oriented arrows, which are drawn in the zone DO in Figure 6.4.

Zone diagrams provide a very compact representation of the effects of intrinsic and extrinsic quantities on voltammograms for a particular reaction mechanism. They also play an important role in simplifying complex mechanisms by a suitable choice of experimental conditions (Section 6.5.7).

Many zone diagrams for important electrochemical reaction mechanisms have been compiled in the literature (e.g., in [7]). Complex examples related to electrocatalytic reactions can be found in [23].

6.3 When Is a Model Mechanism Plausible?

Using experiments and qualitative considerations as described in Section 6.2, we formulate a hypothesis in the form of a mechanistic model of the reaction steps at the electrode, in the diffusion layer, and, with respect to chemical reaction steps, inside the bulk phase of the electrolyte. Before we continue with advanced, quantitative interpretations, we will show how to confirm whether this model is consistent and plausible.

To explain the phenomena observed in the experiment, generally the model must be *as complex as necessary* on the one hand, but *as simple as possible* on the other ("Ockham's razor"[67] or "principle of parsimony"). Thus, superfluous reaction steps must be omitted.

Alternative mechanistic hypotheses must be ruled out, if necessary, by additional experiments, sometimes requiring a reexamination of the system under modified conditions or with a new set of control parameters. We will discuss further methods that are used for plausibility testing toward the end of this section.

6.3.1 Simulation (Modeling)

A simulation is the mathematical reproduction (modeling) of a natural process and the solution of the corresponding equations, which is usually done numerically and digitally [24].

Q: What does "numerical" and "digital" mean here? Is it not the same, and are there also "nonnumerical" simulations?

A: Numerical means, "based on numbers". A numerical solution for an equation or a system of equations results from *calculations* with numbers. In contrast, simulations in which real electronic circuits are assembled and varied until they exhibit the same behavior as an electrochemical cell would be nonnumerical. In the precomputer era, such approaches were tested to understand electrode reactions. Other examples, sometimes used for didactic purposes, are networks of containers and tubes, where the flow of liquid is associated with the electric current. The solution of the equation(s) is experimentally determined by *measurement* of a physical quantity, which is not necessarily equal, but rather analogous to the desired result. For example,

[67] William of Ockham, ca. 1288–1347, English medieval philosopher, theologian, and writer.

the liquid flow behaves analogously to the electric current. "Digital" means that the solution (especially of partial differential equations) is not only carried out with numbers, but also stepwise (*discretely*). Nowadays, such a task is executed effectively by means of a digital computer controlled by an algorithm. This contrasts with simulations by means of analog computers, where electric circuits are used for computational operations (basic arithmetics, integration). Such a process would be *continuous*. Analog computers have been supplanted almost exclusively by digital computers today. In terms of epistemology, simulation is considered a third way of gaining knowledge, complementing theory building and experimentation [25].

Electrochemical simulations serve the electrochemist at different levels of understanding and complexity, and address different questions. For example, the electronic structure of adsorbates is studied by solving the Schrödinger equation[68] (first principle or *ab initio* approach), possibly using approximation methods. The results are very accurate but limited to small systems. The motions of molecules and ions in the pores of a polymer membrane, that is, systems with many atoms, become accessible only with significant simplifications considering statistical properties of restricted movements, each with its own probabilities (called "Monte Carlo" method" after the location of a famous casino). Finally, it is common to abandon the notion of individual particles altogether and to calculate properties like the concentration instead, which can only be defined meaningfully for ensembles in a continuous medium. Therefore, such calculations are called continuum simulations. Of particular importance in molecular electrochemistry is the simulation of continuous transport processes near interfaces coupled to the conversion of the transported species at the electrode and in the electrolyte according to kinetic rate laws. We refer to this class of models as reaction-transport models. They can be extended to include other physical processes such as the generation and conduction of heat, possibly calculating the concentrations somewhat less precisely (and less expensively). Such calculations are very important, for example, for the design of technical batteries and fuel cells.

Simulations of reactions and transport processes include – as far as required for the mechanistic model – the steps listed in Section 6.1, especially diffusion to and from the electrode, heterogeneous ET, and coupled chemical reaction steps. The selection of an appropriate physicochemical model is the initial step in a simulation study (Figure 6.5). The further workflow is discussed below.

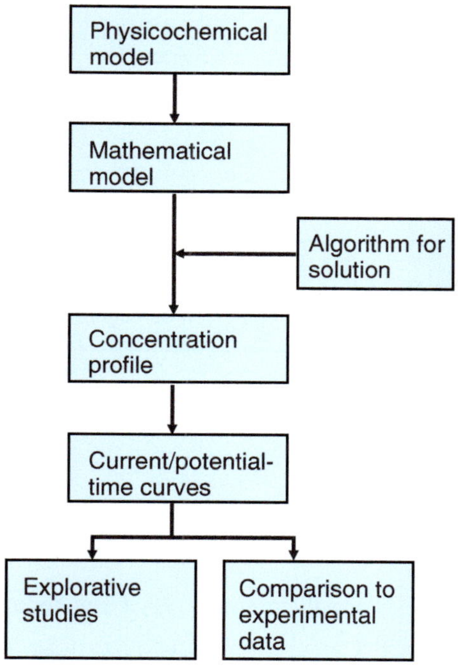

Figure 6.5 Flow chart detailing the steps of a simulation study in electrochemistry.

Building a physicochemical model always simplifies the real situation. For example, convection and migration are usually neglected, although these phenomena are present in a real cell in addition to diffusion. Also, it is often assumed that the electrode has an infinite extension, so that the movement of the redox-active molecules can be simplified as being solely perpendicular to the electrode (planar diffusion, Section 4.5.2). Depending on requirements, however, such restrictions can or must be abandoned – usually with an increase in mathematical complexity.

Let us illustrate these concepts with a simple example, the EC mechanism [Eq. (6.12)]. Species A is assumed to be oxidizable. The electrode is a flat disk with a radius of several mm and is inserted into a solution at rest. For experiments lasting up to approximately one minute, planar diffusion conditions apply, and convection can be neglected. A single space coordinate, x, suffices to describe this situation. Let the electrochemical cell be so large that the solution volume is practically unlimited.

[68] Erwin Rudolf Josef Alexander Schrödinger, 1887–1961, Austrian physicist, professor in Jena, Stuttgart, Breslau (today Wrocław in Poland), Zurich, Berlin, Oxford, Graz, Dublin, and Vienna. Nobel Prize 1933.

Therefore, we call the diffusion *semi-infinite* (Section 5.5.1). Let the supporting electrolyte concentration exceed the concentration of A c^*_A by a factor of 100. Then we do not have to consider migration. At the beginning of the experiment, only A should be present, while B and C are absent.

In the mathematical model, we describe diffusion by partial differential equations (PDEs) derived from Fick's second law [Eq. (4.120)]. We write for the starting compound A (Section (4.5):

D_A – Diffusion coefficient of A [$cm^2\ s^{-1}$]
$c_A(x, t)$ – Concentration of A [$mol\ cm^{-3}$] as a function of space and time
x – Distance from the electrode [cm], space coordinate
t – Time [s]

$$\frac{\partial c_A(x,t)}{\partial t} = D_A \frac{\partial^2 c_A(x,t)}{\partial x^2} \tag{6.22}$$

The space coordinate extends from the electrode surface (located at $x = 0$) into the electrolyte. The time at the beginning of the experiment is $t = 0$, which is the start of the potential scan.

For B, we must consider that this species is not only transported by diffusion, but that its concentration c_B is reduced by the homogeneous following reaction to form C [as shown by the negative sign in Eq. (6.23)]. If we assume a first-order rate law with rate constant k, we find:

D_B, D_C – Diffusion coefficients of B and C [$cm^2\ s^{-1}$]
$c_B(x, t)$, $c_C(x, t)$ – Concentrations of B and C [$mol\ cm^{-3}$]
k – Homogeneous first-order rate constant [s^{-1}]

$$\frac{\partial c_B(x,t)}{\partial t} = D_B \frac{\partial^2 c_B(x,t)}{\partial x^2} - k\, c_B(x,t) \tag{6.23}$$

The term after the minus sign corresponds to the kinetic rate law and describes the removal of B by the chemical reaction.

On the other hand, the concentration of C is increased by the chemical reaction, where the term kc_B with a positive sign describes the effect mathematically:

$$\frac{\partial c_C(x,t)}{\partial t} = D_C \frac{\partial^2 c_C(x,t)}{\partial x^2} + k\, c_B(x,t) \tag{6.24}$$

Equations (6.22) to (6.24) are valid for all x and all t within the diffusion layer. In order to find a unique solution to such a system of coupled PDEs, initial and boundary conditions must be provided. For time $t = 0$, the initial conditions are shown in Table 6.10. We denote the initial concentration of A in the bulk phase by c^*_A, which is usually determined by the initial amount of the starting compound dissolved in a certain volume of the electrolyte solution before the beginning of the experiment. The expressions for c_B and c_C in Eqs. (6.26) and (6.27) show the absence of species B and C before the onset of the potential scan.

Table 6.10 Initial conditions for the simulation of cyclic voltammograms.

$c_A(t = 0) = c^*_A$	(6.25)
$c_B(t = 0) = 0$	(6.26)
$c_C(t = 0) = 0$	(6.27)

For selected positions along the x-coordinate, the boundary conditions in Table 6.11 apply. For instance, the concentrations remain at the value of the initial concentrations for all t at a large distance from the electrode ($x \to \infty$). This is a consequence of the assumption of semi-infinite diffusion: mass transport has no significant effect for $x \to \infty$. Mathematically, boundary conditions, which define values of the desired function $c(x,t)$ *directly*, are referred to as Dirichlet[69] boundary conditions.

Table 6.11 Boundary conditions for the simulation of cyclic voltammograms.

$c_A(x \to \infty) = c^*_A$	(6.28)
$c_B(x \to \infty) = 0$	(6.29)
$c_C(x \to \infty) = 0$	(6.30)

On the other hand, we consider the Nernst equation as a boundary condition at the electrode surface ($x = 0$),

$E(t)$ – Time-dependent electrode potential [V]
$E^{\circ\prime}$ – Formal potential [V] of the redox couple $A \ll B + ne^-$
$E_{NF} = RT/(nF)$ – Nernst factor [V]

$$E(t) = E^{\circ\prime} + E_{NF} \ln \frac{c_B(x=0)}{c_A(x=0)} \tag{6.31}$$

which links the concentrations of the redox-active species A and B at $x = 0$ to the potential $E(t)$ at a given time defined by the excitation or input signal of the experiment, that is, a triangular potential ramp for CV.

Other boundary conditions concern the material flux at the electrode: Any A that reaches the electrode and is converted into B by the ET reaction must diffuse away

[69] Peter Gustav Lejeune Dirichlet, 1805–1859, German mathematician, professor in Berlin and Göttingen.

from the electrode as B. In the present model, as a consequence, the material fluxes of A (to the electrode) and B (from the electrode to the bulk of the electrolyte) at $x = 0$ must have the same absolute value with opposite sign [Fick's first law, Eq. (4.110)]. Boundary conditions that include the *derivative* of the desired quantity are called Neumann[70] boundary conditions:

$$D_{\mathrm{A}}\left(\frac{\partial c_{\mathrm{A}}(x,t)}{\partial x}\right)_{x=0} = -D_{\mathrm{B}}\left(\frac{\partial c_{\mathrm{B}}(x,t)}{\partial x}\right)_{x=0} \tag{6.32}$$

The redox-inactive C does not "cross" the electrode surface (i.e., is neither consumed nor generated at the electrode), so its material flux at $x = 0$ is zero:

$$D_{\mathrm{C}}\left(\frac{\partial c_{\mathrm{C}}(x,t)}{\partial x}\right)_{x=0} = 0 \tag{6.33}$$

For quasireversible or irreversible conditions, the diffusional flux of a redox-active species, for example, A, is linked to its ET rate. This can be described, for example, by the Butler–Volmer Equation (3.27). We replace Eq. (6.31) by

$$D_{\mathrm{A}}\left(\frac{\partial c_{\mathrm{A}}(x,t)}{\partial x}\right)_{x=0} = k^{\circ}\left\{c_{\mathrm{A}}(x=0,t)\exp\left(\frac{(1-\alpha)[E(t)-E^{\circ\prime}]}{E_{\mathrm{NF}}}\right) - c_{\mathrm{B}}(x=0,t)\exp\left(\frac{\alpha[E(t)-E^{\circ\prime}]}{E_{\mathrm{NF}}}\right)\right\} \tag{6.34}$$

The two terms on the right-hand side of Eq. (6.34) correspond to the disappearance or formation of A at the electrode as a result of the anodic or the cathodic partial reaction, respectively. The term on the left-hand side describes the concomitant transport by diffusion. The potential $E(t)$ in the exponents on the right-hand side again corresponds to the time-dependent excitation signal. Eq. (6.32) is still valid if Eq. (6.34) is applied.

Equations (6.22) to (6.34) form a complete system of PDEs with initial and boundary conditions and represent the mathematical model.

The equations of the mathematical model are often converted into a normalized, dimensionless form before use, dividing all quantities with units by a normalization quantity expressed in the same unit. The normalization quantity must be defined according to the problem. It is common to denote the normalized quantities with Greek characters or upper-case Latin characters (Table 6.12). Normalization of concentrations is customarily done using the initial concentration of the starting compound A [Eq. (6.37)]. The quantity a [s^{-1}], which is important in the normalization of the time variable [Eq. (6.38)], has already been defined in Eq. (6.6). Furthermore, x_{max} in Eq. (6.39) is the distance from the electrode where the concentration does not change significantly during the entire simulation up to a maximum time t_{max}. The thickness of diffusion layers (Section 4.5) increases proportionally to $\sqrt{(Dt)}$, and reaches a maximum of $\sqrt{Dt_{\mathrm{max}}}$. The quantity x_{max} is usually specified to be larger than $\sqrt{Dt_{\mathrm{max}}}$ by a "safety factor" f_x, for which values of up to 10 are recommended in the literature. The normalized quantities for kinetic parameters (κ, ψ) already used in Section 6.2.5 are also meaningful in the context of simulations, and result from the application of Eqs. (6.37) to (6.39) to equations of the type exemplified by Eqs. (6.23), (6.24), or (6.34).

Table 6.12 Normalized quantities for the simulation of cyclic voltammograms.

$$C = \frac{c}{c_{\mathrm{A}}^{*}} \tag{6.35}$$

$$\tau = at = \frac{vt}{E_{\mathrm{NF}}} \tag{6.36}$$

$$X = \frac{x}{x_{\mathrm{max}}} = \frac{x}{f_x\sqrt{Dt_{\mathrm{max}}}} \tag{6.37}$$

The solution of the system of PDEs requires the determination of the functions $C_{\mathrm{A}}(X,\tau)$, $C_{\mathrm{B}}(X,\tau)$, and $C_{\mathrm{C}}(X,\tau)$ that satisfy the equations for all X and τ, that is, the time-dependent concentration profiles of all three species of our example mechanism. The integration of the system of PDEs is, in general, not possible as an analytical

[70] Carl Gottfried Neumann, 1832–1925, German mathematician, professor in Leipzig.

solution, but can only be approximated computationally. Several numerical algorithms are commonly used for this purpose in molecular electrochemistry [24]. They include:

- Conversion into systems of equations, which are solvable with less numerical effort, for example, by Laplace transformation (Excursion 4-3), so-called semianalytical approach.
- Approximation of the differential quotients in Eqs. (6.22) to (6.24) and Eqs. (6.32) to (6.34) by difference quotients from which we obtain systems of ordinary equations. Space and time coordinates as well as concentration changes are decomposed into finite steps. This discretization of the continuous equations results in a finite difference method. The concentration profiles are replaced by step-like functions or functions composed of linear pieces. Several algorithms exist for the solution of the resulting extensive systems of often linear equations [26].
- Solution based on orthogonal collocation, where the profiles $C(X,\tau)$ are replaced by polynomials such that the differential equations are always exactly satisfied at certain space coordinates, but only approximated in between.
- Approximation in delimited domains ("elements") of the spatial coordinate X by an appropriately defined function and minimization of the integrated deviations between the approximate and the exact solution over the elements to zero yields a rather effective spatial discretization (finite element method). Particularly flexible numerical techniques are obtained if the spatial discretization is coupled to a second algorithm, which integrates along the time axis; examples for the latter are the so-called "Rosenbrock methods".[71]

Further details have been compiled in [24].

The finite difference and finite element methods can be formulated in a very general way and allow the treatment of a large number of different models, for example, diverse types of following reactions with their rate laws. The general description of reaction models enables software modules to input reaction mechanisms in an intuitive manner as in Eq. (6.12), which are subsequently translated into the appropriate systems of PDEs with a few additional specifications [27–29]. This allows the users to apply such advanced mathematical formalisms for solving the system of PDEs without engaging themselves in the mathematically demanding integration of differential equations. Altogether, this has led to the development of software packages called "electrochemical simulators" or "electrochemical problem-solving environments", which can be used without knowledge of a programming language.

The simulation tools are particularly efficient in combination with adaptive algorithms, where the step size of the space and time discretization automatically adapts to the problem under study, even in rather complex situations, including

- discontinuous boundary conditions, for example, in CA with its step-like potential and resulting steep concentration profile near the electrode at time $t = 0$.
- very fast coupled chemical reactions. The corresponding mathematical models are then so-called stiff PDE systems.
- chemical reactions with a reaction order greater than 1, leading to nonlinearities in the PDE system.

In these cases, integration with the approximation methods mentioned above can provide incorrect solutions in which numerical errors dominate, sometimes resulting in temporal or spatial oscillations of the concentration profiles. Unfortunately, the

[71] Howard H. Rosenbrock, 1920–2010, British electrical engineer, professor at University of Manchester.

errors are not always so obvious. For this reason, effective error control mechanisms and built-in error tolerance in the simulation software are of great importance.

Based on the concentration profiles provided by the integration of the system of equations, we obtain the current through the electrode by relatively simple calculation of the material flux in the direction of the electrode surface according to Fick's first law. This can be done in the dimensionless form or, after reverse transformation, with the real quantities associated with units. For the reaction in Eq. (6.12), the following applies in a notation taking units into account:

$$I = nFAD_A \left(\frac{\partial c_A(x,t)}{\partial x} \right)_{x=0} \tag{6.38}$$

In our current example, it is sufficient to consider the flux of A, since the flux of B is equal in magnitude according to Eq. (6.32) and C does not contribute to the current. The current is calculated as a function of time $I = f(t)$. The conversion to $I = f(E)$ is straightforward due to the proportionality of E and t given by

$$v = \frac{dE}{dt} = \frac{\Delta E}{\Delta t}; \quad E = E_{\text{initial}} + v(t - t_{\text{initial}}) \tag{6.39}$$

In this relation, we use the starting values for the respective half-cycle for E_{initial} and t_{initial}. For the scan rate v, we have to consider the algebraic sign "+" for the positive-going half-cycle and "−" for the negative-going half-cycle. For the first half-cycle, we obtain $E = E_{\text{initial}} + v\,t$ with $t_{\text{initial}} = 0$.

How can simulation results be helpful? We distinguish two basic concepts according to Figure 6.5:

- Exploratory use – the simulations illustrate how, for example, the cyclic voltammogram would look if a certain model mechanism operates or if certain external conditions (which may not be easy to realize experimentally) are specified. Because of the complexity of the relationships, especially in transient, time-dependent experiments such as CV, an appropriate prediction can be very informative. Together with the consideration of concentration profiles, which are difficult to access experimentally, the understanding of the dynamics within the diffusion layer is also facilitated. We have already used simulations and concentration profiles in an exploratory manner in this book in Section 5.3.5 when introducing CV.
  Qualitatively, we can evaluate whether the model that we have derived from cyclic voltammograms according to Section 6.2 likely produces the I–E curves found in an experiment. Finally, explorative simulations can be the basis for zone diagrams (Section 6.2.5) or can be used for the derivation of diagnostic criteria (Sections 6.2.3 and 6.2.4).
- Comparison with experimental data – a fit of simulated to experimental curves facilitates the determination of numerical values for the reaction parameters. We discuss this application in Section 6.5.

Excursion 6-2 Electrochemical Simulators

Electrochemical simulators are computer programs that enable chemistry-oriented users to carry out simulations without their own programming effort and without detailed knowledge of the mathematical methods used to solve PDEs. For this purpose, general algorithms have been formulated and were provided with convenient input and output options for reaction mechanisms and parameters.

Such software packages are no longer restricted to only a single, specific problem, but rather to entire problem *classes*, for example, to all cyclic voltammetric experiments defined within the framework of an "input language" for ETs (E) and coupled chemical reaction steps (C) (Figure 6.6).

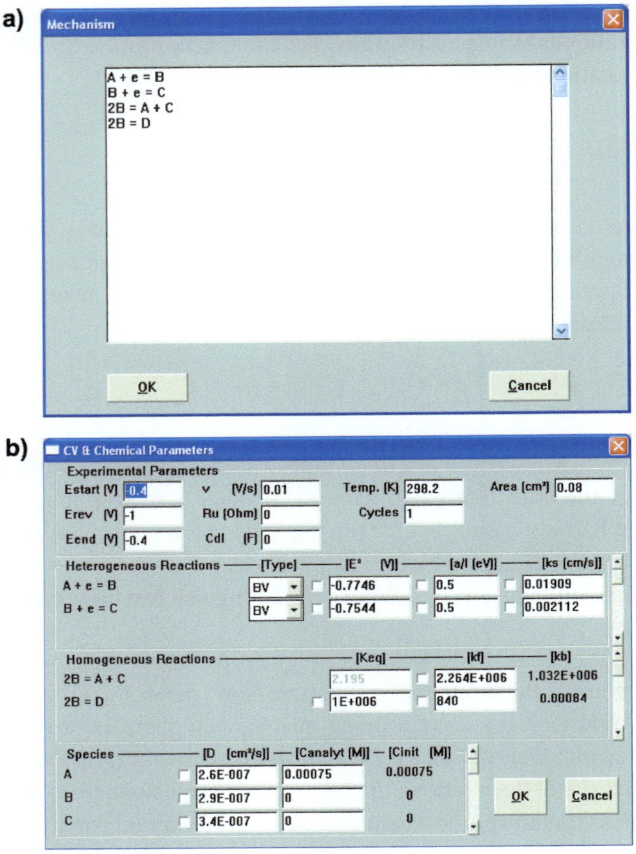

Figure 6.6 Screen shots of DigiSim input windows [32]. a) Definition of a reaction mechanism; b) definition of control parameters (termed "experimental parameters" in the input dialog) and intrinsic quantities for the model mechanism defined in a). The mechanism and input variables are used for analysis of a particular example in Section 6.5.5.

Useful programs offer sound error tolerance with consistency checks of the proposed mechanistic model and its parameters, stability of the numerical solutions, and an appropriate output of the computational results. Note that their support to perform simulations with built-in "basic knowledge" of electrochemistry is very valuable on the one hand, but should not tempt one to blindly trust the algorithms on the other hand.

Besides modules for input and output, an electrochemical simulator needs program units for the translation of the mechanism given in a chemically oriented input language to a mathematical model (compiler) and a solution algorithm (solver). In most cases, the programs are tailored to CV (preferably with a symmetric potential triangle). However, a flexible definition of the excitation signals, analogous to the definition of the reaction mechanisms, has also been published [30].

Apart from the simulators in Table 6.13 specialized for electrochemistry, simulations are also possible with mathematical program libraries such as MATLAB or COMSOL Multiphysics. Some relevant problems with such more general software for mechanistic studies have been pointed out [31]. They have distinct advantages when reaction mechanisms are simulated under observation of additional conditions, for instance, in a complex device.

Table 6.13 Electrochemical simulators.

Name	Description	Website
DigiSim	Commercial Windows program, arguably the first example of an electrochemical simulator as defined above. Its introduction in 1994 was a breakthrough in terms of the formulation of mechanistic models with separate chemical and ET steps [27] and the identification and consideration of so-called "thermodynamically superfluous reaction steps." For a thermodynamically superfluous reaction (TSR), the thermodynamics in terms of equilibrium constants and/or formal potentials is completely determined by other reaction steps. A TSR is part of a thermodynamic cycle.	[32]
DigiElch	Reimplementation and further development of DigiSim, including data fitting and other techniques, for example, impedance spectroscopy	[33]
ELSIM	Package with a more demanding user interface for mathematically inclined users in search of higher flexibility	[34]
EChem++	Open source project in the C++ language for the Linux operating system; allows the definition of complex potential functions to initiate ETs.	[35]

6.3.2 Isolation and Characterization of Intermediates and Products

Further experimental evidence for the plausibility of hypotheses about reaction mechanisms may come from the isolation and structural characterization of intermediates or final products. This compensates for the limitation of electrochemical measurements, which in principle cannot provide direct structural information themselves. The procedure is related to electrolysis (Section 6.6.1) or electrosynthesis (Chapter 17).

It can also be instructive to synthesize suspected intermediates or products separately (even nonelectrochemically) and then study their electrochemistry. The intermediates are no longer formed in the electrochemical cell, and consequently, the overall system is simplified (a similar approach is described in Section 6.5.7).

A special case is the isolation of an ET product (B, without chemical following reaction) by *chemical* oxidation or reduction of the investigated starting compound A. A and B then each give cyclic voltammograms with identical formal potentials. However, the open-circuit potential is on different sides of $E^{\circ\prime}$ on the potential axis, and accordingly, the potential cycle must be started either at potentials located negative or positive of $E^{\circ\prime}$.

Q: Where do we find the open-circuit potential for a cyclic voltammetric experiment?

A: The best way to discuss this is to first define the direction of the redox process: We will consider an oxidation, and the analyte A is the reduced form of the redox couple A/B. Under usual conditions, practically *only* A is present at the beginning of the experiment (Section 6.5.6, selection of the potential range). We assume that the cell is in open-circuit operation: no current can pass through the electrode, and no conversion of A to B occurs. Thus, in the Nernst Equation (3.29), the concentration ratio in the logarithmic term is very small, that is, the second summand in Eq. (3.29) becomes negative. The open-circuit potential E_{OCP} of the electrode is then less positive than $E^{\circ\prime}_{A/B}$ and located to the *left* of $E^{\circ\prime}_{A/B}$ when applying the IUPAC convention (more positive potentials to the right). If we later record a cyclic voltammogram, we will start from the open-circuit potential, where A is predominant, into the positive direction and begin to oxidize A to B.

If, on the other hand, B is the analyte, the sign of the logarithmic term is reversed. The open-circuit potential is more positive than $E^{\circ\prime}_{A/B}$ and is thus located to the *right* of $E^{\circ\prime}_{A/B}$ on the potential axis. Of course, the formal potential has the same value in both cases, and there is only *one* redox potential. We do *not* have separate oxidation and reduction potentials!

6.4 Fundamental Mechanistic Models of Molecular Electrochemistry in Liquid Phases

In this Section, we discuss a selection of important reaction mechanisms in molecular electrochemistry. We extend and systematize the ideas from Section 6.2, yet we are only able to look at a small segment of this diverse field. We review the models mainly in the context of CV using two complementary modes of presentation: Either we consider specific cyclic voltammograms as a function of real, dimensioned reaction parameters (such as in Table 6.2) or I–E curves in abstract form as a function of dimensionless, normalized quantities, as defined in Section 6.2.5. Both approaches have certain advantages and are commonly used in the electrochemical literature.

The cyclic voltammograms shown in this section result from exploratory simulations (Section 6.3.1). Unless otherwise specified, the following parameters were employed: electrode area $A = 1\ \text{cm}^2$, potential scan rate $v = 1\ \text{V s}^{-1}$, temperature $T = 298.2\ \text{K}$, transfer coefficient $\alpha = 0.5$, diffusion coefficient $D = 10^{-5}\ \text{cm}^2\ \text{s}^{-1}$ (assuming that this value is the same for all species in the mechanism), and the initial concentration of the starting compound $c^* = 1\ \text{mmol L}^{-1}$. The formal potential for the electrochemical conversion of the starting compound was set to +0.3 V.

When specific cyclic voltammograms are used in this Section, we usually compare curves for different reaction parameters (e.g., rate constants). Since a rate constant is an intrinsic quantity for a particular chemical system, this means that we compare results expected from different chemical compounds (where other parameters could always change as well). However, there is a relation between these parameters and the timescale as expressed in the respective normalized quantities, for example, in Eqs. (6.13) and (6.21). Therefore, the experimenter is able to adjust the decisive normalized quantities (and thus the resulting curves) in wide ranges by changing the scan rate v as a control parameter.

6.4.1 One-Electron Transfer Reactions Without Separate Chemical Steps

The simplest model mechanism was already used in Section 6.2 as an example: the reversible transfer of an electron at the electrode (reversible ET, E_{rev}, Section 5.3.5). Diffusion is the slowest and therefore rate-determining step, all redox-active molecules arriving at the electrode are immediately converted in the fast heterogeneous ET. Thus, the nernstian equilibrium is always followed *at the electrode surface* ($x = 0$). The ET is much faster than the transport process and cannot be analyzed, as dictated by the bottleneck principle. The criteria of Eqs. (6.2) to (6.5) apply.

If, on the other hand, the ET is much slower than diffusion, it becomes the rate-determining step. Its rate constant k depends on the potential E as used for the derivation of the Butler–Volmer Equation (3.27): The step becomes faster with stronger deviation of E from $E^{0\prime}$. The heterogeneous standard reaction rate constant $k^\circ = k(E^{\circ\prime})$ is used to characterize different analytes. As described by the transfer coefficient α, only a part of the overpotential $E - E^{\circ\prime}$ affects the oxidation or reduction reaction in the Butler–Volmer equation, respectively (Chapter 3). The peaks in the cyclic voltammograms shift to larger overpotentials. A peak is absent on the reverse scan (irreversible process, E_{irr}, Figure 6.7a), because the reverse ET reaction is negligible at the potentials depicted. Strictly speaking, a reverse peak in Figure 6.7a would be observed outside the studied potential range.

Of course, the rate of the ET step (characterized by $k°$) must be related to the experimental timescale. Thus, whether a specific process is considered as irreversible depends on both $k°$ and the potential scan rate v used in CV. Conventionally, the range of irreversibility is defined according to Eq. (6.43).

$$k° = 2 \times 10^{-5} \sqrt{\frac{v}{\text{V s}^{-1}}} \text{ cm s}^{-1} \tag{6.43}$$

Eq. (6.43) holds for $n = 1$, $\alpha = 0.5$, $T = 25\,°\text{C}$, and $D = 10^{-5}$ cm² s⁻¹ (where $D = D_O = D_R$). The equation is sometimes given in dimensionless form with v in V s^{-1}. We have already discussed this practice and the resulting problems when introducing Eq. (6.9).

We have formulated the interplay between $k°$ and v more generally in the discussion of zone diagrams in Section 6.2.5 as a dimensionless parameter ψ [20] or Λ [21] ($\psi = \pi^{-1/2}\,\Lambda$ for $D_O = D_R$). A system is considered irreversible if $\psi \le \pi^{-1/2}\, 10^{-2(1+\alpha)}$. Diagnostic criteria for irreversible voltammograms are listed in Table 6.14.

According to these criteria, the peak shifts by a fixed amount for a 10-fold increase in v. For oxidations, E_{pa} increases, for reductions, E_{pc} becomes more negative, which is why we present two expressions in Eq. (6.40) in Table 6.14. Note that we must also take care to use the correct transfer coefficients for the partial reactions (α or $1 - \alpha$) when applying the criteria in Table 6.14.

The quantity $E_{p/2}$ is the half-peak potential at which half of the peak current is reached. The difference $E_p - E_{p/2}$ in Eq. (6.41) indicates how broad the voltammetric signal appears. Irreversible peaks are much wider than reversible ones.

As in the case of the reversible voltammogram, the peak current I_p is proportional to c^* and $\sqrt{v}$ according to Eq. (6.42). However, the proportionality constant has a different numerical value than in Eq. (6.7) for reversible systems, and the relationship further depends on α or $(1 - \alpha)$. The quantity ΔE_p is not defined because a peak is only observed in the forward scan.

If we treat the kinetics of the ET step according to the Marcus[22]–Hush[72] theory (instead of the Butler–Volmer equation), α becomes potential-dependent and more detailed conclusions can be drawn [7]. Basic features of the Marcus theory will be discussed in Section 12.2.2.

Between the reversible and the irreversible limiting situations (Section 6.2.5, zone diagram), we find a transition region for $8.46 \ge \psi \ge \pi^{-1/2}\, 10^{-2(1+\alpha)}$. We speak of a quasireversible ET (E_{qr}, Figure 6.7b). A reverse signal appears in the voltammograms, but both peaks are shifted away from $E°'$ compared to the voltammogram of a reversible ET (Figure 5.29). In Figure 6.7b, voltammograms with decreasing $k°$ are shown. Starting from the most intense signal (blue curve, reversible ET, approximated by a large rate constant $k° = 10^4$ cm s^{-1}), the peaks move farther and farther apart and become broader and less intense.

The exact shape of the current–voltage curve depends on v, $k°$ as well as α. This is also true for the exact value of $|E_p - E_{p/2}|$ and the shift of E_p with v.

The difference between the anodic and cathodic peak potentials ΔE_p is greater than 58 mV/n and increases with v (Figure 6.8). The peak current also increases with v, but is no longer proportional to $v^{1/2}$. A plot $I_p = f(v^{1/2})$ therefore deviates downward from the straight line that is characteristic of the reversible case (Figure 6.2).

A transition from the reversible to the irreversible case can occur by decreasing $k°$, that is, when comparing *different analytes* as in Figure 6.7b. An analogous change in

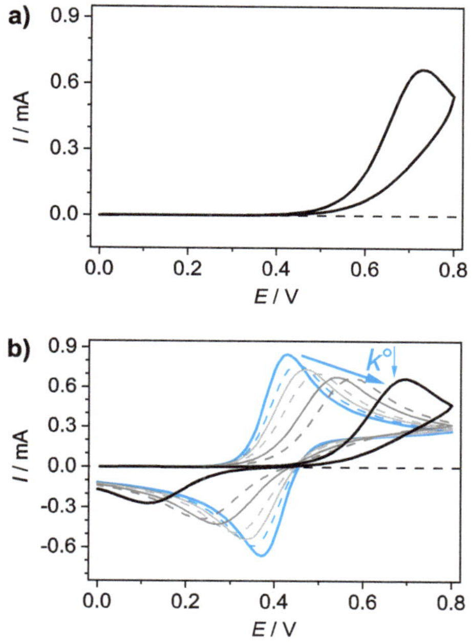

Figure 6.7 Cyclic voltammograms of slow ET reactions. a) Irreversible ET (E_{irr}) with $k° = 10^{-6}$ cm s^{-1}, $E°' = +0.2$ V; b) quasi-reversible ET (E_{qr}) with $k°$/(cm s^{-1}) 10^4, 2×10^{-2}, 5×10^{-3}, 2×10^{-3}, 10^{-3}, 10^{-4}, $E°' = +0.4$ V; $v = 1$ V s^{-1}.

Table 6.14 Diagnostic criteria for irreversible cyclic voltammograms for $n = 1$. $E_{p/2}$ is the potential at which half of the peak current is reached.

$$\frac{\partial E_p}{\partial \log v} = \begin{cases} +1.15\dfrac{RT}{(1-\alpha)F} & \text{(oxidation)} \\ -1.15\dfrac{RT}{\alpha F} & \text{(reduction)} \end{cases} \tag{6.40}$$

$$|E_p - E_{p/2}| = 1.857\frac{RT}{\beta F} \tag{6.41}$$

$$|I_p| = 0.496 F A c^* \sqrt{\frac{D \beta F v}{RT}} \tag{6.42}$$

$$\beta = \begin{cases} 1 - \alpha & \text{(oxidation)} \\ \alpha & \text{(reduction)} \end{cases}$$

[72] Noel Sydney Hush, 1924–2019, Australian theoretical chemist, professor at the University of Sydney after a long tenure in the UK, worked out the quantum mechanical extension of Marcus theory.

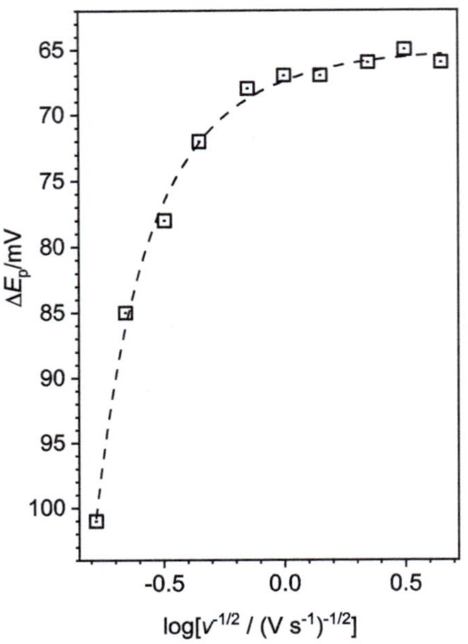

Figure 6.8 Peak potential difference ΔE_p in quasireversible cyclic voltammograms of 0.063 mmol L^{-1} ferrocene + 0.1 mol L^{-1} $NBu_4PF_6 + CH_2Cl_2$. Data from S. Schundelmeier; the graph is analogous to [20], where ΔE_p is plotted against log ψ. Note the proportionality between $v^{-1/2}$ and ψ in Eq. (6.21) for the definition of ψ.

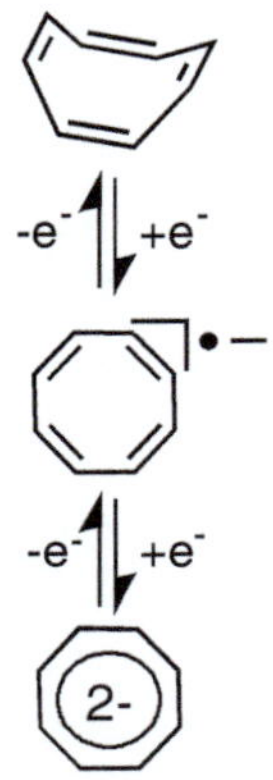

Figure 6.9 Stepwise reduction of cyclooctatetraene (COT) to its aromatic dianion with 10 π electrons.

In a disproportionation reaction, two particles of the same oxidation state react to form one particle in a higher, and one particle in a lower oxidation state. The comproportionation reaction is the reverse of the disproportionation reaction.

K_c – Equilibrium constant of comproportionation [dimensionless]

[A], [B], [C] – Concentrations of A, B, C also c_A, c_B, c_C [mol L^{-1}]

voltammograms is obtained for *a single analyte* by increasing v, because $k°$ and v are linked by means of the dimensionless parameter ψ. However, the current then also increases with v (in contrast to Figure 6.7b).

The chemical reason for a slow ET and the associated high activation energy often lies in the so-called internal or external reorganization energy involving the reorientation of solvent molecules or structural rearrangement of atoms within the analyte (see Section 12.2.2 for more details). A well-studied example is the reduction of the tub-shaped cyclooctatetraene (COT, Figure 6.9) to its planar radical anion, $COT^{\bullet -}$ [36, 37]. The dramatic conformational change manifests itself in a quasireversible first reduction step ($k° = 0.0085\,\text{cm}\,\text{s}^{-1}$) [36]. In contrast, the second reduction step proceeds about 200 times faster than the first one with much smaller changes in the internal coordinates to a likewise planar (and moreover aromatic) dianion COT^{2-} [36].

Irreversibility based on following chemical reactions is discussed in Section 6.4.3.

6.4.2 Multielectron Transfer Reactions Without Separate Chemical Steps

In Section 6.2.4 we have already used multiple ETs as examples of how to draw conclusions about individual mechanistic steps based on diagnostic criteria. Now, the problem of transferring more than one electron will be discussed more systematically.

The case of clearly separated potentials (normal potential order) has already been mentioned. Such a situation is found, for example, for a difference in the formal potentials $\Delta E°'$ larger than 180 mV (Figure 6.10, curve 1). When we consider successively smaller $\Delta E°'$ for the EE mechanism [Eq. (6.17)], we will observe a superposition of the signals and the formation of shoulders (Figure 6.10, curve 2), and finally the appearance of a new pair of peaks with intense, relatively sharp signals (Figure 6.10, curve 4). In the latter case, the second electron is transferred thermodynamically more easily (i.e., with lower energetic effort) than the first one. Voltammograms of this type are observed, for example, when the ETs are associated with significant changes in chemical structure [38]. For $\Delta E°' \leq 0\,\text{V}$, we speak of potential inversion or inverted potential order.

In the intermediate region (Figure 6.10, curves 2 and 3), the Randles–Ševčík Equation (6.7) or (6.9) does not apply. Only in the extreme case of potential inversion, the Randles–Ševčík equation becomes valid again for the peak current, but now with $n = 2$. The current is then multiplied by a factor of $n^{3/2}$ as compared to the case of a single-electron transfer.

The different forms of cyclic voltammograms in the EE mechanism are related to the fact that species A, B, and C in Table 6.8 not only perform ETs with the electrode, but also equilibrate through a homogeneous ET reaction [Eq. (6.44)].

$$A + C \rightleftharpoons 2\,B \qquad (6.44)$$

Eq. (6.44) can be understood from left to right as a comproportionation and *vice versa* as a disproportionation, since A, B, and C represent species with ascending or descending oxidation states. This reaction does not take place at the electrode surface, but in solution close to the interface.

According to the law of mass action, the equilibrium constant is given by Eq. (6.45)

$$K_c = \frac{[B]^2}{[A][C]} \qquad (6.45)$$

We assume that equilibrium [Eq. (6.44)] is reached quickly.

The Nernst equations describing the thermodynamics of the two heterogeneous ETs in Eq. (6.17) for the case of oxidations are

$$E = E^{\circ\prime}_{A/B} + \frac{RT}{F}\ln\frac{[B]}{[A]} \tag{6.46}$$

and

$$E = E^{\circ\prime}_{B/C} + \frac{RT}{F}\ln\frac{[C]}{[B]} \tag{6.47}$$

where B and C represent oxidation products of A. Thus, the formal potentials are linked to K_c:

$$\frac{RT}{F}\ln K_c = E^{\circ\prime}_{B/C} - E^{\circ\prime}_{A/B} \tag{6.48}$$

and therefore

$$K_c = \exp\left[\frac{E^{\circ\prime}_{B/C} - E^{\circ\prime}_{A/B}}{E_{NF}}\right] = \exp\left[\frac{F}{RT}\left(E^{\circ\prime}_{B/C} - E^{\circ\prime}_{A/B}\right)\right] \tag{6.49}$$

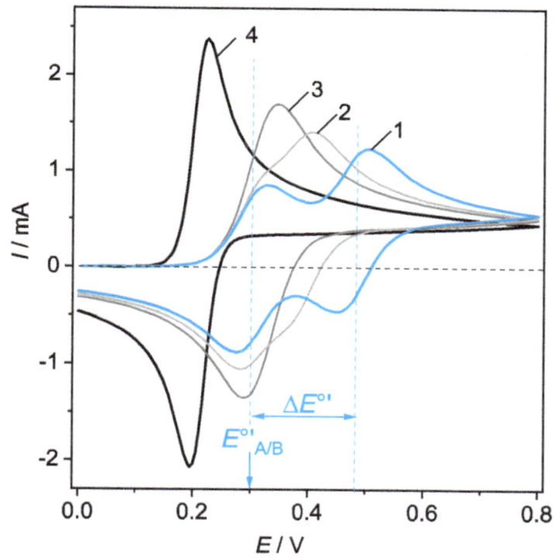

Figure 6.10 Cyclic voltammograms for the EE mechanism for various differences in the two formal potentials $\Delta E^{\circ\prime}$: 1) 180 mV, 2) 90 mV, 3) 35.6 mV, 4) −180 mV; $E^{\circ\prime}_{A/B} = +0.3$ V in all cases.

For normal potential order ($E^{\circ\prime}_{B/C} > E^{\circ\prime}_{A/B}$), this results in $K_c > 1$ and the equilibrium [Eq. (6.44)] is shifted to the right-hand side. Species B, which corresponds to the intermediate oxidation state, is stable with respect to disproportionation. In the case of potential inversion ($E^{\circ\prime}_{B/C} < E^{\circ\prime}_{A/B}$), $K_c < 1$, and the equilibrium Eq. (6.44) is shifted to the left-hand side. Consequently, B disproportionates into A and C. The case $K_c = 1$ corresponds to both formal potentials being equal.

Q: How do we derive Eq. (6.48)?

A: We solve Eqs. (6.46) and (6.47) for the concentration ratios, substitute them into Eq. (6.45), and collect the terms with the formal potentials on the right-hand side.

Let us now consider a molecule A with two identical, noninteracting redox centers (Figure 6.11). This molecule has two independent sites for the ET to occur. For molecule B that has already transferred one electron, there is only *one* further possibility for ET. In contrast, there are two equivalent sites for ET in molecule A (and thus, the probability of ET at A is twice as high as that of an ET at B). The same applies to C. Statistically, the concentration [B] will therefore be twice as large as [A] or [C]: $[A] = [C] = \frac{1}{2}[B]$. Thus, species B appears somewhat stabilized. Substituting the concentration ratio into Eq. (6.45) yields $K_c = 4$ for this case and finally $\Delta E^{\circ\prime} = 35.6$ mV at 25 °C from Eq. (6.48).

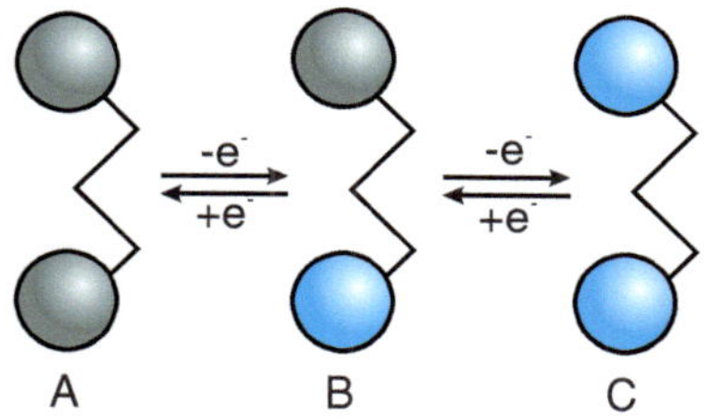

Figure 6.11 Two identical redox centers (spheres) connected by a molecular linker. The colors indicate the redox state.

The superposition of the currents for $A \rightarrow B + e^-$ on the one hand and for $B \rightarrow C + e^-$ on the other hand results in a cyclic voltammogram with a shape corresponding to that of a one-electron transfer reaction with $(E_{p,a} + E_{p,c})/2 = \left(E^{\circ\prime}_{A/B} + E^{\circ\prime}_{B/C}\right)/2$ and doubled peak currents (Figure 6.10, curve 3). The fact that we find $\Delta E^{\circ\prime} > 0$ for this situation (and not $\Delta E^{\circ\prime} = 0$, as one might expect at first glance) is sometimes referred to as "statistical" or "entropic" effect.

Note that the argument is presented for oxidation. If B and C represent reduction products of A, however, an analogous reasoning applies (Exercise 6.1).

The two hypothetical partial curves would be shifted against each other by 35.6 mV because of the positive $\Delta E^{\circ\prime}$. However, the stabilization of B causes $E^{\circ\prime}_{A/B}$ in Eq. (6.46) to be slightly increased, while $E^{\circ\prime}_{B/C}$ in Eq. (6.47) is decreased by the same amount, so that the curves shift toward each other. Consequently, the resulting cyclic voltammogram is also shifted by $\Delta E^{\circ\prime}/2$ on the potential axis with respect to $E^{\circ\prime}_{A/B}$ (selected as $E^{\circ\prime}_{A/B} = +0.3$ V in Figure 6.10).

If the molecule contains two interacting redox centers, various cases can occur. If the redox centers are *identical* and the redox state of *one* center influences the $E^{\circ\prime}$ of the *second* center, peak splitting can occur, which can be easily recognized.

If the second ET requires more energy because of the already converted first center (larger $E^{\circ\prime}_{B/C}$, thermodynamically unfavorable with respect to the first ET), $\Delta E^{\circ\prime}$

increases beyond the entropic effect. Conversely, if the second ET requires less energy than the ET that already occurred at the first center leads to a lower and sometimes even negative $\Delta E^{\circ\prime}$.

The mutual effect of the redox centers is significantly affected by the linker. Conjugated linkers enable an electronic interaction (sometimes called "electronic effect"). Saturated linkers, on the other hand, allow exclusively electrostatic effects (coulombic interaction, effect of the *charge* of one of the redox centers on the other). The difference of the formal potentials $\Delta E^{\circ\prime}$ is used to estimate the extent of the electronic or electrostatic interaction between the centers. Other types of interaction have also been discussed [39]. Dinuclear transition metal complexes such as the Creutz–Taube ion[73] [40] (Figure 6.12) represent an important class of compounds exhibiting such effects. The intensity of the electronic interaction between the centers is classified according to Robin and Day from weak to strong into three classes I to III [41]. In compounds of Robin–Day class III, the interaction is so strong that the ET cannot be assigned to one of the centers at all. In other words, the two redox sites can no longer be distinguished.

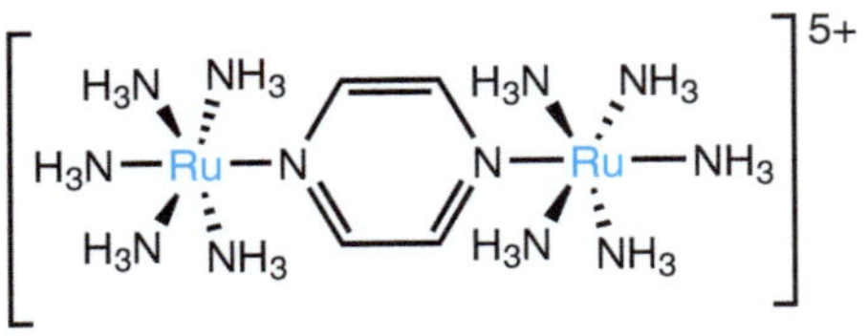

Figure 6.12 Structure of the Creutz–Taube complex.

Nonidentical redox centers in a molecule usually have genuinely different $E^{\circ\prime}$. Moreover, structural changes due to the ETs can change the value of $\Delta E^{\circ\prime}$, often leading to potential inversion.

Dendrimers are polymers with a tree-like branched structure (Greek *dendron*, the tree).

E°_{m} – Standard redox potential of the monomer and all redox-active units in the polymer [V]

$E^{\circ\prime}_{j}$ – Formal potential [V] at which the concentrations of the two species with j and $j-1$ oxidized monomer units are equal, corresponds to the formal potential for redox process $j-1$

n – Number of monomers in the polymer

The concept of two-electron transfers has been extended to n redox centers in a molecule. This case is realized, for example, in polymers (especially in dendrimers) with many, essentially identical redox-active groups. Each group thus has the same standard redox potential E°_{m}, which corresponds to that of the monomer ("m" stands for "monomer"). It is assumed that the redox centers do not interact with each other. However, we have to consider the statistical effect, which we include in the formal potentials for the successive redox steps. It follows for the formal potential of the jth redox process [42]:

$$E^{\circ\prime}_{j} = E^{\circ}_{m} + \frac{RT}{F}\ln\frac{j}{n-j+1} \tag{6.50}$$

As n increases, the formal potentials move closer and closer together. The current–potential curve is formed again as a superposition of the individual voltammetric peaks [42]

$$I(E) = nI_1(E)\sqrt{\frac{D_n}{D_1}} \tag{6.51}$$

Note that an *oxidation process* is treated here. The sign of the logarithmic term in Eq. (6.50) is reversed as compared to Ref [42], because the reference considers a reduction.

where I_1 is the curve of a monomer (*one* single isolated redox center), and the ratio of the diffusion coefficients of the polymer D_n and the monomer D_1 must be taken into account.

Deviations from these theoretical considerations are expected if the ETs are not reversible, interactions between the centers or structural changes occur, or if adsorption effects play a role.

Q: Why is the total current in Eq. (6.51) proportional to n and not to $n^{3/2}$ as in the Randles–Ševčík Equation (6.7)?

A: We are dealing with two different cases (Table 6.15). In the application of the Randles–Ševčík equation with $n = 2$, we assume *simultaneous* transfer of several electrons, as for example, in the case of potential inversion. Here, on the other hand, the electrons are transferred *stepwise*; see also [43].

Table 6.15 Comparison of simultaneous and stepwise ETs.

simultaneous ETs	
$A \rightleftharpoons C + 2e^-$	(6.52)
stepwise ETs	
$A \rightleftharpoons B + e^-$	
$B \rightleftharpoons C + e^-$	(6.53)

[73] Henry Taube, 1915–2005, Canadian-US chemist, professor at Cornell University, the University of Chicago, and Stanford University, Nobel Prize in Chemistry 1983; Carol Creutz, PhD student of H. Taube.

As this discussion shows, the number of ET steps cannot be derived simply by counting the peaks in a cyclic voltammogram.

Many electroanalytical studies of multielectron transfers have been reported in the literature. For example, during the oxidation of *N,N,N′,N′*-tetramethylphenylenediamine (TMPD, Figure 6.13a), a reversible stepwise two-electron transfer to the radical cation ("Wurster's[74] blue") and further to the dication with normal potential order has been observed [44]. The structures of all oxidation states are similar (planar). However, systems with strong structural changes usually show potential inversion as found for the planar aromatic 3,6-bis(dimethylamino)durene (Figure 6.13b), where the substituents at the N-atom and the C-atoms of the ring linked to them are bent out of the plane in the dication (Figure 6.13c) [44].

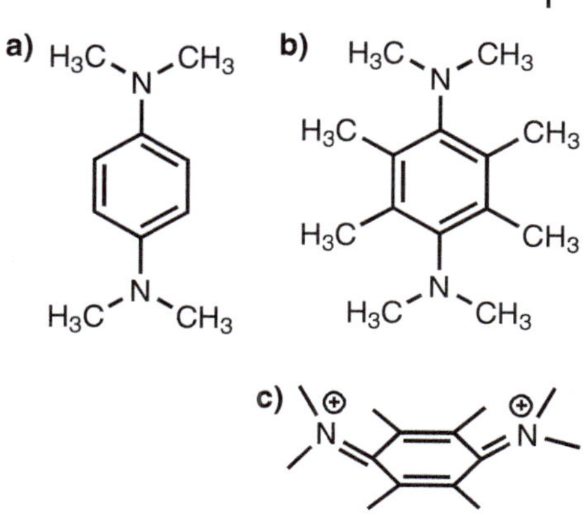

Figure 6.13 Structure of a) *N,N,N′,N′*-tetramethylphenylenediamine (TMPD); b) 3,6-bis(dimethylamino)durene and c) dication of 3,6-bis(dimethylamino)durene.

6.4.3 Coupled Homogeneous Chemical Reactions

Next, we want to analyze mechanisms where chemical reaction steps are coupled to the ET. In Section 6.2.4, we already mentioned the EC mechanism (Table 6.7), where the ET is followed by an irreversible chemical reaction. The variation of the two relevant rate constants $k°$ and k, as well as the control parameter v, led to the zone diagram in Figure 6.4. This will be the basis for the discussion of further mechanisms. Cyclic voltammograms of EC systems are characterized by the decrease in the reverse peak until it completely disappears (Figure 6.14a) when the following reaction becomes fast enough (rate constant k increases). Similarly, the voltammograms change with increasing (longer) timescale of the experiment (v becomes smaller). The reasons for this have already been discussed in Section 6.2.5. The interplay of kinetics and timescale is characterized by the normalized rate constant $\kappa = k/a = RTk/(nF \cdot v)$.

At the same time, the forward peak changes as well: Its intensity increases slightly and shifts (for oxidation) to less positive potentials (Figure 6.14b). For fast following reactions and fast ETs, this shift is 29.6 mV per decade (change in v or k by a factor of 10, resulting in a respective change in κ, area KP of the zone diagram in Figure 6.4).

If the ET becomes slow, the peak broadens until it finally takes on the form of an irreversible signal [45] (zone IR of the zone diagram in Figure 6.4). As an extreme form of this reaction, the ET and the following reaction "merge" into *one* concerted reaction step. If the following reaction is a bond cleavage, this extreme is referred to as dissociative ET [46, 47]. Bond formation may also occur concerted with an ET, for example, in the transfer of electrons and protons. A detailed discussion of this mechanism, which is extremely important in chemistry and biochemistry, can be found in [48, 49]. In these cases, the transfer coefficient α usually deviates considerably from 0.5.

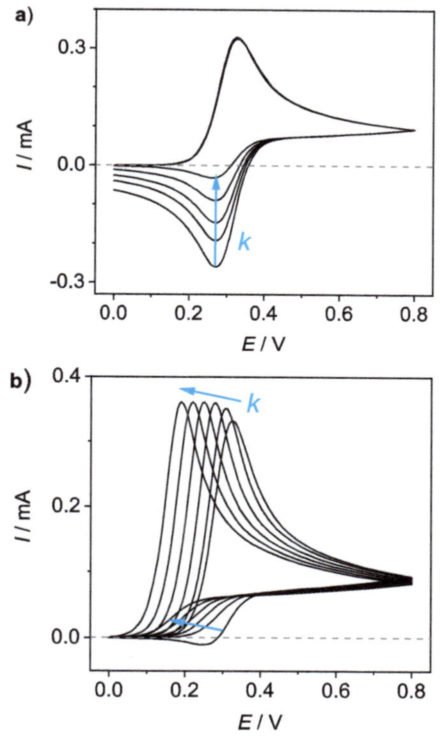

Figure 6.14 Cyclic voltammograms for the EC mechanism with fast ET; effect of the rate of the following reaction; a) $k = 0, 0.5, 1.0, 2.0, 5.0\,\mathrm{s}^{-1}$, in this order the magnitude of the reverse peak decreases; b) $k = 10, 100, 10^3, 10^5, 10^6\,\mathrm{s}^{-1}$, in this order the forward peak shifts to less positive potentials. Other parameters are $A = 1\,\mathrm{cm}^2$, $v = 1\,\mathrm{V\,s}^{-1}$, $T = 298.2\,\mathrm{K}$, $\alpha = 0.5$, $D = 10^{-5}\,\mathrm{cm}^2\,\mathrm{s}^{-1}$ (for all species), and $c^* = 1\,\mathrm{mmol\,L}^{-1}$, $E°' = +0.3\,\mathrm{V}$.

Further variations of the EC mechanism concern the type of the following reaction. If we assume an equilibrium, the overall reaction sequence is referred to as the E$\underline{\text{C}}$ mechanism (Table 6.16). We use two intrinsic quantities (reaction parameters) to characterize the chemical thermodynamics (equilibrium constant K) and kinetics (the sum of the rate constants k_1 and k_{-1} of the forward and reverse reactions):

$$K = \frac{k_1}{k_{-1}} \tag{6.54}$$

The sum of k_1 and k_{-1} is called the "mobility" of the equilibrium. We relate the two rate constants to the experimental timescale by means of a and denote the result as κ [Eq. (6.55)]. A zone diagram is constructed [7] as a function of K and κ.

$$\kappa = \frac{k_1 + k_{-1}}{a} = \frac{E_{\mathrm{NF}}(k_1 + k_{-1})}{v} = \frac{RT(k_1 + k_{-1})}{nFv} \tag{6.55}$$

k_1, k_{-1} – Homogeneous first-order rate constants of the forward and reverse reactions [s^{-1}]

K – Equilibrium constant [dimensionless].

[74] Casimir Wurster, 1854–1913, German chemist.

Table 6.16 E$\underline{\underline{C}}$ mechanism.

First step is oxidation	
$A \xrightleftharpoons{E^{\circ\prime},\, k^{\circ}} B + e^-$	
(identical to $A - e^- \rightleftharpoons B$)	
$B \xrightleftharpoons[k_{-1}]{k_1} C$	(6.56)
First step is reduction	
$A + e^- \xrightleftharpoons{E^{\circ\prime},\, k^{\circ}} B$	
$B \xrightleftharpoons[k_{-1}]{k_1} C$	(6.57)
General formulation	
$A \pm e^- \xrightleftharpoons{E^{\circ\prime},\, k^{\circ}} B$	
$B \xrightleftharpoons[k_{-1}]{k_1} C$	(6.58)

Table 6.17 Mechanism of the radical–radical coupling reaction ($ED_{2,2}$).

First step is oxidation	
$A \xrightleftharpoons{E^{\circ\prime},\, k^{\circ}} B + e^-$	
$2\,B \xrightarrow{k} C$	(6.60)
First step is reduction	
$A + e^- \xrightleftharpoons{E^{\circ\prime},\, k^{\circ}} B$	
$2\,B \xrightarrow{k} C$	(6.61)
General formulation	
$A \pm e^- \xrightleftharpoons{E^{\circ\prime},\, k^{\circ}} B$	
$2\,B \xrightarrow{k} C$	(6.62)

At high mobility (κ large), the concentrations of B and C in the solution immediately follow a change in the electrode potential (similar to the nernstian equilibrium). This limiting case leads to cyclic voltammograms of reversible shape, but shifted on the potential axis by $E_{NF} \ln[1 + K]$. We ultimately observe the equilibrium between A and C and its thermodynamics.

For less mobile equilibria (κ small), the shape of the cyclic voltammogram depends critically on K. If $K \gg 1$, the equilibrium is on the side of C and the voltammogram corresponds to that of an EC mechanism with an irreversible following reaction. For smaller K, on the other hand, the reverse peak changes sensitively with the kinetic parameters of the chemical reaction step [15].

If the reaction order of the following reaction is larger than one, we find two interesting variants:

1) **Electrodimerization.**

Two particles of the primary ET product react with each other in a homogeneous reaction (Table 6.17). If the following reaction consists of a single irreversible reaction step, a second-order rate law applies

$$\frac{d[B]}{dt} = k[B]^2 \tag{6.59}$$

where k is a second-order rate constant with the typical unit L mol^{-1} s^{-1}. Many electrode reactions in organic electrochemistry proceed according to this mechanism, including synthetically important ones. Often, B is an electrochemically generated radical whose dimerization allows a new chemical bond to be formed (radical–radical coupling, Table 6.17). In electrohydrodimerizations the coupling reaction is extended by additional protonation [50, 51].

An alternative mechanism for electrodimerization is the radical–substrate coupling (Table 6.18), which leads to the same product C, albeit after a second ET step. This mechanism can also be regarded as a variant of the ECE mechanism (Section 6.4.4).

Cyclic voltammograms of systems with coupling reactions are concentration dependent (at least at suitable time scales). This is the consequence of second-order kinetics leading to the quadratic concentration term in Eq. (6.59). With increasing starting concentration c^*, the concentration of B also increases, and the following reaction is accelerated. The current-potential curves change accordingly. However, in practice, it is difficult to distinguish between the two alternatives mentioned above. In addition, the reaction steps are often linked to further ET or chemical steps. Radical–radical coupling has been observed for diphenylpolyene radical cations (Figure 6.15) [52], while radical–substrate coupling has been postulated for the technically useful reduction of acrylonitrile [53] (Figure 6.16, Section 17.3.3). There is also evidence that the latter reaction can proceed via radical–radical coupling [54], but under conditions that differ significantly from those of the technical process [55].

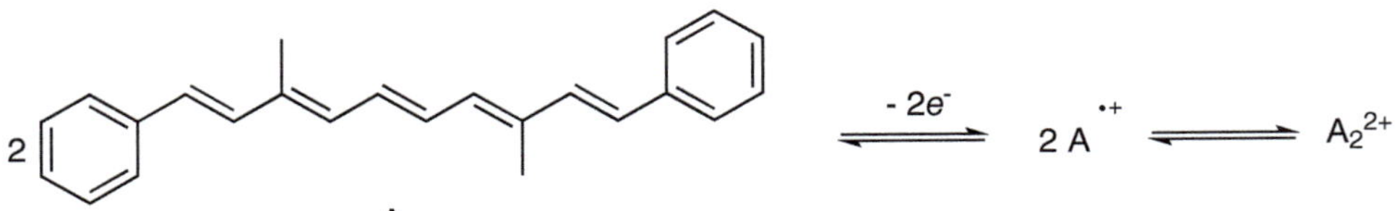

A

Figure 6.15 Oxidation of a diphenylpolyene (selected part of the reaction mechanism according to [52]).

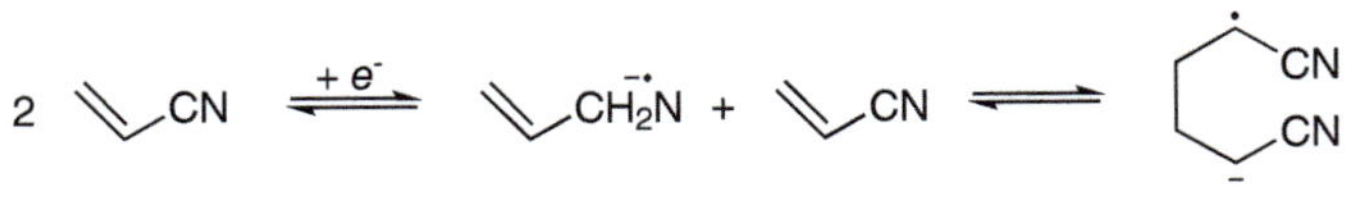

Figure 6.16 Reduction of acrylonitrile (selected part of the reaction mechanism according to [53]).

2) **Reaction with a reagent that is *not* in excess.**

In this case, the primary ET product reacts with an added reagent X (Table 6.19). If the concentration of X in the bulk solution is not very much larger* or even smaller than that of A, a second-order rate law applies

$$\frac{d[B]}{dt} = k[B][X] \tag{6.63}$$

with a homogeneous second-order rate constant k (typical unit: L mol^{-1} s^{-1}). In the absence of the following reaction ($k = 0$), a reversible cyclic voltammogram is obtained (Figure 6.17, curve 1). However, if the following reaction proceeds with a finite k, X may be depleted in the diffusion layer in front of the electrode, and not all particles B can react with the reagent. Cyclic voltammograms with a prepeak (Figure 6.17, curves 2 and 3) are characteristic of this situation. In the example selected for Figure 6.17, $[X]^* = \frac{1}{2}[A]^*$, which defines the size of the prepeak.

The reason for this phenomenon is as follows: initially, little B is formed, and sufficient X is present to allow the chemical reaction in Eq. (6.67) to proceed. As already discussed for the EC mechanism with an irreversible following reaction, a peak occurs that is shifted from $E^{\circ\prime}_{A/B}$ to less positive potentials ("to the left"). As soon as X is depleted, the following reaction can no longer proceed and the peak appears at the usual potential, slightly positive of $E^{\circ\prime}_{A/B}$. As a result, the voltammogram shows a shoulder in the rise of the current (Figure 6.17, curve 2) or even two separate peaks (Figure 6.17, curve 3). In this mechanism, $\kappa = k[A]^*/a$ is a suitable dimensionless parameter. Two extrinsic quantities (control parameters), $[A]^*$ and scan rate (implicit in a), are combined with k.

The reduction of a certain Ni complex in the presence of protons represents a reaction that follows the mechanism in Eq. (6.67) [56].

A prepeak can also occur in mechanisms that involve the adsorption of reactants on the electrode as a reaction step. Therefore, the observation of a prepeak is not a proof of the mechanism in Table 6.19. However, the variation of the intensity of the prepeak with $[X]^*$ provides evidence for the occurrence of this mechanism.

Preceding chemical reaction. Just as the product of an ET can disappear by a chemical reaction, there are mechanisms involving chemical reactions that generate the redox-active initial species from a precursor molecule (CE mechanism [15]). We discuss this for a preceding equilibrium (C̲E) in which the redox-active species A is formed from an electrochemically inert species A′ (Table 6.20). In analogy to the EC̲ mechanism, the intrinsic quantities are the equilibrium constant K and the mobility of the equilibrium (in dimensionless form: κ [Eq. (6.55)]).

Q: Are there also irreversible preceding reactions?

A: Yes, of course. But we only treat the equilibrium case here because, after all, we could let an irreversible preceding reaction take place completely *before* starting the measurement, that is, we would rather wait until all A′ is converted into A. Then the analysis of the data becomes much simpler.

If the equilibrium reaction is slow (κ small), CV maps the equilibrium situation, that is, we obtain an I–E curve corresponding to the equilibrium concentration of A. For the special case of a fast (reversible) ET, the curve has a reversible form (ΔE_p!) with correspondingly reduced peak currents.

At fast equilibrium (κ large), the electrochemical reaction immediately replenishes A that has been removed by ET, so that ultimately all particles A′ react via A to form B. The peak current increases to the value defined by the sum of $[A]^*$ and $[A']^*$ present at the beginning of the experiment. However, the two reaction steps in Eq. (6.68) are then

*"Very much larger" is reached only when the excess is at least 100 times.

Table 6.18 Mechanism of radical–substrate coupling ($ED_{1,2}E$).

First step is oxidation

$$\mathrm{A} \xrightleftharpoons{E_1^{\circ\prime},\, k_1^\circ} \mathrm{B} + e^-$$
$$\mathrm{A} + \mathrm{B} \xrightarrow{k} \mathrm{A{-}B}$$
$$\mathrm{A{-}B} \xrightleftharpoons{E_2^{\circ\prime},\, k_2^\circ} \mathrm{C} + e^- \tag{6.64}$$

First step is reduction

$$\mathrm{A} + e^- \xrightleftharpoons{E_1^{\circ\prime},\, k_1^\circ} \mathrm{B}$$
$$\mathrm{A} + \mathrm{B} \xrightarrow{k} \mathrm{A{-}B}$$
$$\mathrm{A{-}B} + e^- \xrightleftharpoons{E_2^{\circ\prime},\, k_2^\circ} \mathrm{C} \tag{6.65}$$

General formulation

$$\mathrm{A} \pm e^- \xrightleftharpoons{E_1^{\circ\prime},\, k_1^\circ} \mathrm{B}$$
$$\mathrm{A} + \mathrm{B} \xrightarrow{k} \mathrm{A{-}B}$$
$$\mathrm{A{-}B} \pm e^- \xrightleftharpoons{E_2^{\circ\prime},\, k_2^\circ} \mathrm{C} \tag{6.66}$$

Table 6.19 Reaction with reagent X not in excess.

General formulation

$$\mathrm{A} \pm e^- \xrightleftharpoons{E^{\circ\prime},\, k^\circ} \mathrm{B}$$
$$\mathrm{B} + \mathrm{X} \xrightarrow{k} \mathrm{C} \tag{6.67}$$

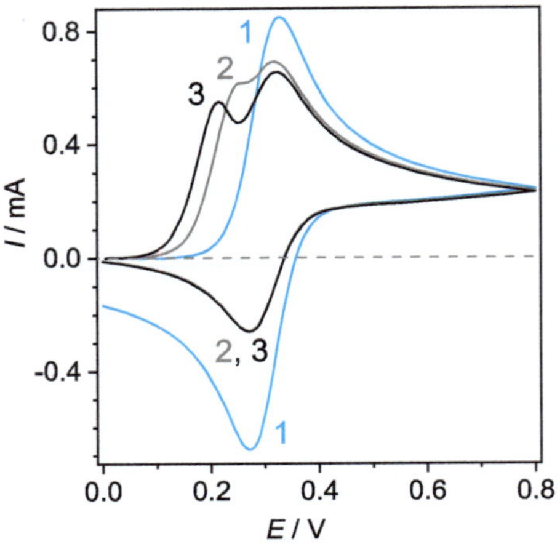

Figure 6.17 Cyclic voltammograms for a system with depletion of the reagent in the diffusion layer; $[A]^* = 1\ \text{mmol}\ \text{L}^{-1}$; $[X]^* = 0.5\ \text{mmol}\ \text{L}^{-1}$; $v = 1\ \text{V}\ \text{s}^{-1}$; k in $\text{L}\ \text{mol}^{-1}\ \text{s}^{-1}$, 1) 0, 2) 10^7, 3) 10^8; A is oxidized. $A = 1\ \text{cm}^2$, $T = 298.2$ K, $\alpha = 0.5$, $D = 10^{-5}\ \text{cm}^2\ \text{s}^{-1}$ (for all species), $E^{\circ\prime} = +0.3$ V.

again thermodynamically combined, and the peak shifts under the control of K by a value of $E_{\text{NF}} \ln[K/(1+K)]$. The exact shape of the cyclic voltammogram is controlled by the timescale, which in turn is varied experimentally by means of v.

There are also cases where both A and A′ undergo an ET at the electrode (Table 6.21). As before, the cyclic voltammetric behavior depends on the timescale of the equilibrium reaction and v, that is, on κ.

At small κ, the equilibrium $\text{A}' \rightleftharpoons \text{A}$ is "frozen". A cyclic voltammogram then shows the two ETs at the formal potentials $E^{\circ}_{\text{A}'/\text{B}'}$ and $E^{\circ\prime}_{\text{A/C}}$ with the respective equilibrium concentrations of A and A′ resulting in two peaks in the forward scan. However, if κ is large, the electrochemically more easily converted species, A′ or A, reacts first. The depleted compound is immediately replenished by the equilibrium, so that in the extreme case, only *one* peak appears in the forward scan, namely that of the more easily converted species. In principle, we can again switch between the two types of behavior by variation of v. This mechanism is usually part of more complex schemes, as we discuss in Section 6.4.5.

6.4.4 Variants of the ECE Mechanism

Two ETs linked by a chemical (equilibrium) reaction as in Table 6.21, constitute a special case of the ECE mechanism, which was already mentioned as an example in Section 6.2.4. The radical–substrate coupling [Eq. (6.65)] can also be understood as an ECE reaction. These examples indicate the importance of such sequences for mechanistic considerations in molecular electrochemistry.

The general mechanism has already been formulated in Table 6.9. The number of intrinsic quantities (i.e., reaction parameters) is larger than in the other mechanisms discussed so far. Therefore, we can only treat some selected cases and the factors that control their behavior. For the following examples, we assume two reversible ETs and first vary the normalized rate constant of the chemical reaction step $\kappa = k/a$ and the difference in the two formal potentials $E^{\circ\prime}_{\text{A/B}}$ and $E^{\circ\prime}_{\text{C/D}}$:

$$\Delta E^{\circ\prime} = E^{\circ\prime}_{\text{C/D}} - E^{\circ\prime}_{\text{A/B}} \tag{6.70}$$

For very small κ, the chemical reaction $\text{B} \rightarrow \text{C}$, which we assume to be irreversible at the moment, has only a negligible effect (Figure 6.18a, curve 1). Species C is not formed in appreciable amounts, and we find only the signal of the redox process $\text{A} - \text{e}^- \rightleftharpoons \text{B}$. However, if κ increases, more and more C is generated, and the peaks for its oxidation as well as the reduction of D back to C on the reverse scan increase in intensity (Figure 6.18a, curves 2–4) until all of B is converted to C during the potential sweep between the two formal potentials. Thus, the second oxidation peak reaches its maximum size (Figure 6.18a, curve 5). Experimentally, the control by means of v is again crucial.

We assumed in Figure 6.18a for the oxidation process that $E^{\circ\prime}_{\text{C/D}} > E^{\circ\prime}_{\text{A/B}}$, that is, $\Delta E^{\circ\prime}$ is positive. Now, we keep a large κ and let the quantity $\Delta E^{\circ\prime}$ decrease until it finally reaches negative values (Figure 6.18b). Curve 1 in Figure 6.18b corresponds to curve 5 in Figure 6.18a: Species B is completely converted to C during the scan between the widely separated $E^{\circ\prime}$ values.

If $\Delta E^{\circ\prime}$ decreases starting from the latter situation, the oxidation peaks begin to overlap (Figure 6.18b, curve 2). Negative $\Delta E^{\circ\prime}$ values are similar to the potential inversion case discussed in Section 6.4.2: The entire reaction from A to D takes place in the single peak observed on the forward scan, with a total of two electrons being transferred. This is evident from the high peak current in Figure 6.18b, curves 3 and 4. In the forward scan, the two cyclic voltammograms are practically indistinguishable, although $\Delta E^{\circ\prime}$ is different. The reverse peak corresponds to the rereduction of D to C. Since the

We discuss here an oxidation as the first ET step in all cases. The second redox step is also an oxidation for the first few examples. Corresponding relations are valid, of course, when the first step is a reduction!

Table 6.20 Reaction according to the CE mechanism.

General formulation	
$\text{A}' \underset{k_{-1}}{\overset{k_1}{\rightleftharpoons}} \text{A}$	
$\text{A} \pm \text{e}^- \overset{E^{\circ\prime},\, k^{\circ}}{\rightleftharpoons} \text{B}$	(6.68)

Table 6.21 Reaction according to the CE(E) mechanism. Both precursors are redox active.

General formulation	
$\text{A}' \underset{k_{-1}}{\overset{k_1}{\rightleftharpoons}} \text{A}$	
$\text{A}' \pm \text{e}^- \overset{E^{\circ}_{\text{A}'/\text{B}}{}',\, k^{\circ}_1}{\rightleftharpoons} \text{B}$	
$\text{A} \pm \text{e}^- \overset{E^{\circ}_{\text{A/C}}{}',\, k^{\circ}_2}{\rightleftharpoons} \text{C}$	(6.69)

chemical reaction step is irreversible, there is no reverse chemical reaction to B. The reduction D → C starts only at potentials negative of $E^{\circ\prime}_{C/D}$. In the case of curve 3 and especially curve 4, this is far negative with respect to the oxidation peak. We observe a correspondingly large peak separation, with the location of the reverse peak depending strongly on $\Delta E^{\circ\prime}$. The low intensity of the reverse peak can be explained as follows: D formed during the forward scan continues to diffuse from the electrode surface to the bulk of the solution during the reverse scan, even when oxidation of A has already ceased (potentials negative with respect to $E^{\circ\prime}_{A/B}$). Species D is depleted, and correspondingly, less D can return to the electrode when the potential is finally sufficient to reduce D.

In the case of curves 1 and 2, the reverse peak is, as expected, close to the oxidation peak for step $C \rightleftharpoons D + e^-$, that is, at potentials of about +0.5 V and +0.4 V, respectively.

Similar to the EE mechanism, we expect a homogeneous comproportionation-disproportionation equilibrium (homogeneous redox reaction) in ECE reactions, now between the species A/D on the one hand and B/C on the other:

$$B + C \rightleftharpoons A + D \tag{6.71}$$

Species B and C have the same oxidation state, whereas it is different for A and D. Accordingly, the equilibrium constant K for Eq. (6.71) is determined by the two formal potentials:

$$K = \frac{[A][D]}{[B][C]} = \exp\left[\frac{\left(E^{\circ\prime}_{A/B} - E^{\circ\prime}_{C/D}\right)}{E_{NF}}\right] = \exp\left[\frac{nF}{RT}\left(E^{\circ\prime}_{A/B} - E^{\circ\prime}_{C/D}\right)\right]. \tag{6.72}$$

Equation (6.72) follows from formulating the two Nernst equations for the ETs and equating E.

The reaction given in Eq. (6.71) does not change the thermodynamics of the system. However, its kinetics can have a considerable influence on the course of the overall reaction, because it represents an alternative to the stepwise transition from A via B and C to finally D. In fact, a further distinction is made according to whether the kinetics of the overall process is first or second order: In the case of a first-order reaction, the transformation $B \rightleftharpoons C$ is rate-determining (DISP1), whereas in the second case (DISP2), this is true for the reaction given in Eq. (6.71) [19]. Thus, the relative rates of these two chemical steps determine whether a DISP1 or a DISP2 reaction is observed. Depending on the kinetic and thermodynamic quantities, the three-dimensional zone diagram shown in Figure 6.19 has been derived.

As in Figure 6.4, the arrows indicate how changes in the parameters affect the behavior of the system: For small mobility of equilibrium $B \rightleftharpoons C$ (denoted as k in Figure 6.19, it is κ in our notation), the system remains in the zone DO (diffusion-controlled one-electron transfer A/B) because no C is formed and therefore all other reactions are blocked. In the range of small k and/or slow disproportionation (characterized by the disproportionation rate constant k_d), we "move" up into the ECE zone as k increases. In Figure 6.19, C^0 is the initial concentration of A, that is, $[A]^*$ in our notation. A small value of $[A]^*$ works against the influence of disproportionation and favors the progress of oxidation C → D at the electrode. Conversely, large k_d or a high concentration $[A]^*$ leads to DISP mechanisms with the formation of D as a result of the homogeneous redox reaction, with DISP2 being the predominant reaction process only in a small segment.

In the previous ECE examples, we have assumed that both ETs proceed "in the same direction", for example, that A and C each represent the *reduced* form of a redox couple. Now, let us consider the case where the directions of the two redox processes

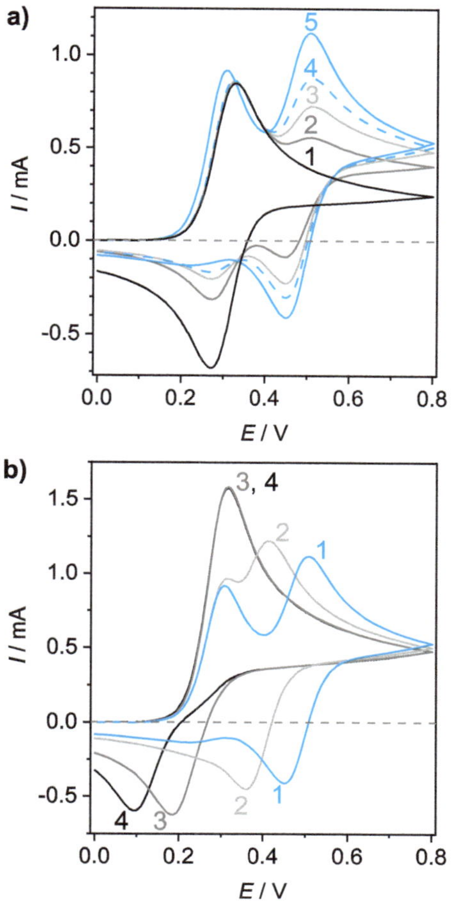

Figure 6.18 Cyclic voltammograms for systems with ECE mechanisms; $E^{\circ\prime}_{A/B} = 0.3$ V, $v = 1$ V s^{-1}, $[A]^* = 1$ mmol L^{-1}. a) Effect of κ, curves 1 to 5; k in s^{-1}: 1) 0, 2) 2, 3) 5, 4) 10, 5) 100; $\Delta E^{\circ\prime} = +180$ mV; b) Effect of $\Delta E^{\circ\prime}$ in mV: 1) +180, 2) +90, 3) −90, 4) −180 mV; $k = 100$ s^{-1}; $A = 1$ cm^2, $T = 298.2$ K, $\alpha = 0.5$, $D = 10^{-5}$ cm^2 s^{-1} (for all species). Curves 3 and 4 in panel b) overlap in the forward scan.

! According to the convention in the literature [57], we write the equilibrium in this case as a disproportionation.

The reaction between B and C is formulated here in a general way as an equilibrium.

Of course, A can also be the oxidized partner and C the reduced partner.

Note the difference in the formulation of the equilibrium constant in Eq. (6.73) compared to Eq. (6.72).

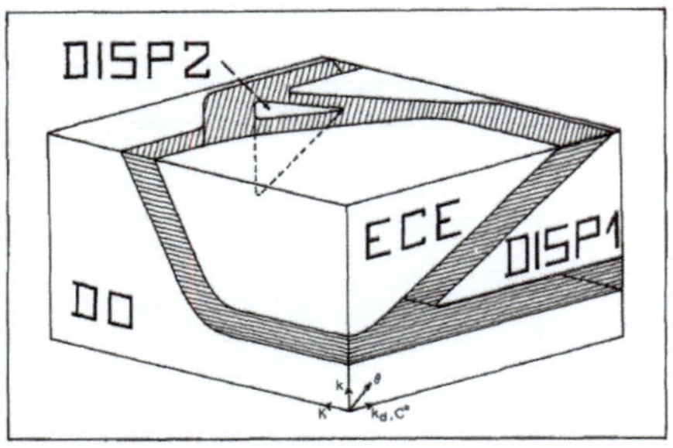

Figure 6.19 Zone diagram for competition between ECE and disproportionation mechanisms, DO is the "diffusion-controlled" zone with the transfer of a single electron. θ corresponds to the timescale of the experiment; the other abbreviations and symbols are explained in the text. Source: Reproduced with permission from [19], Copyright 1979, Elsevier.

Table 6.22 $\vec{\text{E}}\text{C}\overleftarrow{\text{E}}$ Mechanism.

First step is oxidation	
$A \rightleftharpoons B + e^-$	
$B \rightarrow C$	
$C + e^- \rightleftharpoons D$	(6.77)
First step is reduction	
$A + e^- \rightleftharpoons B$	
$B \rightarrow C$	
$C \rightleftharpoons D + e^-$	(6.78)
General formulation	
$A \pm e^- \rightleftharpoons B$	
$B \rightarrow C$	
$C \pm e^- \rightleftharpoons D$	(6.79)

are opposite to each other. This mechanism is denoted as $\vec{\text{E}}\text{C}\overleftarrow{\text{E}}$ sequence [58], here for the case of an oxidation of A and the corresponding reduction of C (Table 6.22). Analyte A is the *reduced* form, whereas C is the *oxidized* form of the respective redox couple. Again, a homogeneous redox reaction is possible, but now between A and C on the one hand (different oxidation states) and B and D on the other hand (same oxidation state).

$$A + C \rightleftharpoons B + D \quad \text{with} \quad K = \frac{[B][D]}{[A][C]} \tag{6.73}$$

Using the Nernst equilibria in Eq. (6.74) and Eq. (6.75)

$$E = E^{\circ\prime}_{A/B} + E_{NF} \ln \frac{[B]}{[A]} = E^{\circ\prime}_{A/B} + \frac{RT}{nF} \ln \frac{[B]}{[A]} \tag{6.74}$$

$$E = E^{\circ\prime}_{D/C} + E_{NF} \ln \frac{[C]}{[D]} = E^{\circ\prime}_{D/C} + \frac{RT}{nF} \ln \frac{[C]}{[D]} \tag{6.75}$$

it follows

$$K = \exp\left[\frac{\left(E^{\circ\prime}_{D/C} - E^{\circ\prime}_{A/B}\right)}{E_{NF}}\right] = \exp\left[\frac{nF}{RT}\left(E^{\circ\prime}_{D/C} - E^{\circ\prime}_{A/B}\right)\right] \tag{6.76}$$

The difference from Eq. (6.72) has interesting consequences: For $E^{\circ\prime}_{D/C} > E^{\circ\prime}_{A/B}$, the equilibrium constant becomes $K > 1$ and the equilibrium Eq. (6.73) is shifted to the right-hand side. Once we generate species C from A by ET via B, it reacts with A to form the final product D of the reaction sequence. This also produces B, which is again converted to C. Thus, a catalytic reaction cycle is initiated with electrons as the chain carriers (ET chain), called ETC reaction). Under these circumstances, the oxidation power of C is sufficient to oxidize the parent compound A.

This mechanism is applied, for example, in the electrochemically triggered $S_{RN}1$ reaction [59] (Section 6.4.7). Importantly, the current can drop to very small values and may even change its sign despite a highly efficient chain reaction occurring. This is due to the catalytic effect of the charge carriers [58, 60]. This mechanism represents the transition to the "square schemes" (Section 6.4.5).

$\vec{\text{E}}\text{C}\overleftarrow{\text{E}}$-type mechanisms are also an instructive example of why it is useful to acquire cyclic voltammograms with more than one complete potential cycle (Figure 6.20).

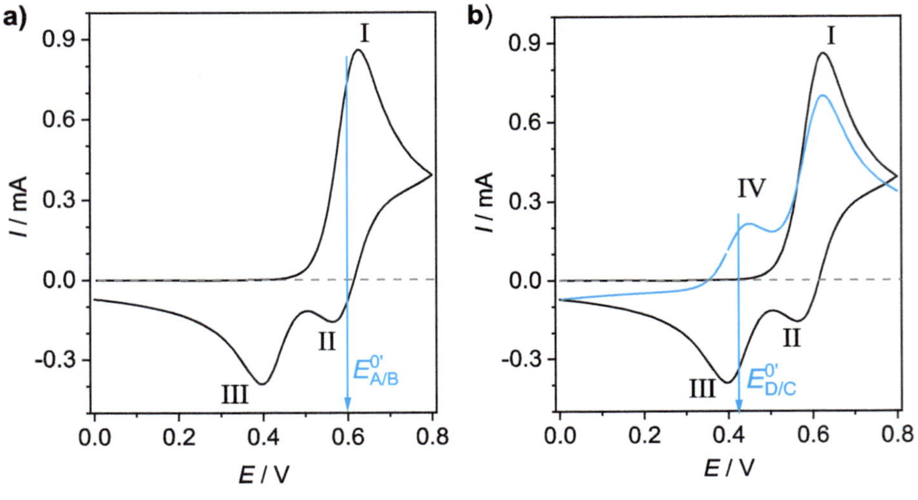

Figure 6.20 Cyclic voltammogram of an $\vec{\text{E}}\text{C}\overleftarrow{\text{E}}$ system with $E^{\circ\prime}_{A/B} = +0.6$ V, $E^{\circ\prime}_{D/C} = +0.42$ V, $k = 5\ s^{-1}$; a) simple potential cycle; b) as a) but with a third, positive going half-cycle. $A = 1\ cm^2$, $T = 298.2$ K, $\alpha = 0.5$, $D = 10^{-5}\ cm^2\ s^{-1}$ (for all species), $v = 1\ V\ s^{-1}$.

Consider again the reaction in Eq. (6.77), but now with $E^{\circ\prime}_{D/C} < E^{\circ\prime}_{A/B}$. Let the rate of the conversion B → C attain an intermediate value. In the first half-cycle, only A can react to B (intense peak I). In the second scan, we see the reverse reaction of the remaining B to A as well as that of C to D (*two* peaks, II and III). The location of peak couple I/II is defined by $E^{\circ\prime}_{A/B}$. Only in the third half-cycle (Figure 6.20b, blue curve) does the ET D → C appear as a new, additional signal IV, so that now $E^{\circ\prime}_{D/C}$ can also be determined from the curve. Peak I has a lower intensity in the third half-cycle because the diffusion layer is depleted of A. The behavior of further $\vec{E}C\vec{E}$ variants is discussed in Section 6.4.7.

Further reading. A detailed study of the ECE mechanism with its many facets can be found in [18].

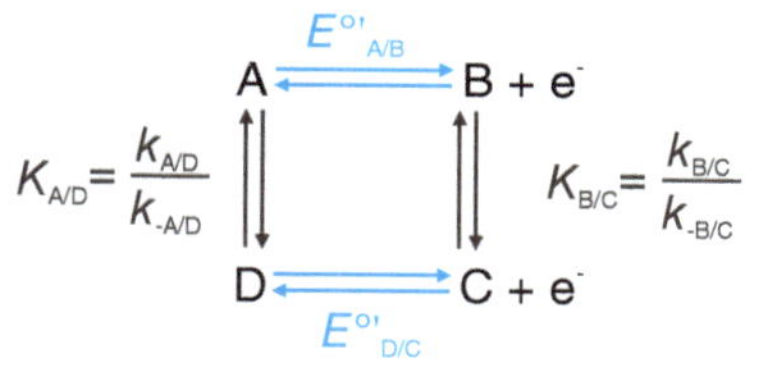

Figure 6.21 Square scheme with horizontal arrangement of ETs and vertical arrangement of chemical equilibria; the homogeneous redox reaction Eq. (6.73) is not shown for clarity.

6.4.5 The Square Scheme and Related Mechanisms

Let us now consider the ECE mechanism in Table 6.22 with the additional assumption that species D is also in equilibrium with A. This is a theoretically and practically very interesting mechanism, the so-called square scheme (Figure 6.21). ET steps (horizontal) and chemical equilibria (vertical) are closely linked. Note that the reaction of B to C in this scheme is also formulated as an equilibrium. Of course, the homogeneous redox reaction [Eq. (6.73)] plays a role too, although this is not explicitly shown for clarity.

In practice, the square scheme is a model for equilibria, such as protonation, ligand exchange, or isomerization, combined with ET [60].

The reaction scheme can be understood based on both thermodynamic and kinetic considerations. The equilibrium constants $K_{A/D}$ and $K_{B/C}$ as well as the formal potentials $E^{\circ\prime}_{A/B}$ and $E^{\circ\prime}_{D/C}$ constitute four contributions to the Gibbs free energy. The Gibbs free energy is a state variable and, therefore, the sum of all $\Delta_R G^{\circ}$-values in the thermodynamic cycle must be zero. Using $\Delta_R G^{\circ} = -RT \ln K$ or $\Delta_R G^{\circ} = -nFE^{\circ\prime}$, one obtains

$$0 = \sum \Delta_R G^{\circ} = -RT \ln K_{A/D} - RT \ln K_{B/C} - nF\, E^{\circ\prime}_{A/B} - nFE\, E^{\circ\prime}_{D/C} \quad (6.80)$$

Passing through the thermodynamic cycle in Figure 6.21, there cannot be any creation or loss of Gibbs free energy. Therefore, if three of the four parameters are fixed, Eq. (6.80) defines the fourth one. The equilibrium constant K_{hom} of the homogeneous redox reaction (6.73) is also related to the other thermodynamic quantities:

$$K_{hom} = K_{A/D}/K_{B/C} \quad (6.81)$$

If the chemical reactions in Figure 6.21 are in equilibrium, this also applies to the homogeneous redox reaction. In electrochemical experiments, however, starting from species A and D (in equilibrium), B and C can only be generated by ET, so that the rate constants and the kinetics of the homogeneous redox reaction affect the shape of the voltammograms. As a result, there is a large number of limiting cases [61], for example, ECE variants as in Eq. (6.77) or catalytic reactions (Section 6.4.6).

Further Reading. A comprehensive overview of the square scheme can be found in [60]. The reaction possibilities become even more complex when several "squares" are linked as in the fence [62], ladder [63], and mesh schemes [64] (Figure 6.22).

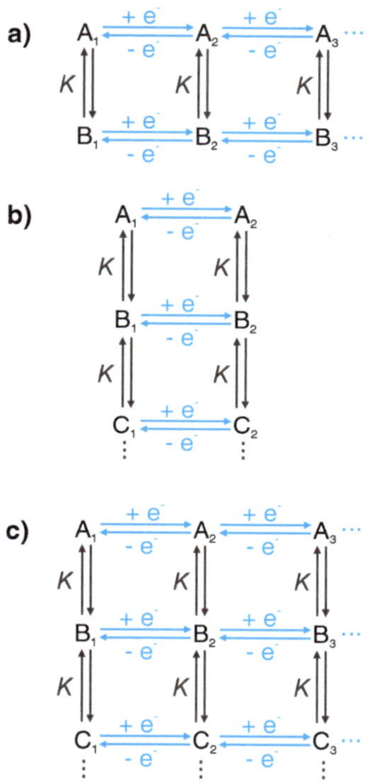

Figure 6.22 a) Fence scheme; b) ladder scheme; and c) mesh scheme. The symbols $+e^-/-e^-$ denote ETs, whereas the symbol K denotes homogeneous equilibrium reactions. The possible homogeneous redox reactions are omitted for clarity.

6.4.6 Catalysis by Electron Transfer

We already know from Section 6.4.4 that electrons can act as chain carriers in catalytic chain reactions, which – depending on kinetic and thermodynamic conditions – represent very efficient reaction processes. We will now deal more systematically with ETs in which a redox-active *molecule* acts as a catalyst. In the context of this book, we refer to such reactions as homogeneous electrocatalysis as opposed to heterogeneous electrocatalysis (Section 12.1.1) in which reaction intermediates are stabilized as adsorbates at the electrode.

Table 6.23 EC' Mechanism.

First step is oxidation	
$A \rightleftharpoons B + e^-$	
$B \xrightarrow{k} A$	(6.84)
First step is reduction	
$A + e^- \rightleftharpoons B$	
$B \xrightarrow{k} A$	(6.85)
General formulation	
$A \pm e^- \rightleftharpoons B$	
$B \xrightarrow{k} A$	(6.86)

Table 6.24 Mechanism of the mediated ET between the electrode and substrate C by mediator A/B.

First step is oxidation	
$A \rightleftharpoons B + e^-$	
$B + C \xrightarrow{k'} A + D$	(6.87)
First step is reduction	
$A + e^- \rightleftharpoons B$	
$B + C \xrightarrow{k'} A + D$	(6.88)
General formulation	
$A \pm e^- \rightleftharpoons B$	
$B + C \xrightarrow{k'} A + D$	(6.89)

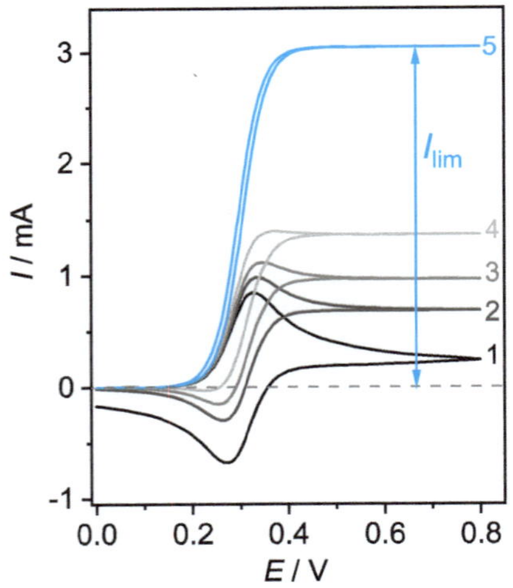

Figure 6.23 Cyclic voltammograms for the EC' mechanism. k in s^{-1} 1) 0, 2) 5, 3) 10, 4) 20, 5) 100 (blue); $v = 1\ V\ s^{-1}$, $A = 1\ cm^2$, $T = 298.2$ K, $\alpha = 0.5$, $D = 10^{-5}\ cm^2\ s^{-1}$ (for all species).

Note that species B can also undergo (undesirable) following reactions.

Table 6.23 shows the simplest, although practically not particularly useful, case of catalysis by ET. The oxidation product B formed from A reacts back to A in a homogeneous reaction according to a first-order rate law with rate constant k. The chemical reaction step takes place in *parallel* to the ET. This mechanism has been used as a limiting case and as a test model for simulation programs [26]. It is sometimes referred to as EC′ (speak EC prime). However, note the critical remarks with respect to this notation, for example, in [23].

The normalized parameter characterizing the homogeneous reaction is given in Eq. (6.82).

$$\kappa = \frac{k}{a} = E_{NF}\frac{k}{v} = \frac{RT}{nF}\frac{k}{v} \quad (6.82)$$

Species A is regenerated in the course of the reaction, which leads to an increase in the current with increasing κ (Figure 6.23). We consider first curve 1, where the chemical reaction has only a negligible effect ($k = 0$; $\kappa = 0$, E mechanism as a limiting case). On the other hand, for large κ, the system enters a state where B is in a steady-state, that is, the concentration of B is constant in time (steady-state approximation, Section 6.2.5; curves 4 and 5 in Figure 6.23). For a chosen timescale (given by a), this implies a large k. For a selected chemical system with fixed k, we achieve a large κ by decreasing v (smaller a). In the intermediate region (curves 2 and 3 in Figure 6.23), the shape of the curve changes continuously, with a concomitantly increasing peak current. At the same time, the current *after* the peak, which usually drops because of the expansion of the diffusion layer, becomes nearly independent of E. Thus, an interesting feature of the resulting voltammograms is the transition to the shape of a sigmoid function (like a stretched “S”). For large κ, the limiting current I_{lim} at the plateau of the S-shaped curve is independent of v and proportional to the root of the rate constant k

$$I_{lim} = nFAc^*\sqrt{Dk} \quad (6.83)$$

which is thus easy to determine.

Now, let the catalyzed chemical reaction occur according to a second-order rate law (Table 6.24). If C is present in large excess over A ($[C]^* \gg [A]^*$), [C] can be assumed constant because the change in [C] is negligible against $[C]^*$. Thus, k can be replaced by $k = k'[C]^*$ in Eq. (6.83), where k' and k are second-order and pseudo-first order rate constants, respectively. This simplification does not take into account a slow decrease in the concentration owing to the consumption of the substrate C. This complication would be important for preparative electrolyses following this mechanism (Section 17.4).

This reaction scheme is realized in the mediated or indirect electrochemical processes. The redox couple A/B reacts at the electrode and transfers an electron to the redox couple C/D. Species C is the “substrate”. The catalyzed ET $C \rightarrow D$ proceeds in homogeneous solution, that is, in a solution layer close to the electrode within the diffusion layer, but at a certain distance from the electrode surface. If the homogeneous reaction is fast, a reaction zone forms with particularly steep concentration gradients. The redox couple A/B acts as a mediator (electron transfer agent) between the electrode and the substrate in solution. Important electrosyntheses discussed in Section 17.4 rely on mediators because energy costs are reduced and selectivity is increased [65].

As an alternative to the redox catalysis formulated in Table 6.24 [66], chemical catalysis is discussed [Table 6.25, Eq. (6.92)] [66]: Here, an intermediate BC forms, which is an adduct of the (oxidized) mediator B and substrate C. Its decay yields the initial form A of the mediator and the product D. In contrast to redox catalysis, a chemical modification (bonding between B and C) is now essential. The explicit ET is "hidden" and occurs within BC. This is also referred to as an *inner sphere* ET, whereas in the redox catalysis discussed above, an *outer sphere* ET occurs [67].

The mechanism in Table 6.24 is of great importance for electrochemical sensors (Section 18.3.3) [68], where the prosthetic group in a redox enzyme can act as a mediator, or such an enzyme can be coupled by ETs on the one hand to a mediator and on the other hand to a substrate ("electroenzymatic reactions") [69]. The enzyme is sometimes immobilized on the electrode [7], which further complicates the mechanism and leads to extremely complex zone diagrams [23]. Among others, this reaction mechanism is also the basis for the electrochemically supported catalytic reduction of CO_2 to CO [23].

Q: What is an enzyme?

A: An enzyme is a catalyst for biological processes and typically consists of one or more polypeptide chains (protein) and an active group (prosthetic group). Enzymes accelerate the reaction of a substrate to a product. An important group of enzymes, the oxidoreductases, catalyzes redox reactions. Their prosthetic group exchanges electrons with the substrate. They are used in enzymatic biosensors, for example, in glucose sensors for diabetics (Section 18.3.3).

Table 6.25 Mechanism of mediated ET with chemical catalysis.

First step is oxidation	
$A \rightleftharpoons B + e^-$	
$B + C \xrightarrow{k'} BC$	
$BC \xrightarrow{k} A + D$	(6.90)
First step is reduction	
$A + e^- \rightleftharpoons B$	
$B + C \xrightarrow{k'} BC$	
$BC \xrightarrow{k} A + D$	(6.91)
General formulation	
$A \pm e^- \rightleftharpoons B$	
$B + C \xrightarrow{k'} BC$	
$BC \xrightarrow{k} A + D$	(6.92)

Enzyme-modified electrodes highlight possible advantages of immobilizing (binding) redox-active molecules directly to the electrode. For such systems, various complex reaction mechanisms are possible. The simplest case has already been discussed in Section 5.5. After the conversion of the entire monolayer, the current drops to zero because reactants cannot be supplied from the solution by diffusion. However, the situation becomes much more complicated when the redox-active particles are deposited as a film, or when diffusing and surface-bound particles are simultaneously present. We will return to such systems in Chapter 13.

6.4.7 Special Phenomena

Finally, we will discuss three striking effects in cyclic voltammograms that provide clues to rather unique mechanistic details. If the current-potential curves for the two scans of a cyclic voltammogram cross (Figure 6.24a and b), we speak of curve crossings.

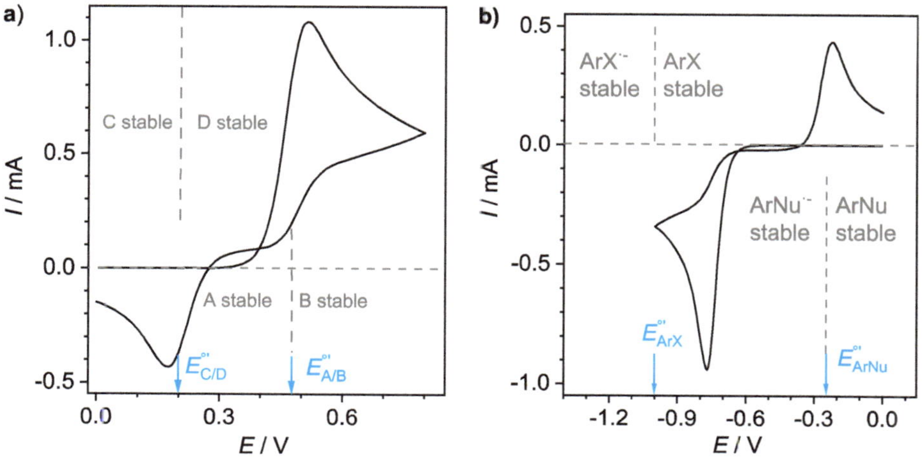

Figure 6.24 Cyclic voltammogram with curve crossing; "stable" refers to the respective nernstian equilibrium. a) ECE mechanism, $E^{o\prime}_{A/B} = 0.48$ V, $E^{o\prime}_{C/D} = +0.2$ V, $k = 10\ s^{-1}$, $[A]^* = 1\ mmol\ L^{-1}$, $v = 1\ V\ s^{-1}$; b) $\vec{E}C\overleftarrow{E}$ mechanism, $E^{o\prime}_{ArX} = -1.0$ V, $E^{o\prime}_{ArNu} = -0.25$ V, $k_1 = k_2 \cdot [Nu^-] = 10^{10}\ s^{-1}$, $k_d = 2 \times 10^4\ L\ mol^{-1}\ s^{-1}$, $[A]^* = 1\ mmol\ L^{-1}$, $v = 1\ V\ s^{-1}$.

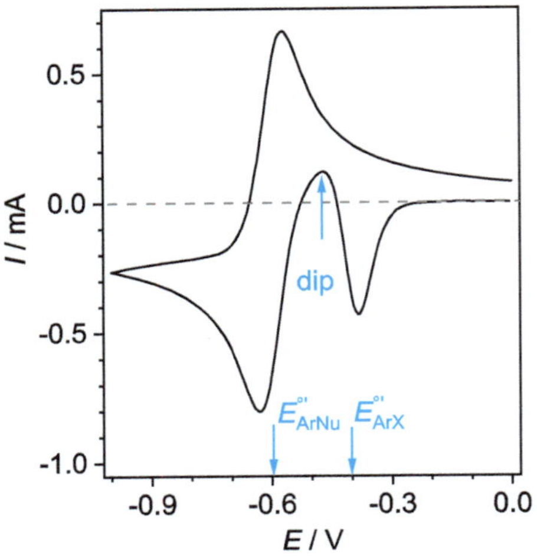

Figure 6.25 Cyclic voltammogram with current dip. $\vec{E}C\overleftarrow{E}$-Mechanism, $E^{o\prime}_{ArX} = -0.4$ V, $E^{o\prime}_{ArNu} = -0.6$ V, $k_1 = k_2$ [Nu$^-$] $= 100\,s^{-1}$, $k_d = 4\times10^6\,L\,mol^{-1}\,s^{-1}$, $[A]^* = 1\,mmol\,L^{-1}$, $v = 1\,V\,s^{-1}$.

Table 6.26 $\vec{E}C\overleftarrow{E}$ Mechanism for the electrochemically induced nucleophilic aromatic substitution ($S_{RN}1$).

First step is reduction	
$ArX + e^- \rightleftharpoons ArX^{\bullet -}$	(6.94)
$ArX^{\bullet -} \xrightarrow{k_1} Ar^{\bullet} + X^-$	(6.95)
$Ar^{\bullet} + Nu^- \xrightarrow{k_2} ArNu^{\bullet -}$	(6.96)
$ArNu^{\bullet -} - e^- \rightleftharpoons ArNu$	(6.97)
$ArX + ArNu^{\bullet -} \overset{K}{\rightleftharpoons} ArX^{\bullet -} + ArNu$	(6.98)

ArX – Starting compound with leaving group X
Nu$^-$ – Nucleophile
ArNu – Substitution product

In contrast to the previous examples, the first step of the $\vec{E}C\overleftarrow{E}$ sequence is a *reduction* in the $S_{RN}1$ reaction.

Current dips lead to peaks in the "opposite" direction (Figure 6.25). Isopotential points may occur when cyclic voltammograms are recorded immediately one after the other, and the resulting curves are superimposed (Figure 6.26).

Curve crossing. What are the reasons for these unusual curve shapes? Curve crossings occur, for example, in the ECE sequences previously discussed (Figure 6.24a) [18]. We assume conditions as in Figure 6.18b with inverted potential order ($\Delta E^{o\prime} = -280$ mV), but consider a slow following reaction B → C (characterized by κ). In this case the product B of the first ET will diffuse away from the electrode into the solution prior to its reaction to C. Thus, C is formed from B in a rather extended diffusion layer. When the potential is reversed (and E is scanned between $E^{o\prime}_{A/B}$ and $E^{o\prime}_{C/D}$), C will still be transported from the diffusion layer to the electrode, where it will be oxidized to D even after the oxidation of A to B has ceased. This occurs because at potentials $E < E^{o\prime}_{A/B}$ (where A is stable), C can still be oxidized to D as long as $E > E^{o\prime}_{C/D}$ ("D stable"). An oxidation current is observed in the voltammograms, which in some cases is even larger than the current in the first half cycle (with only A being present) at the same potentials. The sign of the current changes only when E approaches $E^{o\prime}_{C/D}$ and D having been enriched in the diffusion layer is reduced to C ("C stable"). The crossing of the traces is a result of the combination of these effects. An explanation for the small peak current of the reduction peak has already been given in Section 6.4.4.

A particularly interesting practical example, showing curve crossing, is the electrochemically triggered nucleophilic aromatic substitution ($S_{RN}1$-reaction) (Table 6.26) [70–72]. The basis for this mechanism is an $\vec{E}C\overleftarrow{E}$ sequence. Reduction of an aromatic compound ArX [ET $\vec{E}$, Eq. (6.94)] induces breaking of the bond between a carbon atom in the aromatic ring and the substituent X, which is cleaved off as anion X$^-$ [Eq. (6.95)]. This chemical reaction is characterized by the rate constant k_1. The substitution itself [Eq. (6.96)] proceeds at the radical stage ("R" in the symbol $S_{RN}1$) with a second-order rate constant k_2. However, we assume that the concentration of the nucleophile $[Nu^-]^*$ is much larger than [ArX]. In this case, the term $k_2 \cdot [Nu^-]^*$ is a pseudo-first-order rate constant. The reaction produces the reduced form of the substitution product, which is immediately oxidized at the electrode according to Eq. (6.97) [ET $\overleftarrow{E}$]. Alternatively, the homogeneous redox reaction [Eq. (6.98)] leads to ArNu as well. Its equilibrium constant in Eq. (6.93) results from the difference in the formal potentials of the starting compound and the substitution product.

$$K = \exp\left[\frac{E^{o\prime}_{ArX} - E^{o\prime}_{ArNu}}{E_{NF}}\right] = \exp\left[\frac{F}{RT}\left(E^{o\prime}_{ArX} - E^{o\prime}_{ArNu}\right)\right] \tag{6.93}$$

First, ArX is now assumed to be *more difficult* to reduce than ArNu, $E^{o\prime}_{ArX} < E^{o\prime}_{ArNu}$, which corresponds to the conditions in Figure 6.24b. Upon reduction, at negative potentials, ArX$^-$ is first formed in reaction (6.94) and then the radical anion of the product results according to Eqs. (6.95) and (6.96). This is stable at potentials far negative of $E^{o\prime}_{ArNu}$. Essentially, only $ArNu^{\bullet -}$ is present at the electrode.

Q: Why is ArX already reduced at about −0.75 V in Figure 6.24b, although $E^{o\prime}_{ArX}$ is −1.0 V?

A: The reduction peak in the cyclic voltammogram is shifted from $E^{o\prime}_{ArX}$ to less negative potentials owing to the sequence of fast following reaction steps [Eq. (6.95) and (6.96)]. Please compare this with the peak shift in the EC mechanism in Section 6.4.3.

Q: Then, how can a reduction current still be observed when the potential is scanned back to $E > E^{o\prime}_{ArX}$?

A: Here reaction (6.98) becomes important. This equilibrium is shifted to the left side for the assumed potential order ($K < 1$). If $ArNu^-$ diffuses away from the electrode, it encounters ArX that has not yet been reduced, and despite the preference for the left side of the equilibrium, at least some ArNu is formed. If this species reaches the electrode, it is reduced as long as $E < E^{\circ\prime}_{ArNu}$. The resulting current leads to the curve crossing.

It should further be mentioned that $Ar^{\bullet}$ not only reacts according to Eq. (6.96), but can also be reduced at the electrode or by a homogeneous redox reaction with ArX^- or $ArNu^-$. The resulting anion Ar^- is then protonated to the hydrocarbon ArH, with the solvent or impurities acting as proton donors. ArH represents a side-product of the substitution.

The phenomenon of curve crossing does not only occur in the molecular systems discussed here. It can also indicate delayed phase formation, for example, delayed nucleation in electrodeposition reactions (Figure 9.1).

Current Dips. A completely different situation is the basis for the voltammogram shown in Figure 6.25. Here, ArX is *easier* to reduce than the product ArNu ($E^{\circ\prime}_{ArX} > E^{\circ\prime}_{ArNu}$) [71]. During the voltammetric scan, reactions (6.94) to (6.97) take place in the first reduction peak: ArX is reduced, followed by the two chemical steps leading to substitution. Furthermore, the potential is initially such that generated $ArNu^-$ is immediately oxidized at the electrode as long as $E > E^{\circ\prime}_{ArNu}$. However, if E approaches the potential range in which ArNu becomes less stable according to the Nernst equation,

$$E = E^{\circ\prime}{}_{ArNu^{\bullet-}/ArNu} + \frac{RT}{F} \ln \frac{[ArNu]}{[ArNu^{\bullet-}]} \tag{6.99}$$

we expect a reduction peak according to the reverse reaction in Eq. (6.97). However, $ArNu^-$ formed in the process initiates equilibration (6.98) in the diffusion layer. The order of the formal potentials causes $K > 1$. Thus, the parent compound ArX is reduced to ArX^- at some distance from the electrode without current. A catalytic cycle is initiated that consumes ArX. The ArX concentration near the electrode drops, and the current collapses, resulting in a current dip (Figure 6.25). The current even crosses the zero line, that is, there is a sign inversion caused by the oxidation of $ArNu^-$ reaching the electrode and transferring an electron according to Eq. (6.97).

The electrochemically induced $S_{RN}1$ mechanism is an excellent example of the complex processes that can be relevant when recording and analyzing a cyclic voltammogram.

Isopotential points occur in certain cases when we record the current for several potential cycles and superimpose the resulting voltammograms (Figure 6.26). Under particular circumstances, all these curves intersect in one or more singular points, that is, the current at these potentials has the same value regardless of the cycle.

A prerequisite for this phenomenon is a quantitative chemical transformation of one redox-active species into another, progressing from cycle to cycle. Ultimately, this is a variant of an ECE mechanism [Eq. (6.20)] with a chemical reaction $B \rightarrow C$ that is irreversible or at least far away from the equilibrium state. Side reactions defocus the isopotential point [73]. Figure 6.26 shows three isopotential points (marked with numbered arrows). The total current is composed of two components for the ET of A and C, respectively [73], each resulting from relationships analogous to Eq. (6.35).

$$I = FA \left[D_A \left(\frac{\partial [A]}{\partial x} \right)_{x=0} + D_C \left(\frac{\partial [C]}{\partial x} \right)_{x=0} \right] \tag{6.100}$$

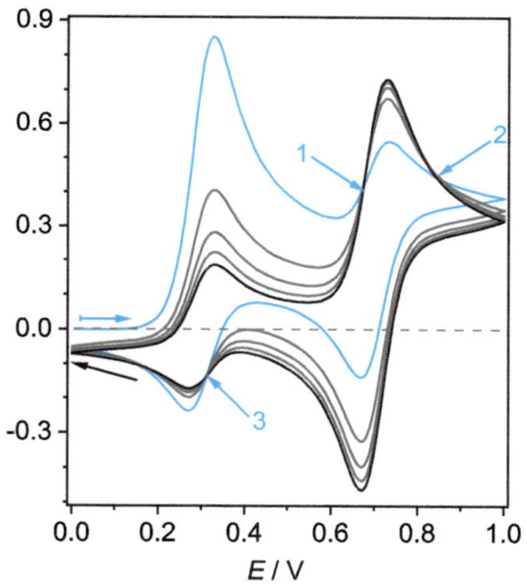

Figure 6.26 Three isopotential points (1), (2), and (3) in five successive cyclic voltammograms for the ECE mechanism shown in Table 6.9 with $E^{\circ\prime}_{A/B} = 0.3$ V, $E^{\circ\prime}_{C/D} = 0.7$ V, $K = 10^2$, $k = 2$ s^{-1}, $[A]^* = 1$ mmol L^{-1}, $v = 1$ V s^{-1}. The first cycle is highlighted in blue, the last one is drawn in black. The arrows not labeled with numbers indicate the scan directions.

The number of electrons transferred is assumed to be one for both A and C. We express the concentrations as a fraction of the respective volume concentration at the beginning of the experiment, $C_X = [X]/[X]^*$.

$$I = FA\left[D_A\left(\frac{\partial C_A}{\partial x}\right)_{x=0}[A]^* + D_C\left(\frac{\partial C_C}{\partial x}\right)_{x=0}[C]^*\right] \tag{6.101}$$

A special case

$$f = D_A\left(\frac{\partial C_A}{\partial x}\right)_{x=0} = D_C\left(\frac{\partial C_A}{\partial x}\right)_{x=0} \tag{6.102}$$

only occurs at certain potentials E of the CV. There,

$$I = FAf([A]^* + [C]^*) \tag{6.103}$$

holds. If the sum $[A]^* + [C]^*$ remains constant, while $[A]^*$ and $[C]^*$ individually change, the current according to Eq. (6.103) is also constant. With each passage of a potential cycle, B is now formed from A by ET, and further reacts chemically to form C. Accordingly, the current caused by C increases, while the current attributable to A decreases. However, the sum of the concentrations of A and C remains constant, and so does I, provided that side reactions are absent.

The situation was modeled by recording cyclic voltammograms of mixtures of two redox-active species, $[Fe(CN)_6]^{3-}$ and $[Ru(NH_3)_6]^{3+}$, while varying the concentration ratio but keeping the sum of the concentrations constant [73].

Isopoints. Phenomenologically similar but of a different origin are the isopoints, which occur when reversible or quasireversible cyclic voltammograms are superimposed that were recorded at different scan rates [74]. An isoalpha point refers to a common point in a series of cyclic voltammograms recorded at constant scan rate but with successive changes of the electrode properties [75].

The examples discussed in Section 6.4 illustrate that increasing complexity of the reaction mechanism causes an increasing diversity of shapes in cyclic voltammograms. The qualitative analysis of the I–E curves in return provides important clues about specific reaction steps and reaction mechanisms.

6.5 Extracting Quantitative Information from Experiments

After a model mechanism has been identified by qualitative observations as described in Section 6.4, a quantitative analysis of the voltammograms allows one to determine the values of reaction parameters (e.g., $k°$, k, K, $E°'$, and D) for individual reaction and transport steps (Table 6.2), for which some procedures are discussed here. In general, we are interested in a determination for *all* steps. This can be very time-consuming for complex reaction mechanisms and, therefore, may require compromises in practice. A consistent set of reaction parameters, which allows one to reproduce the experimental data by simulation over a wide range of control parameter values, provides an indication that the hypothetical mechanism may be correct. For principal reasons, this cannot be solid proof because there might be other model reaction mechanisms that result in a similarly good match. However, failure to obtain consistent reaction parameters or a gross mismatch between experiment and simulation for an assumed reaction mechanism is a firm confirmation that the assumed reaction mechanisms are not correct. Consequently, the mechanistic hypothesis must be modified, and the quantitative analysis must be reattempted. This leads to an iterative process of quantitative and qualitative data analysis. Examples are discussed in Sections 6.5.5–6.5.7.

6.5.1 Quantitative Data Analysis in Mechanistic Molecular Electrochemistry

Quantitative analysis of electrochemical processes proceeds in several steps analogous to other physicochemical measurements (Figure 6.27). Generally speaking, for each quantitative physicochemical experiment (Figure 6.27, left, e.g., measurement of temperature by means of a classical thermometer), data must be determined first (length of the liquid thread in the thermometer capillary), then converted into the desired chemical information (temperature) and finally interpreted in the context of the initial question (e.g., "The boiling point of the studied substance is ....”). As part of the overall procedure, the uncertainty range of the estimated value can also be analyzed.

Strictly speaking, the task is to determine the "numerical value of a parameter". The parameter itself is either known when making an assumption about the reaction mechanisms or must be deduced when analyzing a mechanism not studied before. This is a completely different task than discussed here. However, the quoted phrase leads to very complicated sentences, and the reader will often find a shorthand like "to determine the parameter" in the literature. This book will also make use of this shorthand, but not without indicating its inaccuracy and occasionally spelling out the full meaning.

In a cyclic voltammetric experiment, the data consist of (I, E) or (I, t) data points corresponding to the current at a defined electrode potential or at a certain time. These values in themselves are usually of no direct interest to the experimenter. Instead, they are used to find the numerical values of reaction parameters (intrinsic quantities, the desired chemical information) describing the relevant reaction steps in detail. Finally, this information will be interpreted (here, usually as properties of the molecules or materials used).

In the following, we will consider the extraction of chemical information from the measured data. While such a process is directly "built-into" a liquid thermometer by a numerical and calibrated scale, it requires dedicated effort when analyzing cyclic voltammograms. From the electrochemical data, we want to determine the values of those intrinsic quantities which are relevant for the interpretation [Eq. (6.104)].

$$I = f(E, t) \rightarrow E_i^{\circ\prime}, k_i^{\circ}, \alpha_i, k_j, k_{-j}, K_j, D_k, \ldots \tag{6.104}$$

i, *j*, *k* are index variables for the ET steps, the homogeneous steps, and the substances involved.

We could regard this as a calibration task. However, the conditions are much more complex in the case of CV than in the case of the temperature measurement because ...

- ... we always deal with a multiparameter problem, and even two parameters ($E^{\circ\prime}$ and D) have to be determined for the simplest electrode process with a single diffusion-controlled ET (E_{rev}).
- ... the relationship between the data and the desired parameters is generally nonlinear. We will discuss examples later on.
- ... correlations between the parameters can occur, for example, between rate constants that are linked by the relationship $K = k_1/k_{-1}$.

Owing to the importance of CV in practice, the examples in the following sections present data analysis mainly based on cyclic voltammograms. In principle, the considerations also apply analogously to other voltammetric methods. Sometimes we even need the combination of data from different methods for the determination of all parameter values.

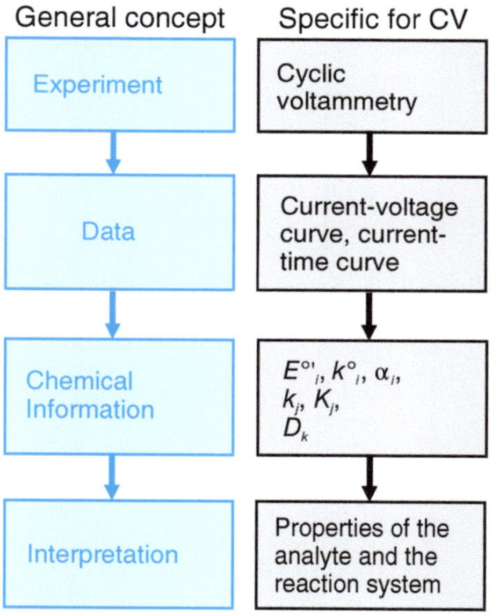

Figure 6.27 Work flow for the quantitative analysis of experimental data; left for a generic experiment, right specified for cyclic voltammetry. Symbols *i*, *j*, and *k* enumerate ET steps, chemical steps, and diffusion coefficients (as a property of the chemical species involved), respectively.

6.5.2 Direct Evaluation of Parameter Values

In some cases, the desired information is obtained directly from the experimental data by simple calculations. A pertinent example is the determination of the formal potential $E^{\circ\prime}$ for a system that is fully reversible under the experimental conditions (Figure 5.29) by applying Eq. (6.105).

$$E^{\circ\prime} = \frac{E_{pa} + E_{pc}}{2} \tag{6.105}$$

Only the potentials of the extreme values in the cyclic voltammogram (peak potentials E_{pa} and E_{pc}) have to be sampled and inserted into Eq. (6.105).

Sometimes, additional quantities are needed for direct evaluation, for which the determination of the diffusion coefficient D provides an example. Assuming a reversible ET process, the Randles–Ševčík Equation (6.7) and the peak currents I_p are used. From

$a = v/E_{NF} = nFv/(RT)$ [s^{-1}]

$$I_p = 0.4463\, nFAc^* \sqrt{Da} \tag{6.106}$$

it follows

$$D = \frac{I_p^2}{(0.4463 nFAc^*)^2 \cdot a} \tag{6.107}$$

However, D can only be calculated directly when the electroactive area of the electrode A, the concentration of the substrate c^* and the stoichiometry number of electrons n are known. The area A can be easily determined by calibration measurements with a standard substance. Data for ferrocene and Eq. (6.106) resolved for A are often used for this purpose. Some diffusion coefficients for ferrocene in common solvents are compiled in Table 6.27 and can be used for area calibrations. For ferrocene, $n = 1$, and only the exact concentration is needed as further information to determine A. We assume A to remain unchanged after the calibration experiment. The stoichiometric number of electrons n for an analyte can often be derived from chemical considerations (oxidation states). If this is not the case, n must be determined by specific experiments (Section 6.6.2) or simultaneously with D (two-parameter problem; Section 6.6.2). Please note that I_p enters quadratically in the result when using Eq. (6.107). Thus, uncertainties in I_p exert a particularly strong effect. In addition, exceptions to the validity of the Randles–Ševčík Equation (6.7) should be noted as discussed in the context of Figure 6.10.

Table 6.27 Diffusion coefficients of ferrocene in nonaqueous electrolyte solutions, supporting electrolyte is 0.1 mol L^{-1} NBu_4PF_6 at 298 K, ferrocene concentration 10 to 25 mmol L^{-1}.

Solvent	$D/(10^{-6}$ cm^2 s$^{-1})$
Propylene carbonate (PC)	4.1
Dimethylformamide (DMF)	10.7
Acetonitrile	25
Dichloromethane (DMC)	22
Dimethyl sulfoxide (DMSO)	4.9

Source: Values according to [76] from voltammetric data; determined as described in Section 6.6.2. The reference contains further detailed values for other experimental conditions.

6.5.3 Data Transformation: Linearization and Semi-integration

Sometimes electrochemical data can only be evaluated after mathematical transformation into a simple, preferably linear relationship. An example is the use of chronocoulometry for the study of dithiocarboxylate oxidation (Figure 6.28). According to Figure 5.1, chronocoulometry belongs to the larger group of potential step methods. The procedure corresponds to chronoamperometry (CA): A potential step is imposed on an electrode in quiescent solution from a potential at which no electrode process takes place to a potential in the range of the diffusion-limited current. A diffusion layer is formed. The initially high current decreases ideally with transport limitation, that is, according to the Cottrell Equation (4.130) with $t^{-1/2}$. The chronoamperometric current is often linearized by plotting it as a function of $1/\sqrt{t}$ (Figure 4.53).

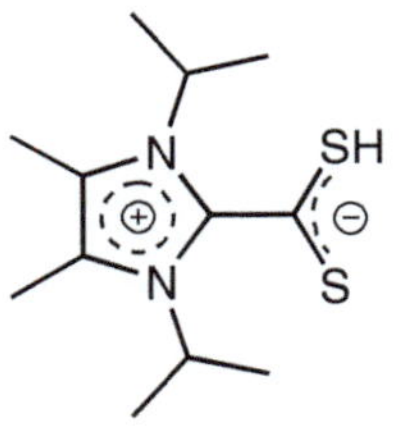

Figure 6.28 Structure of *N,N′*-di(isopropyl)-4,5-dimethylimidazolium-2-dithiocarboxylate.

Chronocoulometry is a variant of CA and records the charge Q rather than the current I. Q is the time integral of I (Figure 6.29a). The integration process smooths out any perturbations in the chronoamperogram caused, for example, by electronic noise. Substitution of Eq. (4.130) into Faraday's first law [Eq. (1.14)] yields

$$Q_F(t) = \int_0^t I_F(t)\mathrm{d}t = \frac{2nFA\sqrt{D}}{\sqrt{\pi}} c^* \sqrt{t} \tag{6.108}$$

A plot of the charge versus $\sqrt{t}$ is also referred to as the Anson plot[75] (Figure 6.29b), which again is a linearization. Often, a double potential step is used in which the initial potential step is immediately followed by a second one back to the initial potential (double potential step chronocoulometry, Figure 6.29a).

[75] Fred C. Anson, 1933–2024, US-American electrochemist, professor at the California Institute of Technology, Pasadena, California, USA.

Only the faradaic current I_F of the diffusing reactants is considered in Eq. (6.108). Let us include other current components that are not associated with the conversion of the analyte at the electrode [Eq. (6.109)]. The ordinate intercept, that is, Q at $t = 0$, corresponds to the sum of Q_{dl} for charging the electrical double layer (Section 4.3.1) and Q_{ads} for converting adsorbed species. The two contributions Q_{dl} and Q_{ads} do not increase with $\sqrt{t}$.

$$Q = Q_F + Q_{ads} + Q_{dl} \tag{6.109}$$

From the slope of the straight line in the Anson plot, corresponding to the prefactor of $\sqrt{t}$ in Eq. (6.108), n, A, c^*, or D can be determined if the other quantities are known. The intercept provides $Q_{dl} + Q_{ads}$ and, if $Q_{ads} = 0$, information about the double-layer capacity.

Semi-integration is a more complex data transformation, related to convolution integrals. The semi-integral $\mathfrak{I}(t)$ (semi = half) is a transformation which – executed twice – results in the integral of the initial function, that is, it is considered "halfway" to integration [77, 78]. This mathematical operation is applied to the current $I(t)$, where u represents a dummy variable.

$$\mathfrak{I}(t) = \frac{1}{\sqrt{\pi}} \int_0^t \frac{I(u)}{\sqrt{t-u}} \mathrm{d}u \tag{6.110}$$

The semi-integral $\mathfrak{I}(t)$ has an interesting property: under certain conditions, its value is independent of the exact excitation signal (the potential waveform applied to the cell) during the experiment [79]. Thus, distortion of the triangular potential waveform in CV due to ohmic voltage drop no longer affects the resulting data. Algorithms for the calculation of $\mathfrak{I}(t)$ can be found in the literature [79]. The semi-integral of a cyclic voltammogram is an S-shaped curve, similar to a polarogram. It is therefore also called a neopolarogram (Figure 6.30b).

The analysis of the neopolarograms can be performed by linearization of the curve (applying a suitable plot) and allows, for example, the characterization of

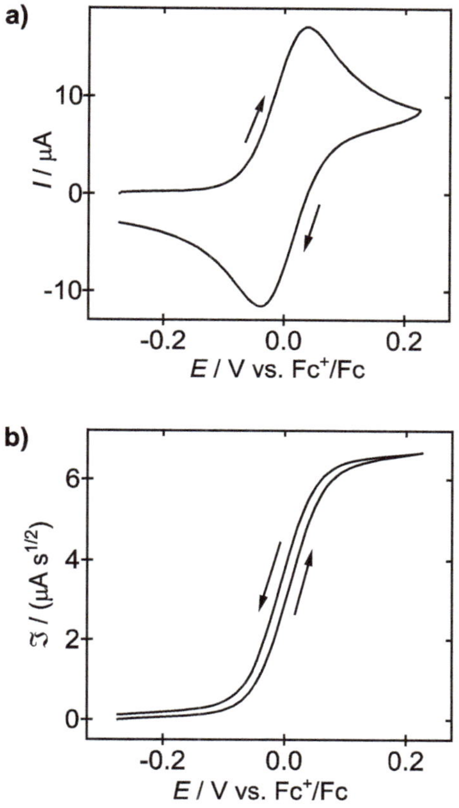

Figure 6.30 a) Cyclic voltammogram and b) corresponding semi-integral (neopolarogram) of a solution of 0.5 mmol L^{-1} ferrocene +0.1 mol L^{-1} NBu_4NPF_6 + acetonitrile; $v = 1\ V\ s^{-1}$. Source: Reproduced with perrmission from [79]. Copyright 2000, American Chemical Society.

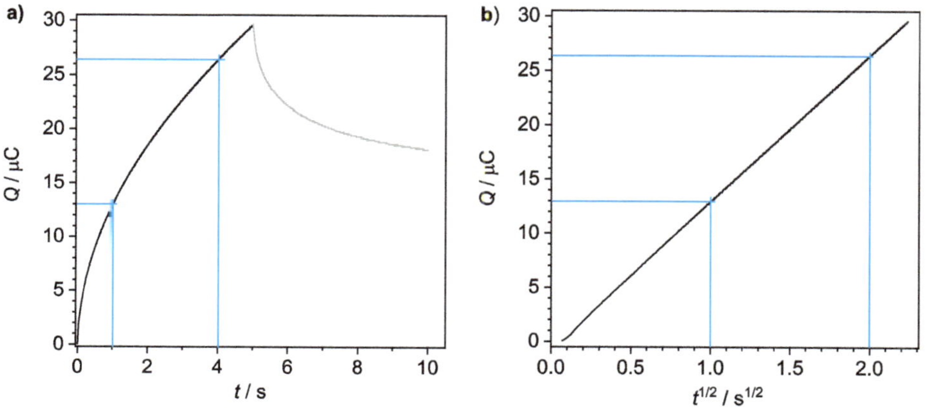

Figure 6.29 a) Double potential step chronocoulogram and b) Anson plot for the first potential step for the dithiocarboxylate shown in Figure 6.28. Solution: 0.24 mmol L^{-1} dithiocarboxylate + 0.2 mol L^{-1} NBu_4PF_6 + tetrahydrofuran (THF); duration of potential steps $\tau = 5$ s, potential step from −1.95 V to −2.55 V versus Fc/Fc^+; glassy carbon electrode. Source: Reproduced with permission from S. Dümmling, B. Speiser, N. Kuhn, G. Weyers; Electrochemical Reduction of *N,N′*-Dialkyl-4,5-dimethylimidazolium-2-dithiocarboxylates to 1,1-Dithiolate Dianions in THF. Steric Modulation of Potential Ordering by Substituents. *Acta Chem. Scand.* **1999**, 53, 876–886, Copyright 1999, Blackwell Munksgaard.

quasireversible ETs [80]. More complex electrode reactions, such as the electrochemically induced cleavage of sulfur–sulfur bonds, have also been studied in detail using the convolution transform [81]. In this case, semi-integration provided values for the heterogeneous ET rate constant $k(E)$ as a function of potential.

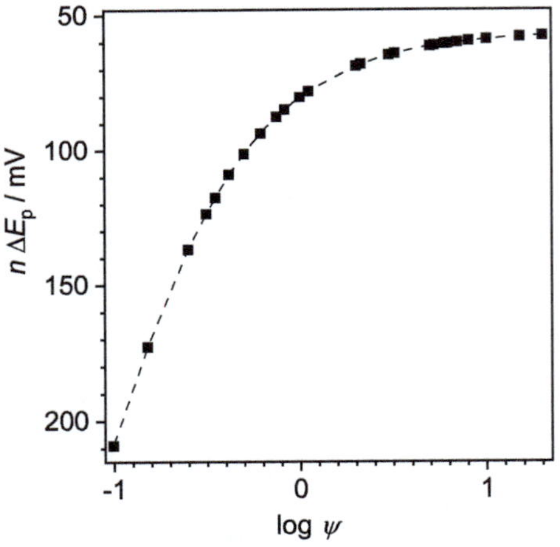

Figure 6.31 Dependence of the peak potential difference on the dimensionless parameter ψ. Plot is analogous to [20]; simulations for the quasireversible ET model, E_{qr}, for $T = 17\,°C$ and $D_O = D_R$.

6.5.4 Use of Working Curves

Sometimes, it is impossible to achieve a linear relationship between experimental data and the desired information. If necessary, we must use nonlinear working curves, most often to analyze features of the experimental voltammograms that are directly related to the variables in the diagnostic criteria.

As an example, we consider the determination of the standard rate constant of ET $k°$ from cyclic voltammograms for quasireversible systems. As shown in Section 6.4.1, the peak potential difference ΔE_p strongly increases from the "reversible value" (about 58 mV for $n = 1$) with increasing scan rate v under mixed control by diffusion and kinetics of the ET step. The interplay between the kinetics and the cyclic voltammetric timescale is summarized by the dimensionless parameter ψ [20] or Λ [21] (Section 6.4.1).

Figure 6.31 shows the relationship between $n\,\Delta E_p$ and ψ [Eq. (6.21)], as obtained from exploratory simulations (Section 6.3.1). This working curve is clearly nonlinear and can be used either graphically or numerically after finding a best-fit curve to the individual data points.

Q: This curve is very similar to the one in Figure 6.8. What is the reason?

A: The dimensionless quantity ψ and the quantity $v^{-1/2}$, calculated from one of our control parameters, are proportional, as can be seen from the definition of ψ in Eq. (6.21).
However, Figure 6.8 was for a particular analyte (ferrocene) under conditions as defined in the figure legend. It corresponds to a specific value of the rate constant. The curve in Figure 6.31 is more general and not restricted to such conditions. It can be used for various analytes with different $k°$.

The experiment typically yields values for ΔE_p, determined at different values of the control parameters (extrinsic quantities) such as v or c^*. If the stoichiometric number of charges n is known, we obtain a corresponding value of ψ for each experimental value of ΔE_p (or $n\,\Delta E_p$) according to the working curve. With v (and thus a), a value for the desired rate constant $k°$ follows from Eq. (6.21). These results are averaged over all ΔE_p. Of course, one must check that the estimated $k°$ values do not systematically depend on the control parameters, for example, on v. This would indicate that the assumed model mechanism, here E_{qr}, is not appropriate.

In the example, the working curve can be conceived as a nonlinear calibration function. Such correlations have been described early in the electrochemical and electroanalytical literature [20] as graphical methods. However, they can also be expressed in terms of a mathematical formula derived by interpolating the results of exploratory simulation. This allows one to find the desired information to be obtained by computer programs and has even been applied to the simultaneous determination of more than one reaction parameter [82]. The two-dimensional working curves are then replaced by three- or higher-dimensional "hypersurfaces". Mathematically, the determination of parameters in this way is a nonlinear optimization problem [83]. Simultaneous use of multiple features of cyclic voltammograms, for example, ΔE_p and I_p, has also been described [84]. However, it must be mentioned that such complex applications of working curves have been performed only in selected cases, and uncertainties in ΔE_p, I_p, etc., may propagate strongly to the obtained values of the reaction parameters.

6.5.5 Global Fitting to Experimental Curves

It is a common technique in the literature to fit simulated cyclic voltammograms to experimental curves over the entire potential range. The agreement between experimental and calculated curves is quite often judged subjectively, but it can also be defined more exactly in a statistical sense, for example, based on minimizing the residual sum of squares (sum of the squared differences between experimental and simulated currents at each potential of the voltammogram for a set of parameters used to calculate the data). Methods of this type are termed "least-squares fitting".

In a typical procedure, we analyze several experimental cyclic voltammograms that are as free from artifacts as possible (*IR* drop, background current, signals from impurities). In principle, fitting could also be applied to a single curve, but then the results have only limited significance. The real benefits of this methodology are in the analysis of *sets* of voltammograms obtained at a *range* of control parameter values. It thus leverages the particular advantage of CV to change the timescale of the experiment by varying v and to modulate the effect of individual reaction steps. The determination of *kinetic* parameters should therefore *always* be based on the analysis of a large number of voltammograms, in particular at different v (preferably also at different c^*, and possibly T).

Q: What are the advantages of fitting multiple voltammograms?

A: The effort required to measure several I–E curves, for example, at different scan rates v is low. However, the different time scales allow the analysis of kinetic processes. In addition, this approach increases the reliability of the results because more data are included in the analysis. Single curves can often be reproduced approximately equally well by several mechanistic models, making it difficult to obtain unambiguous results. This is much less likely when several curves for different values of the extrinsic quantities are considered simultaneously.

From the experimental curves, we qualitatively derive (Sections 6.2.3 and 6.2.4) a plausible reaction sequence as a mechanistic hypothesis (Section 6.3). The hypothesis is used in the framework of an electrochemical simulator (Excursion 6-2) to generate simulated curves. The values of the extrinsic quantities are known as control parameters of the experiments. However, the simulations additionally require values of all intrinsic quantities. Initially, these can be freely chosen within reasonable limits. From this starting combination, we are looking for just that *set of values* that leads to computed curves best matching the experimental curves when the control parameters are varied over the entire accessible range. While the control parameters may vary from curve to curve, the intrinsic quantities are fixed for a given chemical system, say for an entire set of voltammograms recorded at different v. This is true as long as we do not change the composition of the electrolyte or the temperature, for example. For the present discussion, we refrain from doing so. The values of the intrinsic quantities must therefore be identical for all curves.

We may have obtained values of some intrinsic quantities by direct evaluation (Section 6.5.2) or other procedures, as discussed in Sections 6.5.3 and 6.5.4. These should be used here to minimize the number of freely variable parameters. We now observe the agreement between experiments and simulations for a set of values of the intrinsic quantities. Then we vary these values and compute a new set of curves. The goal is to find a better agreement. By repeating this sequence, the fit is improved stepwise or iteratively.

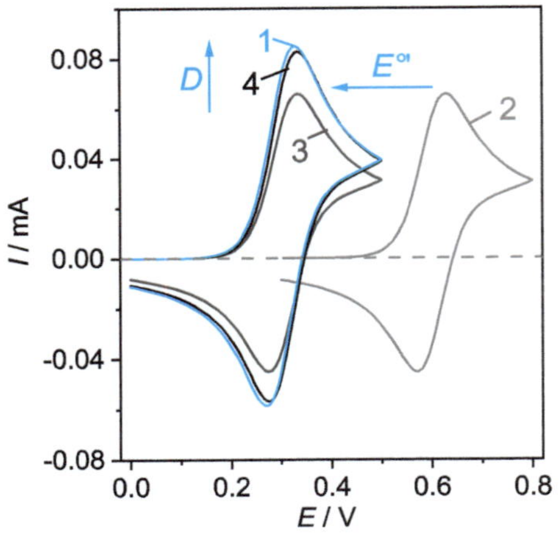

Figure 6.32 Fitting of simulated cyclic voltammograms for the E_{rev} mechanism to an experimental *I–E* curve (1). The parameters used are listed in the table.

Curve	$E^{\circ\prime}$/V	D/(cm^2 s^{-1})
1	experimental ($E^{\circ\prime} = 0.3$ V, $D = 1 \times 10^{-5}$ cm^2 s^{-1})	
2	0.600	6.0×10^{-6}
3	0.305	6.0×10^{-6}
4	0.305	9.5×10^{-6}

Figure 6.32 demonstrates this principle for the simple example of a reversible cyclic voltammogram (assumed mechanism E_{rev}) and the step-by-step determination of values for $E^{\circ\prime}$ and D with only *one I–E* curve. This typical procedure will have to be controlled by human decisions even with today's technical possibilities and cannot be left to an autonomous algorithm. The stoichiometric number of charges is assumed to be $n = 1$ based on the chemical context. Curve 1 (in blue) represents the experimental cyclic voltammogram, to which the model is fitted. For the purpose of this didactic text, this curve was also generated computationally. The fit begins with curve 2. The formal potential $E^{\circ\prime} = 0.6$ V, assumed in the first iteration step, is obviously too positive. Reducing the value to $E^{\circ\prime} = 0.305$ V yields curve 3, where the peak potentials almost match the "experimental" values. However, the simulated peak currents (I_p) are too small. We reason that possibly the diffusion coefficient $D = 6 \times 10^{-6}$ cm^2 s^{-1} used in the simulation of curve 3 is too small. Increasing D to 9.5×10^{-6} cm^2 s^{-1} (curve 4) leads to a calculated voltammogram that practically agrees with the "experimental" one. In a further iteration step, $E^{\circ\prime}$ and D could be further refined with smaller step sizes. For the present example, the *final values* of the intrinsic quantities were deliberately chosen to deviate somewhat from the "correct" values in order to illustrate the principle and to depict curves that are still distinguishable in Figure 6.32. In an actual application, of course, the correct values are not known in advance, and one would aim for the best possible match (and overlap in the figure).

Mathematically, the search for the "best" set of intrinsic quantities is again a nonlinear optimization problem [83]. We usually determine the differences R (residuals) between experimental and simulated current values for all data points in the individual measurements. The sum of the squared residuals, S_{R^2}, is minimized (least-squares fit). A starting set for values of all intrinsic quantities must be specified at the outset. The optimization algorithm checks whether and how these quantities can be changed so that S_{R^2} achieves smaller values. Step by step, a minimum of this sum is approached. It should be noted that S_{R^2} generally has several local minima as a function of the intrinsic quantities, but the search is for the global minimum. To check if the result really corresponds to the global minimum, we run the optimization starting from *different sets* of the values of intrinsic quantities. If these calculations lead to the same minimum, there is a high probability that indeed the global minimum was found. Finally, we check whether the values obtained for the intrinsic quantities are physically and chemically reasonable. For example, minima associated with negative values of a rate constant must be excluded.

Although curve fitting is based on a mathematically sound algorithm, and is usually executed with a computer, this is not a purely mathematical procedure! The expertise of the electrochemist is involved in the reasonable setup of the model, the determination of intrinsic quantity values by independent means, the selection of the possible ranges of values, and the specification of reasonable start parameters. Uncritical use of automatic fitting algorithms in quantitative analysis is discouraged. Especially in cases with a large number of intrinsic quantities, these programs quickly reach their limits [27].

We highly recommend to keep human control for at least the initial adjustment of parameters and the observation of trends. This will provide improved judgment when the assumed mechanism appears to be incorrect and the whole process has to be restarted with another model, or further experiments may be necessary to distinguish between related mechanisms. Current computer programs for electrochemical data analysis are not (yet) sufficiently reliable in this complex decision-making process, although machine learning algorithms might prove to be helpful in the future.

The more complex examples shown in the remainder of this chapter will therefore continue to follow a stepwise approach with intervention by the human experimenter: whenever a reaction step is included in the mechanism, the human specifies values of the intrinsic quantities, varies them step-by-step, and compares the simulation results with the experimental data.

To fit an experimental data set, it may become tempting to increase complexity by adding another reaction step to the mechanism. This process is called "model expansion" [85]. This can improve the fitting quality, and is indeed indicated if that step is chemically reasonable. However, the principle of parsimony (Section 6.3) should always be applied. Moreover, "over-fitting" can occur when too many parameters are fitted. Then, the model may lead to artifacts such as incorrectly estimated reaction parameters. In these cases, additional experimental data can help, for example changing the experimental timescale. On the other hand, it may prove advisable to simplify the model mechanism and eliminate redundant reaction steps ("model reduction") if they do not significantly contribute to the voltammetric behavior [85].

We have already discussed the general advantage of using multiple experimental voltammograms. In this context, a systematic approach is recommended for experimental variation of the control parameters v and c^*. For the Ru^{II} complex of Figure 6.33, selected cyclic voltammograms are arranged as a matrix in Figure 6.34, with v increasing from left to right, and c^* increasing from top to bottom. The Ru complex is reduced at approximately −0.80 V. The reverse peak is intense at large scan rates (Figure 6.34c and f). Any following reactions are excluded under these conditions. In contrast, at slow v (Figure 6.34a and d), the peak current of the reverse peak decreases relative to the forward peak. It does so also as a function of concentration: In Figure 6.34d (higher c^*), the reverse peak is smaller than in Figure 6.34a at lower c^* indicating a following chemical reaction that proceeds faster at higher concentration. From this observation, it must be concluded that there is a following reaction with an order greater than one. By analogy with other aromatic complexes of Ru^{2+} [86], a two-electron reduction to the Ru^0 complex is assumed. Several mechanisms may be consistent with these considerations. It was shown that an EE mechanism with disproportionation and dimerization of the Ru^{1+} intermediate is best suited to describe the data (Table 6.28).

This mechanism was derived by qualitative analysis and then used for simulation. Fitting yielded the simulated curves in Figure 6.34, which reproduce the experiments very well.

All experiments and also most simulations were performed with a potential step size of 1 mV. However, only a subset of the experimental data points is shown in the graphs in order to avoid confusion in the graphical representation and, thus, to make the comparison of the curves obvious.

The values of the intrinsic quantities used will not be discussed in detail here. However, it should be mentioned that the formal potentials of the two ETs in Eqs. (6.111) and (6.112) are inverted by about 20 mV. Thus, the equilibrium (6.113) is shifted to the right-hand side (compare with the discussion in Section 6.4.2).

In such optimizations, we cannot expect complete agreement between the experimental and the simulated curves. The goal is to reproduce as closely as possible the main *features* of the voltammograms, for example, the position and intensity of the peaks as well as their *changes* with the control parameters. In addition, the quality of the fit should be considered over the *entire range* of values for each control parameter. Systematic deviations may be due to missing reaction steps in the mechanism. For example, if the curves match well at high v, but poorly at low v, it may be assumed that a kinetically relevant component of the real mechanism has not been included in the model. Please refer to the explanations in Section 6.5.7.

Table 6.28 EE mechanism for the reduction of the Ru^{II} complex from Figure 6.33.

First step is reduction	
$Ru^{2+} + e^- \rightleftharpoons Ru^+$	(6.111)
$Ru^+ + e^- \rightleftharpoons Ru^0$	(6.112)
$2\,Ru^+ \rightleftharpoons Ru^{2+} + Ru^0$	(6.113)
$2\,Ru^+ \rightleftharpoons Dim$	(6.114)

Ru^{z+} – Ru complex according to Figure 6.33 in the corresponding oxidation state
Dim – Dimer of two Ru units, each with $z = 1$

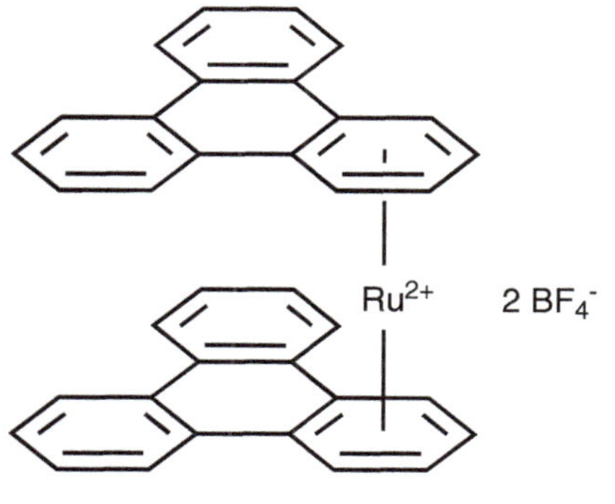

Figure 6.33 Structure of bis(triphenylene) ruthenium-bis(tetrafluoroborate).

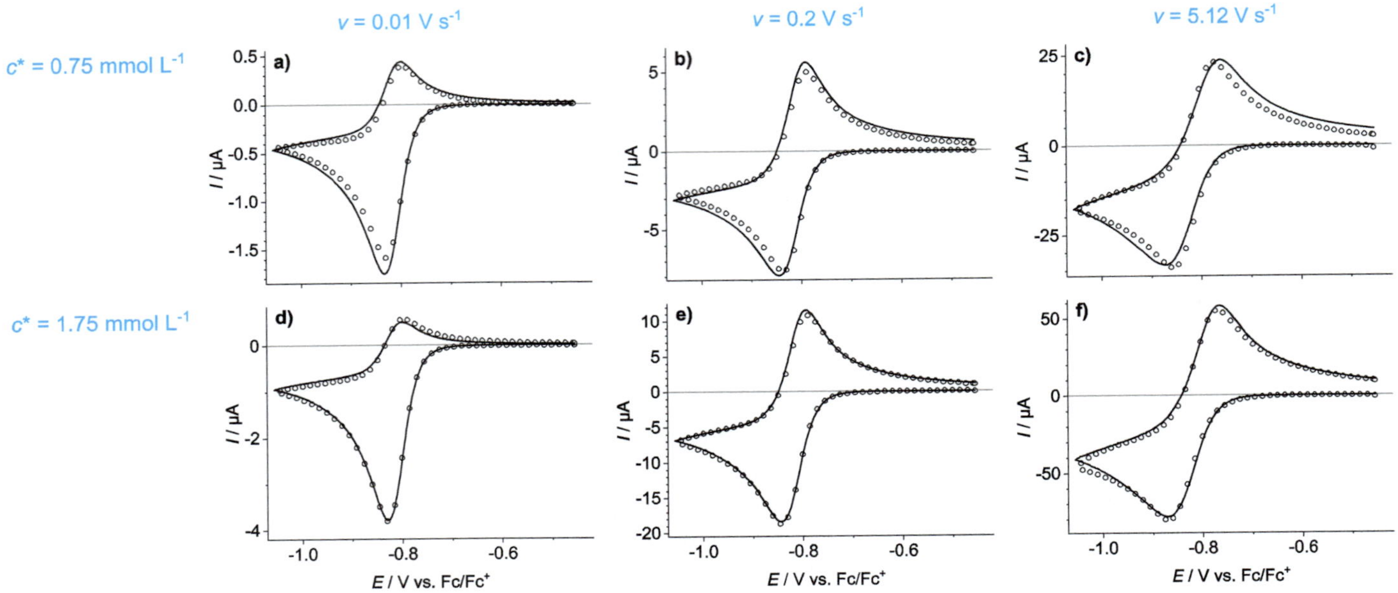

Figure 6.34 Systematic variation of v and c^* for cyclic voltammograms of the Ru complex shown in Figure 6.33. Circles represent selected experimental data, solid lines are simulations; values for c^* and v are given in blue. Electrolyte: 0.1 mol L^{-1} NBu_4PF_6 + propylene carbonate; glassy carbon disk electrode. Source: Data from S. Dümmling; PhD thesis, Eberhard-Karls University Tübingen, 2000.

6.5.6 How Do We Optimize the Quality of Voltammetric Data?

Every experiment is subject to varying degrees of measurement uncertainties, which are statistical or random deviations that can be decreased by averaging. In addition, typical systematic deviations frequently occur in electroanalytical chemistry as well. They are partially due to instrumental problems and must be corrected by dedicated approaches as demonstrated for the mitigation of ohmic potential drop in Excursion 6-1 and the treatment of background currents, especially the double layer charging currents in Chapter 5. Since the quality of the measured data is so decisive for the success of quantitative evaluations in the sense of Sections 6.5.2 to 6.5.5, we will revisit some important experimental details below.

Design of the Experiment. When designing experiments, we first and foremost follow the rule to record voltammograms for various control parameters over as wide a range of parameter values as possible. Single curves are largely worthless for the analyses addressed here. Rather, we usually aim for an ensemble of experimental curves forming a complete data set only in its entirety.

The timescale of the experiment should be changed over several orders of magnitude by changing the potential scan rate v as explained in Sections 6.2.1 and 6.2.2. A practical lower limit of v is imposed by convection and the edge effect in diffusional mass transport, which occurs, for example, in the organic solvent dichloromethane at room temperature at $v < 0.05\,\mathrm{V\,s^{-1}}$. An upper limit of v is given by the rise time of the potentiostat (i.e., the time for the potentiostat to reach a given potential) and the data acquisition rate f (i.e., the frequency by which the analog-to-digital converters can acquire and store I–t or I–E data points from a voltammetric recording). The interval between individual data points (potential step size) should not be larger than 1 mV (Exercise 6.3). In staircase voltammetry (Section 5.3.9) often used in modern instruments, the temporal and potential step sizes are linked through the scan rate $v = \mathrm{d}E/\mathrm{d}t$. For the following example, we assume that the rise time is fast enough to make f the limiting factor. The minimum time step between two data points is then $\mathrm{d}t = 1/f$, yielding $\mathrm{d}E = v\,\mathrm{d}t = v/f$. With typical values of $f = 10^5\,\mathrm{s^{-1}}$ and $\mathrm{d}E = 10^{-3}\,\mathrm{V}$, the maximum *sensible* scan rate amounts to only 100 V s^{-1}, even if a voltammetric instrument nominally provides larger v. Larger scan rates require either specialized instrumentation or compromises with respect to data quality.

Variation of concentration of the analyte can be used to identify higher-order kinetics. A lower limit is given by about 10^{-5} mol L^{-1}, depending on how well the faradaic current can be separated from the background current. An upper limit is about 10^{-3} mol L^{-1}, resulting on the one hand from solubility of the analyte in the electrolyte solution, and on the other hand also from the consideration that the current through the WE should not be larger than some 100 μA to limit the ohmic potential drop. Temperature-dependent measurements are not always easy to realize and are therefore less frequently described in the literature. However, they provide information about entropic contributions to the thermodynamics of the electrode reaction and about the temperature-dependence of kinetics. For example, reactions can be "frozen" at low temperatures. Pressure-dependent experiments are found very rarely. An example of a study under variation of temperature *and* pressure is given in [87].

The experimental curves should be reproduced several times under identical or similar conditions. It is also advisable to reassemble the entire electrochemical cell before repeating the experiment, if this is at all possible.

Ensemble versus Single Curve. For quantitative evaluation of voltammograms, the variation of the control parameters is highly important. It should be emphasized once again that, in particular, v and c^*, potentially also the temperature T, should be varied over a range as large as possible in order to facilitate the data evaluation by curve fitting. It is preferable to use all available experimental curves *simultaneously* for the analysis. For global fitting, as mentioned before, this means finding a set of reaction parameters that provides a satisfactory numerical reproduction for *all* curves of the data set. For an example, see Figure 6.34.

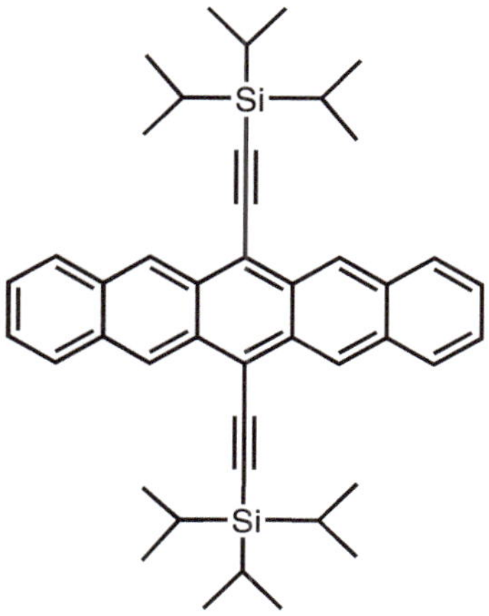

Figure 6.35 Structure of 6,13-bis(triisopropylsilylethynyl)pentacene ("TIPS-pentacene").

Electrolyte Purity. High-purity solvents and supporting electrolytes are an important prerequisite for high-quality experimental data. With relatively small analyte concentrations and a large excess of the other solution components, impurities in the solvent and the supporting electrolyte usually cause additional signals and a significant reduction of the accessible potential window (Sections 5.3.2 and 5.3.3). Furthermore, basic or acidic impurities can react with the analyte or intermediates, forming deprotonated or protonated species, respectively, which themselves may be redox active and cause additional peaks. If data for comparison are not available, the effects may be difficult to detect and, thus, may lead to drastic misinterpretations of the experimental data! The exclusion of oxygen is also essential, because O_2, on the one hand, is a biradical and possibly reacts with starting compounds or (intermediate) products of the electrochemical reaction, and on the other hand shows a pronounced reduction signal in many organic solvents (formation of superoxide $O_2^{\bullet-}$). This reduction peak can overlap with the peak of the analyte in voltammograms, as shown in the example of the reduction of TIPS-pentacene (Figure 6.35), where the reduction of O_2 overlaps with the signals from the analyte in curve 1 of Figure 6.36. Electrochemical cells designed to operate under strict exclusion of oxygen and humidity using vacuum (Schlenk line) techniques and gas-tight designs with dosing options by means of a septum offer advantages in such situations.

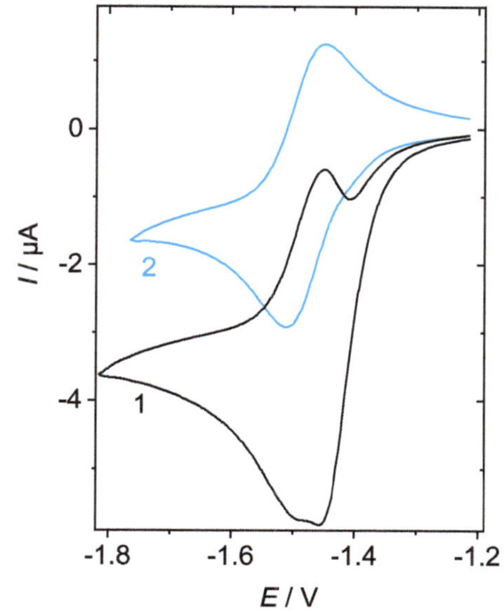

Figure 6.36 Cyclic voltammograms of TIPS-pentacene (Figure 6.35) in 0.1 mol L^{-1} NBu_4PF_6 + dichloromethane; reduction to radical anion, v = 0.1 V s^{-1}, c^* = 0.12 mmol L^{-1}, 1) with and 2) without contamination by oxygen in the electrolyte.

The Background Current I_{bg} in dynamic electrochemical measurements arises from the reactions of impurities, the decomposition of the electrolyte at the electrodes, and the capacitive charging of the double layer. The capacitive current component is intrinsically related to the cell and electrodes used. It cannot be prevented and may even dominate the measured response from which the faradaic current constitutes only a small fraction. In CV, this may occur at large v. In the context of Eq. (4.100), we have shown that the capacitive current I_C in the case of a potential ramp is *proportional* to the scan rate v after a short turn-on period. The double-layer capacitance C_{dl} is the proportionality factor. Thus, I_C increases linearly with v, while the diffusion-limited faradaic current rises only proportionally to $\sqrt{v}$ [Eq. (5.47) and Section 6.2.3]. Software settings in modern instruments may provide some assistance in limiting I_C (Section 5.3.9). For practical application, it is recommended to check carefully that the intended effect is indeed achieved for a particular system under study.

As already mentioned, the current resulting from reactions of impurities must be minimized by diligent sample preparation. However, other effects should also be considered. Electrolytic decomposition of solvent, supporting electrolyte, or electrode material causes sharp current increases at very positive and/or very negative potentials and defines the so-called potential window. The type of electrode affects the potential window by electrolytic dissolution of the material itself or by its electrocatalytic action on the electrochemical conversion of the solvent or supporting electrolyte (e.g., glassy carbon will usually give a wider potential window than platinum). Meaningful data can only be obtained inside this potential range. The potential window is probed by scanning toward positive and negative extremes of the electrode potential *before* adding the analyte. It is recommended to carefully approach the potential range of strong current rise, as this may either irreversibly alter the properties of the electrode surface or form new (reactive) chemical species by ET from the electrolyte components. Such species may inadvertently react with the analyte or reaction intermediates, thus changing the system under study. Thoughtful selection of the potential range is therefore essential. In addition, it is also recommended to change the solution and repolish the electrode after application of potentials outside the potential window.

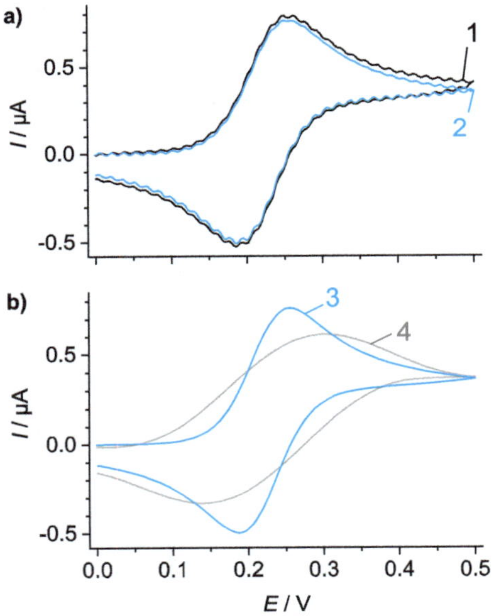

Figure 6.37 Cyclic voltammograms of 0.038 mmol L^{-1} ferrocene in 0.1 mol L^{-1} NBu_4PF_6 + dichloromethane; v = 0.05 V s^{-1}. a) Unprocessed original data (1, black) and curve after subtraction of the background current (2, blue); b) FFT filtering of a superimposed interfering frequency (3, blue) and curve after "over-filtering" (4, gray, see text).

Undesired background currents are advantageously eliminated by a difference technique (Section 6.2.2) where one subtracts data recorded under the same conditions but without the analyte. Figure 6.37a shows the voltammogram of ferrocene with a contribution of double-layer charging current (curve 1). This becomes particularly evident from the broadening of the cyclic voltammogram (difference between the forward and reverse scans) close to the switching potential. After subtracting a background voltammogram recorded in the same electrolyte solution without ferrocene (Figure 6.37a, curve 2), the typical "pointed" shape of the cyclic voltammogram is obtained without the distorting artifact.

Noise and Other Interfering Signals. During measurement, the electrochemical cell, the cable connections (especially to the reference electrode), and connectors act as sensitive antennas that pick up electromagnetic signals from the laboratory environment (from electrical equipment, spectrometers, mobile phones, etc.). Even when the electrochemical cell is operated in a shielding Faraday cage with earth connection, the experimental curve may be superimposed by noise with the mains frequency of 50 Hz or 60 Hz or by noise generated internally in the potentiostat. Software tools in modern potentiostats allow selective filtering of noise as shown in Figure 6.37b. Curve 3 was obtained from the cyclic voltammogram with background current correction (curve 2) by means of the FFT algorithm (fast Fourier[50] transform). However, filtering can also distort the signal when an incorrect selection of the frequency cut-off removes current components that are relevant for the faradaic signal (curve 4, "over-filtering").

Selection of the Potential Range. The potential range that we cover in the experiments must, of course, include the peaks associated with the ETs under investigation. However, this is not a sufficient criterion. Rather, the following considerations should be applied.

The initial potential $E_{initial}$ is selected to cause as little current through the WE as possible. In this way, we ensure that the system is in equilibrium at the beginning of the experiment. If current passes through the cell before starting the experiment, the results can be dramatically different. The easiest way to achieve equilibrium is to select as E_{initial} the so-called open-circuit potential of the electrode (E_{OCP}, defined by $I = 0$). In some devices, this potential is termed the "rest potential". This selection avoids a high (switch-on) current at the start of the experiment. In voltammograms that do not meet this requirement, a significant or even dominating current is already observed at the beginning of the cycle. Modern commercial systems for recording cyclic voltammograms often have a function to determine E_{OCP}.

To obtain an overview of as many redox processes as possible, it is recommended to scan the potential range in the entire accessible electrochemical window with two voltammograms starting from E_{OCP}. The potential cycles should extend from E_{OCP} to more positive *and* to more negative potentials, respectively (Figure 6.38). These traces are sometimes referred to as exploratory voltammograms. A potential program that extends from E_{OCP} first to the chosen positive potential limit, then across the entire potential window to the negative limit, and finally back to E_{initial} (or vice versa) is also common and often useful (Figure 6.38). Subsequently, smaller potential ranges are used to *isolate* certain processes (Section 6.5.7). Separate background current curves must be recorded for each potential range applied and for each scan rate used (Section 6.2.2).

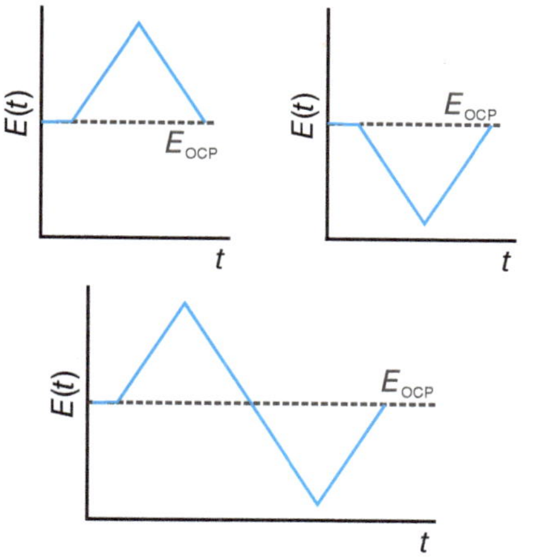

Figure 6.38 Commonly applied excitation functions in CV.

The switching potential E_λ should be located sufficiently far beyond the peak potential E_p. Starting with a distance $|E_\lambda - E_\text{p}| > 35\,\text{mV}/n$, the reverse peak has a shape that is independent of E_λ [15]. However, E_λ still has an influence on the peak potentials and thus on ΔE_p. Only for $|E_\lambda - E_\text{p}| \geq 150\,\text{mV}$ does this effect disappear. In practice, the choice of E_λ is limited by the presence of further peaks (see below, *wave clipping*) or the beginning decomposition of the electrolyte. By clever choice of E_λ, redox steps can often be virtually "separated".

Number of Potential Cycles. For quantitative evaluation, it is recommended to record cyclic voltammograms with well-defined concentration conditions at the beginning of the potential cycle: the concentration of the analyte should be constant throughout the solution and equal to c^* (bulk concentration). It is also typical to ensure insignificantly small product concentrations that may have been formed during previous voltammetric experiments. The establishment of the desired concentrations is achieved by starting at the open-circuit potential and by mixing the solution between the acquisition of two voltammograms to replenish the "concentration scar" (Section 5.3.5) and to provide unambiguous concentration conditions at the beginning of the experiment. This is conveniently achieved by stirring or by bubbling nitrogen or argon. The first potential cycle after mixing is called the initial voltammogram and can advantageously be evaluated with the techniques discussed in this chapter. Only a tiny fraction of the redox-active compound present in the electrolyte is converted in a single voltammogram, and consequently, many hundreds of curves can be recorded in this way without having to replace the electrolyte.

Q: I read in a technical article that the authors waited a few cycles until a "stable" voltammogram developed and then recorded the curve. To me, these curves look somehow more regular. What would be bad about it?

A: In the literature, we find indeed sometimes cyclic voltammograms that were recorded after many preceding potential cycles without mixing the solution. Such voltammograms can be recognized by the fact that the current forms a "closed-loop" and differs from zero at both the negative *and* the

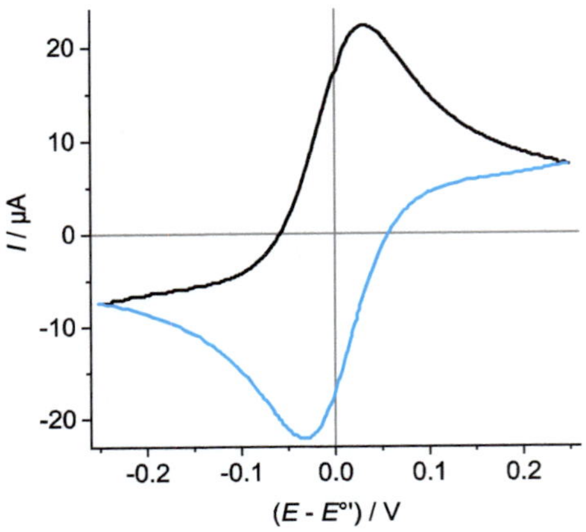

Figure 6.39 Ultimate cyclic voltammogram for the E_{rev} mechanism. If the reduced form constitutes the analyte, the black segment of the curve corresponds to the forward scan, while the blue segment corresponds to the reverse scan. Source: Reproduced with permission from [88]. Copyright 2013, Springer.

positive switching potentials (Figure 6.39). For these voltammograms, E_{initial} and E_{λ} are not evident, and the concentration profiles are not well-defined. Although their closed, symmetric shape may be aesthetically appealing, they are not suitable for quantitative extraction of reaction parameters! There have been attempts to evaluate such ultimate cyclic voltammograms [88, 89]. To date, no particular applications in molecular electrochemistry have been identified for which ultimate voltammograms would be necessary or would offer advantages. Conventional initial voltammograms, with E_{initial} and E_{λ} defined, provide more information in a shorter time and can be compared to a simulation for quantitative understanding. Although it is certainly possible to calculate ultimate cyclic voltammograms by simulations, it might take many cycles and consequently a rather long time in the experiment until the "ultimate" situation is reached [89]. Then, effects such as convection may interfere and disturb the voltammetric experiment, which relies by definition, on a quiescent solution. Figure 6.20 shows that a second potential cycle in addition to the initial voltammogram can still be useful in exceptional cases, for example, in the analysis of the ECE mechanism.

Further Reading. Further useful practical advice for CV can be found in [90].

6.5.7 Hiding and Modifying Reaction Steps

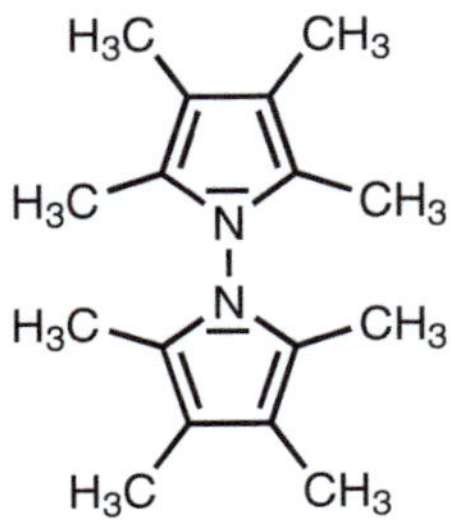

Figure 6.40 Structure of 2,2′,3,3′,4,4′,5,5′-octamethylbipyrrole (OMB).

For particularly complex mechanisms, the techniques of direct evaluation, data transformation, the use of working curves, and global fitting of entire voltammograms may not be sufficient to determine all the desired parameters from the experimental data. In such cases, it may be helpful to hide or modify individual reaction steps by variation of the reaction conditions to simplify the overall reaction and, thus, the voltammograms. We now discuss some of these approaches using the oxidation of the molecule OMB as an example (Figure 6.40) [91]. The mechanism in Figure 6.41 takes into account all electrochemical data and is based on a two-electron oxidation via a radical cation to a dication (top row, $E_1^{\circ\prime}$, $E_2^{\circ\prime}$). The radical cation $OMB^{\bullet+}$ abstracts a hydrogen atom from the solvent dichloromethane (k_2), forming a protonated form of the parent compound $OMB\text{-}H^+$ that reacts to form OMB in an acid–base equilibrium (K_1, k_1). The protonated form can also be oxidized to a radical dication (bottom row, $E_3^{\circ\prime}$), which is strongly acidic and releases a proton to form $OMB^{\bullet+}$ (k_3). The radical dication is also formed from the dication by H-atom abstraction (k_4). All individual steps can be deduced from the cyclic voltammograms. Here, we will particularly emphasize how we use variation of the reaction conditions to support the correct formulation of the mechanism and the estimation of values for the reaction parameters.

Use of Relatively Short Timescales. With short timescales (corresponding to large scan rates v in CV), all chemical reaction steps become hidden. In our example, this is already the case at $v = 1\,\text{V}\,\text{s}^{-1}$ (Figure 6.42). Depending on the potential range, the cyclic voltammograms are fully described by an E_{rev} or an EE mechanism with one or two reversible ETs, respectively. The experimental curves and simulations agree very well. Accordingly, all relevant criteria (Section 6.2.3) are fulfilled.

From Figure 6.42, it is straightforward to determine the two formal potentials ($E_1^{\circ\prime} = 0.491$ V and $E_2^{\circ\prime} = 0.917$ V) by direct evaluation (Section 6.5.2). The two formal potentials correspond to "normal potential ordering". Direct evaluation of the peak currents I_p of the first oxidation process at about 0.5 V gives the diffusion coefficient $D = 1.09 \times 10^{-5}\,\text{cm}^2\,\text{s}^{-1}$. The values for $E_1^{\circ\prime}$, $E_2^{\circ\prime}$, and D were used in the subsequent simulations. They are also necessary for the further steps of the analysis discussed

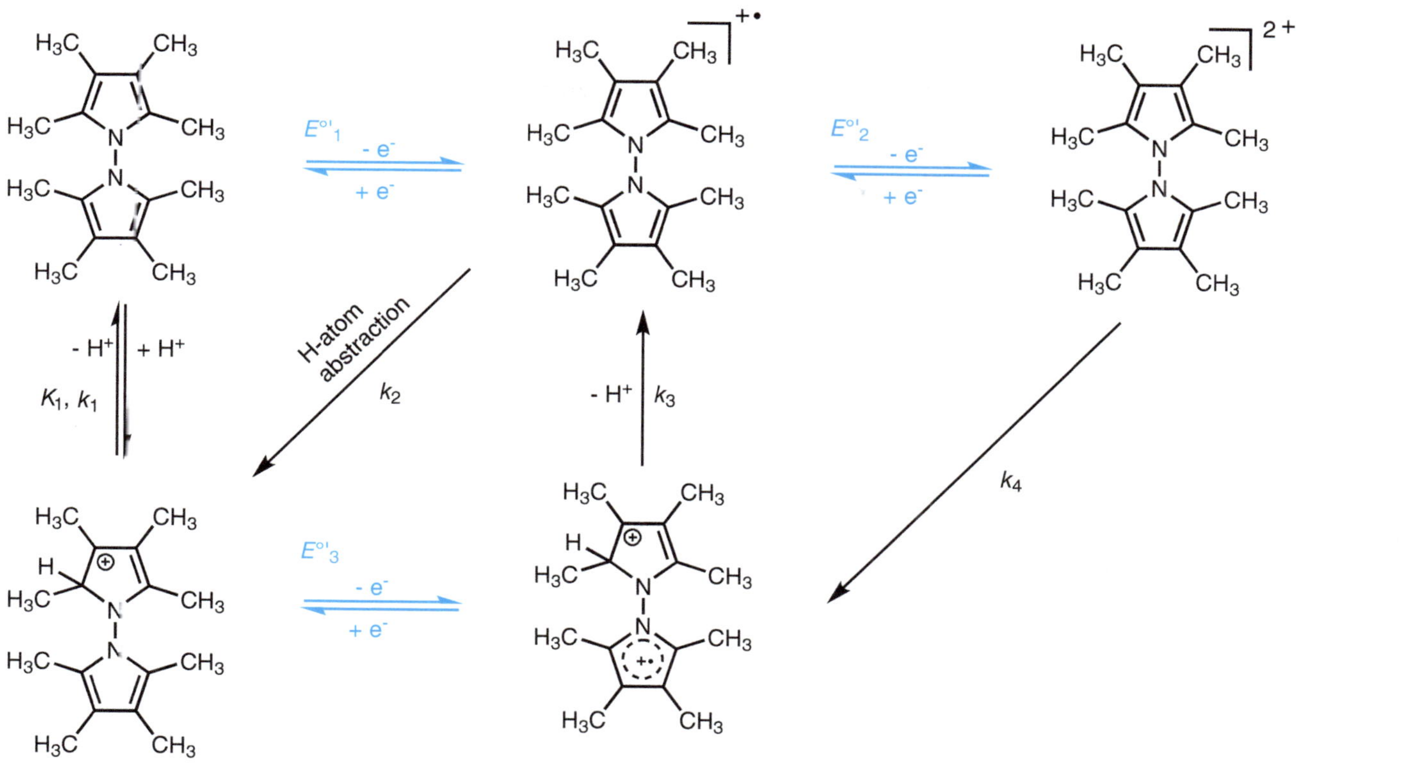

Figure 6.41 Mechanistic hypothesis for the oxidation of OMB in 0.1 mol L^{-1} NBu_4PF_6 + dichloromethane. Source: Adapted from [91].

below. Obviously, all complications from the following reactions are avoided with the scan rate $v = 1\,V\,s^{-1}$ because they do not have a significant impact due to the combination of small reaction rates and a relatively short timescale of the experiment. Thus, all chemical reaction steps were masked in this particular experiment, which helps in the accurate determination of $E_1^{\circ\prime}$, $E_2^{\circ\prime}$, and D.

Reduction of the Potential Range. Individual ETs at the electrode and the related following chemical reactions can be eliminated by restrictive selection of the potential range (wave clipping). If we confine the potential range in Figure 6.42 to $E < +0.8\,V$ (curve 1), the second ET step is no longer observed. Thus, we avoid oxidation of any protonated species that may be present and the associated complications, even at scan rates $v < 1\,V\,s^{-1}$. This is shown in Figure 6.43 for the variation of v.

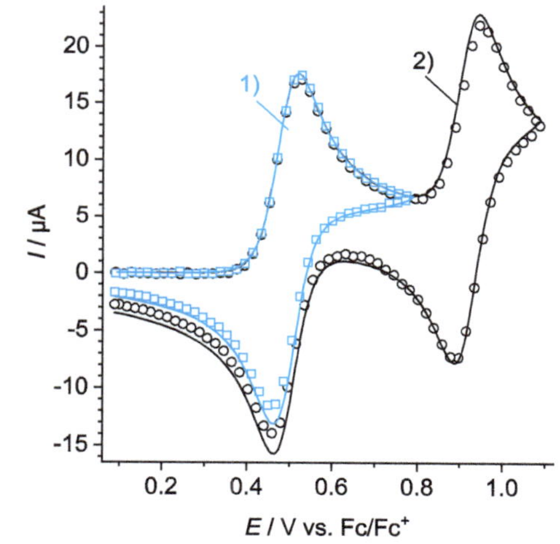

Figure 6.42 Cyclic voltammograms of OMB at $v = 1\,V\,s^{-1}$. Symbols correspond to selected experimental data, $c^* = 0.28\,mmol\,L^{-1}$, lines are simulations for an E_{rev} or an EE mechanism; potentials versus Fc/Fc$^+$. Potential scan ranges are 1) 0.09 V to 0.8 V; 2) 0.09 V to 1.09 V. Source: Adapted from [91].

The simulations (solid lines) for an E_{rev} mechanism (first oxidation assumed to proceed without complications) fit well with the experiments at $v \geq 0.1\,V\,s^{-1}$, if we use the value of $E_1^{\circ\prime}$ found earlier. However, simulated currents for the reverse peak are increasingly exaggerated at scan rates below $0.1\,V\,s^{-1}$ compared to the experimental data (symbols) (Figures 6.43a, b, and c). By taking into account the hydrogen atom abstraction (EC mechanism, Figures 6.43d, e, and f; optimized $k_2 = 0.05\,s^{-1}$), the fitting for $v = 0.05\,V\,s^{-1}$ and $v = 0.1\,V\,s^{-1}$ is significantly improved. On the other hand, especially for $v = 0.02\,V\,s^{-1}$ (Figure 6.43d), the previous deviation is overcompensated. However, if we consider the additional protolysis equilibrium between OMB-H$^+$ and OMB (Figures 6.43g, h, and i; EC mechanism with coupled equilibrium – upper left triangle in Figure 6.41; optimized values are $k_1 = 0.277\,L\,mol^{-1}\,s^{-1}$, $K_1 = 10^{-3}$), good agreement is obtained even at these longer time scales. The reaction parameters were optimized by the experimenter using the fitting procedure (see the comment in Section 6.5.5).

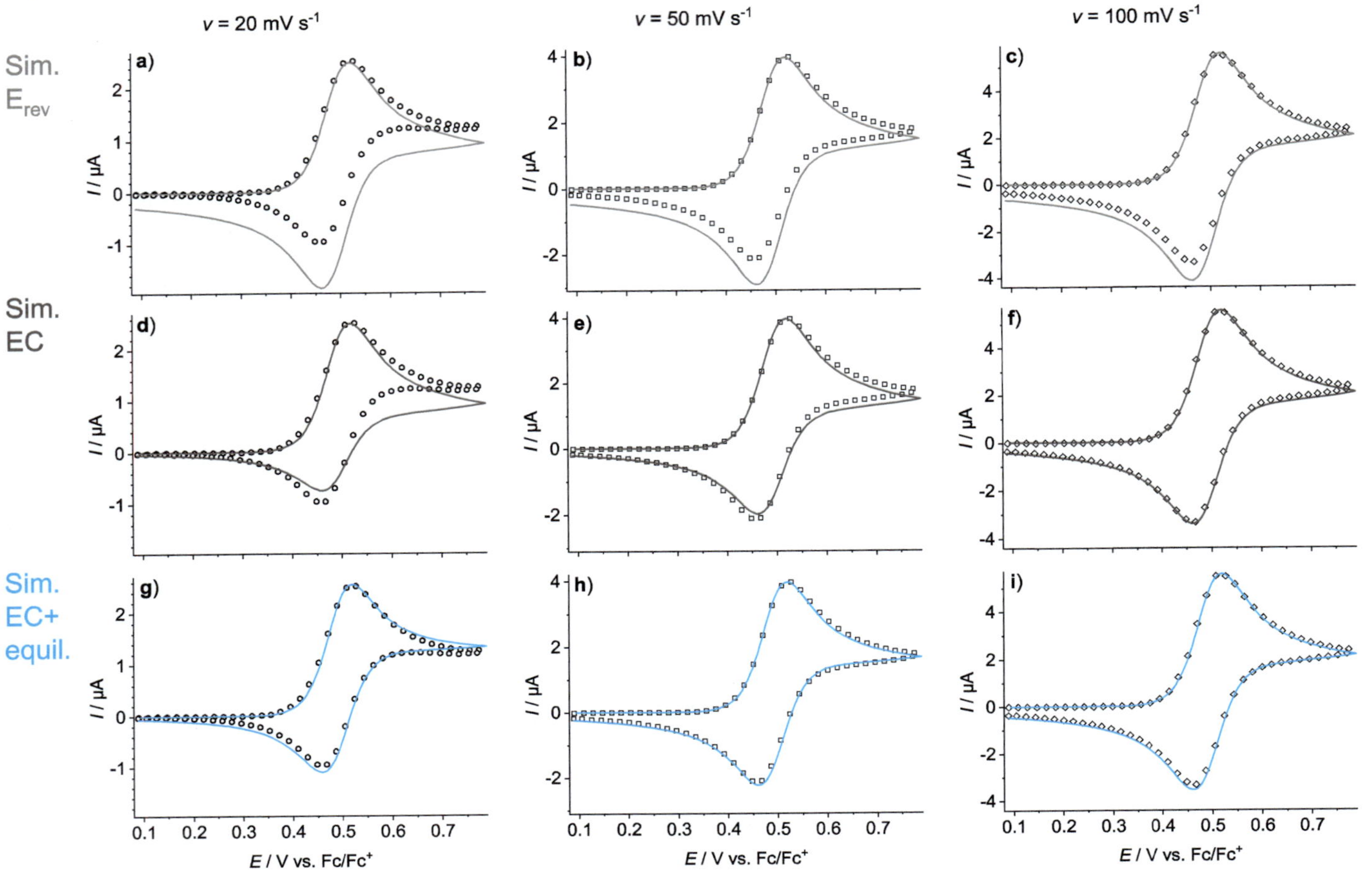

Figure 6.43 Cyclic voltammograms of OMB at $v < 1\,\mathrm{V\,s^{-1}}$ in the potential range of the first oxidation. Symbols correspond to selected experimental data for $c^* = 0.28\,\mathrm{mmol\,L^{-1}}$, lines are simulations for mechanisms as discussed in the text and shown on the left-hand side; $v = 0.02\,\mathrm{V\,s^{-1}}$ (a, d, g), $v = 0.05\,\mathrm{V\,s^{-1}}$ (b, e, h), $v = 0.1\,\mathrm{V\,s^{-1}}$ (c, f, i).

Note that the hydrogen abstraction with k_2 is assumed to be an equilibrium reaction strongly shifted to the side of the product OMB-H$^+$ ($K_2 = 10^{10}$). The protolysis reaction, on the other hand, is described by a second-order rate constant (k_1) and the corresponding equilibrium constant K_1.

For an extended potential range above +0.8 V, the simulations include the peak couple for the second ET (with the previously determined $E_2^{\circ\prime}$), leading to the OMB^{2+} dication. Quantitatively, the agreement between experimental and simulated curves becomes worse at higher potentials (Figure 6.44a, b, and c; mechanism as in the voltammograms of Figure 6.43, bottom line), because the oxidation of OMB-H$^+$ at $E_3^{\circ\prime}$ and the reaction steps with k_3 and k_4 have not yet been included. Therefore, the full mechanism of Figure 6.41 is now taken as the basis for the simulations (Figure 6.44d, e, and f), which adequately reproduces the experiments. Again, the parameters for the chemical reactions were optimized by fitting [91], starting with the previously obtained values for the reactions related to the first oxidation process. The potential $E_3^{\circ\prime}$ must be rather close to $E_2^{\circ\prime}$ to cause the currents of these two ETs to merge and obtain the intense second oxidation peak (Figure 6.43*a* and *d*; $E_3^{\circ\prime} = 0.887\,\mathrm{V}$). The values for $K_3 = 4927$ and $K_4 = 3.2\times10^{10}$ are fixed as these equilibrium constants correspond to "thermodynamically superfluous reactions"(TSR, Excursion 6-2). Only the rate constants $k_3 = 0.5\,\mathrm{L\,mol^{-1}\,s^{-1}}$ and $k_4 = 0.12\,\mathrm{s^{-1}}$ are adjustable here.

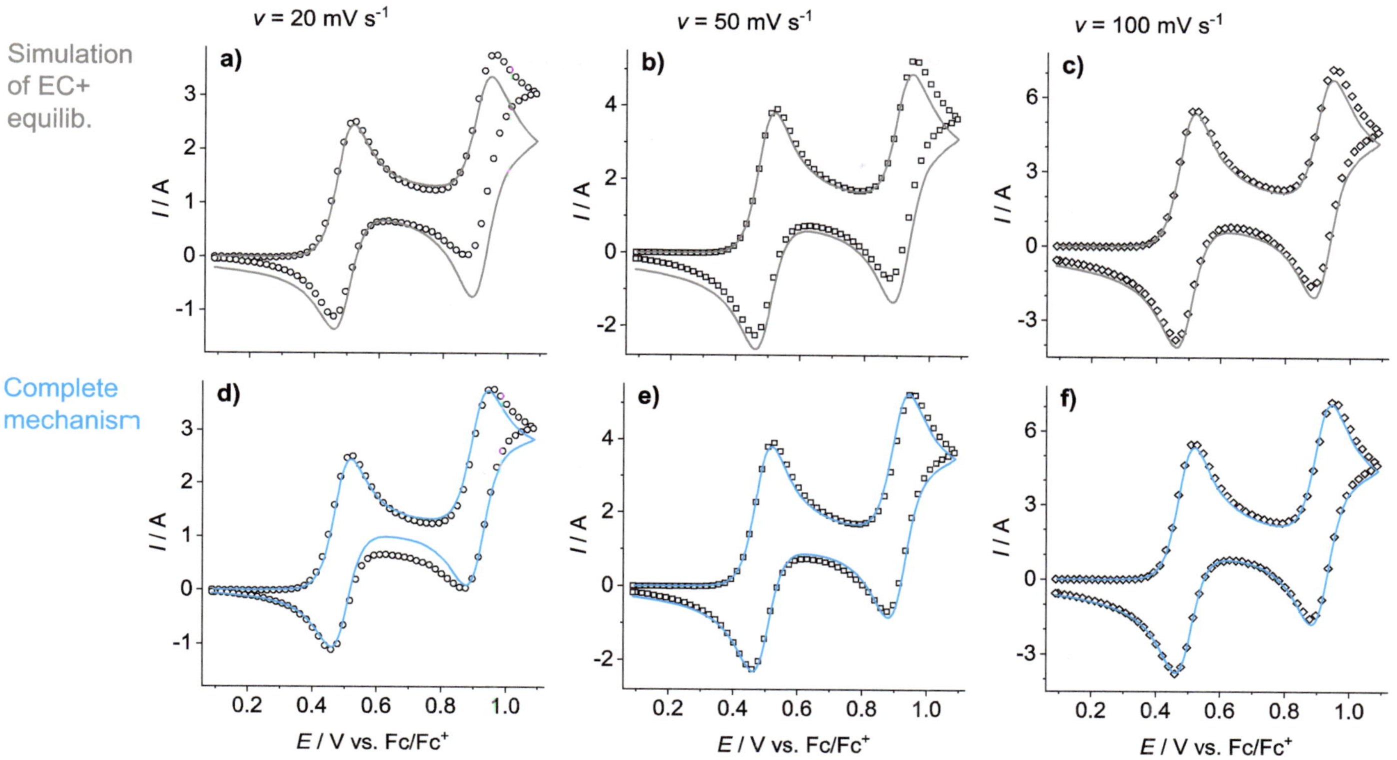

Figure 6.44 Cyclic voltammograms of OMB at $v < 1\,\mathrm{V\,s^{-1}}$, potential range including first *and* second oxidations. Symbols: selected experimental data for $c^* = 0.28\,\mathrm{mmol\,L^{-1}}$, lines: simulations; a), b), and c) mechanism as in the bottom row of Figure 6.43; d), e), and f) complete mechanism as in Figure 6.41.

Modification of the Mechanism by Addition of a Reagent. By adding suitable reagents, a complex reaction scheme can sometimes be modified and possibly simplified. In the solution, OMB is present exclusively as shown in Figure 6.40 because of the small value of K_1. The addition of tetrafluoroboric acid HBF_4 shifts the protolysis equilibrium toward OMB-H^+. With increasing concentration of the acid (up to twice the OMB concentration; Figure 6.45a, b, and c), the first oxidation of OMB becomes almost completely suppressed. In the second oxidation signal, the contributions of two ETs (of OMB-H^+ on the one hand and of the radical cation $OMB^{\bullet+}$ on the other hand) are obvious as evidenced by the shoulder in the I–E curve at +0.82 V. The two peaks overlap owing to the very close potentials of the two ETs, as best shown in Figure 6.45c. The incipient splitting of the peaks occurs because the added acid modifies the kinetics and thermodynamics of those reaction steps involving protons.

The experiments under acidic conditions qualitatively support the mechanism. Furthermore, based on the previously obtained parameters, the simulations also show good agreement in the overall course of the cyclic voltammograms and confirm the previous considerations (Figure 6.45d, e, and f; some numerical values of the reaction parameters were modified again for further optimization).

Freezing of Reaction Steps at Low Temperature. Finally, chemical reaction steps can be slowed down and frozen by lowering the temperature. Examples of this approach can be found in the literature [92–94]. However, this procedure was not used in the study of OMB. The approach is limited by the reduced solubility of the analyte and the supporting electrolyte, by the increased viscosity (decrease in the transport rate) and increased resistance (increased IR drop), as well as by the temperature range in which the electrolyte is in the liquid state.

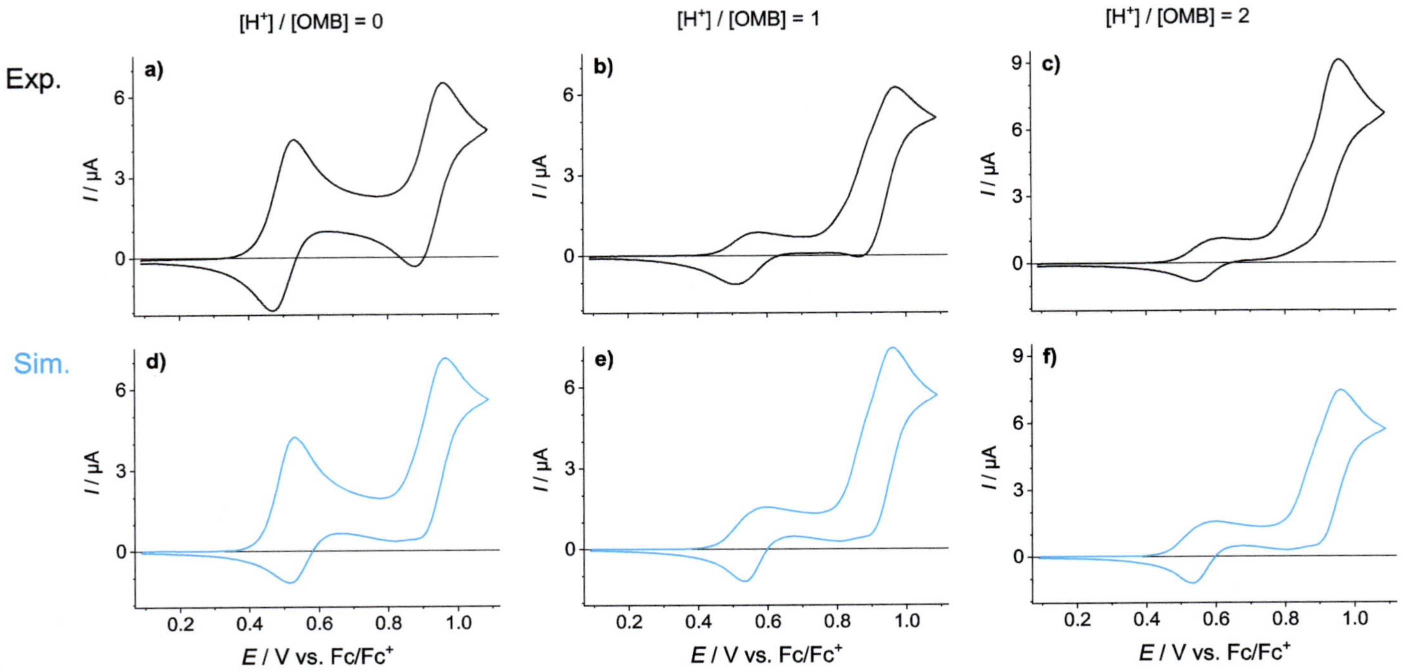

Figure 6.45 Cyclic voltammograms of OMB after addition of HBF_4. Top: experimental data (all data points), $c^* = 0.47$ mmol L^{-1}, $v = 0.02$ V s^{-1}; bottom: simulations with the full mechanism shown in Figure 6.41, adjusted reaction parameters $E_1^{\circ\prime} = 0.502$ V, $E_2^{\circ\prime} = 0.933$ V, $E_3^{\circ\prime} = 0.899$ V, $k_1 = 0.896$ L mol^{-1} s^{-1}, $K_2 = 10^{-5}$, $k_3 = 19$ L mol^{-1} s^{-1}, $k_4 = 0.122$ s^{-1}; values of $K_3 = 50.23$ and $K_4 = 3.8 \times 10^{10}$ are fixed according to TSR considerations.

6.6 Complementary Techniques

The classical electroanalytical methods and the techniques for evaluating the respective experimental data are often supplemented or combined with additional experiments. Some examples will be discussed in the following sections.

6.6.1 Bulk Electrolysis and Coulometry

In a typical electroanalytical experiment, for example, recording a cyclic voltammogram, only a tiny fraction of the initially supplied analyte is actually converted at the electrode. In the bulk of the solution, the initial concentration remains practically unchanged. In contrast, a bulk electrolysis is usually conducted until a macroscopic amount of the substrate is consumed. If we complement such an experiment with the determination of the transferred charge, the experiment is called coulometry.

For electrolysis and coulometry experiments, we use electrodes with large electroactive areas (sheet, mesh, or spiral wire electrodes with an area A in the range of cm^2) as we do for preparative electrolyses (Chapter 17). The solution is agitated by stirring. Large A and stirring minimize the electrolysis time, because the conversion per time increases with A and convection helps to provide a thin diffusion layer and thus rapid mass transport (Nernst diffusion layer model, Figures 4.71 and 5.34). We perform the experiment under potentiostatic or galvanostatic control. In either case, we may electrolyze all of the starting compound (exhaustive electrolysis) or we may add the substrate continuously to the electrolysis cell to keep its concentration constant (constant concentration mode). We will only discuss the former variant.

In the case of exhaustive potentiostatic electrolysis, a potentiostat regulates the electrode potential to a constant value, often 0.2 V above the formal potential determined for example by CV. In this potential range, we typically observe a transport-limited current and the concentration of the reactant at the electrode surface ($x = 0$) approaches zero according to the Nernst equation. The diffusion layer thickness δ is controlled by convection and remains constant. In the Nernst approximation, the concentration within δ increases linearly from $c^* = 0$ at $x = 0$ to c^* at $x = \delta$. The resulting concentration gradient (and hence the current) is proportional to c^*.

Note that other ETs may interfere with the investigated process at this potential.

$$I \propto \frac{\Delta c}{\Delta x} = \frac{c^* - c(x=0)}{\delta} = \frac{c^*}{\delta} \tag{6.115}$$

Since c^* decreases with time t owing to the macroscopically noticeable conversion of the reagent, the current is also a function of time

$$I(t) \propto c^*(t)/\delta \tag{6.116}$$

In simple cases, the decrease in concentration is given by an exponential function similar to a first-order reaction, and we obtain current-time curves as shown in Figure 6.46. In this example, ferrocene is oxidized at a potential in the transport-limited region (here 0.14 V positive of $E^{\circ\prime}$) in a dichloromethane electrolyte. At the beginning of the electrolysis (t = 0 min), a current of 50 mA flows, which continuously decays. The current has dropped to 0.8% (I = 0.43 mA) of the initial value after 228 min at which point the electrolysis is stopped.

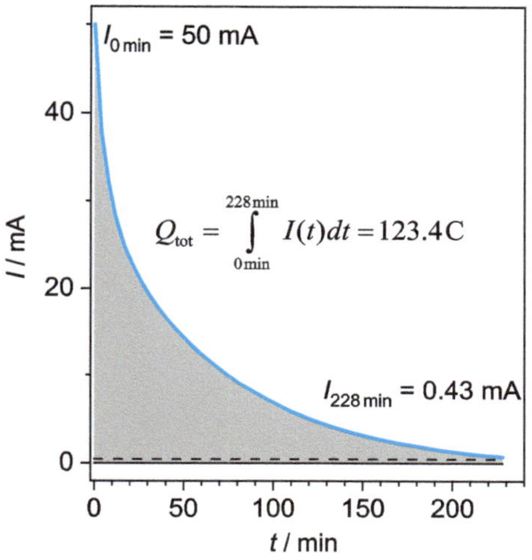

Figure 6.46 Electrolytic oxidation of ferrocene in 0.1 mol L^{-1} NBu_4PF_6 + dichloromethane. Initial amount of ferrocene m = 230.8 mg (1.24 mmol) giving an initial bulk concentration of $c^*(t = 0)$ = 35 mmol L^{-1}, E = 0.14 V versus $E^{\circ\prime}$(Fc/Fc$^+$).

Typically, the decrease in current to values below 1% of its initial value is used as a criterion for terminating the electrolysis, but other values of a so-called "residual current" can be chosen as well. Waiting even longer for the current to decrease below 1% of the initial current is often not sensible because such low currents may be caused by side reactions or impurities. Thus, these low currents do not result from the reaction under study, and their use as a criterion does not enhance the accuracy of the result.

The consumption of the initial compound or an intermediate present in the solution can also serve as a termination criterion, if a suitable detection method is available. Such methods could be spectroscopic methods and also electrochemical methods, for example, CV. In the latter case, voltammograms are recorded intermittently during electrolysis, and the decrease in the current of the reactant peak is monitored.

The charge Q_{tot} transferred during electrolysis is determined as the time integral of the current (Faraday's first law) and corresponds to the gray area in Figure 6.46. According to Faraday's second law, Q_{tot} is proportional to the amount of substrate N or its mass m.

$$Q_{tot} = n \cdot F \cdot N = n \cdot F \cdot m/M \tag{6.117}$$

- Q_{tot} – Transferred charge [C] = [A s]
- n – Stoichiometric number of charges [dimensionless]
- N – Amount of reagent [mol]
- m – Mass of the reagent [g]
- M – Molar mass of the reagent [g mol^{-1}]

Equation (6.117) can be used to determine the stoichiometric number of electrons n. Note that a coulometrically determined value of n can differ from a value determined for the same analyte by CV owing to the much longer timescale of the electrolysis experiment (cf. Section 6.6.2).

The occurrence of background currents can be taken into account for very accurate electrolysis experiments. For example, we could assume that the background current is constant and equal to the current at the end of the experiment (dashed line in Figure 6.46), and then subtract the charge resulting from the electrolysis time and this final current from Q_{tot} (Exercise 6.7).

With galvanostatic reaction control, the current I through the electrolytic cell is kept constant. The charge transferred is then simply proportional to the duration of the electrolysis ($Q_{tot} = I \cdot t$). However, the electrode potential is no longer constant during the experiment. As soon as the initially added compound is strongly depleted in the bulk of

Note that owing to the variation of E, there is a risk that electrode processes other than the one under investigation will influence Q_{tot}, and, for example, higher oxidation states could be formed.

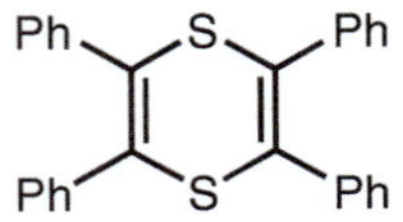

Figure 6.47 Structure of tetraphenyl-1,4-dithiin.

the solution, another species must be converted at the electrode to maintain the set current I. Thus, the electrode potential rises to the electrolysis potential required for this additional reaction, which often corresponds to electrolyte decomposition. Therefore, the endpoint of the experiment can be recognized by an increase in E.

Alternatively, CV can be used to determine complete consumption. This is demonstrated for the oxidation of the dithiin in Figure 6.47 as an example. A cyclic voltammogram of this heterocyclic compound was recorded once every minute (Figure 6.48) and shows a reversible one-electron oxidation process E_{rev}. The voltammogram at the beginning of the experiment ($t = 0$) is the most intense. We apply a constant current $I = 50$ mA. The peak currents decrease linearly with t during electrolysis and thus represent the change in concentration of dithiin during a galvanostatic electrolysis. Per unit time, a certain charge $\Delta Q = I \cdot \Delta t$ is transferred and a corresponding fraction of the substrate is converted. A plot of the peak currents I_p versus time t is linear (Figure 6.48a). Extrapolation of the straight line to $I = 0$ suggests the endpoint for the electrolysis. Given that 0.1 mmol of dithiin was electrolyzed, the total charge results in $n \approx 2$. Apparently, a second electron is removed from the substrate molecule, and a different product is formed in the timescale of the electrolysis experiment, which is much longer than that of CV. The problem of secondary, slower reactions must always be considered when comparing analytical experiments and bulk electrolyses because they proceed on different timescales.

Besides the application of coulometry to the determination of n, exhaustive electrolysis is also a tool for generating macroscopic amounts of products (Chapter 17). In favorable cases, the products can be separated from the supporting electrolyte and from the solvent by conventional methods of preparative chemistry. Isolation and identification provide valuable information about the electrode reaction.

More recently, techniques have been developed in which flow electrolysis cells are combined directly with an analytical detector for product identification. The detector could be a mass spectrometer or a UV–Vis spectrometer. If necessary, a chromatographic separation column can also be integrated into the system (Figure 6.49). The reactant (or mixture of reactants) is injected into the flow system and separated chromatographically either *before* (pathway a) or *after* the electrochemical conversion (pathway b), and finally the components are analyzed. This technique proved to be advantageous for the study of biologically active compounds. The electrolytic cell and the redox reactions taking place model enzymatic redox processes, for example, the formation of metabolites [96].

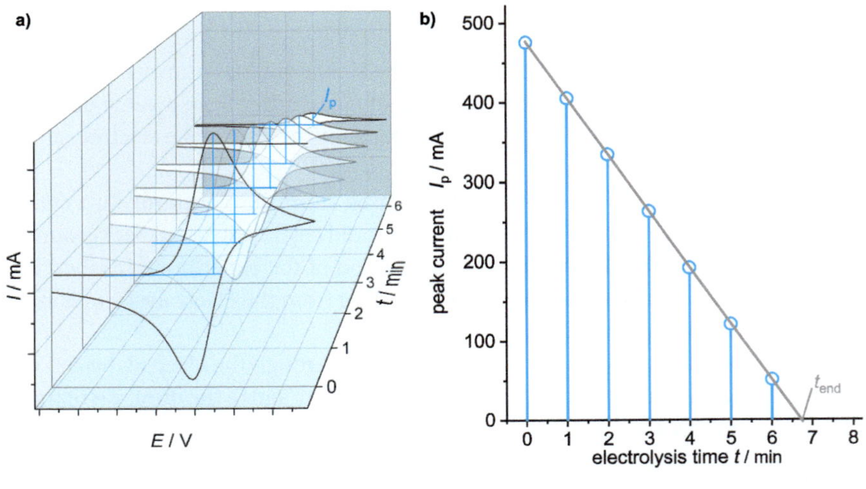

Figure 6.48 a) Development of the galvanostatic electrolysis of the dithiin in Figure 6.47 as observed in cyclic voltammograms recorded at one-minute intervals (schematic). b) The anodic peak currents from a) show the progress of the electrolysis. Idea for the figure inspired by Figure 2.44 in [95].

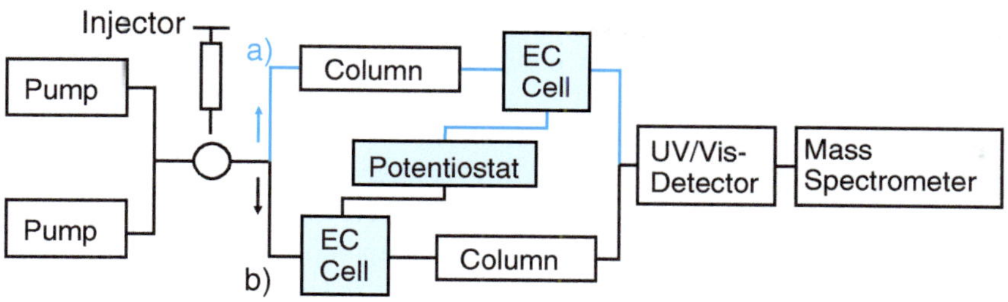

Figure 6.49 Scheme of a flow electrolysis coupled to chromatographic separation columns with a UV–Vis detector and a mass spectrometer. Source: Adapted from [96] with permission, Copyright 2004, Wiley-VCH.

6.6.2 How Can We Determine the Stoichiometric Number of Charges *n* and the Diffusion Coefficient *D D*?

The stoichiometric number of charges n, that is, the number of electrons transferred per analyte molecule, and the diffusion coefficient D represent two important reaction parameters. In principle, we can derive them from current measurements with the methods discussed in Chapter 5, where n naturally affects I directly, while D has an influence through diffusional transport of the redox-active species to the electrode surface. For example, we could use Eq. (6.7) or Eq. (6.9) to evaluate cyclic voltammograms directly. However, this is straightforward only if all other quantities except n (or D) in Eq. (6.7) or Eq. (6.9) are known. For some cases, diffusion coefficients can be estimated from the molar mass [97] of the dissolved compound. Although several assumptions or simplifications limit the application, approximate values for D may be found. However, both n and D often remain unknown and must be determined by means of one of the following methods. The two quantities are accessible in favorable cases by electrochemical or nonelectrochemical experiments, done separately or simultaneously.

Separate Determination of *n*. For an electrochemically reversible process, n can be estimated from ΔE_p in cyclic voltammograms by means of the diagnostic criterion established in Eq. (6.3). The difference of peak potentials is affected by experimental details, but n must be an integer, so the method is quite robust. However, quasi-reversibility, uncompensated potential drop, or closely spaced, superimposed peaks of multiple ETs may cause considerable problems. This also applies to the application of the formula:

$$\frac{I_p/\sqrt{v}}{I(t)\cdot\sqrt{t}} = 4.92\ \mathrm{V}^{-1/2}\cdot\sqrt{n} \tag{6.118}$$

proposed by Malachesky [98]. Here, $I_p\,v^{-1/2}$ is obtained from CV, and $I(t)\,t^{1/2}$ from CA. Eq. (6.118) can only be applied if a) the process is reversible and b) the signals are either clearly separated or the formal potentials are inverted (Section 6.4.2). When multiple peaks with normal potential ordering are superimposed in cyclic voltammograms, the method is unsuitable because the underlying Eq. (6.7) or Eq. (6.9) does not hold. The problem is discussed in detail by Amatore et al. [99]. It is related to the fact that CV uses the potential as the independent variable, and the position and overlap of peaks depend on the mechanism and other factors.

In contrast, coulometric methods can be advantageous in which the charge (i.e., the integral of the current) is directly determined (Section 6.6.1). Based on Faraday's second law, Eq. (6.119) follows from Eq. (6.117) for a complete conversion

$$n = \frac{Q_{tot}M}{F\,m} \tag{6.119}$$

However, coulometry involves bulk electrolysis on a rather long timescale as compared to CV. As already mentioned before, the relevant mechanisms may differ between the two methods (Section 6.6.1). This causes coulometric results and n

determined from the Randles–Ševčík Equation (6.7) to possibly differ as well. In particular, the occurrence of side reactions may prove problematic for bulk electrolysis.

PGSE-NMR is a nuclear magnetic resonance spectroscopic method (PGSE is an acronym for pulse gradient spin echo, and NMR is an acronym for nuclear magnetic resonance)

Separate Determination of *D*. The classical physicochemical literature describes many techniques for the determination of diffusion coefficients. Recently, the so-called pulse gradient spin echo nuclear magnetic resonance (PGSE-NMR) method has been used successfully. It is essential that the molecule contains atomic nuclei that are NMR active. The solution of the analyte is examined in a pulsed magnetic field with a spatial gradient in magnetic field strength [100]. The nuclear spins are "marked" by excitation at a certain point in time and space, and diffuse into a sample region that has a different magnetic field strength due to the field gradient. After a defined delay time, it is interrogated how far the spin-marked molecules have moved on average. The theory of NMR spectroscopy [100] leads to Eq. (6.120) for the signal strength (given by the integrals of the NMR peaks I_g with or I_0 without gradient),

I_g – Integral of the signal *with* applied gradient [dimensionless]
I_0 – Integral of the signal *without* applied gradient [dimensionless]
γ – Gyromagnetic ratio of the observed atomic nucleus [Hz G^{-1}]
δ – Pulse duration for the applied field gradient [s]
G – Field gradient [G cm^{-1}]
Δ – Time between two gradient pulses [s]

$$\ln \frac{I_g}{I_0} = -\left[\gamma^2 \delta^2 G^2 \left(\Delta - \frac{\delta}{3}\right)\right] D \tag{6.120}$$

from which D can be easily determined. Examples that also have relevance to molecular electrochemistry include the study of redox-active "capsules" [101], where the association of capsule fragments was observed via a change in D, or the comparative determination of the diffusion coefficient of the standard ferrocene in a series of solvents [76].

Note, however, that in this experiment we observe "self-diffusion" in a homogeneous solution, whereas molecules diffuse in the concentration gradient of the diffusion layer during an electrode reaction ("transport diffusion"). Only for low concentrations of the analyte do the diffusion coefficients measured for both transport modes become identical. Further information can be found in [102].

With access to modern NMR spectrometers that have appropriate hardware and software for the PGSE technique, the experiment is easy to perform.

Simultaneous Determination of *n* and *D*. Electroanalytical experiments further allow one to determine the two quantities n and D *simultaneously*. In this case, the fact is exploited that different transport conditions prevail on macroelectrodes and microelectrodes with planar and spherical diffusion, respectively. The relationships between n and D in the equations that describe the current are mathematically different for the two cases. Equation (6.7) or (6.9) is valid for a CV experiment at a macroelectrode under conditions of planar diffusion for freely diffusing redox species, provided that artifacts from the edge effect (Figure 4.50d) or the uncompensated IR drop are avoided or are negligible. Here, the peak current I_p is proportional to n and $\sqrt{D}$. Equation (5.109) applies at the microelectrode with hemispherical diffusion under steady-state conditions; the steady-state current I_{ss} is proportional to n and D. Equations (6.7) or (6.9) on the one hand and Eq. (5.109) on the other hand can be combined to determine the two parameters.

I_{ss} – Steady-state current from CV or CA at the microelectrode [A]
I – Current [A] at the macroelectrode at time t [s]
A – Geometric area of the macroelectrode [cm^2]
π – Mathematical constant pi [dimensionless]
r_0 – Radius of the microelectrode [cm]
c^* – Bulk concentration [mol cm^{-3}]
F – Faraday constant [A s mol^{-1}]
D – Diffusion coefficient [cm^2 s^{-1}]
n – Stoichiometric number of charges [dimensionless]

Problems caused by overlapping peaks in cyclic voltammograms can be avoided by analyzing diffusion characteristics using CA. The Cottrell equation describes the behavior for planar diffusion [Eq. (4.130)] in which again n and $\sqrt{D}$ appear. Under conditions of spherical diffusion, a stationary current is obtained, which is independent of the experimental method [Eq. (5.109)]. Combining the equations yields Eq. (6.121) for the case in which both measurements are made in the same solution with bulk concentration c^*.

$$D = \left(\frac{I_{ss}}{I\sqrt{t}}\right)^2 \frac{A^2}{16\pi r_0^2} \tag{6.121}$$

Then n is obtained by transforming Eq. (5.109).

$$n = \frac{I_{ss}}{4F\,Dr_0c^*} \tag{6.122}$$

For this analysis, only the geometric electrode surface area A of the macroelectrode and the radius r_0 of the microelectrode must be known in addition to $I(t)$ and c^*. Under

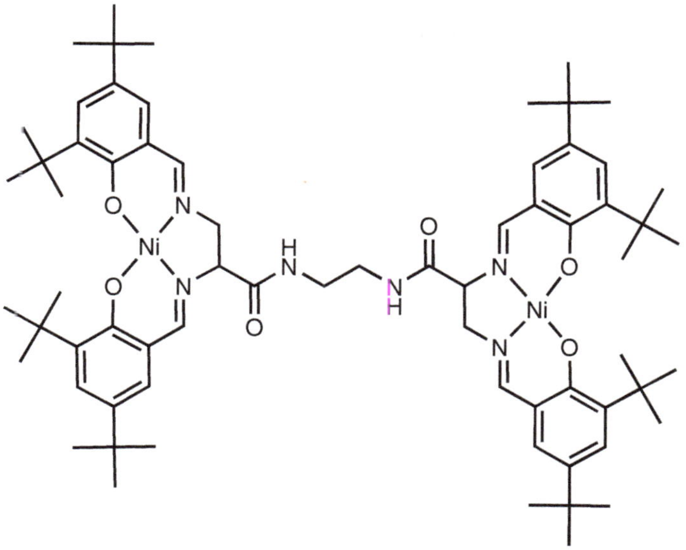

Figure 6.50 Structure of the di-Ni(II) complex μ-{7,7′-[1,1′-(1,2-diaminoethane)-bis-2,3-diiminopropionate-κ N,N′]-bis-3,5-di-*tert*-butyl-salicylidenato-κ O,O′}-di-nickel(II).

favorable circumstances, planar and spherical diffusion conditions can be achieved with the same electrode at short and longer measurement times, respectively. In this case, only one of the two geometrical parameters (A, r_0) must be provided, since the other can be calculated. Even if D is known, this kind of evaluation can be advantageous, because the observation that a value close to an integer is found for n can be used as evidence for the consistency of the data.

As an example, let us consider the dinuclear Ni complex in Figure 6.50 in which both Ni(II) centers might be oxidized. The ligand environment at each Ni atom is formed by a bis(salicylidene)ethylenediamine unit. The cyclic voltammogram in Figure 6.51a shows only a little deviation from the shape of a reversible one-electron process; only ΔE_p is slightly increased to 77 mV at small scan rates as compared to the limit of 58 mV.

The chronoamperogram (Figure 6.52a) and the resulting Cottrell plot (Figure 6.52b) show that the process at the macroelectrode is limited by planar diffusion. The slope of the straight line provides the product $I{\cdot}\sqrt{t}$. Considering also the value of I_{ss} from Figure 6.51b, for experiments in dichloromethane with 0.1 mol L^{-1} NBu_4PF_6 as the supporting electrolyte, the values $n \approx 2.0$ and $D = 6.7 \times 10^{-6}$ cm^2 s^{-1} are obtained.

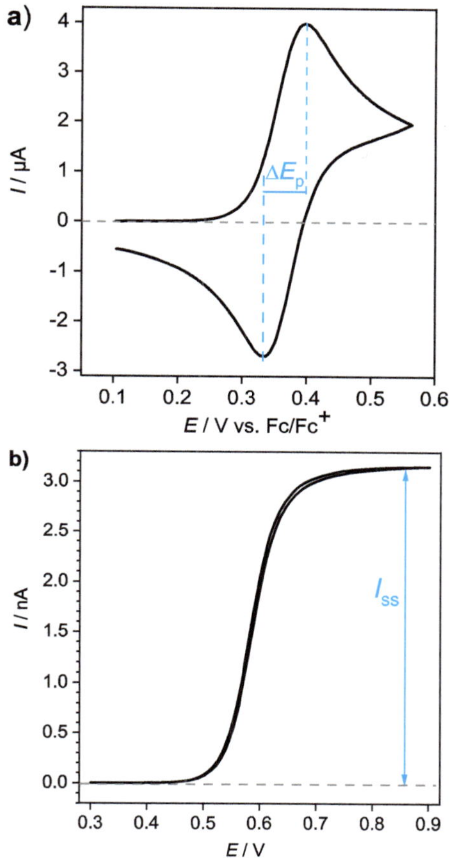

Figure 6.51 Cyclic voltammograms of the di-Ni(II) complex of Figure 6.50. a) At a macroelectrode; $v = 0.1$ V s^{-1}, $c^* =$ 0.15 mmol L^{-1}, $A = 0.064$ cm^2; b) at a microelectrode, $r_0 = 50$ μm, $v = 0.002$ V s^{-1}, $c^* = 0.10$ mmol L^{-1}. Source: Reproduction of part a) with permission from [103], Copyright 2016, Elsevier. Data for part b) were provided by A. Schank (Eberhard Karls University, Tübingen, Germany).

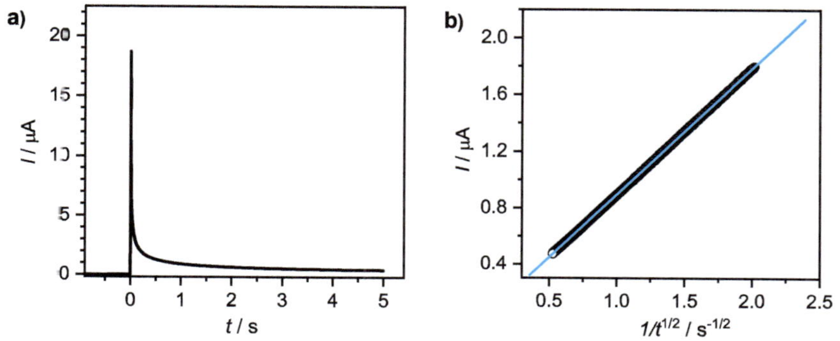

Figure 6.52 a) Chronoamperogram and b) Cottrell plot for the oxidation of the di-Ni(II) complex of Figure 6.50, $c^* = 0.051$ mmol L^{-1} at a macroelectrode with $A = 0.064$ cm^2. Source: Reproduced with permission from [103], Copyright 2016, Elsevier.

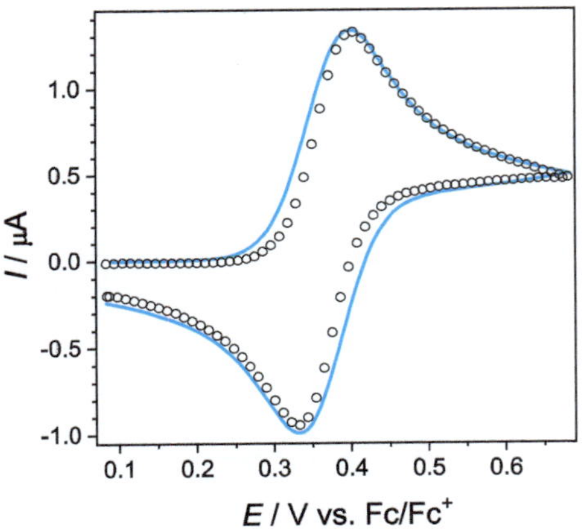

Figure 6.53 Comparison between an experimental cyclic voltammogram of di-Ni(II) complex from Figure 6.50 (circles) and a cyclic voltammogram simulated for the EE mechanism (line); $v = 0.1\,\mathrm{V\,s^{-1}}$, $c^* = 0.051\,\mathrm{mmol\,L^{-1}}$, $A = 0.064\,\mathrm{cm^2}$, $E_1^{\circ\prime} = +0.343\,\mathrm{V}$, $E_2^{\circ\prime} = +0.388\,\mathrm{V}$ versus Fc/Fc$^+$. Source: Reproduced with permission from [103], Copyright 2016, Elsevier.

With these reaction parameters, the cyclic voltammograms can be simulated assuming an EE mechanism with the formal potentials $E_1^{\circ\prime} = +0.343\,\mathrm{V}$ and $E_2^{\circ\prime} = +0.388\,\mathrm{V}$ (versus Fc/Fc$^+$). An exemplary comparison between experimental and simulated data is shown in Figure 6.53. All data are consistent with this mechanistic model. In particular, the peak currents are correctly reproduced. The difference between the formal potentials $\Delta E^{\circ\prime}$ is 45 mV, which is only a few mV higher than the value of 35.6 mV at which an EE voltammogram with a shape identical to the E_{rev} voltammogram with $n = 1$ is expected (see Section 6.4.2). This explains the shape of the experimental voltammogram, which at first glance looks like the voltammogram of a one-electron transfer with a just slightly increased value for ΔE_p compared to the E_{rev} case. Thus, based on $\Delta E^{\circ\prime}$ (Section 6.4.2), the interaction between the two identical Ni centers is small. The complete analysis can be found in [103].

6.6.3 Spectroelectrochemistry

We have already emphasized in Section 6.2.1 that purely electrochemical investigation methods allow very detailed information on reaction mechanisms, but in principle cannot provide chemical structure information. Complementary structural information can only be obtained by combining electrochemical experiments with spectroscopy, for example, by isolating products and separately characterizing them spectroscopically as commonly done in a synthesis laboratory. This approach is possible only for stable compounds and fails if reactions are not quantitative or reactive intermediates are converted faster than the time needed for isolation and characterization. In some of these cases, an analytical system connected directly to an electrochemical reactor, such as that shown in Figure 6.49, can be a great advantage, but in other cases the time needed for transporting the intermediates from the electrochemical cell to the analytical equipment (transfer time) may still be too long to capture short-lived intermediates before they enter a following reaction. A clear proof of the existence of intermediates is often important to confirm or reject a hypothetical reaction mechanism.

Spectroelectrochemical methods comprise the direct coupling of the electrochemical process and spectroscopic/spectrometric detection within the electrochemical cell.

There is a plethora of spectroelectrochemical techniques combining electrochemical with spectroscopy tools. Some are covered in Chapter 8 and concern approaches with which structural information from the solid electrode *surface* is collected simultaneously with performing an electrochemical experiment. Other important combined techniques aim to unravel reaction mechanisms involving adsorbed reaction intermediates. They are important in the context of electrocatalysis, for which *in situ* mass spectrometry coupled to electrochemical experiments is particularly popular (Section 12.4.5). In the following, we focus on detecting molecular species dissolved in solution but located *near* the electrode surface.

There are several approaches for designing a spectroelectrochemical cell depending on electrode materials and other conditions during measurements. Particularly careful considerations must be given to wavelength or spectral range, the optical path length, the compatibility with solvents, the arrangements for maintaining an inert atmosphere, and the relative positions of reference, working, and auxiliary electrodes. First, we consider types of cells for ultraviolet (UV), visible (Vis), and near-infrared (NIR) radiation (Figure 6.54), which are mainly associated with electronic transitions.

In addition, the mid-infrared (IR) spectral range is associated with excitations of molecular vibration. Window materials need to be chosen specifically for the spectral range in which the spectroelectrochemical measurements will be performed.

The measurement cell must be compatible with the spectrometer with respect to wavelength range, shape, and size, as well as time resolution. Usually, one aims to record the entire spectrum rapidly to follow changes in composition during the experiment. For work in the UV–Vis range, diode array spectrometers offer rapid data acquisition with temporal resolution of typically 1 ms (i.e., 1000 spectra per second). Data quality is related to acquisition time. Noise can be suppressed by running the experiment multiple times and taking advantage of signal averaging. The electrochemical cell setup has to meet similar requirements as described elsewhere in the book. Particular care is required with respect to compliance voltage (between WE and Aux) to ensure that potentiostatic operation can be sustained in thin-layer cells where the mass transport in hindered. The long current path causes a large solution resistance so that a high compliance voltage may be required to make the potential adjustment (measured between WE and Ref).

The cell design in Figure 6.54a and b contains an optically transparent electrode (OTE) as the WE. Transparent electrodes can be based on glass slides coated with a thin film of transparent conducting oxides (TCOs). Typical materials are indium-doped tin oxide (ITO) or fluorine-doped tin oxide (FTO). Alternatives are metal meshes (minigrids) based on gold or other noble metals. Minigrids offer the advantages (i) that reactions can be studied at conventional electrode materials, (ii) that the full light spectrum is passing through the pores in the grid, and (iii) the attenuation of light intensity (without absorption by the analyte) is uniform over the entire spectral range and can be calculated simply from the ratio between the projected area of the metal bars of the grid and the total area of the grid. Thin layers of metal evaporated onto the glass slide (< 100 nm) remain transparent, are conductive, and thus, can be used as an OTE. More recently, graphene layers on quartz or glass have been employed as OTEs [104].

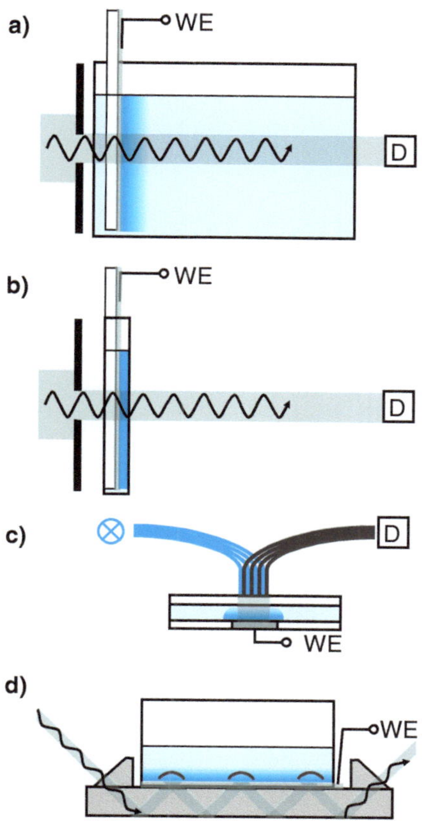

Figure 6.54 Types of spectroelectrochemical cells. a) Use of a transparent electrode with semi-infinite diffusion; b) transparent electrode in a thin layer cell; c) reflection cell with thin layer configuration for fiber optic detection; d) attenuated total reflection (ATR) configuration to probe the diffusion layer close to the electrode surface. The diffusion layer is indicated by blue coloring in all panels; D is the detector.

Spectroelectrochemical Cells with Semi-infinite Solution Layer. Figure 6.54a shows a cross section of a transmission cell with a transparent electrode and the adjacent solution. The solution layer of the cell is much thicker than the diffusion layer near the electrode. In other words, there will always be a part of the solution along the light path that is not affected during the timescale of the experiment (and thus forms a reservoir for supplying reagents to the electrode). Such a situation is called semi-infinite planar diffusion. The Cottrell Equation (4.130) remains valid and can be used to predict the current and the absorption of light. The light beam is attenuated over the entire thickness of the solution layer. However, the products of the electrolysis reside only within the diffusion layer, that is, a small fraction of the entire solution layer. A reference spectrum is recorded and used to account for reflection, scattering losses, and absorption in the transparent electrode material and the electrolyte solution containing the reactant A. We consider a chronoamperometric experiment with a one-electron transfer reaction $A \pm e^- \rightleftharpoons B$ under diffusion control (mechanism E_{rev}). The absorbance (optical density) $\mathcal{A} = \log(\mathcal{I}_0/\mathcal{I})$ can be dissected into additive contributions from two compounds, A and B.

$$\mathcal{A}(\lambda, t) = \varepsilon_A(\lambda) \cdot \int_0^\infty c_A(x, t)\mathrm{d}x + \varepsilon_B(\lambda) \cdot \int_0^\infty c_B(x, t)\mathrm{d}x \tag{6.123}$$

$\mathcal{A}$ – Absorbance or optical density [dimensionless]
ε_i – Molar decadic absorbance coefficient of compound i [mol^{-1} L cm^{-1}]
λ – Wavelength [nm]
c_i – Concentration of substance i [mol L^{-1}]
x – Spatial coordinate along the optical path [cm]
t – Time [s]

There are two limiting cases to be considered based on the absorption of compounds A and B in a given spectral range. If compound B contributes to the absorbance and A does not [$\varepsilon_A(\lambda) = 0$], then the entire signal is caused by the formation of B. The integration of c_B over the distance x yields the amount of B per geometric electrode area $N(B)/A$, which can be obtained from the total faradaic charge Q_F converted in a CA experiment [Eq. (6.108)]. Substitution of $N(B)/A = Q_F/(nFA)$ for the integral over c_B in Eq. (6.123) gives Eq. (6.124)

$$\mathscr{A}(\lambda, t) = \varepsilon_B(\lambda) \cdot 2\sqrt{\frac{D_A}{\pi}} c_A^* \sqrt{t} \tag{6.124}$$

The quantities with the index A describe how fast the starting material diffuses to the electrode and thus how much B is formed. Plotting $\mathscr{A}$ against $t^{1/2}$ should yield a straight line. This is analogous to the Anson plot in Figure 6.29b.

If the condition $\varepsilon_A(\lambda) > 0$ is *not* fulfilled, the spectrum for B is obtained with a strong background from A, because there is a substantial concentration of A in the entire solution layer that contributes to the absorbance, while B is found only in the diffusion layer. Difference spectra can be used to suppress the background. A reference spectrum of the solution of A before electrochemical conversion is taken (Figure 6.55a). Further spectra are taken during the electrochemical conversion $A \pm e^- \rightleftharpoons B$. The difference spectrum (Figure 6.55b) contains negative and positive peaks. Negative peaks are associated with loss of A in a spectral range with $\varepsilon_A(\lambda) > \varepsilon_B(\lambda)$. Positive peaks are associated with the production of B in spectral ranges with $\varepsilon_A(\lambda) < \varepsilon_B(\lambda)$. Difference spectra can be powerful tools in resolving reaction mechanisms and identifying reaction intermediates [105]. Results for further reaction mechanisms studied in this cell are compiled in [105].

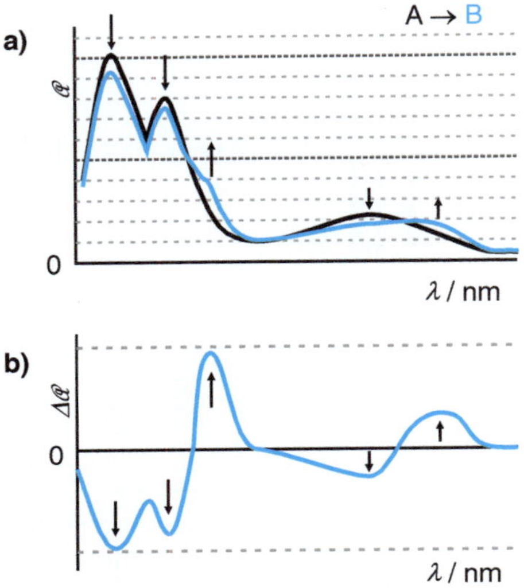

Figure 6.55 Schematic of a) a spectrum of the working solution before electrolysis (black) and during CA (blue); and b) the corresponding difference spectrum between the spectra.

Optically Transparent Thin-Layer Cells. Such cells combine an OTE, reference, and auxiliary electrodes with a thin layer of electrolyte solution (Figures 6.54b and 6.56). The solution layer is so thin that the diffusion layer grows within seconds through the entire electrolyte layer. This allows complete electrolysis of the starting material to the product within the timescale of the electrochemical experiment and also offers a lower background from solvent and electrolyte. A range of examples of electrochemical reactions studied in optically transparent thin-layer electrolysis (OTTLE) cells have been reported [106]. Within this textbook, we will discuss the investigation of redox centers in proteins with OTTLEs in Section 18.2.4.

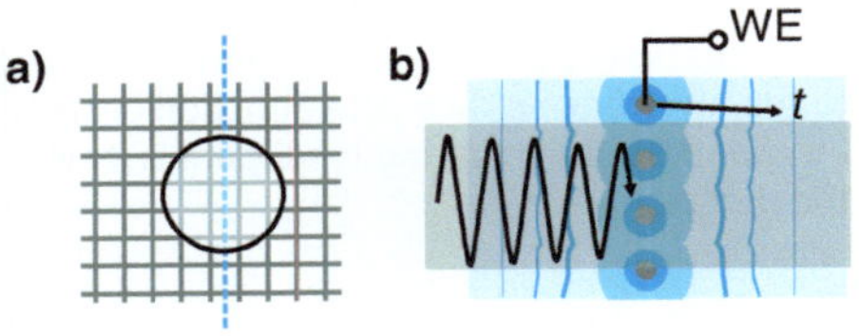

Figure 6.56 Evolution of the diffusion layer around a minigrid electrode; a) view along the optical path; b) cross-sectional view along the dashed blue line in a).

OTTLE cells also have disadvantages. Because the reference and auxiliary electrodes are located outside of the thin solution layer between the OTE and the opposite container wall, the ionic current has to be conducted though this confined geometry of the thin solution layer and may cause a substantial ohmic potential drop (*IR* drop) that is also not uniform for different parts of the WE. Therefore, the current responses can be distorted and the electrode process can become nonuniform (commencing first closer to the auxiliary electrode and then propagating across the WE). Such a situation can also cause instability of the potentiostatic operation, resulting in potential oscillations.

For thin-layer cells, the time dependence of the diffusion process must be taken into account. In CA mode at shorter times, planar semi-infinite diffusion will dominate. The spectral data will become time independent only after exhaustive reaction when the diffusion layer has expanded to cross the cell. The situation with minigrid electrodes is even more complex: planar diffusion around each bar in the grid is followed by cylindrical diffusion around each bar of the grid before switching again to planar diffusion across the entire grid electrode (Figure 6.56b).

The thickness of the diffusion layer can be estimated with Eq. (6.125)

$$\delta_{diff} = \sqrt{D_A t} \tag{6.125}$$

Here, t is the electrolysis time (assuming uniform reactivity across the WE). It is now possible to consider the development of the diffusion layer for different reaction mechanisms, for example, the EC-mechanism (Table 6.7). For a redox transformation of A to B followed by a chemical reaction to C, we assume for simplification that $D = D_A = D_B = D_C$. Spectroelectrochemical measurements are possible at short timescales to investigate the short-lived intermediate B electrochemically and spectroscopically, although the cell design needs to be improved compared to Figure 18.19 [107]. For B to be detected, it has to be formed in the ET step, but should not have reacted further to C. In this case, the diffusion layer of B has the same extension as the reaction layer in which the homogeneous reaction of B to C occurs. For planar diffusion, the diffusion layer thickness depends on the scan rate v. We can express both diffusion layer thickness $\delta_{diff,p}$ and the reaction layer thickness δ_{react} for a reversible redox system.

$$\delta_{diff,p} = \frac{1}{0.446}\sqrt{\frac{E_{NF}D}{v}} \tag{6.126}$$

$\delta_{diff,p}$ – Thickness of the diffusion layer at the peak potential [cm]
E_{NF} – Nernst factor = $RT/(nF)$ [V]
v – Potential scan rate [V s^{-1}]
D – Diffusion coefficient [cm^2 s^{-1}]
δ_{react} – Thickness of the reaction layer [cm]
k – Rate constant of the homogeneous following reaction [s^{-1}]

The following approximation exists for the thickness of the reaction layer δ_{react} [107]:

$$\delta_{react} = \sqrt{\frac{D}{k}} \tag{6.127}$$

Combining Eqs. (6.126) and (6.127) and rearranging for v provides a lower threshold of v that should be applied to optimize the spectroscopic detection of B:

$$v = \frac{1}{0.446^2}E_{NF}k \tag{6.128}$$

It is clear that in many experiments, despite a thin-layer cell with a solution layer of 0.2 to 1 mm, the cell operates in the regime of semi-infinite expansion of the diffusion layer, and one obtains only a noisy spectrum with *one* run. It is therefore useful to repeat the experiment several times. To do so, the "concentration scar" in front of the electrode during the previous run must be removed, preferably by introducing a "fresh" working solution to the thin-layer cell. This is a laborious manual activity for a cell similar to the cell in Figure 18.19. Arrangements such as those in Figure 6.57 offer good possibilities for remedial action here.

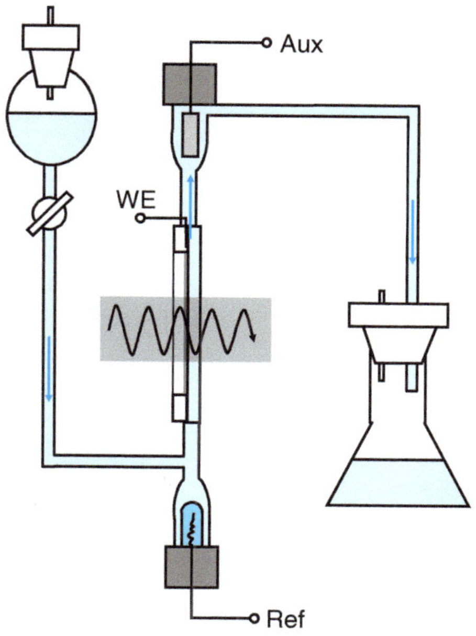

Figure 6.57 Spectroelectrochemical thin-layer cell with gravity flow configuration. Note the positioning of the Aux and Ref electrodes.

Voltammograms obtained in thin-layer cells are often distorted. For small potential scan rates v, conditions may switch from thin-film diffusion (Figure 5.57) to semi-infinite diffusion (Figure 5.29). For higher v, the current increases so much [Eq. (6.7)] that the uncompensated ohmic potential drop $U_{drop} = I \cdot R_{uc}$ causes problems. For spectroelectrochemical thin-layer cells, not only the solution resistance R_E but also the resistance R_{el} within the thin layer of metal oxide at the WE contributes to the ohmic potential drop. R_{uc} causes a deviation of the potential effective at the WE from the set potential of the potentiostat (Section 5.3.1). The effect can be mitigated by:

- Positioning of the reference electrode or Luggin capillary as close as possible to the thin-layer electrode.
- Lowering currents by restricting the electrode area to those regions through which light is passed. This can be achieved, for example, by masking off unnecessary electrode regions.
- Adding a conductive Cu film to the TCO to shorten the conduction path within the transparent WE material to the back contact (Figure 6.58).
- Connecting Ref and Aux electrically via a small external capacitor to suppress oscillations in a cell such as shown in Figure 6.57. The absence of oscillations should be checked by means of an oscilloscope.

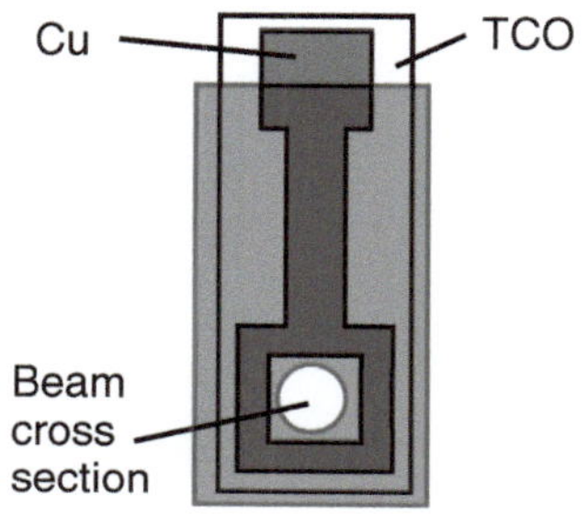

Figure 6.58 Improved electrical contact for an optically transparent electrode. The conduction path within the TCO (with limited conductivity) is decreased.

An elegant spectroelectrochemical cell is based on LIGA microstructures. LIGA is a German acronym for *Lithographie – galvanische Abformung* (lithography – galvanic molding). The transparent electrodes are metal foils with a thickness of typically 100 μm containing exactly parallel channels typically 25 μm in diameter, separated by 10 μm-thick walls. These structures form a minigrid with high transparency, provided the beam is directed parallel to the channel axis and walls of the minigrid.

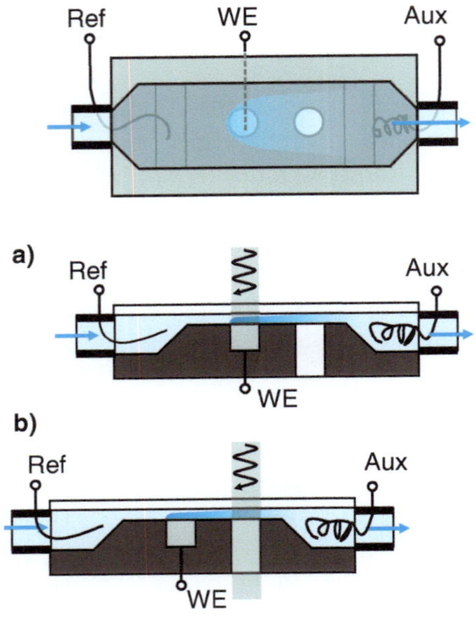

Figure 6.59 Cross section of a UV–Vis–NIR spectroelectrochemical flow cell that can be applied in both a) reflection mode and b) transmission mode.

The diffusion layer forms within the channels through which the light is passed. This significantly improves the effectiveness of the optical detection [107].

Spectroelectrochemical Cells in Reflection Mode. With highly reflective metal WE surfaces, beam reflection (Figure 6.54c) offers an opportunity to combine planar diffusion at nontransparent surfaces and effective interaction with the light beam. Conventional massive metal electrodes can be employed. The thickness of the thin solution layer can be adjusted in these cells with a micrometer mechanism so that the thickness is optimized for the detection of intermediates (see above). The cell illustrated in Figure 6.59 is a flow-through thin-layer cell allowing operation in reflection mode using the position a) and in transmission mode for the downstream detection of products using position b) for the light beam. The time between the electrochemical generation of intermediates and products at the WE and their detection can be adjusted by the linear flow rate and the distance between the WE and the beam position in Figure 6.59b. The incoming working solution always has the same composition and replenishes the reactant near the WE. This allows operation in steady-state regimes and optical data accumulation over prolonged periods of time to improve data quality and sensitivity toward a low concentration of intermediates.

ATR Cells. These cells work on the principle of attenuated total reflection by which an optical beam approaches from an optically dense medium and is completely reflected at an interface to an optically thin medium (Figure 6.54d), provided the incident angle (versus the surface normal) is larger than the critical angle. In spite of the total beam reflection, an evanescent field is formed at the point of incidence and penetrates 1–5 μm into the optically thin medium, which is the sample solution. The exact penetration depth depends on the optical properties and the wavelength. The surface of the ATR crystal can be coated with a thin metal layer to define the WE surface [108]. Products formed at this electrode are within the evanescent field and can absorb radiation energy from the evanescent field as in a conventional transmission cell. The absorption is detected by an attenuation of the reflected beam intensity in the optically dense medium.

Electron Paramagnetic Resonance (EPR) Spectroelectrochemistry. EPR can record signals from reactive radicals or transition metal coordination compounds with unpaired electrons (open shell systems) with high sensitivity. Such compounds and intermediates are commonly produced at electrode surfaces and require detection at concentrations as low as 1 μmol L^{-1} or even below, which can be achieved by prolonged acquisition periods. The spectroelectrochemical cell is inserted into a strong magnetic field, whose field strength is swept. The magnetic field leads to a splitting of spin-degenerate electronic energy levels. Excitation by microwave radiation causes a transition between electron spin energy levels. The resonance produces a characteristic line pattern with information about the type of radical (organic or metal-based) and structural features through the coupling to nuclear spins in the vicinity of the radical site [109, 110].

Further Reading. Reviews are available with further details on instrumentation and signal generation [106–108]. EPR spectroelectrochemistry is specifically covered in [109, 110]. Coupling of electrochemical experiments with NMR spectroscopy as a particularly important method for chemical structure elucidation is summarized in [111]. A further collective book [112] and reviews [113–115] provide information for specific applications including metal–organic coordination materials [116–118], radical dimerization [119], molecular carbon allotropes [120], molecular machines [121], homogeneous catalysts [122], and redox mediators in energy storage [123]. Many of the listed examples also illustrate the importance of combining information from multiple spectroscopic techniques for the study of complex systems. Spectroelectrochemistry is also of fundamental importance in the characterization of conducting polymers [124].

6.7 Summary

Electroanalytical methods, in particular CV, provide qualitative information about reaction mechanisms, as well as quantitative tools to extract equilibrium constants and rate constants from measurements. In this chapter, the general approach for the analysis of complex reaction mechanisms has been illustrated by examples. Although a complete coverage of all relevant examples is not possible here, the significance of such thermodynamic and kinetic parameters should be evident, as they provide the academic foundations for further research. This holds despite the dauntingly high effort required for recording high-quality data (without artefacts), their interpretation, and simulation. Their understanding is also closely linked to effective applications of electrochemical processes in analytics, synthesis, and modern energy storage and conversion.

Test your knowledge

Exercise 6.1: Consider the reduction of two noninteracting identical redox centers in a molecule. Provide rules for the two redox potentials affecting the appearance of cyclic voltammograms. Sketch these voltammograms for the case in which the formal potential is −0.5 V (vs. Fc/Fc^+) for the redox center.

Exercise 6.2: Some equations in molecular electrochemistry differ depending on the direction of the respective electrochemical reaction, that is, whether we deal with an oxidation or a reduction. In the case of CV, this applies to the first part of the potential cycle. Consider Eq. (5.45) for the calculation of the peak current ratio. Which direction of the electron transfer is the formulation written for? Transform the equation into one that applies to the opposite direction. Finally, write down the equation with appropriate symbols such that it can be used for both directions equally.

Exercise 6.3: Let the best time resolution of a potentiostat be 1 μs per data point. What is the maximum potential scan rate v for a linear potential ramp if we want to achieve a potential resolution of at least 1 mV per data point?

Exercise 6.4: Sketch the cyclic voltammogram of a simple reversible one-electron transfer. Define a scale for the current axis, for example, such that the peak current has a value of 10 μA. Now, calculate the *IR* drop for some selected points on the curve (for example, at the foot of the peak, at approximately half the peak current, the peak, and 100 mV after the peak on the forward scan), assuming an uncompensated resistance between the working and the reference electrodes of 2 kΩ. Do so for some selected points on the second part of the curve (reverse scan), too. Sketch the cyclic voltammogram when including the *IR* drop effects. Discuss the relevance of this curve for the qualitative and quantitative data evaluation.

Exercise 6.5: Plot the data from Table 6.27 against the viscosity of the solvents used. Formulate a hypothesis as to whether there might be a correlation and which expression should apply. Does the plot correspond to the hypothesis?

Exercise 6.6: Define and discuss the terms "electrochemical reversibility" and "chemical reversibility" or "irreversibility". How would you verify the presence or absence of these phenomena in a cyclic voltammogram (procedure and criteria)?

Exercise 6.7: Calculate the stoichiometric number of charges n using the numerical values given in Figure 6.46, taking into account a background current (corresponding to the current at the end of the electrolysis). Compare with the value of n determined without this correction and discuss the relevance of the correction in this case.

Exercise 6.8: In Section 6.6.1, we discussed the galvanostatic electrolysis of a dithiin (Figure 6.47) for the coulometric determination of n. Calculate n from the given data ($I = 50\,\text{mA}$, electrolysis time 6.44 min, amount of substance used 0.1 mmol).

Exercise 6.9: Check the units in Eqs. (6.37) through (6.39).

Exercise 6.10: Give Eqs. (6.40) and (6.41) explicitly for the case of an oxidation and a reduction as the electrochemical step and compare. What is the effect on the resulting equations?

Exercise 6.11: Repeat the spectroscopic parameters for the energy scale (wavelength [nm], wave number [cm^{-1}], and energy [eV]) and intensity scale (absorbance, transmission, and absorption) as well as their mutual interconversion.

Exercise 6.12: Carry out the derivation of Eq. (6.123) step by step. Discuss how the optical measurements would change when the mechanism is $A \pm e^- \rightleftharpoons B$, $B \rightarrow C$. In doing so, apply different assumptions for $\varepsilon_C(\lambda)$.

Exercise 6.13: A reaction proceeds according to the EC mechanism (Table 6.7) with a homogeneous rate constant $k = 100\,\text{s}^{-1}$. How large must the potential scan rate v be for B to be detected spectroelectrochemically? How thick would the diffusion layer be in this case at the peak potential? Discuss the resulting challenges and possible solutions for such an experiment.

References

1 D. H. Evans; Voltammetry: Doing chemistry with electrodes. *Acc. Chem. Res.* **1977**, *10*, 313–319.

2 http://www.ise-online.org/divisions/divisions.php (last access 24 Aug 2025).

3 A. J. Bard, L. R. Faulkner, H. S. White; *Electrochemical Methods: Fundamentals and Applications*. 3rd ed., John Wiley & Sons, Hoboken, NJ, 2022.

4 R. G. Compton, C. E. Banks; *Understanding Voltammetry*. 2nd ed., Imperial College Press, London, 2011.

5 R. G. Compton, C. Batchelor-McAuley, E. J. F. Dickinson; *Understanding Voltammetry: Problems and Solutions*. Imperial College Press, London, 2012.

6 R. G. Compton, E. Laborda, K. R. Ward; *Understanding Voltammetry: Simulation of Electrode Processes*, Imperial College Press, London, 2014.

7 J.-M. Savéant; *Elements of Molecular and Biomolecular Electrochemistry*, Wiley, Hoboken, 2006.

8 O. Hammerich, B. Speiser (Eds.); *Organic Electrochemistry*. 5th ed., CRC Press, Boca Raton, 2016.

9 C. P. Andrieux; Terminology and notations for multistep electrochemical reaction mechanisms. *Pure Appl. Chem.* **1994**, *66*, 2445–2450.

10 E. Vieil, G. Cauquis; An alphanumerical notation for electrochemical and photoelectrochemical mechanisms. Propositions and application to various known and characteristic mechanisms: Part I. Electron transfers, chemical reactions and photochemical steps. *J. Electroanal. Chem.* **1983**, *148*, 183–200.

11 A. W. Bott; A comparison of cyclic voltammetry and cyclic staircase voltammetry. *Curr. Sep.* **1997**, *16*, 23–26.

12 C. Batchelor-McAuley, M. Yang, E. M. Hall, R. G. Compton; Correction factors for the analysis of voltammetric peak currents measured using staircase voltammetry. *J. Electroanal. Chem.* **2015**, *758*, 1–6.

13 K. D. Moeller; Combinatorial electrochemistry and miniaturization. in *Organic Electrochemistry* (O. Hammerich, B. Speiser, Eds.). 5th ed., CRC Press, Boca Raton, 2016, pp. 345–369.

14 J. H. Brown; Analysis of two redox couples in a series: An expanded experiment to introduce undergraduate students to cyclic voltammetry and electrochemical simulations. *J. Chem. Educ.* **2016**, *93*, 1326–1329.

15 R. S. Nicholson, I. Shain; Theory of stationary electrode polarography. Single scan and cyclic methods applied to reversible, irreversible, and kinetic systems. *Anal. Chem.* **1964**, *36*, 706–723.

16 E. Eichhorn, A. Rieker, B. Speiser; Numerical method to correct *iR* drop errors in cyclic voltammetric potential data for (quasi)reversible electrode processes. *Anal. Chim. Acta* **1992**, *256*, 243–249.

17 D. H. Evans, K. Hu; Inverted potentials in two-electron processes in organic electrochemistry. *J. Chem. Soc., Faraday Trans.* **1996**, *92*, 3983–3990.

18 S. W. Feldberg; Nuances of the ECE mechanism. III. Effects of homogeneous redox equilibrium in cyclic voltammetry. *J. Phys. Chem.* **1971**, *75*, 2377–2380.

19 C. Amatore, J. M. Savéant; ECE and disproportionation: Part VI. General resolution. Application to potential step chronoamperometry. *J. Electroanal. Chem.* **1979**, *102*, 21–40.

20 R. S. Nicholson; Theory and application of cyclic voltammetry for measurement of electrode reaction kinetics. *Anal. Chem.* **1965**, *37*, 1351–1355.

21 L. Nadjo, J. M. Savéant; Linear sweep voltammetry: Kinetic control by charge transfer and/or secondary chemical reactions: I. Formal kinetics. *J. Electroanal. Chem.* **1973**, *48*, 113–145.

22 L. P. Hammett; *Physical Organic Chemistry*. 2nd ed., McGraw-Hill, New York, 1970, p. 77.

23 J.-M. Savéant; Molecular electrochemistry: Recent trends and upcoming challenges. *ChemElectroChem* **2016**, *3*, 1967–1977.

24 B. Speiser; Application of digital simulation. in *Organic Electrochemistry* (O. Hammerich, B. Speiser, Eds.). 5th ed., CRC Press, Boca Raton, 2016, pp. 205–227.

25 M. Heymann; Modeling reality. Practice, knowledge, and uncertainty in atmospheric transport simulation. *Hist. Stud. Phys. Biol. Sci.* **2006**, *37*, 49–85.

26 D. Britz, J. Strutwolf; *Digital Simulation in Electrochemistry*. 4th ed., Springer, 2016.

27 M. Rudolph, D. P. Reddy, S. W. Feldberg; A simulator for cyclic voltammetric responses. *Anal. Chem.* **1994**, *66*, 589A–600A.

28 L. K. Bieniasz; A reaction compiler for electrochemical kinetics. *Comput. Chem.* **1996**, *20*, 403–418.

29 K. Ludwig, B. Speiser; EChem++ – An object-oriented problem solving environment for electrochemistry. 2: The kinetic facilities of Ecco – a compiler for (electro-)chemistry. *J. Chem. Inf. Comput. Sci.* **2004**, *44*, 2051–2060.

30 K. Ludwig, L. Rajendran, B. Speiser; EChem++ – An object-oriented problem solving environment for electrochemistry. Part 1: A C++ class collection for electrochemical excitation functions. *J. Electroanal. Chem.* **2004**, *568*, 203–214.

31 I. J. Cutress, E. J. F. Dickinson, R. G. Compton; Analysis of commercial general engineering finite element software in electrochemical simulations. *J. Electroanal. Chem.* **2010**, *638*, 76–83.

32 http://www.basinc.com/products/ec/digisim (last access 02 Sep 2025).

33 http://www.elchsoft.com (last access 02 Sep 2025).

34 http://www.cyfronet.krakow.pl/~nbbienia/elsim3ad.html (last access 02 Sep 2025).

35 https://sourceforge.net/projects/echempp (last access 02 Sep 2025).

36 R. D. Allendoerfer, P. H. Rieger; Electrolytic reduction of cyclooctatetraene. *J. Am. Chem. Soc.* **1965**, *87*, 2336–2344.

37 D. H. Evans, K. M. O'Connell; Conformational change and isomerization associated with electrode reactions. in *Electroanalytical Chemistry, Vol. 14* (A. J. Bard, Ed.), Marcel Dekker, New York, 1986, pp. 113–207.

38 K. Lam, W. E. Geiger; Influence of molecular and medium effects on two-electron processes. in *Organic Electrochemistry* (O. Hammerich, B. Speiser, Eds.). 5th ed., CRC Press, Boca Raton, 2016, pp. 395–431.

39 R. Winter; Half-wave potential splittings $\Delta E_{1/2}$ as a measure of electronic coupling in mixed-valent systems: Triumphs and defeats. *Organometallics* **2014**, *33*, 4517–4536.

40 C. Creutz, H. Taube; Binuclear complexes of ruthenium ammines. *J. Am. Chem. Soc.* **1973**, *95*, 1086–1094.

41 M. B. Robin, P. Day; Mixed valence chemistry – A survey and classification. in *Advances in Inorganic Chemistry and Radiochemistry, Vol. 10* (H. J. Emeléus, A. G. Shape, Eds.), Academic Press, New York, 1968, pp. 247–422.

42 J. B. Flanagan, S. Margel, A. J. Bard, F. C. Anson; Electron transfer to and from molecules containing multiple, noninteracting redox centers. Electrochemical oxidation of poly(vinylferrocene). *J. Am. Chem. Soc.* **1978**, *100*, 4248–4253.

43 K. Aoki; Is voltammetric current proportional to the number of transferred electrons for multi-charged ions or to 3/2 power of the number? *Electroanalysis* **2005**, *17*, 1379–1383.

44 K. Hu, D. H. Evans; Electron-transfer reactions accompanied by substantial structural changes: Oxidation of 9,10-bis(dimethylamino)anthracene and 3,6-bis(dimethylamino)durene. *J. Electroanal. Chem.* **1997**, *423*, 29–35.

45 D. H. Evans; Theory for a homogeneous reaction following a quasireversible electrode reaction. *J. Phys. Chem.* **1972**, *76*, 1160–1165.

46 J.-M. Savéant; Dissociative electron transfer. *Adv. Electron Transfer Chem.* **1994**, *4*, 53–116.

47 C. Costentin, M. Robert, J.-M. Savéant; Dissociative electron transfers. in *Organic Electrochemistry* (O. Hammerich, B. Speiser, Eds.). 5th ed., CRC Press, Boca Raton, 2016, pp. 511–530.

48 C. Costentin, M. Robert, J.-M. Savéant; Proton-coupled electron transfers. in *Organic Electrochemistry* (O. Hammerich, B. Speiser, Eds.). 5th ed., CRC Press, Boca Raton, 2016, pp. 481–509.

49 J.-M. Savéant; Concerted proton-electron transfers: Fundamentals and recent developments. *Annu. Rev. Anal. Chem.* **2014**, *7*, 537–560.

50 J. H. P. Utley, R. D. Little, M. F. Nielsen; Reductive coupling. in *Organic Electrochemistry* (O. Hammerich, B. Speiser, Eds.). 5th ed., CRC Press, Boca Raton, 2016, pp. 621–704.

51 H. J. Schäfer; Oxidative coupling. in *Organic Electrochemistry* (O. Hammerich, B. Speiser, Eds.). 5th ed., CRC Press, Boca Raton, 2016, pp. 705–773.

52 A. Smie, J. Heinze; Reversible dimerization of diphenylpolyene radical cations: An alternative to the bipolaron model. *Angew. Chem. Int. Ed. Engl.* **1997**, *36*, 363–367.

53 P. J. Kinlen, C. J. H. King; Mechanistic studies of the electrohydrodimerization of acrylonitrile. *J. Electroanal. Chem.* **1991**, *304*, 133–151.

54 F. Zhou, A. J. Bard; Detection of the electrohydrodimerization intermediate acrylonitrile radical anion by scanning electrochemical microscopy. *J. Am. Chem. Soc.* **1994**, *116*, 393–394.

55 J. H. P. Utley, R. D. Little, M. F. Nielsen; Reductive coupling. in *Organic Electrochemistry* (O. Hammerich, B. Speiser, Eds.). 5th ed., CRC Press, Boca Raton, 2016, p. 629.

56 F. Magno, D. Perosa, G. Bontempelli; Digital simulation of electrochemical processes involving very fast chemical reactions: Part 1. A simple general approach. *Anal. Chim. Acta* **1985**, *173*, 211–217.

57 M. Mastragostino, L. Nadjo, J. M. Savéant; Disproportionation and ECE mechanisms – I. Theoretical analysis. Relationships for linear sweep voltammetry. *Electrochim. Acta* **1968**, *13*, 721–749.

58 S. W. Feldberg, L. Jeftic; Nuances of the $\vec{E}C\overleftarrow{E}$ mechanism. IV. Theory of cyclic voltammetry and chronoamperometry and the electrochemical reduction of hexacyanochromate(III). *J. Phys. Chem.* **1972**, *76*, 2439–2446.

59 K. Chiba, Y. Okada; Electron transfer-catalyzed reactions. in *Organic Electrochemistry* (O. Hammerich, B. Speiser, Eds.). 5th ed., CRC Press, Boca Raton, 2016, pp. 531–555.

60 D. H. Evans; Solution electron-transfer reactions in organic and organometallic electrochemistry. *Chem. Rev.* **1990**, *90*, 739–751.

61 E. Laviron, L. Roullier; The square scheme with the electrochemical reactions at equilibrium. A study by diffusion or thin layer cyclic voltammetry and by scanning potential coulometry. *J. Electroanal. Chem.* **1985**, *186*, 1–15.

62 S. F. Nelsen, G. T. Cunkle, D. H. Evans, K. J. Haller, M. Kaftory, B. Kirste, H. Kurreck, T. Clark; 8,8'-Bi(8-azabicyclo[3.2.1]octane) and its oxidized forms. *J. Am Chem. Soc.* **1985**, *107*, 3829–3839.

63 E. Eichhorn, A. Rieker, B. Speiser; The electrochemical oxidation of [Co^{II}(salen)] in solvent mixtures – an example of a ladder scheme with coupled electron-transfer and solvent-exchange reactions. *Angew. Chem. Int. Ed. Engl.* **1992**, *31*, 1215–1217.

64 E. Laviron; Electrochemical reactions with protonations at equilibrium. Part XIII. Experimental study of the homogeneous electron exchange in quinone/dihydroquinone systems. *J. Electroanal. Chem.* **1986**, *208*, 357–372.

65 E. Steckhan; Indirect electroorganic syntheses – A modern chapter of organic electrochemistry. *Angew. Chem. Int. Ed. Engl.* **1986**, *25*, 683–701.

66 C. P. Andrieux, J. M. Dumas-Bouchiat, J. M. Savéant; Homogeneous redox catalysis of electrochemical reactions: Part I. Introduction. *J. Electroanal. Chem.* **1978**, *87*, 39–53.

67 R. Francke, R. D. Little; Electrons and holes as catalysts in organic electrosynthesis. *ChemElectroChem* **2019**, *6*, 4373–4382.

68 P. Song, A. C. Fisher, J. D. Wadhawan, J. J. Cooper, H. J. Ward, N. S. Lawrence; A mechanistic study of the EC′ mechanism – The split wave in cyclic voltammetry and square wave voltammetry. *RSC Adv.* **2016**, *6*, 70237–70242.

69 C. Kohlmann, S. Lütz; Electroenzymatic synthesis. in *Organic Electrochemistry* (O. Hammerich, B. Speiser, Eds.). 5th ed., CRC Press, Boca Raton, 2016, pp. 1511–1541.

70 C. Amatore, J. Pinson, J. M. Savéant, A. Thiebault; Trace crossings in cyclic voltammetry and electrochemical inducement of chemical reactions: Aromatic nucleophilic substitution. *J. Electroanal. Chem.* **1980**, *107*, 59–74.

71 C. Amatore, J. Pinson, J. M. Savéant, A. Thiebault; Current dips in polarography and cyclic voltammetry associated with electrochemical inducement of chemical reactions: Aromatic nucleophilic substitution. *J. Electroanal. Chem.* **1980**, *107*, 75–86.

72 J.-M. Savéant; *Elements of Molecular and Biomolecular Electrochemistry*, Wiley, Hoboken, 2006, Section 2.5.6, pp. 158ff.

73 A. Fitch, G. J. Edens; Isopotential points as a function of an allowed cross reaction. *J. Electroanal. Chem.* **1989**, *267*, 1–13.

74 H. J. Paul, J. Leddy; Direct determination of the transfer coefficient from cyclic voltammetry: Isopoints as diagnostics. *Anal. Chem.* **1995**, *67*, 1661–1668.

75 P. M. Skitał, P. T. Sanecki, D. Saletnik; Iso points in electrochemistry. Classification and verification by experiment and theory. *Electroanalysis* **2018**, *30*, 1990–2003.

76 J. Janisch, A. Ruff, B. Speiser, C. Wolff, J. Zigelli, S. Benthin, V. Feldmann, H. A. Mayer; Consistent diffusion coefficients of ferrocene in some non-aqueous solvents: Electrochemical simultaneous determination together with electrode sizes and comparison to pulse-gradient spin-echo NMR results. *J. Solid State Electrochem.* **2011**, *15*, 2083–2094.

77 J. C. Imbeaux, J. M. Savéant; Convolutive potential sweep voltammetry: I. Introduction. *J. Electroanal. Chem.* **1973**, *44*, 169–187.

78 K. B. Oldham, J. Spanier; The replacement of Fick's laws by a formulation involving semidifferentiation. *J. Electroanal. Chem.* **1970**, *26*, 331–341.

79 A. M. Bond, K. B. Oldham, G. A. Snook; Use of the ferrocene oxidation process to provide both reference electrode potential calibration and a simple measurement (via semiintegration) of the uncompensated resistance in cyclic voltammetric studies in high-resistance organic solvents. *Anal. Chem.* **2000**, *72*, 3492–3496.

80 K. B. Oldham, J. C. Myland; Extracting parameter values from quasireversible cyclic voltammograms. *J. Solid State Electrochem.* **2012**, *16*, 3691–3693.

81 S. Antonello, K. Daasbjerg, H. Jensen, F. Taddei, F. Maran; Formation and cleavage of aromatic disulfide radical anions. *J. Am. Chem. Soc.* **2003**, *125*, 14905–14916.

82 B. Speiser; Multiparameter estimation: Extraction of information from cyclic voltammograms. *Anal. Chem.* **1985**, *57*, 1390–1397.

83 N. R. Draper, H. Smith; *Applied Regression Analysis*. 3rd ed., Wiley-Interscience, New York, 1998.

84 B. Scharbert, B. Speiser; Chemical information from electroanalytical data. Part 1 – Determination of system parameters for quasi-reversible electron transfer reactions from cyclic voltammetric test data and data for the reduction of cerium (IV) bis(octaethylporphyrinate). *J. Chemomet.* **1988**, *3*, 61–80.

85 L. K. Bieniasz, S. Dümmling, B. Speiser, M. Würde; Use of sensitivity analysis methods in the modelling of electrochemical transients: Part 2. Model expansion and model reduction. *J. Electroanal. Chem.* **1998**, *447*, 173–186.

86 D. T. Pierce, W. E. Geiger; Electrochemical kinetic discrimination of the single-electron-transfer events of a two-electron-transfer reaction: Cyclic voltammetry of the reduction of the bis(hexamethylbenzene)ruthenium dication. *J. Am. Chem. Soc.* **1992**, *114*, 6063–6073.

87 M. Martinez, M.-A. Pitarque, R. van Eldik; Outer-sphere redox reactions of $(N)_5$-macrocyclic cobalt(III) complexes. A temperature and pressure dependence kinetic study on the influence of size and geometry of different macrocycles. *Inorg. Chim. Acta* **1997**, *256*, 51–59.

88 K. B. Oldham; Ultimate cyclic voltammetry: An analytical examination of the reversible case. *J. Solid State Electrochem.* **2013**, *17*, 2749–2756.

89 K. B. Oldham; Multi-scan reversible cyclic voltammetry: Is the ultimate condition approached and, if so, how fast? *J. Solid State Electrochem.* **2013**, *17*, 2757–2769.

90 N. Elgrishi, K. J. Rountree, B. D. McCarthy, E. S. Rountree, T. T. Eisenhart, J. L. Dempsey; A practical beginner's guide to cyclic voltammetry. *J. Chem. Educ.* **2018**, *95*, 197–206.

91 N. Kuhn, H. Kotowski, M. Steimann, B. Speiser, M. Würde, G. Henkel; Synthesis, oxidation and protonation of octamethyl-1,1′-bipyrrole. *J. Chem. Soc. Perkin Trans.* **2000**, *2*, 353–363.

92 C. P. Andrieux, J.-M. Savéant, C. Tardy; Transition between concerted and stepwise dissociative electron transfers. An example of how a change of temperature may trigger a change in mechanism in electrochemical experiments. *J. Am. Chem. Soc.* **1997**, *119*, 11546–11547.

93 W. J. Bowyer, E. E. Engelman, D. H. Evans; Kinetic studies by cyclic voltammetry at low temperatures using microelectrodes. *J. Electroanal. Chem.* **1989**, *262*, 67–82.

94 T. J. Donohoe, D. J. Johnson, R. G. Compton, J. D. Wadhawan; Low temperature electrochemistry as a mechanistic probe for the partial reduction of heterocycles. *Tetrahedron* **2004**, *60*, 5945–5952.

95 O. Hammerich, B. Speiser; Techniques for studies of electrochemical reactions in solution. in *Organic Electrochemistry* (O. Hammerich, B. Speiser, Eds.). 5th ed., CRC Press, Boca Raton, 2016, pp. 97–168.

96 U. Karst; Electrochemistry/mass spectrometry (EC/MS) – A new tool to study drug metabolism and reaction mechanisms. *Angew. Chem. Int. Ed.* **2004**, *43*, 2476–2478.

97 D. F. Valencia, F. J. González; Estimation of diffusion coefficients by using a linear correlation between the diffusion coefficient and molecular weight. *J. Electroanal. Chem.* **2012**, *681*, 121–126.

98 P. A. Malachesky; Correlation of linear sweep voltammetric and chronoamperometric data for *n*-value determinations. *Anal. Chem.* **1969**, *41*, 1493–1494.

99 C. Amatore, M. Azzabi, P. Calas, A. Jutand, C. Lefrou, Y. Rollin; Absolute determination of electron consumption in transient or steady state electrochemical techniques. *J. Electroanal. Chem.* **1990**, *288*, 45–63.

100 E. O. Stejskal, J. E. Tanner; Spin diffusion measurements: Spin echoes in the presence of a time-dependent field gradient. *J. Chem. Phys.* **1965**, *42*, 288–292.

101 K. Moon, A. E. Kaifer; Dimeric molecular capsules under redox control. *J. Am. Chem. Soc.* **2004**, *126*, 15016–15017.

102 J. Kärger; Transport phenomena in nanoporous materials. *ChemPhysChem* **2015**, *16*, 24–51.

103 A. Schank, B. Speiser, A. Stickel; Oxidation electron stoichiometry, diffusion coefficients and formal potentials of two di-nickel-bis(salen) complexes. *J. Electroanal. Chem.* **2016**, *779*, 137–145.

104 C. M. Weber, D. M. Eisele, J. P. Rabe, Y. Liang, X. Feng, L. Zhi, K. Müllen, J. L. Lyon, R. Williams, D. A. Vanden Bout, K. J. Stevenson; Graphene-based optically transparent electrodes for spectroelectrochemistry in the UV–Vis region. *Small* **2010**, *6*, 184–189.

105 T. Kuwana, N. Winograd; Spectroelectrochemistry at optically transparent electrodes: I. Electrodes under semi-infinite diffusion condition. in *Electroanalytical Chemistry*, Vol. 7 (A. J. Bard, Ed.), Marcel Dekker, New York, 1974, pp. 1–78.

106 W. R. Heineman, F. M. Hawkridge, H. N. Blount; Spectroelectrochemistry at opically transparent electrodes: II. Electrodes under thin-layer and semi-infinite diffusion conditions and indirect coulometric titrations. in *Electroanalytical Chemistry*, Vol. 13 (A. J. Bard, Ed.), Marcel Dekker, New York, 1984, pp. 1–113.

107 A. Neudeck, F. Marken, R. G. Compton; UV/Vis/NIR spectroelectrochemistry. in *Electroanalytical Methods – A Guide to Experiments and Applications* (F. Scholz, Ed.), Springer, Berlin, 2002, pp. 167–189.

108 T. E. Keyes, R. J. Forster; Spectroelectrochemistry. in *Handbook of Electrochemistry* (C. G. Zoski, Ed.), Elsevier, Amsterdam, 2007, pp. 591–635.

109 T. M. McKinney; Electron spin resonance and electrochemistry. in *Electroanalytical Chemistry*, Vol. 10 (A. J. Bard, Ed.), Marcel Dekker, New York, 1977, pp. 97–278.

110 P. Rapta, E. Dmitrieva, A. A. Popov, L. Dunsch; *In situ* spectroelectrochemistry of organic compounds. in *Organic Electrochemistry* (O. Hammerich, B. Speiser, Eds.). 5th ed., CRC Press, Boca Raton, 2016, pp. 169–190.

111 R. D. Webster; In situ electrochemical-NMR spectroscopy. Reduction of aromatic halides. *Anal. Chem.* **2004**, *76*, 1603–1610.

112 W. Kaim, A. Klein (Eds.); *Spectroelectrochemistry*, Royal Society of Chemistry, Cambridge, 2008.

113 H. E. Toma, K. Araki; Spectroelectrochemical characterization of organic and metal–organic compounds. *Curr. Org. Chem.* **2002**, *6*, 21–34.

114 W. Kaim, J. Fiedler; Spectroelectrochemistry: The best of two worlds. *Chem. Soc. Rev.* **2009**, *38*, 3373–3382.

115 L. Dunsch; Recent advances in in situ multi-spectroelectrochemistry. *J. Solid State Electrochem.* **2011**, *15*, 1631–1646.

116 B. Sarkar, S. Patra, J. Fiedler, R. B. Sunoj, D. Jandardanan, G. K. Lahiri, W. Kaim; Mixed-valent metals bridged by radical ligand: Fact or fiction based on structure-oxidation state correlations. *J. Am. Chem. Soc.* **2008**, *130*, 3532–3542.

117 N. Kobayshi, H. Lam, W. A. Nevin, P. Janda, C. C. Leznoff, A. B. P. Lever; Electrochemistry and spectroelectrochemistry of 1,8-naphthalene- and 1,8-anthracene-linked cofacial binuclear metallophthalocyanines. New mixed-valence metallophthalocyanines. *Inorg. Chem.* **1990**, *29*, 3415–3425.

118 J. Poppe, M. Moscherrosch, W. Kaim; An unusually weak intervalence transition in a very stable bis-chelate analogue of the mixed-valent Creutz-Taube ion. UV/Vis/Near-IR and EPR spectroelectrochemistry of $[(NH_3)_4Ru(\mu\text{-bptz})Ru(NH_3)]_4^{n+}$ (bptz = 3,6-bis(2-pyridyl)-1,2,4,5-tetrazine; $n = 3\text{-}5$). *Inorg. Chem.* **1993**, *32*, 2640–2643.

119 D. D. Graf, R. G. Duan, J. P. Campbell, L. L. Miller, K. R. Mann; From monomer to π-Stacks. A comprehensive study of the structure and properties of monomeric, π-dimerized, and π-stacked forms of the cation radical of 3′,4′-dibutyl-2,5″-diphenyl-2,2′:5′,2″-terthiophene. *J. Am. Chem. Soc.* **1997**, *119*, 5888–5899.

120 A. A. Popov, N. Chen, J. R. Pinzon, S. Stevenson, L. A. Echegoyen, L. Dunsch; Redox-active scandium oxide cluster inside a fullerene cage: Spectroscopy, voltammetry, electron spin resonance spectroelectrochemical, and extended density functional theory study of $Sc_4O_2@C_{80}$ and its ion radicals. *J. Am. Chem. Soc.* **2012**, *134*, 19607–19618.

121 Z. Zhu, A. C. Fahrenbach, H. Li, J. C. Barnes, Z. Liu, S. M. Dyar, H. Zhang, J. Lei, R. Carmieli, A. A. Sarjeant, C. L. Stern, M. R. Wasielewski, J. F. Stoddart; Controlling switching in bistable [2]catenanes by combining donor-acceptor and radical-radical interactions. *J. Am. Chem. Soc.* **2012**, *134*, 11709–11720.

122 C. W. Machan, M. D. Sampson, S. A. Chabolla, T. Dang, C. P. Bubiak; Developing a mechanistic understanding of molecular electrocatalysts for CO_2 reduction using infrared spectroelectrochemistry. *Organometallics* **2014**, *33*, 4550–4559.

123 J. Wandt, P. Jakes, J. Granwehr, H. A. Gasteiger, R.-A. Eichel; Singlet oxygen formation during the charging process of an aprotic lithium-oxygen battery. *Angew. Chem. Int. Ed.* **2016**, *55*, 6892–6895.

124 E. Dmitrieva, L. Dunsch; How linear is "linear" polyaniline. *J. Phys. Chem. B* **2011**, *115*, 6401–6411.

7

Electrochemical Impedance Spectroscopy

Electrochemical impedance spectroscopy (EIS) is an important methodology for the characterization of kinetics of electrode reactions and performance of electrochemical systems. As a method, EIS is nondestructive and ideal for monitoring status and aging parameters for complete cells or half-cells (electrodes), for example, for fuel cells, batteries, semiconductor electrodes, corroding surfaces, surface coatings, and sensors.

EIS is based on an electrochemical cell (the system under investigation) with an applied DC bias (DC, direct current or applied constant potential) with a superimposed AC signal (AC, alternating current or potential modulation) with small amplitude (to avoid nonlinear responses). The EIS measurement involves sweeping the frequency of the applied modulation f similar to sweeping the frequency of light in a spectrometer, thus the name of the technique. In this picture, terms like absorption and transmission are reflected in impedance and admittance. The response of the electrochemical system is recorded as impedance (the complex electrical resistance) and phase shift between the applied AC voltage and the measured AC current.

The superimposed input signal applied to the electrochemical system drives both chemical reactions and transport processes. In Chapters 3 and 4, we saw that chemical rates and transport processes are limited in rate. For a given input signal with frequency f, we can measure a response signal only if the system is able to respond by moving charges within the associated constant $\tau = 1/(2\pi f)$. By varying f, the effects on different processes (with different time constants) can be probed one after the other. With increasing frequency f, processes in the electrochemical cell can be switched off when the time constant becomes shorter than the time constant of the process in the electrochemical system. Analogously, a high-frequency excitation signal will not excite lower energy transitions in spectroscopy. The physical reality of processes with time constants τ can be mimicked with an equivalent circuit model based on resistors and capacitors representing the same time constant. In Chapter 4.2.2, we employed equivalent circuits in conductometry and for characterizing the electrical double layer (please revisit Sections 4.2.2 and 4.3). In the description of electrical circuits and in linear response theory, we will need complex numbers and complex exponentials. Therefore, this chapter will initially revisit and introduce these mathematical concepts and tools, which are essential for understanding EIS.

Further Reading. This textbook focuses on electrochemical reaction mechanisms and transport phenomena and the associated equivalent circuit diagrams. The applications of EIS and more complex equivalent circuits can be challenging even for experienced electrochemists. Specialized textbooks on EIS [1–4] and textbooks, reviews, and book chapters are recommended for a broader range of examples and a more elaborate mathematical treatment [5–12].

Q: EIS seems very similar to AC voltammetry (ACV) in Section 5.3.9. Do we really need so many new fundamental concepts?

A: For both, EIS and ACV, a combination of DC and AC excitation signals are applied. During ACV experiments, the DC input voltage is scanned and the AC signal has a constant frequency f. In EIS experiments, the DC input signal remains constant, and the frequency of the AC signal is sweeping. Yes, there are really many different terms in electrochemistry, but EIS and ACV refer to different concepts. EIS experiments probe the *time domain*, while voltammetry probes the *potential domain*. At a given potential, EIS can probe into the time constants for chemical and transport processes.

Excursion 7-1 Complex Numbers.

Definition. The set of complex numbers $\mathbb{C}$ [13] can be seen as an expansion of the set of real numbers $\mathbb{R}$, so that for every algebraic equation

$$x^n + a_{n-1}x^{n-1} + \ldots + a_1 x + a_0 = 0 \tag{7.1}$$

with coefficients $a_0, a_1, \ldots a_{n-1}$

there are n complex solutions (or roots). A complex number has two components, a real term and an imaginary term. The imaginary term is based on the **imaginary number i** obtained, for example, when solving the equation $x^2 = -1$.

$$\mathrm{i} = \sqrt{-1} \tag{7.2}$$

Vector Notation. Having a real and an imaginary term, a complex number can be represented in two dimensions (real and imaginary) as a vector in the complex plane. The abscissa represents the real axis, and the ordinate represents the imaginary axis. The complex number z is a vector sum of the real vector a (on the real axis) and the imaginary vector ib (on the imaginary axis) (Figure 7.1).

$$z(a, b) = a + \mathrm{i}b, \quad \text{with} \quad a, b \in \mathbb{R} \tag{7.3}$$

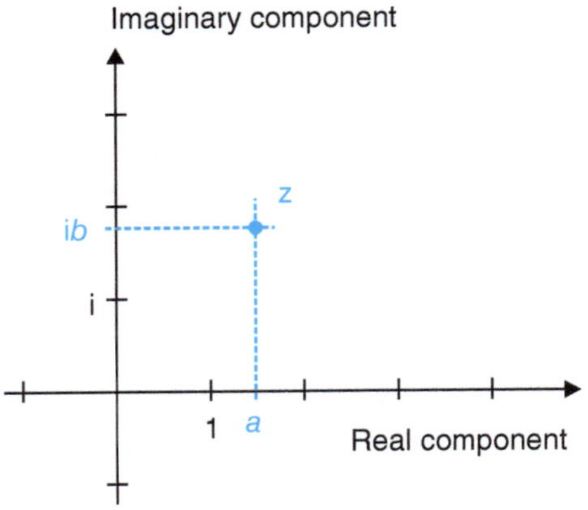

Figure 7.1 Vector representation of a complex number.

The real number a is the real term of z, $a = \mathrm{Re}(z)$. The real number b is the imaginary term of z, $b = \mathrm{Im}(z)$. The set of real numbers $\mathbb{R}$ is a subset of the set of complex numbers $\mathbb{C}$ with $b = 0$.

Polar Coordinates. The number z can also be represented as a point in polar coordinates $z(r, \theta)$ (Figure 7.2). Vector and polar coordinates can be interconverted easily.

$$a = r\cos\theta \tag{7.4}$$

$$\mathrm{i}b = \mathrm{i}r\sin\theta \tag{7.5}$$

$$\theta = \arctan\frac{b}{a} = \tan^{-1}\frac{b}{a} \tag{7.6}$$

$$z(r, \theta) = r(\cos\theta + \mathrm{i}\sin\theta) \tag{7.7}$$

$$r = \sqrt{a^2 + b^2} \tag{7.8}$$

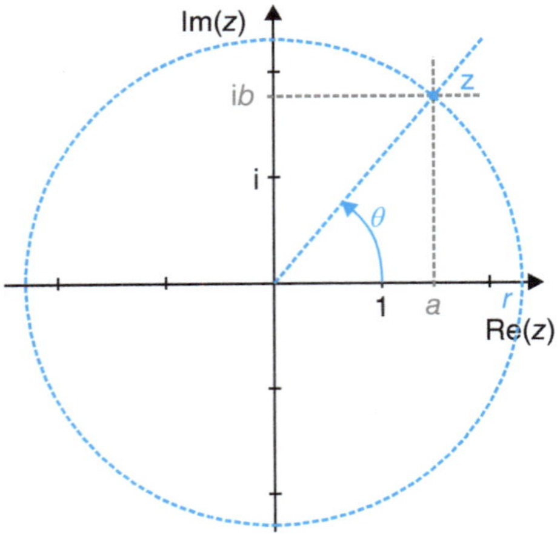

Figure 7.2 Polar coordinates for a complex number.

r is the magnitude of the complex number and always positive. We can state a, b, r, $\theta \in \mathbb{R}$; θ is the phase angle in radians, that is, the arc length on the unit circle.

Euler's Notation. The notation introduced by Euler[76] (Figure 7.3) is very effective and allows access to expressions for complex properties.

$$z = r\mathrm{e}^{\mathrm{i}\theta} = r\exp(\mathrm{i}\theta) \tag{7.9}$$

[76] Leonhard Euler, 1707 (Basel)–1783 (St. Petersburg), Swiss mathematician, professor for mathematics in St. Petersburg, where most of this 866 (!) publications were written. Euler contributed to mathematical notation [e, π, i, Σ, $f(x)$, ...]. Euler's theorem will also be important in Chapter 8.

With Euler's equation, the complex exponential is converted into a sum (the complex number).

$$e^{i\theta} = \cos\theta + i\sin\theta. \tag{7.10}$$

The complex conjugate z^* to a complex number $z = a + ib$ is given by inverting the sign of the imaginary component.

$$z^*(a, b) = a - ib \tag{7.11}$$

When we multiply a complex number z with the complex conjugate z^*, we obtain a real number.

$$z^* \cdot z = (a - ib)(a + ib) = a^2 - (i)^2 b^2 = a^2 - (-1)\,b^2 = a^2 + b^2 \tag{7.12}$$

Therefore, the magnitude r of the complex number z represents the square root of the complex product of z and z^*.

$$r = |z| = \sqrt{z^* \cdot z} = \sqrt{a^2 + b^2} \tag{7.13}$$

You may compare this expression to $|a| = \sqrt{a \cdot a}$ in $\mathbb{R}$, with $|a| \geq 0$, but a being either positive or negative.

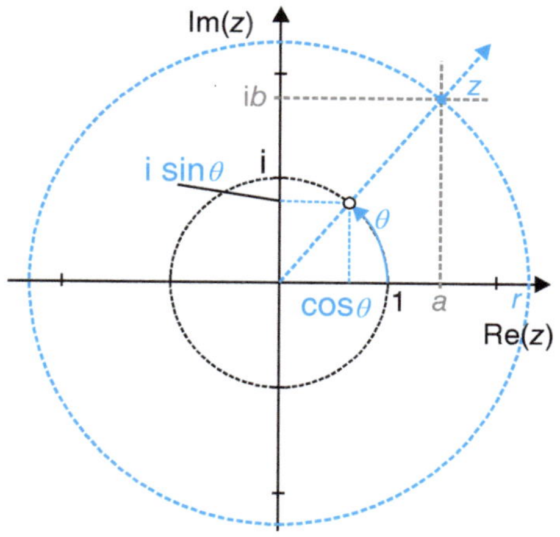

Figure 7.3 Illustration of Euler's notation: The exponential term represents a point on the circle with radius one (black) defining the angle, and r defines the magnitude (blue circle).

Algebra with Complex Numbers. In order to work with complex numbers, we have to establish rules consistent with those for real numbers. For two complex numbers z_1, z_2 with real terms a_1 and a_2 and with imaginary terms ib_1 and ib_2, we can express the following.

Equality: $z_1 = z_2$ only if $a_1 = a_2$ and $b_1 = b_2$.

Addition: $$z_1 + z_2 = (a_1 + a_2) + i\,(b_1 + b_2) \tag{7.14}$$

Multiplication: $$z_1 \cdot z_2 = (a_1 a_2 - b_1 b_2) + i\,(a_1 a_2 + b_2 b_1) \tag{7.15}$$

Division: $$\frac{z_1}{z_2} = \frac{(a_1 a_2 + b_1 b_2) + i(-a_1 b_2 + a_2 b_1)}{a_2^2 + b_2^2} = \frac{(a_1 a_2 + b_1 b_2)}{a_2^2 + b_2^2} + i\frac{(-a_1 b_2 + a_2 b_1)}{a_2^2 + b_2^2} \tag{7.16}$$

Equation (7.17) can be used to express reciprocity $y = 1/z = z^{-1}$ for a complex number z (Figure 7.4):

$$y = z^{-1} = \left(re^{i\theta}\right)^{-1} = \frac{1}{r}e^{-i\theta} \tag{7.17}$$

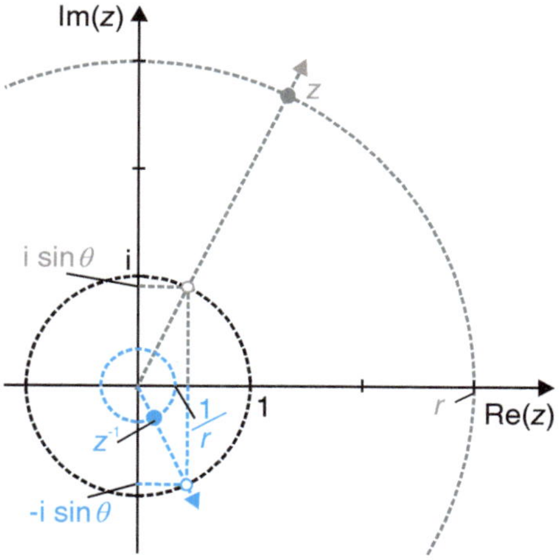

Figure 7.4 Graphical illustration of the reciprocal of a complex number $z = 1.5 + i\,2.598 \approx 3\,\exp(i\,\pi/3)$.

For addition and for multiplication, we retain communicative, associative, and distributive laws equivalent to those for real numbers [$z_1 + z_2 = z_2 + z_1$; $(z_1 + z_2) + z_3 = z_1 + (z_2 + z_3)$; $z_1(z_2 + z_3) = z_1 z_2 + z_1 z_3$, etc.].

7.1 Ohmic Resistors and Ideal Capacitors in AC Circuits

We start by considering the resistance of a circuit element in an AC circuit at a point in time with the associated applied voltage $\dot{E}$. The symbol is introduced to express a dynamic electrode potential measured against a reference electrode.

Often a point symbol indicates a time differential, for example, $\dot{x} = dx/dt$. This is not meant here. Instead, $\dot{E}$ is employed to express the applied potential at a point in time. Therefore, an extra-large dot is used.

ω – Angular frequency (in radians = 2π per cycle) [s^{-1}]
f – Frequency (revolution per time) [Hz] = [s^{-1}]
t – Time [s]
ΔE – Amplitude of AC voltage [V]

$$\dot{E} = \Delta E \sin(\omega t) = \Delta E \sin(2\pi f t) \tag{7.18}$$

For an ohmic resistor R, the ohmic law can be stated by definition as $R = E/I$ and $I = E/R$. We can calculate the time-dependent current $\dot{I}$ with Eq. (7.19).

R – Resistance [Ω] = [V/A]
ΔI – Amplitude in current [A]
$\dot{I}$ – Time-dependent current [A]

$$\dot{I} = \frac{\Delta E}{R}\sin(\omega t) = \Delta I \sin(\omega t) \tag{7.19}$$

For the ohmic resistor, E and I are in phase independent of frequency. In a vector diagram (or phasor diagram), vectors $\dot{E}$ and $\dot{I}$ point in the same direction. Their length represents the amplitudes ΔE and ΔI. Projection of the vectors onto the vertical axis of the phasor diagram gives the real components of $\dot{E}$ and $\dot{I}$. A summary is shown in Figure 7.5, circuit 1.

For the ideal capacitor, the current response is obtained from expressions for charge Q, capacitance C, and voltage E [Eq. (7.21)] by using the first derivative of Faraday's law [Eq. (1.14)].

Q – Charge [A s] = [C]
C – Capacitance [F] = [A s V^{-1}]
E – Voltage or electrode potential [V]

$$Q = CE \tag{7.20}$$

$$\frac{dQ}{dt} = I = C\frac{dE}{dt} = C\frac{d(\Delta E \sin(\omega t))}{dt} = \omega C\ \Delta E \cos(\omega t) \tag{7.21}$$

At this point, it is convenient to introduce the capacitive reactance, $X_C = (\omega C)^{-1}$, in Eq. (7.21). The cosine function is shifted with respect to the sine function by an angle of 90°, called phase shift, that is, $\cos(x) = \sin(x + \pi/2)$.

X_C – Capacitive reactance [V A^{-1}]

$$I = \frac{\Delta E}{X_C}\sin\left(\omega t + \frac{\pi}{2}\right) = \Delta I \sin\left(\omega t + \frac{\pi}{2}\right) \tag{7.22}$$

The comparison of Eqs. (7.19) and (7.22) suggests that the current $\dot{I}$ flowing through a capacitor is shifted by +90° (or $\pi/2$ in radians) with respect to the voltage $\dot{E}$ (Figure 7.5b, circuit 2).

Complex numbers can help expressing impedances and phase shifts effectively. The capacitive reactance (along the ordinate in the vector diagram) is multiplied with $i = \sqrt{-1}$ and represented as an imaginary component. The ohmic resistance is represented along the abscissa as the real component. When expressing the relationship

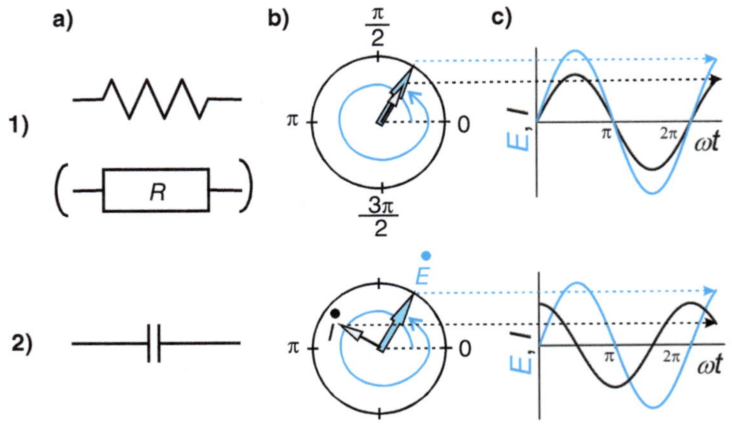

Figure 7.5 Comparison of AC responses for (1) ohmic resistors and (2) ideal capacitors. a) Circuits, b) vector diagrams, and c) time-dependence of $\dot{E}$ and $\dot{I}$. The symbol for the ohmic resistor in a circuit diagram can also be a rectangle depending on which convention is followed.

between current and voltage for an ideal capacitor, we obtain a very compact notation [Eq. (7.23)].

$$\dot{E} = -\mathrm{i}X_C\dot{I} \tag{7.23}$$

This equation is valid for all vector positions in Figure 7.5b, circuit 2 as long as we retain the phase shift at 90° = π/2. A convention for complex plane presentation of these vectors is (i) to place the current vector onto the positive abscissa so that (ii) the vector for voltage points downward (Figure 7.6).

$\dot{E}$ – Vector for voltage (or electrode potential) in the vector diagram [V]
$\dot{I}$ – Vector for current in the vector diagram [A]
i – Imaginary number

Figure 7.6 Vector diagram showing current and voltage for an ideal capacitor with the current vector on the abscissa.

We can now define **impedance Z** as a generalized resistance to an AC current with an ohmic resistance component and a capacitive reactance component.

As a generalization of Ohm's law, we can write

$$\dot{E} = \mathbf{Z}\dot{I} \quad \text{and} \quad \mathbf{Z} = \frac{\dot{E}}{\dot{I}}. \tag{7.24}$$

Generally, the impedance **Z** is frequency-dependent. For an ideal capacitor, we can write $\mathbf{Z} = -\mathrm{i}X_C(\omega)$ and for an ideal resistor, $\mathbf{Z} = R$ (frequency-independent).

7.2 Impedance Measurements with Electrochemical Cells

7.2.1 Instrumentation and Cells

Instrumentation. Classically, EIS data were obtained with a Wheatstone bridge[29] (Figure 4.1C). Both the DC- and the AC-components in the signal needed to be adjusted [6]. With an AC-input voltage applied, resistors and capacitors were adjusted iteratively to match the impedance of the measurement cell [9]. This process has to be repeated frequency-by-frequency with frequencies spanning several orders of magnitude. In order to plot data on a logarithmic scale, frequencies are selected to give the same number of data points for each decade. Although this approach to EIS measurements is very accurate, it is time consuming, and it requires a "stationary" electrochemical cell without any changes in properties during the course of the measurement. Very often, electrochemical cells are nonstationary due to chemical changes or adsorption at electrodes.

Selection of evenly spaced frequencies based on

three points per decade:

1 – 2 – 5 – 10 …

four points per decade:

1 – 1.5 – 2.5 – 5 – 10

five points per decade:

1 – 1.5 – 2.5 – 4 – 6 – 10

etc. (see Figure 7.8a and b).

An elegant approach to adjust the Wheatstone bridge can be based on Lissajous curves.[77] The AC-input signal (potential) is displayed on the x-axis of an oscilloscope and the AC-current response (converted to potential) is displayed on the y-axis. From the Lissajous curves, phase shift and ΔI can be evaluated [1]. Using the sensitivity adjustment, the signals on the screen of the oscilloscope are adjusted to cover the range from −1 to +1 (they are normalized). Then the phase angle is obtained from the Lissajous curve and the setting of the sensitivity for I provides ΔI. The procedure is also executed at each frequency and thus is time consuming.

More modern electrochemical work stations [1, 2] automatically vary frequencies and detect the current output. This can be achieved with a lock-in-amplifier and employing the input signal as a reference signal. The instrument determines the correlation integral with the output signal. This is given as a mathematical series with a constant term and subsequent terms with multiples of the resonance frequency.

[77] Jules Antoine Lissajous, 1822–1880, French physicist.

The constant term is used to determine amplitude and phase shift. There are variations to this procedure [1, 2]. The frequency can be freely varied. This measurement approach is implemented, for example, in instruments from EG&G Princeton Applied Research.

Alternatively, a frequency response analyzer (FRA) can be employed to measure impedance based on correlation of input and output signal (e.g., in instruments by Solartron, PAR-METEK, Metrohm Autolab, Zahner, Scribner). The output signal is multiplied with the sine and cosine version of the input signal, followed by integration, to give real and imaginary terms, respectively. This corresponds to the result of a Fourier analysis. The frequency of the input signal is changed stepwise. For all types of EIS measurements, the validity of data (data all obtained under stationary conditions) can be assessed with the Kramers–Kronig relationships[78] (KKR) [1]. Points that do not satisfy the KKR, can be excluded from further data treatment.

The Fourier transform (FT) approach[50] as routinely employed, for example, in IR- or NMR-spectroscopy lends itself to EIS as well. The input signal has to be a multisine function based on the sum of all input frequencies. The output signal is then dissected by fast FT into a set of lines in the frequency domain. This brings a substantial reduction in measurement time. Each line in the frequency domain can be transformed back and analyzed independently. The resulting decrease in measurement time (good for nonstationary systems) has to be traded against the need for higher input signal amplitudes. The FT methodology requires constant frequency intervals $\Delta\omega$, which leads to an unnecessarily high density of measurement points at higher frequencies when small $\Delta\omega$ intervals are used for low frequencies. The obtained data always follow the KKR, so that this criterion cannot be used to check the data quality and to exclude individual data points from further analysis. The FT-EIS methodology is beneficial particularly for monitoring the evolution of systems with known properties. More specialized approaches to EIS have been proposed based on pulse excitation (analogous to pulse-methods in NMR spectroscopy) [1–6]. However, these methods have so far not been developed into routine EIS measurement tools.

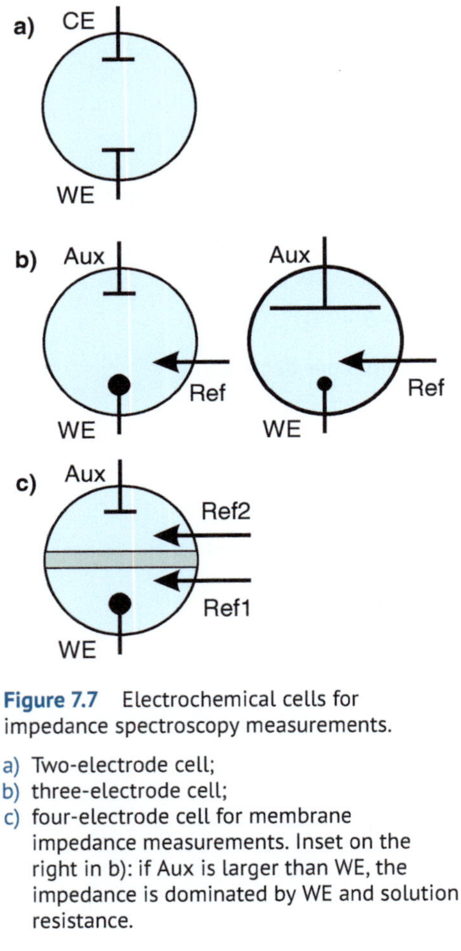

Figure 7.7 Electrochemical cells for impedance spectroscopy measurements.

a) Two-electrode cell;
b) three-electrode cell;
c) four-electrode cell for membrane impedance measurements. Inset on the right in b): if Aux is larger than WE, the impedance is dominated by WE and solution resistance.

Measurement Cells. EIS can be applied to a range of electrochemical cell configurations. In the two-electrode configuration with a working electrode (WE) and a counter electrode (CE), measurements will always return the impedance data $\mathbf{Z}_{\text{cell}}$, which combine contributions from both WE and CE (Figure 7.7a). These can be difficult to dissect as long as the CE is not chosen to be significantly larger in area.

In the three-electrode configuration, a reference electrode (Ref) is introduced and this allows impedance data from the working electrode (WE) and from the auxiliary electrode (Aux) to be determined separately. The cell impedance is then the sum of Z_{WE} and Z_{Aux}. Both contain a contribution from the impedance of the electrolyte (Chapter 5, Figure 5.16). Usually, it is desirable to measure the impedance data for the WE (Figure 7.7b). For good quality data to be obtained, it is recommended to use a large Aux (placed symmetrically to the working electrode) and to position the reference electrode with the help of a Luggin capillary[57] close to the working electrode (but not so close that it blocks diffusional access to the WE). If possible, the measurement is made at the open circuit potential (OCP) of the WE to avoid DC currents in the background. For solution redox systems, it is customary to select equal concentrations for O and R to measure at the equilibrium potential and without current bias. A typical solution redox system for impedance studies is 0.001 mol L^{-1} $K_4[Fe(CN)_6]$ (= R), 0.001 mol L^{-1} $K_3[Fe(CN)_6]$ (= O), and 1 mol L^{-1} Na_2SO_4 (as supporting electrolyte).

[78] Hendrik Anthony Kramers, 1894–1952, Dutch physicist, professor in Utrecht and in Leiden; Ralph de Laer Kronig, 1904–1995, German-US-American physicist, professor at TU Delft.

For the EIS of membranes, both the impedance contributions of the working electrode and the impedance contributions of the counter electrode need to be suppressed. The remaining component, $\mathbf{Z}_{cell}$, can be measured with the four-electrode configuration (WE, Aux, Ref1, Ref2; Figure 7.7c). The impedance is obtained from the potential difference between Ref1 and Ref2 and the current between WE and Aux:

$$\mathbf{Z}_{1,2} = \frac{\dot{E}_{Ref1} - \dot{E}_{Ref2}}{\dot{I}} \tag{7.25}$$

Current Density Distribution. The current density through the working electrode should be uniform to avoid dispersion effects. Current will predominantly flow along the path of lowest impedance, and therefore, symmetry between WE and Aux can help avoid nonuniform current density distributions. Other sources of nonuniformity are linked to nonuniform mass transport and heterogeneous electrode surfaces with nonuniform reaction rates (see Sections 9.3.1 and 9.3.2). Improved uniformity in mass transport can be achieved with rotating disk electrodes, wall-tube electrodes, or rotating cylinder electrodes to improve data quality and simplify the subsequent data analysis/interpretation [1].

7.2.2 Impedance Spectroscopy in Practice

Input Signal. For EIS data to be obtained, a constant DC bias is applied with an AC modulation with a constant amplitude. The modulation frequency ranges over 7 orders of magnitude (typically 0.01–10^5 s^{-1}) and is limited only by time requirements at low frequencies and by cell design at higher frequencies. Within a single EIS experiment, time constants (or rate constants) are determined for a wide range of processes (e.g., surface reactions, pore transport, solution transport, ion transfer, and formation of electrolyte gradients).

The frequency is reported as angular frequency ω [s^{-1}, radians per second] or as AC frequency f [s^{-1}, cycles per second]. In mathematical expressions, the angular frequency ($\omega = 2\pi f$) is convenient and compact. However, the AC frequency f is often given in commercial instruments. In EIS, conversion of frequency units is important for correct data analysis and interpretation.

The analysis of EIS data is usually based on linear response theory which assumes that there are no nonlinear phenomena (rectification). In order to maintain linear response conditions, the input signal amplitude ΔE has to be small. Typical values are $\Delta E = 5$ mV to reflect a peak-to-peak voltage input of 10 mV. Even lower values of ΔE are possible, but lead to losses in the signal-to-noise ratio. A test to explore the validity of data is to vary ΔE. The current amplitude ΔI is proportional to ΔE and, thus, the measured impedance parameters should not depend on ΔE for small variation of ΔE. However, the signal-to-noise ratio might change with ΔE.

Q: The instrument in our laboratory requires an input of E_{eff} (the effective amplitude) rather than the amplitude ΔE. How can I calculate E_{eff} from ΔE?

A: E_{eff} is based on the average input voltage (the square root of the integrated input signal, RMS, *root mean square*) and is convenient when calculating the power. Equation (7.26) can be used to interconvert the two parameters.

$$E_{eff} = \frac{\Delta E}{\sqrt{2}} \tag{7.26}$$

For a desired value of $\Delta E = 5$ mV, we can apply a root mean square input amplitude of $E_{eff} \approx 3.5$ mV.

The quality of EIS data is dependent on measurement control parameters, in particular the integration time (or better the number of integrated cycles of the AC current). An increase in the integration time will improve the signal-to-noise ratio. As a rule of thumb quadrupling the integration time doubles the signal-to-noise ratio. However, the measurement time increases proportional to integration time. Often the electrode process under investigation will not remain stationary over prolonged periods of time.

With a frequency of $f = 0.01$ Hz, the cycle time = $1/f = 100$ s. Integration over four cycles requires 400 s = 6 min 40 s. In contrast, with $f = 100$ kHz, the corresponding integration over four cycles will take 4×10^{-5} s. This is likely to be less than the time for processing the signal and displaying it on a computer screen. Penalties from long integration times are associated with measurements in the lower-frequency range.

EIS with modern instrumentation can appear very user-friendly, but there are pitfalls to be aware of. The measurement data are further treated by fitting to idealized impedance models. In order to obtain conclusive data from the fitting procedure, the EIS data must be of high quality for which an optimized measurement cell and a stationary electrochemical system is very important. Data fitting is then performed based on an equivalent circuit model. Equivalent circuits should not be selected simply to fit, but with a clear vision of each resistor and capacitor providing physical meaning associated with a process in the electrochemical cell. The obtained electrical parameters of the circuit elements should then reveal parameters associated with specific reactions steps, for example, transport and interfacial reaction rates. (The equivalent circuit simulates the behavior of the electrochemical cell, but as such is not an interpretation of the measurement!)

For an assumed equivalent circuit, a data point is calculated for each frequency and compared with the measured data point (least squares fitting). Any mismatch in model and theory is minimized by varying resistance and capacitance values for circuit elements. A good match is required *for all frequencies*. In case the fitting of experimental and model data cannot be achieved, the equivalent circuit *must* be replaced. A good match *suggests* that the assumption made in developing the equivalent circuit may indeed hold for the studied electrochemical cell. However, equivalent circuits are not unique. That is, there are always many equivalent circuits that fit equally well the electrical characteristics of the electrochemical cell. In fact, the calculation and fitting of EIS data is typically conducted on powerful computer programs on personnel computers using mathematical rules for transforming equivalent circuits [1, 2, 14, 15].

In the next section, some commonly employed equivalent circuit models for electrochemical cells are discussed.

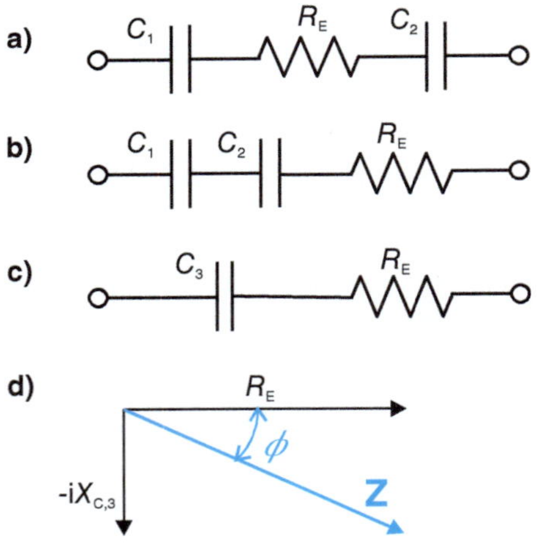

Figure 7.8 Equivalent circuit for a two-electrode cell without faradaic reaction;
a) derived from Figure 4.9;
b) after applying the commutative law to a);
c) equivalent circuit with capacitor C_3 as in Eq. (7.29);
d) as vector diagram.

7.3 Equivalent Circuits for Electrodes and Cells

7.3.1 Impedance for an Electrochemical Cell Without Faradaic Reaction

The equivalent circuit shown in Figure 4.9 was introduced in connection with electrolyte conductivity measurements. Without an electrochemical reaction at the electrode surface, the charge transfer resistance converges to infinity ($R_{CT}[E] \to \infty$). We can simplify the equivalent circuit to that shown in Figure 7.8a. In this series circuit, ohmic and capacitive elements contribute and the sum can be used to derive the total impedance [Eq. (7.27)]. We can use the commutative law and associative law for complex numbers to change the sequence of the circuit elements (Figure 7.8b) and to combine some elements [Eq. (7.28)].

$$\mathbf{Z} = \mathbf{Z}_1 + \mathbf{Z}_E + \mathbf{Z}_2 = -iX_{C,1} + R_E - iX_{C,2} \tag{7.27}$$

$$\mathbf{Z} = R_E - i\left(\frac{1}{\omega C_1} + \frac{1}{\omega C_2}\right) = R_E - i\left(\frac{C_2 + C_1}{\omega C_1 C_2}\right) = R_E - i\frac{1}{\omega C_3} \tag{7.28}$$

$$\frac{1}{C_3} = \left(\frac{1}{C_1} + \frac{1}{C_2}\right) \quad \text{or} \quad C_3 = \frac{C_1 C_2}{C_2 + C_1} \tag{7.29}$$

Note that due to $i^2 = -1$, there are different versions of these equations in textbooks and papers:

$$R_\mathrm{E} - \mathrm{i}\frac{1}{\omega C_3} = R_\mathrm{E} + \frac{1}{\mathrm{i}\omega C_3} \tag{7.30}$$

The total capacitance for this circuit is C_3 [Eq. (7.29)] and allows representation of the entire electrochemical cell by only two elements. This common configuration is also called an RC circuit. Inspection of Eq. (7.29) shows that C_3 is always smaller than either C_1 or C_2. For a combination of two very different capacitors (e.g., a WE with small area and a CE with large area), C_3 is essentially equal to the capacitance of the smaller electrode. This condition is employed when aiming for the characterization of the WE only instead of the entire cell. With this simplification, the total impedance

$$\mathbf{Z}(\omega) = \mathrm{Re}(\mathbf{Z}) - \mathrm{i}\ \mathrm{Im}(\mathbf{Z}) \tag{7.31}$$

can be understood as an RC circuit with $\mathrm{Re}(\mathbf{Z}) = R_\mathrm{E}$ and $\mathrm{Im}(\mathbf{Z}) = X_{C,3}(\omega)$ (Figure 7.8d). The minus sign in Eq. (7.31) is helpful, and as a rule in electrochemical systems, the imaginary component is observed in the negative quadrant of the complex plane. By defining the impedance with a minus sign, both real and imaginary impedance appear in the first (positive) quadrant [6]. The magnitude of the impedance is

$$|\mathbf{Z}| = Z = \sqrt{\mathrm{Re}(\mathbf{Z})^2 + \mathrm{Im}(\mathbf{Z})^2} = \sqrt{R_\mathrm{E}{}^2 + \frac{1}{\omega^2 C_3^2}}. \tag{7.32}$$

For the phase angle ϕ, we can write

$$\tan\phi = \frac{\mathrm{Im}(\mathbf{Z})}{\mathrm{Re}(\mathbf{Z})} = \frac{X_{C,3}}{R_\mathrm{E}} = \frac{1}{\omega C_3 R_\mathrm{E}}. \tag{7.33}$$

In EIS, datasets $\mathbf{Z}(\omega)$ or $\mathbf{Z}(f)$ are acquired, and they can be presented graphically as a function of frequency or in the complex plane (Figure 7.9). Both types of plots are useful and complementary.

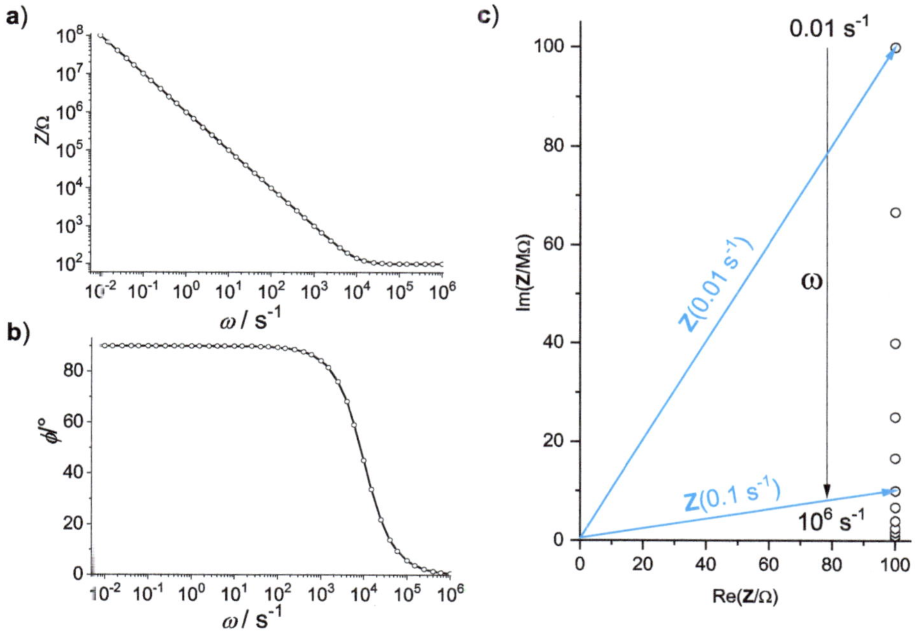

Figure 7.9 Graphical representation of impedance as a function of frequency for an RC circuit.

a) Bode plot for $|Z|$ (logarithmic);
b) Bode plot for the phase angle;
c) Nyquist plot.

Source: Data were simulated for a capacitance of $C_3 = 1\ \mu\mathrm{F}$ and an electrolyte resistance of $100\ \Omega$. The value of C_3 corresponds to a gold electrode with $C_\mathrm{dl} = 20\ \mu\mathrm{F\ cm^{-2}}$ and an area of $A = 5\ \mathrm{mm}^2 = 0.05\ \mathrm{cm}^2$ (assuming a much larger counter electrode; $C_3 = C_\mathrm{dl}\,A \approx C_1$).

In **Bode plots**,[79] the magnitude of impedance $|Z|$ (logarithmic) and the phase angle ϕ are plotted versus the logarithm of angular frequency (Figure 7.8a and b).

In a **Nyquist plot**,[80] the imaginary term Im(**Z**) is plotted versus the real term Re(**Z**) for a range of frequencies (Figure 7.8c). Nyquist plots emphasize low-frequency data.

When discussing impedance spectroscopy data and equivalent circuits, it is important to distinguish intensive (area-independent) and extensive (area-dependent) parameters. In equivalent circuits, the extensive parameters are required. This distinction is not always clear in the literature and can be confusing for new users. With some experience in the area, the correct type of parameter can often be derived based on the context or from the stated units. Here, the area-normalized capacitance (intensive) is indicated with subscript A. In Figure 7.9, C_3 is an extensive capacitance (1 μF). This is linked to the intensive double layer capacitance $C_{\mathrm{dl},A}(\mathrm{Au}) = 20\,\mu\mathrm{F\,cm^{-2}}$ by multiplication with area $A = 2\,\mathrm{mm} \cdot 2.5\,\mathrm{mm} = 5\,\mathrm{mm}^2 = 0.05\,\mathrm{cm}^2$.

7.3.2 Impedance of an Electrochemical Cell with Faradaic Reaction and Without Diffusion Limitation

In the equivalent circuit in Figure 4.9, each electrode is represented by parallel arrangement of an ohmic resistance R_k and a capacitor reflecting the electrical double layer capacitance C_k. This allows the simultaneous occurrence of faradaic current I_F and a capacitive current I_C which we have already discussed as the essential signal components in electroanalytical measurements (Chapter 5). Now, we investigate impedance of the RC parallel circuit. What is the effect of frequency on total impedance and phase angle? This question can be addressed by introducing the admittance.

Real quantity	Complex quantity
Ohmic resistance R	Impedance **Z**
Conductivity G	Admittance **Y**

Figure 7.10 Relationship between transport parameters on the real axis as resistance or conductivity (R, G) and in the complex plane (**Z**, **Y**) as "barrier" (R, **Z**) or as "conduit" (G, **Y**).

Z – Impedance [V A^{-1}] = [Ω]
Z – Magnitude of the impedance [Ω]
ϕ – is the phase angle in radians [1]
i – Complex number

Y – Admittance [A V^{-1}] = [Ω^{-1}]

Admittance Y is the complex conductivity and the reciprocal of impedance.

Figure 7.10 illustrates the relationship between resistivity and conductivity (both real) and the related complex impedance and admittance. Impedance in Euler's notation is

$$\mathbf{Z} = Z\mathrm{e}^{\mathrm{i}\phi} \tag{7.34}$$

The admittance is derived as Eq. (7.35) [see Eq. (7.17), Figure 7.4)].

$$\mathbf{Y} = \frac{1}{\mathbf{Z}} = \frac{1}{Z}\mathrm{e}^{-\mathrm{i}\phi}. \tag{7.35}$$

For a parallel circuit (subscript p), the total admittance $\mathbf{Y}_\mathrm{p}$ can be written as the sum of individual admittances ($\mathbf{Y}_1 + \mathbf{Y}_2$). This is in analogy to Ohm's law for resistors in a DC circuit, where the conductivities are $G_\mathrm{p} = G_1 + G_2$, and the resistivities are connected by $1/R_\mathrm{p} = 1/R_1 + 1/R_2$. For individual resistors or capacitors in the AC circuit, we obtain $\mathbf{Y}_1 = 1/R\,\mathrm{e}^0$ and $\mathbf{Y}_2 = 1/X_\mathrm{c}\,\mathrm{e}^{+\mathrm{i}\pi/2} = C_1\omega\,\mathrm{e}^{+\mathrm{i}\pi/2}$, respectively.

$$\begin{aligned}\mathbf{Y}_\mathrm{p} &= \frac{1}{Z_1}e^{-\mathrm{i}\phi_1} + \frac{1}{Z_2}e^{-\mathrm{i}\phi_2} = \sum_k \frac{1}{Z_k}\; e^{-\mathrm{i}\phi_k} \\ &= \sum_k \left[\frac{1}{Z_k}\cos(\phi_k)\right] - \mathrm{i}\sum_k \left[\frac{1}{Z_k}\sin(\phi_k)\right]\end{aligned} \tag{7.36}$$

[79] Hendrik Wade Bode, 1905–1982, US-American engineer at Bell Laboratories, professor at Harvard University.

[80] Harry Nyquist, 1889, Nilsby (Sweden)–1976, Harlingen (Texas, USA), US-American electronic engineer at AT&T and at Bell Laboratories.

The expressions for $\phi(\omega)$ and for $\mathbf{Y}(\omega)$ are obtained in Euler's notation based on Eqs. (7.6) and (7.8). The impedance is obtained as the reciprocal of the admittance $\mathbf{Y}_p$ ($\mathbf{Z}_p = \mathbf{Y}_p^{-1}$). Figure 7.11 shows the results as Bode plots and as a Nyquist plot. For low frequencies ($\omega \to 0$), the capacitor exhibits a high impedance, and the current flow is predominantly across the resistor R_1. Therefore, we find $Z = R_1 = 100\,\Omega$ in Figure 7.11a for low frequencies. In the Nyquist plot, the data points follow a semicircle pattern with two crossing points of the real axis (abscissa). For $\omega \to 0$, the semicircle touches the real axis at $\mathrm{Im}(\mathbf{Z}) = 0$ and $\mathrm{Re}(\mathbf{Z}) = R_1$. For $\omega \to \infty$, we find $\mathrm{Im}(\mathbf{Z}) = 0$ and $\mathrm{Re}(\mathbf{Z}) = 0$, and the capacitor dominates the impedance. The summit point of the semicircle is the point of maximum $\mathrm{Im}(\mathbf{Z})$. It occurs at the angular frequency $\omega = (C_1R_1)^{-1}$. Based on our knowledge of R_1 and the angular frequency at the summit point, we can calculate C_1.

For real electrochemical cells and electrodes, an additional resistor (the electrolyte resistor R_E) needs to be added in series to the RC parallel circuit. For a two-electrode measurement, the electrolyte resistance is across the full cell (Figure 4.9). For a three-electrode measurement, R_E represents the electrolyte resistance between working electrode and reference electrode (Figure 5.18). In order to calculate the impedance, we first use the parallel component $\mathbf{Z}_P$ with R_1 and C_1 and then add R_E. This can be illustrated in terms of a vector addition of complex numbers in the complex plane (Figure 7.12). Due to R_E being real, the semicircle is shifted along the real axis in the Nyquist plot in Figure 7.11c. The resulting equivalent circuit is the Randles equivalent circuit.[64] This is a fundamental equivalent circuit to describe electrochemical cells.

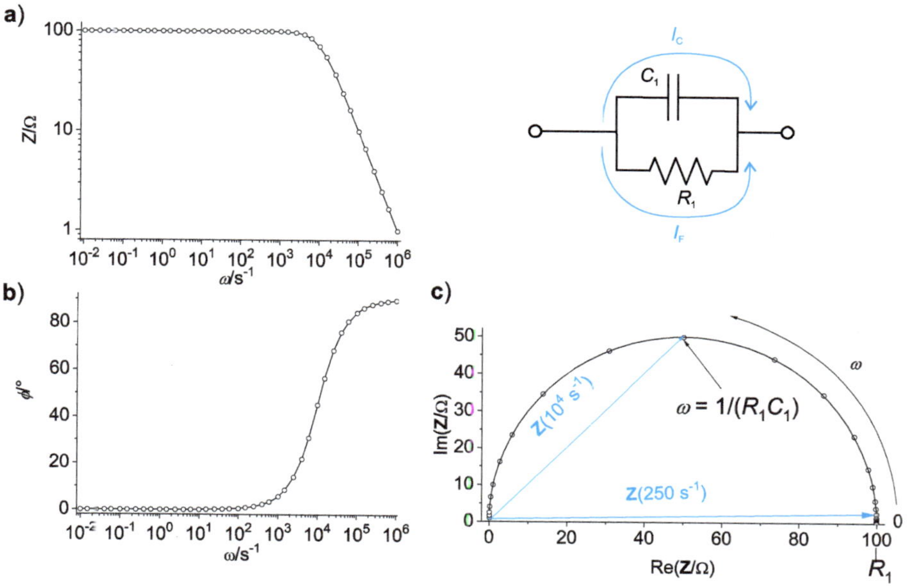

Figure 7.11 Graphical representations of the impedance as a function of angular frequency for an RC parallel circuit with $R_1 = 100\ \Omega$ and $C_1 = 1\ \mu F$.

a) Bode plot for the magnitude of impedance;
b) Bode plot for the phase angle;
c) Nyquist plot.

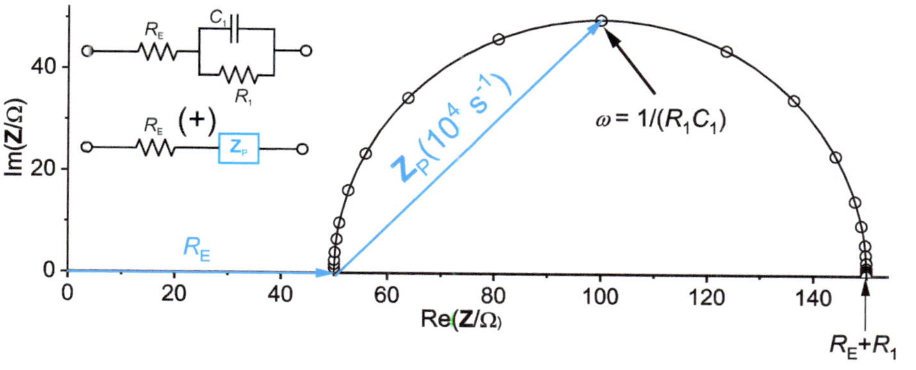

Figure 7.12 Nyquist plot for an electrolyte resistance $R_E = 50\ \Omega$ in series with a parallel RC circuit with $C_1 = 1\ \mu F$ and $R_1 = 100\ \Omega$.

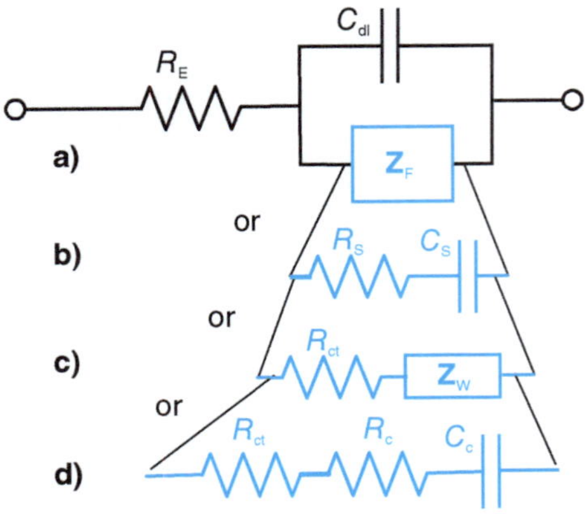

Figure 7.13 Approximations for the faradaic impedance.

a) Generalized equivalent circuit;
b) approximation of $\mathbf{Z}_F$ by R_s and C_s;
c) approximation of $\mathbf{Z}_F$ by R_{CT} and $\mathbf{Z}_W$;
d) approximation of $\mathbf{Z}_W$ by the concentration resistance R_c and the concentration capacitance C_c.

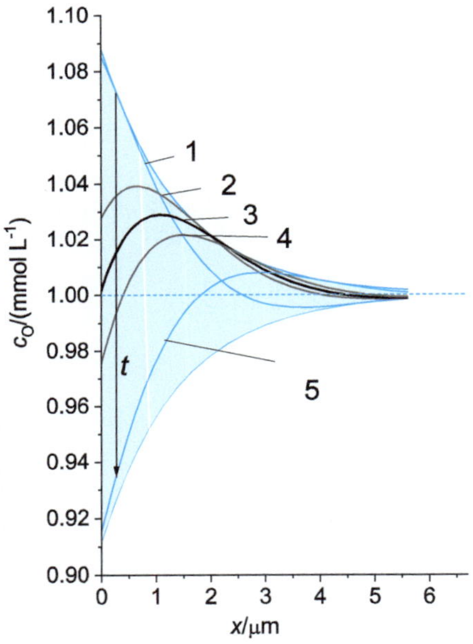

Figure 7.14 Concentration $c(x, t)$ as a function of distance x to the electrode. Shadowed zones represent decaying concentration fluctuations. Curves 1–5 show concentration profiles at specific times within a half-cycle in $E(t)$. Bulk concentration $c_O^* = c_R^* = 1\,\text{mmol L}^{-1}$ (blue dashed line). Local concentrations oscillate by approximately 9% around c^* ($\Delta E = 5$ mV, $n = 1$, reversible system). Graph was calculated according to [9].

7.3.3 Impedance of an Electrochemical Cell with Faradaic Reaction and with Diffusion Limitation

We can now develop equivalent circuit models for more complex cases. In the circuit in Figure 7.12, all elements are ideal capacitors or ohmic resistors. In most cases, the electrolyte resistance (R_E) is close to ideal and the double layer capacitance (C_{dl}) is reasonably close to an ideal capacitor (apart from roughness or heterogeneity effects). However, the resistance for the faradaic current is not ideal and considered here as faradaic impedance $\mathbf{Z}_F$. $\mathbf{Z}_F$ can be dissected into combinations of resistors and capacitors in different ways (equivalent circuits are not unique!). A first approximation for $\mathbf{Z}_F$ can be based on a series of an ohmic resistor R_s [subscript s for serial] and a capacitor C_s (Figure 7.13b).

In order to better describe the faradaic impedance, we can describe two components: the charge transfer resistance R_{CT} associated with electron or ion transfer and a further general impedance contribution called Warburg Impedance[81] $\mathbf{Z}_W$, which is associated with the perturbation of the concentration of reactant species at the electrode surface. The AC-current input signal causes an electrolysis with periodically changing consumption of the reactant. This leads to a concentration wave (damped) propagating from the electrode surface into the solution (Figure 7.14). This concentration perturbation is similar to the concentration changes caused by planar diffusion in cyclic voltammetry (Figure 5.27). This perturbation persists for some time even if the electrolysis at the electrodes stops. We have used the metaphor of a "concentration scar" in Chapter 5 to describe such a situation. For EIS, the amplitude of the concentration change is small. The thickness of the diffusion layer at the electrode–solution interface depends on the frequency f (approx. $\sqrt{DRT/(fF)}$), but usually decays within 10 μm to the bulk value c^*.

Once we have identified a suitable equivalent circuit, the further interpretation of the data can follow the flow chart in Figure 7.15. We derive an expression (here based

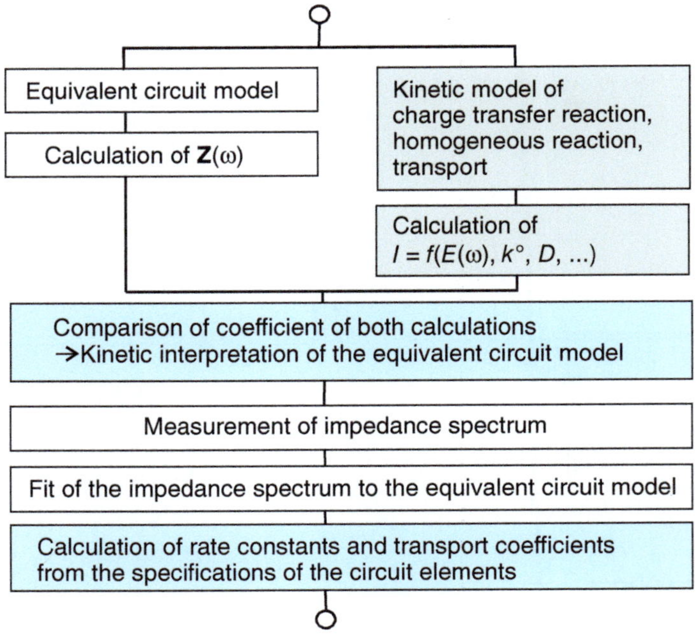

Figure 7.15 Flow chart for the interpretation of data from electrochemical impedance spectroscopy.

---

[81] Emil Gabriel Warburg, 1846–1931, German physicist, professor in Strasbourg, Freiburg, and in Berlin.

on dE/dt) that can subsequently be compared to the corresponding expression derived from the treatment of kinetics and mass transport for electrode reactions. We can identify and assign physical meaning to equivalent circuit parameters based on comparison of coefficients. This approach was used previously when investigating electrolyte conductivity (Section 4.2.3). A detailed derivation can be found in [1, 6, 9] with further examples in [1, 2].

We consider a reversible redox system (R/O) with equal bulk concentrations ($c_O^* = c_R^*$) and with added supporting electrolyte in an electrochemical cell.

$$O + ne^- \rightleftharpoons R \tag{7.37}$$

Electrical Network Expressions. The open circuit potential of the electrode (OCP) is now equal to the formal potential $E^{\circ\prime}$ of the redox couple R/O. There is no bias or background (DC) current. Superimposed on this potential, one applies a modulation in the potential, and an output current is generated.

$$I = \Delta I \sin(\omega t) \tag{7.38}$$

I – Time-dependent current [A]
ΔI – Current amplitude [A]

For a series circuit with R_s and C_s, we can express the voltage, and based on this, the time derivative of the voltage.

$$E = IR_s + \frac{Q}{C_s} \tag{7.39}$$

Q – Charge [A s]
R_s – Series resistor [Ω]
C_s – Series capacitor [C]

$$\frac{dE}{dt} = R_s \frac{dI}{dt} + \frac{I}{C_s} = R_s \cdot \Delta I \omega \cos(\omega t) + \frac{1}{C_s} \Delta I \sin(\omega t). \tag{7.40}$$

Electrochemical Expressions. The potential at the electrode surface (at $x = 0$) is defined by the concentrations for O and R, and the concentrations are linked to the flow of current. Generally, E and its time derivative can be expressed.

$$E = E\left[I, c_O(0,t), c_R(0,t)\right]. \tag{7.41}$$

$$\frac{dE}{dt} = \left(\frac{dE}{dI}\right)_{c_O,c_R} \frac{dI}{dt} + \left(\frac{\partial E}{\partial c_O(0,t)}\right)_{I,c_R} \frac{dc_O(0,t)}{dt} + \left(\frac{\partial E}{\partial c_R(0,t)}\right)_{I,c_O} \frac{dc_R(0,t)}{dt} = R_{CT}\frac{dI}{dt} + \beta_O \frac{dc_O(0,t)}{dt} + \beta_R \frac{dc_R(0,t)}{dt} \tag{7.42}$$

$c_O(0,t)$, $c_R(0,t)$ – Concentrations of O and R at the electrode surface ($x = 0$) [mol L^{-1}]

Each differential coefficient can be evaluated. The time derivative of Eq. (7.39) provides dI/dt. The expression for concentration at the electrode surface requires solving the partial differential equation with boundary conditions applying Laplace transformation [6].

$$\frac{dc_O(0,t)}{dt} = \frac{\Delta I}{nFA}\sqrt{\frac{\omega}{2D_O}}\left[\cos(\omega t) + \sin(\omega t)\right] \tag{7.43}$$

$$\frac{dc_R(0,t)}{dt} = \frac{-\Delta I}{nFA}\sqrt{\frac{\omega}{2D_O}}\left[\cos(\omega t) + \sin(\omega t)\right] \tag{7.44}$$

With the abbreviation σ

$$\sigma = \frac{1}{nFA\sqrt{2}}\left(\frac{\beta_O}{\sqrt{D_O}} - \frac{\beta_R}{\sqrt{D_R}}\right) \tag{7.45}$$

and with Eq. (7.42), we obtain

$$\frac{dE}{dt} = \left(R_{CT} + \frac{\sigma}{\sqrt{\omega}}\right)\Delta I \omega \cdot \cos(\omega t) + \Delta I \sigma \sqrt{\omega} \cdot \sin(\omega t) \tag{7.46}$$

Comparison of Electrical and Electrochemical Expressions. Comparing the coefficients in Eqs. (7.40) and (7.46) provides

$$R_s = R_{CT} + \frac{\sigma}{\sqrt{\omega}} \tag{7.47}$$

$$C_s = \frac{1}{\sigma\sqrt{\omega}} \tag{7.48}$$

The partial differential coefficients R_{CT}, β_O, and β_R are obtained from the partial differential (with respect to c_O and c_R) of the linearized Butler–Volmer Equation (3.38). The linearization is based on a Taylor series ($e^x \approx 1 + x$ close to $x = 0$). This approximation is justified when ΔE and ΔI are very small.

E_{NF} – Nernst factor = $RT/(nF)$ [V]
I_0 – Exchange current [A]

$$\eta = E - E_{eq} = E - E^{\circ\prime} = E_{NF}\left[\frac{I}{I_0} + \frac{c_R(0,t)}{c_R^*} - \frac{c_O(0,t)}{c_O^*}\right] \tag{7.49}$$

$$R_{CT} = \frac{E_{NF}}{I_0} \tag{7.50}$$

$$\beta_O = \frac{E_{NF}}{c_O^*} \tag{7.51}$$

$$\beta_R = -\frac{E_{NF}}{c_R^*} \tag{7.52}$$

Rearranging Eq. (7.47) and inserting into Eq. (7.48) yield the relationship between R_s and R_{CT}:

$$R_s - \frac{\sigma}{\sqrt{\omega}} = R_s - \frac{\sigma\sqrt{\omega}}{\sqrt{\omega}\cdot\sqrt{\omega}} = R_s - \frac{1}{\omega C_s} = R_{CT} = \frac{RT}{nFI_0} \tag{7.53}$$

The power of EIS is the possibility of dissecting complex processes into individual steps and pinpointing one element in the equivalent circuit to extract the desired quantitative value. In this way, many precision measurements of kinetic parameters for fast electron transfer reactions have been performed. Figure 7.13 illustrates that R_E and C_{dl} are measurable with high precision in the absence of O or R. Then, R_s and C_s can be determined (e.g., using the compensation method) to allow determination of R_{CT}. It would be possible to work at a single frequency, but better precision and reproducibility are possible with impedance spectra that also allows consistency checks at different frequencies. Plotting $R_s(\omega)$ and $1/(\omega C_s)$ versus $\omega^{-1/2}$ results in parallel lines with slope σ. The ordinate intercept ($\omega \to \infty$) is R_{CT}. Once we have determined R_{CT}, we can calculate I_0 and the heterogenous standard rate constant $k°$ [Eq. (3.44)].

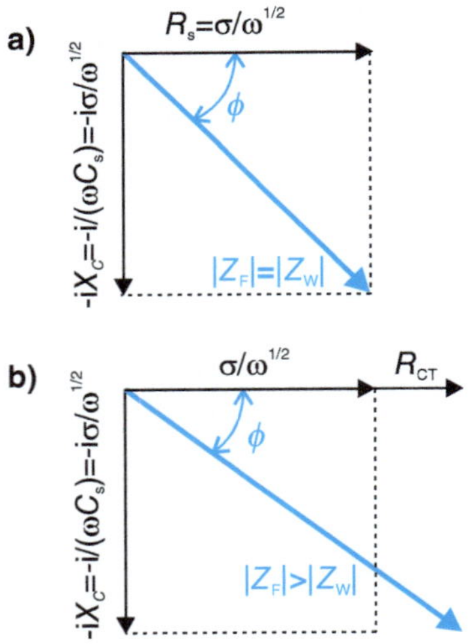

Figure 7.16 Vector diagram for the faradaic impedance for a) a reversible redox system and b) for a quasireversible redox system.

It is interesting to explore the vector diagram for fast electrochemical reactions. For a reversible system, we can set $R_{CT} = 0$; and with Eq. (7.47), we obtain $R_s = \sigma/\omega^{1/2}$. With $C = C_s$, the reactance is $X_C = 1/(\omega C) = \sigma/\omega^{1/2}$ [Eq. (7.48)]. In this case, the real term and the imaginary term are equal, and the phase angle is 45° (Figure 7.16a). For a quasireversible system, R_{CT} has a finite value, and this causes a shift of the data points along the abscissa (real axis) to increase the total impedance. The phase angle will always be lower than 45° (Figure 7.16b).

The Nyquist plot for charge transfer reactions is similar to that in Figure 7.12 but shows an additional linear branch at lower frequencies. This can be attributed to mass transport limitations (planar diffusion) (Figure 7.17a). Depending on the chosen frequency range, the bulk concentrations, the standard rate constant $k°$, and the electrode area A (which affects the exchange current I_0), some features of the Nyquist plot such as mass transport limited or kinetically limited domains) may be missing for a particular system (Figure 7.17b and c). Key parameters associated with the equivalent circuit are obtained from characteristic points in the Nyquist plot. The intersection of

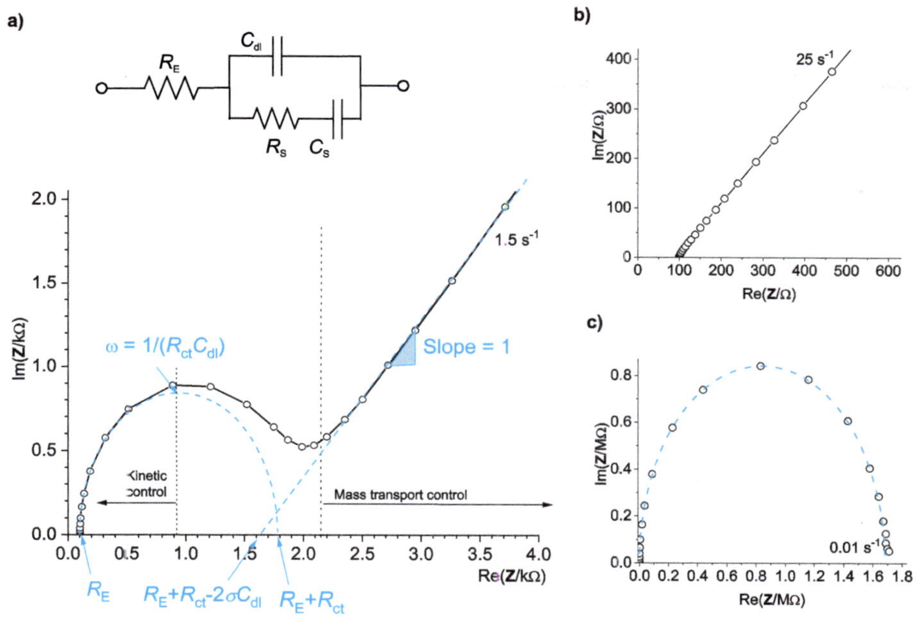

Figure 7.17 Nyquist plot for a charge transfer reaction. a) Incomplete semicircle for a kinetically controlled frequency range and mass transport controlled branch at lower frequencies [parameters $k° = 10^{-4}$ cm s^{-1}, $\alpha = 0.5$, $D_O = D_R = 10^{-5}$ cm^2 s^{-1}, $c_O^* = c_R^* =$ 1 mmol L^{-1} = 10^{-6} mol cm^{-3}, $C_{dl,A} = 20$ µF cm^{-2}, $A = 0.05$ cm^2, $R_E = 100$ Ω]; b) only mass transport controlled range visible [$k° = 0.1$ cm s^{-1}, others as in (a)]; c) only kinetically controlled range visible [$k° = 10^{-7}$ cm s^{-1}, others as in (a)].

the semicircle with the abscissa gives the electrolyte resistance (R_E), the frequency of the summit point [$\omega = (R_{CT}\ C_{dl})^{-1}$], the second intersection point of the semicircle and abscissa ($R_{CT} + R_E$), and the intersecting of the extrapolated linear branch for planar diffusion with the abscissa ($R_{CT} + R_E - 2\,\sigma C_{dl}$).

Data in Figure 7.12 are calculated based on [6]. We obtain $I_0 = j_0\,A$ with j_0 from Eq. (3.33), R_{CT} from Eq. (7.50), σ from Eq. (7.45), R_s from Eq. (7.47), and C_s from Eq. (7.48),

$$O = 1 + \frac{C_{dl,A}A}{C_s} \quad \text{and} \quad P = \omega R_s C_{dl,A}A.$$

Using the quantities O and R, the real and imaginary components can be calculated from.

$$\text{Re}(\mathbf{Z}) = R_E + \frac{R_s}{O^2 + P^2} \quad \text{and} \quad \text{Im}(\mathbf{Z}) = \frac{\frac{P^2}{\omega C_{dl,A}A} + \frac{O}{\omega C_s}}{O^2 + P^2}.$$

7.4 Developing Equivalent Circuits

In practice, electrodes are rough or porous or coated with films, which adds complexity to the ideal mathematical model assumptions. New equivalent circuits can be developed to aid data analysis for more complex electrode systems. The approach below follows that in [1]. A more detailed mathematical treatment can be found in [2].

7.4.1 Partially Blocked Electrodes

In the course of electrochemical processes, electrode surfaces can change or become coated due to oxide formation, formation of electrically insulating polymers, electrode fouling in the sense of formation of precipitates, self-assembled monolayers, or coatings with receptors in biosensors. In the first approximation, we can define a coated fraction of the electrode surface θ with significantly slower charge transfer kinetics. The remaining uncoated fraction $(1 - \theta)$ stays active, retaining fast charge transfer kinetics. In the limit of a substantial coating, the rate of charge transfer approaches zero, resulting in a faradaic impedance $\mathbf{Z}_{F,l}$ approaching infinity. The subscript "l" (for layer) indicates parameters associated with the coating. The equivalent circuit in Figure 7.18 represents a parallel combination of electrode surface regions with a coating and regions without a coating. The total capacitance of the electrode is additive [Eq. (7.54)], which allows the surface coverage to be determined. With the total area A known, and the measured double layer capacitance before coating ($C_{dl,A}$) and after coating a complete monolayer ($C_{l,A}$ for $\theta = 1$), we can use Eq. (7.55) to evaluate intermediate coverages.

C – Capacitance from EIS measurement and analysis as in Section 7.3.3 [μF]
θ – Coverage [dimensionless]
A – Total electrode area [cm^2]
$C_{l,A}$ – Area-normalized capacitance for the coated electrode [$\mu F\ cm^{-2}$]
$C_{dl,A}$ – Area-normalized capacitance for the uncoated electrode [$\mu F\ cm^{-2}$]

$$C = \theta A C_{l,A} + (1 - \theta) A C_{dl,A} \tag{7.54}$$

$$\theta = \frac{A C_{dl,A} - C}{A C_{dl,A} - A C_{l,A}} \tag{7.55}$$

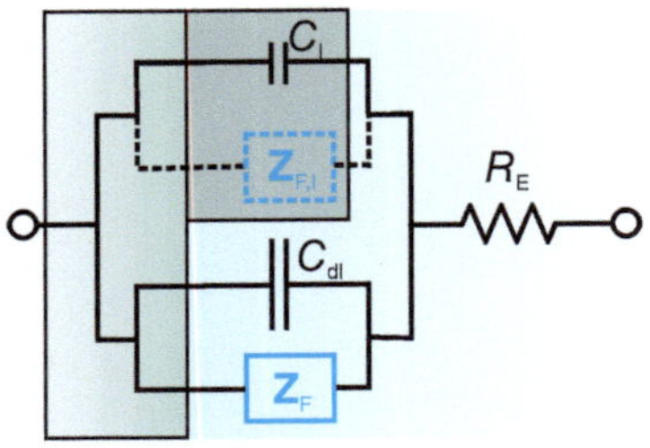

Figure 7.18 Equivalent circuit for a partially blocked electrode. For the coated region, we assume zero faradaic current ($\mathbf{Z}_{F,l} = \infty$). Note that the drawing is not to scale, and the coating is assumed to be very thin.

7.4.2 Electrodes with Porous Coatings

Coatings considered in Section 7.4.1 were assumed to not affect migration-diffusion to the uncoated regions of the electrode surface. However, this assumption is valid only for very thin coatings. For more substantial coatings, we have to consider the additional limitation of charge transport through a porous layer by means of a new impedance term (Figure 7.19). We can introduce a "layer resistance" R_l for the case that charge transfer occurs at the electrode surface and transport occurs through pores in an electrically insulating material (e.g., insulator coatings with defects, porous oxide layers, corroding surfaces). The equivalent circuit (Figure 7.19) is based on a capacitance C_l for the coated region parallel to a series of R_l and parallel C_{dl} and $\mathbf{Z}_F$. Diffusion and transport of molecules in the porous layer can differ from transport in bulk solution, and therefore, the faradaic impedance $\mathbf{Z}_F$ may also differ from the simple case discussed in Section 7.3.3.

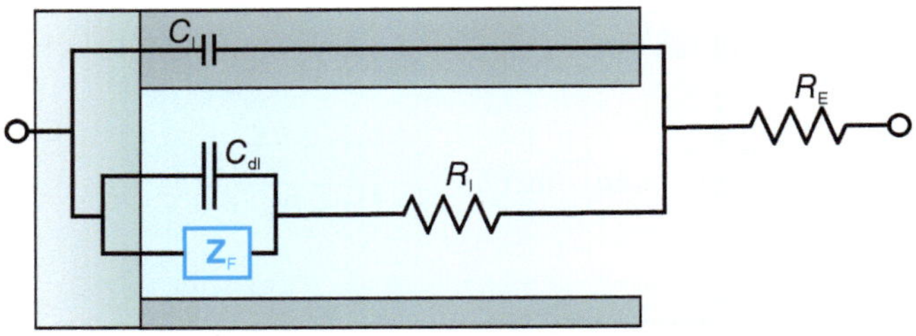

Figure 7.19 Equivalent circuit for an electrode with an electrically insulating porous coating. For the charge transport within the porous coating, an additional resistance R_l has been introduced. For the fully covered regions, we assume absence of faradaic reaction, that is, $\mathbf{Z}_{F,l} = \infty$.

7.4.3 Porous Electrodes

Q: Isn't a porous coating in Section 7.4.2 the same as a porous electrode?

A: No. For porous electrodes, the electrode material itself is porous. In this case, charge transfer reactions occur at the interface of pores and electrolyte. In Section 7.4.2, we assumed a planar electrode (for charge transfer) coated with an insulator coating (such as oxide layers in the context of corrosion, Chapter 14). There are many examples for porous electrodes in batteries and in fuel cells (Chapters 15 and 16). For porous electrodes, we have to consider a combination of electrical conductivity and ion conductivity. Of course, there are even more complex cases of porous electrodes that are coated with porous insulator coatings.

Porosity of electrodes can be very beneficial, for example, by increasing the real surface area while maintaining constant geometric area. Porous electrodes can be produced from powders (Chapter 15) or obtained by partial dissolution of alloys (Section 14.5.7). By mixing particles of different sizes, electrodes can be produced with a distribution of pore diameters and pore lengths. The behavior of porous electrodes with a distribution of pore diameters and pore length can often be approximated by a *single* cylindrical model pore with the characteristic pore length l_{pore} and the characteristic pore radius r_{pore} (Figure 7.20a and b).

Next, we divide the model pore into N_0 ring segments with length $\Delta l_{\text{pore}} = l_{\text{pore}}/N_0$. Each ring segment has an impedance composed, for example, of a Randles component with double layer capacitor and faradaic impedance. The resulting equivalent circuit in Figure 7.20c is called a transmission line model. Transmission line models were originally developed for the description of deep-sea cable impedances (transatlantic telegraph cables) but are now commonly used to describe mixed electrical and ionic charge transport in porous structures. A potential step at such a network is transmitted with a delay from element to element. Thus, it is also called a delay line.

The electrode material could be poorly conducting (e.g., for pressed powders with high grain boundary resistances; see dashed lines in the equivalent circuit). In the derivation below, these resistances are ignored. Charge transfer at the bottom of the pore has also been neglected for simplicity.

$\mathbf{Z}_A$ is a complex function of ω.

In the simplest case, the specific impedances for every ring segment are assumed constant and equal to $\mathbf{Z}_{0,l}$. $\mathbf{Z}_{0,l}$ is obtained from the area-normalized surface impedance $\mathbf{Z}_A$ and the pore circumference $2\pi r_{\text{pore}}$.

$$\mathbf{Z}_{0,l} = \frac{\mathbf{Z}_A}{2\pi r_{\text{pore}}} \tag{7.56}$$

$\mathbf{Z}_{0,l}$ – Specific surface impedance (based on ring width Δl_{pore}) [Ω cm]
$\mathbf{Z}_A$ – Area-normalized surface impedance [$\Omega\ \text{cm}^2$], measured at a planar electrode
r_{pore} – Characteristic pore radius [cm]

However, deeper ring segments experience a higher electrolyte resistance and will therefore contribute less to the current and thus make a different contribution to the impedance. We assume a specific conductivity κ for the electrolyte and obtain the resistance $\rho = 1/\kappa$. Given ρ, the pore cross-sectional area πr_{pore}^2, and the length Δl_{pore} of the segment, we can express the specific electrolyte resistance per segment as

$$R_0 = \rho_{\text{E}} \frac{\Delta l_{\text{pore}}}{\pi r_{\text{pore}}^2} = R_{0,l} \cdot \Delta l_{\text{pore}} \tag{7.57}$$

$R_{0,l}$ – Specific pore resistance [$\Omega\ \text{cm}^{-1}$] = $\rho/\left(\pi r_{\text{pore}}^2\right)$
ρ_{E} – Specific electrolyte resistance [Ω cm]
Δl_{pore} – Length of the pore segment [cm]

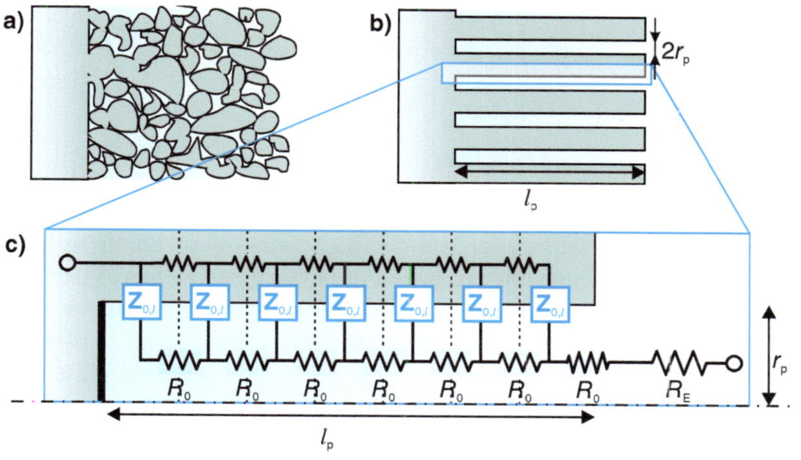

Figure 7.20 Equivalent circuit for a porous electrode.

a) Schematic of a porous electrode;
b) simplification of the pore structure as cylindrical model pore with characteristic length l_{pore} and characteristic radius r_{pore};
c) equivalent circuit for the pore wall.

For a single pore, we obtain the impedance $\mathbf{Z}_{\text{pore}}$ (in approximation)

$\mathbf{Z}_{\text{pore}}$ – Impedance of a single pore [Ω]

$$\mathbf{Z}_{\text{pore}} = \sqrt{R_{0,l}\mathbf{Z}_{0,l}} \coth\left(l_{\text{pore}}\sqrt{\frac{R_{0,l}}{\mathbf{Z}_{0,l}}}\right) \tag{7.58}$$

For an electrode with N_{pore} pores, the impedance $\mathbf{Z}'_{\text{pore}}$ is [16]

N_{pore} – Number of pores [dimensionless]

$$\mathbf{Z}'_{\text{pore}} = R_{\text{E}} + \frac{\mathbf{Z}_{\text{pore}}}{N_{\text{pore}}} = R_{\text{E}} + \sqrt{\frac{\rho \mathbf{Z}_A}{2N_{\text{pore}}\pi^2 r_{\text{pore}}^3}} \coth\left(l_{\text{pore}}\sqrt{\frac{2\rho}{r_{\text{pore}}\mathbf{Z}_A}}\right). \tag{7.59}$$

The Nyquist plot for this type of electrode shows a distorted semicircle dependent on pore length l_{pore} and penetration depth λ of the electrical signal. The penetration depth λ can be expressed as $(\mathbf{Z}_{0,l}/R_0)^{1/2}$. The ratio l_{pore}/λ appears in the coth function in Eq. (7.58) and Eq. (7.59). For $l_{\text{pore}} \ll \lambda$, all of the electrode responds to a potential input. The electrode behaves similar to a planar electrode with the same geometric area. However, for $l_{\text{pore}} \gg \lambda$, the electrical signal is not affected because the additional area of the pore walls are not charged up within the period of the AC signal and the pores behave as if semi-infinite. Increasing the pore depth further does not affect the impedance.

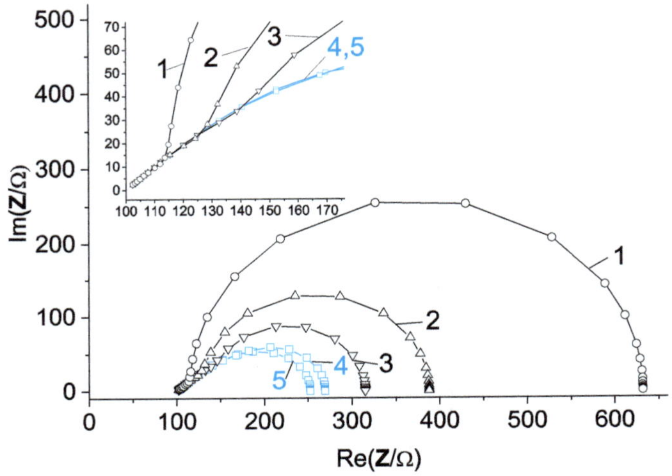

Figure 7.21 Nyquist plot for a porous electrode. The ratio l_p/λ is

1) 0.01 □;
2) 0.02 △;
3) 0.03 ▽;
4) 0.05 □ (blue);
5) 0.5 ■ (blue).

For curves 4 and 5 (blue), the pore depth increases, but the impedance does not change substantially. For a small l_{pore}/λ (= short pores), we obtain a Nyquist plot similar to a semicircle for a planar electrode. Deviations occur at higher frequencies (inset). ($C_{\text{dl},A}$ = 20 µF cm^{-2}, R_{CT} = 100 Ω, r_{P} = 1 µm, N_{pore} = 25 × 10^6, A = 1 cm^2, R_{E} = 100 Ω.)

The treatment of porous electrode structures can be refined based on more complex pore geometries [2] or potential and concentration gradients along the pore [17]. These cases are too complex for analytical equations but can be treated with numerical simulation models. Modern commercial software packages provide tools for this kind of simulations [14].

7.4.4 Constant Phase Elements (CPE)

In many practically relevant applications of EIS, no satisfactory fit of model and experimental data is obtained with equivalent circuits based only on ohmic resistors and ideal capacitors. The reason is the dispersion in physical parameters, which we have already discussed in Sections 7.4.2 and 7.4.3. Simplified relationships are valid only for those parts of the electrode being associated with a *constant* $C_{\text{dl},A}$, and all parts exhibiting a *uniform* rate constant $k(E)$. Furthermore, we require uniform mass transport (diffusion and/or migration) because $c(x = 0)$ and $k°$ (via j_0) also affect R_{CT} [Eqs. (3.33) and (3.44)]. In practice, planar electrodes or hanging mercury drop electrodes satisfy these requirements. For many solid electrodes, there are problems. For polycrystalline electrodes, the surface is characterized by crystal facets with steps, grain boundaries, or defects with varying $k°$. Even an ideal disk electrode is associated with a nonuniform current density distribution under diffusion-controlled conditions due to edge effects. Coatings with adsorbates affect both, k and $C_{dl,A}$, and the coating can be nonuniform. In all these cases, each surface segment could be represented by a distinct equivalent circuit as in Figure 7.8 or with faradaic currents as shown Figure 7.11. For a series C_{dl}–R_E circuit, each of the m surface segments will exhibit a distinct time constant $\tau_m = R_{E,m}C_{dl,m}$ instead of a uniform time constant. This phenomenon is called **dispersion** of time constants. In our example, there are parameters with dispersion (k and C_{dl}), thus the situation is called two-dimensional dispersion (2D dispersion, Figure 7.22).

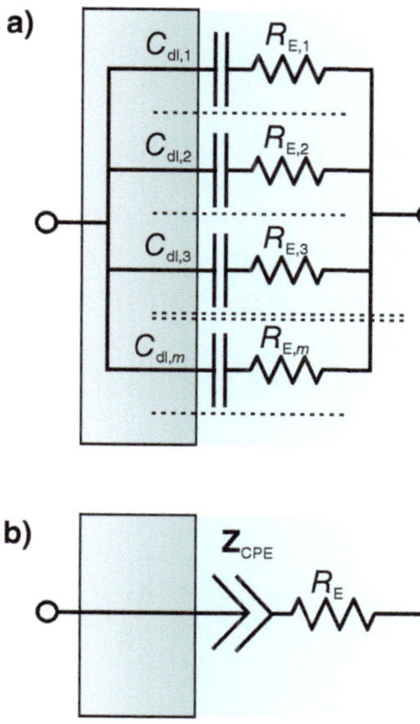

Figure 7.22 Equivalent circuit for an electrode with dispersion in the time constant $\tau_m = C_{dl,m}R_{E,m}$ without faradaic current.

a) Illustration with capacitors and resistors for each surface segment;
b) illustration with *one* constant phase element (CPE).

Even more complicated are cases for rough or porous electrodes with a dispersion in the time constant orthogonal to the electrode surface (3D-dispersion). For a porous electrode, the electrolyte resistance is much larger at the bottom of a pore than at its orifice toward the solution phase. Because the electrolyte resistance influences the time constant τ_m, τ_m varies along the pore.

It is important to note that increasing the number of elements in the equivalent circuit model will always improve the fitting but will not necessarily help understanding the system. By introducing a large number of circuit elements (RC elements in parallel associated with parameters $R_{E,1} \ldots R_{E,m}$, $C_{dl,1} \ldots$, $C_{dl,m}$), good fittings may be possible, but without real insight into the behavior of the system. One approach to improve this situation is based on replacing free fitting parameters ($R_{E,1} \ldots R_{E,m}$, $C_{dl,1} \ldots$, $C_{dl,m}$) with a physically reasonable distribution (e.g., Gaussian distribution, normal distribution, or Pascal triangle approaches to mimic a normal distribution). We could assume that the parameter $\ln(\tau/\tau_0)$ obeys a normal distribution (a bell-shaped distribution with a central $\tau = \tau_0$ and $\ln(\tau/\tau_0) = 0$, Figure 7.23a). Because a normal distribution is applied to the logarithm, the obtained distribution G of τ itself is more complicated (Figure 7.23a and b).

$$G(\tau) = \frac{1}{2\pi\tau} \cdot \frac{\sin\left[(1-\gamma)\pi\right]}{\cosh\left[\gamma \ln\left(\frac{\tau}{\tau_0}\right)\right] - \cos\left((1-\gamma)\pi\right)} \tag{7.60}$$

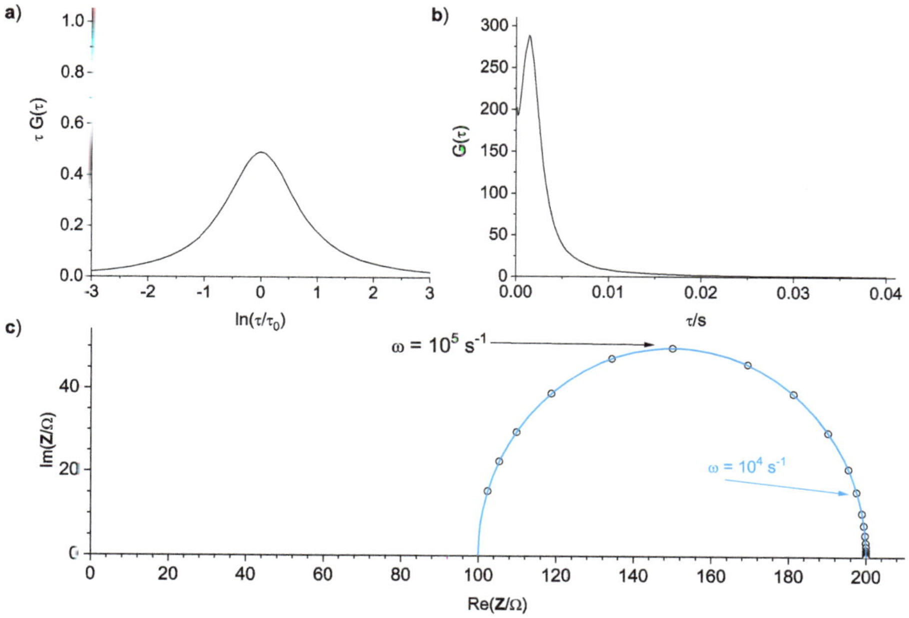

Figure 7.23 Nyquist plot for an electrode with CPE (without faradaic current).

a) Dispersion function of time constants (the abscissa is the natural logarithm of the normalized time constant);
b) dispersion function plotted versus time constant;
c) Nyquist plot for the electrode; for comparison, a Nyquist plot for a Randles circuit is shown in blue (R_E = 100 W, C_{dl} = Q_{CPE} = 1 μFs$^{0.8}$, γ = 0.8, τ_0 = 0.002 s).

Here, γ is a dimensionless factor between 0 and +1. The admittance $\mathbf{Y} = \mathbf{Z}^{-1}$ is obtained by integration over all segments.

$$\mathbf{Y} = \mathbf{Z}^{-1} = \frac{1}{R_E}\left[1 - \int_{\tau=0}^{\infty} \frac{G(\tau)}{1 + i\omega\tau} d\tau\right]. \tag{7.61}$$

Q_{CPE} – Parameter for constant phase element (CPE) [s^{γ} A V^{-1}]

$Q_{CPE,A}$ – Area-normalized parameter for a CPE [s^{γ} A V^{-1} cm^{-2}]

γ – Exponent for a CPE [$0 < \gamma < 1$]

For the impedance of an electrode without faradaic current, the resulting Nyquist plot is shown in Figure 7.23 [based on Eq. (7.62)].

$$\mathbf{Z} = R_E + \frac{1}{(i\omega)^{\gamma} Q_{CPE}} \tag{7.62}$$

Note the similarity of structure in Eq. (7.62) and in Eq. (7.28). The second term in Eq. (7.62) depends on parameters Q_{CPE} and γ. Q_{CPE} is a measure for capacitance. For the limiting case of $\gamma = 1$, Eq. (7.62) is equal to Eq. (7.28). In this case, Q_{CPE} can be identified as the double layer capacitance C_{dl} [μF]. For $\gamma \neq 1$, Q_{CPE} is affected by the dispersion in the surface properties. Q_{CPE} has units of [$s^{(\gamma-1)}$ μF] = [s^{γ} A V^{-1}] = [s^{γ} Ω^{-1} cm^{-2}]. We can area-normalize Q_{CPE} to give $Q_{CPE,A}$ [s^{γ} A V^{-1} cm^{-2}] [analogous to $C_{dl,A}$ in Eq. (7.28)]. The phase of the CPE is always $-\gamma\pi/2 = -(90-\gamma)°$; hence, the name **constant phase element** (CPE). For $\gamma = 0.5$, the CPE behaves like a Warburg element with phase angle 45°. For $\gamma = 0$, the CPE behaves like an ohmic resistor. The CPE can be regarded as a nonideal capacitor with leak currents.

Q: It looks like the CPE data and the semicircle for the Randles circuit are very similar. Do we need the CPE at all here?

A: The plot in Figure 7.23 is misleading! For example, the summit frequencies for the maximum in Im(**Z**) are different. The differences become very clear in the Bode plots (Figure 7.24). When comparing data sets, both the Bode plots for |**Z**| = Z and ϕ, as well as the Nyquist plot must match!

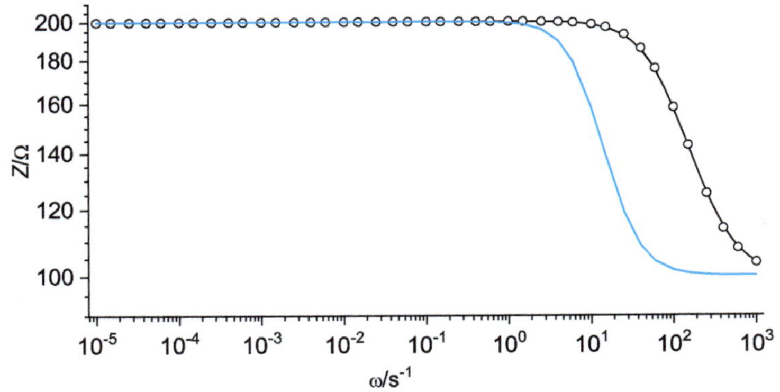

Figure 7.24 Bode plot for total impedance |**Z**| = Z for an electrode with CPE (Figure 7.23). The blue line shows data for a Randles circuit for comparison. The difference in the response of the two types of equivalent circuit models is clearly discernable.

When considering one simple charge transfer reaction (equivalent circuit in Figure 7.25 especially when $\mathbf{Z}_F = R_{CT}$), the impedance with CPE is

$$\mathbf{Z} = R_E + \frac{R_{CT}}{1 + (i\omega)^{\gamma} Q R_{CT}}. \tag{7.63}$$

The application of the CPE often provides good fits with experimental data with only three variable parameters. The CPE is popular, especially in situations when the

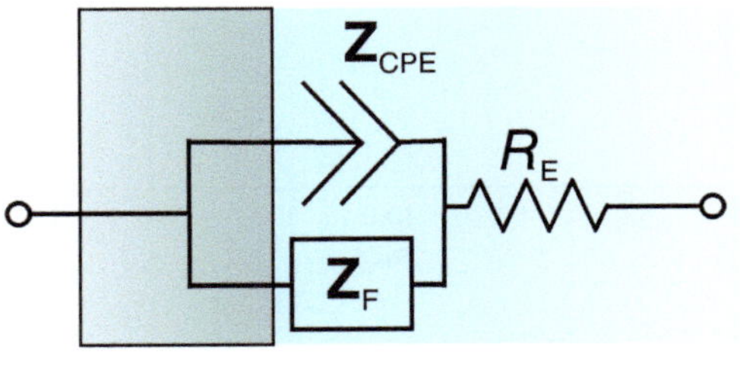

Figure 7.25 Equivalent circuit model for an electrode with distribution of the time constant $\tau_k = C_{dl,k} R_{E,k}$ and faradaic currents.

physical meaning of elements is not important. Even more benefits from the CPE are possible when allowing γ to range from -1 to $+1$. For $\gamma = -1$, the behavior of the CPE is consistent with that of an ideal inductor (coil). Inductive elements are rare in electrochemical cells but possible when intermediates in electrocatalytic reactions adsorb on the electrode surface and inhibit the overall electrode reaction. Further examples are found in more complex corrosion processes.

Uncritical application of the CPE can be problematic. The following arguments make the use of the CPE questionable:

- Investigation of a distribution in τ can be beneficial compared to a fixed τ value. However, the proof for a normal distribution in $\ln(\tau/\tau_0)$ is difficult. Other types of distribution are possible. In those cases, the simplification to Eqs. (7.62) and (7.63) are not sensible.
- A perfect match of experimental data to equivalent circuit fitting is not a sufficient proof for the circuit being valid (but may suggest exactly this). However, other circuits with alternative physical interpretations are possible.

7.5 Summary

EIS provides a widely employed standard test method in electrochemistry to characterize complex electrode and cell reactions. A sine wave input signal (either an input voltage or an input current) superimposed on a DC bias is applied to the cell. Small input amplitudes are employed to assure a linear response from the system, which is recorded as a function of frequency. By selecting an appropriate cell configuration, measurements of a single electrode, of the full cell, or of electrolyte or membrane properties are possible. The huge frequency range during EIS measurements (typically 10^{-2}–10^5 Hz) allows a range of rate constants and time constants to be determined in a single experiment.

Interpretation of EIS data is facilitated with equivalent circuit models based on capacitors and resistors. Particularly relevant is the Randles equivalent circuit model for electrodes with faradaic currents. Based on the Randles circuit, further elements can be added to adapt the model circuit to more complex physical systems. Computer-based analysis (nonlinear fitting) in combination with equivalent circuits allows key parameters (rate constants, double layer capacitance, electrolyte resistivity, diffusion coefficients, etc.) to be extracted. Equivalent circuits are not unique, and it is important to assign physical parameters to avoid meaningless fitting! Finding a fitting equivalent circuit is secondary, but assigning elements in the circuit to reaction and transport parameters provides physically reasonable data interpretation. In industrial applications, EIS provides a tool to monitor and diagnose complex systems (e.g., battery health) which is possible even without a full physical understanding of the data.

Test your knowledge

Exercise 7.1: Derive the reciprocal of the complex number $z(a,b)$ using the trigonometric notation in Eq. (7.16). Compare to Eq. (7.17).

Exercise 7.2: Derive the expression for the magnitude of the impedance Z based on Euler's notation.

Exercise 7.3: Calculate Bode and Nyquist plots in Figure 7.11 and 7.12 with a spreadsheet program for $R_E = 100\ \Omega$, $C_3 = 10\ \mu F\ cm^{-2}$, and $R_1 = 80\ \Omega$.

Exercise 7.4: An electrode immersed in a solution with surfactant is partially coated with a monolayer. The solution also contains supporting electrolyte and a redox system R/O. The heterogeneous standard reaction rate constant for the bare electrode is $k° = 1\ \text{cm s}^{-1}$. The measured rate constant is $k° = 0.05\ \text{cm s}^{-1}$. The solution resistance is 100 Ω. Develop an equivalent circuit model to account for all important elements.

Exercise 7.5: With an impedance analysis software package, illustrate the Nyquist plot and Bode plots for the electrode in Exercise 7.4 for a frequency range $0.01 < \omega/\text{s}^{-1} < 10^5$. For missing parameters, select sensible values such as $D_O = D_R = 10^{-5}\ \text{cm}^2\ \text{s}^{-1}$, $A = 1\ \text{cm}^2$, and $C_{dl} = 20\ \mu\text{F cm}^{-2}$. Alternatively (slightly more difficult, but beneficial), solve this exercise with a spreadsheet software package.

References

1 M. E. Orazem, B. Tribollet; *Electrochemical Impedance Spectroscopy*. 2nd ed., John Wiley & Sons, Hoboken, 2008.

2 A. Lasia; *Electrochemical Impedance Spectroscopy and Its Application*, Springer, New York, 2016.

3 E. Barsoukov, J. R. Macdonald; *Impedance Spectroscopy: Theory, Experiment, and Applications*, John Wiley & Sons, Hoboken, 2018.

4 L. M. Peter; *Electrochemical Impedance Spectroscopy and Related Techniques: From Basics to Advanced Applications*, World Scientific, London, 2024.

5 A. C. Lazanas, M. I. Prodromidis; Electrochemical impedance spectroscopy — A tutorial. *ACS Meas. Sci. Au* **2023**, *3*, 162–193.

6 A. J. Bard, L. R. Faulkner, H. S. White; *Electrochemical Methods - Fundamentals and Applications*. 3rd ed., John Wiley & Sons, New York, 2022, pp. 443–488.

7 H. H. Girault; *Analytical and Physical Electrochemistry*, EPFL Press, distributed by Marcel Dekker, New York, 2004.

8 D. Pletcher, R. Greef, R. Peat, L. M. Peter, J. Robinson; *Instrumental Methods in Electrochemistry*, Horwood Publisher, Chichester, 2001, pp. 251–282.

9 A. Hamnett, C. H. Hamann, W. Vielstich; *Electrochemistry*, 2nd ed., Wiley-VCH, Weinheim, 2008.

10 S. Krause; Impedance methods. in *Encyclopedia of Electrochemistry, Vol. 3. Instrumentation and Electroanalytical Chemistry* (A. J. Bard, M. Stratmann, P. Unwin, Eds.), Wiley-VCH, Weinheim, 2003, pp. 196–229.

11 M. Sluyters-Rehbach, J. H. Sluyter; Sine wave methods in the study of electrode processes. in *Electroanalytical Chemistry, Vol. 4* (A. J. Bard, Ed.), Marcel Dekker, New York, 1970, pp. 1–128.

12 M. Sluyters-Rehbach, J. H. Sluyters; Alternating current and pulse methods. in *Comprehensive Chemical Kinetics, Vol. 26* (C. H. Bamford, R. G. Compton, Eds.), Elsevier, New York, 1986, pp. 203–354.

13 P. G. Francis; *Mathematics for Chemists*, Springer, Dordrecht, 1984.

14 A. Bondarenko, G. A. Ragoisha; EIS spectrum analyser - Analysis and simulation of impedance spectra. http://www.abc.chemistry.bsu.by/vi/analyser/ (last access 22 April 2026), for noncommercial use.

15 S. Fletcher; Tables of degenerate electrical networks for use in the equivalent circuit analysis of electrochemical systems. *J. Eletrochem. Soc.* **1994**, *141*, 1823–1826.

16 R. de Levie; Electrochemical responses of porous and rough electrodes. in *Advances in Electrochemistry and Electrochemical Engineering, Vol. 6* (P. Delahay, Ed.), Interscience, New York, 1967, pp. 329–397.

17 A. Lasia, Porous electrodes in the presence of a concentration gradient. *J. Electroanal. Chem.* **1997**, *428*, 155–164.

8

Surface Electrochemistry

We have emphasized in Chapters 5 and 6 that voltammetric experiments can provide very detailed insights into electrochemical reaction mechanisms, but principally, they cannot resolve the chemical structures of products and intermediates. Therefore, Section 6.6.3 was devoted to spectroelectrochemical methods, which can provide additional information on the structure of *dissolved* reactants in the vicinity of the electrode. This chapter is devoted to the study of the *structure of the solid phase* of a solid–liquid interface, which is decisive for the reactivity of electrodes. Structural features of different length scales, ranging from atomic dimensions to microstructures, may influence the reactivity in different applications.

Below, we provide an overview of surface-analytical methods particularly well-suited for the investigation of electrochemical problems. We emphasize the methods used most frequently. Within this textbook, neither a complete coverage of all methods nor a sufficiently deep presentation of each of the most important methods is possible.

Sample preparation methods are crucial for surface electrochemistry. It is emphasized right from the start that each surface analysis experiment can only be successful if the sample preparation allows it to be.

The requirements for the quality of samples and the cleanliness of electrolyte solutions can be extremely high, and in some cases, meaningful results can only be obtained with single-crystal electrodes (in contrast to poly-crystalline electrodes). Therefore, a full section is devoted entirely to single-crystal electrodes.

Further Reading. Very good descriptions of surface analytical methods without special relation to electrochemistry can be found [1, 2]. Further literature for specific methods is given at the end of each section. Collective works are available on the use of surface analysis in electrochemistry [3, 4]. A specialized textbook on the topic is available [5] with focus on the theoretical foundations of interfacial phenomena at different types of electrodes.

8.1 Overview and Systematics

The most important measurement principles for surfaces analysis are based on

- Spectroscopy (absorption or emission techniques)
- Diffraction and scattering techniques
- Microscopy (based on interaction with radiation or probes), and within the area of electrochemistry most importantly scanning probe microscopies.

ex situ, Latin outside of the original location; *in situ*, Latin at the original location. These antonyms (words of opposite meaning) can have different meanings depending on the field of application. In surface science, in which samples are produced by epitaxy in vacuum and studied subsequently in vacuum, "*in situ*" means that the sample is studied without having left the vacuum. An *in situ* method in electrochemistry aims for the analysis of a solid–liquid interface, which is intact only if the liquid part of the interface is preserved. At the boundary of surface science and surface electrochemistry, these different meanings of "*in situ*" could cause misunderstanding. Compare also "*operando*" methods.

Often electrodes must be removed from solution and transferred to ultrahigh vacuum (UHV) for specialized structural analysis techniques (e.g., for X-ray photoelectron spectroscopy). In the context of electrochemistry, such methods are regarded as *ex situ* methods (Figure 8.1a). When removing an electrode from the electrolyte, the interfacial structures may change, for instance, if organic adsorbates volatilize into the vacuum. Further undesired changes could be the adsorption of molecules from the ambient atmosphere to the electrode or the surface oxidation of metals during the transfer through air. Because of those problems, there is a particular need for *in situ* methods (Figure 8.1b), in which interfacial structures are investigated while the solid part of the electrode remains in contact with the electrolyte solution. Those techniques are not yet widely available for electrochemical problems. In many cases, the use of *in situ* methods requires unavoidable compromises with respect to the kind of samples or the construction of the electrochemical cell. If the cell construction for

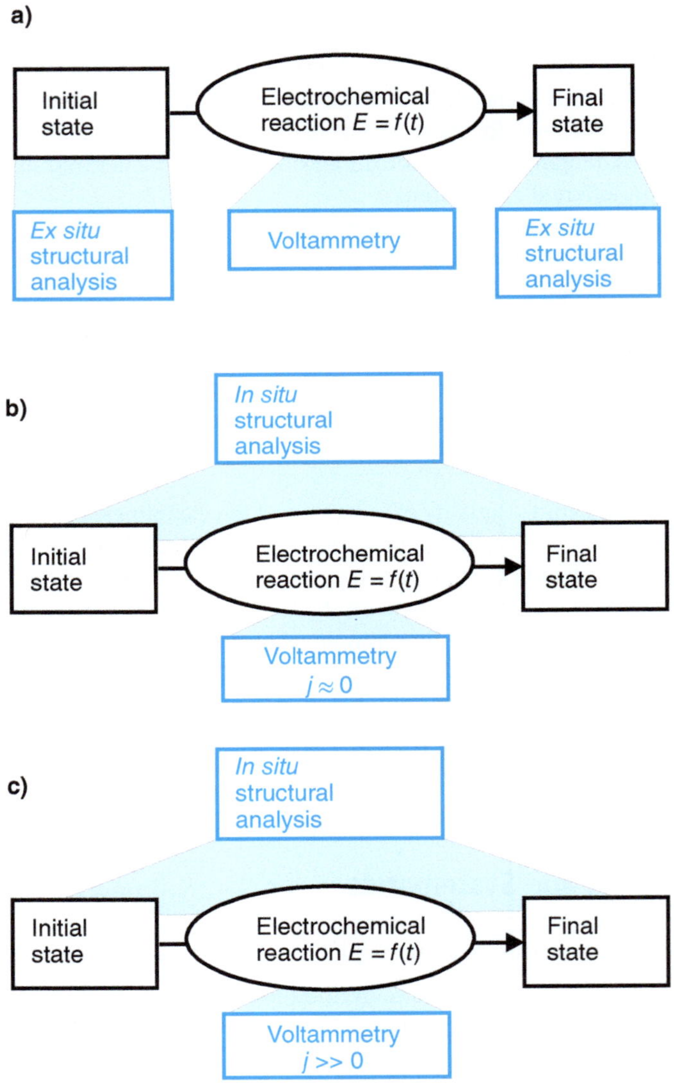

Figure 8.1 Principles for the coupling of electrochemical experiments with spectroscopic characterization methods for electrodes for a) *ex situ*, b) *in situ*, and c) *operando* tools. Analogous relations also exist for microscopies or diffraction experiments instead of a spectroscopic analysis.

an *in situ* experiment contains only a very thin liquid film above the electrode (to allow penetration of radiation), the mass transport to such a thin-layer configuration is significantly diminished and high current densities cannot be achieved. The new term *operando* methods has been coined to distinguish the methods that study the electrode surface not only while in contact with the electrolyte solution but also at relevant current densities (Figure 8.1c).

Further important characteristics for experimental methods are

- the time resolution: How much time must pass so that the method can distinguish a change in surface structure?
- the lateral resolution: How far apart can surface structures be located so that they can be distinguished?
- the information depth: How thick is the surface layer that contributes to the signal?

Below, we will discuss surface spectroscopic methods roughly in the sequence of increasing photon energy. Then, we move to diffraction methods and methods for the detection of mass changes, followed by a discussion of scanning probe techniques. In the spectroscopy literature, it is very common to symbolize the photon energy by E [eV]. We will follow this common practice, although in most parts of the book E [V] stands for the electrode potential. We will use subscripts for distinction only in those sections where there is a risk of confusion.

8.2 Vibrational Spectroscopy at Surfaces

Please first study an introductory text about infrared (IR) spectroscopy (for example [6]), before attempting to study this section. Vibrational spectroscopy provides information about the structure of molecules that can be deduced from information about the energy of transition between different vibrational energy levels. Vibrations occur *between different atoms* within *one* molecule. The energy of the vibrational transition is influenced in a characteristic manner by the chemical structure of the molecule (polarity, bond strength, bond angles, etc.). Information about vibrational transitions can be recorded by IR spectroscopy (Section 8.2.1), Raman spectroscopy[82] (Section 8.2.2), and high-resolution electron energy loss spectroscopy in vacuum (HREELS, Section 8.2.4).

A molecule with N atoms has $3N - 6$ independent normal vibrations ($3N - 5$ for linear molecules). For each normal vibration, all atoms move with the same frequency and phase relative to each other. The complex inner motions of a molecule are a superposition of all normal vibrations. If the vibration changes predominantly the length of a bond, the vibration is called a valence vibration (ν), if the vibration affects mostly bond angles, it is called a deformation vibration (δ). Further distinctions are common according to the symmetry of the vibration with respect to symmetry elements of the molecule [6]. Usually, the vibration can be described with sufficient accuracy by the harmonic oscillator approximation. The frequency of a normal vibration ν_{osc} depends on the force constant k_F and the reduced mass m_r of *both* atoms with the masses m_1 and m_2, that vibrate against each other [Eq. (8.1)].

$$\nu_{osc} = \frac{1}{2\pi}\sqrt{\frac{k_F}{m_r}} \tag{8.1}$$

If ν_{osc} is clearly distinguished from other normal vibrations of the molecule, either by a large k_F (double and triple bonds) or a particularly low m_r (CH, NH, OH, SH

In the classical description, "harmonic oscillator" means that the restoring forces are always proportional to the deflection from the equilibrium position. In reality, this is only a good approximation for small deflections. For a larger deflection, it becomes easier to stretch a molecular bond than to compress it. Compressing would require the extreme case of pressing the positively charged nuclei of the atoms into each other. A better description for large deflections is the anharmonic oscillator. Due to anharmonicity, there are small deviations from the simple cases treated in this section. For most practical analyses, anharmonic spectral features are not important.

ν_{osc} – Vibrational frequency [Hz]
k_F – Force constant [N m^{-1}]
m_r – Reduced mass $= m_1 \cdot m_2/(m_1 + m_2)$

[82]Sir Chandrasekhara Venkata Raman, 1888–1970, Indian physicist, founding president of the Indian Academy of Science, 1930 Nobel Prize in physics.

valence vibrations), the vibration is called a characteristic vibration, because it allows a straightforward conclusion of the functional groups present in the molecule. Those vibrations are also particularly useful for spectroelectrochemistry. Each normal vibration can assume different vibrational energy levels by uptake or release of quantized radiation energy (which are specific energy portions of small, yet very precise value of *one* photon). The energy of vibrational energy states depends on one quantum number v (Latin character v)

E_v – Vibrational energy of the harmonic oscillator [eV]
h – Planck constant[52] [J s]
v – Vibrational quantum number, dimensionless

$$E_v = h\nu_{osc}\left(v + \frac{1}{2}\right), \quad v = \{0, 1, 2, 3, 4 \ldots\} \tag{8.2}$$

The energy difference between the vibrational ground state E_0 and the excited vibrational states E_1, E_2, … is so large for most molecules that, according to Boltzmann statistics[18], almost all molecules occur in their vibrational ground state at room temperature. Thus, the different vibrational spectroscopies observe the transitions from the ground state into excited vibrational energy levels, mostly to the first excited vibrational energy level.

8.2.1 Infrared Reflection Absorption Spectroscopy (IRRAS)

Basics. If electromagnetic radiation interacts with a molecule, the radiation can be absorbed by the molecule only if three conditions are met:

$E_{initial}$ – Energy of the initial state [eV]
E_{final} – Energy of the final state [eV]
ν – Radiation frequency [Hz]

Resonance condition: $E_{final} - E_{initial} = h\nu$ (8.3)

The photon energy equals the energy *difference* between the vibrational levels of *one* molecule.

μ – Dipole moment of the molecule [A s m]
q – Inner coordinate (generalized term for either bond length or bond angle)

Excitation condition: $d\mu/dq \neq 0$ (8.4)

The radiation can only interact with a particular molecular vibration if the dipole moment of the molecule changes during the vibration being considered.

Selection rule: $\Delta v = \pm 1$ (8.5)

The vibrational quantum number v can only change by one (within the model of the harmonic oscillator).

For an anharmonic oscillator, transitions with $\Delta v = \pm 2, \pm 3, \ldots$ are also observed as "overtones", but with strongly reduced intensities compared to $\Delta v = \pm 1$.

Because of $E_{initial} = E_0$ and $\Delta v = 1$, we consider only the transition from E_0 to E_1. Specifically for IR spectroscopy, the energy is recalculated to an energy-proportional quantity, $F_v = E_v/(hc)$. This vibration term has the unit of a wave number $\tilde{\nu}$ [cm^{-1}] (speak nu-tilde), which specifies the number of wave crests within 1 cm in the direction of wave propagation. The same conversion is applied to the photon energy $F = h\nu/(hc) = \nu/(\nu \cdot \lambda) = 1/\lambda = \lambda^{-1}$ [cm^{-1}] on the abscissa of IR spectra.

We will observe the vibrational transitions at

F – Vibration term [cm^{-1}]
c – Speed of light in vacuum [cm s^{-1}]
$\tilde{\nu}$ – Wave number [cm^{-1}]

$$F_1 - F_0 = \tilde{\nu}_{osc}\left[\left(1 + \frac{1}{2}\right) - \left(0 + \frac{1}{2}\right)\right] = \tilde{\nu}_{osc} \tag{8.6}$$

The position of a transition in the spectrum agrees with the oscillator frequency of the corresponding normal vibration. Vibrational transitions are typically observed between 500 and 4000 cm^{-1}, that is, within the mid-IR region. Concomitantly with the change of the vibrational state, there is also a change of the rotational state (as seen in rotation-vibration spectra), which causes a fine structure for spectra of gaseous molecules. In liquid and solid phases, rotational movements of molecules are hindered. Thus, a broadened vibrational band is observed instead of rotational fine structure.

Surface Selection Rule. For the study of electrodes, an IR beam is directed toward the *electrode* (e.g., Au, Ag, Pt, Cu, Al, steel) that can function *as a mirror* for IR radiation. In this situation, a further limitation, the **surface selection rule**, applies to the transitions between vibrational states:

> The excitation of the vibration of an adsorbed molecule can only occur if the **transition dipole moment μ** has a component perpendicular to the reflecting surface.

The quantity μ is a vector. Its direction indicates how the charge distribution within the molecule changes during a vibration. Often, it can be predicted by looking at the motion of atoms during one period of the normal vibration.

The limitation to metals as support for IRRAS can be circumvented by depositing a layer of a few nanometer thickness of the material of interest on an IR mirror, so that adsorbates interact with the material of interest but the underlying support functions as an IR mirror [7]. The surface selection rule is of central importance for IRRAS. Therefore, we would like to discuss it a bit more in detail, but leave out the complex set of mathematical formulas [7–12]. Each impinging photon is characterized by an electric and a magnetic field vector, which are perpendicular to each other and to the propagation direction of the radiation (Figure 8.2). Below, we only discuss the electrical field vector **E**. If the electric field vector $\mathbf{E}_p$ of the incoming radiation is parallel to the plane of incidence (defined by the incoming radiation and the surface normal), we call such radiation parallelly polarized (index p). In case the electrical field vector is oriented perpendicular to the plane of incidence, the radiation is called *s*-polarized (index s from German *senkrecht*). Because the electric field strength is a vectorial quantity, any other orientation of the electric field vector can be considered as a linear combination of $\mathbf{E}_p$ and $\mathbf{E}_s$. This is very convenient for the mathematical treatment. We only need to investigate separately the behavior of the orthogonal components $\mathbf{E}_p$ and $\mathbf{E}_s$ and obtain the behavior of any arbitrary orientation as a linear combination of $\mathbf{E}_p$ and $\mathbf{E}_s$. The task is thus reduced to describing the behavior of $\mathbf{E}_p$ and $\mathbf{E}_s$ in the vicinity of the reflecting surface and the question of which interaction of those field vectors would exist with an adsorbed molecule.

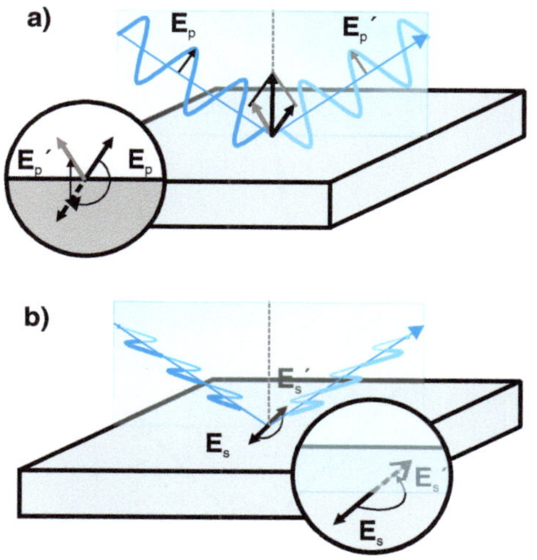

Figure 8.2 Superposition of the a) parallel (p) and b) perpendicular (s) components of the electrical field in the moment of reflection at a highly reflecting interface.

The impinging electric field vectors $\mathbf{E}_p$ and $\mathbf{E}_s$ experience a reflection at the interface and a phase shift, which depends on the angle of incidence and the complex index of refraction of the two phases forming the interface (solution and electrode material with possible layers on it). For metals that are highly reflective, there is a small range of incident angles, in which the phase shift of $\mathbf{E}_p$ and $\mathbf{E}_s$ is 180°.

An adsorbate molecule is exposed to a field strength composed of the vectorial addition of the electrical field vectors of the incident beam **E** and the reflected beam $\mathbf{E}'$. For *p*-polarized radiation ($\mathbf{E} = \mathbf{E}_p$), the field vectors of incident and reflected beams ($\mathbf{E}_p$ or $\mathbf{E}_p'$) form a parallelogram at the reflection point and the resulting field vector is directed perpendicularly to the surface (Figure 8.2a). The magnitude of the vector is larger than that of the incoming or reflected beam. For *s*-polarized radiation ($\mathbf{E} = \mathbf{E}_s$), the field vector is in the plane of the interface. For the special case of a phase shift of 180° at a highly reflective surface, the reflection at the interface and the phase shift result in field vectors $\mathbf{E}_s$ and $\mathbf{E}_s'$ that are of equal magnitude and pointing in opposite directions at the reflection point. Consequently, they cancel out (Figure 8.2b). As a result, there is no electrical field acting parallel to the surface *right at the interface*, while the field perpendicular to the interface is enhanced. If the transition dipole moment μ of the adsorbed molecule has a component perpendicular to the surface, radiation energy can be absorbed into the vibrational transitions. However, no absorption is possible if μ is completely parallel to the surface.

Q: Why should I bother at all with these vectors and dipole moments? I just want to record a spectrum.

A: Exact calculations of light propagation, attenuation, and reflection according to Fresnel's[83] equations and the law of refraction (Snell's law)[84] with the consideration of the complex frequency-dependent refractive indices of the electrode, the adsorbate layer, the electrolyte solution, and the window material are necessary to determine the optimum incident angle for a given material combination [7–12]. The recording of useful spectra can already be prevented by small deviations from the optimum optical arrangement! When attempting IRRAS measurements for a new material system, calculations and construction of a specially adapted electrochemical cell are often necessary. If you feel like dipping a toe in the water of such calculations, you may leap forward to Excursion 8-1.

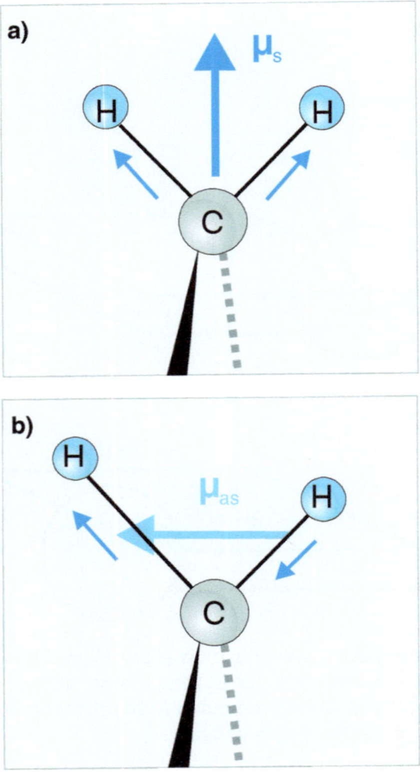

Figure 8.3 Orientation of a) $\mathbf{\mu}_s$ and b) $\mathbf{\mu}_{as}$ of the CH-valence vibration of a CH_2-group relative to the direction of the alkyl chain, which is perpendicular to the plane of the paper (highlighted in light gray).

Let us consider the implications of the surface selection rule using an example that is important for the investigation of self-assembled monolayers of long-chain molecules on electrodes, which serve as model systems for biological membranes [13]. The molecules contain methylene groups $-CH_2-$. This group has two valence vibrations, ν_s and ν_{as}. Both hydrogen atoms move synchronously away from the carbon atom (and again toward it) in the symmetric valence vibration ν_s. For this vibration, $\mathbf{\mu}_s$ is on the bisector of the HCH-bond angle (Figure 8.3a). For the asymmetric valence vibration ν_{as}, $\mathbf{\mu}_{as}$ is located in the plane defined by the carbon atom and the two hydrogen atoms but perpendicular to $\mathbf{\mu}_s$ (Figure 8.3b). In an ordered film, the alkyl chains are "stretched". That is, they are in an all-trans conformation along the C–C-bonds. If such a chain is bound flat on an electrode, $\mathbf{\mu}_s$ is perpendicular to the surface and the maximum intensity for this vibration will be reached for a given amount of adsorbates (Figure 8.4c). In contrast, if the alkyl chain is oriented perpendicular to the surface, $\mathbf{\mu}_s$ as well as $\mathbf{\mu}_{as}$ are parallel to the surface and no absorption is possible for the two vibrations (Figure 8.4a). If the same amount of molecules is available in a random orientation as shown in Figure 8.4b, all orientations of $\mathbf{\mu}_s$ and $\mathbf{\mu}_{as}$ occur with the same probability, and only a few molecules will have the optimum orientation. Thus, the intensity of the spectrum should be in between the two limiting cases. Indeed, the orientation is one-third of the intensity at optimum orientation.

Of course, there is also *one* angle between $\mathbf{\mu}_s$ and the surface normal, at which the ordered arrangement of molecules leads to exactly the same intensity as the random orientation of the same number of absorber molecules. This ("magic") angle amounts to $\theta_{magic} = 54°$ ($\cos^2\theta_{magic} = 1/3$). The experimental spectrum in Figure 8.4b (blue curve) has a lower intensity than the same number of molecules with random orientation. That is, first, some preferential orientation must exist at all, and second, the tilt angle of the chain axis must be between 0° and 54°. A more detailed analysis described in [13] yields a tilt angle of 32° against the surface normal.

[83] Augustin Jean Fresnel, 1788–1827, French physicist and engineer (Fresnel lens, Fresnel integral, Fresnel zone plate, Fresnel's double mirror, Fresnel's number).

[84] Willebrord Snellius, 1580–1626, astronomer and mathematician from the Netherlands. The relation ($\sin\theta_1/\sin\theta_2 = \lambda_1/\lambda_2 = n_2/n_1$) is called Snell's law in the English literature. In fact, this relation has been described by the Arab scientist Ibn Sahl at the court of Bagdad as early as 984.

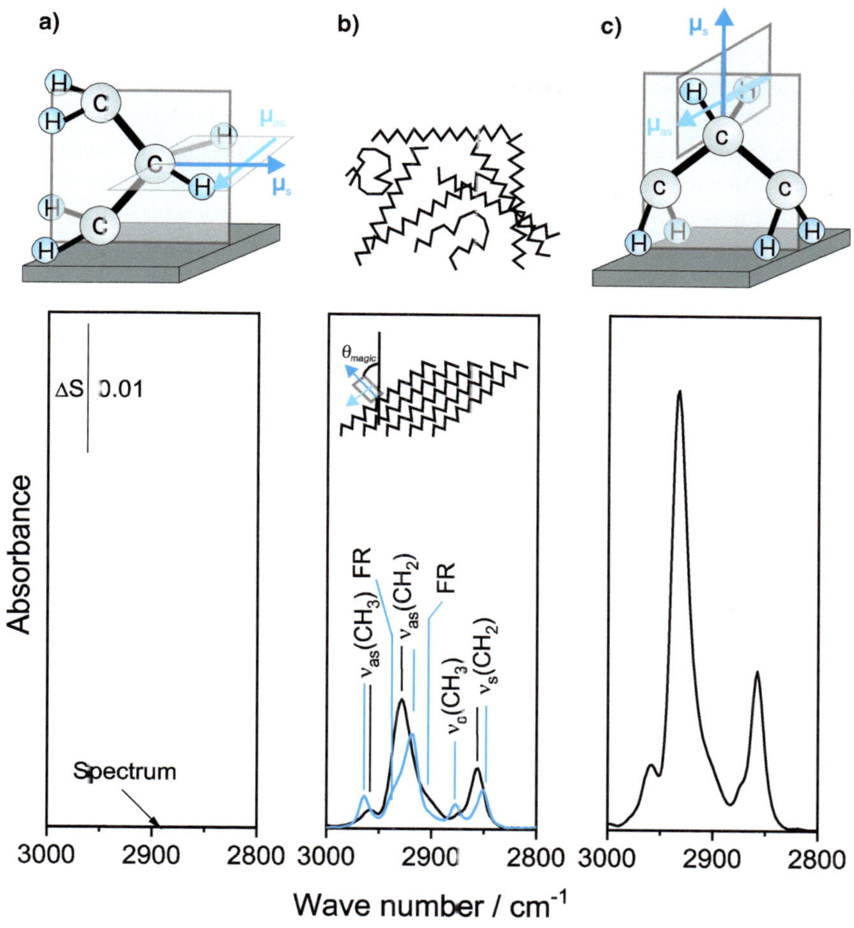

Figure 8.4 Orientation of the transition dipole moment of the valence vibration of the CH_2 group within an ordered monolayer.

a) Calculated absorption band for the orientation of the alkyl chain perpendicular to the electrode surface (μ_s and μ_{as} are parallel to the electrode);
b) calculated intensity (black) for random orientation of the CH_2 groups relative to the electrode or for an ordered arrangement with $\theta = 54°$ between the surface normal with respect to μ_s and μ_{as};
c) calculated intensity for the parallel orientation of the alkyl chain with respect to the electrode (μ_s is perpendicular to the electrode surface).

The experimental spectrum of an ordered *n*-hexadecane thiolate monolayer on Au is shown in blue in panel (b). The alkyl chains have a tilt angle of the chains of 32° against the surface normal. Data were provided by Izabella Brand, University of Oldenburg.

Q: What does FR mean in Figure 8.4b?

A: FR stands for Fermi resonance[85]. This emerges if a normal vibration ($\Delta v = 1$) has a similar frequency like an overtone ($\Delta v = 2, 3, \ldots$) to another vibration. In such a case, the frequencies of both signals are shifted (the signals appear to "repel" each other) and their intensities become similar. The FR at (2890 ± 10) cm^{-1} in Figure 8.4b results from an overtone of the deformation vibration of the CH_2 group $\delta(CH_2)$ at ca. 2×1468 cm^{-1} and $\nu_s(CH_2)$ at 2850 cm^{-1}. The FR at (2935 ± 10) cm^{-1} comes from the overtone of the asymmetric deformation vibration of the CH_3 group $\delta_{as}(CH_3)$ at ca. 2×1455 cm^{-1} and $\nu_s(CH_3)$ at 2880 cm^{-1}. Both FRs are evident in Figure 8.4b as a shoulder of the main peaks.

Q: Why is there a disagreement between calculated and measured peak positions for $\nu_{as}(CH_2)$ in Figure 8.4b?

A: The monolayer is densely packed and reaches an ordered state almost similar to a solid (paraffin crystal). This causes the bands to be narrower and red-shifted. This effect has not been considered in the simple calculation for Figure 8.4 using an isolated molecule.

[85] Enrico Fermi, 1901–1954, Italian-US-American theoretical physicist and nuclear physicist, professor in Florence, Rome; 1938 emigration to the USA because of antisemitic prosecution of his wife and coworkers, significantly contributed to the development of US nuclear weapons; Nobel Prize in physics; his achievements are evident from terms like Fermions, Fermi level, Fermi gas, Fermi's Golden Rule, Fermi–Dirac statistics, Thomas–Fermi theory, Fermi surface.

Background compensation. The application of IR spectroscopy in electrochemistry aims to analyze an adsorbate (often just a monolayer) and thin films on electrode surfaces. IR spectroscopy by itself is not particularly sensitive. The central challenge for IRRAS is the very low signal intensities relative to the background caused by the molecules in the electrolyte [$H_2O(l)$] or the optical path [$H_2O(g)$, $CO_2(g)$]. Consequently, all "trump cards must be played" to increase the signal-to-background ratio. First of all, the control of the gas atmosphere inside the spectrometer and within the external reflection unit limits the amount of trace gases or helps keep them at least constant during the measurement. The intensity advantages of Fourier transform techniques are exploited [14, 15]. There are different approaches for background compensation for IRRAS at electrodes.

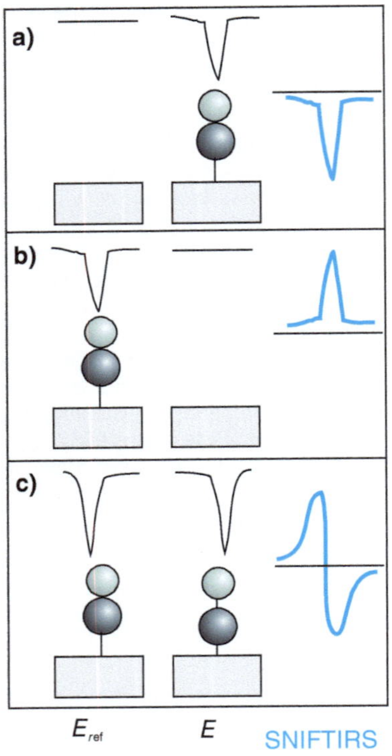

Figure 8.5 Typical signals (blue) in SNIFTIRS experiments and absorption bands at the measurement (E) and reference potential (E_{ref}).

a) molecule is desorbed at E_{ref}, and adsorbed at E;
b) molecule is adsorbed at E_{ref} and desorbed at E;
c) molecule is adsorbed both at E_{ref} and E, but experiences a slight shift of $\tilde{\nu}_{osc}$. This causes the appearance of a bipolar band.

SNIFTIRS. This acronym stands for subtractively normalized interfacial Fourier transform infrared spectroscopy. First, a reference spectrum is taken at an electrode potential E_{ref} at which the molecule of interest does not adsorb. The reference spectrum contains the signals of trace gases in the optical path and of the liquid electrolyte. Next, the potential is switched to a value E, at which the adsorbate is present at the electrode surface. The spectrum is again recorded with identical settings of the spectrometer. The spectrum contains contributions from the adsorption of the trace gases in the optical path, the liquid electrolyte (as in the reference spectrum) and additionally the contribution of the adsorbate at the electrode surface. Correction for the absorbance of the reference spectrum yields the absorbance of the adsorbate. There can be positive, negative, and bipolar bands in the spectrum (Figure 8.5). Molecules that are not adsorbed at the E_{ref} and are adsorbed at E cause a positive band. Molecules that are adsorbed at E_{ref} and are desorbed at E_{ref} give rise to a negative band. Molecules that are adsorbed at both potentials often experience a small shift in the absorption frequency. This leads to a double wave (bipolar band, Figure 8.5c). Figure 8.6a shows schematically the setup of an SNIFTIRS experiment.

PM IRRAS. This abbreviation stands for polarization modulation infrared reflection absorption spectroscopy. It is also recorded using an FT-IR spectrometer with an external reflection unit (Figure 8.6b). One exploits the fact that randomly oriented molecules in the optical path and in the liquid electrolyte absorb *s*- and *p*-polarized radiation with the same probability, while adsorbed molecules cannot absorb *s*-polarized radiation and thus contribute only to the diminution of the intensity of the *p*-polarized radiation. Thus, the difference between the absorbances of *s*- and *p*-polarized radiation represents the absorbance of the adsorbed molecules. Practically, a constant electrode potential is used together with polarized IR radiation. The polarization direction is changed ("modulated") with high frequency by a photoelastic modulator between the two polarization directions. After lock-in amplification of the signals and consideration of the correction functions of the photoelastic modulator

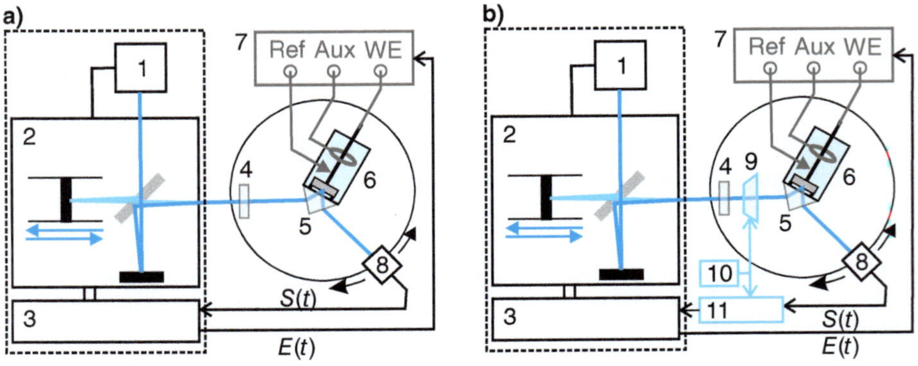

Figure 8.6 Experimental setup for a) SNIFTIRS and b) PM IRRAS for the study of adsorbates on electrodes. 1) IR source, 2) interferometer, 3) power supply, data recording, processing and display of the FT-IR spectrometer (the FT-IR spectrometer is marked by a dashed line), 4) polarizer, 5) CaF_2 prism, 6) electrochemical three-electrode cell, 7) potentiostat, 8) IR detector, 9) photoelastic modulator (for PM IRRAS only), 10) function generator, and 11) lock-in amplifier.

[16, 17], a background-corrected spectrum in absorbance units is obtained, very much like in a common FT-IR instrument. The investigation can be continued at a different potential. This procedure avoids the problematic determination of a suitable reference potential (required for SNIFTIRS) without adsorbates on the surface.

Cell Construction. There are different options for designing the spectroelectrochemical cell. In each case, the liquid electrolyte layer must be very thin (e.g., 1–2 μm), because water (and other polar solvents) are strong IR absorbers. D_2O is often used as a solvent to shift the absorption of the solvent into a noninterfering energy range so as not to affect the analysis of the signal from the adsorbates (large difference between m_r of H_2O and D_2O!). Important features of the cell design are: The optical window is made as a prism to allow adjustment of the incident angle. The window material must be transparent to IR radiation (ZnSe, CaF_2, BaF_2, Si) and at the same time stable in the chosen electrolyte. Especially, the *s*-polarized light must not be reflected at the window–electrolyte interface. This last condition accounts for the much higher intensities that can be achieved using CaF_2 and BaF_2 prisms. (Which electrolyte solutions are compatible with those window materials?) The thickness of the electrolyte layer between the window and the electrode can be adjusted with a micrometer screw. A large electrolyte reservoir located behind the working electrode holds the symmetrically arranged auxiliary electrode (Aux) and a reference electrode (Ref), as well as a tube for introducing an inert gas [16]. A water jacket for connection to a thermostat is quite useful, especially when phase transitions in adsorbate layers are to be expected [18]. The illuminated part of the sample is about 1.5 cm × 1 cm, thus, spatial resolution cannot be achieved with this technique.

Enhancement Mechanisms. Despite those provisions for the instrumental compensation of the background, the signals of adsorbates are often weak. This can be mitigated by adding nanostructures from Ag or Au to the electrode surface. Those structures cause a strong localized enhancement of optical fields at the adsorption sites near the noble metal nanostructures and a higher polarizability of molecules adsorbed to such nanostructures (SEIRA, surface-enhanced infrared absorption).

Application. Typical applications of IRRAS include small polar molecules such as CO and CN^- with absorption in the IR range [19–21]. CO and similar adsorbates play an important role as intermediates and poisons for catalysts in fuel cells [22–24]. In the case of CO, the C–O valence vibration is studied. CO binds with a nonbonding electron pair at the carbon atom to the metal surface. Concomitantly, there is a transfer of electron density from d-orbitals of the metal into antibonding π^*-orbitals between C and O (π back bonding). The bond order between C and O is decreased by the higher electron density in the antibonding orbitals. The lower bond order goes along with a lower force constant for the C–O valence vibration and consequently a lower wave number of the CO vibrational transition compared to gaseous CO [Eq. (8.1)].

The decrease in the force constant of the C–O bond becomes more pronounced the larger the coordination number of the adsorbed CO and the stronger the π back bonding character between the C atom and the surface.

The population of the different adsorption sites depends strongly on the coverage (Table 8.1). Figure 8.7 shows an experimental example. The band position also changes with the applied electrode potential by about 30 cm^{-1} V^{-1} because the extent of π back bonding from the metal to $2\pi^*$-orbitals of CO depends on the charge on the electrode surface [21–28]. This phenomenon is also called the electrochemical shift.

CN^- and related compounds are used as ligands in galvanic baths. Their interaction with surfaces is important for processes like nucleation and layer growth during electrodeposition of metals [19]. IRRAS is also suitable for following the formation of

Table 8.1 Band position for ν(CO) in CO adsorbed at different adsorption sites of metal electrodes. The exact position can vary due to the kind of metal, the solution composition, and most importantly, the applied electrode potential. Therefore, the quoted values serve as approximations only.

CO in the gas phase	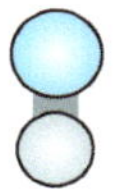	2143 cm^{-1}
Linearly bound CO, on-top		ca. 2100 cm^{-1}
CO on bridging site	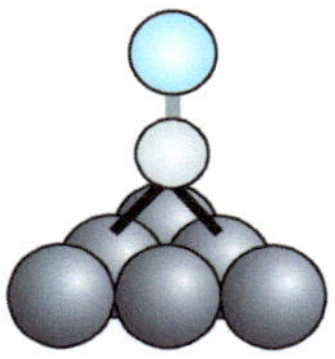	ca. 1900 cm^{-1}
CO at a threefold hollow site		ca. 1850 cm^{-1}

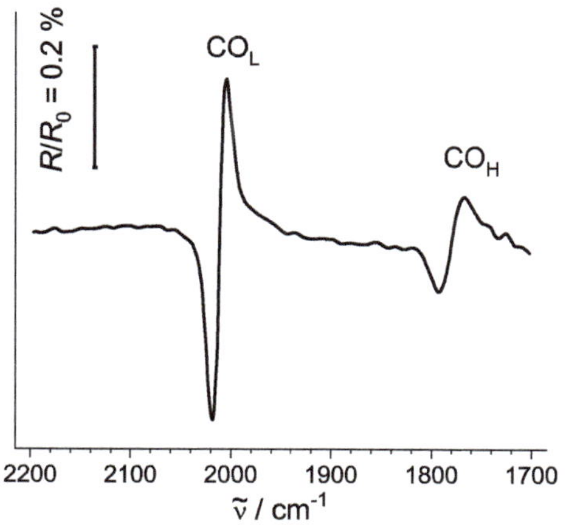

Figure 8.7 *In-situ* FT-IR spectrum of linearly bound CO (CO_L) and CO at a threefold hollow site CO_H at Ru(0 0 0 1) in 0.1 mol L^{-1} $HClO_4$ at 10 °C. The spectrum was recorded at +0.2 V vs. Ag|AgCl and related to the reference spectrum obtained at 0.2 V. Source: Reproduced with permission from [15], Copyright 2003, Wiley-VCH.

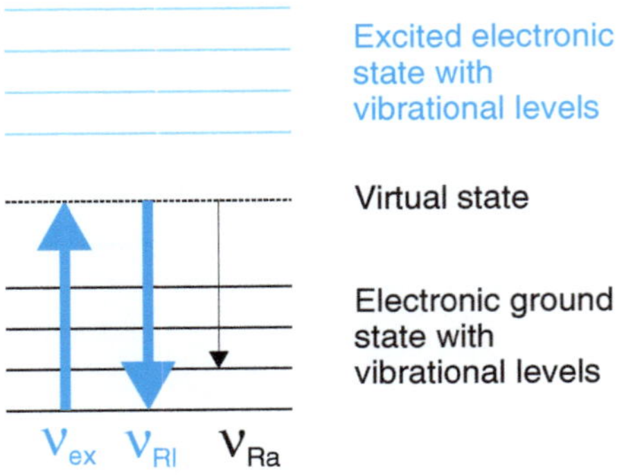

Figure 8.8 Energy transfer during a Raman scattering process; ν_{ex} – frequency (energy) of the excitation, ν_{Rl} – frequency of the Rayleigh scattering radiation, ν_{Ra} – frequency of the Raman scattering radiation.

passive layers [17, 29–31] or to study the adsorption of corrosion inhibitors to metal surfaces [32].

Another important application field for PM IRRAS is biophysical studies at artificial lipid double layers on electrode surfaces. If a selectively deuterated lipid is used for one of the two layers, the characteristic CH_2 and CD_2 vibrations of the molecules in both layers appear in different wavelength ranges of the spectrum, which can be studied separately. The electrode potential allows adjustment of a precise and large electric field strength across the artificial membrane, which mimics the extreme field strength in biomembranes (Excursion 13-1, Section 18.1.1). A quantitative analysis according to the principle in Figure 8.4 yields the supramolecular orientation of the molecular fragments as a function of electrode potential (and thus the electric field strength in the membrane) [13, 18].

8.2.2 Raman Spectroscopy

"Inelastic" means that the energy of the molecule changes during the scattering process.

The Raman effect represents an inelastic scattering process. The incident radiation excites the molecule (or solid) to a virtual state, which is a nonstationary state that cannot be described by *one* set of quantum numbers. From this state, the molecule returns to the electronic ground state. However, the quantum numbers of the vibrational levels for normal vibrations may change by $\Delta v = \{-1, 0, +1\}$. By far, most molecules undergo transitions with $\Delta v = 0$. In this case, the scattered light has the same photon energy $h\nu$ as the excitation light, but a different propagation direction (Rayleigh scattering[86]). A small fraction of the photons in the scattered light appears at lower photon energy (Raman scattering, $\Delta v = +1$). Those spectral lines of the scattered light are called Stokes lines[26]. $\Delta v = -1$ (anti-Stokes lines) is also conceivable; however, most molecules are initially in the vibrational ground state ($v = 0$), so that no further decrease in vibrational energy is possible. Therefore, anti-Stokes lines are practically not observed for vibrational transitions. Decisive for the intensity of Raman lines is the change in the polarizability α during the period of the normal vibration. Consequently, polar bonds tend to have low Raman intensities and nonpolar bonds have higher Raman intensities. This is opposite to the intensities observed in IR absorption lines.

$E_{initial}$ – Initial level [eV]
E_{final} – Final level [eV]
ν_{in} – Frequency of the excitation radiation [Hz]
ν_{out} – Frequency of the Raman line [Hz]

α – Polarizability of the molecule
q – Inner coordinates (generalized for bond angles and bond lengths)

Resonance condition: $E_{final} - E_{initial} = h\nu_{in} - h\nu_{out}$ (8.7)

The energy difference between the Raman line and the excitation radiation corresponds to the energy difference of the involved vibrational levels.

Excitation condition: $d\alpha/dq \neq 0$ (8.8)

The polarizability must be changed during the vibration.

Selection rule: $\Delta v = 1$ (8.9)

The vibrational quantum number can be changed only by one unit during the transition.

A significant advantage for spectroelectrochemistry is the fact that the excitation radiation for Raman is in the visible part of the spectrum. This light that can be passed

[86] John William Rayleigh, 3rd Baron Rayleigh, 1842–1919, English physicist, Nobel Prize in physics 1904. In addition to the description of Rayleigh scattering, his name is associated with Rayleigh distribution, Rayleigh limit, Rayleigh–Taylor instability, Plateau–Rayleigh instability, Rayleigh number, Rayleigh wave etc.

through glass windows and lenses and is not absorbed by solvents, including water. Lasers are the preferred light source to deliver monochromatic light of high intensity with a wavelength appropriate for the spectral range of CCD detectors. A notch filter is used to remove the photons with the excitation wavelength from the scattered light. Such filters are completely opaque within a wavelength range of about 10 nm, while radiation intensity with other wavelengths is essentially unattenuated. A monochromator is used to disperse the light after passing the notch filter with a wavelength similar (but not identical) to the excitation radiation. This radiation is recorded with high spectral resolution. The shift in Raman lines relative to the excitation light is only (500–3000) cm^{-1} (Figure 8.9). Visible light can be focused very well, facilitating analyses with high lateral resolution. The experiment is often directly integrated into an optical microscope.

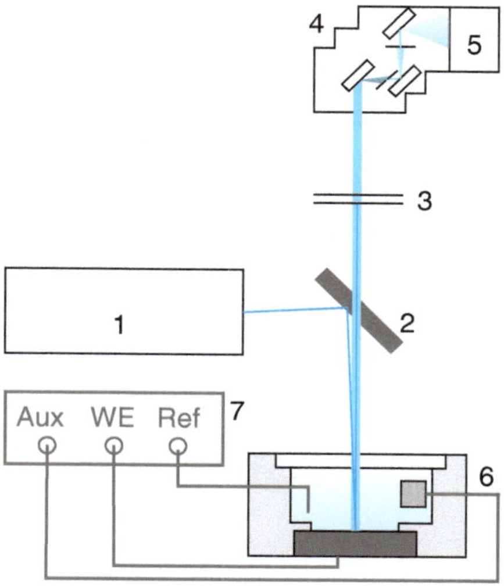

Figure 8.9 Main components of a Raman spectrometer coupled to an electrochemical cell.

1) Laser as excitation source, for example, frequency-doubled neodymium-doped yttrium-aluminum garnet (Nd:YAG) with 532 nm;
2) semitransparent mirror;
3) notch filter for removal of the excitation wavelength;
4) triple monochromator;
5) CCD as detector (charge-coupled device);
6) electrochemical three-electrode cell;
7) potentiostat.

Overall, the Raman effect is a *very weak* effect; only 1 out of 10^{11} incident photons leads to Raman scattering. The method can be used for the investigation of battery materials (Chapter 15). In this case, the electrode consists of a layer of the battery material with a thickness of several micrometers, providing sufficiently high Raman intensity. If the molecules of interest are adsorbed at the interface as a monolayer, enhancement mechanisms are required for their investigation by Raman spectroelectrochemistry. Resonance Raman spectroscopy and surface-enhanced Raman spectroscopy (SERS) provide two powerful options. Resonance Raman spectroscopy can be used for molecules with an absorption in the visible spectral range. The excitation radiation is selected in such a way that its photon energy is close to that of an electronic transition of the molecule of interest (Figure 8.10a). This enhances the interaction between the molecule and the electromagnetic field of the excitation radiation that can lead to enhancement factors of up to 10^6. The effect is suitable for studying molecules in homogeneous solution, for example, molecules within the diffusion layer in the vicinity of an electrode surface.

SERS is possible only for adsorbates on roughened electrodes of some metals, mainly Au, Ag, and Cu. The surfaces are roughened by electrochemical pretreatment (surface oxidation and reduction cycles). The resulting metal nanostructures cause a strong scattering of light and the excitation of plasmons within the metal nanostructures that strongly enhance the local electrical field strength immediately at the metal surface (blue shading in Figure 8.10b). The chemical bond between the adsorbate molecules and the surface provides an additional opportunity to shift electron density between the adsorbate and the metal electrode upon excitation (charge transfer, lens in Figure 8.10b). This enhances the probability of the interaction between a photon and the molecule enormously. Typical enhancement factors are from 10^5 to 10^6. Even larger enhancement factors of up to 10^{14} are possible with nanoparticles of Ag and Au of the optimum size of 80–100 nm. This approach allows even single-molecule spectroscopy of adsorbates. Applications of Raman spectroscopy in the context of electrochemistry comprise aromatic compounds and coordination complexes of transition metals on noble metal electrodes [33]. The restriction to only metals with high Raman signal enhancement can be circumvented by covering the noble metal with a thin film of the metal of interest. In such a situation, the adsorbate binds to the metal of interest, while the underlying noble metal enables the SERS effect [33]. An excellent survey is available about the principles, cell construction, instrumentation, and diverse applications [33].

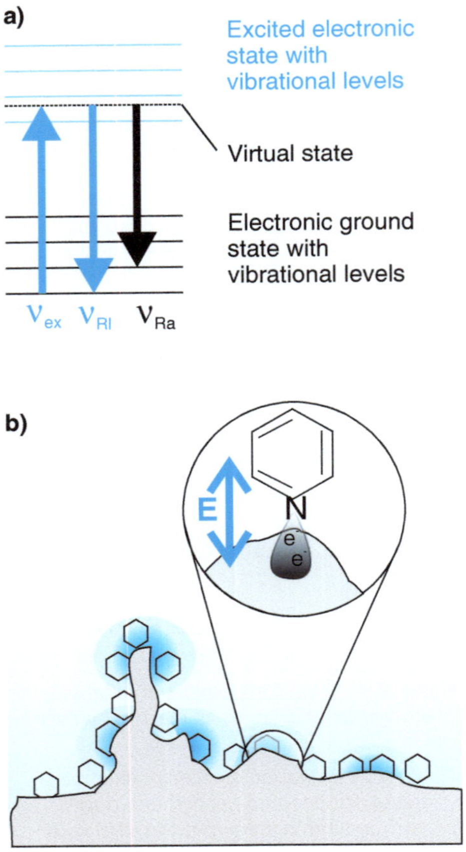

Figure 8.10 Enhancement effect for Raman spectroscopy.

a) Resonance Raman effect and
b) SERS effect.

8.2.3 Exploitation of Nonlinear Optical Effects

Nonlinear optical effects can emerge in optical media that lack a centrosymmetric structure. The broken symmetry can be caused by an interface with strict surface sensitivity to spectroscopic methods, provided that the adjoining phases themselves are centrosymmetric. The techniques require very high light intensities because the excitation probability is only 10^{-12}. The detection of the signal is also instrumentally demanding.

Second harmonic generation (SHG) occurs upon reflection of a high-intensity pulsed laser beam. As a result, the reflected radiation contains photons with double the frequency compared to the incident light. Due to the low intensity of this spectral component, it has to be measured with sensitive detectors. The intensity of signals is often determined by the nonlinear susceptibility of the metal surface, which in turn can be influenced by adsorbates or underpotential metal deposition. As the radiation can be conveniently transmitted through the liquid phase, *in situ* observation of potential-dependent coverages is possible, but chemical identification is rarely achieved [34–38].

For sum frequency generation (SFG), a laser pulse of constant frequency in the visible range is directed to the surface of interest together with another laser in the IR range under a different angle of incidence and with a tunable wavelength [39, 40]. The pulses of the two lasers must be correlated to reach the surface under study *at the same location* and *exactly at the same time*. The reflected light contains a low-intensity component with a frequency that is the sum of the frequencies of the visible and IR lasers (sum frequency signal). This radiation component is detected. If the wave number of the IR laser is equal to a vibrational transition of an adsorbate, the sum frequency signal is significantly enhanced. A spectrum is obtained by scanning the wave number of the IR laser. It contains a background and signatures of the vibrational resonances of molecules at the interface. The signal intensity is proportional to the square of the adsorbate concentration, and it depends on the orientation of adsorbed molecules at the interface. Experimental challenges are linked to bringing the IR radiation through the solvent layer to the interface and adjusting the very short laser pulses to arrive at the same time at the same location at the interface.

8.2.4 Electron Energy Loss Spectroscopy

Electron energy loss spectroscopy (EELS) or high-resolution electron energy loss spectroscopy (HREELS) requires an ultra-high vacuum (UHV) and is thus an *ex situ* method. It is used mainly in connection with electron spectroscopies and electron microscopy (Section 8.4). Nevertheless, it is of interest to study electrode surfaces because it offers higher sensitivity than IRRAS and Raman spectroscopy at interfaces. The sample is exposed to a beam of monoenergetic electrons from an electron gun. The back-scattered electrons are separated according to their energy. The energy difference between the impinging electrons and the scattered electrons corresponds to the energy difference between vibrational energy levels of adsorbates (inelastic scattering). The selection rules allow only the excitation of vibrational modes with total symmetry [1, 41, 42].

8.3 Ellipsometry

In ellipsometry, the reflective sample with an overlayer is illuminated with linearly polarized visible light, which becomes an elliptically polarized beam after reflection. The phase shift and change in the orientation of the polarization are measured to allow conclusions about the layer thickness and complex dielectric function (i.e., the complex refractive index). For electrochemical applications, ellipsometry enables the determination of layer thicknesses during galvanic deposition, the growth of oxide layers, or the formation of conducting polymer films on electrodes. Because visible light is well transmitted through aqueous electrolyte solutions, the method is particularly suited for *operando* measurements. A disadvantage is its inability to provide information about the chemical nature of the film, causing phase shift and orientation change of the polarized light. Furthermore, the quantity of interest (film thickness and refractive index) must be extracted from the raw data by a fitting procedure that, similar to other fitting procedures (Chapter 7), may depend critically on proper models that are applied for the surface structure.

Each photon traveling along the propagating direction can be described by the superposition of two orthogonal waves of equal frequency, from which one is parallelly oriented to the plane of incidence $\mathbf{E}_p = \mathbf{E}_{p,0} \cdot \cos(\omega t)$, while the other one is perpendicularly oriented toward the plane of incidence $\mathbf{E}_s = \mathbf{E}_{s,0} \cdot \cos(\omega t)$. If the phase shift Δ between the two waves is zero, linearly polarized light results from the superposition (Figure 8.11a). The polarization plane can be specified by an angle Ψ with the plane of incidence, which is given by

$$\tan \Psi = \frac{|\mathbf{E}_{s,0}|}{|\mathbf{E}_{p,0}|} \tag{8.10}$$

During reflection and refraction at interfaces, the angular frequency of the two components $\mathbf{E}_s$ and $\mathbf{E}_p$ remains constant, but the amplitudes $|\mathbf{E}_{s,0}|$, $|\mathbf{E}_{p,0}|$, and the phase shift Δ can change. If $\Delta \neq 0$ and $|\mathbf{E}_{s,0}| \neq |\mathbf{E}_{p,0}|$ after reflection, an elliptically polarized wave is formed (Figure 8.11b).

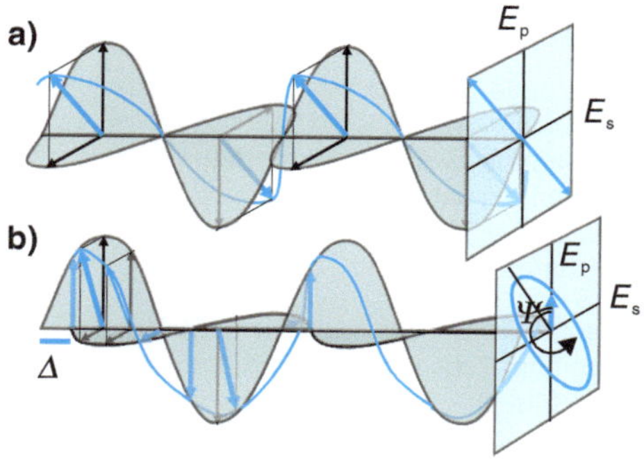

Figure 8.11 Schematic depiction of a) a linearly polarized wave (before reflection) and b) an elliptically polarized wave (after reflection).

Ψ – Polarization angle relative to the plane of incidence (psi) [radian]

$|\mathbf{E}_{s,0}|$ – Amplitude of the alternating electrical field component perpendicular (index "s" from German "senkrecht") to the plane of incidence [V m^{-1}]

$|\mathbf{E}_{p,0}|$ – Amplitude of the alternating electrical field component parallel to the plane of incidence [V m^{-1}]

Δ – Phase shift [radian]

Q: What is circularly polarized light?

A: For circularly polarized light, the field vector resulting from the superposition of $\mathbf{E}_s$ and $\mathbf{E}_p$ rotates around the propagation direction. This corresponds to a phase shift $\Delta \neq 0$ and equal amplitudes of the field components ($|\mathbf{E}_{s,0}| = |\mathbf{E}_{p,0}|$)

A typical ellipsometer comprises a light source with a polarizer (Figure 8.12, (1)). The resulting linearly polarized radiation has a polarization angle somewhere between parallel and perpendicular polarization. During reflection (2) part of the light is reflected, and the other part is refracted (lens in Figure 8.12). Typically, medium 1 in electrochemical applications is the electrolyte solution, medium 2 is the growing film, and medium 3 is the metallic electrode. The quantitative description of the reflection process at such an interfacial layer is complicated because the incident beam (index i) is partially reflected (index r) and partially refracted (index t for transmission) at each interface. The intensity distribution for reflected and transmitted light can be computed by means of Fresnel's equations[83] for each interface (Excursion 8-1). Various reflected waves are superimposed to form an elliptically polarized wave. The compensator (3) is used to change the phase shift to zero. Afterward, a linearly polarized wave propagates further. Its polarization angle is determined in the analyzer (4). For this task, the polarizer is rotating, and the detector records the instantaneous intensity, which is converted into an AC voltage, from which the polarization angle Ψ can be determined. The recording of an individual data point does not take more than 1 ms with contemporary equipment. This short acquisition time permits the recording at several incident angles and polarization directions.

Here Ψ has a different meaning than the wave function Ψ in quantum mechanics or the current function ψ in other parts of this book. The complex refractive index $\hat{n} = n + \mathrm{i}k = n' + \mathrm{i}n''$ should not be confused with the number of transferred electrons n. Danger for confusion also exists for the absorption index k (we use here n'' instead) and the wave vector $\mathbf{k}$ (Chapter 10).

Data points allow the determination of the layer thickness d of medium 2 and the wavelength-dependent optical constants $\hat{n}_1(\omega), \ldots, \hat{n}_m(\omega)$. Because the data evaluation

Figure 8.12 Schematic representation of an ellipsometer.

1) – source of monochromatic and polarized light;
2) – sample (working electrode);
3) – compensator;
4) – rotating analyzer;
5) – detector.

Blue arrow shows the electric field vector. The magnifying lens shows multiple reflections at a thin film.

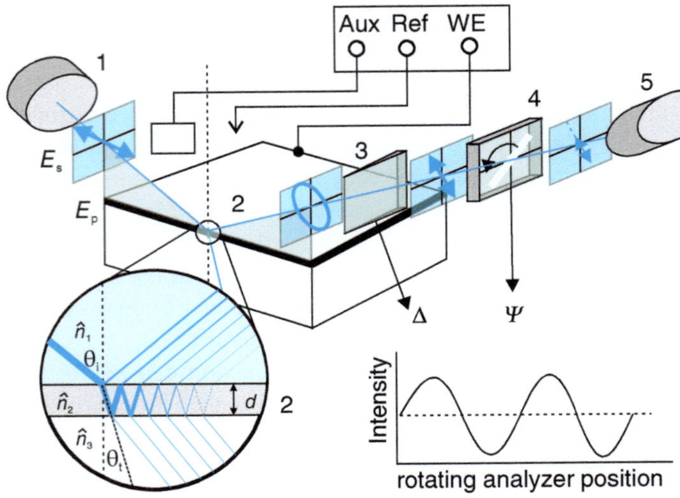

problem is overdetermined (there are more independent measurements than parameters to be determined), least-square fitting procedures are used. Nevertheless, it can happen that several sets of parameters provide a good fit of the experimental data. Therefore, one tries to determine as many of the fitting parameters by independent experiments. One could, for instance, measure the samples without a cover layer or use tabulated optical constants $\hat{n}_1, \dots, \hat{n}_m$. The calculations are executed by dedicated, powerful computer programs that are typically provided together with the ellipsometry instruments. For correct results, it is absolutely necessary to give realistic structural models for the interface layers (roughness, optical constants, etc.). An interesting further development is imaging ellipsometry, which provides information about layer thickness and optical constants for each point on an extended two-dimensional sample by means of a CCD detector.

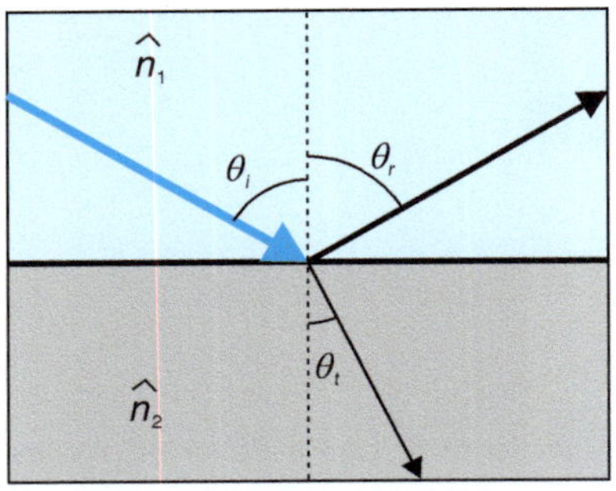

Figure 8.13 Single reflection and refraction at an interface between two media (phases). The schematic is a cross-sectional view through the interface along the incident plane.

θ_i – Incident angle [radian]
θ_r – Reflection angle [radian]

θ_t – Refraction angle [radian]
$\hat{n}_j$ – Complex refraction index for material j [dimensionless]

Excursion 8-1 Fresnel's Equations

Already when discussing infrared reflection absorption spectroscopy and again here for ellipsometry, we realize that the quantitative understanding of the *in situ* spectroelectrochemical methods requires a detailed description of the reflection and refraction processes at solid–liquid interfaces. Therefore, we would like to reexamine the quantitative relations for the single reflection and refraction at an interface (Figure 8.13). This will facilitate the understanding of more specialized literature. We take the incident angle θ_i as a given quantity. It is measured between the incident beam and the surface normal at the reflection point. The reflection angle θ_r is obtained from the law of reflection.

$$\theta_i = \theta_r \tag{8.11}$$

The refraction angle θ_t (index t from transmission) is obtained from Snell's law

$$\hat{n}_1 \sin\theta_i = \hat{n}_2 \sin\theta_t \tag{8.12}$$

The quantities $\hat{n}_1, \hat{n}_2$ [Eq. (8.13)] are complex optical constants of the media 1 and 2. They do depend on the frequency (wavelength) of light and on the temperature. They can be represented as complex numbers (Excursion 7-1). The real term n' describes the ratio between the speed of light in vacuum and in the considered medium $n' = c_{vac}/c_M$. The imaginary term is the absorption index. It is linked to the negative decadic absorbance coefficient α from Lambert–Beer's law[87] $\log I_0/I = \alpha \cdot d$ by the relation $\alpha = 2\omega n''/c_{vac}$.

$$\hat{n}(\omega) = n'(\omega) + in''(\omega) = \frac{c_{vac}}{c_M(\omega)} + i\frac{\alpha c_{vac}}{2\omega} \tag{8.13}$$

The complex refractive index is also related to the permittivity number $\hat{\varepsilon}_r(\omega)$ (also known as dielectric constant or dielectric function) and the magnetic permittivity number $\hat{\mu}_r(\omega)$, which is equivalent to the "conductivity of the medium for electric and magnetic fields":

$$\hat{n}(\omega) = \sqrt{\hat{\varepsilon}_r(\omega) \cdot \hat{\mu}_r(\omega)} \tag{8.14}$$

For nonmagnetic material ($\hat{\mu}_r \approx 1$), this yields

$$\hat{n}(\omega) = \sqrt{\hat{\varepsilon}_r(\omega)} = \sqrt{\varepsilon'(\omega) + i\varepsilon''(\omega)} \tag{8.15}$$

Detectors measure the reflectivity or (reflectance) R; this is the fraction of the incident radiation power that is reflected. R is proportional to the ratio of the squared amplitudes:

$$R = \frac{\text{reflected power}}{\text{incident power}} = \frac{|\mathbf{E}_{0,r}|^2}{|\mathbf{E}_{0,i}|^2} = r_{12}^2 \tag{8.16}$$

Analogous relations hold separately for the s- and p-components of the electromagnetic radiation. The complementary part T, that is the transmittance (transmissivity), is refracted and passes the medium 2, where the power transmittance is different. Therefore, there is a correction term to the squared amplitude ratio t_{12} in Eq. (8.17).

$$T = 1 - R = \frac{\text{transmitted power}}{\text{incident power}} = \frac{n_2 \cos\theta_t |\mathbf{E}_{0,t}|^2}{n_1 \cos\theta_i |\mathbf{E}_{0,i}|^2} = \frac{n_2 \cos\theta_t}{n_1 \cos\theta_i} t_{12}^2. \tag{8.17}$$

Fresnel's equations enable the calculation of R and T from the amplitude ratios r_{kl} and t_{kl}. The indices k and l specify the adjoining phases, whereby k refers to the phase from which the light is coming and l refers to the phase interacting with the incident beam. Because one can consider the polarization directions (p and s) as independent, one can use equations for both components separately. Furthermore, we consider $1 = \sin^2\theta + \cos^2\theta$ and Eq. (8.12). The dependencies of $\hat{n}_j(\omega)$ and $\hat{\mu}_r(\omega)$ have been omitted for clarity.

$$r_{12,p} = \frac{\hat{n}_2 \frac{\mu_{r1}}{\mu_{r2}} \cos\theta_i - \hat{n}_1 \cos\theta_t}{\hat{n}_2 \frac{\mu_{r1}}{\mu_{r2}} \cos\theta_i + \hat{n}_1 \cos\theta_t} = \frac{\hat{n}_2 \frac{\mu_{r1}}{\mu_{r2}} \cos\theta_i - \hat{n}_1 \sqrt{1 - \left(\frac{\hat{n}_1}{\hat{n}_2}\right)^2 \sin^2\theta_i}}{\hat{n}_2 \frac{\mu_{r1}}{\mu_{r2}} \cos\theta_i + \hat{n}_1 \sqrt{1 - \left(\frac{\hat{n}_1}{\hat{n}_2}\right)^2 \sin^2\theta_i}} \tag{8.18}$$

$$t_{12,p} = \frac{2\hat{n}_2 \cos\theta_i}{\hat{n}_2 \frac{\mu_{r1}}{\mu_{r2}} \cos\theta_i + \hat{n}_1 \cos\theta_t} = \frac{2\hat{n}_2 \cos\theta_i}{\hat{n}_2 \frac{\mu_{r1}}{\mu_{r2}} \cos\theta_i + \hat{n}_1 \sqrt{1 - \left(\frac{\hat{n}_1}{\hat{n}_2}\right)^2 \sin^2\theta_i}} \tag{8.19}$$

$$r_{12,s} = \frac{\hat{n}_1 \cos\theta_i - \hat{n}_2 \frac{\mu_{r1}}{\mu_{r2}} \cos\theta_t}{\hat{n}_1 \cos\theta_i + \hat{n}_2 \frac{\mu_{r1}}{\mu_{r2}} \cos\theta_t} = \frac{\hat{n}_1 \cos\theta_i - \frac{\mu_{r1}}{\mu_{r2}} \hat{n}_2 \sqrt{1 - \left(\frac{\hat{n}_1}{\hat{n}_2}\right)^2 \sin^2\theta_i}}{\hat{n}_1 \cos\theta_i + \frac{\mu_{r1}}{\mu_{r2}} \hat{n}_2 \sqrt{1 - \left(\frac{\hat{n}_1}{\hat{n}_2}\right)^2 \sin^2\theta_i}} \tag{8.20}$$

c_{vac} – Speed of light in vacuum [cm s^{-1}]
c_M – Speed of light in the medium [cm s^{-1}]
n' – Real term of the refractive index [dimensionless]
n'' – Imaginary term of the refractive index = absorption index [dimensionless]
α – Absorbance coefficient [cm^{-1}]
ω – Angular frequency [s^{-1}]
i – Imaginary unit
$\hat{\varepsilon}_r(\omega)$ – Permittivity number, dielectric function [dimensionless]
$\hat{\mu}_r(\omega)$ – Magnetic permittivity number [dimensionless]
ε' – Real term of the dielectric function [dimensionless]
ε'' – Imaginary term of the dielectric function = absorption index [dimensionless]

[87] Johann Heinrich Lambert, 1728–1777, born in Mulhouse, at the time in the Swiss Confederacy today in France, position at the Prussian Academy of Sciences in Berlin as mathematician, physicist, philosopher, astronomist, meteorologist. August Beer, 1825–1863, German-Jewish mathematician, chemist and physicist, professor in Bonn.

$$t_{12,s} = \frac{2\hat{n}_1 \cos\theta_i}{\hat{n}_1 \cos\theta_i + \hat{n}_2 \frac{\mu_{r1}}{\mu_{r2}} \cos\theta_t} = \frac{2\hat{n}_1 \cos\theta_i}{\hat{n}_1 \cos\theta_i + \frac{\mu_{r1}}{\mu_{r2}} \sqrt{\hat{n}_2^2 - \hat{n}_1^2 \sin^2\theta_i}} \tag{8.21}$$

The general equations [Eqs. (8.18)–(8.21)] can be simplified if $\mu_{r,1}(\omega) \approx \mu_{r,2}(\omega)$ [Eqs. (8.22) to (8.25)] or if there is no absorption and thus $\hat{n}_j = n'_j$.

$$r_{12,p} = \frac{\hat{n}_2 \cos\theta_i - \hat{n}_1 \cos\theta_t}{\hat{n}_2 \cos\theta_i + \hat{n}_1 \cos\theta_t} = \frac{\hat{n}_2^2 \cos\theta_i - \hat{n}_1 \sqrt{\hat{n}_2^2 - \hat{n}_1^2 \sin^2\theta_i}}{\hat{n}_2^2 \cos\theta_i + \hat{n}_1 \sqrt{\hat{n}_2^2 - \hat{n}_1^2 \sin^2\theta_i}} \tag{8.22}$$

$$t_{12,p} = \frac{2\hat{n}_1 \cos\theta_i}{\hat{n}_2 \cos\theta_i + \hat{n}_1 \cos\theta_t} = \frac{2\hat{n}_1 \hat{n}_2 \cos\theta}{\hat{n}_2^2 \cos\theta_i + \hat{n}_1 \sqrt{\hat{n}_2^2 - \hat{n}_1^2 \sin^2\theta_i}} \tag{8.23}$$

$$r_{12,s} = \frac{\hat{n}_1 \cos\theta_i - \hat{n}_2 \cos\theta_t}{\hat{n}_1 \cos\theta_i + \hat{n}_2 \cos\theta_t} = \frac{\hat{n}_1 \cos\theta_i - \sqrt{\hat{n}_2^2 - \hat{n}_1^2 \sin^2\theta_i}}{\hat{n}_1 \cos\theta_i + \sqrt{\hat{n}_2^2 - \hat{n}_1^2 \sin^2\theta_i}} \tag{8.24}$$

$$t_{12,s} = \frac{2\hat{n}_1 \cos\theta_i}{\hat{n}_1 \cos\theta_i + \hat{n}_2 \cos\theta_t} = \frac{2\hat{n}_1 \cos\theta_i}{\hat{n}_1 \cos\theta_i + \sqrt{\hat{n}_2^2 - \hat{n}_1^2 \sin^2\theta_i}} \tag{8.25}$$

The equations for the amplitude ratios allow the calculation of the distribution of light intensities for *s*- and *p*-polarization for a layer model. For multiple reflection in layered systems (lens in Figure 8.12), the effects can be summarized. The experimentally accessible quantities Δ and Ψ are related to the amplitude factors according to Eq. (8.26).

$$\rho = \frac{r_p}{r_s} = \frac{|\mathbf{E}_{p,0,r}|/|\mathbf{E}^r_{p,0,i}|}{|\mathbf{E}_{s,0,r}|/|\mathbf{E}_{s,0,i}|} = e^{i\Delta} \tan\Psi \tag{8.26}$$

The calculation must be made for the entire light path. They are effectively represented in a matrix formalism, which can be studied in more specialized textbooks. Similar formalisms can be used to calculate the optimum layer thicknesses and materials for IRRAS spectroelectrochemical cells. Such optimizations are a necessary condition for obtaining quality spectroscopic data for complex multilayer systems.

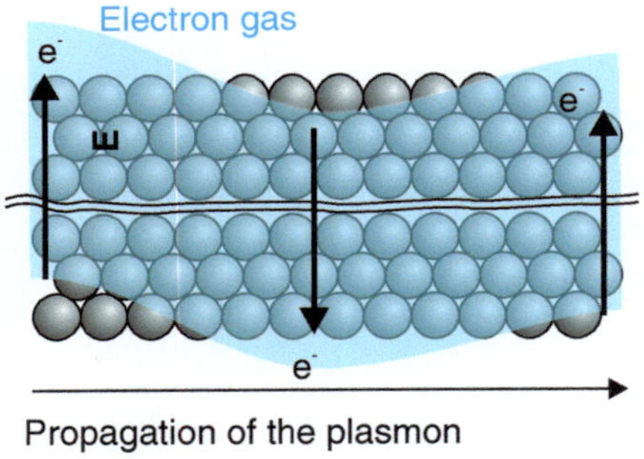

Figure 8.14 Schematic representation of surface plasmon resonance. The alternating electric field causes a displacement of the electron density of the electron gas relative to the atoms on the lattice positions of the metal. This generates a restoring force that pulls the electrons back. However, their relocation is not instantaneous because they possess a mass, and the solid body of the metal exerts a resistance against the motion of the electrons. At a specific frequency of the external field, a resonance occurs between the periodic oscillations of the electron gas and the external electric field. In this situation, energy can be particularly effectively transmitted from the alternating field to the electron gas.

8.4 Surface Plasmon Resonance

Surface plasmon resonance (SPR) has become a standard method in bioanalytics with a range of commercially available instruments. It is also attractive for the characterization of surface-modified gold electrodes (biosensors, organic thin films). Similar to ellipsometry, the method provides *in situ* information about film growth and binding of molecules to the surface; however, without the possibility of chemical identification. The SPR technique is based on the excitation of collective oscillations of the free electron gas in metals versus the atoms on the lattice positions (Figure 8.14). Figure 8.15 illustrates an SPR experiment using the Kretschmann configuration[88], coupled to an electrochemical cell. A *p*-polarized laser beam is directed to the back side of a thin gold film, which concomitantly serves as the working electrode and substrate for the excitation of SPR. The incident angle is varied, and the reflectivity *R* is

[88] Erwin Kretschmann; Die Bestimmung optischer Konstanten von Metallen durch Anregung von Oberflächenplasmaschwingungen. *Z. f. Phys. A* **1971**, *241*, 313–324, doi: 10.1007/BF01395428

measured. Gold is known as an excellent mirror for visible light. Therefore, R is larger than 90% over a large range of angles of incidence. However, the reflectivity drops sharply to almost zero for an incident angle θ_i of ca. 43° (for a laser with a wavelength of 633 nm). In this situation, the frequency of the incident radiation is at resonance with the collective oscillation of the electron gas. If a thin layer is deposited on the gold electrode, for example, by underpotential deposition of a monolayer of a foreign metal (Section 8.9.8), a self-assembled monolayer (Section 13.1.1), or a polymer (Section 13.2), the deposition process can be followed by SPR *in situ* and with excellent time resolution. The plasmons create an evanescent alternating electric field that penetrates about 200 nm into the solution (for excitation with visible light). Surface layers that build up within this evanescent field have different optical constants than gold and exert an influence on the exact SPR frequency via the evanescent field. Direction and frequency can be efficiently described by the wave vector **k**, with a direction pointing in the propagation direction and with a magnitude (visualized as length) $2\pi/\lambda \cdot \varepsilon_0^{1/2} = 2\pi\nu/c \cdot \varepsilon_0^{1/2} = \omega \cdot \varepsilon_0^{1/2}/c$, that is, proportional to frequency (Excursion 10-1). The **k**-vector of the incoming light can be thought of as the combined action of vectors that are aligned parallel ($\mathbf{k}_x$) and perpendicular to the gold surface (Figure 8.15). By changing the incident angle, $\mathbf{k}_x$ can be tuned so that it has the same magnitude compared to the wave vector $\mathbf{k}_{sp}$ of the surface plasmon that propagates parallel to the gold surface. The magnitude is determined by the material properties of the adjoining phases. Equation (8.27) holds for the resonance case.

$$|\mathbf{k}_x| = |\mathbf{k}_{sp}|$$

$$\frac{2\pi}{\lambda}\cdot\sqrt{\varepsilon_0}\sin\theta_r = \frac{2\pi}{\lambda}\sqrt{\frac{\varepsilon_1\varepsilon_2}{\varepsilon_1+\varepsilon_2}} \tag{8.27}$$

θ_r – Incident angle at which resonance occurs [radian]
$\varepsilon_0(\lambda)$ – Complex permittivity of the materials [dimensionless]
$\varepsilon_1(\lambda)$ – Complex permittivity of the metal film on the prism [dimensionless]
$\varepsilon_2(\lambda)$ – Complex permittivity of the electrolyte solution [dimensionless]
λ – Wavelength in free space [m]

Although the wavelength does not appear explicitly in the resonance condition after canceling $2\pi/\lambda$ in Eq. (8.27), the angle $\theta = \theta_r$ at which SPR is observed does depend on the wavelength of the incident light because ε_0, ε_1, and ε_2 are functions of the wavelength.

For the determination of the SPR minimum, one may, in principle, choose one of the following options (Figure 8.16):

1) Use of a fixed wavelength for excitation (laser) and variation of θ to find the minimum of the reflectivity (SPR).
2) Use of a fixed incident angle and measuring the dependence of the reflectivity as function of the variable wavelength of the incident light.
3) Use of a fixed wavelength and fixed incident angle close to, but not exactly at the SPR minimum. One is then observing the intensity of the reflected beam that changes with the formation of complex overlayers.

The third option, in which no optical parts (monochromator or prisms) must be moved is also suited for imaging methods based on SPR in which SPR data are recorded simultaneously from different regions of the sample.

Deposition of cover layers usually leads to a small, yet measurable increase in the angle θ_r, at which SPR is observed. A particular advantage is seen in the fact that the build-up of complex layers, for example, on biosensor surfaces, can be followed step by step because each layer increases the SPR angle slightly (Figure 8.16). SPR has become a routine method in biochemical binding analytics. In this application, angle changes are measured relative to an initial state. However, the exact value of the SPR minimum is influenced by layer thickness and optical constants of all components of the film, the SPR substrate, and the solvent. Those multiple dependencies are typically dealt with by a reference measurement, for instance, at two different positions of one

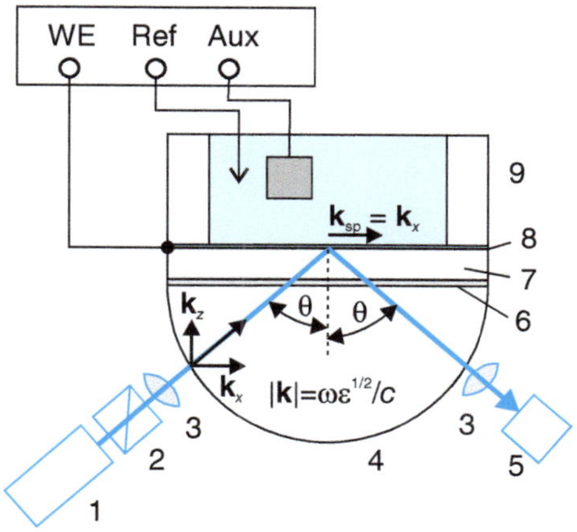

Figure 8.15 Experimental configuration for an SPR instrument coupled to an electrochemical cell illustrated as cross section:
1) laser as light source,
2) polarizer,
3) lenses,
4) hemispherical prism,
5) detector,
6) index-matched contact liquid,
7) cover slide for microscopy (usually 0.17 mm thick),
8) gold film (typically 50 nm thick),
9) cell vessel.
The scheme is not to scale.

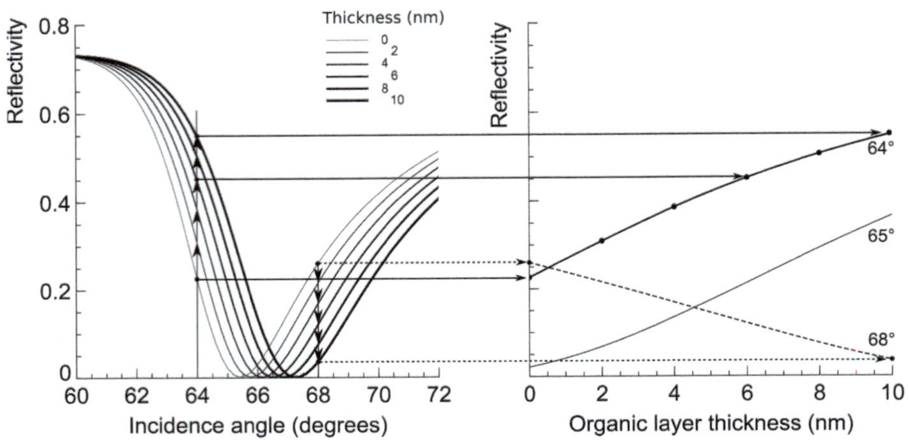

Figure 8.16 Illustration of the development of the SPR signal with increasing layer thickness. This offers an opportunity (*i*) for the detection with a fixed wavelength of the light by variation of the incident angle probing the angle at which the reflection minimum is reached, or (*ii*) following the reflectivity at a fixed angle close to the reflection minimum. The right panel shows the resulting calibration curves for three different angles of incidence at a fixed wavelength of $\lambda = 650$ nm at 22 °C. Measurements at a fixed angle are particularly useful for imaging experiments. The curves were calculated for an adhesion layer of 2 nm Cr ($\hat{n}_{Cr} = 3.105 + 3.327$ i), 50 nm Au ($\hat{n}_{Au} = 0.1685 + 3.138$ i), glass ($\hat{n}_{glass} = 1.615$), organic layer of 0, 2, 4, 6, 8 or 10 nm with ($\hat{n}_L = 1.49$) and water ($\hat{n}_W = 1.3317$). Source: Reproduced with permission from S. Nizamov, V. Scherbahn, M. V. Mirsky; *Sensors Actuators B: Chemical* **2015**, *207*, 740–747, Copyright 2015, Elsevier.

and the same substrate. Patterned layers and microfluidic networks for local delivery of the sample and reference solution are used in this context to ensure that the analyte molecule binds only at one of the two measurement spots.

Because SPR is very sensitive, small temperature changes (e.g., by solution exchange or Joule heating[51] due to an electrolysis current) or minute density changes during a solution exchange cause a change in the SPR signal. This makes precise thermostatic control mandatory in addition to careful preparation of all electrolyte solutions. New method developments aim for improvement of the referencing to utilize the remarkable sensitivity of SPR. In this way, *one* out of the many dependencies of the SPR signal can be followed with high precision while suppressing the effects of all other parameters either by keeping them constant or by compensating for them. Because the thin gold layer is simultaneously the SPR resonator and the working electrode, SPR is especially suited to study modified electrodes and the build-up of multilayer systems [43, 44].

8.5 Spectroscopy of Core Electrons

8.5.1 Primary and Secondary Processes

Spectroscopy involving inner (core) electrons requires excitation energies in the range of X-rays, that is, electromagnetic radiation with a photon energy of 50–10 000 eV. Various methods have been developed using this radiation for the structural characterization of electrode surfaces. All methods have in common that the signals are element-specific, because energy levels of core electrons strongly depend on the atomic number (Mosley's rule[89]). In addition, there is usually a weak dependency on a specific binding situation in which the observed atoms occur (chemical shift).

When X-rays hit a body, primary and secondary processes are triggered that provide the basis for different spectroscopic methods. **Primary processes** lead to an initial vacancy in an electronic core level. In the process of X-ray absorption, an inner electron is excited to an initially unoccupied energy level (Figure 8.17a). The resulting state is a neutral atom with a vacancy in an electronic core level. This process is the basis for

[89] Henry Gwyn Jeffreys Mosley, 1887–1915, British physicist, killed in combat in the battle of Gallipoli (WW I) close to the Dardanelles (today Turkey). The law named after him describes the relation between the wavelength of the characteristic X-rays of an element and its atomic number. At the time, this provided important experimental evidence for the theory of atoms.

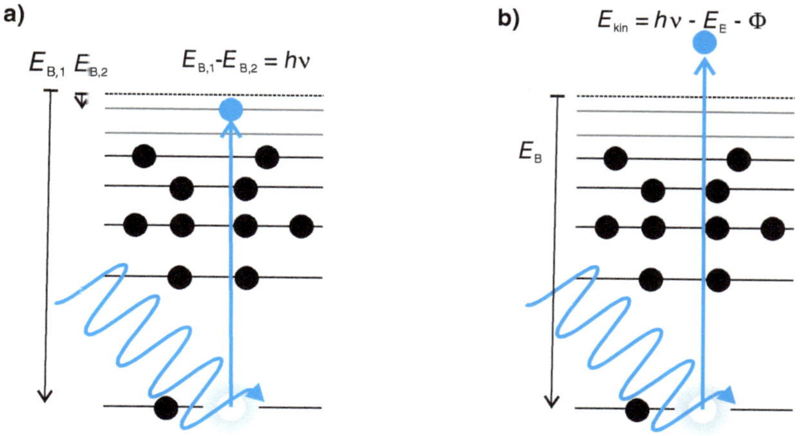

Figure 8.17 Orbital energy diagrams (of one atom of the sample) for primary processes during interaction of X-rays with the sample, a) X-ray absorption, b) photoemission. The vacancies (holes) in the inner shells after the excitation are marked in gray.

X-ray absorption spectroscopy (XAS). Of analytical relevance are the methods X-ray absorption near edge structure (XANES) and extended X-ray absorption fine structure (EXAFS). If higher photon energies are used, the excited electron can be emitted from the atom. This process is called the external photoelectric effect or photoemission (Figure 8.17b). The binding energy E_B of the emitted core electron can be calculated from measured kinetic energy E_{kin} and the photon energy $h\nu$ by rearranging the relation for energy conservation in the photoemission process [Eq. (8.28)]. A correction term Φ accounts for the difference in work functions between the sample and the detector (e.g., effects of charging). The final state is a singly positively charged ion with a vacancy in the core level from which the electron was emitted. This effect is the central process in X-ray photoelectron spectroscopy (XPS).

$$h\nu = E_B + E_{kin} + \Phi \tag{8.28}$$

$h\nu$ – Energy of the X-ray photon [eV]
E_B – Binding energy of the electron in the emitter atom [eV]
E_{kin} – Kinetic energy of the electron after emission [eV]
Φ – Correction term for the difference in work function between sample and detector [eV]

After the primary processes, a sequence of **secondary processes** fills the hole in the inner shell until all the hole resides in the valence shell. They are called secondary processes because they require a hole in an inner shell. Initially, the primary hole may be filled from one of the occupied higher shells. The energy difference between the two electronic states is emitted as radiation. The energy of this radiation is typically within the range of X-rays; thus, the process is called the X-ray emission (Figure 8.18a). Analytical methods using this effect are X-ray fluorescence analysis (XFA, generation of the primary hole by X-rays) and energy-dispersive X-ray analysis (EDX, generation of the primary hole by a focused electron beam). The X-ray emission is in competition with Auger emission[90] (Figure 8.18b), in which the energy difference between the primary hole $E_{B,A}$ and the initial binding energy $E_{B,B}$ of the electron that fills this hole leads to the emission of a second electron which had the binding energy $E_{B,C}$. The kinetic energy of Auger electrons does not depend on photon energy (or the electron energy of the electron beam used to generate the primary hole [Eq. (8.29)]. The spectroscopy based on this effect is called Auger electron spectroscopy (AES). For light elements, Auger electron emission is dominating, for heavy elements, X-ray emission is more effective.

[90] Pierre Victor Auger, 1899–1993, French physicist, professor at the Sorbonne, later General Director of the European Space Research Organization.

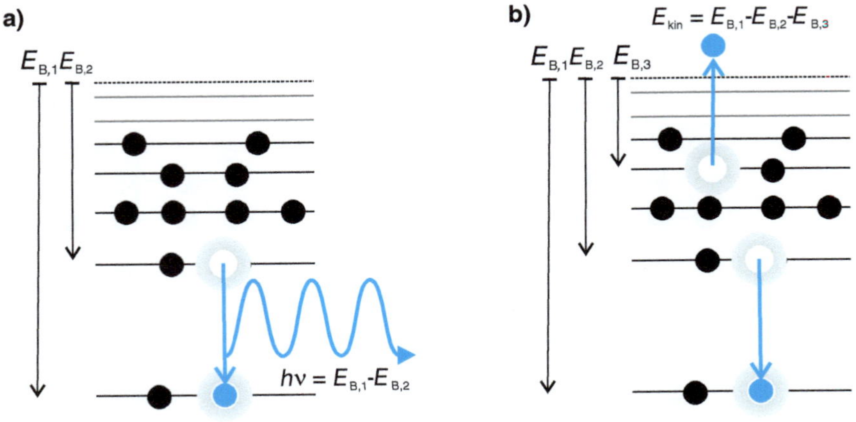

Figure 8.18 Orbital energy diagram (of one atom of the sample) for secondary processes filling a hole in the inner shells; a) X-ray emission and b) Auger electron emission. The states that are not occupied in the ground state are shown as gray lines.

Q: Can I imagine the Auger process as the emission of an X-ray photon that is immediately causing another photoemission process?

A: No, the Auger process is *one* single elementary process leading to the emission of a second electron of the *same* emitter atom without an intermediate occurrence of a photon. If a photon were emitted as an intermediate step, it would be much more likely that this photon would cause the emission of an electron from another, surrounding atom. The result would be two singly ionized atoms. However, the Auger process always leads to a doubly ionized atom because the electron emitted during the Auger process originates from the same atom in which the primary hole was located.

$$E_{kin} = E_{B,1} - E_{B,2} - E_{B,3} \tag{8.29}$$

$E_{B,1}$ – Binding energy of the primary hole [eV]
$E_{B,2}$ – Binding energy of the electron filling the primary hole [eV]
$E_{B,3}$ – Binding energy of the emitted electron [eV]

8.5.2 Electron Spectroscopy

The term "electron spectroscopy" summarizes XPS, AES, and ultraviolet photoelectron spectroscopy (UPS). These techniques are usually conducted under UHV conditions. Recent developments enable the work under near ambient gas pressures (using differential pumping), which now allows, for example, the characterization of gas electrodes (NAP XPS, near-ambient pressure XPS) [45, 46].

Surface Sensitivity. While X-rays can penetrate some micrometers through solids, electrons emitted in XPS, AES, and UPS can move in solids only a few nanometers without loss of energy by colliding with atoms in the sample material. The characteristic quantity for this phenomenon is the attenuation length λ, which typically measures 1–4 nm. Within a distance of λ, only a fraction of 1/e (ca. 37%) of the electrons has retained the original kinetic energy. The complementary fraction of 63% has transferred energy by collisions. After a distance of 2λ, only 13.5% and after 3λ, only 5% of the electrons move with their initial energy E_{kin} according to Eqs. (8.28) or (8.29). This means that almost all electrons that reach the detector without a loss of their initial energy have moved less than 3λ in the solid sample. Electrons that experience an energy loss contribute to a background signal (Figure 8.19a). The precise value of λ depends on the density of the studied material and shows a very strong dependence on E_{kin} of the electron (Figure 8.19b). There are approximate equations derived from experiments with well-defined samples for the estimation of λ in different types of materials [47].

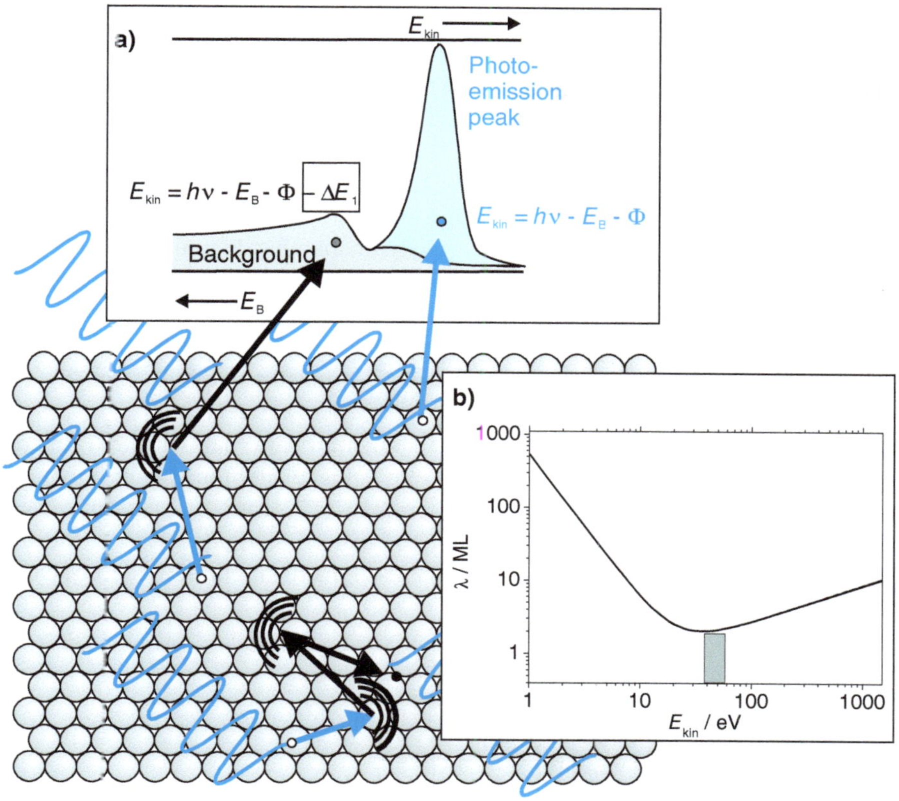

Figure 8.19 Surface sensitivity of XPS. Inset a) shows schematically a photoemission line with its background signal formed by electrons with lower E_{kin} and (thus apparently higher E_B). Those electrons have lost part of their initial kinetic energy. Inset b) shows the dependency of the attenuation length on E_{kin} (note the double logarithmic plot). The energy range with minimum attenuation length and therefore highest surface sensitivity is marked in gray at approximately $E_{kin} \approx 50$ eV.

Because of the short attenuation length, electron spectroscopies are surface-sensitive methods; this means that the information about the sample originates from a very thin layer near the surface of the sample. Just this property is of high interest for the investigation of electrode surfaces. The information depth Λ refers to the range of the sample from which 95% of the detected photoelectrons are emitted. It can be calculated as a function of the emission angle θ (taken relative to the surface normal) according to Eq. (8.30).

$$\Lambda = 3\lambda \cos\theta \qquad (8.30)$$

Sometimes, the already small information depth of XPS is still too large, especially when characterizing monolayers on surfaces that contain the same elements as in the bulk volume (e.g., thiol adsorbate on metal sulfides). Figure 8.20 shows how an increase in θ decreases the information depth. A further option for decreasing the information depth is the selection of a suitable emission line for which the emitted photoelectrons exhibit E_{kin} as close as possible to $E_{kin} \approx 50$ eV, that is, where λ shows a minimum in Figure 8.19b. The best option is the use of synchrotron radiation, a poly-chromatic electromagnetic radiation emitted by electron storage rings, from which a suitable photon energy can be selected by means of a monochromator. The photon energy is selected in such a way that the resulting photoelectrons have a kinetic energy according to Eq. (8.28) of $E_{kin} \approx 50$ eV. The information depth is then only in the range of 3–6 atomic layers. Application examples in the context of electrochemistry can be found in [48–51].

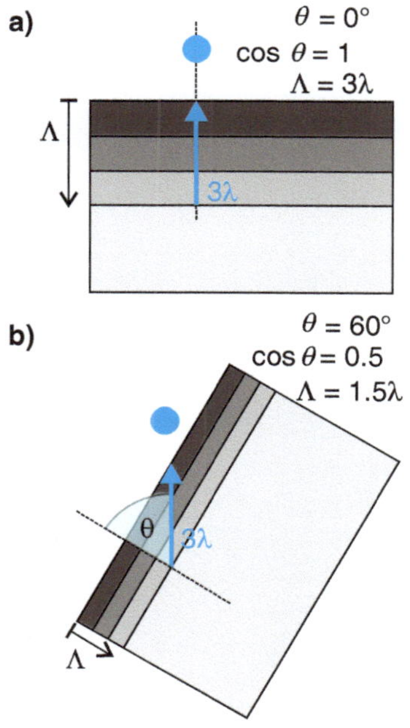

Figure 8.20 Diminution of the information depth by change in the emission angle (tilt of the sample with respect to the entry of the detector during the measurement).

Ultrahigh Vacuum. The small information depth is a blessing for the study of electrode surfaces because only the parts of the solid close to the actual interface contribute to the signal. The high surface sensitivity can also be a curse because every, even low levels of contamination of the surface leads to very strong signals, which can mask signals from the sample of interest. Even carefully prepared samples that must be transported through ambient air always show typical contamination by organic compounds that contain carbon (in different binding states) and oxygen. In order to restrict the influence of such contamination, the measurements are conducted in ultrahigh vacuum (UHV, $p \approx 10^{-9}$ mbar) because the surface would be very quickly covered by adsorbate layers under ambient pressure. The electron energy analyzer (Figure 8.21, parts 5–7) with the detector determine E_{kin} and the intensities of the electrons. It is mounted to the central vacuum chamber (analyzer chamber, Figure 8.21, part 1).

Because the techniques are often used together with one instrument, XPS and AES are also jointly referred to as ESCA (electron spectroscopy for chemical analysis).

Combination of Different Electron Spectroscopies. Modern instruments often provide several excitation sources (X-ray, UV for UPS, electron beam for AES). UPS is important for the investigation of semiconductors and for the determination of the Fermi energy [52]. UPS studies valence electrons for which a simple assignment (as in XPS) is not possible. AES allows a simple assignment to elements similar to XPS, but the determination of chemical states is less straightforward than with XPS. In addition, the excitation of AES by an electron beam tends to alter the chemical nature of the sample more easily than excitation by X-rays in XPS. However, an electron beam can be focused much better than X-rays and, therefore, AES is often used for the analysis of small regions of the sample (e.g., elements of microstructured electronic chips).

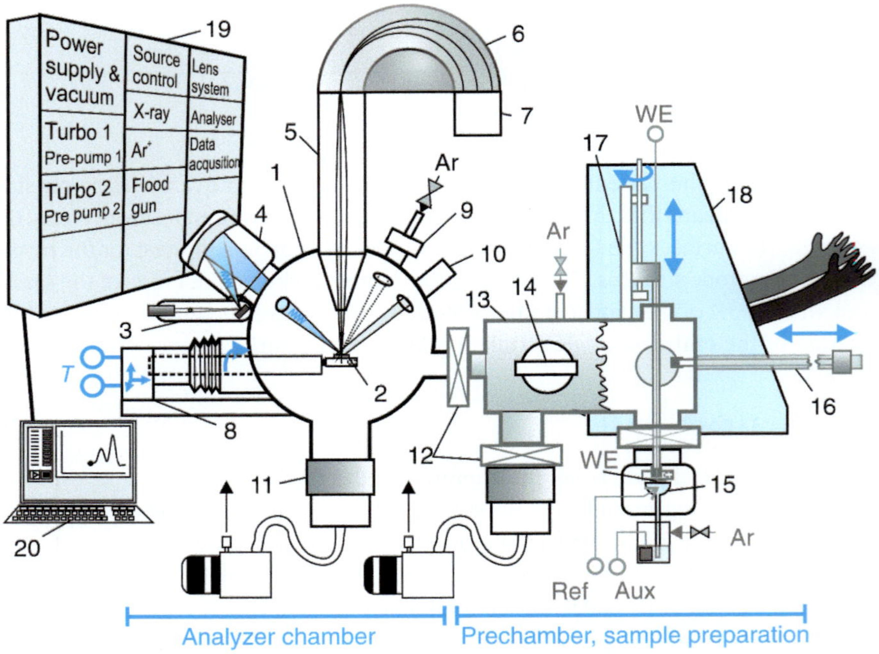

Figure 8.21 Illustration of an XP spectrometer for the investigation of electrodes. The scheme shows the coupling of the electrochemical cell to the system and the coupling of a glove box to the UHV system; 1) analyzer chamber; 2) sample on sample holder; 3) X-ray tube; 4) monochromator; 5) electron lens system; 6) hemispherical analyzer; 7) electron detector (channeltrons, channel plate or delay line detector); 8) sample manipulator with three space directions, rotation of the sample as well as cooling and heating of the sample; 9) Ar ion gun for sputtering of sample layers; 10) flood gun, a source of low energy electrons for charge compensation; 11) turbomolecular pump; 12) gate valve; 13) antechamber or prechamber with option for flooding by Ar (or N_2); 14) opening for introduction of the sample to the UHV system; 15) electrochemical three-electrode cell in a prechamber; 16) transfer rod for transporting samples from the electrochemical cell to the analyzer position; 17) sample manipulator (vertically) with separate electrical contacts; 18) glove box with electrochemical cell in inert environment (not drawn in all details); 19) control electronics; 20) control computer.

The measured intensity is often plotted as the first drivative in AES. For surface analysis with identification of chemical binding states, XPS is the most powerful among the methods. We therefore focus on the XPS method.

Q: Are all XP spectrometers that complicated?

A: The scheme describes different possibilities to attach an electrochemical cell to a UHV instrument. However, there will rarely be two of them at a single XP spectrometer because each of these construction types offers advantages for specific applications but also comes with its own limitations, which will feed back to the entire UHV instrument. In this respect, Figure 8.21 illustrates two basic options, but not a sensible realization of *one* particular instrument. Although some connections of the different components to a *system* are indicated in Figure 8.21, essential details of the vacuum apparatus, a large range of supply lines for power and gases, safety switches, and mechanical scaffolds have been omitted for clarity.

Radiation Sources for XPS. The most frequently used radiation source in the laboratory is an X-ray tube with an Al anode. The emitted X-ray contains a continuum called *bremsstrahlung* and overlaid characteristic X-ray emission lines ($K_{\alpha 3}$, $K_{\alpha 4}$, etc.). This radiation is directed toward a monochromator where the K_{α} line is selected and directed toward the sample [53]. The use of monochromatic X-rays in modern XP spectrometers increases the energy resolution of the spectra, it avoids the appearance of excitation satellites, reduces the intensity of the background, and can be combined with spatial focusing of the X-rays on a spot of 20 μm to 1.2 mm in diameter.

Bremsstrahlung, German, for "deceleration radiation", is a continuum radiation caused by the deceleration of the impinging electrons at the anode. The highest photon energy in the Bremsstrahlung can be calculated from the energy conservation, that is, if all the electric energy used to accelerate the electron in the X-ray tube ($e\,U$) is converted to a single photon with the energy $h\,\nu$. Using the relationship $c = \nu\,\lambda$ yields $\lambda_{min} = h\,c/(e\,U)$.

Modern XP spectrometers recalculate the measured E_{kin} to E_B using Eq. (8.28) and the known X-ray line used in the instrument. If the anode material of the X-ray tube is changed and thus the excitation energy is changed, the photoemission lines appear at the same E_B, while the Auger signals move along the E_B scale. (Stating E_B is not meaningful for Auger lines [Eq. (8.29)]. Although the Auger lines also contain useful information, they are more often an interfering signal.)

Because photoemission leaves behind a hole in a core shell, secondary processes are triggered, including the emission of Auger electrons. The Auger electrons are also captured by the electron analyzer and may overlap with photoemission lines, making their assignment more challenging. The spectrometers often contain a second X-ray source that can help distinguish experimentally Auger lines and photoemission lines. If the X-ray source is changed, $h\nu$ of the characteristic lines changes by a fixed value $h(\nu_1 - \nu_2)$. Thus, E_{kin} of all observed photoemission lines changes by the same amount, while E_{kin} of the Auger lines remains constant because it depends only on the energy of the three participating electron energy levels [Eq. (8.29)]. Another use of the second X-ray source is the *increase* in the information depth. This can be desirable if signals from bulk materials have to be obtained through a thick passive layer. An instructive example is the study of battery materials with their interfacial layers (Section 15.6.3, [54]). A larger excitation energy of the X-rays allows the observation of a larger variety of photoemission lines. Unfortunately, the X-ray lines with higher photon energy tend to exhibit a larger spectral linewidth (Table 8.2) causing a degradation of the spectral resolution in the resulting XP spectra. The spectral resolution of the XP lines is needed for chemical analysis. Those counteracting requirements can be met best by synchrotron radiation. With the inauguration of electron storage rings specifically dedicated to materials research (Table 8.4 in Section 8.5.3) and of setups for electrochemical experiments available at these facilities, the accessibility of synchrotron radiation sources has increased, and it can be predicted that their usage in various areas of electrochemistry will rapidly expand [45, 46, 48–51, 54].

Sample Transfer. From the perspective of electrochemistry, XPS is always an *ex situ* method, because the solid sample must be immersed in the electrolyte and transferred to UHV. During the transfer, the sample may suffer changes compared to its properties

Table 8.2 Survey about laboratory X-ray sources for XPS and UPS.

Line	$h\nu$/eV	Line width/eV	Comment
He I	21.22		UPS
He II	40.8		Second source for UPS
Mg K_α	1253.6	0.7	For differentiation between Auger and XPS lines
Al K_α	1486.6	0.85	Most frequently used excitation source
Zr L_α	2042.4	1.6	
Ag L_α	2984.3	2.6	For a larger energy range and increase in information depth
Ti K_α	4510.9	2.0	
Cr K_α	5417	2.1	

when immersed in an electrolyte solution. These unintended changes are collectively referred to as artifacts:

- Incompletely rinsed electrolyte solution may lead to crystallization of salts once the remaining solvent evaporates in the UHV.
- Molecular adsorbates or layers at the interface may evaporate or sublime from the surface in UHV. This applies, for instance, to physisorbed organic compounds but also to solids like elemental sulfur. The tendency can be counteracted by cooling the sample to the temperature of liquid nitrogen. During cooling, the freezing of moisture from the air must be avoided.
- After removal of the sample from the electrolyte solution, trace components of the ambient air may adsorb to the electrode surface. Particularly delicate in this respect is the transfer of the sample through the liquid–air interface. Hydrophobic organic compounds are collected at this interface because they do not dissolve in the aqueous phase. This contamination layer can be transferred to the sample surface during removal, analogously to the Langmuir–Blodgett transfer[91,92] discussed in Section 13.1.7. The effect can be mitigated by strictly controlled purity of solutions with respect to hydrophobic organic contaminants and by aspirating the surface layer of the liquid reservoir as shown in Figure 13.25.
- Oxygen from the atmosphere can react during the transfer with metals from the sample. As a consequence, a subsequent surface analysis would find an oxide layer (instead of the anticipated metallic surface).

There are technical means to facilitate an artifact-free sample transfer. They offer advantages for certain application areas. Option A is enabled by a prechamber (antechamber) of the UHV system, which allows the insertion of a small electrochemical cell vial after establishing an Ar (N_2) atmosphere under normal pressure. The sample is mounted on a sample holder and is immersed for a short period of time to conduct an electrochemical experiment (Figure 8.21, part 15). After the electrochemical experiment, for example, recording a cyclic voltammogram, the cell vial is lowered, separated from the UHV system, and the prechamber is pumped

[91] Irving Langmuir, 1881–1957, US American physical chemist, PhD in Göttingen, worked in the laboratories of General Electric Company, Nobel Prize in Chemistry 1932; Langmuir probe, Langmuir–Blodgett trough and the Langmuir–Blodget films made using this device as well as Langmuir–Schaefer transfer are named after him (Sections 13.1.6–13.1.7).
[92] Katherine B. Blodgett, 1898–1979, US American physicist, first woman who earned a doctoral degree at the University of Cambridge, worked with Langmuir at General Electric Company.

down to UHV. Subsequently, the sample is transferred to the analyzer chamber and characterized there. The separation from the laboratory atmosphere is close to perfect in this system, but the effort of changing the electrolyte in the little vial may be considerable [55].

If switching between different electrolytes is intended, the electrochemical experiments can be conducted in a glass apparatus that is connected via a flange and a valve to the UHV system (option B). The pretreated sample is first characterized by UHV-based methods, then brought to the Ar- or N_2-filled glass apparatus by linear transfer mechanisms. In the glass apparatus, different electrolyte solutions can be pneumatically transferred into the small vial of the electrochemical cell. The sample is set on the surface of the liquid in this cell (hanging meniscus configuration), and the electrochemical experiment is executed. Afterward, the sample is lifted, rinsed, blown dry and transferred back to the UHV system for characterization. Any spilling of liquids in this cell will necessitate cleaning of the glass apparatus. During this operation, the UHV part is not compromised in any way. However, shielding of the sample against the environment is less perfect than in option A [56].

Even more flexible is the work with a glove box (option C, Figure 8.21, part 18) directly connected to the entry air-lock system of the UHV system. Such systems allow the extraction of highly reactive electrodes from test batteries and their rinsing and mounting to sample holders without access to air or moisture.

Designation of Photoemission Lines. One emission line is observed for each electronic core level of each element contained in the sample within the information depth, provided the excitation energy is high enough to cause a photoemission from this particular energy level. The energy levels are characterized by the principal quantum number $n = \{1, 2, 3, 4\}$ and the orbital angular momentum quantum number $l = \{0, \ldots, n-1\}$. The particular values of l are encoded similarly to the designation of atomic orbitals: s stands for $l = 0$, p for $l = 1$, d for $l = 2$, and f for $l = 3$. Emission line for p-, d-, and f-electrons are split into doublets (spin–orbit coupling) according to the total angular momentum quantum number j, which is obtained as the *absolute value* of the sum of l and the two possible values $+½$ and $-½$ of the spin quantum number m_s. Thus, j can assume only positive values.

The notation of an emission line such as S $2p_{3/2}$ contains the elemental symbol of the emitter atom, defining the atomic number ("S" in this example) followed by a blank space and the principal quantum number n ("2" in the example specifying the L shell). The character "p" stands for the quantum number of the orbital angular momentum $l = 1$ and the subscript "3/2" indicates the total angular momentum number $j = 3/2 = l + ½$. The energetic difference (spin–orbit coupling) between the lines of *one* doublet is roughly constant for one energy level of the *particular* element, such as S 2p, in different compounds. There is a constant intensity ratio between the two components of a doublet (Table 8.3) that follows from the ratio of the degeneracy $g = 2j + 1$ of the involved energy level (there are "exceptions"). Whether the two signals of a doublet appear as separate signals depends on the particular value of the spin–orbit coupling and the energy resolution of the spectrometer, which is influenced by the pass energy used during recording of the spectrum (Figure 8.22).

Designation of Auger Lines. The three-character code is formed from the designations of the principal shells (K, L, M, ...) of the three involved electronic energy levels from Figure 8.18, for example, Cu LMM. Roman numbers are used as subscripts to indicate the electronic sublevels.

Qualitative Analysis. The experimental spectrum represents a superposition of all photoemission lines, of all Auger lines, and the background on the side of lower E_{kin} of each line, according to Figure 8.19 for all elements contained within the information

Table 8.3 Intensity ratio within doublets. Intensity is calculated from the peak *area*.

Line	$g_{j+1} : g_j$	$I_{j+1} : I_j$
$I(p_{3/2})/I(p_{1/2})$	4 : 2	2 : 1
$I(d_{5/2})/I(d_{3/2})$	6 : 4	3 : 2
$I(f_{7/2})/I(f_{5/2})$	8 : 6	4 : 3

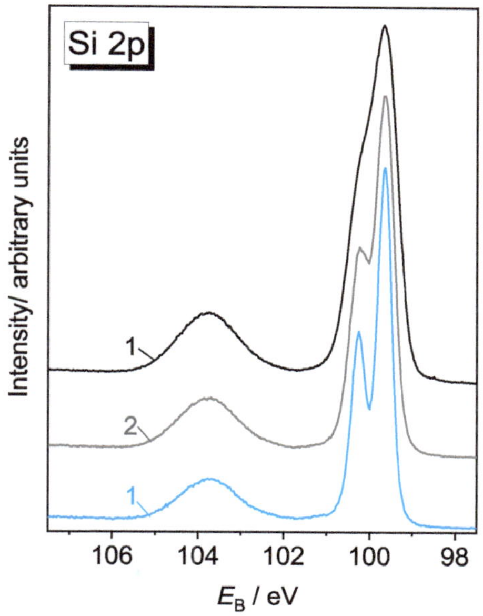

Figure 8.22 Si 2p spectra of a Si wafer with native oxide layer, recorded with different pass energy 1) 5 eV; 2) 18 eV; 3) 30 eV. Only for small pass energies in curve 1, the Si $2p_{3/2}$ and Si $2p_{1/2}$ components appear as distinct signals. At low resolution in curve 3 (but with much higher sensitivity!), only one slightly asymmetric ("skewed") Si 2p signal is recorded. The intensities of each spectrum were normalized to fit in one plot. Source: Data were provided by C. Dosche, Carl von Ossietzky University of Oldenburg.

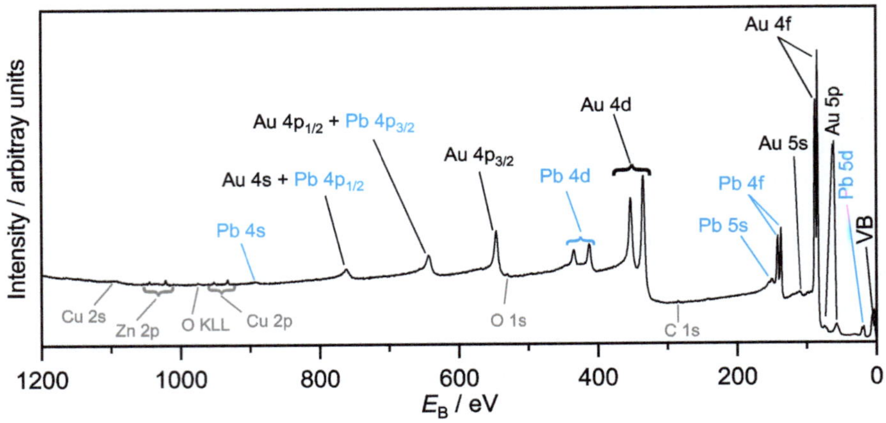

Figure 8.23 Survey spectrum of an Au electrode after underpotential deposition (UPD) of Pb. The lines are annotated in the spectrum. Besides the lines from the Au substrate and the UPD layer, there are small signals for O and C (typically contamination from air) and Cu and Zn (probably contamination from the Pb salt used). Source: Data were provided by C. Dosche, Carl von Ossietzky University of Oldenburg.

depth of the sample. Qualitative analysis starts with the identification of all lines in the survey spectrum (Figure 8.23).

The Chemical Shift. After the assignment of each line in the survey spectrum, the spectral ranges in which signals of interest appear are recorded again with higher resolution (achieved by setting a smaller pass energy), and the resulting lower sensitivity is compensated for by longer acquisition time. An example is shown in Figure 8.22. Chemical identification can be made by comparison of the exact binding energy with reference spectra. Such reference spectra can be taken from the scientific literature, from databases [57, 58] or, preferentially, from in-house measurements of pure compounds with a similar binding energy as the emitter atom in the sample. This comparison is very informative for electrodes because metals contained in the bulk of an electrode material often form a thin oxide layer (corrosion), form special interfacial phases of layers in contact with the electrolyte (Li-ion batteries) or change the effective electron density at the surface by underpotential deposition, adsorption, or alloying. The change in E_B from the binding energy $E_B{}^\circ$ of the pure element is called a chemical shift. The interpretation of chemical shifts must consider several possible reasons for this phenomenon. The experimentally determined binding energy is the energy difference between a charged final state of the emitter atom with $(n-1)$ electrons and the neutral initial state of the emitter atom.

$E_f(n-1)$ – Energy of the photoionized atom [eV]
$E_i(n)$ – Energy of the initial state [eV]

$$E_B = E_f(n-1) - E_i(n) \tag{8.31}$$

According to Koopmans' theorem[93] the experimentally determined value of E_B corresponds to the calculated negative orbital energy ε of the emitted photoelectron ($E_B \approx -\varepsilon$). In this notion, E_B is determined exclusively by the initial state. This approximation equates the energy difference between multielectron systems with n and $n-1$ electrons with a one-electron energy that can be calculated with quantum chemical methods for the wave function of an atom. The exclusive focus on initial state effects neglects the relaxation energy E_r (the other electrons of the emitters and the neighboring atoms' response to the emission process), the correlation energy $d\varepsilon_{cor}$, and relativistic effects $d\varepsilon_{rel}$. A more comprehensive balance is given in Eq. (8.32)

ε – Calculated orbital energy of the emitted electron [eV]
E_r – Relaxation energy [eV]
$\delta\varepsilon_{cor}$ – Contribution of the correlation energy [eV]
$\delta\varepsilon_{rel}$ – Contribution of relativistic effects [eV]

$$E_B = -\varepsilon - E_r - \delta\varepsilon_{cor} - \delta\varepsilon_{rel} \tag{8.32}$$

Surprisingly, Koopmans' theorem provides chemically plausible interpretations in many cases because the contributions of E_r and $\delta\varepsilon_{cor}$ compensate each other, and the contribution of E_r is often a constant for one element. In such a case, the chemical shift

[93] Tjalling C. Koopmans, 1910–1985, US American economist born in the Netherlands. He took on the topic of the theorem during his studies in Utrecht in 1934. He has been working in the USA since 1940 on different subjects and received the Nobel Prize in economics in 1975.

$\Delta E_B = E_B - E_B^\circ$ of a photoemission line versus the emission line of the pure element E_B° has a direct correlation to the effective charge of the emitter atom q_i (and thus to its effective oxidation number) as well as to the charge of the neighboring atoms q_j with their mutual distance r_{ij}.

$$E_B - E_B^\circ = kq_i + \sum_{j \neq i} \frac{q_j}{r_{ij}} \tag{8.33}$$

k – Proportionality constant
q_i – Effective charge number of the emitter atom
q_j – Effective charge of the neighboring atoms
r_{ij} – Distance between emitter and neighboring atoms

This proportionality gives excellent agreement, especially for nonmetals (C, N, O, Si, P, S) with a large range of oxidation states as well as for polymers. This is one of the reasons for the popularity of the method. Because the chemical shift is not very large compared to the line width, signals from different binding states partially overlap. Numerical peak shape analysis ("peak fitting") is broadly used to identify the spectral components in such spectra. This requires mathematical models for the line shape, for the backgrounds and asymmetry (especially for peaks from metallic materials), etc. [59, 60].

Final State Effects. The phenomenon of shake-up satellites is observed, if a certain probability exists, that the absorption of an X-ray photon causes the emission of an electron and concomitantly the excitation of a π-electron from the valence band to a band with antibonding π* character (Figure 8.24). In this case, E_{kin} of the photoelectron will be reduced compared to the simple photoemission process and the emitted photoelectron appears at (apparently) higher E_B. In an electrochemical context, the effect is observed in XP spectra of conducting polymers and of adsorbates containing aromatic moieties.

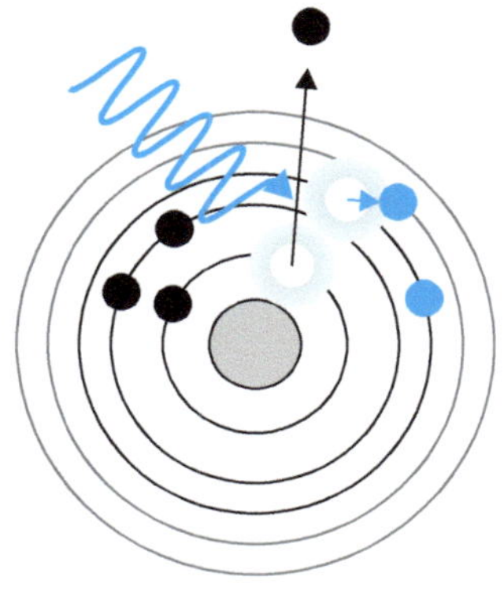

Figure 8.24 Principle for the emergence of *shake-up* satellites. Valence electrons are marked in blue.

Shake-off satellites are observed when the photoemission of a core electron occurs concomitantly with the emission of a second valence electron (Figure 8.25). The photoelectron from the core level has a reduced E_{kin}, showing up at apparently higher E_B in the spectrum.

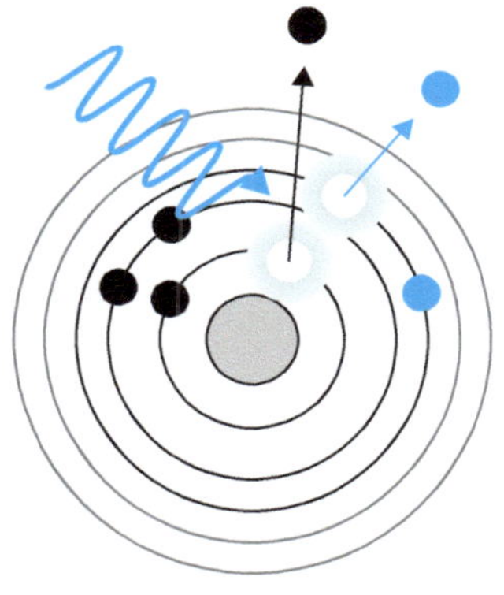

Figure 8.25 Principle of the emergence of a shake-off satellite. Valence electrons are marked in blue.

Photoemissions from emitter atoms in a metallic state often cause asymmetric peaks that are not the consequence of the overlap of different binding states but a result of the concomitant excitation of electrons from the valence band of the emitter to the conduction band (Figure 8.26). Because metals show a continuum of states around the Fermi level E_F (Section 10.1.3), there are many possible values of energy losses for this process.

Figure 8.26 Band diagram of a metal and schematic depiction of the emergence of asymmetric peak shapes from metallic emitter atoms. Valence band is blue, initially unoccupied states in light gray, occupied core levels in black.

Transition metals also show other effects of the final state on E_{kin} and thus on the determined E_B. There can be a significant difference in extra-atomic E_r between metals and ionic compounds of the same metal because the valence electrons in metals are very mobile and respond differently to the primary hole caused by the photoemission process than the holes in an oxide. As an observable result, there can be a strong deviation from Eq. (8.33). For instance, E_B of Cu $2p_{3/2}$ is 932.5 eV for both, metallic copper and Cu_2O. For elemental silver, the Ag 3d lines appear at a higher E_B than that of Ag(I)! If a differentiation is not possible by XPS data, binding states can often be distinguished by the exact E_{kin} of the Auger electrons [61].

A further phenomenon, multiplet splitting, is observed for compounds containing transition metal ions with unpaired d-electrons. Instead of the expected doublet, a larger number of lines results from the photoemission from one and the same energy level. Figure 8.27a shows a Fe $2p_{3/2}$ XP spectrum of poly(vinyl ferrocene). Ferrocene is a low-spin complex with two spin-paired electrons in the three t_{2g} orbitals. A "normal" photoemission doublet results from this electronic structure. After electrochemical oxidation, there is one unpaired electron in the t_{2g} orbitals. The energy of the final states is slightly different depending on the interaction between the electron in the t_{2g} orbital and the spin of the remaining electron in the core level, from which the $2p_{3/2}$ photoelectron was emitted. The calculated intensity distribution for the multiplet is shown as line

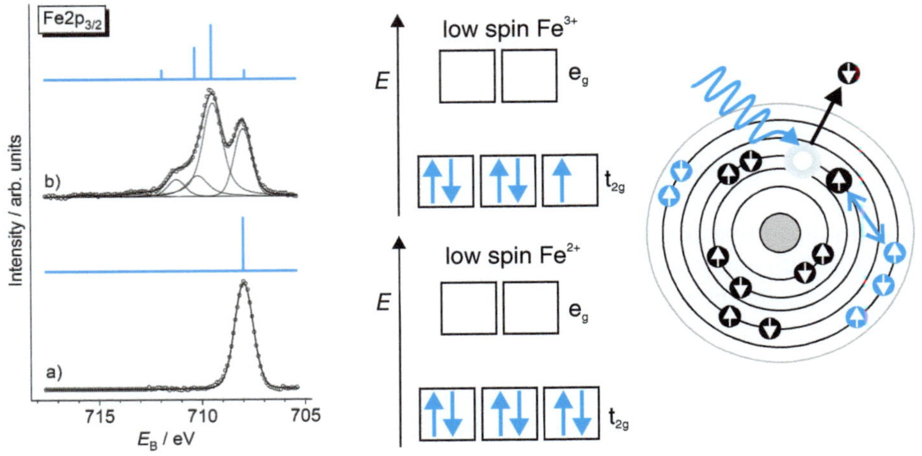

Figure 8.27 Emergence of multiplet splitting for compound of transition metals; a) Fe $2p_{3/2}$ spectrum of poly(vinyl ferrocene); b) Fe $2p_{3/2}$ of electrochemically oxidized poly(vinyl ferrocene). The line spectra in blue are the theoretically expected spectra for comparison. The unpaired electron in b) causes a splitting of the emission line to a multiplet. The background has already been subtracted for this plot. Source: Reproduction of a) and b) with permission from [62], Copyright 2001, Royal Society of Chemistry.

spectra above the experimental spectrum in Figure 8.27b. The significantly increased signal at 708 eV shows clearly that oxidized poly(vinyl ferrocene) (with Fe^{III}) contains significant amounts of nontransformed ferrocene moieties (with Fe^{II}).

Electrons within the conduction bands of electronically conductive materials can be excited to collective oscillations called plasmons. The excitation of plasmons can also be caused by photoelectrons, which results in a series of broad signals on the low kinetic energy side of the XP spectrum. These signals can make the determination of a suitable background difficult and thus complicate quantification. A prominent example is the Al 2s spectrum, which contains the singlet for metallic Al at 119 eV and, in addition, five plasmon loss peaks between 130 and 200 eV on the E_B scale (Figure 8.28). The example also makes clear that plasmon signals can significantly compromise the accuracy of compositional analysis by XPS. There are two plasmon signals at 104 and 121 eV that are excited together with the Al 2p photoemission line at 74 eV (outside of the energy range shown). They appear as additional spectral components or make the establishment of a reasonable background very difficult.

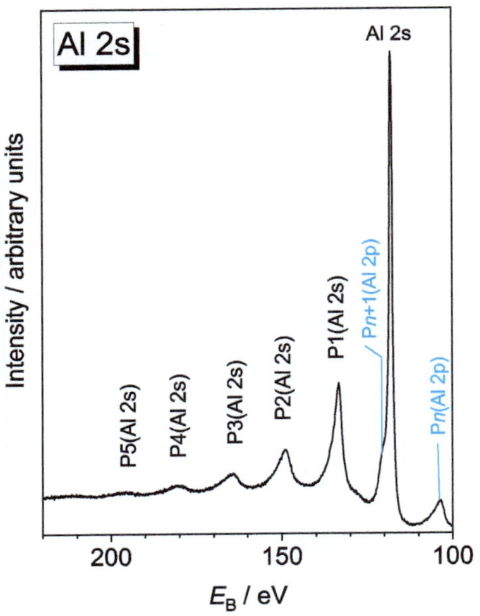

Figure 8.28 Al 2s spectrum of a technical aluminum sample with loss peaks due to the excitation of plasmons together with the photoemission on the side of lower E_{kin} (and apparently higher E_B). Source: Data were provided by C. Dosche, Carl von Ossietzky University of Oldenburg.

Charge Compensation. By emission of electrons, the sample charges up positively. Consequently, more energy is needed for the subsequent photoelectrons to overcome the electrostatic attraction by the positively charged sample in addition to the atomic binding energy. This is not a serious problem for conductive samples because the emitted electrons can be replenished via the electrical contact of the sample to the instrument and eventually to the ground. This is not so easy for insulating samples. In those cases, various charge compensation methods are used. The sample can be exposed to a "shower" of low-energy electrons from a technical electron source (flood gun). In order to compensate for the negative charge of the space around the sample, one often doses concomitantly Ar^+ ions. Modern instruments have achieved a high level of reliability for these procedures, even allowing automatic operation.

In each spectrum, one clearly identifiable line with known binding energy has to be selected as a reference signal. If there is no calibration line like Au $4f_{7/2}$ or Ag $3d_{5/2}$, the C 1s signal from carbon contamination is often employed. This line has a binding energy of 284.8 eV (stated as 285.0 eV in older literature). The position of all other peaks is then measured relative to the reference line, which allows for compensation of smaller charging effects because a uniform charging of a sample shifts all peaks by the same amount.

A feared phenomenon is differential charging, for instance, in composite materials or in thicker oxide layers on metals during corrosion studies. In such samples, the charging as well as the charge compensation by low-energy electrons proceeds differently in different parts of the sample. The measurement of a reference line does not solve the problem (e.g., signals from oxide layer are shifted differently than those of bulk metal).

Quantitative Analysis. The peak area under the XP signal is proportional to the amount of emitter atoms c_i contributing to the signal. Because there is a strong dependence on the depth at which the emitter atoms are located, a concentration determination by XPS must consider further aspects.

The data evaluation starts with the subtraction of the background signals caused by electrons that have lost part of their kinetic energy due to inelastic collisions. As an illustration, background and photoemission signals are differentiated by color in Figure 8.19. Subtraction of the background is typically made with a dedicated computer program that may offer different mathematical background models such as linear, Shirley[94] or Tougard[95] backgrounds. The selection of a sensible background model is of decisive importance for correct peak fitting and quantification. Relevant information must be taken from specialized literature [63]. Partially overlapping peaks are analyzed by peak fitting yielding the intensity of each spectral component. In such a peak fitting procedure, one aims to describe the measured signal by a background and a minimum number of spectral components. The parameters of each component are varied systematically until the computed spectrum agrees with the experimental one. The exact intensity I_{ij} of peak j of the elements i depends on several factors [60, 64]. The most important influences are described by Eq. (8.34).

$$I_{ij} = k \cdot T(E_{\text{kin}}) \cdot L_{ij}(\gamma) \cdot \sigma_{ij}(h\nu) \int_{z=0}^{\infty} c_i(z) \exp\left[\frac{-z}{\lambda(E_{\text{kin}})\cos\theta}\right] dz \tag{8.34}$$

I_{ij} – Background-corrected intensity (peak area) of the line j of the element i [cps] = [counts per second]
k – Summary of further instrumental parameters, which may depend on the settings during recording of the spectrum
$T(E_{\text{kin}})$ – Transmission function [dimensionless]
L_{ij} – Angle-dependent asymmetry [dimensionless]
γ – Angle [degree]
$\sigma_{ij}(h\nu)$ – Atomic excitation cross section for photons $h\nu$ with the orbital j of element i
$c_i(z)$ – Concentration profile of the element i normal to the surface [mol cm^{-3}]
z – Vertical coordinate [nm]
$\lambda(E_{\text{kin}})$ – Mean free path length [nm]
θ – Emission angle relative to the surface normal [degree]
σ'_{ij} – instrument-dependent sensitivity factor

Here, k is a constant that summarizes the effects of the spectrometer such as excitation intensity and acceptance angles of the analyzer, $T(E_{\text{kin}})$ is the transmission function of the analyzer that changes with E_{kin}, L_{ij} is an angle-dependent asymmetry factor for the orbital j of the atom i. The photoionization cross section σ_{ij} specifies the probability of a photoemission from the orbital j of the element i if a photon with the energy $h\nu$ reaches the atom. Values calculated by Scofield are often used for this atomic property [65]. In order to compensate for many other effects that are not known with sufficient precision, one often aims for the determination of concentration *ratios* or mole fractions, rather than the absolute concentration. Some of the lesser-known effects cancel out depending on the structural model. Particularly straightforward is the analysis of amorphous solids, which are considered a homogeneous mixture of several elements (or binding states). This means c_i is not a function of z, which simplifies Eq. (8.34) to Eq. (8.36). For a given excitation energy, for example, Al K$_\alpha$, and a given spectrometer, the factors can be determined for each energy level of each element. They can then be summarized in an instrument-dependent sensitivity factor σ'_{ij} for the excitation energy used [Eq. (8.35)].

$$\sigma'_{j} = \sigma_{ij} L_{ij} \lambda(E_{\text{kin}}) T(E_{\text{kin}}) \tag{8.35}$$

The values of σ'_{ij} are often determined by the instrument manufacturer (with varying precision) and stored in a database that is used by the control and data evaluation software of the XPS instrument [53, 58].

94 D. A. Shirley; *Phys. Rev. B* **1972**, *5*, 4709–4714.
95 S. Tougaard; *Surf. Interface Anal.* **1998**, *11*, 453–472.

Q: Why is there demand for an amorphous solid in the description above?

A: Let us consider a crystalline solid of two elements A and B that are regularly arranged in the crystal lattice. This causes diffraction of the emitted photoelectrons, which changes the angular distribution of photoelectron intensity. For sufficiently large E_{kin}, the intensity will be enhanced along the connecting lines of the lattice components A and B (forward scattering). This effect can cause intensity variations of up to 10%! The effect can be mitigated instrumentally by selecting a lens system with a particularly large acceptance angle for the photoelectrons (averaging over a large angle range). Most modern instruments provide options for large acceptance angles in order to facilitate quantitative determination with a high signal-to-noise ratio and another setting with a small acceptance angle for angle-resolved measurements (see below).

$$\int_0^{\infty} \exp[-az]dz = \frac{1}{a}$$

$$\begin{aligned} I_{ij} &= k \cdot T(E_{kin}) \cdot L_{ij}(\gamma) \cdot \sigma_{ij}(h\nu) \cdot \lambda(E_{kin}) \cdot \cos \cdot c_i \\ &= k \cdot \cos\theta \cdot \sigma'_{ij} \cdot c_i \end{aligned} \tag{8.36}$$

For the same spectrometer settings ($\theta = const$ and $k = const$), the spectra of different elements yield the concentration ratio of two elements A and B, or two binding states of the same element

$$\frac{c_A}{c_B} = \frac{I_{Aj}/\sigma'_{Aj}}{I_{Bk}/\sigma'_{Bk}} \tag{8.37}$$

The mole fraction x_A of the element A [atom-%] out of an arbitrary number Z of elements is then given by Eq. (8.38) when selecting one photoemission line j for each element.

$$x_A = \frac{c_A}{\sum_Z c_Z} = \frac{I_{Aj}/\sigma'_{Aj}}{\sum_Z I_{Zk}/\sigma'_{Zk}} \cdot 100\% \tag{8.38}$$

The colloquial term "atomic percent" is often used for mole fraction. The underlying simple assumption about the structure of a homogeneous mixture of several elements or binding states applies to samples in electrochemistry only in exceptional cases, because we rather expect for electrodes as reactive interfaces the formation of passive layers, adsorbate layers, local nucleation phenomena, the occurrence of nanoparticles as electrocatalysts, and similar complications (Figure 8.29). In all of these cases, the use of Eq. (8.38) will give numerical values, but those values have *no* physical meaning! This remains true even if mole fractions x_A are automatically generated by the instrument control software. Instead, it is necessary to use more advanced structural models and more complicated calculations [60, 64].

Nondestructive Depth Profiling using angle-resolved photoelectron spectroscopy (ARXPS, angle resolved XPS) is a preferred method for the situation depicted in Figure 8.29b that occurs frequently in electrochemistry. The aim is the determination of the thickness d of the overlayer. There are several ways to accomplish this task. In any case, one measures the intensity I_{Sj} (index "S" for substrate) of one element (more precisely of one binding state of an element) occurring only in the substrate using a lens mode of the XP instrument *with small acceptance angle*. In addition, the same signal I°_{Sj} is recorded with the same instrumental settings for a sample without an overlayer. Such a reference sample can be obtained by removing an overlayer by sputtering with Ar^+ ions in the vacuum of the XPS instrument or by not preparing the overlayer. Due to collisions in the overlayer, the signal of the substrate declines exponentially, when photoelectrons pass through the overlayer.

$$I_{Sj} = I^{\circ}_{Sj} \exp\left[\frac{-d}{\lambda(E_{kin})\cos\theta}\right] \tag{8.39}$$

Rearranging yields directly $d = \lambda(E_{kin})\cos\theta \ln(I^{\circ}_{Sj}/I_{Sj})$. The precision can be increased by using several angles for this analysis. Especially for oxide layers of Si and metals with thin compact oxide layers (Al, Ti, Nb, Ta, Zr, …), it is advantageous to also record signals from the overlayer I_{Oj} (index "O" for overlayer) at different angles θ but with otherwise identical instrumental settings. Reference measurements of pure substrates and bulk material of the overlayer provide I°_{Sj} and I°_{Oj}. The intensity data should obey Eqs. (8.40) and (8.41).

$$\frac{I_{Oj}}{I^{\circ}_{Oj}} = 1 - \exp\left[\frac{-d}{\lambda_{O}(E_{kin,Oj})\cos\theta}\right] \tag{8.40}$$

$$\frac{I_{Sj}}{I^{\circ}_{Sj}} = \exp\left[\frac{-d}{\lambda_{O}(E_{kin,Sj})\cos\theta}\right] \tag{8.41}$$

When using photoemission for the same element and the same orbital (e.g., Si 2p from Si and SiO_2), the mean free path length of Si 2p photoelectrons in the overlayer is equal regardless of whether they originate from the substrate or from the overlayer itself [$\lambda_O(E_{kin,O}) \approx \lambda_O(E_{kin,S}$] because their kinetic energies deviate by only a few eV, a negligible quantity for this consideration. The thickness d of the overlayer is then obtained from the slope of the linearization in Eq. (8.42).

$$\ln\left[\frac{I_{Oj}I^{\circ}_{Sj}}{I^{\circ}_{Oj}I_{Sj}} + 1\right] = \frac{-d}{\lambda_{O}(E_{kin,Oj})} \cdot \frac{1}{\cos\theta} \tag{8.42}$$

Even more complicated layer architectures can be treated by assuming a model of stratified layers, calculating the intensities using the concentrations, and calculating the expected intensity attenuation by collisions in the overlayers. The model is iteratively refined until the calculated intensities agree with the measured intensities. As for any other fitting routine, such a procedure may provide several numerically equivalent solutions or show large uncertainties if the experimental data are associated with large uncertainties due to noise in the data, problematic peak fitting to extract intensities from overlapping signal components, or intensity variations for crystalline samples due to forward scattering [66, 67].

Depth Profiles by Sputtering. Angle-resolved XPS (ARXPS) cannot be applied when the thickness of the overlayer is larger than the information depth of XPS. In such cases, the overlayer must be removed in a well-defined process, which is interrupted periodically by recording XP spectra of the sample. The removal of material is often accomplished by a beam of accelerated Ar^+ ions. The time used for sputtering can be correlated to sputter depth using calibration samples of well-defined thickness and composition.

Despite the conceptual simplicity of this procedure, one should remember that there are practical limitations. The impact of Ar^+ ions on the sample may change the chemical state of the bombarded sample. An analysis of the binding state of molecular solids is not sensible after sputtering. Ar^+ ions are implanted into the sample material, which can be recognized by the appearance of an Ar 2p signal after sputtering. The samples are not sputtered completely homogeneously. Instead, the samples are roughened, which can be mitigated to some extent by rotating the sample during sputtering. The removal of material occurs at different rates for different materials (faster for light elements and slower for heavy elements). Material that has been removed from the sample may redeposit on the sample.

The chemical modification by Ar^+ bombardment ruins the chemical selectivity, one of the main advantages of XPS compared to other surface-sensitive techniques. Therefore, intensive efforts are made to improve the sputtering methods. One possibility is to sputter with clusters of 1000–2000 Ar atoms that are all together singly oxidized and

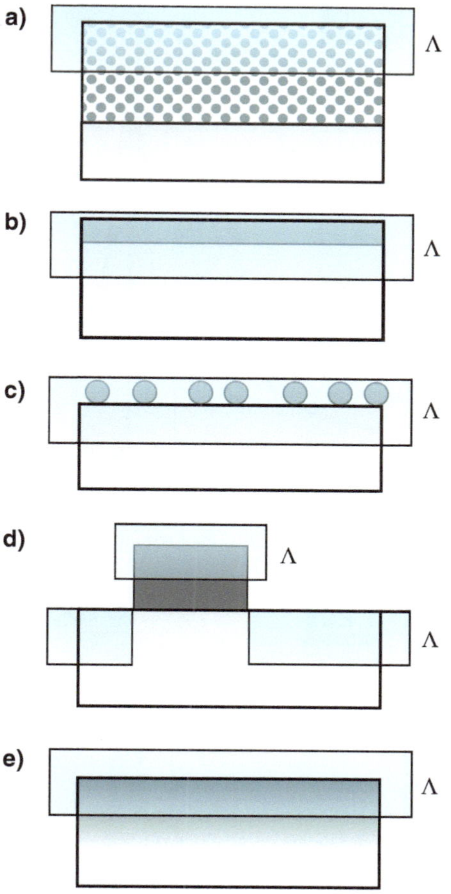

Figure 8.29 Structural models for the spatial distribution of two elements close to an interface with examples from electrochemistry (in brackets). The information depth of an XPS measurement is indicated by the blue transparent band. Elements contained in this region will contribute to the spectrum.

a) homogeneous mixture (polymer layer on an electrode that is thicker than the information depth of XPS);
b) cover layer of constant thickness that is thinner than the information depth of XPS (native passive oxide layer on Si and on metals with a thin and compact oxide layer such as Al, Ti, Ta, Nb, Zr, …);
c) incomplete cover layer of particles that have a smaller diameter than the information depth of XPS (patterned self-assembled monolayers or supported catalyst nanoparticles;
d) incomplete cover layer that is thicker than the information depth of XPS;
e) Concentration gradient close to the surface (some de-alloyed metals, partially discharged battery electrodes).

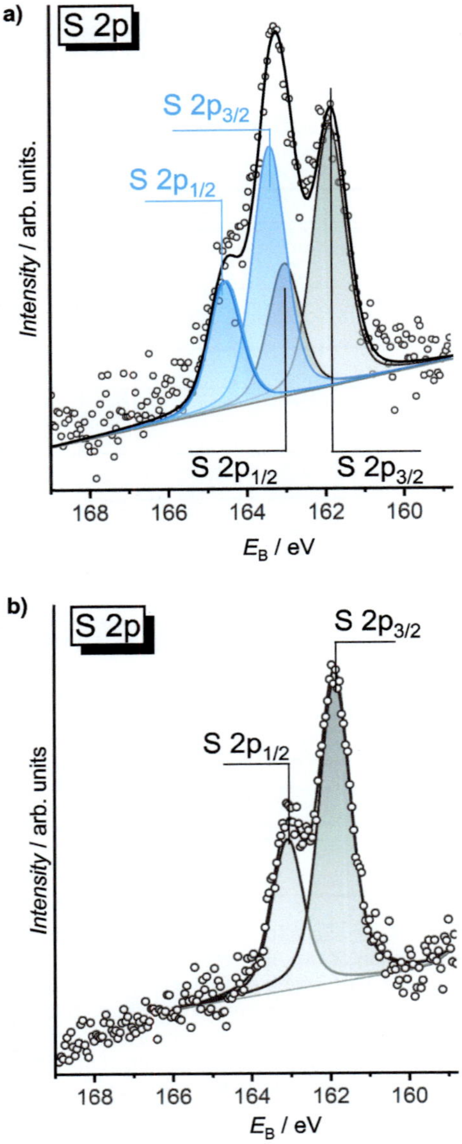

Figure 8.30 S 2p signal of an alkanethiolate monolayer on Au. a) Accumulation of the signal in one region of the sample: There are *two* S $2p_{3/2}$ signals at 161.9 eV and 163.5 eV. b) The accumulations were made on sample regions that had not been irradiated before. There is only *one* S $2p_{3/2}$ signal at 161.9 eV, which proves that the species causing the signal at 163.5 eV (blue in a)) was formed during the measurement. Source: Data were supplied by T. Sieling and A. Markovic, Carl von Ossietzky University of Oldenburg.

then accelerated. If such an Ar^+_{1000} cluster (or perhaps "drop") impacts on the sample, its kinetic energy is distributed among many Ar atoms, causing a gentle removal of organic contamination layers as well as of layers damaged by preceding Ar^+ sputtering without altering significantly the chemical identity of the sample. In return, the sputter rate is very low. Switching between the sputter methods is a suitable approach to combine fast removal of overlayers and the retention of the original chemical state of the sample.

Radiation Damage. Although we have introduced XPS as a destruction-free method for surface analysis, there are situations in which the sample changes during the measurement. Those phenomena are summarized under the term radiation damage. Such changes of the sample are most often caused by low-energy electrons, which have lost so much of their original kinetic energies through inelastic collisions in the sample that their remaining E_{kin} falls into the range that is particularly suitable for chemical modifications, especially for reductions. In addition, the photoionization can cause breaking of covalent bonds and subsequent loss of molecular fragments into the vacuum. Therefore, it is always a good idea to start an XPS investigation with a high-resolution scan of the line presumed to be the most sensitive toward radiation damage. After recording all other spectra, the initial energy range is measured again to look for possible alterations in the spectrum. If no changes in the spectrum can be found, the absence of radiation damage is demonstrated, unless the respective processes are so fast that they have already been *completed* when the first spectrum was recorded. If differences between the first and last spectrum are evident, the affected lines can be measured after different irradiation times. The changes often follow a first-order rate law ($I = I_0 e^{-kt}$). A plot of the logarithm of the intensity versus time yields a straight line. The intersection with the ordinate allows conclusions to be made about the original intensity ($\ln I = \ln I_0\ kt$).

Radiation damage can be limited by optimized measurement protocols. In some cases, this is traded against a lower-than-optimum spectral resolution or lower signal-to-noise ratio. In most cases, high-resolution XP spectra require data accumulation in one energy range. Because modern XP spectrometers focus the X-rays for excitation on the sample, the excitation spot can be moved across the sample. Provided the sample is macroscopic and homogeneous in the horizontal direction, the accumulations can be made at different spots of the sample, which means that each accumulation is made at a region of the sample that has not been hit previously by X-rays. The example in Figure 8.30a shows two S 2p doublets of a self-assembled alkanethiolate monolayer on Au. The spectrum in Figure 8.30a supports the hypothesis that S occurs in two chemically distinct states in such monolayers. However, if the S 2p signal is accumulated in different regions of the same homogeneous sample (Figure 8.30b), it becomes very clear that S is contained in the sample in only one binding state and the second doublet (blue in Figure 8.30a) must be the result of radiation-induced damage.

Further Reading. Specialized monographs about XPS as a method as well as edited books can be recommended for a more extensive coverage of the opportunities and approaches to specific analytical problems [53, 68–70].

8.5.3 X-Ray Absorption Spectroscopy

X-ray absorption spectroscopy (XAS) deals with the absorption of X-rays by matter. The absorption either leads to the excitation of core electrons in previously unoccupied orbitals (Figure 8.17a) or to the emission of core electrons (Figure 8.17b). The absorption is quantified either as the decrease in intensity during passage of an X-ray beam through a sample, through the emission of X-ray fluorescence, or the total yield of emitted electrons. *In situ* operation is possible (which is a difference when compared to electron spectroscopies). A comprehensive survey and methodical introduction into XAS is available on the website xafs.org of the International X-ray Absorption Society.

Within materials, the attenuation of X-ray intensity I_t along the propagation direction x follows the Lambert–Beer law:

$$I_t(x) = I_i \exp[\mu(h\nu)x] \tag{8.43}$$

I_i – Incident intensity [W]
I_t – Transmitted intensity [W]
x – Space coordinate [mm]
μ – Linear absorption coefficient [mm^{-1}]
$h\nu$ – Photon energy [eV]

The X-ray absorption spectrum features absorption edges. If the energy $E = h\nu$ of the incident X-ray photons is smaller than an absorption edge of a particular energy level of one of the elements in the sample, no absorption of that photon is possible by electrons of the respective level. Consequently, X-rays penetrate into the material. If the photon energy is increased, electrons from the respective energy level may first be excited to initially unoccupied electronic states, and with even more photon energy, the electron may be emitted from the atom. In both cases, the photon is absorbed, and the intensity of the X-ray beam is attenuated. If the photon energy is steadily increased, the ionization processes occur with lower efficiency (lower probability) and the X-rays become again more penetrating for the considered material.

An X-ray absorption spectrum is a plot of $\mu(h\nu)$ as a function of the photon energy $E = h\nu$ (Figure 8.31a). For more detailed studies, the range of photon energies is restricted to the surroundings of one specific absorption edge (Figure 8.31b). In this Figure, it is already clear that the recording of such a spectrum requires a continuous

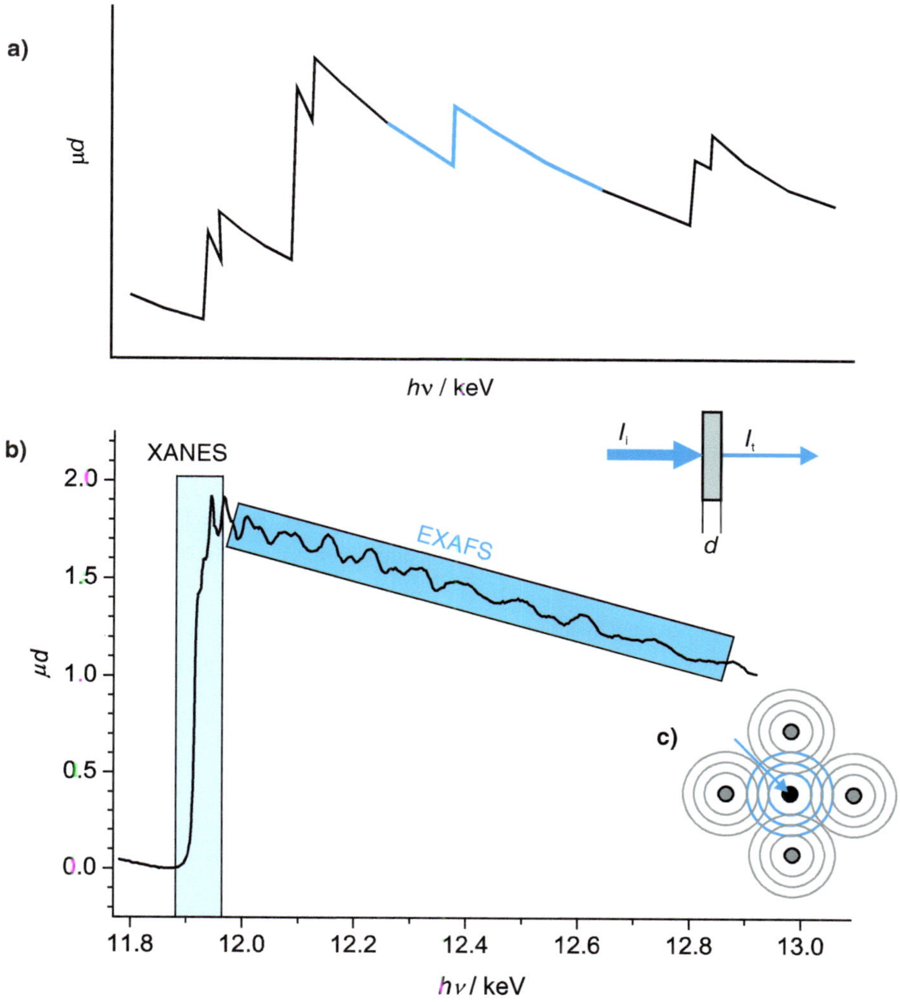

Figure 8.31 X-ray absorption spectrum of the Au L_3 edge;
a) ... in a wide energy range (schematically);
b) ... around the Au L_3 absorption edge. The blue coloration indicates the near-edge range for XANES and the extended range for EXAFS, data from [71].
c) Scattering of emerging electron waves at the neighboring atoms in the sample material.

variation of the photon energy in the range of the absorption edge. This task is efficiently accomplished only by means of synchrotron radiation. Storage rings specifically dedicated to the generation of X-rays for different areas of materials science are operated as national or even international research facilities (Table 8.4). The core of such an installation is an electron storage ring, in which electrons move with high speed in an ultrahigh vacuum and are forced on a curved trajectory. The energy for changing the translational direction is provided by strong magnetic fields. The energy is constantly emitted by the moving electron as highly focused electromagnetic radiation of high intensity and in a large range of photon energies in the tangential direction to the curved path of the electrons. So-called insertion devices allow the emitted electromagnetic radiation to be focused at the specific wavelength ranges needed for a particular experiment. The electromagnetic radiation can be guided in a beam tube to a monochromator. X-ray monochromators use Bragg diffraction[96] at crystals to direct only a particular wavelength to the sample. By rotating the crystal, the photon energy can be varied for the wavelength range that is diffracted in the direction of the sample. Research teams apply for beam time at synchrotron sources at a particular experimental setup (beamline) that is optimized for experiments in a certain photon energy range. Typically, the measurement time is allocated after a competitive review process of applications [72]. At a synchrotron facility, many beamlines work in parallel. Some beamlines are dedicated to electrochemical experiments.

The absorption spectrum is divided into two regions (Figure 8.31b), from which different types of structural information can be extracted. The range of 10–50 eV around the absorption edge is called X-ray absorption near-edge structure (**XANES**)

Table 8.4 Synchrotron radiation sources.

Acronym	Name
ALBA	Barcelona, Spain
ALS	Advanced Light Source (Berkley, CA, USA)
ANKA	Angström source Karlsruhe, Germany
BESSY	Berlin Electron Storage ring for SYnchrotron radiation, Berlin, Germany
CLS	Canadian Light Source, Saskatoon, SK, Canada
DESY	Deutsches Elektronen Synchrotron [German Electron Synchrotron], Hamburg, Germany
ELETRA	Trieste, Italy
ESRF	European Synchrotron Radiation Facility, Grenoble, France
LNLS	Laboratório Nacional de Luz Sincrotron, Campinas, Brazil
MAX-LAB	Lund, Sweden
NSLS	National Synchrotron Light Source, Long Island, NY, USA
SESAME	Synchrotron-light for Experimental Science and Application in the Middle East, Allaan, Jordan
SLS	Swiss Light Source, Villingen, Switzerland
SPring-8	Super Photon Ring 8 GeV, Harima, Japan
SSLS	Singapore Synchrotron Light Source, Singapore
SSRF	Shanghai Synchrotron Radiation Facility, Shanghai, China

[96]William Lawrence Bragg, 1890–1971, British physicist of Australian origin, professor of physics at Manchester and Cambridge, Nobel Prize in physics 1915 together with his father Henry William Bragg.

or near-edge X-ray absorption fine structure (**NEXAFS**). This part is followed by a higher photon energy range (50–330 eV), in which $\mu(h\nu)$ oscillates around a decaying curve. This is called extended X-ray absorption fine structure (**EXAFS**).

XANES. Photons in the near-edge region cause the excitation of electrons from core levels to unoccupied orbitals. The necessary energy difference is not only element-specific but also depends on the oxidation state and the ligand environment of the excited atom, very similar to binding energies in XPS.

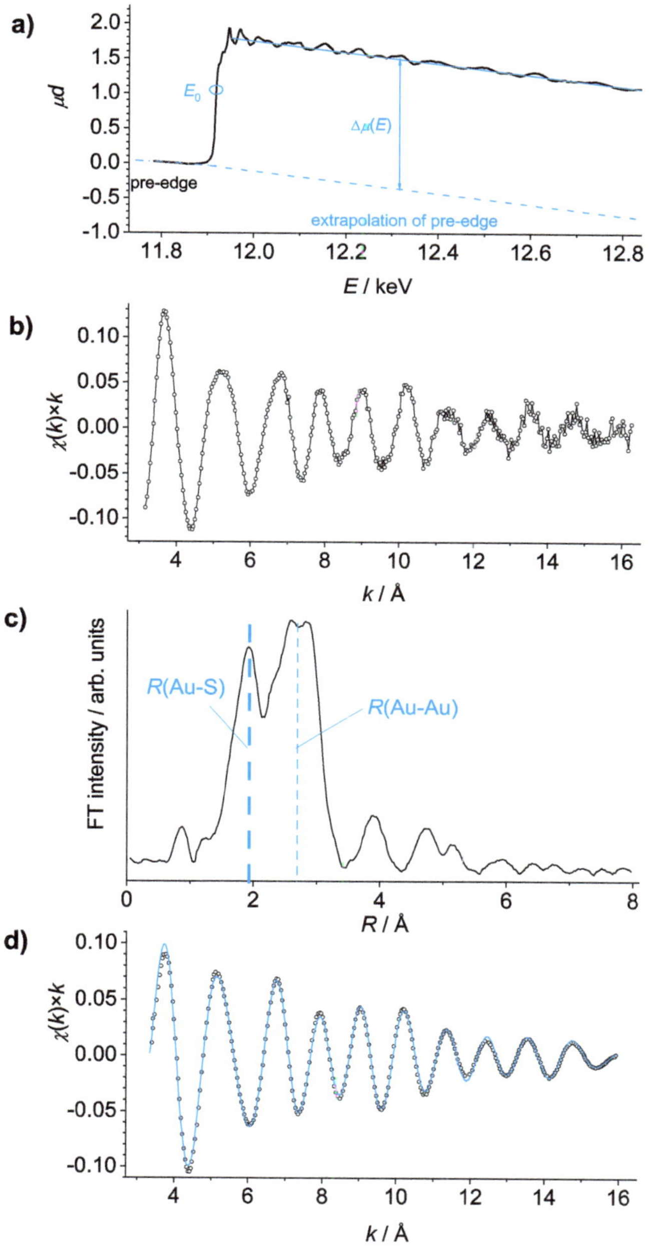

Figure 8.32 Procedure for the evaluation of EXAFS measurements on the example of thiolate-protected gold nanoparticles with an average diameter of 2 nm [and thus a high fraction of surface atoms (!)].

a) High-resolution spectrum of an Au-L_3 absorption edge with labelling of the characteristic edge energy and step height;
b) EXAFS oscillations;
c) Fourier transform of signal in b);
d) Filtered EXAFS signal, to which only scattering between 155 pm and 340 pm contributed (solid line). The calculated EXAFS oscillation for an assumed binding model is shown as open dots. The fit has been achieved with the following parameters:

	Au-Au	Au-S
R [Å]	2.843 ± 0.007	2.34 ± 0.02
N	7.2 ± 0.8	0.8 ± 0.2
σ [Å]	0.078 ± 0.004	0.06 ± 0.02
ΔE_0 [eV]	0.0 ± 0.6	10 ± 2

Source: Reproduced with permission from [71], Copyright 2000, Wiley-VCH.

EXAFS. The intensity oscillations in a larger energy range above the absorption edge in Figure 8.31b are caused by the interference of the emerging electron wave with a scattered fraction of the wave. The interference pattern allows conclusions about the radial distance distributions and the chemical elements surrounding the emitter atom. In contrast to diffraction methods, this type of analysis is possible even when the sample is not crystalline (Section 8.6). For evaluating EXAFS data, the precise position of the absorption edge is determined as the photon energy at which the intensity is 50% of the background-corrected step height. Next, the intensity $\mu_0(h\nu)$ is subtracted from the continuously decaying signal (Figure 8.32a). The remaining function is reminiscent of a damped oscillation (Figure 8.32b). For a better representation, the data on the abscissa are converted to the wave number vector **k** (Excursion 10-1, $|\mathbf{k}| = \sqrt{2m_e h\nu/\hbar}$). The obtained function is subjected to a Fourier transform (FT) and provides the radial distance distribution function (Figure 8.32c). Often, only the first coordination sphere is extracted from this function and back-transformed (inverse FT). The result looks like a smoothed EXAFS oscillation. The last step is a fit procedure for the interpretation of the data. A hypothetical arrangement of the neighboring atoms is assumed (element, number of neighboring atoms N, distance R, Debye–Waller factor[36,97] σ, element-specific data from databases, backscattering amplitude, phase shift, and amplitude factor) and used to calculate the expected EXAFS oscillations. The calculated oscillations are compared to the experimental oscillations. An iterative procedure is used to vary the number of neighboring elements as well as their distances until good agreement between the calculated and measured EXAFS oscillations is reached. If no agreement can be reached, the hypothesis (assumed environment of the emitter atom) has to be rejected, and a new model must be built. If good agreement between calculated and measured EXAFS oscillations is reached, the hypothesis *likely* reflects the real structure. However, it might also be possible that an alternative atomic arrangement could result in a similarly good fit.

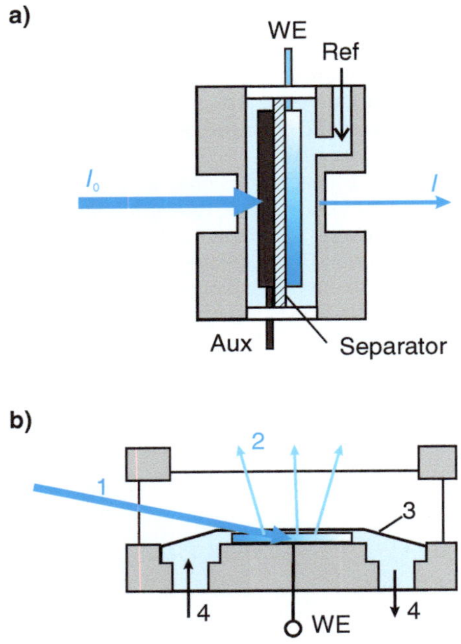

Figure 8.33 Electrochemical cell for *in situ* XAS; a) for transmission mode, b) for fluorescence mode; 1) X-ray from monochromator, 2) fluorescence radiation toward the detector, 3) Kapton foil, 4) electrolyte influx and efflux.

Measurement Cells. Measurement cells are distinguished depending on whether they will be used in fluorescence mode or in transmission mode (Figure 8.33) [73]. Although X-rays penetrate solids, their intensity is attenuated depending on the elements contained in the sample. The attenuation can be calculated for the photon energy range using tabulated data for the elements and their relative content in the sample [72]. The sample to be investigated is placed underneath a thin electrolyte reservoir that can be penetrated by the incident X-ray beam. The radiation enters and leaves the cell through thin window materials of light elements (polymer films, graphite, Si_3N_4). Additional parts may facilitate influx and efflux of the electrolyte solution or the concomitant illumination by visible light (for the investigation of photoelectrodes). XAS is frequently used for the characterization of electrocatalysts, which is summarized for carbon-supported Pt nanoparticles in [74, 75]. References [72, 76, 77] are recommended as an entry for a more general in-depth study of the subject.

8.6 X-Ray Diffraction

XAS should not be confused with X-ray diffraction methods. Similar to X-ray single-crystal diffraction, a monochromatic X-ray beam is directed to the sample. The photon energy is about 12 keV, corresponding to a wavelength $\lambda \approx 0.1$ nm = 100 pm and thus to a typical value for atomic distances or distances d between lattice planes in crystals. X-rays of this energy are diffracted. According to Bragg's equation[96] [$n\lambda = 2d \sin(\theta)$], the intensity is enhanced in specific directions. The diffraction pattern allows conclusions about the position of the scatterers (atoms) in the crystal. The X-ray beam is strongly attenuated even by the thin electrolyte layer. For this reason, grazing incidence X-ray diffraction (GIXRD) is often applied, which uses a shallow

[97] Ivar Waller, 1898–1919, Swedish physicist and crystallographer, professor at the University of Uppsala.

angle between the incident X-rays and the sample surface and a very thin electrolyte layer above the sample (Figure 8.34). The technique is especially suitable for studying structural changes of electrodes while they are in contact with an electrolyte and during current flow. A recent review provides an excellent overview of the physical basics, current developments in instrumentation, and applications for problems in electrochemistry and especially in electrocatalysis [78].

Further Methods. There are a number of more specialized techniques of surface analysis that are applicable to electrodes after transfer to an UHV. Thus, the techniques are used as *ex situ*. The most important techniques are ion scattering spectroscopy (ISS) for the identification of elements at the very surface of a solid and low energy electron diffraction (LEED) for the study of superstructures on monocrystalline electrodes (Section 8.9.3) [1, 79, 80]. Both techniques are more surface-sensitive than X-ray diffraction, because they are based on the diffraction of particles that, other than X-rays, cannot penetrate deeply into solid samples.

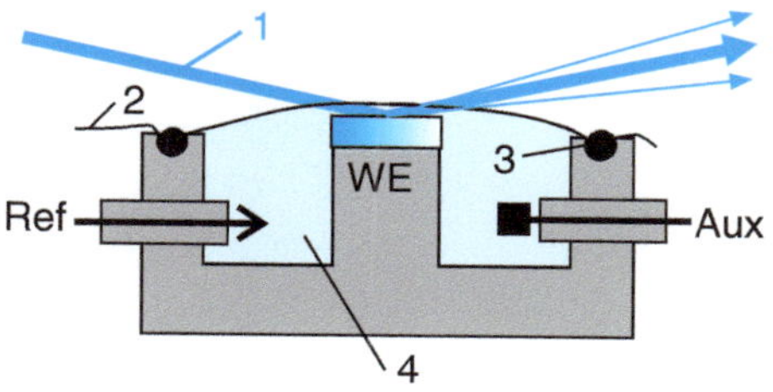

Figure 8.34 Cross-sectional view of a cell for *in situ* X-ray diffraction. The single-crystal working electrode (WE) is placed on a "pole" and is covered by a thin electrolyte (4) layer of 10–20 μm thickness behind a polypropylene film (2). The film is secured by an O-ring (3) pressed into a groove within the cell body. The incident X-ray beam (1) reaches the cell under a grazing angle that is in fact even shallower than schematically depicted here.

8.7 Mass Spectrometry

In the context of electrochemical studies, mass spectrometry is versatile and can be used in different ways. It can be used to identify electrolysis products in the liquid phase (Section 12.4.5). However, here we focus on the discussion of the analysis of the solid phase. Because mass spectrometry is conducted under UHV conditions, electrodes must be removed from the electrolyte solution and transferred to UHV; thus, mass spectrometry is an *ex situ* method. The material at the electrode surface must be transferred into the vacuum and ionized to enable detection in the spectrometer. Secondary ion mass spectrometry (SIMS) is a routine method for this task. The sample is bombarded with a beam of heavy ions (e.g., Cs^+, 15 keV, primary ion). Upon impact of the primary ions, the sample surface is sputtered, and part of the removed atoms and molecular fragments are ionized (secondary ions). The secondary ions are accelerated by an electric field and subsequently separated and detected in the mass spectrometer. Compared to XPS (Section 8.5.2), SIMS is more sensitive to lower concentrations. Because the primary ion beam can be focused and controlled very well, the method allows spatially resolved analysis. The primary ions remove the sample material very effectively, which facilitates depth profiling. Destructive depth profiling is also feasible by means of XPS with a sputter source. However, the relatively light Ar^+ ions, which are preferred in XPS instruments for the removal of contamination layers on the sample, provide only a very slow removal of thicker sample layers compared to Cs^+, typically applied in dedicated SIMS instruments.

The efficient removal of sample material in SIMS, which allows the analysis of three-dimensional element distributions in materials, can be a disadvantage in other situations. The high-energy primary ions are also implanted into the sample, and the binding state of the surface layer is changed. In addition, the sample surface is roughened. As a consequence, atoms from different depths are simultaneously detected.

Because SIMS causes strong fragmentation of molecular solids, more gentle ionization techniques are used as well. This includes thermal desorption mass spectrometry and laser desorption mass spectrometry. Thermal desorption is suitable for the analysis of adsorbates and physiosorbed layers. Laser desorption is used for the analysis of covalently bound thin layers.

8.8 Electrochemical Quartz Crystal Microbalance

Quartz crystal resonators change their piezo-mechanical resonance frequency as a function of the mass deposited onto them. They are routinely used in UHV deposition chambers to follow vapor deposition processes. Around 1980, it was shown that this type of measurement can be applied even when the crystal resonator is in contact with a liquid electrolyte solution. This is even though the liquid strongly damps the mechanical oscillations. One of the electrical contact areas of the crystal resonator can be used simultaneously as the working electrode in a potentiostatic three-electrode cell. This allows simultaneous recording of a voltammogram and the mass changes associated with the electrode processes. The device based on this special variation of the original principle is called an **electrochemical quartz crystal microbalance (EQCM)**. This technique is particularly useful for the study of electrode processes that are associated with a mass change of the solid phase. This includes galvanic deposition processes, the formation of surface layers, and the insertion of ions into conducting polymer films or insertion electrodes. For the latter, voltammetric curves only indicate a net current and do *not* reveal whether an anion was inserted or a cation was ejected. The EQCM can be used to reveal to what extent solvent molecules were transferred together with the ions.

We will cover here the basics of the EQCM. The interpretation of EQCM data is seldom straightforward. Specialized review papers provide a broader coverage of the physical principles and various applications [81–83]. Crystal resonators in contact with a liquid are also an important class of transducers (signal converters) for mass-sensitive biosensors. In this application, the electrical contact pads of the resonator are not needed as the working electrode for an electrochemical cell, and thus, there is more freedom for specialized and more sensitive designs [84]. In this section, we will restrict the discussion to the EQCM in a more narrow sense.

8.8.1 Piezoelectric Effect and Quartz Microbalance

Piezoelectric Effect. Piezoelectric materials generate a voltage difference along specific crystallographic directions when they are deformed by mechanical force. The effect only occurs in acentric crystals. These are crystals that do not have a centrosymmetric space group. The inverse piezoelectric effect describes the mechanical deformation of a material caused by an applied voltage. This is used for actuators in scanning probe microscopy (Section 8.11). For EQCM, flat disks of a quartz crystal that have been cut by the manufacturer at an angle of 32.25° relative to the x-axis of a macroscopic single crystal of α-quartz are almost exclusively used [85]. This preparation is called an AT cut. An advantage of quartz compared to other materials is its chemical stability in most electrolyte solutions and the almost complete absence of temperature-dependent piezoelectric properties of AT-cut crystals in vacuum. Despite this, EQCM measurements are typically temperature-sensitive because the viscoelastic properties of the liquid electrolyte change with temperature! The crystal disks are often processed into round chips with a diameter of 25.4 mm ($2r_Q$, this is 1 in.). Crystals of this size allow clamping the disks to the liquid reservoir in such a way that the mechanical contact to the cell vessel does not compromise the resonator quality as much as for smaller disks. On the other hand, larger disks would be too expensive and more fragile. The thickness defines the resonance frequency.

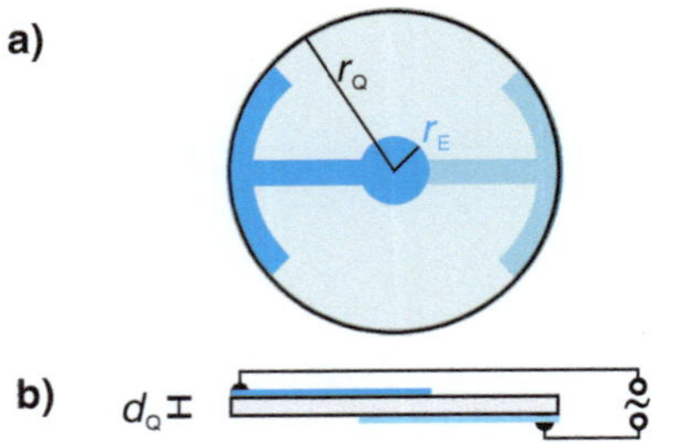

Figure 8.35 a) Electrical contacts to a quartz disk for EQCM in the scale 1:1; b) cross section (not to scale).

Contacting. Thin evaporated metal films on both sides of the disk allow for electrical contact. For the EQCM, a gold layer is employed on top of a Cr layer as an adhesion promoter. The metal layers form a central electrode with a diameter $2r_E = 5.4$ mm. Electrical contact pads of 2 mm width on the front and rear sides of the disk lead to opposing electrodes on the quartz disk (Figure 8.35). Plano-convex resonators are slightly thinner in the center compared to the outer parts. This has the advantage that only mass changes on the working electrode area contribute to the signal. Deposits slightly off the contact electrodes also contribute to the signal for planar disks. Such contributions are not desired for EQCM. The sensitivity and oscillation amplitude are highest in the center of the electrode.

Q: How does the Cr layer function as an adhesion promoter?

A: Cr tends to form stable surface oxides and, therefore, can bind covalently to the surface of quartz (SiO_2). On the other side, it can form a metal bond with the gold layer provided there is no surface oxidation after the deposition of Cr and before the deposition of Au. Due to chemical inertness, Au alone would not adhere well to an oxidic surface like quartz. One problem in using Cr as an adhesion promoter is its tendency to form alloys with Au. Therefore, Cr can appear in very low concentration at the electrode–electrolyte interface and thereby subtly change the properties.

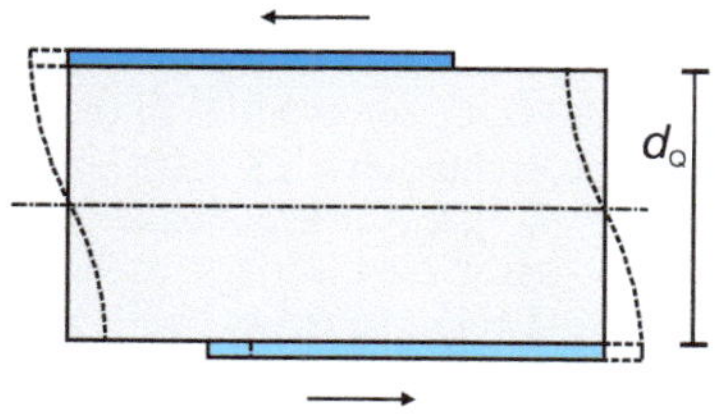

Figure 8.36 Displacement of the crystal in the thickness shear mode. The thickness of the quartz disk, the thickness of the contact electrodes, and the displacement are exaggerated. The displacement is about 10–100 nm, depending on the amplitude of the applied AC voltage.

Resonances. An AC voltage is applied between the two contact pads and causes a periodic displacement of the crystal parallel to the disk surface (Figure 8.36). In addition to this **thickness shear mode** (TSM), there are other vibrations that can couple to the TSM depending on the thickness d_Q of the quartz disk [85]. The suppression of such unwanted coupled modes requires the following relations $2r_Q/d_Q > 50$ and $2r_E/d_Q > 18$. For this reason, the quartz disks should be as thin as possible but at the same time not too fragile. A thickness of $d_Q = 0.334$ mm has turned out to be a good compromise between the two opposing requirements. In resonance at the frequency f_0, there is a standing wave with a node in the center of the disk and a maximum displacement at the surface of the disk (Figure 8.36). The thickness of the disk d_Q is half of the wavelength λ of the acoustic wave. The relationship for the speed of sound in quartz $v_Q = f_0 \cdot \lambda = 3340\ \text{m s}^{-1}$ yields the resonance frequency of the TSM wave:

$$d_Q = \lambda/2$$

$$2d_Q = v_Q/f_0$$

$$f_0 = v_Q/2d_Q = \frac{3340\ \text{m/s}}{2 \cdot 334 \times 10^{-6}\ \text{m}} = 5 \times 10^6\ \text{s}^{-1} = 5\ \text{MHz}. \qquad (8.44)$$

d_Q – Thickness of the quartz disk [m]
f_0 – Resonance frequency of the TSM wave [s^{-1}]
λ – Wavelength of the TSM wave [m]
v_Q – Speed of sound in quartz [m s^{-1}]

Because the sensitivity of the EQCM increases with the resonance frequency f_0 and, on the other hand, because a finite thickness d_Q of the quartz disk is required to ensure mechanical stability, there are variations of the technique that use overtones of uneven order (at $3f_0$, $5f_0$, etc.) in which 1.5 wavelengths fit within d_Q. The corresponding signals are smaller and require dedicated electronics, but the mass sensitivity is higher. Overall, working with 5 MHz is by far the most widespread practice.

Measurement Cells. The quartz disk is typically clamped between two O-rings to the electrolyte reservoir that also accommodates the reference and auxiliary electrodes (Figure 8.37). The gold electrode facing the liquid electrolyte solution is connected to the potentiostat as the working electrode. Simultaneously, the contact pads on the quartz disk are connected to an electrical oscillator. The frequency applied is monitored by a frequency counter, which is controlled together with the potentiostat via a computer. The frequency counter can register a frequency change of 1 Hz on a base frequency of 1 MHz within milliseconds. For more complex measurements at soft films, the frequency counter is replaced by an impedance analyzer or a network analyzer, which are similar to the instrumentation used in electrochemical impedance spectroscopy (EIS). Here, it is the mechanical resonator that is analyzed instead of the current–voltage characteristics of an electrochemical cell in EIS. Furthermore, EQCM usually works at higher frequencies than EIS (MHz instead of kHz).

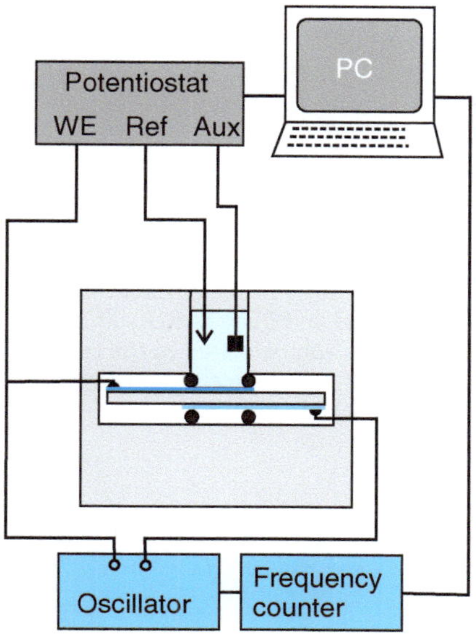

Figure 8.37 Schematic setup of an EQCM measurement cell and its connections to a potentiostat (gray) and the oscillator circuit with frequency counter (blue).

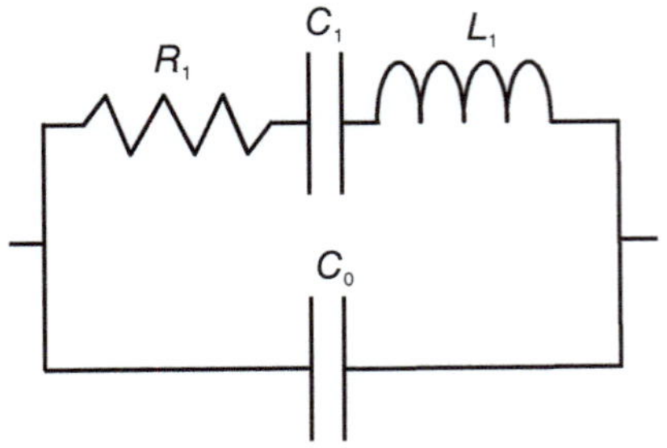

Figure 8.38 Equivalent circuit for a quartz resonator. R_1, dissipation of the mechanical energy ≈ 100 Ω; C_1 corresponds to the energy stored in the oscillation ≈ 23 × 10^{-15} F [1 F = 1 A^2 s^4 kg^{-1} m^{-2}]; L_1 corresponds to the inertia contribution associated with the oscillation ≈ 45 × 10^{-3} H [1 H = 1 kg m^2 A^{-2} s^{-2}]; C_0, electrical capacitance of the quartz crystal between the contact area ≈ 7.3 × 10^{-12} F. Values of the quantities according to [81] for a crystal with air on both sides.

Electrical Properties of an EQCM. The frequency-dependent current between the two contact electrodes can be described by the equivalent circuit in Figure 8.38. For the impedance analysis of the EQCM, one determines the resonance frequency f_s of the mechanical oscillation, and the frequency f_r, for which the phase angle φ between the alternating current I_{AC} and the applied AC voltage becomes zero. The simplified equivalent circuit in Figure 8.39 allows the unequivocal determination of only the reactance $X_e(f)$, the resistance $R_e(f)$, and the admittance **Y** (conductance G). From these quantities, the characteristics given in Figure 8.38 can be derived [81]. Often the relationship $R_1 \approx R_e$ holds if f is slightly larger than f_s ($X_e \approx 0$). C_0 is obtained from a measurement far off f_s (e.g., at 100 kHz). C_1 is calculated by $C_1 = (f_r - f_s)/(2\pi^2 f_s{}^3 C_0 R_1{}^2)$. When knowing R_1, C_1 and C_0, L_1 can be determined by $f_s = \{2\pi[L_1 C_1]^{1/2}\}^{-1}$.

8.8.2 Influences on the Mechanical Resonance Frequency

Mass Loading. The resonance frequency of the quartz resonator shifts after deposition of an additional mass according to the Sauerbrey Equation (8.45)[98]. For practical EQCM work, the prefactors are often summarized into a single sensitivity factor S, which is determined by calibration. The sensitivity value is influenced by the experimental system and by the resonance frequency of individual resonators.

$$\Delta f = -\frac{2kf_0^2}{\sqrt{\rho_Q \mu_Q}} \frac{\Delta m}{A} = S\frac{\Delta m}{A} \tag{8.45}$$

Δf – Shift of the resonance frequency [Hz = s^{-1}]
f_0 – Resonance frequency of the basic mode [Hz]
k – Order of the oscillation (1, 3, …)
ρ_Q – Density of quartz = 2.648 g cm^{-3}
μ_Q – Shear modulus of quartz = 2.947 × 10^{11} g cm^{-1} s^{-2}
$\Delta m/A$ – Area-related mass change [g cm^{-2}]
S – Sensitivity factor [Hz g^{-1} cm^2]

An area-related mass change is detected. This implies that the mass must be evenly distributed on the resonator, because there is a *strong* variation of sensitivity across the surface for locally deposited mass points. Inserting the material properties for quartz and $k = 1$ into Eq. (8.45) yields a sensitivity of 56.6 Hz µg^1 cm^2, that is, an evenly distributed mass of 1 µg cm^2 on the crystal leads to a *decrease* (negative sign) of the resonance frequency by 56 Hz, a readily measurable quantity. Assuming that a shift in resonance frequency down to 1 Hz is measurable, we can calculate the detection limit as 2 × 10^{-8} g cm^{-2}, which corresponds to less than a monolayer! (For comparison: A monolayer of about 1 × 10^9 mol cm^{-2} Cu with M = 63.5 g mol^{-1} corresponds to 6.4 × 10^8 g cm^{-2}.)

Equation (8.45) is valid only in a vacuum and if the deposited mass Δm is stiff and amounts to less than 2% of the mass of quartz itself. Equations for higher mass loading are available. They consider explicitly the acoustic impedances of the deposited material (Z-match method) [81].

Figure 8.39 Simplified equivalent circuit for the analysis of an EQCM. The circuit in Figure 8.38 is approximated by the serial connection of a reactance X_e and ohmic resistor R_e.

Influence of the Electrolyte Solution. If the oscillating crystal is in contact with a liquid, the thickness shear wave propagates into the liquid medium. However, the wave is strongly damped and only a layer of 1 µm of the liquid is actually displaced (Figure 8.40). This small thickness is considerably less when compared to the diffusion layer in typical cyclic voltammetry experiments. Therefore, cyclic voltammograms can be recorded simultaneously with EQCM measurements without forced convection or effects of flow that are observed at rotating disk electrodes in Section 12.4.2. Furthermore, the crystal can store less mechanical energy when in contact with a liquid (instead of a vacuum). Contact with a liquid reduces the quality factor [Q-factor, $Q = (2\pi f_s C_1 R_1)^1$. Attention: do not confuse with charge Q[A s]]. The reduced Q-factor is evident from the broader and flattened resonance curve **Y**(f) in Figure 8.41. The decrease in resonance frequency is proportional to $(\rho_l \eta_l)^{1/2}$ [86].

$$\Delta f = -f_0^{3/2} \frac{\sqrt{\rho_l \eta_l}}{\sqrt{\pi}\sqrt{\rho_Q \mu_Q}} \tag{8.46}$$

ρ_l – Density of the liquid [g cm^{-3}], (Index l – *liquid*)
η_l – Viscosity of the liquid [g cm^{-1} s^{-1}]

[98] Günter Sauerbrey, 1933–2003, professor at the Technical University Berlin.

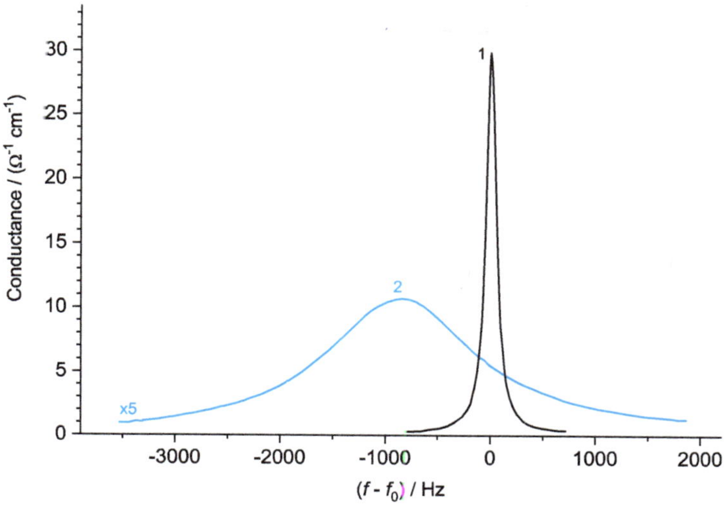

Figure 8.41 Resonance curve of a quartz oscillator in 1) air and 2) water. Curve 2 was multiplied by a factor 5, which means it is actually considerably flatter. Source: Reproduced with permission from [81], Copyright 1991, Marcel Dekker.

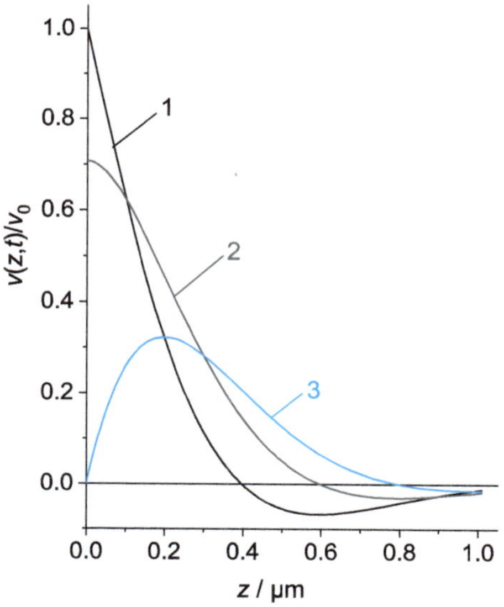

Figure 8.40 Velocity distribution parallel to the surface of a quartz resonator in contact with an aqueous solution. The plot shows the distributions at three points of an oscillation period; 1) maximum velocity at the surface ($t = 0$), 2) intermediate value ($t = 1/(8f_0)$), and 3) reversal point of the oscillation, momentary velocity at the surface is zero ($t = 1/(4f_0)$). Calculated according to [81, 86] with $v(z, t)/v_0 = \exp(-k_1 z) \cdot \cos(k_1 z - 2\pi f_0 t)$, wave propagation constant $k_1 = (\pi f_0 \rho_1/\eta_l)^{1/2}$, density of the liquid $\rho_l = 1$ g cm^{-3}, viscosity of the liquid $\eta_l = 0.01$ g cm^{-1} s^{-1}, $f_0 = 5$ MHz.

Influence of Temperature. In general, a change in temperature acts on f_s via the density ρ_Q and the shear modulus of quartz, as well as via the density and viscosity of the liquid. Especially for quartz crystals with AT cut, the temperature dependence of ρ_Q and μ_Q close to room temperature is very small. Therefore, the temperature dependence of the EQCM signal is mainly a consequence of the changes in the density and the viscosity of the liquid. The changes are indeed significant, as illustrated by the following example for water taken from [81]: Inserting the data $\rho_l(20\,°\text{C}) = 0.99823$ g cm^{-3}, $\rho_l(25\,°\text{C}) = 0.99707$ g cm^{-3}, $\eta_l(20\,°\text{C}) = 0.01002$ g cm^{-1} s^{-1}, $\eta_l(25\,°\text{C}) = 0.008904$ g cm^{-1} s^{-1} with $f_0 = 5$ MHz for quartz into Eq. (8.46), yields a frequency shift of 41 Hz for a temperature shift of 5 K. This value exceeds the frequency shift expected for the deposition of a monolayer of an adsorbate. Therefore, experimentation with EQCM, especially when conducted over longer periods, requires thermostatic control to 0.1 K or even better to 0.02 K. Shorter relative measurements, especially versus a defined initial state, for example, during the recording of a cyclic voltammogram, the temperature drift during the recording remains small. Double resonator experiments with a second reference crystal oscillator in the same solution but not used as an active electrode can help account for temperature drift.

8.8.3 Typical Applications

Calibration of Quartz Crystals. Usually, a layer of Cu, Pb or Ag is electrodeposited on the contact electrode. These reactions proceed with a faradaic efficiency close to 100%. Using the transferred charge Q, the molar mass M, and the stoichiometric number n of transferred electrons per ion, the deposited mass can be calculated by inserting Eq. (1.16) into Eq. (8.45). The accuracy is increased by plotting Δf versus Q at different times of the experiment. This plot should yield a straight line.

$$\Delta f = S\frac{M}{nF}Q \tag{8.47}$$

M – Molar mass of the metal [g mol^{-1}]
n – Stoichiometric number of electrons [dimensionless]
F – Faraday constant [A s mol^{-1}]
Q – Charge [A s = C]
S – Sensitivity [Hz g^{-1}]

To obtain this particular simple relation, the deposited film must be significantly thinner than the quartz crystal so that the displacement inside the deposited layer is essentially equal to the displacement at the surface of the crystal resonator. In addition, the film ideally must be flat because the solution trapped inside the pores would move together with the stiff metal film and thus contribute to the weighted mass.

Deposition of Flat and Stiff Films. Equation (8.47) can also be used when depositing films with unknown molar mass. In this case, the apparent molar mass M' replaces

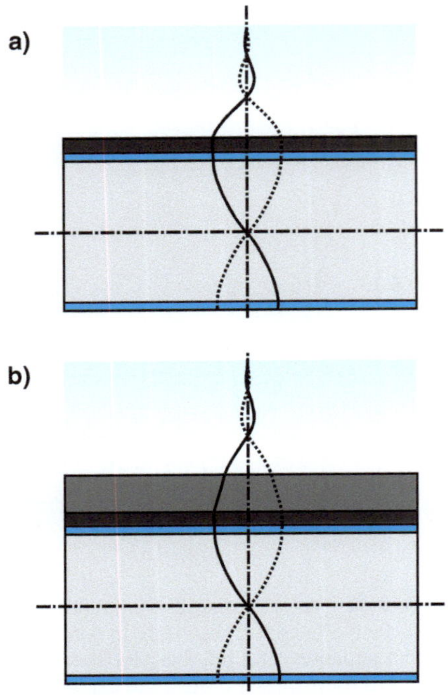

Figure 8.42 Propagation of the thickness shear wave of an EQCM.

a) Coverage with a stiff film in solution – the EQCM operates as mass sensor and the translation of the contact electrode and the film are coupled to that of the quartz surface;
b) coverage by a viscoelastic film in solution – the EQCM operates as a viscoelastic probe. The contribution of a stiff film can also be caused by the entrapment of solution in recessed parts at the surface of the electrode.

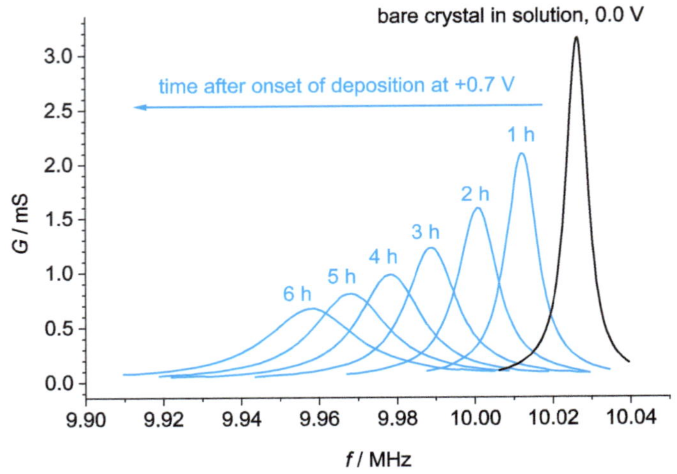

Figure 8.43 Admittance spectra recorded during the electrodeposition of poly(vinyl ferrocene) (PVF) from a PVF solution in dichloromethane (concentration of ferrocene moieties was 2 mmol L^{-1} in 0.1 mol L^{-1} tetrapropylammonium perchlorate in CH_2Cl_2); 10 MHz AT-cut quartz crystal with an Au electrode ($A = 0.23$ cm^2). Source: Reproduced with permission from H. L. Bandey, M. Gonsalves, A. R. Hillman, A. Gilde, S. Bruckenstein; *J. Electroanal. Chem.* **1996**, *410*, 219–227, Copyright 1996, Elsevier.

M in Eq. (8.47). Using a previously determined S, the slope of the function $\Delta f(Q)$ can be used to determine M', that is, the deposited mass per transfer of 1 mol electrons. This procedure provides valuable insights for adsorption processes and underpotential deposition, because the underpotential deposition of metal cations is occasionally connected with the coadsorption of anions, which is not evident from analyzing the transferred charge in a cyclic voltammogram. For investigations at Au electrodes, the contact electrodes can be used directly. Other metals can be deposited as thin layers onto the Au contact electrodes for use as a working electrode.

Roughness and Porosity. Rough and porous layers represent a problem for analysis by means of EQCM, because the solvent contained in the pores will contribute to the determined mass change. If the electrode under study smoothens or roughens without net deposition or net dissolution, EQCM methods can be used to follow the kinetics of morphology changes *in situ*.

Polymer Films. Polymer films represent an important broad field of application for EQCM measurements, which have improved the understanding of the possible mechanisms and competing processes. An extensive review [82] is recommended for a more detailed survey. When interpreting the data, a decision has to be made as to whether the film can be considered as acoustically thin (Figure 8.42a). In this case, the frequency change can be related to a mass change using the Sauerbrey equation. For thicker films, the viscoelastic properties of the film have to be considered (Figure 8.42b). The EQCM facilitates the study of viscoelastic properties even during ongoing electrochemical reactions. The amplitude of the oscillation is rapidly decaying in the electrolyte solution (Figure 8.42, see also Figure 8.40). From the amount of deposited polymer, it is possible to control whether the polymer layer behaves as an acoustically thin or an acoustically thick film. By measuring the complete admittance spectrum of the loaded crystal, both situations can be distinguished experimentally. Figure 8.43 shows the admittance of a crystal with different amounts of poly(vinyl ferrocene), which were obtained by conducting the electrodeposition for different durations.

Q: How can poly(vinyl ferrocene) be deposited electrochemically? The polymer is already synthesized and available as a powder.

A: Indeed, the procedure starts from the chemically synthesized product, which is dissolved in an electrolyte based on dichloromethane (e.g., 0.1 mol L^{-1} $([CH_3-CH_2-CH_2)_4N]^+[ClO_4]^-$ in CH_2Cl_2). The initially neutral ferrocene moieties are potentiostatically oxidized to positively charged ferrocenium moieties (Fc^+). The charged polymer becomes insoluble in the nonpolar solvent CH_2Cl_2 and the polymer is deposited as a film on the electrode surface. In this process, one anion per Fc^+ moiety is deposited into the film to balance the charge. In addition, solvent molecules may be incorporated.

With increasing duration of the deposition process, the resonance frequency decreases (Figure 8.43), which corresponds to a mass increase of the film. Concomitantly, the peak width increases on the frequency axis, that is, the Q-factor of the resonator decreases. This indicates the formation of an acoustically thick film, in which the viscoelastic properties of the film material influence the oscillation. EQCM measurements of acoustically thick films often allow only such qualitative insights. Newer measurement equipment and data evaluation strategies facilitate the determination of mechanical properties of the film. These more complicated approaches use fitting routines for the interpretation of admittance data of electrical or mechanical models of the oscillating system, quite analogous to EIS in Chapter 7. They enrich the available methods for polymer characterization as they provide access to mechanical properties in a frequency domain (>1 MHz) that is not accessible by other methods. Even insulating films can be studied if they are deposited by a self-limiting process on the contact electrodes of a quartz resonator. Typically, these films can be treated as acoustically thin films.

Redox polymers contain a discrete redox-active group that is covalently attached to the polymer backbone, as in poly(vinyl ferrocene). Another important class of polymers that have been studied by EQCM are intrinsically conductive polymers, for example, poly(aniline) or poly(pyrrole), in which an oxidation causes the removal of an electron from an extended π-electron system, by which an electronically conducting polymer chain is formed. For stabilization of the charged polymer chain, an anion is incorporated into the film, a process often termed “doping”. A change in the redox state of the polymer is associated with the incorporation or ejection of ions from/to the (usually concentrated) electrolyte solution that is in contact with the polymer film. Thus, the electrodes are insertion electrodes. Other important classes of compounds that were studied by EQCM are redox-active inorganic coordination polymers such as metal hexacyanometallates, heteropoly acids, and metal–organic frameworks (MOFs).

During redox reactions in deposits, the incorporation or ejection of counter ions occurs coupled to the transfer of solvent and further ions. These processes operate on different timescales toward establishing the thermodynamic equilibrium, in which the electrochemical potentials $\tilde{\mu}_i$ of all transferable species (ions, solvent molecules) are equal in the adjoining phases (polymer film, electrolyte solution). Redox reactions inside the film are accompanied by very significant changes in the activity coefficients of species inside the film. The resulting changes in activities inside the film for equilibration (Eq. 2.11) lead to a gradient of the electrochemical potentials $\tilde{\mu}_i$ which drives the transport processes.

The incorporation of solvent molecules between the polymer chains of an acoustically thin film can be considered as a limiting case of solvent trapped in the pores of a stiff layer that was described above. However, the film thickness and the viscoelastic properties of organic polymers do change when solvent molecules are taken up. This may be a negligible effect for strongly crosslinked and/or thin films, for which a mass determination of the incorporated solvent is possible (example: redox reactions of inorganic electrochromic films of Prussian Blue and of other hexacyanometallates). The data evaluation can be refined when using deuterated solvents, which allows one to ascribe an observed mass change to the incorporation of an ion (no mass change

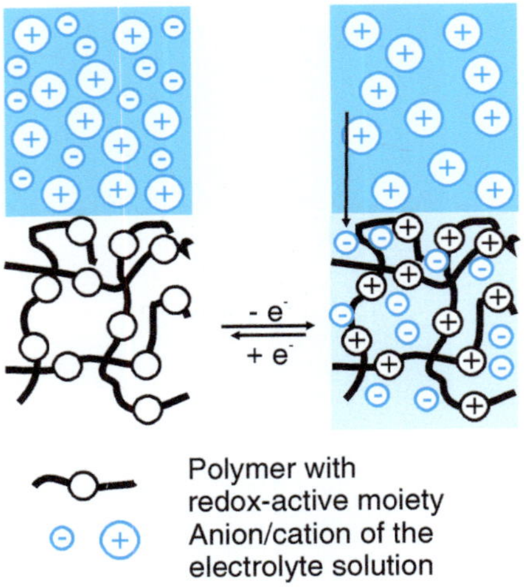

Figure 8.44 Charge compensation during oxidation of a permselective polymer film by incorporation of anions from the electrolyte solution.

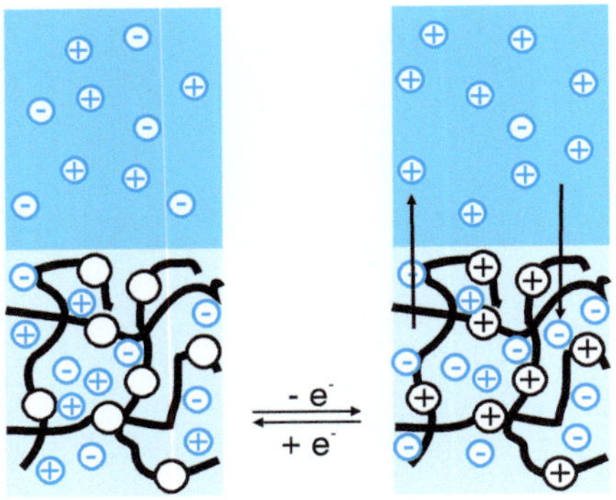

Figure 8.45 Charge compensation during oxidation of a film that is not permselective by incorporation of anions and/or ejection of cations.

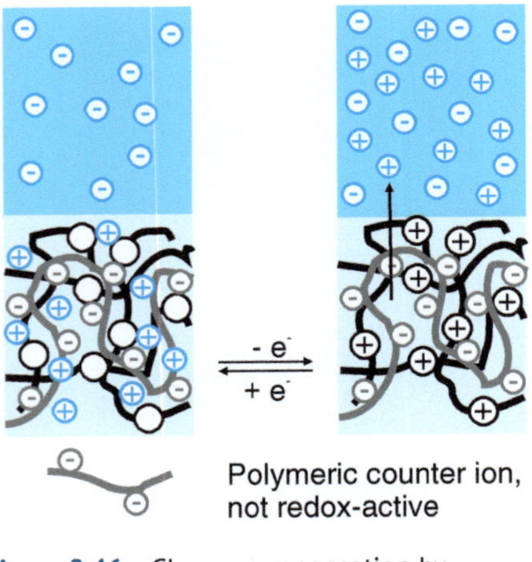

Figure 8.46 Charge compensation by incorporation of cations from the electrolyte upon reduction of a polymer film containing an immobile counter ion (e.g., a polymeric poly-anion).

with a change to deuterated solvent) or a solvent molecule (precise increment of mass change when using a deuterated solvent). Organic polymers often exhibit a thickness increase when taking up solvent molecules (swelling). The solvent acts as a plasticizer, which may change the film properties from an initially acoustically thin film to an acoustically thick film, in which the behavior after solvent uptake is determined by the viscoelastic properties of the swollen film.

Figures 8.44–8.46 show some greatly simplified limiting cases for the incorporation and ejection of ions to and from polymer films. If the polymer film is permselective, that is, only the anion (or only the cation) contained in the electrolyte can pass through the film, the charge compensation upon oxidation of the polymer occurs by uptake of an equivalent amount of anions (Figure 8.44), which is often accompanied by the transfer of a defined amount of solvent molecules. This amount can be determined for acoustically thin films from the ratio of transferred charge and mass change according to Eq. (8.47) using the known molecular weights of the anion and the solvent. In this case, the molecular weight of *one* anion M_X is substituted by the *sum* $M_X + yM_{solv}$ of the molecular weights of the anion and the y concomitantly transferred solvent molecules. Provided M_X and M_{solv} are known, the number y of solvent molecules that move across the polymer-solution interface per transferred elemental charge can be determined.

Films that are not permselective can contain counter ions (for charge compensation) and amounts of the salt dissolved in the electrolyte solution (Figure 8.45). In the literature on ion-exchangers, these are termed "salt contained in the polymer". Extensive literature about ion-exchange resins forms a basis for prediction or interpretation as to whether a particular material will tend to show selective or nonselective ion transfer immersed in a specific electrolyte solution (at a specific pH). If a salt is contained in an oxidized polymer film (Figure 8.45, right), charge compensation upon reduction may occur by ejection of anions into the electrolyte solution and/or uptake of further cations from the electrolyte solution. These processes are generally coupled to the transfer of solvent molecules.

The observation of the reaction by time-dependent EQCM measurements with high time resolution facilitates the disentanglement of the reaction mechanisms controlled by the mobilities of anions and cations in the polymer. For instance, it is possible that the current in the external electrical circuit is initially controlled by the transfer of the faster of the two ions. After decay of the current at the working electrode, ion exchange processes involving the less mobile ion and solvent molecules continue without the flow of a net charge through the working electrode. Such processes would remain undetected in voltammetric or chronoamperometric experiments, because no net current passes through the potentiostat. For a refined analysis, the experiment initially starts with a slow potential scan rate v. Subsequently, the experiment is repeated with an incremental increase in v, thereby sequentially freezing out the slower processes. That is, the experiment is terminated before such a "frozen" process can contribute to a measurable mass change. This approach simplifies the assignment of the interdependent processes. The reaction mechanisms are often represented as scheme-of-cubes (adjoining cubes) analogous to the square schemes in Section 6.4.5. One dimension of the cubes describes the electron transfer, one shows the ion transfer, and the third dimension depicts the solvent transfer.

The possibility to switch the redox state of polymers so that ions are ejected or incorporated has significant applications in the area of biomedicine for controlled delivery of charged pharmaceutically relevant compounds (e.g., salicylates). An extreme case with respect to the different mobilities of anions and cations is shown in Figure 8.46. During the film deposition, a polymeric counter ion was used (Figure 8.46, right). A well-documented example is the redox-active polymer poly(xylyl viologen) (PXV) and poly(styrene sulfonate) (PSS) as polymeric counterion [87]. Upon reduction of the viologen moieties in PXV, the charge neutrality can be maintained only by the mobile counter ions of the electrolyte. In Figure 8.46 this is shown greatly simplified by transfer of cations from the electrolyte only. The real extent of ion transfer processes of anions and cations of the electrolyte solution depends on the timescale and the ratio PXV:PSS.

EQCM experiments have also been essential for explaining the break-in effect and the particular feature of the first potential cycle in multicycle voltammetric experiments of redox-active polymer films. The first voltammogram is often distinctly different from all subsequently recorded voltammograms that are very similar to each other. This effect is particularly evident for poly(vinyl ferrocene) that we have used as an example above. After preparation of the film in CH_2Cl_2 and transfer of the polymer-modified electrode to an aqueous electrolyte, the first oxidation of the ferrocene (Fc) moieties to ferrocenium (Fc^+) moieties causes the uptake of anions and about 5 water molecules per Fc moiety. Upon reduction of the Fc^+ moieties, some of the water molecules remain within the polymer film. In the subsequent second voltammetric cycle, the incorporation of anions is facilitated by the water molecules already contained in the film, and thus the oxidation occurs at a significantly less positive potential than in the first potential cycle.

EQCM measurements have also provided insights into the reaction of insertion materials for batteries. In this case, the viscoelastic properties have to be considered and are by themselves of interest [88].

8.9 Single-Crystal Electrodes

Metallic elements crystallize with very few exceptions in one of the following crystal structures (Table 8.5).

- body-centered cubic packing (bcc)
- face-centered cubic packing (fcc)
- hexagonal close packing of spheres (hcp)

Table 8.5 Crystal structures of important metal.

fcc	Ca, Sr, Rh, Ir, Ni, **Pd**, **Pt**, **Cu**, **Ag**, **Au**, Al, Pb
hcp	Be, Mg, Sc, Y, La, Ti, Zr, Hf, Tc, Re, Ru, Os, Co, Zn, Cd, Tl
bcc	Li, Na, K, Rb, Cs, Ba, V, Nb, Ta, Cr, Mo, W, Fe

Macroscopic pieces of metals and electrode materials are composed of many small metal crystals, crystallites, or grains, which are separated from each other by grain boundaries. Such structures are called **polycrystalline** (Figure 8.47). Multicrystal is a synonym for polycrystal. Within one crystallite, the structure is formed by a three-dimensional periodic repetition of the elementary cell. Such crystallites in metals exhibit different sizes between micrometers and centimeters, depending on the kind of metal and the pretreatment of the metal. Neighboring crystallites have a different orientation of their crystallographic axes in space. If a polycrystal is cut along an arbitrary plane, ground and polished, its surface is formed from facets from various crystallites cut at different angles (Figure 8.47). The two-dimensional periodic arrangement of atoms from cutting a three-dimensional crystal is called a crystal face. The metal atoms exposed at different crystal faces have different arrangements and coordination environments (Section 8.9.1). The different arrangements lead to different binding energies and thus, for instance, to different reactivities for surface-bound reaction intermediates in electrocatalytic reactions or during metal deposition. The different crystal facets can be visualized by metallographic surface preparation and special imaging techniques.

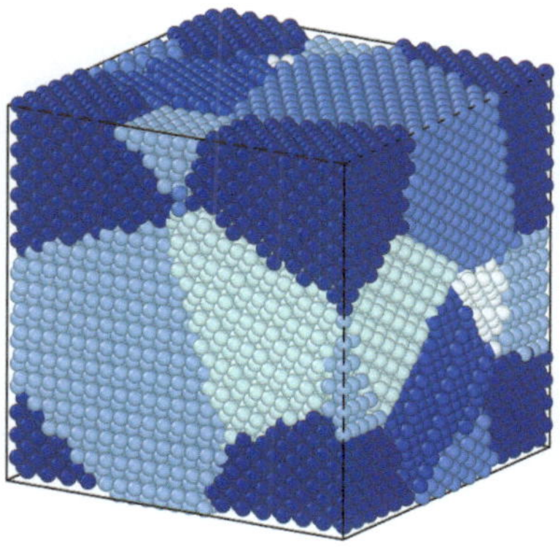

Figure 8.47 Schematic representation of the surface of a polycrystal. Source: The image was generated by the program Atomsk [89] and was kindly provided by Pierre Hirel, University of Lille.

Voltammetric measurements with polycrystalline electrodes provide an integrated or averaged signal from all the crystal facets exposed to the solution. A well-known cyclic voltammogram for polycrystalline Pt electrodes is shown in Figure 5.22. The simultaneous occurrence of signals from different crystal facets impedes the investigation of fundamental structure-reactivity relationships at polycrystalline electrodes.

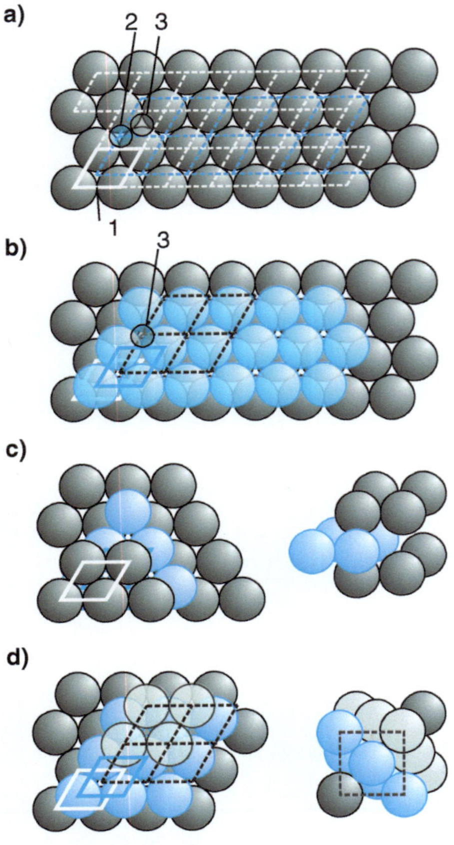

Figure 8.48 Packing of spheres.

a) Hexagonal arrangement of atoms in one layer with indication of the elementary cell (1) and two types of threefold hollow sites (2, 3);
b) second atomic layer with the threefold hollow site for the formation of the fcc lattice (3);
c) hexagonal close packing with stacking order ABA (left top-view, right side-view);
d) third atomic layer in the stacking order ABC (left) and orientation of the atomic layers in the elementary cell of the fcc lattice (right).

Therefore, electrodes with a well-defined surface structure composed of one crystallite only (single crystal) have been used intensively since about 1990. The surface exposed by grinding and annealing such an electrode exhibits a uniform periodic structure over the entire electrode surface. Such single-crystal electrodes facilitate very detailed structure-sensitive analytical techniques. Some methods, such as diffraction methods and scanning probe techniques, provide meaningful data only when using such electrodes. Because of the uniform structure and limited number of different surface sites, voltammetric methods can resolve the reactions at different sites very well, which facilitates an assignment of different processes to specific surface structures. Because of the tremendous importance of single crystals for interfacial electrochemistry, we will discuss them here together with the methods for their characterization.

8.9.1 Structure of Metal Crystals and Their Ideal Surfaces

The metallic bond is characterized by structures with high coordination numbers due to the closest packing of spheres (number of nearest neighboring atoms is 12 in a close packing of spheres as opposed to 2 for the O atom in H_2O). Within a plane, a hexagonal arrangement of spheres forms the closest packing (Figure 8.48a). When going to a second layer, the layer itself has again a hexagonal arrangement. However, it is shifted with respect to the lateral coordinates of the first layer in such a way that each atom of the second layer sits in a threefold hollow site of the first layer (Figure 8.48b). For the third atomic layer, which has a hexagonal arrangement in the plane, there are two nonequivalent types of threefold hollow sites in the second layer. Depending on which of the two nonequivalent sites is occupied by the atoms of the third layer, there are two different stacking sequences of identical packing density. If every atom in the third layer is placed exactly above one atom of the first layer, the stacking order is called ABA, which represents the hexagonal close packing (hcp, Figure 8.48c). In contrast, the stacking order ABC results if the atoms of the third layer are located above unoccupied hollow sites of the first atomic layer as in Figure 8.48d. The fourth atomic layer would then be positioned on the lateral positions of the first layer. This arrangement is the cubic close packing of spheres or the cubic face-centered lattice (fcc, Figure 8.48d). The close-packed atomic layers are oriented perpendicularly to the space diagonal of the cubic lattice (Figure 8.48d). The surface of metal crystals is formed by lattice planes. The **Miller indices**[99] $(h\ k\ l)$ are used to denote particular lattice planes. Parts of a surface with uniform periodic atomic structure are called **facet**. We speak of a **face** if the entire surface shares the same atomic arrangement.

Example Calculation 8.1 Determine the Miller indices of the lattice plane that contains the colored triangle in Figure 8.49.

The unit cell of the crystal shall be characterized by basis vectors $\mathbf{a}_1$, $\mathbf{a}_2$, and $\mathbf{a}_3$. The plane cuts the crystallographic coordinate system at points $4\mathbf{a}_1$, $2\mathbf{a}_2$, and $\mathbf{a}_3$. The reciprocal values are formed from the coordinates of the cutting points with the coordinate system. They show the same proportionality to each other as the unknown Miller indices $(h\ k\ l)$.

$$h : k : l = \frac{1}{4} : \frac{1}{2} : \frac{1}{1} \tag{8.48}$$

The fractions on the right side of Eq. (8.48) are multiplied by the least common multiple of all denominators (= 4). After reduction, one obtains the Miller indices, which

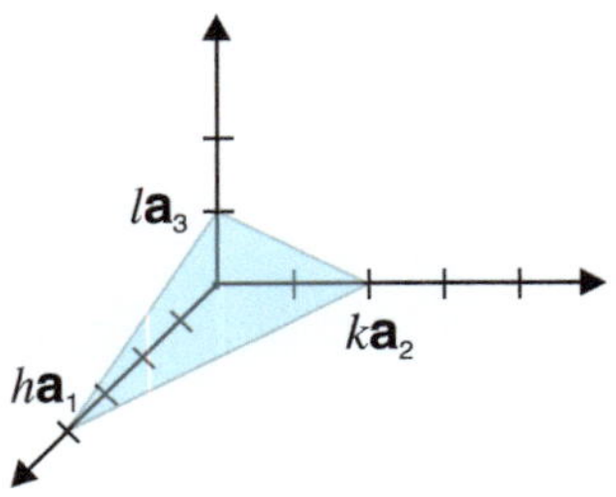

Figure 8.49 Determination of the Miller indices for a lattice plane.

[99] William Hallows Miller, 1801–1880, British mineralogist, crystallographer and physicist, professor of mineralogy at the University of Cambridge.

are given in round brackets and without commas. For this example, the solution is (1 2 4). Often a fixed-width blank space is placed between individual index numbers. This is important if larger index numbers occur to distinguish (1 11 0) from (1 1 10).

$$h : k : l = \frac{4}{4} : \frac{4}{2} : \frac{4}{1} = 1 : 2 : 4 \tag{8.49}$$

Example Calculation 8.2 Determine the location of the lattice plane (1 2 3).

We proceed in the inverse order compared to the Example Calculation 8.1. First, we divide the Miller indices by their least common multiple (= 6) to obtain the reciprocal values of integer numbers having the same proportionality as the Miller indices.

$$h : k : l = 1 : 2 : 3 = \frac{1}{6} : \frac{2}{6} : \frac{3}{6} = \frac{1}{6} : \frac{1}{3} : \frac{1}{2} \tag{8.50}$$

The unknown coordinates of the intersection of the (1 2 3) plane with the crystallographic coordinate system are the reciprocal values of the coefficients in Eq. (8.50), that is, $6\mathbf{a}_1$, $3\mathbf{a}_2$, and $2\mathbf{a}_3$ (Figure 8.50).

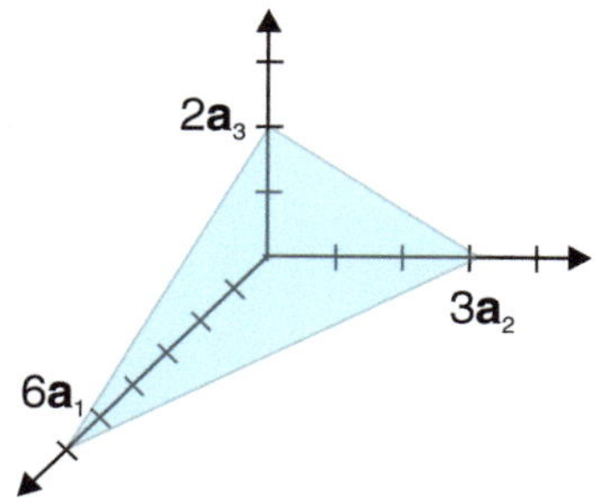

Figure 8.50 Determination of the lattice plane from known Miller indices (1 2 3).

The brackets have a well-defined meaning (Figure 8.51): Round brackets $(h\ k\ l)$ contain the indices of *one* lattice plane as in Figure 8.49. Curly brackets $\{h\ k\ l\}$ denote the *family* of all lattice planes *that are parallel* to the specified lattice plane. Square brackets $[h\ k\ l]$ specify the *direction* within the crystal that is *perpendicular* to the lattice plane $(h\ k\ l)$. The proportionality of the Miller indices to the reciprocal space coordinate turns out to be practically advantageous because the distance d_{hkl} of two neighboring lattice planes within a group of parallel lattice planes $\{h\ k\ l\}$ is experimentally accessible by diffraction experiments using the Bragg equation[96] $n\lambda = 2d_{hkl}\sin\theta$. The distance d_{hkl} is related to the Miller indices and the basis vectors by Eq. (8.51).

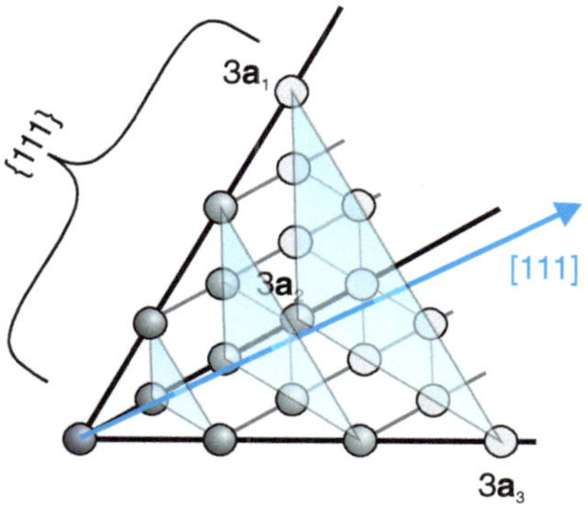

Figure 8.51 Relation between an individual lattice plane (1 1 1), the family of lattice planes {1 1 1}, and the direction [1 1 1] in the crystal.

$$\frac{1}{d_{hkl}^2} = \frac{h^2}{a_1^2} + \frac{k^2}{a_2^2} + \frac{l^2}{a_3^2} \tag{8.51}$$

The subsequent discussion of the crystal faces is restricted to the fcc lattice because the investigation of metal surfaces without a surface oxide layer is restricted to the coinage metals (Cu, Ag Au) as well as Pd and Pt, which all crystallize in the fcc lattice (Table 8.5). For other metals and alloys, it is almost impossible to study single-crystal surfaces in aqueous solution, because the surface will instantaneously be covered by a thin oxide film due to their low redox potential.

Figure 8.52 shows the atomic positions for a selection of crystal faces of an fcc metal lattice. The comparison of the most simple and most common low-index planes (1 0 0), (1 1 0), and (1 1 1) indicates distinct atomic arrangements and adsorption sites for specifically adsorbed ions in the electrical double layer and for intermediates of electrocatalytic reactions. High-index lattice planes can be exposed experimentally by cutting the crystal lattice with a shallow angle with respect to the elementary cell. Steps and kinks are formed in regular patterns and distances (Figure 8.53). When starting out from a (1 1 1) face and changing the direction of the cut in such a way that two indices remain equal, a $(l\ h\ h)$ surface is formed with steps in regular distances (shown in light blue in Figure 8.54). Steps are only one class of surface defects that

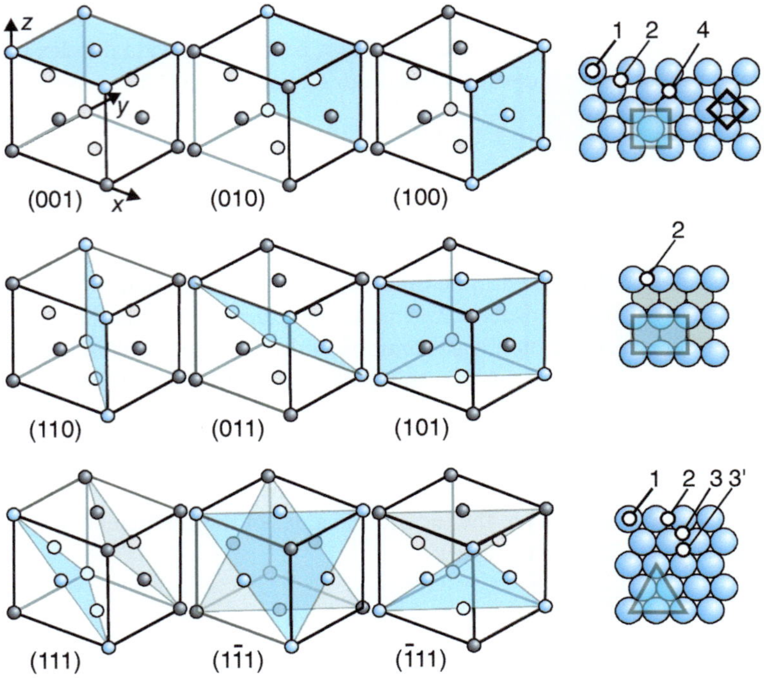

Figure 8.52 Important low-index lattice planes in fcc metals with the adsorption sites shown as white circles in the right part; 1) linearly bound (atop site), 2) bridge site, and 3) threefold hollow site, fourfold hollow site. Places 3 and 3′ of the (1 1 1) plane are distinguished by the atom positions in the second atomic layer. The shading emphasizes the arrangement of atoms in the elementary cell and within a particular lattice plane. The atoms of one elementary cell are connected by lines in the right part of the figure.

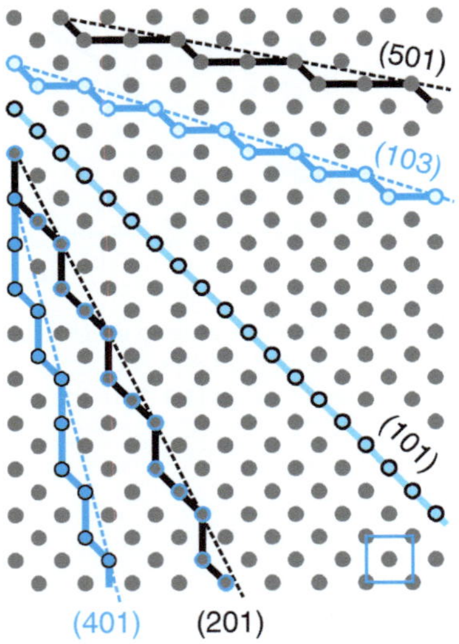

Figure 8.53 Atom positions of high index lattice planes of fcc metals.

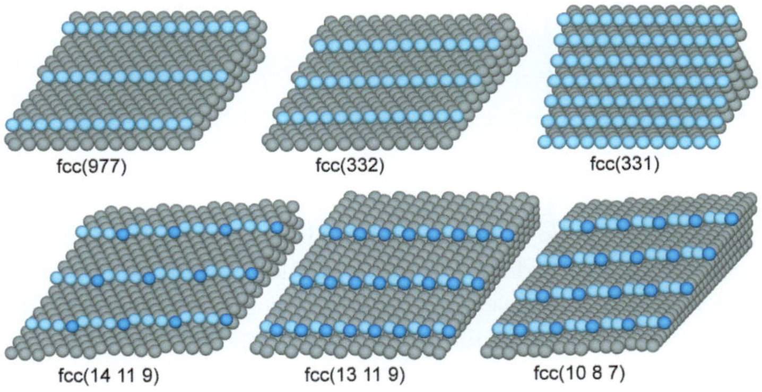

Figure 8.54 Examples of stepped single-crystal surfaces and stepped surfaces with kinks. Atoms at steps and kinks are highlighted by different colors. Source: The Figure was kindly provided by W. Dononelli, University of Oldenburg.

show a particular reactivity. Such surfaces can also be considered as a combination of a terrace of a low-index face of defined width and a monoatomic step. If all three indices are different, the surface exhibits monoatomic kinks or kink sites (Figure 8.54, blue). Kink sites are locations of particularly fast reaction rates during metal dissolution and metal deposition, because the coordination number of a kink site is exactly half of the coordination number in the bulk of the metal crystal. Thus, the activation energy for removing/forming coordinating bonds to solvent molecules or bonds to neighboring metal atoms is low.

8.9.2 What Is a Surface Reconstruction?

Even on a perfectly prepared single-crystal surface, the position of surface atoms often differs from the regular position of the atoms in the bulk of the crystal due to the incomplete coordination environment at the surface. **Surface relaxation** is found if the distance between the first and the second atomic layer is changed compared to the layer distance within the crystal, while the lateral atom positions are those found in the bulk of the crystal. In contrast, **surface reconstruction** also involves modification of the lateral atomic positions with respect to the bulk crystal. Reconstruction is accompanied by a reduction of the interfacial energy, which is the driving force of the process. Reconstruction phenomena are well known from investigations under UHV conditions. At single-crystal electrodes in solution, reconstructions known from UHV experiments can be retained or lifted depending on the electrode potential and the presence of specifically adsorbed anions [90]. As already introduced in Excursion 4-2, scanning tunneling microscopy facilitates the direct imaging of such potential-dependent structures in the electrochemical cell.

The observation of reconstruction phenomena can also be used to investigate the quality and cleanliness of single-crystal electrodes after their preparation. Particularly impressive is the herringbone pattern of the surface reconstruction of Au(1 1 1) electrodes in H_2SO_4 solution (Figure 8.55a and b). At higher electrode potentials, the unreconstructed surface can be observed (Figure 8.55c) before specifically adsorbed bisulfate HSO_4^- ions form a superstructure (Figure 8.55d).

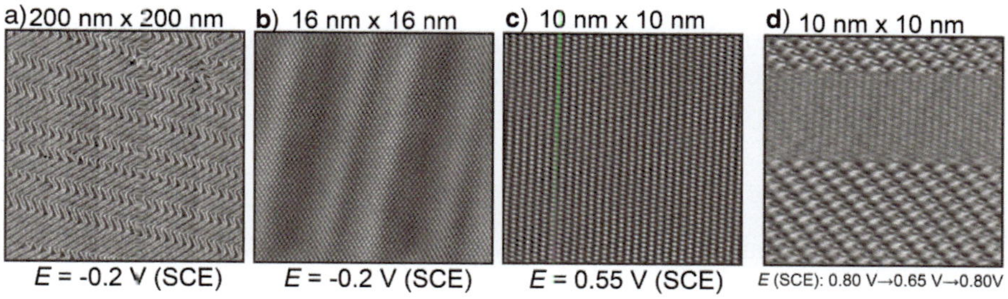

Figure 8.55 Potential-dependent structures of an Au(1 1 1) electrode in 0.1 mol L^{-1} H_2SO_4; a) $(\sqrt{3} \times 22)$-reconstruction at $E = 0.2$ V; b) as a) but smaller image frame; c) unreconstructed surface at $E = +0.55$ V; d) superstructure formed by specifically adsorbed HSO_4^- in the upper and lower part of the image at +0.8 V. A potential step to $E = +0.65$ V in the central part of the image frame exposes the unreconstructed metal surface for calibration purposes. All potentials are given with respect to SCE. Source: Panels a) to c) reproduced with permission from [91], Copyright 1999, Marcel Dekker; panel d) reproduced with permission from A. Cuesta, M. Kleinert, D. M. Kolb; *Phys. Chem. Chem. Phys.* **2000**, *2*, 5684–5690, Copyright 2000, Royal Society of Chemistry.

8.9.3 Which Types of Defects Exist at Surfaces?

First, a distinction is made between surface defects and volume defects. Volume defects traverse through the entire crystal and can also be observed at the surface. A screw dislocation shown in Figure 11.4b is an example of such a defect. We will discuss volume defects systematically in Chapter 11. Even in crystals that are free of volume defects, there are typical surface structures in addition to surface relaxation and surface reconstruction that we may consider as defects. Such defects are particularly important for electrochemical reactions. Their direct imaging by scanning probe methods has contributed significantly to the current understanding of many electrochemical phenomena. Figure 8.57 shows a terrace, an adatom (represented by a cube), an island, a step, a kink site, a hole, and a vacancy.

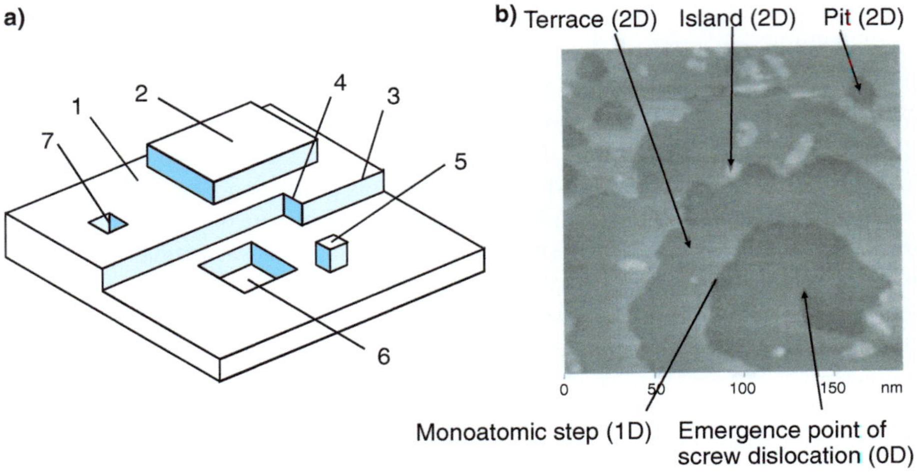

Figure 8.56 Defect structures at surfaces. a) Schematic representation of 1) a terrace, 2) an island, 3) a step, 4) a kink site, 5) an adatom, 6) a hole, and 7) a single atom vacancy; b) ECSTM image of an Au(1 1 1) electrode in 0.005 mol L^{-1} $AgClO_4$ + 0.5 mol L^{-1} $HClO_4$ at $E = +0.2$ V vs. Ag|Ag^+. Source: Panel b) is reproduced with permission from [92], Copyright 1999, Wiley-VCH.

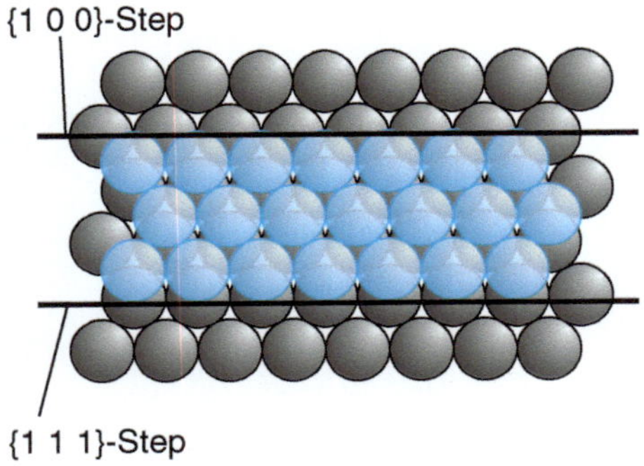

Figure 8.57 Different atomic structures of steps at a (1 1 1) surface of a fcc metal crystal.

There are different reasons for the occurrence of defects. Steps originate from a slight misalignment during cutting and polishing of the macroscopic single crystal during the preparation step. Different to the scheme in Figure 8.57a, steps can have different atomic structures. Figure 8.56 compares two possible step structures of a (1 1 1) surface. The small difference results from the atomic positions in the first and second atomic layer at the step edge.

8.9.4 Preparation of Single-Crystal Electrodes

Single crystals of metals can be obtained by melting and controlled solidification of a small pearl of a highly pure metal at the end of a wire, or crystal growth followed by cutting and polishing of macroscopic metal crystals. Some metals yield single crystals by evaporation onto suitable substrates. Of high importance is the treatment of the surface and its transfer to the electrochemical cell, during which contamination by organic impurities in the ambient air or by some specifically adsorbed anions must be prevented. All chemicals must be of the highest purity. Cell vessels from glass or Teflon are typically treated for several hours with peroxymonosulfuric acid, also known as Piranha solution or Caro's acid. (**Caution**, this compound formed in a mixture of 30% H_2O_2 and concentrated H_2SO_4 can ignite/explode when coming in contact with organic compounds. The amount of solution should be minimized to a few milliliters, and spent solutions must be deactivated immediately. This process requires special safety precautions, a specialized laboratory, and approval by the local occupational safety engineer!) Subsequently, the cell vials are rinsed and boiled several times in ultrapure water. An excellent summary with many further literature references is available online [93].

Pearl-Shaped Single-Crystal Electrodes. Such electrodes of Pt, Au, Ag, Ir, Rh, or Pd can be obtained relatively easily by melting one end of a high-purity noble metal wire in a flame [94–96]. Figure 8.58 shows a resulting electrode from Pt. The surface is composed mainly of (1 1 1) facets with large areas of terraces and monoatomic steps. Other facets can be exposed after orientation and cutting/polishing. The cleaning of the electrode is achieved by heating and quenching in ultrapure water without the need for complicated vacuum equipment.

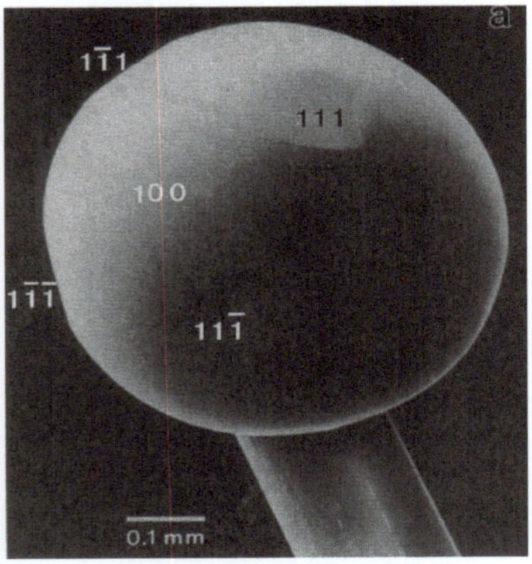

Figure 8.58 Pearl-shaped single-crystal Pt electrode. Source: Reproduced with permission from [80], Copyright 1999, Wiley-VCH.

Film electrodes from Ag and Au can be made by evaporation in a vacuum on suitable flat substrates (silicon wafer, mica, or glass). In the case of Au, a layer of an adhesion promoter such as Cr or Ti is employed between the substrate and Au. The strong affinity of Cr or Ti for oxygen allows a strong adhesion to the oxide surface of the substrate. The fresh metallic layer of the adhesion promoter is immediately covered by the noble metal. This allows the formation of metallic bonds between the adhesion promoter and Au before the surface of the adhesion promoter has the chance to collect oxygen-containing compounds from the residual gas of the vacuum or the ambient environment. Direct evaporation of Au on an oxidic support results in layers that do not adhere to the substrate when exposed to electrolyte solution with applied potentials. The preferential formation of large domains with predominantly (1 1 1)-orientation is promoted by several cycles of heating to a slight red glow and cooling in a stream of Ar or N_2.

A special variation of film electrodes is called template-stripped gold (TSG). First, Au is evaporated *without* an adhesion promoter on mica or on a silicon wafer. Next, a glass plate is glued to the gold film with epoxy resin. Once the gold layer, sandwiched between Si or mica and epoxy, is exposed to a solution, the template (Si or mica) can be peeled off. The result is an almost atomically flat gold film that was formed on the atomically smooth template.

Massive Single Crystals. The largest variety of single-crystal electrodes can be obtained from massive crystals (several millimeters to centimeters). They are produced by slow cooling of a melt in a crucible (disadvantage: the wall of the crucible can influence the crystallization) or by slow and controlled pulling of a seed from a melt (Czochralski process[100]). The crystallization occurs at the meniscus between melt and seed. The disadvantage is the need for elaborate equipment. The crystals are cut by sawing, spark erosion, electrical or chemical erosion. The surfaces are further treated, which is often more difficult than the actual growth of the crystal. First, the crystals are attached to a goniometer and oriented with the help of an X-ray beam. The crystals embedded in a resin can then be polished very cautiously with an accuracy of orientation better than 0.1°. Suitable polishing agents are silicon carbide paper followed by diamond paste, for soft Au crystals also Al_2O_3. Smoothing of the surface is achieved by sputter-heating cycles in UHV instruments. The formation of surface structures and elemental surface composition can be monitored.

Further **treatment of the surface and the recovery of a well-defined surface** between individual measurements starts with annealing in the flame of a Bunsen burner[101] or in a hydrogen flame. The flame has two functions. The oxidizing effect is used for the removal of ubiquitous organic contaminants. The transferred thermal energy increases the mobility of the surface atoms, which facilitates a smoothing of the surface and the healing of small defects. For low-melting metals, too strong heating must be avoided because melting of the crystal would destroy the single crystal. Therefore, the crystal is heated in intervals of 1–2 min up to a weak red glow.

A flame is very effective in the removal of organic contaminants. However, contaminants immediately adsorb again to the metal crystal from the surrounding atmosphere during cooling. This can be prevented by protecting the freshly treated surface. A drop of ultrapure water can be applied to protect and quench the hot metal. This works well for small pearl-shaped single crystals. In contrast, the mechanical tension between the contracting cooled surface and the still hot inner part of a larger massive single crystal may cause damage to the crystal. Therefore, such crystals are carefully cooled in a stream of inert gas. The surface is covered with a drop of water once the temperature of the metal has dropped below 100 °C. This works well for Ag and Au. More reactive

---

[100] Jan Czochralski, 1885–1953, Polish chemist, industrial occupation in Germany, professor at the Technical University of Warsaw.

[101] Robert Wilhelm Eberhard Bunsen, 1811–1899, professor of chemistry in Kassel, Marburg, Breslau (today Wroclaw), Heidelberg.

metals such as Pd, Pt, Rh, and Ir require further precautions. One approach is to heat the single crystals in an inert (O_2-free) atmosphere by resistive or inductive heating. Interestingly, the surface can be protected by an intentional contact with a strongly adsorbing gas like CO. (**Caution**, carbon monoxide CO is an odorless, strongly toxic gas. The work with CO requires special safety and warning equipment and approval by the local occupational safety authorities!). CO occupies all free adsorption sites on the metal surface and, thus, prevents the adsorption of other organic molecules from the gas phase. After transfer to an electrochemical cell, CO can be removed oxidatively in a controlled fashion (CO stripping, Section 12.4.1).

These brief descriptions of procedures have to be adapted for specific metals and sometimes even for specific facets on the crystal surface. Procedural details can be very important for the successful preparation of the intended facets. The collection of procedures in Ref. [93] is recommended as an excellent information source about the different approaches.

8.9.5 Characterization of Single-Crystal Electrodes

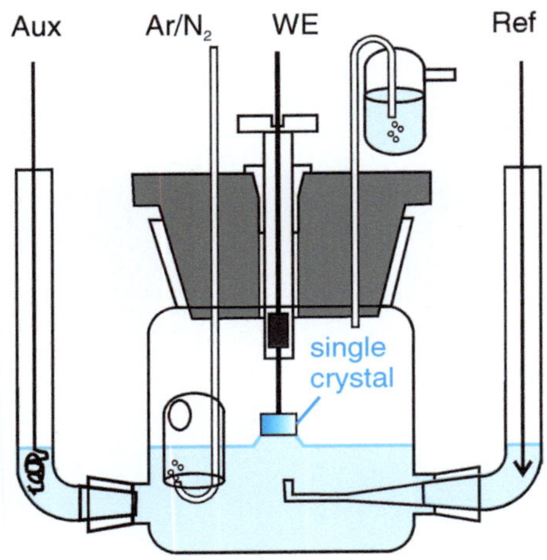

Figure 8.59 Cell body (glass) for the voltammetric experiments at single-crystal electrodes. Source: Adapted from [94].

Single crystal electrodes can be characterized by cyclic voltammetry in the sense of a fingerprint procedure. As discussed earlier regarding the preparation, there are specific requirements with respect to the cleanliness of glassware and solutions. A suitable cell construction is depicted in Figure 8.59. Compared to the cell containers discussed in Chapter 5, the following details are important: The electrode material is in contact with the electrolyte solution only via a liquid meniscus. The part of the solution undergoing purging with an inert gas (N_2, Ar) is separated by a hood to prevent gas bubble agitation or splashing solution to the side walls of the single-crystal electrode. Reference and auxiliary electrodes are placed in separate compartments. The initial potential, usually within the double-layer potential range, has to be applied to the electrode before making contact with the surface of the liquid electrolyte solution. Despite these precautions, the surface reconstruction changes can occur upon initial contact with the electrolyte solution, and this can only be excluded or confirmed by structure-sensitive methods (e.g., ECSTM).

Starting from the double-layer region, a cyclic voltammogram is recorded in an electrolyte without redox systems and without particularly strongly adsorbing anions (e.g., in 0.5 mol L^{-1} H_2SO_4 or 1 mol L^{-1} $HClO_4$). Sometimes, strongly adsorbing anions or cations are intentionally added in order to study their specific adsorption or underpotential deposition (Sections 8.9.7 and 8.9.8). The current signals in single-crystal voltammograms are caused by:

- adsorption/desorption of atomic hydrogen (also called underpotential deposition/stripping),
- adsorption/desorption of hydroxide ions and formation of defined thin oxide films,
- changes of the surface reconstruction upon change in the electrode potential that is connected to a change in the specific double-layer capacitance (thus the current flow) [91],
- specific adsorption/desorption of anions or changes of the superstructure of adsorbates with a concomitant change in the specific double-layer capacitance [97],
- underpotential deposition/stripping of metal ions.

Often, very sharp, needle-like signals are recorded at well-prepared single-crystal electrodes, due to a limited number of energetically uniform surface sites. The cyclic voltammograms of such surfaces have a diagnostic value in the sense of a fingerprint, even if a precise assignment of each signal component is not possible. Figures 8.60–8.62

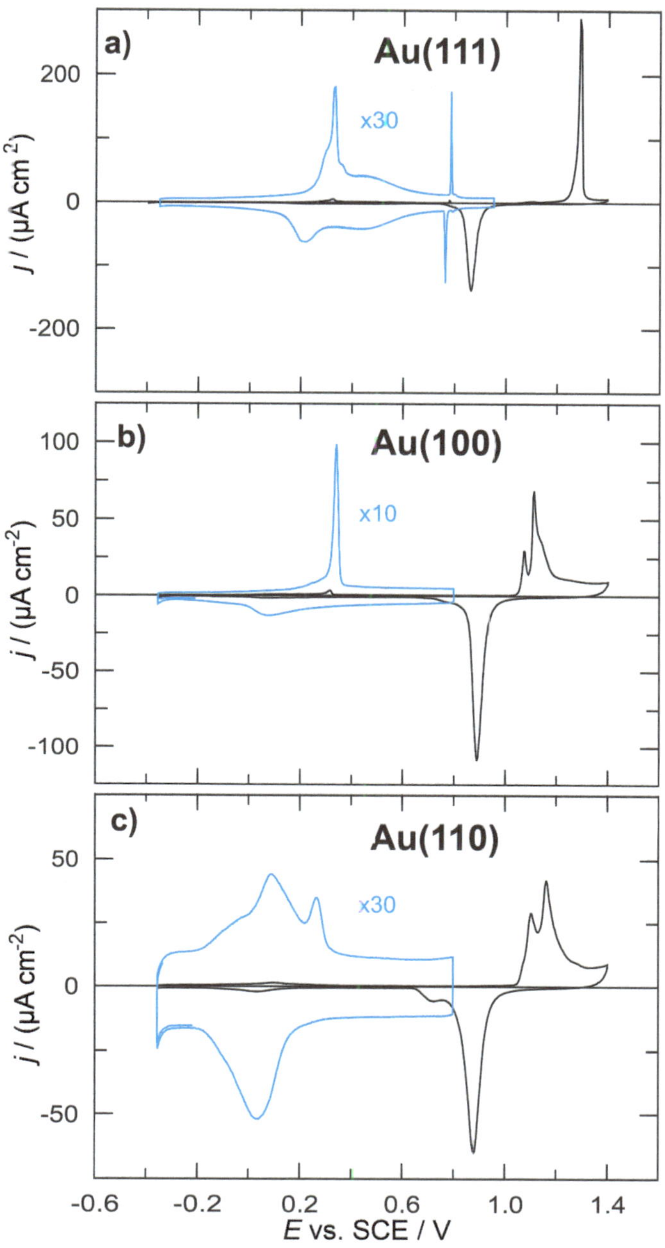

Figure 8.60 Cyclic voltammograms of a) Au(1 1 1), b) Au(1 0 0), c) Au(1 1 0) in 0.1 mol L^{-1} H_2SO_4, $v = 0.01\ V\ s^{-1}$, $T = 298$ K; blue curves are enlarged to better show the range of H-adsorption. Source: Reproduced with permission from [91], Copyright 1999, Marcel Dekker.

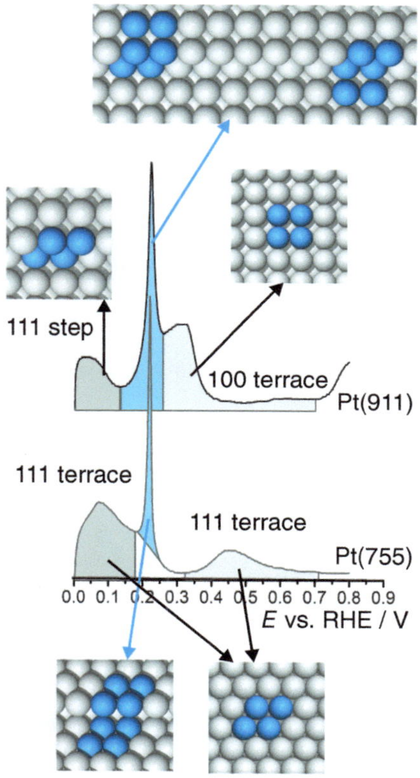

Figure 8.61 Assignment of voltammetric signals of H-adsorption/stripping at stepped single-crystal surfaces. Source: Reproduced with permission from [97], Copyright 2017, Wiley-VCH. The layout was adjusted by the textbook author and V. Climent, University of Alicante.

show examples of such voltammograms of common single-crystal electrodes in pure supporting electrolytes.

The exact assignment of all signals requires the correlation of voltammetric data to the results from structure-sensitive methods that are recorded simultaneously with the recording of voltammograms. Furthermore, it must be considered that steps and kinks can also cause prominent signals that overlay with those of the facet of interest. On the other hand, some faces, as for instance perfectly prepared Pt(1 1 1) electrode, cause a quite unspecific broad signal over a wide potential range (Figure 8.62). This is particularly evident in the comparison of the voltammograms of different stepped single-crystal surfaces in Figure 8.61. Due to the broad and overlapping signals, a determination of the fractional contribution of different facets at polycrystalline

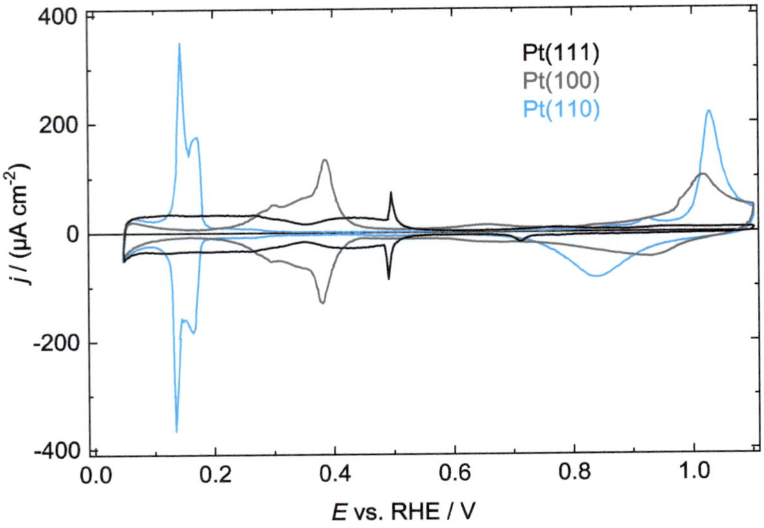

Figure 8.62 Cyclic voltammograms of a) Pt(1 1 1), b) Pt(1 0 0), c) Pt(1 1 0) in 0.1 mol L^{-1} H_2SO_4, $v = 0.05\ V\ s^{-1}$, $T = 298$ K. Source: Reproduced with permission from L. A. Kibler, A. Cuesta, M. Kleinert, D. M. Kolb; *J. Electroanal. Chem.* **2000**, *484*, 73–82, Copyright 2000, Elsevier.

electrodes or metal nanoparticles requires the additional consideration of voltammograms from adsorption/desorption and underpotential deposition/stripping experiments (Section 8.9.8). This process is demonstrated for Pt in Ref. [98].

8.9.6 What Are Superstructures?

If anions, molecules, or metal cations are adsorbed or deposited to defined adsorption sites on metal electrodes (also called "substrate" in this context), these species occupy the energetically most advantageous sites. If the adsorbed monolayer contains exactly the same number of particles as atoms in the top atomic layer of the metal electrode, the arrangement could be that of Figure 8.63a. There are also layers with linearly bound adsorbates, such as H_{ads}–Pt, that show the same number of adsorbate molecules as there are metal atoms in the top layer (Table 8.1, row 2). In many cases, the adsorbate layer contains fewer particles than the topmost layer of the metal electrode. This can be the case if the adsorbate has a larger diameter than the metal atom of the electrode, or if there is considerable electrostatic repulsion between the adsorbed ions that makes a close packing energetically unfavorable. In these cases, the distance between two neighboring adsorbate species is larger than between two metal atoms in the topmost atomic layer of the electrode. There are different possibilities for the packing of the adsorbate relative to the lattice of the electrode. If the adsorbate molecules occupy only the energetically most favorable sites with the smallest just feasible separation (Figure 8.63b), a periodic adsorbate structure is obtained that is different from the atomic structure of the electrode. However, the lattice vectors of the adsorbate structure are connected to the horizontal lattice vectors of the electrodes by rational factors. This is called a commensurate superstructure.

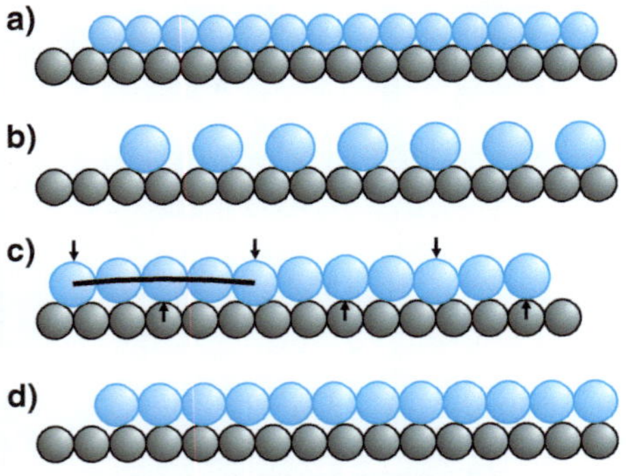

Figure 8.63 Schematic cross section through adsorbate layers of different structure;

a) same arrangement of adsorbate and substrate atoms (commensurate);
b) periodic arrangement of the adsorbate with a rational multiple of the periodicity of the substrate (commensurate);
c) periodic arrangement of higher order, in which only every fourth adsorbate occupies an equivalent adsorption site on the substrate (commensurate of higher order);
d) incommensurate arrangement, in which the lattices of substrate and adsorbate layer are *not* related to each other.

A two-dimensional periodic arrangement of adsorbates with defined periodicity with respect to the crystalline substrate is called a **superstructure**.

There are also examples in which a size difference between adsorbate and substrate is compensated by a structure, in which only every n^{th} adsorbate species occupies an equivalent site with respect to the substrate. Such layers are called commensurate layers of higher order (Figure 8.63c). Furthermore, there are cases in which the lattice of the adsorbates or surface layers is not in a regular relationship to the lattice of the substrate (Figure 8.63d). Such arrangements are called incommensurate.

We have briefly touched on superstructures during the discussion of surface reconstruction in Figure 8.55. There are two independent nomenclatures to specify superstructures. The nomenclature according to Wood[102] follows the Scheme "S(h k l)–($|\mathbf{b}_1|/|\mathbf{a}_1| \times |\mathbf{b}_2|/|\mathbf{a}_2|$) R$y$ A". S stands for the element symbol of the substrate (electrode metal, e.g., Ag, Au, and Pt). A specifies the adsorbate, (h k l) are the Miller indices of the considered face of the substrate, $|\mathbf{b}_1|/|\mathbf{a}_1|$ is the ratio between the absolute amount of the basis vectors of the adsorbate lattice ($\mathbf{b}_1$) and the basis lattice of the metal surface ($\mathbf{a}_1$). $|\mathbf{b}_2|/|\mathbf{a}_2|$ is the corresponding value for the second basis vector along the surface. Ry° specifies a possible rotation of the adsorbate lattice ($\mathbf{b}_1\mathbf{b}_2$) with respect to the substrate lattice ($\mathbf{a}_1\mathbf{a}_2$) as rotational angle y in degrees. When the rotational angle y is zero, the specification is omitted. For further specification, the letters "c" (centered) or "p" (primitive) can precede the superstructure. "c" indicates that the center of the adsorbate unit cell is occupied. This position is unoccupied in a primitive adsorbate lattice. "p" is only specified if there could be a structure with the same lattice vectors and an occupied central position. The advantage of the nomenclature – according to Wood – rests in its intuitive clarity.

However, this nomenclature cannot always be applied, for instance, if the angles between the lattice vectors of substrate and adsorbate lattices differ from each other, or the length ratio between the lattice vectors assumes "exotic" values. In these cases, the more general matrix nomenclature $\mathbf{b} = \mathbf{S}\,\mathbf{a}$ is more suitable. It follows the same rules as the matrix description of coordinate transforms; $\mathbf{a}$ and $\mathbf{b}$ are the lattice vectors, and $\mathbf{S}$ is the 2×2 transform matrix with the matrix elements $\{\{S_{11}, S_{12}\}, \{S_{12}, S_{22}\}\}$. Thus, the adsorbate lattice vectors are given by $\mathbf{b}_1 = S_{11}\mathbf{a}_1 + S_{12}\mathbf{a}_2$ and $\mathbf{b}_2 = S_{21}\mathbf{a}_2 + S_{22}\mathbf{a}_2$.

Superstructures can be particularly well analyzed using low-energy electron diffraction (LEED) [1, 79, 80]. The diffraction pattern of a superstructure with $|\mathbf{b}_1| > |\mathbf{a}_1|$ or $|\mathbf{b}_2| > |\mathbf{a}_2|$ contains *additional* reflexes with *lower* distance. As a reminder: Diffraction patterns show structures in the reciprocal space with the reciprocal lattice vectors $h/\mathbf{a}_1, k/\mathbf{a}_2$. Table 8.6 shows a selection of important superstructures for fcc metals. Further superstructures can be found in [99].

If there is a size difference between adsorbate atoms and the substrate, a commensurate arrangement of higher order can be formed, of which we have discussed a cross section in Figure 8.63c. In ECSTM images, such structures can be recognized by their periodicities that are much larger than the periodicities expected from atomic distances. Such features are also called moiré pattern (from the French word *moire* – shimmering or optical pattern for example on silk textiles), which can emerge from the superposition of dot patterns if they do not have the same periodicity or show a rotation with respect to each other. Figure 8.64 shows a selection of such structures. An experimental example has already been discussed in Figure 4.42.

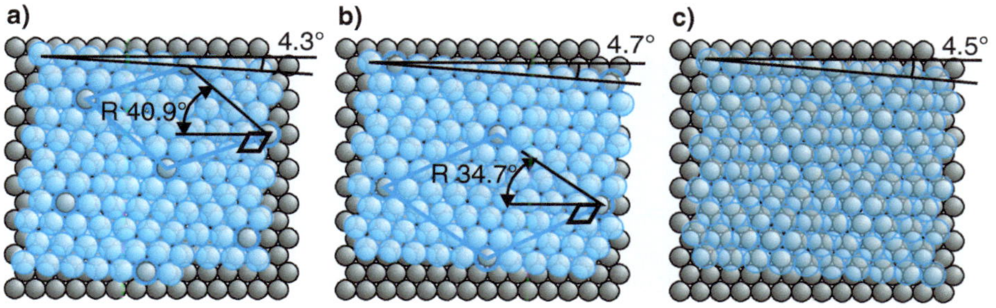

Figure 8.64 Arrangement of adsorbates in complex superstructures.

a) Commensurate structure of higher order with moiré pattern S(1 1 1)–($\sqrt{28} \times \sqrt{28}$) R 40.9° $A_{moiré}$;
b) Commensurate structure of higher order with moiré pattern S(1 1 1)–($\sqrt{37} \times \sqrt{37}$) R 34.7° $A_{moiré}$;
c) Incommensurate structure, in which a hcp adsorbate layer is rotated with respect to the substrate. Adsorbate atoms positioned exactly on top of a substrate atom are highlighted in a) and b).

Figure has been redesigned after [92].

[102] Elizabeth Armstrong Wood, 1912–2006, US American crystallographer and geologist. She led a research program at Bell Telephone Laboratories that eventually led to the discovery of superconductivity and lasers. Cofounder of the American Crystallography Association, which has conveyed the Science Writing Award for understandable science communication named after her since 1997.

Table 8.6 Important superstructures of fcc metal electrode in real space and as diffraction patterns. Adsorbates are drawn in blue and transparent in the real space and as small dots in the diffraction patterns.

Nomenclature	Image in real space	Diffraction pattern (reciprocal space)
S(1 0 0)–(1 × 1) A	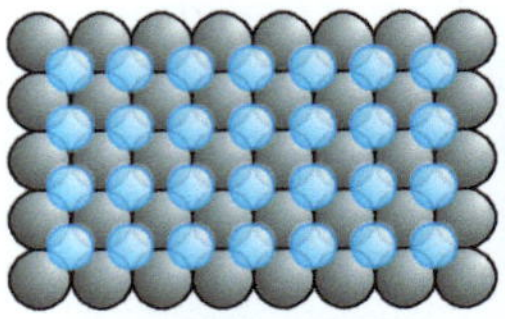	
S(1 0 0)–c(2 × 2) A or S(1 0 0)–($\sqrt{2}$ × $\sqrt{2}$) R 45° A	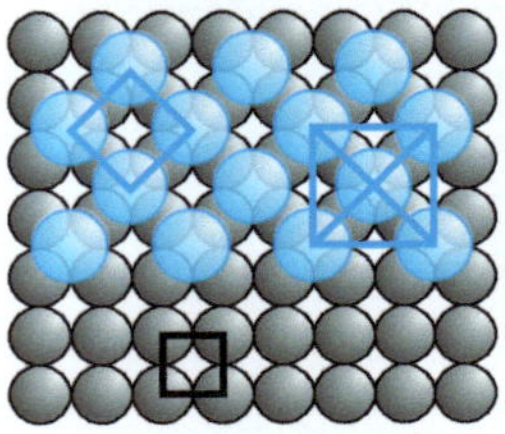	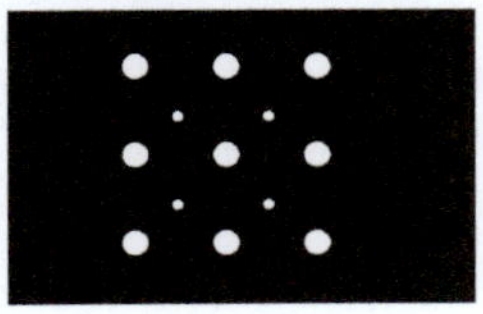
S(1 0 0)–(1 × 2) A	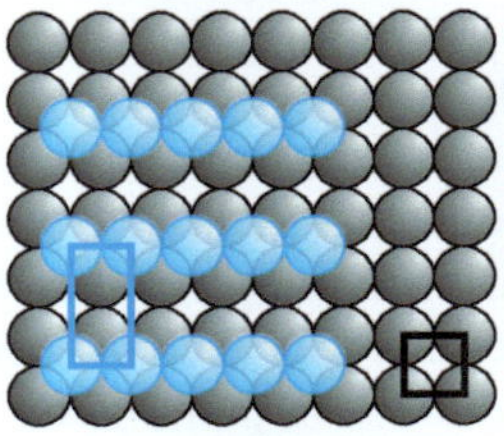	
Domains of S(1 0 0)–(1 × 2) A *and* S(1 0 0)–(2 × 1) A	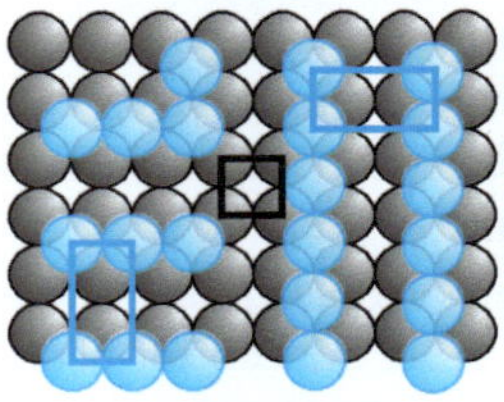	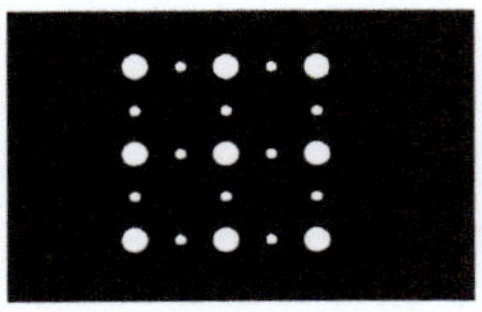
S(1 1 0)–(1 × 2) A	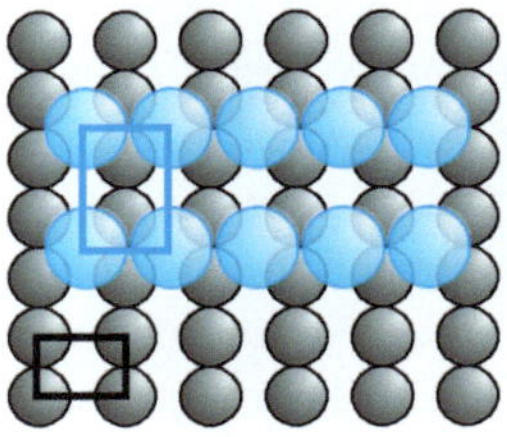	
S(1 1 1)–($\sqrt{3}$ × $\sqrt{3}$) R 30° A	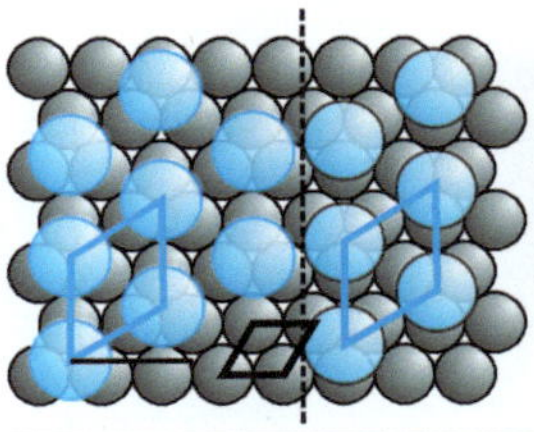	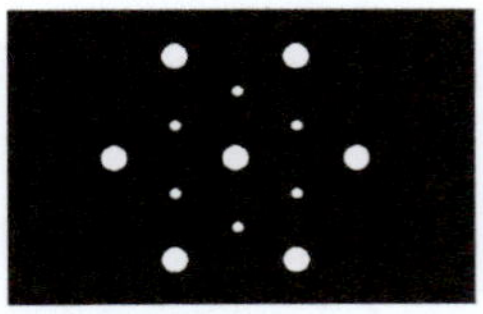

8.9.7 Superstructures of Specifically Adsorbed Anions

Ions with strongly bound solvation shells that include most cations and F^-, ClO_4^-, PF_6^- are retained in the electrical double layer exclusively by electrostatic attraction ("non-specific adsorption"). Thus, these kinds of ions and their arrangement in the double layer do not depend on the chemical nature of the anion. Less strongly solvated ions (Cl^-, Br^-, I^-, CN^-, SCN, S^{2-}, SO_4^{2-}, HSO_4^-) may lose part of their solvation shell upon approach to the electrode interface. Instead, they form a directed bond to the metal surface. The chemical bond between the metal electrode and the anion causes an enrichment of the anion compared to the surface concentration that would be required for charge compensation (see Figure 8.75). The bond between metal and anion is a *directed* bond that depends on a specific arrangement of the anion toward the metal atoms at the electrode surface. Depending on the charge of the surface, different arrangements of the anions can be favored thermodynamically in superstructures. Changes of the electrode potential can cause phase transitions between different superstructures, which can be recognized in cyclic voltammograms at single-crystal electrodes as very sharp current spikes. The example in Figure 8.65 shows the simultaneous occurrence of regions (domains) with a (3×3) or a $(\sqrt{7} \times \sqrt{7})$ R 21.8° superstructure of specifically adsorbed iodide ions on a Pt(1 1 1) surface. For further information and examples, a review paper [100] is recommended.

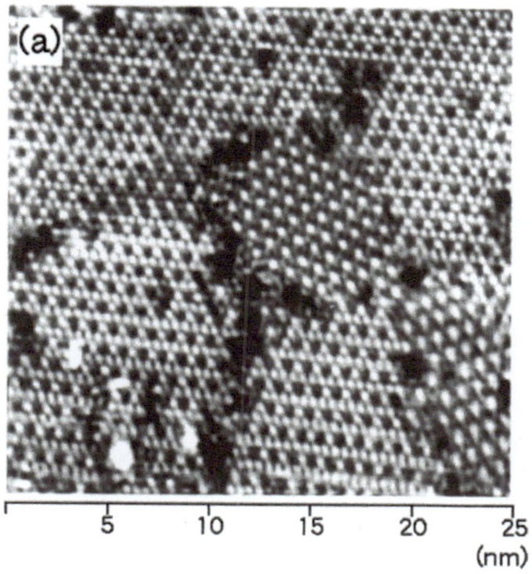

Figure 8.65 ECSTM image of a monolayer of specifically adsorbed I ions on a Pt(1 1 1) electrode in 50 mmol L^{-1} H_2SO_4 + 1 mmol L^{-1} KI at $E = 0.15$ V versus RHE.
Source: Reproduced with permission from J. Inuski, Y. Osawa, M. Wakisaka, K. Sashikata, Y.-G. Kim, K. Itaya; *J. Phys. Chem. B* **1998**, *102*, 3498. Copyright 1998, American Chemical Society.

8.9.8 Superstructure of Metal Adatoms After Underpotential Deposition

Generally, the equilibrium potential of a metal electrode of the first kind is obtained according to Eq. (8.52) (Chapter 2).

$$E_{eq} = E^{\circ}_{M|M^{z+}} + \frac{RT}{zF} \ln\left(\frac{a_{M^{z+}}}{a^{\circ}}\right) \tag{8.52}$$

This equation describes the potential at which the rate of dissolution and the rate of deposition of a metal M are just equal (valid only if the dissolution and deposition are the only processes; Figure 8.66b). The corresponding electrode reaction was noted in Chapter 2 as

$$M \rightleftharpoons M^{z+} + ze^- \tag{8.53}$$

E_{eq} – Equilibrium potential [V]
E° – Standard electrode potential [V]
R – Gas constant = 8.314 A s K^{-1} mol^{-1}
T – Temperature [K]
z – Ionic charge of Me^{z+} (= number of transferred electrons n in discharge reaction) [dimensionless]
$a_{M^{z+}}$ – Concentration activity = $c^* f_c$ [mol L^{-1}]
a° – Standard concentration activity = 1 mol L^{-1}
λ – Partial charge transfer coefficient [dimensionless]

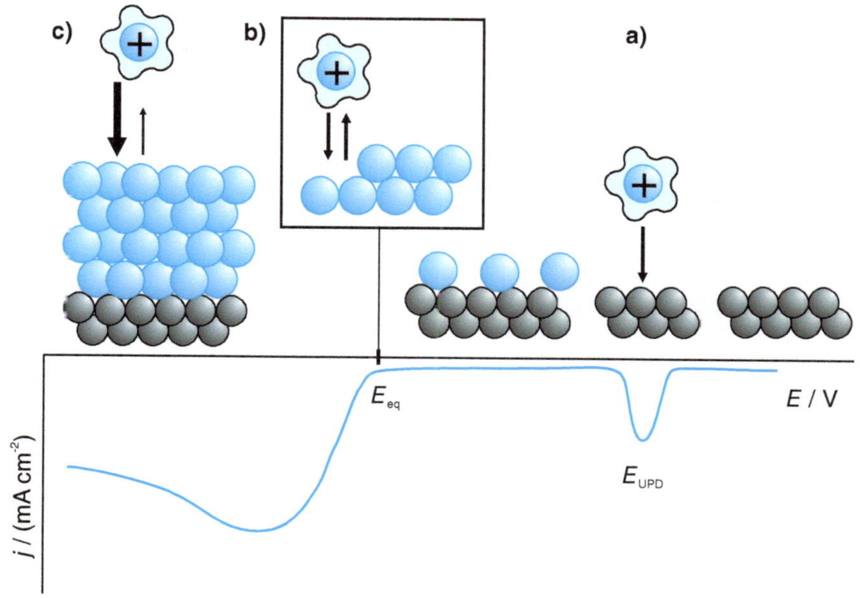

Figure 8.66 Schematic representation of a) the underpotential deposition of M on a foreign substrate S; b) the reaction at a metal ion electrode of the first kind at E_{eq}; and c) the metal deposit on of the metal ion M^+ at an overpotential onto the already deposited atomic layers of the same metal M.

This notation ignores the possibility that a deposition can occur on a substrate S. The corresponding reaction equation would then be more exactly written as

$$\text{S–M}_{\text{ads}}^{(1-\lambda)z+} \rightleftharpoons \text{S} + \text{M}^{z+} + \lambda \cdot z\text{e}^- \tag{8.54}$$

S can be, but does not necessarily need to be, the deposited metal M. We could, for instance, deposit Cu on Au. If the Au substrate is completely covered by a layer of Cu, we will obtain the equilibrium potential according to Eq. (8.52). However, this is *not* expected for the first layer of Cu atoms on the Au substrate because the interaction between Cu atoms and an Au substrate differs from that of Cu atoms with a Cu substrate (Figure 8.66c). Furthermore, the interaction *between* the Cu atoms of the first Cu layer may be different from that within a Cu crystal because the atomic distances are determined by the superstructure formed on the Au substrate, which is not identical to the Cu–Cu distance within a Cu crystal.

There are also options that the deposited ion Me^{z+} is completely discharged to $Me^{(0)}$ ($\lambda = 1$) or that the ion is incorporated into the double layer as a positively charged cation ($\lambda = 0$). The partial charge transfer coefficient λ ($0 \leq \lambda \leq 1$) allows a description of any thinkable situation between these two limiting cases [101, 102]. The product $\lambda \cdot z$ is also called electrosorption valency. A comprehensive description of this phenomenon can be found in [103].

In many cases, the deposition of the first or the first few monolayers of metals on more noble substrates occurs at a potential *positive* of the potential E_{eq} according to Eq. (8.52). Such a deposition is referred to as **underpotential deposition** (UPD, Figure 8.66a). This terminology emphasizes the difference compared to the deposition of thicker metal layers of Me^{z+} on M, for which an overpotential (i.e., a more *negative* electrode potential than E_{eq}) must be applied to drive the reaction (Figure 8.66c).

The UPD process is very important practically as it represents the initial step of a metal deposition on a foreign substrate (Chapter 9). UPD layers can also modify the electrocatalytic properties of surfaces (Chapter 12). The analysis of UPD processes has diagnostic value. Single-crystal electrodes enable the unraveling of complex interdependent relationships between energetic and structural factors for the formation of UPD layers, which are often much more complicated than in Figure 8.66, when specifically adsorbed anions influence the structure formation. Specifically adsorbed anions may form different superstructures on the substrate S and the deposited metal M. This topic has been studied very intensively, and comprehensive texts can be recommended for further reading [92, 104].

Interpretation of UPD Data. Figure 8.67a shows a cyclic voltammogram of an UPD process of Pb^{2+} ions on polycrystalline Ag. The curve is broad with a few shoulders both in the positive-going and the negative-going half-cycles because there are several energetically slightly different adsorption sites on a polycrystalline Ag electrode. Due to the distribution of the overall charge on several (unresolved) peaks, the recorded signal is broad, but much smaller than for a single crystalline Ag(1 1 1) electrode in Figure 8.67b. For an electrode that has been prepared even better by an electrochemical method, the charge is concentrated even more on the peak pair A_2/D_2 (A – adsorption/deposition, D – desorption/stripping). The voltammetric signals have been assigned to reactions at specific sites of the surface by means of electrochemical tunneling microscopy (ECSTM, Section 8.11). The signal pair with A_1 corresponds to the decoration of steps, the signal with A_2 and the highest charge results from filling the terraces with a monolayer of Pb. In this process, a small range at the upper part of a step edge remains uncovered and is filled only when reaching the signal A_3. The

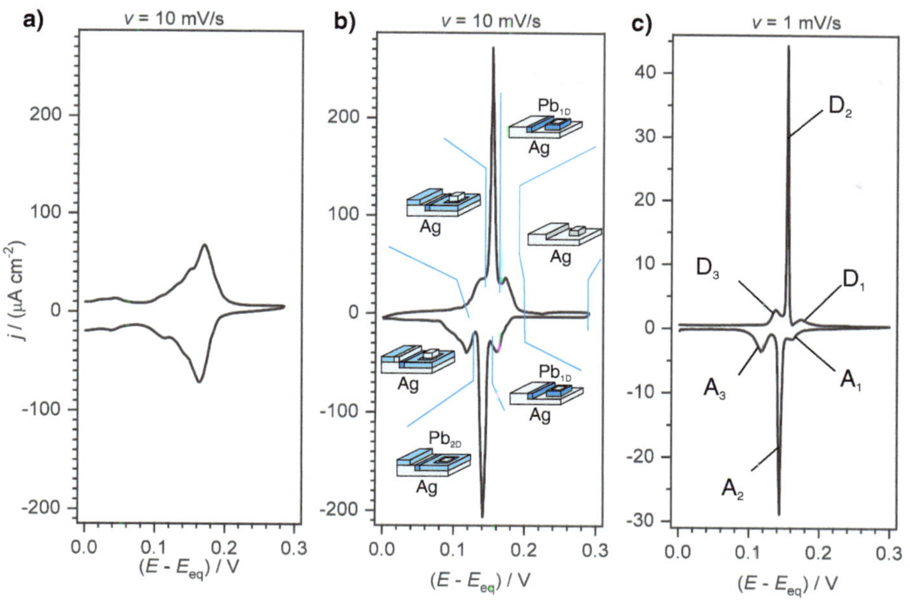

Figure 8.67 UPD processes of Pb on Ag;

a) Ag(poly);
b) chemically polished and cleaned Ag(1 1 1);
c) electrochemically grown and almost perfect Ag(1 1 1).

Measured data for a) and b) in 0.05 mol L^{-1} $Pb(ClO_4)_2$ + 0.5 mol L^{-1} $NaClO_4$ + 0.005 mol L^{-1} $HClO_4$, $v = 0.01\,V\,s^{-1}$, $T = 298$ K; c) recorded in 0.005 mol L^{-1} $Pb(ClO_4)_2$ + 0.5 mol L^{-1} $NaClO_4$ + 0.005 mol L^{-1} $HClO_4$, $v = 0.001\,V\,s^{-1}$, $T = 298$ K.
Source: Data for panels a) and b) from [104], data for panel c) from [92]; design and schematics adapted from [92].

deposition of Pb on an existing monolayer of Pb commences only at $E - E_{eq} = 0\,V$, that is, 0.15 V negative of A_2! Due to the energetic differences between the adsorption sites, cyclic voltammograms of UPD processes are suitable for characterizing the atomic structure of surfaces (such as metal nanoparticles and nanorods) including the contained defect sites, provided the assignment of the different UPD signals is already known from experiments with single-crystal electrodes and there are not too many energetically similar defects on the surface simultaneously.

The interpretation of the voltammetric data from UPD processes can be further complicated compared to conditions in Figure 8.67, because further processes can occur in addition to the deposition/stripping of the two-dimensional adsorbate phase, depending on the particular M and S involved:

- formation of two-dimensional surface alloys (M and S occur together in the topmost atomic layer),
- formation of three-dimensional M–S alloys at $E > E_{eq}$,
- potential-dependent phase transitions between different superstructures of M on S.

Collections of studied systems can be found in the following reviews and books: [105–108]. For Pt electrodes, comprehensive surveys are available [97, 109] covering preparation, voltammetry, and results from UPD experiments and adsorption experiments linked to consideration of the interfacial thermodynamics of the system. Characteristic voltammograms for Au can be found in [110–112].

8.10 A Closer Look at the Electrical Double Layer

In Sections 4.3.2 and 4.3.3, we have discussed different models of the electrical double layer in a qualitative manner. Here, we complement the earlier considerations with an algebraic description. In doing so, we follow comprehensive treatments of the subject (including experimental examples) in [113, 114] and descriptions in other textbooks [115–117]. These sources also provide applications of the system of equations to special electrodes such as interfaces between immiscible electrolyte solutions (ITIES) or semiconductor electrodes [117]. In the context of double-layer theory, the adsorption of neutral compounds is important because such layers may influence the rate of electrochemical reactions, which is broadly exploited in industrial galvanic deposition processes for which modern treatments are available [118, 119]. Below, we begin with a thermodynamic approach that does not presume a particular structural model of the interface, but characterizes its properties by extensive and intensive physical quantities. The application of more general thermodynamic relations allows us to relate these quantities to each other and derive interdependencies.

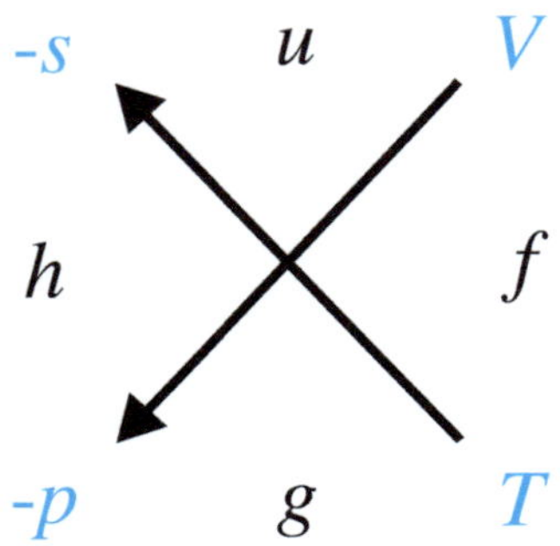

Figure 8.68 The Guggenheim square.

Excursion 8-2 The Guggenheim Square

The Guggenheim square[103] (Figure 8.68) is a mnemonic to get partial derivatives of the thermodynamic potentials Gibbs free energy $g(T, p)$, Helmholtz free energy $f(V, T)$, internal energy $u(V, s)$, and enthalpy $h(p, s)$ with respect to the variables of state s, p, T, and V. The amount of substance N is kept constant in all cases. The sides of the square are formed by one of the thermodynamic potentials (printed in black). Each potential is defined by characteristic state variables (printed in blue at the corners of the square). In order to obtain a partial derivative, we start from one of the thermodynamic potentials X, move to either of the two neighboring state variables Y, for which we seek the derivative and follow the diagonal line toward the result. Next, we need to pay attention to the sign. There is a minus sign left of the state variables s and p. The sign of the expression $\partial X/\partial Y$ is multiplied by -1, if we follow the diagonal in the opposite direction to the arrowhead. In this way, we obtain the following partial derivatives:

$$\left(\frac{\partial u}{\partial s}\right)_{T,N} = \left(\frac{\partial h}{\partial s}\right)_{V,N} = T \tag{8.55}$$

$$\left(\frac{\partial u}{\partial V}\right)_{T,N} = \left(\frac{\partial f}{\partial V}\right)_{T,N} = -p \tag{8.56}$$

$$\left(\frac{\partial h}{\partial p}\right)_{T,N} = \left(\frac{\partial g}{\partial p}\right)_{T,N} = V \tag{8.57}$$

$$\left(\frac{\partial g}{\partial T}\right)_{p,N} = \left(\frac{\partial f}{\partial T}\right)_{V,N} = -s \tag{8.58}$$

There are further relations between the variables of state in the corners of the square:

$$\left(\frac{\partial V}{\partial T}\right)_{p,N} = -\left(\frac{\partial s}{\partial p}\right)_{T,N} \tag{8.59}$$

$$-\left(\frac{\partial p}{\partial T}\right)_{V,N} = -\left(\frac{\partial s}{\partial V}\right)_{T,N} \tag{8.60}$$

The partial derivatives of the thermodynamic potentials with respect to the amount of one substance N are the electrochemical potential of that compound in the system (not contained in the square):

$$\left(\frac{\partial g}{\partial N}\right)_{p,T} = \left(\frac{\partial f}{\partial N}\right)_{V,T} = \left(\frac{\partial h}{\partial N}\right)_{p,T} = \left(\frac{\partial u}{\partial N}\right)_{V,T} = \tilde{\mu} \tag{8.61}$$

[103] Edward Armand Guggenheim, 1901–1970, English physicochemist, professor of chemistry at the University of Reading.

Finally, we need another mnemonic to help us arrange the symbols at the correct positions within the square. There are several of them, such as "good physicists have studied under Very fine Teachers", which starts at the lower side (at 6 o'clock on an analog watch) and then follows the square in clockwise direction. Another one, starting in the top-left corner and going counterclockwise goes like "some hard problems go To finish Very Easy" (here E is used instead of u for the internal energy).

8.10.1 Interfacial Region and Excess Quantities

Let us examine the thermodynamics of a metal–electrolyte interface (Figure 8.69a): For indexing the phases in equations, we use α for the metal and β for the liquid electrolyte phases. A charge on the metal is compensated by an excess of the oppositely charged ion in the electrical double layer. This increases the concentration of the compensating ion above its concentration in the bulk of the electrolyte. To facilitate the treatment of this system with concentrations that depend on the distance to the interface, it is sensible to introduce conceptually a third phase, the so-called interphase with the index σ [Figure 8.69b, note the difference between interphase (a layer of finite thickness) and interface (a mathematical plane)]. We set the thickness of the interphase in such a way that it comprises *at least* the range in which the local concentrations deviate from those of the undisturbed bulk of the adjoining phases α and β. That means this layer must at least have the area of the interface and the thickness of the entire electrical double layer. To be sure, we can set the thickness "generously" a bit larger; this will not change the following treatment. Outside of the interphase, the phases α and β have their ideal, undisturbed composition. This system S (α–σ–β) is now compared to a hypothetical reference system R (Figure 8.69c), in which the same amounts of all substances are present, but the undisturbed phases α and β adjoin each other.

This concept allows us to define an interfacial excess for each quantity by forming the difference of that quantity for the system S and the reference system R. This excess must result only from the interphase in system S. Thus, all excess quantities are indicated by the superscript σ. For instance, the excess amount of Na^+ ions is

$$N^{\sigma}_{\mathrm{Na^+}} = N^{\mathrm{S}}_{\mathrm{Na^+}} - N^{\mathrm{R}}_{\mathrm{Na^+}} \tag{8.62}$$

or in general for the ith substance:

$$N^{\sigma}_{i} = N^{\mathrm{S}}_{i} - N^{\mathrm{R}}_{i} \tag{8.63}$$

If we relate the total excess amount of a substance to the area of the interface, we obtain the interfacial concentration Γ_i.

$$\Gamma_i = \frac{N^{\sigma}_{i}}{A} \tag{8.64}$$

Please note that Γ_i is also an **excess quantity**, it states the surplus or the deficiency (!) of the content relative to the content in the undisturbed bulk phases. Thus, it can also assume a negative value if a substance is depleted close to the interface. The word "excess" is used here as a generalization of "surplus" and "deficiency".

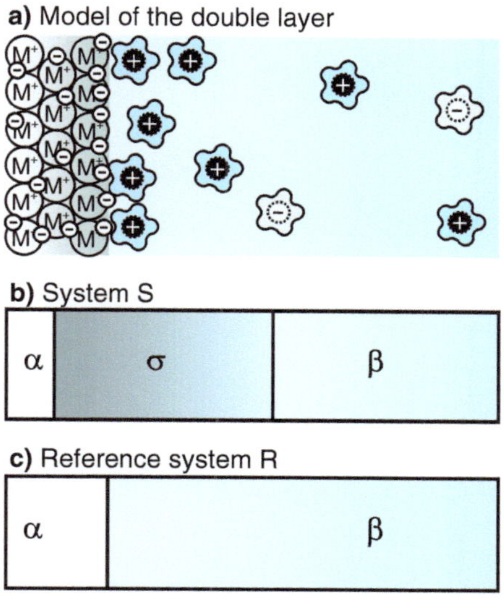

Figure 8.69 Schematic cross section through
a) a metal–solution interface;
b) designation of the phases α, β, and the interphase σ;
c) conceptual reference system in which the undisturbed phases α and β directly adjoin each other.

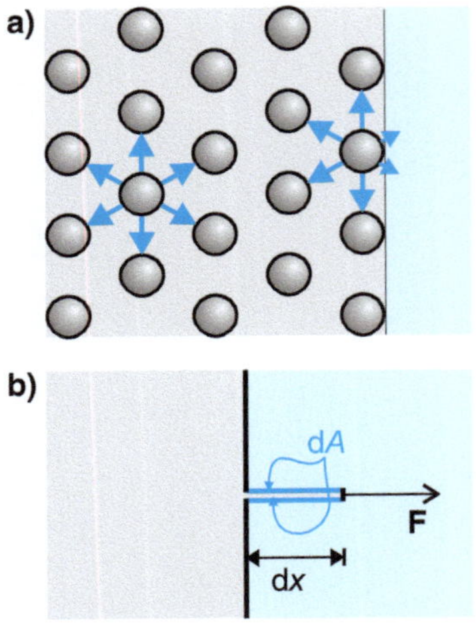

Figure 8.70 Cross section perpendicular to an interface; a) intermolecular forces at the surface and in the bulk of a condensed phase; b) work for enlarging the surface. The schematic is not to scale!

Surface Tension. Atoms and molecules at the surface (s|g, l|g) or an interface (s|s, s|l, l|l) experience a lower stabilizing interaction than the same species within the bulk of the phase (Figure 8.70a). Consequently, there is a net force toward the inner part of the phase acting on each particle located at the interface. Each system is therefore developing toward minimizing the interfacial area, for example, spherical Hg droplets and aerosol droplets form spontaneously as the shape with the least possible surface-to-volume ratio. The enlargement of the area for a fixed amount of substance requires additional energy. This energy is depicted in Figure 8.70b as mechanical work W, provided by the force **F** along the displacement dx. The provided work is proportional to the enlargement dA of the interface and to an intensive quantity γ, the surface tension or interfacial tension. The latter is the partial derivative of the Gibbs free energy g of the system with respect to the interfacial area A. Thus, its unit is [Nm/m^2] = [N m^{-1}].

$$\mathrm{d}W = \mathbf{F}\mathrm{d}x = \gamma \mathrm{d}A = \left(\frac{\partial g}{\partial A}\right)_{T,p,N_i} \tag{8.65}$$

Gibbs Adsorption Isotherm. With consideration of Eq. (8.65) the total Gibbs free energy of the considered system depends on the area of the interface. We can state the extensive quantities of the system S as the sum of the respective quantities for all subsystems. For instance, the total volume V is the sum of the volumes of all subsystems ($V = V^\alpha + V^\sigma + V^\beta$). This relationship also holds for the extensive version of the thermodynamic potentials u, h, g, and f. In this textbook, we indicate the extensive version of those quantities by lowercase symbols and use uppercase symbols U, H, G, and F for the corresponding molar quantities, which are intensive quantities.

At equilibrium, important intensive quantities are constant at all positions within the system, namely

- pressure equilibrium $p = const$, d$p = 0$,
- thermal equilibrium: $T = const$, d$T = 0$,
- electroneutrality of the total system,
- electrochemical equilibrium $\tilde{\mu}_i = const$, d$\tilde{\mu}_i = 0$.

Furthermore, the Gibbs free energy g of the reference system is

$$g^\mathrm{R}(T, p, N_i^\mathrm{R}) = g^{\alpha,\mathrm{R}} + g^{\beta,\mathrm{R}} \tag{8.66}$$

For the system S, g is obtained by summing up g of the three subsystems

$$g^\mathrm{S}(T, p, A, N_i^\mathrm{S}) = g^{\alpha,\mathrm{S}} + g^\sigma + g^{\beta,\mathrm{S}} \tag{8.67}$$

Next, we form the total differentials for g$^\mathrm{R}$ and g$^\mathrm{S}$ considering the equilibrium conditions d$T = 0$ and d$p = 0$.

$$\begin{aligned}\mathrm{d}g^\mathrm{R} &= \left(\frac{\partial g^\mathrm{R}}{\partial T}\right)_{p,N_i} \mathrm{d}T + \left(\frac{\partial g^\mathrm{R}}{\partial p}\right)_{T,N_i} \mathrm{d}p + \sum_i \left(\frac{\partial g^\mathrm{R}}{\partial N_i}\right)_{T,p,N_{j\neq i}} \mathrm{d}N_i^\mathrm{R} \\ &= \sum_i \left(\frac{\partial g^\mathrm{R}}{\partial N_i}\right)_{T,p,N_{j\neq i}} \mathrm{d}N_i^\mathrm{R}\end{aligned} \tag{8.68}$$

$$\begin{aligned}\mathrm{d}g^\mathrm{S} &= \left(\frac{\partial g^\mathrm{S}}{\partial T}\right)_{p,A,N_i} \mathrm{d}T + \left(\frac{\partial g^\mathrm{S}}{\partial p}\right)_{T,A,N_i} \mathrm{d}p + \left(\frac{\partial g^\mathrm{S}}{\partial A}\right)_{T,p,N_i} \mathrm{d}A + \sum_i \left(\frac{\partial g^\mathrm{S}}{\partial N_i}\right)_{T,p,A,N_{j\neq i}} \mathrm{d}N_i^\mathrm{R} \\ &= \left(\frac{\partial g^\mathrm{S}}{\partial A}\right)_{T,p,N_i} \mathrm{d}A + \sum_i \left(\frac{\partial g^\mathrm{S}}{\partial N_i}\right)_{T,p,A,N_{j\neq i}} \mathrm{d}N_i^\mathrm{R}\end{aligned} \tag{8.69}$$

Using the aid from Excursion 8-2, we can immediately substitute the partial derivatives; $\partial g/\partial N_i$ equals the electrochemical potential $\tilde{\mu}_i$, which has to be equal in all phases at equilibrium. We have identified the surface tension $\gamma = \partial g/\partial A$ before in Eq. (8.65). The substitutions yield:

$$dg^R = \sum_i \tilde{\mu}_i dN_i^R \tag{8.70}$$

$$dg^S = \gamma dA + \sum_i \tilde{\mu}_i dN_i^S \tag{8.71}$$

Thus, Eq. (8.72) is the differential of the interfacial excess of the Gibbs free energy.

$$dg^\sigma = dg^S - dg^R = \gamma dA + \sum_i \tilde{\mu}_i \left(dN_i^S - dN_i^R\right) \tag{8.72}$$

Using Eq. (8.63) yields Eq. (8.73).

$$dg^\sigma = \gamma dA + \sum_i \tilde{\mu}_i dN_i^\sigma \tag{8.73}$$

The excess Gibbs free energy g^σ at constant p and T is a function of A and all N_i^σ, which are extensive quantities. Specifically, g^σ depends linearly on those quantities and thus allows the application of Euler's theorem[76] yielding Eq. (8.74).

$$\begin{aligned} g^\sigma &= \left(\frac{\partial g^\sigma}{\partial A}\right)_{T,p,N_i^\sigma} A + \sum_i \left(\frac{\partial g^\sigma}{\partial N_i^\sigma}\right)_{T,p,A,N_{j\neq i}^\sigma} N_i^\sigma \\ &= \gamma A + \sum_i \tilde{\mu}_i N_i^\sigma \end{aligned} \tag{8.74}$$

From Eq. (8.74), we form the total differential Eq. (8.75) and subtract from it Eq. (8.73).

$$dg^\sigma = \gamma dA + \sum_i \tilde{\mu}_i dN_i^\sigma + A d\gamma + \sum_i N_i^\sigma d\tilde{\mu}_i \tag{8.75}$$

$$-dg^\sigma = -\gamma dA - \sum_i \tilde{\mu}_i dN_i^\sigma \tag{8.73}$$

$$0 = A d\gamma + \sum_i N_i^\sigma d\tilde{\mu}_i \tag{8.76}$$

Isolating $d\gamma$ from Eq. (8.76) and substituting the surface concentration $\Gamma_i = N_i^\sigma/A$ from Eq. (8.64) yield the most frequently used form of the **Gibbs adsorption isotherm**.[8]

$$-d\gamma = \sum_i \Gamma_i d\tilde{\mu}_i \tag{8.77}$$

The Gibbs adsorption isotherm is a special form of the **Gibbs–Duhem equation**[8,104]. For a system with only *one* phase, this relationship shows that T, p, and the phase composition cannot be varied independently from each other.

$$0 = sdT - Vdp + \sum_i N_i d\tilde{\mu}_i \tag{8.78}$$

The relationship simplifies further for constant T and p to Eq. (8.79).

$$0 = \sum_i x_i d\tilde{\mu}_i \tag{8.79}$$

$x_{,i}$ – Mole fraction of component i

We will use this relationship later.

[104] Pierre Maurice Marie Duhem, 1861–1916, French physicist and theoretician of science.

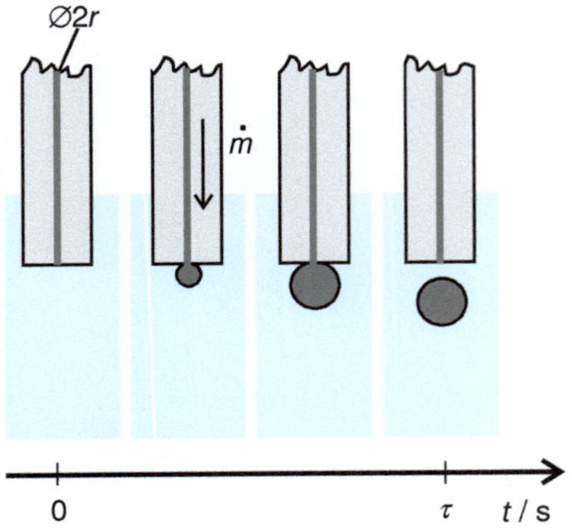

Figure 8.71 Determination of γ by means of the drop time method at a DME.

8.10.2 Excess Charge at Electrodes

We will apply Eq. (8.77) to a particular electrode now [115]. Because of Eq. (2.11) $\tilde{\mu}_i = \mu_i^\circ + RT\ln(a_i/a^\circ) + z_iF\varphi$, we hypothesize that a change in activity/concentration and/or the electrode potential will cause a change in surface tension γ. Such relations have been studied intensively by the classical dropping mercury electrode (DME) because this electrode facilitates the determination of surface tension from the easily accessible lifetime of the Hg droplet (Excursion 8-3, Figure 8.71). Furthermore, a liquid metal is devoid of the complications caused by grain boundaries and surface defects. For the exercise, we choose a polarizable electrode because no charge should be transferred across the electrode interface in the studied potential range, that is, there should be no electrolysis reaction. Hg exhibits a particularly large overvoltage for the hydrogen evolution reaction. Thus, the usable potential range is extended into the negative direction beyond the thermodynamically considered value. The reference and auxiliary electrode is a Ag|AgCl electrode immersed without a diaphragm or membrane in the working solution (thus avoiding the formation of liquid junction potentials). The working solution shall contain an adjustable concentration of KCl and a neutral surfactant Sur. Consequently, the short notation of the cell is

$$\mathrm{Cu|Ni|Hg|K^+}(aq), \mathrm{Cl^-}(aq), \mathrm{Sur}(aq)\mathrm{|AgCl|Ag|Cu'}$$

Excursion 8-3 The Drop Time Method

A classical experimental configuration uses a capillary of radius r through which liquid mercury flows into an electrolyte solution with a constant mass flux $\dot{m} = \mathrm{d}m/\mathrm{d}t$. This is similar to the polarographic experiment discussed in Figure 5.31. The mercury droplets fall down periodically with a drop lifetime τ (Figure 8.71). When the droplet separates from the mercury column, the weight of the drop $G = mg = \dot{m}\tau g$ (g is the gravitational constant [$\mathrm{m^2/s}$]) just exceeds the holding force that is caused by the surface tension γ along the perimeter of the capillary ($2\pi r$). Equating those forces and rearranging yields the surface tension as $\gamma = \dot{m}\tau g/(2\pi r)$. The addition of a surface-active substance will lower γ and thus shorten τ. The DME, which is barely in use anymore for analytical determinations of metal ions (Excursion 5-2), was originally developed exactly for such thermodynamic studies of interfaces. Later, the analytical possibilities were realized, which led to the development of polarography as the first instrumental electroanalytical method.

An alternative and very precise method uses a nulling principle, in which the application of an external pressure keeps the interface between Hg and the electrolyte solution in a specific position. Further information can be found in [116].

The symbols used to denote the different quantities are not uniform in the literature. This can cause confusion. Frequently, the surface tension or interfacial tension is abbreviated by γ or σ, the surface charge density as σ^{M} or q_{M} (index M as general for metal).

For relating the surface tension to the solution composition, we sort the terms in Eq. (8.77) containing the electrochemical potentials according to the compounds for the electrode material, the ions in solution and the neutral substances in the electrolyte solution (here they are distinguished by color for clarity).

$$\begin{aligned}-\mathrm{d}\gamma &= \Gamma_{\mathrm{Hg}}\mathrm{d}\tilde{\mu}_{\mathrm{Hg}} + \Gamma_{\mathrm{e}}\mathrm{d}\tilde{\mu}_{\mathrm{e}}^{\mathrm{Hg}} \\ &\quad + \Gamma_{\mathrm{K^+}}\mathrm{d}\tilde{\mu}_{\mathrm{K^+}} + \Gamma_{\mathrm{Cl^-}}\mathrm{d}\tilde{\mu}_{\mathrm{Cl^-}} \\ &\quad + \Gamma_{\mathrm{Sur}}\mathrm{d}\tilde{\mu}_{\mathrm{Sur}} + \Gamma_{\mathrm{H_2O}}\mathrm{d}\tilde{\mu}_{\mathrm{H_2O}}\end{aligned} \tag{8.80}$$

The different quantities are not independent of each other. There are four intensive quantities for the solid phase (T, p, $\tilde{\mu}_{\mathrm{Hg}}$, $\tilde{\mu}_{\mathrm{e}}^{\mathrm{Hg}}$) and six for the liquid electrolyte phase (T, p, $\tilde{\mu}_{\mathrm{K^+}}$, $\tilde{\mu}_{\mathrm{Cl^-}}$, $\tilde{\mu}_{\mathrm{Sur}}$, $\tilde{\mu}_{\mathrm{H_2O}}$). Altogether, these are 10 intensive variables. However, they are connected to each other by 5 equilibrium conditions ($p_{Hg} = p_{aq}$, $T_{Hg} = T_{aq}$, Gibbs–Duhem equation for Hg and for electrolyte solution, $q_{Hg} = -q_{aq}$). The number of degrees of freedom is therefore $10 - 5 = 5$ or equal to the number of compounds minus (the number of phases minus 2) [Eq. (15.29)]. Additionally, the electrochemical potential of the electrons $\tilde{\mu}_e^{Hg}$ (equivalent to the Fermi level, see Chapter 10) is equal for all metals in electronic contact.

$$\tilde{\mu}_e^{Hg} = \tilde{\mu}_e^{Ni} = \tilde{\mu}_e^{Cu} \text{ and } \tilde{\mu}_e^{Ag} = \tilde{\mu}_e^{Cu'} \tag{8.81}$$

Furthermore, the sum of the electrochemical potentials of the ions is equal to the electrochemical potential of the dissolved salt KCl. This in turn is a neutral compound and thus the electrochemical potential is equal to the chemical potential of KCl.

$$\tilde{\mu}_{K^+} + \tilde{\mu}_{Cl^-} = \tilde{\mu}_{KCl} = \mu_{KCl} \tag{8.82}$$

For the neutral compounds H_2O and Sur, the equality of electrochemical and chemical potentials holds as well.

$$\tilde{\mu}_{H_2O} = \mu_{H_2O}$$
$$\tilde{\mu}_{Sur} = \mu_{Sur} \tag{8.83}$$

The metal of the Hg electrode is a pure liquid phase, for which the activity is equal to the standard activity. Thus, the electrochemical potential is equal to the chemical standard potential, and this relationship does not change when shifting from one set of state variables to another set of state variables.

$$d\tilde{\mu}_{Hg} = d\tilde{\mu}^{\circ}_{Hg} = d\mu^{\circ}_{Hg} = 0 \tag{8.84}$$

Substitution of the relations (8.81) to (8.84) into Eq. (8.80) yields

$$\begin{aligned} -\mathrm{d}\gamma &= \Gamma_e \mathrm{d}\tilde{\mu}_e^{Cu} \\ &+ \Gamma_{K^+} \mathrm{d}\mu_{KCl} - \Gamma_{K^+} \mathrm{d}\tilde{\mu}_{Cl^-} + \Gamma_{Cl^-} \mathrm{d}\tilde{\mu}_{Cl^-} \\ &+ \Gamma_{Sur} \mathrm{d}\mu_{Sur} + \Gamma_{H_2O} \mathrm{d}\mu_{H_2O}. \end{aligned} \tag{8.85}$$

Further simplification can be derived by considering the Ag|AgCl reference electrode that is in equilibrium with one of the electrolyte components (Cl^-). From the reaction equation and consideration of Eq. (8.81) we obtain

$$\mathrm{AgCl}(s) + \mathrm{e}^-{}_{(Ag)} \rightleftharpoons \mathrm{Ag}(s) + \mathrm{Cl}^-(aq) \tag{8.86}$$

$$\tilde{\mu}_{AgCl} + \tilde{\mu}_e^{Cu'} = \tilde{\mu}_{Ag} + \tilde{\mu}_{Cl^-} \tag{8.87}$$

Analogous to the consideration for Eq. (8.84), we obtain for the pure, neutral solid compounds AgCl and Ag $\mathrm{d}\tilde{\mu}_{AgCl} = \mathrm{d}\tilde{\mu}_{Ag} = 0$. With this, Eq. (8.87) leads to

$$\mathrm{d}\tilde{\mu}_e^{Cu'} = \mathrm{d}\tilde{\mu}_{Cl^-} \tag{8.88}$$

Inserting this into Eq. (8.85) provides

$$\begin{aligned} -\mathrm{d}\gamma &= \Gamma_e \mathrm{d}\tilde{\mu}_e^{Cu} - \left[\Gamma_{K^+} - \Gamma_{Cl^-}\right] \mathrm{d}\tilde{\mu}_e^{Cu'} \\ &+ \Gamma_{K^+} \mathrm{d}\mu_{KCl} + \Gamma_{Sur} \mathrm{d}\mu_{Sur} + \Gamma_{H_2O} \mathrm{d}\mu_{H_2O} \end{aligned} \tag{8.89}$$

If we make up the balance for the surface charge density σ^{Hg} on the metallic side of the Hg-electrolyte interface, this excess charge will be brought up exclusively by the *additional* electron density Γ_e in the Hg electrode.

$$\sigma^{Hg} = -F\Gamma_e \quad \text{or} \quad \Gamma_e = \sigma^{Hg}/F \tag{8.90}$$

Attention: According to Eq. (8.64), the interfacial concentrations Γ_i are already defined as an excess quantity!

There must be an equally large charge excess of opposite sign on the solution side (aq) of the interface to satisfy the electroneutrality condition. This charge excess results from the interfacial concentrations of cations and anions.

$$\sigma^{\mathrm{aq}} = -\sigma^{\mathrm{Hg}} = F\left(\Gamma_{\mathrm{K}^+} - \Gamma_{\mathrm{Cl}^-}\right) \quad \text{or} \quad \Gamma_{\mathrm{K}^+} - \Gamma_{\mathrm{Cl}^-} = -\sigma^{\mathrm{Hg}}/F \tag{8.91}$$

In addition, each change in the electrochemical potentials of the Cu cables, Cu and Cu′ must result exclusively from the difference between their (inner) Galvani potentials.

$$\mathrm{d}\tilde{\mu}_{\mathrm{e}}^{\mathrm{Cu}} - \mathrm{d}\tilde{\mu}_{\mathrm{e}}^{\mathrm{Cu}'} = -F\mathrm{d}(\phi^{\mathrm{Cu}} - \phi^{\mathrm{Cu}'}) \tag{8.92}$$

The quantity $\phi^{\mathrm{Cu}} - \phi^{\mathrm{Cu}'}$ is the electrode potential E of the Hg electrode with respect to the Ag|AgCl reference electrode. It can be directly measured or adjusted by a voltage source. With this, one obtains from Eq. (8.89)

$$-\mathrm{d}\gamma = \frac{\cancel{-}\sigma^{\mathrm{Hg}}}{\cancel{F}} \cdot (\cancel{-F}\mathrm{d}E) + \Gamma_{\mathrm{K}^+}\mathrm{d}\mu_{\mathrm{KCl}} + \Gamma_{\mathrm{Sur}}\mathrm{d}\mu_{\mathrm{Sur}} + \Gamma_{\mathrm{H_2O}}\mathrm{d}\mu_{\mathrm{H_2O}}. \tag{8.93}$$

Next, we consider the terms printed in blue that refer to the solution components. It becomes intuitively clear, that they cannot be independent from each other because the increase in Γ_i for one component requires the concomitant decrease in one or more Γ_j of other components to make room for more of species i close to the interface. This fact is described by the Gibbs–Duhem Equation (8.79) for an aqueous solution. It is used to eliminate $\tilde{\mu}_{\mathrm{H_2O}}$ from Eq. (8.93).

$$\begin{aligned} 0 &= x_{\mathrm{KCl}}\,\mathrm{d}\mu_{\mathrm{KCl}} + x_{\mathrm{Sur}}\,\mathrm{d}\mu_{\mathrm{Sur}} + x_{\mathrm{H_2O}}\,\mathrm{d}\mu_{\mathrm{H_2O}} \\ \mathrm{d}\mu_{\mathrm{H_2O}} &= -\frac{x_{\mathrm{KCl}}}{x_{\mathrm{H_2O}}}\mathrm{d}\mu_{\mathrm{KCl}} + -\frac{x_{\mathrm{Sur}}}{x_{\mathrm{H_2O}}}\mathrm{d}\mu_{\mathrm{Sur}} \end{aligned} \tag{8.94}$$

Inserting this result yields for the blue terms in Eq. (8.93):

$$\begin{aligned} \Gamma_{\mathrm{K}^+}\,\mathrm{d}\mu_{\mathrm{KCl}} + \Gamma_{\mathrm{Sur}}\,\mathrm{d}\mu_{\mathrm{Sur}} + \Gamma_{\mathrm{H_2O}}\,\mathrm{d}\mu_{\mathrm{H_2O}} &= \left[\Gamma_{\mathrm{K}^+} - \frac{x_{\mathrm{KCl}}}{x_{\mathrm{H_2O}}}\Gamma_{\mathrm{H_2O}}\right]\mathrm{d}\mu_{\mathrm{KCl}} \\ &+ \left[\Gamma_{\mathrm{Sur}} - \frac{x_{\mathrm{Sur}}}{x_{\mathrm{H_2O}}}\Gamma_{\mathrm{H_2O}}\right]\mathrm{d}\mu_{\mathrm{Sur}}. \end{aligned} \tag{8.95}$$

The expressions in the square brackets are called relative surface excesses, for which we introduce separate symbols:

$$\Gamma_{\mathrm{K}^+(\mathrm{H_2O})} = \left[\Gamma_{\mathrm{K}^+} - \frac{x_{\mathrm{KCl}}}{x_{\mathrm{H_2O}}}\Gamma_{\mathrm{H_2O}}\right] \tag{8.96}$$

$$\Gamma_{\mathrm{Sur}(\mathrm{H_2O})} = \left[\Gamma_{\mathrm{Sur}} - \frac{x_{\mathrm{Sur}}}{x_{\mathrm{H_2O}}}\Gamma_{\mathrm{H_2O}}\right] \tag{8.97}$$

Inserting Eqs. (8.96) and (8.97) into Eq. (8.95) and then into Eq. (8.93) finally gives the **electrocapillary equation**, which we can use directly for interpretating experiments because it contains only measurable or adjustable variables.

$$-\mathrm{d}\gamma = \sigma^{\mathrm{Hg}} \cdot \mathrm{d}E + \Gamma_{\mathrm{K}^+(\mathrm{H_2O})}\,\mathrm{d}\mu_{\mathrm{KCl}} + \Gamma_{\mathrm{Sur}(\mathrm{H_2O})}\,\mathrm{d}\mu_{\mathrm{Sur}} \tag{8.98}$$

In summary, the surface tension will decrease upon an increase in the surface charge density on one side of the double layer because the electrostatic repulsion of the equal charges on one side of the interface counteracts the force resulting from the surface tension.

8.10.3 The Potential of Zero Charge

Figure 8.72 shows the electrocapillary curve for KCl. Rearranging Eq. (8.98) reveals that the surface charge density at the interface can be determined as the slope of the tangent of the electrocapillary curve at constant composition of the electrolyte solution:

$$\sigma^{\mathrm{Hg}} = -\left(\frac{\partial \gamma}{\partial E}\right)_{\mu_{\mathrm{KCl}},\mu_{\mathrm{Sur}}} \tag{8.99}$$

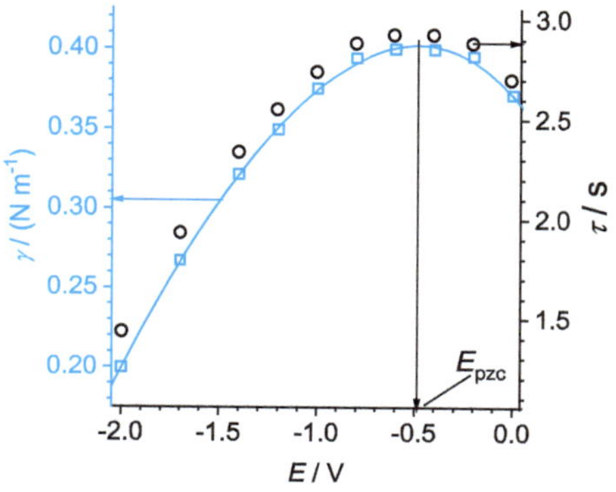

Figure 8.72 Electrocapillary curve of a Hg electrode in 0.1 mol L^{-1} KCl. Source: Adapted from L. Meites; *J. Am. Chem. Soc.* **1961**, *73*, 2035.

The electrocapillary curve is reminiscent of a parabola. It exhibits a maximum, the electrocapillary maximum. Accordingly, the slope $\partial\gamma/\partial E$ is zero at this point in Figure 8.72. That means that the surface charge density is zero at this point as well (Figure 8.73). The potential at which the electrocapillary maximum is observed is called the **potential of zero charge** E_{pzc}.

The second derivative of the surface tension with respect to the electrode potential is the differential double-layer capacitance $C^{\mathrm{d}}_{\mathrm{dl},A}$ [A s V^{-1} cm^{-2}] in Figure 8.74.

$$C^{\mathrm{d}}_{\mathrm{dl},A} = \left(\frac{\partial \sigma^{\mathrm{Hg}}}{\partial E}\right)_{\mu_{\mathrm{KCl}},\mu_{\mathrm{Sur}}} = -\left(\frac{\partial^2 \gamma}{\partial E^2}\right)_{\mu_{\mathrm{KCl}},\mu_{\mathrm{Sur}}} \tag{8.100}$$

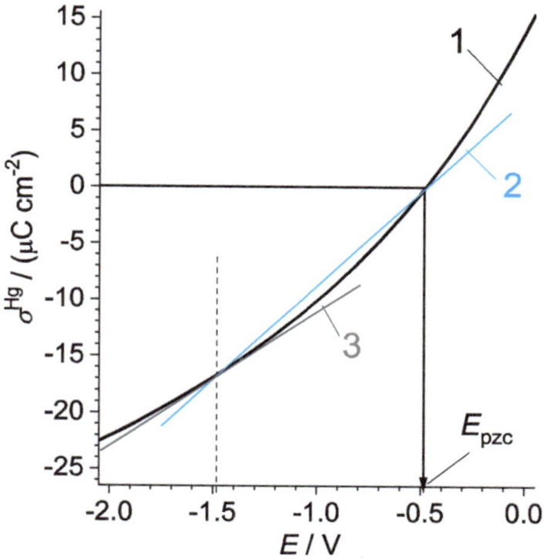

Figure 8.73 Potential-dependent surface charge density of the Hg electrode in 0.1 mol L^{-1} KCl calculated from the data in Figure 8.72 (curve 1). The slope of the secant (2) corresponds to the area-related integral double-layer capacitance $C^{\mathrm{i}}_{\mathrm{dl},A}$ at −1.0 V versus E_{pzc}. The slope of the tangent (3) corresponds to the differential double-layer capacitance $C^{\mathrm{d}}_{\mathrm{dl},A}$ at the same potential.

The superscript d indicates that the symbol refers to the differential capacitance. It corresponds to the slope of the tangent (3) at the function $\sigma^{\mathrm{Hg}}(E)$ at the potential E of interest (Figure 8.73). It is very common to calculate the double-layer capacitance only as an area-related quantity and to use the symbol C_{dl} for it. Here we use a distinct symbol $C^{\mathrm{d}}_{\mathrm{dl},A}$ [A s V^{-1} cm^{-2}] = [F cm^{-2}] to emphasize the difference from the extensive quantity C_{dl} [A s V^{-1}] = [F], which we have used already in Chapter 4. The two different symbols shall highlight the difference between the intensive and extensive quantities that is important for real calculation. In return, we have to accept the somewhat more cumbersome symbol and nomenclature.

In addition to the differential double-layer capacitance, there is also the (area-related) integral double-layer capacitance $C^{\mathrm{i}}_{\mathrm{dl},A}$. Experimentally, this is the charge that flows after a potential jump from E_{pzc} to the considered potential E. $C^{\mathrm{i}}_{\mathrm{dl},A}$ corresponds to the secant (2 in blue) in Figure 8.73 and, thus, forms the integral average value of $C^{\mathrm{d}}_{\mathrm{dl},A}$ within the considered potential range.

$$C^{\mathrm{i}}_{\mathrm{dl},A}(E) = \frac{1}{E - E_{\mathrm{pzc}}} \int_{E_{\mathrm{pzc}}}^{E} C^{\mathrm{d}}_{\mathrm{dl},A}(E)\,\mathrm{d}E \tag{8.101}$$

Q: Is the potential of zero charge linked to the redox potential, so that we reduce at a negatively charged electrode and we oxidize at a positively charged electrode?

A: No, this is not the case! The decision, whether we oxidize or reduce, depends mainly on the standard electrode potential of the redox system and the momentary electrode potential. In Figure 8.72, E_{pzc} = 0.5 V. This does not agree with any of the standard potentials of the components of the system, such as $E°(\mathrm{K|K^+})$, $E°(\mathrm{Hg|Hg_2Cl_2})$, or $E°(\mathrm{Cl|Cl_2})$. In fact, positively charged Cu^{2+} ions can be *reduced* at a *positively charged* Hg electrode at −0.3 V vs. Ag|AgCl, which is +0.2 V vs. E_{pzc}.

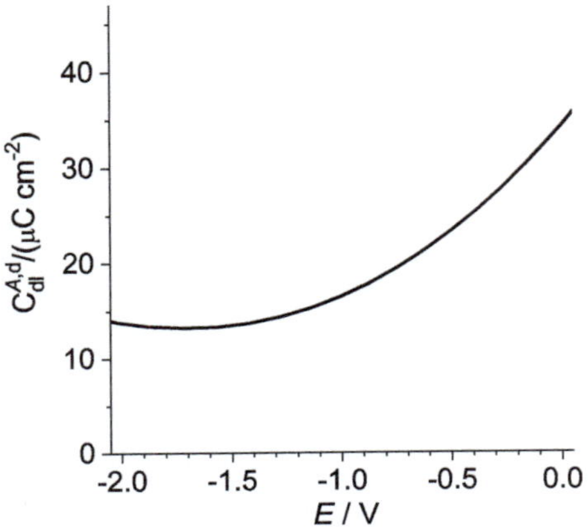

Figure 8.74 Potential-dependent area-related double-layer capacitance of the Hg electrode in 0.1 mol L^{-1} KCl calculated from the data of Figure 8.72.

Solid Electrodes. So far, we have restricted our treatment to liquid metal electrodes because they are particularly well-suited for such experiments. Of course, there is also a potential of zero charge at solid electrodes. However, the measurement will be strongly influenced by grain boundaries at polycrystalline electrodes. There can be different values for E_{pzc} at different crystal facets of the same metal. Therefore, an exact investigation of the interfacial thermodynamics is only sensible when using single-crystal electrodes. Studies are restricted to noble metal electrodes because other electrodes are quickly covered by a thin oxide layer, affecting the interfacial properties strongly. Because the overpotential for the H_2 evolution is lower at most noble metal electrodes compared to Hg, the experimentally accessible potential range for electrocapillary curves is smaller. For the practically important Pt electrode, a recent and comprehensive review is available [109].

In addition to the determination method introduced in Section 4.3.1, the double-layer capacitance can also be investigated by phase-selective AC voltammetry (Section 5.3.9) or, more precisely, by impedance spectroscopy and fitting of the data to an equivalent circuit (Chapter 7). By definite integration of the function $C_{dl,A}(E)$, one obtains the surface charge density of the metal M $\sigma^M(E)$.

$$\sigma^M(E) = \int_{E_{pzc}}^{E} C_{dl,A}^{d}(E)\,dE \tag{8.102}$$

This operation requires an integration constant from an independent measurement, for which E_{pzc} is typically used (often from an independent measurement). A comprehensive description and discussion of the conceptual basis and the experimental procedures is provided in [119]. E_{pzc} can be determined by the **immersion method**. A carefully prepared single-crystal electrode is positioned in a measurement cell (Figure 8.59) above the solution. It is connected as the working electrode to a potentiostat, which is used to apply a constant potential E *before* the electrode is moved down to touch the surface of the electrolyte solution. A transient current $I(t)$ will flow for a short time after immersion of the electrode, which provides the charge to build up the electrical double layer. Integration of the current density $j(t) = I(t)/A$ over the time t provides the excess charge density $\sigma^M(E)$. Based on selected potentials at which the measurements were carried out, the function $\sigma^M(E)$ is interpolated. The zero of the $\sigma^M(E)$ is E_{pzc}.

A particularly important class of modified surfaces is chemisorbed alkane thiolate on Au(1 1 1) electrodes (see also Section 13.1). The chemisorption of alkanethiols can be considered as a charge transfer reaction, which can be reversed by a reductive desorption.

$$\text{Au–S–R} + e^- \rightleftharpoons \text{Au} + {}^-\text{S–R} \tag{8.103}$$

The charge required for this reaction can be determined by integration of chronoamperograms. For this, an initial, or base potential is selected at which the monolayer is stable. After immersion, the electrode is equilibrated at this potential. Then, a potential step to the potential of interest E_c is applied and the charge required for rebuilding the double layer is determined. Afterward, a second potential step is applied to the potential E_{des}, at which the monolayer is completely desorbed. For E_{des}, the absolute excess charge density $\sigma^{Au}(E_{des})$ can be obtained from immersion measurements of an uncovered electrode. With this additional information, the absolute surface excess charge σ^{Au} can be determined for the potential of interest E_c using the charges measured from the chronoamperometric transient. By variation of E_c, the entire potential range can be covered.

For systems with chemisorbed adsorbate, it becomes clear that it is sensible to distinguish between the potential of zero free charge E_{pzfc} and the potential of zero total charge E_{pztc}. E_{pzfc} is the potential at which the solution side of the double layer is not charged, that is, there is no excess of ions, neither of anions nor of cations. E_{pzfc} is obtained by means of the immersion method at monolayer-covered electrodes (*vide supra*). In contrast, E_{pztc} refers not only to the charge of the ions in the double layer, but also includes the charge required to desorb the charged adsorbates according to Eq. (8.103) in the balance. Therefore, E_{pztc} is about 1.0 V more negative than E_{pzfc} for thioglucose and similar compounds on Au(1 1 1) electrodes in 0.1 mol L^{-1} NaF solutions [119]! A collection of E_{pzc} values for different electrodes is available [120].

8.10.4 Surface Excess and Adsorption Isotherms

One can use the electrocapillary equation (8.98) for the Hg electrode to determine further quantities. For instance, the relative surface excess of a substance is the derivative of the surface tension with respect to the chemical potential of the compound. Because of $\mu = \mu^\circ + RT\,\ln(a/a^\circ)$, we need a measurement of the surface tension as a function of the concentration activity ($a = f_c\, c$) of the solution constituents.

$$\Gamma_{K^+(H_2O)} = -\left(\frac{\partial \gamma}{\partial \mu_{KCl}}\right)_{E,\mu_{Sur}} = \frac{-1}{RT}\left(\frac{\partial \gamma}{\partial \ln a_{KCl}}\right)_{E,\mu_{Sur}} \tag{8.104}$$

$$\Gamma_{Sur(H_2O)} = -\left(\frac{\partial \gamma}{\partial \mu_{Sur}}\right)_{E,\mu_{KCl}} = -\frac{1}{RT}\left(\frac{\partial \gamma}{\partial \ln a_{Sur}}\right)_{E,\mu_{KCl}} \tag{8.105}$$

After obtaining $\Gamma_{K^+(H_2O)}$ from Eq. (8.104) for different potentials, the charge balance, $\Gamma_{K^+(H_2O)}$, and $\sigma^{Hg}(E)$ are used to determine $\Gamma_{Cl^-(H_2O)}$. A plot of these quantities is shown in Figure 8.75. In case of KF as electrolyte (Figure 8.75a), anions and cations interact by electrostatic attraction/repulsion (i.e., nonspecifically) with the electrode surface. At potentials negative of E_{pzc}, there is an electron deficiency on the metal side of the double layer that is compensated for by an ion excess on the solution side of the double layer. $\Gamma_{K^+(H_2O)}$ assumes negative values in this potential range, which means that the activity (concentration) of K^+ ions is smaller than in the undisturbed bulk phase of the solution. The situation at potentials positive of E_{pzc} is almost a mirror image of the situation at $E < E_{pzc}$.

The situation is quite different for KBr as an electrolyte (Figure 8.75b). Br^- ions bind chemically (i.e., specific adsorption) to Hg. Therefore, Br^- is more strongly enriched at the Hg-solution interface than expected from electrostatic attraction/repulsion that is effective for KF. For maintaining the charge neutrality, $\Gamma_{K^+(H_2O)}$ must be increased above the activity (concentration) in the bulk phase even at potentials positive of E_{pzc}. Regardless of specific or nonspecific adsorption, at $E = E_{pzc}$, there is the same relative surface concentration of anions and cations.

Neutral solution components would adsorb in the potential range around E_{pzc}. This potential range can be determined by phase-selective AC voltammetry (Figure 5.45).

Practically important in this context are functions that use the bulk concentration c^* of a component i to predict its surface concentration $\Gamma_i = f(c_i^*)$ at equilibrium with $T = const.$ Such functions are called **adsorption isotherms**. The particular shape of an adsorption isotherm allows conclusions about the way a compound interacts with the surface. The determination of the surface concentration is achieved by different methods, such as capacitance measurements, voltammetry of immobilized redox systems (Section 5.5), use of radioactive tracers [121, 122], or spectroscopy of interfaces. The quantities obtained in this way are often only an approximation to the exact, thermodynamically derived quantities according to Eq. (8.97). Therefore, it is common to

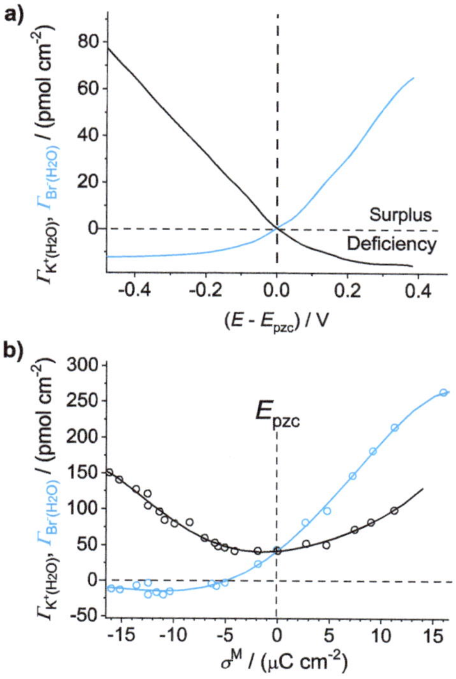

Figure 8.75 Comparison of the relative surface excess at Hg for a) 0.1 mol L^{-1} KF (nonspecific adsorption only) as a function of the electrode potential and b) 0.1 mol L^{-1} KBr as a function of the excess charge density. Source: Data for a) from D. C. Grahame, B. A. Soderberg, *J. Chem. Phys.* **1954**, *22*, 449–4459; data for b) from M. A. V. Devanatham, S. G. Canagaratna, *Electrochim. Acta* **1963**, *8*, 77–85.

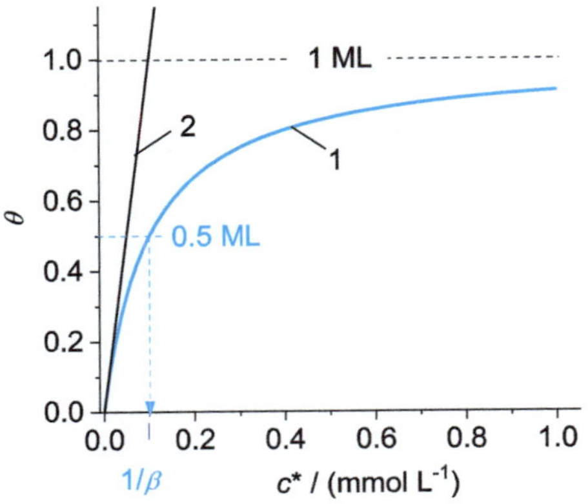

Figure 8.76 Langmuir isotherm (1) with $\beta = 10\,\text{mmol}^{-1}\,\text{L} = 10^4\,\text{mol}^{-1}\,\text{L}$ and Henry isotherm (2) with the same β.

use the coverage $\theta = \Gamma/\Gamma_{max}$, that is, the ratio of the actual interfacial concentration $\Gamma = f(c^*)$ to the maximum possible interfacial concentration Γ_{max} of a monolayer. The coverage is often easier to determine experimentally than Γ itself.

A particularly plausible isotherm is called the Langmuir isotherm[91] (Figure 8.76, curve 1). It can be derived by assuming that

1) The surface has a finite number of adsorption sites.
2) All adsorption sites are structurally and thus energetically uniform and can bind one adsorbate each.
3) The coverage cannot exceed a monolayer.
4) There are no lateral interactions between the adsorbates on the surface.
5) The adsorption is reversible.

Thus, an equilibrium expression is obtained with the adsorption equilibrium constant β $[-\Delta_{ads}G = \text{const} + RT\,\ln(\beta/c°)]$ by equating the rate laws for the adsorption and desorption processes. This expression can be stated in various forms. We use the standard concentration $c° = 1\,\text{mol L}^{-1}$ to obtain a dimensionless quantity before calculating the logarithm from it (Chapter 2). For simplification, we also assume equality of the values for the (amount) concentration c and (amount) concentration activity a_c. This assumption has (silently) be implied for the surface concentration and the corresponding activity, too. We obtain Eq. (8.106) as expressions for the Langmuir isotherm.

$$\Gamma = \Gamma_{max}\frac{\beta c^*}{1+\beta c^*} \quad \text{or} \quad \frac{\Gamma}{\Gamma_{max}} = \theta = \frac{\beta c^*}{1+\beta c^*} \quad \text{or} \quad \frac{\theta}{1-\theta} = \beta c^* \tag{8.106}$$

The adsorption equilibrium constant β usually depends on the electrode potential. The function $\Gamma = f(c^*)$ increases for $c^* \ll 1/\beta$ almost linearly and levels off for larger c^* when the surface is almost entirely covered (Figure 8.76, curve 1). For many practical cases, this describes the function $\Gamma = f(c^*)$ roughly correctly. Furthermore, the derivation of the equation from a physical model is straightforward, making this isotherm very popular for interpreting adsorption data for electrodes where often only one monolayer is bound. If there are several compounds (index j) competing for adsorption at the electrode with adsorption equilibrium constants β_j, the interfacial concentrations are described by Eq. (8.107). Because all β_j's are usually potential-dependent, the situation easily becomes complicated if the electrolyte solution contains several compounds.

$$\frac{\Gamma_i(E)}{\Gamma_{max}} = \theta_i(E) = \frac{\beta_i(E)c_i^*}{1+\sum_j \beta_j(E)c_j^*} \tag{8.107}$$

The assumptions made for deriving the Langmuir isotherm are so restrictive that a more exact description of experimental data requires refined isotherms, for which some of the assumptions made for the Langmuir isotherm are relaxed. The Henry isotherm[55] does not assume a maximum coverage. In this case, the surface concentration is related linearly to the bulk concentration.

$$\theta = \beta c^* \tag{8.108}$$

The isotherm can be applied either if $c^* \ll 1/\beta$ or if there is no defined number of adsorption sites (Figure 8.76, curve 2). The latter applies to the uptake of gases into liquids. However, the latter process is described by the term *absorption* (uptake) as compared to *adsorption* (sticking to). For the uptake of gases in a thin liquid film, for example, in a gas sensor, the concentration of the dissolved gas in the liquid film is a linear function of the partial pressure of the gas ($c_{O_2} = \beta p_{O_2}$, β has a different unit in this case!).

The Temkin isotherm[105] (Figure 8.77, curve 1) is suitable for the practically relevant case of H_2 adsorption on porous Pt. In this example, there is a finite number of adsorption sites. However, different amounts of heat are released during the adsorption of H_2 to structurally different sites (that can be between q_1 and a maximum value q_0). This situation corresponds to different adsorption equilibrium constants between β_0 and β_1 for the different adsorption sites. To simplify the description, the following assumptions are made:

1) For each type of site, the Langmuir isotherm can be applied.
2) Within the interval between q_1 and q_0, all intermediate values occur with the same probability.

For this case, the energetic distribution of the adsorption sites can be summarized by *one* inhomogeneity factor $f = (q_0 - q_1)/(RT)$. For medium coverages $0.2 \leq \theta \leq 0.8$, one obtains Eq. (8.172) that can be used for fitting procedures with the empirical parameters k_1 and k_2.

$$\frac{\Gamma}{\Gamma_{\max}} = \theta = \frac{\ln(\beta_0 c^\circ) + \ln(c^*/c^\circ)}{f} = k_1 + k_2 \ln(c^*/c^\circ) \tag{8.109}$$

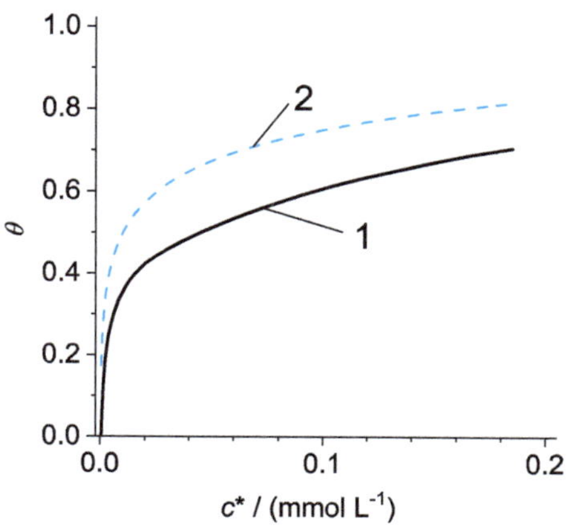

Figure 8.77 Temkin isotherm (1) with $\beta_0 = 10^4$ mol^{-1} L and $\beta_1 = 10^8$ mol^{-1} L. The Langmuir isotherm (2) from Figure 8.76 is shown for comparison.

c° – Standard concentration = 1 mol L^{-1}
β_0 – Lower limit of adsorption equilibrium constants [mol^{-1} L]
f – Inhomogeneity factor
k_1, k_2 – Dimensionless constants

The Frumkin isotherm[106] in Figure 8.78 considers lateral interaction between adsorbed species, for example, the repulsion of equally charged adsorbed ions or strong intermolecular attraction between adsorbed organic molecules. In the case of repulsive interaction, adsorption of dissolved particles to a free surface site becomes less favored if the neighboring sites are already occupied (Figure 8.78, curve 2). In contrast, the adsorption to a free site becomes even more favorable if there are attractive interactions between adsorbates and the surrounding sites are already occupied (Figure 8.78, curve 1). The lateral adsorbate–adsorbate interactions are described by an interaction coefficient, also called Frumkin-coefficient, a or g, which are related to each other by $g = -2a$. They summarize the Gibbs free energy for the interaction of the adsorbate A and the solvent S. Different sign conventions are in use.

$$a = \frac{-\Delta G_{AA} + \Delta G_{AS} - \Delta G_{SS}}{RT} = -\frac{g}{2} \tag{8.110}$$

a, g – Dimensionless coefficients
ΔG_{AA} – Molar Gibbs free energy of adsorbate–adsorbate interaction [kJ mol^{-1}]
ΔG_{AS} – Molar Gibbs free energy of adsorbate–solvent interaction [kJ mol^{-1}]
ΔG_{SS} – Molar Gibbs free energy of solvent–solvent interaction [kJ mol mol^{-1}]

The resulting isotherm is a generalization of the Langmuir isotherm, which is a limiting case of the Frumkin isotherm for $a = g = 0$. The Frumkin isotherm is cited in various slightly different forms, such as

$$\Gamma = \Gamma_{\max} \frac{\beta c^* \exp(2a\theta)}{1 + \beta c^* \exp(2a\theta)} \quad \text{or} \quad \frac{\theta}{1-\theta}\exp(g\theta) = \beta c^* \tag{8.111}$$

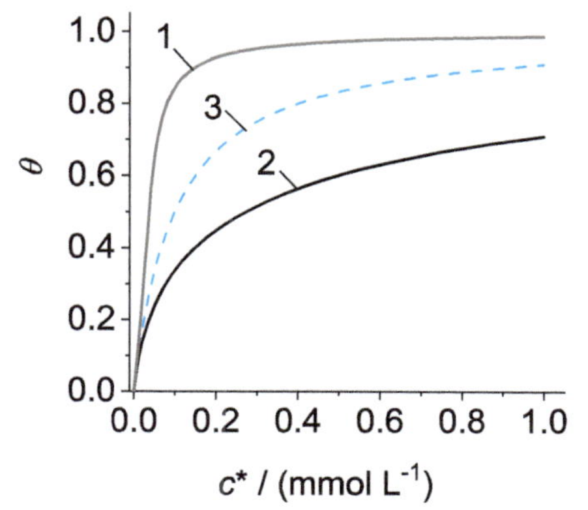

Figure 8.78 Frumkin isotherm with $\beta = 10^4$ mol^{-1} L; 1) $a = +1$ (attractive adsorbate–adsorbate interaction); 2) $a = -1$ (repulsive); 3) $a = 0$ (no adsorbate-adsorbate interaction = Langmuir isotherm from Figure 8.76).

8.10.5 Comparison of Experimental Surface Excess Data and Double-Layer Models

We have already derived models of the structure of the electrical double layer based on qualitative considerations in Sections 4.3.2 and 4.3.3 but largely avoided quantitative descriptions or a comparison of the models with experimental data. Now we aim to make up for that. First, we select experimental data sets that can either

[105] Mikchail Isaakovich Temkin [pronounced tjomkin], 1908 Białystok (today Poland) – 1991 Moscow, Russian physicochemist.
[106] Alexander Naumovich Frumkin, 1895 in Chişinău, today Moldova – 1976 Moscow, Russian electrochemist, Professor in Moscow, founder and director of the Institute of Electrochemistry of the Soviet Academy of Sciences.

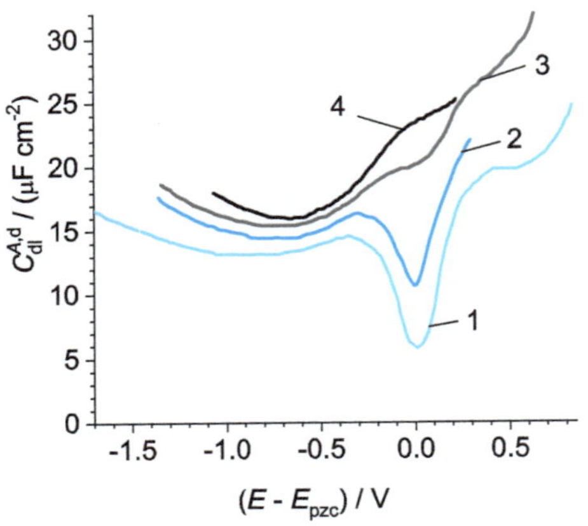

Figure 8.79 Differential double-layer capacitance of the interface between a Hg electrode and NaF solutions at 298 K, $E_{pzc} = -0.48$ V; [NaF] in mol L^{-1} is 1) 0.001, 2) 0.01, 3) 0.1, and 4) 1.0. Source: Data from [113].

confirm or reject the assumption made in the different models. A potential-dependent quantity is the area-related differential double-layer capacitance $C^{d}_{dl,A}(E)$ [μF cm^{-2}] at the Hg electrode in different electrolyte compositions because this quantity can be determined without the disturbing influences of grain boundaries and surface oxide layers (Sections 8.10.2 and 8.10.3). We select an electrolyte salt with ions that do not form specific (chemical) bonds to the electrode metal. This is, of course, an idealized limiting case, but NaF shows close to ideal behavior. The resulting experimental data set is shown in Figure 8.79. The unsymmetrical shape of the curves already informs us that, even in this carefully selected case, there are challenges. Next, we will attempt to derive a function $C^{d}_{dl,A} = f(E, c^*)$ for each double-layer model and compare the result to the experimental data set.

Model of the Compact Double Layer. The occurrence of rigidly arranged counter ions in front of the metal–electrolyte interface is called Helmholtz model[11] (Figure 4.34). The capacitance of this double layer is that of a parallel plate capacitor with the area-related capacitance

$$C_A = \frac{\partial \sigma^{Hg}}{\partial (E - E_{pzc})} = \frac{\varepsilon_r \varepsilon_0}{d} \tag{8.112}$$

ε_r - Relative permittivity of the solvent [dimensionless]

ε_0 - Permittivity of vacuum $\varepsilon_0 = 8.854 \times 10^{-12}$ A s V^{-1} m^{-1} = 8.854×10^{-12} F m^{-1}

where d is the distance between the center of charge of the ions and the electrode surface. Thus the prediction of the model would be $C^{d}_{dl,A}(E, c^*) = const$, which is clearly in contrast to the experimental findings in Figure 8.79 as we already concluded in Chapter 4.

Model of the Diffuse Double Layer. The assumption of a distribution of the counter ions from the surface into the solution (Figure 4.35, Gouy–Chapman layer[40,41]) permits a multilayer coverage of charge-balancing ions and considers the random thermal motion of ions. We consider the distribution of dissolved ions of the type i perpendicular to the electrode surface as a stack of thin solution slabs of constant thickness, in which the same number of ions can be accommodated (same degree of degeneracy for all slabs). Due to electrostatic interaction, the ions in the different slabs have different electrostatic energies $-z_i e(\varphi - \varphi^S)$. The concentration of ions c_i in each slab relative to the undisturbed bulk concentration c_i^* is obtained from the Boltzmann distribution taking the electrostatic energy of the ions in the undisturbed bulk as the "ground state" (c_i [mol L^{-1}] is proportional to the particle number density in particles per cm^3).

e - Elementary charge = 1.602×10^{19} A s

R - Gas constant = 8.314 V A s K^{-1} mol^{-1}

T - Temperature [K]

E_{NF} - Nernst factor = RT/F = 0.0257 V (298 K)

ρ - Charge density [A s cm^{-3}]

$$\frac{c_i}{c_i^*} = \exp\left[\frac{-z_i F(\varphi(x) - \phi_S)}{RT}\right] = \exp\left[\frac{-z_i(\varphi(x) - \varphi^S)}{E_{NF}}\right] \tag{8.113}$$

The charge density ρ is then obtained by summation over the charges of all kinds of ions i at all distances x to the electrode using Eq. (8.114).

$$\rho(x) = \sum_i z_i F c_i(x) = F \sum_i z_i c_i^* \exp\left[\frac{-z_i(\varphi(x) - \varphi^S)}{E_{NF}}\right] \tag{8.114}$$

According to the laws of electrostatics, we can calculate the charge density also from the Poisson equation[39].

$$\rho(x) = -\varepsilon_0 \varepsilon_r \frac{d^2\varphi}{dx^2} \tag{8.115}$$

Equating Eqs. (8.114) and (8.115) yields the Poisson[39]–Boltzmann[18] equation.

$$\frac{d^2\varphi}{dx^2} = \frac{-F}{\varepsilon_0 \varepsilon_r} \sum_i z_i c_i^* \exp\left[\frac{-z_i(\varphi(x) - \varphi^S)}{E_{NF}}\right] \tag{8.116}$$

Further consideration of the Gauss theorem[47] allows the derivation of different quantitative relations that can be tested experimentally. For details, we refer to [115] or textbooks of physics. Here, we only consider the results for a symmetric 1:1 electrolyte (e.g., NaF) consisting of one kind of cation (index $i = 0$) and one kind of anion (index $i = 1$). The following equalities hold: $z_0 = -z_1 = |z|$ and $c_0^* = c_1^* = c^*$.

$$\frac{d\varphi}{dx} = -\left(\frac{8RTc^*}{\varepsilon_0\varepsilon_r}\right)^{1/2} \sinh\left(\frac{|z|(\varphi(x) - \varphi^S)}{2E_{NF}}\right) \tag{8.117}$$

Integration of Eq. (8.117) gives the potential distribution

$$\frac{\tanh\left[\dfrac{|z|(\varphi(x) - \varphi^S)}{4E_{NF}}\right]}{\tanh\left[\dfrac{|z|(\varphi(0) - \varphi^S)}{4E_{NF}}\right]} = \exp\left(\frac{x}{r_D}\right) \tag{8.118}$$

with

$$\frac{1}{r_D} = \left(\frac{2|z|^2F^2c^*}{\varepsilon_0\varepsilon_r RT}\right)^{1/2} \tag{8.119}$$

For small potential differences $\varphi(x) - \varphi^S \approx 0.01$ V, this expression approaches a simple exponential decay function $\varphi(x) - \varphi^S = (\varphi(x) - \varphi^S)\cdot e^{-x/r_D}$ because of $\tanh(y) \approx y$ for $y = |z|E_{NF}(\varphi(x) - \varphi^S)/4 < 0.5$.

For the charge distribution at the Hg electrode σ^{Hg} and the charge distribution in the diffuse double layer σ^S, one obtains

$$\sigma^{Hg} = -\sigma^S = (8\varepsilon_0\varepsilon_r RTc^*)^{1/2} \sinh\left(\frac{|z|(\varphi(0) - \varphi^S)}{2E_{NF}}\right) \tag{8.120}$$

from which the area-related differential double-layer capacitance is calculated by

$$C_{dl,A}^d = \frac{d\sigma^{Hg}}{d(\phi(0) - \phi_L)} = |z|F\left(\frac{2\varepsilon_0\varepsilon_r c^*}{RT}\right)^{1/2} \cosh\left(\frac{|z|(\varphi(0) - \varphi^S)}{2E_{NF}}\right)$$
$$= 228\frac{\mu F}{cm^2}\sqrt{\frac{L}{mol}}|z|\sqrt{c^*}\cosh(19.5\,V^{-1}|z|(\varphi(0) - \varphi_S)), (T = 298\,K) \tag{8.121}$$

Function (8.121) has a minimum at $\varphi(0) - \varphi^S$ (the electrocapillary minimum) with $C_{dl,A,min}^d = |z|F\,(2\varepsilon_0\varepsilon_r c^*/RT)^{1/2}$ (Figure 8.80, curve 1). Indeed, this minimum is observed in the experimental data at low electrolyte concentrations (Figure 8.80, curve 3). However, the minimum vanishes when increasing the electrolyte concentration c^*. Generally, Eq. (8.121) predicts values of $C_{dl,A}^d$ that are too large. Finally, Eq. (8.121) fails to describe the marginal changes in the double-layer capacitance at larger deviations of the applied potential from E_{pzc}. In conclusion, we note that the Gouy–Chapman theory captures some aspects of the double-layer structure correctly but fails in some other aspects. The reasons for failure are the simplifying assumptions made when deriving the theory. One oversimplifying assumption is the description of ions as charge carriers that do *not* possess a volume of their own. Therefore, they can approach the electrode surface arbitrarily close, which leads to an error. Furthermore, the relative permittivity ε_r was assumed to be independent of the distance from the electrode and thus independent of the local composition of the electrolyte. More elaborate theoretical calculations, including the orientation of water molecules close to the charged metal–solution interface, suggest that this assumption does not agree with reality. Furthermore, we assumed that the interaction between the ions in the electrolyte is based on electrostatic forces only, and all chemical interactions can be ignored. However, we already know from Section 4.2.9 (activity coefficients) that this

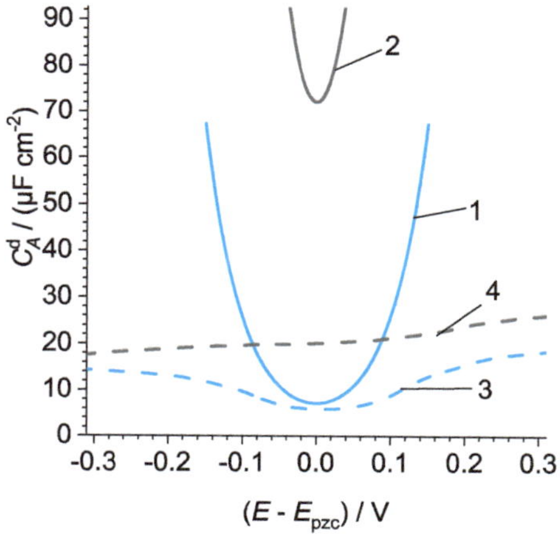

Figure 8.80 Differential double-layer capacitance calculated according to Eq. (8.121) for the Hg electrode immersed in NaF solutions with NaF concentrations of 1) 0.001 mol L^{-1}, 2) 0.1 mol L^{-1}. Parts of the experimental data from Figure 8.79 according to [113] are overlaid for 3) 0.001 mol L^{-1} and 4) 0.1 mol L^{-1} as dashed lines.

simplification does not hold at higher electrolyte concentrations (and thus smaller ion–ion distances) at which specific (chemical) interactions become significant.

Q: What do sinh, tanh, and cosh mean in Eqs. (8.117), (8.118), (8.120), and (8.121)?

A: These functions are called hyperbolic functions. They are the hyperbolic sine function or *sinus hyperbolicus* $y = \sinh(x) = (e^x - e^{-x})/2$ (Figure 8.81a), the hyperbolic cosine function (*cosinus hyperbolicus*) $y = \cosh(x) = (e^x + e^{-x})/2$ (Figure 8.81a) and the hyperbolic tangent function (*tangens hyperbolicus*) $y = \tanh(x) = (e^x - e^{-x})/(e^x + e^{-x})$ (Figure 8.81b). As for the known trigonometric functions, there is also a hyperbolic cotangent function (*cotangens hyperbolicus*) $y = \coth(x) = (e^x + e^{-x})/(e^x - e^{-x})$ (Figure 8.81b). The function $y = \cosh(x)$ can be quite well approximated by the parabola $y = 1 + ½\, x^2$ near the vertex point.

A further analogy to the trigonometric functions is in the structure of the calculus rules for hyperbolic functions. Formally, they can be obtained when taking the calculus rules for trigonometric functions and substituting $\sin(x) \rightarrow \mathrm{i}\sinh(x)$ and $\cos(x) \rightarrow \cosh(x)$ (note the equality of $\mathrm{i}^2 = -1$). With this, one obtains $\sinh^2(x) = [\sinh(x)]^2 = \cosh^2(x) - 1$, $\tanh(x) = \sinh(x)/\cosh(x)$. The inverse functions of the hyperbolic functions are the area functions, that is, $y = \mathrm{arsinh}(x) = \sinh^{-1}(x)$ with $x = \sinh(y)$. The hyperbolic functions occur in solutions of differential equations as in the example above. Further information on those functions can be found in textbooks for mathematics [123].

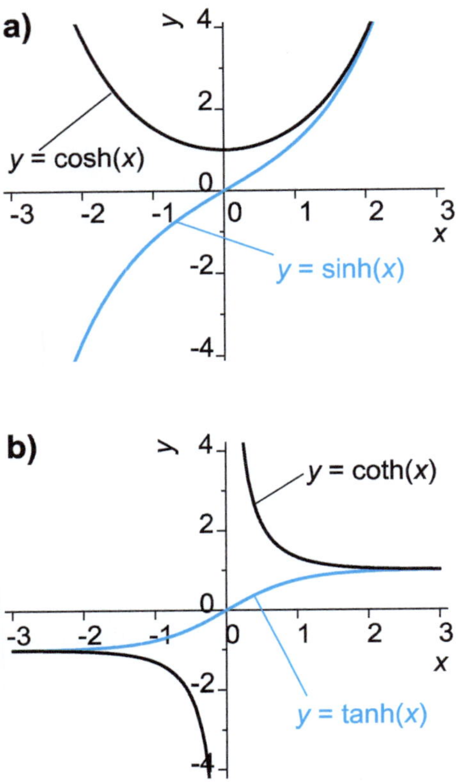

Figure 8.81 Hyperbolic functions.

φ_2 – Potential at the transition between compact and diffuse parts of the double layer

$C_{C,A}$ – Area-related capacitance of the compact part [F cm^{-2}]

$C_{D,A}$ – Area-related capacitance of the diffuse part [F cm^{-2}]

E_{NF} – Nernst factor $= RT/F = 0.0257$ V (298 K)

x_2 – Distance of the outer Helmholtz plane from the electrode [cm]

Combination of a Rigid and Diffuse Double Layer. The model proposed by Stern[107] represents a combination of a compact double layer and a diffuse double layer (Figure 4.35). Therefore, it is also called the Gouy–Chapman–Stern model. Its salient feature is the consideration of the volume of the ions and the minimum distance x_2 of approach following from it. We can imagine x_2 as the radius of the solvated ions. At a distance x_2, there shall be the potential φ_2. When going from the electrode to the solution bulk, the first part is the compact part with the area-related capacitance $C_{C,A}$, in which the potential decays linearly from $\varphi(x = 0)$ to $\varphi_2(x_2)$. It is followed by the diffuse part with the area-related capacitance $C_{D,A}$, in which the potential decays from φ_2 to φ^S. This arrangement is a serial electrical connection of two capacitors, for which the total capacitance, that is, the capacitance of the electrical double layer $C^{d}_{dl,A}(E, c^*)$, is obtained from Eq. (8.122).

$$\frac{1}{C^{d}_{dl,A}(E,c^*)} = \frac{1}{C_{C,A}} + \frac{1}{C_{D,A}(E,c^*)} = \frac{x_2}{\varepsilon_0\varepsilon_r} + \frac{1}{|z|F\left(\dfrac{2\varepsilon_0\varepsilon_r c^*}{RT}\right)^{1/2}\cosh\left(\dfrac{|z|(\varphi_2-\varphi^S)}{2E_{NF}}\right)} \tag{8.122}$$

The smaller of the two capacitances, $C_{C,A}$ and $C_{D,A}$ in Eq. (8.122), determines the overall capacitance. Correspondingly, Figure 8.82 shows a sharp minimum for small electrolyte concentrations at $E = E_{pzc}$ ($\varphi(x = 0) = \varphi^S$). The minimum becomes

[107] Otto Stern (1888 Sohrau, Upper Silesia – 1969 Berkley, USA), 1914–1918 voluntary participation in WW I for the German Realm, professor of physics in Rostock and Hamburg, 1933 emigration to the USA because of repressions in relation to his Jewish ancestry, 1939 US American citizen, Nobel prize of physics 1943 for the discovery of the quantized angular momentum (Stern-Gerlach experiment); also known for his work on zero point energy and the Stern–Volmer equation (photochemical kinetics).

shallower as the electrolyte concentration is increased. In Eq. (8.122), the contribution $C_{D,A}$ grows unlimited, while the total capacitance cannot rise above the concentration-independent and potential-independent contribution $C_{C,A}$. This combination predicts important properties of the function $C^{d}_{dl,A}(E, c^*)$ correctly (Figure 8.82) although significant differences remain. The experimental curves $C^{d}_{dl,A}(E, c^*)$ are inclined for higher concentrations, which can only be captured by further refinement of the double-layer model. Different sizes could be introduced for anions and cations in the compact part of the double layer. This makes the constant value x_2 become a function $x_2(E)$ because either more anions or more cations are contained in the compact part of the double layer as a function of the applied potential. Furthermore, the relative permittivity changes with the ion composition and the field strength in the double layer. Finally, a more specific description would be needed for the ion–ion, ion–solvent molecule, and solvent-molecule–solvent-molecule interactions.

The description of the double layer at practically relevant solid electrodes is even more difficult when compared to the case of the liquid Hg electrode because the partial coverage of the surface by chemisorbed molecules adds another layer of complexity. For Pt electrodes in sulfuric acid, a frequently studied and important experimental system, a review paper is available that takes into account detailed structural analyses of the double layer [97]. A selection of more advanced theoretical treatments of the electrode–solution interface by means of modern theoretical methods can be recommended for further study [124].

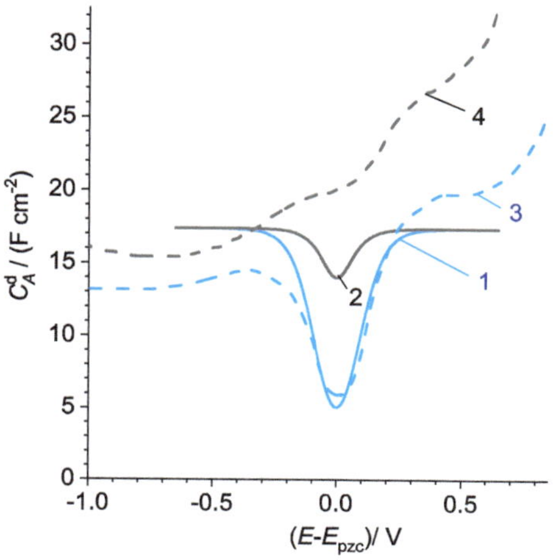

Figure 8.82 Calculated differential double-layer capacitance according to Eq. (8.122) for a Hg electrode immersed in NaF solution of the concentrations 1) 0.001 mol L^{-1}, 2) 0.1 mol L^{-1}. For comparison, sections of the experimental data from Figure 8.79 according to [113] for 3) 0.001 mol L^{-1} and 4) 0.1 mol L^{-1} are overplotted as dashed lines.

8.10.6 How Does the Structure of the Electrical Double Layer Influence Reaction Rates?

Inhibition. Adsorbates on the electrode surface often slow down the electrode reaction because they prevent the approach of the reactants or block the adsorption of intermediates. This effect is called inhibition. A simplistic description of this situation can be based on a linear combination of the fraction for the covered part θ of the electrodes surface and $[1 - \theta]$ for the bare part of the electrode surface:

$$k^\circ = k^\circ_{\theta=0}[1 - \theta] + k^\circ_{\theta=0}\theta \tag{8.123}$$

θ – Coverage $= \Gamma/\Gamma_{max} < 1$

Here, $k^\circ_{\theta=0}$ and $k^\circ_{\theta=1}$ are the standard reaction rate constants of the bare ($\theta = 0$) and completely covered ($\theta = 1$) electrode.

Frumkin Effect.[106] The structure of the double layer can have a very profound influence on the rate of electrochemical reactions if the reactants themselves are charged. We want to consider an irreversible reduction of a component O^z of charge z.

$$O^z + e^- \rightleftharpoons R^{z-1} \tag{8.124}$$

We derived the following expression for the reduction current density j in Chapter 3 for the limiting case when $E \ll E^\circ$.

$$j = -Fk^\circ c_O(0, t) \exp\left[\frac{-\alpha F(E - E^{\circ\prime})}{RT}\right] = -Fk^\circ c_O(0, t) \exp\left[\frac{-\alpha(E - E^{\circ\prime})}{E_{NF}}\right] \tag{8.125}$$

If the reactants are *not* specifically adsorbed, they can approach the electrode surface (at $x = 0$) up to the distance x_2, the distance of the outer Helmholtz plane (OHP, Section 4.3.2, Figure 4.35). At this position, there is the potential $\varphi_2(x_2)$, which is different from the potential φ^S in the solution bulk because part of the potential drop occurs in the diffuse part of the double layer at $x > x_2$ and perhaps some ions are specifically adsorbed. The driving force for the reaction is therefore not the entire potential difference $\varphi^M - \varphi^S$ between the electrode and the solution bulk, but a potential difference diminished by φ_2: $(\varphi^M - \varphi_2) - \varphi^S$. This situation corresponds to an effective electrode potential $E_{eff} = E - \varphi_2$. Even in the absence of a diffusion layer, the equilibrium concentration of O^z differs from that in the solution bulk because a charged component is enriched or depleted in the double layer. Because of electrostatic reasons, the concentration at the position of the OHP is given by

$$c_{O^z}(x_2, t) = c^*_{O^z} \exp\left[\frac{-z\varphi_2}{E_{NF}}\right]. \tag{8.126}$$

Therefore, empirically determined values for $k°$ and exchange current densities j_0 appear to be potential-dependent. They also change with the kind and concentration of the supporting electrolyte, even if the ions of the electrolyte do not appear in the reaction equation. Their concentration determines how strongly the potential decays to the OHP. The composition of the electrolyte is also important for the extent of specific adsorption in the double layer. If one uses the true standard rate constant k°_t (index t for "true"), that is, the rate constant corrected for double-layer effects, expressions for E_{eff} instead of E, and $c_{O^z}(x_2, t)$ instead of $c_O(0,t)$ in Eq. (8.125), one obtains

$$\begin{aligned} j &= -Fk^\circ_t c^*_{O^z} \cdot \exp\left[\frac{-z\varphi_2}{E_{NF}}\right] \cdot \exp\left[\frac{-\alpha(E - \varphi_2 - E^{\circ\prime})}{E_{NF}}\right] \\ &= -Fk^\circ_t c^*_{O^z} \cdot \exp\left[\frac{(\alpha - z)\varphi_2}{E_{NF}}\right] \cdot \exp\left[\frac{-\alpha(E - E^{\circ\prime})_2}{E_{NF}}\right]. \end{aligned} \tag{8.127}$$

If one further assumes that the substitution $c^*_{O^z} \approx c_O(0, t)$ is possible in Eq. (8.125) if there is no diffusion layer, a comparison of coefficients in Eqs. (8.125) and (8.127) yields

$$k^\circ = k^\circ_t \exp\left[\frac{(\alpha - z)\varphi_2}{E_{NF}}\right] \tag{8.128}$$

or

$$k^\circ_t = k^\circ \exp\left[\frac{-(\alpha - z)\varphi_2}{E_{NF}}\right] \tag{8.129}$$

The exponential term in Eq. (8.129) is called the Frumkin correction. A similar treatment is possible for the exchange current density in Eq. (3.33) and gives Eq. (8.130).

$$j_{0,t} = j_0 \exp\left[\frac{-(\alpha - z)\varphi_2}{E_{NF}}\right] \tag{8.130}$$

A prototypical example for which Frumkin effects are observed is the reduction of peroxodisulfate $S_2O_8^{2-}$. The Frumkin effect can have significant effects on electrode kinetics. Further examples are given in [115].

However, the treatment shown above is mainly of conceptual importance and is rarely applied for the evaluation of particular measurements due to some important limitations. Reactions of charged reagents, in which specific adsorption of the reagent does not play any role, constitute an exception rather than the rule. The specific adsorption of electrolyte anions modifies the consideration and complicates the estimation of φ_2. The specific adsorption of electrolyte anions can also cause an inhibition analogous to Eq. (8.124). The different correction terms are estimated best if σ^{Hg} is known from electrocapillary curves. However, Hg electrodes have a reduced practical importance. Further information can be found in [125]. Ref. [126] is recommended as an overview of further advances.

Further Reading for Section 8.10. Modern reviews [127, 128] are available with details on the theoretical description of the electrical double layer and on the processes influenced by it.

8.11 Electrochemical Scanning Tunneling Microscopy (ECSTM) and Electrochemical Scanning Force Microscopy (ECSFM)

Next, our attention turns to high-resolution microscopies that are often applied to study the structure of electrode surfaces.

8.11.1 Microscopy Beyond the Diffraction Limit

The distance *d* between two objects that can just be resolved as two separate objects is defined as the Abbe limit[108] and it corresponds roughly to the wavelength λ of the applied light ($d = \lambda/NA$, NA – numerical aperture of the used optical objective). With violet light ($\lambda = 400$ nm), objects much smaller than 200 nm cannot be resolved. Structures that are important for electrochemical reactivity tend to be even smaller. To observe these much smaller structures, there are two approaches. One approach uses radiation of a smaller wavelength. X-ray microscopy can resolve much smaller structures but requires optical elements of completely different design (lenses, windows, etc.). It has therefore found only limited application. Much more common is **electron microscopy**, which uses high-energy electrons instead of photons to generate the image. In transmission electron microscopy (TEM), electrons are first accelerated by a high voltage (30 000–3 000 000 V) and then penetrate a very thin sample (Figure 8.83). The electrons can be characterized by a de Broglie wavelength[109] $\lambda = hc/eU$. This wavelength is so small that objects with atomic dimensions can be visualized.

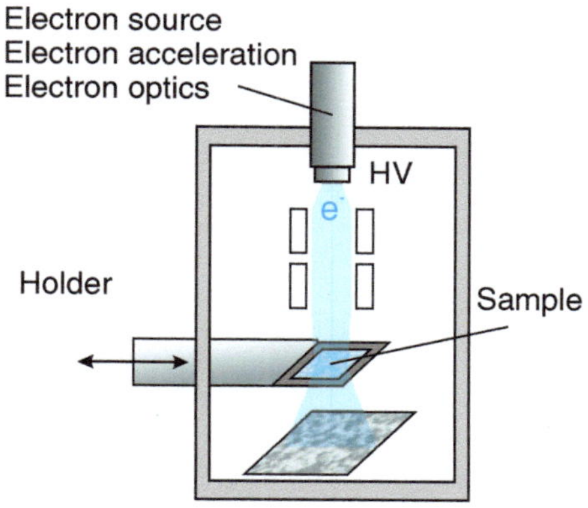

Figure 8.83 Illustration of TEM.

The electrons can be focused as a beam and scanned over the surface of a (thick) sample, which is the principle of scanning electron microscopy (SEM) (Figure 8.84). Secondary electrons are emitted at the point where the focused electron beam impinges on the sample. The intensity of the secondary electrons is measured. The image is composed of the sequentially measured intensities of the secondary electrons as a function of the lateral coordinates of the electron beam. Thicker samples can be studied by SEM, but a thin metallic layer is sometimes required for high-resolution imaging of insulating samples. TEM and SEM require the sample to be placed in a high vacuum, so they are used in electrochemical contexts mostly as *ex situ* methods. Further comprehensive treatments of SEM and TEM are available [129–131].

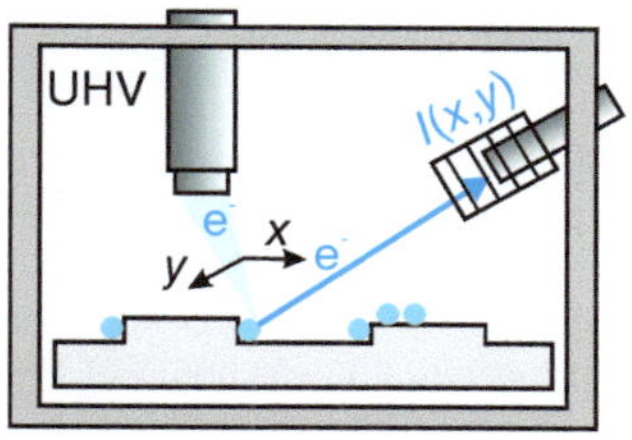

Figure 8.84 Illustration of SEM.

A special TEM technique is investigating an identical location of a sample before and after an electrochemical experiment (IL TEM, identical location TEM). As an example, we discuss the investigation of electrochemical aging mechanisms of fuel cell catalysts in connection with Figure 16.8. Normally, samples for TEM are placed on small metallic grids, which are then transferred into a high vacuum to allow transmission of the electron beam through the very thin slice of the sample (Figure 8.83). Instead of the metallic grid, a sophisticated electrochemical thin-layer cell can be inserted into the electron beam (Figure 8.85). Such liquid cells are now commercially available with several manufacturers and combine an in-plane three-electrode cell on a flat chip with a very thin liquid-filled channel. The chip is covered by a second chip to form the gas-tight thin-layer cell. By means of electrical connections to the flat electrodes and connections of the capillary to a liquid pumping system, electrolyte solutions can be transported to the liquid reservoir of the thin-layer cell. Potentials

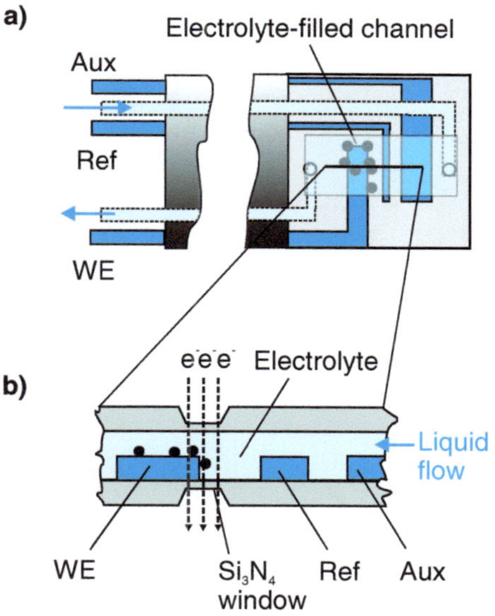

Figure 8.85 General experimental configuration for a TEM sample holder for the *operando* investigation of electrochemical reactions at nanostructures (significantly simplified).

[108] Ernst Karl Abbe, 1840–1905, German physicist, entrepreneur, and social reformer. Together with Carl Zeiss and Otto Schott, he built the scientific and economic foundation of the commercial optical industry.

[109] Luis-Viktor de Broglie (pronounced "de broy"), 1892–1987, French physicist, Nobel prize of physics 1929 for his theory of matter waves (described in his PhD thesis).

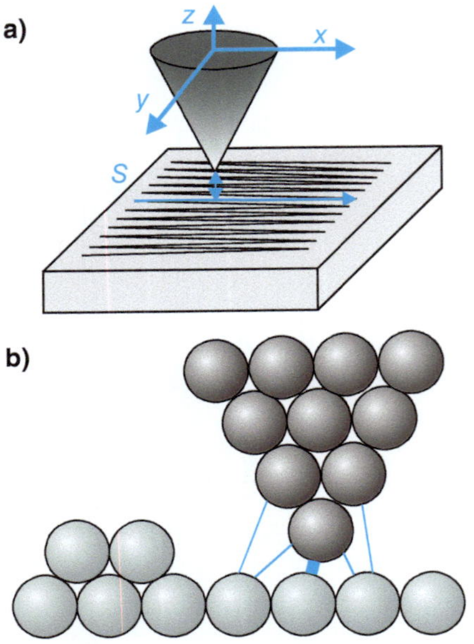

Figure 8.86 General working principle of scanning probe methods. a) General setup and trajectory of the tip above the sample; b) distance-dependent interaction between the tip and the sample, the thickness of the blue line shall indicate the intensity of the interaction causing the signal $S(d)$.

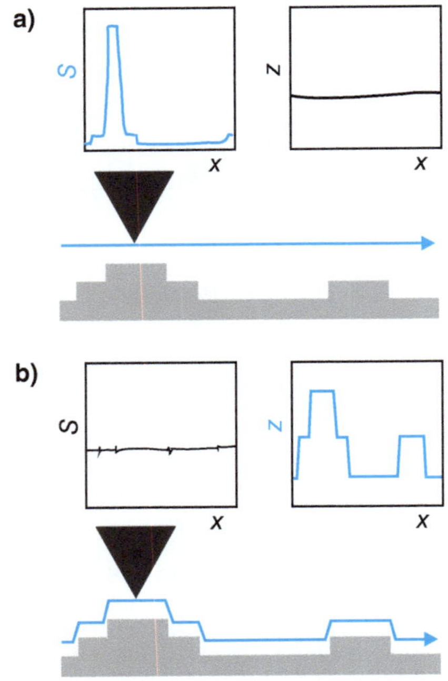

Figure 8.87 General working modes of scanning probe methods a) in constant height mode and b) in constant distance mode.

can be applied with a potentiostat to drive electrochemical reactions during TEM imaging. Above and below the working electrode, the chip is thinned and consists of a thin membrane of 3–8 nm thickness (e.g., silicon nitride). The membrane acts as a "window" for the electron beam. It is electron-transparent but gas-tight. The liquid reservoir with a thickness of 20–200 nm degrades somewhat the achievable resolution, but still facilitates the observation of reactive nanostructures under changing (!) reaction conditions. A significant challenge is finding a compromise between the requirements for meaningful electrochemical experiments and the need for a high-quality TEM measurement. The TEM experiment benefits from an extremely thin liquid reservoir, which makes pumping and sample preparation in the imaged part of the working electrode more challenging. The pumping pressure for the liquid leads to a bulging of the thin Si_3N_4 membranes, which has an adverse effect on the TEM image quality. This can be counteracted by smaller Si_3N_4 windows, which complicates the selection of a suitable viewing area on the working electrode. Attention must be paid to possible artifacts originating from the interaction of the electron beam with the sample and/or the electrolyte solution. In order to limit the electron dosage, a scanning electron beam is often used that transmits the sample (scanning transmission electron microscopy, STEM).

A fundamentally different approach to high-resolution imaging is represented by **scanning probe microscopies** (SPM or SXM). This term refers to an entire "family" of methods that have gained enormous popularity in interfacial physics and chemistry since the beginning of the 1990s. The "X" in SXM is a variable; by inserting "T", "F", or "EC", the abbreviations of important methods are obtained (STM, SFM, SECM). These methods operate differently in detail but share some general principles. The enormous opportunities offered by SXM methods have been used very effectively to solve electrochemical problems, as we have seen in Excursion 4-2 with the example of electrochemical scanning tunneling microscopy (ECSTM). The general principle of all SXM methods is shown in Figure 8.86a. They are characterized by the following features:

1) The signal S is produced at a microscopically small probe, often in the shape of a conical or pyramidal tip.
2) There is an interaction between the probe and the sample, which has a strong distance dependence [$S = f(d$, materials properties)]. This means that the interaction diminishes upon an increase in the distance d. The current between tip and sample in ECSTM is about 1 nA and decreases exponentially with d. There is only 10% of the current when the distance is incremented by 0.1 nm (i.e., roughly the diameter of an atom). If a selected distance provides a current of 1 nA, it means that almost the entire current passes through only *one* atom of the tip and the sample (Figure 8.86b). All other combinations of atoms from the tip and the sample have a mutual distance that is larger by at least one atom diameter than the smallest distance, and thus they contribute a current that is an *order of magnitude* smaller. The strong distance dependence $S(d)$ is the basis of the high resolution achievable by those methods.
3) The probe moves sequentially to the different locations of the sample (Figure 8.86a). The image is thus obtained point-by-point and line-by-line. Structures can only be recognized when the atoms on the surface of the sample change their position so slowly that they are roughly at the same position if the probe comes close to their location again. The image is a collection of line scans. The axis along which the line scans are recorded is called the high-frequency (hf) axis, as compared to the orthogonal low-frequency (lf) axis. In Figure 8.86a, the x-axis is the hf axis and the y-axis is the lf axis. The line-by-line recording of an image frame takes some time. Whether or not a particular scanning probe technique is suitable to solve a particular

problem depends, among other things, on whether the sample can be kept stable over the recording time of an image frame.

4) Scanning probe techniques can be carried out in different modes. The vertical actuator can continuously adjust the vertical position z of the probe in such a way that the signal S remains constant at a preselected set value (constant distance mode). At the same time, the instrument stores the actuator position z for each horizontal coordinate (x, y), which is later used to construct an image (Figure 8.87b with a line scan as an example). Alternatively, the probe can also move in a plane with $z = const.$ In this case, the signal S itself is used for constructing the image (Figure 8.87a, constant height mode). To apply this mode, the sample must be extremely flat so that the probe and the sample do not collide. In this mode, no feedback is necessary to adjust the vertical tip position. Thus, the recording speed can be much higher compared to that for the constant distance mode.

The primary measured data in scanning probe microscopes have the structure in parametric representation $S(t)$, $z(t)$, $x(t)$, $y(t)$. From that, the explicit images $S(x, y)$ or $z(x, y)$ can be formed. Mathematically, the representation $S(x, y)$ can be considered as a curved area in a three-dimensional space. This function of two variables x and y can only be reproduced on a two-dimensional paper plane by different projection methods. Possible projection methods are:

- False color representation, in which the values of S (or z) are translated to a scale of colors or gray shades (Figure 8.88a).
- Isoline plots, in which thin lines connect points that share the same value (or height, Figure 8.88b). This is similar to common topographical maps.
- Parallel projection, in which a surface in three-dimensional space is projected on a two-dimensional flat plane (the paper plane). This is a process well known from arts or photography (Figure 8.88c,d).

We discuss the data representation more extensively because some of the apparently spatial images may cause a strong suggestive effect on the viewer. Biological evolution has selected in humans neuronal image processing capabilities that enable us to recognize effortlessly patterns in complex images (especially if they are of vital interest to us) and even allow us to make very rapid decisions based on such obscured data. However, the scientific use of image data requires us to take a more skeptical, critical, and informed attitude toward features that we might "spot" in data. This attitude has to be built on knowledge about how the data have been acquired and how the image was generated from the primary data. All these steps are prone to artifacts. For instance, we must remain aware that the projection of a topography signal $z(x, y)$ from a constant distance measurement as in Figure 8.87b also depends on the local materials properties because the feedback loop uses the signal $S(d$, materials properties) for adjusting the height z. Therefore, a plot of the height signal $z(x, y)$ is rarely a "purely" topographic image. This is even more so for work in the constant height mode (Figure 8.87a). What appears as "hills" in Figure 8.88 has no topographic reality on the sample because this particular image shows the variation of the local reaction rate on a flat surface! These material properties are of particular interest in understanding electrochemical problems. Therefore, one ongoing development in the area of scanning probe techniques is the design of experiments, experimental conditions, probes, and measurement routines that facilitate the mapping of material properties with no influence of topography. However, this aim was never completely achieved because in the signal $S(d$, materials properties) topography and materials properties are genuinely interlinked.

Apart from image construction (purely mathematical projection), we have to consider modification of the measured data, such as flattening of the background signal or filtering to suppress interfering signals and to enhance the signal of interest.

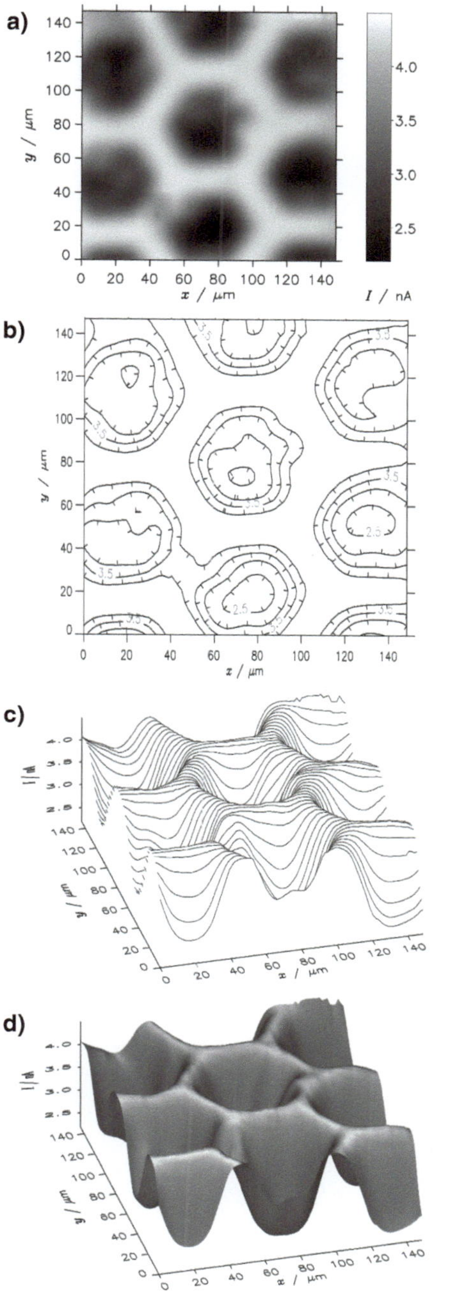

Figure 8.88 Image construction from data of scanning probe techniques. a) False-color or gray-scale image; b) isoline plot; c) parallel projection with individual line scans; and d) parallel projection with a shading calculated as diffuse reflection at the curved surface.

Occasionally, this is problematic because the experimenter selects a filter algorithm that enhances the interesting structures within the measured data set. In extreme cases, this can cause signals to appear in the filtered data set, although they actually did not occur in the original data set. Thus, these signals must be considered as artifacts.

Further unavoidable image distortion may be caused by finite probe size effects. This phenomenon causes a sharp step-like change in a property between two adjacent regions of a sample to result in a gradual transition from one value to another value in the image. This is caused by the fact that a probe of finite size, while passing from one region to the other, simultaneously interacts with both regions, thus causing an intermediate signal. We shall come back to this phenomenon in Section 8.11.7.

The huge success of scanning probe techniques and their broad use today results from the fact that four basic principles (see above) realized for the first time in scanning tunneling microscopy can be used along with other interactions between probe and sample. This facilitates the use of similar instrumental components to develop new methods that record different material properties. Table 8.7 summarizes the most important methods, in which we also include SEM by considering the focused electron beam as a local probe. In the following section, we turn to those scanning probe techniques that can be used directly within electrochemical cells. Of course, scanning probe techniques are also frequently used according to Figure 8.1 as *ex situ* techniques. The sample is taken out of the electrochemical cell after an experiment and can then be studied by almost every scanning probe technique.

New developments aim for combined scanning probe techniques (e.g., SECM-SFM, Kelvin probe-SFM) to obtain from one sample region both topographical information as well as functional properties at the same time (and thus at the same state of the sample) [132].

Finally, the interaction between the local probe and the sample can also alter the sample surface. In the sense of an imaging process, this is not desired because the object of interest is changed during the analysis. However, such local changes can be used purposefully to modify or investigate material surfaces on a very small length scale in a very precise and controlled manner.

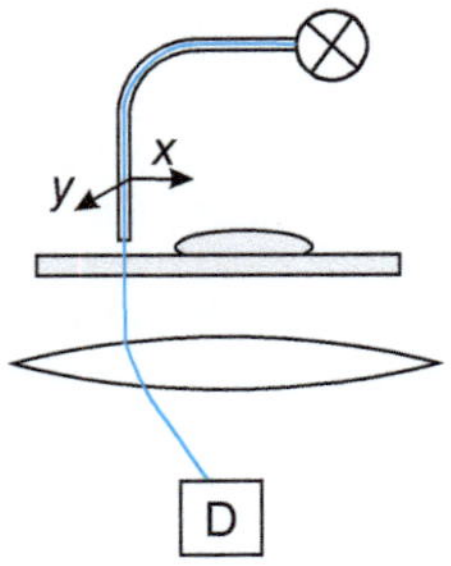

Figure 8.89 Principle of SNOM.

Table 8.7 Important scanning probe techniques. The abbreviations are used in the literature both for the method as well as for the instrument. In this book, the abbreviation is used for the method only.

Abbreviation	Method	Section with more details
ECSTM	Electrochemical scanning tunneling microscopy	Figure 8.86b, Sections 8.11.2–8.11.3
ECSFM	Electrochemical scanning force microscopy	Figure 8.96, Sections 8.11.4–8.11.6
SECM	Scanning electrochemical microscopy	Section 8.12
SPECM	Scanning photoelectrochemical microscopy	Section 8.13.1
SICM	Scanning ion-conductance microscopy	Section 8.13.2
SECCM	Scanning electrochemical cell microscopy	Section 8.13.3
SRET	Scanning reference electrode technique	Section 8.13.4
SVET	Scanning vibrating electrode technique	Section 8.13.4
LEIS	Localized electrochemical impedance spectroscopy	Section 8.13.4
KPM	Kelvin probe microscopy	Section 8.13.5
SEM	Scanning electron microscopy	Figure 8.84
SNOM	Scanning near-field optical microscopy	Figure 8.89

8.11.2 Principle of Scanning Tunneling Microscopy

Scanning tunneling microscopy (STM) allowed for the first time the imaging of atomic arrangements at surfaces in real space. The probe is a metal tip; the sample can be a metal or a semiconductor. The technique is based on the quantum-mechanical tunneling effect, for which there is no comparison in classical mechanics.

The movement of a valence electron in a metallic body, such as the STM tip, can be described as the movement of an electron in a potential well $W(z)$. Figure 8.90a shows the energetic relationships in a simplified fashion. For $z \leq 0$, $W = 0$; and for $z > 0$, W assumes a large and constant value.

In order to contrast with classical mechanics, we can imagine metal spheres rolling back and forth in a well. They cannot leave the well if their total energy is smaller than the height of the energy barrier (Figure 8.90b). In quantum mechanics, the solution of the Schrödinger equation[68] for the same system yields a one-electron wave function Ψ (Psi) that decays exponentially from the position of the energy barrier at $z = 0$ (Figure 8.90c). Therefore, the probability ($<\Psi^* \Psi>$) of finding the particle outside the well (e.g., tunneling) can be calculated from $\Psi(z)$.

$$\Psi(z) = \Psi(0) \cdot \exp[-\beta z] \text{ for } z > 0. \tag{8.131}$$

The quantity β (kappa) is a decay constant [m^{-1}], for which Eq. (8.132) holds.

$$\beta = \frac{\sqrt{2m_e(W(z) - E_e)}}{\hbar} \tag{8.132}$$

Ψ – One-electron wave function
z – Space coordinate perpendicular to the surface [nm]
β – Decay constant of the wave function [nm^{-1}]
m_e – Mass of the electron [kg]
W – Barrier height [eV]
E_e – One-electron energy [eV]
$\hbar$ – Reduced Planck constant = $h/(2\pi)$ [J s]

Typical values are $\beta = 1\ Å^{-1} = 10\ nm^{-1}$. The probability density of the electron decays starting from $z = 0$ for $z > 0$ is proportional to $e^{-2\beta z}$. In other words:

Although the one-electron energy is smaller than the barrier height, there is a small but finite probability to find the electron outside of the potential well (i.e., the STM tip). This small probability vanishes when increasing the distance from the barrier by a few Å (or the size of a few atoms).

If one brings a second electronically conducting body (here the sample) so close to the tip that the probability density of electrons from the tip still has finite values, according to Figure 8.90c, electrons can transfer from the tip to the sample (or *vice versa*). One condition for such a transfer is that the electron must maintain its energy during the transfer and arrive in a previously unoccupied orbital. In other words, no energy exchange is possible with the environment to reach an unoccupied orbital.

To understand this fact, we need to recall that there is a large number of wave functions Ψ_i with their Eigenvalues of energy $E_{e,i}$ as solutions to the Schrödinger equation. The different electron energy levels $E_{e,i}$ of the tip on the left side of Figure 8.91a are so close to each other on the energy scale, that it is not sensible to depict them as discrete energy levels. Instead, one speaks of an energy band. The highest occupied energy level in a metallic body corresponds roughly to the Fermi level E_F. Because of the proportionality $W = eU$, E_F can be stated in the unit volt (V) or electron volt (eV). The gray-filled area in Figure 8.91a shows the position of the occupied electronic states *on the energy axis*. We will return to a more detailed consideration of electronic states in solids in Chapter 10. For now, it is sufficient to realize that it is not practical to make a list of all one-electron energetic states because they have only infinitesimally small energy differences from each other. Instead, one uses the density of states (DOS) $D_S(E_e)$. This is the number of states within a small energy interval. The density of states depends on the particular one-electron energy E_e and the material considered.

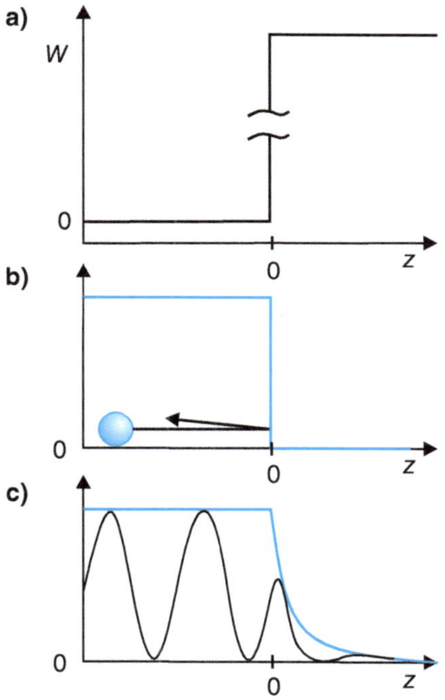

Figure 8.90 Movement of electrons in a potential well. a) Simplified potential profile $W(z)$ perpendicular to the surface of the tip (along the z coordinate in Figure 8.86a); b) probability density of a macroscopic body in classical mechanics; and c) probability density of electrons in the vicinity of the potential barrier (blue) and exemplary one-electron wave function close to the potential barrier (black).

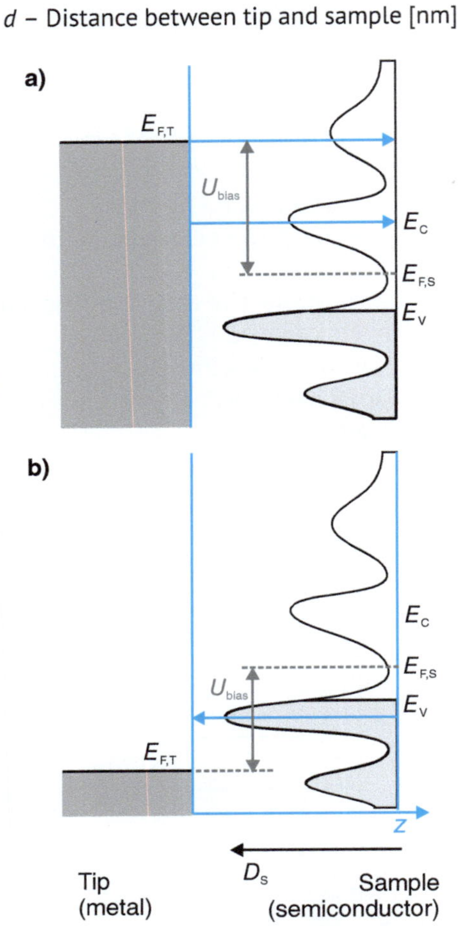

Figure 8.91 a) Schematic representation of the tunneling process from the tip to the sample; b) schematic representation of the tunneling process from the sample to the tip upon reversal of the bias voltage U_{bias}.

If a bias voltage U_{bias} is applied between tip and sample, the Fermi levels of tip and sample are shifted by the value eU_{bias} against each other. This causes a net flux of electrons from the occupied states of the tip into the unoccupied states of the sample (Figure 8.91a). The current is proportional to U_{bias}, the DOS of unoccupied states of the sample $D_S(E_F)$, and a transmission coefficient $T_e(E_e)$, which is proportional to $e^{-2\beta d}$. Thus, the tunneling current I is

$$I = const \cdot U_{bias} \cdot D_S(E_F) \cdot \exp[-2\beta d] \qquad (8.133)$$

Because there is a proportionality between I and U_{bias} in Eq. (8.133), the tunneling contact behaves electrically like an ohmic resistor. In a vacuum, the barrier height W, which enters the calculation of β in Eq. (8.132), amounts roughly to the work function Φ of the metals of the tip and sample. Small decay constants mean a larger tunneling probability. For instance, the exponential term in Eq. (8.131) will have the same value of 1/e for the following combinations of decay constant β and distance d: (0.5 Å^{-1}, 2 Å), (1 Å^{-1}, 1 Å), and (2 Å^{-1}, 0.5 Å). When β in Eq. (8.132) assumes small values, the difference $W - E_e$ has to be as small as possible. This is satisfied for electrons with energy close to the Fermi level E_F (Figure 8.91a). With this consideration, the largest part of the tunneling current is contributed by electrons of the tip with a one-electron energy close to E_F.

Upon reversal of the sign of U_{bias}, the tunneling current occurs from occupied states of the sample with an energy close to E_F to unoccupied states of the tip (Figure 8.91b). This opens up the opportunity to study the local density of states by keeping the position of tip and sample constant but scanning U_{bias}. This working mode is called scanning tunneling spectroscopy (STS). While STS in vacuum plays an important role in surface science, it is rarely used in electrochemical cells, because the potential window, in which no electrochemical reaction interferes with the recording of the tunneling current, is too narrow and thus very limiting.

8.11.3 Electrochemical Scanning Tunneling Microscopy

Figure 8.92 shows schematically the design of an electrochemical cell for tunneling microscopy in solution (see also Figure 4.40 in Excursion 4-2). In the solution, the height of the energy barrier $W(z)$ decreases significantly to about 1 eV and is much smaller than in a vacuum [133]. In addition, electrochemical reactions may occur at the tip and at the substrate. The resulting faradaic current is superimposed on the tunneling current and complicates any exact interpretation of the results. Because the electrochemical reactions at the sample shall be observed under controlled conditions, reactions at the tip must be suppressed as strictly as possible. Experimentally, the bias voltage is set via a bipotentiostatic setup (Figure 8.92a). The tip and the sample are the working electrodes (WE1 and WE2). The potential is controlled versus a reference electrode (Ref) by means of a bipotentiostat. In this way, it is ensured that the potential of the tip is within a potential range in which no faradaic reaction is possible (metal deposition, H_2 evolution, and O_2 reduction). The bias voltage is the difference between the electrode potentials of tip and sample ($U_{bias} = E_{WE1} - E_{WE2} = E_T - E_S$). The currents through the tip and the sample are not necessarily of equal magnitude in this circuit, and an auxiliary electrode (Aux) must deliver the remaining current.

Parasitic currents cannot be prevented in ECSTM. Therefore, the area of the tip is kept very small by an almost complete insulation of the metal surface by a polymer layer that only leaves the very apex of the tip exposed. This is possible because the tunneling current requires only the topmost atoms (Figure 8.92b). An alloy of 90% Pt and 10% Ir is typically used for the tip. The noble metals are stable in a wide potential range and also when exposed to acids. The addition of Ir increases the mechanical stiffness of the wire. In contrast, pure Pt would be too soft.

The mechanical translation of the tip is driven by piezoelectric actuators. They consist of a ceramic piezo material [e.g., lead-zirconium-titanate (PZT from Pb-Zr-Ti)], which responds to an externally applied voltage by a change in length. The ceramic can be manufactured in different shapes and is provided with contact electrodes in such a way that a translation in all three dimensions (x, y, z) is possible. Integrated devices for measuring the translation allow the construction of so-called closed-loop scanners in which the real translocation is measured and the applied voltage is adjusted until the position exactly matches the desired position. This increases the image quality, especially for large image frames, and the handling is simplified.

In order to measure the current of around 1 nA or smaller, preamplifiers are employed, and the experiment is placed into a Faraday cage[3] for electromagnetic shielding. As in other STM experiments, the experiment is so sensitive that thermal expansion can lead to distortion of the results. Therefore, temperature control of the room and thermal insulation may be necessary. It is often necessary to provide a means of mechanical insulation against vibrations from the environment (vibration of the building due to outside traffic, vibrations in the air ventilation system, closing doors, and sometimes acoustic noise). Often, the system works best with a stiff and compact design of the actual measurement head to block low-frequency vibration. The measurement head is attached to a large mass (for instance, a table of stone) and a spring with a low force constant (such as a bungee rope) that block high-frequency vibrations. There are many home-made and commercial solutions.

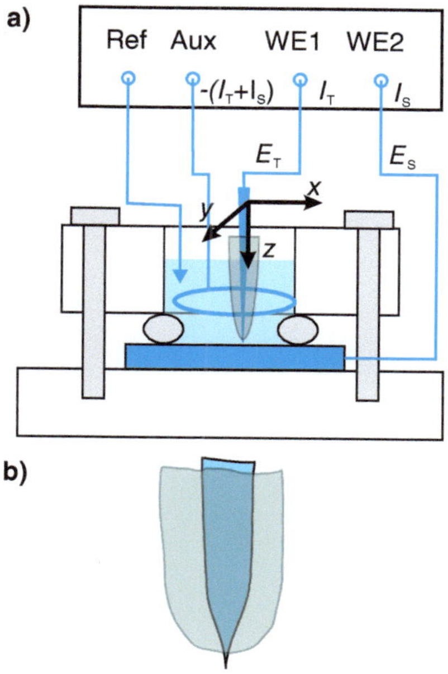

Figure 8.92 a) Illustration of a measurement cell in bipotentiostatic arrangement for an ECSTM experiment; b) enlarged view of the partially coated STM tip. The schematic is not to scale.

The use of a very pure electrolyte solution is critically important for the success of the experiments. This requires substantial effort for the scrupulous cleaning of glassware and measurement cells because impurities attached to them can contaminate the electrolyte solution and adsorb to the surface of the sample and the tip. Due to the extreme sensitivity, this may cause significantly diminished quality of the images and results.

Atomically resolved images can only be obtained from single-crystal surfaces (Section 8.9). If a sample has too many steps and defects, it cannot be avoided that a conically shaped tip establishes tunneling contacts with the surface of the sample at several and changing locations. The result is a coarse topographic image with a lack of detail, which could probably be obtained more easily by scanning force microscopy.

Applications. Typical investigations using ECSTM aim for the analysis of ordered two-dimensional structures on single-crystal surfaces, for example, reconstruction processes as a function of the electrolyte composition and the electrode potential (Section 8.9.2, Figure 8.55, [90]).

Another recurring topic is the analysis of the potential-dependent structure of the electrical double layer, in which layers of specifically adsorbed anions contribute to the image contrast. They form characteristic superstructures, with lattice parameters that are different from those of the electrode metal (Section 8.9.7) [100]. By setting a (desired) tunneling current by specifying the voltage via U_{bias} and the current as the set current, a specific working distance will be enforced. The settings also determine to which orbitals the electrons will be transferred. This can be used sometimes to control whether the metal surface itself is imaged or the arrangement of the anions attached to it [134]. A typical phenomenon is the appearance of a moiré pattern in which slightly larger distances of the specifically adsorbed anions compared to the metal atoms at the surface cause structures with a much larger repeat length (periodicity) than atomic dimensions (Figures 8.63 and 8.64). The larger periodicity emerges because only every n^{th} atom is located on an equivalent adsorption site [135].

The image is formed only by the ions or atoms that change their location significantly more slowly than the recording of the ECSTM image. Thus, many important dynamic structures escape analysis. Low temperatures that are used in a vacuum to enhance image quality by freezing thermal motion are not applicable in ECSTM because the temperature range is limited to the liquid state of the electrolyte (typically with water as solvent). Another option is to increase the recording speed, even up to the rate of a video recording (Video ECSTM). This is achieved in the constant height mode. Video ECSTM has been used for recording and quantitative analysis of trajectories of two-dimensional surface diffusion processes of specifically adsorbed sulfide ions [136].

Organic molecules compete with anions for adsorption to the electrode surface, in which hydrophobic organic molecules are preferentially bound to the electrode surface at potentials close to E_{pzc}. If the organic adsorbates form an *ordered* structure in the adsorbate layer, they are stabilized by adsorbate–adsorbate interaction in addition to the adsorbate-substrate interaction. Such an interaction is typically required to reduce the mobility of organic adsorbates so that they can be imaged by ECSTM. The formation of an ordered structure depends very much on the geometry of the organic molecule, but also on the electrode potential. Transitions from one ordered adsorbate structure via an unordered structure to another ordered structure can be followed by ECSTM. For a complete elucidation of the adsorbate structure, complementary methods (e.g., IRRAS) are often necessary. Surveys with case studies are compiled in [137, 138].

If large adsorbate molecules contain a redox-active moiety, a strong local enhancement of the tunneling current can occur at the redox-active group within the molecule if the formal potential of the redox-active moiety is between the potentials of the tip and the substrate. In this case, the highest occupied molecular orbital (HOMO) and the lowest unoccupied molecular orbital (LUMO) of the redox-active moiety participate in the tunneling process, which reduces the tunneling distance and the barrier height [139].

Metal deposition processes and the resulting topographic features can be observed by ECSTM in all phases of the process [104, 140, 141]. Interesting information can be obtained by imaging while slowly sweeping the electrode potential (Figure 8.93). Each line scan is then recorded at a slightly different potential. This working mode is particularly suited to follow potential-dependent nucleation processes or the advancement of growing terraces. ECSTM has also been used to follow the formation of very thin passivation layers in the context of corrosion processes [142].

Local Surface Modification. STM instruments can be used to induce site-directed local modifications of the substrate surface based on electrochemical reactions. Around the STM tip in air, a thin meniscus of liquid water is often formed, which

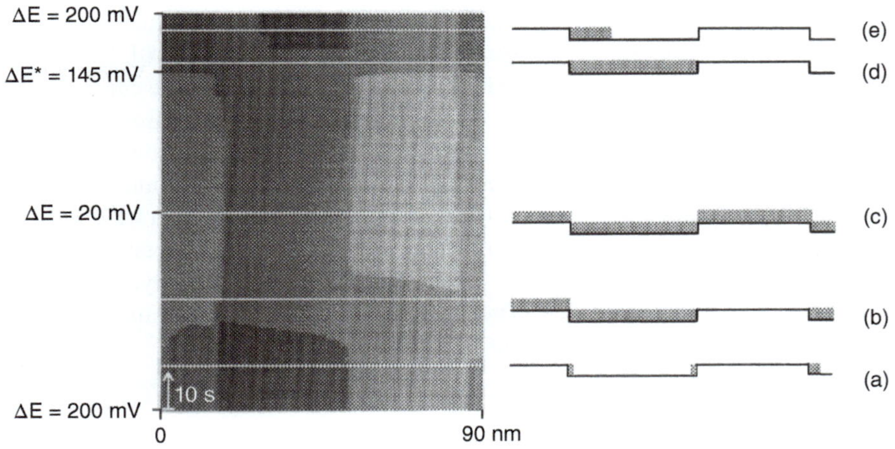

Figure 8.93 ECSTM line scans showing Pb underpotential/stripping on a Ag(1 1 1) electrode as a function of the applied potential during a cyclic voltammogram at the sample between +0.2 V, +0.02 V and +0.2 V and v = 5 mV s^{-1}; solution 10^{-3} mol L^{-1} $Pb(ClO_4)_2$ + 10^{-2} mol L^{-1} $HClO_4$. The schematic cross sections illustrate a) step decoration; b) formation of a 2D Pb-UPD island; c) formation of the Pb-UPD layer also on Ag islands; d) stripping of Pb from the Ag islands; and e) stripping of Pb-UPD layer from a pit in the Ag substrate (Compare also to Figure 8.67). Source: Reproduced with permission from [104], p. 119, Copyright 1996, Wiley-VCH; there reproduced from E. Ammann; Master Thesis, University of Bern 1995.

can be used to induce the local formation of a surface oxide. Another electrochemical mechanism facilitates the placement of metal clusters at the surface of the substrate. First, metal cations from solution are reduced at the STM tip from a diluted solution of the metal cations. The deposited metal can be transferred to the sample as clusters either by a pulsed metal dissolution from the tip under concomitant metal deposition on the substrate [143] or by a mechanical contact triggered by a pulse of the set current [144] (Figure 8.94).

Further Reading. Original communications and review papers [145–152] are recommended for in-depth study of the subject ECSTM.

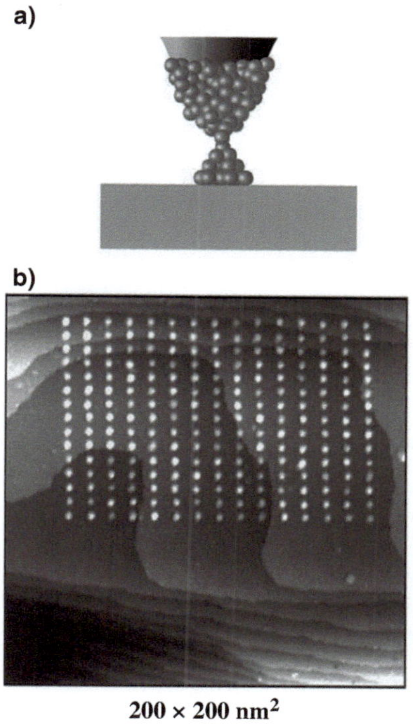

Figure 8.94 Local deposition of a metal cluster; a) principle of the transfer by means of a mechanical contact and b) examples of Cu clusters on an Au(1 1 1) single-crystal electrode. Source: Reproduced with permission from [144], Copyright 2005, Elsevier.

8.11.4 Basic Principles of Scanning Force Microscopy

Instrumentation. Scanning force microscopy (SFM) is based on recording forces that act between the sample and the microscopic tip. SFM has become a routine method in many branches of science and technology. The method allows the study of solid samples regardless of their conductivity. This includes insulating samples like passive layers on electrodes. Therefore, the scope of application is significantly broader than that of STM. SFM can be applied in vacuum, in air, as well as in solution with electrochemical control of the sample potential (ECSFM, electrochemical scanning force microscopy). On the other hand, it is usually much more difficult to obtain atomically resolved images with SFM than with STM. However, SFM can image other interesting material properties such as local adhesion, local friction, local magnetic forces, and local charge.

The tip is produced by methods of silicon technology and typically has a pyramidal shape with an aperture radius in the order of 10 nm. A cantilever carries the tip and connects to a chip that is mounted to the holder (Figure 8.95). The dimensions of the cantilever are in the micrometer range. Because of the small dimension, even a hard material such as silicon or silicon nitride becomes elastically deformable by small forces of 10^{-11} bis 10^{-6} N. The exact force constant of the cantilever is controlled by its dimensions. The cantilever can have the shape of a bar with a rectangular cross section or a V-shape (Figure 8.95). The deflection of the cantilever is proportional to the acting force ($\mathbf{F} = -k_0\ \Delta x$, Δx – is the deflection [nm], that is the displacement from the position without acting force). The very small deflection is measured with the beam of a laser diode that is reflected off the back side of the cantilever (Figure 8.96). A metallic coating on the backside of the cantilever is often used to enhance the reflectivity for the optical detection of deflection. The reflected beam is directed toward a split photodiode. Depending on the beam deflection, the four segments of the photodiode receive a different fraction of the impinging radiation

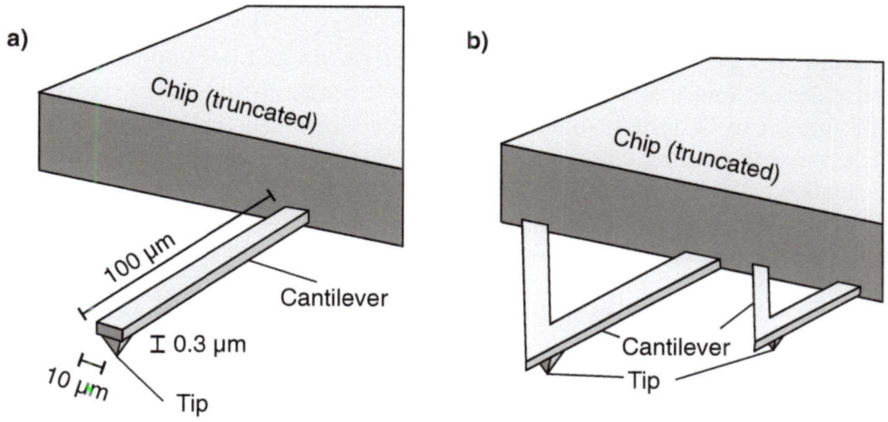

Figure 8.95 Examples of a) a bar-shaped cantilever with rectangular cross section and b) V-shaped cantilever. One chip often holds several cantilevers with different k_0 and ω_0. The schematic is not to scale.

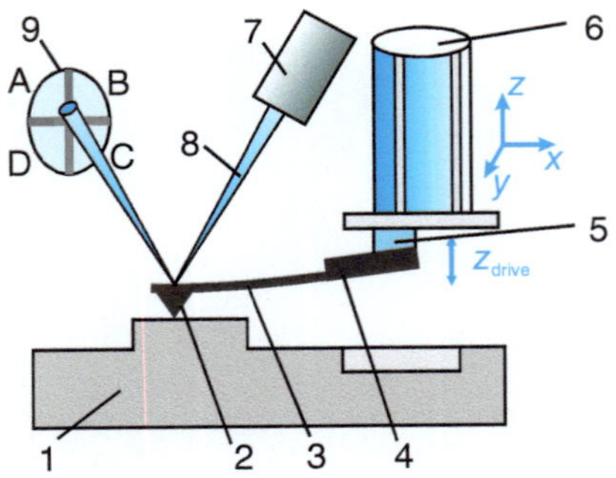

Figure 8.96 Basic experimental configuration of an SFM;

1) sample,
2) tip,
3) cantilever,
4) cantilever chip,
5) drive piezo for dynamic modes,
6) scanner,
7) laser diode,
8) laser beam as pointer (beam is simplified, in real devices the beam is redirected several times by mirrors), and
9) split photodiode. The schematic is not to scale.

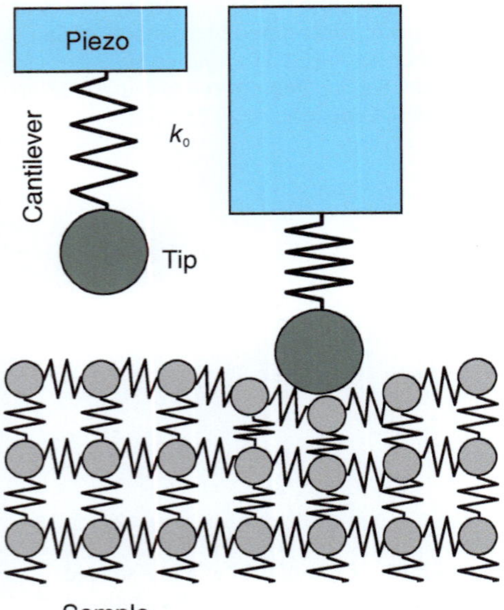

Figure 8.97 Basic model of the interaction between the tip and the sample represented as mass-bearing point-particles that are interconnected by springs. Depending on the relative size of the force constants, the interaction between tip and sample may lead to a deflection of the cantilever only, the deformation of the sample only, or the deflection of cantilever and deformation of sample. The schematic is not to scale.

energy. This leads to different photovoltages at the segments of the photodiode. Those voltages form the primary signal that is further processed and used to construct the image.

Cantilever. The system of cantilever and sample can be thought of as interacting mass points connected by springs (Figure 8.97). The selection of a cantilever with a suitable force constant and resonance frequency is very important for performing successful SFM and ECSFM experiments. For topographic imaging, deformation of the sample should be avoided, which is important and challenging when working with soft organic coatings or biomolecules. In principle, very soft cantilevers (small force constant k_0) would be suitable for this task. However, the thermal energy stored in the cantilever at a temperature T amounts to $\frac{1}{2}k_BT = \frac{1}{2}k_0(\Delta x)^2$ (k_B is the Boltzmann constant). Therefore, the oscillations caused by thermal energy lead to increasing amplitudes Δx with decreasing force constants k_0. This amplitude is not caused or influenced by the sample and results in undesired noise in the images that becomes more important with small k_0. Table 8.8 contains some guideline values for selecting cantilevers.

Table 8.8 Typical cantilevers for experiments in air or in liquids, after [153].

Measurement task	k_0/(N m^{-1})	$f_0 = \omega_0/(2\pi)$/ kHz (in air)	Comment
Contact mode in air or in solutions	<1	10–30	Soft cantilevers, V-shaped cantilevers minimize the lateral deflection, bar-shaped cantilevers for measurement of friction forces.
Intermittent contact mode in air	1.5–3	25–70	Relatively large k_0 in order to overcome capillary forces of a thin water film on the sample in air
Intermittent contact mode in solution	<0.1	20	Because capillary forces do not act and adhesion forces are lower, softer cantilevers may be used than in air
Dynamic mode in air	15–60	130–350	Stiff cantilevers with high Q-factor can overcome the capillary adhesion between tip and sample, moderate amplitude > 5 nm
Dynamic mode in solution	<0.1	30–50	Bar-shaped cantilevers and V-shaped cantilevers with Q-factor above 1 (tips longer than 10 μm, oscillation amplitude < 5 nm)

Metallic coatings on the lower side of the cantilevers and the tip yield conductive cantilevers for measuring local conductivity in addition to topography (C-SFM, conductive scanning force microscopy).

Forces. van der Waals Forces[110] act between the tip and the sample. These are intermolecular forces between at least one dipole at the tip and one at the sample. Such forces can be further differentiated: Dispersion forces (or London forces[111]) act between fluctuating dipoles. They emerge due to the random and *nonuniform* distribution of electron density within a neutral atom at a particular point in time. This favors the opposite distribution of electron density in the neighboring atom. This situation enables electrostatic attraction between neighboring neutral atoms,

[110] Johannes Diderik van der Waals, 1837–1923, Dutch physicist, professor of physics in Amsterdam, Nobel prize in Physics 1910.
[111] Fritz London, 1900–1954, German-US American physicist, 1933 emigration because of prosecution due to his Jewish origin, professor of theoretical chemistry at Duke University.

although their *average* charge density may be completely balanced in space. Dispersion forces decay with the seventh (!) power of the distance d ($|\mathbf{F}| \propto 1/d^7$). The potential energy E_{pot} obtained in this force field follows the relationship $E_{pot} \propto 1/d^6$. The strong distance dependence $E_{pot}(d^6)$ is a prerequisite for the high-resolution images based on dispersion forces.

In addition to fluctuating dipole forces, Keesom forces[112] act between permanent dipoles (molecules with unbalanced average charge distribution) and charged particles or between permanent dipoles and an induced dipole in another particle (Debye forces[36]).

Attractive dipole–dipole forces have to be complemented by repulsive forces. We may picture a contact on the atomic scale when thinking of two atoms that are brought very close together. Initially, van der Waals forces cause an attraction between the two atoms. Upon further approach, the electronic shells of the two atoms would penetrate each other (perhaps with the formation of a chemical bond). If the approach is continued, the repulsive force between two positively charged atomic nuclei will become dominant and has an even stronger distance dependence. The distance dependence of the energy resulting from the van der Waals forces and the counteracting repulsive forces of the two nuclei is often approximated by the Lennard-Jones potential[113] [Eq. (8.134)]. This relationship is shown in Figure 8.98.

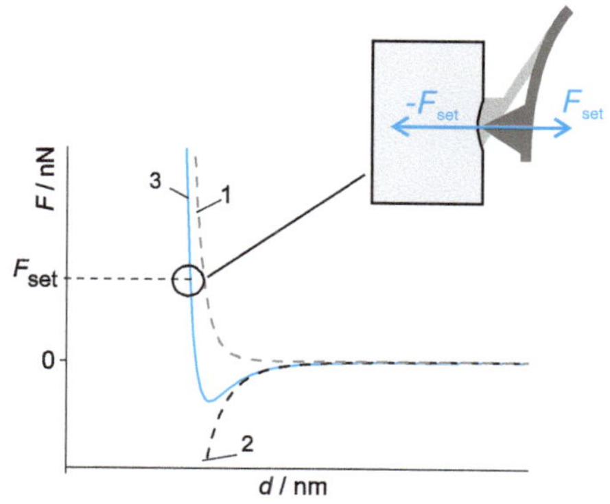

Figure 8.98 Repulsive (1) and attractive forces (2), from which the force–distance curves (3) result. Selecting the setpoint (circle) in the contact mode determines the force F_{set}, which causes a constant deflection of the cantilever and may additionally also deform the sample.

$$E_{pot} = 4E_{min}\left[\left(\frac{d_0}{d}\right)^{12} - \left(\frac{d_0}{d}\right)^{6}\right] \tag{8.134}$$

d – Distance [nm]
d_0 – Distance at $E_{pot} = 0$
E_{min} – Amount of the minimal energy [Nm]

The resulting force is obtained by deriving E_{pot} with respect to the distance to d [Eq. (8.135)].

$$|\mathbf{F}_{tip\text{-}sample}| = -\frac{dE_{pot}}{dd} = \frac{24E_{min}}{d_0}\left[2\left(\frac{d_0}{d}\right)^{13} - \left(\frac{d_0}{d}\right)^{7}\right] \tag{8.135}$$

These forces are also called atomic forces, from which another popular name of the method, atomic force microscopy (AFM), is derived.

8.11.5 Working Modes of Scanning Force Microscopy

The working modes of SFM differ from each other in the types of forces employed for image generation and the way of measuring them. We follow here a more detailed discussion of the forces from Ref. [154] because this knowledge is extremely helpful for the selection of cantilevers, adjustment of measuring parameters, and interpretation of the image data.

Contact mode. The measurement of the force can be accomplished in a static way, in which the cantilever is maintained in a range of repulsive forces by setting a constant deflection and letting the feedback system of the instrument to adjust the vertical position to maintain the desired deflection. In this range, the force–distance curve is very steep (i.e., has a large absolute value of the slope), which facilitates a precise adjustment of the set value for the deflection. Consequently, the tip remains in permanent contact with the sample, thus the name contact mode for this working principle. The advantage of this mode is in the clarity of the setpoint on the force–distance curve (Figure 8.98); the considerable forces are a disadvantage because they act on the sample and may damage it. In addition, the usually used V-shaped cantilever

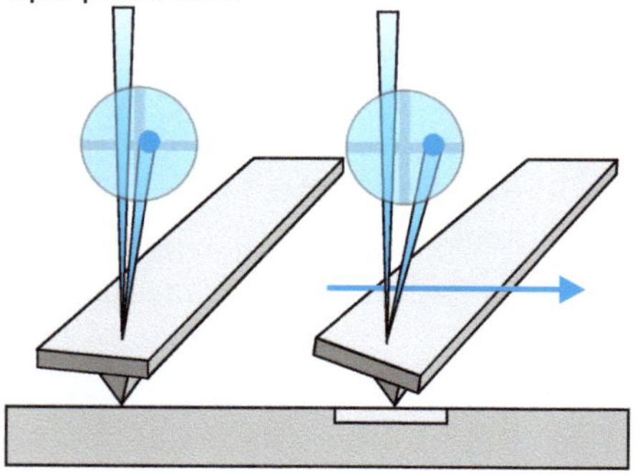

Figure 8.99 Principle of lateral force microscopy.

[112] Willem Hendrik Keesom, 1876–1956, Dutch physicist.
[113] John Edward Lennard-Jones, 1894–1954, British mathematician and physicist, professor of theoretical physics in Bristol and Cambridge, known for the mathematical description of molecular orbitals as a linear combination of atomic orbitals, one of the founding fathers of computational chemistry.

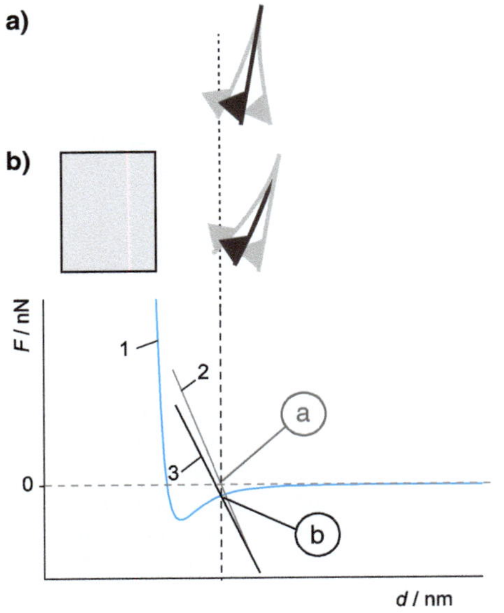

Figure 8.100 Force–distance curve (1) and restoring force of the cantilever without interaction with the sample (a, Curve 2) and resulting force in the case of attractive interaction with the sample (b, Curve 3 = Curve 1 + Curve 2). Schematic drawings in a) and b) depict strongly exaggerated oscillations of the cantilever a) in the field-free space and b) in case of attractive interactions with the sample in the noncontact mode.

ω_0 – Angular frequency of the free oscillation [s^{-1}]
k_0 – Force constant of the cantilevers [N m^{-1}]
μ – Equivalent point mass of the cantilever [kg]

ω – Angular frequency with interaction to the sample [s^{-1}]
k – Modified force constant [N m^{-1}]
d – Tip-sample distance [m]
d_0 – Distance at which $F_{\text{tip-sample}}$ is zero [m]
$\mathbf{F}_{\text{cantilever}}$ – Restoring force of the cantilever [N]
$\mathbf{F}_{\text{tip-sample}}$ – Force between tip and sample [N]
$\mathbf{F}_{\text{total}}$ – Total force [N]

(Figure 8.95b) is also moved laterally, which causes lateral forces on the cantilever in permanent contact with the sample. The lateral forces decrease the quality of the images and may cause increased abrasion at the tip and the sample.

Lateral forces can also be exploited intentionally in lateral force microscopy (LFM) (Figure 8.99). The scanner moves the cantilever laterally and perpendicularly to its main extension. The photodiode not only measures the deflection of the cantilever but also its torsion, which depends on the local friction forces, which in turn are determined by the functional group at the surface of the sample and the tip. This facilitates a material-sensitive imaging, for instance, of areas of a patterned self-assembled monolayer with different terminal groups but identical chain length of the spacer.

Noncontact Mode. In the dynamic mode (also called AC mode), the tip is oscillating during the entire measurement. The image is constructed from small variations in the resonance frequency. The mode is useful when working in the range of attractive forces at larger distances than in the contact mode (Figure 8.98). This mode counteracts unintended alterations of the sample during the imaging process. This comes at the expense of a much more demanding measurement routine. First, there are always two points with the same attractive forces in a force–distance curve like Figure 8.98. Second, the force–distance curve has a smaller absolute slope in this range, which complicates the work of the electronic feedback system. Third, larger working distances are related to a lower lateral resolution because long-range forces are exploited. The oscillation of the usually rectangular cantilever (Figure 8.95a) without interaction with the sample occurs with an angular frequency of

$$\omega_0 = \sqrt{\frac{k_0}{\mu}}. \tag{8.136}$$

The momentary restoring force at the cantilever along the vertical distance coordinate is given by Hook's elasticity law[114] $\mathbf{F}_{\text{cantilever}} = k_0(d - d_0)$. This force has a much steeper gradient than the attractive forces between tip and sample, and it is shown as a gray line in Figure 8.100b. Because $\mathbf{F}_{\text{tip-sample}}$ is negative in the range in which the tip is oscillating above the sample and does not change significantly over the range of the oscillation amplitude of the tip, the total force on the cantilever is $\mathbf{F}_{\text{total}}(d) = \mathbf{F}_{\text{cantilever}}(d) - \mathbf{F}_{\text{tip-sample}}(d)$ in Figure 8.100b. It is approximately linear with a *somewhat flatter slope* than $\mathbf{F}_{\text{cantilever}}(d)$. The flatter slope means that the force constant k (i.e., the deviation of the force with respect to distance $k = \mathrm{d}\mathbf{F}/\mathrm{d}d$) is smaller. In other words, the cantilever appears softer when the tip interacts with the sample than without the presence of the sample. The reduced force constant amounts to

$$k = k_0 - \Delta k = k_0 - \frac{\mathrm{d}\mathbf{F}_{\text{tip-sample}}}{\mathrm{d}d}. \tag{8.137}$$

The smaller force constant influences the resonance frequency of the cantilever $\omega = \sqrt{k/\mu}$. A Taylor series[21] of $\omega(k)$ in the source point (k_0, ω_0) with $\Delta k = \mathrm{d}F/\mathrm{d}d$ leads to

$$\omega \approx \omega_0 + \frac{\mathrm{d}\omega}{\mathrm{d}k}\Delta k = \omega_0\left[1 - \frac{1}{2\mu}\cdot\frac{\mathrm{d}\mathbf{F}_{\text{tip-sample}}}{\mathrm{d}d}\right]. \tag{8.138}$$

Thus, we can conclude:

The measurement of the shift in the resonance frequency of the cantilever in the dynamic working mode is equivalent to recording the gradient $\mathrm{d}F/\mathrm{d}d$ of the force–distance curve.

[114] Robert Hooke, 1635–1703, English polymath active as a scientist and architect, member of the Royal Society.

The measurement of this gradient is usually more sensitive than the measurement of absolute forces and their incremental changes on a large background value. Therefore, it has become the preferred working mode in many applications. We can also conclude that the frequency shift $\omega - \omega_0$ is again proportional to the frequency of the free oscillation. Thus, cantilevers with high resonance frequency tend to facilitate more sensitive measurements.

Intermittent Contact Mode. Working with only attractive forces in air turns out to be difficult. From the atmospheric humidity, a thin moisture film is often deposited on the sample. This thin liquid film can form a meniscus at the tip, which pulls the tip by an additional strong force toward the sample surface (Figure 8.101). A popular working mode uses *very large* oscillation amplitudes to make sure that the tip is pulled out of the thin moisture film during each oscillation period. The approach to the sample is made in a way that the tip briefly (intermittently) contacts the sample surface, thus the name intermittent contact mode. The equivalent expression "tapping mode" was a trade name introduced by the company Digital Instruments, together with a very popular instrument model in the 1990s. In intermittent contact mode, the force $\mathbf{F}_{\text{tip-sample}}$ is relevant only in a small range of the tip movement close to the sample surface. It causes a frequency shift and a phase shift of the oscillation, which can both be used for image construction. Because the tip moves into the distance range, in which repulsive forces dominate, the topography can be recorded. There are much smaller lateral forces compared to those in the contact mode, which drastically decreases artifacts.

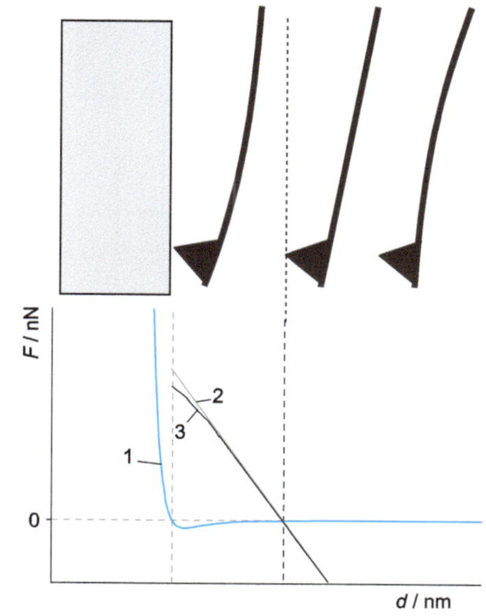

Figure 8.101 Force–distance curve (1) between the tip and the sample, restoring force of the freely oscillating cantilever and (2) the resulting force (Curve 3 = Curve 1 + Curve 2). In the intermittent contact mode, the gradient of the resulting force deviates from the trend in the rest of the curve in the vicinity of the sample surface.

For a more detailed analysis of the behavior of the vibrating cantilever, the mechanical model of a damped, driven oscillator (DDO, Figure 8.102) is helpful. It contains an oscillator and a dashpot in parallel. The dashpot absorbs the mechanical energy, which is constantly resupplied by the driver. In this way, a dynamic steady state can be reached. The damping force $\mathbf{F}_d = b \cdot v$ is proportional to the translation velocity v of the oscillator and to the damping coefficient b. In SFM, the damping is caused by the medium (air or electrolyte solution) surrounding the tip and the sample. Even in a vacuum, some damping occurs due to the inner inelasticity of the material from which the cantilever is made and from the connection between the cantilever and the driver piezo. In addition, the sample takes up some mechanical energy upon each contact to the tip and dissipates this energy more or less well (depending on the stiffness of the sample material). The resonance frequency of the DDO is

$$\omega_0^{\text{DDC}} = \sqrt{\omega_0^2 - \frac{1}{2}\left(\frac{b}{m_r}\right)^2} = \omega_0\sqrt{1 - \frac{1}{2Q^2}}. \tag{8.139}$$

Q – Q-factor [dimensionless]
$\omega_0{}^{\text{DDO}}$ – Resonance frequency of the damped, driven oscillator [s^{-1}]
ω_0 – Resonance frequency without damping and driving [s^{-1}]
b – Damping coefficient [N s m^{-1}]

The dimensionless Q-factor (quality factor)

$$Q = \frac{\mu\omega_0^{\text{DDO}}}{b} \tag{8.140}$$

is a measure of how good the resonator stores mechanical energy. Q increases when the maximum of the resonance curve becomes "sharper" and is inversely proportional to the damping coefficient. Typical values for work in liquids are $Q = 2$ for $f_0 = \omega_0/(2\pi) =$ 40 kHz (in air $Q = 500, f_0 = 300$ kHz). The Q-factor is determined experimentally as the ratio between the resonance frequency and the total width of the resonance curve at $1/\sqrt{2} \approx 70\%$ of the maximum amplitude Δx. At the point of the maximum amplitude Δx, the phase shift ϕ also changes strongly (Figure 8.103).

$$\Delta x = \frac{\Delta x_{\text{drive}} Q}{\sqrt{\left(\frac{\omega}{\omega_0}\right)^2 + Q^2\left[1 - \left(\frac{\omega}{\omega_0}\right)^2\right]^2}} \tag{8.141}$$

Δx_{drive} – Amplitude of the driver [nm]
Δx – Amplitude of the cantilever [nm]
ω – Angular frequency of the drive [s^{-1}]
ω_0 – Angular resonance frequency [s^{-1}]
Q – Q-factor [dimensionless]
ϕ – the phase shift [°].

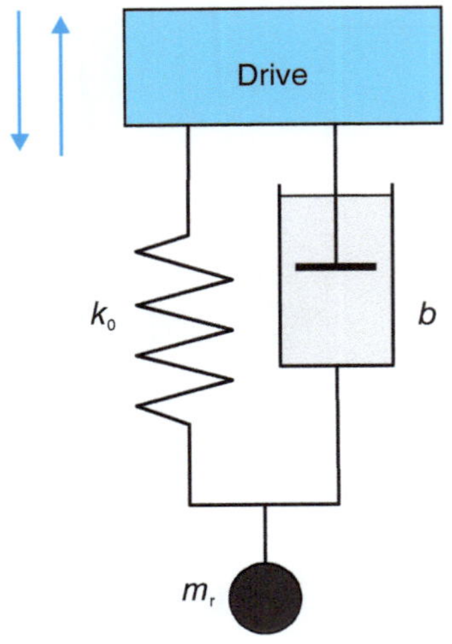

Figure 8.102 Scheme of a damped, driven oscillator (DDO); k_0 is the force constant of the spring; m_r is the reduced mass; b is the force constant of the damping force.

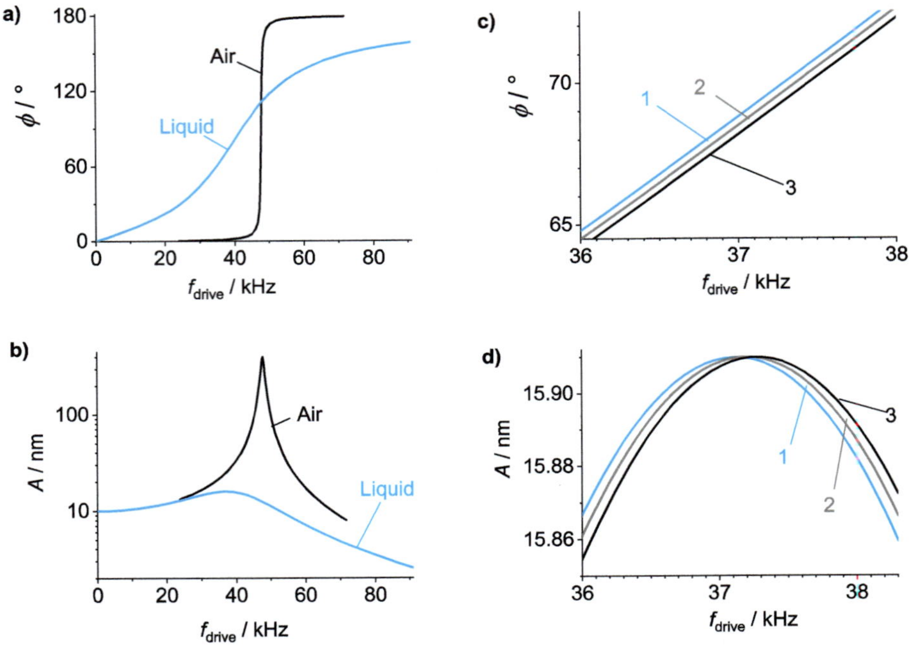

Figure 8.103 Phase shift (a) and amplitude (b) of the cantilever oscillation in air ($f_0 = 47.8$ kHz, $Q = 40$) and in liquid ($f_0^{DDO} = 42.2$ kHz, $Q = 1.5$); c) shift of the resonance frequency and shift of the phase angle for (1) attractive net interaction between tip and sample, (2) no interaction, and (3) repulsive net interaction. Curves were calculated according to Eqs. (8.139), (8.141), and (8.142).

$$\phi = \arctan\left[\frac{\frac{\omega}{\omega_0}}{Q\left[1-\left(\frac{\omega}{\omega_0}\right)^2\right]}\right] \tag{8.142}$$

Material-Sensitive Working Modes. Figure 8.103 shows how the resonance frequency and the phase shift change with additional tip–sample interaction compared to free oscillation. When working with a constant driver frequency, corresponding to the resonance of the free oscillation (Figure 8.104a), and adjusting the working distance by keeping a constant amplitude signal (Figure 8.104b–d), the change in the phase angle ϕ is indicative of the net interaction between tip and sample. Further working modes (implementation depending on the manufacturer) utilize adhesion forces and extract simultaneously local material properties in each run of the tip.

8.11.6 Force–Distance Curves

Understanding the force–distance curves is important to quickly gain an understanding of the mechanical properties of a sample and to use this insight to select appropriate cantilevers and measurement parameters, especially the setpoint.

Figure 8.105a shows a force–distance curve as it is often plotted by the software of SFM instruments for the optimization of the setpoint. This curve looks qualitatively different from the Lennard-Jones potential in Figure 8.98. The reason for the difference is that a different space coordinate is used in both plots for the abscissa. In Figure 8.105a it is the extension of the driver piezo (part 5 in Figure 8.96) and *not* the distance between tip and sample as in the Lennard-Jones potential in Figure 8.98.

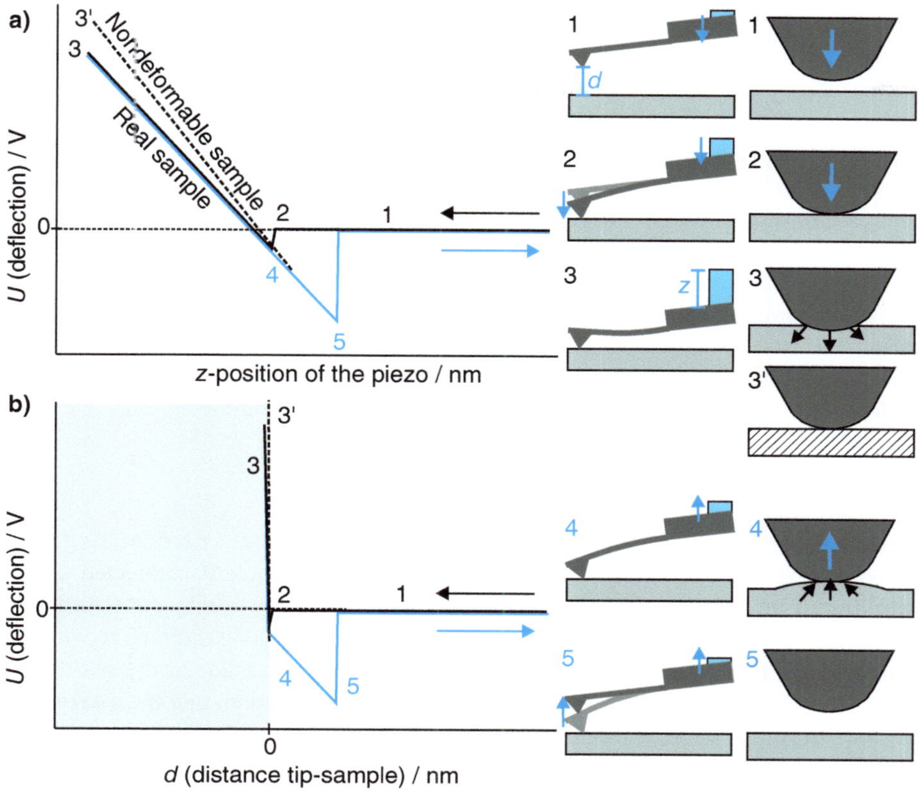

Figure 8.105 Schematic with force–distance curves. a) Plot of deflection versus the extension of the driver piezo (part 5 in Figure 8.96); b) force–distance curves obtained by recalculating the curve from (a) in the contact zone by means of Hook's law. The numbers next to the curve sections correspond to the cantilever movement and the tip–sample interaction in the right part of the figure. The hypothetical curve for a sample without any mechanical deformation is shown in a) and b) and labeled with 3′ in the contact zone. In this case, the cantilever is bent more strongly than in the case of a real, deformable sample.

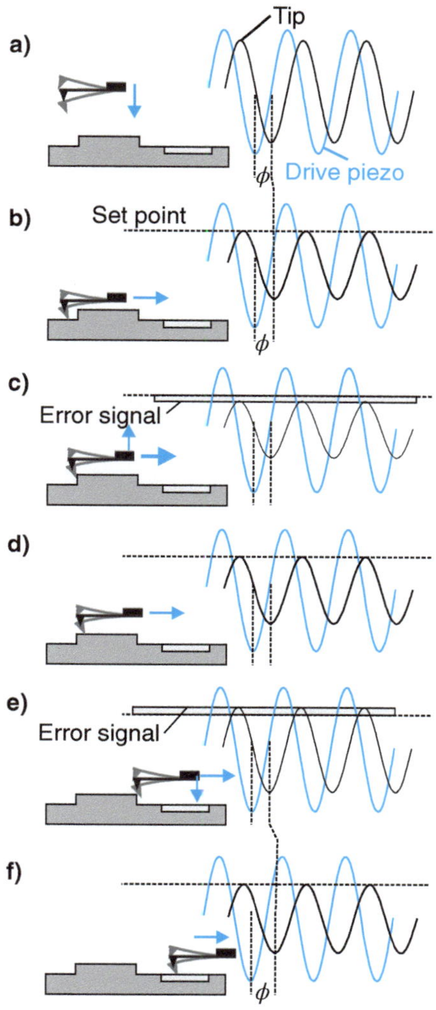

Figure 8.104 Emergence of the material contrast during phase imaging. a) Free amplitude; b) contact, reduction of the amplitude up to the setpoint (strongly exaggerated in the scheme); c) further reduction at an upward step and retraction of the tip; d) stable imaging conditions on an island; e) increase in the amplitude at a downhill step and tracking by the tip movement; and f) change in the phase shift caused by a change in the local viscoelastic properties of the sample.

In section 1 of the curve in Figure 8.105a, the tip–sample distance is so large that the force on the sample is zero and thus the deflection is also zero (measured as a voltage between the upper and lower segments of the split photodiode). In point 2 of the curve, the attractive forces are so large that the cantilever snaps toward the sample. This effect can be considerably enhanced by the water meniscus due to a moisture film on the sample surface. This is in turn diagnosed by the approach curves. In section 3 of the curve, the tip is in mechanical contact with the sample, and a further extension of the driver piezo causes a stronger bending of the cantilever. Due to Hook's law, the bending measured by means of a laser beam reflected on the back side of the cantilever is proportional to the force exerted by the cantilever on the sample. The setpoint is selected within this section of the approach curve in such a way that the determination of the topography is facilitated. This is easier with a stronger bending of the cantilever. On the other hand, the bending should not be so strong that the forces cause damage to the sample. Overall, this means the forces should be as small as necessary. Indeed, it is not only the cantilever that reacts to the force by bending, but also an elastic sample will be indented slightly at the touching point (Figure 8.97). Curve 3′ shows the hypothetical force–distance curve for a nondeformable sample, in which the system responds to the extension of the drive piezo exclusively by bending of the cantilever. After passing through the reversal point, the deflection signal initially follows the signal during approach. However, after passing beyond the point at which the tip snapped to the sample in the forward scan, the withdrawal curve departs from the approach curves. The cantilever bends now toward the sample. At even more contraction of the driver piezo, the bending (and thus the restoring force) will become so strong that the contact between sample and tip is ruptured (section 5). Then, the cantilever continues the backward movement without further bending until the initial point is reached. One can now use Hook's law to recalculate the bending in section 3 of the curve to a sample-probe distance *d*. This yields the plot in Figure 8.105b, in which the curve for section 3 is similar to the Lennard-Jones potential in Figure 8.98.

The force–distance curves contain information about the local mechanical properties of the sample and allow conclusions about the distribution of phases of composite materials on the sample surface. New SFM instruments permit the recording of a complete force–distance curve at each point of the scanning grid and the extraction of different properties after the completion of recording (force volume mode).

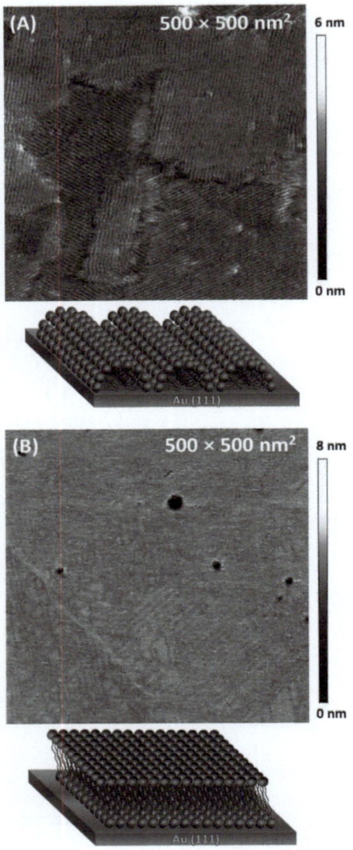

Figure 8.106 Formation of lipid bilayers from DMPC (di-myristoyl phosphatidyl choline) on an Au(1 1 1) electrode by vesicle spreading. ECSFM images show a) the coverage of the surface by rod-like micelles of semicircular cross section at −0.4 V vs. Ag|AgCl|KCl(*sa*) and B) the lipid bilayer formed after a potential step to +0.2 V vs. Ag|AgCl|KCl(*sa*). Source: The figure was kindly provided by Slawomir Sek, University of Warszaw.

8.11.7 Application of ECSFM

Besides using SFM as an *ex situ* method, there are opportunities linked to using SFM directly in an electrochemical cell (called ECSFM). The sample is connected as the working electrode in a three-electrode configuration. The tip is usually not electrically contacted. In order to direct the laser beam through the liquid–gas interface without losing light intensity, the holder for the cantilever chip can be made from glass. There are open-cell and closed-cell designs. When working in solution, one must take into account the much stronger damping of the mechanical oscillator by the liquid medium (lower *Q*-factor).

It is usually much more difficult to reach atomic resolution by means of ECSFM compared to ECSTM. Although studies of adsorption phenomena of simple ions in the electrical double layer are possible [155], they are preferentially conducted by ECSTM. On the other hand, the types of materials that can be studied by ECSFM are much more diverse because, unlike ECSTM, ECSFM does not depend on the conductivity of the sample. ECSFM is used for the characterization of polymer deposits on electrodes, of other organic thin films on electrodes, and of membrane materials [156]. The example in Figure 8.106 shows the potential-controlled formation of a lipid bilayer (as a model of a biological membrane) on an Au(1 1 1) electrode. The lipid is provided as vesicles in solution. At −0.4 V, they form aligned rod-like micelles with semicircular cross section (Figure 8.106a). Only after a potential step to +0.2 V, a lipid bilayer is formed (Figure 8.106b). Further areas of application are the study of corrosion phenomena, in which insulating inclusions can be characterized. The imaging of composite materials is another important area of application for ECSFM, as is the investigation of intercalation processes into battery materials [157]. Attempts have been reported to integrate a microelectrode into the SFM tip to record local electrochemical properties of the sample employing SECM in parallel to SFM imaging [158]. We will cover such experiments in Section 8.12.8.

8.11.8 Artifacts, Image Resolution, and Sample Manipulation in STM and SFM Experiments

The term "artifact" refers to features of the image that have no correspondence to the real structure of the sample. A common error in imaging is caused by multiple tips. During the imaging process, two parts of the tip are simultaneously in contact with the sample. As a result, many structures in the SFM image show a feature reminiscent of a "shadow" that is caused by the interaction with a second tip (Figure 8.107a).

Such artifacts can be distinguished from real structures by rotating the sample by 90° and repeating the image acquisition. If the location of the "shadow" remains the same after the rotation of the sample, it is indeed an artifact (Figure 8.107c). If the shadow rotates with the sample by 90°, the imaged structures are likely to be real and present on the sample (Figure 8.107b).

Deviations between the sample structure and the image are also caused by finite tip size effects. Due to the opening angle of the tip, perpendicular steps on the sample cannot be imaged correctly (Figure 8.108a). The region directly next to the step is not accessible for the tip. Small round and protruding structures (such as nanoparticles) appear in SFM images usually much broader than they are in fact (Figure 8.108b). In these cases, mainly the radius of the tip apex determines the width of the structure in the image. However, the height can be measured with much better accuracy.

Another group of erroneous images is caused by unsuitable settings of the PID controller (proportional-integral-differential) of the SFM instrument. A typical signature is sharp structures on one side and blurred features on the other side of a protruding feature along the scan direction. This can be an indication that the tip loses contact with the sample at downhill steps. In this case, the PID controller must be set to respond faster. However, too fast responses can cause oscillations because the instrument responds to a small deviation between the measured and the set signal by a too strong change in the piezo position.

Finally, the tip may remove material from the sample, which can have different consequences. The removed material from the sample can stick to the tip and degrade the image resolution due to undefined contact points between the tip and the sample. The image can also become unstable because material attaches/detaches to/from the tip. On the sample, the removal of material may cause a crater (Figure 8.109). This effect should be excluded by the following procedure: After acquiring several images from an area of interest of the sample, the scan length is increased (zoom out) and another survey scan is recorded. If an abrasion zone can be recognized in the center of the survey image, the removal of material has at least been detected. It can be avoided by a lower setpoint, a softer cantilever, another imaging mode, or more appropriate settings of the PID controller.

The removal of material can also be done intentionally to obtain useful information. For the thickness determination of soft organic films, an abrasion crater (analogous to Figure 8.109) is formed by imaging with high force and detuned PID settings until the tip reaches the hard substrate (inorganic electrode material). Afterward, a lower force and well-adjusted PID settings are used to record a survey image of the sample topography. The film thickness is obtained as the depth of the crater from a profile across the walls of the abrasion crater under consideration of the ejected material along the line scans.

Other procedures use alterations of the sample for nanofabrication. The SFM tip can locally destroy a thin passivating layer and immediately facilitate an (electro)chemical modification of the uncovered electrode. Such modification may be metal deposition, the formation of a new self-assembled monolayer, or the local generation of an oxide layer.

Electrically conducting SFM tips can induce local reactions using the liquid meniscus to form an electrochemical two-electrode cell between the conductive tip and a conducting sample. More information about local surface modification by SFM and STM, which is often called scanning probe lithography, can be found in a review paper [159].

Further Reading. Information about the theoretical foundations for SFM can be found in a recent textbook [154]. Introductory materials are available from instrument

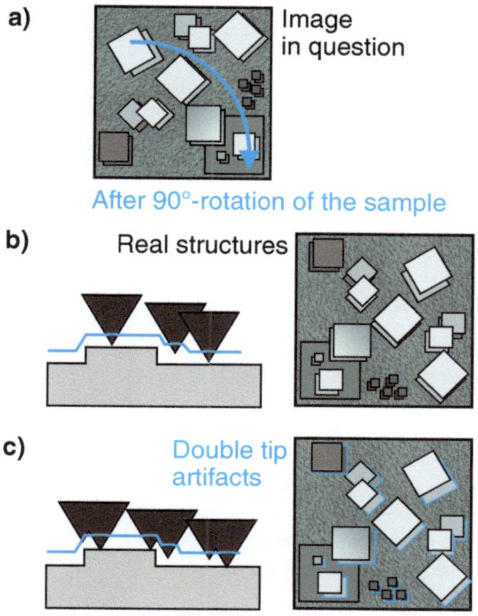

Figure 8.107 Recognition of double tip artifacts; a) image in question; b) image after 90°-rotation of the sample if the structures are real; c) image after 90°-rotation of the sample in case of double tip artifacts.

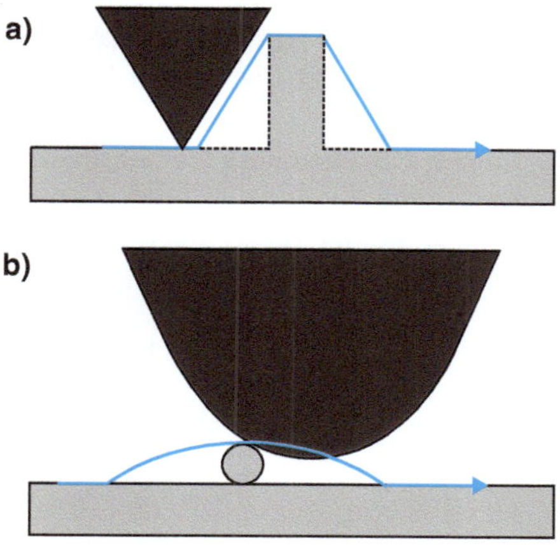

Figure 8.108 Artifacts caused by the finite size of the tip; a) apparent beveling of perpendicular steps; b) imaging of a small object, which appears much broader than it is in reality due to the larger radius of the tip apex. The measured height is imaged correctly. Real sample in gray, trajectory of the tip apex in blue, noncontacted areas of the sample in (a) are shown as a dashed line.

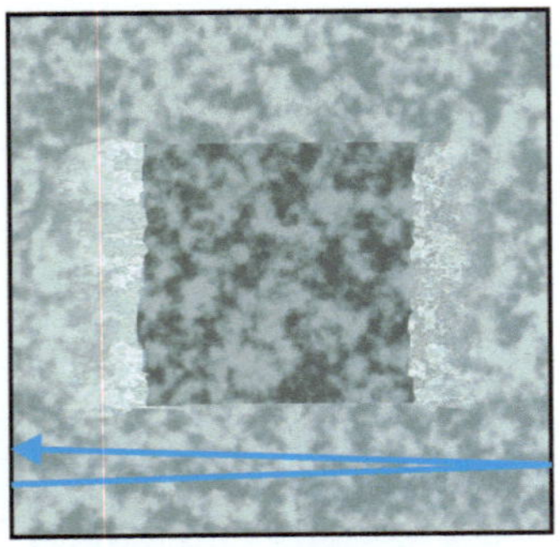

Figure 8.109 Artifacts due to abrasion of the sample (schematically). The region in the center of the image is the range in which the tip moved before recording the survey image. Due to abrasion, the tip dug into the soft film on the hard substrate. The material removed from the square is redeposited left and right at the rim of the abraded region (along the scan direction shown in blue).

It is common practice in the SECM literature to represent dimensionless normalized currents and distances by upper-case characters and to use lower-case symbols for variables that carry a unit. For reasons of consistency within the textbook, we will depart from this and abbreviate the current [nA] as I_T (elsewhere i_T). We will use ψ_T (elsewhere I_T) for the dimensionless (i.e., normalized) current, similar to the current functions introduced in Chapters 5 and 6.

manufacturers [160–163]. Useful experimental procedures for various aspects of SFM and educational experiments of different kinds of samples have been collected in dedicated books [149, 164, 165]. Further collections are devoted to the special considerations when working with SFM in liquids [166]. Reviews are recommended for a deeper coverage of the subject [155].

8.12 Scanning Electrochemical Microscopy (SECM)

Scanning electrochemical microscopy (SECM) has evolved into a very versatile method that is not only used for imaging but can also be applied to fundamental kinetic studies. In this textbook, we discuss only a few principles. Before you read on, please review Section 5.6 about the electrochemistry of microelectrodes.

Further Reading. For a more in-depth coverage, authoritative books [167–169] and review papers [170–172] are recommended. Links to further literature on specific topics will be given below.

8.12.1 Instrumentation and Overview

The experimental configuration for SECM in Figure 8.110 bears many similarities to the ECSTM instrumentation in Figure 8.92. For historical reasons, even many symbols (T for tip, S for sample/substrate) of important variables are equal (e.g., E_T and E_S). Both devices cause a current I_T at a very small electrode while the electrode is moved along the sample surface by means of a positioning system. The potential of the probe electrode and the sample is controlled by a bipotentiostat. While there are several obvious similarities, there are also fundamental differences between the two methods. In ECSTM, the current is caused by tunneling processes of electrons between the tip and the sample. In SECM, the current is caused by an **electrolysis** at the probe electrode. Thus, the current in SECM is faradaic. There are different mechanisms by which the faradaic current at the tip can be influenced by local properties of the sample. We will cover them in the subsequent sections. Several of these mechanisms can also be used to modify the sample surface.

The electrolysis current of the probe is coupled into the electrochemical cell via an auxiliary electrode (Aux in Figure 8.110). This is another major difference to ECSTM, where the current through the tip and the sample are essentially equal but of opposite sign. Because the tip current in SECM does *not* need to pass through the sample, SECM can be applied not only to conducting and semiconducting samples, but also to insulating samples. The tunneling current decays over a distance of a few nanometers. Electrolysis currents at the SECM microelectrode change with the extent of the diffusion layer, usually over distances of tens of micrometers. The exact values depend on the size of the microelectrode. The much lower gradient of the current–distance curve in SECM as compared to ECSTM is the fundamental reason for a lower lateral resolution of SECM.

The potential of the sample can be controlled by a bipotentiostat. However, this is not necessary in all cases. For the investigation of biological samples, catalysts, or membranes, it is sufficient to control only the microelectrode probe potential.

The exact shape of the probes is another difference between SECM and ECSTM. Pointed tips with as small a radius as possible are used in ECSTM to localize the tunneling current. In most SECM experiments, the probe is not a tip but a disk-shaped microelectrode (ME), which was already discussed in Section 5.6. The ME is characterized by its radius r_T (T for tip, although it is not a tip, but a flat disk). The symbol

a is also frequently used in the literature for the disk radius. The insulating sheath is usually made to be in plane with the disk. This is important to achieve good contrast, especially in the feedback mode (Section 8.12.2). The active electrode area should be located in the center of the glass sheath of the radius r_{glass}. It is common to state the most important variables in SECM as dimensionless, normalized quantities. Their general advantages have already been discussed in Chapter 5. Division by the ME radius r_{T} provides a sensible normalization for all distances in SECM:

Radius of the ME $$R_{\text{T}} = r_{\text{T}}/r_{\text{T}} = 1 \tag{8.143}$$

Radius of the active part of the sample $$R_{\text{S}} = r_{\text{S}}/r_{\text{T}} \tag{8.144}$$

Radius of the glass sheath $$R_{\text{g}} = r_{\text{glass}}/r_{\text{T}} \tag{8.145}$$

Working distance $$L = d/r_{\text{T}} \tag{8.146}$$

r_{T} – Radius of the microelectrode [μm]
r_{S} – Radius of the active part of the sample [μm]
r_{glass} – Radius of the glass sheath [μm]
d – Working distance [μm]

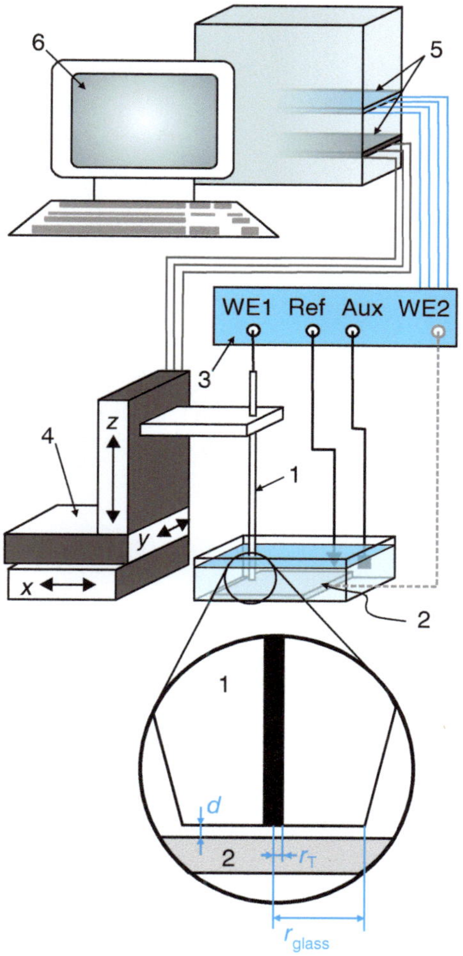

Figure 8.110 Experimental configuration for scanning electrochemical microscopy: 1) microdisk electrode as probe, 2) sample, 3) (bi)potentiostat, 4) positioning system, 5) insertion boards for controlling the potentiostat and the positioning system, and 6) PC for controlling the instrument. The lens shows an enlarged view of the probe and sample with labels of the important variables.

8.12.2 The Feedback Mode

For experiments in the feedback (FB) mode, one redox form of a quasireversible redox couple is added to the supporting electrolyte. This compound is referred to as a mediator in the context of this method. The selection of the mediator depends on the particular problem to be solved. Below, we assume that the reduced form was added. We use R to indicate generically the reduced form of an arbitrary redox couple. R will be oxidized to its oxidized form O at the ME.

$$\text{R} \rightleftharpoons \text{O} + ne^- \tag{8.147}$$

R could be, for instance, ferrocene methanol. According to Eq. (5.109), the diffusion-limited, steady-state current is given by Eq. (8.148) with the symbols from the context of SECM for the electrolysis of R when the ME is in the bulk of the solution, that is, at quasi-infinite distance to the sample surface (thus the index ∞).

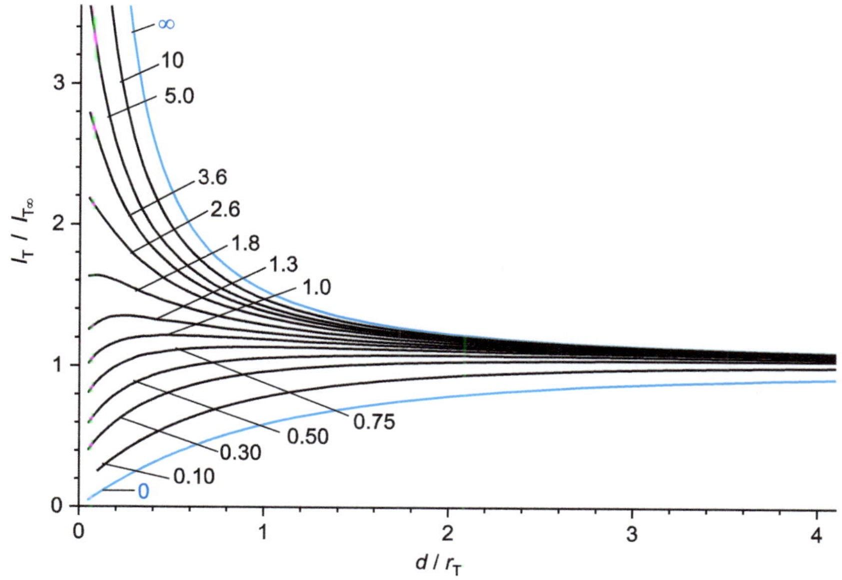

Figure 8.111 Approach curves in the FB mode calculated according to Eqs. (8.151)–(8.156) for $R_{\text{g}} = 5$. The curves are labelled with the normalized heterogeneous rate constants κ for the reaction at the sample. The blue graphs correspond to the limiting cases of an insulating, inert, and impermeable sample ($\kappa = 0$) and a diffusion-controlled mediator regeneration at the sample ($\kappa = \infty$).

$$I_{T,\infty} = g(R_g)nFDr_Tc^* \quad (8.148)$$

$I_{T,\infty}$ – Steady-state probe current [A]
D – Diffusion coefficient of R [$cm^2\ s^{-1}$]
r_T – Radius [cm], sensible conversion of units, for example, 10 μm = 0.001 cm
c^* – Concentration of R [$mol\ cm^{-3}$], sensible conversion of units, for example, 1 mmol $L^{-1} = 10^{-6}$ mol cm^{-3}
g – Geometric factor [dimensionless]

Table 8.9 Values of the geometry-dependent prefactor.

R_g	$g(R_g)$
100	4.000
10	4.072
7	4.108
5	4.156
4	4.200
3	4.272
2	4.440
1.5	4.652
1.2	4.952

Source: C. G. Zoski, M. V. Mirkin; Steady-State Limiting Currents at Finite Conical Microelectrodes. *Anal. Chem.* **2002**, *74*, 1986–1992, Copyright 2002, American Chemical Society.

The geometry-dependent prefactor g goes toward the value 4 as in Eq. (8.148) if $R_g \to \infty$. Table 8.9 shows that $g > 4$ for a thin insulating sheath around the ME, which is preferred for SECM. The current according to Eq. (8.148) amounts to a few nA only. This small value corresponds to a very large current *density,* which is a consequence of the hemispherical diffusion layer (Section 5.6.1, Figure 8.112a). This high current density is not influenced very much when the ME moves slowly (<15 μm s^{-1}) relative to the solution. This facilitates the possibility of recording a quantitatively well-defined electrolysis current at the moving ME. If the ME is located in close distance above an insulating, inert, and impermeable sample, the diffusion of R toward the ME is hindered (Figure 8.112b) and thus I_T decreases as the working distance d is reduced. The plot of $I_T = f(d)$ is called an approach curve and is commonly provided in normalized coordinates $\psi_T = f(L)$. The distance is given as L according to Eq. (8.146). The normalized, dimensionless current is the current function ψ_T according to Eq. (8.149)

$$\psi_T = I_T(d)/I_{T,\infty} \quad (8.149)$$

The graph of the function $\psi_T(L)$ is shown in Figure 8.111 for $\kappa = 0$ corresponding to an insulating, inert, and impermeable sample (Figure 8.112b). The closer the ME is brought to the sample, the stronger the hindrance to the diffusion of R to the ME and the smaller the ME current. Let us stay a moment with the constraining properties of the sample: The sample must not be permeable because otherwise R would simply diffuse through the sample, a situation found for some mesoporous and macroporous solids or for some polymer gels. If the sample is not inert, a strong oxidizer generated at the ME may oxidize the sample material even if it is insulating. Such a situation can be utilized for an intentional and chemically well-defined local surface modification.

If the sample is electronically conducting (thus, it is an electrode itself), the initial form R of the mediator can be regenerated by reversal of the reaction (8.147) as shown in Figure 8.112c. The mediator R can be transported very effectively over the small interelectrode distance from the sample to the ME, where it increases the diffusional flux of R to the ME. The closer the ME comes to the sample, the higher the current at the ME. If the reaction at the sample is diffusion-controlled, the curve "∞" in Figure 8.111 is obtained. The reaction rate at the sample is described by an effective heterogeneous first-order reaction rate constant k_{eff} [cm s^{-1}]. This rate constant can be transformed to a corresponding dimensionless heterogeneous rate constant κ according to Eq. (8.150).

$$\kappa = \frac{k_{eff} r_T}{D} \quad (8.150)$$

In Figure 8.111, the curves "0" for $\kappa = 0$ and the curve "∞" for $\kappa \to \infty$ are the limiting cases. There is a separate distinct approach curve for each of the intermediate κ-values, from which a selection is depicted in Figure 8.111. The information in Figure 8.111 suggests two basic experiments:

1) If the ME moves in a constant distance d_1 horizontally (x, y) over the sample, the different current values $I_T(x, y, d_1)$ indicate different rate constants $k_{eff}(x, y)$ at the surface of the sample. $I_T(x, y)$ can be used for generating an image by one of the options from Figure 8.88. The image can be considered as a map of the local reaction rate constants for the mediator regeneration at the sample.

2) Alternatively, one can record an approach curve $I_T(d)$ at constant lateral coordinates (x, y). Simulation, for example, by the finite element method, allows the calculation of the normalized current $I_T/I_{T,\infty} = \psi_T = f(L, R_g, \kappa)$ for different normalized distances L and parameters R_g and κ. The analytical expressions in Eqs. (8.151)–(8.156) have been found by skillful interpolation of the simulated values and can be used for curve fitting to determine κ from experimental data [173]:

$$\psi(L, \kappa, R_g) = \psi_T^{cond}\left(L + \frac{1}{\kappa}, R_g\right) + \frac{\psi_T^{ins}(L, R_g) - 1}{\left(1 + 2.47R_g^{0.31}L\kappa\right)\left(1 + L^{0.006RG+0.113}\kappa^{-0.0236RG+0.91}\right)} \quad (8.151)$$

The symbols stand for the limiting cases of a completely inert sample with $\kappa = 0$

$$\psi_T^{cond}\left(L + \frac{1}{\kappa}, R_g\right) = \alpha(R_g) + \frac{1}{2\beta(R_g)\xi\left(L + \frac{1}{\kappa}\right)} + \left(1 - \alpha(R_g) - \frac{1}{2\beta(R_g)}\right)\xi\left(L + \frac{1}{\kappa}\right) \quad (8.152)$$

and a sample with diffusion-controlled regeneration of the mediator ($\kappa = \infty$)

$$\psi_T^{ins}(L, R_g) = \frac{\frac{2.08}{R_g^{0.358}}\left(L - \frac{0.145}{R_g}\right) + 1.585}{\frac{2.08}{R_g^{0.358}}(L + 0.0023R_g) + 1.57 + \frac{\ln R_g}{L} + \frac{2}{\pi R_g}\ln\left(1 + \frac{\pi R_g}{2L}\right)} \quad (8.153)$$

The coefficients α, β, ξ summarize the effects of R_g and the rate constant κ:

$$\alpha(R_g) = \ln 2 + \ln 2\left(1 - \frac{2}{\pi}\arccos\left(\frac{1}{R_g}\right)\right) - \ln 2\left[1 - \left(\frac{2}{\pi}\arccos\left(\frac{1}{R_g}\right)\right)^2\right] \quad (8.154)$$

$$\beta(R_g) = 1 + 0.639\left(1 - \frac{2}{\pi}\arccos\left(\frac{1}{R_g}\right)\right) - 0.186\left[1 - \left(\frac{2}{\pi}\arccos\left(\frac{1}{R_g}\right)\right)^2\right] \quad (8.155)$$

$$\xi\left(L + \frac{1}{\kappa}\right) = \frac{2}{\pi}\arctan\left(L + \frac{1}{\kappa}\right) \quad (8.156)$$

The normalized rate constant κ can be obtained by a curve fit of the function $\psi = f(L; R_g, \kappa)$ with the adjustable parameters R_g, κ, r_T, $I_{T,\infty}$ and the position of the surface z_0. L is obtained from $L = (z_0 - z)/r_T$. If κ as well as r_T and D are known, k_{eff} is obtained by rearranging Eq. (8.150). Because the approach curves are not very structured and there are no less than five adjustable parameters, it is advisable to determine some of them by independent experiments and to keep them constant in the fitting procedure afterward. For instance, R_g is approximately obtained from light microscopy, with a refinement after an independent determination of r_T from an approach curve to a microscope slide from glass ($\kappa = 0$). The parameters R_g and r_T are then kept constant during the fit of the approach curve to the sample of interest. A step-by-step instruction of the procedure is available [174].

$\psi_T = I_T/I_{T,\infty}$ – Current function [dimensionless]

$\psi_T^{cond} = I_T^{cond}/I_{T,\infty}$ – Current function for $\kappa = \infty$ [dimensionless]

$\psi_T^{ins} = I_T^{ins}/I_{T,\infty}$ – Current function for $\kappa = 0$ [dimensionless]

κ – Normalized rate constant [dimensionless]

L – Normalized vertical distance, Eq. (8.146) [dimensionless]

R_g – Normalized radius of the insulating sheath, Eq. (8.145) [dimensionless]

α, β, ξ – Dimensionless numerical constants

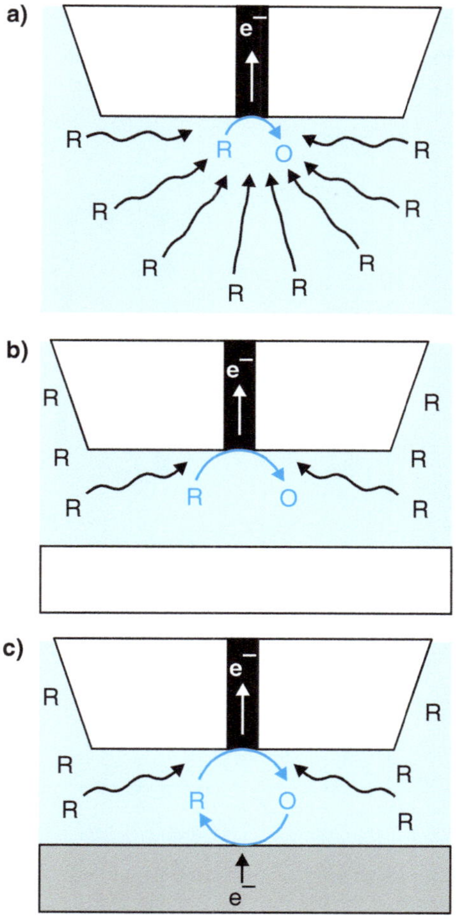

Figure 8.112 SECM FB mode. a) ME in the solution bulk; b) hindered diffusion when the ME approaches an insulating, inert, and impermeable sample ("negative feedback"); and c) mediator regeneration by a heterogeneous reaction at the sample ("positive feedback").

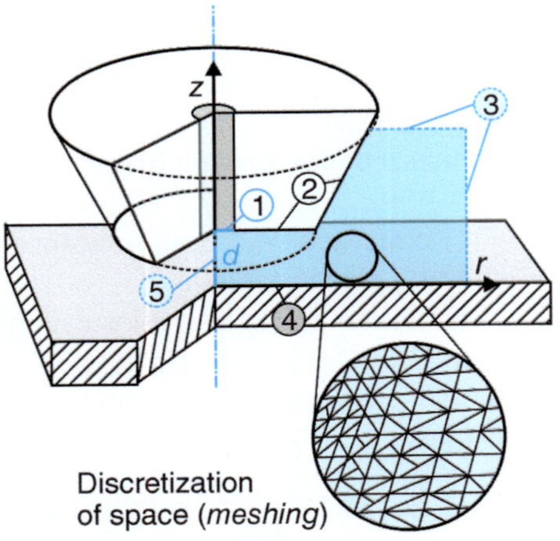

Figure 8.113 Simulation domain for the calculation of SECM signals.

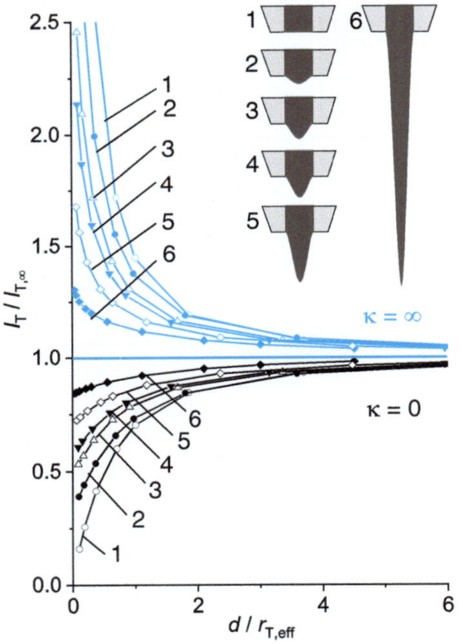

Figure 8.114 The role of the insulating sheath of the ME for the FB mode of SECM. The plot shows normalized approach curves calculated for conical microelectrodes moving toward a conducting sample ($\kappa = \infty$, $I_T/I_{T,\infty} > 1$) and an inert insulating sample ($\kappa = 0$, $I_T/I_{T,\infty} < 1$). $r_{T,eff}$ in the label of the abscissa refers to an effective radius, that is, the radius of a disk-shaped electrode that *would* provide the same diffusion-limited steady-state current in the solution bulk as the considered conical ME, $r_{T,eff} = I_{T,\infty}/(4nFDc^*)$. Source: Reproduced with permission from [175], Copyright 2005, Elsevier.

Excursion 8-4 How Can Currents Be Calculated for Arbitrary Distances?

The partial differential equation to solve is Fick's second law for a microdisk electrode (Excursion 4-3)

$$\frac{\partial[\mathrm{R}]}{\partial t} = D\left\{\frac{\partial^2[\mathrm{R}]}{\partial r^2} + \frac{1}{r}\cdot\frac{\partial[\mathrm{R}]}{\partial r} + \frac{\partial^2[\mathrm{R}]}{\partial z^2}\right\}. \quad (8.157)$$

In contrast to the case in Excursion 4-3, the solution is obtained numerically by discretization of space and time (Figure 8.113). The magnifying lens in this schematic indicates the discretization of space in small triangles (within a cross-sectional plane). Triangles are typically used for the discretization of space in the finite element method (FEM). Furthermore, equality of diffusion coefficients of R and O is assumed for simplification, which allows one to express the concentration of O by $[\mathrm{O}] = [\mathrm{R}]^* - [\mathrm{R}]$. The simulation is iterated until the local concentrations do not change anymore (steady state). From the local concentration in this situation, the current is calculated by first calculating the concentration gradients along a radial axis between 0 and $r = r_T$, followed by integration for a full rotation (2π) around the symmetry axis.

$$I_T = 2\pi nFD\int_{r=0}^{r_T}\frac{\partial[\mathrm{R}]}{\partial z}r\,dr \quad (8.158)$$

The following are the boundary conditions applied for the calculation. Because a diffusion-controlled current is determined, the concentration of R at the surface of the ME (1) is $[\mathrm{R}] = 0$. On the surface of the glass sheath of the electrode (2), the flux perpendicular to the interface is zero $J = 0$. At the domain boundary representing the solution bulk (3), the concentration is equal to the initial bulk concentration $[\mathrm{R}] = [\mathrm{R}]^*$. The rotational symmetry of the arrangement prevents any flux across the symmetry axis (5), that is, $\partial[\mathrm{R}]/\partial r = 0$ for $r = 0$. The result is affected by the radius of the microelectrode r_T, the working distance d, and the kinetics of the reaction

$$\mathrm{O} + ne^- \xrightarrow{k_{\mathrm{eff}}} \mathrm{R}$$

at the surface of the sample. The kinetics are expressed by the boundary condition for the region (4)

$$\frac{\partial[\mathrm{R}]}{\partial z} = -\frac{k_{\mathrm{eff}}}{D}([\mathrm{R}]^* - [\mathrm{R}]). \quad (8.159)$$

Two features of the feedback mode are important. If the microelectrode probe has the shape of a tip as in Figure 8.114, the mediator reaches the ME surface by diffusion from the solution bulk without significant hindrance by the insulating sheath of the ME. For a given radius r_T, $I_{T,\infty}$ increases with increasing height of the conical tip and thus with the electrode area. The recycling of the mediator still occurs at the sample surface, but it provides a relatively smaller contribution to the current at the probe. The contrast between sample regions with $\kappa = 0$ and $\kappa = \infty$ decreases (curves 6 in Figure 8.114). Note that the currents have been normalized in Figure 8.114 using Eq. (8.150) because only in this way can the results of the different probes be assembled into one plot and compared. Thus, the probe must have the shape of a disk or at least have a shape very close to a disk, avoiding the shape of a pointed tip. This is the reason why we avoid the term "tip" for the probe in SECM, although tip and the analogy to ECSTM were the origin of the common index for quantities of the probe, such as E_T, I_T, r_T, etc.

The sample does not necessarily need a connection to an external potential source. In the mediator solution, an OCP is established that is controlled by the formal potential $E^{\circ\prime}$ of the mediator and the concentration ratio [O]/[R]. According to the Nernst

Equation (2.38), the equilibrium potential is negative of $E^{\circ\prime}$ of the mediator redox couple (R/O) because the solution contains almost exclusively R and very little O. Only directly beneath the ME is the situation perturbed by the electrolysis at the ME. If the electrolysis product O arrives at the sample electrode surface, O is immediately reduced. The electrons are provided by a stoichiometrically equivalent amount of R that is oxidized at *different* locations of the sample surface (Figure 8.115a). The sites for the oxidation at the sample surface are distributed over the entire area of the conductive sample. The local concentrations of O and R are not significantly altered by the oxidation of R to O away from the microelectrode. Therefore, the local reduction of O to R at the surface of the sample beneath the microelectrode can be sustained over many hours. The situation corresponds to a differential concentration cell between the position of the ME and the remaining parts of the conductive sample.

This feedback mechanism is only operative if the conductive region below the sample is connected to the bulk phase of the solution by an electron-conducting link that is *not* completely located underneath the insulating sheath of the ME (Figure 8.115b). In case that the conductive region of the sample is of a similar size as the ME and there is no electron-conducting pathway to other conductive parts of the sample surface as in Figure 8.115c, the feedback mechanism is not operative.

The mediator regeneration can be enabled by a heterogeneous electron transfer reaction (in case the sample is an electrode by itself), but also by redox enzymes (Figure 8.116a) that are immobilized to an insulating sample. In this case, the mediator regeneration is coupled to the enzymatic oxidation of a cosubstrate. Further possibilities are the local oxidative etching of the sample (Figure 8.116b) or photoelectrochemical reactions (Figure 8.116c).

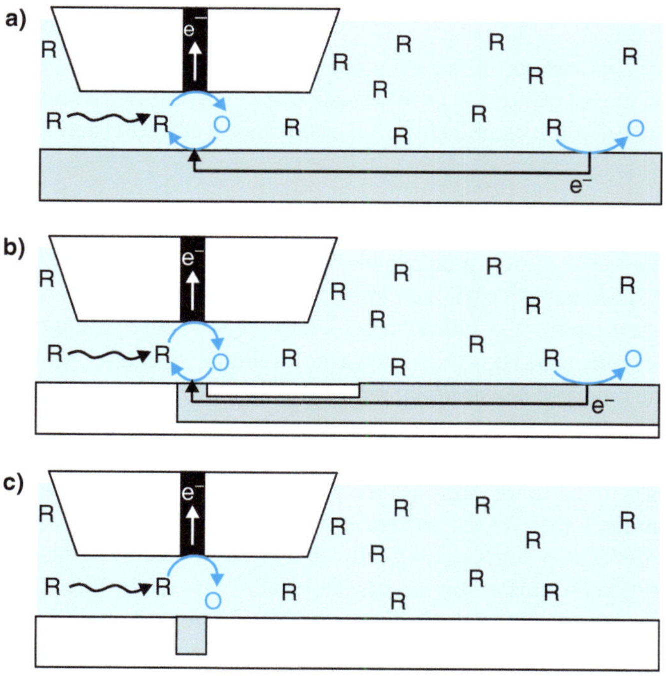

Figure 8.115 Mechanism of the feedback effect at the sample. a) Macroscopic conductive sample surface; b) microscopic sample surface that is connected to a macroscopic conductive region by an electron-conducting link; and c) no feedback effect above a conductive microscopic region of the sample surface without electron-conductive connection to another conductive part of the sample surface.

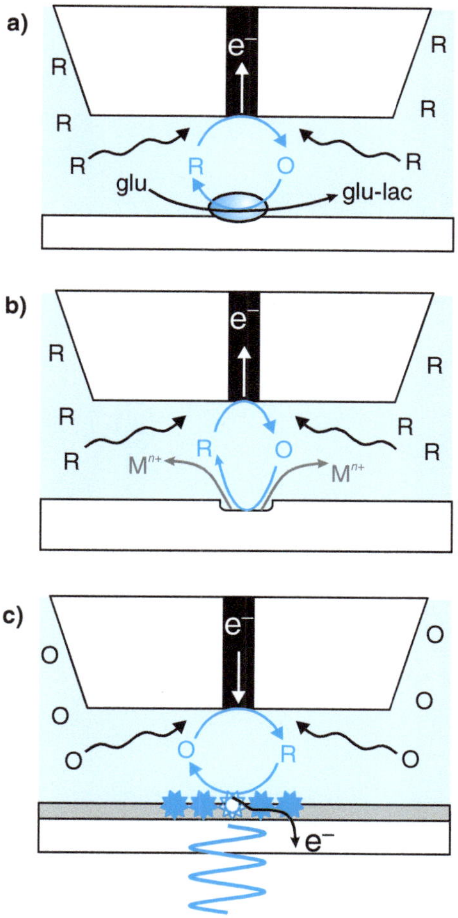

Figure 8.116 Mediator regeneration by a) an immobilized redox enzyme; b) by oxidative etching of the sample; and c) by a photochemical reaction. The schematic is not to scale!

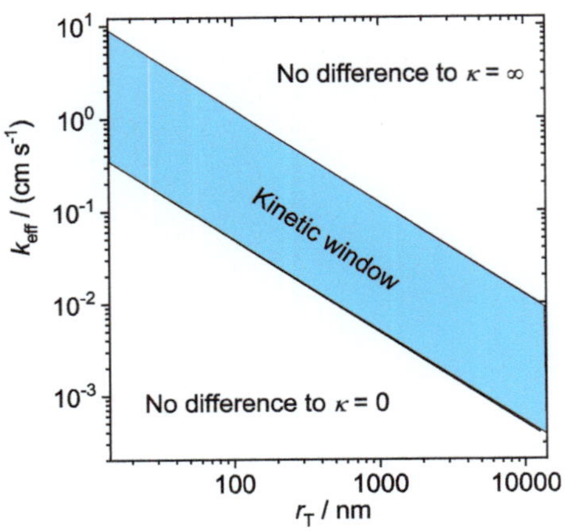

Figure 8.117 Kinetic window. The marked region between $\kappa = 0.03$ and $\kappa = 3.6$ comprises the combinations of rate constants and ME radii that allow a kinetic analysis or a reactivity mapping by SECM.

An important property of the FB mode is that reactions at the sample occur only directly beneath the ME. This is also the reason why the FB mode is particularly well described quantitatively.

Next, we discuss for which range of κ-values the FB mode can be used for a kinetic analysis. There is always a continuous electrolysis at the ME probe. To detect an influence of the sample reaction, the *additional* diffusive flux due to the mediator recycling at the sample must provide a detectable increase in the ME current. This condition is approximately met if $\kappa = 0.03$. For fast reactions, the approach curve must be just distinguishable from the curve for a diffusion-controlled reaction at the sample ($\kappa \to \infty$). This condition is approximately met for the curve with $\kappa = 3.6$. If we plot the heterogeneous rate constants k_{eff} [cm s^{-1}] with units that we obtain from the two conditions for different radii of the ME according to Eq. (8.150), we obtain Figure 8.117. From this plot, we can deduce, for instance, that a sample with a rate constant of k_{eff} = 0.02 cm s^{-1} would cause an approximately diffusion-controlled enhancement of the ME current (positive feedback) if the microelectrode has a radius of 10 μm (10 000 nm). The study of the same sample with a nanoelectrode of r_T = 100 nm would result in an approach curve that cannot be clearly distinguished from an inert and insulating sample! This situation imposes fundamental limits on the miniaturization of the ME for enhancing the lateral resolution in the FB mode despite the fact that the FB mode provides the best lateral resolution among the SECM working modes [176].

Q: Can the FB mode be used if only the oxidized form O of a redox couple is added to the bulk solution?

A; Yes, this is often done. A good example is the mediator $[Ru(NH_3)_6]^{3+}$, which is reduced at the ME to $[Ru(NH_3)_6]^{2+}$ and reoxidized at the sample. The quantitative relations explained before remain valid under reversal of the reaction directions.

Exemplary applications of the FB mode are the investigation of passivating layers (e.g., organic monolayers [177] or thin oxide films [178]), solid electrolyte interphases in batteries [179], photochemically [180] or enzymatically [181] active layers.

8.12.3 Surface Interrogation Mode

The FB mode yields steady-state signals, that is, the signal does not depend on how long the ME is located at a particular location. The situation is different if the mediator reacts at the sample surface with a surface-bound redox-active species. In this case, the mediator recycling (and thus the increased currents) is sustained (as in the FB mode) as long as the mediator can enter into a redox reaction with the surface-bound species on the sample. Once all redox centers at the sample surface that can react with the mediator have been converted, the mediator recycling stops and the microelectrode current decays to the value found for hindered diffusion at the given working distance [182]. The sudden decay of the current is similar to a titration experiment. If the current is recorded chronoamperometrically, that is, as a function of time, the charge for the reaction with the surface-bound species can be obtained by integration of the $I_T(t)$ function after subtraction of the background curve. The background is formed by the current measured above an insulating sample with the same ME probe positioned at the same working distance (hindered diffusion only). This background curve can be conveniently obtained by repeating the potential pulse at the ME *after* the redox centers have been consumed completely in the measurement pulse [183]. In a simple version of the experiment with a macroscopic sample, one cannot be sure about the region of the sample that effectively contributes to the mediator recycling. Furthermore, this area is also changing with the duration of the potential pulse at the ME because the diffusion layer expands. Two experimental "tricks" of the surface interrogation mode of scanning electrochemical microscopy (SI-SECM) overcome this difficulty for the application to modified electrodes as sample [184]:

1) The sample itself will also be prepared as a microelectrode with roughly the same size as the ME probe, which is also called the interrogator electrode in this context. The sample ME and the interrogator ME are brought to a colinear arrangement by line scans in the conventional FB mode.
2) Now the surface-bound redox centers will be brought into a defined redox state by applying a potential pulse to the sample while the interrogator electrode is left at open circuit potential (OCP). The potential pulse at the sample can cause the formation of surface oxides or the adsorption of a compound (Figure 8.118a). Afterward, the connection of the sample to the potentiostat is interrupted (Figure 8.118b) and the sample remains at OCP from now on. Therefore, charge cannot flow from the sample electrode via the external circuit, but can only be exchanged via the sample–electrolyte interface. Concomitantly with setting the sample at OCP, the interrogator electrode is connected to the potentiostat and the chronoamperometric titration experiment is started (Figure 8.118b). The converted charge is obtained by integration of the measured current $I_{T,1}(t)$ and subtraction of the charge obtained from the chronoamperogram $I_{T,ref}(t)$, measured by an analogous procedure but without formation of the adsorbate on the sample as in 8.118c. From the charge difference, the surface concentration of the redox-active centers, adsorbates or surface oxides is calculated using Eq. (8.160).

The effect can be observed at macroscopic electrodes only if the redox-active component is bound to an insulating support. Otherwise, the mediator recycling would happen by reaction with the surface-bound redox centers but also by mediator recycling at a conductive sample according to Figure 8.115a without involvement of the surface-bound redox centers.

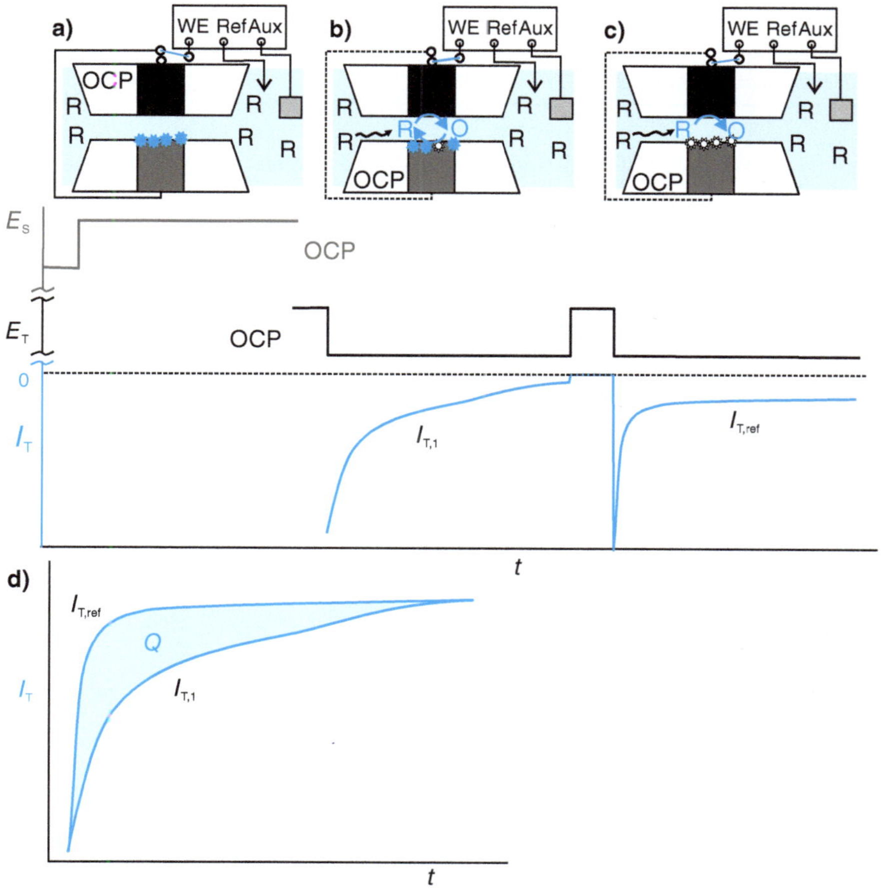

Figure 8.118 Procedure for the surface interrogation mode.

a) Adjustment of the surface concentration of the species of interest at the sample;
b) disconnection of the sample from the potentiostat and connection of the interrogator electrode to the potentiostat with complete conversion of the redox active species at the sample;
c) repetition of the potential pulse at the interrogator without preceding formation of the adsorbates/surface oxides/redox active surface species on the sample; and
d) comparison of the chronoamperograms with and without formation of surface-bound redox species and determination of the charge Q.

The collection efficiency ε has to be considered. The collection efficiency is defined by the fraction of the mediator molecules reacting at the sample that subsequently reach the interrogator electrode and thus contribute to I_T. Typical values of ε in SI-SECM are 0.9 (90%) and, thus, very high.

$$\Gamma_{ads} = \frac{\left| \int_0^{t_e} I_{T,1}(t)dt - \int_0^{t_e} I_{T,ref}(t)dt \right|}{nFA\varepsilon} \tag{8.160}$$

The advantage compared to other adsorbate stripping methods discussed in Section 12.4.1 is the fact that during the determination of the charge of the redox process, no potential excursion of the sample is required, which could alter the surface concentration of the species of interest, for instance, if there is a potential-dependent adsorption equilibrium. In addition, the method avoids the problem of separating double-layer charging effects and the charge from the redox process at the sample as illustrated and discussed in Figure 13.11. Table 8.10 provides a survey of the mediators used for different adsorbates and surface species.

Table 8.10 Mediators and surface species in SI-SECM mode.

Mediator	Adsorbate	Ref.
Ferrocenemethanol, N,N,N′,N′-tetramethyl-*p*-phenylendiamine	H_{ads}|Pt, H_{abs}|Pd	[185]
Br^-	CO_{ads}|Pt	[186]
MV^{2+}, $[Ru(NH_3)_6]^{3+}$, $[Fe^{II}(edta)]^3$, $[Fe(CN)_6]^{3-}$	PtO_x|Pt,	[187]
MV^{2+}, $[Ru(NH_3)_6]^{3+}$	AuO_x|Au	[184]
$[Co(bpy)_3]^{3+}$	OH_{ads}|Me_xO_y	[188]
Fe-triethanolamine complex	$Ni_{1-x}Fe_xOOH$	[189]

Strictly speaking, SI-SECM should not be regarded as microscopy because no image is formed. Instead, the configuration of a microscope is used for a very detailed interfacial analysis without lateral resolution. Despite this consideration, we follow here the common parlance that uses SI-SECM as an established technical term.

Further Reading. The review paper [190] and the original papers from Table 8.10 are recommended for a more detailed explanation of numerical simulation in SI-SECM. The power of SI-SECM can be expanded if a delay period is inserted between the generation pulse applied to the sample and the chronoamperometric detection at the interrogator electrode. During the delay period, surface-bound intermediates of the electrocatalytic reaction can enter into follow-up reactions, which decreases their interfacial concentration [189]. The generation of surface-bound species can also be made by photoelectrochemistry [191] and the concept of SI-SECM is applicable to liquid–liquid interfaces [192].

8.12.4 Generation–Collection Mode

The generation–collection (GC) mode can be conducted in variants such as the sample-generation/tip-collection (SG/TC) and the tip-generation/sample-collection (TG/SC) modes.

The sample-generation/tip-collection SG/TC has similar applications as the FB mode. In the SG/TC mode, the added molecules are first converted/released at the sample and afterward detected at the probe electrode (Figure 8.119a). Similar to the FB mode, the reaction at the sample can also be an (enzyme-)catalyzed [193] or a photoelectrochemical reaction [194]. The SG/TC mode is particularly advantageous for the study of transport processes through membranes (Figure 8.119b). The probe can be an amperometric ME or a potentiometric microsensor (Figure 8.119c) [195]. If the working distance is very small, the product of the reaction at the probe can diffuse to the sample and react there again, which eventually enhances the signal (Figure 8.119a). While this mechanism enhances the contrast in imaging experiments, it is undesirable if fluxes from the sample are determined assuming a formalism that treats the probe as a passive sensor. A passive sensor is a probe that does not alter the concentration of the analyte it detects close to the sample. Strictly speaking, only potentiometric electrodes behave as passive sensors (Figure 8.119c). Under certain conditions, amperometric MEs can approach the behavior of a passive probe. This is achieved if the working distance is so large that a back diffusion of the reaction products from the microelectrode to the sample does not significantly change the signal ($d > 3\ r_T$). Experiments under these conditions have been used extensively to quantify mass transport through pores [196, 197]. For an approach curve over the center of a pore, the probe current is described by Eq. (8.161).

$$I_T = gnFDr_T c_S \frac{2}{\pi} \arctan\left(\frac{r_S}{d}\right) \tag{8.161}$$

The current during a horizontal scan over the center of the pore is given by

$$I_T = gnFDr_T c_S \zeta \tag{8.162}$$

with

$$\zeta = \frac{2}{\pi} \arctan \frac{\sqrt{2} r_S}{\left(\sqrt{(\Delta x^2 + d^2 - r_S^2) + \sqrt{(\Delta x^2 + d^2 - r_S^2)^2 + 4d^2 r_S^2}}\right)} \tag{8.163}$$

g – Geometry-dependent factor from Table 8.9
r_S – Radius of the pore [μm]
d – Vertical working distance [μm]
r_T – Radius of the ME [cm]
D – Diffusion coefficient [$cm^2\ s^{-1}$]
F – Faraday constant [$A\ s\ mol^{-1}$]
c_S – Concentration of the analyte at the pore opening [$mol\ cm^{-3}$]
n – Number of electrons transferred from one analyte molecule
Δx – Horizontal distance of the probe to the center of the pore [μm]

With the availability of two working modes for the same application, the question arises: Which mode should be preferred? There is no simple answer to this question. In the FB mode, the reactions occur only in the immediate vicinity of the probe. Essentially, the electrolysis at the probe generates the reagents that can enter into a reaction at the sample. Therefore, this mode is described quantitatively much more accurately by algebraic equations. In the SG/TC mode, nonstationary diffusion layers will form above the sample because the reaction at the sample starts immediately upon contact of the sample with the working solution. Therefore, the currents will depend on the time that has passed between the contact of the sample with the working solution and the actual recording of the SECM data.

The FB mode always has a "background current" caused by the hindered diffusion of the mediator from the solution bulk to the ME probe. This current is measured even above totally inert and insulating samples. As a background signal is associated with noise, it fundamentally restricts the limit of detection of the FB mode because the contribution of mediator regeneration at the surface of the sample must be detectable against the background current. This limitation does not occur in the SG/TC mode, which gives a clear advantage with respect to sensitivity to the SG/TC mode. However, the diffusion of the reagents to the sample surface can be hindered by the presence of a bulky probe electrode in the SG/TC mode. Such an effect can be evident in SG/TC images when the recorded images show the highest currents close to the edge of active regions but not in their center (doughnut-shaped response from roughly circular active regions). This unwanted effect can be counteracted by increasing the working distance (less hindrance to diffusion). In the case of enzymatic reactions at the sample, an increase in the provided reagent concentration is another option (there

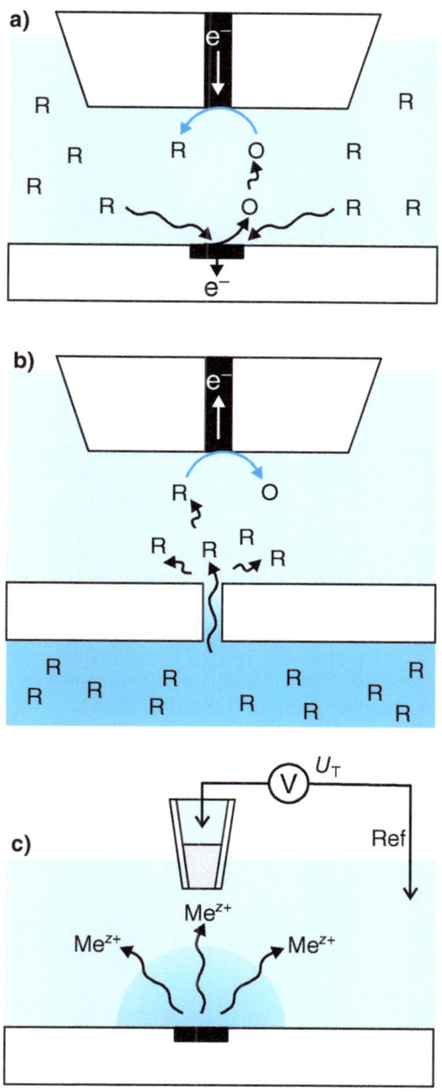

Figure 8.119 Principle of the SG/TC mode; a) electrochemical generation at the sample and amperometric detection at the probe; b) local transport through a pore and amperometric detection at the probe; c) local release of ions and potentiometric detection at the probe.

is no further concentration dependence of the enzymatic reaction rate if the regime of reagent-saturation of Michaelis–Menten kinetics has been reached, cf. Section 18.3.3) [198].

In an enzymatic reaction, the SG/TC mode can be applied to detect the activity of immobilized biotechnologically important enzymes that are not redox enzymes (oxidoreductases), for example, alkaline phosphatase or galactosidase. The success of such an experiment depends on finding an enzyme substrate that is not redox active at the applied probe potential, while one of the products of the enzymatic reaction is redox active at the same potential. A systematic review of the advantages and disadvantages of the FB and the SG/TC mode for the study of enzymatic reaction is available [198].

A typical application is the imaging of localized enzyme activities and the detection of short-lived reaction intermediates very close to their point of formation. This is, for instance, important with respect to oxygen reduction reaction in aqueous and organic solvents [199, 200]. Such studies can also benefit from the use of **tip–substrate voltammetry** (TSV). While a cyclic voltammogram is conducted at the sample, the probe ("tip") is used to detect soluble reaction products released at a specific potential from the sample. The principle of this experiment is reminiscent of the operation of a rotating ring-disc electrode covered in Section 12.4.3. One advantage of TSV is that it does not require a sample in the shape of an embedded disk electrode. Thus, TSV provides new opportunities for the characterization of nanostructured surfaces for which the fabrication of a ring-disk electrode assembly would constitute a formidable additional challenge [201].

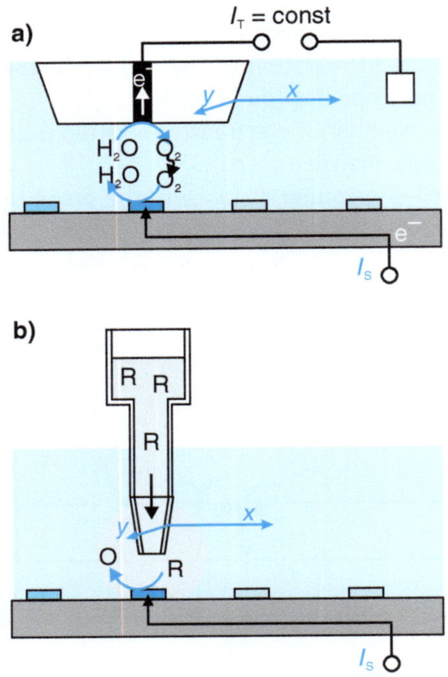

Figure 8.120 Principle of the TG/SC mode for the investigation of catalyst libraries. a) The image is formed by plotting the substrate current I_S as a function of the position of the probe (*x*, *y*). The probe is often operated galvanostatically in order to generate a constant flux of the precursor (O_2 in the example) for the reaction at the sample. b) The constant flux of the reagent can also be achieved by hydrodynamic or diffusive efflux of a reagent-containing solution from a capillary opening.

The tip-generation/sample-collection mode uses the substrate (sample) current to generate the image and the scanning probe is used for the local delivery of reagents for the reaction at the sample (Figure 8.120). The probe can be a ME, generating the reagent by electrolysis, or a capillary providing a reagent solution by efflux into the working solution of the SECM setup. The TG/SC mode is mostly used to study irreversible reactions that are significant in the context of electrochemical energy conversion.

8.12.5 Redox Competition Mode

The redox competition (RC) mode exploits the competition between the probe electrode and the sample for the same reagent (Figure 8.121a). The more active the sample, the more reagent consumed at the sample and correspondingly the less reagent available for reaction at the probe ME, and the ME current decreases. The RC mode

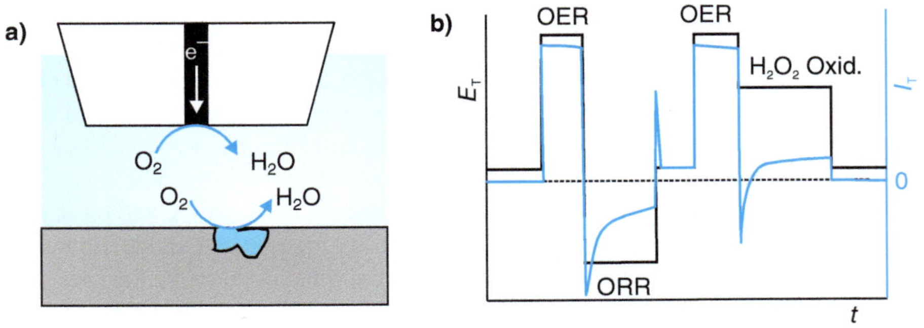

Figure 8.121 Schematic of the redox competition mode and exemplary potential pulse sequence for the study of the oxygen reduction reaction with determination of side products; a) schematic arrangement of the probe and the sample; b) time sequence of the potential pulses (black) and the current response at the ME (blue).

is useful particularly for studying irreversible reactions such as the electrocatalytic oxygen reduction reaction because the product of the reaction (H_2O) is not used for the detection of the activity. A clear disadvantage is the low dynamic range of the method. It is not easy to further differentiate between active, that is, interesting samples because they consume most of the available reagent anyway. The sensitivity can be improved by designing pulse programs that temporarily increase the local reagent concentration and enable the additional detection of by-products (e.g., H_2O_2) (Figure 8.121b) [202].

8.12.6 Direct Mode

The direct mode utilizes the probe ME as the counter or auxiliary electrode for the much larger sample working electrode in a two-electrode or three-electrode setup. Different from the FB mode, the reaction at the probe is not necessarily the reverse reaction of the reaction at the sample. The example in Figure 8.122 illustrates that typically two different redox couples are involved in the reactions at the sample and the ME. The reaction at the sample is localized to the area underneath the ME by a potential drop in solution. This is not a very efficient mechanism to confine the sample area at which the reaction proceeds. For most of the successful examples in the direct mode, an additional mechanism for focusing the reaction at the sample has been exploited. Examples of such mechanisms include the breakup of a passivating layer or a local reagent delivery. The direct mode has been used intentionally only very occasionally, but has to be considered as a possible mode of action in complex experiments involving conductive SFM tips or occurs unintentionally in STM experiments when samples measured in air are covered by an ultrathin film of condensed moisture [204]. Further applications comprise local surface structuring [203, 205, 206].

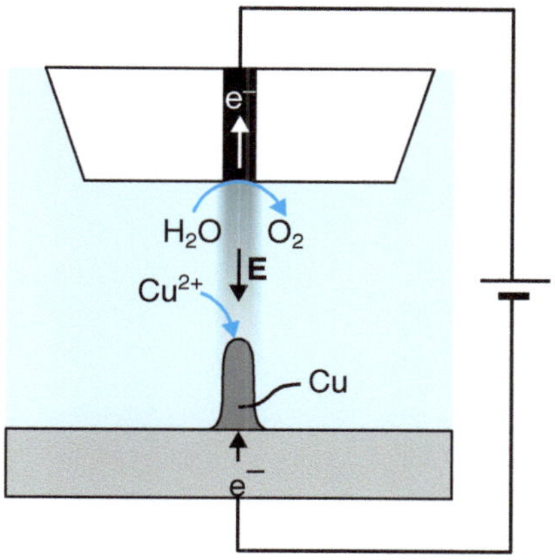

Figure 8.122 Principle of the direct mode for the local deposition of Cu according to [203].

8.12.7 Use of Homogeneous Reactions

Electrochemical "Time-of-Flight" Experiments. SECM instrumentation can also be used as an experimental configuration for measuring the kinetics of a homogeneous reaction occurring in the solution between a microscopic generator electrode (probe) and a detector electrode. The use of the microscope as a tool for physicochemical investigations, apart from the generation of microscopic images, is reminiscent of the SI-SECM in Section 8.12.3. A particularly descriptive terminology is the "time-of-flight" system depicted in Figure 8.123. The probe is used to form a reagent by electrolysis [Eq. (8.164)], that enters into homogeneous follow-up reactions in the interelectrode space, for example, a dimerization of organic radicals or a reaction with solvent molecules [Eq. (8.165)].

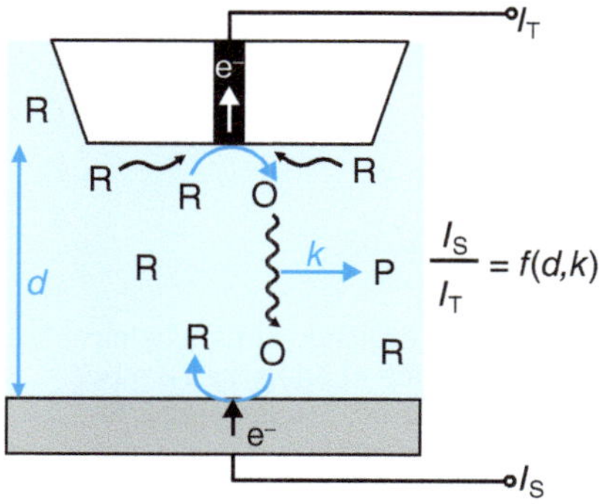

Figure 8.123 Principle of the electrochemical "time-of-flight" experiment for studying the kinetics of homogeneous reactions O → P following an initial electron transfer reaction $R \rightarrow O + e^-$.

Generator electrode: $R \rightarrow O + ne$ (8.164)

Interelectrode space: $O \rightarrow P$ (8.165)

Detector electrode: $O + ne \rightarrow R$ (8.166)

The products of the reaction at the probe diffuse toward the detector electrode and react back to the reagent [Eq. (8.166)]. The transfer coefficient $\varepsilon = I_S/I_T$ between the probe and the sample is experimentally determined with a reversible redox couple R'_{rev}/O'_{rev} that does *not* undergo homogeneous follow-up reactions [Eqs. (8.167) and (8.168)] or calculated by continuum simulation. The transfer coefficient describes the fraction of the product formed at the probe that reaches the detector electrode.

Generator electrode: $R'_{rev} \rightarrow O'_{rev} + ne$ (8.167)

Interelectrode space: No reaction

Detector electrode: $O'_{rev} + me \rightarrow R'_{rev}$ (8.168)

Because the detector electrode is selected to be larger than the probe electrode, the collection efficiency ε approaches unity for R'_{rev}/O'_{rev} and is rather independent of the working distance d. For the redox couple R/O under study, ε does change with d because more time is required for the diffusion between the generator electrode and the collector electrode, which is then also available for the homogeneous follow-up reaction (8.165) to proceed and to decrease the concentration of the product O for its reaction at the detector electrode.

The experimental configuration has some similarity to the use of a rotating ring–disk electrode (RRDE) that we will discuss in Section 12.4.3. The time-of-flight system has the advantages of a much higher collection efficiency and a much lower consumption of reagents, which makes it useful for the mechanistic study of organic compounds [207, 208]. RRDE is commonly used for the investigation of electrocatalytic reactions of small molecules (O_2, H_2, CO_2, CH_3OH).

The term chemical lens refers to the coupling of an SECM imaging mode to a homogeneous reaction. The chemical lens can be used to improve the lateral resolution in imaging and in microstructuring experiments. An example is shown in Figure 8.124 [209]. The reduction of nitrate at the ME causes acidification of the slightly alkaline $[Ag(NH_3)_2]^+$ working solution. The potential of the sample is selected such that it is sufficiently negative to cause the reduction of Ag^+ to $Ag^{(0)}$, but insufficient to deposit $Ag^{(0)}$ from $[Ag(NH_3)_2]^+$ (for explanation, see Section 5.3.12). Only after local acidification in the vicinity of the probe electrode, NH_3 will be protonated, followed by decomposition of the $[Ag(NH_3)_2]^+$ complex and liberation of Ag^+. Because the protonation and decomplexation reactions are very fast, there will be a sharply delimited reaction zone exactly at the position where the diffusional flux of H^+ away from the microelectrode is equal to the flux of $[Ag(NH_3)_2]^+$ toward the reaction zone. The metal deposition occurs at the sample regions that are within the acidified region. A pattern can be formed by the slow movement of the ME (Figure 8.124b).

The confined etchant layer technique (CELT) is related to the idea of the chemical lens. The ME, or even a macroscopic tool electrode of complex shape, is used to generate a chemical etchant. The diffusion layer of this etchant is confined by the homogeneous reaction with a scavenger reagent. This restricts the etching reaction

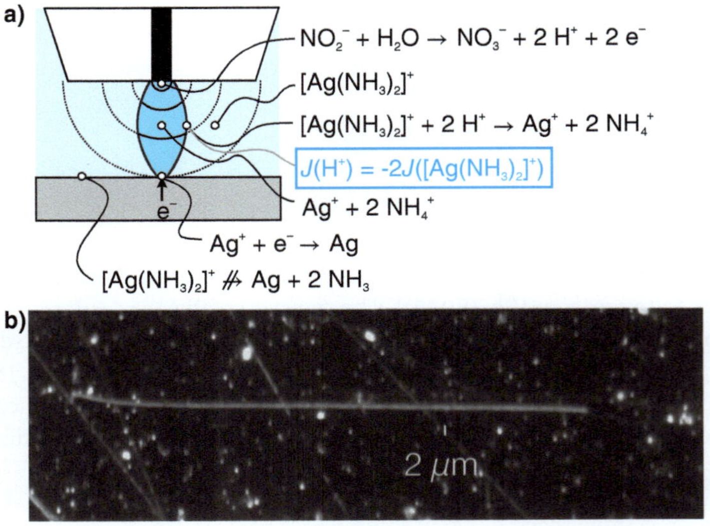

Figure 8.124 Principle of the chemical lens for the deposition of Ag in the FB mode. a) Reaction scheme; b) example for the deposition of an Ag line of 1 μm width on an Au surface. Panel b) is reproduced with permission from [209], Copyright 2005, The Royal Society of Chemistry.

to the workpiece surface that is significantly smaller than the diffusion layer of the etchant in the scavenger-free working solution [210, 211].

A slow homogeneous follow-up reaction can also be used to enhance the image quality in the SG/TC mode [212]. In Section 8.12.4, we have learned that macroscopic sample electrodes cause a steady expansion of their diffusion layers in the SG/TC mode. The reaction product is continuously accumulating in the solution and causes a slow increase in the ME current background during an imaging experiment, which typically takes 30–60 min. This is shown in Figure 8.125a for the detection H_2O_2 above a region with the immobilized enzyme glucose oxidase (GOx). The sample forms H_2O_2 continuously by the GOx-catalyzed oxidation of glucose with O_2 according to Eq. (8.169).

$$\text{Sample: glucose} + O_2 \xrightarrow{[\text{GOx}]} \text{gluconolactone} + H_2O_2 \tag{8.169}$$

Without follow-up reaction, the background current increases in Figure 8.125b from low-left to high-right during the image recording. The addition of the enzyme catalase (Cat) to the working solution enhances the rate of H_2O_2 disproportionation:

$$\text{Solution: } 2\,H_2O_2 \xrightarrow{[\text{Cat}]} O_2 + 2\,H_2O \tag{8.170}$$

In this situation, the homogeneous reaction (8.170) competes with the reaction (8.171) at the ME

$$\text{Probe: } 2\,H_2O_2 \rightarrow O_2 + 2\,e^- + 2\,H^+ \tag{8.171}$$

for the reagent H_2O_2. The conditions (especially the concentration of Cat) must be adapted in such a way that the lifetime of H_2O_2 resulting from the reaction (8.170) is sufficiently long to allow the diffusion of H_2O_2 from the sample to the ME, but too short for a steady accumulation in the solution. This situation is reached by the addition of a very minute amount of the enzyme catalase. In the subsequently recorded image (Figure 8.125c), the background is decreased to a small and constant value. After adding higher amounts of Cat, fewer and fewer H_2O_2 molecules can diffuse across the interelectrode distance without disproportionation, and the signal height is substantially decreased (Figure 8.125c). The plot in Figure 8.125e shows the effect of catalase addition on the background and signal width even more clearly using current–distance profiles along the image diagonals from Figure 8.125b–d.

The possible coupling of homogeneous reactions with the heterogeneous reactions at the sample or the ME under well-defined mass transport conditions underlines the value of SECM methodology as a *chemically selective* microscopy, whereas other scanning probe techniques, such as ECSFM or ECSTM, enable a much better lateral resolution.

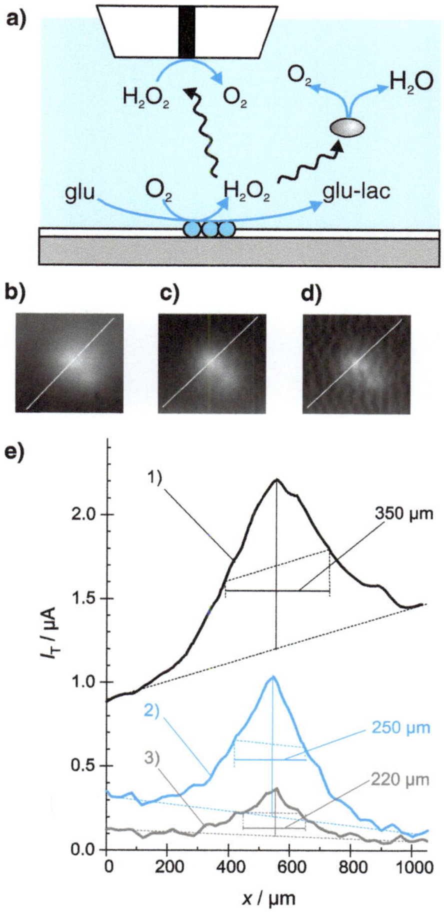

Figure 8.125 Diminution of background currents in the SG/TC mode by consumption of the formed redox-active compound in a homogenously catalyzed reaction. a) Scheme of the glucose oxidation at the sample under formation of H_2O_2; b) to d) gray-scale images for SG/TC detection of H_2O_2 at the ME after addition of catalase suspension (20 mg mL^{-1}, specified in µL suspension per mL working solution); b) 0; c) 0.005; d) 0.03; e) current–distance profiles along the lines in the gray-scale images; assignment of the lines 1) from b); 2 from c); 3) from d). Source: Panels b) to e) are reproduced with permission from [213], Copyright 1997, American Chemical Society.

8.12.8 Possibilities for Improving the Lateral Resolution

There are extensive research efforts to fundamentally improve the lateral resolution of SECM. A first and obvious step is to decrease the size of the ME probe. There is already a range of established methods for the production of such electrodes (Excursion 5-4). As can be deduced from Figure 8.111, a smaller SECM probe also requires a proportionally smaller working distance *d* to achieve a sufficient contrast. A reduced working distance increases the probability of a mechanical collision between the ME probe and the sample that bears the danger of mechanical damage to the sample and/or the probe. For this reason, several approaches have been tested to combine SECM with another scanning probe technique that provides topographical information and allows guiding the ME probe in a small but constant distance over the sample.

Topographic Information from Shear-Force Measurements. In one approach, the ME is excited to a mechanical vibration by a piezoelectric ceramic. The vibration has a small amplitude in the lateral direction to the sample surface. This vibration is damped when the ME approaches the sample to 200–700 nm. The damping of the vibration can be followed by a second piezoelectric ceramic as a decrease in amplitude or as a phase shift [213]. Amplitude and phase do not change a lot within a strongly damping electrolyte solution. Therefore, the experimental challenge is selecting an appropriate resonance frequency of a vibrational spectrum of the experimental setup. The regulation of the shear-force signal becomes additionally challenging when the sample contains regions with very different micromechanical properties, which cause different effects on the vibration of the ME (quite analogous to SFM) [214]. Despite these conceptual challenges, shear-force detection is very useful to improve image quality if only to compensate for a tilt between the sample surface and the scanning plane of the positioning system.

Integration of Electrodes into ECSTM and ECSFM Probes. Microfabrication methods allow integration of amperometric electrodes into tips and cantilevers for scanning probe microscopies (Figure 8.126) using different approaches [158]. The combination with SFM is of particular interest because SFM (as SECM) allows the investigation of conductive as well as of insulating samples. The combination of ECSTM and SECM is also possible [175], but it is restricted to conductive samples. With the miniaturization of the area of the ME of the probe, the kinetic window from Figure 8.117 is shifted to very fast kinetics at the sample that is reached only with a very limited set of sample–mediator combinations. Furthermore, there is a significant impact of finite probe size effects despite the effort to keep the working distance constant [215].

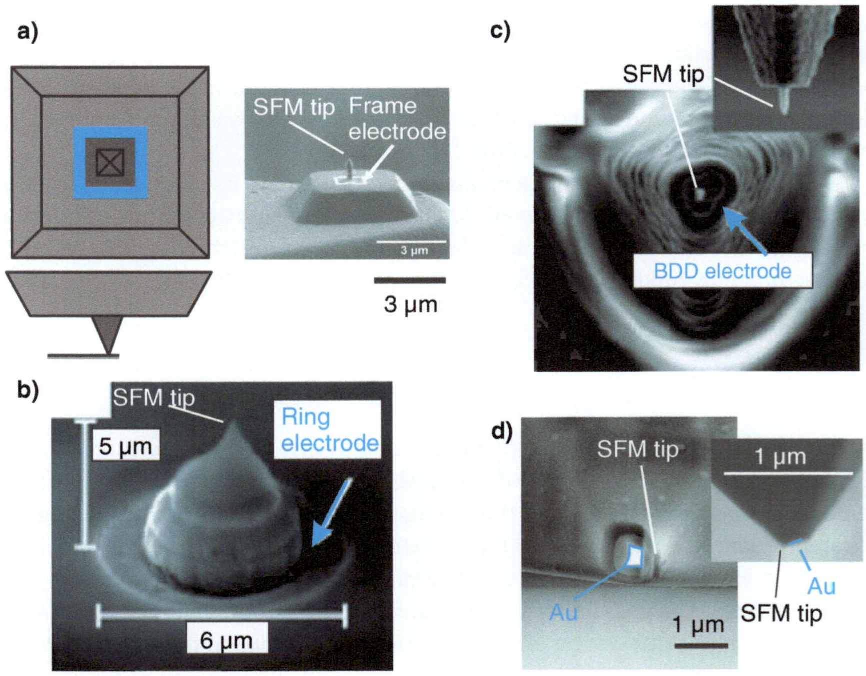

Figure 8.126 Integrated SFM-SECM probes;
a) basic principle of the recessed electrode according to Kranz made as a rectangular frame;
b) recessed ring electrodes produced in a batch process; c) example using a boron-doped diamond (BDD) as electrode material; variant with a nonsymmetric but very small cut of the insulating cover layer on a gold-covered SFM tip.

Source: Reproduced with permission a) from [171], Copyright 2008, Wiley-VCH; b) from [158], Copyright 2014, Royal Society of Chemistry; c) from A. Eifert, W. Smirnov, S. Frittmann, C. Nebel, B. Mizaikoff, C. Kranz; *Electrochem. Commun.* **2012**, *25*, 30–34, Copyright 2012, Elsevier; d) from M. Salomo, S. E. Pust, G. Wittstock, E. Oesterschulze; *Microelectr. Engineer.* **2010**, *87*, 1534–1536, Copyright 2010, Elsevier.

Therefore, the potential for application of combined probes in the FB mode is probably limited. Much more flexible in this respect is the SG/TC mode because it offers a much lower limit of detection. On the other hand, the SG/TC mode has an inherently lower lateral resolution compared to the FB mode. As a consequence, SECM imaging of small active regions depends critically on the preparation of samples, in which active regions occur on the sample surface with a large mutual distance.

8.12.9 Application of Scanning Electrochemical Microscopy

There are broad areas of application of SECM, and a textbook cannot do justice in terms of giving a complete overview. There are important reference sources and review papers [167–172]. Here, we would like to introduce a few important cases of application and provide advice for further reading with respect to these applications. Review papers are also available for general analytical figures of merit, such as resolution, sensitivity, and reproducibility [176, 216].

Corrosion (Chapter 14) is often a localized phenomenon, for which localized characterization methods have a particular value. SECM can be used in combination with SVET and SRET (Section 8.13.4) [217, 218]. Electrochemical energy conversion is one of the important topics of electrochemical research (Chapters 15, 16). Energy conversion systems often use composite electrodes. In such electrodes, the local reaction rates can be much higher than the integral values obtained by averaging the externally measured current (and thus reaction rate) over the entire interfacial area. Very high local current densities can be the cause of degradation phenomena. Their characterization in materials for batteries [219–221], fuel cells [222], and electrolyzers [223] represents an important application of SECM. The SECM experiments can be conducted with freshly prepared electrodes or after their use over a specified time (*post mortem* experiments) to characterize changes occurring during the operation of the device. SECM also allows the study of photoelectrochemical reactions [224, 225], especially because the products formed can be directly determined by tip–substrate voltammetry (TSV) [201, 226, 227].

A further group of applications concerns mass transport across pores and membranes [196, 197, 228] or porous materials, including biological materials like cartilage [229] or dentine [230].

SECM is an important method for determining the activity of immobilized enzymes, for instance on surfaces of biosensors or of microstructured components for assays [231]. Both the SG/TC mode and the FB mode might be useful. While the FB mode allows the investigation of redox enzymes (oxidoreductases) only, the SG/TC mode can also be applied for selected other, technologically important enzymes such as alkaline phosphatase [232] and galactosidase [233].

Positioning of electrodes provides excellent opportunities to study metabolic processes at individual biological cells (Excursion 5-7). SECM setups facilitate different working modes depending on how the selected mediator interacts with the cell metabolism. Hydrophilic mediators dissolved in an aqueous buffer usually cannot pass through a hydrophobic cell membrane. Cells that are attached to a substrate (adherently growing cells) hinder the diffusion of a mediator and allow coarse topographic imaging. In contrast, hydrophobic mediators can pass through the cell membrane, after which they can transfer electrons from/to substrates and enzymes within the cell. This can be exploited for functional imaging of biological cells [234–236]. In addition, the SG/TC mode allows the detection of metabolic products released from biological cells. The redox competition mode has been intensively used with biological cells as the oxygen-consuming sample. For the interpretation of the result, one has to consider that the ME consumes oxygen very efficiently, and this may significantly influence the oxygen metabolism of the biological cell under study [237].

SECM offers very interesting opportunities for chemical surface patterning and local surface modification. Surface modification can be caused by reagents (etchants like Br_2, Fenton's reagent, metal ions from a sacrificial electrode), generated locally at the ME surface, that can be used for the modification of metals, metal oxides, classical semiconductor materials, or organic layers on surfaces. Local surface modification has also been achieved by local perturbation of chemical equilibria, for example, by depletion of a component in solution. Recommended collections and review papers include sections on this topic [171, 238].

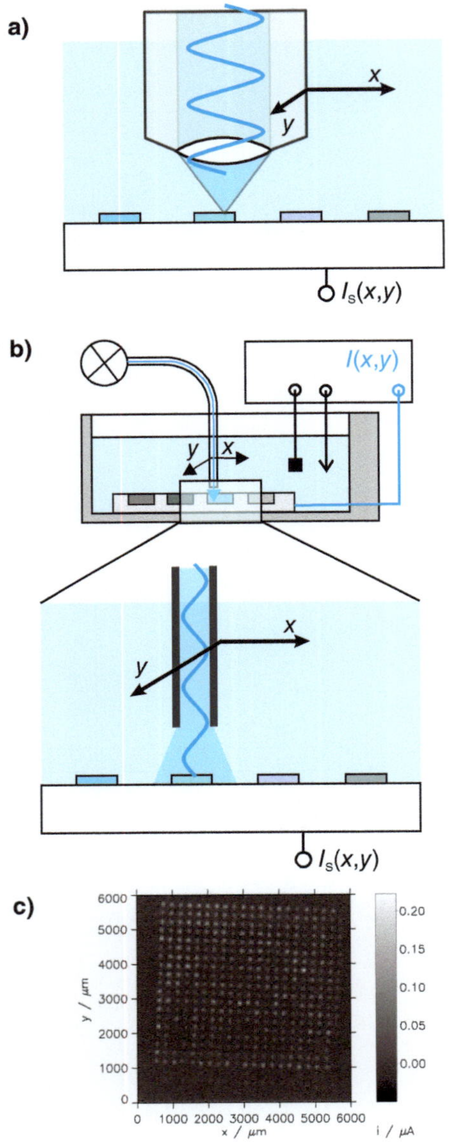

Figure 8.127 Principle of SPECM; a) with a focused (laser) beam and b) with a scanning optical fiber; example for the study of material libraries of photoelectrocatalysts.

8.13 Further Scanning Probe Microscopies

Further scanning probe microscopies are similar to SECM with respect to the experimental configuration and concepts. They are sometimes described in the literature as derivatives of SECM. A current trend in the development of these techniques is the merging of different imaging principles to hybrid or hyphenated techniques, which provide spatially correlated information of different sample properties.

8.13.1 Photoelectrochemical Microscopy

Scanning photoelectrochemical microscopy or scanning photocurrent microscopy (SPECM) has been applied relatively early for the investigation of semiconductor electrodes and photovoltaic devices. The sample is locally illuminated by a scanning laser beam or by a scanning optical fiber. The local illumination can cause a photocurrent at the sample, leading to oxidation of the solvent or of a scavenging reagent. The current from the sample I_S is measured in the external circuit and plotted as a function of the lateral position of the optical fiber or the laser beam (Figure 8.127a). If one regards the photon provided by the laser beam or the optical fiber as a reagent for the reaction at the sample, the analogy to the TG/SC mode of SECM in Figure 8.120 becomes very evident.

SPECM has the advantage that the investigated region can be changed very quickly without considering a finite transfer time for the diffusion of a probe-generated substance from the probe to the sample as in TG/SC-SECM. The working mode is also relatively insensitive to the working distance, which makes the image immune against a slight and sometimes difficult-to-control tilt between the sample and the scanning plane. Due to the low distance dependence, the working distance can be selected so large that a collision between probe and sample can be avoided easily. On the other hand, a very high lateral resolution is difficult to achieve because the space charge layers developed in semiconducting samples blur the response beyond the illuminated area. Due to the possible fast movement of the probe and the low distance dependence, the method is particularly well suited for imaging large samples. Material libraries with photocatalysts for water splitting are a good example of the exploitation of this feature (Figure 8.127b,c). Every spot in the array has another, known overall composition of the elements X, Y, Z, and oxygen. The bright spots exhibit the highest photocurrent for water oxidation and are lead structures for further study and optimization of the material system. For a general discussion of combinatorial approaches, reviews are recommended [239, 240].

As a comparison to other photoelectrochemical methods, Ref. [241] can be recommended. Recently, new approaches have been demonstrated for studying local electrochemical reactions using experiences from single-molecule spectroscopy and optical super-resolution microscopy [240], which enable the identification of preferred reaction sites on nanocrystalline semiconductors.

8.13.2 Scanning Ion-Conductance Microscopy

Scanning ion-conductance microscopy (SICM) was one of the first scanning probe techniques that were developed after the introduction of STM [242]. It is based on the diminution of an ionic current between an inner and outer quasireference electrode through an orifice of a capillary (Figure 8.128a) when the capillary opening approaches a sample surface and the current path is obstructed in the vicinity of the sample. The full potential of the method for contact-free imaging could only be used after instrumental improvements. SICM is of particular advantage for imaging biological objects that have a strong curvature of the surface (a strong corrugation) and would react even upon a gentle mechanical contact with a SFM tip or respond to the laser beam used in most SFM instruments to detect the cantilever bending [243].

An increase in the sensitivity was made possible by the use of modulation techniques in which an external control variable is changed periodically and causes a periodic change in the probe response, that is, the ionic current through the capillary orifice. The frequencies of the external control variable and the probe current are equal. A lock-in amplifier allows for the measurement of such signals more sensitively than DC signals by other equipment. The periodic signals are analyzed using the externally controlled frequency with respect to amplitude and phase shift between the control variable and the probe current. The external control variable can be the working distance d or the voltage U between the inner and outer electrodes.

In the case of a distance modulation (Figure 8.128b), the current decreases when the capillary orifice approaches the sample and increases if it retracts. Thus, the measured signal consists of a DC component and an overlaid AC component. By means of the lock-in amplifier, the AC component is isolated. I_{AC} approaches zero in the bulk phase of the solution and increases when approaching a surface. With a user-defined setpoint and an electronic feedback loop for vertical positioning of the capillary above the sample, I_{AC} can be kept constant in order to maintain a constant working distance, for instance, within a combined scanning probe experiment.

The phenomenon of ionic current rectification can be observed at capillary orifices or working distances that are smaller than the extension of the diffuse part of the electrical double layer at the capillary walls or the surface of the sample. Due to changes in the local concentrations of anions and cations in the double layer compared to the undisturbed bulk solution (Figure 8.129a), the decrease in ionic current is different for anions and cations in the vicinity of the capillary orifice if a significant fraction of the ionic current has to pass through the diffuse part of the electrical double layer of the capillary walls or the sample surface (Figure 8.129b). This rectification effect becomes more prominent at low ionic strength due to the larger extension of the electrical double layer at lower ionic strength (Table 4.4). The rectification effect can also distort topographical images.

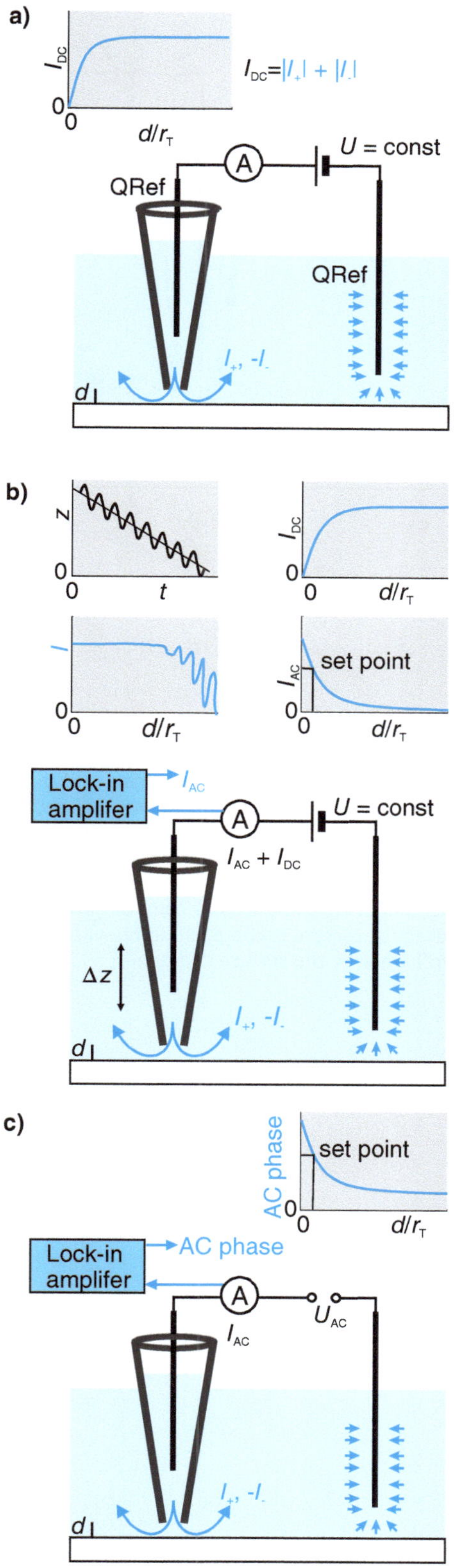

Figure 8.128 a) Variants of SICM; a) DC current; b) distance modulation; and c) voltage modulation. Source: Scheme adapted by book author from [244].

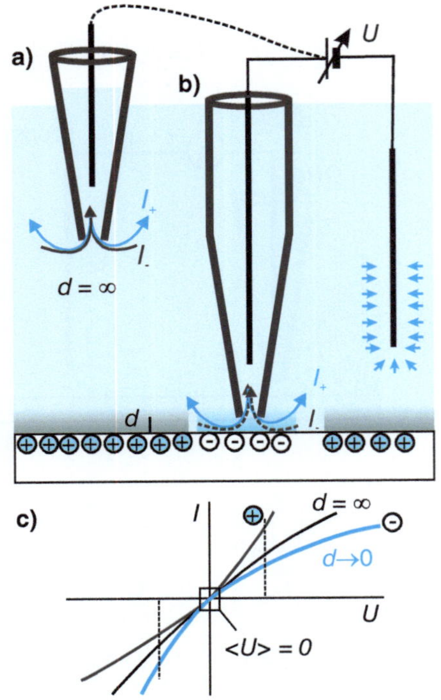

Figure 8.129 Measurement of surface charges by means of SICM. a) Ionic current in the bulk solution; b) additional decrease in the anion current close to a negatively charged sample surface; c) current–voltage curve in the bulk solution and close to the sample surface. Small voltage modulations within the rectangle around the coordinate origin can be used for vertical probe positioning without influence by the surface charge.

Voltage modulation (Figure 8.128c) between the inner and outer quasireference electrodes avoids this problem. The DC component of the excitation signal is set to zero and overlaid with a sine voltage. The ionic current through the capillary is equivalent to I_{AC} through a serial combination of an ohmic resistor and a capacitor (Section 7.3.1). The ohmic resistor corresponds to the migration of ions. An increase in the ohmic resistance in the vicinity of the sample surface causes a change in the phase shift of I_{AC} with respect to the applied AC voltage. The phase shift can then be used as the distance-dependent signal and is independent of the surface charge of the sample. An important advantage is the higher frequency at which voltage modulation can be used compared to distance modulation. The piezoelectric positioners for relatively long travel ranges exhibit a frequency dependence of their extension. Above a critical frequency, for example, 160 Hz, the extension for a given voltage strongly decreases. The higher frequency of the voltage modulation facilitates faster lateral translations of the probe because at each position of the capillary, some periods of the modulation frequency must pass to acquire the distance-dependent signal. If this signal can be obtained faster (at higher modulation frequencies), the lateral movement of the probe can also be made faster. Avoiding a vertical translation of the probe also reduces the mechanical pressure that is transmitted via the liquid electrolyte solution onto the sample surface. Using a zero DC component in the excitation signal prevents the formation of strong electric fields at the capillary orifice that could have a potential influence on biological samples. Because the distance is not changed periodically for the modulation, the average distance between capillary and sample is smaller in the case of a voltage modulation and, thus, should in principle allow higher lateral resolution [244].

By combining the different options, specific information can be obtained. A voltage modulation, as described above, can be used for positioning the capillary orifice in close proximity to the sample surface. From its true vertical position, an undistorted topographic image can be obtained. At this position, the DC component of the voltage between the two quasireference electrodes is varied in order to investigate how the surface charge changes the current–voltage curves compared to the bulk solution (Figure 8.129c). Meanwhile, such measurements have even been demonstrated in physiological buffer solutions (ionic strength $I_c = 0.15\,\text{mol L}^{-1}$). The obtained signals could be recalculated to local surface charges by means of simulation of the involved transport processes. Such local data of surface charges are not accessible by other techniques [244].

Another application is based on observing local reactions that generate or consume ions. The capillary is positioned at a constant distance over the surface using modulation techniques. Subsequently, a cyclic voltammogram $I_S(E_S)$ is recorded at the sample. If a reaction occurs at the sample, there is almost always a change of the *local* ionic composition of the solution (e.g., $N_2H_4 \rightarrow N_2 + 4\,H^+ + e^-$, $Fc \rightarrow Fc^+ + e^-$ etc.) that causes a change in the ionic current at the capillary (I_T) while the cyclic voltammogram is recorded at the sample. This procedure is now repeated at all points of a two-dimensional grid. The experimental data are a four-dimensional dataset that can be visualized in different ways, and from which small video clips are particularly impressive and instructive. They show for each potential of the sample E_S the measured capillary current as a false-color image (x, y, I_T). Often, the onset of a particular reaction at different regions of the sample with different potentials is clearly discernible and can be correlated to topographic features (steps, grain boundaries, etc.) [244].

8.13.3 Scanning Electrochemical Cell Microscopy

The droplet cell is a classical concept to conduct electrochemical experiments on selected regions of a macroscopic sample. The region to be studied is defined by the sample area wetted by a drop suspended from a capillary opening (Figure 8.130a). The sample is the working electrode (WE). The capillary from which the drop is suspended also accommodates the reference (Ref) and auxiliary (Aux) electrodes. The extension of the drop can be controlled by a gasket at the capillary orifice (Figure 8.130b). With this provision, a continuous exchange of the working solution is possible by application of a hydrostatic pressure difference between a solution inlet and outlet. This may be necessary to remove reaction products when studying metal dissolution processes at high current densities. Cells with inlet and outlet can also be used to transfer the solution to a mass spectrometer for chemical analysis. An example is discussed later in Figure 14.36. Such cells facilitate all common dynamic electrochemical techniques and have found application in different variants in the field of corrosion science (Figures 14.14c and 14.27) and even for the restoration of historical objects of art made from silver (Excursion 14-3). Cells with suspended droplets may even allow a scanning movement of the capillary and thus the recording of line scans. Repeated placement of the cell on the sample can be used to scan an array of sample regions, for instance, to characterize gradient material in the combinatorial search for new electrocatalysts or to optimize experimental parameters for electrochemical surface modification. A review about different variants of the droplet cells and their application is available [245].

Recent research has considerably expanded those possibilities by coupling the droplet cell with SICM (Figure 8.130c), called scanning electrochemical cell microscopy (SECCM) [244, 246]. A divided capillary is used, which is called theta capillary because the cross section is reminiscent of the Greek character (Figure 8.131a). Both capillaries contain a silver wire as combined quasireference and counter electrodes (QRCE). A DC voltage U_2 is applied between the two QRCEs, causing an ionic current $I_{+,-}$ through the hanging electrolyte drop. If the drop is placed on the sample surface, it is deformed in shape and this hinders the path for ionic migration and consequently $I_{+,-}$ decreases. A small modulation of the vertical probe position by Δz causes $I_{+,-}$ to follow the modulation, which can be recorded sensitively by a lock-in amplifier. Due to the very sensitive measurement, extremely small capillaries (smaller than the resolution limit of optical microscopy) can be positioned on the surface rapidly, reproducibly, and in an automated fashion. The droplet cell formed with the wetted region of the sample and one of the QRCEs can then be used for a voltammetric characterization. Due to the automation of the procedure and the instantaneous response of the ionic current to the distance modulation, an impressive reduction in imaging time has been demonstrated compared to SECM. This advantage can be used to record movies of the onset of local reactions as a function of the voltage U_1 between the sample and one of the QRCEs. The drop can be placed repeatedly on the surface in an array pattern with a recording of voltammetric curves at each position. Capillaries with four barrels (Figure 8.131b), from which one or two are transformed to a carbon nanoelectrode (Excursion 5-4) provide further flexibility because such probes can be used simultaneously for SICM-like and SECM-like imaging modes [246].

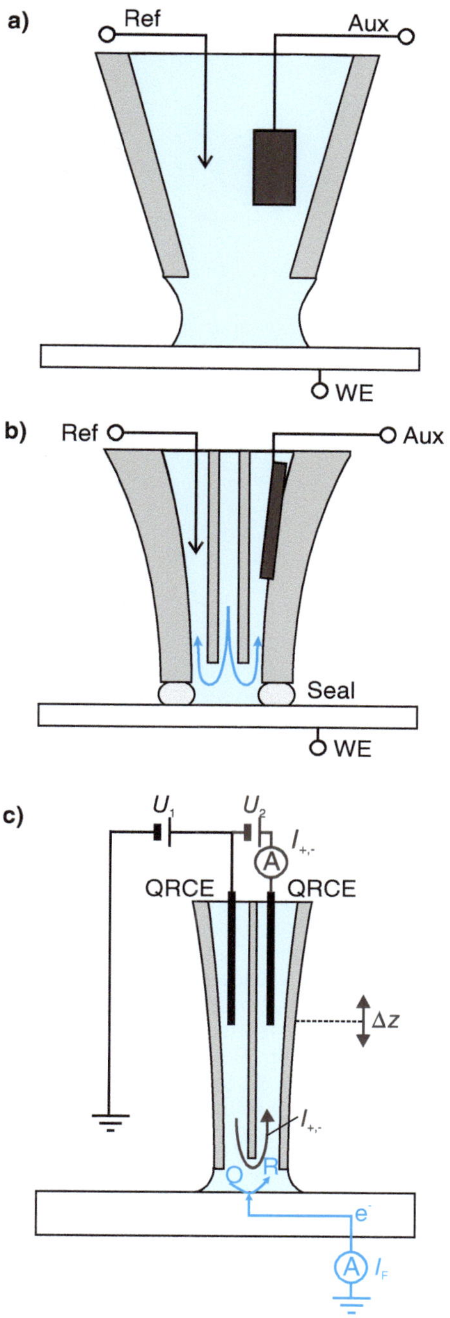

Figure 8.130 Droplet cells (schematic). a) Freely suspended drop; b) drop with gasket and concentrically arranged inlet and outlet for solution exchange; c) freely suspended droplet with distance regulation by SICM. The schematics are not in the same scale. The cells in a) and b) have a diameter between some mm to some tens of μm. The distance regulation in c) permits miniaturization into the nanometer range. The positions of the reference and counter electrodes are usually further up in the capillary than shown in the schematics.

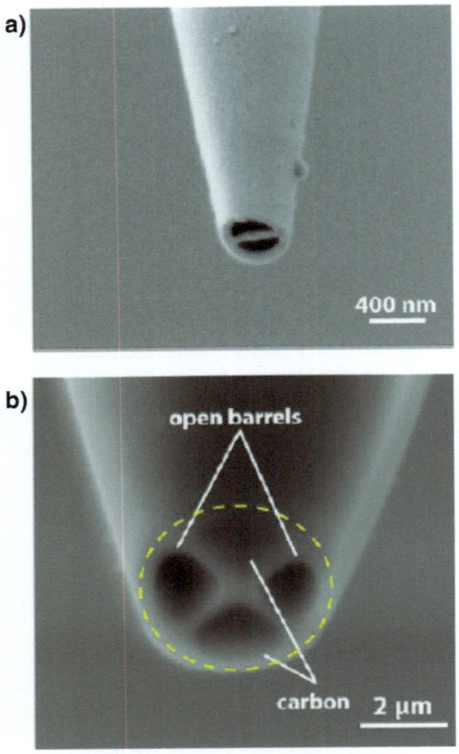

Figure 8.131 Probes for SECCM. a) Divided capillary (theta-capillary); b) quad-barrel capillary with two open channels for positioning by SICM with distance modulation and for SECCM measurements. Two carbon-filled capillaries allow SECM experiments. Source: Panel a) is reproduced with permission from [244], Copyright 2016, American Chemical Society; panel b) is reproduced with permission from B. P. Nadappuram, K. McKelvey, J. C. Byers, A. G. Güell, A. W. Colburn, R. A. Lazenby, P. R. Unwin; *Anal. Chem.* **2015**, *87*, 3566–3573, Copyright 2015, Amercian Chemical Society.

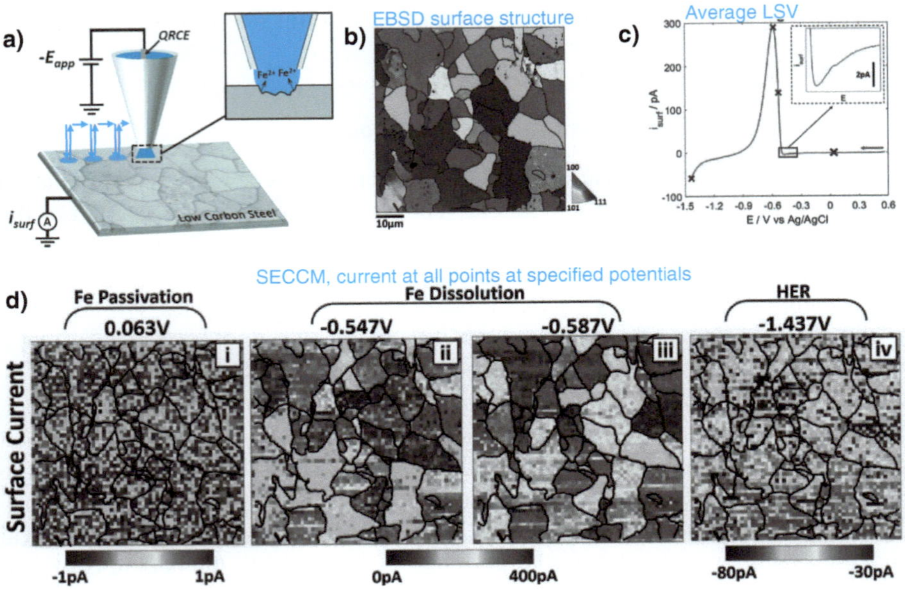

Figure 8.132 SECCM investigation of a steel sample. a) Scheme of the experimental procedure; b) EBSD image of the prevailing crystal orientation of the sample region investigated by SECCM; c) averaged linear sweep voltammograms from all measurement points; d) false-color images of the currents at the selected potentials indicated in panel c). It is clearly visible that the iron dissolution occurs at different rates (current densities) depending on the grain orientation. The grain boundaries have been marked by solid lines. Source: Reproduced with permission from [247], Copyright 2020, Elsevier.

Figure 8.132 illustrates the new possibilities opened up by the new combined techniques [247]. Electron backscatter diffraction (EBSD) was used to determine the orientation of crystallites of a polycrystalline steel sample (Figure 8.132b). The prevailing orientation of each grain can be indicated by a color code. SECCM was then used to record a voltammogram in 5 mmol L^{-1} H_2SO_4 solution at every point of an orthogonal grid. The voltammogram averaged from all the individual voltammograms is shown in Figure 8.132c. The currents at each point of the grid at a particular potential can be converted to a false-color image as shown in Figure 8.132d. At potentials at which the active dissolution of iron is observed, the currents and thus the local reaction rates depend particularly strongly on the grain orientation. Furthermore, voltammograms from the four-dimensional data set can be selected in order to assess the influence of further structural features such as inclusions or grain boundaries.

8.13.4 Scanning Reference Electrode Technique, Scanning Vibrating Electrode Technique, Localized Impedance Spectroscopy, and Ohmic Microscopy

When a local current occurs at an electrode, the resistance of the electrolyte solution causes a potential difference between two points along the current path within the solution (Figure 8.133a). The potential drop $U = \Delta\phi$ is proportional to the local current density, which can be obtained from Eq. (8.172) using Ohm's law $I = U/R$ and the known relationship for the electrical resistance $R = \rho_E l/A$ (ρ_E – specific electrolyte resistance).

$$j = \frac{I}{A} = \frac{U}{AR} = \frac{U}{\cancel{A}\left(\rho\frac{l}{\cancel{A}}\right)} = \frac{U}{\rho_E l} \tag{8.172}$$

This proportionality is the basis for a variety of scanning techniques that measure the local current density via a potential drop within the electrolyte near the electrode interface. The current measurement is nonspecific and captures the local currents independently of the particular interfacial reaction occurring. There are different ways of measuring the potential drop in solution that have been introduced to the literature under different names, but exploit essentially the same principle.

In the scanning reference electrode technique (**SRET**), the opening of a Luggin capillary[57] or a fine Pt wire (as a quasireference electrode, QRE) is scanned above the surface. The potential difference $U = \phi(x, y, z) - \phi_{\text{bulk}}$ between the scanned electrode and a reference electrode at a fixed position in the solution bulk is measured at each grid point (Figure 8.133b). This potential difference can be recalculated to a local current density using Eq. (8.172). There is also a variant in which two Pt wires with a different working distance are scanned together over the surface.

The scanning vibrating electrode technique (**SVET**) measures the potential drop with an electrode oscillating in its position. Often, but not always, the vibration is executed perpendicularly to the sample surface. Because the scanning electrode measures alternately the potential at two different working distances, the recorded signal is an AC voltage with the frequency of the mechanical distance modulation where the amplitude is proportional to the local current density (Figure 8.133c). The use of lock-in amplifiers in SVET allows the measurement of smaller currents than in SRET. SVET is particularly suitable for measuring local reactions at the open circuit potential of the sample, that is, without application of an external voltage source. This situation is important in many corrosion scenarios (Chapter 14), which is also the main application area of SVET.

Localized electrochemical impedance spectroscopy (LEIS) is also related to those techniques. In this case, an AC voltage is applied to the sample as in electrochemical impedance spectroscopy (Chapter 7). The measured signal is not the current I_{AC} averaged over the entire sample surface, but the AC voltage measured between two QRE's that are positioned at different distances from the sample electrode (Figure 8.133d). This AC voltage is proportional to the local alternating current density at the sample. The principle of LEIS is quite different from SECM working modes using AC(AC-SECM), in which the AC voltage is applied to the microelectrode probe [218].

Another related technique is called ohmic microscopy, in which a complex potential program, for example, a cyclic voltammogram, is executed at the sample electrode, and the potential drop between two locations along the current path in solution is measured between two Luggin capillaries. This arrangement allows one to record local voltammograms of surface processes (surface oxidation, chemisorption) similar to those in Figure 5.22, but originating from a small fraction of a macroscopic sample [248].

Further Reading. Reference [249] is recommended as an entry for a more in-depth coverage of SRET, SVET, and LEIS.

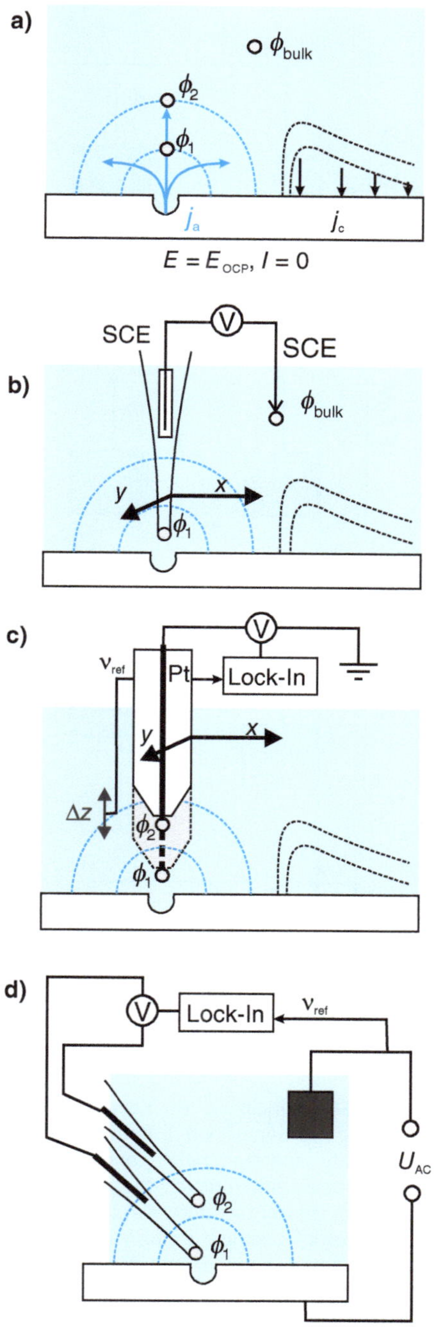

Figure 8.133 Measurement of the local potential drop in solution; a) principle; b) SRET with Luggin capillary; c) SVET with Pt quasireference electrode; and d) LEIS measurement.

8.13.5 Scanning Kelvin Probe Microscopy

Kelvin probe microscopy (KPM) is based on the principle of the Kelvin probe[115] for the measurement of work functions. Two pieces of metal, that are not connected by a wire, are in relation to a uniform vacuum level of electrons by their respective work functions, as shown in Figure 8.134a for the filled valence bands. (We will cover the band

[115] William Thomson, (Baron Kelvin of Largs), 1824–1907, British physicist with a very broad working area. Besides the Kelvin probe, the SI unit of temperature and the Joule–Thomson effect are named after him. Practically important were his contribution to cooling equipment and electrical information transmission.

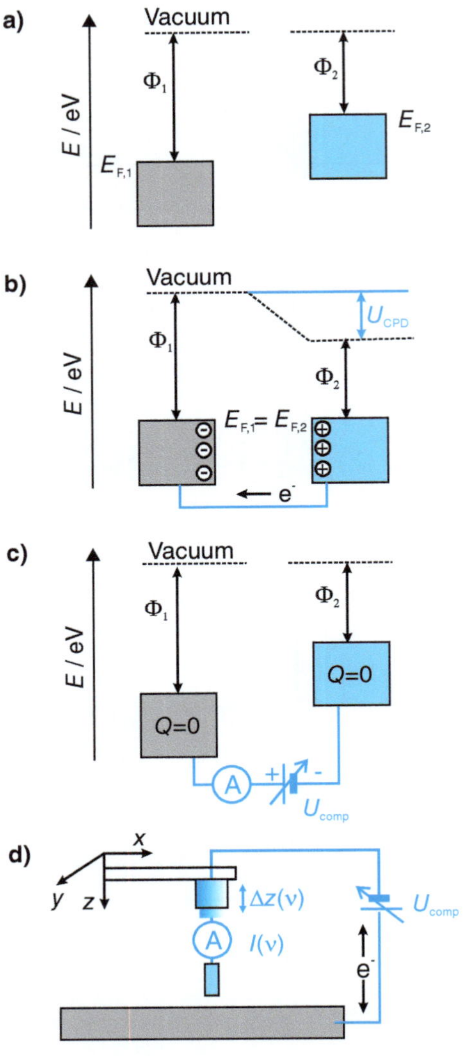

Figure 8.134 Principle of the Kelvin probe. a) Energetic position of two different, electronically separated metals; b) charging of the metals upon electronic contact and formation of the contact potential difference; c) application of a compensation voltage until both metals are neutral; d) design of one metal body as a local probe for a three-dimensional positioning system and distance variation Δz between the metals of the sample and the probe cause I_{AC} that is measured by means of a lock-in amplifier.

structure of solids in more detail in Chapter 10.) The difference between the two work functions Φ_1 and Φ_2 is the contact potential difference U_{CPD}. After connecting the two metal pieces by a wire, electrons move from the metal with a low work function to the metal with a higher work function. The charging of the two metals leads to an equilibrium, at which the charge on the two metal pieces just compensates for the contact potential difference (Figure 8.134b). The contact potential difference can be measured by application of an adjustable compensation voltage U_{comp}, which is required to keep both metal pieces without an excess charge (Figure 8.134c). In this situation, the relation $U_{CPD} = -U_{comp}$ holds. Now, we consider one of the metal pieces as the sample and the other as a probe positioned above the sample (Figure 8.134d). The arrangement is reminiscent of a parallel plate capacitor with area A, variable distance d, and capacitance $C = \varepsilon \cdot \varepsilon_0 \cdot A/d$. For the charge stored in a parallel plate capacitor at a particular voltage, the relation $Q = C \cdot U$ holds. If the distance d between the plates of the capacitor is changed, this changes C, and hence at a fixed voltage U, a current must flow to adjust to the new equilibrium condition. A vertical oscillation of the probe metal causes an alternating current I_{AC} with the frequency of the distance modulation that can be measured very well by means of a lock-in amplifier (Figure 8.134d). During the measurement, U_{comp} is changed continuously. I_{AC} vanishes exactly at the potential when $U_{CPD} = -U_{comp}$ and both metal pieces are uncharged. Only in this situation does a distance variation not cause a current (principle of zero adjustment, which can be made very precisely). If the measurement is carried out at different places on the surface, the measured contact potential differences $U_{CPD}(x, y)$ can be used to generate a false-color image similar to other scanning probe techniques.

The contact potential differences can be recalculated to the local work function provided that a reference sample with known work function Φ_{Ref} is available and a probe is used whose work function does not change. A Ni–Cr alloy is preferentially used for the probe. A reference sample with known work function is Au that is cleaned by Ar^+ sputtering and characterized by photoelectron spectroscopy. However, this procedure is suitable only for UHV instruments because the work function changes when the Au sample is transferred from a vacuum to ambient air and the surface is immediately covered by adsorbates. A reference sample that is more suitable for work under air is HOPG (highly ordered pyrolytic graphite, $\Phi_{Ref} = (4.6\text{–}4.7)$ eV), which binds the typical contaminants from air (H_2O, hydrocarbons) less strongly. In an electrochemical context, the change of the work function rather than its absolute value is used because it provides information about the interesting properties of the sample and independent of a correct referencing.

Electrochemical applications of KPM [132] are found in corrosion science. KPM is then conducted in moist air and can detect changes at buried interfaces [250]. Figure (8.135) shows schematically the KPM response from a progressing delamination frontier between a galvanized workpiece and a polymer film applied for corrosion protection. Starting at a damaged part of the polymer film, the film loses contact with the metal piece, and a small solution volume is encapsulated in which different conditions prevail than at the actually damaged part of the film. Using the contact potential differences, the reactions occurring underneath the intact film and the progressing delamination can be visualized in a length scale of some 20 μm up to several millimeters. For increasing the sensitivity of KPM, it is advantageous to keep a constant working distance above rough and tilted samples [250]. Apart from corrosion protection layers, buried polymer–metal interfaces are also of practical importance for understanding and optimizing advanced adhesive layers in industrial manufacturing.

The principle of KPM can be combined advantageously with the principles of scanning force microscopy (KPFM, Kelvin probe force microscopy) (Section 8.11.5). The probes are metalized cantilevers in this case. The vertical cantilever vibration, often

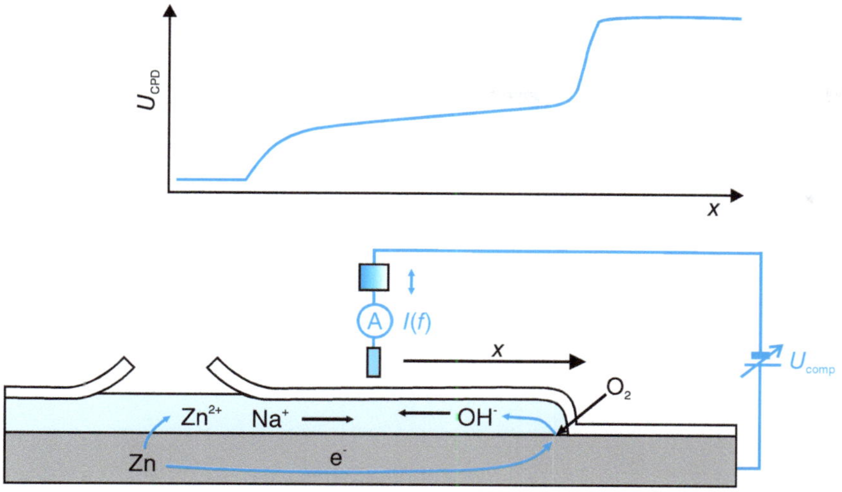

Figure 8.135 Principle of the investigation of the delamination of a polymer film from a metal surface.

used in SFM, is exploited as a periodic distance variation for the KPM measurement. SFM principles are used to maintain a constant average working distance. Miniaturization of the setup enables a much better lateral resolution. Interfaces between metals, semiconductors, and solid or polymeric electrolytes represent samples from an electrochemical context than can be advantageously studied by KPFM [251].

Test your knowledge

Exercise 8.1: Name three methods for vibration spectroscopy at surfaces together with the corresponding excitation conditions. Collect the instrumental basics and the information that those methods can provide.

Exercise 8.2: What is the surface selection rule in IR spectroscopy? Explain the consequences of using the valence vibration of a methylene group ($-CH_2-$) within an alkyl chain in an all-trans conformation, for example, within a self-assembled monolayer with a) perpendicular orientation of the chain and b) tilt of the chain by 45°.

Exercise 8.3: Compare the frequencies of the CO valence vibration of adsorbed carbon monoxide with the CO valence vibration in metal carbonyls [https://en.wikipedia.org/wiki/Metal_carbonyl, last access 01 Jul 2022].

Exercise 8.4: Assume the excitation wavelength of a laser for Raman spectroscopy is $\lambda_{in} = 532$ nm. The vibrational transition is expected at $\tilde{\nu} = 1200\ \text{cm}^{-1}$. At which wavelength would the transition be found in the Raman spectrum?

Exercise 8.5: Explain the advantages and disadvantages of Raman spectroscopy compared to IRRAS for the characterization of electrodes. What would the preferred application areas of both methods be based on this analysis?

Exercise 8.6: Explain the principles of SHG and SFG using a rough sketch.

Exercise 8.7: Collect primary and secondary processes in which X-rays play a role. Design a schematic that illustrates the respective processes. Name a method (with abbreviation and main area of application) that makes use of the respective primary or secondary process.

Exercise 8.8: Why do the doublets observed in XPS have a fixed intensity ratio?

Exercise 8.9: What is the reason for the background in XP spectra? Why must the background be removed before concentration determination? What models for background correction exist? Which model is preferred for which situation?

Exercise 8.10: Which advantage can be used in XPS when changing the excitation energy?

Exercise 8.11: What is the chemical shift in XPS? Use databases and collections of spectra to provide a table of chemical shifts for the important compounds of the following elements: C, Si, S, Ag.

Exercise 8.12: How are chemical shifts in XPS interpreted using Koopmans' theorem? Why is this only a very rough approximation? Use the results of Exercise 8.11 for your explanation.

Exercise 8.13: What is radiation damage in the context of XPS? Explain two mechanisms for radiation damage in XPS experiments.

Exercise 8.14: Which possibilities exist for the determination of the thickness of overlayers at modern XPS spectrometers (with peripheral options)?

Exercise 8.15: Why is synchrotron radiation needed for X-ray absorption spectroscopy?

Exercise 8.16: Why is X-ray absorption spectroscopy in principle useful for *in situ* or *operando* spectroscopy?

Exercise 8.17: Explain the information content of XANES and EXAFS data.

Exercise 8.18: Under which conditions can the Sauerbrey equation be considered as a good description of data from a quartz-crystal microbalance? When is the use of more complicated models compulsory?

Exercise 8.19: Provide a table with the coordination numbers of the atoms near the defects shown in Figure 8.57. Assume a cubic primitive lattice first and then expand the idea for an fcc crystal. Assume the terraces in Figure 8.57 are (1 1 1) faces.

Exercise 8.20: A (1 1 1) face of an Au single crystal is made with a misalignment of 0.1°. What would be the average width of the terraces between two monoatomic steps? The atomic distance in the fcc lattice is 288.4 pm.

Exercise 8.21: a) Use the data in Appendix A-7 to determine the surface concentration of Au atoms on an unreconstructed (1 1 1) surface. Provide the solution in atoms per cm^2 and in mole per cm^2.

b) Determine the surface concentration of an adsorbate on an Au(1 1 1)–$(\sqrt{3} \times \sqrt{3})$ R 30° superstructure. Provide the solution in atoms per cm^2 and in mole per cm^2. Compare the surface concentration of the adsorbate and of the Au atoms.

Exercise 8.22: Collect the methods for the preparation of single-crystal metal electrodes.

Exercise 8.23: What are superstructures? What are moiré patterns? Illustrate your explanation with a schematic for each structural element.

Exercise 8.24: a) How can Pt atoms be arranged along the image frames of Figure 8.65 side by side? How many Pt atoms are approximately covered under the specifically adsorbed anion layer in Figure 8.65?

Exercise 8.25: What does UPD mean? How can UPD phenomena be observed by cyclic voltammetry? Why are UPD peaks larger at single-crystal electrodes than at polycrystalline electrodes of the same material? What would you expect for a comparison of XPS signals from UPD layers from single-crystal and polycrystalline materials?

Exercise 8.26: Prove the equivalence of the two expressions in Eq. (8.111) with $g = -2a$.

Exercise 8.27: Compare the expression of the Debye length in Eq. (8.119) to the expression for the radius of an ionic cloud in diluted electrolyte solution according to Eq. (4.74).

Exercise 8.28: Collect in a table the names, the equations, the assumptions made, and the most important application areas for the different adsorption isotherms.

Exercise 8.29: Determine the Langmuir adsorption isotherm of ubiquinone-10 (UQ) at a Hg electrode in ethanolic 0.2 mol L^{-1} $LiClO_4$ solution from the data provided in the table using a nonlinear curve fit in a spreadsheet application. How large is the coverage for the solution concentration [UQ] = 1.00 μmol L^{-1}? What is the maximum interfacial concentration according to the Langmuir model?

[UQ]*/(μmol L^{-1})	Γ/(mol cm^{-2})
0.177	0.343×10^{11}
0.305	0.526×10^{11}
0.814	0.925×10^{11}
1.53	1.27×10^{11}
2.54	1.44×10^{11}
4.56	1.58×10^{11}
8.08	1.46×10^{11}

Additional challenging exercises: Would the description using a Frumkin isotherm provide a better description of the experimental data of this example?

Exercise 8.30: Show that κ has the unit m^{-1} in Eq. (8.132).

Exercise 8.31: Execute the series expansion to derive Eq. (8.138). Why is the expression an approximation and not an exact equation?

Exercise 8.32: A patterned self-assembled monolayer on gold consists of three regions with *n*-mercaptododecanoic acid (HOOC–$C_{15}H_{30}$SH), *n*-hexanethiol (C_6H_{13}SH), and bare gold. The surface is immersed in an aqueous buffer pH 7.4 and studied with the following scanning probe techniques: STM, SFM in lateral force mode, SFM in intermittent contact mode, SECM, and SICM. Discuss the expected contrasts between the regions for each method using rough sketches. Which optimization of the measurement parameters can you suggest to maximize the contrast?

Exercise 8.33: The enzyme glucose oxidase, which converts glucose according to Eq. (8.169) is immobilized on a polymer fiber with a diameter of 10 μm. Analogously to Eq. (8.169), ferrocenium derivatives can accept electrons from the active center of the enzyme.

Discuss how the activity of the enzyme can be imaged by SECM in the GC, FB, and RC modes. What challenges do you expect for each of those experiments, and how can they be dealt with?

Exercise 8.34: Provide the partial differential equation from Figure 8.113 together with the boundary conditions in normalized quantities according to Eqs. (8.143)–(8.146), (8.149), and (8.150).

Exercise 8.35: For the sequential study of a sample by SECM and SFM, the region of interest is indicated by applied "landing marks" that allow relocation of the region of interest after transfer of the sample from one instrument to the next one. If the substrate of the sample is a passivated Si wafer (Si/SiO_x), the landing marks could be a scratch (or a depression) or an applied contact line from Pt. Discuss which contrast principles of the two methods would be useful to relocate the landing marks in SFM as well as in SECM.

Exercise 8.36: Cyclic voltammetry is used to deposit a film of conducting polypyrrole on the electrode surface. The film is removed together with the electrode from the electrochemical cell in its oxidized state and subsequently investigated by XPS. Would this experiment be regarded as an *ex situ*, *in situ*, or *operando* method? Suggest other *ex situ*, *in situ*, or *operando* methods to study the structural changes of the film during the redox reactions caused by cycling the electrode potential in the electrochemical cell.

References

1 J. C. Vickerman, I. S. Gilmore; *Surface Analysis: The Principal Techniques*, 2nd ed., John Wiley & Sons, Chichester, 2009.
2 J. P. Hornak; *Encyclopedia of Imaging Science and Technology*, 2, John Wiley & Sons, New York, 2002.
3 H. D. Abruña, Ed.; *Electrochemical Interfaces Modern Techniques for In-Situ Interface Characterization*, Wiley-VCH, New York, 1991.
4 A. Wieckowski, Ed.; *Interfacial Electrochemistry, Theory, Experiment and Application*. Marcel Dekker, New York, 1999, pp. 1–966.
5 W. Schmickler, E. Santos; *Interfacial Electrochemistry*. 2nd ed., Springer, Berlin, 2010, pp. 1–272.
6 C. N. Banwell, E. M. McCash; *Fundamentals of Molecular Spectroscopy*, McGraw-Hill, London, 1994.
7 I. Zawisza, G. Wittstock, R. Boukherroub, S. Szunerits; PM IRRAS investigations of thin silica films deposited on gold: Part 1. Theory and proof of concept. *Langmuir* **2007**, *23*, 9303–9309.

8 P. W. Faguy, W. R Fawcett; Infrared reflection-absorption spectroscopy of the electrode/electrolyte solution interface: Optical considerations. *Appl. Spectrosc.* **1990**, *44*, 1309.

9 S. M. Stole, D. D. Popenoe, M. D. Porter; Infrared spectroelectrochemistry: A probe of the molecular architecture of the electrochemical interface. in *Electrochemical Interfaces Modern Techniques for In-Situ Interface Characterization.* (H. D. Abruña, Ed.), Wiley-VCH, New York, 1991, pp. 339–410.

10 R. J. Lipart, B. D. Lamp, M. D. Porter; Specular reflection spectroscopy. in F. M. Mirabella, Ed.; *Modern Techniques in Applied Molecular Spectroscopy*, John Wiley & Sons, New York, 1998.

11 W. N. Hansen; Electric fields produced by the propagation of plane coherent electromagnetic radiation in a stratified medium. *J. Opt. Soc. Am.* **1968**, *58*, 380–390.

12 M. Moskovits; Surface selection rules. *J. Chem. Phys.* **1982**, *77*, 4408–4416. Correction: *J. Chem. Phys.* **1983**, *79*, 1558.

13 I. Brand; *Application of Polarization Modulation Infrared Reflection Absorption Spectroscopy in Electrochemistry*. Monographs in Electrochemistry (F. Scholz, Series Ed.), Springer, Cham, 2020.

14 M. Tasumi, A. Sakamoto (Eds.); *Introduction to experimental infrared spectroscopy: Fundamentals and practical methods*, John Wiley & Sons, Chichester, 2015.

15 P. A. Christensen; In situ infrared spectroelectrochemistry. in *Encyclopedia of Electrochemistry. Vol. 3, Instrumentation and Electroanalytical Chemistry* (A. J. Bard, M. Stratmann, P. R. Unwin, Eds.), Wiley-VCH, Weinheim, 2003, pp. 530–571.

16 V. Zamlynny, J. Lipkowski; Quantitative SNIFTIRS and PM IRRAS of organic molecules at electrode surfaces. in *Advances in Electrochemical Science and Engineering, Vol. 9*, (R. C. Alkire, D. M. Kolb, J. Lipkowski, P. N. Ross, Eds.), Wiley-VCH, Weinheim, 2006, pp. 315–376.

17 R. Wiesinger, U. Schade, C. Kleber, M. Schreiner; An experimental set-up to apply polarization modulation to infrared reflection absorption spectroscopy for improved in situ studies of atmospheric corrosion processes. *Rev. Sci. Instrum.* **2014**, *85*, 064102/1-8.

18 I. Brand; Application of infrared spectroscopy for structural analysis of planar lipid bilayers under electrochemical control. in *Advances in Planar Lipid Bilayers and Liposomes, Vol. 18*, (A. Iglic, C. V. Kulkarni, Eds.), Academic Press, New York, 2013, pp. 21–62.

19 J. F. Foley, C. Korzeniewski, J. L. Daschbach, S. Pons; Infrared vibrational spectroscopy of the electrode–solution interface. in *Electroanalytical Chemistry, Vol. 14*, (A. J. Bard, Ed.), Marcel Dekker, New York, 1986, pp. 309–440.

20 K. Ashley, S. Pons; Infrared spectroelectrochemistry. *Chem. Rev.* **1988**, *88*, 673–695.

21 F. C. Nart, T. Iwasita; In situ FTIR as a tool for mechanistic studies. Fundamentals and application. in *Encyclopedia of Electrochemistry. Vol. 2, Interfacial Kinetics and Mass Transport* (A. J. Bard, M. Stratmann, E. J. Calvo, Eds.), Wiley-VCH, Weinheim, 2003, pp. 243–289.

22 S.-G. Sun; Studying electrocatalytic oxidation of small organic molecules with in-situ infrared spectroscopy. in *Electrocatalysis* (J. Lipkowski, P. N. Ross, Eds.), Wiley-VCH, New York, 1998, pp. 243–290.

23 T. Iwasita, E. Pastor; A Fourier transform infrared spectroscopy approach to double-layer and electrocatalysis problems. in *Interfacial Electrochemistry, Theory, Experiment and Application* (A. Wieckowski, Ed.), Marcel Dekker, New York, 1999, pp. 353–372.

24 A. Rodes, J. M. Pérez, A. Aldaz; Vibrational spectroscopy. in *Handbook of Fuel Cells – Fundamentals, Technology and Applications, Vol. 2*, (W. Vielstich, A. Lamm, H. A. Gasteiger, Eds.), John Wiley & Sons, Chichester, 2003, pp. 191–219.

25 M. J. Weaver, S. Zhou; Adsorbate bonding and reactivity in electrochemistry: Some vibrational spectroscopic links to surface science and interfacial chemistry. in *Interfacial Electrochemistry, Theory, Experiment and Application* (A. Wieckowski, Ed.), Marcel Dekker, New York, 1999, pp. 301–316.

26 A. Rodes, R. Gomez, J. M. Pérez, J. M. Feliu, A. Aldaz; On the voltammetric and spectroscopic characterization of nitric oxide adlayers formed from nitrous acid on Pt(h,k,l) and Rh(h,k,l) electrodes. *Electrochim. Acta* **1996**, *41*, 729–745.

27 K. Kunimatsu, W. G. Golden, H. Seki, M. R. Philipott; Carbon monoxide adsorption on a platinum electrode studied by polarization modulated FT-IRRAS. 1. Carbon monoxide adsorbed in the double-layer potential region and its oxidation in acids. *Langmuir* **1985**, *1*, 245–250.

28 W. G. Golden, D. D. Saperstein, M. W. Severson, J. Overend; Infrared reflection-absorption spectroscopy of surface species: A comparison of Fourier transform and dispersion methods. *J. Phys. Chem.* **1984**, *88*, 574–580.

29 C. Leygraf, J. Hedberg, P. Qiu, J. Hernandez, C. M. Johnson; Molecular in situ studies of atmospheric corrosion. *Corrosion* **2007**, *63*, 715–721.

30 N. Le Bozec, D. Persson, A. Nazarov, D. Thierry; Investigation of filiform corrosion on coated aluminum alloys by FTIR microspectroscopy and scanning Kelvin probe. *J. Electrochem. Soc.* **2002**, *149*, B403–B408.

31 I. Frateur, L. Lartundo-Rojas, C. Méthivier, A. Galtayries, P. Marcus; Influence of bovine serum albumin in sulphuric acid aqueous solution on the corrosion and the passivation of an iron-chromium alloy. *Electrochim. Acta* **2006**, *51*, 1550–1557.

32 M. P. Desimone, G. Grundmeier, G. Gordillo, S. N. Simison; Amphiphilic amido-amine as effective corrosion inhibitor for mild steel exposed to CO_2 saturated solution: Polarization, EIS and PM-IRRAS studies. *Electrochim. Acta* **2011**, *56*, 2990–2998.

33 Z.-Q. Tian, B. Ren; Raman spectroscopy of electrode surfaces. in *Encyclopedia of Electrochemistry. Vol. 3, Instrumentation and Electroanalytical Chemistry* (A. J. Bard, M. Stratmann, P. R. Unwin, Eds.), Wiley-VCH, Weinheim, 2003, p. 572–659.

34 R. M. Corn; In situ second harmonic generation studies of molecular orientation at electrode surfaces. In *Adsorption of Molecules at Metal Electrodes* (J. Lipkowski, P. N. Ross, Eds.), Wiley-VCH, New York, 1992, pp. 391–408.

35 G. L. Richmond; Investigation of electrochemical interfaces by nonlinear optical methods. in *Electrochemical Interfaces Modern Techniques for In-Situ Interface Characterization*. (H. D. Abruña, Ed.), Wiley-VCH, New York, 1991, pp. 265–337.

36 G. L. Richmond; Optical second harmonic generation as in situ probe of electrochemical interfaces. in *Electroanalytical Chemistry, Vol. 18*, (A. J. Bard, Ed.), Marcel Dekker, New York, 1991, pp. 87–180.

37 B. Pettinger, C. Bilger, J. Lipkowski, W. Schmickler; Second harmonic generation anisotropy from single crystalline electrode surfaces. in *Interfacial Electrochemistry, Theory, Experiment and Application* (A. Wieckowski, Ed.), Marcel Dekker, New York, 1999, pp. 373–404.

38 B. Pettinger, C. Bilger, J. Lipkowski; SHG studies on halide adsorption at Au(111) electrodes. in *Solid-Liquid Interphases – Macroscopic Phenomena – Microscopic Understanding. Topic in Applied Physics, Vol. 85*, (K. Wandelt, S. Thurgate, Eds.), Springer Verlag, Berlin, 2003, pp. 223–240.

39 A. Tadjeddine, A. Le Rille; Sum and difference frequency generation at electrode surfaces. in *Interfacial Electrochemistry, Theory, Experiment and Application* (A. Wieckowski, Ed.), Marcel Dekker, New York, 1999, pp. 317–352.

40 S. Ye, K. Uosaki; Sum frequency generation (SFG) evaluation of the chemically modified solid surface. in *Encyclopedia of Electrochemistry, Vol. 10, Modified Electrodes*. (A. J. Bard, M. Stratmann, M. Fujihara, J. R. Rusling, I. Rubinstein, Eds.), Wiley-VCH, Weinheim, 2007, pp. 513–551.

41 N. Batina, D. G. Frank, J. Y. Gui, B. E. Kahn, C.-H. Lin, F. Lu, J. W. McCargar, G. N. Salatia, D. A. Stern, D. C. Zapien, A. T. Hubbard; Oriented adsorption at well-defined electrode surfaces studied by Auger, LEED, and EELS spectroscopy. *Electrochim Acta* **1989**, *34*, 1031–1044.

42 B. A. Sexton; Identification of adsorbed species at metal surfaces by electron energy loss spectroscopy (EELS). *Appl. Phys. A* **1981**, *26*, 1–18.

43 D. G. Hanken, C. E. Jordan, B. L. Frey, R. M. Corn; Surface plasmon resonance measurements of ultrathin organic films at electrode surfaces. in *Electroanalytical Chemistry, Vol. 20*, (A. J. Bard, I. Rubinstein, Eds.), Marcel Dekker, New York, 1998, pp. 141–225.

44 N. Zhang, R. Schweiss, Y. Zong, W. Knoll; Electrochemical surface plasmon spectroscopy – Recent developments and applications. *Electrochim. Acta* **2007**, *52*, 2869–2875.

45 X. Liu, W. Yang, Z. Liu; Recent progress on synchrotron-based in-situ soft X-ray spectroscopy for energy materials. *Adv. Mater.* **2014**, *26*, 7710–7729.

46 S. Axnanda, E. Mao, S. Rani, R. Chang, P. G. Karlsson, M. O. M. Edwards, M. Lundqvist, R. Moberg, P. Ross, Z. Hussain, Z. Liu; Using "tender" X-ray ambient pressure X-ray photoelectron spectroscopy as a direct probe of solid–liquid interface. *Sci. Rep.* **2015**, *5*, 09788/1–12.

47 M. P. Seah, W. A. Dench; Quantitative electron spectroscopy on surfaces: A standard data base for electron ineleastic mean free paths in solids. *Surf. Interf. Anal.* **1979**, *38*, 76–83.

48 A. P. Chandra, A. R. Gerson; The mechanism of pyrite oxidation and leaching: A fundamental perspective. *Surf. Sci. Rep.* **2010**, *65*, 293–315.

49 E. J. Crumlin, E. Mutoro, W. T. Hong, M. D. Biegalski, H. M. Christen, Z. Liu, H. Bluhm, Y. Shao-Horn; In situ ambient pressure X-ray photoelectron spectroscopy of cobalt perovskite surfaces under cathodic polarization at high temperatures. *J. Phys. Chem. C* **2013**, *117*, 16087–16094.

50 J.-P. Veder, R. De Marco, K. Patel, P. Si, E. Gyrgolowicz-Pawlak, M. James, M. T. Alam, M. Sohail, J. Lee, E. Pretzsch, E. Bakker; Evidence for a surface confined Ion-to-electron transduction reaction in solid contact ion-selective electrodes based on poly(3-octylthiophene). *Anal. Chem.* **2013**, *85*, 10495–10502.

51 G. Wittstock, I. Kartio, D. Hirsch, S. Kunze, R. Szargan; Oxidation of galena in acetate buffer investigated by atomic force microscopy and photoelectron spectroscopy. *Langmuir* **1996**, *12*, 5709–5721.

52 G. V. Hansson, R. I. G. Uhrberg; Photoelectron spectroscopy of surface states on semiconductor surfaces. *Surf. Sci. Rep.* **1988**, *9*, 197–292.

53 P. van der Heide; *X-ray Photoelectron Spectroscopy – An Introduction to Principle and Practice*. John Wiley & Sons, Hoboken, 2012.

54 C. Xu, F. Lindgren, B. Philippe, M. Gorgoi, F. Björefors, K. Edström, T. Gustafsson; Improved performance of silicon anode for Li-ion batteries: Understanding the surface modification mechanism of fluorethylene carbonate as an effective electrolyte additive. *Chem. Mater.* **2015**, *27*, 2591–2599.

55 T. Solomun; W. Richtering, H. Gerischer; Electrochemical reactivity of ordered and disordered *n*-GaAs(1 1 0) surfaces. A combined XPS, LEED and electrochemical study. *Ber. Bunsenges. Phys. Chem.* **1987**, *91*, 412–416.

56 J.-M. Abels, D. Hecht, H.-H. Strehblow; A closed specimen transfer for surface analytical application. *Surf. Interface Anal.* **1996**, *24*, 332–338.

57 J. T. Grant; Appendix I. Databases. in *Surface Analysis by Auger and X-ray Photoelectron Spectroscopy*. (D. Briggs, J. T. Grant, Eds.), IM Publications and Surface Spectra Ltd., Charlton, 2003, pp. 869–873.

58 A. V. Naumkin, A. Kraut-Vaas, S. W. Gaarenstroom, C. J. Powell; *NIST X-ray Photoelectron Spectroscopy Database*. https://srdata.nist.gov/xps/ (last access 23 Dec 2024).

59 J. T. Grant; Appendix H. Software for processing AES and XPS data. in *Surface Analysis by Auger and X-ray Photoelectron Spectroscopy*. D. Briggs, J. T. Grant, Eds.), IM Publications and Surface Spectra Ltd., Charlton, 2003, pp. 861–868.

60 S. Hofmann; *Auger and X-Ray Photoelectron Spectroscopy in Materials Science – A User-Oriented Guide*. Springer, Berlin, 2013, pp. 77–204.

61 C. D. Wagner, L. H. Gale, R. H. Raymonds; Two-dimensional chemical state plots: A standardized data set for use in identifying chemical states by X-ray photoelectron spectroscopy. *Anal. Chem.* **1979**, *51*, 466–482.

62 S. Sauter, G. Wittstock, R. Szargan; Localisation of electrochemical oxidation processes in nickel and cobalt hexacyanoferrates investigated by analysis of the multiplet patterns in X-ray photoelectron spectra. *Phys. Chem. Chem. Phys.* **2001**, *3*, 562–569.

63 P. M. Shearwood; Curve fitting in surface analysis and the effect of background inclusion in the fitting process. *J. Vac. Sci. Technol. A* **1996**, *14*, 1424–1432.

64 M. P. Seah; in *Practical Surface Analysis, Vol. 1, Auger and X-ray Photoelectron Spectroscopy*, (D. Briggs, M. P. Seah, Eds.), Wiley, Chichester. 1990, pp. 223–251.

65 J. H. Scofield; Hartree-Slater subshell photoionization cross-sections at 1254 and 1487 eV. *J. Electron Spectr. Rel. Phenom.* **1976**, *8*, 129–137.

66 P. J. Cumpson; Angle-resolved XPS and AES: Depth-resolution limits and general comparison of properties of depth-profile reconstruction methods. *J. Electron Spectrosc. Rel. Phenomen.* **1995**, *73*, 25–52.

67 N. Vandencasteele, F. Reniers; Plasma-modified polymer surfaces: Characterization using XPS. *J. Electron Spectrosc. Rel. Phenomen.* **2010**, *178/179*, 394–408.

68 D. Briggs, J. T. Grant (Eds.); *Surface Analysis by Auger and X-Ray Photoelectron Spectroscopy*. IM Publication and Surface Spectra Ltd, Chichester, 2003.

69 D. Briggs, M. P. Seah (Eds.); *Practical Surface Analysis, Vol. 1 – Auger and X-ray Photoelectron Spectroscopy*, 2nd ed., John Wiley & Sons, Chichester, 1983.

70 S. Hofmann; *Auger and X-Ray Photoelectron Spectroscopy in Materials Science – A User-Oriented Guide*. Springer Verlag, Berlin, 2013.

71 D. Zanchet, B. D. Hall, D. Ugarte; X-ray characterization of nanoparticles, in *Characterization of Nanophase Materials*. (Z. L. Wang, Ed.), Wiley-VCH, Weinheim, 2000, pp. 13–36.

72 A. Minguzzi, P. Ghigna; X-ray absorption spectroscopy in Electrochemistry – From fundamentals to fixed energy X-ray absorption voltammetry. in *Electroanalytical Chemistry, Vol. 27*, (A. J. Bard, C. G. Zoski, Eds.), CRC Press, Boca Raton, 2017, pp. 119–181.

73 L. R. Sharpe, W. R. Heineman, R. C. Elder; EXAFS spectroelectrochemistry. *Chem. Rev.* **1990**, *90*, 705–722.

74 J. McBreen, S. Mukerjee; In situ X-ray absorption studies of carbon-supported Pt and Pt alloy nanoparticles. in *Interfacial Electrochemistry, Theory, Experiment and Application* (A. Wieckowski, Ed.), Marcel Dekker, New York, 1999, pp. 895–914.

75 A. E. Russels, A. Rose; X-ray absorption spectroscopy of low temperature fuel cell catalysts. *Chem. Rev.* **2004**, *104*, 4613–4635.

76 H. D. Abruña; X-ray absorption spectroscopy in the study of electrochemical systems. in *Electrochemical Interfaces Modern Techniques for In-Situ Interface Characterization.* (H. D. Abruña, Ed.), Wiley-VCH, New York, 1991, pp. 1–54.

77 C. H. M. van Oversteeg, H. Q. Doan, F. M. F. de Groot, T. Cuk; In situ X-ray absorption spectroscopy of transition metal based water oxidation catalysts. *Chem. Soc. Rev.* **2017**, *46*, 102–125.

78 O. M. Magnussen, J. Drnec, C. Qiu, I. Martens, J. J. Huang, R. Cattot, A. Singer; In situ and operando X-ray scattering methods in electrochemistry and electrocatalysis. *Chem. Rev.* **2024**, *124*, 629–721.

79 L. R. Clarke; *Surface Crystallography: An Introduction to Low-Energy Electron Diffraction.* John Wiley & Sons, New York, 1985, pp. 1–72.

80 G. Lehmpfuhl, Y. Uchida, M. S. Zei, D. M. Kolb; Electron diffraction and electron microscopy of electrode surfaces. in *Imaging of Surfaces and Interfaces* (J. Lipkowski, P. N. Ross, Eds.), Wiley-VCH, New York, 1999, pp. 57–98.

81 D. A. Buttry; Application of the quartz crystal microbalance to electrochemistry. in *Electroanalytical Chemistry, Vol. 17*, (A. J. Bard, Ed.); Marcel Dekker, New York, 1991, pp. 1–85.

82 A. R. Hillmann; The electrochemical quartz crystal microbalance. in *Encyclopedia of Electrochemistry, Vol. 3. Instrumentation and Electroanalytical Chemistry* (A. J. Bard, M. Stratmann, P. R. Unwin, Eds.), Wiley-VCH, Weinheim, 2003, pp. 230–289.

83 V. Tsionsky, L. Daikhin, M. Urbakh, E. Gileadi; Looking at the metal/solution interface with the electrochemical quartz crystal microbalance: Theory and experiment. in *Electroanalytical Chemistry, Vol. 22*, (A. J. Bard, I. Rubinstein, Eds.), Marcel Dekker, New York, 2004, pp. 1–99.

84 M. Kaspar, H. Stadler, T. Weiß, C. Ziegler; Thickness shear mode resonators ("mass-sensitive devices") in bioanalysis. *Fresenius J. Anal. Chem.* **2000**, *366*, 602–610.

85 https://en.wikipedia.org/wiki/Crystal_oscillator (last access 23 Dec 2024).

86 K. K. Kanazawa, J. G. Gordon; Frequency of a quartz microbalance in contact with liquid. *Anal. Chem.* **1985**, *57*, 1770–1771.

87 G. S. Ostrom, D. A. Buttry; Manipulation of the ionic transport processes during redox cycling for polymer/polymer complex films of poly(styrenesulfonate)/poly(xylylviologen). *J. Phys. Chem.* **1995**, *99*, 15236–15240.

88 N. Shpigel, M. D. Levi, S. Sigalov, L. Daikhin, D. Aurbach; In situ real-time mechanical and morphological characterization of electrodes for electrochemical energy storage and conversion by electrochemical quartz crystal microbalance with dissipation monitoring; *Acc. Chem. Res.* **2018**, *51*, 69–79.

89 P. Hirel; Atomsk: A tool for manipulating and converting atomic data files. *Comput. Phys. Comm.* **2015**, *197*, 212–219.

90 D. M. Kolb; Reconstruction phenomena at metal–electrolyte interfaces; *Prog. Surf. Sci.* **1996**, *51*, 109–173.

91 A. S. Dakkouri, D. M. Kolb; Reconstruction of gold surface. in *Interfacial Electrochemistry, Theory, Experiment and Application* (A. Wieckowski, Ed.), Marcel Dekker, New York, 1999, pp. 151–173.

92 G. Staikov, W. J. Lorenz, E. Budevski; Low-dimensional phases and nanostructures of solid surfaces. in *Imaging of Surfaces and Interfaces* (J. Lipkowski, P. N. Ross, Eds.), Wiley-VCH, New York, 1999, pp. 1–56.

93 L. A. Kibler; *Preparation and Characterization of Noble Metal Single Crystal Electrodes.* Ulm 2003. Available under www.uni-ulm.de/fileadmin/website_uni_ulm/nawi.inst.080/mitarbeiter/Kibler/KiblerSingleCrystals2003.pdf (last access 23 Dec 2024).

94 J. Clavilier, R. Faure, G. Guinet, R. Durand; Preparation of monocrystalline Pt microelectrodes and electrochemical study of the plane surfaces cut in the direction of the {1 1 1} and {1 1 0} planes. *J. Electroanal. Chem.* **1980**, *107*, 211–216.

95 J. Clavilier; Flame-annealing and cleaning technique. in *Interfacial Electrochemistry, Theory, Experiment and Application* (A. Wieckowski, Ed.), Marcel Dekker, New York, 1999, pp. 231–248.

96 V. Climent, J. M. Feliu; Thirty years of platinum single crystal electrochemistry. *J. Solid State Electrochem.* **2011**, *15*, 1297–1315.

97 V. Climent, J. M. Feliu; Surface electrochemistry with Pt single crystal electrodes. in *Advances in Electrochemical Science and Engineering*, *Vol. 17*, (R. C. Alkire, P. N. Bartlett, J. Lipowski, Eds.), Wiley-VCH, Weinheim, 2017, pp. 1–57.

98 J. Solla-Gullón, P. Rodríguez, E. Herrero, A. Aldaz, J. M. Feliu; Surface characterization of platinum electrodes. *Phys. Chem. Chem. Phys.* **2008**, *10*, 1359–1373.

99 R. I. Masel; *Principles of Adsorption and Reaction on Solid Surfaces.* Wiley-Interscience, New York, 1996.

100 O. M. Magnussen; Ordered anion adlayer on metal electrode surfaces. *Chem. Rev.* **2002**, *102*, 679–726.

101 W. Lorenz, G. Salie; Zur Anwendung der Gibbschen Adsorptionsthermodynamik auf Adsorptionsreaktionen mit partiellem Landungsübergang. *Z. Phys. Chem.* **1962**, *219O*, 421–423.

102 W. Plieth; *Electrochemistry for Materials Science*, Elsevier, Amsterdam, 2008, pp. 110–113.

103 W. Schmickler, E. Santos; *Interfacial Electrochemistry*. 2nd ed., Springer Verlag, Berlin, 2010, pp. 59–62.

104 E. Budevski, G. Staikov, W. J. Lorenz; *Electrochemical Phase Formation and Growth – An Introduction to the Initial Stages of Metal Deposition*. Wiley-VCH, Weinheim, 1996, pp. 43–148.

105 Budevski, G. Staikov, W. J. Lorenz; *Electrochemical Phase Formation and Growth – An Introduction to the Initial Stages of Metal Deposition*, Wiley-VCH, Weinheim, 1996, pp. 317–327.

106 G. Staikov, W. J. Lorenz, E. Budevski; Low-dimensional phases and nanostructures of solid surfaces. in *Imaging of Surfaces and Interfaces* (J. Lipkowski, P. N. Ross, Eds.), Wiley-VCH, New York, 1999, p. 24.

107 S. Ye, K. Uosaki; Underpotential deposition. in *Encyclopedia of Electrochemistry. Vol. 1*, (A. J. Bard, M. Stratmann, E. Gileadi, M. Urbakh, Eds.), Wiley-VCH, Weinheim, 2002, pp. 469–512.

108 O. A. Oviedo, L. Reinaudi, S. G. García, E. P. M. Leiva; *Underpotential Deposition – From Fundamentals and Theory to Applications at the Nanoscale*, Springer, Cham, 2016.

109 C. Korzeniewski, V. Climt, J. M. Feliu; Electrochemistry at platinum single crystal electrodes. in *Electroanalytical Chemistry, Vol. 24*, (A. J. Bard, C. Zoski, Eds.), CRC Press, Boca Raton, 2012, pp. 75–169.

110 A. Hamlin; Cyclic voltammetry at gold single-crystal surfaces. Part 1. Behavior at low-index faces. *J. Electroanal. Chem.* **1996**, *407*, 1–11.

111 A. Hamlin; Cyclic voltammetry at gold single-crystal surfaces. Part 2 Behavior at high-index faces. *J. Electroanal. Chem.* **1996**, *407*, 13–21.

112 A. Hamelin; Underpotential deposition of lead on single crystal faces of gold. Part 1. The influence of crystallographic orientation of the substrate. *J. Electroanal. Chem.* **1984**, *165*, 167–180.

113 D. C. Grahame; The electrical double layer and the theory of electrocapillarity. *Chem. Rev.* **1947**, *41*, 441–501.

114 D. M. Mohilner; The electrical double layer. Part I. Elements of double-layer theory. in *Electroanalytical Chemistry, Vol. 1*, (A. J. Bard, Ed.), Marcel Dekker, New York, 1966, pp. 241–409.

115 A. J. Bard, L. R. Faulkner, H. S. White; *Electrochemical Methods – Fundamentals and Applications*. 3rd ed., John Wiley & Sons, Hoboken, 2022, pp. 599–651.

116 N. Eliaz, E. Gileadi; *Physical Electrochemistry: Fundamentals, Techniques, and Applications*. 2nd ed., Wiley-VCH, Weinheim, 2019, pp. 97–122.

117 H. H. Girault; *Analytical and Physical Electrochemistry*, EPFL Press, distributed by Marcel Dekker, Lausanne, 2004, pp. 177–220.

118 D. Bizotto, V. Zamlynny, I. Burgess, C. A. Jeffrey, H.-Q. Li, J. Rubinstein, R. A. Merril, J. Lipkowski, Z. Galus, A. Nelson, B. Pettinger; Amphiphilic and ionic surfactants at electrified interfaces. in *Interfacial Electrochemistry – Theory, Experiment, and Application*. (A. Wieckowski, Ed.), Marcel Dekker, New York, 1999, pp. 405–426.

119 Z. Su, J. Leitch, J. Lipkowski; Measurement of the potentials of zero free charge and zero total charge for 1-thio-β-D-glucose and DPTL modified Au(111) surface in different electrolyte solutions. *Z. Phys. Chem.* **2012**, *226*, 995–1009.

120 E. Lust; Electrical double layers at single-crystals and polycrystalline electrodes. in *Encyclopedia of Electrochemistry, Vol. 1*, (E. Gileadi, M. Urbakh, A. J. Bard, A. Stratmann, Eds.), Wiley-VCH, Weinheim, 2002, pp. 188–224.

121 N. A. Balashova, V. E. Kazarinov; Use of radioactive-tracer method for the investigation of the electric double-layer structure. in *Electroanalytical Chemistry, Vol. 3*, (A. J. Bard, Ed.), Marcel Dekker, New York, 1969, pp. 135–197.

122 G. Horányi; Radiotracer studies of adsorption/sorption phenomena at electrode surfaces. in *Interfacial Electrochemistry – Theory, Experiment, and Application*. (A. Wieckowski, Ed.), Marcel Dekker, New York, 1999, pp. 477–491.

123 K. R. Riley, M. P. Hobeson, S. J. Bence; *Mathematical Methods for Physics and Engineering*. 3rd ed. Cambridge University Press, Cambridge, 2024, pp. 453–459.

124 M. L. Berkowitz, I.-C. Yeh, E. Spohr; Structure of water at the water-metal interface: Molecular dynamics computer simulations. in *Interfacial Electrochemistry – Theory, Experiment, and Application*. (A. Wieckowski, Ed.), Marcel Dekker, New York, 1999, pp. 33–45.

125 G. A. Tsirlina; The role of supporting electrolyte in heterogeneous electron transfer. *J. Solid State Electrochem.* **2017**, *21*, 1833–1845.

126 W. R. Fawcett; Double layer effect on the electrode kinetics of electron and ion transfer reactions. in *Electrocatalysis* (J. Lipkowski, P. N. Ross, Eds.), Wiley-VCH, New York, 1998, pp. 323–371.

127 A. A. Kornyshev, E. Spohr, A. A. Vorotynzsev; Electrochemical interfaces: At the border line. in *Encyclopedia of Electrochemistry, Vol. 1*, (E. Gileadi, M. Urbakh, A. J. Bard, A. Stratmann, Eds.), Wiley-VCH, Weinheim, 2002, pp. 33–132.

128 W. Schmickler; Electrical double layers: Theory and simulations; in *Encyclopedia of Electrochemistry, Vol. 1*, (E. Gileadi, M. Urbakh, A. J. Bard, A. Stratmann, Eds.), Wiley-VCH, Weinheim, 2002, pp. 133–224.

129 K. Shimizu, T. Mitani; *New Horizons of Applied Scanning Electron Microscopy*, Springer, Berlin, 2010.

130 A. Ul-Hamid; *A Beginners' Guide to Scanning Electron Microscopy*, Springer, Cham, 2018.

131 D. B. Williams, C. B. Carter; *Transmission Electron Microscopy, Vol. 1–4*, 2nd ed., Springer, New York, 2009.

132 S. Szunerits, S. E. Pust, G. Wittstock; Multidimensional electrochemical imaging in material science. *Anal. Bioanal. Chem.* **2007**, *389*, 1103–1120.

133 W. Schmickler; Theoretical aspects of the scanning tunneling microscope operating in an electrolyte solution. in *Imaging of Surfaces and Interfaces* (J. Lipkowski, P. N. Ross, Eds.), Wiley-VCH, New York, 1999, pp. 305–338.

134 P. Broekmann, M. Wilms, M. Kruft, C. Stuhlmann, K. Wandelt; In-situ STM investigation of specific anion adsorption on Cu(111). *J. Electroanal. Chem.* **1999**, *467*, 307–324.

135 O. M. Magnussen, B. M. Ocko, J. X. Wang, T. R. Adziz; In-situ X-ray diffraction and STM studies of bromide adsorption on Au(111) electrodes. *J. Phys. Chem.* **1996**, *100*, 5500–5508.

136 T. Tansel, O. M. Magnussen; Video STM studies of adsorbate diffusion at electrochemical interfaces. *Phys. Rev. Lett.* **2006**, *96*, 026101.

137 F. Cunha, N. J. Tao; Surface charge induced order-disorder transitions in an organic monolayer. *Phys. Rev. Lett.* **1995**, *75*, 2376.

138 T. Dretschkow, T. Wandlowski; Structural transitions in organic adlayers – A molecular view. in *Solid-Liquid Interphases – Macroscopic Phenomena – Microscopic Understanding. Topic in Applied Physics, Vol. 85*, (K. Wandelt, S. Thurgate, Eds.), Springer Verlag, Berlin, 2003, pp. 259–321.

139 E. P. Friis, M. H. Thuesen, J. Zhang, J. Ulstrup; Interfacial electron transfer in electrochemistry and in situ scanning tunneling microscopy. in *Interfacial Electrochemistry – Theory, Experiment, and Application.* (A. Wieckowski, Ed.), Marcel Dekker, New York, 1999, pp. 115–129.

140 R. Nichols; Imaging metal electrocrystallization at high resolution. in *Imaging of Surfaces and Interfaces* (J. Lipkowski, P. N. Ross, Eds.), Wiley-VCH, New York, 1999, pp. 99–137.

141 P. Allongue, F. Maroun; Electrodeposition of two-dimensional magnetic nanostructures on single crystal electrode surfaces. in *Electrocrystallization in Nanotechnology* (G. Staikov, Ed.), Wiley-VCH, Weinheim, 2007, pp. 217–241.

142 P. Marcus, V. Maurice; Structure of thin anodic films formed on single-crystal metal surfaces. in *Interfacial Electrochemistry – Theory, Experiment, and Application.* (A. Wieckowski, Ed.), Marcel Dekker, New York, 1999, pp. 541–558.

143 W. Schindler, P. Hugelmann; Localized electrocrystallization of metals by STM tip nanoelectrodes. in *Electrocrystallization in Nanotechnology* (G. Staikov, Ed.), Wiley-VCH, Weinheim, 2007, pp. 117–137.

144 D. M. Kolb, F. C. Simenone; Electrochemical nanostructuring with an STM: A status report. *Electrochim. Acta* **2005**, *50*, 2989–2996.

145 H. Siegenthaler; STM in electrochemistry. in *Scanning Tunneling Microscopy II* (R. Wiesendanger, H. J. Gunterodt, Eds.), Springer Verlag, New York, 1992, pp. 7–49.

146 J. Lipkowski, P. N. Ross, Eds.; *Imaging of Surfaces and Interfaces*, Wiley-VCH, New York, 1999.

147 T. P. Moffat; Scanning tunneling microscopy studies of metal electrodes. in *Electroanalytical Chemistry, Vol. 21*, (A. J. Bard, I. Rubinstein, Eds.), Marcel Dekker, New York, 1999, pp. 211–316.

148 A. J. Bard, F.-R. F. Fan; Application in electrochemistry. in *Scanning Probe Microscopy and Spectroscopy – Theory, Techniques, and Applications*, 2nd ed. (D. A. Bonnell, Ed.), Wiley-VCH, New York, 2001, pp. 421–452.

149 R. J. Colton, A. Engel, J. E. Frommer, H. E. Gaub, A. A. Gewirth, R. Guckenberger, J. Rabe, W. M. Heckl, B. Parkinson; *Procedures in Scanning Probe Microscopies.* John Wiley & Sons, Chichester, 1998.

150 A. A. Gewirth, B. K. Niece; Electrochemical application of in situ scanning probe microscopy. *Chem. Rev.* **1997**, *97*, 1129–1162.

151 K. Itaya; In situ scanning tunneling microscopy in electrolyte solution. *Prog. Surf. Sci.* **1998**, *58*, 121–248.

152 T. P. Moffat; Electrochemical STM. in *Encyclopedia of Electrochemistry, Vol. 3*, (A. J. Bard, M. Stratmann, P. R. Unwin, Eds.), Wiley-VCH, Weinheim, 2003, pp. 393–414.

153 F. Moreno-Herrrero, J. Gomez-Herrero; AFM: Basic concepts. in *Atomic Force Microscopy in Liquid – Biological Application*. (A. M. Baró, R. G. Reifenberger, Eds.), Wiley-VCH, Weinheim, 2012.

154 G. Haugstad; *Atomic Force Microscopy – Understanding Basic Modes and Advanced Application*. John Wiley & Sons, Hoboken, NJ, 2012.

155 J. V. Macpherson; Electrochemical AFM; in *Encyclopedia of Electrochemistry, Vol. 3*, (A. J. Bard, M. Stratmann, P. R. Unwin, Eds.), Wiley-VCH, Weinheim, 2003, pp. 415–443.

156 E. Aleksandrova, R. Hiesgen, K. A. Friedrich, E. Roduner; Electrochemical atomic force microscopy study of proton conductivity in a Nafion membrane. *Phys. Chem. Chem. Phys.* **2007**, *9*, 2735–2743.

157 S. V. Kalinin; N. Balke; Local electrochemical functionality in energy storage materials and devices by scanning probe microscopies: Status and perspectives. *Adv. Mater.* **2010**, *22*, E193–E209.

158 C. Kranz; Recent advancements in nanoelectrodes and nanopipettes und in combined scanning electrochemical microscopy techniques. *Analyst* **2014**, *139*, 336–352.

159 S. Krämer, R. F. Fuierer, C. B. Gorman; Scanning probe lithography. *Chem. Rev.* **2003**, *103*, 4367–4418.

160 Park systems; https://www.parksystems.com/en/learning-center/webinar (last access 23 Dec 2024).

161 Bruker; https://www.bruker.com/en/products-and-solutions/microscopes/materials-afm/afm-webinars.html (last access 23 Dec 2024).

162 Oxford Instruments; https://afm.oxinst.com/webinars/ (last access 23 Oct 2024).

163 Nanosurf; https://www.nanosurf.com/en/resources/webinars (last access 23 Dec 2024).

164 P. Eaton, P. West; *Atomic Force Microscopy*. Oxford University Press, Oxford, 2010.

165 W. C. Sanders; *Atomic Force Microscopy: Fundamentals Concepts and Laboratory Investigations*. CRC Press, Boca Raton, FL, 2020.

166 A. M. Baró, R. G. Reifenberger, Eds.; *Atomic Force Microscopy in Liquid – Biological Application*. Wiley-VCH, Weinheim, 2012.

167 A. J. Bard, M. V. Mirkin (Eds.); *Scanning Electrochemical Microscopy*. 1st ed., Marcel Dekker, New York, 2001.

168 A. J. Bard, M. V. Mirkin (Eds.); *Scanning Electrochemical Microscopy*. 2nd ed., CRC Press, Boca Raton, FL, 2012.

169 A. J. Bard, M. V. Mirkin (Eds.); *Scanning Electrochemical Microscopy*. 3rd ed., CRC Press Boca Raton, FL, 2022.

170 B. R. Horrocks; Scanning electrochemical microscopy, in *Encyclopedia of Electrochemistry, Vol. 3*, (A. J. Bard, M. Stratmann, P. R. Unwin, Eds.), Wiley-VCH, Weinheim, 2003, pp. 444–490.

171 G. Wittstock, M. Burchardt, S. E. Pust, Y. Shen, C. Zhao; Scanning electrochemical microscopy for direct imaging of reaction rates. *Angew. Chem. Int. Ed.* **2007**, *46*, 1584–1617.

172 D. Polcari, O. Dauphin-Ducharme, J. Mauzeroll; Scanning electrochemical microscopy: A comprehensive review of experimental parameters from 1989 to 2015. *Chem. Rev.* **2016**, *116*, 13234–13278.

173 R. Cornut and C. Lefrou; New analytical approximation of feedback approach curves with a microdisk SECM tip and irreversible kinetic reaction at the substrate. *J. Electroanal. Chem.*, **2008**, *621*, 178–184.

174 C. Nunes Kirchner, G. Wittstock; Kinetic analysis of titanium nitride thin films by scanning electrochemical microscopy. in *Electrochemical Sensor Analysis* (S. Alegret, A. Merkorçi, Eds.), *Comprehensive Analytical Chemistry, Vol. 49*, Elsevier, 2007, pp. e363–369 [Another equation for interpolation was used here. The general approach can also be executed with the newer equation of interpolation Eqs. (8.151)–(8.156)].

175 O. Sklyar, T. H. Treutler, N. Vlachopoulos, G. Wittstock; The geometry of nanometer-sized electrodes and its influence on the electrolytic currents and metal deposition process in scanning tunneling and scanning electrochemical microscopy. *Surf. Sci.*, **2005**, *597*, 181–195.

176 G. Wittstock, H. Emons, T. H. Ridgway, E. A. Blubaugh, W. R. Heineman; Development and experimental evaluation of a simple system for scanning electrochemical microscopy. *Anal. Chim. Acta* **1994**, *298*, 285–302.

177 A. Lesch, B. Vaske, F. Meiners, D. Momotenko, F. Cortés-Salazar, H. H. Girault, G. Wittstock; Parallel imaging and chemical template-free patterning of self-assembled monolayers with soft linear microelectrode arrays. *Angew. Chem. Int. Ed.* **2012**, *51*, 10413–10416.

178 S. E. Pust, D. Scharnweber, C. Nunes Kirchner, G. Wittstock; Heterogeneous distribution of reactivity on metallic biomaterials: Scanning probe microscopy studies of biphasic Ti alloy Ti6Al4V. *Adv. Mater.* **2007**, *19*, 878–882.

179 H. Bülter, F. Peters, J. Schwenzel, G. Wittstock; Spatiotemporal changes of the solid electrolyte interphase in lithium-ion batteries detected by scanning electrochemical microscopy. *Angew. Chem. Int. Ed.* **2014,** *53*, 10531–10535.

180 U. M. Tefashe, K. Nonomura, N. Vlachopoulos, A. Hagfeldt, G. Wittstock; Effect of cation on dye regeneration kinetics of N719-sensitized TiO_2 films in acetonitrile-based and ionic-liquid-based electrolytes investigated by scanning electrochemical microscopy. *J. Phys. Chem. C* **2012,** *116*, 4316–4323.

181 W. Nogala, K. Szot, M. Burchardt, F. Roelfs, J. Rogalski, M. Opallo, G. Wittstock; Feedback mode SECM study of laccase and bilirubin oxidase immobilised in a solgel processed silicate film. *Analyst* **2010**, *135*, 2051–2058.

182 C. A. Nijhuis, J. K. Sinha, G. Wittstock, J. Huskens, B. J. Ravoo, D. N. Reinhoudt; Controlling supramolecular assembly of redox-active dendrimers at molecular printboards by scanning electrochemical microscopy. *Langmuir* **2006**, *22*, 9770–9775.

183 M. J. W. Ludden, J. K. Sinha, G. Wittstock, D. N. Reinhoudt, J. Huskens; Control over binding stoichiometry and specificity in supramolecular immobilization of cytochrome c on a molecular printboard. *Org. Biomol. Chem.* **2008**, *6*, 1553–1557.

184 J. Rodriguez-Lopez, M. A. Alpuche-Aviles, A. J. Bard; Interrogation of surfaces for the quantification of adsorbed species on electrodes: Oxygen on gold and platinum in neutral media, *J. Am. Chem. Soc.* **2008**, *130*, 16985–16995.

185 J. Rodriguez-Lopez, A. J. Bard; Scanning electrochemical microscopy: Surface interrogation of adsorbed hydrogen and the open circuit catalytic decomposition of formic acid at platinum, *J. Am. Chem. Soc.* **2010**, *132*, 5121–5129.

186 Q. Wang, J. Rodriguez-Lopez, A. J. Bard; Reaction of Br_2 with adsorbed CO on Pt, studied by the surface interrogation mode of scanning electrochemical microscopy, *J. Am. Chem. Soc.* **2009**, *131*, 17046–17047.

187 J. Rodriguez-Lopez, A. Minguzzi, A. J. Bard; Reaction of various reductants with oxide films on Pt electrodes as studied by the surface interrogation mode of scanning electrochemical microscopy (SI-SECM): Possible validity of a Marcus relationship, *J. Phys. Chem. C* **2010**, *114*, 18645–18655.

188 H. S. Park, K. C. Leonard, A. J. Bard; Surface interrogation scanning electrochemical microscopy (SI-SECM) of photoelectrochemistry at a W/Mo-$BiVO_4$ semiconductor electrode: Quantification of hydroxyl radicals during water oxidation, *J. Phys. Chem. C* **2013**, *117*, 12093–12102.

189 H. S. Ahn, A. J. Bard; Surface interrogation scanning electrochemical microscopy of $Ni_{1-x}Fe_xOOH$ ($0 < x < 0.27$) oxygen evolving catalyst: Kinetics of the "fast" iron sites. *J. Am. Chem. Soc.* **2016**, *138*, 313–318.

190 J. Rodríguez-López; Surface interrogation mode of scanning electrochemical microscopy (SI-SECM): An approach to the study of adsorption and (electro)catalysis at electrodes. in *Electroanalytical Chemistry, Vol. 24*, (A. J. Bard, C. Zoski, Eds.), CRC Press, Boca Raton, FL, 2012, pp. 287–351.

191 D. Zigah, J. Rodríguez-Lopez, A. J. Bard; Quantification of photoelectrogenerated hydroxyl radical on TiO_2 by surface interrogation scanning electrochemical microscopy. *Phys. Chem. Chem. Phys.* **2012**, *14*, 12764–12722.

192 S. Rastgar, G. Wittstock; In situ microtitration of intermediates of water oxidation reaction at nanoparticles assembled at water/oil interfaces. *J. Phys. Chem. C* **2018**, *122*, 12963–12969.

193 C. Zhao, G. Wittstock; Scanning electrochemical microscopy of quinoprotein glucose dehydrogenase. *Anal. Chem.* **2004**, *76*, 3145–3154.

194 F. Conzuelo, K. Sliozberg, R. Gutowski, S. Grützke, M. Nebel, W. Schuhmann; High-resolution analysis of photoanodes for water splitting by means of scanning electrochemical microscopy. *Anal. Chem.* **2017**, *89*, 1222–1228.

195 J. Izquierdo, L. Nagy, I. Bitter, R. M. Souto, G. Nagy; Potentiometric scanning electrochemical microscopy for local characterization of the electrochemical behaviour of magnesium-based materials. *Electrochim. Acta* **2013**, *87*, 283–293.

196 H. S. White, F. Kanoufi; Imaging molecular transport across membranes. in *Scanning Electrochemical Microscopy*. 2rd ed. (A. J. Bard, M. V. Mirkin, Eds.); Marcel Dekker, New York 2012, pp. 233–273.

197 S. Lee, Y. Zhang, H. White, C. C. Harrel, C. R. Martin; Electrophoretic capture and detection of nanoparticles at the opening of a membrane pore using scanning electrochemical microscopy. *Anal. Chem.* **2004**, *76*, 6106–6115.

198 C. Zhao, G. Wittstock; Scanning electrochemical microscopy for detection of biosensor and biochip surfaces with immobilized pyrroloquinoline quinone (PQQ)-dependent glucose dehydrogenase as enzyme label. *Biosensors Bioelectron.* **2005**, *20*, 1277–1284.

199 Y. Shen, M. Träuble, G. Wittstock; Detection of hydrogen peroxide produced during electrochemical oxygen reduction using scanning electrochemical microscopy. *Anal. Chem.* **2008**, *80*, 750–759.

200 P. Schwager, S. Dongmo, D. Fenske, G. Wittstock; Reactive oxygen species formed in organic lithium–oxygen batteries. *Phys. Chem. Chem. Phys.* **2016**, *18*, 10774–10780.

201 B. Konkena, J. Masa, A. J. R. Botz, I. Sinev, W. Xia, J. Koßmann, R. Drautz, M. Muhler, W. Schuhmann; Metallic $NiPS_3$@NiOOH Core–Shell heterostructures as highly efficient and stable electrocatalyst for the oxygen evolution reaction. *ACS Catal.* **2017**, *7*, 229–237.

202 K. Eckhard, W. Schuhmann; Visualisation of O_2 consumption and H_2O_2 formation by means of SECM for the characterisation of fuel cell catalyst activity. *Electrochim. Acta* **2007**, *53*, 1164–1169.

203 E. M. El-Giar, R. A. Said, G. E. Bridges, D. J. Thomson; Localized electrochemical deposition of copper microstructures. *J. Electrochem. Soc.* **2000**, *147*, 586–591.

204 F.-R. F. Fan, A. J. Bard; Imaging of biological macromolecules on mica in humid air by scanning electrochemical microscopy. *Proc. Natl. Acad. Sci.* **1999**, *96*, 14222–14227.

205 O. E. Hüsser, D. H. Craston, A. J. Bard; Scanning electrochemical microscopy: High-resolution deposition and etching of metals. *J. Electrochem. Soc.* **1989**, *136*, 3222–3229.

206 T. Wilhelm, G. Wittstock; Patterns of functional proteins formed by local electrochemical desorption of self-assembled monolayers. *Electrochim. Acta* **2001**, *47*, 275–281.

207 F. Cao, J. Kim, A. J. Bard; Detection of short-lived cation radical intermediate in the electrochemical oxidation of N,N-dimethylaniline by scanning electrochemical microscopy. *J. Am. Chem. Soc.* **2014**, *136*, 18163–18169.

208 F. Zhou, A. J. Bard; Detection of electrohydrodimerization intermediate acrylonitrile radical anion by scanning electrochemical microscopy. *J. Am. Chem. Soc.* **1994**, *116*, 393–394.

209 J. Ufheil, C. Hess, K. Borgwarth, J. Heinze; Nanostructuring and nanoanalysis by scanning electrochemical microscopy (SECM). *Phys. Chem. Chem. Phys.* **2005**, *7*, 3185–3190.

210 Z. Tian, Z. Fen, Z. Tian, X. Zhuo, J. Mu, C. Li, H. Lin, B. Ren, Z. Xie, W. Hu; Confined etchant layer technique for two-dimensional lithography at high resolution using electrochemical scanning tunneling microscopy. *Faraday. Discuss.* **1992**, *94*, 37–44.

211 L. Zhang, X. Z. Ma, M. X. Lin, Y. Lin, G. H. Cao, J. Tang, Z. W. Tian; A comparative study of electrochemical micromachining of *n*-GaAs and *p*-Si by using confined etchant layer technique. *J. Phys. Chem. B* **2006**, *110*, 18432–18439.

212 G. Wittstock, W. Schuhmann; Formation and imaging of microscopic enzymatically active spots on an alkanethiolate-covered gold electrode by scanning electrochemical microscopy. *Anal. Chem.* **1996**, *69*, 5059–5066.

213 B. Ballesteros-Katemann, A. Schulte, W. Schuhmann; Constant-distance mode scanning electrochemical microscopy (SECM) — Part I: Adaptation of a non-optical shear-force-based positioning mode for SECM tips. *Chem. Eur. J.* **2003**, *9*, 2025–2033.

214 U. M. Tefashe, G. Wittstock; Quantitative characterization of shear force regulation for scanning electrochemical microscopy. *C. R. Chimie* **2013**, *16*, 7–14.

215 S. E. Pust, M. Solomo, E. Oesterschulze, G. Wittstock; Influence of electrode size and geometry on electrochemical experiments with combined SECM-SFM probes. *Nanotechnology* **2010**, *21*, 105709/1–12.

216 J. Izquierdo, P. Knittel, C. Kranz; Scanning electrochemical microscopy: An analytical perspective. *Anal. Bioanal. Chem.* **2018**, *410*, 307–324.

217 S. E. Pust, W. Maier, G. Wittstock; Investigation of localized catalytic and electrocatalytic and corrosion reactions with scanning electrochemical microscopy (SECM). *Z. Phys. Chem.* **2008**, *222*, 1463–1517.

218 M. B. Jensen, D. E. Tallman, Application of SECM to Corrosion Studies. in *Electroanalytical Chemistry, Vol. 24*, (A. J. Bard, C. G. Zoski, Eds.), CRC Press, Boca Raton, FL, 2012, pp. 171–286.

219 P. Schwager, H. Bülter, I. Plettenberg, G. Wittstock; Review of local in situ probing techniques for the interfaces of lithium-ion and lithium–oxygen batteries. *Energy Technol.* **2016**, *4*, 1472–1485.

220 L. Danis, S. M. Gateman, C. Kuss, S. B. Schougaard, J. Mauzeroll; Nanoscale measurement of lithium-ion-battery materials using scanning probe techniques. *ChemElectroChem* **2017**, *4*, 6–19.

221 E. Ventosa, W. Schuhmann; Scanning electrochemical microscopy of Li-ion batteries. *Phys. Chem. Chem. Phys.* **2015**, *17*, 28441–28450.

222 W. Schulte, S. Liu, I. Plettenberg, S. Kuhri, W. Lüke, W. Lehnert, G. Wittstock; Local evaluation of processed membrane electrode assemblies by scanning electrochemical microscopy. *J. Electrochem. Soc.* **2017**, *164*, F873–F878.

223 A. R. Zeradjanin, E. Ventosa, A. S. Bondarenko, W. Schuhmann; Evaluation of catalytic performance of gas-evolving electrodes using local electrochemical noise measurements. *ChemSusChem* **2012**, *5*, 1905–1911.

224 I. Schmidt, I. Plettenberg, D. Kimmich, H. Ellis, J. Witt, C. Dosche, G. Wittstock; Spatially resolved analysis of screen printed photoanodes of dye-sensitized solar cells by scanning electrochemical microscopy. *Electrochim. Acta* **2016**, *222*, 735–746.

225 H. Ellis, I. Schmidt, A. Hagfeldt, G. Wittstock, G. Boschloo; Influence of dye architecture of triphenylamine based organic dyes on the kinetics in dye-sensitized solar cells. *J. Phys. Chem. C* **2015**, *119*, 21775–21783.

226 H. Bülter, G. Denuault, S. Mátéfi-Tempfli, M. Mátéfi-Tempfli, C. Dosche, G. Wittstock; Electrochemical analysis of nanostructured iron oxides using cyclic voltammetry and scanning electrochemical microscopy. *Electrochim. Acta* **2016**, *222*, 1326–1334.

227 S. Rastgar, G. Wittstock; Characterization of photoactivity of nanostructured $BiVO_4$ at polarized liquid-liquid interfaces by scanning electrochemical microscopy. *J. Phys. Chem. C* **2017**, *121*, 25961–25948.

228 M. Shen, S. Amemiya; Molecular transport in membranes. in *Scanning Electrochemical Microscopy*. 3rd ed. (A. J. Bard, M. V. Mirkin, Eds.); CRC Press, Boca Raton, FL, 2022, pp. 181–205.

229 M. Gonsalves, A. L. Barker, J. V. Macpherson, P. R. Unwin, D. O'Hare, C. P. Winlove; Scanning electrochemical microscopy as a local probe of oxygen permeability in cartilage. *Biophys. J.* **2000**, *78*, 1578–1588.

230 J. V. Macpherson, P. R. Unwin; Scanning electrochemical microscopy as an in vitro technique for measuring convective flow rates across dentine and the efficacy of surface blocking treatments. *Electroanalysis* **2005**, *17*, 197–204.

231 G. Wittstock; Modification and characterization of artificially patterned enzymatically active surfaces by scanning electrochemical microscopy. *Fresenius J. Anal. Chem.* **2001**, *370*, 303–315.

232 G. Wittstock, K. Yu, H. B. Halsall, T. H. Ridgway, W. R. Heineman; Imaging of immobilized antibody layers with scanning electrochemical microscopy. *Anal. Chem.* **1995**, *67*, 3578–3582.

233 C. Zhao, J. Sinha, C. A. Wijayawardhana, G. Wittstock; Monitoring β-galactosidase activity by means of scanning electrochemical microscopy. *J. Electroanal. Chem.* **2004**, *561*, 83–91.

234 B. Liu, W. Cheng, S. A. Rotenberg, M. V. Mirkin; Scanning electrochemical microscopy of living cells: Part 2. Imaging redox and acid/basic reactivities. *J. Electroanal. Chem.* **2001**, *500*, 590–597.

235 Y. Matsumae, Y. Takahashi, K. Ino, H. Shiku, T. Matsue; Electrochemical monitoring of intracellular enzyme activity of single living mammalian cells by using a double mediator system. *Anal. Chim. Acta* **2014**, *842*, 20–26.

236 C. Du, T. Suduwella, I. Beaulieu, S. B. Schougaard, J. Mauzeroll; Scanning Electrochemical Microscopy of Living Cells. in *Scanning Electrochemical Microscopy*. 3rd ed. (A. J. Bard, M. V. Mirkin, Eds.), CRC Press, Boca Raton, FL, 2022, pp. 297–330.

237 M. Nebel, S. Grützke, N. Diab, A. Schulte, W. Schuhmann; Visualization of oxygen consumption of single living cells by scanning electrochemical microscopy: The influence of the faradaic tip reaction. *Angew. Chem. Int. Ed.* **2013**, *52*, 6355–6338.

238 D. Mandler; Micro- and nanopatterning using scanning electrochemical microscopy. in *Scanning Electrochemical Microscopy*. 3rd ed. (A. J. Bard, M. V. Mirkin, Eds.), CRC Press, Boca Raton, FL, 2022, pp. 379–418.

239 C. M. Hill, S. Pan; SECM techniques for locally interrogating the photocatalytic activity of semiconducting materials for solar driven chemical transformations. in *Scanning Electrochemical Microscopy*. 3rd ed. (A. J. Bard, M. V. Mirkin, Eds.), CRC Press, Boca Raton, FL, 2022, pp. 361–378.

240 G. Wittstock, S. Rastgar, S. Scarabino; Local studies of photoelectrochemical reactions at nanostructured oxides. *Curr. Opin. Electrochem.* **2019**, *13*, 25–32.

241 D. V. Esposito, J. B. Baxter, J. John, N. S. Lewis, T. P. Moffat, T. Ogitsu, G. D. O'Neil, T. A. Pham, A. A. Talin, J. M. Velazquez, B. C. Wood; Methods of photoelectrochemical characterization with high spatial and temporal resolution. *Energy Environ. Sci.* **2015**, *8*, 2863–2885.

242 P. K. Hansma, B. Drake, O. Marti, S. A. C. Gould, C. B. Prater; The scanning ion-conductance microscopy. *Science* **1989**, *243*, 641–643.

243 P. Novak, C. Li, A. I. Shevchuk, R. Stepanyan, M. Caldwell, S. Hughes, T. G. Smart, J. Gorelik, V. P. Ostanin, M. L. Lab, G. W. J. Moss, G. I. Frolenko, D. Klenerman, Y. E. Korchev; Nanoscale live-cell imaging using hoping probe ion conductance microscopy. *Nat. Methods* **2009**, *6*, 279–281.

244 M. Kang, D. Momotenko, A. Page, D. Perry, P. R. Unwin; Frontiers in nanoscale electrochemical imaging: Faster, multifunctional, and ultrasensitive. *Langmuir* **2016**, *32*, 7993–8008.

245 M. M. Lohrengel. A. Moehring, M. Pilarski; Electrochemical surface analysis with the scanning droplet cell. *Fresenius J. Anal. Chem.* **2000**, *367*, 334–339.

246 Y. Takahashi, A. Kumatami, H. Shiku, T. Matsue; Scanning probe microscopy for nanoscale electrochemical imaging. *Anal. Chem.* **2017**, *89*, 342–357.

247 L. C. Yule, V. Shkirskiy, J. Aarons, G. West, B. A. Shollock, C. L. Bentley, P. R. Unwin; Nanoscale electrochemical visualization of grain-dependent anodic iron dissolution from low carbon steel. *Electrochim. Acta* **2020**, *332*, 135267/1-12.

248 I. Plettenberg, G. Wittstock; Combined detection of electrochemical reactions and topographical effects – imaging with scanning ohmic microscopy. *Electrochim. Acta* **2016**, *197*, 318–329.

249 R. S. Lillard; Scanning electrode techniques for investigating near-surface solution current densities. in *Analytical Methods in Corrosion Science and Engineering* (P. Marcus, F. Mansfeld, Eds.), Taylor & Francis, 2006, pp. 571–604.

250 K. Wappner, B. Schoenberger, M. Stratmann, G. Grundmeier; Height regulated scanning Kelvin probe for simultaneous measurement of surface topology and electrode potentials at buried polymer/metal interfaces. *J. Electrochem. Soc.* **2005**, *152*, E114–E122.

251 L. C. S. Pingree, O. C. Reid, D. S. Ginger; Electrical scanning probe microscopy on active organic electronic devices. *Adv. Mater.* **2009**, *21*, 19–28.

Part III

Materials from and for Electrochemistry

9

Electrochemical Deposition of Materials

Perhaps galvanic deposition reactions were the first electrochemical reactions you became familiar with in high school chemistry. Indeed, such reactions are easy to follow visually by the emerging metallic gloss of copper, the formation of a metallic mirror inside a Christmas tree ball, or the mass increase during an electrogravimetric experiment. These processes also have a great economic significance: Many metallic raw materials pass through at least one electrochemical procedure during their processing, for example, electrolytic preparation of Al and other nonnoble metals or the electrolytic refining (purification) of Cu. Galvanic processes also yield decorative coatings, corrosion-retarding layers, or functional metal structures for the electronic industry. Because the raw materials for plating transform almost completely into the final product (usually as a very thin layer), electrochemical coating processes offer significant cost advantages, especially for thin layers of expensive metals. The electrochemical metal deposition is a central reaction for rechargeable batteries. The modern understanding of and know-how about galvanic deposition has evolved since the middle of the nineteenth century, quite often in a very empirical way. Should you expect an easily understood chapter of electrochemistry due to the long history and due to the possibility for visual observation, the book author has to disappoint you: The key in Figure 3.1a has already shown that a metal deposition reaction comprises both the adsorption (with desolvation) as well as the surface diffusion, assembly into the crystal lattice, or the formation of a new growing crystal nucleus on a foreign support. In this chapter, we will systematically cover the individual steps of metal deposition and the means to influence them. We will focus on the deposition of metals (Section 9.1) in this chapter; the deposition of conducting polymers will be covered in Chapter 13. It is helpful to study or refresh the content of Chapters 1–4 and Sections 8.9 and 8.10 before you go on.

The symbol I (or i) is often used for the current *density* in the literature about electrodeposition. In this textbook, we use j for the current density and stay with this usage also in this chapter for reasons of consistency. Furthermore, the symbol z is often used for the stoichiometric number of transferred charges (instead of n in this textbook). The symbol A is often used for the rate constant of nucleation (here k_n). Please also note the frequent use of η, both for the overpotential and the dimensionless faradaic efficiency or coulombic efficiency (here ε_F).

Further reading. There are a number of classical texts about the mechanisms of galvanic deposition [1–7] and further recommended sources, which are more focused on the materials to be deposited [8–10].

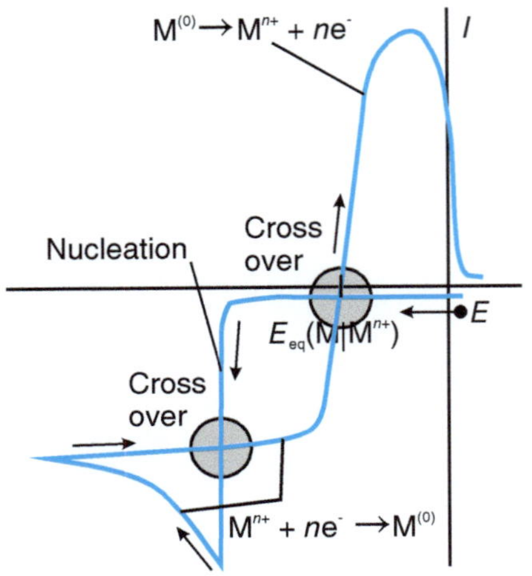

Figure 9.1 Cyclic voltammogram of a metal deposition reaction on a foreign substrate with a large nucleation overpotential. Once nuclei have been formed, metal deposition proceeds at a high rate. In the positive-going half-cycle, the metal deposition occurs up to the equilibrium potential, that is, to more positive potentials than the potential at which nucleation occurred in the negative-going half-cycle. At even more positive potentials, electrolytic metal dissolution is observed. The lines for the negative-going and positive-going half-cycles cross each other at two points.

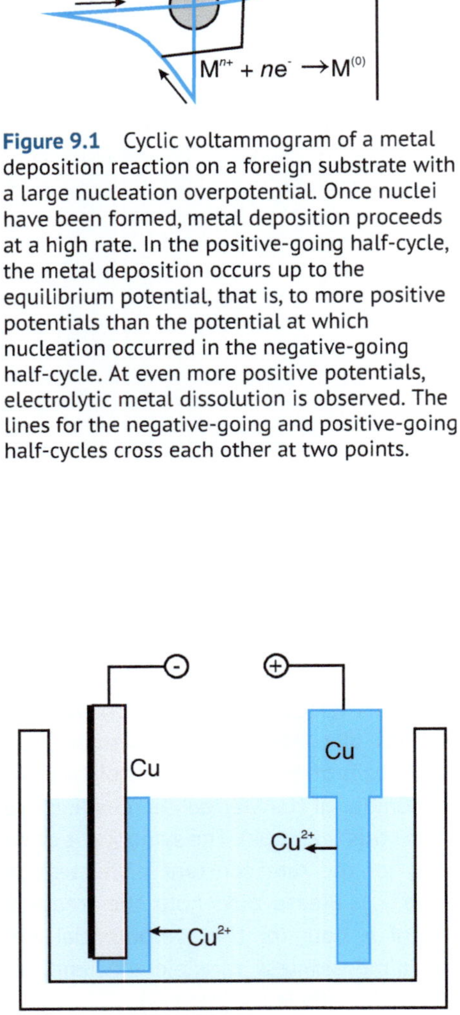

Figure 9.2 Basic principle of electrodeposition. Metal ions are reduced at the workpiece and deposited as a metallic coating. The counterreaction is the electrolytic oxidation of a sacrificial anode, which resupplies the consumed metal cations. Alternatively, the oxygen evolution reaction can occur at the anode. In this case, the galvanic bath must be periodically replenished with the metal cations.

9.1 Which Methods Can Be Used for Electrodeposition of Materials?

9.1.1 Galvanic Deposition

Galvanic deposition involves the reduction of dissolved metal cations $M^{n+} + n\, e^- \rightarrow M^{(0)}$ at an electrode (substrate) connected to an external electrical power supply. Figure 3.1a shows that the solvation shell of the metal cation, composed of solvent molecules Solv (e.g., H_2O), has to be broken up in a stepwise process. Thus, a better representation of the deposition reaction would be

$$[M^{(z+)}(Solv)_m] + \lambda\, z\, e^- \rightarrow [M^{(1-\lambda)\, z+)}(Solv)_k]_{ads} + (m - k)\, Solv \tag{9.1}$$

$$[M^{(1-\lambda)\, z+)}(Solv)_k]_{ads} + (1 - \lambda)\, z\, e^- \rightarrow M^{(0)} + k\, Solv \tag{9.2}$$

The ion adsorbs initially as a partially discharged ion (adatom, $0 < \lambda \leq 1$) with concomitant stripping of a few molecules from the solvation shell. The adatom diffuses on the surface to step edges and finally to a kink site (Section 8.9.8), where it is incorporated into the metal lattice while releasing further molecules from the solvation shell.

If the metal deposition occurs on a foreign substrate, for example, Cu on Fe, a new phase must be formed first. This process is called nucleation and requires more negative potentials (and thus occurs more slowly) than those required for the growth of a metal layer on its own substrate, for example, Cu on Cu. Such retardation due to phase formation processes can be identified in cyclic voltammograms from the crossing over points marked in Figure 9.1.

The deposited mass of the metal can be calculated from the transferred charge and the molar mass using Faraday's law, provided the current efficiency (or faradaic efficiency) ε_F is 100% [Eq. (1.24)]. Very often, this is *not* the case because the deposition of nonnoble metals is accompanied by the hydrogen evolution reaction (HER) due to thermodynamic reasons. The HER not only decreases the faradaic efficiency, but the evolving hydrogen may be incorporated into the deposited metal as an interstitial metal hydride, depending on the kind of metal. We have already touched on such hydrogen insertion reactions in connection with the PdH_x reference electrode in Section 2.4.1. The insertion of hydrogen is undesirable in many cases because it leads to a change in mechanical and electronic properties (hydrogen embrittlement, Section 14.1). Also, the undesirable inclusion of gas bubbles into the growing metal film structure can occur.

In order to avoid the hydrogen evolution during metal deposition of electronegative elements, molten salt electrolytes (e.g., for Al production), aprotic organic electrolyte solutions (Section 15.6.3), or ionic liquids (Section 11.4.2) can be used.

Apart from metals and hydrogen, the electrolytic reduction reaction yields further products, for example, from the anions of the electrolyte. For instance, hypophosphite anions ($H_2PO_2^-$) are incorporated into the metal layer during nickel deposition, where they react with Ni to form Ni_3P during a subsequent heat treatment. This leads to harder coatings (galvanic nickel phosphorous plating).

Alloys can also be obtained by simultaneous electrodeposition of multiple metals (Section 9.5.1). The formation of dispersion layers or composite materials is another example of the electrodeposition of complex materials, in which inert hard materials such as Al_2O_3, WC, SiC, Cr_2C_3, or other solids such as graphite or PTFE are incorporated in the layer of the electrodeposited metal (electrocodeposition, Section 9.5.3). Such functional layers have excellent wear resistance or emergency running characteristics.

In many cases, ions consumed by the electrodeposition can be resupplied by anodic metal dissolution at the counter electrode (Figure 9.2).

9.1.2 Immersion Plating

Immersion plating or immersion deposition exploits a galvanic displacement reaction, in which a more noble metal is spontaneously deposited on a less noble metal and an equivalent amount of the less noble metal is dissolved. Equation (9.3) provides an example.

$$Zn^{(0)} + Cu^{2+} \rightarrow Zn^{2+} + Cu^{(0)} \tag{9.3}$$

We have omitted the solvent molecules associated with the cations, although they are of course involved in the reaction, similar to Eq. (9.1). The more noble metal may form a coating on the less noble metal and, in this way, interrupts the contact between the reductant (the less noble metal) and the solution (Figure 9.3). This makes immersion plating a self-limiting process. Immersion plating plays an important role as an individual step within complex coating technologies.

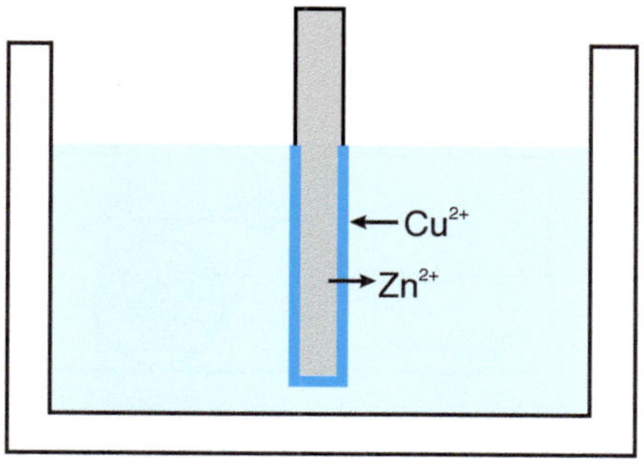

Figure 9.3 Immersion plating is exemplified by a zinc rod immersed in a $CuSO_4$ solution.

9.1.3 Autocatalytic Metal Deposition

Autocatalytic metal deposition is frequently called electroless or currentless deposition. Similarly to immersion plating, it exploits a spontaneous chemical reaction. In this case, the reductant is a chemical reagent dissolved in the electrolyte solution rather than the less noble metal of the workpiece in immersion plating. For thermodynamic reasons, we might expect a spontaneous reaction between the metal cations or precursors to be deposited and the chemical reductant in solution. Although possible in principle, this reaction proceeds very slowly (if at all) due to the absence of nucleation sites for the formation of the new metal phase. In other words, the reaction is inhibited by the absence of nucleation sites. However, at a catalytic surface or at nucleation sites, the reaction can progress at a high rate. The rate is determined by the mixed potential that forms at the nucleation site. It is influenced by both the potential-dependent rate of the reduction of metal cations and the potential-dependent rate of the oxidation of the chemical reductant. We have discussed the formation of mixed potentials in connection with Figure 3.14 and will discuss its significance again in Section 14.3.1 in connection with corrosion. The composition of the electrolyte solution, called "bath" in the context of electrodeposition, offers many opportunities to tune the exact value of the mixed potential and thus to control the rate of the overall plating reaction to give high-quality films and deposits.

A popular example of an autocatalytic metal deposition is the formation of a silver mirror on the glass surface of a reaction vessel containing an ammoniacal silver nitrate solution that is reduced by an aldehyde. Autocatalytic metal deposition has technological importance for the metallization of insulating surfaces, for instance, in the deposition of metallic current paths on printed circuit boards (PCBs), on epoxy resins, or for decorative metal layers on plastic pieces. The following steps can play a role in the process (Figure 9.4): Oxygen-containing functional groups are generated on the surface of the epoxy resin, for instance, by a plasma or a chromate pickle. Those groups bind Sn^{2+} ions to the surface after immersion in a tin(II) salt solution. Next, $Pd^{(0)}$ seeds are formed during an immersion plating under oxidation of the adsorbed Sn^{2+} ions to dissolved Sn^{4+} species.

$$Sn^{2-}(ads) + Pd^{2+}(solv) \rightarrow Sn^{4+}(solv) + Pd^{(0)}(s) \tag{9.4}$$

The $Pd^{(0)}$ seeds are the nucleation sites for the subsequent autocatalytic Cu deposition (reduction) that is the reaction forming the desired conducting metal connections. This reaction is complemented by the oxidation of a dissolved aldehyde according to Eq. (9.5).

$$Cu^{2+} + RCHO + OH^- \rightarrow Cu^{(0)}(s) + RCOO^- + H_2O \tag{9.5}$$

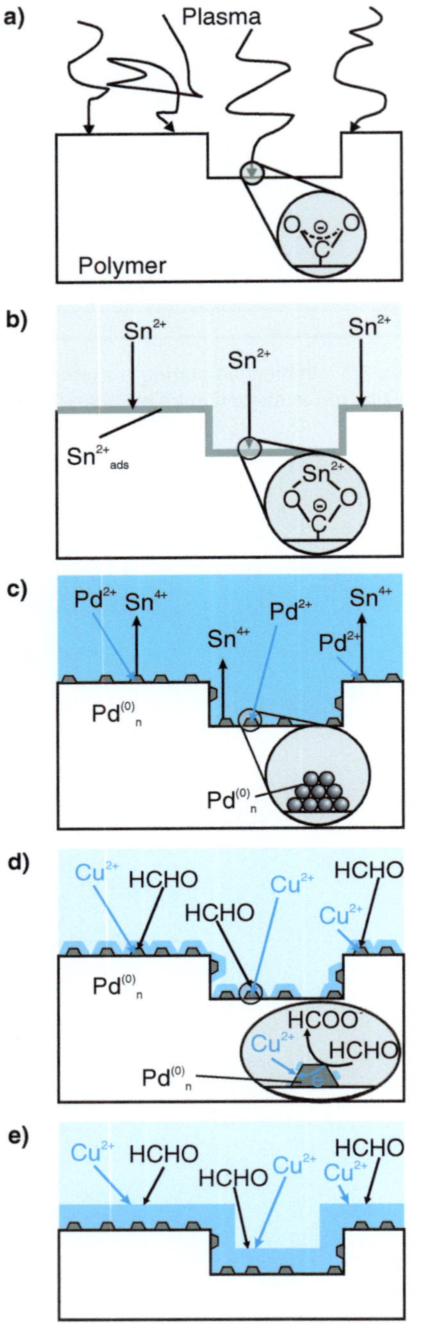

Figure 9.4 Autocatalytic metal deposition on an insulating workpiece. a) Surface activation; b) adsorption of Sn^{2+} ions; c) formation of $Pd^{(0)}$ seeds by immersion plating; d) autocatalytic Cu deposition on the Pd seeds; and e) further Cu deposition on the already formed Cu and formation of a continuous film. The scheme is not to scale.

This reaction continues even if the $Pd^{(0)}$ seeds are completely covered by copper because both Pd and Cu (autocatalysis) promote the oxidation of the reductant.

Autocatalytic metal deposition processes are used mainly for Au, Cu, and Ni_3P. The process does not need a counter electrode nor a connection of the workpiece to an external power supply. This makes it easier to form a layer of uniform thickness on workpieces of complex shape, as well as to coat insulating workpieces. The macrothrowing power and the microthrowing power (Section 9.3.3) of the process are excellent and can be further tuned by the concentration and kinetics of the reducing agent. However, autocatalytic metal deposition can be used technically only for layers of up to 1 μm thickness. Such films can be advantageously used as a conductive support for the subsequent galvanic deposition of the same or another material. The main disadvantage of the autocatalytic metal deposition is the thermodynamic instability of the bath solution. The thermodynamically possible homogeneous reaction of the metal precursor and the reducing agent must be strictly confined to the activated surface. Otherwise, the deposition bath decomposes rapidly in an autocatalytic process (e.g., due to impurities).

Further reading: [11].

9.1.4 Deposition of Semiconductors

The electrochemical deposition of oxides, sulfides, and other compound semiconductors can be accomplished via different routes. First, a metal (e.g., Fe) can be electrodeposited as a thin metal film, which is subsequently oxidized in an oxygen-containing atmosphere at elevated temperatures.

Another process uses the anodization of metals that form passive oxide layers to generate oxide films of defined structure (Section 14.6.1). In this process, no new metal atoms or ions are added to the surface. An additive layer formation, that is, a process with the addition of metal atoms or metal ions to the surface, can be made by means of an electrochemically induced pH shift. An alkaline pH shift of the solution close to the surface of a conducting workpiece can be caused by the following reactions:

$$O_2 + 4\,e^- + 2\,H_2O \rightarrow 4\,OH^- \tag{9.6}$$

$$NO_3^- + 2\,e^- + H_2O \rightarrow NO_2^- + 2\,OH^- \tag{9.7}$$

OH^- ions can then react with the metal cation to form hydroxides that are thermally converted to the corresponding oxides or spontaneously transformed when working in warm electrolyte solutions. An example is the deposition of zinc oxide.

$$Zn^{2+}(aq) + 2OH^- \xrightarrow{80°C} ZnO(s) + H_2O \tag{9.8}$$

A similar principle is used in the electrochemically triggered deposition of sol–gel films, which can be used to form a large variety of functional layers on metal surfaces [12]. The reaction is described by the following equations:

$$\begin{aligned} &Si(OR^1)_4 + R^2_n\,Si(OR^1)_{4-n} + (8-n)\,H_2O \xrightarrow{[OH^-]} \\ &\quad Si(OH)_4 + R^2_n\,Si(OH)_{4-n} + (8-n)\,R^1OH \end{aligned} \tag{9.9}$$

$$Si(OH)_4 + R^2_n\,Si(OH)_{4-n} \longrightarrow (OH)_3Si\text{–}O\text{–}SiR^2_n(OH)_{3-n} + H_2O \tag{9.10}$$

The condensation reaction of the precursors $Si(OR^1)_4$ and $R^2_n\,Si(OR^1)_{4-n}$ in the deposition bath is very slow at pH 7, but can be catalyzed by H^+ or OH^-. The local acidification or an alkaline pH shift can be induced by an electrolysis reaction similar to reactions (9.6) and (9.7). The product of reaction (9.10) undergoes further condensation steps to yield a three-dimensional (3D) cross-linked material containing the groups $-R^2$ that can be used to endow the material with different functional properties [12]. Electrochemically triggered pH shifts are also the principle behind electrophoretic painting by polyelectrolytes (Section 11.6.2). Semiconducting metal chalcogenide layers can be obtained by a chemical deposition that is somewhat reminiscent of autocatalytic metal deposition. Salts and hydroxides of some heavy metals (Ag, Hg, Pb, Cd, etc.) react with a sulfur-containing precursor such as thiourea as described in a simplified way by the following equation:

$$[ML_n]^{2+} + H_2N\text{–}CS\text{–}NH_2 + 2\,OH^- \rightarrow MS + 2\,H_2O + nL + HN{=}C{=}NH \quad (9.11)$$

Coatings of high quality are obtained if crystallization occurs preferentially on the surface of the workpiece to be coated [13].

Electrochemical atomic layer epitaxy starts from a monoatomic underpotential deposition (UPD) layer (e.g., Cd, Pb) on the substrate that is converted in the subsequent step with the chalcogenide-containing precursor (e.g., S^{2-}, $S_2O_3^{2-}$). Next, another layer of the metal cations that would form a salt of low solubility with the sulfur-containing precursor is deposited. By repetition of the cycle, thin films of small and precisely defined thickness can be formed that follow exactly the outer shape of the workpiece (conformal coating) because each individual step is self-limiting [14].

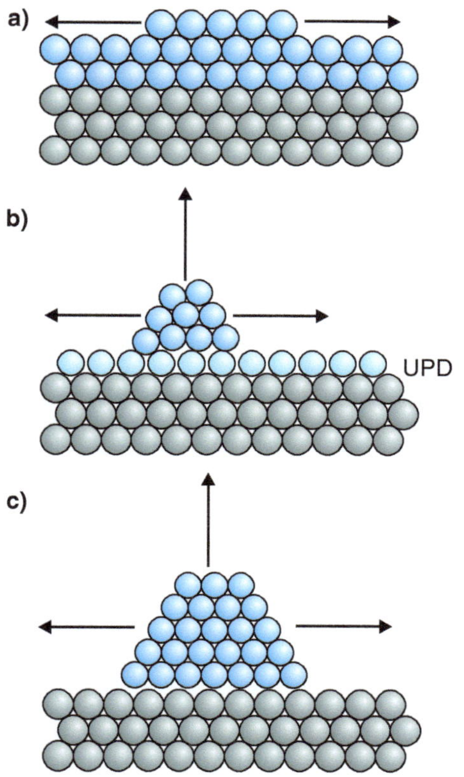

Figure 9.5 Growth modes of thin metal layers on a foreign substrate.

a) Frank–van der Merwe growth (layer growth);
b) Stranski–Krastanov growth (3D islands on an UPD monolayer);
c) Volmer–Weber growth (3D island growth).

9.2 Growth Modes During Metal Deposition

9.2.1 Growth of a Thin Metal Layer on a Foreign Substrate

The first deposition step is called electrocrystallization and comprises the formation of a thin metal film on a workpiece of a different material. Because the metallic bond requires high coordination numbers, only certain positions are favorable for the attachment of a metal atom on the surface of the workpiece. Those sites are determined by the ratio of the lattice spacings between the metal atoms of the substrate S ($d_{0,S}$, from substrate) and the undisturbed structure of the metal M to be deposited ($d_{0,M}$). This ratio is called the crystallographic misfit and is given as a relative quantity $(d_{0,M} - d_{0,S})/d_{0,S}$. Important for the growth mode is also the interaction energy between the metal adatoms M_{ads} to be deposited with its own metal M and with the foreign substrate S.

We have already discussed the fitting of the first metal layer on a foreign substrate in connection with underpotential deposition (UPD) in Section 8.9.8. This process plays an important role here as well for the initial step during metal deposition reactions in cases where the Gibbs free adsorption energy of the metal M is more negative on the substrate S than on its own bulk metal [$\Delta_R G_{ads}(M\text{–}S) < \Delta_R G_{ads}(M\text{–}M)$, $\Delta_R G_{ads} < 0$]. In the case that the crystallographic misfit between M and S is negligible, M grows layerwise on S. This layer growth is also called Frank[116]–van der Merwe[117] growth (Figure 9.5a).

[116] Frederick Charles Frank, 1911–1998, after a study of chemistry, he researched material science problems; professor of physics at the University of Bristol, UK since 1954.
[117] Jan H. van der Merwe, 1922–2016, South African physicist and mathematician, professor at the University of Pretoria.

If there is an UPD layer and a crystallographic misfit, there must be an adaptation of the lattice constant, because otherwise the metal M would grow with the lattice constant $d_{0,S}$ of the substrate material. The deviation of the lattice constants at the interface from the bulk values $d_{0,M}$ and $d_{0,S}$ causes strong internal strain in the material. The modification of the lattice constant has different consequences depending on the magnitude of the crystallographic misfit. Often, the first layer has the structure of an UPD layer, where further layers cover the substrate as small islands only. Within the small islands, the deposited metal M has its own lattice constant $d_{0,M}$. As a result, there is a small strain between the first and the second layer. The strain remains small because the islands cover only a fraction of the substrate surface. The island growth on a monolayer of M is called Stranski[118]–Krastanov[119] growth (Figure 9.5b).

If $\Delta_R G_{ads}(M{-}S) > \Delta_R G_{ads}(M{-}M)$, that is, there is no UPD layer, the crystallographic misfit must be accommodated beginning with the first deposited layer. Therefore, the layer growth starts from 3D islands, which grow laterally and vertically with the lattice spacing $d_{0,M}$ of the deposited metal M. The 3D island growth is called Volmer–Weber growth[19] (Figure 9.5c).

The schematics in Figure 9.5 suggest that the growth of thin metal layers during electrocrystallization is identical to the formation of metal layers by evaporation processes in an ultra-high vacuum (UHV). In a UHV, the work functions of the metals determine the interaction energies. During electrocrystallization, $\Delta_R G_{ads}(M{-}S)$ and $\Delta_R G_{ads}(M{-}M)$ also depend on other electrolyte components and external control variables. The applied electrode potential can shift the adsorption equilibria of ions and neutral molecules. Additives contained in the electrolyte solution compete for the adsorption sites at the substrate surface and can modify $\Delta_R G_{ads}(M{-}S)$ and $\Delta_R G_{ads}(M{-}M)$. This offers many technically important opportunities to tune processes in the galvanic industry. However, this fact also contributes to the complexity of the subject.

9.2.2 Thermodynamics of Nucleation

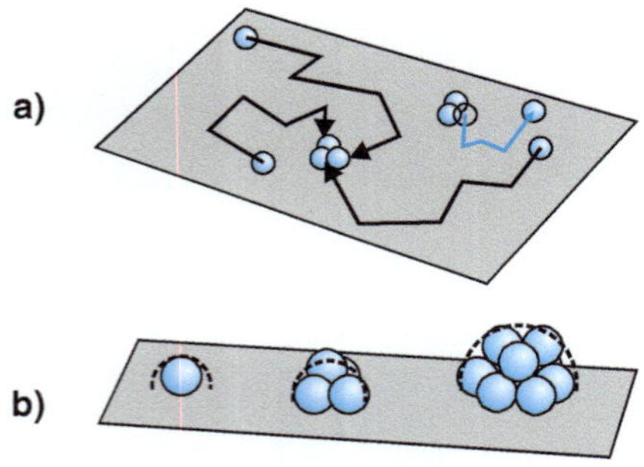

Figure 9.6 Nucleation at surfaces.

a) Encounter of three adatoms and decay of a cluster with subcritical size;
b) Description of three-dimensional nuclei by a hemisphere with the same volume as the respective nucleus.

Adatoms can move on a planar support surface by two-dimensional (2D) diffusion (surface diffusion), provided the thermal energy is sufficiently high to overcome the barrier for a place exchange of the adatom and a component of the electrical double layer. Collisions of several adatoms under the formation of a cluster occur during surface diffusion (Figure 9.6a). Those clusters of adatoms are in a dynamic equilibrium with isolated adatoms. The clusters also disintegrate back to isolated adatoms. As a result, stationary concentrations of adatoms and clusters of different sizes exist (Bodenstein principle[66]). Only from a critical cluster size onward can the cluster keep growing spontaneously. These clusters are called critical nuclei. Smaller clusters are termed subcritical nuclei. The size of a cluster at which the growth changes from a dynamic equilibrium between the cluster and isolated atoms to a spontaneous bulk growth depends on the kind of metal and the experimental parameters. A strongly simplified model can be used to illustrate the principle. We consider a hemispherical cluster of radius r (Figure 9.6b, volume $V = 2\pi r^3/3$, cluster–solution interfacial area $A = 2\pi r^2$). For an estimation of

[118] Iwan Nicolá Stranski, 1897–1979, Bulgarian physical chemist, study of medicine and biology in Vienna, Sofia and Berlin, researcher in Sofia (1926–1929, 1937–1941), Berlin (1929–1930, 1944–1963), Sverdlovsk (today Jekatarinburg, Russia (1935–1936), Breslau (today Wroclaw, 1941–1944). In Berlin, he worked at Kaiser Wilhelm Institute (today the Fritz Haber Institute of the Max Planck Society) and at the Technical University.

[119] Lyobomir Krastanov, 1908–1977, Bulgarian physicist, described the Stranski–Krastanov growth mode in his PhD thesis prepared under guidance of Stranski (1938). Later, he became a geophysicist and President of the Bulgarian Academy of Science (1961–1968).

whether a particular cluster tends to decay or grow spontaneously, we can analyze the Gibbs free energy Δg during the encounter of the adatoms. We consider a bulk contribution Δg_{bulk} and a contribution of the surface Δg_{surf}, coming from the fact that some atoms of the cluster are located at the interface.

$$\Delta g = \Delta g_{bulk} + \Delta g_{surf} \tag{9.12}$$

The contribution of the bulk is calculated from the molar Gibbs free energy of supersaturation $\Delta_{SS}G$ and the number of atoms in the cluster. The number of atoms in the cluster is obtained by dividing the cluster volume V by the molar volume V_m of the deposited metal.

$$g_{bulk} = \frac{-V}{V_m}\Delta_{SS}G = \frac{-2\pi}{3}\frac{\Delta_{SS}G}{V_m}r^3 \tag{9.13}$$

Because of $\Delta_{SS}G < 0$, the bulk contribution is always negative (Figure 9.7, curve 1a). In contrast, the contribution of the interface Δg_{surf} is always positive and is obtained by the area of the hemisphere and the area-related interfacial energy γ.

$$g_{surf} = A\gamma = 2\pi r^2 \gamma \tag{9.14}$$

Figure 9.7 curve 3a shows the result of adding the bulk and surface contributions in the overall Δg. The critical radius r_{crit} of the nucleus is the radius at which the probability for decay is exactly equal to the probability for spontaneous growth. For this size, Δg should not change with an infinitely small increase or decrease in the cluster size. As an algebraic equation, this condition is stated as $d(\Delta g)/dr = 0$. Using this relation, we can determine r_{crit}.

$$\frac{\partial}{\partial r}g_{bulk} = \frac{\partial}{\partial r}\left[\frac{-2\pi}{3}\frac{\Delta_{SS}G}{V_m}r^3 + 2\pi\gamma r^2\right] \overset{!}{=} 0$$

$$0 = -3\frac{2\pi}{3}\frac{\Delta_{SS}G}{V_m}r_{crit}^2 + 2\cdot 2\pi\gamma r_{crit}$$

$$0 = \frac{-\Delta_{SS}G}{V_m}r_{crit} + 2\gamma \tag{9.15}$$

$$r_{crit} = \frac{2\gamma V_m}{\Delta_{SS}G} = \frac{-2\gamma V_m}{nF|\eta|} \tag{9.16}$$

The Gibbs free energy of supersaturation $\Delta_{SS}G = -nF|\eta|$ is obtained directly from the overpotential $\eta = E - E_{eq}$ (similar to Eq. 2.23). E_{eq} is accessible from the Nernst Equation (2.37) for different bulk concentrations of the metal ion. The critical radius decreases with increasing absolute value of the overpotential. Nuclei become supercritical already at smaller cluster sizes (cf. curves 3a and 3b in Figure 9.7). As a consequence, the probability of forming a critical nucleus strongly increases with a higher absolute value of the overpotential, an observation that is also evident from experiments.

Knowing an expression for the critical radius of a nucleus, we can estimate the number n_{crit} of adatoms in a nucleus of critical size. For this purpose, we divide the volume of the nucleus by the volume contribution of a single atom within a bulk metal lattice ($V_{ad} = V_m/N_A$).

$$n_{crit} = \frac{V_{crit}}{V_{ad}} = \frac{2\pi}{3}\left(\frac{-2\gamma V_m}{nF|\eta|}\right)^3\frac{N_A}{V_m} = \frac{16\pi N_A V_m^2\gamma^3}{3(nF|\eta|)^3} \tag{9.17}$$

The general idea of this derivation can be refined and then also yields expressions for 3D nuclei of other geometries:

$$n_{crit} = \frac{8B_{3D}N_A V_m^2\gamma^3}{27(nF|\eta|)^3}. \tag{9.18}$$

$B_{3D} = A^3/V^2$ is a geometry parameter.

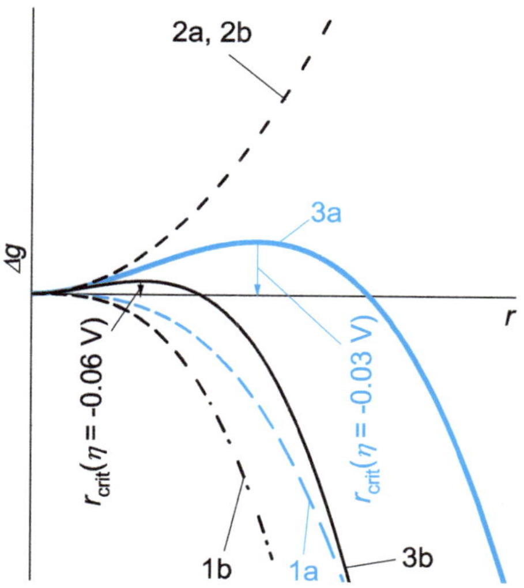

Figure 9.7 Gibbs free energy as a function of the radius of a hemispherical nucleus for overpotentials of −0.03 V (blue) and −0.06 V (black); 1) contribution of the bulk, 2) contribution of the interface, and 3) resulting overall Δg. The critical radius is indicated for both potentials.

$\Delta_{SS}G$ – Molar Gibbs free energy of supersaturation [J mol^{-1}]
V_m – Molar volume [m^3 mol^{-1}]
r_{crit} – Radius of a hemisphere with the same volume as the critical nucleus [m]
V_{crit} – Critical volume of a nucleus [m^3]
n_{crit} – Number of adatoms in a nucleus of critical size
V_{ad} – Volume contribution of an individual adatom in a metal crystal [m^3]
n – Number of transferred electrons per metal ion (here equal to ionic charge) [dimensionless]
F – Faraday constant [A s mol^{-1}]
γ – Interfacial tension [N m^{-1}]
$|\eta|$ – Absolute value of the overpotential [V]
N_A – Avogadro number [mol^{-1}]

B_{3D}, B_{2D} – Geometry parameter [dimensionless]
A – Footprint area of the cluster [m^2]
A_{ad} – Area covered by one adatom [m^2]
P – Perimeter of a 2D-cluster [m]
$\langle\varepsilon\rangle$ – Energy contribution for the formation of a step (per individual atom) [J m^{-1}]. The value is assumed constant or calculated as an average over different geometries. A typical value for Ag is 2×10^{-13} J cm^{-1} = 2×10^{-11} J m^{-1} [5].

For 2D nuclei, the geometry parameter is $B_{2D} = P^2/(4A)$ and is obtained from the perimeter P and the area A_{ad} covered by one adatom [3]. B_{2D} together with the energy contribution $\langle\varepsilon\rangle$ for the generation of a step yields another dependence of n_{crit} on the overpotential.

$$n_{crit} = \frac{B_{2D} N_A \varepsilon^2}{(nF|\eta|)^2} \tag{9.19}$$

Further reading: [15, 16].

9.2.3 Kinetics of Nucleation

We may regard the quantity Δg from Eq. (9.12) as an activation energy of nucleation in an Arrhenius-like [17] approach. From this, we obtain the rate of nucleation $J = (1/A) \cdot dn_{nucl}/dt$ [cm^{-2} s^{-1}]:

J – Rate of nucleation [cm^{-2} s^{-1}]
k_B – Boltzmann constant [J K^{-1}]
R – Gas constant [J K^{-1} mol^{-1}]

$$J = k_1 \cdot \exp\left[\frac{-\Delta g_{crit}}{k_B T}\right] \tag{9.20}$$

Using r_{crit} from Eq. (9.16) yields for Δg:

$$\begin{aligned} \Delta g_{crit} &= \frac{-2\pi}{3}\frac{nF|\eta|}{V_m}\left(\frac{-2\gamma V_m}{nF|\eta|}\right)^3 + 2\pi\gamma\left(\frac{-2\gamma V_m}{nF|\eta|}\right)^2 \\ &= \frac{8\pi\gamma^3 V_m^2}{(nF|\eta|)^2}\left[\frac{-2}{3}+1\right] = \frac{8\pi\gamma^3 V_m^2}{3(nF)^2}\cdot\frac{1}{|\eta|^2} \end{aligned} \tag{9.21}$$

After insertion into Eq. (9.20) and taking the natural logarithm, we obtain

$$\ln J = \ln k_1 + \frac{-8\pi\gamma^3 V_m^2}{3(nF)^2 k_B T}\cdot\frac{1}{|\eta|^2} \tag{9.22}$$

In many cases, a plot of ln J versus $|\eta|^{-2}$ will yield a straight line. Also of interest is a further derivatization of Eq. (9.22) with respect to the overpotential.

$$\frac{\partial(\ln J)}{\partial|\eta|} = (-2)\cdot\frac{-8\pi\gamma^3 V_m^2}{3(nF)^2 k_B T}\cdot\frac{1}{|\eta|^3} \tag{9.23}$$

A comparison of the coefficients with Eq. (9.17) reveals that the slope of a plot of ln J versus $|\eta|^{-3}$ amounts to $(nF/RT)n_{crit}$. From this expression, the number of atoms in a cluster of critical size can be determined.

$$\frac{1}{k_B T} = \frac{N_A}{RT};\quad \frac{1}{(nF)^2} = \frac{1}{(nF)^3}nF$$

Terms from Eq. (9.17) in blue.

$$\frac{\partial(\ln J)}{\partial|\eta|} = \frac{16\pi\gamma^3 V_m^2}{3(nF)^3}\cdot\frac{N_A nF}{RT}\cdot\frac{1}{|\eta|^3} = \frac{nF}{RT}n_{crit} \tag{9.24}$$

In a chronoamperometric experiment (Section 4.5.2), the formation of nuclei and the growth of nuclei occur simultaneously. Therefore, the rate of nucleation cannot be easily extracted from such an experiment. The experiment is conducted with a very short potential pulse, after which the supercritical nuclei are counted. Because the supercritical nuclei are still too small to be identified in microscopy, a double pulse is used. The nucleation occurs during the first short potential pulse to more negative overpotentials. Subsequently, the supercritical nuclei grow at less negative overpotential until they are visible in microscopy. The potential during the growth step is sufficient for the spontaneous growth of nuclei, but it is insufficient for the formation of new supercritical nuclei. Thus, a general rule of phase formation processes is exploited:

> Nucleation requires a higher energy than the subsequent growth of nuclei into bulk material.

Q: Why is this a general principle? Where else can it be observed?

A: The principle applies to all phase formation processes. You may know the problem of boiling retardation from a laboratory course (or from the microwave at home). If a liquid is cautiously heated in a bulb with smooth walls, it is possible to reach liquid temperatures above the boiling point without boiling of the liquid. In this state, a mechanical disturbance of the container or the addition of a porous boiling chip is sufficient to cause a sudden and very strong gas evolution with danger of spills and scalding. The (late) addition of a boiling chip with a rough surface and trapped gas bubbles facilitates the restrained formation of a new phase (here the formation of a small gas bubble). Once gas bubbles are formed, they grow extremely fast because the temperature of the liquid is already well above the boiling point, where the thermodynamically stable state of the substance is gaseous. A similar principle holds for crystallization from supercooled melts in which the nucleation is often retarded. Retarded nucleation processes can be identified in electrochemical experiments, for instance, by crossover points in cyclic voltammograms, as shown in Figure 9.1.

Further reading about other cluster shapes can be found in [17].

Our treatment so far has connected the momentary rate of nucleation J with an activation barrier. This cannot predict how J changes over time. However, this information is essential for the calculation of the transient (time-dependent) deposition current. If there is an identical probability for each point in time that a nucleus is formed on one of the N_0 equivalent sites on the interface, the number of nuclei N at an arbitrary time t is given by Eq. (9.25).

N – Number of nuclei [cm^{-2}]
N_0 – Number of sites suitable for nucleation [cm^{-2}]
k_n – Rate constant of nucleation [s^{-1}]
t – Time [s]

$$N = N_0[1 - \exp(-k_n t)] \tag{9.25}$$

Depending on the magnitude of the rate constant k_n for the nucleation in relation to the growth rate of supercritical nuclei, two limiting cases can be identified (Figure 9.8). They are called "instantaneous nucleation (IN)" and "progressive nucleation (PN)".

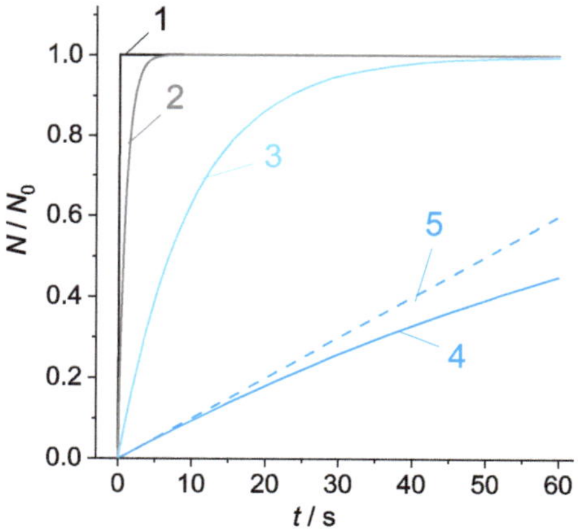

Figure 9.8 Time-dependent development of the number of nuclei as a function of the rate constant of nucleation k_n; 1) instantaneous nucleation, 2) $k_n = 1\ s^{-1}$, 3) $k_n = 0.1\ s^{-1}$, 4) $k_n = 0.01\ s^{-1}$, and 5) linearization according to Eq. (9.26) for $k_n = 0.01\ s^{-1}$.

During IN (large k_n, Figure 9.8, curve 1), N_0 nuclei are formed at the beginning of the deposition process. This number does not change further during the deposition process ($k_n \to \infty, N \approx N_0$). This scenario can be found when the sites for nucleus formation are determined by defects at the surface of the substrate and/or the nucleation is conducted at such a high absolute overpotential that many closely spaced nuclei are formed and all potentially suitable sites for nucleation are instantaneously occupied by a nucleus.

During PN, nuclei are not immediately generated on all potentially suitable sites (Figure 9.8, curves 2–4). The possible sites are rather occupied sequentially by a nucleus while previously formed nuclei continue to grow. As we will see below, this has important implications for the deposition current and the layers obtained during deposition.

We can obtain a simple equation for not too long times with a Taylor expansion[21] for Eq. (9.25), Figure 9.8, curve 5.

$e^x \approx 1 + x$ for values in the proximity of $x_0 = 0$.

$$N \approx N_0\,[1 - \{1 + (-k_n t)\}] = k_n N_0 t \tag{9.26}$$

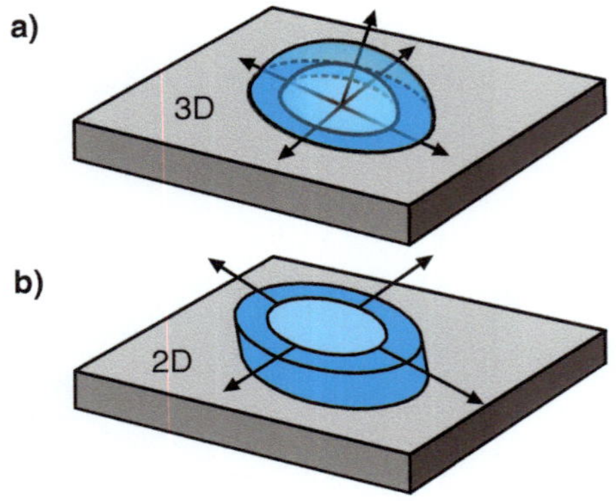

Figure 9.9 a) Three-dimensional growth of a nucleus and b) two-dimensional growth of a nucleus.

V_m – Molar volume [cm^3 mol^{-1}] = molar mass M [g mol^{-1}]/density ρ [g cm^{-3}]
H – Height of a nucleus [cm]
k – Rate constant for the attachment of adatoms to the nucleus [mol cm^{-2} s^{-1}]

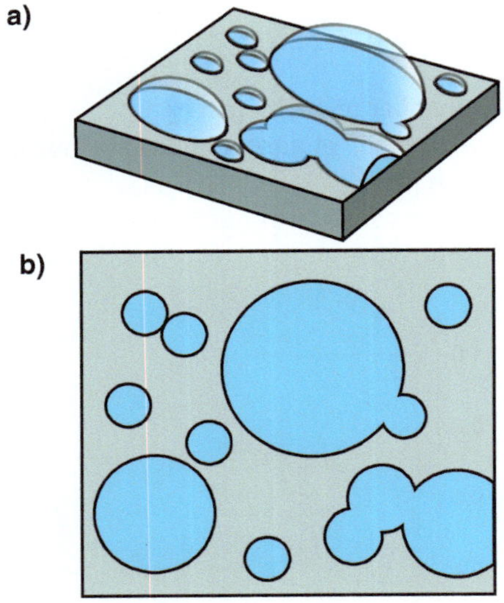

Figure 9.10 Diffusion layers above nuclei; a) as a perspective view and b) as a projection on the substrate surface. Within the regions colored in blue in panel b), the metal cation concentration (and thus the supersaturation) is decreased compared to the undisturbed solution bulk, and thus no new nuclei can be generated.

9.2.4 Growth of Nuclei

The growth of the nuclei and thus the electrolytic current during electrodeposition depends on the rate of the actual charge transfer step and the transport of metal cations to the nuclei. For uniform mass transport to each nucleus, there is a difference between 2D and 3D growth (Figure 9.9). Different electrodeposition currents for an individual nucleus lead to differences in chronoamperometric transients for the different dimensionalities of the growth front.

$$\text{2D growth:}\quad I = \text{const} \cdot t = nFk^2 2\pi h V_m t \tag{9.27}$$

$$\text{3D growth:}\quad I = \text{const} \cdot t^2 = nFk^3 2\pi V_m t^2 \tag{9.28}$$

The transport of the metal cation to the nucleus can be limited by the surface diffusion of adatoms or by the transport of the solvated metal cations in the electrolyte solution. We have already covered mass transport in electrolyte solutions in Chapters 4 and 5. An ensemble of growing nuclei at the surface of the substrate electrode can conceptually be regarded as an array of microscopically small electrodes (Section 5.6.6, Figure 5.83). Initially, a hemispherical diffusion layer forms above each nucleus, which accompanies a very rapid transport of metal cations toward the nucleus and thus a rapid growth of the nucleus. The nuclei grow independently of each other (growth of independent nuclei).

With expansion of the diffusion layer and/or increase in the number of nuclei, the individual diffusion layers start to overlap. This has two consequences: first, fewer metal cations reach an individual nucleus. Thus, the growth of the nuclei slows down, and the deposition current decreases compared to the case of independently growing nuclei. Second, the supersaturation of the solution is decreased in the vicinity of the nuclei (Figure 9.10) by the reduced metal ion concentration within the diffusion layer around the nuclei. According to Eq. (9.13) and the subsequent derivation, a decrease in supersaturation ($nF|\eta|$) reduces the probability that a new supercritical nucleus is formed within an already established diffusion layer. New supercritical nuclei can only emerge on those parts of the surface that are not within an existing diffusion layer around a growing nucleus.

If the individual hemispherical diffusion layers overlap, the current is limited by the emerging planar diffusion toward the surface (Figure 5.84d). As a result of the expanding diffusion layers, the mass transport becomes less effective with ongoing deposition time, and the deposition current decreases. This growth mode is called the growth of interacting nuclei.

The effect of the overlapping diffusion layers can be described by the so-called Avrami Equation[120] (9.29) that was developed for the general case of phase formation processes. For the 2D case in Figure 9.10b, Eq. (9.29) yields the ratio Θ between the coverage obtained with partial overlap of the diffusion layers (blue in Figure 9.10b) and the coverage that would result when the same diffusion layer would not overlap.

$$\Theta = 1 - \exp[-\pi D(8\pi c^* V_m)^{1/2} N(t)t] \tag{9.29}$$

[120] Melvin Avrami, US American metallurgist, made the equation popular in the fields of electrochemistry and metallurgy by three important publications [M. Avrami; *J. Chem. Phys.* **1939**, *7*, 1103–1112; *J. Chem. Phys.* **1940**, *8*, 212–224; *J. Chem. Phys.* **1941**, *9*, 177–184]. In fact, Eq. (9.29) is a special case of the more general problem of phase and microstructure transformations. Avrami used fundamental insights from other scientists. Therefore, the group of equations is also called Johnson–Mehl–Avrami–Kolgomorow equations. Robert Franklin Mehl, 1898–1976, US American metallurgist, Naval Research Laboratory, cofounder of modern materials science; William Austin Johnson, *1913, PhD student of Mehl; Andrey Nikolaevich Kolmogorov, 1903–1987, Russian mathematician, Soviet Academy of Science, important contributions to statistics.

Table 9.1 Growth modes for metal deposition and associated rate laws for 2D/3D growth, instantaneous/progressive nucleation as well as independent/interacting growth of nuclei.

Mechanism	2D	3D
IN + independent growth	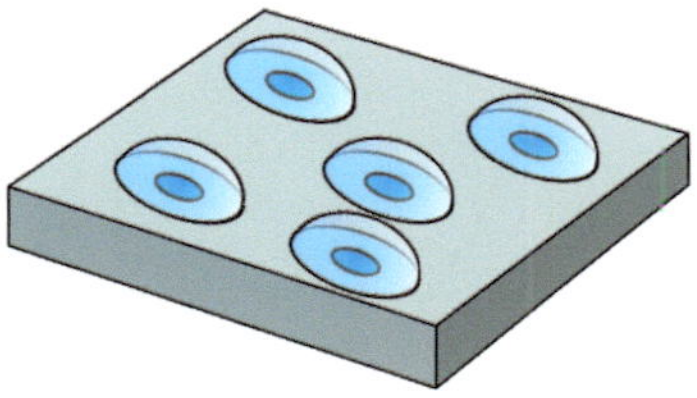 Kinetic control: $$j = 2\pi nFV_{\mathrm{m}}hN_0k^2t \quad (9.30)$$	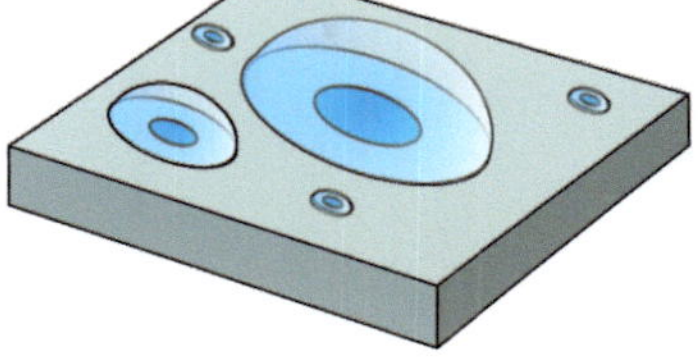 Kinetic control: $$j = 2\pi nFV_{\mathrm{m}}^2N_0k^3t^2 \quad (9.31)$$ Diffusion control: $$j = nF\pi(2Dc^*)^{3/2}V_{\mathrm{m}}^{1/2}N_0t^{1/2} \quad (9.32)$$
PN + independent growth	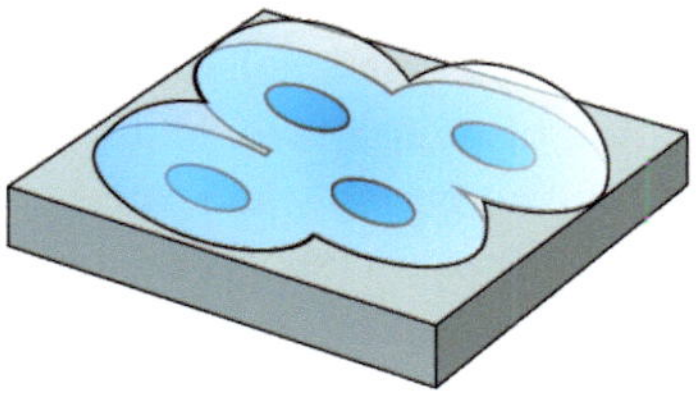 Kinetic control: $$j = \pi nFV_{\mathrm{m}}hk_nN_0k^2t \quad (9.33)$$	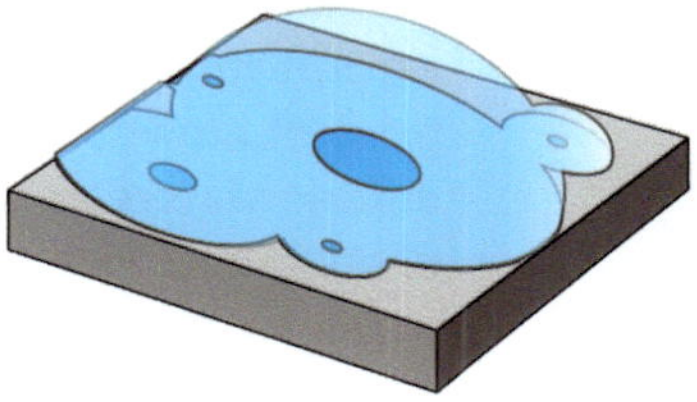 Kinetic control: $$j = \frac{2\pi}{3}nFV_{\mathrm{m}}k_nN_0k^3t^3 \quad (9.34)$$ Diffusion control: $$j = \frac{2\pi}{3}nF(2Dc^*)^{3/2}V_{\mathrm{m}}^{1/2}k_{\mathrm{n}}N_0t^{3/2} \quad (9.35)$$
IN + interacting growth	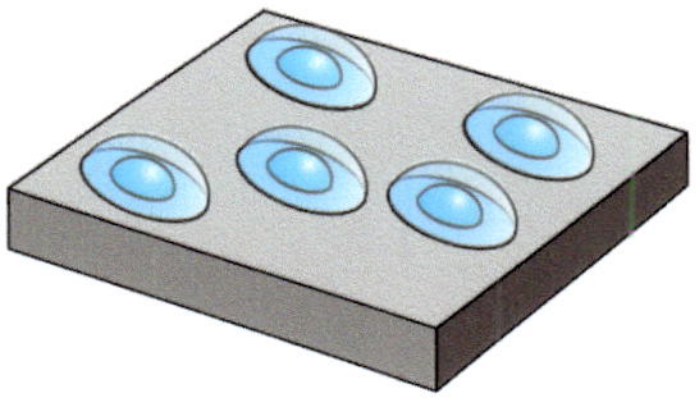 Kinetic control: $$j = 2\pi nFV_{\mathrm{m}}hN_0k^2t \cdot \exp\left[-\pi V_{\mathrm{m}}^2k^2t^2\right] \quad (9.36)$$	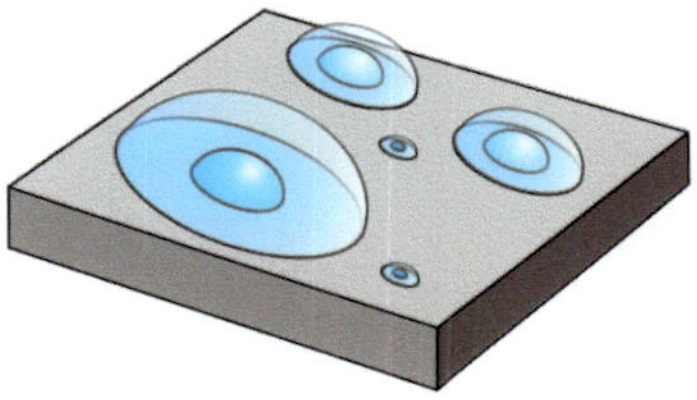 Diffusion control: $$j = nF\left(\frac{D}{\pi t}\right)^{1/2}\{1-\exp[\pi Dk_1N_0t]\} \quad (9.37)$$
PN + interacting growth	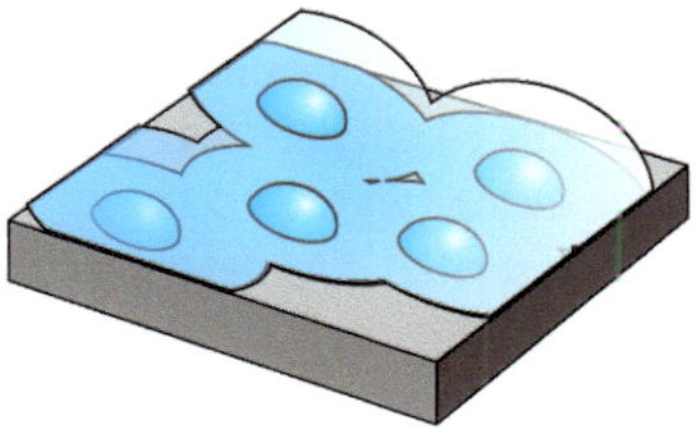 Kinetic control: $$j = \pi nFV_{\mathrm{m}}hk_nN_0k^2t^2 \cdot \exp\left[-\frac{\pi}{3}V_{\mathrm{m}}^2k_nN_0k^2t^3\right] \quad (9.38)$$	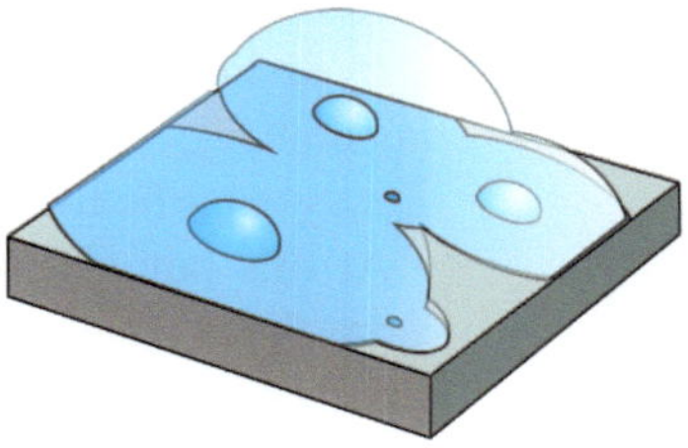 Diffusion control: $$j = nF\left(\frac{D}{\pi t}\right)^{1/2}\left\{1-\exp\left[\pi Dk_1N_0\left(t-\frac{1}{k_{\mathrm{n}}}+\frac{\exp(k_{\mathrm{n}}t)}{k_{\mathrm{n}}}\right)\right]\right\} \quad (9.39)$$

Abbreviations and symbols: IN, instantaneous nucleation; PN, progressive nucleation; 2D, two-dimensional; 3D, three-dimensional.
n, number of transferred electrons per ion [dimensionless]; F, Faraday constant = 96 485 A s mol^{-1}; V_{m}, molar volume [cm^3 mol^{-1}]; M, molar mass [g mol^{-1}]; ρ, density [g cm^{-3}]; c^*, solution bulk concentration [mol cm^{-3}]; D, diffusion coefficient of metal cations [cm^2 s^{-1}]; h, height of a 2D nucleus [cm]; k_{n}, rate constant of nucleation [s^{-1}]; k, rate constant for the attachment of adatoms to the nucleus [mol cm^{-2} s^{-1}]; N_0, number of places for nucleation (e.g., defects [cm^{-2}]); t, time [s]; π, number Pi; $k_1 = (8\pi c^*V_{\mathrm{m}})^{1/2} = (8\pi c^*M/\rho)^{1/2}$).
Source: According to [5, 6, 9].

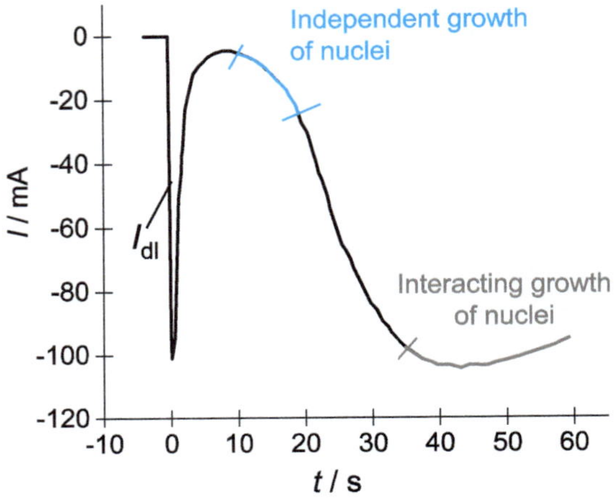

Figure 9.11 Potentiostatic current transient with consideration of the capacitive current and the transition from independent to interacting growth modes of nuclei. Source: Reproduced with permission from H. Bort, K. Jüttner, W. J. Lorenz, G. Staikov, E. Budevski; Underpotential-overpotential transition phenomena in metal-deposition processes. *Electrochim. Acta* **1983**, *28*, 985–991, Copyright 1983, Elsevier.

Figure 9.11 shows how the limiting factors change over the duration of a deposition experiment. Initially, there are only a few nuclei with small diffusion layers and independent growth. Later, after the formation of more nuclei and the expansion of the diffusion layers, the diffusion layers overlap increasingly and cause a transition to planar diffusion with a decreasing efficiency of mass transport and decreasing electrodeposition currents. We already know this behavior from the discussion of the Cottrell equation (Section 4.5.2) and cyclic voltammetry (Section 5.3.5).

Table 9.1 shows the different limiting cases for the growth modes (2D/3D nuclei; IN/PN, independent/interacting growth) with their characteristic time-dependent deposition current densities. This dependency can serve as a diagnostic criterion for the differentiation of growth modes. Further collections for other geometries of nuclei can be found in [3].

The normalized versions of Eqs. (9.37) and (9.39) [5] have received particular importance for 3D IN

$$\left(\frac{I}{I_{\max}}\right)^2 = \frac{3.8181}{t/t_{\max}}\left\{1 - \exp\left[-1.2564\frac{t}{t_{\max}}\right]\right\}^2 \tag{9.40}$$

and for 3D PN

$$\left(\frac{I}{I_{\max}}\right)^2 = \frac{1.2254}{t/t_{\max}}\left\{1 - \exp\left[-2.3367\left(\frac{t}{t_{\max}}\right)^2\right]\right\}^2. \tag{9.41}$$

By dividing the transient current by the maximum current $I_{\max}$ and the corresponding time $t_{\max}$, many constants cancel out in Eqs. (9.40) and (9.41). The resulting expressions are particularly well-suited for curve fitting (Figure 9.12).

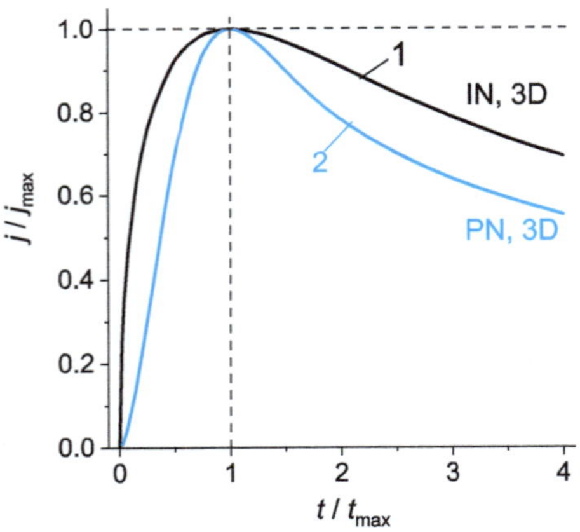

Figure 9.12 Calculated normalized current transients for (1) instantaneous nucleation [Eq. (9.40)] and (2) progressive nucleation [Eq. (9.41)] of three-dimensional nuclei with interacting growth.

At this point, it is necessary to emphasize that the determination of structural data alone from current transients, such as Figure 9.12, does not provide sufficiently sound results. It is better to complement electrochemical measurements with structure-sensitive methods such as electron microscopy or scanning probe techniques (Section 8.11). The recording of current transients is associated with some experimental difficulties: the faradaic current, which has been analyzed above, overlaps with the capacitive current immediately after the potential pulse (I_{dl} in Figure 9.11). This complicates the analysis of current transients, especially at short times. In practically used galvanic bath compositions, the potential-dependent adsorption of additives represents another challenge for the mechanistic analysis.

9.2.5 From Nuclei to a Continuous Film

The formation of a continuous metal film can occur by different mechanisms and result in very different material properties. Those technically important properties can be influenced by the support, the mechanism of nucleation, the overpotential, and by additives in the solution. This textbook can only indicate some trends, while the description of the large variety of material-specific effects is clearly beyond the scope.

Layer growth is relatively lucid because the metal film growth occurs layer-by-layer. Layer growth can be the dominating growth mode within an individual grain of a metal film. In this case, the rate of nucleation is so low that a layer is completed first before the growth of a new layer commences after the formation of a new nucleus on the completed atomic layer. The growth of the next layer can start only after the formation of a new nucleus because the adatoms require a kink site (Section 8.9.8) for integration into the metal lattice. In favorable cases, that is, at low current densities, the layer growth can even be followed from current transients because the current density rises after each nucleation event but decays when an atomic layer is completed and a new nucleus is not yet available to continue the growth of the next layer.

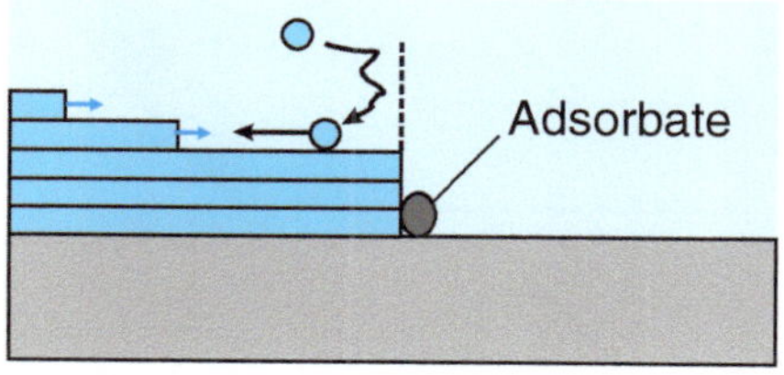

Figure 9.13 Bunching of atomic steps to a macroscopic step by blocking the further growth of the lowest layer by an inhibiting adsorbate (step bunching).

Layer growth can often be observed by microscopy, sometimes even by light microscopy. This is due to the effect of step bunching (Figure 9.13). If impurities or additives adsorb at a step edge, the layer growth can come to a halt at this point. The deposition of the next layers can only advance to the very same location. In this way, an initially atomically high step edge is amplified to a macroscopic step at the surface of the deposited material. Figure 9.14 shows such macrosteps in the vicinity of a screw dislocation.

Figure 9.14 Screw dislocation with macroscopically visible steps during the deposition of Cu from 1 mol L^{-1} $CuSO_4$ + 1 mol L^{-1} H_2SO_4 on Cu(1 0 0) at 298 K. Source: Reproduced with permission from H. Seiter, H. Fischer, L. Albert; *Z. Elektrochemie*. **1959**, *63*, 249–257. Copyright 1959, Wiley-VCH.

The role of surface defects for layer growth becomes particularly evident for a screw dislocation. Screw dislocations are preferred locations for layer growth because adatoms can be easily attached to the metal lattice. In addition, the defect does not disappear with the completion of one layer, but rather, the growth spiral extends laterally (Figure 9.15). The density of screw dislocations and thus the rate of a subsequent electrodeposition may be enhanced by mechanical pretreatment of metal substrates. We will discuss the generation of screw dislocations and other defects in solids in connection with Figure 11.4.

If **3D growth** is prevailing, a continuous film forms when the nuclei contact each other, and further lateral expansion is thus not possible. Only in the case of a single-crystal substrate, the convergence of nuclei may yield a uniform orientation of the crystal of the deposited metal layer, provided the nuclei have a uniform epitaxial orientation with respect to the lattice of the support.

In the case of polycrystalline substrates and epitaxial nuclei, the crystal lattices of the nuclei have different crystallographic orientations determined by the local microstructure of the grains in the support material. If the grains of the substrate have a preferential orientation, this may reappear in the deposited layer. When two nuclei of different orientations converge, a grain boundary is formed. For noncrystalline (amorphous) substrates, for example, glassy carbon, the nuclei show a random orientation.

9.2.6 The Development of the Microstructure

For the further growth of the deposited layer, some trends are important. The growth of the deposited film occurs at different rates along different crystallographic

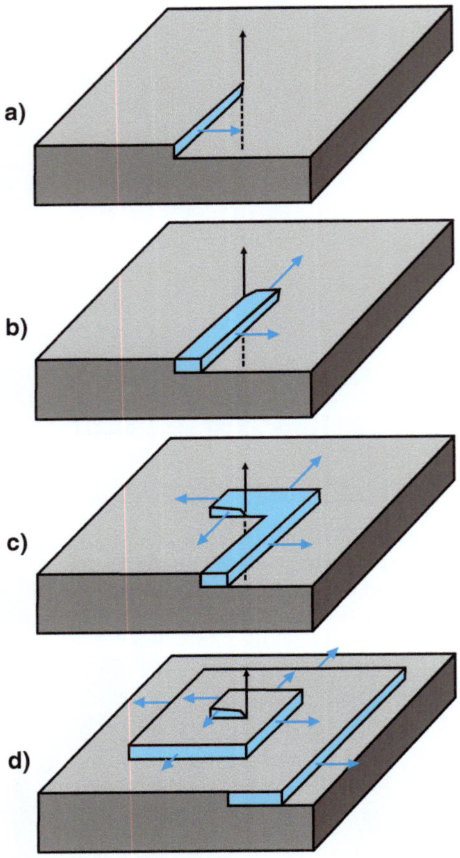

Figure 9.15 Layer growth at a screw dislocation.

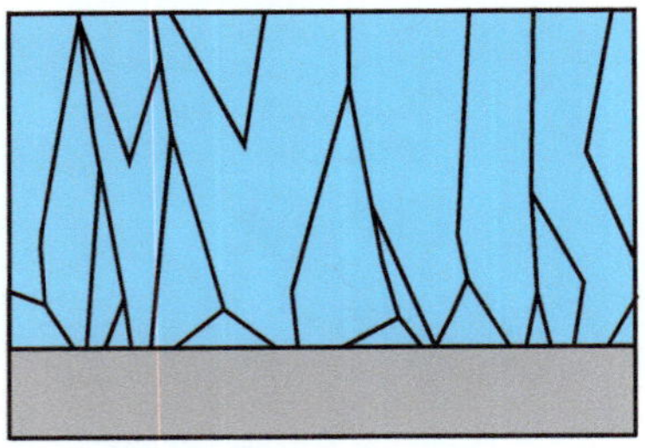

Figure 9.16 Formation of a columnar microstructure due to the different growth rates of individual nuclei. Schematic cross-sectional view.

orientations. Therefore, some grains show a stronger growth of layer thickness than others after the convergence of the nuclei. As a result, a columnar microstructure evolves within the deposited film that originates from those nuclei that grow fastest in the direction perpendicular to the substrate surface (Figure 9.16).

The current density j relative to the exchange current density j_0 and the overpotential are very important for the emerging microstructure of the deposited film. The exchange current density is a function of the momentary area density of kink sites. The overpotential also determines the probability for the formation of new supercritical nuclei. The final microstructure of the growing layer results from the competing influences of those parameters.

Different systems are in use for the classification of the microstructure. We follow here the classification by Fischer [1] and its taxonomy by Winand [18]. This system is strongly simplified but provides a good first orientation for many examples. According to Fischer, the following types of structures can be distinguished.

- The field-oriented isolated crystal type (FI) forms whiskers, dendrites, and crystalline powders.
- The basis reproduction type (BR) has a coarse microstructure initially determined by the substrate.
- The field-oriented texture type (FT) has a fibril-like structure along the electric field lines during the deposition and a preferred crystallographic orientation.
- The unoriented dispersion type (UD) shows extremely small crystallites with a randomly oriented crystal structure.

The Winand diagram (Figure 9.17) is a zone diagram and shows in a simplified manner the dominating film growth modes as a function of the current density normalized to the diffusion-limited current density and the degree of inhibition for the metal deposition. These two factors are *the* main technical means to tune the deposition process (Section 9.3).

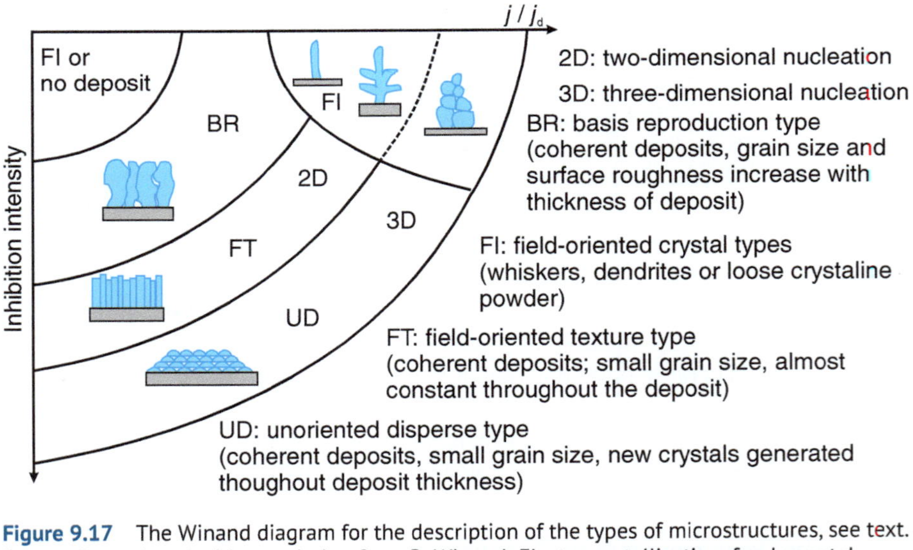

Figure 9.17 The Winand diagram for the description of the types of microstructures, see text. Source: Reproduced with permission from R. Winand; Electrocrystallization: fundamental considerations and application to high current density continuous steel sheet plating. *J. Appl. Electrochemistry*. **1991**, *21*, 377–385, Copyright 1991, Springer.

9.3 How Does the Current Density Influence the Deposition Process?

9.3.1 Relationship Between Current Density and Microstructure

The Winand diagram (Figure 9.17) shows in a qualitative manner the enormous influence of the local current density on the resulting microstructure and thus the technological properties of the deposited metals. We will now consider the most important specifications and laboratory experiments for their characterization in more detail. Figure 9.18 shows the dependence of the overpotential on the applied current density and on the formed microstructures for an acidic copper electrolyte.

The substrates used in the laboratory experiments were smooth and oriented perpendicularly to the electric field lines in the electrochemical cell. With respect to the potential distribution, this is the ideal condition for the deposition of a smooth film of uniform thickness (Figure 9.21a). The average current density $\langle j \rangle$ is easily obtained according to Eq. (9.42) from the total geometric area A of the substrate and the overall current I measured outside the electrochemical cell (in potentiostatic mode). In galvanostatic mode, the overall current is imposed on the cell and is set by the experimenter.

$$\langle j \rangle = I/A \tag{9.42}$$

In the ideal case, the average current density is equal to the local current density at each point of the substrate. Despite the favorable conditions, smooth and bright layers are not automatically obtained depending on the current density, as we have seen qualitatively in Figure 9.17.

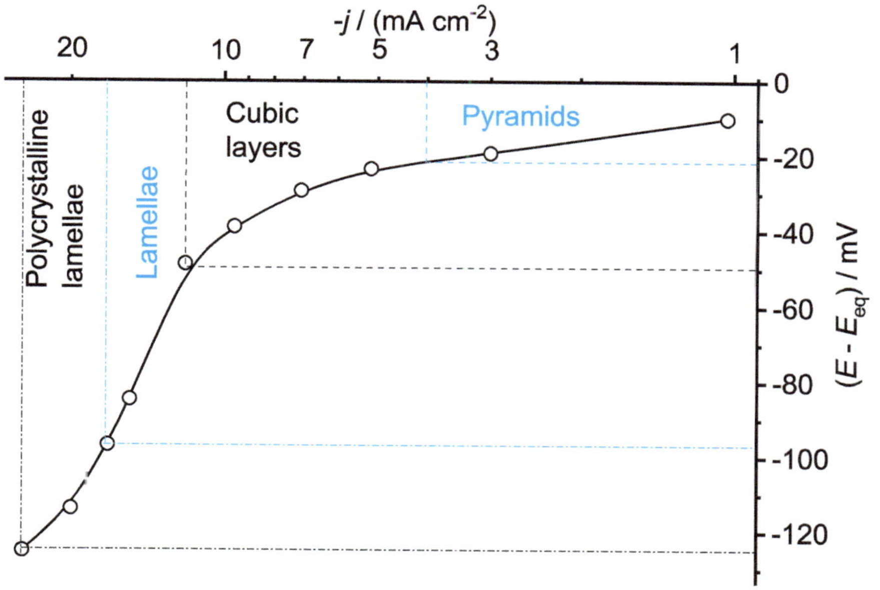

Figure 9.18 Morphologies of electrodeposited Cu at different current densities or overpotentials from 0.5 mol L^{-1} $CuSO_4$ + 5 × 10^{-4} mol L^{-1} glycol + 0.5 mol L^{-1} H_2SO_4 at 298 K. Source: Figure by the textbook author using data from H. Seiter, H. Fischer, L. Albert; *Electrochim. Acta* **1960**, *2*, 97.

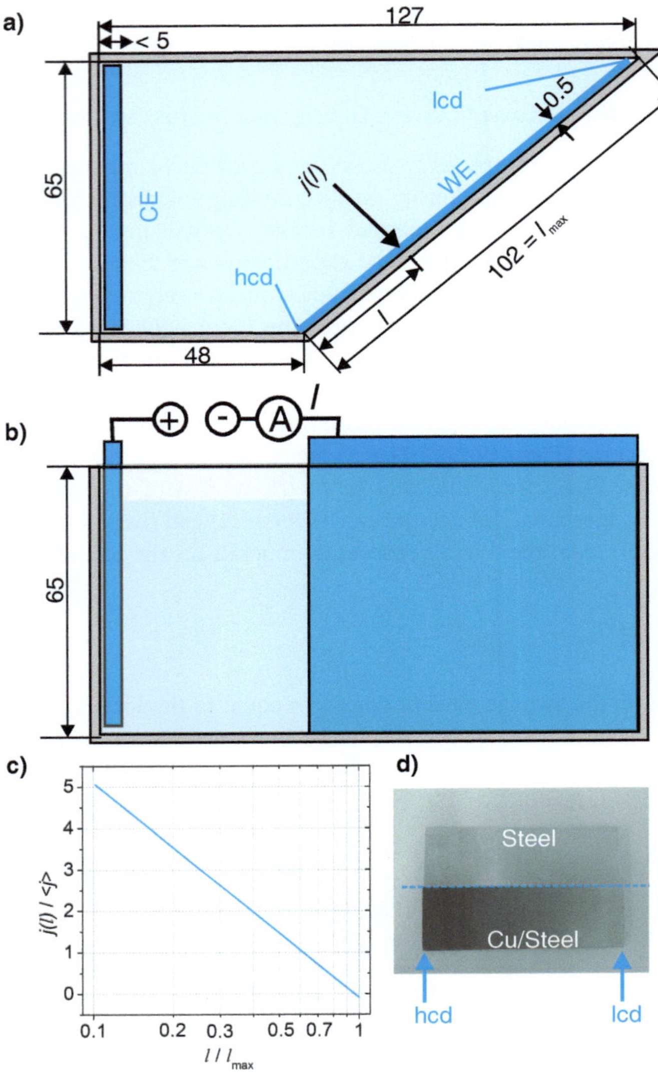

Figure 9.19 a, b) Description of a Hull cell (hcd – high current density, lcd – low current density). The industrial standard DIN 50957 defines two sizes with an electrolyte volume of 0.25 L or 1.00 L. The numbers in the schematic are the length in mm for the 0.25 L-cell. a) Top-view; b) side view; c) ratio between the local current density $j(l)$ and the average current density $\langle j \rangle = I/A$ as a function of the normalized distance l/l_{max}; d) photographic image of copper deposition on a steel sheet in a Hull cell.

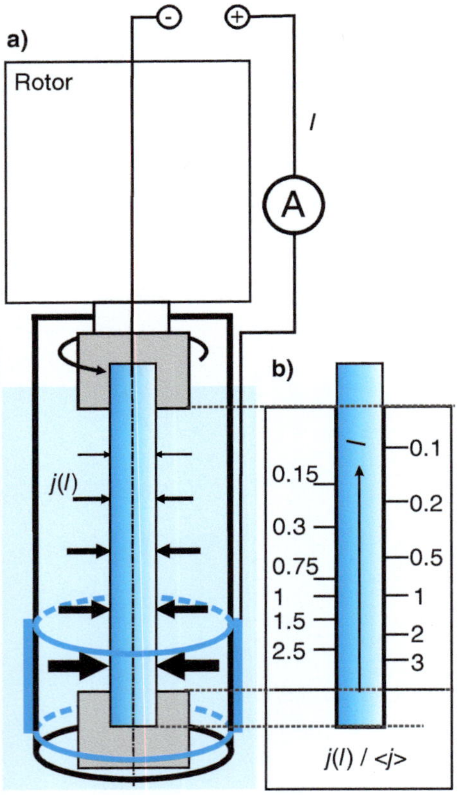

Figure 9.20 Hull cell with rotating cylinder electrode. a) Setup; b) ruler for determining the ratio between the local current density and the average current density along the height of the cylinder electrode.

A Hull cell[121] (Figure 9.19) [19] is used to determine suitable current densities for a particular combination of metal, deposition bath, and substrate in a very efficient way. Instead of conducting many experiments with different current densities, different local current densities are simultaneously applied to *one* substrate. The resulting film quality is judged easily by its optical appearance, in which blackening and shrouds indicate unwanted diminution of the quality. The local current density j that produces a film with a particular optical appearance can be determined from the normalized distance to the high current density (hcd) point using Figure 9.19c and the known average current density $\langle j \rangle = I/A$, *provided* the local current density follows the primary current density distribution (Section 9.3.2). This function is typically provided with the cell in the form of rulers for typical overall currents I.

There are also Hull cells with a rotating cylinder electrode that allow the local current density to vary under practically more relevant convective conditions [20, 21]. The metal deposition occurs on the outer cylindrical surface of the rotating test substrates

[121] Richard. O. Hull (USA), industrial chemist and entrepreneur, published in the 1930s.

(Figure 9.20a). The local current density can be determined from the average current density by means of conversion rulers, as shown in Figure 9.20b.

9.3.2 What Causes Current Density Distributions?

Generally, workpieces have imperfections, indents, and raised regions. Due to the longer distance to the anode, the ohmic drop in the electrolyte is larger between the anode and the indents of the workpiece and so causes a lower local current density (if there are no other counteracting factors). The electrolyte resistance R_E on an area element A of the substrate is obtained from Eq. (4.9) as $R_E = \rho l/A$.

The current density distribution that follows from the potential distribution under current flow in an electrolyte with finite conductivity is called the **primary current density distribution.**

The current density distribution can also result from different kinetics at different regions of the substrate electrode [Butler–Volmer Equation (3.27)]. For this case, we can state a characteristic charge transfer resistance R_{CT} for every current density in Figure 9.18 as

$$R_{CT} = \frac{1}{A} \cdot \frac{\partial \eta}{\partial j}. \tag{9.43}$$

This is analogous to $R = U/I = U/(j \cdot A)$ (Section 3.2.5).

The current density distribution caused by the different kinetics of the charge transfer reaction at different regions of the substrate is called the **secondary current density distribution.**

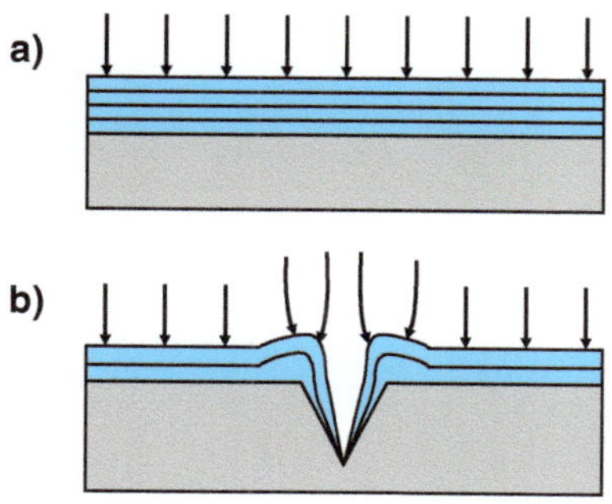

Figure 9.21 Current density distribution and uniformity of deposited layers. a) Ideal surface, on which a smooth coating can grow; b) surface with an indent, indicated field lines, and nonuniform coating.

An uneven distribution of the local current density can also be caused by different diffusive transport of the reactants to different parts of the substrate (Figure 9.21b). In this case, the local current density is

$$j = -nFD\frac{c^* - c_S}{\Delta x}. \tag{9.44}$$

j – Local current density [A cm^{-2}]
n – Stoichiometric number of electrons per ion [dimensionless]
F – Faraday constant [A s mol^{-1}]
D – Diffusion coefficient [cm^2 s^{-1}]
c^* – Bulk solution concentration of metal cations [mol cm^{-3}]
c_S – Local solution concentration of metal cations close to the surface of the workpiece [mol cm^{-3}]
Δx – Thickness of the diffusion layer [cm]

We have derived this approximate equation in Excursion 5-2 for the example of polarography. In a galvanic bath, the convection causes a hydrodynamic boundary layer that determines the extension Δx of the diffusion layer. Within an indent, the thickness of the diffusion layer is larger ($\Delta x_{recess} > \Delta x_{peak}$). Consequently, the deposition current and the resulting layer thickness are smaller within an indent. An initially present indent on the substrate would tend to become deeper if a metallic film is electrodeposited over it.

The current density distribution resulting from different diffusive mass transport to an area element of the substrate is called the **tertiary current density distribution.**

The current density distribution is also important for the sacrificial anode because it should electrolytically dissolve in such a way that it supports maximum utilization of the metal while maintaining the geometric area. To achieve this goal, sacrificial anodes are produced in optimized shapes, in which corners are thickened to account for the higher local dissolution current densities in these parts.

Further reading about the mathematical treatment of current density distributions can be found in [22].

9.3.3 The Wagner Number – Which Current Density Distribution Is the Relevant One?

W_a – Wagner number [dimensionless]
R_{CT} – Charge transfer resistance [Ω]
R_E – Electrolyte resistance [Ω]
A – Geometric electrode area [cm^2]
η – Overpotential [V]
j – Local current density [A cm^{-2}]
ρ_p – Specific resistance [Ω cm]
κ – Specific conductivity = $1/\rho$ [Ω cm^{-1}]
l – Characteristic length (electrode distance) [cm]

The Wagner number[122] W_a is a dimensionless indicator specifying the ratio between the charge transfer resistance and the solution resistance for a particular electrodeposition experiment.

$$W_a = \frac{R_{CT}}{R_E} = \frac{1}{A}\cdot\frac{\partial\eta}{\partial j}\cdot\frac{A}{\rho_E l} = \frac{1}{\rho_E l}\cdot\frac{\partial\eta}{\partial j} = \frac{\kappa}{l}\cdot\frac{\partial\eta}{\partial j}. \tag{9.45}$$

The Wagner number is easily determined by a conductivity measurement and a differentiation of a curve similar to that in Figure 9.18. If $W_a \ll 1$, the solution resistance controls the electrodeposition. The local current density (and thus the local layer thickness) is determined by the cell geometry for this case. If $W_a \gg 1$, R_{CT} and thus the secondary current density distribution is the dominating effect and determines the local thickness of the deposited metal. In the general case, the length l is a characteristic length over which the voltage drop occurs. Please note that the current density in Eq. (9.45) shows only the partial current density for the metal deposition. The overall current resulting from the integration of the local partial current densities may deviate considerably from the measurable current in the external circuit when $\varepsilon_{F,Me}$ is smaller than unity, because the hydrogen evolution reaction (HER) or other side reactions occur. Because the partial currents (reaction rates) of metal deposition and HER show different potential-dependencies (Figure 9.23), a concomitant HER may promote a uniform coating. Raised parts of the workpiece with higher local current densities have a lower faradaic efficiency for metal deposition, and less metal is deposited than expected from the (primary) current density distribution.

It is technically desirable to coat indents and recessed areas with the same layer thickness as raised parts of the workpiece (Figure 9.22a). Such a uniform coating, irrespective of the shape of the workpiece, is called conformal coating and is by itself sufficient to level indents and small recessed areas in the workpiece. In many cases, this aim is achieved even faster by the use of electrolyte additives, called levelers. Levelers adsorb to the protruding parts of the workpiece, thus decreasing the current densities there and enhancing the (relative) current density $j/\langle j\rangle$ in the recessed parts (Figure 9.22b).

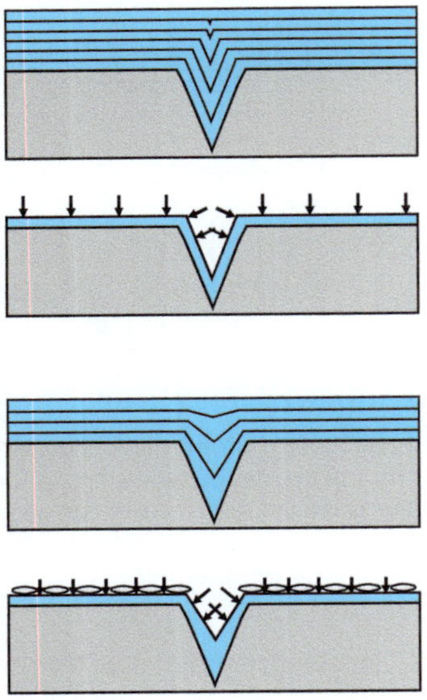

Figure 9.22 Leveling of an indent in a workpiece by a) uniform layer thickness and b) use of a leveler additive.

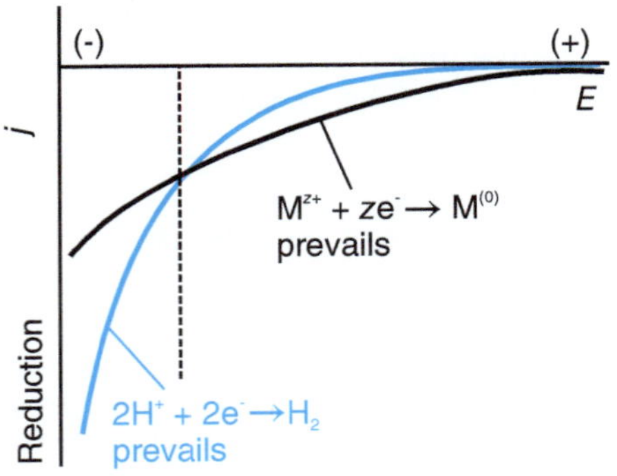

Figure 9.23 Potential-dependence of the current density for metal deposition and the H_2 evolution reaction (schematic).

In this context, the **throwing power T_p** of an electrolyte specifies the ability to yield a uniform thickness of the deposited metal film despite an uneven primary current density distribution.

It is common to distinguish between macrothrowing power and microthrowing power. **Macrothrowing power** describes the ability to produce an approximately even film thickness on complex-shaped workpieces, including the recessed areas. **Microthrowing power** is the ability to deposit metal inside of indents and scratches, that is, arbitrarily distributed microscopic variations of the outer shape. A good macrothrowing power does not automatically go along with a good microthrowing power. The macrothrowing power can be determined quantitatively using the

[122] Carl Wagner, 1901–1976, professor in Jena, Hamburg, Darmstadt, MIT, Göttingen (MPI), one of the founders of solid-state electrochemistry.

Haring–Blum cell[123] (Figure 9.24). The cell consists of a trough with a rectangular cross section. At the two faces, electrically coupled cathodes are inserted. *One* anode made as a net or grid is placed between them. The anode has *unequal* distances to both cathodes. The ratio of the two cathode–anode distances $P_x = l_{lcd}/l_{hcd}$ is formed in a way to be larger than one. A typical value would be $P_x = 5$. The metal deposition is made with this configuration for some time. Afterward, the mass increase of the two cathodes is determined, from which the ratio $M_m = m_{hcd}/m_{lcd}$ of the mass increases is determined. The throwing power T_P is calculated by Eq. (9.46).

$$T_P = \frac{P_x - M_m}{P_x + M_m - 2} \cdot 100\%. \quad (9.46)$$

T_P – Throwing power [dimensionless]
P_x – Ratio between the distances of the cathodes to the anode [dimensionless]
M_m – Mass ratio of the deposited metal on the two cathodes [dimensionless]

Example Calculation 9.1 How large would T_P be for an ideal macrothrowing power in Figure 9.24? How large would T_P be if the electrodeposition process followed exactly the primary current density distribution?

The cell shown in Figure 9.24 has $P_x = l_{lcd}/l_{hcd} = 5$. In the case of an ideal macrothrowing power, the same mass of metal is deposited on both cathodes despite their different distances to the anode. Then $M_m = m_{hcd}/m_{lcd} = 1$. Insertion into Eq. (9.46) yields:

$$T_{max} = \frac{5-1}{5+1-2} \cdot 100\% = 100\%$$

In the case of the primary current density distribution alone determining the mass of the deposited metal, we expect $m_{hcd}/m_{lcd} = l_{lcd}/l_{hcd}$. This means $M_m = P_x = 5$. Inserting into Eq. (9.46) yields

$$T_{min} = \frac{5-5}{5+5-2} \cdot 100\% = 0\%$$

◢

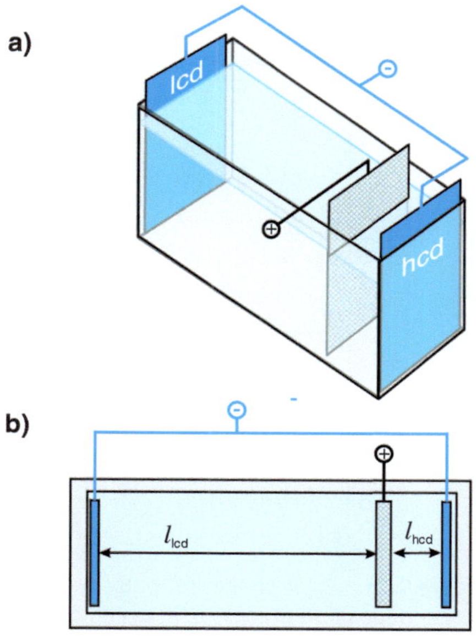

Figure 9.24 Schematic view of a Haring–Blum cell. a) Perspective view and wiring; b) top-view with exemplified dimensions (in mm); lcd, low current density; hcd, high current density.

The requirement for a high throwing power is $W_a \gg 1$. This can be achieved by additives to the electrolyte that enhance the specific conductivity and slow down the charge transfer step.

Further reading: [9].

9.3.4 Pulse Deposition

Many electrodeposition processes are conducted with current pulses (galvanostatic) or voltage pulses (potentiostatic) because the control of the coating process can be improved in these modes. Rectangular pulses (Figure 9.25a) are often used. Metal is deposited during the time t_{on}, and a diffusion layer is formed. In order to avoid a deposition according to the tertiary current density distribution, the deposition is paused for the rest time t_{off}. During this time, the diffusion layer is almost replenished, and there will be almost everywhere at the workpiece the same metal ion concentration when the next deposition pulse commences. A typical concentration profile is shown in Figure 9.26. The rest time t_{off} can also be used to generate a small anodic current

[123] H. E. Haring, W. Blum; *Trans. Am. Electrochem. Soc.* **1923**, *44*, 313.

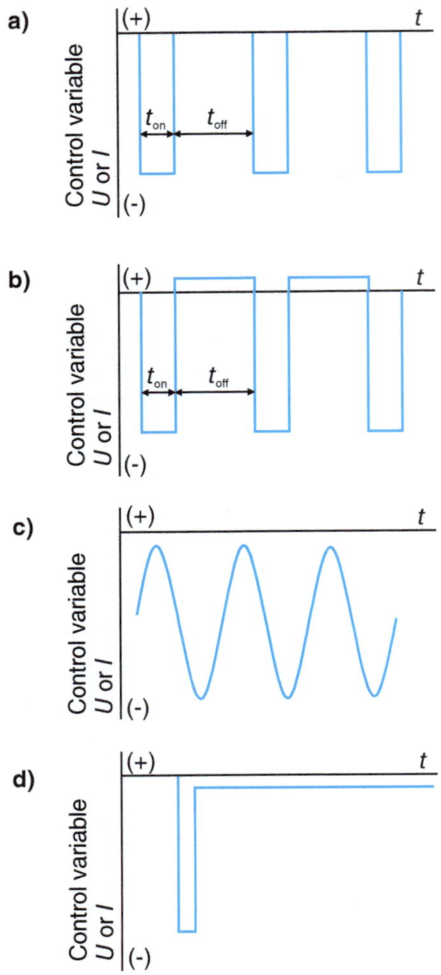

Figure 9.25 Pulse shapes for pulsed electrodeposition processes. a) Rectangular pulses; b) rectangular pulses with anodic dissolution pulse; c) sine function of the control variable; and d) nucleation pulse.

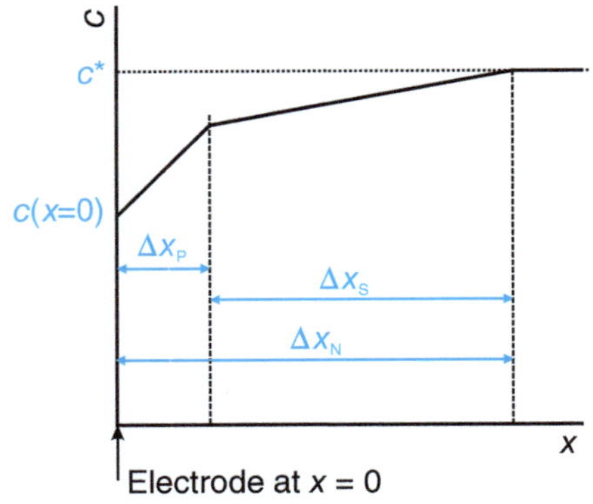

Figure 9.26 Concentration profile during pulsed electrodeposition according to Figure 9.25a. Δx_N, Nernst diffusion layer thickness; Δx_S, steady-state part of the diffusion layer with small changes between the parts of a pulse cycle; Δx_P, diffusion layer thickness with strong concentration changes between the parts of a pulse cycle.

called a periodic reverse pulse (Figure 9.25b). This causes the electrodissolution of a small fraction of the deposited material. Because the transport of the dissolved material is retarded in indents of the workpiece, there will be an enhancement of the metal cation concentration in scratches and microscopic indents relative to the raised parts of the workpiece. This causes more material to be deposited in the recessed areas during the next deposition pulse. In addition, the dissolution current density is higher at unwanted dendrites and protruding structures, which will be leveled periodically by the procedure. For higher overall currents, sine functions of current are used as the control variable instead of rectangular pulses because they can be more easily realized from an electrical engineering point of view (Figure 9.25c). If the amplitude of the sine wave is larger than the mean current, there will be periodic phases of anodic dissolution currents.

The nucleation pulse at the beginning of the deposition procedure in Figure 9.25d aims for generating many nuclei initially and subsequently allowing their growth at lower current densities without the formation of new nuclei.

Pulsed voltages and currents also affect the behavior of surface-active additives, whose adsorption is generally potential-dependent (Section 8.10.4). If an additive X is more strongly adsorbed at E_{on} than at E_{off}, the interfacial concentration Γ_X of the additive X can be depleted during the rest time. For $\Gamma_{X,eq}(E_{on}) < \Gamma_{X,eq}(E_{off})$, the additive X will be accumulated at the interface during t_{off}. If there is more than one additive (competing adsorption), the relative interfacial excess concentration will change during a pulse cycle, offering further possibilities to tune the process at the expense of increasing the complexity of the interacting equilibria and the transient processes.

Further reading: [23].

9.3.5 Influence of Bath Temperature

The temperature of the galvanic bath exerts an influence on almost every step of the metal deposition process. Thus, it has to be kept constant, typically within a 5 K interval. Higher temperatures tend to favor higher deposition rates, a faster dissolution of the sacrificial anode, the diffusive mass transport, and the possibility of working with a low cation concentration of the metal precursor to be deposited. The throwing power must be optimized for each desired temperature range and current density range by adaptation of the bath composition.

9.4 Composition of Galvanic Baths

9.4.1 Basic Types of Galvanic Baths

There are different known bath compositions for each metal to be deposited. Many of them have been published with their main constituents as early as the mid-nineteenth century. The use of additives has become extremely important nowadays in order to adjust the coating process to the requirements of new products or to stricter environmental protection regulations (avoidance of heavy metal waste, cyanide-containing solutions, or Cr(VI) compounds) or in order to achieve process intensification. Often, the particular technological solutions are not disclosed. Within this textbook, we cannot even attempt to provide a representative overview of the large variety of galvanic bath compositions, but refer the interested reader to the literature [10].

Table 9.2 shows a comparison of four baths for the electrodeposition of copper. Below, we will discuss the function of the most important constituents. In an acid

Table 9.2 Composition of galvanic baths for Cu deposition.

Property	Acidic Cu bath	Cu for printed circuit boards	Cyanidic Cu bath	Rochelle electrolyte[a]
Metal source	(160–240) g L^{-1} $CuSO_4 \times 5H_2O$	(20–80) g L^{-1} $CuSO_4 \times 5H_2O$	25 g L^{-1} CuCN	70 g L^{-1} CuCN
Supporting electrolyte	(40–100) g L^{-1} H_2SO_4	(160–200) g L^{-1} H_2SO_4	Complexing agent is also a supporting electrolyte	Complexing agent is also a supporting electrolyte
Complexing agent	–	–	35 g L^{-1} NaCN	80 g L^{-1} NaCN 30 g L^{-1} Na-K-tartrate ($NaKC_4H_4O_6 \times 4\ H_2O$)
Additional salts	(30–150) mg L^{-1} Cl^-	(30–150) mg L^{-1} Cl^-	20 g L^{-1} Na_2CO_3	20 g L^{-1} Na_2CO_3
Additives	Typical are poly(ethylene glycol), organosulfur compounds like thiophene or thiourea Brightener: cresyl violet	Typical are poly(ethylene glycol), organosulfur compounds like thiophen or thiourea Brightener: cresyl violet		Used without additives or with thiosulfate, Se compounds, or alkanethiols
Bath temperature	(20–40) °C to 50 °C	(20–40) °C to 50 °C	(20–30) °C	(50–70) °C
Current density	(0.02–0.10) A cm^{-2}	(0.01–0.04) A cm^{-2}	(0.002–0.005) A cm^{-2}	(0.01–0.05) A cm^{-2}
Typical application	General application, but not for deposition on substrates from electronegative metals	Printed circuit boards	Better throwing power than acid baths, applicable for coating of electronegative metals	Same as for cyanidic bath, but for higher current densities

a) Named after Rochelle salt, that is, potassium-sodium-tartrate.
Source: According to [9, 24].

electrolyte, the addition of sulfuric acid provides a high ionic conductivity (extra conductivity of H^+). Maintaining a low pH is also important to prevent the formation of metal hydroxides of low solubility.

9.4.2 What Is the Function of Complexing Agents?

The addition of complexing agents is necessary when a deposition from acidic electrolytes is not possible, for instance, when substrates from electronegative metals are to be coated that would dissolve in an acidic solution. The complexing agent keeps the precursor metal cations in solution but also shifts the equilibrium potential for the metal deposition toward more negative potentials depending on the complex stability constant (discussed in Section 5.3.12). Important complexing agents are halogenides (Cl^-, Br^-, I^-) and pseudohalogenides (such as cyanide CN^-, thiocyanate CNS^-, and selenocyanate $SeCN^-$), ammonia NH_3, hydrazine N_2H_4, phosphates $P_4O_{12}^{4-}$ and $P_2O_7^{4-}$, sulfate SO_4^{2-}, sulfite SO_3^{2-}, and thiosulfate $S_2O_3^{2-}$. A table of relevant complexes with their stability constants and formal potentials can be found in [24] and in the Appendix, Table A-4.

The addition of a base (Na_2CO_3) ensures that the solution cannot become acidic during the deposition process. This could be very dangerous in the cyanidic bath of Table 9.2 due to the release of extremely toxic and volatile hydrogen cyanide [$CN^-(aq) + H^+(aq) \rightleftharpoons HCN(g)$]. Often there is an equilibrium between complexes of different stoichiometries, such as $[Cu(CN)_2]^-$, $[Cu(CN)_3]^{2-}$, and $[Cu(CN)_4]^{3-}$ in the cyanidic bath from Table 9.2. In order to elucidate the reaction mechanisms and, in particular, the emerging surface-bound intermediates during the electrodeposition process, spectroscopic *operando* methods such as SERS (Section 8.2.2) are indispensable. Such methods provide insights facilitating a rational optimization of process parameters.

9.4.3 What Is the Function of Brighteners, Levelers, Activators, and Wetting Agents?

Brighteners, levelers, surfactants or wetting agents, and activators adsorb to the surface of the workpiece and the growing metal film, thereby decelerating the rate of the galvanic deposition. The adsorption of those additives can be investigated using the methods discussed in Chapter 8 or using voltammetric methods (Chapter 5, AC voltammetry, stripping voltammetry).

However, one should keep in mind that the methods as discussed in Chapter 8 provide adsorption isotherms of individual compounds, which means the interfacial concentration (or coverage) as a function of the solution bulk concentration at equilibrium at a specified temperature and at a specified electrode potential. For understanding the role of additives in a deposition experiment, this is only a brief, yet useful, orientation before an in-depth study. For a deposition experiment, the *dynamic* interfacial concentration of the additives is important because the equilibrium is generally not reached! Typical values for the coverage θ are $0.2 < \theta < 0.8$. The additives are contained in the deposition bath in much lower concentrations than the metal cations (or their complexes) to be deposited.

The function of a leveler (Figure 9.22b) is actually based on the diffusion-controlled adsorption and incomplete coverage of the leveler on metal film. Therefore, the coverage is higher at the protruding parts compared to the recessed parts of the workpiece. In a very simple approach, the deposition rate (i.e., the local current density) is thought to be proportional to the uncovered fraction of the metal electrode surface $j \approx k(1-\theta)$. This explains why medium coverages are required to achieve the leveling effect. If there is no coverage, the growth in the best case is uniform everywhere. This would also be the case if the electrode were almost completely covered. Only if the coverage is significantly lower in the recessed parts of the workpiece than in the elevated parts will the layer growth be faster in the recessed areas. When studying the isotherms in Section 8.10.4, it becomes clear that the transition from an uncovered to an almost completely covered surface is expected within a rather narrow range of bulk concentration. Thus, additive concentrations have to be adapted carefully and must be monitored during the usage of the bath.

Because additives change the rate of deposition for different crystallographic directions in a different way, additives can also change the morphology of the deposited film and thus the macroscopic appearance of the obtained metal layers.

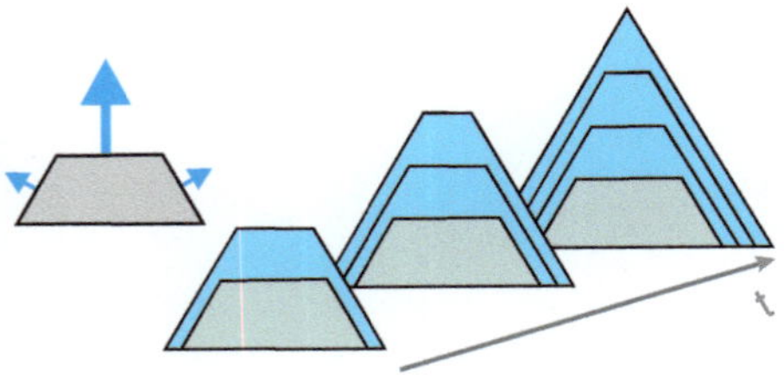

Figure 9.27 Development of crystal morphology by the different growth rates along different crystallographic axes.

> According to Figure 9.27, the external surface of a crystal is determined by the facet with the lowest growth rate (Gibbs[8]-Wulff[124] theorem [25]).

The molecular mechanisms by which the additives influence the deposition process are complicated and are often unknown. Their application is based on empirical knowledge. This situation facilitates only a very limited predictability of new bath formulations. We have already discussed the thermodynamic foundation for the adsorption of organic compounds to electrode surfaces in Section 8.10.

[124] Georgi (Yuri) Viktorovich Wulff, 1863–1925, Russian crystallographer.

Typically, several additives are present simultaneously in electrodeposition processes and compete for adsorption sites at the interface between the workpiece and the electrolyte solution. The **HSAB concept** (hard and soft acids and bases) by Pearson[125] allows qualitative predictions for the tendency of molecules and functional groups of molecules to adsorb on electrode surfaces [26].

According to the HSAB concept, hard acids (or bases) bind preferentially to hard bases (or acids), and soft acids (or bases) prefer soft bases (or acids) as binding partners.

The organic molecules with their nonbinding electron pairs are the Lewis[34] bases, and the metal surface is the electron-deficient Lewis acid. The hard and soft properties are associated with the polarizability of electrons in different binding states. Polarizability describes how easily electron density can be shifted within atoms/molecules by an external electric field.

For metals, the hard/soft character can be predicted from the type of binding electrons, for which the hardness decreases from s-electrons to f-electrons (Figure 9.28). This tendency is based on the larger probability density of s-electrons in the vicinity of the atomic nucleus and the very spread-out probability density of d-electrons and f-electrons.

Hard | s > p > d > f | Soft

Figure 9.28 HSAB-Konzept für Metallionen.

For organic molecules, the hard/soft character is determined by the number of σ-electrons and π-electrons: the polarizability (and soft character) increases with an increasing number of π-electrons. This reasoning can be applied in the same way to fragments of molecules (or functional groups).

When comparing different elements, their electronegativity can be used as a measure of hardness: the higher the electronegativity, the harder the element in the HSAB concept. Consequently, the hardness increases within a row of the periodic table of the elements (from Li to F) and decreases within a main group (from Li to Cs). For transition metals, the HSAB hardness initially increases within a d-block (from Sc to Zn) and decreases within a group (from Cu to Au).

The different facets of a single crystal (Section 8.9.1) can also be compared within the framework of the HSAB concept. For fcc crystals, the (1 1 1) surface has the highest number density of surface atoms and thus the highest HSAB hardness, followed by the (1 0 0) and (1 1 0) facets. Polycrystalline surfaces expose different facets to the solution, and their HSAB hardness is the weighted average of the contributing facets. It is often between that of the (1 0 0) and the (1 1 0) facets. The HSAB character of metal electrodes also depends on the electrode potential. The softest character is observed at the point of zero charge E_{pzc} (Section 8.10.3). The HSAB hardness increases with deviation of the electrode potential from E_{pzc} in either positive or negative direction.

Additives may influence the rate of metal deposition by intercepting different steps within this complex process. Solvated metal ions can be desolvated at a step edge and diffuse along the step edge to be incorporated in the metal lattice at a kink site (step edge ion-transfer mechanisms). Alternatively, they can be bound as an adatom to the surface of a terrace first, then reach a kink site by surface diffusion on the terrace to a step and along the step to be finally incorporated into the metal lattice at the kink site (terrace ion-transfer mechanism). Adsorbed additives can alter the accessibility of the kink sites, influence the rate of surface diffusion, or stabilize/destabilize adatoms.

[125] Ralph G. Pearson, 1919–2022, US-American chemist, professor at the University of California in Santa Barbara.

Nucleation is also influenced by adsorption processes. Let us imagine an adsorbed monolayer of an additive. In most cases, the monolayer will suppress the formation of an adatom and thus also the formation of a nucleus. If the adsorption and desorption rates are high, there will always be adsorbate molecules leaving their place at the surface. In the moment of desorption, the uncovered surface site may allow the formation of a nucleus.

Levelers and Brighteners exert a similar role. Some compounds also fulfill both functions in the sense that levelers act on a length scale corresponding to the extension of the diffusion layer (Figure 9.22), while brighteners act on irregularities on a much smaller length scale (less than 0.5 μm).

A further molecular mechanism is the competing adsorption of the primary brightener with an inhibitor that would otherwise cause step bunching (Figure 9.13) and thus yield coarse-grained metal layers. Important primary brighteners are sulfonamides, sulfonimides, benzene-monosulfonic acid, benzene-disulfonic acid, benzene-trisulfonic acid, alkylsulfonic acids, sulfinic acids, and arylsulfonates.

The action of a primary brightener is often enhanced by a secondary brightener at low concentrations. The combination of both additives allows the formation of mirror-smooth surfaces. Typical secondary brighteners are thiourea, acylthiourea, mercaptoalkyl-sulfonic acids, bis-sulfoalkyl-disulfides, thiocarbazones, thiosemicarbazones, and thiohydanthonin.

The effect of brighteners is often bound to a narrow range of bath temperature and current density. The effect of *different* current densities can be probed effectively in *one* experiment by means of a Hull cell[121] experiment (Figure 9.19).

Wetting agents improve the wetting of the surface and facilitate the detachment of gas bubbles from the surface of the workpiece. The gas bubbles originate from the competing HER or are attached air bubbles after immersion. The surfactants have both hydrophilic moieties that orient toward the aqueous electrolyte and hydrophobic groups pointing toward the metal surface or toward the interior of a gas bubble. The surfactants also promote the dissolution of other additives that would otherwise be barely soluble in aqueous solutions. They prevent or retard the coagulation and the sedimentation of solid dust particles that would otherwise be easily incorporated into the growing metal films. Many wetting agents also inhibit the crystal growth and thus share properties with brighteners. Finally, wetting agents facilitate the emersion of the workpiece from the deposition bath and thus decrease losses of the bath liquid and carryover of the bath solution into subsequently used rinsing facilities or further coating processes.

Additives are only dynamically exchanged between the electrode surface and the electrolyte solution and, thus, do not appear in the reaction equations for the metal deposition reactions. However, additives are slowly depleted in the bath. This necessitates a periodic resupply of additives to the bath. Besides the carryover of additives with the liquid films during emersion of workpieces from the deposition bath, additives can also be incorporated into the growing metal film either in their original form or after reduction. This also alters the layer properties (mechanical properties, conductivity, and corrosion properties). They can also be oxidized at the counter electrode.

Further reading about the action of additives and typical bath compositions can be found in [27–29].

Excursion 9-1 Filling of Structures with Extreme Aspect Ratio

The filling of recessed structures with copper is of considerable importance for different products of the electronic industry. It is also called superconformal film growth. The copper structures are usually used for electrical contacting of electronic chips based on silicon (Figure 9.29a) produced in the Damascene process. Other applications are the contacts between layers of 3D stacks of wafers or between layers of printed circuit boards (PCBs). Such boards have multiple layers in modern products, such as mobile phones. This is necessary to achieve a high density of the attached active elements without crossing the circuit paths (Figure 9.29b). Multilayer architectures require contacts between the layers that are made through small openings within the layers. Their diameters may vary between 40 nm for an individual element within a silicon chip and 80 μm on a modern PCB. Such a "through hole" contact is called "VIA", from vertical interconnect access. The electrical contacts also conduct a considerable fraction of the heat produced by the electronic elements. The planning of their dimensions and paths has to take into account both the heat transport and the electric currents.

Deposition experiments are conducted in test structures with trenches in the required size regime and with the aspect ratio aimed for. The aspect ratio is the ratio between depth and width. The surface of the insulating test structure is first decorated with Cu nuclei (e.g., by a process described in Section 9.1.3). On top of this layer, the actual electrodeposition is carried out, after which cross sections of the test structures are prepared and assessed under a scanning electron microscope. Unwanted features are voids that may be formed when the thickness growth at the vertical walls of the trench structures causes an occlusion of the trench before the void space is filled by the metal to be deposited (Figure 9.30a). Such electrolyte-filled voids not only decrease the electronic conductivity of the vertical interconnect but can also become the reason for early failure of the device. When large current densities pass the connecting structure, copper ions may be dissolved at one side and redeposited at the other side of the void. In this way, the voids migrate through the device. Several initially very small voids may join to a larger void that is trapped under a bond pad or similar structure (Figure 9.30b). Eventually, this may cause an interruption of the current path and thus failure of the entire device.

The electrodeposition in tight slots becomes possible only by the combined action of different additives (Figure 9.31) in an acid copper bath that also contains chloride. Typical baths contain poly(ethylene glycol) (PEG) as a suppressor that forms a film together with chloride ions, which strongly inhibits the Cu deposition. In this film, bis-3-sulfopropyl-disulfide (SPS) is incorporated as an accelerator or brightener. SPS binds to the copper surface under cleavage of the disulfide group. The hydrophilic sulfonic acid groups direct the molecular backbone toward the electrolyte solution and cause SPS to not, or to only a very small extent, be embedded in the growing Cu film. All additives are surface-active compounds whose binding to the surface can be studied using the methods discussed in Chapter 8. On a surface with the film described above, the electrodeposition rate is much smaller than the diffusion-controlled deposition. Therefore, concentration gradients of Cu^{2+} within the slot structure do not play a significant role.

The competing adsorption of the additives is very important for this process. The incorporation of SPS into the inhibiting film causes a strong acceleration of the Cu deposition. The local deposition rate is proportional to the local interfacial

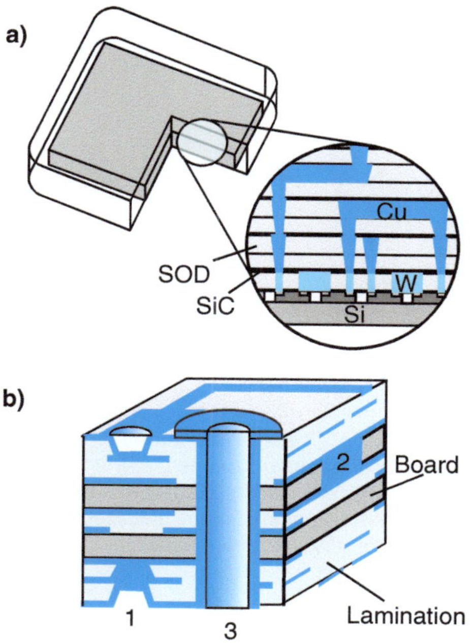

Figure 9.29 Metal deposition into structures with high aspect ratios. a) Contacting of individual electronic circuit elements in an integrated circuit; SOD is the spin on dielectric – insulating layer (SiO_2) made from precursors that can be applied by spin coating; b) cross-sectional view of a multilayer printed circuit board with 1) BMV (blind micro via), 2) conformal metal layer, and 3) through hole metallization. The figure is schematic and not to scale. Copper tracks are in blue.

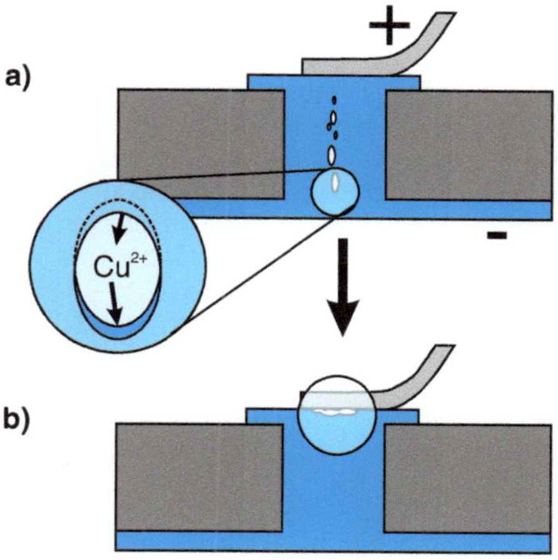

Figure 9.30 a) Schematic representation of voids in vias; b) mechanism for the migration of voids and failure of devices.

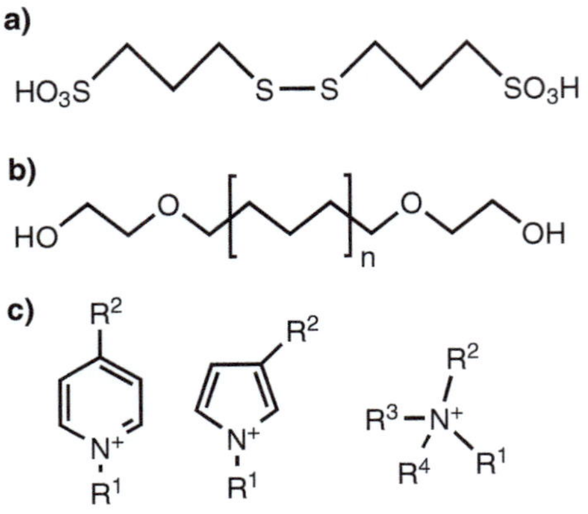

Figure 9.31 Chemical structures of a) bis-3-sulfopropyl-disulfide (SPS), b) poly(ethylene glycol) (PEG), and c) levelers on the basis of pyrimidinium, imidazolium, and quaternary ammonium ions.

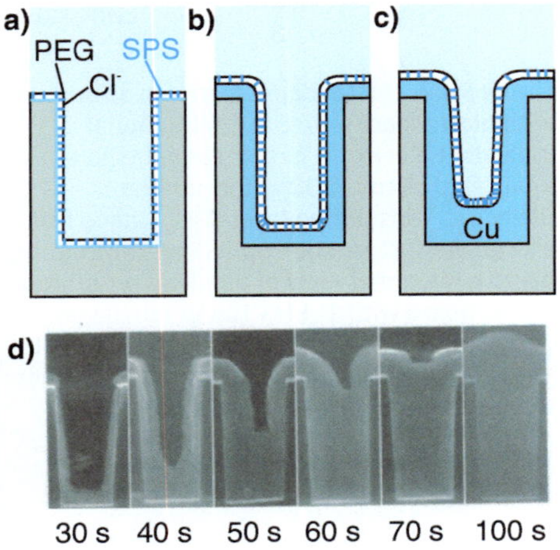

Figure 9.32 a–c) Mechanisms by which SPS is concentrated at the bottom of test structures. a) Before the onset of the electrodeposition; b) formation of zones with enhanced and decreased interfacial SPS concentration; c) filling of the trench structure; and d) experimental example for the time course of the electrodeposition.
Source: Panel d) reproduced with permission from [31], Copyright 2003, Wiley-VCH.

SPS concentration. SPS is contained in the deposition bath in much lower concentrations than PEG. Therefore, the inhibiting film initially contains much less SPS than PEG (Figure 9.32a). The filling of the trench structure can be explained by a model based on curvature-enhanced adsorbate coverage: At the beginning of the electrodeposition, a uniform thickness growth commences. During this process, the surface area increases at the convex curvature at the upper edge of the trench. The competing adsorption equilibria are not established instantaneously, mainly because of the low bulk concentration of SPS in the bath. The inhibiting film at the growing part of the surface contains more chloride and PEG and less SPS than the initial equilibrated film. This causes progressively stronger local retardation of the film growth.

The opposite scenario is encountered at the bottom of the trench. Here, the surface area shrinks due to the deposition. Because SPS molecules do not leave the film once incorporated, the local SPS concentration in the film is enhanced (Figure 9.32b), causing also an enhanced local deposition rate. This further reduces the interfacial area and further increases the local interfacial SPS concentration – the trench is filled from the bottom up (Figure 9.32c).

Because there is a high local interfacial SPS concentration at the original position of the trench, the film continues to grow faster there once it has reached the upper plane of the test structures (Figure 9.32d). This causes a small elevation of the film before the SPS concentration equalizes within the entire inhibiting film. A further fine-tuning of the process for different applications can be achieved by the use of levelers, the adaptation of the current densities, the use of pulsed deposition procedures, the use of convective mass transport, and the optimization of the temperature.

Further reading: Numerous examples are collected in [31], and the main lines of a quantitative description based on data from planar electrodes are outlined in [30, 32].

9.5 Deposition of Complex Materials

The vast majority of technologically used materials are not pure metals, but alloys. Alloys and composites surpass pure metals by a wide margin in important properties such as corrosion resistance (e.g., Zn–Sn, Zn–Cd, and Cr–Ni), heat resistance (Cr–Mn, Cr–W, Cr–Mo, and W–Fe), and dynamic friction properties (Zn–Bi, Ag–Sn, and Pb–Cu), solderability (Sn–Pb, Sn–Bi, and Sn–Sb), magnetic properties (Ni–Co, Ni–Fe, Ni–Fe–Co, Co–W, Co–Pt, and Fe–Pt), electrical contact (Ag–Sb, Au–Pd, and Au–Ni), as well as decorative and protective properties (Ni–Zn, Ni–Au, and Au–Ag). Thus, it is hardly surprising that the electrodeposition of alloys has developed in parallel with the deposition of pure metals that we have already discussed. Composites provide the opportunity to combine the properties of very different classes of materials and to achieve completely new combinations of functional properties. Below, we will identify the peculiarities of the direct electrodeposition of alloys and then turn our attention to multilayer systems and composites.

9.5.1 Direct Galvanic Deposition of Alloys

The direct deposition of alloys yields at *one* deposition potential E_d the alloy of two (or more) elements with the composition A_xB_y in a solution containing the precursors (A^{m+}, B^{n+}, ...) of two (or more) metal cations or metal complexes.

$$x\,A^{m+} + y\,B^{n+} + (x \cdot m + y \cdot n)\,e^- \rightarrow A_xB_y \tag{9.47}$$

In many cases, the HER will occur simultaneously. The total current density j is therefore obtained by summing up three (or more) partial current densities: j_A for A, j_B for B, and j_{HER} for H_2.

$$j = j_A + j_B + j_{HER} \tag{9.48}$$

The partial current densities j_A and j_B have to be similar at the deposition potential if the alloy is to contain the elements in the desired mole fractions $x_A = x/(x + y)$ and $x_B = y/(x + y)$. The partial current densities are determined according to the Butler–Volmer Equation (3.38) by the respective exchange current densities $j_{A,0}$ and $j_{B,0}$ and the equilibrium potentials of the two metal electrodes $E_{eq}(A|A^{m+})$ and $E_{eq}(B|B^{n+})$. For an aqueous electrolyte of only one precursor (A^{m+} or B^{n+}) and a constant E_d, the overpotential of the metals A and B is $\eta_A = E_d - E_{eq}(A|A^{m+})$ and $\eta_B = E_d - E_{eq}(B|B^{n+})$, respectively. Due to the exponential term in the Butler–Volmer equation, no alloy deposition would be expected if the standard redox potentials of the two metals are too different from each other.

However, the partial current densities *cannot* usually be approximated from current density curves in a solution containing only a precursor for one metal component. Instead, another approach is used: Deposition experiments are conducted under steady-state conditions at different deposition potentials. For each potential, the following steps are executed for the evaluation of the results:

1) The partial current density j_{HER} for HER is subtracted from the total current density j. Data for j_{HER} are calculated from a volumetric measurement of the evolved H_2. The total mass of the deposited alloy can be obtained from a quartz crystal microbalance (Section 8.9.2).
2) Next, the mole fractions x_A and x_B in the obtained alloy are determined spectroscopically.
3) The partial current density j_A is then calculated by Eq. (9.49).

$$j_A = \frac{(j - j_{HER})mx_A}{mx_A + nx_B}. \tag{9.49}$$

Then, j_B is obtained from Eq. (9.50)

$$j_B = j - j_{HER} - j_A \tag{9.50}$$

Figure 9.33 shows the partial current densities for alloy A–B. Because the contribution of j_A and j_B to the total current changes with potential, the exact composition of the alloy changes with the deposition potential (Figure 9.33, curve 4 for x_A).

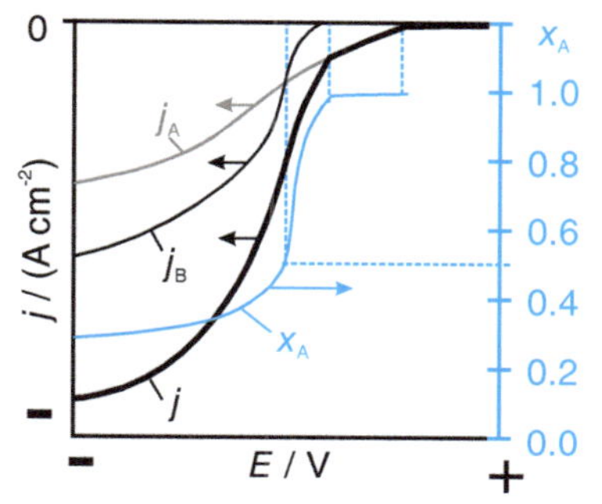

Figure 9.33 Current density–potential curves for the galvanic deposition of an alloy of metals A and B (schematically). 1) Overall current density corrected for j_{HER}; 2) partial current density j_A for the deposition of A; 3) partial current density j_B for the deposition of B; and 4) mole fraction x_A of the element A in the alloy as a function of the electrode potential (right ordinate).

Figure 9.34 shows the comparison of current density–potential curves for the deposition of the pure metals in an equivalent electrolyte containing either the precursor for A or the precursor for B only. In this case, the curve for the more noble metal A is shifted during alloy deposition in Figure 9.33 to more negative values (overpolarization of the deposition), while the partial current density for the alloy deposition of the less noble component B is shifted positively compared to the deposition of pure B (depolarization of the deposition). The following factors must be considered as possible reasons for the shifts:

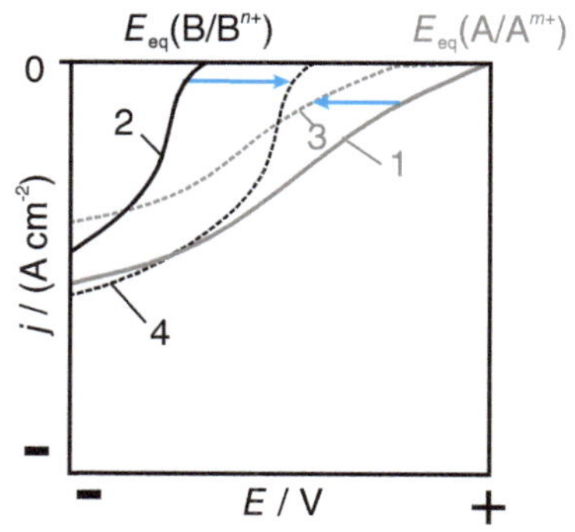

Figure 9.34 Current density–potential curves for the pure components (1) A and (2) B in an electrolyte containing *only* the precursor for A or *only* the precursor for B with an adapted ionic strength. The curves for the alloy deposition from Figure 9.33 are shown for comparison as dashed lines (3, 4).

- The mole fraction activities of the metal atoms are not one and are reduced compared to pure metals.
- The surface of the alloy can have a different structure of the electrical double layer, which also causes changes in the surface charge and the adsorbed solution components. Especially, the adsorption of hydrogen changes with the elements at the electrode surface.

- With the occurrence of different kinds of atoms at the electrode surface, the rates change for surface diffusion and for incorporation of adatoms into the metal lattices at kink sites compared to the pure metals.

Some of these factors can be captured by a statistical treatment of the problem that takes into account, for instance, the probability of the incorporation of an atom A or B at a kink site made by the element B (or A) and connects this to the potential-dependent residence times of A and B at a surface site. This treatment is described in [24] for binary and tertiary alloys.

9.5.2 Which Conditions Are Required for Direct Deposition of Alloys

There are further boundary conditions for the direct deposition of alloys besides the considerations discussed in Section 9.5.1.

Phase Diagrams of Alloys. Two metallic elements are miscible in any ratio only in exceptional cases, such as for the Au-Ag system. In most cases, an alloy phase is only formed within a certain range of mole fractions of the constituting elements. If the elemental composition of a melt or the overall atomic ratio of electrodeposited metals is outside of the stability range of an alloy phase, several phases with different element compositions and structures (arrangements of the atoms in the metallic lattice) can form simultaneously during solidification or electrocrystallization. Such solids are called multiphase alloys. The phases can be different if made by solidification of a melt or by electrocrystallization (Ag–Pb). The compositions are often similar. In a few cases, electrocrystallization allows the formation of supersaturated solid solutions of metals that would form multiphase alloys during the solidification of a melt with the same overall composition (Cu–Ag, Cu–Pb, Cu–Ni at low polarization). Due to the large variety of phase diagrams of alloys, generalizations that we have to make here for reasons of brevity must be considered with caution. A systematic introduction of the structural types of alloys with relation to alloy deposition can be found in [33].

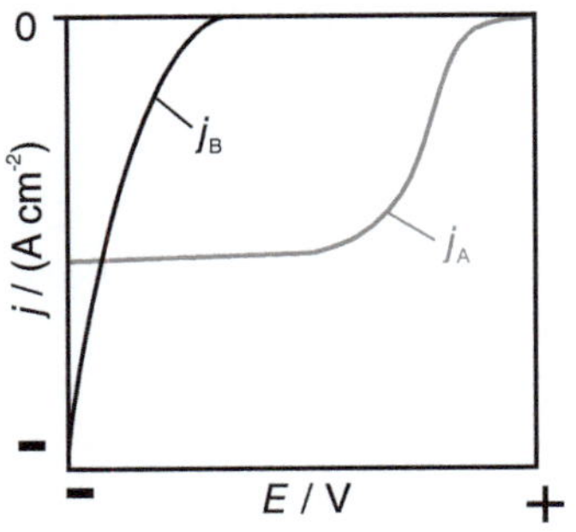

Figure 9.35 Alloy deposition at diffusion-limited conditions for the more noble element (schematic).

Adjustment of the Partial Current Densities by Means of Diffusion Limitation of One Component. If it is not possible to adjust the deposition potential of the precursors for the alloy by a complexing agent (see below), the content of the more noble element A can be limited by exploiting its limited transport by diffusion (Figure 9.35, normal codeposition). It must be ensured that the consumption of both components A and B occurs in the right ratio because a concentration variation of the more noble component immediately affects its content in the deposited alloy. In addition, there is a tendency for dendrite formation in diffusion-controlled metal depositions that can be counteracted by the addition of surface-active additives.

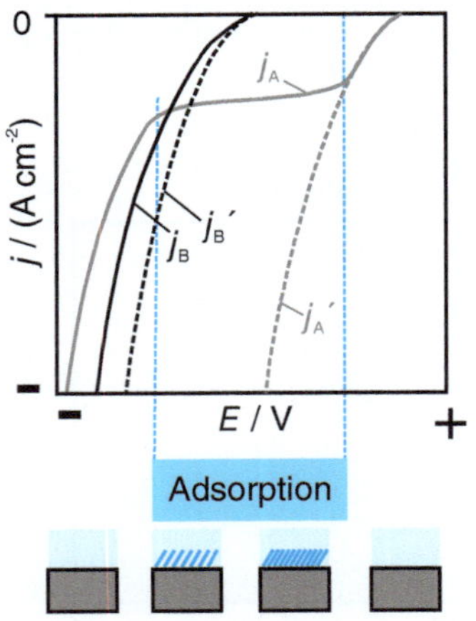

Figure 9.36 Alloy deposition is enabled by the selective inhibition of the deposition of the more noble element. j_A and j_B are the partial current densities in the presence of the additive, j_A' and j_B' are the partial current densities in the absence of the additive given for comparison (schematic).

Adjustment of Partial Current Densities by Selective Inhibition of the Deposition of the More Noble Metal. Sometimes an additive is able to selectively decrease the partial current density of the more noble element A (Figure 9.36) and thus adjust it to that of the less noble element B (abnormal codeposition).

Adjustment of the Partial Current Densities by Complexation. We consider the deposition of brass (Cu–Zn) as an example. The difference of the standard potentials of Cu|Cu^+ and Zn|Zn^{2+} amounts to about 1.3 V. Despite this large difference, the direct deposition of brass is possible provided that the complexation with CN^- decreases the concentration (concentration activity) of the *free* metal ions of the more noble element Cu much more than that of the less noble element Zn (see data in marginal column). A further condition is the *fast response of the complexation equilibria* (reversibility). This causes the immediate replenishment of metal ions consumed in the electrodeposition process by dissociation of some coordination complexes. Despite the very low concentration of free metal ions, a diffusion limitation is avoided in this way. The complexation greatly shifts the equilibrium potential in the negative direction. The derivation has already been provided in Section 5.3.12.

$E°(Cu|Cu^+) = +0.52$ V vs. SHE
$Cu^+ + 3\,CN^- \rightleftharpoons [Cu(CN)_3]^{2-}$
Brutto stability constants
$\beta = 10^{28}$ mol^{-3} L^3

$E°(Zn|Zn^{2+}) = -0.76$ V vs. SHE
$Zn^{2+} + 4\,CN^- \rightleftharpoons [Zu(CN)_4]^{2-}$
Brutto stability constants
$\beta = 10^{17}$ mol^{-4} L^4

Induced Electrodeposition and Induced Codeposition. Some metals, for example, W and Mo, cannot be electrodeposited from molybdate (MoO_4^{2-}) and tungstate (WO_4^{2-}) solutions. Instead, oxide layers of low quality are obtained. However, alloys of both metals can be electrodeposited with nickel as the second element from citrate-containing electrolytes.

To a lesser extent, variation of the solution concentration of the precursors for A and B, temperature control, and pulse programs can be used to achieve an alloy deposition.

9.5.3 Deposition of Multilayers and Composites

Besides the deposition of alloys, desired properties of a coating can also be achieved by local variation of the composition within the coating. The embedding of small particles of particularly hard materials (e.g., ceramics and boron nitride) into the growing metal film (electrocodeposition) increases the abrasion resistance, for example, of decorative coatings. We call such materials **composites** because the embedded particles are of a completely different binding type (e.g., partial ionic bond in a metal oxide) than the metallic bond in the surrounding metal matrix (Figure 9.37a).

Another variation is **multilayer systems** that are obtained by alternating electrodeposition of metal films in baths of different precursors (Figure 9.37b). Besides a periodic change of the deposition bath, which is not very attractive for industrial processes, multilayer formation can also be achieved in *one* electrolyte system, behaving similarly to Figure 9.35 by switching the deposition potential. At one potential, only the more noble element A is deposited, and an alloy with the less noble element B as the dominating component is obtained at a significantly more negative potential.

Thermal posttreatment of layered systems leads to a diffusive mixing of the two elements A and B, which can be used for making alloy coatings that are not accessible by simultaneous deposition. Some layered systems have their own significant technological importance, for example, layers exhibiting a giant magnetoresistance (GMR). They consist of alternating layers of a ferromagnetic material and a nonmagnetic material. The electrical resistance of those layered systems perpendicular to the orientation of the layers is a function of the magnetization of the layers. Depending on the thickness of the layers, very small magnetic fields are sufficient to change the magnetization of the thin layers, which in turn causes a change in the electrical resistance across the layer stack. This principle is used in reading heads of magnetic storage devices (hard drives). The first systems used a stack sequence of Fe–Cr–Fe with a layer thickness of 3 nm. Currently, Ni–Cu–Ni layers are preferred. Despite a smaller GMR effect, they have significant advantages due to better stability.

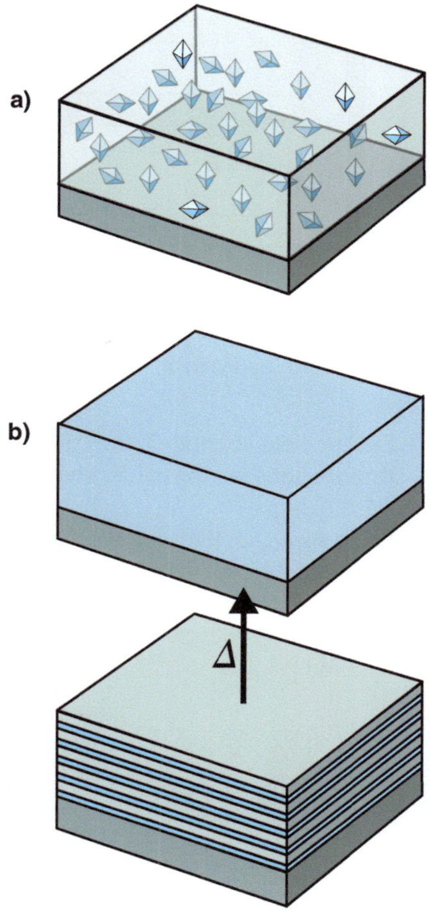

Figure 9.37 Multilayer and composite films. a) Embedding of ceramic particles into a thin composite film; b) multilayer film with the possibility of alloy formation during a thermal treatment after electrodeposition.

9.6 Electrochemistry of Bipolar Electrodes

9.6.1 Fundamentals

A bipolar electrode is characterized by the simultaneous occurrence of anodic and cathodic reactions at both ends of the electrode within an external electric field.

A classic example of the simultaneous occurrence of anodic and cathodic reactions at the same electrode is local corrosion reactions in which the local dissolution of the

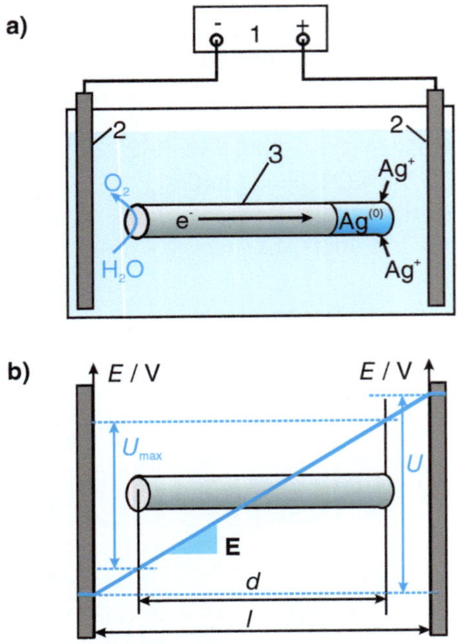

Figure 9.38 a) Electrochemical cell with a bipolar electrode; 1) external power source; 2) feeder electrodes; 3) bipolar electrode; b) potential distribution in a).

E – Electric field strength [$V\ cm^{-1}$]

l – Distance between the feeder electrodes [cm]

U – Externally applied voltage between the feeder electrodes [V]

d – Length of the rod [cm]

U_{max} – Maximum voltage between the end of the rod [V]

metallic workpiece is balanced by the reduction of oxygen at a much larger surface. We will discuss these important processes in Chapter 14. While corrosion reactions are spontaneous reactions, bipolar electrodes find increasing applications in cells with enforced reactions that require an input of energy ($\Delta_R G > 0$), but in which the bipolar electrode itself is *not* physically connected to a power source. We discuss those electrodes here because the principles of bipolar electrochemistry can be used for metal deposition. Currently, many more applications are being explored [34].

Figure 9.38 shows a simple setup with a bipolar electrode. The external voltage U is applied between two feeder electrodes and causes an electrolysis current at them. An electric field $\mathbf{E} = U/l$ is established between the feeder electrodes, as we have discussed in Sections 4.2.1 and 4.6. The rod-shaped bipolar electrode of length d in Figure 9.38a has a uniform potential, but the potential of the surrounding electrolyte solution changes over the length of the rod (Figure 9.38b) by U_{max}.

$$U_{max} = \mathbf{E}\, d \tag{9.51}$$

The voltage can be sufficient to cause different electrolysis reactions at both ends of the bipolar electrode. In a solution with a silver salt, this could be the electrodeposition of silver

$$Ag^+ + e^- \rightarrow Ag,\ E_1^{\circ\prime} = +0.80\ V \tag{9.52}$$

balanced by the anodic oxygen evolution reaction (OER).

$$6H_2O \rightarrow O_2 + 4e^- + 4H_3O^+, E_2^{\circ\prime} = +1.23\ V \tag{9.53}$$

Both reactions can occur simultaneously at *one* rod that is *not* connected to an external power source, provided that condition (9.54) holds.

$$U_{max} \geq E_2^{\circ\prime} - E_1^{\circ\prime} \tag{9.54}$$

Actually, we need to account also for the overpotentials of both reactions, $\eta_1 < 0$ and $\eta_2 > 0$, because a measurable current is achieved only for $E > E^{\circ\prime}$ (Chapter 3). With this, we find the condition (9.55).

$$U_{max} \geq \left(E_2^{\circ\prime} + \eta_2\right) - \left(E_1^{\circ\prime} + \eta_1\right) \tag{9.55}$$

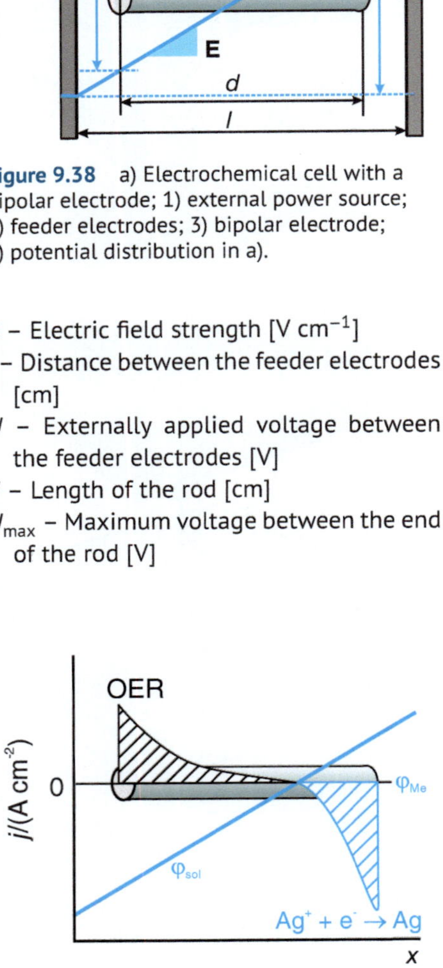

Figure 9.39 Distribution of the current density along the rod length.

The partial current densities can be estimated for each point along the rod by the Butler–Volmer equation, taking the local difference of the Galvani potentials $\Delta\varphi = \varphi_{Me} - \varphi_{sol}$ as the driving force in the exponential term of the Butler–Volmer equation. The constant potential of the rod spontaneously adjusts to the value at which the total anodic current I_a equals the total cathodic current I_c. I_a and I_c are obtained by integration over the local current densities [Eq. (9.56)] along the rod with radius r as schematically shown in Figure 9.39.

$$I_a = 2\pi r \int_0^l j_a(x)\,dx \quad \text{and} \quad I_c = 2\pi r \int_0^l j_c(x)\,dx \tag{9.56}$$

The current within the bipolar electrode $I_{be} = |I_a| = |I_c|$ can be measured by cutting the rod in the middle and electrically connecting both parts via an ammeter (Figure 9.40a). It is interesting to compare the current I_{be} within the bipolar electrode to the total current I_{tot} between the two feeder electrodes. The ratio

$$\varepsilon_{be} = \frac{I_{be}}{I_{tot}} \tag{9.57}$$

represents an efficiency number that we could name as bipolar current efficiency in analogy to similar quantities in other areas of electrochemistry. In the example, ε_{be} is smaller than one because only a fraction of I_{tot} passes through the bipolar electrode while another fraction is conducted through the electrolyte solution surrounding the rod. This part is called bypass current I_{bp}. Setups with $|I_{bp}| > 0$ are called open bipolar setups, and the equivalent circuit in Figure 9.40b illustrates the situation and makes clear the most important means to optimize ε_{be}. One approach would be to increase the electrolyte resistance around the bipolar electrode, for instance, by embedding the bipolar electrode in a less conductive matrix. We will discuss an example later in connection with Figure 9.41.

I_{be} – Current through the bipolar electrode (absolute value) [A]

I_{tot} – Current between the two feeder electrodes (absolute value) [A]

ε_{be} – Bipolar current efficiency [dimensionless]

In principle, we can also attempt to decrease the resistance of the bipolar electrode itself, which is essentially determined by the impedances at both ends of the bipolar electrode. Each of the two impedances is composed in a complicated manner of capacities and resistors, as discussed in Chapter 7. Among them, the potential-dependent charge-transfer resistances of the anodic side and the cathodic side make the largest contribution and can be greatly influenced by the deposition of electrocatalysts.

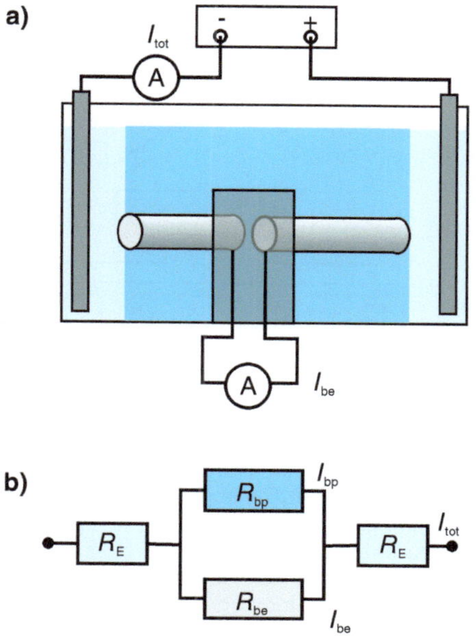

Figure 9.40 a) Separated bipolar electrode with the possibility of current measurement; b) simplified equivalent circuit of a bipolar electrode. R_E, electrolyte resistance; R_{be}, resistance of the bipolar electrode; R_{bp}, electrolyte resistance in the vicinity of the bipolar electrode (in the bypass).

9.6.2 Cell Setups with Bipolar Electrodes and Their Applications

Below we discuss some exemplary applications of bipolar electrodes in order to illustrate how the fundamentals introduced in Section 9.6.1 help to comprehend processes and phenomena for quite different systems. Figure 9.41 shows a **bipolar nanoelectrode**, which is embedded in a capillary. Such electrodes can be prepared in different ways. For a conventional operation, the embedded wire must be contacted electronically, which can be complicated due to the fragility of the thin metal wire. This difficulty can be circumvented if the metal wire is operated as a bipolar electrode between two feeder electrodes, one inside and the other outside of the capillary. Provided there is a leak-free embedding of the nanoelectrode into the capillary, the total current equals the current through the bipolar electrodes ($\varepsilon_{be} = 1$). Setups with $I_{bp} = 0$ are called closed bipolar setups; examples can be found in [36, 37].

Figure 9.42 shows a cell for the deposition of metals and polymers on bipolar electrodes. The objects to be coated are embedded within a gel-type electrolyte and placed in the middle compartment. The gel prevents the rotation of the objects and maintains their initial orientation with respect to the electric field. This setup can be used to produce Janus particles [35]. These are objects that are coated differently on the sides facing the two feeder electrodes. The name is derived from the Greek deity Janus, who is typically depicted with two faces looking backward and forward. Figure 9.43 shows another option for the continuous coating of rod-like objects (e.g., carbon nanotubes) using an apparatus for capillary electrophoresis (Excursion 4-4 and Figure 4.62) [38]. The objects to be coated are provided in the anolyte together with a metal salt. They are transported with the electroosmotic flow through the capillary. Only during passage through the capillary do the objects act as bipolar electrodes. Due to the small capillary diameter, R_{bp} is dramatically increased compared to a bulk cell. Within the capillary, a metal deposition occurs on the rod-like objects only on one side. Such microobjects have been used as self-propelled microswimmers. The propulsion is generated by a catalytic gas evolution reaction (e.g., $2H_2O_2 \rightarrow O_2 + 2H_2O$) at the metal-coated side of the rod. Impressive videos of such autonomous swimmers can be found on the internet. Perspectively, microswimmers with other chemical propulsion mechanisms and loaded with pharmaceuticals could be used to recognize cancer cells circulating in the bloodstream, attach to them via recognition molecules, and discharge their drug cargo to destroy them while minimizing many side effects on other organs of the patient.

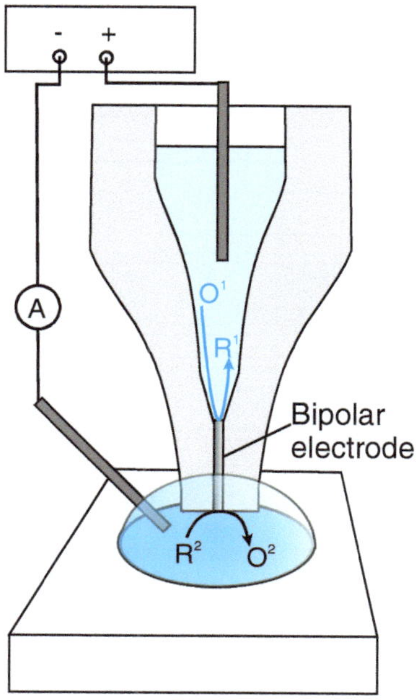

Figure 9.41 Micro- or nanoelectrode as a bipolar electrode.

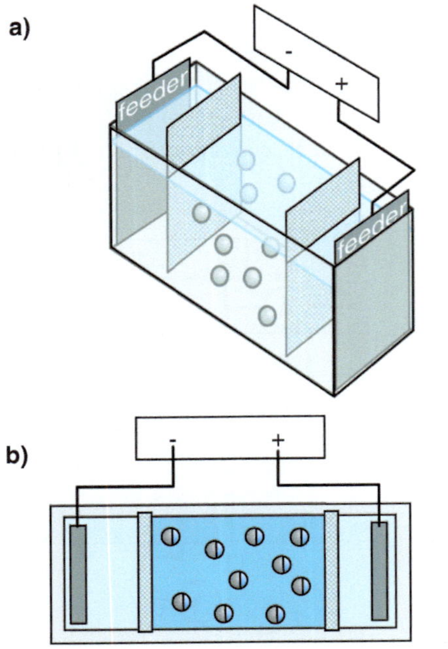

Figure 9.42 a) Cell for bipolar electrodeposition. The objects to be coated are held in a gel-type electrolyte in the middle compartment between two separators.
b) Top-view of the cell with Janus particles with metal deposition on only one side, according to [35], not to scale.

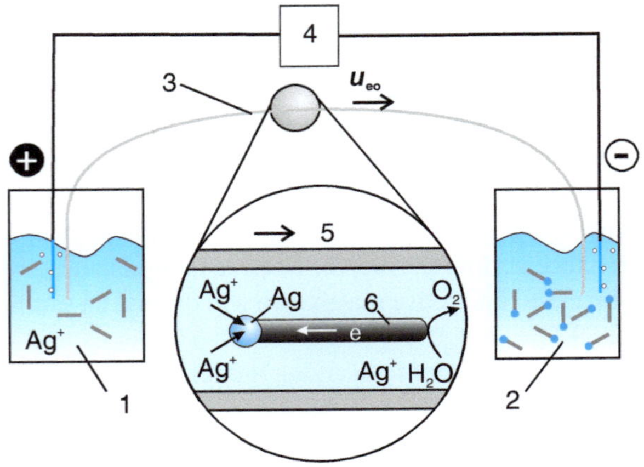

Figure 9.43 Bipolar electrodeposition by means of an electrophoresis instrument.
1 – Anolyte, 2 – catholyte, 3 – capillary, 4 – high voltage power source, 5 – sectional view along the longitudinal axis of the capillary, and 6 – carbon nanotube. The schematic is not to scale! Source: Adapted from [38].

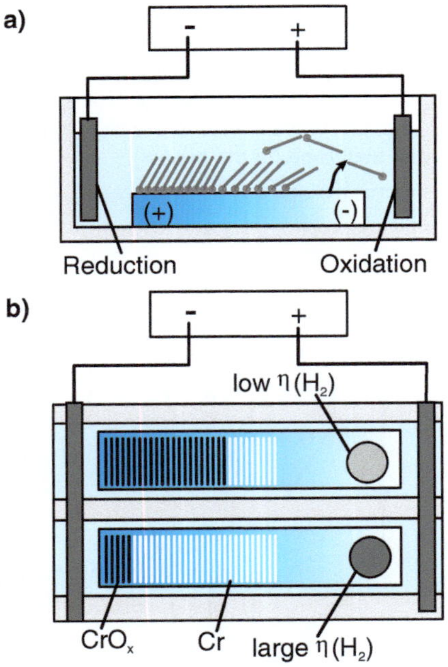

Figure 9.44 a) Generation of gradient materials at a bipolar electrode and b) metal oxidation for the optical indication of the intensity of the hydrogen evolution reaction (or oxygen reduction reaction), making use of the oxidation of Cr to CrO_x. Source: a) Adapted from [39]; b) adapted from [40].

The generation of gradient materials represents another application in which a potential-dependent surface modification is made within a region of the bipolar electrode. Examples are the formation of a mixed self-assembled monolayer (Section 13.1.5) with systematically varied composition of two adsorbates (Figure 9.44a) [39] or visually readable microtests for the electrocatalytic activity (Figure 9.44b) [40]. These tests are important for the combinatorial development of electrocatalysts (Chapter 12). At the cathodic side of the bipolar electrode, H_2 is formed (HER) or O_2 is reduced (ORR). These reactions require different overpotentials depending on the electrocatalyst used. The aim is the discovery of new materials with a particularly low overpotential for the target reaction. Such a material leaves a higher fraction of the externally imposed voltage U_{max} between the two feeder electrodes for the anodic partial reaction of the bipolar electrode. A suitable anodic reaction for optical indication is the conversion of metallic Cr to chromium oxides. Thus, an active electrocatalyst will cause the conversion of more bar-shaped ruler ticks of the test stripe in Figure 9.44b. The conversion starts at the outermost bar and progresses to the inner bars. This decreases the driving force for the bipolar reaction. The overpotential of the target reaction (HER or ORR) can be directly obtained from the number of converted bars.

Interestingly, bipolar electrodes have been used in large-scale industrial processes since the 1970s. Figure 9.45a shows a stack arrangement of bipolar electrodes. Because the electrodes are inserted with an insulating and impermeable connection to the side walls of the cell container, there is no bypass current, and the total current is passed through each of the bipolar electrodes ($I_{bp} = 0$ and $I_{tot} = I_{be}$) [41]. We will also encounter such a cell configuration when discussing the setup of fuel cells in Chapter 16. Larger surface area can be realized with bipolar fixed-bed reactors (Figure 9.46) [42]. The volume between the feeder electrodes is filled with a bed of conducting particles. Each particle is covered with an insulating net that prevents the electronic conduction between neighboring particles [43]. Thus, each particle acts as a bipolar electrode, maximizing the space–time yield of the reactor.

Further Reading. Besides the references provided for the examples discussed, a review [34] is recommended as a systematic introduction.

Test your knowledge

Exercise 9.1: Demonstrate the equivalence of Eqs. (9.17) and (9.18) for a hemispherical nucleus.

Exercise 9.2: Calculate the expected value for the critical size n_{crit} of 3D clusters of hemispherical, cubic, and pyramidal shapes according to Eq. (9.18) for the following absolute values of the overpotential: 10, 20, 30, 40, and 50 mV. Use as representative material constants those of Ag(1 0 0)/$AgNO_3$ with $\gamma = 1 \times 10^{-6}$ J cm^{-2}. Advice: Derive specific equations for the different shapes of the nuclei, then use a spreadsheet for the actual numerical operations.

Exercise 9.3: Similar to Exercise 9.2, calculate the critical cluster size for circular and square-shaped 2D nuclei. Advice: Use Eq. (9.19) and the specific value for silver $<\varepsilon> = 2 \times 10^{-13}$ J cm^{-1} [5].

Exercise 9.4: The two cathodes in a Haring–Blum cell (Figure 9.24) have a distance of 20 cm. The wetted area of the electrodes amounts to 10 cm × 7.5 cm. The anode is placed at a distance of 15 cm from one of the cathodes. In an electrodeposition experiment in 2 mol L^{-1} $CuSO_4$ solution with additives at 30 °C, the lcd cathode exhibits a mass increase of 0.2 g while the hcd cathode exhibits a mass increase of 0.4 g. Calculate the throwing power of the electrolyte. Is the calculated value the macrothrowing power of the microthrowing power?

Exercise 9.5: Inform yourself about the chemical structure of the organic additives listed in Section 9.4.2. Collect in a table the systematic names, possible trivial names, and the Lewis formulas of the compounds.

Exercise 9.6: Calculate with the data from the margin column in Section 9.5.1 the equilibrium potentials of the following electrodes:

$$\text{Cu}|1\ \text{mol L}^{-1}\ [\text{Cu(CN)}_3]^{2-}, 2\ \text{mol L}^{-1}\ \text{CN}^-$$

$$\text{Zn}|1\ \text{mol L}^{-1}\ [\text{Zn(CN)}_4]^{2-}, 2\ \text{mol L}^{-1}\ \text{CN}^-$$

Compare the shift of both potentials against the standard redox potentials of Cu|Cu^+ and Zn|Zn^{2+}, respectively.

Exercise 9.7: A production line for the electrolytic tinning of steel strips (tin plate) with a thickness of 0.1 mm and a width of 60 cm continuously processes coils of steel strips with a mass of 20 t. It generates a coating of 2 g of tin per m^2. The sheet runs through the galvanic bath with a linear velocity of 10 m s^{-1}. [ρ(β-Sn) = 7.31 g cm^{-3}, ρ(steel) = 7.86 g cm^{-3}].

a) How large is the applied mass of tin per cm^2?
b) How thick is the applied tin coating?
c) How large is the consumption of tin per day (24 h uninterrupted operation)?
d) Which minimum current is required for the production line, provided the current yield is 100% and there are no further losses?
e) How many 20 t coils of steel strips can be coated in one day?

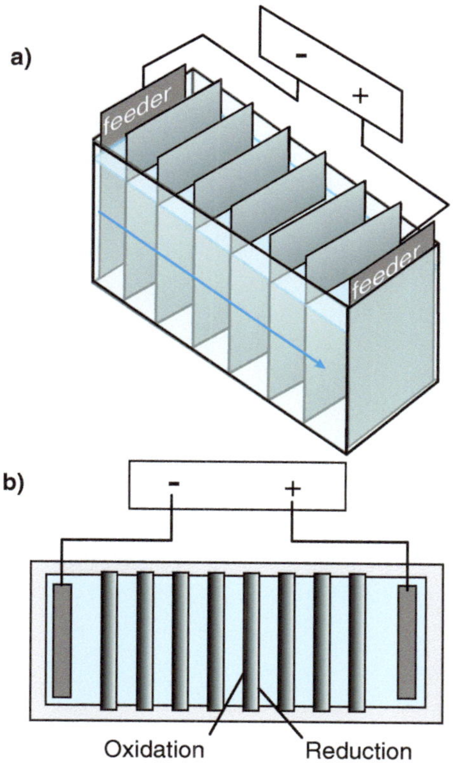

Figure 9.45 a) Bipolar electrochemical reactor and b) top-view. Source: Adapted from [41].

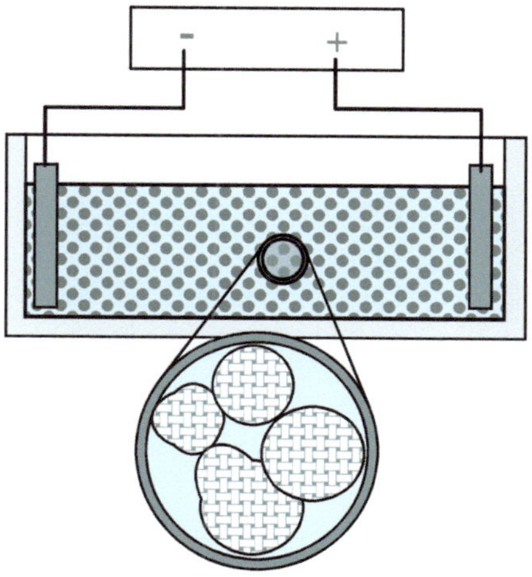

Figure 9.46 Sectional view across a bipolar electrochemical fixed-bed reactor.

Exercise 9.8: A cell for bipolar electrochemistry has two feeder electrodes with a mutual distance of 10 cm, between which a voltage of 50 V is applied. Ag will be deposited on one side of a rod-shaped microobject with a diameter of 65 nm from an Ag^+-containing solution, while OER will occur at the opposite side of the microobject. The overpotential for Ag deposition will be −50 mV, and the overpotential for the OER will be +330 mV. How long must the rods be in order to observe a successful Ag deposition on one side of the objects?

References

1. H. Fischer; Electrocrystallization of metals under ideal and real conditions. *Angew. Chem. Int. Ed.* **1969**, *8*, 108–119.
2. J. A. Harrison, H. R. Thirsk; Fundamentals of metal deposition. in *Electroanalytical Chemistry, Vol. 5* (A. J. Bard, Ed.), Marcel Dekker, New York, 1971, pp. 67–148.
3. E. Budevski, G. Staikov, W. J. Lorenz; *Electrochemical Phase Formation and Growth – An Introduction to the Initial Stages of Metal Deposition*, VCH, Weinheim, 1996.
4. N. Kanani; *Electroplating: Basic Principles, Processes and Practice*, Elsevier, Oxford, 2004.
5. M. Paunovic, M. Schlesinger; *Fundamentals of Electrochemical Deposition*. 2nd ed., John Wiley & Sons, Hoboken, NJ, 2006.
6. B. R. Sharifker, J. Mostany; Electrochemical nucleation and growth, in *Encyclopedia of Electrochemistry, Vol. 2* (A. J. Bard, M. Stratmann, E. J. Calvo, Eds.), Wiley-VCH, Weinheim, 2003, pp. 512–540.
7. G. Staikov (Ed.); *Electrocrystalization in Nanotechnology*, Wiley-VCH, Weinheim, 2007.
8. S. S. Djokić (Ed.); *Electrodeposition in Theory and Practice (Modern Aspects of Electrochemistry), Vol. 48*, Springer, New York, 2010.
9. Y. Gamburg, G. Zangari; *Theory and Practice of Metal Electrodeposition*, Springer, New York, 2011.
10. M. Schlesinger, M. Paunovic (Eds.); *Modern Electroplating*. 5th ed., John Wiley & Sons, Hoboken NJ, 2010.
11. S. S. Djokić, P. Cavalotti; Electroless deposition: Theory and applications. in *Modern Aspects of Electrochemistry, Vol. 48: Electrodeposition* (R. E. White, G. Vayenas, S. S. Djokić, Eds.), Springer, New York, 2010, pp. 251–289.
12. A. Walcarius, D. Mandler, J. A. Cox, M. Collison, O. Lev; Exciting new directions in the intersection of functionalized sol–gel materials with electrochemistry. *J. Mater. Chem.* **2005**, *15*, 3663–3689.
13. D. Lincot, M. Froment, H. Cachet; Chemical deposition of chalcogenide thin films from solution. in *Advances in Electrochemical Science and Engineering* (R. C. Alkire, D. M. Kolb, Eds.), Wiley-VCH, Weinheim, 1999, pp. 165–235.
14. J. L. Stickney, T. L. Wade, B. H. Flowers, R. Vaidyanathan, U. Happek; Electrodeposition of compound semiconductors by electrochemical atomic layer epitaxy (EC-ALE). in *Encyclopedia of Electrochemistry, Vol. 1* (A. J. Bard, M. Stratmann, E. Gileadi, M. Urbakh, Eds.), Wiley-VCH, Weinheim, 2002, pp. 513–560.
15. E. Budevski, G. Staikov, W. J. Lorenz; *Electrochemical Phase Formation and Growth – An Introduction to the Initial Stages of Metal Deposition*, VCH, Weinheim, 1996, pp. 157–163.
16. Y. Gamburg, G. Zangari; *Theory and Practice of Metal Electrodeposition*, Springer, New York, 2011, pp. 97–122.

17 E. Budevski, G. Staikov, W. J. Lorenz; *Electrochemical Phase Formation and Growth – An Introduction to the Initial Stages of Metal Deposition*, VCH, Weinheim, 1996, pp. 163–199.

18 R. Winand; Electrodeposition of metals and alloys – New results and perspectives. *Electrochim. Acta* **1994**, *39*, 1091–1105.

19 D. R. Gabe, G. D. Wilcox; The Hull cell (tutorial note 10). *Trans. Inst. Met. Finish.* **1993**, *71*, 71–73.

20 C. Madore, M. Matlosz, D. Landolt; Experimental investigation of the primary and secondary current distribution in a rotating cylinder Hull cell. *J. Appl. Electrochem.* **1992**, *22*, 1155–1160.

21 C. Madore, D. Landolt; The rotating cylinder Hull cell: design and application. *NASF Surf. Technol. White Papers* **2015**, *79*, 10–20.

22 J. Newman, K. E. Thomas-Alyea; *Electrochemical Systems.* 3rd ed., John Wiley & Sons, New York, 2004, pp. 419–430

23 W. E. G. Hansal, S. Roy; *Pulse Plating. Eugen G. Leuze Verl. KG., Bad Saulgau 2012.* Open access https://www.hanser-elibrary.com/doi/pdf/10.12850/9783874803236.fm (last access 28 Dec. 2024).

24 W. Plieth; *Electrochemistry for Materials Science*, Elsevier, Amsterdam, 2008.

25 E. Budevski, G. Staikov, W. J. Lorenz; *Electrochemical Phase Formation and Growth – An Introduction to the Initial Stages of Metal Deposition*, VCH, Weinheim, 1996, pp. 152–156.

26 W. Plieth; Additives in the electrocrystallization process. *Electrochim. Acta* **1992**, *37*, 2115–2121.

27 W. Plieth; *Electrochemistry for Materials Science*, Elsevier, Amsterdam, 2008, pp. 220–228.

28 M. Paunovic, M. Schlesinger; *Fundamentals of Electrochemical Deposition.* 2nd ed., John Wiley & Sons, Hoboken, NJ, 2006, pp. 117–198.

29 N. Kanani; *Electroplating: Basic Principles, Processes and Practice*, Elsevier, Oxford, 2004, pp. 55–85.

30 T. P. Moffat, D. Wheeler, D. Josell; Superconformal film growth. in *Electrocrystallization in Nanotechnology* (G. Staikov, Ed.), Wiley-VCH, Weinheim 2007, pp. 96–114.

31 T. P. Moffat, J. E. Bonevich, W. H. Huber, A. Stanishevsky, D. R. Kelly, G. R. Stafford, D. Josell; Superconformal electrodeposition of copper in 500–90 nm features. *J. Electrochem. Soc.* **2000**, *147*, 4524–4535.

32 T. P. Moffat, D. Wheeler, D. Josell; Superconformal film growth. in *Advances in Electrochemistry and Electrochemical Engineering, Vol. 10* (R. C. Alkire, D. M. Kolb, P. N. Ross, K. Lipkowski, Eds.), Wiley-VCH, Weinheim, 2008, pp. 107–189.

33 Y. Gamburg, G. Zangari; *Theory and Practice of Metal Electrodeposition*, Springer, New York, 2011, pp. 211–221.

34 L. Bouffier, D. Zigah, N. Sojic, A. Kuhn; Recent advances in bipolar electrochemistry. in *Electroanalytical Chemistry, Vol. 27* (A. J. Bard, C. G. Zoski, Eds.), CRC Press, Boca Raton, 2017, pp. 27–117.

35 G. Loget, J. Roche, A. Kuhn; True bulk synthesis of Janus objects by bipolar electrochemistry. *Adv. Mater.* **2012**, *24*, 5111–5116.

36 J. P. Guerette, S. M. Oja, B. Zhang; Coupled electrochemical reactions at bipolar microelectrodes and nanoelectrodes. *Anal. Chem.* **2012**, *84*, 1609–1616.

37 J. P. Guerette, S. J. Percibal, B. Zhang; Fluorescence coupling for irect imaging of electrochemical heterogeneity. *J. Am. Chem. Soc.* **2013**, *13*, 855–861.

38 C. Warakulwit, T. Nguyen, J. Majimel, M.-H. Delville, V. Lapeyere, P. Garrigue, V. Ravaine, J. Limtrakul, A. Kuhn; Dissymmetric carbon nanotubes by bipolar electrochemistry. *Nano Lett.* **2008**, *8*, 500–504.

39 C. Ulrich, O. Andersson, L. Nyholm, F. Björefors; Formation of molecular gradients on bipolar electrodes. *Angew. Chem. Int. Ed.* **2008**, *47*, 3034–3036.

40 S. E. Fosdick, S. P. Berglund, C. B. Mullins, R. M. Crooks; Parallel screening of electrocatalyst candidates using bipolar electrochemistry. *Anal. Chem.* **2013**, *85*, 2493–2499.

41 C. Comninellis, E. Plattner, P. Bolomey; Estimation of current bypass in a bipolar electrode stack from current–potential curves. *J. Appl. Electrochem.* **1991**, *21*, 415–418.

42 K. Kusukabe, S. Morooka, Y. Kato; Current path and electrolysis efficiency in bipolar packed-bed electrodes. *J. Chem. Eng. Jpn.* **1986**, *19*, 43–47.

43 J. J. Byerley, K. Enns; Discrete particulate bipolar reactor. US Patent 4517067, 1985.

10

Semiconductor Electrochemistry

This chapter provides an overview of the electrochemistry of semiconductors, which not only plays a role in the processing of semiconductor materials in the electronics industry (electrochemical polishing, deposition of interconnects in semiconductor integrated circuits), but also forms the basis for current research on the conversion of light into electrical or chemical energy. Since most metal oxides are semiconductors, reactions at passivating oxide layers of metals in the context of corrosion are also based on the special electrochemical properties of semiconductors. Before discussing electrochemical reactions at semiconductors, we briefly introduce important concepts in solid-state physics necessary for interdisciplinary communication and for the study of more advanced literature. The comparison of semiconductors and metals also allows us to gain a deeper understanding of electron transfer reactions at metal electrodes. Electrochemical reactions are always interfacial reactions. Therefore, the interfacial properties of semiconductors (in addition to their bulk properties) are especially important. The interfacial properties of semiconductors are characterized by structural and electronic defects, which are much more difficult to describe theoretically than the bulk material of crystals.

Further Reading. Pictorial representations of the crystal structures of semiconductors can be found in textbooks on inorganic chemistry [1] or solid-state physics [2–4]. The textbooks on solid-state physics discuss in more detail the electronic structure and spectroscopic methods for characterizing semiconductors. Specialized introductions to semiconductor electrochemistry can be found in [5–11]. Experimental principles for the determination of important characteristics of semiconductor electrodes are summarized in [12] using the example of water splitting.

10.1 Structure and Binding in Semiconductors

10.1.1 Crystal Structure of Semiconductors

Many inorganic semiconductors are crystalline solids in which the bonding between atoms has significant covalent character. As covalent bonds are directed bonds, lower coordination numbers are found in semiconductors compared to nondirected metallic bonds in metal crystals (e.g., Na metal) or nondirected ionic bonds in typical ionic crystals (e.g., NaCl). In typical semiconductors, each atom is surrounded by four neighboring atoms in a tetrahedral arrangement, for example, silicon occurs in the diamond structure (Table 10.1). The four neighboring atoms are Si atoms. In compound semiconductors such as GaAs, we often encounter the sphalerite (zinc blende) structure. Here, the four neighbors of a Ga atom are the As atoms in a tetrahedral arrangement;

Table 10.1 Crystal structures of semiconductors.

Structure type	Examples
Diamond	Si, Ge, $C_{Diamond}$
Sphalerite	ZnS, SiC, GaP, GaAs, InP, CdS
Wurtzite	SiC, ZnS, CdS, ZnO
Rutile	TiO_2
Anatase	TiO_2
Pyrite	FeS_2
Layered chalcogenide	SnS_2, MoS_2, WSe_2
Rock salt (NaCl)	PbS

conversely, each As atom is surrounded by four Ga atoms in a tetrahedral arrangement. The sphalerite lattice for GaAs consists of interpenetrating face-centered cubic lattices of As and Ga atoms, with both lattices shifted with respect to each other in all spatial directions. Compound semiconductors can occur in more than one structural type (modification). Thus, ZnS exists in the sphalerite structure or in the wurtzite structure. In the wurtzite structure, interpenetrating hexagonal close-packed sublattices of Zn and S are present, also with a tetragonal arrangement of nearest neighbors. Other crystal types occur in compounds with stoichiometries AB_2 and AB_3.

10.1.2 Electronic Structure of Semiconductors

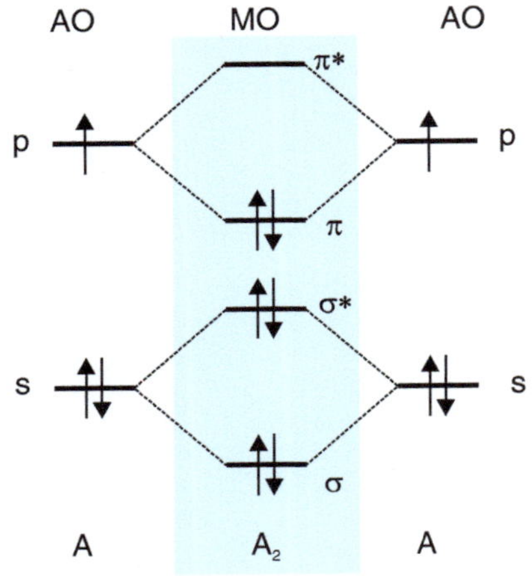

Figure 10.1 Principle of formation of MOs (highlighted in light blue) from AOs for a molecule A_2. Only two selected atomic orbitals are shown.

The "overlapping of the bands" occurs on the energy axis. States of different bands (i.e., of different types) have the same energy. In other contexts, chemists often speak of the "overlap of orbitals". This usually refers to an overlap on the spatial axis, which can be calculated as an overlap integral of two wave functions and is qualitatively outlined in pictorial representations as interpenetration of orbital lobes.

Energy bands. The electronic structure (i.e., the distribution of possible states with a one-electron energy E_e [eV] on the energy axis) can be explained by analogy. In atom A, there are atomic orbitals (AOs, which are one-electron wave functions) corresponding to defined one-electron energies. If a molecule A_2 is formed from two atoms A, the properties of the electrons in the molecule A_2 can be described in terms of molecular orbitals (MOs) formed by a linear combination of similar AOs of the two atoms (Figure 10.1). One of the MOs corresponds to a lower E_e. This MO is called *bonding*. The second MO corresponds to an increased E_e compared to the initial AOs, and this is called *antibonding*. After the MOs have been calculated, the electrons can be "distributed" over the MO energy levels according to population rules. In this process, starting with the lowest E_e, the MO energy levels are each filled with two (energetically degenerate) electrons with opposite spin orientations. If both orbitals of one type (bonding and antibonding MOs) are filled, there is no net energy difference (for a first approximation) compared to the case of fully filled AOs. Therefore, it is usually sufficient to consider only the valence electrons to describe the bonding. As a simplification, the inner electrons are assumed to be fully localized at the atoms.

If we apply this approach to a homonuclear crystalline solid (e.g., Si_n, $n \approx 10^{20}$ atoms, corresponding to 46 mg Si), 10^{20} AOs of the same type from the 10^{20} atoms give rise to 10^{20} MOs of the same type. These 10^{20} MOs distribute over an energy range of about 1 eV. If we now try to estimate the energetic difference between two of these MOs (1 eV/10^{20}), we get such a tiny energy difference that it is no longer experimentally accessible. It is therefore more useful to think of the energy range between the MO of lowest energy (lower band edge) and the MO of highest energy (upper band edge) as an **energy band** of continuous states. The occupation of the bands with electrons is again determined according to the population rules. Different situations can now occur because bands formed from unoccupied AOs and occupied AOs overlap in different ways on the energy axis (Figure 10.2).

Band Structure Plots and Band Gap at a Glance. Contrary to what Figure 10.2 suggests, the number of valence electrons, the extent of overlap of the orbitals, and the coordination number cannot be influenced by the experimenter for a given substance. Therefore, it is useful to plot the bands for each substance in a *separate* plot as in Figure 10.3. The vertical axis is the one-electron energy E_e. The axis is often not drawn in but is implied. The horizontal axis can often be taken to be a spatial coordinate perpendicular to the surface (Section 10.2). We can identify in each plot a **valence band (VB)**, that is, the energetically highest band containing electrons. It corresponds to the highest occupied molecular orbital (HOMO) of a molecule in Figure 10.2. Analogously, there is the **conduction band (CB)**. This is the unoccupied band, which is lowest in energy and corresponds to the lowest unoccupied molecular orbital (LUMO) in molecules.

Metals and Semiconductors. In metals, there are occupied *and* unoccupied states at the energy of the highest occupied states either because a band is only partially

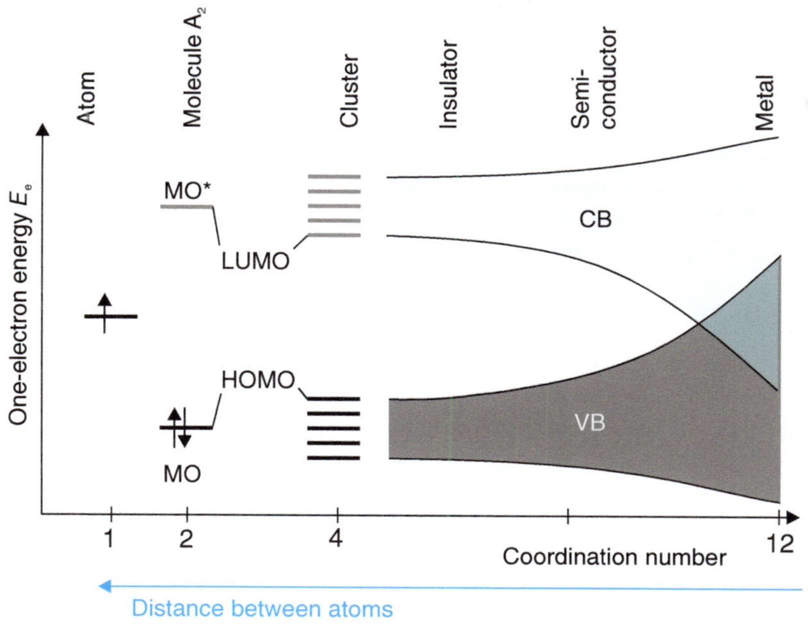

Figure 10.2 Change of the "bandwidth" of energetic states of valence electrons during the transition from atoms via diatomic molecules and clusters to solids (insulators, semiconductors, conductors). The scheme is oversimplified. Occupied states in dark gray, unoccupied states in light gray.

filled (3s band in Na, Figure 10.3a) or because of overlap of occupied and unoccupied bands (Figure 10.3b). The bands extend over the entire spatial region. If we apply an electric field to the metallic body, the electrons can move in the electric field because they can move from an occupied electronic state to a previously unoccupied state of similar energy without a change in energy. Because of the applied electric field, the energy of inherently similar states will decrease along the electric field. This enables the directional charge transport during such transitions. For semiconductors, the energetic states also extend over the entire body. In semiconductors, the VB is completely filled. VB and CB are separated by an energy region in which there are no states (Fig. 10.3c).

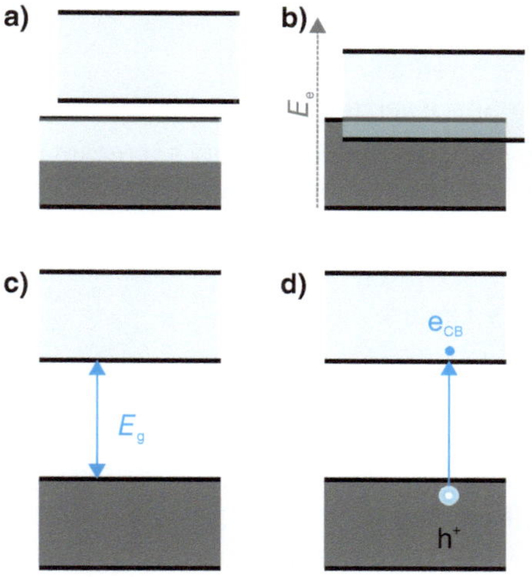

Figure 10.3 Band structure plots of.

a) a metal with a half-filled valence band (Na);
b) a metal with overlapping bands (Mg);
c) an intrinsic semiconductor (Si) in the electronic ground state; and
d) an intrinsic semiconductor after generation of an electron–hole pair. Occupied states are shown in dark gray, empty states are shown in light gray.

We refer to the energy range between VB and CB as **band gap** E_g [eV]. Synonyms are "energy gap" or "forbidden band".

Since all states are occupied in the semiconductor up to the highest energy of the VB, called valence band maximum E_V or valence band edge, the contained electrons cannot perform any net motion. Only when electrons are excited from the VB into the CB (Figure 10.3d) can they move from state to state and cause a macroscopic charge transport. The remaining unoccupied electronic state in the VB is called a **hole** (h^+). The hole can accept another electron of the same energy from the VB, again creating an unoccupied state elsewhere. Therefore, in an electric field, the position of the hole will shift in a directional manner. The hole reacts with an inertia to the acceleration by the electric field. It can be conceived as a particle (quasi-particle), to which particle properties, such as a mass m_h, can be assigned. Instead of describing the motion of all electrons involved, each sequentially filling a hole, it is sufficient to describe the motion of *one* hole, which is much simpler.

Intrinsic Semiconductors and Insulators. Excitation of electrons from the VB into the CB can occur by absorption of thermal energy or by absorption of a photon with $h\nu > E_g$ (internal photoelectric effect). Semiconductors that contain charge carriers only in the form of electron–hole pairs (e_{CB}–h_{VB}^+ pairs) after excitation are called **intrinsic semiconductors,** or undoped semiconductors. Examples are pure Si and pure Ge. In contrast, **extrinsic semiconductors**, or doped semiconductors, are those in which the number of charge carriers is increased by the addition of impurities (doping) (Section 10.1.4). Solids with $E_g > 5$ eV are called **insulators**.

Excursion 10-1 Wave Functions in Solids

Here, we will derive the notion of energy bands from a somewhat more detailed analysis of the quantum mechanical basis of chemical bonds.

Free Electron Gas (Jellium Model). In this simplest case, the wave function of the valence electrons are derived for the situation when the electrons move between the ions of the solid. The ions are formed by the atomic nuclei and the core electrons. The charge is balanced by the valence electrons moving between the ions. The charge of ions and valence electrons balances in such a way that the valence electrons inside the solid are not subjected to a potential gradient [$V(\mathbf{r}) \approx 0$]. This is a very rough approximation, which at best has meaning for alkali metals. However, the procedure can be illustrated relatively easily here. The one-electron Schrödinger equation[68] has to be solved and can be stated as

$$\left(\frac{\hbar}{2m_e}\nabla^2 - E_{e,j}\right)\Psi_j = 0 \tag{10.1}$$

V – Potential energy [eV]
$\mathbf{r}$ – Space vector [m, cm]
$\hbar = h/(2\pi)$ – Reduced Planck's constant [J s]
m_e – Mass of an electron [kg]
$E_{e,j}$ – One-electron energy [eV]
Ψ_j – One-electron wave function
∇^2 – Laplace operator $= \frac{\partial^2}{\partial x^2} + \frac{\partial^2}{\partial y^2} + \frac{\partial^2}{\partial z^2}$
$\mathbf{k}$ – Angular wave vector $\{k_x, k_y, k_z\}$ [cm^{-1}]
i – Imaginary unit
e^{ix} – Euler's representation of a complex number, see Excursion 7-1 Complex Numbers
a – Lattice constant in a cubic crystal [cm]

To solve this equation, we select a sinusoidal wave in the space $\Psi \propto e^{-i\mathbf{kr}}$. We require periodicity in a cube of edge length a, so we let $\Psi(x, y, z) = \Psi(x + a, y, z) = \Psi(x, y + a, z) = \Psi(x, y, z + a)$. This must hold when inserting the solution approach $\Psi \propto e^{i\mathbf{kr}} = \exp(i\mathbf{kr})$ in a spatial direction, for example, for x: $\exp(ik_x x) = \exp[ik_x(x + a)] = \exp(ik_x x) \cdot \exp(ik_x a)$. This gives $\exp(ik_x a) = 1$, which is satisfied if $k_x a = n_x \cdot 2\pi$ ($n_x = 0, 1, 2, \ldots$). k_x is the magnitude of the angular wave vector (in the direction of the x-axis), defined as $|\mathbf{k}| = 2\pi/\lambda$, which for this problem can only take the values given by Eq. (10.2).

$$k_m = \frac{2\pi}{a}n_m \quad \text{with} \quad m = \{x, y, z\} \quad \text{and} \quad n_m = \{0, 1, 2, \ldots\} \tag{10.2}$$

The angular wave number vector or k-vector has several physical meanings. We will encounter it repeatedly and recall its meanings here:

1) The vector direction is the propagation direction of the wave. The magnitude is proportional to the reciprocal of the lattice constant $1/a$. We can associate the vector directly with the reciprocal lattice in a solid. This reciprocal lattice is directly accessible experimentally by diffraction techniques.
2) After normalization to the fundamental domain, the wave function and the energy eigenvalues are obtained.

$$\Psi_j = \frac{1}{\sqrt{a^3}}e^{i\mathbf{kr}} \tag{10.3}$$

$$E_{e,j} = \frac{\hbar}{2m_e}\mathbf{k}^2 = \frac{\hbar}{2m_e}\left(\frac{2\pi}{a}\right)^2 (n_x^2 + n_y^2 + n_z^2) \tag{10.4}$$

The expression $r^2 = x^2 + y^2 + z^2$ is the analytical description of a sphere with radius r. Thus, we can first characterize the one-electron energies $E_{e,j}$ via the square of the angular wave vector $\mathbf{k}^2$ and, in turn, illustrate $\mathbf{k}^2$ as a sphere.

3) For electrons in field-free space (i.e., $V[\mathbf{r}] \approx 0$), the one-electron energy is obtained as $E_e = \mathbf{p}^2/(2m_e)$, from which the momentum is obtained as $\mathbf{p} = \hbar\mathbf{k}$. Therefore, the k-vector is also proportional to the momentum vector. Because of the population rules, we need one energy level $E_{e,j}$ to accommodate two spin-degenerate electrons. In an alkali metal, the most energetic electron at $T = 0$ K just occupies the Fermi level[85] E_F, to which we assign the corresponding angular wave vector $\mathbf{k}_F$. We can imagine $\mathbf{k}_F$ as the radius of a sphere, the so-called Fermi sphere, with the volume $V = 4\pi\mathbf{k}_F{}^3/3$.

In summary, the meanings of the k-vector are as follows:

$$\mathbf{k} = \mathbf{p}/\hbar \quad \text{(momentum vector)} \tag{10.5}$$

$$\mathbf{k} = 2\pi/\lambda \quad \text{(angular wave vector)} \tag{10.6}$$

$$\mathbf{k} = n\pi/a \quad \text{(reciprocal lattice)} \tag{10.7}$$

$\mathbf{k}_F$ – Wave vector corresponding to the Fermi energy [cm^{-1}]

N_e – Number of valence electrons within a unit cell [dimensionless]

n_e – Concentration of valence electrons, often called electron density, number of valence electrons per space unit [cm^{-3}]

$\mathbf{p}$ – Momentum [kg m s^{-1}]

Electron Density and Density of States. Since only two spin-degenerate electrons can share an identical set of quantum numbers (n_x, n_y, n_z), that is, occupy the same state, we see from Eq. (10.2) that a state occupies a fixed volume of $(2\pi/a)^3$ in the Fermi sphere. We can express the electron density n_e [m^{-3}, cm^{-3}] of an alkali metal in terms of the number of valence electrons N_e per unit cell of the lattice $n_e = N_e/a^3$. The Fermi level can be estimated by comparing the volume of the Fermi sphere with the space required by the N_e twofold-degenerate valence electrons in $\mathbf{k}$-space:

$$\frac{4\pi}{3}|\mathbf{k}_F|^3 = \frac{N_e}{2}\left(\frac{2\pi}{a}\right)^3$$

$$|\mathbf{k}_F|^3 = \left(3\pi^2\frac{N_e}{a^3}\right)$$

$$|\mathbf{k}_F| = \left(3\pi^2 n_e\right)^{1/3} \tag{10.8}$$

$$E_F = \frac{\hbar^2}{2m_e}|\mathbf{k}_F|^2 = \frac{\hbar^2}{2m_e}\left(3\pi^2 n_e\right)^{2/3} \tag{10.9}$$

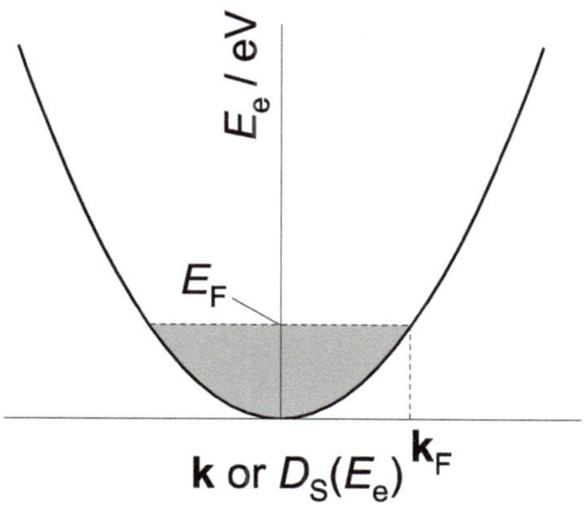

Figure 10.4 One-electron energy as a function of **k** or the density of states along the x coordinate for a free electron within a metal.

The number of states $Z_S(E_e)$ up to energy E_e is the density of states in space and, because of spin degeneracy, is equal to half the electron density $n_e(E_e)$ that can be accommodated up to E_e. Substituting E_e for E_F and $n_e(E_e)$ for n_e into E. (10.9) yields

$$Z_S(E_e) = \frac{n_e(E_e)}{2} = \frac{1}{6\pi^2}\left(\frac{2m_e E_e}{\hbar^2}\right)^{3/2} \tag{10.10}$$

$Z_S(E_e)$ – Number of states within a volume element [cm^{-3}]

Another important quantity is the **density of states** ($D_S(E_e)$), the number of states in an energy interval and space element:

$$D_S(E_e) = \frac{\partial Z_S(E_e)}{\partial E_e} = 2\pi\left(\frac{2m_e}{h}\right)^{3/2}\sqrt{E_e} \tag{10.11}$$

$D_S(E_e)$ – Density of states, number of states within an energy range [cm^{-3} eV^{-1}]

h – Planck's constant [J · s, eV · s] = $2\pi\,\hbar$

Rearranging the equation gives the parabola $E_e = const \cdot [D_S(E_e)]^2$, in which we can plot the population with electrons up to $E_e = E_F$ for $T = 0$ K (Figure 10.4).

The Wigner–Seitz Cell and the Brillouin Zone. One describes the periodicity of the crystal structures by a translation vector **R**, which is formed as a linear combination of the basis vectors $\mathbf{a}_1, \mathbf{a}_2, \mathbf{a}_3$, and integer coefficients m_1, m_2, m_3.

$$\mathbf{R} = m_1\mathbf{a}_1 + m_2\mathbf{a}_2 + m_3\mathbf{a}_3 \tag{10.12}$$

One can represent a unit cell with lattice atoms on the corners of a cube. In contrast, the more practical Wigner–Seitz cell[126] is based on a lattice atom located in the center of the unit cell. The side faces of the unit cell run perpendicularly through the center of the connecting lines to the neighboring atoms (Figure 10.5).

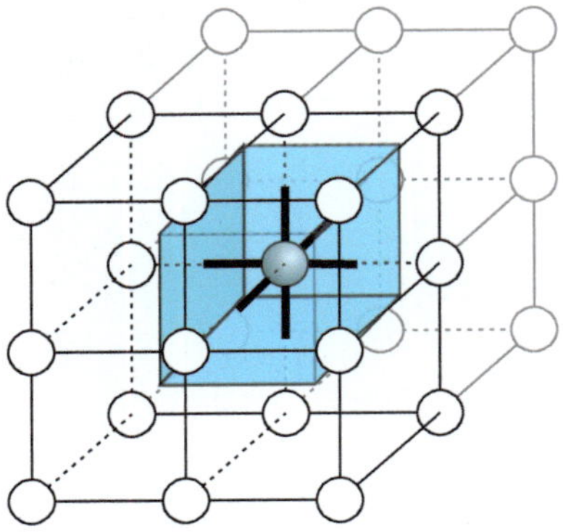

Figure 10.5 Illustration of the Wigner–Seitz cell in a cubic lattice.

We already discussed the meaning of **k**-space as a vector in the reciprocal lattice space. The translation vector **K** of the reciprocal lattice is

$$\mathbf{K} = n_1\mathbf{b}_1 + n_2\mathbf{b}_2 + n_3\mathbf{b}_3 \tag{10.13}$$

The Wigner–Seitz cell of the reciprocal lattice consisting of the basis vectors $\mathbf{b}_1$, $\mathbf{b}_2$, and $\mathbf{b}_3$ is called the Brillouin zone.[127] The following relationships hold for the basis vectors of the real space lattice and the reciprocal lattice:

$$a_i b_j = 2\pi \cdot \delta_{ij} \tag{10.14}$$

$$\delta_{ij} = \begin{cases} 1, & i = j \\ 0, & i \neq j \end{cases} \tag{10.15}$$

$$a_i \perp b_j, b_k \tag{10.16}$$

$$b_i \perp a_j, a_k. \tag{10.17}$$

This yields the following relationship for **R** and **K**:

$$\mathbf{R} \cdot \mathbf{K} = 2\pi \left(n_1 m_1 + n_2 m_2 + n_3 m_3\right) = 2\pi N \tag{10.18}$$

with the integer numbers $n_1, n_2, n_3, m_1, m_2, m_3$, and N. All lattice points **R** that satisfy the equation $\mathbf{R} \cdot \mathbf{K} = 2\pi N$ for a given N lie on a lattice plane perpendicular to **K**. This allows the condition for Bragg reflection[96] in crystals to be formulated in a particularly compact way.

$$\mathbf{k}' - \mathbf{k} = \mathbf{K} \tag{10.19}$$

k – Wave vector of the incipient wave [cm^{-1}]
k′ – Wave vector of the reflected wave [cm^{-1}]
K – Translation vector in the reciprocal lattice [cm^{-1}]

If a wave with wavelength λ ($\mathbf{k} = 2\pi/\lambda$) travels perpendicularly toward a lattice plane, the wave number vector of the reflected wave is $\mathbf{k}' = -2\pi/\lambda$ (direction reversal). **K** is then $2\pi/\lambda - (-2\pi/\lambda) = 4\pi/\lambda = Nb_1$. b_1 is related to the spacing of the lattice elements a_1 in the relationship $a_1 b_1 = 2\pi$. Substitution for $N = 1$ yields $4\pi/\lambda = 2\pi/a_1$ or $2/\lambda = 1/a_1$ or $\lambda/2 = a_1$. In this case, reflection occurs when half the wavelength corresponds to the lattice spacing a_1, so the angular wave vector is $\mathbf{k} = 2\pi/2a_1 = \pi/a_1$.

Bound Electrons in the Crystal Lattice. The electronic properties of semiconductors cannot be fully explained by the free electron gas. The valence electrons are also bound to the lattice atoms (tight binding model), and the bonding between

[126] Eugen Paul Wigner, 1902–1995, US-American physicist of Hungarian origin, Nobel Prize for Physics 1963, also made important contributions to the Manhattan project; Frederick Seitz, 1911–2008, US-American physicist. President of the Rockefeller University and President of the National Academy of Science of the USA, 1962–1969.
[127] Léon Nicolas Brillouin, 1889–1969, French-US-American physicist.

the atoms has a strongly covalent character in many semiconductors. Thus, we would have to combine very many AOs, from which we would get a large number of MOs for the crystal. We want to trace here in a very abbreviated way the concept of solid-state MOs, the Bloch functions.[128] The stationary Schrödinger equation to be solved is

$$\left(\frac{\hbar}{2m_e}\nabla^2 + \mathbf{V}(\mathbf{r}) - E_{e,j}\right)\Psi_j = 0 \tag{10.20}$$

The operator $\mathbf{V(r)}$ describes the interactions of the considered electron with the nuclei and with all other electrons. Because of the periodicity of the crystal lattice, it must be equal at all equivalent positions in the crystal: $\mathbf{V(r)} = \mathbf{V(r + R)}$. The Bloch functions satisfy this eigenvalue equation and have the structure

$$\Psi_j(\mathbf{k}, \mathbf{r}) = e^{i\mathbf{kr}} \cdot u_j(\mathbf{k}, \mathbf{r}) \tag{10.21}$$

where u_j is a periodic function with the same periodicity as the crystal lattice. This means $u_j(\mathbf{k}, \mathbf{r}) = u_j(\mathbf{k}, \mathbf{r} + \mathbf{R})$. The term $e^{i\mathbf{kr}}$ is the plane wave, which already appeared in the free electron gas model. For the exact determination, we can use a perturbation theory for which we separate the operator $\mathbf{V}$ into contributions for each translation vector in reciprocal space:

$$\mathbf{V}(\mathbf{r}) = \sum_{m \neq 0} \{\mathbf{V}[\mathbf{K}(m_1, m_2, m_3) \cdot \exp(i\mathbf{K}(m_1, m_2, m_3)\,\mathbf{r})]\} \tag{10.22}$$

The perturbation is generally quite small – except when the $\mathbf{k}$-vector is near a lattice point of the reciprocal lattice. There, the perturbation is significant because the wave is reflected according to Eq. (10.19). The one-electron energy with the effective electron mass m_e^* and the perturbation term V results in

$$E_{e,j} = \frac{\hbar}{2m_e^*}\mathbf{k}^2 \pm |V(\mathbf{K} - \mathbf{k})| \tag{10.23}$$

This function is shown in Figure 10.6a. From the second derivative $\mathrm{d}^2E/\mathrm{d}|\mathbf{k}|^2$, the effective mass m_e^* can be determined. High m_e^*, that is, electrons that are difficult to

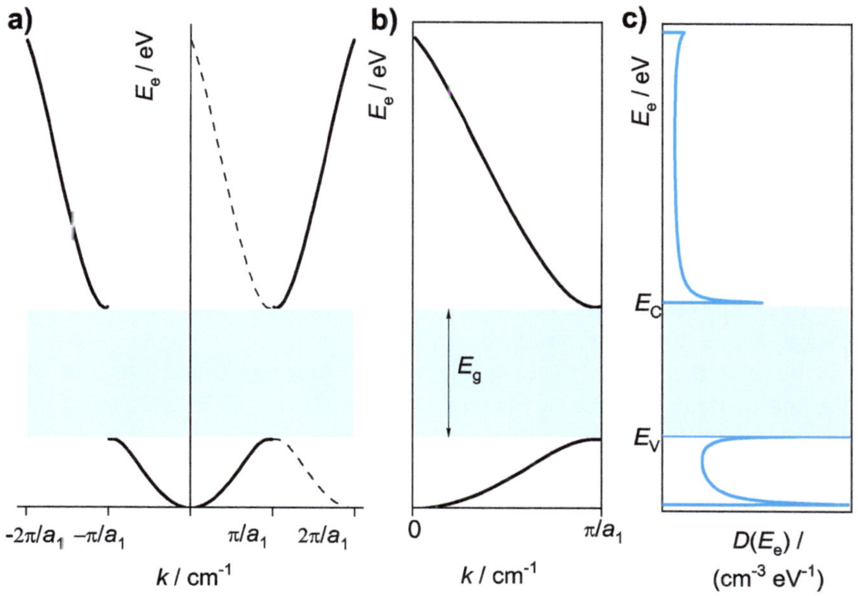

Figure 10.6 One-electron energy as a function of the wave vector for $j = 1$ along the direction of the k_x-vector;

a) bound electron;
b) modified plot of a) in which the graph was mirror-imaged at the straight line $k_x = \pi/a_1$;
c) Density of states for the band structure shown in b).

[128] Felix Bloch, 1905–1983, Swiss-US-American physicist, professor of theoretical physics at Stanford, Nobel Prize of Physics 1952 for the discovery of nuclear magnetic resonance.

accelerate, result from strongly curved bands, whereas slightly curved bands lead to electrons that are easy to accelerate (low m_e^*).

No energy eigenvalues and associated wave functions exist for the region marked in light blue on the E_e axis in Figure 10.6. Therefore, this region is a gap where no energy states exist. Of particular importance is the case, where the lower band is the VB and the upper band is the CB. The gap between the VB and CB is often referred to as the **band gap E_g** of a substance. In the following, we often use simplified representations where we consider only the upper edge of the VB E_V and the lower edge of the CB E_C.

Diagrams, as in Figure 10.6a, are the starting point for the construction of **band structure plots**, which accommodate as much information as possible into one figure. The plot in Figure 10.6a is symmetrical to the ordinate, so the part for negative **k**-vectors contains no new information and has been omitted in Figure 10.6b. In addition, the graph in the range $\pi/a_1 < k < 2\pi/a_1$ of Figure 10.6a was mirrored at the straight line $E_e = \pi/a_1$ and contains, for the considered example, only two curves in Figure 10.6b. This procedure makes sense because graphs with different points of origin overlap (periodicity in k-space is $b_1 = 2\pi/a_1$). In addition, graphs along different spatial directions within the Brillouin zone are combined into *one* plot to accommodate more information. The scheme in Figure 10.7 explains the procedure. Examples in the literature often appear even more complex, because many bands are shown in one diagram.

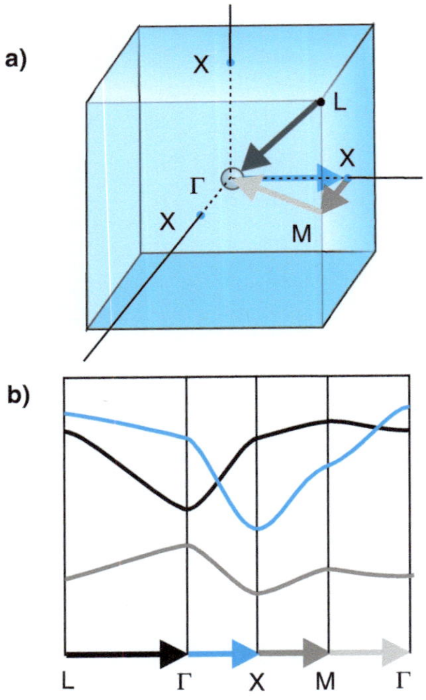

Figure 10.7 Construction of band structure plot along different directions in space.

a) Brillouin zone of a primitive cubic crystal.
b) Schematic band structure plot. The different sections on the abscissa are color coded and correspond to the vectors in panel a).

Consideration of Overlapping of AOs. From a chemical point of view, the construction of band structure plots using the overlap of the AOs in a solid is also illustrative; it is demonstrated here using a linear arrangement of four atoms (a very small section of a crystal) [6]. An orbital of the solid is energetically lower (thus possesses stronger bonding character), the fewer nodal planes the orbital has. Nodal planes are formed when neighboring AOs are combined with opposite signs. Let us first consider spherically symmetric s-AOs (Figure 10.8a). A positive sign of an orbital lobe is indicated by gray shading. For the combination of s orbitals with equal signs ($|\mathbf{k}| = k = 0$), there are no nodal planes. With one nodal plane, all AOs to the left of the plane have the same sign. The same applies to all AOs to the right of the nodal plane. Further combinations are obtained with more nodal planes. With four atoms considered, there are four possible combinations of the s-AOs in a linear arrangement. For more than four atoms and a three-dimensional arrangement in the crystal, the number of possible combinations naturally increases very rapidly. The energy changes with the **k**-vector as shown in Figure 10.8b.

The p_x orbitals are aligned in the direction of the atomic chain (Figure 10.9a). If all AOs are combined with positive signs, then there is a nodal plane between each of the atoms (thus, the highest energy). If the p_x AOs are alternately aligned, there is the optimal overlap and thus, the lowest energy (Figure 10.9a and d). For the p_y and p_z orbitals, the energy changes qualitatively resemble those of the s orbital (Figure 10.9b, c, e, and f).

If we plot the energy for a variation of $|\mathbf{k}|$ between 0 and π/a_1, we obtain the energy range covered by each orbital type (Figure 10.9e and f). Differences between s orbitals on the one hand and p_y and p_z orbitals on the other hand become apparent here, since the overall overlap is larger for s orbitals and thus the influence of the nodal planes on E_e is increased. Figure 10.9e and f also show how a band gap is created, depending on whether or not a gap remains on the energy axis between the largest values of the s band and the smallest values of the p_x band.

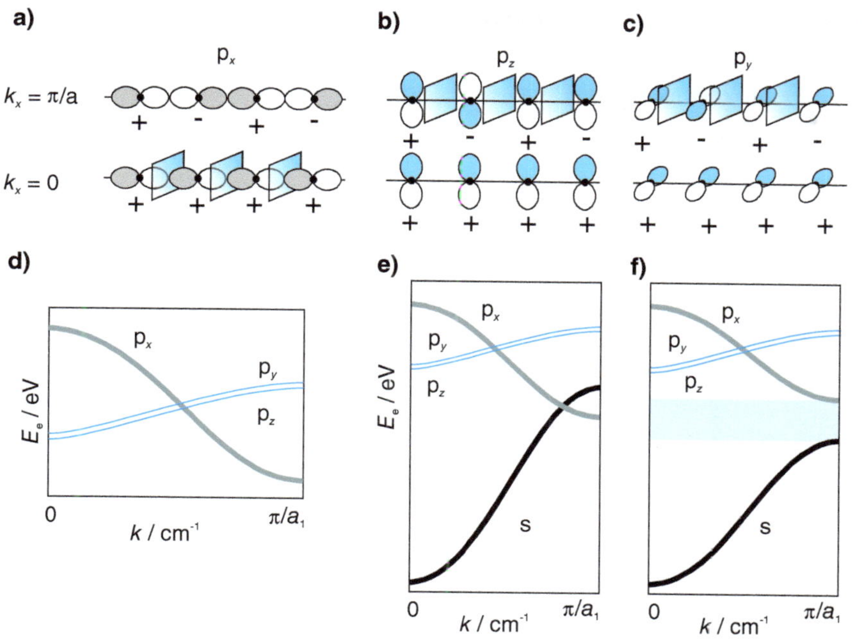

Figure 10.9 Combination of p-AOs to solid-state MOs shown for the example of a linear chain of four atoms.

a) Scheme for p_x orbitals;
b) scheme for p_y orbitals;
c) scheme for p_z orbitals;
d) orbital energies of the p orbitals as function of the wave vector within the first Brillouin zone;
e) combined plots of s orbitals and p orbitals with overlap of the s and p_x band;
f) combined graphs of s orbitals and p orbitals with formation of a band gap (shown in light blue).

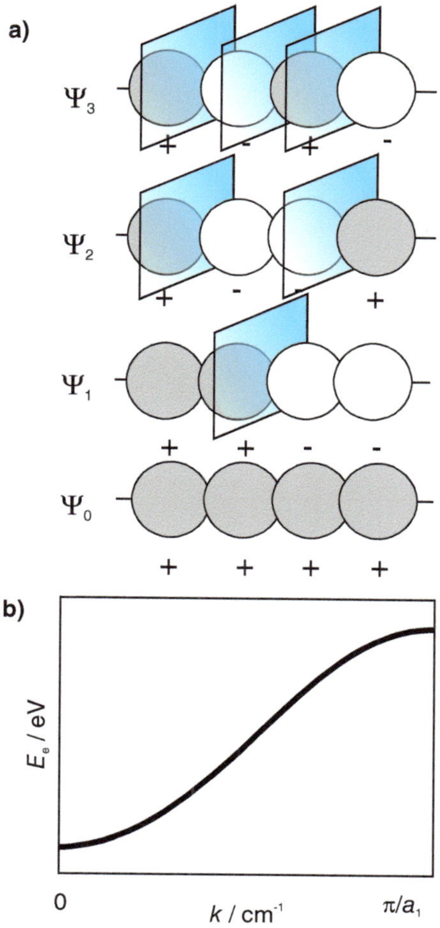

Figure 10.8 Combination of s-AOs to solid-state MOs.

a) Scheme for four s-orbitals;
b) Change of orbital energy as a function of the angular wave vector within the first Brillouin zone.

From the schematic band structure plot, we can qualitatively derive the density of states. To do this, we divide the wave vector axis and the energy axis into small increments, since each state occupies a constant volume in **k**-space. We then "count off" how many increments occur in **k**-space to sweep over one increment on the energy axis. Qualitatively, the trend is shown in Figure 10.6c and illustrates very well the different widths of the resulting energy bands.

Direct and Indirect Transitions. Band gaps and band structures are determined from spectroscopic experiments, which we cannot discuss here [2, 5]. The difference between direct and indirect optical transitions is important for applications exploiting electronic excitations with light. In the schematic representation of the band structure in Figure 10.9f, the maximum of the s-band (VB) is exactly below the minimum of the p_x-band (CB). In this case, the transition with the lowest energy runs "perpendicular" in the diagram (example GaAs). Excitation of the electron is therefore possible without simultaneous momentum transfer. Such a transition is called a **direct transition**. We speak of **indirect transitions** when the maximum of the VB and the minimum of the CB are not at the same point of the Brillouin zone (Figure 10.10, example Si). In order to absorb energy, momentum must be transferred at the same time so that the electron can pass from the VB to the CB. Conceptually, direct and indirect transitions in solids are linked to allowed and forbidden transitions in molecules, respectively.

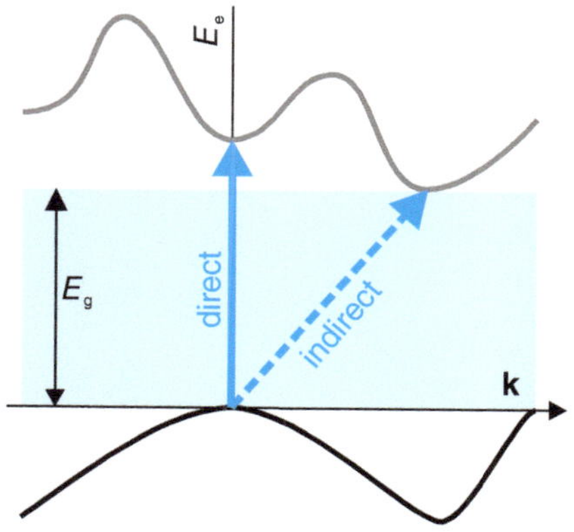

Figure 10.10 Schematic representation of direct and indirect transitions of electrons between VB and CB of Si, for which the minimum of the CB is not located at the same point in the Brillouin zone as the VB maximum.

10.1.3 The Fermi Level

The **Fermi level** E_F [eV] is the one-electron energy level at which the occupation probability of an electronic state is just 50%.

This energy plays an important role in all considerations of semiconductor electrochemistry and is closely related to the distribution of valence electrons among the available energy levels.

Population at $T = 0$ K. At zero absolute temperature and in darkness, no excitation processes are possible. In this case, the electrons are filled into the VB in a metal, for example, Na (Figure 10.11a, hashed area). The energy of the highest occupied level is E_F. At almost identical energy, unoccupied levels also exist. For an intrinsic semiconductor, at $T = 0$ K, the VB is completely filled (Figure 10.11b, hashed area, occupation probability = 100%) and the CB is completely empty (occupation probability 0%). The energy E_F, for which the occupation probability is 50%, lies approximately in the middle of the band gap (Figure 10.11b, also see Excursion 10-2).

Population at $T \gg 0$ K. At higher temperatures, thermal electron excitation processes are possible. In metals, an electron is excited from a state $E_e < E_F$ to a state $E_e > E_F$. Since the thermal energy ($k_B T$) is small relative to the width of the bands and there are unoccupied states only at $E_e > E_F$, only electrons with energies close to E_F can be excited. This results in a population density as in the blue shaded area in Figure 10.11a. For an intrinsic semiconductor, excitation of some electrons occurs from the VB into the CB (Figure 10.11b, blue shaded area). In the plot, the number of electrons in the CB is greatly exaggerated. We will calculate the concentration of e_{CB} and h^+ in more detail in Excursion 10-2.

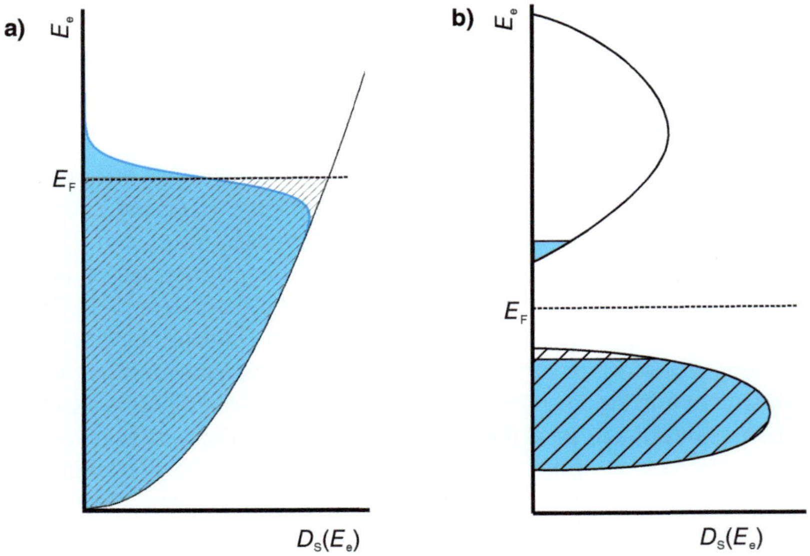

Figure 10.11 Distribution of electrons a) in a metal and b) in an intrinsic semiconductor at $T = 0$ K (hashed) and at high temperature (shaded blue). The amount of e_{CB} is strongly exaggerated.

Q How is the blue curve in Figure 10.11a calculated?

A The shape of the blue curve in Figure 10.11a represents the product of density of states $D_S(E_e) = const \cdot E_e$ and population probability according to Fermi[85]–Dirac statistics[129] $f(E_e) \approx 1/\{1 + \exp[(E_e - E_F)/(k_B T)]\}$. This makes sense as long as E_F and the chemical potential of the electron have the same value ($T \ll 10^5$ K). Then, the chemical potential of the electron is equal to E_F, which applies to all relevant situations in electrochemistry. The curves in Fig. 10.11b are shown linear in first approximation, but should also be sigmoidal when evaluated more carefully.

Excursion 10-2 How to Estimate the Charge Carrier Concentration and the Fermi Level for Undoped Si at Room Temperature?

We follow here the argument used in Ref. [5]. The equilibrium state for valence electrons in the semiconductor is E_F. Electrons in the CB have a higher energy than E_F. We obtain the concentration n_e [cm^{-3}] of these electrons by integrating the number of occupied states above the lowest energy of the CB, called conduction band minimum E_C:

$$n_e = \int_{E_C}^{\infty} D_S(E_e)\, f(E_e)\, dE_e \tag{10.24}$$

$D_S(E_e)$ – Density of states [cm^{-3} EV^{-1}]
$f(E_e)$ – Fermi–Dirac-distribution function (dimensionless)
E_C – One-electron energy at the conduction band minimum [eV]

Since this integral cannot be solved, we have to use approximate solutions. Electrons must be treated according to the Fermi–Dirac statistics. However, this converges into the (simpler) Boltzmann statistics[18] for $(E_e - E_F) \gg k_B T$.

$$f(E_e) \approx 1/\left\{\exp\left[(E_e - E_F)/(k_B T)\right]\right\} = \exp\left[-(E_e - E_F)/(k_B T)\right] \tag{10.25}$$

We calculate $k_B T$ as a measure of the thermal energy at 298 K in units of eV. We convert the units by recalling that 1 eV = 1.60 × 10^{-19} J or 1 = 1 eV/1.60 × 10^{-19} J. Because we can always multiply an expression with 1 without changing the result, such a procedure represents a "very safe" way of dealing with unit conversion, especially if situations become even more complicated.

$$\begin{aligned} k_B T &= 1.38 \times 10^{-23}\,\mathrm{J\,K^{-1}} \cdot 1 \cdot 298\,\mathrm{K} \\ &= 1.38 \times 10^{-23}\,\mathrm{J\,K^{-1}} \cdot \frac{1\,\mathrm{eV}}{1.60 \times 10^{-19}\,\mathrm{J}} \cdot 298\,\mathrm{K} \\ &= 0.0257\,\mathrm{eV} \end{aligned} \tag{10.26}$$

Thus, $k_B T$ is much smaller than E_g, which justifies the use of the Boltzmann statistics.

Instead of integrating the density of states from E_C to ∞, we can use a single value D_C as approximation, which corresponds to the density of states within a few $k_B T$ above E_C:

D_C – Density of states [cm^{-3}] within a few $k_B T$ above E_C
m_e^* – Effective mass of the valence electron in the field of the remaining cations of the crystal lattice [kg]. For simplification it is assumed here that $m_e^* = m_e$ (mass of the electron in the field-free space).

$$D_C = \frac{2\left(2\pi m_e^* k_B T\right)^{3/2}}{h^3} \approx 5 \times 10^{19}\ \mathrm{cm^{-3}} \tag{10.27}$$

$$n_e = D_C \exp\left[\frac{-(E_C - E_F)}{k_B T}\right] \tag{10.28}$$

[129] Paul Adrien Maurice Dirac, 1902–1984, British theoretical physicist, professor in Cambridge, since 1970 in Tallahassee, Florida, USA. Nobel prize in Physics 1933 together with E. Schrödinger, established many of the common formalisms still in use today including the Ket-Bra notation (<Ψ|), the Dirac functional ($\delta_{ij} = 0$ for $i \neq j$ and $\delta_{ij} = 1$ for $i = j$), theory of holes, theoretical prediction of anti-matter, and explanation of the spin.

In a similar way, we obtain the hole concentration p_h [cm^{-3}] (usually referred to as "hole density") from the following approach:

p_h – Hole concentration [cm^{-3}]
E_V – One-electron energy at the valence band maximum [eV]

$$p_h = \int_{-\infty}^{E_V} D_S(E_e)\,(1 - f(E_e))\,dE_e. \tag{10.29}$$

After application of corresponding simplifications made for the electrons in the CB, the results for holes in the VB are given by Eqs. (10.30) and (10.31).

D_V – Density of state within a few k_BT below the valence band maximum [cm^{-3}]
m_h^* – Effective mass of the hole [kg]

$$D_V = \frac{2(2\pi m_h^* k_B T)^{3/2}}{h^3} \tag{10.30}$$

$$p_h = D_V \exp\left[\frac{-(E_F - E_V)}{k_B T}\right] \tag{10.31}$$

In an intrinsic semiconductor, h^+ and e_{CB} are formed and recombine, as approximately represented by the following "reaction equations":

$$e_{VB} + h\nu \rightarrow e_{CB} + h^+ \tag{10.32}$$

$$e_{CB} + h^+ \rightarrow e_{VB} + \text{heat} \tag{10.33}$$

In a photoluminescence process, h^+ and e_{CB} recombine to emit light and heat

$$e_{CB} + h^+ \rightarrow e_{VB} + h\nu' + \text{heat} \tag{10.34}$$

Because of electroneutrality, the concentrations of e_{CB} and h^+ are equal to the intrinsic carrier concentration $n_{int} = n_e = p_h$. The quantity n_{int}^2 is obtained by multiplying Eqs. (10.28) and (10.31) and substituting the relation $-(E_C - E_F) - (E_F - E_V) = (-E_C + E_V) = -E_g$.

$$n_{int}^2 = n_e p_h = D_C D_V \exp\left[\frac{-E_g}{k_B T}\right] \tag{10.35}$$

Q: Provided $n_e = p_h$ for intrinsic semiconductors, isn't then $n_e \cdot p_h$ something like an ionic product or a solubility product?

A: Yes, that is indeed the case! We can take $n_e \cdot p_h = n_{int}^2$ as an equilibrium constant and treat it like an equilibrium constant in many considerations. The analogies to the ionic product of water are particularly striking. At neutral pH, H_2O has an ("intrinsic") conductivity that is very low due to autoprotolysis. Addition of HCl ("doping") increases [H^+] and decreases [OH^-]. However, since the total carrier concentration [H^+] + [OH^-] increases, the conductivity of the solution increases. Corresponding considerations can be made for the addition of NaOH.

With the simplification $m_h^* = m_e^* = m_e$ and the material values for Si ($E_g = 1.1$ eV), we obtain $n_e = p_h = 2.5 \times 10^{10}$ cm^{-3} for $T = 298$ K. This number looks very large. However, we have to consider it in relation to the number of Si atoms in 1 cm^3:

N_A – Avogadro number [mol^{-1}]
N_{Si} – Amount of Si atoms [mol]
ρ_{Si} – Density Si, 2.336 g cm^{-1}
M_{Si} – Molar mass Si, 28 g mol^{-1}

$$\frac{\sqrt{n_{int}^2}}{\frac{N_A N_{Si}}{V}} = \frac{\sqrt{n_{int}^2}}{\frac{N_A m_{Si}}{V\ M_{Si}}} = \frac{\sqrt{n_{int}^2}}{\frac{N_A \rho_{Si}}{M_{Si}}} = \frac{2.5 \times 10^{10}\ \text{cm}^{-3}}{5.0 \times 10^{22}\ \text{cm}^{-3}} = 5 \times 10^{-13} \tag{10.36}$$

This consideration immediately reveals that excitation occurs only for a very small fraction of all Si atoms. The representation in Figure 10.11b is therefore a gross exaggeration! The specific conductivity results from the knowledge of the charge carrier mobilities of e_{CB} and h^+ according to Eq. (4.31).

After equating Eqs. (10.28) and (10.31), we can solve for E_F.

$$E_F = \frac{E_C + E_V}{2} + \frac{k_B T}{2} \ln\left(\frac{D_V}{D_C}\right) = \frac{E_C + E_V}{2} + \frac{k_B T}{2} \ln\left(\frac{m_h^*}{m_e^*}\right)^{3/2} \tag{10.37}$$

As we suspected before, E_F lies approximately in the middle of E_C and E_V or in the middle of the band gap. The deviation from the center is very small because of $k_B T \ll E_g$ [Eq. (10.26)] and disappears completely when $m_e^* = m_h^*$.

Relationship to Thermodynamic Quantities. Later, we will consider reactions of dissolved reactants at semiconductor electrodes, for which it is important to establish a relation between the Fermi level and the electrochemical potential.

The Fermi level of a solid phase α is the electrochemical potential $\tilde{\mu}_e^\alpha$ of electrons in the phase α:

$$E_F^\alpha = \tilde{\mu}_e^\alpha = \mu_e^\alpha + e\varphi^\alpha. \tag{10.38}$$

$\tilde{\mu}_e^\alpha$ – Electrochemical potential of the electrons in the phase α [eV]; say as "mu-tilde-e-in-alpha"
μ_e^α – Chemical potential of the electrons in the phase α [eV]
e – Elemental charge [e]
φ^α – Inner potential of the phase α [V]

It is common to state E_F in eV, although a conversion into other energy units is of course possible. In Chapter 2, for example, we have expressed μ in J mol^{-1}.

Conversion of the Potential Scales. The units eV and J mol^{-1} can be easily converted into each other [analogous to Eq. (10.26)], but this does not solve the problem of the different reference states. E_F usually refers to the level of an electron in a field-free vacuum (vacuum level E_{vac}). In metals and semiconductors, a bound electron has a lower energy. Therefore, the numerical values of E_{vac} on the vacuum scale are negative (Figure 10.12, left). The electrochemical potential scale E_{SHE} (Figure 10.12, right) refers to the SHE. The direction of the two scales is opposite. Small absolute values on the vacuum scale mean that the electron is strongly reducing, which corresponds to

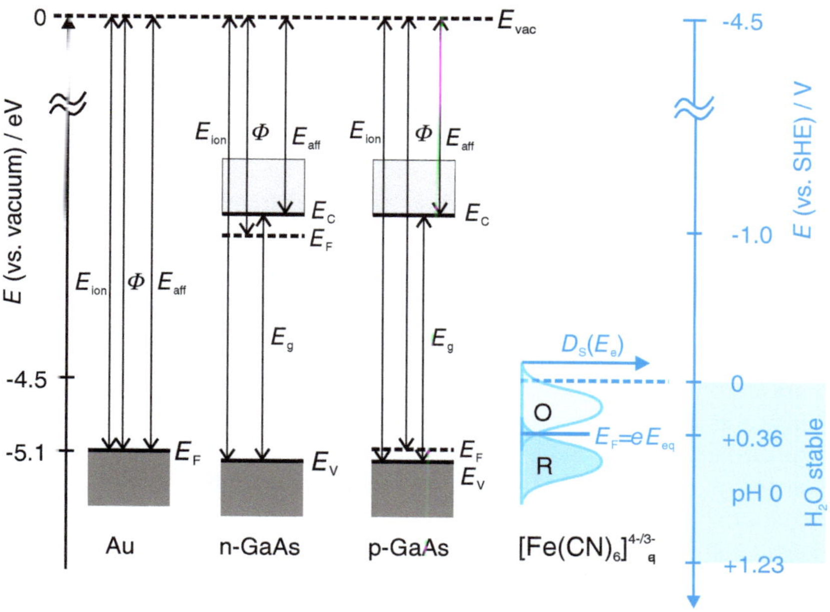

Figure 10.12 a) Comparison of vacuum energy scale and electrochemical potential scale measured against SHE for metals, n-type semiconductors, p-type semidonductors, and a dissolved redox couple. The examples are Au, n-GaAs, p-GaAs, and $[Fe(CN)_6]^{4-/3-}$. For the dissolved redox couple, equal concentrations of the oxidized (O) and reduced (R) forms were assumed, [O] = [R]. For [O] ≠ [R], the curves for $D_S(E_e)$ have a different magnitude. The energy of R and O differ by the number of electrons and by the reorganization of the ligand shell when the charge of the central ion is increased (reorganization energy). The distribution of the energy is not caused by different energies within an band structure, but by the fluctuation of the energy during molecular vibrations in the dissolved reagent. The stability region of H_2O at pH 0 (!) has been highlighted on the potential axis. The work function Φ, the electron affinities E_{aff}, and the ionization energies E_{ion} (Section 10.2.1) are marked as well.

negative values on the SHE scale. There are different suggestions for the conversion [5, 7]. Many authors recommend the relation by Lohmann [13] for $T = 298$ K:

E_{vac} – Energy of an electron in a field-free vacuum [eV]

E_{SHE} – Voltage with respect to the standard hydrogen electrode [V]

$$E_{vac} = -e \cdot E_{SHE} - (4.5 \pm 0.1)\ \text{eV} \tag{10.39}$$

The current IUPAC recommendation [14] is:

$$E_{vac} = -e \cdot E_{SHE} - (4.44 \pm 0.02)\ \text{eV} \tag{10.40}$$

The paper [14] also contains recommendations on conversions for organic electrolyte solutions and a comprehensive discussion of the conceptual and experimental procedures to relate the two scales. Figure 10.12 contains the work function Φ, the electron affinities E_{aff}, and the ionization energies E_{ion} (Section 10.2.1).

Q: This sounds quite confusing. How am I supposed to orient myself?

A: You can orient yourself on the vacuum scale with the help of a simple mnemonic. Electrons as particles “roll” downhill (from more reducing to less reducing) like balls on the band structure plots. We can think of the holes as “bubbles” in the VB otherwise filled with electrons. Holes rise upward like bubbles, from strongly oxidizing states to less strongly oxidizing states. This is important for electrochemistry because at interfaces, the bands are bent (Section 10.2.1), with the change in local energy occurring over micrometers, or many atomic distances. The representation of bent bands in Section 10.2.1 is therefore *not* a band structure within a Brillouin zone!

10.1.4 Doping of Semiconductors

Many semiconductors are called doped semiconductors or extrinsic semiconductors because a noticeable conductivity is only created by the deliberate addition (doping) of foreign elements. If Si atoms are exchanged for As atoms at some positions in the silicon crystal, each of these doping atoms brings an additional valence electron (Figure 10.13c). The electron energy level E_D (in the ground state) is far above E_V but close to E_C (by about 0.05 eV, i.e., about 2 k_BT at room temperature) and is largely localized at the As atom. Since the energetic distance to E_C is very small and there are many unoccupied energy levels at E_C, thermal energy can easily excite this extra electron into the CB. The energy levels in the CB are delocalized. So once the electron is excited, it can contribute to the conductivity. In addition, of course, the generation of e_{CB}–h^+ pairs still proceeds to a very small extent as in the intrinsic semiconductor (Fig. 10.13b). Knowing the dopant concentration n_D, we can now easily determine the concentration of conduction band electrons n_e and of holes p_h (Example Calculation 10.1). Typically, $n_D = 5 \times 10^{16}$ cm^{-3}, that is, one in a million Si atoms has been replaced. The subscript “D” comes from “donor”. While $n_e = p_h$ applies to undoped Si, the concentration of negative charge carriers predominates with As doping.

These comparative statements always refer to mobile charge carriers. Of course, As-doped Si is electrically neutral to the outside!

Semiconductors that have significantly more mobile negative than positive charge carriers are called ***n*-type semiconductors**.

The charge carriers that are clearly in excess in a material are called **majority charge carriers**.

The charge carriers that occur in the material in a significantly lower density (concentration) than the majority charge carriers are called **minority charge carriers**.

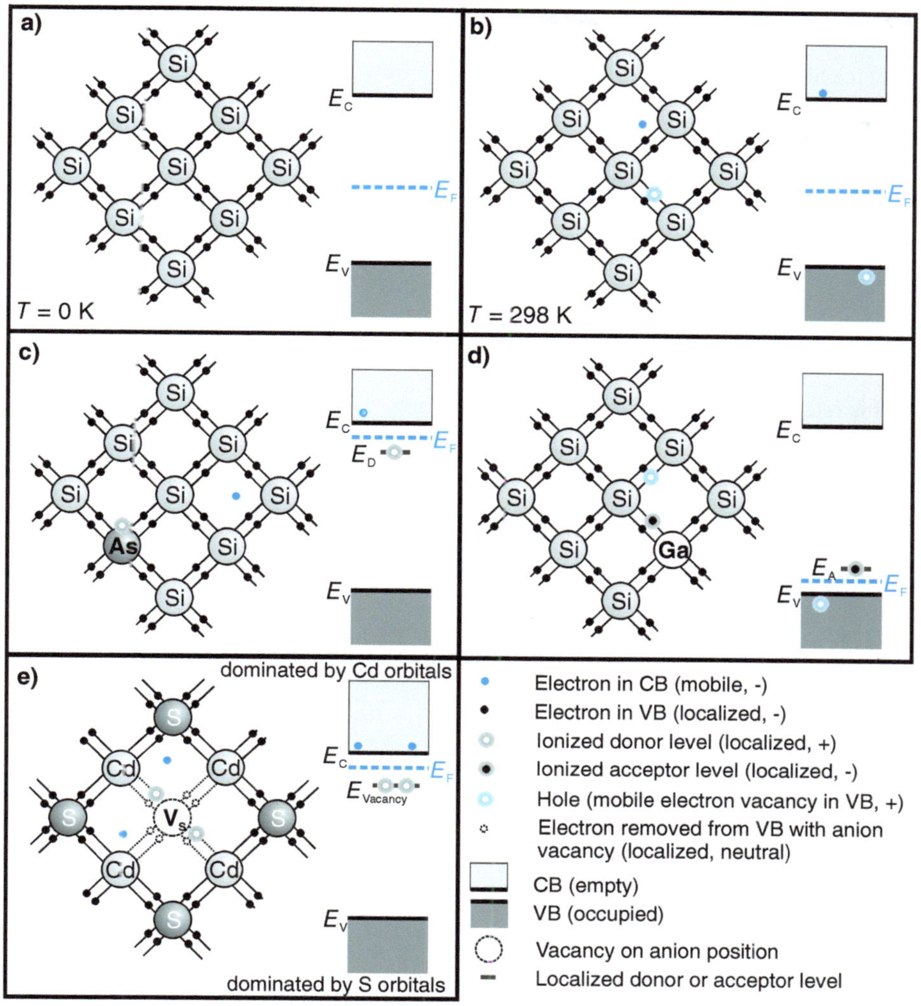

Figure 10.13 Comparison of a) pure Si at 0 K, b) pure Si at 298 K, c) *n*-type Si, d) *p*-type Si, and e) CdS_{1-x} with a vacancy on an S-position.

The two-dimensional representation of the crystal structure serves only to illustrate the bonding situation. The spatial arrangement of the atoms in Si corresponds to the diamond structure, for CdS to the zincblende structure. Moving charge carriers are highlighted in blue. **Unlike in chemical structure formulas, a dash symbolizes an electron level in the VB, *not* an electron *pair*!** A dot in the spaces represents an electron in the CB (nonbonding).

The opposite situation arises when the Si crystal is doped with Ga atoms (Figure 10.13d). Here, too, a concentration of $n_A = 5 \times 10^{16}$ cm^{-3} is typical. The subscript "A" comes from "acceptor". The Ga atoms have only three valence electrons and therefore cannot contribute an electron to one of the four bonds with the surrounding Si atoms. Energetically, this situation corresponds to a localized (nonmobile) unoccupied state at $T = 0$ K, which is slightly above E_V. At $T > 0$ K, electrons can be excited from the VB into the localized acceptor orbitals. A vacancy (hole) remains in the VB, which is mobile and contributes to the conductivity (Figure 10.13d). Because of the very large number of electrons in the VB and the small energetic difference of the acceptor levels E_A to E_V, almost all acceptor levels are "filled" with a (nonmobile) electron. Thus, the charge is now predominantly transported by holes (Example Calculation 10.1). The generation of e_{CB}–h^+ pairs still occurs to a very small extent, as was the case in intrinsic semiconductors.

Semiconductors in which holes are the majority charge carriers are called ***p*-type semiconductors**.

The Fermi level lies between E_D and E_C for *n*-type semiconductors (Figure 10.13c) and between E_V and E_A for *p*-type semiconductors (Figure 10.13d), as a first approximation, in the middle in each case.

Example Calculation 10.1 Determine the majority and minority carrier concentrations for Si, doped a) with 1×10^{16} cm^{-3} As and b) with 1×10^{16} cm^{-3} Ga.

Unknown: n_e, p_h

Given:

$$n_{int} = 2.5 \times 10^{10} \text{ cm}^{-3}$$

$$n_D = 1 \times 10^{16} \text{ cm}^{-3} \text{ (a)}$$

$$n_A = 1 \times 10^{16} \text{ cm}^{-3} \text{ (b)}$$

$$n_{int}^2 = n_e \cdot p_h$$

This relationship also holds for doping, but now $n_e \neq p_h$.

a) $n_e \approx n_D = 1 \times 10^{16}$ cm^{-3} (Almost all donor atoms have donated an electron to the CB.)

$$p_h = n_{int}^2 / n_e = \left(2.5 \times 10^{10} \text{ cm}^{-3}\right)^2 / \left(1 \times 10^{16} \text{ cm}^{-3}\right)$$

$$p_h = 6.25 \times 10^4 \text{ cm}^{-3}$$

b) $p_h \approx n_A = 1 \times 10^{16}$ cm^{-3} (Almost all acceptor atoms have accepted an electron from the VB.)

$$n_e = n_{int}^2 / p_h = \left(2.5 \times 10^{10} \text{ cm}^{-3}\right)^2 / \left(1 \times 10^{16} \text{ cm}^{-3}\right)$$

$$n_e = 6.25 \times 10^4 \text{ cm}^{-3}$$

In Si with a doping concentration of 1×10^{16} cm^{-3}, the majority carriers have a density (concentration) of 1×10^{16} cm^{-3} and the minority carriers have a density of 6×10^4 cm^{-3}. Thus, the density ratio between majority and minority carriers is $1.6 \times 10^{11} = 160\,000\,000\,000$!

In compound semiconductors (oxides, sulfides), the bonds have a partially ionic character. The compounds are often nonstoichiometric, that is, they can give up a small part of the anions, for example, CdS → CdS_{1-x} + x S in Figure 10.13e. As a result, a vacancy is produced at an S^{2-} site (V_S) in the crystal lattice. If an S *atom* leaves an S^{2-} lattice position, the two valence electrons of the original S^{2-} *anion* remain in the semiconductor. As they cannot find a place in the VB, they must assume states in the CB. Thus, the anion vacancy acts similar to a donor atom.

10.1.5 The Quasi-Fermi Level

The previous considerations apply only to semiconductors in electronic equilibrium, that is, in the dark. Only in this case can the energy of e_{CB} and h^+ be described reasonably by *one* Fermi level E_F (Figure 10.14a). When a semiconductor is irradiated with light of suitable energy ($h\nu > E_g$), e_{CB}–h^+ pairs are continuously generated. At the same time, e_{CB} and h^+ recombine with heat release. A steady-state nonequilibrium state is established. "Steady state" means that the concentrations of e_{CB} and h^+ are constant in time. "Nonequilibrium" means that $n_e \cdot p_h > n_{int}^2$. Thus, the charge carrier concentrations can no longer be calculated from the Fermi level E_F at equilibrium. It has proved convenient to define separate quasi-Fermi levels for electrons ($E_{F,n}$) and for holes ($E_{F,p}$) (Figure 10.14b). If we substitute $E_{F,n}$ and $E_{F,p}$ for E_F in Eqs. (10.28) and (10.31), respectively we obtain the carrier concentrations under illumination. The following relations apply:

$$E_{F,n} = E_C - \ln\left(\frac{D_C}{n_e}\right) \tag{10.41}$$

$$E_{F,p} = E_V - \ln\left(\frac{D_V}{p_h}\right) \tag{10.42}$$

D_V – Density of states within a few k_BT below the upper edge of the VB [cm^{-3}]
D_C – Density of states within a few k_BT above the lower edge of the CB [cm^{-3}]
E_C – One-electron energy at the lower edge of the CB [eV]
E_V – One-electron energy at the upper edge of the VB [eV]
$E_{F,n}$ – Quasi-Fermi level of electrons [eV]
$E_{F,p}$ – Quasi-Fermi level of holes [eV]

This situation has some important consequences for electrochemical reactions at illuminated semiconductor electrodes [5]. Let us make this clear with an *n*-type semiconductor. Optical excitation ($h\nu \rightarrow e_{CB} + h^+$) increases the number of e_{CB} and h^+ by the same absolute amount ($\Delta n_e = \Delta p_h$). Because the equilibrium concentration of carriers (in the dark) is very different ($n_{e,eq} \gg p_{h,eq}$, Example Calculation 10.1a), the relative increase in carrier concentrations is dramatically different. It is $\Delta n_e \ll n_{e,eq}$ and thus $n_e = n_{e,eq} + \Delta n_e \approx n_{e,eq}$. In contrast, $\Delta p_h \gg p_{h,eq}$, and thus, under illumination $p_h = p_{h,eq} + \Delta p_h \approx \Delta p_h$. If we insert these values into Eqs. (10.41) and (10.42), it becomes clear that the quasi-Fermi level of the electrons $E_{F,n}$ will not differ significantly from E_F, whereas for the holes a clear lowering of $E_{F,p}$ compared to E_F is to be expected (Figure 10.14b). How large this lowering turns out to depend on Δp_h, that is, on the absorption probability, the irradiation intensity, and the probability for recombination processes. However, since the penetration depth of electromagnetic radiation is small due to absorption forming e_{CB}–h^+ pairs, a large shift of $E_{F,p}$ occurs only near the semiconductor surface and drops off toward the interior of the semiconductor (Figure 10.14c). Since electrochemical reactions occur precisely at the surface, the concept of quasi-Fermi levels plays an important role in explaining reactions at illuminated semiconductor electrodes. They correspond to the electrochemical potentials of charge carriers under illumination.

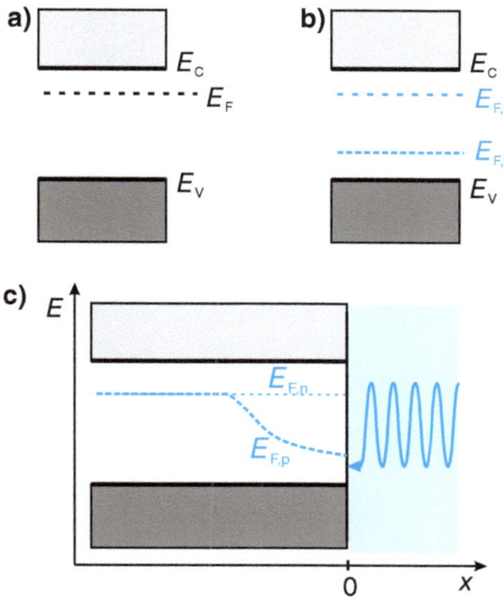

Figure 10.14 Energy diagrams of an *n*-type semiconductor
a) in the dark (equilibrium);
b) with quasi-Fermi levels under electronic excitation; and
c) under illumination of the semiconductor surface.

10.2 Semiconductor Electrodes

10.2.1 Interfaces at Semiconductors and the Space Charge Region

Semiconductor to Vacuum Interface. Let us first consider the surfaces of semiconductors in a vacuum. The representation in Figure 10.15 is strongly reminiscent of the simplified band structure diagrams in Figure 10.13. We can take the horizontal axis (x) as a spatial axis perpendicular to the interface. The vertical axis is an energy axis with the vacuum level E_{vac} as the zero point. For intrinsic semiconductors such as pure Si, E_F is approximately in the middle of the band gap. The distance from E_F to the vacuum level is the work function Φ. The electron affinity E_{aff} corresponds to the energy released when an electron is transferred from the vacuum to a free level of

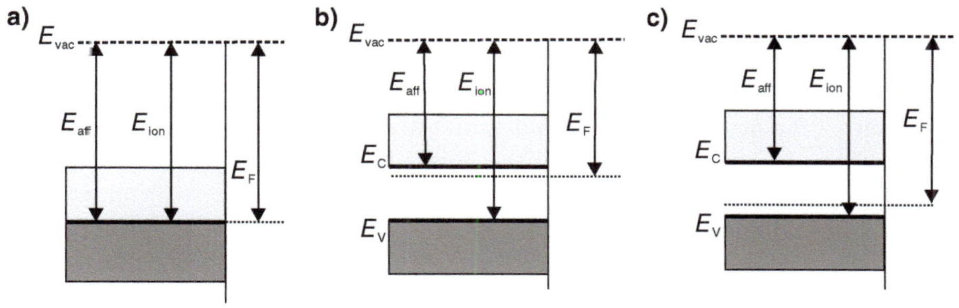

Figure 10.15 Representation of the band structure on the surface of
a) a metal,
b) an *n*-type semiconductor,
c) a *p*-type semiconductor relative to the vacuum level.

the solid (E_C). The ionization energy E_{ion} is the minimum energy required to transfer an electron from an occupied level (E_V) to the vacuum. For a metal, E_F, E_{aff}, and E_{ion} are equal in magnitude.

Surface States. Figure 10.15 shows a highly simplified situation, because further states exist at the surface of each semiconductor, which are collectively called surface states. Their energy E_s can also be within the band gap. Surface states change the band structure diagram at the surface (Figure 10.16). These states arise because the periodic crystal structure is broken at the surface. The lack of coordination of the surface atoms is partially compensated for by surface reconstruction, but this is also accompanied by altered energy levels (intrinsic surface states). Another cause of surface states is adsorbed molecules and ions (extrinsic surface states), which also play a role in semiconductor–electrolyte interfaces.

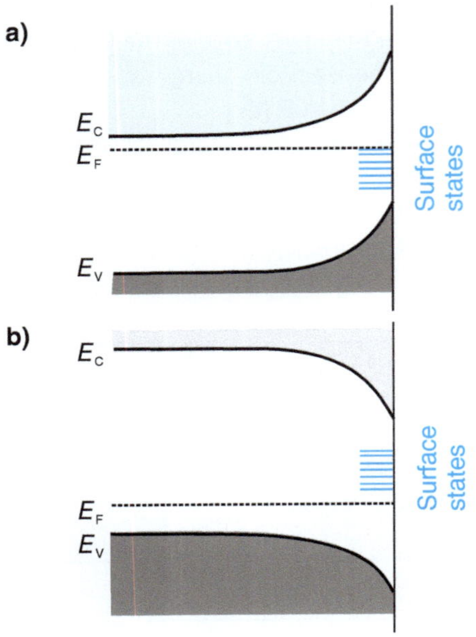

Figure 10.16 Influence of surface states on the formation of the space charge region.

a) in an *n*-type semiconductor and
b) in a *p*-type semiconductor.

In *n*-type semiconductors, the gap from E_S to E_V is much smaller than the band gap E_g, and therefore, surface states are easier to fill. However, to maintain electronic equilibrium, the number of electrons in the CB must be lowered because of $n_e \cdot p_h = const$ ($T = const$). This corresponds to an increase in the band edges at the surface (Figure 10.16a).

An analogous diagram can be constructed for *p*-type semiconductors. Due to the occupation of the surface states, there are more electrons in the CB at the surface and correspondingly fewer holes in the VB, which leads to a downward bending of the bands (Figure 10.16b).

Defect States. In addition to energy states due to disorder of the lattice at the surface, semiconductors can also contain defects in the bulk. For example, semiconducting oxides, which are significant for corrosion processes or in photoelectrochemical applications, contain a large number of perturbations of the crystal structure also inside the material, which are associated with new (localized) energy states in the band gap, which are called gap states. They affect the charge carrier mobility. Let us make this clear with the example shown in Figure 10.17. After optical excitation of an e_{CB}–h^+ pair, the e_{CB} can only move over a short distance before it "falls" into a localized and energetically lower state (trap state) and thus loses mobility. However, it can be "lifted" back into the CB from the trap state by thermal excitation and thus regain mobility. Transport of charge carriers occurs via a sequence of stays in traps and thermal excitation processes. Moreover, the band edges are "smeared out" in the case of optical excitations because there is an appreciable number of states in the band gap that can be the starting point or target of optical transitions. The charge carrier mobility changes very strongly with light intensity. If very many electrons are excited, almost all trap states in the band gap are filled. If the states are occupied, they cannot act as traps anymore for other e_{CB} (Figure 10.17b). The e_{CB} that are further formed by light absorption then exhibit a higher mobility. Perturbations in the crystal structure at the surface or in the bulk volume are always preferred centers for recombination processes ($e_{CB} + h_{VB} \rightarrow e_{VB}$ + heat).

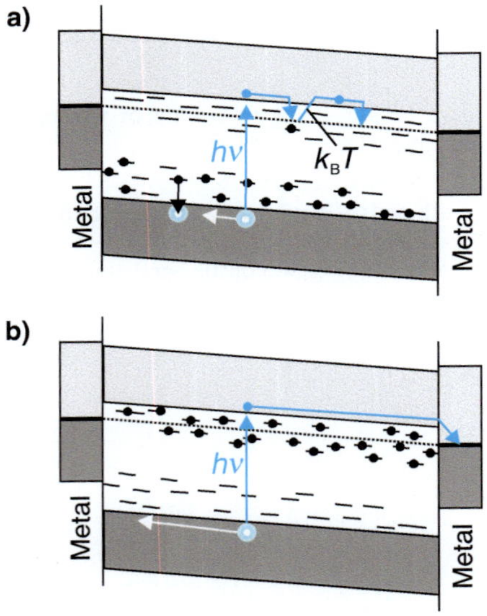

Figure 10.17 Influence of defect states on charge transport through a semiconductor with applied voltage bias: a) at low concentration of e_{CB} and b) at high concentration of e_{CB} (e.g., at high illumination intensity).

For clarity, the contact metal was chosen in a way to avoid the formation of space charge regions at the metal–semiconductor interfaces (see below).

Semiconductor–metal interfaces are determinant for a number of electronic devices and can serve as a starting point for describing semiconductor–electrolyte interfaces. They are important for contacting semiconductor electrodes. Figure 10.18 compares the band structure before/after contact for metals with different work functions. The left part of the figure shows the materials before contact formation, and the right part of the figure shows the band structure after contact formation. We assume thermodynamic equilibrium, where the Fermi levels of the metal and the semiconductor equalize with concurrent bending of the bands in the semiconductor. In Figure 10.18b, some electrons are transferred from the metal to the semiconductor. As a result, the metal becomes positively charged and the semiconductor becomes negatively charged. While the excess charge in the metal remains immediately at

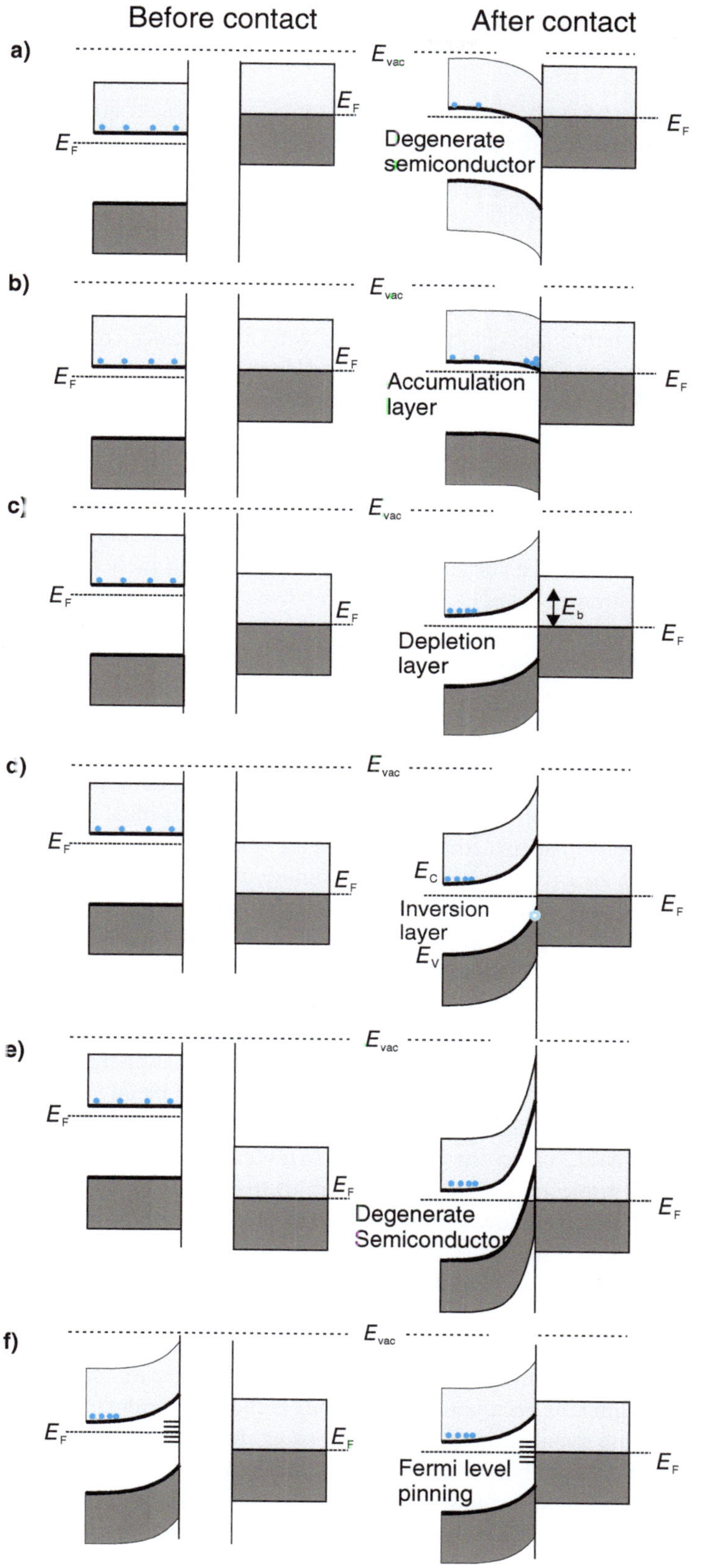

Figure 10.18 Band structures of *n*-type semiconductor–metal interfaces for metals with different work functions. Left diagrams before the contact, right diagrams after the contact;

a) formation of a degenerate semiconductor;
b) formation of an accumulation layer;
c) formation of a depletion layer;
d) formation of an inversion layer;
e) formation of a degenerate semiconductor;
f) Fermi-level pinning at an *n*-type semiconductor with a large density of surface states.

the surface and the entire interior of the metal is field-free, the excess charge in the semiconductor is distributed over a *layer* of finite thickness near the interface.

The area near the surface of a semiconductor where excess charge is distributed is called the **space charge region.**

Q That sounds to me like something I have read before in Section 4.3 with the electrical double layer in electrolyte media. Is this a coincidental similarity?

A: This is well observed! For both the metal–semiconductor interface and the metal–electrolyte interface, the metal with a high concentration of mobile charge carriers faces a phase with much lower conductivity (i.e., lower charge carrier concentration and/or charge carrier mobility). Therefore, it is not possible to fully compensate for the excess charge in a very thin layer, and the zone of unequal charge distribution is distributed over a larger region in the phase with lower conductivity.

The excess charge creates a local electric field in the semiconductor that modifies the electron energies near the surface, changing the occupation probability of electronic states near the surface. Here, E_C is lowered relative to E_F, and mobile charge carriers (here e_{CB}) are enriched at the surface (accumulation layer). In Figure 10.18c, a metal with a larger work function is present. During contact, some CB electrons e_{CB} of the semiconductor are transferred into the metal, causing the surface to be depleted of e_{CB} (depletion layer). For the transport of an electron from the metal into the semiconductor, an energy barrier E_b must be overcome. This barrier represents an additional resistance and should be avoided when contacting semiconductor electrodes.

If the work function of the metal is even larger, the bands must bend so extremely that E_V has a smaller gap to E_F than E_F to E_C at the surface of the *n*-type semiconductor (Figure 10.18d). At E_F, the occupation probability of an electronic state is just 50%. Thus, it is more likely in this case to find an h^+ than an e_{CB} as a charge carrier at the surface of the *n*-type semiconductor. The role of majority and minority charge carriers reverses at the surface. Therefore, we call this region the inversion layer.

With even higher work function of the chosen metal, E_F passes through the strongly bent VB of the semiconductor after equilibration (Figure 10.18e). At the surface, both occupied and unoccupied states then exist (as in a metal) in the energetic vicinity of E_F. The semiconductor surface behaves accordingly similar to a metal. Analogously, if the work function of the metal is very small, the CB may be bent downward after contact to such an extent that E_F passes through the CB. In this case, too, we would find metal-like behavior of the semiconductor surface (Figure 10.18a). We refer to semiconductors in situations such as those shown in Figure 10.18a and e as degenerate semiconductors.

If a large density of surface states is present in the *n*-type semiconductor, the equilibrium adjustment upon contacting often takes place primarily by rearranging the charge state of the surface states. Even before contact is made with the metal, E_F passes through the surface states (Figure 10.18f, left). Since the occupation probability there is about 50% in the example, further electrons can be accepted or released without significantly changing the occupation in the bands of the semiconductor. For this reason, the band bending remains approximately constant after contact (Figure 10.18f, right). Thus, the barrier height is determined only by the surface states. This situation, where the Fermi level of the semiconductor is "stuck", is called "Fermi level pinning".

Energy Levels in Electrolytes with Redox Systems. To understand charge transfer reactions of dissolved redox systems at semiconductor electrodes, it is convenient to translate the thermodynamic considerations from Chapter 2 for a redox

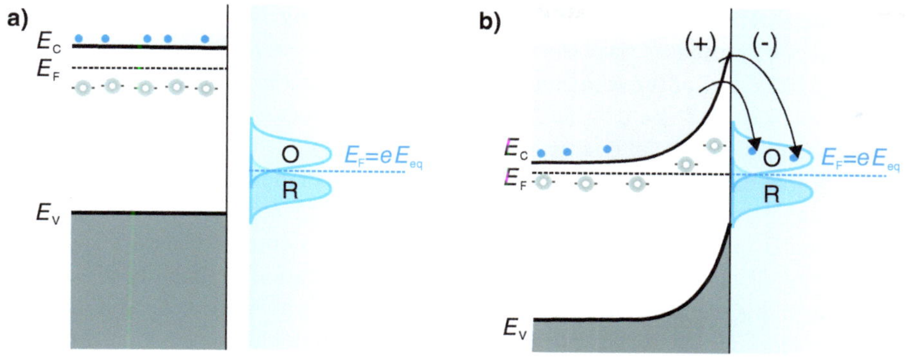

Figure 10.19 Adjustment of an equilibrium between an *n*-type semiconductor and a Fe^{2+}/Fe^{3+} solution:

a) before contact;
b) after contact and equilibration.

pair R $\rightleftharpoons$ O + e^- (e.g., Fe^{2+}/Fe^{3+}) into the band model. We think of the redox equilibrium Fe^{2+}/Fe^{3+} as a distribution of electronic "states" of the dissolved iron species. The Fe^{2+} ions represent the occupied states, and the Fe^{3+} ions correspond to the unoccupied states (Figure 10.19a). Due to the different interactions of ions with their ligands and solvation shells (reorganization energy), the average energy of the Fe^{2+} states is lower than that of the Fe^{3+} states. However, the energy is not exactly constant but is modulated by molecular vibration during which the distance of the ligands and solvent molecules to the central ion changes periodically. Thus, the energy of the "states" changes in time; the solution species have *only one energy level* (and not a band), but the precise energy of this level changes with the molecular vibration. The energy distribution corresponds to a bell-shaped curve with a width between 0.3 and 0.6 eV. The bell-shaped curves for R and O intersect at one point. At this point, the probability that a state is occupied or unoccupied is equal. Thus, this point corresponds to the Fermi level of the redox system in solution. We can interpret this as the electrochemical potential of an electron distributed in the solution between Fe^{2+} (more generally R) and Fe^{3+} (O). After conversion (eV → V, reference to vacuum → reference against SHE), E_F corresponds to the Galvani potential difference at a Fe^{2+}/Fe^{3+} redox electrode. The concentration of Fe^{3+} and Fe^{2+} ions is proportional to the height of the bell curves. Thus, in the example in Figure 10.19, both redox forms would be present in the solution in equal concentrations. By changing the height of one of the bell curves, the position of the intersection point on the E axis would shift – but only very slightly, as predicted by the Nernst equation.

Note similarities and differences of (measurable) electrode potentials and (nonmeasurable) Galvani potential differences that we have already discussed in connection with Eq. (2.110).

Q: This I cannot follow. If an e^- occupies the "unoccupied state" in the Fe^{3+}, surely that becomes an Fe^{2+}, and according to Figure 10.19a, it then occupies a lower energy level than Fe^{3+}.

A: This is correct. As soon as we "fill" an e^- into the unoccupied states (Fe^{3+}), Fe^{2+} is formed. However, the drop in energy is not directly caused by the "filling in" of the e^-, but is due to the change in interactions of the less charged Fe^{2+} with the ligand and solvent environment. We would have to think of the "filling" of the e^- into the unoccupied state as transferring the e^- with the ligand and solvent environment fixed. While this is not practically possible, it can be simulated with a quantum chemical computer program as a conceptually very useful model. The amounts of converted reactants at the interfaces are so small that they are easily balanced by mass transport processes. Thus, the solution concentrations are only imperceptibly changed by the "filling" of some electrons into some Fe^{3+}-ions.

Semiconductor Electrodes at Equilibrium. If we bring the *n*-type semiconductor into contact with the Fe^{2+}/Fe^{3+}electrolyte solution at the energetic situation depicted in Figure 10.19, e_{CB} of the semiconductor will transfer to the "unoccupied states" of the solution (i.e., $Fe^{3+} + e^-_{CB} \rightarrow Fe^{2+}$) until the semiconductor has become charged to such an extent that the net charge transfer comes to an end (We already discussed in Chapter 2 that the turnover necessary for this is very small in relation to the number of particles in chemical systems!!). The positive excess charge in the semiconductor distributes into a space charge region. The resulting local potential changes the position of the energy levels near the surface and thus the occupation probability of states at the surface. This manifests itself in the band diagram by a bending of E_C and E_V. This is a depletion layer analogous to Figure 10.18c, since there are fewer mobile charge carriers at the surface than inside the field-free semiconductor.

Analogous considerations can be made when bringing a *p*-type semiconductor into contact with an electrolyte solution (Figure 10.20). In this case, a few electrons from Fe^{2+} will pass into the semiconductor and be taken up there by holes in the VB ($Fe^{2+} + h^+ \rightarrow Fe^{3+}$). The net transfer comes to a halt when the semiconductor has become sufficiently negatively charged. The excess charge is distributed in a space charge region where the number of holes is less than in the volume of the semiconductor. Again, a depletion zone (of h^+) would form.

Although the above illustration correctly represents the rough picture, it should be noted that, in the case of semiconductor electrodes, the establishment of thermodynamic equilibrium is often prevented by corrosion of the semiconductor material with the formation of oxide layers or by very slow electron transfer processes.

10.2.2 The Flat Band Potential

Analogous to Section 10.2.1, we can ask how the band structure at a semiconductor-electrolyte interface behaves when we apply a voltage across the interface by means of a potentiostat. In this case, we do not need a redox couple in the solution to charge the semiconductor. Rather, charge carriers flow away across the external contact. Assuming an ohmic contact of the semiconductor electrode (Section 10.2.5), the external cell voltage drops between the interior of the semiconductor electrode and the interior of the (metallic) counter electrode. An electrical double layer is present at the counter electrode, just as we discussed in Section 4.3.2. We now have to consider the details of how the potential drops at the semiconductor–electrolyte interface. Because we can freely vary the voltage of the cell with a potentiostat, we can set any excess charge in the semiconductor without replacing a redox pair R/O. Qualitatively, similar band structure diagrams result on the semiconductor side as in the right column of Figure 10.18. At exactly *one* potential, no excess charge will occur at all in the semiconductor – this corresponds to a situation *between* Fig. 10.18b and c.

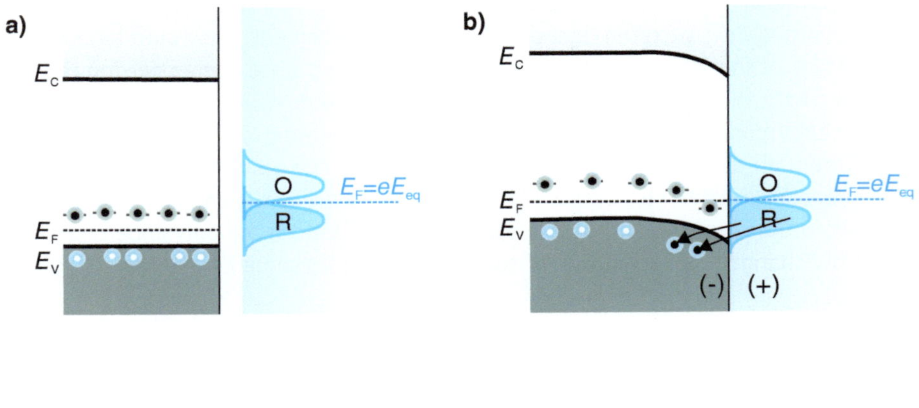

Figure 10.20 Adjustment of an equilibrium between a *p*-type semiconductor and a Fe^{2+}/Fe^{3+} solution:
a) before contact;
b) after contact and equilibration.

The potential, at which there is no excess charge at a semiconductor–electrolyte interface, is called the **flat band potential** E_{fb}, because in this case, no band bending occurs. E_{fb} corresponds to the potential of zero charge E_{pzc} at metal electrodes that was discussed in Section 8.10.3.

If we apply a more negative potential than E_{fb} to an n-type semiconductor, the bands "bend downward" and e_{CB} accumulate at the surface (accumulation layer, analogous to Figure 10.18b). Even a degenerate semiconductor may be formed as in Figure 10.18a. If the potential applied to an n-type semiconductor becomes more positive (more oxidizing) than E_{fb}, the bands "bend upward". The surface becomes depleted in e_{CB}. A depletion layer forms analogous to Figure 10.18c. At even more positive potentials, an inversion layer forms analogous to Figure 10.18d, or it may even happen that E_F is below E_V (analogous to Figure 10.18e). In this case, the occupation probability in the normally filled VB at the surface is 50%. At the surface, many occupied *and* unoccupied states are present side by side on the energy scale, similar to what is typical for metals (degenerate semiconductor).

Analogous considerations can be made for p-type semiconductors, where the terms enrichment and degeneracy are applied to h^+ as the majority charge carriers:

$E > E_{fb}$	Accumulation of h^+, band bending upward
$E = E_{fb}$	no excess charge, no band bending
$-E_g < (E - E_{fb}) < 0$	depletion of h^+, band bending downward
$(E - E_{fb}) < -E_g$	strong band bending downward, at the surface $E_F > E_C$ (degenerate semiconductor)

Whether all the cases shown in Figure 10.18 really occur in the experiment depends on the band gap, the solvent used, and the electrolysis or corrosion reactions of the semiconductor.

The space charge region is associated with a measurable area-related electrical capacitance $C_{sc,A} = f(E)$ that is analogous to the electrical double layer capacitance (Sections 4.3 and 8.10). Starting from the Poisson equation, the derivation is done for an n-type semiconductor using the band bending $\Delta\phi_{sc} = -(E - E_{fb})$, the dopant concentration n_D, and the temperature-dependent charge carrier densities in the volume of the semiconductor [5]. We obtain a linearized relation $C_{sc,A}^{-2} = a_0 + a_1 E$, which is called the **Mott–Schottky equation**.[130]

$$\frac{1}{C_{sc,A}^2} = \frac{2}{e\varepsilon_0\varepsilon_r n_D}\left(E - E_{fb} - \frac{k_B T}{e}\right) = \left(\frac{2r_D}{\varepsilon_0\varepsilon_r}\right)^2\left(\frac{eE}{k_B T} - \frac{eE_{fb}}{k_B T} - 1\right). \qquad (10.43)$$

$C_{sc,A}$ – Area-related capacity of the space charge region [A s V^{-1} cm^{-2}]
e – Elemental charge [A s]
ε_0 – Dielectric constant of a vacuum = 8.8542×10^{-12} A s V^{-1} m^{-1}
ε_r – Relative dielectric constant of the semiconductor [dimensionless]
n_D – Dopant concentration [cm^{-3}]
E – Applied potential [V versus SHE]
E_{fb} – Flat band potential [V versus SHE]
k_B – Boltzmann constant [eV K^{-1}]
T – Temperature [K]
r_D – Debye length [cm] = $[\varepsilon_0\varepsilon_r k_B T/(2N_D e^2)]^{1/2}$

The Debye length[36] r_D (Section 4.2.8) is calculated according to Eq. (4.74), using the number of charge carriers ($\approx$ dopant concentration $n_D = 10^{17}$ cm^{-3}) in the volume of the semiconductor instead of the number of ions per cm^3. Typical values for r_D are 10^{-6} cm = 10 nm [5].

[130] Walter H. Schottky, 1886–1976, German physicist, professor of theoretical physics in Rostock and physicist at Siemens AG; Sir Nevill Francis Mott, 1905–1996, English physicist, professor of physics in Bristol and Cambridge, Nobel prize in physics 1977 for his work on the electronic structure of magnetic and disordered solids.

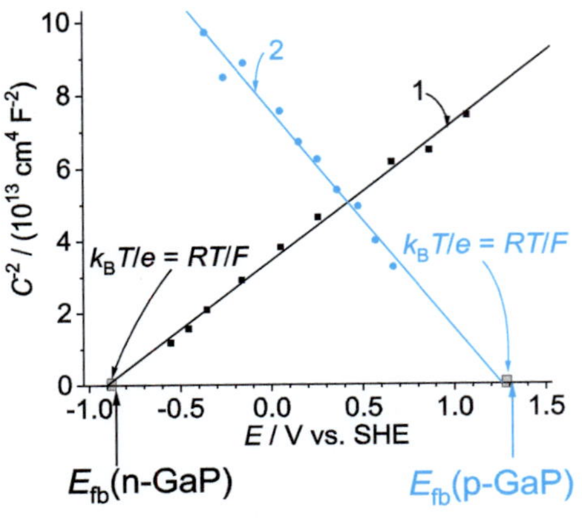

Figure 10.21 Example of Mott–Schottky plot for 1) *n*-type GaP and 2) *p*-type GaAs in 0.05 mol L^{-1} H_2SO_4. Data from [15]. The position of the Fermi levels and band edges of analogous *n*-type GaAs and *p*-type GaAs on the different potential scales is shown in Figure 10.12.

Mott–Schottky plots (Figure 10.21) are made for the potential range where a depletion zone occurs. An extrapolation of the straight line to $C_{sc,A}^{-2} = 0$ (intersection with the E axis) yields the flat band potential E_{fb} after correction by $k_BT/e = 0.0257$ V at $T = 298$ K. From the slope, the donor concentration n_D is accessible (for flat surfaces). The position of the band edges E_C and E_V can be estimated from E_{fb} together with E_g (e.g., from spectroscopic data) and n_D, which are very important for the prediction of electrochemical reactions on semiconductors. Figure 10.22 shows such band edges for important semiconductors. The results from Mott–Schottky plots are susceptible to errors due to the influence of surface states. Therefore, we measure $C_{sc,A}$ at different frequencies and check whether they yield equal values for E_{fb} and n_D. In addition, it should be noted that the space charge region with capacitance C_{sc} and the electric double layer with capacitance C_{dl} represent two capacitors connected in series with the total capacitance $C^{-1} = C_{sc}^{-1} + C_{dl}^{-1}$. In this situation, we can determine C_{sc} only if $C_{sc} \ll C_{dl}$, which is usually the case for charge carrier concentrations smaller than 10^{20} cm^{-3}.

Flat Band Potential and pH. In aqueous solutions, E_{fb} is often pH-dependent or, more generally, dependent on chemisorption of ions that can lead to surface charging. We consider this, taking TiO_2 as an example, where a protonation equilibrium exists for OH groups bound at the surface to Ti ions (Ti_s). The following reactions must be considered (index s for surface).

$$\equiv Ti_s\text{-}OH_2^+ \rightleftharpoons \equiv Ti_s\text{-}OH + H^+ \tag{10.44}$$

$$\equiv Ti_s\text{-}OH \rightleftharpoons \equiv Ti_s\text{-}O^- + H^+ \tag{10.45}$$

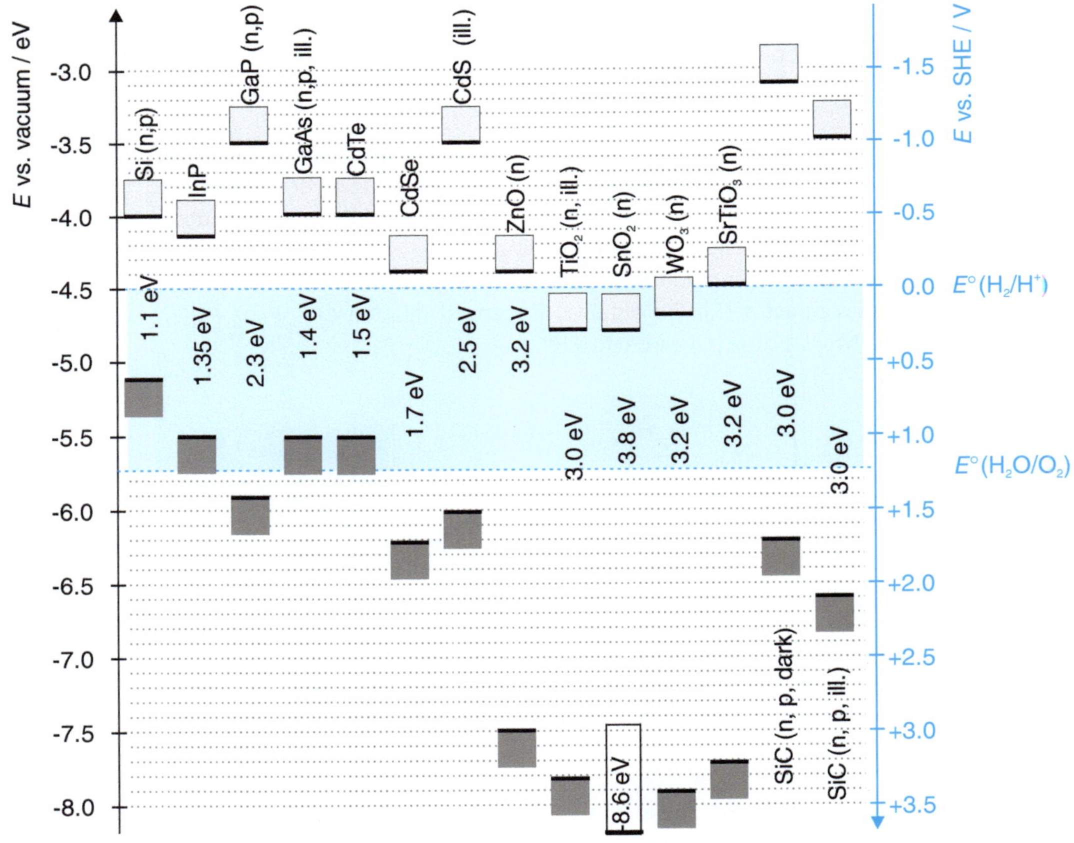

Figure 10.22 Approximate position of band edges on the energy scale for important semiconductors. Data for Si, InP, CdTe at pH 1 from [11], other compounds from [5] at pH 0; n, *n*-type semiconductor; p, *p*-type semiconductor; ill., under illumination; dark, in the dark. The stability range of H_2O at pH 0 is highlighted in color.

Increasing the pH will lead to an increasing deprotonation of the surface and thus to a negative charge. E_{fb} must be more negative than for a neutral surface for the bands to show no bending at this negative surface charge. The reverse holds if the surface is positively charged at low pH values. Adsorption of anions acts similarly to a pH increase, and adsorption of cations acts similar to a pH decrease. The influence of the pH of the surrounding medium on E_{fb} plays an important role, for example, in the biocompatibility of metallic implant materials that are covered by a passive oxide layer (e.g., Ti|TiO_2) [16].

10.2.3 Current–Voltage Curves at Semiconductor Electrodes in the Dark

At metal and carbon electrodes, the current depends on the concentration of reactants at the electrode surface $c_O(x = 0)$, $c_R(x = 0)$, and a potential-dependent rate constant $k(E)$ [cm s^{-1}] according to a (pseudo)-first order rate law. We were able to explain the potential dependence with the Butler–Volmer equation [Eqs. (3.27) and (3.38)]. The carrier concentration in the metal did not appear in the equations. We now realize that this is because there are large numbers of occupied *and* unoccupied states at energies around E_F in metals. Therefore, the carrier concentrations hardly change with the applied potential. The carrier concentration enters $k(E)$ as a constant. In the case of semiconductor electrodes, the situation is quite different. If a redox system that has energy levels within the band gap is to react, only the majority charge carriers are available in a moderately doped semiconductor in the dark. In these situations, reductions can be carried out at *n*-type semiconductors. However, no holes are available for oxidation reactions. Conversely, oxidations (but no reductions) can take place on *p*-type semiconductors.

Q: So, the electrode behaves somewhat like a diode then?
A: Yes, this is indeed a good comparison for reactions in the dark.

The concentration of charge carriers in the space charge region ($n_{e,sc}$ bzw. $p_{h,sc}$ [cm^{-3}]) depends strongly on the potential (Figure 10.18):

$$p_{h,sc} = N_A \exp\left[\frac{E - E_{fb}}{E_{NF}}\right] \tag{10.46}$$

$$n_{e,sc} = N_D \exp\left[\frac{-(E - E_{fb})}{E_{NF}}\right]. \tag{10.47}$$

Both quantities must be explicitly considered in the corresponding rate laws:

$$I_a = nFA\, k'_a(E)\, p_{h,sc}(E)\, c_R(x = 0) \tag{10.48}$$

$$I_c = nFA\, k'_c(E)\, n_{e,sc}(E)\, c_O(x = 0) \tag{10.49}$$

We obtain the rate constants k'_a and k'_c in an analogous way to the Butler–Volmer Equations (3.27) and (3.38).

In the depletion layer, $n_{e,sc} < N_D$ or $p_{h,sc} < N_A$. As we will see from Exercise 10.7 (at the end of this Chapter), $C_{sc} \ll C_{dl}$ holds in most realistic situations. Therefore, almost all of the voltage drop between the interior of the semiconductor electrode and the interior of the solution occurs in the space charge region (but not in the electrical double layer). Consequently, in this region, a plot of ln$I$ versus E yields the slope $1/E_{NF}$. In the context of a Tafel plot (Figure 3.12), this would result in a transfer coefficient $\alpha = 1$. Only when E has been shifted to the point where a degenerate semiconductor is present does the influence of $k'_a(E)$ or $k'_c(E)$ become important. The corresponding Tafel plot yields $\alpha \approx 0.5$, as in most reactions at metal electrodes.

I_a (I_c) – Anodic (cathodic) current [A]
n – Number of transferred electrons per reactant molecule (dimensionless)
F – Faraday constant [A s mol^{-1}]
k'_a (k'_c) – Anodic (cathodic) rate constant for second-order rate law [cm^4 s^{-1}]
$c_R(x = 0)$ ($c_O(x = 0)$) – Concentration of R (O) at the electrode surface [mol cm^{-3}]
$p_{h,sc}$ ($n_{e,sc}$) – Concentration of h^+ (e_{CB}) in the space charge region [cm^{-3}]
N_A (N_D) – Concentration of acceptors (donors) [cm^{-3}]
E_{NF} – Nernst factor = RT/F
E_{fb} – Flat band potential [V]
$E - E_{fb}$ – Band bending [V]

These idealized relationships are rarely found so clearly in experiments, because corrosion processes of the semiconductor, effects of the internal electrical resistance in the semiconductor electrode, and reactions over surface states may occur in the same potential range as the reaction with dissolved redox systems.

The current–voltage curves for an n-type semiconductor differ greatly depending on whether the formal potential of the redox couple $E^{\circ\prime}(\text{R}'/\text{O}')$ is less than or greater than E_{fb}. For $\text{E}^{\circ\prime}(\text{R}'/\text{O}') < E_{\text{fb}}$, an accumulation layer is present near the formal potential, and the curves differ little from those at metal electrodes (for comparison, see Figure 10.23a, curve 3 shown for Pt). On the other hand, if $E(\text{R/O}) > E_{\text{fb}}$, the reaction cannot start appreciably until an accumulation layer is formed at $E < E_{\text{fb}}$, and CB electrons e_{CB} are available at the interface. For the oxidation reaction, the currents remain very small until the band bending becomes so large that E_{F} at the interface is near the upper edge of the VB E_{V} (Figure 10.23b, curve 1).

For a p-type semiconductor, the curve at $E^{\circ\prime}(\text{R}'/\text{O}') > E_{\text{fb}}$ differs little from that at metal electrodes (Figure 10.24b, curves 1 and 3). If $E^{\circ\prime}(\text{O/R})$ is within the band gap, oxidation currents do not occur until the holes accumulate on the surface at $E > E_{\text{fb}}$. Reductions can only be carried out at very negative potentials when E_{F} is close to the lower edge of the CB (E_{C}), and CB electrons e_{CB} become available at the interface (Figure 10.24a, curve 1).

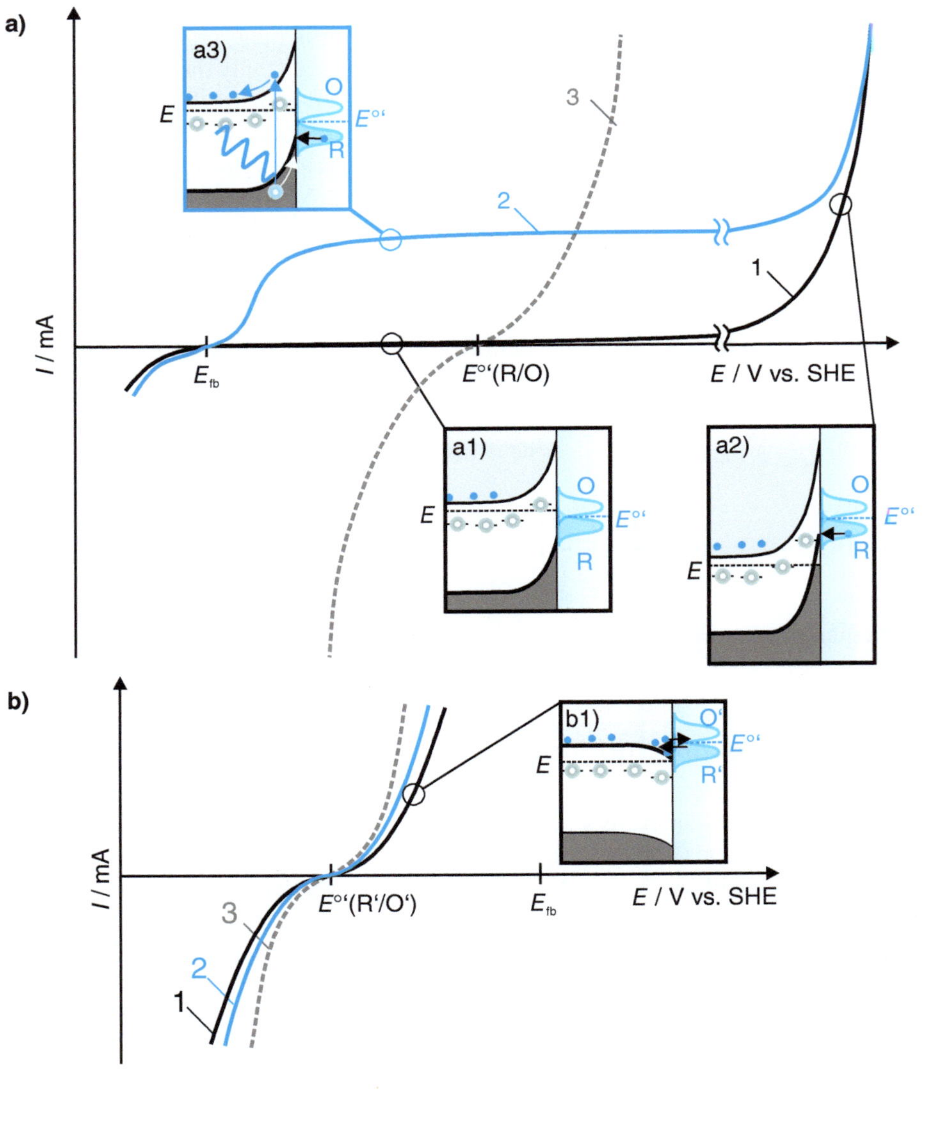

Figure 10.23 Current–voltage curves of an n-type semiconductor electrode in a solution with [R] = [O] or [R′] = [O′]:

a) $E^{\circ\prime}(\text{R/O}) > E_{\text{fb}}$;
b) for a redox system R′/O′ with $E^{\circ\prime}(\text{R}'/\text{O}') < E_{\text{fb}}$;

Individual curves:

1) semiconductor electrode in the dark,
2) semiconductor electrode under illumination with $h\nu > E_{\text{g}}$,
3) for comparison, current–voltage curve of R/O or R′/O on metal or glassy carbon electrodes.

Mass transport limitations are not considered (corresponds to experiments on rotating electrodes). The insets show the band bending at the respective potentials; the symbols correspond to those in Figure 10.13.

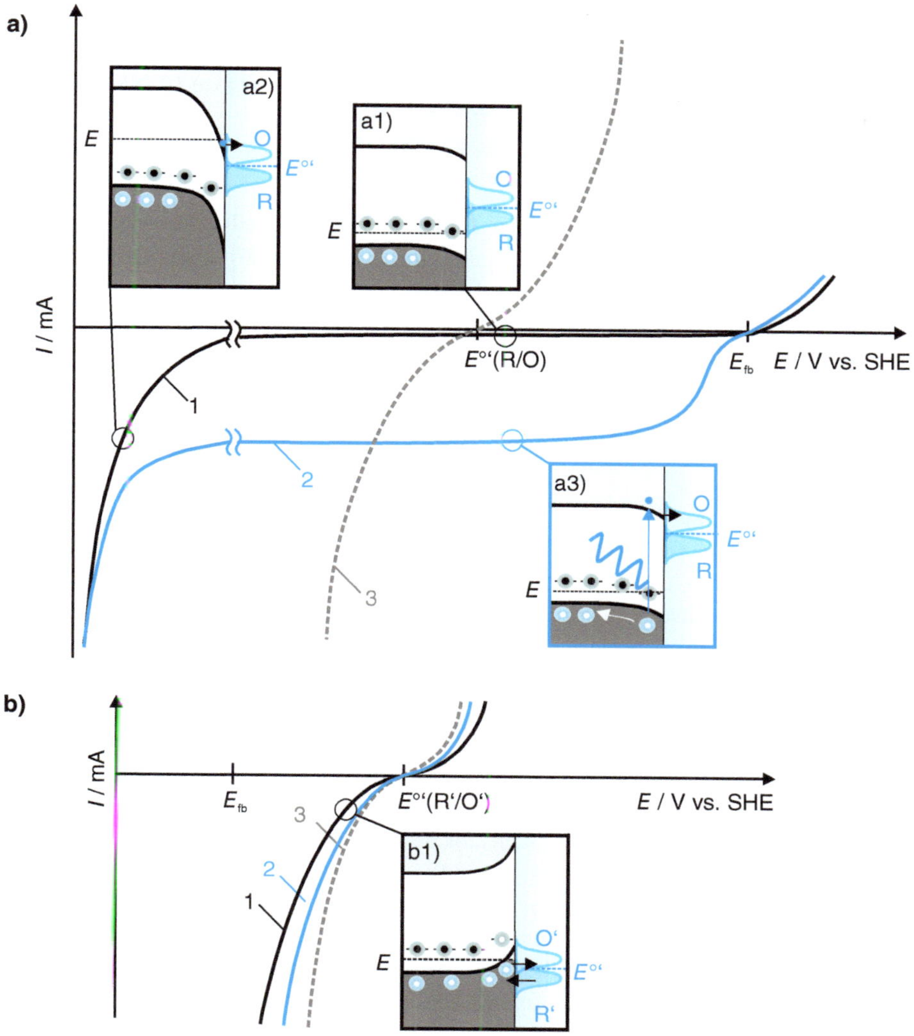

Figure 10.24 Current–voltage curves of a *p*-type semiconductor electrode in a solution with [R] = [O]:

a) with $E^{\circ\prime}$(R/O) < E_{fb};
b) for a redox system R′/O′ with $E^{\circ\prime}$(R′/O′) > E_{fb};

Individual curves:

1) semiconductor electrode in the dark,
2) semiconductor electrode under illumination with $h\nu > E_g$,
3) for comparison, current–voltage curve of R/O or R′/O on metal or glassy carbon electrodes.

Mass transport limitations are not considered (corresponds to experiments on rotating electrodes). The insets show the band bending at the respective potentials; the symbols correspond to those in Figure 10.13.

10.2.4 Photocurrents at Semiconductor Electrodes

If semiconductor electrodes are irradiated with photons $h\nu > E_g$, e_{CB}–h^+ pairs can be generated. Inside the semiconductor, e_{CB} and h^+ recombine with heat generation because the carrier concentration after excitation exceeds the thermodynamically determined constant carrier product $n_e \cdot p_h$. However, if an e_{CB}–h^+ pair is generated in the space charge region of an *n*-type semiconductor, or gets there by diffusion before recombination, the electric field accelerates the electron into the volume of the semiconductor and the hole toward the surface. The h^+ is available as a minority charge carrier for oxidation reactions. We see in Figure 10.23b, curve 2, that oxidation reactions under illumination start at potentials *negative* to $E^{\circ\prime}$(R/O) of the redox couple!!! The energy for the oxidation at such low potentials comes from the energy of the absorbed photons. The principal possibility for capturing light energy with semiconductor electrodes by using separated light-generated charge carriers (e_{CB}, h^+), and storing the energy permanently in chemical compounds (thus mimicking photosynthesis), represents a major driving force for further research in the field of semiconductor electrochemistry (Section 10.3).

Similarly, the absorption of a photon in a *p*-type semiconductor, polarized to form a depletion layer, can bring e_{CB} to the interface. They can enter reduction reactions at potentials *positive* of the formal potential of the redox couple (Figure 10.24a, curve 2).

10.2.5 Experimental Considerations

Preparation and Contacting of the Electrodes. Semiconductor electrodes are mostly used as single-crystal electrodes in the form of wafers (Si) or small blocks. The electrodes must then be provided with an ohmic contact on the back of the electrode. Special materials (Table 10.2) must be used (not required for metal electrodes!) to avoid creating additional barriers to electron transport (such as those shown in Fig. 10.18). Ohmic contacting methods for semiconductor electrodes are either based on a combination of materials as shown in Figure 10.18b, in which almost no barrier exists, or they use methods to increase the doping at the contact point to such an extent that barrier-free charge transfer becomes possible. The contact to the semiconductor electrode is made by two leads. By measuring the resistance at various applied voltages between the two leads, it can be proven that a low-resistance ohmic contact was indeed established. The electrically contacted electrode can then be integrated into an electrochemical cell, for example, by pressing it against an O-ring to seal the solution reservoir. Alternatively, the semiconductor electrode can be embedded into a polymer, and the exposed front surface used as an electrode. Preparation is particularly challenging when a semiconductor is used as a rotating disk electrode (RDE) to establish well-defined mass transport conditions. The semiconductor must be processed into circular disks using special techniques before embedding [5].

Semiconducting metal oxides are often formed spontaneously or are deposited directly onto the metal. Powdered oxides for photoelectrochemical cells are deposited from slurries and thermally treated (sintered).

Specialized Cells. For wafers, there is a range of cell designs in Figure 10.25 that allow fast sample exchange and also provide illumination either from the solution side (front side illumination) or from the back (back side illumination). Further information on cutting and contacting Si wafers with different crystal orientations can be found in Ref. [17].

Pretreatment of the Electrodes. Polishing processes cannot be used for the pretreatment of single-crystal electrodes because they would destroy the crystalline structure at the surface. Therefore, etching processes are preferred, whereby hydrofluoric acid (HF) is often used. **Caution:** HF is very *corrosive*, even to glassware, and therefore, requires the use of containers made of fluorinated plastics. HF is also *very toxic*. It can be absorbed into the body through the respiratory tract, digestive tract, or skin and

Table 10.2 Methods for contacting silicon wafers according to [17] with advantages (A) and disadvantages (D).

$n_D > 10^{19}$ cm^{-3}	Contacting with a metal clamp
$n_D < 10^{19}$ cm^{-3}	GaIn eutectic applied to backside with sandpaper; A: fast, D: possible contamination of equipment for handling semiconductors
$n_D < 10^{19}$ cm^{-3}	Application of a metal film on the back side and heat treatment; D: possible contamination of equipment for handling semiconductors
$n_D < 10^{19}$ cm^{-3}	Doping of the back side with $n_D > 10^{19}$ cm^{-3} and contacting with metal clamp, A: avoidance of contamination, optically transparent

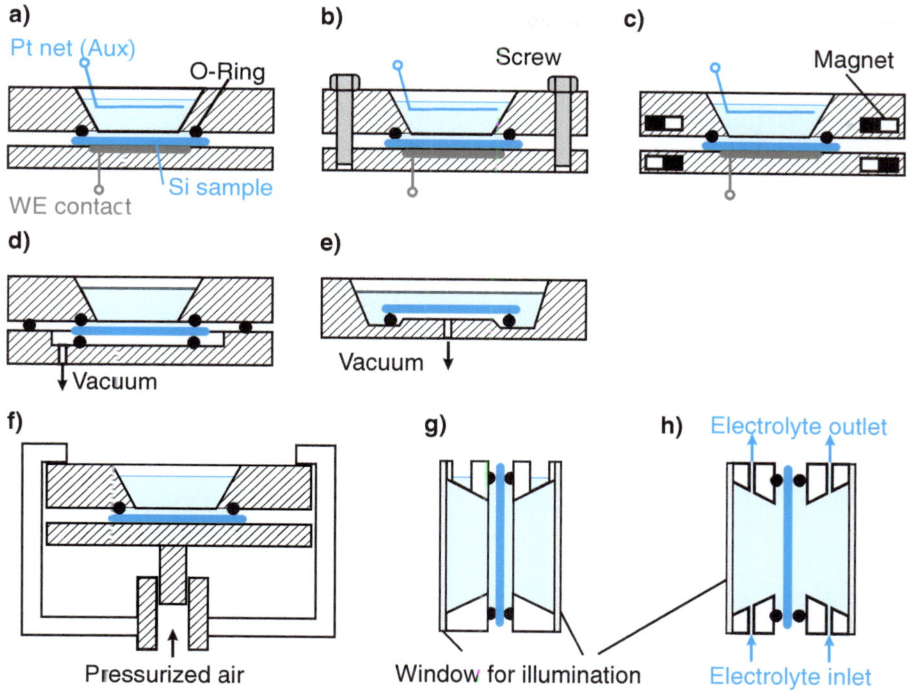

Figure 10.25 Cross sections through electrochemical cells designed for semiconductor wafers. a) Sealing via O-ring by the weight of the cell body; b) pressing by screws; c) pressing by integrated permanent magnets; d) and e) suction by vacuum; f) pressing by pneumatically operated stamp; g) vertically arranged cell; and h) closed cell with inlet and outlet tubing. Source: Adapted from [17], Copyright 2002, Wiley-VCH.

leads to slow/certain death. Concentrated solutions develop considerable vapor pressure. Therefore, in addition to the self-evident use of a fume hood, further personal protective equipment and precautions must be taken when working with HF in case of spilled solutions!

10.2.6 Photoelectrochemical Methods for the Characterization of Semiconductor Electrodes

In this section, we will discuss the principles of some measurement techniques that are used specifically for semiconductor electrodes. Further information on these techniques can be found in [18, 19].

Photocurrent measurements examine the current difference I_{ph} between the current under illumination I_{illum} and in darkness I_{dark} (Figure 10.23b). I_{illum} and I_{dark} can be elegantly determined by linear sweep voltammetry (LSV, Section 5.3.5) with *one* measurement, during which the sample is *intermittently* illuminated (chopped illumination). The current then jumps back and forth between I_{illum} and I_{dark} (Figure 10.26). However, a steady-state current is established only after some time (usually several seconds). Immediately after the onset of illumination and also after switching off the light source, a characteristic overshoot of the current occurs, resulting from the interaction of reaction and recombination processes (Figure 10.26). Figure 10.26a shows a typical band structure diagram of an n-type semiconductor with surface states and a depletion zone. Immediately after the onset of illumination, a hole current density j_h sets in. This can be calculated according to the Gärtner Equation[131] (10.50) and is shown in Fig. 10.26b-1.

$$j_h = \frac{I_h}{A} = qI_0\left(1 - \frac{\exp(-\alpha r_D)}{1 + \alpha L_h}\right) \quad (10.50)$$

I_h, (j_h) – Hole current (hole current density) [A] ([A cm^{-2}])
A – Electrode area [cm^2]
q – Correction factor for losses by reflection ($0 < q < 1$), dimensionless
I_0 – Light intensity of illumination [W cm^{-2}]
α – Absorption coefficient [cm^{-1}]
r_D – Thickness of the space charge region [cm]
L_h – Hole diffusion length [cm]

[131] Wolfgang W. Gärtner; Depletion-Layer Photoeffects in Semiconductors. *Phys. Rev.* **1959**, *116*, 84–87.

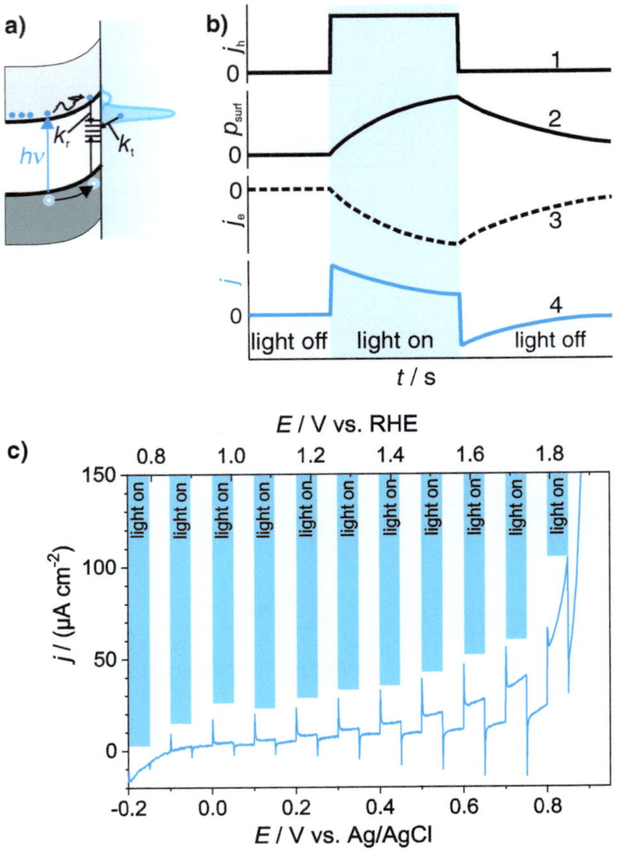

Figure 10.26 Photocurrent measurements on *n*-type semiconductors.

a) Band structure diagram under illumination according to [20];
b) temporal behavior according to [20] of
   1) h^+ current density,
   2) surface concentration of h^+,
   3) e_{CB} current density,
   4) net current density;
c) experimental photocurrent measurement with slow change of electrode potential at an iron oxide–zinc oxide electrode in 0.1 mol L^{-1} NaOH solution, kindly provided by D. Taffa, Carl von Ossietzky University Oldenburg.

The hole current leads to an enrichment of h^+ in the surface states. The instantaneous concentration of holes at the surface p_{surf} [cm^{-3}] is shown in Figure 10.26b-2. Now, charge transfer occurs from these holes to reactants in solution (with the heterogeneous rate constant k_t [cm s^{-1}]) or recombination with the majority charge carriers, which diffuse slowly (depletion zone) to the surface states. The process can be described by a rate law of pseudo-first order with a rate constant k_r [cm s^{-1}]. When the hole concentration p_{surf} increases, the rate of recombination also increases. This rate corresponds to an electron current density towards the surface (Figure 10.26b-3). In the external circuit, the measurable net current density is $j = j_h + j_e$ (note the sign of $j_h + j_e$, Figure 10.26b-4), and we observe the characteristic overshoot as j_e is delayed after the onset of illumination, but also decays slowly after the illumination is turned off. In contrast, j_h follows the light intensity immediately. From the decaying part of the curve $j(t)$ the sum $k_r + k_t$ can be determined. The ratio of the steady-state current (or the steady-state current density) to the current (or the current density) immediately after the onset of illumination is $k_t/(k_r + k_t)$.

Steady-state photocurrents show a characteristic dependence on the light intensity (number of photons per second and per area) and the energy of a single photon. These relationships differ in a characteristic way according to whether the absorption is a direct or an indirect transition (Figure 10.10). Direct transitions usually have much larger absorption coefficients.

I_{ph} – Photo current [A]
I_0 – Light intensity [W cm^{-2}]
$h\nu$ – Photon energy [eV]
$const_1$ – Proportionality constant [eV V^{-2} cm^{-4}]
$const_2$ – Proportionality constant [(eV)$^{-1/2}$ $V^{-1/2}$ cm^{-1}]

$$\text{Direct transition:} \quad \left(\frac{I_{ph}\cdot h\nu}{I_0}\right)^2 = const_1\left(h\nu - E_g\right) \tag{10.51}$$

$$\text{Indirect transition:} \quad \left(\frac{I_{ph}\cdot h\nu}{I_0}\right)^{1/2} = const_2\left(h\nu - E_g\right) \tag{10.52}$$

If the photocurrent is proportional to the carrier concentration, plots according to Eqs. (10.51) or (10.52) allow the identification of direct or indirect transitions. In principle, E_g can also be determined. This opportunity is indeed used in cases where optical absorption measurements are impossible.

Impedance spectroscopy (Chapter 7) is widely used in the study of semiconductor electrodes [21] with modulation of the electrode potential in the dark or under illumination. I_{ph} changes with the frequency of this modulation, but exhibits a phase shift with respect to the excitation. The data analysis often begins with the establishment of an equivalent circuit to describe the semiconductor–electrolyte interface. By fitting the data over the experimentally accessible frequency range at different potentials or light intensities, the characteristic quantities such as C_{sc} and resistance are determined for the charge carrier transport in the semiconductor, for charge carrier recombination, or for solution reactions. Further details can be found in the literature [22].

Intensity-modulated photocurrent spectroscopy (IMPS) can be regarded as a variant of impedance spectroscopy in which the illumination intensity applied to an electrode is periodically modulated (instead of the potential). In this way, for example, the transit time of photogenerated charge carriers can be determined for complex electrode structures [18, 19, 22].

Photopotential Transients. The semiconductor electrode is irradiated at open circuit until a steady state is established between carrier generation and recombination. In this state, the product $n_e \cdot p_h$ exceeds the equilibrium value. In Section 10.1.5, we have introduced the quasi-Fermi levels to describe the different electrochemical potentials for electrons and holes in such a situation. The measured photovoltage results from the shift of the quasi-Fermi level with respect to the equilibrium Fermi level. If illumination is abruptly terminated, carrier generation ceases immediately, while carrier recombination and interfacial reactions continue for some time. The latter leads to a gradual decrease in photovoltage. If the measurement is performed with a high-impedance measurement setup, the decrease must be due to charge carrier recombination in the semiconductor electrode (possibly involving surface states). If the solution contains a reactant for the transfer of charge carriers, the photovoltage usually decays much faster due to the additional decay channel for charge carriers by the interfacial reaction. The rate constants for these processes can be determined from a curve fit.

Intensity-modulated photovoltage spectroscopy (IMVS) is a variant of open circuit photovoltage measurement with frequency-dependent modulation of light intensity. It is used to determine the average lifetime of generated charge carriers before their recombination in the electrode or at its interface [18, 19, 22, 23]. Thus, in combination with IMPS, the mean free path length of charge carriers can also be determined. This is the path that a free charge carrier travels on average before it recombines.

10.3 Application of Semiconductor Electrodes

In the semiconductor industry, the electrochemistry of silicon plays a central role in surface treatments and in the production of porous silicon, which has special luminescence properties [17, 24, 25]. In addition to this application, semiconducting metal oxides play a crucial role in corrosion processes (Chapter 14).

Here, we will mainly focus on photochemical applications where light absorbed by semiconducting electrodes is used to perform an electrochemical reaction. There are three basic types of photoelectrochemical cells (Table 10.3). In photovoltaic cells, light energy is converted into useful electrical energy. In photosynthetic cells, light energy is used to drive an endergonic reaction and to produce high-energy products

Table 10.3 Types of photoelectrochemical cells according to [7].

Energetics	*n*-Type semiconductor | electrolyte | metal	*p*-Type semiconductor | electrolyte | metal
Photovoltaic cell $\Delta_R G > 0$ (no reaction possible without light; with light there is no net reaction because O and R are symmetrically generated and consumed at both electrodes of the cell)	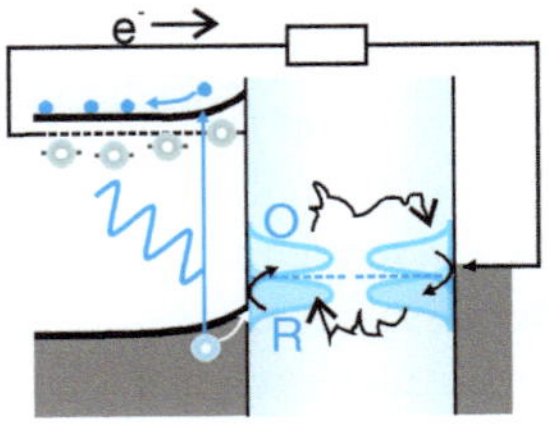	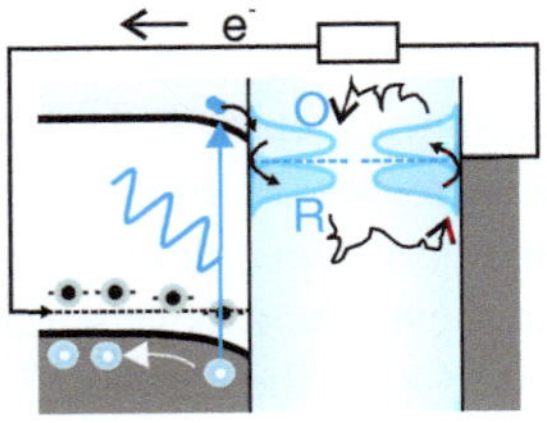
Photosynthetic cell $\Delta_R G > 0$ (no reaction possible without light, divided cell)	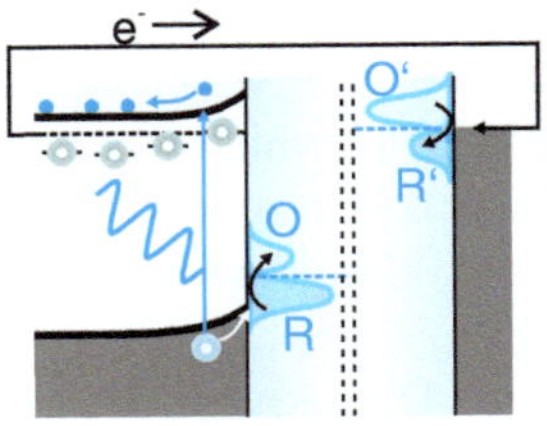	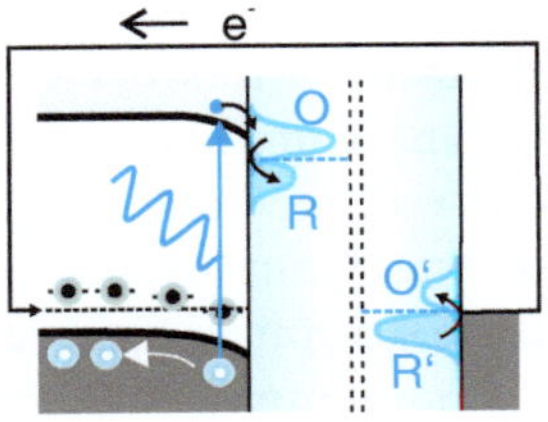
Photocatalytic cell $\Delta_R G < 0$ (reaction is thermodynamically possible even without light – but proceeds very slowly; light energy is used to activate the reaction that otherwise would have a larger overpotential; divided cell)	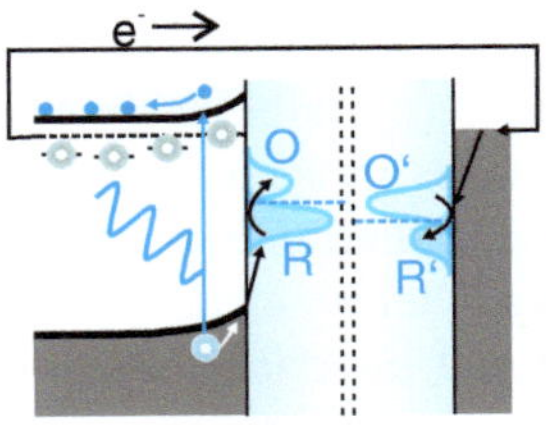	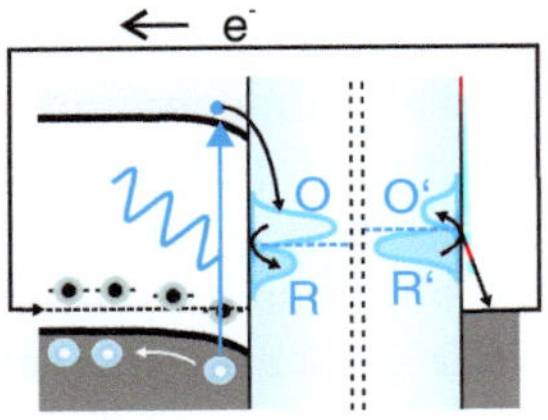

("chemical fuels"), for example, by splitting H_2O into H_2 and O_2 (water splitting). In photocatalytic cells, the reaction is not endergonic, and light energy is used to provide the activation energy, which is difficult to apply otherwise in the dark. The term "catalysis" is not used correctly here, because additional energy is absorbed, and the absorption of light corresponds more to external heating in conventional chemical reactions.

10.3.1 Mineralization of Organic Pollutants in Wastewater

The purification of wastewater from organic substances is an important stage in wastewater treatment. Substances that are often not amenable to biological degradation (halogenated organic compounds, some pesticides, dyes) must be chemically degraded. Reactions on semiconductor particles may be suitable for this purpose. TiO_2 is mainly used (*n*-type semiconductor). Because of the large band gap, this semiconductor is stable against photooxidation, which is otherwise a preferred anodic reaction. The use of particles massively increases the available surface area of the semiconductor electrode, and the use in a suspension or on a coated electrode surrounded by a thin layer of wastewater reduces the transport problem and difficulty of achieving homogeneous illumination that would exist on solid electrodes or cells. Therefore, these processes can be used at low concentrations of the substances to be degraded, where other processes are uneconomical.

Special features of semiconductor particles when compared to solid semiconductor electrodes of the same material do not result only from the larger surface-to-volume ratio. If the dimensions of the semiconductor become smaller than about 2 nm, quantum size effects become clearly apparent. We can distinguish between thin films (quantum wells, one dimension <2 nm), quantum wires (two dimensions <2 nm), and quantum dots (three dimensions <2 nm). In these cases, the electronic structure of the semiconductor is greatly changed compared to a bulk crystal. The band gaps become larger and the width of the bands decreases as the particles become smaller. This can be used to adapt absorbers to specific wavelengths of light. This is important for the applications considered here insofar as it also allows adapting the positions of the band edges to the requirements of the targeted reactions.

Suspensions of larger particles also behave differently from electrodes made of the same solid bulk material. There are two reasons for this: First, the e_{CB}–h^+ pairs can only be permanently separated if the charge carriers react at an interface or enter an external circuit. If this does not happen within the lifetime of the charge carriers, the e_{CB}–h^+ pairs recombine with heat generation and without the desired photochemical reaction (Figure 10.27a). In particles, the distances that e_{CB} and h^+ have to travel to arrive at the interface are often shorter than the average transport distances within the lifetime τ (about $\sqrt{2D\tau}$). This fact increases the probability that the generated charge carriers will reach the interface. Second, we would initially expect the depletion layer in the particle to prevent e_{CB} from reaching the interface. However, this is not the case because after the transfer of some h^+ at the interface (i.e., the oxidation of an electron donor, e.g., the wastewater component to be mineralized), the negative charge of the particle counteracts the band bending (Figure 10.27b). If the particles are smaller than the extent of the space charge region (specified as Debye length r_D) in the solid electrode, band bending does *not* occur. In such cases, the charge carriers move in the semiconductor particle exclusively by diffusion (Figure 10.27c) and not field-driven. Oxidation and reduction reactions must occur *simultaneously* on each particle; otherwise, the particle becomes so charged that no further reactions are possible. To accelerate the interfacial reactions, noble metal clusters are often attached to the semiconductor particles as catalysts.

The investigation of photoelectrochemical reactions in suspensions of semiconductor particles is only useful with inert (irreversible) redox couples R/O and R′/O′, since otherwise the thermodynamically expected back reaction $R' + O \rightarrow O' + R$ would occur immediately.

The **mineralization reactions** of organic compounds are complicated, irreversible, multistep reactions; in many cases, they are not yet fully elucidated. The target reaction is usually the total oxidation to CO_2 and H_2O. This can be accomplished by h^+ transfer to the organic compound. For TiO_2, a hole in the VB corresponds

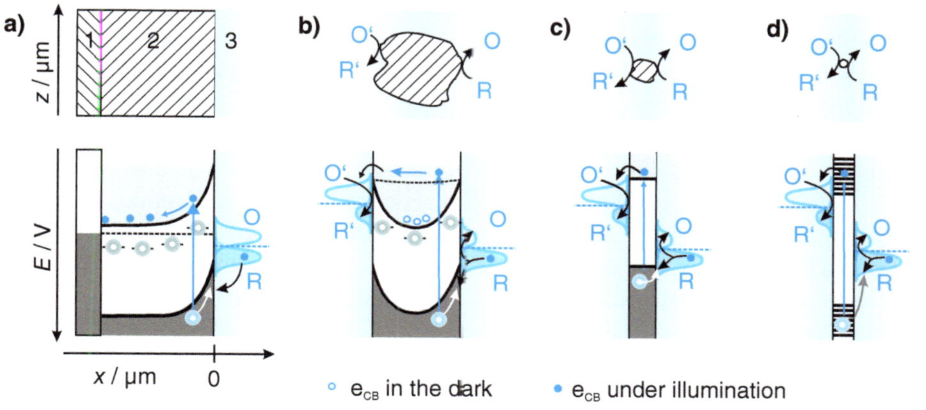

Figure 10.27 Comparison of the charge transfer processes at

a) macroscopic electrodes;
b) on "large" particles ($d > r_D$);
c) on "small" particles ($d < r_D$);
d) at a quantum dot with discrete energy levels.

R is the donor and O′ is the acceptor.

1) – Metal contact,
2) – Semiconductor layer,
3) – Electrolyte.

The schemes are not to scale!!!

to a potential of +2.7 V vs. SHE, a potential at which most organic compounds are destroyed. However, these reactions are slow, and the probability of finding the organic compound adsorbed on the semiconductor electrode in a suitable configuration in a dilute solution is low. Therefore, it is often assumed that water first reacts with h^+ to form hydroxyl radicals $OH^•$, which then convert the organic compounds to CO_2 and H_2O in subsequent homogeneous follow-up reactions. However, TiO_2 as a widely available semiconductor absorbs only UV light ($\lambda \leq 414$ nm, $E_g = 3.0$ V). Thus, most photons in the visible light spectrum are not usable. Therefore, efforts are directed at introducing other elements into the oxide to shift the absorption into the visible region of the spectrum and thereby make a larger fraction of the irradiated light energy usable.

Water splitting is another important reaction that is currently being investigated very intensively, as it promises access to regeneratively produced fuels [12, 26–33]. This is a photosynthetic reaction in the sense of Table 10.3, since $\Delta_R G > 0$. There are variants with massive electrodes (advantage: products accrue separately) [20], or in suspensions (advantage: large surface area, but explosive oxyhydrogen gas accrues as a product mixture). Some reactions proceed under absorption of two photons per separated charge carrier pair ("Z-scheme", analogous to natural photosynthesis) [34]. Review papers illustrate current research directions focused on finding fundamentally new materials from abundant elements, combining different materials at the nanometer scale [35], experimentally analyzing and modeling the local processes at different time and length scales [36], as well as optimizing the solid–liquid interface.

Further Reading: A systematic review of other photoelectrochemical conversions is provided by [37].

10.3.2 Dye-Sensitized Solar Cells as Example for Photovoltaic Cells

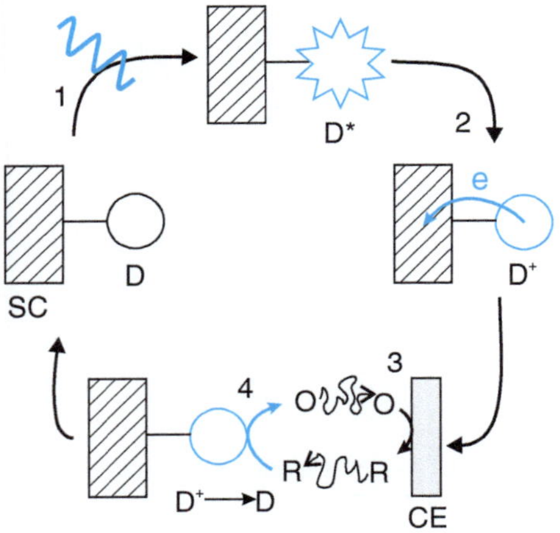

Figure 10.28 Reaction sequence in a dye-sensitized solar cell;

1) light absorption by D adsorbed on the semiconductor SC;
2) injection of the electron into the CB;
3) reduction of the mediator at the CE;
4) reduction of the photooxidized dye D^+.

Principle. TiO_2 can be the basis for a photoelectrochemical solar cell. However, only photons with $\lambda \leq 414$ nm, that is, only a very small fraction of the visible light, would be absorbed in a solid TiO_2 electrode. A decisive improvement is possible with sensitizers. These are dyes D that adsorb at the semiconductor–electrolyte interface and have a high absorbance coefficient in the visible range. After absorption of a photon ($D + h\nu \rightarrow D^*$), an electron from the LUMO of the excited dye D^* is injected into the CB of the semiconductor within less than a picosecond ($D^* \rightarrow D^+ + e_{CB}$). The remaining oxidized dye is reduced by a redox mediator (R/O) from solution ($D^+ + R \rightarrow D + O$). Thus, D is regenerated and can subsequently absorb another photon. The redox mediator is reduced again after its diffusion to the counter electrode ($O + e^- \rightarrow R$). Figure 10.28 shows the reactions of the dye in terms of a chemical reaction. However, the energetic relationships and the direction of the processes are not clear from this illustration.

The reactions are usually shown in the context of the band model (Figure 10.29). In the band model, we immediately recognize that the energetic LUMO–HOMO difference from D is smaller than E_g of the semiconductor. The dye should have a broad absorption band to absorb light over the widest possible wavelength range. HOMO and LUMO should "enclose" the CB edge E_C of the semiconductor to allow efficient injection of the electron from the LUMO of the dye into the CB. The maximum possible voltage is determined by the difference between E_F of the semiconductor (more precisely, the quasi-Fermi level of electrons under illumination) and the equilibrium potential of the redox mediator, which is a quasireversible redox couple (R/O). The choice of an appropriate redox couple is important in conjunction with its overpotentials. Significant features are the increase in lifetime and diffusion length of the generated e_{CB} and the location of the quasi-Fermi levels relative to the redox potential of the mediator, as they determine the achievable cell voltage. A detailed discussion of

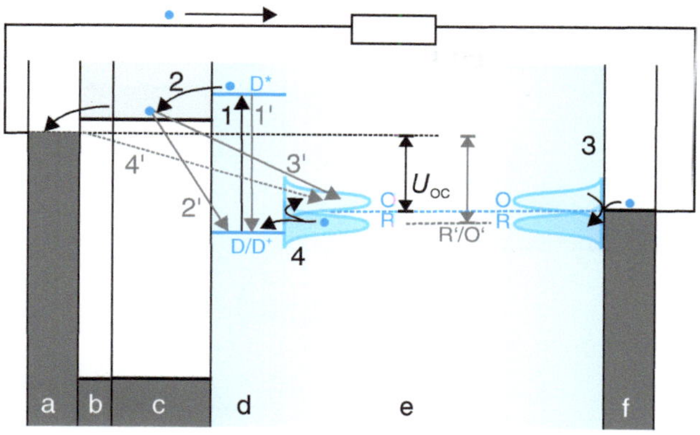

Figure 10.29 Reactions in a dye-sensitized solar cell in the band model with two different redox mediators R/O and R′/O′.

The desired reactions are indicated by black arrows: 1) light absorption; 2) electron injection into the CB; 3) reduction of the mediator at the CE; 4) regeneration of the photooxidized dye D^+.

Undesired reactions are indicated by gray arrows: 1′) luminescence; 2′) recombination by transfer of e_{CB} to D^+; 3′) recombination by transfer of e_{CB} to the oxidized mediator O; 4′) recombination by transfer of electrons from backside contact to O.

Components: a) thin conductive oxide layer; b) blocking layer to shield the back contact from the electrolyte; c) nanoporous semiconductor layer; d) dye on the surface of the semiconductor; e) electrolyte layer; f) counter electrode.

important characterization methods and modeling of these cells based on determined quantities can be found in [38].

Experimental Realization. The surface area of the semiconductor electrode must be as large as possible to accommodate many dye molecules per geometric area. To prevent side reactions (e.g., photocorrosion of the semiconductor), semiconductors with a large band gap are advantageous. In addition, organic solvents are used because there are fewer side reactions in them than in water (oxidation/reduction of water with the formation of reactive compounds). Dye-sensitized solar cells (DSCs) based on TiO_2 nanoparticles have gained particular popularity [39, 40]. A glass coated with a transparent thin conductive metal oxide layer, that is referred to as [transparent conducting oxide (TCO), for example, indium-doped tin oxide (ITO) or fluorine-doped tin oxide (FTO)] serves as a substrate for the photoanode. It is first covered with a dense blocking layer of compact TiO_2 with a thickness of less than one micrometer. This layer prevents direct contact of the mediator solution with the back contact. Without this layer, electrons from the back contact could react directly with the oxidized mediator component O.

A suspension of TiO_2 nanoparticles is then applied; the particles are cemented together with a sintering process. Since nanoparticles are used, the band bending in this material is low (Figure 10.27c). A pore volume remains between the nanoparticles, which provides a large internal surface area where many dye molecules can be adsorbed. Ruthenium complexes with derivatized bipyridyl ligands have been established as dyes (Figure 10.30a). These compounds often contain several carboxylic acid groups, which cause the dyes to adsorb on the TiO_2 surface. Additional substituents provide an asymmetric charge distribution that increases the absorption coefficient and aids in the injection of the electron into the CB. In order to avoid the use of the rare heavy metal Ru, organic dyes, often with the triphenylamine structural motif, have been intensively studied (Figure 10.30b,c).

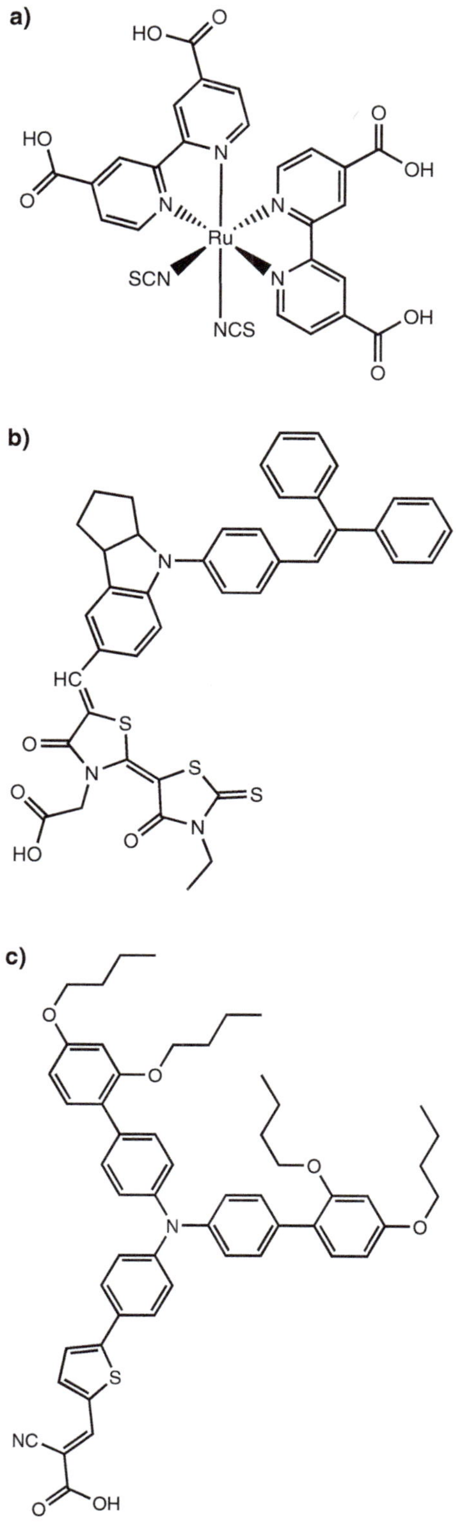

Figure 10.30 Structures of dyes used as sensitizers in DSCs.

a) N719 as an example of a Ru dye;
b) D35; and
c) D149 as examples of organic dyes.

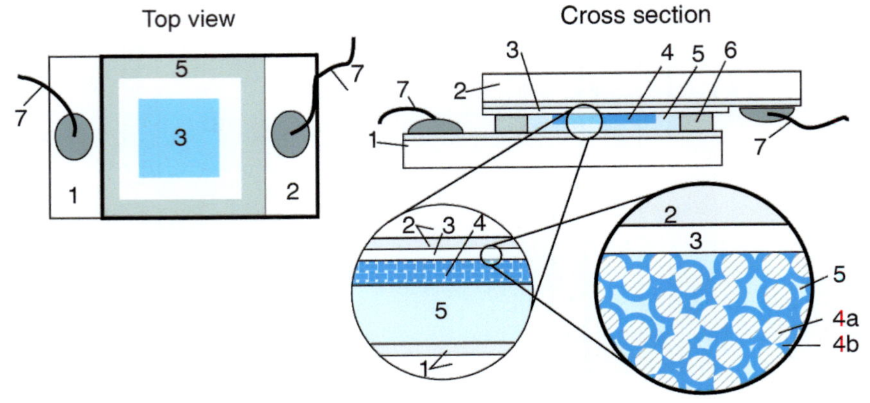

Figure 10.31 Setup of a dye-sensitized solar cell:
1) glass substrate with transparent conductive oxide layer as counter electrode (modified at the surface);
2) photoanode with transparent conductive oxide layer;
3) blocking layer;
4) nanostructured semiconductor (4a) with adsorbed dye (4b);
5) electrolyte with mediator;
6) sealing;
7) lead and contact.

After the deposition of the dye molecules, the photoanodes are assembled and sealed with a counter electrode to form a thin-layer cell. The counter electrode can be a second electrode made of TCO modified with Pt nanoparticles or a conductive polymer to increase the reaction rate of the mediator reduction. Finally, the electrolyte solution is filled in that facilitates the dye regeneration (Figure 10.31). The most commonly used mediator is the iodide–triiodide system [Eq. (10.53)], since it can be dissolved in organic solvents in high concentrations.

$$3I^- \rightleftharpoons I_3^- + 2e^- \tag{10.53}$$

Due to the complex follow-up reactions, the system has quite different reaction rates for oxidation and reduction. This is advantageous here for suppressing recombination processes.

Recombination processes can occur in competition with the process outlined in Figure 10.28. They are indicated by gray arrows in Figure 10.29. It is crucial for the efficiency of the solar cell that these recombination processes be suppressed. The transfer of an electron from the backside contact to the oxidized form of mediator O is inhibited by the blocking layer. The transfer of e_{CB} from the TiO_2 to O can be modulated by coadsorbates, but always poses a problem. The reaction of e_{CB} with the photooxidized form of the dye D^+ can be influenced by the asymmetric structure of the dye.

Specifications. The performance of DSCs can be compared by diagnostic parameters. A current–voltage curve is often plotted for standardized illumination (Figure 10.32). At $I = 0$, no current flows through the external circuit. The cell voltage measured then is the open-circuit photovoltage U_{OCV}. This is a measure of the maximum driving force the cell can deliver. On the other hand, if the cell voltage becomes zero, the current corresponds to a short-circuit current I_{SC} (short-circuit photocurrent) across the external circuit. It characterizes how many electrons can be moved when there is no external (but only internal) resistance to the current flow and is a measure of the overall reaction rate of the cell. A hypothetical maximum power is the product of U_{OCV} and I_{SC}. However, there are limitations to the reaction process such that the maximum power is $P_{max} = I_{P,max} \cdot U_{P,max}$, which can be read from plots of $U(I) \cdot I$ versus I or $U(I) \cdot I$ versus U (analogous to Figure 2.33). U_{OCV} and I_{SC} span a rectangle in Figure 10.32, as do $I_{P,max}$ and $U_{P,max}$. The ratio of the areas of these rectangles is the ratio of the measured maximum power and the hypothetical maximum power. This ratio is called the fill factor *FF*. The closer the fill factor is to one, the fewer electrons recombine, resulting in smaller internal voltage drops:

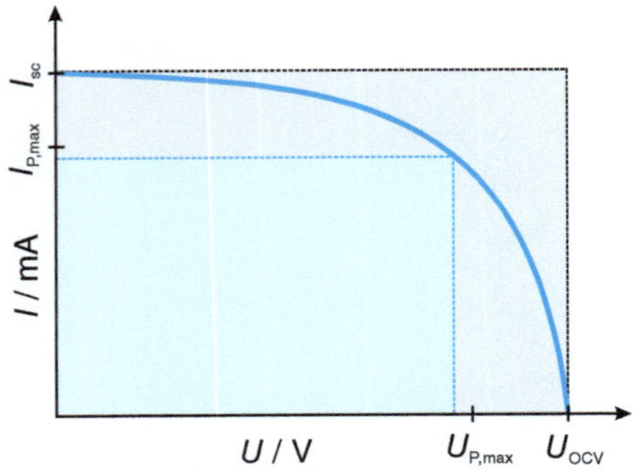

Figure 10.32 Schematic current–voltage curve of a dye-sensitized solar cell with fill-factor analysis at maximum power $P_{max} = I_{P,max}\ U_{P,max}$.

$$FF = \frac{I_{P,max} \cdot U_{P,max}}{I_{SC} \cdot U_{OCV}} \tag{10.54}$$

Another important parameter for the practical use of a cell is its efficiency ε, which describes the ratio of the electrical power output to the radiated power. For efficient DSCs, ε is between 10 and 14%. Solar cells with a single junction (single junction cells) have a fundamentally limited efficiency, which results from the spectral distribution in sunlight and E_g or the HOMO–LUMO energy difference of the dye D. If the energy of a photon is lower than the optical transition, the photon will not be absorbed. If the energy is greater than E_g, the photon can be absorbed forming a high-energy charge carrier pair, but this pair gives off its excess energy in thermal relaxation processes until its energy equals E_g or the HOMO–LUMO difference. Only this remaining energy can be used in an external electrical circuit. The maximum possible efficiency is 33% (the Shockley–Queisser[132] limit) for a solar cell with an optical transition of about 1.4 eV. Since ideal behavior is assumed, this maximum possible efficiency is independent of a particular material system [41].

Other important parameters are the lifetime of the cells, which depends on the stability of the dyes, but also on the sealing of the cells (evaporation of the solvent, exclusion of O_2), as well as the availability of the raw materials used (life cycle analysis) and the manufacturing costs. Especially for the latter two aspects, DSCs give hope for more favorable alternatives compared to established solar cell technologies [38].

Research. Different approaches to further development address almost all levels of this integrated functional system [42]. On the one hand, semiconducting oxides can also be deposited by electrochemical processes using structure-directing agents in nanoporous and crystalline form [43]. This eliminates the need for an energy-intensive thermal sintering process in manufacturing and potentially shortens the time for a DSC to achieve a positive energy balance (energy payback time = usable energy delivered over time minus energy to manufacture). Ru-containing dyes would require mandatory recycling if used on a large scale, as Ru is an element with very limited availability and is also toxic as a heavy metal. Therefore, there are intensive efforts to use purely organic dyes. These also require other redox mediators, for example, based on $[Co(bpy)_3]^{2+/3+}$. The replacement of the I^-/I_3^- redox system is also desirable for other reasons, since it is corrosive and can react with the organic solvents or sealants. Another advantage of the $[Co(bpy)_3]^{2+/3+}$ system is the slightly higher U_{OCV}, indicated in gray in Figure 10.29. Since the redox potential of $[Co(bpy)_3]^{2+/3+}$ is more positive than that of I^-/I_3^-, less energy is converted to heat during the regeneration of D^+. However, there is also a smaller difference of Gibbs free energy available for activation of the electron transfer process to the photooxidized dye. The current records for the efficiencies of DSCs and other types of solar cells are available on the Internet [44].

A derivative approach is taken by so-called all-solid-state DSCs, where the redox mediator has been replaced by a solid organic hole conductor. These cells exhibit fewer stability problems but have the problem of often greatly increased recombination. Perovskite solar cells have been developed from these cells in recent years. Instead of the dye, an organic–inorganic hybrid semiconductor of $[CH_3{-}CH_2{-}NH_3{}^+]Pb^{II}I_3$ or similar compounds with ABX_3 structure are used here as an absorber between the contact layers. The contact layers allow the selective passage of electrons from the CB of the perovskites or holes from the VB of the perovskites. The improvements in cell efficiencies achieved in a short period of time give rise to high hopes, although stability problems and safeguards against the introduction of lead compounds into the environment remain to be solved [45–47].

---

[132] William Bradford Shockley Jr., 1912–1989, US American physicist, worked at Bell Laboratories and later as professor at Stanford University, Nobel prize in physics 1956 for work on semiconductors and the discovery of the transistor; Hans-Joachim Queisser, *1931, German physicist, worked with Shockley, than with Bell Laboratories and became professor in Frankfurt and founding director of the Max Planck Institute of Solid State Research in Stuttgart.

Test your knowledge

Exercise 10.1: Calculate the minimum value that a band gap of an intrinsic semiconductor must have to prevent electron–hole pair formation by visible light (390 nm < λ < 800 nm).

Exercise 10.2: Using the relationships discussed in Chapter 4, calculate the specific conductivity of undoped silicon with the mobilities for CB electrons $u_e = 1350\ \text{cm}^2\ \text{s}^{-1}\ \text{V}^{-1}$ and holes $u_h = 480\ \text{cm}^2\ \text{s}^{-1}\ \text{V}^{-1}$. Use the estimated carrier concentrations from Example Calculation 10.1.

Exercise 10.3: Compare the conductivity of undoped Si and Si with dopant concentrations of 10^{13}, 10^{15}, 10^{17}, and $10^{20}\ \text{cm}^{-3}$ with that of aqueous HCl solutions with [HCl]/(mol L^{-1}) of 0.001 and 1, respectively. Use a spreadsheet program and assume the conductivity coefficient for HCl to be $f_\lambda = 1$. Find the carrier mobilities in Exercise 10.2 and Table 4.3. Document the conversion of units accurately. Discuss the impact of the made simplifying assumption on the validity of the comparison.

Exercise 10.4: Prepare sketches analogous to Figure 10.18 for the contact of a *p*-type semiconductor with a metal, taking into account various possible work functions of the metal.

Exercise 10.5: Make sketches analogous to Figure 10.18 for the band bending of an *n*-type semiconductor when its potential is adjusted via a potentiostat. Show a typical situation for an accumulation layer, a depletion layer, an inversion layer, and a degenerate semiconductor.

Exercise 10.6: Make sketches analogous to Figure 10.18 for the band bending of a *p*-type semiconductor when its potential is adjusted via a potentiostat. Show a typical situation for an accumulation layer, a depletion layer, an inversion layer, and a degenerate semiconductor.

Exercise 10.7: Compare the area-related capacitances and extension (as Debye length) for the space charge region in Si with dopant concentrations of 10^{13}, 10^{15}, 10^{17}, $10^{20}\ \text{cm}^{-3}$ and the electrochemical double layer of Si with a 1–1 electrolyte of the concentrations of 0.001 mol L^{-1} and 1 mol L^{-1}, respectively. Use a spreadsheet program, $T = 298$ K, $\varepsilon_r(H_2O) = 80$, and $\varepsilon_r(\text{Si}) = 12$. Document the conversion of units accurately.

Exercise 10.8: Table 10.4 shows the ratio of the transient photopotential E_{ph} relative to the maximum photopotential $E_{ph,max}$ at a *p*-GaAs electrode in H_2SO_4. Make a plot of the time-dependent photovoltage and determine the decay constant from a fit to a first-order rate law. Use the result to determine the lifetime of the charge carriers. Which charge carriers are we dealing with here?

Table 10.4 Photopotential transients of *p*-GaAs in H_2SO_4.

t/ms	0.0	0.5	1.0	1.5	2.0	2.5	3.0	3.5	4.0
$E_{ph}/E_{ph,max}$	1	0.69	0.54	0.43	0.37	0.31	0.28	0.25	0.23

Source: Data from [6].

Exercise 10.9: Table 10.5 shows sections of a current–voltage curve of a dye-sensitized solar cell measured at room temperature and standardized illumination. Plot the current–voltage curve using a spreadsheet program. Determine the short-circuit current, the open-circuit voltage, the fill factor, and the maximum delivered electrical power.

Table 10.5 Section of a current–voltage curve.

E/V	0.048	0.154	0.362	0.468	0.547	0.574	0.586
I/mA	9.041	8.973	8.699	8.493	8.219	8.014	7.945
E/V	0.609	0.626	0.640	0.662	0.693	0.714	0.758
I/mA	7.740	7.602	7.397	6.438	5.342	4.041	1.986

Exercise 10.10: Explain what factors determine the open circuit voltage using the solution to Exercise 10.9. Which quantities determine the short-circuit current?

References

1 D. F. Shriver, P. W. Atkins; *Inorganic Chemistry*. 4th ed., W. H. Freeman & Co, 2006.

2 N. W. Ashcroft, N. D. Mermin; *Solid State Physics*. Int. ed., 21st print, Saunders College Publishing, New York, 1995.

3 C. Kittel, P. Mc Euen; *Introduction to Solid State*. 9th ed., Global edition, John Wiley & Sons, Hoboken, NJ, 2018.

4 H. Ibach, H. Lüth, *Solid State Physics: An Introduction to Principles of Materials Science*. 3rd ed., Springer Verlag, Berlin, 2003.

5 R. Memming; *Semiconductor Electrochemistry*, 2nd ed., Wiley-VCH, Weinheim, 2015.

6 W. Plieth; *Electrochemistry for Material Science*, Elsevier, Amsterdam, 2008, Chapters 2 and 9.

7 A. J. Bard, L. R. Faulkner, H. S. White; *Electrochemical Methods – Fundamentals and Application*. 3rd ed., John Wiley & Sons, New York, 2022, Ch. 20.

8 H. Gerischer; The impact of semiconductors on the concepts of electrochemistry. *Electrochim. Acta* **1990**, *35*, 1677–1699.

9 S. R. Morrison; *Electrochemistry at Semiconductor and Oxidized Metal Electrodes*, Plenum Press, New York, 1980.

10 W. Jaegermann; The semiconductor/electrolyte interface: A surface science approach. in *Modern Aspects of Electrochemistry, Vol. 30* (R. E. White, B. E. Conway, J. O'M. Bockris, Eds.), Plenum Press, New York, 1996, pp. 1–185.

11 K. Rajeshwar; Fundamentals of semiconductor electrochemistry and photoelectrochemistry. in *Encyclopedia of Electrochemistry, Vol. 6 Semiconductor Electrodes and Photoelectrochemistry* (S. Licht, A. J. Bard, M. Stratmann, Eds.), Wiley-VCH, Weinheim, 2002, pp. 1–54.

12 Z. Chen, H. N. Dinh, E. Miller (Eds.); *Photoelectrochemical Water Splitting – Standards, Experimental Methods, and Protocols*. Springer Brief in Energy. Springer, New York, 2013 (www.escience.cn/system/file?fileId=69496 (last access 25 Dec 2018).

13 F. Lohmann; Fermi-Niveau und Flachbandpotential von Molekülkristallen aromatischer Kohlenwasserstoffe. *Z. Naturforsch.* **1967**, *22a*, 843.

14 S. Trasatti; The absolute electrode potential: An explanatory note. *Pure Appl. Chem.* **1986**, *58*, 955–966.

15 R. Memming; Mechanism of the electrochemical reduction of persulfates and hydrogen peroxide. *J. Electrochem. Soc.* **1969**, *116*, 785–790.

16 R. Thull; Tissue–implant interaction. in *Metals as Biomaterials* (J. A. Helsen, H. J. Breme, Eds.), John Wiley & Sons, Chichester, 1998, pp. 291–315.

17 V. Lehmann; *Electrochemistry of Silicon – Instrumentation, Science, Materials and Applications*, Wiley-VCH, Weinheim, 2002.

18 L. M. Peter; Dynamic aspects of semiconductor photoelectrochemistry. *Chem. Rev.* **1990**, *90*, 753–769.

19 L. M. Peter, D. Vanmaeckelbergh; Time and frequency resolved studies of photoelectrochemical kinetics. in *Advances in Electrochemical Science and Engineering, Vol. 6* (R. C. Alkire, H. Gerischer, D. M. Kolb, C. W. Tobias, Eds.), Wiley-VCH, Weinheim, 1999, pp. 77–163.

20 L. M. Peter, K. G. U. Wijayantha, A. A. Tahir; Kinetics of light-driven oxygen evolution at α-Fe_2O_3 electrodes. *Faraday Discuss.* **2012**, *155*, 309–322.

21 J. Halme, P. Vaherma, K. Miettunen, P. Lund; Device physics of dye sensitized solar cells. *Adv. Energy Mater.* **2010**, *22*, E210–E234.

22 J. Halme; Linking optical and electrical small amplitude pertubation techniques for dynamic performance characterization of dye solar cells. *Phys. Chem. Chem. Phys.* **2011**, *13*, 12435–12446.

23 G. Schlichthörl, S. Y. Huang, J. Sprague, A. J. Frank; Band edge movement and recombination kinetics in dye-sensitized nanocrystalline TiO_2 solar cells: A study by intensity modulated photovoltage spectroscopy. *J. Phys. Chem. B* **1997**, *101*, 8141–8155.

24 L. Canham (Ed.); *Handbook of Porous Silicon.* 2nd ed., Springer, Cham, 2018.

25 M. J. Sailor; *Porous Silicon in Practice: Preparation, Characterization and Applications*, Wiley-VCH, Weinheim, 2012.

26 A. J. Bard, M. A. Fox; Artificial photosynthesis: Solar splitting of water to hydrogen and oxygen. *Acc. Chem. Res.* **1995**, *28*, 141–145.

27 F. E. Osterloh; Inorganic materials as catalysts for photoelectrochemical splitting of water. *Chem. Mater.* **2008**, *20*, 35–54.

28 M. Woodhouse, B. A. Parkinson; Combinatorial approaches for the identification and optimization of oxide semiconductors for efficient solar photoelectrolysis. *Chem. Soc. Rev.* **2009**, *38*, 197–210.

29 M. G. Walter, E. L. Warren, J. R. McKone, S. W. Boettcher, Q. Mi, E. A. Santori, N. S. Lewis; Solar water splitting cells. *Chem. Rev.* **2010**, *110*, 6446–6473.

30 L. Vayssieres (Ed.); *On Solar Hydrogen & Nanotechnology*, John Wiley & Sons, Singapore, 2009.

31 I. Roger, M. A. Shipman, M. D. Symes; Earth-abundant catalysts for electrochemical and photoelectrochemical water splitting. *Nat. Rev. Chem.* **2017**, *1*, 0003:1–11.

32 R. van de Krol, B. A. Parkinson; Perspectives on the photoelectrochemical storage of solar energy. *MRS Energy Sustain.* **2017**, *4*, E13.

33 K. Sivula, F. Le Formal, M. Grätzel; Solar water splitting: Progress using hematite (α-Fe_2O_3) photoelectrodes. *ChemSusChem* **2011**, *4*, 432–449.

34 K. Maeda; Z-Scheme water splitting using two different semiconductor photocatalysts. *ACS Catal.* **2013**, *3*, 1486–1503.

35 H. Chen, S. Yang; Hierarchical nanostructures of metal oxides for enhancing charge separation and transport in photoelectrochemical solar energy conversion systems. *Nanoscale Horiz.* **2016**, *1*, 96–108.

36 D. V. Esposito, J. B. Baxter, J. John, N. S. Lewis, T. P. Moffat, T. Ogitsu, G. D. O'Neil, T. A. Pham, A. A. Talin, J. M. Velazquez, B. C. Wood; Methods of photoelectrochemical characterization with high spatial and temporal resolution. *Energy Environ. Sci.* **2015**, *8*, 2863–2885.

37 H. Kisch; *Semiconductor Photocatalysis*, Wiley-VCH, Weinheim, 2015.

38 L. M. Peter; Characterization and modeling of dye-sensitized solar cells. *J. Phys. Chem. C* **2007**, *111*, 6601–6612.

39 B. O'Reagan, M. Grätzel; A low-cost, high-efficiency solar cell based on dye-sensitized colloidal titanium dioxide films. *Nature* **1991**, *353*, 737–740.

40 M. Grätzel; Recent advances in sensitized mesoporous solar cells. *Acc. Chem. Res.* **2009**, *42*, 1788–1798.

41 W. Shockley, H. J. Queisser; Detailed balance limit of efficiency of *p-n*-junction solar cells. *J. Appl. Phys.* **1961**, *32*, 510–519.

42 L. M. Peter; The Grätzel cell: Where next? *J. Phys. Chem. Lett.* **2011**, *2*, 1861–1867.

43 T. Yoshida, J. Zhang, D. Komatsu, S. Sawatani, H. Minoura, T. Pauporté, D. Lincot, T. Oekermann, D. Schlettwein, H. Tada, D. Wöhrle, K. Funabiki, M. Matsui, H. Miura, H. Yanagi; Electrodeposition of inorganic/organic hybrid thin films. *Adv. Funct. Mater.* **2009**, *19*, 17–43.

44 D. Levi; Conversion efficiencies of best research solar cells worldwide from 1976 to present for various photovoltaic technologies. Efficiencies determined by certified agencies/laboratories. https://www.nrel.gov/pv/cell-efficiency.html (last access 22 Dec. 2023).

45 H. J. Snaith; Perovskite: The emergence of a new era of low-cost, high-efficiency solar cells. *J. Phys. Chem. Lett.* **2013**, *4*, 3623–3630.

46 M. Grätzel; The light and shade of perovskite solar cells. *Nat. Mater.* **2014**, *13*, 838–842.

47 Y. Yuan, J. Huang; Ion migration in organometal trihalide perovskite and its impact on photovoltaic efficiency and stability. *Acc. Chem. Res.* **2016**, *49*, 286–293.

11

Solid-State Electrolytes and Other Unconventional Electrolyte Systems

In Section 5.2.8, we have already dealt with the lambda probe, in which two platinum electrodes in contact with gases and an O^{2-}-ion conductor form a galvanic cell. Yet, we have not dealt with the mechanisms of ionic conduction in solids. We want to make up for this, as solid ionic conductors have a great and growing importance not only for sensor technology, but also for corrosion processes on passivated metals (Chapter 14), batteries (Chapter 15), fuel cells (Chapter 16), and inorganic membranes for gas separation. When discussing conductivity mechanisms in Section 11.1, we will see that point defects in crystalline solids play a central role in ion transport. The defect concentration can be calculated from equilibrium considerations and adjusted by doping. The Kroeger–Vink notation represents the defects relative to the perfect lattice. From the description of these point defects, we will see that the concepts of solid-state chemistry can be viewed as a generalization of the chemistry of both electrolyte solutions and semiconductors (Chapter 10). The thermodynamics of defect structures have immediate consequences for the mechanisms of ionic conduction in solids (Section 11.2). Analogous to the electrochemical double layer, boundary layers are also formed in ion-conducting solids in which the concentration of defects deviates from the defect concentration inside the solid electrolyte. Such boundary layers open up the interesting possibility of selectively increasing the conductivity (Section 11.3).

In Section 11.4, we will discuss electrochemical cells with solid electrolytes that are used in the laboratory to characterize such solid electrolytes. After considering conductivity via defects in solids, we will apply these concepts to some unconventional electrolyte systems: molten salts and ionic liquids, ion-conducting glasses and polymer electrolytes. Ion-conducting glasses and polymer electrolytes find applications, for example, in batteries and as glass electrodes in analytical chemistry.

Further Reading. Most electrochemistry textbooks contain a chapter on solid-state electrolytes [1–4]. In addition, some review articles can also be recommended [5–7]. Specialized textbooks [8] and edited volumes [9, 10] address solid-state electrolytes and their applications more broadly. Interfaces of immiscible electrolyte solutions are interesting systems for special analytical applications [11, 12]. Ionic liquids represent an intensively studied research area for which edited volumes [13–16] and review articles [17–19] are available.

11.1 Why Do Crystalline Solids Contain Defects?

11.1.1 Types of Point Defects and Kröger–Vink Notation

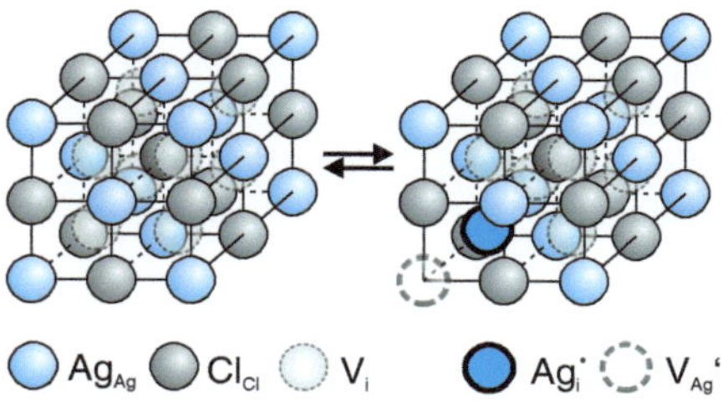

Figure 11.1 Formation of a point defect in an AgCl crystal with the corresponding Kroeger–Vink notation.

Consider a crystal of AgCl crystallizing in the NaCl structure (Figure 11.1). If all ions were firmly locked into their lattice positions, an ideal crystal structure would be formed. However, such a crystal would not be able to conduct an ionic current and could therefore not be used as an electrolyte in electrochemical cells.

Q: In Chapter 10, we discussed how conductivity in solids is made possible by the excitation of electrons from the valence band (VB) into the conduction band (CB). This should also be the case for AgCl, regardless of whether an ideal or nonideal crystal is present.

A: That is correct. However, here we are analyzing solids that take up the function of an electrolyte in an electrochemical cell. According to Section 1.2, these materials must have *ionic* conductivity. This is linked to the presence of mobile defects. Substances with ionic conductivity can also have an *electronic* conductivity, which arises according to the mechanisms discussed in Chapter 10. The simultaneous presence of ionic and electronic conductivity is called *mixed conductivity*. The ratio of ionic to electronic conductivity is of great importance in selecting materials for many applications. We will return to this problem later.

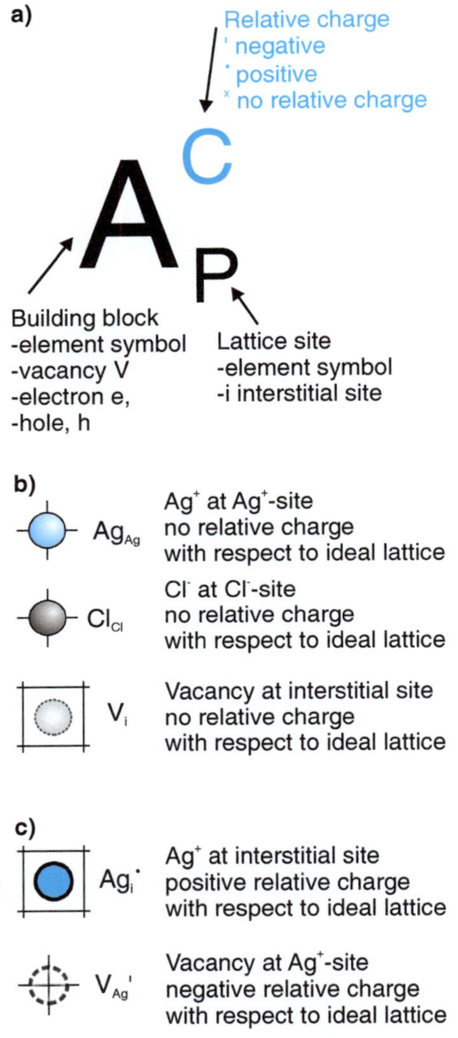

Figure 11.3 Kroeger–Vink notation of point defects.
a) General scheme;
b) example for the ideal structure in Figure 11.1, left; and
c) example for the point defects in Figure 11.1, right.

Between the regular structural units of the crystal, there are interstitial sites V_i. One way to form a **point defect** can be imagined as follows: An Ag^+ ion transfers from the regular lattice position to an interstitial site (Ag_i^+) with the uptake of energy (Figure 11.1). A vacancy V_{Ag} remains at the Ag^+ site in the regular lattice. Sometimes the symbol for the vacancy is also □, which is especially useful when the element symbol V for vanadium also occurs.

We can therefore imagine a real, crystalline solid as a "sum" of the ideal structure and charged point defects. After rearranging the "addition equation", the point defects, which are decisive for the electrochemical properties of the solids, can be understood as "difference" or, expressed more precisely, as "deviation" of the real structure from the ideal crystal structure (Figure 11.2). This concept also underlies the **Kroeger–Vink notation**[133] of the point defects (Figure 11.3). We can formulate the formation of a point defect in AgCl as a chemical equilibrium:

$$Ag_{Ag} + V_i \rightleftharpoons V'_{Ag} + Ag_i^{\cdot} \tag{11.1}$$

The relative charge indicates the charge difference between a building unit in the real crystal and the building unit at the specific site in the ideal crystal.

On the left side of Eq. (11.1), there are only elements of the ideal crystal structure. Therefore, the silver ion on a regular site in the overall neutral AgCl crystal carries no *relative* charge. To distinguish the charge of individual crystal elements and the

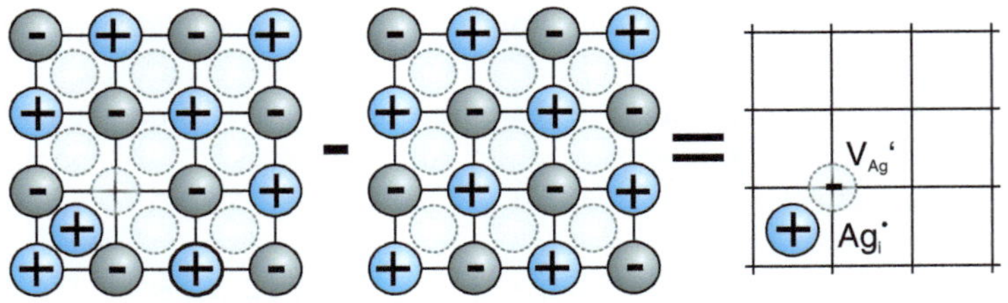

Figure 11.2 Defect structures as a deviation ("difference") of the real structure from the ideal crystal structure. For reasons of clarity, the three-dimensional lattice from Figure 11.1 has been simplified and drawn as a two-dimensional lattice.

[133] Ferdinand Anne Kroeger, 1915–2006, Dutch chemist, worked at Philips and at the University of Southern California; Hendrik Jan Vink. 1915–2009, Dutch physicist, worked at Philips in Eindhoven, member of the Dutch Royal Academy of Sciences.

differential charge with respect to the ideal crystal structure, the notation ′ is preferred for negative and • for positive *relative* charges in the Kroeger–Vink notation. Therefore, on the right-hand side of the equation, a vacancy on a lattice site for Ag^+ ions carries a negative relative charge to the lattice building unit Ag^+. This is because, a charged lattice building block Ag^+ is missing in the vacancy compared to the ideal lattice and the charge compensation must be applied by another charge carrier. In this case, it is the positively charged Ag^+ ion on the interstitial site. The interstitial site is indicated by the subscript "i" and the positive charge of this defect by "•". If necessary, a zero relative charge is stated by "x". The symbols on the left side of Eq. (11.1) correspond to the ideal crystal structure (Ag^+ on an Ag^+ site and a vacancy on the interstitial site). We could therefore "omit" these symbols and instead use Eq. (11.2):

$$\text{Null} \rightleftharpoons V'_{Ag} + Ag^{\bullet}_{i} \tag{11.2}$$

Equation (11.2) immediately shows us a basic principle in the formation of point defects: There must be a balanced charge. That is, point defects always arise in pairs. In addition, there must be a balanced elemental composition, as in all chemical reactions. It is not so obvious from the shorthand in Eq. (11.2) as it is in Eq. (11.1). Finally, there must also be a balance of lattice sites. This can be seen from the subscripted symbols for the lattice sites in Eq. (11.1): On the left as well as on the right side of Eq. (11.1), the lattice sites $_{Ag}$ and $_{i}$ appear once each.

A crystal M^+X^- can have the following point defects: The positioning of a cationic lattice unit M^+ in an interstitial space is called Frenkel defect[134] (a lattice disorder).

$$M_M + V_i \rightleftharpoons V'_M + M^{\bullet}_i \tag{11.3}$$

The analogous arrangement of an anionic lattice unit X^- in an interstitial site is called anti-Frenkel defect and is expressed as

$$X_X + V_i \rightleftharpoons V^{\bullet}_X + X'_i \tag{11.4}$$

Both ion species can also be disordered simultaneously. In the case of Schottky defects[130] (e.g., in NaCl), one Na^+ and one Cl^- are removed from the structure, leaving vacancies on the Na^+ sites V'_{Na} and on the Cl^- sites $V^{\bullet}_{Cl}$.

$$M_M + X_X \rightleftharpoons MX + V'_M + V^{\bullet}_X \tag{11.5}$$

In case of anti-Schottky defects, additional anions and cations are incorporated on interstitial sites, such as Pb^{2+} and O^{2-} in the yellow lead(II) oxide.

$$MX + 2V_i \rightleftharpoons M^{\bullet}_i + X'_i \tag{11.6}$$

Table 11.1 gives an overview of these and other defect types that will be discussed in Section 11.1.2.

11.1.2 Defects in Solids as a Generalization of the Chemistry of Electrolyte Solutions and the Electronic Structure of Semiconductors

It is clear from Section 11.1.1 that defects are created by the disordering of lattice elements without the contamination of the solid by foreign atoms. Such defects are called

[134]Yakov Il'ich Frenkel, 1894–1952, Russian-Soviet physicist, Physical-Technical Institute St. Petersburg, also introduced the hole in semiconductors as a type of defect.

Table 11.1 Overview about point defects and comparison with ions in aqueous solution and charge carriers in semiconductors.

Two-dimensional schematic representation	**Reaction equation and equilibrium condition**
Autoprotolysis of water (Chapter 4) 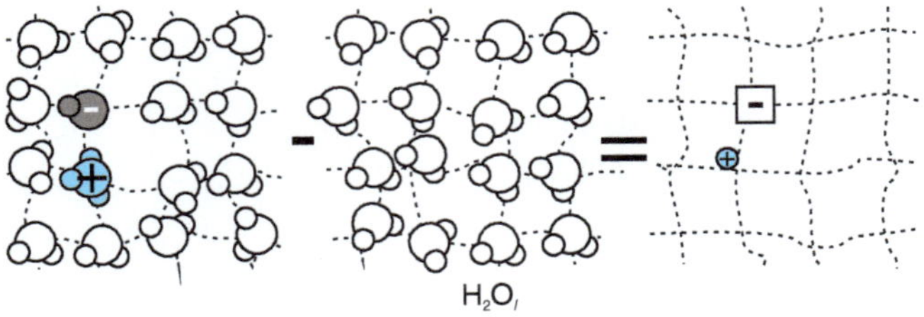	$2H_2O \rightleftharpoons H_3O^+ + OH^-$ (11.7) $K_W = [H_3O^+]\cdot[OH^-]$ (11.8)
Electronic band-to-band excitation in semiconductors (Chapter 10) 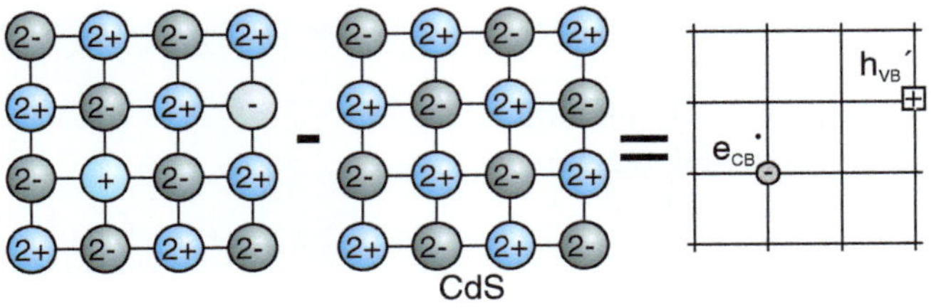	$\text{Null} \rightleftharpoons h^{\bullet} + e'_{CB}$ (11.9) $K_B = const\cdot \exp\left[\frac{-E_g}{k_B T}\right] = [h^{\bullet}]\cdot[e']$ (11.10)
Frenkel defect 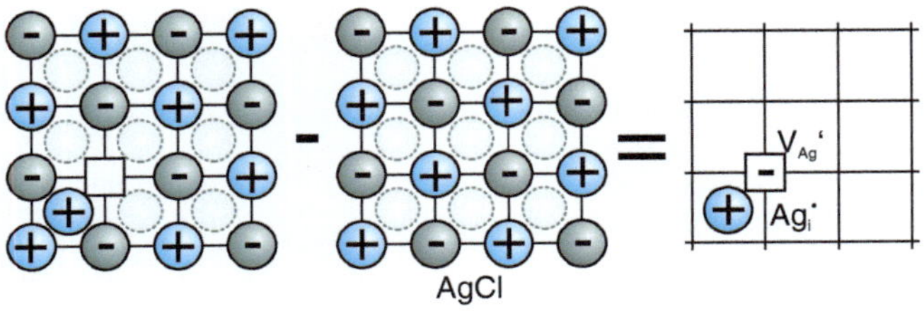	$M_M + V_i \rightleftharpoons V'_M + M_i^{\bullet}$ (11.11) $K_F = [M_i^{\bullet}]\cdot[V'_M]$ (11.12)
Anti-Frenkel defect 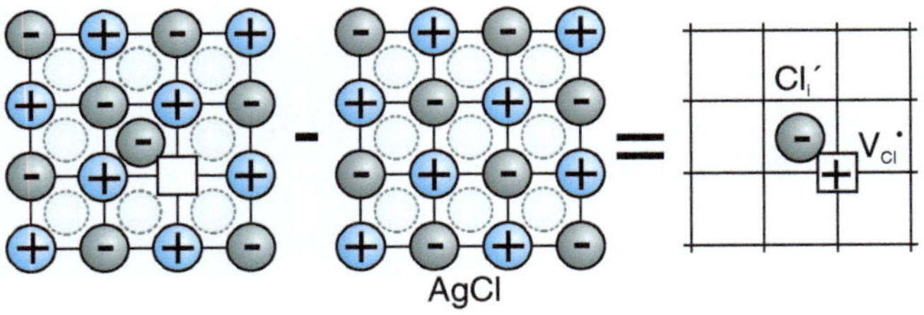	$X_X + V_i \rightleftharpoons V_X^{\bullet} + X'_i$ (11.13) $K_{AF} = [X'_i]\cdot[V_X^{\bullet}]$ (11.14)
Schottky defect 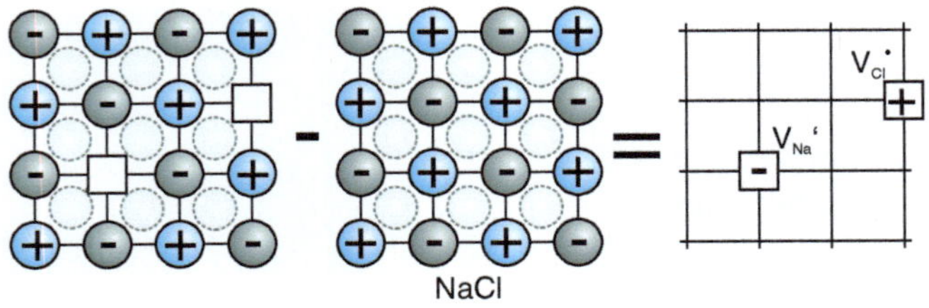	$M_M + X_X \rightleftharpoons MX + V'_M + V_X^{\bullet}$ (11.15) $K_S = [V'_M]\cdot[V_X^{\bullet}]$ (11.16)

(Continued)

Table 11.1 (Continued)

Two-dimensional schematic representation	Reaction equation and equilibrium condition	
Anti-Schottky defect		
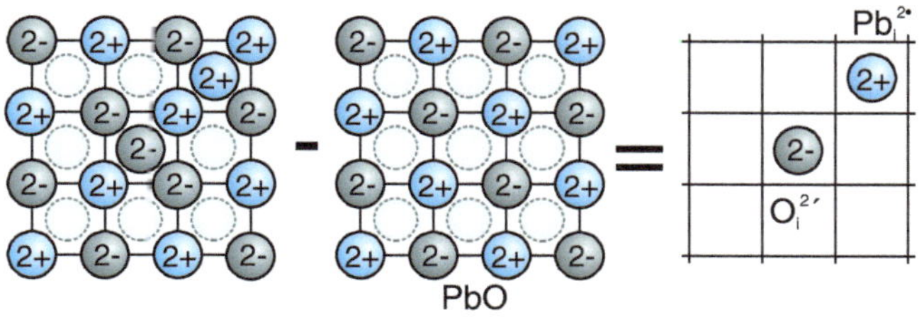	$MX + 2\,V_i \rightleftharpoons M_i^{\bullet} + X_i'$ $K_{AS} = [M_i^{\bullet}] \cdot [X_i^{\bullet}]$	(11.17) (11.18)
Reaction with gas atmosphere from $X_{2,g}$		
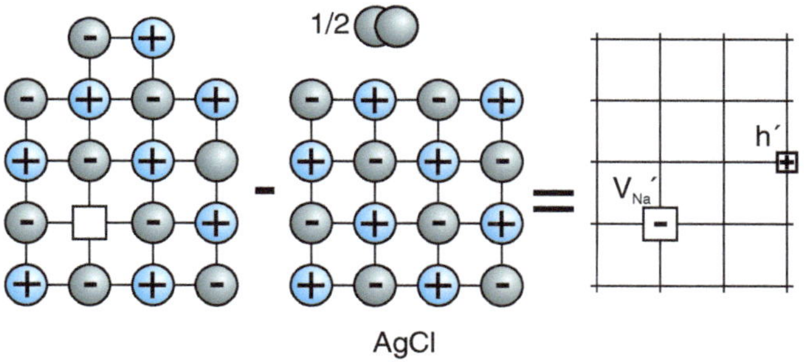	$\frac{1}{2}X_2(g) + M_M \rightleftharpoons MX + V_M' + h^{\bullet}$ $K_X = \frac{[h^{\bullet}]\cdot[V_M']}{\sqrt{p_{X_{2,g}}}}$	(11.19) (11.20)
Doping with higher valent cation		
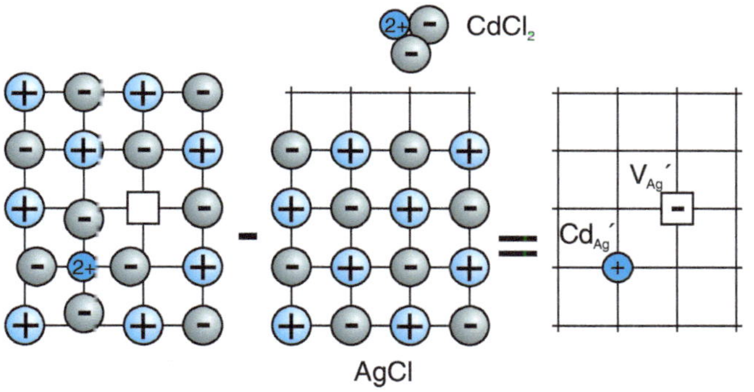	$NX_2 + 2M_M \rightarrow N_M^{\bullet} + V_M' + 2MX$	(11.21)

intrinsic defects. Let us now compare the situation with an aqueous electrolyte solution. In pure water, autoprotolysis produces a few charge carriers:

$$2H_2O \rightleftharpoons H_3O^+ + OH^- \tag{11.22}$$

By subtracting the ideal structure of water, we could write the corresponding equation as

$$\text{Null} \rightleftharpoons H_i^{\bullet} + V_H' \tag{11.23}$$

We can imagine the "ideal structure" of water as a "lattice" of building units "$\{H^+\}$–$\{O^{2-}\}$–$\{H^+\}$". An OH^- ion is then a vacancy on an H^+ site of the ideal structure. An oxonium ion (H_3O^+) can then be thought of as an additional H^+-ion that must be accommodated between the water molecules of the ideal structure. From these considerations, it is clear that Eq. (11.23) represents an unusual notation of the simplified notation of the autoprotolysis equilibrium $H_2O \rightleftharpoons H^+ + OH^-$. However, there are also conceptual differences between solid electrolytes and liquid electrolyte solutions.

Since the ideal lattice positions and interstitial sites are not defined in aqueous electrolyte solutions, autoprotolysis can be understood both as a rearrangement of a proton (Frenkel disorder) or as additional H^+ and OH^- introduced (Schottky defects).

We can imagine an oxide semiconductor as a crystal of metal cations M^{m+} and oxide ions O^{2-}. Therefore, the valence band (VB) has predominantly the character of oxygen 2p orbitals, and the empty conduction band has the character of s orbitals of the metal ion. Excitation of the semiconductor can be thought of as an internal redox reaction in which an electron is transferred from O^{2-} to M^{m+} and can be represented by

$$O^{2-} + M^{m+} \rightleftharpoons O^- + M^{(m-1)+} \tag{11.24}$$

In the relative notation, we can understand this reaction as the emergence of two electronic defects, an additional electron at the metal e′ (= e_{CB}) and a vacancy at the oxygen, which we described in Chapter 10 as hole $h^{\bullet}$ (= h_{VB}). The notation in Eq. (11.25) correctly indicates no localization of the electronic defects.

$$\text{Null} \rightleftharpoons h^{\bullet} + e'_{CB} \tag{11.25}$$

The similarities with respect to the "intrinsic" charge carriers also cause the similarity of the corresponding equilibrium expression (compare with ionic product and with Section 10.1.3).

11.1.3 Interactions Between Neighboring Phases

In the case of Schottky defects, a Na^+- and a Cl^--ion must leave the structure, for example, as a NaCl molecule into an adjacent gas phase. On the other hand, a PbO molecule, for example, must be incorporated into the PbO crystal as $Pb_i^{2\bullet}$ and $O_i^{2\prime}$ in the case of anti-Schottky defects. This requires an interface between the crystal and an adjacent gas phase. Such interfaces open up further possibilities for the formation of defects.

Consider a crystal of M^+ and X^- with the possibility of ionic and electronic defects, where deviations from ideal stoichiometry can also occur as a result of interaction with a gas phase. If the gas phase contains the molecule X_2, X^- can be incorporated in a vacant X site:

$$\frac{1}{2}X_2(g) + V_X^{\bullet} + e' \rightleftharpoons X_X \tag{11.26}$$

In the process, electrons are bound and, because of the coupled equilibria (analogous to aqueous electrolyte solutions), the number of positive charge carriers must increase (such as $h^{\bullet}$ or $M_i^{\bullet}$). We can also indicate this by other formulations of the equilibrium reaction (Table 11.1):

$$\frac{1}{2}X_2(g) + M_M + e' \rightleftharpoons MX + V'_M \tag{11.27}$$

$$\frac{1}{2}X_2(g) + M_M \rightleftharpoons MX + V'_M + h^{\bullet} \tag{11.28}$$

Let us now investigate what happens during the incorporation of a divalent cation N^{2+} namely the (mostly irreversible) incorporation of an impurity:

$$NX_2 + 2M_M \rightarrow N_M^{\bullet} + V'_M + 2MX \tag{11.29}$$

We refer to such defects as **extrinsic defects**. The incorporation of the divalent N^{2+} on an M^+ site must be compensated by the formation of a vacancy on an M site (Table 11.1). The formation of this vacancy leads to the formation of a bivalent N by coupling with the equilibrium (11.5) to the reduction in the concentration of anion vacancies $V_X^{\bullet}$. This can also be described by the expression

$$NX_2 + V_X^{\bullet} + M_M \rightarrow N_M^{\bullet} + X_X + MX \tag{11.30}$$

Taking into account the equilibrium (11.26), it is clear that the incorporation of N^{2+} increases the e′ concentration:

$$NX_2 + M_M \rightarrow MX + \frac{1}{2}X_2 + N_M^{\cdot} + e' \qquad (11.31)$$

In summary, an important conclusion emerges:

The incorporation of an impurity with an effective positive charge causes both a decrease in the concentration of all positive defects and an increase in the concentration of all negative defects.

Let us return to the example of AgCl: doping with $CdCl_2$ leads to the incorporation of Cd as $Cd_{Ag}^{\cdot}$, to an increase in the concentration of silver vacancies V_{Ag}' and to an increase in the concentration of conduction band electrons e′, while the concentration of Ag^+ ions on interstitial sites and the concentration of holes decrease. Doping with Ag_2S after incorporation as S_{Cl}' leads to an increase in concentrations of $V_{Cl}^{\cdot}$, $h^{\cdot}$ and of $Ag_i^{\cdot}$ as well as to a decrease in the concentration of e′.

Excursion 11-1 Yttrium-Stabilized Zirconia (YSZ)

Yttrium-doped zirconium dioxide is an important ionic conductor for O^{2-} ions (see the lambda probe in Section 5.2.8 and SOFC in Section 16.4.1). A typical material is obtained by mixing a molar fraction of 8% Y_2O_3 with ZrO_2 at high temperatures (8YSZ, the number 8 stands for the molar fraction of Y_2O_3). The stoichiometrically correct reaction equation is

$$92\ ZrO_2 + 8\ Y_2O_3 \rightleftharpoons Zr_{92}Y_{16}O_{208} \qquad (11.32)$$

However, this equation does not reflect the formation of the oxygen vacancies, which are crucial for the application as an ionic conductor. We will write the reaction in the Kroeger–Vink notation. The trivalent Y^{3+}-ion is incorporated into the lattice on Zr sites. Since only 3 O^{2-} ions (instead of 4 O^{2-} for 2 Zr^{4+}) are required for charge compensation of 2 Y^{3+}-ions, one oxygen vacancy must occur together with 2 Y^{3+}. If we now formulate the reaction as the deviations from the ideal structure, we get

$$Y_2O_3 + 2\ Zr_{Zr} + O_O \rightleftharpoons 2\,Y_{Zr}' + V_O^{2\cdot} + 2\,ZrO_2 \qquad (11.33)$$

We express the balances of elements, lattice space, and charge.

Balance of elemental constituents: left side 2Y, 2Zr, 4O; right side 2Y, 2 Zr, 4O; => balanced.

Balance of lattice sites: left side 2 Zr sites, 1 O site; right side 2 Zr sites, 1 O site. [Only the sites in the ZrO_2 lattice count, not the sites in the added (Y_2O_3) or removed crystals (ZrO_2 on the right)].

Balance of charge: Left side = 0; right side 2 $[Y_{Zr}']$ + $[V_O^{2\cdot}]$ = 0

Thus, the sum formula $Zr_{92}Y_{16}O_{208}$ must contain 8 vacancies on O^{2-} sites for 16 Y^{3+}. This results in

$$\frac{[V_O]}{[O_O]} = \frac{8}{208} = \frac{1}{26} \qquad (11.34)$$

Thus, every 27th oxygen site remains unoccupied. We could also give the molecular formula as $Zr_{92}Y_{16}O_{208}\{V_O\}_8$. In general, the formula would be $Zr_{1-x}Y'_{2x}O_{2+x}$ $\{V_O{}^{2\cdot}\}_x$. The color highlighting is to clarify the derivation from ZrO_2. This notation is uncommon in the literature and is given here for illustrative purposes only.

We should also ask why technical materials are produced with just 8 mol-% Y_2O_3. Pure ZrO_2 exists in three modifications. At room temperature, it exists as a Baddelyt with a monoclinic structure and a coordination number (CN) of Zr of 7. Above 1170 °C, the tetragonal structure with CN = 8 is preferred, and at temperatures above 2370 °C, the cubic structure (CN = 8) is the most stable. During the transition from the monoclinic to the tetragonal structure, a decrease in volume of 7% occurs, which leads to cracking of the sintered bodies. Dopants can not only change the conductivities, but also stabilize certain phases. For example, the incorporation of divalent (e.g., Ca^{2+}) and trivalent (e.g., Y^{3+}) metal ions with contents ≥8 mol% of metal ions can stabilize the cubic structure, where the activation energy for ion migration is lowest. Considering the phase diagrams and the mechanical properties correctly represents another important criterion for material synthesis.

As an illustration, we can remember that the ionic conductivity of 8YSZ at 700 °C corresponds approximately to the conductivity of a 0.1 mol L^{-1} NaCl solution at room temperature. This translates to a microscopic notion of each O^{2-} ion jumping 10^8 times per second into an adjacent oxygen vacancy [20]!

11.1.4 Complex Defect Structures

In addition to point defects, more complex defect structures also occur. The simplest are spatial **associations of** oppositely charged defects (analogous to ion pairs in solution). The driving force for their formation is the electrostatic attraction of the effective charges:

$$V'_M + V^{\bullet}_X \rightleftharpoons (V_M V_X)^X \quad (11.35)$$

The two neighboring voids can also be interpreted as a preliminary stage of pore formation.

Associations between electronic and ionic defects can also occur, which are particularly significant when their energy is within the band gap:

$$V'_M + h^{\bullet} \rightleftharpoons V^X_M \quad (11.36)$$

$$V^{\bullet}_X + e' \rightleftharpoons V^X_{Cl} \quad (11.37)$$

$$N^{\bullet}_M + e' \rightleftharpoons N^X_M \quad (11.38)$$

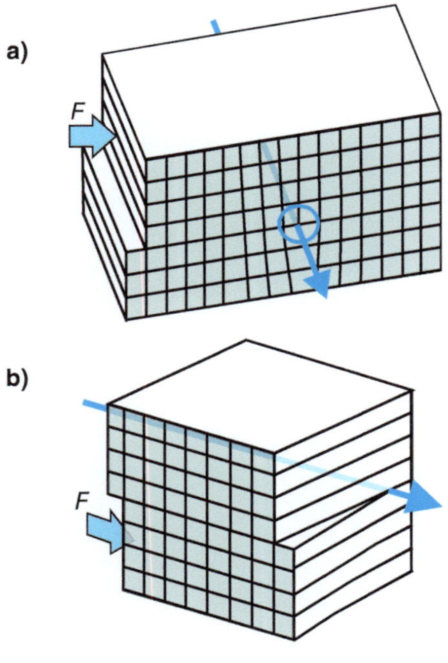

Figure 11.4 Formation of a) steps and b) screw dislocations due to mechanical impact.

They often attract further defects, leading to the formation of extensive **defect clusters**. They are important, especially in solid electrolytes with a high concentration of defects.

In crystals, there are also typical extended defects. Dislocations occur when parts of a crystal are displaced with respect to another part. Special forms are the edge dislocation and screw dislocation in Figure 11.4, which are not only significant for the mechanical properties, but also for the formation of growth lines in deposition processes. These defects run through the crystal as line defects.

The effects considered above can also occur in single crystals. However, most solids consist of several crystallites (grains) which have the same composition but a different orientation of the crystal axes. Where two grains with different orientations of the crystal axes meet, a **grain boundary** is formed. Grain boundaries therefore represent plane defects. Depending on how strongly the crystal axes of the adjacent grains are tilted against each other and which lattice constants occur, different types of grain boundaries are formed (Figure 11.5). A good introduction can be found in [8].

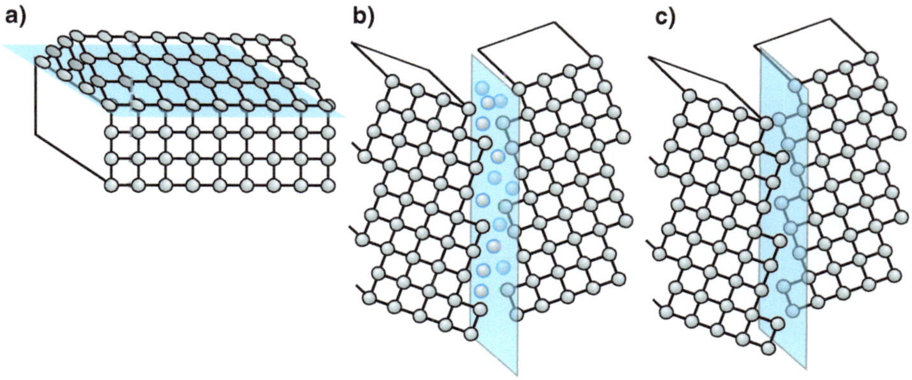

Figure 11.5 Examples for grain boundaries;

a) coherent interface (lattice points of one grain are also lattice points of the other grain);
b) grain boundary with amorphous interlayer; and
c) grain boundary without interlayer and without relaxation.

11.1.5 Thermodynamics of Point Defects

In Section 11.1.1, we established that the formation of point defects takes place under energy input. We now want to consider why a certain, nonzero equilibrium concentration of defects $N_{d,eq}$ occurs in crystals. We want to understand the energetic principles rather than the different types of defects and the diversity of materials. More detailed considerations of this can be found in more specialized textbooks [8]. For simplicity, we consider the formation of vacancies in a single-element crystal (e.g., diamond) of 1 mol atoms. Let the concentration of vacancies be so small that they cannot interact with each other (this corresponds to "infinite dilution"). In Figure 11.6, a section of the crystal is drawn schematically in two dimensions to simplify the principle. However, the considerations outlined below also apply to three-dimensional arrangements of lattice elements.

To determine the equilibrium concentration, we need to find an expression for the Gibbs free energy G and its first derivative with respect to the number of defects N_d. Thus, the problem is an "accounting problem" in which we account for ergonic "costs" and "gains". The starting point is the molar Gibbs free energy of the ideal lattice G_{ideal}.

In the real lattice (with G_{real}), further "cost contributions" arise from breaking the four bonds of an atom inside the crystal (Figure 11.6). This atom is not released into the gas phase, but is attached to the surface, recovering a Gibbs free energy equivalent of two bonds. In fact, the situation is even more complicated. The lattice elements around the created vacancy change their position slightly (they relax) to minimize the energetic cost of the vacancy inside the crystal. In addition, the vibrational frequencies of the lattice units change near the defect. For vacancies, the vibrational frequency usually decreases, while it increases when interstitial sites are occupied. Let the energy input related to a defect be $\Delta_d g$. Multiplied by the number of defects N_d in a mole of substance, there is a positive contribution $\Delta_d G$ to the change in G_{real}:

$$\Delta_d G = N_d \cdot \Delta_d g = N_d(\Delta_{bd} g + \Delta_{vib} g) \tag{11.39}$$

$\Delta_d G$ – Gibbs free energy contribution for the creation of defects in the crystal [kJ mol^{-1}]

N_d – Number of defects per mole of lattice elements [mol^{-1}]

$\Delta_d g$ – Gibbs free energy contribution for one defect [kJ]

$\Delta_{bg} g$ – Gibbs free energy contribution due to bond breaks for one defect [kJ]

$\Delta_{vib} g$ – Gibbs free energy contribution due to the change of vibration frequencies at the defect location for one defect [kJ]

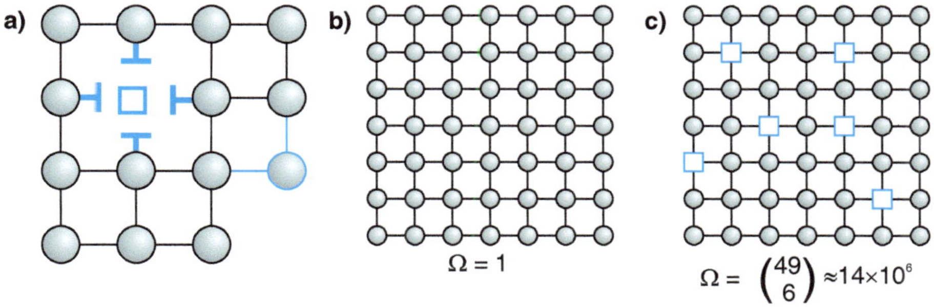

Figure 11.6 Schematic two-dimensional representation of the formation of point defects in a crystal of an element. a) Balance of broken bonds and newly formed bonds; b) and c) balance based on increasing the number of possible microstates.

N_{CN} – Coordination number at the defect location [dimensionless]

$\Delta_{bg}g_{EE}$ – Gibbs free energy contribution due to bond breaking one element–element bond [kJ]

T – Temperature [K]

k_B – Boltzmann constant = 1.38×10^{-26} kJ mol^{-1}

$\widetilde{\nu}$ – Vibrational wave number in an ideal crystal [cm^{-1}]

$\widetilde{\nu}_d$ – Vibrational wave number for the neighboring atoms of the defect site [cm^{-1}]

Rough estimates can provide approximations for $\Delta_{bg}g$ and $\Delta_{vib}g$ in the manner outlined above.

$$\Delta_{bd}g = \frac{1}{2}N_{CN} \cdot \Delta_{bd}g_{E-E} \tag{11.40}$$

$$\Delta_{vib}g = N_{CN}k_BT \cdot \ln\frac{\widetilde{\nu}_d}{\widetilde{\nu}} \tag{11.41}$$

Inserting realistic values in Eq. (11.41) shows that this contribution is negligible at low temperatures compared to $\Delta_{bd}g$, but assumes significant values at 1000 K, that is, at temperatures that play a role in the synthesis and application of solid electrolytes.

The contribution Δ_dG is positive and cannot explain the occurrence of finite equilibrium defect concentrations ($N_{d,eq} > 0$). For this to happen, a negative contribution (entropic) must be added, which arises from the dramatic increase in microstates. Consider a two-dimensional crystal of 49 lattice points: If all lattice points are occupied, there is exactly *one* realization (one microstate, $\Omega_{ideal} = 1$) for it. [Since the building units are indistinguishable, swapping two of the same building units does not result in a distinguishable microstate]. If we now distribute six blanks among the lattice places (as in the lottery 6 out of 49), this results in $\Omega_{real} = \binom{49}{6} = 49!/[6!(49-6)!] = 49{\cdot}48{\cdot}47{\cdot}46{\cdot}45{\cdot}44/(1\ 2\ 3\ 4\ 5\ 6) \approx 14$ million possibilities (i.e., 49 over 6). The gain in configuration entropy ΔS_{conf} amounts to

$$\Delta S_{conf} = k_B \cdot \ln\frac{\Omega_{real}}{\Omega_{ideal}} = k_B \cdot \ln \Omega_{real} \tag{11.42}$$

S_{conf} – Configuration entropy [kJ K^{-1} mol^{-1}]

$\Delta_{conf}G$ – Contribution to the change in molar Gibbs free energy from the configuration entropy [kJ mol^{-1}]

Ω_{real} – Number of microstates in a real crystal [dimensionless]

Ω_{ideal} – Number of microstates in an ideal crystal = 1

N_A – Avogadro's number = 6.022×10^{23} mol^{-1}

The contribution $\Delta_{conf}G = -T\Delta S_{conf}$ is thus an energetic measure of the greater probability of disordered states in the considered ensemble of particles on lattice sites. In total, the molar Gibbs free energy is given for the real crystal by

$$G_{real} = G_{ideal} + \Delta_dG + \Delta_{conf}G \tag{11.43}$$

For an ensemble with large numbers of particles, Stirling's[135] approximation can be used to calculate factorials. With $\ln(N!) \approx N\ln(N) - N$ we obtain

$$\ln \Omega_{real} \approx N_A \ln\frac{N_A}{N_A - N_d} - N_d \ln\frac{N_d}{N_A - N_d} \approx -N_d \ln\frac{N_d}{N_A} \tag{11.44}$$

Because of $N_d \ll N_A$, the second (blue) term dominates the result of Eq. (11.44), and the contribution of the configuration to the change in Gibbs free energy is a negative term of

$$\Delta_{conf}G = -k_BT\ln\Omega_{real} = k_BTN_d\ln\frac{N_d}{N_A} \tag{11.45}$$

μ_d – Chemical potential of defects [kJ mol^{-1}]

μ_d° – Standard chemical potential of defects [kJ mol^{-1}]

x_d – Mole fraction of the defects [dimensionless]

$x_{d,eq}$ – Equilibrium mole fraction of the defects [dimensionless]

R – Gas constant = N_Ak_B = 8.314×10^{-3} kJ mol^{-1} K^{-1}

$\Delta_{bg}g$ – Gibbs free energy contribution for bond breaking for one defect [kJ]

$\Delta_{vib}g$ – Gibbs free energy contribution due to change of vibrational frequencies at the defect location for one defect [kJ]

The difference in Gibbs free energy $\Delta_{real}G = G_{real} - G_{ideal}$ is shown by the two contributions Δ_dG and $\Delta_{conf}G$ in Figure 11.7 for two temperatures. The minimum of $\Delta_{real}G$ shifts to higher values of the defect concentration N_d with increasing temperature. N_d can also be expressed as mole fraction $x_d = N_D/N_A$. At this minimum, the chemical potential of the defects is zero and is represented by

$$\Delta\mu_d = N_A\frac{\partial G}{\partial N_d} \approx \mu_d^\circ + RT\ln x_d = 0 \tag{11.46}$$

We can derive the equilibrium mole fraction $x_{d,eq}$ from this equilibrium condition, which corresponds to a Boltzmann approximation.[18]

$$x_{d,eq} \approx \exp\left(\frac{-\mu^\circ}{RT}\right) \approx \exp\left(\frac{-(\Delta_{bd}g + \Delta_{vib}g)}{k_BT}\right) = 0 \tag{11.47}$$

[135] James Stirling, 1692–1770, Scottish mathematician.

Depending on the value of $\Delta_{bg}g$, the mole fractions x_d can be very small even at high temperatures. Nevertheless, these intrinsic defects in solids, just like the H_3O^+- and OH^--ions in aqueous solutions, are of enormous importance for reactions and transport processes.

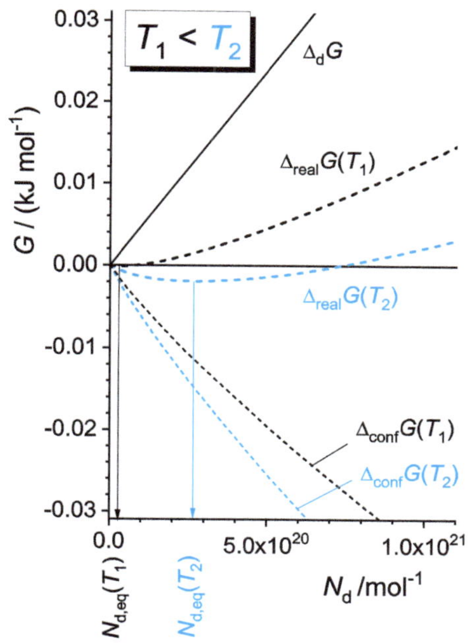

Figure 11.7 Contributions to the molar Gibbs free energy of an element crystal as a function of the number of defects per mole of atoms.

11.2 Ionic Conductivity in Solids

11.2.1 How Do Ions Move in Solids?

The following mechanisms can be considered for the transport of ions in solids:

- A lattice element moves to a vacancy. This creates a vacancy on another lattice site. Similarly to hole conduction in semiconductors (Chapter 10), it is useful to think of this process as the migration of the vacancy (Figure 11.8a).
- Ions can move from an interstitial site to an adjacent interstitial site (Figure 11.8b).
- An ion on an interstitial site can displace an ion from a regular lattice site and take its place. The ion that was originally on the lattice site then occupies an interstitial site (Figure 11.8c).

For any movement of ions in a crystal, an activation barrier E_A must be overcome. We therefore expect that with increasing temperature, not only does the concentration of defects increase (Figure 11.7), but also their mobility. The specific conductivity κ often follows an exponential temperature dependence, which is represented by

$$\kappa \approx a \cdot \exp\left(\frac{-E_A}{k_B T}\right) \tag{11.48}$$

where E_A can be understood as the height of the activation barrier or activation energy. In the literature, the symbol σ is often used instead of κ for the specific conductivity and the symbol μ for the ion mobility. For consistency and because of the risk of confusion between the symbol μ for the chemical potential and μ for the mobility, we stick to the symbols introduced in Chapter 4.

For a more detailed consideration, we resort to Eq. (4.31), but consider only *single* mobile ionic species i, for which we obtain the specific conductivity as

$$\kappa = c_i |z_i| u_i F \tag{11.49}$$

Inserting the Einstein Equation (4.113) $D_i = RTu_i/(|z_i|F)$, which describes the relationship between ion mobility and diffusion coefficients, and the T-dependence of $D_i = D_{i,0} \exp[E/(RT)]$ leads to

$$\begin{aligned} \kappa &= c_i|z_i|F\frac{|z_i|FD_i}{RT} = \frac{c_i|z_i|^2F^2}{R}\cdot\frac{1}{T}D_i \\ &= \frac{c_i|z_i|^2F^2D_{i,0}}{R}\cdot\frac{1}{T}\cdot\exp\left[\frac{E}{RT}\right] \\ \kappa T &= \frac{c_i|z_i|^2F^2D_{i,0}}{R}\cdot\exp\left[\frac{E}{RT}\right] \end{aligned} \tag{11.50}$$

κ – Specific conductivity [S cm^{-1}] = [Ω^{-1} cm^{-1}]
c_i – Defect concentration [mol cm^{-3}]
$|z_i|$ – Charge number (dimensionless)
F – Faraday constant = 96 485 A s mol^{-1}
u_i – Charge carrier mobility of the defect i [cm^2 s^{-1} V^{-1}]
R – Gas constant = 8.314 V A s K^{-1} mol^{-1}
D_i – Diffusion coefficient of defect type i [cm^2 s^{-1}]

The linearization of Equation (11.50) should yield the straight line $\ln(\kappa T) = a_0 + (E/R)\, T^{-1}$. However, in this relation the exponential temperature dependence dominates. Therefore, a simplified linearization $\ln(\kappa) = b_0 + (E/R)\, T^{-1}$ is sufficient for many systems or in restricted temperature ranges. On the other hand, even more complicated dependencies are found for measurements over large temperature ranges for most solid-state ionic conductors, since the dominant conduction mechanisms can change with temperature and contributions from electronic conductivity and

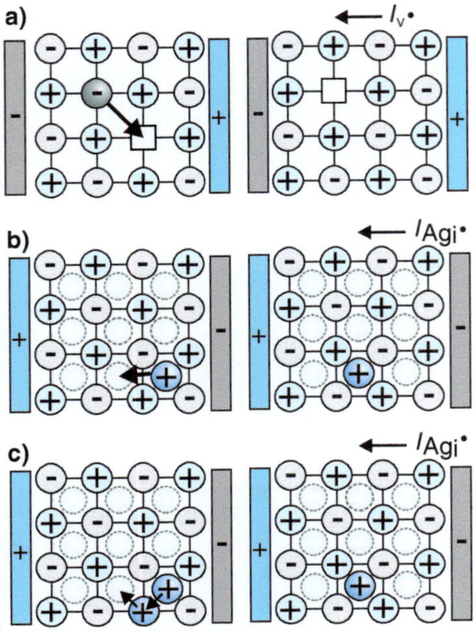

Figure 11.8 Ion conductance in solids:

a) migration of a vacancy in the electric field;
b) migration of ions to interstitial sites; and
c) displacement of a lattice element by an ion at an interstitial site.

The three-dimensional crystal is reproduced as a two-dimensional arrangement for a simplified illustration of the principle.

from temperature-dependent reactions with neighboring phases (both with a different T-dependence) play a role. The T-dependence of specific conductivity has high diagnostic value. More detailed information and examples can be found in [3, 8]. A detailed derivation of the relations from the generalized flow equations of irreversible thermodynamics is provided in [6, 8]. The term $D_{0,i}$ in the preexponential factor is related to the frequency with which ions attempt to overcome the activation barrier.

Particularly high ionic conductivities can be expected if the interstitial sites are readily accessible to an ion species, as in the case for Ag^+ in silver halides. The high polarizability of Ag^+, that is, the ability to shift electron density within the ion, facilitates interstitial movement and stabilizes defects. Another way to stabilize defects is to have a high concentration of vacancies on regular lattice sites, as we discussed in Excursion 11-1 using YSZ as an example. Finally, at higher temperatures, the defect concentration can increase so much that all ions of one species can move almost freely across low-energy barriers between regular lattice positions and interstitial sites. We can conceive of this state as the liquid state of the anion or cation sublattice, a state that is called structural disorder and is associated with very high ionic conductivities (super ionic conductivity). An example of this is the α-AgI.

11.2.2 How Is Conductivity Measured in Solids?

In the simplest case, the conductivity of solid electrolytes can be measured in an electrochemical cell, which is prepared by vapor deposition of metal contacts on opposite sides of the solid electrolyte. As with liquid electrolytes, we apply an AC voltage of sufficiently high frequency to prevent chemical conversions from occurring at the interfaces. The AC voltage causes a current flow in the electrolyte, which in turn leads to the charging of an electrical double layer. Before charging is complete, polarity reversal occurs. The specific conductivity κ measured in this way is the *sum of* the contributions of all mobile charge carriers, which is thus composed of the contribution of the electronic conductivity κ_{el} and the ionic conductivity κ_{ion}:

$$\kappa = \kappa_{el} + \kappa_{ion} = \kappa_{el} + F \sum_{i} c_i |z_i| u_i \tag{11.51}$$

The electronic conductivity is caused by the electrons in the conduction band and the holes. All defects i contribute to the ionic conductivity in proportion to their concentration, charge, and mobility according to Eq. (11.49).

There are two ways to determine the conductivity contributions of individual charge carriers. If the frequency dependencies of the charge carriers differ significantly, a differentiation can be made using impedance spectroscopy. However, direct current measurements are frequently possible in conjunction with suitable contact layers, which allow the transport of only one type of charge carrier. Figure 11.9 shows the measurement of the specific conductivity of silver(I) sulfide. Ag_2S has a mixed conductivity, which can be determined by means of an AC experiment. If we cover the Ag_2S sample on both sides with a layer of AgI (a pure Ag^+-ion conductor), the electronic contact between Ag_2S and the electronically conducting feeder electrodes (made of Ag) is prevented. In a DC experiment (Figure 11.9), current flow in the Ag_2S electrolyte can now only occur through Ag^+ ions. In addition, silver deposition occurs at the negative electrode and silver dissolution at the positive electrode. These reactions can be demonstrated experimentally (Excursion 11-2). Metal deposition can be associated with overvoltages. Silver dissolution can also influence the contact resistances, which are also included in the measurement. These effects at the contacts can be estimated by measurements on Ag_2S samples of different thicknesses. Four-point measurements

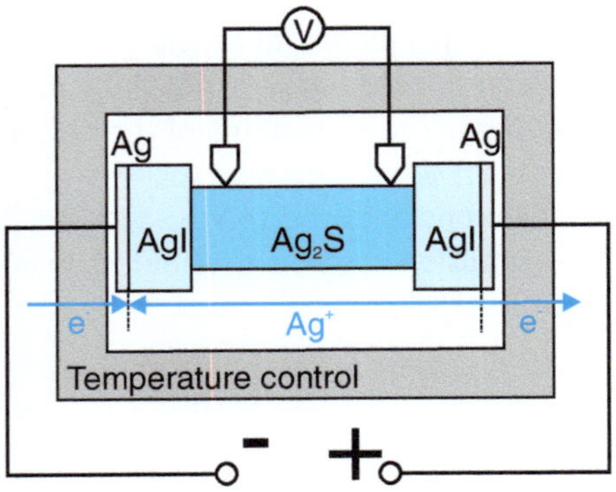

Figure 11.9 Cell for measuring the Ag^+ ionic conductivity in Ag_2S.

represent another experimental possibility, for which two more contact electrodes are applied along the current path on the Ag_2S sample. One measures the potential difference $U = R\,I$ between the inner contact electrodes without current, while recording the current I via the outer contact electrodes.

Excursion 11-2 Ag^+ Ionic Transport in a Solid Electrolyte with Mixed Conductivity

The Ag^+ conductivity and electronic conductivity of Ag_2S can be visualized in an experiment (Figure 11.10). One places a densely pressed tablet of Ag_2S on a silver foil and presses this arrangement, with the silver foil down onto an ordinary laboratory hot plate. Shortly after switching on the hot plate, silver precipitates become visible above 176 °C on the surface of the Ag_2S tablet. The shape of the silver precipitates depends on the exact experimental conditions, in particular on the preparation of the Ag_2S tablet. However, thin single-crystalline wires, so-called whiskers, often appear.

The following reaction takes place at the interface between the Ag foil and the Ag_2S tablet:

$$Ag \rightleftharpoons Ag^{+}_{i,Ag_2S} + e_{CB,Ag_2S} \qquad (11.52)$$

The transport of Ag^+ and e^- through the Ag_2S tablet occurs along the gradients of their electrochemical potentials $\tilde{\mu}$. At the Ag_2S|air interface, elemental silver is formed again by reversing Eq. (11.52). The emerging Ag phase "pushes" itself into the gas phase due to the ongoing formation of "new" silver.

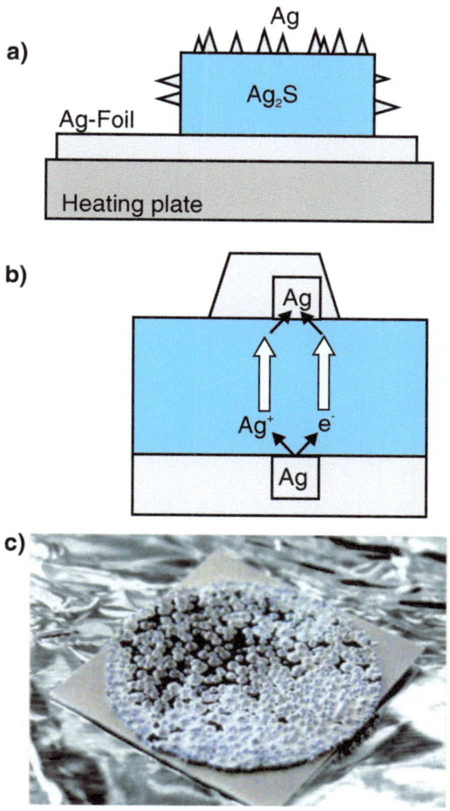

Figure 11.10 Demonstration of mixed conductivity of Ag_2S: a) experimental setup; b) schematic of the reactions and processes taking place; and c) optical image of Ag precipitates. Source: Reproduced with permission from E. Mutoro et al. [20], Copyright 2008, Wiley-VCH.

This example can be used to identify experimental problems in solid-state electrochemistry. Reactions at the interfaces lead to changes in the shape of the solid phase, which are accompanied by changes in the contact areas and changes in volume. Unlike a liquid electrolyte, a solid electrolyte cannot adapt to such shape changes. If substances are to be continuously converted from an adjacent gas phase (SOFC in Section 16.4.1), they must be able to pass through a porous contact electrode to the phase boundary between the electronic conductor and the ionic conductor. It is possible to contact solid electrolytes by liquids, such as liquid sodium in Na–S batteries (Table 15.1), which contain molten Na and S in contact with solid β-Al_2O_3. β-Al_2O_3 has an elemental composition between $Na_2O \cdot 5\,Al_2O_3$ and $Na_2O \cdot 11\,Al_2O_3$.

The choice of contact electrodes suitable for a wide temperature range is a major challenge in solid-state electrochemistry.

As a result of very high defect concentrations at grain boundaries of a solid, it may be that the observed conductivity is significantly influenced by these internal interfaces. Often, grain boundary conductivity and bulk conductivity can both be observed simultaneously in electrochemical impedance measurements. Therefore, it is important to use structurally well-characterized solid electrolytes for the determination of conductivity data.

Another experimental approach is tracer diffusion. For this purpose, an oxide-ion conductor is brought into contact with a $^{18}O_2$ atmosphere. After incorporation of the oxygen isotope into the solid, its diffusion coefficient can be determined, for example, by mass spectrometry. When comparing the transport quantities, correction factors must be considered, since only the diffusion of $^{18}O^{2-}$ is determined in tracer experiments while the diffusion of all defects is determined by electrochemical measurements [6].

Nuclear magnetic resonance (NMR) spectroscopy provides another approach to study point defects and their motion in solids. This requires special NMR instrumentation. Often samples have to be enriched with NMR-active isotopes (e.g., ^{17}O) to achieve sufficient sensitivity. In addition to observing the chemical shift and its assignment to individual defect states, relaxation times can also be measured and transport mechanisms inferred from their change. Another possibility is to label atomic spins in an inhomogeneous magnetic field at a specific position of the sample and to characterize the relaxation of this label based on its translocation [21, 22].

11.2.3 Important Ion-Conducting Solids

Figure 11.11 shows the temperature-dependent specific conductivities of important material classes discussed below [23].

Oxide-Ion Conductors. Ion conduction occurs by the movement of O^{2-} from lattice positions to oxygen vacancies in the lattice. These vacancies are created in mixed oxides by adding low-valent cations (e.g., YSZ, Excursion 11-1). Another variant is the formation of vacancies under partial reduction of the metal cations (CeO_{2-x} = "$Ce^{III}{}_{x/2}Ce^{IV}{}_{1-x/2}O_{2-x}$"), which leads to mixed conductivity. This partial reduction can also occur in contact with reducing gases, thus limiting the range of application.

Bi_2O_3 has a very high O^{2-}-conductivity at temperatures above 729 °C. At this temperature, a phase transition to δ-Bi_2O_3 occurs. δ-Bi_2O_3 can be stabilized by doping with cations of other rare earth metals [e.g., $(Bi_2O_3)_{0.75}(Y_2O_3)_{0.25}$]. These compounds are not stable in reducing atmospheres, which significantly limits their application in electrochemistry.

Oxides of the perovskite type ABO_3, which are derived from lanthanum gallate $LaGaO_3$, are currently studied intensively as solid electrolytes because they are expected to be able to operate solid oxide fuel cells (Section 16.4) at moderate temperatures as low as 650–800 °C. The oxide-ion conductivity can be increased by partially substituting La(III) with divalent ions of alkaline earth metals, and Ga(III) with Mg(II), for example, $La_{1-x}Sr_xGa_{1-y}Mg_yO_{3-(x+y)/2}$ (LSGM).

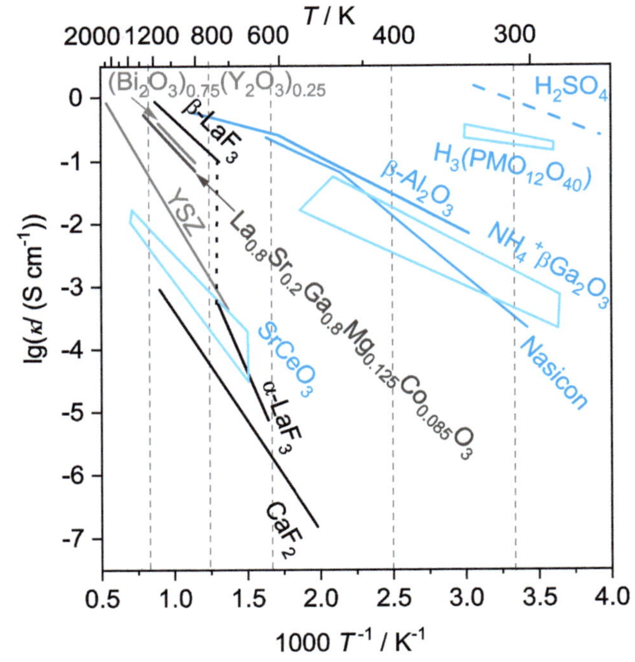

Figure 11.11 Temperature-dependent specific conductivities of some ionic conductors. Cationic conductors are shown in blue, anionic conductors in gray. Sulfuric acid is shown for comparison. Source: Adapted from [3, 23].

Excursion 11-3 Influence of Foreign Metal Ions on the Corrosion of Oxide-Covered Metals

The application of many metals is only possible due to the presence of a passive oxide layer. This layer slows down the thermodynamically expected oxidation of the metals in an oxygen-containing atmosphere (Section 14.3.2). The robustness of the oxide layer depends on many factors. If the molar volume of the metal oxide differs from that of the metal, the oxide layer forms islands or cracks, allowing continuing contact to the oxidant [$O_2(g)$]. Therefore, corrosion progresses rapidly. If, on the other hand, a dense oxide layer forms with a similar molar volume of the oxide layer and the metal, corrosion can only proceed if this oxide layer has ionic conductivity. In this case, a chemical diffusion of the *metal* occurs, that is, a rectified migration of M^{n+} and ne^- analogous to that shown in Figure 11.10. In this case, the continuous formation of new oxide occurs at the interface between the oxide layer and the corrosion medium. Alternatively, chemical diffusion of *oxygen*, that is, the opposing movement of O^{2-} and $2e^-$ into the oxide layer, occurs. Thus, new oxide forms at the interface between the metal and the oxide film. Corrosion of metals with a dense oxide film requires mixed conductivity under the corrosion conditions. If either ionic conductivity or electronic conductivity is negligible in the oxide at ambient temperature, as in the case of Al_2O_3 and oxides of other so-called valve metals (Ti, Zr, Ta, Nb) at room temperature, atmospheric corrosion is unlikely to occur. However, this picture can change at higher temperatures when ionic conductivity increases sharply (high-temperature corrosion).

A mixed conductivity is assumed for zinc oxide, where e'_{CB} and $Zn_i^{2\cdot}$ on interstitial sites are the mobile charge carriers, whose concentration can be strongly influenced by homogeneous doping. According to the rule (from Section 11.1.3) that the incorporation of an effectively positively charged dopant causes a reduction in the concentration of all positive defects (and *vice versa*), it is possible to predict that the addition of lithium to zinc will increase the corrosion rate because the oxide contains Li'_{Zn}-defects with a negative effective charge. Conversely, low levels of aluminum in zinc will greatly reduce the corrosion rate because the resulting oxide layer contains positively charged defects ($Al^{\cdot}_{Zn}$ or $Al^{\cdot}_i$) that lower the concentration of mobile charge carriers.

Fluoride-Ion Conductors. A class of F^--conducting solids is derived from fluorite CaF_2. This structure acquires F^--conductivity by partial substitution of Ca^{2+} by Na^+. β-PbF_2 exhibits an even higher conductivity, which can be further increased by substitution (e.g., $PbSnF_4$ or $Pb_{1-x}Bi_xF_{2+x}$). Another class of materials is derived from LaF_3 in the tysonite structure, which already exhibits low conductivity at room temperature, so that crystal disks are used as membranes in ion-selective electrodes for fluoride (Section 5.2.5).

Silver-Ion and Copper-Ion Conductors. Ag(I) and Cu(I) halides occur at room temperature in the β-phase with a wurtzite structure. At higher temperatures (e.g., 147 °C for AgI, 430 °C for CuI), a phase transition occurs to the cubic α-phase, which is accompanied by an increase in the ionic conductivity by three orders of magnitude that is then comparable to aqueous H_2SO_4 solutions! The increase in conductivity is due to structural disorder in which many cation sites are unoccupied and the easily polarizable cations move between the anions to the vacant lattice sites.

Sodium-Ion Conductors. A group of compounds referred to as β-Al_2O_3 and β″-Al_2O_3 are formed from Na_2O and Al_2O_3 in various mixing ratios, such as $Na_2O \cdot 11\ Al_2O_3$ (or $NaAl_{11}O_{17}$) or $Na_2O \cdot 5.33\ Al_2O_3$. These typically nonstoichiometric compounds have a layered structure that implies much higher conductivity along the layers than perpendicular to the layers. The conductivity already reaches 0.01 S cm^{-1} at room temperature. The β″-phase exhibits much higher conductivity than the β-phase and shows sufficiently high values at about 300 °C for application in Na–S batteries (Table 15.1). The compound is also stable to liquid Na. Another class of Na^+-ion conductors is called NASICON (*Na superionic conductors*) and represents a solid solution of $NaZr_2(PO_4)_3$ and $Na_4Zr_2(SiO_4)_3$. The maximum conductivity is observed for the composition $Na_{1+x}Zr_2P_{3-x}Si_xO_{12}$, $x = 2$ and also does not exhibit preferential direction. It is due to only partial occupation of the Na^+ sites.

Lithium-Ion Conductors. Li^+-ion conductors are of obvious importance for lithium-ion batteries (Section 15.6). Very different requirements arise depending on the function of the material. Active materials, that is, the compounds that store the charge in batteries, must have mixed conductivity so that both electrons from the external circuit and the working ion from the electrolyte phase can be transported in the solid active material. The electrolyte, on the other hand, must have an exclusively ionic conductivity to prevent a short circuit between the positive and negative electrodes. Replacing solutions of Li salts in (liquid) organic solvents, which show strong volume expansion during heating and evaporation and are flammable when in contact O_2, by a solid Li^+-ion conductor could significantly improve the safety properties of Li-ion batteries. Lithium halides, after doping, can achieve conductivities already at room temperature (e.g., 10^{-5} S cm^{-1} at 298 K for LiI with 1 mol% CaI_2). Layered Li_3N and oxidic compounds such as Li_4SiO_4 and Li_4GeO_4 represent starting points for the development of solid Li^+-ion conductors, which can be further improved by doping. One such product is LISICON ($Li_4ZnGe_4O_{16}$, 0.1 S cm^{-1} at 598 K).

Solid proton conductors are of great importance for modern technological developments in the field of fuel cells, electrolyzers, sensors, mass separation, and electrochromic materials. We have already explained the proton conductivity of Nafion in Figure 2.30. Many heteropolyacids like $H_3[PMo_{12}O_{40}] \cdot 29\ H_2O$ also show a high proton conductivity of 0.1 S cm^{-1} already at room temperature. As in the case of Nafion, this is linked to the presence of water in the structure, and therefore, collapses when the material dries out. β-Al_2O_3 or β″-Al_2O_3 can acquire proton conductivity after ion exchange of Na^+ for H_3O^+ or NH_4^+. Another compound of this class has the composition $(NH_4)(H_3O)_{2/3}Mg_{2/3}Al_{31/3}O_{17}$ (10^{-4} S cm^{-1} at 298 K). Ion exchange can proceed even more readily when using the Ga analogs of β-Al_2O_3 or of β″-Al_2O_3 (Figure 11.11). Perovskite-type oxides such as $SrCeO_3$ with partial substitution of trivalent cations (5% in Y^{3+}, Yb^{3+}, Nd^{3+}, etc.) on Ce(IV) sites are *p*-type semiconductors at elevated temperatures in air. In the presence of H_2 or H_2O, they develop proton conductivity at elevated temperatures as the holes are consumed by reaction with the reducing atmosphere:

$$H_2(g) + 2h^+ \rightleftharpoons 2H_i^+ \tag{11.53}$$

and

$$H_2O(g) + O_O + V_O^{2\cdot} \rightleftharpoons 2OH_O^{\cdot} \quad (\text{or } \text{“}2O_O + 2H_i^{\cdot}\text{”}) \tag{11.54}$$

Further Reading. The development of new solid electrolytes for lithium-ion batteries is currently receiving significant attention. A current overview can be found in [24].

Excursion 11-4 Resistive Random Access Memory (RRAM) with Solid Ionic Conductors

A random access memory (RAM) is a rewritable memory device. A new development in this field uses memory cells in which nanometer-sized metallic filaments are reversibly formed and dissolved within nanoseconds in a solid ionic conductor (Figure 11.12). The switching state is nonvolatile, that is, it is retained without a separate energy supply. Each memory cell consists of an oxidizable electrode (e.g., Ag) covered with a 30–50 nm thick layer of a suitable ionic conductor. This ionic conductor must have a high electronic resistance. Based on the above considerations, a suitable material for this purpose would be AgI. However, in semiconductor technology, amorphous metal oxides (Ta_2O_5), perovskites [$(Ba, Sr)TiO_3$], dielectrics such as Al_2O_3, Gd_2O_3, and higher chalcogenides are attractive, because they are compatible with other processes and structures used in the semiconductor industry. The ionic conductor is covered by an inert electrode (Ni, W, Pt).

A minimum voltage is required for the deposition of Ag filaments. Below this voltage, the resistance is very high. With the formation of filaments, the electronic resistance decreases sharply, and a current can flow even at low voltages. Thus, two easily readable states exist with the values 0 and 1 (Figure 11.12). By reversing the voltage, the connection of the Ag filaments to the Ag electrode can be broken and the resistance at the readout voltage can be increased significantly. By miniaturizing the cells, high storage densities are possible. This principle can also be applied to other material combinations.

Further reading: [25, 26].

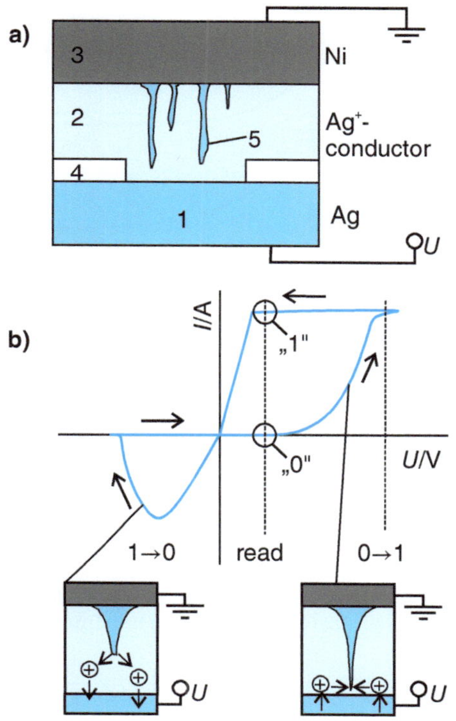

Figure 11.12 Principle of a nonvolatile RRAM memory cell. a) Principle structure; 1) Ag electrode, 2) Ag^+-ion conductor, 3) counter electrode (Ni), 4) insulation layers for lateral structuring, 5) Ag filaments; b) cyclic voltammogram (characteristic I–V curve) with marking of the voltages for writing and reading.

11.3 Boundary Layers in Solid Electrolytes

11.3.1 How Can We Generalize the Insights About the Electrical Double Layer and Space Charge Region?

In the treatment of the electrical double layer at interfaces between an electronic conductor and an electrolyte solution, we have already found that deviations from the volume concentration of the charge carriers do exist, but they are confined to small spatial regions near interfaces (Sections 4.3 and 8.10). The equilibrium condition is based on the local constancy of the electrochemical potentials of ions, electrons, and defects ($\widetilde{\mu}_{M^+}$, $\widetilde{\mu}_{X^-}$, $\widetilde{\mu}_{e^-}$). The local constancy of a quantity means that its derivative with respect to a spatial coordinate x is zero. As introduced in Chapter 4, we choose the spatial coordinate x to be perpendicular to the phase boundary. We also normalize the coordinate to the Debye length[36] r_D, meaning we state the distance in units of the Debye length. The following Eq. (11.55) applies to the electrochemical potential of a component k.

$$0 = \frac{\partial}{\partial x}\widetilde{\mu}_k = \frac{\partial}{\partial x}[\mu_k + z_k F\varphi] \tag{11.55}$$

Thus, it is possible that local variations of the chemical potential are compensated by a gradient of the Galvani potential φ:

$$\frac{\partial}{\partial x}\left[\mu_k^\circ + RT\ln a_k\right] = -\frac{\partial}{\partial x}[z_k F\varphi] \tag{11.56}$$

Since the standard term μ_k° is location-independent, the local composition must change via the activity a_k in the case of local potential variations, for example, when the neighboring phase A is charged. We express the composition of the phase by the

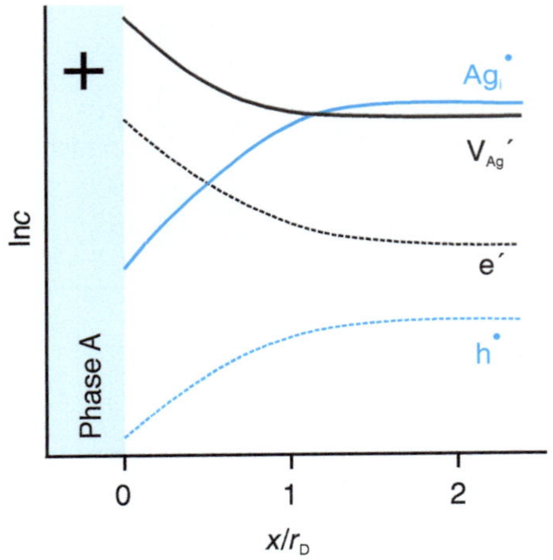

Figure 11.13 Local concentrations of metal ions on interstitial sites, vacancies on cation sites, mobile electrons, and holes in AgCl near the interface when the neighboring phase A is positively charged. Source: Adapted from [8].

concentration activity, but assume for simplicity that the activity coefficient of dilute defects is unity in complete analogy to the ideal solution model ($c_k = a_k$):

$$RT\frac{\partial \ln c_k}{\partial x} = -z_k F\frac{\partial \varphi}{\partial x} \tag{11.57}$$

The integration of Eq. (11.57) between the volume with the volume concentration c_k^* and the potential φ^* and the location x near the interface yields

$$\left[\frac{c_k(x)}{c_k^*}\right]^{1/z_k} = \exp\left[\frac{-F(\varphi(x) - \varphi^*)}{RT}\right] \tag{11.58}$$

The right-hand side of Eq. (11.58) is a term that is independent of the type of defect considered. Thus, it is affected in the same way for each mobile ionic species. Only the charge number of the defect $z_k = \{-2, -1, 0, 1, 2\}$ in the exponent of the left-hand side and the concentration of the considered defect in the bulk phase c_k^* lead to a distinction. The term $c_k(x)/c_k^*$ is a dimensionless concentration, indicating the concentration relative to the concentration in the undisturbed bulk phase. We discussed the advantage of such normalized quantities in Chapter 4. Figure 11.13 shows the local concentrations of mobile cations on interstitial sites, vacancies on cation sites, mobile electrons, and holes. Note the logarithmic ordinate.

Example Calculation 11.1 Calculate the relative defect concentration at the surface of AgCl at room temperature when the potential difference between the interface and the interior of the phase is 250 mV.

Unknown: $c_k(0)/c_k^*$

Given:

$$\begin{aligned}\varphi(0) - \varphi^* &= -0.25\ \text{V}\\ z_k &= 1\\ T &= 298\ \text{K}\\ R &= 8.314\ \text{J K}^{-1}\ \text{mol}^{-1} = 8.314\ \text{V A s K}^{-1}\ \text{mol}^{-1}\\ F &= 96\,485\ \text{A s mol}^{-1}\end{aligned}$$

$$\begin{aligned}\left[\frac{c_k(x)}{c_k^*}\right] &= \exp\left[\frac{-z_k F(\varphi(x) - \varphi^*)}{RT}\right]\\ &= \exp\left[\frac{(-1)\cdot 96\,485\ \text{A s mol}^{-1}\cdot(-0.25\ \text{V})}{8.314\ \text{V A s K}^{-1}\ \text{mol}^{-1}\cdot 298\ \text{K}}\right]\\ &= \text{e}^{-9{,}736}\\ &\approx \frac{1}{17000}.\end{aligned}$$

Thus, the concentration of interstitial ions is reduced by 4 orders of magnitude, while the concentration of vacancies is increased by 4 *orders of magnitude.*

The formalism of solid-state electrochemistry represents a generalization of both, the chemistry of electrolyte solutions and the consideration of electronic charge carriers in semiconductors. Accordingly, the information can also be plotted analogously to the band diagrams introduced in Chapter 10. The Fermi level E_F corresponds to the electrochemical potential and denotes the energy at which the probability of occupying a state is 0.5 (50%). We define the standard electrochemical potential $\widetilde{\mu}_k^\circ$ of component k as

$$\widetilde{\mu}_k = \widetilde{\mu}_k^\circ + RT\ln c_k = \mu_k^\circ + z_k F\varphi + RT\ln c_k \tag{11.59}$$

From the changes of the local electrochemical potential in Figure 11.14, we can construct a diagram (Figure 11.15) analogous to band diagrams in semiconductors as discussed in Section 10.2.1. The Fermi level E_F is replaced by the electrochemical potential $\widetilde{\mu}_{M^+} = \mu_{M^+} + z_k F\varphi$. Analogous to the change in a "conduction band minimum", the energy of a cation on interstitial sites increases as it approaches a positively charged neighboring phase. Accordingly, the "valence band maximum", which corresponds to the energy of a vacant lattice site, bends upward. At the interface, it approaches the "Fermi energy". This means that the probability for the occurrence of a vacancy on a cation lattice site approaches 50%.

The exact concentration profiles are given by the Poisson Equation[39] (11.60):

$$\frac{d^2\varphi(x)}{dx^2} = \frac{\sum_k (z_k F c_k(x))}{\varepsilon_r \varepsilon_0} \tag{11.60}$$

Taking into account the electroneutrality condition for the volume $\left(\sum_k (z_k F c_k^*) = 0\right)$, the Poisson–Boltzmann[18] Equation (11.61) is obtained.

$$\frac{d^2(\varphi(x) - \varphi^*)}{dx^2} = \frac{F}{\varepsilon_r \varepsilon_0} \sum_k \left(z_k c_k^* \cdot \exp\left[\frac{z_k F(\varphi(x) - \varphi^*)}{RT} \right] \right) \tag{11.61}$$

Influence on the Conductivity. As with the ionic product of water or the concentration of electrons and of holes in a semiconductor, the *product* of the concentration of ions on interstitial sites and the concentration of vacancies is a (temperature-dependent) constant in solid-state ionic conductors. What is not constant, however, is the *sum of* these concentrations, which is, according to Eq. (11.62), decisive for the specific conductivity:

$$\kappa = \sum_k (|z_k| F u_k c_k) \tag{11.62}$$

Thus, by increasing *one* charge carrier concentration in the layer near the interface, the specific conductivity parallel to the interface can be greatly enhanced. Such effects can be exploited, for example, in gas sensors, in which the absorption of a gas leads to an increase in the concentration of mobile defects at the surface and can be measured as a change in conductivity along the surface.

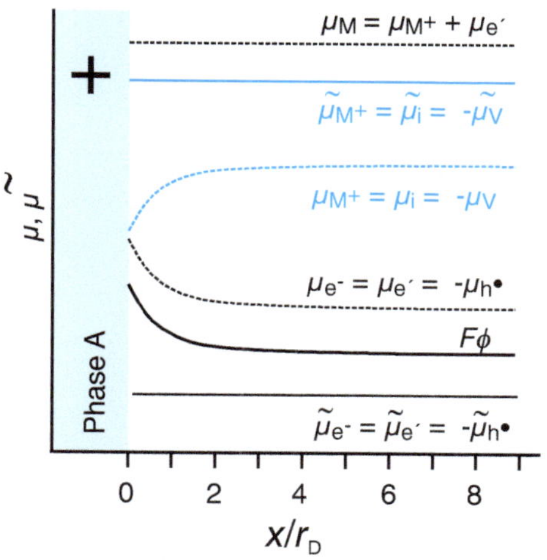

Figure 11.14 Map of electric, chemical, and electrochemical potentials for metal ions on interstitial sites and of mobile electrons in AgCl. Source: Adapted from [8].

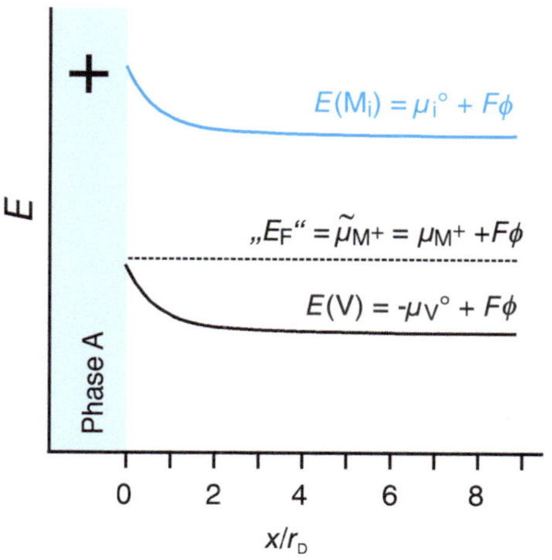

Figure 11.15 "Band structure diagram" of AgCl when applying a potential between the interface and bulk. Source: Adapted from [8].

11.3.2 Increase of Conductivity in Nanocrystalline Solids

The increase in conductivity parallel to the interface can only be measured directly in a few systems. However, the effect can be greatly increased by using nanoscopic surfaces with a correspondingly large surface-to-volume ratio.

As an example, let us again consider AgCl with conductivity from $Ag_i^{\cdot}$ and V'_{Ag}. An increase in conductivity can be achieved by doping the entire crystal with $CdCl_2$ (Table 11.1). This procedure is called homogeneous doping because Cd^{2+} is uniformly distributed throughout the AgCl phase.

Heterogeneous doping refers to a process in which the ion conductor is brought in contact with another, often nanocrystalline phase. As an example, when an AgCl crystal is brought into contact with Al_2O_3 particles, Ag^+ ions are bound to the nonbonding electron pairs of the oxide, a process that proceeds approximately according to Eq. (11.63)

$$[Al_2O_3]\text{-}OH + Ag_{Ag} \rightleftharpoons [Al_2O_3]\text{-}OHAg^+ + V'_{Ag} \tag{11.63}$$

The adsorption of Ag^+-ions at the interface leaves vacancies in the near-surface region of the AgCl crystal, while the number of $Ag_i^{\cdot}$ defects is reduced compared to the bulk phase. There is a decrease in conductivity by interstitial sites (i) due to the decrease in the concentration of $Ag_i^{\cdot}$ defects. Often this effect is overcompensated for by the extreme increase in vacancy concentration and the resulting vacancy conductivity (v). Such a local change of the dominant conductivity mechanism from i to v conduction is analogous to a *p*–*n* transition in semiconductors! The effect can be exploited to synthesize new materials with macroscopically altered conductivities. For example, finely dispersed Al_2O_3, which itself has no ionic conductivity and a very high melting point, can be introduced into an AgCl melt. On cooling, the Al_2O_3 particles accumulate on the surfaces of the AgCl crystallites, increasing the conductivities at the grain boundaries and leading to a reduction in the size of the AgCl crystallites. Excess Al_2O_3 is deposited at the grain boundaries only at contents above 10% Al_2O_3. In such a situation, Al_2O_3 blocks the conductivity *between* the AgCl crystallites, thus the macroscopic conductivity collapses if the overall content in Al_2O_3 is too high.

Reference [27] provides a systematic account of this approach.

11.4 Electrochemistry in Molten Salts and Ionic Liquids

Molten salts, oxides, and bases often exhibit ionic conductivity because of mobile charge carriers. The conductivity mechanisms and applications vary greatly, depending on the melting temperature and the dominant nature of chemical bonds. Compounds that exist in the solid state in an ionic lattice with a low contribution of covalent bonding (alkali halides, alkaline earth halides, as well as hydroxides, nitrates, carbonates, and sulfates) are largely dissociated in the liquid state (analogous to strong electrolytes), forming complexes of varying structure. Substances that have a substantial contribution of covalent bonding in the solid state are present in solution as undissociated molecules, analogous to weak electrolytes (e.g., $AlCl_3$). In addition, salts, oxides, and hydroxides can dissolve in excess of another molten salt, forming a molten solution.

Currently, ionic liquids receive particular attention. They consist of molecular ions with an *asymmetric structure* in which the *charge is strongly delocalized*. They occur in the liquid state already below 100 °C or even at room temperature.

11.4.1 Alkali Metal Halides

Melts of alkali metal halides MeX such as NaCl and LiCl are ionic. The specific conductivities of their melts are correspondingly high (Table 11.2). Their viscosities and thermal conductivities are comparable to aqueous electrolyte solutions. However, the conductivity mechanism can only be explained with major oversimplifications. Unlike in aqueous electrolyte solutions, there is no solvent to separate the ions from each other. Therefore, strong interionic interactions are to be expected. Moreover, these are chemically specific. Complexes such as $[MeX_4]^{3-}$ and $[XMe_4]^{3+}$ with coordination number 4 are expected to occur in the melt. However, these complexes can also combine to form larger aggregates. The conduction mechanism is similar to that in a highly disordered ion-conducting solid. First, a single ion Me^+ or X^- must detach from the complex, analogous to the occupation of an interstitial site by an ion in an ionic lattice. This is followed by site-to-site movement in between complex ions in the melt. The movement ends with the "capture" of a single ion by a "vacancy" complex ion.

The temperature dependence of conductivity follows an Arrhenius expression [Eq. (11.48)], where the activation energy corresponds to the energy barrier for the movement of a single ion from one place to the next.

Table 11.2 Melting points T_m, measurement temperature T, specific conductivity κ, molar conductivity Λ, transference number of cations t_+, and activation energy E_A in Arrhenius expression for the temperature dependence of κ.

Compound	T_m/°C	T/°C	κ/(S cm^{-1})	t_+	Λ/(S cm^{-1} mol^{-1})	E_A/(kJ mol^{-1})
LiCl	614	620	5.87	0.75	183	5.2
NaCl	800	850	4.75	0.62	113.5	9.2
KCl	790	800	2.19	0.62	103.5	15
RbCl	715	715	1.17	0.58	94	16.7
CsCl	646	660	1.14	0.64	66.7	15
$MgCl_2$	712	730	1.05	0.48	28.8	15.9
$CaCl_2$	772	800	2.00	0.48	64	19.1
$SrCl_2$	873	900	1.98	0.26	55.7	18.1
$PbCl_2$	500	550	1.69	0.24	53	7.2
0.1 M KCl(*aq*)	—	18	0.0112	0.49	112	11.3

Source: Adapted from [28].

When comparing different alkali metal halides, it is noticeable that, in contrast to aqueous solutions, the conductivity decreases with increasing crystallographic radius (LiCl > NaCl > KCl > RbCl) since no solvation shell with a hydrodynamic radius is formed, which in the case of aqueous solutions overcompensates for the increasing crystallographic radii (Figure 4.21).

As in aqueous electrolyte solutions, two inert electrodes of equal size are used to measure the conductivity. However, since the conductivities are relatively high, the electrodes are connected via a thin channel filled with the molten salt. This ensures that the current in the cell is limited by the resistance in the melt.

Ion-conducting molten salts are used in important technical processes. For example, molten carbonates incorporated into a solid matrix form the electrolyte of the molten carbonate fuel cell (Section 16.5). One advantage of conducting the reaction in molten salts is the usually very high exchange current densities of 1–10 A cm^{-2}, because the high temperatures are sufficient for thermal activation of the reaction even for high activation barriers.

Another important application is the technical preparation of reactive metals that cannot be deposited from aqueous solutions. The most important of these processes is molten salt electrolysis for the preparation of aluminum and alkaline earth metals. However, halide melts are very corrosive, which can limit the effectiveness of electrolysis. For instance, in a $CaCl_2$ molten-salt electrolysis, the metal just deposited at the cathode reacts again with the electrolyte, approximately according to the equation $Ca + CaCl_2 \rightleftharpoons Ca \cdot CaCl_2$. The product can be thought of as a solvated Ca atom or a solvated electron with a cationic countercharge. These mobile charge carriers can also reach the anode and be oxidized there. This shuttling between anode and cathode leads to a current flow without net conversion or turnover in the cell and thus to reduced current yield (faradaic efficiency). Salt additives, which promote the formation of aggregates, can push back this effect. The corrosiveness of the molten salts requires a material for the cell container that can withstand both the high temperatures and the corrosive conditions.

Finally, wetting of the electrode materials with the melt also presents a technical challenge because, on the one hand, the surface tension of the melts is higher than that of aqueous electrolyte solutions, which makes complete wetting of solid electrodes difficult. Second, the formation of gaseous electrolysis products (e.g., Cl_2) can cause a thin layer of gas to form between the anode and the melt. This increases the cell voltage required to maintain a given current density to such an extent that sparking can occur at the electrodes.

11.4.2 Ionic Liquids

Ionic liquids are compounds and homogeneous mixtures which are in the liquid state below 100 °C but are composed exclusively of ionic constituents.

In a narrower definition, only those compounds are considered that are in the liquid state at room temperature (room-temperature ionic liquids, RTIL). Such compounds have been known for a long time. Low-melting chloroaluminates play a role of stable solvents in nuclear fuel reprocessing, but were only suitable for special applications because of their sensitivity to moisture. Around the year 2000, work on the role of anions in polymer electrolytes (Section 11.7.1) based on polyethylene oxide made it clear that certain anions can prevent the crystallization of ionic compounds. The combination of such anions with cations having similar structural features then leads to a very large number of liquid ionic compounds with new, attractive properties. These compounds can be handled under typical laboratory conditions. The physical properties of ionic liquids can be found in databases [13, 29]. Subsequently, intensive research efforts were directed toward various aspects of electrochemistry in ionic liquids, which have already been systematically reviewed [19].

Let us first look at a prototypical ionic liquid from which we can identify essential properties. A typical anion is bis(trifluoromethylsulfonamyl)imide (TFSI, Figure 11.16a). The extremely electron-withdrawing CF_3–SO_2 moiety and the existence of alternative resonance structures favor a distribution of the electron density of the imide nitrogen across almost the entire anion, so that local charge density differences are largely avoided. As a result, these anions are not very reactive and coordinate metal ions only weakly. The S–N–S bonding is flexible, with multiple

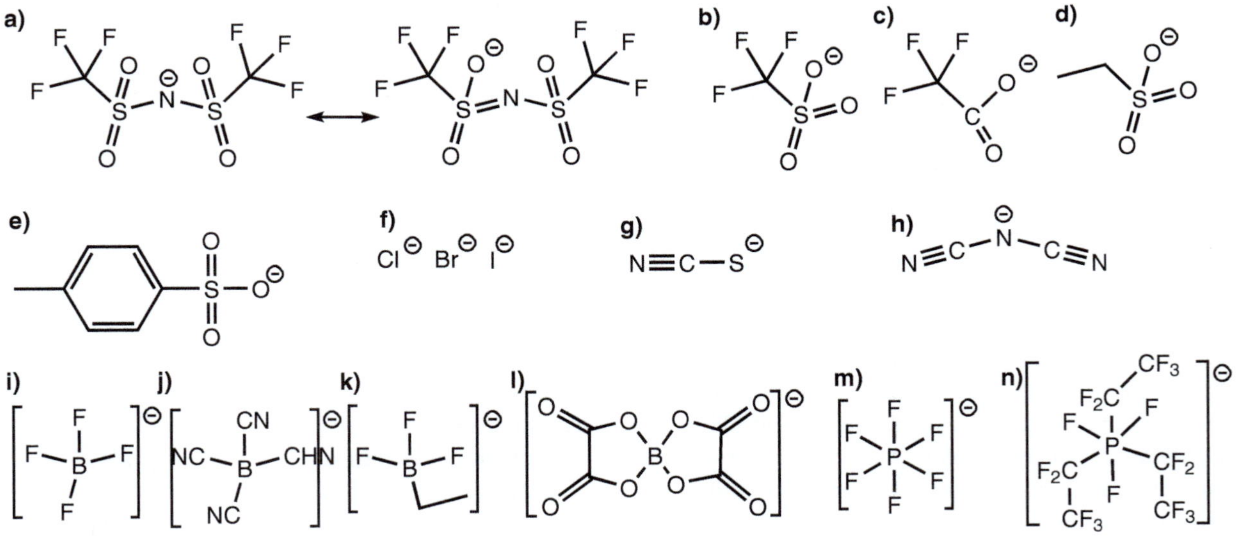

Figure 11.16 Anions in RTILs: a) two resonance structures of bis(trifluoromethylsulfonyl)imide (TFSI or Tf_2N); b) trifluoromethanesulfonate; c) trifluoroacetate; d) ethanesulfonate; e) *p*-toluenesulfonate (tosylate); f) halides; g) thiocyanate ("rhodanide"); g) dicyanimide; i) trifluoroborate; j) tetracyanoborate; k) ethyl trifluoroborate; l) bisoxalylborate (BOB); m) hexafluorophosphate; and n) trifluoro-triperfluoroethyl-phosphate.

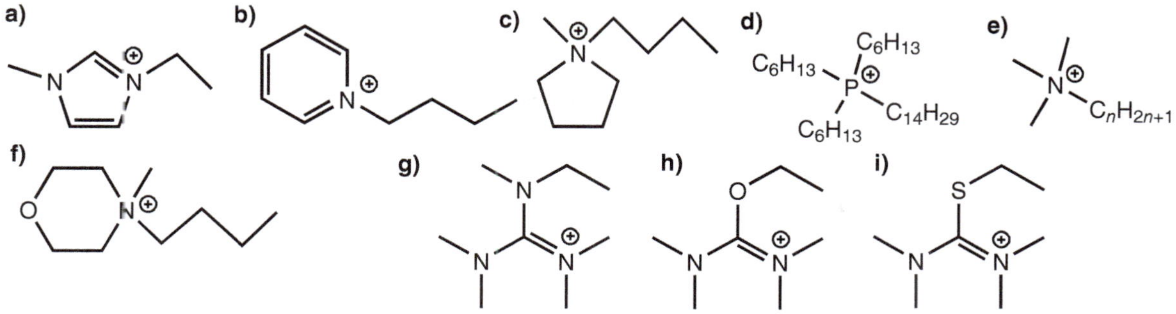

Figure 11.17 Cations in RTILs: a) *N*-ethyl-*N*′-methylimidazolium; b) *N*-butylpyridinium; c) *N*-butyl-*N*-methylpyrrolidinium; d) trihexyl(tetradecyl)phosphonium; e) alkyltrimethylammonium; f) *N*-butyl-*N*-methylmorpholinium; g) ethylpentamethylguanidinium; h) *O*-ethyl-*N*,*N*,*N*′,*N*′-tetramethyluronium (derived from urea, uronium); and i) *S*-ethyl-*N*,*N*,*N*′,*N*′-tetramethylthiouronium.

conformations that have only small energy differences and small barriers for transition. Similar properties are found for cations in ionic liquids (Figure 11.17). Here we can consider the 1-ethyl-3-methylimidazolium cation as a typical example (Figure 11.17a). The heteroaromatic ring system distributes the overall positive charge. The two alkyl substituents of different chain lengths have many energetically similar conformations. The different space requirements of the two alkyl chains also make space-filling packing difficult. All in all, this is a compound that is unlikely to solidify!

The preparation of very clean ionic liquids is difficult due to the lack of possibility for recrystallization. Other separation methods also work only to a limited extent, and because of their ionic nature, these compounds have very low, often negligible, vapor pressure. Thus, they cannot be distilled. Protocols for the preparation of ionic liquids that meet the purity requirements of electrochemistry are now established.

The low vapor pressure opens up some new application areas. For example, ionic liquids can be used as lubricants. In electrochemistry, they are advantageous when thin and stable electrolyte films are needed in contact with a gas atmosphere, such as in gas sensors or in metal–air batteries. The low vapor pressure prevents a gas atmosphere above an ionic liquid from accumulating organic molecules. As a result, ionic liquids – unlike most organic solvents – are not flammable or are very difficult to ignite. Because of their low volatility, valuable substances can be easily separated from ionic liquids by distillation, and the ionic liquids themselves can be recycled with essentially no loss of mass. However, the claim to be "green" solvents based on this property must be evaluated on a case-by-case basis, taking into account also the toxicity of the ions *and* their decomposition products (HF, HCN, …), especially in the case of slow decomposition reactions with water.

For electrochemistry, it is particularly significant that these liquids can serve as both a solvent for other substances and as a conductive electrolyte. The theoretical description of the conductivity of ionic liquids must consider that, in contrast to aqueous or organic electrolyte solutions, there is no solvent to enclose and separate the ions from each other. Therefore, the limiting case of dilute solutions, where long-range electrostatic forces dominate, is not realistic for ionic liquids. The assumptions made in Chapter 4 that ions can be described as spheres of a certain size and charge density with a bound solvent environment and with a hydrodynamic radius do *not* apply to ionic liquids. Indeed, one of their central characteristics is precisely that their ions deviate strongly from the spherical shape. This poses considerable challenges for the theoretical description and prediction of their properties [17].

In particular, our notion of the structure of the electrical double layer from Section 4.3 (rigid and diffuse layer, superimposed by specific adsorption of individual ions) is challenged, since the dense packing of one type of ion leads to large repulsive forces between equally charged ions due to the absence of a dielectric solvent between them. New structural models are needed for the double-layer structure in ionic liquids, which are very important for understanding reactivity and properties. Theoretical approaches as well as coupled electrochemical and structure-sensitive methods (Chapter 8) are used for this purpose [17, 30, 31].

Due to the diffuse charge distribution in ionic liquids, their constituting ions are often difficult to oxidize or to reduce. Therfore, ionic liquids often have a very wide usable potential window of 4 V, which corresponds to that of high-purity organic electrolyte solutions.

Although they are composed exclusively of ions, many ionic liquids cannot be mixed with water. However, they quickly absorb traces of water from the environment, much like dried organic solvents. As in organic electrochemistry, this poses a problem because water limits the usable potential range or reacts with the materials under study or with electrochemically generated intermediates.

Ionic liquids are attractive alternatives to organic electrolyte solutions [32], which they can outperform in their dissolving power for polar compounds and, especially, for ionic compounds. This results in new electrochemical applications such as the deposition of reactive metals, which are either deposited at such negative potentials that intense hydrogen evolution occurs in aqueous electrolytes, or whose precursors form stable compounds with water ($\{SiO_2\}$, Al_2O_3) and make deposition of the elemental metal impossible. Here, ionic liquids open new access routes for the deposition of refractory metals and semiconductors [15, 18]. In addition, the morphology of the obtained films can be controlled by the chosen ionic liquid, which can be attractive, for example, for the deposition of reactive metals in rechargeable batteries [19].

One disadvantage compared to organic solvents is the high viscosities, which can be 1–3 orders of magnitude higher than those of conventional solvents. This significantly slows down mass transport processes. For a given cation, the viscosity is very much determined by the nature of the anions in the ionic liquid. Strong charge delocalization in the anion, as in TFSI, favors relatively low viscosity. Short and branched alkyl chains of the cation favor low viscosity.

The high viscosity of ionic liquids is a major factor affecting the specific conductivities. They range from 0.1 to 20 mS cm^{-1}. Although ionic liquids are composed only of ions, these values lag behind those of aqueous electrolyte solutions by at least an order of magnitude. For example, the specific conductivity of a solution containing 29.4 mass % KOH used in alkaline batteries is 540 mS cm^{-1}. In general, the water content in ionic liquids (as impurities or as a deliberate addition) leads to an increase in specific conductivity. While these conductivities are too low for some applications, they are generally sufficient for experiments traditionally conducted in organic electrolyte solutions (Chapter 6). Similarly, the equipment is essentially transferable. One difficulty is selecting a reference electrode. Although the commonly used quasireference electrodes (Pt or Ag wires) are convenient, they do not allow comparison between different experiments. Therefore, reversible redox couples such as metallocenes are used to compare potentials [32–34]. Attempts have also been made to construct special reference electrodes for ionic liquids [35, 36].

As solvents for mechanistic studies of organic and organometallic compounds, ionic liquids often behave similarly to organic electrolyte solutions. However, special features of some ionic liquids should be noted. For example, some cations, such as *N*-ethyl-*N*′-methylimidazolium, possess a C–H acidity, which leads to the rapid reaction of electrochemically generated products and can suppress other subsequent reactions (e.g., dimerization).

Interestingly, many enzymes retain their activity in ionic liquids. This opens up the possibility of monitoring the activity of immobilized enzymes in ionic liquids by electrochemical measurements. In this case, the ionic liquids can be applied as a thin film or droplet similar to an electrode modification (Chapter 13). The studies can also be performed entirely in ionic liquids [32].

11.4.3 Deep Eutectic Solvents

Deep eutectic solvents (DESs) are eutectic salt melts consisting of several components with a melting point below 100 °C or even close to room temperature. The best-known DES is formed by mixing and slightly heating one mass fraction of choline chloride (Figure 11.18a, melting point $T_{m,A}$ = 575 K) and two mass fractions of urea (Figure 11.18b, $T_{m,B}$ = 406 K). Eutectic mixtures of two substances A and B have melting points that are significantly *lower than* those of the pure substances A and B (Figure 11.19, for the example, T_e = 285 K). In DES, component A is an asymmetric ammonium, phosphonium, or sulfonium salt, usually a chloride. Component B is a hydrogen bond donor that interacts with the anion of salt A. Because of these intermolecular interactions among the components, many local structures of similar energies exist, preventing crystallization in a periodic ionic lattice. Occasionally, a strong oxoacid is also added to the DES to increase ionic conductivity. DES can dissolve many metal salts.

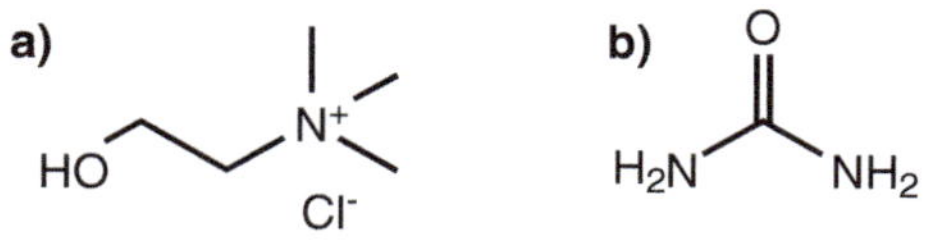

Figure 11.18 Structure of a) choline chloride and b) urea.

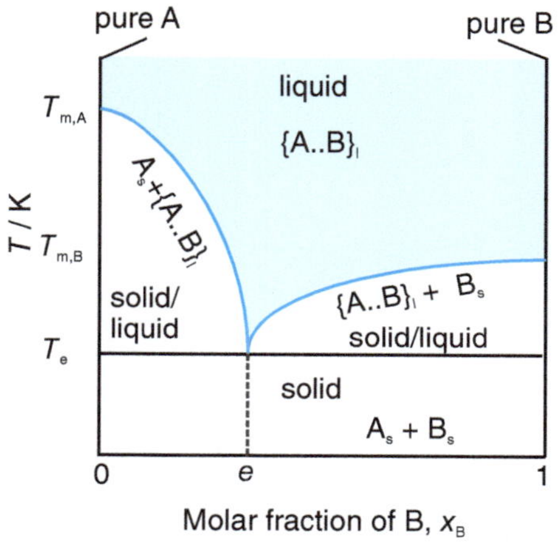

Figure 11.19 Schematic phase diagram of a eutectic mixture of two components A and B.

Unlike ionic liquids, DESs consist of more than just two kinds of ions and can be obtained from relatively simple starting materials without the effort of separating by-products. Therefore, they appear suitable for large-scale applications. In addition to these fundamental differences, many other properties of DES show broad analogies to ionic liquids for electrochemical applications: DES are viscous and have negligible vapor pressure, which implies low flammability. Their electrochemical potential window is somewhat smaller than that of ionic liquids, but allows for deposition of base metals at high current yields (i.e., without parallel hydrogen evolution or other reductive side reactions). The solubility of many metal salts is also interesting for electrochemical polishing processes, where the electrolytic dissolution of protruding metal parts results in a smoothing of workpieces (Section 14.6.2).

DESs do not contain fluorine-containing components, and the starting materials are toxicologically safe. Together with the low vapor pressure, these properties make them environmentally benign solvents. Recent review articles [37, 38] provide insight into these new electrolyte systems of ever-increasing importance.

11.5 Conductivity in Glasses

Glasses are amorphous solids that have a near-range order but no far-range order typical of crystals. Most glasses are obtained by solidification of a melt, which occurs faster than crystallization of the solid. Thus, we can view a glass as a liquid in which the order of the molecular environment is preserved (near-range order), but nonuniform orientation of the building blocks leads to a very high number of defects (lack of far-range order). These defects provide vacancies for the movement of ions. The conductivities are often selective (exploited in the use of glasses for pH- and Na^+-selective electrodes, Section 5.2).

Very different conductivities can result in the glasses depending on the elements present. Simple binary glasses of LiO_2 and SiO_2 show conductivities of about 10^{-6} S cm^{-1} at room temperature. Sulfur analogs such as Li_2S-B_2S_3 or Li_2S-SiS_2 exhibit hundreds of times higher conductivities (10^{-4} S cm^{-1}) because the sulfide ion is more polarizable. Other cation-conducting (Ag^+) glasses are $(AgI)_{1-x} \cdot (Ag_2MoO_4)_x$ ($0.02\ S\ cm^{-1}$) and $AgI \cdot Ag_2SeO_4$ ($0.01\ S\ cm^{-1}$). Anion-conducting glasses also exist, for example, the system $ZrF_4 \cdot BaF_4$ with a fluoride conductivity of $10^{-6}\ S\ cm^{-1}$ at 200 °C.

11.6 Conductivity in Polymers

Polymers can exhibit considerable conductivity and are very important for some applications. *Electronically conductive polymers* serve as functional layers on electrodes (Section 13.2). Here, we will only deal with those polymers that exhibit exclusively *ionic conductivity* and can therefore be used as electrolytes in electrochemical cells.

11.6.1 Polymer Electrolytes and Gel Electrolytes

Many polymers can be mixed with salts when nonbonding electron pairs on heteroatoms can complex alkali metal cations. A practically important **polymer electrolyte** is poly(ethylene oxide), (Figure 11.20) whose helically arranged chains can coordinate Li^+-ions, for example, from $LiClO_4$. Since not all possible coordination sites are occupied by Li^+, place changes of Li^+ can occur, which are supported by a certain mobility of the polymer chains. The composition of the polymer electrolyte is given as the number n of ethylene oxide units per formula unit of $LiClO_4$, for example, $PEO_n\ LiClO_4$. The perchlorate ions stabilize the chain structure and balance the charge. The conductivity increases with the proportion of chains that are in an amorphous state. The transition between predominantly crystalline structures and amorphous state occurs at the glass transition temperature T_g, which serves as a reference point for the temperature dependence of the ionic conductivity, often given by Eq. (11.64), which is also known as the Vogel–Fulcher–Tammann equation.[136]

$$\kappa = \frac{A}{\sqrt{T}} \exp\left[\frac{-E_A}{R(T - T_g)}\right] \tag{11.64}$$

A – Constant depending on free carrier concentration [$S\ cm^{-1}\ K^{0.5}$]
T – Temperature [K]
T_g – Glass transition temperature [K]
E_A – Apparent activation energy for place change [$J\ mol^{-1}$]
R – Gas constant [$J\ K^{-1}\ mol^{-1}$]

The conductivity can be increased by the addition of larger anions or of plasticizers, which are incorporated between the polymer chains and thus reduce the size of crystalline regions and increase the proportion of amorphous regions. We speak of **gel electrolytes** when only gel-like areas are present with plasticizers intercalated between the polymer chains. Gels can also be formed when the solution of a salt in an organic solvent is absorbed by a polymer matrix. We already discussed an example in Excursion 5-1 (Preparation of an ion-selective membrane).

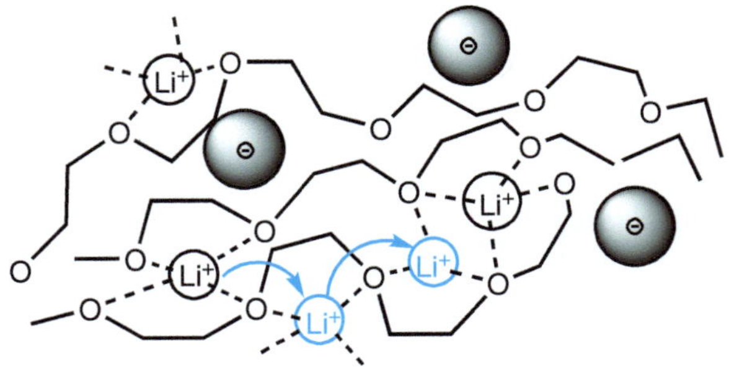

Figure 11.20 Coordination of Li^+ ions in poly(ethylene oxide).

[136] H. Vogel; *Physikal. Zeitschr.* **1921**, *22*, 645; Gordon Scott Fulcher, 1884–1971; Gustav Tammann, 1861–1938, German-Baltic chemist, professor in Göttingen.

Important applications of polymer electrolytes with Li^+-ion conductivity are lithium-ion batteries in which these electrolytes separate the anode from the cathode mechanically and electronically, whilst connecting them as an ionic conductor (Section 15.6).

11.6.2 Polyelectrolytes

Polyelectrolytes are polymers containing a large number of ionic groups in the polymer chain or in side chains with charges compensated for by counter ions. Another subdivision, not consistently applied in the literature, considers ionomers as copolymers that have a hydrophobic backbone into which an adjustable number of charged monomers can be incorporated. A typical ionomer example is Nafion (Figure 2.30). The actual polyelectrolytes are distinguished from ionomers because *each* monomer unit carries a charge as, for example, in poly(styrene sulfonate). The concentration of charged groups in polyelectrolytes reaches 5–6 mol L^{-1}. Other frequently occurring charged groups in polyelectrolytes are $-NH_3^+$ and $-COO^-$. Depending on solubility and counterions, these polymers can be processed into very different membrane materials or coatings.

Below, we consider some applications: Polyions can strongly bind multivalent counterions due to the chelate effect. In solution, polyion chains can be reversibly crosslinked via cations and thus precipitated from solutions. If the polymer strands are previously converted into networks with a low degree of cross-linking between the chains, gel-like materials result with interstices that are easily accessible to solvents and ions. If the charge of a polyanion is initially compensated for by monovalent cations, these can be exchanged for divalent cations according to Eq. (11.65):

$$2\,-SO_3Na + Ca^{2+} \rightleftharpoons \{-SO_3\}_2Ca + 2Na^+ \tag{11.65}$$

This is the basic principle of ion exchangers, which are widely used, for example, in water treatment. In the laboratory, they are found in supply systems for ultrapure water. Once all Na^+-ions have been exchanged, the capacity of the ion exchanger is exhausted. However, it can be regenerated by treating with a concentrated NaCl solution, whereby the reaction of Eq. (11.65) is reversed. The same principle applies to the retention of polyvalent ions (e.g., pollutants such as heavy metal ions) by humic matter. Humic matter is a mixture of natural or artificial substances that are formed, for example, during the degradation of plant residues. They contain linked benzene rings with phenolic OH groups and carboxylate groups. Many aluminosilicates such as zeolites form another class of ion exchangers. They are natural inorganic substances, but are also synthesized for technical applications.

Ion Exchange Chromatography uses ion exchangers for the analytical separation of ions. The exchanger material is packed into a chromatographic column that is inserted into a high-pressure liquid chromatography system. Because the ions of the sample are retained on the column to a different extent and are displaced by the ions of the eluent at different rates, a specific retention of the analyte ions of the sample occurs that can be used for identification.

Ion Exchange Membranes are ion-exchange materials processed into thin films or membranes. The membranes can be placed between two electrodes and thus can also be used as the sole electrolyte, such as in the polymer electrolyte membrane fuel cell (Sections 2.4.4 and 16.2). In other processes, membranes separate the anode and cathode compartments, or they are the key elements for electrochemically driven separation processes. All these processes make use of the permselectivity of the membranes, that is, the exclusive permeability for the small counterions as opposed to the immobility of the oppositely charged ions covalently anchored to the polymer chains or the impermeability for larger ions.

Ion-exchange membranes are also key components in the large-scale electrolysis of NaCl by the membrane process (Excursion 17-1).

Electrophoretic Deposition. Polyelectrolytes from weak acids or bases exhibit pH-dependent solubility in water. For example, at high pH values, carboxylate groups are present in the deprotonated form ($-COO^-$) and at low pH values, amino groups occur as a charged form ($-NH_3^+$). With the charged groups along the polymer chain, the polyelectrolytes are soluble in water. However, if the pH shifts, the ionic groups of the polymer can become neutral groups ($-COOH$ or $-NH_2$), which is accompanied by a loss of solubility. If the pH shift is triggered by electrolysis on a metallic workpiece, the polyelectrolytes deposit as a film directly on the workpiece (Figure 11.21), where the film thickness is self-limiting because the resulting film reduces the reaction rate of the electrolysis that causes the necessary pH shift. Therefore, even curved surfaces, cavities, or edges can be coated uniformly. Thermal posttreatment allows the double bonds or reactive groups contained in the polymer chains to connect the individual polymer chains, thus forming a stable cross-linked film. This process is used on a large scale in the coating of entire car bodies [39, 40].

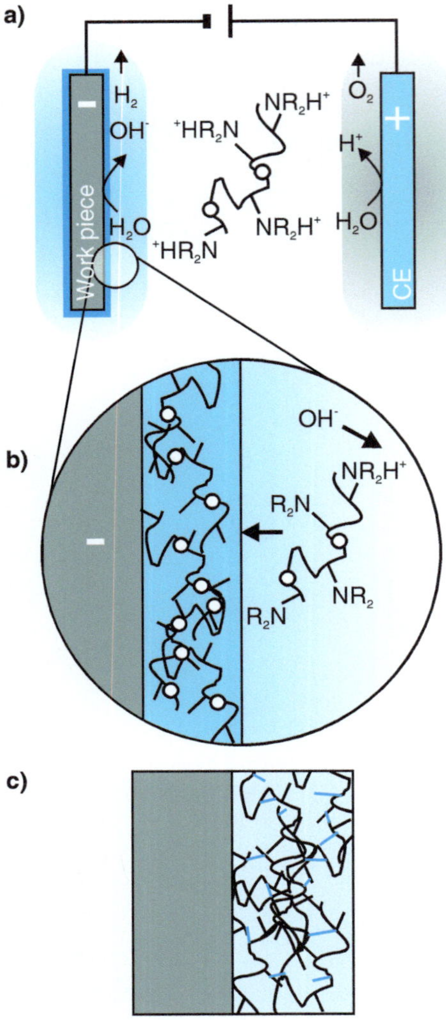

Figure 11.21 Principle of cathodic electrophoretic deposition.

a) Water electrolysis increases the pH at the cathode.
b) The polymers with ammonium groups and reactive groups (○) are neutralized by OH^- near the cathode and precipitate as a wet film on the workpiece.
c) Reactive groups in the polymer are thermally cross-linked (blue lines) to form a permanently stable film.

The deposition bath may also contain pigments and other additives. The pH shift at the counter electrode (CE) is buffered by organic acid anions.

Excursion 11-5 Electrokinetic Membrane Separation Processes

Electrodialysis uses an electric field applied across a stack of alternating anion and cation exchange membranes (Figure 11.22). The membranes can be passed either by cations only ($t_+ = 1$) or by anions only ($t_- = 1$), so that the ion content in the solutions between the membranes is alternately depleted or enriched. Electrodialysis is suitable for the production of drinking water from brackish water.

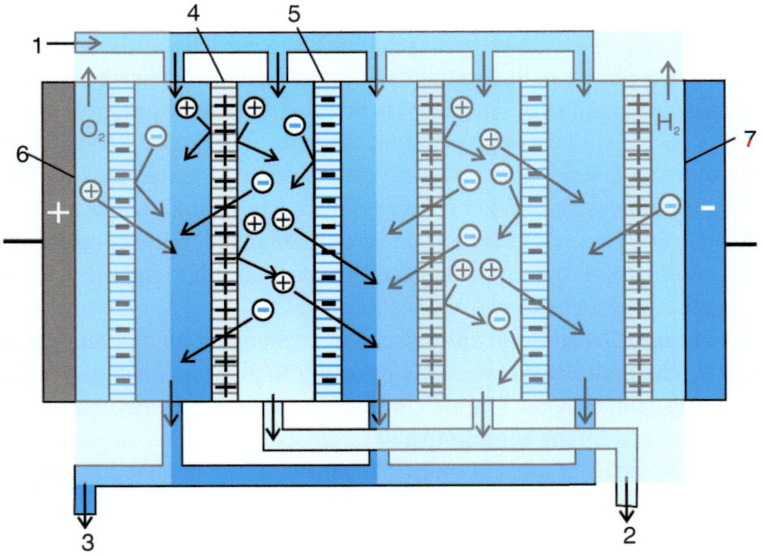

Figure 11.22 Principle of electrodialysis. 1) Feed solution, 2) dilute (purified water), 3) concentrate, 4) anion exchange membrane (fixed positive charges in the polymer), 5) cation exchange membrane (fixed negative charges in the polymer), 6) anode ($2\,H_2O \rightarrow O_2 + 4\,e^- + 4\,H^+$), 7) cathode ($4\,H_2O + 4\,e^- \rightarrow 2\,H_2 + 4\,OH^-$). The stacking sequence of technical units is indicated with only two units. A single unit is highlighted.

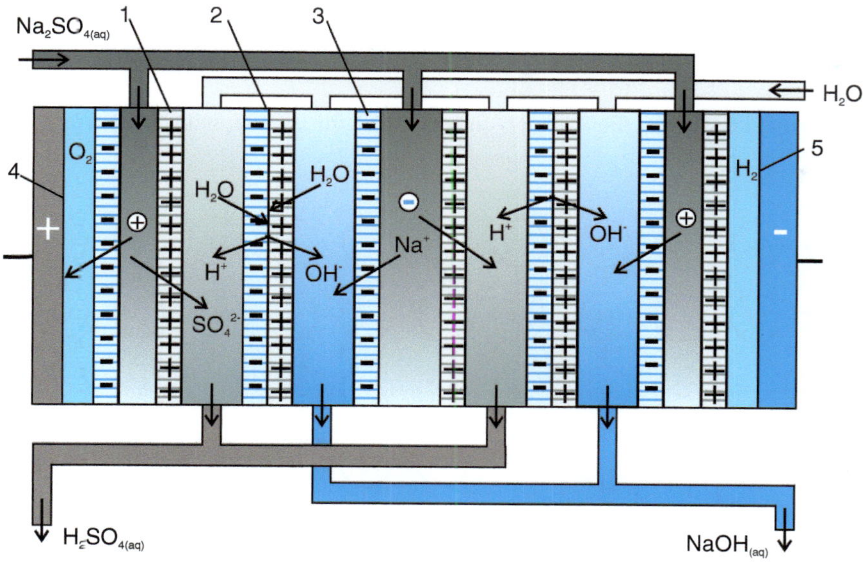

Figure 11.23 Principle of electrodialysis with a stack of bipolar membranes using the preparation of H_2SO_4 and NaOH from Na_2SO_4 solution as example. 1) Cation-exchange membrane, 2) bipolar membrane, 3) anion-exchange membrane, 4) anode ($2\ H_2O \rightarrow O_2 + 4\ e^- + 4\ H^+$), and 5) cathode ($4\ H_2O + 4\ e^- \rightarrow 2\ H_2 + 4\ OH^-$). Technical systems use a stacking sequence with far more than the two cells shown here.

In a modified version, a bipolar membrane is used, which is a direct combination of an anion and a cation exchange membrane (Figure 11.23). Neither anions nor cations can pass through the bipolar membrane. When a current is forced from the outside, H^+ will leave the membrane in the direction of the cathode and OH^- will leave the membrane in the direction of the anode. The (neutral) water consumed in this process is replenished from the surrounding media. With a clever arrangement of the liquid spaces, acids, and bases can be obtained in a single process. The energy efficiency is significantly improved compared to pure electrolysis processes. This becomes clear when we realize that Figure 11.23 represents only a section of a periodic stacking sequence. Thus, for each charge converted at the outer electrodes, N equivalents of acids and bases are also formed at the N bipolar membranes. The voltage drop in electrodialysis on bipolar membranes is thus essentially determined by the membranes (which is similar in magnitude to that in membrane electrolysis), while the losses due to overpotentials of the charge transfer reactions at the outer electrodes do not significantly enter into the balance of the reactor with a stack of bipolar membranes. In contrast, the energy loss due to the overpotential is a major limitation of pure electrolysis processes with a single membrane per electrode couple.

Further Reading. Reviews on electrokinetic separation for remediation, substance recovery, and water treatment are recommended [41–43].

Test your knowledge

Exercise 11.1: Using Chapter 4, compile a table with definitions, symbols, units, and equations for the determination of the following quantities: conductivity, resistivity, specific conductivity, molar conductivity, molar conductivity at infinite dilution, transference number.

Exercise 11.2: Using textbooks of inorganic chemistry or solid-state physics, review the structure of face-centered cubic, hexagonal, or body-centered cubic lattices.

Exercise 11.3: Give the Kroeger-Vink notation for the following structures.

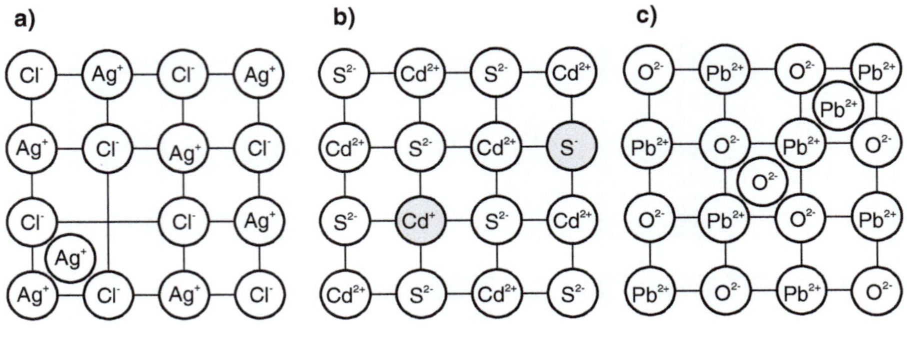

Exercise 11.4: What are intrinsic and extrinsic defects in solid-state electrolytes? Use one sketch for each of the two terms in your explanation.

Exercise 11.5: Formulate the incorporation of Ca^{2+} on Zr sites into ZrO_2 as a reaction equation using the Kroeger–Vink notation.

Exercise 11.6: Formulate the incorporation of Na^+ on Ca sites into CaF_2 as a reaction equation using the Kroeger–Vink notation.

Exercise 11.7: Give the relative charges for the following defects:

a) partial incorporation of MnO_4^- on perchlorate sites in $KClO_4$,

b) incorporation of Cu^{2+} on silver sites in AgCl.

Exercise 11.8: Calculate the data for Figure 11.7 using a spreadsheet program.

Exercise 11.9: What is structural disorder? What influence does it have on the specific conductivity?

Exercise 11.10: Draw an electrical circuit diagram of a four-point measurement to determine the conductivity of a solid electrolyte.

Exercise 11.11: Why is AC voltage measurement *not* usually used to determine the specific conductivity of solid electrolytes (unlike for liquid electrolyte solutions in Chapter 4)?

Exercise 11.12: Present three applications of solid-state electrolytes. Determine which specific requirements must be met by the ionic conductors in these applications.

Exercise 11.13: Tabulate the main classes of ion-conducting solids with relevant temperature ranges, specific conductivities, and possible applications.

Exercise 11.14: Why are two salts often used simultaneously in molten electrolytes?

Exercise 11.15: Find out about molten-salt electrolysis for the large-scale production of aluminum (anodic reaction, cathodic reaction, electrolyte, setup of electrolysis cell, reaction control).

Exercise 11.16: What are ionic liquids? Name and explain the advantages and disadvantages

- compared to aqueous electrolyte solutions
- compared to organic electrolyte solutions
- compared to molten salt electrolytes.

Exercise 11.17: What are glasses? Why can they have ionic conductivity?

Exercise 11.18: Explain three ways in which polymers can acquire ionic conductivity. To which of these three classes does the product Nafion belong?

Exercise 11.19: Why does the ionic conductivity of many systems increase with temperature in a similar way as the rate constant of chemical reactions in the Arrhenius equation? (For comparison, the electronic conductivity of a metal wire decreases with increasing temperature!)

References

1. W. Plieth; *Electrochemistry for Material Science*, Elsevier, Amsterdam, 2008, pp. 6–24.
2. H. Rickert; *Electrochemistry of Solids. An Introduction*, Springer Verlag, Berlin, 1982.
3. U. Stimming, H. Tu; Solid-state electrochemistry. in *Fundamentals of Electrochemistry*. 2nd ed. (V. S. Bagotsky, Ed.), John Wiley & Sons, Hoboken, 2006, pp. 419–447.
4. C. H. Hamann, A. Hamnett, W. Vielstich; *Electrochemistry*. 2nd ed., Wiley-VCH, Weinheim, 2007, pp. 381–396.
5. H. Rickert; Solid ionic conductors: Principles and applications. *Angew. Chem. Int. Ed. Engl.* **1978**, *90*, 37–46.
6. J. Maier; Defect chemistry: Composition, transport, and reactions in solid state; Part I: Thermodynamics. *Angew. Chem. Int. Ed.* **1993**, *32*, 313–335.
7. J. Maier; Defect chemistry: Composition, transport, and reactions in solid state; Part II: Kinetics. *Angew. Chem. Int. Ed.* **1993**, *32*, 528–542.
8. J. Maier; *Physical Chemistry of Ionic Materials: Ions and Electrons in Solids*. 2nd ed., John Wiley & Sons, Hoboken, 2023.
9. P. J. Gellings, H. J. M. Bouwmeester (Eds.); *The CRC Handbook of Solid State Electrochemistry*, CRC Press, Boca Raton, 1997.
10. V. V. Kharton (Ed.); *Solid State Electrochemistry I: Fundamentals, Materials and Their Application*, Wiley-VCH, Weinheim, 2009.
11. H. H. Girault; Electrochemistry at liquid–liquid interfaces. in *Electroanalytical Chemistry, Vol. 23* (A. J. Bard, C. G. Zoski, Eds.), CRC Press, Boca Raton, 2010, pp. 1–104.
12. D. W. M. Arrigan, G. Herzog, A. D. Scanlon, J. Strutwolf; Bioanalytical application of electrochemistry at liquid–liquid microinterfaces. in *Electroanalytical Chemistry, Vol. 25* (A. J. Bard, C. G. Zoski, Eds.), CRC Press, Boca Raton, 2014, pp. 105–178.

13 S. Zhang, X. Lu, Q. Zhou, X. Li, X. Zhang, S. Li; *Ionic Liquids: Physicochemical Properties*, Elsevier, Oxford, 2009.

14 H. Ohno (Ed.), *Electrochemical Aspects of Ionic Liquids*, John Wiley & Sons, New York, 2005.

15 F. Endres, D. MacFarlane, A. Abbott; *Electrodeposition from Ionic Liquids*, John Wiley & Sons, New York, 2008.

16 A. A. J. Torriero (Ed.); *Electrochemistry in Ionic Liquids, Vol. 1: Fundamentals, Vol. 2: Applications*, Springer Verlag, Heidelberg, 2015, pp. 1–623.

17 M. F. Fedorov, A. A. Kornyshev; Ionic liquids and electrified interfaces. *Chem. Rev.* **2014**, *114*, 2978–3036.

18 F. Endres; Ionic liquids: Solvents for the electrodeposition of metals and semiconductors. *ChemPhysChem* **2002**, *3*, 144–154.

19 M. Armand, F. Endres, D. R. MacFarlane, H. Ohno, B. Scrosati; Ionic liquid materials for the electrochemical challenges of the future. *Nat. Mater.* **2009**, *8*, 621–629.

20 E. Mutoro, B. Luerßen, J. Janek; Grenzflächen fester Ionenleiter. *Chem. Unserer Zeit* **2008**, *42*, 80–90.

21 C. V. V. Kharton, Ed. V. Chandran, P. Heitjans; Solid-state NMR studies of lithium ion dynamics across material classes. in *Annual Reports on NMR Spectroscopy, Vol. 89* (G. A. Webb, Ed.), Elsevier, Amsterdam, 2016, pp. 1–102.

22 J. Habasaki, C. León, K. L. Ngai; *Dynamics of Glassy, Crystalline and Liquid Ionic Conductors: Experiments, Theories, Simulations. Topics in Applied Physics, Vol. 123*, (M. S. Dresselhaus, Y. P. Lee, P. M. Ossi, Eds.), Springer, Cham, 2017.

23 T. Kudo; Survey of types of solid electrolytes. in *The CRC Handbook of Solid State Electrochemistry* (P. J. Gellings, H. J. M. Bouwmeester, Eds.), CRC Press, Boca Raton, 1997, pp. 195–221.

24 Z. Zhang, Y. Shao, B. Lotsch, Y.-S. Hu, H. Li, J. Janek, L. F. Nazar, C.-W. Nan, J. Maier, M. Armand, L. Chen; New horizons for inorganic solid state ion conductors. *Energy Environ. Sci.* **2018**, *11*, 1945–1976.

25 M. N. Kozicki, C. Gopalan, M. Balakrishnan, M. Park, M. Mitkova; Non-volatile memory based on solid electrolytes. in *Proc. IEEE Non-Volatile Memory Technology Symposium*, 2004, https://doi.org/10.1109/NVMT.2004.1380792. (last access Oct. 16, 2022, open access).

26 I. Valov, E. Linn, S. S. Tappertzhofen, S. Schmelzer, J. van den Hurk, F. Lentz, R. Waser; Nanobatteries in redox-based resistive switches require extension of memristor theory. *Nat. Commun.* **2013**, *4*, 1771.

27 J. Maier; Nanoionics: Ionic charge carriers in small systems. *Phys. Chem. Chem. Phys.* **2009**, *11*, 3011–3022.

28 V. S. Bagotsky; *Fundamentals of Electrochemistry*, John Wiley & Sons, Hoboken, 2006, p. 132.

29 National Institute of Standard and Technology; Standard Reference Database #147. http://ilthermo.boulder.nist.gov/ (last access Oct. 16, 2022).

30 R. Wen, B. Rahn, O. M. Magnussen; Potential-dependent adlayer structure and dynamics at the ionic liquid/Au(111) interface: A molecular-scale in situ video-STM study. *Angew. Chem. Int. Ed.* **2015**, *54*, 6062–6066.

31 T. Sieling, I. Brand; In situ spectroelectrochemical investigation of potential-dependent changes in an amphiphilic imidazolium-based ionic liquid film on the Au(111) electrode surface. *ChemElectroChem* **2020**, *7*, 3233–3243.

32 P. Hapiot, C. Lagrost; Electrochemical reactivity in room-temperature ionic liquids. *Chem. Rev.* **2008**, *108*, 2238–2264.

33 P. de Vreese, K. Haerens, E. Matthijs, K. Binnemans; Redox reference systems in ionic liquids. *Electrochim. Acta* **2012**, *76*, 242–248.

34 A. A. J. Torriero, J. Sunarso, P. C. Howlett; Critical evaluation of reference systems for voltammetric measurements in ionic liquids. *Electrochim. Acta* **2012**, *82*, 60–68.

35 B. Huber, B. Roling; Development of a Ag/Ag^+ micro-reference electrode for electrochemical measurements in ionic liquids. *Electrochim. Acta* **2011**, *56*, 6569–6572.

36 D. Cicmil, S. Anastasova, A. Kavanagh, D. Diamond, U. Mattinen, J. Bobacka, A. Lewenstam, A. Radu; Ionic liquid-based, liquid-junction-free reference electrode. *Electroanalysis* **2011**, *23*, 1881–1890.

37 E. L. Smith, A. P. Abbott, K. S. Snyder; Deep eutectic solvents and their application. *Chem. Rev.* **2014**, *114*, 11060–11082.

38 L. I. N. Tomé, V. Baião, W. da Silva, C. M. A. Brett; Deep eutectic solvents for the production and application of new materials. *Appl. Mater. Today* **2018**, *10*, 30–50.

39 G. E. F. Brewer; Electrophoretic painting. *J. Appl. Electrochem.* **1983**, *13*, 269–275.

40 L. Besra, M. Liu; A review on fundamental applications of electrophoretic deposition (EPD). *Prog. Mater. Sci.* **2007**, *52*, 1–61.

41 C. Huang, T. Xu; Electrodialysis with bipolar membranes for sustainable development. *Environ. Sci. Technol.* **2006**, *40*, 5233–5243.

42 M. A. Alkhadra, X. Su, M. E. Suss, H. Tian, E. N. Guyes, A. N. Shocron, K. M. Conforti, J. P. de Souza, N. Kim, M. Tedesco, K. Khoiruddin, I. G. Wenten, J. G. Santiago, T. A. Hatton, M. Z. Bazant; Electrochemical methods for water purification, ion separation, and energy conversion. *Chem. Rev.* **2022**, *122*, 13547–13635.

43 V. Montiel, V. García-García, E. Epósito, J. M. Ortiz, A. Aldaz; Membrance processes, electrodialysis. in *Encyclopedia of Applied Electrochemistry* (G. Kreysa, K.-I. Ota, R. F. Savinell, Eds.), Springer Verlag, New York, 2014, pp. 1224–1229.

12

Electrocatalysis

In this chapter, we will investigate electrocatalytic reactions, that is, electrode reactions with a rate strongly depending on the electrode material. An important application of electrocatalysis is in the development of fuel cells (Chapter 16). Fuel cells require accelerated reactivity for intrinsically inert compounds such as O_2 or methanol. In the context of fuel cell research, many experimental and theoretical studies have been carried out on electrocatalytic reactions, for which a systematic understanding is currently beginning to emerge. Water electrolysis to generate green H_2 is also based on electrocatalytic reactions. Electrocatalytic reactions involving small molecules (H_2O, NH_3, H_2, O_2, N_2, CO_2) are currently topical to ensure that in the future, excess solar or wind energy can be converted into storable chemical energy carriers.

The importance of electrocatalytic reactions goes far beyond their application in electrochemical energy conversion. Oxygen reduction or hydrogen evolution are the reduction reactions during corrosion processes. These take place at the same time as oxidative metal dissolution causes damage. The retardation or suppression of these electrocatalytic processes represents an important starting point in corrosion protection (Section 14). The large-scale production of chlorine (Excursion 17-1) is carried out on electrocatalytic electrodes, as is the production of other chemicals, for example, H_2O_2.

Finally, electrocatalytic reactions may also play a crucial role in future technological breakthroughs addressing global challenges. From the textbook author's perspective, these include the discovery of a technology to close the carbon cycle through the reduction of carbon dioxide or the direct production of nitrogenous compounds from atmospheric nitrogen. The large-scale use of atmospheric nitrogen for electrochemical production of fertilizers (for agriculture to ensure food supply for a growing world population) could, in the not-so-distant future, replace the energy-consuming Haber[56]–Bosch[137] process to avoid CO_2 emissions.

For reasons of space, we must limit ourselves in this chapter to the presentation of important basic principles and methods to study electrocatalytic reactions. You should be familiar with the contents of Chapters 1–3 and Sections 8.9–8.11 before reading this chapter. A review of Sections 10.1.2 and 10.1.3 on the electronic structure of solids is also recommended.

Further literature. Some general overviews [1–7] can be recommended on the subject of electrocatalysis. Further literature on particular aspects are given in the respective sections.

[137] Carl Bosch, 1874–1940, German chemist and engineer. Leading role in the development of high-pressure syntheses at BASF company.

12.1 Concepts and Definitions

12.1.1 Catalysis and Electrocatalysis

According to Wilhelm Ostwald,[31] a catalyst is a substance that accelerates the rate of a chemical reaction compared to a reaction without a catalyst, but it itself does not undergo any *chemical* change.

The corresponding phenomenon is called catalysis.

Q: Why is the emphasis on *chemically* unchanged important?
A: When the catalyst forms an intermediate bond with the reactants, atoms can, for example, be released from a crystal and then redeposited again at another location. The external shape of the catalyst may therefore change.

In homogeneous catalysis, the catalyst and all the reactants are present in the *same* phase (e.g., dissolved in a solvent). In heterogeneous catalysis, the catalyst is present in a *separate* phase (e.g., a solid catalyst within a gaseous or liquid reaction mixture). Transferring these definitions to electrochemical reactions is not straightforward due to multiple functions of the electrode surface (in causing oxidation, reduction, and chemical processes).

We can approach the definition in a different way. We first consider the effect of a catalyst at the molecular level. For this purpose, we plot the change in Gibbs free energy G of the reacting substances as a function of a reaction coordinate as shown in Figure 12.1 (compare to Figure 3.5). The energetically most favorable path can be seen running along the reaction coordinate. The rate constant (and thus the reaction rate) with otherwise constant state variables is given by an Arrhenius-type expression[17] with $k = A' \exp[-\Delta G^{\neq}/(RT)]$ [Eq. (3.14)]. In the uncatalyzed case (Figure 12.1, curve 1), the Gibbs free energy of activation is relatively large because the transition state is energy-rich. Accordingly, the probability (associated with k) that two colliding particles will form products is low.

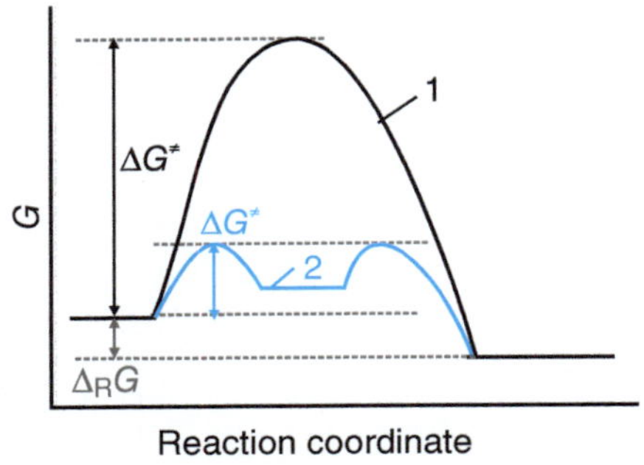

Figure 12.1 Energetic landscape of a reaction 1) without catalyst and 2) with catalyst and with an intermediate forming on the way to the product.

In catalytic reactions, intermediates are stabilized by specific interactions with a catalyst, thus lowering the activation energy and increasing the rate constant.

The difference in the Gibbs free energy of reaction $\Delta_R G$ remains unaffected, but the energy of the transition state $\Delta G^{\neq}$ is significantly reduced compared to the uncatalyzed case (Figure 12.1, curve 2). Because of the exponential relationship between $\Delta G^{\neq}$ and k, even a small decrease in $\Delta G^{\neq}$ leads to a significant increase in k. This approach can be applied in the case of electrochemical reactions.

Electrocatalytic reactions lead within at least one reaction step to a temporary and specific binding of reaction intermediates to the electrode or to another substance close to the electrode. The electrode or catalytic compound remains chemically unchanged after the reaction.

In the literature, electrocatalysis in the narrower sense is often understood only as a reaction in which the reaction intermediates bind to the electrode itself. In analogy to catalysis in general, we will refer to this situation in this textbook as heterogeneous electrocatalysis. Accordingly, in homogeneous electrocatalysis, the reaction intermediates would be stabilized by complexation to solutes in the immediate vicinity of the electrode. These reactive complexes can then undergo electron transfer, for example, by mediators (Section 6.4.6). Similar principles are used in modified electrodes (Section 13.4.8) and in amperometric biosensors (Section 18.3.3). In the scheme for the nomenclature of electron transfer reactions, such reactions are designated as EC′ (pronounced ec-prime). An example is the reduction of $Fe^{(III)}$ in the presence of H_2O_2:

$$\text{At the electrode: } Fe^{(III)} + e^- \rightarrow Fe^{(II)} \tag{12.1}$$

$$\text{In solution: } 2\,Fe^{(II)} + H_2O_2 + 2\,H^+ \rightarrow 2\,Fe^{(III)} + 2\,H_2O \tag{12.2}$$

A homogeneous catalyst can be located close to the electrode surface by a molecular spacer. The situation is then analogous to heterogenized or immobilized catalysts in which a discrete molecular reaction center is bound to a solid phase. Although the catalyst is no longer dissolved in the reaction mixture, the reaction mechanism can be characterized and quantified using methods employed for dissolved catalysts.

Q: Are there any electrode reactions that are *not* catalyzed following this definition?

A: That's a very good question! According to the key diagram in Figure 3.1, surface-bound intermediates *can* occur in both ion transfer reactions and in electron transfer (ET) reactions. However, there are some redox pairs for which the interconversion of the redox forms depends very little, if at all, on the electrode material used. These are outer sphere redox systems such as $[Ru(NH_3)_6]^{2+/3+}$, Fc/Fc^+, $[IrCl_6]^{3-/2-}$. The ET then proceeds similarly to a tunneling process between the electrode and the redox system, which is usually located near the electrode with the ligand and solvation shell intact. Figure 12.2 shows a comparison of two exemplary reactions at a microelectrode for which we expect a sigmoidal cyclic voltammogram (Section 5.6.1). The oxidation of hydrazine $N_2H_4$is an electrocatalytic reaction with at least one adsorbed intermediate. Consequently, the reaction is suppressed by chemisorption of a thiol (Figure 12.2a, curve 2) because the thiol occupies the surface sites that are required for the N_2H_4 oxidation reaction. In the case of the oxidation of ferrocenemethanol, adsorption on the electrode is not required for the reaction to proceed. Therefore, it is not affected by the chemisorption of the short-chain thiol (Figure 12.2b). Thus, we can conclude that the oxidation of ferrocenemethanol is not an electrocatalytic reaction.

Note that for a reaction proceeding under diffusion-controlled conditions (such as the oxidation of ferrocenemethanol), it is generally not possible to determine whether the reaction could be further accelerated by changing the electrode material, since it is already limited by mass transport.

Against the background of the concept of stabilized intermediates, Sabatier's principle[20] immediately becomes plausible. It states that the catalyst must have an optimum binding strength to the reaction intermediates. If the bond is too weak, the activation energy is not lowered sufficiently; if the bond is too strong, the intermediates remain bound to the catalyst and a continuous reaction is not possible. In electrochemistry, this principle can be illustrated by volcano plots (Figure 3.9). We will discuss them in Section 12.3.

In this chapter, we will limit our discussion to heterogeneous electrocatalysis.

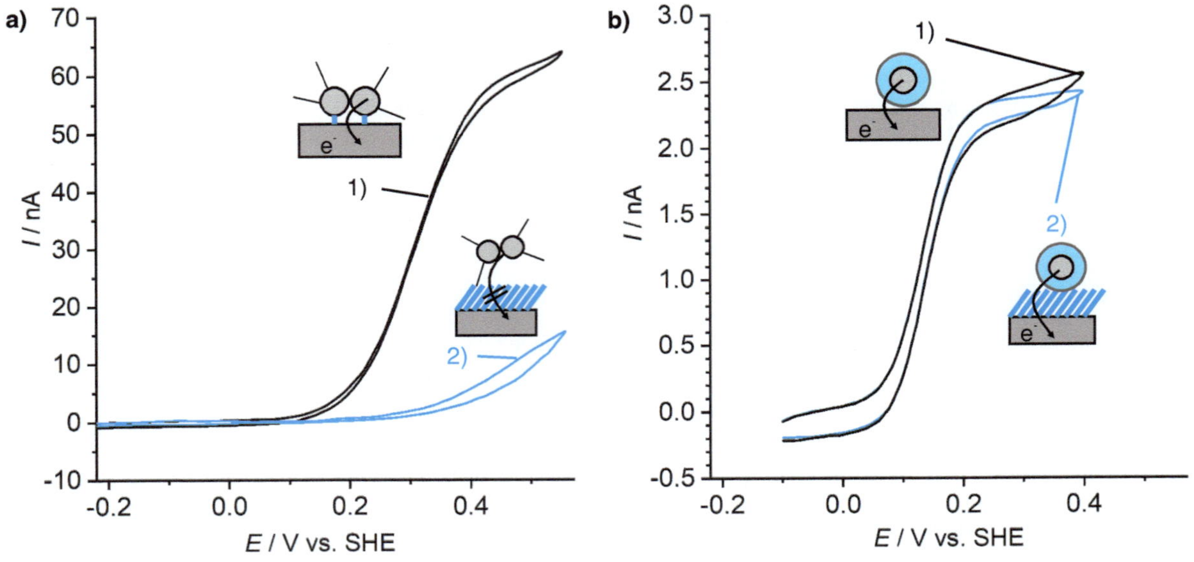

Figure 12.2 Cyclic voltammograms for a) oxidation of hydrazine N_2H_4 at an Au microelectrode as an example of an electrocatalytic reaction; b) oxidation of ferrocenemethanol at an Au microelectrode as an example of a noncatalyzed reaction. Curve 1 is the reaction at an uncovered Au microelectrode; curve 2 is for reaction after adsorption of 3-mercaptopropionic acid onto the gold. In the case of the electrocatalytic reaction, N_2H_4 oxidation is suppressed. The surface sites are not available for adsorption of the intermediate. The oxidation of ferrocenemethanol is not affected. Source: Idea and measured data from A. J. Bard; Inner sphere heterogeneous electrode reactions. Electrocatalysis and photocatalysis: The challenge. *J. Am. Chem. Soc.* **2010**, *132*, 7560–7567. Schematics, rescaling according to IUPAC convention, and formatted by the textbook author. Details: a) 12 mmol L^{-1} N_2H_4 in 50 mmol L^{-1} phosphate buffer; b) 1 mmol L^{-1} ferrocene methanol in 50 mmol L^{-1} $NaClO_4$; adsorption of 3-mercaptopropionic acid from a 50 μmol L^{-1} solution.

12.1.2 Activity and Selectivity

The **activity of a catalyst** refers to its ability to accelerate the reaction rate compared to the uncatalyzed case.

Q: In Chapter 2 and elsewhere in this textbook, we have used activity a to denote an effective or corrected concentration (e.g., $a_c = f_c c$ [mol L^{-1}]). This is not the same, is it?

A: No, you are absolutely right! "Activity" is another example of the occasional ambiguity of terms in electrochemistry. However, the concept of activity of a catalyst is as well-established a concept as that of effective concentration in the thermodynamics of mixtures.

Most reagents can be oxidized or reduced to various products. Consider, for example, the oxygen reduction reaction (ORR):

$$O_2 + e^- \rightarrow O_2^{-\bullet} \quad (12.3)$$

$$O_2 + 2\,e^- + 2\,H^+ \rightarrow H_2O_2 \quad (12.4)$$

$$O_2 + 4\,e^- + 4\,H^+ \rightarrow 2H_2O \quad (12.5)$$

A catalyst that accelerates all possible reaction paths equally is usually of little technical use. In a fuel cell, for example, a lot of H_2O_2 would then be generated, which would greatly accelerate the corrosion and degradation of cell components. What is desirable is a catalyst that accelerates the reaction (12.5), but *does not* enhance the rates of reactions (12.3) and (12.4).

By **selectivity** we understand the ability of an electrocatalyst to accelerate only *one* of several possible oxidation or reduction reactions of a reactant and thus to influence the composition of the resulting product mixture.

Figure 12.3 illustrates the effect of increased activity and selectivity. Figure 12.3b shows an increase in activity compared with Figure 12.3a without improvement in selectivity. We obtain the same product mixture in a shorter time, or can achieve the same conversion per unit time with less catalyst. For example, this would allow the noble metal content in a fuel cell to be lowered. However, the distribution of the reaction products remains the same. Thus, if a corrosive, toxic, or otherwise undesirable by-product should occur, its share in the conversion of a given mass of the starting material in Figure 12.3 will remain unchanged.

If, on the other hand, the selectivity can be increased, the composition of the product mixture is changed (Figure 12.3c). Now, the desired target product is predominantly formed. This is often more important (but more difficult to achieve) than increasing the activity. With improved selectivity, the conversion required for an application can still be improved with the use of a larger amount of catalyst.

During the development of new electrocatalysts, the bonding between intermediates and the catalysts is modified. Activity and selectivity are both affected. Both are also temperature-dependent. Thus, the selectivity can also be influenced by reaction control.

The determination of quantitative parameters for activity and selectivity will be discussed in Section 12.4.

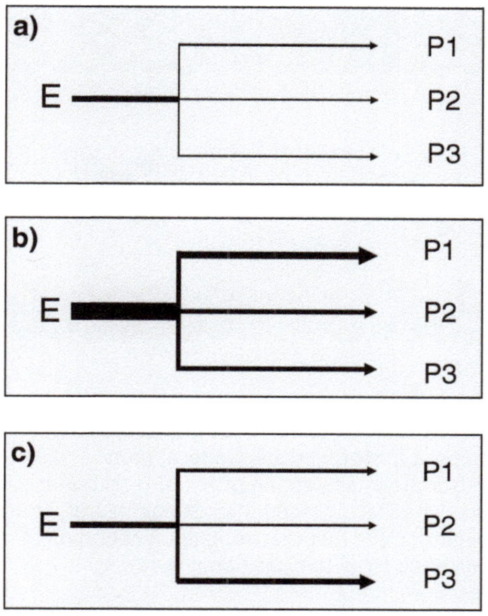

Figure 12.3 Illustration of activity and selectivity. The thickness of the reaction arrows corresponds to the rate constant for each reaction pathway: a) assumed uncatalyzed case; b) increase in activity without change in selectivity; c) increase in selectivity for product P3 with small increase in activity (E – reactant; P1, P2, P3 – different products).

12.1.3 Important Elementary Steps Involving Adsorbed Intermediates

We will discuss the concepts of heterogeneous electrocatalysis, primarily using the hydrogen electrode as an example. In this electrode reaction, hydrogen is either oxidized to give two protons or formed by the reduction of two protons. Depending on the reaction medium, the solvent water and autoprotolysis products (OH^-/H_3O^+) can be involved.

$$\text{Acidic solution: } 2\,H_3O^+ + 2e^- \rightleftharpoons H_2 + 2\,H_2O \tag{12.6}$$

$$\text{Alkaline solution: } 2\,H_2O + 2e^- \rightleftharpoons H_2 + 2\,OH^- \tag{12.7}$$

Frequently, reaction equations such as Eq. (12.6) are written in an abbreviated space-saving notation without explicit solvent and with H_3O^+ abbreviated as H^+. Equation (12.6) would then read as

$$2\,H^+ + 2e^- \rightleftharpoons H_2 \tag{12.8}$$

For the representation of reaction mechanisms on solid surfaces, we also need a symbol for a free surface site, where adsorption of a compound or intermediate may occur. We use an open square (□). However, note that this does not indicate the geometry of the adsorption site (compare with Table 8.1).

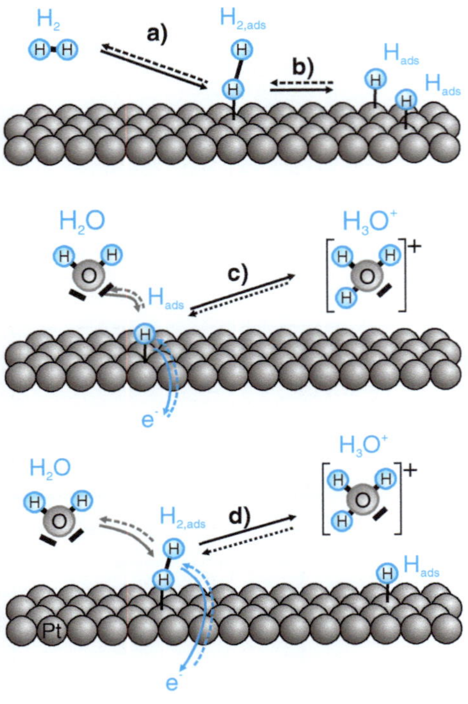

Figure 12.4 Illustration of the elementary steps at a hydrogen electrode. a) (non-dissociative) adsorption of H_2; b) H–H bond breaking (Tafel reaction); c) oxidation and hydration (Volmer reaction); d) oxidation of adsorbed H_2 with hydration (Heyrovský reaction).

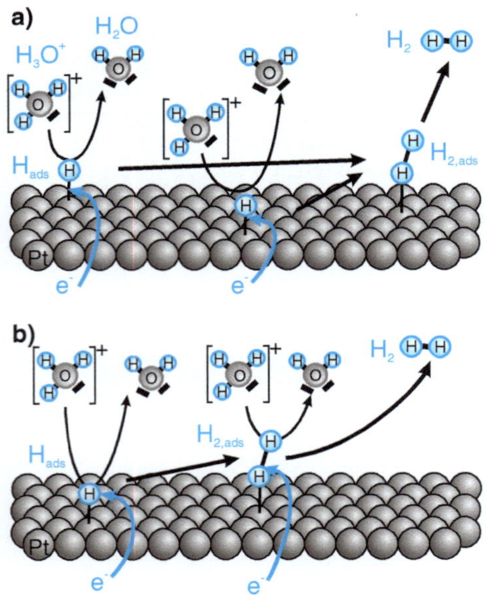

Figure 12.5 Schematic comparison of a) the Volmer–Tafel mechanism and b) the Volmer–Heyrovský mechanism of the hydrogen evolution reaction.

The hydrogen evolution reaction (HER) begins with the reduction of a proton and the binding of the product to a free surface site. The reaction is called the Volmer reaction[19].

$$H_3O^+ + \square + e^- \rightleftharpoons H_{ads} + H_2O \tag{12.9}$$

In an alkaline solution, the proton is supplied by water, and OH^- is released.

$$H_2O + \square + e^- \rightleftharpoons H_{ads} + OH^- \tag{12.10}$$

Two atomic hydrogen adsorbates can combine to form one molecule of $H_{2,ads}$. The reverse would be the splitting of a hydrogen molecule. This reaction is called the Tafel reaction[16].

$$2\,H_{ads} \rightleftharpoons H_{2,ads} + \square \tag{12.11}$$

Please note the difference between $H_{2,ads}$ and H_{ads}! For illustration, the reactions are shown graphically in Figure 12.4. It is also conceivable that the atomically adsorbed hydrogen formed according to Eq. (12.9) can react with another proton (from H_3O^+ in solution) and uptake another electron according to Eq. (12.12) (Heyrovský reaction[58]).

$$H_{ads} + H_3O^+ + e^- \rightarrow H_{2,ads} + H_2O \tag{12.12}$$

In an alkaline solution, the Heyrovský reaction results in

$$H_{ads} + H_2O + e^- \rightarrow H_{2,ads} + OH^- \tag{12.13}$$

The desorption of an intact molecule H_2 from a surface

$$H_{2,ads} \rightleftharpoons H_2 + \square \tag{12.14}$$

then results in the complete conversion to the H_2 molecule. Thus, to form a molecule of H_2 at an electrode, at least two reaction steps involving adsorbed intermediates are required. These mechanisms are referred to as the Volmer–Tafel mechanism [combination of (12.9), (12.11), and (12.14)] or as the Volmer–Heyrovský mechanism [combination of (12.9), (12.12), and (12.14)] and are illustrated in Figure 12.5.

Although the reaction steps shown here are specific to the hydrogen electrode, analogous reaction steps exist in other types of electrocatalytic reactions.

12.2 Theoretical Treatment of Electron Transfer Reactions

In the following section, we will continue the considerations from Chapter 3 to ultimately understand how the internal structures of reactants, products, and electrode surfaces in a solvent environment affect the rate of heterogeneous electron transfer (ET) reactions. However, the complexity of this problem requires the introduction of some very broad assumptions in order to facilitate computational solutions. Experimentally determined kinetic parameters can therefore usually only be compared to empirical estimates, but not (yet) compared to calculated values from first principles. Therefore, the benefit of the following considerations, which are the core of the Marcus theory[22] [8], is in the conceptual interpretation of *trends* in the kinetics of electrochemical reactions. Its importance lies in the formalization of the *decisive* influencing factors in electron transfer. Extensive efforts exist in the research community to refine this theory – first, to increase the accuracy in calculating reactivity data, and second, to incorporate the quantum mechanical treatment of electrons. This approach, referred to as the Marcus–Hush[72] theory, can be applied to ET processes at electrodes as well as to ET between redox-active compounds in homogeneous solution. We will attempt to sketch the crucial components of the theory for ET at electrodes. More in-depth treatments can be found in more specialized textbooks [11, 12].

12.2.1 Tunneling Processes of Electrons

Tunneling processes of electrons were touched upon in the discussion of scanning tunneling microscopy (STM) in Section 8.11.2. In essence, the tunneling process describes the change of location of an electron between two bound states through an energy barrier. The electron overcomes the barrier *without* exchanging energy. Accordingly, these processes *cannot* be accelerated by adding thermal energy. The tunneling process occurs with a probability that depends exponentially on the distance d over which the charge transfer occurs.

$$\text{Tunneling probability} \propto \exp(-\beta d) \tag{12.15}$$

The factor β is a decay constant and summarizes the dependence of the tunneling probability on the height of the energy barrier and the medium between the two locations. Example values for electron tunneling can be found in Table 12.1. For the treatment of ET theories, the tunneling probability is expressed as the transmission coefficient T_e.

$$T_e(d) = T_e^\circ \exp(-\beta d) \tag{12.16}$$

Table 12.1 Decay constant for the tunneling process.[a)]

System	β/Å^{-1}
Between two Pt electrodes in vacuum	2.4
Between metals through saturated hydrocarbon chains	1–1.2
Between metals through π-conjugated molecules.	0.4–0.6

Source of the data: [12].

a) In addition to measuring β, estimations can be made using the following equation:

$\beta = \dfrac{2\sqrt{2m_e\phi}}{\hbar}$ with

$m_e = 9.1 \times 10^{-28}$ g (mass of the electron)

$\phi = 5.7$ eV (work function of Pt),

$\hbar$ is the reduced Planck constant $= h/(2\pi) = 1.05 \times 10^{-34}$ J s $= 6.6 \times 10^{-16}$ eV s.

12.2.2 Electron Transfer Reactions Follow the Franck–Condon Principle – Marcus Theory

An ET reaction consists not only of a shift in the location of electrons between the reactants and the metal of the electrode, but also involves a change in the nuclear coordinates of the reactants and their surrounding solvent molecules that form the solvation shell. According to the Franck–Condon principle[24,25] (familiar from lectures on UV–Vis spectroscopy), we must assume that the change of electronic states (or more generally, the shift in electron density) is much faster than the movement of atomic nuclei in the reactant. In other words:

> The shift in electron density occurs at stationary nuclear coordinates.

This realization allows us to separately consider energies for nuclear motion and electronic motion. We will illustrate the consequences for ET reactions using the example of a seemingly very simple ET reaction with $n = 1$ of a transition metal complex of the metal Me^{z+} with six ligands L and a solvation shell (*aq*). Practical examples would be

$$[Ru(NH_3)_6]^{3+}(aq) + e^-_{(electrode)} \rightleftharpoons [Ru(NH_3)_6]^{2+}(aq) \tag{12.17}$$

$$[Fe(H_2O)]_6^{2+}(aq) \rightleftharpoons [Fe(H_2O)_6]^{3+}(aq) + e^-_{(electrode)} \tag{12.18}$$

This approach can also be used to describe reactions of molecules, such as the one-electron reduction of dioxygen to the superoxide radical anion in dry dimethyl sulfoxide (DMSO).

$$O_2(DMSO) + e^-_{(electrode)} \rightleftharpoons O_2^{\bullet-}(DMSO) \tag{12.19}$$

The energy of the reactants depends on the nuclear coordinates, which change periodically because of molecular vibrations, whereby potential energy and kinetic energy of the nuclear motion are interconverted. Because of the zero-point vibrational energy, this vibration also takes place in the energetic ground state. We can now describe the potential energy as a function of the nuclear coordinates. Here, it is useful to use a set $\mathbf{q}$ of m internal coordinates $(q_1, \ldots q_i, \ldots q_m)$ instead of Cartesian coordinates (x, y, z) for each atomic nucleus of the reactant and the solvation shell. These internal coordinates are, for example, bond lengths or bond angles. For the oxidized and the reduced forms of the reacting redox couple, there is one set of inner coordinates each, that is, $\mathbf{q}_R = \{q_{R,1}, q_{R,2}, ..., q_{R,m}\}$ and $\mathbf{q}_O = \{q_{O,1}, q_{O,2}, ..., q_{O,m}\}$, at which the potential energy has a minimum and thus corresponds to the equilibrium conformation. For the rest of our considerations, we make the highly simplifying assumption that we can describe the change in the potential energy of the molecular system in terms of just *one* of the internal coordinates (q). This could be, for example, the distance of a ligand from the central atom in examples (12.17) and (12.18) or the O–O bond length in example (12.19).

Note that the equilibrium distance $q°$ of this coordinate changes during an electron transfer reaction, that is, it takes the value q_R for the reduced form and the value q_O for the oxidized form.

This change occurs because a more positively charged central ion exerts a stronger attraction to the ligands and the dipoles of the solvent, or during the reduction of O_2 an electron enters an antibonding molecular orbital, thereby reducing the bond order and increasing the bond length.

We now seek an expression for the Gibbs free energy of activation $\Delta G^{\neq}$ as a function of the change in nuclear coordinates for an ET reaction. One can express the contribution of the potential energy to $\Delta G^{\neq}$ during a vibration around the equilibrium position using the harmonic oscillator model with sufficient accuracy for many molecular systems (Section 8.2). We use the force constant k_F and the displacement $\Delta q = q - q_R$ or $\Delta q = q - q_O$ (as in the treatment of infrared spectroscopy of molecules).

$$G_{\text{pot,vib}} = \frac{1}{2}k_F \cdot (\Delta q)^2 \tag{12.20}$$

When drawing the functions

$$G_R(q) = G_{\text{pot,vib,R}}(q) + \Delta G° = \frac{k_F}{2}(q - q_R)^2 + \Delta_R G° \tag{12.21}$$

and

$$G_O(q) = G_{\text{pot,vib,O}}(q) = \frac{k_F}{2}(q - q_O)^2 \tag{12.22}$$

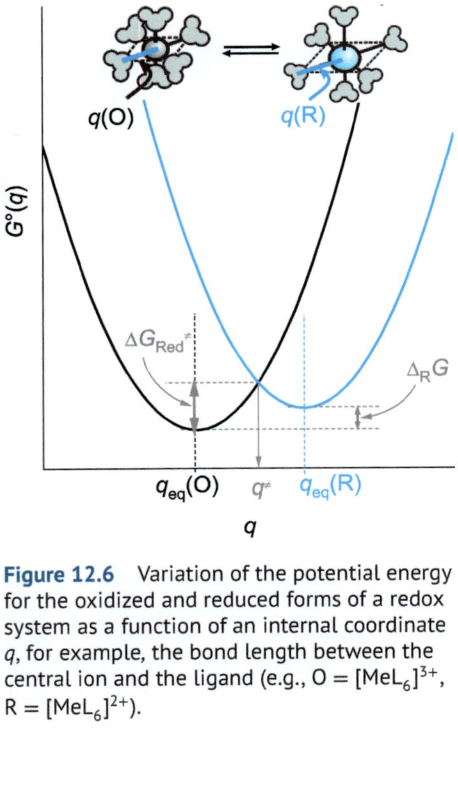

Figure 12.6 Variation of the potential energy for the oxidized and reduced forms of a redox system as a function of an internal coordinate q, for example, the bond length between the central ion and the ligand (e.g., O = $[MeL_6]^{3+}$, R = $[MeL_6]^{2+}$).

for the oxidized and reduced form, one obtains the two parabolas in Figure 12.6. In the case of homogeneous reactions, the energetic difference between the minima at q_R and q_O of the two curves is $\Delta_R G°$. For electrode reactions, $\Delta_R G°$ must have contributions from the Gibbs free energy for the reacting electron at a given electrode potential E relative to the standard electrode potential $E°$ [$\Delta_R G° = -F(E - E°)$]. The applied potential can move the position of the parabolas vertically relative to each other.

To identify the transition state, we need to find the inner coordinate $q^{\neq}$, at which G_{pot} *does not* differ for both oxidation states. At this value of the coordinate $q = q^{\neq}$ an electron transfer may occur without change in the total Gibbs free energy of the molecular system, that is, without exchanging energy between the reactants and the environment. Of course, such energy exchange processes occur naturally, but they are upstream or downstream of electron transfer because they are much slower than an elementary ET process. In Figure 12.6, we can see the transition state as the intersection of the blue and black parabolas. To determine $q^{\neq}$, we equate Eqs. (12.21) and (12.22) and solve for $q = q^{\neq}$. [For simplicity, we assume that the force constant k_F does not change during the reaction.]

$$\frac{k_F}{2}(q^{\neq} - q_R)^2 + \Delta G° = \frac{k_F}{2}(q^{\neq} - q_O)^2$$

$$q^{\neq} = \frac{q_R + q_O}{2} + \frac{\Delta_R G°}{k_F(q_R - q_O)} \tag{12.23}$$

The Gibbs free energy of activation for the reduction is then

$$\Delta G^{\neq}_{\mathrm{Red}} = G_{\mathrm{O}}(q^{\neq}) - G_{\mathrm{O}}(q_{\mathrm{O}}) = \frac{k_{\mathrm{F}}}{2}(q^{\neq} - q_{\mathrm{O}})^2 \tag{12.24}$$

Inserting Eq. (12.23) into Eq. (12.24) results in

$$\begin{aligned}\Delta G^{\neq}_{\mathrm{Red}} &= \frac{k_{\mathrm{F}}}{2}\left[\frac{q_{\mathrm{R}} + q_{\mathrm{O}}}{2} + \frac{\Delta_{\mathrm{R}}G^{\circ}}{k_{\mathrm{F}}(q_{\mathrm{R}} - q_{\mathrm{O}})} - \frac{2q_{\mathrm{O}}}{2}\right]^2 \\ &= \frac{k_{\mathrm{F}}(q_{\mathrm{R}} - q_{\mathrm{O}})^2}{8}\left[1 + \frac{4\Delta_{\mathrm{R}}G^{\circ}}{k_{\mathrm{F}}(q_{\mathrm{R}} - q_{\mathrm{O}})^2}\right]^2\end{aligned} \tag{12.25}$$

Furthermore, we substitute

$$\lambda = \frac{k_{\mathrm{F}}}{2}(\Delta q)^2 = \frac{k_{\mathrm{F}}}{2}(q_{\mathrm{R}} - q_{\mathrm{O}})^2 \tag{12.26}$$

The parameter λ is called **reorganization energy** and indicates how much the potential energy would change if the oxidized form is expanded *without ET* to the equilibrium structure of the reduced form. With the substitution in Eq. (12.26), we obtain for Eq. (12.25)

$$\Delta G^{\neq}_{\mathrm{Red}} = \frac{\lambda}{4}\left(1 + \frac{\Delta_{\mathrm{R}}G^{\circ}}{\lambda}\right)^2 = \frac{(\lambda + \Delta_{\mathrm{R}}G^{\circ})^2}{4\lambda} \tag{12.27}$$

leading to Eq. (12.28) for the cathodic reaction at the electrode with $\Delta_{\mathrm{R}}G^{\circ} = e(E - E_{\mathrm{eq}}) = e\eta$.

$$\Delta G^{\neq}_{\mathrm{Red}} = \Delta G^{\neq}_{\mathrm{c}} = \frac{\lambda}{4}\left(1 + \frac{e\eta}{\lambda}\right)^2 = \frac{(\lambda + e\eta)^2}{4\lambda} \tag{12.28}$$

Q: Why do we write $e(E - E_{\mathrm{eq}})$ now? In Chapters 2 and 3 we have always used $F(E - E_{\mathrm{eq}})$?

A: This is just a simple unit conversion. In the case of $F(E - E_{\mathrm{eq}})$, we get the energy in J mol^{-1}, that is, for one mole of formula conversions. In the case of $e(E - E_{\mathrm{eq}})$, we get the energy in eV for a single formula turnover. This turns out to be more convenient here because we obtain some energy quantities directly from quantum chemical calculations in eV. Moreover, band structure diagrams, which we already know from Chapter 10 and which we will need in this chapter, are plotted in units of eV.

In addition, there are other contributions from G_{w} for the oxidized and reduced forms, often referred to as work terms. They may include contributions for the electrical energy of a charged reactant near a charged electrode or contributions for the binding energies within the ligands. The Gibbs free energy is then

$$G_{\mathrm{pot}} = \frac{1}{2}k_{\mathrm{F}} \cdot (\Delta q)^2 + G_{\mathrm{w}} \tag{12.29}$$

and the Gibbs free energy of activation for the reduction at an electrode is

$$\Delta G^{\neq}_{\mathrm{c}} = \frac{\lambda}{4}\left(1 - \frac{e\eta - G_{\mathrm{w,O}} + G_{\mathrm{w,R}}}{\lambda}\right)^2. \tag{12.30}$$

Q: So far we have considered only *one* coordinate. But a typical complex like $[Ru(NH_3)_6]^{3+}$ already has a lot of internal coordinates, and then there are also coordinates for the surrounding solvent molecules. How should we deal with this?

A: First, the formal description for *one* essential internal coordinate is helpful in understanding, at least qualitatively, the behavior of redox systems (Section 12.2.3). In order to include other influencing variables, the following procedure is useful:

First, the reorganization energy can be divided into an inner and outer part.

$$\lambda = \lambda_i + \lambda_o \tag{12.31}$$

However, the terms "inner" and "outer" do not refer to the concept of redox reactions in the inner and outer coordination spheres. To determine the inner reorganization energy, we sum the contributions of all m normal vibrations of the reactant. If the normal vibrations of the reacting molecule are harmonic (and therefore independent of each other), the energy contributions of all vibrations are additive. Each of them has its own force constant $k_{F,j}$ and a normal coordinate q_j, thus

$$\lambda_i = \sum_{j=1}^{m} \frac{k_{F,j}}{2}(q_{R,j} - q_{O,j})^2. \tag{12.32}$$

For the consideration of the external reorganization energy, that is, the contribution of the solvent environment, we treat the solvent as a continuum characterized by a static and optical dielectric constant ($\varepsilon_0\varepsilon_s$, $\varepsilon_0\varepsilon_{op}$). In addition, the distance R between a charge of the reactant near the electrode and its image charge in the electrode is important. This value is somewhere between the distance of the reactants to the electrode (in case of complete shielding of the charge by counterions) and the double of this distance in case of absence of shielding. In addition, the radius r_O of the reactant occurs (assumed to be a sphere). Under these assumptions, we obtain for the external reorganization energy

$$\lambda_{o,el} = \frac{e^2}{8\pi\varepsilon_0}\left(\frac{1}{r_O} - \frac{1}{R}\right)\left(\frac{1}{\varepsilon_{op}} - \frac{1}{\varepsilon_s}\right). \tag{12.33}$$

In a homogeneous reaction O1 + R2 $\rightleftharpoons$ R1 + O2 between two redox systems O1 and R2, the following equation holds:

$$\lambda_{o,ex} = \frac{e^2}{4\pi\varepsilon_0}\left(\frac{1}{2r_{O1}} + \frac{1}{2r_{R2}} - \frac{1}{d}\right)\left(\frac{1}{\varepsilon_{op}} - \frac{1}{\varepsilon_s}\right) \tag{12.34}$$

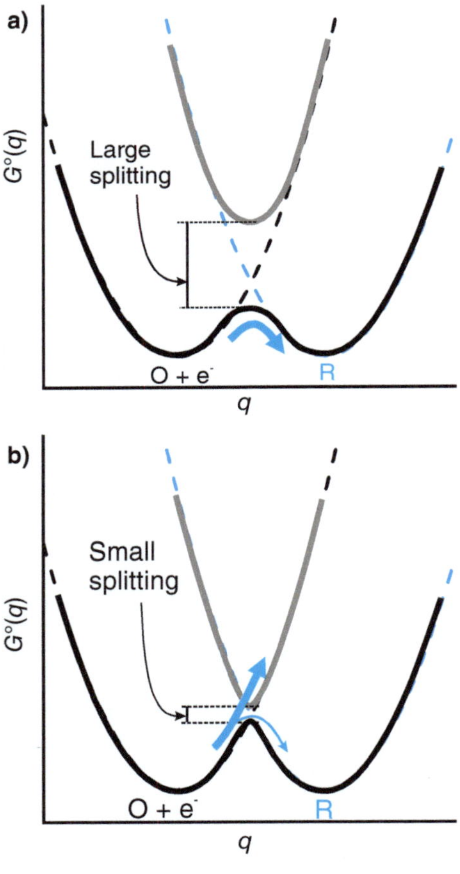

Figure 12.7 Qualitative depiction of the potential energy with a) strong electronic coupling between electrode and reactant (adiabatic reaction) and b) with weak coupling between reactant and electrode (nonadiabatic reaction); $E = E^{\circ\prime}$.

12.2.3 From Reorganization Energy to Rate Constants

During an electrode reaction, the complex O constantly changes its internal coordinates by oscillations. If the inner coordinates of the transition state are reached, that is, those coordinates where the potential energies of O and R are equal – tunneling of the electron to the electrode can occur with a certain probability. The rate constant is

$$k_{Red} = A \exp\left(\frac{-\Delta G^{\neq}_{Red}}{k_B T}\right) = K_{P,O} \nu_n T_e \exp\left(\frac{-\Delta G^{\neq}_{Red}}{k_B T}\right). \tag{12.35}$$

The preexponential factor from Section 3 can now be interpreted more clearly: $K_{P,O}$ is an *equilibrium* constant indicating the concentration of molecules at a distance from the electrode suitable for the reaction relative to the volume concentration (precursor equilibrium constant). For reactions at electrodes, $K_{P,O} = \Gamma_O / c^*_O$ [cm]. ν_n is the nuclear frequency factor [s^{-1}], which indicates how many attempts per second are made to overcome the energy barrier. This quantity can often be related to vibrational frequencies and solvent motions. The dimensionless factor T_e is then obtained as the transmission factor according to Section 12.2.1.

When the distance d of the molecule from the electrode has an optimum value for electron transfer, T_e approaches one. The reaction in this case is called adiabatic electron transfer. More generally, in ET theory, the extent of electronic coupling between reactant and electrode is called adiabaticity. Thus, the highly simplified Figure 12.6 has to be refined, because in the case of strong electronic coupling with the electrode, the parabolas of Figure 12.6 split into two separate curves. Along the lower curve, a smooth transition from O to R can be made. We can imagine the upper curve as an electronically excited state separated from the lower curve by an "energy gap".

The effects on an ET reaction depend on the "splitting" between the two curves. If the splitting is larger than the thermal energy ($>k_B T$), then a reactant molecule whose internal coordinates are moving toward the transition state crosses the barrier smoothly ($T_e = 1$, Figure 12.7a, adiabatic reaction). On the other hand, if the splitting is smaller than $k_B T$, the reactant molecule in the transition state can perform an ET with a certain probability $T_e < 1$ or move along the excited state potential curve without ET (Figure 12.7b, nonadiabatic electron transfer reaction.)

12.2.4 What Are the Predictions of Marcus Theory?

Symmetry Factor. The value of Marcus theory lies above all in the possibility to explain and predict trends for experimentally observable quantities. For instance, the symmetry factor α for a one-electron reaction is obtained from Marcus theory as

$$\alpha = \frac{1}{e}\frac{\partial \Delta G_c^{\neq}}{\partial E} = \frac{1}{2} + \frac{e\eta}{2\lambda} \tag{12.36}$$

e – Elementary charge
$\Delta G^{\neq}$ – Gibbs free energy of activation per molecule [eV], calculations in J mol^{-1} are possible, then e is to be replaced by the Faraday constant F.
E – Electrode potential [V]
η – Overpotential ($E - E_{eq}$) [V]

with the prediction that for small overpotentials $\alpha \approx 0.5$, but for larger overpotentials – unlike the assumption in the derivation of the Butler–Volmer equation in Chapter 3, α is no longer a constant. Further refinements are possible if working terms are included [12].

Comparison of Homogeneous and Heterogeneous ET Reactions. Homogeneous exchange reactions (self-exchange), for which Marcus first developed his theory, lend themselves to comparison with charge transfer reactions at electrodes. Let us consider an ET between Fe complexes $[FeL_6]^{2+/3+}$ (L = ligand). To follow the reaction, one of the redox forms is labeled radioactively or followed by magnetic resonance techniques (ESR, NMR) (marked with *).

$$[FeL_6]^{2+} + [Fe^*L_6]^{3+} \xrightarrow{k_{ex}} [FeL_6]^{3+} + [Fe^*L_6]^{2+} \tag{12.37}$$

The reorganization energies for the electrode reaction and the homogeneous exchange reaction differ only in the external reorganization energies. A comparison of Eq. (12.33) and Eq. (12.34) yields Eq. (12.38) with the reasonable simplifications $r_O = r_{O1} = r_{R2} = r$ and $R = d = 2\,r$.

$$\lambda_{el} = \frac{\lambda_{ex}}{2} \tag{12.38}$$

For an exchange reaction, $\Delta_R G^\circ = 0$ holds which simplifies Eq. (12.27) to $\Delta G^{\neq}_{Red} = \lambda_{ex}/4$. For ET at an electrode at the standard potential $E = E^\circ$, the heterogeneous rate constant is equal to the standard heterogeneous reaction rate constant $k_{el} = k^\circ$, and $\Delta G_c^{\neq} = \lambda_{el}/4$ applies according to Eq. (12.28). This yields

$$\sqrt{\frac{k_{ex}}{A_{ex}}} = \frac{k^\circ}{A_{el}}. \tag{12.39}$$

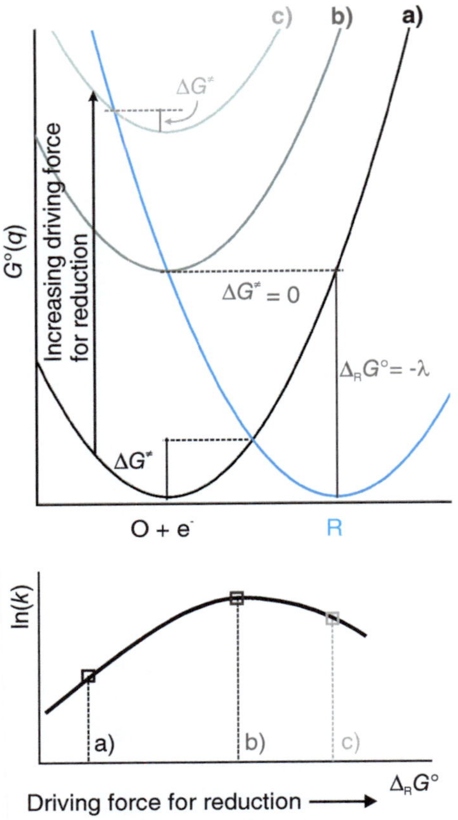

Figure 12.8 Prediction of activation barrier for electron transfer reactions as a function of the change in reaction Gibbs free energy $\Delta G°$ as the driving force of the reaction.

a) normal region, activation barrier decreases with increasing driving force;
b) minimum barrier and maximum rate constant at $\Delta_R G° = -\lambda$; and
c) inverted region, where barrier increases and rate constant decreases with further increase in driving force.

Structure of the Reactants. Moreover, from the consideration that the relations $\Delta G^{\neq}_{\text{Red}} = \lambda_{\text{ex}}/4$ and $k° = k_a = k_c$ hold for $E = E°$, it follows directly that large values of $k°$ are to be expected when the structure of O and R are very similar, whereas redox reactions with significant changes in bond lengths and bond angles tend to proceed more slowly. The molecular size r_O affects the external reorganization energy [Eq. (12.33)] to the extent that larger molecules show lower overall differences in solvation energies than smaller molecules and therefore tend to have higher $k°$.

Influence of the Solvent. The influence of the solvent is noticeable in the contribution to the external reorganization energy λ_o and additionally in the preexponential factor. Here, the longitudinal relaxation time τ_L enters, which is approximately proportional to the viscosity η. The nuclear frequency factor ν_n in Eq. (12.35) is inversely proportional to τ_L andη. This can be seen experimentally, among other things, in the decrease in $k°$ values in aqueous solutions when sucrose is added to increase the viscosity without any significant effect on the solvent environment of charged reactants and the associated λ_o.

Inverted Region. Normally, we expect an ET reaction to proceed faster as the driving force $-\Delta_R G°$ increases, because the activation barrier to be overcome decreases at the same time (Figure 12.8a). However, this does not hold if the intersection $q^{\neq}$ of the parabolas $G_O(q)$ and $G_R(q)$ is not between q_O and q_R, but at $q^{\neq} < q_O, q_R$ (Figure 12.8c) or at $q^{\neq} > q_O, q_R$. The barrier is smallest, and thus the reaction rate constant k is largest, when $\Delta_R G° = -\lambda$ (Figure 12.8b). Experimentally, this prediction has been confirmed for photochemically triggered ET reactions [8]. The existence of the inverted region is plausible because at a strongly negative $\Delta_R G°$ a lot of released energy has to be distributed among the vibrational modes of the reactants, for which the probability decreases when $|-\Delta_R G°| > \lambda$. Moreover, in these cases, the reaction often no longer takes an adiabatic course. Reactions with strongly negative $\Delta_R G°$ are important for the field of electrogenerated chemiluminescence, one of the most sensitive electrochemical detection techniques [9, 10].

At metal electrodes, there is no prospect of being able to observe the inverted region. The previous considerations were based on the assumption that the ET between the molecule and the metal always occurs with electrons in levels close to the Fermi level E_F of the metal electrode. If we attempt to set parameters so that ET would occur in the inverted region and involve energy levels close to E_F in the metal electrode, there would exist occupied/unoccupied states below/above E_F in the metal that could donate/accept an electron with vanishing barrier height. Dissipation of energy would then occur in the metal instead of dissipation in the reactant (compare with Figure 8.91).

12.2.5 What Role Does the Electrode Material Play in the Speed of the Electrode Reaction?

When treating the kinetics of ET reactions according to Marcus theory, the material of the electrode is not considered, since the kinetics depends only on the reorganization energy of the dissolved reactant. One approach to also include the electronic properties of the electrode material is to consider the overlap of the electronic states of the electrode and the boundary orbitals of the dissolved reactants as an essential parameter for the kinetics of the reaction. This is based on the insight, as it was a starting point for the Marcus theory – that the ET occurs so rapidly that the molecular system cannot move at the moment of transfer and cannot absorb or release energy during the ET. Therefore, reduction always occurs by the transfer of an electron from an occupied state of the electrode to an unoccupied orbital of the reactant of the *same* energy, and oxidation occurs by the transfer of an electron from an occupied orbital of the reactant to an unoccupied state of the electrode material of the *same* energy. Decisive for the reaction rate are then both the number of states of equal energy available in the electrode and the energy states (orbitals) of the reactant involved in the ET (Excursion 10-1).

The density of states $D_S(E_e)$ is the number of energetically equivalent electronic states per volume [eV^{-1} cm^{-3}] or the number of states per surface element of the electrode [eV^{-1} cm^{-2}].

In Chapter 10, we used $D_S(E_e)$ when discussing semiconductor electrodes for which this theoretical approach was introduced, in particular, by the work of Gerischer.[23] The importance of this theoretical approach is obvious for semiconductor electrochemistry because in a semiconductor electrode, either only electrons or only holes are involved in the redox reactions, depending on the doping level and the applied electrode potential. Near the Fermi level E_F, there is a band gap where only extremely few energy states exist and are available for ET reactions. We will limit ourselves to qualitative consideration in this book, and refer to other textbooks [11, 12] for a derivation of equations describing the effects quantitatively.

12.3 The Role of Intermediates for Electrocatalysis

In metals, occupied and unoccupied states are always present in large density near the Fermi level, so that this factor is usually not limiting for the course of ET reactions. Nevertheless, a very strong dependence on the electrode material is observed for many reactions, such as the hydrogen evolution reaction (HER) (Figure 3.9). In contrast, other reactions, such as the oxidation of ferrocene, show an approximately equal heterogeneous standard reaction rate constant $k°$ for different electrode materials (Figure 12.2b). The key here is the presence of adsorbed intermediates in the reaction. Therefore, we will organize the following section according to the number of adsorbed intermediates. The explanations follow approaches in [6, 7].

12.3.1 One-Electron Transfer Without Intermediate

We consider the relationship between the Gibbs free energy of activation $\Delta G^{\neq}$ and the reorganization energy λ according to Eq. (12.28) and insert $\lambda = \lambda_i + \lambda_o$ according to Eq. (12.31).

$$\Delta G_c^{\neq} = \frac{(\lambda_i + \lambda_o + e\eta)^2}{4(\lambda_i + \lambda_o)} \tag{12.40}$$

$\Delta G^{\neq}$ – Gibbs free energy of activation per molecule [eV]
e – Elementary charge
η – Thermodynamic driving force, here specifically the overpotential ($E - E_{eq}$) [V]
λ_i – Internal reorganization energy [eV]
λ_o – External reorganization energy [eV]

As long as there are no strong interactions with the substrate (i.e., no intermediate is formed) and a sufficient number of electronic states are present in the electrode material in the energy range around E_F, λ_i is the same for these reactions on different electrode materials. This is also true for λ_o. Thus, the type of electrode material has no significant influence on the reaction rate (reactions in the outer sphere).

12.3.2 Two-Electron Transfer with One Intermediate

When two electrons are transferred to a molecule, the electrons can be transferred either concertedly (at the same time) or sequentially (one after the other). According to Marcus theory, in the first case, the activation energy is four times higher than in the case of two sequential ETs. Thus, it is energetically advantageous to store the charge in an intermediate state (called intermediate).

For reactions with an intermediate, a “three-level problem” occurs according to the nomenclature of ET theory, for example, O/intermediate/R. If the intermediate is an adsorbed intermediate, as in HER, the energy level of this intermediate has a crucial importance for the activation energy and thus for the rate of the overall reaction.

Let us now consider as an example the Volmer–Heyrovský mechanism[19,58] of HER in detail [combination of Eqs. (12.9), (12.12), and (12.14)], or in a simplified version

$$H^+ + \square + e^- \rightleftharpoons H_{ads} \text{ with } E^\circ\left(H/H^+_{ads}\right),\ n_1 = 1 \tag{12.41}$$

$$H_{ads} + H^+ + e^- \rightleftharpoons H_2 + \square \text{ with } E^\circ(H_{ads}, H^+/H_2),\ n_2 = 1 \tag{12.42}$$

From Figure 12.9, the relationship between the standard potentials of the two elementary reaction steps and the standard potential of the overall reaction 2 H^+ + 2 e^- $\rightleftharpoons$ H_2 ($n = 2$) becomes clear. Since by definition $E^\circ(H^+/H_2) = 0$, the following relation must hold:

$$E^\circ(H^+|H_2) = \frac{E^\circ(H^+/H_{ads}) + E^\circ(H_{ads}, H^+/H_2)}{2} = 0 \tag{12.43}$$

Excursion 12-1 The Quantum Chemical Hydrogen Electrode

The equilibrium potential of a redox reaction involving an adsorbate can be calculated if a suitable standard state is agreed upon. According to Nørskov et al. [13], this is possible in the following way: For the reaction (12.41), equilibrium is reached if the following holds true:

$$\Delta G(H_{ads}) = \Delta G(H^+ + e^-) \tag{12.44}$$

For this purpose, it is agreed that the right side of the equation becomes zero when $E = 0$ V vs. SHE. For any other potential, $\Delta G(H^+ + e^-)$ is then eE because only the Gibbs free energy of the electron in the metal is affected by the potential change, not the Gibbs free energy of the (hydrated) proton (computational hydrogen electrode). In fact, quantum chemical calculations using density functional theory (DFT) show that binding energies of H_{ads} do not depend significantly on the applied potential nor on the electric field. However, this is only true if the dipole moment of the bond between H_{ads} and the electrode surface is approximately zero and the electrosorption valence (Section 8.9.8) of the reaction (12.41) is close to one.

We obtain the standard electrode potential for the adsorption reaction from ΔG_{ads} at the standard state

$$E^\circ(H^+/H_{ads}) = \frac{-\Delta G^\circ(H_{ads})}{e} \tag{12.45}$$

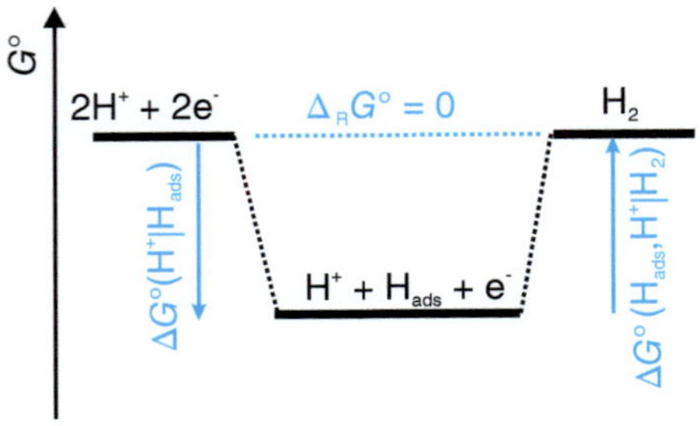

Figure 12.9 Derivation of Eq. (12.43) from the consideration of the path independence of $\Delta_R G^\circ$ for the Volmer–Heyrovský mechanism (G is a state variable) at $E = 0$ V vs. SHE.

$\Delta G^\circ(H_{ads})$ is the Gibbs free energy of adsorption of H_{ads} relative to H_2 (by definition = 0) at a sensible standard state, such as at a coverage $\theta = 0.25$ on the (1 1 1) surface of the metal under consideration. Thus, $E_1{}^\circ(H^+/H_{ads}) > 0$, if adsorption is energetically favorable [$\Delta G(H_{ads}) < 0$]. We have learned about situations in which an adsorbate is (reductively) bound to the surface at a more positive potential than the standard potential, namely underpotential deposition (UPD) in Section 8.9.8. The Eq. (12.43) is zero for the total reaction only when Eq. (12.46) is valid for the Heyrovský reaction as the second step in the Volmer–Heyrovský mechanism.

$$E^\circ(H^+/H_{ads}) = -E^\circ(H_{ads}, H^+/H_2) \tag{12.46}$$

Thus, if hydrogen adsorption is thermodynamically favored, H_2 formation in the Heyrovský step *must be* thermodynamically unfavorable (Figure 12.9).

According to Norskov et al. [13], the minimum "thermodynamic overpotential" η_T of the overall reaction can be regarded as the potential difference between the thermodynamically most unfavorable partial reaction and the standard electrode potential of the overall reaction. If the formation of H_{ads} is thermodynamically favorable, the thermodynamic overpotential is $\eta_T = E°(H_{ads},H^+/H_2) - E°(H^+/H_2)$ for the considered reaction. The least favorable reaction step is also called the potential-determining step.

Q: Is the potential-determining step the same as the *rate-determining* step (RDS) that we already talked about in Chapter 3?

A: No, conceptually it's something different. However, it's not so rare that the potential-determining step is also the RDS. This is because the thermodynamic driving force $\Delta_R G$ of a reaction and its Gibbs free energy of activation $\Delta G^{\neq}$ often – but not always – correlate. Often, a linear relationship can be observed for the same reaction of structurally related compounds. The relationship between kinetic and thermodynamic quantities assumed for such cases is called the linear free energy relationship (LFER), also known as the Bell–Evans–Polyani postulate[138] or Brønsted–Evans–Polyani[138] postulate. This postulate also underlies our considerations. The reaction studied here is the HER, the variation being in the type of electrode metal chosen. We can determine potential-determining steps for any electrode potential. However, the potential-determining step at any potential need not be the RDS at the potential at *which the reaction actually occurs.*

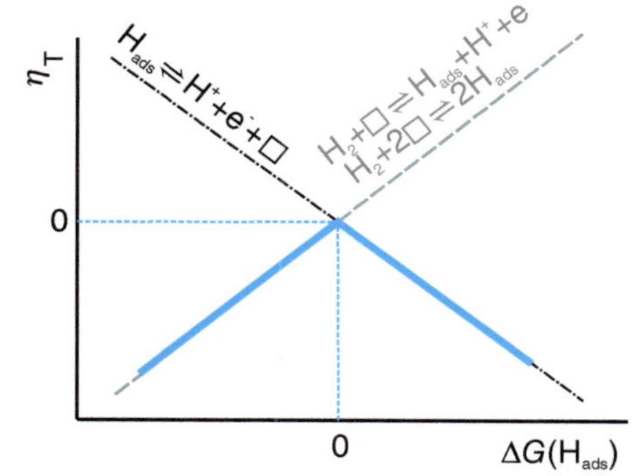

Figure 12.10 Schematic volcano plot for the HER. $E°(H_{ads},H^+/H_2)$ and $E°(H^+/H_{ads})$ are accessible from calculations and plotted as a function of $\Delta G°(H_{ads})$ (dashed lines). The resulting thermodynamic overpotential η_T is plotted as a thick blue line and corresponds to the *more negative* of the two potentials.

We can now ask what properties the ideal electrocatalyst must possess for the HER according to the Volmer–Heyrovský mechanism: It is the one with the smallest value for η_T. The smallest possible value for η_T is zero, which can be obtained by considering Eq. (12.47).

$$\eta_T = E°(H_{ads}, H^+/H_2) = E°(H^+/H_{ads}) = -\Delta G°(H_{ads})/e = 0 \tag{12.47}$$

Any electrocatalyst with $\Delta G°(H_{ads}) \neq 0$ thus has a $\eta_T > 0$, since either the Volmer reaction or the Heyrovský reaction is thermodynamically unfavorable. When both tendencies are balanced, η_T reaches its optimal value – an example of the classical Sabatier principle. This example can also be used to trace the origin of the volcano plots. In Eq. (12.47), we see that essential kinetic properties depend on only *one* quantity, here $\Delta G°(H_{ads})$. Such quantities are called descriptors. Figure 12.10 shows the dependences $E°(H^+/H_{ads}) = -\Delta G°(H_{ads})/e$ and $E°(H_{ads},H^+/H_2) = -E°(H^+/H_{ads}) = +\Delta G°(H_{ads})/e$. The thermodynamic overpotential η_T, the smaller of these two potentials, is highlighted as a thick blue line.

The same analysis can be performed for the hydrogen oxidation reaction. In this case, identical relationships are obtained for $E°(H^+/H_{ads})$ and $E°(H_{ads},H^+/H_2)$. η_T is then the larger of the two potentials, and the volcano plot has the appearance of a "funnel" instead (Figure 12.11).

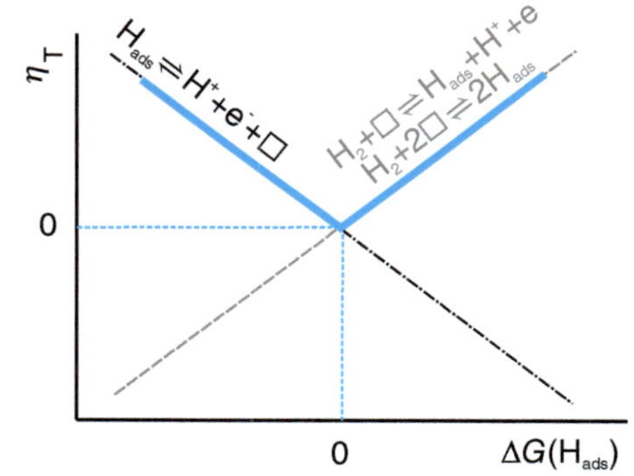

Figure 12.11 Schematic "volcano" plot for the HOR, which has the shape of a "funnel". $E°(H_{ads},H^+/H_2)$ and $E°(H^+/H_{ads})$ are accessible from quantum chemical calculations and plotted as a function of $\Delta G°(H_{ads})$ (dashed lines). The resulting thermodynamic overpotential η_T is plotted as a thick blue line and corresponds to the *more positive* of the two potentials.

---

[138] Johannes Nicolaus Brønsted, 1879–1947, Danish physical chemist, professor in Copenhagen, also known for the acid–base theory named after him; Michel Polanyi, 1891–1976, grew up in Hungary, initially studied medicine and worked as a medical officer in World War I. In the 1920s, he became a professor of physical chemistry in Berlin. In 1933, he emigrated to the United Kingdom, where he became a professor of physical chemistry in Manchester. He was the first to use contour diagrams to represent transition states (as in Figure 3.4). The explanation of the transfer coefficient in Section 3.2.4 is essentially due to him, as is the interpretation of the plastic deformability of metals by dislocations. After World War II, he became a professor of social sciences. His son John Polanyi *1929 also worked in the field of reaction dynamics and was awarded the Nobel Prize in Chemistry in 1986; M. G. Evans, coworker of Polanyi.

The Volmer–Tafel mechanism can be subjected to an analysis in the same way that yields an identical volcano plot, because this analysis does not distinguish between the details of the two mechanisms with sufficient precision [6].

The key insight from the analysis is that on each side of the volcano plot, a different reaction step of the overall reaction is the potential-determining step.

In the experimentally determined volcano plot in Figure 3.9, the exchange current density was plotted on the ordinate (instead of η_T in Figures 12.10 and 12.11). Because of various shortcomings, such as the inclusion of oxide-covered valve metals, caution must always be exercised when interpreting volcano plots (Section 3.2.3). Experimental studies that closely resemble the analysis shown here use well-defined single crystals on which, for example, a monolayer of palladium has been deposited. Because of the different spacing of the adsorption sites (determined by the metal of the support) the spacing of the Pd atoms in the monolayer is also affected, the band structure of the electrode can be varied without changing the metal. The procedure allows fine tuning of $\Delta G°(H_{ads})$ while maintaining otherwise equal binding partners for the intermediates. The highest reaction rate is actually found for the system where $-\Delta G°(H_{ads})$ is close to zero [14].

12.3.3 Transfer of More Than Two Electrons with the Participation of More Than One Adsorbed Intermediate

We want to apply the approach in Section 12.3.2 to the analysis of the oxygen reduction reaction (ORR)

$$O_2 + 4\,H^+ + 4\,e^- \rightleftharpoons 2\,H_2O, \quad E°(O_2/H_2O) = +1.23\ V \tag{12.48}$$

and the oxygen evolution reaction (OER). Both reactions have significant technological importance.

Thermodynamics of the Overall Reaction. For the following considerations, it is convenient to define some abbreviations. We consider liquid water as our reference state and define all other standard molar Gibbs free energies and potentials relative to it:

$$eE°(O_2/H_2O) = \frac{\Delta G°(O_2) - 2\Delta G°(H_2O)}{4} = C_O = 1.23\ eV \tag{12.49}$$

The constant C_O corresponds to a thermodynamically easily accessible constant with which the following ideas can be compactly formulated. A rearrangement of Eq. (12.49) gives the standard molar Gibbs free energy of O_2 (relative to H_2O).

$$\Delta G°(O_2) = 4C_O = 4.92\ eV \tag{12.50}$$

Intermediates. In this reaction, possible intermediates are OOH_{ads}, OH_{ads}, and $O_{2,ads}$. First, oxygen is bound with the uptake of one electron as a superoxide-like intermediate

$$O_2 + H^+ + e^- + \square \rightleftharpoons OOH_{ads} \tag{12.51}$$

For this purpose, we can specify the standard equilibrium potential:

$$eE°(O_2/OOH_{ads}) = \Delta G°(O_2) - \Delta G°(OOH_{ads}). \tag{12.52}$$

The overpotential of this step relative to the total reaction is obtained using the substitution in Eq. (12.49) as shorthand

$$\begin{aligned} e\eta_1 &= \Delta G°(O_2) - \Delta G°(OOH_{ads}) - eE°(O_2/H_2O) \\ &= \Delta G°(O_2) - \Delta G°(OOH_{ads}) - C_O. \end{aligned} \tag{12.53}$$

Upon uptake of a further electron, two surface-bound OH groups are formed from the superoxide-like intermediate by cleavage of the O–O bond:

$$OOH_{ads} + H^+ + e^- + \square \rightleftharpoons 2\,OH_{ads} \tag{12.54}$$

$$eE^\circ(OH_{ads}/H_2O) = \Delta G^\circ(OH_{ads}) - \Delta G^\circ(H_2O) \tag{12.55}$$

$$e\eta_3 = \Delta G^\circ(OH_{ads}) - \Delta G^\circ(H_2O) - C_O. \tag{12.56}$$

The reduction of OH_{ads} finally yields water:

$$OH_{ads} + H^+ + e^- \rightleftharpoons H_2O + \square \tag{12.57}$$

$$eE^\circ(OH_{ads}/H_2O) = \Delta G^\circ(OH_{ads}) - \Delta G^\circ(H_2O) \tag{12.58}$$

$$e\eta_3 = \Delta G^\circ(OH_{ads}) - \Delta G^\circ(H_2O) - C_O \tag{12.59}$$

As long as we consider the same adsorbed intermediates, we obtain these results even with slight differences in the reaction mechanism. Analogous to Eq. (12.43), a boundary condition links the potentials of the partial reactions to the standard potential of the overall reaction.

$$\begin{aligned} E^\circ(O_2|H_2O) &= \frac{E^\circ(O_2/OOH_{ads}) + E^\circ(OOH_{ads}/OH_{ads}) + 2E^\circ(OH_{ads}/H_2O)}{4} \\ &= 1.23\,\text{V}. \end{aligned} \tag{12.60}$$

Requirements for the Optimum Catalyst. The optimum catalyst would be the one for which all the overpotentials of the partial reactions become zero with respect to the total reaction. This condition is fulfilled if we insert into Eqs. (12.53), (12.56), and (12.59) the following Gibbs free energies of adsorption:

$$\Delta G^\circ(OH_{ads}) = C_O = 1.23\ \text{eV} \tag{12.61}$$

$$\Delta G^\circ(OOH_{ads}) = 3\ C_O = 3.69\ \text{eV} \tag{12.62}$$

For adsorbed atomic oxygen O_{ads} possibly occurring as an intermediate with two bonds between the oxygen atom and the surface, the following relation is obtained:

$$\Delta G^\circ(O_{ads}) = 2\ C_O = 2.46\ \text{eV} \tag{12.63}$$

Scaling Relationships for the ORR. Quantum mechanical calculations of adsorbates on (1 1 1) surfaces of various metals [16] showed that the Gibbs free adsorption energies of the intermediates O_{ads}, OH_{ads}, and OOH_{ads} *do not* change independently. If $\Delta G^\circ(O_{ads})$ is particularly large on one metal, $\Delta G^\circ(OH_{ads})$ and $\Delta G^\circ(OOH_{ads})$ will also assume particularly large values compared to other metals. With some approximations, we find the following scaling relationships for the (1 1 1) surfaces of all metals.:

$$\Delta G^\circ(OH_{ads}) \approx 0.5\ \Delta G^\circ(O_{ads}) + K_{OH} \tag{12.64}$$

$$\Delta G^\circ(OOH_{ads}) \approx 0.5\ \Delta G^\circ(O_{ads}) + K_{OOH}. \tag{12.65}$$

The factor 0.5 in Eqs. (12.64) and (12.65) accounts for the fact that O_{ads} is linked to the metal electrode by two bonds, and OOH_{ads} and OH_{ads} are linked by one bond each. The parameters in the equations also include the correction terms for stabilization by water and are approximately $K_{OH} = 0.04$ eV and $K_{OOH} = 3.33$ eV.

The scaling relationships allow one to simplify the considerations of the catalysts insofar as all energies can be represented as a function of *one* decisive parameter, the so-called descriptor [here $\Delta G^\circ(O_{ads})$]. Figure 12.12 shows a calculated volcano plot

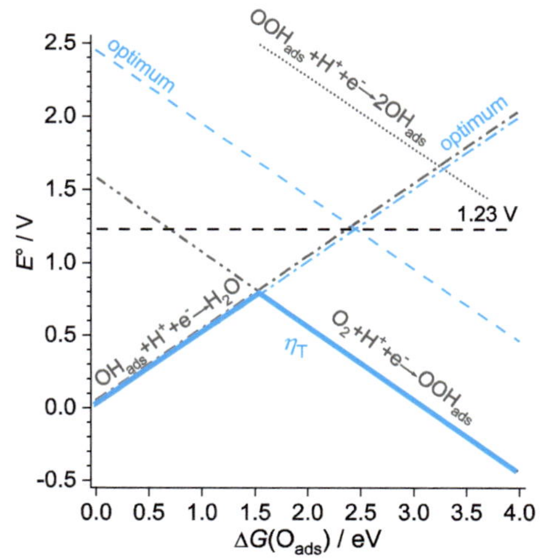

Figure 12.12 Thermodynamic volcano plot for the ORR. Equilibrium potentials of the reactions (12.51), (12.54), and (12.57). Depending on $\Delta G^\circ(O_{ads})$, the quantity η_T (thick blue line) results from Eq. (12.51) or Eq. (12.57). Reaction (12.54) is never the potential-determining reaction, no matter what value $\Delta G^\circ(O_{ads})$ has. Source: Adapted from [6], with permission, Copyright 2011, Elsevier.

for the ORR that depends on only *one* parameter, although several ET steps and several intermediates are involved. The minimum thermodynamic overpotential η_T (thick blue curve) *never becomes zero* – that is, the optimum catalyst defined according to Eqs. (12.61) and (12.62) cannot exist at all! The reason for this lies in the scaling relationship, which states that the free adsorption energies of O_{ads}, OH_{ads}, and OOH_{ads} cannot be varied independently of each other. The scaling relationship fixes the difference $\Delta G°(OOH_{ads}) - \Delta G°(OH_{ads})$ to the amount $K_{OOH} - K_{OH} \approx 3.29$ eV, whereas the condition for the optimal electrocatalyst requires a difference of $2C_O = 2.46$ eV. This conclusion from the calculations is consistent with experimental observations insofar as even the best-known ORR electrocatalysts have a considerable overpotential. In order to describe the best possible electrocatalyst that can be practically implemented, we look for the maximum of η_T, which is located at the intersection of the lines for the reactions (12.51) and (12.57). At this point, η_T amounts to 0.41 V, which corresponds to $(K_{OOH} - K_{OH} - 2C_O)/(2e)$. In fact, this agrees quite well with experimental values of about 0.3 V for the currently best ORR electrocatalysts (alloys such as Pt–Ni, Pt–Co, and Pt–Y). Significantly better-performing electrocatalysts can only be expected from materials or classes of materials for which the described scaling relationships would *not* apply. Although such materials have not yet been discovered, their existence and thus the possibility of their discovery cannot be excluded, even if it seems unlikely from the perspective of current knowledge.

Let us consider some more limitations of the previous analysis: the scaling relationships [Eqs. (12.64) and (12.65)] apply only to the (1 1 1) surfaces of the metals because other surface terminations or entirely different catalysts, such as transition metal centers in a coordination environment of N and C atoms, have different scaling relationships. In this case, the slope in the relationships would probably not vary much after all, because it is mainly determined by the ratio of the bond orders of OH_{ads} and O_{ads} to the catalyst surface. The changes would mainly affect K_{OOH} and K_{OH}. It is not so much the absolute values as the *difference* between K_{OOH} and K_{OH} that exerts the decisive influence on η_T. Thus, even if K_{OOH} and K_{OH} were to change, if their difference remained constant, there would be no fundamentally different prediction for the best possible catalyst!

A limitation of this analysis is the lack of conclusions about the selectivity of the ORR. In aqueous solutions, this can also lead to the usually undesirable side product H_2O_2.

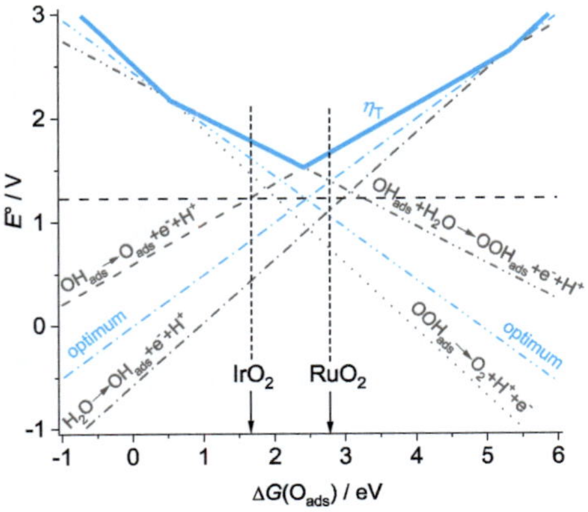

Figure 12.13 Thermodynamic volcano plot for the OER and equilibrium potentials for the partial reactions (12.66)–(12.69) of the OER. The adsorbate energies for IrO_2 and RuO_2 are also plotted. Source: Adapted from [6], with permission, Copyright 2011, Elsevier.

Scaling Relationships for the OER. The effects of different scaling relationships are made very clear by a comparison between OER and ORR. Indeed, the same considerations for the thermodynamics of the overall reaction and for the intermediates apply to both reactions. First, water binds to the electrode

$$H_2O + \square \rightleftharpoons OH_{ads} + H^+ + e^- \tag{12.66}$$

followed by the oxidation to O_{ads}

$$OH_{ads} \rightleftharpoons O_{ads} + H^+ + e^-. \tag{12.67}$$

The next step could be the formation of the superoxide-like intermediate

$$O_{ads} + H_2O \rightleftharpoons OOH_{ads} + H^+ + e^- \tag{12.68}$$

from which oxygen is then finally released

$$OOH_{ads} \rightleftharpoons O_2 + H^+ + e^- + \square. \tag{12.69}$$

The fact that the OER proceeds in reverse sequence to the ORR does not affect the thermodynamic considerations.

However, the oxygen evolution proceeds at metal oxides, because all metals are covered with a more or less thick oxide film at OER potentials. Quantum chemical calculations result in *different* scaling relationships [17] at these metal oxides compared to those at the metal surfaces relevant for the ORR:

$$\Delta G^{\circ}(\mathrm{OH_{ads}}) \approx 0.61\,\Delta G^{\circ}(\mathrm{O_{ads}}) + K_{\mathrm{OH}} \tag{12.70}$$

$$\Delta G^{\circ}(\mathrm{OOH_{ads}}) \approx 0.64\,\Delta G^{\circ}(\mathrm{O_{ads}}) + K_{\mathrm{OOH}} \tag{12.71}$$

with $K_{\mathrm{OH}} \approx -0.58$ eV und $K_{\mathrm{OOH}} \approx +2.40$ eV. The slopes of the scaling relationships are similar (ORR: 0.5 and OER: 0.6), but the K values are about 1 eV more negative for the OER at oxide surfaces. In contrast, the difference $K_{\mathrm{OOH}} - K_{\mathrm{OH}}$ is quite similar (ORR: 3.29 eV, OER: 2.98 eV). With these scaling relations, the standard potentials of all substeps can be calculated as a function of $\Delta G^{\circ}(\mathrm{O_{ads}})$ (Figure 12.13). As in the discussion above, the thick blue line in Figure 12.13 represents the thermodynamic overpotential η_{T} for every conceivable value of $\Delta G^{\circ}(\mathrm{O_{ads}})$, that is, the connecting line with the highest E°. It again corresponds to a volcano plot (in the shape of a "funnel" as we called it in the previous section). $\Delta G^{\circ}(\mathrm{O_{ads}})$ for IrO_2 and for RuO_2 are also plotted. The two oxides are effective catalysts for the OER. According to Figure 12.13, IrO_2 binds oxygen slightly too weakly and RuO_2 a little too strongly to reach the theoretical optimum η_{T}.

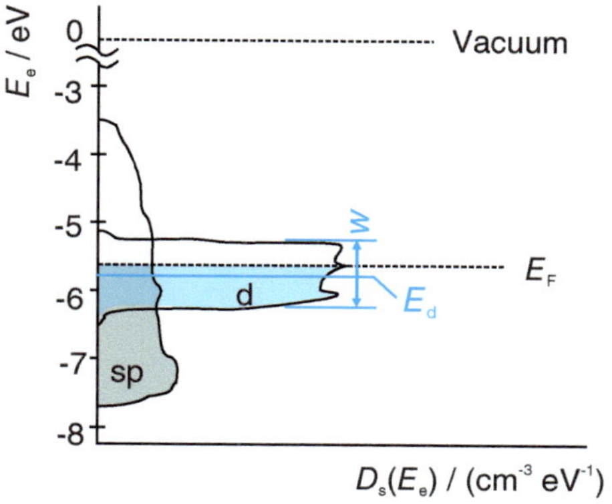

Figure 12.14 Schematic of the density of states and occupation of sp- and d-bands in metals. (The fuzziness of the occupation around E_F at $T > 0$ K is not visible in this scale.)

12.3.4 How Can We Rationalize the Different Adsorption Energies of Intermediates on Transition Metals?

After the previous section, the question arises *why* the scaling relationships exist at all, that is, why the Gibbs free energies of adsorption change in the same way for different adsorbates on different metals or metal oxides. For this purpose, we have to compare the electronic structures (Sections 10.1.2 and 10.1.3) of different metals in terms of adsorbate interactions.

The energy bands, which are important for the electronic structure, are formed by the valence electrons of the same subshells. Their notation is derived from the subshells: The 3d band is formed by overlapping the 3d orbitals of many atoms in a metal crystal. An important representation of the electronic structure are band structure plots (Figure 12.14), where the ordinate is the one-electron energy E_e and the abscissa is the density of states $D_S(E_e)$. We speak of "broad bands" if they cover a large interval w on the ordinate (E_e-axis). This is the case when the involved atomic orbitals overlap strongly, which applies to s- and p-electrons. They form a common broad sp-band. The probability density of finding the electron at a specific position is smeared over a large spatial range, as represented by the electron gas model in an even greater simplification. The d-orbitals of the metals do not overlap as much, thus resulting in a narrower d-band as schematically depicted in Figure 12.14 and more localized electronic states.

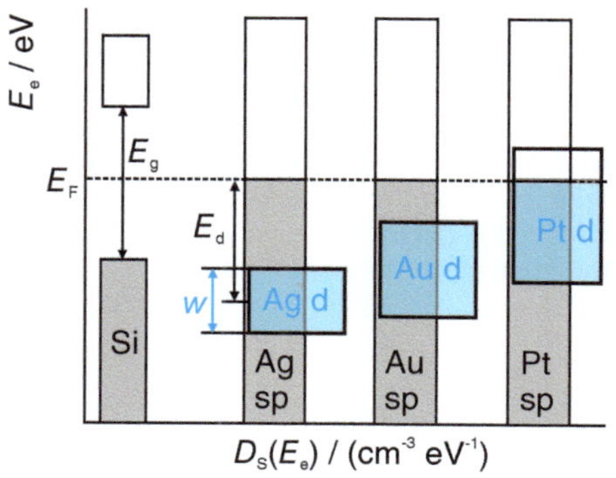

Figure 12.15 Comparison of the strongly schematized sp- and d-bands of the intrinsic semiconductor Si with those of the typical electrode metals Ag, Au, and Pt. The occupation of the bands is indicated by the color shading. (The blurring of the occupation around E_F at $T > 0$ K is not apparent at this scale.) Idea of the figure according to [11].

The energy bands are occupied up to the Fermi level E_F (Section 10.1.3), although at temperatures $T > 0$ K, the thermal excitation leads to a fuzzy occupation around E_F (Figure 10.11a). The various metals now differ from each other by how the sp- and d-bands are located relative to E_F and how wide the d-bands are. This is shown in Figure 12.15 for three typical electrode metals. For a quick comparison, only the width w of the d-band is important. The position of the d-band is strongly simplified by the center of the d-band E_d.

If an adsorbate interacts with a broad band of the metal, the level of the adsorbate is greatly broadened compared to the isolated molecule. The result is weak chemisorption. There are no large energy gains of the overall system in the bound state (Figure 12.16). On the other hand, a splitting of the levels into a "bonding" and

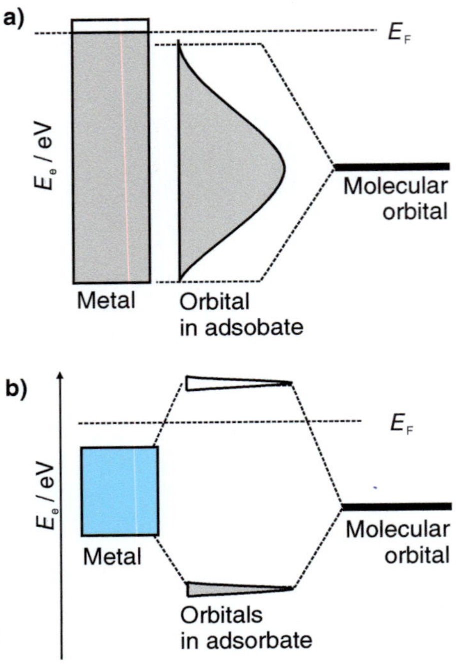

Figure 12.16 Interaction of an adsorbate level with a) a broad d-band of a metal and b) with a narrow d-band.

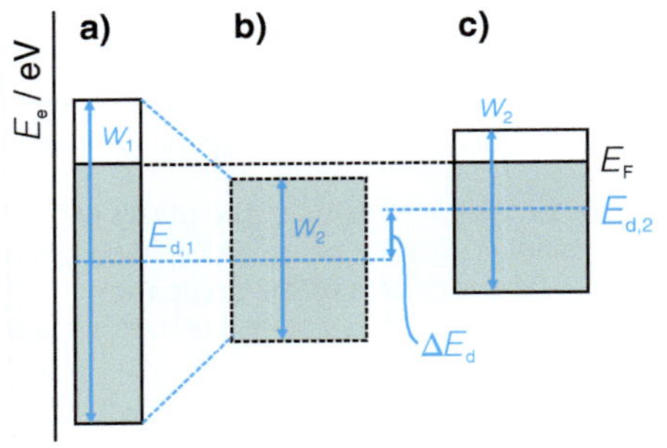

Figure 12.17 Effect of stretching a surface on the position of the d-band. a) Band structure diagram in the unstretched state; b) (notional) reduction of the bandwidth by increasing the atomic spacing; and c) reduction of the bandwidth *and* increase in E_d, required to keep the occupation state constant.

an "antibonding" adsorbate level occurs with only slight broadening when the orbitals of the molecule interact with a narrow d-band (Figure 12.16b). If only the binding adsorbate level is occupied, the result is strong chemisorption. The illustrations in Figure 12.16 depict the limiting cases; in fact, intermediate levels also occur in which the binding adsorbate level and the antibinding adsorbate level are connected by a "band" with a low density of states. The position of the adsorbate levels depends on the position of the d-band of the metal relative to E_F, which can be described in simplified terms by the energy difference between E_F and E_d.

This concept allows the prediction of trends for the electrocatalytic effect of noble metals, alloys, and surface modifications. For example, consider the consequences of the band structures of the metals for the mechanism of HER (hydrogen evolution reaction) from Section 12.1.3: In the sp-metals, the d-band is fully occupied and is overall below E_F. On these metals, adsorption of a proton is energetically unfavorable because the energy released is insufficient to compensate for the desolvation energy of the proton. The inverse of the reaction in Eq. (12.9) is therefore the rate-determining step. In contrast, the Tafel reaction (recombination of from H_{ads} to $H_{2,ads}$) and its desorption occur without much barrier. Therefore, the conversion proceeds according to the Volmer–Tafel mechanism with the adsorption of H^+ as RDS.

At Cu and Ag electrodes, the subsequent reaction of H_{ads} occurs preferentially by the Heyrovský mechanism [Eq. (12.12)]. Experimentally, the two limiting cases can be distinguished by chronoamperometric measurements. In both cases, coverage by H_{ads} builds up after the onset of the reaction. In the case of the Volmer–Tafel mechanism, such coverage slows down the RDS, which depends on the concentration of free surface sites. In the case of the Volmer–Heyrovský mechanism, the reduction current initially increases as the rate of RDS [Eq. (12.12)] increases with increasing H_{ads} coverage.

On transition metals with partially filled d-bands, the effects on the rate of the HER are more complicated, and usually, both mechanisms occur simultaneously, with one or the other mechanism dominating in the overall turnover depending on the applied potential, the facet of the metal crystal considered, or the pH. These are precisely the metals of interest for electrocatalysis and require detailed consideration.

As an example, let us analyze the effect of increased spacing of metal atoms (strain) on the surface. In Figure 12.17a, a highly simplified d-band is shown filled up to E_F. If the atomic distance at the surface of the metal increases, the overlap of the d-orbitals is reduced, and the d-band therefore becomes narrower (Figure 12.17b) and is then completely below E_F. In order for the occupation of states of the d-band to remain unchanged (there are no more electrons in the metal), the d-band must be raised compared to the undisturbed metal (Figure 12.17c). The center of the d-band thus shifts in the direction of E_F.

If an adsorbate binds to the undisturbed metal surface, both the binding and antibinding adsorbate states can be below E_F (Figure 12.18a). Thus, *both* adsorbate states are occupied, and at most a low net binding energy results. With the stretched surface, as a result of the elevation of the d-band center, the probability increases that more antibonding adsorbate states lie above E_F and therefore remain unoccupied (Figure 12.18b). This results in a stronger binding of the adsorbate at the stretched than at the undisturbed metal surface.

Various methods have indeed succeeded in implementing such concepts experimentally, such as mechanical stretching of metal films [18, 19], buildup of metal films from only one or only a few atomic layers on a substrate with larger bond distances [14], or alloy formation [20–22] each of which confirmed the computationally predicted trends [23].

Thus, it is the narrower and more structured d-band that is important for electrocatalytic action (as in catalysis in general).

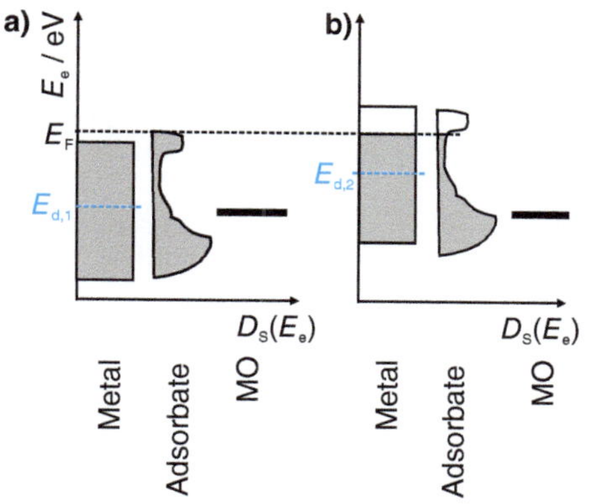

Figure 12.18 Effect of increasing the d-band center on the occupation of adsorbate levels. a) All adsorbate levels are below E_F and thus are occupied and b) part of the adsorbate levels with antibonding character are above E_F and remain unoccupied.

Further Reading. Further considerations on the topics listed in Section 12.3 for different reactions can be found in an anthology [24]. The d-band model and explanations of electrocatalytic reaction data derived from it are comprehensively presented in [11, 25, 26]. A critical discussion of the possibilities and limitations of the volcano curves is provided in [27]. An instructive comparison on the application of DFT methodology in heterogeneous catalysis, electrocatalysis, and for insertion electrodes is provided by [28]. An overview of modeling electrocatalytic reactions on different size and timescales can be found in [24, 29].

12.4 Experimental Investigation of Electrocatalytic Reactions

12.4.1 How Do We Determine the Effective Surface Area?

Measurable Quantities. In the case of electrocatalytic reactions, the measurement of reaction rates (central to catalysis) can be carried out by determining the current density [Eq. (1.23)]. Many metals that exhibit good electrocatalytic properties, such as Pt, are very expensive. Therefore, realizing the largest possible surface area per mass of the noble metal is important. This is achieved by means of nanometer-sized Pt particles, which are deposited onto a support such as carbon (Figure 2.31). The support material ensures the electronic contact to the current collector and simultaneously stabilizes the nanoparticles against agglomeration.

When comparing different catalyst preparations, the question arises to which area the measured current should be related. The current density j_{geom} refers to the current per geometric area and is of great importance for the design of technical devices. The area used as a reference value can be identified with the unit A/cm^2_{geom}. However, the current density referred to the geometric area can vary substantially even when the same catalyst material is used. This area of the electrode can be covered with a large or small mass of the catalyst. Therefore, to make a statement about the quality of a catalyst, the current is related to the area of the Pt (j_{Pt}) and usually represented by the units A/cm^2_{Pt}. The effective surface area of Pt is accessible from electrochemical measurements.

The quantity related to the Pt surface area is an important specification for the intrinsic activity of the electrocatalyst material used, but it does not provide information on how efficiently this material was processed into particles. For the same current density j_{Pt} $[A/cm^2_{Pt}]$ and the same mass of Pt used, many small particles would provide a much larger total current, and thus be more beneficial than a few larger particles. Therefore, the Pt surface area related to the mass of Pt used is another important quantity. This quantity is usually referred to as the electrochemically active surface area (ECSA) or, better, as specific surface area

$$A_{ECSA} = \frac{A_{Pt}}{m_{Pt}} \left[m^2_{Pt} g^{-1}_{Pt}\right] . \tag{12.72}$$

The activity of an electrode is the kinetically limited current related to the mass of Pt

$$j_{m,Pt} = \frac{i_{kin}}{m_{Pt}} \quad \left[A\, mg^{-1}_{Pt}\right] \tag{12.73}$$

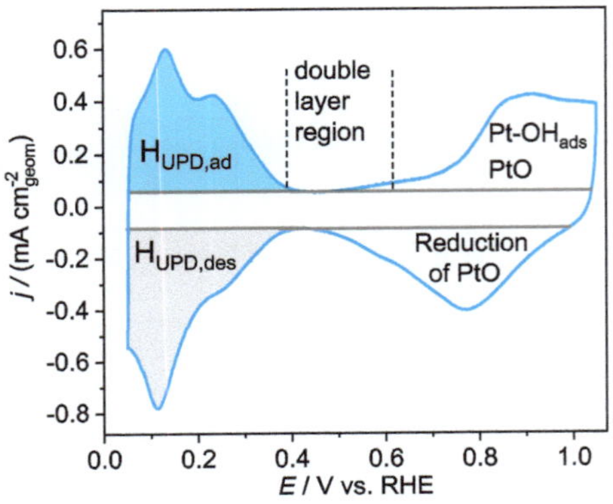

Figure 12.19 Integration of the H_{ads} and the H_{des} peaks of a supported Pt electrocatalyst. Measurement of the voltammogram in 0.1 mol L^{-1} $HClO_4$ at $v = 0.05$ V s^{-1}. Source: Data were provided by Marius Gollasch and Julian Lorenz (DLR Institute for Networked Energy Systems Oldenburg and Carl v. Ossietzky University Oldenburg).

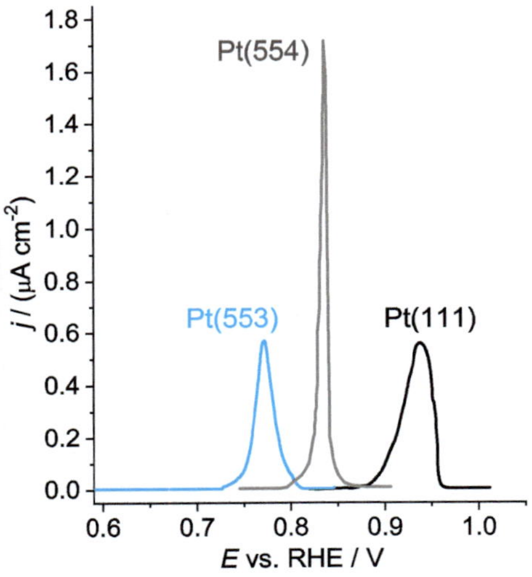

Figure 12.20 Stripping voltammogram for the oxidation of CO adsorbed on different Pt single crystal electrodes; $v = 0.05$ V s^{-1}. CO_{ads} is oxidized at lower potentials on steps than on the smooth Pt(1 1 1) surfaces. The stepped surfaces (combination of steps and smooth terraces between steps) show only *one* sharp signal at low potentials rather than two signals for steps and terraces. This shows that the CO molecules adsorbed on the terraces diffuse very rapidly to the steps, where they are converted. Source: Reproduced with permission from [35], Copyright 2009, John Wiley & Sons.

Preparation of Test Electrodes. A disc electrode made of glassy carbon is cleaned by polishing and rinsing in distilled water. Then one prepares a suspension of the catalyst particles from the powdered catalyst with precisely known mass and precious metal content in water, isopropanol, and a portion of Nafion. After a drying step, Nafion binds the powder particles to the electrode. The exact proportions vary depending on the catalyst. The following values are therefore only guidelines and should be optimized according to literature values (H_2O: 4 mL, isopropanol 1 mL, Pt: 5 mg, Nafion[R] 20 μL of a commercially available 5% solution). Ultrasonic treatment of the suspension, preferably with a rod-shaped ultrasonic horn or a high-power ultrasonic bath, produces a good dispersion. Using a pipette, 5–25 μL of the suspension is placed on the active surface of the glassy carbon electrode, depending on the size of the electrode. The suspension is then left to dry on the glassy carbon electrode. Heated air or slight rotation of the droplet can be helpful to achieve uniform coverage of the surface. From the volumes used and the mass of the catalyst (Pt plus carbon), the mass fraction of Pt in the catalyst and the area of the disk electrode, the mass loading $[\mu g_{Pt}\, cm^{-2}_{geom}]$ can be calculated.

Determination of the Active Surface Area of Pt by Hydrogen Adsorption/Desorption. For this purpose, adsorption/desorption of hydrogen is used, which is often also referred to as underpotential deposition (UPD) of hydrogen. A cyclic voltammogram is recorded with the prepared electrode in 0.5 mol L^{-1} H_2SO_4 or 0.1 mol L^{-1} $HClO_4$ at a scan rate v of 0.05–0.5 V s^{-1} between 0.05 and 1.2 V vs. RHE (Figure 12.19). It is useful to use a potentiostat with an analog ramp generator for this purpose (Section 5.3.9), since staircase ramps, which are generated by many digital instruments, can introduce distortions in the signals of surface-bound redox systems that are significant, especially for the hydrogen UPD signals. To correct for the contribution of the electric double layer charging in an approximate way, we apply a baseline that is the tangent to the current at about 0.45 V vs. RHE and parallel to the potential axis. The charge for the UPD peak is then given by Eq. (5.96) [note v!]. From the charge, we calculate the Pt surface area using the value of 210 μC $cm^{-2} \approx 2 \times 10^{-4}$ A s cm^{-2} for a complete coverage of the surface with H_{ads} [30].

$$A_{Pt} = Q(H_{ads})/2.1 \times 10^{-4}\ \text{A s cm}^{-2} \qquad (12.74)$$

For other noble metals with hydrogen UPD, the same procedure is used, but with different numerical values for the adsorption/stripping of a complete monolayer.

Surface Area Determination by Reduction of an Oxide Film. For precious metals, the charge for the reduction of a thin oxide film formed in a preceding potential scan up to a well-defined upper potential can be used to determine the active surface area. For example, the charge for the reduction of surface gold oxides on polycrystalline gold is 386 μC cm^{-2} for a potential scan up to the anodic current minimum between the formation of the surface oxides and the onset of the OER (Burshtein minimum[139]) [32].

Surface Area Determination from the Double Layer Capacitance. In some cases, the specific double-layer capacitance can be used to estimate the surface area, but not for supported electrocatalysts, because there the support material (such as carbon) would also contribute to the double-layer capacitance.

CO Stripping. This measurement principle is based on the adsorption of a monolayer of carbon monoxide on Pt (or other noble metals) and the subsequent CO oxidation. **Caution:** *CO is a dangerous respiratory poison, because it is odorless. Therefore, professional safety equipment and gas warning sensors, which are subject to regulations, must be used when working with CO.* From the charge of the CO oxidation peak, the

[139] А. А. мичри, А. Г. Пщеничников, Р. Х. Бурштейн; Элекрохимия **1972**, *8*, 364–366 (A. A. Mitchri, A. G. Pshchenichnikov, R. K. Burshtein; Elektrokhimiya **1972**, *8*, 364–366).

amount of CO molecules and thus the area of the noble metal electrode can be determined. However, it must be taken into account that the oxidative removal of CO from an adsorption site not only requires two electrons per CO molecule, but additionally, the double layer capacitance changes drastically, and electrons must flow to charge up the double layer with its increased capacitance [31]. Therefore, a correction for the double-layer capacitance must be subtracted from the charge of the stripping peak before Faraday's law is used to determine the amount of CO molecules and thus the number of adsorption sites (see also Section 13.1.2 and Figure 13.11).

In addition, the potential at which CO oxidation occurs depends very strongly on the adsorption site and the type of metal. Thus, qualitative statements can also be derived from the CO stripping experiments. A distinction can be made between linearly bound CO, CO on bridge sites on Pt and CO in threefold hollow sites on Pt(1 1 1) terraces (Table 8.1). CO stripping experiments provide information on the presence of the individual facets on nanoparticle catalysts. An example of the differentiation of the individual single crystal facets based on CO stripping is shown in Figure 12.20.

Further Reading. Data for other surfaces can be found in the reviews [32–34] and summaries of experimental findings on CO stripping in [2, 23, 30, 35]. Analyses of the surface structure of Pt nanoparticles and polycrystalline Pt in comparison to single crystal surfaces by adsorption of H, CO, and underpotential deposition of metals are summarized in [36, 37].

12.4.2 Determination of Catalyst Activity by Means of the Rotating Disk Electrode

Rotating disk electrodes (RDE) and rotating ring–disk electrodes (RRDE) are examples of electrode systems with forced convection. There is a relative motion between the solution and the electrode surface during voltammetry at such electrodes (hydrodynamic methods). Convection accelerates the mass transport. Diffusion dominates transport only in the hydrodynamic boundary layer. The thickness of this layer can be adjusted by the velocity of the electrode surface versus the solution (i.e., the rotation rate). The convection is kept constant during a measurement. This results in steady-state voltammograms, which are easier to evaluate than the voltammograms with transient diffusion layers (transient voltammograms, Figure 5.27) that were used extensively in Chapter 6. At electrodes with forced convection, intermediates can be detected in systems with two working electrodes (e.g., RRDE), where the products of the reaction at the first electrode (disk) are transported with the solution flow to the second electrode (ring) and detected there.

In addition to the RDE and RRDE, systems with forced convection comprise flow channels with electrodes inserted into the channel walls, rotating cylinders (Figure 9.20), vibrating electrodes, wall-jet cells (Figure 5.51b), electrodes under the influence of ultrasound, and others. A disadvantage of hydrodynamic systems is the difficulty in producing systems with precisely controlled mass transport conditions, because small deviations from the ideal geometry have a strong effect. In addition, the mathematical description is also demanding because not only do the partial differential equations for diffusion have to be solved, but also those for convection.

We will discuss only the results for the RDE and RRDE because they are the most popular for characterizing electrocatalysts. This may be because it is comparatively easy to coat the catalysts, as described in Section 12.4.1, onto the disk electrode in a reproducible manner. In addition, the RDE is one of the so-called uniformly accessible electrodes. This means that the local current density is the same at all points of the surface, which greatly simplifies the data treatment [38, 39].

Figure 12.21 shows the experimental design of an RDE. During rotation, the liquid is accelerated radially by the electrode surface and causes a liquid flow to compensate for

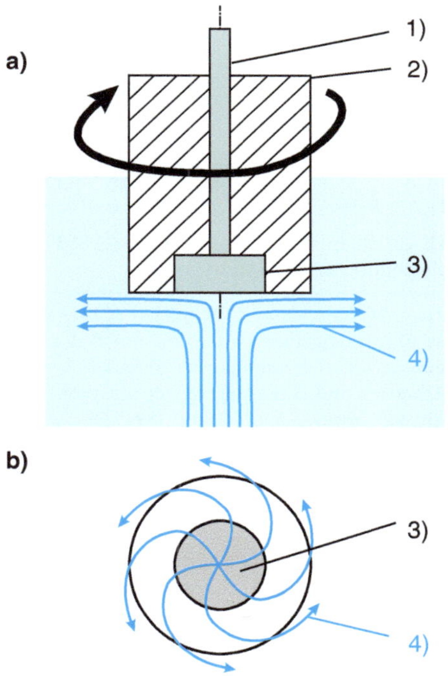

Figure 12.21 Setup of a rotating disk electrode; a) cross-sectional view and b) top view.

1) Mechanical connection to the rate-controlled motor and electrical connection to the potentiostat;
2) Insulating electrode sheath;
3) Working electrode;
4) Trajectories of liquid motion (streamlines of convection).

The important electrical contacting of the rotating cylinder is not detailed.

I_L – Levich current [A]
n – Stoichiometric number of charges [dimensionless]
F – Faraday constant [A s mol^{-1}];
A – Geometric electrode area [cm^2]
D – Diffusion coefficient [cm^2 s^{-1}]
ω – Angular frequency [s^{-1}]
ν – Kinematic viscosity [cm^2 s^{-1}]
c^* – Volume concentration [mol cm^{-3}]
Δx – Thickness of the Nernst[10] diffusion layer [cm]

Table 12.2 Kinematic viscosities of some electrolyte solutions at 298 K (25 °C).

Solution 0.1 mol L^{-1} supporting electrolyte/solvent	ν/(cm^{2}s^{-1})
TEAP/H_2O	0.009132
TEAP/D_2O	0.01028
KCl/H_2O	0.008844
TEAP/CH_3CN	0.004536
TEAP/DMSO	0.01896
TEAP/Pyridine	0.009518
TEAP/DMF	0.008971
TEAP/*N*,*N*-dimethylacetamide	0.01067
TEAP/hexamethyl-phosphoric acid triamide	0.03530

Source: M. Tsushima, K. Tokuda, T. Ohsaka; Use of hydrodynamic chronocoulometry for simultaneous determination of diffusion coefficients and concentrations of dioxygen in various media. *Anal. Chem.* **1994**, *66*, 4551–4556, Copyright 1994, American Chemical Society. TEAP stands for $[(C_2H_5)_4N^+]ClO_4^-$.

I_{tot} – Total current, experimentally measured quantity [A]
I_K – Kinetically limited current [A]

the displaced solution. Immediately in front of the electrode, there is a hydrodynamic boundary layer in which the liquid moves together with the electrode (a quiescent layer in first approximation). The relative motion between the electrode and the liquid layer becomes zero only for molecules directly at the rotating surface. Transport of reactants through this layer is dominated by diffusion, which is the limiting step for mass transport. The hydrodynamic boundary layer becomes thinner, and thus the mass transport more intense, the higher the rotation rate of the electrode, which is expressed as angular frequency ω [s^{-1}] = $2\pi f$ [s^{-1}] (f rotational frequency = number of revolutions per second or revolutions per minute RPM). Analogous to the procedure in the preceding chapters, we now first investigate the transport-limited current I_L for a reversible system at the RDE as a function of angular frequency and other quantities.

$$I_L = 0.62\, nFAD^{2/3}\omega^{1/2}\nu^{-1/6}c^* \qquad (12.75)$$

Equation (12.75) is called **Levich equation**[140] and I_L is the mass transport-limited Levich current. In addition to known quantities, the Levich equation contains the kinematic viscosity ν, which is the ratio of viscosity to density. A typical value for aqueous solutions is 0.01 cm^2 s^{-1}. Other values are given in Table 12.2. From a comparison of coefficients with the simple model of the diffusion-limited current at a steady-state diffusion layer $I_d = nFA(D/\Delta x)c^*$ [Eq. (5.48) from Excursion 5-2], the mass transport coefficient can be expressed as

$$m_t = \frac{D}{\Delta x} = 0.62\, D^{2/3}\omega^{1/2}\nu^{-1/6}. \qquad (12.76)$$

Next, the thickness Δx of the stationary diffusion layer (the Nernst diffusion layer) can be obtained from

$$\Delta x = 1.61\, D^{1/3}\omega^{-1/2}\nu^{1/6}. \qquad (12.77)$$

The shape of the current–voltage curve is obtained analogously to Eqs. (5.52) and (5.53) with $k'/k = (D_R/D_O)^{2/3}$, where D_R and D_O are the diffusion coefficients of the oxidized and reduced redox forms, respectively.

For the more general case, where both mass transport and charge transfer kinetics can be limiting, the total current I_{tot}, is given by the **Koutecký–Levich equation**[141].

$$\frac{1}{I_{tot}} = \frac{1}{I_K} + \frac{1}{I_L} \qquad (12.78)$$

For such steady-state current–voltage curves, we obtain an expression for the kinetically limited current by transforming Eq. (12.78):

$$I_K = \frac{I_{tot}I_L}{I_L - I_{tot}} \qquad (12.79)$$

Inserting Eq. (12.75) into Eq. (12.78) leads to the following expression that holds specifically for the RDE:

$$\frac{1}{I_{tot}} = \frac{1}{I_K} + \frac{1}{0.62\, nFAD^{2/3}\nu^{-1/6}c^*}\cdot\frac{1}{\omega^{1/2}}. \qquad (12.80)$$

[140] Venjan (Benjamin) Grigorievich Levich, 1917–1987, founder of electrochemical hydrodynamics, was one of the first Jewish scientists to emigrate from the Soviet Union in 1978, first to Israel, then to the United States, with the support of an international solidarity campaign.
[141] Jaroslaw Koutecký, 1922–2005, Czech theoretical electrochemist, professor at Charles University in Prague, from 1973 on at the Free University of Berlin.

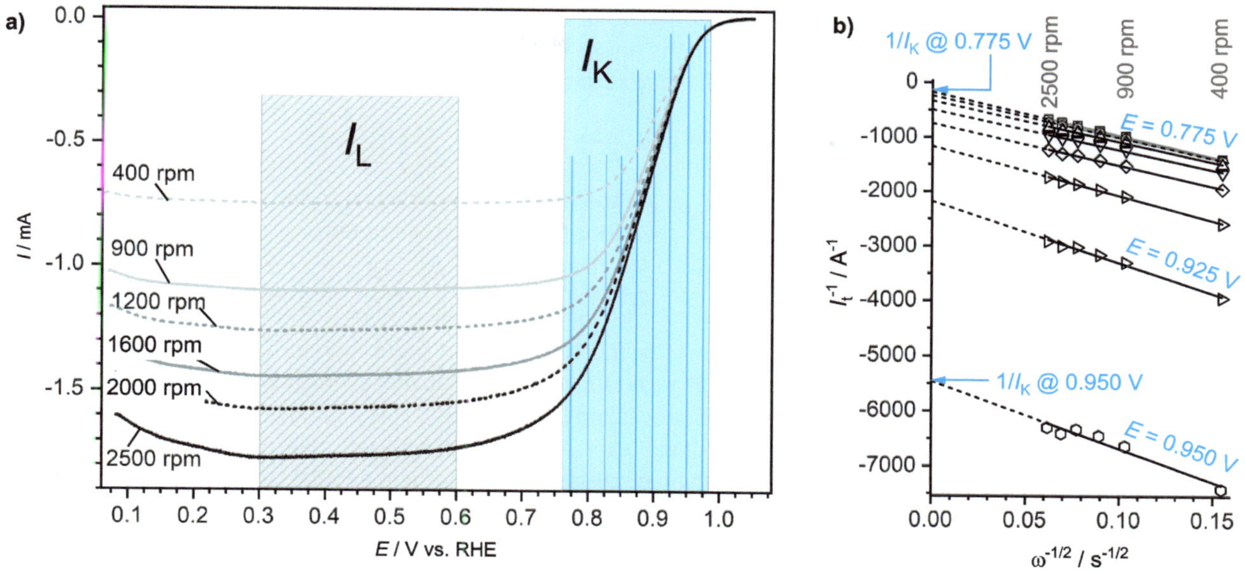

Figure 12.22 a) Current–voltage curves for oxygen reduction at a glassy carbon RDE coated with carbon-supported Pt nanoparticles. The transport-limited current (Levich current) occurs in the gray-shaded area. The vertical blue lines were used to read the currents for the construction of the Koutecký–Levich plot in b) to determine the kinetically limited current. Details: supported Pt catalyst in 0.1 mol L^{-1} $HClO_4$, $v = 0.005$ V s^{-1}, potential values are corrected for ohmic drop (Figure 12.25). Source: Data were provided by Marius Gollasch and Julian Lorenz (DLR Institute for Networked Energy Systems Oldenburg and Carl v. Ossietzky University Oldenburg).

Curves for different angular frequencies of the RDE are shown in Figure 12.22a. The kinetically limited current for each electrode potential E can be determined from the predominantly kinetically controlled region of such a dataset. To do this, we determine the experimental current I_{tot} for each ω at potentials approximately 0.02 V apart. Equation (12.80) can be considered as the equation of a straight line in the form $y = a_0 + a_1 x$, where $y = 1/I_{tot}$, $a_0 = 1/I_K$ and $x = 1/\omega^{1/2}$ (Figure 12.22). From the intercept of the linear fits of the data and the ordinate, I_K can be read directly. Alternatively, Eq. (12.79) can be used in spreadsheet applications.

Q: I do not really understand Eq. (12.78). Doesn't this mean that the experimentally measured total current I_{tot} is always slightly smaller than the smaller of I_L and I_K. Does this make sense? After all, the diffusion-controlled current is measured directly at very negative potentials.

A: The exact meaning of I_L and I_K is complicated. Since there are similar relationships in many electrochemical problems, it is worthwhile to consider again Eq. (12.78), but from a different perspective. I hope to make the relationship better understandable with this. According to Figure 3.1b, we can regard mass transport and charge transfer as two steps of a consecutive reaction. I_L and I_K are the *limiting* currents of the two steps, these are those currents that *would* occur if the other process did not contribute at all to limiting the total current. Figure 12.23 shows a short form of the scheme in Figure 3.1b together with a very primitive equivalent circuit, with which Eq. (12.78) can be made plausible. Let the total voltage U_{tot} be applied to the circuit. The voltage drops U_1 and U_2 occur at the variable resistors R_1 and R_2. The total current I_{tot} in the series circuit is equal to the currents through each of the two resistors $I_{tot} = I_1 = I_2$ [i.e., quite different from the form of Eq. (12.78)!]. The resistance can be expressed as $R_i = U_i/I_i$ or rearranged $1/I_i = R_i/U_i$ for each of the two resistors (i = [1, 2]). The limiting currents $I_{i,\mathrm{lim}}$ of each step are then the currents that would occur when the

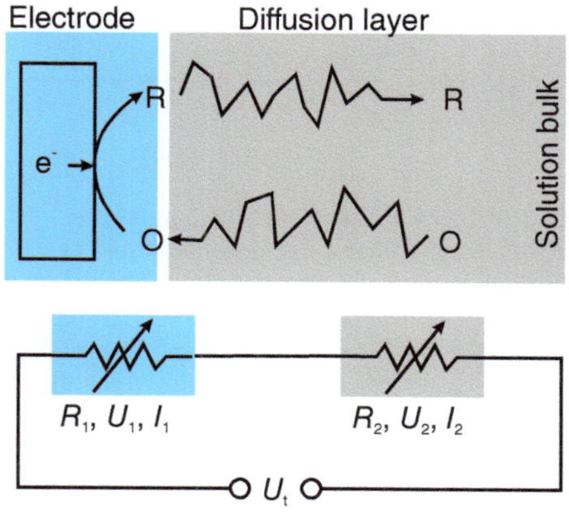

Figure 12.23 Analogy for explanation of the limiting currents for two steps in a consecutive reaction.

other resistor does *not* contribute to the current limitation, that is, $R_{j \neq i} \rightarrow 0$. In this case, the entire voltage drops across R_i, and U_i is equal to U_{tot}. This gives the *limiting* currents at R_1 and R_2 as

$$\frac{1}{I_{1,lim}} = \frac{R_1}{U_{tot}} \text{ and } \frac{1}{I_{2,lim}} = \frac{R_2}{U_{tot}} \tag{12.81}$$

$I_{1,lim}$ and $I_{2,lim}$ are therefore no longer equal to the total current! If we now add the two equations from (12.81), we obtain

$$\frac{1}{I_{1,lim}} + \frac{1}{I_{2,lim}} = \frac{R_1}{U_{tot}} + \frac{R_2}{U_{tot}} = \frac{R_1 + R_2}{U_{tot}} = \frac{R_{tot}}{U_{tot}} = \frac{1}{I_{tot}}, \tag{12.82}$$

and thus exactly Eq. (12.78). I_L and I_K are therefore not the actual currents occurring at a certain potential, but those which would be observed if all other steps proceeded infinitely fast. This analogy can simplify derivations of formal kinetic expressions for *steady-state* measurements of consecutive reactions in many situations and became popular, among others, due to the work of Savéant and coworkers [41]. Another example of this approach can be found in Section 13.4.7 of this book.

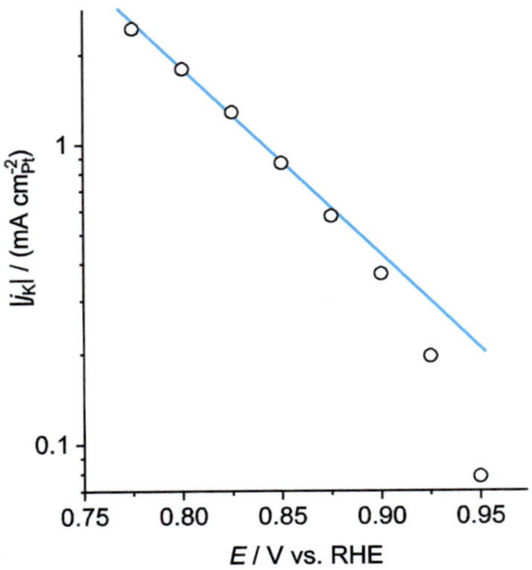

Figure 12.24 Tafel plot constructed from j_K-values of the ORR at an RDE with supported Pt catalyst in 0.1 mol L^{-1} $HClO_4$. Source: Data calculated from Figure 12.22, potential values are corrected for the uncompensated ohmic drop (Figure 12.25).

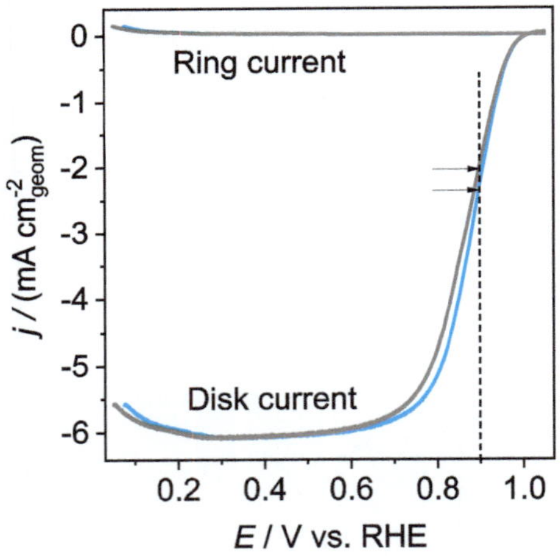

Figure 12.25 Effect of correction of the ohmic drop on the current density read at 0.9 V vs. RHE in an RRDE measurement of ORR on a supported Pt catalyst in 0.1 mol L^{-1} $HClO_4$, $v = 0.005$ V s^{-1}, $f = 2500$ U min^{-1}, $R = 19.5\ \Omega$ at $E = 0.5$ V; gray – raw data, blue data after IR correction. Source: Data were provided by Marius Gollasch and Julian Lorenz (DLR Institute for Networked Energy Systems Oldenburg and Carl v. Ossietzky University Oldenburg).

The procedure described above is used in particular for the characterization of catalysts for the **oxygen reduction reaction** (ORR) on Pt-containing catalysts. The I_K-values determined from Eq. (12.80) can then be related to the electrochemically active surface area A_{ECSA} or to the mass of Pt used. Another parameter is the activity, where the current at a potential of +0.90 V vs. RHE is related to the mass or A_{ECSA} of the Pt used. A Tafel plot $\ln(I_K)$ versus E [or E versus $\ln(j_K)$] can then be constructed from the determined values $I_K(E)$ (Figure 12.24).

Two more details are important when performing the experiment: for exact values, a correction of the uncompensated ohmic drop (*IR* correction) should be made (Section 5.3.1). To do this, we measure the uncompensated electrolyte resistance R_E before the measurement (e.g., by electrochemical impedance spectroscopy) and use a spreadsheet program to calculate a corrected electrode potential $E_{corr} = E - IR_E$ for each applied voltage (I – current with proper sign) [45]. The correction after the measurement is possible here due to the steady-state nature of the current. The effect of the *IR* correction is shown in Figure 12.25. Alternatively, the compensation capabilities of the potentiostats (Section 5.7.2) can be used [46], but potential oscillation due to overcompensation must be avoided [45].

Unlike what we discussed in Chapter 5, we begin the LSV in Figure 12.22 at a potential at which the diffusion-controlled current occurs, that is, at about 0.1 V vs. RHE, and then change the potential toward positive values at about 0.005 V s^{-1} until no current flows. This is possible because it does not matter from which side a certain potential value is approached for stationary measurements. The advantage of this procedure is that the measurements begin from the double-layer region of the Pt electrode, that is, when no specific adsorbates occur on Pt. At more positive potentials, the surface is first covered with OH_{ads} and then with a thin oxide layer, which affects the kinetics of the ORR. If we start the LSV in the region of the oxide-covered surface, we can expect the ORR current to start only after the retarded reduction of the oxide layer, which typically occurs in *one* reduction peak as in Figure 5.22.

The studies are carried out in an electrolyte that does not specifically adsorb to the Pt surface. In particular, traces of the halide ions should be avoided [47]. Therefore, hydrogen electrodes (such as in Figure 2.21) or a chloride-free electrode of the second type, for example, $Hg|Hg_2SO_4|K_2SO_4(sa)$ or Hg|HgO in alkaline solutions are used as reference electrodes. The studies are often carried out in suprapure 0.5 mol L^{-1} H_2SO_4

because this electrolyte provides a similar condition as Nafion (Figure 2.30). However, HSO_4^- adsorbs somewhat to Pt. Therefore, some authors prefer diluted suprapure $HClO_4$ solutions for activity determinations.

12.4.3 Determination of Selectivity by Means of a Rotating Ring–Disk Electrode

The selectivity of ORR can be characterized using the rotating ring–disk electrode (RRDE). The experimental system of the RRDE is shown in Figure 12.26. A ring-shaped electrode, usually made of Pt, concentrically surrounds the disk. Ring and disk can be operated with a bipotentiostat as two independent working electrodes against common reference and auxiliary electrodes. In this case, the reduction of O_2 runs simultaneously at the disk in a two-electron reduction [Eq. (12.4)] to H_2O_2 and in a four-electron reduction [Eq. (12.5)] to H_2O. H_2O_2 as the soluble product of the two-electron reduction is flushed with the flowing solution from the generation site (the disk) for detection at the ring. The ring electrode is operated at a potential at which H_2O_2 is oxidized under diffusion-controlled conditions. However, not all H_2O_2 molecules formed at the disk reach the ring, because some H_2O_2 molecules also diffuse into the solution. The *fraction* of molecules reaching the ring is called the transfer ratio or collection efficiency ε_C (often also N). The collection efficiency ε_C can be determined by carrying out a test reaction on the disk, in which only one stable product is formed, from which a fraction ε_C will be detected at the ring. The oxidation of $[Fe(CN)_6]^{4-}$ can be used for this purpose.

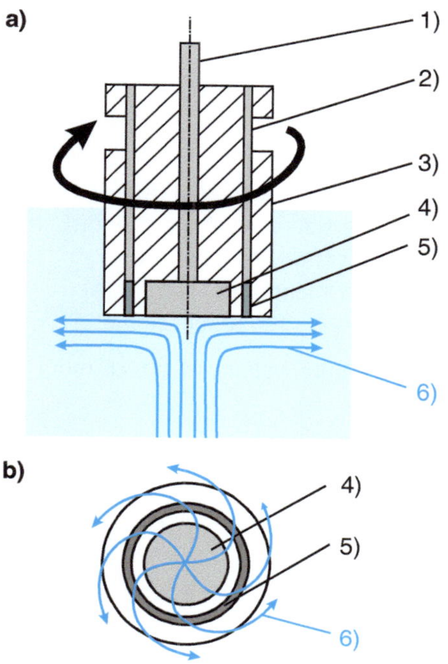

Figure 12.26 Setup of an RRDE.
a) cross-sectional and b) top view.

1) Mechanical connection to the rate-controlled motor and electrical connection between the disk and the bipotentiostat;
2) Electrical connection between ring and WE2 of the bipotentiostat;
3) Insulating electrode sheath;
4) Disc electrode (WE1);
5) Ring electrode mostly made of Pt (WE2);
6) Trajectories of liquid flow (streamlines of convection).

$$\text{Disc: Current } I_D \text{ from } [Fe(CN)_6]^{4-} \rightarrow [Fe(CN)_6]^{3-} + e^- \tag{12.83}$$

$$\text{Ring: Current } I_R \text{ from } [Fe(CN)_6]^{3-} + e^- \rightarrow [Fe(CN)_6]^{4-}\ e^- \tag{12.84}$$

The collection efficiency is then

$$\varepsilon_C = \frac{I_R}{I_D} \tag{12.85}$$

ε_C depends only on the geometry of the RRDE,[142] not on the rotation rate. Equation (12.85) states that the ring current I_R is caused by the fraction ε_C of all

[142] An exact hydrodynamic treatment of the problem shows that the collection efficiency depends only on the ratios of the radii of the RRDE [Y. B. Ivanov, V. G. Levich; *Dokl. Akad. Nauk U.S.S.R* **1959**, *126*, 1029–1032]. Other formulations are available, all with the same conclusion [W. J. Albery, S. Bruckenstein; *Trans. Faraday Soc.* **1966**, *62*, 1920–1931].

$$\varepsilon_C = \frac{0.8}{3}\left[\frac{1}{2}\ln\frac{\left[1-\frac{3}{4}\left(\frac{r_1}{r_2}\right)^3\right]+\left[\left(\frac{r_3}{r_2}\right)^3-1\right]}{\left\{\left[1-\frac{3}{4}\left(\frac{r_1}{r_2}\right)^3\right]^{1/3}+\left[\left(\frac{r_3}{r_2}\right)^3-1\right]^{1/3}\right\}^3}\right.$$
$$\left.+\sqrt{3}\arctan\frac{2\left[\left(\frac{r_3}{r_2}\right)^3-1\right]^{1/3}-\left[1-\frac{3}{4}\left(\frac{r_1}{r_2}\right)^3\right]^{1/3}}{\sqrt{3}\left[1-\frac{3}{4}\left(\frac{r_1}{r_2}\right)^3\right]^{1/3}}+0.9069\right]$$

where r_1 is the disk radius [cm], r_2 is the inner radius of the ring [cm], and r_3 is the outer radius of the ring [cm].

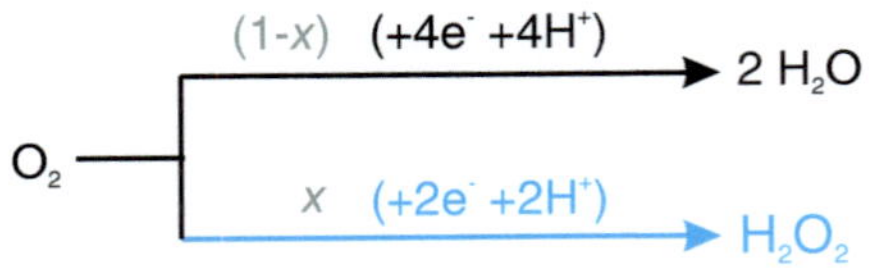

Figure 12.27 Simplified reaction scheme for the ORR.

I_D – Experimental current at the disk [A]
$I_{D,2}$ – Partial current at the disk from the two-electron reduction [A]
$I_{D,4}$ – Partial current at the disk from four-electron reduction [A]
I_R – Experimental ring current [A]
ε_C – Transfer ratio [dimensionless]

The absolute values of all currents are used.

particles that reacted at the disk, while the fraction $1 - \varepsilon_C$ diffuses into the solution and does not reach the ring.

For the ORR, the disk current $I_D = I_{D,2} + I_{D,4}$ is composed of the contributions of the two-electron reduction $I_{D,2}$ and the four-electron reduction $I_{D,4}$. We now ask what fraction x of the O_2 molecules reacted at the disk according to the two-electron reaction (Figure 12.27). Taking Faraday's laws into account, the quantities of substances N formed for both partial reactions are obtained from the steady-state currents because of $It = nNF$. After reduction, the fraction x can be calculated using Eq. (12.86).

$$x(H_2O_2) = \frac{|I_{D,2}|/2}{(|I_{D,4}|/4) + (|I_{D,2}|/2)} = \frac{2|I_{D,2}|}{|I_{D,4}| + 2|I_{D,2}|} = \frac{2|I_{D,2}|}{|I_{D,4}| + |I_{D,2}| + |I_{D,2}|} = \frac{2|I_{D,2}|}{|I_D| + |I_{D,2}|} \tag{12.86}$$

We can calculate the initially unknown partial current for the two-electron reduction in Eq. (12.86) from Eq. (12.85) ($I_{D,2} = I_R/\varepsilon_C$). Substituting this into Eq. (12.86) yields the working equation

$$x(H_2O_2) = \frac{2\,|\,I_R\,|/\varepsilon_C}{|\,I_D\,| + (\,|\,I_R\,|\,/\varepsilon_C)}. \tag{12.87}$$

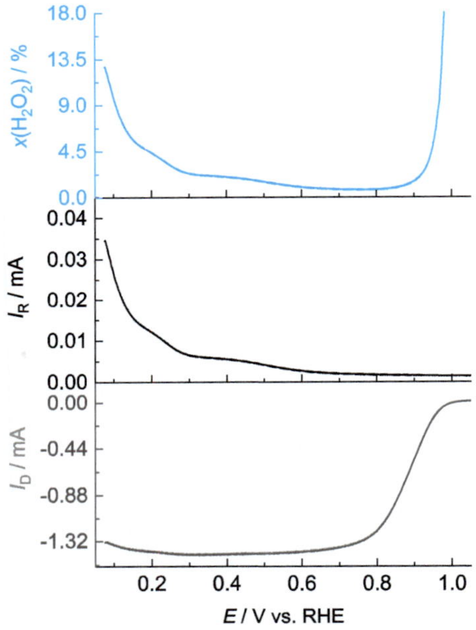

Figure 12.28 Determination of selectivity with linear sweep voltammograms of ORR at an RRDE with supported Pt catalyst on a glassy carbon disk in 0.1 mol L^{-1} $HClO_4$, $v = 0.005$ V s^{-1}, $f = 1600$ U min^{-1}. Potentials are *IR* corrected. Source: Data were provided by Marius. Gollasch and Julian Lorenz (DLR Institute for Networked Energy Systems Oldenburg and Carl v. Ossietzky University Oldenburg).

Figure 12.28 shows linear sweep voltammograms recorded with a ring–disk electrode for a Pt catalyst. Note the different axis scaling for I_D and I_R. For comparison, I_D and I_R are shown with uniform scaling of the current axes in Figure 12.25.

Further Reading for Sections 12.4.2. and 12.4.3. The derivation of the equations for convective mass transport is summarized in [40]. A comparative consideration for other hydrodynamic methods, such as the wall-jet cell or the channel electrodes, can be found in [42]. We already discussed the generation of hydrodynamic flow in micro-analysis systems in Excursion 4-4. Examples for the use of RDE and RRDE to study ORR can be found in [43, 44]. Detailed methodological recommendations regarding RDE and RRDE studies on electrocatalysts can be found in [37, 45–50].

12.4.4 Electrochemical Methods for the Characterization of Electrocatalysts with Improved Mass Transport

RDE and RRDE are used to increase the mass transport to electrocatalysts. Despite the convective mass transport, the realized current densities are much lower than in technical devices, for example, fuel cells. This is due to the fact that O_2 in a fuel cell (Chapter 16) is transported to the electrocatalysts via the gas phase with higher concentrations and about 100 times higher diffusion coefficients. Therefore, significantly higher current densities are obtained in fuel cells than in experiments with RDE or RRDE. Moreover, concentration differences occur *within the* catalyst layers deposited on a RDE or RRDE, which affect the comparability of measurements, especially for very active catalysts [51].

From these limitations, it follows that newly manufactured catalysts cannot be adequately characterized by RDE experiments alone as to whether they actually enable the desired high current densities (i.e., high power output) of a fuel cell. In contrast, the construction of a complete membrane electrode assembly (MEA) requires larger amounts of material, which is not possible with newly fabricated complex nanostructures (such as those shown in Figure 12.36) from which initially only a small amount may be available. Therefore, some new characterization methods have been developed in recent years, which represent an intermediate step between RDE experiments on the one hand and the construction of a complete laboratory cell with a MEA on the other hand. The basic principle is that O_2 is transported to the catalyst via the gas phase, which increases the rate of mass transport. The catalyst is manufactured as a thin layer and is simultaneously in contact with a proton-conducting phase.

In the floating electrode [52], the support structure consists of a polycarbonate track-etch membrane. These are commercially available thin polymer films in which breaks within the polymer chains have been created by bombardment with particle radiation. The molecular defects occur along the straight trajectories of the high-energy particles. The line of defective polymer chains can be expanded in a subsequent chemical process to form nanometer-sized pores of uniform diameter of about 400 nm. The polymer film with the pores is coated with gold on one side and placed with the gold side onto an aqueous electrolyte solution. The uncoated side faces the gas phase. The gold coating forms the electrical connection to Pt nanoparticles on carbon supports placed in the pore openings on the solution side. In this situation, effective transport of O_2 through the gas-filled pore to the catalysts can occur, and on the other hand, H^+ can also be effectively transported from the electrolyte side.

In other variants, a small piece of a Nafion membrane (i.e., much smaller than would be required to produce a single cell) separates the gas space from a macroscopic liquid reservoir. The catalyst is applied or pressed onto the Nafion layers on the side facing the gas atmosphere. Significantly higher current densities can be achieved with this arrangement than with RDE experiments [53, 54].

12.4.5 How Does Differential Electrochemical Mass Spectrometry (DEMS) Work?

Mass spectrometry (MS) is an excellent technique for obtaining information on chemical structure of reaction products, which is not accessible from current–voltage measurements (Figure 8.1). We already mentioned SIMS and desorption mass spectrometry in Section 8.7, which are used to analyze elemental composition and molecular adsorbates on the electrode after emersion from the electrolyte solution.

Integral Determination of Gaseous Reaction Products. A different methodology is needed to identify the electrolysis products that do not adsorb to or are not deposited on the electrode. For this purpose, the confined gas space above an electrolyte solution can be sampled periodically for gaseous or high vapor pressure products (Figure 12.29). In this case, the gas phase above the solution contains molecules that were formed at the electrode at different reaction times (or at different potentials). However, with products building up, the analysis result is integrated over time, which does not allow the detection of products formed at a specific electrode potential in a voltammetric experiment.

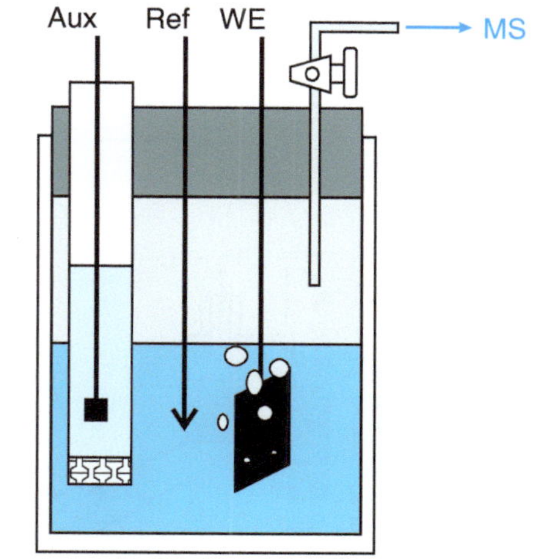

Figure 12.29 Principle of sampling of the gas atmosphere above an electrolyte solution. The auxiliary electrode is separated so that the products formed there are not detected.

The determination of dissolved reaction products is possible, for example, if the products flow continuously past an electrode in a microfluidic channel arrangement. This solution flow is fed into a mass spectrometer. Electrospray ionization is suitable for this purpose, for example, for larger molecules or for mechanistic studies [55]. We discuss the use of ionization in an inductively coupled plasma (ICP-MS) using Figure 14.36 for an application in corrosion research. Capillary systems with one end coupled directly to the inlet of a mass spectrometer and the other end scanned over a sample also facilitate local studies, for example, of products formed at different locations in a materials library [56–58].

DEMS. For electrocatalytic reactions, a different mode of operation finds diverse applications. The central component of the system is a porous membrane made of poly-tetrafluoroethylene (Teflon®, about 75 μm thick). The structure and function of this membrane correspond to those of a Gore-Tex membrane in functional outdoor clothing. The membrane separates the liquid electrolyte from the high vacuum of the

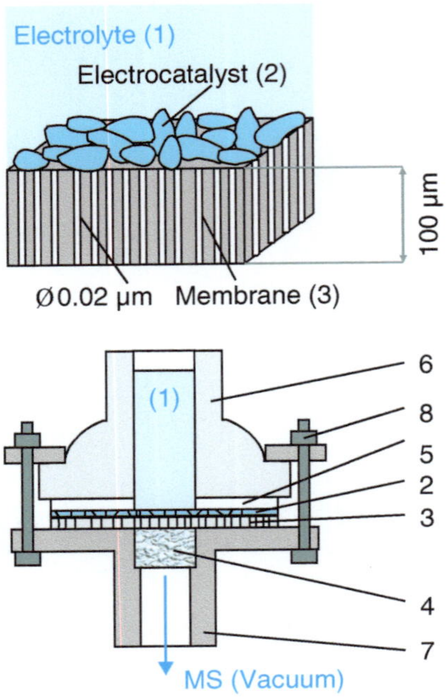

Figure 12.30 Separation of the electrolyte (1) from the vacuum of a mass spectrometer by a porous hydrophobic membrane (3) coated on one side with a porous electrode material (2). Other parts: 4) steel frit as mechanical support for the membrane; 5) spacer and gasket; 6) cell body made of glass; 7) flange of the vacuum apparatus; 8) connection of the glass and metal components.

mass spectrometer (Figure 12.30). The *liquid* water cannot wet the inside of the membrane because of the hydrophobic material and the narrow pores (about 20 nm in diameter). Thus, the liquid cannot escape into the vacuum. However, compounds with a vapor pressure (including the solvent H_2O) can pass through the pores of the membrane as *gaseous* molecules and thus enter the ionization chamber of the mass spectrometer. The electrochemical cell, on the one hand, and the vacuum system of the mass spectrometer, on the other hand, are now arranged around this membrane. The system is designed in such a way that the time for the transfer of volatile product molecules from the working electrode to the mass spectrometer is as short as possible.

In the first realization of DEMS, the electrode material was applied directly to the electrolyte-facing side of the membrane (Figure 12.30). The membrane was mechanically stabilized by a steel frit and connected to the inlet of the mass spectrometer [59]. The transfer time was about 20 s. Today DEMS systems achieve transfer times of no more than 0.2 s, making it possible to detect product formation during the run of a cyclic voltammogram with scan rates up to 50 mV s^{-1}. Here, the mass spectrometer ion current I_i is proportional to the transient electrolysis current I_F for the given product at a given mass-to-charge ratio (m/z). Therefore, the mode of operation is referred to as differential electrochemical mass spectrometry (DEMS), in distinction to the integrating mode of operation shown in Figure 12.29.

On the mass spectrometer side, decisive improvements were achieved with the introduction of a differential pumping scheme along the trajectory of the molecules in the mass spectrometer and the use of inexpensive, compact, and fast quadrupole mass spectrometers (Figure 12.31). Although this type of spectrometer has a low mass resolution (hence its designation as a mass filter), it is sufficient for mostly small gaseous electrolysis products such as CO_2, CO, methyl formate, and formaldehyde.

On the electrochemical cell side, various designs have been developed [60, 61], which, among other things, allow the use of single-crystal electrodes. These enable particularly detailed insights into structure-reactivity relationships for electrocatalytic reactions (Figure 12.32). The cell allows solution exchange and can also be operated under conditions of continuous electrolyte flow. The small electrolyte volume is also suitable when working with small amounts of expensive isotope-labeled substances to elucidate reaction mechanisms [62]. On the other hand, the limited electrolyte volume may cause rapid depletion of the reactants. In a further development is a DEMS cell with two parts (Figure 12.33). The supply of the reagent to the working electrode is made by a wall jet arrangement (Figure 5.51b). The effluent from the wall-jet cell is

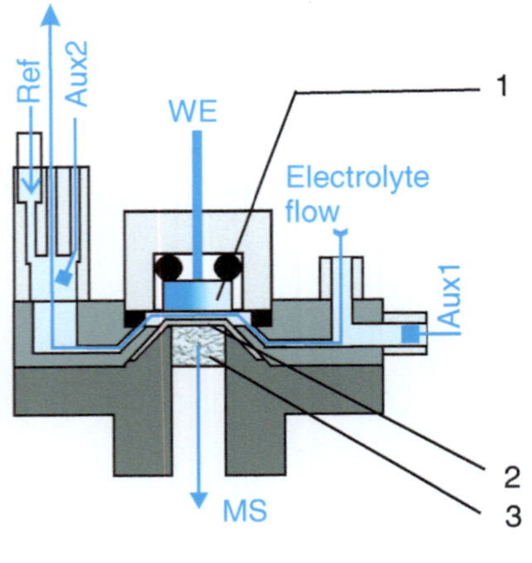

Figure 12.32 DEMS cell for a solid single crystal electrode 1) as working electrode (WE); 2) hydrophobic porous membrane; and 3) steel frit. Source: Adapted from [61].

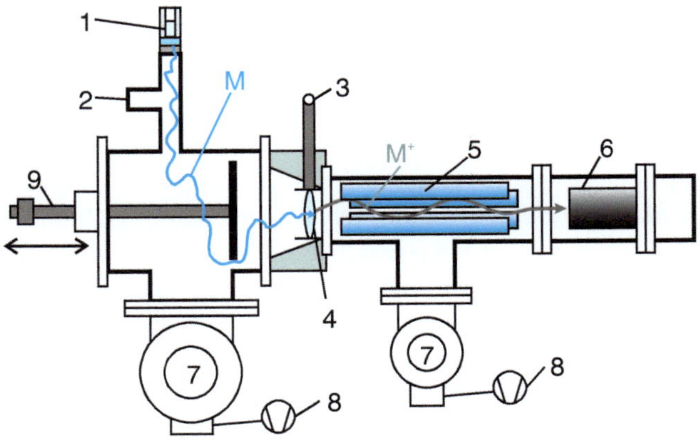

Figure 12.31 Mass spectrometer for DEMS. 1) Electrochemical cell; 2) inlet for calibration; 3) direct inlet; 4) ion source; 5) quadrupole elements; 6) detector; 7) turbomolecular pump; 8) rotary pump; and 9) transfer rod.

then directed to the membrane in a very compact arrangement that separates the electrolyte from the vacuum system (Figure 12.33). It was also possible to integrate illumination for the study of photoelectrochemical reactions in such cells [63].

Calibration. All DEMS cells require calibration for the transfer time, that is, for the delay between the electrochemical conversion and the detection of the products in the mass spectrometer. For each mass number, the sensitivity of the mass spectrometric detection must also be determined because the ionization of different molecules occurs with different efficiencies, and in each cell, only a portion of the products formed at the working electrode are transferred to the spectrometer. For this purpose, considerations of the transfer ratio (collection efficiency) are suitable (similar to the collection efficiency for the RRDE). For calibration, reactions are used with a known amount of products, for example, the oxidation of a monolayer of CO to CO_2. Since the reactant is adsorbed, for example, on Pt, the amount of product is exactly defined. After calibration and for less defined reactions, the ion current of the mass spectrometer can be used to calculate the fraction of the electrolysis current, leading to the formation of a specific product.

Example. The oxidation of methanol in acidic solution at a flat Pt electrode can lead to CO_2 ($m/z = 44$) or formic acid. Formic acid reacts with excess methanol to form methyl formate (methyl formate, $m/z = 60$). Figure 12.34 shows the typical cyclic voltammogram of methanol oxidation in a DEMS cell. As with other alcohols and with small organic molecules, the oxidation current drops to almost zero when an oxide film is formed on the Pt electrode at $E > 0.8$ V, preventing electrocatalysis at the *metal* surface. After the reduction of the Pt surface oxides in the negative-going potential scan, an oxidation current commences (Figure 12.34a). However, it is not possible to determine from the voltammograms which products are formed. The mass traces recorded in parallel for the products CO_2 ($m/z = 44$, Figure 12.34b) and $HCOOCH_3$ ($m/z = 60$, Figure 12.34c) show that both products are formed independently of each other.

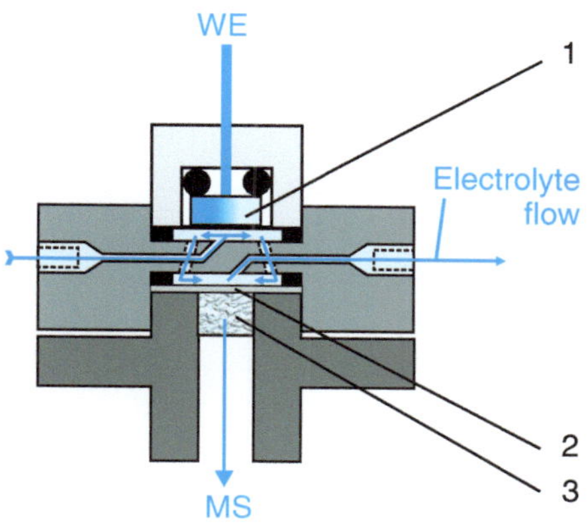

Figure 12.33 Two-part DEMS cell for a solid single-crystal electrode 1) with separate transfer cell; 2) hydrophobic, porous membrane, 3) steel frit. Source: Adapted from [61].

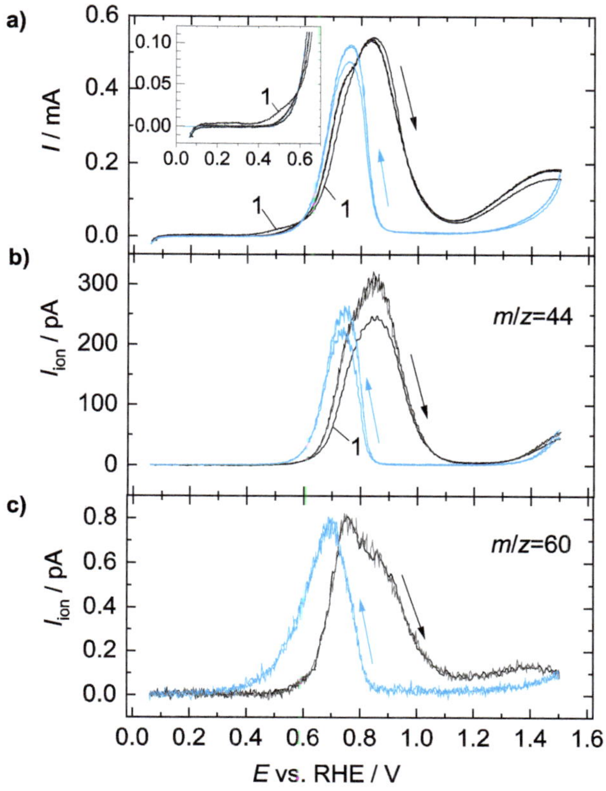

Figure 12.34 Simultaneously recorded cyclic voltammogram and mass spectrometer ion currents for $m/z = 44$ (CO_2) and $m/z = 60$ ($HCOOCH_3$) on flat polycrystalline Pt in 0.1 mol L^{-1} methanol +0.5 mol L^{-1} $HClO_4$, $v = 0.01$ V s^{-1}, electrolyte flow rate 5 μL s^{-1}. The first of three potential cycles is labeled "1". Source: Reproduced with permission from [61], Copyright 2004, American Society of Mass Spectrometry.

In particular, it can be seen from the mass trace $m/z = 60$ that formic acid and methyl formate are generated at lower potentials than CO_2. Chronoamperometric experiments with simultaneous mass spectrometric detection of the products are also possible [61].

Further Reading: Reviews of the DEMS approach and the information that can be gained from it are summarized in [60, 61].

12.5 Important Electrocatalytic Reactions

Practically important reactions are mainly reactions of small molecules that cannot be activated at all without binding to a surface. These include, in particular, the evolution of H_2 (HER), O_2 (OER), Cl_2 (CER), the reductions of O_2 (ORR) and CO_2, the oxidation of H_2 (HOR), and the oxidation reactions of small organic molecules such as methanol or formic acid. We discussed HER and HOR when introducing basic principles of heterogeneous electrocatalysis in Section 12.3 and we will therefore not revisit these reactions in the present section. N_2 reduction represents an even greater challenge than O_2 reduction because a triple bond must be broken in N_2. So far, this can only be achieved with specially prepared electrode materials and is not yet mature for a larger-scale technical application [64].

12.5.1 The Oxygen Reduction Reaction (ORR) and the Oxygen Evolution Reaction (OER)

We have discussed ORR and OER in Section 12.3.3 as an example of a reaction with more than one adsorbed intermediate. We assumed chemically plausible intermediates and were able to derive important predictions about the suitability of electrocatalyst surfaces from the adsorption energies of key intermediates. In this section, we will revisit these reactions because of their great importance. But now we will focus on the sequence of possible reaction steps and the possibilities for optimizing catalysts.

The ORR and OER are of particular importance for electrochemical energy conversion as reactions to balance the oxidation of fuels in fuel cells and in the electrolytic generation of H_2, respectively. Since ORR and OER proceed at high overpotentials, progress toward a possible hydrogen economy depends much more on the improvement of electrocatalysts for these reactions than on catalytic electrodes for H_2 oxidation or H_2 evolution! ORR has further applications in sensor technology. It plays an important role in the corrosion of metals. The OER is also important as a balancing reaction during electrochemical deposition processes.

The reactions shown in Figure 12.35 are conceivable depending on the intermediates that occur and the reactions between them. As a dissociative mechanism of ORR, we can consider a reaction path in which the O–O bond is broken directly upon chemisorption of O_2, followed by a stepwise reaction of O_{ads} (Figure 12.35a). The reverse of this sequence would be the associative mechanism of OER. The associative mechanism of ORR (Figure 12.35b) first leads to OOH_{ads}, followed by cleavage into O_{ads} and OH_{ads}. The peroxo-mechanism (Figure 12.35c) involves the initial formation of OOH_{ads}, followed by further reduction to an adsorbed peroxide $HOOH_{ads}$. Only then does the O–O bond breaking occur.

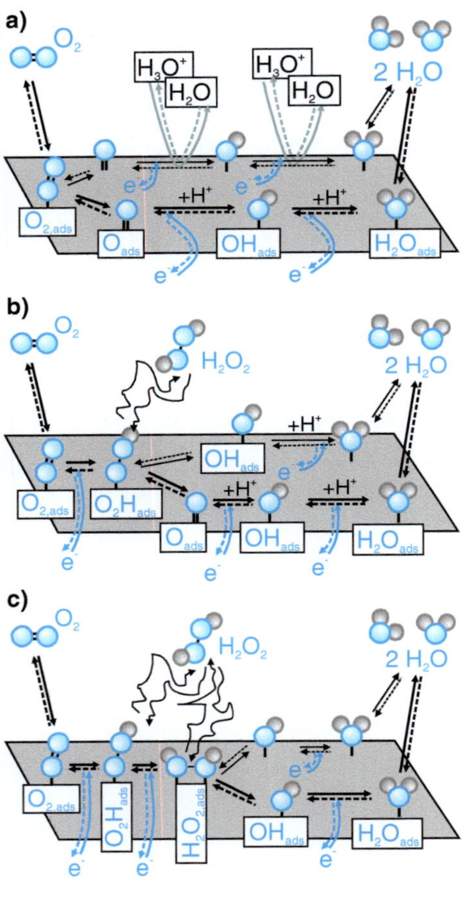

Figure 12.35 Possible mechanisms of ORR (——) and OER (---); a) dissociative mechanism; b) associative mechanism; and c) peroxo-mechanism. For clarity, the involvement of dissolved H_3O^+/H_2O is indicated only for ORR in the foreground of panel a).

The ideal catalyst for **ORR** must have a good balance for the adsorption of OOH_{ads} and OH_{ads}. Among the pure metals, Pt best meets these requirements, but it is still not in the optimum reactivity range when compared to quantum-chemically calculated volcano plots [13]. Since the volcano plots represent semilogarithmic orders, even small differences on the ordinate correspond to very large differences in reactivity. Even different facets of a Pt crystal show divergent activities [43]. This opens up the possibility of increasing activity by optimizing the structure of the catalysts, and the following approaches have been used [20–22, 66, 67]:

- Pt is deposited only as a monolayer on a substrate metal by a galvanic exchange reaction or an UPD process.
- Pt is enriched from a Pt-containing alloy by heat treatment on the surface.
- A surface alloy is produced in which a monolayer of another metal is first deposited by UPD and then exchanged with the surface atoms by heat treatment.
- Electrochemical or chemical dissolution of the less noble element of a Pt-containing alloy. The remaining Pt forms a layer of a few atoms on a core, which still contains the less noble element (Figure 12.36). The base metal influences the reactivity of the Pt layer by either electronic effects and/or strain or compression.

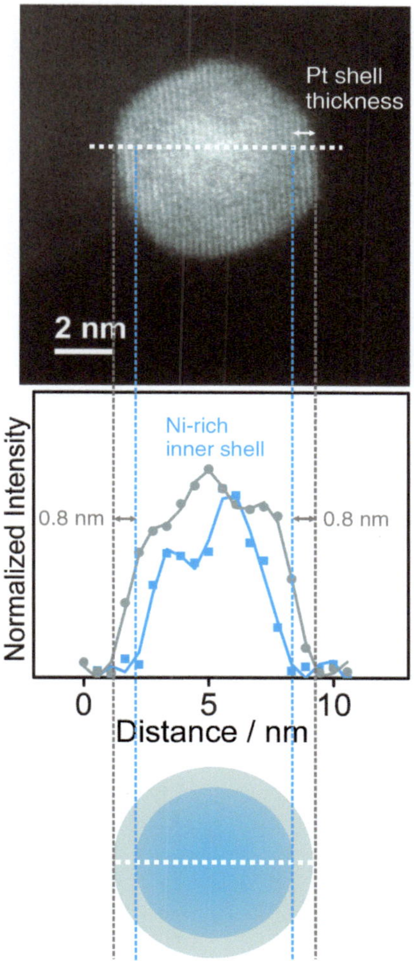

Figure 12.36 HAADF-STEM images of core–shell nanoparticles obtained by dealloying $PtNi_3$ nanoparticles. The Pt-rich shell and the Ni-rich core can be seen particularly well in the two-dimensional intensity diagram along the dotted line (HAADF-STEM – high angular annular dark field scanning transmission electron microscopy). Source: Reproduced with permission from L. Gan, C. Cui, S. Rudi, P. Strasser; Core–shell and nanoporous particle architectures and their effect on the activity and stability of Pt ORR electrocatalysts. *Top. Catal.* **2014**, *57*, 236–244. Copyright 2014, Springer.

The selectivity of the ORR is subject to a complex dependence on the surface structure of the catalyst. Furthermore, the presence of specifically adsorbed ions (such as Cl^-) affects the mechanism. These ions compete with the intermediates for adsorption sites, favoring formation of the undesirable and corrosive by-product H_2O_2 [47]. This leads to the recommendations given at the end of Section 12.4.2 for the experimental characterization of this reaction.

The dependence on Pt, which has a low abundance and can be supplied on Earth at only about 270 tons (2010) per year, with an additional 30% from recycling [65], poses a major problem for mass production. Therefore, considerable research effort is dedicated to using more abundant (and therefore less expensive) materials, possibly at the cost of lower activity. For example, oxides of transition metals can be good catalysts for ORR, provided that alkaline solutions are used. However, there are no membranes so far for alkaline electrolytes that are as stable and efficient as Nafion under acidic conditions (Sections 16.1, 16.3.3). Additionally, these alkaline membranes must exhibit tolerance to CO_2 in the air.

Another class of intensively studied noble metal-free catalysts is nitrogen-coordinated metal centers in carbon materials, in which transition metal ions such as Co and Fe are coordinated in a square planar arrangement and can thus stabilize further adsorbates. Molecular model compounds include cobalt-phthalocyanine, cobalt-tetramethoxy-phenyl-porphyrin, and cobalt-tetraazanulene. Important structural motifs of these catalysts are pyridinic N atoms incorporated into the carbon layers. They have an XPS binding energy of 398.6 eV. The C atoms around the active site act as a catalyst support material and conduct the electrons. To incorporate the N atoms, the edges of the graphitic material must be accessible or must be created. There is some dispute about the importance of transition metals for these catalysts. Since some materials exhibit catalytic activity even after the transition metals are removed, it is reasonable to assume that the transition metals are only necessary to support the incorporation of an aromatic system containing N atoms into the support structure. However, it is analytically difficult to provide evidence for a completely metal-free material, so some measured activities are probably still caused by residual transition metal content [68].

It is interesting to compare the best catalysts for ORR and those for OER: Since OER proceeds at much more positive potentials than ORR, all materials, including noble metals, are covered with an oxide layer at the working potential [34]. The limitations here result from too strong adsorption of OOH_{ads} or too weak binding of O_{ads}. The highest activities are achieved by oxides with the rutile, spinel, or perovskite structure, with IrO_2 and RuO_2 having the highest activities. The activity can be further increased by mixed oxides such as TiO_2/RuO_2 or Ta_2O_5/IrO_2. The structure-reactivity relationships for OER are so far less well understood than for ORR, since the methodological approach of performing reactivity studies on structurally well-characterized single-crystal electrodes is not possible for oxide surfaces.

In addition to the binding of intermediates, other properties such as surface effects on nucleation and detachment of gas bubbles are important for technical applications, because they affect local resistivity and influence the process window in which the materials may be used. There are important differences between ORR and OER. Since O_2 occurs only in low concentrations in aqueous solution, it is possible to completely convert the reactant in the outer layers of a technical electrocatalyst. This is different for OER with the starting material water, which permeates the entire pore space. For OER, a structure is required and becomes a design criterion that facilitates the transport of O_2 out of the pore space and the preferential nucleation of bubbles in places where detachment is possible.

In addition to the activity, the **stability of the catalysts** is significant. In recent years, the systematic understanding of catalyst degradation mechanisms has developed decisively through the use of new investigation methods. We will discuss more details in the context of fuel cells (Section 16.2.4) and for the coupling of droplet cells with online elemental analysis (Figure 14.36), which allowed demonstration of the dissolution of Pt catalysts when changing from an oxide-covered to a metallic surface. In addition to the changes at the Pt surface, the oxidation of the carbon support material is also an important degradation process.

Further Reading. An overview of basic and application-oriented aspects of ORR and OER is provided in [66]. A comparison of different catalyst systems can be found in [69] and reviews on N-coordinated metal centers as electrocatalysts are available in [70–72]. The development of engineered catalysts continues to be inspired by new insights into the mechanisms of enzymatic ORR, which is also of interest for biofuel cells (Section 18.3.4) [73–75]. A broad overview of electrocatalysis in alkaline solutions is provided in [76].

12.5.2 Oxidation of Carbon Monoxide, Formic Acid, Methanol, and Other Alcohols

The oxidation of compounds with one carbon atom takes place stepwise through the sequential cleavage of all, mostly inert C–H bonds in these compounds. For example, the direct oxidation of methane at room temperature is very difficult or impossible. The oxidation of methanol usually proceeds via formaldehyde to formic acid. The further oxidation to CO_2, according to

$$\mathrm{HCOOH} \rightleftharpoons \mathrm{CO_2} + 2\,\mathrm{H^+} + 2\mathrm{e^-} \tag{12.88}$$

can be dissected into two steps

$$\mathrm{HCOOH} \rightleftharpoons \mathrm{HCOO_{ads}} + \mathrm{H^+} + \mathrm{e^-} \tag{12.89}$$

$$\mathrm{HCOO_{ads}} \rightleftharpoons \mathrm{CO_2} + \mathrm{H^+} + \mathrm{e^-}. \tag{12.90}$$

Unlike the case in the HER, two different fragments (CO_2 /H^+) are involved here. The following applies for the optimum catalyst [6]:

$$\Delta G^\circ(\mathrm{HCOO_{ads}}) = \frac{\Delta G^\circ(\mathrm{HCOOH}) + \Delta G^\circ(\mathrm{CO_2})}{2} \tag{12.91}$$

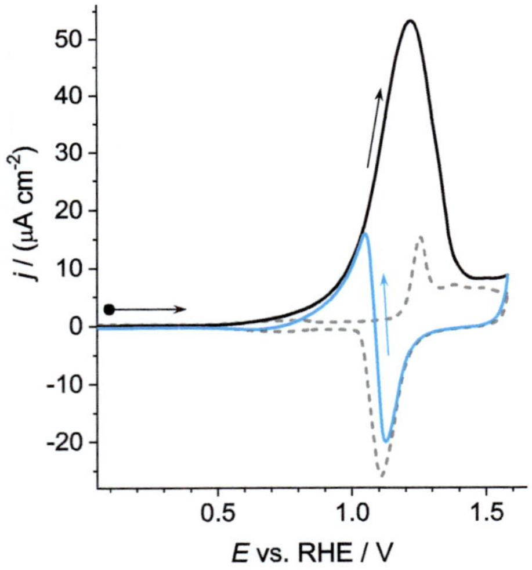

Figure 12.37 Cyclic voltammogram for the oxidation of 1 mol L^{-1} methanol in 0.1 mol L^{-1} KOH on nanoporous Au. The cyclic voltammogram in methanol-free 0.1 mol L^{-1} KOH solution is given as a dashed curve. Source: Data were kindly provided by A. R. Silva Olaya, Carl v. Ossietzky University Oldenburg.

Figure 12.37 shows the cyclic voltammogram for the oxidation of methanol in alkaline solution on nanoporous gold. In the positive-going half-cycle, the current collapses at $E > 1.2$ V vs. RHE because the electrode is covered with a thin oxide layer that inhibits methanol oxidation. *Reduction* of the surface gold oxides occurs in the negative-going potential scan at $E = 1.1$ V vs. RHE, which immediately turns on the *oxidation* of methanol. The current in the outer circuit is then the *differential current* from the reduction of the surface oxides of the metal and the oxidation of the alcohol.

This gives rise to the change in sign of the current in the negative-going potential scan often seen in cyclic voltammograms of alcohols and carbohydrates. The degree of this effect may vary depending on details of the surface structure, the scan rate, and the concentration of the reactants. The exact course of the reaction can be further influenced by underpotential deposition of metal layers in many ways. The peak current in these voltammograms is usually not diffusion-controlled, but determined by the kinetics of the surface reaction. For example, a methanol concentration of 1 mol L^{-1} was used in Figure 12.37. Otherwise, concentrations between 0.0001 and 0.01 mol L^{-1} are more common for cyclic voltammetry.

The oxidation of higher alcohols usually proceeds up to the stage of carboxylic acids without any particular impediments. For further oxidation, the C–C bond would have to be cleaved, which is particularly difficult but highly desirable for the energetic utilization of renewable raw materials (especially for bioethanol for fuel cells) and has therefore led to corresponding research efforts.

The oxidation of CO proceeds via two different intermediates CO_{ads} and OH_{ads}, which cannot be converted into each other. Therefore, more complicated conditions apply here than in the case of ORR. It is crucial that, at the working potential, both intermediates are stabilized at the surface and that they can meet each other. This occasionally occurs, but only at surfaces that have distinctly different functionalities. CO_{ads} is then bound to one subarea, OH_{ads} to another. By surface diffusion, the adsorbates reach the boundary between the subregions and react with each other.

The oxidation of CO has practical significance in the context of fuel cells. CO acts as a catalyst poison in low-temperature polymer electrolyte membrane fuel cells (PEMFC). However, since the formation of CO as an intermediate or by uptake as a trace gas cannot be completely ruled out (e.g., from the exhaust gases of other vehicles with internal combustion engines), an important task is to increase the CO tolerance of fuel cells, which can be achieved if their CO oxidation activity is improved.

Further Reading. Appropriate chapters in the collective works [2, 24, 77, 78] can be recommended.

12.5.3 Carbon Dioxide Reduction Reaction

The reduction of CO_2 to a versatile fuel or building blocks for chemical synthesis is of historical importance to complete the carbon cycle using a technical process. The electrochemical reduction of CO_2 is possible on copper electrodes, albeit at very large overpotentials and therefore with very low energy efficiency. Other suitable metals are lead and tin. It is crucial that the metals have a high overpotential against hydrogen evolution, because the HER represents here an unwanted competing reaction and reduces the faradaic efficiency (current yield) of the CO_2 reduction. Generally, CO_2 reduction yields a product mixture requiring elaborate further processing, a difficulty of its own.

Possible two-electron products are CO, HCOOH, or oxalic acid (HOOC–COOH) with formation of a C–C bond. CO and oxalic acid themselves are not useful fuels; CO could possibly find use as an additive to syngas for the Fischer–Tropsch reaction to build longer-chain hydrocarbons. The most promising product is formic acid, which so far can only be produced electrochemically with high overvoltages. In nature, the enzyme formate dehydrogenase exists, which can realize a reversible conversion $HCOOH \rightleftharpoons CO_2 + 2\,H^+ + 2\,e^-$ when it is immobilized at an electrode [15]. This finding gives rise to the hope that it is ultimately possible to produce such an important catalyst also by purely technical means [79–81].

12.5.4 Chlorine Evolution Reaction

The chlorine evolution reaction (CER) proceeds according to the overall reaction

$$2Cl^- \rightleftharpoons Cl_2 + 2e^- \tag{12.92}$$

with the standard electrode potential

$$E^\circ(Cl_2/Cl^-) = \frac{\Delta_R G^\circ}{2e} = \frac{\Delta G^\circ(Cl_2) - 2\Delta G^\circ(Cl^-)}{2e} \tag{12.93}$$

We can formulate the mechanism with two steps and analyze it analogously to the case of the HER.

$$Cl^- \rightleftharpoons Cl_{ads} + 2e^- \tag{12.94}$$

$$Cl^- + Cl_{ads} \rightleftharpoons Cl_2 + 2e^- \tag{12.95}$$

For the optimal catalyst, $\Delta G(Cl_{ads}) = \Delta G(Cl_2)/2$. This means that the adsorbed chlorine atom is ideally bound to the electrode surface as strongly as in a Cl_2 molecule [6]. In practical implementation, however, it is particularly important to increase the selectivity with respect to the competing oxygen evolution reaction, which proceeds at approximately the same potential.

Further reading: [82–84].

Test your knowledge

Exercise 12.1: For a cube of Pt with a 1 cm edge, calculate the area and the area per mass of Pt if the same amount of Pt atoms were contained in smaller cubic particles with 1 mm, 0.1 mm, 10 μm, 1 μm, 0.1 μm, 20 nm, 10 nm, 5 nm, and 2 nm edge length. For each cube size, also calculate the number of atoms per cube. Document the equations for the calculations and perform the actual calculations in a spreadsheet program.

Exercise 12.2: For which of the following two-electron transfer reactions in aprotic organic solvents would you expect a larger standard reaction rate constant? Justify your hypothesis using Marcus theory.

$$O_2 + e^- \rightleftharpoons O_2^{\bullet -}$$
$$Ar + e^- \rightleftharpoons Ar^{\bullet -} (Ar = \text{Anthracene})$$

Exercise 12.3: Show the equivalence of the expressions in Eqs. (12.23), (12.25), (12.27), and (12.28).

Exercise 12.4: A 25 mg initial mass of a Pt catalyst on carbon with a platinum content of 27% (by mass) is suspended in 4.0 mL of water with 1.0 mL of isopropanol and 20 μL of a Nafion solution [5% (by mass) in ethanol]. Of this solution, 12 μL was applied to a 6 mm-diameter disc electrode. What is the mass loading (mass of Pt per geometric electrode area)?

Exercise 12.5: In Section 12.4.2, we introduced the rotating disk electrode (RDE) as a uniformly accessible electrode. Which other electrodes would have similar properties and under which conditions? For which electrode geometries would you anticipate a nonuniform diffusion layer? How can these considerations be verified experimentally?

Exercise 12.6: In Table 12.2, the viscosity of a 0.1 mol L^{-1} KCl solution in water is given as 0.008844 $cm^2\ s^{-1}$. However, many textbooks recommend the value of 0.01 $cm^2\ s^{-1}$ as an approximation. How large is the relative error in the calculation of the Levich current when the approximation is used instead of the exact value?

Exercise 12.7: The blue shaded areas in Figure 12.19 are $\int_{0.06V}^{0.40V} I(E)\mathrm{d}E = 2.41 \times 10^{-5}\mathrm{A} \cdot \mathrm{V}$ for H desorption (stripping) and $\int_{0.40V}^{0.06V} I(E)\mathrm{d}E = -2.51 \times 10^{-5}\mathrm{A} \cdot \mathrm{V}$ H adsorption The charge for a monolayer of H on Pt is $210\,\mu\mathrm{C\,cm}_{\mathrm{Pt}}^{-2}$. Further details of the experiment are given in the caption of Figure 12.19 and in the Data Collection below the exercises. Calculate the charge converted for H adsorption and H desorption, the Pt area [cm^2], the specific electrochemically active surface area $[\mathrm{m}^2\,\mathrm{g}_{\mathrm{Pt}}^{-1}]$ and the Pt loading $[\mathrm{g\,cm}_{\mathrm{geom}}^{-2}]$.

Exercise 12.8: Calculate the area-related and mass-related activity of the catalyst from the data in the Data Collection below the exercises. Use Eq. (12.79) to determine the kinetically limited current at 0.900 V vs. RHE.

Exercise 12.9: Create a Koutecký–Levich plot for all potentials using the information in the Data Collection below the exercises, and determine the Tafel line from it. Use a spreadsheet program for the calculation.

Exercise 12.10: Use the information from the Data Collection below to calculate the fraction x of oxygen molecules that are not converted to H_2O, but to H_2O_2 at 0.9 V and 1600 rpm.

Exercise 12.11: List all theoretically possible electrochemical oxidation reactions for methanol. Also think of possible products that can be formed from two different intermediates.

Exercise 12.12: Perform the analysis of the chlorine evolution reaction from Section 12.5.4 in the same way as for the HER.

Data collection for Exercises 12.7–12.10. An RRDE with a geometric area of 0.2376 cm^2 for the disk and a collection efficiency of 38.3% was modified with 124 µL of a catalyst ink, which dried completely on the disk electrode under slight rotation to form a uniform film. To prepare the ink, 36 mg of the catalyst powder [Pt nanoparticles on carbon support, 40% (*m*/*m*) Pt content] was dispersed with 2.7 mL of H_2O, 0.9 mL of isopropanol, and 14.4 µL of a 5% (*m*/*m*) Nafion solution using an ultrasonic horn.

In the RRDE measurements, the following values were determined as the average of three LSV measurements at the corrected potentials $E_{corr} = E - I_D R$ ($R = 19.2\ \Omega$).

f rpm	I_D/mA Mean value between 0.3 V and 0.6 V	I_D/mA @0.775 V	I_D/mA @0.800 V	I_D/mA @0.825 V	I_D/mA @0.850 V	I_D/mA @0.875 V	I_D/mA @0.900 V	I_D/mA @0.925 V	I_D/mA @0.950 V
400	−0.7468	−0.7170	−0.7037	−0.6703	−0.6105	−0.5151	−0.3883	−0.2539	−0.1349
900	−1.0968	−1.0291	−0.9924	−0.9291	−0.8175	−0.6609	−0.4811	−0.3040	−0.1507
1200	−1.2554	−1.1577	−1.1055	−1.0232	−0.8856	−0.7077	−0.5115	−0.3175	−0.1551
1600	−1.4354	−1.2901	−1.2257	−1.1159	−0.9501	−0.7550	−0.5394	−0.3322	−0.1579
2000	−1.5891	−1.3752	−1.2892	−1.1628	−0.9870	−0.7714	−0.5517	−0.3340	−0.1561
2500	−1.7549	−1.4972	−1.3924	−1.2401	−1-0397	−0.8157	−0.5783	−0.3449	−0.1588

The ring current at 1600 rpm and $E = 0.900$ V vs. RHE is $I_R = +0.00158$ mA.

References

1 J. Lipkowski, P. N. Ross (Eds.); *Electrocatalysis*, Wiley-VCH, New York, 1998.

2 E. Herrero, J. M. Feliu, A. Aldaz, W. Vielstich; Kinetics and mechanisms of selected electrochemical processes. in *Encyclopedia of Electrochemistry, Vol. 2* (A. J. Bard, M. Stratmann, E. J. Calvo, Eds.), Wiley-VCH, Weinheim, 2003, pp. 441–511.

3 C. A. Luca, N. M. Markovic; Structure relationships in electrochemical reactions. in *Encyclopedia of Electrochemistry, Vol. 2* (A. J. Bard, M. Stratmann, E. J. Calvo, Eds.), Wiley-VCH, Weinheim, 2003, pp. 291–359.

4 G. Kokkinidis; The influence of ad-atom adsorption on the reaction rates and mechanisms. in *Encyclopedia of Electrochemistry, Vol. 2* (A. J. Bard, M. Stratmann, E. J. Calvo, Eds.), Wiley-VCH, Weinheim, 2003, pp. 391–417.

5 A. Wieckowski, E. R. Savinova, C. G. Vayenas (Eds.); *Catalysis and Electrocatalysis at Nanoparticle Surfaces*, Marcel Dekker, New York, 2003.

6 M. T. M. Koper; Thermodynamic theory of multi-electron transfer reactions: Implications for electrocatalysis. *J. Electroanal. Chem.* **2011**, *660*, 254–260.

7 M. T. M. Koper, Catalysis of redox reactions. in *Comprehensive Inorganic Chemistry II: From Elements to Application, Vol. 8* (J. Reedijk, K. Poeppelmeier, Eds.), Elsevier, Oxford, 2013, pp. 459–474.

8 R. A. Marcus; Electron transfer reactions in chemistry. Theory and experiment (Nobel Lecture). *Angew. Chem. Int. Ed.* **1993**, *32,* 1111–1121.

9 A. J. Bard (Ed.); *Electrogenerated Chemiluminscence*, New York, Marcel Dekker, 2004.

10 J. L. Delaney, C. F. Hogan, W. Shen; Electrogenerated chemiluminescence detection in paper-based microfluidic sensors. *Anal. Chem.* **2011**, *83*, 1300–1306.

11 W. Schmickler, E. Santos*; Interfacial Electrochemistry*. 2nd ed., Springer Verlag, Berlin, 2010.

12 A. J. Bard, L. R. Faulkner, H. S. White; *Electrochemical Methods: Fundamentals and Applications*. 3rd ed., John Wiley & Sons, New York, 2022, pp. 142–168.

13 J. K. Nørskov, J. Rossmeisl, A. Logadottir, L. Lindqvist; Origin of the overpotential for oxygen reduction at a fuel-cell cathode. *J. Phys. Chem. B* **2004**, *108*, 17886–17892.

14 J. Greeley, J. K. Nørskov, L. A. Kibler, A. M. El-Aziz, D. M. Kolb; Hydrogen evolution over bimetallic systems: Understanding the trends. *ChemPhysChem* **2006**, *7*, 1032–1035.

15 T. Reda, C. M. Plugge, N. J. Abram, J. Hirst; Reversible interconversion of carbon dioxide and formate by an electroactive enzyme. *Proc. Natl. Acad. Sci. U.S.A.* **2008**, *105*, 10654–10658.

16 J. Rossmeisl, A. Logadottir, J. K. Nørskov; Electrolysis of water on (oxidized) metal surfaces. *Chem. Phys.* **2005**, *319*, 178–184.

17 J. Rossmeisl, Z.-W. Qu, H. Zhu, G.-J. Kroes, J. K. Nørskov; Electrolysis of water on oxide surfaces. *J. Electroanal. Chem.* **2007**, *607*, 83–89.

18 J. Weissmüller; Electrocapillarity of solids and its impact on heterogeneous catalysis. in *Electrocatalysis: Theoretical Foundation and Model Experiments* (R. C. Alkire, L. A. Kibler, D. M. Kolb, J. Lipkowski, Eds.), Wiley-VCH, Weinheim, 2013, pp. 163–219.

19 J. Weissmüller; Adsorption-strain coupling at solid surfaces; *Curr. Opin. Chem. Eng.* **2019**, *24*, 45–53.

20 V. Stamenkovic, B. S. Mun, K. J. J. Mayrhofer, P. N. Ross, N. M. Markovic; Effect of surface composition on electronic structure, stability and electrocatalytic properties of Pt-transition metal alloys: Pt-skin versus Pt-skeleton surfaces. *J. Am. Chem. Soc.* **2006**, *128*, 8813–8819.

21 Y. Bing, H. Liu, L. Zhang, D. Ghosh, J. Zhang; Nanostructured Pt-alloy electrocatalysts for PEM fuel cell oxygen reduction reaction. *Chem. Soc. Rev.* **2010**, *39*, 2184–2202.

22 M. Özaslan, F. Hasché, P. Strasser; Pt-based core-shell catalyst architectures for oxygen fuel cell electrodes. *Phys. Chem. Lett.* **2013**, *4*, 3273–3291.

23 N. M. Marković, P. N. Ross; Surface science studies of model fuel cell electrocatalysts. *Surf. Sci. Rep.* **2002**, *45*, 117–229.

24 M. T. M. Koper (Ed.); *Fuel Cell Catalysis: A Surface Science Approach*, John Wiley & Sons, Hoboken, 2009.

25 B. Hammer, J. K. Nørskov; Theoretical surface science and catalysis – Calculations and concepts. *Adv. Catal.* **2000**, *45*, 71–129.

26 B. Hammer, J. K. Nørskov; Why is gold the noblest of all the metals? *Nature* **1995**, *376*, 238–240.

27 A. R. Zeradjanin, J.-P. Grote, G. Polymeros, K. J. J. Mayrhofer; A critical review on hydrogen evolution electrocatalysis: Re-exploring the volcano-relationship. *Electroanalysis* **2016**, *28*, 2256–2269.

28 K. S. Exner; Recent advancement towards closing the gap between electrocatalysis and battery science communities: The computational lithium electrode and activity-stability volcano plots. *ChemSusChem* **2019**, *12*, 2330–2344.

29 J. E. Mueller, D. Fantauzzi, T. Jacob; Multiscale modeling of electrochemical systems. in *Electrocatalysis: Theoretical Foundation and Model Experiments* (R. C. Alkire, L. A. Kibler, D. M. Kolb, J. Lipkowski, Eds.), Wiley-VCH, Weinheim, 2013, pp. 1–73.

30 D. Chen, Q. Tao, L. W. Liao, S. X. Liu, Y. X. Chen, S. Ye; Determining the active surface area for various platinum electrodes. *Electrocatalysis* **2011**, *2*, 207–219.

31 V. Climent, J. M. Feliu; Thirty years of platinum single crystal electrochemistry. *J. State Solid Electrochem.* **2011**, *15*, 1297–1315.

32 G. Tremiliosi-Filho, L. H. Dall'Antonia, G. Jerkiewicz; Limit to extent of formation of the quasi-two-dimensional oxide state on Au electrodes. *J. Electroanal. Chem.* **1997**, *422*, 149–159.

33 S. Trasatti, O. A. Petrii; Real surface area measurements in electrochemistry. *Pure Appl. Chem.* **1991**, *63*, 711–734.

34 B. E. Conway; Electrochemical oxide film formation at noble metals as a surface chemical process. *Prog. Surf. Sci.* **1995**, *49*, 331–452.

35 M. T. M. Koper, S. C. S. Lai, E. Herrero; Mechanisms of the oxidation of carbon monoxide and small organic molecules at metal electrodes. in *Fuel Cell Catalysis: A Surface Science Approach* (M. T. M. Koper, Ed.), John Wiley & Sons, Hoboken, 2009, pp. 159–207.

36 J. Solla-Gullón, P. Rodríguez, E. Herrero, A. Aldaz, J. M. Feliu; Surface characterization of platinum electrodes. *Physiol. Chem. Phys.* **2008**, *10*, 1359–1373.

37 K. J. J. Mayrhofer, D. S. Strmcnik, B. B. Blizanac, V. R. Stamenkovic, M. Arenz, N. M. Markovic; Measurement of oxygen reduction activities via rotating disc electrode method: From Pt model surfaces to carbon-supported high surface area catalysts. *Electrochim. Acta* **2008**, *53*, 3181–3188.

38 J. Newman; Current distribution on a rotating disk below the limiting current. *J. Electrochem. Soc.* **1966**, *113*, 1235–1241.

39 W. J. Albery, M. L. Hitchman, *Ring–Disc Electrodes*, Calderon Press, Oxford, 1971.

40 A. J. Bard, L. R. Faulkner, H. S. White; *Electrochemical Methods: Fundamentals and Applications.* 3rd ed., John Wiley & Sons, Hoboken, NJ, 2022, pp. 411–442.

41 C. P. Andrieux, J. M. Savéant; Kinetics of electrochemical reactions mediated by redox polymer films: Irreversible cross-exchange reactions: Formulation in terms of characteristics currents for stationary techniques. *J. Electroanal. Chem.* **1982**, *134*, 163–166.

42 P. R. Unwin; Dynamic electrochemistry as a quantitative probe of interfacial physicochemical processes. *J. Chem. Soc. Faraday Trans.* **1998**, *94*, 3183–3195.

43 N. M. Marković, H. A. Gasteiger, P. N. Ross; Oxygen reduction on platinum low-index single-crystal surfaces in sulfuric acid solution: Rotating ring–Pt(*hkl*) disk studies. *J. Phys. Chem.* **1995**, *99*, 3411–3415.

44 U. A. Paulus, T. J. Schmidt, H. A. Gasteiger, R. J. Behm; Oxygen reduction on a high-surface area Pt/Vulcan carbon catalyst: A thin-film rotating ring–disk electrode study. *J. Electronal. Chem.* **2001**, *495*, 134–145.

45 C. Wei, R. R. Rao, J. Peng, B. Huang, I. E. L. Stephens, M. Risch, Z. J. Xu, Y. Shao-Horn; Recommended practices and benchmark activity for hydrogen and oxygen electrocatalysis in water splitting and fuel cells. *Adv. Mater.* **2019**, *31*, 1806296, 1–24.

46 D. van der Vliet, D. S. Strmcnik, C. Wang, V. R. Stamenkovic, N. M. Markovic, M. T. M. Koper; On the importance of correcting for the uncompensated ohmic resistance in model experiments of the oxygen reduction reaction. *J. Electroanal. Chem.* **2010**, *647*, 29–34.

47 T. J. Schmidt, U. A. Paulus, H. A. Gasteiger, R. J. Behm; The oxygen reduction reaction on a Pt/Carbon fuel cell catalyst in the presence of chloride anions; *J. Electroanal. Chem.* **2001**, *508*, 41–47.

48 K. Shinozaki, J. W. Zack, R. M. Richards, B. S. Pivovar, S. S. Kocha; Oxygen reduction reaction measurement on platium electrocatalysts utilizing rotating disk electrode technique. I. Impact of impurities, measurement protocols and applied correction. *J. Electrochem. Soc.* **2015**, *162*, F1144–F1158.

49 K. Shinozaki, J. W. Zack, Svitlana Pylypenko, B. S. Pivovar, S. S. Kocha; Oxygen reduction reaction measurement on platium electrocatalysts utilizing rotating disk electrode technique. II. Influence of ink formulation, catalyst layer uniformity and thickness. *J. Electrochem. Soc.* **2015**, *162*, *12*, F1384–F1396.

50 S. S. Kocha, K. Shinozaki, J. W. Zack, D. J. Myers, N. N. Mariuki, T. Nowicki, V. Stamenkovic, Y. Kang, D. Li, D. Papgeorgopoulos; Best practice and testing protocols for benchmarking ORR activity of fuel cell electrocatalysts using rotating disk electrodes. *Electrocatalysis* **2017**, *8*, 366–374.

51 S. Martens, L. Asen, G. Ercolano, F. Dionigi, C. Zalitis, A. Hawkins, A. Martinez Bonastre, L. Seidl, A. C. Knoll, J. Sharman, P. Strasser, D. Jones, O. Schneider; A comparison of rotating disc electrode, floating electrode technique and membrane electrode assembly measurements for catalyst testing. *J. Power Sources* **2018**, *392*, 274–284.

52 C. M. Zalitis, D. Kramer, A. R. Kucernak; Electrocatalytic performance of fuel cell reactions at low catalyst loading and high mass transport. *Phys. Chem. Chem. Phys.* **2013**, *15*, 4329–4340.

53 M. Inaba, A. Westergard Jensen, G. W. Sievers, M- Escudero-Escribano, A. Zana, M. Arenz; Benchmarking high surface area electrocatalysts in a gas diffusion electrode: Measurement of oxygen reduction activities under realistic conditions. *Energy Environ. Sci.* **2018**, *11*, 988–994.

54 K. Ehelebe, D. Seeberger, M. T. Y. Paul, S. Thiele, K. J. J. Mayrhofer, S. Cherevko; Evaluating electrocatalysts at relevant currents in a half-cell: The impact of Pt loading on oxygen reduction reaction. *J. Electrochem. Soc.* **2019**, *166*, F1259–F1268.

55 M. Abonnenc; L. Qiao, B. Liu, H. H. Girault; Electrochemical aspects of electrospray and laser desorption/ionization for mass spectrometry. *Annu. Rev. Anal. Chem.* **2010**, *3*, 321–354.

56 A. D. Modestov, S. Srebnik, O. Lev, J. Gun; Scanning capillary microscopy/mass spectrometry for mapping spatial electrochemical activity of electrodes. *Anal. Chem.* **2001**, *73*, 4229–4240.

57 K. Jambunathan, S. Jayaramann, A. C. Hillier; A multielectrode electrochemical and scanning differential mass spectrometry study of methanol oxidation on electrodeposited Pt_x Ru_y. *Langmuir* **2004**, *20*, 1856–1863.

58 D. Momotenko, L. Qiao, F. Cortés-Salazar, A. Lesch, G. Wittstock, H. H. Girault; Electrochemical push-pull scanner with mass spectrometry detection. *Anal. Chem.* **2012**, *84*, 6630–6637.

59 S. Bruckenstein, R. R. Gadde; Use of a porous electrode for in situ mass spectrometric determination of volatile electrode reaction products. *J. Am. Chem. Soc.* **1971**, *93*, 793–794.

60 B. Bittins-Cattaneo, E. Cattaneo, P. Königshoven, W. Vielstich; New developments in electrochemical mass spectrometry. in *Electroanalytical Chemisty, Vol. 17* (A. J. Bard, Ed.), Marcel Dekker, New York, 1991, pp. 181–220.

61 H. Baltruschat; Differential electrochemical mass spectrometry. *J. Am. Soc. Mass Spectrom.* **2004**, *15*, 1693–1706.

62 N. A. Anastijevic, J. Heitbaum; On the hydrogen evolution during the electrochemical oxidation of aldehydes at Ib metals. *Electrochim. Acta* **1993**, *38*, 1067–1072.

63 P. Bogdanoff, P. Friebe, P. Alonso-Vante; A new inlet system for differential electrochemical mass spectroscopy applied to the photocorrosion of *p*-InP(111) single crystals. *J. Electrochem. Soc.* **1998**, *145*, 576–582.

64 H. Iriwan, S. Z. Andersen, X. Zhang, B. J. Comer, J. Barrio, P. Chen, A. J. Medford, I. E. L. Stephens, I. Chorkendorff, Y. Shao-Horn; Methods for nitrogen activation by reduction and oxidation. *Nat. Rev. Methods Primers* **2022**, *1*, 56.

65 D. R. Wilburn; Global Exploration and Production Capacity for Platinum-Group Metals From 1995 Through 2015. Scientific Investigation Reports 2012-5164, US Department of the Interior and U.S. Geological Survey. Reston, Virginia 2012. https://pubs.usgs.gov/publication/sir20125164 (last access 10 Aug 2024).

66 I. Katsounaros, S. Cherevko, A. R. Zeradjanin, K. J. J. Mayrhofer; Oxygen electrochemistry as a cornerstone for sustainable energy conversion. *Angew. Chem. Int. Ed.* **2014**, *53*, 102–121.

67 I. E. L. Stephens, A. S. Bondarenko, U. Grønberg, J. Rossmeisl, I. Chorkendorff; Understanding the electrocatalysis of oxygen reduction on platinum and its alloys. *Energy Environ. Sci.* **2012**, *5*, 6744–6772.

68 J. Masa, W. Xia, M. Muhler, W. Schuhmann; On the role of metals in nitrogen-doped carbon electrocatalysts for oxygen reduction. *Angew. Chem. Int. Ed.* **2015**, *54*, 10102–10120.

69 H. A. Gasteiger, N. M. Marković; Just a dream – or future reality? *Science* **2009**, *324*, 48–49.

70 Y. P. Zhu, C. Guo, Y. Zheng, S.-Z. Qiao; Surface and interface engineering of noble-metal-free electrocatalysts for efficient energy conversion processes. *Acc. Chem. Res.* **2017**, *50*, 915–923.

71 L. Lin, Q. Zhu, A.-W. Xu; Noble-metal-free Fe-N/C catalysts for highly efficient oxygen reduction reaction under both alkaline and acidic conditions. *J. Am. Chem. Soc.* **2014**, *136*, 11027–11033.

72 U. Tylus, Q. Jia, K. Strickland, N. Ramaswamy, A. Serov, P. Atanassov, S. Mukerjee; Elucidating oxygen reduction active sites in pyrolyzed metal-nitrogen coordinated non-precious-metal electrocatalyst systems. *J. Phys. Chem. C* **2014**, *118*, 8999–9008.

73 J. A. Cracknell, K. A. Vincent, F. A. Armstrong; Enzymes as working or inspirational electrocatalysts for fuel cells and electrolysis. *Chem. Rev.* **2008**, *108*, 2439–2461.

74 S. C. Barton, J. Gallaway, P. Atanassov; Enzymatic biofuel cells for implantable and microscale devices. *Chem. Rev.* **2004**, *104*, 4867–4886.

75 V. Artero, M. Chavarot-Kerlidou, M. Fontecave; Splitting water with cobalt. *Angew. Chem. Int. Ed.* **2011**, *123*, 7238–7266.

76 Y. Yang, C. R. Peltier, R. Zeng, R. Schimmenti, Q. Li, X. Huang, Z. Yan, G. Potsi, R. Selhorst, X. Lu, W. Xu, M. Tader, A. V. Soudackov, H. Zhang, M. Krumov, E. Murray, P. Xu, J. Hitt, L. Xu, H.-Y. Ko, B. G. Ernst, C. Bundschu, A. Luo, D. Markovich, M. Hu, C. He, H. Wang, J. Fang, R. A. DiStasio, Jr, L. F. Kourkoutis, A. Singer, K. J. T. Noonan, L. Xiao, L. Zhuang, B. S. Pivovar, P. Zelenay, E. Herrero, J. M. Feliu, J. Suntivich, E. P. Giannelis, S. Hammes-Schiffers, T. Arias, M. Mavrikakis, T. E. Mallouk, J. D. Brock, D. A. Muller, F. J. DiSalvo, G. W. Coates, H. D. Abruña; Electrocatalysis in alkaline media and alkaline membrane-based energy technologies. *Chem. Rev.* **2022**, *122*, 6117–6321.

77 W. Vielstich; CO, formic acid, and methanol oxidation in acid electrolytes – Mechanisms and electrocatalysis. in *Encyclopedia of Electrochemistry, Vol. 2* (J. Calvo, A. J. Bard, M. Stratmann, Eds.), Wiley-VCH, Weinheim, 2003, pp. 466–511.

78 N. S. Marinkovic, M. Li, R. R. Adzic; Pt-based catalysts for the electrochemical oxidation of ethanol. *Top. Curr. Chem.* **2019**, *377*: 11, 1–39.

79 Y. Hori; Electrochemical CO_2 reduction on metal electrodes. in *Modern Aspects of Electrochemistry, Vol. 42.* (C. Vayenas, Ed.), Springer, New York, 2008, pp. 89–189.

80 J. He, N. J. J. Johnson, A. Huang, C. P. Berlinguette; Electrocatalytic alloys for CO_2 reduction. *ChemSusChem* **2018**, *11*, 48–57.

81 S. Nitopi, E. Bertheussen, S. B. Scott, X. Liu, A. K. Engstfeld, S. Horch, B. Seger, I. E. L. Stephens, K. Chan, C. Hahn, J. K. Nørskov, T. F. Jaramillo, I. Chorkendorff; Progress and perspectives of electrochemical CO_2 reduction on copper in aqueous electrolyte. *Chem. Rev.* **2019**, *119*, 7610–7672.

82 R. K. B. Karlsson, A. Cornell; Selectivity between oxygen and chlorine evolution in the chlor–alkali and chlorate process. *Chem. Rev.* **2016**, *116*, 2982–3028.

83 K. S. Exner, J. Anton, T. Jacob, H. Over; Controlling selectivity in the chlorine evolution reaction over RuO_2-based catalysts. *Angew. Chem. Int. Ed.* **2014**, *53*, 11032–11035.

84 K. S. Exner; Beyond the thermodynamic-based material-screening concepts: Kinetic scaling relations exemplified by the chlorine evolution reaction over transition-metal oxides. *Electrochim. Acta* **2020**, *334*, 13555.

13

Modified Electrodes

In this chapter, we will consider electrodes that have been *intentionally* coated with a material of a different chemical composition to give them special properties. Such desired properties include, for example, electrocatalytic activity for electrosynthetic conversions or for sensors, enrichment or attachment of components from a solution, electrochromic properties for displays, or passivation of reactive surfaces. Following the above definition, we could also consider adsorption of organic compounds (Section 8.10), underpotential deposition (Section 8.9), and the production of a passivating oxide film in the context of corrosion protection (Chapter 14) as an electrode modification. In the extremely broad field of modified electrodes, both in terms of substance classes and applications, we want to concentrate on those layers that differ fundamentally from the underlying electrode in terms of the elements they contain and their bonding forms, such as an organic polymer film on a metal electrode.

A metal electrode exhibits electronic conductivity. If the coating of the electrode is extremely thin, that is, up to 1 nm (or 10 C–H bond lengths), electron transfer (ET) between the metal and a reagent in solution can take place through the thin layer (Figure 13.1a). For thicker films, this is no longer possible. Here, the film itself must have a conductivity for ions and/or electrons so that an electrode reaction can take place. Figure 13.1b shows a possible reaction scheme for a film with mixed conductivity. Figure 13.1c schematically depicts an electronically insulating polymer film that swells in contact with an electrolyte solution and then becomes ion-conducting so that ET reactions are possible at the electrode–film interface. Without conductivity, the film has a passivating effect and reactions do not take place, even if the reaction is expected based on thermodynamic reasoning.

We will use some examples to investigate how the electrochemical behavior of modified electrodes changes during the transition from thin, highly ordered layers to molecular-scale thick films. The underlying principles can be applied to many other material systems, such as organic–inorganic hybrid materials.

To understand this chapter, it is recommended to review Chapters 1–4 as well as Sections 5.3.5 and 5.5. For the characterization of the films, the methods discussed in Chapter 8 are employed. Therefore, the reader should be familiar with Chapter 8.

Further Reading. To deepen understanding of the subject, textbook chapters on modified electrodes [1–4], review articles [5–10], and encyclopedic compilations [11–15] are available. For special aspects, further sources are listed at the end of each respective section.

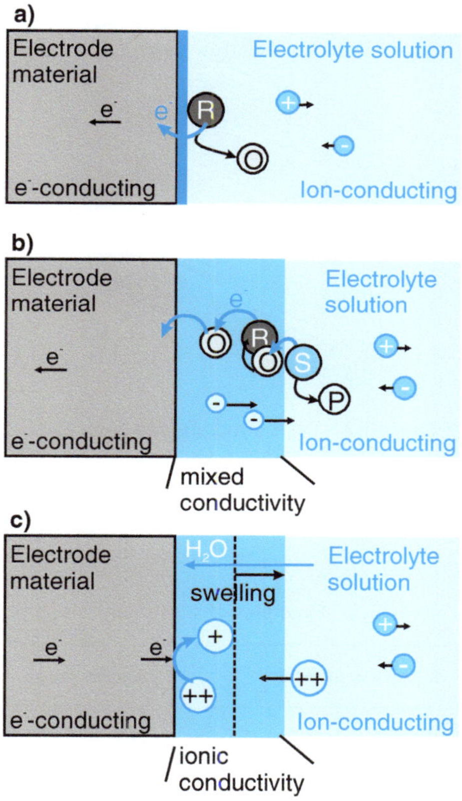

Figure 13.1 Electrochemical reactions on modified electrodes.

a) Tunneling of electrons through thin films;
b) reactions on films with electronic *and* ionic conductivity;
c) passage of ions through a swollen, electronically insulating polymer film and ET reaction at the electrode–film interface.

13.1 Monolayer Systems

Most organic compounds tend to adsorb from aqueous solutions onto electrode surfaces, where they form disordered adsorbate films according to the principles described in Section 8.10. These films can slow down or inhibit electrode reactions. If the adsorbed molecules are themselves redox active, their redox chemistry can be observed. The appearance of the voltammograms then depends on how strong the adsorption of the reactants and products is and whether diffusing molecules react at the electrode in addition to the adsorbed molecules (Section 13.5).

The binding of redox systems to surfaces is of particular interest if the redox system is covalently immobilized at the electrode surface via a linker so that it can be transferred together with the electrode body into an electrolyte system in which the redox system alone would not dissolve at all. Thus, it could not be investigated electrochemically. Such targeted modifications of electrode surfaces are particularly successful with the aid of ordered monolayers. Since thermal motion counteracts the spontaneous formation of an ordered molecular system, there must be either special molecular structures allowing energetically favorable interaction between the adsorbates in the ordered layer that overcome the tendency toward disorder (Section 13.1.1) or work must be performed during the preparation of the system to enforce the formation of ordered systems (Section 13.2.1).

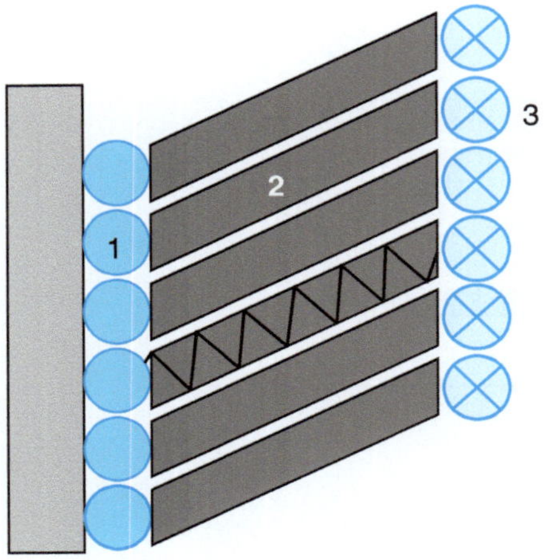

Figure 13.2 Basic structure of a SAM.
1) Adhesive group;
2) spacer;
3) terminal group.

13.1.1 Self-Assembled Monolayers

Self-assembled monolayers (SAMs) form when molecules with an anchor group or head group bind chemically to a solid surface (in our case, an electrode), and the spacers attached to the anchor group allow for a tight packing that also includes the terminal group or tail group (Figure 13.2). Some examples of SAM systems are summarized in Table 13.1. Suitable combinations of anchor group and electrode material provide a kind of "construction kit" or "molecular Lego(R)", with which very different functions can be realized. The anchor groups can be combined according to uniform principles (of organic chemistry) with spacers of different lengths and designs, and with a range of terminal groups. The most suitable anchor groups for many metallic electrodes are thiols and disulfides. Some other sulfur-containing functional groups have also been used occasionally. Molecules with two sulfur atoms (e.g., lipoic acid) are

Table 13.1 Examples of self-organized monolayers.

Substrate surface	Anchor group	Bond	Spacer
Au, Ag, Cu, Pd, GaAs, InP	HS–R, R–S–S–R	Au–SR	
SiO_2, glass	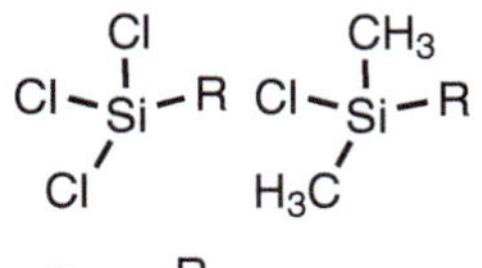	Si–O–Si	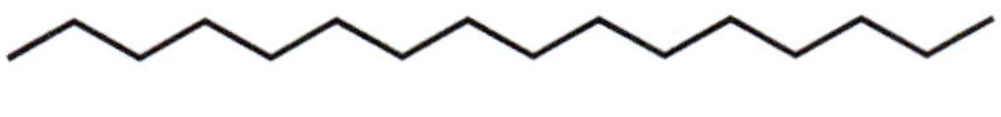 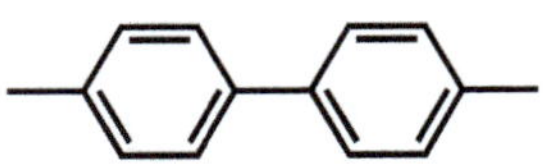
Si|SiH	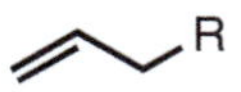	Si–R	
ZrO_2, TiO_2, In_2O_3/SnO_2 (ITO)	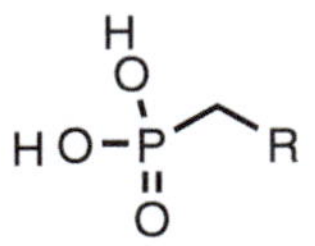	M^{n+} $[O_3P–R]^{2-}$	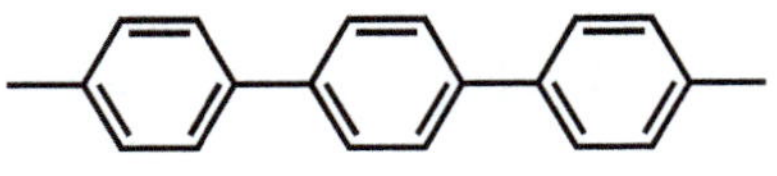
Al_2O_3	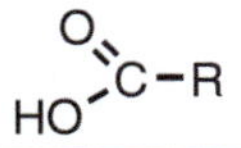	$M^{n+}[OOC–R]^-$	

interesting because they often form more stable monolayers. Similar experiments have been conducted with selenium analogs. Oxide surfaces can be modified with silanes. Hydrogen-terminated silicon surfaces can be modified with terminal alkenes. Phosphonates bind to most transition metal oxides such as transparent conducting oxides (TCO). Suitable spacers are unbranched chain-like molecules such as alkanes, perfluorinated alkanes, and diphenyl or *p*-terphenyl groups. The terminal group determines the desired function. Common functions are the introduction of a charge [$-SO_3^-$, $-COO^-$, $(-N^+CH_3)_3$], reactive groups allowing covalent binding of other molecules (e.g., using click chemistry [16]), or groups that can be activated electrochemically [17] or photochemically [18] for binding or interaction with biological cells or macromolecules. Layers that suppress cell adhesion and protein adsorption (e.g., oligoethylene glycol) are also important in certain applications [19]. The diversity of applications forces us to present the formation principles of SAMs using a few particularly well-studied examples and to leave a further in-depth treatment of the applications to self-study. Monolayers have been used for some time in the large-scale processing of minerals by flotation. For example, flotation of sulfidic ores can be described as an electrochemical process, in which surface oxidation starts by injecting O_2 gas bubbles into a suspension of the ground metal sulfides. This process is accompanied by adsorption of sulfur-containing hydrophobic compounds (e.g., xanthates called "collector" in this context). They adsorb to the surface of sulfide minerals controlled by the potential, which is increased slowly by the addition of oxygen to the system. The adsorbate film renders the surface of the mineral particle hydrophobic and causes ore particles to attach to the rising gas bubbles that carry them to the surface of the liquid. Ore particles can then be skimmed off as a froth [20].

13.1.2 Alkanethiolates on Gold and Silver

Preparation. Alkanethiolate SAMs on gold are particularly popular, probably because of the ease of preparation. We can obtain them by immersing a *fresh* gold surface into an approximately 10^{-3} mol L^{-1} solution of an alkanethiol or dialkyl disulfide in ethanol or isopropanol for 12–24 h.

Q: What does "fresh" mean in this context?

A: "Fresh" is of course somewhat colloquial here. It refers to a surface onto which no other substances have previously been adsorbed, that is, on which the entire area is still available for the binding of alkanethiols. Often, this can only be achieved with electrodes made by vapor deposition processes that are immersed in the adsorption solution within seconds after retrieval from the vacuum deposition chamber. The short exposure time to ambient air is important to prevent other molecules from the gas phase from binding tightly to the electrode during the transfer. Particularly in laboratories, where thiols are handled, trace concentrations of these compounds are always present in the air. Even these trace gases can be sufficient to prevent the formation of clean SAMs. Further manufacturing and purification procedures are described in Section 3.9.4. For a discussion of further details, see [7].

SAM Formation Reaction. We can describe the reactions of alkanethiol RSH by the following reaction equations, which show that SAM formation itself can be understood as a redox reaction.

$$\text{R–SH} + \text{Au}_n^{(0)} \rightleftharpoons [\text{RS}^-]\left[\text{Au}^+\text{–Au}_{n-1}^{(0)}\right] + \frac{1}{2}\text{H}_2 \tag{13.1}$$

In the presence of oxygen, the electrons can be consumed by the oxygen reduction reaction:

$$\text{R–SH} + \text{Au}_n^{(0)} + \frac{1}{2}\text{O}_2 \rightleftharpoons [\text{RS}^-]\left[\text{Au}^+\text{–Au}_{n-1}^{(0)}\right] + \frac{1}{2}\text{H}_2\text{O}_2. \quad (13.2)$$

Especially in the case of thiols with a complex structure, it is often easier to synthesize and store the corresponding dialkyl disulfide, since thiols easily oxidize to the corresponding disulfides in an oxygen-containing atmosphere. The dialkyl-disulfides react according to

$$\text{R–SS–R} + \text{Au}_{2n}^{(0)} \rightleftharpoons 2\ [\text{RS}^-]\left[\text{Au}^+\text{–Au}_{n-1}^{(0)}\right] \quad (13.3)$$

and also lead to monolayers, which are indistinguishable from those after reactions (13.1) and (13.2). Reports stating the opposite are likely due to artifacts in the structural analyses performed by low-energy electrons (cf. Figure 8.30).

Insights into the structure of the monolayers were obtained by experiments with asymmetric disulfides R–SS–R′, such as $HO–(CH_2)_{10}–SS–(CH_2)_{10}–CF_3$. After chemisorption, both terminal groups are present in equal amounts in the SAM; thus, both alkyl residues must be incorporated into the monolayer. If these SAMs are immersed again in solutions of thiols or disulfides, a partial exchange of the adsorbates occurs (Figure 13.3), with the smaller thiols causing much faster substitution than the larger disulfides. In this process, the thiolate with the trifluoromethyl end group of the original SAM exchanges for $HS–(CH_2)_{10}–CN$ about 1000 times faster than the thiolate with the hydroxyl group. This is only possible if the two parts of the original disulfide R–SS–R′ are present in the SAM as separate, independent units RS^- and $R'S^-$ [8].

Kinetics of Self-Assembly. The adsorption of thiols from millimolar alcoholic solutions occurs in two steps. In the first step, 80–90% of the final coverage is reached within a few minutes. This process follows the Langmuir mechanism ($R + \square \rightleftharpoons R_{ads}$) and is usually diffusion-controlled. The process takes about 1 min in a solution containing 1 mmol L^{-1} thiol and about 100 min in a 1 μmol L^{-1} solution as a rough estimate. In a second, much slower step, the alkyl groups present in solution mostly with a random conformation change to an *all-trans conformation* by rotation around the C–C bonds, which stabilize the monolayer by van der Waals forces[110] between the chains of *different* molecules. Because of the many possible conformations, this process takes much longer (12–24 h) than the attachment of thiols to the surface.

Energy Contributions During the Formation of the SAM. The energy contributions consist of a contribution for the attachment of the head group (Example Calculation 13.1) and an additional contribution for the interaction between the chains. The second contribution increases with increasing chain length of the alkanethiols.

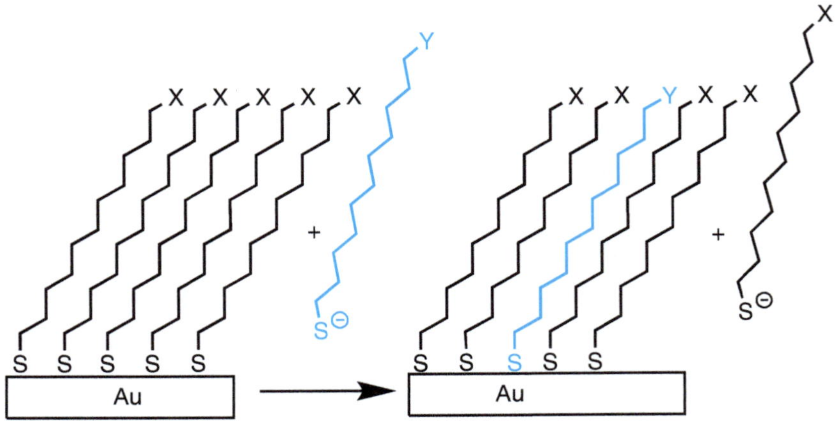

Figure 13.3 Schematic of a substitution reaction of alkanethiolates on gold.

Example Calculation 13.1 The energies for bond cleavage in SAM are Au–S: +167 kJ mol^{-1}, H_2: +435 kJ mol^{-1}, RS–H: +365 kJ mol^{-1}, RS–SR: +233 kJ mol^{-1}. From these values, estimate the molar enthalpy of reaction for the formation of a SAM from R–SH and R–SS–R [21].

For reaction (13.1), energy must be provided to cleave the S–H bond, but energy is released in the formation of the RS–Au and H–H bonds:

$$\begin{aligned}\Delta H(13.1) &= \Delta H(\text{RS–H}) - \Delta H(\text{RS–Au}) - \frac{1}{2}\cdot\Delta H(\text{H–H})\\ &= (365 - 167 - \frac{1}{2}\cdot 435)\ \text{kJ mol}^{-1}\\ &= -19.5\ \text{kJ mol}^{-1}\ \text{(exothermic)}\end{aligned}$$

$$\begin{aligned}\Delta H(13.2) &= \frac{1}{2}\cdot\Delta H(\text{RS–SR}) - \Delta H(\text{RS–Au})\\ &= \left(\frac{1}{2}\cdot 233 - 167\right)\ \text{kJ mol}^{-1}\\ &= -50.5\ \text{kJ mol}^{-1}\ \text{[more exothermic than reaction (13.1)]}\end{aligned}$$

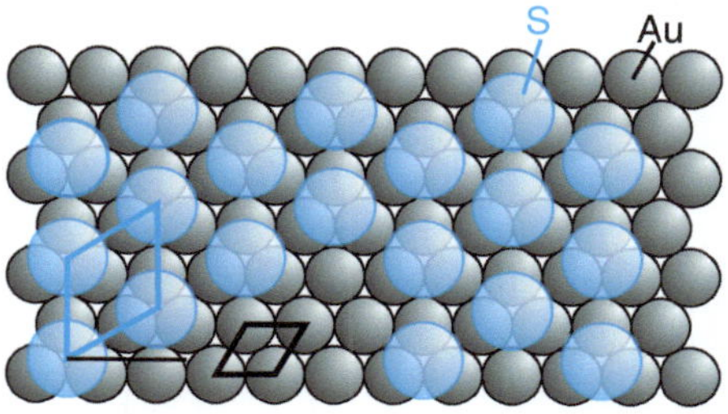

Figure 13.4 Superstructure of sulfur atoms (blue) of an alkanethiolate SAM on Au(1 1 1). Au atoms in gray.

The Structure of the SAM on Au(1 1 1). The sulfur atoms occupy threefold hollow sites on the Au(1 1 1) surface and form a hexagonal superstructure with a unit cell length $\sqrt{3}$ times the length of the unit cell of Au atoms on the surface. The two lattices are rotated by 30° with respect to each other. The short notation is $\sqrt{3}\times\sqrt{3}$R30° (Figure 13.4).

In this superstructure, the distance between two adsorbates is slightly larger than the van der Waals diameter of the alkyl chains. The distance between the alkyl chains can be reduced by tilting the chains with respect to the surface normal (Figure 13.5). The optimum chain spacing exists at a tilt angle of 32°. At this angle, each additional CH_2 group in the chain causes an increase in film thickness of 600 pm or 0.6 nm.

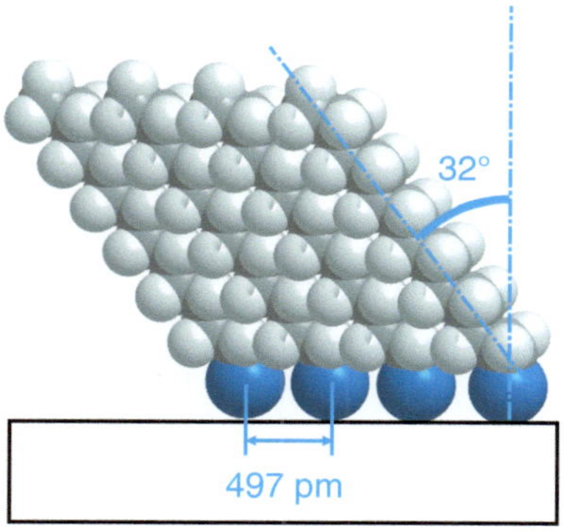

Figure 13.5 Schematic cross section through a SAM at the experimentally observed tilt angle.

The tilt of the chains causes further structural peculiarities. For example, the angle of the end groups changes systematically between alkyl chains with an even and odd number of C atoms (odd–even effect, Figure 13.6). This can be significant if the function desired for an application depends on the orientation of the terminal group.

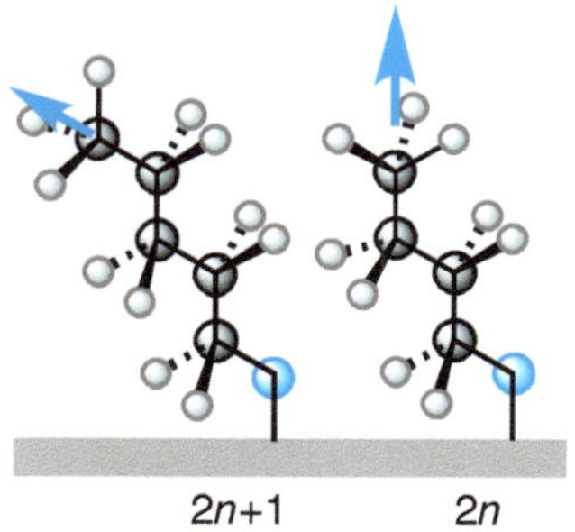

Figure 13.6 Influence of alkyl chains with even and odd numbers of carbon atoms on the orientation of the terminal group.

Defects in SAMs. Alkanethiolate SAMs on Au exhibit defects at different structural levels. Domain boundaries result from chain tilting (Figures 13.7 and 13.9d). If the molecules align together during self-assembly, the tilt of the chains occurs in different directions at different points on the surface but is uniform within a region called a domain (Figure 13.7a). When the domains grow together, molecules with different chain orientations eventually meet at the domain boundary (Figure 13.7b).

Other defects exist at different structural levels and can be visualized by scanning tunneling microscopy (STM) as shown in Figure 13.9.

- Gauche defects are deviations from the preferred conformation (Figure 13.8). Instead of all-trans conformations, some alkyl chains adopt a gauche conformation.
- Molecular defects are those in which a different molecule is incorporated into the layer, for example, an alkanethiol with a different chain length, a different terminal group, or, for example, a thiophenol (Figure 13.9e, f).
- When the thiols are packed on the surface, different superstructures can meet, leaving one row free (Figure 13.9b).
- Etch pits are formed during self-assembly by dissolving some Au atoms out of the surface with the support of the thiols (trace analytical detection of Au in the adsorption solution). The resulting pits are typically one atomic layer deep (detection by STM) and refilled with adsorbed thiols (Figure 13.9a).

These defects affect the electrochemical reactions at SAMs in different ways.

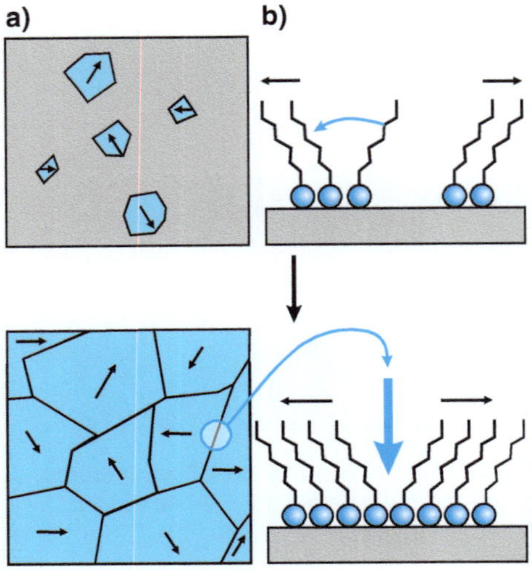

Figure 13.7 Formation and structure of domain boundaries in alkanethiolate SAM on Au. a) Top view and b) side view. The black arrows indicate the direction of chain tilt.

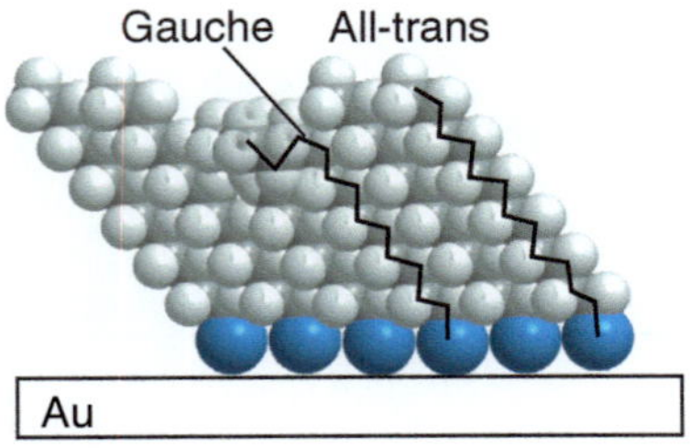

Figure 13.8 Schematic of a gauche defect in an alkanethiolate SAM on Au.

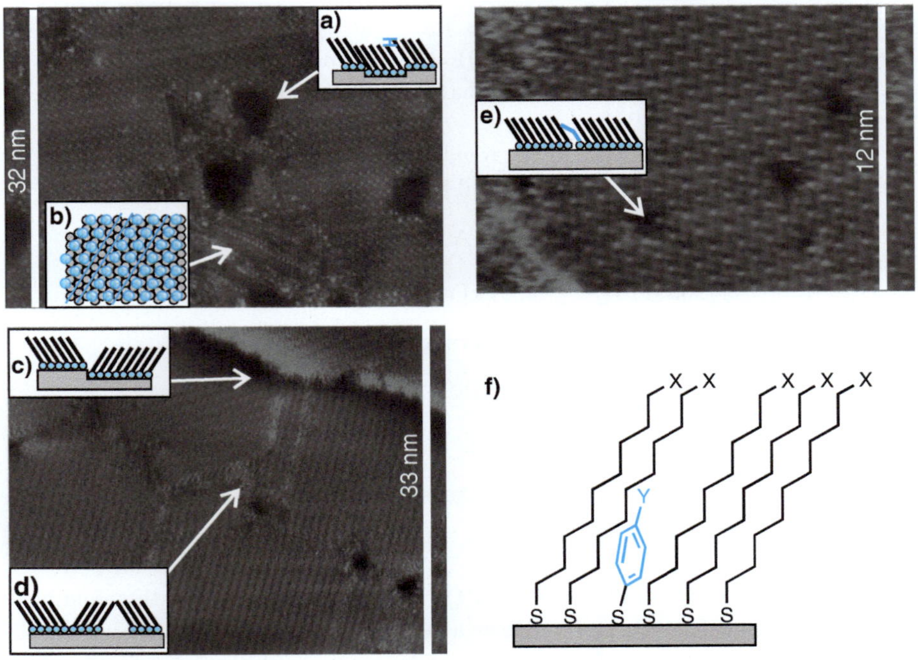

Figure 13.9 STM images of alkanethiolate monolayers with different defects. a) Etch pits; b) missing row; c) monoatomic step with domain boundary; d) domain boundaries; e) molecular point defect; f) example of molecular point defect. Source: Panels (a–e) reproduced with permission from [22], Copyright 2010, Royal Society of Chemistry.

Comparison of the Structure of Alkanethiolate SAMs on Ag(1 1 1) and Au(1 1 1). It is very instructive for the understanding of self-assembly processes at interfaces to compare SAMs on Ag and Au [8]. Some of their properties are listed in Table 13.2. Although both, Ag and Au, have a face-centered cubic lattice with an almost identical lattice constant, SAMs on the two metals have entirely different structures! What are the reasons for the differences between them? If we calculate the adsorption energy of alkanethiolates on different locations on Au(1 1 1) and Ag(1 1 1) surfaces, the difference between the most favorable site and the least favorable site is 25.1 kJ mol^{-1} for Au and only 13.8 kJ mol^{-1} for Ag. Iodide ions, which also chemisorbs strongly on Au and on Ag, occupy the most energetically favorable sites on both metals with a $\sqrt{3} \times \sqrt{3}$R30° superstructure. The sulfur atoms of a SAM on Au are present in the same superstructure, but this does not correspond to the maximally close packing of the alkyl chains. Therefore, the alkyl chains are tilted, and this tilt allows for additional stabilization through interchain interactions. If more molecular chains were to be accommodated on the surface, the S atoms would have to "pay" an energetic "penalty" for occupying energetically less favorable sites of 25 kJ mol^{-1}, a "price" that would not be compensated for by the additional interactions between alkyl chains. In contrast, an alkanethiolate SAM on Ag accommodates significantly more chains on the surface. In this case, the additional interchain interactions can bring up the lower energetic "penalty" of 13.8 kJ mol^{-1} for occupying the less favorable adsorption sites, resulting in a completely different superstructure.

Because the supramolecular structure of self-assembled systems depends on many small, competing interactions, even small changes to the system can lead to large structural changes.

Table 13.2 Comparison of the structure of alkanethiolate SAMs on Au(1 1 1) and Ag(1 1 1). Data from [8].

Feature	SAM on Au(1 1 1)	SAM on Ag(1 1 1)
Atomic distance of the metal	289 pm	288 pm
Superstructure	$\sqrt{3}\times\sqrt{3}$R30°	$\sqrt{7}\times\sqrt{7}$R10.9°
	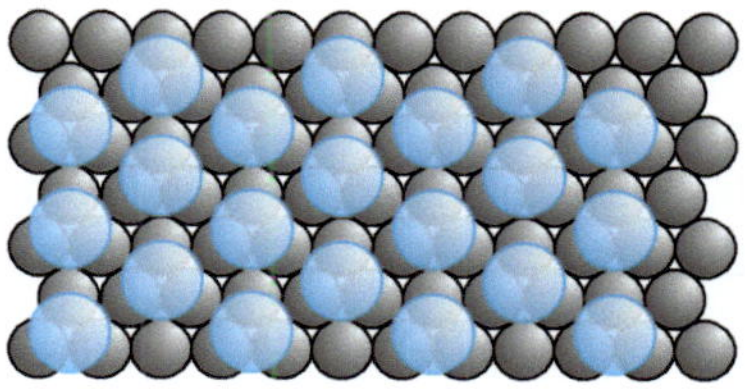	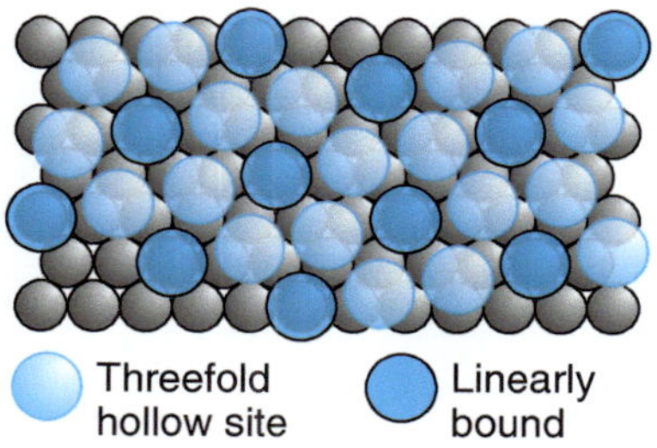
S–S distance	479 pm	441 pm, smaller than chain spacing in kerosene wax
Chain inclination	32°	≈ 0°
Adsorption sites	Threefold hollow site	Two nonequivalent places, one of them 50 pm higher
Surface reconstruction	None	Reconstructed
For comparison: superstructure of chemisorbed I^-	$\sqrt{3}\times\sqrt{3}$R30°	$\sqrt{3}\times\sqrt{3}$R30°
Energy of the adsorption sites (schematic)	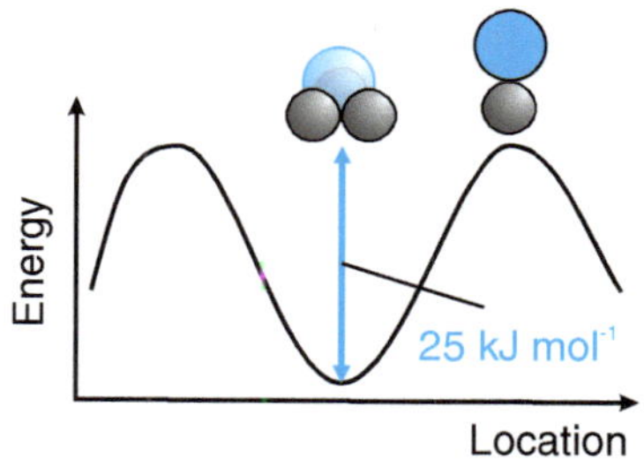	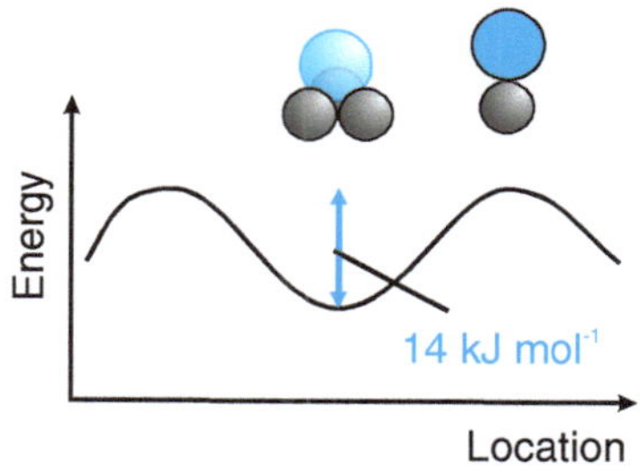

Electrochemical Properties of Alkanethiolate SAMs. The main effect of an alkanethiolate SAM is its blocking effect on other redox systems. The electron transfer between the gold electrode and outer sphere redox systems in solution is noticeably inhibited when the alkyl chain contains more than eight C atoms. The longer the chain, the stronger the inhibition because longer chains favor the all-trans conformation in the adsorbate molecules, which is also a requirement for dense packing and which increases layer thickness. As described earlier, SAMs always contain defects, and thus, electron transfer (ET) reactions can also occur to a small extent at a fully covered electrode at point defects (pinholes) or domain boundaries. The extent to which a defect compromises the blocking behavior of a SAM depends on the polarity of the redox system and on any charge on the terminal group of the SAM. Polar redox systems are usually more strongly inhibited than nonpolar ones. Inhibition of electron transfer is stronger when the terminal group and the redox system carry a charge of the same sign. Observation of the blocking properties allows a quick assessment of the quality of a SAM, which may also deteriorate, for example, with storage in air or

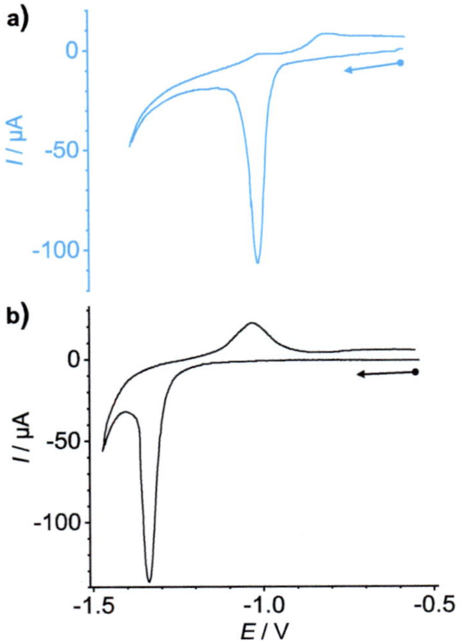

Figure 13.10 Cyclic voltammograms for reductive desorption of a) *n*-hexanethiol and b) *n*-octadecanethiol in 0.5 mol L^{-1} KOH, $v = 0.1\ V\ s^{-1}$. Source: Data from [24] reformatted according to IUPAC convention by the book author.

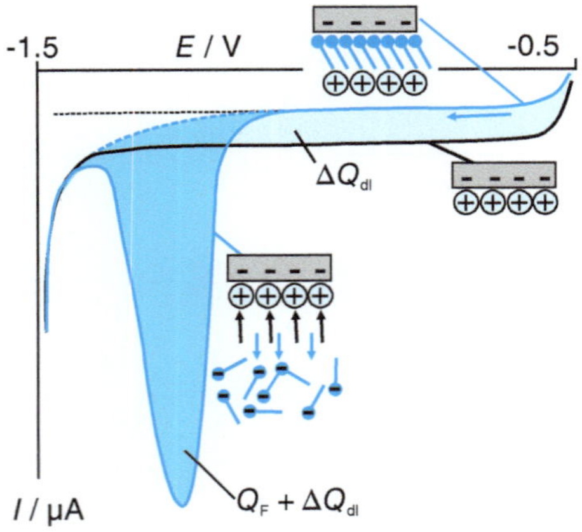

Figure 13.11 Schematic of the charge contributions for the electrode reaction [Eq. (13.4)] and the current flow due to the change in $C_{dl,A}$.

with continued experimentation. The behavior of an ensemble of point defects can be explained with the zone diagram in Figure 5.87.

SAMs decrease the double-layer capacitance of electrodes. The longer the alkyl chain of the SAM, the lower the double-layer capacitance, which is to be expected according to the model of the double layer. (Section 8.10, $C_{dl,A} = \varepsilon_0\ \varepsilon_r/d$, where d is the thickness of the SAM). Experimentally, the dependence $1/C_{dl,A} \propto d \propto n$ could be confirmed for alkyl chains with more than $n = 10$ C atoms [23]. Measurements are made either via EIS (Chapter 7) or from the double-layer charging current in cyclic voltammograms (Figure 4.32).

The SAM itself can be converted electrochemically. At potentials negative of about 1 V vs. Ag|AgCl, reductive desorption takes place according to Eq. (13.4).

$$\text{Au–S–R} + e^- + M^+ \rightleftharpoons \text{Au}^{(0)} + \text{R–S}^- + M^+. \tag{13.4}$$

The reaction equation contains the cations M^+, because the reaction proceeds differently depending on the cation. Clear signals are obtained in 0.5 mol L^{-1} KOH solution (Figure 13.10). The exact desorption potential shifts in the negative direction with increasing chain length. In mixed monolayers, separate signals can be obtained for the desorption of individual alkanethiolate molecules. It has been suggested that the charge for reductive desorption of about 90 μC cm^{-2} can be used for a quantitative determination of thiols [24]. However, as already mentioned for CO stripping in connection with Figure 12.20, one must take into account that not only the charge Q_F for the redox process according to Eq. (13.4), but also the double-layer charge ΔQ_{dl} is transported to the interface during desorption. Since the double-layer capacitance also changes massively during the removal of the SAM (Figure 13.11), the charge must be transferred to the interface that the uncovered electrode would otherwise store during a potential change from the potential of zero charge to the desorption potential. The capacitance of the SAM-covered electrode is much lower than that of the uncovered electrode, so for the same change in potential, a much smaller charge is stored at the SAM-covered electrode. While the double-layer charge ΔQ_{dl} in CO desorption in Section 12.4.1 is only a relatively small correction term, ΔQ_{dl} is a large fraction in SAM desorption, which also depends on the chain length of the alkanethiolate [25].

At potentials between +0.5 V and +1.0 V vs. Ag|AgCl in 0.5 mol L^{-1} KOH, alkanethiols on gold are irreversibly oxidized and then desorbed. Based on the charge of 289 μC cm^{-2}, the presumed oxidation products are sulfinic acids.

$$\text{Au–S–R} + 4OH^- \rightarrow \text{Au}^{(0)} + \text{R–SO}_2^- + 3e^- + 2H_2O. \tag{13.5}$$

Accurate analysis is complicated by the surface oxidation of Au that occurs at similar potentials.

> Cleanly prepared SAMs from alkanethiolates exhibit excellent blocking behavior even on rough and polycrystalline Au electrodes, but they are desorbed at negative potentials and decomposed at positive potentials.

Further Reading. References [2, 7–9, 22, 26] are recommended as further literature with a different focus.

13.1.3 Silanes on Oxide Surfaces

Silanization with trichloro-alkylsilanes or triethoxy-alkylsilanes is frequently used for surface modification of oxidic surfaces, such as glass, quartz, mica, Al_2O_3, surface-oxidized silicon Si|SiO_2, germanium Ge|GeO_2, or transparent conductive oxides (TCO), such as fluorine- or indium-doped SnO_2. In this process, the silanes react with OH groups on the surface with the elimination of HCl or ethanol. The chemical flexibility is very high with respect to the terminal group. Two possible idealized structures are shown in Figure 13.12. However, many more structures exist, since trichlorosilanes can form dimers already in solution, which subsequently attach to the surface and point toward the solution as a branched chain or cross-link further in three dimensions (Figure 13.13). Therefore, it is difficult to prepare structurally uniform monolayers of silanes [27]. The unwanted binding of oligomers to the surface can be circumvented by silanization with monochloro-dimethyl-alkylsilanes containing just one Cl-atom necessary for binding to the surface. These molecules can still dimerize, but the dimer has no further Cl-atom to bind to the surface.

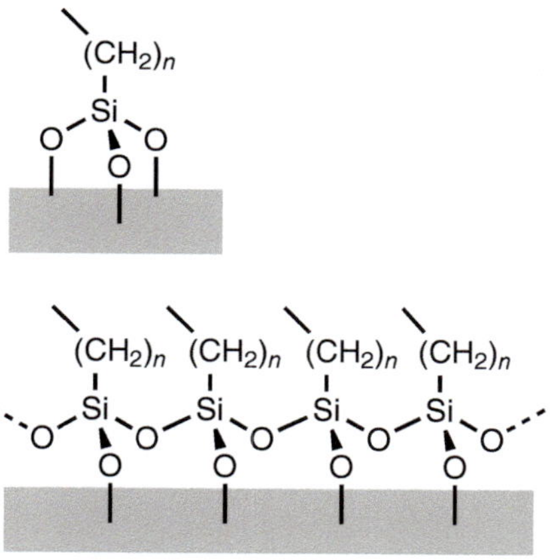

Figure 13.12 Idealized structures of silane layers on oxide surfaces.

13.1.4 Reduction of Aryldiazonium Salts

Another very flexible way of modifying surfaces can be achieved by the chemical or electrochemical reduction of aryldiazonium salts in aqueous or organic electrolyte solutions. The resulting radical intermediates can bind to various carbon electrodes or metal electrodes (Figure 13.14a). A particular advantage is that the coating can be started by a negative-going potential scan. A large irreversible reduction peak can be observed in the first potential scan (Figure 13.15). In further potential cycles, this reduction peak becomes smaller and smaller, because the surface is gradually

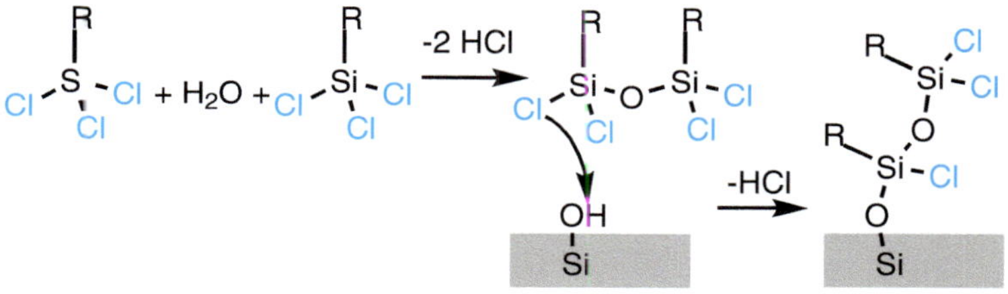

Figure 13.13 Realistic structures of silane layers.

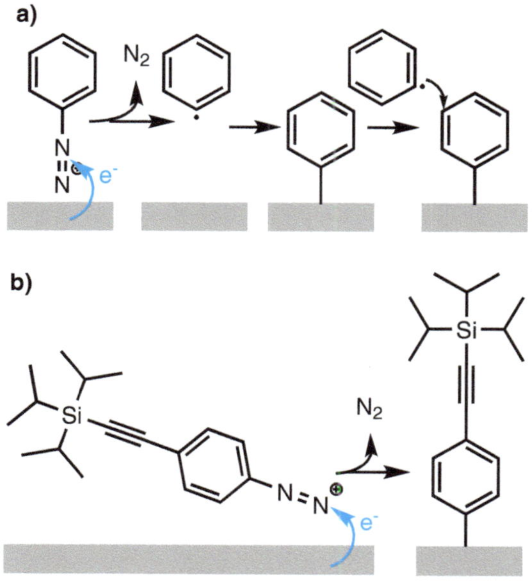

Figure 13.14 a) Schematic of film formation by the reduction of aryldiazonium salts; b) example of an aryldiazonium ion with sterically demanding substituents that forms only a monolayer and can be deprotected for further modifications.

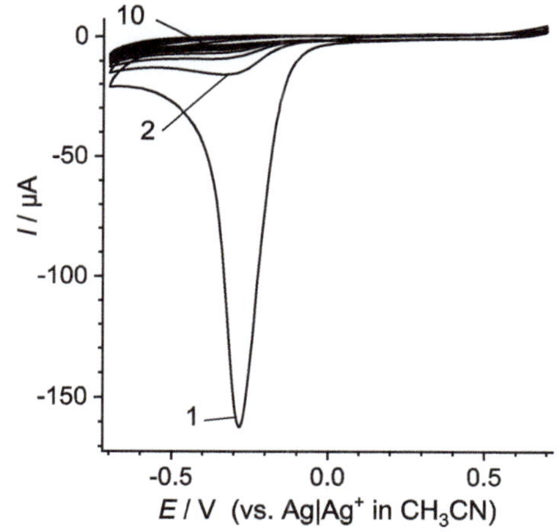

Figure 13.15 Ten consecutive cyclic voltammograms for the reduction of 10^{-2} mol L^{-1} 4-(tris-isopropyl)-silyl-ethenyl-benzenediazonium tetrafluoroborate (Figure 13.14b) on ITO in 10^{-2} mol L^{-1} tetrabutyl-ammonium hexafluorophosphate in acetonitrile; $v = 0.05$ V s^{-1}. Data were kindly provided by Sweety Ann Samuel, Carl von Ossietzky University Oldenburg.

passivated by the growth of the organic film. Thus, the progress of the film formation can be monitored via the current. The modification is self-limiting.

A possible disadvantage is that the radicals formed during the reduction react nonspecifically with both the substrate surface and molecular fragments already bound to the electrode (Figure 13.14a). This results in branched cauliflower-like structures instead of homogeneous molecular monolayers. Multilayer formation can be prevented by using diazonium salts with sterically demanding substituents (Figure 13.14b).

The structural diversity of the layers is large and there are many possible terminal groups. Because of the additional flexibility in terms of the electrodes to be coated, the application scope is broad.

Further Reading. References [28–31] can be recommended for further reading.

13.1.5 Adjustment of Surface Properties with SAMs

Chemisorbed monolayers offer the possibility to control surface properties such as wettability, electrical charge, adhesion of biological cells, and other properties via the terminal group of the SAM independently of the type of substrate electrode material. SAMs facilitate anchoring reactive groups to the surface for further modifications. This approach is also referred to as surface engineering.

Mixed Monolayers. In order to control properties, such as the density of functional groups in a continuous way, it is convenient to use mixed monolayers. These monolayers contain two different adsorbate molecules. It is important that both adsorbate molecules have a chemically similar spacer, so that ordered packing is possible in the mixture. In the case of incompatible spacers, place exchange reactions of the adsorbates, particularly at elevated temperatures, may cause the formation of domains containing only one type of adsorbate molecule. To avoid this effect, mixed monolayers of HS–$(CH_2)_n$–X and HS–$(CH_2)_n$–Y with the same spacer and different, small end groups X (e.g., $-CH_3$) and Y (e.g., –OH, –COOH, –CN) are preferable. These mixed monolayers exhibit a statistical distribution of terminal groups.

In mixed monolayers, the ratio of the interfacial concentrations of the adsorbates in the monolayer usually does not agree with the concentration ratio in the adsorption solution [$\Gamma(X)/\Gamma(Y) \neq c(X)/c(Y)$].

Figure 13.16 Schematic cross section through a mixed monolayer with a large functional terminal group and a shorter, methyl-terminated diluent. In the close-packed part of the layer, the molecules are rigid, while the protruding part of the longer building blocks exhibits a flexible conformation. Source: Modified according to [32].

The required composition of the adsorbate solution $c(X)/c(Y)$ to obtain a desired ratio of terminal groups $\Gamma(X)/\Gamma(Y)$ at the surface must be determined by a series of experiments (i.e., empirically).

Often, the terminal group of interest is significantly larger than the cross section of the spacer. A densest packing of the spacers as a prerequisite for highly ordered layers is not possible simultaneously with a densest packing of the end groups. In these cases, the molecule with the end group of interest is often embedded into a monolayer that additionally contains a "diluent". The diluent is a molecule with a similar *but shorter* spacer (Figure 13.16). In the mixed monolayer, the end groups of interest protrude from the close-packed monolayer. We discuss a corresponding example in Excursion 13-1, Figure 13.34. This approach is also suitable if the functional terminal group of the monolayer is to interact with proteins or biological cells and must therefore exhibit a flexible conformation.

To prepare the layers, adsorption is performed in two steps. First, an incomplete monolayer is formed with the longer chain molecule containing the functional group of interest (often the more expensive component) by keeping the concentration low and/or the adsorption time short. The parts of the gold surface that initially remain free are backfilled with the shorter chain diluent in a second adsorption step.

Patterned Monolayers. A range of advanced applications requires the deposition of microscopic patterns onto an initially uniform electrode surface. These include DNA arrays in bioanalytics, material libraries, or components in molecular electronics. SAM systems offer excellent opportunities to obtain such structures with relatively simple laboratory resources.

Microcontact Printing. Particularly popular for this purpose is microcontact printing (MCP), for which the special properties of poly(dimethyl)siloxane (PDMS) are exploited. This polymer, which is both, transparent and flexible, can be cured (i.e., cross-linked) on surfaces. The finished polymer precisely molds details of the surface down to the nanometer scale. Structured glass or silicon surfaces are suitable as mold templates (masters). Less stable, but easier to produce, are templates made of cured photoresists on smooth silicon wafers (Figure 13.17). The ratio of lateral structure widths to depths of the structures must be between 0.2 and 2 (Figure 13.18). To facilitate detachment of the cured polymer, the surface of the master is often modified with a fluorinated trichlorosilane as a release agent (by keeping it in an evacuated desiccator for 12 h together with an open bottle of the liquid silane, from which a small amount of silane passes from the liquid via the gas phase to the mold). The master is placed in a Petri dish, and the prepared mixture of PDMS precursor and hardener is poured over it (Figure 13.19). After placing the Petri dish in a desiccator, application of negative pressure can facilitate the escape of bubbles from the highly viscous polymer precursor mixture. After curing, the polymer block can be removed from the master. The molded micro- and nanostructures are retained, but the molded part is flexible and can conform to other surfaces. This makes it an ideal material for a miniaturized stamping process.

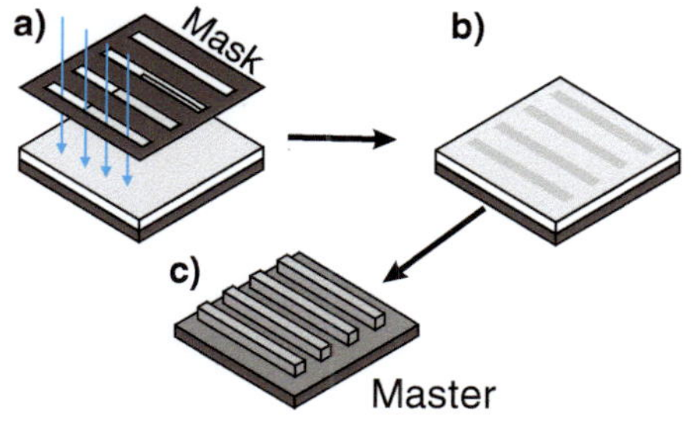

Figure 13.17 Production of a master for microcontact printing. a) Si wafer coated with photoresist; b) wafer with exposed photoresist; c) wafer after removal of the unexposed parts of the photoresist layer. This structure serves as a template (master) for casting the actual stamp.

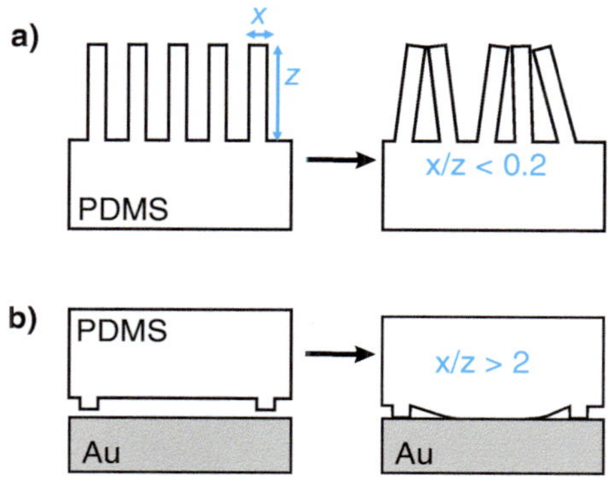

Figure 13.18 Cross sections of MCP stamps with improper aspect ratios. a) Structures that are too slender coalesce; b) between structures that are too flat, the retracted regions sag. Source: Modified after [9].

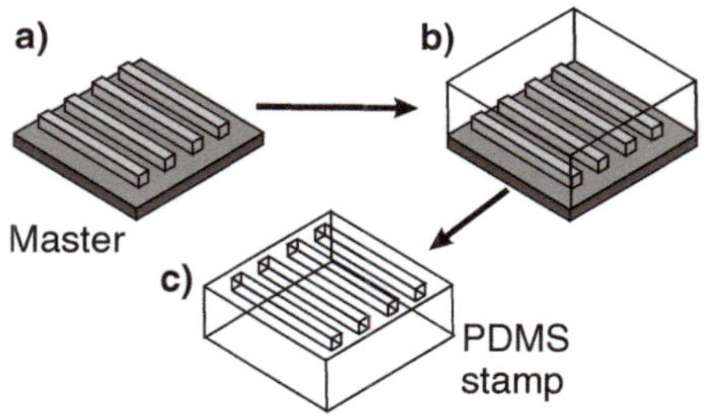

Figure 13.19 Molding of the stamp. a) Master; b) a mixture of PDMS precursor and hardener is poured over the master; c) released stamp made of PDMS.

The "ink" used is an approximately 1 mol L^{-1} ethanolic solution of an alkanethiol, which is spread as a small drop onto the surface of the PDMS stamp. One waits until the solvent has evaporated *completely*. (The stamp looks completely dry again.) However, the alkanethiol remains on the surface of the stamp as a film invisible to the eye. Now, the stamp is placed on a clean gold surface and a SAM will form at the contact points as alkanethiol from the film on the surface of the PDMS stamp reacts with the Au surface (Figure 13.20). The contact time should be about 30 s. For shorter contact times, the quality of the monolayer is unsatisfactory (see Section 13.1.2, kinetics of SAM formation). Longer contact times should also be avoided, because then areas of the gold, which are not in mechanical contact with the stamp, are coated with alkanethiol because of transport via the gas phase (Figure 13.20).

The parts of the Au surface not initially occupied during a printing process are often filled with a second alkanethiol (backfilling), which passivates the Au surface and stabilizes the pattern. Printing processes with multiple passes using different alkanethiols or rotating the stamps can be used to build flexible molecular architectures. An example is shown in Figure 13.21. In addition, variations of the methodology exist, which are summarized under the name soft lithography. They are used, for example, in the construction of microelectrochemical systems [9].

The soft lithographic processes are examples of parallel structuring processes in which large surface areas are modified in a single operation using a complex template. The advantage of a relatively high speed is offset by the low flexibility in terms of variations of the generated patterns. A new template must be generated for each new pattern. For research where pattern optimization is a concern, it is sometimes more appropriate to use a sequential process in which areas of the surface are modified one at a time using a tip (Section 8.11.8). A new pattern can be programmed relatively easily.

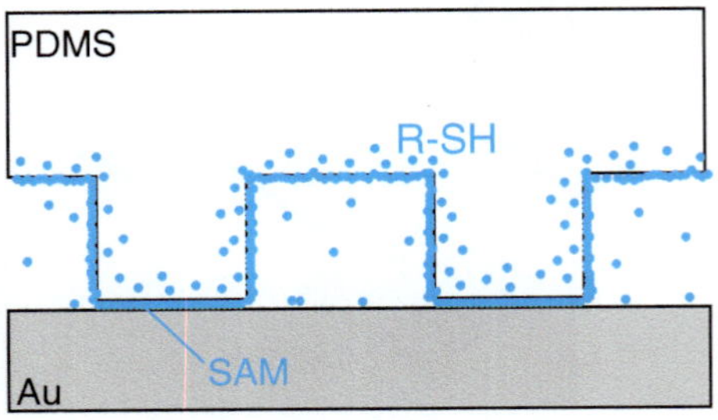

Figure 13.20 MCP stamp during contact with the Au substrate (cross section).

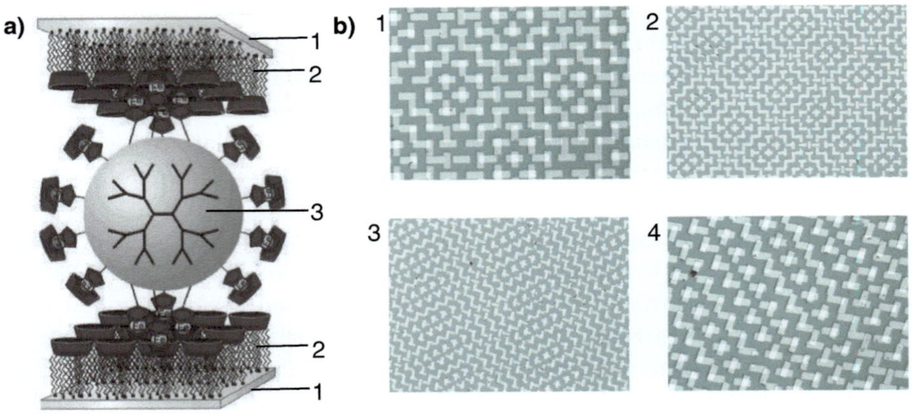

Figure 13.21 A combination of MCP and microtransfer printing of a complex monolayer with attached gold layer is used for the fabrication of devices for molecular electronics.

a) Schematic of the molecular units bridging the overlapping gold layers; 1) Au substrate; 2) SAM with β-cyclodextrin (which is a ring-shaped oligosaccharide that has a binding pocket for noncovalent bonding of, for example, ferrocene groups); 3) ferrocene-decorated dendrimer;
b) Optical image (the reflective thin gold layers printed on each other with different angles are bright); 1) 90°, 2) 85°, 3) 80°, and 4) 75°.

Source: Reproduced with permission from C. A. Nijhuis, J. ter Maat, S. Z. Bisri, M. H. H. Weusthof, C. Salm, J. Schmitz; B. J. Ravoo, J. Huskens, D. N. Reinhoudt; *New J. Chem.* **2008**, *32*, 652–661, Copyright 2008, The Royal Society of Chemistry.

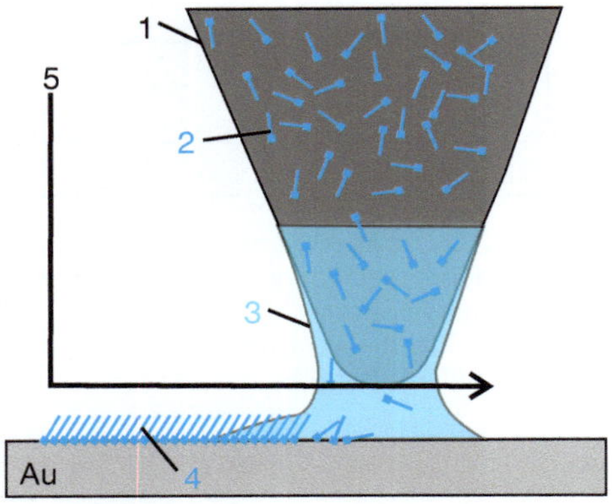

Figure 13.22 Principle of dip-pen nanolithography.

1) Tip of a scanning force microscope;
2) alkanethiol molecules;
3) meniscus of water and solvent;
4) SAM;
5) trajectory of the tip movement.

Nanoshaving. In "nanoshaving", a monolayer is removed by abrasion with the tip of a scanning force microscope (Section 8.11). The exposed surface of the Au substrate is then filled with another alkanethiol, which may be contained in the working solution [33]. An analogous procedure is possible with exposure of the Au surface by localized electrochemical desorption (Section 13.1.2) [34].

In **dip-pen nanolithography (DPN)**, the tip of a scanning force microscope is wetted with a solution of an alkanethiol and, when in contact with the sample, transfers a thin liquid film containing the alkanethiol to the substrate, similar to a classical dip-pen (Figure 13.22). The "ink" on the tip can be renewed by a microfluidic arrangement integrated into the tip or by dipping the tip into a reservoir. This technique is of particular interest because not only can molecular reagents be transferred but also nanoparticles, nanotubes, and even whole biological cells.

Further information can be found in review articles on the MCP [9], on DPN [35–37], on the use of alkanethiol SAMs in sensors [38], and on layers formed by reduction of aryldiazonium salts [29].

13.1.6 Langmuir Layers

Langmuir films[91,143] are two-dimensional (2D) arrangements of amphiphilic molecules at the interface between a liquid (subphase) and a gas atmosphere. These

[143] The design of the Langmuir trough, used today in a much more refined form, can be traced back to the German autodidact Ms. Agnes Pockels, 1862–1935, who first observed the phenomenon in 1880 on the surface of dishwater. She used the simplest of means (e.g., metal buttons instead of Wilhelmy plates) to construct a device known as a "sliding trough". She reported the discovery in 1890 to Sir Rayleigh, who arranged for its publication in the prestigious journal *Nature*. It was only after this that Agnes Pockels was taken seriously by the established scientific community in her home country, Germany, and she was given scientific work opportunities at the Technische Hochschule Braunschweig and an honorary doctorate in 1932.

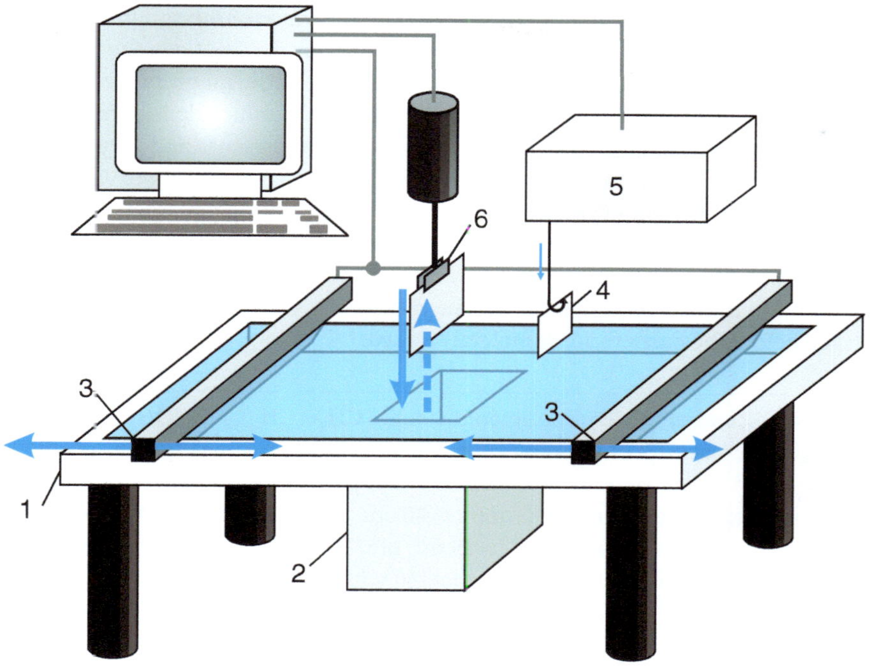

Figure 13.23 Schematic of a Langmuir trough. 1) PTFE trough; 2) well for holding substrates in transfer experiments; 3) movable barriers; 4) Wilhelmy plate for measuring film pressure; 5) electronic balance for force measurement; 6) holder for substrates on a movable arm.

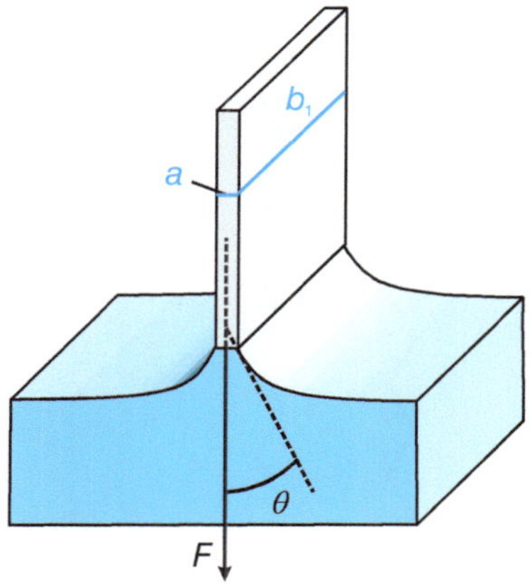

Figure 13.24 Measuring the interfacial tension with the Wilhelmy plate. Either a roughened Pt sheet or a small piece of filter paper is used as the plate. For both materials, $\theta \approx 0°$. The force F on the Wilhelmy plate is given by $F = u\gamma \cos\theta$, where $u = 2\,(a + b_1)$ is the length of the contact line between the plate and the liquid surface. The film pressure $\Pi = \gamma - \gamma_0$ is the difference in the interfacial tensions of the film γ and pure subphase γ_0. If $\gamma_0 \ll \gamma$, additionally the thickness a of the platelet is negligible compared to the width b_1 and the surface is ideally wettable ($\theta = 0°$), the film pressure is obtained from $\Pi \approx F\,/\,(2b_1)$.

films are best observed in a shallow trough of polytetrafluoroethylene (PTFE, Teflon) as shown in Figure 13.23. Movable barriers that are parallel to the short sides of the trough allow a surface film to be pushed together along the longer side of the trough. A Wilhelmy plate[144] is used to track the film order (Figure 13.24). The interfacial tension γ is measured and the film pressure $\Pi = \gamma - \gamma_0$ is obtained as the difference of the interfacial tensions with (γ) and without (γ_0) addition of the amphiphilic molecule. Often, the trough contains a well in the center to hold substrates during transfer experiments (Section 13.1.7) and a clamp for substrates on a motorized actuator arm. In addition to this classic design for research laboratories, more complex setups are also available to coat a large number of substrates in a pilot production [39].

Preparation of Langmuir Films. The trough is meticulously cleaned, first with hydrophobic solvents, then with water. Subsequently, the subphase is filled into the trough at the maximum distance of the barriers. The subphase is either pure water or an aqueous electrolyte solution. The trough is filled so that the liquid meniscus is clearly visible above the rim of the trough (Figure 13.25a). Then the barriers are pushed together, and the film pressure is measured as a function of the distance between the barriers. This should result in a completely flat line. However, this will often not be the case initially because hydrophobic organic matter from the ambient air accumulates on the surface of the aqueous subphase and is pushed together by the barriers ("monolayer of disordered dirt"). Such layers greatly interfere with sensitive experiments. However, the impurities can be elegantly aspirated with a pipette connected to a water aspirator pump. To do this, the pipette tip is briefly placed on the liquid surface several times with the barriers closed (Figure 13.25b), and the blank curve is then recorded again until Π only changes by less than 1 mN m^{-1} over the entire displacement path.

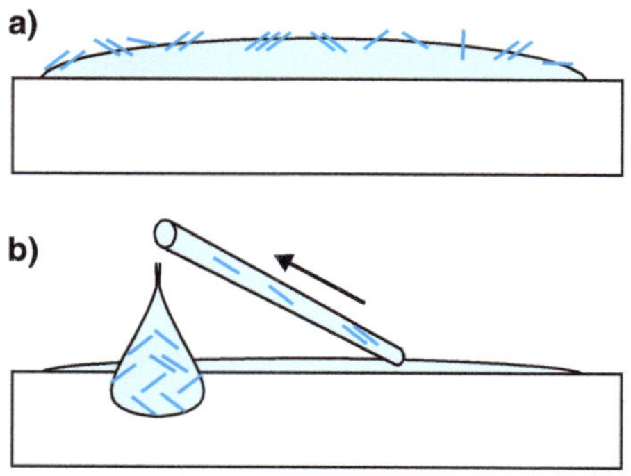

Figure 13.25 a) Side view of the filled trough with surface-active contaminants (schematic) and b) aspiration of the top liquid layers with surface-active contamination from the liquid–gas interface.

Then, the amphiphilic molecule (or mixture of molecules) is dissolved in a volatile, hydrophobic solvent (e.g., $CHCl_3$) that is immiscible with the subphase. As a

[144] Ludwig Wilhelmy, 1812–1864, German scientist best known for the first quantitative studies of chemical kinetics.

microliter-volume of this solution is placed on the surface of the subphase, the complete evaporation of the solvent takes about 10 min for $CHCl_3$. Only the molecule of interest remains at the interface. The hydrophilic part of the amphiphilic molecule is well solvated by the subphase, but the other part of the molecule is poorly solvated. In addition, the amphiphiles must have both low vapor pressure and very low overall solubility in the subphase. Lipids (Figure 13.26) or long-chain carboxylic acids are prototypical amphiphiles from which Langmuir films are prepared. The concentration of the molecules in the organic stock solution must be such that, after application of the solutions with the barriers open, there is such a large distance between the amphiphiles that no film pressure can be measured. A densely packed monolayer is formed before the barriers are completely closed.

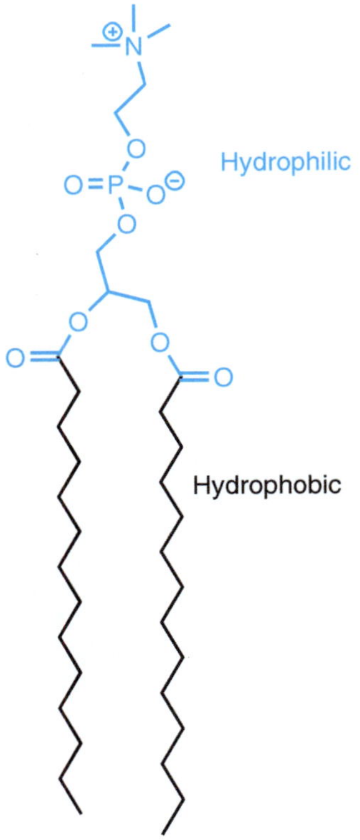

Figure 13.26 Hydrophilic and hydrophobic moieties of 1,2-dimyristoly-sn-glycero-3-phosphocholine (DMPC).

Example Calculation 13.2 For arachidic acid [CH_3–$(CH_2)_{18}$–COOH, M = 312.54 g mol^{-1}], dense packing of the molecules is achieved approximately at a specific area of A' = 20 Å^2 per molecule. The Langmuir trough has a width of 8 cm. When fully open, the distance between the barriers is 40 cm. The stock solution was obtained by dissolving 50 mg of arachidic acid in 25 mL of $CHCl_3$. What volume of this solution must be applied to obtain the densest packing with the barrier half closed?

To be calculated: V (the added liquid volume)

given:

$$b_2 = 8 \text{ cm} = 0.08 \text{ m}$$
$$l_{max} = 40 \text{ cm} = 0.40 \text{ m}$$
$$l = l_{max}/2 = 0.20 \text{ m} \quad \text{(Distance at half-closed barrieres)}$$
$$m = 50 \text{ mg} \quad \text{(Weighted mass)}$$
$$M = 312.54 \text{ g mol}^{-1}$$
$$V_{tot} = 25 \text{ mL} \quad \text{(Total volume of stock solution)}$$
$$A' = 20 \text{ Å}^2 = 20 \times 10^{-20} \text{ m}^2 \quad \text{(Area per molecule in compressed state)}$$
$$N_A = 6.022 \times 10^{23} \text{ mol}^{-1} \quad \text{(Avagadro constant)}$$

The following applies to the added volume V_{stock} of the stock solution:

$$V = \frac{N}{c},$$
$$N = \frac{A}{A' N_A} = \frac{l\, b_2}{A' N_A},$$
$$c = \frac{c_m}{M} = \frac{m}{V_{tot} M},$$
$$V = \frac{l\, b_2}{A' N_A} \cdot \frac{V_{tot} M}{m}$$
$$= \frac{0.2 \text{ m} \cdot 0.08 \text{ m} \cdot \left(\frac{10^{10} \text{Å}}{1 \text{ m}}\right)^2 \cdot 25 \cancel{\text{m}}\text{L} \cdot 312.54 \frac{\text{g}}{\text{mol}}}{20 \text{ Å}^2 \cdot 6.022 \times 10^{23} \text{ mol}^{-1} \cdot 50 \cancel{\text{m}}\text{g}}$$
$$= 2.056 \times 10^{-5} \text{ L} = 20.6 \text{ μL}$$

N – Added amount [mol]
A – Area at half-closed barriers [Å^2]
c_m – Mass concentration of stock solution [g/L]

About 20 μL of the stock solution must be applied so that a tightly packed monolayer is present with half-closed barriers. In order for the Langmuir isotherm to commence at negligible film pressure, about twice the specific area per molecule must be available at the beginning of the experiment (in this example 40 Å^2).

The barriers are then slowly pushed together (20 mm min^{-1}). The hydrophobic molecules are bound to the surface of the subphase, and therefore, they can neither evaporate nor dissolve in the subphase. The subphase flows under the barriers as they move, causing the amphiphilic molecules to be slowly squeezed together. The changing state of molecules within the surface film is often given with respect to the average area A' for one molecule, which is obtained from the known amount N of the applied molecules and the momentary area between the barriers.

$$A' = \frac{A}{N \cdot N_A} = \frac{b_2 l}{N_A c V}. \tag{13.6}$$

A' – Area per molecule [m^2]
A – Trough surface area between barriers with distance l and width b_2 [m^2]
c – Concentration of amphiphilic molecules in the stock solution [mol L^{-1}]
V – Volume of the applied stock solution [L]
N – Added amount of the amphiphilic molecule [mol]
N_A – Avogadro number [mol^{-1}].

Isotherms. The Wilhelmy plate (Figure 13.24) is used to measure the interfacial tension, which is plotted as a function of the area between the barriers available for one molecule. We call the $\Pi - A'$ curve the film pressure isotherm. Figure 13.27 shows the isotherm for compression of arachidic acid at room temperature. In this example, several states of the monolayer are successively passed with decreasing A'. They are the 2D equivalents of the states of matter, gaseous–liquid–solid, and correspond to the degree of order of the arachidic acid molecules in the film. A transfer of the film could take place at 40 mN m^{-1}, when a packing is present that is equal to that in a solid. At even higher compression, the film collapses with monolayers sliding over each other like ice floes.

The transitions between the order states are also influenced by the subphase, for example, by salt addition to the aqueous phase. For other amphiphilic molecules, the transitions cannot be seen as clearly as in Figure 13.27. For a mixture of three lipids in Figure 13.28a, one cannot see the steep rise in Π at very small A', which corresponds to the solid-like state. However, the isotherm has different slopes in the broad range for the liquid-like state. This becomes clearer when we calculate the compression modulus from the isotherms (inset in Figure 13.28).

$$C_s = -A' \left(\frac{\partial \Pi}{\partial A'} \right)_{T,p,n_i} \tag{13.7}$$

Π – Film pressure [mN m]$^{-1}$
C_s – (2D) compression modulus [mN m^{-1}], reciprocal of 2D compressibility. Compressibility is the relative change in volume or area per change in pressure.

Based on the compression modulus, the liquid-expanded phase (LE) and the liquid-condensed phase (LC) can be distinguished in the example (Table 13.3). Which order states are attained depends on the structure of the film-forming molecules. The inspection of space-filling models of the molecules can often convey good insight into whether a highly ordered packing is geometrically conceivable at all. Further information on packing possibilities of lipids can be found in [40].

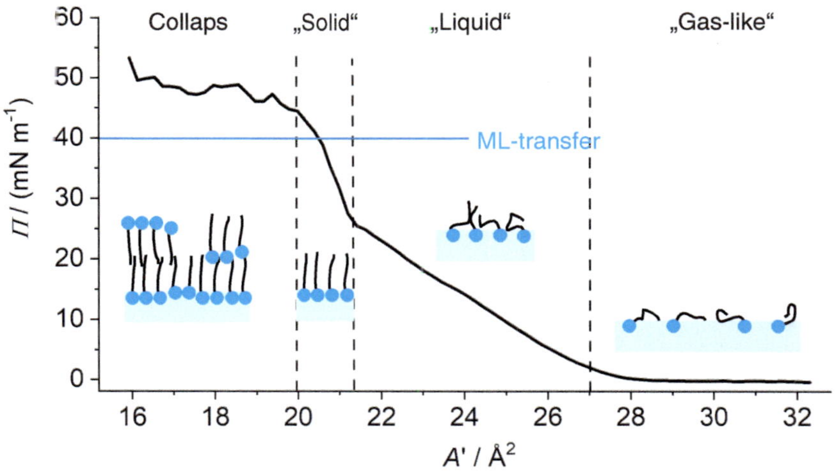

Figure 13.27 Compression isotherm of arachidic acid ($C_{19}H_{39}COOH$) on water, T = 298 K. The schematic shows cross sections of the 2D arrangements of the molecules on the water surface for 1) the gas-like arrangement, 2) the liquid-like arrangement, 3) the solid-like arrangement, and 4) the collapse of the 2D film.

Table 13.3 Typical compression modulus values for the order states in Langmuir isotherms.

Phase	C_s/(mN m^{-1})
LE phase	12–50
LC phase	100–250
Solid state	Until 2000

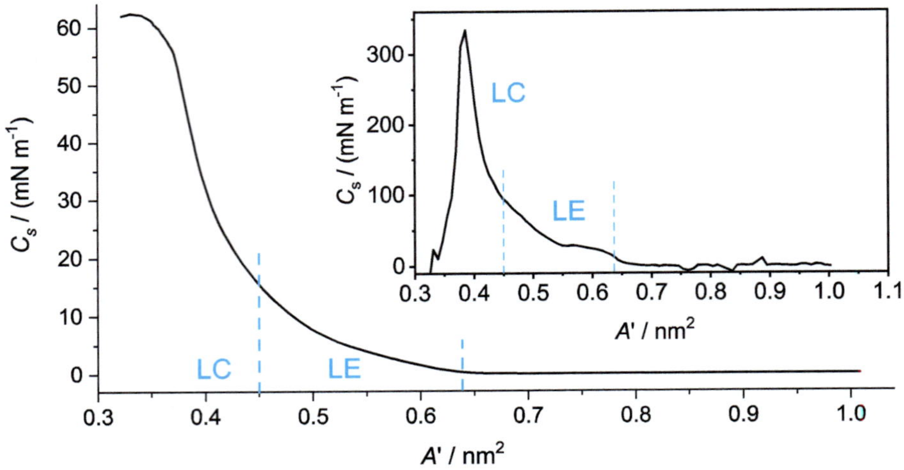

Figure 13.28 Film pressure isotherm of a mixture of DMPC (Figure 13.26) with 30% cholesterol and 20% of the glycosphingolipid GD1a on water. The image inset shows the compression modulus, in which the phase transitions can be observed more clearly than in the film pressure isotherm.
Source: Data were provided by Izabella Brand, Carl von Ossietzky University Oldenburg.

Using film pressure isotherms at different temperatures, temperature-dependent 2D phase transitions can also be characterized. Other effects such as domain formation (2D equivalent of liquid segregation) can occur in layers of two or more surface-active compounds and can be particularly well characterized by Brewster angle microscopy[145] [41, 42]. Overall, 2D equivalents can be found for almost all of the diverse structural and phase transitions in liquid and solid mixtures.

13.1.7 Film Transfer Techniques

In the Langmuir–Blodgett[91,92] (LB)- and Langmuir–Schaefer[146] (LS) transfer techniques, a Langmuir film is first compressed to a predetermined film pressure as described in Section 13.1.6. By applying mechanical work, the molecules within the film are arranged with respect to each other. This preordered film is then transferred to a solid substrate, for example, an electrode.

In the **Langmuir–Blodgett transfer process**, the substrate to be coated is pulled through the film. In the case of a hydrophilic substrate, it moves from the aqueous subphase through the film so that the hydrophilic head groups of the monolayer interact with the hydrophilic substrate (Figure 13.29a). In the case of hydrophobic substrates, the substrate moves from the atmosphere through the film into the subphase, allowing the hydrophobic moieties to interact with the substrate (Figure 13.29b). The barriers of the Langmuir trough are controlled to maintain a constant film pressure during the transfer process, that is, to push the surface film. If a portion of the monolayer is transferred to the substrate, the barriers will move toward each other, causing a complete monolayer to be always present near the substrate.

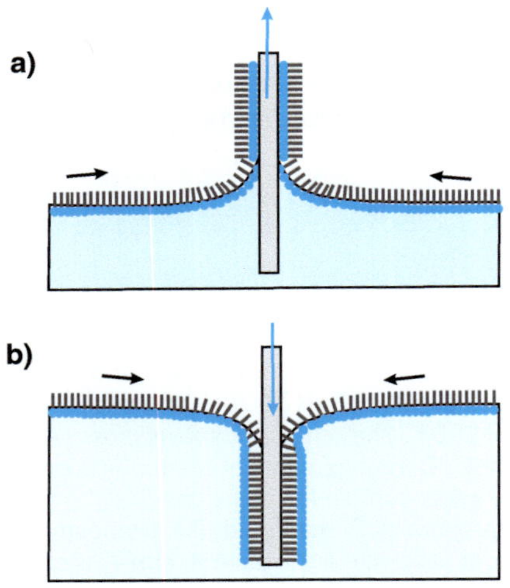

Figure 13.29 Principle of Langmuir–Blodgett transfer a) for hydrophilic substrates and b) for hydrophobic substrates.

Based on the barrier positions recorded during a transfer experiment, the success of the transfer experiment can be assessed. To do this, we form the transfer ratio τ from the change in liquid surface area between the barriers $\Delta A_L = b_2\,\Delta l$ (under compression, ΔA_L, $\Delta l < 0$) and the area of the substrate with width b_3 and the vertical displacement distance of the substrate Δz.

$$\tau = -\frac{\Delta A_L}{A_S} = -\frac{b_2\,\Delta l}{b_3\,\Delta z} \tag{13.8}$$

[145]David Brewster, 1781–1868, British scientist, inventor, historian of science, and science organizer in Scotland.
[146]Vincent J. Schaefer, 1906–1993, American chemist and meteorologist, worked with Langmuir at General Electric on protein films and other topics.

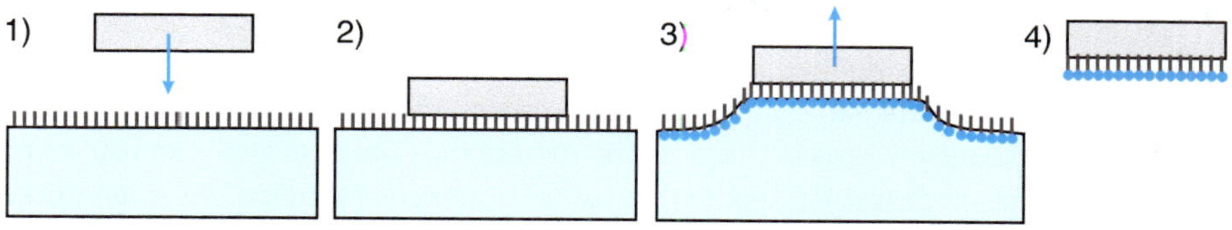

Figure 13.31 Methodology for the Langmuir–Schaefer transfer.

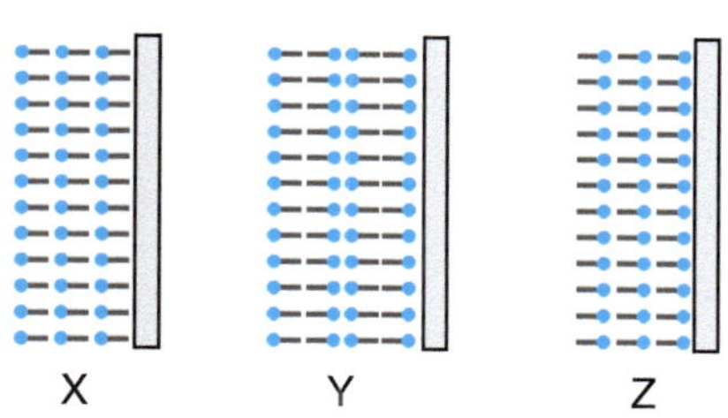

Figure 13.30 Classification of multilayer deposition with LB or LS technique. The dots indicate the hydrophilic part of the amphiphilic molecule.

If $\tau > 0$, the Langmuir film has been transferred from the liquid surface to the substrate; if $\tau < 0$, the film transfers from the substrate surface back to the liquid surface and the barriers move away from each other. An ideal transfer is observed when $\tau = 1$.

The transfer procedure can be repeated several times. Different orientations of the molecules within the film can be set by varying the transfer direction. They are referred to as X-, Y-, and Z-multilayers (Figure 13.30). However, it will not be possible to produce X- or Z-type multilayers for every molecule.

In the **Langmuir–Schaefer transfer technique** (also called horizontal touch technique), the substrate is approached horizontally to the compressed film and a region of the surface layer with the area A_S is "excised" out of it (Figure 13.31).

In **vesicle spreading**, the amphiphilic molecules are preorganized in a double-layered spherical micelle. Similar to biological membranes, molecules can also be embedded in this spherical micelle that protrude from one or both sides out of the double layer. If these vesicles come into contact with an electrode, they may rupture and form a planar double layer on the electrode (Figure 13.32).

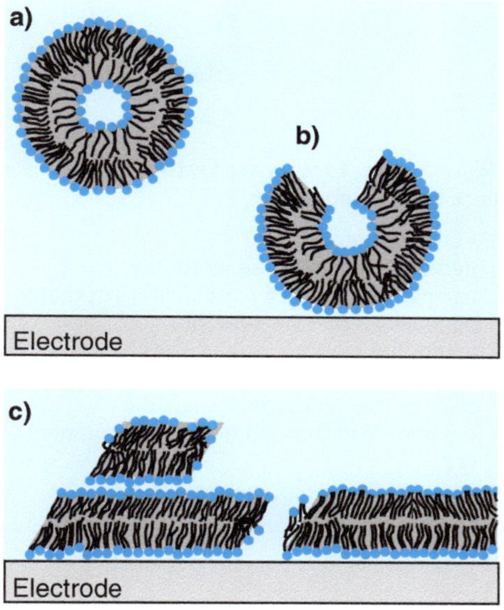

Figure 13.32 Vesicle-spreading. a) Vesicles in solution; b) contact with surface, rupture and spreading of vesicles; c) cross-sectional view of a bilayer on a substrate with pore defect and partial multilayer formation.

All monolayer transfer techniques are suitable for coating electrodes with structurally defined supramolecular films. The main components are chain-like molecules, which are crucial for the internal order. Redox-active or photochemically active molecules can be intercalated and oriented between the chains in an *ordered* manner, which can then be studied or manipulated by electrochemical methods. These layers are of particular interest in bioelectrochemistry for the construction of model systems that can be used to study the biophysical properties of lipid bilayers (Excursion 13-1).

A disadvantage of these systems for electrochemical applications compared to SAMs (Sections 13.1.1–13.1.5) is that the films held on the surface without chemical bonding are displaced from the surface by anions or cations of the electrical double layer at electrode potentials that deviate significantly from the potential of zero charge. The potential-dependent stability range can be investigated, for example, with the aid of tensammetry (Figure 5.45).

Further Reading. Detailed reports on transfer techniques [2, 12] and review articles on their applications in electrochemistry [39] are recommended.

Excursion 13-1 Model Systems for Biological Membranes.

Biological membranes that enclose cells or cell organelles consist of lipid bilayers. They separate storage reservoirs for metabolic products and compartments in which specific metabolic pathways occur (Section 18.1). In most cases, there is a potential difference across the membrane, which can be regarded as an "intermittent storage" for metabolic energy. However, cell membranes are not only barrier layers, but rather a platform to which many proteins are anchored for recognition as well as for conduction and regulation of substance fluxes. The functional properties of these membrane proteins can be studied outside of cells if they are embedded in a lipid bilayer. Many of these proteins play a crucial role in the regulation of cellular processes. G protein-coupled receptors are prominent examples of important targets for therapeutic agents.

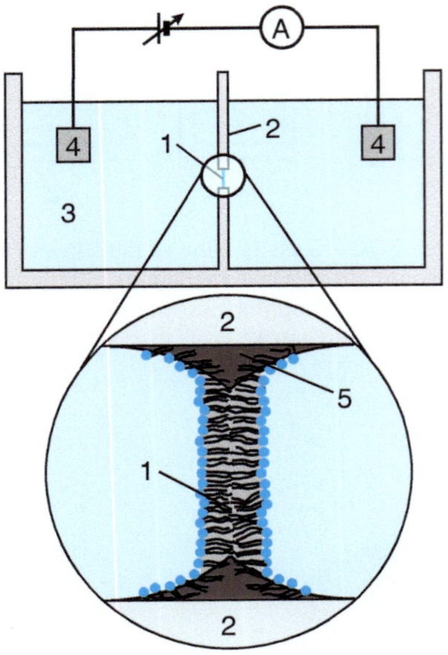

Figure 13.33 Experiment for the study of black lipid membranes.

1) Lipid membrane;
2) hydrophobic wall to separate the compartment with a pore of special shape;
3) aqueous electrolyte solution;
4) feeder electrodes;
5) remnants of the hydrophobic solvent used to apply the lipid, which seals the membrane to the wall of the measurement cell.

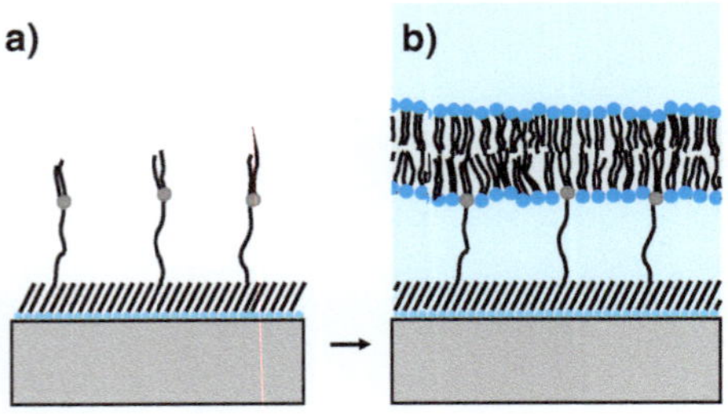

Figure 13.34 Tethered lipid bilayers with short-chain spacers. a) Mixed SAM with some lipids covalently linked to a long-chain thiol and b) completed lipid bilayer.

Therefore, there is great interest in studying such membranes under physicochemically precisely defined conditions, and without exposing these membranes to the unmanageable variety of influencing factors that occur in living cells. Electrodes lend themselves as carriers for the membranes because they can also read out or impose electrical signals. In this context, sensors are being developed that present a simplified membrane on a planar surface, which can be used to study the effect of drugs, for example, on binding processes to or transport processes across membranes. These two aspects, that is, the effort to decipher the functioning of biological systems at the supramolecular level (knowledge gain) and the opportunity to use such systems to rapidly quantify the efficacy of a large number of potential therapeutics under defined conditions (application), constitute an effective, long-term motivation for the further development of model membranes on electrodes. An essential question in this context is how accurately a model membrane reflects the composition and selected functions of the biological membranes of interest – and, most importantly, how well the model membranes support the structural and functional characterization of the membrane. With these points in mind, let us look at some of the model systems.

Vesicles enclose an inner solution volume and thus resemble cell compartments (Figure 13.32). In addition, proteins can also be incorporated into the vesicle wall. However, manipulation of the potential across the membrane is not readily possible because inserting an electrode into the interior of the vesicle always causes disruption of the vesicle wall.

Planar black lipid membranes (Figure 13.33) offer a remedy. They are built up via a hole in a very hydrophobic partition between two compartments with electrolyte solutions [43]. Although transport processes can be analyzed very well, structural characterization is extremely difficult. In addition, the asymmetry of real lipid bilayers is difficult or impossible to reproduce in these systems. In addition, there is the problem that the solvent, in which the lipids are applied, must not be completely removed, since a solvent residue is necessary to bond the membrane to the much thicker wall of the measurement cell.

SAMs provide a structurally well-defined attachment to the electrode, on which complex molecular architectures can be built on the solution side (Figure 13.16). However, the individual molecules of a SAM are firmly anchored to the substrate. Therefore, the individual membrane components do not exhibit lateral mobility, which is, however, a typical and important property of biological membranes. Although potentials can be adjusted across a SAM, transport processes cannot be investigated because the SAM is directly bound to the metal electrode.

Supported lipid bilayers can be formed asymmetrically by a combination of a Langmuir–Blodgett transfer (Figure 13.29a) followed by a Langmuir–Schaefer transfer (Figure 13.31). They sandwich an approximately 1 nm-thin layer of the electrolyte solution between the support material and the lipid double layer. However, this is too thin to realistically reproduce the particular distribution of ions on the membrane surface. In addition, there is a risk that the layers will detach from the electrode – especially during potential excursions.

This limitation can be overcome with tethered lipid bilayers (Figure 13.34). In tethered bilayers, some of the lipid building blocks are chemically functionalized with a chain-like spacer and bound, for example, to a gold electrode like individual molecules of a SAM. After completion of the lipid bilayer by nonfunctionalized lipid molecules, a membrane is formed which encloses a solution volume with a defined thickness between the membrane and the gold electrode. The thickness can even be varied by using spacer molecules of different lengths. After embedding ion channels, transport processes can be studied very precisely, as well as the effects of pharmaceuticals on such processes. Moreover, similar to SAMs or lipid bilayers, these layers can also be characterized structurally using spectroscopic methods (IRRAS, SPR, EQCM, Chapter 8).

If polymers are used as spacers, the solution reservoir between the polymer chains can also mimic the attachment of the membrane to the cytoskeleton (Figure 13.35).

Further Reading. A systematic overview on biological and biomimetic membranes can be found in [44–46], on anchored bilayers in [43, 47], about sensors developed from such membranes in [48].

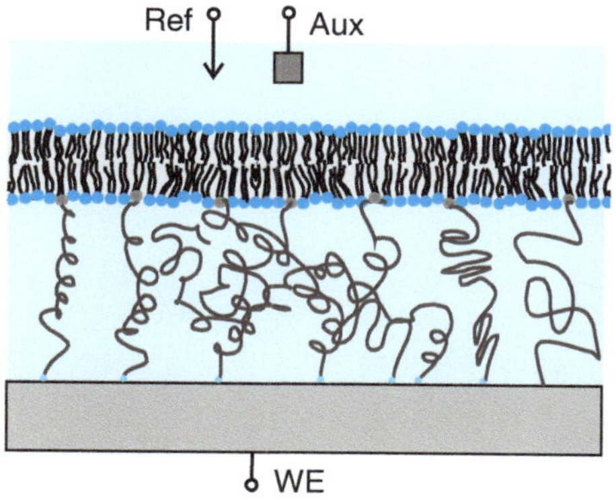

Figure 13.35 Lipid bilayers anchored to a polymer cushion that serves as a spacer.

13.2 Polymer Layers on Electrodes

Polymer layers can be applied to electrodes in a variety of ways. In many cases, techniques are used that have reached a high degree of technical maturity in the semiconductor industry, such as spin coating [49]. Material printing processes, which were originally used for transferring ink onto paper, are becoming increasingly important for coating a wide variety of substrates with functional materials. These are also widely used outside of electrochemical sciences. At this point, we will limit ourselves to processes that produce structurally defined layers at the molecular level and/or processes that utilize electrochemical principles for layer formation.

13.2.1 Layer-by-Layer Deposition of Polyelectrolytes

Layer-by-layer (LbL) adsorption of polyelectrolytes starts from a charged surface, for example, a gold electrode modified with a self-assembled sulfonate-terminated monolayer (Figure 13.36a). This surface is alternately immersed in a solution of polyanions or polycations. Polyanions and polycations are grouped under the term polyelectrolytes, in which a polymeric molecule carries multiple charges of the same type within a chain. We have already learned about Nafion as a polyanion in Section 2.4.4. Further examples are given in Figure 13.37. The charge in polyelectrolytes is compensated by oppositely charged, low molecular-weight ions.

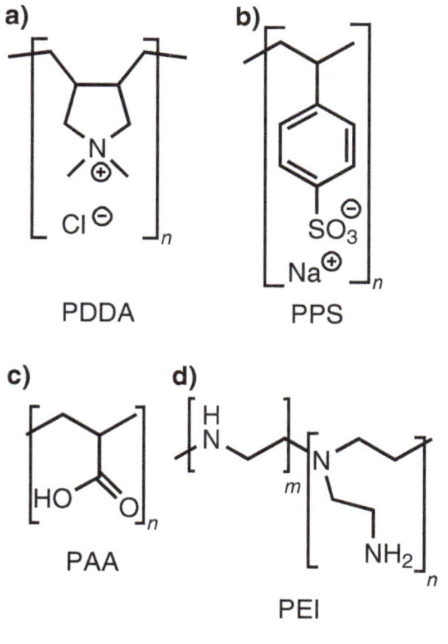

Figure 13.37 Polyelectrolytes for the construction of LbL layers. a) Polydiallyldimethyl-ammonium chloride (PDDA); b) poly-styrene-sulfonate (PSS); c) polyacrylic acid (PAA); d) polyethyleneimine (PEI). PAA and PEI are weak electrolytes that are charged only at a suitable pH.

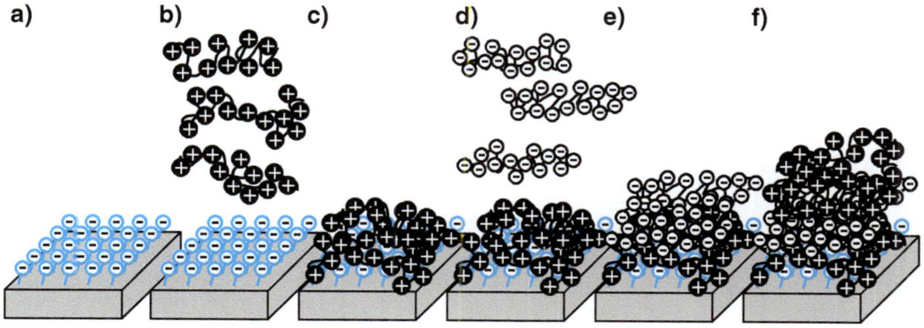

Figure 13.36 Layer-by-layer absorption of polyelectrolyte layers.

a) Negatively charged surface;
b) adsorption of polycations;
c) adsorbed polycations with reversal of surface charge;
d) adsorption of polyanions;
e) adsorbed multilayer with negative total charge;
f) further alternating adsorption of polycations with reversal of surface charge in each adsorption step.

On a negatively charged surface, polycations will arrange themselves to compensate for the surface charge (Figure 13.36b). In this case, one polycation compensates for several charges on the surface. However, because of the chain-like structure, it is not possible for all charges of the polymer chain to participate in charge compensation. It is possible to imagine the polycation being in a coiled conformation with charged side groups pointing outward in all directions. Therefore, only a fraction of the ionic groups will come into contact with the surface and compensate for the charges. Since the molecule contains other charged groups, the surface charge will not only be compensated for, but even overcompensated for, that is, the initially negatively charged surface will have a positive charge after adsorption of a polycation (Figure 13.36b). This circumstance also causes a self-limitation of the adsorption process, since further polycations are repelled from a positively charged surface (Figure 13.36c).

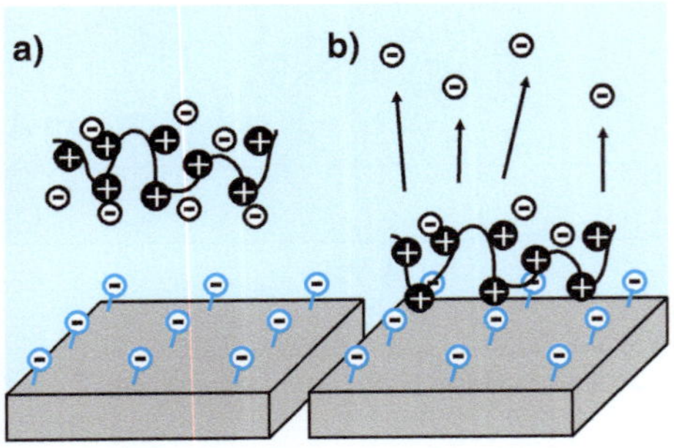

Figure 13.38 Increase in entropy during the formation of a layer-by-layer (LbL) film by releasing low molecular-weight ions. In addition, solvent molecules are also released from the hydration shell of the polyion, which are not shown here for clarity, but contributes to the increase in entropy.

The overcompensation of the charge and the self-limitation lead to a well-defined increase in layer thickness. As a result of the charge reversal, an analogous process can be initiated after a rinsing step by adsorption of a polyanion (Figure 13.36d) with the result that the surface again carries a negative excess charge (Figure 13.36e). Alternating the adsorption of polycations and polyanions causes a defined incremental increase in layer thickness in each adsorption step (Figure 13.36f). The driving force for the buildup and cohesion of the layers is the increase in entropy, which results from the release of many low molecular-weight counterions from the electrical double layer of the substrate and the ion cloud of the polyelectrolyte (Figure 13.38).

Although the thickness increase per cycle is constant, this does not automatically mean that each polymer molecule lies exactly within one layer. Rather, the layers interpenetrate each other and thus form a stable polymer matrix.

LbL films are suitable for flexible electrode modifications because nanoparticles (which carry excess charge to stabilize the suspension) and, especially, proteins can also be embedded in the film instead of one of the two polymeric polyelectrolytes. The films usually have ionic conductivity and a high water content, both of which are conducive to conversions in films. Depending on the film structure, redox centers in proteins can be arranged close enough to each other to allow electron transfer between them [50].

Further literature can be found in an edited book [15] and in original communications.

13.2.2 Insulating Polymers

Insulating polymers are used as blocking layers for the passivation of electrodes. Of interest, here are those layers that are produced directly on the electrode by electrochemical reactions. These allow very small structures to be passivated in a site-directed manner [51]. A prototypical example is polyphenol (or poly-oxyphenylene), which is

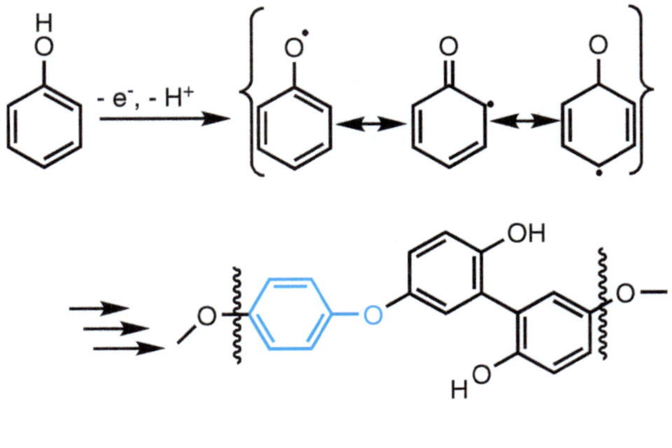

Figure 13.39 Structure evolution of polyphenols.

formed by oxidation of phenol. The initially formed radicals are linked preferentially via the para position, with some branching in the ortho position (Figure 13.39). The resulting polymer is deposited on the electrode. Because it has neither electronic nor ionic conductivity, even a film with a thickness of a few nanometers shields the electrode from the solution and the current is interrupted (electrode fouling). This is an undesirable process in techniques such as electrochemical detection in liquid chromatography, where one cannot choose whether the sample contains phenols or not and the film formed by the detection of phenols changes the sensitivity to the compound(s) of interest.

13.2.3 Ion-Exchangers and Redox Polymers

Films of ion-exchange resins (e.g., Nafion, Figure 2.30) have ionic conductivity, which enables their use, for example, as membrane electrolytes in fuel cells. As a thin layer on an electrode, they can serve other functions as well.

Since ion-exchangers only absorb oppositely charged low molecular-weight ions, they can be used as a barrier layer to increase selectivity in (bio)sensors (Section 18.3.3). For example, a cation-exchanger such as Nafion acts as a barrier layer for the ascorbate anion. Ascorbate can be oxidized electrochemically. Since it occurs in body fluids, it represents an interfering substance in the determination of other oxidizable metabolites [52].

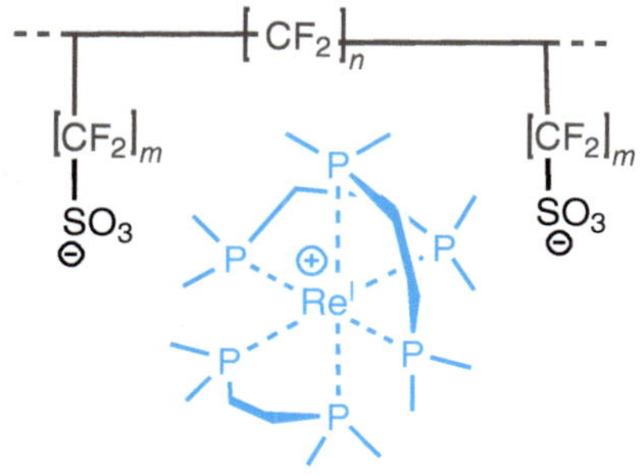

Figure 13.40 Structure of a Nafion film into which $[Re(dmpe)_3]^+$ ions were enriched [dmpe-1,2-bis-(dimethylphosphino)ethane].

Ion-exchangers can enrich oppositely charged ions. Hydrophobic cations are preferentially enriched in Nafion, having an extremely hydrophobic polymer backbone and immobilized sulfonate groups (Figure 13.40). Interesting hydrophobic cations are, for example, complexed transition metal ions that exhibit a rich redox chemistry. After ion-exchange, the loaded polymer has ionic conductivity and in addition, redox activity resulting from the conversion of the absorbed transition metal complexes.

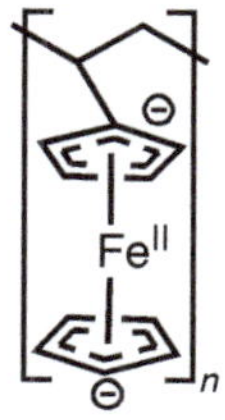

Figure 13.41 Structure of polyvinyl ferrocene.

In addition to these polymers, there are also the so-called redox polymers. They contain covalently bonded redox-active groups in the side chains, but do not exhibit ion-exchange properties because of the balanced charge. An example is polyvinyl ferrocene (Figure 13.41), which we have already discussed in connection with the quartz crystal microbalance in Section 8.8.3. The polymer acquires a charge only by oxidation of the ferrocene groups. This charge must be compensated for by low molecular-weight ions from the solution. We could also say that such films exhibit both ionic and electronic (i.e., a mixed) conductivity only after the initial oxidation (break-in), which we have noted in Section 8.8.3 as a prerequisite for complete conversions of the film.

Other polymers, as shown in Figure 13.42, contain both charged side chains favoring the intercalation of water and low molecular weight ions between the polymer chains as well as side chains with ligands for the complexation of redox-active transition metals. Therefore, they possess electronic conductivity and ionic conductivity with stable anchoring of redox centers. Such polymers are important in biosensors as mediators (electron transfer agents or redox shuttles) between the shielded redox centers of an enzyme and the electrode (Excursion 18-2).

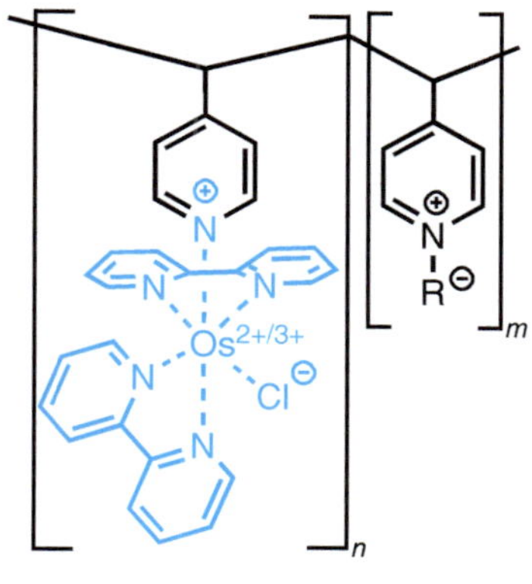

Figure 13.42 Structure of an osmium-bipyridyl complex coordinated to the pyridine side chain of a polymer.

Further reading on the examples discussed here is given in [53, 54]. Background information on redox polymers is provided in [55]; these compounds have become particularly important as matrices and mediators for biosensors. Reference [56] is recommended as an overview about redox polymers and polyelectrolyte layers.

13.2.4 Intrinsically Conductive Polymers

For many polymer materials, electronic conductivity can only be achieved by processing them into a composite with conductive carbon fibers, carbon black, or metal particles.

In contrast, intrinsically conducting polymers (ICP) exhibit electronic conductivity without the addition of other substances.

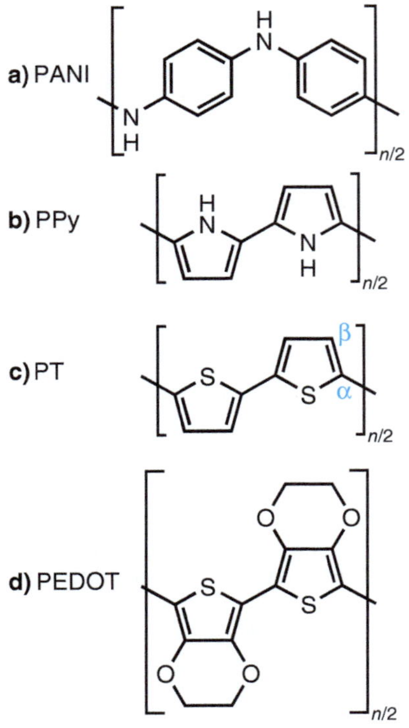

Figure 13.43 Molecular structure of

a) polyaniline (PANI);
b) polypyrrole (PPy);
c) polythiophene (PT);
d) poly-diethoxy-thiophene (PEDOT).

These polymers have chains of conjugated double bonds. The most important representatives of this class of materials for electrochemical applications are derived from polyaniline (PANI), polypyrrole (PPy), and polythiophene (PT) (Figure 13.43).

In the neutral state, the conductivities of these polymers correspond to those of inorganic semiconductors. The polymers are relatively easy to oxidize while the atomic arrangement remains intact. The positive charge of the polymer chain is compensated for by the incorporation of anions into the polymer (Figure 13.44). By analogy with inorganic semiconductors, this process is also known as "doping". In the oxidized state, the conductivity is increased by several orders of magnitude, almost reaching the conductivity of metals; hence, these materials are also called "synthetic metals". The electrochemical oxidation of ICPs occurs over a wide range of potentials as it becomes increasingly difficult to remove further electrons from the already positively charged polymer chain. Under favorable conditions, 25–30% of the monomer units can be *reversibly* oxidized. If oxidation occurs even further, one speaks of overoxidation. This leads to bond rearrangements within the polymer chain, disrupting the conjugated domains of the polymer, resulting in irreversible loss of conductivity. Therefore, overoxidation should be avoided if the conductivity of the polymers is required for the application. In addition to oxidation, the conductivity of some ICPs can also be increased by reduction.

The reversibility of the oxidation-reduction cycles is significantly influenced by chemical side reactions (e.g., with O_2), which attack the β-position of the rings or the heteroatoms. In the technically important polymer PEDOT (Figure 13.43d), the β-positions of each monomer unit are already substituted, which stabilizes the structure chemically.

The structural and functional diversity of this class of materials can be tuned by the following options:

- Details of material deposition
- Use of different counterions A^-
- Use of substituted monomers
- Formation of copolymers from different monomers

Film formation. The coating of electrodes with ICP can be achieved in different ways. For example, a chemically prepared polymer [57] can be dissolved in an organic solvent, and this solution can be drop-casted onto an electrode or applied to it by spin coating. After evaporation of the solvent, a film remains on the electrode.

Figure 13.44 Redox reactions of PT with intercalation of anions (A^-).

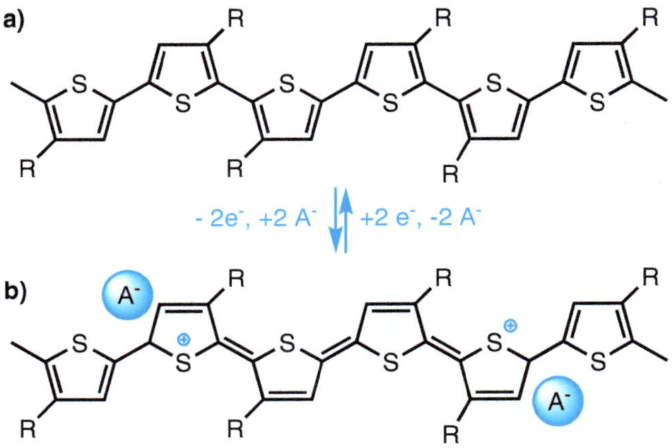

Most ICPs can be produced by electropolymerization directly on the electrode, whereby only the electrochemically active surface is coated and not any insulating sheath of the electrode body. The mechanism of electropolymerization has been intensively studied and has been the subject of some controversy [58]. Today, the mechanism shown for PT in Figure 13.45 is generally accepted [57] and also applicable to PPy. The reaction starts with the oxidation of the monomer to the radical cation (Figure 13.45a). Two radical cations recombine to form a dication that is stabilized by donating two protons to form the dimer (Figure 13.45b,c). The dimer has a larger conjugated π-electron system than the monomer. Therefore, it is more easily oxidized to a radical cation than the monomer (Figure 13.45d). This systematic relationship between increasing chain length of the oligomers and their decreasing redox potential can be seen in lines 3 and 10–12 of Table 13.4. The radical cation of the dimer can recombine with radical cations of the monomer or other oligomers (Figure 13.45e). The resulting higher oligomers are in turn even more easily oxidized, allowing the reaction to continue. A crucial feature for film formation is the fact that the solubility of the oligomers decreases with increasing chain length. Thus, the oligomers will precipitate out of solution at a certain chain length. The reactions shown in Figure 13.45 require repeated oxidations of the monomers and the oligomers at the electrode. Homogeneous radical recombination reactions take place in solution, but this is also confined to the immediate vicinity of the electrode. Therefore, the resulting oligomers precipitate on the electrode.

The morphological appearance of the polymer films can be very diverse and is determined by nucleation and growth processes on the electrode surface. Often, irregular, adjacent domains grow laterally and vertically in competition with each other. The resulting morphology is often called the "cauliflower structure" (Figure 13.46). On the electrode surface, the oligomers continue to grow or cross-link. The layer growth can come to a halt when the polymer layer has reached a thickness at which the oxidation potential of the monomers at the polymer film–electrolyte interface is no longer reached, taking into account the voltage drop across the polymer film.

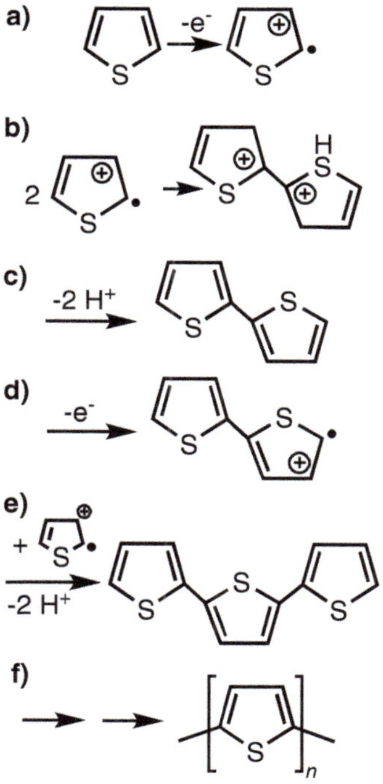

Figure 13.45 Formation of ICP by coupling of radical cations using PT as an example.

Table 13.4 Electrochemical properties of monomers, from which ICP can be formed. Compilation according to [57].

Monomer	Oxidation potential/V	Reference electrode	Source
Aniline	1.4	SCE	[59]
Pyrrole	1.2	SCE	[59]
Thiophene	2.07	SCE	[59]
3,4-Ethyldioxythiophene	1.39	Fc/Fc^+	[60]
2-(3-Thienyl)-ethyl acetate	1.4	$Ag|AgClO_4|CH_3CN$	[61]
3-Methylthiophene	1.86	SCE	[59]
3-Bromothiophene	2.1	SCE	[59]
3-Cyanothiophene	2.46	SCE	[59]
3-Nitrothiophene	2.69	$Ag|AgClO_4|CH_3CN$	[62]
Bithiophene	1.31	SCE	[59]
Terthienyl	1.05	SCE	[59]
Pentathienyl	0.43	SCE	[59]

Figure 13.46 Morphology of PPy. Source: Reproduced with permission from C. Dbiemme-Chouvy, A. Fakjry, F. Pillier; Electrosynthesis of polypyrrole nano/microstructures using an electrogenerated oriented polypyrrole nanowire array as framework. *Electrochim. Acta* **2018**, *268*, 66–72, Copyright 2018, Elsevier.

According to these considerations, it is plausible that many factors of the experiment influence film formation. These include the material and, if applicable, the crystallographic orientation of the electrode on which the oligomers precipitate and arrange themselves. Furthermore, all solution components (anions, cations, solvents) exert a strong influence as they compete with the monomers and oligomers for adsorption sites at the electrode surface, and their adsorption affects the reaction rates of the reaction steps. In particular, the anions, which are incorporated into the growing polymer film as counterions during oxidation, and their interactions with the electrode material and the developing polymer chain significantly influence the film morphology.

An external control variable in the deposition experiments is the potential or current density. Films from ICP can be obtained potentiostatically, potentiodynamically, galvanostatically, or galvanodynamically. The optimization of these excitation techniques must take into account the following aspects. The applied potential or the potential that arises in the case of galvanostatic operation must be positive enough to oxidize the monomers at a high rate, but remain below the potential at which overoxidation of the polymer commences. It is also important that the bimolecular recombination reaction of the radical cations takes place at a high reaction rate. As a second-order reaction, this depends very much on the concentration of radical cations. This transient concentration ultimately depends on the formation rate of the radical cations from the monomers. If the reaction rate slows down during deposition due to mass transport limitations, a radical cation formed at the electrode may not immediately recombine with other radical cations in the immediate vicinity of the electrode. If the recombination occurs at too great a distance from the electrode, the formed oligomer may diffuse away from the electrode or convert into a radical cation only after a further diffusion process to the electrode surface. Such a scenario greatly reduces the efficiency of polymer formation or prevents it altogether. Therefore, more complex conditions are employed in polymer deposition, such as repetitive potential cycles or pulse sequences [63]. If the deposition is carried out by executing repetitive cyclic voltammograms, the film growth can be monitored very well via an incremental current increase in each cycle (Figure 13.47). With pulse sequences, the formation of radical cations is periodically interrupted, which allows the initial concentration of monomers at the electrode to be restored. With the following pulse, a high concentration of radical cations is again formed at the electrode, which can recombine efficiently.

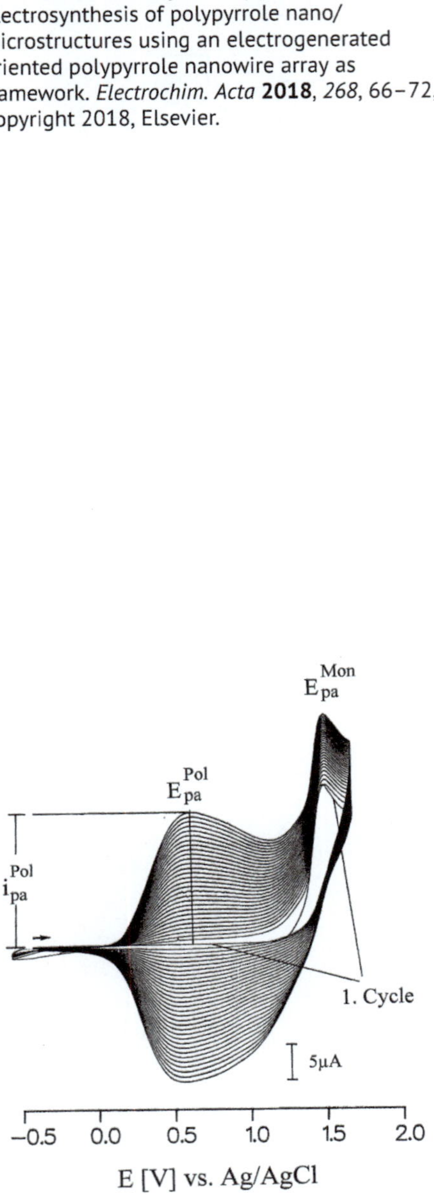

Figure 13.47 Cyclic voltammograms for the electropolymerization of pyrrole to polypyrrole. The first 30 cycles are shown, the first cycle is labeled; 0.1 mol L^{-1} Bu_4NPF_6 in acetonitrile, $v = 0.2$ V s^{-1}. E_{pa}^{Mon} – anodic peak potential for the monomer oxidation, E_{pa}^{Pol} – anodic peak potential for the polymer oxidation, i_{pa}^{Pol} – anodic peak current for the polymer correlating with the amount of polymer deposited. Source: Reproduced with permission from J. Heinze, K. Hinkelmann, M. Land, *DECHEMA Monogr.* **1988**, *112*, 75–87, Copyright 1988, DECHEMA. The plot was adapted to the IUPAC convention.

Copolymers. In a solution with two monomers, A and B, polymerization can lead to copolymers if A and B have similar oxidation potentials. The relative solution concentrations [A]/([A] + [B]) and [B]/([A] + [B]) often do not correspond to the resulting content of monomers A and B in the copolymer film. The decisive factor for the type of copolymer is the mutual reactivity of the radical cations. If the reactivity of the radical cations $A^{+\bullet}$ and $B^{+\bullet}$ is approximately equal, the result is a random copolymer (–A–B–B–A–B–A–B–A–). If the reactivity between $A^{+\bullet}$ and $B^{+\bullet}$ is much greater than that between $A^{+\bullet}$–$A^{+\bullet}$ and $B^{+\bullet}$–$B^{+\bullet}$, an alternating copolymer (–A–B–A–B–A–B–) is formed. In contrast, block copolymers (–A–A–A–...–A–B–B–...–B–B–B–) are formed when the recombination rate of $A^{+\bullet}$–$A^{+\bullet}$ and $B^{+\bullet}$–$B^{+\bullet}$ is higher than that of $A^{+\bullet}$ and $B^{+\bullet}$. The recombination rates can be lowered by sterically demanding substituents. These are often functional units themselves or contain reactive groups for subsequent covalent attachment of functional molecules. Often, these substituted monomers can only be incorporated into an ICP in dilute form.

Adhesion. ICP films cannot be formed on nonnoble metals by electrooxidation because the nonnoble metals dissolve at positive potentials. However, films can be formed on nonnoble metals if the films are anchored to the metal with a self-assembled monolayer, whose terminal group is a monomer. This terminal group is then incorporated into the polymer film during *chemical* polymerization (Figure 13.48), which supports the formation of well-adherent layers that are used, for example, in corrosion protection.

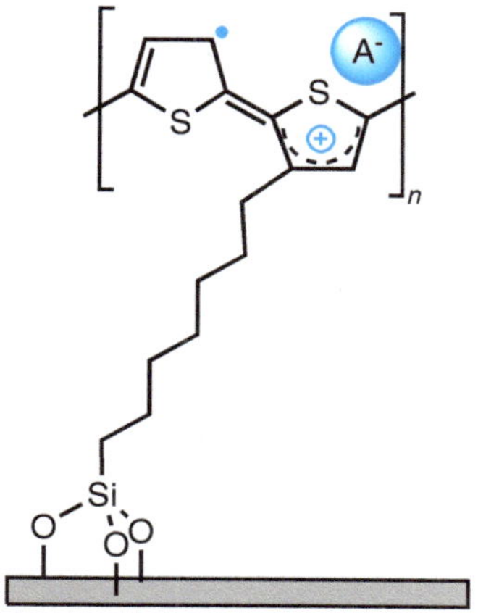

Figure 13.48 PT-film with a SAM as an adhesion promoter.

Redox reactions of ICPs are accompanied by the release and incorporation of anions, cations, and solvent molecules. In Section 8.8.3, we have discussed the use of the electrochemical quartz crystal microbalance (EQCM) to elucidate these processes. Using the simultaneous recording of the converted charge and the mass changes, one obtains the apparent mass change M_{app} per transferred electron. The mass change is composed of the contribution of x anions with molar mass M_{an}, of y cations ($y\,M_{cat}$) and of the exchanged solvent molecules ($z\,M_{solv}$).

$$M_{app} = xM_{an} + yM_{cat} + zM_{solv} \tag{13.9}$$

By varying the size of the ions provided in the electrolyte, the type of the exchanged ion can be influenced. From the comparison of experiments in solvents with different degrees of deuteration, the number of exchanged solvent molecules can be determined.

Conduction Mechanisms. The explanation of the electronic conductivity of ICP is difficult. Different models have been used for the interpretation, but they usually reproduce only some of the experimental observations at a time [58]. One interpretation arises from the polaron and bipolaron model. Polarons and bipolarons are quasi-particles. They describe multielectron states that can move along a polymer chain or within a crystal. This motion can be formally described as that of charge carriers, hence the name quasi-particles. In neutral PPy, the band gap is about 3.2 V (Figure 13.49a), which can be estimated from the spacing of oxidation and reduction signals in cyclic voltammograms of ICP films [64]. After oxidation, new electronic states are formed whose energetic position can be determined by spectroelectrochemical methods [65, 66]. In this model, the formation of the radical cation and the intercalation of an anion corresponds to the formation of a polaron (Figure 13.49b). The positive charge due to the "missing electron" is more likely to be near the anion than at any other point in the chain. The anion and the electron deficit are mobile together. In the energy diagram, they correspond to states within the band gap.

Upon further oxidation, a second electron can be removed (Figure 13.49c). Here, the distance between the two electron-deficient states along the polymer chain is not arbitrary because some distances can be better stabilized by the π-electron system.

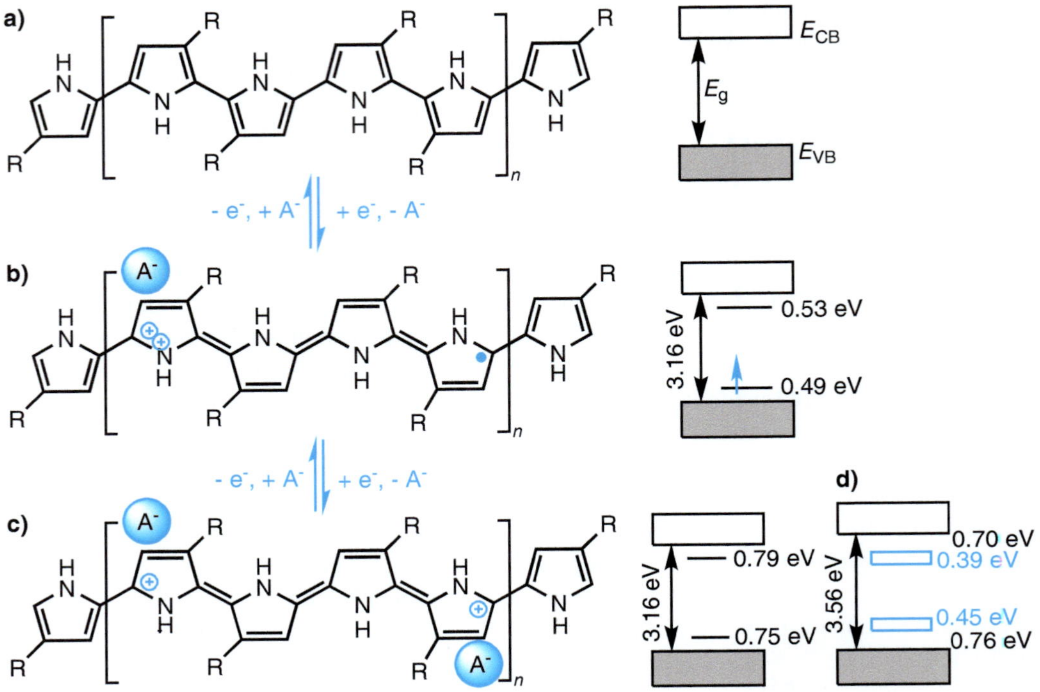

Figure 13.49 Molecular structures for PPy and schematic energy diagrams.

a) Neutral polymer;
b) radical cation or polaron;
c) radical dication or bipolaron;
d) band diagram for the bipolaron states broadened to a band at higher oxidation state.

Source: Modified according to [57].

Accordingly, the two anions are positioned. This state is called bipolaron. It lies within the band gap and has a smaller energetic difference than in polarons.

The further oxidation of the polymer can be seen as the formation of further bipolarons on the polymer chain. In the energy diagram, this leads to an overlap of the bipolaronic states to two energy bands with a much smaller energy gap than the band gap in the neutral polymer (Figure 13.49d). However, this analogy to inorganic semiconductors also has limitations [58] because the conjugated regions in ICPs have a very limited length of only 5–20 monomer units. Experimentally, the spectra of ICPs do *not indicate* the formation of energy bands when comparing oligomers and polymer [58].

In the case of PANI, the conditions are somewhat different from polymers formed from heterocycles, since protonation reactions occur in addition to redox processes and greatly affect conductivity (Figure 13.50). The fully reduced, yellowish form can be converted by partial oxidation (formally one electron for two monomer units) to the green emeraldine base, which has a mixture of benzoid and quinoid moieties. However, it is only when the quinoid structures are protonated by an acid (e.g., HCl) that the electronically conductive emeraldine salt is formed.

Further oxidation leads to the violet pernigraniline base or the pernigraniline salt, which is composed only of quinoid structural units.

The conductivity of the emeraldine salt becomes more plausible when looking at the resonance structures in the right part of Figure 13.51. Here, all N atoms are equivalent, but two positive formal charges have to be distributed among four monomer units and compensated by counter anions A^-. These charges can move along the chain without changing the connectivity of the involved atoms. The charge carrier, that is, the mobile

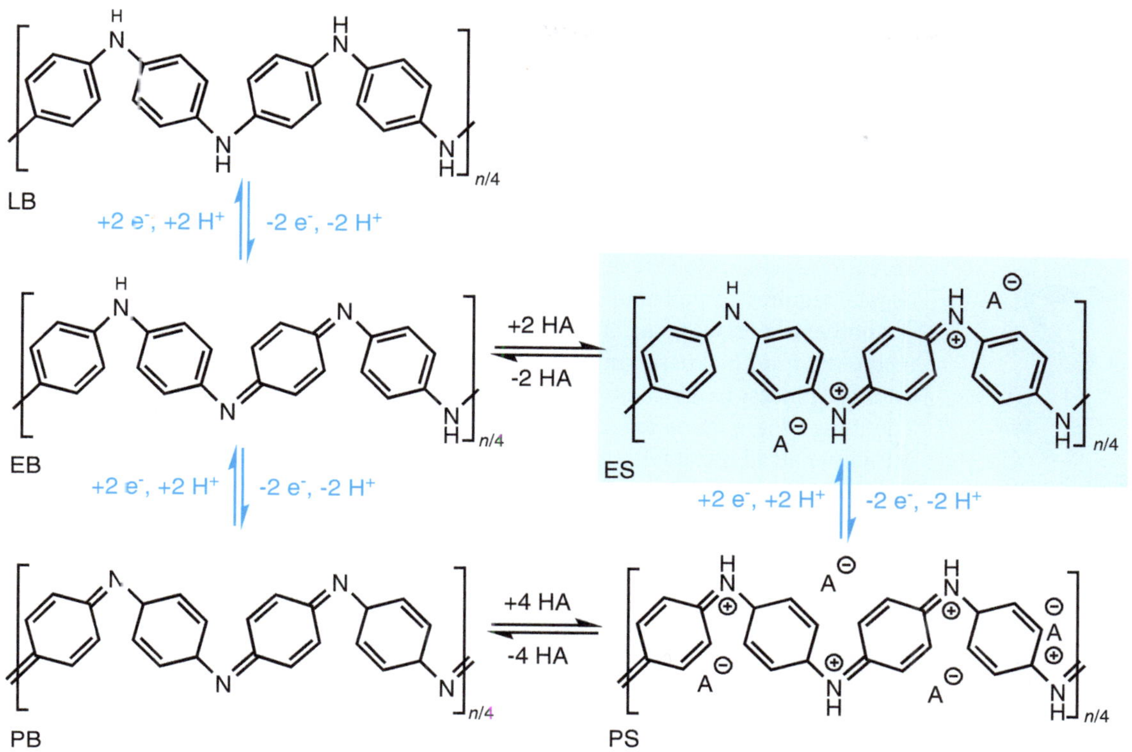

Figure 13.50 Molecular structure and redox processes in PANI.

LB – leucoemeraldine base;
EB – emeraldine base;
PB – pernigraniline base;
ES – emeraldine salt;
PS – pernigraniline salt.

Only ES has a high electronical conductivity.

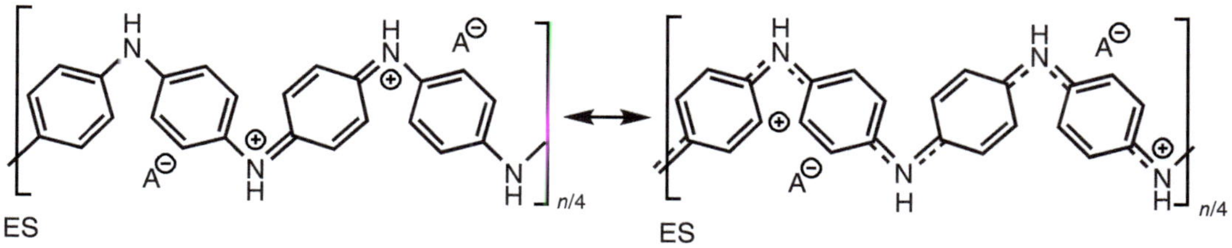

Figure 13.51 Resonance structures of the emeraldine salt from PANI.

charge deficit of the chain, is "smeared" over several monomer units with compensation by an anion. This state is referred to as a polaron.

Further Reading. A comprehensive overview of various ICPs can be found in [67]. The formation mechanisms and the conductivity mechanism are discussed in depth in [58, 68]. An overview and further literature can be found in [67] on the application of ICPs for electrochromic devices, membranes and ion exchangers, corrosion protection, chemo- and biosensors, energy storage, artificial muscles and electrocatalysis. Reference [69] is recommended as an introduction to the use of ICPs for photovoltaic applications and light-emitting diodes (LEDs).

13.3 Films of Organic–Inorganic Hybrid Materials

13.3.1 Ion-Exchangers and Oxide Films

Zeolites and clays are naturally occurring or synthetically produced hydrous aluminosilicates. Zeolites form three-dimensional (3D) spatial lattices, while layered arrangements occur in clays. The mutual arrangement of these layers can be stabilized by treatment with polyoxoanions. The materials can also exhibit catalytic activity and are cation-exchangers that can accumulate redox-active cations. Processing into films usually requires a small amount of a polymeric binder [70].

Another important group of oxidic films is formed by the anodization of valve metals, where the film thickness can be adjusted by the applied voltage. Of particular interest is the process in which nanoporous structures are formed directly (Section 14.6.1). The films adhere to the substrate without a binder. Many of these films can incorporate either alkali metal ions or protons into the oxide lattice when the transition metal ions change valence. Well-known in this context are tungsten bronzes, in which the transition W(VI)→W(V) is associated with the intercalation of H^+ and an appearance of an intense color, which is used for electrochromic windows [71].

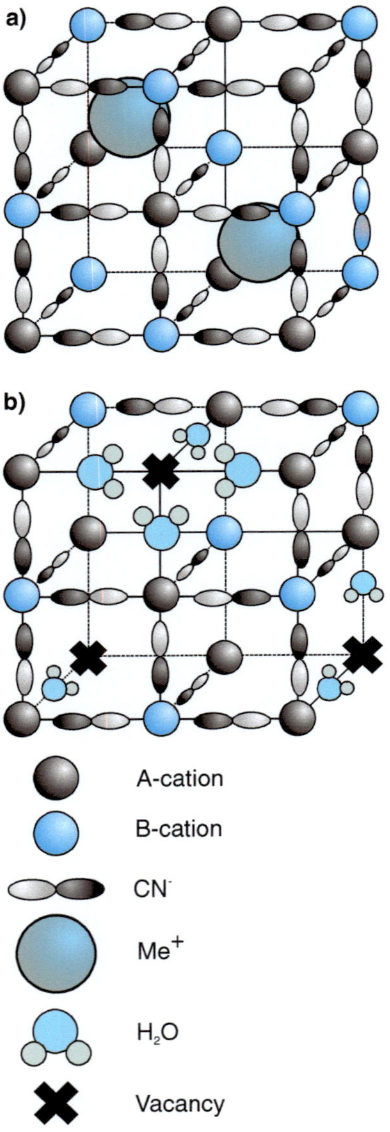

Figure 13.52 Molecular structures of metal hexacyanometallates. a) Alkali metal ions compensate for the charge, example $KFe^{III}[Fe^{II}(CN)_6]$; b) without alkali metals (example: $Fe_4{}^{III}[Fe^{II}(CN)_6]_3 \times mH_2O$). For clarity, ligands have not been drawn in the second layer.

13.3.2 Polyoxometallates, Polycyanometallates, Conductive Coordination Network Compounds

Polyoxymetallates are the anions of heteropolyacids formed by condensation of phosphates with oxoanions of tungsten or molybdenum (e.g., $P_2W_{17}MoO_{62}K_6$). They are molecularly uniform in structure, exhibit a rich redox chemistry due to valence changes of the transition metals, and possess electrocatalytic properties [72].

Films of metal hexacyanometallates (MHCMs) have the general formula $Me_xA_y[B(CN)_n]_z$ (Me – alkali metal, A, B – transition metal). They can be deposited electrochemically using different procedures. The principle here is usually a valence change of one of the two transition metals. This results in a compound with a lower solubility product leading to crystallization of the compound on the electrode surface. The prototypical compound is Prussian Blue ($Fe^{III}{}_4[Fe^{II}(CN)_6]_3 \times mH_2O$). The transition metals are located at the corners of a cubic lattice and are connected along the edges by cyanide as the bridging ligand (Figure 13.52). Alkali metals and NH_4^+ can move in the channels. When the transition metals undergo valence changes by an electron transfer, the transfer of alkali metals causes charge compensation, thus MHCM films are insertion electrodes. MHCMs are compounds with mixed conductivity, which is a prerequisite for the electrochemical conversion of thicker films. MHCMs occur in two structures. In the structure shown in Figure 13.52a, *all* nodes of the lattice are occupied by a transition metal ion. A possible negative charge excess due to the ligands is balanced by intercalated alkali metal ions. In Figure 13.52b, vacancies are present on the nodes in a regular arrangement, through which the charge is balanced without intercalation of the alkali metal ions. In this case, water can replace the cyanide ligand in the vicinity of the vacancies. Only in case of valence changes must other ions be intercalated for charge compensation. In many materials, the structural motifs of Figure 13.52a,b occur together in a solid solution.

Applications include redox-active MHCM films as switchable ion-exchangers, as solid-state contacts for ion-selective electrodes, as electrochromic films, as active battery materials and as electrocatalysts.

Further reviews on the electrochemistry of metal hexacyanometallates can be found in [73, 74].

Coordination networks compounds or metal–organic frameworks (MOF) are currently attracting a lot of attention in research because of potential applications as sorption materials, as active layers in sensors and in photoelectrochemical energy converters or as electrocatalysts. All of these applications require electrical conductivity of the framework compounds and stability in electrochemically useful solvents. Many MOFs cannot meet these requirements. However, a 3D framework compound with molecularly precisely defined and modifiable structure would be highly desirable for many electrochemical applications. A recent summary of the state of research, also with reference to electrochemical experiments, can be found in [75].

13.4 Voltammetry at Modified Electrodes

13.4.1 Overview

In this section, we consider the electrochemical reactions of modified electrodes *after* their complete preparation. Because of the variety of modifications, reactions, and applications, it is useful to first systematize the possible effects and then discuss prototypical situations in more detail. In real films, with their unavoidable structural defects, we must always expect the different phenomena to occur simultaneously. Supplementary structural investigations using the methods discussed in Chapter 8 are therefore indispensable to narrow down the variety of situations in question before quantitative data are obtained using electrochemical methods.

Figure 13.53 shows some scenarios for the turnover of the redox centers in monolayers or thin films. Here, the redox centers may be anchored to the supporting electrode but essentially surrounded by the liquid solvent (Figure 13.53a). In these cases, the formal potentials are usually within 100 mV of those of the same redox system diffusing in

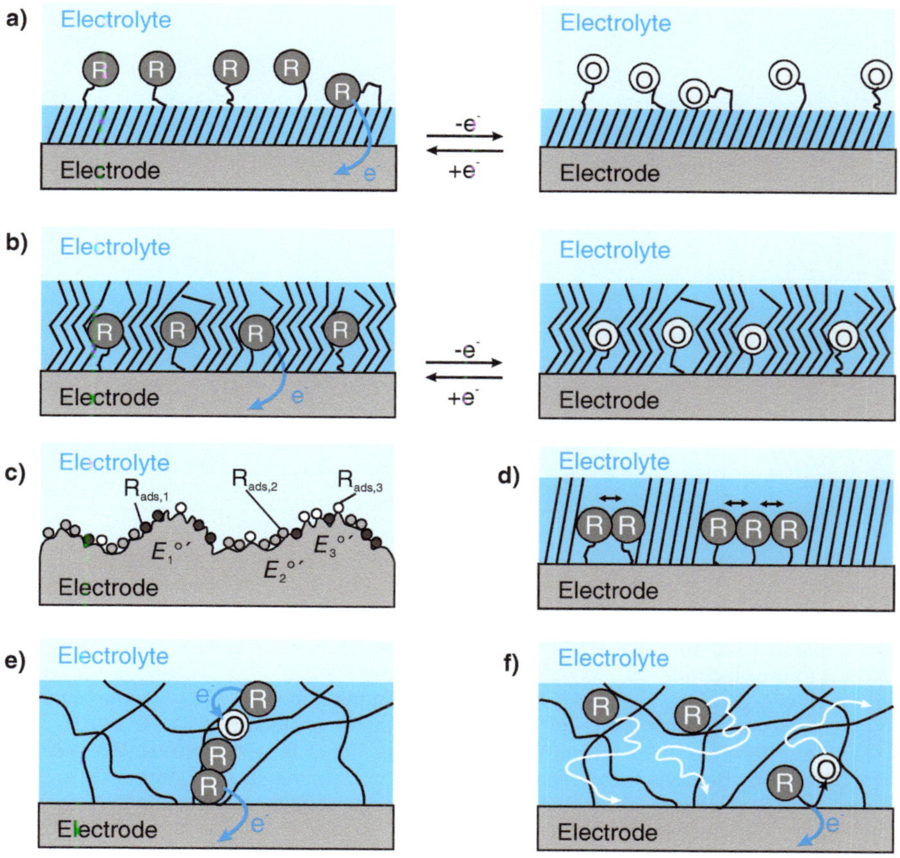

Figure 13.53 Electrochemical conversion of redox centers in monolayers and thin films (shown as cross sections).

a) Isolated immobilized redox centers surrounded by solvent;
b) immobilized redox centers embedded in the film;
c) identical redox centers on nonequivalent adsorption sites, resulting in different redox potentials;
d) interacting redox centers in a thin film;
e) film or thin layer with electron transfer between redox centers in the film;
f) film or thin layer with physical diffusion of the redox centers in the film.

the solution. However, redox centers may also be embedded in a matrix or thin film, which may be considered as a different solvent environment (Figure 13.53b). This can manifest itself in a significant shift in the voltammetric signals if the redox forms are stabilized differently in the film than in the solvent. Analogously, we can also classify the relative stabilization or destabilization of adsorbates by bond formation to the support. In Figure 13.53a and b, all redox centers are present on equivalent binding sites or adsorption sites and are therefore energetically equivalent. However, on real solid-state electrodes, adsorption sites are often present that stabilize the redox centers differently. Consequently, we deal with a distribution of redox centers behaving similarly but not identically (Figure 13.53c). Even when all adsorption sites are equivalent, interaction *between* neighboring redox-active adsorbates can cause activity coefficients to deviate substantially from unity, thus causing changes in the energetics of reactions with coverage or film loading (Figure 13.53d). For thicker films or thin-layer cells, in which not all redox centers perform a direct charge transfer to the supporting electrode, the transport of charges can occur by electron transfer (ET) or hopping between adjacent redox centers (Figure 13.53e) or by physical diffusion (mass transport) of the redox centers in the film and/or thin layer (Figure 13.53f).

The variety of scenarios is further increased if we include the conversion of a substance S dissolved in the electrolyte at the modified electrode (Figure 13.54), which is the relevant situation, for example, in electrocatalytic electrodes for electrosynthesis, energy conversion, and sensing. Here, S can be converted directly at the interface between the supporting electrode and the film if the film is extremely thin (Figure 13.54a) or permeable to S (Figure 13.54b). Despite the coverage by an insulating film, conversion may also occur at the electrode–solution interface if the

The word **"substrate"** is used in the literature with two very different meanings. On the one hand, the term is used in the sense of a carrier or platform for the electrode, onto which a film or monolayer is coated. This corresponds to the vocabulary of surface science and coating technology (e.g., in Section 13.1.1). On the other hand, the dissolved reagent S that is converted at a modified electrode is also referred to as "substrate" in Figure 13.54. This corresponds to its use in biochemistry and catalysis. Since the two fields merge here and in Chapter 18, the vocabulary also overlaps. In this textbook, the use of "substrate" in the sense of reagent of a reaction is preferred, but a complete elimination of the alternative meaning is deliberately omitted in order to prepare readers for the reception of journal literature, in which both meanings occur in parallel *without* warning, sometimes in the same document!

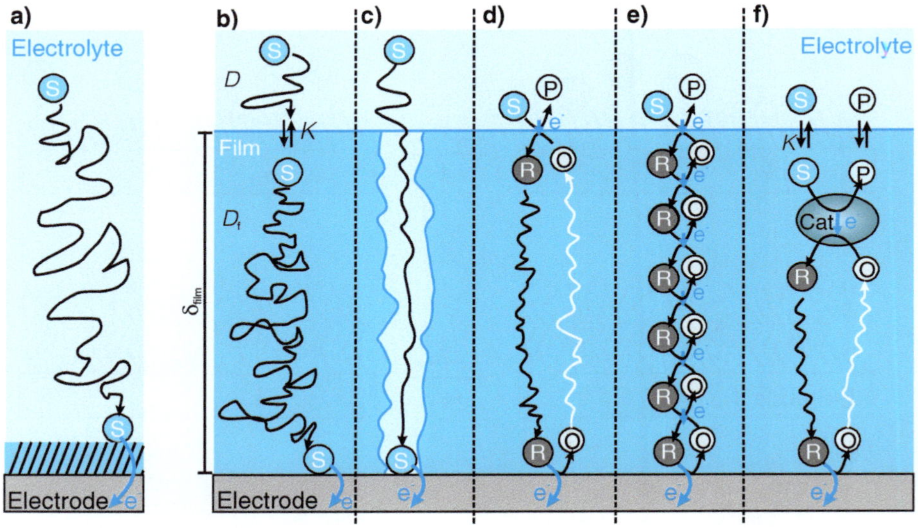

Figure 13.54 Electrochemical conversion of a dissolved reactant S → P at a modified electrode.

a) Direct ET (only for monomolecular films);
b) accumulation and diffusion of S in the film, ET at electrode–film interface;
c) access to the electrode–solution interface via pinholes;
d) reaction of S with a redox system R/O bound within the film at the film–solution interface and charge transfer between redox centers and electrode analogous to Figure 13.53f;
e) reaction of S with a redox system R/O bound in the film at the film–solution interface and charge transfer between redox centers and the electrode analogous to Figure 13.53e;
f) reaction of S with the mediator R/O catalyzed by a catalyst (Cat) bound within the film. Here with film diffusion of the mediator.

film has pinholes (Figure 13.54c). Catalytic conversion of S with a redox system O/R bound within the film can occur at the film–solution interface. For a continued conversion charge transport must occur by physical diffusion of O and R in the film (Figure 13.54d) or by ET between closely spaced redox centers (Figure 13.54e). In many applications, such as enzymatic biosensors, the reaction $S + O \rightarrow P + R$ needs a catalyst embedded in the film (Figure 13.54f).

13.4.2 How Do the Thermodynamics of Redox Reactions Change When Redox Centers Are Embedded in a Film?

We consider here the adsorption of the oxidized (O) and reduced (R) forms of a redox system at an electrode:

$$O_{solv} \rightleftharpoons O_{ads}, \tag{13.10}$$

$$R_{solv} \rightleftharpoons R_{ads}. \tag{13.11}$$

If only energetically equivalent adsorption sites exist at the electrode, adsorbate–adsorbate interactions will be negligible, and interfacial concentrations $\Gamma_{O,sa}$ and $\Gamma_{R,sa}$ (index sa for saturated, that is, usually a monolayer) will be maximum, we can describe the adsorbate concentrations according to the Langmuir competitive adsorption model of O and R given by Eq. (8.106) with the adsorption equilibrium constants β_O and β_R:

Γ_O, Γ_R – Interfacial concentrations [mol cm^{-2}]

$\Gamma_{O,sa}$, $\Gamma_{R,sa}$ – Maximum interfacial concentrations [mol cm^{-2}]

$c_O(x = 0)$, $c_R(x = 0)$ – Solution concentrations of O and R at the electrode ($x = 0$) [mol L^{-1}]

β_O, β_R – Adsorption equilibrium constants of O and R [L mol^{-1}]

b_O, b_R – Adsorption coefficients of O and R ($\Gamma_{O,s}\,\beta_O$, $\Gamma_{R,s}\,\beta_R$) [L cm^{-2}].

$$\Gamma_O = \frac{\Gamma_{O,sa}\beta_O c_O(x=0)}{1 + \beta_O c_O(x=0) + \beta_R c_R(x=0)}, \tag{13.12}$$

$$\Gamma_R = \frac{\Gamma_{R,sa}\beta_R c_R(x=0)}{1 + \beta_O c_O(x=0) + \beta_R c_R(x=0)}. \tag{13.13}$$

Analogously to the formal potential $E^{\circ\prime}$, where $c_O(x = 0) = c_R(x = 0)$ holds for a nernstian redox system, we can now define a formal potential $E^{\circ\prime}_{ads}$ for the adsorbed redox system, where $\Gamma_O = \Gamma_R$ holds. Division of Eq. (13.12) by Eq. (13.13) yields

$$\frac{\Gamma_O}{\Gamma_R} = \frac{\Gamma_{O,sa}\beta_O c_O(x=0)}{\Gamma_{R,sa}\beta_R c_R(x=0)}, \tag{13.14}$$

$$\frac{c_O(x=0)}{c_R(x=0)} = \frac{\Gamma_O \cdot \Gamma_{R,sa}\beta_R}{\Gamma_R \cdot \Gamma_{O,sa}\beta_O}. \tag{13.15}$$

Substituting this into the Nernst equation [Eq. (3.19)] gives

$$E = E^{\circ\prime} + \frac{RT}{nF}\ln\frac{c_O(x=0)}{c_R(x=0)}$$

$$= E^{\circ\prime} + \frac{RT}{nF}\ln\frac{\Gamma_O \cdot \Gamma_{R,sa}\beta_R}{\Gamma_R \cdot \Gamma_{O,sa}\beta_O} = E^{\circ\prime} + \frac{RT}{nF}\ln\frac{\Gamma_{R,sa}\beta_R}{\Gamma_{O,sa}\beta_O} + \frac{RT}{nF}\ln\frac{\Gamma_O}{\Gamma_R} \tag{13.16}$$

or

$$E = E^{\circ\prime}_{ads} + \frac{RT}{nF}\ln\frac{\Gamma_O}{\Gamma_R}. \tag{13.17}$$

The difference $E^{\circ\prime}_{ads} - E^{\circ\prime}$ is therefore

$$E^{\circ\prime}_{ads} - E^{\circ\prime} = +\frac{RT}{nF}\ln\frac{\Gamma_{R,s}\beta_R}{\Gamma_{O,s}\beta_O} = -\frac{RT}{nF}\ln\frac{\Gamma_{O,s}\beta_O}{\Gamma_{R,s}\beta_R} = -\frac{RT}{nF}\ln\frac{b_O}{b_R}. \tag{13.18}$$

The terms b_O and b_R are also called adsorption coefficients. In the case of $E^{\circ\prime}_{ads} \approx E^{\circ\prime}$, the two adsorption coefficients are equal, for example, because O and R are adsorbed to the same extent and their maximum interfacial concentrations are equal (i.e., equal space requirements for adsorbed O and R).

However, if one of the redox forms is adsorbed more strongly than the other, one of the two redox forms is more stabilized than the other. Therefore, $E^{\circ\prime}_{ads}$ is shifted from $E^{\circ\prime}$. More or less energy is then required for the electrochemical conversion between R_{ads} and O_{ads} than for the conversion between R_{solv} and O_{solv}. These energy differences correspond to shifts in the voltammetric signals on the potential axis.

Considerations of the different stabilization of the redox forms O_{ads} and R_{ads} in the film can also be applied to the difference between the sketches in Figure 13.53a and Figure 13.53b. If there is a high density of redox centers in a film, the change in charge converted in the electron transfer reaction must be balanced by counterions. In these cases, a shift of the formal potential by 0.059 V/n (at 298 K) is often observed when the concentration of the conductive electrolyte increases by a factor of 10. Apart from such unspecific effects of ionic strength, specific effects can be distinguished in which the formal potential responds to the concentration of *one* counterion species only. Such effects may be caused for instance by ion-pairing.

For the interpretation of voltammetric signals, there are different consequences depending on the system and the objective: In the mechanistic investigation of redox reactions as in Chapter 6, it is possible that the substances of interest adsorb at the electrode. This is noticeable in the voltammogram by so-called prepeaks (stronger stabilization of the products) or postpeaks (stronger stabilization of the reagents). Generally, one attempts to avoid such effects in mechanistic studies by an appropriate choice of a suitable solvent and electrode material. In addition, by choosing a slow scan rate v, the signal from the reactants that continuously diffuse to the electrode can be enhanced compared to the finite amount of adsorbed reactants. A detailed and quantitative treatment of these situations is given in [1].

If β_O and β_R are very large, extremely small solution concentrations are sufficient to form a complete monolayer at the electrode surface. In this situation, only the immobilized redox centers R_{ads} and O_{ads} contribute to the signal, since the concentrations of O_{solv} and R_{solv} are below the detection limits. This situation is often encountered in redox-active SAMs. In addition, the transfer of the electrode from the adsorption solution to electrolyte solution without O_{solv} and R_{solv} can ensure that only immobilized redox systems react in a voltammetric experiment. We will deal with the voltammetry of such systems in the following section.

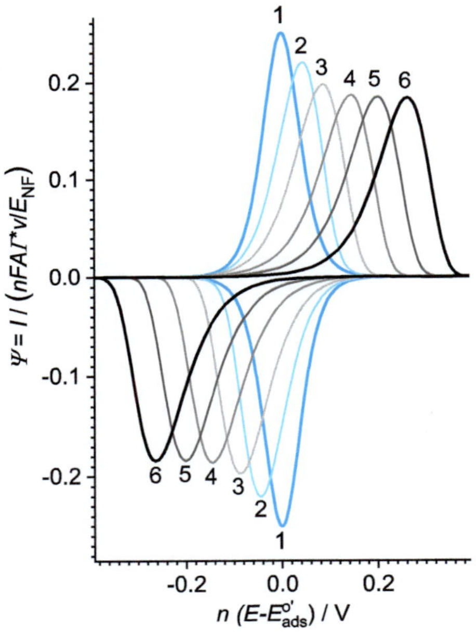

Figure 13.55 Normalized cyclic voltammograms of immobilized redox systems calculated according to [76] with the normalized rate constant κ of 1) ∞ (reversible case); 2) 0.30; 3) 0.10; 4) 0.030; 5) 0.010; 6) 0.003. $\alpha = 0.5$, $T = 298$ K. The plot shows the faradaic current only. Source: Adapted from [76].

13.4.3 Electrode Kinetics of Immobilized Redox Systems

We discussed the voltammograms for an ideal immobilized redox system in Section 5.5 (please review). For these systems, we assumed that there are no redox-active species in the solution ($c_O = c_R = 0$), that the total concentration Γ^* of O_{ads} and R_{ads} at the electrode is constant ($\Gamma^* = \Gamma_O + \Gamma_R$), that the rate constant of the conversion is quasi-infinite, and therefore, that Eq. (13.17) is satisfied at all times (nernstian or reversible system). These assumptions also imply that in the potential range considered, the adsorption equilibrium constants *do not* depend on the electrode potential. The resulting voltammogram (Figure 13.55, curve 1) can be calculated according to Eq. (5.93) and is mirror symmetric with respect to the potential axis and with respect to the line intersecting the potential axis at the peak potential. Other diagnostic criteria can be obtained from Eq. (5.93).

Q: I tried very hard in the lab course, but the peaks of the immobilized redox system are hardly visible in my experimental curves.

A: The curves in Figure 13.55 show only the faradaic current. In experimental curves, the capacitive current appears in addition to the faradaic current. Depending on the total interfacial concentration Γ^* of the immobilized redox system, the faradaic current may be relatively small compared to the capacitive current. Figure 13.56 shows an example of a calculated curve considering the capacitive current according to Eq. (4.100). In some experiments, a "blank curve" of the capacitive current without immobilized redox system can be subtracted (as background signal) from the measured curve to obtain voltammograms as in Figure 13.55 for a simpler evaluation. However, it must be noted that the coating of the electrodes usually also changes the capacitive current significantly! (see also Figure 13.11).

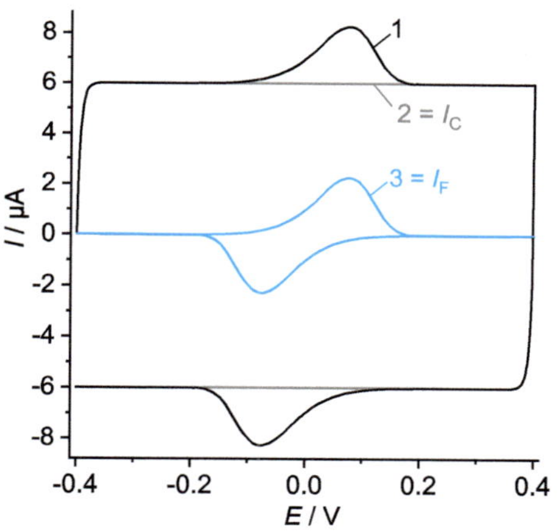

Figure 13.56 Calculated cyclic voltammogram of (1) an immobilized redox system with consideration of the capacitive current (2) according to Eq. (4.100). The faradaic current (3) is obtained from curve (1) minus curve (2). Parameters: $E^{\circ\prime}_{ads} = 0$ V, $k^\circ_s = 5$ s^{-1}, $n = 1$, $\alpha = 0.5$, $\Gamma^* = 1 \times 10^{-11}$ mol cm^{-2} (about 10% of a monolayer), $T = 298$ K, $v = 1$ V s^{-1}, $A = 0.3$ cm^2, $C_{dl,A} = 20$ μC cm^{-2}, and $R_E = 1000$ Ω.

We now want to investigate how the voltammograms change when we relax these strong constraints step-by-step. First, we consider a finite standard rate constant of the interfacial charge transfer reaction k°_s [s^{-1}]. The term "standard" means here that the (potential-dependent) rate constant is meant at the formal potential, that is, $k^\circ_s = k_s(E = E^{\circ\prime}_{ads})$. As with the cyclic voltammetry of dissolved redox systems, the appearance of the voltammograms does not depend solely on k°_s, but rather on the relationship between k°_s and the timescale of the experiment set by v. This combined influence can be captured by a normalized dimensionless rate constant κ (the symbol m is also often used in the literature):

$$\kappa = \frac{k^\circ_s}{\text{normalized scan rate}} = \frac{k^\circ_s}{\frac{v}{E_{NF}}} = E_{NF}\frac{k^\circ_s}{v} = \frac{RT}{nF}\frac{k^\circ_s}{v}. \tag{13.19}$$

For the general case of the voltammogram, that is, for arbitrary values of κ, no analytical expression can be given, but the curves can be determined numerically. For $\kappa < 0.08$ (irreversible case), approximate analytical expressions are available. The corresponding formulas can be found in [6, 76]. Some simulated cyclic voltammograms are shown in Figure 13.55. As κ becomes smaller, the anodic and cathodic peak potentials split. For a given redox system, that is, for $k^\circ_s = const$, κ can be varied by changing v. The resulting change in peak splitting $\Delta E_p = E_{p,a} - E_{p,c}$ can then be used to determine κ and finally k°_s (see Example Calculation 13.3).

Note that the shape of these voltammograms also applies to thin-layer voltammetry when the scan rate is so low that concentration gradients such as those in Figure 5.58b are not significant.

Table 13.5 Anodic and cathodic peak potentials in cyclic voltammograms of an immobilized redox system (n = 2).

v/(V s^{-1})	$E_{p,a}$/V	$E_{p,c}$/V
0.010	0.652	0.639
0.020	0.655	0.637
0.050	0.660	0.636
0.100	0.668	0.630
0.200	0.678	0.626
0.300	0.684	0.622
0.400	0.689	0.619
0.600	0.696	0.607
0.900	0.704	0.597
2.00	0.732	0.570
3.00	0.749	0.557
5.00	0.766	0.537
7.00	0.781	0.520
10.0	0.796	0.510

Reference electrode is SCE. Data from A. Markovic, PhD thesis, Carl von Ossietzky University Oldenburg 2019.

Example Calculation 13.3 Determine the standard rate constant of an immobilized redox system from the peak potentials in cyclic voltammograms given in Table 13.5.

Solution:
To solve this problem, we need to know the stoichiometric number n of transferred electrons per molecule, which is often known from the chemical context. If n must be determined experimentally, we can analyze the peak width at low scan rates. The cyclic voltammogram should then look approximately like the one in Figure 13.55, curve 1 or Figure 5.57, from which n can be determined from the full peak width at half peak height (FWHM) using Eq. (5.98).

In the next step, we make a case distinction with respect to the peak splitting $\Delta E_p = E_{p,a} - E_{p,c}$. If $n\Delta E_p > 0.2$ V applies, the peak potentials $E_{p,a}$ and $E_{p,c}$ [76] are described by:

$$\begin{aligned} E_{p,c} &= E^{\circ\prime}_{ads} - \frac{E_{NF}}{\alpha} \ln \frac{\alpha}{|\kappa|} \\ &= E^{\circ\prime}_{ads} - \frac{2.3E_{NF}}{\alpha} \log \frac{\alpha v}{E_{NF} k^{\circ}_{s}} \\ &= E^{\circ\prime}_{ads} + \frac{2.3E_{NF}}{\alpha} \log \frac{E_{NF} k^{\circ}_{s}}{\alpha} - \frac{2.3E_{NF}}{\alpha} \log v \end{aligned} \tag{13.20}$$

and

$$\begin{aligned} E_{p,a} &= E^{\circ\prime}_{ads} + \frac{E_{NF}}{1-\alpha} \ln \frac{1-\alpha}{|\kappa|} \\ &= E^{\circ\prime}_{ads} + \frac{2.3E_{NF}}{1-\alpha} \log \frac{(1-\alpha)v}{E_{NF} k^{\circ\prime}} \\ &= E^{\circ\prime}_{ads} - \frac{2.3E_{NF}}{1-\alpha} \log \frac{E_{NF} k^{\circ}_{s}}{1-\alpha} + \frac{2.3E_{NF}}{1-\alpha} \log v. \end{aligned} \tag{13.21}$$

The two equations can be regarded as straight-line equations $y = a + bx$ [$y = E_p - E^{\circ\prime}_{ads}$; $x = \log(v)$]. Thus, we plot the values from Table 13.5 against log(v) and draw two straight lines through the points, to which $n\Delta E_p > 0.2$ V applies (Figure 13.57). This diagram is also called trumpet plot (because of its similarity to the silhouette of the bell of the musical instrument). After extrapolation of the fitting lines, we determine the intersections with the horizontal line at $E - E^{\circ\prime}_{ads}$. For the scan rates v_a and v_c of the intersections, the following relationships hold:

E_{NF} – Nernst factor = $RT/(nF)$ [V]
k°_{s} – Standard rate constant of the immobilized redox system [s^{-1}]
α – Transfer coefficient [dimensionless]
v_a, v_c – Scan rates at the intersection of the straight fitting line in Figure 13.57 with the horizontal line at $E = E^{\circ\prime}_{ads}$ [V s^{-1}]

$$\frac{E_{NF} k^{\circ}_{s}}{\alpha} = v_c \quad \text{and} \quad \frac{E_{NF} k^{\circ}_{s}}{1-\alpha} = v_a. \tag{13.22}$$

For the example, we find $\log[v_a/(\text{V s}^{-1})] = -0.6999$ and $\log[v_c/(\text{V s}^{-1})] = -0.5963$. The two relationships in Eq. (13.22) form a system of equations with two unknowns k°_{s} and α. After solving, we obtain $\alpha = v_a/(v_a + v_c) = 0.44$ and $k^{\circ}_{s} = v_c\ \alpha/E_{NF} = v_a(1 - \alpha)/E_{NF} = 8.6\ s^{-1}$. (In the case $\alpha = 0.5$, the two fitting lines should intersect at *one* point with the horizontal line at $E = E^{\circ\prime}_{ads}$.)

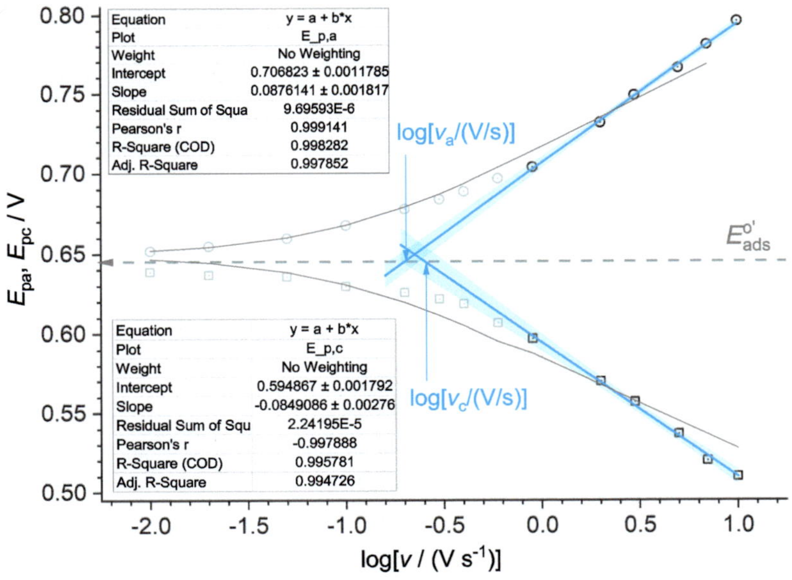

Figure 13.57 Trumpet plot for the data in Table 13.5.

It is important to note the measurement uncertainties, which can be estimated from the 95% confidence intervals in Figure 13.57 and which do affect k_s° and α. Other systematic sources of error include the uncompensated ohmic resistance, which leads to an enlargement of ΔE_p (as discussed in Chapter 6 for other voltammetric experiments). In connection with Figure 5.41, we have already discussed in detail possible deviations of the voltammograms when using staircase voltammetry in digital potentiostats as an approximation of a true analog potential ramp. This has to be considered, especially when analyzing the peak shape (e.g., when determining n).

Alternatively, k_s° can be determined directly from ΔE_p–v data sets using a spreadsheet application [76]:

$$\log k_s^\circ = \alpha \log(1-\alpha) + (1-\alpha)\log\alpha - \log\frac{E_{NF}}{v} - \alpha(1-\alpha)\frac{\Delta E_p}{2.3E_{NF}}. \tag{13.23}$$

This procedure does not require knowledge or estimation of $E^{\circ\prime}_{ads}$, but cannot make use of the averaging associated with the fitting lines in Figure 13.57. In addition, the transfer coefficient α is unknown. In principle, α can be determined from the shape of the voltammograms [76]. Because the peak shape is often slightly changed by further influences, this procedure is greatly affected by corresponding uncertainties. Therefore, $\alpha \approx 0.5$ is often assumed. This is reasonable because the shape of the voltammograms and the resulting further values do not differ very much over the range $0.3 < \alpha < 0.7$.

If $n\Delta E_p < 0.2$ V, we use the working curve shown in Figure 13.58 to read $1/\kappa$ for $n\Delta E_p$ of each cyclic voltammogram (blue arrows). Knowing v, we then obtain a value for k_s° from each voltammogram. We can then average the k_s° values for different v. Analogous curves for other values of α can be found in [76], but $\alpha \approx 0.5$ is often assumed. A more practical approach than fiddling with the working curve is to fit the tabulated data from [76] with a polynomial function and then use this function to calculate $1/\kappa$ column-wise in a spreadsheet program. The interpolation itself can also be carried out relatively easily with spreadsheet programs. For example, the function

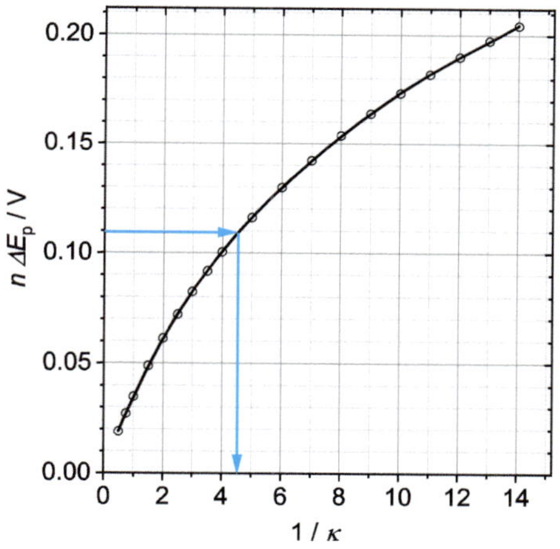

Figure 13.58 Working curve for the determination of the standard rate constant of immobilized redox systems for $n\Delta E_p < 0.2$ V. Source: Data from [76] for $\alpha = 0.5$, $T = 298$ K, $1/\kappa = (nFv)/(RTk_s^\circ)$.

$$y = 0.01442656 + 23.3593x + 131.97x^2 + 293.129x^3 - 499.544x^4 + 5711.95x^5 \tag{13.24}$$

$y = 1/\kappa = [nF/(RT)] - [v/k_s^\circ]$
$x = n\Delta E_p$

describes the data set in Figure 13.58 very well (swapping abscissa and ordinate). Other interpolations are also possible. Overall, all the determinations from individual $n\Delta E_p$–κ^{-1} data points tend to be less accurate than those made from the trumpet plot because no averaging is used. We can also use Eqs. (13.23) and (13.24) for a nonlinear fit with E°_{ads}, α and k_s° as adjustable parameters to interpolate all values from Table 13.5 (e.g., using Excel Solver). This yields the gray lines in Figure 13.57 with $E^\circ_{ads} = 0.650$ V, $\alpha = 0.51$ and $k_s^\circ = 3.0$ s^{-1}.

◢

The study of electron transfer (ET) kinetics of redox centers anchored to SAMs allows the determination of very large rate constants across monolayers of defined molecular structure. This is difficult for diffusing redox systems due to the limiting influence of mass transport or the existence of defects in SAMs. By varying the length d and the type of spacer in the SAM, the distance dependence of ET rate constants through different molecular structures can be investigated. The distance dependence of the rate constant $k_s^\circ(d)$ often follows a logarithmic dependence.

$$k_s^\circ(d) = k^\circ(d=0)\cdot\exp(-\beta d). \tag{13.25}$$

The decay constant usually has values of $\beta \approx 1$ Å^{-1} = 10 nm^{-1} = 0.01 pm^{-1}. Therefore, such a transfer can only occur when the monolayer is less than 1.5 nm thick.

A biological lipid bilayer with a thickness of 3–4 nm already leads to complete blockage of an electrode or to the separation of reaction compartments in biological cells in which different redox reactions can occur independently of each other.

For very fast redox reactions, redox-active monolayers are also investigated by chronoamperometry [1, 7].

13.4.4 How Do Adsorbate–Adsorbate Interactions Influence the Voltammograms?

An essential prerequisite for the Langmuir isotherm is the absence of lateral adsorbate-adsorbate interactions, which, however, often occur in densely packed monolayers. If we want to analyze the effects of such interactions (schematically depicted in Figure 13.53d), we need a formalism that is still mathematically manageable but can reveal the essential changes. For this purpose, it is convenient to describe the adsorption equilibrium by a Frumkin isotherm[106] (Section 8.10.4), which extends the Langmuir isotherm by allowing for the lateral adsorbate-adsorbate interactions [1, 6, 77]. There are three parameters a_{OO}, a_{RR}, and a_{OR}, which describe the pairwise interactions between O_{ads}–O_{ads}, R_{ads}–R_{ads}, and O_{ads}–R_{ads}. Positive values represent attractive interactions and negative values represent repulsive interactions. The parameter

Γ^* – Total interfacial concentration $\Gamma_O + \Gamma_R$ [mol cm^{-2}]
E_{NF} – $RT/(nF)$ – Nernst factor [V]
v – Potential scan rate [V s^{-1}]
θ_O, θ_R – (Fractional) coverages by O_{ads} and R_{ads} [dimensionless].
$\theta_t = \theta_O + \theta_R$ – Total (fractional) coverage [dimensionless]
n_{solv} – Number of solvent molecules displaced from the interface per O_{ads} or R_{ads}

$$g = a_{OO} + a_{RR} - 2\,a_{OR} \tag{13.26}$$

summarizes these pairwise interactions. In addition, we have to consider the number n_{solv} of solvent molecules that are displaced from the interface by O_{ads} or R_{ads}. The cyclic voltammogram can then be represented by

$$I = nFA\Gamma^* \cdot \frac{v}{E_{NF}} \cdot \psi = nFA\Gamma^* \cdot \frac{v}{E_{NF}} \cdot \frac{\theta_R(1-\theta_R)}{1 - 2n_{solv} g \theta_t \theta_R (1-\theta_R)}. \tag{13.27}$$

The potential change is coupled in Eq. (13.17) to Γ_i (i = O, R) and consequently to the coverages $\theta_i = \Gamma_i / \Gamma_{s,i}$.

If $g > 0$, the attractions O_{ads}–O_{ads} and R_{ads}–R_{ads} are significantly larger than that between different redox states; that is, the mixing of O_{ads} and R_{ads} is energetically unfavorable. The peaks in the voltammograms in Figure 13.59 are then much narrower than those for the reference system without lateral interaction ($g = 0$). This is plausible because if R_{ads} were oxidized, the first O_{ads} would be surrounded only by R_{ads}, and therefore, an additional energy contribution must be provided because of the unfavorable lateral interaction O_{ads}–R_{ads}. However, once O_{ads} is formed, further conversion of R_{ads} to O_{ads} is favored because of the attractive O_{ads}–O_{ads} interaction. Continuing this trend, we can think of the process as being similar to a nucleation or a delayed phase formation process, where the reaction starts later than would be expected thermodynamically. But once initiated, the process may abruptly take effect. This corresponds to very sharp peaks that do not have a bell shape (Figure 13.59, curve 6).

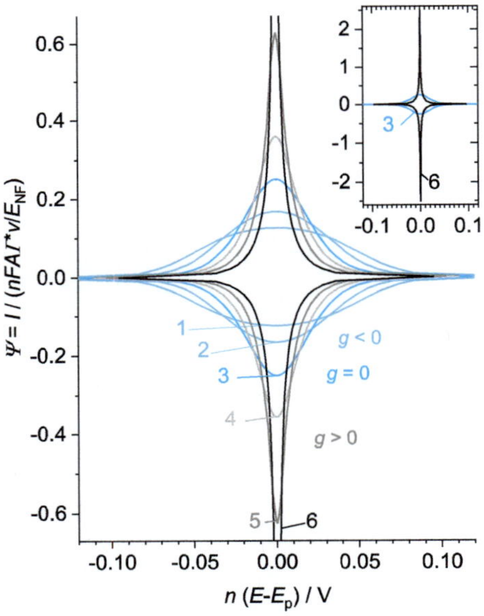

Figure 13.59 Calculated normalized reversible voltammograms of immobilized redox systems with lateral adsorbate–adsorbate interaction corresponding to a Frumkin isotherm for different values of $n_{solv}\,\gamma\theta_t$ of 1) −2; 2) −1; 3) 0 (corresponding to the Langmuir isotherm); 4) 0.6; 5) 1.2; 6) 1.8. Capacitive currents were *not* considered. The peak occurs at $E_p = E^\circ_{ads} + E_{NF}\,\theta_t n_{solv}(a_{RR} - a_{OO})$. All cases are reversible voltammograms ($\kappa \to \infty$).

Conversely, at $g < 0$, the R_{ads}–O_{ads} interaction is more favorable than that between the same redox forms. The formation of the first O_{ads} is favored in an environment of R_{ads}, but the formation of the pure monolayers of R_{ads} or of O_{ads} is then less favorable, and the peaks are broader (with the same area under the curves) than in the absence of lateral interaction.

It should be emphasized here that the lateral adsorbate–adsorbate interaction is not the only possible cause of peak broadening. Often, peak broadening results from energetically different adsorption sites, which occur on almost all solid electrodes (Figure 13.53c). In such cases, energetically nonequivalent adsorbates coexist at the surface and have somewhat different formal potentials according to Eq. (13.16). The system must then be described by a distribution of formal potentials. Different adsorption sites and different orientations of the molecules at the interface can also cause different values of k_s° (kinetic dispersion [78]). This effect as well as mass transport limitations within thicker films also cause peak broadening.

13.4.5 Electron Transport in Thick Films

Electrochemical conversion of thick films with multilayers of redox centers requires a mechanism by which electrons are passed through the film to convert all redox centers. In most cases, the matrix of the film (e.g., the polymer backbone) has no electronic conductivity. Intrinsically conductive polymers are an exception (Section 13.2.4), but we do not consider them here. In other polymer systems, the redox centers are often transition metal ions in a coordination environment bound to an ion-exchange polymer (Figure 13.40). In these systems, there are two mechanisms for electron transport, often occurring in parallel. Together, they are also referred to as "electron diffusion" since they can be described by analogous formulas. First, the redox centers can physically diffuse between the polymer chains by changing places (Figure 13.60). The corresponding transport diffusion coefficient D_f (index f from film) usually differs from the diffusion coefficient D of the same transition metal complex in the liquid electrolyte phase.

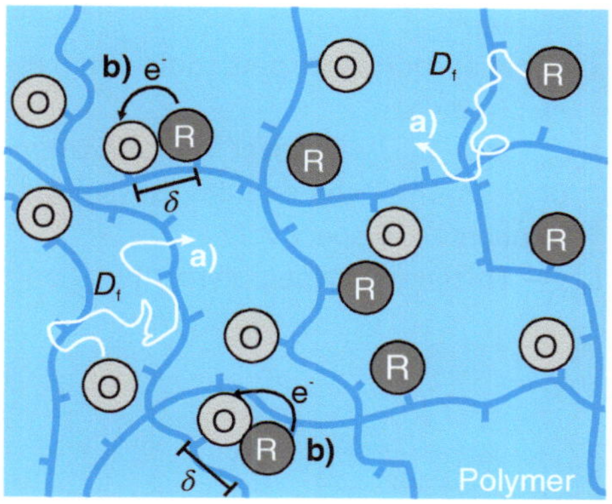

Figure 13.60 Transport of electronic charge in polymer films by a) physical diffusion of redox centers (mass transport) and b) electron transfer between redox centers.

However, the electrons can also change their position by electron transfer (ET) between neighboring redox centers (electron hopping, Figure 13.60). This process is also possible if the redox centers in redox polymers are covalently anchored to the polymer backbone or integrated into a coordination network (as in Figure 13.41 and Figure 13 42). Even then, mobile ions in the film are necessary for charge compensation. It is the movement of counterions that must occur and that can also be limiting for the overall reaction rate. In ET, the rate of electron conduction depends on the mutual distance of the redox centers δ, on the total concentration of redox centers in the film $c^* = c_O + c_R$, and on a bimolecular rate constant k. In addition, a geometry factor $b = 1/2$ for one-dimensional (1D) diffusion, $b = 1/4$ for the 2D, or $b = 1/6$ for the 3D case has to be considered.

The combined contributions of the two processes give an apparent diffusion coefficient D_{CT} (index CT from charge transfer).

$$D_{CT} = D_f + b\,k\,\delta^2\,c^*. \tag{13.28}$$

D_{CT} – Apparent charge transport coefficient [cm^2 s^{-1}]

D_f – Mass transport diffusion coefficient of redox centers in the polymer film [cm^2 s^{-1}]

b – Geometry factor [dimensionless]

k – Distance-dependent bimolecular rate constant for the reaction of oxidized and reduced redox centers (exchange reaction)

δ^2 – Squared distance between redox centers in the film [cm^2]

c^* – Total concentration of redox centers in the film [mol cm^{-3}]

Equation (13.28) is also called the Dahms–Ruff equation.[147]

Further Reading. A detailed discussion of conductivity in polymer films and specifically in redox polymers, including experimental methods for its determination, can be found in [14].

13.4.6 Reactions at Blocked Electrodes

In the following sections, we will turn to the conversion of a substance S (substrate) dissolved in the electrolyte solution at the modified electrode. In the first case, the electrode coating is nonconductive and impermeable to S, as it is desirable for coatings

[147] This transport model was first proposed for transport in solutions, see H. Dahms; Electron conduction in aqueous solution. *J. Phys. Chem.* **1968**, *72*, 362–364; I. Ruff; Experimental evidence of electronic conduction in aqueous solution. *Electrochim. Acta* **1970**, *15*, 1059–1061. The approach was then extended to redox polymers, F. B. Kaufmann, M. B. Engler; Solid-state spectroelectrochemistry of crosslinked donor bound polymer films. *J. Am. Chem. Soc.* **1979**, *101*, 547–549.

that serve as a barrier. However, defects are present in real systems so that S can reach the substrate electrode via cracks or pinholes in the coating. The defects act like an ensemble of randomly arranged microelectrodes and thus can be analyzed with the formalism developed for this type of electrode. The appearance of the corresponding voltammograms has already been discussed in Section 5.6.6.

The zone diagram in Figure 5.87 should be highlighted in this context. Qualitatively different voltammograms are obtained in the individual zones because different processes are limiting. The axes of the zone diagram are formed by normalized quantities, into which structural properties of the modified electrode (degree of coverage, size, and spacing of defects) and characteristic parameters of the experiment (scan rate) are incorporated. Zone diagrams are often used as well to describe catalytically active electrodes (Section 13.4.7).

13.4.7 Reactions at Electrodes with a Permeable Film

We consider here a film of thickness l_{film} (such as hydrogels or films with a pore structure such as Nafion), where the reactant S can diffuse within the film with a diffusion coefficient D_{f}. A distribution equilibrium is established at the interface between the film and the electrolyte solution:

$$K = \frac{c_{\text{S,film}}(x = l_{\text{film}})}{c_{\text{S,sol}}(x = l_{\text{film}})}. \tag{13.29}$$

K – Nernst partition coefficient [dimensionless]

x – Spatial coordinate perpendicular to the electrode [cm]

$c_{\text{S,film}}$ and $c_{\text{S,sol}}$ – Local concentrations of S in the film and in the solution [mol cm^{-3}], respectively

l_{film} – Thickness of the polymer film [cm]

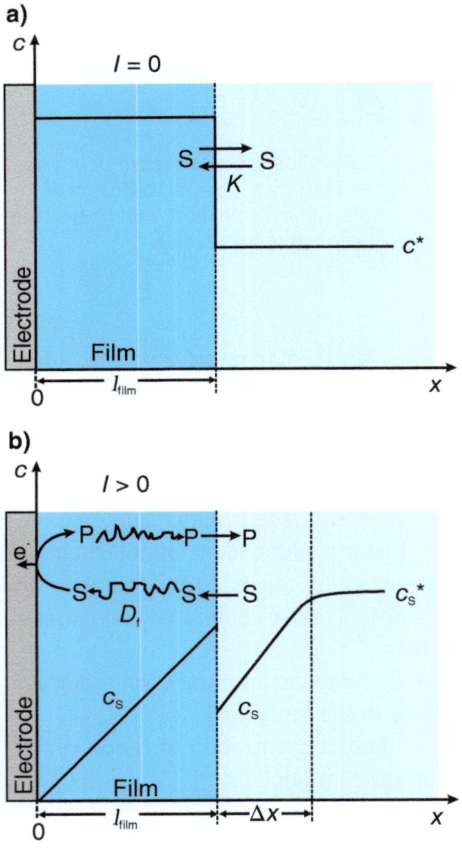

Figure 13.61 Concentration profiles of a substance S at an electrode covered with a permeable film. a) At equilibrium (without net conversion of S and thus without mass transport); b) with conversion of S and thus with mass transport at steady state. The distribution coefficient is $K = 2$.

Figure 13.61a shows the concentration profile at equilibrium, that is, in the absence of a net interfacial reaction and thus in the absence of mass transport. As a result of the partition equilibrium, the concentration of S in the film is greater than in the solution for $K > 1$. Thus, S is extracted from the solution into the film.

If a conversion of S takes place at the electrode–film interface (Figure 13.61b), we can analyze the current particularly well under steady-state conditions, for example, at a rotating electrode (RDE, Section 12.4.2). Here, the extent of the diffusion layer in solution is limited to the hydrodynamic boundary layer of thickness Δx. Within the film, the thickness of the diffusion layer is equal to the film thickness l_{film}. In a steady state, the concentration profiles are linear. To study the limitations of the process, we need to convert the transport and reaction rates to a uniform comparable quantity. In electrochemical systems, the current or the equivalent current, in which the particle fluxes are converted to electrical currents, is suitable for this purpose.

For the reaction at the electrode, two processes connected in series must be considered:

1) Diffusion of S in the electrolyte phase through the diffusion layer of thickness Δx with diffusion coefficient D and yielding the equivalent current I_1.
2) Diffusion of S through the polymer film of thickness l_{film} with diffusion coefficient D_{f} yielding the equivalent current I_2.

Furthermore, we use the generalization that we had already noted in the discussion of the Koutecký–Levich equation [Eq. (12.78)]:

> The reciprocal current of subsequent reactions is the sum of the reciprocal limiting currents of all subprocesses.

We have already used this consideration in the study of electrocatalytic reactions in Section 12.4.2, especially in Figure 12.23. Please repeat this section, if necessary, before continuing, because it is of fundamental importance for the further considerations in this chapter.

For the experimentally measured diffusion-limited current I_d, the following relation applies:

$$\frac{1}{I_d} = \frac{1}{I_{1,lim}} + \frac{1}{I_{2,lim}}. \tag{13.30}$$

We find the equivalent currents I_1 and I_2 analogous to the derivation of the diffusion-limited current in polarography [Eq. (5.47)]:

$$|I_1| = nFAD\frac{c^*_{S,sol} - c_{S,sol}(x = \delta_{film})}{\Delta x}, \tag{13.31}$$

$$|I_2| = nFAD_f\frac{c_{S,film}(x = l_{film}) - 0}{l_{film}}. \tag{13.32}$$

Δx – Thickness of the Nernst diffusion layer [cm]
l_{film} – Thickness of the polymer film [cm]
D – Diffusion coefficient of S in the solution [cm^2 s^{-1}]
D_f – Diffusion coefficient of S in the polymer film [cm^2 s^{-1}]

We use here the absolute amount of the currents $|I_i|$ as a reminder that the sign of I is to be derived from the current direction and the applicable sign convention, but not from the equations themselves. Substituting Eq. (13.29) into Eq. (13.32) gives

$$|I_2| = nFAD_f\frac{Kc_{S,sol}(x = l_{film})}{l_{film}}. \tag{13.33}$$

We obtain the *limiting* partial currents $I_{1,lim}$ and $I_{2,lim}$ if we search for the maximum possible value for each partial current. For process 1), this means that $c_{S,sol}(x = l_{film}) = 0$; and for process 2), it means $c_{S,sol}(x = l_{film}) = c^*_{S,sol}$, that is, no concentration gradient occurs in the solution at all. As in the discussion of Figure 12.23, it is also important here to understand the quantities $I_{1,lim}$ and $I_{2,lim}$ as purely conceptual quantities, each of which is larger than the total current I_d and, therefore, cannot be achieved experimentally.

$$|I_{1,lim}| = nFAD\frac{c^*_{S,sol}}{\Delta x} \tag{13.34}$$

$$|I_{2,lim}| = nFAD_f\frac{Kc^*_{S,sol}}{l_{film}} \tag{13.35}$$

Substituting this into Eq. (13.30) yields

$$\frac{1}{I_d} = \frac{\Delta x}{nFADc^*_{S,sol}} + \frac{l_{film}}{nFAD_fKc^*_{S,sol}}. \tag{13.36}$$

With this equation, we can now predict the effect of various variables. For example, the current at the modified electrode will still be proportional to the volume concentration $c^*_{S,sol}$. However, if we succeed in producing films with different film thicknesses, I_d will be influenced by a new parameter D_fK/l_{film}.

13.4.8 Conversion at an Electrode with a Catalytically Active Film

In many cases, the actual motivation for modifying electrodes in a targeted manner is the production of catalytically active electrodes. The most important commercial products of this type are enzyme electrodes [79] for measuring the glucose concentration in blood (Excursion 18-2). However, the literature also contains many examples of nonenzymatic conversions of organic reactants [80, 81], or of inorganic reactions [82, 83]. According to the nomenclature from Section 12.1.1, these systems are

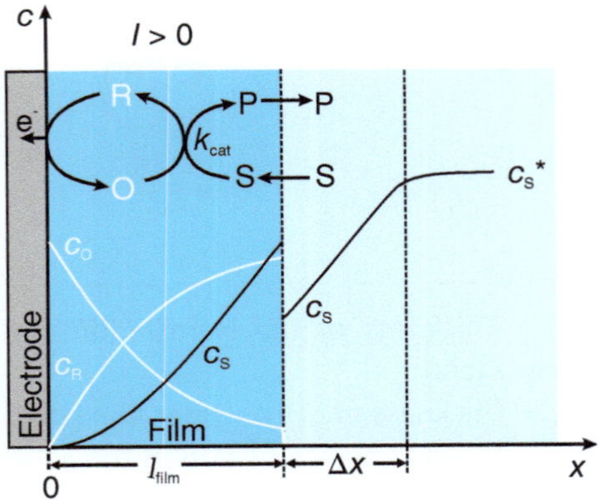

Figure 13.62 Concentration profiles of a substance S and of a mediator at an enzyme-modified electrode (schematic according to [1]).

regarded as homogeneous electrocatalytic systems. The following sequential processes may occur at such modified electrodes, although in special systems some of these processes may be absent (Figure 13.62):

1) Diffusion of S in the electrolyte phase (Section 13.5.6)
2) Extraction of S into the film (Section 13.5.6), possible passage through pores at the film–solution interface [1]
3) Diffusion of S in the film (Section 13.5.6)
4) Reaction with the catalyst
5) Reaction of the catalyst with the redox mediator
6) Charge transport by means of the redox mediator (Section 13.5.5)
7) Conversion of the redox mediator at the interface between the supporting electrode and the film (Section 13.5.2)

An exact solution to the problem is only possible by solving the respective coupled differential equations under consideration of the boundary conditions, because very complex concentration profiles of the involved substances can arise within the film and in the solution adjacent to the electrode. A simplified analysis is possible in principle via the respective limiting currents. However, without further simplifications, the analysis has only limited value because many of the parameters that occur are not known or cannot be easily determined. Experimentally, the following parameters can be determined. In the absence of the reactant S, the reaction of the mediator and the charge transport in the film can be studied, and then after the addition of S, the catalytic conversion can be determined. It is helpful if the transport in the solution phase can be varied in a defined way, for example, using a rotating disk electrode.

Furthermore, the choice of concentrations of S, the choice of catalyst and mediator, and studies at different film thicknesses can be used to induce limiting situations that can be represented in zone diagrams. Examples of this procedure can be found in [1, 84]. A systematic introduction to the modeling of these systems is given in [85, 86]. The value of this analysis is in a conceptual understanding that points the way toward rationally guided optimization of practically applicable systems (such as biosensors) and, alongside this, also highlights the fundamental limitations that cannot be overcome by even the most assiduous trial and error experimentation. In this context, there are also studies that build such catalytic systems with a very well-defined structure in order to better constrain some of the unknown parameters or to be able to neglect some of the processes listed above [87]. This increases the accuracy of the subsequent analysis of the remaining subprocesses.

13.4.9 Electrochemically Triggered Film Modifications

In Section 13.1, we learned about monolayer systems as attractive electrode modifications for anchoring functional molecules to them in a molecularly defined manner. Many organic preparative reactions are suitable for this purpose. However, since the desired products cannot be separated from undesired by-products in monolayers, such reactions that proceed very rapidly and form only one product under mild conditions are preferred.

A special possibility is offered by surfaces that are able to chemically couple functional molecules from the solution only *after* an electrochemical reaction of the monolayer ("activation"). This comes with the great advantage that the reaction can be started at an *externally controllable* time or, if electrode arrays are used, can be carried out *site-selectively* at one of many electrodes.

As an example, we will consider hydroquinone-terminated monolayers (Figure 13.63). The hydroquinone group can be converted into the corresponding quinone by oxidation. As a carbonyl compound, the quinone can then form so-called Schiff bases[148] [88]. As an electrophilic alkene, it can enter into a Diels–Alder cycloaddition[149] [89] or can be used for conjugate addition of thiols [90]. Since the hydroquinone/quinone redox pair usually has a different formal potential than the adducts formed, the reaction progress can be well monitored with cyclic voltammetry via the decrease in the hydroquinone/quinone peak pair and the increase in the adduct peak pair.

As a variation, the processes can also be carried out locally if the oxidation is caused by a chemical oxidant generated locally at a positionable microelectrode, or if the oxidation is confined to the wetting area of a small droplet in a scanning droplet cell [91]. In addition to these build-up methods, monolayers can also be patterned by local etching methods. If an etchant such as Br_2 or OBr^- is generated locally at a microelectrode [92] or applied through a microchannel system [88], patterns can be "written" into monolayers.

Further Reading. In addition to the specific references given above, Ref. [93] is recommended for an overview of temporally and spatially defined surface modifications.

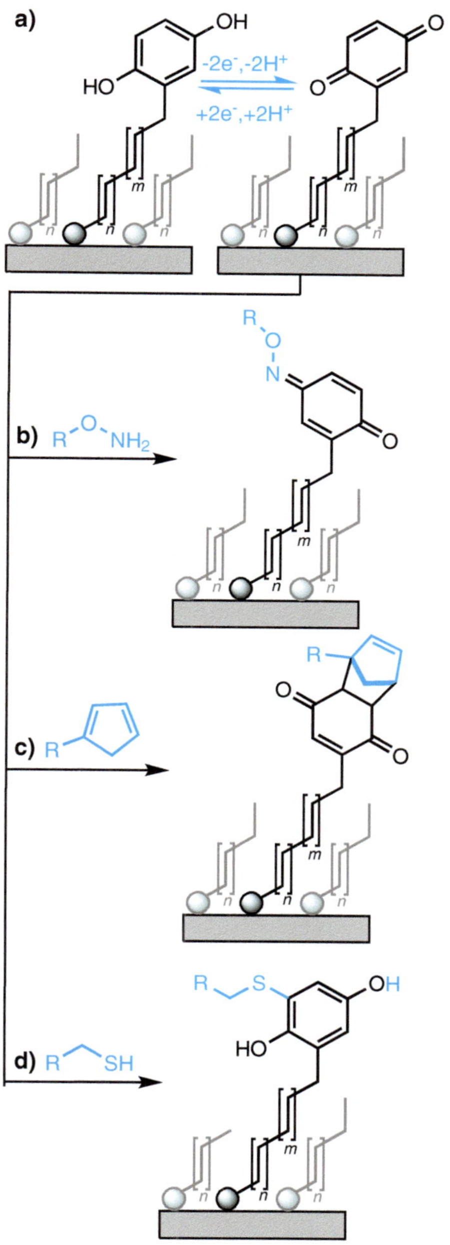

Figure 13.63 Electrochemical activation of hydroquinone-terminated monolayers for subsequent chemical coupling of functional groups R.

a) Oxidation to quinone;
b) formation of a Schiff base;
c) Diels–Alder cycloaddition;
d) conjugate addition.

Test your knowledge

Exercise 13.1: Calculate the interfacial concentration of alkanethiols on Au(1 1 1) for the superstructure shown in Figure 13.4.

Exercise 13.2: Suggest a way to experimentally verify the orientation of a methyl end group shown in Figure 13.6 using the characterization methods presented in Chapter 8.

Exercise 13.3: Prepare sketches of reversible cyclic voltammograms of adsorbed redox systems with lateral adsorbate–adsorbate interactions analogous to Figure 13.59 for the following parameters: $\theta_t = 1, (\kappa \to \infty), E^{0\prime}_{ads} = 0$ V, and the interaction parameters from the table below.

Curve	a_{OO}	a_{RR}	a_{OR}
1	0	0	0
2	1	1	1
3	1	0.6	0.8
4	0.6	1	0.8
5	1	1	0.8
6	1	1	1.8
7	0.6	1	1

Include the case without adsorbate–adsorbate interactions in each sketch for comparison.

[148] Hugo Schiff, 1834–1915, German-Italian chemist.
[149] Otto Diels, 1876–1954, and his student Kurt Alder, 1902–1958, were German chemists who jointly received the Nobel Prize in 1950.

Exercise 13.4: Determine the standard rate constant for the surface-bound redox couple $R_{ads} \rightleftharpoons O_{ads} + 2e^-$ from the peak potentials of cyclic voltammograms. [Ref = SCE.]

v/(V s^{-1})	E_{pa}/V	E_{pc}/V
0.03	−0.085	−0.095
0.06	−0.083	−0.097
0.10	−0.084	−0.096
0.20	−0.082	−0.098
0.30	−0.083	−0.099
0.60	−0.080	−0.102
1.0	−0.078	−0.100
2.0	−0.073	−0.107
4.0	−0.051	−0.127
8.0	−0.039	−0.143
12.0	−0.031	−0.019

Exercise 13.5: Using the data from Example Calculation 13.3, perform a curve fit using a spreadsheet program with k_s, α, and $E^{\circ\prime}_{ads}$ as adjustable parameters. Discuss different possibilities for this.

References

1 A. J. Bard, L. R. Faulkner, H. S. White; *Electrochemical Methods – Fundamentals and Applications*. 3rd ed., John Wiley & Sons, New York, 2022, pp. 755–818.

2 A. Ulman; *An Introduction to Ultrathin Organic Films – From Langmuir-Blodgett to Self-Assembly*, Academic Press, San Diego, 1991.

3 R. J. Forster, T. E. Keyes, J. G. Vos; *Interfacial Supramolecular Assemblies*, John Wiley & Sons, Chichester, 2003.

4 A. E. Kaifer, M. Gómez-Kaifer; *Supramolecular Electrochemistry*, Wiley-VCH, Weinheim, 1999.

5 R. W. Murray; Chemically modified electrodes. in *Electroanalytical Chemistry, Vol. 13* (A. J. Bard, Ed.), Marcel Dekker, New York, 1984, pp. 191–368.

6 E. Lavrion; Voltammetric methods for the study of adsorbed species. in *Electroanalytical Chemistry, Vol. 12* (A. J. Bard, Ed.), Marcel Dekker, New York, 1982, pp. 53–157.

7 H. O. Finklea; Electrochemistry of organized monolayers of thiols and related molecules on electrodes. in *Electroanalytical Chemistry, Vol. 19* (A. J. Bard, I. Rubinstein, Eds.), Marcel Dekker, New York, 1996, pp. 109–335.

8 A. Ulman; Formation and structure of self-assembled monolayers. *Chem. Rev.* **1996**, *96*, 1533–1554.

9 Y. Xia, G. M. Whitesides; Soft lithography. *Angew. Chem. Int. Ed.* **1998**, *37*, 550–575.

10 R. C. Alkire, D. M. Kolb, J. Lipkowski, P. N. Ross (Eds.); *Advances in Electrochemical Science and Technology, Vol. 11 – Chemically Modified Electrodes*, Wiley-VCH, Weinheim, 2009.

11 W. Knoll (Ed.); *Handbook of Biofunctional Surfaces*, Pan Stanford Publ. Pte. Ltd. Singapore, 2013.

12 G. Roberts (Ed.); *Langmuir–Blodgett Films*, Plenum Press, New York, 1990.

13 P. Ceroni, A. Credi, M. Venturi; *Electrochemistry of Functional Supramolecular Systems*, John Wiley & Sons, Hoboken, NJ, 2010.

14 M. E. G. Lyons; *Electroactive Polymer Electrochemistry, Vol. 1 & 2,* Springer Science + Business Media, New York, 1994, 1996.

15 G. Decher, J. B. Schlenoff (Eds.); *Multilayer Thin Films*, Wiley-VCH, Weinheim, 2003.

16 B. Guan. S. Ciampi, G. Le Saux, K. Gaus, P. J. Reece, J. J. Gooding; Different functionalization of the internal and external surfaces in mesoporous materials for biosensing applications using "click" chemistry. *Langmuir* **2011**, *27*, 328–334.

17 W.-S. Yeo, M. Mrksich; Electroactive self-assembled monolayers that permit orthogonal control over adhesion of cells to patterned substrates. *Langmuir* **2006**, *22*, 10816–10820.

18 W. S. Dillmore, M. N. Yousaf, M. Mrksich; A photochemical method for patterning the immobilization of ligands and cells to self-assembled monolayers. *Langmuir* **2004**, *20*, 7223–7231.

19 S. Herrwerth, W. Eck, S. Reinhardt, M. Grunze; Factors that determine the protein resistance of oligoether self-assembled monolayers – Internal hydrophilicity, terminal hydrophilicity, and lateral packing density. *J. Am. Chem. Soc.* **2003**, *125*, 9359–9366.

20 Y. Hu, W. Sun, D. Wang; *Electrochemistry of Flotation of Sulphide Minerals*, Tsinghua University Press, Beijing and Springer, Berlin, 2009.

21 J. B. Schlenoff, M. Li, H. Ly; Stability and self-exchange in alkanethiol monolayers. *J. Am. Chem. Soc.* **1995**, *117,* 12528.

22 C. Vericat, M. E. Vela, G. Benitez, P. Carro and R. C. Salvarezza; Self-assembled monolayers of thiols and dithiols on gold: New challenges for a well-known system. *Chem. Soc. Rev.*, **2010**, *39*, 1805–1834.

23 M. D. Porter, T. B. Bright, D. Allara, C. E. D. Chidsey; Spontaneously organized molecular assemblies. 4. Structural characterization of *n*-alkyl thiol monolayers on gold by optical ellipsometry, infrared spectroscopy, and electrochemistry. *J. Am. Chem. Soc.* **1987**, *109*, 3559–3568.

24 C. A. Widrig, C. Chung, M. D. Porter; The electrochemical desorption of *n*-alkanethiol monolayers from polycrystalline Au and Ag electrodes. *J. Electroanal. Chem.* **1991**, *310*, 335–359.

25 T. W. Schneider, D. A. Buttry; Electrochemical quartz crystal microbalance studies of adsorption and desorption of self-assembled monolayers of alkyl thiols on gold. *J. Am. Chem. Soc.* **1993**, *115*, 12391–12397.

26 M. Buck; Structure, electrochemistry and applications of self-assembled monolayers of thiols. in *Advances in Electrochemical Science and Engineering, Vol. 11* (R. C. Alkire, D. M. Kolb, J. Lipkowski, P. N. Ross, Eds.), Wiley-VCH, Weinheim, 2009, pp. 197–255.

27 A. Glaser, J. Foiser, G. Friedbacher, H. Hoffmann; Low-temperature investigation of the growth mechanism of alkylsiloxane self-assembled monolayers. *Anal. Bioanal. Chem.* **2004**, *379*, 653–657.

28 J. Pinson, F. Podvorica; Attachment of organic layers to conductive or semiconductive surfaces by reduction of diazonium salts. *Chem. Soc. Rev.* **2005**, *34*, 429–439.

29 J. A. Belmont, C. Bureau, M. M. Chehimi, S. Gam-Derouich, J. Pinson; Patents and industrial application of aryl diazonium salts and other coupling agents. in *Aryl Diazonium Salts* (M. M. Chehimi, Ed.), Wiley-VCH, Weinheim, 2012, pp. 309–321.

30 A. Berisha, M. M. Chehimi, J. Pinson, F. I. Podvorica; Electrode surface modification using dianzonium salts. in *Electroanalytical Chemistry, Vol. 26* (A. J. Bard, C. G. Zoski, Eds.), CRC Press, Boca Raton, 2016, pp. 1–71.

31 Y. R. Leroux, P. Hapiot; Nanostructured monolayers on carbon substrates prepared by electrografting of protected aryldiazonium salts. *Chem. Mater.* **2013**, *25*, 489–495.

32 J. P. Folkers, P. E. Laibinis, G. M. Whitesides; Self-assembled monolayers of alkanethiols on gold: Comparisons of monolayers containing mixtures of short- and long-chain constituents with methyl and hydroxymethyl terminal groups. *Langmuir* **1992,** *8*, 1331–1341.

33 S. Xu, G.-Y. Liu; Nanometer-scale fabrication by simultaneous nanoshaving and molecular self-assembly. *Langmuir* **1997**, *13*, 127–129.

34 T. Wilhelm, G. Wittstock; Patterns of functional proteins formed by local electrochemical desorption of self-assembled monolayers. *Electrochim. Acta* **2001**, *47*, 275–281.

35 R. D. Piner, J. Zhu, F. Xu, S. Hong, C. A. Mirkin; "Dip-pen" nanolithography. *Science* **1999**, *283*, 661–663.

36 K. Salaita, Y. Wang, J. Fragala, R. A. Vega, C. Liu, C. A. Mirkin; Massively parallel dip-pen nanolithography with 55 000-pen two-dimensional arrays. *Angew. Chem. Int. Ed.* **2006,** *45*, 7220–7223.

37 J. Kim, Y.-H. Shin, S.-H. Yun, D.-S. Choi, J.-H. Nam, S. R. Kim, S.-K. Moon, B. H. Chung, J.-H. Lee, J.-H. Kim, K.-Y. Kim, K.-M. Kim, J. H. Lim; Direct-write patterning of bacterial cells by dip-pen nanolithography. *J. Am. Chem. Soc.* **2012**, *134*, 16500–16503.

38 J. J. Gooding, F. Mearns, W. Yang, J. Liu; Self-assembled monolayers into the 21st century: recent advances and applications. *Electroanalysis* **2003**, *15*, 81–96.

39 M. Sakomura, M. Fujihira; Langmuir–Blodget (LB) films on electrodes-photoelectrochemistry. in *Encyclopedia of Electrochemistry, Vol. 10* (A. J. Bard, M. Stratmann, M. Fujihira, J. F. Rusling, I. Rubinstein, Eds.), Wiley-VCH, Weinheim, 2007, pp. 171–202.

40 D. M. Small; Lateral chain packing in lipids and membranes. *J. Lipid Res.* **1984**, *25*, 1490–1500.

41 https://en.wikipedia.org/wiki/Brewster_angle_microscope (last accessed Oct. 19, 2022).

42 P. Dynarowicz-Latka, J. Minones Jr., O. Conde, M. Casas, E. Irbarngoeray; BAM studies on the penetration of amphotericin B into lipid mixed monolayers of cellular membranes. *Appl. Opt. Surface Sci.* **2005**, *246*, 334–341.

43 Wikipedia: https://en.wikipedia.org/wiki/Model_lipid_bilayer; (last access 28 Feb 2026).

44 T. Hianik; Biological membranes and membrane mimics. in *Bioelectrochemistry – Fundamentals Experimental Techniques and Applications* (P. N. Bartlett, Ed.), John Wiley & Sons, Chichester, 2008, pp. 87–156.

45 H. Ti Tien; Membrane electrochemistry. in *Encyclopedia of Electrochemistry, Vol. 9* (G. S. Wilson, A. J. Bard, M. Stratmann, Eds.), Wiley-VCH, Weinheim, 2002, pp. 511–557.

46 P. L. Yeagle (Ed.); *The Structure of Biological Membranes.* 2nd ed., CRC Press, Boca Raton, 2005.

47 S. M. Schiller; Biomimetic systems: The tethered bilayer lipid membrane. in *Handbook of Biofunctional Surfaces* (W. Knoll, Ed.), Pan Stanford Publ. Pte. Ltd., Singapore, 2013, pp. 439–683.

48 B. A. Cornell; Ion channel biosensors. *Handbook of Biosensors and Biochips, Vol. 1* (R. S. Marks, D. C. Cullen, I. Karube, C. R. Lowe, H. H. Weetall, Eds.), Wiley, Hoboken, 2007, pp. 321–334.

49 M. C. Petty; Organic thin film deposition techniques. in *Encyclopedia of Nanoscience and Nanotechnology, Vol. 8*, American Scientific Publishers, 2004, pp. 295–304.

50 M. K. Beissenhirtz, F. W. Scheller, W. F. M. Stöcklein, D. G. Kurth, H. Möhwald, F. Lisdat; Electroactive cytochrome *c* multilayers within a polyelectrolyte assembly. *Angew. Chem. Int. Ed.* **2004**, *43*, 4357–4360.

51 M. Ferreira, H. Varela, R. M. Torresi, G. Tremiliosi-Filho; Electrode passivation caused by polymerization of different phenolic compounds. *Electrochim. Acta* **2006**, *52*, 434–442.

52 E. I. Iwuoha, M. R. Smyth; Polymer-based amperometric biosensors. in *Electroactive Polymer Electrochemistry, Vol. 2* (M. E. G. Lyons, Ed.), Springer Science + Business Media, New York, 1996, pp. 297–328.

53 Z. Hu, C. J. Seliskar, W. R. Heineman; Voltammetry of $[Re(DMPE)_3]^+$ at ionomer-entrapped composite-modified electrodes. *Anal. Chem.* **1998**, *70*, 5230–5236.

54 A. Heller; Electrical wiring of redox enzymes. *Acc. Chem. Res.* **1990**, *23*, 128–134.

55 D. M. Kelly, J. G. Voss; Osmium and ruthenium poly(pyridyl) redox polymers as electrode coatings. in *Electroactive Polymer Electrochemistry, Vol. 2* (M. E. G. Lyons, Ed.), Springer Science + Business Media, New York, 1996, pp. 173–232.

56 M. Tagliazucchi, E. J. Calvo; Electrochemically active polyelectrolyte-modified electrodes. in *Electrochemical Science and Engineering, Vol. 11* (R. C. Alkire, D. M. Kolb, J. Lipkowski, P. N. Ross, Eds.), Wiley-VCH, Weinheim, 2009, pp. 57–115.

57 W. Plieth; *Electrochemistry for Materials Science*, Elsevier, Amsterdam, 2008, pp. 323–363.

58 J. Heinze, Electrochemistry of conducting polymers. in *Organic Electrochemistry*, 5th ed. (O. Hammerich, B. Speiser, Eds.), CRC Press, Boca Raton, 2016, pp. 1571–1604.

59 A. F. Diaz; Chemically modified electrodes and conducting polymers. in *Organic Electrochemistry – An Introduction and a Guide*, 3rd ed. (H. Lund, M. Baizer, Eds.), Marcel Dekker, New York, 1991, p. 1377.

60 M. Dietrich, J. Heinze, G. Heywang, F. Jonas; Electrochemical and spectroscopic characterization of polyalkylenedioxythiophenes. *J. Electroanal. Chem.* **1994**, *369*, 87–92.

61 F. Li, W. J. Albery; A novel mechanism of electrochemical deposition of conducting polymers: Two-dimensional layer-by-layer nucleation and growth observed for poly(thiophene-3-acetide acid). *Electrochim. Acta* **1992**, *37*, 393–401.

62 R. J. Waltmann, J. Bargon; Electrically conducting polymers: A review of the electropolymerization reaction, of the effects of chemical structure on polymer film properties, and of applications toward technology. *Can. J. Chem.* **1986**, *64*, 76–95.

63 W. Schuhmann, C. Kranz, H. Wohlschläger, J. Strohmeier; Pulse technique for the electrochemical deposition of polymer films on electrode surfaces. *Biosens. Bioelectron.* **1997**, *12*, 1157–1167.

64 U. Janakiraman, K. Doblhofer, C. Fischmeister, A. B. Holmes; Electrogenerated chemiluminescence in poly(dibutoxyphenylenevinylene) coatings. *J. Phys. Chem. B* **2004**, *108*, 14368–14373.

65 P. Rapta, A. Neudeck, A. Petr, L. Dunsch; In situ EPR/UV-VIS spectroelectrochemistry of polypyrrole redox cycling. *J. Chem. Soc. Faraday Trans.* **1998**, *94*, 3625.

66 R. Holze; Spectroelectrochemistry of conducting polymers. in *Handbook of Advanced Electronic and Photonic Materials and Devices, Vol. 8 – Conducting Polymers* (H. S. Nalwa, Ed.), Academic Press, 2001, pp. 209–301.

67 G. Inzelt; *Conducting Polymers – A New Era in Electrochemistry*, Springer, Berlin, 2008.

68 J. Heinze, B. A. Fronatana-Uribe, S. Ludwigs; Electrochemistry of conducting polymers – Persistent models and new concepts. *Chem. Rev.* **2010**, *110*, 4724–4771.

69 C. Zhan, G. Yu, Y. Lu, Y. Wang, E. Wujcik, S. Wei; Conductive polymer nanocomposites: a critical review of modern advanced devices. *J. Mater. Chem. C*, **2017**, *5*, 1569–1585.

70 A. Fitch; Clay-modified electrodes: A review. *Clay Miner.* **1990**, *38*, 391–400.

71 Z. Hai, Z. Wei, C. Xue, H. Xu, F. Verpoort; Nanostructured tungsten oxide thin film devices: From optoelectronics and ionics to ionotronics. *J. Mater. Chem. C* **2019**, *7*, 12968–12990.

72 R. Wlodarczyk, M. Chojak, K. Miecznikowski, A. Kolary, P. J. Kulesza, R. Marassi; Electroreduction of oxygen at polyoxometallate-modified glassy carbon-supported Pt nanoparticles. *J. Power Sources* **2006**, *159*, 802–809.

73 U. Schröder, F. Scholz; The solid-state electrochemistry of metal octacyanomolybdates, octycyanotungstates and hexacyanoferrates explained on the basis of dissolution and reprecipitation reactions, lattice structures, and crystallinities. *Inorg. Chem.* **2000**, *39*, 1006–1015.

74 A. A. Karyakin; Prussian blue and its analogues: Electrochemistry and analytical applications. *Electroanalysis* **2001**, *13*, 813–819.

75 J. Liu, C. Wöll; Surface-supported metal–organic framework thin films: Fabrication methods, applications, and challenges. *Chem. Soc. Rev.* **2017**, *46*, 5730–5770.

76 E. Lavrion; General expression of the linear potential sweep voltammogram in the case of diffusionless electrochemical systems. *J. Electroanal. Chem.* **1979**, *101*, 19–28.

77 E. Laviron; The use of linear potential sweep voltammetry and of a.c. voltammetry for the study of the surface electrochemical reaction of strongly adsorbed systems and redox modified electrodes. *J. Electroanal. Chem.* **1979**, *100*, 263–270.

78 G. K. Rowe, M. T. Carter, J. N. Richardson, R. W. Murray; Consequences of kinetic dispersion on the electrochemistry of an adsorbed redox-active monolayer. *Langmuir* **1995**, *11*, 1797–1806.

79 P. N. Bartlett, J. Cooper; Application of electroactive polymers in bioelectrochemistry and bioelectronics. in *Electroactive Polymer Electrochemistry, Vol. 2* (M. E. G. Lyons, Ed.), Springer Science + Business Media, New York, 1996, pp. 233–267.

80 T. Kuwana, T. Osa, Y. Kashiwagi; Organic electrocatalysis. in *Encyclopedia of Electrochemistry, Vol. 10* (A. J. Bard, M. Stratmann, M. Fujihara, I. Rubinstein, J. F. Rusling, Eds.), Wiley-VCH, Weinheim, 2007, pp. 695–705.

81 D. Leech; Analytical application of polymer-modified electrodes. in *Electroactive Polymer Electrochemistry, Vol. 2* (M. E. G. Lyons, Ed.), Springer Science + Business Media, New York, 1996, pp. 269–296.

82 N. Kobayshi; Oxygen electrocatalysis. in *Encyclopedia of Electrochemistry, Vol. 10* (A. J. Bard, M. Stratmann, M. Fujihara, I. Rubinstein, J. F. Rusling, Eds.), Wiley-VCH, Weinheim, 2007, pp. 687–694.

83 B. Keita, L. Nadjo; Electrocatalysis of inorganic chemicals by chemically modified electrodes. in *Encyclopedia of Electrochemistry, Vol. 10* (A. J. Bard, M. Stratmann, M. Fujihara, I. Rubinstein, J. F. Rusling, Eds.), Wiley-VCH, Weinheim, 2007, pp. 695–705.

84 J. Leddy, A. J. Bard, J. T. Maloy, J. M. Saveant; Kinetics of film-coated electrodes: The effect of a finite mass transfer rate of substrate across the film-solution interface at steady state. *J. Electroanal. Chem.* **1985**, *187*, 205–227.

85 P. N. Bartlett, C. S. Toh, E. J. Calvo, V. Flexer; Modeling biosensor responses. in *Bioelectrochemistry – Fundamentals Experimental Techniques and Application* (P. N. Bartlett, Ed.), John Wiley & Sons, Chichester, 2008, pp. 267–325.

86 M. E. G. Lyons; Modeling of biosensor interfaces. in *Handbook of Biosensors and Biochips*, *Vol. 1* (R. S. Marks, D. C. Cullen, I. Karube, C. R. Lowe, H. H. Weetall, Eds.), Wiley, Hoboken, 2007, pp. 291–319.

87 A. Anne, C. Demaille, J. Moiroux, J.-M. Savéant, C. Bourdillon; Electron transfer and transport in ordered enzyme layers. in *Encyclopedia of Electrochemistry, Vol. 10* (A. J. Bard, M. Stratmann, M. Fujihara, I. Rubinstein, J. F. Rusling, Eds.), Wiley-VCH, Weinheim, 2007, pp. 763–813.

88 D. Dutta, A. Pulsipher, M. N. Yousaf; Selective tethering of ligands and proteins to a microfluidically patterned electroactive fluid lipid bilayer array. *Langmuir*, **2010**, *26*, 9835–9841.

89 M. N. Yousaf, B. T. Houseman, M. Mrksich; Turning on cell migration with electroactive substrates. *Angew. Chem. Int. Ed.* **2001**, *40*, 1093–1096.

90 M. Shamsipur, S. H. Kazemi, A. Alizadeh, M. F. Mousavi, M. S. Workentin; Self-assembled monolayers of a hydroquinone-terminated alkanethiol onto gold surface. Interfacial electrochemistry and Michael-addition reaction with glutathione. *J. Electroanal. Chem.* **2007**, *610*, 218–226.

91 J. Clausmeyer, J. Henig, W. Schuhmann, N. Plumeré; Scanning droplet cell for chemoselective patterning through local electroactivation of protected quinone monolayers. *ChemElectroChem* **2014**, *15*, 151–156.

92 A. Lesch, B. Vaske, F. Meiners, D. Momotenko, F. Cortéz-Salazar, H. H. Girault, G. Wittstock; Parallel imaging and template-free patterning of self-assembled monolayers with soft linear microelectrode arrays. *Angew. Chem. Int. Ed.* **2012**, *51*, 10413–10416.

93 F. Meiners, I. Plettenberg, J. Witt, B. Vaske, A. Lesch, I. Brand, G. Wittstock; Local control of protein binding and cell adhesion by patterned organic thin films. *Anal. Bioanal. Chem.* **2013**, *405*, 3673–3691.

Part IV

Applications

14

Corrosion and Corrosion Protection

Corrosion has been an important phenomenon since prehistoric times, leading to the deterioration of functional properties of technical products through interaction with the environment and thus posing an immense safety, economic, and ecological challenge (Figure 14.1). In particular, localized damage to load-bearing structures and pipelines is feared, as it can lead to sudden failure of technical systems and serious accidents.

In this textbook, we limit our consideration to corrosion phenomena of metallic materials. However, corrosion can also occur on mineral construction materials, ceramics, and polymers. For electrochemical corrosion of metals, we discuss the scientific principles, the relationships to other subfields of electrochemistry, and some important measurement techniques. It is recommended to review Chapters 2 and 3 on thermodynamics and kinetics of electrochemical reactions first, especially Sections 2.3.5 and 3.2.7, as this chapter builds on them.

Figure 14.1 Heavily corroded ship. It is estimated that 3–4% of the world's gross national product is destroyed by corrosion. The release of metal ions from the corroding metals or pollutants from coatings used for corrosion protection poses an ecological problem.

From a thermodynamic point of view, most metals and semimetals should spontaneously convert to their oxides under atmospheric conditions and in aqueous solutions (Section 14.2). The fact that we are surrounded by a large number of metallic commodities and that silicon wafers can be handled in the open atmosphere is due to their protection by thin oxide layers. This delays the oxidative attack on the materials but usually does not prevent it completely (Section 14.3). This phenomenon is called *passivation*. Passivation can break down in certain corrosive media, whereupon the workpiece is attacked. In the discussion of corrosion protection measures in Section 14.4 and in the presentation of typical corrosion scenarios in Section 14.5, we give priority to systematic understanding over the compilation of all possible details.

The dissolution of metals can be used for targeted shaping of workpieces, for example, for the production of nanostructured materials and for processing workpieces by electrochemical machining (Section 14.6).

Further Reading. Most textbooks on electrochemistry contain chapters on corrosion, but these are sometimes very brief and limited to the core messages [1, 2]. An easy introduction to the study of corrosion problems, even without prior knowledge of chemistry or electrochemistry, is provided in [3]. Advanced considerations of corrosion are a central field of materials science with existing textbooks or dedicated chapters in edited volumes [4] as well as extensive literature on the subject [5–8]. The application of modern measurement techniques for corrosion research is summarized in collective works [7–12]. Further literature on specific aspects is referred to in the respective sections of this chapter.

14.1 What Is Corrosion?

Corrosion is part of a hierarchical system of material damage (Figure 14.2). In the case of excessive thermal or mechanical stress, a workpiece is destroyed because it was either inadequately designed or the load applied was outside the specified application range. This is contrasted with the area of wear and tear, where deterioration of the material properties occurs under normal operating conditions. The term wear refers specifically to mechanical impact. This is contrasted with electrochemical, chemical, or physical attack, which is summarized under the term corrosion. In practice, an important distinction is made between corrosion with and without additional mechanical stress.

While the scheme in Figure 14.2 seems quite clear, it is often not clear when damage is inspected which cause was decisive, for example, when several forms of damage occur simultaneously. However, a traceable assignment of damage to a cause can be very important in order to clarify liability and warranty claims. Corrosion processes play a role in very different branches of industry and are linked to other, very different phenomena. Just think of such different areas like shipbuilding, microelectronics, and medical implants that face significant and important corrosion issues. These different, practically relevant areas have given rise to industry-specific terminologies. On the other hand, there is also a need for *uniform* terms, uniform characterization methodology, and uniform test procedures for corrosion resistance. There is no universally accepted definition of corrosion. A compilation of definitions [16] illustrates that some of them are more general and some are more restrictive with respect to the materials and processes. Probably the most widely accepted definitions were established by the

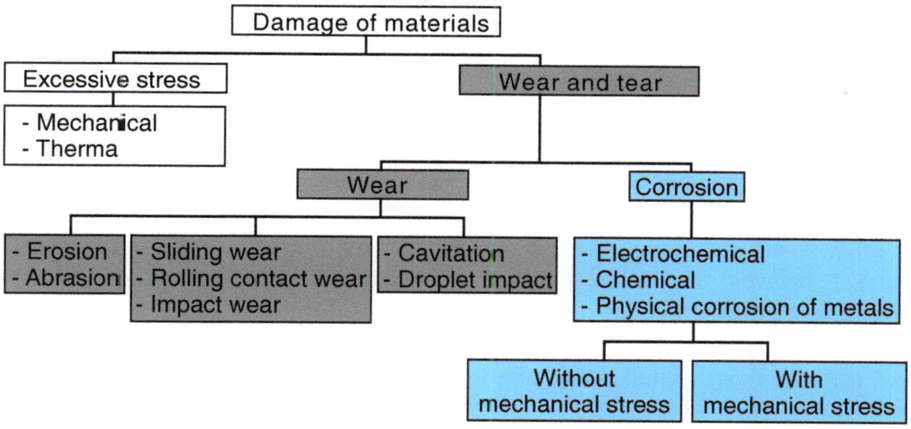

Figure 14.2 Corrosion as a specific kind of material according to [13].

National Association of Corrosion Engineers (NACE/ASTM G193) and by the International Standard Organization located in Switzerland (ISO 8044). They contain the following elements:

> **Corrosion** is the *process* of physicochemical interaction of a material (mostly metals) with its environment that causes a *measurable change* in the properties of the material or its environment.
>
> **Corrosion phenomenon** refers to the measurable *change* due to corrosion.

Examples: Mass decrease, discoloration, roughness increase, as well as current and voltage oscillations.

> The term **corrosion damage** refers to the impairment of the *function* of a metallic component resulting from the corrosion phenomenon.

Examples: Rusting of screws, increased flow resistance in pipelines, and increased sliding friction of a bicycle chain.

> **Corrosion failure** refers to a *complete loss of function* of a technical system due to corrosion phenomena.

Example: Breakage of components.

These definitions already make it clear that not every corrosion process automatically leads to corrosion damage. Rather, corrosion is unavoidable in many situations. The gradual removal of material can, for example, be taken into account in the design of components. Here, functional impairment only occurs when corrosion proceeds unexpectedly rapidly in some places. Sometimes corrosion is even desired, for instance, in bioresorbable metallic implants that dissolve when the presence of the implant is not required anymore. This avoids a second surgery for removing a temporary metallic implant.

Socio-economic Relevance. Metallic components must be extracted by energy-intensive processes from ores, which usually occur in nature as oxides, sulfides, or carbonates. During extraction and processing, there are considerable health risks and adverse effects on the environment (mining and mining consequences, tailings, leaching of problematic substances, SO_2- and CO_2-emission, as well as accumulation

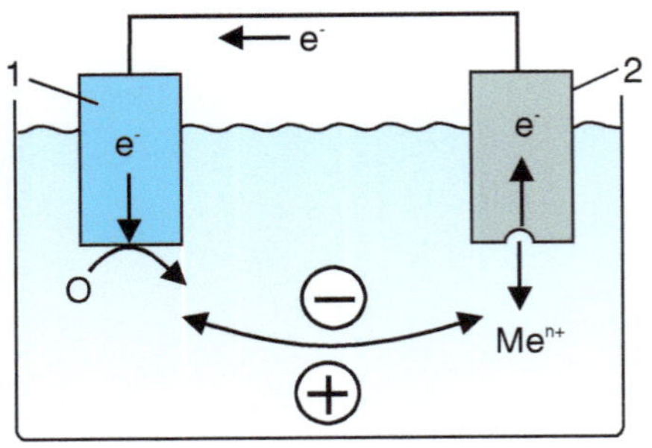

Figure 14.3 Prerequisites for electrochemical corrosion. 1) Cathode; 2) anode with an electron-conducting and an ion-conducting current path between them.

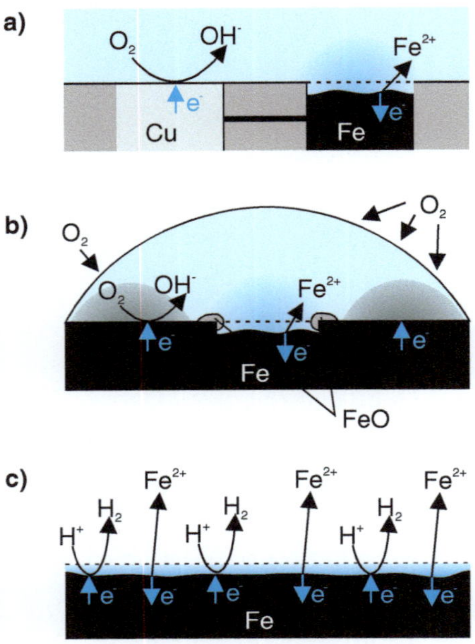

Figure 14.4 Examples of different spatial distributions of anodic and cathodic reactions.

a) Complete separation of the partial reactions in contact corrosion of Cu and Fe.
b) Corrosion of Fe under a drop of salt water. The distribution of anodic and cathodic partial reactions is caused mainly by a gradient in the concentration of dissolved oxygen and the resulting pH gradient. O_2 is taken up from the atmosphere into the droplet and reduced to OH^- at the Fe-electrolyte solution interface near the Fe/water/air three-phase boundary. At the same time, Fe^{2+} goes into the solution at the center of the wetted surface and diffuses outward from there. If the Fe^{2+} meets the inward migrating OH^- ions, $Fe(OH)_2$ precipitates, which eventually leads to a rust ring.
c) Uniform or general corrosion of iron in acids; both partial reactions take place on the same surface elements.

and spread of large quantities of auxiliary materials). It should therefore be a general principle of sustainable management to be particularly careful and economical with metallic materials. However, with the exception of some precious metals, metals are not thermodynamically stable in an oxygen-containing environment and should therefore convert back into oxides or salts without special precautions. Different estimates suggest that about a quarter to a third of the annual production of iron is destroyed by corrosion processes! In value terms, the cost of corrosion in developed countries is equivalent to 3–4% of the gross national product, of which 15–30% would probably be avoidable [14–18]. Corrosion damage requires continuous replacement of damaged parts. In addition to the immediate costs, intangible damage is caused by dissatisfied customers. Switching to more expensive materials leads to higher costs and the use of rare metals. The same applies to surface treatments such as paints or coatings. The consideration of corrosion processes in the planning and design phase also leads to higher development costs. Finally, during corrosion, metal ions are released from the workpiece into the environment, where they can cause consequential damage. Also, the used protective coatings can be a source of problematic emissions into the environment (e.g., organo-tin compounds from old marine coatings to retard fouling now pollute the seas). Finally, accidents with high personal injury and property damage are possible if reinforcement steels in load-bearing concrete structures, pipelines, or pressurized gas containers suddenly fail.

Prerequisite of Electrochemical Corrosion. The most common corrosion phenomenon is the removal of the metallic material from the surface of the workpiece. The metal ions deposit on the surface in the form of oxidic coatings (rust) or go into solution as dissolved ions. This form of corrosion is called **external corrosion**. The oxide coatings on the surfaces form a passive layer, which can retard further metal dissolution. By its nature, metal dissolution is an oxidation reaction. Since the valence electrons remain in the metal, a reduction reaction must concomitantly consume the excess electrons created by metal dissolution for the corrosion process to proceed further. Frequently, the metal dissolution and the reduction reaction take place at spatially separated areas of *one* workpiece. Electron transport must be enabled by an electron-conducting junction between the two reaction sites. In addition, both sites must be connected ionically in the surrounding corrosion medium. We had identified these features in Chapter 1 as the general characteristics of an electrochemical reaction. Therefore, corrosion reactions of this type are summarized as **electrochemical corrosion**. Thus, for electrochemical corrosion, **four conditions** must be met (Figure 14.3):

- an anodic partial reaction (usually metal dissolution),
- a cathodic partial reaction,
- an electron-conducting connection between the two reaction sites,
- an ion-conducting connection between the two reaction sites.

The analysis of any corrosion problem begins with the identification of these four conditions, which also point to the main approaches for corrosion protection. Since we can think of the reactions as sequence of reaction steps, the slowest of the four processes of oxidation, reduction, electron conduction, or ion conduction determines the rate of the overall corrosion process. It should be noted, however, that the areas available for oxidation and reduction, for example, can be of different sizes, which decisively influences the resulting corrosion phenomena (Figure 14.4).

External corrosion is opposed by internal corrosion, in which changes occur not only at the surface but also inside the workpiece. This includes the absorption of elemental hydrogen (from an H_2 atmosphere or after the reduction of dissolved H^+) into the interstitial sites of the metals with the formation of insertion hydrides. These metal hydrides have completely different mechanical properties (elasticity, density, and so on) than the actual metals. Thus, cracking and embrittlement of the material can occur (hydrogen embrittlement). Other reactive gases, liquid metals such as Hg, or metals with high vapor pressures (e.g., Cd at $T > 150\ ^\circ C$) can also penetrate deep into the material at grain boundaries and cause changes there. These reactions responsible for internal corrosion are summarized under the term metal-physical corrosion.

For the sake of completeness, chemical corrosion should be mentioned. It occurs without the need for an ion-conducting phase and without the formation of charged interfaces. For metallic materials, the reaction of magnesium with halogenated hydrocarbons RX to form Grignard compounds[150] RMgX is an example of chemical corrosion. Chemical corrosion also includes many forms of damage to polymers and ceramics, which we will not consider further.

The corroding workpiece together with the surrounding medium (or media such as drops of a salt solution in moist air) forms the corrosion system, which is a thermodynamically open system (Figure 14.5). Open systems in the thermodynamic sense allow an exchange of material and energy with the environment to reach thermodynamic equilibrium. The components of the corrosion system are characterized by various parameters. From them, the parameters of the workpiece can be influenced

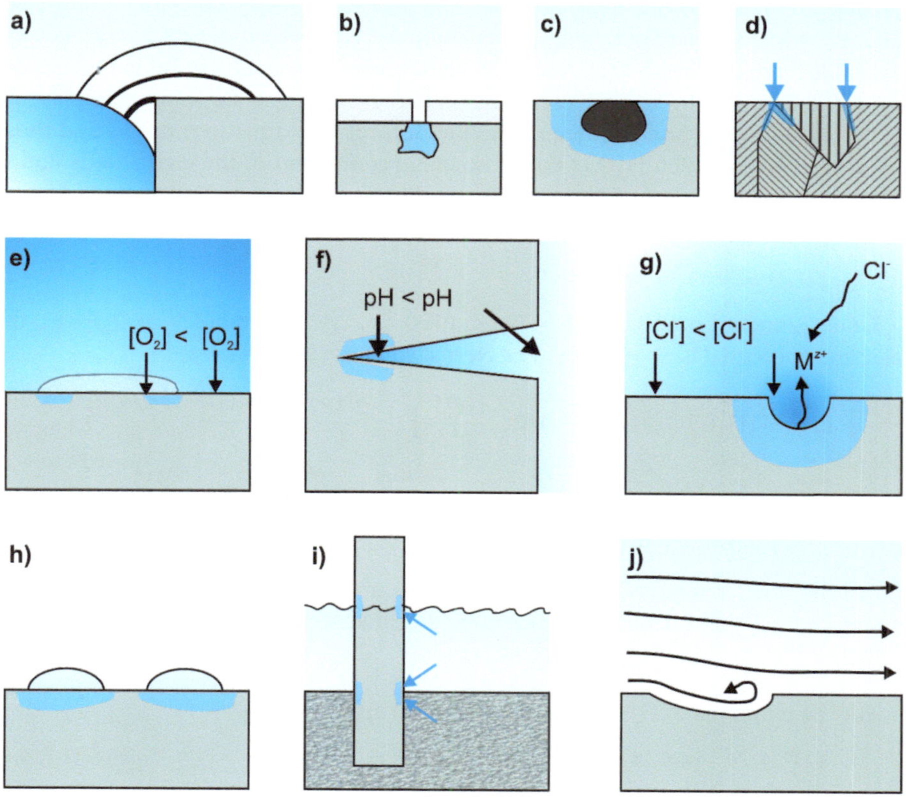

Figure 14.5 The corrosion system decisively influences the speed of corrosion. Influences from the work piece result from a) the electrical contact of different metals; b) cracks in coatings or passive layers; c) inclusions in the material; d) grain boundaries. The corrosion medium can promote corrosion through the occurrence of concentration gradients of e) O_2; f) H^+; g) Cl^-; h) concentration gradient often caused by the special conditions at three-phase boundaries at drops; i) transitions from gas to liquids and from liquid to soil; or j) due to different convective conditions.

[150] Victor Grignard, 1871–1935, French chemist, professor at Nancy, Nobel Prize in Chemistry 1912.

during material selection and manufacturing and are often also known very precisely (Figure 14.5a to d). Less well known and, moreover, not constant in time (transient) are the properties of the medium (Figure 14.5e to j). It should be mentioned that corrosion damage can also occur to the medium. Released metal ions can render water in pipelines or food in packaging unusable. Metal ions often have a catalytic effect and can thus change other ingredients and reduce the shelf life of products (e.g. fretting corrosion involving oxidation of lubricants).

14.2 The Pourbaix Diagram – The Thermodynamics of Corrosion

We have already dealt with electrolyte equilibria involving metal ions in Chapters 1 and 2 and found that by coupling the redox equilibrium with solubility and protolysis equilibria, many metal ions can form solid coatings on electrodes with solution components. Especially in the case of transition metal oxides with multiple oxidation states, there are many possibilities for the formation of solid hydroxides, oxides, and oxyhydroxides, depending on pH. In order to predict which products can form at all from a thermodynamic point of view, potential–pH diagrams or Pourbaix diagrams[151] have proven useful. These are zone diagrams that give the stability range of certain compounds as a function of potential (on the ordinate) and pH (on the abscissa). They are available for many metals [21]. A selection of sources can be found in Table 14.1. Below, we will follow step-by-step the construction of these diagrams for some examples.

Water. First, we probe different combinations of E and pH in order to find those combinations at which H_2O is the most stable compound of the elements H and O (Figure 14.6). Toward positive potentials, the stability range is limited by the oxidation of water according to

H^+ stands as an abbreviation for H_3O^+.

$$2H_2O \rightleftharpoons O_2 + 4e^- + 4H^+ \tag{14.1}$$

Under standard conditions ($T = 298$ K, pH 0, i.e., $\{H^+\} = 1$ mol L^{-1}, $p(O_2) = p°$, the electrode potential is $E° = 1.23$ V. The Nernst equation is as follows:

$$\begin{aligned} E &= E°(H_2O/O_2) + \frac{2.3RT}{4F}\log\left(\frac{\{H^+\}}{a°}\right) + \frac{2.3RT}{4F}\log\frac{p(O_2)}{p°} \\ &= E°(H_2O/O_2) - 2.3 \cdot E_{NF} \cdot pH + \frac{2.3E_{NF}}{4}\log\frac{p(O_2)}{p°} \\ &\approx 1.23\,V - 0.059\,V\,pH(T = 298\,K, p(O_2) = p°) \end{aligned} \tag{14.2}$$

Since the relationship $p(O_2) = p°$ holds for electrolysis with release of O_2 above the solution, Eq. (14.2) further simplifies to the linear equation (14.3):

$E_{NF} = RT/F$ – Nernst factor [V]
$2.3E_{NF} = 2.3RT/F = 0.059$ V for $T = 298$ K
$pH = -\log(\{H^+\}/a°)$
$\{H^+\}$ – Oxonium ion activity [mol L^{-1}]
$a°$ – Standard activity for dissolved ions = 1 mol L^{-1}
R – Ideal gas constant = 8.314 A s V mol^{-1} K^{-1}
F – Faraday constant = 96 485 A s mol^{-1}
$\log(p°/p°) = \log(1) = 0$

$$E = 1.23\,V - 0.059\,V\,pH \tag{14.3}$$

For the reduction of H^+ to form H_2, Eq. (14.4) applies

$$2H^+ + 2e^- \rightleftharpoons H_2 \tag{14.4}$$

With a constant $p(H_2) = p°$ above the solution, another linear equation (14.5) is obtained:

$$\begin{aligned} E &= E°(H_2/H^+) + \frac{2.3E_{NF}}{2}\log\left(\frac{\{H^+\}}{a°}\right) - \frac{2.3E_{NF}}{2}\log\frac{p(H_2)}{p°} \\ &\approx 0\,V - 0.059\,V\,pH\ (T = 298\,K). \end{aligned} \tag{14.5}$$

[151] Marcel Pourbaix (pronounced "Poorbejh"), 1904 (Myshega, Russia)–1998 (Belgium), Belgian chemist.

Equations (14.3) and (14.5) correspond to lines (1) and (2) in Figure 14.6. These lines are equilibrium lines, that is, they connect points where O_2 is in equilibrium with H_2O, or H_2 is in equilibrium with H_2O at different pH values. They frame the stability range of water. Both lines run parallel at a distance of 1.23 V. They are inclined because charge transfer is associated with the consumption of protons or hydroxide ions.

Inclined straight lines in the Pourbaix diagram indicate electrode reactions with turnover of protons or hydroxide ions. The slope results from the ratio between the protons (hydroxide ions) and electrons converted in a formula conversion times the Nernst factor for decadic logarithm $2.3E_{NF} = 2.3RT/F \approx 0.059$ V at $T = 298$ K.

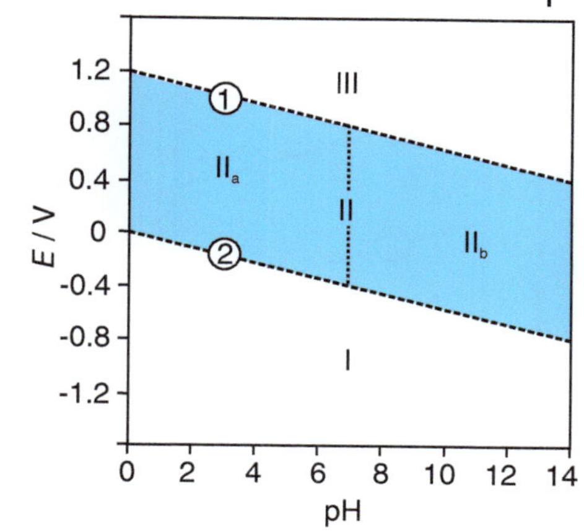

Figure 14.6 Pourbaix diagram for water. Line (1) corresponds to Eq. (14.3), line (2) corresponds to Eq. (14.5)).

a) I – Stability range of H_2;
b) II – stability range of H_2O (marked in blue), II_a – acidic, II_b – alkaline;
c) III – stability range of O_2.

Apart from the field of corrosion, this consideration also leads to illustrative conclusions in other areas. For water electrolysis, the potentials of the electrodes (against SHE) must lie outside the blue-shaded area in Figure 14.6. For an H_2–O_2 fuel cell, on the other hand, the potentials of both electrodes lie within the shaded region. Small corrections in the range of 10 mV are necessary due to deviating partial pressures $p(O_2) = 0.21\,p°$ in air.

Magnesium. Next, we consider a simple Pourbaix diagram of a metal, namely magnesium (Figure 14.7). A Mg electrode has a potential over a very wide pH range that depends only on the activity of the dissolved magnesium ions $\{Mg^{2+}\}$. In the Pourbaix diagrams, the equilibrium lines are given by *convention* for $\{M^{n+}\} = 1 \times 10^{-6}$ mol L^{-1} = 1 μmol L^{-1}. This is a reasonable agreement, since a uniform corrosion current density of 10 μA cm^{-2} can achieve approximately this concentration close to the electrode surface. Differences between numerical values of concentrations and activities are generally negligible at such low concentrations. Higher concentrations sometimes occur in localized corrosion processes, that is, very rapid metal dissolution in a small area of a large electrode. Therefore, equilibrium lines for further ion activities $\{M^{n+}\}$ are occasionally given, usefully in logarithmic gradation as illustrated in Figure 14.7 for 1 mmol L^{-1} and 1 mol L^{-1}. The curves are labeled $-\log(\{Mg^{2+}\}/a°)$ [analogous to pH = $-\log(\{H^+\}/a°)$]. Curve 3 in Figure 14.7 is parallel to the abscissa (pH axis), because the electrode reaction does not involve the conversion of protons.

Q: How can the mentioned value $\{M^{n+}\} = 1 \times 10^{-6}$ mol L^{-1} be estimated?

A: Try by assuming a steady-state with an activity of $\{M^{n+}\}^* = 0$ in the bulk phase and an unknown activity $\{M\}^{n+}{}_{x=0}$ at the electrode surface. Assume a diffusion coefficient of $D = 5 \times 10^{-6}$ cm^2 s^{-1} and the presence of a Nernst diffusion layer of thickness Δx (i.e., linear concentration profile). Using Fick's first law [Section 4.4.2, adjust Eq. (5.45)], you can either specify $\{M^{n+}\}_{x=0} = 1 \times 10^{-6}$ mol L^{-1} and calculate Δx, or set Δx to a reasonable value and calculate $\{M^{n+}\}_{x=0}$. For an estimate of Δx, you may, for example, look at the simulated concentration profiles of a cyclic voltammogram in Figure 5.22. In each case, the results should be within reasonable orders of magnitude. Handling the units becomes easier in the cgs system (gram–centimeter–second), for which it is best to first convert the concentrations to mol cm^{-3} = mol mL^{-1}.

$$E = E°(Mg/Mg^{2+}) + \frac{2.3E_{NF}}{2}\log\left(\frac{\{Mg^{2+}\}}{a°}\right)$$

$$\approx -2.37\,\text{V} + \frac{0.059\text{V}}{2}\log\left(\frac{\{Mg^{2+}\}}{a°}\right) \qquad (14.6)$$

Equilibrium lines in the Pourbaix diagram, which run *parallel* to the abscissa (pH axis) correspond to electrode reactions without participation of protons or hydroxide ions.

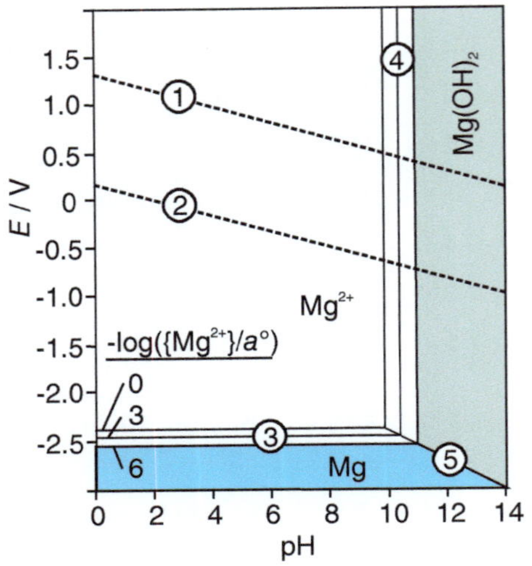

Figure 14.7 Pourbaix diagram of magnesium. The dashed lines mark the stability region of H_2O as in Figure 14.6. The numbered equilibrium lines correspond to the following equations:

1) Eq. (14.3);
2) Eq. (14.5);
3) Eq. (14.6) for $\{Mg^{2+}\}/(mol\ L^{-1}) = 10^{-6}, 10^{-3}$, and 1;
4) Eq. (14.10) for $\{Mg^{2+}\}/(mol\ L^{-1}) = 10^{-6}$, 10^{-3}, and 1;
5) Eq. (14.11).

The area for immunity is shaded in blue, and the area of possible passivity is shaded in gray.

In alkaline solution, Mg^{2+} reacts with hydroxide ions:

$$Mg^{2+}(aq) + 2OH^-(aq) \rightleftharpoons Mg(OH)_2(s) \tag{14.7}$$

The pH value at which equilibrium exists with a particular $\{Mg^{2+}\}$ in solution is obtained from the solubility product:

$$\{Mg^{2+}\} = K_L/\{OH^-\}^2$$

$$\{OH^-\} = \sqrt{\frac{K_L}{\{Mg^{2+}\}}} \tag{14.8}$$

$$pOH = -\log\sqrt{\frac{K_L}{\{Mg^{2+}\}}} = \frac{1}{2}(\log K_L - \log\{Mg^{2+}\}) = \frac{1}{2}(pK_L - pMg) \tag{14.9}$$

$$pK_W = pH + pOH = 14$$

$$pH = pK_W - pOH = pK_W - \frac{1}{2}(pK_L - pMg) \approx 14 - \frac{1}{2}(10.74 - 6) = 11.63 \tag{14.10}$$

The line array 4 runs perpendicular to the pH axis and parallel to the potential axis up to the point of intersection with lines of the line array 3.

Equilibrium lines in the Pourbaix diagram, which run *perpendicular* to the pH axis, belong to reactions without charge transfer at the electrode.

From the Pourbaix diagram in Figure 14.7, we can already draw some important conclusions about the reactions of *pure* Mg with water. There is no overlap between the stability ranges of water and metallic Mg. Only if an external potential of $E < -2.72$ V is imposed on the Mg electrode, Mg cannot dissolve in water for thermodynamic reasons. Under these conditions, we speak of **immunity**. [However, water would be electrolyzed to H_2 at these potentials.] Above the potential $E > -2.72$ V, dissolution of Mg can be expected and, at the same time, H_2 can be formed from water. For example, consider $E = -1.5$ V: At this potential, H_2O forms H_2 and Mg converts to Mg^{2+}. The metal should corrode. Finally, at pH > 12, there is a range where Mg dissolves, but the oxidation product $Mg(OH)_2$ deposits as a solid coating on the Mg electrode. This layer can suppress the progress of the corrosion reaction if it is insulating for electrons and OH^- and does not allow O_2 to pass. In this case, we speak of **passivity**. In this region, the redox equilibrium of Mg|Mg^{2+} is coupled with the solubility equilibrium given by Eq. (14.7). We obtain the activity $\{Mg^{2+}\}$, which we must insert into Eq. (14.6), from the mass action law [Eq. (14.8)]. If we take into account the ionic product of the water, we obtain for line (5) in Figure 14.7:

$$\begin{aligned} E &= E°(Mg/Mg^{2+}) + \frac{2.3E_{NF}}{2}\log\left(\frac{K_L}{\{OH^-\}^2\ a°}\right) \\ &= E°(Mg/Mg^{2+}) + \frac{2.3E_{NF}}{2}pK_L + 2.3E_{NF}\cdot pOH \\ &= E°(Mg/Mg^{2+}) + 2.3E_{NF}\left(\frac{pK_L}{2} + pK_W\right) - 2.3E_{NF}\cdot pH \end{aligned} \tag{14.11}$$

For $T = 298$ K, we obtain

$$\begin{aligned} E &\approx -2.37V + 0.059V\left(\frac{-10.74}{2} + 14\right) - 0.059V\cdot pH \\ &\approx -1.861\,V - 0.059V\cdot pH \end{aligned} \tag{14.12}$$

In the context of corrosion, **immunity** denotes a state (*E*, pH, and *T*), in which a workpiece cannot dissolve for thermodynamic reasons. **Passivity** denotes a state in which, for thermodynamic reasons, a conversion of the metal is to be expected, but this reaction is prevented or extremely slowed down by a dense surface layer.

Aluminum also forms a dense oxide layer that protects the metal from further attack. This is the basis for the wide use of this metal as a lightweight construction material. However, $Al(OH)_3$ has amphoteric properties. Therefore, we need to consider other equilibrium lines in Figure 14.8. At dissolved metal species concentrations of 10^{-6} mol L^{-1}, $T = 298$ K, we obtain the line array (1) for the Al electrode of the first kind in the acidic environment.

$$E = E^{\circ}(Al/Al^{3+}) + \frac{2.3\,E_{NF}}{3}\log\left(\frac{\{Al^{3+}\}}{a^{\circ}}\right)$$
$$= -1.67\,V + 0.019\,V \cdot (-6) = -1.788\,V \tag{14.13}$$

At higher pH values, hydroxides or oxides precipitate. The most stable one is hydrargillite ($Al_2O_3 \cdot 3H_2O$). The line array (2) is for the equilibrium (14.14).

$$2\,Al^{3+} + 3\,H_2O \rightleftharpoons Al_2O_3(s) + 6\,H^{+} \tag{14.14}$$

results from

$$3\,pH = -\log(\{Al^{3+}\}/a^{\circ}) + pK_a \tag{14.15}$$

$pK_a = 5.70, K_a = \{H^+\}^3/\{Al^{3+}\}$

$$pH = \left(\frac{-\log\frac{\{Al^{3+}\}}{a^{\circ}} + pKa}{3}\right) \approx \frac{6.00 + 5.70}{3} = 3.9 \tag{14.16}$$

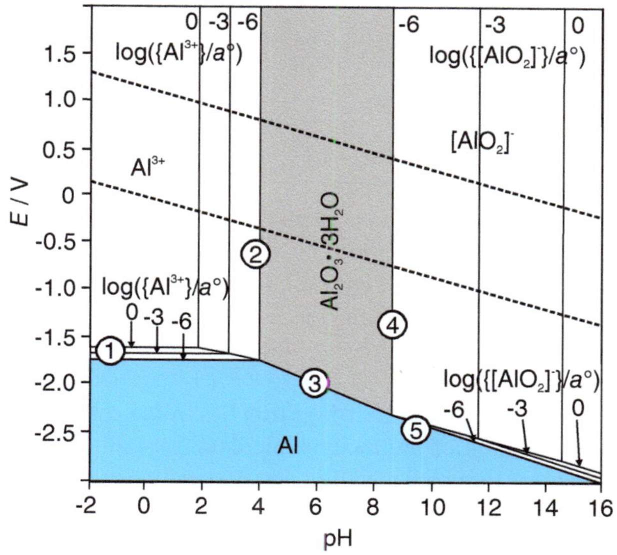

Figure 14.8 Calculated Pourbaix diagram for aluminum. The dashed lines mark the stability region of H_2O. The numbered equilibrium lines correspond to the following equations:

1) Eq. (14.13);
2) Eq. (14.16);
3) Eq. (14.17);
4) Eq. (14.20);
5) Eq. (14.23).

The area of immunity is shaded in light blue; the area of possible passivity is shaded gray.

For the Al electrode of the second kind, we can solve Eq. (14.15) for log({Al^{3+}}/a°) and substitute the result into Eq. (14.13) [line (3)].

$$E = E^\circ(\mathrm{Al/Al^{3+}}) + \frac{2.3E_{\mathrm{NF}}}{3}(\mathrm{p}K_\mathrm{a} - 3\,\mathrm{pH})$$
$$\approx -1.558\,\mathrm{V} - 0.059\mathrm{V}\cdot\mathrm{pH} \tag{14.17}$$

Unlike Mg, the oxide layers dissolve again in strongly alkaline solutions:

$$\mathrm{Al_2O_3 + 2\,OH^- \rightleftharpoons 2\,AlO_2^- + H_2O} \tag{14.18}$$

Let us formulate this reaction as the reaction of the acidic Al_2O_3 with water,

$$\mathrm{Al_2O_3 + 2\,H_2O \rightleftharpoons 2\,AlO_2^- + 2\,H^+} \tag{14.19}$$

$\mathrm{p}K_\mathrm{a} = 14.6,\ K_\mathrm{a} = \{\mathrm{H^+}\}\cdot\{\mathrm{AlO_2^-}\} = \{\mathrm{AlO_2^-}\}\cdot K_\mathrm{W}/\{\mathrm{OH^-}\}$

resulting in line (4) for $\{\mathrm{AlO_2^-}\} = 10^{-6}$ mol L^{-1}.

$$\mathrm{pH} = \mathrm{p}K_\mathrm{a} + \log\frac{\{\mathrm{AlO_2^-}\}}{a^\circ} = 14.6 + (-6) = 8.6 \tag{14.20}$$

We obtain the equilibrium potential of the Al electrode in alkaline solutions if we combine Eqs. ((14.19)) and ((14.14)):

$$\mathrm{Al^{3+} + 2\,H_2O \rightleftharpoons AlO_2^- + 4\,H^+} \tag{14.21}$$

$\mathrm{p}K_\mathrm{a} = \mathrm{p}K_{\mathrm{a},1} + \mathrm{p}K_{\mathrm{a},2} = 5.70 + 14.6 = 20.3 = 4\,\mathrm{pH} - \log\left(\{\mathrm{AlO_2^-}\}/\{\mathrm{Al^{3+}}\}\right)$

For the electrode reaction:

$$\mathrm{Al + 2\,H_2O \rightleftharpoons AlO_2^- + 4\,H^+ + 3\,e^-} \tag{14.22}$$

with the concentration of 10^{-6} mol L^{-1} for the dissolved Al species at $T = 298$ K, line array (5) is then obtained from the following:

$$E = E^\circ(\mathrm{Al/Al^{3+}}) + \frac{2.3E_{\mathrm{NF}}}{3}\log\left(\frac{\{\mathrm{Al^{3+}}\}}{a^\circ}\right)$$
$$= E^\circ(\mathrm{Al/Al^{3+}}) + \frac{2.3E_{\mathrm{NF}}}{3}\left[\log\left(\frac{\{\mathrm{Al^{3+}}\}}{a^\circ}\right) + \mathrm{p}K_\mathrm{a} - 4\cdot\mathrm{pH}\right]$$
$$\approx -1.67\,\mathrm{V} + \frac{0.059\,\mathrm{V}}{3}(-6 + 20.3) - \frac{4}{3}0.059\,\mathrm{V}\cdot\mathrm{pH}$$
$$\approx -1.388\,\mathrm{V} - 0.07866\,\mathrm{V}\cdot\mathrm{pH} \tag{14.23}$$

Pourbaix diagrams become even more complicated when there are different oxidation states for metal ions and metal oxides.

Iron is a typical example of these metals and the most important construction metal. At an iron electrode, Fe^{2+} ions are formed first, which can be further oxidized to Fe^{3+}. The Pourbaix diagram is shown in Figure 14.9, but we refer to the literature for the derivation of the individual equilibrium lines [22].

Table 14.1 summarizes other sources of annotated Pourbaix diagrams in readily available sources.

Table 14.1 Sources for annotated Pourbaix diagrams.

Element	Literature
Mg	[2, 5, 7, 13, 19]
Al	[2, 6, 7, 9, 13, 19]
Ti	[5–7, 9, 13, 19]
Ta	[6]
Cr	[6]
Fe	[2, 4, 6, 7, 13, 19, 20, 22]
Mn	[19]
Ni	[7]
Co	[23]
Sn	[5]
Zn	[5–7, 10, 13, 19]
Cu	[1, 7, 9, 13, 19]
Pd	[6, 13, 24]
Ag	[6, 24]
Au	[9, 24]
Pt, Ir, Rh, Ru	[24]

The Importance of the Pourbaix Diagrams lies in the compact and descriptive summary of a large amount of thermodynamic data. Like all thermodynamic data, these values can only tell us which reactions will *not* occur. For example, in the immunity range, metals will indeed not dissolve. However, material damage in this region can still occur if metal hydrides are formed as a result of hydrogen evolution (gaseous metal hydrides for Sn or hydrogen embrittlement for Fe). Hydride formation reactions are not considered in the discussion of the simplified Pourbaix diagrams in this section. However, if a reaction is possible according to the Pourbaix diagram, this does not mean that it also proceeds at a high rate. We have already seen in the discussion of the volcano curve for the hydrogen evolution reaction (HER) in Chapter 3 that one and the same thermodynamically allowed electrochemical reaction can occur with extremely

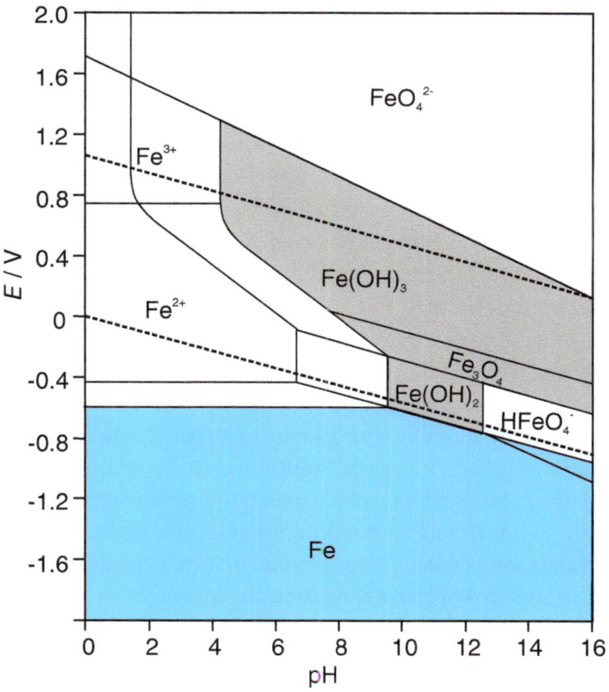

Figure 14.9 Calculated Pourbaix diagram for iron with dissolved iron species concentrations of 1 mol L^{-1} and 10^{-6} mol L^{-1} at $T = 298$ K according to [22]. Dashed lines mark the stability range of H_2O. The area of immunity is shaded in light blue, and the area of possible passivity is shaded gray.

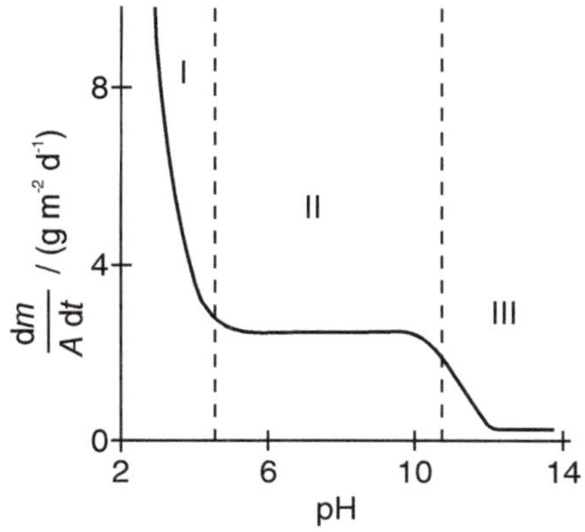

Figure 14.10 Plot of the pH-dependent corrosion rate of unalloyed steel in H_2O under. Source: Reproduced with permission from [13], Copyright 2001, Wiley-VCH.

I – Acid corrosion;
II – Oxygen corrosion;
III – Passivity.

different reaction rates. Finally, a Pourbaix diagram is always valid only for a particular temperature.

For most metals, the hydrogen line is above the range of immunity. This means that uncovered metals should dissolve in acids. If the HER dominates as a cathodic partial reaction, we speak of **acid corrosion**. With increasing pH, the activity of hydrogen ions decreases, and the oxygen reduction reaction (ORR) dominates as a reductive partial reaction. Accordingly, this corrosion regime is called **oxygen corrosion**. Acid corrosion and oxygen corrosion can occur simultaneously. Under air $[p(O_2) = 0.21\,p°]$ at normal pressure and room temperature, the equilibrium concentration of $O_2(aq)$ in solution is only about 0.25×10^{-3} mol L^{-1}. If oxygen corrosion becomes the dominant process it is slower than acid corrosion under acidic conditions because of the low O_2 concentration in solution. At even higher pH, an oxide or hydroxide film forms on Fe, imparting passivity and extremely slowing down the corrosion rate (Figure 14.10). For amphoteric oxides, the corrosion rate increases again in very alkaline solutions, so that metal species go into solution as metalates (Figure 14.11). Pourbaix diagrams cannot indicate how well a special oxide film protects the metal. Only the thermodynamic stability range with respect to pH can be well estimated. A further complication arises when other oxidic phases have to be considered. For example, the Pourbaix diagram for iron becomes even more complex if the oxides Fe_3O_4 and Fe_2O_3 are also considered [20]. A word of caution is in order here: the author found many small differences when comparing Pourbaix diagrams from recent publications for the same metals. This is due not only to the consideration of different phases but also to the use of different values for the physicochemical properties, which are known with only limited accuracy for some reactions. In addition, misprints and transpositions easily creep into the complex diagrams.

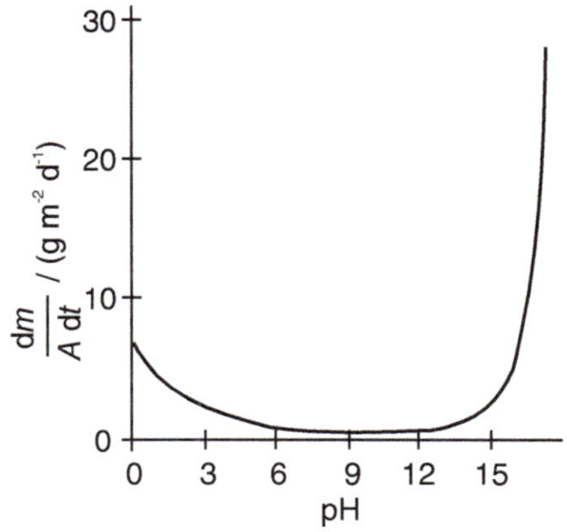

Figure 14.11 Plot of the pH-dependent corrosion rate of Al in H_2O under air. In alkaline conditions, corrosion increases again due to the formation of soluble aluminates. Source: Reproduced with permission from [13], Copyright 2001, Wiley-VCH.

Only for nobler metals such as Cu, the hydrogen line is within the immunity range of the metal. Therefore, copper should not corrode in an oxygen-free environment, but oxygen corrosion is possible in the presence of O_2. This conclusion, which is clear in itself, can be seemingly refuted in practice if copper also corrodes in oxygen-free media. This is possible when the presence of complexing agents shifts the equilibrium electrode potential strongly into the negative direction (Section 5.3.12). Taking this situation into account, there are attempts to establish potential-pH diagrams also for important other media with defined concentrations of complexing agents. For seawater with a high Cl^- concentration, several new solid phases (CuCl, $CuCl_2$, 3 $Cu(OH)_2$–$CuCl_2$), and dissolved Cu species ($[CuCl_2^-]$, $[CuCl_4^{2-}]$, etc.) would then have to be considered.

Pourbaix diagrams apply to pure metals and, in this way, do not capture technically important materials such as alloys or multiphase materials. However, some reasonable hypotheses are also possible for alloys. From looking at the Pourbaix diagrams for Fe and Cr as an example, it can be expected that Fe will dissolve out of chromium-containing steels in acidic solutions until predominantly Cr-containing compounds remain at the surface. The expected passivity range of Cr extends much further to low pH values than that of Fe, so for stainless steels, a significant extension of the passivity range toward low pH values can also be expected [6]. In addition, there are attempts to establish Pourbaix diagrams for alloys as well [25]. Alloy corrosion is nevertheless a complex subarea of electrochemical corrosion, which we can only touch upon within this textbook.

14.3 Kinetics of Corrosion Processes and Passivity

14.3.1 Evans Diagrams for Uniform Corrosion

Besides the fact that the previously discussed Pourbaix diagrams do not give any information about the rate of a corrosion reaction, they also cannot tell us which potential will occur at all on a workpiece in contact with an electrolyte solution. We call this potential, at which no current flows via an external circuit, (free) corrosion potential E_{cor} (synonyms: rest potential, free potential, open circuit potential). By its nature, it is not an equilibrium potential, but a mixed potential (Sections 2.3.5 and 3.2.7).

Q: Why do we use the term "corrosion potential" here when we have used "open circuit potential" in Chapters 2 and 3?

A: There are perhaps two reasons for this. One is that it is simply common to talk about the potential for corrosion in this context. You can see from this example that communication across disciplinary boundaries is important but not always easy. Additionally, this wording emphasizes that – unlike in Figure 3.13 – the corrosion potential is far removed from the equilibrium potential E_{eq} of the metal. Thus, we cannot assume a potential-determining ion species here either!

The anodic partial reaction is the oxidation of the metal:

$$Fe \rightleftharpoons Fe^{2+} + 2e^- \tag{14.24}$$

The reductive partial reaction is the HER:

$$2H^+ + 2e^- \rightleftharpoons H_2 \tag{14.25}$$

We can easily determine E_{cor} by combining the workpiece under investigation with a reference electrode, immersing both electrodes in the corrosion medium and measuring the potential difference under open-circuit conditions. The condition that applies here is that all anodic and cathodic currents are equal in magnitude:

$$\sum_{i=1}^{n} |I_{a,i}| = \sum_{j=1}^{m} |I_{c,j}|$$

$$\sum_{i=1}^{n} I_{a,i} = -\sum_{j=1}^{m} I_{c,j} \tag{14.26}$$

In the example chosen, we are dealing with only two partial reactions ($m = n = 2$). To illustrate this, we plot the current–voltage curves according to the Butler–Volmer equation for $H_2|H^+$ and $Fe|Fe^{2+}$ in a diagram (Figure 14.12). The solid lines are the sum of anodic and cathodic currents for each redox pair. The equilibrium potentials $E_{eq}(Fe|Fe^{2+})$ and $E_{eq}(H_2|H^+)$ depend on the solution concentrations present and therefore do not exactly match the standard electrode potentials $E°(Fe|Fe^{2+})$ and $E°(H_2|H^+)$. This would require the presence of standard activities ($a°$ bzw. $p°$) of the reactants. The corrosion potential E_{cor} is now the potential where $I(Fe|Fe^{2+}) = -I(H_2|H^+)$ in Figure 14.12a, that is, the zero of the curve $j(E)$. The absolute value of the current is the corrosion current $I_{cor} = |I(Fe|Fe^{2+}, E_{cor})| = |I(H_2|H^+, E_{cor})|$, which results in the corrosion current density $j_{cor} = I_{cor}/A$.

In the case of uniform corrosion, also called general corrosion, equal absolute currents of the partial reactions cause equal absolute values of *current densities* ($|j_a| = |j_c|$). In the case of other forms of corrosion, this is not the case and the *current densities* can, in principle, vary greatly. Such phenomena of *local corrosion* are discussed in Sections 18.5.2 and 18.5.3.

Because both partial reactions take place on a macroscopic piece of iron, we *cannot* measure the corrosion current from the outside.

The corrosion rate is experimentally most easily accessible via the mass loss determined by weighing. We obtain the area-related mass loss rate $\dot{m} = dm/dt$, which in turn is linked to the corrosion current density:

$$\frac{\dot{m}}{A} = \frac{dm}{A\,dt} = \frac{M}{nFA}\frac{dq}{dt} = \frac{M}{nF} j_{cor} \tag{14.27}$$

The thickness change Δx of the workpiece results from the volume change of the workpiece $\Delta V = A\,\Delta x$ and the density $\rho = \Delta m/\Delta V$ provided that the material removal is uniform (!):

$$\Delta x = \frac{\Delta m}{A\rho} \tag{14.28}$$

It can also be expressed using the corrosion current density j_{cor}:

$$\dot{x} = \frac{dx}{dt} = \frac{M}{nF\rho} j_{cor} \tag{14.29}$$

It is also immediately apparent from Figure 14.12a that E_{cor} and j_{cor} can change as conditions in the corrosion cell change. For instance, a gradual decrease in pH would cause a shift of E_{cor} in the negative direction, together with a decrease in j_{cor}. In order to predict the corrosion conditions of a system for a wide range of conditions from a limited number of measurements, the so-called **Evans diagrams**[152] are suitable, in which the decadic logarithm of the absolute value of the current density for the electrode reaction is plotted against the electrode potential (semilog plot, Figure 14.12b). The logarithmic plot has the advantage that current densities of very different magnitudes can be compared in one representation. The use of absolute values is necessary because the logarithm is only defined positive numbers. Each individual curve in Figure 14.12a results in a Tafel plot in Figure 14.12b. The linear sections of the curves are important. After extrapolation, we find the point [E_{cor}, log(j_{cor})] as the intersection of the branches for metal dissolution and hydrogen evolution.

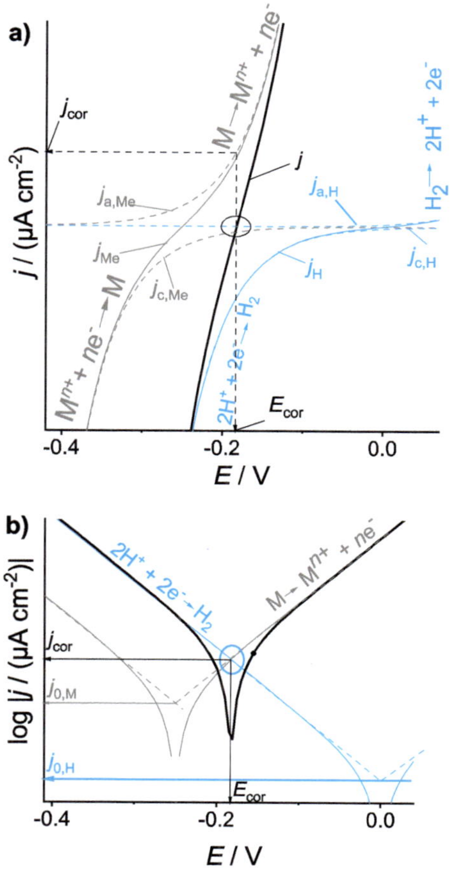

Figure 14.12 Schematic of partial current densities for acid corrosion of iron a) with linear scaling for current density and b) with logarithmic scaling for current density.

[152] Ulick Richardson Evans, 1889–1980, English electrochemist, professor at the University of Cambridge, author of highly influential textbooks (1923, 1948) in which the electrochemical nature of corrosion of metals was put on a scientific footing.

Excursion 14-1 Construction of Evans Diagrams from Polarization Curves

In practice, the current–voltage curves of the individual partial reactions are not always accessible. In these cases, one records a polarization curve of the electrode with more than one reacting redox couple. In order to keep the conditions comparable, standardizations by international technical organizations exist for this purpose [27].

First, a test specimen must be prepared as an electrode from the material under study. In particular, the electrode surface should be defined, and its area should be known so that one can later determine the current densities. In addition, all surface elements should be subject to similar mass transport conditions. Figure 14.13 shows some possibilities for inserting samples with different external shapes into the electrochemical cell.

The electrodes prepared in this way are inserted as a working electrode (WE) into a cell setup for corrosion measurements (Figure 14.14). Different cell designs are common for this purpose. Usually, these cells hold a significantly larger volume of electrolyte solution than the analytical cells that we have covered in Chapters 5 and 6. This prevents significant changes in the composition of the liquid electrolyte during the measurement period, even at quite high corrosion currents and corresponding generation of metal cations. A reference electrode (Ref) and an auxiliary electrode (Aux) complete the three-electrode arrangement. Since relatively large currents can flow, it is important to measure the potential directly in front of the working electrode. For this purpose, a Haber–Luggin capillary[56,57] (see Section 5.3.2) is positioned close to the working electrode. To stabilize the potential measurements, a Pt wire with a capacitor can be run parallel to the Haber–Luggin capillary (Figure 5.24). A thermometer and a bath for

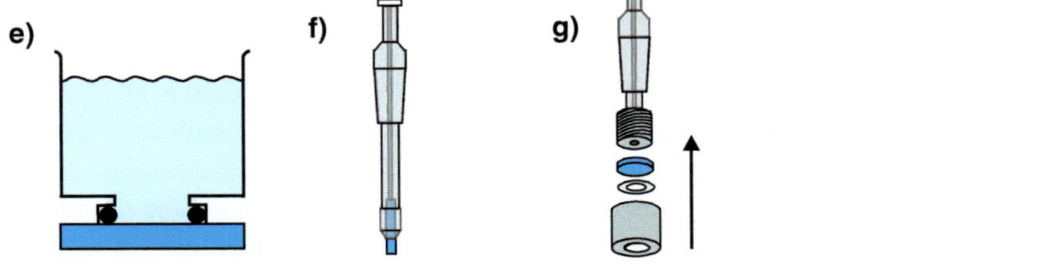

Figure 14.13 Schematic of different electrode shapes for corrosion tests. a) Flag-shaped foils, exposed on both sides; b) wire loop; c) specimen partially masked with cover lacquer; d) specimen embedded in epoxy resin and ground; e) liquid reservoir sealed with O-ring and placed on specimen; f) cylindrical specimen held with pinch seal; g) electrode holder for disk-shaped specimens. In each case, the contact point of the electrical lead wires shall be sealed against the electrolyte. The cells are completed with reference and auxiliary electrodes. Compilation according to [28].

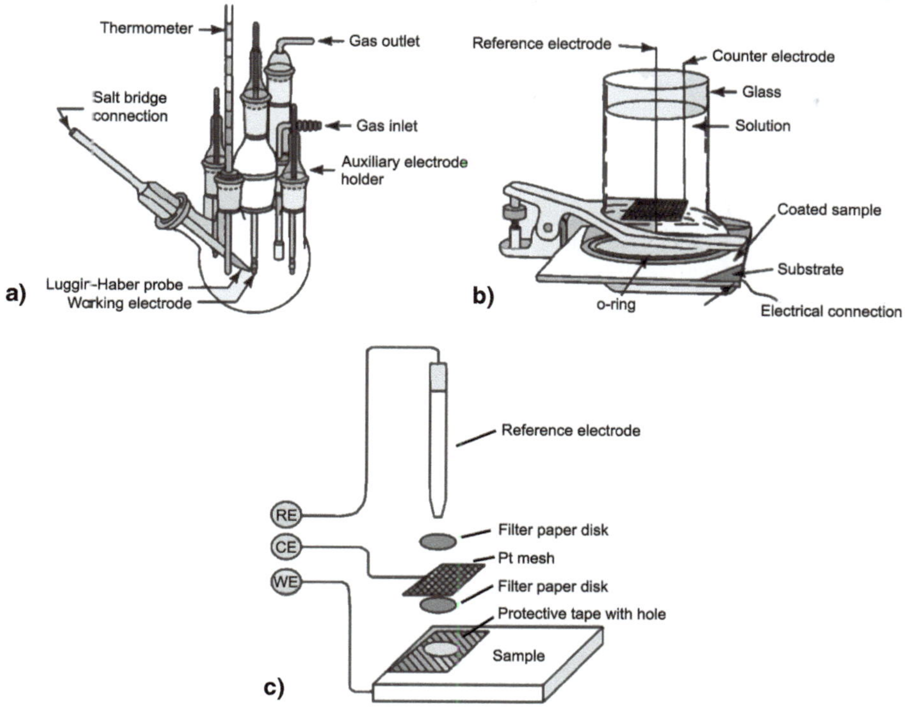

Figure 14.14 Schematic of typical cells for recording polarization curves and other corrosion measurements. a) Corrosion cell for macroscopic specimens; b) cell with clamping mechanisms for disk-shaped specimens; c) droplet cell. Source: Reproduced with adaptation from [28], Copyright 2003, Wiley-VCH.

temperature control of the cell are useful. If many samples of uniform geometry are to be examined, designs with rapid clamping mechanisms are advantageous. Droplet cells are another possibility. They are used in different sizes with different application directions. Here, the area wetted by a hanging drop of the electrolyte defines the area of the working electrode. The reference and auxiliary electrodes are located in the liquid reservoir of the capillary. They allow a fast sample change, but also local investigations on specific regions of a much larger sample. However, for the construction of Evans diagrams for uniform corrosion of electrodes, defined conditions with respect to T and the gas atmosphere in cells are generally preferred (see Figure 14.14a).

After assembling the measuring cell and rinsing the solution with Ar or N_2, one first measures the free corrosion potential as the open-circuit potential. Starting from this potential, one records two linear sweep voltammograms, one in the negative direction and one in the positive direction. They are plotted as absolute current values in semi-log representation (Figure 14.15a). For both voltammograms, about 0.25 V can be considered a reasonable potential range. The potential scan rate v is only 0.1 to 1 mV s^{-1}, since one mainly wants to capture a steady-state situation and suppress capacitive currents to a large extent. Due to the exponential relationship between current and voltage, the measured current can easily vary over several orders of magnitude; thus, a potentiostat with automatic switching of the current ranges is recommended. Standardized procedures and test specimens allow validation of the experimental setup [27]. Deviations from the idealized curves may occur if cover layers build up or degrade during the potential scan, whereby the state of the electrode surface is changed during the measurement of the polarization curve.

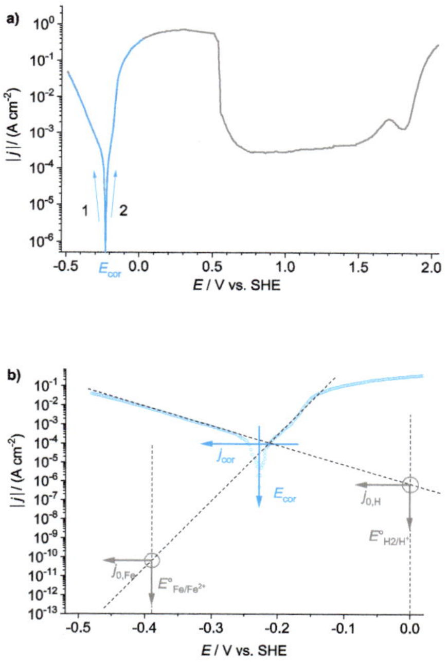

Figure 14.15 Construction of the Tafel plot for the corrosion of iron in 0.5 mol L^{-1} H_2SO_4. a) Measured polarization curve (gray). The potential region of interest is highlighted in blue; b) semilogarithmic plot of the region is highlighted in blue. The dashed straight lines are the extrapolation of the linearized sections of the potential regions between 0.05 and 0.25 V of the corrosion potential. Their intersections with the tabulated standard potentials $E°(Fe|Fe^{2+})$ and $E°(H_2|H^+)$ provide an estimate of the exchange current densities of the two partial reactions, which are usually very different. Source: Data from [28] simplified and rescaled by the textbook author.

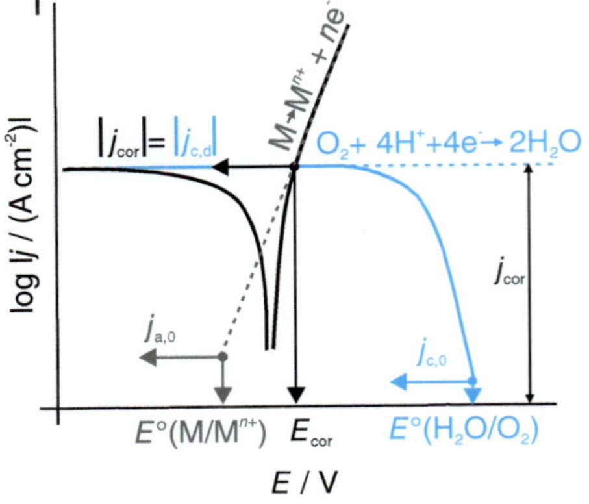

Figure 14.16 Evans diagram for metal dissolution limited by the transport of oxygen for the cathodic partial reaction.

The Tafel plots of the Evans diagram can be described by Eqs. (14.30) and (14.31):

$$\log I_a = \log I_{a,0} + \frac{E - E°(M|M^+)}{b(M|M^+)} \tag{14.30}$$

$$\log I_a = \log I_{a,0} + \frac{E - E°(M|M^+)}{b(M|M^+)} \tag{14.31}$$

Those straight lines are plotted as dashed lines, together with a section of the polarization curve in Figure 14.15b.

The net current is then calculated as follows:

$$\begin{aligned} I &= I_a - |I_c| \\ &= I_{a,0} \cdot 10^{\frac{E-E^{\circ}(M|M^+)}{b(M|M^+)}} - |I_{c,0}| \cdot 10^{\frac{E-E^{\circ}(H_2|H^+)}{b(H_2|H^+)}} \\ &= I_{a,0}\exp[\frac{2.3(E - E°(M|M^+))}{b(M|M^+)}] - |I_{c,0}| \cdot \exp[\frac{2.3(E - E°(H_2|H^+))}{b(H_2|H^+)}] \end{aligned} \tag{14.32}$$

This equation is called **Wagner–Traud equation**[122,153]. At first glance, it resembles the Butler–Volmer Equation (3.27). The main difference is that the anodic and cathodic partial currents originate from *different* redox pairs in the Wagner–Traud relation (Figure 14.15b). Therefore, different slopes of the Tafel lines $b(M|M^{n+})$ and $b(H^+|H_2)$, two exchange currents $j_{0,a}$ and $j_{0,c}$, and two standard electrode potentials $E°(M|M^{n+})$ and $E°(H_2|H^+)$ are also necessary (unlike in the Butler–Volmer equation).

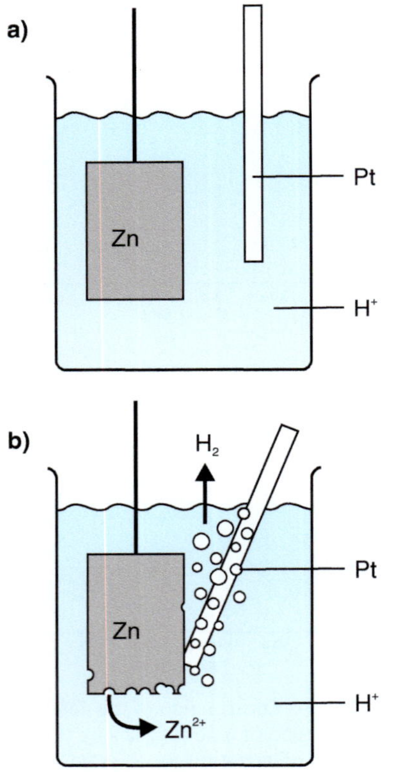

Figure 14.17 Observation of a) free corrosion of Zn in dilute acids and b) galvanic corrosion when a Zn sheet and a Pt sheet are in contact.

Evans diagrams can also be used to predict the corrosion potential and corrosion rate under the condition of limitation by the mass transport of one reactant. This often occurs, for example, in neutral media, where oxygen corrosion dominates. The transport-limited current density for the ORR is drawn as a horizontal line (Figure 14.16). The intersection of this line with the straight line of the Tafel plot for the metal dissolution reaction yields E_{cor} and j_{cor}.

Evans diagrams can also be used to estimate the behavior in galvanic corrosion. The example in Excursion 14-2 is particularly impressive, because here the difference between free corrosion and galvanic corrosion can be visually observed without measuring equipment due to the bubble formation that occurs.

Excursion 14-2 Prediction of Galvanic Corrosion upon Contact of Two Different Metals

Qualitative Observation. If a zinc sheet is immersed in a dilute hydrochloric acid, Zn should actually dissolve under vigorous H_2 development according to the difference of the standard electrode potentials $E°(Zn|Zn^{2+})$ and $E°(H_2|H^+)$. However, only few hydrogen bubbles grow slowly on the Zn sheet. If the zinc sheet is touched in the same acidic solution with a Pt sheet or Pt rod, vigorous H_2 evolution commences on the Pt sheet and the Zn sheet dissolves correspondingly rapidly (Figure 14.17). The cause of this rapid reaction is *not* the higher cell voltage of the galvanic element $Zn|Zn^{2+}$, $Pt^{2+}|Pt$. This element is not effective here at all because the solution at the beginning of the experiment contains neither Zn^{2+} nor Pt^{2+}. Rather, the reason is the extremely different exchange current density for the redox pair $H_2|H^+$ at Zn and Pt (volcano plot in Figure 3.9).

Quantitative Treatment. For this purpose, we use polarization curves recorded on Zn and Pt electrodes of the same area in the acid and plot them in a common coordinate system (Figure 14.18). Extrapolation of the linear parts of the curves

[153] Carl Wagner; Wilhelm Traud; On the interpretation of corrosion processes by the superposition of electrochemical partial processes and on the potential formation at mixed electrodes. *Z. Elektrochemie* **1938**, *44*, 391–402.

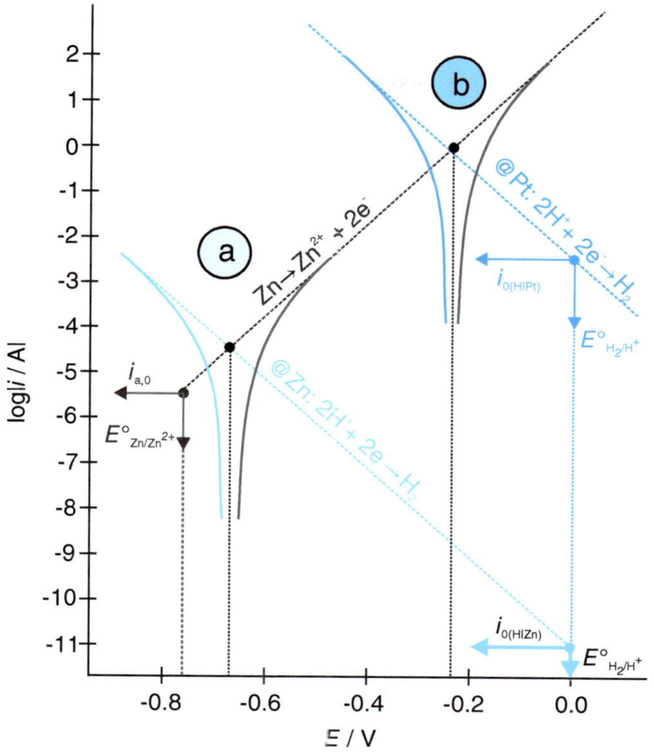

Figure 14.18 Construction of the Evans diagram for a) acid corrosion of Zn and b) galvanic corrosion of Zn in contact with Pt in dilute acids. Situations a) and b) correspond to the corresponding panels in Figure 14.17.

yields E_{cor} and j_{cor} of the Zn electrode [point (a)]. For the Pt electrode, we obtain the data for the hydrogen electrode, $i_0(Pt, H_2|H^+)$ and $E°(H_2|H^+)$, since Pt is an inert electrode under the experimental conditions and is not converted. When Pt contacts Zn in acid, the corrosion potential is obtained as the point (b) at which the magnitude of the partial current for Zn dissolution equals that of the HER at Pt. We obtain this point by extrapolating the measured polarization curves. The current for Zn dissolution in contact with Pt is about five orders of magnitude larger than the free corrosion current of Zn [point (a)]. In this experiment, we must consider the exchange currents (rather than the current densities) because the areas of the Pt and Zn electrodes may be very different. However, different electrode areas play only a minor role due to the acceleration of the corrosion reaction by several orders of magnitude.

14.3.2 Passivity

If we observe polarization curves at larger positive potentials, the current does not behave as expected from Tafel plots (Figure 14.19). For example, consider iron at pH < 2, for example, in 0.5 mol L^{-1} H_2SO_4. The Pourbaix diagram in Figure 14.9 suggests active iron dissolution. In fact, the polarization curve shows a steep increase in current between −0.2 and +0.1 V vs. SCE for anodic dissolution. According to the Butler–Volmer equation, we should actually expect a further increase. However, as is often the case in nature, an exponential increase in a quantity cannot be sustained indefinitely. Also here, the current for active iron dissolution levels off to a limiting current. In this region, the electrode is covered by a high concentration of solvated

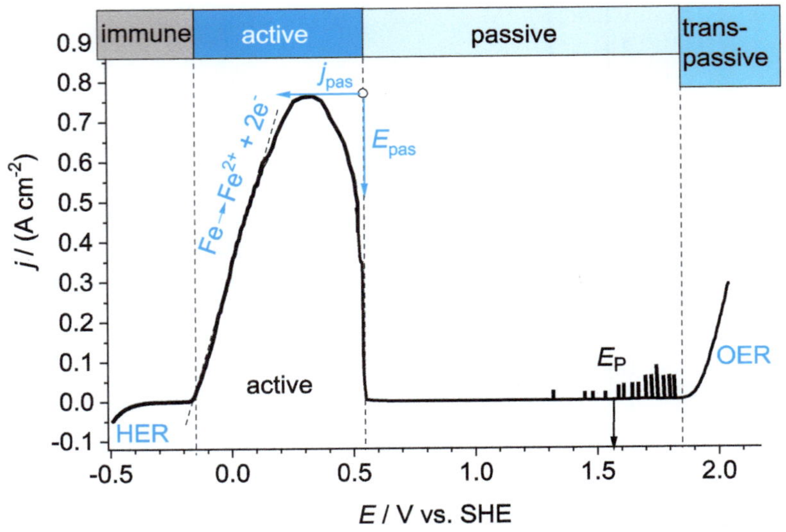

Figure 14.19 Schematic polarization curve of iron in 0.5 mol L^{-1} H_2SO_4 with linear scaling of the current density axis. The regions of immunity, active metal dissolution, passivity due to the formation of dense oxide layers, and transpassivity with the oxygen evolution reaction (OER) are highlighted. Pitting corrosion can occur at potentials greater than E_P.

metal ions and anions from the solution, so that further reaction is not possible until positive metal cations have diffused away from the electrode or additional anions have been transported in to balance the charge. Metal dissolution and removal of the reaction products reach a steady state. The potential region with the current rise and the plateau region together form the region of active metal dissolution.

However, if the passivation potential E_{pas} is exceeded at the passivation current density j_{pas}, the current drops by several orders of magnitude into the range of μA cm^{-2} and remains at this level until about +1.7 V. We call this potential range the passive range. In this potential range, the metal cations form an invisible, very thin (about 1 nm) pore-free surface layer with anions of the solution. The remaining low current density j results from further dissolution of the metal. For this reaction, metal cations must be removed from the metal lattice and incorporated into the oxide lattice. In many cases, the compounds in the surface layer are semiconductors with low cation conductivity. Therefore, the cations must be transported through the solid by the high field strengths in the oxide layer (electric field = potential difference/distance). In the case of free corrosion, the ORR usually takes place as a cathodic partial reaction within this potential range. For this to occur, molecular oxygen must either reach the metal–oxide interface or electrons must be transported through the surface layer to the interface between the surface layer and the surrounding medium. This is possible because the passive layers usually have low electronic conductivity, which is higher than the cation conductivity. Further reactions take place after absorption of electromagnetic radiation with the formation of electron–hole pairs in the semiconducting oxides. A comprehensive overview of the complex subject of semiconductor electrochemistry is provided in Chapter 10.

At even higher potentials, the current increases again because water electrolysis starts with the formation of O_2 (or, in media containing Cl^-, Cl_2 formation). At the same time, oxidation of O^{2-} ions in an oxidic surface layer can start with concomitant dissolution of the surface layer. This potential range is called the transpassive range. Another possible reaction at potentials greater than E_P is pitting corrosion (index P). At locations where the surface layer is locally disturbed, very rapid metal dissolution can occur, while a large fraction of the electrode area is available for the slower cathodic reaction (Section 14.5.3).

In the experiment in Figure 14.19, the potential was imposed externally, and the passive state was established when j_{pas} and E_{pas} were exceeded. In Figure 14.20, we see that j_{pas} and E_{pas} can also be set under the conditions of free corrosion by means of a strong oxidizing acid in the corrosion medium. In this case, concentrated HNO_3 causes a higher potential and thus stable passivation (Figure 14.20a). In contrast, j_{pas} and E_{pas} are not reached in diluted HNO_3 after disturbance of the passive layer, causing active dissolution of the metal piece (Figure 14.20b).

Caution: Concentrated nitric acid causes severe burns; be sure to wear protective gloves and goggles; work behind a protective screen!

In order to understand this initially surprising observation, we consider the polarization curve from Figure 14.15a (semilogarithmic representation). The key question to predict metal dissolution is the identification of conditions under which a cathodic partial reaction can deliver a cathodic partial current to accept the electrons available from metal dissolution. Schematically, we can imagine four possible Tafel lines for cathodic partial reactions, indicated as thin lines in Figure 14.21. If the cathodic partial reaction follows line 1, it lies above the anodic polarization curve of the metal for all possible combinations of j–E. The cathodic reaction can always sustain a metal dissolution at potentials higher than E_{pas} and current densities greater than j_{pas}. This is also true if the exchange current density $j_{0,c}$ of the cathodic reaction is very high, but its standard potential is lower (curve 2). j_{pas} can always be balanced by the cathodic partial reaction in this case, too. The cases for which curves 1 and 2 apply result in stable passivity, since the passive state occurs under all possible conditions of the workpiece. The case of curve 3 results in unstable passivity. There is more than one intersection of the anodic polarization curve of the metal with the Tafel plot of the cathodic partial

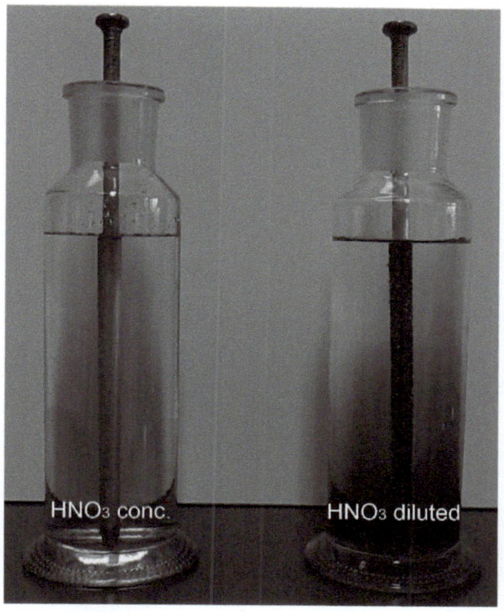

Figure 14.20 Illustration of the passivation of an iron nail. a) The iron nail does not corrode in concentrated HNO_3. When the iron nail is removed from the HNO_3 solution and carefully rinsed, a thin oxide film remains and initially passivates the nail under the liquid film of dilute HNO_3. b) The nail corrodes rapidly in diluted HNO_3. This can be observed visually by the formation of a dark layer of corrosion products and coloration of the solution. By immersion in conc. HNO_3, the corrosion products can be removed, and the passive layer can be regenerated.

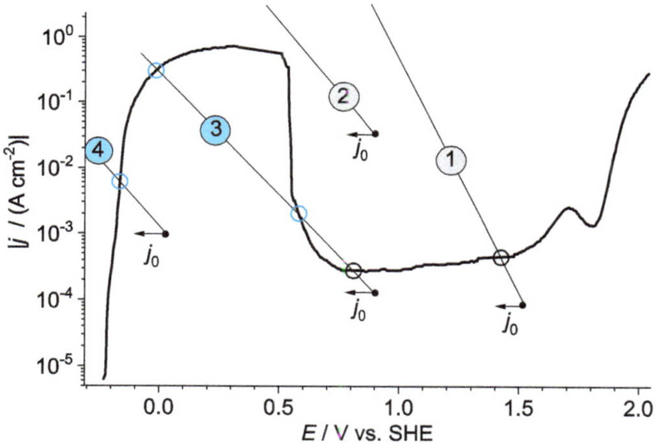

Figure 14.21 Schematic polarization curve of iron in 0.5 mol L^{-1} H_2SO_4 with logarithmic current density axis. The plot also contains four possible Tafel lines for the cathodic partial reaction in a corrosion process.

1) Strong oxidizing agent present. The polarization curve of iron is intersected in the region of stable passivity.
2) Oxidizing agent with high exchange current that exceeds j_{pas} (stable passivity) under all conditions.
3) The Tafel line of the cathodic partial reaction intersects the polarization curve at several points. Depending on the boundary conditions, active metal dissolution or unstable passivity can occur.
4) Weak oxidizing agent present. The Tafel line of the cathodic partial reaction always intersects the polarization curve of iron in the region of active metal dissolution.

reaction. One intersection is in the active region, one is in the transition between the active and passive regions, and another intersection is in the passive region. If such a system is slightly disturbed, its behavior can change the passive state to the active state and remain there. Curve 4 corresponds to a situation in which no passivity is to be expected, since the only intersection point between the polarization curve of the workpiece and the Tafel plot of the cathodic reaction occurs in the active potential range of the metal.

In conclusion, a stronger oxidizing agent is more suitable than a weak one for achieving stable passivity [i.e., permanent inhibition (slowdown, retardation) of metal oxidation]!

An extensive overview of the topic provides a comparison of different materials and discusses in detail the relationship to the semiconducting properties of thin oxide films [26].

14.4 Principal Approaches for Corrosion Protection

The term corrosion protection covers a range of measures used to slow down the rate of metal dissolution. Since metals reach their thermodynamic equilibrium in the form of oxides or salts, corrosion can never be completely avoided, but it can be retarded considerably.

Cathodic protection is the application of a potential at which metal dissolution is greatly slowed down, and instead the HER is the dominant reaction at the workpiece (Figure 14.22b). The externally applied potential shifts the state of the workpiece in the Pourbaix diagram from the region of active metal dissolution to the region of immunity (Figure 14.22a). The external potential can be set in two ways. Either a current source is used to force electrolysis between the workpiece and an inert counter electrode, with oxidation (of water, OER) occurring at the counter electrode and hydrogen evolution (HER) occurring at the workpiece. Figure 14.22c shows, as a blue curve, the Evans diagram of the workpiece to be protected with the free corrosion potential E_{cor}. Applying an external voltage shifts E_{cor} in the negative direction to E'_{cor} and accelerates the cathodic reaction, that is, HER on the workpiece. The rate of metal dissolution decreases exponentially with the potential, and the rate of HER increases (logarithmic scales in Figure 14.22b to d!). When using an external voltage source, the protection is achieved only under consumption of electrical energy.

Alternatively, the potential shift can also be induced by connection to a sacrificial anode made of a less noble metal (e.g., Al, Mg, and Zn for iron workpieces). The sacrificial anode and the workpiece form a galvanic couple, in which the cathodic partial reaction proceeds at the workpiece (HER) and the metal dissolution occurs at the sacrificial anode. We can predict the corrosion potential E'_{cor} from Figure 14.22d as the intersection of the linear section of the Tafel plots for the metal dissolution of the sacrificial anode (gray) and the HER on the workpiece (blue). A possible disadvantage of cathodic corrosion protection is the HER on the workpiece to be protected. Depending on the kind of metal, the developed hydrogen can be incorporated into the metal lattice and cause hydrogen embrittlement. Sacrificial anodes are slowly dissolving and, thus, must be replaced at regular maintenance intervals. Typical applications of cathodic corrosion protection are in the underwater area of ships and in protection of pipelines. The size and sensible placement of sacrificial or counter electrodes around the workpiece are very important in order to achieve an approximately uniform current density distribution at the workpiece [29].

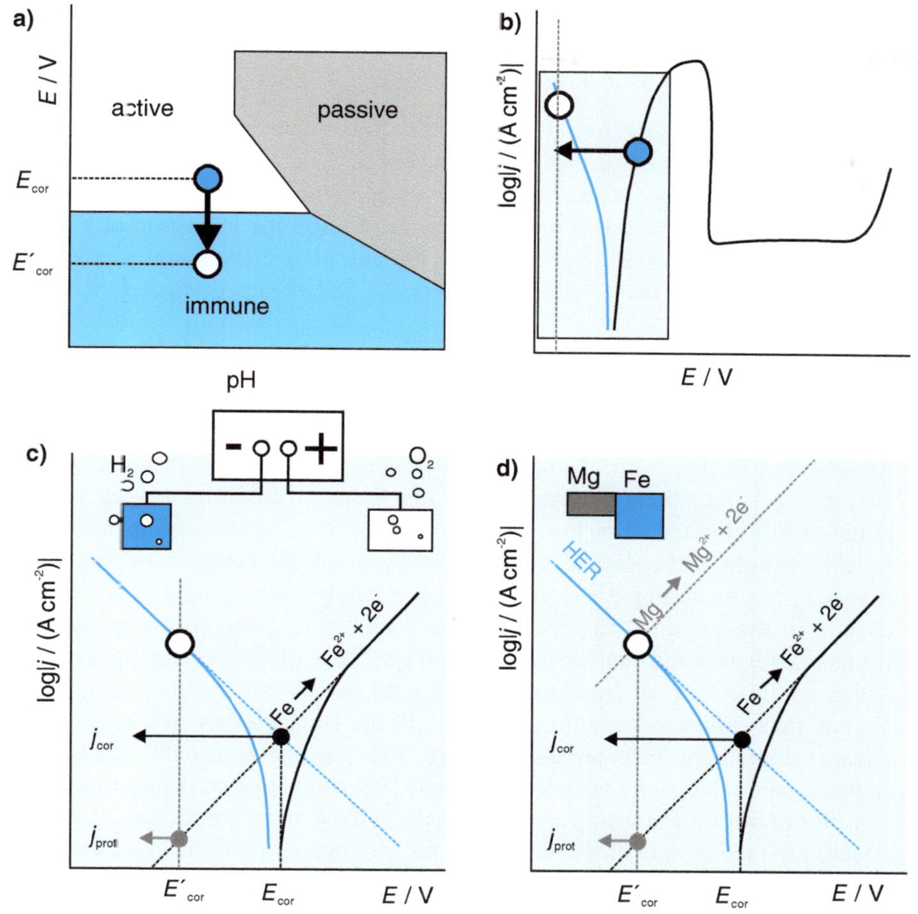

Figure 14.22 Cathodic protection.

a) Potential shift to the region of immunity in the simplified Pourbaix diagram of iron;
b) polarization curve of iron with potential shift induced by cathodic protection;
c) realization of the potential shift by an external voltage source causing H_2 evolution (HER) on the iron workpiece;
d) potential shift due to the formation of a galvanic couple with a less noble metal (Mg), which dissolves while H_2 is formed on the iron workpiece.

For scarificial electrodes and the use of a potential source, the current density for metal dissolution of the workpiece is reduced by several orders of magnitude (note the logarithmic scale of the ordinate in b) to d)).

Anodic protection is based on increasing E_{cor} of passivating metals so that E_{pas} and j_{pas} are exceeded and the workpiece is kept within the potential range of stable passivity. This can be achieved by imposing a potential from a potentiostat between the workpiece and counter electrodes, where the HER and/or the ORR proceed as a cathodic reaction. Although comparatively complex instrumentation is required, energy consumption is low due to the low current density in the passive state. Alternatively, the potential can also be caused by changes in the composition of the corrosion system. This can be achieved by the following measures:

- Especially, valve metals form extremely stable passive layers (e.g., Al, Ti, Ta, and Zr). Anodic surface treatment of these materials during production can give them permanent protection, depending on the application field.
- Oxidizing agents can be added to the corrosion medium, which then lead to the formation of passive layers. One example is chromate CrO_4^{2-} and dichromate $Cr_2O_7^{2-}$, but their replacement is pending because of their carcinogenicity

(i.e., cancer-causing property). The development of equivalent, ecologically, toxicologically, and economically acceptable alternatives is a current area of research.

The addition of small amounts of metals, which are catalytically active for the HER, can raise the current density in free corrosion above j_{pas}, thereby aiding passivation.

Use of alloys. Alloys can have a very strong influence on the corrosion rate of oxide-covered metals. First, the alloyed metal can ensure the formation of a dense oxide layer. If a dense oxide layer is already present, alloy constituents can further reduce the ionic conductivity in the oxide layer. We had already discussed this effect for zinc in Excursion 11-3.

Modification of the corrosion medium by means of additives is another useful approach, especially for closed systems such as cooling circuits. The effect of the additives can be based on different principles:

- **Removal of Problematic Components**. Acidic media can be alkalized by base additions, so that thicker passive layers are formed. Molecular oxygen can be removed from the media by purging with N_2. Reducing agents such as hydrazine N_2H_4 or sodium sulfite Na_2SO_3 chemically reduce molecular oxygen. In both cases, oxygen corrosion is suppressed. Ion exchangers can be used to reduce the total ion content in the medium. The resulting low conductivity between the anodic and cathodic regions of the metallic construction can further limit the corrosion rate.
- Other additives act as **inhibitors**. This term refers mostly to organic compounds that adsorb at potentials near E_{cor} to the surface of the workpiece (e.g., nitrogen-containing heterocycles for Cu). The adsorbates usually inhibit the cathodic partial reaction of corrosion because they compete with the intermediates of HER or ORR for the adsorption sites on the surface of the workpiece. The metal dissolution reaction can also be inhibited, for instance, by phosphate, which forms passivating films with Fe^{3+}.

Surface coatings represent the most widely applied approach to corrosion protection, often in conjunction with other measures. The immediate effect is to physically separate the workpiece from the corrosion medium (breaking the ion-conducting connection of the anode and the cathode in Figure 14.3). Coatings may be of a more noble metal that protects the underlying workpiece. Very common are coatings with less noble metals that form passive layers (e.g., Zn on Fe). Finally, there is the large field of organic polymers (in the form of varnishes) or hydrophobic organic molecules (as oil films for short-term protection). The question of durability is crucial to the selection of coatings. Damage mechanisms vary extremely depending on the type of coating, the material, and the corrosion medium but cannot be discussed in detail here. Another important issue is the consequences of minor damage during use (e.g., a small scratch) on the progress of the corrosion. If not chosen properly, localized corrosion can be triggered here (Section 14.5.3), leading to such a rapid onset of corrosion damage that it is not even noticed during the usual maintenance intervals and, thus, can then have catastrophic consequences (failure of load-bearing mechanical structures and piping). On the other hand, coatings can also act as sacrificial anodes and continue to protect the workpiece even in the case of minor damage (e.g., Zn on Fe).

14.5 Important and Typical Corrosion Phenomena

In the following, we present some important corrosion scenarios. Different perspectives exist here. For example, steel manufacturers would probably be interested in the corrosion properties of their product in different corrosion media. On the other hand, a designer who has to select materials or combinations of materials is more likely to ask about the corrosion properties of a wide range of materials in a particular corrosion medium (soils, seawater, moist air, cooling or steam cycles, chemical reactors, or concrete), taking into account the additional consideration of mechanical loads. An extensive literature exists on these practical issues. The character of this textbook forces a selection, with which we illustrate the variety of corrosion phenomena and their significance in very different application areas. A systematic and complete treatment of the subject is not possible here. This is already evident from the fact that there are 3000 commercial grades of steel today! In a particular application, different corrosion phenomena and materials can occur simultaneously and influence each other. Therefore, it is sometimes not possible to separate the individual corrosion scenarios in practical cases as neatly as suggested by the prototypical examples selected for this textbook.

14.5.1 Uniform Corrosion

Uniform corrosion or general corrosion is an "ideal case" of corrosion, in which microscopic regions of metal dissolution are uniformly mixed with regions for the cathodic reaction on the surface of the workpiece (Figure 14.4c). This form of corrosion is relatively predictable and can therefore be taken into account in design. We will discuss the dangerous forms of localized corrosion starting from Section 14.5.3. In fact, even uniform surface corrosion very rarely leads to completely uniform material removal because workpiece geometries, inhomogeneities in the workpiece, surface layer formations, or variations in the corrosion medium ($[O_2]$, T, flow, and so on, Figure 14.5) lead to different reaction rates.

Atmospheric corrosion represents a special case of uniform corrosion. At a relative humidity of more than 60%, a thin film of water forms on the surfaces, which is the ion-conducting corrosion medium. On unalloyed steels, the anodic reaction is iron dissolution.

$$Fe \rightleftharpoons Fe^{2+} + 2e^- \tag{14.33}$$

The Fe^{2-} ions react with water to form hydroxide and release protons.

$$Fe^{2+} + 2\,H_2O \rightleftharpoons Fe(OH)_2 + 2\,H^+ \tag{14.34}$$

In the presence of O_2, further oxidation of the iron-II-hydroxide to iron-III-oxyhydroxides occurs.

$$4\,Fe(OH)_2 + O_2 \rightleftharpoons 4\,FeO(OH) + 2\,H_2O \tag{14.35}$$

Attention, all reaction equations in this chapter are strongly simplified!

The iron oxides and oxide hydrates do not have a uniform composition and can best be described by the formula $FeO(OH) \cdot n\ H_2O$. They form a reddish-brown colored loose layer known as the typical "rust". Other hydrated cations also react acidically, for example, Cr^{3+}.

$$[Cr(H_2O)_6]^{3+} + H_2O \rightleftharpoons [Cr(H_2O)_5(OH)]^{2+} + H_3O^+ \tag{14.36}$$

The cathodic reaction is the ORR.

$$O_2 + 4e^- + 2\,H_2O \rightleftharpoons 4\,OH^- \tag{14.37}$$

So we recognize:

> Anodic reactions lead to acidification of the corrosion medium, and cathodic reactions lead to alkalization of the corrosion medium.

In uniform corrosion, the released H^+ and OH^- neutralize immediately because the anodic and cathodic microregions are immediately adjacent and evenly distributed

Table 14.2 Atmospheric corrosion of various metals.

Iron and steel
Formation of red-brown rust $FeO(OH) \cdot n\,H_2O$;
Removal of 10–100 μm a^{-1}, often with exponential decrease due to increasing thickness of the oxide layer;
Cu, Cr, and P slow down corrosion by forming denser top layers.
Zinc
Formation of basic zinc carbonates and chlorides;
Uniform removal of 2–3 μm a^{-1}, with air contamination up to 15 μm a^{-1};
With aggressive media, formation of white surface layers (white rust);
Chromating delays initial corrosion and white rust formation.
Aluminum and aluminum materials
Formation of very dense and durable passive layers of Al_2O_3;
Removal rate initially of 2–5 μm a^{-1}, then decreasing and transitioning to localized forms of corrosion;
Some Al alloys (Al–Zn–Mg and Al–Zn–Mg–Cu) may show film corrosion.
Copper
Formation of a compact passive layer of basic carbonates and/or sulfates (depending on air impurities);
Removal rate of 0.2–2 μm a^{-1}.
Tin and tin alloys (e.g., solder alloys)
Highly resistant to atmospheric corrosion if all corrosion-promoting processing chemicals such as soldering fluxes have been removed.

Source: Data from [13].

over the surface (Figure 14.4c). Atmospheric corrosion depends strongly on environmental conditions. Particularly severe exposure occurs when thin condensate films form and disappear several times. At the moment, when a very thin condensate film is present, very effective entry of O_2 can occur. The solubility of O_2 from the atmosphere at normal pressure is only 0.2 mmol L^{-1} in aqueous solutions. Therefore, thick water films limit the transport of O_2 to the workpiece surface. However, over very thin water films, transport can be very rapid due to the large concentration gradient (gradient $\approx \Delta c/\Delta x$, Section 4.4), causing unexpectedly high corrosion rates. Significant acceleration of atmospheric corrosion can occur from air pollutants, especially sulfur dioxide (SO_2) emitted in classic industrial areas. Dust particles can also accelerate corrosion. For example, adhering soot particles contribute to the acceleration of corrosion as permanent microscopic cathodes. Any deposits on the workpiece can affect the porosity and density of the formed rust layer, which decisively influences the further progress of corrosion. An overview of atmospheric corrosion of various metals is given in Table 14.2.

Excursion 14-3 Restoration of Art Objects with an Electrolytic Pen

The corrosion of metallic objects also poses a challenge for the conservation of artistic and archaeological objects. Although electrolytic methods had already been used for some time to reduce oxidic coatings, instrumental measurement procedures only found their way into the repertoire of conservation departments of large museums since around 1990 [35]. The possibility of precisely controlling the potential with potentiostats permits very gentle cleaning of art objects and makes such techniques valuable for preserving cultural heritage. An instructive example is provided by the cleaning of tarnished silver objects. The dark deposits known as silver tarnish consist of Ag_2S and AgCl, which form on the surfaces from Ag^+ and gaseous sulfur compounds and chloride (e.g., from sweat). Traditional methods of cleaning involve mechanical removal (rubbing) with simultaneous chemical dissolution with thiourea. Such procedures remove materials from the art object and thus lead to an irreversible alteration of the objects to be preserved. Electrochemical reduction with a sacrificial Al anode (wrapping the art object in Al foil and immersion in $Na_2CO_3/NaHCO_3$ buffer) can result in uncontrolled reactions. For example, if the potential is too negative, hydrogen evolution can occur in addition to the desired reduction of Ag_2S, leading to the deposition of porous and finely dispersed silver, which does not meet modern conservatory requirements. Scientists from the Rijksmuseum Amsterdam (The Netherlands) and the Haute École de Conservation-Restauration Arc in Neuchâtel (Switzerland) developed "Pleco", an electrolytic pen. The entire project is documented in an exemplary way in Open Source to enable museums worldwide to use the procedure with a minimum financial investment. The name of the device is derived from a catfish species that cleans algae from aquarium glass. In principle, the pen is a droplet cell sealed with a soft, porous, and moistened diaphragm (Figure 14.23). The electrolyte is circulated by a pump and removes the substances released from corrosion layers. This eliminates the risk of alloying elements dissolving in one place and redepositing in another, which is a problem when macroscopic art objects are immersed [35]. The electrolytic pens, needed in small batches, can be effectively manufactured with 3D printers and are made available to restoration departments worldwide. Further information can be found in [36–38].

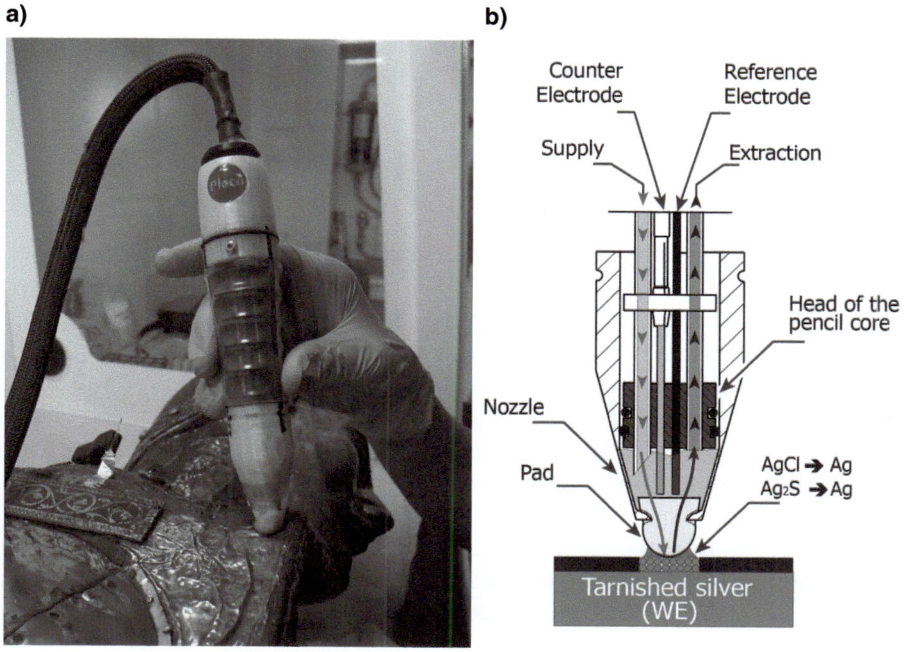

Figure 14.23 Electrolytic pen "Pleco".

a) Photographic image. Source: Photo by Abbaye de Saint-Maurice, St Maurice, CH;
b) Schematic sectional view with the flow pattern of the electrolyte. The figure is kindly provided by Haute École de Conservation-Restauration Arc, Neuchâtel, CH.

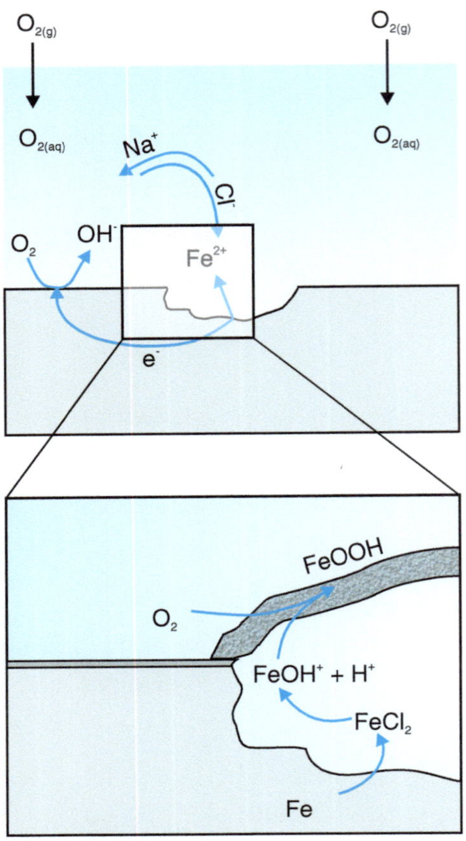

Figure 14.24 Cross-sectional view of a shallow corrosion pit.

Attention, all reaction equations in this chapter are significantly simplified!

Uniform Corrosion in Water. The uniform corrosion of iron and steels in water is limited by the cathodic partial reaction. In acidic solutions, this is the HER, and in neutral and alkaline media, this is the ORR (see Figure 14.10 and related discussion in Section 14.2). The rate of HER is very strongly influenced by alloying elements and metal physical parameters, because HER is particularly dependent on the electrode material (volcano plot, Figure 3.9). If, on the other hand, ORR dominates in neutral and alkaline solutions as a cathodic and simultaneously limiting partial reaction, the transport of oxygen to the workpiece controls the corrosion rate. This transport is mainly influenced by the influx of O_2 and the local flow velocity. In closed water circuits, the corrosion rate can therefore be very low after consumption of the dissolved O_2 originally present.

14.5.2 Corrosion in Salt Solutions – An Example of the Transition to Local Forms of Corrosion

In water (e.g., in a condensate film, Section 14.5.1), there is a low ion concentration and consequently a low electrical conductivity (Section 4.2). For a corrosion reaction to start under these conditions, the anodic and cathodic microregions must be located close together. In salt solutions with higher ionic conductivity, this restriction is lifted, and thus, the anodic and cathodic regions can be spatially separated permanently. In most cases, further requirements must be fulfilled such as crevices, surface layers, or differentially ventilated areas for local corrosion to actually occur, usually in the form of shallow pit formation. We speak of shallow pit formation when hollowed-out areas in the workpiece have a larger diameter than depth.

If iron or steel corrodes according to Eq. (14.33), anions (e.g., Cl^-) migrate to the anodic site to balance the charge. There, they form $FeCl_2$. Chloride salts often exist in a gel-like consistency and hydrolyze with the release of H^+:

$$FeCl_2 + H_2O \rightleftharpoons Fe(OH)^+ + 2Cl^- + H^+ \tag{14.38}$$

and $FeO(OH) \cdot n\,H_2O$ is formed, which is deposited in the form of poorly soluble rust pustules on the anodic sites.

$$4Fe(OH)^+ + O_2 + 2H_2O \rightleftharpoons 4FeO(OH) + 4H^+ \tag{14.39}$$

Within these rust pustules, a much more acidic solution is present than on the rest of the workpiece [Eq. (14.34)]. The more acidic medium within the rust pustules prevents the formation of a passive layer on the metal and may even cause the transition to much faster acid corrosion. The higher conductivity of the solution now allows the cathodic regions to be located relatively far away on the workpiece, so that the alkalization occurring there according to Eq. (14.37) does not immediately lead to neutralization at the anodic regions. Rather, wide areas of the workpiece may be covered with a thin passive layer by alkalization. The passive layer slows down metal dissolution there. Consequently, the anodic regions, once encapsulated, may be permanently preserved (Figure 14.24). In these cases, we speak of **element formation**. Since the cathodic reaction is usually limiting, a very unequal area ratio between anodic and cathodic regions can result. The local current densities are inversely proportional to the respective areas. The situation can be compared to the action of forces on a mechanical lever (Figure 14.25). Shallow pit formation in heating and cooling circuits can be mitigated by treating the aqueous corrosion medium in the following way:

- Reduction of conductivity by means of desalination.
- Prevention of poorly soluble deposits by water softening.
- Blocking of adsorption sites at the surface of the workpiece by inhibitors. The blocked sites are then not available for HER and ORR.

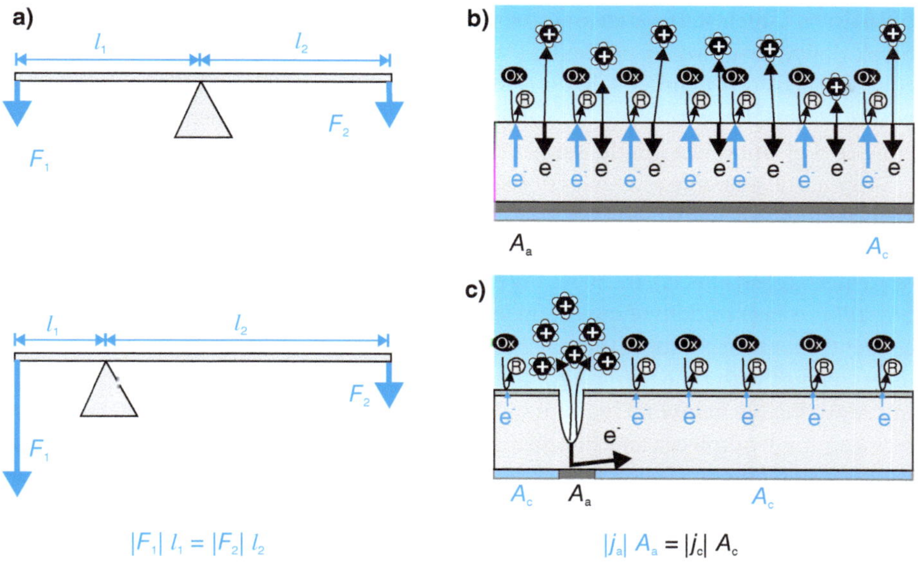

Figure 14.25 Lever system in corrosion reactions.

a) Mechanical lever;
b) "Lever system" for uniform corrosion;
c) "Lever system" for local corrosion.

The cations of the corroding workpiece marked with ⊕ for space reasons will mostly be higher charged cations (e.g., Fe^{2+}). In b) and c), the anodic regions are indicated as a gray stripe and the cathodic regions as a blue stripe below the respective sketch. These regions overlap in b).

14.5.3 Pitting Corrosion

Pitting corrosion occurs when the depth of the hollowed areas of the workpiece exceeds its diameter, or the hollowed areas severely undercut the workpiece surface (Figure 14.26b). While shallow pit formation is mainly connected to active metal dissolution, pitting corrosion is closely associated with the occurrence of passive layers, for example, on high-alloy steels as well as on Al, Cu, Ni, Ti, Ta, and alloys of these metals. In general, passive layers greatly reduce the corrosion rate. However,

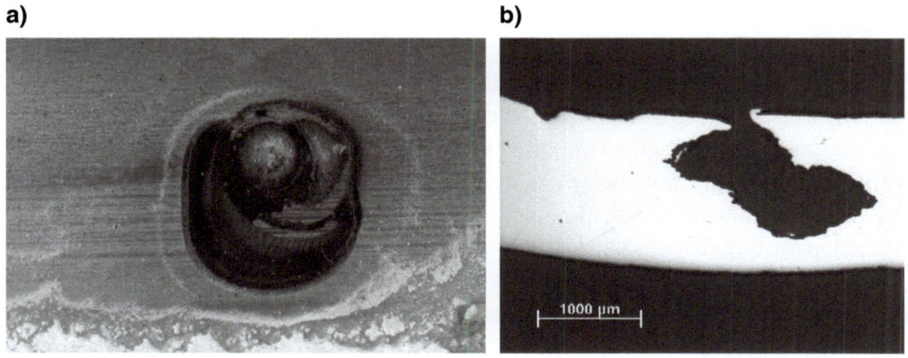

Figure 14.26 Examples of pitting corrosion on the inside of a steel pipeline (1.4401). a) The top view shows a rust-brown pustule, from which the extent of the damage is not directly visible; b) cross-sectional view of the pipe (light) through a pitting pit (dark). The undercutting of the inner surface and the near breakthrough to the outer pipe surface are clearly visible. To the left of the pitting pit approximately above the 1000 µm scale bar, a shallow pit can be observed that quickly repassivated. Photographs by Johann Wilhelm Erning, Federal Institute for Materials Research and Testing (BAM) Berlin.

passivity is a metastable state and therefore, under certain conditions, passivity can break down, resulting in particularly rapid *local* attack, which can have dangerous consequences (cf. Figure 14.26a and b!). Therefore, experimental investigations on pitting corrosion are not only crucial for selecting materials but also for checking the condition of workpieces in the service phase.

Despite considerable research efforts, the mechanisms of pitting corrosion have not yet been fully elucidated [30–32]. This is due to the *diversity of the materials* under consideration in combination with the manifold *additional stresses* occurring in their use (e.g., mechanical effects), the *variety of corrosion media*, and the *deviations of the locally prevailing concentrations* of important ions in the corrosion medium from their volume concentrations. In addition, the stoichiometry and stratification of the complex passive layers change with time (aging). All these make it difficult to perform the experiments in a reproducible manner. Finally, it is difficult to study small sections of the interfaces between the workpieces and the corrosion medium, while the workpiece remains in the corrosion medium (*operando* studies). Since about 1990, considerable progress has been achieved in the local *in situ* observation of electrodes using complex investigation techniques, which are summarized in more detailed reviews [12]. Finally, pitting corrosion represents a stochastic process, in which large differences in the behavior of individual pits occur.

Preconditions. Pitting corrosion also requires the presence of "aggressive" ions acting on a passive film at a potential within the potential range of pitting corrosion. Ions are aggressive with respect to corrosion if they react with the dense oxide films to form colloidal or gel-like products that have poorer barrier properties than the original oxide film. Aggressive ions include primarily the halides (F^-, Cl^-, Br^-, and I^-) and also pseudohalides (especially SCN^-). HSO_4^- and ClO_4^- can initiate pitting corrosion under certain conditions. Among the aggressive ions, Cl^- is the most important one since it occurs in almost all technically relevant corrosion media (seawater, road salt in winter for lowering the melting point, and body fluids). Therefore, we want to limit the discussion to this ion.

Phases of Pitting Corrosion. Pitting corrosion is usually divided into different phases:

- Local passive film breakdown (pit nucleation);
- onset of temporary pitting corrosion (pit initiation);
- stable progressive pitting corrosion (stable pitting);
- repassivation.

One can observe the different stages of pitting corrosion by a closer look at current–voltage curves (or polarization curves). An example is shown in Figure 14.27a; we have already discussed a simplified version of such curves in Figure 14.19, where we marked the pitting potential E_P in the passive region. Already before E_P, the current increases temporarily. What may look like noise is in fact the result of metastable pitting corrosion (Figure 14.27b-4).

If one plots the polarization curves with the positive-going and the negative-going half-cycles, one sees that the onset of stable pitting corrosion requires a potential $E_{P,n}$ (index n of nucleation, Figure 14.27b, 5). This potential is usually somewhat more positive than the potential $E_{P,r}$, where repassivation of the corrosion pit occurs in the negative-going scan (Figure 14.27b, 6). The difference between $E_{P,n}$ and $E_{P,r}$ is due to hysteresis effects, which are typical for electrochemical experiments in general. As an example, we have already discussed hysteresis due to the superposition of interfacial reaction and transport processes using the example of cyclic voltammograms of dissolved compounds in Figure 5.27 in a completely different area of electrochemistry.

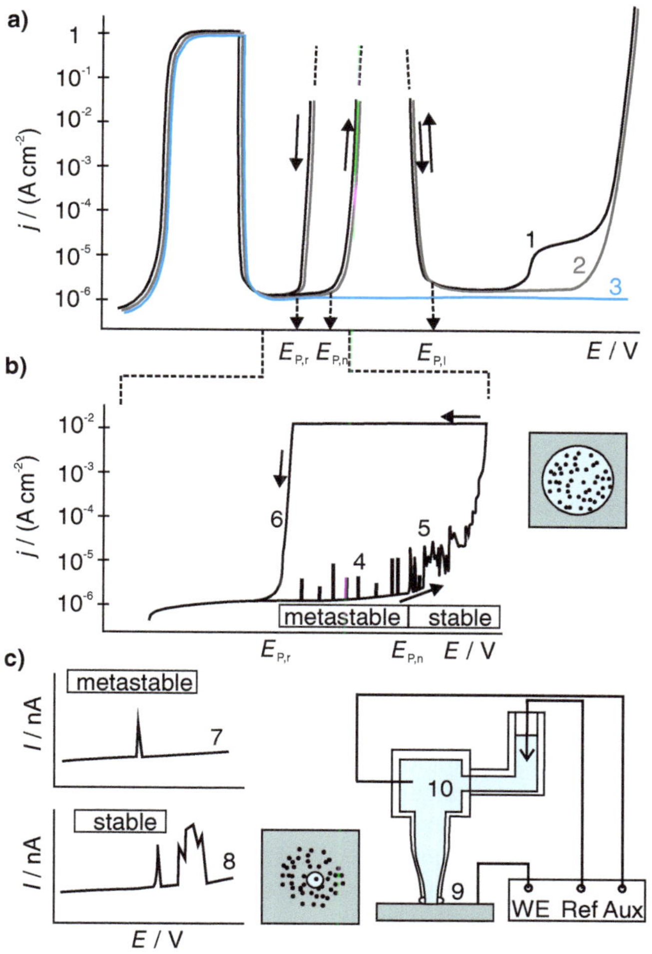

Figure 14.27 Schematic representation of current–potential curves with pitting corrosion.

a) For 1) Cr and Ni, 2) Fe, and 3) valve metals (Al, Ta, Zr, and Hf) without oxygen evolution in the transpassive range;
b) Schematic cyclic voltammogram of stainless steel in NaCl solution in the region of metastable pitting corrosion (4), stable pitting corrosion (5), and repassivation (6);
c) Schematic current–voltage curves of microcells. If the cross-sectional area of the capillary is very small, it may be possible to track the activation of individual metastable pits (7) or stable pits (8) as a function of potential (or time). The microcell consists of a capillary with a seal at the capillary end (9) connected to an electrolyte-filled holder (10) that accommodates the reference and auxiliary electrodes.
d) Depending on the design, the capillary openings have a diameter between 100 nm and 1000 μm.

In pitting corrosion, the following factors contribute to hysteresis:

- The formation of a stable pit requires time, during which the potential scan further advances, before the pitting corrosion becomes detectable by a current increase in the measured curves.
- During pitting corrosion, corrosion products and aggressive ions accumulate inside the pit. E_P typically decreases with increasing concentration of such aggressive ions, and thus, repassivation is shifted in a negative direction by a locally higher concentration of aggressive ions [31].

At an even larger potential $E_{P,I}$, pitting corrosion is inhibited (Figure 14.27a). $E_{P,I}$ can be lowered by inhibiting ions (e.g., NO_3^- and ClO_4^-) [31].

Using micro- and nanoelectrochemical capillary cells (Figure 14.27c), currents have been measured for individual corrosion pits. The currents are in the femtoampere (fA) range for nanometer-sized pits and in the picoampere (pA) range for micrometer-sized pits. These low, transient currents are less of a problem for practical applications because such corrosion pits repassivate very rapidly. However, for mechanistic investigations, they form an important starting point. For example, it was possible to identify MnS inclusions in steel as a possible trigger for the formation of corrosion pits [33]. With a positive going potential scan, the current flowing during the pitting events increases and persists for a longer time until finally stable pitting corrosion commences at $E_{P,n}$. *Individual* corrosion pits may also repassivate in this potential region (Figure 14.27c-8), but the number of active corrosion pits at any time on a

1 fA = 10^{-15} A

1 pA = 10^{-12} A

Depending on the quality of the setup, it may not be possible to distinguish such small currents on macroscopic samples from the general noise of the measurement curve, and thus, they may not be detected.

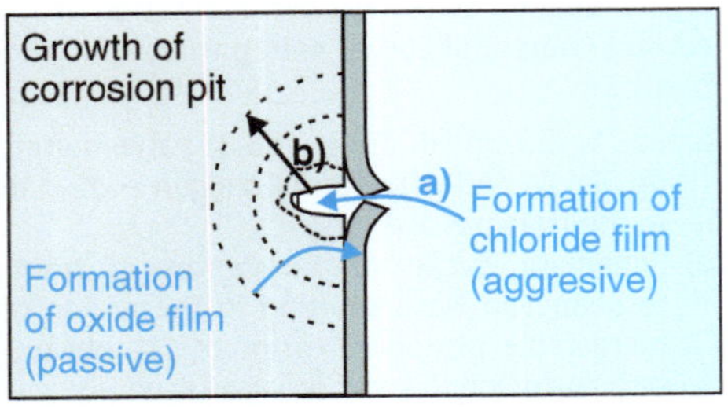

Figure 14.28 Film-breaking mechanism. a) Access of aggressive ions at a defect in the passive layer with b) subsequent pitting corrosion.

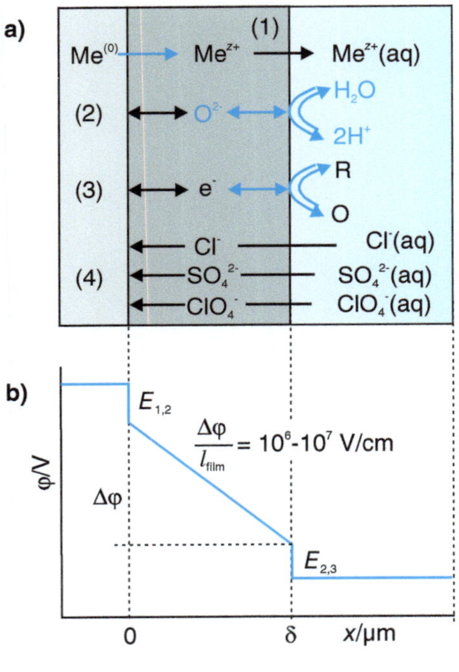

Figure 14.29 Penetration mechanism.

a) Schematic cross section through the oxide film showing the transport processes (black) and reactions (blue): 1) corrosion, 2) film formation, 3) redox reactions, and 4) penetration of aggressive ions;
b) Representation of local potential and electric field in a passive film.

macroscopic sample is so large that a significantly increased total current results (Figure 14.27b). Observation of individual corrosion pits with microelectrochemical capillary cells allows the statistical treatment of individual pitting events and their association with local structural features on the surface of the specimen [33, 34].

Passivity Breakdown. The initiation of pitting corrosion is preceded by a local breakdown of the passive layer. There are three models for this, which can also occur side-by-side in corrosion scenarios [30]. In the film-breaking mechanism, defects in the passive layers allow aggressive ions to reach the unprotected metal surface. From there, metal dissolution progresses under the protective oxide layer (Figure 14.28). In this process, the formation of a new passive oxide film competes with the formation of a chloride film, which promotes the growth of the corrosion pit.

In the penetration mechanism, Cl^- ions reach the boundary between the metal and the oxide film under the influence of a very strong electric field in the thin oxide film (Figure 14.29). However, Cl^- was detected by surface-sensitive spectroscopies only in the outer layers of thicker passive films when those films had been formed in chloride-free media and then transferred to the chloride-containing corrosion medium. Finally, the adsorption mechanism describes a local decrease in film thickness due to the attack of adsorbed aggressive ions (Figure 14.30). Adsorbed anions may cause an increase in cation mobility in the oxide film and/or complexation of cations by the electrolyte anions. The decreasing thickness of the oxide film locally increases the electric field strength and leads to increased dissolution, but also to further migration of aggressive anions to the site of high current density, that is, the areas with a locally thinned oxide layer. Once the passive layer is completely dissolved, active metal dissolution occurs in the corrosion pit. Metal dissolution in the active and passive states can be distinguished experimentally, for example, using a rotating ring–disk electrode (RRDE) to determine the valence state of the released Fe ions. In active metal dissolution, Fe^{2+} ions are released, while Fe^{3+} ions dissolve out of the passive film in metal dissolution via a passive layer.

Growth of the Corrosion Pit. Once the passive layer has been degraded, the further fate of a corrosion pit depends on many factors. Using potentiostatic and galvanostatic pulse experiments in conjunction with imaging techniques, it has been shown that only a fraction of less than 1% of many nanometer-sized nuclei for a corrosion pit mature into a stably growing pit. The vast majority of pit nuclei quickly passivate again. The presence of aggressive ions is critical to the growth of corrosion pits. By forming a thin salt film, aggressive ions prevent repassivation by a new oxide film. Local current densities of more than 100 A cm^{-2} have been determined in growing corrosion pits [31]. Such high current densities should trigger immediate repassivation unless the presence of aggressive ions prevents the formation of a passive film. Their effect is to significantly increase the passivation current density. Therefore, the workpiece can continuously corrode at a uniform potential. Metal dissolution proceeds at high local anodic current densities in the corrosion pit, while most of the workpiece exhibits only a very low passive current density at the same potential.

Repassivation. The time required for repassivation of a corrosion pit heavily depends on its depth. This determines how long it takes for the aggressive anions to diffuse out of the pit after a solution exchange or a potential change.

Avoidance of Pitting Corrosion. The most important measures begin with the avoidance of high Cl^- concentrations. Flow can be used to prevent deposition of non-passivating layers and promote repassivation of the corrosion pits. Finally, materials can be selected so that for a known corrosion potential and a given temperature of the corrosion medium, the passivation potential is exceeded, but the pitting potential is not exceeded.

Excursion 14-4 Electrochemical Noise Analysis (ENA)

Potentiostatic and galvanostatic polarization methods impose a potential or a current on the specimen from the outside that deviates from the free corrosion potential E_{cor} or from the free corrosion current I_{cor}. This is a suitable method in the laboratory to record the current–voltage curves. However, it is less suitable for monitoring or field testing (containers and pipelines in service and steel structures). This is where the electrochemical noise analysis (ENA) has found its place in recent years. Different electrode arrangements are possible, but for a field test, the arrangement in Figure 14.31 is particularly advantageous. Two apparently identical electrodes are connected as working electrodes via a zero-resistance ammeter (ZRA). In addition, the voltages are measured against a noise-free reference electrode. If a metastable corrosion pit with a transient high anodic current occurs on one of the workpieces, a current also flows through the ZRA, and the free corrosion potential changes for a short time. $E_{cor}(t)$ and $I(t)$ are recorded, the standard deviations $\sigma(E_{cor})$, $\sigma(I)$ are calculated from them, and a noise resistance R_n is defined.

$$R_n = \frac{\sigma(E_{cor})}{\sigma(I)} \tag{14.40}$$

A lower R_n indicates that a small fluctuation of E_{cor} causes a large fluctuation of I. Additional kinetic information is accessible if the functions $E_{cor}(t)$ and $I(t)$ are acquired with high time resolution and then subjected to a fast Fourier transform (FFT) to obtain $E_{cor}(f)$ and $I(f)$ from which the spectral noise resistance $R_{sn}(f)$ is determined.

$$R_{sn}(f) = \frac{E_{cor}(f)}{I(f)} \tag{14.41}$$

The limit value of R_{sn} for $f \to 0$ allows a conclusion about the corrosion resistance of the system. However, caution must be exercised in the quantitative analysis of the noise measurements because there can be other sources of noise besides a corrosion process (electronic noise and thermal noise). Formation of metastable corrosion pits is only one of several causes for fluctuating corrosion resistance. Other causes include concentration fluctuations of reacting particles, fluctuations in surface morphology, variations in the electrochemically active surface (e.g., bubble formation and detachment), and variations in flow velocity. A comprehensive introduction to ENA can be found in [39].

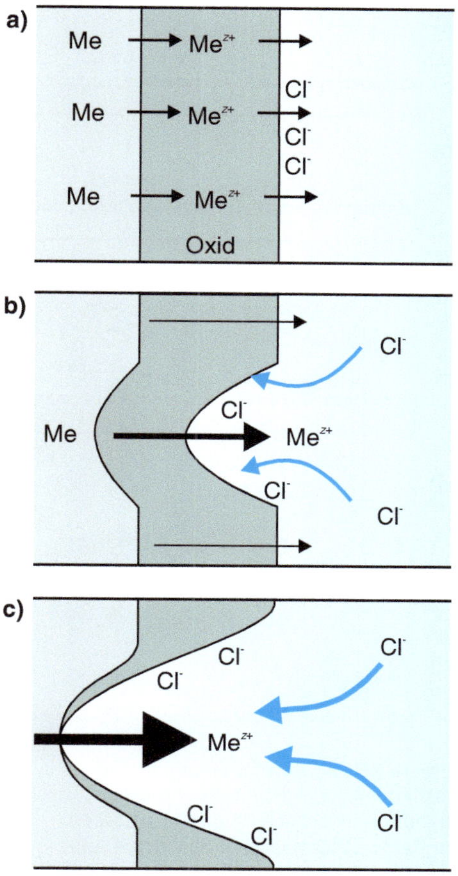

Figure 14.30 Adsorption mechanism. a) Locally increased concentrations of aggressive ions decrease the thickness of the oxide film. b) This increases the local electric field strength across the film compared to the surrounding regions. The resulting local increase in passive current density attracts further aggressive ions by migration to the thinned site to balance the charge. c) When the oxide film is completely degraded, a corrosion pit is formed.

Crevice Corrosion is a form of locally enhanced corrosion in crevices caused by concentration elements. The corrosion reaction changes the local solution composition because the concentration balance with the rest of the corrosion medium is delayed in the crevice. Thus, the oxygen influx may be slowed down compared to the rest of the workpiece. As a result, concentration elements build up, whereby the anodic partial reaction occurs preferentially at the walls of the crevice. Hydrolysis of the hydrated metal cations acidifies the medium in the crevice [Eqs. (14.34) and (14.36)], which further accelerates local corrosion. The corrosive attack finally leads to shallow pit formation or pitting in the crevice. This form of corrosion occurs at crevices that are present on the workpiece due to its construction (joints). Even a joint between a metal and an insulator can trigger crevice corrosion. Approaches to avoid crevice corrosion are proper selection of material and permanent closure of construction-related crevices by polymers.

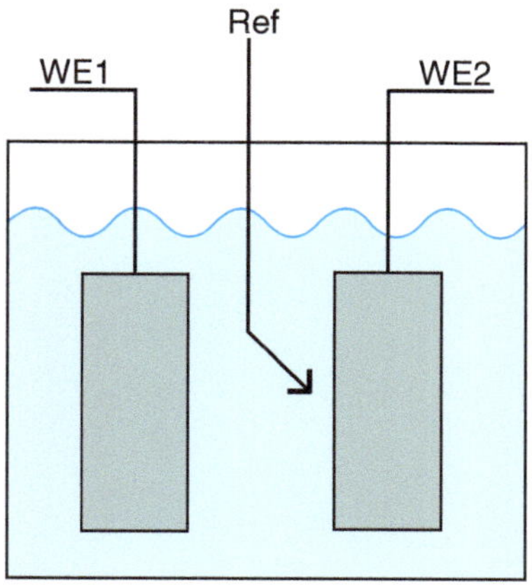

Figure 14.31 Electrode arrangement for electrochemical noise analysis. Two identical working electrodes are connected to a noise-free reference electrode.

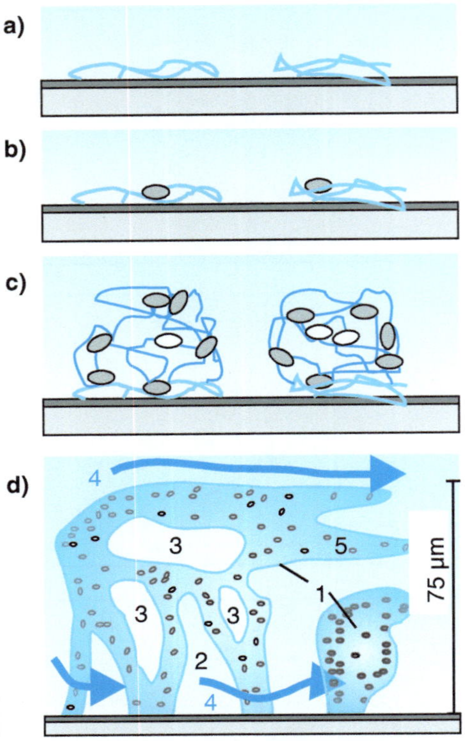

Figure 14.32 Formation of a biofilm (schematic). a) Adsorption of proteins; b) adhesion of bacteria on the adsorbed proteins; c) formation of an extracellular matrix and division of the bacteria; d) cross-sectional view through a fully grown biofilm. Typically, mushroom-shaped cell aggregates (1) shield an inner solution volume (2) from the medium. Here, significantly different local concentrations prevail than in the solution or in the outer region of the biofilm. The cell aggregates may contain solution-filled cavities (3). A hydrodynamic flow (4) leads to the formation of streamers (5, i.e., elongated extensions of cell aggregates) from which cells detach and contribute to the proliferation of the biofilm, in addition to other mechanisms.

14.5.4 Galvanic Corrosion

Galvanic elements can form when dissimilar metals are connected by an electrically and an ionically conductive path. Figure 14.17 shows a particularly striking example. In practice, galvanic corrosion leads to accelerated dissolution of the less noble metal if the potential difference between the metals is more than about 50 mV. Therefore, only materials with similar electrode potentials should be selected. The decisive factor is the open-circuit potential in the corrosion medium (which is a mixed potential by nature) but *not* the standard electrode potential of the metal electrode. An initial orientation for material selection is provided by tabulated values for important structural materials that exhibit similar open-circuit potentials in relevant corrosion media. As shown in Excursion 14-2, the corrosion current densities at the free corrosion potential must also be taken into account, since they determine the dissolution rate.

14.5.5 Microbiologically Induced Corrosion

Microbiologically induced corrosion (MIC) occurs because microorganisms living in the corrosion media colonize solid surfaces as their preferred habitat. Some bacterial species can also use metal dissolution as an energy-producing process for their cellular respiration. MIC usually requires the formation of a biofilm. The development of a biofilm [40] commences with the adsorption of biomacromolecules followed by the adhesion of cells. After cell division, the cells form an extracellular polymeric matrix that connects the cells to each other and to the surface as a porous gel-like film (Figure 14.32). With the incorporation of corrosion products or poorly soluble salts, biofilms become porous, brittle solids. The interstices of the film serve to supply nutrients to the film-forming cells. In those cavities, conditions similar to those of crevice corrosion prevail. The corrosive attack is intensified by the release of water-soluble organic metabolites, which act as complexing agents to accelerate the corrosion. Within a biofilm, microorganisms exhibit significantly increased resistance to mechanical or chemical attack and are then difficult to remove. Even a dead biofilm, analogous to a crevice, promotes the formation of differential concentration elements. Therefore, corrosion protection measures start with the prevention of initial biofilm formation. The choice of materials offers only a few starting points on the materials side. Smoothing of surfaces can be advantageous. However, it is more important to eliminate substances in the corrosion medium that are essential for the growing bacteria including bioavailable carbon sources as well as NO_3^- and SO_4^{2-}. Biocidal coatings can be useful, but their broader ecological impact must be evaluated before deployment. Enzymes can be used to detach or dissolve biofilms.

14.5.6 Stress Corrosion Cracking and Corrosion Fatigue

Stress corrosion cracking and corrosion fatigue occur on metals in the passive state and are examples of the concomitant impact of mechanical and chemical stress on a workpiece. In stress corrosion cracking, the workpiece is stretched by the mechanical stress, so that the passive layer ruptures locally and is not immediately repassivated in the corrosion medium. This can lead locally to very rapid material removal, which in turn further reduces the resistance of the workpiece to mechanical deformation. Cracks in the material can then propagate along grain boundaries (intergranular corrosion) or traverse across a grain (intragranular corrosion). Corrosion fatigue occurs with repetitive mechanical loads. Concomitant hydrogen embrittlement, which changes the mechanical properties of the affected materials, may enhance the effects of stress corrosion cracking and corrosion fatigue. From a practical point of view, this form of corrosion of load-bearing components is particularly dangerous because it usually sets in after long service lives and without any previously detectable material removal.

There is a whole range of other forms of attack in which mechanical and chemical attacks are amplified (drop impact, erosion in flowing liquids and suspensions, and vibrations).

14.5.7 Corrosion of Alloys and Dealloying

Most construction materials are alloys. To date, a complete understanding of the corrosion behavior of alloys has not been developed [41]. On the one hand, this is due to the very large variety of solid-state structures (with their defects!) and compositions, for which a systematic consideration would alone fill a separate textbook. On the other hand, there is a lack of *in situ* investigation methods to observe the processes for these complex materials under corrosion conditions. In other cases, such techniques are available in principle, but their application is not only very expensive but also requires specially prepared model samples (such as single crystals).

A basic distinction can be made between single-phase and multiphase alloys. In single-phase alloys, all crystallites of the metal structure have the same atomic composition. The passive layers differ from grain to grain only in the orientation of the crystal axes to the surface. However, many high-performance alloys are multiphase alloys. The melt of the alloy-forming elements forms crystallites of different elemental compositions (phases) during solidification. The kinetics and thermodynamics of corrosion processes may differ for the different phases of such alloys, resulting in the formation of galvanic elements. However, the corrosion phenomena also depend very much on the oxide layers on the phases, which can have very different properties due to different elemental compositions of the respective phases. Inclusions of carbides, borides, sulfides, and carbon further influence or change the corrosion behavior. We must limit the considerations to a selective discussion of single-phase systems but will come back to a particularly instructive study of multiphase materials in Excursion 14-5.

The individual crystallites of a single-phase alloy are separated from each other by grain boundaries. The structure of the metal lattice is disturbed at a grain boundary to the bulk phase (cf. Figure 11.5), and trace elements can accumulate at the grain boundary during solidification. It would therefore be expected that corrosion mainly attacks these areas, where the thermodynamically most stable metal structure is disturbed. In general, however, this process is not observed, probably because the disturbed zone is small and the metal structure at the surface is always disturbed to some extent anyway as a result of coatings and machining. If the grain boundaries are widened due to metallurgical or mechanical effects, intergranular corrosion can occur, with the removal of material close to the grain boundary. The scenario exhibits transitions to pitting corrosion and stress corrosion cracking. By dissolution of the region near the grain boundary, the metal grains lose their mechanical cohesion [13].

In a model consideration, we can assume that the metals A and B of a single-phase alloy A_xB_y are in equilibrium with the solution at a potential E_{eq} [Eqs. (14.42) and (14.43)] provided that the Nernst Equation (14.44) is satisfied for both metals. In contrast to the considerations in Section 2.2.3, we need to use the activity term $a_x = f_x x$ of the fixed mole fraction activity of the metal.

$$A_xB_y \rightleftharpoons A^{n+} + ne^- \tag{14.42}$$

$$A_xB_y \rightleftharpoons B^{m+} + me^- \tag{14.43}$$

$$E_{eq} = E^\circ_{A|A^{n+}} + \frac{RT}{nF}\ln\frac{\{A^{n+}\}}{f_{x,A^{(0)}}x(A^{(0)})} = E^\circ_{B|B^{m+}} + \frac{RT}{mF}\ln\frac{\{B^{m+}\}}{f_{x,B^{(0)}}x(B^{(0)})} \tag{14.44}$$

To achieve equilibration of an alloy electrode with a solution, both the contents of the dissolved metal ions (expressed by $\{A^{n+}\}$ and $\{B^{m+}\}$) and the mole fractions of the metal atoms in the alloy must change. However, the latter is only possible if mass transport of the metal atoms by diffusion is possible in the alloy. Liquid amalgams with A as the less noble metal component are among the few systems that allow equilibration of the concentrations in the metal phase. Moreover, if the difference in standard redox potentials between A and B is very large, the less noble component A alone determines the potential of the alloy electrode. We have encountered such an example in the discussion of the stripping voltammetry of Zn^{2+}, Cd^{2+}, and Pb^{2+} at Hg electrodes (Section 5.3.10) – but it has no meaning in the context of material damage.

In the much more common cases of very slow diffusion of metal atoms in the metal lattice, the surface of the alloy A_xB_y is depleted in A and enriched in B. Equilibrium then exists between the solution and the near-surface layer, which in extreme cases may be a *thin surface layer of the nearly pure more noble element B* on the alloy A_xB_y. In fact, the relationships are far more complicated, since the free enthalpy of mixing of the metals in the alloy must also be taken into account [42]. In addition, alloys can occur in different modifications (crystal structures), with the element contents stabilizing one or the other crystal structure. If the elemental composition changes in the corrosion process, this can lead to destabilization of one modification. This is often accompanied by changes in volume and mechanical properties, so that shrinkage with cracking can cause the corrosion process to propagate deep into the material. An example is the corrosion of ε-brass (Zn–Cu), where Zn dissolves and porous γ-brass is formed depending on the overpotential. Shrinkage causes cracks, through which corrosion progresses into the interior of the workpiece.

If dissolution of both metals occurs because the corrosion potential is positive from the standard potential of the nobler component, the metals usually dissolve at different rates, which also causes a change in elemental composition near the surface. The slower-dissolving metal accumulates at the surface until it dominates the corrosion process. Fe–Cr alloys represent a technologically important example of this.

Alloy corrosion is usually undesirable, but it can also be utilized for the preparation of nanoporous metals that have a large surface area per mass of the metal. An example is Raney nickel,[154] first described in 1927, which is obtained by selective dissolution of Al from a Ni–Al alloy. It is used as a hydrogenation catalyst. Meanwhile, dealloying has been recognized as a general method for the preparation of highly reactive metals. Some of the most active catalysts for oxygen reduction are obtained by dealloying Cu–Pt nanoparticles [43]. Nanoporous gold (Figure 14.33) has received particular attention. This material is formed during the dealloying of Ag–Au or Cu–Au alloys and can be obtained as macroscopic bodies (monoliths) traversed by a network of pores and ligaments (ribbons) of adjustable size between about 40 and 200 nm [44]. The dealloying process can be caused by free corrosion in an oxidizing acid (HNO_3) or by imposing a potential (or current) using a potentiostat (or galvanostat).

A number of conditions must be met to form a macroscopic nanoporous metal:

- The standard potentials of the two metals must be sufficiently different for selective dissolution of the less noble element to occur.
- The less noble element must be present in a content above the parting limit. Otherwise, a continuous film of the more noble element is formed and prevents the penetration of the corrosion front into the bulk of the material.

[154] Murray Raney, 1885–1966, trained as a mechanical engineer, Raney was active as a teacher, sales representative in the distribution of electricity and paints, engineer of steam engines, and ultimately in the hydrogenation of vegetable oils for the production of margarine. The company he founded in 1950 still produces Raney nickel under a new corporate name (as of 2016).

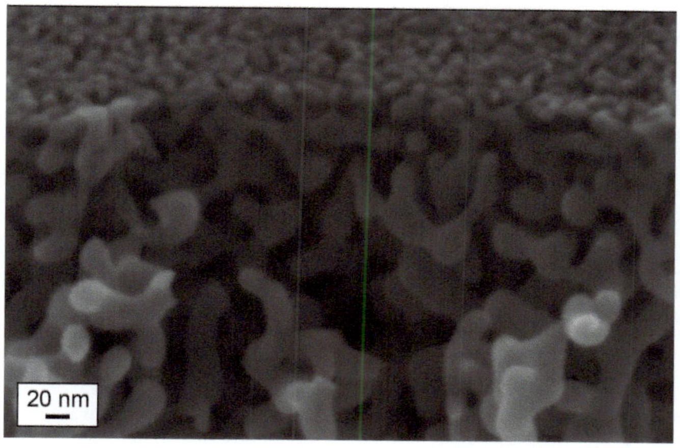

Figure 14.33 Scanning electron microscope image of a cross-sectional view through nanoporous gold obtained by potentiostatic dealloying of $Au_{75}Au_{25}$ in 1 mol L^{-1} $HClO_4$. Source: Reproduced with permission from M. Graf, B. Roschning, J. Weissmüller; Nanoporous Gold by Alloy Corrosion: Method-Structure-Property Relationships. *J. Electrochem. Soc.* **2017**, *164*, C194–C200, Copyright 2017, The Electrochemical Society.

- The corrosion potential (generated by free corrosion or imposed by a potentiostat) must be more positive than a critical potential E_c. E_c is positive from the standard electrode potential of the less noble element and also depends on its content. At this potential, the gain in free energy due to solvation of the resulting metal ions is equal to the energy for breaking the metal lattice *plus* the energy for creating a new surface when a single atom or a few atoms of the less noble element are dissolved out of a lattice plane of the more noble element.
- The atoms of the more noble element must have sufficient mobility (by surface diffusion of atoms) to form ligaments of a defined size by minimizing the surface energy and relieving mechanical stresses in the material. During this process, deeper layers of the less noble element are exposed to the corrosion medium. The surface mobility depends on the temperature, the anions in the electrolyte, and the electrode potential. For example, in the case of Au, Cl^- ions increase the surface mobility of Au atoms. Higher surface mobility leads to coarser ligaments. The ligaments can also be coarsened by thermal posttreatment.

Excursion 14-5 From Glass-Forming Steel to Multiphase Alloys – Structure and Passivity

Stainless steel is usually an Fe–Cr alloy that does not form a visible (i.e., reddish-brown) rust layer or etch pits in a humid atmosphere or in an electrolyte solution because the surface is protected by a 4 nm-thick layer of chromium oxides. Stainless steel requires a minimum chromium content of 12–13% by mass. The passivating chromium oxide layer exists as a composite of crystalline and amorphous components. The structural analysis of such small spatial regions represents a challenge even today. A complete correlation of the microscopic structure and composition to macroscopic corrosion properties has not yet been achieved.

The importance of amorphous structures for passivity was elucidated in a particularly elegant manner using a variety of state-of-the-art structure-sensitive analytical and electrochemical methods for an exemplary set of materials [45]. The authors cooled a melt with the composition $Fe_{50}Cr_{15}Mo_{14}C_{15}B_6$ so rapidly that no crystallization occurred and an amorphous glassy state was obtained (glass-forming steel). Interestingly, passivity is observed in the amorphous state already at 4–5 mass% Cr. Of particular interest for a fundamental understanding is that this glass-forming steel can be transformed into a multiphase crystalline alloy by mild heat treatment at 550, 620, 650, 670, and 800 °C in a well-defined manner. With atom probe tomography (APT), it is possible to observe directly how nanometer-sized regions of different elemental composition emerge after

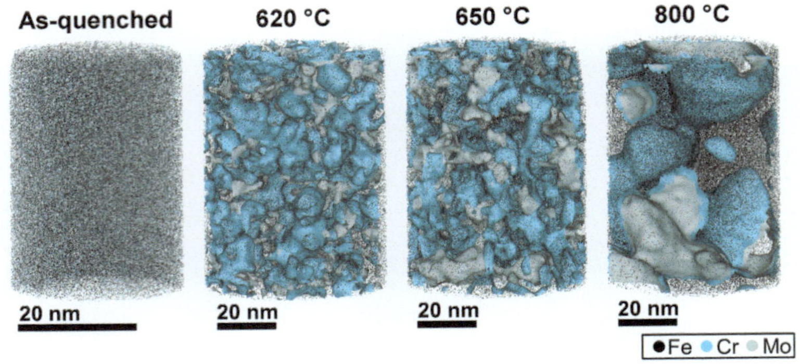

Figure 14.34 Formation of nanoscale phases during annealing of glassy $Fe_{50}Cr_{15}Mo_{14}C_{15}B_6$ from APT data. Temperatures of the heat treatment (for 20 min) are indicated above the columns. The meaning of the false colors in the image is as follows: As quenched sample Fe = black, Cr = blue, Mo = dark gray; 620 °C and 650 °C, areas of equal concentration (iso-concentration surfaces) were colored gray for 22 at% Mo and blue for 16 at% Cr; for 800 °C, the gray isoconcentration surface corresponds to 32 at% Mo and the blue isoconcentration surface corresponds to 15 at% Cr. The figure was kindly provided by Jazmin Duarte and Frank Uwe Renner. Source: Adapted from [45], Copyright 2013, American Association for the Advancement of Science.

a specific heat treatment. Figure 14.34 shows a uniform distribution of elements for a quenched sample. As the temperature in the heat treatment step is increased, small crystalline regions of different compositions form, embedded in an amorphous matrix. With heat treatment at even higher temperatures, the crystalline regions grow at the expense of the amorphous matrix, eventually forming three-dimensional intertwined spatial regions. The exact elemental compositions in the separating phases of the alloy are shown in Table 14.3.

APT is based on the detachment of single ions from a very sharp metallic tip of the sample material in a strong electric field (field ion microscope). The electric field accelerates the ions toward a matrix detector. From the time between detachment of the ions from the tip and their impact on the detector, the mass can be inferred, and from the location of the impact, the original position in the sample material can be determined. This requires that the sequence of detection events be evaluated using fast measurement and computation techniques. The combined measurement data yield a three-dimensional reconstruction of elemental distributions in electrically conducting sample materials [46, 47].

The formation of phases with different elemental compositions has very strong effects on the dissolution behavior. Figure 14.35 shows linear sweep voltammograms of pure Cr, pure Mo (blue), the amorphous alloy $Fe_{50}Cr_{15}Mo_{14}C_{15}B_6$ (black), and the same alloy after heat treatments for 20 min at different temperatures. Three regions a), b), and c) are marked on the voltammograms. In region a), oxidation currents due to metal dissolution occur in curves 2 and 5. In the potential

Table 14.3 Representative concentrations of Fe, Cr, and Mo (from APT measurements) in the individual phases of the alloy $Fe_{50}Cr_{15}Mo_{14}C_{15}B_6$ after heat treatment at different temperatures for 20 min.

T [°C]	Phase	Fe[at%]	Cr[at%]	Mo[at%]
quenched	Amorphous matrix[a)]	59	19	22
620	Cr-rich	61	26	13
	Matrix	58	19	23
650	Cr-rich	65	23	12
	Mo-rich	44	9	47
	Matrix	59	19	22
670	Cr-rich	65	23	12
	Mo-rich	44	8	48
800	Cr-rich	65	22	13
	Mo-rich	42	7	51

a) Corresponds to the average elemental composition of the alloy.
Source: Adapted from [45].

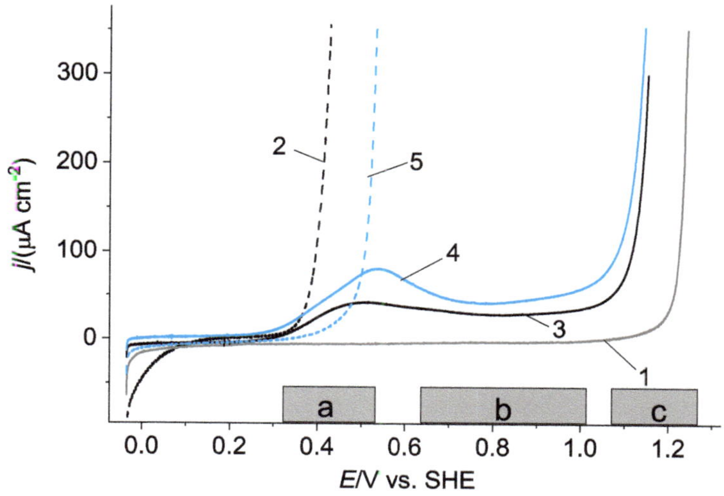

Figure 14.35 Linear sweep voltammetry of alloys with 2 mV s^{-1} in 0.1 mol L^{-1} H_2SO_4 of 1) pure Cr, 2) pure Mo, 3) the amorphous alloy as prepared, 4) and 5) after heat treatment of the alloy at (4) 620 °C, (5) 800 °C. Raw data were kindly provided by Frank Uwe Renner, Universiteit Hasselt, Diepenbeek, Belgium. Source: Adapted from [45], Copyright 2013, American Association for the Advancement of Science.

region b), there is passive behavior in curves 1, 3, and 4, where a lower steady-state current results as the driving force increases. The peak between regions a) and b) marks the transformation of the unstable passive film into a more stable passive film. In region c), the stable passive film breaks down even for curves 1, 3, and 4. First, it can be observed that pure Mo dissolves at about 0.4 V, while pure Cr resists dissolution up to 1.2 V. The amorphous alloy dissolves at about 1.1 V. Although this potential is 0.1 V lower than that for pure Cr, the amorphous alloy behaves much more similar to Cr than to Mo. With the formation of a Mo-rich crystalline phase during heat treatment, the dissolution behavior becomes similar to that of Mo (compare curves 2 and 5). Since the Mo-rich phase permeates the entire specimen as a three-dimensional, interconnected phase, dissolution does not stop after dissolution of a near-surface layer.

Using a microdroplet cell (Figure 14.36), the metal dissolution can be studied in a flow-through system. This has the advantage that the solution can be continuously fed to an inductively coupled plasma source with attached mass spectrometer (ICP-MS). This allows the determination of the elemental concentration in the corrosion medium and its correlation to the composition of the solid sample material. This procedure makes it readily apparent that metal dissolution depends on the formation of the multiphase alloy with spatial and surface regions of different elemental composition and that the dissolution process of the multiphase alloy is similar to that of Mo. If no multiphase crystalline structure is present, dissolution of the near-surface layers quickly generates a passive layer, which consists essentially of remaining Cr oxides that cover and passivate the entire sample.

Although this particular alloy has little commercial significance to date, its investigation is important. Only special materials, which allow a systematic adjustment of critical properties, facilitate new investigation strategies for complex materials using instrumentation of the highest resolution. In this context, it is not only the development of the measuring instruments itself that is crucial but also their use to understand the complexity of the materials and their changes under different conditions. In this respect, this study [45] is groundbreaking for the future treatment of corrosion problems of complex alloys. It also highlights the crucial influence

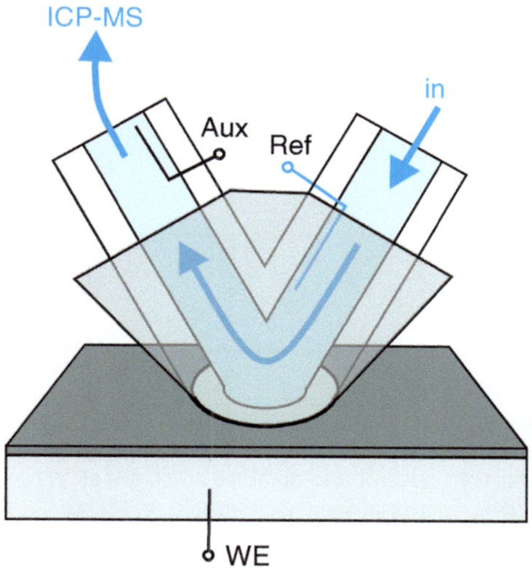

Figure 14.36 Setup of a flow cell according to [45] with which the corrosion medium is continuously fed to an ICP-MS. In addition, it is advantageous that only a small area of the sample is used for the measurement so that several linear sweep voltammograms can be recorded on one specimen.

that thermal treatments can have on the corrosion properties of alloys. We must refrain from discussing the important consequences of mechanical treatment on material properties here (rolling, drawing, forging, etc.).

High-performance alloys can hardly be underestimated in their importance for technological and, ultimately, social developments. Just think of the designation of cultural epochs (Bronze Age and Iron Age) and the importance of noncorroding lightweight alloys for aerospace and energy-saving road and rail vehicles. Noncorroding biocompatible implants for temporary or permanent replacement of bone, as well as stents for keeping blood vessels open, contribute significantly to increasing life expectancy. Interestingly, alloys that dissolve slowly and in a controlled manner without releasing toxic metal ions are a current area of research to identify resorbable implant materials, which could lead to significant therapeutic benefits.

14.6 Deliberate Metal Dissolution

14.6.1 Preparation of Nanoporous Metal Oxide Layers

All valve metals form compact and smooth oxide layers that protect the underlying metal from further dissolution, making them suitable for use in valves. Valve metals include Al, Ti, Zr, Hf, Ta, Nb, W, and many of their alloys. Oxide films on these metals can be formed with high control by anodization, that is, an anodic treatment (as opposed to native oxide films, which are formed spontaneously by corrosion in air or aqueous solutions). An interplay of different processes controls the growth of these oxide films (Figure 14.37). At the metal–oxide interface, metal atoms are oxidized.

$$\mathrm{Me} \rightarrow \mathrm{Me}^{z+} + ze^- \tag{14.45}$$

Under the influence of the electric field, the cations migrate toward the oxide–electrolyte interface. At the interface, cations undergo hydrolysis:

$$\mathrm{Me}^{z+} + \mathrm{H_2O} \rightarrow \mathrm{MeO}_{z/2} + z\mathrm{H}^+ \tag{14.46}$$

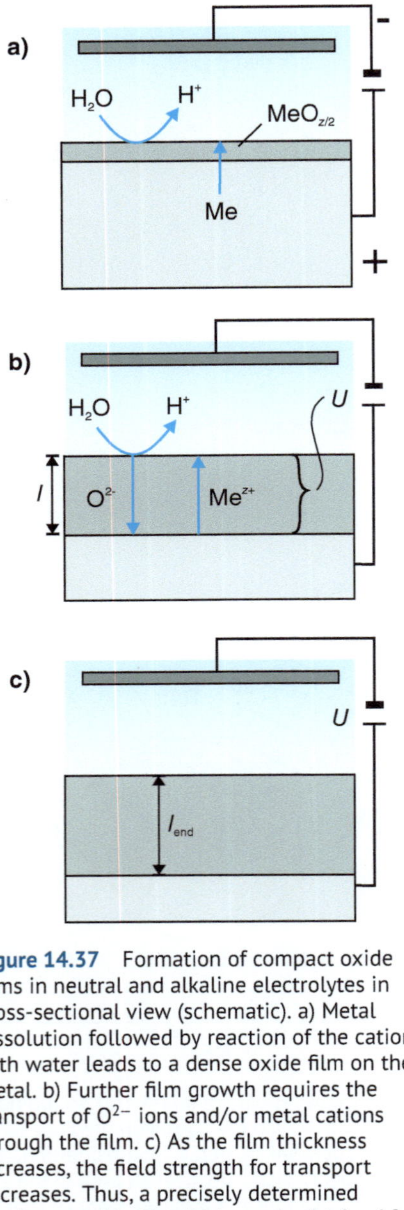

Figure 14.37 Formation of compact oxide films in neutral and alkaline electrolytes in cross-sectional view (schematic). a) Metal dissolution followed by reaction of the cations with water leads to a dense oxide film on the metal. b) Further film growth requires the transport of O^{2-} ions and/or metal cations through the film. c) As the film thickness increases, the field strength for transport decreases. Thus, a precisely determined maximum oxide film thickness is obtained for each anodizing voltage.

The O^{2-} ions formed in this process are incorporated into the oxide lattice, in which they have mobility. Depending on whether Me^{z+} or O^{2-} ions have greater mobility, new oxide forms predominantly at the metal–oxide or the oxide–electrolyte interface. Since the oxide has a lower conductivity than the electrolyte solution, almost all of the cell voltage drops across the oxide film. The prevailing field strength $\mathbf{E} = U/l$ is the driving force for the migration of ions in the solid oxide. During oxidation under constant anodizing voltage U in a neutral or alkaline electrolyte, the increase in oxide film thickness l leads to a decrease in electric field strength $\mathbf{E}$ and thus to slower transport of the ions in the solid oxide. This results in a slower growth rate of the oxide film thickness, which finally comes to a complete stop. The achieved final film thickness l is proportional to U.

It has been known for a long time that a porous oxide layer with a thickness of 5–25 μm forms when Al is anodized in acid electrolytes (Figure 14.38). The pores in this oxide can be filled with foreign substances (e.g., a dye). When the oxide layer is treated in boiling water, Al_2O_3 absorbs water and increases its volume. The increase in volume compresses the oxide, and the pore openings close. The added substance, for example, a dye, is permanently encapsulated in the oxide film, which facilitates the use as passive coatings with an added decorative function (Eloxal process for "electrolytic oxidation of aluminum") [48].

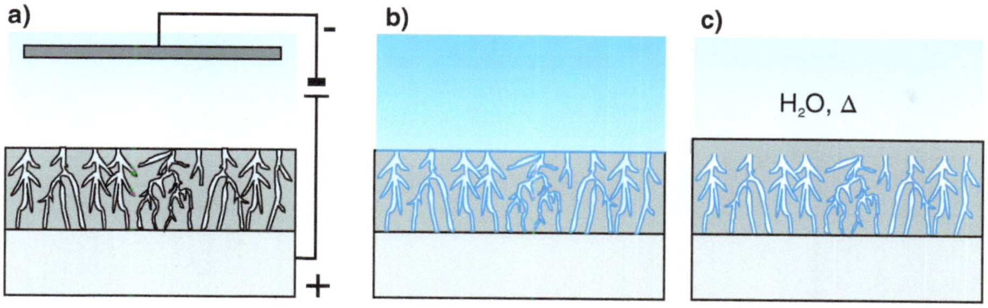

Figure 14.38 a) Formation of a porous oxide layer in acid electrolytes in cross section; b) adsorption of a dye on the walls of the pores; and c) sealing of the pores (schematic is not to scale).

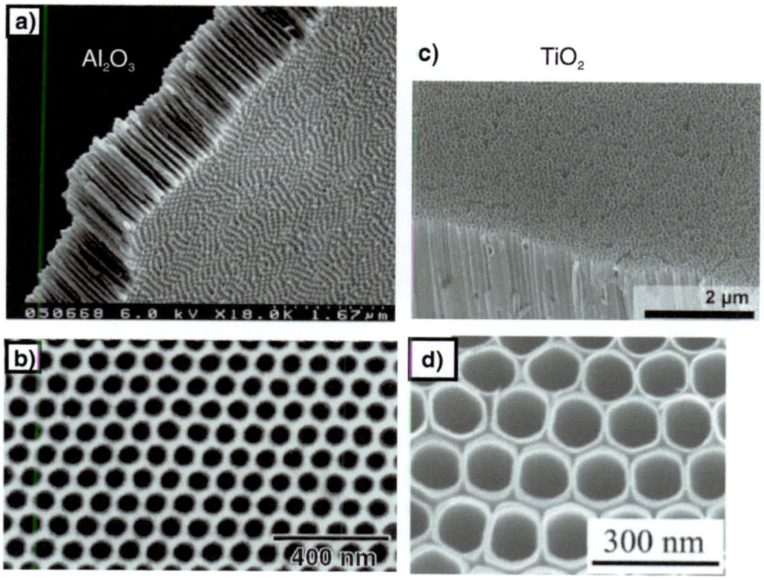

Figure 14.39 Examples of highly ordered arrangements of nanopores in Al_2O_3 (a and b) and TiO_2 (c and d). Sources: a) Reproduced with permission from M. Kang, S. Yu, N. Li, C. R. Martin; *Small* **2005**, *1*, 69, Copyright 2005, Wiley-VCH; b) reproduced with permission from H. Masuda in *Ordered Porous Nanostructures and Applications*. (R. B. Wehrspohn, Ed.), Springer, New York, 2005, p. 39, Copyright 2005, Springer; c) reproduced with permission from S. P. Albu, P. Schmuki; *Phys. Status Solidi RRL* **2010**, *4*, 151–153, Copyright 2010, Wiley-VCH; d) reproduced with permission from J. M. Macak, S. P. Albu, P. Schmuki; *Phys. Status Solidi RRL*, **2007**, *1*, 181–183, Copyright 2007, Wiley-VCH.

In 1995, Masuda and Fukuda discovered [49] that under certain conditions, highly ordered arrays of pores can be formed (Figure 14.39a). Similar highly ordered arrays have later been prepared from other metal oxides with a large variation of the structural parameters. For some metals, densely packed nanotubes form preferentially or exclusively in this process. Ti can be considered prototypical for the formation of nanotubes (Figure 14.39c,d). The great interest in such easy access to controlled nanostructures triggered intensive research efforts. Nevertheless, not all observed phenomena in the generation of nanopores and nanotubes can be explained by a complete and unified concept so far. We will only trace some important trends here. More detailed reviews can be found in [50–53].

Starting from a disordered arrangement of pores (Figure 14.40a), highly ordered arrays of nanopores or nanotubes can form if the electrolyte slowly dissolves the resulting oxide (Figure 14.40b). This can be achieved for Al_2O_3 by an acidic electrolyte. This is insufficient for TiO_2, and an F^--containing compound is added to the electrolyte.

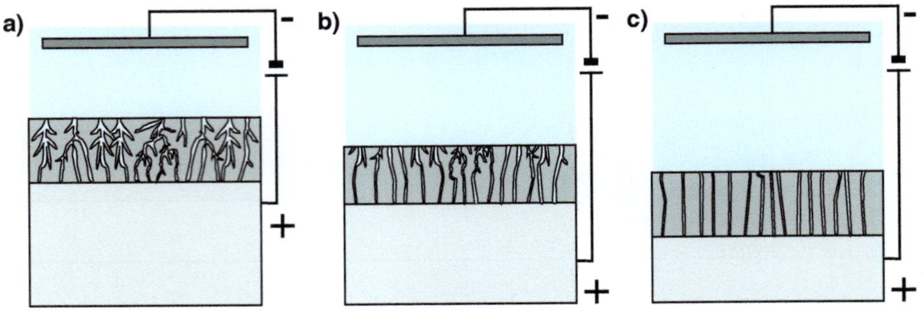

Figure 14.40 Formation of a highly ordered array of nanopores or nanotubes in electrolytes that can dissolve the resulting oxide (schematic is not to scale).

Fluoride ions attack TiO_2 to form a fluoro-complex with Ti^{4+}. This attack occurs particularly at low pH values, that is, at the bottom of the pore [Eq. 14.47].

$$TiO_2 + 6\,F^- + 4\,H^+ \rightarrow [TiF_6]^{2-} + 2\,H_2O \qquad (14.47)$$

The process starts with the formation of disordered pores (Figure 14.40a). During the further growth of the pores into the bulk of the metal, pores continue to grow that have a very regular distance to each other while the initially formed porous oxide slowly dissolves (Figure 14.40b and c). The result is a layer with a very regular arrangement of pores or tubes.

Important for the mechanism of pore formation is the increase in volume during oxide formation (compared to the metal), which leads to mechanical stress in the oxide layer. The interplay of these factors initially causes random small variations in layer thickness that self-reinforce and result in pore growth (Figure 14.41a and b). The highest field strength, and thus, the fastest ion transport in the solid oxide occurs at the bottom of the pores, where the thickness of the oxide layer is lowest. There, the new oxide is formed at the highest rate. Due to the pH prevailing there, the oxide at the bottom of the pore is also dissolved particularly rapidly, so that the pore grows further and further into the material (Figure 14.41b). The increase in volume of the oxide layer relative to the metal causes the oxide at the bottom of the pore to move sideways toward the vertically extending pore walls. The continuous reformation and dissolution of the oxide results in a highly ordered nanoporous oxide film that continues to grow under steady-state conditions (Figure 14.41c). This means that the pores continue to grow at the same rate as they dissolve on the solution side.

The transition from the formation of compact oxide layers to the growth of a nanoporous oxide layer can be easily recognized by the maximum b) in current–time plots during the anodizing process (Figure 14.41d).

In the case of steady-state pore growth, only those pores "survive", which have an optimum spacing with an approximately hexagonal arrangement. Various "tricks" have been described to significantly improve the regularity of the arrangement. For example, anodizing can be performed in two steps by chemically dissolving a primarily formed oxide layer and then starting a new anodizing process on the preetched metal surface. Stamping processes (stamps themselves originate from an anodizing process) can be used to locally damage compact oxide films to induce the subsequent anodizing attack under pore-forming conditions at the predamaged regions. This allows patterns to be created that no longer show only a hexagonal arrangement. The regular patterns can even extend across grain boundaries.

Anions of the electrolyte are also incorporated into the resulting oxide (Figure 14.42a). They are transported in the solid oxide by migration. Depending on their mobility

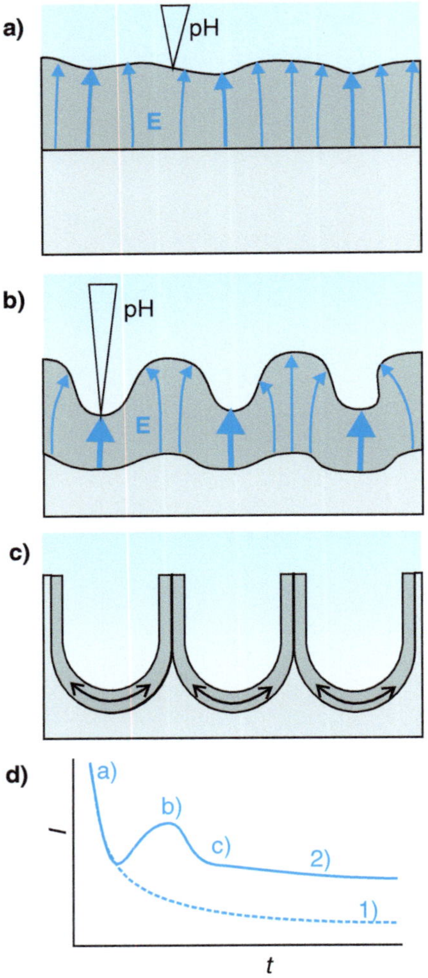

Figure 14.41 Schematic of the formation of highly ordered nanopore arrays according to [53]. a) Random variations in the oxide film thickness lead to variation of the electric field strength in the film. b) The increased dissolution rate in regions of high field strength leads to pore formation in the metal specimen. c) A steady state is reached between metal dissolution at the bottom of the pores and the slow dissolution of the oxide at the pore walls. The pH is reduced within the pores. d) Anodizing curves for (1) the formation of compact oxide films and (2) porous oxide films.

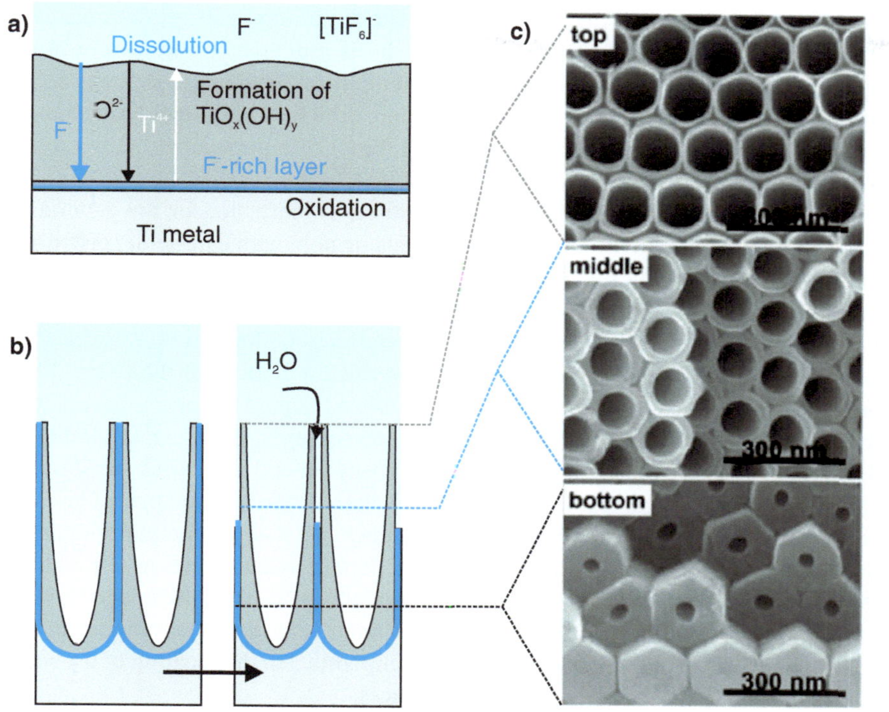

Figure 14.42 TiO_2 nanotubes from steady-state growth; a) schematic representation of fluoride enrichment at the metal-oxide interface; b) preferential dissolution of the fluoride-rich parts of the oxide in the electrolyte; c) electron micrographs of nanotube cross sections at the solution-side pore exit (top), in the middle of the tubes (middle), and at their bottom (bottom). The approximate position is indicated in scheme b). Source: Panel c) reproduced with permission from [53], Copyright 2014, American Chemical Society.

relative to O^{2-}, increased concentration of the electrolyte anion occurs in specific zones of the oxide film. In particular, F^- ion migrates faster than O^{2-}, and therefore, accumulates at the metal–oxide interface, while quite pure oxide is present at the oxide–electrolyte interface.

Whether nanopores or nanotubes are formed depends on the solubility of the oxide phase, which is enriched with the anions of the electrolyte. This layer is exposed in its cross section at the outer surface of the oxide film, where it is subject to dissolution by the electrolyte (Figure 14.42b). If the anion-rich oxide layer is more readily dissolved than the pure oxide, the anion-enriched layer is dissolved out faster than the metal oxide. In this case, as found for Ti, free-standing nanotubes are formed (Figure 14.42c, d).

The particular advantage of nanoporous metal oxide films is the precise control of the structural parameters [51]. In this context, pure and stress-free metal samples as starting material are a prerequisite for regular pore arrangements. Further factors play a role, such as temperature, potential scan rate, final applied potential, composition and pH of the electrolyte, as well as the viscosity of the solution. Finally, aqueous or organic electrolyte solutions are considered. The applied voltage determines the spacing at which the pores or nanotubes form. Here, the distance between the centers of adjacent pores (10–250 nm) is proportional to the applied anodizing voltage (1–45 V). The lengths of the pores and tubes, and thus the aspect ratio, are initially controlled by the duration of the anodizing process, but only until the stage of steady-state pore growth is reached, after which the obtained pore depth does not change significantly. Under these conditions, the maximum achievable pore length is largely determined by the dissolution rate of the formed oxide. The dissolution can be retarded, causing the pores to be elongated if the transport of F^- is hindered (in the case of TiO_2 by low temperatures or viscous solvents). Further control of the pore shape is achieved by changing the anodization voltage during pore growth. This allows branched pores, tubes with intermediate segments ("bamboo tubes"), or multilayer structures with different properties to be created.

The nanopores and nanotubes are widely used as templates for the generation of other nanomaterials [55]. For example, metals or polymers can be deposited inside them. If electrodeposition is to be performed, the valve metal can be deposited on a thin gold layer as a back contact. Afterward, anodization is performed until the pores reach the gold layer. Defined nanostructures can then be formed by galvanic deposition onto the exposed portion of the gold layer at the bottom of the pores. Usually, pulsed electrodeposition is applied in order to allow the pore volume to be replenished with metal ions between deposition pulses. Subsequently, the metal oxide layer can be chemically dissolved so that the electrodeposited metal structures stand up like a "brush" of free nanowires. It is also possible to dissolve the residual parent metal after pore formation in order to obtain a free-standing metal oxide membrane with continuous pores. These can be used, for example, in photoreactors in flow-through mode [54]. Other applications are directly aimed at exploiting the material properties of TiO_2, for example, for photocatalysis, dye-sensitized solar cells, electrochromic displays, or control of cell growth on surfaces of medical implants. Further coating of the surfaces allows the creation of superhydrophobic surfaces.

The principles of nanopore and nanotube generation are transferable to other metals and alloys: Pores are formed on Ta and Nb (as on Al), while Zr and Hf form tubes (as Ti) [52].

Porous silicon is another important material obtained by electrochemical processes (see Section 10.3).

14.6.2 Electrochemical Machining

Electrochemical machining (ECM) refers to a whole class of machining processes used to change the external shape of workpieces made of metals or semiconductors by anodic dissolution [56–59]. These processes are usually much slower than metal cutting processes such as drilling, turning, and milling that dominate the metalworking industry. ECM methods are therefore used whenever metals are to be machined that are too brittle for mechanical processing (e.g., titanium components in jet engines), that are too small to absorb the forces occurring during mechanical machining (e.g., shear foils for electric shavers), that would be destroyed by the heat generated during mechanical processing or laser machining, that have machining points in the workpiece that are mechanically inaccessible (internal holes and connecting channels in nozzles), or that have complex shapes and require defined surface roughness to be produced in one and the same operation (microsystems technology and medical technology).

Figure 14.43 shows the basic principle of the ECM. The workpiece (2) is connected as the anode for oxidative metal dissolution at extreme anodic current densities of 10–300 A cm^{-2}. The cathode is the tool (1), whose shape is the negative of the areas to be removed from the workpiece. An aqueous electrolyte (5) is forced through a working gap with a thickness of 20–800 μm at linear flow rates of 5–50 m s^{-1}. The electrolyte must carry away the metal salts produced by metal dissolution and provide sufficient ionic conductivity to sustain such a high current density. The supply line for the electrolyte can be integrated into the mold (4) or can also be provided via other supply lines.

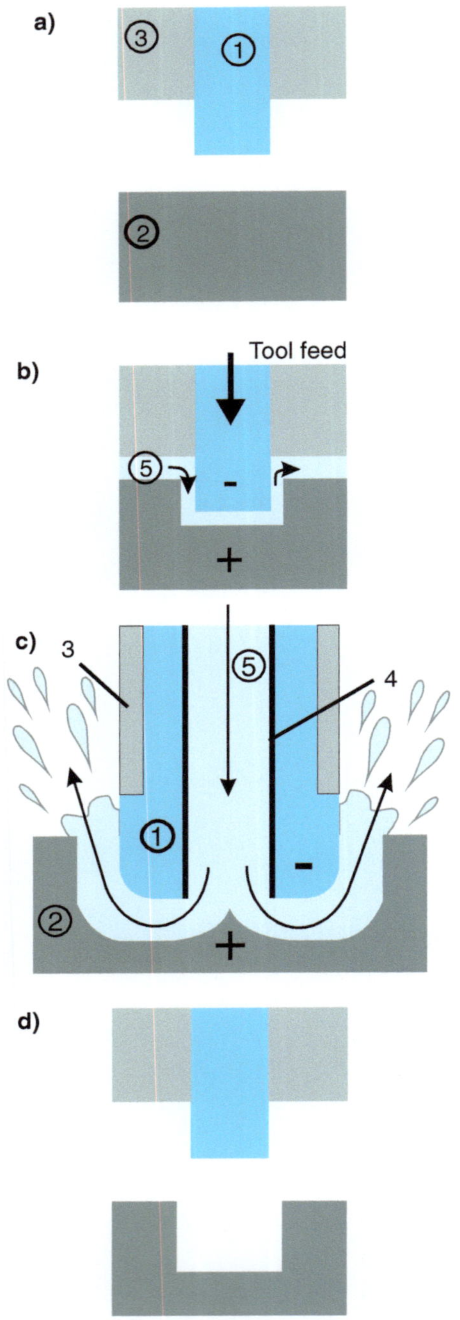

Figure 14.43 Basic principle of ECM sinking.

a) Before machining;
b) during ECM with electrolyte flow from one side of the machined area;
c) during ECM with electrolyte flow through the ECM tool electrode;
d) after ECM.

Components: 1) tool (cathode), 2) workpiece (anode), 3) insulating sheath, 4) supply channel for electrolyte, and 5) electrolyte.

Influencing Parameters. In the following, we will look at some of the parameters influencing the ECM process. We cannot discuss the entire diversity of materials here, and thus, have to limit the discussion to examples for ECM of steel.

ECM machining can be performed with DC voltage or voltage pulses. When working with direct current (DCECM, direct current electrochemical machining), the mains voltage is first transformed to the working voltage and then rectified. The local current densities for metal dissolution result from the working voltage and the local resistance between the tool and the workpiece. For the latter, the working distance is the determining control variable. If the working distance is locally too large, no significant material removal takes place. If the distance is too small, unwanted sparkovers or short circuits can occur. Today, modern ECM systems mostly use pulsed voltages or currents (Figure 14.44), with pulse lengths in the ms to µs range (PECM, pulsed electrochemical machining). Material removal occurs only during the current pulses. The pauses between the pulses allow removal of the reaction products and the generated Joule heat[51]. Thus, a defined initial state is ensured before each pulse (compare also with pulsed electrodeposition in Section 9.3.4).

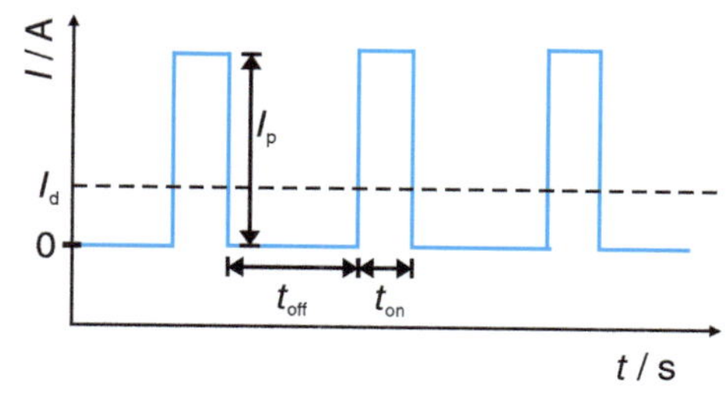

Figure 14.44 Schematic of the transient current at the workpiece during PECM.

The duty cycle γ is the ratio of the pulse time t_{on} to the total cycle time $(t_{on} + t_{off})$. Thus, the pulse current I_p results in the average current I_{av} and the total charge Q during machining.

$$\gamma = \frac{t_{on}}{t_{on} + t_{off}} \tag{14.48}$$

$$I_{av} = \frac{t_{on}}{t_{on} + t_{off}} I_p = \gamma I_p \tag{14.49}$$

$$Q = I_{av} t_{tot} = I_{av} N_p (t_{on} + t_{off}) \tag{14.50}$$

γ – Duty cycle [dimensionless]
T_{on} – Pulse duration [ms]
t_{off} – Idle time [ms]
I_p – Pulse current [A]
I_{av} – Mean current [A]
Q – Total charge [C] = [A s]
t_{tot} – Duration of PECM processing [s]
N_p – Number of current pulses [dimensionless]

The current yield or faradaic efficiency ε_F is the ratio between the actually achieved material removal (measured as mass loss Δm by weighing before and after processing) and the mass removal calculated according to Faraday's law Δm_{theo}.

$$\varepsilon_F = \frac{\Delta m}{\Delta m_{theo}} = \frac{\Delta m\, nF}{QM} = \frac{\Delta m\, nF}{I_{av} t_{tot} M} \tag{14.51}$$

ε_F – Current yield (faradaic efficiency) [dimensionless]
Δm – Mass loss due to weighing [g]
Δm_{theo} – Mass loss calculated according to Faraday's law [g]
n – Reaction valency (number of transferred electrons per metal ion) [dimensionless]
M – Molar mass of the metal [g mol^{-1}]

The reaction valency n is often, but not always, known. If mixed valence dissolution occurs, that is, simultaneous formation of Fe^{2+} and Fe^{3+}, the dominant valence must be selected and it must be taken into account in the interpretation that systematic deviations for ε_F are to be expected due to mixed valence dissolution.

Even shorter voltage pulses in the nanosecond range make it possible to achieve submicrometer precision in material removal [60]. At such short pulses, only the double layer is recharged at those parts of the workpiece surface that have the smallest distance to the tool electrode. The time for recharging the double layer is determined by the time constant of the RC element formed by the double layer capacitance and the electrolyte resistance.

$$\tau = R_E C_{dl} = \rho_E d C_{dl} / A \tag{14.52}$$

τ – Time constant [s]
R_E – Electrolyte resistance [Ω] = [V A^{-1}]
ρ_E – Electrolyte resistivity [Ω cm]
d – Working distance [cm]
C_{dl} – Double-layer capacitance [C V^{-1}] = [A s V^{-1}]
A – Area [cm^2]

The electrical double layer is only formed where $\tau < t_{on}$. Without a charged double layer, the extreme field strengths are not present at the surface, which are necessary for a charge transfer reaction to occur. Only where the working distance d is very small are R_E-values encountered that allow charging the double layer within an ns pulse (Figure 14.45). However, it is not easy to realize such short pulses with electronic devices because the devices themselves have internal capacitances due to different assemblies and cables, resulting in a time constant for the electronic device itself, below which voltage changes cannot be exactly applied.

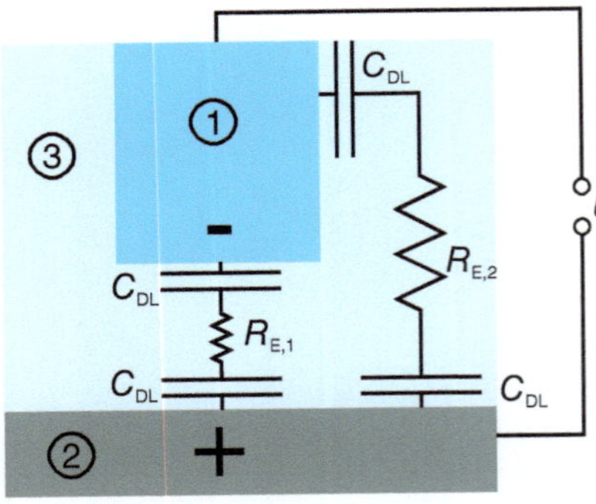

Figure 14.45 ECM with ultrashort voltage pulses that allow charging of the double layer only at the points with the smallest working distance. 1) Tool; 2) workpiece; and 3) electrolyte solution.

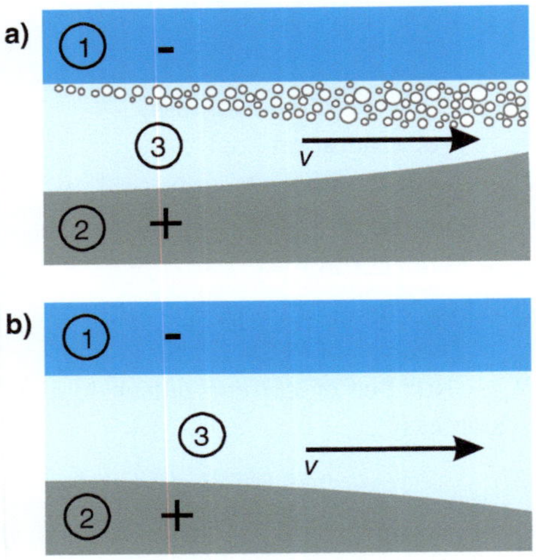

Figure 14.46 Change in material removal rate along the flow direction *v* in the working gap. a) Slower removal at the end of the gap due to increasing resistance caused by gas bubbles; b) higher removal rate at the end of the gap due to reduced resistance caused by temperature increase. 1) Tool; 2) workpiece; and 3) electrolyte solution.

The flow or convection of the electrolyte exerts a very important influence on the ECM process, since the local resistance between the workpiece and the tool determines the local dissolution rate. In the gap between the tool and the workpiece, the resistivity changes by two processes (Figure 14.46). As H_2 gas bubbles form at the cathode, the resistivity of the electrolyte increases along the working gap. The increased resistivity tends to cause less material removal at the end of the working gap (Figure 14.46a). This effect can be further enhanced if O_2 or Cl_2 is formed at the anode (workpiece) as a side reaction. At the same time, the electrolyte heats up as it flows through the gap due to the Joule heat of the ECM process. Ionic conductivity increases with temperature increase and tends to cause an increased machining rate at the end of the working gap. Although the two effects are opposite, they do not automatically cancel out each other.

Electrolyte and Electrode Reaction. The electrolyte must be low cost and have a high solubility to achieve conductivities of 50–300 mS cm^{-1}. These requirements, together with physiological safety and low disposal problems, make NaCl and $NaNO_3$ solutions the most commonly used electrolytes for steels. Other electrolytes are used in special applications. Because of their aggressiveness (dissolution of the metal even without a current pulse), the use of acids is limited to special applications, such as the electrochemical drilling of nickel-based alloys in turbine construction. Here, they enable the dissolved metal cations to be removed from the narrow working gap with a high aspect ratio.

According to the classification of corrosion scenarios, the anions of the electrolyte exert a decisive influence on the mechanisms of ECM. Cl^- ions are called activating anions because they attack the passive layers and favor a transition to active metal dissolution in steel. Accordingly, much higher machining rates can be achieved with NaCl than with $NaNO_3$. The current yield is close to 100% for iron dissolution at a reaction valency of $n = 2$.

$$Fe \rightleftharpoons Fe^{2+} + 2e^- \qquad (14.53)$$

At current densities greater than the limiting current density in the transpassive region, direct formation of Fe^{3+} or further oxidation of Fe^{2+} also occurs, which can be seen from the decrease in faradaic efficiency calculated with $n = 2$.

$$Fe \rightleftharpoons Fe^{3+} + 3e^- \qquad (14.54)$$

$$Fe^{2+} \rightleftharpoons Fe^{3+} + e^- \qquad (14.55)$$

In addition, the O_2 evolution reaction at the workpiece must be considered.

$$2H_2O \rightleftharpoons O_2 + 4H^+ + 4e^- \qquad (14.56)$$

The thermodynamically possible Cl_2 evolution reaction occurs only at very high current densities and correspondingly high overpotentials. It is not observed, for example, at $j < 64$ A cm^{-2} in 4.5 mol L^{-1} NaCl solution [61].

$$2Cl^- \rightleftharpoons Cl_2 + 2e^- \qquad (14.57)$$

Q: Why is Fe^{3+} not reduced again at the tool electrode, similar to redox cycling or SECM?

A: The reduction of Fe^{3+} or Fe^{2+}, which is conceivable, is not possible because of the strong electrolyte flow. In the time required for diffusion/migration between the anode and the cathode, the solution element of the electrolyte solution has already left the working gap (see Exercise 14.8).

Hydrogen is formed at the tool (the cathode).

$$2H_2O + 2e^- \rightleftharpoons H_2 + 2OH^- \qquad (14.58)$$

Since the entire surface is active, the local current density is distributed only according to the local electrolyte resistance. This leads to material removal from a rather large area of the workpiece even at very low local current densities. As a consequence, the machining precision is not very high.

NO_3^-, on the other hand, is one of the passivating anions since it favors the formation of stable passive layers. At these passive layers, O_2 forms as the most important by-product according to Eq. (14.56). Up to about 10 A cm^{-2}, this is even the dominant reaction. The faradaic efficiency for iron dissolution increases significantly only at current densities greater than 20 A cm^{-2}. The anodic reactions are the formation of hematite (Fe_2O_3) or magnetite (Fe_3O_4) or their interconversion.

$$2\,Fe + 3\,H_2O \rightleftharpoons Fe_2O_3 + 6\,H^+ + 6e^- \tag{14.59}$$

$$3\,Fe + 4\,H_2O \rightleftharpoons Fe_3O_4 + 8\,H^+ + 8e^- \tag{14.60}$$

$$2\,Fe_3O_4 + H_2O \rightleftharpoons 3\,Fe_2O_3 + 2\,H^+ + 2e^- \tag{14.61}$$

In addition, dissolved Fe^{2+} ions can be oxidized to Fe^{3+} by O_2 or NO_3^- without transferring charges into the external circuit.

At the cathode, the reduction of nitrate to different compounds takes place, for example, according to the following equations:

$$NO_3^- + H_2O + 2e^- \rightleftharpoons NO_2^- + 2\,OH^- \tag{14.62}$$

$$NO_3^- + 7\,H_2O + 8e^- \rightleftharpoons NH_4OH + 9\,OH^- \tag{14.63}$$

By coupling the material removal to high current densities, the material removal is limited to the regions with particularly high local current densities, while the regions with low current densities form O_2 only. This leads to a higher machining accuracy at a slower machining rate than that obtained in NaCl electrolytes.

Modeling. In order to machine a desired shape by ECM, the design of the tool electrode and also the process parameters must be adjusted. Because of the many influencing parameters, this is not a trivial task. In addition to iterative and lengthy test series, modeling the process on a computer represents an approach to this challenge [59]. The accuracy of the simulation is essentially determined by the fact that the critical parameters of ECM in a particular situation are well reproduced in the model, but that the model can still be handled numerically. Several models of varying complexity are in use to meet these conflicting requirements in as many situations as possible. In the electrostatic model, only the stronger material removal at high local current densities at protruding areas is considered. Influences of flow conditions and the different mechanisms of active and passive metal dissolution are ignored. The model of the compact salt layer assumes a passivated metal surface at which only O_2 is formed. Metal cations are generated only at high current densities and immediately form a compact salt layer with anions because the solubility product of the salt is exceeded. This salt layer inhibits O_2 evolution but allows anodic metal dissolution. The electrostatic model and the compact salt layer model predict surface leveling in every case, but this is not always observed (cf. pitting corrosion, Section 14.5.3). The model of the double salt layer assumes the presence of an inner compact and an outer porous salt layer. The viscous film model replaces the assumption of a salt layer with a viscous product film that is formed by the metal cations with anions migrating to the workpiece (anode). The viscous film is depleted in water due to water oxidation at the anode. Therefore, cations cannot be solvated completely. The viscous film slows down mass transport processes that become limiting for the overall machining process. The shorter transport path through the film at protruding parts of the workpiece causes a leveling of

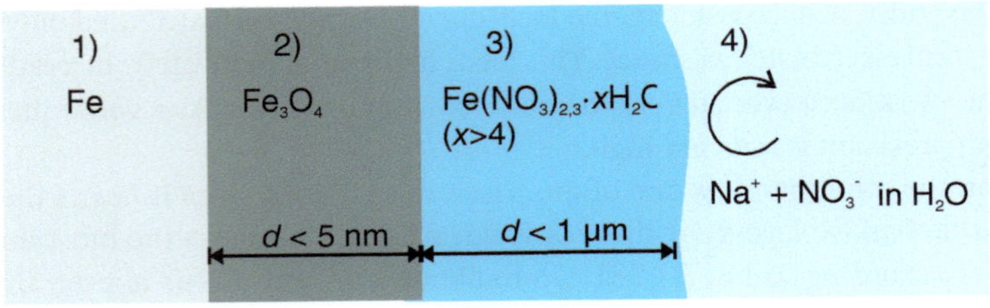

Figure 14.47 Schematic diagram of the oxide layer and the viscous polishing film during ECM of steel in a $NaNO_3$ electrolyte according to [62]. 1) Steel matrix; 2) oxide layer; 3) viscous polishing film; and 4) electrolyte solution with turbulent flow.

the surface. In NO_3^- electrolytes, this model can be refined by assuming an oxide film between the metal and the viscous polishing film [62]. The polishing film consists of iron nitrate hydrates, $Fe(NO_3)_3 \cdot 9H_2O$ and $Fe(NO_3)_2 \cdot 6H_2O$. The two compounds melt at only 47 and 60.5 °C, respectively, and are therefore present in a highly viscous state under ECM conditions (Figure 14.47).

Test your knowledge

Exercise 14.1: Consider the galvanic corrosion of Cu and Fe in Figure 14.3. Assume that this reaction scheme proceeds for a) two Cu plates connected to an Fe rivet; b) two Fe plates connected to a Cu rivet. Let the surface area of the plates be much larger than the surface area of the rivets. What damage pattern do you expect in a) and b)?

Exercise 14.2: Using Figure 14.8 and Figure 14.9, explain the concepts of immunity and passivity using Al and Fe as examples.

Exercise 14.3: Show that the equilibrium line does not change from Eq. (14.23) when the associated electrode reaction is formulated with hydroxide ions as reactants.

$$Al + 4OH^- \rightleftharpoons AlO_2^- + 2H_2O + 3e^- \tag{14.64}$$

Exercise 14.4: Show that the statement "anodic reactions lead to acidification of the corrosion medium, cathodic reactions lead to alkalization of the corrosion medium" is true even when acid corrosion dominates.

Exercise 14.5: Using the data in the volcano plot (Figure 3.9), explain why small inclusions of Pd in Zn can greatly increase the rate of acid corrosion of Zn.

Exercise 14.6: Using an Evans diagram, explain how the deposition of soot-containing dust on sheet metal can lead to the acceleration of atmospheric corrosion.

Exercise 14.7: In an ECM process, a pulse current of 10 A is applied for 10 ms. This is followed by an idle time of 30 ms before the next pulse. In total, one processing step for deburring takes 23.5 s. Calculate the average current. What would be the theoretical removed mass of iron when oxidized to Fe^{2+}?

Exercise 14.8: Estimate the time for diffusion of Fe^{3+} across a working gap of 50 μm and compare this to the residence time of a bulk element at a linear flow rate in the gap of 10 m s^{-1}. Make reasonable assumptions for the missing material data and discuss the limitations of these estimates.

References

1 D. Pletcher; *A First Course in Electrode Processes.* The Electrochemical Consultancy, Hampshire 1991, ISBN 0951730703, pp. 209–232.

2 N. Eliaz, E. Gileadi; *Physical Electrochemistry: Fundamentals, Techniques and Applications*; 2nd ed., Wiley-VCH, Weinheim, 2019, pp. 227–322.

3 A. Groysman; *Corrosion for Everybody*; Springer, Dordrecht, **2010**.

4 H. Kaesche; *Corrosion of Metals – Physicochemical Principles and Current Problems.* Springer, Berlin **2003**.

5 R. W. Revie (Ed.); *Uhlig's Corrosion Handbook.* 3rd ed., John Wiley & Sons, Hoboken, **2011**.

6 E. McCafferty; *Introduction to Corrosion Science*; Springer, New York, **2010**.

7 A. J. Bard, M. Stratmann, G. S. Frankel (Eds.); *Encyclopedia of Electrochemistry, Vol. 4*, Wiley-VCH, Weinheim, **2003**.

8 N. Perez; *Electrochemistry and Corrosion Science.* 2nd ed., Springer, Switzerland, **2016**.

9 R. G. Kelly, D. W. Shoesmith, R. G. Buchheit; *Electrochemical Techniques in Corrosion Science and Engineering*, CRC Press, New York, **2002**.

10 P. Marcuse (Ed.); *Corrosion Mechanisms in Theory and Practice.* 2nd ed., Marcel Dekker, New York, **2002**.

11 E. McCafferty; *Surface Chemistry of Aqueous Corrosion Processes.* Springer, New York, **2015**.

12 P. Marcus, F. Mansfeld (Eds.); *Analytical Methods in Corrosion Science and Engineering*, CRC Press, Bota Raton, **2006**.

13 K.-H Trostmann; *Korrosion – Ursachen und Vermeidung*, Wiley-VCH, Weinheim, **2000**.

14 J. Kruger; Cost of metallic corrosion. in *Uhlig's Corrosion Handbook*, 3rd ed. (R. W. Revie, Ed.), John Wiley & Sons, Hoboken, 2011, pp. 15–20.

15 E. D. Verink; Economics of corrosion. in *Uhlig's Corrosion*, 3rd ed. (R. W. Revie, Ed.), John Wiley & Sons, Hoboken, 2011, pp. 21–30.

16 M. Lannuzzi; Corrosion Definitions. 2013, https://www.aboutcorrosion.com/2013/12/15/corrosion-definition/ (last access 05 Oct **2023**)

17 G. Koch, M. P. H. Bringers, N. G. Thompson, Y. P. Virmani, J. H. Payer; *Corrosion Costs and Preventive Strategies in the United States, FHWA and NACE reports, FHWA-RD-01-156*, 2002, accessed via impact.nace.org/documents/ccsupp.pdf (last access 22 Oct. 2022). FHWA – Federal Highway Administration; NACE – National Association of Corrosion Engineers, now an international professional society.

18 G. Schmitt; Global Needs for Knowledge Dissemination, *Research, and Development in Materials Deterioration and Corrosion Control, White Paper.* World Corrosion Organization, May **2009**, accessed at www.corrosion.org (last access 22 Oct **2022**).

19 E. Ghali, *Corrosion Resistance of Aluminum and Magnesium Alloys*, Wiley, Hoboken, NJ, **2010**.

20 E. D. Verink; Simplified procedure for constructing pourbaix diagrams. in *Uhlig's Corrosion Handbook*, 3rd ed. (R. W. Revie, Ed.), John Wiley & Sons, Hoboken, 2011, pp. 93–101.

21 M. Pourbaix: *Atlas of the Electrochemical Equilibria in Aqueous Solutions.* Pergamon Press, Oxford **1966**.

22 G. S. Frankel, D. Landolt; Fundamentals of corrosion. in *Encyclopedia of Electrochemistry, Vol. 4* (A. J. Bard, M. Stratmann, G. S. Frankel, Eds.), Wiley-VCH, Weinheim, 2003, pp. 9–24.

23 D. Powell, J. Cortez, E. K. Melloni; A laboratory exercise introducing students to the Pourbaix of cobalt. *J. Chem. Educ.* **1984**, *64*, 165–167.

24 M. Schalenbach, O. Kasian, M. Ledendecker, F. D. Speck, A. M. Mingers, K. J. J. Mayrhofer, S. Cherevko; The electrochemical dissolution of noble metals in alkaline media. *Electrocatalysis* **2018**, *9*, 153–161.

25 W. T. Thomson, M. H. Kaye, C. W. Bale, A. D. Pelton; Pourbaix diagrams for multielement systems. in *Uhlig's Corrosion Handbook*, 3rd ed. (R. W. Revie, Ed.), John Wiley & Sons, Hoboken, 2011, pp. 103–109.

26 J. W. Schultze, A. W. Hassel; *Passivity of metals, alloys and semiconductors*; in *Encyclopedia of Electrochemistry, Vol.* 4 (A. J. Bard, M. Stratmann, G. S. Frankel, Eds.), Wiley-VCH, Weinheim, 2003, pp. 216–270.

27 G5-94; Standard reference test method for making potentiostatic and potentiodynamic anodic polarization measurements; *Annual Book of ASTM Standards*, Vol. 3.02, ASTM International, Philadelphia, PA, **2000**, p. 57, www.astm.org/standards/G5. ASTM = American Society for Testing and Materials, founded in **1898**.

28 G. S. Frankel; M. Rohwerder, A. J. Bard, M. Stratmann, G. S. Frankel; Electrochemical techniques for corrosion. in *Encyclopedia of Electrochemistry, Vol. 4* (A. J. Bard, M. Stratmann, G. S. Frankel, Eds.), Wiley-VCH, Weinheim, 2003, pp. 687–723.

29 J. H. Fitzgerald; Engineering of cathodic protection system. in *Uhlig's Corrosion Handbook*, 3rd ed. (R. W. Revie, Ed.), John Wiley & Sons, Hoboken, 2011, pp. 1001–1011.

30 H.-H. Strehblow; Mechanisms of pitting corrosion; in *Corrosion Mechanisms in Theory and Practice.* 2nd ed. (P. Marcuse, Ed.), Marcel Dekker, New York, 2002. pp. 243–285.

31 H.-H. Strehblow, A. J. Bard, M. Stratmann, G. S. Frankel; Pitting corrosion, in *Encyclopedia of Electrochemistry, Vol. 4* (A. J. Bard, M. Stratmann, G. S. Frankel, Eds.), Wiley-VCH, Weinheim, 2003, pp. 308–343.

32 H. Böhni; Localized corrosion of passive metals, in *Uhlig's Corrosion Handbook*, 3rd ed. (R. W. Revie, Ed.), John Wiley & Sons, Hoboken, 2011, pp. 157–169.

33 H. Böhni, T. Suter, A. Schreyer; Micro- and nanotechniques to study localized corrosion. *Electrochim. Acta* **1995**, *40*, 1361–1368.

34 T. Suter, H. Böhni; Microelectrodes for corrosion studies in microsystems. *Electrochim. Acta* **2001**, *47*, 191–199.

35 C. Degrigny; Use of electrochemical techniques for the conservation of metal artifacts: A review. *J. Solid State Electrochem.* **2010**, *14*, 353–361.

36 https://www.he-arc.ch/conservation-restauration/formation-continue/selective-cleaning-of-tarnished-silver-based-artefacts-using-local-electrochemistry/ (last access 01 Mar 2026).

37 C. Degrigny, R. Jeanneret, D. Witschard; Local cleaning with the Pleco electrolytic pencil of the Tarnished Saint Candide Reliquary Head at the Treasury of Saint-Maurice Abbey, Valais (Switzerland). *e-Preserv. Sci.* 2015, *12*, 20–27. http://www.morana-rtd.com/e-preservationscience/2015/ePS_2015_a4_Degrigny.pdf (last access 22 Oct **2022**).

38 C. Degrigny, R. Jeanneret , D. Witschard, C. Baudin, G. Bussy, H. Carrel; A new electrolytic pencil for the local cleaning of silver tarnish, *Stud. Conserv.*, 2016, *61*, 162–173. 10.1179/2047058415Y.0000000015 (last access 22 Oct **2022**).

39 F. Huet, P. Marcus, F. Mansfeld; Electrochemical noise technique. in *Analytical Methods in Corrosion Science and Engineering* (P. Marcus, F. Mansfeld, Eds.) CRC Taylor & Francis, Boca Raton, 2006, pp. 507–570.

40 J. Lennox, A. Cunningham, G. James, P. Stewart, R. Gerlach, G. Geesey et al.; www.hypertextbookshop.com/biofilmbook/v004/r003/ (last access 22 Oct **2022**).

41 H. Kaiser, G. A. Eckstein, A. J. Bard, M. Stratmann, G. S. Frankel; Corrosion of alloys. in *Encyclopedia of Electrochemistry, Vol. 4* (A. J. Bard, M. Stratmann, G. S. Frankel, Eds.), Wiley-VCH, Weinheim, 2003, pp. 156–186.

42 K. E. Heusler; Multicomponent electrodes. *Electrochim. Acta* **1996**, *41*, 411–418.

43 P. Strasser, S. Koh, T. Anniyev, J. Greeley, K. More, C. Yu, Z. Liu, S. Kaya, D. Nordlund, H. Ogasawara, M. F. Toney, A. Nilsson; Lattice-strain control of the activity in dealloyed core-shell fuel cell catalysts. *Nat. Chem.* **2010**, *2*, 454–460.

44 G. Wittstock, M. Bäumer, W. Dononelli, T. Klüner, L. Lührs, C. Mahr, L. V. Moskaleva, M. Oezaslan, T. Risse, A. Rosenauer, A. Staubitz, J. Weissmüller, A. Wittstock; Nanoporous gold: From structure evolution to functional properties in catalysis and electrochemistry. *Chem. Rev.* **2023**, *123*, 6716–6792.

45 M. J. Duarte, J. Klemm, O. Klemm, K. J. J. Mayrhofer, M. Stratmann, S. Borodin, A. H. Romero, M. Madinehei, D. Crespo, J. Serrano, S. S. A. Gerstl, P. P. Choi, D. Raabe, F. U. Renner; Element-resolved corrosion analysis of stainless-type glass-forming steels. *Science* **2013**, *341*, 372–376.

46 T. F. Kelly, M. K. Miller; Atom probe tomography. *Rev. Sci. Instrum.* **2007**, *78*, 031101.

47 https://www.mpie.de/2890898/atom_probe_tomography (last access 22 Oct **2022**).

48 R.S. Alwitt; Anodizing. in *Electrochemistry Encyclopedia*, The Electrochemical Society; https://knowledge.electrochem.org/encycl/art-a02-anodizing.htm (last access 12 Jan **2025**).

49 H. Masuda, K. Fukuda; Ordered metal nanohole arrays made by a two-step replication of honeycomb structures of anodic alumina. *Science*, **1995**, *268*, 1466–1468.

50 H. Masuda; Highly ordered nanohole arrays in anodic porous alumina, in *Ordered Porous Nanostructures and Application* (R. B. Wehrsporn, Ed.), Springer Science and Business Media, New York, 2005, Ch. 3, pp. 37–56.

51 G. D. Sulka; Highly ordered anodic porous alumina formation by self-organized anodization, in *Nanostructured Materials in Electrochemistry* (A. Eftekhari, Ed.), Wiley-VCH, 2008, pp. 1–116.

52 A. Ghicov, P. Schmuki; Self-ordering electrochemistry: A review on growth and functionality of TiO_2 nanotubes and other self-aligned MO_x structures. *Chem. Commun.* **2009**, 2791–2808.

53 K. Lee, A. Mazare, P. Schmuki; One-dimensional titanium dioxide nanomaterials: Nanotubes. *Chem. Rev.* **2014**, *114*, 9385–9454.

54 S. P. Albu, A. Ghicov, J. M. Macak, R. Hahn, P. Schmuki; Self-organized, free-standing TiO_2 nanotube membrane for flow-through photocatalytic applications. *Nano Lett.* **2007**, *7*, 1286–1289.

55 C. R. Sides, C. R. Martin; Deposition into Templates, in *Electrochemistry at the nanoscale* (P. Schmuki, S. Virtanen, Eds.) Springer, New York, 2009, pp. 279–320.

56 M. Datta, D. Harris; Electrochemical micromachining: An environmentally friendly, high speed processing technology. *Electrochim. Acta* **1997**, *42*, 3007–3013.

57 M. Datta; Anodic dissolution of metals at high rates. *IBM J. Res. Dev.* **1993**, *37*, 207–226.

58 M. Datta, D. Landolt; Pit and flow streak formation during high rate dissolution of nickel. *J. Electrochem. Soc.* **1982**, *129*, 1889–1895.

59 A. D. Davydov, V. M. Volgin; Electrochemical machining; in *Encyclopedia of Electrochemistry, Vol. 5* (A. J. Bard, M. Stratmann, D. D. Macdonald, P. Schmuki, Eds.), Wiley-VCH, Weinheim, 2007, pp. 809–854.

60 R. Schuster, V. Kirchner , P. Allongue, G. Ertl; Electrochemical micromachining. *Science* **2000**, *289*, 98–101.

61 K. W. Mao; ECM study in a closed-cell system: I. $NaClO_3$. *J. Electrochem. Soc.* 1971, *118*, 1870–1876; K. W. Mao; ECM study in a closed-cell system: II. NaCl, $NaClO_4$, and $NaNO_3$. *J. Electrochem. Soc.* **1971**, *118*, 1876–1879.

62 M. Lohrengel; Pulsed electrochemical machining of $NaNO_3$: Fundamentals and new aspects. *Mater. Manufact. Proc.* **2005**, *20*, 1–8.

15

Electrochemical Energy Storage

We have discussed fuel cells and the lead–acid battery as examples of galvanic cells at the end of Chapter 2. The conversion of fuel or material is linked to the conversion/storage of energy in all (electro)chemical reactions, but batteries, fuel cells, and electrochemical capacitors are optimized specifically for this purpose rather than for obtaining a new substance.

Electrochemical energy conversion systems are of outstanding importance in many technical applications, often having enabled them in the first place. Electrochemical energy sources in the form of Volta's piles[9] were the first sources of stable electricity in the laboratory and enabled fundamental insights into electricity and the discovery of new metallic elements by galvanic deposition. The Daniell cell[6] (Figure 1.8) was of great importance as energy source for telegraphic communication with the Morse apparatus.[155] The use of battery power for propulsion was explored in the early history of automobiles but was then superseded by the widespread use of internal combustion engines that, however, required the starter battery. Battery-powered vehicles "survived" in niche applications such as vehicles for storage facilities (e.g., forklifts) or "milk floats". Nowadays, electromobility is at the core of public and private research and development efforts worldwide. The reasons are obvious: reserves of fossil fuel are fundamentally limited and, in the long run, may be even more needed as feedstock for chemical production. On the other hand, the Earth's atmosphere has a limited and probably already exhausted capacity to take up CO_2 as an end product of combustion processes without tipping the climate in a dangerous way. In addition, emissions from combustion processes include compounds like soot and oxides of nitrogen (NO_x), causing health hazards, especially in urban areas.

Fuel cells initially attracted attention as an energy source in military submarines and on space missions. They could become a central element of a future, sustainable, and emission-neutral hydrogen economy, provided hydrogen can be produced with renewable energy sources in an efficient way.

The introduction of lithium-ion technology since around 1990 has enabled the widespread use of cell phones and other portable appliances. It has been the subsequent improvement of battery performance that made possible today's smartphones (whose computing power exceeds that of data centers in the 1970s) with batteries, whose size and mass have dramatically decreased when compared to early two-way radiotelephone apparatuses. The permanent availability of information in many formats associated with the widespread introduction of smartphones has triggered major social and cultural changes, with no end point in sight.

[155] Samuel Finley Breese Morse, 1791–1872, professor of painting, sculpture, and drawing; as inventor, he created the first telegraph and the Morse alphabet.

In the following two chapters, the focus will be on fundamentals, measurement methods, and materials using the example of electrochemical energy conversion devices. Before continuing, you should review Sections 2.4.4 and 4.3.

Further Reading. The great variety of special batteries and the considerable history of this branch of electrochemistry have led to a great diversity of publications, but also to a certain fragmentation in the presentation and terminology. For a more in-depth study of the subject, some review articles are recommended [1–7]. A precise and quick overview of a wide range of battery systems is provided by encyclopedias and compilations on the web [8, 9]. Further recommended sources on specific aspects are referred to in the respective sections.

15.1 Overview of Electrochemical Energy Conversion Systems

Batteries and fuel cells supply electrical energy by converting chemical energy into electrical energy at *interfaces*. Electrochemical capacitors are considered electrochemical energy conversion devices with a conversion of material only in the sense of interfacial restructuring. Starting at the interface, ions and electrons are transported separately. The electrons move in the external circuit, the ions in the electrolyte.

Batteries are thermodynamically closed systems in which the materials of the electrodes are converted in redox reactions. The storage of energy and the conversion thus take place within the same system. The reacting materials are also called active materials. It is common to divide the electrodes into anodes and cathodes according to their function in the *discharge reaction*. This is conceptually clear for primary cells (i.e., nonrechargeable cells). In the context of secondary cells, the function of anode and cathode is reversed during the charging reaction. Such secondary cells are also called accumulators or rechargeable cells. To avoid the possible confusion with the designation of anode and cathode in a rechargeable cell, we prefer the designations negative and positive electrodes, which do not interchange during charging and discharging. The redox couple (R/O) with the higher electrode potential (more positive) reacts at the positive electrode. Here, O symbolizes a strongly oxidizing agent. At the negative electrode, the redox couple (R′/O′) reacts with the lower redox potential, and R′ is a strongly reducing agent. Consider the discharging reaction of a lithium-ion battery as an example: The strongly electronegative lithiated graphite will "push" electrons into the external circuit, so graphite is the negative electrode. During recharging, a very negative potential must be applied to graphite from the outside to enforce the uptake of lithium ions into the graphite lattice. It is still the more negative of the two battery electrodes.

In the literature on secondary cells, the negative (positive) electrode is often referred to as the anode (cathode) – even when it comes to the charging reaction in which the charge transfer reaction is reversed, and it acts as the cathode (anode) according to the definition in Section 1.4.

In addition to the active materials, the two electrodes and the cell usually contain passive materials, which are *not* converted but contribute to the total mass and volume of the cell. These include the current collectors, which "collect" the electronic current from the active material and guide it to the contact point with the external circuit. The active materials are usually powders that are processed together with a binder to form a paste that is applied to the current collector. To adjust the electronic conductivity, the paste is supplemented with a conductive additive. An internal electrical short circuit, which in this context would occur via an electron-conducting shunt between the positive and negative electrodes within the electrochemical cell, is prevented by a separator (or spacer). There are different designs such as plastic frames in lead–acid batteries or thin layers of porous fabric-like films that are soaked with the electrolyte solution and placed directly between the two electrodes of a lithium-ion cell.

The electrolyte contains a solvent (H_2O in the case of lead–acid batteries, organic carbonates in the case of lithium-ion batteries) and an acid or base or salt dissolved in it (H_2SO_4 in the case of lead–acid batteries, $LiPF_6$ in the case of lithium-ion batteries). The electrolyte can also be an ion-conducting solid. It is important that the electrolyte has a high, but *exclusively ionic* conductivity. It can contain further additives to adjust the properties of the battery.

The term **battery** covers the entire device and may include one or more electrically interconnected galvanic cells.

Fuel cells are thermodynamically open systems. The active materials are combustible gases or liquids (fuels) as reducing agents and oxygen from the air as oxidizing agent. The fuels are kept in a storage tank from where they are fed to a reaction unit (in this case, the fuel cell). Since oxygen is usually taken from open air, only the fuel needs to be stored. During operation, the components of the fuel cell are not converted themselves (apart from undesirable degradation processes). The electrolyte is usually a gastight but ion-conducting thin polymer or ceramic membrane on which the catalytically active electrodes are deposited. The membrane and electrodes form the membrane electrode assembly (MEA). The MEA is covered on both sides by a porous gas diffusion layer (GDL). The gases are distributed to the catalytically active centers through the pore network of the GDL. The GDL also provides the electrical contact with the bipolar plate. A bipolar plate provides the electrical connection between the anode of one cell and the cathode of the next cell. It contains channel structures called flow field for the inflow of the gases. Fuels can be produced from natural gas or from electrolysis processes. Electrolytic hydrogen is expensive because of the electricity costs *and* the investment costs for the electrolysis reactor. Temporary surpluses of renewable energies and increasing taxation of CO_2 emissions may change the cost structure. The effort to avoid CO_2 emissions will foreseeably lead to changes in the legislation and consumer behavior. These trends are drivers for intensive research efforts to use hydrogen and other low molecular weight compounds as energy storage molecules or energy vectors. In this context, electrolysis processes receive renewed attention. If a fuel cell can also perform electrolysis, we speak of a unitized regenerative fuel cell. For fuel cells, as for batteries, we also speak of single cells. Several interconnected cells are called a stack. We will discuss fuel cells and water electrolyzers in Chapter 16.

In addition to these two basic types of energy conversion systems, there are hybrid forms in which the advantages of batteries and fuel cells are combined. For example, **metal–air cells** use gaseous oxygen from the air as an oxidizing agent (as in a fuel cell), but achieve much higher cell voltages by employing metals as reducing agents instead of H_2 (Section 15.7). They can therefore be regarded as a hybrid of a battery and a fuel cell. However, such hybrids not only combine the advantages of batteries and fuel cells, but also the properties that are problematic from an application perspective. For example, dendrite-free metal deposition must be achieved in rechargeable metal–air batteries and, at the same time, a catalytically active electrode must be available for the oxygen reduction reaction and the oxygen evolution reaction.

Redox flow batteries (RFBs) use redox-active compounds, for example, V^{2+}/V^{3+} and VO^{2+}/VO_2^+, in liquid electrolytes as fuels (Section 15.8). The liquid starting materials are pumped from reservoirs into the electrochemical cell (the reactor), and the products are collected at the outlet. The reactor is conceptually similar to a fuel cell, which also runs on fluid fuels. If we consider the storage tank and the feed lines as part of the system, the starting materials are contained in the system too (hence the name battery). The particular advantage of this cell is the possibility of scaling the storage capacity (energy) separately from the power output (energy per time). For a given cell chemistry, the size of the tanks determines the stored energy, and the size of the reaction unit primarily determines the possible power output.

Fluid phase (fluid) – Summary term for liquid and gaseous phases

Condensed phases – Summary term for solid and liquid phases

Electrochemical capacitors store energy primarily by charging the electrical double layer. Thus, they are not necessarily galvanic cells. Such devices are covered in Section 15.9 together with hybrid forms, in which oxidation and reduction reactions, similar to those in a very small battery, contribute to energy storage. Such devices are sometimes called "electrochemical supercapacitors" or just "supercaps".

The many combinations of different electrodes result in a rather complex scheme for classifying all electrochemical energy conversion systems (Figure 15.1). Overall, batteries dominate in applications. Although over a thousand battery systems have been described in the literature, batteries produced in large numbers have been limited to a few cell types. In the context of this textbook, we will limit ourselves to presenting the basic principles and development potential using selected examples for aqueous and nonaqueous cell chemistries. The use of fuel cells in the Apollo and Gemini space missions triggered great euphoria or hopes, but so far, only a few large-scale applications have developed on Earth. Electrochemical capacitors are used in the electronics industry, but are otherwise found only in niche applications. Use in vehicles to recover braking energy or to cover short distances (e.g., in buses) is in the phase of field-testing.

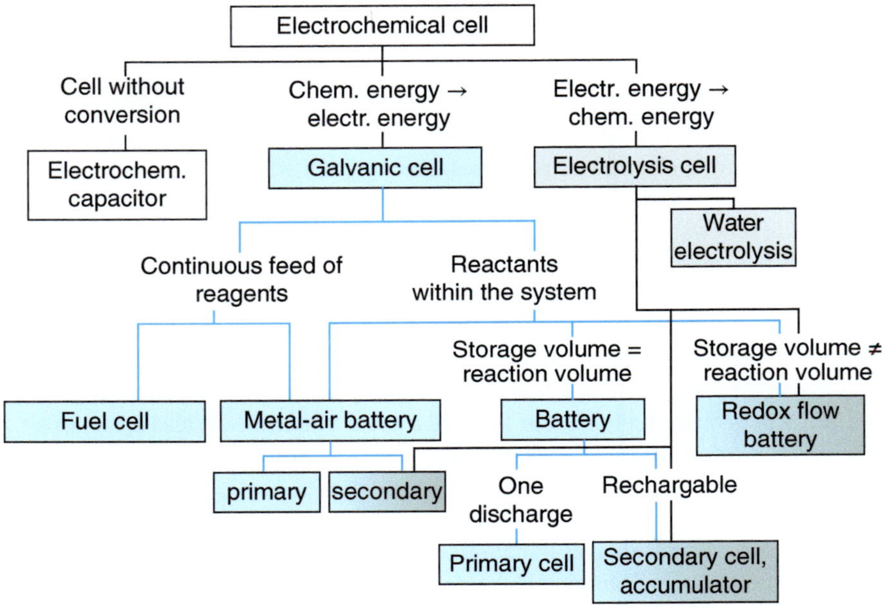

Figure 15.1 Family tree of electrochemical energy conversion systems. Accumulators (rechargeable batteries) are galvanic cells when delivering energy and electrolytic cells when charging, thus their shaded coloration in the scheme.

15.2 Figures of Merit for Batteries and Battery Materials

15.2.1 Capacity and Energy

Attention. Distinguish symbols and units. Symbols are set in italics.

W – Work, energy [W h], [W s], [J]
P – Power = $I(t)\ U(t) = W/t$ [W]

The most important parameters are the capacity Q [A h], the energy W [W h], and the power P [W]. The (mass-)specific energy $W_m = W/m$ is obtained by dividing the stored energy by the mass m of the battery or the fuel cell with a tank [W h kg^{-1}]. The (mass-)specific power [W kg^{-1}] is the work delivered or absorbed per time divided by m: $P_m = P/m = W/(t\ m) = I\ U/m$. Analogously, the energy and power density can be formed by referencing the quantity of interest to the volume of the cell (W_V, P_V). Sometimes these quantities are also called "volumetric energy density" and "volumetric power density," although this can be regarded as an undesired tautology. Depending on the application, the energy density or the specific energy is the more critical quantity.

One obtains the nominal capacity Q_N directly as the product of the amount of material N (mass m/molar mass M) used in the battery, the reaction valency n, and the Faraday constant F. We obtain the specific capacity by relating Q_N to the mass ($Q_{N,m}$) or to the volume ($Q_{N,V}$) (Example Calculation 15.1). For cells, those quantities must be based on the electrode with the smaller capacity.

Very often, nominal capacity is represented by the symbol $C_N = Q_N/m$. We avoid this usage to mitigate the possible confusion with capacitance $C = Q/U$ (cf. Section 15.9).

Q: Why should the capacities of the positive or negative electrode not be equal? If they are equal, the highest specific energy of the cell could be achieved.

A: That's right. In fact, the active masses of positive and negative electrodes are often not exactly balanced, because it should be predetermined which of the two electrodes limits the discharge process. For example, some electrodes should not be fully discharged, because the crystal structure would be irreversibly changed. This can be prevented by using a substoichiometric amount of material from the oppositely charged electrode and by accepting a reduced specific energy.

The capacity can also be measured/calculated for an individual electrode. We determine the theoretical energy density from the capacities and the discharge voltages (Example Calculation 15.2). Either the difference of the standard potentials, the open circuit voltage at the beginning of the discharge or, with more realistic results, the actual mean discharge voltage can be used as cell voltage. Using the (mass) densities from tabulated data, we obtain the theoretical energy density (Example Calculation 15.3). However, these quantities represent only upper bounds, which means that the actual values can never be larger. More realistic estimates are obtained if we also include the mass of other components of the battery. However, this usually requires detailed knowledge of the internal structure and the materials used. For example, water for dissociation of the acid, excess acid for conductivity in the discharged state, and the mass of the case (housing of the battery), current collectors, and spacers must be considered for a lead–acid battery in addition to the active masses (see Example Calculation 15.2). In addition, only 35–60% of the active mass is actually converted. Overall, the practical energy density for this type of battery is only 18–25% of the theoretically possible energy density [1]! This is about 30 W h kg^{-1}, which means that a battery with a total weight of 1 kg can supply a 60 W lamp with energy for 30 min.

Example Calculation 15.1 Specific capacity of a negative graphite electrode of a lithium-ion battery

We consider the mass of graphite to be charged up to the maximum lithiation level LiC_6. This yields

$$M(C_6) = 72\ \text{g mol}^{-1} = 0.072\ \text{kg mol}^{-1}\ (\nu = 1)$$

$$n = 1$$

The usual unit for the specification of specific capacities is A h kg^{-1} = mA h g^{-1}. For the conversion of the units, the author recommends the method of the "unity factor". Here, the expression is multiplied by one, which is the quotient of one and the same quantity, given in different units. We choose the units so that the desired unit is in the numerator and the unit that is to "disappear" is in the denominator. Here this would be 1 = 1000 g/1 kg and 1 = 1 h/3600 s. In this way, the units cancel, and the calculation remains correct and comprehensible in the units. Dimensionless constants, which have a correct meaning only in a certain context, can be avoided in this way.

$Q_{N,m}$ – Specific capacity; 1 [A s g^{-1}] = 0.278 [mA h g^{-1}] = 0.278 [A h kg^{-1}]
ν_i – Stoichiometric coefficients of component i
M_i – Molar mass of component i [g mol^{-1}]
F – Faraday constant = 896 485 A s mol^{-1}
n – Reaction valency [dimensionless]

$$\begin{aligned}Q_{N,m} &= \frac{nF}{\sum \nu_i M_i}\\ &= \frac{1 \cdot 96\,485\ \text{A s mol}^{-1}}{72\ \text{g mol}^{-1}}\\ &= 1340\ \text{A s g}^{-1}\frac{1\ \text{h}}{3600\ \text{s}} \times \frac{1000\ \text{g}}{1\ \text{kg}}\\ &= 372\ \text{A h kg}^{-1}\end{aligned} \tag{15.1}$$

We obtain the volume-based capacity by multiplying by the density of graphite $\rho = 2.25\ \text{g cm}^{-3} = 2.25\ \text{kg L}^{-1}$.

$Q_{N,V}$ – Volume-related capacity or "charge density" [A s cm^{-3}], [mA h cm^{-3}] = [A h L^{-1}]
ρ – Density [g cm^{-3}] = [kg L^{-1}] = [t m^{-3}]

$$\begin{aligned}Q_{N,V} &= \rho Q_{N,m}\\ &= 2.25\ \text{kg L}^{-1} \cdot 372\ \text{A h kg}^{-1}\\ &= 837\ \text{A h L}^{-1}\end{aligned} \tag{15.2}$$

◢

Example Calculation 15.2 Calculation of the theoretical specific energy of a lead–acid battery

We first calculate the nominal capacity from the molar masses of the reactants and the reaction valency. Here, we consider two molecules of H_2SO_4 for a formula conversion.

$$\text{Pb}(s) + \text{PbO}_2(s) + 4\,\text{H}^+ + 2\,\text{SO}_4^{2-} \rightarrow 2\,\text{PbSO}_4(s) + 2\,\text{H}_2\text{O} \tag{15.3}$$

$n = 2$
$M(\text{Pb}) = 207\ \text{g mol}^{-1}$
$M(\text{PbO}_2) = 239\ \text{g mol}^{-1}$
$M(\text{H}_2\text{SO}_4) = 98\ \text{g mol}^{-1}$

$Q_{N,m}$ – Specific capacity; 1 [A s g^{-1}] = 0.278 [mA h g^{-1}] = 0.278 [A h kg^{-1}]
ν_i – Stoichiometric coefficient of component i
M_i – Molar mass of component i [g mol^{-1}]
F – Faraday constant
n – Reaction valency [dimensionless]

$$\begin{aligned}Q_{N,m} &= \frac{nF}{\sum \nu_i M_i}\\ &= \frac{2 \cdot 96485\ \text{A s mol}^{-1}}{(207 + 239 + 2 \cdot 98)\,\text{g mol}^{-1}}\\ &= 300.6\ \text{A s g}^{-1}\end{aligned} \tag{15.4}$$

To convert the units to A h kg^{-1} = mA h g^{-1} we use the method of the "unity factor" (see Example 15.1).

$$\begin{aligned}Q_{N,m} &= 300.6\ \text{A s g}^{-1} \cdot \frac{1\ \text{h}}{3600\ \text{s}} \cdot \frac{1000\ \text{g}}{1\ \text{kg}}\\ &= 83.5\ \text{A h kg}^{-1}\end{aligned} \tag{15.5}$$

This value can be made more realistic by taking into account the necessary mass of excess water, the incomplete mass utilization, and the masses of current collectors, separators, and the cell container. We obtain the maximum possible energy density if we use the open circuit cell voltage of 1.96 V from Example 2.15 as a basis.

W_m – Specific energy [W h kg^{-1}] = [mW h g^{-1}]

$$\begin{aligned}W_m &= \frac{W}{m} = \frac{UIt}{m} = U\frac{It}{m} = U\frac{nF}{\sum \nu_i M_i} = UQ_{N,m}\\ &= 1.96\ V \cdot 83.5\ \text{A h kg}^{-1}\\ &= 164\ \text{W h kg}^{-1}\end{aligned} \tag{15.6}$$

This value is much larger than the experimental value, not only because of the assumptions made when calculating the specific nominal capacity. The voltage used as a basis also corresponds to the maximum possible voltage. However, the voltage will deviate to varying degrees from the thermodynamically calculated value depending on the current.

Example Calculation 15.3 Calculation of the theoretical energy density of a lead–acid battery

To calculate the energy content per volume, we again start from the molar masses of the reactants in the charged state and convert the molar masses with the densities ρ into the volumes of the reactants. We assume pure sulfuric acid as electrolyte. Analogous to ignoring the solvent water, we also refrain from considering other cell components and incomplete mass utilization. In addition, we do not verify whether the active masses occupy a larger volume in the charged or in the discharged state.

$\rho(\mathrm{Pb}) = 11.34\ \mathrm{g\ cm^{-3}}$

$\rho(\mathrm{PbO_2}) = 9.87\ \mathrm{g\ cm^{-3}}\ (\alpha - \mathrm{PbO_2})$

$\rho(\mathrm{H_2SO_4}) = 1.84\ \mathrm{g\ cm^{-3}}$

$$\begin{aligned} W_V &= \frac{W}{V} = \frac{UIt}{V} = U\frac{It}{V} = U\frac{nF}{\sum \frac{\nu_i M_i}{\rho_i}} = UQ_{\mathrm{N},V} \\ &= 1.96\ \mathrm{V} \cdot \frac{2 \times 96\,485\ \mathrm{A\ s\ mol^{-1}}}{\left(\frac{207}{11.34} + \frac{239}{9.87} + \frac{98}{1.87}\right)\frac{\mathrm{g\ mol^{-1}}}{\mathrm{g\ cm^{-3}}}} \\ &= 1.96\ \mathrm{V} \cdot \frac{192\,970\ \mathrm{A\ s\ mol^{-1}}}{(18.25 + 24.21 + 104.81)\frac{\mathrm{g\ mol^{-1}}}{\mathrm{g\ cm^{-3}}}} \\ &= 1.96\ \mathrm{V} \cdot 1310\ \mathrm{A\ s\ cm^{-3}} \cdot \frac{1\ \mathrm{h}}{3600\ \mathrm{s}} \cdot \frac{1000\ \mathrm{cm^3}}{1\ \mathrm{L}} \\ &= 1.96\ \mathrm{V} \cdot 364\ \mathrm{A\ h\ L^{-1}} \\ &= 713\ \mathrm{W\ h\ L^{-1}} \end{aligned} \tag{15.7}$$

$Q_{\mathrm{N},V}$ – Volume-related capacity or "charge density" [A s cm^{-3}], [mA h cm^{-3}] = [A h L^{-1}]

ν_i – Stoichiometric coefficients of component i

M_i – Molar mass of component i [g mol^{-1}]

ρ_i – Density of component i [g cm^{-3}] = [kg L^{-1}] = [t m^{-3}]

F – Faraday constant

n – Reaction valency [dimensionless]

To measure the practical energy density, we record a discharge curve. Here, the delivered voltage is followed at a constant current (CC) and plotted against time (Figure 15.2). The area under the curve multiplied by the constant current yields the energy, which still has to be related to the total mass (or volume) of the system [Eqs. (15.8) and (15.9)].

$$W_m = \frac{W}{m} = \frac{I}{m}\int_{t=0}^{t_{\mathrm{end}}} U(t)\mathrm{d}t \tag{15.8}$$

$$W_V = \frac{W}{V} = \frac{I}{V}\int_{t=0}^{t_{\mathrm{end}}} U(t)\mathrm{d}t \tag{15.9}$$

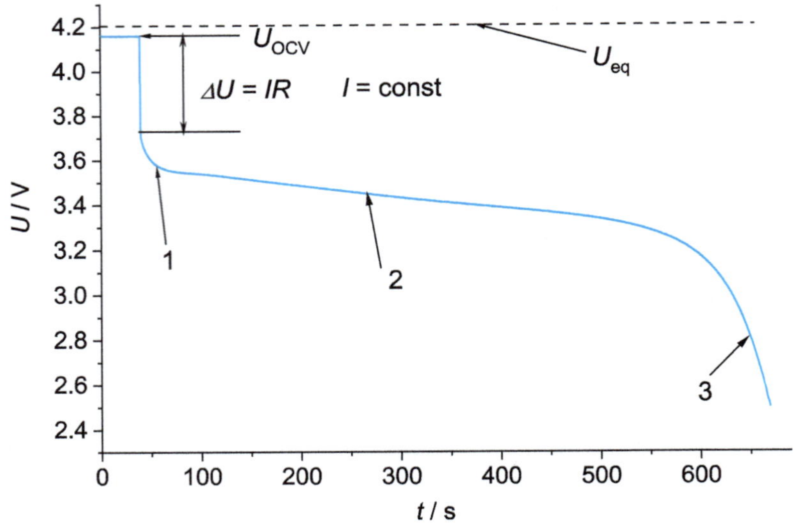

Figure 15.2 Discharge curve of a battery. The small difference between the thermodynamic equilibrium voltage U_{eq} and the open circuit voltage U_{OCV} results from the formation of a mixed potential due to side reactions. With onset of the discharge current, there is a voltage drop across the internal resistances within the battery. Further voltage drops are added to this drop at various stages of discharge (1, 2). When nearly all the convertible material is consumed, the voltage collapses, and the battery reaches the end of its life (3). Raw data were kindly provided by P. Das, EWE Research Center NEXT ENERGY.

15.2.2 Power and Ragone Plot

Initially, it may appear to be arbitrary which current is selected for the experiment in Figure 15.2. If the current is doubled, the discharge time should be halved, but the specific energy should not change. However, this is only a very rough approximation. In fact, at high discharge rates, the cell voltage drops sharply or the capacity drops, because the material utilization decreases. Since the discharge rate naturally depends not only on the current I, but also on the nominal capacity Q_N, it is common to state the discharge rate as the so-called C-rate C_r. This is the ratio of the discharge current I and the nominal capacity Q_N. I and Q_N are also related by the theoretical discharge time t_d ($Q_N = I\, t_d$). The C-rate is usually given in h^{-1}, although a unit is often not specified.

> The C-rate indicates what proportion of the theoretical capacity is charged or discharged within one hour at a given current.

C_r – C-rate [A/(A h)] = [h^{-1}]
t_d – Discharge time [s], [h]
Q_N – Nominal capacity calculated from stoichiometry [A s], [A h], etc.

$$C_r = \frac{I}{Q_N} = \frac{1}{t_d} \tag{15.10}$$

A C-rate of 1 (h^{-1}) means that the cell is discharged in one hour, a C rate of 2 (h^{-1}) means that the discharge process takes only 0.5 h = 30 min. A C-rate of 0.25 (h^{-1}) denotes a charging or discharging process lasting 4 hours.

From what has been said, it is already clear that a purely mathematical estimation (prediction) of the specific power or the power density is more difficult than the prediction of the specific energy or the energy density. For this purpose, information about the kinetics of both electrode reactions must be included.

An illustrative comparison of energy storage systems is provided by a Ragone plot[156], in which the specific power is plotted against the specific energy in a double logarithmic or semilogarithmic coordinate system (Figure 15.3). The plot may contain additional diagonal lines (isochrons). They connect points with equal times for complete charging or complete discharging. The exact appearance of the diagrams depends on the particular energy storage systems used as a basis and changes with further technical development of the cells.

[156] David Vincent Ragone, 1930–2023, U.S. metallurgist and physical chemist, professor at the University of Michigan, Carnegie Mellon University in Pittsburgh, Dartmouth College, Case Western Reserve University in Cleveland, MIT, industrial appointment at General Atomics.

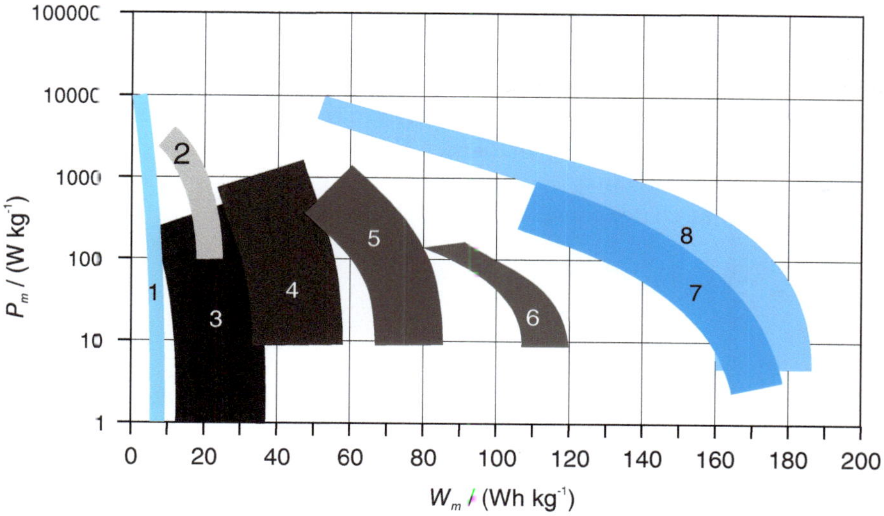

Figure 15.3 Ragone plots for different electrochemical energy storage devices. 1) Electrochemical capacitors (low spec. Energy, high spec. power); 2) lead–acid battery; 3) spiral lead–acid battery; 4) Ni–Cd; 5) Ni-metal hydride; 6) Na–$NiCl_2$ (ZEBRA); 7) Li–polymer; and 8) Li-ion with optimization either toward high specific energy or toward high specific power.

The energy storage devices occupy very different areas in the Ragone plot and, thus, best serve different types of applications. Electrochemical capacitors can absorb and provide high power but have only a small capacity. They are therefore suitable for the rapid absorption of excess energy or for supplying a short-term high-energy demand (e.g., photo flash devices). Batteries can absorb more energy, but cannot release it as quickly (e.g., for a more constant flow of electricity in clocks or digital devices). This trend continues with fuel cells with even larger specific energies and even smaller specific powers. These are perhaps suitable as a power supply for remote settlements or for emergency power generators. Since no type of electric energy conversion system can cover the requirements of all applications and sometimes not even those of a single application (Section 2.4.4), different types of power storage systems are often interconnected electrically. Internal combustion engines (diesel and gasoline engines) appear on a "sweet spot" in the Ragone plot at high energy density and high power density. This is one reason why it is challenging to introduce electric vehicles across the board. The very high characteristics of internal combustion engines result, among other things, from the very large entropy production (distribution of the total combustion products and waste heat in the atmosphere). At the same time, this also constitutes the ultimate reason for the large environmental impact that this technology causes in the long term. A detailed discussion of fuel cells and of batteries to replace internal combustion engines for vehicles can be found in review papers tracing the evolving perspectives and experiences [6, 7].

15.2.3 Efficiency of Energy Storage Devices

To calculate the efficiency of energy storage devices, we relate different quantities of the same dimension, for example, the electric charge during discharging to that during charging. Therefore, these quantities hardly play a role for primary cells. The efficiencies ε (with indices) can be expressed as dimensionless ratios between zero and one or as percentages. The difference in efficiencies from 100% is caused by loss processes. In Eq. (1.24), we discussed the faradaic efficiency $\varepsilon_{F,l}$ (or current efficiency) and in Eq. (2.161) as well as Eq. (2.162) the maximum electrical efficiency $\varepsilon_{el,max}$ of a fuel cell. These are examples of efficiency values. We need to further specify these for electrochemical energy conversion devices and link them to other influencing factors.

! Frequently, the formula symbol η is found in the literature for efficiency quantities, but we avoid this because of the possible confusion with overvoltage η, which is also significant within the same context.

Although some losses are unavoidable in principle according to the second law of thermodynamics (Example 2.16), the analysis of the different loss processes represents an important starting point for the further development of energy conversion devices, because it identifies the meaningful fields of action. The efficiencies are also important input variables for techno-economic assessments [10] and (ecological) system analyses. They form an important intersection between research in the laboratory or in a pilot plant and technical implementations, which we do not elaborate on further here.

ε_C – Charge efficiency or coulombic efficiency [dimensionless]

$|Q_d|, |Q_c|$ – Absolute amounts of the charge for discharging (index d of discharge) and charging (index c for charge) [A s], [A h], etc.

I_d, I_c – Charge and discharge currents, indices as for Q_d, Q_c [A]

t, t_d, t_c – Time in general, discharge time, charging time, indices as for Q_d, Q_c [s], [h]

The **coulombic efficiency** ε_C is the ratio of the electric charge delivered during discharge Q_d to the charge converted during charging Q_c.

$$\varepsilon_C = \left|\frac{Q_d}{Q_c}\right| = \left|\frac{\int_0^{t_d} I_d(t)\mathrm{d}t}{\int_0^{t_c} I_c(t)\mathrm{d}t}\right| \tag{15.11}$$

A quantity charge factor is also used, which is the reciprocal of ε_C.

The values of ε_C are different for batteries. Deviations from 100% are caused by side reactions, which can occur depending on the cell chemistry and operation mode (!). Side reactions reduce the faradaic efficiency of the main reaction $\varepsilon_{F,l}$, because they consume part of the charge flowing over the external circuit. If a storage device shows no side reactions, then $\varepsilon_C = 100\%$ holds. Electrochemical double-layer capacitors achieve values nearly as high ($\varepsilon_C > 99\%$) because the accumulation of ions in the electrical double layer is exploited here for charge storage without chemical conversion. Lithium-ion batteries (Section 15.6) also achieve ε_C between 90 and 100% after they have been passed through the initial charge–discharge cycles (formation).

Water electrolysis is a possible side reaction for almost all batteries with aqueous electrolytes (Example Calculation 2.15). Let us assume that we charge a lead–acid battery with $\varepsilon_{F,1} = 94\%$, that is, 94% of the passed charge is used to convert $PbSO_4$, but a fraction of 6% would be due to water electrolysis. During discharge with $\varepsilon_{F,2} = 95\%$, 95% of the charge Q_{stored} in the form of Pb and PbO_2 is delivered to the external circuit (and 5% leads to further water electrolysis during discharge). Thus,

Q_{stored} – Charge stored after charging [A s], [A h]

$$Q_{stored} = \varepsilon_{F,1}\, Q_c, \tag{15.12}$$

$$Q_d = \varepsilon_{F,2}\, Q_{stored} = \varepsilon_{F,2}\, \varepsilon_{F,2}\, Q_c, \tag{15.13}$$

$$\varepsilon_C = \frac{Q_d}{Q_c} = \frac{\varepsilon_{F,2}\, \varepsilon_{F,1}\, Q_c}{Q_c} = \varepsilon_{F,2}\, \varepsilon_{F,1} = 94\% \cdot 95\% = 89\%. \tag{15.14}$$

When considering the thermodynamics (formal potential), one could expect that water electrolysis would be the dominating reaction in the lead–acid battery. To reduce the rate of water electrolysis to an acceptable level, some "tricks" are necessary (Section 15.5). Depending on the design, the slight gas evolution in some lead–acid batteries is even desirable, since the rising gas bubbles cause a beneficial convective circulation of the electrolyte. However, the extent of the side reactions is by no means constant. Instead, it usually changes with the temperature (rule of thumb: temperature increase by 10 K accelerates side reactions by a factor of two). The temperature also depends significantly on the current. In addition, the proportion between main and side reactions changes with the state of charge of the battery (SoC); SoC = 1 corresponds to the fully charged battery, SoC = 0 denotes the fully discharged state. If the SoC approaches one, the charge can no longer be converted at all in the main reaction, and therefore, a higher proportion of the charge flows into water electrolysis. For accurate measurements, it is necessary to specify ε_C for a given SoC. In addition, ε_C is significantly influenced by the age of the battery. If corrosion protection layers are thinned or added inhibitors for water electrolysis are depleted, the current yield for the main reaction is further reduced. Higher C-rates require higher currents, which are only achieved at higher voltages, which accelerate side reactions.

Using the example of the lead–acid battery, it becomes clear that we must expect a loss of the accumulated charge if the battery is stored at open circuit voltage. The self-discharge (or standby loss) is about 0.1% of the charge per day in standby at the beginning of the lifetime. The significance of self-discharge processes varies depending on the cell chemistry. Since self-discharge depends on the storage time, we determine the discharge capacity immediately after charging. The obtained efficiency is also called round-trip efficiency or cycling efficiency.

Although the specification of ε_C is not meaningful for fuel cells, it can certainly be determined for a fuel cell; in this case, this quantity indicates how much charge relative to the quantity of fuel used enters the external circuit. However, the deviations from 100% depend mainly on the fuel utilization rate of only 60–80% for the single pass. This low value makes fuel recirculation or postcombustion necessary. Another loss mechanism is crossover of some fuels such as methanol through the membrane.

The **voltaic efficiency** ε_V is the ratio of the average discharge voltage to the average charge voltage:

$$\varepsilon_V = \frac{\langle U_d \rangle}{\langle U_c \rangle} \tag{15.15}$$

ε_V – Voltaic efficiency [dimensionless]
$\langle U_d \rangle$, $\langle U_c \rangle$ – Average discharge (d) or charge (c) voltage [V]

Typically, the numerical values of ε_V for SoC ranges are determined with CC charging and CC discharging. Voltage losses occur in all energy conversion devices due to ohmic losses at internal electrical resistances in the electrolyte, membranes, and separators. Therefore, even an electrochemical double-layer capacitor will not achieve ε_V of 100%. In addition to internal electrical resistances, an overvoltage has the effect of activating the reactions. For the charging reaction to proceed sufficiently fast, a much higher voltage must be applied than would be expected according to the standard electrode potential (Chapter 3, Butler–Volmer equation, Figure 3.8). During discharging, only a voltage lower than the difference of the standard electrode potentials is expected. Less energy can be harvested for the same reasons.

The **energy efficiency** ε_{el} (or electrical efficiency) is the ratio of the electrical energy W_d (= useful energy) released during discharging and the energy W_c expended for charging. For CC charging and CC discharging, ε_{el} is the product of ε_C and ε_V, so that all losses discussed above are included in the energy efficiency:

$$\varepsilon_{el} = \left| \frac{W_d}{W_c} \right| = \frac{|I_d| \langle U_d \rangle t_d}{|I_c| \langle U_c \rangle t_c} = \varepsilon_C \varepsilon_V \tag{15.16}$$

ε_{el} – Energy efficiency [dimensionless]
$\langle U_d \rangle$, $\langle U_c \rangle$ – Average discharge and charge voltages [V]
$|I_d|$, $|I_c|$ – Absolute amounts of discharge and charge currents [A]
t_d, t_c – Discharge and charging time [s], [h]

For a fuel cell, we use the enthalpy $\Delta_R H$ of the fuels for the energy expended. For fuel cells operated at constant current, this gives the electrical efficiency:

$$\varepsilon_{el} = \frac{|-U_d|}{\frac{|\Delta_R H|}{nF}}. \tag{15.17}$$

$|-U_d|$ – Absolute value of the harvested voltage [V]
$|\Delta_R H|$ – Absolute value of the molar enthalpy of the reaction [J mol^{-1}] = [V A s mol^{-1}]
n – Reaction valency, number of electrons for the conversion of one molecule of fuel [dimensionless]
F – Faraday constant [A s mol^{-1}]

For $\Delta_R H$, we only take the reaction enthalpy of the *converted* fuels (cf. Example Calculation 2.16). This makes sense if complete conversion is ultimately achieved by fuel recirculation.

In addition, it is necessary to specify the system boundaries because batteries and fuel cells may contain other components such as gas processing, electrical converters, pumps, and water or battery management systems (BMS) that consume some of the released electrical energy and thus reduce the energy provided to the outside.

15.2.4 Other Figures of Merit

Depending on the application area, there are further specifications for batteries, of which the **lifetime** is perhaps the most important. For primary batteries, the shelf life in the charged state is particularly important. It depends on the rate of self-discharge.

For rechargeable batteries, the cycle life is significant. It indicates the number of charge and discharge cycles the battery can perform without violating one or more criteria that are necessary for functional performance, for example, a minimum remaining capacity. The service life (or calendar life) is the time the battery is capable of delivering sufficient energy and refers to a specific application, for example, powering a wristwatch. The end of service life is determined by the exhaustion of the stored capacity.

Secondary cells cannot be recharged an unlimited number of times because, in addition to the main reaction of the battery, the materials and components undergo gradual changes, which are also referred to as aging. As in real life, these are the changes that inevitably occur even with good maintenance (topping up water in lead–acid batteries, following operational procedures, using the humidification regime for fuel gases in a fuel cell, etc.). The most important causes of aging are corrosion or degradation of cell components, including membranes and separators. Other irreversible changes in cell properties are due to changes in the microstructure of the materials. For example, particles can fracture during charging and discharging due to the change in volume. Particles (especially very small metal particles) can be subject to Ostwald ripening[31], in which large particles within a particle population grow at the expense of smaller particles. Particles may lose electrical contact with the current collector because they have cleaved off from another particle.

In addition, the current collector or particles may become covered by an insulating layer of corrosion products. Equally important as the contact to the current collector is contact to the electrolyte, which can be compromised as a result of volume expansion and contraction, particularly in the case of solid or gel electrolytes. Drying out or loss of electrolyte due to electrolyte decomposition can also reduce the contact area of the electrodes to the electrolyte. Finally, the gradual formation of passive layers occurs in many cases, slowing down the cell reaction. As a result, the proportion of current converted into side reactions increases. That means, more current flows to charge a given amount of active material. This additional current flow for side reactions leads to more heating, which further accelerates side reactions. Consequently, the efficiency of the cell continues to decrease. We will return to the aging of lead–acid batteries in Section 15.5.2 and to the aging of lithium-ion batteries in Section 15.6.5.

In summary, aging of secondary cells is described by their "state of health" (SoH). However, there is no definition for this in the sense of a formula. Rather, the manufacturers of BMSs adapt the calculation to the respective application in order to indicate, for practical purposes, when a battery must be replaced. Measured variables can be used with different weightings for this purpose: the internal resistance (or impedance), the capacity compared to the capacity of the new battery, ε_V, the self-discharge rate, and the ability to absorb a charge.

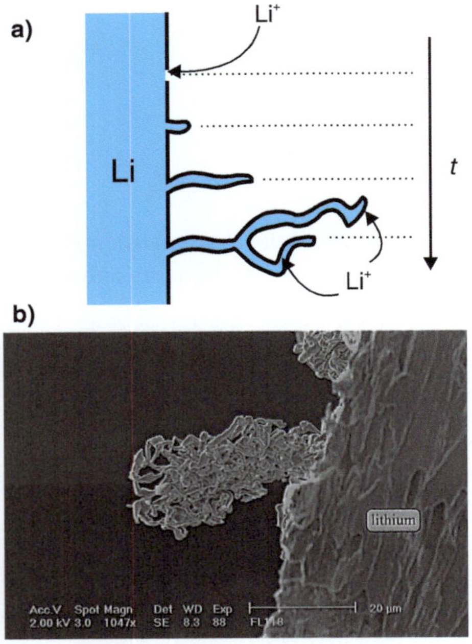

Figure 15.4 Galvanic deposition of mossy and dendritic coatings of Li. a) Schematic; and b) electron micrograph of a Li dendrite. Source: Reproduced with permission from F. Orsini, A. Du Pasquir, B. Beaudoin, J. M. Tarascon, M. Trentin, N. Langenhuizen, E. de Beer, P. Notten; In-Situ Scanning Electron Microscopy (SEM) Observation of Interfaces within Plastic Lithium Batteries. *J. Power Sources* **1998**, *76*, 19–29, Copyright 1998, Elsevier.

Another important aspect is **safety**. Since batteries contain energy that, when released uncontrolled, for example, in the event of a short circuit, is fundamentally associated with risks. Additional safety risks arise with batteries due to possible gas development with leakage or bursting of sealed cells, leakage of corrosive and toxic chemicals after damage to the cell housing. Above all, risks of lithium-ion batteries are associated with flammable organic electrolytes in combination with metals such as Li, which tend to form dendrites or mossy deposits during charging. Dendrites are small thread-like or tree-like extensions (Figure 15.4). If these thin metallic filaments short-circuit the positive and negative electrodes, the stored charge flows off rapidly via this metallic junction with strong heat generation. As a result of strong local heating, battery components decompose [possibly producing O_2 in the cell (Section 15.6.4)] and the electrolyte evaporates. The gas generation leads to pressure buildup and cell bursting, whereupon the flammable cell components come into contact with atmospheric oxygen and can burn. Impressive videos of such processes can be found on the internet. Cheap lithium-ion batteries as used in model sports are particular vulnerable to such incidents [11–14].

Batteries with (nonflammable) aqueous electrolytes are devoid of the risks associated with oxygen contact to the electrolyte. However, the water electrolysis may generate an oxyhydrogen detonating gas in the vicinity of electrical charging infrastructure that may generate sparks. Despite the better safety features, storing energy is never free of hazards. Such risks can be reduced by precise adherence to the operating regime or its adaptation to the state of the cell (SoC, SoH) by means of BMS.

Additional safety risks lie in the charging infrastructure, such as cable fires in charging stations for forklift trucks, etc. The latter safety risks, though important, lie outside the battery and will not be considered further here.

15.3 Battery Classification

Although only about a dozen of the more than a thousand batteries proposed in the research literature have achieved commercial significance, there is a great variety of battery products. (Have you perhaps also had the experience that after painstakingly searching for a battery it did not fit into a camera or other device?) However, this fit mainly refers to the external form, which can conceal different chemical systems. We want to try sorting out the abundance of variants a little bit by giving an overview and encourage further reading in monographs, edited books, and review papers [8, 15].

15.3.1 Classification According to the Type of Active Material

Except for RFBs, which we discuss separately in Section 15.8, all batteries are based on the conversion of solid active masses. There are various ways of doing this: The negative electrodes are usually made of metals, which dissolve during discharge and donate electrons to the current collector. It is important to note that electronegative metals do not react spontaneously with the solvent (at least not continuously), which would be expected since most base metals are stronger reducing agents than H_2. However, the rate of hydrogen evolution reaction on important battery metals such as lead or zinc exhibits a particularly large overpotential. As a result, contrary to the expectation from the electrochemical series, the metal ion is deposited first and hydrogen is formed only at even much more negative potentials. Alkaline electrolytes are also suitable for suppressing the hydrogen evolution reaction.

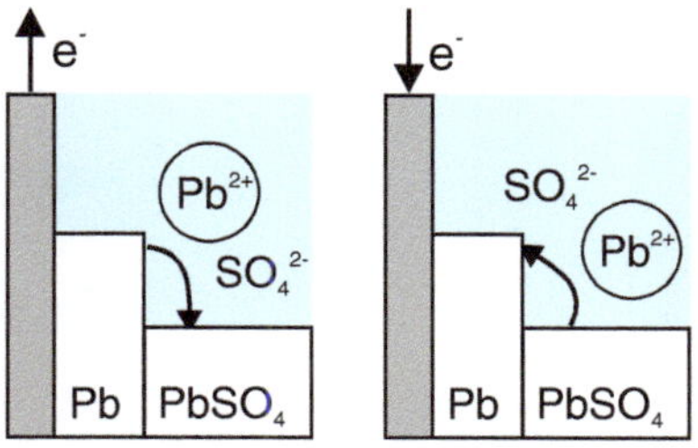

Figure 15.5 Deposition of smooth coatings on a Pb|$PbSO_4$ electrode according to the solution–precipitation mechanism.

Additionally, it is required for recharging that the metal is not just deposited but obtained as a smooth coating. Many metals such as Li or Zn tend to yield rough, porous or dendritic layers (Figure 15.4) [16, 17]. As described in Section 15.2.4, this creates the risk of a short circuit during recharging. Many metals are therefore not suitable (or at least difficult) for secondary cells. The situation is different for lead in contact with sulfuric acid, which forms a metal electrode of the second kind. The concentration of Pb^{2+} is kept very low by the simultaneous solubility equilibrium of $PbSO_4$ which is deposited on the Pb electrode. The associated mechanism is called solution–precipitation mechanism. It also works in the reverse direction (Figure 15.5). Since smooth coatings are formed, this electrode is also suitable for secondary elements.

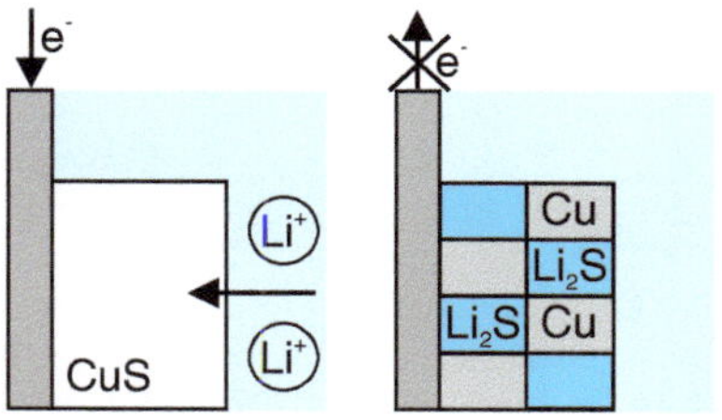

Figure 15.6 Displacement reaction of Cu^{2+} with formation of separate solid phases (Cu and Li_2S) upon reduction (discharge) of CuS as the positive electrode.

An example of a displacement reaction is found in the CuS electrode as a positive electrode. Upon reduction, metallic copper is deposited and Li^+ forms a salt with sulfide S^{2-}. Both, Cu and Li_2S, form separate, stable solid phases, so the reaction cannot be reversed (Figure 15.6).

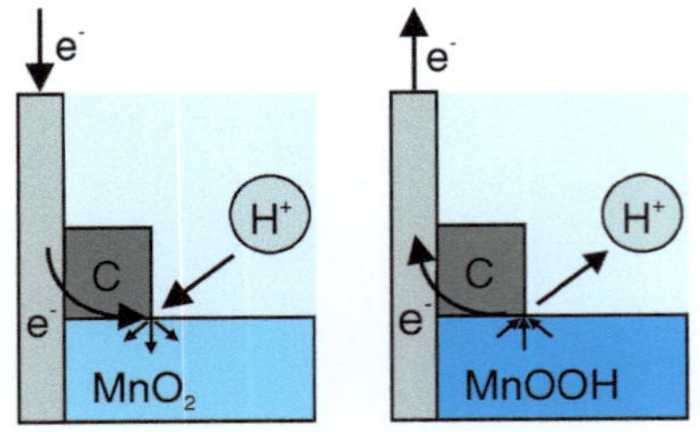

Figure 15.7 Insertion electrode: reduction of MnO_2 at the three-phase boundary between electron conductor (C), MnO_2, and liquid electrolyte with simultaneous insertion of H^+.

Intercalation reactions or insertion reactions are used for many positive electrodes. Manganese dioxide MnO_2 contains Mn in oxidation state IV. Reduction to oxidation state III becomes possible when the charge is compensated by insertion of H^+ or Li^+ into the crystal lattice (insertion electrode, Figure 15.7). The inserted ions are mobile within the solid (Chapter 11) because they can move in channels of the crystal structure. However, these oxides exhibit a very low electronic conductivity. Therefore, it is not readily possible to transport electrons from the current collector to distant MnO_2 particles. The necessary ionic *and* electronic conductivity can be adjusted by admixing electronically conductive carbon powders to the ion-conductive powder of the transition metal oxide.

Electrodes made of one and the same active material can be optimized for high energy density *or* high power density. Because of this possibility, each type of battery occupies a more or less large *area in* the Ragone plot. The basic principle can be explained with the help of Figure 15.2. The active material is processed with a conductive additive of carbon black and an insulating binder to form a composite. If the proportion of the conductive additive is increased, the electronic conductivity increases. If the pore volume is increased, the ionic conductivity increases because the electrolyte fills the pore volume. Because the conductivity increases, the *IR* drop decreases. The *IR* drop is one of the major limitations for high power density because it decreases the discharge voltage and increases the charge voltage. This loss is proportional to the current (Figure 15.2). Therefore, increasing the conductivity enhances the power ($P = U I$) that a battery can deliver during discharge or take up during charging. However, the increased proportion of conductive additive and pore volume comes at the expense of the proportion of active material in the volume and mass of the electrode. A given volume of positive electrode can store less energy if it contains more conductive additive and less active material. If the energy density is to be maximized, as much active material as possible must be accommodated in a given volume element. Hence the pore volume and the proportions of conductive additive and binder must be minimized for this purpose. In turn, this will increase the resistance *R*. Although more energy can be stored, the increased *IR* drop will reduce the power delivered during discharge and increase the power required during charging.

Q: How can Li^+ conductivity within solid active materials be affected by the amount of liquid electrolyte in the pores of the composite electrode?

A: The ionic conductivity within the crystals of the active material is of course not affected by the conductivity of the liquid phase and must be controlled by the choice of material and synthesis conditions (particle sizes, vacancies, Chapter 11). However, if too little pore space is available, ionic transport in the pores of the composite electrode may contribute significantly to the overall resistance of the cell. This represents a starting point for optimization for high power densities.

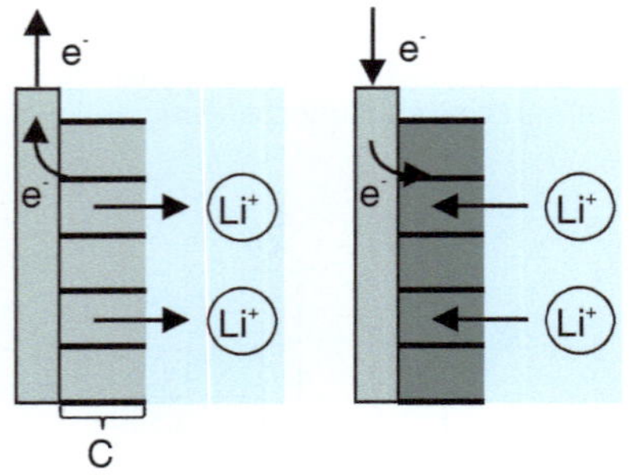

Figure 15.8 Insertion electrode: intercalation of Li^+ between the layers of a graphite crystal.

Lithium-ion batteries (LIBs) consist of two insertion electrodes. In addition to a transition metal oxide as positive electrode, graphite is used for the negative electrode (Figure 15.8). The great advantage of insertion electrodes is the gradual change in stoichiometry without phase transitions. This means that the charging and discharging processes is reversible in many cases. It is already clear here that an essential optimization task is the adjustment of the capacity and power density of the positive and negative electrodes.

Excursion 15-1 Construction and Characterization of a Positive Electrode for Lithium-ion Batteries

Here we want to trace the construction and characterization of a composite electrode made of $LiFePO_4$, looking at the presented characteristics and their calculation in everyday laboratory work. All data and graphics were kindly provided by P. Das, EWE Research Center NEXT ENERGY, Oldenburg.

Solid components of the composite are 1.68 g $LiFePO_4$, 0.12 g polyvinylidene difluoride (PVDF) as binder and 0.2 g carbon black as conductive additive. They are mixed with 5 mL of the solvent *N*-methylpyrrolidone (NMP, Figure 15.9, density $\rho = 1.03\ \text{g mL}^{-1} \approx 1\ \text{g mL}^{-1}$), $m(\text{NMP}) = V(\text{NMP}) \cdot \rho(\text{NMP}) \approx 5$ g). Here, the solvent is first stirred with PVDF for 2 h at 50 °C and then the carbon black is added and stirred for another 30 min. Finally, $LiFePO_4$ is added and stirred for an additional 3 h. The solid fraction x_s is important for the processability of the resulting slurry.

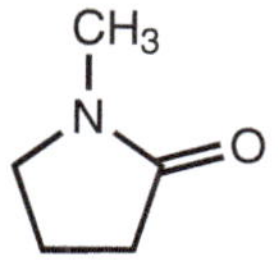

Figure 15.9 *N*-Methylpyrrolidone (NMP).

$$x_s = \frac{m(\text{LiFeSO}_4) + m(\text{PVDF}) - m(\text{C})}{m(\text{LiFeSO}_4) + m(\text{PVDF}) + m(\text{C}) + m(\text{NMP})} = \frac{(1.68 + 0.12 + 0.2)\,\text{g}}{(1.68 + 0.12 + 0.2 + 5)\,\text{g}} = 0.285 = 28.5\% \quad (15.18)$$

Such a slurry has special rheological properties that allow the material to be applied to an Al foil as current collector using a doctor blade process. During drying, sedimentation of the particles must not occur. This would have the effect that larger particles would be found in the finished composite electrode preferentially at the current collector and that the particles would become finer and finer toward the surface. This is not desired. Rather, all components in the composite should be approximately statistically distributed. For this purpose, the slurry should have a fairly high viscosity at rest, but this viscosity should be reduced by shear movements, thus enabling uniform application [18, 19]. In the laboratory test, the slurry is applied with a doctor blade (Figure 15.10, gap height 150 µm, linear speed 1 mm s^{-1}). Drying is carried out at 60 °C in vacuum. The porosity is influenced by pressing or rolling (calendering). This also establishes the electrical contact between the particles.

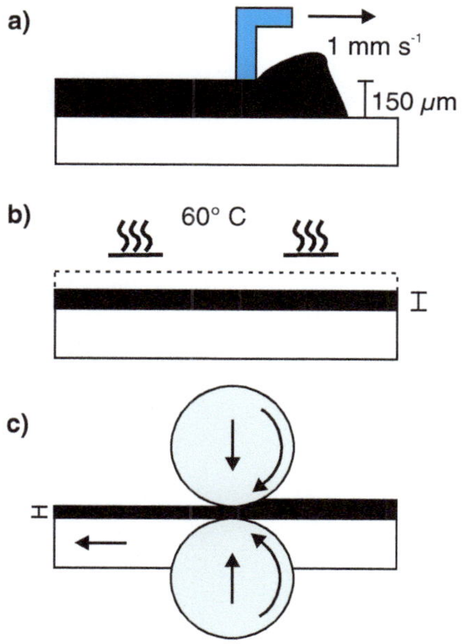

Figure 15.10 Schematic of the coating process for a positive electrode. a) Application of the slurry using a doctor blade; b) volume change during drying; and c) calendering.

After drying of the electrodes, they are assembled into a cell. Swagelok cells are a particularly flexible type of research cell (Figure 15.16), for which 11 mm-diameter discs are punched out from the coated current collectors with a hollow punch. The mass of the material applied can be determined by differential weighing (masses of coated minus uncoated metal foils). In the following example the mass difference is $m = 1.259$ g. A separator made of glass fibers is placed on the positive electrode and is impregnated with the electrolyte containing 1 mol L^{-1} $LiPF_6$ in a mixture of one volume part ethylene carbonate and one volume part ethyl methyl carbonate (Table 15.4). To characterize the positive electrode, we place a foil of lithium metal as a counter electrode on the separator, with a capacitance much higher than the positive electrode under investigation. In many laboratory cells, a lithium wire can also be introduced as a reference electrode. This allows the potentials of the positive and negative electrodes to be measured individually.

To test the positive electrode, the cell is subjected to charging and discharging cycles at constant current. The current can be varied between cycles. Because the lithium counter electrode is not limiting, the results are essentially a characteristic of the $LiFePO_4$ electrode. To set a specific C rate C_r, we calculate the maximum theoretical capacity of the weighed active mass:

$$\begin{aligned} Q_N &= mQ_{N,m} = m\frac{nF}{\sum \nu_i M_i} \\ &= 1.259\ \text{g} \cdot \frac{1 \cdot 96\,485\ \text{A s mol}^{-1}}{(6.94 + 55.84 + 30.97 + 64)\,\text{g mol}^{-1}} \\ &= 1.259\ \text{g} \cdot 611.6\ \text{A s g}^{-1} \cdot \frac{1\ \text{h}}{3600\ \text{s}} \\ &= 1.259\ \text{g} \cdot 0.170\ \text{A h g}^{-1} \\ &= 0.214\ \text{A h} \end{aligned} \tag{15.19}$$

Initial charging is performed at $C_r = 0.2\ \text{h}^{-1}$ (referred to $Q_N = 0.214$ A h) until the cell voltage reaches a value of 4 V. According to Eq. (15.10), the set current is therefore

$$\begin{aligned} I &= C_r Q_N \\ &= 0.2\,\text{h}^{-1} \cdot 0.214\ \text{A h} = 0.0428\ \text{A}. \end{aligned} \tag{15.20}$$

Afterwards, the cell is switched to potentiostatic charging at an applied voltage of 4 V and charging is continued until the current has dropped to 0.1 C_r ($0.1\ \text{h}^{-1}\ Q_N$). After a pause of 15 min, the cell is discharged at $C_r = 0.2\ \text{h}^{-1}$ or a discharge current of 0.0428 A to a final cell voltage of 0.28 V. From the time $t = 3$ h 40 min 22.8 s = 3.678 h required for discharge, the actual capacity is obtained (Figure 15.11):

$$\begin{aligned} Q_m &= \frac{Q}{m} = \frac{I_d t_d}{m} \\ &= \frac{0.0428\ \text{A} \cdot 3.673\ \text{h}}{1.2589\ \text{g}} = \frac{0.1572\ \text{A h}}{1.2589\ \text{g}} \\ &= 1.249\ \text{mA h g}^{-1} \end{aligned} \tag{15.21}$$

The experimental charge Q and the derived specific capacity C_m are used as the basis for the C rates in all further tests. Three tests are performed: (1) The capacity is determined by charging and discharging the cell at regular intervals between other tests at $C_r = 0.2\ \text{h}^{-1}$ (Figure 15.13). The rapid discharge behavior (rate capability) is investigated by charging at $C_r = 0.5\ \text{h}^{-1}$ and discharging at $0.5\ \text{h}^{-1}$, $1\ \text{h}^{-1}$, $2\ \text{h}^{-1}$, and $5\ \text{h}^{-1}$. The discharge time is used to determine Q according to Eq. (15.20) and the capacity at the set discharge current (Figure 15.12). (2) Similarly, the fast charging capability can be tested by charging at different currents and keeping a moderate discharge current constant (e.g., $C_r = 0.5\ \text{h}^{-1}$).

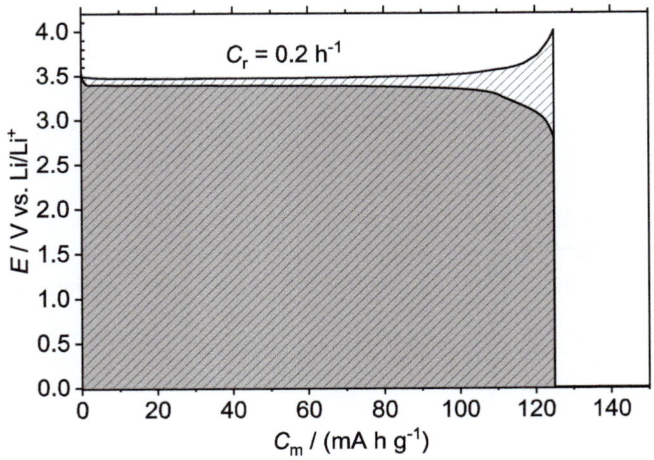

Figure 15.11 Charging and discharging curves of a cell $Li|LiPF_6|Li_xFePO_4$ during slow discharge at a constant current of 0.0428 A (corresponding to $C_r = 0.2\ \text{h}^{-1}$). The gray area under the discharge curve corresponds to the energy extracted from the cell, and the hatched area corresponds to the required energy for charging [Eq. (15.23)]. Source: Raw data were kindly provided by P. Das, EWE Research Center NEXT ENERGY Oldenburg.

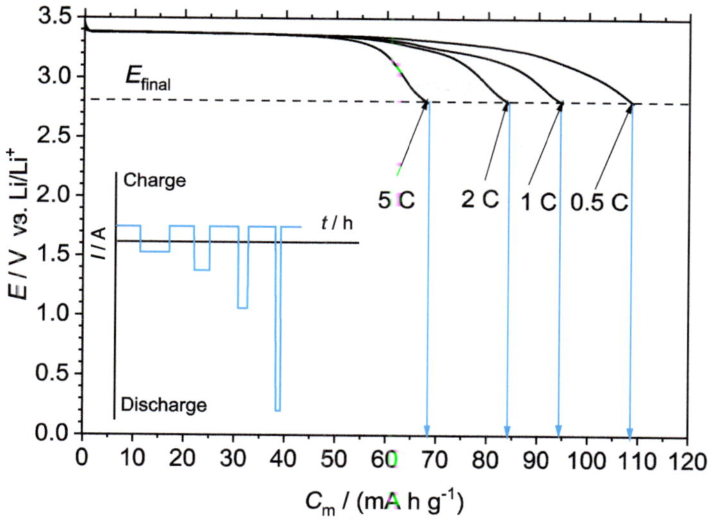

Figure 15.12 Discharge curves of a cell $Li|LiPF_6|Li_xFePO_4$ after charging at $C_r = 0.5\ h^{-1}$ and discharging at the indicated discharge rates. Raw data were kindly provided by P. Das, EWE Research Center NEXT ENERGY Oldenburg.

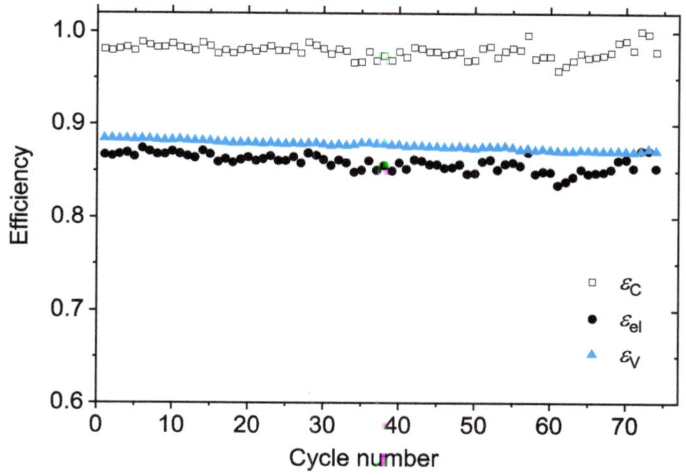

Figure 15.13 Efficiency of a cell $Li|LiPF_6|Li_xFePO_4$ over 75 cycles with $C_r = 0.5\ h^{-1}$. Raw data were kindly provided by L. Komsiyska, EWE Research Center NEXT ENERGY Oldenburg.

(3) The coulombic efficiency is the ratio of the discharge capacity and the capacity during charging.

$$\varepsilon_C = \left|\frac{Q_d}{Q_c}\right| = \frac{|I_d|\, t_d}{|I_c|\, t_c} \tag{15.22}$$

The energy efficiency is obtained from capacity, charging voltage and discharging voltage

$$\varepsilon_{el} = \left|\frac{W_d}{W_c}\right| = \frac{|I_d| \int_0^{t_d} U_d \mathrm{d}t}{|I_c| \int_0^{t_c} U_c \mathrm{d}t} \tag{15.23}$$

The voltage efficiency then results from charge and energy efficiency:

$$\varepsilon_V = \frac{\varepsilon_{el}}{\varepsilon_C} \tag{15.24}$$

The efficiencies are summarized for 75 cycles in Figure 15.13.

15.3.2 Primary and Secondary Cells

From what has been described above, it is already clear that there are many possibilities to combine two different electrodes to obtain a primary cell, but there are considerably fewer systems for secondary cells. A selection of commercially important primary and secondary cells is shown in Table 15.1. For reasons of cost and because of their high

Table 15.1 Major commercial battery systems.

No.	Name	Reaction at negative electrode	Reaction at positive electrode	Electrolyte	*U*/V	Theoretical specific energy[d)]/ (Wh kg^{-1})	Practical specific energy[e)]/ (Wh kg^{-1})	Applications, remarks
	Primary cells							
1	Zn–Mn cell	$Zn + 2\,H_2O \rightarrow Zn(OH)_2 + 2\,H^+ + 2\,e^-$	$2\,MnO_2$ (EMD) + $2\,H^+ + 2e^- \rightarrow 2\,MnO(OH)$	$ZnCl_{2,(aq)}$ or $ZnCl_{2,(aq)} + NH_4Cl_{(aq)}$	1.5	290	75	Especially cheap application where performance and reliability are less important (portable radio)
2	Alkaline cell	$Zn + 4\,OH^- \rightarrow [Zn(OH)_4]^{2-} + 2e^-$ $Zn + 2\,OH^- \rightarrow Zn(OH)_2 + 2e^-$ $Zn + 2\,OH^- \rightarrow ZnO + H_2O + 2e^-$	$2\,MnO_2$ (EMD) + $2\,H^+ + 2e^- \rightarrow 2\,MnO(OH)$	$KOH_{(aq)}$	1.5	290	75	Somewhat more expensive than Zn-Mn cell, higher power density, applications with increased electrical power (flashlight)
3	Zn–Ag cell	$Zn + 4\,OH^- \rightarrow [Zn(OH)_4]^{2-} + 2e^-$ $Zn + 2\,OH^- \rightarrow Zn(OH)_2 + 2e^-$ $Zn + 2\,OH^- \rightarrow ZnO + H_2O + 2e^-$	$Ag_2O + Zn^{2+} \rightarrow 2\,Ag + ZnO + 2e^-$	$KOH_{(aq)}$	1.6	209	100–150	Small electronic devices (wrist watches, calculators), large-scale batteries for military applications
4	Li–MnO_2 cell	$Li \rightarrow Li^+ + e^-$	MnO_2 (EMD) + $Li^+ + e^- \rightarrow LiMnO_2$	$LiClO_4$ in PC/DME[a)] CF_3SO_3Li in PC/DME	3.0	856	230	Good shelf life, wide temperature range, stable voltage when discharged
5	Li–thionyl chloride cell	$Li \rightarrow Li^+ + e^-$	$2\,SOCl_2 + 4\,Li^+ + 4e^- \rightarrow 4\,LiCl + S + SO_2$	$LiAlCl_4$ in $SOCl_2$	3.6	1431	260	Excellent characteristics in wide temperature range, low self-discharge, application in medical devices, and sensor technology

6	Li–$Ag_2V_4O_{11}$ cell	$Li \rightarrow Li^+ + e^-$	$Ag_2V_4O_{11} + xe^- \rightarrow x\,Ag^0 + Li_xAg_{2-x}V_4O_{11}$ $(x < 2)$ $Li_xV_4{}^VO_{11} + y\,Li^+ + ye^- \rightarrow Li_{x+y}V_4{}^{(V,IV,III)}O_{11}$	LiAsF in PC/DME[a)]	2.7–3.2	281	270	Implanted pacemakers, high current pulses, reliability and safety for years, discharge curve with multiple plateaus allowed, safe diagnosis of the state of charge, also coupled with Li:$[CF]_x$
7	Li–poly monofluorocarbon	$Li \rightarrow Li^+ + e^-$	$-[CF_x]_n- + x\,ne^- + x\,n\,Li^+ \rightarrow C_n + x\,n\,LiF$ $(0.5 < x < 1.2)$	$LiBF_4$ or $LiPF_6$ in organic solvent	2.8–3.3	2386	1280	Wide temperature range, long storage stability. Memory protection in computer technology, security systems, and devices in motor vehicles, medical technology, and military use.
8	Zn–air cell	$Zn + 2\,OH^- \rightarrow ZnO + H_2O + 2e^-$	$0.5\,O_2 + H_2O + 2e^- \rightarrow 2\,OH^-$	KOH in a gel	1.65	663 (incl. O_2 and OH^-) 1352 (Zn only)	146	Practical discharge voltage is about 1.3 V, because of the activation of oxygen reduction
	Secondary cells							
9	Lead–acid battery	$Pb + HSO_4^- \rightleftharpoons PbSO_{4(s)} + H^+ + 2e^-$	$PbO_2 + 2e^- + 3\,H^+ + HSO_4^- \rightleftharpoons PbSO_{4(s)} + 2\,H_2O$	$H_2SO_{4,(aq)}$	2.04	167	40	Cheap and reliable, all raw materials sufficiently available, relatively high power output, and safe. Starter battery in motor vehicles, traction battery for forklifts, etc., uninterruptible power supply for telecommunication infrastructure, high recycling rate is necessary and actually achieved.

(Continued)

Table 15.1 (Continued)

No.	Name	Reaction at negative electrode	Reaction at positive electrode	Electrolyte	U/V	Theoretical specific energy[d)]/ (Wh kg^{-1})	Practical specific energy[e)]/ (Wh kg^{-1})	Applications, remarks
10	Ni–Cd cell	$Cd + 2\,OH^- \rightleftharpoons Cd(OH)_{2(s)} + 2e^-$	$2\,NiOOH + 2e^- \rightleftharpoons 2\,Ni(OH)_2 + 2\,OH^-$	$KOH_{(aq)}$	1.29	208	49	Dissolution-deposition Cd electrodes, positive insertion electrode for H$^+$, higher power density than Pb-acid cell, high cycle number, maintenance-free. Use of Cd is problematic for ecotoxicological reasons, application of power tools.
11	NiMH cell[b)]	$LaNi_5H_{6.7} \rightleftharpoons 6.7\,H^+ + 6.7\,e^-$	like (10)	$KOH_{(aq)}$	1.2–1.4	210	40–110	Increasingly replaces Cd-Ni cells, high short-term power density up to 1900 W kg^{-1}, but poor long-term power density, poor characteristics at low temperatures. Inexpensive rechargeable cells for small devices up to cars with hybrid drive.
12	ZEBRA cell[c)]	$2\,Na_{(l)} \rightleftharpoons 2\,Na^+ + 2e^-$	$NiCl_2 + 2\,Na^+ + 2e^- \rightleftharpoons Ni + 2\,NaCl$	$\beta\text{-}Al_2O_{3(s)}$, $NaAlCl_{4(l)}$	2.58	787	120	Working temperature 300 °C, positive electrode in porous Ni framework structure. Electrolyte is solid Na-ion conductor$^+$
13	Na–S cell	$2\,Na_{(l)} \rightleftharpoons 2\,Na^+ + 2e^-$	$3\,S_{(l)} + 2\,Na^+ + 2e^- \rightleftharpoons Na_2S_3$	$\beta\text{-}Al_2O_{3(s)}$	2.1	755	226	1 mm thick solid-state ion-conductor as electrolyte and separator, high cycle number. Must be heated during rest periods, tests for stationary large storage tanks.

14	LIB	$C_6Li_x \rightleftharpoons C_6 + x\,Li^+ + xe^-$	$Li_{1-x}Co^{III,IV}O_2 + x\,Li^+ + xe^- \rightleftharpoons LiCo^{III}O_2$ ($x < 0.7$)	EC and other organic carbonates, $LiPF_6$	3.9	493	230	Moderate energy density 530 Wh L^{-1}, but very high specific power 1500 W kg^{-1} (low density of components), fast charging, low memory effect. Capable of deep discharge up to defined end point, including safety hazards during subsequent charging. Battery management system mandatory
15	LIB	$C_6Li_x \rightleftharpoons C_6 + x\,Li^+ + xe^-$	$(Li_{1-x-y}Ni^{II}{}_y)Ni^{III,IV}O_2 + x\,Li^+ + xe^- \rightleftharpoons (Li_{1-y}Ni^{II}{}_y)Ni^{III}O_2$ ($y \approx 0.12$)	EC and other organic carbonates, $LiPF_6$	3.8	498	150	Improved safety characteristics compared to 14, high discharge rate possible
16	LIB	$C_6Li_x \rightleftharpoons C_6 + x\,Li^+ + xe^-$	$Li_{1-x}[Mn_{2-y}{}^{III/IV}Ni_y]O_4 + x\,Li^+ + xe^- \rightleftharpoons Li[Mn_{2-y}{}^{III}Ni_y]O_4$ ($y \approx 0.5, x < 0.8$)	EC and other organic carbonates, $LiPF_6$	4.7	419	140	High-voltage battery, requires ion-conducting interphase also at the positive electrode
17	LIB	$C_6Li_x \rightleftharpoons C_6 + x\,Li^+ + xe^-$	$Fe^{III}PO_4 + Li^+ + e^- \rightleftharpoons LiFe^{II}PO_4$	EC and other organic carbonates, $LiPF_6$	3.4	396	100	Particularly good safety properties, but problems due to poor conductivity of $LiFePO_4$
18	Vanadium redox flow battery	$V^{2+} \rightleftharpoons V^{3+} + e^-$	$VO_2^+ + e^- + 2\,H^+ \rightleftharpoons VO^{2+} + H_2O$	2 mol L^{-1} H_2SO_4	1.4	33	20–30	Low energy density of only 15–30 Wh L^{-1}

a) PC, propylene carbonate; DME, 1,2-dimethoxyethane.
b) Nickel metal hydride cell; M represents A–B type alloy, where A represents a mixture (La, Ce, Pr, Nd) of elements resembling La, and B represents elements resembling nickel (Ni, Co, Mn, Al).
c) Zero Emission Battery Research Activity.
d) Theoretical specific energy calculated from the active masses without conductive electrolyte (the amount of which can be minimized in principle). In the case of the VRFB, a solution concentration of 2 mol L^{-1} was taken as a basis, since a limit for solubility exists here in principle. In the case of the lead–acid cell, the sulfate ions entering the product were taken into account.
e) The practical specific energy varies according to the product, and the best values evolve according to the maturity of the product.

conductivity (extra conductivity Section 4.2.6) and stability, H_2SO_4 or KOH are used for aqueous electrolytes. A larger potential window is required for lithium-ion batteries; thus, electrolytes are prepared from one or more organic solvents.

Finally, RFBs should be considered, which are only useful as rechargeable systems (Section 15.8). Researchers also use suspensions as a fluid storage medium – especially for large-scale storage – with materials that have good charge storage at suitable potentials but are not suitable for certain reasons as an active material in composite electrodes for secondary cells.

15.3.3 Classification According to Form Factors and Connection Parameters

To build efficient batteries, positive and negative electrodes and the electrolyte with the largest possible interface (determines I and thus the power) and active mass (determines Q_N) must be accommodated in the smallest possible volume. Quite obviously, the arrangement as shown in the schematics of the introductory chapters of this book (e.g., Figure 1.8) is unsuitable for this purpose. Instead, it is convenient to fabricate the electrodes as thin plates or foils and then stack them (Figure 15.14a). If the current collectors are flexible, a large-area electrode can also be rolled up into a "Swiss role" (Figure 15.14b). A cylindrical roll or a flat coil (Figure 15.14c) can be used depending on the design of the housing, which must be able to fit into a finished device.

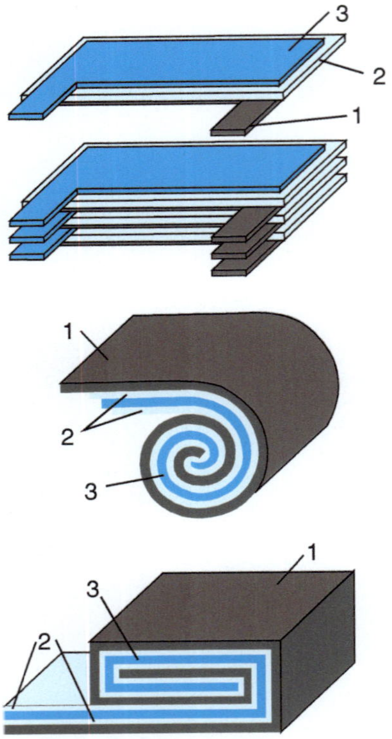

Figure 15.14 Shapes of packaging in battery construction. a) Stack; b) cylindrical coil; and c) prismatic coil. The positive electrode (3) is in blue, the negative electrode (1) in dark gray, and the separator (2) in light blue.

The packed cells are placed in housings with a shape and composition depending on the end product. You are familiar with electronic devices with the round cell. Their sizes are standardized to achieve compatibility between different manufacturers. The form factor is indicated by a five-digit number (Figure 15.15). For an AA cell, the first two digits (14) are derived from the diameter in mm and the other three digits (500) from the length in units of 0.1 mm. Prismatic coils and stacked cells can be placed in a flexible enclosure of metal foil called a "soft pack" or "pouch bag". Another type of enclosure is prismatic boxes with hard sidewalls closed by a lid ("hardcase"). Finally, coin cells are used in various sizes for devices with low energy demand. The different variants of cases and packaging may contain chemically different galvanic cells. It is therefore not possible to draw any conclusions about the cell chemistry from the design of the finished battery! The individual cells are connected in parallel or in series. In the case of a series connection, the voltages of the individual cells add up. The capacity of the battery is determined by the discharge of the single cell with the lowest capacity. The parallel connection increases the current. The power output is therefore increased at a constant nominal voltage. For larger consumers (e.g., electric vehicles), several cells are combined to form battery systems. Specific problems include the selection of mutually compatible individual cells (the weakest link determines the strength of the entire chain) and the consideration of various occupational health and safety problems regarding the electrical and chemical safety during assembly, testing, storage, transport, and repair. For these important topics, we can only refer to the literature [20].

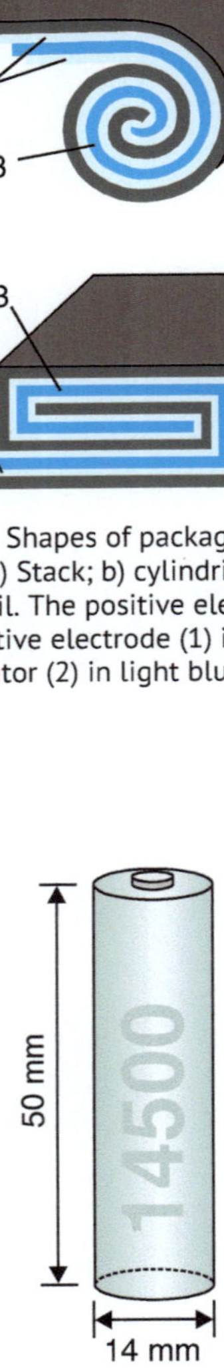

Figure 15.15 Cylindrical AA cell. The shape factor 14 500 is derived as follows: 14 is the diameter in mm, and 500 is the length in units of 0.1 mm.

In a research laboratory dealing with cell chemistry, small amounts of substances are to be tested followed by chemical analysis of the electrode materials after use. For the lithium-ion batteries that are currently being studied particularly intensively, the cells must be shielded against atmospheric oxygen, nitrogen and humidity and thus require a sealed housing or enclosure. Commonly used laboratory cells include coin cells (Figure 15.16a) in which a single cell is typically placed in an embossed metal mold and sealed. The pouch bag cell (Figure 15.16b) is also easy to handle in the laboratory. The cell stack is sealed in a bag made of polymer-coated aluminum foils (like a gastight coffee bag). Particularly flexible are Swagelok cells (Figure 15.16). The latter uses connecting elements for laboratory gas supply systems from a manufacturer of the

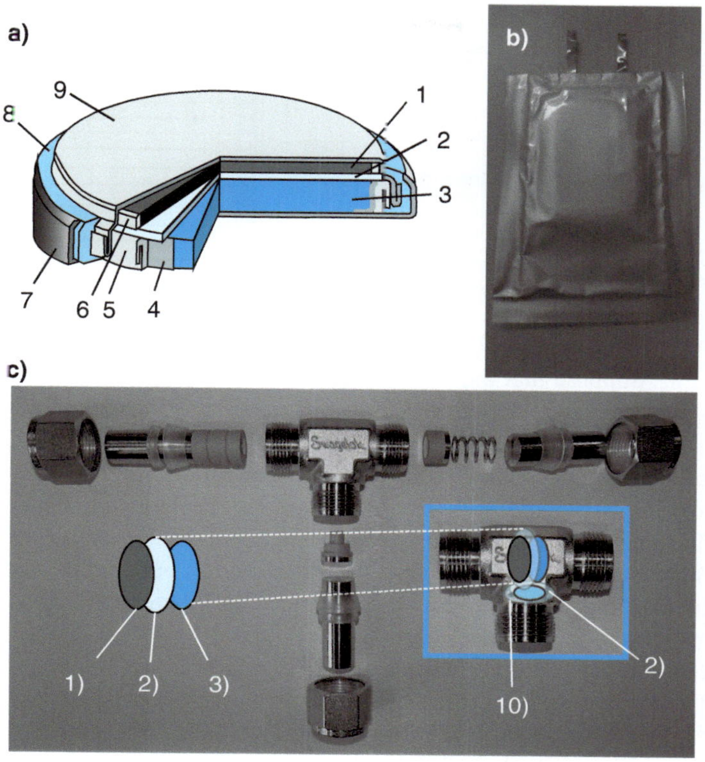

Figure 15.16 Designs of batteries that are particularly useful for research laboratories. a) Coin cell; b) pouch bag cell ("coffee bag" cell); and c) Swagelok cell.

Components of the coin cell or Swagelok cell: 1) Li (negative electrode), 2) electrolyte-impregnated separator, 3) MnO_2 (positive electrode), 4) ring for mechanical support, 5) spacer ring, 7) housing and positive terminal, 8) polymer sealing, 9) metal cap and negative terminal, and 10) Li reference electrode.
Source: Photo (b) by H. Bülter, Carl von Ossietzky University Oldenburg.

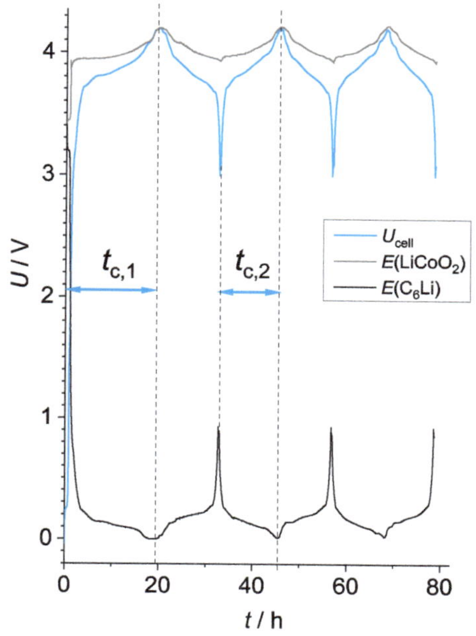

Figure 15.17 Recording of charging and discharging curves using a three-electrode cell. Raw data was kindly provided by L. Komsiyska, EWE Research Center NEXT ENERGY Oldenburg.

same name to hold the electrodes. Via a three-way connector, a reference electrode can be inserted in order to track the potential of each electrode separately (Figure 15.17). This allows relating characteristic changes in cell voltage to processes at one of the two battery electrodes, which greatly facilitates the interpretation of the curves.

15.3.4 Current Density Distribution in Batteries

The composite electrode materials in battery electrodes are typically porous and exhibit ionic conductivity *and* electronic conductivity. This is necessary because the electrons are transported to the current collector, and ions from the electrolyte must first pass through the pore network to all points of the active material in order to react there or, as in the case of intercalation electrodes, to be further transported within the solid active material. Although the electrode materials themselves often show both kinds of conductivity, their numerical value can vary over several orders of magnitude. The current will always take the path with the lowest resistance. The resulting local variation in current density is called primary current density distribution (Section 9.3.2). It can be calculated using Kirchhoff's law[157] if all resistance values are known. As a consequence, chemical conversions occur along the current path, while other parts of the electrode are not involved in the electrode reaction. An extreme situation is shown in

[157] Gustav Robert Kirchhoff, 1824–1887, German physicist, professor in Heidelberg and Berlin.

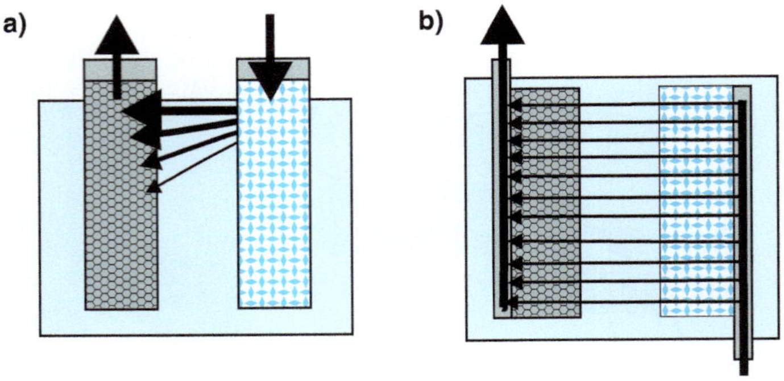

Figure 15.18 Effect of current path on mass utilization. Current collector in light gray, positive electrode in blue pattern, negative electrode in light gray pattern. a) Low electronic conductivity of the positive electrode with low mass utilization and b) optimization of the current path via the current collector for high mass utilization.

Figure 15.18a. Suppose that the ionic conductivity of the electrolyte is high, but the electronic conductivity of the positive electrode material is low. In this case, the current path at the positive electrode goes from the current collector via the shortest path in the active material into the electrolyte right at the immersion line of the electrode in the solution. The material will only react there, and the fraction of material utilized (mass utilization) for energy conversion would be low. At the negative electrode (with assumed higher conductivity), the effects are not so strong, but even there, the current density distribution is not uniform. The situation can be improved by (i) extending the current collector far into the active material and also by (ii) having the connection to the external circuit on opposite sides of the electrode stack (Figure 15.18b). Now, approximately equal resistances are present on each current path (shown as lines in the Figure), since all current paths traverse the same thickness of the three resistive media (positive electrode, electrolyte, negative electrode).

15.4 Zinc–Manganese Batteries as Examples of Primary Cells

In the following, we will discuss a few important batteries to illustrate how the individual cell components interact with each other and therefore mandate a material selection taking into account such interactions.

Zinc–manganese cells use metallic zinc as the reducing agent, which dissolves during discharge. MnO_2 is the oxidizing agent. Carbon is used as current collector of the positive electrode, thus the (common) name zinc–carbon cell. There are different technical realizations such as the Leclanché cell[158] (since 1866!) and the alkaline cell; both are primary cells. In the Leclanché cell, the reaction products of the negative Zn electrode go into solution as Zn^{2+} or deposit on the electrode as $Zn(OH)_2$, depending on the pH:

$$Zn + 2H_2O \rightarrow Zn(OH)_2 + 2H^+ + 2e^-, \tag{15.25}$$

$$Zn \rightarrow Zn^{2+} + 2e^-. \tag{15.26}$$

Zinc has a standard electrode potential of −0.76 V vs. SHE. Thus, we could expect (for thermodynamic reasons) hydrogen to be formed at this electrode in a corrosion reaction without electrons entering the external circuit [Eq. (15.27)].

$$Zn + 2\,H_2O \rightarrow Zn(OH)_2 + H_2(g) \tag{15.27}$$

Without special precautions, this reaction indeed occurs, consuming Zn as one of the active materials and reducing the cell capacity. In addition, the gaseous reaction product leads to pressure buildup that can cause the cell to burst. Consequently, the

[158] Georges Leclanché, 1839–1882, French physical chemist.

reaction rate of the hydrogen evolution reaction must be reduced. Historically, the first mitigation strategy was the addition of up to a few percent HgO to the zinc electrode. HgO is reduced and forms an amalgam with Zn. At amalgams, the exchange current density for the hydrogen evolution reaction is very low (cf. volcano plot in Figure 3.9). Therefore, the reaction mentioned in Eq. (15.27) did not play a role. However, Hg from such cells entered the waste. This is particularly problematic because primary cells are disposed of after a short period of use. Modern primary cells are completely mercury-free. Instead of HgO addition, Zn is alloyed, and corrosion inhibitors are added to the electrolyte.

The positive electrode, consisting of MnO_2 with a standard potential of 1.1 V vs. SHE, is discharged in two steps. In a first step, Mn(IV) is reduced to Mn(III) by taking up an electron with simultaneous insertion of a proton into the oxide lattice:

$$2\,MnO_2 + 2\,H^+ + 2e^- \rightarrow 2\,MnO(OH). \tag{15.28}$$

The solid reaction product MnO(OH) forms *one* joint mixed phase with remaining starting material MnO_2, which can have a continuously adjustable composition.

In the second reduction step, Mn^{2+} is generated and dissolves in the electrolyte. Thus, the starting material and the reaction product form two different phases. The consideration of phases is very important for battery electrodes and for the resulting properties. The basic principle is described by the Gibbs' phase rule[8] relating the number of independent chemical compounds N_c (molecular species, components), the number of phases N_P, and the number N_F of independent variables of state of the system. Here, the independent state variables are temperature T, pressure p, and cell voltage U. In a more general sense, N_F denotes the number of degrees of freedom of the system. The phase rule states

$$N_F = N_C - N_P + 2. \tag{15.29}$$

N_F – Number of degrees of freedom (p, T, U) [dimensionless]
N_C – Number of chemical components in the reaction system [dimensionless]
N_P – Number of phases [dimensionless]

Thus, for the first discharge step, $N_C = 2$ [MnO_2, MnO(OH)] in one solid phase (insertion electrode with variable composition, $N_P = 1$), resulting in $N_F = 3$ according to Eq. (15.29). Thus, in addition to p and T, the third state variable U can also vary in the system. This can be seen in the left part of the discharge curve in Figure 15.19: The voltage of a zinc–manganese cell drops almost uniformly, which is due to the positive electrode. (The negative Zn electrode shows a much smaller drop, see Nernst equation). During operation, the voltage decay is not desirable for many applications, but it also offers a potential advantage: By measuring the cell voltage, the SoC can be estimated at any time, and appropriate indications can be given for the necessary replacement of the battery.

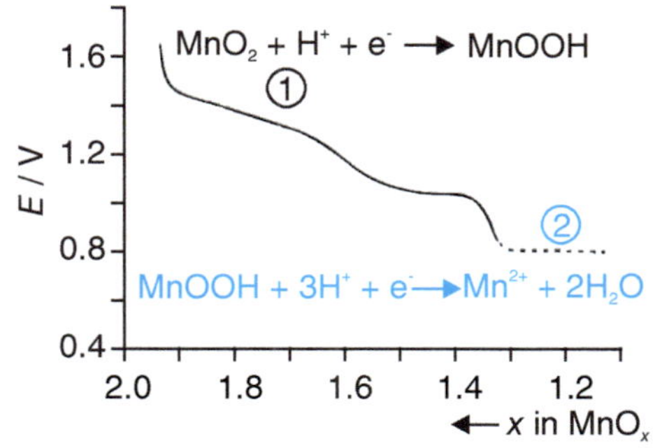

Figure 15.19 Schematic discharge curve of a Zn–MnO_2 cell.

For the second discharge step from MnO(OH) to Mn^{2+} (dissolved in the electrolyte phase), $N_C = 2$ (MnO(OH), Mn^{2+}), $N_P = 2$ (insertion electrode and aqueous electrolyte), resulting in $N_F = 2$ according to Eq. (15.29). If p, and T are fixed, the system has no further degree of freedom, U therefore remains approximately constant at about 0.8 V.

In two-phase reactions, the electrode potential remains approximately constant; in single-phase reactions, the voltage changes during the discharge reactions.

While Leclanché worked with natural manganese dioxide, modern zinc–manganese cells use electrolytic manganese dioxide (EMD), which enables better performance data. The electrolyte is a gel-like paste of $ZnCl_2$ or $ZnCl_2$ with NH_4Cl and some water. The Zn electrode is made in the shape of a cup fitting into the housing of the cell,

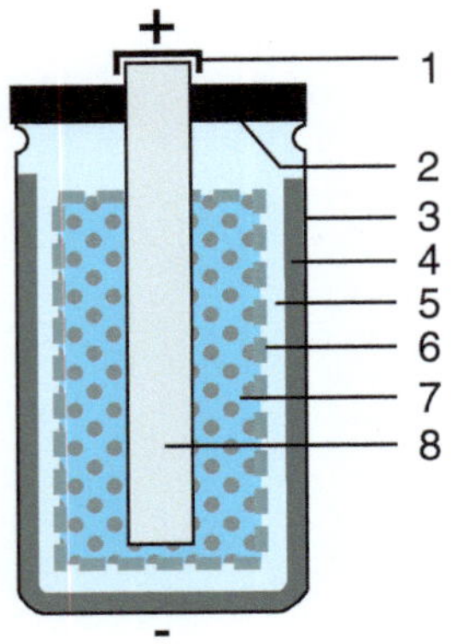

Figure 15.20 Schematic cross-sectional view of a Leclanché cell (Zn–MnO_2 cell).

1) Pole cap;
2) cover and gasket;
3) metal cup (current collector of negative electrode);
4) Zn cup (negative electrode);
5) gel electrolyte;
6) separator;
7) MnO_2 and graphite powder (positive electrode);
8) graphite rod (current collector of positive electrode).

which is also the current collector. The positive electrode is formed by MnO_2, which is mixed with carbon black to increase the electronic conductivity. The composite is arranged around a central carbon rod as a current collector (Figure 15.20).

The alkaline battery uses a similar design. Concentrated KOH is used as the electrolyte, which allows higher currents because of the extra conductivity of OH^-. The MnO_2 electrode largely corresponds to Figure 15.20. Zinc is alloyed and used in the form of a powder, which enables larger currents due to larger interfacial area. The powder is added to a gel and filled into a housing and current collector of Ni-plated steel sheet, providing increased mechanical stability. In the alkaline electrolyte, different reactions take place at the Zn electrode depending on the SoC:

$$Zn + 4OH^- \rightarrow [Zn(OH)_4]^{2-} + 2e^-, \qquad (15.30)$$

$$Zn + 2OH^- \rightarrow Zn(OH)_2 + 2e^-, \qquad (15.31)$$

$$Zn + 2OH^- \rightarrow ZnO + H_2O + 2e^-. \qquad (15.32)$$

The alkaline cells are somewhat more expensive than the Zn–Mn cells. Interestingly, the cheaper and less powerful Zn–Mn cells dominate the market in Europe and Japan, while in the USA, 80% of batteries sold for the same range of applications are alkaline cells. This illustrates that consumers do not only evaluate purely technical or economic criteria when deciding between competing products (otherwise, the decision should be similar).

Other primary cells for niche applications (Table 15.1) are Zn–Ag_2O cells with alkaline electrolytes, which are used in wristwatches, pocket calculators, etc. The Zn electrode works as discussed. The Ag_2O electrode is an example of the use of a displacement reaction (Figure 15.6).

Lithium is not only used for secondary batteries, but also for primary cells with particularly high energy but moderate power density. The use of lithium metal requires an electrolyte that cannot be reduced in contact with the highly electronegative Li metal. Organic electrolyte solutions are used for this purpose instead of aqueous solutions. Nevertheless, decomposition reactions cannot be ruled out, but in some systems, the so-called SEI (solid electrolyte interphase) is formed, whose properties are discussed in Section 15.6.3. The Li metal electrode can be combined with positive electrodes of transition metal compounds (MnO_2, FeS_2, AgV_2O_4).

The lithium–thionyl chloride cell is an example. It contains a lithium electrode, a current collector made of carbon, and $LiAlCl_4$ in thionyl chloride $SOCl_2$. Here, it is not a solid substance that is reduced, but the solvent of the electrolyte. The reduced discharge products are SO_2 (soluble in the electrolyte), elemental sulfur S, and LiCl. S and LiCl deposit as solid phases on the current collector. The discharge reaction is complete when the current collector is completely covered. Because of their excellent operating parameters, even at low temperatures, these systems are used in automotive electronics, smoke detectors, military equipment, and transponders.

Li–AgV_2O_4 cells are preferred for cardiac pacemakers. We discuss the Zn–air cell in Section 15.7.

15.5 Lead–Acid Accumulator as Rechargeable Battery with Aqueous Electrolyte

15.5.1 Overview

The lead–acid accumulator is (still) the most important commercial rechargeable battery (accumulator, secondary cell) worldwide in terms of installed capacity, weight, and sales value. This is due to its use as a starter battery (SLI, starter-lighter-ignition) in vehicles with internal combustion engine. The starter battery provides the energy for the starting process, the ignition, and for the onboard electronics (car radio, various controls, window regulators, ...). Its use for powering the vehicle (traction battery) itself is limited to niche applications such as forklifts, which cannot travel far from a central charging station and, moreover, can make use of the relatively large mass of the battery as a counterweight. We have already considered the basic features of cell chemistry in Section 2.4.4 and build on this here. The automotive starter battery is in many respects adapted to this particular application. For information on this, we must refer to literature [23]. Here, we only want to highlight some special aspects that are also instructive for the design or for the evaluation of other battery systems.

It was quite a long way from the identification of a suitable cell reaction to today's starter battery. If we were to simply immerse a lead plate and a plate of a PbO_2 composite into H_2SO_4 solution, the equilibrium cell voltage would be measured as the value that we calculated in Example Calculation 2.15. However, when current is passed to provide work, the voltage would quickly decline because the electrode area would be too small to provide useful power output. The reaction product $PbSO_4$ would not be deposited on the electrodes but would sink to the bottom of the cell container. Thus, it would no longer be available for recharging, and the capacity would decrease significantly with each discharging–charging cycle.

From this consideration, it is clear that the active materials must be used in the form of **porous electrodes** (Section 15.3.3). The pores must be well wetted by the electrolyte to achieve complete impregnation. For this purpose, a suspension of particles of the active masses is placed in a metallic frame, the current collector. A current collector must conduct the electronic current between the external circuit and the particles of the active material. It also provides mechanical stability to the electrode. Electron conduction is not trivial, especially because $PbSO_4$ is an insulator. Therefore, other cell components in the porous electrodes must provide electronic conductivity even in the discharged state when most metallic Pb has been converted to insulating $PbSO_4$. This implies that these electrodes must not be completely discharged, otherwise recharging would be impossible. Therefore, the mass utilization is restricted to only 35–60%. The recharging reaction for $PbSO_4$ is accomplished by the solution–precipitation mechanism. For this, it is important that solid $PbSO_4$ is in close spatial proximity to the sites consuming Pb^{2+}, so that the locally undersaturated solution immediately causes dissolution of $PbSO_4$. At larger distances, local consumption would have an effect only after the formation of a diffusion layer (Section 4.5.2) between the Pb electrode and the location of solid $PbSO_4$. Due to the macroscopic distances in a battery, the transport via such a concentration gradient would be slow and, thus, would prevent the delivery/uptake of large currents.

15.5.2 Boon and Bane of Side Reactions

In addition to the actual cell reaction [Eq. (15.3)], other reactions take place in the battery that cause a slow change in performance (aging): O_2 evolves at the positive electrode:

$$2\,H_2O \rightarrow O_2 + 4\,H^+ + 4e^- \tag{15.33}$$

This reaction would be expected according to the electrochemical series at less positive potentials than the formation of PbO_2.

$$Pb \rightarrow Pb^{2+} + 2e^- \tag{15.34}$$

$$Pb + 2\,H_2O \rightarrow PbO_2 + 4\,H^+ + 4e^- \tag{15.35}$$

However, the oxygen evolution reaction (OER) is an electrocatalytic reaction that occurs on different metals at very different rates. Pb and PbO_2 have a very low exchange current density for the OER. Therefore, the battery can be charged in the first place. In addition, the dissolution of the current collectors of Pb and co-alloyed elements according to Eq. (15.34) is thermodynamically possible, but extremely slowed down by passive layers.

At the negative electrode, O_2 can be converted to water in the oxygen reduction reaction (ORR) and H_2 can be formed by the hydrogen evolution reaction (HER):

$$O_2 + 4H^+ + 4e^- \rightarrow 2H_2O, \tag{15.36}$$

$$2H^+ + 2e^- \rightarrow H_2. \tag{15.37}$$

These two reactions are thermodynamically expected at less negative potentials than the deposition of lead. However, both reactions are associated with a very large overpotential at the materials used.

Overall, the cell voltage of a lead–acid cell is 2.04 V and thus significantly greater than the difference of the standard electrode potentials of the oxygen and hydrogen electrodes (1.23 V). The exchange current density j_0 for the OER and HER can be reduced to such an extent only by special materials and thus facilitates the operation of this cell with water as solvent! In general, the experience of over 120 years of battery research shows:

A cell voltage of 2 V is the highest achievable cell voltage in electrochemical energy converters with aqueous electrolytes.

Q: A starter is specified with 12 V, isn't it? How does that work?

A: The 12 V is produced by connecting 6 single cells, each with 2 V, in series. Each single cell can, in turn, contain several pairs of plates connected in parallel to deliver more power.

During galvanostatic charging, the charging voltage changes from slightly above 2 V to the final voltage of 2.6 V. The charging voltage is the cell voltage required to pass the set charging current through the battery. The final charging voltage is taken as an indicator to abort the charging process. The voltage does not increase evenly, but in steps. One of these is at 2.4 V, from where HER commences at the negative electrode and OER at the positive electrode (Figure 15.21). Taken together, this corresponds to

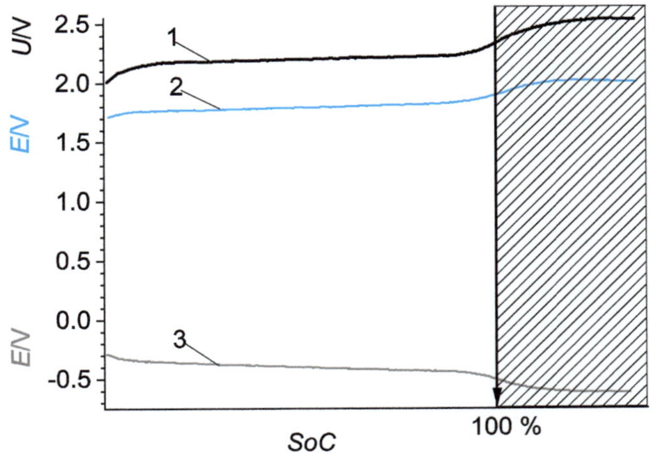

Figure 15.21 Galvanostatic charging ($I = const$) of a lead–acid cell. 1) Cell voltage; 2) potential of the $PbO_2|PbSO_4$ electrode; 3) potential of the $Pb|PbSO_4$ electrode. The transition from the charging reaction to water electrolysis can be read from the development of the cell voltage and the electrode potentials with the state of charge (SoC), which is also roughly proportional to the charging time.

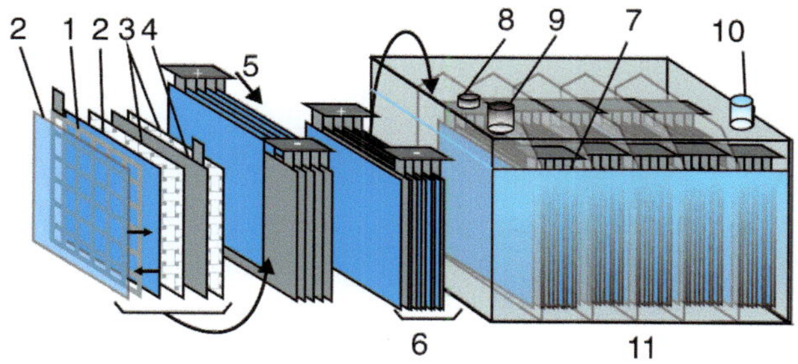

Figure 15.22 Schematic structure of a lead–acid battery with immersed plates (flooded cell).
1) Metal grid as current collector;
2) active mass of positive electrode (PbO_2);
3) porous separators or spacers;
4) negative Pb electrode on metal grid as current collector;
5) parallel connection of several single cells;
6) stack of single cells;
7) H_2SO_4 solution as electrolyte;
8) valve;
9) negative terminal;
10) positive terminal of the battery assembled from serially connected stacks (6);
11) case.

water electrolysis, with a corresponding water consumption, which had to be replenished at maintenance intervals in classical lead–acid batteries. However, the rising gas bubbles also have a positive effect: they mix the sulfuric acid electrolyte very effectively. Without this "stirring," the battery tends to stratify the electrolyte (acid stratification). As H_2SO_4 is either consumed or formed in the cell reaction, the density of the electrolyte is constantly changing during operation. The resulting density gradients cause heavier, more concentrated solutions to sink to the bottom of the cell and lighter, less concentrated acid to rise. As a result, after some time, different solution concentrations are present at one and the same electrode plate, which results in different equilibrium electrode potentials (and thus overpotentials) for the electrode reactions involved. This has the undesirable consequence that the reactions proceed at different rates depending on the immersion depth.

In flooded cells, the electrodes are immersed in liquid electrolyte (Figure 15.22), and the resulting gases escape into the atmosphere under normal pressure. In valve-regulated lead–acid cells (VRLA, Figure 15.23), gas leakage is largely avoided, and water consumption is minimized to such an extent that the batteries remain practically maintenance-free for their entire service life. This requires the following measures: 1) The electrolyte is immobilized in a porous separator. 2) The electrodes are no longer suspended in the electrolyte in the form of plates, but are pressed against

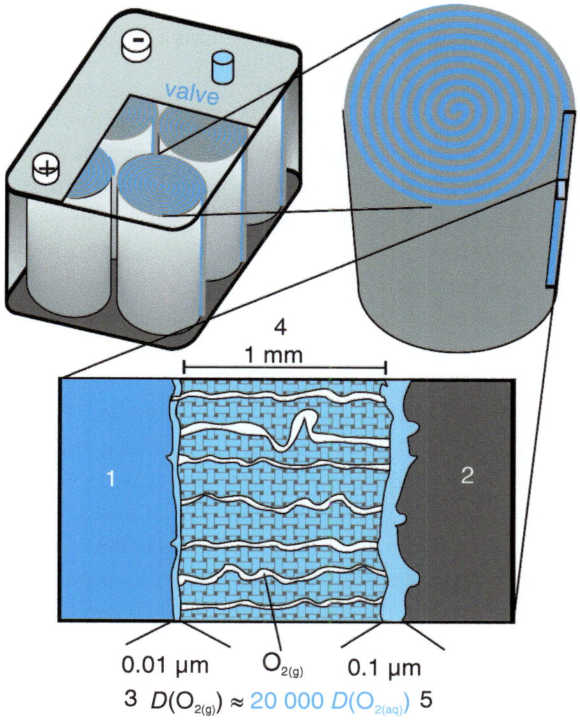

Figure 15.23 Schematic structure of a VRLA cell with enlarged sections of the coiled cell and a schematic cross section through the H_2SO_4-impregnated separator.
1) Positive electrode;
2) negative electrode;
3) liquid electrolyte layer on the negative electrode;
4) H_2SO_4-soaked separator interspersed with gas-filled channels (O_2) facilitating (fast diffusion of O_2);
5) liquid electrolyte layer on the positive electrode.

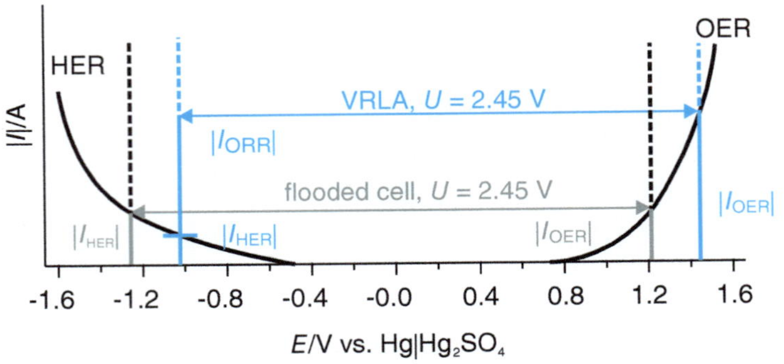

Figure 15.24 Contributions of side reactions to the faradaic current in lead–acid cells after onset of gas evolution reactions at a cell voltage of about 2.45 V.

a) For the flooded cell with immersed plate electrodes and a movable sulfuric acid electrolyte in gray.
b) For a VRLA cell, in which the electrolyte with microscopic gas volume is contained in a separator and gas leakage is prevented during normal operation. Part of the O_2 produced by OER at the positive electrode is rereduced at the negative electrode (ORR), thereby reducing the amount of H_2 produced in the HER (and thus reducing water consumption).

Note: The absolute currents are plotted to facilitate the comparison of anodic and cathodic partial currents.

the electrolyte-impregnated separator, for example, in the form of a roll. 3) Gas leakage is prevented by a valve that only releases a small amount of gas at excessively high pressures and then closes again.

In a VRLA cell, the OER is the dominant side reaction at the positive electrode according to Eq. (15.33). The gas cannot escape and diffuses very rapidly in the separator through microscopic gas channels to the negative electrode [$D_{gas} \approx 2 \times 10^4\ D_{sol}$ (!)]. There, the ORR proceeds according to Eq. (15.36), forming H_2O again. Due to the ORR at the negative electrode, the charging current is already reached at less negative potentials, and thus, the current yield for the HER is reduced according to Eq. (15.37). The situation is depicted in Figure 15.24. Due to the oxygen cycle, the excess energy introduced is converted into heat, and the composition of the battery is no longer irreversibly altered.

The density represents an easily accessible specification for the H_2SO_4 content in the electrolyte. Corresponding tables allow conversion to a (amount) concentration or to molality.

Irreversible sulfation is another side reaction that can limit the cycle number. Under certain conditions (continued high currents, very deep discharge, and a density of the acid above 1.24 g cm^{-3}), undesired large crystals precipitate besides the wanted finely grained $PbSO_4$-crystallites. The large crystals cannot be dissolved during the subsequent charging process, and the capacity of the battery decays.

15.5.3 Selected Details of the Technical Implementation

In starter batteries for passenger cars, the current collectors consist of an alloy of lead with calcium. For other applications, alloys with antimony and tin are also used. In recent decades, important development work has been carried out to adapt the alloy composition to the specific application. Important criteria are processability, mechanical stability, electrochemical stability during multiple charging–discharging cycles, corrosion reactions, and the formation of H_2 and O_2 (and thus water consumption). The introduction of Pb–Ca alloys for current collectors, for example, has inhibited water electrolysis to the point where periodic replenishment of water is no longer necessary. Grids are cast or stamped out of sheets, or they are used as stretch metal (a perforated metal sheet that is then pulled apart). The active masses are then introduced into these grids. The masses consist of 30% elemental Pb and PbO powder, which is partially reacted with dilute H_2SO_4 to form $PbSO_4$. Water accounts for 50% by volume, which escapes during the subsequent drying process, leaving voids with hydrophilic walls. After drying, the plates are charged against each other. This process is called formation. The actual energy-storing compounds Pb and PbO_2 only form during this process. The plates are placed in a polypropylene case with a separator (electronic separation of the plates while maintaining an ion-conducting connection through a porous plastic film).

It is important that the porosity is also maintained over several charging and discharging cycles. This is particularly significant because the volume of the resulting

$PbSO_4$ is 3 times larger than that of Pb and 1.5 times larger than that of PbO_2. Therefore, the pores in the active masses narrow during discharge. In addition, the Pb grains increase in size with multiple charging and discharging. To counteract this, macromolecular compounds (lignosulfonic acid and lignosulfates) are added to the active mass for the negative electrode, which adsorb to the surface of the Pb particles and thus prevent or retard particle growth.

The properties are further optimized with additives to the active masses and to the electrolyte. Components with contents of less than 0.02% are considered impurities, contents of 0.02–2% are referred to as additives, and contents of more than 2% are referred to as components. Addition of $BaSO_4$ to the active masses acts as a crystallization nucleus and promotes the formation of many small $PbSO_4$ crystals instead of a few large crystals. This creates the conditions for high total currents over a large surface area. H_2SO_4 with defined purity is used for the electrolyte. Critical impurities are, on the one hand, metal cations (Pt, Pd, Fe, Co, and Ni), because they lower the overvoltage for the HER when deposited at the negative electrode, thus accelerating self-discharge and increasing H_2O consumption. Ions are also problematic when they promote corrosion of the active materials and the current collectors. Additives include phosphoric acid H_3PO_4 (up to 0.8%) because it favorably affects the growth of PbO_2 crystals and increases the overvoltage for the HER and OER by adsorption on the electrodes.

Conclusion. The lead–acid battery provides the maximum feasible cell voltage of 2.0 V with an aqueous electrolyte. This is achieved by expedient selection of all materials to maximize overpotentials for the OER and HER. Despite the high density of the active materials, the lead–acid cell holds its own in important fields of application because of its robustness and low price. Other secondary cells with aqueous electrolytes (Table 15.1) such as Cd–Ni systems [24–26] and nickel-metal hydride systems [24, 26–29], on the other hand, are more affected by competition from the ever-improving lithium-ion batteries. Batteries with nickel hydroxide electrodes exhibit memory effect [30]. Although the aqueous systems have fundamental advantages in terms of safety characteristics, which are particularly effective for larger batteries, lithium-ion technology is also gaining ground in these areas.

Excursion 15-2 Memory Effect.

The term memory effect is used to describe the reversible loss of capacity of a battery after repeated partial discharge. If the accumulator is fully charged after partial discharge (shallow discharge), it does not release all the stored energy when discharging. This effect is particularly strong in alkaline batteries with Ni as the positive electrode [Ni–Cd, Ni-metal hydride (Ni-MH)]. Most simple devices such as mobile tools, shavers, and electric toothbrushes switch off the consumer at a specified output voltage during discharge. This cut-off voltage value is used as a simple indicator of complete discharge without actually determining the actual SoC. If the discharge voltage is lowered by internal processes (see below), the end point of discharge defined by the set cut-off voltage is reached after a shorter period of use; the battery appears to have lost some of its capacity. In most cases, the original capacity can be restored by (multiple) complete discharges (complete in the sense of stoichiometry) and subsequent complete charging. This may have to be done in a special charger, because simple devices only cycle the batteries between two fixed cut-off voltages. Determining the SoC solely on the basis of cell voltages is inaccurate – a problem that plays a major role for electric traction in particular, because the remaining range would be determined from an inaccurate value.

There are several causes for the memory effect, which cannot be easily read in detail from charging–discharging curves. In Ni–Cd and Ni-MH cells, the effect is attributed to the formation of γ-NiOOH as a by-product in a separate phase during partial overcharging [30, 31]. Discharge of this phase occurs with different voltage characteristics than those of the actual battery materials, leading to premature detection of the discharge end point.

Contrary to popular belief, lithium-ion batteries can also exhibit memory effects. For example, electrodes made of $LiFePO_4$ exhibit a memory effect due to the occurrence of two phases during cycling [32]. At a given stoichiometry (SoC), the occurrence of the two phases on different grains depends on the history. The formation of the two phases leads to slightly different voltage profiles in the discharge curves. However, the discharge curves are generally very flat in two-phase systems (Section 15.4, Gibbs' phase rule). Thus, small deviations in the discharge voltage in this part of the discharge curve cause large deviations in the estimation of the SoC.

15.6 Lithium-Ion Batteries

The term lithium-ion battery (LIB) refers to an entire class of batteries in which both the positive and negative electrodes are Li^+-intercalation electrodes. Compared to lead–acid batteries, LIB is a very recent technology. These batteries were first introduced into the scientific literature around 1972. It took another 20 years of basic research to establish cells with *two* intercalation electrodes between which the Li^+ ions "move back and forth" (rocking chair technology). It was not until the early 1990s that the first LIBs appeared in consumer products of the company Sony. This example clearly illustrates the persistent, continuous, and also very cost-intensive research efforts required to help a brilliant idea in the laboratory achieve a breakthrough in practical application. LIBs work with organic electrolyte solutions, which have a significantly larger potential window than aqueous electrolytes. Due to the larger cell voltage, a significantly greater specific energy is achieved. However, these batteries are also much more expensive than systems with aqueous electrolytes. They were therefore initially only able to establish themselves in high-priced, portable electronic devices such as cameras, cell phones, and laptops. Mobile communication devices connected to the internet triggered technical and social developments that certainly represent the most noticeable overall societal impact, for which basic electrochemical research of the last three decades set the stage. The development of LIB was recognized in 2019 with the Nobel Prize in Chemistry.[159]

A smartphone that we use every day now (2025) as a digital assistant (or play-mate) has more computing power than large-scale data centers had in the 1970s and can even compete with the power of a server 25 years ago. (There was already the internet and email back then, too.)

Further development of LIB technology is highly competitive and aims to open up new applications for this type of battery. These include, above all, the use as a traction battery for passenger vehicles and storage of regenerative but intermittent electrical energy from photovoltaic and wind power plants. In addition to costs, safety is a much more critical aspect than with the battery types discussed so far. As a relatively young technology, there are initially very many different variants of LIBs in terms of the materials used, their processing into electrodes, and the combination of components into cells. We can only mention a few prototypical examples here to make

[159] John Bannister Goodenough, 1922–1923, US-American physicist and materials scientist, professor at the University of Texas at Austin; M. Stanley Wittingham, born 1941, British-US American solid-state chemist and electrochemist, since 1988 professor at Binghamton University (NY) after an industrial career; Akira Yoshino, born 1948, since 1972 active in industrial companies such as Asahi Kasei, A&T Battery in senior research positions.

clear the relationship between laboratory experiments, which is the working domain of electrochemists, and commercial cells. Further reviews are available for a more in-depth study of the topic [21, 22].

Nowadays, we understand LIB as a whole group of batteries that use Li^+ as the working ion moving between a negative intercalation electrode (mostly graphite) and a positive transition metal oxide intercalation electrode.

15.6.1 Basic Structure

Figure 15.25 shows the basic structure of a LIB. The negative electrode consists of a graphite material to which binder and conductivity additives have been added and which is applied to both sides of a thin, surface-treated Cu foil as a current collector. The current collector has high electrical conductivity, but also good thermal conductivity for heat dissipation. It also provides mechanical strength to the electrode and allows it to be processed into different shapes. The positive electrode has a similar structure. Here, the current collector is made of Al. This valve metal is coated with a thin passive layer of Al_2O_3 or AlF_3, which prevents the dissolution of Al at the potentials of the positive electrode $E \gg E(Al/Al^{3+})$. Al also has good thermal and electrical conductivity. A transition metal oxide is deposited on both sides as the active material (Table 15.1).

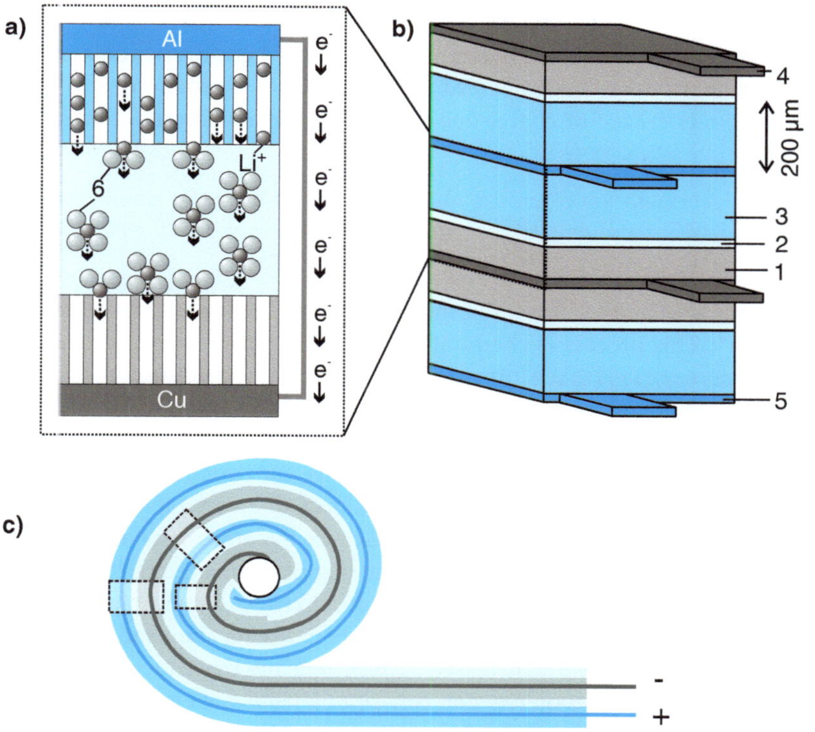

Figure 15.25 Basic structure of a LIB. a) Section of a single cell; b) stacking sequence of the cell components with current collectors coated on both sides; and c) assembly of the cell as a cylindrical coil. The position of the single cell, as shown in a), is marked with dashed rectangles in b) and c). Components: 1) negative electrode (lithiated graphite), 2) electrolyte-impregnated separator, 3) positive electrode ($LiCoO_2$), 4) current collector for the negative electrode (Cu), and 5) current collector for the positive electrode (Al). Source: Parts a) and b) adapted from H. Bülter, PhD thesis, Carl von Ossietzky University Oldenburg, 2016.

This is held together by a binder to form a slurry, to which carbon materials are added to enhance the electronic conductivity. Both electrodes are kept apart by a thin separator. The pores of the separator take up the liquid or gel-like Li^+-containing organic electrolyte. The electrolyte also fills the pore volume of the positive and negative electrodes. The use of porous electrodes increases the area of the solid–liquid interface and thus allows higher overall currents (thus higher power). The cell is sealed gastight from the environment to prevent the entry of O_2 and H_2O. The cells are assembled in the discharged state in glove boxes (filled with anhydrous and O_2-free Ar) and charged before the first use.

Lithium is one of the few elements that also violently reacts with molecular nitrogen.

15.6.2 The Negative Electrode

The supposedly direct way to large energy densities is to use very electronegative metals with low density (Li, Na, or K) or high ionic charges (Mg, Al, or Zn). Some of these metals either form very stable oxide films or react very strongly with water or with the polar organic solvents after removal of the protective oxide films in the sense of a corrosion reaction that would cause self-discharge in a battery (the solvents must be polar in order to dissolve or complex cations). The layers formed from the decomposition products usually passivate the electrodes, so that no useful batteries can be obtained. Li in particular forms a special protective layer, the solid electrolyte interphase (SEI), when in contact with some *organic* solvents. This layer is electronically insulating but is ionically conductive for Li^+. In contrast to many other parts in this book, we are concerned here with a three-dimensional phase between the active electrode material and the liquid electrolyte, thus the name *inter-ph-ase*. This should be distinguished from *interface*, which denotes a two-dimensional plane between two adjoining phases. The formation of the SEI has enabled primary cells to be produced with metallic Li because it limits the self-discharge due to a corrosion reaction with the electrolyte.

> The practical operation of LIBs is made possible by a **kinetic stabilization** of the electronegative metal *beyond the thermodynamically predicted limits.*

For secondary batteries, however, the Li metal must be recovered by electrodeposition of Li^+. This process is chemically possible in organic solvents. However, the deposition does not lead to the formation of smooth Li layers; instead, structures with a large surface area (high-surface area lithium) are formed, for example, dendritic structures (Figure 15.4). If dendrites grow from the negative to the positive electrode, they may form an electronic connection between positive and negative electrodes, and thus, an internal short circuit. Such a short circuit leads to a very rapid discharge during which the thin dendrites heat up rapidly and cause the organic solvents to evaporate. The formed gas causes the buildup of pressure that may burst the cell case. The heated (energy-rich) cell components may ignite upon contact with atmospheric oxygen, and "the doom takes its course." Even without a short circuit, the dendrites are a major hazard because of the greatly increased surface area and the associated opportunities for a chemical decomposition reaction between Li metal and electrolyte components. Extensive efforts to suppress dendrite growth by additives have been unsuccessful so far or have resulted in other intolerable effects on cell properties [33].

The breakthrough for secondary LIB came only with the use of insertion electrodes. A number of materials can be considered for this purpose, which, in addition to graphite, include transition metal oxides, lithium alloys with Sn, Sb, and Si, chalcogenides, nitrides, and polymers (Figure 15.26). Since the host lattice of these compounds contributes to weight and volume but does not itself convert charge,

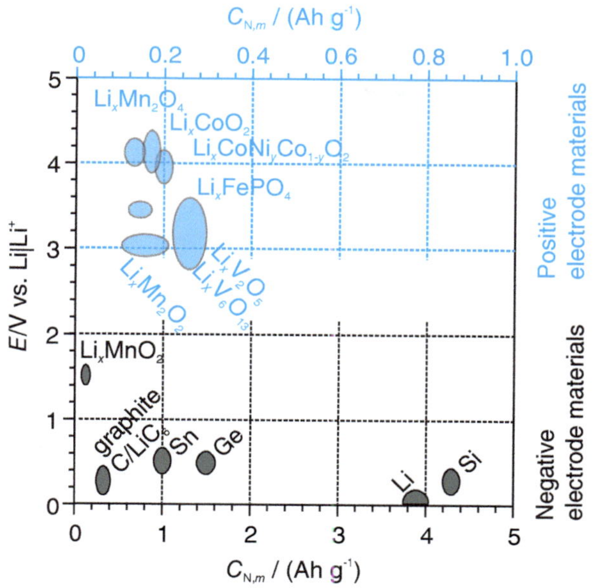

Figure 15.26 Voltage ranges and specific capacities of lithium intercalation materials and lithium alloys. Color-shading is gray for negative electrode materials and blue for positive electrode materials. Source: Adapted from [34, 35].

the performance data for these electrodes are, as expected, lower than those for metallic Li. However, the performance data for metallic Li are dramatically reduced by the very fact that a Li electrode must never be completely dissolved if subsequent redeposition is to occur. From this point of view, other alloy electrodes come close to the capacity of metallic Li. The redox potentials are also important for the energy content. Here, lithiated graphite offers a redox potential similar to that of metallic lithium. In addition, graphite exhibits high dimensional stability during cycling, has its own electronic conductivity, and consists of a common (abundant) element. The ionic transport of Li^+ within graphite is moderately fast. A clear disadvantage is the safety, which could only be increased incrementally by a multitude of measures (see below).

An essential aspect of the insertion materials and of the alloy electrodes is the volume expansion during the incorporation of Li^+ or Li. The volume expansion as well as the capacity for Li^+ varies greatly among the materials. An extreme example is silicon. It can, in principle, form alloys with Li up to the stoichiometric composition of $Li_{22}Si_5$ (theoretically 3570 mA h g^{-1}) and thus exceeds the theoretical specific capacity of graphite (372 mA h g^{-1}) by a factor of ten. However, the lithiation of Si is accompanied by a significant volume expansion of 270%. The much smaller volume expansion during lithiation of graphite can already lead to the cleavage of graphite particles under unfavorable conditions (Figure 15.27). With the much more pronounced volume change of silicon electrodes, macroscopic crystalline silicon electrodes, such as silicon wafers, suffer such extreme internal stresses that the material crumbles after only a few charging–discharging cycles, that is, parts of the electrode lose mechanical adhesion and thus also electronic contact to the current collector. Current research is therefore based on nanoscale silicon materials in which the mechanical stresses during Li^+ insertion are reduced because the volume expansion of the entire particle is more uniform.

Graphite, the most commonly used material, can incorporate Li^+ in varying amounts. Different domains then form in graphite with the approximate stoichiometries C_6Li, $C_{12}Li$, or $C_{18}Li$ [36]. The volume expansion in the fully lithiated state is only 10%. Although this is significantly lower than for Si, the volume changes and stress in the material during charging/discharging can lead to the cleavage of components of the active material, which then no longer participate in the electrochemical

Figure 15.27 Graphite particles with a conductive binder. a) After formation of the SEI and b) after charging–discharging cycles at 10 °C (high current at low temperatures), leading to mechanical damage of the particles. Source: Reproduced from K. Takahashi, V. Srinivasan; *J. Electrochem. Soc.* **2015**, *162*, A635–A645.

reactions during subsequent charging–discharging cycles (Figure 15.27). Accordingly, the capacity of the electrode slowly decreases – it "ages". Degradation processes in secondary cells are due to various causes, but the periodic buildup of mechanical stress in the materials is one of the main causes.

The Li^+-ions do not arrange themselves randomly in the graphite lattice. This would also be energetically unfavorable, since the individual graphene layers would have to "bend" around the randomly incorporated Li^+-ions. Within a layer, the Li^+ ions are located above and below the center of a six-membered ring of the neighboring graphene layer. Between the "occupied" rings, one six-membered ring remains unoccupied (Figure 15.28a). This superstructure (Section 8.9.6) of Li^+-ions is itself hexagonal. If we consider the center-to-center distance of adjacent six-membered carbon ring structures as the base unit a_0, the distance of the Li^+-ions is $\sqrt{3}\,a_0$. The superstructure is therefore also called the $\sqrt{3} \times \sqrt{3}$ structure. When passing through different stoichiometries during partial discharge, one layer is always completely emptied. Thus, every second layer then remains occupied by Li^+, then every third layer, then every fourth layer, and so on. This phenomenon is called staging.

The individual stages are characterized by a suffix k, which indicates the number of graphene layers between the layers of Li^+ (Figure 15.28b). The different stages can be investigated during charging using X-ray diffraction and Raman spectroscopy and correlated with the potential plateaus of a charging curve (Figure 15.29a). Clear transitions at different potentials occur in structurally well-defined materials. In most cases, however, two stages are present next to each other. This then implies a fixed potential. Only at the beginning, each layer is occupied by a very small amount of Li^+. If the first Li^+-ions are incorporated in stage 2, this state is called a liquid-like structure (L):

$$\text{Random occupation} \rightleftharpoons \text{Stage 4}, \quad E = 0.21\ \text{V vs. Li}|\text{Li}^+ \tag{15.38}$$

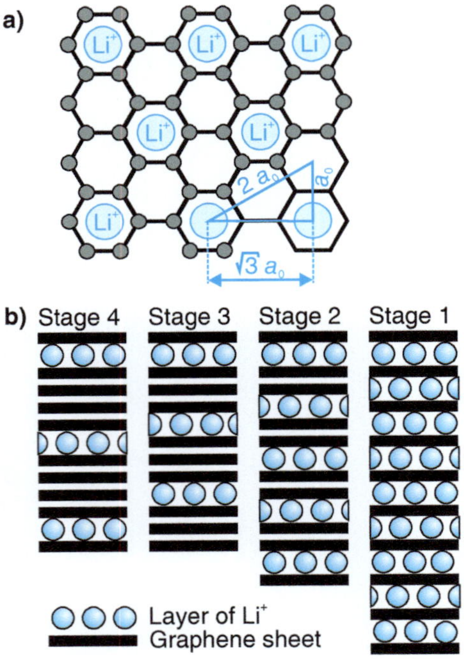

Figure 15.28 Structure of Li graphite intercalation compounds (LiGICs). a) Top view of a stage 4 layer (LiC_6); and b) cross section through the stack of graphene layers with different lithiation levels or stacking sequences (staging).

$$\text{Stage 4} \rightleftharpoons \text{Stage 2L} \tag{15.39}$$

$$\text{Stage 2L} \rightleftharpoons \text{Stage 2},\ E = 0.12\ \text{V vs. Li}|\text{Li}^+ \tag{15.40}$$

$$\text{Stage 2} \rightleftharpoons \text{Stage 1},\ E = 0.09\ \text{V vs. Li}|\text{Li}^+ \tag{15.41}$$

The different lithiation levels can also be easily visualized [42]: LiC_6 shines golden, LiC_{12} is red, and LiC_{18} is blue. The stages can also be seen in a detailed analysis of the charging curves.

Excursion 15-3 From a Charging Curve to a Differential Capacity Curve.

For charging and discharging tests, an experimental cell is usually used in which the material to be tested (here graphite) is combined with an excess of Li metal as the counter electrode. An additionally installed Li wire can be used as the reference electrode for separately measuring the potentials of both electrodes. The battery is charged at a constant current (here −0.19 mA) and discharged at a constant current (here +0.19 mA, corresponding approximately to $C_r = 0.1\ h^{-1}$). Battery testers typically store their measured values in tabulated form as in Table 15.2.

The plot of $U(t)$ versus $Q(t)$ yields the charging curve (Figure 15.29a). Here, Q is on the abscissa, and because of $I = const$ and $Q = I\ t$, Q is proportional to time t. To obtain a meaningful plot, we should subtract the charge irreversibly consumed in the first charging cycle (Section 15.6.3) from the subsequent further charging–discharging curves. The individual charge stages are clearly recognizable as potential stages in Figure 15.29a. The representation of the stages becomes even clearer if we plot Q during charging or discharging at a certain potential E. To achieve this, one differentiates the function $Q(E) = f(E)$ with a spreadsheet program. If this does not work technically, one may form the difference quotient $\Delta Q/\Delta U = (Q_{i+1} - Q_i)/(U_{i+1} - U_i)$. However, the section in Table 15.2 contains consecutive points with equal $U(t)$. $\Delta Q/\Delta U$ is not defined for $\Delta U = 0$. In addition, we have to expect instrumental noise in the measurement of $U(t)$. Altogether, it

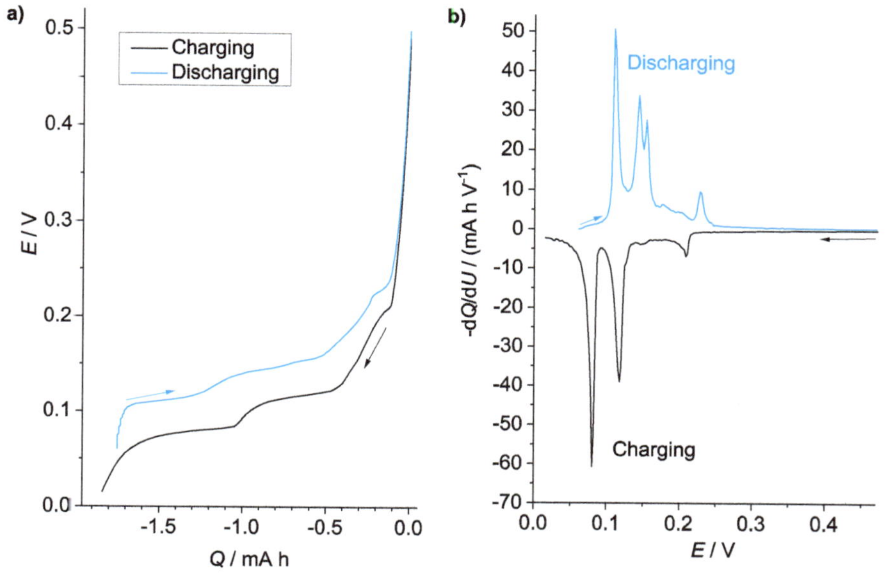

Figure 15.29 a) Charging and discharging curves of a graphite electrode against Li metal and b) differential capacitance curves constructed from them. The different charging levels can be clearly identified. Raw data were kindly provided by L. Komsiyska, EWE Research Center NEXT ENERGY Oldenburg.

Table 15.2 Section of tabulated raw data of a battery cycling test.

t/h	E/V	I/A	Q/Ah
18.79224	0.16059	1.89969×10^{-4}	-8.51827×10^{-4}
18.7928	0.16059	1.89969×10^{-4}	-8.51721×10^{-4}
18.79336	0.16059	1.89969×10^{-4}	-8.51615×10^{-4}
18.79391	0.16078	1.89969×10^{-4}	-8.5151×10^{-4}
18.79447	0.16059	1.90009×10^{-4}	-8.51404×10^{-4}
18.79502	0.16078	1.89969×10^{-4}	-8.51299×10^{-4}
18.79558	0.16059	1.90009×10^{-4}	-8.51193×10^{-4}
18.79613	0.16078	1.89969×10^{-4}	-8.51088×10^{-4}
18.79669	0.16078	1.89969×10^{-4}	-8.50982×10^{-4}
18.79724	0.16078	1.89969×10^{-4}	-8.50877×10^{-4}
18.7978	0.16078	1.89969×10^{-4}	-8.50771×10^{-4}
18.79836	0.16078	1.89969×10^{-4}	-8.50666×10^{-4}
18.79891	0.16078	1.89969×10^{-4}	-8.5056×10^{-4}
18.79947	0.16078	1.89969×10^{-4}	-8.50454×10^{-4}
18.80002	0.16078	1.89969×10^{-4}	-8.50349×10^{-4}
18.80058	0.16078	1.89969×10^{-4}	-8.50243×10^{-4}
…	…	…	…
18.86724	0.16231	1.90009×10^{-4}	-8.37578×10^{-4}
18.8678	0.1625	1.89969×10^{-4}	-8.37473×10^{-4}
18.86836	0.1625	1.89969×10^{-4}	-8.37367×10^{-4}
18.86891	0.1625	1.90009×10^{-4}	-8.37262×10^{-4}
18.86947	0.1625	1.89969×10^{-4}	-8.37156×10^{-4}
18.87002	0.1625	1.89969×10^{-4}	-8.3705×10^{-4}
18.87058	0.1625	1.89969×10^{-4}	-8.36945×10^{-4}
18.87113	0.1625	1.89969×10^{-4}	-8.36839×10^{-4}
18.87169	0.16269	1.89969×10^{-4}	-8.36734×10^{-4}
18.87224	0.1625	1.90009×10^{-4}	-8.36628×10^{-4}
18.8728	0.16269	1.89969×10^{-4}	-8.36523×10^{-4}
18.87336	0.16269	1.90009×10^{-4}	-8.36417×10^{-4}

The data were recorded every 5 s. The data used in Table 15.3 are marked in light blue.

becomes clear that very noisy curves would result from simple numerical differentiation of the charging curve. The following "trick" offers a way out: One sorts selected points of the original data set into a new table. The selected points are those where the potential changes by at least 2 mV (examples are indicated by light-blue shade in Table 15.2).

This results in significantly smaller data sets with larger differences in charge between neighboring measured values and no possibility for $\Delta U = 0$ between adjacent points (Figure 15.29b). However, the measured values are no longer evenly distributed on the time axis. At some potentials, the constant current must flow longer to cause a 2-mV change in potential. Thus, the capacitance is increased at this potential. The capacitance at a given potential is called differential capacitance. One obtains its numerical value by numerical differentiation of Q in the

Table 15.3 Selected data from Table 15.2.

t/h	*E*/V	*I*/A	*Q*/Ah
18.79224	0.16059	1.89969×10^{-4}	-8.51827×10^{-4}
18.87169	0.16269	1.89969×10^{-4}	-8.36734×10^{-4}

Data were copied from Table 15.2 to Table 15.3 when the value of $E(t)$ had changed by at least 2 mV compared to the previously copied value.

data set of Table 15.3 with respect to U (spreadsheet programs or difference quotient). In this way, a much smoother curve is obtained than from the original data in Table 15.2. The point density at $\Delta U = 2$ mV is quite sufficient to differentiate between charging processes. If one plots $-\mathrm{d}Q/\mathrm{d}U$ against U, the peaks for charging the graphite (reduction) point downward and for discharging (oxidation) point upward, as we are used to from cyclic voltammograms following the IUPAC recommendation. A combined plot of differential capacity during charging and discharging (Figure 15.29a) is then reminiscent of a cyclic voltammogram and can also be read analogously with respect to the involved ion transfer processes.

There are different approaches to compensate for the volume expansion of the intercalation materials using design of the materials and their processing into electrodes. First, the active materials are used in the form of powders, which are processed with binders to form a slurry. The pore spaces between the particles in the pastes allow volume expansion of the particles. A small particle can also compensate better for internal stresses than a macroscopic particle. In addition, the mass transport of Li^+ in the solid (!) leads more quickly to a homogeneous distribution of the guest ions.

Q: How can the transport of Li^+ be measured in solids?

A: Various methods of NMR spectroscopy are used for this purpose. The Li nucleus is NMR-active. In a magnetic field, an electromagnetic pulse can change the spin state of the Li atomic nuclei and label them in this way. Different methods can then be used to observe how these nuclei move. For example, their spatial distribution is measured at different times in an inhomogeneous magnetic field (pulsed field gradient, PFG). In addition, conclusions about the transport velocities can be drawn from the measurement of relaxation times.

The results from NMR measurements can be compared with data obtained from impedance spectroscopy, where the movement of Li^+ in the solid can be *one* of the observed processes.

Further literature on the application of NMR spectroscopy to this problem can be found in [43].

Using smaller particles also increases the available surface area of the insertion electrode for a given mass of the active material. While typical particle sizes for graphite intercalation electrodes range between 10 and 25 μm, this size would be much too large for silicon alloy electrodes. In this case, it makes sense to use nanoscale materials, which would further increase the surface area of the electrode. Due to the strong volume expansion during charging, the SEI is repeatedly stressed, and the electrolyte comes into contact with the strongly reducing negative electrode material. The electrolyte is gradually consumed by the subsequent reactions. Serious safety issues can result from incomplete passivation under certain operating conditions.

Further approaches attempt to match the volume expansion of the negative electrode during charging and the simultaneous volume contraction of the positive electrode materials. Further design measures are necessary to cope with the volume expansion of the active materials in a rigid cell case.

Metallates can also be used as negative electrodes. For example, $Li_4Ti_5O_{12}$ (LTO) has a potential of 1.5 V vs. Li|Li$^+$. This is not sufficiently negative to reduce solvent molecules and electrolyte components. Therefore, there is no gradual conversion of the electrolyte and formation of the SEI during operation. As a result, the safety and service life of the battery are significantly increased. However, the specific capacity is only 160 mA h g^{-1}, which is less than half the capacity of graphite. For the same mass, the specific energy of the resulting cells is lowered not only by the lower specific capacity, but also by the voltage decreased by 1.5 V compared to that of graphite. However, the absence of an SEI allows faster charging and discharging of the $Li_4Ti_5O_{12}$ electrode, which can be further accelerated by using $Li_4Ti_5O_{12}$ nanoparticles (shorter ion transport pathways in the solid). A comparison of $Li_4Ti_5O_{12}$, graphite, and silicon as possible electrode materials shows the optimization possibilities in LIBs in terms of capacity, energy density, power density, or safety. As in many other life situations, top values in one parameter require compromises in other criteria. Thus, tuning the properties of LIBs for specific applications remains an important area of applied research. Because of the good trade-off between price, stability, and performance, graphite remains by far the most widely used negative electrode material. Surface modifications, heat treatment, and fabrication of special carbon materials are the subject of intensive research [36].

15.6.3 The Electrolyte and the Solid Electrolyte Interphase

Electrolytes in lithium-ion batteries (LIBs) are solutions, polymers, or solids of complex composition. They must exhibit high conductivity over the entire temperature range of interest for a particular application. For instance, −40 °C to +80 °C is a typical requirement for automotive applications. The electrolyte must be stable over many charging–discharging cycles and must not react chemically with the materials of the negative electrode (strongly reducing) and the positive electrode (strongly oxidizing). This can usually be achieved only by forming protective (passivating) layers, which vary in effectiveness and stability depending on the choice of electrolyte. Therefore, the operation characteristics have to be adapted to preserve the protective layers. The electrolyte is also crucial for the safety of the battery. Ecological and economic aspects must be taken into account during production, use, and disposal/recycling.

The diversity of these requirements already suggests that there is not just *one* electrolyte, but that electrolytes must be optimized for specific applications [37, 38]. In the following, we will outline some important principles, focusing on liquid electrolytes. Liquid electrolytes contain as essential components a Li salt, an organic solvent (mixture), and additives to adjust special properties. The complex properties of electrolytes are still *not fully predictable*. Because of the wide variety of combinations, the make-up of the electrolyte from the components for a LIB in a specific product still requires trial-and-error optimization, in which combinatorial experimentation is also used.

Suitable Li$^+$-containing salts dissolve in high concentrations in the organic solvent and do not tend to form ion pairs. This property is likely found when the negative charge of the anion is distributed over a larger molecular anion. These ions are referred to as weakly coordinating anions. These ions without a center of concentrated negative partial charge are not able to form a coordinative bond with positive ions. At the same time, the anions should not be too large (and thus too heavy), so that the total weight of the cell does not become too large. Among all Li salts, only $LiPF_6$ is used in commercial LIBs today, as it provides a reasonable balance between the various requirements. It is stable up to almost 5 V vs. Li|Li$^+$ and thus also suitable for high-voltage batteries. Solutions with organic carbonates (see below) reach conductivities of 8–12 mS cm^{-1} at 1 mol L^{-1}. $LiPF_6$ also contributes to the passivity of the Al current collector of the positive electrode even at potentials above 3 V vs. Li|Li$^+$. The production process of $LiPF_6$ has been continuously improved. High-purity $LiPF_6$ has been commercially available only since the 1980s. This aspect is absolutely essential for a practical application, since otherwise impurities would also be introduced into the cell along with the Li salt and would remain there because the battery is a closed system. Since the LIB is a kinetically stabilized system, all impurities can potentially lift the protective effect for the electrodes and thus lead to a shortening of the service life.

However, $LiPF_6$ also has disadvantages. For example, it is part of the equilibrium

$$LiPF_6(s) \rightleftharpoons LiF(s) + PF_5(g), \tag{15.42}$$

which shifts to the right side at higher temperatures. Although organic solutions stabilize $LiPF_6$, thermal decomposition sets in more strongly above 70 °C. The released PF_5 is a strong Lewis acid[34] and triggers further reactions such as a slow polymerization of the solvent molecules, which can be visually recognized by a yellow coloration of the solution. The PF bond is susceptible to hydrolysis. PF_6^- and PF_5 react with even trace amounts of water to form hydrofluoric acid HF, which in turn undergoes further reactions. The end products of these reactions are found as components of the SEI or pass into the gas space of the battery:

$$LiPF_6(sol) + H_2O \rightleftharpoons LiF + 2\,HF + POF_3. \tag{15.43}$$

With the aim of reducing the susceptibility to hydrolysis, other electrolyte salts are being worked on. Promising examples in Figure 15.30 are Li-bis(trifluoromethylsulfonyl)imide (LiTFSI) and Li-(bisfluorosulfonyl)imide (LiFSI). With comparable conductivity, LiFSI and LiTFSI exhibit thermal stability up to 200 °C (contributing to the safety of LIBs). In contrast, corrosion of the Al current collector of the positive electrode causes problems, as its passive layer in these electrolytes is incomplete and not sufficiently stable over a wide potential range. This example illustrates a general problem in the optimization of batteries:

> A change in one cell component often necessitates optimization of other cell components because of the many interactions involved.

More complex fluoroalkyl phosphates such as LiFAP [$LiPF_3(C_2F_5)_3$, Figure 15.30] have a higher stability against hydrolysis than $LiPF_6$ but are also much more complicated to produce (thus, more expensive). The complete elimination of fluorine-containing conducting salts is desirable from an ecological point of view and can be achieved with Li-bis(oxalato)borate (LiBOB, Figure 15.30). This salt also supports the formation of the SEI and the formation of passivating layers of $AlBO_3$ on the current collectors of the positive electrode. Although the compound is also sensitive to hydrolysis, no HF is formed during the reaction. However, the achievable conductivity is only half of that in $LiPF_6$, because the solubility of LiBOB is significantly lower in organic solvents than that of $LiPF_6$. LiBOB is therefore mainly used as an electrolyte additive.

Solvents must dissolve lithium salts in high concentrations while stabilizing the ions (not the ion pairs!). Organic carbonates, esters, and ethers have been used so far. Ether and carbonyl oxygen atoms in the molecular structure promote complexation of Li$^+$.

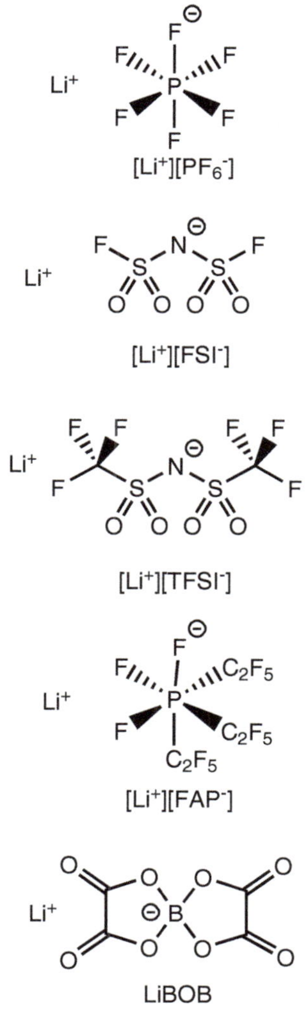

Figure 15.30 Chemical structure of some lithium salts.

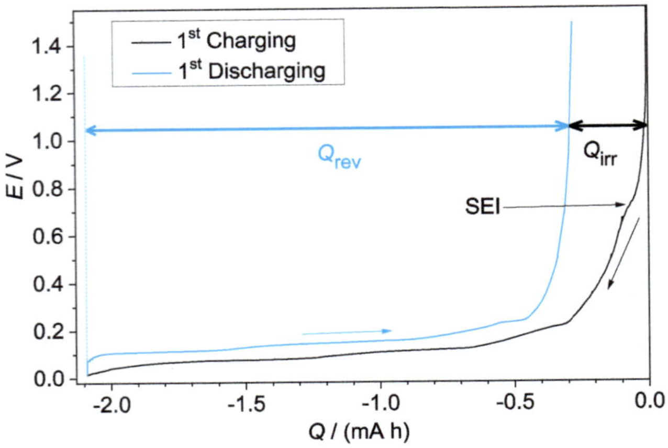

Figure 15.31 Comparison of reversible charge for lithiation and delithiation, and irreversible charge for SEI formation during the first charging.

A large relative dielectric constant ε_r is also a prerequisite to form solutions of molecularly dispersed ions. Mass transport is faster, the lower the viscosity η, also at low temperatures. To achieve this, the solvent must be at least liquid over the temperature range in which the cell shall be operated. This implies a low melting point T_m, and a high boiling point T_b. For safety reasons, the ignition temperature T_i should be as high as possible. Table 15.4 shows a selection of the solvents used. Protic solvents are not suitable because of the formation of H_2 at the negative electrode.

In most cases, solvent mixtures are used instead of pure solvents, which reduces the melting points (lowering of the melting point when a foreign component is added). By mixing cyclic and open-chain carbonates, a balance can be made between ε_r (determines solubility of electrolyte salts) and η (determines ion mobility and diffusion coefficients). Other essential criteria are the properties of the solvents in the formation of the SEI, which we briefly discuss below.

EC is reduced at potentials negative of 1.5 V vs. Li|Li$^+$ [Eq. (15.44)]. This reaction occurs when the LIB is charged for the first time. This is why the converted charge during the first charging is larger than in all subsequent charging–discharging cycles (Figure 15.31). Moreover, Li$^+$-ions are incorporated into the outer layers of the graphite lattice *with* their solvation shell of EC as ternary graphite intercalation compounds (GICs) with low Li content (Figure 15.33a). If more Li$^+$ is incorporated [Eqs. (15.39) to (15.41)], stronger negative potentials are required at which EC is reduced. Organic decomposition products are formed, which either escape as gases or are at least partially insoluble in the electrolyte and contribute to the formation of the solid electrolyte interphase (SEI), which is deposited on the electrode.

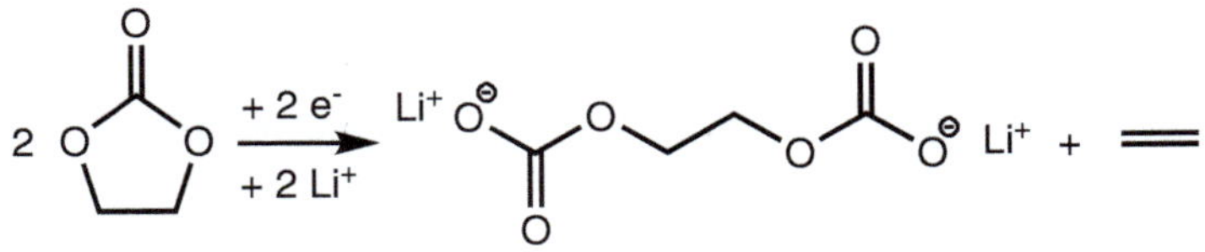

(15.44)

The charge consumed to reduce EC is called irreversible capacity Q_{irr}, because it cannot be recovered in the subsequent discharge step (Figure 15.31). Dilithium ethylene dicarbonate, as an essential component of the SEI, can be thought of as a solid that can coordinate Li$^+$ at the carboxylate groups and pass Li$^+$ from one coordination site to the next (Figure 15.32b). Once a stable SEI is formed, it prevents solvated Li$^+$ from being incorporated into the graphite lattice (Figure 15.32a). Upon entering the SEI, the Li$^+$ strips off its ligand shell, and the coordination is taken over by the components of the SEI instead. The SEI also separates the liquid electrolyte from the strongly reducing electrode material. This prevents or at least extremely retards the solvent reduction. The SEI can be regarded as a selective *protective layer*.

Table 15.4 Structure and properties of important solvents for LIB.

Solvent	Structure	ε_r η/cP	T_m/°C T_b/°C T_i/°C	Comment
Organic carbonates				
Ethylene carbonate (EC)	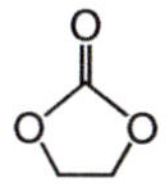	90 (40 °C) 1.9 (40 °C)	36 247–249 160	great ε_r, mediocre η; solid at room temperature
Propylene carbonate (PC)	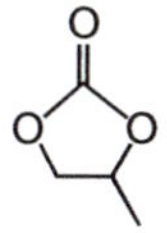	65 2.53	−48 242 135	great ε_r, mediocre η
Dimethyl carbonate (DMC)	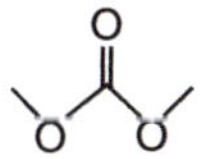	3.1 0.59	2–4 90 15	mediocre ε_r, low η
Diethyl carbonate (DEC)	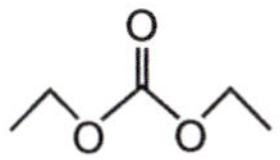	2.8 0.75	−43 125–129 33	mediocre ε_r, low η
Ethyl methyl carbonate (EMC)	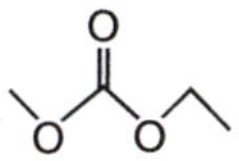	3.0 0.65	−55 108 23	mediocre ε_r, low η
Ethers				
Dimethoxy ethane (DME)	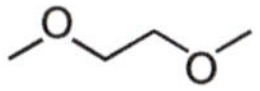	7.2 0.46	−58 84 0	Admixture to improve properties at low temperatures
Tetrahydrofuran (THF)		−7.4 0.46	−108 65–66 −17	
Esters				
Ethyl acetate (EA)	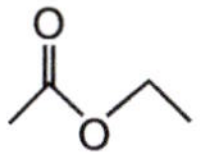	6.0 0.45	−83 77 −4	low viscosity → high conductivity; not stable against 4 V vs. Li|Li$^+$
Methyl butyrate (MB)	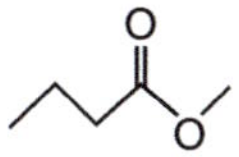	5.6 0.6	−84 102 11	

ε_r – relative dielectric constant; η – viscosity at 25 °C (unless otherwise stated); T_m – melting point; T_b – boiling point; T_i – ignition point.
Source: Adapted from [39].

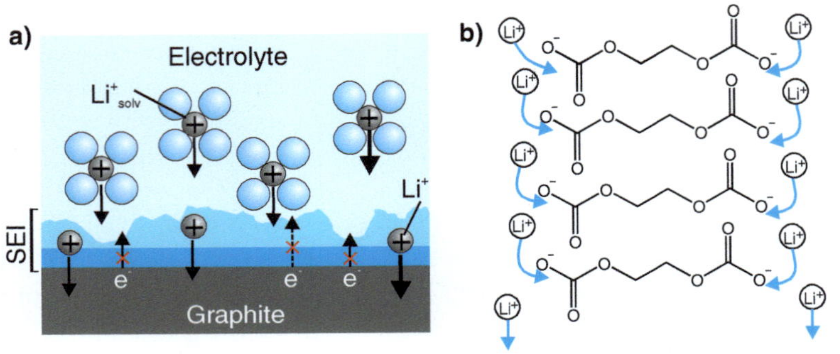

Figure 15.32 Emergence of ionic conductivity in a layer of dilithium ethylene dicarbonate; this layer is electronically insulating. Source: Adapted from H. Bülter, PhD thesis, Carl von Ossietzky University Oldenburg, 2016.

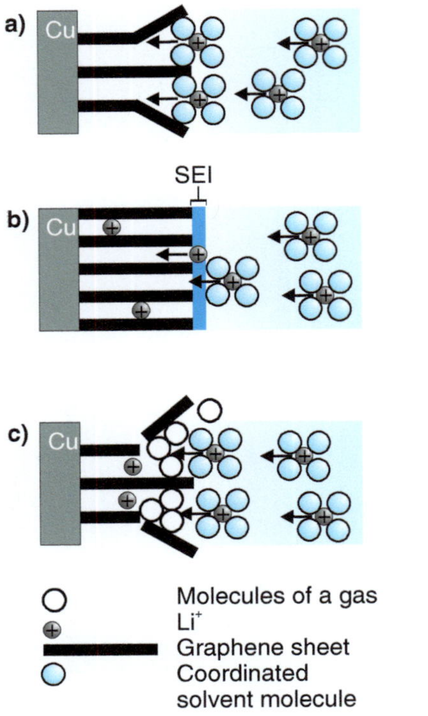

Figure 15.33 Mechanism for exfoliation of graphite.

a) Formation of solvated ternary graphite intercalation compound (GIC) with widening of the distance between graphene layers at $E < 1.5$ V vs. Li|Li^+.
b) Upon reduction of the GIC, a stable SEI is formed, which acts like a membrane for further Li^+ and leads to stripping of the solvation shell before incorporation into the graphite lattice (example: ethylene carbonate).
c) During the reduction of GIC, gases are formed which eventually break up the layer structure of graphite (exfoliation); protection is not achieved (e.g., propylene carbonate).

Source: Schemes adapted from H. Bülter, PhD thesis, Carl von Ossietzky University Oldenburg, 2016.

PC is slightly more stable against reduction than EC and is suitable for low-temperature applications due to its lower melting point. The ternary GICs with PC penetrate deeper into the graphite lattice and initially reach higher concentrations before they are finally reduced. The gas evolution that then starts leads to exfoliation (Figure 15.33c); the graphene layers and the crystallite crumble.

Due to the excellent properties for the formation of an SEI, almost every electrolyte contains EC. To reduce the viscosity, open-chain carbonates are usually added. These also enter into different decomposition reactions. Therefore, the SEI has a very complex composition (ROLi, Li-alkyl carbonates, di-Li oxalate). This is further complicated by the fact that inorganic components are also found in the SEI (LiF, LiOH, Li_2O, Li_2CO_3, ...), although it is partly debatable whether these compounds are an original component of the SEI under operating conditions or are formed only during the transfer of the sample to analytical instruments or during sample preparation procedures such as sputtering with Ar^+. The SEI exhibits a coarse stratification in which inorganic components are preferentially found on the graphite surface, while the organic components tend to occur on the side facing the electrolyte. However, electron micrographs also show very clearly a composition that varies in parallel with the surface. The SEI also changes as the battery is used. Its thickness increases. As a result, charging and discharging processes slow down, and the electrolyte is consumed. In addition, organic components of the SEI dissolve in the liquid electrolyte. Thus, the SEI is *dynamic*. The structural and functional properties of the SEI also depend very much on the parameters (T, I) of the formation process.

Although SEI has been intensively studied for many years [40, 41], a comprehensive understanding of this important boundary phase is still lacking. This is due, among other things, to the fact that some of the techniques that are useful to study it must operate in an ultra-high vacuum (UHV), which is likely to cause changes to the SEI (dissolution of some components during rinsing, sublimation in UHV) or the techniques cannot capture the lateral inhomogeneities of the layers.

Additives are added to the electrolyte to improve specific properties (Figure 15.34), for example, vinyl carbonate (VC), vinyl ethylene carbonate (VEC), or fluoroethylene carbonate (FEC). VC forms a thin polymer-like film on the electrode under reductive conditions and thus improves the stability of the SEI.

Safety risks in lithium-ion batteries arise primarily from overcharging, combined with strong heat generation. This leads to the decomposition of battery components, increased dissolution of the SEI, subsequent reduction of solvent, gas evolution, inflation, mechanical deformation of the cell case, and finally the rupture of the cell case with possible ignition of the energy-rich components upon access of atmospheric O_2. Design measures aim for avoiding these risks and include active cooling, safety valves,

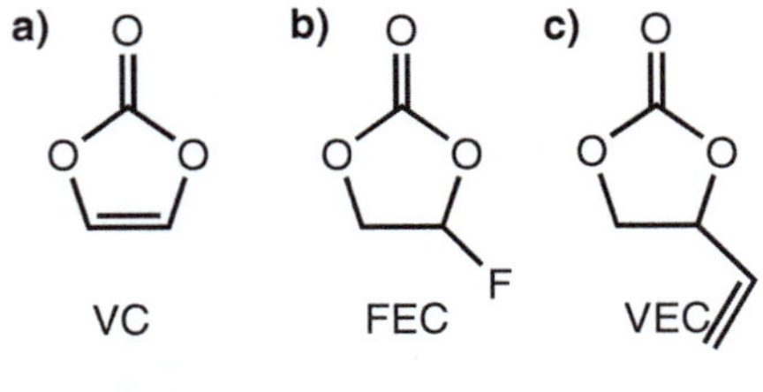

Figure 15.34 Chemical structure of typical additives to improve SEI; a) vinyl carbonate; b) fluoroethyl carbonate; and c) vinyl ethyl carbonate.

and an electronic BMS. The BMS continuously analyzes the state of the battery and adjusts parameters of the charging and discharging processes. Internal safety measures are safety additives. Biphenyl, for example, is a shutdown additive. It polymerizes at +4.5 V vs. Li|Li^+, thereby forming an insulating film on the positive electrode that interrupts the ionic current between positive and negative electrodes. It also releases gas that activates the safety valve. The safety additive permanently disables the battery, which prevents accidents but requires replacement of the affected battery or individual cell.

Another concept is the use of redox mediators (redox shuttles). At high potentials, the mediator transfers electrons to the positive electrode (i.e., is oxidized itself), diffuses to the negative electrode, and accepts electrons there (Figure 15.35a). Like a ferry or shuttle, the mediator transports electrons from the negative to the positive electrode without further conversion or damage to the active materials. An examples of a redox mediator is 2,5-dialkyl-1,4-dialkoxybenzenes (Figure 15.35b).

Specially designed separators contribute significantly to safety. Most separators consist of stretched polymer films with pores. Particularly instructive in this context are three-layer separators whose middle layer of polyethylene (PE) has a lower melting point than the two outer layers of polypropylene (PP). If the cell overheats, the inner layer melts (Figure 15.36). This closes the pores and irreversibly shuts down the ionic current. However, the two outer layers retain their shape and ensure the separation of the positive and negative electrodes. In contrast, a single melting layer would allow the particles of the positive and negative electrodes to be pushed through the molten polymer layer, thus causing an internal short circuit.

Alternative Electrolytes. Polymeric and gel electrolytes provide ionic conduction and separation of positive and negative electrodes. They are often derived from polyethylene glycol because the ether functions coordinate Li^+ and support conductivity. Li salts with weakly coordinating anions (LiTFSI and LiFSI, Figure 15.30) are added to the gel. Gel electrolytes are cross-linked polymer electrolytes that take up an electrolyte between the polymer chains. They achieve higher conductivity than pure polymer electrolytes. Both gel and polymer electrolytes tend to have lower conductivity than liquid electrolytes, but have the advantage of better compensation of vibrations and volume changes of the active materials (compared to a hard separator). On the other hand, they suppress dendrite formation better than a liquid electrolyte.

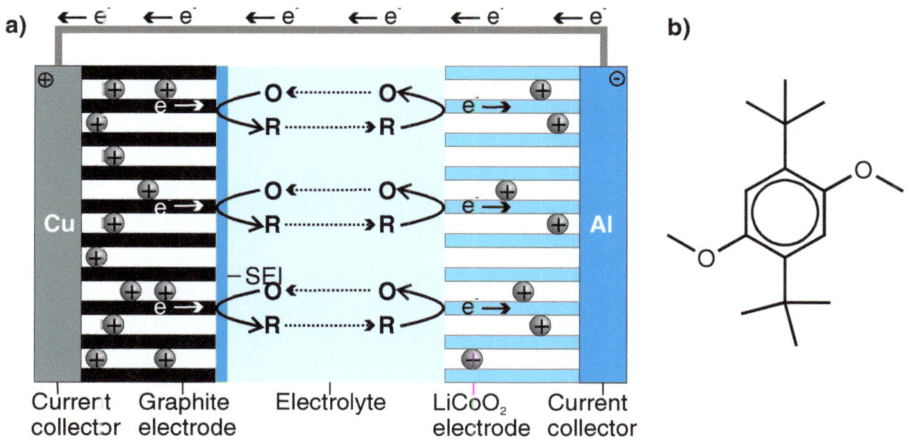

Figure 15.35 a) Operation of redox mediators as overcharge protection additives and b) chemical structure of 2,5-di-*tert*-tertiarybutyl-1,4-di-methoxy-benzene (corresponds to compound R in panel (a)). Source: Panel a) adapted from H. Bülter, PhD thesis, Carl von Ossietzky University Oldenburg, 2016.

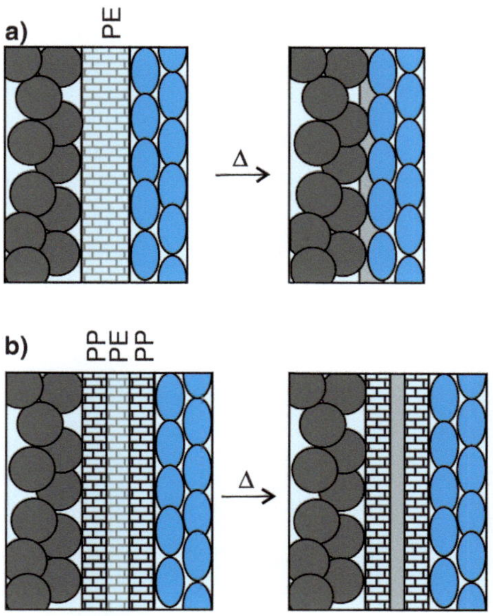

Figure 15.36 Function of a) a single-layer separator and b) a three-layer separator (shut-down separator).

Particularly stable in terms of mechanics and chemical processes are solid electrolytes (Chapter 11). These are solids that conduct Li^+ already at room temperature. However, solid electrolytes for LIBs are still under development. Overall, it is expected that solid electrolytes will tolerate only low currents and that the fabrication of the diverse solid–solid interfaces will be associated with major challenges such as retaining conduction paths during volume expansion of one phase.

15.6.4 Positive Electrode Materials

Li-containing transition metal oxides are used as positive electrodes, which undergo valence changes of the transition metal ions and intercalate or deintercalate Li^+ to maintain charge neutrality. It is important that the crystal structure does not change significantly over a wide compositional range. Here, we can only explain the general trends of the structure–property relationships and the resulting research approaches on the basis of a few typical materials.

The historically first material was lithium cobalt oxide ($Li_{1-x}CoO_2$, LCO). In the discharged state with $x = 0$, Co is in the valence state +3. In the charged state, some of the cobalt atoms are converted to the valence state +4 and Li^+-ions leave the material ($x > 0$, Table 15.1, line 13). This is possible because the crystal [of α-$NaFeO_2$-structure, (R-3 m)] is built up of alternating layers of $Co^{III/IV}$ and Li^+ ions (**layered oxide materials**). Each cation is octahedrally surrounded by O^{2-} ions. This structure exhibits a cubic close packing of O^{2-}. Li^+ and $Co^{III/IV}$ are arranged as in a NaCl structure (Figure 15.37). Li^+ is mobile between the Co layers. However, only a maximum of 70% of Li^+ can be removed from the structure (corresponding to 4.2 V vs. Li|Li^+, 140–150 mA h g^{-1}). If the charging process is continued beyond this threshold, oxidation of O^{2-} ions starts from about 4.6 V vs. Li|Li^+ on and the released oxygen O_2 reacts violently, possibly even explosively, with the organic electrolyte. A further disadvantage of these materials is that cobalt is relatively rare and therefore expensive. This prompted the investigation of layered oxides of the much more common elements Ni and Mn.

Lithium nickel oxide ($Li_{1-x}NiO_2$, LNO) has a higher useful capacity (200 mA h g^{-1}) than LCO with an analogous structure (Figure 15.37). Because of the similar ionic radii

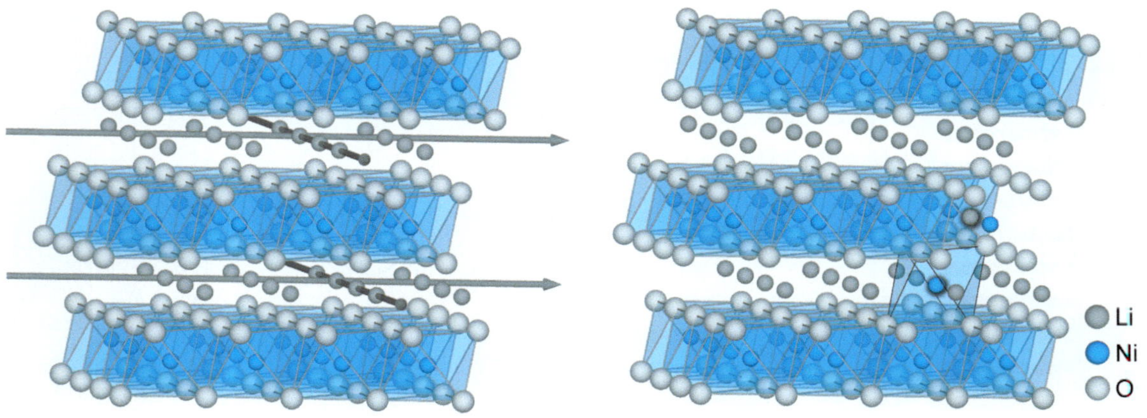

Figure 15.37 Layered structure of $LiNiO_2$ in the discharged state (left) and with defects (right), where Ni^{2+} ions occupy Li^+ sites and Li^+ ions occupy Ni^{2+} sites $(Li_{1-x}Ni_x)(Ni_{1-x}Li_x)O_2$. The directions for Li^+ transport are plotted as vectors. Source: Wilke Dononelli, Carl von Ossietzky Universität Oldenburg, with permission of Wilke Dononelli.

of Li^+ and Ni^{2+}, there is a tendency for Ni^{2+} to be incorporated on the lattice sites of Li^+ during synthesis or during discharge. One could therefore describe the compound by the molecular formula $(Li_{1-x-y}Ni^{II}_y)Ni^{III/IV}O_2$. As with LCO, overcharging LNO leads to the dreaded release of oxygen from the structure. Mixed occupancy of Li^+ sites in LNO can be suppressed by partial substitution (<20%) of Ni by Co. This increases not only the useful capacity but also the ionic and electronic conductivity. The release of O_2 is suppressed in the commercial product nickel-cobalt-aluminum oxide (NCA, $Li[Ni_{0.8}Co_{0.15}Al_{0.05}]O_2$) by the incorporation of Al. Further stability improvements are possible by partial substitution of the anions.

Nickel is not harmless ecologically. Although it is contained in a battery in a closed system for which 100% recycling is conceivable in principle, extremely serious environmental damage occurs during the mining of nickel [49]. It would therefore be desirable to use the far less problematic Mn in the form of its layered oxide $Li_{1-x}MnO_2$. However, for this stoichiometric composition, two phases exist, that is, two crystal structures of identical stoichiometry. During discharge at the composition $Li_{0.5}MnO_2$, a phase transition can occur from the layered oxide to the thermodynamically more stable spinel phase $LiMn_2O_4$, which is associated with a voltage drop during the discharge reaction. A solid solution $Li_{1-x}[Ni_{0.5}{}^{II/III/IV}Mn_{0.5}{}^{IV}]O_2$ can stabilize the layered structure over a wide stoichiometric range. Ni^{2+} ions are partially incorporated on Li^+ sites, which means that capacity losses have to be accepted. The incorporation of Ni^{2+} on Li^+ sites can, in turn, be reduced by the incorporation of Co. This development led to the commercial electrode material nickel manganese cobalt oxide (NMC, $Li_{1-x}[Ni_{0.33}{}^{II}Mn_{0.33}{}^{IV}Co_{0.33}{}^{III}]O_2$ in discharged state, $x < 66\%$, 160 mA h g^{-1}). The incorporation of Co also increases conductivity, allowing for greater current densities and power densities.

Starting from the previously mentioned spinel phase of lithium manganese oxide [$LiMn_2O_4$, spinel LMO, (Fd-3 m), Figure 15.38], intensive developments took place for modified **spinel oxides** as materials for positive electrodes. In spinels, a cubic close packing of O^{2-} is present. The octahedral cavities contain the Mn^{3+} ions, and the tetrahedral cavities contain Li^+. The difficulty is the poor cyclization behavior of

The stoichiometry of commercial battery materials is often specified by the abbreviation of the compound followed by three integer digits, as exemplified in Table 15.5.

Table 15.5 Abbreviation for commercial battery materials with different stoichiometries.

Abbreviation	Chemical composition
NMC333	$Li[Ni_{0.33}Mn_{0.33}Co_{0.33}]O_2$
NMC532	$Li[Ni_{0.5}Mn_{0.3}Co_{0.2}]O_2$
NMC622	$Li[Ni_{0.6}Mn_{0.2}Co_{0.2}]O_2$
NMC811	$Li[Ni_{0.8}Mn_{0.1}Co_{0.1}]O_2$

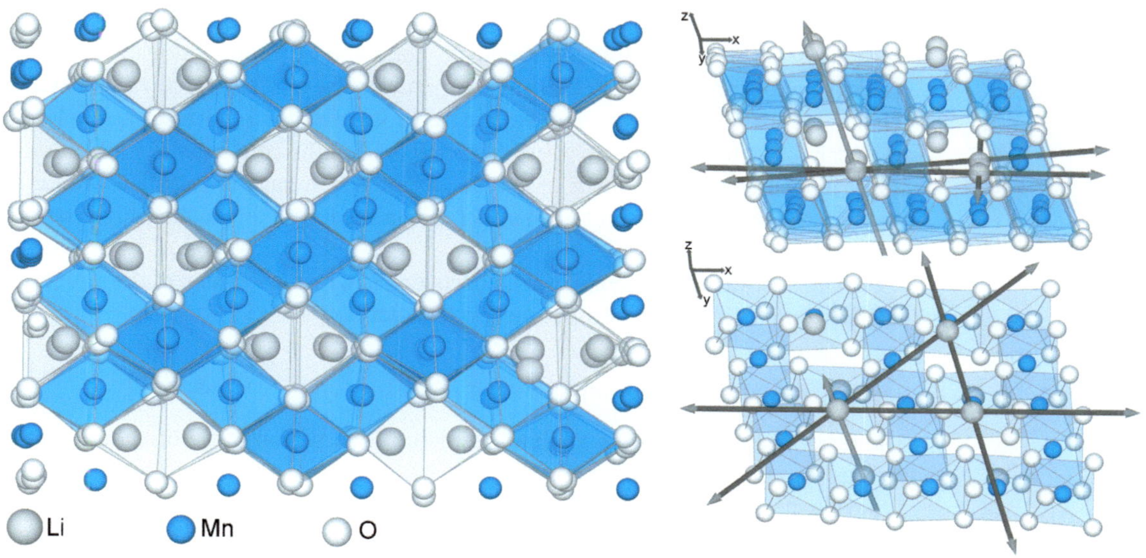

Figure 15.38 Spinel-structure of $LiMn_2O_4$ (discharged state). Left, view along the channels for the Li^+ diffusion; right, two views in which the channel axes are indicated by vectors. Source: Wilke Dononelli, Carl von Ossietzky Universität Oldenburg, with permission of Wilke Dononelli.

simple $LiMn_2O_4$. Moreover, under the influence of HF traces in the cell, Mn^{2+} ions dissociate from the positive electrode, migrate to the negative electrode, where they cause damage the SEI (see below). Discharge can occur at a level of about 4 V vs. Li|Li^+ only to 80% of the nominal capacity. At 3 V, Li^+ can also be absorbed into octahedral cavities. This leads to a 6.5% increase in unit cell size and stresses in the material that accelerate degradation and capacity fading. Stability can be improved by selective doping during material synthesis. In this process, metal ions are incorporated into the octahedral cavities, corresponding to the molecular formula $Li[Mn_{2-y}M_y]O_4$ (M = Li, Cr, Co, Ni, Cu). Spinel LMO can also release O_2. The solubility of Mn^{2+} is reduced by coating.

Of particular interest are the so-called high-voltage spinels, in which voltages of 4.7 V are obtained by substituting about 25% of the Mn^{3+}-ions (for $Li[Ni_{0.5}Mn_{1.5}]O_4$). The specific energy increases proportionally with the cell voltage. However, these voltages are already in a range where the electrolyte decomposes not only reductively but also by oxidation processes. Therefore, these positive electrode materials require a special ion-conducting layer for protection (passivation) similar to the negative electrodes. These layers are thinner than the SEI on the negative electrodes and far less intensively studied. They are referred to as cathode electrolyte interphase (CEI). At the cathode, reduction of the active material takes place during discharge. However, the potential at which this reaction occurs is so extremely positive that the electrolyte is oxidized at the same time! (Hence an oxidation at the "cathode".) Additives can favorably influence the layer formation.

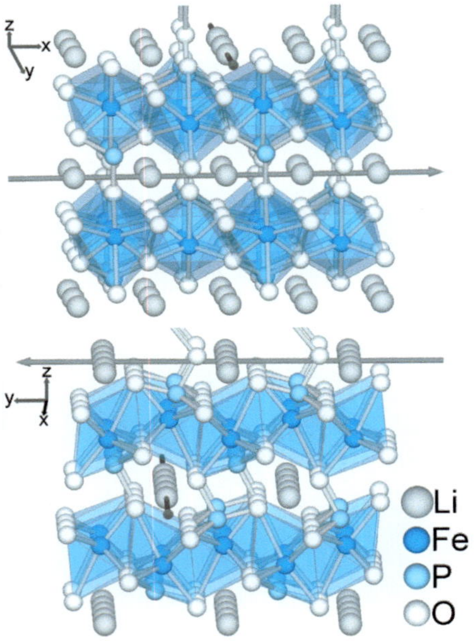

Figure 15.39 Two views of the olivine structure of $LiFePO_4$ (discharged state). The transport directions for Li^+ are indicated by vectors. Source: Wilke Dononelli, Carl von Ossietzky Universität Oldenburg, with permission of Wilke Dononelli.

The third important class of materials is formed by phosphates with lithium iron phosphate ($Li_{1-x}FePO_4$, LFP) as the most important example. It occurs in the **olivine structure** (Pnma, Figure 15.39). The O-atoms form a slightly distorted hexagonal close packing of spheres. The octahedral cavities contain $Fe^{II/III}$- and Li^+-ions. The P-atoms occupy the tetrahedral cavities and form a phosphate ion with the adjacent O-atoms. The cavities occupied by Li^+-ions touch at one edge and facilitate one-dimensional transport of Li^+. The material occurs in two very similar phases, which transform into each other by diffusion of Li^+ with small volume changes. This implies a very flat charge and discharge voltage profile. The delithiated material is very stable, even at elevated temperatures. However, this is offset by low conductivity of $Li_{1-x}FePO_4$. In order to still build operational cells, the size of the particles is reduced to about 500 nm, thus shortening the distances that electrons and Li^+ must travel in the solid. The particles are also coated with a thin, electronically conductive carbon coating to ensure the transport of electrons from all sides of the particle. The olivine structure also allows the complete or partial replacement of the transition metal ions with, for example, Co or Ni. Thus, high-voltage materials can be prepared. If the conductivity problems are solved and CEI formation can be promoted, the interest in these materials, as well as in the high-voltage spinels, could further increase in the future due to their high specific energy.

Further literature is recommended to cover a detailed discussion of the structural and functional diversity of positive electrode materials [44–48].

15.6.5 Aging of Lithium-Ion Batteries

The term aging refers to the gradual decrease in the performance of batteries with increasing number of charging cycles or with increasing lifetime (calendar aging). Typically, it is characterized by measuring the capacities and efficiencies over many cycles. Further diagnostic methods, for example, using electrochemical impedance spectroscopy, are the subject of current developments [50]. The aging of LIB is caused by several processes that influence each other. Depending on the application, aging has a different significance. For example, the LIB of a cell phone may only have to withstand three years with 3×365 charging cycles at a tolerable residual capacity of 50%. The lifetime of such a cell phone battery would not be acceptable for an electric vehicle or a battery storage plant next to a solar power plant that has a lifetime of 10 or 25 years. The elucidation of aging mechanisms (and countermeasures based on them) represents a current area of research with the goal of achieving increased lifetimes for long-life products [51]. We have already discussed some processes above; we intend to summarize and systematize the individual aspects here.

At the negative carbon-based electrodes, slow self-discharge can occur if the SEI is not perfectly formed, is damaged, or partially dissolves in the electrolyte. The latter is favored by increased temperature because the solubility of the SEI components in the electrolyte increases with temperature. An increase in temperature, especially above 80 °C, can also accelerate further side reactions, with the result that the consumption of the electrolyte progresses until finally the amount of dissolved Li salts becomes insufficient for the operation of the cell, and the cell fails. The resistance of the SEI is critically affected by the conditions under which it was formed. Especially important are the current density and temperature during initial charging. At low temperatures or with aged cells, deposition of metallic lithium rather than intercalation can occur on graphite (Figure 15.40). This unwanted process is favored by slowing the diffusion of Li^+ in the graphite, for instance, at low temperature. Metallic lithium not only poses the risk of triggering an internal short circuit via dendrites, but also immediately covers itself with an SEI under consumption of electrolyte.

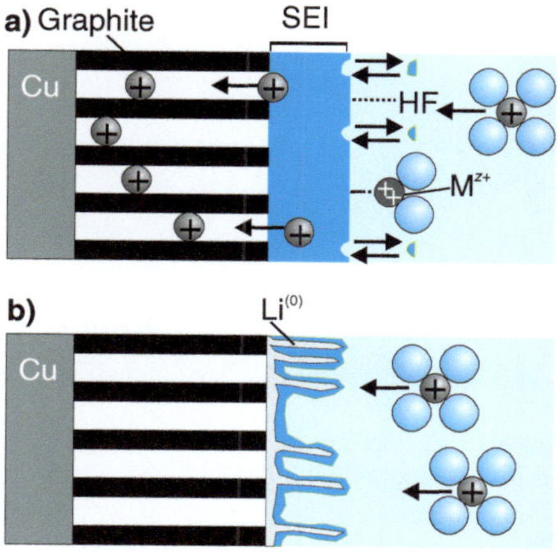

Figure 15.40 Changes of the SEI. a) Dissolution of SEI components and b) electrodeposition of Li metal on graphite. Adapted from H. Bülter, PhD thesis, Carl von Ossietzky University Oldenburg, 2016.

In addition, the graphite particles or the binder of the composite can also change, or the contact to the current collector can be disturbed. However, these processes are less important than those at the SEI.

At the positive electrode, the following processes occur with different weighting depending on the special electrode material:

- Structural changes of the lithium metal oxide, especially the formation of new phases and partial dissolution (oxygen release, Mn^{2+} release). Dissolution/redeposition as a different phase is also conceivable. The Mn^{2+} release leads to catalytic decomposition reactions at the SEI after transport to the negative electrode.
- Volume changes of the active material cause stresses in the material, which lead to fracturing of the particles of the active material. As a result, particles lose contact with the electron-conductive carbon black.
- The decomposition of binder material and of conductive additives can disrupt the functionality of the composite or interrupt the electrical contact to the current collector. This is particularly significant in the case of high-voltage materials. Here, oxidative decomposition of the electrolyte also becomes an important factor.

In addition to aging of the cell components, the overall cell design influences aging. In particular, gradients in temperature and state of charge accelerate aging. Coil cells are more susceptible to the formation of gradients than stacked cells. On the other hand, the edges of the electrodes in stacked cells experiences higher current densities (edge effect) providing multiple starting points for degradation processes, such as Li metal deposition.

15.7 Metal–Air Batteries

Metal–air cells directly combine an electronegative metal (Li, Na, Mg, Al, Zn, and Fe) as the negative electrode with the oxygen electrode as the positive electrode. The use of gaseous oxygen from air promises a large saving of weight in the charged state. However, the fundamental attractiveness of this concept is offset by some major challenges. These are the prevention…

- …of passivation of the metal electrode by dense oxide films during discharge.

- ...of corrosion of the metal electrode due to hydrogen evolution.
- ...of side reactions with other components of ambient air (CO_2, N_2, trace gases; the cell is not a closed system!).
- ...of the evaporation of the solvent of the electrolyte.

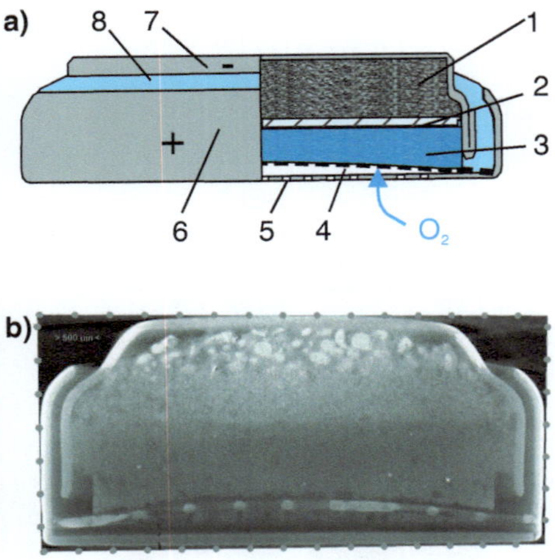

Figure 15.41 Design of a zinc–air primary cell for hearing aids.
a) Schematic.
1) Zn powder impregnated with electrolyte;
2) electrolyte-impregnated separator;
3) carbon powder as cathode for ORR;
4) gas diffusion layer;
5) perforation for air access;
6) housing (positive pole);
7) cap (negative pole);
8) sealant.
b) Cross section of a discharged Zn–air cell reconstructed from computed microtomographic measurements with X-rays. Source: Image by L. Komsiyska, EWE Research Center NEXT ENERGY Oldenburg.

Under these conditions, only primary zinc–air cells (Figure 15.41) have reached commercialization so far. They are used for hearing aids because their characteristics are ideally suited to this application. The high specific capacity allows operation without too frequent changes at a low weight (wearing comfort). The relatively low toxicity of Zn is beneficial to limit the impact on the environment because discarded cells may be damaged, and complete recycling may not be realistic for very small cells at widely distributed end users.

At the negative electrode, Zn is in contact with an alkaline solution. Although Zn should react under hydrogen evolution for thermodynamic reasons, this reaction is very slow in alkaline solutions, so that the cells can even be stored quite well with exclusion of air. The reaction at the negative electrode can be formulated as a multistep reaction, which ultimately leads to ZnO as a solid discharge product:

$$Zn + 4OH^- \rightleftharpoons [Zn(OH)_4]^{2-} + 2e^-, \tag{15.45}$$

$$[Zn(OH)_4]^{2-} \rightarrow ZnO + H_2O + 2OH^-. \tag{15.46}$$

At the oxygen electrode, O_2 is converted to OH^-:

$$\frac{1}{2}O_2 + H_2O + 2e^- \rightarrow 2OH^- \tag{15.47}$$

In alkaline solution, this reaction can be accelerated by a series of nonprecious metal catalysts. The electrode is a gas-diffusion electrode (GDE, Figure 15.41, part 3) made of a carbon material (similar to a GDE in a fuel cell). It must allow access for gaseous oxygen in the Zn–air cell, that is, the liquid electrolyte must not flood the entire pore volume, which is achieved by coating it with a perfluorinated polymer that cannot be wetted by the electrolyte. However, the liquid electrolyte must also reach the pore volume. Therefore, only the side facing the air is fluorinated. The cells are manufactured in the form of a coin cell. The GDE is located at the bottom of the housing where it is supplied with oxygen via air holes. The maximum power can be adjusted with the number of holes. Between production and use, the air holes are sealed with a foil. Under these conditions, the cell is also stable in storage.

Such systems would be particularly attractive as rechargeable cells, especially with lithium metal [52–54]. In organic solvents, a much higher cell voltage and thus energy density could be achieved. Secondary metal–air systems are currently an important area of research [55–57]. For their practical implementation, the following challenges have to be mastered:

- In the case of organic solvents, reactions of the intermediates of the oxygen reduction and oxygen evolution reaction must not attack the electrolyte and the solid electrode materials.
- In the case of aqueous systems, the release of oxygen from the oxides must occur without electrolyzing water.
- There is also a risk of electrolysis of the solvent as a side reaction with many organic electrolytes. Solvents that form a stable SEI and also do not decompose at the positive electrode are not yet known. They should also have a low vapor pressure.
- The metals in question have a very strong tendency to form dendrites during deposition. This must be suppressed by appropriate charging programs and by using three-dimensionally structured and dimensionally stable electrodes or by electrolytes with additives.

15.8 Vanadium Redox Flow Battery

RFBs use a liquid electrolyte containing a redox couple as active masses. The solutions are stored in reservoirs and fed into a reaction unit, which consists of two inert, porous electrodes at which electron transfer reactions take place (Figure 15.42). The principal advantage of this type of battery is the separate scalability of capacity and power. If the power is to be increased, the area of the inert electrodes in the reaction unit must be increased. On the other hand, an increase in capacity requires an increase in reservoir volume. In addition, the cell can switch very flexibly between charging and discharging operations. Therefore, this type of battery appears particularly attractive for the intermediate storage of electrical energy from solar and wind power plants.

The most important flow battery is the vanadium redox flow battery (VRFB). The following reaction takes place at the negative electrode:

$$V^{2+} \rightleftharpoons V^{3+} + e^-. \tag{15.48}$$

At the positive electrode, dioxovanadium-V-ions or pervanadyl ions (VO_2^+) and oxovanadium-IV-ions or vanadyl ions ($V^{IV}O^{2+}$) are converted into each other:

$$VO_2^+ + e^- + 2H^+ \rightleftharpoons VO^{2+} + H_2O \tag{15.49}$$

The half-cells are separated by a cation exchange membrane. Although this membrane should only allow passage of H^+, other vanadium-containing cations can also pass through the membrane (cross over). The solutions contain 1.6 mol L^{-1} vanadium species and about 2 mol L^{-1} H_2SO_4 as electrolyte, to which H_3PO_4 and HCl are added as additives.

Important side reactions are hydrogen evolution by self-discharge,

$$2V^{2+} + 2H^+ \rightarrow 2V^{3+} + H_2, \tag{15.50}$$

or the HER as a competing reaction during charging at the negative electrode. O_2 and CO_2 can evolve at the positive electrode as a result of corrosion of the electrode material. Chemical self-discharge occurs due to crossover processes and the HER. The VRFB has the great advantage that these processes reduce the energy efficiency but do not endanger the stability of the system. During the next charging operation, the vanadium species react to form the storage form typical of the half-cell. It may be necessary to transfer solutions between the tanks of the positive or negative electrodes during maintenance cycles to restore the same total capacity in both half-cells.

RFBs require a technical periphery, for example, for circulating the electrolytes in the reservoirs and monitoring against leakage. For this reason, a technical battery causes energy consumption even in the idle state.

The main disadvantage of the VRFB is its low energy density. Therefore, the VRFB can be considered for applications where the space and the mass of the storage are only minor issues, but robustness is important, for example, for the storage of variable (intermittent) renewable energy. Research approaches [58] to increase the energy density consist of the use of...

- ...nonaqueous electrolytes to increase the cell voltage;
- ...use of electrolyte additives to increase conductivity (HCl);
- ...use of hybrid concepts in which at least one active material is not present in the form of dissolved ions, but as a gas or as suspended solid particles;
- ...use of additional solid storage materials as boosters.

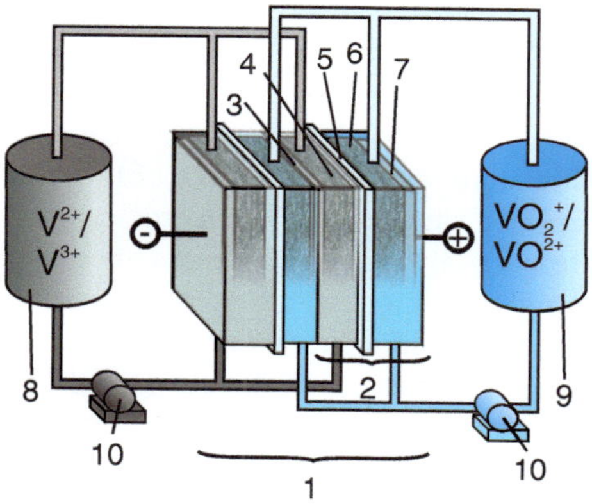

Figure 15.42 Schematic diagram of a vanadium redox flow battery.

1) Reaction unit;
2) single cell;
3) single cell with bipolar plate;
4) graphite felt as a negative electrode;
5) membrane;
6) graphite felt as a positive electrode;
7) current collector for positive electrode;
8) supply reservoir for V^{2+}/V^{3+} electrolyte;
9) storage reservoir for VO_2^+/VO^{2+} electrolyte;
10) pumps for pumping the electrolyte solutions.

For clarity, a stack with only two single cells is shown.

The symbol C for capacitance should not be confused with capacity, which is often represented by C in the literature (in this book, Q_m or Q_V for specific capacity and volumetric capacity, respectively). Capacity refers to the charge stored per mass or per volume, which is physically different from the capacitance that specifies the charge stored per voltage change. Similarly, the unit farad F = C/V = A s/V should not be confused with the symbol $F \approx 96\,485$ A s mol^{-1} for the Faraday constant. Despite these potential confusions from the use of the symbol C for capacity as well as for capacitance, it is very frequently used in the literature.

C – Capacitance [F] = [A s V^{-1}]
Q – Charge [C] = [A s]
U – Voltage [V]
ε_r – Relative dielectric constant [dimensionless]
ε_0 – Vacuum electric permittivity [F m^{-1}]
A – Area of one plate of the capacitor [m^2]
d – Distance between parallel plates [m]

15.9 Electrochemical Capacitors

15.9.1 Overview

The use of capacitors (or condensers) in the form of the Leyden jar (1745), a glass jar coated inside and outside with a thin silver layer, predates the use of batteries for charge storage. Capacitors store energy in the form of an electric field (rather than in energy-rich compounds as batteries do). As no chemical conversions are involved in the basic charge storage mechanisms, capacitors tend to be long-lived. Many modern capacitors use additional charge storage mechanisms to enhance performance, which results in a large variety of materials and detailed mechanisms. The historical Leyden jar represents a simple electrostatic capacitor in which two electron-conducting sheets are separated by a dielectric (insulating material). Capacitors are characterized by their capacitance C.

$$C = \frac{Q}{U} \tag{15.51}$$

For a parallel-plate capacitor with two plates of the same area A, the capacitance can be calculated from Eq. (15.52)

$$C = \varepsilon_r \varepsilon_0 \frac{A}{d} \tag{15.52}$$

The stored energy is

$$W_{\text{el}} = \int_0^U Q dU = \int_0^U C \cdot U\, dU \approx \frac{1}{2} C U^2 \;\text{ for }\; C \neq f(U) \tag{15.53}$$

and the energy density is obtained by inserting Eq. (15.52):

$$W_{\text{el},A,d} = \frac{W_{\text{el}}}{A \cdot d} = \frac{\varepsilon_r \varepsilon_0}{2d^2} U^2 \tag{15.54}$$

For a given capacitor, the energy density increases strongly with increasing U and decreasing distance d between the plates. However, each capacitor has a limit for high voltage, known as the breakdown voltage. The capacitance can also be maximized by increasing the area of the metal sheets within a given volume and using a dielectric material with large ε_r. The optimization of these aspects leads to very similar designs of the devices as shown in Section 15.3.3 for batteries. Compared to batteries, capacitors have a lower specific energy or energy density but a higher specific power or power density, that is, they can take up or release energy much faster than batteries, as indicated in the Ragone plot in Figure 15.3.

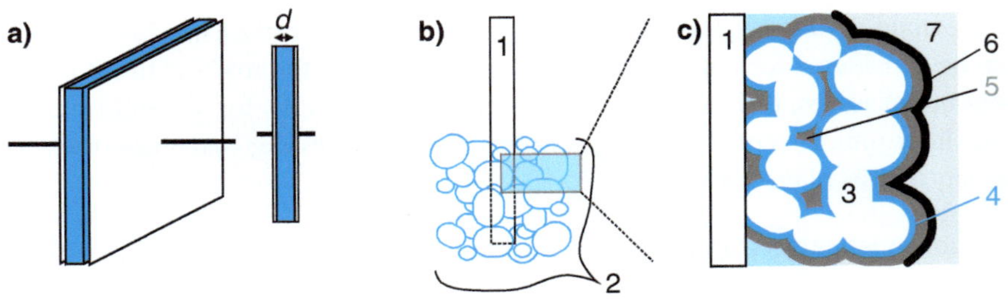

Figure 15.43 Schematic diagram of a) an electrostatic capacitor with two parallel plates and b) an electrolytic capacitor; and c) schematic cross section of the electrolytic Ta/Ta_2O_5 capacitor; 1) Ta wire as current collector; 2) sintered Ta body; 3) metallic Ta particle; 4) layer of Ta_2O_5 as dielectric; 5) pore filling MnO_2 as electrolyte; 6) graphite layer; 7) Ag current collector.

Electrostatic capacitors are traditionally made of two metal foils that are separated by an oil-impregnated paper as both a separator and a dielectric medium between the capacitor plates (Figure 15.43a). Improvements have been achieved by optimizing the dielectric to reduce the distance between the plates (without lowering the breakdown voltage) and by selecting materials with high dielectric constants ε_r, for example, polymer films (polypropylene $\varepsilon_r \approx 2.2$; polyester such as Mylar™ $\varepsilon_r \approx 3$) and mica ($\varepsilon_r \approx 7$). The charge is stored physically, which allows particularly high frequencies f in the gigahertz range for charge and discharge. As the charge storage does not involve any phase or compositional changes, electrostatic capacitors have a very long lifetime.

Electrolytic capacitors are typically made from a sintered pellet of Ta or Nb metal powder or an etched Al foil. Those metals can be coated with a compact insulating oxide layer through an electrolytic oxidation process discussed in Section 14.6.1 (thus the name for this type of capacitors, which should not be confused with electrochemical capacitors discussed below and named after the charge storage mechanism). The nanometer-thick insulating oxide serves as the dielectric layer with a particularly high relative dielectric constant (Ta_2O_5 $\varepsilon_r \approx 26$, Nb_2O_5 $\varepsilon_r \approx 41$, Al_2O_3 $\varepsilon_r \approx 8$). The second electrode must conformally contact the dielectric oxide layer for which different electrolytes are used, that is, the second conducting phase must be able to penetrate the pores in the sintered pellet. This can be achieved by forming a so-called "solid electrolyte" directly in the pores of the sintered oxide-coated metal pellet, for example, by MnO_2 (obtained by pyrolysis) or conducting polymers. In this context, "solid electrolytes" refers to a material of mixed ionic and electronic conductivity. So-called "wet capacitors" use a liquid electrolyte solution, for example, H_2SO_4 solution for Ta/Ta_2O_5 or organic electrolytes based on dimethylformamide, dimethylacetamide, or γ-butyrolactone for Al/Al_2O_3, filling the pores by capillary action.

The charge stored in an electrolytic capacitor can be 10 times larger than that of an electrostatic capacitor of equivalent size. However, charging or discharging rates are limited by the electrolyte resistances R_E [$f = (CR_E)^{-1} \approx$ 10–20 kHz]. The breakdown voltage is defined by the potentials of transpassive currents of those electrodes in the selected electrolyte and depends on the thickness of the oxide layer. An important specification is the breakdown voltage per thickness of the oxide layer ranging between 400 and 1000 V μm^{-1}. Further details can be found in [59].

Electrochemical capacitors store charge in the electrical double layer formed at the interface between a porous electrode material in direct contact with an electrolyte solution, as depicted in Figure 4.8. One may think of the porous electrode material as one plate of the capacitor, the solvent molecules of the liquid electrolyte as an ultrathin dielectric layer, and the nonspecifically adsorbed ions in the outer Helmholtz layer[11] as the second electrode of a capacitor. Thus, the "dielectric layer" is reduced to one layer of the solvent, and thus, much higher capacitances can be expected than from electrolytic capacitors that use a nanometer-thick insulating oxide as the dielectric layer.

The physical device is formed by *two* porous electrodes (Figure 4.35) of the same type (symmetrical cell) or different types (asymmetrical cell) [60]. Both electrodes are connected by a liquid electrolyte layer with a resistance R_E. Thus, the circuit diagram of an electrochemical capacitor contains two double layer capacitances in series with the ohmic resistance R_E (Figure 7.8), and the total capacitance of the device C is smaller than the smaller of the two double layer capacitances ($C^{-1} = C_{dl,1}{}^{-1} + C_{dl,2}{}^{-1}$). The most common porous electrode materials are different forms of activated carbons [61] with specific areas in the range of A_m = 1000 m^2 g^{-1}. With a typical capacitance of one flat electrode in a concentrated electrolyte solution of up to C_A = 50 μF cm^{-2}, a coarse estimate of the specific capacitance C yields $C_{dl,1,m} = A_m\ C_A = 1 \times 10^7$ cm^2 g^{-1} · 50 × 10^{-6} F cm^{-2} = 500 F g^{-1}. For a capacitor with two such electrodes, the capacitance of the device would be $C = \frac{1}{2}C_{dl,1} = 250$ F. Twice the active mass of a single electrode is required to store the charge. Therefore, the specific capacity would be $C_m \approx C/(2m_1) \approx \frac{1}{4}\, C_{dl,1,m}$ and thus about one quarter of that of the single electrode! Practical electrochemical capacitors exceed the capacitance of regular capacitors by a factor of up to 10^4 and thus form a link between the specifications of conventional capacitors and batteries. The specifications of electrochemical capacitors can be improved by incorporating further charge storage mechanisms that will be discussed below.

It is highly recommended to review Sections 4.2, 4.3, 8.10, and Chapter 7 for the basics on which the following section is built.

The electrical behavior of a porous electrode is described by a transmission line model (or delay line, Section 7.4.3 and Figure 7.20c), that is, each segment of the pore is discharged over a different electrolyte resistance depending on the pore diameter and the distance to the outer surface. This causes electrochemical capacitors to respond more slowly to a change in applied voltage than electrostatic or electrolytic capacitors.

15.9.2 Charge Storage Mechanisms of Electrochemical Capacitors

Faradaic processes obey Faraday's law [Eq. (1.16)] and are associated with charge transfer across the electrode–electrolyte interface.

When discussing the determination of the electrical double layer in Figure 4.32, we have already noted that the charging of the electrical double layer should result in a cyclic voltammogram similar to Figure 4.32a, but it is often overlaid by faradaic processes at the surface of the electrode material resulting in additional small currents and capacitances as in Figure 4.32b that are collectively summarized as "pseudocapacitance". In many respects, pseudocapacitive behavior shares many features with true capacitive behavior, for example, the proportionality of current and scan rate holds for both types of processes [Eqs. (4.100) and (5.93)]. These pseudocapacitive processes provide an additional contribution to the overall capacitance and thus constitute a resource for enhancing the energy density of electrochemical capacitors. Devices exploiting true capacitive behavior and pseudocapacitive behavior can be viewed as hybrids of capacitors and batteries [62]. In fact, a clear distinction between the possible mechanisms is often blurred and has led to a variety of nomenclature and classification proposals. A clear understanding and nomenclature would be desirable to apply sensible metrics for the characterization and comparison of devices and results [63–65]. The term "supercapacitor" or "supercap" is often used in this field. It originated from a commercial product (marketed in the 1970s by NEC and as "ultracapacitors" in the 1980s by Pinacle Research Institute) and was later used for a variety of devices with quite different charge storage mechanisms [66]. Despite its appealing brevity, it is recommended to avoid this term in a scientific context because it lacks a universally accepted and clear definition.

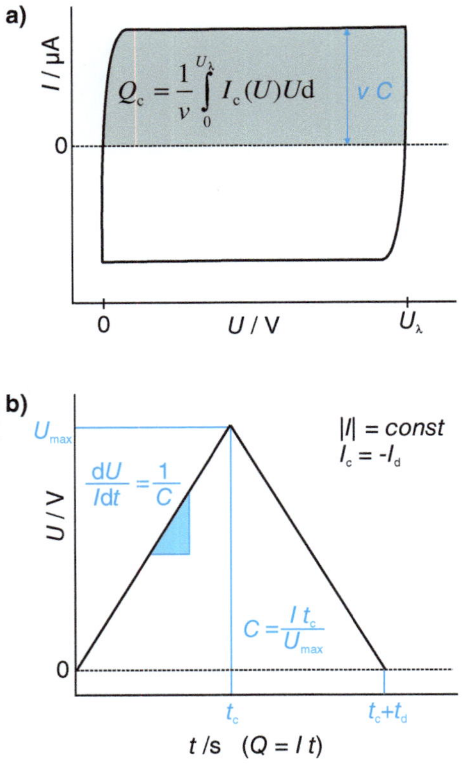

Figure 15.44 a) Schematic cyclic voltammogram of an ideal electrical double layer capacitor (EDLC) and b) voltage versus charge (time) plot for galvanostatic charging and discharging of an ideal EDLC at (|I| = const).

Electrical double layer capacitors (EDLCs) store the charge by reversible adsorption of electrolyte ions on the surface of the electrode. The current–voltage curves are almost rectangular (Figure 15.44a). The galvanostatic charge–discharge curves (a plot of voltage U over time t or charge $Q = I\,t$ at constant current I) yields a saw-toothed straight line (Figure 15.44b), that is, the cell voltage changes linearly as charge is taken up or delivered by the device ($U = (1/C) \cdot Q$). This is in stark contrast to charge–discharge curves of batteries characterized by plateaus in a U versus Q plot (cf. Figure 15.11). It means that charge is taken up or delivered at a nearly constant cell voltage U in batteries because U is determined primarily by the difference in the formal potentials of the redox couples of the positive and negative electrodes ($U \approx E_1^{\circ\prime} - E_2^{\circ\prime}$). The charging process in batteries is often associated with a phase change of the materials. Such phase change processes are one of the reasons for the limited cycle life. Phase formation typically depends on nucleation, a process that can be rate limiting as we have discussed in the example of metal deposition in Sections 9.1.1 and 9.2.3. Thus, batteries charge and discharge more slowly than electrochemical capacitors.

The capacitance of EDLCs depends primarily on the specific surface area of the electrodes. Porous nanostructured carbon materials offer such high specific surface area with high electronic conductivity at low price which makes them clearly the dominating material for this application.

Another important aspect is the ability of the pores to actually accommodate the electrolyte ions. As the pore diameters shrink, the matching between the characteristics of the electrode material and the electrolyte becomes more important [61]. With pore sizes smaller than 1 nm, the pore cannot accommodate all solvated ions, and the performance may depend on the size, charge, and shape of the cations and anions of the electrolyte. An instructive example is the cycling behavior of carbon materials in aqueous electrolytes [67]. Carbons with an average pore size of 0.58 nm showed almost the expected rectangular voltammogram in aqueous $MgSO_4$ electrolyte. Decreasing the average pore size to 0.51 nm leads to a rectangular cyclic voltammogram in the same $MgSO_4$ electrolyte, but with a greatly diminished capacitance because anions and cations of the electrolyte cannot be accommodated in the majority of the pores. Electrolytes with ions, from which one can be accommodated in the pores and the other cannot, result in skewed voltammograms where the capacitance is not constant in varying potential ranges. A comparison of the pore sizes at which such effects are observed with the diameter of solvated ions suggests that the optimum performance in many electrolytes is achieved with (partial) desolvation of ions.

The maximum voltage of the EDLC represents another important parameter for energy density of the capacitor. Of course, the maximum voltage is limited by electrolytic decomposition of the solvent, the electrolyte ions, or the carbon material. Organic electrolytes (similar to batteries) and ionic liquids are explored to widen the potential window. The conductivity of the electrolyte is important to allow fast charging and discharging. It is influenced by the type of ions in the electrolyte, their concentration, the degree of dissociation of the electrolyte, and the viscosity of the electrolyte. Clearly, the choice of electrolyte is a parameter to tune specific power, specific energy, and rate capability of EDLCs [61].

However, there are further limitations caused by saturation of the pores with electrolyte ions. Upon increase in the applied voltage, pores in the material are continuously filled with solvated ions to compensate for the excess of electronic charge on the pore walls. Once the pores are nearly filled with solvated ions, no additional charge can be accommodated even if the voltage is further increased. This leads to a drop in capacitance at higher voltages [61]. Increasing the surface area goes along with thinner pore walls. Eventually, the pore walls become so thin that they are not able to screen the ionic charge completely, and repulsive forces between ions of the same charge will become limiting to the charge that can be accommodated in the double layer [68].

In addition to charge storage in the electrical double layer, the following mechanisms are discussed for **pseudocapacitive behavior**: Pseudocapacitive contributions by adsorption are caused by underpotential metal deposition (Figure 8.67a), formation of adsorbed hydrogen on noble metal electrodes (Figures 8.60 to 8.62), or formation of surface oxide layers (Figure 12.19). Different adsorption sites may cause several closely separated peak pairs as found prominently for hydrogen underpotential deposition on polycrystalline Pt (Figure 12.19).

Pseudocapacitance by surface redox processes is expected to follow the behavior of immobilized redox systems with a clear voltammetric signal centered at the formal potential of the surface redox couple (Figure 5.57). Even if the redox process occurs in a film of finite thickness, the behavior may follow this trend provided the film thickness l_{film} is significantly smaller than the diffusion length during the critical time of the experiment $(D_{film} t_c)^{1/2}$. Such surface-bound redox couples can be redox-active groups at the edges of graphene sheets in carbon materials. In those cases, the redox potential of the surface moieties may vary (depending on the structure and the surrounding groups) and lead to a response in cyclic voltammetry that is much broader on the potential axis than expected from the idealized voltammogram in Figure 5.57.

Some materials such as hydrated RuO_2 in acidic aqueous electrolytes [80] exhibit an almost rectangular cyclic voltammogram over a wide range of potentials. The redox process can be described by Eq. (15.55) [80].

$$Ru^{IV}O_x(OH)_y + \delta H^+ + \delta e^- \quad Ru^{III}O_{x-\delta}(OH)_{y+\delta} \tag{15.55}$$

There are several special features that enable such a pseudocapacitive response over a large potential range. The material has nano-sized domains of RuO_2 surrounded by water with an optimum water content of 0.58 mol water per mole RuO_2. This provides a high electronic conductivity via interconnected RuO_2 particles and a high proton conductivity via the bound water molecules. The nanosized architecture also provides a large internal surface area with a high fraction of Ru atoms exposed to the water shell. The combination of capacity and structural data suggests that only a surface layer equivalent to one unit cell of RuO_2 is involved in the redox transition from Ru(IV) to Ru(III). Due to the high electronic and proton conductivity, the charge is delocalized, causing a capacitor-like charging process. Although RuO_2 is a rare, expensive, and also heavy material, the redox reaction of thin films of this material has been the subject of some controversy in the field, which is traced in [65]. It remains a reference material for comparison to other materials explored for broad practical applications. MnO_2 has also been studied as a pseudocapacitive material. Due to the lower conductivity and the missing hydration shell, the capacity is equivalent to only 20% of the near-surface Mn atoms [65]. K^+ showed a better rate performance than Li^+ and Na^+, probably due to a faster desolvation as a result of lower charge density in K^+ (cf. Figure 4.21) [65].

A thin-layer voltammetric response may also be found if thin layers of redox-active liquid electrolyte solutions containing I^-/I_3^-, V^{2+}/VO^{2+}, quinones/hydroquinones, etc. are electrolyzed in divided cells. In this case, the mechanism is similar to that of RFBs with a minimized liquid reservoir, that is, the electrolyte in the pores and between the electrode and the separator.

Pseudocapacitance by ion insertion occurs in transition metal oxides in which small cations such as Li^+ can move in channel structures. Examples are monoclinic $TiO_2 + x\,Li^+ + x\,e^- \rightleftharpoons Li_xTiO_2$ ($x < 1.25$) or orthorhombic and tetragonal $Nb_2O_5 + x\,Li^+ + x\,e^- \rightleftharpoons Li_xNb_2O_5$ with $x \leq 2$. These processes occur at potentials that are not sufficiently negative for the formation of a solid electrolyte interphase that would retard the ion transfer process. The oxide is a nanometer-thick layer (as opposed to micrometer-sized particles in batteries) to avoid transport limitations in the solid materials that would compromise the rate capability [65].

Other materials that cause pseudocapacitive behavior are thin layers of intrinsically conducting polymers (Section 13.2.4) such as polyaniline, polypyrrole, and polythiophene and their derivatives [66].

Whether the response appears more capacitive or more diffusive may depend on the timescale of the experiment, for example, the frequency in electrochemical impedance spectroscopy or the scan rate in cyclic voltammetry. From a set of cyclic voltammograms recorded for the same cell at different scan rates, the relative contributions can be extracted by assuming that the current at each potential is a linear combination of diffusive and capacitive contributions whose relative contributions may change with potential and with scan rate.

$$I(U,v) = k_1(U)\sqrt{v} + k_2(U)v \tag{15.56}$$

Rearranging yields a linear equation, which is convenient to determine $k_1(U)$ and $k_2(U)$ from linear regression.

$$\frac{I(U,v)}{\sqrt{v}} = k_1(U) + k_2(U)\sqrt{v} \tag{15.57}$$

As the pseudocapacitive behavior is caused by a faradaic reaction, it is common to specify the capacitance related to mass or volume. Similar to batteries, the pseudocapacitance is not uniform over the potential range, as it should be for a pure EDLC [63].

15.9.3 Characterization of Electrochemical Capacitors

Initially, the electrode materials are characterized by cyclic voltammetry in three-electrode cells to determine the maximum possible electrode potential and the capacitance. After assembling two electrodes to a capacitor, the characterization is usually completed by recording galvanostatic charging and discharging curves.

Testing of Electrode Materials. A small amount of weighed material is processed with binder and carbon black to an electrode similar to the description of the processing of battery materials in Excursion 15-1. A slurry is prepared by mixing the powder of the electrode material, a binder (e.g., PTFE or sodium carboxymethylcellulose (NaCMC) in combination with styrene-butadiene rubber (SBR)), a conductive agent (e.g., carbon black), and a small amount of solvent. Typical mass fractions of the solids for research cells are 80% active material, 10% binder, and 10% conductive additive. In commercial capacitors, the formulations are further optimized to reach 96–98% of the active mass (at the expense of the mass fraction of binder and conductive additives). The selection of the solvent depends on the choice of the binder (organic solvents for PTFE and water for NaCMC and SBR). The slurry is coated on a current collector (Al or stainless steel foils). Subsequently, the slurry is dried under elevated temperature and calendered to decrease the contact resistance between the particles in the composite and to adjust the porosity of the electrode layer.

The first aim is to identify the electrochemical stability window and to estimate the specific capacitance. The electrode is connected to a potentiostat as working electrode in a three-electrode setup with auxiliary and reference electrodes and tested by cyclic voltammetry at moderately fast scan rates of 5 or 10 mV s^{-1}. It is important to use the electrolyte solution that is to be used later in the device because the possible potential range depends on both the electrode material and the electrolyte solution. The initial potential is set to the open-circuit potential E_{OCP} of the electrode material in the specific electrolyte solution. The potential range is explored by incrementing the positive switching potentials $E_{\lambda,+}$ in small steps, for example, 50 mV, and determining the coulombic efficiency $\varepsilon_C = |Q_d|/|Q_c|$ for each $E_{\lambda,+}$ from the ratio of charges in the forward and the reverse scan (Figure 15.45a). This ratio is unity (100%) for an ideal capacitor. Values of ε_C below 100% indicate that some charge was consumed by irreversible faradaic reactions. The maximum $E_{\lambda,+}$ is identified as the potential at which ε_C remains larger than 99% (or 97%). The threshold value is somehow a subjective criterion and therefore must be precisely specified in a report. Even at $\varepsilon_C = 99\%$ some irreversible reactions (degradation) will inevitably occur, but rapid detrimental changes are prevented.

The negative potential limit is identified in an analogous manner with the same electrode (for symmetrical capacitors) or the same mass of another electrode material (for asymmetric capacitors). Voltammograms are initiated at E_{OCP} with a negative-going scan to $E_{\lambda,-}$, which is decremented by about 50 mV from cycle to cycle, and determining $\varepsilon_C = |Q_d|/|Q_c|$ (Figure 15.45b). $E_{\lambda,-}$ is the potential when ε_C is just above 99% (97%). Once $E_{\lambda,+}$ and $E_{\lambda,-}$ are known, the corresponding charges Q_+ (at $E_{\lambda,+}$) and Q_- (at $E_{\lambda,-}$)

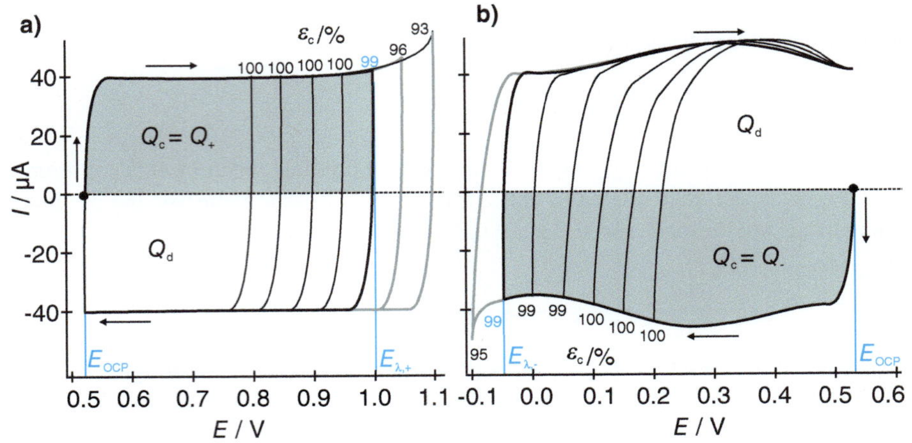

Figure 15.45 Determination of the potential window (schematic). a) Potentiodynamic excursion toward the positive limit; and b) potentiodynamic excursions toward the negative limit. The determination of the stored charge from integration of one half-cycle is indicated.

are determined and used for balancing. A more refined version of this approach with further references is provided in [69].

A second method for the determination of the potential limits is the voltage-hold test. The electrode is charged in the three-electrode setup to the presumed maximum potential. It is then left at open circuit. Slow self-discharge will occur due to faradaic reactions, and this can be followed by recording the open-circuit potential E_{OCP} as a function of time t. The potential decay will be faster when irreversible faradaic reactions occur above the potential limits and will be slow within the suitable potential limits. Furthermore, the shape of the decay curve provides hints about the self-discharge mechanism. If the stored charge reacts with an impurity in the electrolyte under diffusion-controlled conditions, E_{OCP} changes linearly with $t^{1/2}$. If the reaction responsible for self-discharge involves a compound of high concentration (electrolyte salt or solvent, electrode material), E_{OCP} changes linearly with $\ln(t)$. The voltage-hold test also allows estimation of long-term stability. Examples and discussion can be found in [70].

$C_{m,+}$, $C_{m,-}$ – Capacitances [F g^{-1}] for charging the material in the three-electrode cell to $E_{\lambda,+}$ or $E_{\lambda,-}$

$|Q_+|$, $|Q_-|$ – Absolute amounts of charge [C] stored between E_{OCP} and $E_{\lambda,+}$ or $E_{\lambda,-}$ for the same mass of electrode material in the three-electrode setup

m'_+, m'_- – Masses [g] of material for the positive (negative) electrode in the balanced two-electrode setup

$|Q'_+|$, $|Q'_-|$ – Absolute amounts of charge [C] stored between E_{OCP} and $E_{\lambda,+}$ or $E_{\lambda,-}$ of the balanced full cell in the two-electrode setup

m_+, m_- – Masses [g] of the positive and negative electrode materials used in the three-electrode tests

$E_{\lambda,+}$, $E_{\lambda,-}$ – Maximum positive (or minimum negative) switching potentials [V] in the three-electrode test

U_{max} – Maximum voltage [V] of the two-electrode test cell

Balancing Positive and Negative Electrodes. Even if the very same electrode has been used to determine the charges Q_+ and Q_- stored between E_{OCP} and the limits of the stability range $[E_{\lambda,+}, E_{\lambda,-}]$, Q_+ and Q_- are expected to differ because the stability range of electrode and electrolyte is usually not symmetric with respect to E_{OCP}. The final full cell will be charged by storing equal amounts of charge on the positive electrode and on the negative electrode (but this may cause a different change in the electrode potential on both). Charge balancing is necessary before assembling a complete capacitor. Otherwise, the capacitor will be quickly destroyed by irreversible faradaic reactions at the electrode that reaches the limiting potential first. Using the specific capacitance (charge per voltage and mass) at the potential limits $C_{m,+} = |Q_+(E_{\lambda,+})|/m$ and $C_{m,-} = |Q_-(E_{\lambda,-})|/m$ from the three-electrode measurements, the balancing condition for charges $|Q'_+| = |Q'_-|$ in the full cell is obtained by Eq. (15.58)

$$
\begin{aligned}
|Q'_+| &= |Q'_-| \\
m'_+ C_{+,m} E_{\lambda,+} &= m_-{}' C_{-,m} E_{\lambda,-} \\
\frac{m'_+}{m'_-} = \frac{C_{m,-} E_{\lambda,-}}{C_{m,+} E_{\lambda,+}} &= \frac{|Q_-|\, m_+}{m_-\, |Q_+|}
\end{aligned}
\tag{15.58}
$$

Equation (15.58) is further simplified if the same mass $m = m_+ = m_-$ of positive and negative electrode materials is used in the determination of the potential limits. In this case, the ratio of the balanced masses is the reverse ratio of the charges in the forward potential scan from E_{OCP} to $E_{\lambda,+}$ ($E_{\lambda,-}$).

$$\frac{m'_+}{m'_-} = \frac{|Q_-|\,m}{m\,|Q_+|} = \frac{\left|\frac{1}{v}\int_{E_{OCP}}^{E_{\lambda,-}} I_d(E)dE\right|}{\left|\frac{1}{v}\int_{E_{OCP}}^{E_{\lambda,+}} I_d(E)dE\right|} = \frac{\left|\int_{E_{OCP}}^{E_{\lambda,-}} I_d(E)dE\right|}{\left|\int_{E_{OCP}}^{E_{\lambda,+}} I_d(E)dE\right|} \tag{15.59}$$

The ratio of the integrals in Eq. (15.59) corresponds simply to the area under the forward scans between [E_{OCP}, $E_{\lambda,+}$] and [E_{OCP}, $E_{\lambda,-}$] in the cyclic voltammograms of both electrodes.

The maximum voltage window of the electrochemical capacitor is obtained as

$$U_{max} = E_{\lambda,+} - E_{\lambda,-} \tag{15.60}$$

The cell is assembled from the positive and negative electrodes with the balanced mass loadings and the electrolyte-soaked separator.

Characterization of Full Cells. The whole cell is made from the electrodes with a mass ratio according to Eq. (15.58) and a porous electrolyte-soaked separator in a research cell, for example, coin cell, pouch bag cell, or Swagelok cell (Figure 15.16). The electrolyte is the same as used in the determination of the potential window above. The positive electrode is connected as the working electrode (WE), and the ports for auxiliary (Aux) and reference electrodes (Ref) are jointly connected to the negative electrode. The behavior of the devices is then studied by voltammograms between 0 V and a positive value. For cycling devices without a reference electrode port, the positive electrode is the working electrode and the negative electrode is the counter electrode (CE) connected to the Ref *and* to the Aux ports of a potentiostat/galvanostat.

For research cells, it is very helpful to have an additional reference electrode between the positive and the negative electrode, allowing the potentials of the positive and negative electrodes to be recorded separately (cf. to corresponding approaches for batteries in Figure 15.17). In this case, the positive electrode is used as WE, the negative electrode as Aux, and the reference electrode as Ref. The following tests are usually conducted:

1) Verification of the voltage limits by incrementation of the applied maximum voltage U_{max} in cyclic voltammograms similar to Figure 15.46 and monitoring of ε_C.
2) Verification of the shape of the voltammograms at increasing scan rates. Figure 15.46a shows selected voltammograms at different scan rates v. At this scaling, the shape of the curves for small v cannot be clearly seen. Normalizing the current by the scan rate in Figure 15.46b shows that there is an irreversible reaction in curve 1 ($v = 0.002$ V s^{-1}) at voltages larger than 0.5 V. Curve 4 ($v = 0.5$ V s^{-1}) exhibits a clear deviation from the expected almost rectangular shape of the cyclic voltammogram, in Curve 2 and a decrease in capacitance compared to lower scan rates ($I / v = I\, dt/dU = dQ/dU$). This is due to significant ohmic losses caused by internal resistances limiting the rate capability of the capacitor.
3) Determination of the capacitance $C = |I \cdot t_d/U_{max}|$ by galvanostatic charging–discharging curves at different current densities of 1 A per gram of active material of one electrode and for cell voltages between 0 and U_{max}. In addition, the obtained U versus t plots provide indications on the limitations by cell resistance (Figure 15.47a) and the contribution of pseudocapacitance (Figure 15.47b). In the case of pseudocapacitance, the differential capacitance $C = dQ/dU$ varies over the potential range and affects the stored energy, for which it is important to know at which voltage the additional charge is stored. This test is much more informative when a reference electrode is used to obtain the electrode potentials in addition to the cell voltage. In any case, the report must clearly state to which mass the specific quantities were related (active/total mass of one/both electrodes).

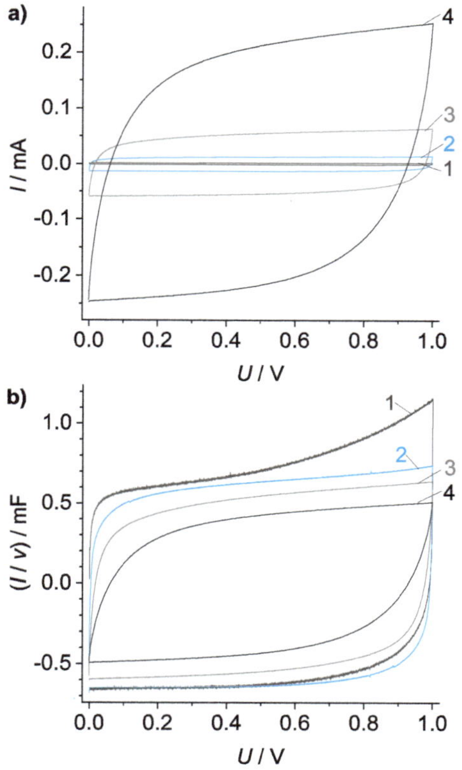

Figure 15.46 Cyclic voltammetry of an assembled electrochemical capacitor at different scan rates; a) plots of currents versus applied voltage; and b) plots of current normalized by the scan rate. Biomass-derived carbon electrodes in aqueous 0.5 mol L^{-1} $MgSO_4$; scan rates are 1) 2 mV s^{-1}, 2) 20 mV s^{-1}, 3) 100 mV s^{-1}, and 4) 500 mV s^{-1}. Data were kindly provided by Anita Cymann-Sachajdak, Gdańsk University of Technology, Poland.

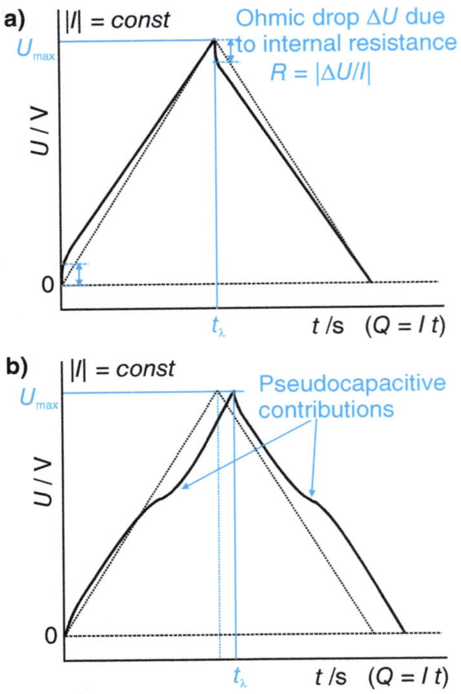

Figure 15.47 a) Schematic galvanostatic charging–discharging curves illustrating a) the effect of internal resistance (voltage is changing without significant charge stored) and b) pseudocapacitive contributions (additional charge is stored by a faradaic reaction with lower change in potential than for an ideal EDLC).

4) Characterization of the internal resistance as the serial resistance from electrochemical impedance spectroscopy (EIS, Chapter 7) at high frequencies. Under these circumstances, the equivalent circuit can be simplified to a serial combination of a resistor and a capacitor, as shown in Figure 15.48. The serial resistance depends on the electrolyte, but also on the internal resistance within the electrode material, that is, the used materials *and* the interparticle resistance in the composite electrode materials, and thus on the way the electrode was prepared. A more elaborate analysis of entire impedance spectra can provide information about pore structures of the electrode materials [71–73].
5) Calculation of the capacitance, the specific capacitance, the stored energy, the specific energy, the maximum power, and the maximum specific power from the obtained data. Specific quantities can be obtained by relating the quantity to the active masses of both electrodes or to the total mass of the device, as shown in Table 15.6. The equations below refer to the masses of the active materials. The approximations refer to the case when $m_+ = m_-$ and $C_{m+} = C_{m,-}$. There is a dispute about a sensible rating of the power [74]. The frequently cited equation $P_{max} = U_{max}{}^2/(4R_s)$ is only valid for a matched load, in which 50% of the stored energy is converted to work and 50% is dissipated as heat. Such a 50%-energy efficiency ε_W is almost unusable in most situations. The equation in Table 15.6 is for a pulse discharge at 75% of U_{max}, removing only a small fraction of the stored energy [75]. In any case, the maximum power output scales with U^2_{max}/R_s.
6) Determination of the lifetime of the capacitor by recording at least 100 000 galvanostatic charging–discharging cycles at a current of 1 A g^{-1} with intermittent EIS determination of the series resistance. The lifetime is reported as the number of completed cycles before one of the following conditions is met:
   i) The series resistance increases by 100% of the initial value.
   ii) The capacitance C drops to 80% of its initial value.
7) Checking the lifetime and capacitances at reduced (−40 °C) and elevated temperatures (up to 65 °C). The properties of the devices often deviate from those at room temperature due to different ion mobilities in the electrolyte solution and different rates of side reactions.

Commercial EDCLs are expected to have a cycle lifetime of over 1 000 000 cycles. Their testing is specified by an international standard IEC 62391-1:2022 [74, 76, 77]. The considerations above show that many relative specifications can be calculated for an electrochemical capacitor and a battery. It is very important to state exactly which

Table 15.6 Typical derived specification of electrochemical capacitors (device level).

Quantity	Absolute quantity	Specific quantity
Capacity or charge [C], [C g^{-1}]	$Q = I_d t_d$	$Q_m = \dfrac{Q_d}{m_+ + m_-}$
Capacitance [F], [F g^{-1}]	$C = \dfrac{Q_d}{U_{max}} = \left(C_+^{-1} + C_-^{-1}\right)^{-1} \approx \dfrac{C_+}{2}$	$C_m = \dfrac{C}{m_+ + m_-} = \dfrac{Q_d}{U_{max}(m_+ + m_-)} \approx \dfrac{C_{m,+}}{4}$
Energy [Ws], [Ws g^{-1}]	$W_{el} = \int_0^{U_{max}} C\,U\,dU \approx \frac{1}{2} C U^2_{max}$	$W_{el,m} = \dfrac{W_{el}}{m_+ + m_-} \approx \dfrac{CU^2_{max}}{2(m_+ + m_-)}$
Power [W], [W g^{-1}]	$P = \dfrac{9}{16}(1 - \varepsilon_W)\dfrac{U^2_{max}}{R_s}$ $\varepsilon_W = 1 - \dfrac{I_d R_s}{U} = 1 - \dfrac{I_d R_s}{0.75\,U_{max}}$	$P_m = \dfrac{P}{m_+ + m_-}$

quantity was used for the calculation and to report the primary data. For instance, two capacitors are reported, one with a specific capacitance calculated with respect to the total mass of one electrode and one specified with respect to the mass of the active material of one(two) electrode(s). A comparison of the materials (or devices) is possible only if the method of the calculation was exactly and transparently documented. Substantial efforts are needed to establish a generally accepted and followed practice of reporting material and device specifications [82].

Further Reading. The classical text on electrochemical aspects of capacitors is a book by Conway [78], which has been complemented by more recent collective books about the basics and characterization of electrochemical capacitors [79]. A selection of recent reviews is recommended for an overview of the very dynamic development of the field that currently experiences an enormous output of scientific publications from different disciplines [62, 65]. Theoretical aspects of the charge storage mechanisms are covered in [81]. The reader is referred to reviews regarding specific material aspects (and matching of different materials), which are critical for the advancement of the field [61, 65, 79–81, 83–85].

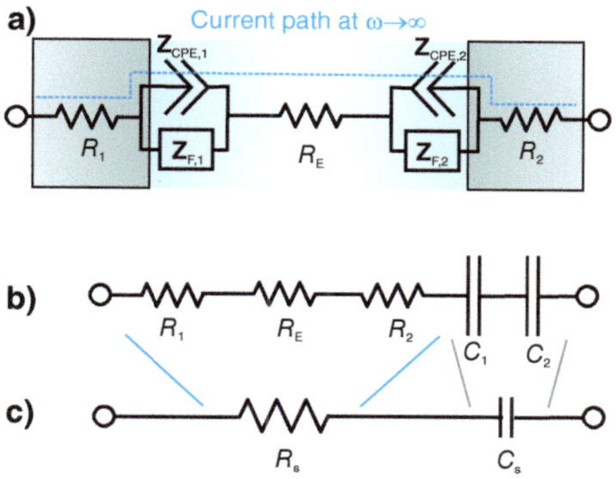

Figure 15.48 Simplification of the equivalent circuit of an electrochemical capacitor. a) The two porous electrodes according to Figure 7.25 are combined in series with the electrolyte resistance R_E and the internal resistances (R_1, R_2) of the two electrodes. For high frequencies, the resistance of the constant phase elements (CPE) become much lower than the almost constant faradaic impedance. Hence, the AC current passes almost entirely via the capacitor (dashed blue line). b) Rearrangement of the sequence of the circuit elements with omission of the faradaic impedances $\mathbf{Z}_{F,1}$ and $\mathbf{Z}_{F,2}$ and replacement of the constant phase elements by simple capacitances; and c) replacements using the serial resistance $R_s = R_E + R_1 + R_2$ and the serial capacitance $C_s^{-1} = C_1^{-1} + C_2^{-1}$. Compare also to Figure 4.9.

Test your knowledge

Exercise 15.1: Calculate the energy efficiency of a hydrogen storage system, assuming electrolysis of water with an energy efficiency of 90% and conversion of the hydrogen formed in a fuel cell with an electrical energy efficiency of 65%. Compare these values with corresponding values for a lead–acid battery or a lithium-ion battery (literature search). What are the advantages of hydrogen as an energy storage material?

Exercise 15.2: Calculate the theoretical specific capacity of $LiCoO_2$,

a) assuming complete release of Li^+ from the host lattice,
b) assuming that only 70% of Li^+ ions can be released without damaging the host lattice.

Exercise 15.3: How many grams of graphite as negative electrode material must be combined with 1 g of $LiCoO_2$ as positive electrode material to form a battery so that the capacity of the negative electrode is 105% of the capacity of the positive electrode? Assume a discharge of up to 70% Li^+ from the $LiCoO_2$ host lattice.

Exercise 15.4: Using the results from Exercise 15.3, calculate the theoretical specific energy of a lithium-ion battery with $LiCoO_2$. Research any missing quantities.

Exercise 15.5: Using the Ragone plot, discuss a reasonable application of electrochemical capacitors and lithium-ion batteries.

Additional Exercise: Research characteristic data for fuel cells and include them in the discussion.

Exercise 15.6: Use Figure 15.11 to estimate the voltage efficiency and energy efficiency of the battery under investigation. Proceed in such a way that you first assume a constant charging voltage and a constant discharging voltage and, in a refinement step, approximate the voltage at high states of charge by a linear function in each case.

Exercise 15.7: Give reasons why batteries can charge or discharge different amounts of charge at different charging currents (rate capability). Describe how this property can be characterized. What is the significance of this property for the practical use of batteries?

Exercise 15.8: Explain why metallic lead can be deposited in a lead–acid battery even though Pb|Pb^{2+} has a more negative standard electrode potential than H_2|H^+.

Exercise 15.9: What is the central reaction in a lead–acid cell? What are the important side reactions that occur? How do the side reactions influence the application properties of the cell?

Exercise 15.10: Propose an electrical circuit of single cells of a lead–acid battery to provide a voltage of (24 ± 1) V at the poles.

Exercise 15.11: Using an appropriate schematic sketch of a lithium-ion battery, explain why its operation is also known as rocking chair technology.

Exercise 15.12: Compile the chemical properties of lithium. Derive occupational safety measures from this when working with metallic lithium. In particular, find out how fires of metallic lithium can be extinguished.

Exercise 15.13: Compare the appearance and information content of a cyclic voltammogram (Chapter 5) with a (cyclic) differential capacitance curve (Figure 15.29a).

Exercise 15.14: Why do electrolytes for lithium-ion batteries usually contain ethylene carbonate?

Exercise 15.15: Why are solvent mixtures preferred in electrolytes for lithium-ion batteries instead of pure solvents?

Exercise 15.16: Explain the origin of the SEI and its importance to the safety and lifetime of a lithium-ion battery.

Exercise 15.17: Explain the term "formation" in the context of lithium-ion batteries.

Exercise 15.18: Explain the advantages of a three-layer PP/PE/PP separator.

Exercise 15.19: Compile possible materials for the positive electrode in lithium-ion batteries. Explain their functional advantages and disadvantages.

Exercise 15.20: Explain the safety hazard of overcharging a $LiCoO_2$ electrode.

Exercise 15.21: Why do lithium-ion batteries "age"?

Exercise 15.22: Compose the reaction equations for metal–air batteries using Zn, Li, and Mg.

Exercise 15.23: Calculate the capacity and specific energy of a vanadium redox flow battery from the concentrations of the electrolytes and the cell voltage. Evaluate these data and compare them with specifications of batteries for a cell phone, for a cordless screwdriver and for a passenger car. Take data from data sheets (e.g., websites of internet retailers) or make reasonable estimates.

Exercise 15.24: Why are increased research efforts being devoted to redox flow batteries despite their relatively low energy density?

Exercise 15.25: A new bio-waste derived carbon material shall be characterized as active material for an electrochemical capacitor. The specific area was determined from a BET isotherm (Brunauer–Emmett–Teller) as 1252.5 $cm^2\ g^{-1}$. A slurry is prepared with 1200 mg carbon material as active material, 150 mg PTFE as binder, 150 mg carbon black as conductive additive, and 750 μL of *N*-methyl pyrrolidone. The slurry is coated on one side of an Al foil that has a mass of 27 $mg\ cm^{-2}$. After complete drying, 100 cm^2 of the coated foil has a total mass of 3.980 g. A disk is cut from the foil with an area of 1 cm^2 and tested in a three-electrode cell with an Ag|AgCl|3 mol L^{-1} KCl reference electrode for the potential limits by cyclic voltammetry between E_{OCP} and $E_{\lambda,+}$ or $E_{\lambda,-}$ at $v = 0.01\ V\ s^{-1}$. The obtained raw data are summarized in the table below:

Cycle number	$E_{initial} = E_{OCP}$/ V vs. Ag|AgCl	E_{λ}/V vs. Ag|AgCl	Q_c/C	Q_d/C
1	0.345	0.645	0.4816896	0.48024453
2	0.345	0.695	0.5677056	0.56543478
3	0.345	0.745	0.65536	0.6504448
4	0.345	0.795	0.7446528	0.73422766
5	0.345	0.845	0.835584	0.8197079
6	0.345	0.895	0.9281536	0.90123715
7	0.345	0.045	−0.516096	−0.5150638
8	0.345	−0.005	−0.6078464	−0.6060229
9	0.345	−0.055	−0.7012352	−0.6963266
10	0.345	−0.105	−0.7962624	−0.789096
11	0.345	−0.155	−0.909312	−0.8956723
12	0.345	−0.205	−1.0182656	−0.9877176

Determine the positive and negative potential limits at which the coulombic efficiencies remain equal or above 99%. *Determine* the specific capacitances of the electrode at both potential limits (relative to the *active* mass).

Calculate from the tabulated data the charge balancing for a symmetric capacitor: The positive electrode shall be made exactly as in this single electrode test. Derive the solid mass (active materials + binder + conductive additive) of the slurry that must be coated on 100 cm^2 for the negative electrode to achieve charge balancing. Use the same mass fractions as in the single electrode tests.

Predict for the resulting symmetric capacitor the maximum cell voltage, the capacitance, the specific capacitance (with respect to the active masses of both electrodes), the stored energy at the maximum voltage, and the specific energy (relative to the active masses of both electrodes).

Exercise 15.26: A symmetrical capacitor has been built from activated carbon with a geometric electrode area of 1 cm^2 for each electrode in KOH solution. The tests showed that the maximum voltage is 0.8 V, solid masses of positive and negative electrodes are 2.06 $mg\ cm^{-1}$ each (80% of the mass is activated carbon). The cell shall be characterized by galvanostatic charging–discharging curves at 1 A g^{-1} (related to the active mass of one electrode). The resulting curve is shown below.

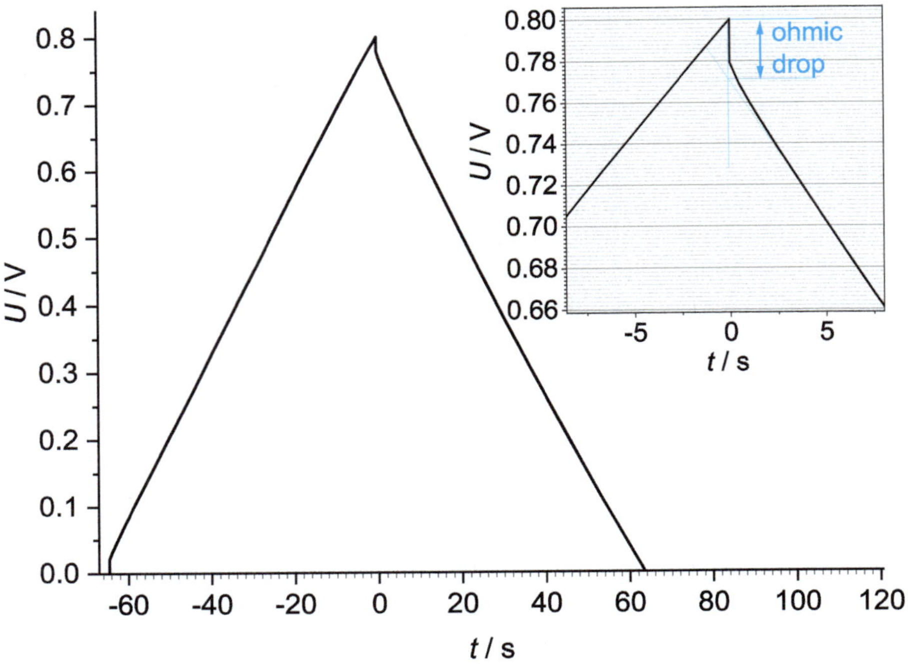

Calculate the current to set the galvanostat at.

Calculate from the graph the charges for charging and discharging, the coulombic efficiency, the serial resistance, the capacitance and specific capacitance of the cell, the stored energy, the specific energy, the energy efficiency at the selected current density, the power, and the specific power of the cell. The specific quantities should be related to the active masses of both electrodes.

Estimate the specific capacitance of a single electrode under the assumption that the specific capacitances of positive and negative electrodes are equal and constant over the entire potential range.

References

1 M. Winter, R. J. Brodd; What are batteries, fuel cells and supercapacitors? *Chem. Rev.* **2004**, *104*, 4245–4269.

2 M. Winter, B. Barnett, K. Xu; Before Li ion batteries. *Chem. Rev.* **2018**, *118*, 11433–11456.

3 F. Wu, J. Maier, Y. Yu; Guidelines and trends for next generation rechargable lithium and lithium-ion batteries. *Chem. Soc. Rev.* **2020**, *49*, 1569–1614.

4 A. Dutta, S. Mitra, M. Basak, T. Banjeree; A comprehensive review on batteries and supercapacitors: Development and challenges since their inception. *Energy Storage* **2023**, *5*, e339.

5 Z. Zhu, T. Jiang, M. Ali, Y. Meng, Y. Jin, Y. Cui, W. Chen; Rechargable batteries for grid scale energy storage. *Chem. Rev.* **2022**, *122*, 16610–16751.

6 O. Gröger, H. A. Gasteiger, J. P. Suchsland; Review – Electromobility: Batteries or fuel cells? *J. Electrochem. Soc.* **2015**, *162*, A2605–A2622.

7 W. Zhang, X. Fang, C. Sun; The alternative path for fossil fuel: Electric vehicles for hydrogen fuel cell vehicles. *J. Environ. Manage.* **2023**, *341*, 118019.

8 J. Grache (Ed.); *Encyclopedia of Electrochemical Power Sources, Vol. 1–7*, 2nd ed., Elsevier, Amsterdam, 2025.

9 I. Buchmann; *Batteries in a Portable World: A Handbook on Rechargeable Batteries for Non-Engineers.* 4th ed., Battery University, 2016, Individual chapters on the web at https://batteryuniversity.com/articles/ (last access 08 Mar 2026).

10 P. Mock, S. A. Schmidt; Techno-economic assessment of hydrogen derivaties. in *Encyclopedia of Electrochemical Power Sources, Vol. 1*, 2nd ed. (J. Garche, Ed.). Elsevier, Amsterdam, 2025, pp. 565–585.

11 Institut für Schadensverhütung; https://www.youtube.com/watch?v=dYq75w9WBJM (last access 08 Mar 2026).

12 R. Spotnitz, J. Franklin; Abuse behavior of high-power, lithium-ion cells. *J. Power Sources* **2002**, *113*, 81–100.

13 Swiss Federal Laboratories for Materials Testing and Research (EMPA); https://www.youtube.com/watch?v=nr1dSkOv-fs (last access 08 Mar 2026).

14 FuelCon AG; https://www.youtube.com/watch?v=wQZvQjIFcmk (last access 08 Mar 2026).

15 C. Daniel, J. O. Besenhard (Eds.); *Handbook of Battery Materials, Vol. 1 and 2*, 2nd expanded. ed., Wiley-VCH, Weinheim, 2011.

16 M. Dollé, L. Sannier, B. Beaudoin, M. Trentin, J.-M. Tarascon; Live scanning electron microscope observation of dendrite growth in lithium/polymer cells. *Electrochem. Solid-State Lett.* **2002**, *5*, A286–A289.

17 H. Kim, G. Jeong, Y. Kim, J. Kim, C. Parke, H. Sohn; Metallic anodes for next generation secondary batteries. *Chem. Soc. Rev.* **2013**, *42*, 9011–9034.

18 J. Goodwin, R. Hughes; *Rheology for Chemists: An Introduction*, 2nd ed.; Royal Society of Chemistry, London, 2008.

19 W. Bauer, N. Nötzel; Rheological properties and stability of NMP based cathode slurries for lithium ion batteries. *Ceram. Int.* **2014**, *40*, 4591–4598.

20 R. Korthauer (Ed.); *Lithium-Ion Batteries: Basics and Application*, Springer, Berlin, 2018.

21 C. Julien, A. Mauger, A. Vijh, K. Zaghib; *Lithium Batteries – Science and Technologies*, Springer, Heidelberg, 2016.

22 G. E. Blomgren; The development and future of lithium ion batteries. *J. Electrochem. Soc.* **2017**, *164*, A5019–A5025.

23 J. Garche (Ed.); *Encyclopedia of Electrochemical Power Sources*; *Vol. 3*, 2nd ed. Elsevier, Amsterdam, 2025, pp. 35–584.

24 N. Eliaz, E. Gileadi, *Physical Electrochemistry: Fundamentals, Techniques, and Applications.* 2nd ed., Wiley-VCH, Weinheim, 2019, pp. 398–402.

25 P. Bernard, P. Kurzweil; Nickel batteries: Nickel–cadmium. in *Encyclopedia of Electrochemical Power Sources, Vol. 3*, 2nd ed. (J. Garche, Ed.), Elsevier, Amsterdam, 2025, pp. 459–481.

26 J. McBreen; Nickel hydroxides. in *Handbook of Battery Materials. Vol. 1*, 2nd ed. (C. Daniel, J. O. Besenhard, Eds.), Wiley-VCH, Weinheim, 2011, pp. 149–165.

27 B. Hariprakash, A. K. Shukla, S. Venugoplan; Nickel-metal hydride: Overview. in *Encyclopedia of Electrochemical Power Sources, Vol. 3*, 2nd ed. (J. Garche, Ed.), Elsevier, Amsterdam, 2025, pp. 674–681.

28 P. H. L. Notten, M. Latroche; Nickel-metal hydrides. in *Encyclopedia of Electrochemical Power Sources, Vol. 3*, 2nd ed. (J. Garche, Ed.), Elsevier, Amsterdam, 2025, pp. 682–700.

29 J. H. Reilly; Metal hydride electrodes. in *Handbook of Battery Materials, Vol. 1*, 2nd ed. (C. Daniel, J. O. Besenhard, Eds.), Wiley-VCH, Weinheim, 2011, pp. 239–268.

30 Y. Sato; Memory effect. in *Encyclopedia of Electrochemical Power Sources, Vol. 3*, 2nd ed. (J. Garche, Ed.), Elsevier, Amsterdam, 2025, pp. 729–746.

31 Y. Sato, S. Takeuchi, K. Kobayakawa; Cause of the memory effect observed in alkaline secondary batteries using nickel electrode. *J. Power Sources* **2001**, *93*, 20–24.

32 T. Sasaki, Y. Ukyo, P. Novak; Memory effect in lithium-ion battery. *Nat. Mater.* **2013**, *12*, 569–575.

33 K. Brandt, P. Kurzweil; Li-metal battery: Overview. in *Encyclopedia of Electrochemical Power Sources, Vol. 4*, 2nd ed. (J. Garche, Ed.), Elsevier, Amsterdam, 2025, pp. 122–140.

34 M. Winter, J. O. Besenhard; Lithiated carbon. in *Handbook of Battery Materials, Vol. 1*, 2nd ed. (C. Daniel, J. O. Besenhard, Eds.), Wiley-VCH, Weinheim, 2011, pp. 433–478.

35 M. Osiak, H. Geaney, E. Armstrong, C. O'Dwyer; Structuring materials for lithium-ion batteries: advancements in nanomaterial structure, composition, and defined assembly on cell performance. *J. Mater. Chem. A* **2014**, *2*, 9433–9460.

36 M. Inaba; Li-ion battery: Negative electrodes: Graphite and carbon. in *Encyclopedia of Electrochemical Power Sources, Vol. 4*, 2nd ed. (J. Garche, Ed.), Elsevier, Amsterdam, 2025, pp. 159–170.

37 T. R. Jow, K. Xu, O. Borodin, M. Ue (Eds.); *Electrolytes for Lithium and Lithium-Ion Batteries*. (Modern Aspects of Electrochemistry, *Vol. 58*, R. E. White, C. G. Vayenas, Eds.), Springer, New York, 2014.

38 E. M. Erickson, E. Markevich, G. Salitra, D. Sharon, D. Hirshberg, E. de la Llave, I. Shterenberg, A. Rosenmann, A. Frimer, D. Aurbach; Review—Development of advanced rechargeable batteries: A continuous challenge in the choice of suitable electrolyte solutions. *J. Electrochem. Soc.* **2015**, *162*, A2424–A2438.

39 C. Hartnig, M. Schmidt; Electrolyte and conducting salts, in *Lithium-Ion Batteries: Basics and Application*. (R. Korthauer, Ed.), Springer, Berlin, 2018, pp. 59–74.

40 P. Verma, P. Maire, P. Novák, A review of the features and analyses of the solid electrolyte interphase in Li-ion batteries, *Electrochim. Acta* **2010**, *55*, 6332–6341.

41 X.-B. Cheng, R. Zhang, C.-Z. Zhao, F. W. J.-G. Zhang, Q. Zhang; A review of solid electrolyte interphases on lithium metal anode. *Adv. Sci.* **2016**, *3*, 1500213/1–20.

42 A. M. Dimiev, G. Ceriotti, N. Behabtu, D. Zakhidov, M. Pasquali, R. Saito, J. M. Tour; Direct real-time monitoring of stage transition in graphite intercalation compounds. *ACS Nano* **2013**, *7*, 2773–2780.

43 C. V. Chandran, P. Heitjans; Solid-state NMR studies of lithium ion dynamics across material classes. in *Annual Reports on NMR Spectroscopy, Vol. 89* (G. A. Webb, Ed.), Elsevier, Amsterdam, 2016, pp. 1–102.

44 C. Graf; Cathode materials for lithium-ion batteries. in *Lithium-Ion Batteries: Basics and Application*. (R. Korthauer, Ed.), Springer, Berlin, 2018, pp. 29–41.

45 T. Ohzuku, R. J. Brodd; An overview of positive-electrode materials for advanced lithium-ion batteries. *J. Power Sources* **2007**, *174*, 449–456.

46 N. Nitta, F. Wu, J. T. Lee, G. Yushin; Li-ion battery materials: Present and future. *Mater. Today* **2015**, *18*, 252–264.

47 A. Manthiram, T. Muraliganth; Lithium intercalation cathode materials for lithium-ion batteries. in *Handbook of Battery Materials, Vol. 1*, 2nd ed. (C. Daniel, J. O. Besenhard, Eds.), Wiley-VCH, Weinheim, 2011, pp. 343–375.

48 H. Arai, A. Ikezawa, R. Kanno, M. Okada, Y. Makimura, K. Zaghid, M. V. Reddy, F. Gendron, C. M. Julien, J. B. Goodenough, M. Wohlfahrt-Mehrens, P. Axmann, A. W. Xiao; Lithium secondary batteries: Li-ion battery: Positive electrode; in *Encyclopedia of Electrochemical Power Sources, Vol. 4*, 2nd ed. (J. Grache, Ed.), Elsevier, Amsterdam, 2025, pp. 218–328.

49 G. M. Mudd; Global trends and environmental issues in nickel mining: Sulfides versus laterites. *Ore Geol. Rev.* **2020**, *38*, 9–25.

50 I. A. J. Gordon, S. Grugeon, H. Takenouti, B. Tribollet, M. Armand, C. Davoisne, A. Debart, S. Laruelle; Electrochemical impedance spectroscopy response study of a commercial graphite-based negative electrode for Li-ion batteries as a function of the cell state of charge and aging. *Electrochim. Acta* **2017**, *223*, 63–73.

51 M. Wohlfahrt-Mehrens, T. Waldmann; Lithium secondary batteries: Li-ion battery: Aging mechanisms and lifetime prediction. in *Encyclopedia of Electrochemical Power Sources, Vol. 4*, 2nd ed. (J. Garche, Ed.), Elsevier, Amsterdam, 2025, pp. 412–425.

52 K. M. Abraham, Z. Jiang; A polymer electrolyte-based rechargeable lithium/oxygen battery. *J. Electrochem. Soc.* **1996**, *143*, 1–5.

53 P. G. Bruce, S. A. Freunberger, L. J. Hardwick, J.-M. Tarascon; Li–O_2 and Li–S batteries with high energy storage. *Nat. Mater.* **2012**, *11*, 19–29.

54 G. Girishkumar, B. McCloskey, A. C. Luntz, S. Swanson, W. Wilcke; Lithium–air battery: Promise and challenges. *J. Phys. Chem. Lett.*, **2010**, *1*, 2193–2203.

55 B. Scrosatti, K. M. Abraham, W. V. Schalkwijk, J. Hassoun (Eds.), *Lithium Batteries*. Wiley, Hoboken, NJ, 2013.

56 N.-S. Choi, Z. Chen, S. A. Freunberger, X. Ji, Y.-K. Sun, K. Amine, G. Yushin, L. F. Nazar, J. Cho, P. G. Bruce; Challenges facing lithium batteries and electrical double-layer capacitors. *Angew. Chem. Int Ed.* **2012**, *51*, 9994–10024.

57 P. Adelhelm, P. Hartmann, C. L. Bender, M. Busche, C. Eufinger, J. Janek; From lithium to sodium: Cell chemistry of room temperature sodium–air and sodium–sulfur batteries. *Beilstein J. Nanotechnol.* **2015**, *6*, 1016–1055.

58 G. L. Soloveichik, Flow batteries: Current status and trends. *Chem. Rev.* **2015**, *115*, 11533–11558.

59 Keyword 'Electrolytic Capacitors' on Wikipedia; https://en.wikipedia.org/wiki/Electrolytic_capacitor (last access 08 Mar 2026).

60 Y. Shao, M. F. El-Kady, J. Sun, Y. Li, Q. Zhang, M. Zhu, H. Wang, B. Dunn, R. B. Kaner; Design and mechanisms of asymmetric supercapacitors. *Chem. Rev.* **2018**, *118*, 9233–9280.

61 F. Béguin, V. Presser, A. Balducci, E. Frąckowiak; Carbons and electrolytes for advanced supercapacitors. *Adv. Mater.* **2014**, *26*, 2219–2251, 2283.

62 M. R. Lukatskaya, B. Dunn, Y. Gogostky; Multidimensional materials and devices architectures for future hybrid energy storage. *Nat. Commun.* **2016**, *7*, 12647.

63 P. Simon, Y. Gogostky, B. Dunn; Where do batteries end and supercapacitors begin? *Science* **2014**, *343*, 1210–1211.

64 T. Brousse, D. Bélanger, J. W. Long; To be or not to be pseudocapacitive. *J. Electrochem. Soc.* **2015**, *162*, A5185–A5189.

65 S. Fleischmann, J. B. Mitchell, R. Wang, C. Zhan, D.-e. Jian, V. Presser, V. Augustyn; Pseudocapacitance: From fundamental understanding to high power energy storage materials. *Chem. Rev.* **2020**, *120*, 6738–6782.

66 T. Pandolfo, V. Ruiz, S. Sivakkumar, J. Nerkar; General properties of electrochemical capacitors. in *Supercapacitors: Materials Systems, and Applications* (F. Béguin, E. Frąckowiak, Eds.) Wiley-VCH, Weinheim, 2013, pp. 69–109.

67 L. Eliad, G. Salitra, A. Soffer, D. Aurbach; Ion sieving effects in the electrical double layer of porous carbon electrodes: Estimating effective ion size in electrolyte solutions. *J. Phys. Chem. B* **2001**, *105*, 6880–6887.

68 K. Fic, A. Platek, J. Piwek, E. Frąckowiak; Sustainable materials for electrochemical capacitors. *Mater. Today* **2018**, *21*, 437–454.

69 P. Ruschhaupt, S. Pohlmann, A. Varzi, S. Passerini; Determining realistic electrochemical stability windows of electrolytes for electrical double-layer capacitors. *Batteries Supercaps* **2020**, *3*, 698–707.

70 A. Laheäär, A. Arenillas, F. Béguin; Change of self-discharge mechanism as a fast tool for estimating long-term stability of ionic liquid based supercapacitors. *J. Power Sources* **2018**, *396*, 220–229.

71 H.-K. Song, H.-Y. Hwang, K.-H. Lee, L. H. Dao; The effect of pores size distribution on the frequency dispersion of porous electrodes. *Electrochim. Acta* **2000**, *45*, 2241–2257.

72 P. L. Taberna, P. Simon, J. F. Fauvarque; Electrochemical characteristics of impedance spectroscopy studies of carbon–carbon supercapacitors. *J. Electrochem. Soc.* **2003**, *150*, A292–A300.

73 H. D. Yoo, J. H. Jang, J. H. Ryu, Y. Park, S. M. Oh; Impedance analysis of porous carbon electrodes to predict rate capability of electric double layer capacitors. *J. Power Sources* **2014**, *267*, 411–430.

74 A. Bruke; Testing of electrochemical capacitors. in *Supercapacitors: Materials Systems, and Applications*. (F. Béguin, E. Frąckowiak, Eds.), Wiley-VCH, Weinheim, 2013, pp. 437–471.

75 A. Bruke, M. Miller; The power capability of ultracapacitors and lithium ion batteries for electric and hybrid vehicle applications. *J. Power Sources* **2011**, *196*, 514–522.

76 H. Yang; A comparative study of supercapacitor capacitance characterization methods. *J. Energy Storage* **2020**, *29*, 101316.

77 International Electrotechnical Commission; IEC 62391-1-1: 2020 Fixed Double-Layer Capacitors for Use in Electric and Electronic Equipment – Part 1: Generic Specifications. https://webstore.iec.ch/en/publication/66557 (last access 08 Mar 2026).

78 B. E. Conway; *Electrochemical Supercapacitors: Scientific Fundamentals and Technological Applications*. Kluwer Academic Publisher and Plenum Publisher, New York, 1999.

79 F. Béguin, E. Frąckowiak (Eds.) *Supercapacitors: Materials Systems, and Applications*. Wiley-VCH, Weinheim, 2013.

80 S. Ardizzone, G. Fregonara, S. Trasatti; "Inner" and "outer" active surface of RuO_2 electrodes. *Electrochim. Acta* **1990**, *35*, 263–267.

81 J. Wu; Understanding the electric double-layer structure, capacitance, and charging dynamics. *Chem. Rev.* **2022**, *122*, 10821–10859.

82 Editorial; The path to accurate reporting. *Nat. Energy* **2024**, *9*, 1175–1176.

83 T.-C. Liu, W. G. Pell, B. E. Conway; Behavior of molybdenum nitrides as materials for electrochemical capacitors. *J. Electrochem. Soc.* **1998**, *145*, 1882–1888.

84 J. Deng, W. Hu, E. Paek, D. Mitlin; Review of hybrid ion capacitor: From aqueous to lithium to sodium. *Chem. Rev.* **2018**, *118*, 6457–6498.

85 M. Boota, Y. Gogostky; MXene-conducting polymer asymmetric pseudocapacitors. *Adv. Energy Mater.* **2019**, *9*, 1802917.

16

Fuel Cells

We have already dealt very briefly with fuel cells (FCs) in Section 2.4.4. In Example Calculation 2.16, we deduced that, in principle, they can convert fuels into work particularly efficiently. They are not limited by the Carnot cycle[14] that applies to internal combustion engines (ICE). Further advantages of fuel cells over ICEs are their high energy density, low acoustic noise, and low emissions of pollutants. We have also considered current limitations and the basic design of important fuel cell types. Chapter 12 focused on the electrocatalytic reactions of small molecules, which are important for fuel cells. In Section 15.1, we placed fuel cells and batteries in the context of electrochemical energy conversion devices. Now that you may have refreshed these basics, we will look at selected more advanced aspects of fuel cell technology.

Further Reading: There are a number of monographs and review articles that are suitable for in-depth study of individual aspects, and a small selection is listed at the end of this chapter [1–4]. Biofuel cells are very interesting type of fuel cell that we will cover in the context of bioelectrochemistry in Section 18.3.5.

16.1 Overview

Contrary to what the name fuel *cell* may suggest, an FC is not just an electrochemical cell but a complete small chemical power plant. It typically contains several individual galvanic cells connected in series (as is the case within batteries). The series connection increases the voltage supplied to the external circuit to a technically usable level. Individual cells are usually installed in a stack and, together with the end plates, form the actual reaction unit of the fuel cell system, which also includes peripheral devices for gas processing, gas humidification, fuel recirculation, and cooling (Figure 2.31). Pressure is exerted on the stack via the end plates using tie rods in order to minimize contact resistance between the layers involved.

The central problem within fuel cells is the slowness of reactions (oxidation/reduction) of low-molecular-weight compounds, which generally only take place with the help of a catalyst. To convert O_2 to H_2O, four electrons and four protons must be transferred to a small molecule. This can only be achieved at practically useful rates if the intermediates are stabilized by adsorption on catalyst surfaces. So far, precious metals have been used as catalyst for this purpose. Only they are sufficiently stable in the presence of an acidic electrolyte, such as a Nafion proton-conducting membrane (Figure 2.30). Metal oxides of less noble metals would dissolve under these conditions. Their use is possible in alkaline electrolytes, which are currently attracting a great deal of attention [5]. For alkaline electrolytes, an anion-conducting membrane is required to separate the anode and the cathode. Unfortunately, anion-conducting membranes are not yet available with sufficient ion conductivity and chemical stability compared

Table 16.1 General advantages and disadvantages of fuel cells compared to other energy conversion systems such as batteries and internal combustion engines (ICEs).

Advantages
– High theoretical efficiency compared to ICE
– High energy density compared to batteries
– Low noise emission compared to ICE
– Low pollutant emissions compared to ICE
Disadvantages
– Complex system structure
– Precious metals as electrocatalysts
– Ion-selective membrane required
– Poisoning of electrocatalysts by trace gases (CO)
– Degradation of system components under operating conditions
– Complex mass and heat fluxes with formation of gradients
– Challenges in hydrogen storage
– Narrow range for operating temperature

to those of Nafion. Ceramic membranes can conduct oxide ions O^{2-} at temperatures around 800 °C and are used in high-temperature solid oxide fuel cells (SOFCs).

The precious metal catalysts used in low-temperature fuel cells are also quickly "poisoned" by traces of other gases such as carbon monoxide CO. These catalyst poisons adsorb more strongly to the surface sites of the catalysts than the reactants. The sites are then no longer available for catalysis. Table 16.1 summarizes the basic advantages and disadvantages of FCs.

The most important fuel for FC is hydrogen. Currently, the most cost-effective way to obtain H_2 is from natural gas. Depending on its origin, natural gas is a mixture of methane CH_4, higher gaseous hydrocarbons, and trace gases. For the sake of clarity, we will only follow the conversion of CH_4 below. For FC systems, autothermal reforming is preferred, in which part of the CH_4 is oxidized to CO with the release of heat, and part reacts with H_2O with the heat uptake to form CO and H_2:

$$2\,CH_4 + O_2 \rightarrow 2\,CO + 4\,H_2 \tag{16.1}$$

$$CH_4 + H_2O \rightarrow CO + 3\,H_2. \tag{16.2}$$

Carbon monoxide (CO) is a poison for most catalytic converters, which becomes so strongly bound to the catalyst surface that hydrogen or oxygen can no longer bind. CO must therefore be removed from the fuel gas. This is done by the water-gas shift reaction:

$$CO + H_2O \rightleftharpoons CO_2 + H_2. \tag{16.3}$$

As this reaction is an equilibrium reaction, 0.2–2% CO remains in the gas flow, which is removed by gas adsorption systems. The remaining hydrogen with a purity of >99.99% can then be stored or transported via pipelines.

The exact requirements regarding the residual CO content vary depending on the FC type. While special measures are required for the polymer electrolyte membrane fuel cell (PEMFC, 80 °C) to tolerate just traces of ubiquitous CO in the air (e.g., when an FC vehicle is used in urban traffic), a solid oxide fuel cell (SOFC, 800 °C) can be operated directly with natural gas.

The higher the operating temperature, the higher the CO concentrations in the fuel gas that can be tolerated by a fuel cell.

Storage of H_2 in the liquified state is possible only in cooled tanks (at 20 K), which is very expensive for decentralized terrestrial applications. It is therefore usually stored in pressurized gas tanks, where the weight of the tank and the low density of the compressed H_2 are severe limitations to energy density. The exact requirements can vary greatly depending on the particular application. For example, it is conceivable that H_2 for FCs in passenger vehicles can be refueled from large-scale industrial production based on natural gas. It would also be conceivable to fill up directly with natural gas instead. This would allow much simpler storage and the use of existing pipeline systems. In this case, H_2 has to be produced in a small reformer unit within the car. However, current fleet trials (2024) with passenger cars rely on hydrogen filling stations.

Terrestrial applications = Applications on Earth under ambient conditions as opposed to applications in space technology (Section 16.7)

Electrolytic H_2 generation for FCs is actually attractive due to the high purity of the H_2 produced in this way. However, due to the high investment and energy costs, this method only plays a role in niche markets. However, this scenario could change as a result of the increased use of renewable energies and accompanying regulatory measures.

16.2 Polymer Electrolyte Membrane Fuel Cell (PEMFC)

16.2.1 Membrane

The polymer electrolyte membrane fuel cell (PEMFC) is currently the preferred FC for decentralized applications, particularly for electric vehicles, following its first use in NASA's Gemini space missions and many further development steps. Specific targets, for example, for use in passenger cars, can be found in strategy papers of the major industrialized nations [6]. The electrolyte membrane is the central element in both senses of the word. The membrane consists of a perfluorinated polymer with side chains whose ends carry sulfonic acid groups. The best-known product is Nafion from DuPont de Nemours. However, there are also other chemically similar products from other manufacturers. Polymers with ionic groups are collectively called ionomers. The combination of the hydrophilic sulfonic acid groups on the one hand and the perfluorinated, extremely hydrophobic chains of the polymer on the other hand leads to a structure with water-filled cavities in a water-repellent polymer matrix that are interconnected by pores and lined with stationary sulfonate groups (Figure 2.30). The molecular weight is 10^5–10^6 g mol^{-1}. One mole of sulfonate groups is contained in approximately 1000–1500 g of Nafion (the so-called equivalent weight M_{eq} [g]). The membranes are characterized by a three-digit number. The first two digits correspond to one hundredth of the equivalent weight, and the third digit to the thickness d of the membrane in mils (milli-inches).

The abbreviation PEMFC is also spelled out as *proton-exchange membrane fuel cell*. Both terms refer to the same system.

Q: What is a mil? Why is such a unit used?

A: An inch equals 2.54 cm or 25.4 mm. A mil is a shortend form for milli-inch, that is 1/1000 of an inch or 25.4 μm. It is also abbreviated as thou (thousandth of an inch). So let's calculate the parameters for the typical membranes (Table 16.2). The inch is a unit of measure for length in the British Imperial System, which, for historical reasons, is commonly used in the United States except in a scientific context. The shortened form 'mil' for 0.001 inch should not be confused with the common colloquial clipped form 'mil' for millimeter!

Table 16.2 Parameters of some important Nafion membranes.

	Nafion 112	**Nafion 117**
M_{eq}/g	11 00	11 00
d/inch	0.002	0.007
d/cm	0.00508	0.01778
d/mm	0.0508	0.1778
d/μm	51	178

These membranes can selectively conduct cations (only H^+ is present here). The conductivity requires the presence of liquid water in the pores of the membrane. Therefore, the operating temperature must not fall significantly below 0 °C or rise above 100 °C. The water content of Nafion may vary depending on the pretreatment and process conditions. The specific conductivity of Nafion is roughly proportional to the water content. This is expressed by the λ-value [mol g^{-1}], which is the ratio of the amount of H_2O to the equivalent weight M_{eq}. The internal structure of Nafion and the volume of Nafion change with the water content [7, 8]. Water losses due to evaporation must be compensated for by humidifying the reactant gases. The typical operating temperature is 80 °C.

In the electrochemical cell, the membrane acts as an ion-conducting electrolyte. It exhibits an ohmic resistance toward the current that is proportional to the thickness of the membrane [Eq. (16.4)]. According to Ohm's law, the resistance results in an internal voltage drop ΔU, which reduces the voltage that can be supplied by the FC to the external electrical circuit under load.

$$\Delta U = RI = \rho \frac{l}{A} I = \rho j l \qquad (16.4)$$

R – Diaphragm or membrane resistance [Ω]
I – Current [A]
l – Thickness of the membrane [cm]
j – Current density [A cm^{-2}]
A – Area [cm^2]
ρ – Specific resistance [Ω cm]

To minimize the membrane resistance, a membrane of a very small thickness would be desirable. However, the membrane must also separate the gas spaces of the reactants O_2 and H_2 from each other in a gastight manner. Partial mixing of the gases reduces efficiency, as O_2 and H_2 would then react to water on the same catalyst. This reaction would generate heat, but the electrons would not move via the external circuit. The typical values for membrane thickness given in Table 16.2 are a compromise with respect to these opposing requirements.

16.2.2 Electrodes and Bipolar Plate

The electrodes are applied to both sides of the membrane. At the electrodes, charges are transferred. The ionic current passes through the membrane, and the electrons flow via the current collector, the external circuit, and the electrical appliance. Both the oxidation

$$2\ H_2(g) \rightarrow 4\ H^+(\text{Nafion}) + 4\ e^-(\text{Carbon}) \tag{16.5}$$

and the reduction

$$O_2(g) + 4\ H^+(\text{Nafion}) + 4e^-(\text{Carbon}) \rightarrow 2\ H_2O_{(l)} \tag{16.6}$$

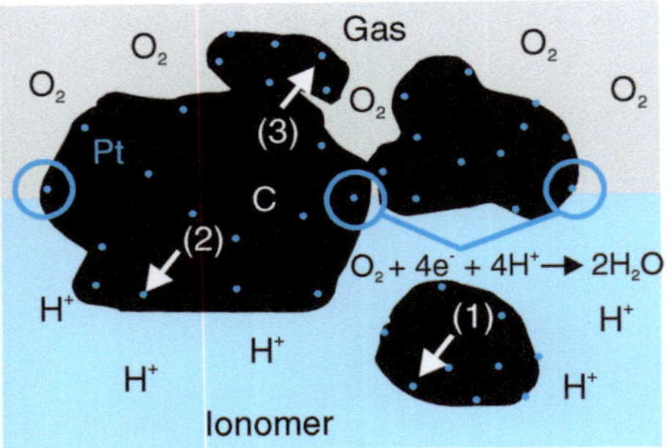

Figure 16.1 Three-phase boundary between gas phase, carbon-supported Pt catalyst, and ion-conducting membrane (schematic). Only the catalyst particles marked with a circle contribute significantly to the total current. Other particles are cut off from the electron-conducting path (e.g., 1), the gas flow (e.g., 2) or the ion-conductor (e.g., 3).

require a catalyst. Pt is the standard catalyst material. In order to reduce the amount of costly Pt, it is deposited in the form of particles with a diameter of 2–5 nm on an electron-conducting porous (gas permeable) graphitic support with specific surface areas of 25–140 $m^2\ g^{-1}$. Such electrodes are called gas diffusion electrodes. A Pt loading of 0.05–0.5 mg cm^{-2} is necessary for the hydrogen oxidation reaction. The oxygen reduction reaction is kinetically significantly slower and therefore requires a higher catalyst loading of 0.2 mg cm^{-2}. The porous structure of the carrier allows the entry of gaseous reactants. In order to keep the particles of the carrier material together and to bring H^+ to the catalyst particles, the particles of the support with attached catalyst nanoparticles are first processed into a suspension with dissolved Nafion and then applied to the membrane by spraying. This layer is called the catalyst layer (CL, thickness 5–30 μm). Ultimately, the reaction can only take place efficiently on the Pt nanoparticles that are located at or near the three-phase boundary between ionomer, gas phase, and catalyst on the electron-conducting support (Figure 16.1).

In order to maintain the activity of the catalyst despite the residual traces of CO contained in the H_2 gas stream, other elements – in particular Ru – are added, which increases the CO tolerance to around 50 ppm. Even higher CO tolerances are possible through "air bleeding", in which a small amount of air (2–10%) is added to the fuel gas. However, this is done at the expense of system efficiency because it also consumes H_2 without electrons reaching the external circuit.

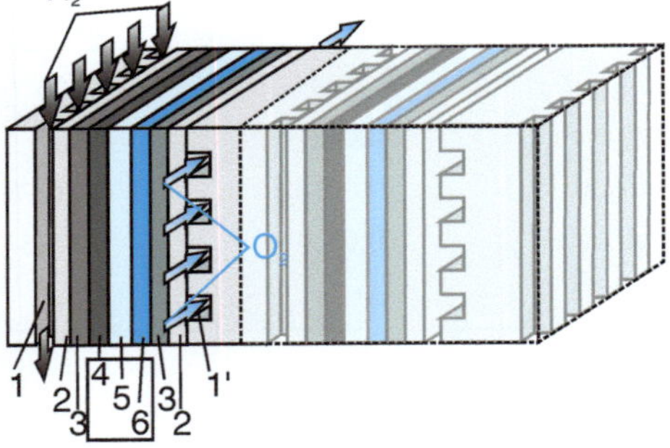

Figure 16.2 Single cell within a cell stack consisting of:

1) Bipolar plate with flow field for H_2;
2) Gas diffusion layers;
3) Microporous layers;
4) Catalyst layer of the anode;
5) Polymer electrolyte membrane;
6) Catalyst layer at the cathode;

1′) Bipolar plate with flow field for O_2 (or air). The actual electrochemical cell consisting of anode, electrolyte, and cathode is highlighted by a black rectangle. The layer sequence is then repeated. The figure is not to scale.

The membrane itself, together with the two catalyst layers applied to both sides of the membrane, is called membrane electrode assembly (MEA). The MEA is covered on both sides by a microporous layer (MPL, 10–30 μm thick) and a gas diffusion layer (GDL, 200–300 μm thick). The MPL consists of graphite particles without a catalyst. The GDL is formed by a carbon fleece. Both layers ensure a uniform distribution of the incoming gas to the catalyst particles, even under the protruding areas of the flow field (see below). MPL and GDL also ensure electronic contact between the catalyst-loaded electrode and the flow field that serves as a bipolar current collector. The surface of the GDL is treated to suppress the condensation of product water, which would lead to flooding of the electrode. If possible, product water should condense in the channels of the flow field, where it can be blown out as droplets.

The GDL is in direct mechanical and electrical contact with the bipolar plate. The bipolar plate connects the anode of one individual cell with the cathode of the next individual cell in an electrical series connection (cf. Figure 9.45). It also contains recessed channels, in which the gases are fed to the electrodes (Figure 16.2). This channel system is called flow field. Its layout is optimized in order to achieve a uniform conversion and uniform temperature distribution over the entire electrode at as many points of the MEA as possible. This is not a trivial task due to the decreasing H_2 and O_2 contents along the course of the channel. The flowing gas also acts as a coolant.

The cell stack is pressed together between end plates with tie rods. The gas inlet and gas outlet are integrated into the end plates, which transport the gas to the flow fields on either side of each individual bipolar plate.

16.2.3 Characteristic Curve of a Fuel Cell

We have already discussed the characteristic performance curve of a fuel cell using Figure 2.33. The construction of this curve is explained in more detail in Figure 16.4.

Open Circuit Voltage. At $I = 0$, the equilibrium voltage corresponds to the difference between the standard electrode potentials $U_{eq} \approx E°(H_2/H^+) - E°(H_2O/O_2) = 1.23$ V. Even if the cell is operated with air [$p(O_2) = 0.21p°$] instead of pure O_2, the value does not change significantly (Exercise 2.8). In fact, at $I = 0$, a significantly lower open circuit voltage U_{OCV} is measured. This open circuit voltage *is not* an equilibrium voltage, but results from mixed potentials (Sections 2.3.5 and 14.3.1). Contrary to popular belief, many polymers exhibit a clearly measurable permeability to gases. The fact that the decrease of U_{OCV} compared to U_{eq} or $U°$ depends, among other things, on the thickness of the polymer electrolyte membrane indicates that H_2 diffuses through the membrane at a low rate and is oxidized at the Pt particles of the cathode (whereby O_2 is reduced at the same time) without electrons flowing through the external circuit.

Cell Voltage under Current. When observing the voltage upon increase of the current delivered to an appliance (electrical load), the potential initially drops steeply at low current density ($j < 0.2$ A cm^{-2} in Figure 16.3, curve for ORR). This voltage drop is described by the Butler–Volmer equation, for example, Eq. (3.44) in the simplified form for $E > 1$ V in Figure 16.4. To calculate the working curve, we must calculate the voltage–current density plots for the hydrogen oxidation reaction (HOR) and the oxygen reduction reaction (ORR) separately, giving the overpotentials of the anodic and cathodic partial reactions η_a and η_c for a specific current density. For both electrode reactions, the relevant parameter exchange current $j_{0,a}$ and $j_{0,c}$ results from the intrinsic heterogeneous standard reaction rate constants k_a°, k_c° (depending on the material and the exact surface conditions of the catalyst) and the surface area of the electrode. The electrochemically active surface area (ECSA) of the Pt catalyst particles is decisive here. The geometric surface area and the surface area of the carbon particles, on the other hand, play no role. In order to make the best possible use of the scarce Pt, it is therefore advantageous to apply the Pt in the form of small nanoparticles. While the HOR is quite fast (i.e., a small change in voltage causes a large change in current), the ORR is a rather slow reaction. Therefore, more Pt is usually applied to the cathode side of an FC. Essentially, the kinetics of ORR determine the shape of the working curve in the range $j < 0.2$ A cm^{-2} in Figure 16.5. The enumerated sections of the voltage–current density plot are

1) Calculated equilibrium cell voltage,
2) Reduction in voltage due to activation of the electron transfer reaction,
3) Further reduction of the cell voltage due to the ohmic loss in the membrane ($\Delta U = I R$),
4) Breakdown of the cell voltage due to limitations in the mass transport of reactants to the electrocatalyst surface.

Influence of the Membrane Resistance. According to Figure 3.8, it would be expected that after the onset of a current in the FC, the current would rise very steeply without any significant further change in voltage. This would correspond to a plateau in the voltage–current density plot in Figure 16.5. However, this is *not* the case. Instead, the voltage in Section 3 of the working curve falls approximately *linearly*. This is due to the ohmic resistances within the circuit, which counteract an uninhibited charge transport. Supply cables do not usually play a significant role here. In contrast, the specific conductivities of the carbon materials of the GDL, MPL, and CL have a greater effect. The influence of the electrical contact resistances between bipolar plates and

You can follow these explanations using an Excel worksheet, which can be found at www.wiley-vch.de/ISBN9783527347216.

Note that U is plotted on the ordinate and I on the abscissa in Figure 2.33 and in Figures 16.3 and 16.5, that is, exactly the opposite of the Butler–Volmer curves in Figure 16.4 and the voltammograms in Chapters 5 and 6!

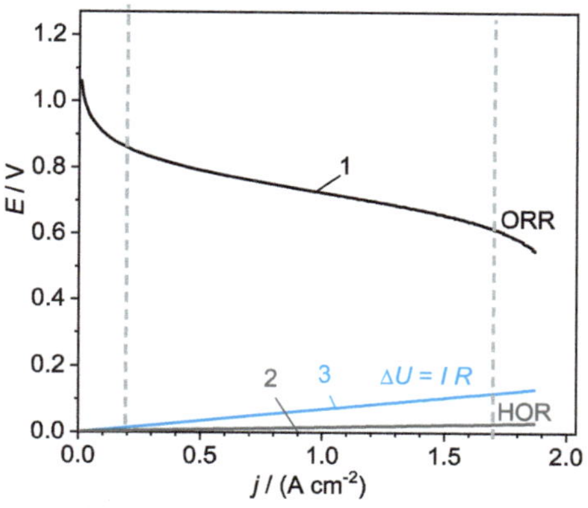

Figure 16.3 Voltage losses in a fuel cell (schematic). 1) Cathode for ORR; 2) anode for HOR; and 3) ohmic voltage drop IR in the fuel cell (due to membrane electrolyte and contact resistances).

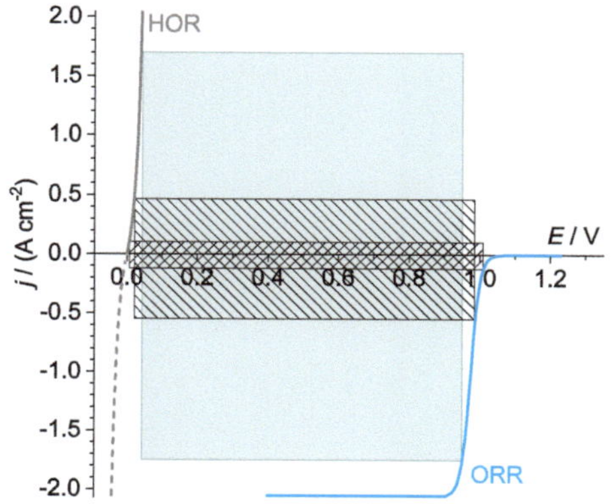

Figure 16.4 Schematic Butler–Volmer curves for H_2 oxidation and O_2 reduction with three marked working points. They are characterized by a certain current or a certain current density. At the working point, there is a balance of total currents $I_a = -I_c$. Only if the electrode areas (here the geometric area) of the anode and cathode are equal, there is an equality of the current densities, and $j_a = -j_c$ holds, too.

U – Cell voltage [V]
U_{OCV} – Open circuit voltage [V] (results from the mixed potentials of both electrodes)
I – Current [A]
R – Membrane resistance [Ω] (and other internal resistances)
η_a, η_c – Overvoltages at anode and cathode [V]

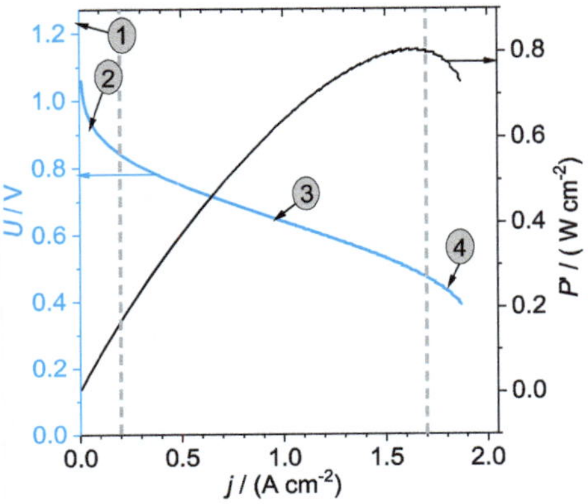

Figure 16.5 Schematic voltage–current density plot (blue) after addition of all voltage losses from Figure 16.3 and characteristic curve for area-related power output $P' = j \cdot U$. The reference value is the geometric area. Explanation of the enumerated Sections is in the text.

the GDL is expressed, for example, in the dependence of the working curve on the contact pressure. The ohmic resistance of the polymer electrolyte membrane according to Eq. (16.4) is the most important contribution. It depends on the specific material of the membrane and the water content, and increases proportionally to the thickness of the membrane. Overall, the working curve in Figure 16.5 is calculated as follows:

$$U = U_{OCV} - IR - \eta_a(I) - \eta_c(I). \tag{16.7}$$

Mass-Transport Limitation. In the range $j > 1.5\ \text{A cm}^{-2}$ in Figure 16.5, the voltage drops sharply over a narrow range of current density (section 4) because an insufficient flux of reactant molecules reaches the electrode surface. This could be due to an almost complete conversion of H_2 and O_2. However, this is not usually the case. As we explained in detail in Chapter 5 using cyclic voltammetry, it is not the complete conversion of the reagent but the limited rate of reactant transport that causes the limitation. This area of the curve is therefore referred to as the mass-transport region. When exactly mass transport limitations occur in a specific cell is influenced, for example, by the pore structure of the individual layers and also by the distribution of liquid water in the cell. At the anode, H^+ ions enter the membrane and entrain their hydration shell during transport. At the cathode, H^+ ions are consumed and water is produced. If the water condenses in the CL or MPL, the droplets block gas access to the catalysts. The complex transport processes within a single cell are illustrated in Figure 16.6.

Simulation of Fuel Cells. An exact quantitative description of the FC is considerably more complicated than in the spreadsheet provided with this textbook because the transport processes are highly temperature-dependent and the temperature distribution is uneven both in the fuel cell stack and in each individual cell (Figure 16.6). In order to calculate a characteristic curve with greater accuracy than in Figures 2.33 and 16.5, it is therefore not sufficient to refine simple expressions similar to those in the spreadsheet program; rather, the mutual influence of the individual processes on each other must also be simulated.

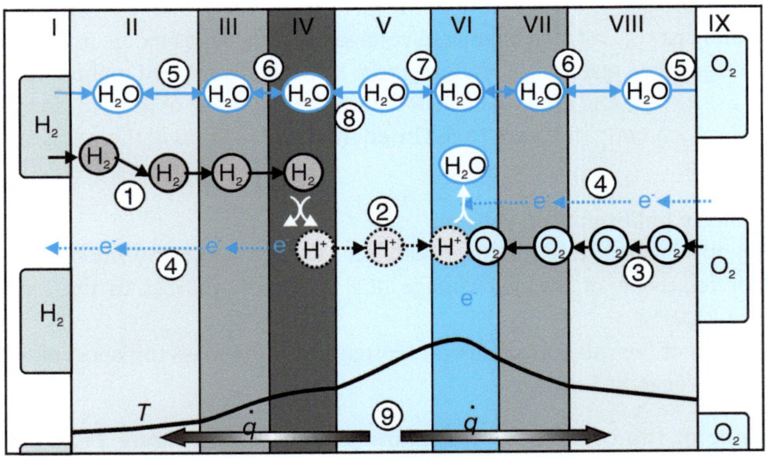

Figure 16.6 Schematic diagram of the cross section of a single cell with the transport processes and distribution of temperature T.
Components: I – flow field for H_2; II – GDL of the anode; III – MPL of the anode; IV – CL of the anode; V – membrane; VI – CL of the cathode; VII – MPL of the cathode; VIII – GDL of the cathode; IX – flow field for O_2.
Processes: 1) H_2 diffusion; 2) diffusion and migration of H^+; 3) diffusion of O_2; 4) electron conduction; 5) diffusion, evaporation and convection of gaseous and liquid H_2O; 6) movement of $H_2O(l)$ due to capillary forces; 7) electroosmotic flow; 8) back diffusion of H_2O; 9) heat transfer to dissipate the reaction heat from anode and cathode.

At this point, we would like to point out once again a fundamental problem that we have already mentioned in the simulation of cyclic voltammograms (Section 6.3.1) and impedance spectra (Section 7.3.3): Although the agreement of simulations with experimental data is a necessary condition for the agreement of the model assumptions with reality, it is not a sufficient condition. The characteristic curve of an FC has a few characteristic features and can be described fairly well by several sets of such parameters, given the large number of parameters. Therefore, in order to validate such computer simulations, an attempt is made to capture many experimental parameters and achieve their correct reproduction by the computer model in as wide a range of operating conditions as possible. After this warning, it should also be pointed out that validated computer models are a prerequisite for installing a specific FC in a product. This is the only way to predict the behavior of the overall system. Therefore, no responsible engineer can be persuaded to actually use an FC for a product if such a model is not available.

A comprehensive compilation of the most important considerations for the simulation of fuel cells can be found in [9, 10].

Excursion 16-1 Electrochemical Impedance Spectra of a Polymer Electrolyte Membrane Fuel Cell (PEMFC)

Compared to the characteristic curves discussed above, more detailed information can be obtained by recording and evaluating electrochemical impedance spectra (EIS, Chapter 7). An AC voltage of small amplitude (5 mV) and variable frequency is superimposed on the DC voltage (either U_{OCP} or the operating voltage occurring at a certain resistance of the external load). In addition to the purely ohmic resistances, which cause a shift along the real axis in the Nyquist plot (Figure 16.7), the processes taking place in the cell have characteristic time constants that result from a characteristic capacitance and a resistance that counteracts charging and discharging ($\tau[\text{s}] = R[\Omega = \text{V A}^{-1}] \cdot C[\text{F} = \text{A s V}^{-1}]$). This corresponds to the representation of these processes by a capacitor C connected in parallel with a resistor R (Figure 7.11). If the frequency of the AC voltage $f = \omega/(2\pi)$ is significantly lower than τ, the current can only flow via R, as the frequency-dependent resistance [$X_C = 1/(\omega C)$] of C is very high. However, as ω increases, the resistance across C decreases and a larger proportion of the current flows across it; thus, the total impedance decreases. In this way, increasing ω enables step-by-step additional current paths. If the time constants are sufficiently different, each process leads to an approximated semicircle in the Nyquist plot. However, it is possible that semicircles overlap in such a way that fewer semicircles are recognizable. The plot starts on the *right* of the diagram at low frequencies (0.1–1 Hz) and ends on the *left* at high frequencies (10–100 kHz).

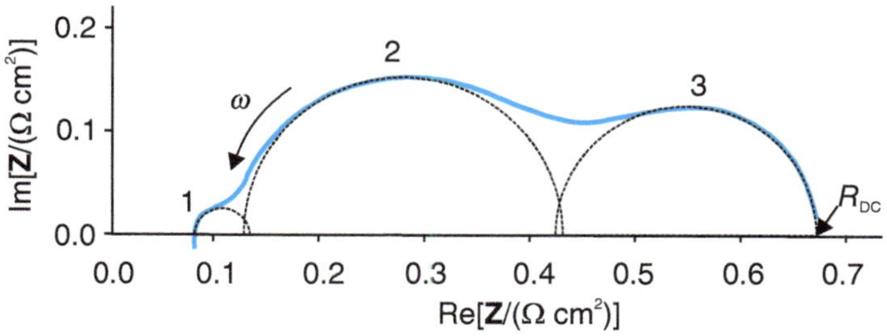

Figure 16.7 Schematic Nyquist plot of a fuel cell with a surface area of 1 cm^2.

Nyquist diagrams with three semicircles are very typical for fuel cells. The circle (1) in Figure 16.7 at high frequencies results from the resistance of the electrolyte and the contact resistances in the setup. In solid electrolyte fuel cells, the resistance is very high and strongly temperature-dependent. With polymer membrane electrolytes, the resistance is primarily dependent on humidification. Such variations in the test parameters can be used to assign physical processes to individual parts of the impedance spectrum and to set up meaningful models. If these models are available and validated, the condition of a fuel cell can be diagnosed during operation. This is a great advantage, because even when recording characteristic curves, the current density (and thus the reaction rate) and the power consumption (and thus also the heat dissipation) must be varied. If the external supply lines contain electrical conductors with very low resistance, inductive contributions can occur in the impedance spectra of fuel cells.

The middle arc (2) in Figure 16.7 is caused by the electrode processes. Often, only the parameters of the limiting (slower) partial reaction, usually the ORR, can be determined from the curve. This part of the EIS data can be influenced by varying the catalyst loading. The low-frequency part of the EIS data (3) in Figure 16.7 is related to mass transport. It can also be formed as a rising straight line. This section of the curve could be influenced, for example, by varying the partial pressures of the gases or the flow velocity. The intersection of this semicircle with the real axis corresponds to the DC resistance $R_{DC} = U_{DC}/I_{DC}$, which would be measured in a working curve at the applied cell potential. Condensed reaction water has a strong influence on the mass transport. It covers the electrodes, and the mass transport of O_2 can no longer take place through the gas phase, but must take place in the liquid phase, in some cases with diffusion coefficients that are orders of magnitude lower.

Fuel cells contain porous electrodes in which not all parts have the same electrical connection and the same supply of reactants. As explained in Section 7.4.3, distorted semicircles are therefore often observed. In addition, a distribution of reaction rates must often be taken into account. For these reasons, EIS of fuel cells is usually modeled using constant-phase elements (CPE) (Section 7.4.4).

Diagnostic methods based on EIS are currently being intensively investigated in order to obtain information about the internal condition of fuel cells during operation and to optimally adapt the operating conditions accordingly [11, 12].

16.2.4 Aging of Fuel Cells

By "aging" we mean the irreversible decline in the performance data of a fuel cell over the course of its use. Aging can be caused by various processes, the exact explanation of which is the subject of current research. This is made more difficult by the fact that a highly integrated chemical system such as a compact FC makes access for sampling or *operando* spectroscopy more difficult. In addition, the intermediates in question are usually short-lived and cannot be detected outside the cell. Since the FC is stable over short periods of time, accelerated aging protocols are used for laboratory tests in which the FC or its components are exposed to much more drastic conditions than a real FC. As a result, analytically detectable changes in the FC components occur after hours or days instead of thousands of operating hours.

Membrane. Although Nafion and similar ionomers are very stable compounds, they are subject to slow degradation in the FC, which can be analyzed by monitoring the continuous release of fluoride ions. A very probable cause is the attack of reactive oxygen species on the membrane. Reactive oxygen species are secondary products of hydrogen peroxide H_2O_2, which is formed during the two-electron reduction of O_2.

In principle, such a reaction could take place at the cathode. However, it is also conceivable that a small amount of O_2 diffuses through the membrane and is reduced to H_2O_2 at the anode. H_2O_2 can form hydroxyl radicals $HO^{\bullet}$ in a homolytic reaction. In accelerated aging tests, this reaction is accelerated by Fe^{2+} ions in a homogeneous reaction (Fenton reagent)[160], and membranes are directly exposed to this solution. The hydroperoxyl radical $HOO^{\bullet}$ is formed from $HO^{\bullet}$ as a secondary product:

$$H_2O_2 + Fe^{2+} \rightarrow HO^{\bullet} + OH^- + Fe^{3+} \tag{16.8}$$

$$H_2O_2 + HO^{\bullet} \rightarrow H_2O + HOO^{\bullet}. \tag{16.9}$$

These reactive intermediates probably attack the fluorinated polymers at the chain ends first. These consist of carboxyl groups, which are incorporated into the polymer in very small quantities with the polymerization initiator. The degradation can be written in simplified form [13]:

$$R{-}CF_2{-}COOH + HO^{\bullet} \rightarrow R{-}CF_2^{\bullet} + CO_2 + H_2O \tag{16.10}$$

$$R{-}CF_2^{\bullet} + HO^{\bullet} \rightarrow R{-}CF_2OH \tag{16.11}$$

$$R{-}CF_2OH \rightarrow R{-}COF + HF \tag{16.12}$$

$$R{-}COF + H_2O \rightarrow R{-}COOH + HF. \tag{16.13}$$

Catalyst and Catalyst Support. As mentioned in Chapter 12, CO and sulfur-containing compounds can adsorb on the surface of the Pt catalyst. These adsorption sites are then no longer available for the FC reactants and the activity decreases. In addition, the catalyst nanoparticles change during operation. Modern transmission electron microscopy (TEM) methods can be used to follow the shape evolution of individual catalyst nanoparticles. For this purpose, a layer of carbon is prepared on a measuring grid for TEM, which is then decorated with catalyst nanoparticles. Potential cycles are executed with these electrodes, for example, between +0.4 V vs. RHE (ORR) and +1.4 V vs. RHE (no ORR). This corresponds to a start/stop cycle in real FC operation and is also a form of accelerated aging. Using landing marks on the TEM grid, it is possible in favorable cases to examine exactly the same carbon particles with attached catalyst nanoparticles before and after the electrochemical experiment (identical location TEM, Figure 16.8).

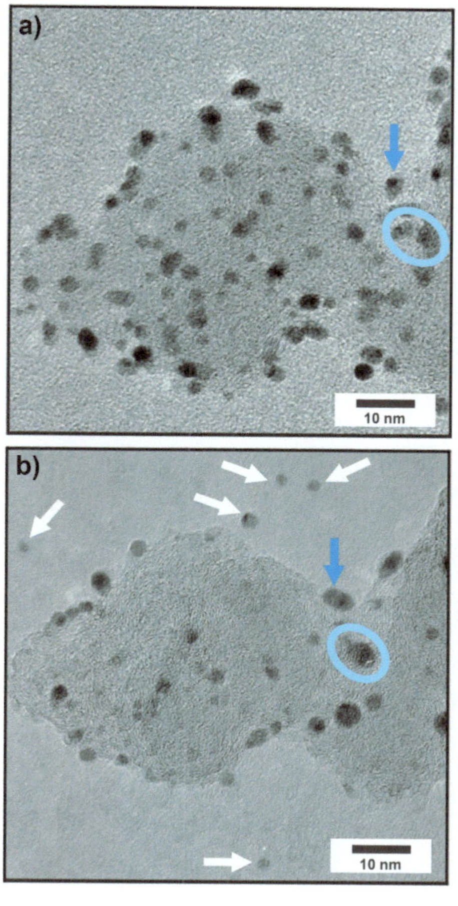

Figure 16.8 TEM images of identical Pt nanoparticles on carbon; a) before and b) after potential cycles between +0.4 V and +1.4 V vs. RHE. Source: Reproduced with permission from [14], Copyright 2013, The Electrochemical Society.

The atoms in the metal particles are much more mobile than we usually imagine. In particular, when an oxide layer is formed on the Pt particles in a start–stop cycle and then reduced again, the Pt atoms on the surface are very mobile for a short time. Three mechanisms can be distinguished to cause changes in the particle size and shape:

- The dissolution and redeposition of metal atoms leads to Ostwald ripening[31] in which larger particles grow at the expense of smaller ones. The driving force behind this process is the reduction of the Gibbs surface energy (Figure 16.8b, blue arrow).
- If particles come into contact with each other, coalescence occurs, that is, a combination of several particles with a reduction in the total surface area (Figure 16.8b, light blue ellipse).
- Nanoparticles marked by white arrows in Figure 16.8b have lost contact with the carbon particle. This can happen if the catalyst catalyzes the oxidation of the carbon support at high positive potentials. This reaction is to be expected in any case from a thermodynamic point of view, but is usually quite slow.

The detachment of Pt ions and their redeposition within the Nafion membrane was also demonstrated.

[160] Henry John Horstman Fenton, 1854–1929, British chemist.

The gradual oxidation of the carbon support not only causes a loss of electronic contact with some catalyst particles. It can also lead to the gradual collapse of the pore structure, which hinders the mass transport of the reactants and products and also increases the electrical contact resistance between the particles and other cell components.

16.3 Types of Polymer Electrolyte Membrane Fuel Cells

16.3.1 Direct Methanol Fuel Cell and Direct Ethanol Fuel Cell

The use of $H_2(g)$ as a fuel requires considerable effort for storage and transportation. Lower alcohols, for example, methanol and ethanol, have a much higher energy density than $H_2(g)$. In addition, liquids in fuel tanks or cartridges are much easier to handle. The standard cell voltage $U° = 1.21$ V of a methanol–O_2 cell is only slightly lower than that of $U°(H_2/O_2) = 1.23$ V. These advantages have led to intensive efforts to use methanol and other lower alcohols directly as fuel, that is, without an upstream reforming process, hence the name direct methanol fuel cell (DMFC). The cells contain Nafion or Nafion-like proton exchange membranes with Pt-containing catalysts. Methanol is supplied to the anode side in gaseous form or as an approximate 3 mol L^{-1} solution in liquid water.

Besides the mentioned advantages, there are also some serious disadvantages. The oxidation of methanol is much slower than that of H_2, which means that a substantial overvoltage must also be accepted on the anode with the resulting lower power output (20–50 mW cm^{-2}). Another problem is the methanol crossover. Crossover of methanol occurs to a much greater extent than the previously discussed H_2 transfer. Methanol is transported to the cathode by diffusion and electroosmotic flow in connection with the transport of protons through the membrane (see Figure 4.64 for a detailed explanation of the electroosmotic flow). At the cathode, methanol is oxidized by consuming O_2 without electrons being transferred to the external circuit. This leads to the formation of a mixed potential (lowering of the open circuit voltage compared to $U°$), a reduction in the fuel utilization rate (lower charge transport in the external circuit per amount of fuel used), and increased oxygen transport (occurrence of O_2 transport limitations even at lower current densities). The intermediates of methanol oxidation (CO) occupy active sites on the catalyst surfaces and reduce the exchange current density. From a practical point of view, it may be even more serious that the incomplete oxidation of methanol CH_3OH, may lead to the emission of formaldehyde HCHO, which has been classified as carcinogenic.

The reaction at the cathode is the ORR, which we have already considered in Sections 12.3.3 and 12.5.1. Methanol oxidation takes place at the anode via several adsorbed intermediates (>Pt should indicate a Pt surface atom):

$$>\mathrm{Pt} + \mathrm{CH_3OH} \rightarrow >\mathrm{Pt{-}CH_2OH} + \mathrm{H^+} + \mathrm{e^-} \tag{16.14}$$

$$>\mathrm{Pt{-}CH_2OH} \rightarrow >\mathrm{Pt{-}CH{=}O} + 2\,\mathrm{H^+} + \mathrm{e^-} \tag{16.15}$$

$$>\mathrm{Pt{-}CH{=}O} \rightarrow >\mathrm{Pt{-}C{\equiv}O} + \mathrm{H^+} + \mathrm{e^-}. \tag{16.16}$$

Adsorbed CO cannot be further oxidized without the presence of an oxygen-containing compound. This reaction occurs via adsorbed hydroxyl groups, which only form on Pt at higher potentials:

$$>\mathrm{Pt} + \mathrm{H_2O} \rightarrow >\mathrm{Pt{-}OH} + \mathrm{H^+} + \mathrm{e^-} \tag{16.17}$$

$$>\mathrm{Pt{-}C{\equiv}O} + >\mathrm{Pt{-}OH} \rightarrow 2 >\mathrm{Pt} + \mathrm{CO_2} + \mathrm{H^+} + \mathrm{e^-}. \tag{16.18}$$

The step in Eq. (16.18) requires adsorbed CO and OH on neighboring sites. However, since CO adsorbs more strongly, there is a tendency for all sites to be gradually covered with CO, and the entire reaction comes to a standstill. By incorporating another metal, in particular Ru, into the catalyst particles, the kinetics are significantly improved. Ru binds OH at lower potentials and does not bind CO as strongly:

$$>\mathrm{Ru} + \mathrm{H_2O} \rightarrow >\mathrm{Ru{-}OH} + \mathrm{H^+} + \mathrm{e^-} \tag{16.19}$$

$$>\mathrm{Pt{-}C{\equiv}O} + >\mathrm{Ru{-}OH} \rightarrow >\mathrm{Pt} + >\mathrm{Ru} + \mathrm{CO_2} + \mathrm{H^+} + \mathrm{e^-}. \tag{16.20}$$

Nevertheless, methanol oxidation is much slower than H_2 oxidation; this can be counteracted by a higher catalyst loading on the anode (2–10 mg cm^{-2} in DMFC instead of 0.05–0.5 mg cm^{-2} in PEMFC with H_2).

Methanol transport in membranes can be mitigated with modified membrane materials [15]. Perfluorinated polymers with solid fillers, polymer blends, and completely new polymers, for example, polybenzimidazoles with sulfonated polyether ether ketones, are used for this purpose. Nafion-type polymers are also used as binders in CL. The advantage here is that H_2O and methanol dissolve quite well in these binders in order to supply all catalyst particles with reactants (cf. Figure 16.1) [16–18].

The oxidation of ethanol and other organic substrates with more than one carbon atom is even more complicated [19–21], as a C–C bond has to be cleaved. So far, this has not been achieved reliably, and therefore ethanol is usually oxidized incompletely to ethanal (acetaldehyde, CH_3CHO). This means that only two of the maximum possible 12 e^-/12 H^+ per CH_3CH_2OH molecule are converted. Despite these existing limitations, the direct ethanol fuel cell is an important long-term development goal, as ethanol can be obtained entirely from renewable raw materials (i.e., CO_2-neutral).

16.3.2 High-Temperature Polymer Electrolyte Membrane Fuel Cell

The tolerance for CO increases with increasing temperature. At 80 °C, Pt can only tolerate a few ppm CO (1 ppm = 0.000 1%). Increasing the operating temperature to 200 °C would make it possible to tolerate 2% CO in the fuel gas and thus drastically reduce the cost of fuel pretreatment. Nafion-type membranes conduct cations in the pores only when liquid water is present. Conductivity is irreversibly lost when pore water dries out. This defines the upper temperature limit of PEMFCs. Smaller increases in the temperature window are possible if the pore structure is stabilized by other proton conductors. By these means, the operating temperature can be increased up to a maximum of 145 °C.

One possible approach to achieving even higher temperatures of 150–220 °C and thus increased CO tolerance is to switch to polymer membranes with higher-boiling electrolyte systems, hence the name high-temperature polymer electrolyte membrane fuel cell (HT-PEMFC). In principle, finely dispersed $Zr(HPO_4)_2$, heteropolyacids (e.g., $H_4[W_{12}SiO_{40}]$), metal oxides, for example, TiO_2, SiO_2, Al_2O_3, or solid acids such as $KH(PO_3H)$ can be considered. These compounds change into a disordered phase with high proton conductivity at elevated temperatures.

Currently, the most important system is based on polybenzimidazole (PBI) or its derivatives impregnated with phosphoric acid H_3PO_4. The high conductivity is achieved by protons that are transported in this system according to a mechanism similar to the Grotthuss mechanism[32] for the extra conductivity of H^+ in aqueous solutions. The challenges are to ensure sufficient mechanical strength of these membranes after the uptake of large amounts of phosphoric acid and to suppress the leaching of H_3PO_4 in temperature ranges in which liquid water is present, for example, immediately after starting the cell. An overview of the basic principles and the state of the art can be found in the literature [22, 23].

16.3.3 PEMFC Free of Precious Metals

The catalysts for PEMFC are made of Pt or Pt alloys, metals that are only available in limited quantities on Earth, which determines the cost of FC to a considerable extent [24]. It is therefore obvious to look for alternative catalysts that contain less or even no precious metals [21, 25]. Enzymes such as bilirubin oxidase and laccase have very high activities for the oxygen reduction reaction and contain several copper ions. This proves that it is possible in principle to find ORR catalysts without precious metals. However, for FC applications, the catalysts must also be stable in combination with a porous support and an ion-conducting membrane, as well as under different operating conditions.

Alkaline Membranes. A whole series of transition metal oxides of common elements such as Fe, Co, and Mn exhibit quite high activities for the ORR. This can be further increased by mixed oxides containing other transition metals. However, these oxides dissolve in the acidic environment of a Nafion-type membrane and can therefore only be used with alkaline membranes. In this case, the working ion is OH^-, that is, the ion that is transported through the electrolyte. OH^--conducting membranes are less well developed than Nafion-type membranes and have a lower conductivity. As a result, these cells can only deliver a lower power per unit area of membrane. In addition, contact with CO_2 from the air poses the risk of solid carbonates being deposited in the membrane and reducing the conductivity. However, significant improvements have recently been achieved in the properties of such membranes [26–28].

Nitrogen-Coordinated Metal Centers. Another approach is to use transition metal ions as Co and Fe, which are coordinated in a square planar arrangement with four nitrogen atoms of an aromatic ligand [29–31]. These structural motifs are often incorporated into thermally treated carbon materials. The C atoms around the active center assume the function of a support material and conduct the electrons. To incorporate the N atoms, edges of the graphitic material must be accessible or created. For further details, refer to Section 12.5.1 and the literature cited above.

16.4 Solid Oxide Fuel Cell (SOFC)

The solid oxide fuel cell (SOFC) operates at temperatures of approximately 700–900 °C, at which certain metal oxides have a high conductivity for oxide ions $\{O^{2-}\}$ as was discussed in Chapter 11. In addition, the high temperature makes the catalysts insensitive to poisoning, so that the cell can be operated directly with CH_4, CO, and H_2. No precious metals are required for the catalyst because the absence of liquid water means that corrosion reactions of metal oxides do not play a significant role. The SOFC also has a membrane electrode assembly (MEA) in which an O^{2-}-conducting solid membrane represents the electrolyte. This results in particularly complicated requirements for the materials. This concerns the mechanical stability of the MEA at widely varying temperatures, which requires the thermal expansion coefficients of the different materials to be matched with each other. In addition, the electrodes themselves must be porous so that the gases can reach the reaction sites. The electrolyte membrane, on the other hand, must be gastight. Finally, there are special requirements for chemical stability, as the reducing fuel gases meet an oxidic solid at around 1000 °C on the anode side. On the cathode side, the electrodes are in contact with an oxygen atmosphere at high temperatures. The layers of the MEA should be as thin as possible in order to reduce the ohmic resistance. However, this also results in low mechanical stability. For this reason, one of the layers usually has to act as a support layer to give the entire MEA the necessary mechanical strength. A distinction is made between metal-supported cells (MSCs), electrolyte-supported cells (ESCs), and anode-supported cells (ASCs).

Further Reading. Relevant material properties can be found in [32, 33].

16.4.1 Solid Oxide Membrane

The central component of the SOFC is a 5–200 μm thick solid oxide membrane. The best-known ion-conducting oxide for this application is yttria-stabilized zirconia (YSZ, Excursion 11-1). Y_2O_3 can stabilize the tetragonal (3 mol-% Y_2O_3) or the cubic phase (at >8 mol-% Y_2O_3) of ZrO_2 over a wide temperature range. The content of the doping oxide (Y_2O_3) given as a mole fraction [in mol-%] is often stated as a numerical prefix. For the two examples, this would result in 3YSZ and 8YSZ. During doping, Y^{3+} occupies sites of Zr^{4+} in the ZrO_2 lattice. This introduces two negative charges per formula unit Y_2O_3. To maintain electroneutrality, vacancies ($V_{O''}$) are created in the O lattice, in which one O^{2-} ion is missing and two positive charges are introduced. The doping can be formally formulated as a substitution reaction (Excursion 11-1):

yttria – Y_2O_3
yttrium – yttrium metal
alumina – Al_2O_3
zirconia – ZrO_2
titania – TiO_2
etc.

$$Y_2O_3 \xrightarrow{ZrO_2} 2\,Y'_{Zr} + 3\,O_O + V_{O''} \qquad (16.21)$$

O^{2-} ions can move from their original position to the vacancies at elevated temperatures and thus transport charge in the solid. It is important for this application that the solid has negligible electronic conductivity and is gastight, so that oxygen and the fuel gas are reliably separated.

The conductivity at a certain temperature also depends on the trace elements in the starting materials and the processing of the membrane. In particular, the resistance at the grain boundaries must be reduced by heat treatment.

Alternative solid electrolytes are based on CeO_2, which is doped with Y, Sm, or Gd ($Ce_{1-x}Me_xO_{2-x/2}$). Sufficiently high conductivities are already achieved at 600 °C. Above 700 °C, however, the *n*-type conductivity increases to such an extent that the solid can no longer be used as an electrolyte. However, the mechanical stability of CeO_2 membranes is very low, and the thermal expansion coefficient differs from that of ZrO_2. Consequently, the membrane cannot be combined with the same electrodes as it is possible for YSZ because the materials expand differently during heating/cooling, causing the formation of micro-gaps, which ultimately leads to failure of the MEA.

Solid solutions of lanthanum gallates with the formula $LaGaO_3$, for example, lanthanum-strontium-magnesium gallate (LSMG, $La_{1-x}Sr_xGa_{1-y}Mg_yO_{3-d}$, $x = 0.1$–0.2, $y = 0.15$–0.2) have even higher conductivities. These compounds are present in a perovskite structure. La^{3+} is partially replaced by Sr^{2+}, Ca^{2+}, or Ba^{2+}; and Mg^{2+} ions are incorporated into the Ga^{3+} sublattice. Traces of other cations can further increase the ionic conductivity without increasing the electronic conductivity too much.

In addition to the high price of Ga and the toxicity of LSMG materials, the main disadvantage of this material is phase instability at $T > 800\,°C$. Under these conditions, Ga is removed from the perovskite lattice and forms another phase. In addition, the electrolyte reacts under oxidizing conditions with electrode materials that also have a perovskite structure, and also with metallic Ni.

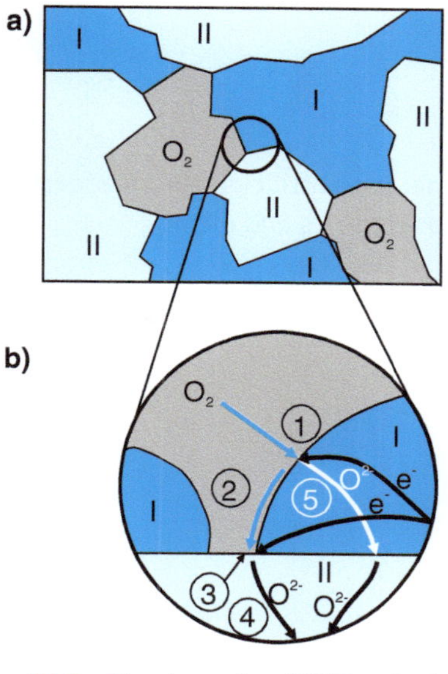

Figure 16.9 Structure of an SOFC cathode.

a) Cross section with indication of the three-phase boundary between the gas phase with O_2: I) electrolyte (e.g., YSZ) and II) cathode material (e.g., LSM).

b) Enlarged representation of the three-phase boundary with indication of the following processes: 1) adsorption and dissociation; 2) surface diffusion of O_{ads}; 3) reduction of O_{ads} at the three-phase boundary with e^- from the cathode material; 4) incorporation of O^{2-} into solid electrolyte and migration of O^{2-} in the solid electrolyte; 5) additional reduction at the surface of the cathode material and transport of O^{2-} to the cathode–electrolyte interface with opposite transport of e^- in the cathode material.

16.4.2 SOFC Cathode Materials

The most important cathode materials are lanthanum strontium manganite $La_{1-x}Sr_xMnO_3$ (LSM) and lanthanum calcium manganite $La_{1-x}Ca_xMnO_3$ (LCM), which are present in a perovskite structure. These materials exhibit good catalytic activity for ORR and chemical and thermomechanical compatibility with YSZ. Other metal ions can be present as dopants. If the cathode materials only had electronic conductivity, the reaction would take place exclusively at the three-phase boundary between the gas atmosphere, the solid electrolyte, and the electrode material (Figure 16.9, steps 1–4). After adsorption (1), oxygen diffuses along the surface of the electron conductor to the three-phase boundary (2), where it is incorporated into the solid electrolyte as O^{2-} after the actual reduction step (3) and transported further there by migration.

$$O_2 + 4e^- + 2V_{O''} \rightarrow 2O^{2-} \tag{16.22}$$

The cathode can have a very complicated structure. For example, a CeO_2 layer separates the electrolyte from Fe- or Co-containing perovskites in the actual cathode material. LSM has a low ionic conductivity. Thus, O^{2-} can also be transported in the cathode material itself to the cathode–YSZ interface and incorporated there. The transport of O^{2-} is synchronized with an opposite transport of e^- to the cathode–gas interface to reduce O_2 there. The cathode materials with mixed electronic and ionic conductivity thus dramatically increase the size of the surface area actually involved in the electrode reaction.

In order to utilize the effect in the best way, a mixture of LSM and YSZ powders is often used, which is processed into a composite. Graded compositions are also applied, that is, the ratio between LSM and YSZ changes within the layer. The achievable characteristics are significantly influenced by the processing, for example, the methods used to apply the initial suspensions (screen printing, spraying) and the sintering technique (T, co-firing). This can also lead to the formation of new phases, which have a strong influence on the electrode properties.

16.4.3 SOFC Anode Materials

The anodes are usually made of metallic nickel. However, as the thermal expansion coefficient is much greater than that of YSZ, nickel must be processed with a ceramic to form a composite, which is called cermet. The metallic particles form a continuous electron-conducting phase, and the ceramic phase of stabilized ZrO_2 or CeO_2 forms an O^{2-}-conducting phase. Finally, interconnected cavities should allow gas transport into the interior of the cermet and thus increase the area of the three-phase boundary (Figure 16.10).

Cermet is an artificial word made up of the words *ceramics* and *metal*.

Various reactions take place at this complex electrode: H_2 conversion to water vapor [Eq. (16.23)], the direct oxidation of CO [Eq. (16.24)], the water-gas shift reaction [Eq. (16.25)] and the internal reforming of hydrocarbons [Eq. (16.26)].

$$H_2(g) + O^{2-} \rightarrow H_2O(g) + V_{O''} \tag{16.23}$$

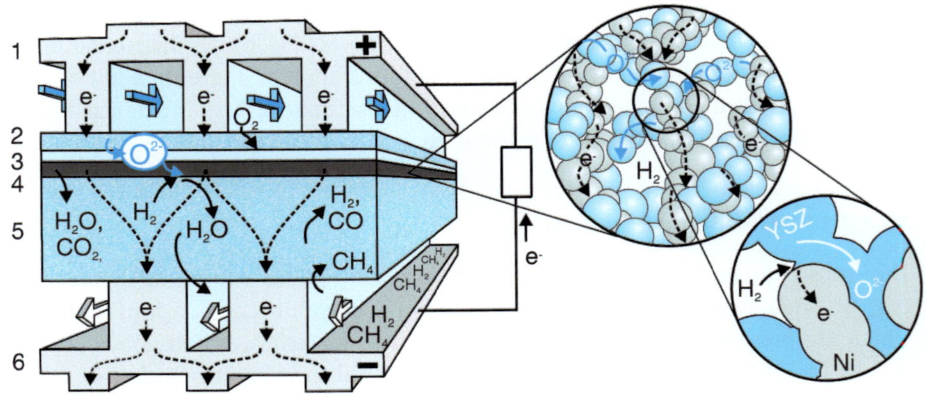

Figure 16.10 Cross section through a single cell of an SOFC. The specific example is an anode-supported MEA with a planar design. The enlargements show the continuous ionic and electronic conduction paths and the pore system running through the anode. 1) Cathode interconnect with air flow channel; 2) cathode; 3) electrolyte; 4) anode catalyst layer; 5) porous anode support; 6) anode interconnect with fuel flow channel.

$$CO + O^{2-} \rightarrow CO_2 + V_{O''} \tag{16.24}$$

$$CO + H_2O(g) \rightarrow CO_2 + H_2 \tag{16.25}$$

$$CH_4 + H_2O(g) \rightarrow CO + 3\,H_2 \tag{16.26}$$

The great advantage of the SOFC is its insensitivity to poisoning of the catalyst by CO. However, the water vapor content in the fuel gas must be sufficiently high; otherwise, carbon will be deposited on the electrode by the Boudouard equilibrium[161]

$$2\,CO(g) \rightleftharpoons CO_2(g) + C(s) \tag{16.27}$$

or by cracking of methane

$$CH_4 \rightarrow C + 2\,H_2. \tag{16.28}$$

Deposited carbon blocks the electrodes and the cavities. Carbon deposition can be reduced by adding, for example, Mo and Au to the cermet. Another problem arises from the presence of H_2S and other organic sulfides in natural gas, which adsorb on the nickel surfaces and deactivate them. Therefore, natural gas must be desulfurized before use in the SOFC, though the requirements are far less stringent than for PEMFCs.

However, the advantage of greater tolerance with regard to fuels is offset by some important disadvantages. Due to the large difference between ambient temperature and operating temperature, the materials of an SOFC undergo substantial thermal expansion If the thermal expansion coefficients of the different materials are unequal, the components suffer deformation. This must occur evenly in the solid layer systems to prevent internal stresses and damage (thermomechanical compatibility). In addition, the sintering and operating temperatures are so high that direct solid-state reactions between the adjoining solids can also occur. For example, when sintering electrolytes made of ZrO_2 and cathodes made of $La_{1-x}Sr_xMnO_3$ at temperatures above 1350 °C, zirconates are formed ($La_2Zr_2O_7$), which have insulating properties and act as a barrier layer. Even when operating at higher current densities, La_2O_3 can initially form in a complex mechanism, which then reacts further with ZrO_2 to form a barrier of $La_2Zr_2O_7$. The need to avoid these solid-state reactions at the interfaces between the different oxides in an SOFC at high temperatures (chemical compatibility) further restricts the wide range of materials and possible processing technologies during production.

The requirements for the electrode materials are porosity, chemical stability, electronic conductivity, thermomechanical compatibility, and catalytic activity.

16.4.4 From Membrane Electrode Assembly to SOFC Cells and Systems

There is a whole range of variations for combining anode, electrolyte, and cathode materials to a MEA. MEAs are produced in different ways that also imply different routes for assembling cells and systems. A central problem with solid electrolytes is their low specific conductivity. Any contribution to lowering the resistance (or equivalently to lowering the operating temperature) is an important contribution to increasing the practicality of the systems. The easiest way to reduce the internal electrical resistance of the cells is to minimize the thickness of the layers. However, this also causes a reduction in mechanical stability. For this reason, *one* layer is usually designed in such a way that it makes the most significant contribution to mechanical

---

[161] Octave Leopold Boudouard, 1872–1923, French chemist, professor at the Conservatoire National des Arts et Métiers, Paris from 1914.

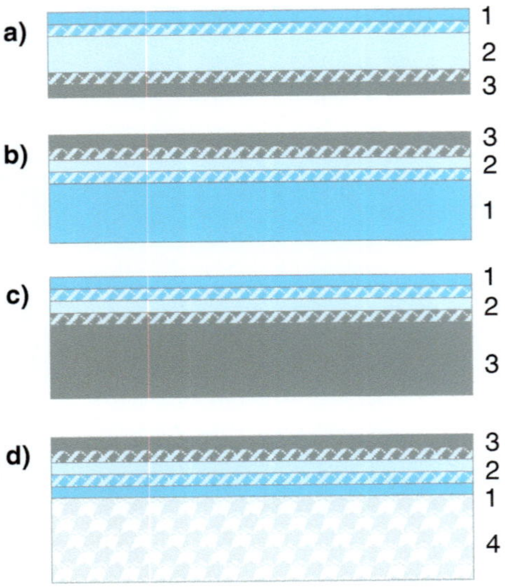

Figure 16.11 Cross section through differently supported MEAs.

a) Electrolyte-supported;
b) cathode-supported;
c) anode-supported;
d) MEA is applied to a porous ceramic or porous metallic substrate.

1) Anode (e.g., Ni cermet); 2) electrolyte (e.g., YSZ); 3) cathode (e.g., LSM/YSZ); 4) porous ceramic or metal alloy support structure.

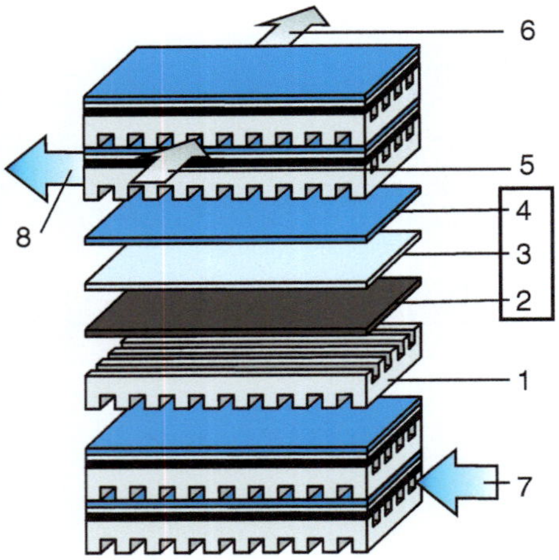

Figure 16.12 SOFC in planar design.
1) Bipolar plate; **2) anode; 3) solid electrolyte; 4) cathode;** 5) influx of fuel gas; 6) CO_2 and H_2O; 7) influx of air; 8) O_2-depleted air.

strength. This is then also the first layer to which the other layers are applied during the manufacturing process (Figure 16.11). Accordingly, metal-supported cells (MSCs), electrolyte-supported cells (ESCs), and anode-supported cells (ASCs) can be distinguished. A separate layer of porous metal or a ceramic can act as a carrier, to which the anodes are applied.

Planar (Figure 16.12) or tubular designs (Figure 16.13) can be considered for the shape of the individual cell.

Cells of planar design are similar in principle to the stacking sequence of the individual cells in a PEMFC (Figure 16.2). A membrane electrode assembly (MEA) is installed between two interconnects (connecting plates) (Figure 16.12). Each interconnect touches an anode on one of its sides and the cathode of the next single cell on its other side, thus it functions as a bipolar plate. Interconnects must have high electronic conductivity, be gastight, and have no ionic conductivity. In addition – and this is a major challenge – they must be chemically stable, even though a hot reducing gas atmosphere passes on one side and a hot oxidizing gas atmosphere flows past the other. In order to direct the air and fuels evenly to all points of the MEA, the connecting plates have gas channels. The layout of these channels must be optimized in order to supply all parts of the MEA with gas as evenly as possible. The decreasing fuel and oxygen content along the length of the channel can lead to locally different current densities (i.e., reaction rates), which are also associated with differences in local heat production. These may reinforce existing temperature gradients caused by heat transfer to the environment. However, different temperatures in the stack also result in different conductivities of the solid electrolytes, which can further increase the differences in the reaction rates. The optimization of flow rates, heat dissipation, and uniform reactant supply using the principles of counter-flow (of air and fuel gases), cross-flow, and co-flow is therefore an important field of optimization, which is supported by continuum simulations. In this context, it should be mentioned that planar cells have been developed with both approximately square MEAs and circular MEAs. The latter are supplied with fuel from a gas channel located in the center of the disk.

Stainless, chromium-containing, ferritic steels for operating temperatures up to 800 °C are suitable as a material for connecting plates. They form a chromium-containing oxide layer that passivates the metal in the reactive gas flows. However, these layers release gaseous chromium species under operating conditions that can deposit on the electrodes and deactivate the cathode in particular. The extent of this effect varies significantly between different passive layers. The release of chromium compounds can be reduced by heat-resistant oxidic coatings (e.g., lanthanum chromite and manganese chromite). Lanthanum chromite can also be used to make the entire interconnect board. It is a *p*-type semiconductor whose conductivity can be increased by doping with Ca, Mg, Sr on the La^{3+} or Cr^{3+} sites and can be used at higher temperatures than metallic interconnect plates.

This design requires a seal of the gas flows against each other, because otherwise the fuel utilization would decrease or air and fuel gases would partially mix and react directly with each other at the high temperatures with further heat release. This would probably lead to local damage to the cell materials. The gastight sealing of materials subjected to a temperature regime from ambient to 1000 °C in reactive atmospheres is an important technical challenge and is one of the main obstacles to the mass application of this technology. Seals can be made of precious metals, composites of precious metals and mica, or glass-ceramics. The latter soften and "glue" the individual layers during production at temperatures above the operating temperature.

However, the main advantage of the planar design lies in simpler, easily automated production processes and the high power densities of up to 1.8 W cm^{-2} that can be achieved. The most important problems lie in the permanent sealing of the gas spaces, maintaining planarity across all production steps, the electrical contact to the current collector, and the mechanical stability of the MEA during temperature changes.

The tubular shape of the individual cell avoids the problem of sealing the gas spaces (Figure 16.13). The MEA is designed as a ceramic tube made of LSM that is closed on one side. Typical diameters are 2 cm with a length of 150 cm. YSZ is deposited as the electrolyte and a Ni–YSZ cermet as the anode and a strip of lanthanum chromite as the connecting element. The air is fed in via a concentrically arranged feed tube made of Al_2O_3 until just before the closed end of the tube. From there, air flows along the inner wall back to the open end of the tube. The fuel gas flows around the outside of the tube. Several tubes are then electrically connected in parallel and in series. The avoidance of seals leads to very stable operation, proven on large-scale demonstrators. Their main disadvantage is the low power density of around 0.2 W cm^{-2}. Newer designs address this weak point and combine flow channels arranged over a large area to form plate-shaped components with a cross section that resembles corrugated cardboard torn apart.

Among fuel cells, the SOFC offers the greatest flexibility in terms of fuels and particularly high energy efficiency, especially if the heat can also be used. This is facilitated by the fact that the waste heat is generated at a high temperature level. An important application prospect is combined heat power (CHP) for residential buildings with outputs of 1–10 kW, especially in areas without access to the electricity grid. One of the problems is that the proportion of electrical energy released is too low compared to thermal energy for many applications. Excess heat must therefore be dissipated at the cost of energy losses or, if the heat is fully utilized, part of the electrical energy must be drawn from an electricity grid.

SOFC systems based on tubular single cells with outputs of 0.4–250 kW and desulfurized natural gas as fuel were tested as power plants for electricity production. Particularly high efficiencies with regard to the generation of electrical energy were achieved when combined with a gas turbine [34].

Although SOFCs have also been tested for portable applications, the fast operational readiness required for these applications cannot be achieved as well with SOFCs as with PEMFCs. This is particularly true for traction energy in vehicles. Prototypes for additional power sources for vehicles (APU, auxiliary power unit) have been tested.

Intensive research is also being carried out into new solid electrolytes that achieve high conductivity even at lower temperatures and could therefore alleviate many of the material problems of current SOFC designs [35].

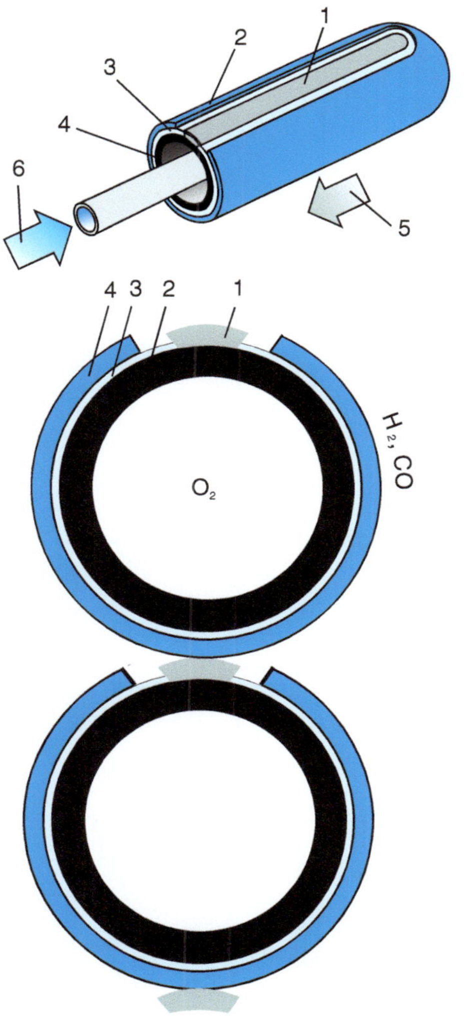

Figure 16.13 SOFC in tubular design. Structure of a single cell: 1) interconnect (electronically conductive connection, for example, Ni, between the cathode of one cell and the anode of the neighboring cell); **2) anode; 3) electrolyte; 4) cathode;** 5) fuel gas flow; 6) air flow.

16.5 Molten Carbonate Fuel Cell

We will now turn to fuel cells that work with liquid electrolytes. The first example is a molten carbonate fuel cell (MCFC) [36, 37]. This cell operates at around 600–650 °C. Like the SOFC, it has the advantage of CO tolerance and shares with it the disadvantages of very complicated material problems. It uses H_2 and CO as fuel gases, which either come from an external reformer or are produced in an internal reforming process.

Structure. The electrolyte consists of a melt of two carbonates, Li_2CO_3 and K_2CO_3 (eutectic mixture 62:38 mol%, better properties at normal pressure) or Li_2CO_3 and Na_2CO_3 (60:40 mol%, better stability of the cathode at higher pressures). The use of salt mixtures decreases the melting point. After assembly of the stack, the molten salt is absorbed with slow heating in a porous ceramic made of α-$LiAlO_2$ during a conditioning phase. The porous ceramic layer is only 200–500 μm thick in order to minimize the ohmic voltage drop. Overall, the electrolyte resistance allows power densities of

0.16 W cm^{-2}. A porous anode made of sintered Ni–Cr and Ni–Al alloys with a thickness of 400–800 μm and a porosity of 55–75% is directly placed next to the electrolyte. The cathode consists of porous nickel metal, which transforms into NiO during operation in the oxygen-containing atmosphere. NiO has a low solubility in carbonate melts, but this is sufficient for NiO to dissolve, migrate through the electrolyte, and deposit on the anode in the form of dendrites. They can ultimately lead to internal short circuits in the cell. This can be mitigated by the following measures:

- The solubility is only half as high in Li_2CO_3/Na_2CO_3 compared to Li_2CO_3/K_2CO_3.
- The addition of basic oxides (CaO, SrO, BaO) further reduces the solubility.
- Operation under normal pressure and low CO_2 partial pressure in the cathode compartment.
- Thick electrolyte matrix to increase the diffusion path length (at the expense of greater internal resistance).

The carbonate melt partially penetrates the pore network of the electrode. In order to equalize the wetting between the different porous materials, the capillary pressure must be set for the largest flooded pores with the diameters d_a, d_{el}, d_c (index "a" for anode, "el" for electrolyte, and "c" for cathode). The interfacial tension γ and the respective wetting angle θ play a role here. Equation (16.29) must then apply.

γ – Interfacial tension [N m^{-1}] = [J m^{-2}]
d – Diameter of the largest filled capillaries [m]
θ – Contact angle between molten salt and pore wall material [°]

$$\frac{\gamma_c \cos\theta_c}{d_c} = \frac{\gamma_a \cos\theta_a}{d_a} = \frac{\gamma_{el} \cos\theta_{el}}{d_{el}} \tag{16.29}$$

The individual cells consisting of anode, matrix with molten salt, and cathode are connected by interconnects (bipolar connecting plates, Figure 16.14). The connecting plates are made of surface-coated stainless steel, whose corrugated structure also defines the channels for the gas flow (cross-flow principle). Tie rods press the stack together. The gas flows are guided via laterally mounted plates with a gas distribution system.

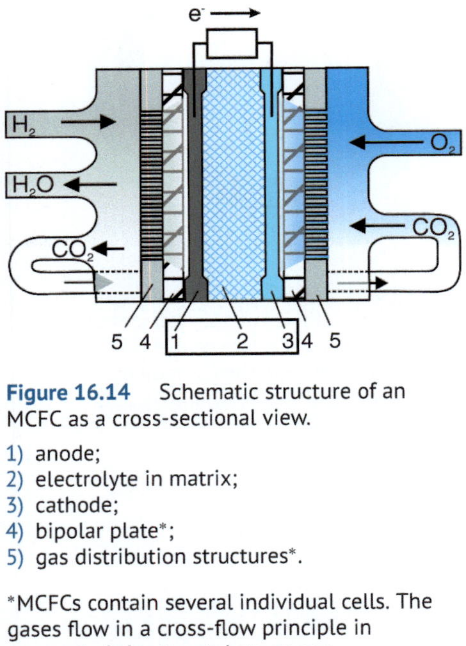

Figure 16.14 Schematic structure of an MCFC as a cross-sectional view.

1) anode;
2) electrolyte in matrix;
3) cathode;
4) bipolar plate*;
5) gas distribution structures*.

*MCFCs contain several individual cells. The gases flow in a cross-flow principle in corrugated sheet metal structures.

Function. In the reformer, natural gas (mainly CH_4) is first reformed with steam [Eq. (16.26)]. Then, H_2 and CO are oxidized at the anode. This produces CO_2, consuming carbonate ions from the electrolyte.

$$H_2 + CO_3^{2-} \rightarrow H_2O + CO_2 + 2e^- \tag{16.30}$$

$$CO + CO_3^{2-} \rightarrow 2\,CO_2 + 2e^- \tag{16.31}$$

H_2O is condensed out of the gas flow, and the remaining combustible gases are burned. The heat can be used to heat the reformer. The hot combustion gases, which contain CO_2, are mixed with O_2 and fed to the cathodes of the cell stack. This forms carbonate, which enters the carbonate melt.

$$\frac{1}{2}O_2 + CO_2 + 2e^- \rightarrow CO_3^{2-} \tag{16.32}$$

In the overall balance, 1 mol of CO_2 must be converted at the anode and cathode for the conversion of a charge of 2 F. In addition to the dependencies of the cell voltage of an H_2–O_2 fuel cell, the cell voltage [Eq. (16.33)] of an MCFC also exhibits a dependency on the CO_2 partial pressures in the anode (a) and cathode (c) compartments.

U – Cell voltage [V]
$E°$ – Standard electrode potential [V]
$RE°$ – Universal gas constant = 8.314 J K^{-1} mol^{-1}
$TE°$ – Absolute temperature [K]
$FE°$ – Faraday constant [A s mol^{-1}]
$pE°$ – Partial pressure [Pa] = [N m^{-2}]

$$U = \left[E^\circ_{O_2/H_2O} - E^\circ_{H^+/H_2}\right] + \frac{RT}{2F}\ln\frac{p(H_2)p(O_2)^{1/2}}{p(H_2O)} + \frac{RT}{2F}\ln\frac{p(CO_2)_c}{p(CO_2)_a} \tag{16.33}$$

Conclusion. Overall, the functionality of the MCFC was demonstrated in several examples in important industrialized countries. These are large-scale stationary generators with outputs of around 0.3 to 3 MW. However, the complex technology is not yet cost-competitive with thermal power plants, especially if the latter are combined with gas turbines.

16.6 Phosphoric Acid Fuel Cell (PAFC)

The phosphoric acid fuel cell (PAFC) uses pure phosphoric acid as an electrolyte at around 200 °C [38, 39]. At elevated temperatures, this acid allows proton conductivity even without the solvent water, as H_3PO_4 itself undergoes significant autoprotolysis above 150 °C:

$$H_3PO_4 \rightleftharpoons H_2PO_4^- + H_4PO_4^+ \tag{16.34}$$

This allows the protons produced at the anode to be transported to the cathode and simultaneously carry the electrical charge in the electrolyte. The operating temperature is limited by the vapor pressure of the phosphoric acid at temperatures above 220 °C. The electrolyte is contained in a porous matrix so that it can be installed in stacked arrangements with the two gas diffusion electrodes. The matrix consists of silicon carbide powder (chemically inert), which is processed with polytetrafluoroethylene (PTFE) to a porous thin layer. The electrodes consist of supported electrocatalysts (Pt/C), which are also processed with PTFE to a thin porous layer. The average pore radius of the electrolyte matrix is smaller than that of the porous active electrode layer in order to retain the liquid electrolyte in the carrier matrix and prevent massive leakage into the gas diffusion electrode, with blockage of the electrode surfaces and the transport channels for the gases. The active electrode layer is applied to a hydrophobized gas diffusion layer. The matrix for the liquid electrolyte is applied directly to the two porous electrodes. The liquid electrolyte in the completely filled pores of the electrolyte matrix ensures that the gas spaces of the anode and cathode are sealed. Bipolar plates made of graphite connect neighboring individual cells. The reactants reach the catalyst layer via the integrated flow field, and the products are removed. There are also channels for a liquid coolant.

The cathode carries out the oxygen reduction reaction at a Pt-containing catalyst according to Eq. (16.6). The achievable current densities are lower than on Pt in contact with Nafion at lower temperatures, as phosphate adsorbs on Pt and therefore the reaction proceeds more slowly. Significantly higher Pt loadings are required at the cathode (0.5–1 mg cm^{-2}) than at the anode (0.1–0.25 mg cm^{-2}). The anode is mainly operated with H_2 from natural gas or methanol. Although the PAFC is significantly less sensitive to CO due to the higher temperature, operation with natural gas still requires upstream reforming and gas purification in order to largely remove CO and sulfur-containing compounds from the fuel gas feed. Unconverted combustion gases can be used to generate heat. Due to the complex PAFC periphery, relatively large plants seem to be the most sensible application scenario for PAFCs. Accordingly, prototypes with an output of 10–500 kW were developed and operated between 1960 and 2010 [38, 39]. Despite various optimizations in the production of the reactors, it was repeatedly confirmed how complicated a permanently stable operation management is due to the highly corrosive liquid electrolyte layer at elevated temperatures, the complex fluid management (e.g., removal of the reaction water), the gradual loss of catalytic activity, and the complex measurement and operation control technology. For example, the cathode must be protected from corrosive, moist O_2 during shutdown. For decentralized combined heat and electricity supply, PAFCs have so far been significantly less economical than gas turbines.

16.7 Alkaline Fuel Cell (AFC)

Alkaline fuel cells (AFCs) were one of the first fuel cells to reach technical maturity; they became particularly well known as a result of NASA's missions with Apollo spacecraft and space shuttles [40, 41]. The anodic and cathodic partial reactions can be described by Eqs. (16.35) and (16.36).

$$2\,H_2 + 4\,OH^- \rightarrow 4\,H_2O + 4e^- \tag{16.35}$$

$$O_2 + 2\,H_2O + 4e^- \rightarrow 4\,OH^- \tag{16.36}$$

Although the overall cell reaction $2\,H_2 + O_2 \rightarrow 2\,H_2O$ leads to the formation of water as in the PEMFC, there is an important difference in that the OH^- ion is transported through the electrolyte. The oxygen reduction reaction is faster in alkaline solutions. As a result, the highest cell efficiencies can be achieved with AFCs. Cells can also be produced without the use of precious metals of the platinum group. Under the acidic conditions of the PEMFC, other metals and metal oxides in question dissolve (with the exception of tungsten carbide). Finally, in some AFC designs, water management proves to be simpler than with PEMFCs. In addition, the cold start behavior is very good.

After the Gemini missions with PEMFCs, NASA opted for the AFC. The 31 electrodes in each of the three stacks were circular with a diameter of 20 cm and a thickness of 2.5 mm. They operated at 200–230 °C with 85% KOH as the liquid electrolyte. With a mass of 109 kg, each stack was specified for an output of 1.5 kW [40].

These advantages are offset by serious disadvantages (otherwise there would now be many more AFCs). The alkaline electrolytes absorb CO_2 from the air (natural content 0.03 mol%).

$$CO_2 + 2\,OH^- \rightleftharpoons CO_3^{2-} + H_2O \tag{16.37}$$

$$CO_2 + OH^- \rightleftharpoons HCO_3^{2-} \tag{16.38}$$

$$CO_2 + 2\,KOH \rightleftharpoons K_2CO_3 + H_2O \tag{16.39}$$

The carbonate ions form poorly soluble carbonates with metal ions, which clog the pores or drastically reduce the conductivity of the electrolyte. For this reason, AFCs are mainly operated with pure H_2 and pure O_2. This expense is not so significant for space missions, but has so far been a decisive obstacle for terrestrial applications. Due to their design, AFCs deliver a lower power density than PEMFCs despite their high efficiency. For this reason, they are currently not considered for vehicle propulsion. However, as AFCs open up important development directions in the long term, we want to take a closer look at the components and system design.

Design of the Single Cell. The alkaline electrolyte can be arranged in three different ways between the porous anode and the porous cathode. As already discussed for the PAFC, it can be held in a porous matrix (Figure 16.15a). This design is particularly simple, but comes at the cost of the serious disadvantage that the electrolyte cannot be changed, and therefore, carbonates that have been absorbed cannot be removed. The product water and the reaction heat must be removed via the flowing gases.

The fuel cells of the NASA space shuttles were AFCs with a 35–45% KOH solution as electrolyte, which was contained in an asbestos felt. The anode was a carbon powder loaded with 10 mg cm^{-2} Pt and Pd. Teflon served as a binder. The cathode consisted of 90% Au and 10% Pt. The individual cells were combined in a bipolar arrangement to form a stack. There were three modules on board, which produced electrical energy and water for the crew and as a coolant. During a 7-day flight, 750 kg of H_2 and O_2 were converted. The progress compared to the AFC in the Apollo program was impressive: with a smaller total volume and a weight reduced by 18 kg, a stack with 12 kW delivered eight times more power [40]!

Another cell design uses a circulating liquid electrolyte that flows between the porous electrodes (Figure 16.15b). A 6 mol L^{-1} solution of KOH is typically employed. The circulating electrolyte can be treated or completely replaced, whereby impurities, carbonates, and product water are removed. The electrolyte can also be drained to suppress the uptake of CO_2 during rest phases of the cell. Finally, it can also be used advantageously for heat dissipation.

In addition, an alkaline anion exchange membrane can be used as an electrolyte. It consists of solid polymers that carry positive functional groups on the side chains, usually quaternary ammonium groups such as $R{-}N^{\oplus}{-}(CH_3)_3$. The stationary positive charges on the polymer backbone are balanced by mobile anions. There are no alkali metal ions in this system. Therefore, CO_2 cannot lead to the precipitation of solid carbonates after reaction to carbonate (CO_2-tolerant cells). This facilitates operational management. In addition, membranes prevent gas transfer between the two gas spaces

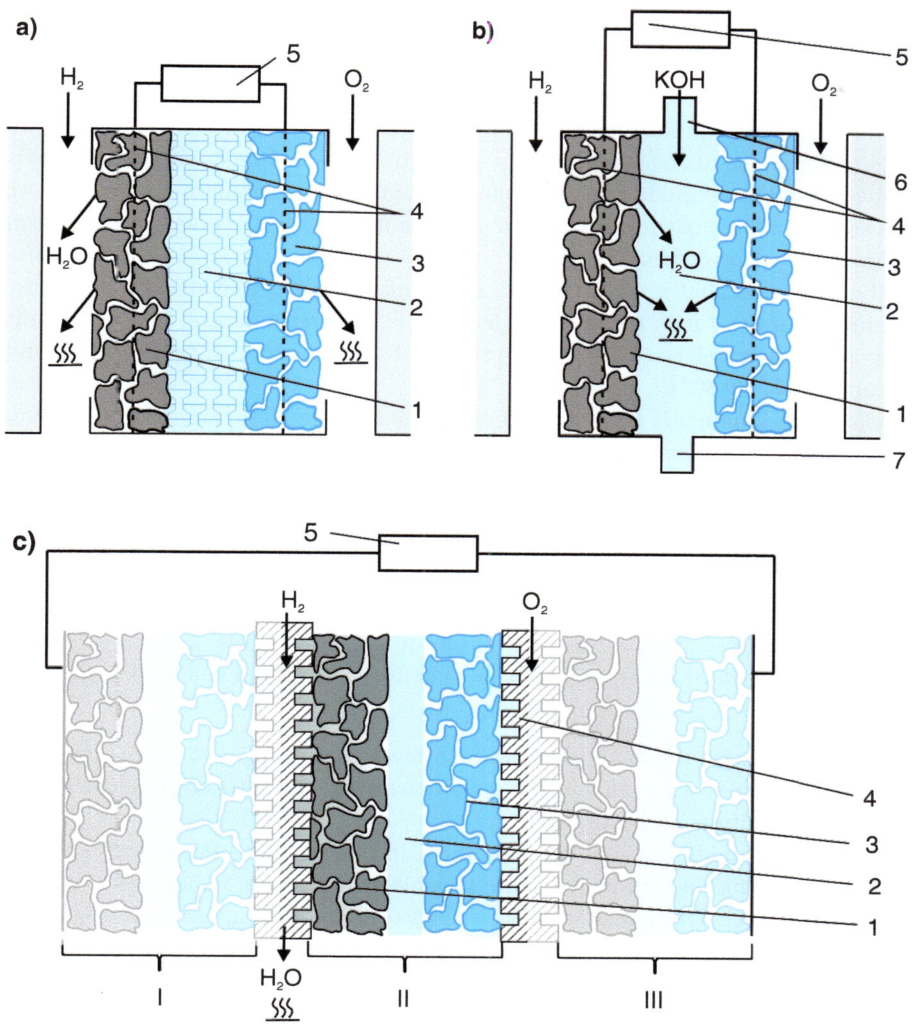

Figure 16.15 Design variants of single cells for AFCs.
a) Liquid electrolyte retained in matrix;
b) Circulating electrolyte;
c) Alkaline polymer membrane.

Components:

1) Anode;
2) Electrolyte;
3) Cathode;
4) Current collector (a, b) or bipolar plate (c);
5) External consumers;
6) Feed of KOH solution (b);
7) Outlet for KOH solution (b).

The symbol ≋ indicates which fluid flow predominantly dissipates the reaction heat.

more reliably than thin liquid films. Membranes can be made much thinner than liquid electrolyte layers, which reduces the internal cell resistance. There is no problem with the leakage of electrolyte through the porous membranes as in other cell designs due to the absence of liquid electrolyte solutions.

After the membrane has been synthesized, the mobile counterions are usually halide ions, which are exchanged for hydroxide ions OH^- by treatment with NaOH or KOH solutions. Although this concept initially appears to be completely analogous to the PEMFC with Nafion membranes, there are some significant differences.

- The ion mobility of hydroxide ions in most media, including polymer electrolyte membranes, is significantly lower than the mobility of H^+.
- In addition, the dissociation constant of perfluorinated sulfonic acids ($pK_a \approx -1$) is five orders of magnitude greater than that of quaternary ammonium salts ($pK_b \approx +4$). These two circumstances result in a significantly lower ionic conductivity, which limits the power output of the cell.
- Hydroxide ions are strong nucleophiles that attack the nitrogen atom of the ammonium group and can, for example, lead to cleaving off methanol $CH_3{-}OH$. This would cause accelerated degradation of the membrane, especially at operating temperatures of 60–80 °C.
- Finally, an ionomer is needed to effectively bind the catalyst particles to the membrane so that they are easily accessible to the gaseous reactants and are in electronic contact with the current collector. Simultanelusly, they must be connected to the alkaline membrane by ion-conducting compounds. In PEMFCs, suspensions of polymeric perfluorinated sulfonic acids fulfill this task. An equivalent component for AFCs does not yet exist. Therefore, less effective alternatives must be used. These consist of Nafion dispersions, that is, a cation exchange material that combines the metal particles and carbon particles to form a gas-permeable electrode. Reactive monomers that form a cross-linked polymer after application and mixing are other alternatives that are still being sought.

Overall, intensive efforts are being made to develop improved anion exchange membranes. Once these are available, the potential advantages of this technology can be used for applications on Earth [26, 27].

Catalysts. In addition to the metals of the platinum group, porous silver for oxygen reduction and nickel for the anode are suitable catalysts, too. Experiments are also being carried out with transition metal oxides or even with carbon materials doped with nitrogen. They can be obtained, for example, by pyrolysis of phthalocyanines.

Structure of the Stacks. Stacks for AFCs can be monopolar or bipolar. In the monopolar design, a gas chamber is surrounded on both sides by electrodes of the same type (Figure 16.16a). The anode of one cell is electrically connected to the cathode of the neighboring cell via an electronically conductive connection. The electrons must therefore be transported laterally through the porous electrode in order to reach the contact point. This is achieved by metal grids within the electrodes, similar to the current collectors in batteries. Nevertheless, an uneven current density distribution and the associated loss of efficiency must be expected.

In a bipolar arrangement of the individual cells of a stack, an electrically conductive plate connects adjacent anodes and cathodes (Figure 16.16b). The supply channels for the gases are typically embedded in these plates (flow field).

AFCs can be operated with gas flow systems. As the product water can be discharged with the flowing electrolyte, particularly with circulating electrolytes, AFCs can also be used with closed gas reservoirs (dead-end operation) for H_2 and O_2, which are only subjected to a constant gas pressure. The advantage of this is that it is easy to ensure full fuel utilization even with changing power outputs. With these cells, periodic flushing must be carried out to ensure that the electrolyte is completely returned since it escapes to a small extent through the porous electrodes.

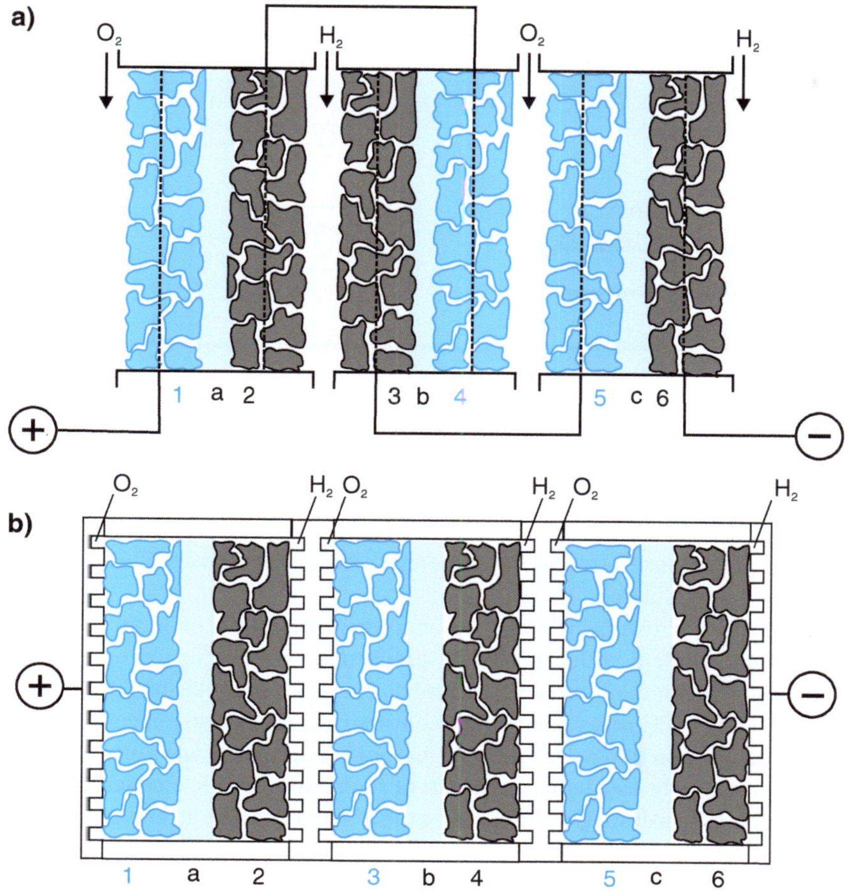

Figure 16.16 Comparison of a) the monopolar arrangement of single cells with b) the bipolar arrangement of single cells in alkaline fuel cells.
In the monopolar arrangement, the series connection of the electrodes and membranes is (+)-1-a-2-4-b-3-5-c-6-(−); in the bipolar arrangement, the sequence is (+)-1-a-2-3-b-4-5-c-6-(−).

Alternative Fuels. In addition to the classic H_2/O_2 systems, variants of the AFC can also be considered for alternative fuels, which could possibly be obtained as chemical energy carriers in the future by using temporary surplus renewable energies [42]. They include, for example, sodium borohydride $NaBH_4$, an adduct of ammonia and borane $NH_3 \rightarrow BH_3$, hydrazine N_2H_4 [43], ammonia NH_3, and lower alcohols, that is, methanol CH_3OH and, above all, bioethanol CH_3CH_2OH [19–21, 41].

16.8 Water Electrolysis

Water electrolysis is the reverse reaction to Eqs. (16.5) and (16.6) in the fuel cell, and aims for efficient production of hydrogen gas H_2 as an energy carrier. It is therefore not surprising that many of the design principles and materials that we have already discussed for the fuel cell can also be found in water electrolyzers.

Under the current economic boundary conditions, the large-scale production of H_2 from natural gas with the emission of CO_2 according to the reactions in Eqs. (16.1) to (16.3) is significantly cheaper than water electrolysis. For this reason, water electrolysis only plays a role in niche markets, for example, when surplus electrical energy from hydroelectric power plants is to be utilized, production of especially pure hydrogen in pilot plants, or isotope enrichment of deuterium is the main aim. However, many forecasts assume that this assessment will change significantly in the near future if the

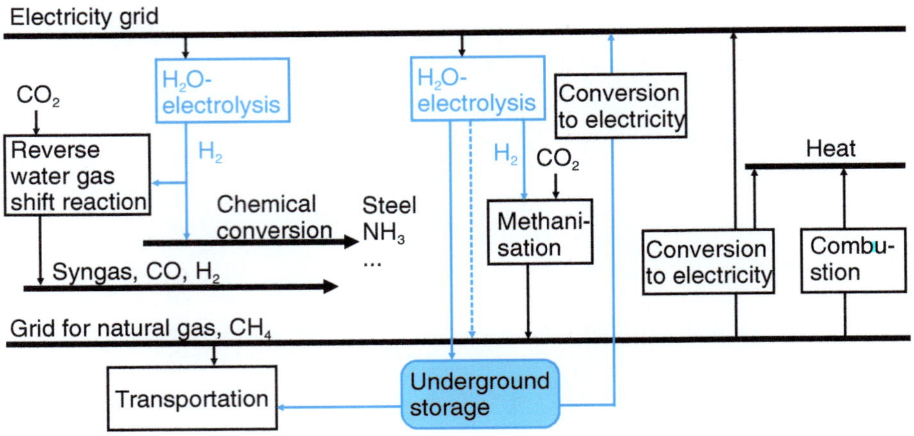

Figure 16.17 Possible use of electrolytic H_2 production in an energy system with a high proportion of fluctuating renewable electrical energy sources.

emission of greenhouse gases becomes more expensive through certificate trading or is prevented by regulation.

The production of hydrogen from wind and solar power would offer options for operating an energy system with a very high proportion of renewable energy and high security of supply, even in times of calm and darkness ("dark doldrums"). Because H_2 can be used in a variety of ways (chemical conversion, steel industry, reconversion into electricity, conversion into other synthetic fuels), there are many ways of balancing out fluctuations in supply and demand. As part of such a strategy, H_2 enables the existing grids for electricity, natural gas, and heating to be coupled (H_2 as an energy vector, sector coupling, Figure 16.17). When required, H_2 generated from natural gas could be fed into the process chains in order to buffer periods of particularly high demand or particularly low supply of electrolytic H_2. Technological problems with the use of H_2 include the high demands on the tightness of containers and pipelines, as well as the risk of an explosive oxyhydrogen reaction if H_2–O_2 mixtures form unintentionally.

Although water electrolyzers with different technologies are marketed commercially by various suppliers, further research efforts are needed in the context of sustainable energy systems to ensure that water electrolyzers can be operated safely on much larger scales than today, and probably with changing load profiles. Furthermore, it must be assessed, how the (expensive) hydrolytic hydrogen can be used most efficiently to avoid CO_2 emissions on an economic scale.

16.8.1 Technical Realization of Water Electrolysis

As with fuel cells, different reactor types are used for water electrolysis. They differ significantly in terms of material, reaction medium, operating temperature, and pressure of the H_2 formed (Figure 16.18). Operation is galvanostatic.

A Polymer Electrolyte Membrane Electrolyzer (PEMEL) is constructed in the same way as the PEMFC (Figure 16.2) and is operated at 20–100 °C. At its heart is a membrane electrode assembly, which comprises a 50–200 μm thick Nafion membrane coated with catalysts on both sides. Carbon-supported Pt nanoparticles are used for the cathode. For the anodes, the particularly rare element iridium must be used, whose oxide film (IrO_2) catalyzes the OER. Due to the high potentials, Ti-coated stainless steel must be used as the material for the bipolar plates. This material is much more stable against oxidation, but also much more expensive than carbon materials preferred in fuel cells. Water is fed to the anode, and H^+ ions pass through the membrane. As with

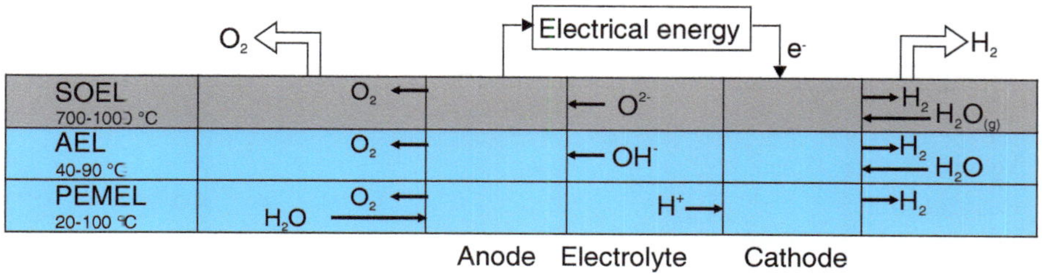

Figure 16.18 Comparison of different electrolysis processes. PEMEL – polymer electrolyte membrane electrolysis; AEL – alkaline electrolysis; SOEL – solid oxide electrolysis. See also Figure 2.28.

fuel cells, several individual cells are connected to form a stack (Figure 16.2) [44, 45]. The cells can tolerate load changes, which is important when using fluctuating solar and wind energy.

Alkaline Electrolysis (AEL) works at 40–90 °C in a KOH electrolyte. Here, water is added on the cathodic side, and OH^- ions pass through the membrane. This electrolyzer offers the possibility of using abundant transition metal oxides as catalysts, which is necessary if such plants are to be built with huge capacities. However, the space–time yields in these reactors have so far been much lower than in PEMEL. The electrolyte is a 20–40% (*m*/*m*) KOH solution, which is held in a felt mat of glass-reinforced polyphenylene sulfide with perforated sheets of Ni-coated steel adjoining as electrodes. Other electrode materials are steel, NiS, Raney nickel[154] [46, 47]. The systems supply H_2 at pressures of up to 30 bar. Tolerating load changes is a challenge for this cell type.

Solid oxide electrolyzer (SOEL) carries out the reactions at 700–1000 °C and, like the SOFC, utilizes the O^{2-}-ion conductivity of certain oxides. Here, all reactants and products are gaseous. Water vapor is supplied on the cathode side. As with the SOFC, the reactor can have a planar or tubular design (Figures 16.12 and 16.13). A typical cell structure consists of a 100–250 μm thick layer of yttrium-stabilized zirconium oxide (YSZ) as the electrolyte, to which a 100 μm thick Ni-cermet layer is applied as the cathode and approximately 40 μm of $LaMnO_3$ or Sr-doped $LaFeO_3$ with YSZ as the anode [48].

16.8.2 Operation of a Water Electrolysis System

In Figure 16.19, the cell voltage is plotted as a function of the applied current. The practical cell voltage is always greater than the thermodynamically expected cell voltage because overvoltage contributions at the anode, the cathode, and in the membrane lead to efficiency losses. We obtain the reversible voltage U_{rev} of a single cell from the molar Gibbs energy.

$$U_{rev} = \frac{\Delta_R G(T, p, x_i)}{nF} \tag{16.40}$$

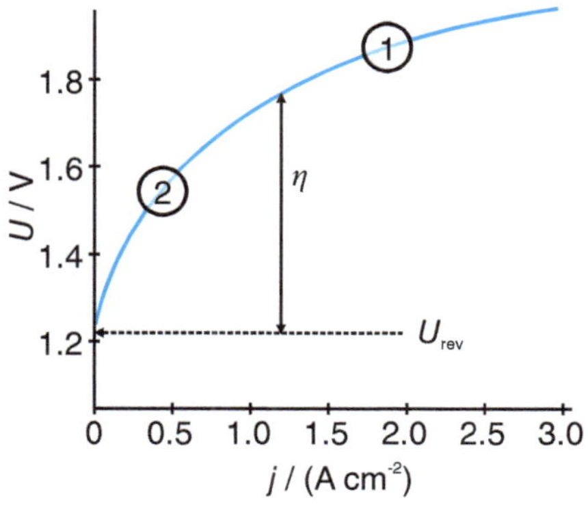

Figure 16.19 Schematic working curve of a PEMEL with two operating points. For details on the operating points, see Table 16.3.

U_{rev} decreases with increasing temperature. The voltage exceeding this value is irreversibly lost. As with fuel cells, the specific energy demand and the efficiency can be derived directly from the cell voltage. The area-related volume flow is directly linked to the current density. Let us consider the consequences using the example of operating points 1 and 2 marked in Figure 16.19. The same volume flow $\dot{V}n, H_2$ of 250 $m^3\,h^{-1}$ H_2 is to be generated at both operating points. The symbol $\dot{V}n, H_2$ indicates the value as a time derivative ($\dot{V} = dV/dt$) by the dot above the symbol. The index "n" indicates that the volume was converted to the standard conditions of 100 kPa and 0 °C. This exact

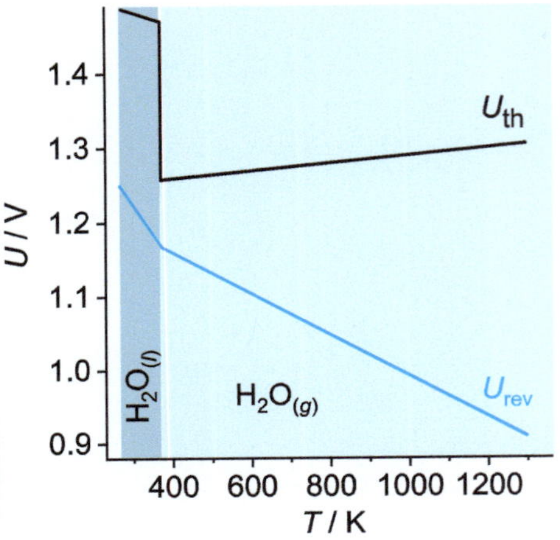

Figure 16.20 Temperature dependence of the reversible and thermoneutral voltages of a water electrolysis cell.

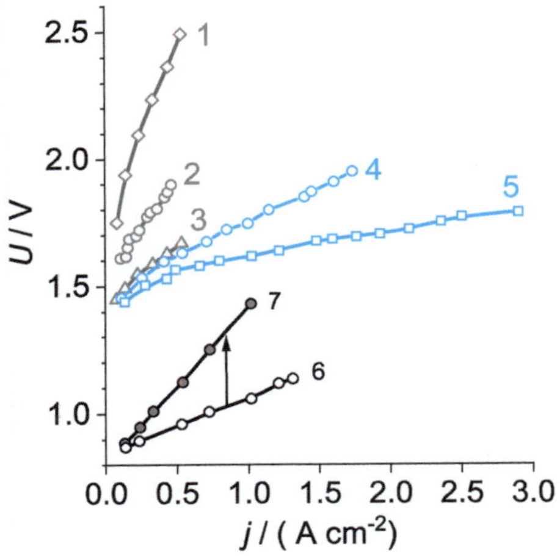

Figure 16.21 Comparison of the characteristic curves of electrolyzers. 1) to 3) different development stages of AELs from conventional systems to advanced laboratory models; 4) and 5) different development stages of PEMEL; 6) SOEL at the start of operation; 7) SOEL at the end of the operating period. Source: Data from [49].

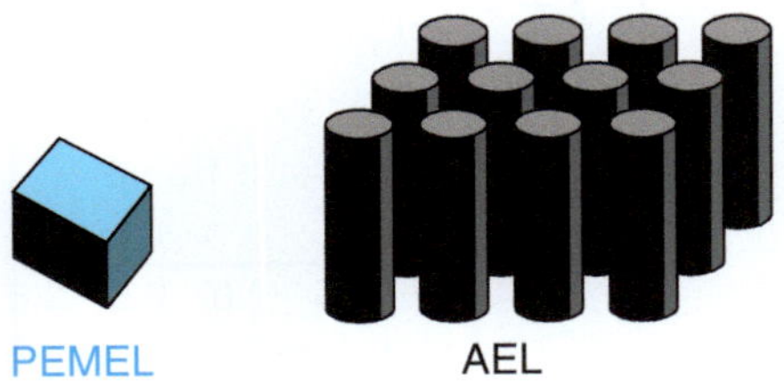

Figure 16.22 Schematic comparison of the reactor sizes at the same H_2 volume flow for a PEMEL and an AEL.

Table 16.3 Comparison of the characteristics of an electrolytic cell at operating points 1 and 2 in Figure 16.19.

Size	Operating point 1	Operating point 2
Volume flow $\dot{V}n$, $H_2/(m^3\,h^{-1})$	250	250
Cell area A/m^2	27	109
Electrical power P_{el}/MW	1	0.86
Efficiency/%	69.4	80.8
Costs	Low investment costs	Low operating costs

notation is often replaced in technology by the specification 250 $Nm^3\,h^{-1}$, where "N" indicates standard conditions. We do not use this notation here for reasons of consistency. A comparison of the two operating points is shown in Table 16.3 and leads to the conclusion:

Higher current densities lead to lower energy efficiencies but to more compact reactors.

Another important consideration is the required heat supply or cooling for an electrolyzer. The thermoneutral voltage of a single cell can be calculated from the molar reaction enthalpy $\Delta_R H(T, p, x_i)$:

$$U_{th} = \frac{\Delta_R H(T,p,x_i)}{nF} \tag{16.41}$$

Below U_{th}, that is, at low current densities, heat must be constantly supplied to the electrolytic cell, for example, by utilizing waste heat from other processes. On the other hand, heat must be constantly dissipated from the reactor at $U > U_{th}$. The exact values of U_{rev} and U_{th} depend on the operating temperature and show a discontinuity in the transition from liquid to gaseous water as the reagent (Figure 16.20).

Figure 16.21 shows a comparison of the polarization curves of the various electrolyzers. These polarization curves translate into very different designs. The schematic size comparison in Figure 16.22 shows, for example, the enormous difference in size between a PEMEL reactor and an AEL reactor of the same power and H_2 production per time.

Further Reading. Comprehensive overviews with a technology comparison are available [49–51].

Test your knowledge

Exercise 16.1: Compare the theoretical efficiencies of a PEMFC (with H_2 as fuel gas at 80 °C) and an SOFC (with H_2 as fuel gas at 800 °C). The oxidizing agent in each case is O_2 from the air at standard pressure.

Exercise 16.2: A PEMFC operates with an electrical output of 1 kW at a single cell voltage of 0.7 V. What quantity of reaction water must be discharged from the system?

Exercise 16.3: A pressurized gas tank for H_2 has a capacity of 100 L and is filled with 100 MPa. How long can a fuel cell be operated with this amount of H_2 when the power output is 50 kW and the electrical efficiency is 65% (100% fuel utilization)?

Exercise 16.4: A single cell of a PEMFC operates with a current of 1 A at a single cell voltage of 0.7 V. The membrane has a specific resistance of 8.77 Ω cm^{-2} and a geometric electrode area of 4 cm^2 for both the anode and the cathode. Calculate the power drop due to the current flow through the membrane.

Exercise 16.5: Make a simplified sketch of an MEA of a PEMFC. Label all components. Indicate which processes take place in the individual parts. What quantitative relationships describe the respective processes?

Exercise 16.6: Make a simplified sketch of an MEA of an SOFC. Label all components. Indicate which processes take place at the individual parts. What quantitative relationships describe the respective processes?

Exercise 16.7: Estimate how much Pt would be required if all cars registered in your home country (alternatively in the European Union, in the United States, or in Japan) were equipped with a modern PEMFC?

Note: Create an Excel worksheet for your estimate in which the assumptions made are documented and the figures are stored (and can also be changed). Try to obtain the most accurate figures possible for the required variables from official government statistics. If this is not possible, make reasonable assumptions and document them. Try out different Pt contents of the fuel cells and average performance data of the cars. Possible starting values could be, for example, required power output 80 kW, power-related Pt content of the fuel cell of 0.125 g/kW.

Expand the estimate (1): How much Pt would have to be provided annually through mining operations or recycling if every car were replaced by a new one every 10 years?

Extend the estimate (2): How large would the additional annual demand be if a populous country such as the P.R. China were to aim for the same level of fuel cell passenger cars (per capita) within 10 years as in the country you are considering? Compare the estimated masses of Pt with the mass of Pt currently traded on the world market (new production 268 tons (2010) plus about 30% (80 tons) from recycling [24]).

Exercise 16.8: Explain possible aging processes in PEMFCs. Propose experimental strategies to demonstrate the occurrence of the individual aging processes in a test cell.

Exercise 16.9: Name the advantages and disadvantages of an SOFC compared to a PEMFC. For which applications does a PEMFC seem more suitable? For which applications would the SOFC be more favorable?

Exercise 16.10: Using Figure 16.13, explain the serial electrical connection of individual cells in the tubular design of the SOFC. *Note*: Draw a sketch and draw the current path.

Exercise 16.11: Discuss the development potential of the alkaline fuel cell with anion exchange membrane. What are the basic advantages of the system? What are the current limitations? Suggest research directions to overcome the existing limitations. What fundamental difficulties are to be expected?

Further Reading

1 K. Kordesch, G. Simander; *Fuel Cells and Their Application*, Wiley-VCH, Weinheim, 1996.

2 K. Sundmacher; Fuel cell engineering: Towards the design of efficient electrochemical power plants. *Ind. Eng. Chem. Res.* **2010**, *49*, 10159–10182.

3 M. Cappadonia, U. Stimmig, K. V. Kordesch, J. C. Tambasco de Oliveira, P. Schmuki, M. Eikerling, A. A. Kornyshev, A. A. Kulikovsky, S. Gottesfeld; Fuel cells, in *Encyclopedia of Electrochemistry, Vol. 5 – Electrochemical Engineering* (A. J. Bard, M. Stratmann, D. D. Macdonald, P. Schmuki, Eds.), Wiley-VCH, Weinheim, 2007, pp. 429–661.

4 J. Garche (Ed.); *Encyclopedia of Electrochemical Power Sources, Vol. 5 and 6*, 2nd ed., Elsevier, Amsterdam, 2025.

5 Y. Yang, C. R. Peltier, R. Zeng, R. Schimmenti, Q. Li, X. Huang, Z. Yan, G. Potsi, R. Selhorst, X. Lu, W. Xu, M. Tader, A. V. Soudackov, H. Zhang, M. Krumov, E. Murray, P. Xu, J. Hitt, L. Xu, H.-Y. Ko, B. G. Ernst, C. Bundschu, A. Luo, D. Markovich, M. Hu, C. He, H. Wang, J. Fang, R. A. DiStasio, Jr., L. F. Kourkoutis, A. Singer, K. J. T. Noonan, L. Xiao, L. Zhuang, B. S. Pivovar, P. Zelenay, E. Herrero, J. M. Feliu, J. Suntivich, E. P. Giannelis, S. Hammes-Schiffers, T. Arias, M. Mavrikakis, T. E. Mallouk, J. D. Brock, D. A. Muller, F. J. DiSalvo, G. W. Coates, H. D. Abruña; Electrocatalysis in alkaline media and alkaline membrane-based energy technologies. *Chem. Rev.* **2022**, *122*, 6117–6321.

6 https://energy.gov/eere/fuelcells/hydrogen-and-fuel-cell-technologies-office (last access 26 Jan 2025).

7 G. Gebel; Structural evolution of water swollen perfluorosulfonated ionomers from dry membrane solutions. *Polymer* **2000**, *41*, 5829–5838.

8 K. A. Mauritz, R. B. Moore; State of understanding Nafion. *Chem. Rev.* **2004**, *104*, 4535–4586.

9 M. Eikerling, A. A. Kornyshev, A. A. Kulikovsky; Physical modeling of fuel cells and their components. in *Encyclopedia of Electrochemistry, Vol. 5 – Electrochemical Engineering* (A. J. Bard, M. Stratmann, D. D. Macdonald, P. Schmuki, Eds.), Wiley-VCH, Weinheim, 2007, pp. 447–543.

10 P. P. Mukherjee, G. Wang, C.-Y. Wang; Direct simulation of polymer electrolyte fuel cell catalyst layers, *Modern Aspects of Electrochemistry*, *Vol. 40* (R. E. White, C. G. Vayenas, A. E. Gambola-Aldeco, Eds.), Springer, New York, 2007, pp. 285–341.

11 S. M. Rezaei Niya, M. Hoorfar; Study of proton exchange membrane fuel cells using electrochemical impedance spectroscopy technique – A review. *J. Power Sources* **2013**, *240*, 281–293.

12 Z. Tang, Q.-A. Huang, Y.-J. Wang, F. Zhang, W. Li, A. Li, L. Zhang, J. Zhang; Recent progress in the use of electrochemical impedance spectroscopy for the measurement, monitoring, diagnosis and optimization of proton exchange membrane fuel cell performance. *J. Power Sources* **2020**, *468*, 228361/1–26.

13 V. Sethuraman; Fuel cells: Polymer-electrolyte membrane fuel cells: Life-limitations. in *Encyclopedia of Electrochemical Power Sources, Vol. 5*, 2nd ed. (J. Garche, Ed.), Elsevier, Amsterdam, 2025, pp. 289–308.

14 A. Zana, J. Speder, M. Roefzaad, L. Altmann, M. Bäumer, M. Arenz; Probing degradation by IL-TEM: The influence of stress test conditions on the degradation mechanism. *J. Electrochem. Soc.* **2013**, *160*, F608–F615.

15 V. Neburchilov, J. Martin, H. Wang, J. Zang; A review of polymer electrolyte membranes for direct methanol fuel cells. *J. Power Sources* **2007**, *169*, 221–238.

16 K. Scott, A. K. Shukla; Direct methanol fuel cells: Fundamentals, problems perspectives, *Modern Aspects of Electrochemistry, Vol. 40* (R. E. White, C. G. Vayenas, A. E. Gambola-Aldeco, Eds.), Springer, New York, 2007, pp. 127–227.

17 B. L. Garcia, J. W. Weidner; Review of direct methanol fuel cell, *Modern Aspects of Electrochemistry, Vol. 40* (R. E. White, C. G. Vayenas, A. E. Gambola-Aldeco, Eds.), Springer, New York, 2007, pp. 228–284.

18 A. S. Arico, S. Srinivasan, V. Antonucci; DMFCs: From fundamental aspects to technology development. *Fuel Cells* **2001**, *1*, 133–161.

19 A. Brouzgou, A. Podias, P. Tsiakaras; PEMFCs and AEMFCs directly fed with ethanol: A current status comparative review. *J. Appl. Electrochem.* **2013**, *43*, 119–136.

20 E. H. Yu, U. Krewer, K. Scott; Principles and materials aspects of direct alkaline alcohol fuel cells. *Energies* **2010**, *3*, 1499–1523.

21 K. I. Ozoemena; Nanostructured platinum-free electrocatalysts in alkaline direct ethanol fuel cells: Catalyst design, principles and application. *RSC Adv.* **2016**, *6*, 89523–89550.

22 R. Zeis; Materials and characterization techniques for high-temperature polymer electrolyte membrane fuel cells. *Beilstein J. Nanotechnol.* **2015**, *6*, 68–83.

23 Y. Liu, W. Lehnert, H. Janßen, R. C. Samsun, D. Stolten; A review of high-temperature polymer electrolyte membrane fuel-cell (HT-PEMFC)-based auxiliary power units for diesel-powered road vehicles. *J. Power Sources* **2016**, *311*, 91–102.

24 D. R. Wilburn; Global Exploration and Production Capacity for Platinum-Group Metals From 1995 Through 2015. Scientific Investigation Reports 2012–5164, US Department of the Interior and U.S. Geological Survey. Reston, Virginia, 2012. https://pubs.usgs.gov/publication/sir20125164 (last access 10 Aug 2024).

25 H. A. Gasteiger, N. M. Markovic, Just a dream – Or future reality? *Science* **2009**, *324*, 48–49.

26 J. R. Varcoe, P. Atanassov, D. R. Dekel, A. M. Herring, M. A. Hickner, P. A. Kohl, A. P. Kucernak, W. E. Mustain, K. Nijmeijer, K. Scott, T. Xu, L. Zhuang; Anion-exchange membranes in electrochemical energy systems. *Energy Environ. Sci.* **2014**, *7*, 3135–3191.

27 D. R. Dekel; Review of cell performance in anion exchange membrane fuel cells. *J. Power Sources* **2017**, *375*, 158–169.

28 S. Gottesfeld, D.-R. Dekel, M. Page, C. Bae, Y. Yan, P. Zelenay, Y. S. Kim; Anion exchange membrane fuel cells: Current status and remaining challenges. *J. Power Sources* **2018**, *375*, 170–184.

29 Y. P. Zhu, C. Guo, Y. Zheng, S.-Z. Qiao; Surface and Interface engineering of noble-metal-free electrocatalysts for efficient energy conversion processes. *Acc. Chem. Res.* **2017**, *50*, 915–923.

30 L. Lin, Q. Zhu, A.-W. Xu; Noble-metal-free Fe-N/C catalysts for highly efficient oxygen reduction reaction under both alkaline and acidic conditions. *J. Am. Chem. Soc.* **2014**, *136*, 11027–11033.

31 U. Tylus, Q. Jia, K. Strickland, N. Ramaswamy, A. Serov, P. Atanassov, S. Mukerjee; Elucidating oxygen reduction active sites in pyrolyzed metal-nitrogen coordinated non-precious-metal electrocatalyst systems. *J. Phys. Chem. C* **2014**, *118*, 8999–9008.

32 X. D. Zhou, S. C. Singhal, K. V. Galloway, N. M. Sammes, B. R. Roy, M. Kusnezoff, L. G. J. de Haart, M. Spiller, R. J. Kee, A. M. Colclasure, H. Zhou, Z. Yang, S. C. Singhal. R. Peters. L. Blum, E. Riensche, A. Weber, S. Zang, S. Yu, K. Wang,

S. B. Beale, F. Kunz; Fuel cells – Solid oxide fuel cells. in *Encyclopedia of Electrochemical Power Sources, Vol. 5*, 2nd ed. (J. Garche, Ed.), Elsevier, Amsterdam, 2025, pp. 622–784.

33 A. J. Jacobson; Materials for solid oxide fuel cells. *Chem. Mater.* **2010**, *22*, 660–674.

34 J. Brouwer; *Gas Turbine Handbook*. https://www.netl.doe.gov/carbon-management/turbines/handbook (last access 31 Oct 2022).

35 E. D. Wachsman, K. T. Lee; Lowering the temperature of solid oxide fuel cells. *Science* **2011**, *334*, 935–939.

36 K. Kordesch, G. Simander; *Fuel Cells and Their Application*, Wiley-VCH, Weinheim, **1996**, pp. 111–133, 196–204.

37 C. Yuh, A. Hilmi, R. Venkataraman, M. Farooque; Fuel cells: Molten carbonate fuel cells. in *Encyclopedia of Electrochemical Power Sources, Vol. 5* (J. Garche, Ed.), Elsevier, Amsterdam, 2025, pp. 557–621.

38 K. Kordesch, G. Simander; *Fuel Cells and Their Application*, Wiley-VCH, Weinheim, 1996, pp. 93–111, 207–241, 276–278.

39 M. Saquib, A. Sharma, A. C. Bhosale, B. Deka, P. Kurzweil, T. Kuwabara, J. Müller-Hülstede, H. Schmies, D. Schonvogel, P. Wagner, T. Murahashi, S. Kanuri, T. Patterson; Fuel cells: Phosphoric acid fuel cells. in *Encyclopedia of Electrochemical Power Sources, Vol. 5*, 2nd ed. (J. Garche, Ed.), Elsevier, Amsterdam, 2025, pp. 460–556.

40 K. Kordesch, G. Simander; *Fuel Cells and Their Application*, Wiley-VCH, Weinheim, 1996, pp. 54–72.

41 G. Muhler, N. Carboni, M. A. Navarra, V. Hacker, S. D. Fraser, K. Kordesch, S. P. Jiang, Q. Li; Fuel cells: Alkaline fuel cells. in *Encyclopedia of Electrochemical Power Sources, Vol. 5*, 2nd ed. (J. Garche, Ed.), Elsevier, Amsterdam, 2025, pp. 411–459.

42 El Hao, X. Wang, U. Krewer, L. Li, K. Scott; Direct oxidation alkaline fuel cells: From materials to systems. *Energy Environ. Sci.* **2012**, *5*, 5668–5680.

43 A. Serov, C. Kwak; Direct hydrazine fuel cell. *Appl. Catal., B* **2010**, *98*, 1–9.

44 E. Rasten, G. Hagen, R. Tunold; Electrocatalysis in water electrolysis with solid polymer electrolyte. *Electrochim. Acta* **2003**, *48*, 3945–3952.

45 M. Carmo, D. L. Fritz, J. Mergel, D. Stolten; A comprehensive review on PEM water electrolysis. *Int. J. Hydrogen Energy* **2013**, *38*, 4901–4934.

46 K. Zeng, D. Zhang; Recent Progress in alkaline water electrolysis for hydrogen production and applications. *Prog. Energy Combust. Sci.* **2010**, *36*, 307–326.

47 S. Marini, P. Salvi, P. Nelli, R. Pesenti, M. Villa, M. Berrettoni, G. Zangari, Y. Kiros; Advanced alkaline water electrolysis. *Electrochim. Acta* **2012**, *82*, 384–391.

48 P. A. Lessing; Materials for hydrogen generation via water electrolysis. *J. Mater. Sci.* **2007**, *42*, 3477–3487.

49 T. Smolinka, E. T. Ojong, J. Garche; Hydrogen production from renewable energies – Electrolyzer technologies. in *Electrochemical Energy Storage for Renewable Sources and Grid Balancing* (P. T. Mosley, J. Garche, Eds.), Elsevier, Amsterdam, 2015, pp. 103–128.

50 M. Chatenet, B. G. Pollet, D. R. Dekel, F. Dionigi, J. Deseure, P. Millet, R. D. Braatz, M. Z. Bazant, M. Eikerling, I. Staffell, P. Balcombe, Y. Shao-Horn, H. Schäfer; Water electrolysis: From textbook knowledge to the latest scientific strategies and industrial developments. *Chem. Soc. Rev.* **2022**, *51*, 4583–4762.

51 A. Makhsoos, B. G. Pollet, D. S. Falcão, D. Santos, A. M. F. R. Pinto, R. B. Ferreira, M. Müller, I. Galkina, F. Scheepers, F. Lohmann-Richters, S. Ramakrishnan, M. Mamlouk, S. A. Grigoriev, S. K. Sampangi, L. Röntzsch, E. Hoppe, M. Riegraf, R. Costa, K. A. Friedrich; Electrolyzer. in *Encyclopedia of Electrochemical Power Sources, Vol. 6*, 2nd ed. (J. Garche, Ed.), Elsevier, Amsterdam, 2025, pp. 1–122.

17

Electrosyntheses in the Laboratory and Industry

In all chemical reactions, the formation of products is directly linked to the conversion of different forms of energy. Specifically, electrical energy and chemical energy are interconverted in electrochemical reactions. In Chapters 15 and 16, chemical transformations in batteries and fuel cells were discussed, associated with the storage and release of energy. In contrast, the preparative aspect of the reaction, that is, the formation of useful products (polymers, pharmaceuticals, intermediates, etc.) is the focus of electrosynthesis treated here. In Chapter 9, the galvanic plating of metals, one of the classical topics in electrochemistry, was discussed as a way of producing solid metallic objects and coatings using electrochemistry. Galvanic plating processes depend on the interplay of nucleation, growth, and transport rates to control the grain size, grain orientation, and quality of the product obtained in *solid form*.

In this chapter, electrochemical processes will be considered in which the product is obtained initially in *fluid form*, that is, dissolved in a solvent, as a gas or a solid dispersed in the electrolyte. The electrolysis products must be separated from the components of the electrolyte solution and side products. In many cases, further processing steps are necessary before the electrolysis products become part of a more complex commercial product. The special case of hydrogen production from water electrolysis was discussed in Section 16.8 because hydrogen is an energy vector and water electrolysis technologies are closely linked to fuel cell technologies. In contrast, electrosynthesis as treated in Chapter 17 is associated with electroorganic syntheses and the transformation of organic compounds (avoiding stoichiometric oxidants or reductants) into valuable products. Prerequisite to the study of this chapter is an understanding of organic chemistry and transformations, as well as Chapters 1–4 and, in particular, Chapter 6 on mechanisms. Further links to materials in Chapters 12, 13, and 16 will be indicated.

Further Reading. Comprehensive details on laboratory and industrial electrosyntheses are provided in [1–8]. Reviews on classic inorganic electrosyntheses beyond the coverage of chlor–alkali electrolysis in this text can be found in [9–12]. Up-to-date reviews on electrosynthesis are recommended for further study [13–18]. More specifically, the investigation of electroorganic reaction mechanisms is covered in Chapter 6 and review articles [19]. Further literature suggestions will be provided below for specific topics.

17.1 Special Features of Electrosynthetic Reactions

Electrosynthetic reactions can be compared to other preparative methods in organic and organometallic chemistry, but we will also contrast electrosynthesis and electrochemical experiments linked to physical and analytical chemistry in Sections 17.1.1–17.1.5.

17.1.1 Comparison of Conventional Chemical Synthesis and Electrosynthesis

With the invention of powerful AC generators (alternating current) based on technology from electrical motors in 1866 by Werner Siemens[28], batteries were replaced as the most important electrical power sources. Batteries were too expensive and of limited power for use in electrosynthesis apart from research laboratories. The AC generator in combination with a rectifier produced DC (direct current) with high power to drive electrolytic processes such as metal winning (Zn and Al) and refining (Cu). New processes were invented for the production of chlorine and chlorate, which remain as the worldwide predominant electrosynthetic processes to this day.

Electrochemical technology is uniquely important in the synthesis of basic inorganic chemicals, including very strong oxidants (F_2 [20, 21], Cl_2 [10, 21], H_2O_2 [10, 21], and $NaClO_3$ [21, 22]), and reductants (alkali and alkaline earth metals, Al, and refractory metals [10, 23–27]). Often, there are no chemical reagents to achieve these transformations, or only very expensive nonelectrochemical options (e.g., for Cl_2 and Al). These processes operate on a huge scale in specialized reactors, under carefully controlled conditions, and with high economic impact (e.g., production of nonnoble metals in molten salt electrolysis, Section 11.4). Because each type of reaction requires very specific conditions and reactors, the concepts cannot be transferred easily to other reactions. In contrast, the chlor–alkali electrolysis (Excursion 17-1) is an example of an established process, where small changes in the reactor concept still allow substantial improvements in energy efficiency.

The situation is different for preparative organic electrochemistry, which remained in the shadow of large-scale inorganic electrolytic processes for a long time. In a review published in 1896, Löb [28] suggested that organic electrosyntheses were to be "composed of largely disjunct facts" (i.e., unpredictability) due to the associated complex homogeneous following reactions after the electron transfer (ET) step. After more than a century of research, the fundamental understanding of electrosynthetic processes has been established. In the 1970s, numerous new electrochemical reactions were developed for functional group transformations (FGT) and C–C bond formation. However, large-scale industrial processes have remained scarce, with the hydrodimerization of acrylonitrile to adiponitrile (for polymer synthesis) in the Monsanto process[162] probably being the most prominent example. The advantages of electroorganic syntheses are evident:

- continuous choice of oxidation or reduction potential applied to the electrode (to fine-tune product type and yield),
- excellent atom-economy for the main reaction by exploiting electrons as reagents and by minimizing waste production (no chemical oxidizing or reducing reagents),
- precise control of reaction rates by external power source (safe reaction engineering),
- typically mild reaction conditions with low/ambient temperatures and mild pH to generate reactive reaction intermediates in their electronic ground state.

There are disadvantages to electroorganic syntheses as well: electrosyntheses require electrical power with losses due to overvoltages and resistivity, especially in organic solvents with lower ionic conductivity. Typically, 150–200% of the thermodynamically required energy has to be supplied. This makes processes less competitive when compared to thermal activation: Electrical energy is more expensive compared to thermal energy, which can be provided by coupling exothermic and endothermic processes in a chemical plant. Electrochemical reactions occur only at interfaces and, therefore,

[162] Monsanto, American chemical company, 2018 bought by Bayer AG, developed the electrolytic adiponitrile synthesis in the 1960s.

require mass transport. Reactors have to be designed with a large electrode area in as small as possible total volume. This makes electrochemical reactors more complex and more expensive when compared to conventional reactors. Reactions occur in the liquid phase with associated slow mass transport processes when compared to gas phase processes. Thus, electrochemical syntheses operate with substantially lower space–time yields compared to catalytic gas-phase reactors. As a result, electrosynthetic H_2 production (water electrolysis) cannot compete with a heterogeneous catalysis converting natural gas [Eqs. (16.1)–(16.3)]. This argument might change when electric power from renewable sources becomes more competitive (e.g., during supply peaks where it has to be used somehow to maintain grid stability), or when costs for the emission of greenhouse gases (CO_2 and CH_4) are factored in.

Research in electroorganic synthesis has been reenergized by the vision of many chemical processes being electrified in the future. Currently, the focus is on high-value pharmaceuticals and fine chemicals rather than on commodity chemicals. For multistep synthesis, electrochemical methods offer lower reagent consumption associated with lower waste production, lower number of reaction steps, and exquisite control in the formation of reactive intermediates. These advantages balance out energy costs for fine chemicals. Remaining barriers to wider adoption of electroorganic synthesis are a lack of specialized reaction ware and potentially a lack of skilled operators for preparative electrosynthesis [14, 15]. When planning electrosynthetic transformations, the link from mass transport to reaction mechanism (Chapter 6) is crucial. This difficulty is less important in the currently preferred metal–organic reactions. The recent trend toward microfluidic and flow-reactor syntheses provides a deeper appreciation for mass transport in reaction control under flow (flow chemistry) in laboratories focusing on preparative transformations. The need for a supporting electrolyte and the separation of product and electrolyte strongly impact the choice of solvents and processes. The price of the salts that are soluble in organic solvents is substantial and often mandates recycling the supporting electrolyte to lower cost and waste.

In the following sections, concepts for electrochemical cell designs are reviewed specifically for application in electroorganic synthesis. Based on the reaction mechanisms discussed in Chapter 6, exemplary reactions will be discussed that are particularly attractive in electrosynthesis.

17.1.2 Power Sources and Reaction Control

The most important peculiarity of electrosynthesis compared to conventional organic synthesis is the use of a DC power supply to directly drive the reaction. The electrical power is supplied in the form of a constant applied voltage (or a constant applied current). For many laboratory-scale synthetic procedures, a laboratory power source is sufficient with a constant, adjustable voltage (or current) output. For the study of the reaction mechanism and the optimization of the synthesis, an electrochemical workstation is more practical that allows two-electrode operation with a galvanostat or three-electrode operation with a potentiostat as power sources. The potentiostat should have a compliance voltage of 30 V and provide current from the milliampere range to approximately 1–2 A for gram-scale transformations. A compliance voltage of 30 V or 100 V is higher than that typically used in analytical or physicochemical instrumentation. Although requiring higher driving voltages and currents, electrosynthetic processes are often less demanding in terms of equipment, and costs tend to be lower than for analytical instrumentation. For analytical and physicochemical instrumentation, low-current conditions lead to a small ohmic voltage drop[27] $\Delta U = I \cdot R$ in cables, in the electrolyte, and in diaphragms or separators (e.g., $I \cdot R = 1\ \mu A \cdot 1000\ \Omega = 1\ mV$). In contrast, higher currents in a poorly conducting organic electrolyte media can cause a

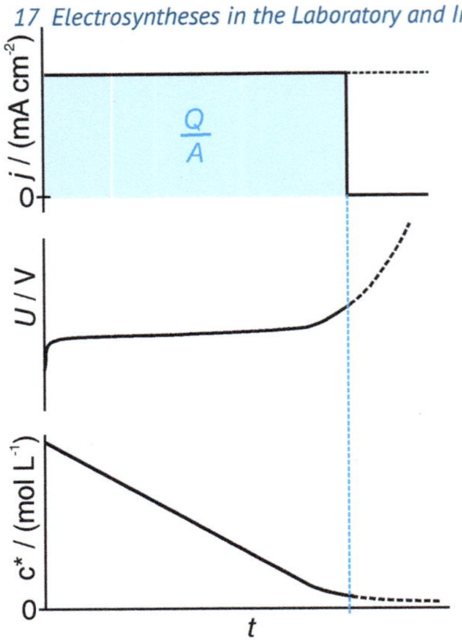

Figure 17.1 Illustration of current density (fixed) and cell voltage (WE versus Aux) versus time with decreasing concentration of the starting material over time during galvanostatic electrolysis. The end point is defined either by (i) a fixed amount of stoichiometric charge (time) or (ii) a cut-off voltage (indicating that the transport of reagent molecules to the electrode surface is insufficient to maintain the faradaic current of the target reaction, causing the cell voltage to increase and unwanted parasitic reactions to commence).

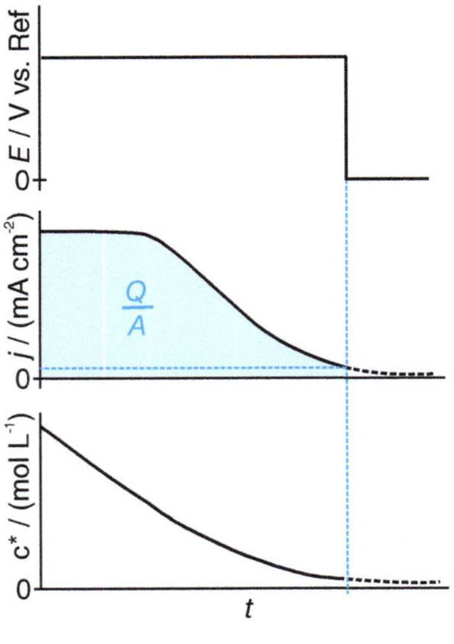

Figure 17.2 Illustration of the potential (fixed, WE versus Ref), the resulting current, and the concentration of starting materials during potentiostatic electrolysis.

substantial voltage drop in preparative electrochemical cells, for example, $I \cdot R = 10$ mA $\cdot$ 1000 Ω = 10 V. The power source has to overcome this ohmic voltage drop in addition to delivering the thermodynamic equilibrium voltage $U = -\Delta_R G/(nF)$ and the overpotentials at the anode and the cathode (Chapter 3). Thus, the applied cell voltage is substantially higher during preparative electrolysis compared to the difference in anodic and cathodic formal potentials for a particular cell reaction.

When using the galvanostatic electrolysis mode (constant current), processes at the anode and the cathode are driven at a constant rate (Figure 17.1). The cell voltage is continually adjusted by the galvanostat in such a way that it provides the driving force for the set current to pass through all components of the cell. During the course of electrolysis processes, the cell voltage usually stabilizes and remains relatively constant. The benefit of galvanostatic control is in the defined reaction progress. The current must be set to a value that can be achieved with the provided reactant concentration and mass transport conditions. Otherwise, the cell voltage will be increased until another (parasitic) process provides the preset current. Good mass transport based on effective stirring is desirable. Eventually, mass transport of the reactant to the electrodes can become rate-limiting toward the end of the electrolysis, and the cell voltage then increases (ultimately to the solvent or electrolyte decomposition voltage). Therefore, the end point of the electrolysis needs to be selected to avoid parasitic processes and the formation of side products.

When employing potentiostatic control during electrolysis, the selectivity for specific products can be substantially higher. This is due to the potential at the working electrode (WE) remaining constant versus a fixed reference electrode (Ref, Section 2.4.1 and Table 2.6) throughout the electrolysis experiment (Figure 17.2). The cell voltage (WE versus Aux) is continuously adjusted by the potentiostat to maintain the constant working electrode potential (Sections 5.3.1 and 5.7.1). The resulting current is automatically adjusted to reflect the decreasing rate of mass transport without any danger of side product formation or solvent electrolysis. The end point of electrolysis is typically set when the current has decayed to below 5% of the initial value, that is, at 95% conversion with 5% of the starting material remaining. For a higher conversion, the required time for electrolysis increases substantially (exponentially).

A further option is electrolysis with constant cell *voltage* (that is, the voltage between the anode and the cathode) based on a two-electrode electrolysis cell (WE versus CE) with a power source delivering a constant voltage. When optimizing the experiment with a potentiostat, this situation can be simulated by operating the WE versus a CE and connecting this simultaneously to the connectors for Aux (CE) and Ref at the potentiostat. Apart from simplicity in terms of instrumentation, a constant cell voltage electrolysis does not offer any benefits, because the overall cell voltage drops over many resistive components in the cell, and the important voltage drop at the interface between WE and solution is difficult to control (see discussion for Figure 17.8).

For large electrolysis cells in the laboratory or technical electrolyzers, galvanostatic reaction control is preferred due to high costs for high-power potentiostats.

A systematic approach to the design of electrosynthetic processes can be based on (i) using cyclic voltammetry (Chapter 6) to elucidate the formal potentials, the effects from the supporting electrolyte, and the current densities in a reactor. Then, (ii) the electrolysis is performed potentiostatically on a small scale, followed by structural analysis of the products. Only then, (iii) it is appropriate to work at a higher scale with larger electrodes/reactors (galvanostatic) using the optimal current density determined with the smaller electrolysis cell.

In order to achieve reproducibility in electrolysis experiments, it is essential to consider and report (i) the concentration of starting materials, (ii) the nature and concentration of the supporting electrolyte, (iii) any additives (acids, bases), (iv) the type and area of the electrodes, (v) the applied potential or current, and (vi) the electrolysis time [29]. Ideally, all parameters are recorded, for example, during galvanostatic control, the cell voltage and the working electrode potential versus reference, similar to battery testing (Figure 15.17). For processes under potentiostatic control, the current must be recorded versus time. This allows conversion to charge versus time plots [galvanostatic $Q = I \cdot t$, potentiostatic Eq. (1.14)], which allow faradaic yields (or the faradaic efficiency ε_F) and n (the stoichiometric number of electrons) to be determined. These parameters are essential for reaction optimization (Section 6.6.1).

Q: I have seen descriptions like "1.1 F/mol" in literature reports. What does it mean?

A: The parameter is based on the proportionality of the amount of converted reagent and the transferred charge. For example, for the conversion of 1 mol reagent, a charge of 1.1 · 96 485 A s = 1.1 · 96 485 C is required. This is a way to express the *effective* stoichiometric number of electrons, that is, including parasitic reactions. Although in common use, formally, there is a problem with using "F/mol". The Faraday constant F (set in italics) has the units "A s mol^{-1}" or "C mol^{-1}". The conversion F/mol then leads to the unit "A s mol^{-2}", which is incorrect. This also does not become right when considering the Faraday constant as a unit (set in upright font). In doing so, there is an even more problematic potential confusion with the unit Farad for capacitance, with the symbol F. The unit Farad is defined as the charge transferred when applying a voltage of 1 V to a capacitor: 1 Farad = 1 F = 1 A s V^{-1}. This is very different from 1 F = 96 485 A s mol^{-1}!

This is the reason for discontinuing the historical notation "F/mol" in recent literature [8]. This approach is also recommended by the textbook authors. We have seen similar problems in the correct handling of units in connection with the "C-rate" employed in battery technology (Section 15.2.2).

17.1.3 How to Design a Laboratory Electrosynthesis Cell

Undivided cells. A minimal cell design is shown in Figure 17.3: both the anode and the cathode are immersed in the same electrolyte. A simple glass vial with a polymer lid [resistant to solvent, e.g., from polyether ether ketone (PEEK) or fluorinated polymer materials] is adapted with ports for anode, cathode, reference electrode, and reagent addition. Shown are metal plate electrodes with geometric area $A \approx 2\,w \cdot h$ (w = width, h = height in solution, thickness ignored). The reference electrode should be positioned between the anode and the cathode. A magnetic stirring bar is employed for effective agitation that enhances interfacial mass transport.

Changes in the cell design can lead to considerable improvements in performance depending on reaction conditions. Some reactions require strongly alkaline environments where glass is not appropriate. The optimization of reactor materials has been discussed in the literature [30]. Often reactions require an inert atmosphere. The flow of inert gas in and out as well as gas flow due to gas-evolving processes associated with the electrochemical/chemical reactions occurring in the cell need to be considered. The cell lid design/mechanism and potential dangers from pressure build-up need to be taken into account.

The geometry chosen for the anode and the cathode in Figure 17.3 is poor because it results in a nonuniform current density distribution (Figure 17.4a) across the electrodes that could even change the reaction mechanism. Integration of the electrodes into the cell walls of the electrolysis reactor can avoid extreme changes in current density that can occur between the front and back sides for immersed plate electrodes

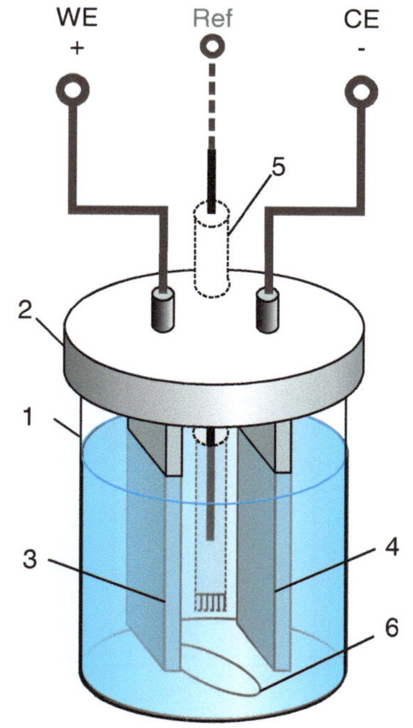

Figure 17.3 Construction of an undivided electrolysis cell for oxidation. 1) Cell container; 2) cell lid; 3) anode (WE); 4) cathode (CE); 5) optional reference electrode for potentiostatic electrolysis; and 6) magnetic stirring bar.

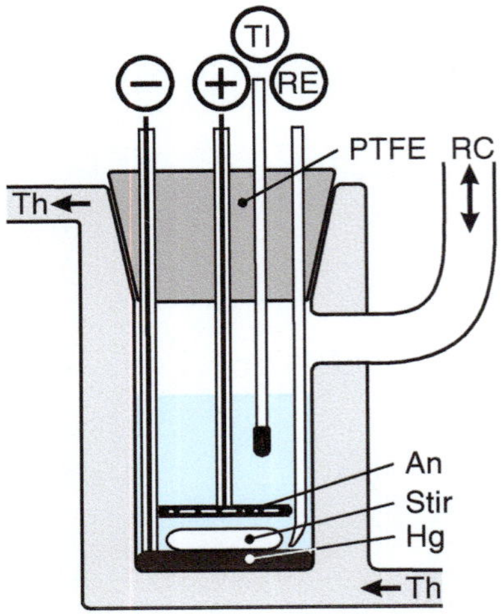

Figure 17.5 Undivided cell with thermostat. An – Pt-gauze anode; Hg – Hg cathode; Stir – magnetic bar stirrer; RE – reference electrode with Haber–Luggin capillary; TI – temperature probe; Th – influx and outflux of fluid from thermostat; (+) and (−) – current connectors; PTFE – stopper made from perfluorinated polymer; RC – reflux condenser. Source: Reproduced with permission from [31]. Copyright 2004, Wiley-VCH.

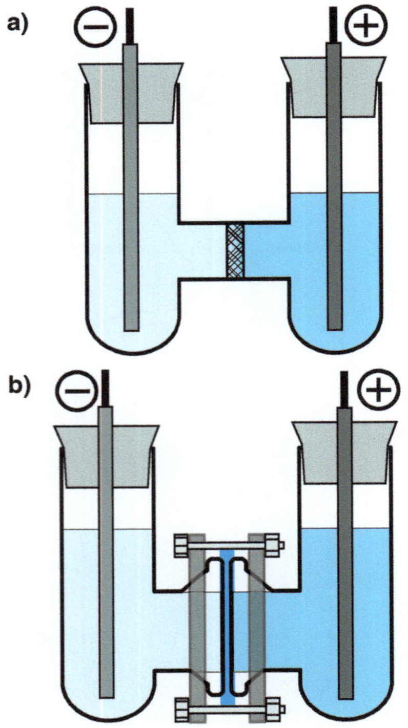

Figure 17.6 H-cells with a) a glass frit and b) an exchangeable polymer membrane as an ion-conductive separator to avoid convective/diffusional mixing of the anolyte and the catholyte.

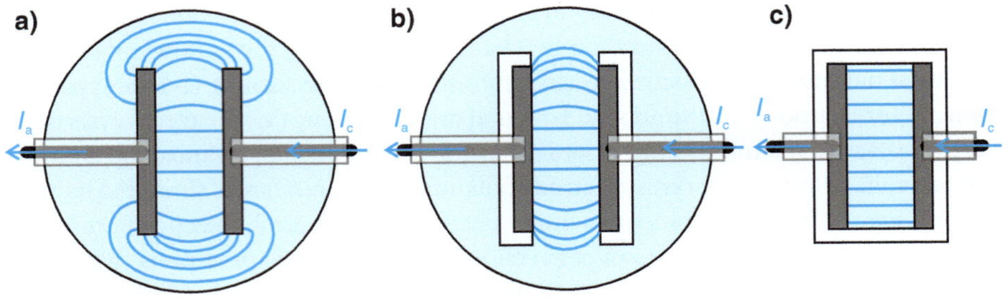

Figure 17.4 Field lines in electrochemical cells. The density of field lines is proportional to the local current density. a) Top view of the cell in Figure 17.3; b) design based on front-side electrolysis with edge effects; c) design with improved uniformity in current density. Source: Adapted from [30].

(Figure 17.4b). There remains a current density distribution due to "edge effects" (higher current densities at edges), which can be avoided by the cell design shown in Figure 17.4c. However, the design in Figure 17.4c could lead to disappointing results because, without additional measures, it is detrimental to convective mass transport, which is a crucial parameter for electrosynthetic reactions. The location of the reference electrode has to be optimized considering the potential distribution in the electrolysis cell (see Figure 5.18). The reference electrode should be close to the working electrode without affecting the potential and current density distribution. Practically, Haber–Luggin capillaries[56,57] are employed to optimize the reference electrode position (Figure 5.24). More practical tips are provided in the literature [30, 31].

Many reactions require temperature control to allow the formation of reactive intermediates or to remove reaction heat (or Joule heating effects[51]) by a reflux condenser. An optimized cell design for a temperature-controlled electrosynthesis cell is shown in Figure 17.5. It contains a mercury pool cathode.

Divided cells. In electrosyntheses, products generated at the WE often react further at the CE. CE products (acids, bases, O_2, and metal ions) could interfere with the reaction at the WE or contaminate the product. In particular, reversible or quasireversible reactions (Chapter 6) would lead to charge shuttling without net product formation. In these cases, the electrolyte at the anode (anolyte) and at the cathode (catholyte) needs to be divided into half-cells separated by an ion-conducting membrane or a porous separator, which is a very common cell design in practical electrochemical reactors. This setup prevents the mixing of the anolyte and catholyte but permits an ionic current between both reservoirs. This concept was introduced in Chapter 1 for the Daniell cell (Figure 1.8) using a salt bridge. For electrosynthetic cells, the electrical resistance of the ion-conducting connections should be minimized to facilitate high current (= high conversion rate). The H-cell design (Figure 17.6a) is based on two beaker-shaped compartments connected via a tube that may include a membrane flange (reminiscent of the character "H").

The barrier between the anolyte and the catholyte can eliminate convective mixing and is called a separator. Separators can be made from porous materials (diaphragms), which allow transport of ions and small neutral molecules, or they can be made from ion-exchange membranes, which restrict transport to specific ions (e.g., protons). Improved performance can be achieved with two membranes in series at the cost of an increased cell voltage (energy loss). The solution between these two membranes can be continuously replaced for further performance improvement. This can be necessary for anolyte and catholyte systems with very different compositions (e.g., different pH values). In such systems, the driving force for diffusive transport across the separator is large, and stable conditions are maintained by continuously replenishing the solution between the two separators.

The separator needs to be chemically robust toward the anolyte and the catholyte and to form a good seal with the reactor flange. Pores in the separator need to be wetted/flooded by the electrolyte (to form the ion-conducting connection between the anolyte and the catholyte) without trapping air or gases, which would increase resistance.

In the past, diaphragms were often based on filamentous asbestos (and similar silicate fibers), but health concerns about asbestos have stopped this practice. For laboratory-based electrosyntheses, glass frits are universally approved. Glass frits can be obtained in a range of porosities, and they can be permanently sealed into glass reactors (Figure 17.6a). Similar porous diaphragms can be based on unglazed porcelain or calcined clay. They can be universally applied, except for alkaline solutions, in which these ceramic materials are not stable. In addition, adsorption of OH^- ions to the pore walls or the deprotonation of surface OH-groups in alkaline solutions forms surface charges on the pore walls, causing electroosmotic flow (Section 4.6.3 and Figure 4.64). This flow can drag solvent molecules from one half-cell to the other half-cell, which can detrimentally affect electrolytic processes.

The cell resistance increases significantly with increasing thickness of separators composed of glass frits. Alternatively, porous films based on polymers such as materials employed in batteries (Section 15.6), for example, polyethylene (PE) or polypropylene (PP) can be advantageous. Commercial materials such as hydrophobic GoreTex® provide porous films that are wettable with an aqueous electrolyte solution only after impregnation with an organic solvent, followed by transfer to more hydrophilic solvents. These flexible polymer membranes can be mounted into the glass flange (Figure 17.6b) and then easily removed and renewed.

Separators based on ion-exchange membranes can be semipermeable to allow the transport of either cations or anions only. A good example is Nafion® (Figure 2.30), which is employed in fuel cell systems (Section 16.2.1). As is the case of undivided cells, inlets for inert gas, a reference electrode, or sampling/monitoring are useful in divided electrosynthesis cells, too.

For exploratory investigations, the concept of the quasi-divided cell is beneficial [17]. This cell is based on a simple design and operated in galvanostatic mode with a WE much larger in area compared to the CE. With an equal current through both electrodes, the current density through the smaller CE is substantially higher. The current at the CE cannot be achieved by diffusion-controlled transport of the reagent, and thus, solvent or electrolyte decomposition will also occur at the CE. The back reaction of the products from the WE does not play a role in this situation. Laboratory cell electrolysis based on the quasi-divided configuration can be employed for volumes up to several 100 mL.

Flow Cells. Scale-up from small batch devices is best achieved with flow reactors with well-defined transport (convection, migration, and diffusion). Flow cells can be operated as divided, quasi-divided, or undivided cells. Often, the benefits of the parallel plate configuration for the anode and the cathode are exploited with a uniform current density distribution (Section 9.3.2), as shown with the electric field lines in Figure 17.4c. Convection parallel to the electrode surface and perpendicular to the current path improves transport in these cells. However, it should be noted that the flow over the electrode surface introduces a nonuniform concentration distribution in the cell along the flow direction. Therefore, reactions may be different at the leading and the trailing edges. A typical design is illustrated in Figure 17.7. Electrodes can also be pressed directly onto thin spacer layers (gasket) to give well-defined and adjustable interelectrode gaps.

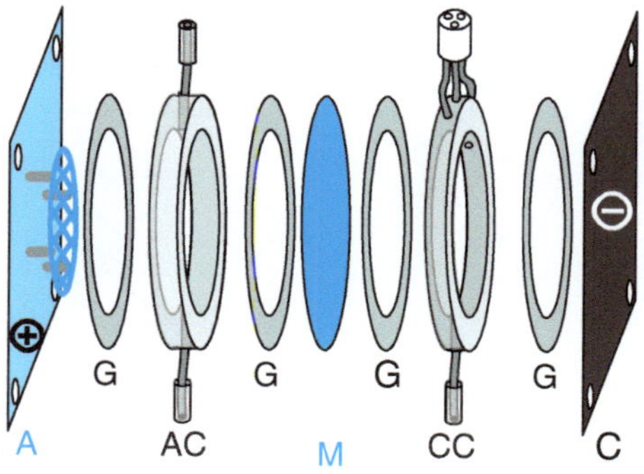

Figure 17.7 Flow cell as a parallel plate reactor. A – Anode, with metal mesh; G – seal rings; AC – frame for the anode compartment with inlets and outlets; M – membrane (diaphragm); CC – frame for the cathode compartment with gas outlet (H_2); C – cathode. Connections for current are at the four corners of the anode and the cathode. The stack is completed with end plates as shown in Figure 2.31a. Source: Reproduced with permission from [31], Copyright 2004, Wiley-VCH.

A practical advantage of flow cells is the choice of operational modes. Exploratory tests can be conducted after filling the electrolyte solution with the reagent without convection to establish basic operational parameters such as current or potential settings. With slow-flow conditions, a complete electrolysis can be achieved for a reservoir with a larger volume of the electrolyte solution with the reagent. Batch operation is possible by pumping the solution several times through the cell (recirculation) with faster flow rates similar to the operation of a redox-flow battery (compare Figure 15.42). A review summarizes flow cell laboratory reactors [32]. Flow reactor electrosynthesis is undergoing rapid development in conjunction with microreactor technology, in which preparative steps such as mixing and heterogeneous catalysis steps are conducted in continuous operation using a sequence of miniaturized flow reactors. Upscaling of flow synthesis procedures is replaced by up-numbering to maintain perfect reaction control (in terms of all important parameters such as heat flow, mass flow, and electricity flow) [33].

17.1.4 Reactors for Industrial Electrosynthesis

Continuous operation is desirable for industrial-scale processes. With this in mind, we can discuss cell designs based on the simple flow cell in Figure 17.7. For large-scale processes, it is important to increase the energy efficiency (lowering electricity costs) and the space–time yield. An important measure is the decrease in the voltage drop between the anode and the cathode that depends on the resistance of the electrolyte and can be minimized by reducing the interelectrode gap. Figure 17.8 illustrates the voltage drop along the current path between the parallel plate electrodes.

In order to achieve a more uniform convective transport, a channel pattern can be embedded into the electrodes, which are positioned with a spacer. Multiple cells are combined into **stacks**. In this configuration, the anode and the cathode are operated

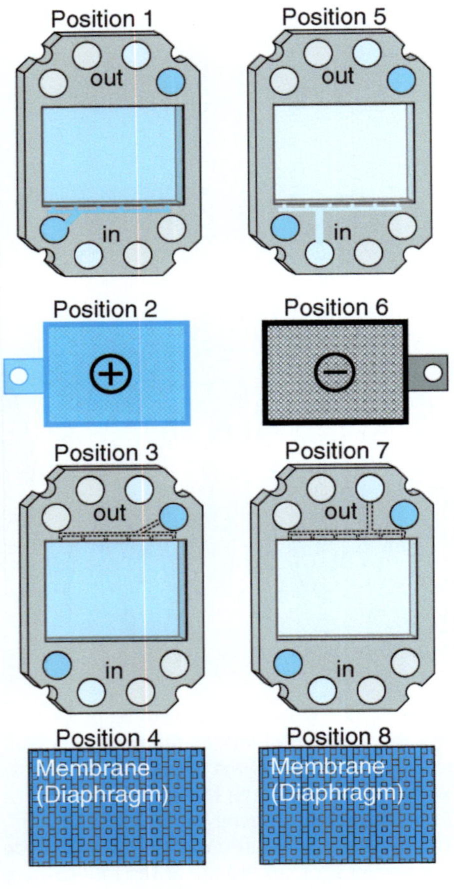

Figure 17.9 Components for a stack of divided electrolysis cells with different anolyte and catholyte. Positions in the figure refer to the sequence in the stack in Figure 17.10.

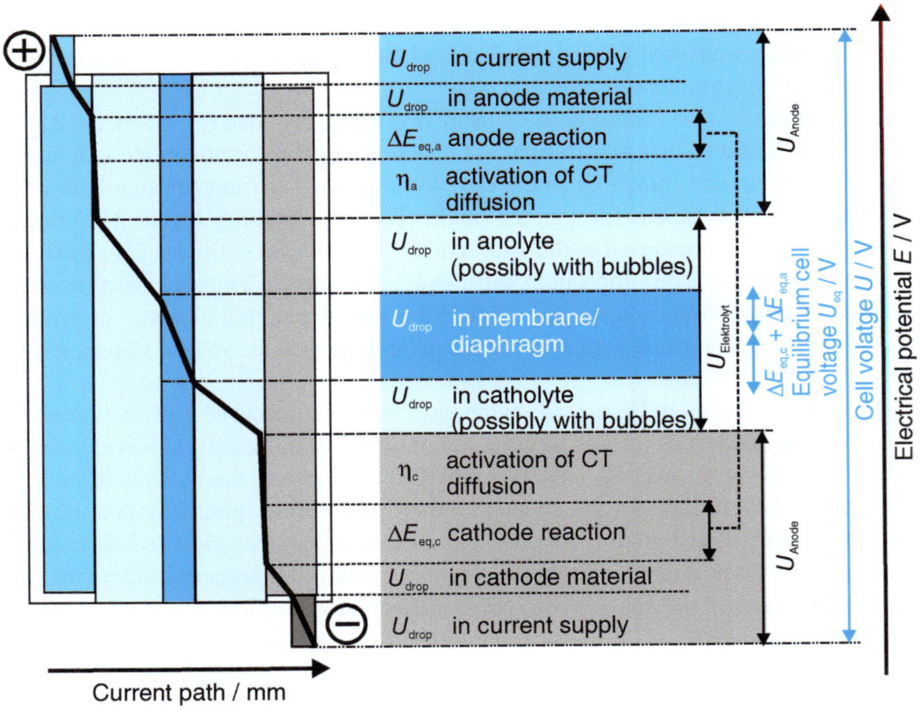

Figure 17.8 Schematic illustrating the voltage drop along the current path in a parallel plate electrolyzer. CT – charge transfer. Source: Reproduced with permission from [31]. Copyright 2004, Wiley-VCH. The schematic is not to scale!

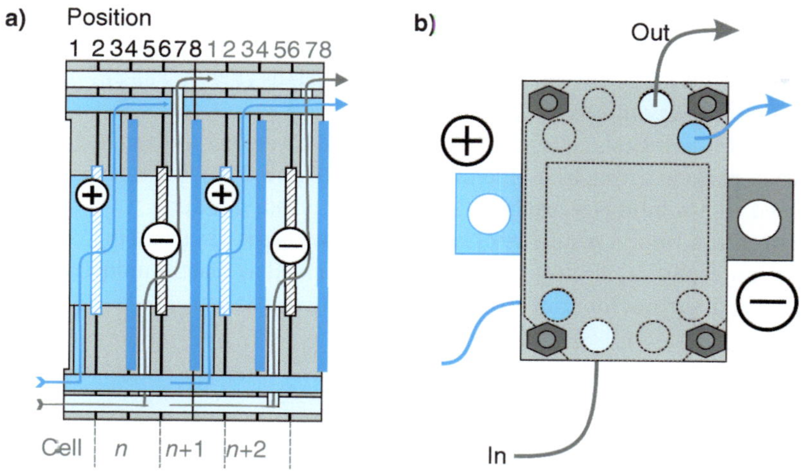

Figure 17.10 Electrolysis cell stack. a) Sequence of elements in Figure 17.9 for a cut into the stack (cross section along the stack direction); b) connection scheme (view along the stack direction) for mass flux toward anodes (blue) and cathodes (gray) and for electricity flux (+ and −). The frame allows further inlets and outlets (for gas diffusion electrodes or thermostating in more complex cells). Components to seal the cells are not shown (drawing not to scale).

as bipolar electrodes – electrically in a series circuit similar to that in a fuel cell – with wire connections only at the end plates (Figures 9.45 and 16.15). Alternatively, the stack can be based on alternating anodes and cathodes with front and back performing the same reaction (Figures 15.14 and 16.16a). The mode of operation is monopolar and corresponds electrically to a parallel circuit. This operation mode requires high currents with lower applied voltages.

Kits of stack elements (Figure 17.9) based on parallel plates can be assembled flexibly in a sequence as required in a particular process (Figure 17.10). Cell frames allow thermostatic control to maintain the reaction temperature. Information about stack design for electrolysis and its operation can be found in YouTube videos [34].

A different approach to process optimization is realized in **capillary gap cells** (Figure 17.11). This type of cell has been employed in the industrial synthesis of aromatic aldehydes. Electrodes are based on circular graphite disks with a central hole. The disks are stacked with spacers defining 1 mm gaps. The stack is operated in the bipolar mode with the top of each disk as the anode and the bottom as the cathode. The electrolyte solution passes radially from the central hole to the outside. The solution can be circulated to optimize yields. The small gap of typically 1 mm between the anode and the cathode allows electrosynthesis in poorly conducting organic electrolyte solutions [35]. Capillary gap cells can be operated as divided cells with the electrolyte penetrating through carbon cloth electrodes. In this case, carbon cloths for the anode and the cathode are separated with ionomer separators that act as electrolyte (cf. fuel cells). A considerable benefit in this case is the absence of electrolyte salts in the solution phase and, consequently, the avoidance of further process steps for separation and recycling of dissolved electrolyte salts [36].

Swiss Roll Cell. Large anode and cathode areas combined with a small electrolyte volume and uniform current density distribution are achieved with cylindrical assemblies of the anode, the cathode, and the separator (similar to battery design, Figure 15.14b). This type of reactor is called a **Swiss roll cell** (Figure 17.12) based on Swiss technology and with obvious links to Swiss cuisine [37]. A nickel mesh anode (coated with NiOOH) and a steel mesh cathode are separated by a PP net and wound around a

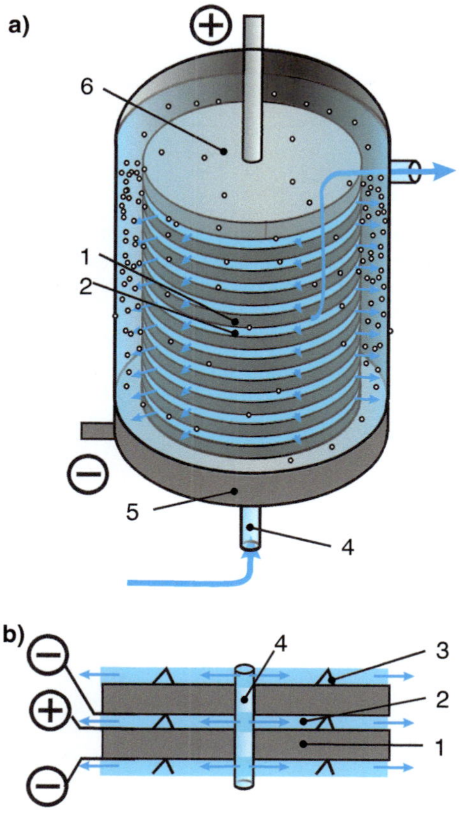

Figure 17.11 a) Schematic of a capillary gap cell; b) cross-sectional view of a stack with bipolar electrodes.

1) Bipolar electrode;
2) electrolyte-filled capillary gap;
3) electrically insulating spacer;
4) inlet for the electrolyte into the capillary gaps;
5) negative end plate;
6) positive end plate with contacts. The formation of gas bubbles is indicated. Drawing not to scale.

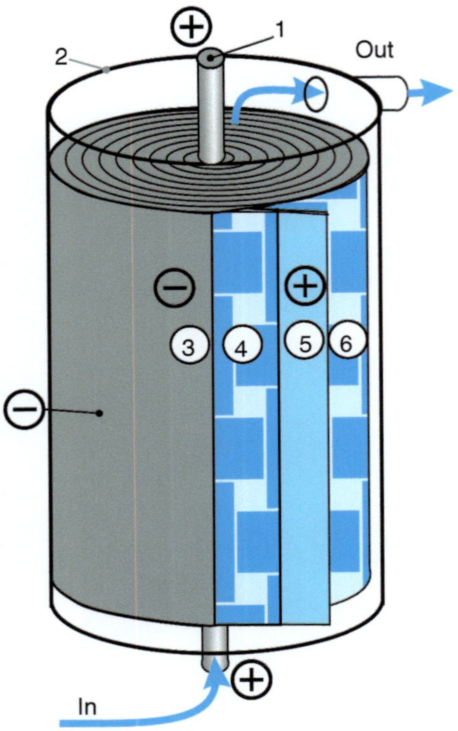

Figure 17.12 Swiss roll cell. 1) Central current connector; 2) steel reactor housing with current connector; 3) steel mesh cathode; 4) polypropylene net as a separator; 5) anode based on mesh of Ni|NiOOH; 6) polypropylene net as separator.

central nickel rod (the current collector for the anode). The steel mesh makes electrical contact with the outer steel housing. Swiss roll reactors with 1 m diameter provide 200 m^2 electrode area with the electrolyte flowing axially (Figure 17.12). An important application is the synthesis of vitamin C [37].

Other electrosynthesis cells have been developed based on fluidized bed technology, with each particle acting as a bipolar electrode (Figure 9.46). Mass transport can be enhanced with rotating electrodes [9] or power ultrasound [38].

Further Reading. A summary of laboratory flow cells is discussed in [39]. An introduction into technical electrosynthesis reactors, as well as reactor modeling and optimization with further literature, is provided in [40].

17.1.5 Electrode Materials

The selection of electrode material, the solvent, and the electrolyte (in combination!) is crucial for developing a successful electrosynthetic reaction. Here, the selection criteria are discussed to rationalize the details given in literature examples.

For electrode materials such as Pt, Ni, or glassy carbon, there is a choice of porous or flat electrodes. Porous electrodes provide a high true electrode area for a given geometric electrode area (Figure 4.33). A high true electrode area in porous structures can help minimize current density (per true surface area) to limit the required overpotential for kinetically controlled processes. This requires sufficient mass transport into (and out of) the porous structure. Flow-through porous electrodes offer advantages in this respect due to convective transport.

An important class of electrode processes is based on gas evolution, either producing gases as the main product (H_2, Cl_2, and CO_2) or as a product at the counter electrode (H_2, and O_2). Often gases form as undesirable side products. Gas bubble formation on the electrode surface (blocking) can be detrimental, but gas bubbles forming in solution can help enhance mass transport and mixing close to the electrode surface. Gaseous products are easy to separate. Some gases can form explosive mixtures that must be prevented by technical means. Large amounts of gas increase the electrolyte resistivity and lead to energy losses. Gas evolution processes are often electrocatalytic reactions and, therefore, extremely sensitive to the choice of electrode material. A high exchange current density is desirable, for example, for the hydrogen evolution reaction (HER, Figure 3.9). A high exchange current density is equivalent to a low reaction overpotential (see Butler–Volmer equation) at a given current density. A high exchange current density is associated with high (electro)catalytic activity (Chapter 12) due to stabilization of adsorbed reaction intermediates.

In industrial electrosynthesis, gaseous starting materials are beneficial for electrode processes at counter electrodes (Excursion 17-1). Effective gas transformations require gas-diffusion electrodes (GDE) similar to those in fuel cells (Chapter 16).

Electrode materials have to be inert, that is, during storage and operation in an electrolyte solution, there should be no corrosion. Exceptions are sacrificial anodes from Zn, Al, or Mg, which dissolve cleanly to give inert side products when employed as the counter electrode for electroreduction reactions at the WE. The metal dissolution occurs at a negative potential that does not cause reoxidation of the products formed at the working electrode. Thus, the use of sacrificial electrodes allows undivided cells to be employed. Such cells require the concomitant formation of anions or the consumption of cations, typically by consumption of H^+ in the target reaction. In some cases, a sacrificial anode is employed to give useful products at the WE, for example, when producing metal–organic compounds from the electrode metal [41].

Platinum and noble metals/alloys provide noncorroding and electrocatalytically active electrode materials that are commonly employed in spite of considerable costs.

Platinum cathodes are highly effective for HER in acidic media. At the same time, the use of platinum as a cathode to form another reduction product suffers from the competing HER process. Noble metals are employed as noncorroding anodes with high activity, although in aqueous media, adsorption of reaction intermediates and passive oxide layer formation have to be considered (see methanol oxidation, Figure 12.34 and Figure 12.37). Noble metal alloys provide a higher tolerance toward catalyst poisoning (e.g., a Pt–Ru catalyst provides improved CO tolerance in fuel cells when compared to Pt). For reproducibility of electrocatalytic processes, the use and composition of alloys need to be carefully documented (down to impurity levels in noble metals). A methodology for enhancing catalytic activity by enhancing the true electrode area is based on the deposition of porous metal layers. Platinum surfaces can be coated with platinum in dendritic structures (platinum black) to give highly porous surfaces.

Nickel is in many ways similar to platinum, but much less costly compared to the electrocatalytic noble metals. Nickel foams are employed in HER in alkaline solutions. When employed as an anode in alkaline electrolyte, a layer of NiOOH is formed with good catalytic reactivity and selectivity. Nickel is often used as a gas-evolving CE (for H_2 or O_2). Nickel can be employed in the form of Raney Nickel[154] to give highly porous and reactive electrodes.

Iron offers a low overpotential for HER and is therefore an economical option for H_2-evolving CEs. Corrosion of iron can be problematic (in the absence of applied negative bias). The application of stainless steel can prevent corrosion with only a minor increase in HER overpotential.

Lead is relatively affordable and offers a low HER exchange current density (Figure 3.9). Lead electrodes can be superior as the WE in electrochemical reductions. At a positive applied potential, a layer of insoluble PbO_2 is formed. Pb(IV) is known as a strong oxidant, and therefore, lead anodes are employed in a range of technical applications (e.g., ozone formation). Films of PbO_2 and Pb itself are mechanically unstable/brittle. Therefore, lead alloys are used similar to their use in battery technology. Lead oxide can be used in the form of films on mechanically stable substrates, for example, in commercially developed $PbO_2|TiO_2|Ti$ anodes.

Mercury has been employed for many electrosynthetic reductions for decades due to its extremely low exchange current density for the HER (mercury does not form bonds to hydrogen) and the ease of renewing the electrode surface of the liquid metal. The toxicity of mercury and the formation of metal–organic mercury products have all but eliminated mercury from the laboratory and industry (even traces of Hg have to be eliminated). With special measures and a lot of care, today mercury electrodes are still in use in electroanalysis (detection of trace metals in seawater) and as amalgam electrodes on zinc or copper.

Carbon is supplied in many forms and employed in many electrosynthetic processes. Plate and cylinder electrodes based on pressed graphite powder offer low cost and low electrocatalytic activity (i.e., avoiding HER). The specific properties of carbon material as electrodes can vary significantly. These electrodes are commonly employed in electrosynthesis. Porosity of carbon electrodes has to be considered, and in some cases, polymer impregnation is applied to provide gas/liquid-proof materials. Variants of carbon electrodes are formed from carbon cloths, carbon mats, or carbon composites with electrical conductivity.

A high-quality electrode material is glassy carbon (or reticulated vitreous carbon) (Figure 5.19). Glassy carbon is more expensive and more difficult to machine but offers high corrosion resistance, mechanical strength, high electrical conductivity, and high reproducibility (many laboratories use Tokai glassy carbon to allow interlaboratory comparison of results). Glassy carbon is polished to a mirror finish (very flat) and has very low electrocatalytic activity. Reticulated vitreous carbon can be obtained as porous three-dimensional electrode materials for flow-through operation.

A special type of carbon electrode is boron-doped diamond (BDD) [42]. Diamond, an exceptionally mechanically strong and electrically insulating material, becomes highly conductive by doping mainly with boron. Layers of BDD are grown on substrates (Ti, Si) to give chemically and electrochemically robust electrodes with a wide potential window in aqueous media (due to low exchange current densities for H_2 and O_2 evolution). These electrodes can be operated at extreme potentials (e.g., for anodic generation of radicals such as $OH^\bullet$) without major corrosion or O_2-evolution contributions. These reactive radicals allow degradation of pollutants or waste, even of polymers (electrochemical incineration). Commercial applications have emerged in water treatment, as well as in some preparative electrosynthetic processes [42]. For educational purposes, BDD electrodes can be employed in laboratory experiments [43].

For industrial applications, inert substrates are often coated with highly catalytic (and often expensive) materials. There are many examples that cannot be discussed in detail here. The use of titanium as a substrate coated with Ir/Ru oxides leads to technically important dimensionally stable anodes (DSA®). The usually passive TiO_2 surface layer is modified with electrically conducting and catalytically active Ti–Ru oxides with high efficiency and stability, for example, in industrial chlorine evolution [44]. The term DSA implies low corrosion rates and extended operational life time, although some loss of noble metal still occurs. The addition of Ir oxides to DSAs provides highly active electrodes for oxygen evolution.

Ebonex® is a hard ceramic bulk material made from substoichiometric titanium oxide Ti_nO_{2n-1} ($3 < n < 10$). This material offers electrical conductivity and low electrocatalytic activity. Ebonex can be employed as a substrate for DSAs [45].

A systematic review is available on the selection of suitable electrode materials for many electroorganic reactions [46].

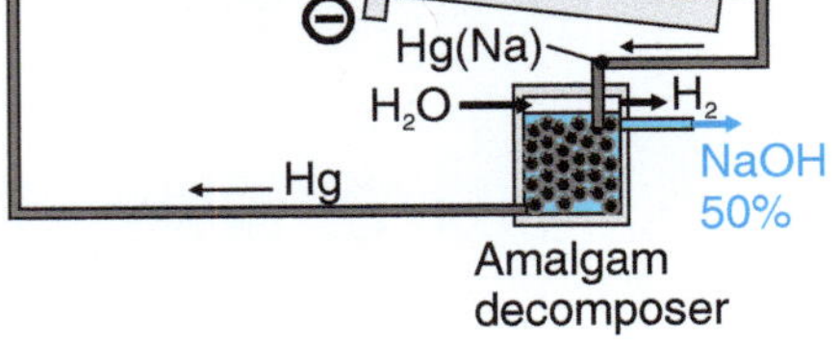

Figure 17.13 Oversimplified illustration of the amalgam process.

Excursion 17-1 Chlor–Alkali Electrolysis

The industrial production of chlorine and sodium hydroxide was developed in the 1880s initially for providing NaOH for soap and glass manufacturing. With the development of industrial organic chemistry, chlorine became the more valuable product of the electrolysis process. The production of polyvinylchloride (PVC) offered a mass market, but many other intermediates also require chlorine that are subsequently converted to chlorine-free final products. From 2005 to 2019, chlorine production in the EU alone was 9 000 000–11 800 000 t per year [47]. The chlor–alkali electrolysis (also called chlor–alkali process, brine electrolysis, and chlorine and caustic technology) is by volume the most significant electrochemical process and is linked to many other industrial processes. The extensive study and optimization have had a major impact on the energetic and environmental footprint of many other value generation chains and supply chains.

The central reaction is expressed in Eq. (17.1).

$$2\,Na^+ + 2\,Cl^- + 2\,H_2O \rightarrow Cl_2 + 2\,NaOH + H_2 \quad (17.1)$$

The anodic half-reaction is the chlorine evolution reaction (CER):

$$2\,Cl^- \rightarrow Cl_2 + 2e^-. \quad (17.2)$$

In competition with chlorine evolution, the oxygen evolution reaction (OER) needs to be considered:

$$2\,H_2O \rightarrow O_2 + 4\,H^+ + 4e^-. \quad (17.3)$$

The current yield of the undesired OER can be minimized by low electrocatalytic activity and thus low exchange current density for this reaction.

The cathodic half-cell reaction varies depending on the type of process. A classic reaction was the reduction of Na^+ to Na amalgam (on mercury electrodes):

$$Na^+ + e^- + xHg \rightarrow Na{\cdot}Hg_x. \tag{17.4}$$

Thermodynamically, we would expect HER

$$2H_2O \rightarrow H_2 + 2OH^- \tag{17.5}$$

or the oxygen reduction reaction (ORR)

$$O_2 + 4e^- + 2H_2O \rightarrow 4OH^- \tag{17.6}$$

to proceed. If cathodically formed hydroxide reaches the anode, Cl_2 can disproportionate to hypochlorite, which further react to form chlorate:

$$Cl_2 + 2OH^- \rightarrow ClO^- + Cl^- + H_2O \tag{17.7}$$

$$2\,HOCl + OCl^- \rightarrow ClO_3^- + 2\,Cl^- + 2\,H^+ \tag{17.8}$$

Raw materials required for the chlor–alkali electrolysis are rock salt or sea salt, from which trace elements and impurities have to be removed.

In the **amalgam process**, electrolysis is performed in an undivided cell in a monopolar configuration. A liquid mercury cathode is present as a liquid pool flowing slowly at the bottom of the slightly inclined cells (Figure 17.13). HER can be avoided on Hg due to the low exchange current density of the HER at Hg electrodes (Figure 3.9), making possible the reduction of Na^+ at a potential of −2.71 V. The electrolyte is depleted from 310 to 270 g L^{-1} NaCl without strongly affecting the thermodynamic conditions along the electrolysis cell. The concentration of NaCl electrolyte leaving the cell is replenished before feeding it into the electrolysis cell again.

The presence of impurities and trace metals can lead to the formation of other types of amalgams with a higher HER exchange current density. The resulting H_2–Cl_2 gas mixtures lower current efficiency and, more importantly, lead to an explosion hazard (H_2–Cl_2 gas mixtures can detonate).

The Na-amalgam is reacted with water to recover Hg, H_2, and NaOH. Due to the extremely low exchange current density for this reaction, the amalgam is brought in contact with graphite balls as decomposers that contain catalytically active Fe or Ni oxides (or Mo/W carbides):

$$\text{Amalgam}|H_2O: \quad Na{\cdot}Hg_x \rightarrow Hg + Na^+ + e^- \tag{17.9}$$

$$\text{Graphite with catalyst: } H_2O + 2e^- \rightarrow H_2 + 2OH^- \tag{17.10}$$

The recovered mercury is recycled, and NaOH is further processed. Alternatively, Na amalgam can be used as a reducing reagent for the production of other products such as sodium alcoholates [Eq. (17.11)] needed in large quantities for organic alkoxylation reactions.

$$\text{Graphite with catalyst: } ROH + 2e^- + 2\,Na^+ \rightarrow H_2 + 2\,RONa \tag{17.11}$$

The anode for chlorine production is located close to the Hg-pool cathode. Initially, graphite electrodes were employed, but they corrode and, therefore, have to be continuously supplied. From 1972 onward, they have been replaced by DSAs (Ti | $TiO_2{\cdot}RuO_2$). Although they are more expensive, they offer lower total costs due to a lifetime of up to seven years without any change in shape. These electrodes

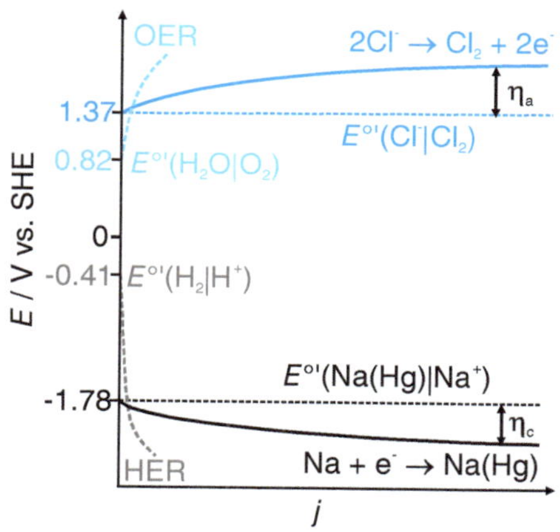

Figure 17.14 Schematic of potential–current density plots for reactions during the amalgam process.

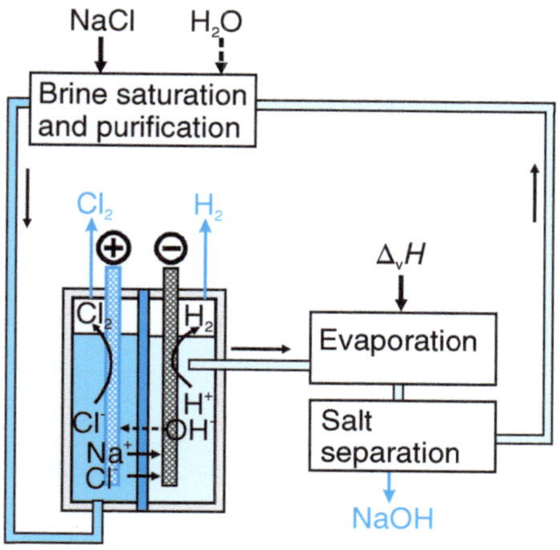

Figure 17.15 Schematic cross-sectional view of a single cell for the diaphragm process.

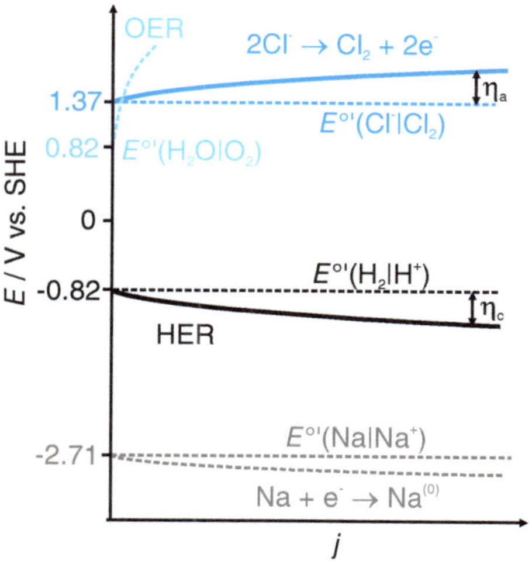

Figure 17.16 Potential–current density plots for the diaphragm process and the membrane process.

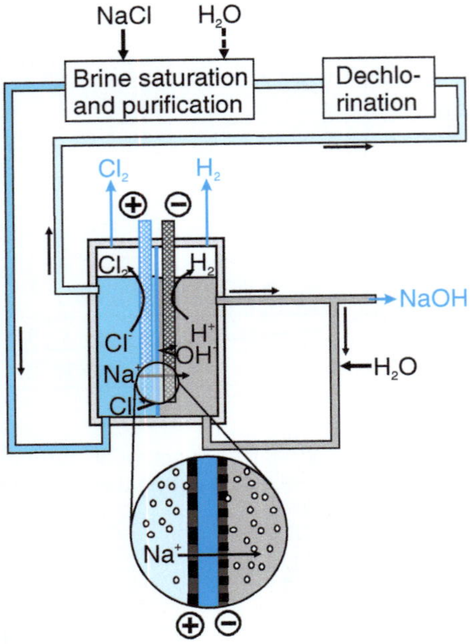

Figure 17.17 Schematic cross section of a single cell for a membrane process. The inset shows the two mesh electrodes pressed on the membrane (zero-gap configuration).

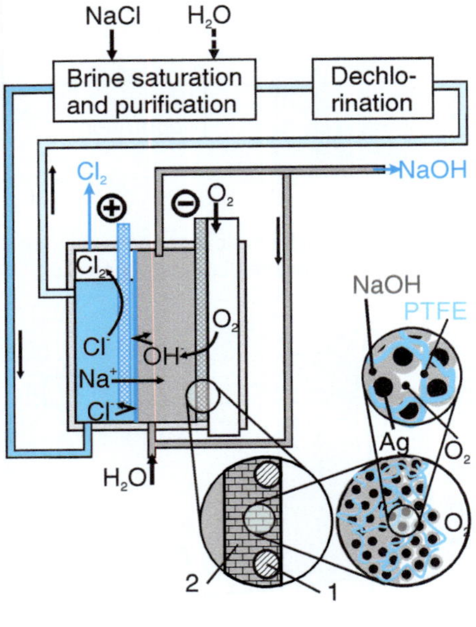

Figure 17.18 Schematic cross section of the single cell for the membrane process with the oxygen-depolarized cathode. The insets show enlarged views of the cathode with a stainless-steel current collector (1) and the gas-diffusion electrode (2) based on a composite of Ag particles and an extremely hydrophobic PTFE binder. The structure of the GDE supports the *partial* permeation of pore space by NaOH solution, maximizing the extension of the three-phase boundary region NaOH|Ag|O_2.

also benefit from the low exchange current density for OER. Electrode potentials for the most important partial reactions in the amalgam process are summarized in Figure 17.14.

Although electrolysis plants are sealed, a small amount of the toxic Hg is released from the reactor with product streams (e.g., approx. 1 g Hg per t Cl_2). This makes the amalgam process problematic and unsustainable.

The **diaphragm process** is based on divided cells (upright) in unipolar circuits (Figure 17.15). Traditionally, asbestos diaphragms were employed with a thickness of 5 mm, leading to health hazards during the processing of the diaphragms, during installation of the reactor, and during maintenance. Today, composite materials made from PTFE and zirconium oxide are employed. The anode is a DSA, and the cathode is based on stainless steel, Ni-coated steel, or iron. Anodes and cathodes are made from metal meshes or stretched perforated metal. The diaphragm is permeable to all ions. The concentrated electrolyte is fed to the anolyte and moves through the diaphragm driven by gravity. At the cathode, hydrogen gas is produced and collected from the reactor without mixing with Cl_2. The potential–current density plot (Figure 17.16) shows that hydrogen evolution dominates over sodium production. The catholyte contains NaOH and NaCl, and separation is achieved by evaporation of water and NaCl precipitation (energy-intensive). The resulting 50% NaOH solution still contains 1% NaCl.

Problems arise due to the diffusion of OH^- into the anolyte against the electrolyte flow. This leads to unwanted processes in Eqs. (17.7) and (17.8). The lower energy efficiency without compensating advantages causes a low economic competitiveness that makes the diaphragm process unviable.

In the **membrane process**, the diaphragm is replaced by a cation exchange membrane (Nafion). The cation-conducting membrane allows only Na^+-ion transport. A DSA is employed as the anode; the cathode consists of stainless steel or Ni-coated steel. Both anode and cathode are made from meshes and are pressed onto the membrane (Figure 17.17). This so-called zero-gap technology minimizes electrolyte resistance. Gases formed at the three-phase junction polymer electrolyte–electrode–liquid electrolyte can escape through the open anode and cathode structures. Liquid transport through the membrane is slow, and therefore, both the anolyte and the catholyte are supplied separately. The anolyte is enriched with purified brine, and gaseous Cl_2 is extracted. Purity requirements for brine are high. Trace earth alkaline metals cause deposition of hydroxides on the membrane, thus blocking membrane pores. The catholyte based on NaOH is recycled. A product is 33% concentrated NaOH with only 0.007% NaCl, from which a fraction is removed as product and replaced with pure water.

In the **membrane process with oxygen-depolarized cathode** (ODC), the ORR at a gas-diffusion electrode (GDE) is used as the cathodic reaction, replacing the HER (Figure 17.18). The principle is roughly equivalent to that of the cathode in alkaline fuel cells. In order to achieve sufficiently high current densities, a porous electrode from Ag with a Teflon binder is employed with a feed of pure oxygen. Thermodynamically, ORR occurs at more positive potentials than HER (Figure 17.19). Therefore, the cell voltage is considerably lower. The resulting energy savings are in the range of 30%. However, hydrogen as a beneficial side product is not obtained. The oxygen reduction proceeds in the three-phase boundary region in partially gas-filled GDE pores (Figure 17.18). Only there can the reactant from the gas phase simultaneously react with electrons from the electrical conductor (Ag) and ions from the membrane (Nafion).

The upright cell construction leads to a gradient in hydrostatic pressure along the 1 m-high GDE. Without precautions, this would cause operational problems (leaking electrolyte at the bottom and bubbling gas into the liquid electrolyte at the upper parts). Gas pockets (Figure 17.20) are introduced to rebalance the pressure within a smaller section vertically along the GDE. Another technical solution is based on the flow of electrolyte along a vertical textile (falling liquid film) to avoid hydrostatic pressure gradients.

Summary. The comparison of chlor–alkali electrolysis processes demonstrates the potential to reduce the ecological footprint of this energy-demanding technology. Improvements are possible because the energy consumption still remains considerably higher than the thermodynamically defined minimum, that is, the equilibrium cell voltage $U_{rev} = -\Delta_R G/(nF)$. A complete analysis needs to take into account purification and separation processes. On the other hand, system-level benefits (e.g., hydrogen and NaOH feeding into other processes, specific energy consumption) do not result in a sudden change of the production plants. These strong path dependencies for large-scale processes are connected to the huge investments required and the long pay-back times (typically 25 years). Furthermore, a change in the technology of chlor–alkali electrolysis requires changes in other coupled processes, which are linked to further huge investment. The supply issue is especially critical during the transition period from one plant technology to another one. The timescales for innovation are illustrated by considering that ODCs were introduced in the 1950s, even before the availability of cation-exchange membranes. The membrane processes emerged in the 1970s. Today, with most processes using the membrane electrolysis approach, there are still some old-fashioned processes in use [47].

Further Reading. A comprehensive description with many details of chlorine production can be found in [48]. Review articles are in [49–51]. Principles and processes for chlor–alkali electrolysis apply to a wider range of technologies, for example, the electrolysis of dilute HCl from polyurethane production [52].

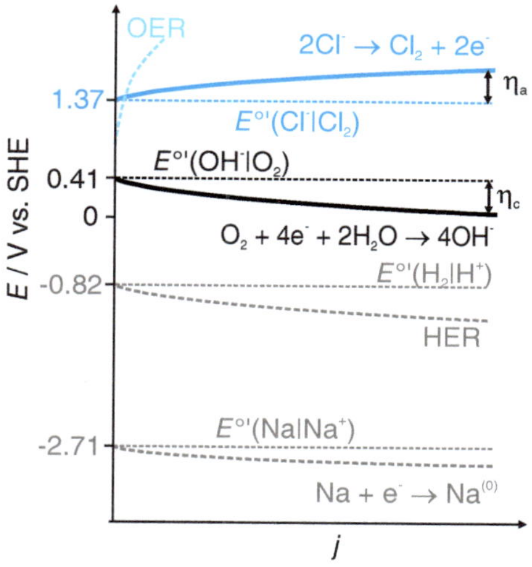

Figure 17.19 Potential–current density plots for reactions in the membrane process with an oxygen-depolarized electrode as the cathode.

17.1.6 Electrolyte Composition

Electrolyte solutions require a solvent that can dissolve the reagent and a salt for ionic conductivity. There are several criteria/parameters for the selection of solvent and electrolyte:

- proton activity;
- potential window for the anode and cathode materials;
- dielectric constant and polarity; solubility of the electrolyte salt; ion-pairing phenomena (Figure 4.15), causing lower ionic conductivity;
- reagent solubility;
- accessible temperature range; vapor pressure; viscosity;
- toxicity;
- cost.

The proton activity of the solvent and electrolyte is important, especially in the catholyte. A high proton activity will make the HER the dominating cathodic process in almost every case. Although unwarranted at the working electrode, HER can be desirable as the counter electrode process. Proton activity is also important for the stabilization of reaction intermediates, especially for radical ions. The solvent can act as a proton donor or a proton acceptor. The reaction mechanism in the reaction layer at the electrode surface can be controlled by changing the proton activity. For example, proton activity controls the formation of the desired product in the

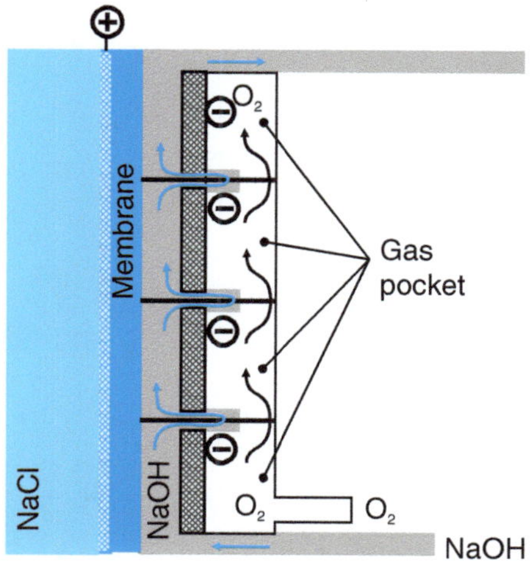

Figure 17.20 Principle of O_2 supply via gas pockets autonomously adjusting the O_2 pressure to the hydrostatic pressure along the vertical extension of the GDE.

adiponitrile synthesis (Section 17.3.3). Bases are often added if the deprotonation of an anodically formed intermediate is rate-limiting.

Protic solvents provide a high ionic conductivity. Commonly employed acidic solutions are sulfuric acid, fluorosulfonic acid, trifluoroacetic acid, and acetic acid. For electrochemical fluorinations, hydrogen fluoride HF is employed. HF is extremely toxic and can lead to death. Solutions in water and methanol should be used only under special safety precautions and with special equipment because HF can attack glass equipment.

Examples of neutral protic solvents are water and methanol. Hexafluoroisopropanol (Figure 17.21) has a very low nucleophilicity and is particularly suited for the study of radical cations.

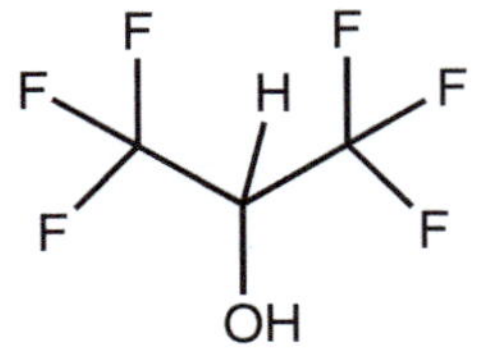

Figure 17.21 Structure of 1,1,1,3,3,3-hexafluoro-2-propanol.

Examples for basic protic solvents are liquid ammonia, liquid methylamine, and ethylenediamine.

In many cases, synthetic electroorganic reactions require dry, polar, and aprotic solvents [Figures 17.22–17.24; acetonitrile, dimethylformamide (DMF), dimethylsulfoxide (DMSO), propylene carbonate (Table 15.4), pyridine, *N*-methylpyrrolidone (Figure 15.9), nitrobenzene, nitromethane, dichloromethane (DCM), and methylene chloride], as discussed in Section 5.3.3. Experimental procedures for solvent purification, recycling, and analysis of water content are provided in [30]. Summaries of procedures and recipes with examples are provided in [53–56]. The book chapter [30] discusses choices of supporting electrolytes and reference electrodes for different types of solvents.

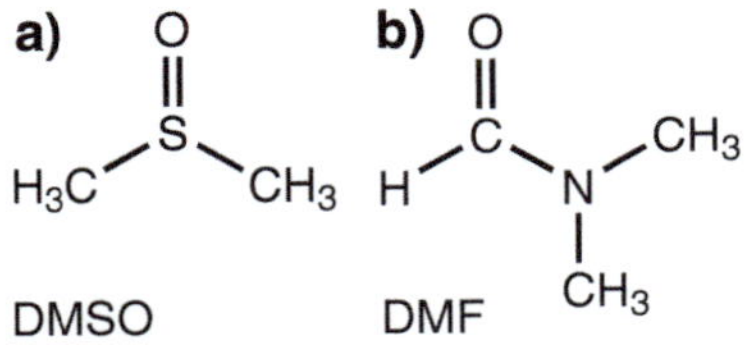

Figure 17.22 Molecular structure for a) dimethylsulfoxide and b) dimethylformamide.

Supporting electrolytes should be electrochemically inert in the potential range employed for the electrochemical synthesis. For organic solvents, supporting electrolytes need to be identified with sufficient solubility and minimal ion-pair formation (Figure 4.15). Commonly, tetraalkylammonium halogenides, hexafluorophosphates, tetrafluoroborates, or perchlorates are employed. The negative limit of the potential window is determined by the reduction potential of the tetraalkylammonium cation. In some cases, metal cations such as Li^+ or Na^+ can be employed ($LiClO_4$ or $NaClO_4$). Mg^{2+} cations can stabilize radical ions in aprotic solvents to slow down following reactions [30].

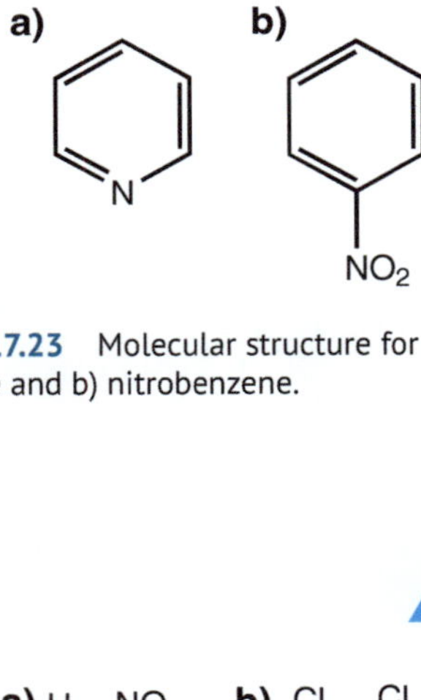

Figure 17.23 Molecular structure for a) pyridine and b) nitrobenzene.

Tetrafluoroborates are preferable to perchlorates, which can be explosive (!) when removing the solvent during workup and recycling of solvent and electrolyte. Trifluoromethylsulfonates (triflates, Figure 17.25a) and their products are not explosive. Aromatic sulfonates such as *p*-toluenesulfonate (tosylate, Figure 17.25b) can act as phase transfer reagents and make water-insoluble reagents water-soluble, especially in combination with a tetraalkylammonium cation. Carboxylates act as both the electrolyte and the reagent in the Kolbe electrolysis[163] (Section 17.3.1). A more detailed discussion of electrolyte salts with literature references for stability and purification can be found in [30]. Weakly coordinating anions with highly delocalized charge play a special role in broadening the accessible potential window toward positive potentials and preventing ion-pair formation. These anions shield positive charges of intermediates to a much lower extent than other ions and also show weaker interactions with anodically generated electrophiles [30, 57].

New electrolyte systems are based, for example, on ionic liquids (Section 11.4.2) or water-in-salt systems, for example, based on tetraalkylammonium-*p*-tolenesulfonate [30]. Further unconventional electrolyte systems are liquified gases (NH_3 and SO_2) or

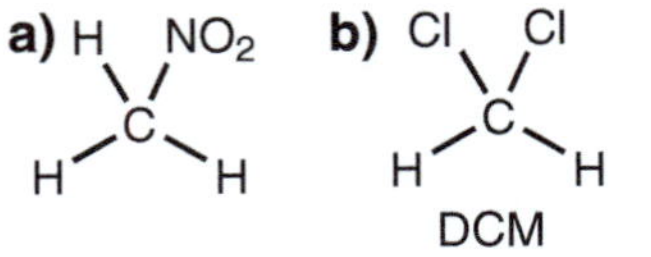

Figure 17.24 Molecular structure for a) nitromethane and b) dichloromethane.

[163] Hermann Kolbe, 1818–1884, German chemist, professor in Marburg and in Leipzig, known also for the first synthesis of acetylsalicylic acid (Aspirin®). He was part-owner of a salicylic acid plant in Radebeul, where the first pharmaceuticals were produced on an industrial scale.

supercritical fluids (H_2O, NH_3, and CO_2) [30]. Emulsions and nanoemulsions provide a way to employ highly hydrophobic substrates in ionically conducting liquids. In liquid two-phase systems, substrates and further reactants can be in separate phases to allow the following chemical reaction steps at the liquid–liquid phase boundary. More compositional variables open up new opportunities for optimization of syntheses, but at the same time, this increases complexity. The state-of-the-art is summarized in recent reviews [38]. Ultrasound has been introduced into reactions to improve mass transport, particularly in viscous media such as ionic liquids [38].

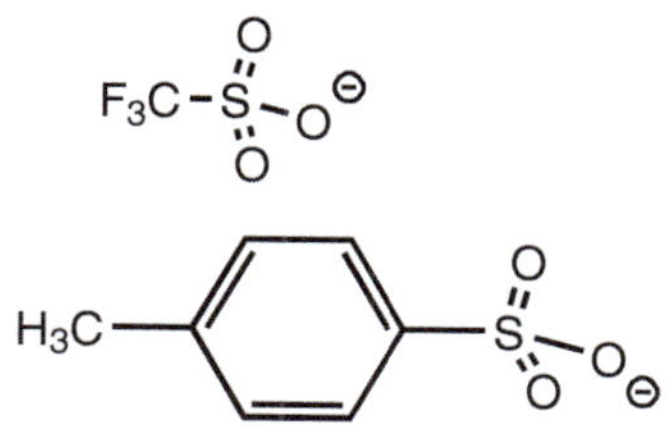

Figure 17.25 Molecular structure of a) trifluoromethylsulfonate (triflate) and b) *p*-toluenesulfonate (tosylate).

17.2 Organic Electrosyntheses – An Overview

Electrosyntheses for the transformation of organic and metal–organic molecules can be grouped based on details of the reaction mechanisms, often dependent on reaction conditions. This implies that with new insights about the reaction mechanism, the association with one group must be revised. A particular reaction may also belong to more than one group, as will be shown below.

There are four criteria for electrosyntheses:

- type of reaction intermediates (Table 17.1),
- reaction mechanism (Table 17.2),
- starting material (Table 17.3), and
- reaction concepts (Table 17.4).

Specialized textbooks [7] and handbooks [8] are based on all four criteria and can be useful for further study before accessing primary literature sources.

In organic electrosynthetic processes, a multitude of reaction steps occur near the electrode surface but within the solvent environment of the reaction layer. Thus, they

Table 17.1 Electrosynthetic processes categorized based on reaction intermediates.

Intermediate	Syntheses	Ref.
Radicals (often C-localized)	C–C coupling in Kolbe electrolysis	[59, 60]
	Nucleophilic aromatic substitution $S_{RN}1$ (cf. Section 6.4.7)	[61]
	Pinacol formation by reductive coupling of ketones	[62]
Cations (often carbenium cations)	Hofer–Moest[164] (non-Kolbe) electrolysis	[60]
	Chinoid spirolactones by phenol oxidation	[63]
	Substitution at oxonium and acyliminium ions in cation pools	[64]
Anions (often carbanions)	Reduction of ketones to secondary alcohols	[62]
	Reduction of α,ω-dihaloalkanes to give strained ring systems via intramolecular nucleophilic substitution	[65]
Radical ions	Electrohydrodimerization of acrylonitrile (Monsanto process[162]; cf. Sections 6.4.3 and 17.3.3)	[66, S. 629]
	(4 + 2)-Cycloadditions [note the comment on the electrocatalytic Diels–Alder reaction[149]]	[67, 68]

Comment: During cycloaddition, two reagent molecules are fused to give a ring system. According to the IUPAC recommendations [69], the numbers in round brackets, for example, (4 + 2), provide the *number of atoms* of the two reagents contributing to ring formation, while numbers in square brackets, for example, [4 + 2], denote the *number of electrons* provided by the two reagents in the pericyclic process. Note that a (4 + 2)-cycloaddition can be based on a [4 + 1]- or a [3 + 2]-pericyclic reaction [67]. The use of the term "electrocatalytic Diels–Alder reaction" is problematic due to the classic Diels–Alder reaction being defined as a concerted [4 + 2]-cycloaddition.

[164] H. Hofer; M. Moest; Ueber die Bildung von Alkoholen bei der Elektrolyse fettsaurer Salze; *Liebigs Annalen der Chemie* **1902**, *323*, 284–323.

Table 17.2 Electrosynthetic processes categorized based on reaction mechanisms.

Mechanism	Syntheses	Lit.
Nucleophilic substitution at a saturated C-atom	Electrochemical generation of nucleophiles: ring formation for α,ω-dihaloalkanes	[65]
	Electrochemical generation of carbenium ions for S_N1: Hofer–Moest reaction (non-Kolbe products)	[60]
Nucleophilic substitution at aromatic C-atoms	Electrochemically induced $S_{RN}1$-reaction (cf. Section 6.4.7)	[61]
Dimerization	Kolbe electrolysis (Section 17.3.1)	[59, 60]
Cycloaddition	(4 + 2); diene/dienophile reaction to give Diels–Alder[149] products	[67, 68]
	(2 + 2); cyclobutane formation by cross-coupling of enolethers and electron-poor alkenes	[70]

Table 17.3 Electrosynthetic processes categorized based on functional groups.

Functional group	Syntheses	Lit.
Carbonic acids	Kolbe electrolysis (Section 17.3.1)	[59, 60]
Disulfides	S–S bond formation, e.g., cystine → cysteine (industrial process)	[71]
Nitro compounds	Reduction to nitrosoaromatics, phenylhydroxylamines, anilines, etc.	[72]
Phenols	Oxidation to quinones, quinol derivatives, and C–C-coupling products	[73]

are homogeneous reaction steps. As a fundamental feature of all electrosynthetic reactions, at least one electron transfer step has to occur right at the electrode surface (i.e., heterogeneous reaction step), generating the reactive intermediates, which start a cascade of homogeneous reaction steps. Usually, the reaction intermediates are formed in the electronic ground state. Reaction intermediates formed at the electrode surface are identical in reactivity when compared to intermediates formed, for example, by adding redox reagents. Many known classic redox-driven organic synthetic methods can be "translated" into cleaner electrosynthetic processes. The knowledge about the reactivity of those intermediates is critically important for the optimization of electrosyntheses [58].

Further categories in the electroorganic process can be based on the reaction mechanism.

For the design of syntheses, specific reaction pathways for functional groups are important. A selection of cases with redox-active functional groups is provided in Table 17.3.

Finally, reactions can be categorized based on electrochemical criteria and reaction control (Table 17.4). More examples for these criteria will be discussed in Sections 17.3–17.5.

Table 17.4 Electrosynthetic processes categorized based on reaction concepts.

Reaction concept	Syntheses	Lit.
Anodic oxidation	Kolbe[163] electrolysis and Hofer–Moest[164] reaction	[59, 60]
Cathodic reduction	Reduction of carbonyl compounds and their derivatives	[62]
Umpolung (inversion of the reactivity pattern of a functional group)	Coupling of electron-rich (nucleophilic) olefins by means of oxidation of one reagent to the corresponding radical cation	[74]
Paired electrolysis	Protected homo-allylalcohols from three-component synthesis (convergent paired electrolysis)	[75]
Mediator-based or indirect electrolysis	Oxidation of dithianes to aldehydes (deprotection) with triarylaminium-radical cations	[76]
Flow-through electrolysis	Intermediate formation by means of *o*-quinone oxidation and their Diels–Alder coupling[149] with dienophiles	[77]
Miniaturized electrolysis	α-Alkoxylation of cyclic secondary amines (example of combinatorial electrosynthesis)	[78]

17.3 Direct Transformations

Following Table 17.4, some reactions will now be discussed in more detail. These examples are not exhaustive or complete and rather serve as illustrations. For further in-depth study, literature and textbook sources are recommended.

Direct electrosyntheses are triggered by electron transfer (ET) between the electrode and the reagent (substrate).

The initial ET can lead to either neutral radicals, radical anions, or radical cations. These can be highly reactive and follow known homogeneous reaction pathways in the following reaction steps. Thus, electrosynthetic processes are dominated by the chemistry of radicals, which by nature tend to be not very selective. Therefore, care is required to ensure selectivity in such radical reactions. Indirect or mediated processes will be investigated in Section 17.4.

Direct electrosyntheses applies to a multitude of functional group transformations (FGT) (Table 17.3). Of particular interest are reactions with high selectivity and potential to replace several traditional reagent-based synthetic steps. An instructive example is the oldest known type of organic electrosynthesis, the Kolbe electrolysis[163]. This type of electrosynthetic reaction still gains importance for example in the transformation of renewable biomolecules and provides an example for reaction control.

17.3.1 Kolbe Electrolysis

The one-electron oxidation of carboxylates at 2.1–2.8 V vs. SHE causes formation of alkyl- or arylcarbonyloxy radicals. Loss of CO_2 then leads to the formation of carbon-centered radicals R•. This radical offers a plethora of reaction pathways (Figure 17.26). In the Kolbe electrolysis, the recombination of two radicals R• produces the alkanes R–R. Thus, it is a C–C-coupling reaction suitable for constructing carbon backbones starting from organic carboxylic acids. Typically, a concentrated solution (0.5 mol L^{-1}) of the carboxylic acid in methanol is employed, from which 5% to 10% are deprotonated by addition of a base such as Na-methanolate, pyridine, triethanolamine, or a solid phase base for a simple separation. Suitable anode materials

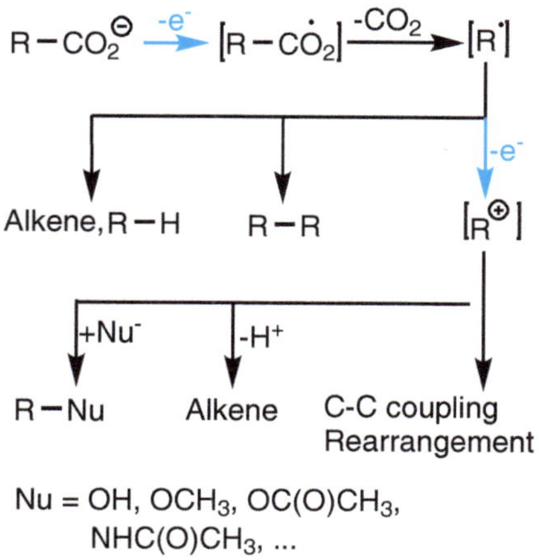

Figure 17.26 Possible following reactions after one- or two-electron oxidation of carboxylic acids according to [30].

are Pt-sheets or Pt-nets. Sometimes glassy carbon, graphite, or BDD is employed, but this can change the reaction pathway. The counter electrode process is the HER [59]. Thus, the cathode material must be electrocatalytically active, and Pt is a good choice here. Galvanostatic conditions are applied with current densities of 250 mA cm^{-2} in undivided cells. The electrolysis continues until the pH changes to more alkaline (when the carboxylic acid has been consumed).

The reaction pathway is linked to the reaction conditions. The addition of a base pushes the deprotonation equilibrium for the carboxylic acid toward the carboxylate, which is then oxidized. Carboxylates as ionic species enhance conductivity in the solution. Only a part of the reagent molecules is deprotonated. When carboxylates are consumed in the ET step, they are immediately replenished from the reservoir of undissociated carboxylic acid, reacting rapidly with free base. At the cathode, HER occurs to remove protons (generated at the anode) from the solution and to balance the overall pH. Both the anode and the cathode produce gases, which ensure the removal of products without back reaction. The reaction of radical $R^{\bullet}$ is chemically irreversible without back reaction. Therefore, undivided cells can be employed. Divided cells are necessary only in those cases when the carboxylic acid reagent contains additional functional groups sensitive to reduction. High current densities are associated with positive potentials at the anode and correspondingly high coverage of the anode with carboxylate anions. This decreases the coverage by the solvent molecules, which, together with a high radical concentration, favors the radical coupling reaction to form Kolbe products. Lower current densities lead to lower current yields (Section 12.5.2) because the fraction of current spent on methanol oxidation increases. There can be problems due to insufficient solubility of coupling products, which then block the anode surface. Solvent mixtures, mechanical/ultrasonic cleaning, temperature variation, or pulse programs allow *in situ* anode cleaning and continuous Kolbe electrolysis.

Particularly useful are Kolbe cross-coupling reactions based on two distinct carboxylic acids, R^1COOH and R^2COOH. The anodic electrode reaction

$$4\,R^1COO^- + 4\,R^2COO^- \rightarrow R^1{-}R^1 + 2\,R^1{-}R^2 + R^2{-}R^2 \tag{17.12}$$

produces homo-coupling products $R^1{-}R^1$ and $R^2{-}R^2$ and the desired cross-coupling product $R^1{-}R^2$. Using a 10-fold excess of the lower-cost reagent, mainly the desired $R^1{-}R^2$ product is obtained (besides a lot of one homo-coupling product). In spite of the unavoidable formation of one homo-coupling product, the Kolbe cross-coupling process is versatile and powerful as a synthetic step. The process has been used in the synthesis of fatty acids, pheromones, chiral intermediates, C-glycosides, and nonnatural amino acids [59]. Kolbe electrolysis allows systematic carbon chain extension reactions:

$$2\,RO_2C{-}(CH_2)_n{-}CO_2^- \xrightarrow[-2\,e^-]{-2\,CO_2} RO_2C{-}(CH_2)_{2n}{-}CO_2R \tag{17.13}$$

$$R^1{-}CO_2^- + {}^-O_2C{-}(CH_2)_n{-}CO_2R^2 \xrightarrow[-2\,e^-]{-2\,CO_2} R^1{-}(CH_2)_n{-}CO_2R^2 \tag{17.14}$$

The resulting esters can be deprotected.

Electrolysis of carboxylates in the presence of alkenes allows the radical addition of $R^{\bullet}$ to the alkene (Figure 17.27). The radical intermediate can react with another $R^{\bullet}$ or undergo dimerization. Disproportionation and H-abstraction are further pathways of the C-centered radicals. The H-abstraction reaction (e.g., with the solvent) leads to a functional group interconversion from R–COOH to R–H.

Important is the radical addition of $CF_3^{\bullet}$ (from CF_3COO^-) to alkenes, conjugated dienes, or aromatic precursors. Intramolecular addition leads to ring-forming reactions. Ring formation is favored by lower current densities (lower radical concentrations) [60].

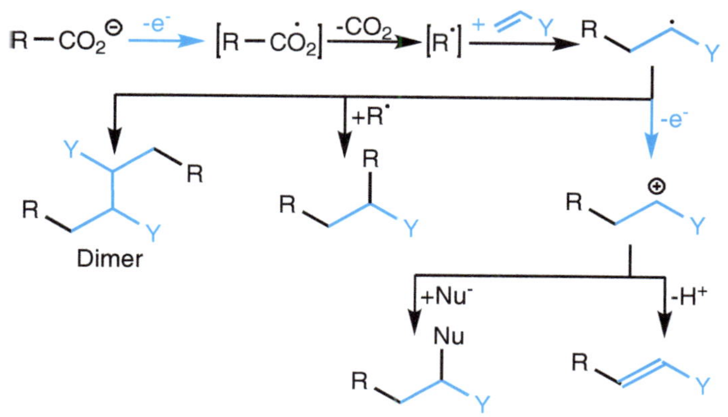

Figure 17.27 Products from Kolbe electrolysis in the presence of an alkene. Source: Adapted from [60].

A promising recent development is the use of a rapid polarity inversion of the applied cell voltage between two similar electrodes [79]. Thus, the target product is formed alternatingly at one the two electrodes. This prevents the acidification around the anode, which in turn can cause protonation of the carboxylic acid and associated selectivity problems in the Kolbe synthesis. Electrolysis with rectangular voltage pulses substantially broadens the synthetic scope of the Kolbe reaction and facilitates the use of carbon electrodes (instead of Pt) [79]. This example of using specifically defined waveforms to control diffusion and reaction layers in the vicinity of the electrode and to prevent electrode passivation is a common concept in several areas of electrochemistry (e.g., Figures 5.52, 9.25, 14.44) and may also hold a broader promise for electrosynthesis.

17.3.2 Non-Kolbe Electrolysis

The transfer of a second electron during Kolbe electrolysis in Figure 17.26 leads to carbenium cations R^+ as reactive intermediates. Based on [60], the reaction pathway can be switched between Kolbe and non-Kolbe products by the variables given in Table 17.5.

The carbenium cation R^+ obtained after the second ET step can react with a diverse range of nucleophiles – as long as the carbenium cation is sufficiently stable. Methoxylation occurs in methanol, acetoxylation in acetic acid, and acetamidation in wet acetonitrile (Figure 17.28). Stabilization of carbenium cations is possible by mesomeric structures, especially in the presence of heteroatoms such as O, N, or S. In these cases, reactions provide acetoxylation or methoxylation products cleanly (Figure 17.29).

Carbenium cations can transform into alkenes depending on structural features in the reactions shown in Figure 17.30 [60].

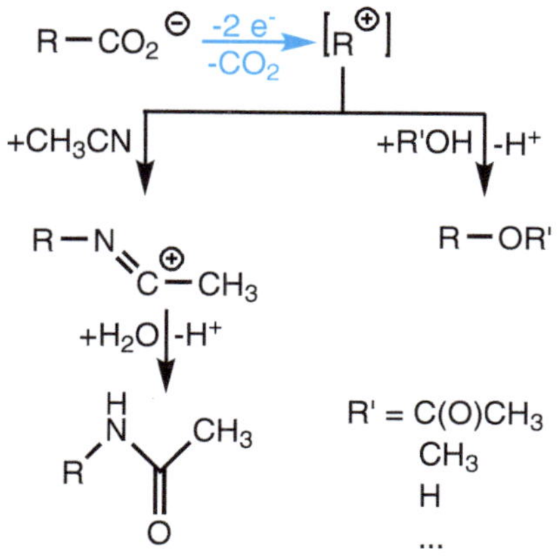

Figure 17.28 Alkoxylation and acetamidation as non-Kolbe reaction pathways. Source: Adapted from [60].

Table 17.5 Experimental parameters in electrodecarboxylation reactions.

Parameter	One-electron process	Two-electron process
Current density	High, 0.25–1 A cm^{-2}	Low
Electrode material	Pt	Soft graphite
Electrolyte solution	Acidic Water, protic solvents (e.g., CH_3OH), mixtures of water and protic solvents	Alkaline Polar aprotic solvents (e.g., pyridine), mixtures of water and polar aprotic solvents, protic solvents
Additives	None	ClO_4^-, SO_4^{2-}, HCO_3^-
Temperature	<50 °C	Not important

Source: Adapted from [60].

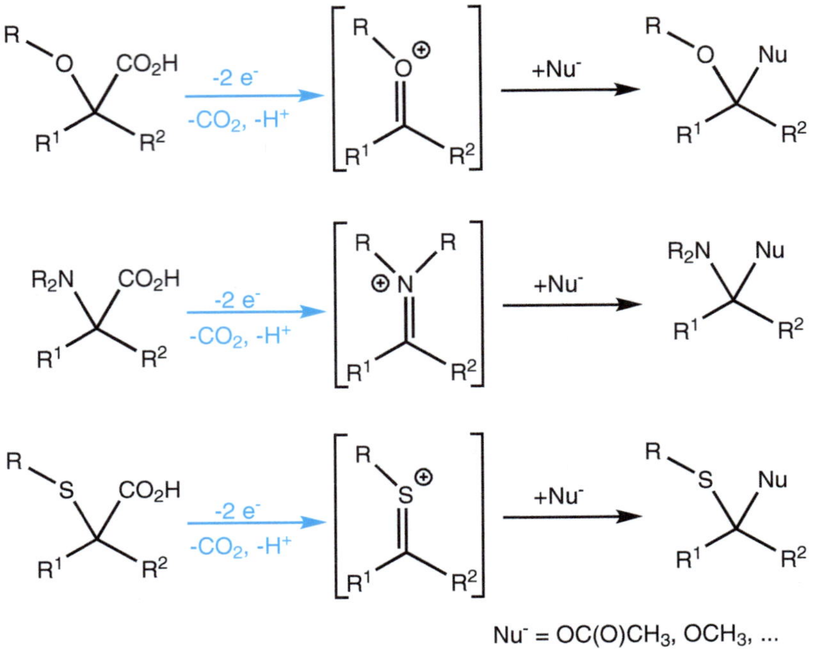

Figure 17.29 Stabilization of non-Kolbe carbenium cations by heteroatoms and acetoxylation or methoxylation reactions. Source: Adapted from [60].

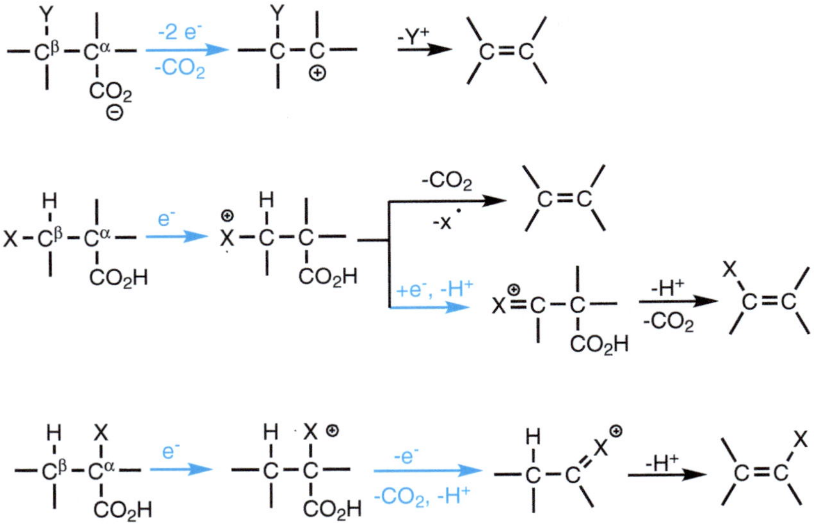

Figure 17.30 Alkene formation from non-Kolbe carbenium cations. Source: Adapted from [60].

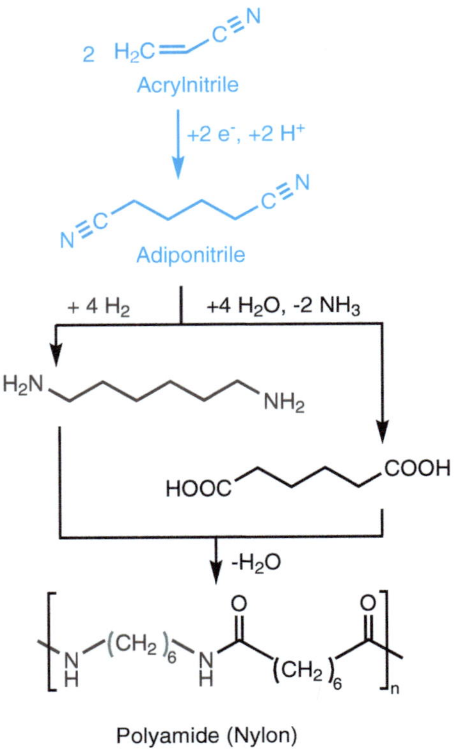

Figure 17.31 Electrosynthesis of adiponitrile during industrial Nylon-6,6 manufacture.

A further carbenium cation intermediate can form from the binding of the primary radical to an alkene, followed by a second ET (Figure 17.27). The new carbenium cation is electronically stabilized by the electron-donating substituent Y. The cation either reacts with a nucleophile Nu⁻ or produces an alkene with the release of a proton. Often, both products are observed.

17.3.3 Reductive Coupling Reactions

In addition to anodic C–C-coupling reactions, there are many reductive coupling reactions [66], including the commercially important and well-studied example of the reductive hydrodimerization in the electrochemical synthesis of adiponitrile (Figure 17.31, Monsanto process[162]). The process requires an alkene with an electron-withdrawing group (EWG) attached to the double bond. The one-electron reduction of the alkene leads to several reaction pathways (Figure 17.32).

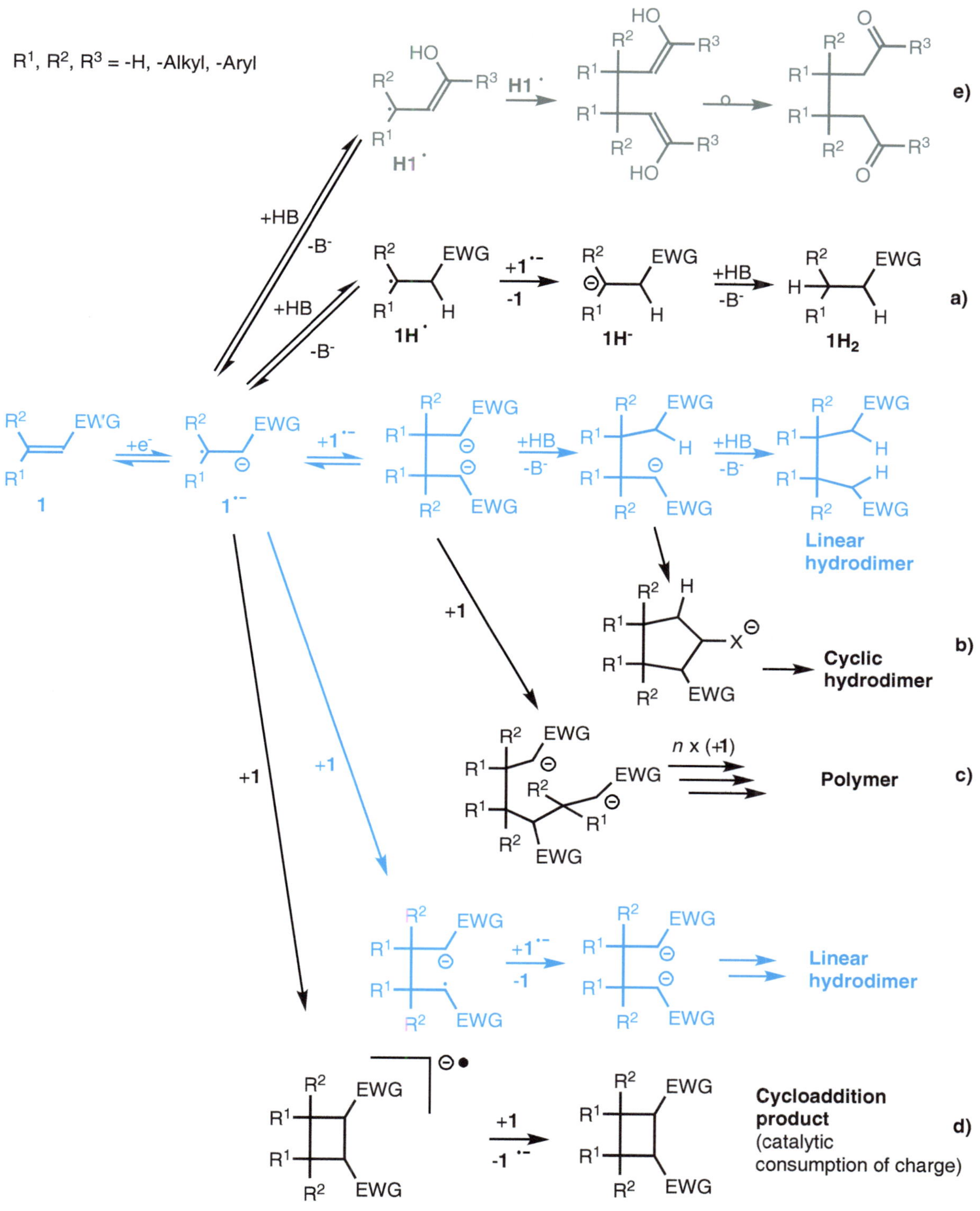

Figure 17.32 Formation of the radical anion and following reaction pathways. Source: Adapted from [66].

Two reaction pathways shown in blue in Figure 17.32 for radical–radical and radical–substrate coupling (cf. Section 6.4.3) lead to the desired linear hydrodimer (LHD). There are several other competing reaction pathways:

a) Protonation, transfer of another electron, followed by protonation to give the hydrogenation product **1H$_2$**.
b) Intramolecular addition instead of the second protonation gives a cyclic hydrodimer. Symbol X in Figure 17.32 denotes a substituent formed from the EWG in the reagent.
c) Coupling of two radical anions and initiation of polymerization of the alkene by the dianionic dimer.
d) Reaction of the radical anion with the reagent and electron transfer to the reagent. Formally, this represents a cycloaddition. However, the catalytic role of the electron in this case becomes obvious when considering that only 0.1 *F* charge has been transferred. This reaction pathway can be regarded as "catalysis by electrons". Provided that there is a suitable order of the redox potentials for the ET steps, this leads to an electron transfer chain reaction (ETC, Section 6.4.4).
e) For ET to ketones or aldehyde functional groups in protic solvents (Figure 17.32 in gray), the radical anion can be O-protonated, followed by radical dimerization, forming the linear coupling product.

The formation of the cyclic hydrodimer (b) is favored over a second protonation when the EWG is a carbonyl, carboxylate, or nitrile group. The polymerization pathway (c) is favored in dry aprotic solvents. The addition of proton donors can suppress polymerization. The experimental investigation of reaction pathways toward linear hydrodimers is problematic (Chapter 6) and requires carefully controlled conditions that are different than those in the synthetic reactor. A combination of electroanalytical know-how and application of microelectrodes (Section 5.6) was used to mimic mass transport conditions in a technical flow reactor with a linear flow rate of typically 1–2 m s^{-1} [81].

Challenges in adiponitrile electrosynthesis are suppressing H_2 evolution (HER) with very negative potentials at the cathode and simultaneously providing sufficient proton activity for the desired product to form. Furthermore, the hydrogenation pathway also needs to be suppressed (Figure 17.32a). As discussed in Section 17.1, there are several optimization steps required. Today, undivided cells are employed with a bipolar circuit (4 V cell voltage, 0.25 A cm^{-2}, 52 °C). Electrodes are made from stainless steel with the cathode further coated with Pb or Cd (to suppress HER, cf. Figure 3.9). The interelectrode gap is 2 mm only. The electrolyte is a 0.8 mol L^{-1} Na_2HPO_4 solution with added hydrophobic cation salts [tetraethylammonium salts or hexamethylene-bis(ethyl-dibutyl-ammonium)-hydrogenphosphate, Figure 17.33a]. The additive has three functions:

Figure 17.33 Molecular structure of organic additives for electrosynthesis of adiponitrile.
a) Hexamethylene-bis(ethyl-dibutyl-ammonium) cation;
b) ethylenediamine-tetraacetic acid (EDTA);
c) triethanolamine (TEA).

1) Emulsion formation in water, with the reagent in the phase containing the bis-ammonium cation. Emulsification increases the apparent reagent concentration in the electrolyte.
2) Lowering surface concentrations of water and HPO_4^{2-} at the cathode.
3) Inhibiting metal plating (due to Fe(III) from corrosion processes).

Particularly interesting is the effect of the hydrophobic cation on the cathode reactivity. With highly negative applied potentials at the cathode, charge-compensating cations are drawn into the double layer and onto the electrode surface. During electrolysis, that is, under operating conditions, a thin film of cations forms on the cathode, replacing the polar and protic solvent. The ET to acrylic acid occurs in this thin hydrophobic film. This favors the subsequent dimerization because of the depleted local concentration of solvent and proton sources. Electric field-driven repulsion of the dianion from the cathode into the bulk solution allows rapid protonation to the adiponitrile product. The use of the bis-ammonium cation allows a lower concentration of the additive. This in turn minimizes side products from anodic ammonium oxidation. When using tetraalkylammonium salts, a divided cell was required, while an undivided cell can be employed with bis-ammonium cations. The latter has a much higher energy efficiency due to lower ohmic losses.

Further additives [81] are complexing agents ethylenediamine-tetraacetic acid (EDTA, 0 01 mol L^{-1}) and triethanolamine (TEA, 0.003 mol L^{-1}) (Figure 17.33b,c). These are required to mask Fe(III), Pb(II), or Cd(II) cations formed by corrosion of the electrodes. In particular, deposition of iron [Fe(III) to Fe(0)] at the cathode would substantially increase the exchange current density for HER and therefore lower current yields for adiponitrile. Complexation of cations also prevents the formation of solid hydroxides. (Reduction with high current densities leads to alkaline conditions at the cathode despite buffering due to consumption of two H^+ per molecule of adiponitrile.) The electrolyte also contains 0.1 mol L^{-1} borax as a buffer and corrosion inhibitor.

With optimization, electrochemical adiponitrile synthesis is possible with 90% yield. The atom economy is excellent (low waste). In spite of this, the most economic production of adiponitrile is based on Ni-catalyzed addition of HCN to butadiene [82]. Costs for this process depend mainly on the price of natural gas. The electrosynthesis requires propene that is more expensive due to its use in other processes.

Further Reading. There are many resources [66–78, 80–82] and textbooks with information about the electrosynthesis of adiponitrile [10, 11]. Multiple variants of reductive dimerization processes with other functional groups have been reviewed [8]. An emerging area of interest is the reductive coupling of aromatic halogenides [15].

17.4 Indirect Transformations

Direct electrosyntheses (Section 17.3) by interfacial ET at one of the reagents produce highly reactive intermediates with multiple following reaction pathways and sometimes problematic polymer film formation at the electrode surface that inhibits further ET reactions. Some substrates require exceedingly high overpotentials (due to bond breaking) to drive ET at the cost of selectivity because another group can also react at the applied potential. Indirect electrosynthesis can be employed to alleviate these problems.

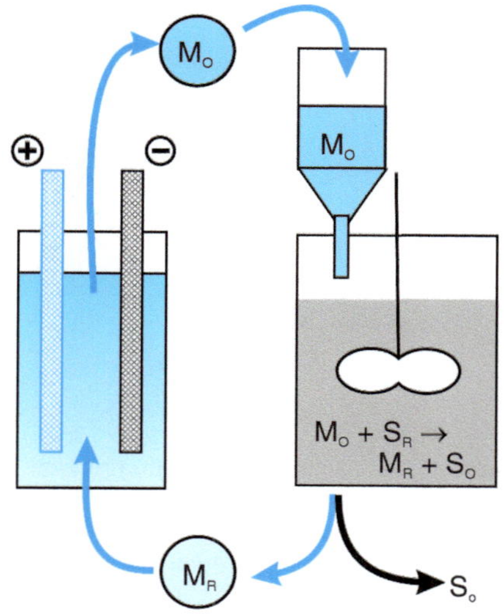

Figure 17.34 Indirect electrooxidation with chemical oxidation of the reagent and electrochemical regeneration of the oxidant.

Indirect electrosyntheses are driven by heterogeneous ET between the electrode and a **mediator** (electron shuttle). The mediator then drives a homogeneous redox process, converting the reagent (substrate) by oxidation or reduction.

The concept of the redox mediator has been covered in the discussion of SECM, dye-sensitized solar cells, modified electrodes, and glucose sensors. Thus, it is an important general principle of modern electrochemistry with many applications in electrosyntheses.

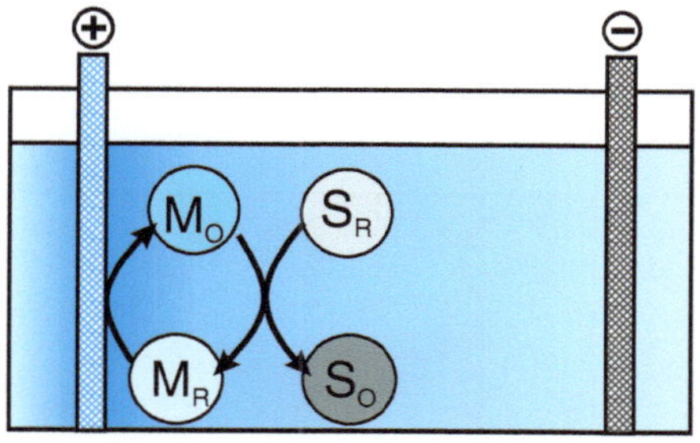

Figure 17.35 Indirect electrooxidation where the electrooxidized mediator homogeneously reacts with the substrate in the electrochemical cell.

17.4.1 Variants of Indirect Electrolyses

Mediated electrosynthetic processes can be achieved in various ways. One option is to react the reagent with the oxidizer (or reductant), followed by electrochemical (re-)generation of the oxidizer (or reductant) in a separate reactor (Figure 17.34). The two reactors can be connected via flow [76].

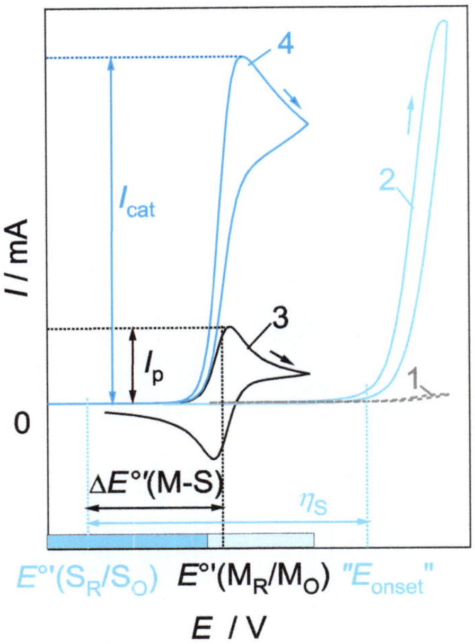

Figure 17.36 Schematic of voltammograms for a mediated oxidation of substrate S_R. 1) Supporting electrolyte; 2) direct oxidation of S_R; 3) voltammogram for mediator M_R without S_R; 4) voltammogram with catalytic wave for oxidation of S_R with catalytic amounts of M_R. Source: Data provided by Robert Francke, Leibniz Institute for Catalysis e.V. (LIKAT), Rostock, Germany.

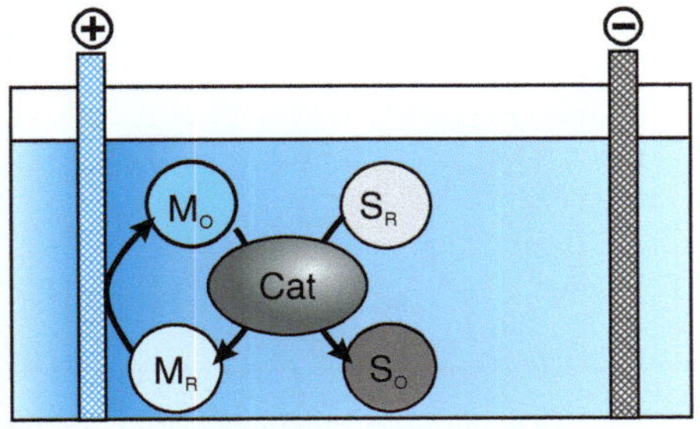

Figure 17.37 Indirect electrooxidation with a homogeneously catalyzed reaction between the oxidized mediator and the substrate.

Alternatively, the substrate can be added to the electrochemical cell to allow the reaction *in situ* in the vicinity of the electrode (Figure 17.35). This mode of operation requires very low exchange current densities for the substrate to avoid direct electrolysis (i.e., a higher onset-potential, E_{onset}, for direct electrolysis than for mediator ET, Figure 17.36, curve 2). The parameter E_{onset} is not fully defined. We had discussed this empirical quantity at the beginning of the book in the context of Figure 2.1. The thermodynamically defined formal potential $E°'$ for a redox pair S_R/S_O is often unknown and experimentally inaccessible. For instance, the reactivity of radical products can be too high for reversible or quasireversible voltammograms to be obtained because the reactive intermediates react too fast to be converted in the reverse scan of a cyclic voltammetry experiment. This prevents the determination of $E°'$.

Mediators should have a high exchange current density to operate at low overpotentials (Figure 17.36, curve 3). The mediator reacts with the substrate (reagent) by an EC′-mechanism (Section 6.4.6, Table 6.24). This reaction happens within the reaction layer in the vicinity of the electrode surface. The mediator is regenerated by the chemical reaction with the substrate. This reaction limits the extension of the mediator diffusion layer near the electrode and causes large concentration gradients ($\partial[M_R]/\partial x$) and $\partial[M_O]/\partial x$) for M_R and M_O, which are associated with a higher current I_{cat}. Often, the voltammetric response is called "catalytic wave" (Figure 17.36, curve 4). The mediator diffuses rapidly between the electrode and the reaction layer and thus undergoes multiple oxidation–reduction cycles. Therefore, catalytic amounts of mediator are sufficient.

For indirect oxidations, the formal potential of the mediator is positive relative to the formal potential of the substrate to provide the driving force for the homogeneous ET (see blue stripe over abscissa in Figure 17.36). For some reactions, the formal potential for the mediator appears to be up to 0.6 V below that of the substrate (light blue stripe over abscissa in Figure 17.36). Here, the driving force for the reaction originates from a fast following chemical step that removes one reaction partner from the equilibrium [76].

Processes such as those in Figure 17.35 can be regarded as "catalysis by ET". They are often called "electrocatalysis" or "redox catalysis". In order to distinguish such processes from heterogeneous catalytic electrode processes proceeding via intermediates adsorbed and stabilized at solid electrode surfaces (e.g., in fuel cells or water electrolysis), this textbook prefers the terminology homogeneous electrocatalysis (Sections 6.4.6 and 12.1.1).

Homogeneous reactions between the mediator and the substrate can be controlled with a cocatalyst (Figure 17.37). Examples of this are homogeneous copper or palladium catalysts. Processes of this nature can be considered as double-mediated, often with very high selectivity [83]. Stereochemically difficult transformations are possible [84], and electrochemically inert substrates, such as CO_2, can be transformed [85]. Very closely related and important are electroenzymatic syntheses [16, 86]. The enzyme introduces high stereoselectivity. Sometimes enzymes require very expensive biochemically produced cofactors, which can be recycled electrochemically like a mediator, making the entire process viable. Achieving this is by no means easy, for example, for $NADH/NAD^+$, one of the most important biochemical cofactors. The electrochemistry of $NADH/NAD^+$ is discussed in more detail in Section 18.2.1.

In some cases, multiphase media are beneficial for indirect electrosynthesis, for example, for mediators solubilized into a thin liquid film on the electrode surface. The reaction then occurs at the liquid–liquid phase boundary between the hydrophobic liquid film and the aqueous solution with the hydrophilic substrate. In the extreme case, the thickness of this layer is reduced to a monolayer surface modification of the solid electrode surface (Section 13.4.8) – a concept equivalent to heterogenization of a molecular catalyst. Mediators or catalysts chemically bound to beads can be used with soluble mediators and facilitate ease of separation and reuse.

17.4.2 Mediator Choice

The selection of the mediator is crucial for successful electrosynthesis. The following criteria apply:

- stability of both redox forms (no following chemical steps apart from ET to the substrate);
- suitable formal potential $E^{\circ\prime}$ to drive ET to the substrate;
- fast kinetics for heterogeneous ET at the electrode;
- fast kinetics for homogeneous redox processes with the substrate.

These criteria allow screening and selection of suitable mediators by mechanistic studies discussed in Chapter 6. Stability and formal potential are obtained by cyclic voltammetry of the mediator in the appropriate electrolyte. The heterogeneous standard reaction rate constant k° can be determined for the ET step between the mediator and the electrode. There are data collections for mediators in the literature [83, 85–89].

Tables 17 6 and 17.7 summarize important redox mediators for electrosynthesis. The formal potential for organic mediators is systematically adjusted by substitution with electron-donating or electron-withdrawing groups, for example, for triphenylamines [76, 87, 91] (Table 17.8), TEMPO [89], or quinones.

Table 17.8 Redox potentials of substituted triphenylamines.

Mediator	X	Y	Z	E°/V vs. SHE
	Br	H	H	1.30
	Br	Br	H	1.75
	Br	Br	Br	1.96

Source: [7], Copyright 2015, John Wiley & Sons.

Table 17.6 Selection of mediators for important oxidation reactions.

Name/acronym	Reaction at the electrode (examples)
TEMPO (2,2,6,6-tetramethyl-piperidine-N-oxyl)	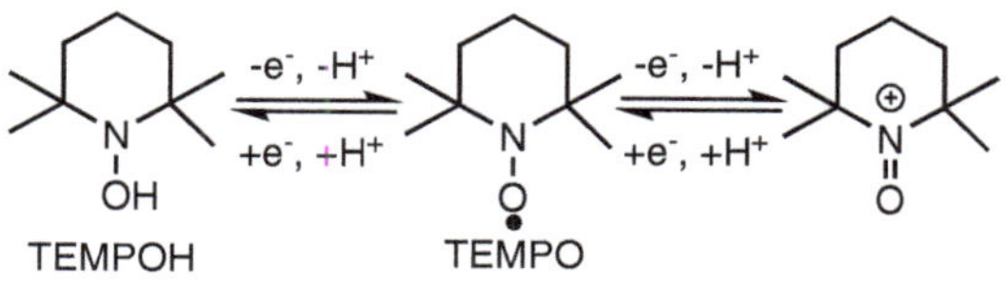 Example: Oxidation of alcohols to give carbonyl compounds [87], Pd-catalyzed cross-coupling of terminal alkynes and arylborates.
PINO (phthalimide-nitroxyl) NHPI (N-hydroxyphthalimide)	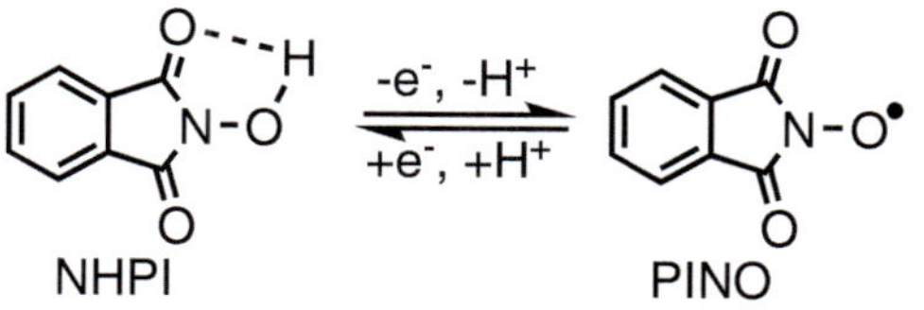 Examples: Oxidation of alcohols [89], oxygenation of allylic compounds [90].
Triarylamine $(R{-}Ph)_3N/(R{-}Ph)_3N^{\bullet+}$	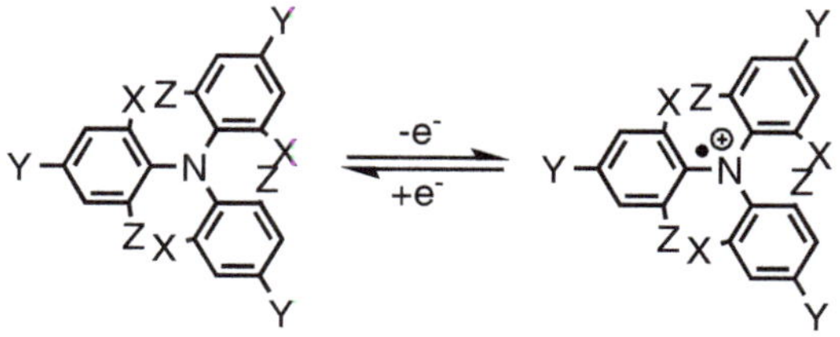 Example: Dimerization of E-anethol (1-methoxy-4-(1-E-propenyl)-benzene [87].

Table 17.6 (Continued)

Name/acronym	Reaction at the electrode (examples)
Halogens (e.g., Br^-/Br^+ or Br^-/Br_2)	$2Br^- \underset{+2e^-}{\overset{-2e^-}{\rightleftharpoons}} Br_2 \xrightarrow[-H_2O]{+2OH^-} Br^- + OBr^-$ $Br^- \underset{+2e^-}{\overset{-2e^-}{\rightleftharpoons}} Br^+$ Example: 2,5-dimethoxylation of solid phase-immobilized furanes [87], overview [91].
Iodate/periodate	$IO_3^- \underset{+2e^-}{\overset{-2e^-}{\rightleftharpoons}} IO_4^-$
Metalorganic complexes $Ru^{2+/4+}$, $Co^{2+/3+}$, $Mn^{2+/3+}$, $Ni^{2+/3+}$, $Ce^{3+/4+}$, $Cr^{3+/6+}$, $Os^{6+/8+}$	Examples: Oxidation of toluene to benzaldehyde with Mn^{3+}/Mn^{4+}, oxidation of toluene to benzoic acid with $Cr^{3+}/^{6+}$ [7]; overview [76].
Hypervalent iodoarenes	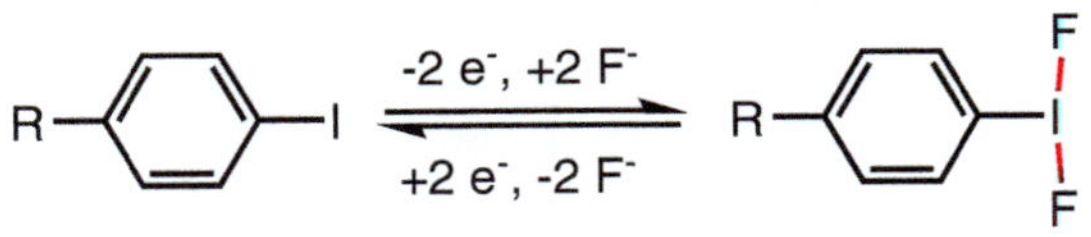 Examples: Indirect anodic fluorination of 2-pyrimidylsulfides with electrochemically generated iodoarene-difluorides [87], anodic vicinal difluorination of olefins [92].
DDQ (2,3-dichloro-5,6-dicyano-1,4-benzoquinone)	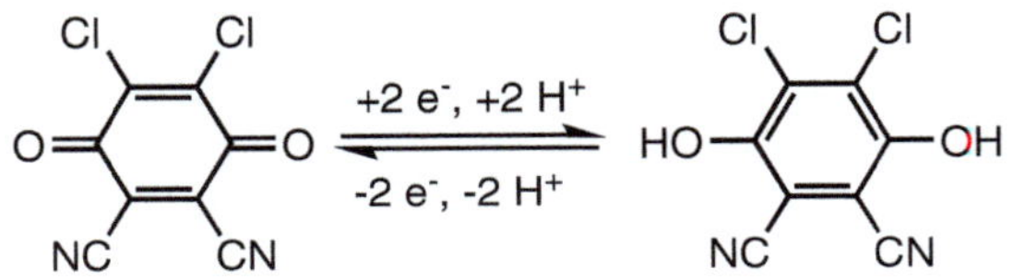 Example: Selective functionalization of sidechains in naphthalenes [94]. The oxidized form is added and then anodically regenerated.
Sulfide	R^1–S–R^2

The initial redox state is listed first, followed by the activated redox form that reacts with the substrate.

The rate constant for the homogeneous redox reaction has to be determined for each substrate individually as described in Section 6.4.6. A step-by-step recipe is described in Example Calculation 18.1. The procedure is applicable for indirect electrosynthesis with small adjustments [85, 87].

Further Reading. Overview papers [83, 85–89] provide a general [76, 91] and a systematic introduction into indirect electrosynthesis.

Table 17.7 Selection of mediators for important reduction reactions.

<table>
<tr><th>Name/acronym</th><th>Reaction at the electrode (examples)</th></tr>
<tr><td>Methylviologen</td><td>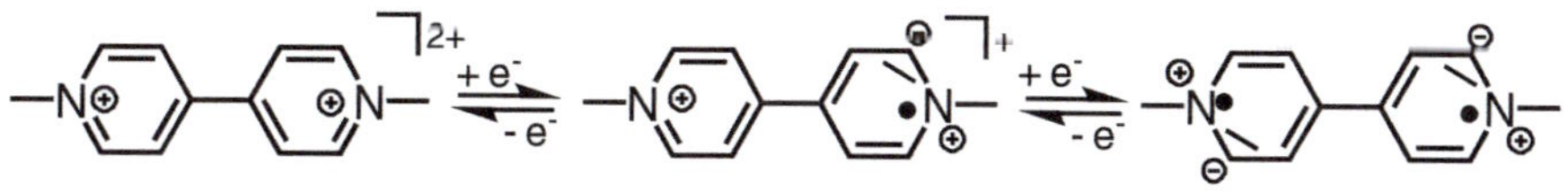
Example: Indirect electroenzymatic reduction of NAD^+ to NADH [76].</td></tr>
<tr><td>Polycyclic aromatic hydrocarbons</td><td>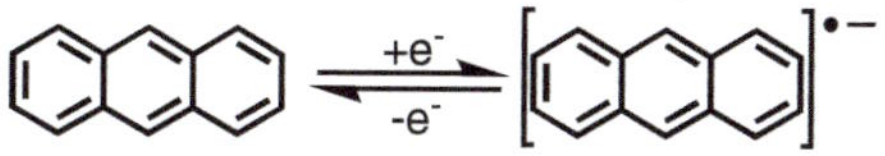
Example: Generation of radical anions from bromoalkanes [87].</td></tr>
<tr><td>Fullerene C_{60}</td><td>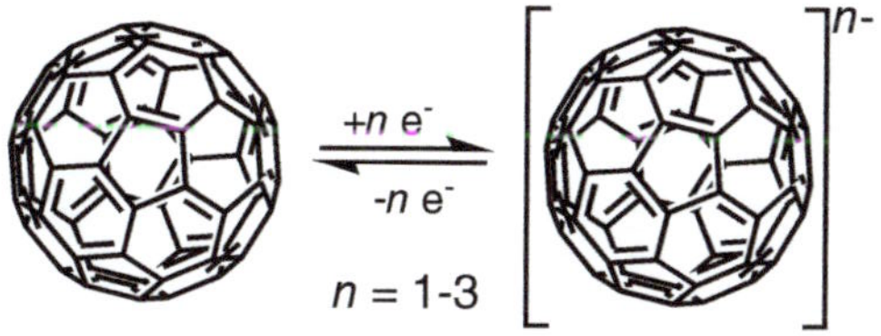
Example: Reduction of vic-dihalogenides and perfluoroalkylhalogenides [93].</td></tr>
<tr><td>Phthalonitrile</td><td>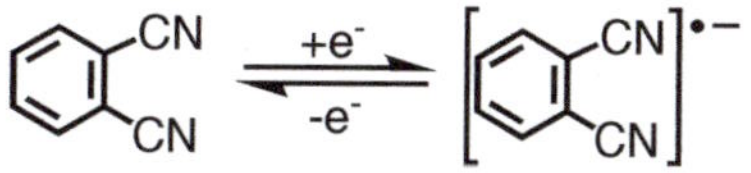
Example: Cathodic cross-coupling of 1-iodo-2-trifluoromethyl-benzene with substituted 1H-imidazoles [87].</td></tr>
<tr><td>Metalorganic complexes, metal salts
$Ni^{2+/(0)}$, $Pd^{2+/(0)}$, $Mo^{2+/(0)}$, $Co^{2+/+}$, $Cr^{3+/2+}$, $Ti^{3+/2+}$, $Sn^{4+/2+}$, $Sn^{2+/(0)}$, $Mg^{2+/(0)}$</td><td>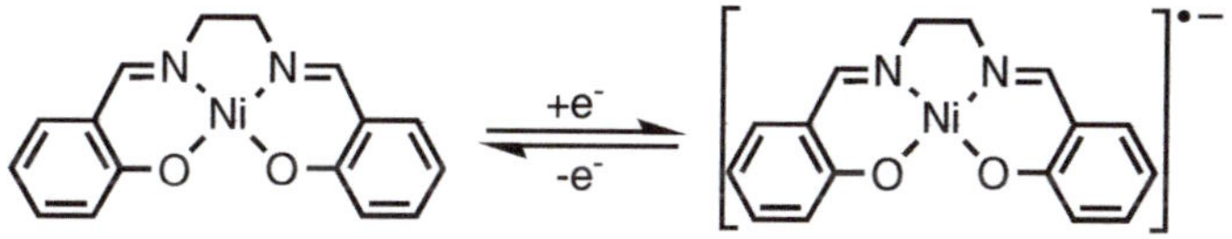
Examples: Intramolecular cyclization of electron-poor olefins with nickel(II)–salen complexes as the mediator [87]; reductive $Pd^{(0)}$- or $Ni^{(0)}$-catalyzed coupling of arylhalogenides [7, 15]; reductive dehalogenation of vic-dibromides with Co-I–salen [7].</td></tr>
</table>

The initial redox state is shown first, followed by the electrochemically activated form of the mediator for reaction with the substrate.

17.5 Examples of Electroorganic Syntheses with Reaction Control

In the following section, we will discuss some important means for reaction control in electrosynthesis.

17.5.1 Cation Pool Method

Preparative targets can often be realized with a cationic intermediate R^+ reacting with a nucleophile Nu^-. Cations are often unstable with short lifetimes, so it is attractive to generate the intermediate cation R^+ chemically or electrochemically directly in a solution of Nu^-, where it immediately reacts to form R–Nu. However, the formal potential for R/R^+ might be positive of that for Nu^-/Nu (Figure 17.38). This situation can be overcome with an elegant method developed since 1990 by Yoshida et al. [95].

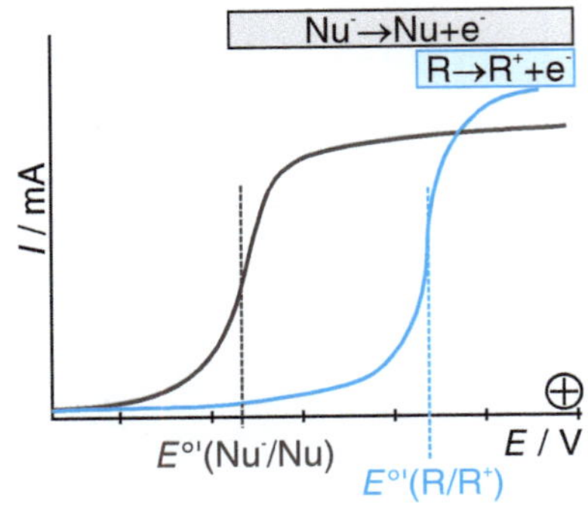

Figure 17.38 Commonly observed sequence of formal potentials for cationic redox intermediates R/R^+ and oxidation of nucleophiles Nu^-/Nu. Electrochemical generation of R^+ would also oxidize the nucleophile Nu^-.

The **cation-pool methodology** is based on preparing R^+ in preelectrolysis, usually in a divided cell. The cation R^+ is generated and kept at a low temperature (−72 °C). After completion of the electrolysis and switching off the electrochemical cell, the nucleophile Nu^- is added to the preelectrolyzed electrolyte solution (the cation pool), where R^+ and Nu^- react rapidly to form R–Nu (Figure 17.39).

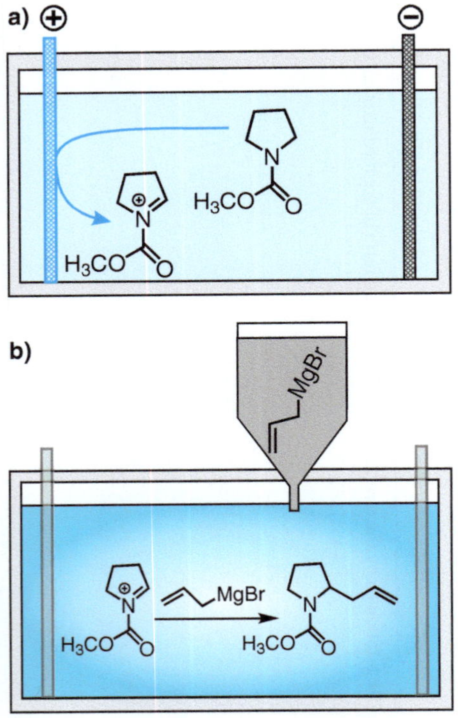

Figure 17.39 Cation-pool methodology based on a) electrolytic generation of *N*-acyliminium cations and b) subsequent reaction with allyl-magnesium-bromide without electrolysis [96].

A particular challenge is finding electrolytes and solvents suitable for low-temperature work. Solidification and ion pairing would make the synthetic process impossible and must be prevented. Some electrolyte systems have been developed, in which *N*-acyliminium ions can be handled at temperatures up to 0 °C [95]. The use of the cation pool idea in microflow reactor systems reduces the time over which the reactive cation must be stabilized to the transfer time between its electrochemical generation and the mixing with the substrate stream (Figure 17.40). By minimizing the interelectrode gap in flow cells (or by employing interdigitated microband array electrodes), work with low electrolyte concentration and less polar solvents is possible. Microflow reactor systems provide flexibility and new opportunities, for example, for combinatorial flow synthesis [64].

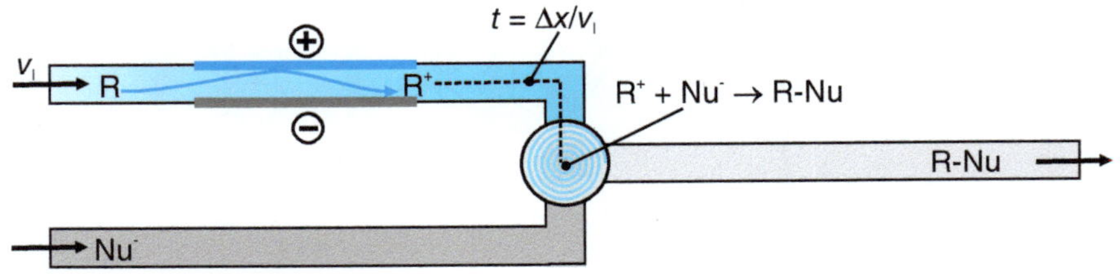

Figure 17.40 Principle of the cation-flow methodology according to [94].

The cation-pool methodology has been fruitful and applied for *N*-acyliminium, carbenium, carboxonium, iodonium, sulfonium, and silyl cations, with the following reactions such as alkylation, allylation, reductive dimerization, cross-coupling, arylthiolation, alkane oxidation, or halogenation [16, 17, 95].

17.5.2 Paired Electrosynthesis

Commonly, the electrosynthetic product of interest is generated either at the anode or at the cathode. Although there also has to be a product formation at the counter electrode to allow a current through the electrochemical cell, the energy demand and the products of the reaction at the counter electrode are often ignored, at least in laboratory electrolysis. Paired electrosynthesis in contrast offers a methodology to minimize loss of energy and/or production of waste.

Syntheses based on the combined or simultaneous use of cathodic and anodic reaction products are called **paired electrosynthesis**.

In a way, the chlor–alkali process (Excursion 17-1) can be considered as an example of a paired electrosynthesis to produce Cl_2 at the anode and NaOH at the cathode from the electrolyte salt NaCl. Conceptually, the term "paired electrosynthesis" is associated more with organic transformations where there is a very clear distinction between neutral reagents and supporting electrolyte salts. Different types of paired electrosynthesis have been developed.

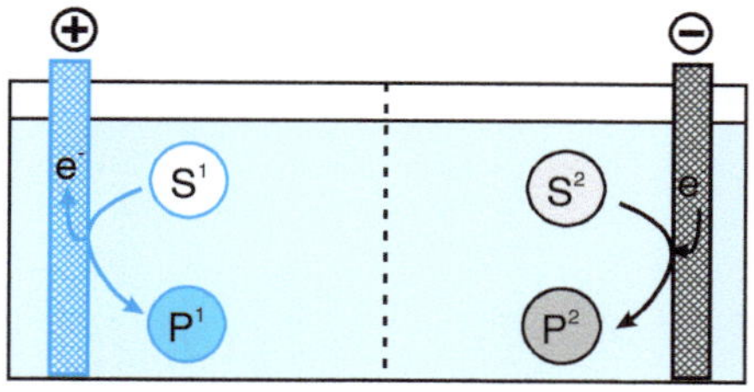

Figure 17.41 Schematic for a parallel paired electrosynthesis.

Parallel paired electrolysis (two reagents, two products). Two reagents (substrates S^1 and S^2) are used with an anodic reaction $S^1 - e^- \rightarrow P^1$ and a cathodic reaction $S^2 + e^- \rightarrow P^2$ running in parallel (Figure 17.41). A divided cell configuration is employed to prevent mixing and homogeneous following reactions between S^1, P^2, P^1, and S^2. A representative technical process developed by BASF is the parallel synthesis of *t*-butylbenzaldehyde-dimethylacetal and phthalide (Figure 17.42) [82]. Hydrolysis of the acetal leads to the product *t*-butylbenzaldehyde. This acetal synthesis offers an example of the selective oxidation of a methyl group at an aromatic ring, which is elegant via electrosynthesis. Traditionally, this conversion would require chlorination and further steps or oxidation with transition metal reagents; both approaches have a low atom economy and bear the risk of overoxidation to carboxylic acids instead of yielding the aldehyde. The electrosynthetic reaction occurs in methanol to immediately produce the ether and the acetal, which are quite stable against overoxidation. Technically, this electrosynthesis is performed in an undivided capillary gap cell (Figure 17.11). Methanol liberated at the cathode is used again at the anode [82]. The current yield (the amount of product per stoichiometric charge transfer) reaches 190% [97] due to the formation of products simultaneously at the anode and the cathode.

Divergent paired electrolysis (one reagent, two products). When using only one reagent, it is possible to generate two products, at the anode and the cathode in parallel (Figure 17.43). There are a limited number of sensible reagents for such a process, that is, the approach is of limited scope. For instance, the reaction of glucose with

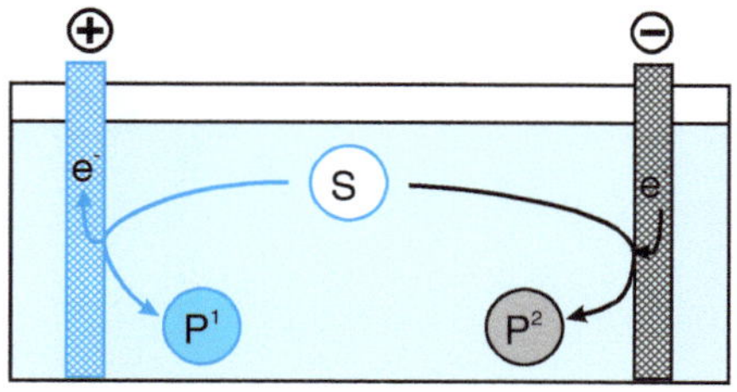

Figure 17.43 Schematic depiction of a divergent paired electrosynthesis.

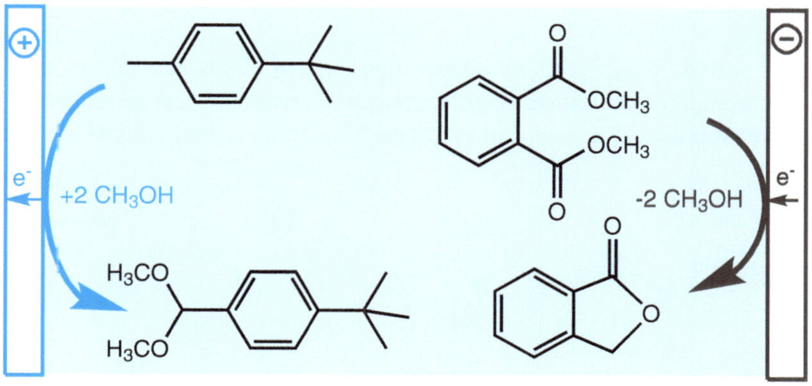

Figure 17.42 Schematic for the parallel paired electrosynthesis of *t*-butylbenzaldehyde-dimethylacetal and phthalide.

anodically generated hypobromite BrO^- produces gluconic acid, while glucose is converted to sorbitol at the Raney Nickel[154] cathode [98]. The reaction can be considered as an electrocatalytic hydrogenation because nickel hydride is formed on the cathode surface and acts as a catalytic intermediate.

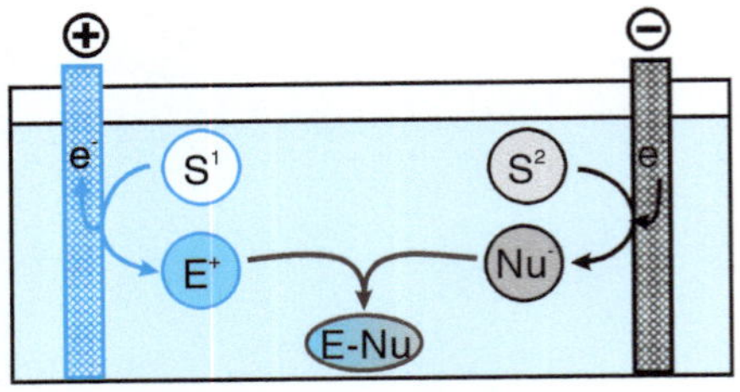

Figure 17.44 Schematic for a convergent paired electrosynthesis.

Convergent paired electrolysis (two reagents, two intermediates, one final product). The anodically generated electrophile E^+ can be paired with cathodic production of the nucleophile Nu^-. A following homogeneous reaction of E^+ and Nu^- then gives E–Nu (Figure 17.44). An example of this reaction type (Figure 17.45) is the anodic generation of benzyl cations coupled with the cathodic formation of nitro-substituted carbanions. The process is performed in an undivided cell in perchlorate electrolyte.

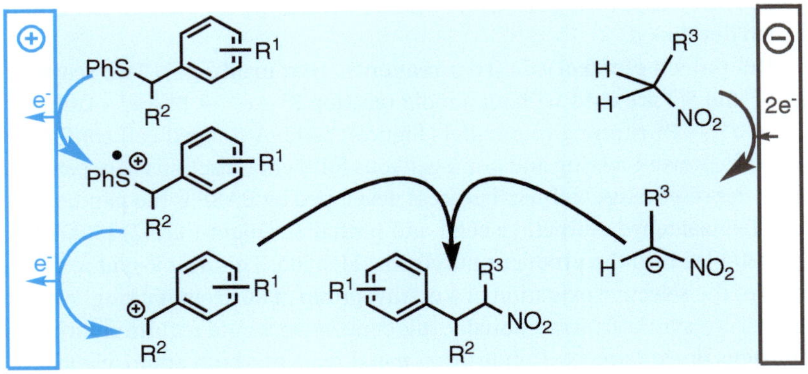

Figure 17.45 Simplified reaction scheme for a benzylic nitroalkylation in a convergent paired electrolysis according to [99].

Linear paired electrolysis (one reagent, anodic and cathodic reactions produce the same product with up to 200% current yield).

Q: This looks very suspicious. How can the oxidation and the reduction of the same reagent deliver the same product?

A: This may seem unusual, but it can be achieved with mediators and molecular oxygen as reagent. At the cathode, O_2 molecules are *reduced* to the highly reactive oxidant H_2O_2. H_2O_2 is then able to perform oxidations (Figure 17.46) of, for example, WO_4^{2-} to WO_5^{2-}. This pertungstate allows further oxidations similar to the process at the anode based on Br_2.

Linear paired electrosynthesis allows high current yields because each electron passing through the cell causes two electrons to be removed from the reagent.

Further Reading. Reviews and examples for paired electrosynthesis are available [17, 99].

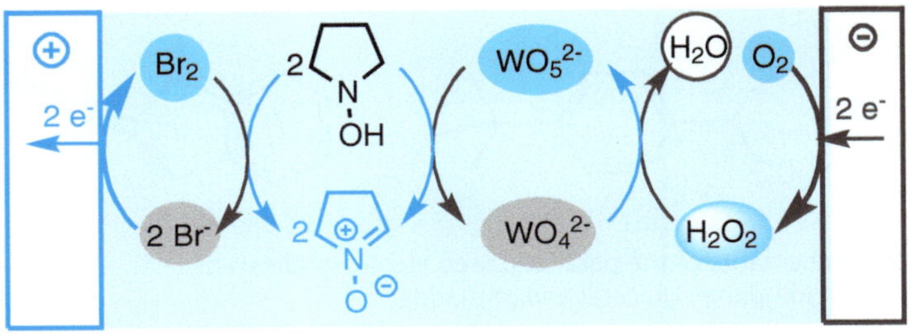

Figure 17.46 Linear paired electrolysis for the oxidation of hydroxylamine according to [17].

17.5.3 Redox Tags and Electrochemically Initiated Ring Formation (Annulation)

This chapter is concluded with the discussion of some electrosynthetic processes that do not exhibit a formal change in oxidation state in the substrate. Reaction equations written out appear to be without ET. Ring formation reactions or annulations provide important examples for such reactions. These are called redox-neutral reactions and have recently been reviewed systematically [101]. For example, a heterogeneous one-electron oxidation can be employed to initiate radical reactions under mild conditions. During this reaction, intramolecular ET stabilizes the radical and the transfer of one electron from the substrate to the product initiates a radical chain reaction. The radical chain reaction (ETC mechanism) causes transformation with only small amounts of charge transferred between the substrate and the electrode. Alternatively, the reduction of the product radical cation could occur at the electrode. This type of reaction mechanism is categorized here as the $\vec{E}C\overleftarrow{E}$ mechanism (Table 6.3 and Section 6.4.4). The chain reaction dominates for high substrate concentrations. Similar to processes discussed in Section 17.4.2, substoichiometric transfer of charge occurs and current yields are higher than 100% (called superstoichiometric current yield).

Important for this type of reaction is a formal potential for the substrate (assuming oxidation) that is lower compared to the formal potential of the product. Homogeneous ET is required from the intermediate back to the substrate. In order to satisfy both conditions, the redox properties of the reagent must be fine-tuned.

> Fine-tuning of electronic properties is possible with substituents, which act as **redox tags**.

The concept of redox tags has wider significance beyond the syntheses discussed in this section and represents a conceptual approach that is increasingly often used in the field of electroorganic chemistry [102–105].

Electrochemically Initiated (2 + 2)-Cycloaddition. A model reaction for this class of transformations is the (2 + 2)-cycloaddition of enolethers with terminal olefins to cyclobutanes [16, 104] (Figure 17.47). This reaction begins with a one-electron oxidation of the enolether. The resulting radical cation binds to the olefin (present in excess). Stabilization of the intermediate occurs via intramolecular ET from the anisyl moiety to form the cyclobutane ring while simultaneously generating a persistent aromatic radical cation. The radical cation oxidizes the enolether substrate by a homogeneous ET. This reaction forms the cyclobutane as the final product and propagates the chain reaction. The product is obtained with 94% yield and superstoichiometric current yield. The aromatic substituent attached to the olefin is initially an electron donor and then an electron acceptor (redox tag). Variation of the structure of the redox tag reveals how sensitive the reaction is to the electronic properties of the redox tag. The reaction occurs with electron-rich aromatic tags (aromatic substitution with 1–2 methoxy groups or with 1–3 methyl groups). The cycloaddition occurs with increased efficiency when the formal potential of the redox tag decreases. However, the formal potential of the redox tag must always be larger than that of the enol; otherwise, the redox tag cannot oxidize the enol in the first step. Based on these criteria, redox-tags fail with three methoxy groups or electron-rich nitrogen- or sulfur-containing groups attached to the aromatic ring.

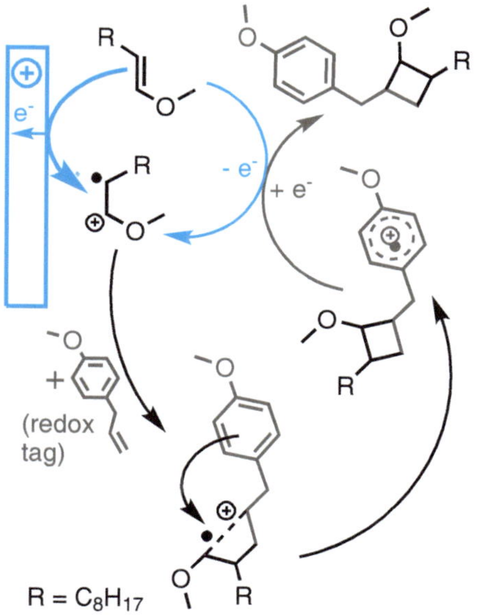

Figure 17.47 Schematic for the electrochemically initiated formal (2 + 2)-cycloaddition according to [104].

For the nomenclature of cycloaddition reactions, see notes in Table 17.1.

Electrochemically Initiated Olefin Metathesis. Phenyl rings without substitution are unsuitable as redox tags (Figure 17.48). The reaction produces the cyclobutyl-radical cation similar to Figure 17.47, but there is no intramolecular ET. Instead, the cyclobutyl-radical cation undergoes cleavage into the starting materials

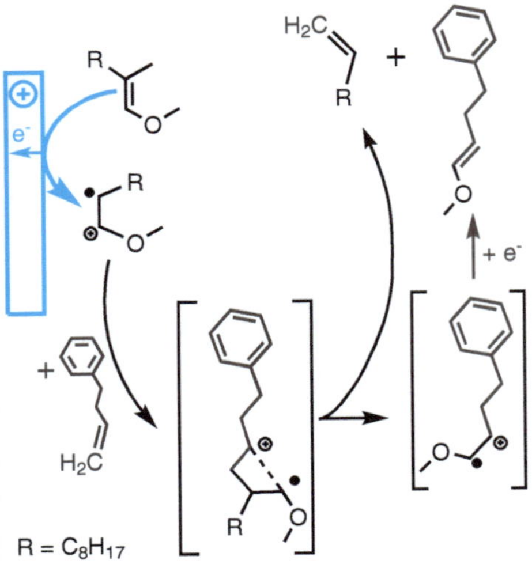

Figure 17.48 Schematic for olefin metathesis with an inefficient redox tag according to [103].

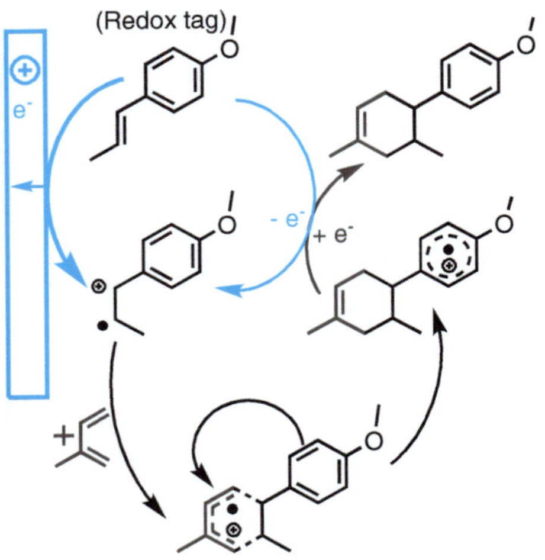

Figure 17.49 Schematic for an electrochemically initiated (4 + 2)-cycloaddition according to [16].

or metathesis products (Figure 17.48) [103, 105]. Yields of metathesis products can be increased from 48% to 79% [16] by using cyclohexyl instead of the phenyl groups that are shown in Figure 17.48.

Electrochemically Initiated (4 + 2)-Cycloaddition. Diels–Alder reactions[149] between dienes and dienophiles are commonly employed in organic synthesis. The electrochemically initiated (4 + 2)-cycloaddition starts with a one-electron transfer, causing an umpolung or polarity inversion of a nucleophilic substrate into an electrophilic intermediate. This intermediate then enters into cycloaddition with the diene that would not be possible otherwise. The redox tag is again crucial because the cationic intermediate with the cyclohexyl ring in Figure 17.49 has to be stabilized by an intramolecular one-electron transfer. With a 4-methoxyphenyl redox tag, the reaction occurs quantitatively with a charge of only 0.1 *F*, while no reaction occurs with an unsubstituted phenyl ring in the position of the redox tag.

Test your knowledge

Exercise 17.1: Consider a mixed Kolbe electrolysis with a molar ratio of the two starting carboxylic acids of 1:1, 1:10, and 1:100. What are the resulting statistical yields for symmetric and asymmetric coupling products?

Exercise 17.2: Based on literature reports [10, 11], compare process characteristics for various adiponitrile electrosyntheses in terms of energy efficiency and yield.

Exercise 17.3: What are the selection rules for pericyclic (concerted) cycloadditions (Woodward–Hoffmann rules[165])? Why do these rules fail in predicting stereospecific processes under conditions for electrochemical Diels–Alder reactions?

Exercise 17.4: Suggest a method to couple the two olefinic double bonds in 5,5-dimethyl-2,7-nonadiene dicarboxylate-dimethylester by ring formation to a cyclopentane derivative. Hint: One of the double bonds requires umpolung. Consider whether these double bonds are electron-rich or electron-poor and how this affects reactivity.

Exercise 17.5: Which electrosynthesis approaches allow current yields of higher than 100%. Identify two types of processes.

References

1 H. Lund, M. M. Baizer; *Organic Electrochemistry. An Introduction and a Guide.* 3rd ed., Marcel Dekker, New York, 1991.

2 D. Pletcher, F. C. Walsh; *Industrial Electrochemistry*, 2nd ed., Springer Science & Business Media, 2013.

3 N. L. Weinberg, B. V. Tilak (Eds.); *Technique of Electroorganic Synthesis. Scale-up and Engineering Aspects*, John Wiley & Sons, New York, 1982.

4 S. Torii; *Electroorganic Synthesis. Methods and Application.* Monographs in Modern Chemistry, *Vol. 15 and 16*, Kodansha Ltd., Tokyo and VCH-Verlagsgesellschaft, Weinheim, 1985.

5 A. J. Bard, M. Stratmann, H. J. Schäfer (Eds.); *Encyclopedia of Electrochemistry, Vol. 8*, Wiley-VCH, Weinheim, 2004.

[165] Robert Burns Woodward, 1917–1979, American organic chemist, received the Nobel Prize of chemistry in 1965; Roald Hoffmann, born 1937 in Złoczów (Poland, today located in Ukraine), US American chemist, received the Nobel Prize of chemistry in 1981.

6 S. Torii; *Electroorganic Reductions Synthesis*, Wiley-VCH, Weinheim, 2006.

7 T. Fuchigami, S. Inagi, M. Atobe; *Fundamentals and Applications of Organic Electrochemistry: Synthesis, Materials, Devices*, John Wiley & Sons, Chichester, 2015.

8 O. Hammerich, B. Speiser (Eds.); *Organic Electrochemistry*, 5th ed., CRC Press, Boca Raton, 2016.

9 F. Hine; *Electrode Processes and Electrochemical Engineering*, Plenum Press, New York, 1985.

10 V. M. Schmidt; *Electrochemical Process Engineering*, Wiley-VCH, Weinheim, 2006.

11 C. H. Hamann, A. Hamnett, W. Vielstich; *Electrochemistry*, 2nd ed. Wiley-VCH, Weinheim, 2007, pp. 301–337.

12 G. Kreysa, K.-I. Ota, R. Savinell (Eds.); *Encyclopedia of Applied Electrochemistry*, Springer, New York, 2014.

13 M. Yan, Y. Kawamata, P. S. Baran; Synthetic organic electrochemical methods since 2000: On the verge of a renaissance. *Chem. Rev.* **2017**, *117*, 13230–13319.

14 M. Yan, Y. Kawamata, P. S. Baran; Synthetic organic electrochemistry: Calling all engineers. *Angew. Chem. Int. Ed.* **2018**, *57*, 4149–4155.

15 A. Wiebe, T. Gieshoff, S. Möhle, E. Rodrigo, M. Zirbes, S. R. Waldvogel; Electrifying organic synthesis. *Angew. Chem. Int. Ed.* **2018**, *57*, 5594–5619.

16 S. Möhle, M. Zirbes, E. Rodrigo, T. Gieshoff, A. Wiebe, S. R. Waldvogel; Modern electrochemical aspects for the synthesis of value-added organic products. *Angew. Chem. Int. Ed.* **2018**, *57*, 6018–6041.

17 G. Hilt; Basic strategies and types of applications in organic electrochemistry. *ChemElectroChem* **2020**, *7*, 395–405.

18 C. Kingston, M. D. Palkowitz, Y. Takahira, J. C. Vantourout, B. K. Peters, Y. Kawamata, P. S. Baran; A survival guide for the "electro-curious". *Acc. Chem. Res.* **2020**, *53*, 72–83.

19 E. C. R. McKenzie, S. Hosseini, A. G. Coutro Petro, K. K. Rudman, B. H. R. Gerroll, M. S. Mubarak, L. A. Baker, R. D. Little; Versatile tools for understanding electrosynthetic mechanisms. *Chem. Rev.* **2022**, *122*, 3292–3335.

20 F. Marken; The electrochemistry of halogens. in *Encyclopedia of Electrochemistry. Vol. 7*, (A. J. Bard, M. Stratmann, F. Scholz, C. J. Pickett, Eds.), Wiley-VCH, Weinheim, 2006, pp. 273–302.

21 C. H. Hamann, T. Röpke, P. Schmittinger; Electrowinning of other inorganic compounds. in *Encyclopedia of Electrochemistry. Vol. 5,* (A. J. Bard, M. Stratmann, D. D. Macdonald, P. Schmuki, Eds.), Wiley-VCH, Weinheim, 2007, pp. 267–315.

22 A. Cornell; Chlorate synthesis cells and technology. in *Encyclopedia of Applied Electrochemistry* (G. Kreysa, K.-I. Ota, R. F. Savinell, Eds.), Springer, New York, 2014, pp. 181–187.

23 J. Aromaa, J. E. Ewans, D. P. Ziegler; Electrowinning of metals. in *Encyclopedia of Electrochemistry. Vol. 5,* (A. J. Bard, M. Stratmann, D. D. Macdonald, P. Schmuki, Eds.), Wiley-VCH, Weinheim, 2007, pp. 159–265.

24 G. M. Haarberg; Aluminum smelter technology. in *Encyclopedia of Applied Electrochemistry* (G. Kreysa, K.-I. Ota, R. F. Savinell, Eds.), Springer, New York, 2014, pp. 59–65. https://doi.org/10.1007/978-1-4419-6996-5_452.

25 P. Taxil; Refractory metal production by molten salt electrolysis. in *Encyclopedia of Applied Electrochemistry* (G. Kreysa, K.-I. Ota, R. F. Savinell, Eds.), Springer, New York, 2014, pp. 1801–186; https://doi.org/10.1007/978-1-4419-6996-5_456.

26 G. M. Haarberg; Magnesium smelter technology. in *Encyclopedia of Applied Electrochemistry* (G. Kreysa, K.-I. Ota, R. F. Savinell, Eds.), Springer, New York, 2014, pp. 1210–1213; https://doi.org/10.1007/978-1-4419-6996-5.

27 G. M. Haarberg; Alkali and alkaline earth metal production by molten salt electrolysis. in *Encyclopedia of Applied Electrochemistry* (G. Kreysa, K.-I. Ota, R. F. Savinell, Eds.), Springer, New York, 2014, pp. 21–25; https://doi.org/10.1007/978-1-4419-6996-5_451.

28 W. Löb; Unsere Kenntnisse in der Elektrolyse und Elektrosynthese organischer Verbindungen. *Z. Elektrochem.* **1895**, *2*, 293–308.

29 S. B. Beil, D. Pollok, S. R. Waldvogel; Reproducibility in electroorganic synthesis – myths and misunderstandings. *Angew. Chem. Int. Ed.* **2021**, *60*, 14750–14759.

30 J. Jörissen, B. Speiser; Preparative electrolysis on the laboratory scale. in *Organic Electrochemistry*, 5th ed. (O. Hammerich, B. Speiser, Eds.), CRC Press, Boca Raton, 2016, pp. 263–330.

31 J. Jörissen; Practical aspects of preparative scale electrolysis. in *Encyclopedia of Electrochemistry, Vol. 8*, (A. J. Bard, M. Stratmann, H. J. Schäfer), Wiley-VCH, Weinheim, 2004, pp. 25–72.

32 D. Pletcher, R. A. Green, R. C. D. Brown; Flow electrolysis cells for the synthetic organic chemistry laboratory. *Chem. Rev.* **2018**, *118*, 4573–4591.

33 M. Atobe, H. Tateno, Y. Matsumura; Application of flow microreactors in electrosynthetic processes. *Chem. Rev.* **2018**, *118*, 4541–4572.

34 YouTube: https://www.youtube.com/watch?v=C8UNt3d5_3Q&feature=emb_logo (last access 15 Mar 2026); https://www.youtube.com/watch?v=4f-ZafVaMf8&feature=emb_logo (last access 16 Mar 2026).

35 F. Beck; *Elektroorganische Chemie.* Grundlagen und Anwendungen. Verl. Chemie, Weinheim, 1974.

36 E. Steckhan, T. Arns, W. R. Heineman, G. Hilt, D. Hoormann, J. Jörissen, L. Kröner, B. Lewall, H. Pütter; Environmental protection and economization of resources by electroorganic and electroenzymatic syntheses. *Chemosphere* **2001**, *43*, 63–73.

37 P. M. Robertson, P. Berg, H. Reimann, K. Schleich, P. Seiler; Application of “swiss-roll” electrolysis cell in vitamin-C production. *J. Electrochem. Soc.* **1983**, *130*, 591–596.

38 F. Marken, J. D. Wadhawan; Multiphase methods in organic electrosynthesis. *Acc. Chem. Res.* **2019**, *52*, 3325–2238.

39 M. Elsherbini, T. Wirth; Electroorganic synthesis under flow conditions. *Acc. Chem. Res.* **2019**, *52*, 3287–3296.

40 E. M. Stuve; Electrochemical reactor design and configurations. *Encyclopedia of Applied Electrochemistry* (G. Kreysa, K.-I. Ota, R. F. Savinell, Eds.), Springer, New York, 2014; https://doi.org/10.1007/978-1-4419-6996-5_326.

41 J.-I. Yoshida, T. Nokami, S. Suga; Organometallic compounds. in *Organic Electrochemistry*, 5th ed. (O. Hammerich, B. Speiser, Eds.), CRC Press, Boca Raton, 2016, pp. 1357–1392.

42 S. R. Waldvogel, S. Mentizi, A. Kirste; Boron-doped diamond electrodes for electroorganic chemistry. in *Radicals in Synthesis III* (M. Heinrich, A. Gansäuer, Eds.), *Top. Curr. Chem.* **2012**, *320*, Springer. 1–32.

43 https://www.condias.de/Produkte/Synthese_StarterKit (last access 16 Mar 2026).

44 N. Ž. Nikolić, V. Panić; Electrocatalysis of chlorine evolution. in *Encyclopedia of Applied Electrochemistry* (G. Kreysa, K.-I. Ota, R. F. Savinell, Eds.), Springer, New York, 2014, pp. 411–417.

45 F. C. Walsh, R. G. A Wills; The continuing development of Magnéli phase titanium sub-oxides and Ebonex® electrodes. *Electrochim. Acta* **2010**, *55*, 6342–6351.

46 D. M. Heard, A. J. J. Lennox; Electrode materials in modern organic electrochemistry. *Angew. Chem. Int. Ed.* **2020**, *59*, 18866–18884.

47 Review on statistics of chlorine production and its geographic distribution are published annually by the European interest group EuroChlor of chlorine producers. https://www.chlorineindustryreview.com (currently accessible under the menu item "Download Review"; last access 20 Jun 2024).

48 T. F. O'Brien, T. V. Bommaraju, F. Hine; *Handbook of Chlor–Alkali Technology, Vol. I–V*; Springer, New York, 2005.

49 J. Jörissen; Chlorine and caustic technology. in *Encyclopedia of Applied Electrochemistry* (G. Kreysa, K.-I. Ota, R. F. Savinell, Eds.), Springer, New York, 2014, pp. 187–207.

50 T. V. Bommaraju, P. J. Orosz, E. A. Sokol; Brine electrolysis. *Electrochemistry Encyclopedia*. https://knowledge.electrochem.org/encycl/art-b01-brine.htm (last access 16 Mar 2026).

51 Wikipedia: Chloralkali process; https://en.wikipedia.org/wiki/Chloralkali_process#Manufacturer_associations (last access 16 Mar 2026).

52 J. Jörissen; Hydrochloric acid electrolysis. in *Encyclopedia of Applied Electrochemistry* (G. Kreysa, K.-I. Ota, R. F. Savinell, Eds.), Springer, New York, 2014, pp. 1030–1035.

53 D. T. Sawyer, J. L. Roberts Jr.; *Experimental Electrochemistry for Chemists*, Wiley, New York, 1974.

54 C. K. Mann; Nonaqueous solvents for electrochemical use. in *Electroanalytical Chemistry, Vol. 3*, (A. J. Bard Ed.), Marcel Dekker, New York, 1969, pp. 57–134.

55 K. Izutsu; *Electrochemistry in Nonaqueous Solvents*. Wiley-VCH, Weinheim, 2002, pp. 287–311.

56 S. Creager; Solvents and supporting electrolytes. in *Handbook of Electrochemistry* (C. G. Zosky, Ed.), Elsevier, Amsterdam, 2007, pp. 57–72.

57 W. E. Geiger; Electrolytes based on weakly coordinating anions: An advance in anodic molecular electrochemistry. in *Electroanalytical Chemistry, Vol. 25*, (A. J. Bard, C. G. Zoski, Eds.), CRC-Press, Boca Raton, 2014, pp. 179–222.

58 K. D. Moeller; Anodic olefin coupling reactions: A mechanism driven approach to the development of new synthetic tools. *ECS Interface* **2016**, *25*, 53–59.

59 H. J. Schäfer; Oxidative coupling. in *Organic Electrochemistry*, 5th ed. (O. Hammerich, B. Speiser, Eds.), CRC Press, Boca Raton, 2016, pp. 705–773.

60 H. Tanaka, M. Kuroboshi, S. Torii; Oxidation of carboxylic acid derivatives. in *Organic Electrochemistry*, 5th ed. (O. Hammerich, B. Speiser, Eds.), CRC Press, Boca Raton, 2016, pp. 1267–1307.

61 J. Pinson, J. M. Savéant; Electrochemically induced aromatic nucleophilic substitution. *J. Am. Chem. Soc.* **1978**, *100*, 1506–1510.

62 J. Ludvík; Reduction of aldehydes, ketones, and azomethines. in *Organic Electrochemistry*, 5th ed. (O. Hammerich, B. Speiser, Eds.), CRC Press, Boca Raton, 2016, pp. 1201–1247.

63 A. Hutinec, A. Ziogas, M. El-Mobayed, A. Rieker; Spirolactones of tyrosine: Synthesis and reaction with nucleophiles. *J. Chem. Soc. Perkin Trans.* **1998**, *1*, 2201–2208.

64 K. D. Moeller; Combinatorial electrochemistry and miniaturization. in *Organic Electrochemistry*, 5th ed. (O. Hammerich, B. Speiser, Eds.), CRC Press, Boca Raton, 2016, pp. 345–369.

65 M. R. Rifi; Electrochemical preparation of anticyclones and other strained cycloalkanes. *J. Am. Chem. Soc.* **1967**, *89*, 4442–4445.

66 J. H. P. Utley, R. D. Little, M. F. Nielsen; Reductive coupling. in *Organic Electrochemistry*, 5th ed. (O. Hammerich, B. Speiser, Eds.), CRC Press, Boca Raton, 2016, pp. 621–703.

67 A. Ozaki, Y. Yamaguchi, Y. Okada, K. Chiba; Bidirectional access to radical cation Diels–Alder reactions by electrocatalysis. *ChemElectroChem* **2017**, *4*, 1852–1855.

68 Y. Imada, N. Shida, Y. Okada, K. Chiba; A novel thermomorphic system for electrocatalytic Diels–Alder reactions. *Chin. J. Chem.* **2019**, *37*, 557–560.

69 IUPAC; *Compendium of Chemical Terminology*, 2nd ed. (the "Gold Book"). Compiled by A. D. McNaught, A. Wilkinson. Blackwell Scientific Publications, Oxford, 1997. Keyword "Cycloaddition"; Online version (2019) created by S. J. Chalk. https://goldbook.iupac.org/terms/view/C01496 (last access 16 Mar 2026).

70 K. Chiba, T. Miura, S. Kim, Y. Kitano, M. Tada; Electrocatalytic intermolecular olefin cross-coupling by anodically induced formal [2+2] cycloaddition between enol ethers and alkenes. *J. Am. Chem. Soc.* **2001**, *123*, 11314–11315.

71 E. Fritz-Langhals; German Patent DE 100 39 171 A 1, 2002.

72 O. Hammerich; Reduction of nitro compounds and related substrates. in *Organic Electrochemistry*, 5th ed. (O. Hammerich, B. Speiser, Eds.), CRC Press, Boca Raton, 2016, pp. 1149–1200.

73 R. Francke, T. Quell, A. Wiebe, S. R. Waldvogel; Oxygen-containing compounds. in *Organic Electrochemistry*, 5th ed. (O. Hammerich, B. Speiser, Eds.), CRC Press, Boca Raton, 2016, pp. 981–1033.

74 K. D. Moeller; Intramolecular anodic olefin coupling reactions: Using radical cation intermediates to trigger new umpolung reactions. *Synlett* **2009**, 1208–1218.

75 G. Hilt; Convergent paired electrolysis for the three-component synthesis of protected homoallylic alcohols. *Angew. Chem. Int. Ed.* **2003**, *42*, 1720–1721.

76 E. Steckhan; Indirect electroorganic syntheses – a modern chapter of organic electrochemistry. *Angew. Chem. Int. Ed.* **1986**, *25*, 683–701.

77 T. Noël, Y. Cao, G. Laudadio; The fundamentals behind the use of flow reactors in electrochemistry. *Acc. Chem. Res.* **2019**, *52*, 2858–2869.

78 A. K. Yudin, T. Siu; Combinatorial electrochemistry. *Curr. Opin. Chem. Biol.* **2001**, *5*, 269–272.

79 Y. Hioki, M. Costantini, J. Griffin, K. C. Harper, M. Prado Merini, B. Nissl, Y. Kawamata, P. S. Baran; Overcoming the limitations of Kolbe coupling with waveform-controlled electrosynthesis. *Science* **2023**, *380*, 81–87.

80 M. C. Leech, K. Lam; Electrosynthesis using carboxylic acid derivatives: New tricks for old reactions. *Acc. Chem. Res.* **2020**, *53*, 121–134.

81 M. Watson, D. Pletcher, D. Sopher; A microelectrode study of competing electrode reactions in the commercial process for the hydrodimerization of acrylonitrile to adiponitrile. *J. Electrochem. Soc.* **2000**, *147*, 3751–3758.

82 N. Aust; Organic electrochemistry, industrial aspects. in *Encyclopedia of Applied Electrochemistry* (G. Kreysa, K.-I. Ota, R. F. Savinell, Eds.), Springer, New York, 2014, pp. 1392–1397.

83 F. Wang, S. S. Stahl; Electrochemical oxidation of organic molecules at lower overpotential: Accessing broader functional group compatibility with electron-proton-transfer mediators. *Acc. Chem. Res.* **2020**, *53*, 561–574.

84 A. R. Amundsen, E. N. Balko; Preparation of chiral diols by the osmium-catalysed, indirect anodic oxidation of olefins. *J. Appl. Electrochem.* **1992**, *22*, 810–816.

85 R. Francke, B. Schille, M. Roemelt; Homogeneously catalyzed electroreduction of carbon dioxide – methods, mechanisms, and catalysts. *Chem. Rev.* **2018**, *118*, 4631–4701.

86 C. Kohlmann, S. Lütz; Electroenzymatic synthesis. in *Organic Electrochemistry*, 5th ed. (O. Hammerich, B. Speiser, Eds.), CRC Press, Boca Raton, 2016, pp. 1511–1541.

87 K. Chiba, Y. Okada; Electron transfer-catalyzed reactions. in *Organic Electrochemistry*, 5th ed. (O. Hammerich, B. Speiser, Eds.), CRC Press, Boca Raton, 2016, pp. 531–555.

88 T. Fuchigani, S. Inagi; Organic electrosynthesis. in *Fundamentals and Applications of Organic Electrochemistry: Synthesis, Materials, Devices*. (T. Fuchigami, S. Inagi, M. Atobe), John Wiley & Sons, Chichester, 2015, pp. 83–90.

89 J. E. Nutting, M. Raffa, S. S. Stahl; Tetramethylpiperidine *N*-oxyl (TEMPO), phthalimide *N*-oxyl (PINO) and related *N*-oxyl species: Electrochemical properties and their use in electrocatalytic reactions. *Chem. Rev.* **2018**, *118*, 4834–4885.

90 E. J. Horn, B. R. Rosen, Y. Chen, J. Tang, K. Chen, M. D. Eastgate, P. S. Baran; Scalable and sustainable electrochemical allylic C–H oxidation. *Nature* **2016**, *533*, 77–81.

91 R. Francke, R. D. Little; Redox catalysis in organic electrosynthesis: Basic principles and recent developments. *Chem. Soc. Rev.* **2014**, *43*, 2492–2521.

92 S. Doobary, A. T. Sedikides, H. P. Caldora, D. L. Poole, A. J. J. Lennox; Electrochemical vicinal difluorination of alkenes: Scalable and amendable to electron-rich substrates. *Angew. Chem.* **2020**, *132*, 1171–1176., *Angew. Chem. Int. Ed.* **2020**, *59*, 1155–1160.

93 T. Fuchigami, M. Kasuga, A. Konno; Electro-organic chemistry of fullerenes. Part 1. Indirect cathodic reduction of *vic*-dihalides and perfluoroalkyl halides using C_{60} as mediator. Cyclic voltammetric study and preparative scale electrolysis. *J. Electroanal. Chem.* **1996**, *411*, 115–119.

94 J. H. P. Utley, H. G. Rozenberg; Electroorganic reactions. Part 57. DDQ mediated anodic oxidation of 2-methyl- and 2-benzylnaphtalenes. *J. Appl. Electrochem.* **2003**, *33*, 525–532.

95 J.-I. Yoshida, A. Shimizu, R. Hayashi; Electrogenerated cationic reactive intermediates: The pool methods and further advances. *Chem. Rev.* **2018**, *118*, 4702.

96 S. Suga, M. Okajima, J.-I. Yoshida; Reaction of an electrogenerated 'iminium cation pool' with organometallic reagents. Direct oxidative α-alkylation and -arylation of amine derivatives. *Tetrahedron Lett.* **2001**, *42*, 2173–2176.

97 R. Breinbauer, M. Peters; Reduction of carboxylic acids and derivatives. in *Organic Electrochemistry*, 5th ed. (O. Hammerich, B. Speiser, Eds.), CRC Press, Boca Raton, 2016, pp. 1249–1265.

98 K. Park, P. N. Pintauro, M. M. Baizer, K. Nobe; Flow reactor studies of the paired electro-oxidation and electroreduction of glucose. *J. Electrochem. Soc.* **1985**, *132*, 1850–1855.

99 C. A. Paddon, M. Atobe, T. Fuchigami, P. He, P. Watts, S. J. Haswell, G. J. Pritchard, S. D. Bull, F. Marken; Towards paired and coupled electrode reactions for clean organic microreactor electrosynthesis. *J. Appl. Electrochem.* **2006**, *36*, 617–634.

100 S. Kim, R. Uchiyama, Y. Kitano, M. Tada, K. Chiba; Benzylic nitroalkylation by paired electrolysis of benzylsulfides in nitroalkanes. *J. Electroanal. Chem.* **2001**, *507*, 152–156.

101 R. Francke, R. D. Little; Electrons and holes as catalysts in organic electrosynthesis. *ChemElectroChem* **2019**, *6*, 4373–4382.

102 Y. Okada, K. Chiba; Redox-tag process: Intramolecular electron transfer and its broad relationship to redox reaction in general. *Chem. Rev.* **2018**, *118*, 4592–4630.

103 Y. Okada, K. Chiba; Electron transfer-induced four-membered cyclic intermediate formation: Olefin cross-coupling vs. olefin metathesis. *Electrochim. Acta* **2011**, *56*, 1037–1042.

104 Y. Okada, A. Nishimoto, R. Akaba, K. Chiba; Electron-transfer-induced intermolecular [2 + 2] cycloaddition reactions based on the aromatic "redox tag" strategy. *J. Org. Chem.* **2011**, *76*, 3470–3476.

105 M. Arata, T. Miura, K. Chiba; Electroactive formal [2+2] cycloaddition reaction between anodically activated enyloxy benzene and alkenes. *Org. Lett.* **2007**, *9*, 4347–4350.

18
Bioelectrochemistry

Bioelectrochemistry applies the concepts of electrochemistry and its experimental methods to the study of living systems, or uses biological systems or biomolecules in technical systems. Examples for the latter, applied aspect of bioelectrochemistry are redox enzymes isolated from biological cells that are used in sensors, biofuel cells, and electroenzymatic syntheses.

Charge transfer processes at interfaces such as biomembranes play a crucial role in photosynthesis, respiration, and transmission of signals in the nervous system, on which life on Earth is directly dependent. Individual areas of bioelectrochemistry are closely linked to other scientific disciplines, such as biophysics, electrophysiology, biochemistry, cell biology, the ecology of biofilms, and biotechnology. The various areas of bioelectrochemistry develop in close exchange with these disciplines.

This chapter on bioelectrochemistry has been placed at the end of the textbook, which may suggest less importance when compared to "tangible" topics such as corrosion or batteries. However, this is by no means the case. On the contrary, bioelectrochemistry in the form of Luigi Galvani's "frog leg experiments"[5] (Figure 18.1) was historically at the beginning of electrochemistry as a science. This inspired the development of the first laboratory power sources (batteries), which became fundamentally important for the further development of chemistry and physics in the nineteenth century. The topic of bioelectrochemical processes in cells, tissues, and in health is still a growing research topic.

Figure 18.1 Monument to Luigi Galvani[5] in front of the historic building of the University of Bologna, Italy. The frog legs are commemorated on the open page of his book. Source: Photo by the book author.

Today, one of the seven Divisions of the International Society of Electrochemistry deals exclusively with bioelectrochemistry, and within the Electrochemical Society, one of the thirteen Divisions is devoted to Organic and Biological Electrochemistry. For all these reasons, it makes sense to conclude our journey through electrochemistry with a return to one of its cornerstones. Thus, after discussing the combination of electron transfer (ET) and ion transfer (IT) processes in the energy metabolism of biological cells, we will turn our attention to the question of how such reactions provide inspiration for the development of technical systems. In doing so, we will encounter important principles that we are already familiar with from previous sections of the book. The parallels that emerge perhaps best document the way in which the high degree of integration of biological cells was and is a source of inspiration for the development of modern technologies.

If you have studied the previous chapters carefully, you will have noticed that we have already presented some aspects of bioelectrochemistry in Excursions 5-6, 5-7, 13-1, and in Section 13.4. In addition to the basics in Chapters 1 to 4, this chapter also builds on Chapters 5, 6, and 13. Therefore, it is recommended to revisit them before studying this chapter.

Further Reading. Basic introductions to bioelectrochemistry can be found in collective books [1, 2]. A comprehensive multivolume survey is available [3]. For the popular subfield of biosensor technology, there are basic textbooks [4] and a series of

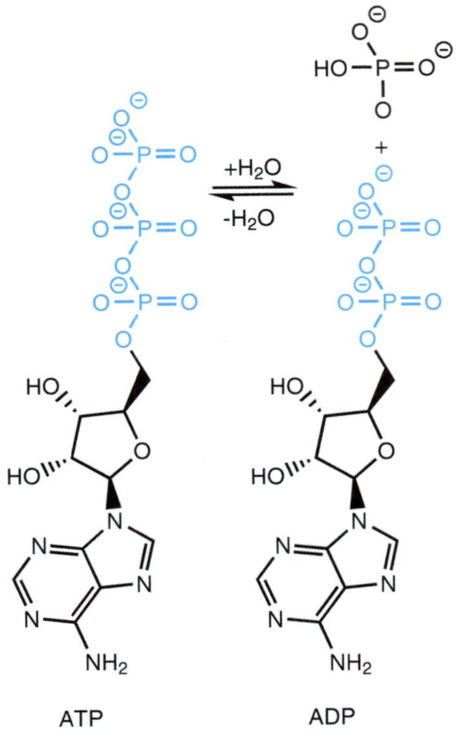

Figure 18.2 Structure of adenosine triphosphate (ATP) and adenosine diphosphate (ADP).

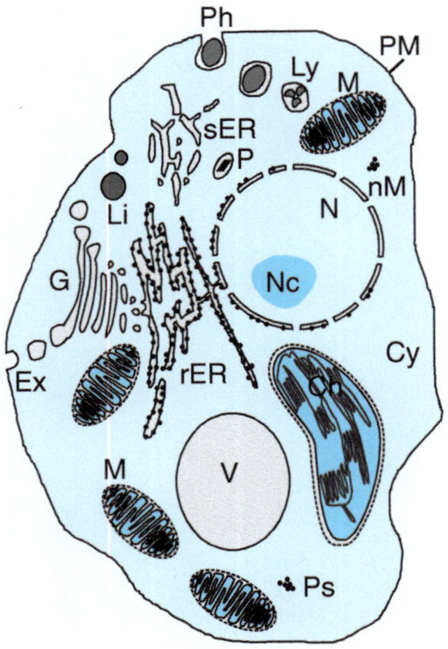

Figure 18.3 Components of an eukaryotic cell. The assignment to the compartments is listed in Table 18.1. Cy, cytoplasm; PM, plasma membrane; N, nucleus; Nc, nucleolus; nM, nuclear membrane; Ch, chloroplast; M, mitochondria; sER, smooth endoplasmic reticulum; rER, rough endoplasmic reticulum; G, Golgi apparatus; Ex, exocystosis; Ph, phagocytosis; Ly, lysosome; V, vacuole; P, peroxisome; Ps, polysome; Li, lipid body.

comprehensive anthologies [5, 6]. References to further reading on specific aspects are provided at the end of the relevant sections.

18.1 Coupling of Ion Transfer and Electron Transfer Processes in Photosynthesis and in Cellular Respiration

Living organisms are hierarchically structured, complex systems. This means that although they consist of a relatively small number of recurring building blocks, their countless number of possible connections and resulting interactions at different levels of integration produce an almost incomprehensible variety of functions at the higher structural levels. We also find this basic principle in complex technical systems such as computers or complex machinery [7]. From a thermodynamic point of view, living organisms are dissipative structures that constantly export entropy into the environment in order to maintain an internal order far from thermodynamic equilibrium [8, 9]. This primarily occurs through energy metabolism, with which "highly concentrated" energy is distributed into the environment in many "small portions". Despite the diversity of the living world, evolution has produced only a few realizations of this central function. The "energy currency" used in all cells is adenosine triphosphate (ATP), which is converted to the lower-energy adenosine diphosphate (ADP) by splitting off a phosphate residue, thereby driving other metabolic processes (Figure 18.2). The energy source for synthesizing the energy-rich ATP is the difference in electrochemical potentials (Chapter 2) of protons across a membrane. This driving force is built up via very complicated ET processes driven either by the absorption of light (photosynthesis) or the oxidation of energy-rich substances (cellular respiration). We concentrate here on general principles that connect the various living organisms. Precisely these proven general principles, some of which we have already encountered in the previous chapters, give us hope that a solution does indeed exist for some of the major unsolved technical problems, such as the efficient reduction of O_2 in neutral solutions (Chapters 12 and 16).

18.1.1 Biomembranes – All in One: "Circuit board" for Functional Units and Short-Term Energy Storage Device

Biological cells represent the lowest hierarchical level of living nature, in which the essential life functions of metabolism, reproduction, sensitivity, mutability, growth, and limitation are realized (elementary organism). Simple bacterial cells (prokaryotes) have only an outer cell membrane. All other cell components are dissolved in the cytoplasm without further subdivision. The cells of all higher organisms (fungi, plants, animals, humans) share a further subdivision of the cell interior, especially they contain a cell nucleus. The higher organisms are collectively referred to as eukaryotes. Here, we will take a closer look at the mitochondria (respiration) and the chloroplasts of plants (photosynthesis). Both are important in energy metabolism. An overview of other cell components can be found in Figure 18.3 with the assigned functions in Table 18.1. Mitochondria and chloroplasts evolved from originally independent prokaryotic cells into components of the eukaryotic host cell through endosymbiosis. They have their own genome, for example, mitochondrial deoxyribonucleic acid (mt-DNA). Both organelles form special reaction compartments within a cell, in which different reaction conditions prevail. This enables different reactions to occur simultaneously, very much like distinct reactions in different flasks of a chemical laboratory.

Table 18.1 Important functions of the cell compartments.

Compartment	Function
Cytoplasm	Cy, intracellular movement, metabolism, protein synthesis
Plasma membrane	PM, Ph, Ex; demarcation from and connection to the environment, stimulus reception, selective transport (here without details)
Cell nucleus	N, Nc, nM; DNA replication, transcription, RNA processing
Endoplasmic reticulum	rER with polysomes Ps; protein synthesis, sER, lipid synthesis
Golgi apparatus	Polysaccharide synthesis, export of metabolic products
Mitochondria	M, cellular respiration
Plastids, e.g., chloroplasts	Ch, photosynthesis
Lytic compartment	Ly, V, Li; digestion, storage, disposal
Microbodies, e.g., peroxisomes	Special tasks, including detoxification of reactive oxygen species

The abbreviations refer to Figure 18.3.

The biomembrane [10] represents the boundary between the cell and its environment. Its structure is derived from a lipid bilayer (Figure 13.33). In addition to phosphoglycerolipids (e.g., DMPC, in Figure 13.26), glycoglycerolipids, and cholesterol, the membrane also contains proteins (Figure 18.4). Biomembranes always surround their interior completely and, thus, separate the cytoplasm or a plasmatic phase from an outer, nonplasmatic or extraplasmatic phase. Accordingly, we refer to the two sides of the membrane as the P-side and E-side (Figure 18.4). This structural principle is common to all cells and cell organelles and is called the **compartmentalization rule**:

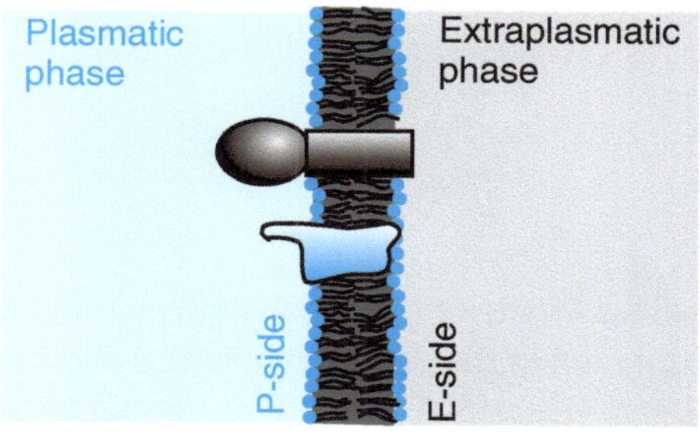

Figure 18.4 A biomembrane is asymmetric and separates a plasmatic phase (P) from an extraplasmatic phase (E).

A biological membrane always separates a plasmatic from a nonplasmatic phase.

For this reason, biomembranes are structurally and functionally asymmetrical (unlike the model membrane in Figure 13.33!). In order to separate the plasmatic and extraplasmatic phases, biomembranes must always be self-contained. Therefore, free-standing lamellae do not exist, but stacked invaginations do. In the membrane, the lipids are oriented with the hydrophobic molecular parts toward the inside of the membrane. Although this construction principle is uniform everywhere in nature, there is a great deal of variability in the exact composition of biomembranes, both between different organisms and within one and the same organism.

Let us now take a closer look at the contents of the cell organelles using the example of the mitochondria. The interior of the mitochondrion is a plasmatic phase that must not mix with the cytoplasm and is therefore surrounded by a membrane. The membrane has invaginations to increase the surface area. However, according to the compartmentalization rule (Figure 18.4), the outside of this membrane must be adjacent to an extraplasmatic phase, that is, it cannot directly adjoin the cytoplasm of the cell. Instead, the mitochondrion as a whole is surrounded by another membrane, the outer membrane. The outer and inner membranes of the mitochondria enclose a thin layer of an extraplasmatic phase. Analogous conditions are found in the chloroplasts. These structures can be marveled at in high-resolution electron microscopy images reproduced in textbooks on cell biology [11, 12].

Generally, a difference between two electrical potentials is called voltage. However, it is not common to call the potential difference across a membrane a "voltage". Instead, the difference between two potentials is often called just "potential" in this context. This somewhat sloppy usage collides with the nomenclature commonly used for discussing the origin of cell voltages in electrochemistry (see Chapter 2 of this textbook). We therefore use "potential difference" to remain close to the common usage in this field, but make a distinction between "potential" and "potential difference" as in Chapter 2.

The biomembranes not only contain lipids and various membrane-bound transport proteins, which ensure selective transport, but also a large number of proteins for recognition processes and redox processes. These proteins are called peripheral proteins if they are functionally associated with the membrane but can be easily detached from it. Integral membrane proteins are linked to the hydrophobic parts of the membrane-forming lipids by hydrophobic interactions. They can often only be isolated from the membranes using detergents, in which case they may lose their structure and function. The analysis and biotechnological utilization of isolated integral membrane proteins is therefore much more challenging than that of dissolved proteins. Often, about 20% of the protein is located within the hydrophobic part of the membrane, and the remaining part protrudes from one or both sides of the membrane. Since the lipid bilayers of the membrane are fluid, all components can perform lateral movements (fluid mosaic model [14]). The average distance between two proteins in the membrane is smaller than in a highly concentrated protein solution! Due to the orientation of the proteins along the surface normal of the membrane and their small and variable distance, functional aggregates of several proteins can form very quickly, which are of enormous importance for charge transfer and charge translocation processes to be discussed here.

Due to the hydrophobic interior, only small hydrophobic molecules can dissolve inside the membrane and diffuse through the membrane along a concentration gradient. Ions and polar molecules require special protein structures that enable their specific transport through the membrane. The transport is referred to as passive or catalyzed diffusion if it can only take place up to concentration equilibrium. Active transport, on the other hand, also takes place against a concentration gradient (ion pumps) under the consumption of energy.

The term "proton motive force" is common in the area of biophysics and has been selected in analogy to the outdated term "electromotive force" for the equilibrium voltage of a galvanic cell. Both quantities refer to a *voltage* (with unit V) and *not to a force* (with unit N). The proton motive force describes the voltage acting on protons on two sides of a membrane, which originates from a difference in the Galvani potential (membrane voltage or membrane potential) and the difference in pH.

If a specific type of ion is transported through a membrane, a Galvani potential difference builds up across the membrane, as we have already discussed using the example of the Donnan potential (difference)[13] in Figure 2.16. The Donnan potential difference is an equilibrium voltage. However, biological membranes are not in equilibrium, but in a steady state, which is why the description of the membrane potential differences is more complicated and often requires further approximations. It is particularly important to understand that the difference in the electrochemical potential $\Delta\tilde{\mu}_{H^+}$ of the oxonium ions ($H_3O^+ \equiv H^+$) and the associated motive force, also known as proton motive force ($pmf = \Delta\tilde{\mu}_{H^+}/F$), is capable of performing work, in particular driving the endergonic synthesis of ATP (Section 18.1.4). We calculate $\Delta\tilde{\mu}_{H^+}$ from the expressions for the electrochemical potential of H^+ for the plasmatic (superscript P) and the extraplasmatic phase (E) on both sides of the membrane:

$$\Delta\tilde{\mu}_{H^+} = \Delta\tilde{\mu}^{P}_{H^+} - \Delta\tilde{\mu}^{E}_{H^+} \tag{18.1}$$

Substituting Eq. (2.11) $(\tilde{\mu}^{\alpha}_{H^+} = \tilde{\mu}^{\alpha\circ}_{H^+} + RT\,\ln\left(a^{\alpha}_{H^+}/a^{\circ}\right) + z_{H^+}F\varphi^{\alpha})$ with (α = P, E) and the ionic charge $z_{H^+} = 1$ and taking into account that the *standard* chemical potential $\mu^{\circ}_{H^+}$ of H^+ and the standard activity a° are the same in both phases, yields

$$\Delta\tilde{\mu}_{H^+} = RT\,\ln\frac{a^{P}_{H^+}}{a^{E}_{H^+}} + F(\varphi^{P} - \varphi^{E}). \tag{18.2}$$

After conversion to the decadic logarithm and taking into account that pH = $-\log a^{\alpha}_{H^+} = -2.3\ln a^{\alpha}_{H^+}$, it gives

$$\Delta\tilde{\mu}_{H^+} = -2.3\,RT\,\Delta\text{pH} + F\Delta\varphi. \tag{18.3}$$

The proton motor force (pmf) is then at 298 K

$$pmf = \frac{\Delta\tilde{\mu}_{H^+}}{F} = -\frac{2.3RT}{F}\Delta\text{pH} + \Delta\varphi = 0.059\,\text{V}\cdot\Delta\text{pH} + \Delta\varphi. \tag{18.4}$$

$\Delta\tilde{\mu}_{H^+}$ – Difference in the electrochemical potentials of H^+ on the P and E side of the membrane [J mol^{-1} K^{-1}]

$a^{\alpha}_{H^+}$ – Amount concentration activity of H^+ of the phase α = P or E [mol L^{-1}]

Φ^{α} – Galvani potential of the phase α = P or E [V]

F – Faraday constant [A s mol^{-1}]

R – Gas constant [J K^{-1} mol^{-1}]

T – Absolute temperature [K]

Table 18.2 Example for ion concentrations on the P and E side of a frog muscle cell.

Ion	c_P/(mmol L^{-1})	c_E/(mmol L^{-1})
Na^+	9	120
K^+	140	2.5
Cl^-	3.4	121

Source: Data from H. Kleinig, P. Sitte; *Zellbiologie*. 2nd ed., Gustav Fischer Verlag, Jena, 1986, p. 54.

Table 18.3 Contributions to *pmf* in mitochondria- and thylakoid membranes.

Quantity	Mitochondrion	Thylakoid membrane
ΔpH	1	4
0.059 V ΔpH	60 mV	240 mV
$\Delta\varphi$	160 mV	0
pmf	220 mV	240 mV

Source: Adapted from [13].

From Eq. (18.4), we can see that the *pmf* is fed by two contributions, namely a pH difference across the membrane and a difference $\Delta\varphi$ between the internal potentials of the two adjacent phases. The latter is determined by the distribution of *all* ionic species, especially Na^+, K^+, and Cl^-, across the membrane (Table 18.2). The unequal concentrations of the ions on both sides of the membrane reach a steady state through selective transport processes. The relative contributions of ΔpH and $\Delta\varphi$ differ between the mitochondria and the thylakoid membrane of chloroplasts (Table 18.3).

These apparently low membrane voltages are based on a charge shift of 1–10 μmol L^{-1} at a total ion concentration of 100–200 mmol L^{-1}. However, the membrane potential difference of $\Delta\varphi = 0.22\,\text{V}$ drops across the hydrophobic part of the membrane with a thickness d of only $3\,\text{nm} = 3 \times 10^{-9}\,\text{m}$. This results in an electric field strength of $\mathbf{E} = \Delta\varphi/d \approx 7 \times 10^7\,\text{V m}^{-1}$. The membrane is therefore very efficient at building up extremely high field strengths. A slightly higher membrane voltage would already lead to breakdown – a process that is specifically used in electroporation to introduce substances into biological cells.

18.1.2 How Is the Proton Motive Force Generated During Cellular Respiration?

Through cellular respiration, the complete conversion of 1 mol of glucose to H_2O and CO_2 releases a molar Gibbs energy of $\Delta_R G° = -2870\,\text{kJ mol}^{-1}$, which is immediately used to synthesize up to 32 mol of ATP. In eukaryotic cells, 1 mol glucose is first converted to 2 mol pyruvate, 2 mol ATP, and 2 mol NADH, the reduced form of β-nicotinamide adenine dinucleotide (NAD^+), in the cytoplasm. The molecular structures of the biological redox systems involved are summarized in Section 18.2. Pyruvate and NADH are used in the mitochondria to form up to 30 mol of ATP using oxygen. Mitochondria in the cells can therefore be considered as the biological equivalent of fuel cells. In a human organism with an energy consumption of 116 W and a membrane potential of 0.2 V, this corresponds to a current flow of 500 A across all mitochondrial membranes [15]! In the mitochondria, three interlinked processes take place simultaneously via diffusing reactants: In the plasma of the mitochondria, pyruvate is converted in the citrate cycle to CO_2 with the formation of reduction

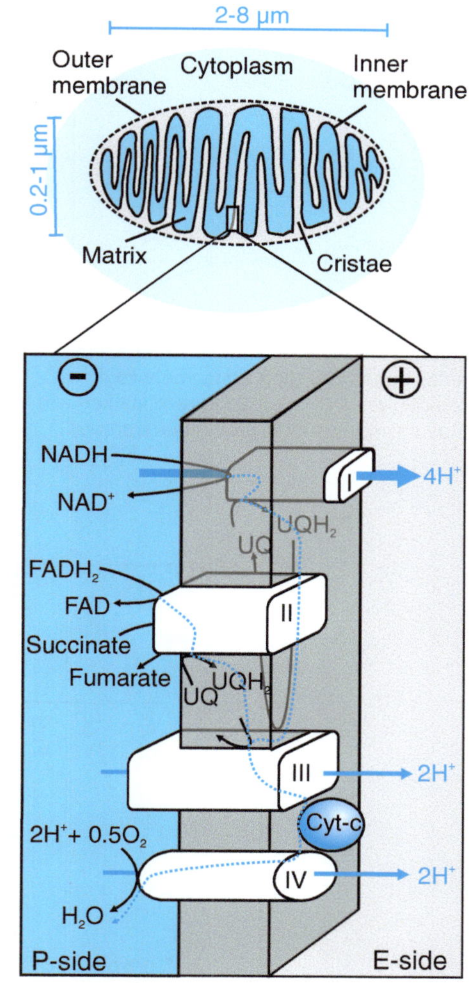

Figure 18.5 Overview of the components of the electron transport chain at the inner mitochondrial membrane. The translocation of the protons is indicated by blue arrows, and the path of the electrons is indicated by dotted blue arrows.

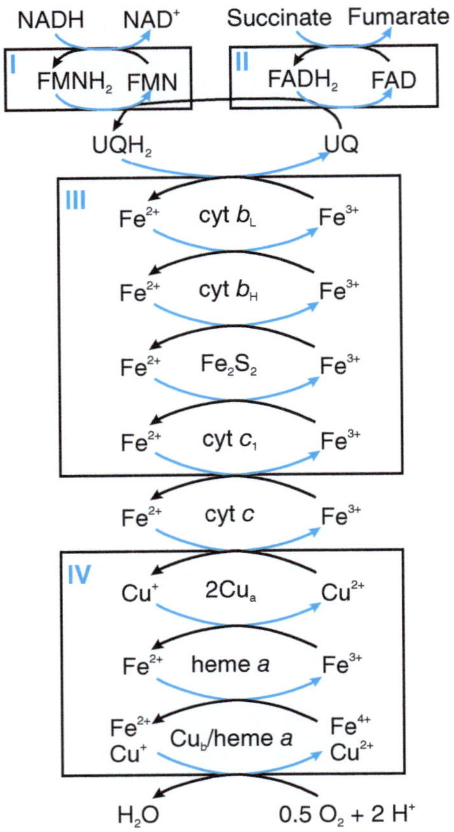

Figure 18.6 Overview of the interlocking redox cycles and their assignment to complexes I to IV. Oxidations are marked with blue arrows, reductions with black arrows.

equivalents in the form of NADH and $FADH_2$, the dehydrogenated or fully reduced form of flavin adenine dinucleotide (FAD). NADH and $FADH_2$ transfer electrons to the electron transport chain at the inner mitochondrial membrane (Figure 18.5). The electrons then undergo a multistep process until they finally reduce O_2 to H_2O. During this process, H^+ ions are simultaneously shifted from the P side of the inner mitochondrial membrane to the E side. The resulting *pmf* then drives ATP synthesis at the membrane-bound enzyme F_0F_1-ATPase (Section 18.1.4).

From the perspective of bioelectrochemistry, the processes in the electron transport chain and their coupling with ATP synthesis are of particular interest (oxidative phosphorylation). Here, we only consider the example of eukaryotic cells. However, prokaryotes are also of interest in biotechnology, especially when it comes to extremophilic organisms that can live at high temperatures, in the presence of toxic substances, or at high salt levels. In biochemistry textbooks, you will find many different illustrations of the processes in the electron transport chain, with a focus on particular aspects. In Figure 18.5, the four large membrane-bound protein complexes are particularly easy to recognize, each of which transports protons across the membrane. The oxidation of NADH takes place at complex I, the oxidation of succinate and $FADH_2$ at complex II. NADH comes from the citrate cycle, which takes place in the plasma of the mitochondria, but also in the cytoplasm of the cell. NADH is transported across the inner mitochondrial membrane by special shuttle proteins.

From these separate starting points, further processes take place with a common final pathway (which follows general principles of metabolic processes). Between complexes I and II on the one hand and complex III on the other, the electrons are transported by the ubiquinone/ubiquinol redox pair (UQ/UQH_2, Figure 18.11, also known as coenzyme Q). Due to the long hydrophobic side chain, UQ can only move within the lipid bilayer and act as a mediator (electron shuttle). This principle has been adapted from nature (biomimetics) for use in a variety of electrochemical applications and measurement methods with completely different substances and for different purposes (in this book: SECM in Section 8.12, dye-sensitized solar cells in Section 10.3.2, modified electrodes in Section 13.4.8, indirect electrosynthesis in Section 17.4, glucose sensors in Section 18.3.3). Electrons are transferred from complex III to cytochrome *c* (cyt *c*),

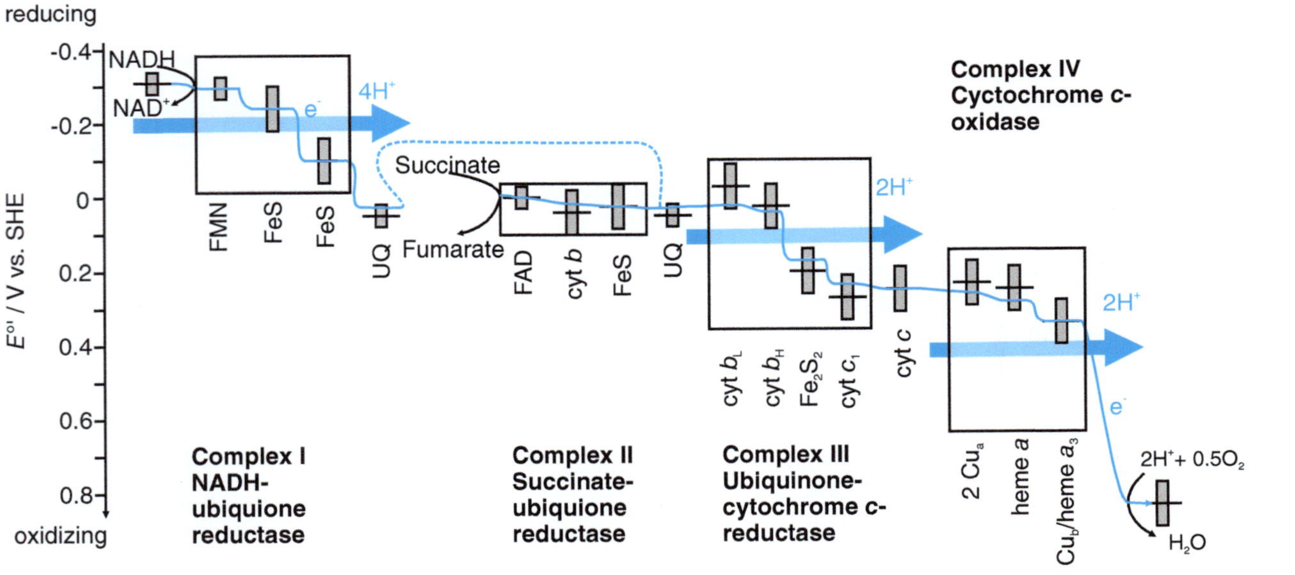

Figure 18.7 Overview of the energetic position of the redox systems in the electron transport chain of mitochondria. For each redox system, the rectangle indicates the potential range, which is covered by a variation of the concentration ratio [O]/[R] between 1:100 and 100:1.

a redox protein on the E side of the inner mitochondrial membrane. From there, electrons are transferred to complex IV, and from there to O_2 as the terminal electron acceptor. In each redox center, coupled redox cycles take place (Figure 18.6) in which the reaction rates are influenced by the difference in redox potentials, the reorganization energy during the redox transition (Marcus theory[22] Section 12.2.2), the geometric distance, and the type of bridging chemical structures (Section 12.2.1).

The path of the electrons is linked to the active transport of H^+ across the membrane to build up the *pmf*. For the electron transport chain to function properly, the redox potentials of the involved systems must be finely graded (Figure 18.7). The potential differences must be just large enough for the ETs to proceed quickly and to provide the driving force for proton transport from the P to the E side of the membrane. If the difference were even greater, the excess energy would be lost. This is a typical optimization problem that also occurs in technical systems. The required gradation of redox potentials in the electron transport chain results from the use of a few basic types of redox systems (heme group in cytochromes, FeS clusters, and coordinated Cu ions). By varying the substituents, the redox potentials are fine-tuned, which corresponds to a basic principle that is also used in the optimization of technical redox mediators (examples in Table 17.8 and Section 18.3.4).

18.1.3 How Is the Proton Motive Force Generated in Photosynthesis?

In photosynthesis in the chloroplasts of green plants, the energy for the generation of the *pmf* is provided by the absorption of light. The optically excited electrons are spatially removed from the absorbing chromophore via a redox chain. During the gradual change in their energy, as in cellular respiration, H^+ is transported from the P-side through the membrane to the E-side.

The chloroplasts contain three important membranes (Figure 18.8). The outer membrane, which separates the chloroplast from the cytoplasm of the plant cell, contains many channels (porins) that allow easy passage of metabolic products. The inner membrane controls the selective transport of metabolic products into the interior of the chloroplast (the stroma), but – unlike the mitochondria – is not the site of energy conversion. Light absorption and the formation of *pmf* take place at the thylakoid membrane. This encloses an extraplasmatic phase, the lumen, and is folded several times, like a deflated air mattress that has been folded several times. The folded areas are called grana and show a great variety of shapes in nature.

The energy conversion (Figure 18.9) starts with the absorption of a photon in the light-harvesting complex (LHC) of Photosystem II (PS II). PS II is a large protein complex with several subunits and the associated LHC and a photoreaction center, where O_2 is formed. The LHC consists of several chlorophyll *a* (Chl *a*), chlorophyll *b* (Chl *b*), and carotenoid dyes. The different dyes broaden the absorption band of the LHC so that a larger spectral portion of the sunlight can be utilized. In addition, the aggregated dye molecules act as light collectors in which the absorbed energy is conducted to one of the two chlorophyll P_{680} molecules of the photoreaction center through a very fast resonant energy transfer. The actual charge separation takes place at these two P_{680} units. The excited electron is transferred to pheophytin (chlorophyll, in which two H^+ replace the central Mg^{2+} ion). From there, the electron is transferred to the quinone acceptor molecule on the stroma side of the thylakoid membrane in a spatially and energetically aligned ET chain within PS II (Figure 18.8) with two quinones, Q_A and Q_B.

The photooxidized $P_{680}{}^+$ molecules obtain electrons from the oxidation of water

$$2H_2O \rightarrow O_2 + 4H^+ + 4e^- \tag{18.5}$$

in the photoreaction center of PS II. As a result of the associated release of the 4 H^+, the lumen is acidified. The mechanistically complicated transfer of $4e^-$ for a molecule O_2

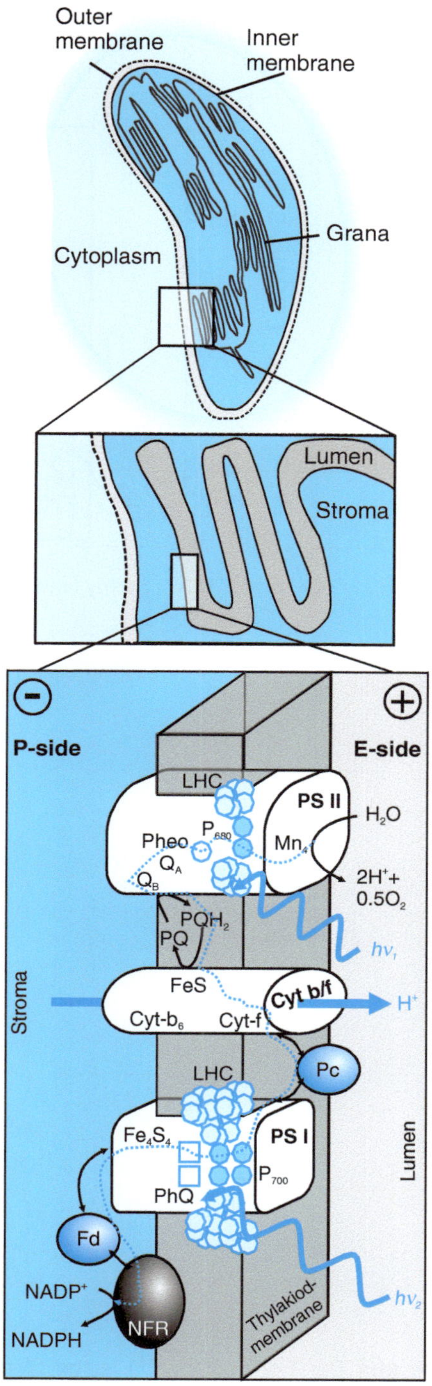

Figure 18.8 Schematic representation of a chloroplast with an enlarged view of the thylakoid membrane. The translocation of the protons is marked with blue arrows. The path of the electrons is traced with a dotted line. The concentrations of PS II, the cyt *bf* complex, and PS I show a large lateral variation in the thylakoid membrane, so that electrons must be transported over greater distances via the plastoquinone pool.

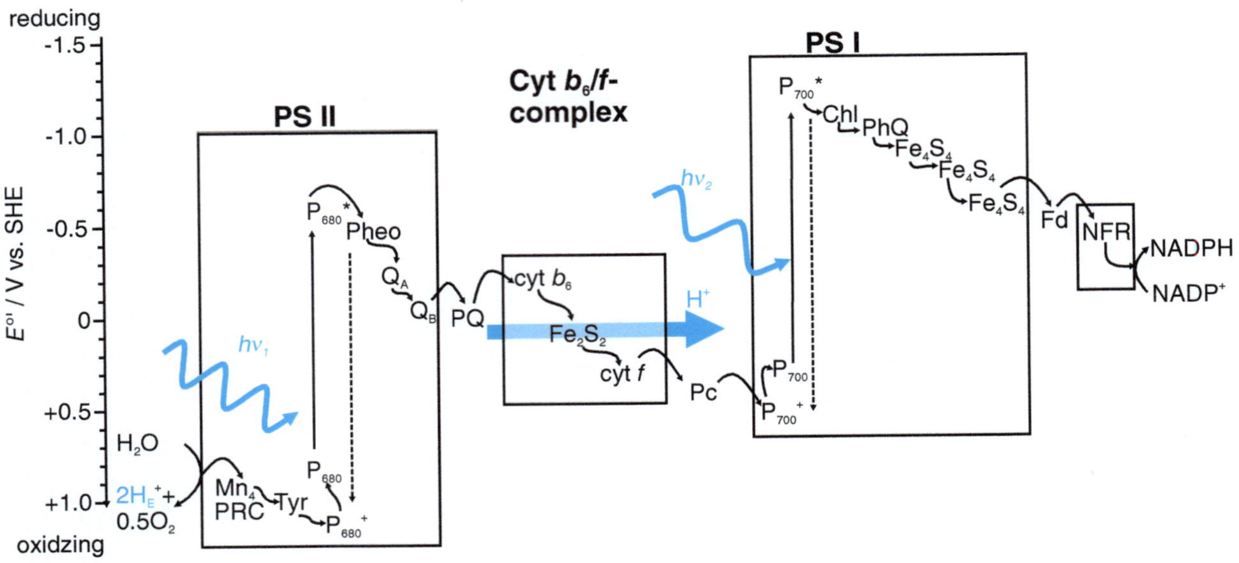

Figure 18.9 Energy landscape in the ET processes of photosynthesis.

(cf. Chapter 12) is made possible by intermediate storage of the electrons in a cluster with 4 Mn ions, which undergoes a catalytic cycle with 4 redox states and in this way couples the one-electron ET reactions at $P_{680}/P_{680}{}^{+}$ with the four-electron oxidation in Eq. (18.5).

On the stroma side of PS II, plastoquinones (PQs) are reduced to the corresponding hydroquinones (PQH_2). The PQs comprise several structurally related quinones (Figure 18.11). These hydrophobic molecules can diffuse laterally in the lipid bilayer of the thylakoid membrane and play an analogous role in photosynthesis as UQ does in cellular respiration. They transport the electrons and protons to cytochrome *bf* (cyt *bf*), where they are transferred to a plastocyanin (Pc). This process is sometimes, but not always, associated with the pumping of 2 H^+. The cyt *bf* complex contains two parallel redox chains, involving an Fe_2S_2 cluster on which PQH_2 is oxidized. The electrons are transferred to Pc, which thus plays a similar role to cyt *c* in the mitochondria.

The electrons on the lumen side of the membrane are transferred from Pc to the PS I. PS I is larger than PS II and, with 96 chlorophyll units, contains significantly more leaf pigment than PS II. The LHC is in two layers and contains 90 of the chlorophyll units and approximately 20 carotenoid pigment molecules. The photon energy absorbed there is somewhat lower and the corresponding wavelength longer than the energy absorbed by PS II. The energy is passed within the LHC by resonance energy transfer to a pair of chlorophyll P_{700} units in the center of the photoreaction center. After excitation, an electron is transferred from there to ferredoxin (Fd) on the stroma side of the thylakoid membrane via an electron transport chain within PS I, which consists of four further chlorophyll units, two phylloquinones, and three Fe_4S_4 clusters. Fd is a small soluble protein (11 kDa) with an Fe_2S_2 cluster that can perform a one-electron transfer. After diffusion to the stroma side of the thylakoid membrane, Fd donates an electron at the membrane-bound enzyme ferredoxin-NADP oxidoreductase for the reduction of $NADP^+$ to NADPH according to Eq. (18.6). NADPH, the reduced form of β-nicotinamide adenine dinucleotide phosphate ($NADP^+$), serves as a universal reduction equivalent and is structurally and functionally similar to NADH (Section 18.2).

$$NADP^+ + H^+ + 2\,Fd \rightarrow NADPH + 2\,Fd^+ \tag{18.6}$$

The unit Da (from Dalton[166]) is the term commonly used in the English literature for the atomic mass unit ($1\,u = 1.6605 \times 10^{-27}$ kg). The mass of a molecule can be determined directly, for example, using mass spectrometry. This quantity with the corresponding unit is often used in biochemistry to characterize the molar mass ("molecular weight") *M* of biomacromolecules; 1 kDa corresponds to a molar mass of 1 kg mol^{-1}.

[166] John Dalton, 1766–1844, English naturalist and teacher.

This also involves a translation from the one-electron processes at Fd to the two-electron processes at $NADP^+/NADPH$.

In PS I, the photooxidized ${P_{700}}^+$ units in the photoreaction center are reduced by Pc.

18.1.4 How Is the Gradient in Proton Activity Used for ATP Synthesis?

The proton motive force (*pmf*) (Sections 18.1.2 and 18.1.3) is used in the enzyme F_0F_1-ATP synthase, or ATPase for short, to build up the energy-rich ATP, which serves as "fuel" for all endergonic processes in the cell.

$$ADP^{(3-)} + H_2PO_4^- + x H_E^+ \rightarrow ATP^{(4-)} + H_2O + x H_P^+ \tag{18.7}$$

The formation of ATP corresponds to a standard reaction enthalpy of $\Delta_R H° = +30.5\ kJ\ mol^{-1}$ and, under physiological concentrations, to a reaction enthalpy of $\Delta_R H = +52\ kJ\ mol^{-1}$ [16]. The "daily production" of a human is approximately 40 kg ATP (!) and is immediately used up as it is formed. The resulting steady-state concentrations in the cytoplasm are about 2.5 mmol L^{-1} ATP and 0.25 mmol L^{-1} ADP. The CF_0CF_1-ATPase (C for chloroplasts) in the chloroplasts is not discussed here separately because it is similar to the ATPase in the mitochondria.

The ATPase is a particularly fascinating enzyme that is an example of a "molecular machine" in which the energy of the *pmf* is first converted into a mechanical rotary motion and then into the chemical energy of the ATP. This "machine" consists of the two basic units F_0 and F_1, which in turn are composed of several subunits (F_0 from a to c, F_1 from α to ε, Figure 18.10). The F_0 unit is firmly attached to the membrane. Ten to fourteen c subunits form a membrane-spanning tube, the proton channel. The a and b units, which hold the detachable F_1 unit to the P-side of the membrane, are bound to the sides of this tube. The F_1 unit consists of a "wheel bearing", which is made up of three alternating α and β subunits and is locked to the proton channel via the F_0b-unit. This "wheel bearing" is located on the F_1 γ-unit, which represents a rotatable "axle". This axle is connected to the ion channel consisting of F_0c-units via the F_1ε-subunit.

The rotation of the F_1γ-unit in the "wheel bearing" leads in each F_1β-unit alternatingly to 1) binding of ADP, 2) formation of ATP, 3) release of ATP, similar to processes on a "camshaft" of an engine. The exact function of this enzyme was elucidated by a combination of precise structural analyses, nuclear magnetic resonance (NMR) experiments, and single-molecule fluorescence experiments [13]. The motor mechanism is particularly easy to recognize in animations, but is still the subject of research [17].

ATP produced in the mitochondria is exported to the cytoplasm via the adenine nucleotide translocator.

Further Reading. More detailed overviews of the bioenergetics of respiration and photosynthesis, including processes in prokaryotes, can be found in [3 vol. 2, 13]. For a more detailed consideration of cell anatomy and metabolic processes, textbooks on cell biology or biochemistry are recommended, for example, [11, 12, 18].

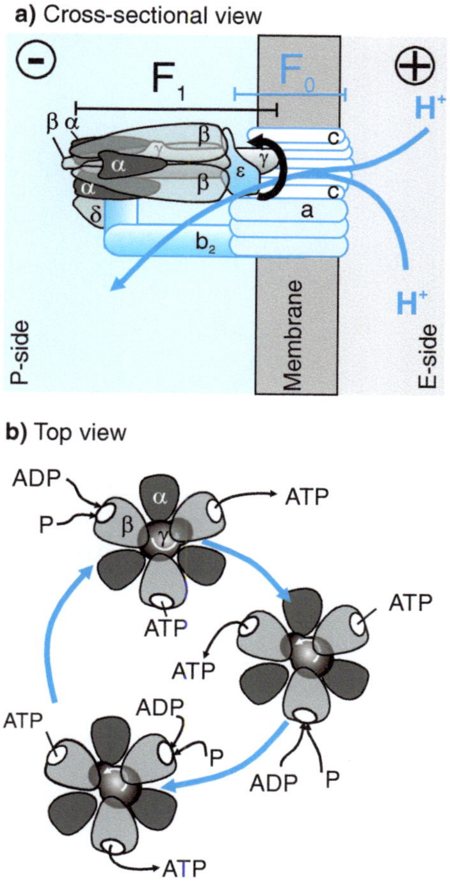

Figure 18.10 a) Schematic structure of the F_0F_1-ATPase; b) formation of ATP at the F_1β-units upon rotation of the F_1γ-unit. P, inorganic phosphate.

18.2 The Electrochemical Characterization of Biomolecules

In this Section, we will look at how the redox systems, presented with their biological functions in Section 18.1, can be characterized outside a biological cell using electrochemical methods, in particular cyclic voltammetry (CV, Section 5.3.5, Chapter 6).

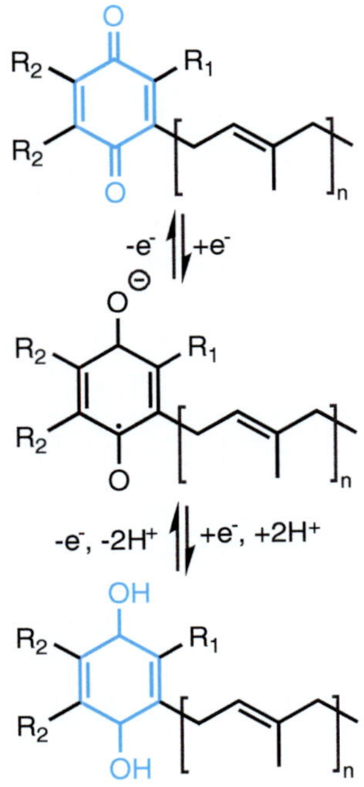

	R_1	R_2	n	$E^{\circ\prime}$/V
UQ	$-CH_3$	$-OCH_3$	4-10	+0.10
PQ	-H	$-CH_3$	6-9	+0.08
MQ	$-CH_3$		3-4	+0.07

Figure 18.11 Redox reactions of biological quinones. UQ, ubiquinone; PQ, plastoquinone; MQ, menaquinone.

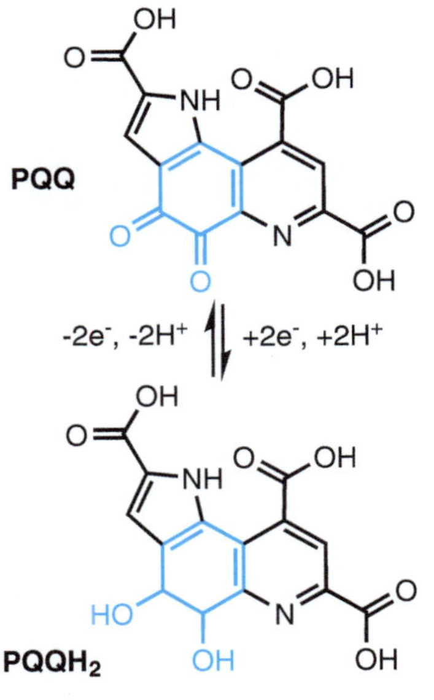

Figure 18.12 Redox reactions of PQQ – pyrroloquinoline quinone.

18.2.1 Low-Molecular Weight Redox Centers

Quinones. Ubiquinone (UQ), plastoquinone (PQ), and menaquinone (MQ) (Figure 18.11) can be reduced to the corresponding hydroquinones by accepting $2e^-$ and $2H^+$. Their formal potential is pH-dependent and changes by 0.059 V per pH unit at 298 K [Eqs. (2.95) to (2.97)]. By taking up H^+ on the P side and releasing protons on the E side of the membrane, the quinones are directly involved in the formation of the *pmf*. As we saw in Chapter 6, the transfer of electrons and protons occurs stepwise (scheme of squares, Figure 6.19a), whereby the semiquinone radical is formed as an intermediate. In aqueous solutions, the intermediate is sometimes difficult to observe, as the proton transfer reactions take place extremely rapidly. However, the quinones in the biological cell do not react at all in an aqueous system, as they are always dissolved in the hydrophobic part of the lipid bilayer. This is ensured by the side chain of isoprene units, whereby the number of isoprene units varies between organisms. For example, UQ has 10 isoprene units with 50 carbon atoms in humans, hence the name UQ_{10} or coenzyme Q_{10}. The existence of the semiquinone is functionally important, as the quinones act as mediators between the two-electron redox centers of the flavins and the one-electron centers such as the heme groups.

The side chains on the aromatic ring are important to prevent side reactions analogous to those in Figure 13.63. They also influence the formal potential to a small extent by shifting the electron density at the redox-active group (Figure 18.11).

Pyrroloquinoline quinone (PQQ, methoxatin, Figure 18.12) is another redox-active prosthetic group, which is of interest in a series of biotechnologically interesting redox enzymes [$E^{\circ\prime} = +0.116$ V (vs. SHE, pH 7, 59 mV/pH)].

Flavins. Flavin adenine dinucleotide (FAD) and flavin mononucleotide (FMN, Figure 18.13) are also $2e^-$-$2H^+$-redox systems with a formal potential $E^{\circ\prime} = -0.21$ V (vs. SHE at pH 7). As with UQ, the pH dependence of $E^{\circ\prime}$ is 59 mV/pH at $T = 298$ K. The structural difference between FAD and FMN is irrelevant for the redox reaction, as the additional phosphate, ribose, and adenosine residues in FAD are present at a large distance and without a conjugated link to the redox-active group. In most cases, FAD and FMN occur as redox centers bound to proteins. The incorporation into a protein environment can cause a considerable shift in $E^{\circ\prime}$ of the redox center, exactly according to the same principles that we have discussed in Section 13.4.2 for the embedding of redox centers in films. In the case of flavins, it should be noted that the relative dielectric constant ε_r inside a typical protein is about 5, whereas in aqueous

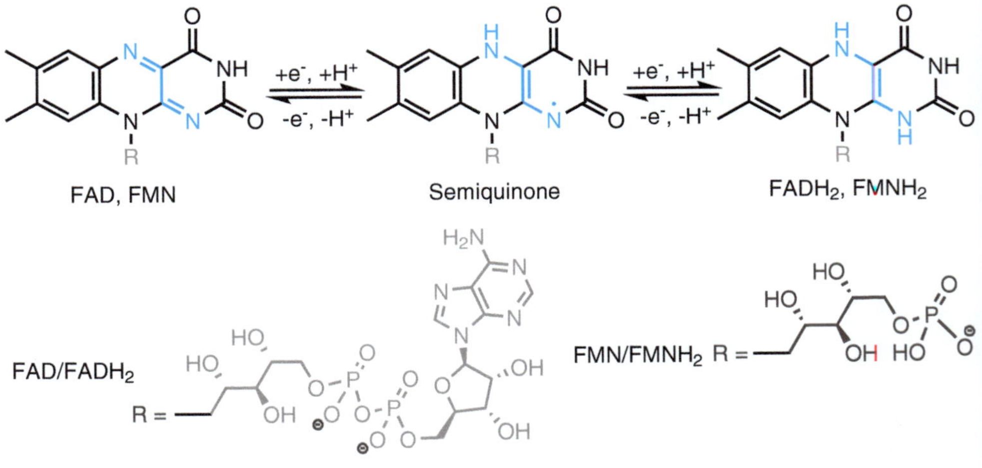

Figure 18.13 Redox reactions of flavins. FAD, flavin adenine dinucleotide; FMN, flavin mononucleotide.

solutions it is close to $\varepsilon_r(H_2O) = 78$, that is, there is a very significant difference. Furthermore, $E^{\circ\prime}$ of the protein-bound flavins are significantly shifted compared to the free flavins, especially by the π-stacking of aromatic amino acid residues of the surrounding protein above and below the aromatic ring system of the flavins. The shift can be as large as −0.47 V for the FMN-containing bacterial ET protein flavodoxin [13, 19].

NADH and NADPH. β-nicotinamide adenine dinucleotide (NAD^+) and β-nicotinamide adenine dinucleotide phosphate ($NADP^+$) differ only by a phosphate residue on the ribose ring of adenosine (Figure 18.14). This modification is located too far from the redox-active nicotinamide group to influence the formal potential. Thus, both compounds have an equal formal potential of −0.320 V (vs. SHE, pH 7), which is why we write $NAD(P)^+$ in the following discussion when referring equally to both compounds. The additional phosphate group influences the binding to different proteins and therefore has an important biological function.

The reduction of $NAD(P)^+$ represents a $2e^-{-}1H^+$ reaction. During the reaction, a radical intermediate has not been detected. This reaction thus corresponds to the transfer of a hydride ion (H^-), which is a strong reducing agent in chemistry, for example, in the well-known reagent $LiAlH_4$. Accordingly, NAD(P)H is the universal reduction equivalent in biological cells. However, it is essential that NAD(P)H is inert with respect to direct oxidation by O_2 because O_2 is ubiquitous in biological cells. Inert means that the reaction with O_2 does not occur at all or occurs sufficiently slowly that the ET from NADH to O_2 via the complicated ET chain in Figures 18.5 and 18.7 is much faster and involves displacement work for protons, avoiding a direct reaction of 2 NADH + O_2. The inertness of O_2, which we have already discussed in detail in the context of electrocatalysis (Chapter 12) and fuel cells (Chapter 16), also contributes to the efficiency of this mechanism.

NAD(P)H, as a soluble reduction equivalent, would represent a very interesting starting point for many bioelectrochemical applications. However, all conventional, non-specifically modified electrode materials produce only irreversible and poorly reproducible voltammograms with a high overpotential of about 1 V [20]. This is because either $NAD(P)^+$ or NAD(P)H adsorbs strongly on the electrodes and blocks the ET for other molecules, and because the radicals formed as intermediates dimerize very quickly to form inhibiting polymer films on the electrodes (electrode fouling).

NAD(P)H oxidation is only successful if a mediator is used, which chemically oxidizes NAD(P)H to $NAD(P)^+$ near the electrode and is then in turn oxidized at the electrode by a heterogeneous ET (Figure 18.15, mediator principle). Various mostly heterocyclic organic molecules have been tested as mediators for NAD(P)H oxidation, together with different approaches for integrating these mediator molecules with a solid electrode to form a system. The main challenge in integration, for example, for use in sensors, is that the mediator should oxidize NAD(P)H in a homogeneous or quasi-homogeneous reaction without diffusing away from the electrode or being washed out during operation over longer periods of time.

Further Reading. An early detailed overview of the electrochemistry of cofactors can be found in [21]. The state of research on the problem of NAD(P)H oxidation is summarized in several comprehensive overviews [22, 23].

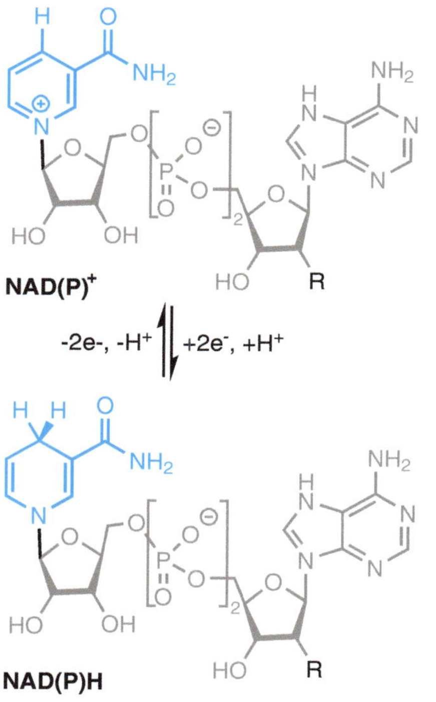

NAD^+/NADH	R = -OH
$NADP^+$/NADPH	R = $-O-PO_3H^-$

Figure 18.14 Redox reactions of NAD^+ (β-nicotinamide dinucleotide) and $NADP^+$ (β-nicotinamide adenine dinucleotide phosphate).

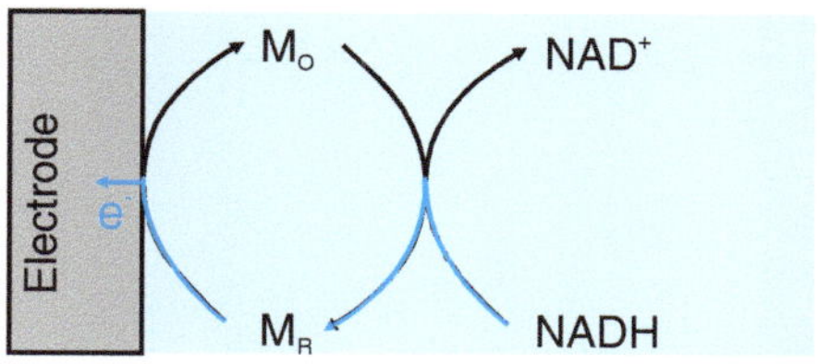

Figure 18.15 NADH oxidation with a mediator (M_O, oxidized form; M_R, reduced form). The path of the electrons is shown in blue.

18.2.2 Metal-Containing Redox Centers

Figure 18.6 clearly shows that most redox reactions in the electron transport chain of the mitochondria occur via valence changes of transition metal ions, which are contained in the proteins as nonpeptide groups. Such groups are called prosthetic groups or cofactors. FAD, FMN, and PQQ from Section 18.2.1 also belong to this category.

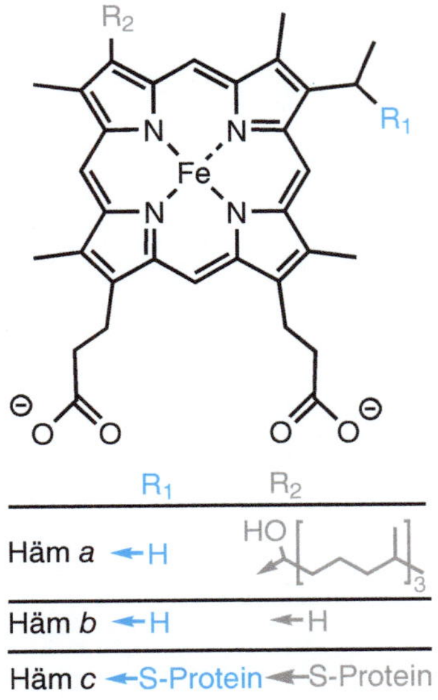

Figure 18.16 Structure of heme groups. S – sulfur atom of a cysteine side chain.

From the large variety of such redox centers, we will only take a closer look at the heme groups here (Figure 18.16). The spelling heme or haem occur as synonyms. A glance at the electron transport chains discussed in Section 18.1 makes it clear that these heme groups are found at very different points in the electron transport chain. Therefore, they must have different redox potentials despite being of a similar structure. This is made possible by the different binding of the prosthetic groups to the polypeptide chains of the proteins containing them. In heme *c*, which occurs in cyt *c*, the linkage is via S atoms of the amino acid cysteine, whereas it is via a hydrocarbon chain in heme *b*. As the electron-withdrawing S atoms are localized close to the π-electron system of the porphyrin, a shift in electron density affects the entire ring system and thus also the central Fe ion, whose redox potential shifts in a characteristic manner. In addition, the formal potential of the heme group is influenced by two axial ligands, which are located in front of or behind the drawing plane (not shown in Figure 18.16). These axial ligands are side chains of histidine or methionine of the surrounding protein, which interact with the central metal via N or S atoms. The resulting formal potentials of the $Fe^{2+/3+}$ redox transition vary between −0.3 V vs. SHE when two histidine side chains are coordinated up to +0.4 V vs. SHE when the heme *c* group is coordinated between histidine and methionine side chains [24]. The electron-withdrawing S-atoms in the thioether bridges at the periphery of the porphyrin and the axially coordinated methionine and the shielding of the Fe center from the polar solvent H_2O stabilize the less charged Fe^{2+} redox form of the central Fe ion and thus increases $E^{\circ\prime}$. Another possible variation in $E^{\circ\prime}$ results from the interaction of the central Fe ion with charged amino acid residues in the interior or on the surface of the surrounding protein [25, 26]. Through site-directed mutagenesis on the protein myoglobin, which contains heme *c* as a prosthetic group, it was demonstrated how large the shift in the redox potential of the heme *c* group is when a specific amino acid is substituted within a genetically modified protein [27]. These principles for the variation of $E^{\circ\prime}$ are also found in other redox-active metal centers of proteins.

Cu-ions can be coordinated, for example, by histidine, cysteine, methionine, or glutamate residues. In these cases, stabilization of the less charged Cu^+ form can also be achieved by shielding the metal center from the polar solvent H_2O, which can be further enhanced by distorting the coordination environment toward a tetragonal geometry. Many enzymes contain several Cu centers, which have different functions and can also be coordinated differently [28, 29].

A large group of proteins contains clusters of Fe and S atoms in different ratios and geometries, which in turn influence their redox potentials. The basic construction principle consists of embedding a small section of an FeS lattice in a coordination environment of histidine and cysteine residues of the surrounding protein. Depending on the number of Fe ions, these clusters can accept and release a variable number of electrons (multielectron redox centers) [30].

18.2.3 Difficulties in the Electrochemical Characterization of Proteins

The most important building blocks of proteins are linearly connected amino acids, most of which are stable over a very wide potential range. Only with this stability can they fulfill their biological functions in an oxygen-containing environment. The electrochemical oxidation of some amino acids, namely of methionine, tryptophan, and tyrosine within proteins, is only possible under special conditions and is sometimes even sequence-specific [31, 32].

In the following, we will discuss the electrochemistry of proteins and, more specifically, the electrochemistry of prosthetic groups (or cofactors) within proteins. The aim of the electrochemical experiments is to characterize the redox reactions of these groups when they are linked to the polypeptide chains of the protein. This is important because it influences the kinetics and thermodynamics of these redox centers (Sections 18.2.1 and 18.2.2).

Redox proteins comprise redox-active enzymes (biocatalysts) and redox-active proteins without a catalytic function in the living cell (e.g., hemoglobin as an O_2 transporter). However, these proteins can also have catalytic activity outside the cell for individual reactions. Enzymes are categorized using a numerical classification system (EC number for Enzyme Commission number) into seven classes according to the catalyzed reaction (1 = oxidoreductases, 2 = transferases, 3 = hydrolases, 4 = lyases, 5 = isomerases, 6 = ligases, 7 = translocases). The term redox enzymes is therefore synonymous with oxidoreductases, the largest enzyme class in terms of numbers. A further subdivision is made according to the substrates and cosubstrates. The final number is a consecutive number. For example, the technologically important enzyme glucose oxidase (GOx) has the EC number 1.1.3.4.

Do not confuse the EC number of the proteins with the EC nomenclature of electrochemical reaction mechanisms from Chapter 6!

The cofactors of the enzymes can be bound to the protein structure to varying degrees (covalently or through intermolecular forces). An enzyme from which the cofactor has been removed is called an apoenzyme. The enzyme with prosthetic groups is called a holoenzyme.

$$\text{Holoenzyme} = \text{Apoenzyme} + \text{Prosthetic group(s)} \tag{18.8}$$

Many proteins also contain polysaccharides on the periphery of the protein structure (glycoproteins).

Q: Why have such complicated structures evolved with natural evolution if it is just to speed up a simple oxidation reaction?

A: Only these complicated structures enable the selectivity of the biochemical reactions, which in turn is a prerequisite for the fact that anabolic and catabolic reactions (build-up and degradation of complicated compounds) can take place side by side in the same reaction volume. If we look at the cofactors from Sections 18.2.1 and 18.2.2, UQ reacts rather easily with various other redox systems. It can therefore only be part of a selective electron transport chain because it is embedded in the lipid bilayer and thus shielded from the compounds dissolved in the aqueous plasma. NAD(P)H, on the other hand, is so inert that it does not react with many redox systems in the absence of a catalyst, despite the presence of a substantial thermodynamic driving force. The prosthetic groups themselves are often not particularly selective when isolated. Therefore, a limited set of such cofactors can be used in redox enzymes for a large variety of functions. In these cases, the selectivity only arises because the prosthetic group is hidden deep within the protein structure. The peptidic part of the protein forms a binding pocket through which only very specific substrates can reach the prosthetic group and then react quickly.

The first attempts to investigate the electrochemistry of proteins using voltammetric methods similar to those used for organic and inorganic substances led to frustrating results, as the signals were often not reproducible and also very weak, did not appear at all or – if they did appear – changed within a very short time.

Due to the embedding of the prosthetic groups in the protein structure, the distance between the cofactor and the electrode is too great for a fast ET even when the peptidic part of the protein touches the electrode surface (Figure 18.17a). If at all, we might expect ETs to proteins only for relatively small redox proteins such as cyt *c* (Figure 18.17c). Another problem is the low concentration of redox centers. Given the very large molar masses of the proteins, a large protein mass must be dissolved in order to achieve a concentration of the prosthetic group that can be meaningfully investigated using voltammetric methods. Membrane proteins are particularly difficult to isolate and lose their function without membrane fragments. However, the electrode can be blocked by surface-active membrane fragments. In addition, the side chains of many proteins adsorb very strongly on the usual electrode materials such as Hg, Au, and Pt. This distorts the three-dimensional protein structure (denaturation). The environment of the prosthetic group and thus the redox properties are often also changed (Figure 18.17b). Once a layer of damaged proteins is adsorbed on the electrode, diffusing proteins can no longer reach the electrode (electrode fouling). In the case of proteins with redox centers near the protein surface, the orientation of the proteins is also important. With a random orientation, many of the redox centers would point away from the electrode and would not be able to perform ET (Figure 18.17d).

Q: If the electrochemical investigation of redox proteins is so difficult, how could the redox potentials within the electron transport chain be determined?

A: Various "tricks" were used for this. We have already seen the use of mediators in the discussion of NAD(P)H, which transport electrons like a ferry between the electrode and the prosthetic group hidden in the protein structure. Mediators can be used in the context of titrations, which are monitored optically, or in voltammetric experiments. The process is known as mediated electron transfer (MET). Targeted electrode modifications also allow some enzymes to bind in an oriented fashion in such a way that their redox centers can perform an ET with the electrode. We refer to this process as direct electron transfer (DET).

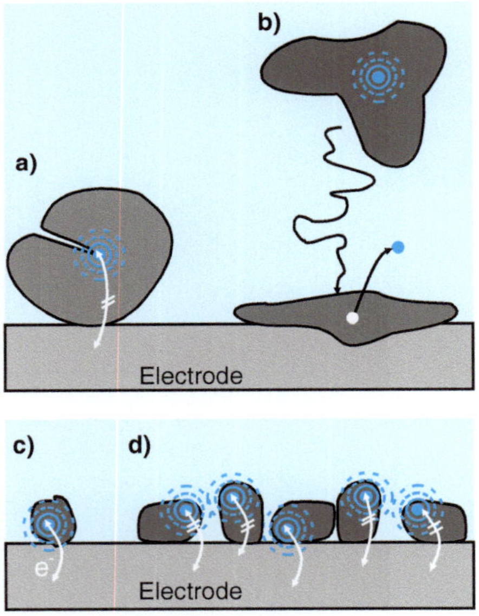

Figure 18.17 Difficulties in the voltammetric analysis of redox-active proteins: a) Redox centers are shielded by polypeptide chains so that the distance to the electrode is too great for an ET. b) On contact with the electrode, individual amino acid residues bind to the electrode surface and thus distort the structure of the protein, possibly with loss of the prosthetic group. c) ET is only possible with small redox proteins. d) If the proteins are adsorbed with random orientation, ET can only occur in some orientations. The prosthetic group is shown in blue, and the dashed circles indicate the range of the ET.

The use of DET and MET is of considerable technological importance. In order to understand which enzymes can possibly perform a DET, it is useful to differentiate between intrinsic and extrinsic redox proteins [33].

In the case of intrinsic redox proteins (Figure 18.18a), the reaction with the oxidizing agent and the reducing agent takes place at redox centers that are localized in the protein structure at one location and are usually shielded. An ET with an electrode can then only take place if one of the following conditions is met:

- The catalytic center is located near the surface of the protein.
- The enzyme can deform reversibly and temporarily bring the catalytic center close to the electrode.
- The electrode protrudes into the redox protein with extensions such as nanoparticles or nanotubes.
- The ET between the prosthetic group and the electrode takes place via artificially created electron transport pathways that do not exist in nature.

Extrinsic redox proteins usually work together with other redox proteins in the biological cell to transport electrons, for example, ubiquinone-cytochrome *c* reductase and cyt *c* in Figures 18.5 and 18.7. Extrinsic redox proteins must contain a redox center near their "docking site" to the second redox protein to facilitate ET to that second protein

(Figure 18.18b). There is a chance of observing the DET of extrinsic redox proteins at an electrode if it is possible to bind the docking site close to (< 1.5 nm) an electrode. Since these enzymes often have a second binding site for the second substrate, they have an internal electron transport chain, like ubiquinone-cytochrome *c* reductase, in which various redox centers are arranged at defined distances from each other.

With the rapidly increasing availability of high-resolution protein structures, the structures of more and more proteins can be viewed in freely accessible databases on the internet [34]. This textbook deliberately refrains from reproducing these structures, as their clarity often only becomes apparent when the structures are rotated on the screen, individual components are faded in and out, or partial structures are displayed in different resolutions. Please make yourself familiar with these possibilities!

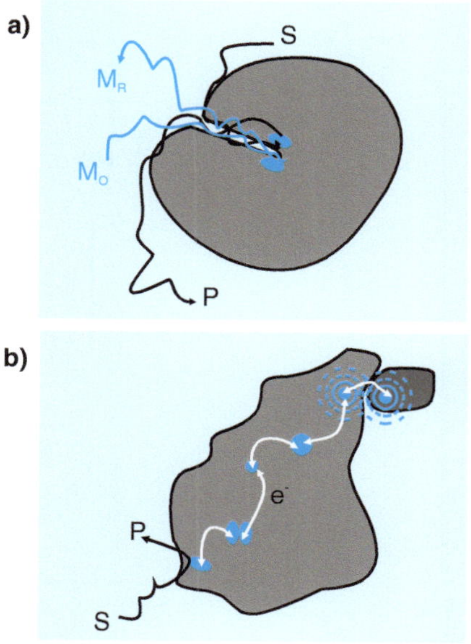

Figure 18.18 Schematic representation of the reactions on a) intrinsic and b) extrinsic redox proteins.

18.2.4 Determination of the Formal Potential by Potentiometric Redox Titration on Optically Transparent Thin-Layer Electrodes

The basic principle is a potentiometric titration, in which the charge is first transferred from an electrode to a mediator (usually referred to as a mediator titrant in the relevant literature), and then the concentration ratio of the oxidized and reduced forms of the redox protein ($[E_O]$, $[E_R]$) is measured spectroscopically. This is possible because the redox-active groups are often also chromophores whose spectra differ significantly between the oxidized and reduced states. The experiment is carried out with an optically transparent thin-layer electrode (OTTLE, Figure 18.19). The mediator acts as a redox buffer and can adjust the redox potential of the solution by approximately $\pm 2 \cdot 2.3E_{NF} = \pm 4.6\ RT/(n_M F)$ around its own formal potential $E^{\circ\prime}(M_R/M_O)$. Different potentials E_{appl} are successively applied in the range of $E^{\circ\prime}(M_R/M_O)$. After a waiting period to allow the conversion of the entire solution volume between the two slides of the OTTLE (approx. 5 min, depending on the thickness of the OTTLE), an electrochemical equilibrium is reached for the mediator with the Nernst equation.

$$E_{appl} = E^{\circ\prime}(M_R/M_O) + \frac{2.3RT}{n_M F} \log \frac{[M_O]}{[M_R]} \tag{18.9}$$

By plotting E_{app} versus $\log([M_O]/[M_R]$, $E^{\circ\prime}(M_R/M_O)$ can be determined as the ordinate intercept and the stoichiometric number of electrons of the mediator n_M from the slope.

The experiment is then repeated after the addition of the redox protein. Depending on the concentration ratio of protein and mediator, it is necessary to wait longer to reach equilibrium because the homogeneous redox reaction

$$\text{In solution: } (n_E/n_M)\, M_O + E_R \rightleftharpoons (n_E/n_M)\, M_R + E_O \tag{18.10}$$

requires the transfer of more charge in the heterogeneous reaction

$$\text{At the transparent electrode: } M_R \rightleftharpoons M_O + n_M\, e^- \tag{18.11}$$

compared to the conversion of the mediator alone. After equilibration, the spectra are recorded (Figure 18.20) and corrected for the spectra of the mediators at the same potentials (inset in Figure 18.20). The concentration ratio $[E_O]/[E_R]$ of the two redox forms of the proteins can be obtained from the spectra. In this situation, two redox equilibria are simultaneously established, and hence, the following condition holds:

$$\begin{aligned} E_{appl} &= E^{\circ\prime}(M_R/M_O) + \frac{2.3RT}{n_M F} \log \frac{[M_O]}{[M_R]} \\ &= E^{\circ\prime}(E_R/E_O) + \frac{2.3RT}{n_E F} \log \frac{[E_O]}{[E_R]}. \end{aligned} \tag{18.12}$$

A linear regression of E_{app} versus $\log([E_O]/[E_R]$ yields n_E and $E^{\circ\prime}(E_R/E_O)$.

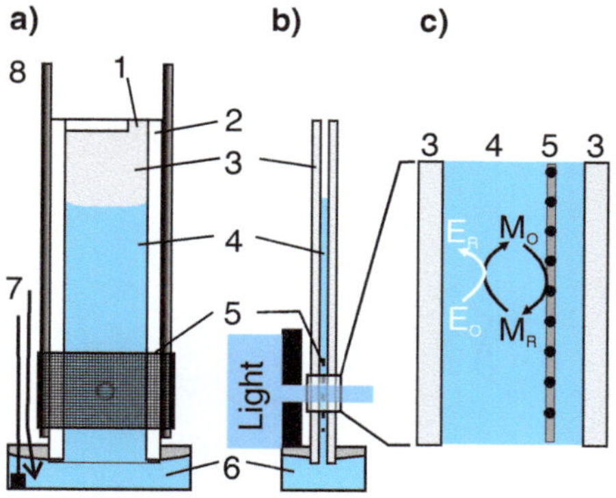

Figure 18.19 Structure of a simple cell using an optically transparent thin-layer electrode. a) View in the direction of light propagation; b) view perpendicular to the direction of light propagation; c) enlarged section of b).

1) Point for aspirating solution;
2) Teflon tape as a spacer;
3) microscope slide;
4) working solution;
5) optically transparent electrode (gold mini-grid or thin conductive oxide;
6) container for solution;
7) Aux and Ref electrodes;
8) wire of current collector for the OTTLE.

The working solution is drawn into the approximately 0.2 mm-thick gap between the two slides by capillary action. The light beam should be constricted to the center of the transparent electrode using a mask to eliminate edge effects. Image adapted from [35].

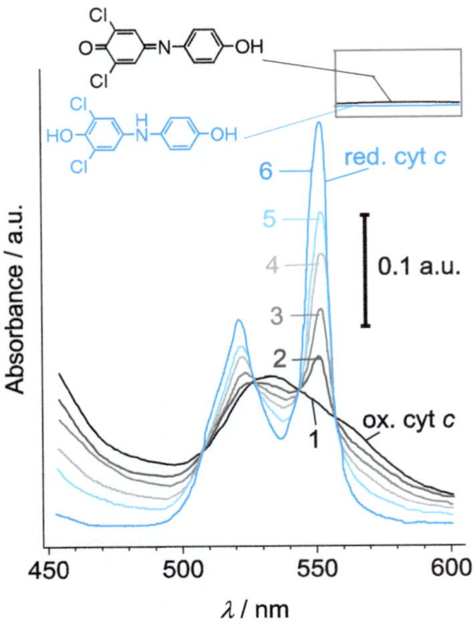

Figure 18.20 Example for a potentiometric titration of **0.5 mmol L^{-1}** cyt *c* with **0.01 mmol L^{-1}** 2,6-dichlorophenol indophenol; E_{appl}/mV vs. SCE is 1) −250, 2) −10, 3) +10, 4) +30, 5) +50, 6) +250. The inset shows the converted spectra of the mediator at the concentration used in the range of maximum absorbance. Source: Data from [35] were replotted by the textbook author.

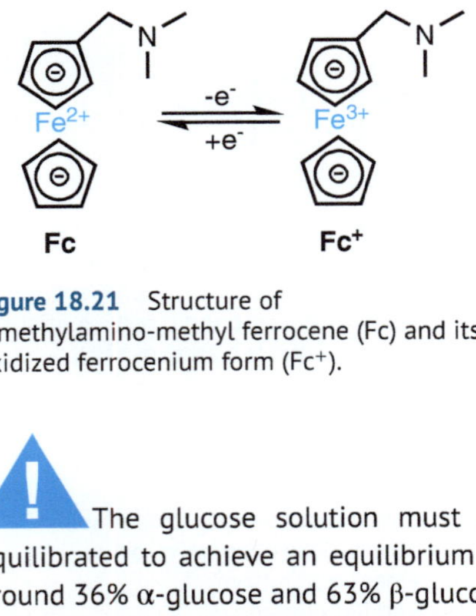

Figure 18.21 Structure of dimethylamino-methyl ferrocene (Fc) and its oxidized ferrocenium form (Fc^+).

! The glucose solution must be equilibrated to achieve an equilibrium of around 36% α-glucose and 63% β-glucose before addition to the electrochemical cell. This process is relatively slow and can be followed with simple means in a classic lab course experiment on chemical kinetics.

The possible applications of the method depend on the availability of mediators with formal potentials spread over as wide a range as possible. This is because an effective potential control by the mediator is only possible if the formal potentials $E^{\circ\prime}(M_O/M_R)$ and $E^{\circ\prime}(E_R/E_O)$ differ by less than $2.3E_{NF}$. The mediators must also

- be chemically stable in both redox forms,
- perform a reversible or quasireversible ET with the electrode,
- react quickly with the redox protein in reaction (18.10), but *must not* permanently bind to the redox protein.
- A low optical absorption of the mediator is also desirable.

If there is significant absorption of the mediator, the concentration ratio between mediator and redox protein can be used to optimize the conditions for quantitative spectroscopic determination.

Further Reading. In addition to a simple example in [35], further overviews are available [36–37]. An update can be found in [38] with a particular focus on working under anaerobic conditions, which is essential for a number of redox enzymes.

18.2.5 Mediated Electron Transfer with Catalytic Turnover of Dissolved Enzymes

For amperometric biosensors that work with a mediator, a catalytic reaction of the enzyme with a substrate takes place continuously in addition to the reactions (18.10) and (18.11). As an example, we consider the enzyme GOx, which contains $FAD/FADH_2$ as a prosthetic group and uses β-D-glucose as substrate (S). A suitable mediator would be dimethylamino-methyl ferrocene (Fc, Figure 18.21) and its oxidized ferrocenium form (Fc^+). The substrate is oxidized to D-glucono-1,5-lactone ($C_6H_{10}O_6$, P). The mechanism of the electrode reaction is:

$$\text{At the elctrode: } 2\,\text{Fc} \rightleftharpoons 2\,\text{Fc}^+ + 2\text{e}^- \tag{18.13}$$

$$\text{In solution: } 2\,\text{Fc}^+ + \text{GOx-FADH}_2 \xrightarrow{k} 2\,\text{Fc} + \text{GOx-FAD} + 2\,\text{H}^+ \tag{18.14}$$

$$\text{In solution: GOx-FAD} + \text{S} \rightarrow \text{GOx-FADH}_2 + \text{P} \tag{18.15}$$

"In solution" in this scheme means that the reaction is homogeneous, but takes place in the immediate vicinity of the electrode (in the reaction layer). In reaction (18.14), Fc^+ replaces oxygen as the natural electron acceptor (O_2 + GOx-$FADH_2$ → H_2O_2 + GOx-FAD).

These and similar reaction schemes are often illustrated as in Figure 18.22. The reaction (18.15) prevents equilibrium from being reached in the solution space. Instead, a continuous conversion takes place, which under certain conditions leads to a steady state and can be observed voltammetrically. Curve 1 in Figure 18.23 shows the reversible cyclic voltammogram of Fc, which we already know from the discussion of cyclic voltammetry in Chapter 5. The voltammogram does not change in the presence of glucose, as glucose has a very high overpotential at the glassy carbon electrode (GCE) used.

After adding GOx to the solution (curve 2 in Figure 18.23), the FAD centers of GOx oxidize glucose very quickly while being reduced to $FADH_2$. Fc^+ diffuses away from the electrode and reacts with GOx-$FADH_2$ after a short diffusion path from the electrode. It is then reduced to Fc in reaction (18.14). Fc can then be oxidized again at the electrode in reaction (18.13). If the relative concentrations of the reactants and the scan rate v are set as in Figure 18.23 so that all Fc^+ molecules are converted back to Fc within a thin reaction layer, a steady-state current is recorded (plateau), which depends on the reaction rate constant k'' between the mediator (Fc^+) and the enzyme (GOx-$FADH_2$).

The half-wave potential of the sigmoidal cyclic voltammogram (curve 2 in Figure 18.23) corresponds to $E°'(M_O/M_R)$ of the mediator and is *not* related to the formal potential of the $FAD/FADH_2$ redox center of the enzyme.

For oxidation of the $FADH_2$ centers $E°'(M_O/M_R) = E°'(Fc/Fc^+) = +0.16$ V vs. SCE must be more positive than the formal potential of the $FAD/FADH_2$ group bound in the protein, which amounts to −0.33 V (vs. SCE, pH 7) in GOx. Thus, the difference between the two formal potentials of the mediator and the prosthetic group is considerable, and so is the driving force of reaction (18.14).

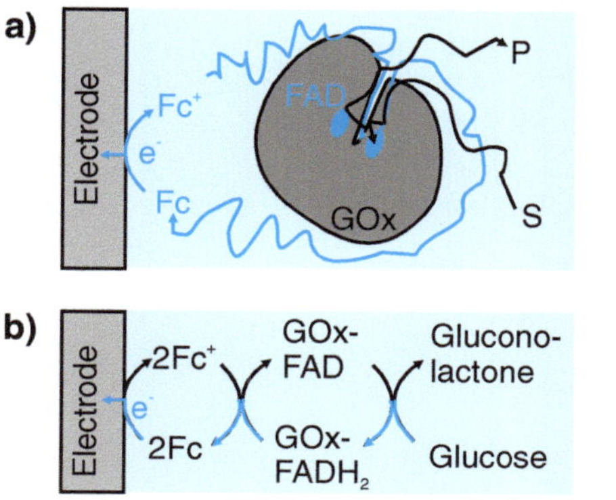

Figure 18.22 Different schematic representations of the mediated electron transfer of glucose oxidase with catalytic turnover. a) Emphasis on the spatial arrangement; b) emphasis on the redox cycles of the ferrocene mediator (Fc/Fc^+) and the prosthetic group ($FAD/FADH_2$) of the enzyme. The path of the electrons is highlighted in blue in b).

Example Calculation 18.1 Determine the bimolecular reaction rate constant k'' for reaction (18.14) from the cyclic voltammograms in Figure 18.23.

Three stock solutions were used:
Stock solution A: 0.050 mol L^{-1} glucose in 0.1 mol L^{-1} phosphate buffer pH 7
Stock solution B: $c_{B,Fc} = 0.005$ mol L^{-1} Fc in solution A
Stock solution C: 8 mg glucose oxidase ($M = 186\,000$ g mol^{-1}) in solution A
Working solution for curve 1:
Combine $V_A = 2$ mL solution A and $V_B = 0.5$ mL solution B.
Working solution for curve 2:
Combine $V_A = 1.6$ mL solution A, $V_B = 0.5$ mL solution B, and $V_C = 0.4$ mL solution C.

We determine the concentrations of Fc and GOx in both working solutions.
Fc in working solution 1:

$$c_{Fc,1} = \frac{c_{B,Fc} \cdot V_B}{V_A + V_B} = \frac{0.5 \times 10^{-3}\ \mathrm{L} \cdot 0.005\ \mathrm{mol\ L^{-1}}}{(2 + 0.5) \times 10^{-3}\ \mathrm{L}} = 0.001\ \mathrm{mol\ L^{-1}}$$

Fc in working solution 2:

$$c_{Fc,2} = \frac{c_{B,Fc} \cdot V_B}{V_A + V_B + V_C} = \frac{0.5 \times 10^{-3}\ \mathrm{L} \cdot 0.005\ \mathrm{mol\ L^{-1}}}{(1.6 + 0.5 + 0.4) \times 10^{-3}\ \mathrm{L}} = 0.001\ \mathrm{mol\ L^{-1}}$$

Thus $c_{Fc,1} = c_{Fc,2}$, which is a necessary condition for the application of the working curves below. Next, we obtain the amount concentration of GOx in the working solution 2 from the weighted mass, the molar mass, the prepared volume $V_{\text{stock solution}}$ of stock solution C, and the dosing scheme:

$$c_{GOx,2} = \frac{c_{C,GOx} \cdot V_C}{V_A + V_B + V_C} = \frac{\dfrac{m_{\text{weighted, GOx}}}{M(\mathrm{GOx}) \cdot V_{\text{stock solution}}} \cdot V_C}{V_A + V_B + V_C} = \frac{\dfrac{8 \times 10^{-3}\mathrm{g}}{186000\ \mathrm{g\ mol^{-1}} \cdot 1 \times 10^{-3}\ \mathrm{L}} \cdot 0.4 \times 10^{-3}\ \mathrm{L}}{(1.6 + 0.5 + 0.4) \times 10^{-3}\ \mathrm{L}} = 6.882 \times 10^{-6}\ \mathrm{mol\ L^{-1}} \quad (18.16)$$

Thus, the concentration of GOx (the catalyst) is much lower than that of the mediator ($c_{GOx,2} \ll c_{Fc,2}$). After the addition of glucose, each molecule of GOx reacts many

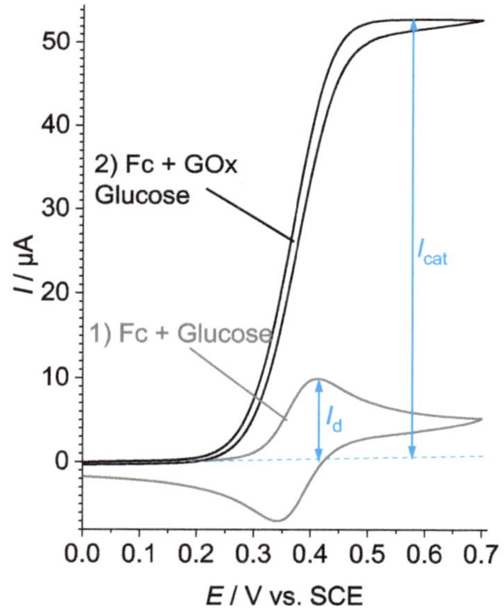

Figure 18.23 Cyclic voltammograms of 1) 1 mmol L^{-1} dimethylamino-methylferrocene (Fc) + 50 mmol L^{-1} glucose; 2) 1 mmol L^{-1} Fc + 50 mmol L^{-1} glucose +0.007 mmol L^{-1} glucose oxidase; GCE, phosphate buffer pH 7.0, $v = 50$ mV s^{-1}, $T = 298$ K.

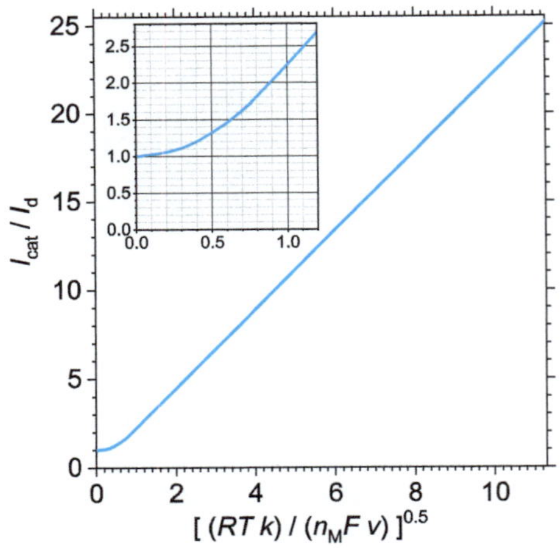

Figure 18.24 Working curve for determining the homogeneous rate constant. Source: Adapted from [39].

times with the mediator (catalytic turnover). Only this catalytic turnover enables current amplification and thus electrochemical observation of the homogeneous catalytic reaction.

We take the peak current $I_d = 9.5$ μA from curve 1 in the solution of Fc and GOx (no glucose) and the catalytically controlled current I_{cat} = 52 μA from curve 2 obtained in a solution of Fc, GOx, and glucose. We use a working curve (Figure 18.24) for the EC′ mechanism (Section 6.4.6) to determine a normalized rate constant $\kappa = RT k/(n_M F v) = 6.035$ of (pseudo)first order from the ratio $I_{cat}/I_d = 5.47$.

The current ratio $I_{cat}/I_d = 5.47$ falls within the linear range given by

$$\frac{I_{cat}}{I_d} = 2.2282\sqrt{\kappa} = 2.2282\sqrt{\frac{RT}{n_M F}\frac{k}{v}} \tag{18.17}$$

By rearranging, we obtain $k = 11.75\ s^{-1}$ for the example. The second-order rate constant is then $k'' = k/c_{GOx,2} = 1.7 \times 10^6\ L\ mol^{-1}\ s^{-1}$.

Although we can determine k from just two voltammograms, it is more accurate to measure at several v in both solutions (e.g., at v = 0.01, 0.02, 0.05, and 0.25 V s^{-1}). This results in several I_{cat}/I_d values, for which $\kappa = (RTk)/(nFv)$ is then determined either using Eq. (18.17) for $I_{cat}/I_d > 2.3$ or graphically from the inset in Figure 18.24 for $I_{cat}/I_d \leq 2.3$. We then plot these determined $(RTk)/(nFv)$ values over $1/v$.

$$\frac{RT}{n_M F}\cdot\frac{k}{v} = m\frac{1}{v}, \tag{18.18}$$

By linear regression, we determine the slope m [V s^{-1}], from which we obtain k [s^{-1}] according to

$$k = m\frac{n_M F}{RT}. \tag{18.19}$$

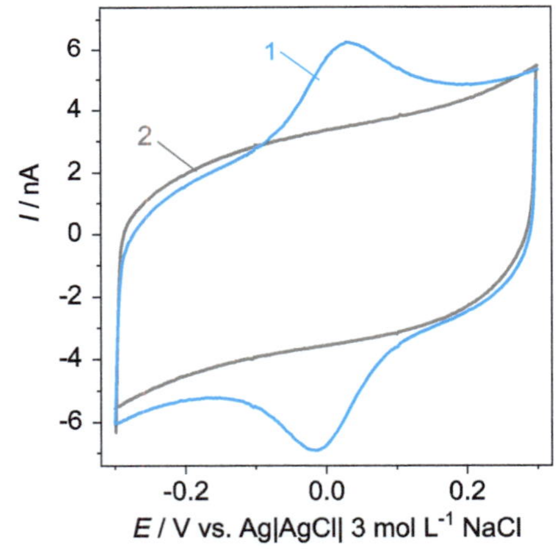

Figure 18.25 Cyclic voltammograms of 1) cytochrome *c* (cyt *c*) adsorbed on a monolayer-covered Au electrode, 2) monolayer-covered Au electrode without cyt *c*; working solution is 0.1 mol L^{-1} phosphate buffer, v = 0.05 V s^{-1}. Source: Reproduced with permission from S. Dongmo, J. Leyk, C. Dosche, C. Richter-Landsberg, U. Wollenberger, G. Wittstock; Electrogeneration of $O_2^{\bullet-}$ and H_2O_2 Using Polymer-modified Microelectrodes in the Environment of Living Cells. *Electroanalysis* **2016**, *28*, 2400–2407. Copyright 2016, Wiley-VCH.

18.2.6 Direct Electron Transfer with Immobilized Proteins

While the voltammetric investigation of many intrinsic redox proteins, and thus many redox enzymes, is only possible with mediators, an increasing number of reports are available on voltammetric studies of extrinsic redox proteins without mediators. Such proteins are investigated as surface-bound protein films in the sense of modified electrodes (Chapter 13). The success of this approach depends in particular on the choice of the electrode material, a suitable electrode pretreatment, or electrode modification, which must be matched to the protein. For example, the voltammogram of cyt *c* (Figure 18.25) was obtained by adsorbing the protein onto a mixed self-assembled alkanethiolate monolayer (SAM) on a gold electrode. Interestingly, this electrode is also catalytically active for the detection of H_2O_2 and $O_2^{\bullet-}$ [40]. Figure 18.25 shows both the cyclic voltammograms of the SAM without and with adsorbed cyt *c*. In the literature, the voltammograms are often shown after subtraction of the capacitive current (cf. Figure 13.56, curve 3). This results in voltammograms that correspond to the voltammograms of immobilized redox systems (Section 13.4.3; largely symmetrical peaks, after the peak, the current drops to the baseline, $I_p \propto v$). The quantities $E^{\circ\prime}$, k_s°, and n can be determined as discussed in Section 13.4.3 and by using the trumpet plot (Example Calculation 13.3).

Other possibilities of attaching proteins to an electrode in an *oriented* manner exploit their interaction with a metal oxide or their embedding in Langmuir–Blodgett films (Sections 13.1.5 and 13.1.7). A particularly large number of proteins have been successfully studied on pyrolytic graphite in which the oxidized edges of the graphene layers

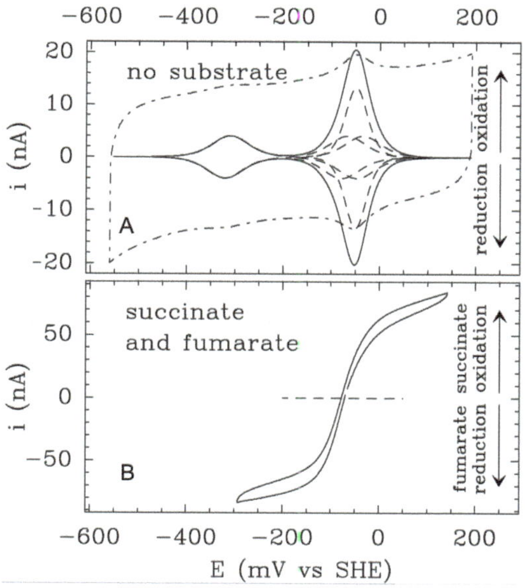

Figure 18.26 A) Noncatalytic voltammograms of a film of fumarate reductase (FrdAB) on a pyrolytic graphite electrode; (· – ·) unprocessed, unscaled (!) voltammogram, (—) voltammogram after subtraction of capacitive current, (- -) components after curve fit; v = **10 mV s**$^{-1}$, T = 293 K, pH 7, A = 0.03 cm^2. B) Voltammograms during catalytic conversion of a film of FrdAB at a rotating electrode; v = **1 mV s**$^{-1}$, T = 293 K, pH 7, ω = 3000 U min^{-1}, [succinate] = 1.2 mmol L^{-1}, [fumarate] = 0.004 mmol L^{-1}. Source: Reproduced with permission from [41], Copyright 2001, American Chemical Society.

come into contact with the solution (edge plane pyrolytic graphite, see lens inset in Figure 5.19). Protein adsorption is often improved by inclusion in polycations.

Figure 18.26A shows the cyclic voltammogram of fumarate reductase from *E. coli*, which contains an FAD and three Fe–S clusters as prosthetic groups [41]. In the unprocessed voltammogram (· – ·), the signals can only be detected as weak waves because the surface concentration of redox centers on the electrode is very low. The signals for the redox reactions (—) can be clearly recognized only after subtracting a polynomial baseline. The redox signals originate from the four redox centers. The prominent peak at −50 mV vs. SHE can be assigned to the FAD group (2e$^-$), the other peaks at −35 mV ($[2Fe2S]^{+/2+}$), −67 mV ($[3Fe4S]^{0/+}$), and −310 mV ($[4Fe4S]^{+/2+}$) to the three Fe–S clusters (1e$^-$). One signal is clearly separated from the others on the potential axis, while the signals of the other two clusters overlap with that of the FAD group. The dashed lines (--) indicate the result of a curve fit for these signals. By adding the substrates fumarate and succinate, the catalytic currents can be observed at a lower scan rate, where the enzyme reacts with fumarate or succinate and the active center of the enzyme is regenerated by ET with the electrode, so that the reaction proceeds continuously. In order to avoid mass transport limitations of fumarate or succinate, Figure 18.26B was recorded at a rotating electrode.

In Figure 18.26B and similar experiments, the catalytic current commences at the formal potential of the redox center that is decisive for this reaction (here FAD/FADH$_2$).

DET has many advantages over MET for the development of biosensors and biofuel cells. Consequently, there are intensive efforts to find enzymes and suitable electrode modifications that enable DET. However, it is not so easy to clearly demonstrate DET for enzymes. In an exemplary and thorough study, Bartlett and Al-Lolage [42] showed that, contrary to the claims in over one hundred publications, DET is *not* possible on native GOx. The literature reports mentioned were based on unrecognized influences of insufficiently purified starting materials. If DET is to be detected on enzymes, *all of* the following criteria must be met:

- The DET must be demonstrated with highly purified enzyme preparations, because otherwise cofactors can be released from other proteins, adsorb directly onto the electrode, and cause misleading signals that are similar to those of the enzyme under investigation without substrate addition.

- In addition to the DET, the catalytic reaction must be demonstrated as shown in Figure 18.26A. The catalytic current must start at the formal potential of the prosthetic group, as shown in Figure 18.26B.
- The causal relationship between current response and enzymatic activity should be verified using alternative substrates or enzyme inhibitors. For example, by using L-glucose instead of the enzyme substrate D-glucose, it is possible to test whether the small current changes to be evaluated in these experiments are due to purely physical effects such as a change in viscosity, substrate adsorption, or a nonenzymatic conversion. By varying the substrates for which the enzyme under investigation has different activities, it is possible to test whether the current response changes in the same way as the activity.

A reasonable hypothesis as to whether a DET is to be expected can be made by looking at the *biological* function of the enzymes and the data available in databases [34] for many enzymes. Only if the enzymes are extrinsic redox proteins that interact *in vivo* with other proteins or biopolymers during ET is DET plausible at an electrode surface designed accordingly. This applies, for example, to the biotechnologically important enzymes laccase, horseradish peroxidase (HRP), bilirubin oxidase (BOD), and cellobiose dehydrogenase. For intrinsic redox enzymes, on the other hand, skepticism regarding the possibility of DET is appropriate [42].

One interesting possibility is the construction of an artificial ET chain, in which a cofactor is covalently bound to the electrode surface via a suitable redox-active group. The corresponding apoenzyme can then bind to an electrode modified in this way via the cofactor (reconstitution of the holoenzyme). In the reconstituted holoenzyme, electrons can be transported from the redox center, which is not accessible *per se*, to the electrode [43].

Further Reading. Various review articles illustrate the possibilities of attaching proteins to electrodes to investigate the DET of the cofactors they contain [44–47]. The aforementioned article [42] is highly recommended with regard to proving DET.

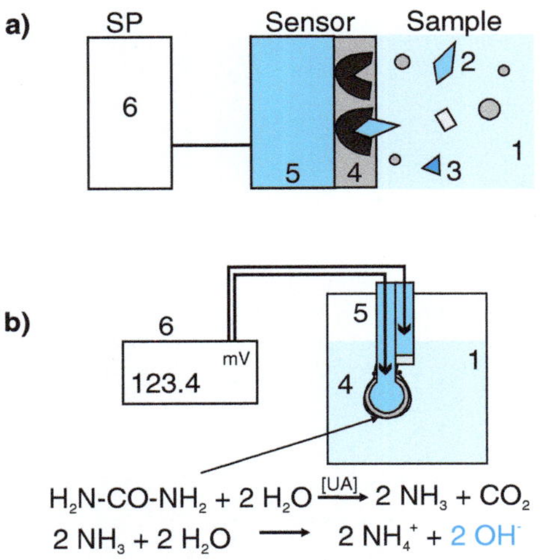

Figure 18.27

a) General principle of a biosensor. Sample (1) with analyte (2), interfering substance (3), sensor with recognition element (4), transducer (5), and signal processing device (SP, 6) for amplification, recording, and display.
b) Example of a potentiometric urea sensor with the enzyme urease (UA) as a recognition element; sample with urea (1), UA cross-linked to form a gel (4), and applied behind a dialysis membrane attached to the combination pH electrode (5) as a transducer. The cell voltage is displayed with the pH meter (6) and converted into a urea concentration after calibration.

For the exact structure of the combination pH electrode, see Figure 5.5.

18.3 Electrochemical Bioanalytics

Electrochemical bioanalytics, as an important application for electrochemistry, is based on the principles presented in Sections 18.1 and 18.2. They are used for the development of inexpensive analysis systems for complex analytes in medical diagnostics, quality assurance in food production, drug development, and other monitoring tasks. The basic idea is to combine an electrode as an instrumentally simple, but less selective transducer with a selective biological or biomimetic recognition elements (REs). The term "recognition element" (RE) is intended as a generalization to describe the biochemical components of sensors and test arrangements with which the analyte, that is, the substance to be determined, is recognized via a specific interaction among many other molecules in the sample (the matrix). Depending on the analytical problem, the RE can be an enzyme, an antibody (Ab), a specific sequence of single-stranded DNA (ssDNA), or another receptor molecule. In bioanalytics, a high selectivity is crucial, as molecules usually have to be determined in a matrix that contains many other molecules that have only a slightly different molecular structure but a completely different biological function or effect.

From an application perspective, the use of electrochemical detection methods often competes with optical and mass-spectrometric methods. The advantages of electrochemical methods are particularly evident when low-cost solutions or highly miniaturized systems are required. (Optical detection depends on the length of the optical path in the sample, which limits its miniaturization.)

18.3.1 Electrochemical Biosensors, Biochips, and Assays

The combination of REs with electrodes can be implemented in different ways.

Biosensors represent a direct spatial coupling of a biological recognition element and an electrochemical transducer for the qualitative or quantitative analysis of gaseous or dissolved compounds (Figure 18.27).

Direct spatial coupling facilitates highly integrated and therefore very small systems. However, it also has disadvantages: If biological molecules are bound to an electrode – we say "immobilized" – there is a tendency for them to slow down (or inhibit, Section 8.10.6) the desired electrode reaction. It is possible that the biological component and the inorganic electrode material have very different lifetimes. If one of the two components fails, the entire system may have to be discarded. Other possible combinations of REs and transducers avoid these problems, but often do not have the necessary precision when determining individual values.

In **biochips**, the biological recognition elements are bound to a portable, exchangeable solid carrier and, after interaction with the gaseous or liquid sample, transferred to an external analytical instrument to read out the signals (Figure 18.28).

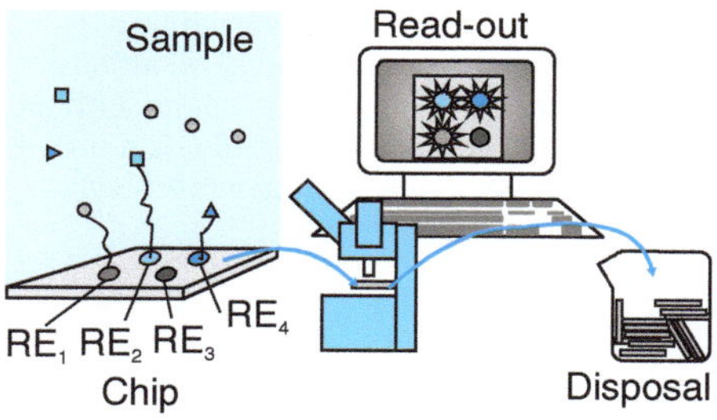

Figure 18.28 Principle of a biochip. Several recognition elements, RE$_1$ to RE$_4$, on an exchangeable carrier react with one or more analytes and are read out on one external instrument. The schematic is not to scale.

The chips are often about the size of coins. After analysis, the chip is discarded, and the read-out instrument can process the next chip. Several different REs can be immobilized on one chip in the form of an array so that several analytes can be detected simultaneously.

The even more general term **assay** refers to analysis methods, in which biological REs are often (but not always) bound to the walls of microtiter plates, capillaries, or micrometer-sized polymer beads. After a complex, often automated sample treatment, the read-out is carried out using electrochemical or optical methods in a relatively simple apparatus (e.g., simple compared to a high-resolution mass spectrometer).

The wide variety of biosensors described in the literature to date can be further subdivided according to various criteria, for example, according to the analyte (glucose, urea, creatine sensors), which is the most important information for the user. The REs used (enzymes, antibodies, DNA sequences, aptamers, molecularly imprinted polymers, organelles of biological cells or tissue) offer a further differentiation option. In the same way, sensors can be divided into optical, electrochemical, mass-sensitive, or thermal sensors according to the transducer. The electrochemical biosensors of interest here can be divided into potentiometric, amperometric, and impedimetric sensors according to the electrochemical detection principles (Chapters 5 and 7, Table 18.4). In order to understand the functional principle of the sensors, it is most informative to assign a sensor either to the group of metabolism sensors (Figure 18.29a) or bioaffinity sensors (Figure 18.29b), because the categorization of the biological recognition principle leads to direct conclusions about the transduction principles that can be used, which in turn are linked to characteristic advantages and disadvantages.

Table 18.4 Important electrochemical transducers.

Transducer	Explanation
Ion-selective electrode (ISE)	Potentiometry, $U \propto \ln\{X\}$, $I = 0$ (Section 5.2, example in Figure 18.27b)
Ion-selective field effect transistor (ISFET)	Charge accumulated at the interface switches current between source and drain $I_{SD} \propto \ln\{X\}$
Polarizable electrode	Amperometry, $E = const$, $I = f(t)$, example in Figure 18.35; voltammetry, $E = f(t)$, $I = f(E)$
Polarizable electrode	Impedimetry, $I = f(\omega)$; capacitive sensors, $C = f(E)$, $C = f(\omega)$

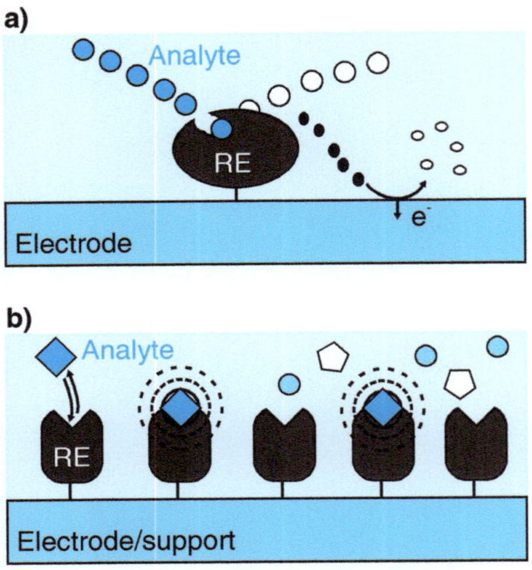

Figure 18.29 Principles of a) a metabolism sensor, in which many analyte molecules are continuously bound and converted by a recognition element (RE); and b) an affinity sensor, in which the analyte is bound without chemical conversion. The metrological task is to quantify the bound analyte molecules.

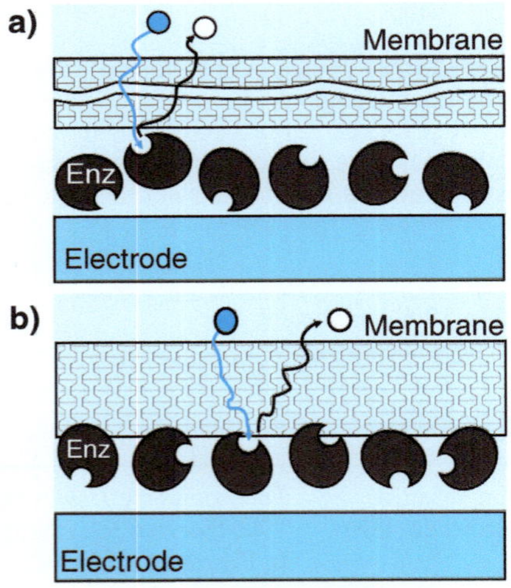

Figure 18.30 Immobilization of enzymes with semipermeable membranes. a) Retention behind a membrane; b) enzyme molecules bound to a membrane. The sketches are not to scale! The diameter of an enzyme molecule is a few nm, and the thickness of membranes is 10–500 μm.

Metabolism sensors (Figure 18.29a) continuously convert the analyte. The REs are therefore enzymes. Because *one* enzyme molecule can convert many analyte molecules in a short time, many more product molecules are available for the amperometric detection by the electrochemical transducer than the number of immobilized enzyme molecules.

In **bioaffinity sensors** (Figure 18.29b), each binding site reacts only *once* with an analyte molecule. Therefore, each detection element can only bind a small integer number of analyte molecules (two for antibodies, one for DNA sequences). The maximum number of analyte molecules interacting with the transducer can, therefore, be only as large as the number of binding sites in the biological REs. An electrochemical conversion of the extremely small quantity of bound analyte molecules often does not provide a sufficiently large current. The binding process can be detected, for example, via the changes in the electrochemical double layer using impedance spectroscopy, which responds very sensitively to this. However, it must be ensured that nonspecific changes in the double layer due to adsorption of other sample components without the involvement of the REs can be easily distinguished from the specific binding of the analyte to the RE that is actually of interest. In general, nonspecific binding events of matrix elements pose a major difficulty for all affinity sensors. In many cases, this problem is solved by using labeled substances (Section 18.3.4).

18.3.2 Immobilization of Enzymes or Other Biological Recognition Elements

The immobilization of the enzymes or biological REs shall ensure their permanent anchoring to the electrode or chip surface. It is also important to maintain the recognition function as much as possible, that is, not to change the actual binding site and its accessibility. The simplest method is the adsorption of the enzyme on the electrode (or on a specially selected monolayer on the electrode), as in the example used for Figure 18.26. This method is mainly applied for basic or explorative investigations [41], but does not offer sufficient stability for commercial sensors.

Another option is to retain the enzyme behind a membrane (Figure 18.30a). Although this procedure is simple, the membrane represents an additional diffusion barrier for the analyte, which reduces the sensitivity and prolongs the response time. The enzyme can also be bound to the membrane (Figure 18.30b), which offers the advantage that the enzymes can simply be renewed with a new membrane if the enzyme has lost its activity. The sensor can thus be regenerated.

Enzymes and other REs can also be bound to the electrode by cross-linking (Figure 18.31a). A bifunctional reagent, such as glutaraldehyde, links two protein molecules together. Because each protein is bound multiple times, an insoluble network of interconnected protein molecules is formed. In the case of glutaraldehyde, the coupling can also take place near the channel to the active center so that it is no longer passable for the substrate. Therefore, these procedures often need to be optimized in order to maintain enzymatic activity.

Cross-linking must be distinguished from approaches in which the REs are covalently bound to the electrode (Figure 18.31b), for example, grafted to a self-assembled monolayer (Figure 13.16) [48].

Another immobilization approach often used in the sense of modular building blocks is the biotin–avidin bond. Biotin is a medium-sized organic molecule for which the macromolecular protein avidin has four binding pockets. The binding pockets have a binding constant of 10^{15} mol^{-1} L. This noncovalent bond is therefore almost as stable as a covalent linkage. Biotin can be covalently bound to the RE and avidin to the surface of the transducer. When the biotin-labeled RE meets the avidin-modified surface, a defined construct is formed, similar to the coupling of two railroad carriages.

Processes in which the enzymes are embedded in a polymer film are widely used (Figure 18 32). We have already discussed different variants in Chapter 13. By electropolymerization in the presence of an enzyme (Figure 13.47), the latter can be incorporated into the resulting film [49–52]. A particular advantage is that the coating only takes place on the active electrode surface, and microstructures can be specifically addressed. Locally defined depositions can also be achieved with photopolymerizable polymers.

At a suitable pH, the enzyme is either a polyanion or a polycation and can be reacted with a polymeric counterion by layer-by-layer absorption (Figure 13.36) to form well-defined coatings. Also suitable for embedding are electrophoretic deposition (Section 11.6.2) or sol–gel layers [Eqs. (9.9) and (9.10)], whose deposition is triggered by a pH change caused by an electrolysis at the electrode to be coated [52].

Hydrogels in which the enzyme molecules are embedded in a relatively wide-meshed hydrophilic polymer are of the greatest practical importance [53]. This polymer can also contain the redox mediator in the side chains, as in Figure 13.42, so that reagent addition is not required for sensor operation. The water-soluble analytes can reach the enzymes almost unhindered through the wide-meshed, solvent-filled polymer network.

Another immobilization technique is the embedding of enzymes in carbon paste electrodes. Carbon pastes are suspensions of carbon particles and a hydrophobic oil, for example silicone oil [54]. The cofactors of the enzyme can also be included in the pastes (Figure 18.33). The carbon particles provide electrical conductivity. By extruding the paste, enzyme molecules that were not previously exposed to the solution can be brought to the surface, thereby regenerating the sensor surface. This approach can be extended so that pastes with curable components yield mechanically stable electrodes after a screen-printing process. Variations of the principle include, for example, the use of carbon nanotubes or graphene materials, which, due to their high degree of dispersion, can reach very close to the active center of the enzyme and have an enormous specific surface area.

Further Reading. Comparative reviews [55, 56] are recommended in addition to the literature examples for the individual methods cited above.

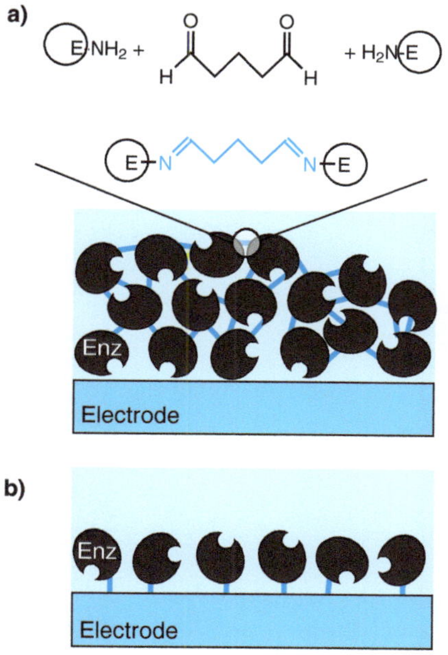

Figure 18.31 Binding of proteins (E) by
a) cross-linking or
b) grafting to the surface.

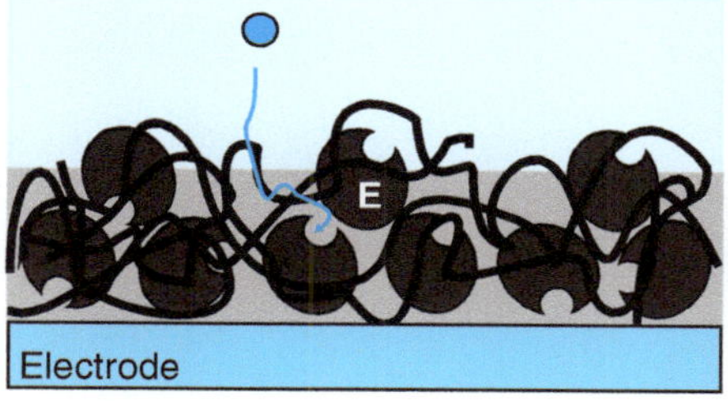

Figure 18.32 Embedding of enzymes (E) in polymer films, which can be accomplished by electropolymerization, layer-by-layer adsorption, electrophoretic deposition, or sol–gel processes.

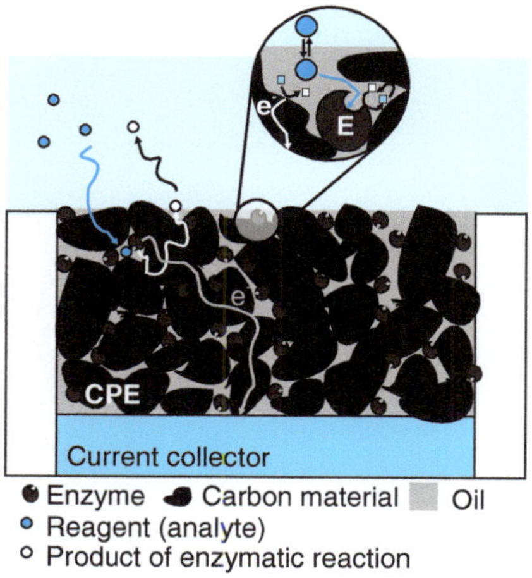

Figure 18.33 Embedding enzymes in a carbon paste electrode.

18.3.3 Amperometric Enzyme Electrodes for the Determination of Glucose

As an example of an electrochemical metabolism sensor, we will discuss the amperometric glucose sensor. According to Figure 18.22, $2e^-$ and $2H^+$ are transferred to FAD at GOx during the conversion of glucose (substrate, S) to gluconolactone (product, P).

$$S + \text{GOx-FAD} \rightarrow P + \text{GOx-FADH}_2 \tag{18.20}$$

In a further step, $2e^-$ and $2H^+$ are passed on to O_2 to form H_2O_2,

$$O_2 + \text{GOx-FADH}_2 \rightarrow H_2O_2 + \text{GOx-FAD}. \tag{18.21}$$

The enzyme molecule can then convert another glucose molecule. In addition to GOx, glucose can be oxidized by glucose dehydrogenases (GDHs), which also have biotechnological significance and are used in sensors. Dehydrogenases differ in the cofactor, which can be either PQQ, NAD^+, or FAD (PQQ-GDH, NAD-GDH, or FAD-GDH).

The recognition reaction is Eq. (18.20). This can now be coupled with the electrochemical detection in different ways. For this purpose, either the consumption of O_2 or the formation of H_2O_2 can be determined amperometrically. However, this procedure has some disadvantages: Firstly, it must be ensured that the O_2 concentration

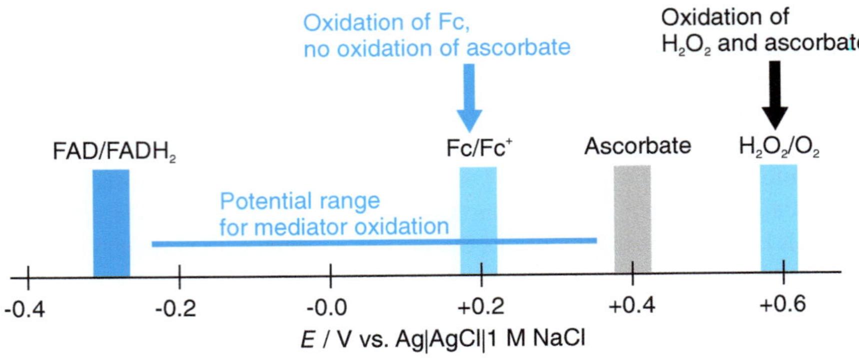

Figure 18.34 Relationship between the peak potentials for the oxidation of the FADH$_2$ center within GOx, a ferrocene redox mediator (Fc/Fc$^+$), H_2O_2, and ascorbate as an example for an interfering compound. The potentials shown correspond approximately to the anodic peak potentials in cyclic voltammograms. Depending on the compound, the peak potentials may be affected by considerable overpotentials and thus cannot be compared directly to thermodynamically defined equilibrium potentials.

present in the sample, or more precisely the product of the O_2 concentration and O_2 diffusion coefficient, is somewhat greater than the corresponding value for glucose, because otherwise the response of the sensor would be limited by the availability of O_2.

The oxidation of H_2O_2 takes place in neutral solutions at a potential of around +0.6 V vs. Ag|AgCl and is associated with a considerable overpotential. At the detection potential, other components of body fluids, such as ascorbate, urate, and cysteine, are also oxidized (Figure 18.34). The current from the oxidation of these interfering compounds then distorts the sensor signal.

These problems could be overcome by using mediators, in particular ferrocene derivatives (Section 18.2.5). These compounds have a very low overpotential at many electrodes and react rapidly with the enzyme. The detection potential must therefore only be slightly positive of the formal potential of the ferrocene derivative, so that interfering molecules are not yet oxidized (Figure 18.34). With the mediated biosensors, glucose sensors for self-monitoring have become widely available for diabetes patients (Excursion 18-2). The mediator reaches the FAD/FADH$_2$ centers of the enzyme by diffusion in the area close to the electrodes. It is therefore possible that the mediator is washed out of the sensor into the sample, which is unproblematic for use as a disposable sensor, but prevents its use as an implantable sensor.

The challenge of developing a sensor without leachable components was overcome by using redox hydrogels (Figure 13.42) in which transition metal complexes act as mediators. These complexes are covalently bound to a polymer chain and therefore cannot be washed out.

For GDH, $2e^-$ and $2H^+$ are first transferred to PQQ and from there to a mediator or a redox hydrogel. An additional advantage of GDH is that a competing transfer of electrons from PQQ to O_2 does not occur or occurs much more slowly than to mediators.

Excursion 18-1 Fabrication of a Simple Glucose Biosensor

A simple enzyme electrode can be produced by encapsulating GOx in an anodic electrophoretic paint (Section 11.6.2). The sensor can be subsequently characterized. The procedure is based on lab course instructions by W. Schuhmann (Ruhr University Bochum). More detailed information can be found in reference [57].

Preparation of the Sensor. The electrophoretic deposition is based on reducing the solubility of the polyanionic polymer through an electrochemically induced acidification (pH reduction) in the vicinity of the working electrode of the sensor. The oxygen evolution reaction ($2\ H_2O \rightarrow 2\ O_2 + 4\ H^+ + 4\ e^-$) reduces the pH and causes protonation of the polyanionic polymer molecules, whereby the solubility decreases and leads to the deposition of the polymer on the electrode under encapsulation of GOx. Approximately 70 μL mL^{-1} anodic electrophoretic paint and 4 mg mL^{-1} GOx are introduced. The deposition is carried out at a Pt-WE by 1 to 10 potential pulses to 2.2 V vs. Ag|AgCl|3 mol L^{-1} KCl over 0.5–1 s followed by a 5 s pause at 0 V. The quality of the sensor can be influenced by varying the pulse sequence. Due to the cost of the enzyme, it makes sense to produce the sensor in a small-volume cell container.

Recording the Calibration Curve. A stock solution of 1 mol L^{-1} glucose is prepared in phosphate buffer pH 7.4 and kept for at least 8 h to adjust the anomeric equilibrium between the α-D-glucose and β-D-glucose. The measuring cell is filled with $V_{buffer} = 20$ mL of air-saturated phosphate buffer, pH 7.4. The solution is gently agitated by a magnetic stir bar. The coated Pt electrode is connected as the working electrode (WE) to a potentiostat. A Ag|AgCl-Ref and a Pt wire as Aux complete the three-electrode cell. A potential of +0.6 V vs. Ag|AgCl|3 mol L^{-1} KCl is applied to the WE, and the current is recorded every 0.1 s. The recorded time series $I(t)$ is called the sensogram (Figure 18.35). The magnetic stirrer is switched off to read the measured value. Then, glucose stock solution is added ($V_{add} = 20$ μL, $c_{stock} = 1$ mol L^{-1}). The solution is thoroughly mixed by means of the magnetic stir bar. After stopping the magnetic stirrer, the approximately steady-state current is read and followed by the addition of another aliquot of the glucose stock solution. Due to the expected nonlinear calibration plot, it is advisable to increase the added volumes of the glucose stock solution concentration so that the current is incremented by an approximately constant value after each addition of stock solution.

Construction of the Calibration Plot. The glucose concentration present at each point in time is calculated from the additions of the glucose stock solution, taking into account the increase in total solution volume in the electrochemical cell.

$$c_i^* = \frac{\sum_{i=1}^{k} c_{stock} V_{add,i}}{V_{buffer} + \sum_{i=1}^{k} V_{add,i}} \tag{18.22}$$

The glucose consumption by the sensor itself, on the other hand, is neglected as it is very small compared to the total amount of glucose in the working solution (Section 5.3.5). The calibration plot of the steady-state current versus glucose concentration is shown in Figure 18.36.

GOx only converts β-D-glucose that is present in solution in equilibrium with α-D-glucose and as the open-chain form. This equilibrium must first be established after dissolving solid glucose. If the equilibrium is not established, the amount of β-D-glucose in the analysis solution changes despite a constant total glucose concentration.

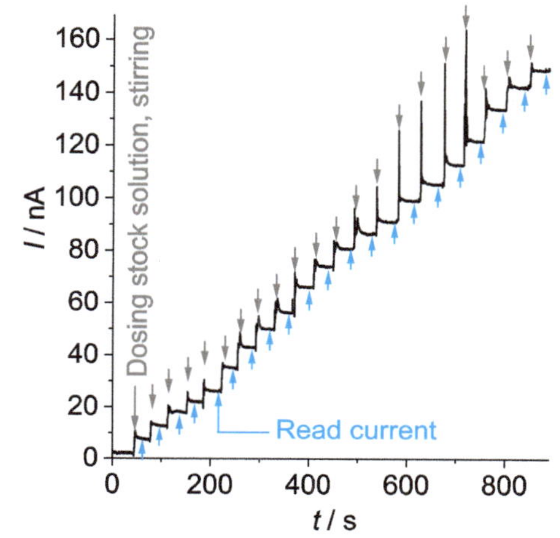

Figure 18.35 Sensogram of a glucose sensor. Data were kindly provided by S. Schmidt and W. Schuhmann, Ruhr University Bochum.

Calibration Function. The calibration function of an amperometric enzyme electrode is often similar to that shown in Figure 18.36 and can be represented by Eq. (18.23).

$$I(c^*) = \frac{I_{max}}{\frac{K'_M}{c^*} + 1} = \frac{I_{max}}{K'_M + c^*} c^* \tag{18.23}$$

If the analyte concentration c^* is significantly greater than the apparent Michaelis–Menten constant,[167] the term K'_M/c^* in the denominator of Eq. (18.23) can be neglected compared to one and $I(K'_M/c^*)$ tends toward I_{max}. If, on the other hand,

$I(c)$ – Concentration-dependent current of the sensor [A];

I_{max} – Maximum current at very high analyte concentrations [A];

c^* – Amount concentration of the analyte in the bulk of the sample [mol L^{-1}];

K'_M – Apparent Michaelis–Menten constant [mol L^{-1}].

[167] Leonor Michaelis, 1875–1949, German-Jewish-American biochemist and physician, professor in Berlin, from 1922 in Nagoya, from 1926 at Johns Hopkins University in Baltimore; Maud Leonora Menten, 1879–1960, Canadian physician, worked with Michaelis in Berlin from 1912 to 1916, later professor of pathology in Pittsburgh.

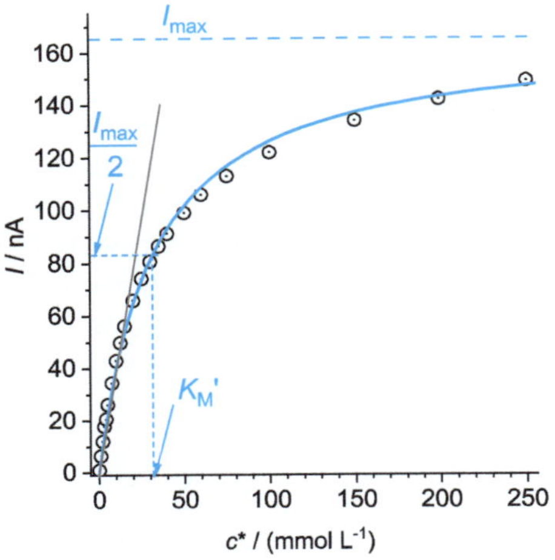

Figure 18.36 Calibration plot for the glucose sensor from the sensogram in Figure 18.35. The blue line is a fit to Eq. (18.23). The gray line is the calibration function for the linear range of the sensor for the concentration range $c \ll K'_M$.

$c^* \ll K'_M$, one in the denominator can be neglected with respect to K'_M/c^*. In this range, the sensor signal is approximately linearly dependent on the concentration and represents the preferred working range of the sensor.

$$I(c^*) \approx \frac{I_{max}}{K'_M} c^* \quad \text{for } c^* \ll K'_M \tag{18.24}$$

The slope of the calibration plot $b = \partial I/\partial c^* = I_{max}/K'_M$ is the sensitivity of the sensor. The limit of detection (LOD) c_{LOD} is the concentration at which the signal exceeds the noise of the background signal or the blank value (recorded as the standard deviation $S_{c,BV}$ when measuring a sample without analyte several times) by a factor of 3 $[c_{LOD} = 3S_{c,BV}/b = 3S_{c,BV}/(K'_M/I_{max})]$.

K'_M is referred to as the *apparent* Michaelis–Menten constant because the actual Michaelis–Menten constant K_M is determined for dissolved enzymes. K_M is a property of the enzyme in relation to the respective substrate. However, the enzyme is immobilized in a sensor. The concentration of the analyte in the volume phase K'_M/c^* is *not* present in the enzyme film. Rather, the local concentration in the film is modified by a diffusive mass transport and possibly by a distribution equilibrium of the analyte between the sample solution and the enzyme film, as shown in Figure 13.62. Therefore, K'_M is not only influenced by the enzyme and the kind of substrate, but also by mass transport characteristics and other properties of the sensor.

The Michaelis–Menten expression for two substrates [glucose and O_2, or glucose and mediator (O)] contains two concentration terms and two Michaelis–Menten constants $K_{M,glu}$ and $K_{M,O}$ [58], one or both of which may be negligible under certain conditions.

A complete physicochemical understanding of the calibration function and traceability of the variables occurring in it to clearly defined parameters of the enzyme or the enzyme electrode requires solution of the coupled differential equations for this system [59–61]. Such solutions are helpful for identifying possible improvements of the sensor. However, the effort is not necessary to simply use the calibrated sensors for concentration determination.

Adaptation of the Sensor Characteristics. In order to turn a sensor for a lab course experiment into a commercial product, further important developmental steps must be carried out. Here, we want to look at the possibilities for adapting the analytical characteristics of the sensor to the analytical problem to be solved. We cannot go into the extremely important aspect of the approval of sensors in the medical field, which follows similar strict protocols as the approval of a new drug.

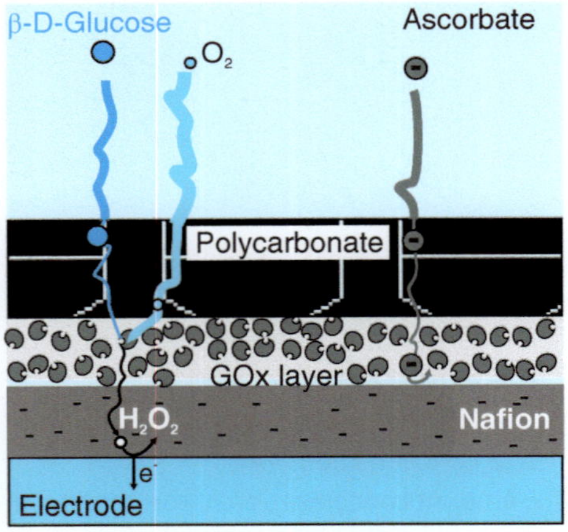

Figure 18.37 Effect of membranes on interferences. Polyanionic membranes exclude anionic interferences. Membranes with a size exclusion effect favor the transport of small molecules (O_2, H_2O_2) relative to larger ones. The widths of the trajectories indicate the relative values of the diffusion coefficients.

The influence of interfering substances in glucose sensors that oxidize H_2O_2 can be reduced by polyanionic membranes (e.g., from Nafion, cf. Figure 2.30), which are placed between the enzyme-containing layer and the electrode (Figure 18.37). H_2O_2 can pass through this layer, but anionic interferences such as ascorbate or urate cannot. Neutral interferences, such as cysteine, cannot be stopped by such a membrane. The additional use of neutral membranes with a size exclusion effect, such as cellulose acetate or polycarbonate, can help here because the small H_2O_2 molecule can pass through these membranes much better than larger molecules.

In many samples, the concentration range of interest is greater than K'_M, that is, c^* lies in a range in which the calibration function (Figure 18.36) is distinctly nonlinear or substrate saturation occurs. In this case, it is possible to dilute the sample, but this requires handling of the liquid sample, which makes the overall procedure more complex. A membrane placed between the enzyme layer and the sample solution (polycarbonate membrane in Figure 18.37) can help here by slowing down the transport of glucose to the enzyme layer. In this case, K'_M increases because the steady-state analyte concentration in the enzyme layer is lowered, although the analyte concentration in the sample remains unchanged by the additional transport resistance between the solution and the enzyme layer.

Catalytic amplification mechanisms can be used for the case in which the sensitivity is too low or the LOD is too high. This scenario is not relevant for typical glucose determinations, but is found in immunoassays or in the detection of contaminants in environmental analysis. Figure 18.38 shows how the structure of an enzyme electrode with PQQ-GDH can be used for the trace determination of phenol. The sensor contains two enzymes, PQQ-GDH and tyrosinase. Tyrosinase oxidizes phenols in two $2e^-$–$2H^+$ steps to 1,2-quinones [62]. The electrons are transferred to O_2. The reduction in O_2 concentration is recorded as a measurement signal at a Pt electrode. As the substrate is converted twice, four electrons are transferred from each analyte molecule instead of two, as in most other electrode reactions. However, the decisive increase in sensitivity is achieved by the fact that the reaction product 1,2-quinone is a good electron acceptor for PQQ-GDH. If glucose is added to the sample *as a reagent*, electrons are transferred from glucose via PQQ-GDH to 1,2-quinone, which is converted back to 1,2-hydroquinone and can thus react again with tyrosinase [63]. The fact that PQQ-GDH does not transfer electrons to O_2 is important for the correct function of the sensor.

Sensors with multiple enzymes can also be used to expand the range of analytes that can be determined by reacting analytes of interest in multiple steps, with only one coupled to an electrochemical reaction, or by having two enzymatic reactions competing with each other [64].

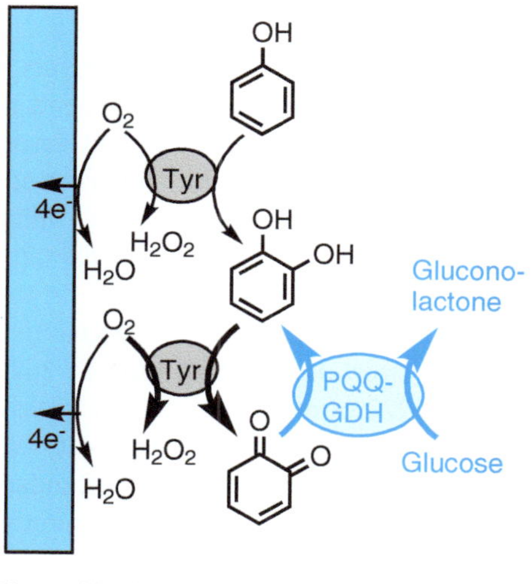

Figure 18.38 Mechanism of signal amplification for the detection of phenol. The consumption of dissolved O_2 is recorded at the electrode. The analyte phenol is converted twice by the enzyme. Additionally, 1,2-hydroquinone is regenerated by PQQ-dependent GDH when an excess of glucose is added as a reagent, because 1,2-quinone is an effective mediator for PQQ-GDH. Tyr, tyrosinase. Source: Scheme according to [63].

Excursion 18-2 Development of Blood Glucose Monitoring

For a healthy person, the concentration of glucose in blood is between 4.2 and 5.2 mmol L^{-1}. Deviations from this concentration range can cause acute and long-term damage to health. Since around 1974, amperometric glucose sensors have been used in large clinical laboratories [68]. The sensors consist of a replaceable membrane coated with GOx and a Pt electrode as the working electrode (WE). Additional membranes provide interference shielding analogous to Figure 18.37. Diluted blood samples are used, in which the glucose turnover of the red blood cells has been stopped, and the samples can therefore be analyzed within a period of about one day, which greatly facilitates the organization of the workflow compared to an immediate measurement.

Another scenario for the use of glucose biosensors is their application as a portable device for intensive care units, doctors' surgeries, and emergency services (bedside monitoring). These devices use sensors with membranes that adhere firmly to the electrode and measure in undiluted blood. Between measurements, the sensor is cleaned with a few microliters of rinsing solution and can be used for around 30 days or 1000 samples [68].

The regulation of blood glucose concentration is impaired in diabetes mellitus. This is why many patients have to monitor and regulate their blood glucose concentration themselves. The sensors described above are not suitable for the daily self-monitoring of patients because patients cannot safely perform optimum cleaning, regeneration, and storage of a sensor in a domestic environment. For this reason, only so-called disposable sensors are used at home. The sensors are discarded after one measurement. This application requires the sensors to be manufactured with extremely low tolerances, as the sensors have to be calibrated by the manufacturer and not by the user. In addition, extremely cost-effective production is required so that the use of several sensors per day is affordable for healthcare systems.

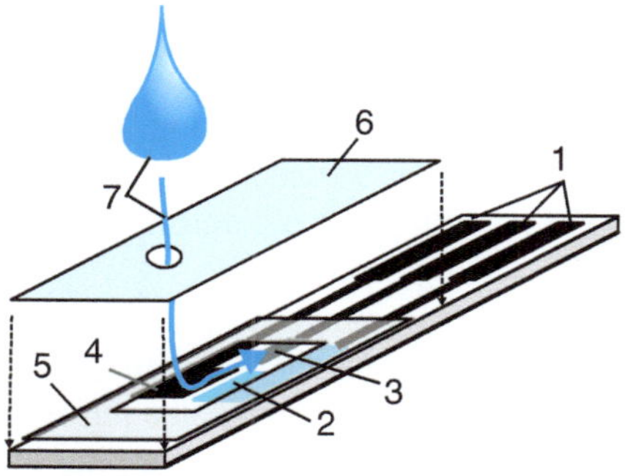

Figure 18.39 Schematic structure of a disposable sensor for self-monitoring of diabetic patients.

1) Contact pads;
2) WE with GOx and mediator and possibly other polymers;
3) Ag|AgCl-Ref;
4) Aux;
5) spacer with a flow channel, which absorbs the sample with a hydrophilic network;
6) cover foil with dosing opening;
7) drop of capillary blood and flow path of the sample.

Diabetes mellitus

The term diabetes mellitus [65] covers a range of metabolic diseases that lead to chronic hyperglycemia. If left untreated, elevated blood sugar levels lead to serious damage to the blood circulation, which results in kidney failure and nerve damage, and even necessitate amputations. In the United States, 11.6% of the population suffered from diabetes mellitus in 2021 [66]. In 2022, about 25% of the US healthcare costs were spent on the treatment of diabetes mellitus, including secondary diseases [67]! There are differences in the incidence worldwide, but it is generally increasing.

Depending on the cause and severity of the disease, treatment includes various measures ranging from a balanced diet to regular injections of the hormone insulin. The success of lifelong treatment increases when avoiding even short-term elevated levels in blood glucose concentrations. However, an overdose of insulin can cause hypoglycemia, which can lead to unconsciousness and thus represents an acutely life-threatening situation (hypoglycemic shock). By closely monitoring the blood glucose level over time, food intake and insulin doses can be adjusted to the current metabolic situation. The more closely this monitoring is carried out, the greater the chance of mimicking the effect of healthy biological regulation of the blood glucose level.

The disease means a significant loss of quality of life for affected patients and poses major economic challenges for healthcare systems due to the high incidence of the disease. However, the high incidence is also a very strong incentive for the development of new therapeutic approaches in which the electrochemical determination of glucose concentration using biosensors plays a major role and is being continuously refined.

The screen printing process is central to the production of such sensors [69] by which the paste for producing the electrodes is applied to the substrate through a wire mask. After closing some of the meshes, patterns can be transferred to the substrate via the remaining open meshes. The necessary storage stability is achieved by airtight welding of the sensors with the addition of a desiccant [70].

The capillary blood sample is obtained by puncturing tissue with a good blood supply, such as the fingertip. These parts of the body contain many nerve endings, which is why the puncture is painful. Therefore, a key development goal has been to reduce the volume of the required blood sample in order to be able to use smaller, less painful puncture procedures. The first test strips at the end of the 1980s required 10–20 μL of blood, which was drawn into a fluid channel on a test strip (Figure 18.39). The amount of the blood sample that reaches the electrodes is precisely adjusted through this channel, regardless of the size of the drop of blood collected. The electrodes of the electrochemical cell, which were produced using screen printing, are adjacent to the channel. As the electrodes are arranged flat on the support material, this arrangement is called coplanar. The electrochemical cell consists of a WE, an auxiliary electrode (Aux), and a reference electrode (Ref). Aux and Ref are sometimes combined in one electrode. Additional electrodes are used to monitor a minimum filling level of the channel. The ion-conducting connection between the electrodes is only established when a sufficient volume of blood is taken up.

In addition, the coating of the WE contains the mediator and the enzyme GOx. Analogous to Figure 18.23, the oxidation current of the mediator was measured amperometrically at the WE. The response depended on how quickly the mediator is reduced by the enzymatic reaction with glucose. By using the mediators, the detection potential could be lowered to such an extent that most interfering substances did not elicit a signal at this low potential [71]. By reducing the size, it was possible to lower the necessary volume to 3–5 μL by the mid-1990s. The products of the various companies differed according to the mediators used, the enzyme (GOx, FAD-GDH, PQQ-GDH, NAD-GDH), and additional strategies to exclude interfering substances [68].

The following problem had to be solved for further miniaturization: In the first sensors, the mediators diffused freely in the liquid channel. However, if WE and Aux are too close to each other, the mediator can be oxidized at the WE and be reduced at the Aux without this signal being caused by the presence of glucose in the sample. The solution to this problem was to "wire" the enzymes with redox polymers (Figure 13.42). Here, the $Os^{2+/3+}$ redox centers can only move in a small volume around their anchoring site within the polymer film due to conformational changes of the polymer chains. However, this restricted mobility is sufficient to take up electrons from the redox centers of the proteins and pass them on from Os complex to Os complex to the WE [68]. This scientific discovery made it possible to reduce the necessary sample volume to about 0.3 μL or 300 nL (Figure 18.40) and thus to use less well-perfused and therefore less painful puncture sites than the fingertip. In addition to this development, which is roughly outlined here, many small development steps were taken on all components of the sensor system. From an electrochemical point of view, the fine-tuning of the Os polymer is particularly interesting. By varying the ligands around the $Os^{2+/3+}$ ions, the redox potential of the complex can be "adjusted" (see also Figure 18.41 below). In this way, interferences can be excluded even better, and the mediator regeneration at the combined auxiliary reference electrode (labeled as combined counter-reference electrode in Figure 18.41) can be completely suppressed [53].

An important change in the electrochemical measurement technique was also implemented, which is not apparent from just looking at the structure of the sensor. Until then, the enzyme electrodes were operated amperometrically, that is, the current, which is proportional to the conversion per time, is measured. The current depends on the glucose concentration in the solution, but also on the mass transport conditions at the WE, and also, on the temperature. As the sample volume and channel dimensions are reduced, it becomes increasingly difficult to control these parameters with sufficient accuracy. Therefore, the transition was made to a coulometric measurement principle in which the glucose contained in the tiny sample is completely consumed during a few seconds, and the equivalent charge transferred to the mediator is determined. This procedure also leads to improved linearity. For the result of the coulometric analysis to become available to the user after the very short analysis time required for this application, tiny sample volumes are a necessity.

With the reduction in sample volumes, the problem arose that this small drop could no longer be visually perceived by older patients, which posed challenges in positioning the opening of the inflow channel next to the blood drop. The solution was to integrate the sampling and the sensor into one structure. This approach also improves therapy for other patients, as it allows the glucose concentration to be continuously monitored and thus creates the conditions for better regulation of the blood glucose level, which can further reduce the likelihood of serious secondary diseases. The measurement is carried out in the subcutaneous interstitial fluid (under the skin). The subcutaneous glucose concentration in the steady state is linked to the blood glucose concentration via a proportionality constant and follows the blood glucose concentration with a time delay (lag time), which must be taken into account when interpreting the measured values during a rise or fall in concentration. The sensor has the shape of a large button and is attached to the inside of the upper arm. There is a 5 mm long puncture cannula on the underside through which the tissue fluid enters the sensor [53]. The electrodes are placed on a small carrier plate. The sensor also contains a small battery to operate the electrochemical cell. The sensor sends the measured values at periodic intervals (around 15 min) to a device about the size of a cell phone, which alerts the patient to impending hyperglycemia or hypoglycemia with warning signals. The continuously available data allows patients and doctors to tailor food intake and insulin doses even better to the individual patient.

The artificial pancreas

The long-term goal of an artificial pancreas has been pursued for many years. An implantable sensor continuously records the blood glucose level and automatically triggers insulin administration via a micropump. The experience gained from sensor development means that this goal no longer seems so attractive today for broad application. Since such a system can easily put patients in life-threatening situations due to technical defects, including possible hacker attacks on the computer or transmission electronics used in it, a high and therefore costly technical effort would be necessary to meet the strict approval requirements. Under the conditions of a financially feasible mass deployment of such systems, two separate systems would appear to make more sense, in which the patient must first confirm a suggested insulin dosage. Such systems necessitate powerful microbatteries for sensors and pumps, whose requirements in terms of capacity, maximum power, storage stability, weight, and volume have not yet been fully met. Yet another challenge for electrochemists!

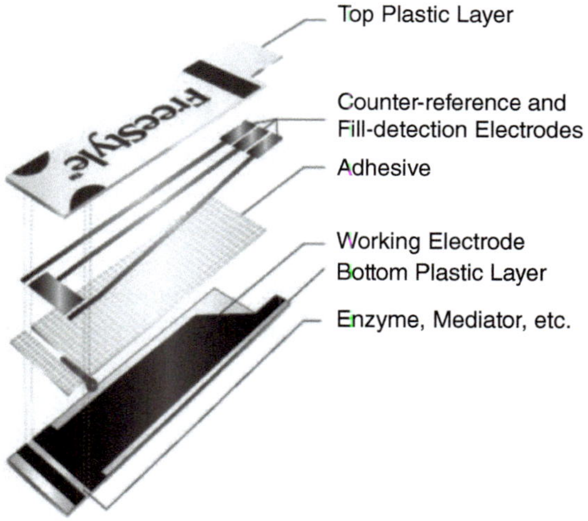

Figure 18.40 Layered structure of a coulometric single-use sensor. The enzyme is enclosed in an Os polymer and thus mechanically and electrically connected to the WE. Due to the absence of a freely diffusing mediator, the distance to the combined auxiliary and reference electrode (labeled "counter-reference" in this figure) and thus the volume of the sample channel can be reduced. Source: Reproduced with permission from [68], Copyright 2008, American Chemical Society.

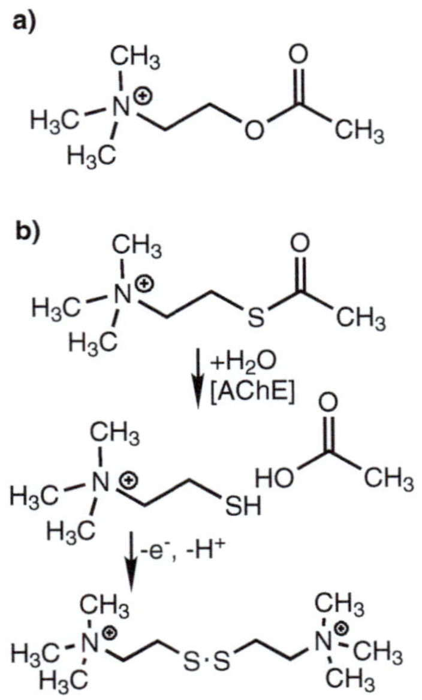

Figure 18.41 Structure of a) acetylcholine, b) acetylthiocholine, and the conversion to disulfide of thiocholine.

18.3.4 Enzymatic Sensors for Inhibitors

The principle of inhibitor sensors shall be briefly discussed. An enzyme electrode is used here, for example with the enzyme acetylcholinesterase (AchE). This enzyme cleaves the neurotransmitter acetylcholine (Figure 18.41a) and thus renders it ineffective immediately after transmission of the nerve impulse. A whole range of substances inhibit the activity of AchE and therefore act as neurotoxins. Those compounds include the insecticide parathion and paraoxon; organic phosphoric acid esters in the chemical warfare agents such as Sarin, Soman, Tabun, and Novichok; and the pharmaceuticals donepezil, rivastigmine, and galantamine for the symptomatic treatment of Alzheimer's disease.

The inhibitor sensors quantify the activity of the immobilized enzyme AchE, which decreases in the presence of inhibitors. The activity determination can be performed in a bienzymatic system, in which acetylcholine is first cleaved into acetate and choline by the enzyme AchE, and in a second step, choline is oxidized with the enzyme choline oxidase [72]. Alternatively, an artificial substrate, acetylthiocholine (Figure 18.41b), can be used, which is cleaved into thiocholine and acetate by AchE. Thiocholine can be oxidized electrochemically to the corresponding disulfide [73].

In each case, a constant amount of the enzyme substrate (acetylcholine or acetylthiocholine) is added to the sample as a reagent. It is observed whether the sensor response is reduced by the effect of any inhibitor that may be present (screening, monitoring) [72, 73]. A sample identified as suspicious in this way must be further analyzed using more complex analytical methods to determine the content and chemical identity of the inhibitor. Inhibitor sensors based on AchE are used to monitor drinking water.

Further Reading. Many textbooks on analytical chemistry contain chapters with a systematic overview of the determination of glucose [74], including a comparison with competing, nonelectrochemical methods. A coherent introduction that traces the development of glucose measurement specifically in the context of diabetes can be found in [68, 75]. The discovery of Os polymers and their applications for glucose measurements are very clearly described in connection to the nonelectrochemical aspects in [53]. A discussion of the commercialization and product development of blood glucose measurements can be found in more detail in [70, 74]. The particular challenges for implantable sensors (*in vivo* sensors) are described in [76, 77].

Several overviews are dedicated to multienzyme sensors [78, 79] or sensors with other enzyme classes [23, 80]. All these sources contain extensive bibliographies with references to original work.

18.3.5 Biofuel Cells

The progress in coupling enzymatic catabolic reactions to electrodes raises hope that the released Gibbs energy can be converted into usable electrical energy as in a fuel cell using highly active biological catalysts. Depending on the biological components used, different types of biofuel cells can be distinguished, which can be further subdivided according to the combination with a second electrode and the arrangement with or without a membrane.

The enzymatic biofuel cells use immobilized enzymes at the anode and cathode. They would be desirable, for example, as an energy supply for cardiac pacemakers, insulin pumps, or implantable sensors. Particularly attractive is the use of enzymes that carry out DET or at least work without freely diffusing mediators and are electronically connected to the electrode via an Os redox hydrogel, as in Figure 18.42. In these cases, the biofuel cell does not require a membrane to separate the anolyte and catholyte. In contrast, freely diffusing mediators (which can also be biological metabolic products)

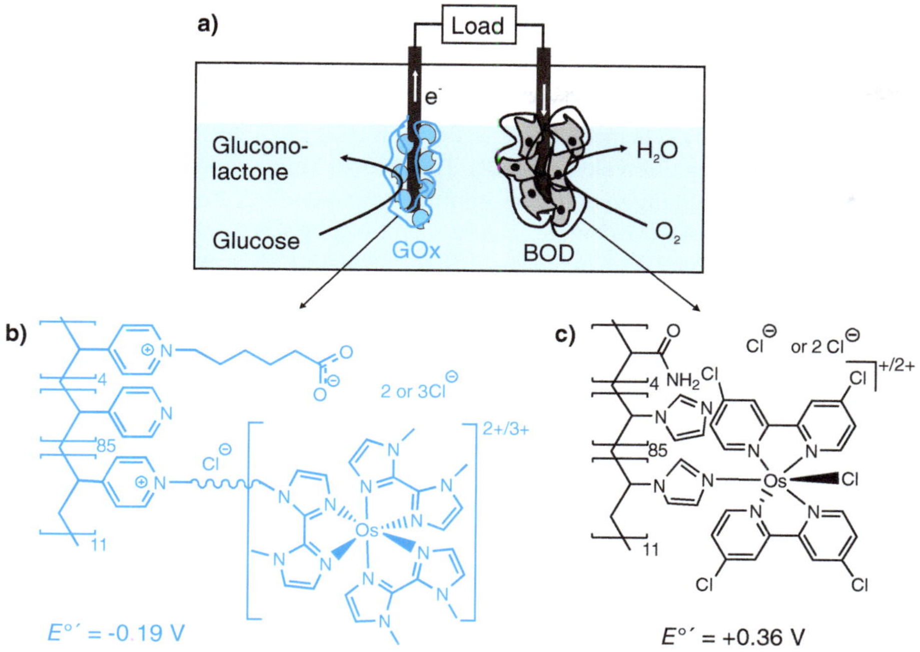

Figure 18.42 a) Structure of a membrane-free enzymatic biofuel cell with GOx at the anode and bilirubin oxidase (BOD) at the cathode; b) Os complex for the inclusion of GOx at the anode with $E^{\circ\prime}$ = −0.19 V vs. Ag|AgCl|1 mol L^{-1} NaCl; c) Os polymer for the inclusion of BOD at the cathode with $E^{\circ\prime}$ = +0.36 V vs. Ag|AgCl|1 mol L^{-1} NaCl. Source: Adapted from [81].

short-circuit the electrodes and reduce the electrical power output. Enzymatic biofuel cells without a membrane enable extensive miniaturization [81]. However, this is also accompanied by a limitation of the power output [82]. For this reason and due to their limited long-term stability, these cells have not yet found their way into commercial products.

In microbial fuel cells (Figure 18.43), the oxidation of organic compounds takes place at an electrode coated with a biofilm (Figure 14.32). Here, the enzymes are embedded in the protective environment of a biological cell. Bacterial communities consisting of different microorganisms can be specifically selected by the growth conditions so that they prefer the metabolic pathways that are advantageous for the biofuel cell, that is, using the anode instead of chemical oxidizers. This approach attracted considerable attention following the discovery that some bacteria are not only able to exchange electrons with the environment via soluble metabolic products, but also have electrically conductive extensions (pili) with which they can transfer electrons over greater distances to other microorganisms or to the electrode [83]. Electrodes with biofilms of such microbial communities can be combined with a robust and large-area electrode for oxygen reduction. Microbial fuel cells could improve the energy balance of wastewater treatment, for which a significant part of the available electrical energy is currently used (aeration and circulation of the clarification tanks). However, the organic substances contained in the wastewater represent a considerable energy content that cannot be directly utilized technically. Another application would be energy sources at measuring points that are very difficult to access, such as in the deep sea [84]. Despite impressive relative increases in performance, the characteristics for an economically viable use of microbial fuel cells have not yet been achieved.

Further Reading. Review articles on enzymatic [82–86] and microbial [83, 87] fuel cells and related technologies [88, 89] are recommended for further study of the topic.

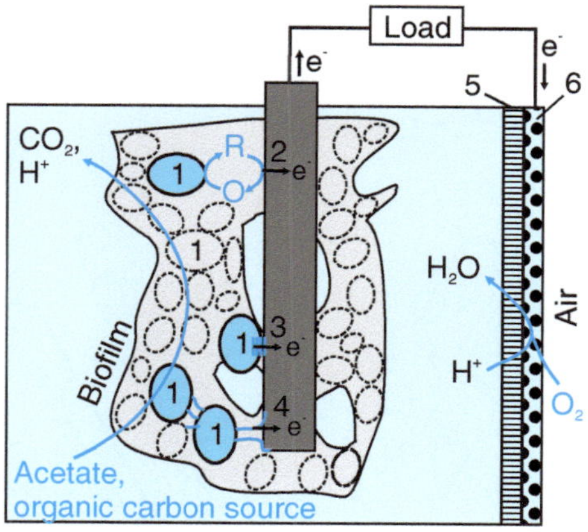

Figure 18.43 Structure of a microbial fuel cell. The microorganisms (1) within the biofilm around the anode are indicated as ellipses. For the cells highlighted in color, the possibilities of electrical coupling to the electrode are shown. 2) Mediated electron transfer (MET) via redox-active metabolic products, 3) DET via membrane-bound cytochromes, and 4) ET chains via electronically conductive pili between cells and between cells and anode. In the example, a cation exchange membrane (5) is used to separate the gas diffusion electrode (6, cathode). The sketch is not to scale.

18.3.6 Electrochemical Immunoassays

Immunoassays are based on the selective recognition of an antigen (Ag) by an antibody (Ab) directed against it (for explanations of the terminology, see Excursion 18-3). From the perspective of chemical analysis, the Ag is most often the analyte and the Ab is the biological recognition element (RE). The reverse configuration is also possible, for example, when aiming to detect an immune response of a patient to a pathogen. The formation of an Ag–Ab complex can be quantified by a mass action law with the binding constant K [mol^{-1} L], analogously to the formation of inorganic coordination compounds.

$$K = \frac{[\text{Ag–Ab}]}{[\text{Ag}][\text{Ab}]} \tag{18.25}$$

The binding constants of Abs used in immunoassays are often 10^8–10^{10} mol^{-1} L. As each binding site can only be occupied once by an Ag, the corresponding sensors belong to the group of bioaffinity sensors. The transducer must be able to detect this one-time binding event.

Excursion 18-3 Monoclonal Antibodies

Antibodies (also known as immunoglobulins) are produced in the B-cells of mammals as part of the immunological response to a foreign macromolecular substance – the Ag. Each B-cell produces only one type of Ab, which differs from the antibodies produced by other B-cells. There are several classes of Abs (G, A, M, D, E), of which immunoglobulin G (IgG) is the most important for immunoassays. IgG consists of four polypeptide chains (Figure 18.44) and is a relatively large protein with a total weight of 150 kDa. The chains are linked to each other by disulfide bridges (–S–S–), and the globular domains within the chains are also stabilized by disulfide bridges. Two of the four polypeptide chains are called heavy chains (H). They contain four domains, three of which have a relatively constant (C) amino acid sequence, and one domain has a highly variable (V) sequence. The light chains (L) contain one C-domain of constant sequence and one V-domain of variable sequence. One light and one heavy chain form an "arm", the Fab fragment (fragment of antigen binding). The actual binding site (paratope) for the Ag is located at the end of the Fab fragment and is formed by the V-domains of the heavy and the light chains. This is where the N-terminus of both chains is located. The C_H2 and C_H3 domains of the heavy chains form a third arm, the Fc fragment (constant fragment), which is important in the living organism for triggering a complex immune response, but is not involved in the recognition of the Ag.

Antibodies against a specific substance are obtained by injecting the macromolecular Ag into a test animal. Medium-sized organic molecules (such as pharmaceuticals and pesticides) can act as haptens. Haptens can only trigger an immune response and thus the Ab formation after binding to a suitable macromolecular carrier. After the immune response, the animal's serum is obtained, and the immunoglobulins are isolated. The serum contains a large number of different Abs, which are constantly produced by all mammals. After a targeted immunization, many but not all immunoglobulins are directed against the injected Ag. In their entirety, these Abs are referred to as polyclonal antibodies. These Abs attack various structures of the Ag, called epitopes, but exhibit different binding strengths. Although such sera are easy to obtain in small quantities, they can hardly be used for immunoassays, because reimmunization of the same or a different animal provides serum again, but its Abs have different binding properties. This would not be reproducible as a RE in a sensor.

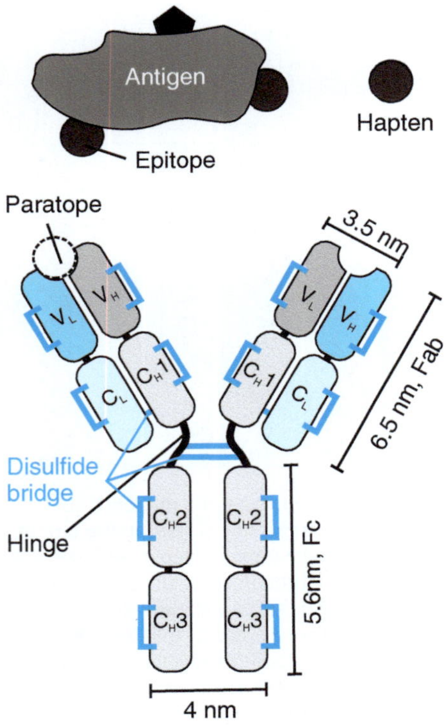

Figure 18.44 Structure of an IgG antibody. Abbreviations and terms are explained in the text.

To produce monoclonal antibodies, spleen cells from the immunized test animal are fused with cultured myeloma cells. Myeloma cells are a continuously dividing, immortal cell line. The fused cells (hybridoma cells) are tested to see whether they contain both the genetic information of a spleen cell and that of the myeloma cell and form Abs against the desired target molecule. If this is the case, the cells are separated in several culture steps to such an extent that a culture vessel only contains daughter cells of a single hybridoma cell – that is, all cells carry the same genetic information, and therefore, form one molecularly uniform Ab, a monoclonal Ab. This Ab is characterized in terms of its binding properties and other criteria. An immunization experiment can result in several monoclonal Abs, but it is also possible that no stable cell line is created that forms a high-affinity Ab. Hybridoma cells that have a stable Ab with a high binding constant to the target molecule are cultivated on a large scale. Monoclonal antibodies can then be isolated from the culture supernatant as molecularly uniform macromolecules, which can then be commercially obtained for a long time.

The description of the procedure has been greatly simplified here. Many new developments have refined the method. Additional site-directed mutagenesis enables the formation of recombinant antibodies with binding specificities that are otherwise unattainable.

Further Reading. Textbooks on biochemistry [18] and molecular immunology [90] provide further details on the biochemistry of the immune response, the production of monoclonal Abs, and immunodiagnostic procedures.

Labeled reagents are used in most assays. The label is indicated with a "*" in chemical equations. There are different **formats** for how the Ag, the Ab, and possible labeled reagents react with each other and how a signal is generated. In homogeneous immunoassays, Ab and Ag react with each other in the solution; in heterogeneous immunoassays, one binding partner – usually the Ab – is immobilized on the surface of a solid phase. The solid phase can be the inside of a microtiter plate, the inside of a capillary, the surface of paramagnetic polymer particles (beads), or in rare cases, the electrode itself. The use of capillaries or beads allows the surrounding solution to be flushed to the electrode within a miniaturized cell. The support surface for the biorecognition element can be renewed or discarded, while the electrode remains usable. Heterogeneous immunoassays make it particularly easy to separate the Ag–Ab complex from the sample matrix and from excess reagent. The disadvantage is the requirement that all Ag molecules must first be transported to the container wall in order to be able to react with an immobilized Ab. This is particularly problematic for trace analysis. Surface-modified magnetic particles combine the advantages of both approaches. They can be dosed as a liquid suspension and mixed with the sample while stirring so that Ag in the entire sample can react with the Ab on the surface of the particles. The particles can then be retained with a magnet while the unreacted matrix components are removed.

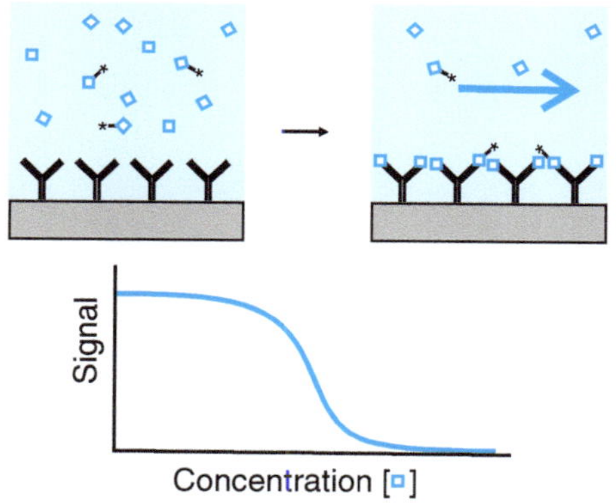

Figure 18.45 Schematic of a competitive heterogeneous immunoassay.

According to the role of the labeled reagents, a further distinction is made in competitive immunoassays, sandwich immunoassays, and titration assays. In a competitive assay (Figure 18.45), the Ag (as analyte) competes with the labeled Ag* (as added "standard") for a limited number of binding sites of the Ab. The higher the Ag content, the lower the signal caused by the Ag* label. In order for the measurement to be evaluated, the total amount of Ag and Ag* must be sufficient to saturate all binding sites. Therefore, particularly low detection limits cannot be achieved with this format. In the sandwich immunoassay (Figure 18.46), the immobilized primary Ab retains the Ag, whereby not all binding sites are saturated. In a second recognition reaction, a secondary, labeled Ab* detects the Ag previously retained on the surface. However, this

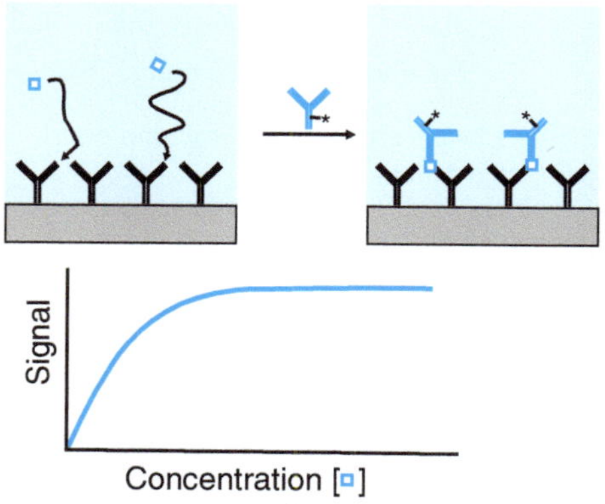

Figure 18.46 Schematic of a sandwich immunoassay.

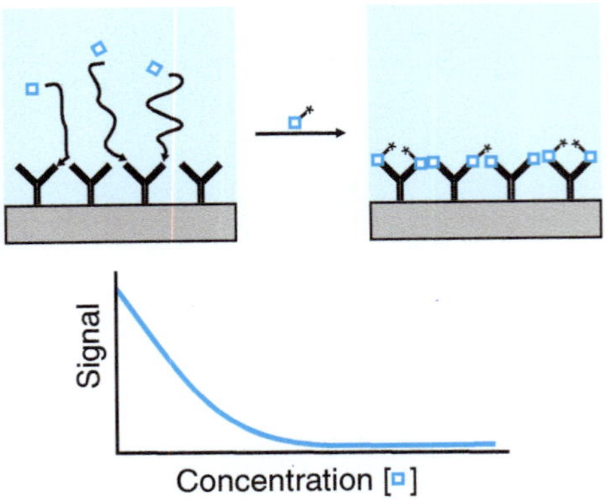

Figure 18.47 Schematic of a titration immunoassay.

is only possible if the Ag has at least two different epitopes (Excursion 18-3) and these are used by the two Ab. In a titration assay (Figure 18.47), the Ag reacts first with the immobilized Ab. The sites left free in the first binding step are then filled with Ag*, whereby it must be ensured that Ag* cannot displace the previously bound Ag during this reaction.

Radioactive labels were initially used; they can be detected with excellent sensitivity and do not significantly change the binding properties of Ag* compared to Ag due to their small size. Due to the complex storage, handling, and disposal, radioactive markers are not practical for mass applications. A variety of other markers were therefore explored, of which fluorescent markers and enzymes (E) are the most important. Enzyme labels form product molecules at a high rate in the detection step of the assay after the addition of a working solution optimized for the enzyme label. We will limit ourselves here to assays, in which the products of the enzymatic reaction are detected electrochemically.

First of all, it should be remembered that the enzymes (E) themselves are macromolecules. Thus, the labeled Ag is a conjugate of the analyte Ag and a macromolecular label (Ag* = Ag–E). It is not unlikely that the binding and transport properties differ slightly between Ag–E and Ag. This difference is taken into account by calibration procedures. The electrochemical detection of the products formed has several advantages. Since electrochemistry is an interfacial technique, the LOD does not depend on the optical path length (in contrast to optical detection methods) and is therefore particularly suitable for integration into small, highly integrated cartridges and chips, as we have already discussed for the enzymatic glucose sensors. The detection electronics can be designed cost-effectively for precisely specified applications. Electrochemical detection can also be used in turbid or light-scattering solutions.

A labeling enzyme should have the following properties:

- a high turnover number, which indicates the maximum number of product molecules formed per time;
- a low Michaelis–Menten constant;
- high stability under storage and assay conditions;
- high availability of the enzyme in its pure form;
- electrochemical detection methods for the products of the enzyme reaction;
- compatibility of the optimal working conditions of the enzyme (pH, ionic strength, buffer composition) with the conditions for the Ag–Ab reaction.

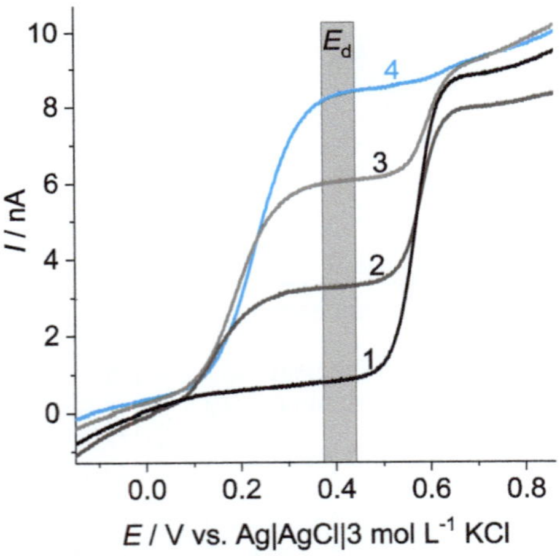

Figure 18.48 Linear sweep voltammetry (LSV) of 2 mmol L^{-1} *p*-aminophenyl-β-D-galactopyranoside in phosphate buffer (pH 7.0) on a Pt microelectrode after different reaction times following the addition of 40 μg mL^{-1} galactosidase. 1) Before enzyme addition; 2) 10 min; 3) 20 min; 4) 30 min (corresponding to almost complete conversion to *p*-aminophenol); $v = 0.1$ V s^{-1}. E_d indicates a suitable detection potential for an electrochemical immunoassay. Figure made by bookauthor using data from C. Zhao, J. K. Sinha, C. A. Wijayawardhana, G. Wittstock, *J. Electroanal. Chem.* **2004**, *561*, 83–91.

These conditions are approximately fulfilled by the enzymes alkaline phosphatase (ALP), β-galactosidase (Gal), glucose oxidase (GOx) and peroxidases, especially horseradish peroxidase (HRP). Most electrochemical immunoassays use one of these four enzymes (Table 18.5).

Let us examine Gal in more detail. This enzyme cleaves galactose from a substrate (Table 18.5); the catalyzed reaction is therefore not a redox reaction (just like the cleavage of phosphate by ALP). For amperometric or voltammetric detection, a substrate *p*-aminophenyl-β-galactopyranoside (PAPG) is used. This substrate is converted to galactose and *p*-aminophenol (PAP) by Gal. Figure 18.48 shows the linear sweep voltammograms at a microelectrode after different reaction times for the Gal-catalyzed conversion of PAPG to PAP. Only PAPG was present when curve 1 was recorded. In contrast, PAP is almost exclusively available at the time when curve 4 was recorded. Oxidation of the substrate PAPG requires potentials above +0.5 V vs. Ag|AgCl|3 mol L^{-1} KCl, whereas PAP can be detected at lower potentials of +0.3 V. For the analysis, the detection potential E_d is selected between 0.35 V and 0.43 V, so that PAP is oxidized to *p*-quinone imine (PQI), but PAPG is not converted. For each PAP molecule formed and converted at the electrode, $2e^-$ enter into the external circuit.

Table 18.5 Reactions of the most important labeling enzymes for electrochemical enzyme immunoassays.

Enzyme	Enzyme reaction	Electrochemical detection
ALP EC 3.1.3.1	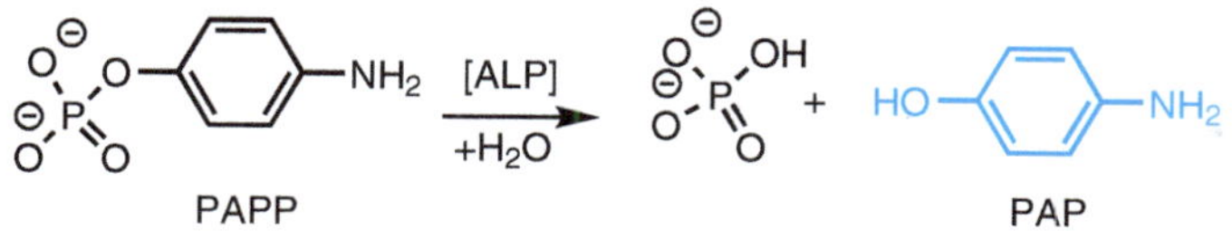	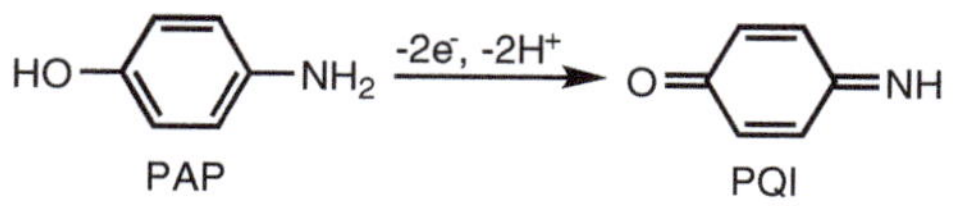
Gal EC 3.2.1.93	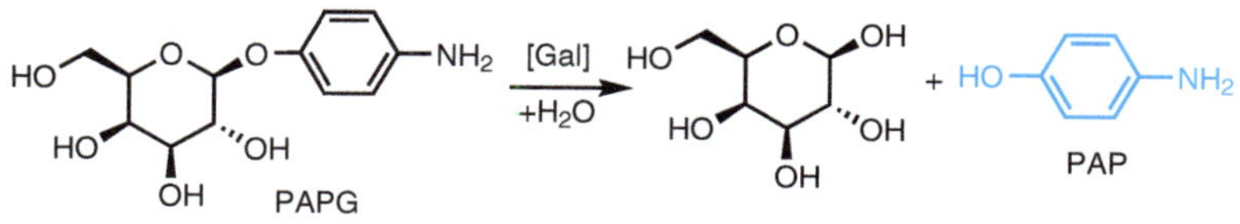	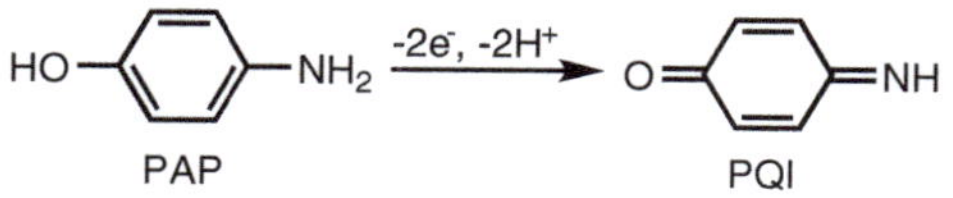
HRP EC 1.11.1.7	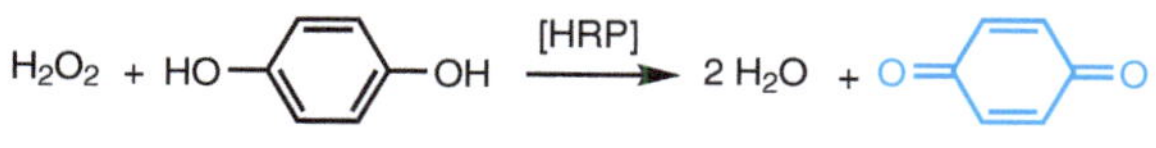	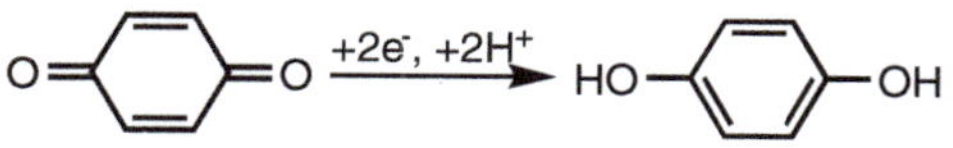
GOx EC 1.1.3.4	Glucose + O_2 $\xrightarrow{[GOx]}$ Gluconolactone + H_2O_2	H_2O_2 $\xrightarrow{-2e^-, -2H^+}$ O_2

The sensitivity of the detection can be further increased by these strategies:

- PQI/PAP is oxidized and reduced several times at an interdigital electrode array (cf. Figure 5.88) [91, 92].
- PQI/PAP or phenol formed from *p*-phenylphosphate and *o*-quinone/1,2-hydroquinone serves as a mediator for a multienzyme system consisting of glucose dehydrogenase and tyrosinase, as in Figure 18.38. The recorded signal is then the oxygen consumption of this system. Record detection limits have been achieved with such sensors [63].

In both cases, more than $2e^-$ are transferred to the external circuit for each phenol or PAP molecule formed at the enzyme label.

The enzyme labels facilitate a large increase in sensitivity. The need for labeling and the possible associated change in the binding properties of Ab or Ag stimulated the search for label-free detection schemes with capacitive and impedimetric methods. Different measuring principles are used to determine how the distribution of charged species changes in the electrical double layer of the electrode during Ag–Ab binding (Figure 18.49). From Section 8.10, we know that the electrical double layer can be compared in a very simplified way to a plate capacitor with the capacitance $C = \varepsilon_0\, \varepsilon_r\, A/d$ [Eq. (4.18)]: One plate is the solid electrode and the other "plate" is formed by the ions of the compact double layer, which can approach the electrode up to distance d (Figure 18.49). By binding the Ab, d is increased and the relative dielectric constant changes, causing C to decrease. The measurement task is now to reliably detect the additional change in capacitance when the Ag is bound to the Ab. Especially with small Ags and incomplete saturation of the binding sites, it is not easy to achieve sufficient sensitivity because the nonspecific adsorption also leads to small signal changes that can mask the specific Ag–Ab binding. A change in coverage is also detected during impedimetric detection.

The **nonspecific adsorption** represents a fundamental problem for heterogeneous immunoassays. This refers to the adsorption of Ag, Ag*, or matrix components to the

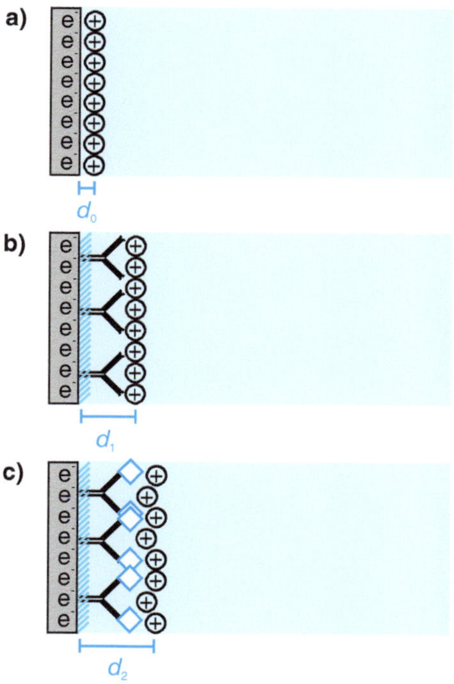

Figure 18.49 Capacitive detection of the Ag–Ab bond. a) Schematic and highly simplified representation of the double layer at the uncovered electrode; b) decrease in capacitance by increasing the distance between the electrode and the ions in the compact double layer from d_0 to d_1 after immobilization of the Ab; c) further increase in distance to d_2 and corresponding decrease in capacitance upon formation of the Ag–Ab complexes.

surfaces without formation of Ag–Ab complexes. The molecules bound in this way contribute to the signal or can interfere with the formation of other Ag–Ab complexes. Therefore, assay protocols usually contain precautions to suppress nonspecific adsorption, such as solvent additives or special surface coatings [93].

Further Reading. Please refer to review articles on electrochemical immunoassays [94–96] for information on the exact measurement procedure, calibration strategies, and sample preparation. In addition to Ab, artificial REs such as molecularly imprinted polymers [97, 98] or aptamers [99, 100] are used. [Aptamers are DNA or ribonucleic acid (RNA) oligomers, in which loops with very diverse structures are formed by the hybridization of segments within a single strand. These can be selected in vitro in such a way that they bind specific regions of proteins and other molecules, similar to an Ab.]

18.3.7 Electrochemical DNA Assays and DNA Chips

DNA and RNA are linear macromolecules, in which nucleotides are linked to the bases (A, adenine; G, guanine; C, cytosine; T, thymine; in RNA, U, uracil instead of T) via phosphate residues. DNA and RNA can occur in different conformations. In B-DNA, the most common helical conformation of double stranded DNA (dsDNA), there are two complementary strands in opposite directions in a double helix, which is held together by hydrogen bonds between the base pairs A–T (A–U) and G–C. The DNA can be split or "melted" into two single-stranded DNA molecules (ssDNA) by heating. The base-specific association of two ssDNA molecules into one dsDNA molecule is referred to as hybridization.

The bases within the DNA strand can be both oxidized and reduced. Specific electrochemical signals can also be obtained from chemically damaged DNA or from compounds that intercalate in DNA [101, 102].

Base pairing means that a sequence of nucleotides on one strand results in a complementary sequence on the second strand of the dsDNA. The complementary strand can be synthesized using an ssDNA strand as template both *in vivo* and *in vitro*. Oligomeric DNA segments can be produced sequence-specifically in the laboratory. Using the polymerase chain reaction (PCR), individual DNA segments can be selected from a digested cell sample and amplified exponentially. In addition, further synthetic modifications to DNA oligomers are possible, which dramatically expand the possibilities for biotechnological applications (genetic engineering).

The sequence of nucleotides in a DNA segment forms the basis of the genetic code. This information is important, for example, in the detection of bacterial or viral infections, adjustment of therapies to drug-resistant bacterial strains, diagnosis of tumor diseases, determination of predispositions to genetic diseases and initiation of prophylactic measures, for proof of identity in forensics, as well as the detection of genetically modified organisms in food.

The possibilities of using **ssDNA as a biological recognition element** for the complementary strand in bioaffinity sensors or assays are somewhat similar to the approach in immunoassays (Section 18.3.5) [103, 104]. However, there are also important differences:

- DNA and RNA are polyanionic molecules. This influences their interactions with surfaces regardless of the sequence [105].
- ssDNA strands bind to each other even if there are some base mismatches. Mismatches decrease the binding strength. As a result, melting occurs at a lower temperature than with completely complementary strands. In order to exclude the binding of ssDNA with few mismatches during the recognition reaction, hybridization must be carried out under restrictive conditions (increased temperature, precisely adjusted ionic strength, and possible application of electric fields).

- In addition to B-DNA, other conformations are possible. In particular, ssDNA and RNA can form complex structures by hybridizing complementary regions within *the same* strand. This can be used for completely new recognition and transduction principles [106].
- An ssDNA strand can bind to different areas on different other ssDNA strands. As a result, large and also quite complex, three-dimensional structures can be spontaneously assembled (DNA origami), which can, for example, block a flow of ions through nanopores [107] or decompose a blocking biopolymer film on electrodes when hydrolytic enzymes are bound to the origami structure. The process can be tracked via the onset of a current [108].

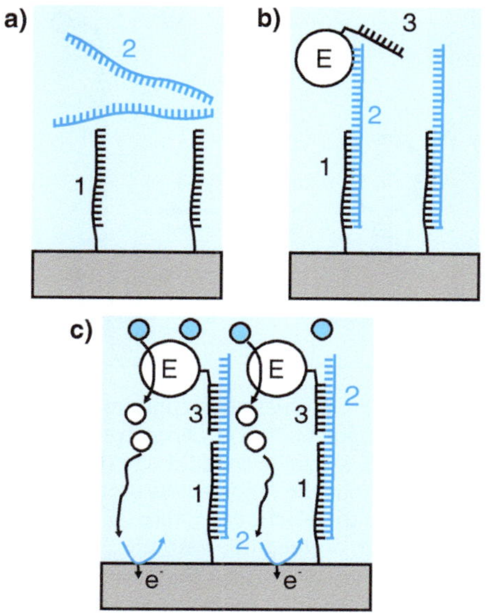

Figure 18.50 Electrochemical DNA hybridization assay in sandwich format. a) A surface-bound probe (1) interacts with the target DNA [(2), the analyte] of the sample; b) the bound target sequence is labeled with an enzyme-ssDNA conjugate (3); c) electrochemical readout of the labeled double strands.

In applications that require cost-effective, portable, and rapid analytical methods (i.e., without PCR), electrochemical detection methods may have advantages. During the completion of this book, the need for such analytical methods became particularly evident in the context of mitigating the COVID-19 pandemic when the testing capacities in large centralized laboratories and the logistics of transmitting samples and results were repeatedly overwhelmed, even in developed countries.

Without claiming to be exhaustive, we would like to present some assay formats below that illustrate these statements.

Sandwich Assays. The recognition molecule (often referred to as a probe in the context of DNA sensors) is immobilized on the surface of the chip or electrode, for example, in the form of a mixed self-assembled monolayer (cf. Section 13.1.5). The DNA segment to be determined (i.e., the analyte or target DNA) is attached to the surface by a specific base-pairing with the probe DNA. Some parts of the target DNA protrude beyond the probe. A labeled ssDNA is bound to the protruding part of the target DNA in a sequence-specific manner (Figure 18.50). The label can be a labeling enzyme from Table 18.5, whose reaction products are detected electrochemically during the detection step (readout).

Another approach is the enzymatic formation of an insoluble substance that is deposited as a layer on the electrode and thereby blocks it. The signal during the readout step is a decreasing current of an added redox probe.

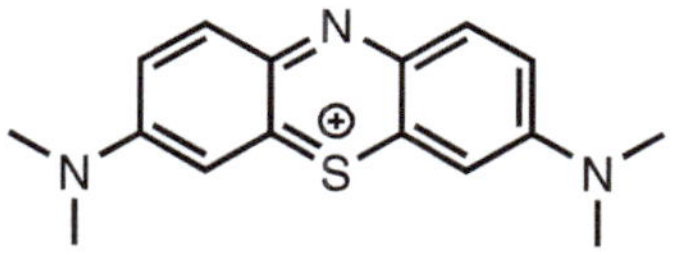

Figure 18.51 Structure of methylene blue.

Enzyme labels on an ssDNA strand greatly increase the size compared to an ssDNA with several 10 base pairs. However, enzymes are much more unstable than DNA, so enzyme labeling can cause stability problems. Attempts are therefore being made to establish label-free detection of the hybridization reaction in DNA assays by impedimetric or capacitive measurements on electrodes [105].

Redox Labels. Hybridization can also be detected by redox labeling. There are various possibilities for this. For example, redox-active groups, which have fast redox kinetics and are converted at potentials at which the DNA bases do not react, can be bound to the end of the strand. A frequently used group is methylene blue (Figure 18.51). Planar aromatic compounds can insert (intercalate) between the stacked base pairs of a double strand without covalently binding to a single strand. By linking an intercalating molecule with a redox-active group, as shown in Figure 18.55, labeling reagents can be obtained that intercalate into a double strand after it has been formed, thus making the hybridization reaction electrochemically detectable. Another labeling method is based on the binding of redox-active cations to the negatively charged DNA backbone, which can be enhanced by fitting into the large groove of B-DNA. A clever choice of probes can ensure that the redox-active group is bound in the vicinity of the electrode after the formation of the construct consisting of immobilized probe, target DNA, and the second soluble probe, as depicted in Figure 18.52 [109].

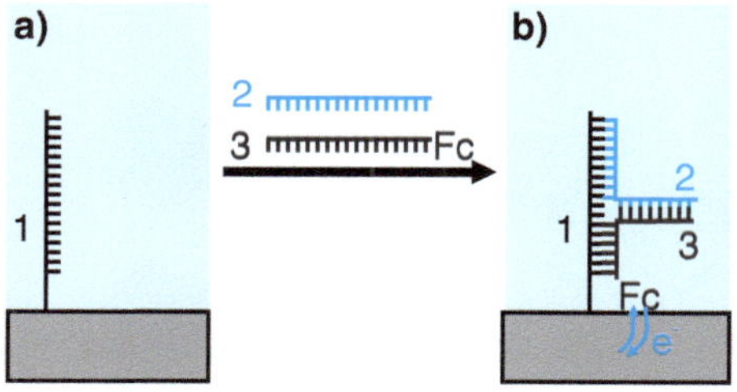

Figure 18.52 DNA sandwich assay with redox labeling. a) Probe (1) is bound to a gold electrode and hybridized with the target DNA (2) and a ferrocene (Fc)-conjugated second probe (3); b) after hybridization, the Fc group is positioned close to the surface and can undergo fast ET. Source: Adapted from [109].

In order to avoid the sensitivity problems of cyclic voltammetry when investigating a redox-active monolayer, it is advisable to use alternating current voltammetry

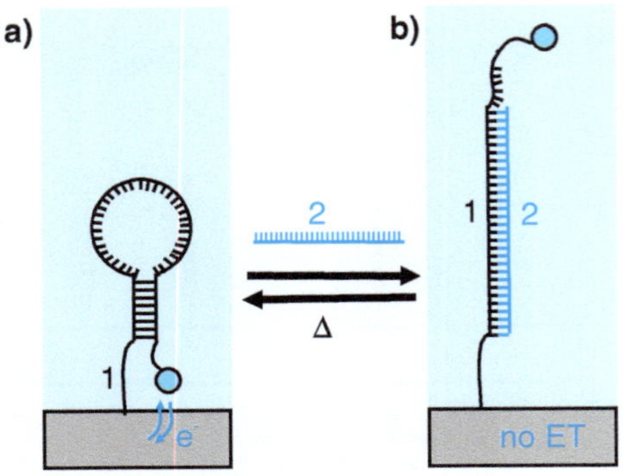

Figure 18.53 Evaluation of conformational changes to detect a hybridization event. The immobilized, single-stranded probe (1) forms a loop that brings the redox-active group at its end close to the electrode surface. Hybridization of the probe with the complementary target DNA (2) breaks up the loop and removes the redox-active group from the electrode so that no ET is possible. Source: Schematic figure adapted from [106].

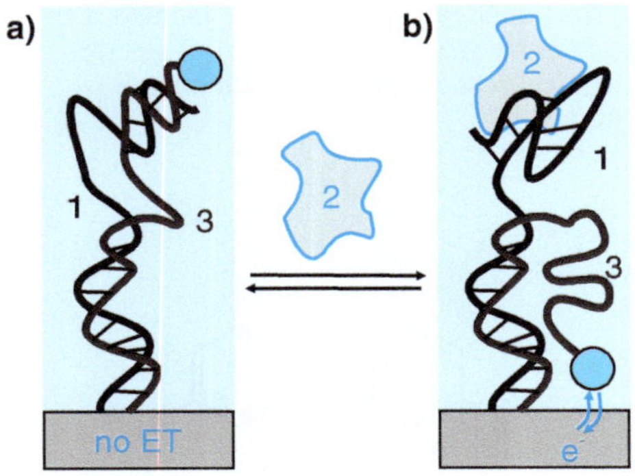

Figure 18.54 Utilization of conformational changes to detect the binding of an aptamer to thrombin. Source: Schematic according to [111].

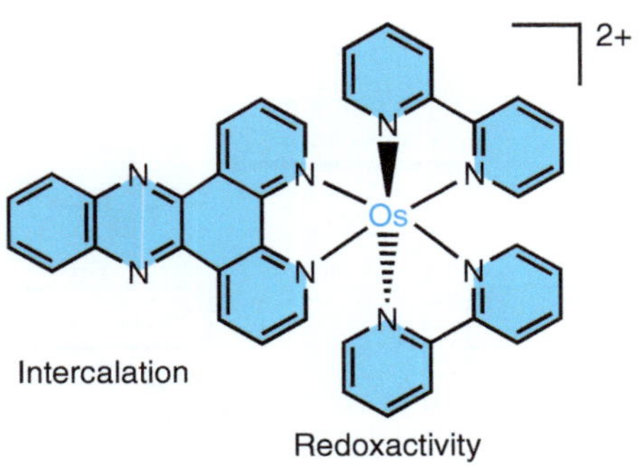

Figure 18.55 Structure of $[Os(bpy)_2dppz]^{2+/3+}$ (dppz stands for dipyrido[3,2-a : 2,3-c]-phenazine).

(ACV, Figure 5.44) or square wave voltammetry (SQWV, Figure 5.38) for detection. Electrons pass several times between the electrode and the redox label if the label is in the vicinity of the electrode and the electrode potential corresponds to the formal potential of the redox-active moiety.

Use of Conformational Changes for Detection. Redox labels can only exchange electrons with the electrode if they are at a distance of around 1 nm from the electrode. This fact can be exploited when a loop-forming, redox-labeled probe DNA changes into a stretched conformation through hybridization, in which the redox-active label is removed from the electrode (Figure 18.53) [106, 110]. Consequently, no ET is possible in the stretched conformation.

Switching by conformational changes is a detection principle that can also be used in other assays [111, 112]. Figure 18.54 shows an aptamer sensor. The analyte is thrombin, an important protein in the blood coagulation cascade. The RE is an aptamer formed *in vitro* against thrombin. Chemically, the aptamer is an ssDNA strand. This aptamer is immobilized on an electrode together with a second strand that carries a redox label. The sequence of the labeled strand is selected in such a way that it hybridizes with the aptamer at some points, thereby removing the label from the electrode (Figure 18.54a). However, the bond between the two strands is not very strong. In the presence of thrombin, the aptamer binds more strongly to thrombin. As a result, the end of the second strand is released and becomes mobile so that the redox-active label reaches the electrode and causes a signal (Figure 18.54b).

Homogeneous assay formats can also be combined very advantageously with electrochemical detection principles (Figure 18.56) [113]. The signal measured is the electrolysis current of an intercalating redox probe $[Os(bpy)_2dppz]^{2+/3+}$ (Figure 18.55) at a printed coplanar three-electrode electrochemical cell (Figure 18.56). The screen-printed sensor can potentially be as inexpensive as a glucose sensor. The redox label can only intercalate into dsDNA. However, association with such a large molecule greatly reduces the diffusion coefficient of the redox label and, hence, the current in the SQWV (Figure 5.38) decreases. In the course of the assay, the target DNA is copied multiple times using the loop-mediated isothermal amplification (LAMP) method [114]. In contrast to PCR, LAMP does not require separation of dsDNA by multiple, precise temperature-programmed heating–cooling cycles (and, thus, is much faster and requires less equipment). Instead, a polymerase is used in LAMP that displaces the second strand and complements the first to a new double strand. In this way, the test can be carried out in 50 μL samples. If the digested biological cells of the sample contain the target DNA, more and more dsDNA is formed into which the redox probe intercalates. Once the precursors for DNA replication have been used up, the experiment comes to a standstill. A measure of the amount of target DNA originally present in the sample is the time it takes for the SQWV peak current to fall to a certain level (approximately 10–40 min). The current does not have to be measured continuously in a reaction vessel, but only from time to time. Therefore, *one* potentiostat can alternately monitor several electrochemical cells (multiplexing), which opens up the possibility of high-throughput procedures such as those required for mass testing in mitigating a pandemic.

18.4 Investigation of Ionic Currents Across Membranes

In Section 18.1, we have already discussed the great importance of differences in the ionic composition on the P and E sides of biomembranes in the context of bioenergetics. Apart from biochemical energy conversion, the differences in ion concentrations are particularly important for the conduction of stimuli along axons of nerve cells, which is made possible by voltage-gated selective ion channels in the cell membranes. The patch clamp technique (Section 18.4.1) contributed significantly to the elucidation of their function. Recently, some of the measurement principles developed in this process have been incorporated into new electroanalytical systems for other problems, a selection of which is discussed in Section 18.4.2.

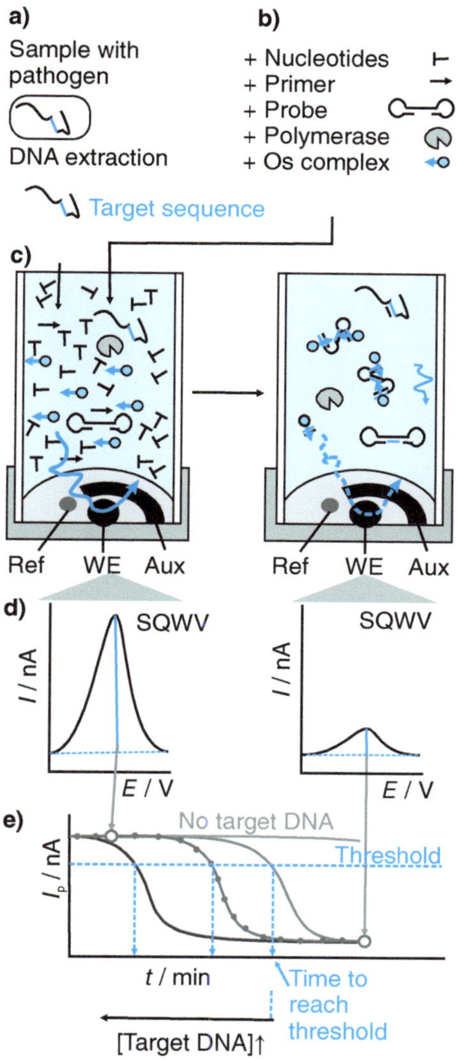

Figure 18.56 Homogeneous DNA assay with electrochemical detection and loop-mediated isothermal amplification (LAMP). (a) DNA with the target sequence typical for a specific pathogen is extracted. The sample (a) is placed in a miniaturized cell tube (c) with a reagent mixture (b) ($V = 50$ μL). The electrodes of the three-electrode cell are located at the bottom of the tube. dsDNA is formed only when the target sequence is amplified (c, right). The electrochemical signal (d) decreases if a larger fraction of the Os complexes (as redox probe molecule, Figure 18.55) is associated with the formed dsDNA due to the reduced diffusion coefficient. The reaction time (e) required for the current of the probe molecule to fall below a threshold value is evaluated. Source: Schematic according to [113].

18.4.1 Characterization of Individual Ion Channels by Means of the Patch Clamp Method

Electrophysiology and biophysics have long used pulled pipettes with an internal Ag|AgCl electrode to measure potential differences across biomembranes (often also referred to as "membrane potential"). The techniques and instrumentation used for this purpose have been increasingly refined, albeit largely independently of developments in other areas of electrochemistry. As a result, a completely independent terminology and associated conventions have been established. For example, measuring ranges for most potentiostats are specified in current per output voltage (Excursion 5-8). However, electrophysiological workstations use the reciprocal value, that is, output voltage per current. Table 18.6 shows examples of the conversion between the two values. The current convention follows the electrical current convention, according to which the positive current direction is defined along the movement of the cations from the positive to the negative electrode (and thus *contrary to* the IUPAC definition followed in this textbook). The explanations below aim to relate the terms used in electrophysiology to the terms used elsewhere in this book to facilitate the use of the methodical achievements and the available instrumentation of electrophysiology, which is very valuable for some new developments in other areas of electrochemistry. An in-depth presentation of the fundamentals of electrophysiology is beyond the scope of this book and can be found in specialized textbooks [115].

The probe electrode is a glass or quartz capillary pulled to a very fine end and filled with a chloride-containing electrolyte. A silver-silver chloride electrode (Ag|AgCl) is inserted into the unchanged opening of the capillary. This device is used to probe the

Table 18.6 Conversion of sensitivity factors (measurement ranges) between conventional potentiostats and patch clamp amplifiers.

I_{max}	Potentiostat, S/(A/V)	Patch clamp amplifier, S'/(V/A)
10 μA	1 μA/V	1 V/μA
1 μA	100 nA/V	10 V/μA
100 nA	10 nA/V	100 V/μA
10 nA	1 nA/V	1 V/nA
1 nA	100 pA/V	10 V/nA
500 pA	50 pA/V	20 V/nA
250 pA	25 pA/V	40 V/nA
100 pA	10 pA/V	100 V/nA
10 pA	1 pA/V	1 V/pA
1 pA	100 fA/V	10 V/pA
100 fA	10 fA/V	100 V/pA
10 fA	1 fA/V	1 V/fA

The typical analog output voltages for devices are ±10 V, ±4 V, or ± 1 V. The examples are shown for an analog voltage range of ±10 V. The units can also be shown differently, for example, 100 V/nA = 0.1 V/pA. Conversion can be carried out using the method of "unity factor" (e.g., multiplication by 1 = 1 nA/1000 pA).

local electrical potential at the capillary opening, similar to the use of a Luggin capillary in Figure 5.24. In some experiments, bath electrodes are also used in which a larger quantity of AgCl are used in form of a pellet. Depending on the specific implementation, the AgCl pellet can be separated from the solution by an agar plug that is permeable to ions. The design of these electrodes is therefore reminiscent of various types of reference electrodes that have already been discussed elsewhere in this book. The Ag|AgCl reference electrode in the three-electrode cell in other Sections of this book is currentless (or the current is negligibly small compared to the current at the WE). However, the Ag|AgCl electrodes are often used in electrophysiology to quickly convert a (very small) electronic current in the metallic Ag lead into an ionic current in solution by the interfacial reaction:

$$e^- + AgCl \rightleftharpoons Ag + Cl^- \tag{18.26}$$

This reaction is preferred for this purpose because of its reversibility (i.e., high reaction rate and absence of side reactions).

Pulled capillaries can be routinely produced with openings of 1 μm or smaller. This allows the pipette opening to be placed by means of micromanipulators and under observation with an optical microscope close to selected biological cells or even close to substructures of cells. The pipettes can also be used to penetrate the cell membrane and thus carry out electrical measurements on the P-side of the membrane.

There are different measurement programs that can be run either between two (or more) Ag|AgCl electrodes inside pulled pipettes or between one Ag|AgCl electrode inside a pipette and an Ag|AgCl bath electrode. It is advisable to classify the measurement programs according to the signal-response relationship, analogously to the treatment in Section 5.3. The measurement with a **current clamp** corresponds to a galvanostatic operation (Figure 18.57). Typically, an ionic current is fed into a cell via a pipette (i.e., forced), and the resulting change in the potential difference across the membrane is recorded between another pipette and the bath electrode. From the perspective of biology, the experiment corresponds to the situation after a stimulus was taken up at a synapse.

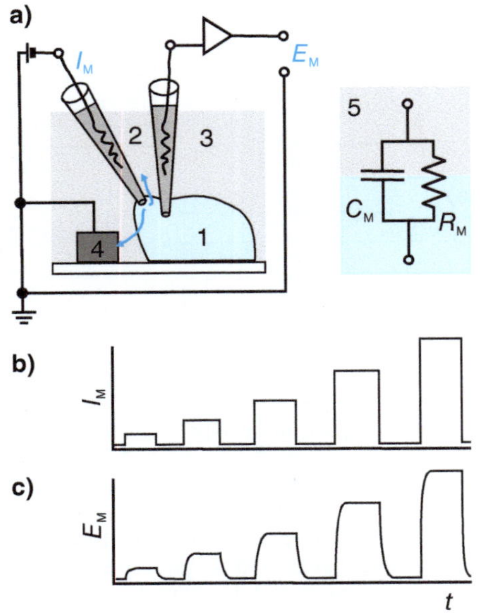

Figure 18.57 Principle of the current clamp measurement.

a) A current I_M is passed through the current-carrying electrode (2) into the biological cell (1) for a certain period of time. The resulting change in the potential difference across the membrane E_M ("membrane potential") is recorded between a second electrode (3) and the bath electrode (4).
b) The experiment is repeated several times with other currents.
c) The behavior of the membrane corresponds to that of an RC element or a capacitor with leakage current.

For a **voltage clamp** measurement, a defined potential difference is applied between the P side of the cell membrane and the extracellular medium by means of Ag|AgCl capillary electrodes inserted into the cell and the external bath electrode (Figure 18.58). The resulting current between another capillary electrode (3) inside the cell and the bath electrode (4) is measured at a fixed potential difference across the membrane. This corresponds to a potentiostatic operation and chronoamperometry in other areas of electrochemistry. Since diffusion potentials can also occur at the pipettes, the potential difference between the potential-measuring electrode (2) and the bath electrode (4) is first adjusted, and afterward the capillary electrode is inserted into the cell. Because it is much easier to position *one* pipette in a biological cell (instead of two Figure 18.58a), a two-electrode circuit is often used, in which the current-carrying pipette (designated as "Input" or "Signal Input") is used with the bath electrode as a common reference and auxiliary (CE, or Ref-Aux). The electrode connector for the bath electrode is often labeled as "Ground" or "GND" in electrophysiology instruments.

The basic elements of circuits with operational amplifiers, which appear in Figures 18.57 and 18.58, have already been discussed in Excursion 5-8. Since the currents are very small, the corresponding devices work with preamplifiers (head stages) that are mounted in the immediate vicinity of the electrode in order to achieve the lowest possible noise with very fast switching of the current. In addition to the properties of the electronic amplifier according to Eq. (4.14), the rate at which such a switching of the object under study can be measured is determined by the time constant τ of the measuring arrangement, which depends on the capacitance of the membrane and the electrolyte resistance.

The variety of electrophysiological experimental setups is further increased by the fact that the current clamp and voltage clamp methods can be combined with different positioning of the pipettes and different biological cells. The size of the biological cells or organelles is an important boundary condition because the pipette tips must be positioned close to or within these structures, and the experiment may also require the precise positioning of a second pipette on or in the biological cell. For example, pipettes used for studying the giant axons of squid [116] have a diameter of up to one mm and an even greater length, while pipets for studying mitochondria have dimensions of only 0.5 to 1 μm. Except for unusually large cells, the positioning of micropipettes requires experience and manual skill. However, there is also a need for electrophysiological measurements in routine applications, for example, in the development of pharmaceuticals, which often act via ion channels. Extensive efforts are, therefore, being made to automate ion transport measurements or to transfer them to planar structures (chips, arrays) [115].

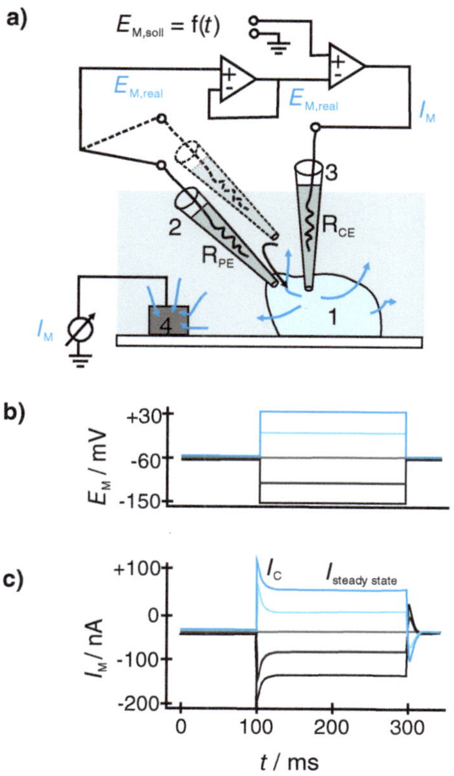

Figure 18.58 Principle of the voltage clamp measurement on a whole cell.

a) After adjusting the potentials between the potential-measuring electrode (2) and the bath electrode (4) outside the cell, electrode 2 and the current-carrying electrode (3) are inserted into the cell.
b) Different potentials are set one after the other between electrodes 2 and 4. The control circuit causes an ion current to flow into the cell via the current-carrying electrode until the actual value of the membrane potential difference matches the set value.
c) Like a chronoamperogram on a microelectrode, the current flowing during a potential pulse contains a capacitive component I_C and a steady-state component, which balances the charges flowing through the ion channels of the cell.

Pure lipid bilayers are very good insulators and do not allow ions to pass through, even at high field strengths (Excursion 13-1). Only ion channels enable ion transfer (IT) across biomembranes. Ion channels are complex transmembrane proteins that have a hydrophilic channel inside through which the ions – driven by the concentration gradient (Table 18.2), or more precisely the difference in electrochemical potentials (Section 2.2.1) – are transported between the P and E sides of the membrane. The permeability of many ion channels is specific for only one type of ion (Na^+, K^+, Ca^{2+}, Cl^-, …) and is voltage-gated, that is, the conductivity of the channel depends on the transmembrane potential difference (voltage). There are different ion channels with different switching properties for the same type of ion [116]. At a membrane potential difference (voltage) specific for a particular channel protein, the channel switches between an open and a closed state at a specific rate due to a conformational change. According to the Hodgkin–Huxley model[168], conduction *within* a nerve cell results from the interaction of fast voltage-dependent Na^+ channels, slow voltage-dependent K^+ channels, and the different Na^+ and K^+ concentrations present in the resting state (Table 18.2) on the P and E side of the membrane [116]. Local stimulation, for example, at a synapse, locally disturbs (depolarizes) the membrane potential, which leads to the opening of the Na^+ channels and results in an influx of Na^+ into the cell. The K^+ channels open more slowly. The subsequent outflow of K^+ from the cell counteracts the change in potential and leads to repolarization. Once the resting potential is reached, the channels close again. As depolarization and repolarization occur at different speeds, the excitation propagates along the axon in only one direction. Stimulation is transmitted *between* nerve cells via the synaptic cleft by neurotransmitters (Excursion 5-6).

The conduction of an excitation along an axon can be recorded using an *extracellular measurement* with two micropipettes (Figure 18.59). The electrolyte resistance opposes the ionic current with a resistance $R_E = U/I$. If a potential difference or voltage U is now measured at two points along the current path, this is proportional to the local ionic current ($I = U/R_E$).

Sakmann and Neher[169] achieved a decisive methodological breakthrough in the elucidation of the function of voltage-dependent ion channels by introducing the **patch**

[168] Andrew Fielding Huxley, 1917–2012, British biophysicist and physiologist; Alan Lloyd Hodgkin, 1914-1998, British physiologist and biophysicist; Huxley and Hodgkin were awarded the Nobel Prize in Physiology and Medicine in 1963.
[169] Bert Sakmann, born 1942, German physician and physiologist, Director of the Max Planck Institute for Medical Research in Heidelberg; Erwin Neher, born 1944, German biophysicist, Director of the Max Planck Institute for Biophysical Chemistry in Göttingen. Sakmann and Neher received the Nobel Prize for Physiology and Medicine in 1991.

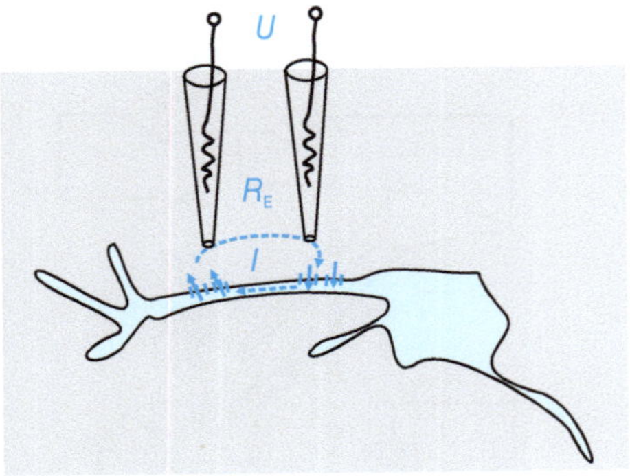

Figure 18.59 Principle of extracellular recording on an axon.

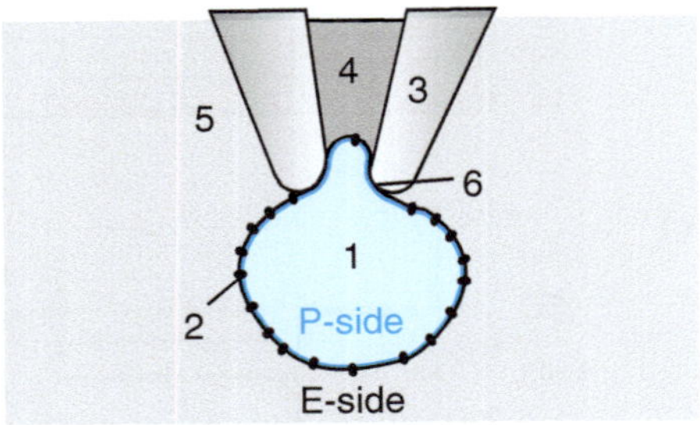

Figure 18.60 Patch clamp with a cell (1) in the on-cell configuration. After placing the pipette orifice (3) on the cell membrane with the ion channels (2), the inner solution (4) is isolated via a resistance >1 GΩ (gigaseal, 6) from the bath solution (5) by aspiration of the solution inside the pipette. The current between the electrode inside the pipette and the bath electrode is limited by the ion channel insulated by the membrane patch.

Figure 18.61 a) Whole cell configuration with relatively small cells after destruction of the membrane patch by a voltage or suction pulse; b) removal and reformation of the membrane in a Ca^{2+}-containing medium; c) cell-free examination of the membrane patch, with the E-side of the membrane facing the bath solution (outside-out patch).

clamp technique. This approach made it possible to measure the switching behavior of individual ion channels [117]. The basic idea is to electrically isolate a small membrane segment (the patch) by placing a very small pipette orifice on the cell membrane and to measure a current that flows through this tiny membrane segment. It is statistically expected that a segment of this size contains either no, only one, or only very few ion channels, which change their switching state depending on the transmembrane voltage. In order to measure the current flowing through a single ion channel, the seal between the pipette and the surrounding membrane must be so good that, in the absence of ion channels, there is a resistance *R* between the Ag|AgCl electrode in the pipette and the external Ag|AgCl bath electrode in the range of >1 GΩ ($10^9\ \Omega = 0.05\ V/5 \times 10^{-11}\ A$). In laboratory jargon, this situation is called "gigaseal" (= sealing with gigaohm resistance) and can be achieved by first placing a clean, freshly pulled pipette on the membrane surface and then sucking the membrane section into the pipette orifice by applying a small negative pressure (Figure 18.60). The connection between the cell membrane and the pipette opening achieved in this way is surprisingly mechanically stable, so that the cells can be transferred to another working solution using the pipette. Since the currents through a single ion channel are very small, they can be measured in a two-electrode arrangement between the Ag|AgCl electrode inside the pipette and the bath electrode.

By modifying the procedure, different arrangements of the membrane patch relative to the pipette orifice can be realized: A larger negative pressure in the pipette can destroy the membrane patch and create a free passage between the inner solution of the pipette and the cytoplasm of the cell (Figure 18.61a), which allows whole-cell measurements (whole-cell configuration). In addition, substances can be exchanged between the pipette and the cytoplasm, which is used for the targeted dosing of metabolites, inhibitors, or fluorescence indicators into individual biological cells (microinjection).

By withdrawing the pipette (Figure 18.61b) after formation of a gigaseal, a membrane patch can be excised out of the cell membrane while retaining the electrical seal on the capillary orifice. Depending on the details of the procedure, either the P-side (Figure 18.61c) or the E-side (Figure 18.62) of the membrane faces the inside of the pipette. Voltage clamp experiments can then be carried out with isolated membrane patches. Such experiments showed that all voltage-dependent ion channels exhibit two switching states only, that is, an "on-off" or "all-or-nothing" behavior (Figure 18.63). Each individual channel behaves stochastically, that is, its voltage-dependent switching state can only be specified with probabilities. The behavior of the potential difference across the membrane and the ionic currents as found experimentally in whole cells and as described by differential equations in the Hodgkin–Huxley model results as an ensemble average of the behavior of many stochastically reacting individual ion channels.

Patch clamp measurements are one of the first examples of single entity measurements, which were later expanded, for instance, by single-molecule spectroscopy and other approaches to single entity electrochemistry (Section 5.5.3) [119, 120]. This ongoing development makes it possible to identify subsets of the objects under investigation in complex systems that are otherwise masked during ensemble averaging and thus would remain unrecognized.

Further Reading. References [116, 118] are recommended as an introduction to the biophysics and electrophysiology of stimulus conduction with an overview of the large group of ion channels. Some textbooks and manuals from device manufacturers provide a systematic overview of the electrochemical and electronic principles of the patch clamp technique [115, 121].

18.4.2 Electroanalytical Methods Using Artificial Nanopores

The success of single-channel measurements by means of the patch clamp technique has stimulated method developments to solve other analytical questions by monitoring ionic currents through isolated, artificial pores. We have already discussed examples in the context of scanning probe techniques in Sections 8.13.2 and 8.13.3.

Another classic application is the Coulter counter[170], which is used in medical laboratories to determine the number of blood cells. A direct current (DC) is observed between two electrochemical half-cells that are connected via a pore (Figure 18.64). The suspension with the blood cells is moved between the half-cells by a pump. Each time a cell enters the pore, there is less cross section available for charge transport by ions, and the current between the two half-cells breaks down as long as the cell passes through the pore. The frequency of these current dips correlates with the particle concentration (cell concentration) in the examined sample; the size of the current dip is a measure of the size of the particle (the blood cell) and therefore also allows differentiation between the blood cells. Red blood cells and platelets can be determined in one run; the number of white blood cells is determined after lysis of the other two cell types.

The principle of the Coulter counter can also be used to analyze suspensions of charged nanoparticles if pores with significantly smaller orifices are used. The accuracy of the size determination can be further increased if the polarity of the voltage is inverted after a current dip. This causes a single nanoparticle to be driven back into the pore after passing through it. Thus, the measurement is repeated, which improves the accuracy of the size determination significantly [122].

Using the example of electrophoretic processes, we have already discussed in Section 4.6.3 how the presence of an electric field parallel to a charged surface of a container leads to the coupling of different transport processes. Charged surfaces form an electrical double layer whose thickness can exceed the radius of very small capillary orifices in dilute electrolyte solutions (cf. Table 4.4). In these cases, either the transport of cations or that of anions or correspondingly charged nanoparticles is favored. The current–voltage curves, which are linear for large pores, exhibit curvatures for small pores. This effect is known as voltage-rectified currents and is reminiscent of the behavior of semiconductor diodes. It can occur on conical nanopores with charged walls (Figure 18.65). Depending on the material, the charge of the insulating capillary wall can either be introduced via a synthetic surface modification or arise spontaneously from the dissociation of surface groups or an uneven charge distribution on the surface (cf. Figure 4.65). The occurrence of an electroosmotic flow (EOF, Section 4.6.3) is coupled to the rectification effect at nanopore orifices. This can potentially be used for very precise flow control in microfluidic and nanofluidic devices of highly integrated analytical systems [123]. In addition, these effects must also be considered when describing transport processes in other nanoporous materials (Figures 14.33 and 14.39).

In addition to the use of pore orifices created by nanofabrication processes or precisely controlled etching processes, another approach is to first seal a larger pore with a lipid bilayer (Figure 13.33) and then incorporate purposefully selected, biotechnologically produced, pore-forming proteins into the bilayer membrane. In addition to ion channels as pore-forming proteins, there is a variety of other proteins with pores of different diameters [124] that can be further modified chemically or by genetic engineering. This allows the interaction between the pore wall and an analyte molecule passing through the pore to be chemically selective, so that the time for passage through the

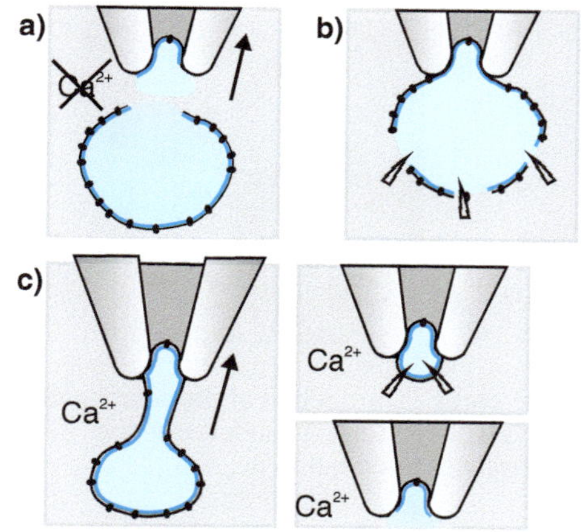

Figure 18.62 Preparation of the inside-out patch of a membrane starting from the on-cell configuration in Figure 18.60. a) A membrane patch can be torn out of the cell by removing the pipette in Ca^{2+}-free solution; b) mechanical or chemical destruction of the aspirated cell; c) withdrawal of the pipette in Ca^{2+}-containing solution leads to the formation of a vesicle, which is subsequently destroyed chemically or mechanically. Finally, the E side of the membrane faces the bath solution.

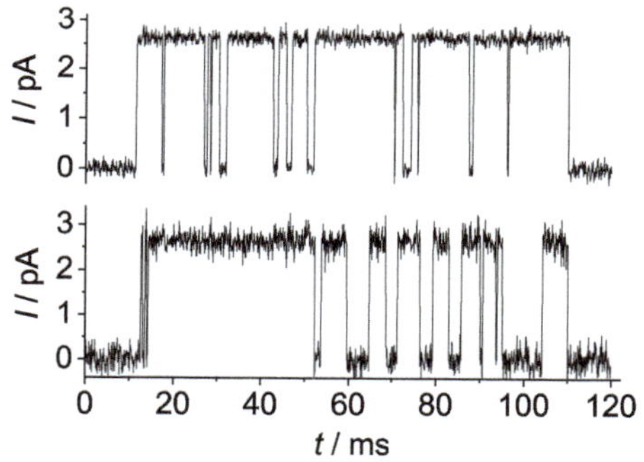

Figure 18.63 Voltage clamp recording on an isolated ion channel. Source: Dataset was kindly provided by HEKA Elektronik, Lambrecht, Germany.

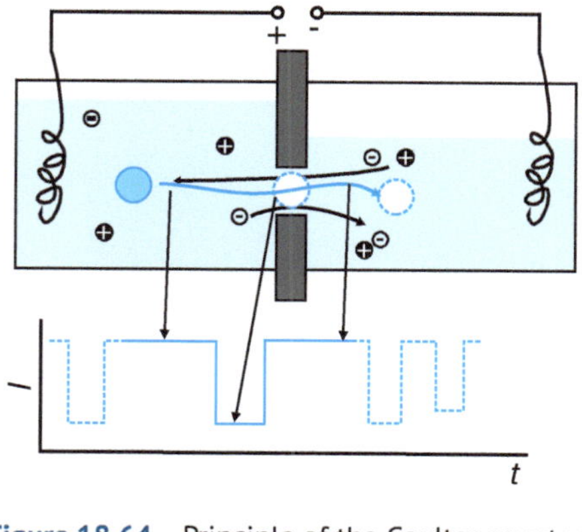

Figure 18.64 Principle of the Coulter counter for determining the number of blood cells (blue circle). The schematic is not to scale.

[170] Wallace H. Coulter, 1913–1998, US engineer and entrepreneur. The company he founded, Coulter Electronics, is now part of Beckman Coulter.

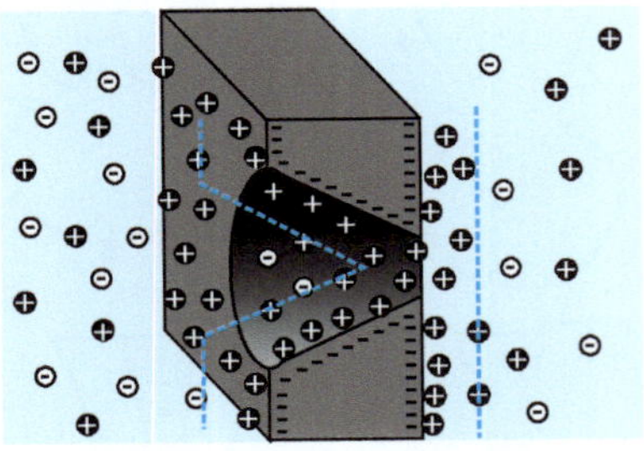

Figure 18.65 Distribution of the anions and cations within a conical nanopore with a negatively charged wall.

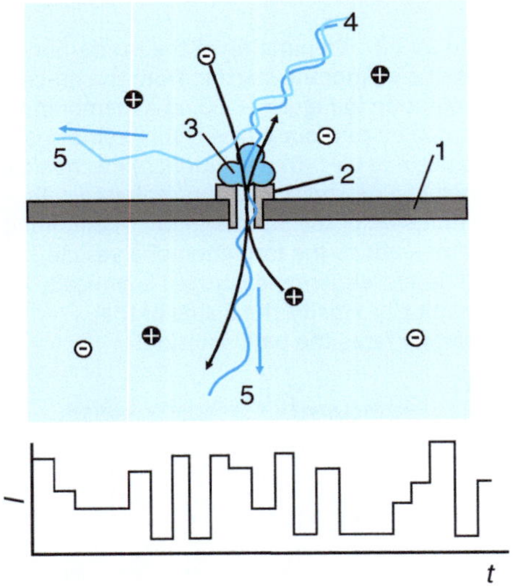

Figure 18.66 Reading DNA sequence information via the ionic current fluctuation during the passage of a DNA fragment through a specifically selected nanopore. 1) Separation membrane; 2) nanopore (e.g., a specially modified channel protein); 3) motor protein; 4) dsDNA; 5) ssDNA.

channel can be used for chemical identification. The frequency of the single-molecule passages allows the concentration to be determined.

Of particular interest in this context is the possibility of reading out a DNA sequence while a single DNA molecule passes through a pore specifically selected for this application [125]. This technology has been developed into commercial products with the aid of various improvements to the pores used, the reagents used, and above all, bioinformatics [126, 127]. The basic idea is that a motor protein holds dsDNA at the entrance of a specially selected pore, ensures that the DNA strands are disentangled, and releases one of the two individual strands into the pore at a precisely defined speed (Figure 18.65). The ion current through the pore adjusts to different plateaus depending on which base is currently passing through the pore. This portable analysis can now provide valuable services, for example, in the fight against infectious diseases in areas that are difficult to access [128]. Modifications of the method can also elucidate further functional details of nucleic acids [129] (Figure 18.66).

Further Reading., The anthology [130] and a selection of special articles [121, 123, 131] are recommended for an overview of the topic.

18.5 Review and Outlook

The consideration of the highly integrated and miniaturized analytical devices with nanopores is particularly suitable to conclude the textbook because it emphasizes some basic principles that the author has followed in the conception of this book: Electrochemistry, even in the simplest experiments, deals with a problem of systems chemistry, in which the interaction of different chemical compounds (reactant, supporting electrolyte, solvent, electrode material) and the use of different integration principles is much more important than the properties of a single substance. This interdependence is a key difficulty for students to enter the field. On the other hand, integration enables great technological breakthroughs, which we have witnessed over the last two decades: Lithium-ion batteries made possible the digital networking of the world, the blood glucose sensors significantly improved the quality of life of millions of diabetes patients and, perhaps in the future, individualized diagnostics and treatment of infectious diseases can be provided at costs that can be borne by a community thanks to portable diagnostic devices.

And what might come next? Many electrochemists hope that efforts to develop sustainable energy conversion processes with electrocatalysts free of precious metals can make a significant contribution to the decarbonization of the economy. In contrast to the examples listed above, we are talking about large, even gigantic scales here if measurable effects are to be achieved on a global level. However, this challenge also represents a problem of systems chemistry, the solution of which will likely require the use of ion-conducting solids, clever wet-chemical coating processes, prevention of corrosion of new construction and functional materials, possibly the integration of organic and inorganic functional materials, and the use of the entire range of electrochemical sensor technology and characterization methods. This constellation applies not only to such "hard nuts", but also to the many smaller research and development projects in electrochemistry.

If this book was able to introduce students and academics from different disciplines to the extensive original literature on the sub-areas while keeping the system aspects in mind, the author and his colleagues would consider their task fulfilled.

Test your knowledge

Exercise 18.1: What are intrinsic and extrinsic redox proteins? What are their biological functions?

Exercise 18.2: Using examples for electrochemical experiments from Chapter 18, explain the similarities and differences between the state of chemical equilibrium and a steady state in general.

Exercise 18.3: Compile a table of biological $1e^-$ and $2e^-$ redox systems with the corresponding structural formulas.

Exercise 18.4: Look at the molecular structure of glucose oxidase in [34]. Alternatingly show and hide H_2O, the prosthetic groups, the polypeptides, and the carbohydrates. Try to show the position of the prosthetic groups relative to the surface of the enzyme using a rough sketch. Make a well-founded hypothesis as to whether MET and/or DET is to be expected for this enzyme.

Exercise 18.5: Analyze Figure 18.26. Make a rough sketch of Figure 18.26B. Add to the coordinate system the signals (without capacitive currents) that would arise if the experiment were repeated with the adsorbed enzymes on a rotating disk electrode in a solution without fumarate and succinate. The concentrations of fumarate and succinate used in Figure 18.26B are different, but the limiting currents are approximately equal. What conclusion can you draw from this observation?

Exercise 18.6: Explain the difference between the Michaelis–Menten constant K_M and the apparent Michaelis–Menten constant K'_M. How are both quantities determined?

Exercise 18.7: How would the calibration curve in Figure 18.36 change if the sensor were produced on an electrode with half the surface area? How would the curve change if the accessibility of the enzyme for glucose were improved during embedding? Explain your ideas using semiquantitative rough sketches.

Exercise 18.8: The measurement values in Figure 18.36 are

$c/(\text{mol L}^{-1})$	I/A	$c/(\text{mol L}^{-1})$	I/A
0.00	9.83×10^{-10}	2.50×10^{-2}	7.42×10^{-8}
9.99×10^{-4}	6.25×10^{-9}	3.00×10^{-2}	8.08×10^{-8}
1.97×10^{-3}	1.21×10^{-8}	3.50×10^{-2}	8.65×10^{-8}
2.99×10^{-3}	1.79×10^{-8}	4.00×10^{-2}	9.12×10^{-8}
3.99×10^{-3}	2.06×10^{-8}	5.00×10^{-2}	9.91×10^{-8}
4.99×10^{-3}	2.61×10^{-8}	6.00×10^{-2}	1.06×10^{-7}
7.51×10^{-3}	3.44×10^{-8}	7.50×10^{-2}	1.13×10^{-7}
1.00×10^{-2}	4.30×10^{-8}	1.00×10^{-1}	1.22×10^{-7}
1.25×10^{-2}	4.99×10^{-8}	1.50×10^{-1}	1.34×10^{-7}
1.45×10^{-2}	5.62×10^{-8}	1.99×10^{-1}	1.42×10^{-7}
2.00×10^{-2}	6.60×10^{-8}	2.50×10^{-1}	1.49×10^{-7}

Determine the maximum current and the apparent Michaelis–Menten constant. Use a spreadsheet program and try a nonlinear fit and a fit with a suitable linearization. Provide the result in nA and mmol L^{-1}.

Exercise 18.9: The experiment in Figure 18.36 should be repeated with a different electrode. 20 mL of buffer solution are added to the cell container. A stock solution of 1 mol L^{-1} glucose in buffer is available. Calculate a dosing scheme (addition of stock solution to the buffer solution in the cell vessel) that achieves approximately the concentrations in the Table in Exercise 18.8. Note that small concentration steps should be realized first, followed by larger concentration changes due to the expected nonlinear calibration plot. The change in the total volume must also be taken into account. Use a spreadsheet program and properly document the units.

Exercise 18.10: Blood glucose concentrations in an electrochemical laboratory are usually given in mmol L^{-1}. In clinical chemistry, however, it is established practice to state these values in mg/dL. Convert the standard range of (4.2–5.2) mmol L^{-1} into mg/dL.

Exercise 18.11: Describe the process of electrochemical immunoassays in more detail (coating of the solid surface of the carrier, dosing of the sample, incubation, rinsing steps, detection, etc.). Note which processes require an incubation time and how the length of the incubation time affects the signal.

Exercise 18.12: Discuss which assay format from Figures 18.45–18.47 is most suitable for the determination of very low concentrations.

Exercise 18.13: Explain why the use of ACV or SQWV is preferable over CV for electrochemical DNA assays or sensors.

References

1 P. N. Bartlett (Ed.); *Bioelectrochemistry – Fundamentals, Experimental Techniques and Applications*. John Wiley & Sons, Chichester, 2008.

2 G. S. Wilson; A. J. Bard, M. Stratmann (Eds.); *Encyclopedia of Electrochemistry – Vol. 9 Bioelectrochemistry*. Wiley-VCH, Weinheim, 2002.

3 S. R. Caplan, I. R. Miller, G. Milazzo (Eds.); *Bioelectrochemistry: Principles and Practice. Vol. 1–5*, Birkhäuser, Basel, 1995.

4 B. Eggins; *Biosensors – An Introduction*. Wiley-Teubner, Chichester, 1996.

5 F. W. Scheller; F. Schubert, J. Fedrowitz (Eds.); *Frontiers in Biosensorics, Vol. 1* and 2. Birkhäuser Publications, Basel, 1997.

6 R. S. Marks, D. C. Cullen, I. Karube, C. R. Lowe, H. H. Weetall; *Handbook of Biosensors and Biochips, Vol. 1* and *2*. John Wiley & Sons, Chichester, 2007.

7 A. J. Bard; *Integrated Chemical Systems – A Chemical Approach to Nanotechnology*. Wiley-Interscience, New York, 1995.

8 I. Prigogine; *Introduction to Thermodynamics of Irreversible Processes*. Wiley Interscience, New York, 1967.

9 D. Walz, S. R. Caplan; Nonequilibrium thermodynamics and kinetics. in *Bioelectrochemistry: Principles and Practice. Vol.* 1, Birkhäuser, Basel, 1995, pp. 1–48.

10 P. L. Yeagle (Ed.); *The Structure of Biological Membranes*. 2nd ed. CRC Press, Boca Raton, 2005.

11 B. Alberts, A. Johnson, J. Lewis, D. Morgan, M. Raff, K. Roberts, P. Walter, J. Wilson, T. Hunt: *Molecular Biology of the Cell*. 6th ed. Garland Science, Taylor and Francis Group, New York, NY, 2015.

12 G. M. Cooper, R. E. Hausman; *The Cell: A Molecular Approach*. 3rd ed., ASM Press, Washington, DC, 2004.

13 P. N. Bartlett; Bioenergetics and biological electron transport; in *Bioelectrochemistry – Fundamentals, Experimental Techniques and Applications*. (P. N. Bartlett, Ed.) John Wiley & Sons, Chichester, 2008, pp. 1–37.

14 Wikipedia; https://en.wikipedia.org/wiki/Fluid_mosaic_model (last access 18 Mar 2026).

15 P. Rich; The cost of living. *Nature* **2003**, *421*, 583–583.

16 R. E. Blankenship; *Molecular Mechanisms of Photosynthesis*. Blackwell Science, Oxford, 2002.

17 Wikipedia; https://en.wikipedia.org/wiki/ATP_synthase (last access 18 Mar 2026).

18 J. M. Berg, J. L. Tymoczko, G. J. Gatto, L. Stryer; *Stryer Biochemistry*, 9th ed., MacMillan International, New York, NY, 2019.

19 J. Sancho; Flavodoxins: sequence, folding, binding, function and beyond. *Cell. Mol. Life Sci.* **2006**, *63*, 855–864.

20 Z. Samec, P. J. Elving; Anodic oxidation of dihydronicotineamide adenine dinucleotide at solid electrodes: Mediation by surface species. *J. Electroanal. Chem.* **1983**, *144*, 217–234.

21 L. Underwood, R. W. Burnett; Electrochemistry of biological compounds. in *Electroanalytical Chemistry, Vol. 6* (A. J. Bard, Ed.), Marcel Dekker, New York, 1973, pp. 1–85.

22 L. Gorton; Electrochemistry of NAD(P)$^+$/NAD(P)H. in *Encyclopedia of Electrochemistry - Vol. 9 Bioelectrochemistry*. (G. S. Wilson; A. J. Bard, M. Stratmann, Eds.), Wiley-VCH, Weinheim, 2002, pp. 67–143.

23 L. Gorton, P. N. Bartlett; NAD(P)-based biosensors. in *Bioelectrochemistry – Fundamentals, Experimental Techniques and Applications*. (P. N. Bartlett, Ed.), John Wiley & Sons, Chichester, 2008, pp. 157–198.

24 J. J. R. Fráusto da Silva, R. J. P. Williams; *The Biological Chemistry of the Elements*. Oxford University Press, Oxford, 1993.

25 A. G. Mauk, G. R. Moore; Control of metalloprotein redox potentials: What does site-directed mutagenesis of hemeproteins tell us? *J. Bio. Inorg. Chem.* **1997**, *2*, 119–125.

26 G. Battistuzzi, M. Borsari, M. Sola; Medium and temperature effects on the redox chemistry of cytochrome *c*. *Eur. J. Inorg. Chem.* **2001**, *12*, 2989–3004.

27 R. Varadaranjan, T. E. Zewert, H. B. Gray, S. G. Boxer; Effects of buried ionizable amino acids on the reduction potential of recombinant myoglobin. *Science* **1989**, *243*, 69–72.

28 V. V. Vu, S. T. Ngo; Copper active site in polysaccharide monooxygenases. *Coord. Chem. Rev.* **2018**, *368*, 134–157.

29 A. G. Sykes (Ed.) *Advances in Inorganic Chemistry, Vol.* 36, Elsevier, Amsterdam, 1991.

30 F. A. Armstrong; Voltammetric investigation of iron–sulfur clusters in proteins. in *Electroanalytical Methods for Biological Materials*. (A. Brajter-Toth, J. Q. Chambers, Eds.), Marcel Dekker, New York, 2002, pp. 143–194.

31 D. J. Walton, P. G. Richards, J. Hepinstall, B. Coles; Electrosynthetic modification of proteins: Electrooxidation of methionine and tryptophan in hen egg-white lysozyme. *Electrochim. Acta* **1997**, *42*, 2285–2294.

32 A. H. B. Dourado, F. C. Pastrián, S. I. C. de Torresi; The long and successful journey of electrochemically active amino acids. From fundamental adsorption studies to potential surface engineering tools. *Ann. Braz. Acad. Sci.* **2018**, *90* (1 Suppl. 1), 607–630.

33 L. H. Gua, H. A. O. Hill; Direct electrochemistry of proteins and enzymes. in *Advances in Inorganic Chemistry, Vol. 36* (A. G. Sykes, Ed.), Elsevier, 1991, 7 pp. 341–373.

34 RCSB PDB https://www.rcsb.org/3d-view/1GPE/0 (last access 18 Mar 2026).

35 W. R. Heineman, B. J. Norris, J. F. Goelz; Measurement of enzyme $E^{\circ\prime}$ values by optically transparent thin layer electrochemical cells. *Anal. Chem.* **1975**, *47*, 79–84.

36 R. Szentirmay, P. Yeh, T. Kuwana; Evolution of mediator-titrants for indirect coulometric titration of biocomponents. in *ACS Symposium Series, Vol.* 38, American Chemical Society., Washington, DC, 1977, pp. 143–169.

37 W. R. Heineman, F. M. Hawkridge, H. N. Blount; Spectroelectrochemistry at optically transparent electrodes. II electrodes under thin-layer and semi-infinite diffusion conditions and indirect coulometric titrations. in *Electroanalytical Chemistry, Vol. 13* (A. J. Bard, Ed.), Marcel Dekker, New York, 1984, pp. 1–113.

38 J. Pellet, M. Stankovich; Potentiometric measurements of proteins. in *Encyclopedia of Electrochemistry – Vol. 9 Bioelectrochemistry.* (G. S. Wilson; A. J. Bard, M. Stratmann, Eds.), Wiley-VCH, Weinheim, 2002, pp. 485–509.

39 R. S. Nicholson, I. Shain; Theory of stationary electrode polarography: Single scan and cyclic methods applied to reversible, irreversible, and kinetic systems. *Anal. Chem.* **1964**, *36*, 706–723.

40 B. Ge, F. Lisdat; Superoxide sensor based on cytochrome *c* immobilized on a mixed-thiol SAM with a new calibration method. *Anal. Chim. Acta* **2002**, *454*, 53–64.

41 C. Léger, K. Heffron, H. R. Pershad, E. Maklashina, C. Luna-Chavez, G. Cecchini, B. A. Ackrell, F. A. Armstrong; Enzyme electrokinetics: Energetics of succinate oxidation by fumarate reductase and succinate dehydrogenase. *Biochemistry* **2001**, *40*, 11234–11245.

42 P. N. Bartlett, F. A. Al-Lolage; There is no evidence to support literature claims of direct electron transfer (DET) for native glucose oxidase (GOx) at carbon nanotubes or graphene. *J. Electroanal. Chem.* **2018**, *819*, 26–37.

43 B. Willner, I. Willner; Reconstituted redox enzymes on electrodes: From fundamental understanding of electron transfer at functionalized electrode interfaces to biosensor and biofuel cell applications. in *Bioelectronics – From Theory to Applications* (I. Willner, E. Katz, Eds.), Wiley-VCH, Weinheim, 2005, pp. 35–97.

44 F. A. Armstrong; Electrochemistry of hydrogenases. in *Electroanalytical Chemistry, Vol.* 25 (A. J. Bard, C. G. Zoski, Eds.), CRC Press, Boca Raton, 2014, pp. 33–104.

45 J. A. Cracknell, K. A. Vincent, F. A. Armstrong; Enzymes as working or inspirational electrocatalysts for fuel cells and electrolysis. *Chem. Rev.* **2008**, *108*, 2439–2461.

46 F. A. Armstrong; Voltammetry of proteins. in *Encyclopedia of Electrochemistry – Vol. 9 Bioelectrochemistry.* (G. S. Wilson; A. J. Bard, M. Stratmann, Eds.), Wiley-VCH, Weinheim, 2002, pp. 11–49.

47 J. F. Rusling, B. Wang, S.-e. Yun; Electrochemistry of redox enzymes. in *Bioelectrochemistry – Fundamentals, Experimental Techniques and Applications.* (P. N. Bartlett, Ed.), John Wiley & Sons, Chichester, 2008, pp. 40–85.

48 T. Lötzbeyer, W. Schuhmann, H.-L. Schmidt; Electron transfer principles in amperometric biosensors: Direct electron transfer between enzymes and electrodes surface. *Sensors Actuators B Chem.* **1996**, *33*, 50–54.

49 Z. Zhang, H. Liu, J. Deng; A glucose biosensor based on immobilization of glucose oxidase in electropolymerized *o*-aminophenol film on platinized glassy carbon electrode. *Anal. Chem.* **1996**, *68*, 1632–1638.

50 M. Trojanowicz, O. Geschke, T. Krawczyński vel Krawcyk, K. Cammann; Biosensors based on oxidases immobilized in various conducting polymers. *Sensors Actuators B Chem.* **1995**, *28*, 191–199.

51 C. Kranz, H. Wohlschläger, H.-L. Schmidt, W. Schuhmann; Controlled electrochemical preparation of amperometric biosensors based on conducting polymer multilayers. *Electroanalysis* **1998**, *10*, 546–552.

52 D. Guschin, H. Shkil, W. Schuhmann; Electrodeposition polymers as immobilization matrices in amperometric biosensors: Improved polymer synthesis and biosensor fabrication. *Anal. Bioanal. Chem.* **2009**, *395*, 1693–1706.

53 A. Heller, B. Feldman; Electrochemistry in diabetes management. *Acc. Chem. Res.* **2010**, *43*, 963–973.

54 L. Gorton; Carbon paste electrodes modified with enzymes, tissues and cells; *Electroanalysis* **1995**, *7*, 23–45.

55 K. Habermüller, M. Mosbach, W. Schuhmann; Electron-transfer mechanisms in amperometric biosensors. *Fresenius J. Anal. Chem.* **2000**, *366*, 560–568.

56 W. Schuhmann; Amperometric enzyme biosensors based on optimized electron-transfer pathway and non-manual immobilization procedures. *Rev. Mol. Biol.* **2002**, *81*, 425–441.

57 C. Kurzawa, A. Hengstenberg, W. Schuhmann; Immobilization method for the preparation of biosensors based on pH shift-induced deposition of biomolecule-containing polymer films. *Anal. Chem.* **2002**, *74*, 355–361.

58 D. T. Pierce, P. R. Unwin, A. J. Bard; Scanning electrochemical microscopy. 17. Studies of enzyme-mediator kinetics for membrane- and surface-immobilized glucose oxidase. *Anal. Chem.* **1992**, *64*, 1795–1804.

59 P. N. Bartlett, J. Cooper; Application of electroactive polymers in bioelectrochemistry and bioelectronics. in *Electroactive Polymer Electrochemistry, Vol.* 2 (M. E. G. Lyons, Ed.), Springer Science + Business Media, New York, 1996, pp. 233–267.

60 P. N. Bartlett, C. S. Toth, E. J. Calvo, V. Flexer; Modeling biosensor response. in *Bioelectrochemistry – Fundamentals, Experimental Techniques and Applications.* (P. N. Bartlett, Ed.), John Wiley & Sons, Chichester, 2008, pp. 267–325.

61 M. E. G. Lyons; *Modeling Biosensor Interfaces; in Handbook of Biosensors and Biochips, Vol.* 1, John Wiley & Sons, Chichester, 2007, pp. 291–319.

62 M. G. Peter, U. Wollenberger; Phenol-oxidizing enzymes: Mechanisms and application in biosensors. in *Frontiers in Biosensorics, Vol. 1* (F. W. Scheller; F. Schubert, J. Fedrowitz, Eds.); Birkhäuser Verlag, Basel, 1997, pp. 64–82.

63 C. G. Bauer, A. V. Eremenko, E. Ehrentreich-Förster, F. F. Bier, A. Makower, H. B. Halsall, W. R. Heineman, F. W. Scheller; Zeptomole-detecting biosensor for alkaline phosphatase in an electrochemical immunoassay for 2,4-dichlorophenoxyacetic acid. *Anal. Chem.* **1996**, *68*, 2453–2458.

64 C. Ziller, J. Lin, P. Knittel, L. Friedrich, C. Andronescu, S. Pöller, W. Schuhmann, C. Kranz; Poly(benzoxazine) as immobilization matrix for miniaturized ATP and glucose biosensors. *ChemElectroChem* **2017**, *4*, 864–871.

65 Wikipedia; https://en.wikipedia.org/wiki/Diabetes (last updated 18 Mar 2026).

66 American Diabetes Association; https://diabetes.org/about-diabetes/statistics/about-diabetes (last access 18 Mar 2026).

67 E. D. Parker; J. Lin, T. Mahoney, N. Ume, G. Yang, R. A. Gabbay, N. A. ElSayed, R. R. Bannuru; Economic costs of diabetes in the US in 2022. *Diabetes Care* **2024**, *47*, 26–43.

68 A. Heller, B. Feldman; Electrochemical glucose sensors and their application in diabetes management. *Chem. Rev.* **2008**, *108*, 2482–2505.

69 Wikipedia; https://en.wikipedia.org/wiki/Screen_printing (last access 18 Mar 2026).

70 J. M. Montornes, M. S. Vreecke, I. Katakis; Glucose biosensors. in *Bioelectrochemistry – Fundamentals, Experimental Techniques and Applications.* (P. N. Bartlett, Ed.), John Wiley & Sons, Chichester, 2008, pp. 199–217.

71 A. E. G. Cass, G. Davis, G. D. Francis, H. A. O. Hill, W. J. Aston, I. J. Higgins, E. V. Plotkin, L. D. L. Scott, A. P. F. Turner; Ferrocene-mediated enzyme electrode for amperometric determination of glucose. *Anal. Chem.* **1984**, *56*, 667–671.

72 M. Bernabei, C. Cresimini, M. Mascini, G. Palleschi; Determination of organophosporous and carbamic pesticides with a choline and acetylcholine electrochemical biosensor. *Anal. Lett.* **1991**, *24*, 1317–1331.

73 T. Neufeld, I. Eshkenazi, E. Cohen, J. Rishpon; A micro flow injection electrochemical biosensor for organophosporous pesticides. *Biosens. Bioelectron.* **2000**, *15*, 323–329.

74 D. C. Harris; *Quantitative Analytical Chemistry*. 10th ed., Macmillan Learning, New York, 2019.

75 E. A. H. Hall; Glucose measurement within diabetes via "traditional" electrochemical biosensors. in *Handbook of Biosensors and Biochips, Vol.* 2, John Wiley & Sons, Chichester, 2007, pp. 1112–1129.

76 P. Vadgama, M. Schoenleber; in vivo applications: Glucose monitoring, fuel cells. in *Bioelectrochemistry – Fundamentals, Experimental Techniques and Applications*. (P. N. Bartlett, Ed.), John Wiley & Sons, Chichester, 2008, pp. 443–466.

77 A. Heller; Implanted electrochemical glucose sensors for the management of diabetes. *Annu. Rev. Biomed. Eng.* **1999**, *1*, 153–175.

78 U. Wollenberger, F. Schubert, D. Pfeiffer, F. W. Scheller; Enhancing biosensor performance using multienzyme systems. *Trends Biotechnol.* **1993**, *111*, 255–262.

79 F. W. Scheller, U. Wollenberger; Enzyme electrodes. in *Encyclopedia of Electrochemistry – Vol. 9 Bioelectrochemistry*. Wiley-VCH, Weinheim, 2002, pp. 431–459.

80 U. Wollenberger, F. Lisdat, A. Rose, K. Streffer; Phenolic biosensors; in *Bioelectrochemistry – Fundamentals, Experimental Techniques and Applications*. (P. N. Bartlett, Ed.), John Wiley & Sons, Chichester, 2008, pp. 219–248.

81 N. Mano, F. Mao, A. Heller; Characteristics of a miniature compartment-less glucose-O_2 biofuel cell and its operation in a living plant. *J. Am. Chem. Soc.* **2003**, *125*, 6588–659.

82 S. Shleev; Quo Vadis, implanted fuel cell? *ChemPlusChem* **2017**, *82*, 522–539.

83 B. E. Logan, R. Rossi, A. Ragab, P. E. Saikaly; Electroactive microorganisms in bioelectrochemical system. *Nat. Rev. Microbiol.* **2019**, *17*, 307–319.

84 L. M. Tender, C. E. Reimers, H. A. Stecher III, D. E. Holmes, D. R. Bond, D. A. Lowy, K. Pilobello, S. J. Fertig, D. R. Lovley; Harnessing microbially generated power on the seafloor; *Nat. Biotechnol.* **2002**, *20*, 821–825.

85 A. Heller; Minature biofuel cells. *Phys. Chem. Chem. Phys.* **2004**, *6*, 209–216.

86 M. Grattieri, S. D. Minteer; Self-powered biosensors. *ACS Sens.*, **2018**, *3*, 44–53.

87 C. Santoro, C. Arbizzani, B. Erable, I. Ieropoulos; Microbial fuel cells: From fundamentals to application. A review. *J. Power Sources* **2017**, *356*, 225–244.

88 U. Schröder, F. Harnisch, L. T. Angenet; Microbial electrochemistry and technology: Terminology and classification. *Energy Environ. Sci.* **2015**, *8*, 513–519.

89 F. Harnisch, L. F. M. Rosa, F. Kracke, B. Virdis, J. O. Krömer; Electrifying white biotechnology: Engineering and economic potential of electricity-driven bio-production. *ChemSusChem* **2015**, *8*, 758–766.

90 K. M. Murphy, C. Weaver. L. J. Berg; *Janeway's Immunobiology*, 10th ed., W. Norton & Company Inc., New York, NY, 2022.

91 K. Aoki, M. Morita, O. Niwa, H. Tabei; Quantitative analysis of reversible diffusion-controlled currents of redox soluble species at interdigitated array electrodes under steady-state conditions. *J. Electroanal. Chem. Interfacial Electrochem.* **1988**, *256*, 269–282.

92 U. Wollenberger, M. Paeschke, R. Hintsche; Interdigitated array microelectrodes for the determination of enzyme activities. *Analyst* **1994**, *119*, 1245–1249.

93 G. Wittstock, S. H. Jenkins, H. B. Halsall, W. R. Heineman; Continuing challenges for the immunoassay field. *Nanobiology* **1998**, *4*, 153–162.

94 C. A. Wijayawardhana, H. B. Halsall, W. R. Heineman; Electrochemical immunoassay. in *Encyclopedia of Electrochemistry –Vol. 9 Bioelectrochemistry*. (G. S. Wilson; A. J. Bard, M. Stratmann, Eds.), Wiley-VCH, Weinheim, 2002, pp. 145–174.

95 N. Ronkainen-Matsuno, J. H. Thomas, H. B. Halsall, W. R. Heineman; Electrochemical immunoassays moving into the fast lane. *Trends Anal. Chem.* **2002**, *21*, 213–225.

96 A. Warsinke, A. Benkert, F. W. Scheller; Electrochemical immunoassays. *Fresenius J. Anal. Chem.* **2000**, *366*, 622–634.

97 K. Haupt, K. Mosbach; Molecularly imprinted polymers and their use in biomimetic sensors. *Chem. Rev.* **2000**, *100*, 2495–2504.

98 F. W. Scheller, X. Zhang, A. Yarman, U. Wollenberger, R. E. Gyurcsányi; Molecularly imprinted polymer based electrochemical sensors for biopolymers. *Curr. Opin. Electrochem.* **2019**, *14*, 53–59.

99 H. Sun, Y. Zu; A highlight of recent advances in aptamer technology and its application. *Molecules* **2015**, *20*, 11959–11980.

100 P. Dauphine-Ducharme, K. Yang, N. Arroyo-Currás, K. L. Ploense, Y. Zhang, J. Gerson, M. Kurnik, T. E. Kippin, M. N. Stojanovic, K. W. Plaxco; Electrochemical aptamer-based sensor for improved therapeutic drug monitoring and high-precision, feedback-controlled drug delivery, *ACS Sens.* **2019**, *4*, 2832–2837.

101 E. Paleček, M. Fojta, F. Jelen, V. Vetterl; Electrochemical analysis of nucleic acids. in *Encyclopedia of Electrochemistry – Vol. 9 Bioelectrochemistry*. (G. S. Wilson; A. J. Bard, M. Stratmann, Eds.), Wiley-VCH, Weinheim, 2002, pp. 365–429.

102 A. M. Oliveira-Brett; Electrochemical DNA assays. in *Bioelectrochemistry – Fundamentals, Experimental Techniques and Applications*. (P. N. Bartlett, Ed.), John Wiley & Sons, Chichester, 2008, pp. 411–442.

103 J. Wang; Electrochemical DNA biosensors. in *Electroanalytical Methods for Biological Materials*. (A. Brajter-Toth, J. Q. Chambers, Eds.), Marcel Dekker, New York, 2002, pp. 27–41.

104 V. Ruiz-Valdepeñas Montiel, E. Povedano, E. Vargas, R. M. Torrente-Rodríguez, M. Pedrero, A. Julio Reviejo, S. Campuzano, J. M. Pingarrón; Comparison of different sensing strategies for the development of highly sensitive electrochemical nucleic acid biosensors using neither nanomaterials nor nucleic acid amplification. *ACS Sens.* **2018**, *3*, 211–221.

105 M. Gebala, W. Schuhmann; Understanding properties of electrified interfaces as a prerequisite for label-free DNA hybridization detection. *Phys. Chem. Chem. Phys.* **2012**, *14*, 14933–14942.

106 C. Fan, K. W. Plaxco, A. J. Heeger; Electrochemical interrogation of conformational changes as a reagentless method for sequence-specific detection of DNA. *Proc. Natl. Acad. Sci. USA* **2003**, *100*, 9134–9137.

107 R. Wei, T. G. Martin, U. Rant, H. Dietz; DNA origami gatekeeper for solid state nanopores. *Angew. Chem. Int. Ed.* **2012**, *51*, 4864–4867.

108 D. Fapyane, J. S. Nielsen, E. E. Ferapontova; Electrochemical enzyme-linked sandwich assay with a cellulase label for ultrasensitive analysis of synthetic DNA and cell-isolated RNA. *ACS Sens.* **2018**, *3*, 2104–2111.

109 Y. Thang, Y. Wang, H. Wang, J.-H. Jiang, G.-L. Shen, R.-Q. Yu, J. Li; Electrochemical DNA biosensor based on the proximity-dependent surface hybridization assay. *Anal. Chem.* **2009**, *81*, 1982–1987.

110 A. A. Lubin, B. V. S. Hunt, R. J. White, K. W. Plaxco; Effects of probe length, probe geometry, and redox-tag placement on the performance of electrochemical E-DNA sensor. *Anal. Chem.* **2009**, *81*, 2150–2158.

111 Y. Xiao, B. D. Piontek, K. W. Plaxco, A. J. Heeger; A reagentless signal-on architecture for electronic, aptamer-based sensors via target-induced strand displacement. *J. Am. Chem. Soc.* **2005**, *127*, 17900–17991.

112 N. E. Odgen, M. Kurnik, C. Parolo, K. W. Plaxco; An electrochemical scaffold sensor for rapid syphilis diagnosis. *Analyst* **2019**, *144*, 5277–5283.

113 A. Martin, K. B. Grant, F. Stressmann, J.-M. Ghigo, D. Marchal, B. Limoges; Ultimate single-copy DNA detection using real-time electrochemical LAMP. *ACS Sens.* **2016**, *1*, 904–912.

114 T. Notomi, H. Okayama, H. Masubuchi, T. Yonekawa, K. Watanbe, N. Amino, T. Hase; Loop-mediated isothermal amplification of DNA. *Nucl. Acids Res.* **2000**, *28*, e63/i–vii.

115 J. Rettinger, S. Schwarz, W. Schwarz; *Electrophysiology, Basics, Modern Approaches and Application*. Springer International, Switzerland, 2016.

116 B. Hille; *Ion Channels of Excitable Membranes*, 3rd ed, Sinauer Associates, Sunderland, MA, 2001.

117 E. Neher; Ion channels for communication between and within cells (Nobel Lecture). *Angew. Chem. Int. Ed.* **1992**, *31*, 824–829.

118 B. Sakmann, E. Neher (Eds.); *Single-Channel Recording*. 2nd ed., Springer, Boston, MA, 1995.

119 Y.-L. Ying, J. Wang, A. R. Leach, Y. Jiang, R. Gao, C. Xu, M. A. Edwards, A. D. Pendergast, H. Ren, C. K. T. Weatherly, W. Wang, P. Actis, L. Mao, H. S. White, Y.-T. Long; Single-entity electrochemistry at confined sensing interfaces. *Sci. Chin.* **2020**, *63*, 589–618.

120 M. Copsey, Ed. Single entity electrochemistry. *Faraday Discuss.* **2016**, *193*, 1–558.

121 *The Axon Guide – Electrophysiology and Biophysics Laboratory Techniques*, 4th ed., Molecular Devices, San Jose, CA, 2020. https://www.moleculardevices.com/en/assets/user-guide/dd/cns/axon-guide-to-electrophysiology-and-biophysics-laboratory-techniques (last access 18 Mar 2026).

122 M. A. Edwards, S. R. German, J. E. Dick, A. J. Bard, H. S. White; High-speed multipass coulter counter with ultrahigh resolution. *ACS Nano* **2015**, *9*, 12274–12282.

123 W. J. Lan, M. A. Edwards, L. Luo, R. T. Perera, X. Wu, C. R. Martin, H. S. White; Voltage-rectified current and fluid flow in conical nanopores. *Acc. Chem. Res.* **2016**, *49*, 2605–2613.

124 Y.-L. Ying, Y.-T. Long; Nanopore-based single-biomolecule interfaces: From information to knowledge. *J. Am. Chem. Soc.* **2019**, *141*, 15720–15729.

125 M. Jain, H. E. Olsen, B. Paten, M. Akeson; The Oxford nanopore MinIon: Delivery of nanopore sequencing to the genomics community. *Genome Biol.* **2016**, *17*, 239/1–11.

126 D. R. Garalde et al.; Highly parallel direct RNA sequencing in an array of nanopores. *Nat. Meth.* **2017**, *15*, 201–206.

127 https://nanoporetech.com/about-us/history (last access 18 Mar 2026).

128 Y. Erlich; A vision for ubiquitous sequencing. *Genome Res.* **2015**, *25*, 1411–1416.

129 J. T. Simpson, R. E. Workman, P. C. Zuzarte, M. David, L. J. Dursi, W. Timp; Detection of DNA cytosine methylation using nanopore sequencing. *Nat. Meth.* **2017**, *14*, 407–410.

130 J. B. Edel, T. Albrecht (Eds.); *Engineered Nanopores for Bioanalytical Applications*. Elsevier, Amsterdam, 2013.

131 K. Fu, P. W. Bohn; Nanopore electrochemistry: A Nexus for molecular control of electron transfer reactions. *ACS Cent. Sci.* **2018**, *4*, 20–29.

Appendix

Table A-1 Standard electrode potentials against SHE in aqueous solution at 298 K in alphabetical order.

Electrode reaction		$E°$/V
$Ag^+ + e^-$	$\rightleftharpoons Ag$	+0.80
$Ag^{2+} + e^-$	$\rightleftharpoons Ag^+$	+1.98
$Al^{3+} + 3e^-$	$\rightleftharpoons Al$	−1.66
$Au^+ + e^-$	$\rightleftharpoons Au$	+1.69
$Au^{3+} + 3e^-$	$\rightleftharpoons Au$	+1.40
$Ba^{2+} + 2e^-$	$\rightleftharpoons Ba$	−2.91
$Be^{2+} + 2e^-$	$\rightleftharpoons Be$	−1.85
$Bi^{3+} + 3e^-$	$\rightleftharpoons Bi$	+0.20
$Br_2 + 2e^-$	$\rightleftharpoons 2\ Br^-$	+1.09
$BrO^- + H_2O + 2e^-$	$\rightleftharpoons Br^- + 2\ OH^-$	+0.76
$Ca^{2+} + 2e^-$	$\rightleftharpoons Ca$	−2.87
$Cd(OH)_2 + 2e^-$	$\rightleftharpoons Cd + 2\ OH^-$	−0.81
$Cd^{2+} + 2e^-$	$\rightleftharpoons Cd$	−0.40
$Ce^{3+} + 3e^-$	$\rightleftharpoons Ce$	−2.48
$Ce^{4+} + e^-$	$\rightleftharpoons Ce^{3+}$	+1.61
$Cl_2 + 2e^-$	$\rightleftharpoons 2Cl^-$	+1.36
$ClO^- + H_2O + 2e^-$	$\rightleftharpoons Cl^- + 2\ OH^-$	+0.89
$ClO_4^- + 2\ H^+ - 2e^-$	$\rightleftharpoons ClO_3^- + H_2O$	+1.23
$ClO_4^- + H_2O + 2e^-$	$\rightleftharpoons ClO_3^- + 2\ OH^-$	+0.36
$Co^{2+} + 2e^-$	$\rightleftharpoons Co$	−0.28
$Co^{3+} + e^-$	$\rightleftharpoons Co^{2+}$	+1.81
$Cr^{2+} + 2e^-$	$\rightleftharpoons Cr$	−0.91
$Cr_2O_7^{2-} + 14\ H^+ + 6e^-$	$\rightleftharpoons 2\ Cr^{3+} + 7\ H_2O$	+1.33
$Cr^{3+} + 3e^-$	$\rightleftharpoons Cr$	−0.74
$Cr^{3+} + e^-$	$\rightleftharpoons Cr^{2+}$	−0.41
$Cs^+ + e^-$	$\rightleftharpoons Cs$	−2.92
$Cu^+ + e^-$	$\rightleftharpoons Cu$	+0.52

(Continued)

Table A-1 (Continued)

Electrode reaction		$E°/V$
$Cu^{2+} + 2e^-$	$\rightleftharpoons Cu$	+0.34
$Cu^{2+} + e^-$	$\rightleftharpoons Cu^+$	+0.16
$F_2 + 2e^-$	$\rightleftharpoons 2\,F^-$	+2.87
$Fe^{2+} + 2e^-$	$\rightleftharpoons Fe$	−0.44
$Fe^{3+} + 3e^-$	$\rightleftharpoons Fe$	−0.04
$Fe^{3+} + e^-$	$\rightleftharpoons Fe^{2+}$	+0.77
$[Fe(CN)_6]^{3-} + e^-$	$\rightleftharpoons [Fe(CN)_6]^{4-}$	+0.36
$2\,H^+ + 2e^-$	$\rightleftharpoons H_2$	0 (by definition for all temperatures)
$2\,H_2O + 2e^-$	$\rightleftharpoons H_2 + 2\,OH^-$	−0.83
$2\,HBrO + 2\,H^+ + 2e^-$	$\rightleftharpoons Br_2 + 2\,H_2O$	+1.60
$2\,HClO + 2\,H^+ + 2e^-$	$\rightleftharpoons Cl_2 + 2\,H_2O$	+1.63
$H_2O_2 + 2\,H^+ + 2e^-$	$\rightleftharpoons 2\,H_2O$	+1.78
$H_4XeO_6 + 2\,H^+ + 2e^-$	$\rightleftharpoons XeO_3 + 3\,H_2O$	+3.0
$Hg_2^{2+} + 2e^-$	$\rightleftharpoons 2\,Hg$	+0.79
$Hg^{2+} + 2e^-$	$\rightleftharpoons Hg$	+0.86
$2\,Hg^{2+} + 2e^-$	$\rightleftharpoons Hg_2{}^{2+}$	+0.92
$I_2 + 2e^-$	$\rightleftharpoons 2\,I^-$	+0.54
$I_3{}^- + 2e^-$	$\rightleftharpoons 3\,I^-$	+0.53
$In^+ + e^-$	$\rightleftharpoons In$	−0.14
$In^{2+} + e^-$	$\rightleftharpoons In^+$	−0.40
$In^{3+} + 2e^-$	$\rightleftharpoons In^+$	−0.44
$In^{3+} + 3e^-$	$\rightleftharpoons In$	−0.34
$In^{3+} + e^-$	$\rightleftharpoons In^{2+}$	−0.49
$K^+ + e^-$	$\rightleftharpoons K$	−2.93
$La^{3+} + 3e^-$	$\rightleftharpoons La$	−2.52
$Li^+ + e^-$	$\rightleftharpoons Li$	−3.05
$Mg^{2+} + 2e^-$	$\rightleftharpoons Mg$	−2.36
$Mn^{2+} + 2e^-$	$\rightleftharpoons Mn$	−1.18
$Mn^{3+} + e^-$	$\rightleftharpoons Mn^{2+}$	+1.51
$MnO_2 + 4\,H^+ + 2e^-$	$\rightleftharpoons Mn^{2+} + 2\,H_2O$	+1.23
$MnO_4^- + 8\,H^+ + 5e^-$	$\rightleftharpoons Mn^{2+} + 4\,H_2O$	+1.51
$MnO_4^- + e^-$	$\rightleftharpoons MnO_4^{2-}$	+0.56
$MnO_4^{2-} + 2\,H_2O + 2e^-$	$\rightleftharpoons MnO_2 + 4\,OH^-$	+0.60
$Na^+ + e^-$	$\rightleftharpoons Na$	−2.71
$Ni^{2+} + 2e^-$	$\rightleftharpoons Ni$	−0.23
$NiOOH + H_2O + e^-$	$\rightleftharpoons Ni(OH)_2 + OH^-$	+0.49
$NO_3^- + 2\,H^+ + e^-$	$\rightleftharpoons NO_2 + H_2O$	+0.80
$NO_3^- + 4\,H^+ + 4e^-$	$\rightleftharpoons NO + 2\,H_2O$	+0.96
$NO_3^- + H_2O + 2e^-$	$\rightleftharpoons NO_2{}^- + 2\,OH^-$	+0.10
$O_2 + 2\,H_2O + 4e^-$	$\rightleftharpoons 4\,OH^-$	+0.40

(Continued)

Table A-1 (Continued)

Electrode reaction		$E°/V$
$O_2 + 4\,H^+ + 4e^-$	$\rightleftharpoons 2\,H_2O$	+1.23
$O_2 + e^-$	$\rightleftharpoons O_2{}^-$	−0.56
$O_2 + H_2O + 2e^-$	$\rightleftharpoons HO_2{}^- + OH^-$	−0.08
$O_3 + 2\,H^+ + 2e^-$	$\rightleftharpoons O_2 + 2\,H_2O$	+2.07
$O_3 + H_2O + 2e^-$	$\rightleftharpoons O_2 + 2\,OH^-$	+1.24
$Pb^{2+} + 2e^-$	$\rightleftharpoons Pb$	−0.13
$Pb^{4+} + 2e^-$	$\rightleftharpoons Pb^{2+}$	+1.67
$Pt^{2+} + 2e^-$	$\rightleftharpoons Pt$	+1.20
$Pu^{4+} + e^-$	$\rightleftharpoons Pu^{3+}$	+0.97
$Ra^{2+} + 2e^-$	$\rightleftharpoons Ra$	−2.92
$Rb^+ + e^-$	$\rightleftharpoons Rb$	−2.93
$S + 2e^-$	$\rightleftharpoons S^{2-}$	−0.48
$S_2O_8^{2-} + 2e^-$	$\rightleftharpoons 2\ SO_4^{2-}$	+2.05
$Sn^{2+} + 2e^-$	$\rightleftharpoons Sn$	−0.14
$Sn^{4+} + 2e^-$	$\rightleftharpoons Sn^{2+}$	+0.15
$Sr^{2+} + 2e^-$	$\rightleftharpoons Sr$	−2.89
$Ti^{2+} + 2e^-$	$\rightleftharpoons Ti$	−1.63
$Ti^{3+} + e^-$	$\rightleftharpoons Ti^{2+}$	−0.37
$Ti^{4+} + e^-$	$\rightleftharpoons Ti^{3+}$	0.00
$Tl^+ + e^-$	$\rightleftharpoons Tl$	−0.34
$U^{3+} + 3e^-$	$\rightleftharpoons U$	−1.79
$U^{4+} + e^-$	$\rightleftharpoons U^{3+}$	−0.61
$V^{2+} + 2e^-$	$\rightleftharpoons V$	−1.19
$V^{3+} + e^-$	$\rightleftharpoons V^{2+}$	−0.26
$Zn^{2+} + 2e^-$	$\rightleftharpoons Zn$	−0.76

Source 1: P. W. Atkins, J. de Paula; *Physikalische Chemie*. 4th ed., Wiley-VCH, Weinheim, 2006, p. 1117; and
Source 2: E.-G. Jäger, K. Schöne, G. Werner; *Elektrolytgleichgewichte und Elektrochemie: Arbeitsbuch 5*. 3rd ed., Deutscher Verlag für Grundstoffindustrie, Leipzig, 1977, pp. 273–276.

Table A-2 Formal potentials of important reversible molecular redox couples at 298 K.

Redox system	$E°'$/V vs. SHE, 298 K, pH 7
N,*N*′-Dimethyl-*p*-phenylendiamine	0.374
1,4-Benzoquinone	0.274
N,*N*,*N*′,*N*′-Tetramethyl-*p*-phenylendiamine	0.264
2,6-Dichlorophenol-indophenol	0.211
1,2-Napthoquinone-4-sulfonsäure	0.211
Toluene blue	0.109
Phenazinium-methylsulfate	0.054
Cresyl blue	0.053
Methylene blue	0.005
Tetramethyl-*p*-benzoquinone	0.011
Indigo-disulfonate	−0.131
Safranine T	−0.295
Neutral red	−0.331
1,1′-Dibenzyl-4,4′-bipyridyl salt	−0.356
1,1′-Dimethyl-4,4′-bipyridyl dichloride	−0.436
$[Ru(bpy)_3]^{2+/3+}$	1.266
$[Ru(phen)_3]^{2+/3+}$	1.226
$[Fe(bpy)_3]^{2+/3+}$	1.068
$[IrCl_6]^{3-/2-}$	0.886
$[W(CN)_8]^{3-/2-}$	0.504
$[Co(edta)]^{2-/-}$	0.374
$[Fe(edta)]^{2-/-}$	0.114
$[Ru(NH_3)_6]^{2+/3+}$	0.045
$[Cr(bpy)_3]^{2+/3+}$	−0.256
Co-sepulcrat$^{2+/3+}$	−0.306

The potentials are listed versus NHE in the original source. They have been converted for this table to SHE.
Source 1: M. L. Fultz, R. A. Durst; Mediator compounds for the electrochemical study of biological redox systems: a compilation. *Anal. Chim. Acta* **1982**, *140*, 1–18; and
Source 2: Selected data from M. D. Arning, S. D. Minteer; Electrode potentials. in *Handbook of Electrochemistry* (C. G. Zoski, Ed.); Elsevier, Amsterdam, 2007, pp. 813–827.

Table A-3 Negative decadic logarithm of solubility products (pK_{SP} values) of salts with low solubility.

Compound	pK_{SP}	Temperature / °C
$AgCl$	9.75	25
$AgCl$	8.88	50
$AgBr$	12.28	25
AgI	16.08	25
$AgCN$	14.2	25
$AgSCN$	12.0	25
$AgBrO_3$	4.28	25
$AgIO_3$	7.52	25
$AgOH$	7.9	18
$AgOOCCH_3$	2.40	25
AgN_3	8.54	25
Ag_2CO_3	11.2	25
Ag_2CrO_4	11.95	25
$Ag_2Cr_2O_7$	6.7	25
$Ag_2C_2O_4$	11.30	25
Ag_2S	49.6	25
Ag_3SO_4	4.80	25
Ag_3PO_4	17.8	25
Ag_3AsO_4	19	25
Ag_3AsO_3	18.5	25
Ag_3VO_4	6.3	20
$Al(OH)_3$	32.7	25
$Ba(IO_3)_2$	9.2	25
BaF_2	5.76	25
$Ba(IO_3)_2$	9.2	25
$BaCO_3$	9.26	25
BaC_2O_4	6.79	18
$BaCrO_4$	9.93	25
$BaSO_4$	9.96	25
$BaSO_4$	9.70	25
$Be(OH)_2$	18.6	25
$[BiO]Cl$	6.15	25
Bi_2S_3	97	25
CaF_2	10.40	25
$Ca(IO_3)_2$	6.13	18
$Ca(OH)_2$	5.26	18
$CaCO_3$ (calcite)	8.18	25
CaC_2O_4	8.07	25
$CaCrO_4$	1.64	18
$CaSO_4$	4.32	25
$Ca[HPO_4](H_2O)_2$	6.57	25

(Continued)

Table A-3 (Continued)

Compound	pK_{SP}	Temperature / °C
$Ca_3(PO_4)_2$	26.0	25
$Cd(OH)_2$	13.9	18
$CdCO_3$	13.9	25
CdS (yellow)	26.1	25
CdS (red)	28	25
$Ce(IO_3)_3$	9.5	25
$Ce_2(C_2O_4)_3$	28.4	25
$Co(OH)_2$	16	25
$CoCO_3$	12.0	25
CoS	22.1	(22…26)
$Cr(OH)_3$	30	25
CuCl	6.0	20
CuBr	7.4	20
CuI	11.3	20
CuSCN	10.8	18
Cu_2S	48.0	25
$Cu(IO_3)_2$	7.13	25
$Cu(OH)_2$	19.3	25
$CuCO_3$	9.86	25
CuC_2O_4	7.54	25
$CuCrO_4$	5.44	25
CuS	35.1	(37…44)
$Fe(OH)_2$	15.3	18
$FeCO_3$	10.6	20
FeC_2O_4	6.7	25
FeS	17.3	(18…21)
$Fe(OH)_3$	37.4	18
$[Hg_2]Cl_2$	17.88	25
$[Hg_2]Br_2$	20.9	25
$[Hg_2]I_2$	27.9	25
$[Hg_2](CN)_2$	39.3	25
$[Hg_2]CO_3$	16	25
$[Hg_2]CrO_4$	8.70	25
$[Hg_2]SO_4$	6.17	25
HgS (black)	51.8	25
HgS (red)	52.4	25
$KClO_4$	2.05	25
$K_2[PtCl_6]$	5.85	25
$La(IO_3)_3$	11.21	25
$La(OH)_3$	20	25
$La_2(C_2O_4)_3$	27.7	25
Li_2CO_3	0.5	25

(Continued)

Table A-3 (Continued)

Compound	pK_{SP}	Temperature / °C
MgF_2	8.18	25
$Mg(OH_2)$	10.74	25
$MgCO_3(H_2O)_3$	4.25	25
MgC_2O_4	4.1	25
$MgNH_4PO_4$	12.6	25
$Mn(OH)_2$	14.2	25
$MnCO_3$	10.1	25
MnS (green)	12.6	25
MnS (pink)	9.6	25
MnS	15	25
$NaHCO_3$	2.9	25
$Ni(OH)_2$	13.8	25
$NiCO_3$	6.87	25
NiS	20.7	(21…26)
PbF_2	7.5	9
$PbCl_2$	4.77	25
$PbBr_2$	4.4	25
PbI_2	8.06	25
$Pb(OOCH)_2$	6.70	25
$Pb(IO_3)_2$	12.59	25
$PbCO_3$	13.5	18
PbC_2O_4	10.56	18
$PbCrO_4$	13.75	25
$PbSO_4$	7.80	25
PbS	26.6	(27…29)
$RaSO_4$	10.37	20
$RbClO_4$	2.4	25
$Sn(OH)_2$	25.3	25
SnS	25.0	(25…28)
SrF_2	8.54	25
$SrCO_3$	9.96	25
SrC_2O_4	7.25	18
$SrCrO_4$	4.44	18
$SrSO_4$	6.55	25
TlCl	3.72	25
TlBr	5.41	25
TlI	7.24	25
$TlIO_3$	5.66	25
TlN_3	3.66	25
Tl_2CrO_4	12.01	25
$Zn(OH)_2$	17	25
α-ZnS	23.8	25
β-ZnS	21.6	(22…25)

Source: E.-G. Jäger, K. Schöne, G. Werner; *Elektrolytgleichgewichte und Elektrochemie: Arbeitsbuch 5*. 3rd ed., Deutscher Verlag für Grundstoffindustrie, Leipzig, 1977, pp. 283–284.

Table A-4 Formation constants of selected coordination complexes.

Complex	Temperature / °C	Ionic strength / mol L^{-1}	log *K*	log *β*
$[Cu(NH_3)]^{2+}$	30	0.5…5	4.15	4.15
$[Cu(NH_3)_2]^{2+}$	30	0.5…5	3.50	7.65
$[Cu(NH_3)_3]^{2+}$	30	0.5…5	2.89	10.54
$[Cu(NH_3)_4]^{2+}$	30	0.5…5	2.13	12.67
$[Cu(NH_3)]^{+}$			6.18	6.18
$[Cu(NH_3)_2]^{+}$			4.69	10.87
$[Ag(NH_3)]^{+}$	30	0.5…5	3.20	3.20
$[Ag(NH_3)_2]^{+}$			3.83	7.03
$[Zn(NH_3)_4]^{2+}$				9.46
$[Cd(NH_3)_4]^{2+}$				7.12
$[Ni(NH_3)_4]^{2+}$				7.95
$[Ni(NH_3)_6]^{2+}$				8.90
$[AlF_4]^{-}$	25	0.5		17.75
$[AlF_5]^{2-}$	25	0.5	1.62	19.37
$[AlF_6]^{3-}$	25	0.5	0.47	19.84
FeF_3	25	0.5		12.06
$[FeF_5]^{2-}$	25	0.5		15.40
$[Zn(OH)_4]^{2-}$	25			15.44
$[Al(OH)_4]^{-}$			−1	32
$[AgCl_2]^{-}$	25	0		4.75
$[AgCl_4]^{3-}$	25	5		5.32
$[AuCl_4]^{-}$	18			21.30
$[HgI_4]^{2-}$	25	0.5		29.83
$[CdI_4]^{2-}$	25	5		6.50
$[PbI_4]^{2-}$	25	0		3.85
$[Fe(CN)_6]^{4-}$	18			35
$[Fe(CN)_6]^{3-}$	18			42
$[Co(CN)_6]^{4-}$				19.5
$[Co(CN)_6]^{3-}$				48
$[Ni(CN)_4]^{2-}$				15.5
$[Cu(CN)_4]^{3-}$	25	0		30.3
$[Zn(CN)_4]^{2-}$	25			16.9
$[Cd(CN)_4]^{2-}$	25	3		18.45
$[Hg(CN)_4]^{2-}$	25			41.4
$[Au(CN)_2]^{-}$				38
$[Ag(CN)_2]^{-}$	18	0.3		21.1
$[Ag(SCN)_2]^{-}$	25	2.2		7.6
$[Ag(SCN)_4]^{3-}$	25	2.2		10.1
$[Hg(SCN)_4]^{2-}$	25	0.3		21.3
$[Ag(S_2O_3)]^{-}$	20	0		8.8
$[Ag(S_2O_3)_2]^{3-}$	20	0		13.5
$[Ni(en)_2]^{2+}$	30	1.2		14.1

(Continued)

Table A-4 (Continued)

Complex	Temperature / °C	Ionic strength / mol L^{-1}	log *K*	log *β*
$[Ni(en)_3]^{2+}$	30	1.2		18.6
$[Cu(en)_2]^{2+}$	30	0.5		19.6
$[Zn(en)_2]^{2+}$	30	1		10.4
$[Ag(en)_2]^{+}$	18			7.84
$[Cd(en)_2]^{2+}$	30	0.5		10.0
$[Cd(en)_3]^{2+}$	30	0.5		12.1
$[Co(en)_3]^{3+}$	30	1		48.7
$[Fe(phen)_3]^{2+}$				21.3
$[Fe(dipyr)_3]^{2+}$	25	0.33		17.58
$[Cu(dipyr)_2]^{2+}$	25	0.1		14.2
$[Cu(dipyr)_3]^{2+}$	25	0.1		17.85
$[Cu(citr)]^{-}$	25	0.5		14.2
$[Cu(OH)_2(citr)_2]^{-}$	25			18.77
$[Cu(OH)(citr)]^{2-}$	25			16.35
$[Cu(OH)_2(citr)_3]^{6-}$	25			19.3
Ca-Erio T				5.4
Mg-Erio T				7.0

Source: E.-G. Jäger, K. Schöne, G. Werner; *Elektrolytgleichgewichte und Elektrochemie: Arbeitsbuch 5.* 3rd ed., Deutscher Verlag für Grundstoffindustrie, Leipzig, 1977, pp. 281–282.

Table A-5 Activity coefficients of important electrolyte solutions.

Substance		Molality *b* [mol in 1000 g water]										
		0.001	0.005	0.01	0.05	0.1	0.5	1.0	2.0	3.0	4.0	5.0
$AgNO_3$	f_m					(0.734)	0.536	0.429	0.316	0.252	0.210	0.181
	f'_m		0.925	0.896	0.787	0.717	0.501	0.390				
$AlCl_3$	f_m					(0.337)	0.331	0539				
$Al(ClO_3)_3$	f'_m	0.783	0.620	0.533	0.350	0.299	0.258					
$Al_2(SO_4)_3$	f_m					(0.035)	0.0143	0.0175				
$BaBr_2$	f_m		0.800	0.740	0.594		0.443	0.511				
$BaCl_2$	f_m	0.881	0.774	0.716	0.564	0.499	0.392	0.388				
BaI_2	f'_m	0.907	0.839	0.798	0.686	0.634						
$Ba(NO_3)_2$	f_m	0.882	0.772	0.705	0.517	0.433						
$Ba(OH)_2$	f_m		0.773	0.712	0.526	0.443						
$Be(NO_3)_2$	f'_m	0.885	0.762	0.694	0.541	0.478						
$BeSO_4$	f'_m	0.754	0.534	0.426	0.222	0.157						
$CaCl_2$	f_m	0.883	0.783	0.730	0.589	0.531	0.447	0.505				
CaI_2	f'_m	0.885	0.779	0.717	0.589	0.538						
$Ca(NO_3)_2$	f'_m	0.885	0.78	0.71	0.55	0.48	0.34	0.31				
CaS_2O_3	f_m	0.754	0.540	0.446	0.267	0.208	0.122	0.111				

(Continued)

Table A-5 (Continued)

Substance		Molality b [mol in 1000 g water]										
		0.001	**0.005**	**0.01**	**0.05**	**0.1**	**0.5**	**1.0**	**2.0**	**3.0**	**4.0**	**5.0**
$CdBr_2$	f'_m		0.527	0.432	0.249	0.179	0.072					
	f_m	0.85	0.65	0.50	0.23	0.17	0.08	0.06				
$CdCl_2$	f'_m		0.648	0.567	0.380	0.289	0.130					
	f_m	0.755	0.569	0.475	0.277	0.206	0.093	0.061				
CdI_2	f'_m	0.70	0.47	0.36	0.15	0.094	0.032	0.021				
	f_m		0.56	0.40	0.14	0.092	0.031	0.02				
$Cd(NO_3)_2$	f'_m	0.906	0.835	0.800	0.696	0.653	0.60	0.61				
$CdSO_4$	f_m	0.754	0.540	0.432	0.227	0.166	0.067	0.045	0.032	0.033		
$CeCl_3$	f_m					(0.309)	0.264	0.342	0.847			
$CoCl_2$	f'_m	0.900	0.806	0.751	0.618	0.567	0.524	0.628				
$CrCl_3$	f_m					(0.331)	0.314	0.481				
$Cr(NO_3)_3$	f_m					(0.319)	0.291	0.401				
$Cr_2(SO_4)_3$	f_m					(0.046)	0.019	0.021				
CsBr	f'_m		0.926	0.898	0.795	0.733	0.564	0.483				
	f_m					0.754	0.603	0.538	0.486	0.465	0.457	0.453
$CsCH_3CO_2$	f_m					0.799	0.762	0.802	0.950	1.145		
CsCl	f_m		0.942	0.896	0.795	0.739	0.598	0.534	0.495	0.478	0.473	0.474
CsF	f'_m	0.982	0.963	0.952	0.913	0.892	0.851	0.874				
CsI	f_m					0.754	0.599	0.533	0.470	0.434		
$CsNO_3$	f_m		0.924	0.894	0.780	0.707	0.528	0.422				
CsOH	f_m					0.795	0.739	0.771				
Cs_2SO_4	f'_m	0.894	0.806	0.752	0.587	0.494						
$CuBr_2$	f'_m	0.896	0.812	0.768	0.667	0.634	0.659					
$CuCl_2$	f_m	0.888	0.783	0.723	0.577	0.518	0.416	0.43				
$CuSO_4$	f_m	0.74	0.53	0.41	0.209	0.149	0.061	0.041				
$EuCl_3$	f_m					(0.318)	0.276	0.371	0.995			
$FeCl_2$	f'_m	0.895	0.804	0.752	0.624	0.580	0.568	0.668				
$FeCl_3$	f'_m	0.80	0.65	0.59	0.47	0.41	0.35	0.42				
HBr	f_m			0.91	0.84	0.81	0.78	0.87				
HCl	f_m	0.965	0.928	0.904	0.830	0.796	0.757	0.809	1.009	1.136	1.762	2.38
$HClO_4$	f_m					0.803	0.769	0.823	1.055	1.448	2.08	3.11
HI	f_m		0.927	0.902	0.822	0.818	0.839	0.963	1.356	2.015		
HIO_3	f'_m	0.961	0.908	0.865	0.691	0.580	0.294	0.186				
HNO_3	f_m	0.965	0.927	0.902	0.823	0.785	0.715	0.720	0.793	0.909		
H_2SO_4	f_m	0.837	0.646	0.543		0.379	0.221	0.186				
KBO_2	f'_m		0.930	0.906	0.818	0.763	0.582	0.495				
KBr	f_m	0.965	0.927	0.903	0.822	0.777	0.665	0.625				
$KBrO_3$	f'_m		0.926	0.896	0.795	0.729						
KCH_3CO_2	f_m					0.796	0.751	0.783	0.910	1.086		
KCN	f'_m		0.926	0.900	0.809	0.755	0.617	0.615				

(Continued)

Table A-5 (Continued)

Substance		Molality b [mol in 1000 g water]										
		0.001	0.005	0.01	0.05	0.1	0.5	1.0	2.0	3.0	4.0	5.0
K_2CO_3	f_m	0.892	0.807	0.745	0.576	0.497	0.357	0.327				
KCl	f_m	0.965	0.927	0.902	0.818	0.771	0.655	0.611				
$KClO_3$	f_m	0.967	0.932	0.0907	0.813	0.755	0.568					
$KClO_4$	f_m	0.965	0.924	0.895								
KF	f_m	0.971	0.949	0.934	0.881	0.848	0.741	0.710	0.658	0.705	0.779	
$K_4Fe(CN)_6$	f'_m	0.650	0.447	0.360	0.189	0.134	0.062					
$K_3Fe(CN)_5$	f'_m	0.785	0.618	0.547	0.365	0.291	0.155					
KI	f_m	0.965	0.927	0.905	0.837	0.799	0.706	0.680				
KNO_3	f_m					0.739	0.545	0.443	0.333	0.269		
KOH	f_m		0.924	0.898	0.805	0.754	0.666	0.675				
KH_3PO_4	f_m					0.731	0.529	0.421				
KSCN	f'_m		0.926	0.902	0.820	0.771	0.629	0.555				
	f_m					0.769	0.646	0.559	0.556	0.538	0.529	0.524
K_2SO_4	f_m	0.889	0.781	0.715	0.529	0.441						
$LaCl_3$	f_m					(0.314)	0.266	0.342	0.847			
$La(NO_3)_3$	f'_m	0.792	0.630	0.551	0.380	0.317						
LiBr	f_m	0.966	0.932	0.909	0.842	0.810	0.783	0.848				
	f_m					0.796	0.753	0.803	1.015	1.341	1.897	2.74
$LiCH_3CO_2$	f_m					0.784	0.700	0.689	0.729	0.798	0.877	
LiCl	f_m	0.963	0.921	0.895	0.819	0.782	0.729	0.761				
	f_m					0.790	0.739	0.774	0.921	1.156	1.510	2.02
$LiClO_3$	f'_m	0.967	0.933	0.911	0.842	0.810	0.769	0.808				
$LiClO_4$	f'_m	0.967	0.935	0.915	0.853	0.825	0.821	0.913				
	f_m					0.812	0.808	0.887	1.158	1.582	2.18	
LiF	f'_m	0.965	0.922	0.889								
LiI	f'_m		0.95	0.94	0.91	0.90	0.92					
	f_m					0.815	0.824	0.910	1.198	1.715		
$LiNO_3$	f_m	0.966	0.930	0.904	0.834	0.798	0.743	0.765				
	f_m					0.788	0.726	0.743	0.835	0.966	1.125	1.310
LiOH	f_m					0.760	0.617	0.554	0.513	0.494	0.481	
$MgCl_2$	f'_m	0.891	0.800	0.751	0.627	0.577	0.540	0.659				
$Mg(NO_3)_2$	f_m	0.882	0.771	0.712	0.554	0.508	0.443	0.496				
$MgSO_4$	f'_m				0.262	0.195	0.091	0.067	0.042			
$MnCl_2$	f'_m	0.892	0.790	0.731	0.594	0.543	0.490	0.568				
$MnSO_4$	f_m	0.780	0.621	0.536	0.333	0.247	0.110	0.080				
	f_m					(0.150)	0.064	0.044	0.035	0.038	0.048	
NH_4Br	f'_n	0.964	0.901	0.870	0.780	0.733	0.617	0.572				
NH_4Cl	f_m	0.961	0.911	0.880	0.790	0.742	0.620	0.574				
NH_4I	f'_m	0.962	0.917	0.889	0.804	0.760	0.646	0.600				
NH_4NO_3	f_m	0.959	0.912	0.882	0.783	0.726	0.558	0.471				

(Continued)

Table A-5 (Continued)

Substance		Molality *b* [mol in 1000 g water]										
		0.001	**0.005**	**0.01**	**0.05**	**0.1**	**0.5**	**1.0**	**2.0**	**3.0**	**4.0**	**5.0**
NaBr	f_m	0.966	0.934	0.914	0.844	0.807	0.726	0.717				
	f_m					0.782	0.697	0.687	0.731	0.812	0.929	
$NaBrO_3$	f_m	0.967	0.934	0.911	0.826	0.775						
$NaCH_3CO_2$	f_m					0.791	0.735	0.757	0.851	0.982		
Na_2CO_3	f'_m	0.891	0.791	0.729	0.565	0.488	0.281					
NaCl	f_m	0.966	0.929	0.906	0.828	0.786	0.688	0.664				
	f_m					0.778	0.681	0.657	0.668	0.714	0.783	0.874
$NaClO_3$	f'_m	0.966	0.930	0.905	0.819	0.769	0.621	0.544				
	f_m					0.772	0.645	0.589	0.538	0.515		
$NaClO_4$	f'_m	0.966	0.929	0.904	0.821	0.773	0.640	0.576				
	f_m					0.775	0.668	0.629	0.609	0.611	0.626	0.649
NaF	f'_m		0.926	0.900	0.807	0.752	0.615					
	f_m					0.765	0.632	0.573				
NaI	f_m	0.966	0.935	0.917	0.866	0.841	0.808	0.844				
	f_m					0.787	0.723	0.736	0.820	0.963		
$NaIO_3$	f_m		0.924	0.895	0.784	0.714						
$NaNO_3$	f_m					0.762	0.617	0.548	0.478	0.437	0.408	0.386
NaOH	f_m			0.905	0.815	0.772	0.678	0.668				
	f_m					0.766	0.690	0.678	0.709	0.784	0.903	1.077
Na_2HPO_4	f'_m	0.885	0.771	0.706	0.530	0.441						
NaH_2PO_4	f_m					0.744	0.563	0.468	0.371	0.320	0.293	0.276
NaSCN	f_m					0.787	0.715	0.712	0.744	0.804	0.897	
Na_2SO_3	f'_m	0.891	0.808	0.757	0.610	0.530	0.327	0.24				
Na_2SO_4	f_m	0.887	0.778	0.714	0.536	0.453						
Na_2SiO_3	f_m				0.50	0.41	0.23	0.18				
$NdCl_3$	f_m					(0.310)	0.264	0.344	0.867			
$NiCl_2$	f'_m	0.900	0.807	0.753	0.619	0.567						
$Ni(NO_3)_2$	f_m	0.90	0.805	0.76	0.63	0.58	0.515	0.58				
$NiSO_4$	f'_m	0.764	0.561	0.455	0.246	0.180	0.078	0.056				
$PbCl_2$	f_m	0.802	0.660	0.584								
$Pb(NO_3)_2$	f'_m	0.885	0.763	0.687	0.464	0.373	0.168	0.112				
$PrCl_3$	f_m					(0.311)	0.262	0.338	0.825			
RbBr	f'_m		0.929	0.905	0.818	0.765	0.621	0.556				
	f_m					0.763	0.632	0.578	0.536	0.520	0.514	0.515
$RbCH_3CO_2$	f_m					0.796	0.755	0.792	0.933	1.126		
RbCl	f_m		0.927	0.903	0.816	0.765	0.638	0.585	0.546	0.536	0.538	0.546
RbF	f'_m		0.940	0.924	0.871	0.839	0.756	0.729				
RbI	f_m					0.762	0.629	0.575	0.533	0.518	0.515	0.517
$RbNO_3$	f'_m		0.297	0.902	0.792	0.720						
	f_m					0.734	0.534	0.430	0.321	0.257	0.216	

(Continued)

Table A-5 (Continued)

Substance		Molality *b* [mol in 1000 g water]										
		0.001	0.005	0.01	0.05	0.1	0.5	1.0	2.0	3.0	4.0	5.0
$ScCl_3$	f_m					(0.320)	0.298	0.443				
$SmCl_3$	f_m					(0.314)	0.271	0.362	0.940			
$SnCl_2$	f_m	0.809	0.624	0.512	0.283	0.233						
$SrCl_2$	f_m	0.898	0.814	0.759	0.616	0.558	0.455	0.496				
SrI_2	f'_m	0.93	0.88	0.85	0.76	0.73	0.82	1.11				
$Sr(NO_3)_2$	f'_m	0.895	0.82	0.77	0.62	0.54	0.34	0.27				
$Th(NO_3)_4$	f_m					(0.279)	0.189	0.207	0.326	0.486	0.647	0.791
$TlCH_3CO_2$	f_m					0.750	0.589	0.515	0.444	0.405	0.376	0.354
$TlNO_3$	f_m		0.922	0.890	0.765	0.680						
UO_2Cl_2	f'_m	0.93	0.88	0.85	0.75	0.70	0.57					
$UO_2(NO_3)_2$	f'_m	0.912	0.845	0.801	0.674	0.615	0.554	0.658				
UO_2SO_4	f'_m	0.754	0.533	0.420	0.214	0.150	0.060					
YCl_3	f_m					(0.314)	0.278	0.385	1.136			
$ZnCl_2$	f'_m	0.891	0.797	0.746	0.62	0.580	0.490	0.465				
	f_m	0.881	0.767	0.708	0.556	0.502	0.376	0.325				
$Zn(NO_3)_2$	f_m	0.89	0.79	0.75	0.62	0.57	0.49	0.56				
$ZnSO_4$	f_m	0.700	0.477	0.387		0.144	0.060	0.043	0.035	0.041		

f_m – Stoichiometric molality activity coefficient at 298 K.
f'_m – Stoichiometric molality activity coefficient close to the freezing temperature of the solution.
Source: Jean d'Ans, E. Lax, C. Synowietz; *Taschenbuch für Chemiker und Physiker, Vol. 1*, Springer, 1967.

Table A-6 pK_L values for solubility of gases in water at 298 K.

Gas absorption process	pK_L
$CO_2(g) \rightleftharpoons CO_2(aq)$	1.4
$Cl_2(g) \rightleftharpoons Cl_2(aq)$	1.2
$N_2(g) \rightleftharpoons N_2(aq)$	3.2
$NH_3(g) \rightleftharpoons NH_3(aq)$	−4.67
$O_2(g) \rightleftharpoons O_2(aq)$	2.86
$H_2S(g) \rightleftharpoons H_2S(aq)$	0.99

Source: E.-G. Jäger, K. Schöne, G. Werner; *Elektrolytgleichgewichte und Elektrochemie: Arbeitsbuch 5.* 3rd ed., Deutscher Verlag für Grundstoffindustrie, Leipzig, 1977, p. 284.

Table A-7 Properties of important metals for electrodes.

Metal	Lattice type	a / pm	$2r$ / pm	ρ / (g cm^{-3})	T_m / °C
Cu	fcc				
Ag	fcc	408.64	289.0		961.9
Au	fcc	407.86	288.4		1064.4
Ni	fcc				
Pd	fcc	388.98	275.1		1544
Pt	fcc	392.31	277.4		1772
Rh	fcc	380.43	269.0		1966
Ir	fcc	383.94	271.5		2410

a – Lattice constant, $2r$ – atomic diameter, ρ – density, T_m – melting point.
Source: L. A. Kibler; *Preparation and Characterization of Noble Metal Single Crystal Electrodes*, Ulm, 2003. Available under www.uni-ulm.de/fileadmin/website_uni_ulm/nawi.inst.080/mitarbeiter/Kibler/KiblerSingleCrystals2003.pdf (last access 06 Nov 2022).

Table A-8 Properties of single-crystal electrodes.

Facet	φ / eV	E_{pzc} / V vs. SCE	C_{dl} / (A s cm^{-2})	γ / (μJ cm^{-2})
Ag				
Ag(1 1 1)	4.74	−0.695		61.8
Ag(1 0 0)	4.64	−0.865		70.3
Ag(1 1 0)	4.52	−0.975		75.6
Au				
Au(1 1 1) ($\sqrt{3} \times 22$)	5.20–5.31[a]	0.32		
Au(1 1 1) (1 × 1)		0.23		
Au(1 0 0)–(hex)	5.00–5.47[a]	0.30		
Au(1 0 0)–(1 × 1)		0.08		
Au(1 1 0)–(1 × 2)	4.8–5.37[a]	−0.04		
Au(1 0 0)–(1 × 1)		−0.02		

φ – Work function; E_{pzc} – potential of zero charge vs. SCE, γ – surface energy.
[a]Values scatter, probably because crystals with different surface reconstructions have been measured.

Source: L. A. Kibler; *Preparation and Characterization of Noble Metal Single Crystal Electrodes*, Ulm, 2003. Available under www.uni-ulm.de/fileadmin/website_uni_ulm/nawi.inst.080/mitarbeiter/Kibler/KiblerSingleCrystals2003.pdf (last access 06 Nov 2022).

The table contains a selection of references for single crystal electrodes. It should be noted that some properties depend not only on the metal but also on the electrolyte solution with which it is in contact. The data were taken from existing compilations that give references to the original literature. Only the consideration of methodical details allows a sensible discussion of deviation of the cited numerical values.

Table A-9 Thermodynamic data for elements and inorganic compounds.

	M/ (g mol^{-1})	$\Delta_F H°$/ (kJ mol^{-1})	$\Delta_F G°$/ (kJ mol^{-1})	S_m°/ (J K^{-1} mol^{-1})[a]	$C_{p,m}^\circ$/ (J K^{-1} mol^{-1})
Aluminum					
Al(s)	26.98	0	0	28.33	24.35
Al(l)	26.98	+10.56	+7.20	39.55	24.21
Al^{3+}(aq)	26.98	−531	−485	−321.7	
Al_2O_3(s, α)	101.96	−1675.7	−1582.3	50.92	79.04
$Al(OH)_3$(s)^{t}	78.0	−308	−304	189.5	99.2
$Al(NO_3)_3 \cdot 9\,H_2O$(s)^{b}	375.1	1676	−1582	50.9	79.04
Antimony					
Sb(s)	121.75	0	0	45.69	25.23
Sb_2O_3(s)^{b}	291.5	−1417.1	−1250.8	265.3	209.2
Sb_2O_5(s)^{b}	323.5	−1007.5	−864.7	125.1	117.7
Sb_2S_3(s)^{b}	339.7	−157.7	181.6	156.1	123.2
Arsenic					
As(s, α)	74.92	0	0	35.1	24.64
As_2O_3(s)^{b}	197.8	−656.8	−575	107.1	95.65
As_2O_5(s)^{b}	229.8	−924.9	−782.4	105.4	116.5
Barium					
Ba(s)	137.34	0	0	62.8	28.07
Ba^{2+}(aq)	137.34	−537.64	−560.77	+9.6	
BaO(s)	153.34	−553.5	−525.1	70.43	47.78
$BaCl_2$(s)	208.25	−858.6	−810.4	123.68	75.14
Beryllium					
Be(s)	9.01	0	0	9.50	16.44
$Be(OH)_2$(s)^{b}	42.03	−165		−210	71.1
BeO(s)^{b}	25.0	−1196	−1088	90	88
Bismuth					
Bi(s)	208.98	0	0	56.74	25.52
$Bi(OH)_3$(s)^{b}	260.0	−577.8	−497.3	151	114
Bi_2O_3(s)^{b}	496.0	−395.3	−311.1	86.2	
Boron					
B(s)^{b}	10.8	0	0	5.86	11.1
B_2O_3(s)^{b}	69.6	−1254	−1184	80.8	62.8
H_3BO_3(s)^{b}	61.8	−1094	−968.8	88.7	81.34
Bromine					
Br_2(l)	159.82	0	0	152.23	75.689
Br_2(g)	159.82	+30 907	+3.110	245.46	36.02
Br^-(aq)	79.91	−121.55	−103.96	+82.4	−141.8
HBr(g)	90.92	−36.40	−53.45	198.70	29.142

(Continued)

Table A-9 (Continued)

	M/ (g mol^{-1})	$\Delta_F H°$/ (kJ mol^{-1})	$\Delta_F G°$/ (kJ mol^{-1})	$S°_m$/ (J K^{-1} mol^{-1})[a]	$C°_{p,m}$/ (J K^{-1} mol^{-1})
Cadmium					
$Cd(s, \gamma)$	112.40	0	0	51.76	25.98
$Cd^{2+}(aq)$	112.40	−75.90	−77.612	−73.2	
$Cd(OH)_2(s)$[b]	146.4	−561.5	−473.8	93	
$CdO(s)$	128.40	−258.2	−228.4	54.8	43.43
$CdCO_3(s)$	172.41	−750.6	−669.4	92.5	
$CdSO_4(s)$[b]	208.5	−934.4	−823.9	123.05	99.62
Cesium					
$Cs(s)$	132.91	0	0	85.23	32.17
$Cs^+(aq)$	132.91	−258.28	−292.02	+133.05	−10.5
Calcium					
$Ca(s)$	40.08	0	0	41.42	25.31
$Ca^{2+}(aq)$	40.08	−542.83	−553.58	−53.1	
$CaO(s)$	56.08	−635.09	−604.03	39.75	42.80
$CaCO_3$(*s*, calcite)	100.09	−1206.9	−1128.8	92.9	81.88
$CaCO_3$(*s*, aragonite)	100.09	−1207.1	−1127.8	88.7	81.25
$CaF_2(s)$	78.08	−1219.6	−1167.3	68.87	67.03
$CaCl_2(s)$	110.99	−795.8	−748.1	104.6	72.59
$CaBr_2(s)$	199.90	−682.8	−663.6	130	
Carbon (for organic compound, see Table A.10)					
C(*s*, graphite)	12.011	0	0	5.740	8.527
C(*s*, diamond)	12.011	+1.895	+2.900	2.377	6.113
$CO(g)$	28.011	−110.53	−137.17	197.67	29.14
$CO_2(g)$	44.010	−393.51	−394.36	213.74	37.11
$CO_2(aq)$	44.010	−413.80	−385.98	117.6	
$H_2CO_3(aq)$	62.03	−699.65	−623.08	187.4	
$HCO_3^-(aq)$	61.02	−691.99	−586.77	+91.2	
$CO_3^{2-}(aq)$	60.01	−677.14	−527.81	−56.9	
$CS_2(l)$	76.14	+89.70	+65.27	151.34	75.7
$HCN(g)$	27.03	+135.1	+124.7	201.78	35.86
$HCN(l)$	27.03	+108.87	+124.97	112.84	70.63
$CN^-(aq)$	26.02	+150.6	+172.4	+94.1	
Cerium					
$Ce(s)$[b]	140.1	0	0	64	26.9
$CeCl_3(s)$[b]	246.5	−1057.9	−983.9	171.5	
$CeO_2(s)$[b]	172.1	−1088.3	−1025.5	62.3	61.63
$Ce(SO_4)_2(s)$[b]	332.2	−2343	−2123	210.7	

(Continued)

Table A-9 (Continued)

	M/ (g mol^{-1})	$\Delta_F H°$/ (kJ mol^{-1})	$\Delta_F G°$/ (kJ mol^{-1})	$S°_m$/ (J K^{-1} mol^{-1})[a]	$C°_{p,m}$/ (J K^{-1} mol^{-1})
Chlorine					
$Cl_2(g)$	70.91	0	0	223.07	33.91
$Cl^-(aq)$	35.45	−167.16	−131.23	+56.5	−136.4
HCl(g)	36.46	−92.31	−95.30	186.91	29.12
HCl(aq)	36.46	−167.16	−131.23	56.5	−136.4
Chromium					
Cr(s)	52.00	0	0	23.77	23.35
CrO(s)[b]	68.0	−570.3	−500.7	124.7	91.8
$Cr_2O_3(s)$[b]	152.0	−1140	−1059	81.2	118.8
$CrO_3(s)$[b]	100.0	−590.4	−513	73.2	
$CrO_4^{2-}(aq)$	115.99	−881.15	−727.75	+50.21	
$Cr_2O_7^{2-}(aq)$	215.99	−1490.3	−1301.1	+261.9	
Cobalt					
Co(s)[b]	58.9	0	0	30.04	24.66
$Co(OH)_2(s)$[b]	92.9	−102.1	−07.5	158.2	
CoO(s)[b]	74.9	−868.2	−761.9	113.4	138
Copper					
Cu(s)	63.54	0	0	33.150	24.44
$Cu^+(aq)$	63.54	+71.67	+49.98	+40.6	
$Cu^{2+}(aq)$	63.54	+64.77	+65.49	−99.6	
$Cu_2O(s)$	143.08	−168.6	−146.0	93.14	63.64
CuO(s)	79.54	−157.3	−129.7	42.63	42.40
$CuSO_4(s)$	159.60	−771.36	−661.8	109	100.0
$CuSO_4 \cdot H_2O(s)$	177.62	−1085.8	−918.11	146.0	134
$CuSO_4 \cdot 5\,H_2O(s)$	249.68	−2279.7	−1879.7	300.4	280
Deuterium					
$D_2(g)$	4.028	0	0	144.96	29.20
HD(g)	3.022	+0.318	−1.464	143.80	29.196
$D_2O(g)$	20.028	−249.20	−234.54	198.34	34.27
$D_2O(l)$	20.028	−294.60	−243.44	75.94	84.35
HDO(g)	19.022	−245.30	−233.11	199.51	33.81
HDO(l)	19.022	−289.89	−241.86	79.29	
Fluorine					
$F_2(g)$	38.00	0	0	202.78	31.30
$F^-(aq)$	19.00	−332.63	−278.79	−13.8	−106.7
HF(g)	20.01	−271.1	−273.2	173.78	29.13
Gold					
Au(s)	196.97	0	0	47.40	25.42
$Au(OH)_3(s)$[b]	248.0	−477.8	−349.8	121	

(Continued)

Table A-9 (Continued)

	M/ (g mol^{-1})	$\Delta_F H°$/ (kJ mol^{-1})	$\Delta_F G°$/ (kJ mol^{-1})	$S°_m$/ (J K^{-1} mol^{-1})[a]	$C°_{p,m}$/ (J K^{-1} mol^{-1})
Hydrogen (see also Deuterium)					
$H_2(g)$	2.016	0	0	130.684	28.824
$H^+(aq)$	1.008	0	0	0	0
$H_2O(s)$	18.015			37.99	
$H_2O(l)$	18.015	−285.83	−237.13	69.91	75.291
$H_2O(g)$	18.015	−241.82	−228.57	188.83	33.58
$H_2O_2(l)$	34.015	−187.78	−120.35	109.6	89.1
Iodine					
$I_2(s)$	253.81	0	0	116.135	54.44
$I_2(g)$	253.81	+62.44	+19.33	260.69	36.90
$I^-(aq)$	126.90	−55.19	−51.57	+111.3	−142.3
$HI(g)$	127.91	+26.48	+1.70	206.59	29.158
Iron					
$Fe(s)$	55.85	0	0	27.28	25.10
$Fe^{2+}(aq)$	55.85	−89.1	−78.90	−137.7	
$Fe^{3+}(aq)$	55.85	−48.5	−4.7	−315.9	
$Fe_3O_4(s$, magnetite)	231.54	−1118.4	−1015.4	146.4	143.43
$Fe_3O_3(s$, hematite)	159.69	−824.2	−742.2	87.40	103.85
$FeS(s, \alpha)$	87.91	−100.0	−100.4	60.29	50.54
$FeS_2(s)$	119.98	−178.2	−166.9	52.93	62.17
Lithium					
$Li(s)$	6.94	0	0	29.12	24.77
$Li^+(aq)$	6.94	−278.49	−293.31	+13.4	68.6
Lead					
$Pb(s)$	207.19	0	0	64.81	26.44
$Pb^{2+}(aq)$	207.19	−1.7	−24.43	+10.5	
$PbO(s$, gelb)	223.19	−217.32	−187.89	68.70	45.77
$PbO(s$, rot)	223.19	−218.99	−188.93	66.5	45.81
$PbSO_4(s)$[b]	303.3	−723.4	−606.2	211.3	149.9
$PbO_2(s)$	239.19	−277.4	−217.33	68.6	64.64
Magnesium					
$Mg(s)$	24.31	0	0	32.68	24.89
$Mg^{2+}(aq)$	24.31	−466.85	−454.8	−138.1	
$MgO(s)$	40.31	−601.70	−569.43	26.94	37.15
$MgCO_3(s)$	84.32	−1095.8	−1012.1	65.7	75.52
$MgCl_2(s)$	95.22	−641.32	−591.79	89.62	71.38

(Continued)

Table A-9 (Continued)

	M/ (g mol^{-1})	$\Delta_F H°$/ (kJ mol^{-1})	$\Delta_F G°$/ (kJ mol^{-1})	$S°_m$/ (J K^{-1} mol^{-1})[a]	$C°_{p,m}$/ (J K^{-1} mol^{-1})
Manganese					
$Mn(s)$[b]	54.9	0	0	31.76	26.32
$Mn(OH)_2(s)$[b]	88.9	−996		117	
$MnO(s)$[b]	70.9	−385.1	−363.3	61.5	44.1
$Mn_2O_3(s)$[b]	157.87	−1387.6	−1282.9	154.8	139.3
$MnO_2(s)$[b]	86.9	−957.5	−879.9	110.5	107.5
Mercury					
$Hg(l)$	200.59	0	0	76.02	27.983
$Hg^{2+}(aq)$	200.59	+171.1	+164.40	−32.2	
$Hg_2^{2+}(aq)$	401.18	+172.4	+153.52	+84.5	
$HgO(s)$	216.59	−90.83	−58.54	70.29	44.06
$Hg_2Cl_2(s)$	472.09	−265.22	−210.75	192.5	102
$HgCl_2(s)$	271.50	−224.3	−178.6	146.0	
$HgS(s$, black)	232.65	−53.6	−47.7	88.3	
Molybdenum					
$Mo(s)$[b]	95.9	0	0	28.9	23.78
$MoO_2(s)$[b]	127.9	−589.1	−533.2	46.3	55.98
$MoO_3(s)$[b]	143.9	−744.6	−666.8	77.7	75.02
$MoS_2(s)$[b]	160.1	−234.8	−225.2	63.2	63.5
Nickel					
$Ni(s)$[b]	58.7	0	0	29.98	26.07
$Ni(OH)_2(s)$[b]	92.7	−543.5	−458.3	79	
$NiO(s)$[b]	74.7	−239.7	−211.4	37.99	44.31
$NiSO_4 \cdot 7\,H_2O(s)$[b]	280.8	−2997.4	−2463.3	378.9	364.6
$NiS(s)$[b]	90.7	−79	−76.9	52.97	47.11
Nitrogen					
$N_2(g)$	28.013	0	0	191.61	29.125
$NO(g)$	30.01	+90.25	+86.55	210.76	29.844
$N_2O(g)$	44.01	+82.05	+104.20	219.85	38.45
$NO_2(g)$	46.01	+33.18	+51.31	240.06	37.20
$N_2O_4(g)$	92.1	+9.16	+97.89	304.29	77.28
$N_2O_5(s)$	108.01	−43.1	+113.9	178.2	143.1
$N_2O_5(g)$	108.01	+11.3	+115.1	355.7	84.5
$HNO_3(l)$	63.01	−174.10	−80.71	155.60	109.87
$HNO_3(aq)$	63.01	−207.36	−111.25	146.4	−86.6
$NO_3^-(aq)$	63.01	−205.0	−108.74	+146.4	−86.6
$NH_3(g)$	17.03	−46.11	−16.45	192.45	35.06
$NH_3(aq)$	17.03	−80.29	−26.50	111.3	
$NH_4^+(aq)$	18.04	−132.51	−79.31	+113.4	79.9
$NH_2OH(s)$	33.03	−114.2			

(Continued)

Table A-9 (Continued)

	M/ (g mol^{-1})	$\Delta_F H°$/ (kJ mol^{-1})	$\Delta_F G°$/ (kJ mol^{-1})	$S°_m$/ (J K^{-1} mol^{-1})[a]	$C°_{p,m}$/ (J K^{-1} mol^{-1})
$HN_3(l)$	43.03	+264.0	+327.3	140.6	
$HN_3(g)$	43.03	+294.1	+328.1	238.97	43.68
$N_2H_4(l)$	32.05	+50.63	+149.43	121.21	98.87
$NH_4NO_3(s)$	80.04	−365.56	−183.87	151.08	139.3
$NH_4Cl(s)$	53.49	−314.43	−202.87	94.6	84.1
Oxygen					
$O_2(g)$	31.999	0	0	205.138	29.355
$O_3(g)$	47.998	+142.7	+163.2	238.93	39.20
$OH^-(aq)$	17.007	−229.99	−157.24	−10.75	−148.5
Phosphorus					
P(s, weiß)	30.97	0	0	41.09	23.840
$PCl_3(l)$	137.33	−319.7	−272.3	217.1	
$PCl_5(s)$	208.24	−443.5			
$H_3PO_3(s)$	82.00	−964.4			
$H_3PO_3(aq)$	82.00	−964.8			
$H_3PO_4(l)$	94.97	−1266.9			
$H_3PO_4(aq)$	94.97	−1277.4	−1018.7	−222	
$PO_4^{3-}(aq)$	94.97	−1277.4	−1018.7	−221.8	
$P_4O_{10}(s)$	283.89	−2984.0	−2697.0	228.86	211.71
$P_4O_6(s)$	219.89	−1640.1			
Potassium					
K(s)	39.10	0	0	64.18	29.58
$K^+(aq)$	56.11	−252.38	−283.27	+102.5	21.8
KOH(s)	58.10	−424.76	−379.08	78.9	64.9
KF(s)	74.56	−576.27	−537.75	66.57	49.04
KCl(s)	119.01	−436.75	−409.14	82.59	51.30
KBr(s)	166.01	−393.80	−380.66	95.90	52.30
KI(s)	166.01	−327.90	−324.89	106.32	52.93
Silicon					
Si(s)	28.09	0	0	18.83	20.00
$SiO_2(s, \alpha)$	60.09	−910.94	−856.64	41.84	44.43
Silver					
Ag(s)	107.87	0	0	42.55	25.351
$Ag^+(aq)$	107.87	+105.58	+77.11	+72.68	21.8
AgBr(s)	187.78	−100.37	−96.90	107.1	52.38
AgCl(s)	143.32	−127.07	−109.79	96.2	50.79
$Ag_2O(s)$	231.74	−31.05	−11.20	121.3	65.86
$AgNO_3(s)$	169.88	−129.39	−33.41	140.92	93.05

(Continued)

Table A-9 (Continued)

	M/ (g mol^{-1})	$\Delta_F H°$/ (kJ mol^{-1})	$\Delta_F G°$/ (kJ mol^{-1})	S°_m/ (J K^{-1} mol^{-1})[a]	$C^\circ_{p,m}$/ (J K^{-1} mol^{-1})
Sodium					
Na(*s*)	22.99	0	0	51.21	28.24
$Na^+(aq)$	22.99	−240.12	−261.91	+59.0	46.4
NaOH(*s*)	40.00	−425.61	−379.49	64.46	59.54
NaCl(*s*)	58.44	−411.15	−384.14	72.13	50.50
NaBr(*s*)	102.90	−361.06	−348.98	86.82	51.38
NaI(*s*)	149.89	−287.78	−286.06	98.53	52.09
Sulfur					
S(*s*, α, rhombic)	32.06	0	0	31.80	22.64
S(*s*, β, monocline)	32.06	+0.33	+0.1	32.6	23.6
$S^{2-}(aq)$	32.06	+33.1	+85.8	−14.6	
$SO_2(g)$	64.06	−296.83	−300.19	248.22	39.87
$SO_3(g)$	80.06	−395.72	−371.06	256.76	50.67
$H_2SO_4(l)$	98.08	−813.99	−690.00	156.90	138.9
$H_2SO_4(aq)$	98.08	−909.27	−744.53	20.1	−293
$SO_4^{2-}(aq)$	96.06	−909.27	−744.53	+20.1	−293
$HSO_4^-(aq)$	97.07	−887.34	−755.91	+131.8	−84
$H_2S(g)$	34.08	−20.63	−33.56	205.79	34.23
$H_2S(aq)$	34.08	−39.7	−27.83	121	
$HS^-(aq)$	33.072	−17.6	+12.08	+62.08	
$SF_6(g)$	146.05	−1209	−1105.3	291.82	97.28
Tin					
Sn(*s*, β)	118.69	0	0	51.55	26.99
$Sn^{2+}(aq)$	118.69	−8.8	−27.2	−17	
SnO(*s*)	134.69	−285.8	−256.9	56.5	44.31
$SnO_2(s)$	150.69	−580.7	−519.6	49.0	52.59
Zinc					
Zn(*s*)	65.37	0	0	41.63	25.40
$Zn^{2+}(aq)$	65.37	−153.89	−147.06	−112.1	46
ZnO(*s*)	81.37	−348.28	−318.30	43.64	40.25

[a]Standard entropies of ions can have positive or negative values because they are relative values with respect to $H^+(aq)$.

[b]Compounds marked by an asterisk were taken from Source 2 (Altmann et al., 1987) and added to the data from Source 1 (Atkins et al., 2021).

Source 1: P. W. Atkins, J. de Paula, J. J. Keeler; *Physikalische Chemie,* 6th ed., Wiley-VCH, Weinheim, 2021, pp. 1132–1137 and

Source 2: R. Altmann, G. Brandes, O. Regen, J. Schneider; *Chemisch-technische Stoffwerte: Eine Datensammlung* 2nd ed., Deutscher Verlag für Grundstoffindustrie, Leipzig, 1987.

Table A-10 Thermodynamic data for organic compounds.

	M/ (g mol^{-1})	$\Delta_F H°$/ (kJ mol^{-1})	$\Delta_F G°$/ (kJ mol^{-1})	$S_m^°$/ (J K^{-1} mol^{-1})[a]	$C_{p,m}^°$/ (J K^{-1} mol^{-1})	$\Delta_C H°$/ (kJ mol^{-1})
Hydrocarbons						
CH_4(*g*), methane	16.04	−74.81	−50.72	186.26	35.31	−890
C_2H_2(*g*), ethyne	26.04	+226.73	+209.20	200.94	43.93	−1300
C_2H_4(*g*), ethene	28.05	+52.26	+68.15	219.56	43.56	−1411
C_2H_6(*g*), ethane	30.07	−84.68	−32.82	229.60	52.63	−1560
C_3H_6(*g*), propene	42.08	+20.42	+62.78	267.05	63.89	−2058
C_3H_6(*g*), cyclopropene	42.08	+53.30	+104.45	237.55	55.94	−2091
C_3H_8(*g*), propane	44.10	−103.85	−23.49	269.91	73.5	−2220
C_6H_6(*l*), benzene	78.12	+49.0	+124.3	173.3	136.1	−3268
C_6H_6(*g*), benzene	78.12	+82.93	+129.72	269.31	81.67	−3302
C_6H_{12}(*l*), cyclohexane	84.16	−156	+26.8	204.4	156.5	−3920
$C_6H_5CH_3$(*g*), methylbenzene (toluene)	92.14	+50.0	+122.0	320.7	103.6	−3953
C_8H_{18}(*l*), octane	114.23	−249.9	+6.4	361.1		−5471
C_8H_{18}(*l*), *iso*-octane	114.23	−255.1				−5461
$C_{10}H_8$(*s*), naphthalene	128.18	+78.53				−5157
Alcohols and phenols						
CH_3OH(*l*), methanol	32.04	−238.66	−166.27	126.8	81.6	−726
CH_3OH(*g*), methanol	32.04	−200.66	−161.96	239.81	43.89	−764
C_2H_5OH(*l*), ethanol	46.07	−277.69	−174.78	160.7	111.46	−1368
C_2H_5OH(*g*), ethanol	46.07	−235.10	−168.49	282.70	65.44	−1409
C_6H_5OH(*s*), phenol	94.12	−165.0	−50.9	146.0		−3054
Carbonic acids, hydroxy acids, and esters						
HCOOH(*l*), formic acid	46.03	−424.72	−361.35	128.95	99.04	−255
CH_3COOH(*l*), acetic acid	60.05	−484.5	−389.9	159.8	124.3	−875
CH_3COOH(*aq*), acetic acid	60.05	−485.76	−396.46	178.7		
$CH_3CO_2^-$(*aq*), acetate	59.05	−486.01	−369.31	+86.6	−6.3	
$(COOH)_2$(*s*), oxalic acid	90.04	−827.2			117	−254
C_6H_5COOH(*s*), benzoic acid	122.13	−385.1	−245.3	167.6	146.8	−3227
$CH_3CH(OH)COOH$(*s*), lactic acid	90.08	−694.0				−1344
$CH_3COOC_2H_5$(*l*), acetic acid ethylester	88.11	−479.0	−332.7	259.4	170.1	−2231
Alkanes and alkanones						
HCHO(*g*), methanal	30.03	−108.57	−102.53	218.77	35.40	−571
CH_3CHO(*l*), ethanal	44.05	−192.30	−128.12	160.2		−1166
CH_3CHO(*g*), ethanal	44.05	−166.19	−128.86	250.3	57.3	−1192

(Continued)

Table A-10 (Continued)

	M/ (g mol^{-1})	$\Delta_F H°$/ (kJ mol^{-1})	$\Delta_F G°$/ (kJ mol^{-1})	S°_{m}/ (J K^{-1} mol^{-1})[a]	$C^{\circ}_{p,m}$/ (J K^{-1} mol^{-1})	$\Delta_C H°$/ (kJ mol^{-1})
Polysaccharides (sugars)						
$C_6H_{12}O_6(s)$, α-D-glucose	180.16	−1274				−2808
$C_6H_{12}O_6(s)$, β-D-glucose	180.16	−1268	−910	212		
Nitrogen compounds						
$CO(NH_2)_2$ urea	60.06	−333.51	−197.33	104.60	93.14	−632
$CH_3NH_2(g)$, methylamine	31.06	−22.97	+32.16	243.41	53.1	−1085
$C_6H_5NH_2(l)$, aniline	93.13	+31.1				−3393
$CH_2(NH_2)COOH(s)$, glycine	75.07	−532.9	−373.4	103.5	99.2	−969

[a] Standard entropies of ions can have positive or negative values because they are relative values with respect to $H^+(aq)$.
Source: P. W. Atkins, J. de Paula, J. J. Keeler; *Physikalische Chemie*. 6th ed., Wiley-VCH, Weinheim, 2021, pp. 1131–1132 (selection).

Table A-11 Greek alphabet.

$A\ \alpha$	alpha
$B\ \beta$	beta
$\Gamma\ \gamma$	gamma
$\Delta\ \delta$	delta
$E\ \epsilon$	epsilon
$Z\ \zeta$	zeta
$H\ \eta$	eta
$\Theta\ \theta$	theta
$I\ \iota$	iota
$K\ \kappa$	kappa
$\Lambda\ \lambda$	lambda
$M\ \mu$	mu
$N\ \nu$	nu
$\Xi\ \xi$	xi
$O\ o$	omicron
$\Pi\ \pi$	pi
$P\ \rho$	rho
$\Sigma\ \sigma$	sigma
$T\ \tau$	tau
$Y\ \upsilon$	upsilon
$\Phi\ \phi$	phi
$X\ \chi$	chi
$\Psi\ \psi$	psi
$\Omega\ \omega$	omega

Table A-12 Selection of fundamental physical constants.

Constant	Symbol	Numerical value	SI unit
Avogadro constant	N_A	$6.022\,140\,76 \times 10^{23}$	mol^{-1}
Planck constant	h	$6.626\,070\,15 \times 10^{-34}$	$J \cdot s$
Speed of light in vacuum	c	$2.997\,924\,58 \times 10^{8}$	$m \cdot s^{-1}$
Elementary charge	e	$1.602\,176\,634 \times 10^{-19}$	C
Bohr radius (= length unit for atomic sizes)	a_0	$5.291\,772\,109\,03 \times 10^{-11}$	m
Faraday constant	F	$9.648\,533\,21\,2 \times 10^{4}$	$C \cdot mol^{-1}$
Boltzmann constant	k_B	$1.380\,649 \times 10^{-23}$	$J \cdot K^{-1}$
Molar gas constant	R	$8.314\,462\,618$	$J \cdot mol^{-1} \cdot K^{-1}$
Bohr magneton	μ_B	$9.274\,010\,078\,3 \times 10^{-24}$	$A \cdot m^2$
Nuclear magneton	μ_N	$5.050\,783\,746\,1 \times 10^{-27}$	$A \cdot m^2$
Vacuum electric permittivity	ε_0	$8.854\,187\,812\,3 \times 10^{-12}$	$A^2\,s^4\,kg^{-1}\,m^{-3} =$ $[A\,s\,V^{-1}\,m^{-1} = F\,m^{-1}]$
Electron mass	m_e	$9.109\,383\,701\,5 \times 10^{-31}$	kg
Proton mass	m_p	$1.672\,621\,923\,69 \times 10^{-27}$	kg
Neutron mass	m_n	$1.674\,927\,498\,04 \times 10^{-27}$	kg
Rydberg constant	R_∞	$1.097\,373\,156\,816\,0 \times 10^{7}$	m^{-1}

Source: E. Tiesinga, P. J. Mohr, D. B. Newell, B. N. Taylor; CODATA Recommended Values of the Fundamental Physical Constants: 2018. *J. Phys. Chem. Ref. Data* **2021**, *50*, 033105. https://doi.org/10.1063/5.0064853.

Index

B

E

K

L

M

Q

R